To the Student

As you begin the school year and your AP® Calculus course, you may feel anxious about the many theorems, definitions, problems, and pages that you see in your book. Don't worry. Your concerns are normal. You can learn it all by May.

We wrote this calculus book with you in mind. We began with our college calculus book, and then adapted it to the AP® Calculus curriculum. If you pay attention in class, keep up with your homework, and read and study what is in this book, you will build the knowledge and skills needed to master calculus and to be successful on the AP® Calculus exam.

Here are some ways to use the book to your benefit.

Read Carefully

When you get busy, it is tempting to skip the reading and go right to the problems. Don't! This book has a large number of examples with clear, precise explanations that help you learn how to logically solve mathematics problems. Reading the book with a pencil in hand and doing the worked examples along with us will help you develop a clearer understanding beyond simple memorization.

Read before class, not after. If you do, the words will sound familiar as your teacher talks, and you can ask questions about anything you don't understand. You'll be amazed at how much more you'll get out of class if you read first.

But don't stop there. Getting a 5 on the exam requires understanding of the concepts. Learn the definitions and the theorems. Do the suggested NOW WORK Problems and as many of the other problems at the end of each section as you can, including the AP® Practice and Retain Your Knowledge Problems. The AP® Review Problems at the end of most chapters have video clips that guide you through a solution. The Break It Down feature models the strategy you can use to solve problems on the exam. The AP® Cumulative Review Problems help you to build up your proficiency across chapters. Be sure not to fall behind, and seek help as soon as you need it.

Use the Features of the Book

In a classroom, we use many methods to communicate. These methods, in a book, are called "features." The features in *Calculus for the AP® Course* serve several purposes. They include:

- Objectives that organize the material,
- Need to Review? and Recall boxes that provide a just-in-time review of material you have previously encountered,
- Break It Down features that apply a stepped-out method for solving AP® Practice Problems,
- Chapter Review, Review Exercises, AP® Review Problems, Retain Your Knowledge AP® Problems, and AP® Cumulative Review Problems, plus two complete exams:
  - AP® Practice Exam: Calculus AB
  - AP® Practice Exam: Calculus BC

The Alumni Profile feature that opens each chapter spotlights how some of our friends and colleagues have used the strategies learned in AP® Calculus in their professional lives. These features will help prepare you for quizzes, classroom tests, and ultimately the AP® Exam.

To make this process easier, on pages xix to xxix we provide a brief overview of the features and how they appear in the textbook. Spend a few minutes looking it over before diving into the course. Then use the features as you read the book, do the problems, and study for exams.

Please do not hesitate to contact us with any questions, suggestions, or comments that might improve this text. We look forward to hearing from you. We hope that you enjoy learning calculus and do well—not only in AP® Calculus but in all your studies.

Best wishes,

Mike   *Kathleen*

Michael Sullivan and Kathleen Miranda

CALCULUS

for the AP® Course

Fourth Edition

Michael Sullivan
Chicago State University

Kathleen Miranda
State University of New York,
Old Westbury

bedford, freeman & worth
publishers
New York

Program Director: Yolanda Cossio
Senior Program Manager: Raj Desai
Executive Development Editor: Lisa Samols
Development Editor: Andrew Sylvester
Senior Media Editor: Justin Perry
Associate Media Editor: Michael Emig
Editorial Assistant: Calyn Liss
Director, High School Marketing: Janie Pierce-Bratcher
Marketing Manager: Thomas Menna
Marketing Assistant: Brianna DiGeronimo
Senior Director, Content Management Enhancement: Tracey Kuehn
Executive Managing Editor: Michael Granger
Manager, Publishing Services: Edward Dionne
Senior Workflow Project Manager: Paul Rohloff
Production Supervisor: Robert Cherry
Director of Design, Content Management: Diana Blume
Senior Design Services Manager: Natasha A. S. Wolfe
Interior Design: Maureen McCutcheon
Cover Design: John Callahan
Art Manager: Matthew McAdams
Illustration Coordinator: Janice Donnola
Illustrations: Network Graphics
Icon (people around table in a blue circle, for group work): appleuzr/Getty Images
Executive Manager, Permissions: Christine Buese
Rights and Billing Editor: Alexis Gargin
Senior Director of Digital Production: Keri deManigold
Lead Media Project Manager: Jodi Isman
Composition: Lumina Datamatics, Inc.
Printing and Binding: Transcontinental

ISBN 978-1-319-45342-8

Library of Congress Control Number: 2023943081

Printed in Canada.

1 2 3 4 5 6   29 28 27 26 25 24

Acknowledgments

Acknowledgments and copyrights appear on the same page as the text and art selections they cover; these acknowledgments and copyrights constitute an extension of the copyright page.

AP® is a trademark registered by the College Board, which is not affiliated with, and does not endorse, this product.

Bedford, Freeman & Worth Publishers
120 Broadway, New York, NY 10271
bfwpub.com/catalog

What Is Calculus?

Calculus is a part of mathematics that evolved much later than other subjects. Algebra, geometry, and trigonometry were developed in ancient times, but calculus as we know it did not appear until the seventeenth century.

The first evidence of calculus has its roots in ancient mathematics. For example, in his book *A History of* π, Petr Beckmann explains that Greek mathematician Archimedes (287–212 BCE) "took the step from the concept of 'equal to' to the concept of 'arbitrarily close to' or 'as closely as desired'... and reached the threshold of the differential calculus, just as his method of squaring the parabola reached the threshold of the integral calculus."* But it was not until Sir Isaac Newton and Gottfried Wilhelm Leibniz, each working independently, expanded, organized, and applied these early ideas, that the subject we now know as calculus was born.

Although we attribute the birth of calculus to Newton and Leibniz, many other mathematicians, particularly those in the eighteenth and nineteenth centuries, contributed greatly to the body and rigor of calculus. You will encounter many of their names and contributions as you pursue your study of calculus.

But what is calculus?

The simple answer is: calculus models change. Since the world and most things in it are constantly changing, mathematics that explains change becomes immensely useful.

Calculus has two major branches, differential calculus and integral calculus. Let's take a peek at what calculus is by looking at two problems that prompted the development of calculus.

The Tangent Problem—The Basis of Differential Calculus

Suppose we want to find the slope of the line tangent to the graph of a function at some point $P = (x_1, y_1)$. See Figure 1(a). Since the tangent line necessarily contains the point P, it remains only to find the slope to identify the tangent line.

Suppose we repeatedly zoom in on the graph of the function at the point P. See Figure 1(b). If we can zoom in close enough, then the graph of the function will look approximately linear. Then we can choose a point Q on the graph of the function different from the point P, and use the formula for slope.

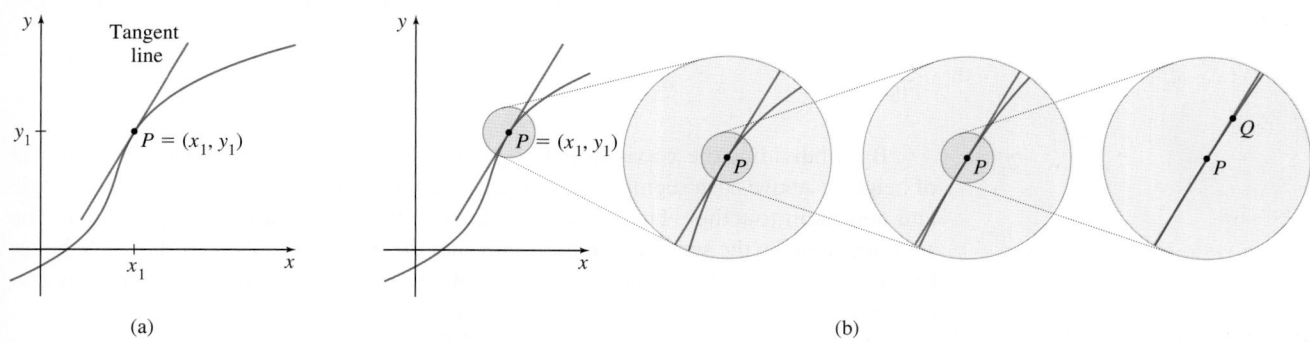

(a)   (b)

Figure 1

Repeatedly zooming in on the point P is equivalent to choosing a point Q closer and closer to the point P. Notice that as we zoom in on P, the line connecting the points P and Q, called a *secant line*, begins to look more and more like the tangent line to the graph of the function at the point P. If the point Q can be made as close as we please to the point P, without equaling the point P, then the slope of the tangent line $m_{\tan}$ can be found. This formulation leads to differential calculus, the study of the derivative of a function.

* Beckmann, P. (1976). *A History of* π (3rd ed., p. 64). New York: St. Martin's Press.

The derivative gives us information about how a function changes at a given instant and can be used to solve problems involving velocity and acceleration; marginal cost and profit; and the rate of change of a chemical reaction. Derivatives are the subjects of Chapters 2 through 5.

The Area Problem—The Basis of Integral Calculus

If we want to find the area of a rectangle or the area of a circle, formulas are available. See Figure 2. But what if the figure is curvy, but not circular as in Figure 3? How do we find this area?

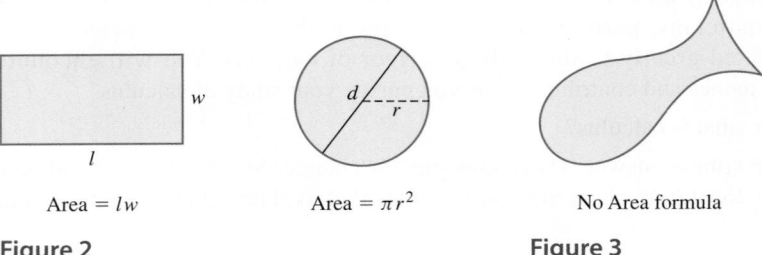

Area = lw Area = πr^2 No Area formula

Figure 2   **Figure 3**

Calculus provides a way. Look at Figure 4(a). It shows the graph of $y = x^2$ from $x = 0$ to $x = 1$. Suppose we want to find the area of the shaded region.

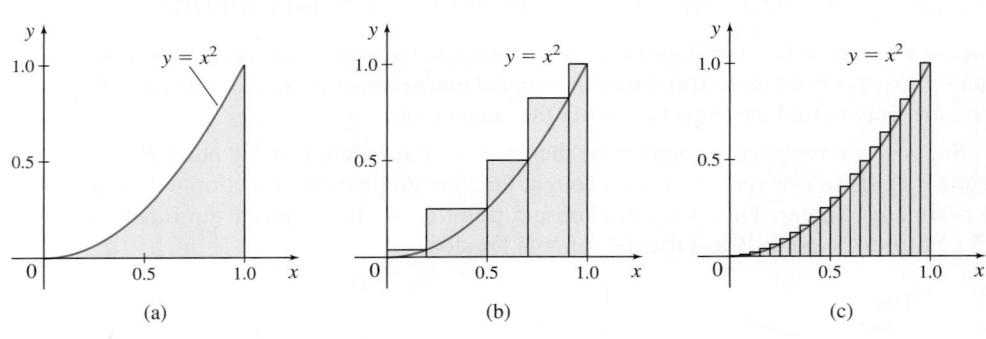

Figure 4

By subdividing the x-axis from 0 to 1 into small segments and drawing a rectangle of height x^2 above each segment, as in Figure 4(b), we can find the area of each rectangle and add them together. This sum approximates the shaded area in Figure 4(a). The smaller we make the segments of the x-axis and the more rectangles we draw, the better the approximation becomes. See Figure 4(c). This formulation leads to integral calculus, and the study of the integral of a function.

Two Problems—One Subject?

At first differential calculus (the tangent problem) and integral calculus (the area problem) appear to be different, so why call both of them calculus? The *Fundamental Theorem of Calculus* establishes that the derivative and the integral are related. In fact, one of Newton's teachers, Isaac Barrow, recognized that the tangent problem and the area problem are closely related, and that derivatives and integrals are inverses of each other. Both Newton and Leibniz formalized this relationship between derivatives and integrals in the *Fundamental Theorem of Calculus.*

Brief Contents | *Calculus*

Contents | *Calculus*

Limits and Continuity 80

Michael Collins, Apollo 11, NASA

Ayhan Altun/Getty Images

Unit 4

Fotografias Jorge León Cabello/Getty Images

Unit 5

Yellow Dog Productions/Getty Images

LINE FORMS TO THE RIGHT

Unit 6, Part 1

Brian Handy/500px/Getty Images

Unit 6, Part 2

tilialucida/Shutterstock

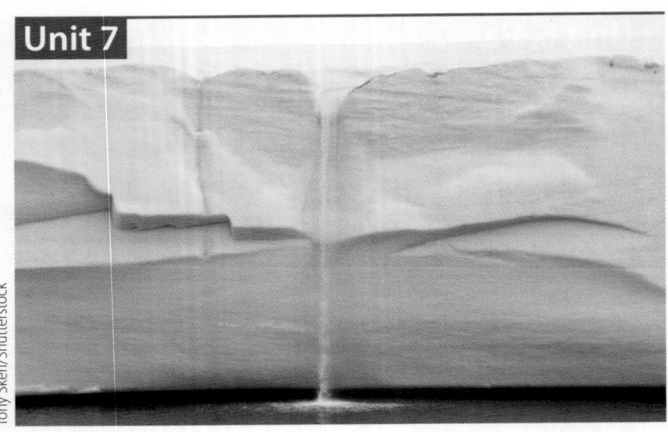

Unit 7

Tony Skerl/Shutterstock

Unit 8

David R. Frazier Photolibrary, Inc. / Alamy Stock Photo

Unit 9
paffy/Shutterstock

Unit 10
kelvin38/Shutterstock

About the Authors

Michael Sullivan

Michael Sullivan, Emeritus Professor of Mathematics at Chicago State University, received a Ph.D. in mathematics from the Illinois Institute of Technology. Before retiring, Mike taught at Chicago State for 35 years, where he honed an approach to teaching and writing that forms the foundation for his textbooks. His publications include *Finite Mathematics, College Algebra, Algebra and Trigonometry, Precalculus, Business Calculus, Engineering Calculus,* and *AP® Calculus.*

Mike is a member of the American Mathematical Association of Two Year Colleges, the American Mathematical Society, the Mathematical Association of America, and the Textbook and Academic Authors Association. Mike serves on the governing board of TAA and represents TAA on the board of the Authors Coalition of America, a consortium of 22 author/creator organizations in the United States. Several of his texts have received TAA's McGuffey Longevity Award. In 2007, he was awarded the TAA Lifetime Achievement Award and in 2023 the Alumni Award for Excellence from St. Ignatius College Prep in Chicago.

His influence in the field of mathematics extends to his four children: Kathleen, who teaches college mathematics; Michael III, who also teaches college mathematics, and who is his coauthor on two precalculus series; Dan, who is a sales director for a college textbook publishing company; and Colleen, who teaches middle-school and secondary school mathematics. Twelve grandchildren round out the family.

Mike would like to dedicate this text to his four children, 12 grandchildren, and future generations.

Kathleen Miranda

Kathleen Miranda, Ed.D. from St. John's University, is an Emeritus Associate Professor of the State University of New York (SUNY), where she taught for 25 years. Kathleen is a recipient of the prestigious New York State Chancellor's Award for Excellence in Teaching, and she particularly enjoys teaching mathematics to underprepared and fearful students. Kathleen has served as director of Curriculum and Assessment Development at SUNY Old Westbury.

In addition to her extensive classroom experience, Kathleen has worked as accuracy reviewer and solutions author on several mathematics textbooks, including Michael Sullivan's *Brief Calculus* and *Finite Mathematics.* Kathleen's goal is to help students unlock the complexities of calculus and appreciate its many applications.

Kathleen has four children: Edward, a plastic surgeon in San Francisco; James, an anesthesiologist in Philadelphia; Kathleen, a chemical engineer who directs a biologics division at a major pharmaceutical firm; and Michael, a corporate strategy specialist and entrepreneur.

Kathleen would like to dedicate this text to her children and grandchildren.

Content Resources Team, 4th Edition

Karen Hyers

Content Advisor and Author, Teacher's Edition

Karen has worked in high school and collegiate mathematics education for 35 years and has taught AP® Calculus for more than 25 years. She is an AP® Calculus consultant and mentor for College Board and an AP® exam reader and leader. Karen received the 2009 Presidential Award for Excellence in Mathematics and Science Teaching. She is the past president of Minnesota's Council of Teachers of Mathematics and has served on the Board of Directors for NCSM: Math Ed Leadership, the Council of Presidential Awardees in Math, and SciMath^{MN}. She is a regular presenter at state, regional, and national math education conferences. Her recent action research focuses on enriching the student experience through collaborative learning, self-reflection, and critical thinking.

Ben Hedrick

Content Advisor and Solutions Author, Solutions Manual

Ben Hedrick holds a bachelor of science degree in mathematics and statistics from The Pennsylvania State University, a master of arts in teaching degree from Duke University, and a Ph.D. in Mathematics Curriculum and Instruction from Stanford University. He has taught calculus at the high school, college, and university levels, as well as English as a second language in Japan at the elementary and middle school levels and elementary mathematics in Finland as part of a Fulbright Distinguished Award in Teaching grant. He previously worked at College Board as the course lead for AP® Calculus and AP® Statistics.

Eliel Gonzalez

Content Advisor and Accuracy Reviewer, Student Edition, Teacher's Edition, and Solutions Manual

Eliel Gonzalez served as a mathematics educator in Massachusetts for over three decades, retiring in 2022. He holds a bachelor of science degree in mathematics from Purdue University (2021 Distinguished Education Alumnus Award) as well as a master in education degree from the University of Massachusetts. His primary duties included teaching AP® Calculus, honors calculus, and precalculus, as well as being involved with Math Counts and/or HS Math League every year of his career. Furthermore, Eliel has served the College Board as K-12 professional development project manager in mathematics and AP® Annual Conference Steering Committee member. Eliel will soon start teaching AP® Precalculus workshops to complement his over two-decade tenure as an AP® Calculus teacher trainer.

Content Resources Team, 3rd Edition

James Bush

Waynesburg University, PA — Content Advisor, Solutions Manual

James Bush, Ph.D., is a Professor of Mathematics at Waynesburg University.

Erica Chauvet

Waynesburg University, PA — Content Advisor, Teacher's Edition

Erica has over 10 years of experience teaching AP® Calculus in a high school setting and currently teaches at Waynesburg University.

Brent Ferguson

Princeton High School, NJ — Content Advisor, Teacher's Edition

Brent has taught AP® Calculus AB and BC in private and public schools in the Midwest, and on the West and East Coasts.

Owen G. Glennon

Marist High School, Chicago, IL — Content Advisor, Solutions Manual

Owen has taught mathematic at Marist for all 48 years of his teaching career. He is the director of the Marist Honors Math Program and a moderator of the math team.

Joshua Newton

Dallas Independent School District, Dallas TX — Content Advisor, Teachers Edition

Joshua has more than 10 years of experience teaching AP® Calculus in the high-school setting.

Acknowledgments

Content Advisory Board for the Fourth AP® Edition

Thank you to our advisory board members for their helpful direction throughout the development of this book. The content advisory board understands the needs of the AP® Calculus teacher and student. They provided valuable guidance on how to make the content relevant, engaging, and appropriate for the AP® high-school classroom.

Kimberly A. Foltz, *Indiana Academy for Science, Mathematics, and Humanities*

Eliel Gonzalez (retired), *East Longmeadow High School, MA*
Ben Hedrick
Theresa Horvath, *McKinney High School, McKinney, TX*
Karen Hyers, *Tartan High School, Oakdale, MN*
Teresita Lemus, *School for Advanced Studies, West Campus, FL*

Siena Morse, *Orange Vista High School, Perris, CA*
Larry Peterson, *Northridge High School, Layton, UT*
Anthony J. Record, *Avon High School, IN*
Dixie Ross, *Pflugerville High School, TX*
William Semus, *Cherry Hill High School East, NJ*
Allen Wolmer, *Technology Based Educational Support*

Content Advisory Board for the Third AP® Edition

We would also like to thank the advisory board members who participated in the development of the previous edition.

John Aversa, *Catholic Memorial School, West Roxbury, MA*
Joseph Brandell, *St. Mary's Preparatory, Orchard Lake, MI*
Mikki Castor, *Frontier High School, Mansfield, TX*
Brent Ferguson, *Princeton High School, Princeton, NJ*

Jon Kawamura, *West Salem High School, Salem, OR*
Joshua Newton, *Dallas Independent School District, Dallas, TX*
William Roloff, *Lake Park High School, Roselle, IL*

Reviewers of the AP® Editions

We would like to thank the following reviewers and consultants for sharing their expertise about the AP® Calculus course with us. Their advice has helped us to identify and carefully address the needs of AP® Calculus students and teachers:

Howard Alcosser, *Diamond Bar High School, Diamond Bar, CA*
Robert Arrigo, *Scarsdale High School*
John Aversa, *Catholic Memorial School, West Roxbury, MA*
Joseph Brandell, *West Bloomfield High School, West Bloomfield, MI*
James Bush, *Waynesburg University*
Mikki Castor, *Mansfield ISD, Mansfield, TX*
Erica Chauvet, *Waynesburg University, Waynesburg, PA*
Julie Morrisett Clark, *Hollins University*
Brent Ferguson, *Princeton High School, Princeton, NJ*

Owen G. Glennon, *Marist High School, Chicago, IL*
Paul Gruber, *Elizabeth High School, Elizabeth, NJ*
John Jensen, *Rio Salado College*
Jon Kawamura, *West Salem High School, Salem, OR*
Stephen Kokoska, *Bloomsburg University of Pennsylvania*
Brendan Murphy, *John Bapst Memorial High School*
Joshua Newton, *Yvonne A. Ewell Townview Magnet Center, Science and Engineering Magnet, Dallas, TX*
John Quintrell, *Montgomery High School*
William Roloff, *Lake Park High School, Roselle, IL*

Reviewers of the Second College Edition

Kent Aeschliman, *Oakland Community College*
Anthony Aidoo, *Eastern Connecticut State University*
Jason Aran, *Drexel University*
Alyssa Armstrong, *Wittenberg University*
Mark Ashbrook, *Arizona State University*
Beyza Aslan, *University of North Florida*
Nick Belloit, *Florida State College at Jacksonville*
Robert W. Benim, *University of Georgia*
Dr. Benkam B. Bobga, *University of North Georgia, Gainesville Campus*
Light Bryant, *Arizona Western College*
James Bush, *Waynesburg University*

Debra Carney, *Colorado School of Mines*
David Chan, *Virginia Commonwealth University*
Jeffrey Cohen, *El Camino College*
Roberto Diaz, *Fullerton College*
Tim Doyle, *De Paul University*
Deborah A. Eckhardt, *St. Johns River State College*
Steven Edwards, *Kennesaw State University*
Greg Fry, *El Camino College*
Sami M. Hamid, *University of North Florida*
Karl Havlak, *Angelo State University*
Beata Hebda, *University of North Georgia*
Victor Kaftal, *University of Cincinnati*

Wei Kitto, *Florida State College at Jacksonville*
Richard C. Le Borne, *Tennessee Technological University*
Frank Lynch, *Eastern Washington University*
Filix Maisch, *Oregon State University*
John R. Metcalf, *St. Johns River State College*
Val Mohanakumar, *Hillsborough Community College*
Tom Morley, *Georgia Institute of Technology*
Alfred Mulzet, *Florida State College at Jacksonville*
Shai Neumann, *Eastern Florida State College*
Mike Nicholas, *Colorado School of Mines*

Jon Oaks, *Macomb Community College*
Stan Perrine, *Georgia Gwinnett College*
Katherine Pinzon, *Georgia Gwinnett College*
Elise Price, *Tarrant County College*
Brooke P. Quinlan, *Hillsborough Community College*
Mahbubur Rahman, *University of North Florida*
Vicki Sealey, *West Virginia University*
Plamen Simeonov, *University of Houston-Downtown*
Bruce Thomas, *Kennesaw State University*
H. Joseph Vorder Bruegge, *Hillsborough Community College*

Student Contributors to the College Editions

Because the student is the ultimate beneficiary of this material, we would be remiss if we did not mention their input. We would like to thank students from the following schools who have provided us with feedback and suggestions for exercises throughout the text:

Barry University
Catholic University
Colorado State University
Murray State University

University of North Texas
University of South Florida
Southern Connecticut State University
Towson State University

Aside from commenting on exercises, a number of the applied exercises in the text were contributed by students. Although we were not able to use all of the submissions, we would like to thank all of the students from the following schools who took the time and effort to participate in this project:

Boston University
Bronx Community College
Idaho State University
Lander University
Minnesota State University

Millikin University
University of Missouri—St. Louis
University of North Georgia
Trine University
University of Wisconsin—River Falls

Content Reviewers of the First College Edition

James Brandt, *University of California, Davis*
William Cook, *Appalachian State University*
Mark Grinshpon, *Georgia State University*
Karl Havlak, *Angelo State University*
Steven L. Kent, *Youngstown State University*

Stephen Kokoska, *Bloomsburg University of Pennsylvania*
Larry Narici, *St. John's University*
Frank Purcell, *Twin Prime Editorial*
William Rogge, *University of Nebraska, Lincoln*

Exercise and Accuracy Checkers of the First College Edition

Nick Belloit, *Florida State College at Jacksonville*
Jennifer Bowen, *The College of Wooster*
Brian Bradie, *Christopher Newport University*
James Bush, *Waynesburg University*
Deborah Carney, *Colorado School of Mines*
Julie Clark, *Hollins University*
Adam Coffman, *Indiana University-Purdue University, Fort Wayne*
Alain D'Amour, *Southern Connecticut State University*
John Davis, *Baylor University*
Amy Erickson, *Georgia Gwinnett College*
Judy Fethe, *Pellissippi State Community College*
Tim Flaherty, *Carnegie Mellon University*
Melanie Fulton
Jeffery Groah, *Lone Star College*
Peter Lampe, *University of Wisconsin-Whitewater*
Donald Larson, *Penn State Altoona*

Denise LeGrand, *University of Arkansas, Little Rock*
Serhiy Levkov and his team
Benjamin Levy, *Wentworth Institute of Technology*
Roger Lipsett, *Brandeis University*
Aihua Liu, *Montclair State University*
Joanne Lubben, *Dakota Weslyan University*
Betsy McCall, *Columbus State Community College*
Will Murray, *California State University, Long Beach*
Vivek Narayanan, *Rochester Institute of Technology*
Nicholas Ormes, *University of Denver*
Chihiro Oshima, *Santa Fe College*
Vadim Ponomarenko, *San Diego State University*
John Samons, *Florida State College at Jacksonville*
Ned Schillow, *Carbon Community College*
Mark Smith, *Miami University*
Timothy Trenary, *Regis University*

Kiryl Tsishchanka and his team
Marie Vanisko, *Carroll College*
David Vinson, *Pellissippi Community College*

Bryan Wai-Kei, *University of South Carolina, Salkehatchie*
Lianwen Wang, *University of Central Missouri*
Kerry Wyckoff, *Brigham Young University*

Additional applied exercises, as well as many of the chapter projects that open and close each chapter of this text, were contributed by a team of creative individuals. We would like to thank all of our exercise contributors for their work on this vital and exciting component of the text:

Wayne Anderson, *Sacramento City College (Physics)*
Allison Arnold, *University of Georgia (Mathematics)*
James Bush, *Waynesburg University (Mathematics)*
Erica Chauvet, *Waynesburg University (Mathematics)*
Kevin Cooper, *Washington State University (Mathematics)*
Owen G. Glennon, *Marist High School, Chicago, IL*
Adam Graham-Squire, *Highpoint University (Mathematics)*

Sergiy Klymchuk, *Auckland University of Technology (Mathematics)*
Eric Mandell, *Bowling Green State University (Physics and Astronomy)*
Eric Martel, *Millikin University (Physics and Astronomy)*
Rachel Renee Miller, *Colorado School of Mines (Physics)*
Kanwal Singh, *Sarah Lawrence College (Physics)*
John Travis, *Mississippi College (Mathematics)*
Gordon Van Citters, *National Science Foundation*

Reviewers and Advisors of the First College Edition

Marwan A. Abu-Sawwa, *University of North Florida*
Jeongho Ahn, *Arkansas State University*
Martha Allen, *Georgia College & State University*
Roger C. Alperin, *San Jose State University*
Weam M. Al-Tameemi, *Texas A&M International University*
Robin Anderson, *Southwestern Illinois College*
Allison W. Arnold, *University of Georgia*
Mathai Augustine, *Cleveland State Community College*
Carroll Bandy, *Texas State University*
Scott E. Barnett, *Henry Ford Community College*
Emmanuel N. Barron, *Loyola University Chicago*
Abby Baumgardner, *Blinn College, Bryan*
Robert W. Bell, *Michigan State University*
Nicholas G. Belloit, *Florida State College at Jacksonville*
Daniel Birmajer, *Nazareth College of Rochester*
Justin Bost, *Rowan-Cabarrus Community College—South Campus*
Laurie Boudreaux, *Nicholls State University*
Alain Bourget, *California State University at Fullerton*
David Boyd, *Valdosta State University*
Jim Brandt, *Southern Utah University*
Light R. Bryant, *Arizona Western College*
Kirby Bunas, *Santa Rosa Junior College*
Dietrich Burbulla, *University of Toronto*
Shawna M. Bynum, *Napa Valley College*
Joe Capalbo, *Bryant University*
Mindy Capaldi, *Valparaiso University*
Luca Capogna, *University of Arkansas*
Jenna P. Carpenter, *Louisiana Tech University*
Nathan Carter, *Bentley University*
Vincent Castellana, *Craven Community College*
Stephen Chai, *Miles College*
E. William Chapin, Jr., *University of Maryland, Eastern Shore*
Sunil Kumar Chebolu, *Illinois State University*
William Cook, *Appalachian State University*
Sandy Cooper, *Washington State University*
David A. Cox, *Amherst College*
Mark Crawford, *Waubonsee Community College*
Charles N. Curtis, *Missouri Southern State University*
Larry W. Cusick, *California State University, Fresno*
Rajendra Dahal, *Coastal Carolina University*
Ernesto Diaz, *Dominican University of California*

Robert Diaz, *Fullerton College*
Geoffrey D. Dietz, *Gannon University*
Della Duncan-Schnell, *California State University, Fresno*
Deborah A. Eckhardt, *St. Johns River State College*
Karen Ernst, *Hawkeye Community College*
Mark Farag, *Fairleigh Dickinson University*
Kevin Farrell, *Lyndon State College*
Walden Freedman, *Humboldt State University*
Md Firozzaman, *Arizona State University*
Kseniya Fuhrman, *Milwaukee School of Engineering*
Douglas R. Furman, *SUNY Ulster Community College*
Tom Geil, *Milwaukee Area Technical College*
Jeff Gervasi, *Porterville College*
William T. Girton, *Florida Institute of Technology*
Darren Glass, *Gettysburg College*
Gisèle Goldstein, *University of Memphis*
Jerome A. Goldstein, *University of Memphis*
Lourdes M. Gonzalez, *Miami Dade College*
Pavel Grinfeld, *Drexel University*
Mark Grinshpon, *Georgia State University*
Jeffrey M. Groah, *Lone Star College, Montgomery*
Gary Grohs, *Elgin Community College*
Paul Gunnells, *University of Massachusetts, Amherst*
Semion Gutman, *University of Oklahoma*
Christopher Hammond, *Connecticut College*
James Handley, *Montana Tech*
Alexander L. Hanhart, *New York University*
Gregory Hartman, *Virginia Military Institute*
Ladawn Haws, *California State University, Chico*
Mary Beth Headlee, *State College of Florida, Manatee-Sarasota*
Janice Hector, *De Anza College*
Anders O. F. Hendrickson, *St. Norbert College*
Shahryar Heydari, *Piedmont College*
Max Hibbs, *Blinn College*
Rita Hibschweiler, *University of New Hampshire*
David Hobby, *SUNY at New Paltz*
Michael Holtfrerich, *Glendale Community College*
Keith E. Howard, *Mercer University*
Tracey Hoy, *College of Lake County*
Syed I. Hussain, *Orange Coast College*
Maiko Ishii, *Dawson College*

Nena Kabranski, *Tarrant County College*
William H. Kazez, *The University of Georgia*
Michael Keller, *University of Tulsa*
Eric S. Key, *University of Wisconsin, Milwaukee*
Raja Nicolas Khoury, *Collin College*
Michael Kirby, *Colorado State University*
Alex Kolesnik, *Ventura College*
Natalia Kouzniak, *Simon Fraser University*
Ashok Kumar, *Valdosta State University*
Geoffrey Laforte, *University of West Florida*
Tamara J. Lakins, *Allegheny College*
Justin Lambright, *Anderson University*
Carmen Latterell, *University of Minnesota, Duluth*
Glenn W. Ledder, *University of Nebraska, Lincoln*
Namyong Lee, *Minnesota State University, Mankato*
Aihua Li, *Montclair State University*
Rowan Lindley, *Westchester Community College*
Matthew Macauley, *Clemson University*
Filix Maisch, *Oregon State University*
Heath M. Martin, *University of Central Florida*
Vania Mascioni, *Ball State University*
Philip McCartney, *Northern Kentucky University*
Kate McGivney, *Shippensburg University*
Douglas B. Meade, *University of South Carolina*
Jie Miao, *Arkansas State University*
John Mitchell, *Clark College*
Val Mohanakumar, *Hillsborough Community College*
Catherine Moushon, *Elgin Community College*
Suzanne Mozdy, *Salt Lake Community College*
Gerald Mueller, *Columbus State Community College*
Kevin Nabb, *Moraine Valley Community College*
Bogdan G. Nita, *Montclair State University*
Charles Odion, *Houston Community College*
Giray Ökten, *Florida State University*
Kurt Overhiser, *Valencia College*
Edward W. Packel, *Lake Forest College*
Joshua Palmatier, *SUNY College at Oneonta*
Teresa Hales Peacock, *Nash Community College*
Chad Pierson, *University of Minnesota, Duluth*
Cynthia Piez, *University of Idaho*
Joni Burnette Pirnot, *State College of Florida, Manatee-Sarasota*
Jeffrey L. Poet, *Missouri Western State University*
Shirley Pomeranz, *The University of Tulsa*
Elise Price, *Tarrant County College, Southeast Campus*

Harald Proppe, *Concordia University*
Michael Radin, *Rochester Institute of Technology*
Jayakumar Ramanathan, *Eastern Michigan University*
Joel Rappaport, *Florida State College at Jacksonville*
Marc Renault, *Shippensburg University*
Suellen Robinson, *North Shore Community College*
William E. Rogge, *University of Nebraska, Lincoln*
Yongwu Rong, *George Washington University*
Amber Rosin, *California State Polytechnic University*
Richard J. Rossi, *Montana Tech of the University of Montana*
Bernard Rothman, *Ramapo College of New Jersey*
Dan Russow, *Arizona Western College*
Adnan H. Sabuwala, *California State University, Fresno*
Alan Saleski, *Loyola University Chicago*
Brandon Samples, *Georgia College & State University*
Jorge Sarmiento, *County College of Morris*
Ned Schillow, *Lehigh Carbon Community College*
Kristen R. Schreck, *Moraine Valley Community College*
Randy Scott, *Santiago Canyon College*
George F. Seelinger, *Illinois State University*
Rosa Garcia Seyfried, *Harrisburg Area Community College*
John Sumner, *University of Tampa*
Geraldine Taiani, *Pace University*
Barry A. Tesman, *Dickinson College*
Derrick Thacker, *Northeast State Community College*
Millicent P. Thomas, *Northwest University*
Tim Trenary, *Regis University*
Pamela Turner, *Hutchinson Community College*
Jen Tyne, *University of Maine*
David Unger, *Southern Illinois University–Edwardsville*
William Veczko, *St. Johns River State College*
James Vicknair, *California State University, San Bernardino*
Robert Vilardi, *Troy University, Montgomery*
Klaus Volpert, *Villanova University*
David Walnut, *George Mason University*
James Li-Ming Wang, *University of Alabama*
Jeffrey Xavier Watt, *Indiana University–Purdue University Indianapolis*
Qing Wang, *Shepherd University*
Rebecca Wong, *West Valley College*
Carolyn Yackel, *Mercer University*
Hong Zhang, *University of Wisconsin, Oshkosh*
Qiao Zhang, *Texas Christian University*
Qing Zhang, *University of Georgia*

Special thanks to Kaleigh Marie O'Hara, University of Virginia, Masters Degree Program, for research and advice.

Finally, we would like to thank the editorial, production, and marketing teams at Bedford, Freeman & Worth whose support, knowledge, and hard work were instrumental in the publication of this text: Yolanda Cossio, Lisa Samols, Raj Desai, Andrew Sylvester, Calyn Liss, Justin Perry, and Michael Emig, in Editorial; Edward Dionne, Paul Rohloff, Tracy Kuehn, Natasha Wolfe, Janice Donnola, Matthew McAdams, Christine Buese, and Alexis Gargin on the Project Management, Design, and Production teams; and Thomas Menna, Janie Pierce-Bratcher, and Brianna DiGeronimo on the Marketing team. Thanks as well to the diligent group at Lumina Datamatics, Inc., led by Aravinda Doss, for their expert composition, and to Ron Weickart at Network Graphics for his skill and creativity in executing the art program.

Michael Sullivan
Kathleen Miranda

Student's Guide to This Book

The next page begins a tour that shows you how this book has been set up to help you learn calculus and do well on the AP® Calculus Exam. Take a few minutes to read the following pages to see how you can get the most from this book and your AP® Calculus course.

. . . .

Looking for the videos and other helpful resources or digital options?

Go to

bfwpub.com/APCalculus4e or log into your

Achieve **course**

Conceptual understanding and applied practice, delivered when you need it.

CHAPTER

2

The Derivative and Its Properties

Michael Collins, Apollo 11, NASA

The Apollo Lunar Module
"One Giant Leap for Mankind"

On May 25, 1961, in a special address to Congress, U.S. president John F. Kennedy proposed the goal "before this decade is out, of landing a man on the Moon and returning him safely to the Earth." Roughly eight years later, on July 16, 1969, a Saturn V rocket launched from the Kennedy Space Center in Florida, carrying the *Apollo 11* spacecraft and three astronauts—Neil Armstrong, Buzz Aldrin, and Michael Collins—bound for the Moon.

The *Apollo* spacecraft had three parts: the Command Module with a cabin for the three astronauts; the Service Module that supp[...] propulsion, electrical power, oxygen, and wate[...] Moon. After its launch, the spacecraft traveled [...] orbit. Armstrong and Aldrin then moved into t[...] flat expanse of the Sea of Tranquility. After mo[...] the surface of the Moon crawled into the Luna[...] Command Module, which Collins had been pil[...] then headed back to Earth, where they splashe[...]

In 2022, NASA initiated the Artemis I prog[...] crewed Orion mission and eventually for the re[...] the Moon and then on to Mars in the 2030s.

Explore some of the physics at work that allowe[...] Lunar Module to the Moon's surface in the Chapt[...]

Chapters align with the AP® Calculus units.
Every chapter boldly displays the book's chapter title as well as the unit title in the AP® course description. This way, you know that the book covers the material that you need to know to prepare for the exam.

Immediate engagement.
Each chapter opens with a brief story that relates the calculus covered in it to real-world scenarios in fields such as biology, engineering, environmental sciences, technology, and space travel.

The project in Chapter 2 is an exploration of how physics is used to maneuver the Lunar Module.

CHAPTER 2 PROJECT

Michael Collins, Apollo 11, NASA

The *Apollo* Lunar Module

This Project may be done individually or as part of a team.

The Lunar Module (LM) was a small spacecraft that detached from the *Apollo* Command Module and was designed to land on the Moon. Fast and accurate computations were needed to bring the LM from an orbiting speed of about 5500 ft/s to a speed slow enough to land it within a few feet of a designated target on the Moon's surface. The LM carried a 70-lb computer to assist in guiding it successfully to its target. The approach to the target was split into three phases, each of which followed a reference trajectory specified by NASA engineers.* The position and velocity of the LM were monitored by sensors that tracked its deviation from the preassigned path at each moment. Whenever the LM strayed from the reference trajectory, control thrusters were fired to reposition it. In other words, the LM's position and velocity were adjusted by changing its acceleration.

The reference trajectory for each phase was specified by the engineers to have the form

$$r_{\text{ref}}(t) = R_T + V_T t + \frac{1}{2} A_T t^2 + \frac{1}{6} J_T t^3 + \frac{1}{24} S_T t^4 \qquad (1)$$

The end-of-chapter project, suitable for group or individual work, takes over where the opening story leaves off. Answer the questions to shed light on how the calculus techniques you have learned can be applied in these different fields.

Carefully constructed presentation to guide you through complex material.

Alumni Profiles.
Each chapter opens with a brief testimonial from friends and colleagues of the author team describing how their experience with AP® Calculus has shaped their professional lives.

Ryan Murphy
Athlete

I am a professional swimmer and four-time Olympic gold medalist. The knowledge of velocity and acceleration that I learned in AP® Calculus has helped me better understand and develop the strategy behind my backstroke.

Chapter 2 opens by returning to the tangent problem to find an equation of the tangent line to the graph of a function f at a point $P = (c, f(c))$. Remember in Section 1.1 we found that the slope of a tangent line is a limit,

$$m_{\tan} = \lim_{x \to c} \frac{f(x) - f(c)}{x - c}$$

This limit is one of the most significant ideas in calculus, the *derivative*.

In this chapter, we introduce interpretations of the derivative, treat the derivative as a function, and consider some properties of the derivative. By the end of the chapter, you will have a collection of basic derivative formulas and derivative rules that will be used throughout your study of calculus. ∎

2.1 Rates of Change and the Derivative

AP® EXAM TIP

BREAK IT DOWN: As you work through the problems in this chapter, remember to follow these steps:

Step 1 Identify the underlying structure and related concepts.

Step 2 Determine the appropriate math rule or procedure.

Step 3 Apply the math rule or procedure.

Step 4 Clearly communicate your answer.

On page 231, see how we've used these steps to solve Section 2.2 AP® Problem 12 on page 191.

OBJECTIVES *When you finish this section, you should be able to:*

1 Find equations for the tangent line and the normal line to the graph of a function (p. 168)

2 Find the rate of change of a function (p. 169)

3 Find average velocity and instantaneous velocity (p. 170)

4 Find the derivative of a function at a number (p. 173)

In Chapter 1, we discussed the tangent problem: *Given a function f and a point P on its graph, what is the slope of the tangent line to the graph of f at P?* See Figure 1, where ℓ_T is the tangent line to the graph of f at the point $P = (c, f(c))$.

The tangent line ℓ_T to the graph of f at P must contain the point P. Since finding the slope requires two points, and we have only one point on the tangent line ℓ_T, we reason as follows.

Suppose we choose any point $Q = (x, f(x))$, other than P, on the graph of f. (Q can be to the left or to the right of P; we chose Q to be to the right of P.) The line containing the points $P = (c, f(c))$ and $Q = (x, f(x))$ is a secant line of the graph of f. The slope $m_{\sec}$ of this secant line is

$$m_{\sec} = \frac{f(x) - f(c)}{x - c} \qquad (1)$$

Figure 2 shows three different points Q_1, Q_2, and Q_3 on the graph of f that are successively closer to the point P, and three associated secant lines ℓ_1, ℓ_2, and ℓ_3. The closer the points Q are to the point P, the closer the secant lines are to the tangent

Objectives describe core knowledge and key skills.
Every section starts with objectives that are repeated as headlines to reinforce ideas. Objectives are linked to the worked examples and their clear, succinct explanations.

Easy-to-follow signposts.
Headlines reinforce the order of the coverage and the objectives.

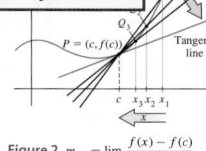

the secant line.

Figure 2 $m_{\tan} = \lim\limits_{x \to c} \dfrac{f(x) - f(c)}{x - c}$

AP® EXAM TIP

Problems on the exam often ask about the tangent line.

RECALL Two lines, neither of which is horizontal, with slopes m_1 and m_2, respectively, are perpendicular if and only if

$$m_1 = -\frac{1}{m_2}$$

① Find Equations for the Tangent Line and the Normal Line to the Graph of a Function

THEOREM Equation of a Tangent Line

If $m_{\tan} = f'(c)$ exists, then an equation of the tangent line to the graph of a function $y = f(x)$ at the point $P = (c, f(c))$ is

$$y - f(c) = f'(c)(x - c)$$

The line perpendicular to the tangent line at a point P on the graph of a function f is called the **normal line** to the graph of f at P.

THEOREM Equation of a Normal Line

An equation of the normal line to the graph of a function $y = f(x)$ at the point $P = (c, f(c))$ is

$$y - f(c) = -\frac{1}{f'(c)}(x - c)$$

provided $f'(c)$ exists and is not equal to zero. If $f'(c) = 0$, the tangent line is horizontal, the normal line is vertical, and the equation of the normal line is $x = c$.

Easy-to-grasp titles.
Worked EXAMPLE titles underscore the objectives and tell you the scope of the explanation.

CALC CLIP

EXAMPLE 1 Finding Equations for the Tangent Line and the Normal Line

(a) Find the slope of the tangent line to the graph of $f(x) = x^2$ at the point $(-2, 4)$.

(b) Use the result from (a) to find an equation of the tangent line at the point $(-2, 4)$.

(c) Find an equation of the normal line to the graph of f at the point $(-2, 4)$.

(d) Graph f, the tangent line to f at $(-2, 4)$, and the normal line to f at $(-2, 4)$ on the same set of axes.

Solution

(a) At the point $(-2, 4)$, the slope of the tangent line is

RECALL One way to find the limit of a quotient when the limit of the denominator is 0 is to factor the numerator and divide out common factors.

$$f'(-2) = \lim_{x \to -2} \frac{f(x) - f(-2)}{x - (-2)} = \lim_{x \to -2} \frac{x^2 - (-2)^2}{x + 2} = \lim_{x \to -2} \frac{x^2 - 4}{x + 2}$$

$$= \lim_{x \to -2} (x - 2) = -4$$

Clear explanations supported by rigorous examples.

Support in everyday language.
Use **In Words** to translate the complex formulas, theorems, proofs, rules, and definitions into everyday language to ease your understanding of them.

IN WORDS The derivative of the sum of two differentiable functions equals the sum of their derivatives. That is, $(f + g)' = f' + g'$.

Access to practice.
After working through an EXAMPLE, try the NOW WORK Practice Problems and **AP® Problems** at the end of the section to master the concepts.

3 Differentiate the Sum and the Difference of Two Functions

We can find the derivative of a function that is the sum of two functions whose derivatives are known by adding the derivatives of each function.

THEOREM Sum Rule

If two functions f and g are differentiable and if $F(x) = f(x) + g(x)$, then F is differentiable and

$$F'(x) = f'(x) + g'(x)$$

Proof If $F(x) = f(x) + g(x)$, then

$$F(x + h) - F(x) = [f(x + h) + g(x + h)] - [f(x) + g(x)]$$
$$= [f(x + h) - f(x)] + [g(x + h) - g(x)]$$

So, the derivative of F is

$$F'(x) = \lim_{h \to 0} \frac{[f(x + h) - f(x)] + [g(x + h) - g(x)]}{h}$$
$$= \lim_{h \to 0} \frac{f(x + h) - f(x)}{h} + \lim_{h \to 0} \frac{g(x + h) - g(x)}{h} \quad \text{The limit of a sum is the sum of the limits.}$$
$$= f'(x) + g'(x) \qquad \blacksquare$$

In Leibniz notation, the Sum Rule takes the form

$$\frac{d}{dx}[f(x) + g(x)] = \frac{d}{dx}f(x) + \frac{d}{dx}g(x)$$

EXAMPLE 4 Differentiating the Sum of Two Functions

Find the derivative of $f(x) = 3x^2 + 8$.

Solution
Here f is the sum of $3x^2$ and 8. So, we begin by using the Sum Rule.

$$f'(x) = \frac{d}{dx}(3x^2 + 8) = \frac{d}{dx}(3x^2) + \frac{d}{dx}8 = 3\frac{d}{dx}x^2 + 0 = 3 \cdot 2x = 6x \qquad \blacksquare$$
$$\underset{\text{Sum Rule}}{\uparrow} \qquad \underset{\substack{\text{Constant Multiple} \\ \text{Rule}}}{\uparrow} \qquad \underset{\substack{\text{Simple} \\ \text{Power Rule}}}{\uparrow}$$

NOW WORK Problem 7 and AP® Practice Problem 6.

Information-rich examples.
Worked EXAMPLES provide step-by-step instruction. Look for annotations in blue that show you what formula or reasoning is involved in solving the problem.

CALC CLIP
EXAMPLE 7 Differentiating an Expression Involving $y = e^x$

Find the derivative of $f(x) = 4e^x + x^3$.

Solution
The function f is the sum of $4e^x$ and x^3. Then

$$f'(x) = \frac{d}{dx}(4e^x + x^3) = \frac{d}{dx}(4e^x) + \frac{d}{dx}x^3 = 4\frac{d}{dx}e^x + 3x^2 = 4e^x + 3x^2 \qquad \blacksquare$$
$$\underset{\text{Sum Rule}}{\uparrow} \qquad \qquad \uparrow \qquad \qquad \uparrow$$

NOW WORK Problem 25 and AP® Practic...

Calculus in our world.
Applied EXAMPLES examples show how calculus is beneficial and relevant to a wide variety of fields and endeavors.

Extra support when you need it.
EXAMPLES marked with the Calc Clip button ▶ are supported by short video clips in 🐟 Achieve that walk you through each step in the process of solving a similar problem.

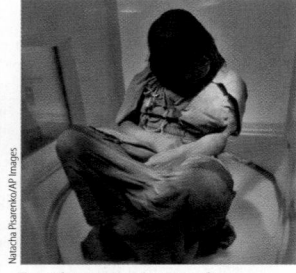

The perfectly preserved mummy of La Doncella, a 15-year-old girl, is displayed in a museum in Salta, Argentina.

Natacha Pisarenko/AP Images

EXAMPLE 4 Application: Carbon-14 Dating

All carbon on Earth contains some carbon-14, which is radioactive and exists in a fixed ratio with some nonradioactive carbon-12. When a living organism dies, the carbon-14 begins to decay at a fixed rate. The formula $P(t) = 100e^{-0.000121t}$ gives the percentage of carbon-14 present at time t years. Notice that when $t = 0$, the percentage of carbon-14 present is 100%. When the preserved bodies of 15-year-old La Doncella and two younger children were found in Argentina in 2005, 93.5% of the carbon-14 remained in their bodies, indicating that the three had died about 550 years earlier.

(a) What is the rate of change of the percentage of carbon-14 present in a 550-year-old fossil?

(b) What is the rate of change of the percentage of carbon-14 present in a 2000-year-old fossil?

Solution
(a) The rate of change of P is given by its derivative

$$P'(t) = \frac{d}{dt}\left(100e^{-0.000121t}\right) = 100\left(-0.000121e^{-0.000121t}\right) = -0.0121e^{-0.000121t}$$
$$\uparrow$$
$$\frac{d}{dx}e^{u(x)} = e^{u(x)}\frac{du}{dx}$$

Built-in support where and when you need it most.

RECALL gives a quick refresher of key results used in theorems, definitions, and examples.

NEED TO REVIEW? shows you where in the book to find concepts that you want to refresh. Each of these is on-the-spot help to send you to the right information.

AP® EXAM TIPS offer advice and tips on how to succeed on the exam.

CAUTION and NOTES warn you about potential misconceptions and pitfalls.

ORIGINS point out the life story and key discoveries of people who helped in the development of calculus.

RECALL If the functions f and g are inverses and if $f(a) = b$, then $g(b)$ exists, and $g(b) = a$.

NEED TO REVIEW? Inverse trigonometric functions are discussed in Section P.7, pp. 63–67.

AP® EXAM TIP

The exam often uses arcsin x instead of $\sin^{-1} x$, arccos x instead of $\cos^{-1} x$, and so on. Be familiar with both notations.

CAUTION Keep in mind that $f^{-1}(x)$ represents the inverse function of f and not the reciprocal function.

That is, $f^{-1}(x) \neq \dfrac{1}{f(x)}$.

NOTE Alternatively, we can use the Derivative of an Inverse Function. If $y = \sin^{-1} x$, then $x = \sin y$, and

$$\frac{dy}{dx} = \frac{1}{\dfrac{dx}{dy}} = \frac{1}{\cos y}$$

provided $\cos y \neq 0$.

ORIGINS Logarithmic differentiation was first used in 1697 by Johann Bernoulli (1667–1748) to find the derivative of $y = x^x$. Johann, a member of a famous family of mathematicians, was the younger brother of Jakob Bernoulli (1654–1705). He was also a contemporary of Newton, Leibniz, the French mathematician Guillaume de l'Hôpital, and the Japanese mathematician Seki Takakazu (1642–1708).

EXAMPLE 2 Finding an Equation of a Tangent Line

The function $f(x) = x^3 + 2x - 4$ is one-to-one and has an inverse function g. Find an equation of the tangent line to the graph of g at the point $(8, 2)$ on g.

Solution

The slope of the tangent line to the graph of g at the point $(8, 2)$ is $g'(8)$. Since f and g are inverse functions, $g(8) = 2$ and $f(2) = 8$. Now use (1) to find $g'(8)$.

$$g'(8) = \frac{1}{f'(2)} \qquad g'(y_0) = \frac{1}{f'(x_0)}; \ x_0 = 2; \ y_0 = 8$$

$$= \frac{1}{14} \qquad f'(x) = 3x^2 + 2; \ f'(2) = 3 \cdot 2^2 + 2 = 14$$

Now use the point-slope form of an equation of a line to find an equation of the tangent line to g at $(8, 2)$.

$$y - 2 = \frac{1}{14}(x - 8)$$

$$14y - 28 = x - 8$$

$$14y - x = 20$$

The line $14y - x = 20$ is tangent to the graph of g at the point $(8, 2)$. ∎

NOW WORK Problem 51 and AP® Practice Problems 7 and 11.

② Find the Derivative of the Inverse Trigonometric Functions

Table 1 lists the inverse trigonometric functions and their domains.

TABLE 1 The Domain of the Inverse Trigonometric Functions

f	Restricted Domain	f^{-1}	Domain
$f(x) = \sin x$	$\left[-\dfrac{\pi}{2}, \dfrac{\pi}{2}\right]$	$f^{-1}(x) = \sin^{-1} x$	$[-1, 1]$
$f(x) = \cos x$	$[0, \pi]$	$f^{-1}(x) = \cos^{-1} x$	$[-1, 1]$
$f(x) = \tan x$	$\left(-\dfrac{\pi}{2}, \dfrac{\pi}{2}\right)$	$f^{-1}(x) = \tan^{-1} x$	$(-\infty, \infty)$
$f(x) = \csc x$	$\left(-\pi, -\dfrac{\pi}{2}\right] \cup \left(0, \dfrac{\pi}{2}\right]$	$f^{-1}(x) = \csc^{-1} x$	$|x| \geq 1$
$f(x) = \sec x$	$\left[0, \dfrac{\pi}{2}\right) \cup \left[\pi, \dfrac{3\pi}{2}\right)$	$f^{-1}(x) = \sec^{-1} x$	$|x| \geq 1$
$f(x) = \cot x$	$(0, \pi)$	$f^{-1}(x) = \cot^{-1} x$	$(-\infty, \infty)$

To find the derivative of $y = \sin^{-1} x$, $-1 \leq x \leq 1$, $-\dfrac{\pi}{2} \leq y \leq \dfrac{\pi}{2}$, we write $\sin y = x$ and differentiate implicitly with respect to x.

$$\frac{d}{dx} \sin y = \frac{d}{dx} x$$

$$\cos y \cdot \frac{dy}{dx} = 1 \qquad \frac{d}{dx} \sin y = \cos y \frac{dy}{dx} \text{ using the Chain Rule.}$$

$$\frac{dy}{dx} = \frac{1}{\cos y}$$

EXAMPLE 4 Finding Derivatives Using Logarithmic Differentiation

Find y' if $y = \dfrac{x^2}{(3x - 2)^3}$.

Solution

It is easier to find y' if we take the natural logarithm of each side before differentiating. That is, we write

$$\ln y = \ln\left[\frac{x^2}{(3x - 2)^3}\right]$$

and simplify using properties of logarithms.

$$\ln y = \ln x^2 - \ln(3x - 2)^3$$

$$= 2 \ln x - 3 \ln(3x - 2)$$

To find y', use implicit differentiation.

$$\frac{d}{dx} \ln y = \frac{d}{dx}[2 \ln x - 3 \ln(3x - 2)]$$

$$\frac{y'}{y} = \frac{d}{dx}(2 \ln x) - \frac{d}{dx}[3 \ln(3x - 2)] \qquad \frac{d}{dx} \ln y = \frac{1}{y} \frac{dy}{dx} = \frac{y'}{y}$$

$$\frac{y'}{y} = \frac{2}{x} - \frac{9}{3x - 2}$$

$$y' = y\left[\frac{2}{x} - \frac{9}{3x - 2}\right] = \left[\frac{x^2}{(3x - 2)^3}\right]\left[\frac{2}{x} - \frac{9}{3x - 2}\right] \qquad ∎$$

Practice, Practice, Practice: Hone your skills with exercises at every turn.

Start with the **Concepts and Vocabulary** to check your comprehension of the section's main points.

Skill Building problems help you to develop computation skills and the ability to select the best approach to solve a problem.

Once you've mastered computational skills, tackle the **Applications and Extensions** problems, which are applied or extend the concepts of the section…

…and the **Challenge** problems—more difficult, thought-provoking extensions of the section material — often combine concepts learned in previous chapters.

Throughout the section-level problem sets, a **Group Work** icon 🔗 identifies problems selected by the authors as potential candidates for group projects or assignments.

2.3 Assess Your Understanding

Concepts and Vocabulary

[194] 1. $\dfrac{d}{dx}\pi^2 = $ _____ ; $\dfrac{d}{dx}x^3 = $ _____ .

2. When n is a positive integer, the Simple Power Rule states that $\dfrac{d}{dx}x^n = $ _____ .

3. **True or False** The derivative of a power function of degree greater than 1 is also a power function.

4. If k is a constant and f is a differentiable function, then $\dfrac{d}{dx}[kf(x)] = $ _____ .

5. The derivative of $f(x) = e^x$ is _____ .

6. **True or False** The derivative of an exponential function $f(x) = a^x$, where $a > 0$ and $a \neq 1$, is always a constant multiple of a^x.

Skill Building

In Problems 7–26, find the derivative of each function using the formulas of this section. (a, b, c, and d, when they appear, are constants.)

[195] 7. $f(x) = 3x + \sqrt{2}$ 8. $f(x) = 5x - \pi$

9. $f(x) = x^2 + 3x + 4$ 10. $f(x) = 4x^4 + 2x^2 - 2$

11. $f(u) = 8u^5 - 5u + 1$ 12. $f(u) = 9u^3 - 2u^2 + 4u + 4$

13. $f(s) = as^3 + \dfrac{3}{2}s^2$ 14. $f(s) = 4 - \pi s^2$

15. $f(t) = \dfrac{1}{6}(t^6 - 5t)$ 16. $f(x) = \dfrac{1}{8}(x^8 - 5x^2 + 2)$

Applications and Extensions

39. **Slope of a Tangent Line**   An equation of the tangent graph of a function f at $(2, 8)$ is $y = -5x + 12$. What i

40. **Slope of a Tangent Line**   An equation of the tangent function f at $(3, 4)$ is $y = \dfrac{4}{3}x + 1$. What is $f'(3)$?

41. **Tangent Line**   Does the tangent line to the graph of y at $(1, 1)$ pass through the point $(2, 5)$?

42. **Tangent Line**   Does the tangent line to the graph of y at $(1, 1)$ pass through the point $(2, 5)$?

[170] 43. **Respiration Rate**   A human being's re (in breaths per minute) is given by $R =$ where p is the partial pressure of carbo Find the rate of change in respiration wh

44. **Instantaneous Rate of Change**   The volume V of the right circular cylinder height 5 m and radius r m shown in the figure is $V = V(r) = 5\pi r^2$. Find the instantaneous rate of change of the volume with respect to the radius when $r = 3$ m.

Want to master calculus concepts? Assess Your Understanding exercises appear at the end of each section. Selected answers appear in the back of the book.

19. $f(x) = \dfrac{x^3 + 2x + 1}{7}$ 20. $f(x) = \dfrac{1}{a}(ax^2 + bx + c)$

21. $f(x) = 4e^x$

[199] 2  $f(x) = x - \ln x$

[198] 25. $f(u) = 5\ln u - 2e^u$

In Problems 27–32, find e

27. $\dfrac{d}{dt}\left(\sqrt{3}\,t + \dfrac{1}{2}\right)$

29. $\dfrac{dA}{dR}$ if $A(R) = \pi R^2$

[195] 31. $\dfrac{dV}{dr}$ if $V = \dfrac{4}{3}\pi r^3$

In Problems 33–36:
(a) Find the slope of the tangent line to the graph of each function f at the indicated point.
(b) Find an equation of the tangent
(c) Find an equation of the normal
(d) Graph f and the tangent line an in (b) and (c) on the same set of

[197] 33. $f(x) = x^3 + 3x - 1$ at $(0, -1)$

35. $f(x) = e^x + 5x$ at $(0, 1)$

Easy cross-references back to text.
In the practice problems, look for the **red icon** with page number that directs you back to the corresponding worked example in the chapter text.

Icons identify problems that require a **graphing calculator** 🔗 or **computer algebra system** [CAS].

In Problems 49 and 50, for each function f:
(a) Find $f'(x)$ by expanding $f(x)$ and differentiating the polynomial.
[CAS] (b) Find $f'(x)$ using a CAS.
(c) Show that the results found in parts (a) and (b) are equivalent.

49. $f(x) = (2x - 1)^3$ 50. $f(x) = (x^2 + x)^4$

Challenge Problem

60. Another way of finding the derivative of $y = \sqrt[n]{x}$ is to use inverse functions. The function $y = f(x) = x^n$, n a positive integer, has the derivative $f'(x) = nx^{n-1}$. So, if $x \neq 0$, then $f'(x) \neq 0$. The inverse function of f, namely, $x = g(y) = \sqrt[n]{y}$, is defined for all y if n is odd and for all $y \geq 0$ if n is even. Since this inverse function is differentiable for all $y \neq 0$, we have

$$g'(y) = \dfrac{d}{dy}\sqrt[n]{y} = \dfrac{1}{f'(x)} = \dfrac{1}{nx^{n-1}}$$

Since $nx^{n-1} = n\left(\sqrt[n]{y}\right)^{n-1} = ny^{(n-1)/n} = ny^{1-(1/n)}$, we have

$$\dfrac{d}{dy}\sqrt[n]{y} = \dfrac{d}{dy}y^{1/n} = \dfrac{1}{ny^{1-(1/n)}} = \dfrac{1}{n}y^{(1/n)-1}$$

Use the result from above and the Chain Rule to prove the formula

---

106. **Student Approval**   Professor Miller's student approval rating is modeled by the function $Q(t) = 21 + \dfrac{10\sin\frac{2\pi t}{7}}{\sqrt{t} - \sqrt{20}}$, where $0 \leq t \leq 16$ is the number of weeks since the semester began.

(a) Find $Q'(t)$.
(b) Evaluate $Q'(1)$, $Q'(5)$, and $Q'(10)$.
(c) Interpret the results obtained in (b).
🔗 (d) Use technology to graph $Q(t)$ and $Q'(t)$.
(e) How would you explain the results in (d) to Professor Miller?

Source: Mathematics students at Millikin University, Decatur, Illinois.

Review, Review, Review: Gauge your progress.

Summaries organized for ease of use.
The **Chapter Review** briefly recaps the main ideas and key concepts of the chapter.

Things to Know
contains a detailed list of definitions, formulas, and theorems with page references so they can be found easily in the chapter.

Chapter Review

THINGS TO KNOW

2.1 Rates of Change and the Derivative

- **Definition** Derivative of a function f at a number c

$$\textbf{Form (1)} \quad f'(c) = \lim_{x \to c} \frac{f(x) - f(c)}{x - c}$$

provided the limit exists. (p. 173)

Three Interpretations of the Derivative

- *Geometric* If $y = f(x)$, the derivative $f'(c)$ is the slope of the tangent line to the graph of f at the point $(c, f(c))$. (p. 173)

- *Rate of change of a function* If $y = f(x)$, the derivative $f'(c)$ is the rate of change of f with respect to x at c. (p. 173)
- *Physical* If the signed distance s from the origin at time t of an object moving on a line is given by the position function $s = f(t)$, the derivative $f'(t_0)$ is the velocity of the object at time t_0. (p. 173)

2.2 The Derivative as a Function
- **Definition of a derivative function**

$$\textbf{Form (2)} \quad f'(x) = \lim_{h \to 0} \frac{f(x + h) - f(x)}{h}$$

provided the limit exists. (p. 179)

The Objectives table displays section-by-section lists of the objectives and the worked examples of the chapter with page references. It also includes references to **Review Exercises** and **AP® Review Problems** that pertain to each objective.

OBJECTIVES

Preparing for the **AP® Exam** — AP® Review Problems

Section	You should be able to ...	Examples	Review Exercises	AP® Review Problems
2.1	1 Find equations for the tangent line and the normal line to the graph of a function (p. 168)	1	67–70	7, 10
	2 Find the rate of change of a function (p. 169)	2, 3	1, 2, 73 (a)	6
	3 Find average velocity and instantaneous velocity (p. 170)	4, 5	71(a), (b); 72(a), (b)	12
	4 Find the derivative of a function at a number (p. 173)	6–9	3–8, 75	5, 11
2.2	1 Define the derivative function (p. 179)	1–3	9–12	2, 13
	2 Graph the derivative function (p. 181)	4, 5	9–12, 15–18	
	3 Identify where a function is not differentiable (p. 182)	6–8	13, 14, 75	4
	4 Explain the relationship between differentiability and continuity (p. 184)	9, 10	13, 14, 75	4
2.3	1 Differentiate a constant function (p. 192)	1		
	2 Differentiate a power function; the simple power rule (p. 192)	2, 3	19–22	
	3 Differentiate the sum and the difference of two functions (p. 195)	4–6	23–26, 33, 34, 40, 51, 52, 67	6, 8, 12
	4 Differentiate the exponential function $y = e^x$ and the natural logarithm function $y = \ln x$ (p. 197)	7, 8	44, 45, 53, 54, 56, 59, 69	6, 7, 9

Review Exercises offer an opportunity to return to the key concepts of the chapter for each objective.

REVIEW EXERCISES

In Problems 1 and 2, use a definition of the derivative to find the rate of change of f at the indicated numbers.

1. $f(x) = \sqrt{x}$ at **(a)** $c = 1$ **(b)** $c = 4$
(c) c any positive real number

2. $f(x) = \dfrac{2}{x - 1}$ at **(a)** $c = 0$ **(b)** $c = 2$
(c) c any real number, $c \neq 1$

In Problems 3–8, use a definition of the derivative to find the derivative of each function at the given number.

3. $F(x) = 3x + 6$ at -2
4. $f(x) = 8x^2 + 1$ at -1
5. $f(x) = 3x^2 + 5x$ at 0
6. $f(x) = \dfrac{3}{x}$ at 1
7. $f(x) = \sqrt{4x + 1}$ at 0
8. $f(x) = \dfrac{x + 1}{2x - 3}$ at 1

In Problems 9–12, use a definition of the derivative to find the derivative of each function. Graph f and f' on the same set of axes.

9. $f(x) = x - 6$
10. $f(x) = 7 - 3x^2$
11. $f(x) = \dfrac{1}{2x^3}$
12. $f(x) = \pi$

17. Use the information in the graph of $y = f(x)$ to sketch the graph of $y = f'(x)$.

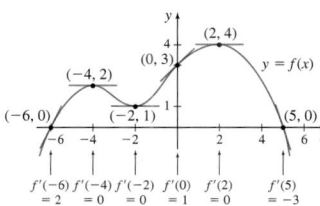

$$\begin{array}{ccccccc} f'(-6) & f'(-4) & f'(-2) & f'(0) & f'(2) & & f'(5) \\ = 2 & = 0 & = 0 & = 1 & = 0 & & = -3 \end{array}$$

18. Match the graph of $y = f(x)$ with the graph of its derivative.

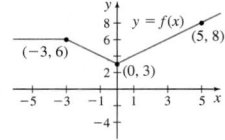

XXV

Sharpen the strategies needed to tackle AP® problems.

$$\frac{\Delta y}{\Delta x} = \frac{f(x) - f(c)}{x - c} \qquad x \neq c$$

where Δy is the change in y and Δx is the change in x.

The derivative of f at c, $f'(c)$, if it exists, equals the limit of the average rate of change of f as x approaches c. That is,

$$f'(c) = \lim_{x \to c} \frac{f(x) - f(c)}{x - c} \qquad x \neq c$$

An average rate of change measures the change in f from c to x; the derivative measures change at c.

NEED TO REVIEW? The derivative and interpretations of the derivative are discussed in Section 2.1, pp. 168–173.

1 Interpret a Derivative as the Slope of a Tangent Line to the Graph of a Function

AP® EXAM TIP

BREAK IT DOWN: As you work through the problems in this chapter, remember to follow these steps:

Step 1 Identify the underlying structure and related concepts.

Step 2 Determine the appropriate math rule or procedure.

Step 3 Apply the math rule or procedure.

Step 4 Clearly communicate your answer.

On page 329, see how we've used these steps to solve Section 4.2 AP® Practice Problem 9(a) on page 300.

EXAMPLE 1 Interpreting the Derivative as the Slope of a Tangent Line

(a) Find the derivative of $f(x) = x \cos x + x - 1$.
(b) Find the slope of the tangent line to the graph of f at the point $(0, f(0))$.
(c) Find an equation of the tangent line to the graph of f at the point $(0, f(0))$.

Solution

(a) Use the Product Rule.

$$f(x) = x \cos x + x - 1$$
$$f'(x) = -x \sin x + \cos x + 1$$

Stepped-out approach.

The authors have constructed a simple step-by-step framework that you can use when solving the AP® Practice Problems. This AP® Exam Tip also connects you to the BREAK IT DOWN feature at the end of the chapter, where the authors have applied these steps to one of the section-level problems.

Free-Response Question

PAGE 295

9. A roofer's 13-meter ladder is placed against the wall of a building with its base on level ground. The top of the ladder slips down the wall as the bottom of the ladder slips away from the building at a constant rate of 5 m/s.

   (a) At what rate is the top of the ladder moving when it is 5 m from the ground?

   (b) At what rate is the area of the triangle formed by the ladder, the wall, and the ground changing when the top of the ladder is 5 m from the ground?

   (c) If θ is the angle formed by the ladder and the ground, what is the rate of change in θ when the top of the ladder is 5 m from the ground?

See the BREAK IT DOWN on page 329 for a stepped out solution to AP® Practice Problem 9(a).

Break It Down

Chapter 4 • R

Preparing f

Let's take a closer look at **AP® Practice Problem 9 part (a)** from Section 4.2 on page 300.

9. A roofer's 13-meter ladder is placed against the wall of a building with its base on level ground. The top of the ladder slips down the wall as the bottom of the ladder slips away from the building at a constant rate of 5 m/s.

(a) At what rate is the top of the ladder moving when it is 5 m from the ground?

Step 1	Identify the underlying structure and related concepts.	There are rates changing with respect to time: the rate the top down the wall, and the rate the bottom of the ladder moves aw is a **related rates** question.
Step 2	Determine the appropriate math rule or procedure.	When possible, the first step in solving a related rates problem picture (see the figure in Step 3). The ladder is 13 m long. Nex variables needed for the given situation. Let h denote the heig ladder from the ground and let x denote the distance the botto

the wall. As the ladder slides down the wall, $\frac{dh}{dt}$ equals the rat
is changing and $\frac{dx}{dt}$ equals the rate at which the distance from
We are told that $\frac{dx}{dt} = 5$ m/s and are asked to find the rate of change of the
height $\left(\frac{dh}{dt}\right)$ when $h = 5$ m.

Step 3	Apply the math rule or procedure.	Together, the ladder, the height h, and the base x form a right triangle. Using the Pythagorean Theorem, we have $h^2 + x^2 = 13^2$. When $h = 5$ m, we find $x^2 = 169 - 25 = 144$, so $x = 12$ m.

The distance x and the height h are each changing over time t, so we will take the derivative of the above expression with respect to t to obtain an equation involving the derivatives of h and x.

$$\frac{d}{dt}(h^2 + x^2) = \frac{d}{dt}(13^2)$$

$$2h \frac{dh}{dt} + 2x \frac{dx}{dt} = 0$$

When $h = 5$ m, $x = 12$ m and $dx/dt = 5$ m/s. After dividing out the 2's, we have

$$5 \frac{dh}{dt} + (12)(5) = 0$$

$$\frac{dh}{dt} = -12$$

Step 4	Clearly communicate your answer.	(a) When $x = 5$ m, the height of the ladder is *decreasing* at a rate of -12 m/s.

Methodical problem solving.

In the BREAK IT DOWN feature at the end of each chapter, you can follow a detailed solution to one of the section-level AP® Practice Problems using the steps outlined in the chapter opening AP® Exam Tip.

AP® Exam prep every step of the way.

Immediate reinforcement of skills.
Each section covering content that may appear on the exam includes a comprehensive selection of **AP® Practice Problems**. Practicing all year will pay off when you are more confident taking the exam.

Continual practice.
At the end of each section, you are continually prompted to reinforce prior knowledge through the **Retain Your Knowledge AP® Practice Problems**, which present three multiple-choice questions and one free-response question drawing from concepts covered in prior chapters.

Become a savvy test-taker.
Every chapter ends with **AP® Review Problems** that include multiple-choice problems for exam prep all year as well as open-ended questions to build your skills at answering free-response questions.

More test prep, more comprehensive practice.
Starting with Chapter 2, every chapter ends with an **AP® Cumulative Review Practice Problem** set. These sets reinforce concepts across the course. They prepare you for the AP® exam by keeping the most important concepts and techniques fresh in your mind.

AP® Practice Problems

Multiple-Choice Questions

1. The function $f(x) = \begin{cases} x^2 - ax & \text{if } x \le 1 \\ ax + b & \text{if } x > 1 \end{cases}$, where a and b are constants. If f is differentiable at $x = 1$, then $a + b =$

   (A) -3 (B) -2 (C) 0 (D) 2

2. The graph of the function f, given below, consists of three line segments. Find $\lim\limits_{h \to 0} \dfrac{f(3+h) - f(3)}{h}$.

   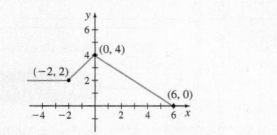

   (A) -1 (B) $-\dfrac{2}{3}$ (C) $-\dfrac{3}{2}$ (D) The limit does not exist.

5. If $f(x) = |x|$, which of the following statements about f are true?

   I. f is continuous at 0.
   II. f is differentiable at 0.
   III. $f(0) = 0$.

   (A) I only (B) III only
   (C) I and III only (D) I, II, and III

6. The graph of the function f shown in the figure has horizontal tangent lines at the points $(0, 1)$ and $(2, -1)$ and a vertical tangent line at the point $(1, 0)$. For what numbers x in the open interval $(-2, 3)$ is f not differentiable?

   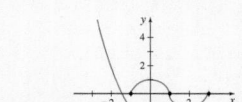

Retain Your Knowledge

Multiple-Choice Questions

1. The function $f(x) = e^x + x$ has an inverse function g. Then, $g'(1) =$

   (A) $\dfrac{1}{2}$ (B) 2 (C) $\dfrac{1}{e+1}$ (D) $\dfrac{1}{e}$

2. Find y' for $e^x \sin y - x^2 \cos y = x$.

   (A) $y' = \dfrac{1 + 2x \cos y - e^x \sin y}{x^2 \sin y - e^x \cos y}$

   (B) $y' = \dfrac{1 - 2 \cos y - e^x \sin y}{x^2 \sin y + e^x \cos y}$

   (C) $y' = 1 + \dfrac{2x \cos y - e^x \sin y}{x^2 \sin y + e^x \cos y}$

   (D) $y' = \dfrac{1 + 2x \cos y - e^x \sin y}{x^2 \sin y + e^x \cos y}$

3. If $f(x) = x^4 \ln(2x^2 + 3)$, then $f'(1)$ equals

   (A) $\dfrac{4}{\ln 5} + 4 \ln 5$ (B) $4 + 4 \ln 5$

   (C) $\dfrac{4}{5} + 4 \ln 5$ (D) $\dfrac{1}{5} + 4 \ln 5$

Free-Response Question

4. The resistance R, in ohms, of a wire of radius x cm is given by the formula $R(x) = \dfrac{0.0048}{x^2}$. The radius x is given by $x = 0.991 + 10^{-5}T$, where T is the temperature in Kelvin. Determine how R is changing with respect to T when $T = 320$ K.

AP® Review Problems: Chapter 2

Multiple-Choice Questions

1. If $f(x) = \sec x$, then $f'\left(\dfrac{\pi}{4}\right) =$

   (A) $\dfrac{\sqrt{2}}{2}$ (B) 2 (C) 1 (D) $\sqrt{2}$

2. If a function f is differentiable at c, then $f'(c)$ is given by

   I. $\lim\limits_{x \to c} \dfrac{f(x) - f(c)}{x - c}$
   II. $\lim\limits_{x \to c} \dfrac{f(x+h) - f(x)}{h}$
   III. $\lim\limits_{h \to 0} \dfrac{f(c+h) - f(c)}{h}$

   (A) I only (B) III only
   (C) I and II only (D) I and III only

3. If $y = \dfrac{3}{}$, then $\dfrac{dy}{} =$

4. The graph of the function f is shown below. Which statement about the function is true?

   (A) f is differentiable everywhere.
   (B) $0 \le f'(x) \le 1$, for all real numbers.
   (C) f is continuous everywhere.
   (D) f is an even function.

5. The table displays select values of a differentiable function f. What is an approximate value of $f'(2)$?

x	1.996	1.998	2.002	2.004
$f(x)$	3.168	3.181	3.207	3.220

   (A) 6.5 (B) 0.154 (C) 0.013 (D) 1.5

AP® Cumulative Review Problems: Chapters 1–2

Multiple-Choice Questions

1. $\lim\limits_{x \to 4} \dfrac{x - 4}{4 - x} =$

   (A) -4 (B) -1 (C) 0 (D) The limit does not exist.

2. $\lim\limits_{x \to 0} \dfrac{3x + \sin x}{2x} =$

   (A) 0 (B) 1 (C) 2 (D) The limit does not exist.

3. Let h be defined by
   $h(x) = \begin{cases} f(x) \cdot g(x) & \text{if } x \le 1 \\ k + x & \text{if } x > 1 \end{cases}$
   where f and g are both continuous at all real numbers. If $\lim\limits_{x \to 1} f(x) = 2$ and $\lim\limits_{x \to 1} g(x) = -2$, then for what number k is h continuous?

   (A) -5 (B) -4 (C) -2 (D) 2

5. Suppose the function f is continuous at all real numbers and $f(-2) = 1$ and $f(5) = -3$. Suppose the function g is also continuous at all real numbers and $g(x) = f^{-1}(x)$ for all x. The Intermediate Value Theorem guarantees that

   (A) $g(c) = 2$ for at least one c between -3 and 1.
   (B) $g(c) = 0$ for at least one c between -2 and 5.
   (C) $f(c) = 0$ for at least one c between -3 and 1.
   (D) $f(c) = 2$ for at least one c between -2 and 5.

6. The line $x = c$ is a vertical asymptote to the graph of the function f. Which of the following statements cannot be true?

   (A) $\lim\limits_{x \to c} f(x) = \infty$ (B) $\lim\limits_{x \to \infty} f(x) = c$
   (C) $f(c)$ is not defined. (D) f is continuous at $x = c$.

Need more help?
Every **AP® Review Problem** is accompanied by a short video clip ▶ in ♨ Achieve that guides you through the solution. Let an experienced AP® Calculus teacher help you when you need it most.

Prepare and practice for the AP® Calculus exams.

AP® Practice Exam: Calculus AB

Preparing for the AP® Exam

Section 1: Multiple Choice, Part A
A calculator may not be used for Part A.

1. If $f(x) = e^{4x} + \sin(2x)$, then $f'(0) =$
   (A) 1   (B) 2   (C) 4   (D) 6

2. If $\lim_{x \to 1} f(x) = 3$, then $\lim_{x \to 1} \frac{(x^2 - 1)f(x)}{x - 1} =$
   (A) 0   (B) 4   (C) 6   (D) Does not exist.

3. The graph of the function f is shown below. Which of the following statements is false?

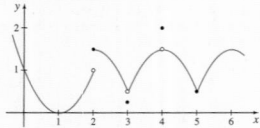

   (A) $\lim_{x \to 2^-} f(x) = f(2)$
   (B) The function f is discontinuous at $x = 3$.
   (C) $\lim_{x \to 4} f(x)$ exists.
   (D) The function f is continuous at $x = 5$.

4. Let f be the function $f(x) = x^3 - 15x^2 - 1800x + 2000$. On which interval is the function f both decreasing and concave down?
   (A) $x < -20$ (B) $-20 < x < 5$
   (C) $5 < x < 30$ (D) $x > 30$

5. If $y = e^{2x}\cos(3x)$, then $\frac{dy}{dx} =$
   (A) $-6e^{2x}\sin(3x)$
   (B) $e^{2x}(2\cos(3x) - 3\sin(3x))$
   (C) $e^{2x}(2\cos(3x) + 3\sin(3x))$
   (D) $e^{2x}(\cos(3x) - \sin(3x))$

6. Let f be the function defined below. For what value of k is f continuous at $x = 0$?
$$f(x) = \begin{cases} \dfrac{\sin(7x)}{2x} & \text{for } x < 0 \\ k + 2\ln(x + e^{x+1}) & \text{for } x \geq 0 \end{cases}$$
   (A) $-\frac{3}{2}$ (B) -1 (C) $\frac{3}{2}$ (D) $\frac{7}{2}$

8. The function f is differentiable and its derivative is continuous on the interval $(-4, 5)$. The table below lists several values of f and f' in the interval.

x	-3	-1	0	3
$f(x)$	-16	$\frac{2}{3}$	2	2
$f'(x)$	3	-1	0	15

Then $\int_{-1}^{3} f'(x)\, dx =$
   (A) -16 (B) $\frac{4}{3}$

9. If $y = \sqrt{4x + 6e^{\tan x}}$,
   (A) $\sqrt{4 + 6e^{\tan x}\sec^2}$
   (C) $\frac{2 + 3e^{\tan x}}{\sqrt{4x + 6e^{\tan x}}}$

10. The function f is conti differentiable on $1 < x$ then the Mean Value T
   (A) f is linear on the
   (B) $f'(c) = 10$ for at
   (C) $f'(c) = 0$ for at le
   (D) $f(c) = 30$ for at

11. The Riemann sum $\frac{1}{20}$ approximation for whi
   (A) $\int_0^2 e^{x/20}\, dx$
   (C) $\frac{1}{20}\int_0^2 e^x\, dx$

12. What is the area of the the graph of $y = 1 + e$
   (A) $2 - 2e^{-6}$ (B)
   (C) $\frac{7}{2} - \frac{1}{2}e^{-6}$ (D)

Section 2: Free Response, Part A
A graphing calculator is required for Part A.

1. The figure below shows the graphs of
$r_1(\theta) = 2 + 3\sin\theta + 4\cos\theta$ and $r_2(\theta) = 2 + 6\cos\theta$
on the interval $0 \leq \theta \leq \frac{\pi}{2}$. Let S be the shaded region bounded by the two graphs, the x-axis, and the y-axis. The two curves intersect at point P.

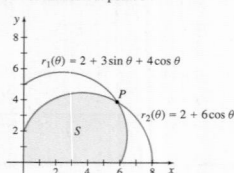

$r_1(\theta) = 2 + 3\sin\theta + 4\cos\theta$
$r_2(\theta) = 2 + 6\cos\theta$

   (a) Write, but do not evaluate, an expression involving one or more integrals that gives the area of S.
   (b) Find the angle θ in the interval $0 \leq \theta \leq \frac{\pi}{2}$ that corresponds to the point on the curve
$r_1(\theta) = 2 + 3\sin\theta + 4\cos\theta$
that is furthest from the origin. Justify your answer.
   (c) The radial distance between the two curves changes for $0 \leq \theta \leq \frac{\pi}{2}$. Suppose θ changes at a constant rate of

Section 2: Free Response Part B
No calculator is allowed for Part B.

3. Water is dripping from a faucet into a 2-gallon bucket that has height h 24 in. The height of the water in the bucket at time t, $0 \leq t \leq 60$, can be modeled by a differentiable function h, where h is measured in inches and t in minutes. The table below shows the height of the water at select times t.

t (minutes)	0	10	20	30	40	50	60
h (inches)	0	9.4	15.2	18.6	20.8	22.0	22.8

   (a) Using the data above, approximate $h'(25)$. Show the computations and explain the result, including the units.
   (b) Approximate $\int_0^{60} h(t)\, dt$ with a trapezoidal sum with three intervals of equal length. Use the result to calculate $\frac{1}{60}\int_0^{60} h(t)\, dt$, and interpret the meaning of the result. Be sure to use correct units.
   (c) If the function that models h is $h(t) = 24 - 24e^{-0.05t}$, determine the concavity of h. Does the trapezoidal sum found in (b) overestimate or underestimate the height of the water in the bucket?
   (d) Is there a time t between $t = 40$ and $t = 60$ minutes at which $h'(t) = 0.10$? Justify your answer.

4. Let f be a function defined on the closed interval $0 \leq x \leq 16$ with $f(0) = -3$. The graph of f', the derivative of f, consists of 4 line segments as shown below. Also shown is the graph of $y = \frac{4}{x}$ which intersects the graph of f' at points P and Q.

AP® Practice Exam: Calculus BC

Preparing for the AP® Exam

Section 1: Multiple Choice, Part A
A calculator may not be used for Part A.

1. $\int \frac{3x^2 - 7x - 4}{x - 3}\, dx =$
   (A) $\frac{3}{2}x^2 + 2x + 2\ln|x - 3| + C$
   (B) $\frac{3}{2}x^2 + 2x - 10\ln|x - 3| + C$
   (C) $\frac{1}{2}(3x + 2)^2 + 2\ln|x - 3| + C$
   (D) $3x + 2 + 2\ln|x - 3| + C$

2. $\int_0^{x^2} e^{3t}\, dt =$
   (A) $\frac{1}{3}(e^{3x^2} - 1)$ (B) $3(e^{3x^2} - 1)$
   (C) $2xe^{3x^2}$ (D) $\frac{1}{3}(2x)(e^{3x^2} - 1)$

3. What is the slope of the tangent line to the graph of $y = e^{6x} - \sin(4x)$ at $x = 0$?
   (A) 0   (B) 1   (C) 2   (D) 10

4. $\lim_{x \to 0} \frac{2e^{5x} - 5e^{2x} + 3}{x^2 - 2\cos(2x) + 2} =$
   (A) 0   (B) $\frac{3}{4}$ (C) $\frac{3}{2}$ (D) 3

5. Which of the following series converges?
   I. $\sum_{k=1}^{\infty} \frac{1}{k^{2/3}}$
   II. $\sum_{k=1}^{\infty} \frac{k}{k^3 + 5k + 1}$

7. Find y' at the point $(3, 6)$ on the graph of $y^2 = 2x^2 + xy$.
   (A) $\frac{1}{6}$ (B) $\frac{6}{5}$ (C) $\frac{7}{4}$ (D) 2

8. The function f has a second derivative given by $f''(x) = x^2(x - 1)\sqrt{x + 1}$. At what values of x does f have a point of inflection?
   (A) 0 only   (B) 1 only
   (C) 0 and 1 only   (D) -1, 0, and 1

9. If the function f is continuous and if $F'(x) = f(x)$ for all real numbers x, then $\int_2^5 f(3x + 2)\, dx$ equals
   (A) $3F(5) - 3F(2)$ (B) $\frac{1}{3}F(5) - \frac{1}{3}F(2)$
   (C) $\frac{1}{3}F(17) - \frac{1}{3}F(8)$ (D) $3F(17) - 3F(8)$

10. What is the sum of the series $\sum_{k=1}^{\infty} \frac{3^{k+2}}{5^{k+1}}$?
   (A) $\frac{3}{2}$ (B) 2   (C) $\frac{5}{2}$ (D) $\frac{27}{10}$

11. The base of a solid S is the region enclosed by the graph of $y = \sqrt{x}$, the line $x = 3$, the line $x = 1$, and the x-axis. If the cross sections of S perpendicular to the x-axis are squares, then the volume of S is
   (A) 2   (B) $2\sqrt{3}$ (C) 4   (D) $\frac{4}{3}(3\sqrt{3} - 1)$

12. Let f be the function given by $f(x) = x^x$. If three subintervals of equal length are used, what is the value of the Left Riemann sum approximation for $\int_{1.2}^{1.8} x^x\, dx$?
   (A) $5.0(1.2^{1.2} + 1.4^{1.4} + 1.6^{1.6})$
   (B) $0.2(1.2^{1.2} + 1.5^{1.5} + 1.8^{1.8})$
   (C) $0.2(1.2^{1.2} + 1.4^{1.4} + 1.6^{1.6})$
   (D) $0.2(1.4^{1.4} + 1.6^{1.6} + 1.8^{1.8})$

13. The volume of a box with a square top and bottom is to be k cubic inches. If a minimum amount of cardboard is to be used to construct the box, what must be the area, in square inches, of the top of the box?
   (A) $\sqrt[3]{k}$ (B) $2\sqrt[3]{k}$ (C) $\sqrt[3]{k^2}$ (D) $4\sqrt[3]{k^2}$

14. Let $P(x) = a + b(x - 1) + c(x - 1)^2 + d(x - 1)^3$ be the third-degree Taylor polynomial for the function f about $x = 1$. A table of values for $f^{(k)}(1)$ is given below. Find $c + d$.

k	0	1	2	3
$f^{(k)}(1)$	4	4	-2	3

Practice, read, and learn online anywhere with 📖 Achie√e.

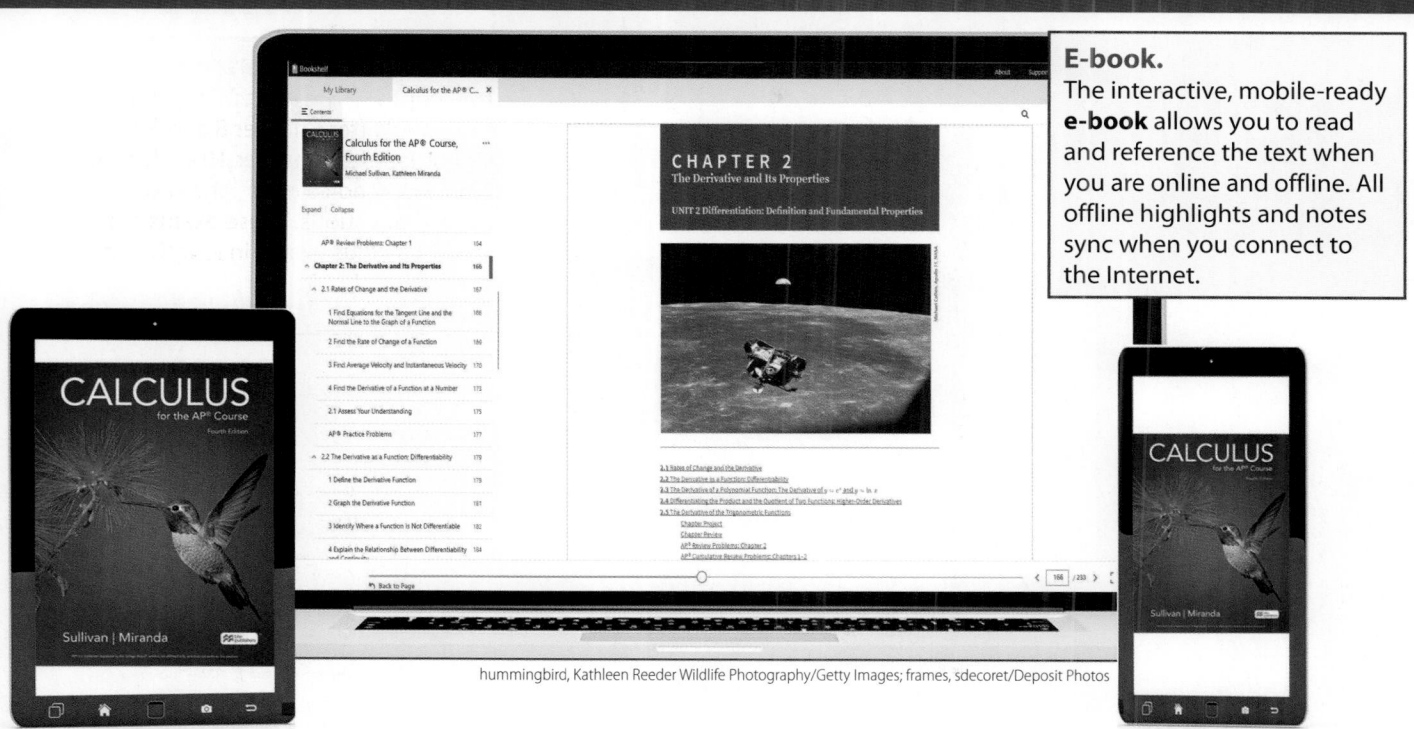

E-book.
The interactive, mobile-ready **e-book** allows you to read and reference the text when you are online and offline. All offline highlights and notes sync when you connect to the Internet.

hummingbird, Kathleen Reeder Wildlife Photography/Getty Images; frames, sdecoret/Deposit Photos

📖 Achie√e
bfw publishers

Homework system
Read, learn, and practice with 📖 Achie√e. Self-grading practice problems from the book with guided feedback based on common misconceptions help you learn as you practice. Students who use this system find that it significantly improves their understanding and helps them build essential foundations.

Get extra help when you need it by accessing CalcClips and AP® Review Problem videos in 📖 Achie√e.

CALCULUS

for the AP® Course

Preparing for Calculus

Museum of Science and Industry, Chicago/Getty Images

Until now, the mathematics you have encountered has centered mainly on algebra, geometry, and trigonometry. These subjects have a long history, well over 2000 years. But calculus is relatively new; it was developed less than 400 years ago.

Calculus deals with change and how the change in one quantity affects other quantities. Fundamental to these ideas are functions and their properties. For example, Foucault's Pendulum, pictured above and invented by 19th century physicist Leon Foucault, demonstrates the Earth's rotation—its angular velocity is a function of the sine of the pendulum's latitude over time. (Essentially, the plane of the pendulum remains at a fixed position while the universe moves!)

In Chapter P, we discuss many of the functions used in calculus. We also provide a review of techniques from precalculus used to obtain the graphs of functions and to transform known functions into new functions.

Your teacher may choose to cover all or part of this chapter. Regardless, throughout the text, you will see the **NEED TO REVIEW?** marginal notes. They reference specific topics, often discussed in Chapter P.

Raj
Program Manager

AP® Calculus taught me how to strategically think through a problem—to understand the question, apply appropriate strategies, and evaluate outcomes. As a textbook editor, I have to see the big picture and manage the moving parts to help create quality content solutions.

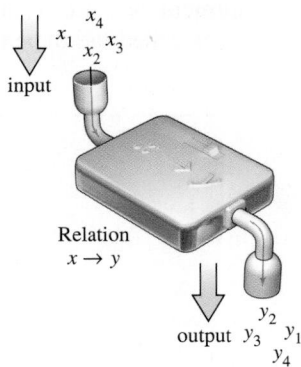

Figure 1

P.1 Functions and Their Graphs

OBJECTIVES *When you finish this section, you should be able to:*

1 **Evaluate a function (p. 4)**
2 **Find the difference quotient of a function (p. 5)**
3 **Find the domain of a function (p. 6)**
4 **Identify the graph of a function (p. 7)**
5 **Analyze a piecewise-defined function (p. 9)**
6 **Obtain information from or about the graph of a function (p. 9)**
7 **Use properties of functions (p. 11)**
8 **Find the average rate of change of a function (p. 13)**

Often there are situations where one variable is somehow linked to another variable. For example, the price of a gallon of gas is linked to the price of a barrel of oil. A person can be associated with her phone number(s). The volume V of a sphere depends on its radius R. These are examples of a **relation**, a correspondence between two sets called the **domain** and the **range**. If x is an element of the domain and y is an element of the range, and if a relation exists from x to y, then we say that y **corresponds to** x or that y **depends on** x, and we write $x \rightarrow y$. It is often helpful to think of x as the **input** and y as the **output** of the relation. See Figure 1.

Suppose an astronaut standing on the moon throws a rock 20 meters up and starts a stopwatch as the rock begins to fall back down. If x represents the number of seconds on the stopwatch and if y represents the altitude of the rock at that time, then there is a relation between time x and altitude y. If the altitude of the rock is measured at $x = 1, 2, 2.5, 3, 4,$ and 5 seconds, then the altitude is approximately $y = 19.2, 16.8, 15, 12.8, 7.2,$ and 0 meters, respectively. This is an example of a relation expressed **verbally**.

The astronaut could also express this relation **numerically**, **graphically**, or **algebraically**. The relation can be expressed by a table of numbers (see Table 1) or by the set of ordered pairs $\{(0, 20), (1, 19.2), (2, 16.8), (2.5, 15), (3, 12.8), (4, 7.2), (5, 0)\}$, where the first element of each pair denotes the time x and the second element denotes the altitude y. The relation also can be expressed visually, using either a graph, as in Figure 2, or a map, as in Figure 3. Finally, the relation can be expressed algebraically using the equation

$$y = 20 - 0.8x^2$$

TABLE 1	
Time, x (in seconds)	**Altitude, y** (in meters)
0	20
1	19.2
2	16.8
2.5	15
3	12.8
4	7.2
5	0

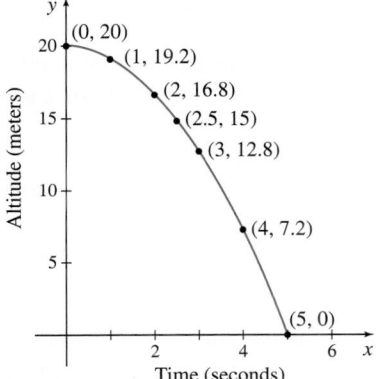

Figure 2

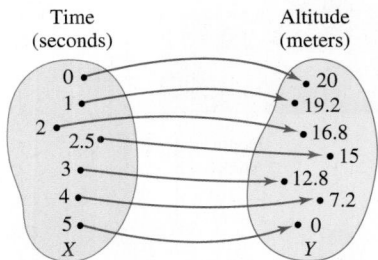

Figure 3

NOTE Not every relation is a function. If any element x in the set X corresponds to more than one element y in the set Y, then the relation is not a function.

In this example, notice that if X is the set of times from 0 to 5 seconds and Y is the set of altitudes from 0 to 20 meters, then each element of X corresponds to one and only one element of Y. Each given time value yields a **unique**, that is, exactly one, altitude value. Any relation with this property is called a *function from X into Y*.

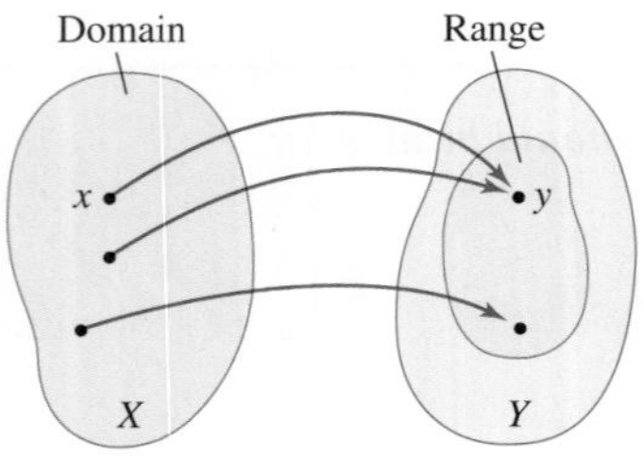

Domain   Range

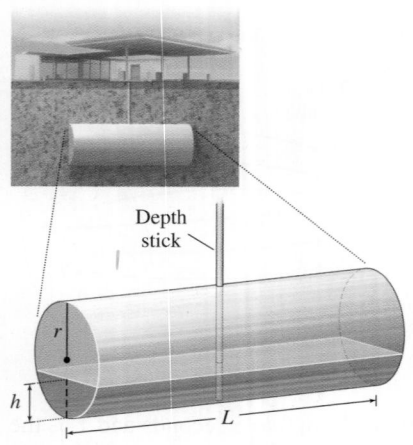

Figure 4

DEFINITION Function

Let X and Y be two nonempty sets.* A **function** f from X into Y is a relation that associates with each element of X exactly one element of Y.

The set X is called the **domain** of the function. For each element x in X, the corresponding element y in Y is called the **value** of the function at x, or the **image** of x. The set of all the images of the elements in the domain is called the **range** of the function. Since there may be elements in Y that are not images of any x in X, the range of a function is a subset of Y. See Figure 4.

1 Evaluate a Function

Functions are often denoted by letters such as f, F, g, and so on. If f is a function, then for each element x in the domain, the corresponding image in the range is denoted by the symbol $f(x)$, read "f of x." $f(x)$ is called the **value of f at x**. The variable x is called the **independent variable** or the **argument** because it can be assigned any element from the domain, while the variable y is called the **dependent variable** because its value depends on x.

EXAMPLE 1   **Evaluating a Function**

For the function f defined by $f(x) = 2x^2 - 3x$, find:

(a) $f(5)$ **(b)** $f(x+h)$ **(c)** $f(x+h) - f(x)$

Solution

(a) $f(5) = 2 \cdot 5^2 - 3 \cdot 5 = 50 - 15 = 35$

(b)   The function $f(x) = 2x^2 - 3x$ gives us a rule to follow. To find $f(x+h)$, expand $(x+h)^2$, multiply the result by 2, and then subtract the product of 3 and $(x+h)$.

$$f(x+h) = 2(x+h)^2 - 3(x+h) = 2(x^2 + 2hx + h^2) - 3x - 3h$$

In $f(x)$, replace x by $x+h$

$$= 2x^2 + 4hx + 2h^2 - 3x - 3h$$

(c) $f(x+h) - f(x) = [2x^2 + 4hx + 2h^2 - 3x - 3h] - [2x^2 - 3x] = 4hx + 2h^2 - 3h$ ∎

NOW WORK   Problem 13.

EXAMPLE 2   **Finding the Amount of Gasoline in a Tank**

A gas station stores its gasoline in an underground tank that is a right circular cylinder lying on its side. The volume V of gasoline in the tank (in gallons) is given by the formula

$$V(h) = 40h^2 \sqrt{\frac{96}{h} - 0.608}$$

where h is the height (in inches) of the gasoline as measured on a depth stick. See Figure 5.

(a) If $h = 12$ inches, how many gallons of gasoline are in the tank?

(b) If $h = 1$ inch, how many gallons of gasoline are in the tank?

Depth stick

r

h

L

Figure 5

*The sets X and Y will usually be sets of real numbers, defining a **real function**. The two sets could also be sets of complex numbers, defining a **complex function**, or X could be a set of real numbers and Y a set of vectors, defining a **vector-valued function**. In the broad definition, X and Y can be any two sets.

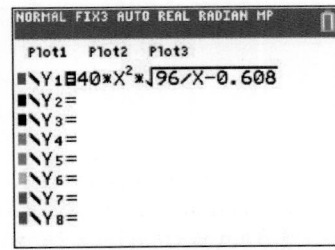

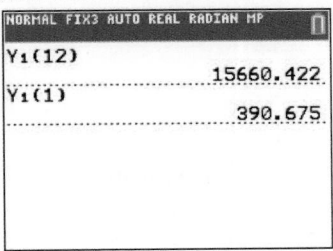

Figure 6

Solution

(a) We evaluate V when $h = 12$.

$$V(12) = 40 \cdot 12^2 \sqrt{\frac{96}{12} - 0.608} = 40 \cdot 144\sqrt{8 - 0.608} = 5760\sqrt{7.392} \approx 15{,}660$$

There are about 15,660 gallons of gasoline in the tank when the height of the gasoline in the tank is 12 inches.

(b) Evaluate V when $h = 1$.

$$V(1) = 40 \cdot 1^2 \sqrt{\frac{96}{1} - 0.608} = 40\sqrt{96 - 0.608} = 40\sqrt{95.392} \approx 391$$

There are about 391 gallons of gasoline in the tank when the height of the gasoline in the tank is 1 inch. ∎

Figure 6 shows the answers obtained with a graphing calculator.

Implicit Form of a Function

In general, a function f defined by an equation in x and y is said to be given **implicitly**. If it is possible to solve the equation for y in terms of x, then we write $y = f(x)$ and say the function is given **explicitly**. For example,

Implicit Form	**Explicit Form**
$x^2 - y = 6$	$y = f(x) = x^2 - 6$
$xy = 4$	$y = g(x) = \dfrac{4}{x}$

In calculus, we sometimes deal with functions that are defined implicitly but cannot be expressed in explicit form. For example, if the function with independent variable x and dependent variable y is defined by $\sin(xy) = xy - 2y + \cos x - \sin y$, there is no method to solve for y and express the function explicitly.

❷ Find the Difference Quotient of a Function

An important concept in calculus involves working with a certain quotient. For a given function $y = f(x)$, the inputs x and $x + h$, $h \neq 0$, result in the images $f(x)$ and $f(x + h)$. The quotient of their differences

$$\frac{f(x+h) - f(x)}{(x+h) - x} = \frac{f(x+h) - f(x)}{h}$$

with $h \neq 0$, is called the *difference quotient of f* at x.

DEFINITION Difference Quotient

The **difference quotient** of a function f at x is given by

$$\boxed{\dfrac{f(x+h) - f(x)}{h} \qquad h \neq 0}$$

The difference quotient is used in calculus to define the derivative, which is used in applications such as the velocity of an object and optimization of resources.

When finding a difference quotient, it is necessary to simplify the expression in order to divide out the h in the denominator, as illustrated in the next example.

EXAMPLE 3   Finding the Difference Quotient of a Function

Find the difference quotient of each function.

(a) $f(x) = 2x^2 - 3x$ (b) $f(x) = \sqrt{x}$

Solution

(a)

$$\frac{f(x+h) - f(x)}{h} \underset{\substack{\uparrow \\ f(x+h) = 2(x+h)^2 - 3(x+h)}}{=} \frac{[2(x+h)^2 - 3(x+h)] - [2x^2 - 3x]}{h}$$

$$= \frac{2(x^2 + 2xh + h^2) - 3x - 3h - 2x^2 + 3x}{h} \qquad \text{Simplify.}$$

$$= \frac{2x^2 + 4xh + 2h^2 - 3x - 3h - 2x^2 + 3x}{h} \qquad \text{Distribute.}$$

$$= \frac{4xh + 2h^2 - 3h}{h} \qquad \text{Combine like terms.}$$

$$= \frac{h(4x + 2h - 3)}{h} \qquad \text{Factor out } h.$$

$$= 4x + 2h - 3 \qquad \text{Divide out the factor } h.$$

(b)

$$\frac{f(x+h) - f(x)}{h} = \frac{\sqrt{x+h} - \sqrt{x}}{h} \qquad f(x+h) = \sqrt{x+h}$$

$$= \frac{\sqrt{x+h} - \sqrt{x}}{h} \cdot \frac{\sqrt{x+h} + \sqrt{x}}{\sqrt{x+h} + \sqrt{x}} \qquad \text{Rationalize the numerator.}$$

$$= \frac{(\sqrt{x+h})^2 - (\sqrt{x})^2}{h(\sqrt{x+h} + \sqrt{x})} \qquad (A-B)(A+B) = A^2 - B^2$$

$$= \frac{h}{h(\sqrt{x+h} + \sqrt{x})} \qquad \begin{array}{l}(\sqrt{x+h})^2 - (\sqrt{x})^2 \\ = x + h - x = h\end{array}$$

$$= \frac{1}{\sqrt{x+h} + \sqrt{x}} \qquad \text{Divide out the factor } h. \qquad ■$$

NOW WORK   Problem 23.

---

③ Find the Domain of a Function

In applications, the domain of a function is sometimes specified. For example, we might be interested in the population of the United States from 1900 to 2020. The domain of the function is time, in years, and is restricted to the interval [1900, 2020]. Other times the domain is restricted by the context of the function itself. For example, the volume V of a sphere, given by the function $V = \dfrac{4}{3}\pi R^3$, makes sense only if the radius R is greater than 0. But often the domain of a function f is not specified; only the formula defining the function is given. In such cases, the **domain** of f is the largest set of real numbers for which the value $f(x)$ is defined and is a real number.

EXAMPLE 4   Finding the Domain of a Function

Find the domain of each of the following functions:

(a) $f(x) = x^2 + 5x$ (b) $g(x) = \dfrac{3x}{x^2 - 4}$

(c) $h(t) = \sqrt{4 - 3t}$ (d) $F(u) = \dfrac{5u}{\sqrt{u^2 - 1}}$

Solution

(a) Since $f(x) = x^2 + 5x$ is defined for any real number x, the domain of f is the set of all real numbers.

(b) Since division by zero is not defined, $x^2 - 4$ cannot be 0, that is, $x \neq -2$ and $x \neq 2$. The function $g(x) = \dfrac{3x}{x^2 - 4}$ is defined for any real number except $x = -2$ and $x = 2$. So, the domain of g is the set of real numbers $\{x \mid x \neq -2, x \neq 2\}$.

(c) Since the square root of a negative number is not a real number, the value of $4 - 3t$ must be nonnegative. The solution of the inequality $4 - 3t \geq 0$ is $t \leq \dfrac{4}{3}$, so the domain of h is the set of real numbers $\left\{ t \mid t \leq \dfrac{4}{3} \right\}$ or the interval $\left(-\infty, \dfrac{4}{3} \right]$.

(d) Since the square root is in the denominator, the value of $u^2 - 1$ must be not only nonnegative, it also cannot equal zero. That is, $u^2 - 1 > 0$. The solution of the inequality $u^2 - 1 > 0$ is the set of real numbers $\{u \mid u < -1\} \cup \{u \mid u > 1\}$ or the set $(-\infty, -1) \cup (1, \infty)$. ∎

NEED TO REVIEW? Solving inequalities is discussed in Appendix A.1, pp. A-6 to A-8.

NEED TO REVIEW? Interval notation is discussed in Appendix A.1, pp. A-5 to A-6.

If x is in the domain of a function f, we say that f **is defined at** x, or $f(x)$ **exists**. If x is not in the domain of f, we say that f **is not defined at** x, or $f(x)$ **does not exist**. The domain of a function is expressed using inequalities, interval notation, set notation, or words, whichever is most convenient. Notice the various ways the domain was expressed in the solution to Example 4.

NOW WORK   Problem 17.

④ Identify the Graph of a Function

In applications, often a graph reveals the relationship between two variables more clearly than a table or an equation. For example, Table 2 shows the average price of gasoline at a particular gas station in the United States for the years 1995–2021. If we plot these data using year as the independent variable and price as the dependent variable, and then connect the points (year, price) we obtain Figure 7.

TABLE 2 Average retail price of gasoline

Year	Price	Year	Price	Year	Price
1995	1.16	2004	1.90	2013	3.58
1996	1.25	2005	2.31	2014	3.44
1997	1.24	2006	2.62	2015	2.52
1998	1.07	2007	2.84	2016	2.25
1999	1.17	2008	3.30	2017	2.53
2000	1.52	2009	2.41	2018	2.81
2001	1.46	2010	2.84	2019	2.60
2002	1.39	2011	3.58	2020	2.17
2003	1.60	2012	3.68	2021	3.01

Source: U.S. Energy Information Administration

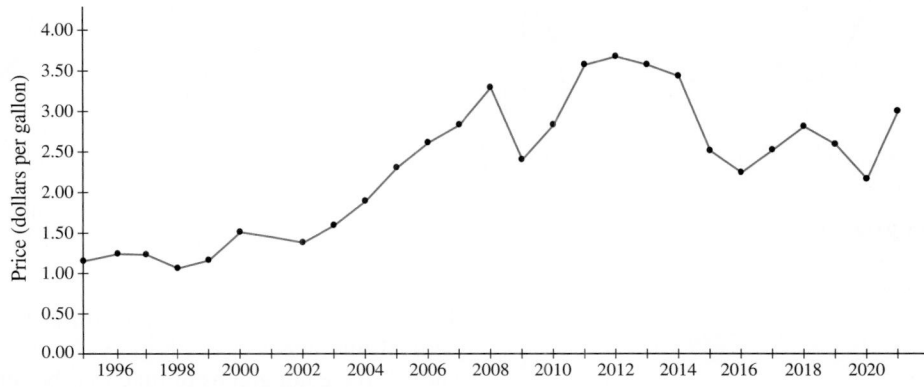

Source: U.S. Energy Information Administration

Figure 7 Graph of the average retail price of gasoline.

In Figure 7 (p. 7), the graph shows that for each date on the horizontal axis there is only one price on the vertical axis. So, the graph represents a function, although the rule for determining the price from the year is not given.

When a function is defined by an equation in x and y, the **graph of the function** is the set of points (x, y) in the xy-plane that satisfy the equation.

But not every collection of points in the xy-plane represents the graph of a function. Recall that a relation is a function only if each element x in the domain corresponds to exactly one image y in the range. This means the graph of a function never contains two points with the same x-coordinate and different y-coordinates. Compare the graphs in Figures 8 and 9. In Figure 8 every number x is associated with exactly one number y, but in Figure 9 some numbers x are associated with three numbers y. Figure 8 shows the graph of a function; Figure 9 shows a graph that is not the graph of a function.

NEED TO REVIEW? The graph of an equation is discussed in Appendix A.3, pp. A-17 to A-19.

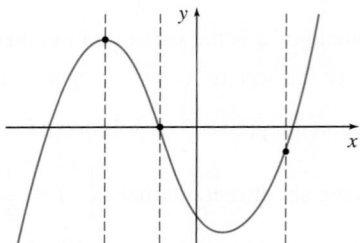

Figure 8 Function: Exactly one y for each x. Every vertical line intersects the graph in at most one point.

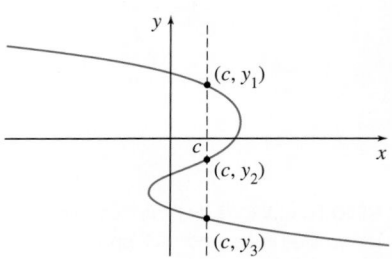

Figure 9 Not a function: $x = c$ has three ys associated with it. The vertical line $x = c$ intersects the graph in three points.

For a graph to be a *graph of a function*, it must satisfy the *Vertical-line Test*.

NOTE The phrase "if and only if" means the statements on each side of the phrase are equivalent. That is, they have the same meaning.

THEOREM Vertical-line Test

A set of points in the xy-plane is the graph of a function if and only if every vertical line intersects the graph in at most one point.

EXAMPLE 5   **Identifying the Graph of a Function**

Which graphs in Figure 10 represent the graph of a function?

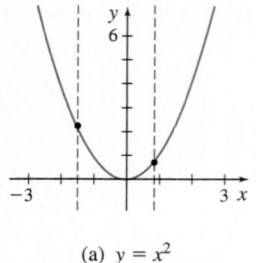

(a) $y = x^2$

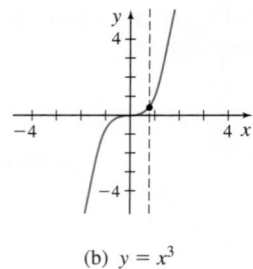

(b) $y = x^3$

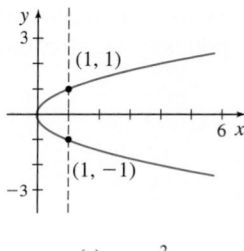

(c) $x = y^2$

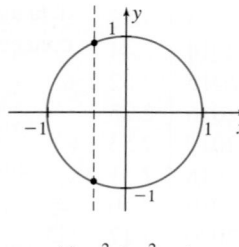

(d) $x^2 + y^2 = 1$

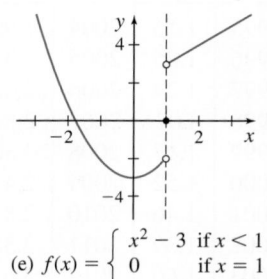

(e) $f(x) = \begin{cases} x^2 - 3 & \text{if } x < 1 \\ 0 & \text{if } x = 1 \\ x + 2 & \text{if } x > 1 \end{cases}$

Figure 10

Solution

The graphs in Figures 10(a), 10(b), and 10(e) are graphs of functions because every vertical line intersects each graph in at most one point. The graphs in Figures 10(c) and 10(d) are not graphs of functions because there is a vertical line that intersects each graph in more than one point. ∎

NOW WORK   Problems 31(a) and (b).

Notice that although the graph in Figure 10(e) represents a function, it looks different from the graphs in (a) and (b). The graph consists of two pieces plus a point and they are not connected. Also notice that different equations describe different pieces of the graph. Functions with graphs similar to the one in Figure 10(e) are called *piecewise-defined functions*.

⑤ Analyze a Piecewise-Defined Function

Sometimes a function is defined differently on different parts of its domain. For example, the *absolute value function* $f(x) = |x|$ is actually defined by two equations: $f(x) = x$ if $x \geq 0$ and $f(x) = -x$ if $x < 0$. These equations are usually combined into one expression as

$$f(x) = |x| = \begin{cases} x & \text{if} \quad x \geq 0 \\ -x & \text{if} \quad x < 0 \end{cases}$$

Figure 11 shows the graph of the absolute value function. Notice that the graph of f satisfies the Vertical-line Test.

When a function is defined by different equations on different parts of its domain, it is called a **piecewise-defined** function.

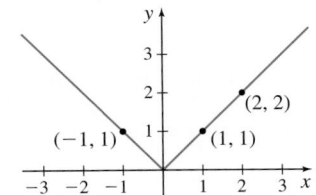

Figure 11 $f(x) = |x|$

EXAMPLE 6   Analyzing a Piecewise-Defined Function

The function f is defined as

$$f(x) = \begin{cases} 0 & \text{if} \ x < 0 \\ 10 & \text{if} \ 0 \leq x \leq 100 \\ 0.2x - 10 & \text{if} \ x > 100 \end{cases}$$

(a) Evaluate $f(-1)$, $f(100)$, and $f(200)$.

(b) Graph f.

(c) Find the domain, range, and the x- and y-intercepts of f.

Solution

(a) $f(-1) = 0$; $f(100) = 10$; $f(200) = 0.2(200) - 10 = 30$

(b) The graph of f consists of three pieces corresponding to each equation in the definition. The graph is the horizontal line $y = 0$ on the interval $(-\infty, 0)$, the horizontal line $y = 10$ on the interval $[0, 100]$, and the line $y = 0.2x - 10$ on the interval $(100, \infty)$, as shown in Figure 12.

(c) f is a piecewise-defined function. Look at the values that x can take on: $x < 0$, $0 \leq x \leq 100$, $x > 100$. We conclude the domain of f is all real numbers. The range of f is the number 0 and all real numbers greater than or equal to 10, which, in set notation, is written as $\{0\} \cup \{x \,|\, x \geq 10\}$. The x-intercepts are all the numbers in the interval $(-\infty, 0)$; the y-intercept is 10. ∎

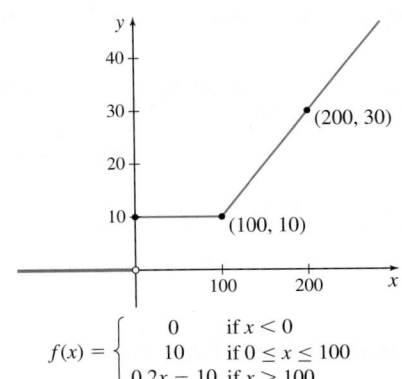

$$f(x) = \begin{cases} 0 & \text{if} \ x < 0 \\ 10 & \text{if} \ 0 \leq x \leq 100 \\ 0.2x - 10 & \text{if} \ x > 100 \end{cases}$$

Figure 12

RECALL x-intercepts are numbers on the x-axis at which a graph touches or crosses the x-axis.

NOW WORK   Problem 33.

---

⑥ Obtain Information from or about the Graph of a Function

The graph of a function provides a great deal of information about the function. Reading and interpreting graphs is an essential skill for calculus.

EXAMPLE 7   Obtaining Information from the Graph of a Function

The graph of $y = f(x)$ is given in Figure 13 on page 10. (x might represent time and y might represent the distance of the bob of a pendulum from its *at-rest* position. Negative values of y would indicate that the bob is to the left of its at-rest position; positive values of y would mean that the bob is to the right of its at-rest position.)

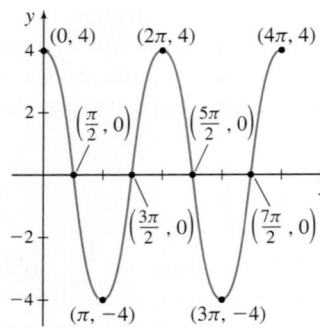

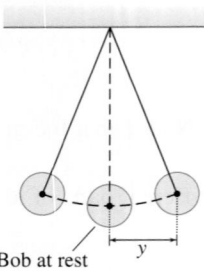

Bob at rest

Figure 13

RECALL Intercepts are points at which a graph crosses or touches a coordinate axis.

(a) What are $f(0)$, $f\left(\dfrac{3\pi}{2}\right)$, and $f(3\pi)$?

(b) What is the domain of f?

(c) What is the range of f?

(d) List the intercepts of the graph.

(e) How many times does the line $y = 2$ intersect the graph of f?

(f) For what values of x does $f(x) = -4$?

(g) For what values of x is $f(x) > 0$?

Solution

(a) Since the point $(0, 4)$ is on the graph of f, the y-coordinate 4 is the value of f at 0; that is, $f(0) = 4$. Similarly, when $x = \dfrac{3\pi}{2}$, then $y = 0$, so $f\left(\dfrac{3\pi}{2}\right) = 0$, and when $x = 3\pi$, then $y = -4$, so $f(3\pi) = -4$.

(b) The points on the graph of f have x-coordinates between 0 and 4π inclusive. The domain of f is $\{x \mid 0 \le x \le 4\pi\}$ or the closed interval $[0, 4\pi]$.

(c) Every point on the graph of f has a y-coordinate between -4 and 4 inclusive. The range of f is $\{y \mid -4 \le y \le 4\}$ or the closed interval $[-4, 4]$.

(d) The intercepts of the graph of f are:

$$(0, 4), \left(\frac{\pi}{2}, 0\right), \left(\frac{3\pi}{2}, 0\right), \left(\frac{5\pi}{2}, 0\right), \text{ and } \left(\frac{7\pi}{2}, 0\right).$$

(e) Draw the graph of the line $y = 2$ on the same set of coordinate axes as the graph of f. The line intersects the graph of f four times.

(f) Find points on the graph of f for which $y = f(x) = -4$; there are two such points: $(\pi, -4)$ and $(3\pi, -4)$. So $f(x) = -4$ when $x = \pi$ and when $x = 3\pi$.

(g) $f(x) > 0$ when the y-coordinate of a point (x, y) on the graph of f is positive. This occurs when x is in the set $\left[0, \dfrac{\pi}{2}\right) \cup \left(\dfrac{3\pi}{2}, \dfrac{5\pi}{2}\right) \cup \left(\dfrac{7\pi}{2}, 4\pi\right]$. ∎

NOW WORK   Problems 37, 39, 41, 43, 45, 47, and 49.

EXAMPLE 8   **Obtaining Information about the Graph of a Function**

Consider the function $f(x) = \dfrac{x+1}{x+2}$.

(a) What is the domain of f?

(b) Is the point $\left(1, \dfrac{1}{2}\right)$ on the graph of f?

(c) If $x = 2$, what is $f(x)$? What is the corresponding point on the graph of f?

(d) If $f(x) = 2$, what is x? What is the corresponding point on the graph of f?

(e) What are the x-intercepts of the graph of f (if any)? What point(s) on the graph of f correspond(s) to the x-intercept(s)?

Solution

(a) The domain of f consists of all real numbers except -2; that is, the set $\{x \mid x \ne -2\}$. The function is not defined at -2.

(b) When $x = 1$, then $f(1) = \underset{\substack{\uparrow \\ x=1}}{\dfrac{1+1}{1+2}} = \dfrac{2}{3}$. The point $\left(1, \dfrac{2}{3}\right)$ is on the graph of f; the point $\left(1, \dfrac{1}{2}\right)$ is not on the graph of f.

(c) If $x = 2$, then $f(2) = \underset{\substack{\uparrow \\ x=2}}{\dfrac{2+1}{2+2}} = \dfrac{3}{4}$. The point $\left(2, \dfrac{3}{4}\right)$ is on the graph of f.

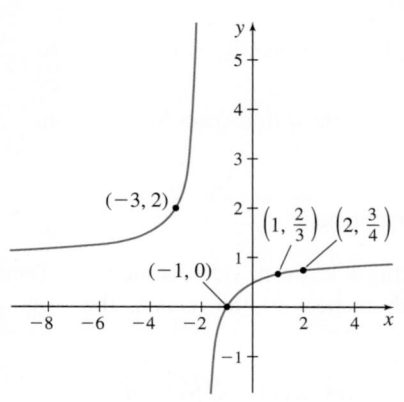

Figure 14 $f(x) = \dfrac{x+1}{x+2}$

(d) If $f(x) = 2$, then $\dfrac{x+1}{x+2} = 2$. Solving for x, we find

$$x + 1 = 2(x+2)$$
$$x + 1 = 2x + 4$$
$$x = -3$$

The point $(-3, 2)$ is on the graph of f.

(e) The x-intercepts of the graph of f occur when $y = 0$. That is, they are the solutions of the equation $f(x) = 0$. The x-intercepts are also called the **real zeros** or **roots** of the function f.

The real zeros of the function $f(x) = \dfrac{x+1}{x+2}$ satisfy the equation $x + 1 = 0$, so $x = -1$. The only x-intercept is -1, so the point $(-1, 0)$ is on the graph of f. ∎

Figure 14 shows the graph of f.

NOW WORK   Problems 55, 57, and 59.

7 Use Properties of Functions

One of the goals of calculus is to develop techniques for graphing functions. Here we review some properties of functions that help obtain the graph of a function.

DEFINITION Even and Odd Functions

A function f is **even** if, for every number x in its domain, the number $-x$ is also in the domain and

$$\boxed{f(-x) = f(x)}$$

A function f is **odd** if, for every number x in its domain, the number $-x$ is also in the domain and

$$\boxed{f(-x) = -f(x)}$$

For example, $f(x) = x^2$ is an even function since

$$f(-x) = (-x)^2 = x^2 = f(x)$$

Also, $g(x) = x^3$ is an odd function since

$$g(-x) = (-x)^3 = -x^3 = -g(x)$$

NEED TO REVIEW? Symmetry of graphs is discussed in Appendix A.3, p. A-19.

See Figure 15 for the graph of $f(x) = x^2$ and Figure 16 for the graph of $g(x) = x^3$. Notice that the graph of the even function $f(x) = x^2$ is symmetric with respect to the y-axis and the graph of the odd function $g(x) = x^3$ is symmetric with respect to the origin.

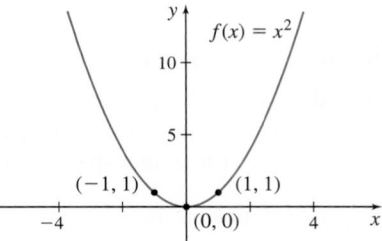

Figure 15 The function $f(x) = x^2$ is even. The graph of f is symmetric with respect to the y-axis.

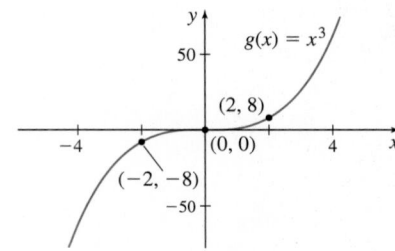

Figure 16 The function $g(x) = x^3$ is odd. The graph of g is symmetric with respect to the origin.

THEOREM Graphs of Even and Odd Functions

- A function is even if and only if its graph is symmetric with respect to the y-axis.
- A function is odd if and only if its graph is symmetric with respect to the origin.

EXAMPLE 9   Identifying Even and Odd Functions

Determine whether each of the following functions is even, odd, or neither. Then determine whether its graph is symmetric with respect to the y-axis, the origin, or neither.

(a) $f(x) = x^2 - 5$ **(b)** $g(x) = \dfrac{4x}{x^2 - 5}$ **(c)** $h(x) = \sqrt[3]{5x^3 - 1}$

(d) $F(x) = |x|$ **(e)** $H(x) = \dfrac{x^2 + 2x - 1}{(x - 5)^2}$

Solution

(a) The domain of f is $(-\infty, \infty)$, so for every number x in its domain, $-x$ is also in the domain. Replace x by $-x$ and simplify.

$$f(-x) = (-x)^2 - 5 = x^2 - 5 = f(x)$$

Since $f(-x) = f(x)$, the function f is even. So the graph of f is symmetric with respect to the y-axis. Figure 17 shows the graph of f using a graphing calculator.

(b) The domain of g is $\{x \mid x \neq \pm\sqrt{5}\}$, so for every number x in its domain, $-x$ is also in the domain. Replace x by $-x$ and simplify.

$$g(-x) = \frac{4(-x)}{(-x)^2 - 5} = \frac{-4x}{x^2 - 5} = -g(x)$$

Since $g(-x) = -g(x)$, the function g is odd. So the graph of g is symmetric with respect to the origin. Figure 18 shows the graph of g using a graphing calculator.

(c) The domain of h is $(-\infty, \infty)$, so for every number x in its domain, $-x$ is also in the domain. Replace x by $-x$ and simplify.

$$h(-x) = \sqrt[3]{5(-x)^3 - 1} = \sqrt[3]{-5x^3 - 1} = \sqrt[3]{-(5x^3 + 1)} = -\sqrt[3]{5x^3 + 1}$$

Since $h(-x) \neq h(x)$ and $h(-x) \neq -h(x)$, the function h is neither even nor odd. The graph of h is not symmetric with respect to the y-axis and not symmetric with respect to the origin.

(d) The domain of F is $(-\infty, \infty)$, so for every number x in its domain, $-x$ is also in the domain. Replace x by $-x$ and simplify.

$$F(-x) = |-x| = |-1| \cdot |x| = |x| = F(x)$$

The function F is even. So the graph of F is symmetric with respect to the y-axis.

(e) The domain of H is $\{x \mid x \neq 5\}$. The number $x = -5$ is in the domain of H, but $x = 5$ is not in the domain. So the function H is neither even nor odd, and the graph of H is not symmetric with respect to the y-axis or the origin. ∎

NOW WORK   Problem 61.

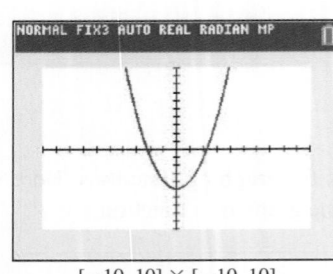

$[-10, 10] \times [-10, 10]$

Figure 17 $f(x) = x^2 - 5$

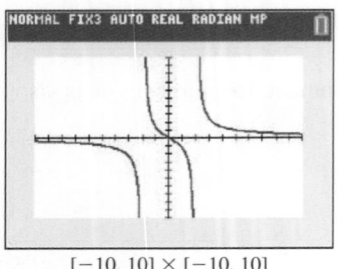

$[-10, 10] \times [-10, 10]$

Figure 18 $g(x) = \dfrac{4x}{(x^2 - 5)}$

Another important property of a function is to know where it is increasing or decreasing.

DEFINITION

A function f is **increasing** on an interval I, if, for any choice of x_1 and x_2 in I, with $x_1 < x_2$, then $f(x_1) < f(x_2)$.

   A function f is **decreasing** on an interval I, if, for any choice of x_1 and x_2 in I, with $x_1 < x_2$, then $f(x_1) > f(x_2)$.

   A function f is **constant** on an interval I, if, for all choices of x in I, the values of $f(x)$ are equal.

Notice in the definition for an increasing (decreasing) function f, the value $f(x_1)$ is *strictly* less than (*strictly* greater than) the value $f(x_2)$. If a nonstrict inequality is used, we obtain the definitions for *nondecreasing* and *nonincreasing* functions.

DEFINITION

A function f is **nondecreasing** on an interval I, if, for any choice of x_1 and x_2 in I, with $x_1 < x_2$, then $f(x_1) \le f(x_2)$.

   A function f is **nonincreasing** on an interval I, if, for any choice of x_1 and x_2 in I, with $x_1 < x_2$, then $f(x_1) \ge f(x_2)$.

IN WORDS From left to right, the graph of an increasing function goes up, the graph of a decreasing function goes down, and the graph of a constant function remains at a fixed height. From left to right, the graph of a nondecreasing function never goes down, and the graph of a nonincreasing function never goes up.

Figure 19 illustrates the definitions. In Chapter 5, we use calculus to find where a function is increasing or decreasing or is constant.

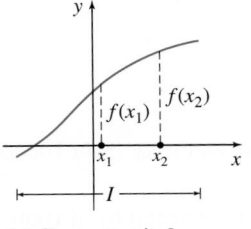
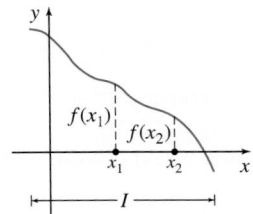
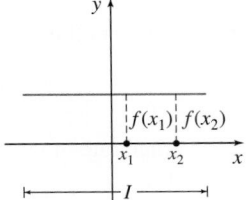
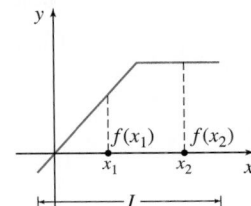
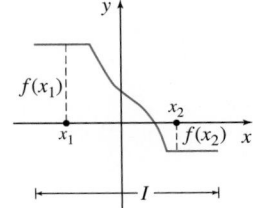

(a) For $x_1 < x_2$ in I, $f(x_1) < f(x_2)$; f is increasing on I

(b) For $x_1 < x_2$ in I, $f(x_1) > f(x_2)$; f is decreasing on I

(c) For all x in I, the values of f are equal; f is constant on I

(d) For $x_1 < x_2$ in I, $f(x_1) \le f(x_2)$; f is nondecreasing on I

(e) For $x_1 < x_2$ in I, $f(x_1) \ge f(x_2)$; f is nonincreasing on I

Figure 19

NOW WORK Problem 51.

---

8 Find the Average Rate of Change of a Function

The average rate of change of a function plays an important role in calculus. It provides information about how a change in the independent variable x of a function $y = f(x)$ causes a change in the dependent variable y. We use the symbol Δx, read "delta x," to represent the change in x and Δy to represent the change in y. Then by forming the quotient $\dfrac{\Delta y}{\Delta x}$, we arrive at an *average rate of change*.

DEFINITION Average Rate of Change

Suppose the domain of a function $y = f(x)$ is an interval l. If a and b, where $a \ne b$, are in l, the **average rate of change of** f from a to b is defined as

$$\boxed{\dfrac{\Delta y}{\Delta x} = \dfrac{f(b) - f(a)}{b - a} \qquad a \ne b}$$

NOTE $\dfrac{\Delta y}{\Delta x}$ represents the change in y with respect to x.

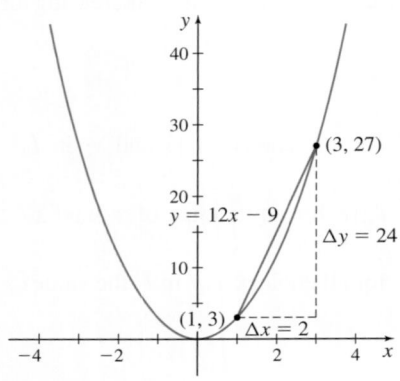

Figure 20 $f(x) = 3x^2$

EXAMPLE 10   Finding the Average Rate of Change

Find the average rate of change of $f(x) = 3x^2$:

(a) From 1 to 3   **(b)** From 1 to x, $x \neq 1$

Solution

(a) The average rate of change of $f(x) = 3x^2$ from 1 to 3 is

$$\frac{\Delta y}{\Delta x} = \frac{f(3) - f(1)}{3 - 1} = \frac{27 - 3}{3 - 1} = \frac{24}{2} = 12$$

See Figure 20.

Notice that the average rate of change of $f(x) = 3x^2$ from 1 to 3 is the slope of the line containing the points $(1, 3)$ and $(3, 27)$.

(b) The average rate of change of $f(x) = 3x^2$ from 1 to x is

$$\frac{\Delta y}{\Delta x} = \frac{f(x) - f(1)}{x - 1} = \frac{3x^2 - 3}{x - 1} = \frac{3(x^2 - 1)}{x - 1}$$

$$= \frac{3(x - 1)(x + 1)}{x - 1} = 3(x + 1) = 3x + 3$$

provided $x \neq 1$. ∎

NOW WORK   Problem 65.

EXAMPLE 11   Analyzing a Cost Function

CALC CLIP

The weekly cost C, in dollars, of manufacturing x lightbulbs is

$$C(x) = 7500 + \sqrt{125x}$$

(a) Find the average rate of change of the weekly cost C of manufacturing from 100 to 101 lightbulbs.

(b) Find the average rate of change of the weekly cost C of manufacturing from 1000 to 1001 lightbulbs.

(c) Interpret the results from parts (a) and (b).

Solution

(a) The weekly cost of manufacturing 100 lightbulbs is

$$C(100) = 7500 + \sqrt{125 \cdot 100} = 7500 + \sqrt{12,500} \approx \$7611.80$$

The weekly cost of manufacturing 101 lightbulbs is

$$C(101) = 7500 + \sqrt{125 \cdot 101} = 7500 + \sqrt{12,625} \approx \$7612.36$$

The average rate of change of the weekly cost C from 100 to 101 is

$$\frac{\Delta C}{\Delta x} = \frac{C(101) - C(100)}{101 - 100} \approx \frac{7612.36 - 7611.80}{1} = \$0.56$$

(b) The weekly cost of manufacturing 1000 lightbulbs is

$$C(1000) = 7500 + \sqrt{125 \cdot 1000} = 7500 + \sqrt{125,000} \approx \$7853.55$$

The weekly cost of manufacturing 1001 lightbulbs is

$$C(1001) = 7500 + \sqrt{125 \cdot 1001} = 7500 + \sqrt{125,125} \approx \$7853.73$$

The average rate of change of the weekly cost C from 1000 to 1001 is

$$\frac{\Delta C}{\Delta x} = \frac{C(1001) - C(1000)}{1001 - 1000} \approx \frac{7853.73 - 7853.55}{1} = \$0.18$$

(c) Part (a) tells us that the unit cost of manufacturing the 101st lightbulb is \$0.56. From (b) we learn that the unit cost of manufacturing the 1001st lightbulb is only \$0.18. The unit cost of manufacturing the 1001st lightbulb is less than the unit cost of manufacturing the 101st lightbulb. ■

NOW WORK   Problem 73.

The Secant Line

The average rate of change of a function has an important geometric interpretation. Look at the graph of $y = f(x)$ in Figure 21. Two points are labeled on the graph: $(a, f(a))$ and $(b, f(b))$. The line containing these two points is called a **secant line**. The slope of a secant line is

$$m_{\text{sec}} = \frac{f(b) - f(a)}{b - a} \underset{\substack{\uparrow \\ h = b - a}}{=} \frac{f(a + h) - f(a)}{h}$$

Notice that the average rate of change of f from a to b equals the difference quotient of f at a equals the slope of a secant line.

> **THEOREM Slope of a Secant Line**
>
> The average rate of change of a function f from a to b equals the slope of the secant line containing the two points $(a, f(a))$ and $(b, f(b))$ on the graph of f.

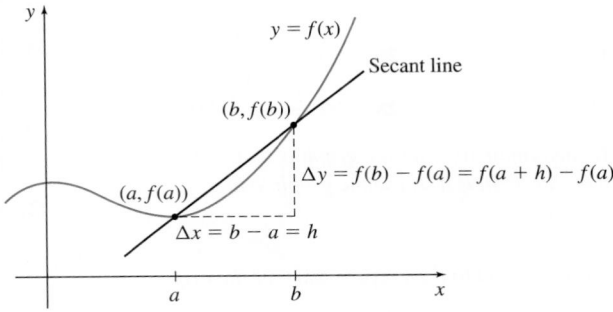

Figure 21 A secant line of the function $y = f(x)$.

P.1 Assess Your Understanding

Concepts and Vocabulary

1. If f is a function defined by $y = f(x)$, then x is called the _____ variable and y is the _____ variable.

2. *True or False* The independent variable is sometimes referred to as the argument of the function.

3. *True or False* If no domain is specified for a function f, then the domain of f is taken to be the set of all real numbers.

4. *True or False* The domain of the function $f(x) = \dfrac{3(x^2 - 1)}{x - 1}$ is $\{x \mid x \neq \pm 1\}$.

5. *True or False* A function can have more than one y-intercept.

6. A set of points in the xy-plane is the graph of a function if and only if every _____ line intersects the graph in at most one point.

7. If the point $(5, -3)$ is on the graph of f, then $f(\text{\_\_\_\_}) = \text{\_\_\_\_}$.

8. Find a so that the point $(-1, 2)$ is on the graph of $f(x) = ax^2 + 4$.

9. *Multiple Choice* A function f is
[**(a)** increasing, **(b)** decreasing, **(c)** nonincreasing, **(d)** nondecreasing, **(e)** constant]
on an interval I if, for any choice of x_1 and x_2 in I, with $x_1 < x_2$, then $f(x_1) < f(x_2)$.

10. *Multiple Choice*   A function f is
[(a) even, (b) odd, (c) neither even nor odd]
if for every number x in its domain, the number $-x$
is also in the domain and $f(-x) = f(x)$.

11. *True or False*   Even functions have graphs that are symmetric
with respect to the origin.

12. The average rate of change of $f(x) = 2x^3 - 3$ from 0 to 2 is
_____.

Practice Problems

In Problems 13–16, for each function find:

(a) $f(0)$ (b) $f(-x)$ (c) $-f(x)$
(d) $f(x+1)$ (e) $f(x+h)$

[PAGE 4] 13. $f(x) = 3x^2 + 2x - 4$ **14.** $f(x) = \dfrac{x}{x^2+1}$

15. $f(x) = |x| + 4$ **16.** $f(x) = \sqrt{3-x}$

In Problems 17–22, find the domain of each function.

[PAGE 7] 17. $f(x) = x^3 - 1$ **18.** $f(x) = \dfrac{x}{x^2+1}$

19. $v(t) = \sqrt{t^2 - 9}$ **20.** $g(x) = \sqrt{\dfrac{2}{x-1}}$

21. $h(x) = \dfrac{x+2}{x^3 - 4x}$ **22.** $s(t) = \dfrac{\sqrt{t+1}}{t-5}$

In Problems 23–28, find the difference quotient of f. That is, find

$$\frac{f(x+h) - f(x)}{h}, h \neq 0$$

[PAGE 6] 23. $f(x) = -3x + 1$ **24.** $f(x) = \dfrac{1}{x+3}$

25. $f(x) = \sqrt{x+7}$ **26.** $f(x) = \dfrac{2}{\sqrt{x+7}}$

27. $f(x) = x^2 + 2x$ **28.** $f(x) = (2x+3)^2$

In Problems 29–32, determine whether the graph is that of a function by using the Vertical-line Test. If it is, use the graph to find
(a) *the domain and range*
(b) *the intercepts, if any*
(c) *any symmetry with respect to the x-axis, y-axis, or the origin.*

29. 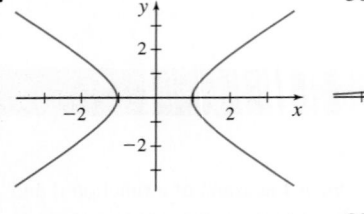   **30.**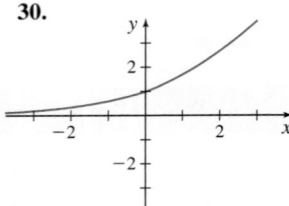

[PAGE 8] 31. 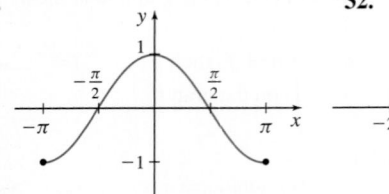   **32.**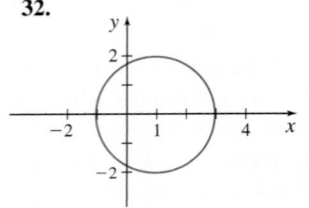

In Problems 33–36, for each piecewise-defined function:
(a) *Find $f(-1)$, $f(0)$, $f(1)$, and $f(8)$.*
(b) *Graph f.*
(c) *Find the domain, range, and intercepts of f.*

[PAGE 9] 33. $f(x) = \begin{cases} x+3 & \text{if } -2 \leq x < 1 \\ 5 & \text{if } x = 1 \\ -x+2 & \text{if } x > 1 \end{cases}$

34. $f(x) = \begin{cases} 2x+5 & \text{if } -3 \leq x < 0 \\ -3 & \text{if } x = 0 \\ -5x & \text{if } x > 0 \end{cases}$

35. $f(x) = \begin{cases} 1+x & \text{if } x < 0 \\ x^2 & \text{if } x \geq 0 \end{cases}$

36. $f(x) = \begin{cases} \dfrac{1}{x} & \text{if } x < 0 \\ \sqrt[3]{x} & \text{if } x \geq 0 \end{cases}$

In Problems 37–54, use the graph of the function f given below to answer the questions.

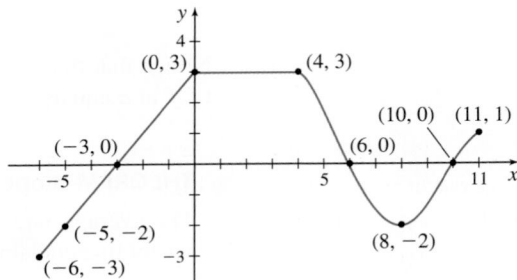

[PAGE 10] 37. Find $f(0)$ and $f(-6)$.

38. Is $f(3)$ positive or negative?

[PAGE 10] 39. Is $f(-4)$ positive or negative?

40. For what values of x is $f(x) = 0$?

[PAGE 10] 41. For what values of x is $f(x) > 0$?

42. What is the domain of f?

[PAGE 10] 43. What is the range of f?

44. What are the x-intercepts?

[PAGE 10] 45. What is the y-intercept?

46. How often does the line $y = \dfrac{1}{2}$ intersect the graph?

[PAGE 10] 47. How often does the line $x = 5$ intersect the graph?

48. For what values of x does $f(x) = 3$?

[PAGE 10] 49. For what values of x does $f(x) = -2$?

50. On what interval(s) is the function f increasing?

[PAGE 13] 51. On what interval(s) is the function f decreasing?

52. On what interval(s) is the function f constant?

53. On what interval(s) is the function f nonincreasing?

54. On what interval(s) is the function f nondecreasing?

In Problems 55–60, answer the questions about the function

$$g(x) = \frac{x+2}{x-6}$$

55. What is the domain of g?

56. Is the point $(3, 14)$ on the graph of g?

57. If $x = 4$, what is $g(x)$? What is the corresponding point on the graph of g?

58. If $g(x) = 2$, what is x? What is(are) the corresponding point(s) on the graph of g?

59. List the x-intercepts, if any, of the graph of g.

60. What is the y-intercept, if there is one, of the graph of g?

In Problems 61–64, determine whether the function is even, odd, or neither. Then determine whether its graph is symmetric with respect to the y-axis, the origin, or neither.

61. $h(x) = \dfrac{x}{x^2 - 1}$

62. $f(x) = \sqrt[3]{3x^2 + 1}$

63. $G(x) = \sqrt{x}$

64. $F(x) = \dfrac{2x}{|x|}$

65. Find the average rate of change of $f(x) = -2x^2 + 4$:
  (a) From 1 to 2 **(b)** From 1 to 3
  (c) From 1 to 4 **(d)** From 1 to x, $x \neq 1$

66. Find the average rate of change of $s(t) = 20 - 0.8t^2$:
  (a) From 1 to 4 **(b)** From 1 to 3
  (c) From 1 to 2 **(d)** From 1 to t, $t \neq 1$

In Problems 67–72, the graph of a piecewise-defined function is given. Write a definition for each piecewise-defined function. Then state the domain and the range of the function.

67.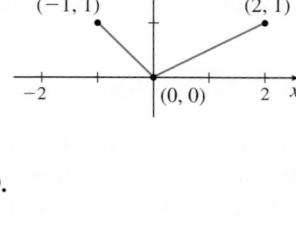

68.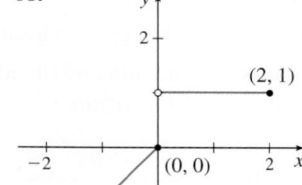

69.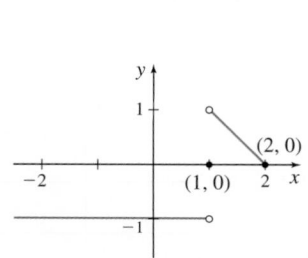

70.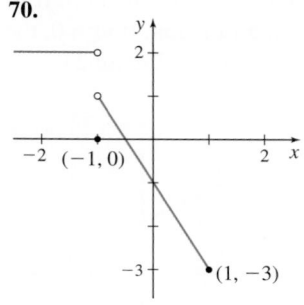

71. **72.**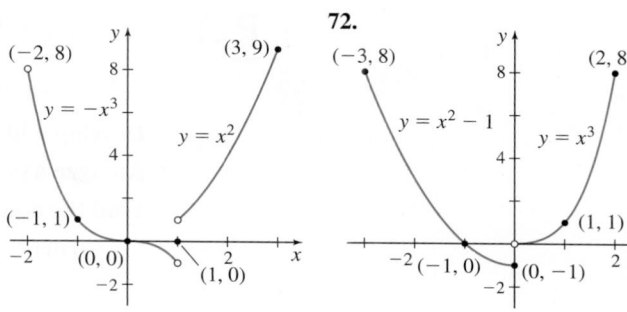

73. The monthly cost C, in dollars, of manufacturing x road bikes is given by the function

$$C(x) = 0.004x^3 - 0.6x^2 + 250x + 100{,}500$$

(a) Find the average rate of change of the cost C of manufacturing from 100 to 101 road bikes.

(b) Find the average rate of change of the cost C of manufacturing from 500 to 501 road bikes.

(c) Interpret the results from parts (a) and (b) in the context of the problem.

74. The graph below represents the distance d (in miles) that Kevin was from home as a function of time t (in hours). Answer the questions by referring to the graph. In parts (a)–(g), how many hours elapsed and how far was Kevin from home during this time?

(a) From $t = 0$ to $t = 2$ **(b)** From $t = 2$ to $t = 2.5$
(c) From $t = 2.5$ to $t = 2.8$ **(d)** From $t = 2.8$ to $t = 3$
(e) From $t = 3$ to $t = 3.9$ **(f)** From $t = 3.9$ to $t = 4.2$
(g) From $t = 4.2$ to $t = 5.3$

(h) What is the farthest distance that Kevin was from home?
(i) How many times did Kevin return home?

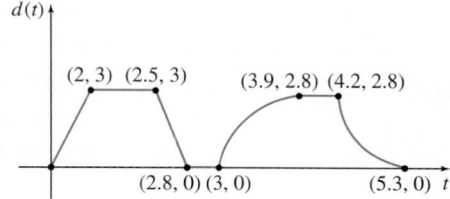

75. Population as a Function of Age The function

$$P(a) = 0.025a^2 - 6.212a + 359.571$$

represents the population P (in millions) of Americans that were a years of age or older in 2018.

(a) Identify the independent and dependent variables.

(b) Evaluate $P(20)$. Interpret $P(20)$ in the context of this problem.

(c) Evaluate $P(0)$. Interpret $P(0)$ in the context of this problem.

Source: U.S. Census Bureau

P.2 Library of Functions; Mathematical Modeling

OBJECTIVES *When you finish this section, you should be able to:*

1 Develop a library of functions (p. 18)
2 Analyze a polynomial function and its graph (p. 21)
3 Find the domain and the intercepts of a rational function (p. 23)
4 Construct a mathematical model (p. 24)

When a collection of functions have common properties, they can be "grouped together" as belonging to a **class of functions.** Polynomial functions, exponential functions, and trigonometric functions are examples of classes of functions. As we investigate principles of calculus, we will find that often a principle applies to all functions in a class in the same way.

1 Develop a Library of Functions

Most of the functions in this section will be familiar to you; several might be new. Although the list may seem familiar, pay special attention to the domain of each function and to its properties, particularly to the shape of each graph. Knowing these graphs lays the foundation for later graphing techniques.

Constant Function $\boxed{f(x) = A \qquad A \text{ is a real number}}$

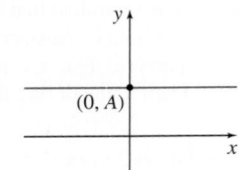

Figure 22 $f(x) = A$

The domain of a **constant function** is the set of all real numbers; its range is the single number A. The graph of a constant function is a horizontal line whose y-intercept is A; it has no x-intercept if $A \neq 0$. A constant function is an even function. See Figure 22.

Identity Function $\boxed{f(x) = x}$

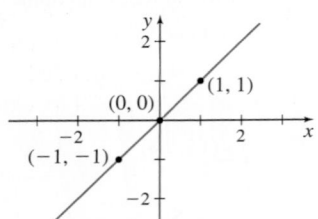

Figure 23 $f(x) = x$

The domain and the range of the **identity function** are the set of all real numbers. Its graph is the line through the origin whose slope is $m = 1$. Its only intercept is $(0, 0)$. The identity function is an odd function; its graph is symmetric with respect to the origin. It is increasing over its domain. See Figure 23.

Notice that the graph of $f(x) = x$ bisects quadrants I and III.

The graphs of both the constant function and the identity function are straight lines. These functions belong to the class of *linear functions:*

Linear Functions $\boxed{f(x) = mx + b \qquad m \text{ and } b \text{ are real numbers}}$

NEED TO REVIEW? Equations of lines are discussed in Appendix A.3, pp. A-20 to A-22.

The domain of a **linear function** is the set of all real numbers. The graph of a linear function is a line with slope m and y-intercept b. If $m > 0$, f is an increasing function, and if $m < 0$, f is a decreasing function. If $m = 0$, then f is a constant function, and its graph is the horizontal line, $y = b$. See Figure 24.

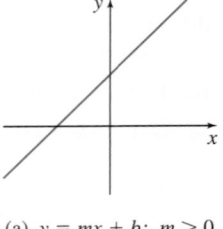

(a) $y = mx + b$; $m > 0$

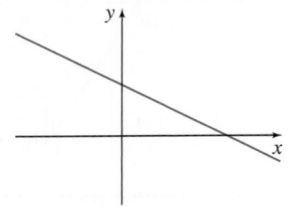

(b) $y = mx + b$; $m < 0$

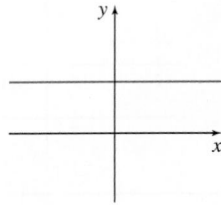

(c) $y = mx + b$; $m = 0$

Figure 24

In a *power function*, the independent variable x is raised to a power.

| **Power Functions** | $f(x) = x^a$ a is a real number |

Below we examine power functions $f(x) = x^n$, where $n \geq 1$ is a positive integer. The domain of these power functions is the set of all real numbers. The only intercept of their graph is the point $(0, 0)$.

$f(x) = x^n$, $n \geq 1$, an odd integer

If f is a power function and n is a positive odd integer, then f is an odd function whose range is the set of all real numbers, and the graph of f is symmetric with respect to the origin. The points $(-1, -1)$, $(0, 0)$, and $(1, 1)$ are on the graph of f. As x becomes unbounded in the negative direction, f also becomes unbounded in the negative direction. Similarly, as x becomes unbounded in the positive direction, f also becomes unbounded in the positive direction. The function f is increasing over its domain.

The graphs of several odd power functions are given in Figure 25.

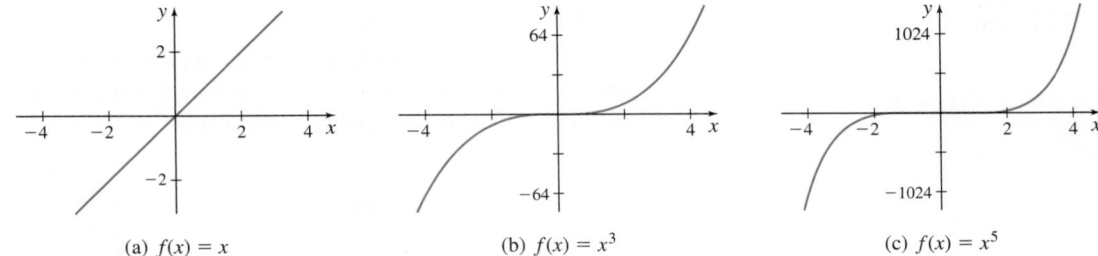

(a) $f(x) = x$ (b) $f(x) = x^3$ (c) $f(x) = x^5$

Figure 25

$f(x) = x^n$, $n \geq 2$, an even integer

If f is a power function and n is a positive even integer, then f is an even function whose range is $\{y \mid y \geq 0\}$. The graph of f is symmetric with respect to the y-axis. The points $(-1, 1)$, $(0, 0)$, and $(1, 1)$ are on the graph of f. As x becomes unbounded in either the negative direction or the positive direction, f becomes unbounded in the positive direction. The function is decreasing on the interval $(-\infty, 0]$ and is increasing on the interval $[0, \infty)$.

The graphs of several even power functions are shown in Figure 26.

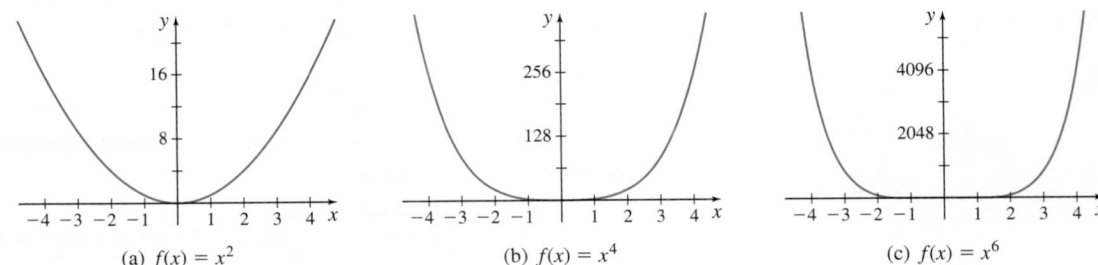

(a) $f(x) = x^2$ (b) $f(x) = x^4$ (c) $f(x) = x^6$

Figure 26

Look closely at Figures 25 and 26. As the integer exponent n increases, the graph of f is flatter (closer to the x-axis) when x is in the interval $(-1, 1)$ and steeper when x is in interval $(-\infty, -1)$ or $(1, \infty)$.

| **The Reciprocal Function** | $f(x) = \dfrac{1}{x}$ $x \neq 0$ |

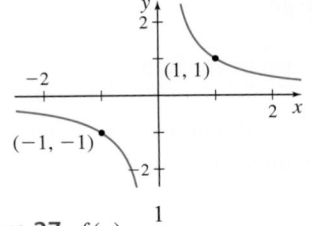

Figure 27 $f(x) = \dfrac{1}{x}$

The domain and the range of the **reciprocal function** (the power function $f(x) = x^a$, $a = -1$) are the set of all nonzero real numbers. The graph has no intercepts. The reciprocal function is an odd function so the graph is symmetric with respect to the origin. The function is decreasing on $(-\infty, 0)$ and on $(0, \infty)$. See Figure 27.

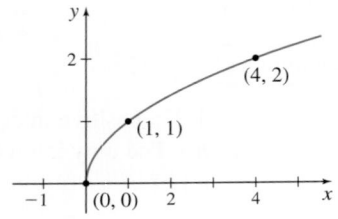

Figure 28 $f(x) = \sqrt{x}$

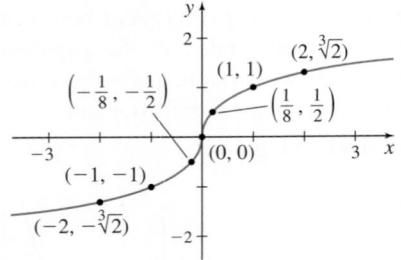

Figure 29 $f(x) = \sqrt[3]{x}$

Root Functions $\boxed{f(x) = x^{1/n} = \sqrt[n]{x} \quad n \geq 2 \text{ is a positive integer}}$

Root functions are also power functions. If $n = 2$, $f(x) = x^{1/2} = \sqrt{x}$ is the **square root function**. The domain and the range of the square root function are the set of nonnegative real numbers. The intercept is the point $(0, 0)$. The square root function is neither even nor odd; it is increasing on the interval $[0, \infty)$.

See Figure 28.

For root functions whose index n is a positive even integer, the domain and the range are the set of nonnegative real numbers. The intercept is the point $(0, 0)$. Such functions are neither even nor odd; they are increasing on their domain, the interval $[0, \infty)$.

If $n = 3$, $f(x) = x^{1/3} = \sqrt[3]{x}$ is the **cube root function**. The domain and range of the cube root function are all real numbers. The intercept of the graph of the cube root function is the point $(0, 0)$.

Because $f(-x) = \sqrt[3]{-x} = -\sqrt[3]{x} = -f(x)$, the cube root function is odd. Its graph is symmetric with respect to the origin. The cube root function is increasing on the interval $(-\infty, \infty)$. See Figure 29.

For root functions whose index n is a positive odd integer, the domain and the range are the set of all real numbers. The intercept is the point $(0, 0)$. Such functions are odd so their graphs are symmetric with respect to the origin. They are increasing on the interval $(-\infty, \infty)$.

Absolute Value Function $\boxed{f(x) = |x|}$

The **absolute value function** is defined as the piecewise-defined function

$$f(x) = |x| = \begin{cases} x & \text{if } x \geq 0 \\ -x & \text{if } x < 0 \end{cases}$$

or as the function

$$\boxed{f(x) = |x| = \sqrt{x^2}}$$

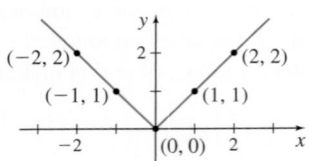

Figure 30 $f(x) = |x|$

The domain of the absolute value function is the set of all real numbers. The range is the set of nonnegative real numbers. The intercept of the graph of f is the point $(0, 0)$. Because $f(-x) = |-x| = |x| = f(x)$, the absolute value function is even. Its graph is symmetric with respect to the y-axis. See Figure 30.

NOW WORK   Problems 11 and 15.

The **floor function**, also known as the **greatest integer function**, is defined as the largest integer less than or equal to x:

$$\boxed{f(x) = \lfloor x \rfloor = \text{largest integer less than or equal to } x}$$

The domain of the floor function $\lfloor x \rfloor$ is the set of all real numbers; the range is the set of all integers. The y-intercept of $\lfloor x \rfloor$ is 0, and the x-intercepts are the real numbers in the interval $[0, 1)$. The floor function is constant on every interval of the form $[k, k+1)$, where k is an integer, and is nondecreasing on its domain. See Figure 31.

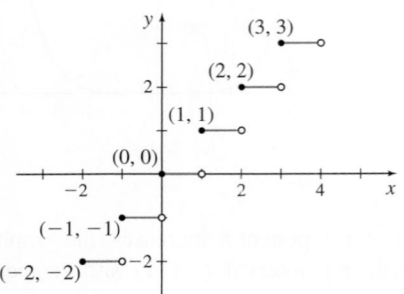

Figure 31 $f(x) = \lfloor x \rfloor$

IN WORDS The floor function can be thought of as the "rounding down" function. The ceiling function can be thought of as the "rounding up" function.

The **ceiling function** is defined as the smallest integer greater than or equal to x:

$$\boxed{f(x) = \lceil x \rceil = \text{smallest integer greater than or equal to } x}$$

The domain of the ceiling function $\lceil x \rceil$ is the set of all real numbers; the range is the set of integers. The y-intercept of $\lceil x \rceil$ is 0, and the x-intercepts are the real numbers

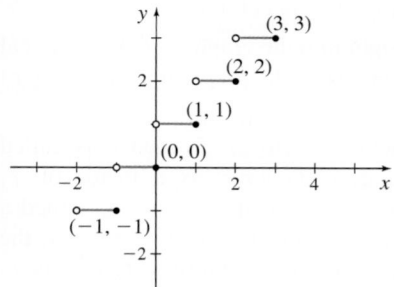

Figure 32 $f(x) = \lceil x \rceil$

in the interval $(-1, 0]$. The ceiling function is constant on every interval of the form $(k,\ k + 1]$, where k is an integer, and is nondecreasing on its domain. See Figure 32.

For example, for the floor function $\lfloor 5 \rfloor = 5$ and $\lfloor 4.9 \rfloor = 4$, but for the ceiling function $\lceil 5 \rceil = 5$ and $\lceil 4.9 \rceil = 5$.

The floor and ceiling functions are examples of **step functions**. At each integer, the function has a *discontinuity*. That is, at integers the function jumps from one value to another without taking on any of the intermediate values.

② Analyze a Polynomial Function and Its Graph

A **monomial** is an expression of the form ax^n, where $a \neq 0$ is a real number and $n \geq 0$ is an integer. *Polynomial functions* are formed by adding a finite number of monomials.

DEFINITION Polynomial Function

A **polynomial function** is a function of the form

$$f(x) = a_n x^n + a_{n-1} x^{n-1} + \cdots + a_1 x + a_0$$

where $a_n,\ a_{n-1},\ \ldots,\ a_1,\ a_0$ are real numbers and n is a nonnegative integer. The domain of a polynomial function is the set of all real numbers.

If $a_n \neq 0$, then a_n is called the **leading coefficient of** f, and the polynomial function is said to be of **degree** n.

The constant function $f(x) = a_0$, where $a_0 \neq 0$, is a polynomial function of degree 0. The constant function $f(x) = 0$ is the **zero polynomial function** and has no degree. Its graph is the x-axis, that is, the horizontal line containing the point $(0, 0)$.

If the degree of a polynomial function is 1, then it is a linear function of the form $f(x) = mx + b$, where $m \neq 0$. The graph of a linear function is a straight line with slope m and y-intercept b.

Any polynomial function f of degree 2 can be written in the form

$$f(x) = ax^2 + bx + c$$

where a, b, and c are constants and $a \neq 0$. The square function $f(x) = x^2$ is a polynomial function of degree 2. Polynomial functions of degree 2 are also called **quadratic functions**. The graph of a quadratic function is known as a **parabola** and is symmetric about its **axis of symmetry**, the vertical line $x = -\dfrac{b}{2a}$. Figure 33 shows the graphs of typical parabolas and their axes of symmetry. The x-intercepts, if any, of a quadratic function satisfy the quadratic equation $ax^2 + bx + c = 0$.

NEED TO REVIEW? Quadratic equations and the discriminant are discussed in Appendix A.1, pp. A-3 to A-4.

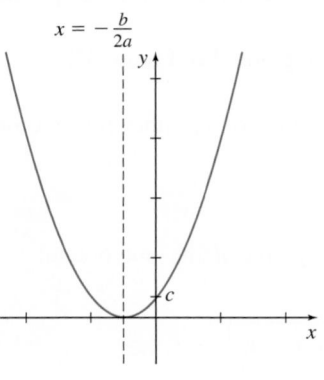

$f(x) = ax^2 + bx + c, a > 0$
$b^2 - 4ac = 0$
One x-intercept

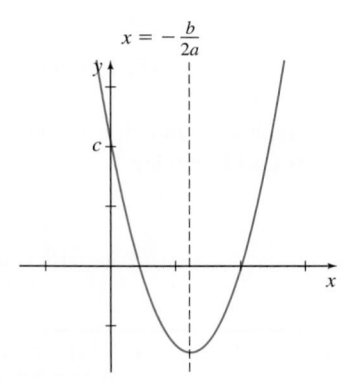

$f(x) = ax^2 + bx + c, a > 0$
$b^2 - 4ac > 0$
Two x-intercepts

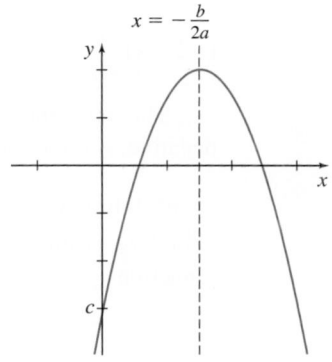

$f(x) = ax^2 + bx + c; a < 0$
$b^2 - 4ac > 0$
Two x-intercepts

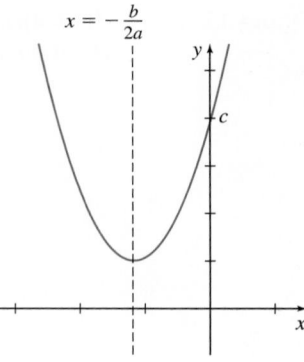

$f(x) = ax^2 + bx + c; a > 0$
$b^2 - 4ac < 0$
No x-intercepts

Figure 33

NOTE Finding the zeros of a polynomial function f is easy if f is linear, quadratic, or in factored form. Otherwise, finding the zeros can be difficult.

Graphs of polynomial functions have many properties in common.

The zeros of a polynomial function f give insight into the graph of f. If r is a real zero of a polynomial function f, then r is an x-intercept of the graph of f, and $(x - r)$ is a factor of f.

If $(x - r)$ occurs more than once in the factored form of f, then r is called a **repeated** or **multiple zero of** f. In particular, if $(x - r)^m$ is a factor of f, but $(x - r)^{m+1}$, where $m \geq 1$ is an integer, is not a factor of f, then r is called a **zero of multiplicity** m **of** f. When the multiplicity of a zero r is an even integer, the graph of f will touch (but not cross) the x-axis at r; when the multiplicity of r is an odd integer, then the graph of f will cross the x-axis at r. As Figure 34 illustrates, when the graph crosses the x-axis, the sign of $f(x)$ changes, and when the graph touches the x-axis, the sign of $f(x)$ remains the same.

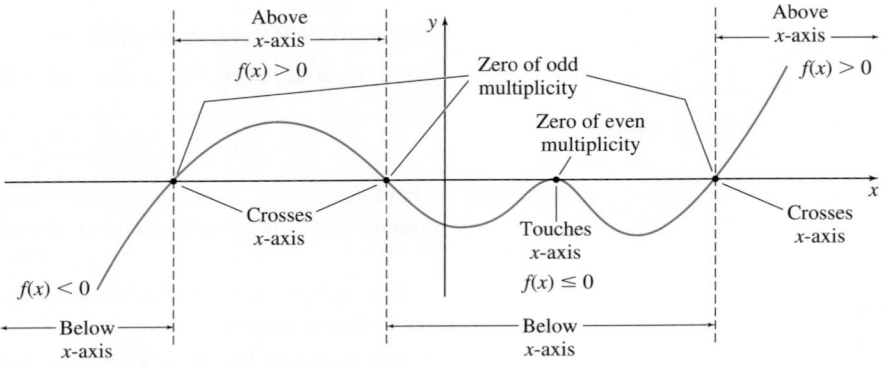

Figure 34 The graph of a polynomial function.

EXAMPLE 1   Analyzing the Graph of a Polynomial Function

For the polynomial function $f(x) = x^2(x - 4)(x + 1)$:

(a) Find the x- and y-intercepts of the graph of f.

(b) Determine whether the graph crosses or touches the x-axis at each x-intercept.

(c) Plot at least one point to the left and right of each x-intercept and connect the points to obtain the graph.

Solution

(a) The y-intercept is $f(0) = 0$.
The x-intercepts are the zeros of the function: 0, 4, and -1.

(b) 0 is a zero of multiplicity 2; the graph of f touches the x-axis at 0. The numbers 4 and -1 are zeros of multiplicity 1; the graph of f crosses the x-axis at 4 and -1.

(c) Since $f(-2) = 24$, $f\left(-\dfrac{1}{2}\right) = -\dfrac{9}{16}$, $f(2) = -24$, and $f(5) = 150$, the points $(-2, 24)$, $\left(-\dfrac{1}{2}, -\dfrac{9}{16}\right)$, $(2, -24)$, and $(5, 150)$ are on the graph. See Figure 35. ∎

Figure 35 $f(x) = x^2(x - 4)(x + 1)$
$= x^4 - 3x^3 - 4x^2$

The behavior of the graph of a function for large values of x, either positive or negative, is referred to as its **end behavior**.

End Behavior

For large values of x, either positive or negative, the graph of the polynomial function

$$f(x) = a_n x^n + a_{n-1} x^{n-1} + \cdots + a_1 x + a_0$$

resembles the graph of the power function

$$y = a_n x^n$$

TABLE 3

x	$f(x)$	$y = -2x^3$
10	-1494	-2000
100	$-1,949,904$	$-2,000,000$
500	$-248,749,504$	$-250,000,000$
1000	$-1,994,999,004$	$-2,000,000,000$

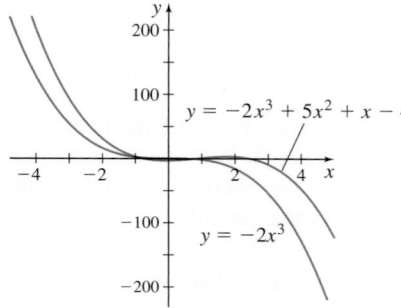

Figure 36

For example, for large values of $|x|$, the end behavior of the graph of the polynomial function $f(x) = x^4 - 3x^3 - 4x^2$, shown in Figure 35, resembles the graph of $y = x^4$.

As another example, if $f(x) = -2x^3 + 5x^2 + x - 4$, then the graph of f will behave like the graph of $y = -2x^3$ for very large values of x, either positive or negative. We can see that the graphs of f and $y = -2x^3$ "behave" the same by considering Table 3 and Figure 36.

NOW WORK Problem **21**.

3 Find the Domain and the Intercepts of a Rational Function

The quotient of two polynomial functions p and q is called a *rational function*.

DEFINITION Rational Function

A **rational function** is a function of the form

$$R(x) = \frac{p(x)}{q(x)}$$

where p and q are polynomial functions and q is not the zero polynomial. The domain of R is the set of all real numbers, except those for which the denominator q is 0.

If $R(x) = \dfrac{p(x)}{q(x)}$ is a rational function, the real zeros, if any, of the numerator, which are also in the domain of R, are the x-intercepts of the graph of R.

EXAMPLE 2   **Finding the Domain and the Intercepts of a Rational Function**

CALC CLIP

Find the domain and the intercepts (if any) of each rational function:

(a) $R(x) = \dfrac{2x^2 - 4}{x^2 - 4}$ (b) $R(x) = \dfrac{x}{x^2 + 1}$ (c) $R(x) = \dfrac{x^2 - 1}{x - 1}$

Solution

(a) The domain of $R(x) = \dfrac{2x^2 - 4}{x^2 - 4}$ is $\{x \mid x \neq -2; x \neq 2\}$. Since 0 is in the domain of R and $R(0) = 1$, the y-intercept is 1. The zeros of R are solutions of the equation $2x^2 - 4 = 0$ or $x^2 = 2$. Since $-\sqrt{2}$ and $\sqrt{2}$ are in the domain of R, the x-intercepts are $-\sqrt{2}$ and $\sqrt{2}$.

(b) The domain of $R(x) = \dfrac{x}{x^2 + 1}$ is the set of all real numbers. Since 0 is in the domain of R and $R(0) = 0$, the y-intercept is 0, and the x-intercept is also 0.

(c) The domain of $R(x) = \dfrac{x^2 - 1}{x - 1}$ is $\{x \mid x \neq 1\}$. Since 0 is in the domain of R and $R(0) = 1$, the y-intercept is 1. The x-intercept(s), if any, satisfy the equation

$$x^2 - 1 = 0$$
$$x^2 = 1$$
$$x = -1 \quad \text{or} \quad x = 1$$

Since 1 is not in the domain of R, the only x-intercept is -1. ∎

Figure 37 shows the graphs of $R(x) = \dfrac{x^2 - 1}{x - 1}$ and $f(x) = x + 1$. Notice the "hole" in the graph of R at the point $(1, 2)$. Also, $R(x) = \dfrac{x^2 - 1}{x - 1}$ and $f(x) = x + 1$ are not the same function: their domains are different. The domain of R is $\{x \mid x \neq 1\}$ and the domain of f is the set of all real numbers.

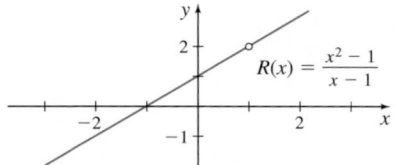

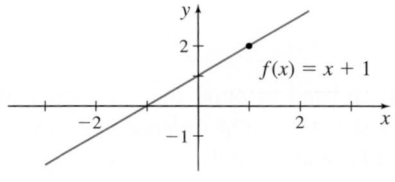

Figure 37

NOW WORK Problem **27**.

Algebraic Functions

Every function discussed so far belongs to a broad class of functions called **algebraic functions**. A function f is called algebraic if it can be expressed in terms of sums, differences, products, quotients, powers, or roots of polynomial functions. For example, the function f defined by

$$f(x) = \frac{3x^3 - x^2(x + 1)^{4/3}}{\sqrt{x^4 + 2}}$$

is an algebraic function. Functions that are not algebraic are called **transcendental functions**. Examples of transcendental functions include the trigonometric functions, exponential functions, and logarithmic functions, which are discussed in the later sections of this chapter.

4 Construct a Mathematical Model

Problems in engineering and the sciences often can be solved by using mathematical models that involve functions. To build a model, verbal descriptions must be translated into the language of mathematics by assigning symbols to represent the independent and dependent variables and then finding a function that relates the variables.

EXAMPLE 3   **Constructing a Model from a Verbal Description**

A liquid is poured into a container in the shape of a right circular cone with radius 4 meters and height 16 meters, as shown in Figure 38. Express the volume V of the liquid as a function of the height h of the liquid.

Solution

The formula for the volume of a right circular cone of radius r and height h is

$$V = \frac{1}{3}\pi r^2 h$$

The volume depends on two variables, r and h. To express V as a function of h only, we use the fact that a cross section of the cone and the liquid form two similar triangles. See Figure 39.

Corresponding sides of similar triangles are in proportion. Since the cone's radius is 4 meters and its height is 16 meters, we have

$$\frac{r}{h} = \frac{4}{16} = \frac{1}{4}$$

$$r = \frac{1}{4}h$$

Then

$$V = \frac{1}{3}\pi r^2 h \underset{\substack{\uparrow \\ r = \frac{1}{4}h}}{=} \frac{1}{3}\pi \left(\frac{1}{4}h\right)^2 h = \frac{1}{48}\pi h^3$$

So $V = V(h) = \dfrac{1}{48}\pi h^3$ expresses the volume V as a function of the height of the liquid. Since h is measured in meters, V will be expressed in cubic meters. ■

NOW WORK   Problem 29.

In many applications, data are collected and are used to build the mathematical model. If the data involve two variables, the first step is to plot ordered pairs using rectangular coordinates. The resulting graph is called a **scatter plot**.

Scatter plots are used to help suggest the type of relation that exists between the two variables. Once the general shape of the relation is recognized, a function can be chosen whose graph closely resembles the shape in the scatter plot.

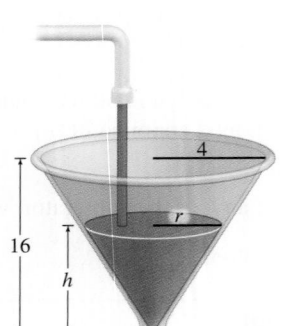

Figure 38

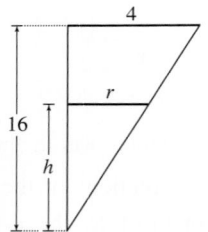

Figure 39

NEED TO REVIEW? Similar triangles and geometry formulas are discussed in Appendix A.2, pp. A-14 to A-15.

EXAMPLE 4 **Identifying the Shape of a Scatter Plot**

Determine whether you would model the relation between the two variables shown in each scatter plot in Figure 40 with a linear function or a quadratic function.

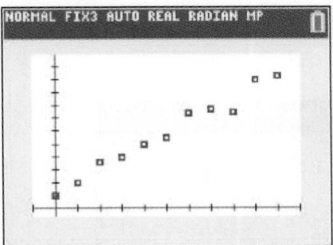 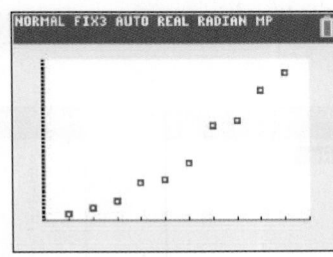 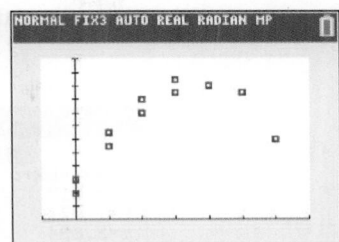 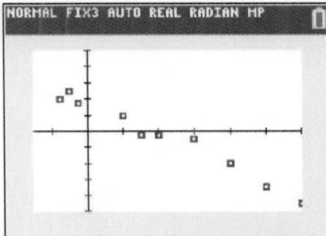

Figure 40

Solution

For each scatter plot, we choose a function whose graph closely resembles the shape of the scatter plot. See Figure 41.

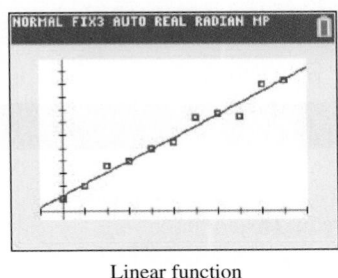 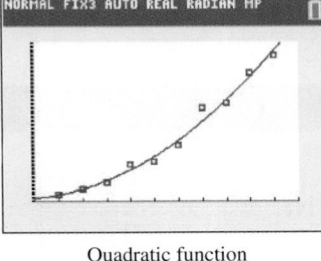 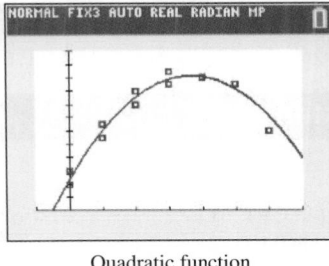 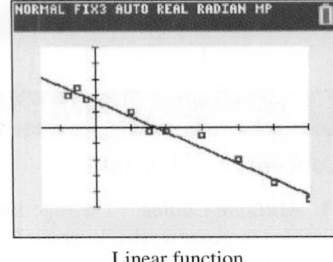

Linear function   Quadratic function   Quadratic function   Linear function

Figure 41

∎

EXAMPLE 5 **Building a Function from Data**

The data shown in Table 4 measure crop yield for various amounts of fertilizer:

(a) Draw a scatter plot of the data and determine a possible type of relation that may exist between the two variables.

(b) Use technology to find the function of best fit to these data.

Solution

(a) Figures 42 and 43 show the data and a scatter plot. The scatter plot suggests the graph of a quadratic function.

TABLE 4

Plot	Fertilizer, x (Pounds/100 ft^2)	Yield (Bushels)
1	0	4
2	0	6
3	5	10
4	5	7
5	10	12
6	10	10
7	15	15
8	15	17
9	20	18
10	20	21
11	25	20
12	25	21
13	30	21
14	30	22
15	35	21
16	35	20
17	40	19
18	40	19

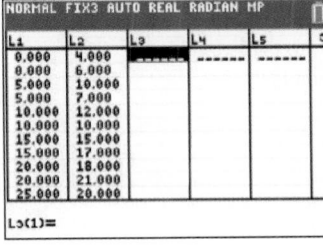

Figure 42

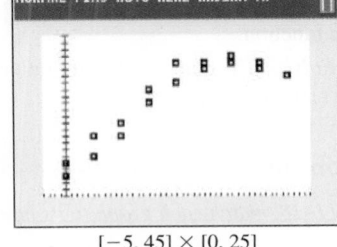

$[-5, 45] \times [0, 25]$

Figure 43

(b) The graphing calculator screen in Figure 44 shows that the quadratic function of best fit is

$$Y(x) = -0.017x^2 + 1.077x + 3.894$$

where x represents the amount of fertilizer used and Y represents crop yield. The graph of the quadratic model is illustrated in Figure 45.

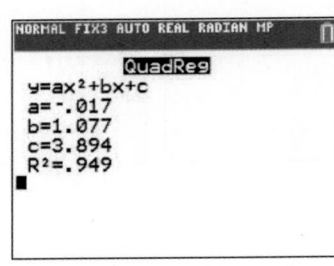

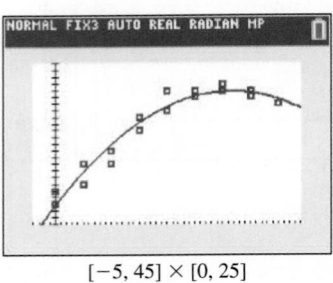

$[-5, 45] \times [0, 25]$

Figure 44   Figure 45

NOW WORK   Problem 31.

P.2 Assess Your Understanding

Concepts and Vocabulary

1. **Multiple Choice** The function $f(x) = x^2$ is [(a) increasing, (b) decreasing, (c) neither] on the interval $[0, \infty)$.

2. **True or False** The floor function $f(x) = \lfloor x \rfloor$ is an example of a step function.

3. **True or False** The cube function is odd and is increasing on the interval $(-\infty, \infty)$.

4. **True or False** The cube root function is decreasing on the interval $(-\infty, \infty)$.

5. **True or False** The domain and the range of the reciprocal function are all real numbers.

6. A number r for which $f(r) = 0$ is called a(n) _____ of the function f.

7. **Multiple Choice** If r is a zero of even multiplicity of a function f, the graph of f [(a) crosses, (b) touches, (c) doesn't intersect] the x-axis at r.

8. **True or False** The x-intercepts of the graph of a polynomial function are called real zeros of the function.

9. **True or False** The function $f(x) = \left[x + \sqrt[5]{x^2 - \pi}\right]^{2/3}$ is an algebraic function.

10. **True or False** The domain of every rational function is the set of all real numbers.

Practice Problems

In Problems 11–18, match each graph to its function:

A. Constant function   B. Identity function
C. Square function   D. Cube function
E. Square root function   F. Reciprocal function
G. Absolute value function H. Cube root function

PAGE 20 11.

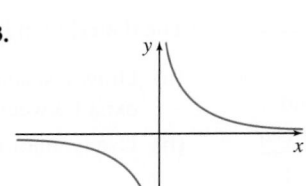

12.

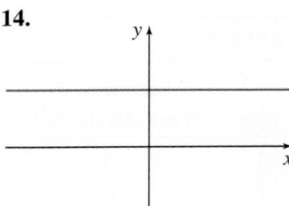

13.

14.

PAGE 20 15.

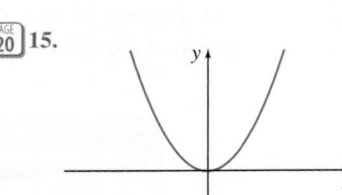

16.

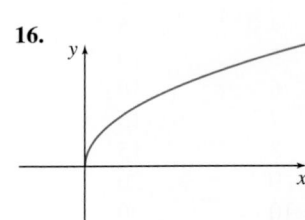

17.

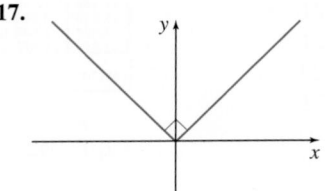

18.

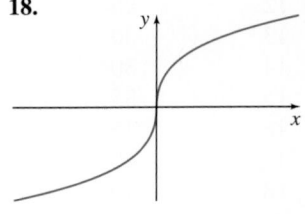

19. If $f(x) = \lfloor 2x \rfloor$, find

  (a) $f(1.2)$ **(b)** $f(1.6)$ **(c)** $f(-1.8)$

20. If $f(x) = \left\lceil \dfrac{x}{2} \right\rceil$, find

  (a) $f(1.2)$ **(b)** $f(1.6)$ **(c)** $f(-1.8)$

In Problems 21 and 22, for each polynomial function f:
(a) List each real zero and its multiplicity.
(b) Find the x- and y-intercepts of the graph of f.
(c) Determine whether the graph of f crosses or touches
    the x-axis at each x-intercept.
(d) Determine the end behavior of the graph of f.

21. $f(x) = 3(x - 7)(x + 4)^3$

22. $f(x) = 4x(x^2 + 1)(x - 2)^3$

In Problems 23 and 24, decide which of the polynomial functions in the list might have the given graph. (More than one answer is possible.)

23. **(a)** $f(x) = -4x(x - 1)(x - 2)$

  (b) $f(x) = x^2(x - 1)^2(x - 2)$

  (c) $f(x) = 3x(x - 1)(x - 2)$

  (d) $f(x) = x(x - 1)^2(x - 2)^2$

  (e) $f(x) = x^3(x - 1)(x - 2)$

  (f) $f(x) = -x(1 - x)(x - 2)$

24. **(a)** $f(x) = 2x^3(x - 1)(x - 2)^2$

  (b) $f(x) = x^2(x - 1)(x - 2)$

  (c) $f(x) = x^3(x - 1)^2(x - 2)$

  (d) $f(x) = x^2(x - 1)^2(x - 2)^2$

  (e) $f(x) = 5x(x - 1)^2(x - 2)$

  (f) $f(x) = -2x(x - 1)^2(2 - x)$

In Problems 25–28, find the domain and the intercepts of each rational function.

25. $R(x) = \dfrac{5x^2}{x + 3}$

26. $H(x) = \dfrac{-4x^2}{(x - 2)(x + 4)}$

27. $R(x) = \dfrac{3x^2 - x}{x^2 + 4}$

28. $R(x) = \dfrac{3(x^2 - x - 6)}{4(x^2 - 9)}$

29. Constructing a Model   The rectangle shown in the figure has one corner in quadrant I on the graph of $y = 16 - x^2$, another corner at the origin, and corners on both the positive y-axis and the positive x-axis. As the corner on $y = 16 - x^2$ changes, a variety of rectangles are obtained.

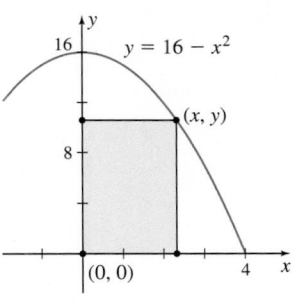

  (a) Express the area A of the rectangles as a function of x.

  (b) What is the domain of A?

30. Constructing a Model   The rectangle shown in the figure is inscribed in a semicircle of radius 2. Let $P = (x, y)$ be the point in quadrant I that is a vertex of the rectangle and is on the circle. As the point (x, y) on the circle changes, a variety of rectangles are obtained.

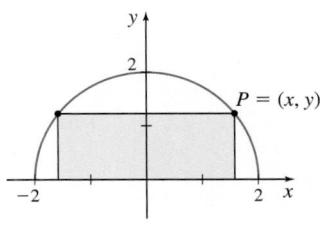

  (a) Express the area A of the rectangles as a function of x.

  (b) Express the perimeter p of the rectangles as a function of x.

31. Height of a Ball   A ballplayer throws a ball at an inclination of 45° to the horizontal. The following data represent the height h (in feet) of the ball at the instant that it has traveled x feet horizontally:

Distance, x	20	40	60	80	100	120	140	160	180	200
Height, h	25	40	55	65	71	77	77	75	71	64

  (a) Draw a scatter plot of the data with x as the independent variable. Comment on the type of relation that may exist between the two variables.

  (b) Use technology to verify that the quadratic function of best fit to these data is

$$h(x) = -0.0037x^2 + 1.03x + 5.7$$

  Use this function to determine the horizontal distance the ball travels before it hits the ground.

  (c) Approximate the height of the ball when it has traveled 10 feet.

32. Geometry

  (a) A wire of length x feet is bent into the shape of a circle. Express the circumference C and area A of the circle as a function of x.

  (b) A wire of length x feet is bent into the shape of a square. Express the perimeter P and area A of the square as a function of x.

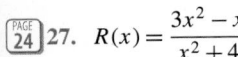

P.3 Operations on Functions; Graphing Techniques

OBJECTIVES *When you finish this section, you should be able to:*

1 Form the sum, difference, product, and quotient of two functions (p. 28)
2 Form a composite function (p. 29)
3 Transform the graph of a function with vertical and horizontal shifts (p. 32)
4 Transform the graph of a function with compressions and stretches (p. 33)
5 Transform the graph of a function by reflecting it about the *x*-axis or the *y*-axis (p. 34)

1 Form the Sum, Difference, Product, and Quotient of Two Functions

Functions, like numbers, can be added, subtracted, multiplied, and divided. For example, the polynomial function $F(x) = x^2 + 4x$ is the sum of the two functions $f(x) = x^2$ and $g(x) = 4x$. The rational function $R(x) = \dfrac{x^2}{x^3 - 1}$, $x \neq 1$, is the quotient of the two functions $f(x) = x^2$ and $g(x) = x^3 - 1$.

DEFINITION Operations on Functions

If f and g are two functions, their sum, $f + g$; their difference, $f - g$; their product, $f \cdot g$; and their quotient, $\dfrac{f}{g}$, are defined by

- Sum: $(f + g)(x) = f(x) + g(x)$
- Difference: $(f - g)(x) = f(x) - g(x)$
- Product: $(f \cdot g)(x) = f(x) \cdot g(x)$
- Quotient: $\left(\dfrac{f}{g}\right)(x) = \dfrac{f(x)}{g(x)}$, $g(x) \neq 0$

For every operation except division, the domain of the resulting function consists of the intersection of the domains of f and g. The domain of a quotient $\dfrac{f}{g}$ consists of the numbers x that are common to the domains of both f and g, but excludes the numbers x for which $g(x) = 0$.

EXAMPLE 1 Forming the Sum, Difference, Product, and Quotient of Two Functions

Let f and g be two functions defined as

$$f(x) = \sqrt{x - 1} \quad \text{and} \quad g(x) = \sqrt{4 - x}$$

Find the following functions and determine their domain:

(a) $(f + g)(x)$ **(b)** $(f - g)(x)$ **(c)** $(f \cdot g)(x)$ **(d)** $\left(\dfrac{f}{g}\right)(x)$

Solution

The domain of f is $\{x \mid x \geq 1\}$, and the domain of g is $\{x \mid x \leq 4\}$.

(a) $(f + g)(x) = f(x) + g(x) = \sqrt{x - 1} + \sqrt{4 - x}$. The domain of $(f + g)(x)$ is the intersection of the domains of f and g. That is, the domain of $(f + g)(x)$ is the closed interval $[1, 4]$.

(b) $(f - g)(x) = f(x) - g(x) = \sqrt{x - 1} - \sqrt{4 - x}$. The domain of $(f - g)(x)$ is the intersection of the domains of f and g. That is, the domain of $(f - g)(x)$ is the closed interval $[1, 4]$.

(c) $(f \cdot g)(x) = f(x) \cdot g(x) = (\sqrt{x - 1})(\sqrt{4 - x}) = \sqrt{-x^2 + 5x - 4}$. The domain of $(f \cdot g)$ is the intersection of the domains of f and g. That is, the domain of $(f \cdot g)(x)$ is the closed interval $[1, 4]$.

(d) $\left(\dfrac{f}{g}\right)(x) = \dfrac{f(x)}{g(x)} = \dfrac{\sqrt{x - 1}}{\sqrt{4 - x}} = \dfrac{\sqrt{-x^2 + 5x - 4}}{4 - x}$. The domain of $\left(\dfrac{f}{g}\right)(x)$ is the intersection of the domains of f and g except where $g(x) = 0$, that is, except for $x = 4$. The domain of $\left(\dfrac{f}{g}\right)(x)$ is the half-open interval $[1, 4)$. ∎

NOW WORK Problem 11.

② Form a Composite Function

Suppose an oil tanker is leaking, and your job requires you to find the area of the circular oil spill surrounding the tanker. You determine that the radius of the spill is increasing at a rate of 3 meters per minute. That is, the radius r of the spill is a function of the time t in minutes since the leak began, and can be written as $r(t) = 3t$.

For example, after 20 minutes the radius of the spill is $r(20) = 3 \cdot 20 = 60$ meters. Recall that the area A of a circle is a function of its radius r; that is, $A(r) = \pi r^2$. So, the area of the oil spill after 20 minutes is $A(60) = \pi (60^2) = 3600\pi$ square meters. Notice that the argument r of the function A is itself a function, and that the area A of the oil spill is found at any time t by evaluating the function $A = A(r(t))$.

Functions such as $A = A(r(t))$ are called *composite functions*.

Another example of a composite function is $y = (2x + 3)^2$. If $y = f(u) = u^2$ and $u = g(x) = 2x + 3$, then substitute $g(x) = 2x + 3$ for u, to get the original function:

$$y = f(u) = f(g(x)) = (2x + 3)^2$$
$$\underset{u = g(x)}{\uparrow} \quad \underset{g(x) = 2x + 3}{\uparrow}$$

In general, suppose that f and g are two functions and that x is a number in the domain of g. By evaluating g at x, we obtain $g(x)$. Now, if $g(x)$ is in the domain of the function f, we can evaluate f at $g(x)$, obtaining $f(g(x))$. The correspondence from x to $f(g(x))$ is called *composition*.

DEFINITION Composite Function

Given two functions f and g, the **composite function**, denoted by $f \circ g$ (read "f composed with g"), is defined by

$$\boxed{(f \circ g)(x) = f(g(x))}$$

The domain of $f \circ g$ is the set of all numbers x in the domain of g for which $g(x)$ is in the domain of f.

Figure 46 illustrates the definition. Only those numbers x in the domain of g for which $g(x)$ is in the domain of f are in the domain of $f \circ g$. As a result, the domain of $f \circ g$ is a subset of the domain of g, and the range of $f \circ g$ is a subset of the range of f.

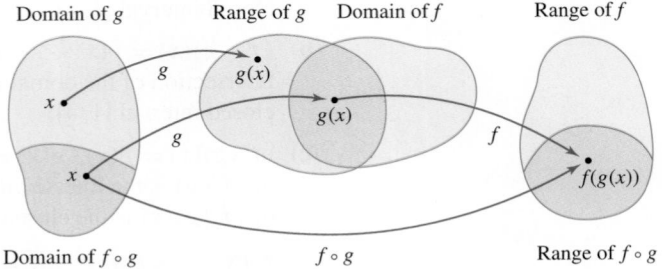

Figure 46

EXAMPLE 2   **Evaluating a Composite Function**

Suppose that $f(x) = \dfrac{1}{x+2}$ and $g(x) = \dfrac{4}{x-1}$.

NOTE The "inside" function g, in $f(g(x))$, is evaluated first.

(a) $(f \circ g)(0) = f(g(0)) = f(-4) = \dfrac{1}{-4+2} = -\dfrac{1}{2}$ $g(x) = \dfrac{4}{x-1}; \; g(0) = -4$

(b) $(g \circ f)(1) = g(f(1)) = g\left(\dfrac{1}{3}\right) = \dfrac{4}{\dfrac{1}{3}-1} = \dfrac{4}{-\dfrac{2}{3}} = -6$
$$f(x) = \dfrac{1}{x+2}; \; f(1) = \dfrac{1}{3}$$

(c) $(f \circ f)(1) = f(f(1)) = f\left(\dfrac{1}{3}\right) = \dfrac{1}{\dfrac{1}{3}+2} = \dfrac{1}{\dfrac{7}{3}} = \dfrac{3}{7}$

(d) $(g \circ g)(-3) = g(g(-3)) = g(-1) = \dfrac{4}{-1-1} = -2$
$$g(x) = \dfrac{4}{x-1}; \; g(-3) = \dfrac{4}{-3-1} = -1$$

A composite function can also be evaluated using technology. See Figure 47 for the answers to Example 2 using a graphing calculator.

NOW WORK   Problem 15.

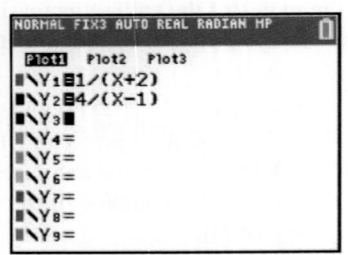

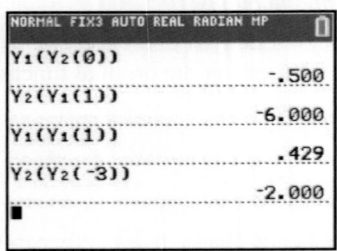

Figure 47

EXAMPLE 3   **Finding the Domain of a Composite Function**

Suppose that $f(x) = \dfrac{1}{x+2}$ and $g(x) = \dfrac{4}{x-1}$. Find $f \circ g$ and its domain.

Solution

$$(f \circ g)(x) = f(g(x)) = \dfrac{1}{g(x)+2} = \dfrac{1}{\dfrac{4}{x-1}+2} = \dfrac{x-1}{4+2(x-1)} = \dfrac{x-1}{2x+2}$$

To find the domain of $f \circ g$, first note that the domain of g is $\{x \mid x \neq 1\}$, so we exclude 1 from the domain of $f \circ g$. Next note that the domain of f is $\{x \mid x \neq -2\}$, which means $g(x)$ cannot equal -2.

To determine what additional values of x to exclude, we solve the equation $g(x) = -2$:

$$\frac{4}{x-1} = -2 \qquad\qquad g(x) = -2$$

$$4 = -2(x-1)$$

$$4 = -2x + 2$$

$$2x = -2$$

$$x = -1$$

We also exclude -1 from the domain of $f \circ g$.

The domain of $f \circ g$ is $\{x \mid x \neq -1, x \neq 1\}$. ∎

We could also find the domain of the function $f \circ g$ by first finding the domain of g: $\{x \mid x \neq 1\}$. So, exclude 1 from the domain of $f \circ g$. Then looking at $(f \circ g)(x) = \dfrac{x-1}{2x+2} = \dfrac{x-1}{2(x+1)}$, notice that $x \neq -1$, so we exclude -1 from the domain of $f \circ g$. Therefore, the domain of $f \circ g$ is $\{x \mid x \neq -1, x \neq 1\}$.

NOW WORK Problem 23.

In general, the composition of two functions f and g is not commutative. That is, $f \circ g$ almost never equals $g \circ f$. For example, in Example 3,

$$(g \circ f)(x) = g(f(x)) = \frac{4}{f(x)-1} = \frac{4}{\dfrac{1}{x+2}-1} = \frac{4(x+2)}{1-(x+2)} = -\frac{4(x+2)}{x+1}$$

Functions f and g for which $f \circ g = g \circ f$ are discussed in Section P.4.

Some techniques in calculus require us to "decompose" a composite function. For example, the function $H(x) = \sqrt{x+1}$ is the composition $f \circ g$ of the functions $f(x) = \sqrt{x}$ and $g(x) = x + 1$.

▶ CALC CLIP

EXAMPLE 4 Decomposing a Composite Function

Find functions f and g so that $f \circ g = F$ when:

(a) $F(x) = \dfrac{1}{x+1}$ **(b)** $F(x) = \left(x^3 - 4x - 1\right)^{100}$ **(c)** $F(t) = \sqrt{2-t}$

Solution

(a) If we let $f(x) = \dfrac{1}{x}$ and $g(x) = x + 1$, then

$$(f \circ g)(x) = f(g(x)) = \frac{1}{g(x)} = \frac{1}{x+1} = F(x)$$

(b) If we let $f(x) = x^{100}$ and $g(x) = x^3 - 4x - 1$, then

$$(f \circ g)(x) = f(g(x)) = f\left(x^3 - 4x - 1\right) = \left(x^3 - 4x - 1\right)^{100} = F(x)$$

(c) If we let $f(t) = \sqrt{t}$ and $g(t) = 2 - t$, then

$$(f \circ g)(t) = f(g(t)) = f(2-t) = \sqrt{2-t} = F(t)$$ ∎

Although the functions f and g chosen in Example 4 are not unique, there is usually a "natural" selection for f and g that first comes to mind. When decomposing a composite function, the "natural" selection for g is often an expression inside parentheses (as in part (b)), in a denominator (as in part (a)), or under a radical (as in part (c)).

NOW WORK Problem 27.

③ Transform the Graph of a Function with Vertical and Horizontal Shifts

At times we need to graph a function that is very similar to a function with a known graph. Often techniques, called **transformations**, can be used to draw the new graph.

First we consider *translations*. **Translations** shift the graph from one position to another without changing its shape, size, or direction.

For example, suppose f is a function with a known graph, say, $f(x) = x^2$. If k is a positive number, then adding k to f adds k to each y-coordinate, causing the graph of f to **shift vertically up** k units. On the other hand, subtracting k from f subtracts k from each y-coordinate, causing the graph of f to **shift vertically down** k units. See Figure 48.

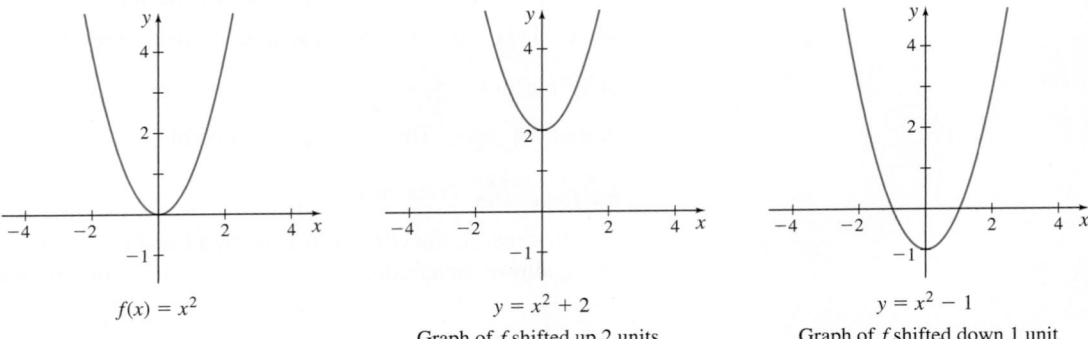

$f(x) = x^2$

$y = x^2 + 2$
Graph of f shifted up 2 units

$y = x^2 - 1$
Graph of f shifted down 1 unit

Figure 48

So, adding (or subtracting) a positive number to a function shifts the graph of the original function vertically up (or down). Now we investigate how to shift the graph of a function right or left.

Again, suppose f is a function with a known graph, say, $f(x) = x^2$, and let h be a positive number. To shift the graph to the right h units, subtract h from the argument of f. In other words, replace the argument x of a function f by $x - h, h > 0$. The graph of the new function $y = f(x - h)$ is the graph of f **shifted horizontally right** h units. On the other hand, if we replace the argument x of a function f by $x + h, h > 0$, the graph of the new function $y = f(x + h)$ is the graph of f **shifted horizontally left** h units. See Figure 49.

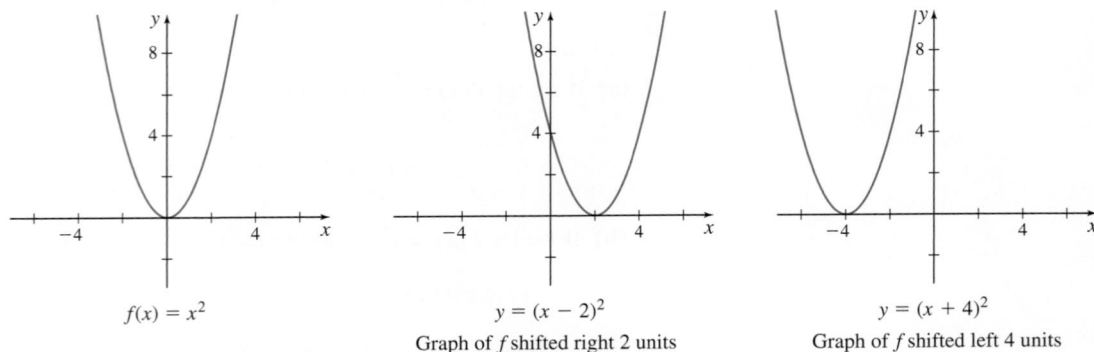

$f(x) = x^2$

$y = (x - 2)^2$
Graph of f shifted right 2 units

$y = (x + 4)^2$
Graph of f shifted left 4 units

Figure 49

The graph of a function f can be moved anywhere in the plane by combining vertical and horizontal shifts.

EXAMPLE 5   **Combining Vertical and Horizontal Shifts**

Use transformations to graph the function $f(x) = (x + 3)^2 - 5$.

Solution

Graph f in steps:

- Observe that f is basically a square function, so first graph $y = x^2$. See Figure 50(a).
- Replace the argument x with $x + 3$ to obtain $y = (x + 3)^2$. This shifts the graph of f horizontally to the left 3 units, as shown in Figure 50(b).
- Finally, subtract 5 from each y-coordinate, which shifts the graph in Figure 50(b) vertically down 5 units, and results in the graph of $f(x) = (x + 3)^2 - 5$ shown in Figure 50(c).

Notice the points plotted in Figure 50. Using key points can be helpful in keeping track of the transformation that has taken place.

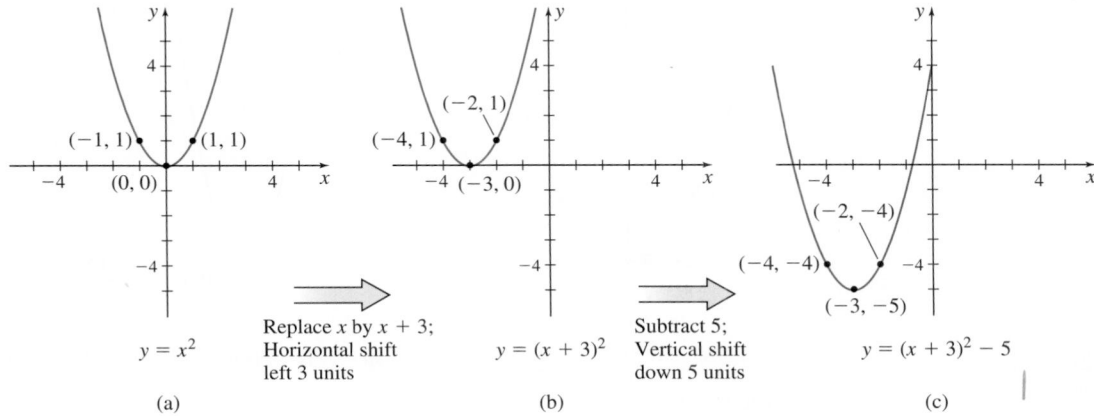

(a)   Replace x by $x + 3$; Horizontal shift left 3 units   (b)   Subtract 5; Vertical shift down 5 units   (c)

$y = x^2$ $y = (x + 3)^2$ $y = (x + 3)^2 - 5$

Figure 50

NOW WORK   Problem 39.

4 Transform the Graph of a Function with Compressions and Stretches

When a function f is multiplied by a positive number a, the graph of the new function $y = af(x)$ is obtained by multiplying each y-coordinate on the graph of f by a. The new graph is a **vertically compressed** version of the graph of f if $0 < a < 1$ and is a **vertically stretched** version of the graph of f if $a > 1$. Compressions and stretches change the proportions of a graph.

For example, the graph of $f(x) = x^2$ is shown in Figure 51(a). Multiplying f by $a = \dfrac{1}{2}$ produces a new function $y = \dfrac{1}{2}f(x) = \dfrac{1}{2}x^2$, which vertically compresses the graph of f by a factor of $\dfrac{1}{2}$, as shown in Figure 51(b). On the other hand, if $a = 3$, then multiplying f by 3 produces a new function $y = 3f(x) = 3x^2$, and the graph of f is vertically stretched by a factor of 3, as shown in Figure 51(c).

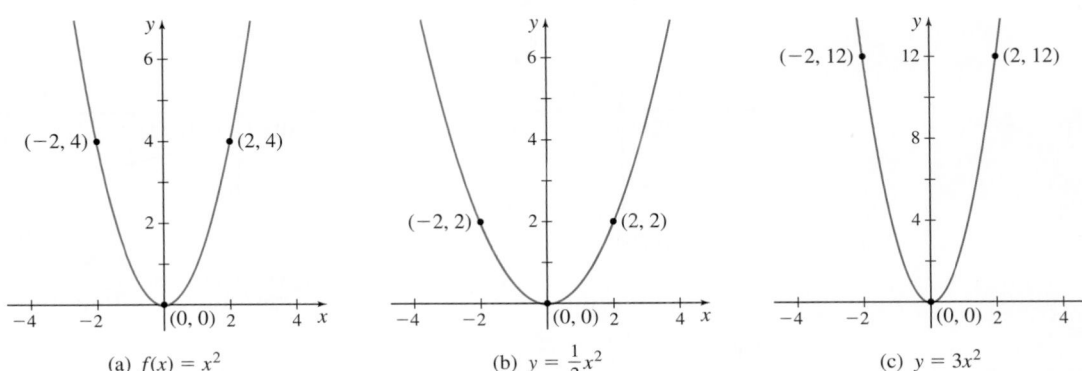

(a) $f(x) = x^2$ (b) $y = \dfrac{1}{2}x^2$ (c) $y = 3x^2$

Graph of f is vertically compressed   Graph of f is vertically stretched

Figure 51

If the argument x of a function f is multiplied by a positive number a, the graph of the new function $y = f(ax)$ is a **horizontal compression** of the graph of f when $a > 1$, and a **horizontal stretch** of the graph of f when $0 < a < 1$.

For example, the graph of $y = f(2x) = (2x)^2 = 4x^2$ is a horizontal compression of the graph of $f(x) = x^2$. See Figures 52(a) and 52(b). On the other hand, if $a = \dfrac{1}{3}$, then the graph of $y = \left(\dfrac{1}{3}x\right)^2 = \dfrac{1}{9}x^2$ is a horizontal stretch of the graph of $f(x) = x^2$. See Figure 52(c).

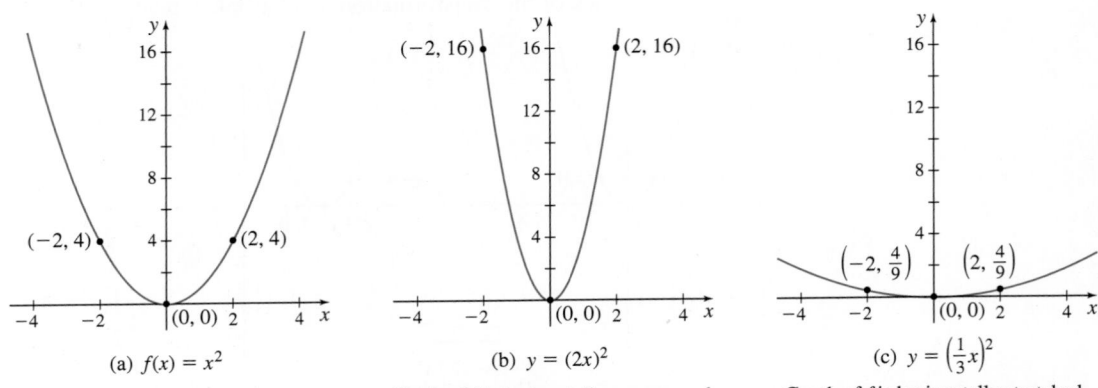

(a) $f(x) = x^2$

(b) $y = (2x)^2$

Graph of f is horizontally compressed

(c) $y = \left(\dfrac{1}{3}x\right)^2$

Graph of f is horizontally stretched

Figure 52

⑤ Transform the Graph of a Function by Reflecting It about the x-Axis or the y-Axis

The third type of transformation, reflection about the x- or y-axis, changes the orientation of the graph of the function f but keeps the shape and the size of the graph intact.

When a function f is multiplied by -1, the graph of the new function $y = -f(x)$ is the **reflection about the x-axis** of the graph of f. For example, if $f(x) = \sqrt{x}$, then the graph of the new function $y = -f(x) = -\sqrt{x}$ is the reflection of the graph of f about the x-axis. See Figures 53(a) and 53(b).

If the argument x of a function f is multiplied by -1, then the graph of the new function $y = f(-x)$ is the **reflection about the y-axis** of the graph of f. For example, if $f(x) = \sqrt{x}$, then the graph of the new function $y = f(-x) = \sqrt{-x}$ is the reflection of the graph of f about the y-axis. See Figures 53(a) and 53(c). Notice in this example that the domain of $y = \sqrt{-x}$ is all real numbers for which $-x \geq 0$, or equivalently, $x \leq 0$.

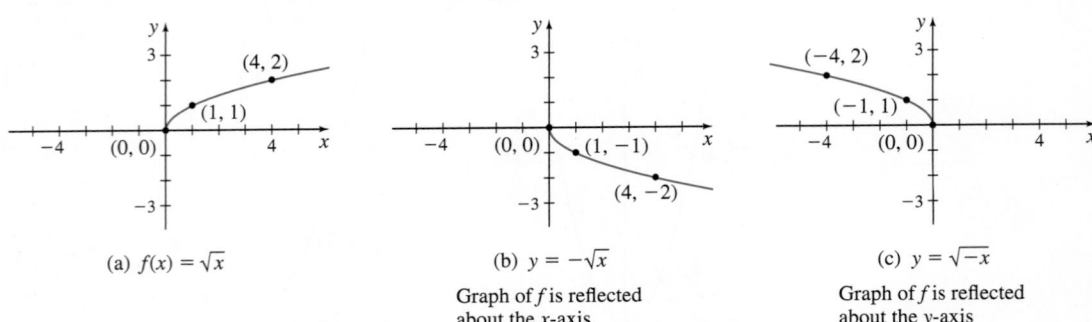

(a) $f(x) = \sqrt{x}$

(b) $y = -\sqrt{x}$

Graph of f is reflected about the x-axis

(c) $y = \sqrt{-x}$

Graph of f is reflected about the y-axis

Figure 53

EXAMPLE 6 **Combining Transformations**

Use transformations to graph the function $f(x) = \sqrt{1-x} + 2$.

Solution

We graph f in steps:

- Observe that f is basically a square root function, so begin by graphing $y = \sqrt{x}$. See Figure 54(a).
- Now replace the argument x with $x + 1$ to obtain $y = \sqrt{x+1}$, which shifts the graph of $y = \sqrt{x}$ horizontally to the left 1 unit, as shown in Figure 54(b).
- Then replace x with $-x$ to obtain $y = \sqrt{-x+1} = \sqrt{1-x}$, which reflects the graph about the y-axis. See Figure 54(c).
- Finally, add 2 to each y-coordinate, which shifts the graph vertically up 2 units, and results in the graph of $f(x) = \sqrt{1-x} + 2$ shown in Figure 54(d).

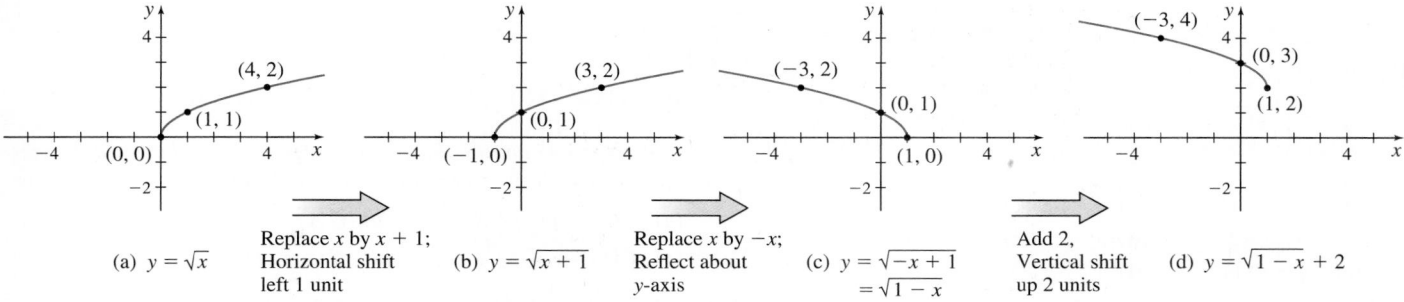

(a) $y = \sqrt{x}$ Replace x by $x + 1$; Horizontal shift left 1 unit   (b) $y = \sqrt{x+1}$ Replace x by $-x$; Reflect about y-axis   (c) $y = \sqrt{-x+1}$ $= \sqrt{1-x}$ Add 2, Vertical shift up 2 units   (d) $y = \sqrt{1-x} + 2$

Figure 54

NOW WORK Problem 43.

P.3 Assess Your Understanding

Concepts and Vocabulary

1. If the domain of a function f is $\{x \mid 0 \le x \le 7\}$ and the domain of a function g is $\{x \mid -2 \le x \le 5\}$, then the domain of the sum function $f + g$ is _____.

2. **True or False** If f and g are functions, then the domain of $\dfrac{f}{g}$ consists of all numbers x that are in the domains of both f and g.

3. **True or False** The domain of $f \cdot g$ consists of the numbers x that are in the domains of both f and g.

4. **True or False** The domain of the composite function $f \circ g$ is the same as the domain of $g(x)$.

5. **True or False** The graph of $y = -f(x)$ is the reflection of the graph of $y = f(x)$ about the x-axis.

6. **True or False** To obtain the graph of $y = f(x + 2) - 3$, shift the graph of $y = f(x)$ horizontally to the right 2 units and vertically down 3 units.

7. **True or False** Suppose the x-intercepts of the graph of the function f are -3 and 2. Then the x-intercepts of the graph of the function $y = 2f(x)$ are -3 and 2.

8. Suppose that the graph of a function f is known. Then the graph of the function $y = f(x - 2)$ can be obtained by a(n) _____ shift of the graph of f to the _____ a distance of 2 units.

9. Suppose that the graph of a function f is known. Then the graph of the function $y = f(-x)$ can be obtained by a reflection about the _____-axis of the graph of f.

10. Suppose the x-intercepts of the graph of the function f are -2, 1, and 5. The x-intercepts of $y = f(x + 3)$ are _____, _____, and _____.

Practice Problems

In Problems 11–14, the functions f and g are given. Find each of the following functions and determine their domain:

(a) $(f + g)(x)$ **(b)** $(f - g)(x)$

(c) $(f \cdot g)(x)$ **(d)** $\left(\dfrac{f}{g}\right)(x)$

11. $f(x) = 3x + 4$ and $g(x) = 2x - 3$

12. $f(x) = 1 + \dfrac{1}{x}$ and $g(x) = \dfrac{1}{x}$

13. $f(x) = \sqrt{x+1}$ and $g(x) = \dfrac{2}{x}$

14. $f(x) = |x|$ and $g(x) = x$

In Problems 15 and 16, for each of the functions f and g, find:

(a) $(f \circ g)(4)$ **(b)** $(g \circ f)(2)$

(c) $(f \circ f)(1)$ **(d)** $(g \circ g)(0)$

15. $f(x) = 2x$ and $g(x) = 3x^2 + 1$

16. $f(x) = \dfrac{3}{x+1}$ and $g(x) = \sqrt{x}$

In Problems 17 and 18, evaluate each expression using the values given in the table.

17.

x	-3	-2	-1	0	1	2	3
$f(x)$	-7	-5	-3	-1	3	5	7
$g(x)$	8	3	0	-1	0	3	8

(a) $(f \circ g)(1)$ **(b)** $(f \circ g)(-1)$

(c) $(g \circ f)(-1)$ **(d)** $(g \circ f)(1)$

(e) $(g \circ g)(-2)$ **(f)** $(f \circ f)(-1)$

18.

x	-3	-2	-1	0	1	2	3
$f(x)$	11	9	7	5	3	1	-1
$g(x)$	-8	-3	0	1	0	-3	-8

(a) $(f \circ g)(1)$ **(b)** $(f \circ g)(2)$

(c) $(g \circ f)(2)$ **(d)** $(g \circ f)(3)$

(e) $(g \circ g)(1)$ **(f)** $(f \circ f)(3)$

In Problems 19 and 20, evaluate each composite function using the graphs of $y = f(x)$ and $y = g(x)$ shown in the figure below.

19. (a) $(g \circ f)(-1)$ **(b)** $(g \circ f)(6)$

(c) $(f \circ g)(6)$ **(d)** $(f \circ g)(4)$

20. (a) $(g \circ f)(1)$ **(b)** $(g \circ f)(5)$

(c) $(f \circ g)(7)$ **(d)** $(f \circ g)(2)$

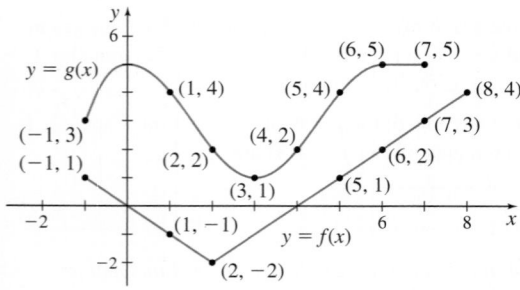

In Problems 21–26, for the given functions f and g, find:

(a) $f \circ g$ **(b)** $g \circ f$ **(c)** $f \circ f$ **(d)** $g \circ g$

State the domain of each composite function.

21. $f(x) = 3x + 1$ and $g(x) = 8x$

22. $f(x) = -x$ and $g(x) = 2x - 4$

23. $f(x) = x^2 + 1$ and $g(x) = \sqrt{x - 1}$

24. $f(x) = 2x + 3$ and $g(x) = \sqrt{x}$

25. $f(x) = \dfrac{x}{x-1}$ and $g(x) = \dfrac{2}{x}$

26. $f(x) = \dfrac{1}{x+3}$ and $g(x) = -\dfrac{2}{x}$

In Problems 27–32, find functions f and g so that $f \circ g = F$.

27. $F(x) = (2x + 3)^4$ **28.** $F(x) = \left(1 + x^2\right)^3$

29. $F(x) = \sqrt{x^2 + 1}$ **30.** $F(x) = \sqrt{1 - x^2}$

31. $F(x) = |2x + 1|$ **32.** $F(x) = |2x^2 + 3|$

In Problems 33–46, graph each function using the graphing techniques of shifting, compressing, stretching, and/or reflecting. Begin with the graph of a basic function and show all stages.

33. $f(x) = x^3 + 2$ **34.** $g(x) = x^3 - 1$

35. $h(x) = \sqrt{x - 2}$ **36.** $f(x) = \sqrt{x + 1}$

37. $g(x) = 4\sqrt{x}$ **38.** $f(x) = \dfrac{1}{2}\sqrt{x}$

39. $f(x) = (x - 1)^3 + 2$ **40.** $g(x) = 3(x - 2)^2 + 1$

41. $h(x) = \dfrac{1}{2x}$ **42.** $f(x) = \dfrac{4}{x} + 2$

43. $G(x) = 2|1 - x|$ **44.** $g(x) = -(x + 1)^3 - 1$

45. $g(x) = -4\sqrt{x - 1}$ **46.** $f(x) = 4\sqrt{2 - x}$

In Problems 47 and 48, the graph of a function f is illustrated. Use the graph of f as the first step in graphing each of the following functions:

(a) $F(x) = f(x) + 3$ **(b)** $G(x) = f(x + 2)$

(c) $P(x) = -f(x)$ **(d)** $H(x) = f(x + 1) - 2$

(e) $Q(x) = \dfrac{1}{2}f(x)$ **(f)** $g(x) = f(-x)$

(g) $h(x) = f(2x)$

47.

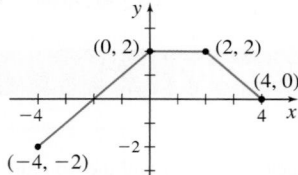

48.

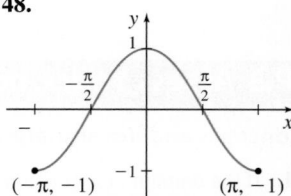

49. Period of a Pendulum   The period T (in seconds) of a simple pendulum is a function of its length l (in meters) defined by the equation

$$T = T(l) = 2\pi\sqrt{\dfrac{l}{g}}$$

where $g \approx 9.8$ meters/second^2 is the acceleration due to gravity.

(a) Use technology to graph the function $T = T(l)$.

(b) Now graph the functions $T = T(l + 1)$, $T = T(l + 2)$, and $T = T(l + 3)$.

(c) Discuss how adding to the length l changes the period T.

(d) Now graph the functions $T = T(2l)$, $T = T(3l)$, and $T = T(4l)$.

(e) Discuss how multiplying the length l by 2, 3, and 4 changes the period T.

50. Suppose $(1, 3)$ is a point on the graph of $y = g(x)$.

(a) What point is on the graph of $y = g(x + 3) - 5$?

(b) What point is on the graph of $y = -2g(x - 2) + 1$?

(c) What point is on the graph of $y = g(2x + 3)$?

P.4 Inverse Functions

OBJECTIVES *When you finish this section, you should be able to:*

1. Determine whether a function is one-to-one (p. 37)
2. Determine the inverse of a function defined by a set of ordered pairs (p. 38)
3. Obtain the graph of the inverse function from the graph of a one-to-one function (p. 39)
4. Find the inverse of a one-to-one function defined by an equation (p. 40)

1 Determine Whether a Function Is One-to-One

By definition, for a function $y = f(x)$, if x is in the domain of f, then x has one, and only one, image y in the range. If a function f also has the property that no y in the range of f is the image of more than one x in the domain, then the function is called a *one-to-one function*.

> **DEFINITION** One-to-One Function
>
> A function f is a **one-to-one function** if any two different inputs in the domain correspond to two different outputs in the range. That is, if $x_1 \neq x_2$, then $f(x_1) \neq f(x_2)$.

IN WORDS A function is not one-to-one if two different inputs in the domain correspond to the same output.

Figure 55 illustrates the distinction among one-to-one functions, functions that are not one-to-one, and relations that are not functions.

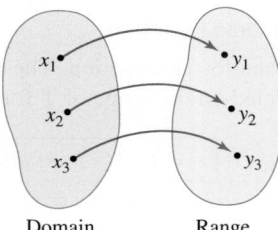
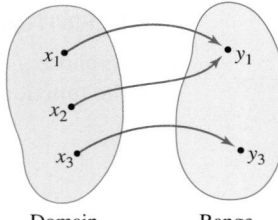
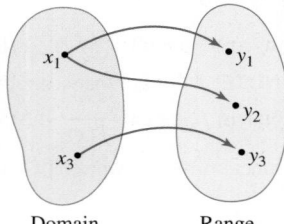

Domain Range	Domain Range	Domain Range
(a) One-to one function: Each x in the domain has one and only one image in the range	(b) Not a one-to-one function: y_1 is the image of both x_1 and x_2	(c) Not a function: x_1 has two images, y_1 and y_2

Figure 55

If the graph of a function f is known, there is a simple test called the *Horizontal-line Test,* to determine whether f is a one-to-one function.

> **THEOREM** Horizontal-line Test
>
> The graph of a function in the xy-plane is the graph of a one-to-one function if and only if every horizontal line intersects the graph in at most one point.

EXAMPLE 1 Determining Whether a Function Is One-to-One

Determine whether each function is one-to-one:

(a) $f(x) = x^2$ **(b)** $g(x) = x^3$

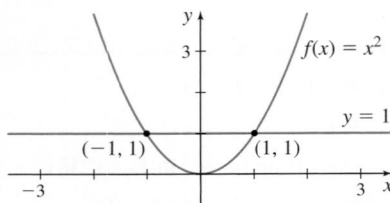

A horizontal line intersects the graph twice; f is not one-to-one

Figure 56

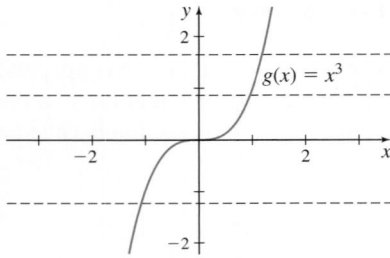

Every horizontal line intersects the graph exactly once; g is one-to-one

Figure 57

NOTE f^{-1} is not the reciprocal function. That is, $f^{-1}(x) \neq \dfrac{1}{f(x)}$. The reciprocal function $\dfrac{1}{f(x)}$ is written $[f(x)]^{-1}$.

Solution

(a) Figure 56 illustrates the Horizontal-line Test for the graph of $f(x) = x^2$. The horizontal line $y = 1$ intersects the graph of f twice, at $(1, 1)$ and at $(-1, 1)$, so f is not one-to-one.

(b) Figure 57 illustrates the Horizontal-line Test for the graph of $g(x) = x^3$. Because every horizontal line intersects the graph of g exactly once, it follows that g is one-to-one.

∎

NOW WORK   Problem 9.

Notice that the one-to-one function $g(x) = x^3$ also is an increasing function on its domain. Because an increasing (or decreasing) function always has different y-values for different x-values, a function that is increasing (or decreasing) on an interval is also a one-to-one function on that interval.

THEOREM One-to-One Function

- A function that is increasing on an interval I is a one-to-one function on I.
- A function that is decreasing on an interval I is a one-to-one function on I.

Suppose that f is a one-to-one function. Then to each x in the domain of f, there is exactly one image y in the range (because f is a function); and to each y in the range of f, there is exactly one x in the domain (because f is one-to-one). The correspondence from the range of f back to the domain of f is also a function, called the *inverse function of f*. The symbol f^{-1} is used to denote the inverse function of f.

DEFINITION Inverse Function

Suppose f is a one-to-one function. The **inverse function of** f, denoted by f^{-1}, is the function defined on the range of f for which

$$x = f^{-1}(y) \quad \text{if and only if} \quad y = f(x)$$

We discuss how to find inverses for three representations of functions: (1) sets of ordered pairs, (2) graphs, and (3) equations. We begin with finding the inverse of a function represented by a set of ordered pairs.

2 Determine the Inverse of a Function Defined by a Set of Ordered Pairs

If the one-to-one function f is a set of ordered pairs (x, y), then the inverse function of f, denoted f^{-1}, is the set of ordered pairs (y, x).

EXAMPLE 2   Finding the Inverse of a Function Defined by a Set of Ordered Pairs

Find the inverse of the one-to-one function:

$$\{(-3, -5), (-1, 1), (0, 2), (1, 3)\}$$

State the domain and the range of the function and its inverse.

Solution

The inverse of the function is found by interchanging the entries in each ordered pair. The inverse function is

$$\{(-5, -3), (1, -1), (2, 0), (3, 1)\}$$

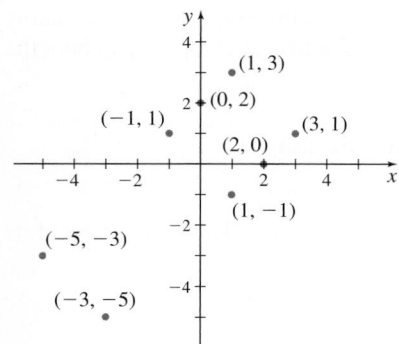

Figure 58

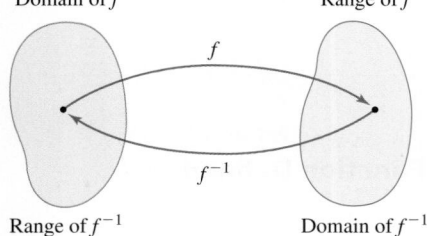

Figure 59

The domain of the function is $\{-3, -1, 0, 1\}$; the range of the function is $\{-5, 1, 2, 3\}$. The domain of the inverse function is $\{-5, 1, 2, 3\}$; the range of the inverse function is $\{-3, -1, 0, 1\}$. ■

Figure 58 shows the one-to-one function (in blue) and its inverse (in red).

NOW WORK Problem 19.

Remember, if f is a one-to-one function, it has an inverse function f^{-1}. See Figure 59. Based on the results of Example 2 and Figure 58, two properties of a one-to-one function f and its inverse function f^{-1} become apparent:

$$\boxed{\text{Domain of } f = \text{ Range of } f^{-1} \qquad \text{Range of } f = \text{ Domain of } f^{-1}}$$

The next theorem provides a means for verifying that two functions are inverses of one another.

THEOREM

Given a one-to-one function f and its inverse function f^{-1}, then

- $(f^{-1} \circ f)(x) = f^{-1}(f(x)) = x$ where x is in the domain of f
- $(f \circ f^{-1})(x) = f(f^{-1}(x)) = x$ where x is in the domain of f^{-1}

EXAMPLE 3 **Verifying Inverse Functions**

Verify that the inverse function of $f(x) = \dfrac{1}{x-1}$ is $f^{-1}(x) = \dfrac{1}{x} + 1$. For what values of x is $f^{-1}(f(x)) = x$? For what values of x is $f\left(f^{-1}(x)\right) = x$?

Solution

The domain of f is $\{x \mid x \neq 1\}$ and the domain of f^{-1} is $\{x \mid x \neq 0\}$. Now

$$f^{-1}(f(x)) = f^{-1}\left(\frac{1}{x-1}\right) = \frac{1}{\left(\dfrac{1}{x-1}\right)} + 1 = x - 1 + 1 = x, \quad \text{provided } x \neq 1$$

$$f(f^{-1}(x)) = f\left(\frac{1}{x}+1\right) = \frac{1}{\left(\dfrac{1}{x}+1\right)-1} = \frac{1}{\dfrac{1}{x}} = x, \quad \text{provided } x \neq 0 \qquad ■$$

NOW WORK Problem 15.

3 Obtain the Graph of the Inverse Function from the Graph of a One-to-One Function

Suppose (a, b) is a point on the graph of a one-to-one function f defined by $y = f(x)$. Then $b = f(a)$. This means that $a = f^{-1}(b)$, so (b, a) is a point on the graph of the inverse function f^{-1}. Figure 60 shows the relationship between the point (a, b) on the graph of f and the point (b, a) on the graph of f^{-1}. The line segment containing (a, b) and (b, a) is perpendicular to the line $y = x$ and is bisected by the line $y = x$. (Do you see why?) The point (b, a) on the graph of f^{-1} is the reflection about the line $y = x$ of the point (a, b) on the graph of f.

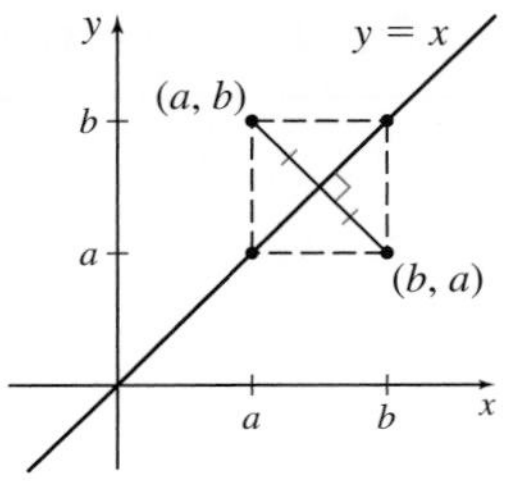

Figure 60

THEOREM Symmetry of Inverse Functions

The graph of a one-to-one function f and the graph of its inverse function f^{-1} are symmetric with respect to the line $y = x$.

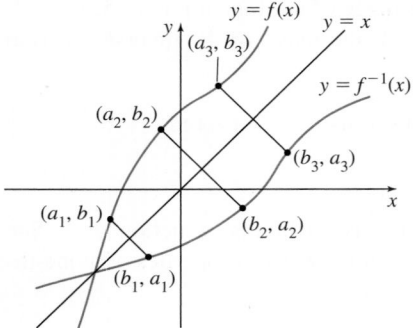

Figure 61

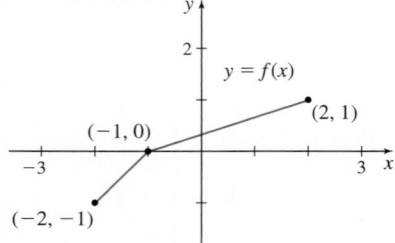

Figure 62

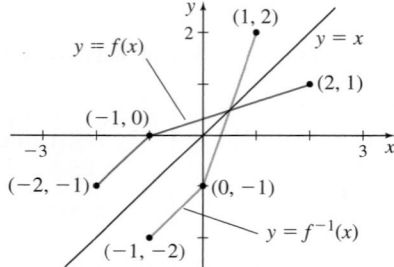

Figure 63

We can use this result to find the graph of f^{-1} given the graph of f. If we know the graph of f, then the graph of f^{-1} is obtained by reflecting the graph of f about the line $y = x$. See Figure 61.

EXAMPLE 4   Graphing the Inverse Function from the Graph of a Function

The graph in Figure 62 is that of a one-to-one function $y = f(x)$. Draw the graph of its inverse function.

Solution

Since the points $(-2, -1)$, $(-1, 0)$, and $(2, 1)$ are on the graph of f, the points $(-1, -2)$, $(0, -1)$, and $(1, 2)$ are on the graph of f^{-1}. Using these points and the fact that the graph of f^{-1} is the reflection about the line $y = x$ of the graph of f, draw the graph of f^{-1}, as shown in Figure 63.   ∎

NOW WORK   Problem 27.

❹ Find the Inverse of a One-to-One Function Defined by an Equation

Since the graphs of a one-to-one function f and its inverse function f^{-1} are symmetric with respect to the line $y = x$, the inverse function f^{-1} can be obtained by interchanging the roles of x and y in f. If f is defined by the equation

$$y = f(x)$$

then f^{-1} is defined by the equation

$$x = f(y) \qquad \text{Interchange } x \text{ and } y$$

The equation $x = f(y)$ defines f^{-1} *implicitly*. If the implicit equation can be solved for y, we will have the *explicit* form of f^{-1}, that is,

$$y = f^{-1}(x)$$

Steps for Finding the Inverse of a One-to-One Function

Step 1 Write f in the form $y = f(x)$.

Step 2 Interchange the variables x and y to obtain $x = f(y)$. This equation defines the inverse function f^{-1} implicitly.

Step 3 If possible, solve the implicit equation for y in terms of x to obtain the explicit form of f^{-1}: $y = f^{-1}(x)$.

Step 4 Check the result by showing that $f^{-1}(f(x)) = x$ and $f(f^{-1}(x)) = x$.

EXAMPLE 5   Finding the Inverse Function

CALC CLIP

The function $f(x) = 2x^3 - 1$ is one-to-one. Find its inverse.

Solution

We follow the steps given above.

Step 1 Write f as $y = 2x^3 - 1$.

Step 2 Interchange the variables x and y.

$$x = 2y^3 - 1$$

This equation defines f^{-1} implicitly.

Step 3 Solve the implicit form of the inverse function for y.

$$x + 1 = 2y^3$$
$$y^3 = \frac{x+1}{2}$$
$$y = \sqrt[3]{\frac{x+1}{2}} = f^{-1}(x)$$

Step 4 Check the result.

$$f^{-1}(f(x)) = f^{-1}(2x^3 - 1) = \sqrt[3]{\frac{(2x^3 - 1) + 1}{2}} = \sqrt[3]{\frac{2x^3}{2}} = \sqrt[3]{x^3} = x$$

$$f(f^{-1}(x)) = f\left(\sqrt[3]{\frac{x+1}{2}}\right) = 2\left(\sqrt[3]{\frac{x+1}{2}}\right)^3 - 1 = 2\left(\frac{x+1}{2}\right) - 1$$
$$= x + 1 - 1 = x \qquad \blacksquare$$

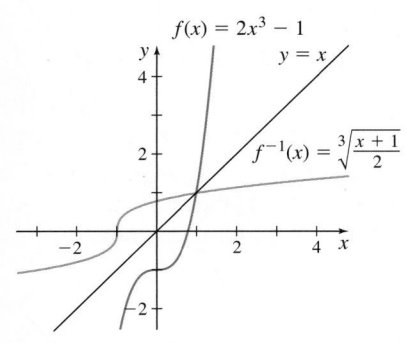

$f(x) = 2x^3 - 1$

$y = x$

$f^{-1}(x) = \sqrt[3]{\frac{x+1}{2}}$

Figure 64

See Figure 64 for the graphs of f and f^{-1}.

NOW WORK   Problem 31.

If a function f is not one-to-one, it has no inverse function. But sometimes we can restrict the domain of such a function so that it becomes a one-to-one function. Then on the restricted domain the new function has an inverse function.

EXAMPLE 6   **Finding the Inverse of a Domain-Restricted Function**

Find the inverse of $f(x) = x^2$ if $x \geq 0$.

NEED TO REVIEW? Principal roots are discussed in Appendix A.1, p. A-9.

Solution

The function $f(x) = x^2$ is not one-to-one (see Example 1(a)). However, by restricting the domain of f to $x \geq 0$, the new function f is one-to-one, so f^{-1} exists. To find f^{-1}, follow the steps.

Step 1 $y = x^2$, where $x \geq 0$.

Step 2 Interchange the variables x and y: $x = y^2$, where $y \geq 0$. This is the inverse function written implicitly.

Step 3 Solve for y: $y = \sqrt{x} = f^{-1}(x)$. (Since $y \geq 0$, only the principal square root is obtained.)

Step 4 Check that $f^{-1}(x) = \sqrt{x}$ is the inverse function of f.

$$f^{-1}(f(x)) = \sqrt{f(x)} = \sqrt{x^2} = |x| = x, \qquad \text{where } x \geq 0$$
$$f(f^{-1}(x)) = [f^{-1}(x)]^2 = [\sqrt{x}]^2 = x, \qquad \text{where } x \geq 0 \qquad \blacksquare$$

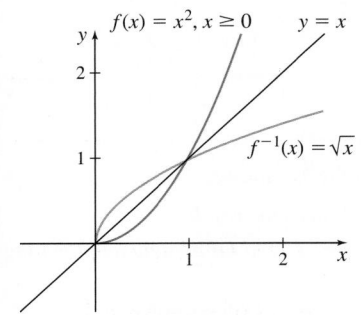

$f(x) = x^2, x \geq 0$

$y = x$

$f^{-1}(x) = \sqrt{x}$

Figure 65

The graphs of $f(x) = x^2$, $x \geq 0$, and $f^{-1}(x) = \sqrt{x}$ are shown in Figure 65.

NOW WORK   Problem 37.

P.4 Assess Your Understanding

Concepts and Vocabulary

1. **True or False** If every vertical line intersects the graph of a function f at no more than one point, f is a one-to-one function.

2. If the domain of a one-to-one function f is $[4, \infty)$, the range of its inverse function f^{-1} is _____.

3. **True or False** If f and g are inverse functions, the domain of f is the same as the domain of g.

4. **True or False** If f and g are inverse functions, their graphs are symmetric with respect to the line $y = x$.

5. **True or False** If f and g are inverse functions, then $(f \circ g)(x) = f(x) \cdot g(x)$.

6. **True or False** If a function f is one-to-one, then $f\left(f^{-1}(x)\right) = x$, where x is in the domain of f.

7. Given the graph of a collection of points (x, y), explain how you would determine if it represents a one-to-one function $y = f(x)$.

8. Given the graph of a one-to-one function $y = f(x)$, explain how you would graph the inverse function f^{-1}.

Practice Problems

In Problems 9–14, the graph of a function f is given. Use the Horizontal-line Test to determine whether f is one-to one.

PAGE 38 **9.**

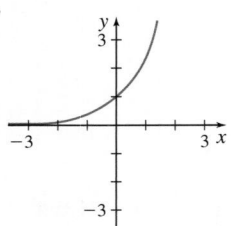

10.

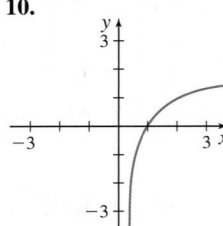

11.

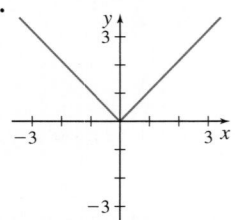

12.

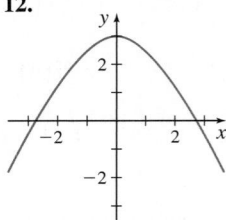

13.

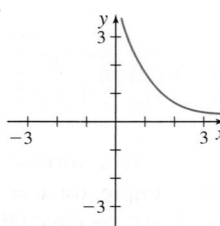

14.

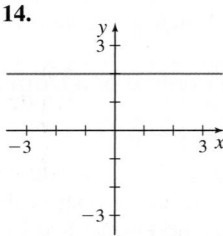

In Problems 15–18, verify that the functions f and g are inverses of each other by showing that $(f \circ g)(x) = x$ and $(g \circ f)(x) = x$.

PAGE 39 **15.** $f(x) = 3x + 4; \ g(x) = \dfrac{1}{3}(x - 4)$

16. $f(x) = x^3 - 8; \ g(x) = \sqrt[3]{x + 8}$

17. $f(x) = \dfrac{1}{x}; \ g(x) = \dfrac{1}{x}$

18. $f(x) = \dfrac{2x + 3}{x + 4}; \ g(x) = \dfrac{4x - 3}{2 - x}$

In Problems 19–22, (a) determine whether the function is one-to-one. If it is one-to-one, (b) find the inverse of each function. (c) State the domain and the range of the function and its inverse.

PAGE 39 **19.** $\{(-3, 5), (-2, 9), (-1, 2), (0, 11), (1, -5)\}$

20. $\{(-2, 2), (-1, 6), (0, 8), (1, -3), (2, 8)\}$

21. $\{(-2, 1), (-3, 2), (-10, 0), (1, 9), (2, 1)\}$

22. $\{(-2, -8), (-1, -1), (0, 0), (1, 1), (2, 8)\}$

In Problems 23–28, the graph of a one-to-one function f is given. Draw the graph of the inverse function. For convenience, the graph of $y = x$ is also given.

23.

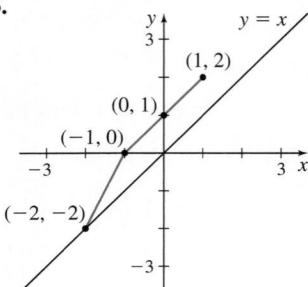

24.

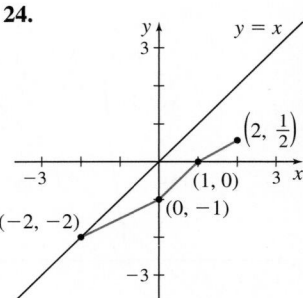

25.

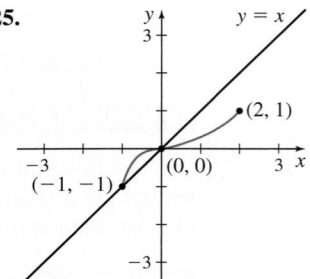

26.

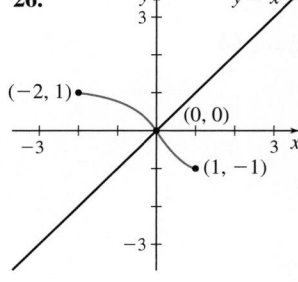

PAGE 40 **27.**

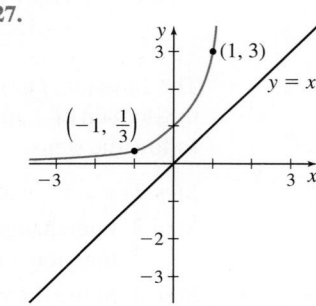

28.

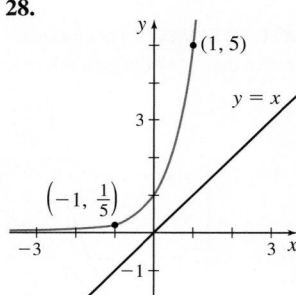

In Problems 29–38, the function f is one-to-one.

(a) *Find its inverse function and check the result.*

(b) *Find the domain and the range of f and the domain and the range of f^{-1}.*

29. $f(x) = 4x + 2$ **30.** $f(x) = 1 - 3x$

PAGE 41 **31.** $f(x) = \sqrt[3]{x + 10}$ **32.** $f(x) = 2x^3 + 4$

33. $f(x) = \dfrac{1}{x - 2}$ **34.** $f(x) = \dfrac{2x}{3x - 1}$

35. $f(x) = \dfrac{2x + 3}{x + 2}$ **36.** $f(x) = \dfrac{-3x - 4}{x - 2}$

PAGE 41 **37.** $f(x) = x^2 + 4, \ x \geq 0$ **38.** $f(x) = (x - 2)^2 + 4, \ x \leq 2$

P.5 Exponential and Logarithmic Functions

OBJECTIVES *When you finish this section, you should be able to:*

1 Analyze an exponential function (p. 43)

2 Define the number e (p. 46)

3 Analyze a logarithmic function (p. 47)

4 Solve exponential equations and logarithmic equations (p. 50)

Here we begin our study of transcendental functions with the exponential and logarithmic functions. In Sections P.6 and P.7, we investigate the trigonometric functions and their inverse functions.

① Analyze an Exponential Function

The expression a^r, where $a > 0$ is a fixed real number and $r = \dfrac{m}{n}$ is a rational number, in lowest terms with $n \geq 2$, is defined as

$$a^r = a^{m/n} = \left(a^{1/n}\right)^m = \left(\sqrt[n]{a}\right)^m$$

So, a function $f(x) = a^x$ can be defined so that its domain is the set of rational numbers. Our aim is to expand the domain of f to include both rational and irrational numbers, that is, to include all real numbers.

Every irrational number x can be approximated by a rational number r formed by truncating (removing) all but a finite number of digits from x. For example, for $x = \pi$, we could use the rational numbers $r = 3.14$ or $r = 3.14159$, and so on. The closer r is to π, the better approximation a^r is to a^π. In general, we can make a^r as close as we please to a^x by choosing r sufficiently close to x. Using this argument, we can define an *exponential function* $f(x) = a^x$, where x includes all the rational numbers and all the irrational numbers.

> **DEFINITION** Exponential Function
>
> An **exponential function** is a function that can be expressed in the form
>
> $$\boxed{f(x) = a^x}$$
>
> where a is a positive real number and $a \neq 1$. The domain of f is the set of all real numbers.

CAUTION Be careful to distinguish an exponential function $f(x) = a^x$, where $a > 0$ and $a \neq 1$, from a power function $g(x) = x^a$, where a is a real number. In $f(x) = a^x$, the independent variable x is the *exponent*; in $g(x) = x^a$, the independent variable x is the *base*.

The base $a = 1$ is excluded from the definition of an exponential function because $f(x) = 1^x = 1$ (a constant function). Bases that are negative are excluded because $a^{1/n}$, where $a < 0$ and n is an even integer, is not defined.

Examples of exponential functions are $f(x) = 2^x$, $g(x) = \left(\dfrac{2}{3}\right)^x$, and $h(x) = \pi^x$.

Consider the exponential function $f(x) = 2^x$. The domain of f is all real numbers; the range of f is the interval $(0, \infty)$. Some points on the graph of f are listed in Table 5. Since $2^x > 0$ for all x, the graph of f lies above the x-axis and has no x-intercept. The y-intercept is 1. Using this information, plot some points from Table 5 and connect them with a smooth curve, as shown in Figure 66.

The graph of $f(x) = 2^x$ is typical of all exponential functions of the form $f(x) = a^x$ with $a > 1$, a few of which are graphed in Figure 67.

TABLE 5

x	$f(x) = 2^x$	(x, y)
-10	$2^{-10} \approx 0.00098$	$(-10, 0.00098)$
-3	$2^{-3} = \dfrac{1}{8}$	$\left(-3, \dfrac{1}{8}\right)$
-2	$2^{-2} = \dfrac{1}{4}$	$\left(-2, \dfrac{1}{4}\right)$
-1	$2^{-1} = \dfrac{1}{2}$	$\left(-1, \dfrac{1}{2}\right)$
0	$2^0 = 1$	$(0, 1)$
1	$2^1 = 2$	$(1, 2)$
2	$2^2 = 4$	$(2, 4)$
3	$2^3 = 8$	$(3, 8)$
10	$2^{10} = 1024$	$(10, 1024)$

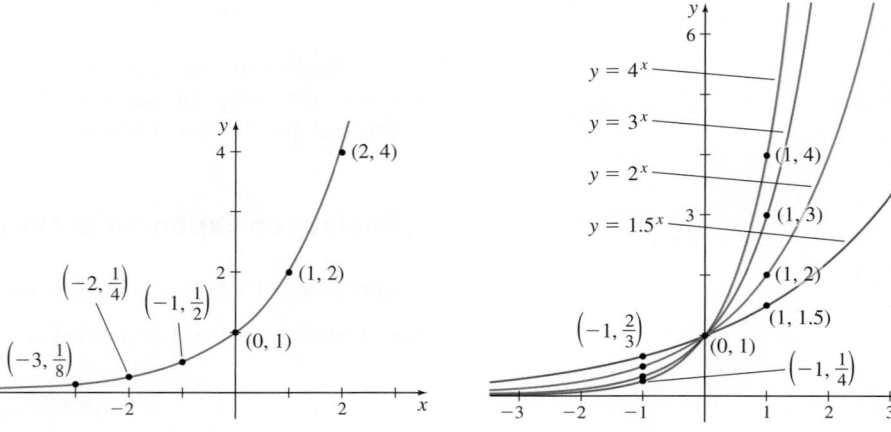

Figure 66 $f(x) = 2^x$

Figure 67 $f(x) = a^x; a > 1$

Notice that every graph in Figure 67 lies above the x-axis, passes through the point $(0, 1)$, and is increasing. Also notice that the graphs with larger bases are steeper when $x > 0$, but when $x < 0$, the graphs with larger bases are closer to the x-axis.

All functions of the type $f(x) = a^x$, $a > 1$, have the following properties:

Properties of an Exponential Function $f(x) = a^x, a > 1$

- The domain is the set of all real numbers; the range is the set of positive real numbers.
- There are no x-intercepts; the y-intercept is 1.
- The exponential function f is increasing on the interval $(-\infty, \infty)$.
- The graph of f contains the points $\left(-1, \dfrac{1}{a}\right)$, $(0, 1)$, and $(1, a)$.
- $\dfrac{f(x+1)}{f(x)} = a$
- Because $f(x) = a^x$ is a function, if $u = v$, then $a^u = a^v$.
- Because $f(x) = a^x$ is a one-to-one function, if $a^u = a^v$, then $u = v$.

NEED TO REVIEW? The laws of exponents are discussed in Appendix A.1, p. A-9.

THEOREM Laws of Exponents

If u, v, a, and b are real numbers with $a > 0$ and $b > 0$, then

$$a^u \cdot a^v = a^{u+v} \qquad \frac{a^u}{a^v} = a^{u-v} \qquad \left(a^u\right)^v = a^{uv} \qquad (ab)^u = a^u \cdot b^u \qquad \left(\frac{a}{b}\right)^u = \frac{a^u}{b^u}$$

For example, we can use the Laws of Exponents to show the following property of an exponential function:

$$\frac{f(x+1)}{f(x)} = \frac{a^{x+1}}{a^x} = a^{(x+1)-x} = a^1 = a$$

EXAMPLE 1 Graphing an Exponential Function

Graph the exponential function $g(x) = \left(\frac{1}{2}\right)^x$.

Solution

Begin by writing $\frac{1}{2}$ as 2^{-1}. Then

$$g(x) = \left(\frac{1}{2}\right)^x = (2^{-1})^x = 2^{-x}$$

Now use the graph of $f(x) = 2^x$ shown in Figure 66 and reflect it about the y-axis to obtain the graph of $g(x) = 2^{-x}$. See Figure 68.

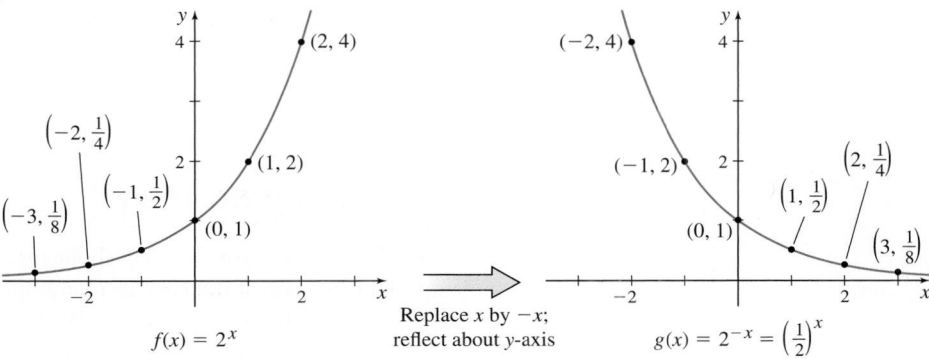

Replace x by $-x$;
reflect about y-axis

$f(x) = 2^x$ $g(x) = 2^{-x} = \left(\frac{1}{2}\right)^x$

Figure 68

NOW WORK   Problem 19.

The graph of $g(x) = \left(\frac{1}{2}\right)^x$ in Figure 68 is typical of all exponential functions that have a base between 0 and 1. Figure 69 illustrates the graphs of several more exponential functions whose bases are between 0 and 1. Notice that the graphs with smaller bases are steeper when $x < 0$, but when $x > 0$, these graphs are closer to the x-axis.

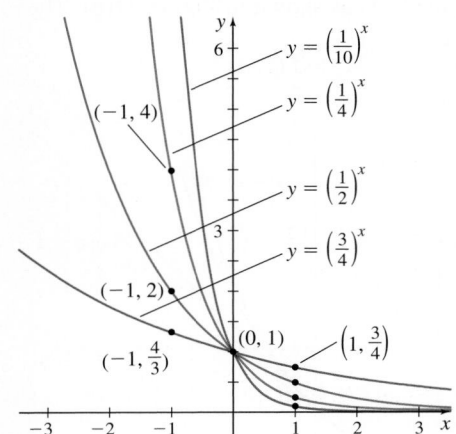

Figure 69 $f(x) = a^x, 0 < a < 1$

Properties of an Exponential Function $f(x) = a^x, 0 < a < 1$

• The domain is the set of all real numbers;
  the range is the set of positive real numbers.

• There are no x-intercepts; the y-intercept is 1.

• The exponential function f is decreasing on the interval $(-\infty, \infty)$.

• The graph of f contains the points $\left(-1, \frac{1}{a}\right)$, $(0, 1)$, and $(1, a)$.

• $\dfrac{f(x+1)}{f(x)} = a$

• Because $f(x) = a^x$ is a function, if $u = v$, then $a^u = a^v$.

• Because $f(x) = a^x$ is a one-to-one function, if $a^u = a^v$, then $u = v$.

EXAMPLE 2 **Graphing an Exponential Function Using Transformations**

Graph $f(x) = 3^{-x} - 2$ and determine the domain and range of f.

Solution

We begin with the graph of $y = 3^x$. Figure 70 shows the steps.

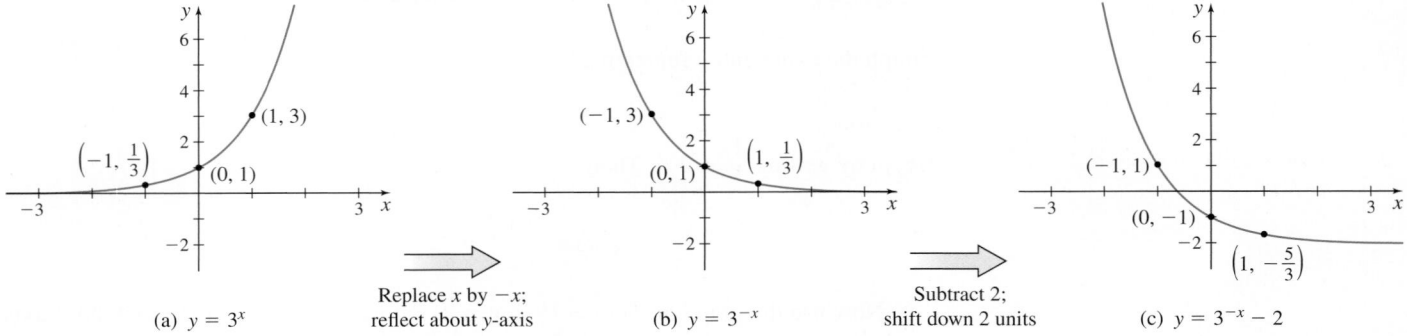

(a) $y = 3^x$ Replace x by $-x$; reflect about y-axis   (b) $y = 3^{-x}$ Subtract 2; shift down 2 units   (c) $y = 3^{-x} - 2$

Figure 70

The domain of f is all real numbers; the range of f is the interval $(-2, \infty)$. ∎

NOW WORK Problem 27.

2 Define the Number e

In earlier courses, you learned about an irrational number called e. The number e is important because it appears in many applications and because it has properties that simplify computations in calculus. We use a geometric approach to define e.

Consider the graphs of the functions $y = 2^x$ and $y = 3^x$ in Figure 71(a), where we have carefully drawn lines that just touch each graph at the point $(0, 1)$. (These lines are *tangent lines*, which we discuss in Chapter 1.) Notice that the slope of the tangent line to $y = 3^x$ is greater than 1 (approximately 1.10) and that the slope of the tangent line to the graph of $y = 2^x$ is less than 1 (approximately 0.69). Between these graphs there is an exponential function $y = a^x$, whose base is between 2 and 3, and whose tangent line to the graph at the point $(0, 1)$ has a slope of exactly 1, as shown in Figure 71(b). The base of this exponential function is the number e.

NEED TO REVIEW? The slope of a line is discussed in Appendix A.3, pp. A-20 to A-22.

imageBROKER / Alamy

ORIGINS The irrational number we call e was named to honor the Swiss mathematician Leonhard Euler (1707–1783), who is considered one of the greatest mathematicians of all time. Euler was encouraged to study mathematics by his father. Considered the most prolific mathematician who ever lived, he made major contributions to many areas of mathematics. One of Euler's greatest gifts was his ability to explain difficult mathematical concepts in simple language.

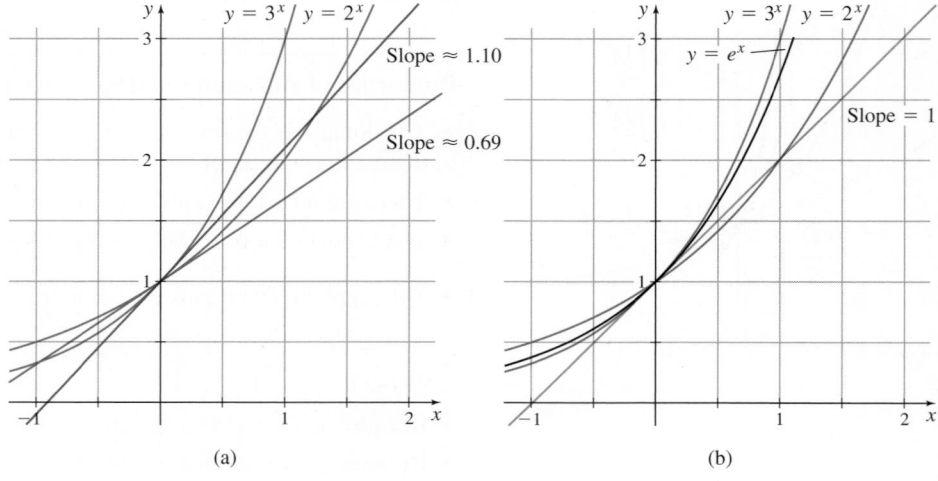

(a)   (b)

Figure 71

DEFINITION The number e

The **number** e is defined as the base of the exponential function whose tangent line to the graph at the point $(0, 1)$ has slope 1. The function $f(x) = e^x$ occurs with such frequency that it is usually referred to as *the* **exponential function**.

The number e is an irrational number, and in Chapter 3, we show $e \approx 2.71828$.

③ Analyze a Logarithmic Function

NEED TO REVIEW? Logarithms and their properties are discussed in Appendix A.1, pp. A-10 to A-12.

Recall that a one-to-one function $y = f(x)$ has an inverse function that is defined implicitly by the equation $x = f(y)$. Since an exponential function $y = f(x) = a^x$, where $a > 0$ and $a \neq 1$, is a one-to-one function, it has an inverse function, called a *logarithmic function,* that is defined implicitly by the equation

$$x = a^y \qquad \text{where } a > 0 \quad \text{and} \quad a \neq 1$$

DEFINITION Logarithmic Function

The **logarithmic function with base** a, where $a > 0$ and $a \neq 1$, is denoted by $y = \log_a x$ and is defined by

$$\boxed{y = \log_a x \quad \text{if and only if} \quad x = a^y}$$

The domain of the logarithmic function $y = \log_a x$ is $x > 0$.

IN WORDS A logarithm is an exponent. That is, if $y = \log_a x$, then y is the exponent in $x = a^y$.

In other words, if $f(x) = a^x$, where $a > 0$ and $a \neq 1$, its inverse function is $f^{-1}(x) = \log_a x$.

If the base of a logarithmic function is the number e, then it is called the **natural logarithmic function**, and it is given a special symbol, **ln** (from the Latin, *logarithmus naturalis*). That is,

$$\boxed{y = \ln x \quad \text{if and only if} \quad x = e^y}$$

Since the exponential function $y = a^x$, $a > 0, a \neq 1$ and the logarithmic function $y = \log_a x$ are inverse functions, the following properties hold:

- $\log_a(a^x) = x$ for all real numbers x
- $a^{\log_a x} = x$ for all $x > 0$

Because exponential functions and logarithmic functions are inverses of each other, the graph of the logarithmic function $y = \log_a x$, $a > 0$ and $a \neq 1$, is the reflection of the graph of the exponential function $y = a^x$, about the line $y = x$, as shown in Figure 72.

Based on the graphs in Figure 72, we see that

$$\boxed{\log_a 1 = 0 \qquad \log_a a = 1 \qquad a > 0 \text{ and } a \neq 1}$$

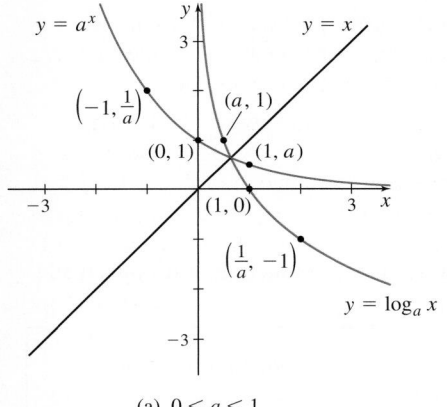

(a) $0 < a < 1$

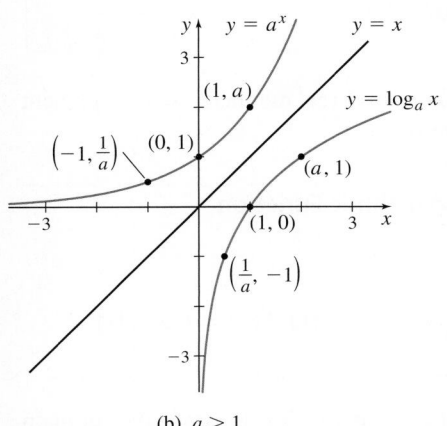

(b) $a > 1$

Figure 72

EXAMPLE 3 Graphing a Logarithmic Function

Graph:

(a) $f(x) = \log_2 x$ **(b)** $g(x) = \log_{1/3} x$ **(c)** $F(x) = \ln x$

Solution

(a) To graph $f(x) = \log_2 x$, graph $y = 2^x$ and reflect the graph about the line $y = x$. See Figure 73(a).

(b) To graph $g(x) = \log_{1/3} x$, graph $y = \left(\dfrac{1}{3}\right)^x$ and reflect it about the line $y = x$. See Figure 73(b).

(c) To graph $F(x) = \ln x$, graph $y = e^x$ and reflect it about the line $y = x$. See Figure 73(c). ∎

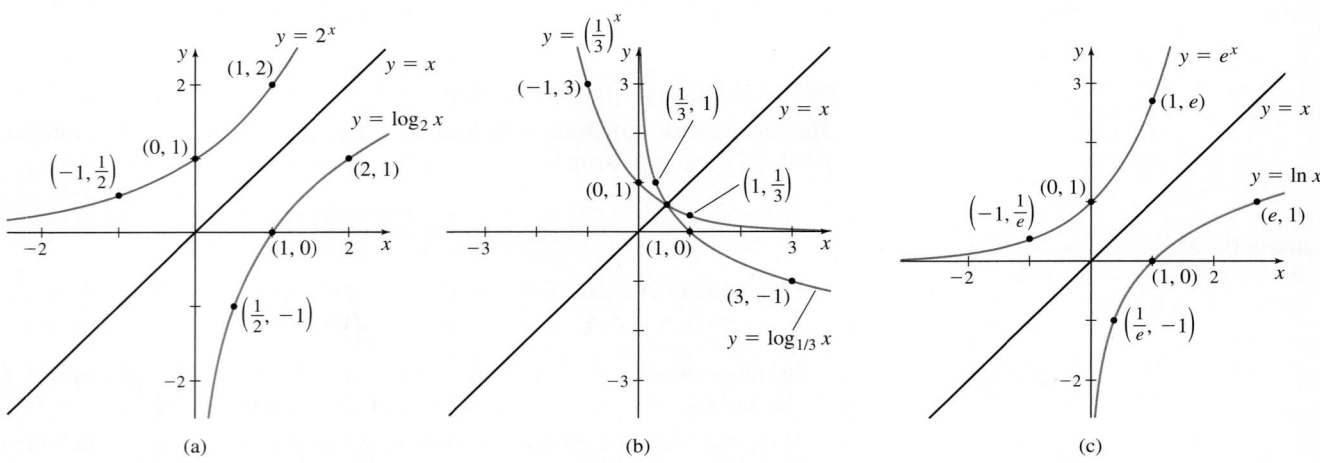

Figure 73

NOW WORK   Problem 39.

Since a logarithmic function is the inverse of an exponential function, it follows that:

> • Domain of the logarithmic function = Range of the exponential function
> $= (0, \infty)$
> • Range of the logarithmic function = Domain of the exponential function
> $= (-\infty, \infty)$

The domain of a logarithmic function is the set of *positive* real numbers, so the argument of a logarithmic function must be greater than zero.

EXAMPLE 4 Finding the Domain of a Logarithmic Function

Find the domain of each function:

(a) $F(x) = \log_2(x+3)$ **(b)** $g(x) = \ln\left(\dfrac{1+x}{1-x}\right)$ **(c)** $h(x) = \log_{1/2}|x|$

Solution

NEED TO REVIEW? Solving inequalities is discussed in Appendix A.1, pp. A-6 to A-8.

(a) The argument of a logarithm must be positive. So to find the domain of $F(x) = \log_2(x+3)$, we solve the inequality $x + 3 > 0$. The domain of F is $\{x \mid x > -3\}$.

(b) Since $\ln\left(\dfrac{1+x}{1-x}\right)$ requires $\dfrac{1+x}{1-x} > 0$, we find the domain of g by solving the inequality $\dfrac{1+x}{1-x} > 0$. Since $\dfrac{1+x}{1-x}$ is not defined for $x = 1$, and the solution to the equation $\dfrac{1+x}{1-x} = 0$ is $x = -1$, we use -1 and 1 to separate the real number line into three intervals $(-\infty, -1)$, $(-1, 1)$, and $(1, \infty)$. Then we choose a test number in each interval, and evaluate the rational expression $\dfrac{1+x}{1-x}$ at these numbers to determine if the expression is positive or negative. For example, we chose the numbers -2, 0, and 2 and found that $\dfrac{1+x}{1-x} > 0$ on the interval $(-1, 1)$. See the table on the left. So the domain of $g(x) = \ln\left(\dfrac{1+x}{1-x}\right)$ is $\{x \mid -1 < x < 1\}$.

(c) $\log_{1/2}|x|$ requires $|x| > 0$. So the domain of $h(x) = \log_{1/2}|x|$ is $\{x \mid x \neq 0\}$. ∎

Interval	Test Number	Sign of $\dfrac{1+x}{1-x}$
$(-\infty, -1)$	-2	Negative
$(-1, 1)$	0	Positive
$(1, \infty)$	2	Negative

NOW WORK Problem 31.

Properties of a Logarithmic Function $f(x) = \log_a x, a > 0$ **and** $a \neq 1$

- The domain of f is the set of all positive real numbers; the range is the set of all real numbers.
- The x-intercept of the graph of f is 1. There is no y-intercept.
- A logarithmic function is decreasing on the interval $(0, \infty)$ if $0 < a < 1$ and increasing on the interval $(0, \infty)$ if $a > 1$.
- The graph of f contains the points $(1, 0)$, $(a, 1)$, and $\left(\dfrac{1}{a}, -1\right)$.
- Because $f(x) = \log_a x$ is a function, if $u = v$, then $\log_a u = \log_a v$.
- Because $f(x) = \log_a x$ is a one-to-one function, if $\log_a u = \log_a v$, then $u = v$.

See Figure 74 for typical graphs of $y = \log_a x$ with base a, $a > 1$; see Figure 75 for graphs of $y = \log_a x$ with base a, $0 < a < 1$.

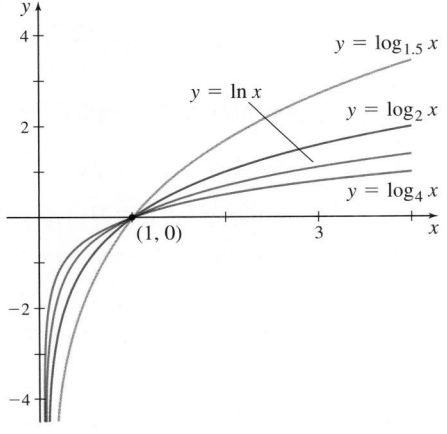

Figure 74 $y = \log_a x, a > 1$

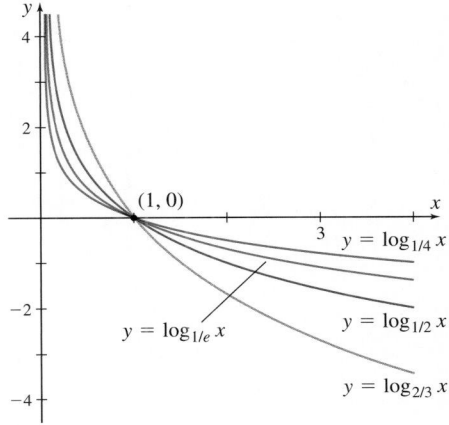

Figure 75 $y = \log_a x, 0 < a < 1$

4 Solve Exponential Equations and Logarithmic Equations

Equations that involve terms of the form a^x, $a > 0$, $a \neq 1$, are referred to as **exponential equations**. For example, the equation of the form $5^{2x+3} = 5^x$ is an exponential equation.

The one-to-one property of exponential functions

IN WORDS In an exponential equation, if the bases are equal, then the exponents are equal.

$$\boxed{\text{if } a^u = a^v \quad \text{then } u = v} \tag{1}$$

can be used to solve certain kinds of exponential equations.

EXAMPLE 5 Solving Exponential Equations

Solve each exponential equation:

(a) $4^{2x-1} = 8^{x+3}$ **(b)** $e^{-x^2} = (e^x)^2 \cdot \dfrac{1}{e^3}$

Solution

(a) Begin by expressing both sides of the equation with the same base so we can use the one-to-one property (1).

$$4^{2x-1} = 8^{x+3}$$
$$\left(2^2\right)^{2x-1} = \left(2^3\right)^{x+3} \qquad 4 = 2^2, \, 8 = 2^3$$
$$2^{2(2x-1)} = 2^{3(x+3)} \qquad (a^r)^s = a^{rs}$$

Now set the exponents equal to each other and solve for x.

$$2(2x - 1) = 3(x + 3) \qquad \text{If } a^u = a^v, \text{ then } u = v.$$
$$4x - 2 = 3x + 9 \qquad \text{Simplify.}$$
$$x = 11 \qquad \text{Solve.}$$

The solution is 11.

(b) Use the Laws of Exponents to obtain the base e on the right side.

$$(e^x)^2 \cdot \frac{1}{e^3} = e^{2x} \cdot e^{-3} = e^{2x-3}$$

As a result,

$$e^{-x^2} = e^{2x-3}$$

Now set the exponents equal to each other and solve for x.

$$-x^2 = 2x - 3 \qquad \text{If } a^u = a^v, \text{ then } u = v.$$
$$x^2 + 2x - 3 = 0$$
$$(x + 3)(x - 1) = 0$$
$$x = -3 \quad \text{or} \quad x = 1$$

The solution set is $\{-3, 1\}$. ∎

NOW WORK Problem 49.

To use the one-to-one property of exponential functions, each side of the equation must be written with the same base. Since for many exponential equations it is not possible to write each side with the same base, we need a different strategy to solve such equations.

EXAMPLE 6   **Solving Exponential Equations**

Solve the exponential equations:

(a) $10^{2x} = 50$ **(b)** $8 \cdot 3^x = 5$

Solution

(a) Since 10 and 50 cannot be written with the same base, we write the exponential equation as a logarithm.

$$10^{2x} = 50 \quad \text{if and only if} \quad \log 50 = 2x$$

RECALL Logarithms to the base 10 are called **common logarithms** and are usually written without a subscript. That is, $x = \log_{10} y$ is written $x = \log y$.

Then $x = \dfrac{\log 50}{2}$ is an exact solution of the equation. Using a calculator, an approximate solution is $x = \dfrac{\log 50}{2} \approx 0.849$.

(b) It is impossible to write 8 and 5 as a power of 3, so we write the exponential equation as a logarithm.

$$8 \cdot 3^x = 5$$

$$3^x = \frac{5}{8}$$

$$\log_3 \frac{5}{8} = x$$

Now use the change-of-base formula to obtain an approximate solution using a calculator.

NEED TO REVIEW? The change-of-base formula, $\log_a u = \dfrac{\log_b u}{\log_b a}$, $a \neq 1, b \neq 1$, and u positive real numbers, is discussed in Appendix A.1, p. A-12.

$$x = \log_3 \frac{5}{8} = \frac{\ln \frac{5}{8}}{\ln 3} \approx -0.428 \qquad \blacksquare$$

Alternatively, we could have solved each of the equations in Example 6 by taking the natural logarithm (or the common logarithm) of each side. For example,

$$8 \cdot 3^x = 5$$

$$3^x = \frac{5}{8}$$

$$\ln 3^x = \ln \frac{5}{8} \qquad \text{If } u = v, \text{ then } \log_a u = \log_a v.$$

$$x \ln 3 = \ln \frac{5}{8} \qquad \log_a u^r = r \log_a u$$

$$x = \frac{\ln \frac{5}{8}}{\ln 3}$$

NOW WORK   Problem 51.

Equations that contain logarithms are called **logarithmic equations**. Care must be taken when solving logarithmic equations algebraically. In the expression $\log_a y$, remember that a and y are positive and $a \neq 1$. Be sure to check each apparent solution in the original equation and to discard any solutions that are extraneous.

Some logarithmic equations can be solved by changing the logarithmic equation to an exponential equation using the fact that $x = \log_a y$ if and only if $y = a^x$.

EXAMPLE 7 Solving Logarithmic Equations

Solve each equation:

(a) $\log_3(4x - 7) = 2$ **(b)** $\log_x 64 = 2$

Solution

(a) Change the logarithmic equation to an exponential equation.

$$\log_3(4x - 7) = 2$$
$$4x - 7 = 3^2 \qquad \text{Change to an exponential equation.}$$
$$4x - 7 = 9$$
$$4x = 16$$
$$x = 4$$

Check: For $x = 4$, $\log_3(4x - 7) = \log_3(4 \cdot 4 - 7) = \log_3 9 = 2$, since $3^2 = 9$.
The solution is 4.

(b) Change the logarithmic equation to an exponential equation.

$$\log_x 64 = 2$$
$$x^2 = 64 \qquad \text{Change to an exponential equation.}$$
$$x = 8 \quad \text{or} \quad x = -8 \qquad \text{Solve.}$$

The base of a logarithm is always positive. As a result, -8 is an extraneous solution, so we discard it. Now we check the solution 8.

Check: For $x = 8$, $\log_8 64 = 2$, since $8^2 = 64$.
The solution is 8. ∎

NOW WORK Problem 57.

The properties of logarithms that result from the fact that a logarithmic function is one-to-one can be used to solve some equations that contain two logarithms with the same base.

CALC CLIP **EXAMPLE 8** Solving a Logarithmic Equation

Solve the logarithmic equation $2 \ln x = \ln 9$.

Solution

Each logarithm has the same base, so

$$2 \ln x = \ln 9$$
$$\ln x^2 = \ln 9 \qquad r \log_a u = \log_a u^r$$

Now use the fact that logarithmic functions with the same base are one-to-one.

$$x^2 = 9 \qquad \text{If } \log_a u = \log_a v, \text{ then } u = v.$$
$$x = 3 \quad \text{or} \quad x = -3$$

We discard the solution $x = -3$ since -3 is not in the domain of $f(x) = \ln x$. The solution is 3. ∎

NOW WORK Problem 61.

Although the equations in the previous examples were relatively easy to solve, this is not generally the case. Many solutions to exponential and logarithmic equations need to be approximated using technology.

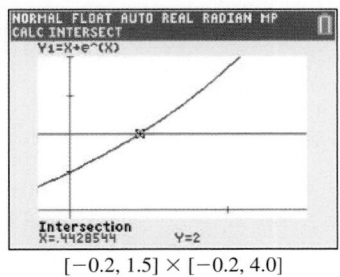

Figure 76

$[-0.2, 1.5] \times [-0.2, 4.0]$

EXAMPLE 9 **Approximating the Solution to an Exponential Equation**

Solve $x + e^x = 2$. Express the solution rounded to three decimal places.

Solution

We can approximate the solution to the equation by graphing the two functions $Y_1 = x + e^x$ and $Y_2 = 2$. Then we use graphing technology to approximate the intersection of the graphs. Since the function Y_1 is increasing (do you know why?) and the function Y_2 is constant, there will be only one point of intersection. Figure 76 shows the graphs of the two functions and their intersection. The graphs intersect when $x \approx 0.4428544$, so the solution of the equation is 0.443 rounded to three decimal places. ∎

NOW WORK Problem 65.

P.5 Assess Your Understanding

Concepts and Vocabulary

1. The graph of every exponential function $f(x) = a^x$, $a > 0$ and $a \neq 1$, contains the three points: _____ , _____ , and _____ .

2. *True or False* The graph of the exponential function $f(x) = \left(\dfrac{3}{2}\right)^x$ is decreasing.

3. If $3^x = 3^4$, then $x =$ _____ .

4. If $4^x = 8^2$, then $x =$ _____ .

5. *True or False* The graphs of $y = 3^x$ and $y = \left(\dfrac{1}{3}\right)^x$ are symmetric with respect to the line $y = x$.

6. *True or False* The range of the exponential function $f(x) = a^x$, $a > 0$ and $a \neq 1$, is the set of all real numbers.

7. The number e is defined as the base of the exponential function f whose tangent line to the graph of f at the point $(0, 1)$ has slope _____ .

8. The domain of the logarithmic function $f(x) = \log_a x$ is _____ .

9. The graph of every logarithmic function $f(x) = \log_a x$, $a > 0$ and $a \neq 1$, contains the three points: _____ , _____ , and _____ .

10. *Multiple Choice* The graph of $f(x) = \log_2 x$ is [(a) increasing, (b) decreasing, (c) neither].

11. *True or False* If $y = \log_a x$, then $y = a^x$.

12. *True or False* The graph of $f(x) = \log_a x$, $a > 0$ and $a \neq 1$, has an x-intercept equal to 1 and no y-intercept.

13. *True or False* $\ln e^x = x$ for all real numbers.

14. $\ln e =$ _____ .

15. Explain what the number e is.

16. What is the x-intercept of the function $h(x) = \ln(x + 1)$?

Practice Problems

17. Suppose that $g(x) = 4^x + 2$.

   (a) What is $g(-1)$? What is the corresponding point on the graph of g?

   (b) If $g(x) = 66$, what is x? What is the corresponding point on the graph of g?

18. Suppose that $g(x) = 5^x - 3$.

   (a) What is $g(-1)$? What is the corresponding point on the graph of g?

   (b) If $g(x) = 122$, what is x? What is the corresponding point on the graph of g?

In Problems 19–24, the graph of an exponential function is given. Match each graph to one of the following functions:

   (a) $y = 3^{-x}$ *(b)* $y = -3^x$ *(c)* $y = -3^{-x}$

   (d) $y = 3^x - 1$ *(e)* $y = 3^{x-1}$ *(f)* $y = 1 - 3^x$

19.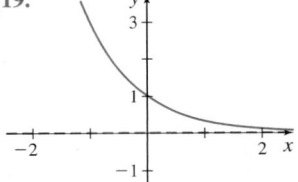

20.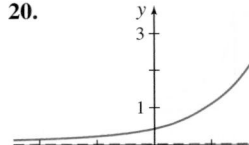

21.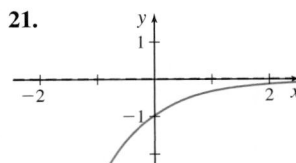

22.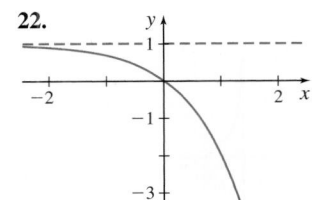

23.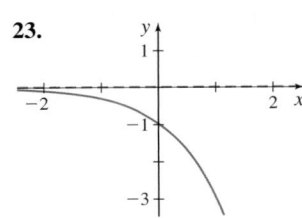

24.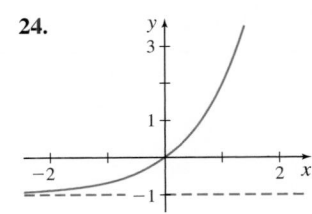

In Problems 25–30, use transformations to graph each function. Find the domain and range.

25. $f(x) = 2^{x+2}$

26. $f(x) = 1 - 2^{-x/3}$

PAGE 46 **27.** $f(x) = 4\left(\dfrac{1}{3}\right)^x$

28. $f(x) = \left(\dfrac{1}{2}\right)^{-x} + 1$

29. $f(x) = e^{-x}$

30. $f(x) = 5 - e^x$

In Problems 31–34, find the domain of each function.

PAGE 49 **31.** $F(x) = \log_2 x^2$

32. $g(x) = 8 + 5\ln(2x + 3)$

33. $f(x) = \ln(x - 1)$

34. $g(x) = \sqrt{\ln x}$

In Problems 35–40, the graph of a logarithmic function is given. Match each graph to one of the following functions:

(a) $y = \log_3 x$ **(b)** $y = \log_3(-x)$

(c) $y = -\log_3 x$ **(d)** $y = \log_3 x - 1$

(e) $y = \log_3(x - 1)$ **(f)** $y = 1 - \log_3 x$

35.
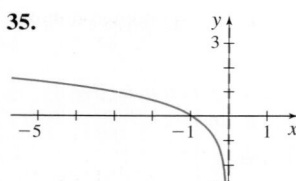

36.
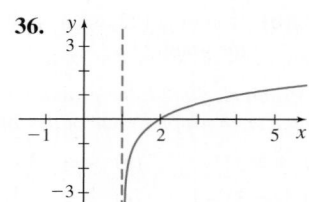

37.
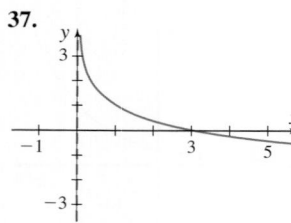

38.
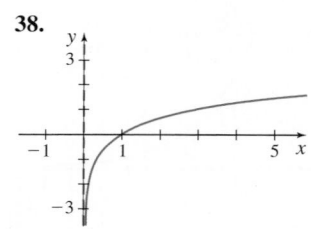

PAGE 48 **39.**
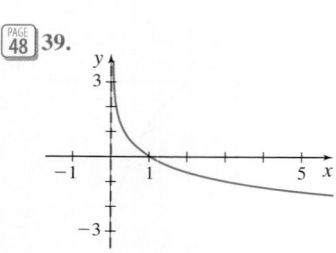

40.
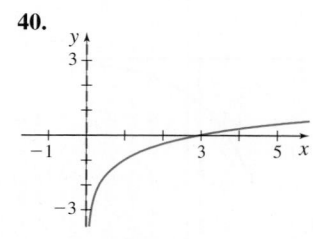

In Problems 41–44,

(a) Find the domain of f.

(b) Graph f.

(c) From the graph of f, determine the range of f.

(d) Find f^{-1}, the inverse of f.

(e) Use f^{-1} to find the range of f.

(f) Graph f^{-1}.

41. $f(x) = \ln(x + 4)$

42. $f(x) = \dfrac{1}{2}\log(2x)$

43. $f(x) = 3e^x + 2$

44. $f(x) = 2^{x/3} + 4$

45. How does the transformation $y = \ln(x + c)$, $c > 0$, affect the x-intercept of the graph of the function $f(x) = \ln x$?

46. How does the transformation $y = e^{cx}$, $c > 0$, affect the y-intercept of the graph of the function $f(x) = e^x$?

In Problems 47–62, solve each equation.

47. $3^{x^2} = 9^x$

48. $5^{x^2+8} = 125^{2x}$

PAGE 50 **49.** $e^{3x} = \dfrac{e^2}{e^x}$

50. $e^{4x} \cdot e^{x^2} = e^{12}$

PAGE 51 **51.** $e^{1-2x} = 4$

52. $e^{1-x} = 5$

53. $5(2^{3x}) = 9$

54. $0.3(4^{0.2x}) = 0.2$

55. $3^{1-2x} = 4^x$

56. $2^{x+1} = 5^{1-2x}$

PAGE 52 **57.** $\log_2(2x + 1) = 3$

58. $\log_3(3x - 2) = 2$

59. $\log_x\left(\dfrac{1}{8}\right) = 3$

60. $\log_x 64 = -3$

PAGE 52 **61.** $\ln(2x + 3) = 2\ln 3$

62. $\dfrac{1}{2}\log_3 x = 2\log_3 2$

 In Problems 63–66, use technology to solve each equation. Express the answer rounded to three decimal places.

63. $\log_5(x + 1) - \log_4(x - 2) = 1$ **64.** $\ln x = x$

PAGE 53 **65.** $e^x + \ln x = 4$ **66.** $e^x = x^2$

67. **(a)** If $f(x) = \ln(x + 4)$ and $g(x) = \ln(3x + 1)$, graph f and g on the same set of axes.

   (b) Find the point(s) of intersection of the graphs of f and g by solving $f(x) = g(x)$.

   (c) Based on the graph, solve $f(x) > g(x)$.

68. **(a)** If $f(x) = 3^{x+1}$ and $g(x) = 2^{x+2}$, graph f and g on the same set of axes.

   (b) Find the point(s) of intersection of the graphs of f and g by solving $f(x) = g(x)$. Round answers to three decimal places.

   (c) Based on the graph, solve $f(x) > g(x)$.

P.6 Trigonometric Functions

OBJECTIVES *When you finish this section, you should be able to:*

1. Work with properties of trigonometric functions (p. 55)
2. Graph the trigonometric functions (p. 56)

NEED TO REVIEW? Trigonometric functions are discussed in Appendix A.4, pp. A-30 to A-35.

AP® EXAM **TIP**

In calculus, radians are used to measure angles, unless degrees are specifically mentioned.

1 Work with Properties of Trigonometric Functions

Table 6 lists the six trigonometric functions and the domain and range of each function.

TABLE 6

Function	Symbol	Domain	Range
sine	$y = \sin x$	All real numbers	$\{y \mid -1 \le y \le 1\}$
cosine	$y = \cos x$	All real numbers	$\{y \mid -1 \le y \le 1\}$
tangent	$y = \tan x$	$\left\{x \mid x \ne \text{ odd integer multiples of } \dfrac{\pi}{2}\right\}$	All real numbers
cosecant	$y = \csc x$	$\{x \mid x \ne \text{ integer multiples of } \pi\}$	$\{y \mid y \le -1 \text{ or } y \ge 1\}$
secant	$y = \sec x$	$\left\{x \mid x \ne \text{ odd integer multiples of } \dfrac{\pi}{2}\right\}$	$\{y \mid y \le -1 \text{ or } y \ge 1\}$
cotangent	$y = \cot x$	$\{x \mid x \ne \text{ integer multiples of } \pi\}$	All real numbers

An important property common to all trigonometric functions is that they are *periodic*.

DEFINITION Periodic Function

A function f is called **periodic** if there is a positive number p with the property that whenever x is in the domain of f, so is $x + p$, and

$$\boxed{f(x + p) = f(x)}$$

If there is a smallest number p with this property, it is called the **(fundamental) period** of f.

The sine, cosine, cosecant, and secant functions are periodic with period 2π; the tangent and cotangent functions are periodic with period π.

THEOREM Period of Trigonometric Functions

• $\sin(x + 2\pi) = \sin x$	• $\cos(x + 2\pi) = \cos x$	• $\tan(x + \pi) = \tan x$
• $\csc(x + 2\pi) = \csc x$	• $\sec(x + 2\pi) = \sec x$	• $\cot(x + \pi) = \cot x$

Because the trigonometric functions are periodic, once the values of a trigonometric function over one period are known, the values over the entire domain are known. This property is useful for graphing trigonometric functions.

The next result, also useful for graphing the trigonometric functions, is a consequence of the even-odd identities, $\sin(-x) = -\sin x$ and $\cos(-x) = \cos x$. From these, we have

$$\tan(-x) = \frac{\sin(-x)}{\cos(-x)} = \frac{-\sin x}{\cos x} = -\tan x \quad \sec(-x) = \frac{1}{\cos(-x)} = \frac{1}{\cos x} = \sec x$$

$$\cot(-x) = \frac{1}{\tan(-x)} = \frac{1}{-\tan x} = -\cot x \quad \csc(-x) = \frac{1}{\sin(-x)} = \frac{1}{-\sin x} = -\csc x$$

THEOREM Even-Odd Properties of the Trigonometric Functions

- The sine, tangent, cosecant, and cotangent functions are odd, so their graphs are symmetric with respect to the origin.
- The cosine and secant functions are even, so their graphs are symmetric with respect to the y-axis.

NEED TO REVIEW? The values of the trigonometric functions for select numbers are discussed in Appendix A.4, pp. A-32 to A-33.

TABLE 7

x	$y = \sin x$	(x, y)
0	0	$(0, 0)$
$\dfrac{\pi}{6}$	$\dfrac{1}{2}$	$\left(\dfrac{\pi}{6}, \dfrac{1}{2}\right)$
$\dfrac{\pi}{2}$	1	$\left(\dfrac{\pi}{2}, 1\right)$
$\dfrac{5\pi}{6}$	$\dfrac{1}{2}$	$\left(\dfrac{5\pi}{6}, \dfrac{1}{2}\right)$
π	0	$(\pi, 0)$
$\dfrac{7\pi}{6}$	$-\dfrac{1}{2}$	$\left(\dfrac{7\pi}{6}, -\dfrac{1}{2}\right)$
$\dfrac{3\pi}{2}$	-1	$\left(\dfrac{3\pi}{2}, -1\right)$
$\dfrac{11\pi}{6}$	$-\dfrac{1}{2}$	$\left(\dfrac{11\pi}{6}, -\dfrac{1}{2}\right)$
2π	0	$(2\pi, 0)$

② Graph the Trigonometric Functions

To graph $y = \sin x$, use Table 7 to obtain points on the graph. Then plot these points and connect them with a smooth curve. Since the sine function has a period of 2π, continue the graph to the left of 0 and to the right of 2π. See Figure 77.

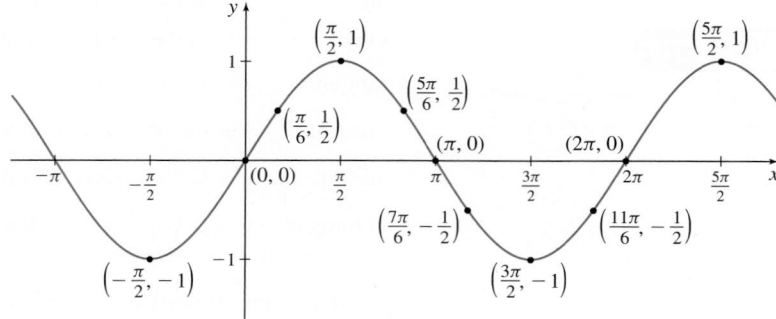

Figure 77 $f(x) = \sin x$

Notice the symmetry of the graph with respect to the origin. This is a consequence of $f(x) = \sin x$ being an odd function.

The graph of $f(x) = \sin x$ illustrates some facts about the sine function.

Properties of the Sine Function $f(x) = \sin x$

- The domain of f is the set of all real numbers.
- The range of f consists of all real numbers in the closed interval $[-1, 1]$.
- The sine function is an odd function, so its graph is symmetric with respect to the origin.
- The sine function has a period of 2π.
- The x-intercepts of f are $\ldots, -2\pi, -\pi, 0, \pi, 2\pi, 3\pi, \ldots$; the y-intercept is 0.
- The maximum value of f is 1 and occurs at $x = \ldots, -\dfrac{3\pi}{2}, \dfrac{\pi}{2}, \dfrac{5\pi}{2}, \dfrac{9\pi}{2}, \ldots$; the minimum value of f is -1 and occurs at $x = \ldots, -\dfrac{\pi}{2}, \dfrac{3\pi}{2}, \dfrac{7\pi}{2}, \dfrac{11\pi}{2}, \ldots$.

The graph of the cosine function is obtained in a similar way. Locate points on the graph of the cosine function $f(x) = \cos x$ for $0 \le x \le 2\pi$. Then connect the points with a smooth curve, and continue the graph to the left of 0 and to the right of 2π to obtain the graph of $y = \cos x$. See Figure 78.

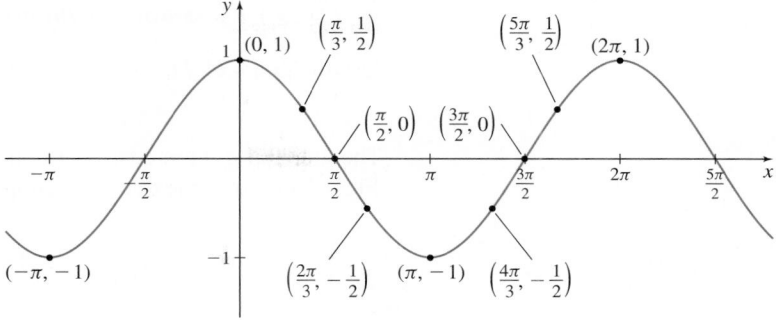

Figure 78 $f(x) = \cos x$

The graph of $f(x) = \cos x$ illustrates some facts about the cosine function.

Properties of the Cosine Function $f(x) = \cos x$

- The domain of f is the set of all real numbers.
- The range of f consists of all real numbers in the closed interval $[-1, 1]$.
- The cosine function is an even function, so its graph is symmetric with respect to the y-axis.
- The cosine function has a period of 2π.
- The x-intercepts of f are $\dots, -\dfrac{3\pi}{2}, -\dfrac{\pi}{2}, \dfrac{\pi}{2}, \dfrac{3\pi}{2}, \dfrac{5\pi}{2}, \dots$; the y-intercept is 1.
- The maximum value of f is 1 and occurs at $x = \dots, -2\pi, 0, 2\pi, 4\pi, 6\pi, \dots$; the minimum value of f is -1 and occurs at $x = \dots, -\pi, \pi, 3\pi, 5\pi, \dots$.

Many variations of the sine and cosine functions can be graphed using transformations.

EXAMPLE 1   Graphing Variations of $f(x) = \sin x$ Using Transformations

Use the graph of $f(x) = \sin x$ to graph $g(x) = 2 \sin x$.

Solution

Notice that $g(x) = 2f(x)$, so the graph of g is a vertical stretch of the graph of $f(x) = \sin x$. Figure 79 illustrates the transformation.

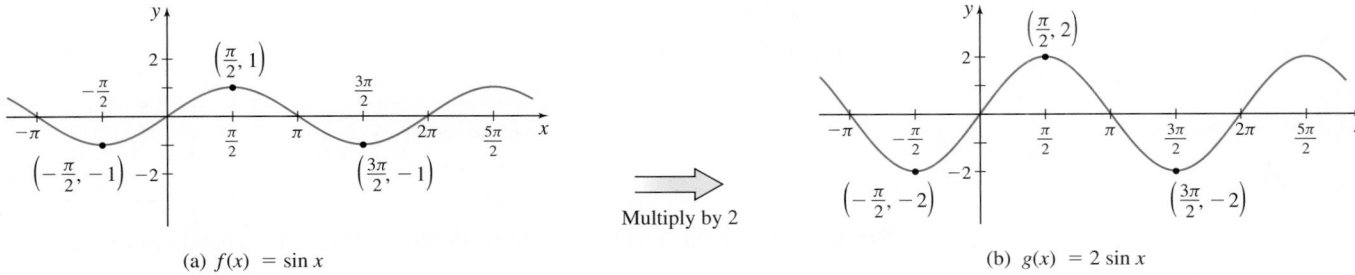

(a) $f(x) = \sin x$ (b) $g(x) = 2 \sin x$

Figure 79

Notice that the values of $g(x) = 2 \sin x$ lie between -2 and 2, inclusive.

In general, the values of the functions $f(x) = A \sin x$ and $g(x) = A \cos x$, where $A \neq 0$, will satisfy the inequalities

$$-|A| \le A \sin x \le |A| \qquad \text{and} \qquad -|A| \le A \cos x \le |A|$$

respectively. The number $|A|$ is called the **amplitude** of $f(x) = A \sin x$ and of $g(x) = A \cos x$.

EXAMPLE 2   Graphing Variations of $f(x) = \cos x$ Using Transformations

Use the graph of $f(x) = \cos x$ to graph $g(x) = \cos(3x)$.

Solution

The graph of $g(x) = \cos(3x)$ is a horizontal compression of the graph of $f(x) = \cos x$. Figure 80 shows the transformation.

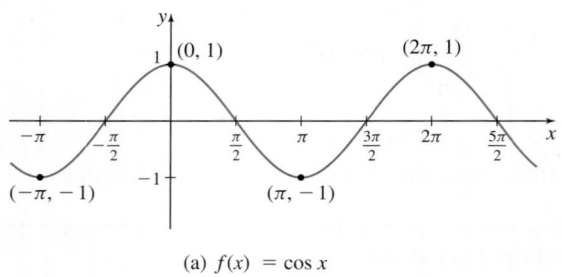

(a) $f(x) = \cos x$

Replace x by $3x$;
horizontal compression
by a factor of $\frac{1}{3}$.

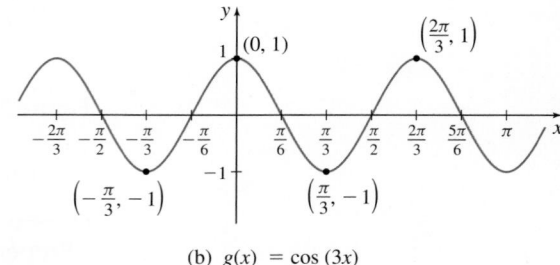

(b) $g(x) = \cos(3x)$

Figure 80

From the graph, we notice that the period of $g(x) = \cos(3x)$ is $\dfrac{2\pi}{3}$.

NOW WORK   Problems 27 and 29.

In general, if $\omega > 0$, the functions $f(x) = \sin(\omega x)$ and $g(x) = \cos(\omega x)$ have period $T = \dfrac{2\pi}{\omega}$. If $\omega > 1$, the graphs of $f(x) = \sin(\omega x)$ and $g(x) = \cos(\omega x)$ are horizontally compressed and the period of the functions is less than 2π. If $0 < \omega < 1$, the graphs of $f(x) = \sin(\omega x)$ and $g(x) = \cos(\omega x)$ are horizontally stretched, and the period of the functions is greater than 2π.

One period of the graph of $f(x) = \sin(\omega x)$ or $g(x) = \cos(\omega x)$ is called a **cycle**.

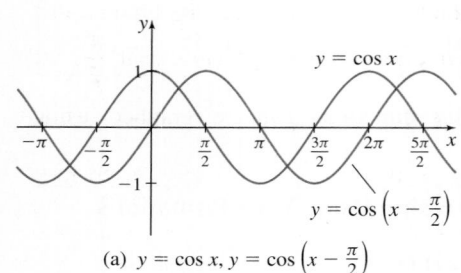

(a) $y = \cos x, \ y = \cos\left(x - \frac{\pi}{2}\right)$

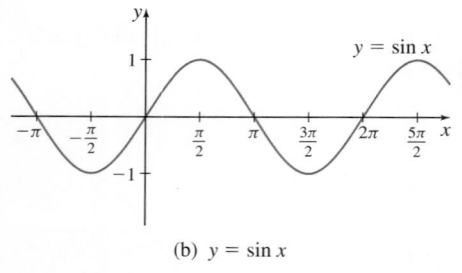

(b) $y = \sin x$

Figure 81

NEED TO REVIEW? Trigonometric identities are discussed in Appendix A.4, pp. A-33 to A-36.

Sinusoidal Graphs

If we shift the graph of the function $y = \cos x$ to the right $\dfrac{\pi}{2}$ units, we obtain the graph of $y = \cos\left(x - \dfrac{\pi}{2}\right)$, as shown in Figure 81(a). Now look at the graph of $y = \sin x$ in Figure 81(b). Notice that the graph of $y = \sin x$ is the same as the graph of $y = \cos\left(x - \dfrac{\pi}{2}\right)$.

Figure 81 suggests that

$$\boxed{\sin x = \cos\left(x - \frac{\pi}{2}\right)} \tag{1}$$

Proof To prove this identity, use the difference formula to expand $\cos\left(x - \dfrac{\pi}{2}\right)$.

$$\cos\left(x - \frac{\pi}{2}\right) = \cos x \cos\frac{\pi}{2} + \sin x \sin\frac{\pi}{2} = \cos x \cdot 0 + \sin x \cdot 1 = \sin x \qquad \blacksquare$$

Because of this relationship, the graphs of $y = A\sin(\omega x)$ or $y = A\cos(\omega x)$ are referred to as **sinusoidal graphs**, and the functions and their variations are called **sinusoidal functions**.

THEOREM Amplitude and Period

For the graphs of $y = A \sin(\omega x)$ and $y = A \cos(\omega x)$, where $A \neq 0$ and $\omega > 0$,

$$\bullet \ \text{Amplitude} = |A| \qquad \bullet \ \text{Period} = T = \frac{2\pi}{\omega}$$

▶ **EXAMPLE 3** **Finding the Amplitude and Period of a Sinusoidal Function**

CALC CLIP

Determine the amplitude and period of $y = -3 \sin(4\pi x)$.

Solution

Comparing $y = -3 \sin(4\pi x)$ to $y = A \sin(\omega x)$, we find that $A = -3$ and $\omega = 4\pi$. Then,

$$\text{Amplitude} = |A| = |-3| = 3 \qquad \text{Period} = T = \frac{2\pi}{\omega} = \frac{2\pi}{4\pi} = \frac{1}{2}$$ ■

NOW WORK Problem 33.

The function $y = \tan x$ is an odd function with period π. It is not defined at odd multiples of $\frac{\pi}{2}$. So, we construct Table 8 for $0 \leq x \leq \frac{\pi}{3}$. Then we plot the points from Table 8, connect them with a smooth curve, and reflect the graph about the origin, as shown in Figure 82. To investigate the behavior of $\tan x$ near $\frac{\pi}{2}$, we use the identity $\tan x = \frac{\sin x}{\cos x}$. When x is close to, but less than, $\frac{\pi}{2}$, $\sin x$ is close to 1 and $\cos x$ is a positive number close to 0, so the ratio $\frac{\sin x}{\cos x}$ is a large, positive number. The closer x gets to $\frac{\pi}{2}$, the larger $\tan x$ becomes. Figure 83 shows the graph of $y = \tan x$, $-\frac{\pi}{2} < x < \frac{\pi}{2}$. The graph of $y = \tan x$ is obtained by repeating the graph in Figure 83. See Figure 84.

TABLE 8

x	$y = \tan x$	(x, y)
0	0	$(0, 0)$
$\dfrac{\pi}{6}$	$\dfrac{\sqrt{3}}{3}$	$\left(\dfrac{\pi}{6}, \dfrac{\sqrt{3}}{3}\right)$
$\dfrac{\pi}{4}$	1	$\left(\dfrac{\pi}{4}, 1\right)$
$\dfrac{\pi}{3}$	$\sqrt{3}$	$\left(\dfrac{\pi}{3}, \sqrt{3}\right)$

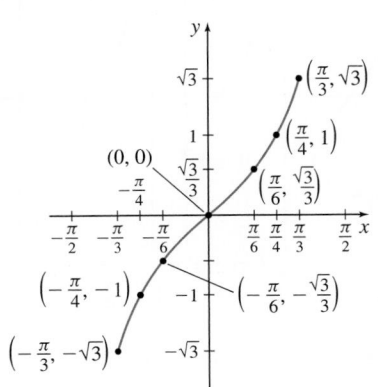

Figure 82 $y = \tan x$, $-\dfrac{\pi}{3} \leq x \leq \dfrac{\pi}{3}$

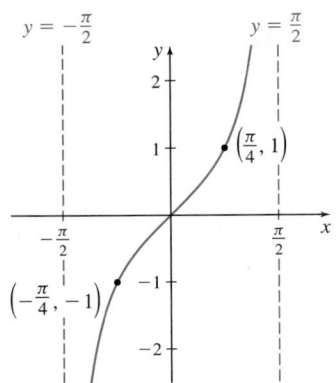

Figure 83 $y = \tan x$, $-\dfrac{\pi}{2} < x < \dfrac{\pi}{2}$

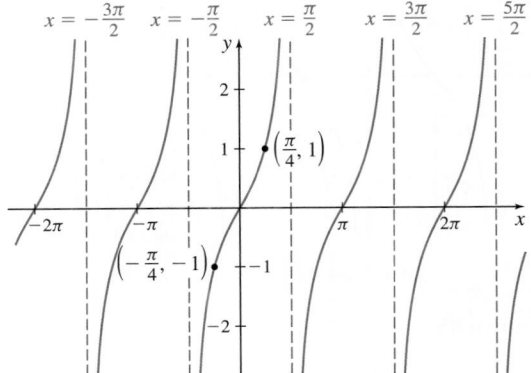

Figure 84 $y = \tan x$, $-\infty < x < \infty$,

$x \neq$ odd multiples of $\dfrac{\pi}{2}$.

The graph of $f(x) = \tan x$ illustrates the following properties of the tangent function:

Properties of the Tangent Function $f(x) = \tan x$

- The domain of f is the set of all real numbers, except odd multiples of $\dfrac{\pi}{2}$.

- The range of f is the set of all real numbers.

- The tangent function is an odd function, so its graph is symmetric with respect to the origin.

- The tangent function is periodic with period π.

- The x-intercepts of f are $\ldots, -2\pi, -\pi, 0, \pi, 2\pi, 3\pi, \ldots$; the y-intercept is 0.

EXAMPLE 4   Graphing Variations of $f(x) = \tan x$ Using Transformations

Use the graph of $f(x) = \tan x$ to graph $g(x) = -\tan\left(x + \dfrac{\pi}{4}\right)$.

Solution

Figure 85 illustrates the steps used in graphing $g(x) = -\tan\left(x + \dfrac{\pi}{4}\right)$.

- Begin by graphing $f(x) = \tan x$. See Figure 85(a).

- Replace the argument x by $x + \dfrac{\pi}{4}$ to obtain $y = \tan\left(x + \dfrac{\pi}{4}\right)$, which shifts the graph horizontally to the left $\dfrac{\pi}{4}$ unit, as shown in Figure 85(b).

- Multiply $\tan\left(x + \dfrac{\pi}{4}\right)$ by -1, which reflects the graph about the x-axis and results in the graph of $y = -\tan\left(x + \dfrac{\pi}{4}\right)$, as shown in Figure 85(c).

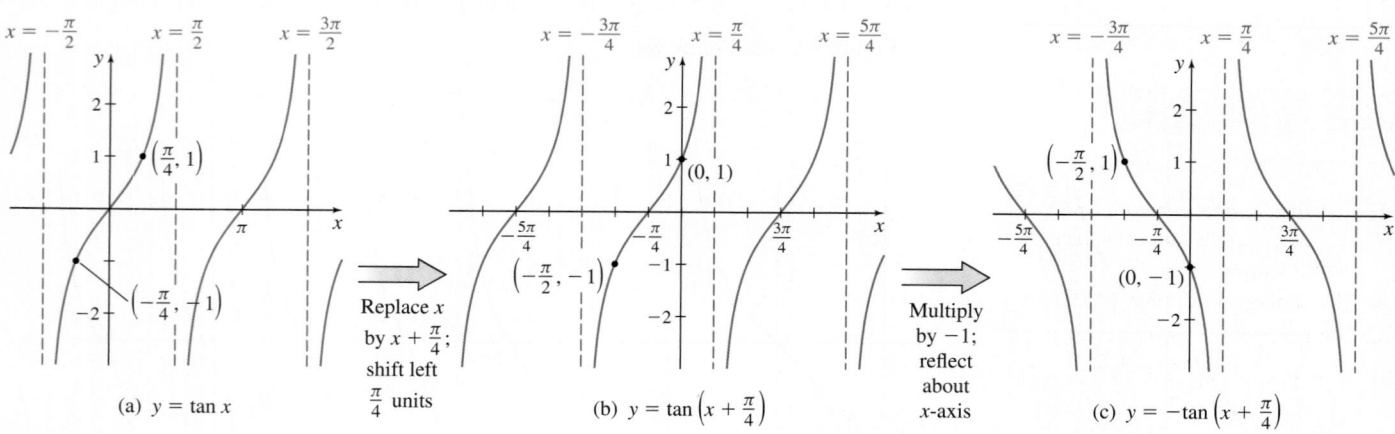

Figure 85

NOW WORK   Problem 31.

The graph of the cotangent function is obtained similarly. $y = \cot x$ is an odd function, with period π. Because $\cot x$ is not defined at multiples of π, graph y on the interval $(0, \pi)$ and then repeat the graph, as shown in Figure 86.

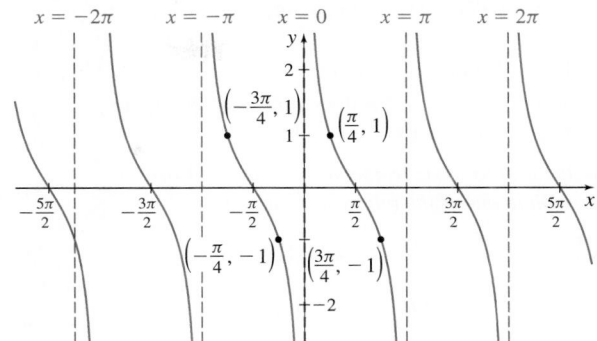

Figure 86 $f(x) = \cot x, \ -\infty < x < \infty, x \neq$ multiples of π.

The cosecant and secant functions, sometimes referred to as **reciprocal functions**, are graphed by using the reciprocal identities

$$\csc x = \frac{1}{\sin x} \quad \text{and} \quad \sec x = \frac{1}{\cos x}$$

The graphs of $y = \csc x$ and $y = \sec x$ are shown in Figures 87 and 88, respectively.

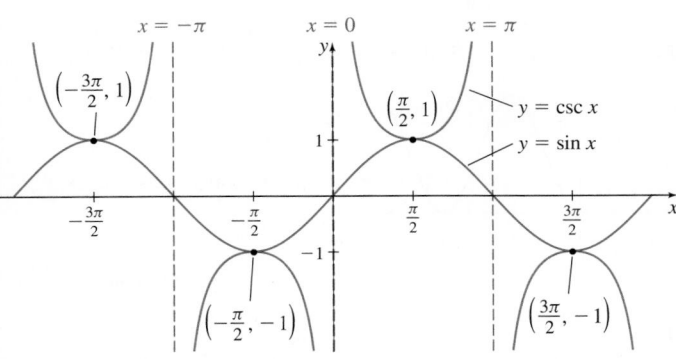

Figure 87

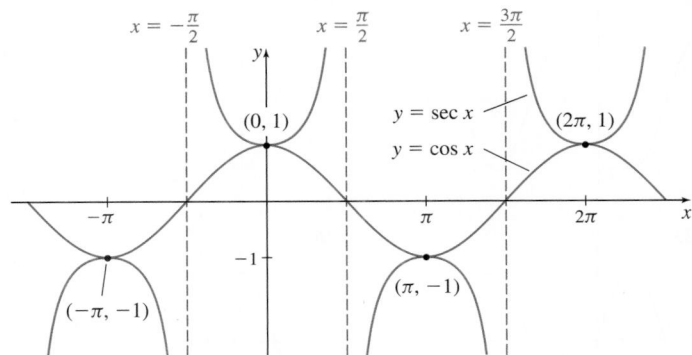

Figure 88

P.6 Assess Your Understanding

Concepts and Vocabulary

1. The sine, cosine, cosecant, and secant functions have period _____; the tangent and cotangent functions have period _____.

2. The domain of the tangent function $f(x) = \tan x$ is _____.

3. The range of the sine function $f(x) = \sin x$ is _____.

4. Explain why $\tan\left(\frac{\pi}{4} + 2\pi\right) = \tan\frac{\pi}{4}$.

5. **True or False** The range of the secant function is the set of all positive real numbers.

6. The function $f(x) = 3\cos(6x)$ has amplitude _____ and period _____.

7. **True or False** The graphs of $y = \sin x$ and $y = \cos x$ are identical except for a horizontal shift.

8. **True or False** The amplitude of the function $f(x) = 2\sin(\pi x)$ is 2 and its period is $\frac{\pi}{2}$.

9. **True or False** The graph of the sine function has infinitely many x-intercepts.

10. The graph of $y = \tan x$ is symmetric with respect to the _____.

11. The graph of $y = \sec x$ is symmetric with respect to the _____.

12. Explain, in your own words, what it means for a function to be periodic.

Practice Problems

In Problems 13–16, use the even-odd properties to find the exact value of each expression.

13. $\tan\left(-\frac{\pi}{4}\right)$

14. $\sin\left(-\frac{3\pi}{2}\right)$

15. $\csc\left(-\frac{\pi}{3}\right)$

16. $\cos\left(-\frac{\pi}{6}\right)$

In Problems 17–20, if necessary, refer to a graph to answer each question.

17. What is the y-intercept of $f(x) = \tan x$?
18. Find the x-intercepts of $f(x) = \sin x$ on the interval $[-2\pi, 2\pi]$.
19. What is the smallest value of $f(x) = \cos x$?
20. For what numbers x, $-2\pi \leq x \leq 2\pi$, does $\sin x = 1$? Where in the interval $[-2\pi, 2\pi]$ does $\sin x = -1$?

In Problems 21–26, the graphs of six trigonometric functions are given. Match each graph to one of the following functions:

(a) $y = 2\sin\left(\dfrac{\pi}{2}x\right)$ (b) $y = 2\cos\left(\dfrac{\pi}{2}x\right)$

(c) $y = 3\cos(2x)$ (d) $y = -3\sin(2x)$

(e) $y = -2\cos\left(\dfrac{\pi}{2}x\right)$ (f) $y = -2\sin\left(\dfrac{1}{2}x\right)$

21.
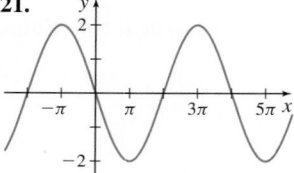

22.
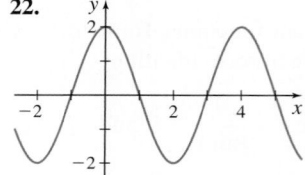

23.
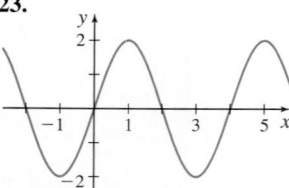

24.
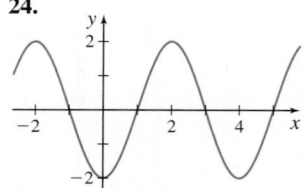

25.
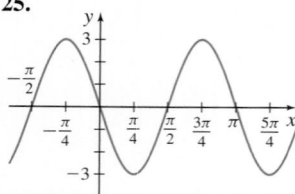

26.
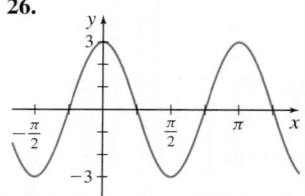

In Problems 27–32, graph each function using transformations. Be sure to label key points and show at least two periods.

27. $f(x) = 4\sin(\pi x)$
28. $f(x) = -3\cos x$
29. $f(x) = 3\cos(2x) - 4$
30. $f(x) = 4\sin(2x) + 2$
31. $f(x) = \tan\left(\dfrac{\pi}{2}x\right)$
32. $f(x) = 4\sec\left(\dfrac{1}{2}x\right)$

In Problems 33–36, determine the amplitude and period of each function.

33. $g(x) = \dfrac{1}{2}\cos(\pi x)$ 34. $f(x) = \sin(2x)$

35. $g(x) = 3\sin x$ 36. $f(x) = -2\cos\left(\dfrac{3}{2}x\right)$

In Problems 37 and 38, write the sine function that has the given properties.

37. Amplitude: 2, Period: π 38. Amplitude: $\dfrac{1}{3}$, Period: 2

In Problems 39 and 40, write the cosine function that has the given properties.

39. Amplitude: $\dfrac{1}{2}$, Period: π 40. Amplitude: 3, Period: 4π

In Problems 41–48, for each graph, find an equation involving the indicated trigonometric function.

41.
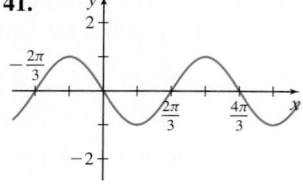

Sine function

42.
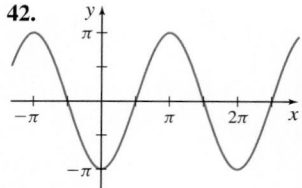

Cosine function

43.
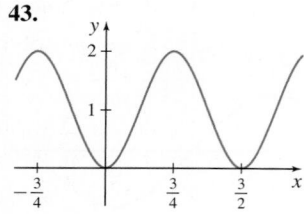

Cosine function

44.
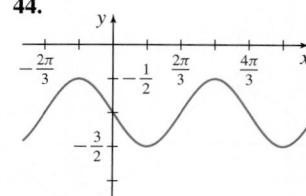

Sine function

45.
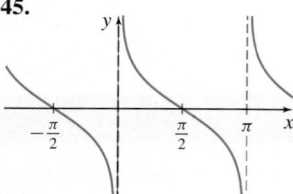

Cotangent function

46.
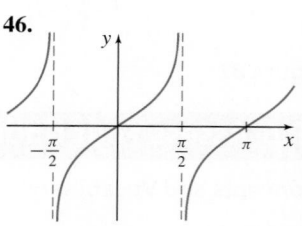

Tangent function

47.
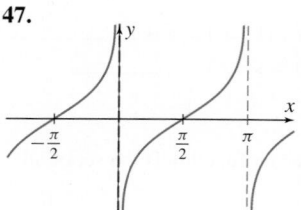

Tangent function

48.
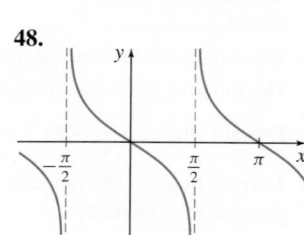

Cotangent function

P.7 Inverse Trigonometric Functions

OBJECTIVES *When you finish this section, you should be able to:*

1 **Define the inverse trigonometric functions (p. 63)**

2 **Use the inverse trigonometric functions (p. 67)**

3 **Solve trigonometric equations (p. 67)**

In Section P.4, we discussed inverse functions, and we concluded that if a function is one-to-one it has an inverse function. We also observed that if a function is not one-to-one, it may be possible to restrict its domain so that the restricted function is one-to-one. For example, the function $y = x^2$ is not one-to-one. However, if we restrict the domain to $x \geq 0$, the function is one-to-one.

Other properties of a one-to-one function f and its inverse function f^{-1} that we discussed in Section P.4 are summarized below.

Properties of a Function f and Its Inverse Function f^{-1}

- $f^{-1}(f(x)) = x$ for every x in the domain of f.
- $f(f^{-1}(x)) = x$ for every x in the domain of f^{-1}.
- Domain of f = range of f^{-1} and range of f = domain of f^{-1}.
- The graph of f and the graph of f^{-1} are reflections of one another about the line $y = x$.
- If a function $y = f(x)$ has an inverse function, the implicit equation of the inverse function is $x = f(y)$.
- If we can solve $x = f(y)$ for y, we obtain the explicit form $y = f^{-1}(x)$.

We now investigate the inverse trigonometric functions. In calculus the inverses of the sine function, the tangent function, and the secant function are used most often. So we define these inverse functions first.

1 Define the Inverse Trigonometric Functions

Although the trigonometric functions are not one-to-one, we can restrict the domains of the trigonometric functions so that each new function will be one-to-one on its restricted domain. Then an inverse function can be defined on each restricted domain.

Figure 89 shows the graph of $y = \sin x$. Because every horizontal line $y = b$, where b is between -1 and 1, inclusive, intersects the graph of $y = \sin x$ infinitely many times, it follows from the Horizontal-line Test that the function $y = \sin x$ is not one-to-one.

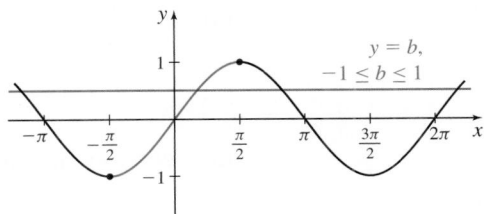

Figure 89 $y = \sin x$, $-\infty < x < \infty$, $-1 \leq y \leq 1$

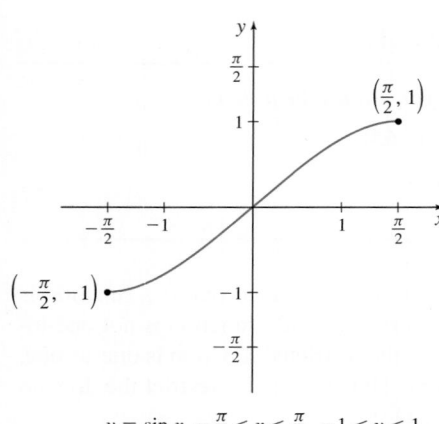

$$y = \sin x, \; -\tfrac{\pi}{2} \le x \le \tfrac{\pi}{2}, \; -1 \le y \le 1$$

Figure 90

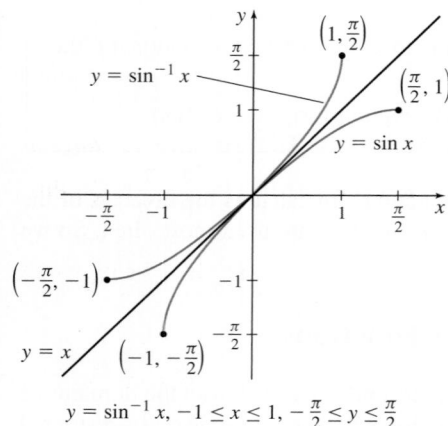

$$y = \sin^{-1} x, \; -1 \le x \le 1, \; -\tfrac{\pi}{2} \le y \le \tfrac{\pi}{2}$$

Figure 91

The Inverse Sine Function

If the domain of the function $f(x) = \sin x$ is restricted to the interval $\left[-\dfrac{\pi}{2}, \dfrac{\pi}{2}\right]$, the function $y = \sin x$, $-\dfrac{\pi}{2} \le x \le \dfrac{\pi}{2}$, is one-to-one and has an inverse function. See Figure 90.

To obtain an equation for the inverse of $y = \sin x$ on $-\dfrac{\pi}{2} \le x \le \dfrac{\pi}{2}$, we interchange x and y. Then the implicit form of the inverse is $x = \sin y$, $-\dfrac{\pi}{2} \le y \le \dfrac{\pi}{2}$. The explicit form of the inverse function is called the *inverse sine* of x and is written $y = \sin^{-1} x$.

DEFINITION Inverse Sine Function

The **inverse sine function**, denoted by $y = \sin^{-1} x$, or $y = \arcsin x$, is defined as

$$y = \sin^{-1} x \quad \text{if and only if} \quad x = \sin y$$

$$\text{where } -1 \le x \le 1 \quad \text{and} \quad -\frac{\pi}{2} \le y \le \frac{\pi}{2}$$

CAUTION The superscript $^{-1}$ that appears in $y = \sin^{-1} x$ is not an exponent but the symbol used to denote the inverse function. (To avoid confusion, $y = \sin^{-1} x$ is sometimes written $y = \arcsin x$.)

The domain of $y = \sin^{-1} x$ is the closed interval $[-1, 1]$, and the range is the closed interval $\left[-\dfrac{\pi}{2}, \dfrac{\pi}{2}\right]$. The graph of $y = \sin^{-1} x$ is the reflection of the restricted portion of the graph of $f(x) = \sin x$ about the line $y = x$. See Figure 91.

Because $y = \sin x$ defined on the closed interval $\left[-\dfrac{\pi}{2}, \dfrac{\pi}{2}\right]$ and $y = \sin^{-1} x$ are inverse functions,

$$\bullet \quad \sin^{-1}(\sin x) = x \quad \text{if } x \text{ is in the closed interval } \left[-\frac{\pi}{2}, \frac{\pi}{2}\right] \qquad (1)$$

$$\bullet \quad \sin(\sin^{-1} x) = x \quad \text{if } x \text{ is in the closed interval } [-1, 1] \qquad (2)$$

EXAMPLE 1 Finding the Values of an Inverse Sine Function

Find the exact value of:

(a) $\sin^{-1} \dfrac{\sqrt{3}}{2}$ **(b)** $\sin^{-1}\left(-\dfrac{\sqrt{2}}{2}\right)$ **(c)** $\sin^{-1}\left(\sin \dfrac{\pi}{8}\right)$ **(d)** $\sin^{-1}\left(\sin \dfrac{5\pi}{8}\right)$

Solution

(a) We seek an angle whose sine is $\dfrac{\sqrt{3}}{2}$. Let $y = \sin^{-1} \dfrac{\sqrt{3}}{2}$. Then by definition, $\sin y = \dfrac{\sqrt{3}}{2}$, where $-\dfrac{\pi}{2} \le y \le \dfrac{\pi}{2}$. Although $\sin y = \dfrac{\sqrt{3}}{2}$ has infinitely many solutions, the only number in the interval $\left[-\dfrac{\pi}{2}, \dfrac{\pi}{2}\right]$, for which $\sin y = \dfrac{\sqrt{3}}{2}$, is $\dfrac{\pi}{3}$. So $\sin^{-1} \dfrac{\sqrt{3}}{2} = \dfrac{\pi}{3}$.

(b) We seek an angle whose sine is $-\dfrac{\sqrt{2}}{2}$. Let $y = \sin^{-1}\left(-\dfrac{\sqrt{2}}{2}\right)$. Then by definition,

$\sin y = -\dfrac{\sqrt{2}}{2}$, where $-\dfrac{\pi}{2} \le y \le \dfrac{\pi}{2}$. The only number in the interval $\left[-\dfrac{\pi}{2}, \dfrac{\pi}{2}\right]$, whose

sine is $-\dfrac{\sqrt{2}}{2}$, is $-\dfrac{\pi}{4}$. So $\sin^{-1}\left(-\dfrac{\sqrt{2}}{2}\right) = -\dfrac{\pi}{4}$.

(c) Since the number $\dfrac{\pi}{8}$ is in the interval $\left[-\dfrac{\pi}{2}, \dfrac{\pi}{2}\right]$, we use (1).

$$\sin^{-1}\left(\sin \dfrac{\pi}{8}\right) = \dfrac{\pi}{8}$$

(d) Since the number $\dfrac{5\pi}{8}$ is not in the closed interval $\left[-\dfrac{\pi}{2}, \dfrac{\pi}{2}\right]$, we cannot use (1).

Instead, we find a number x in the interval $\left[-\dfrac{\pi}{2}, \dfrac{\pi}{2}\right]$ for which $\sin x = \sin \dfrac{5\pi}{8}$.

Using Figure 92, we see that $\sin \dfrac{5\pi}{8} = y = \sin \dfrac{3\pi}{8}$. The number $\dfrac{3\pi}{8}$ is in the

interval $\left[-\dfrac{\pi}{2}, \dfrac{\pi}{2}\right]$, so

$$\sin^{-1}\left(\sin \dfrac{5\pi}{8}\right) = \sin^{-1}\left(\sin \dfrac{3\pi}{8}\right) = \dfrac{3\pi}{8} \quad \blacksquare$$

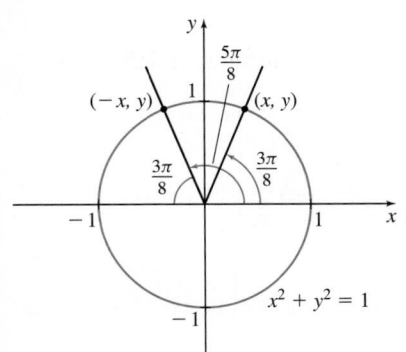

Figure 92

NOW WORK   Problems 9 and 21.

The Inverse Tangent Function

The tangent function $y = \tan x$ is not one-to-one. To define the *inverse tangent function*, we restrict the domain of $y = \tan x$ to the open interval $\left(-\dfrac{\pi}{2}, \dfrac{\pi}{2}\right)$. On this interval, $y = \tan x$ is one-to-one. See Figure 93.

Now use the steps for finding the inverse of a one-to-one function, listed on page 40:

Step 1 $y = \tan x$, $-\dfrac{\pi}{2} < x < \dfrac{\pi}{2}$.

Step 2 Interchange x and y, to obtain $x = \tan y$, $-\dfrac{\pi}{2} < y < \dfrac{\pi}{2}$, the implicit form of the inverse tangent function.

Step 3 The explicit form is called the *inverse tangent* of x, and is denoted by $y = \tan^{-1} x$.

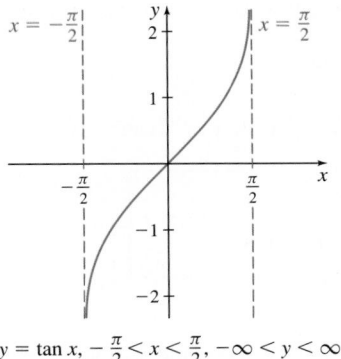

$y = \tan x, \ -\dfrac{\pi}{2} < x < \dfrac{\pi}{2}, \ -\infty < y < \infty$

Figure 93

DEFINITION Inverse Tangent Function

The **inverse tangent function**, symbolized by $y = \tan^{-1} x$, or $y = \arctan x$, is

$y = \tan^{-1} x$	if and only if	$x = \tan y$
where $-\infty < x < \infty$	and	$-\dfrac{\pi}{2} < y < \dfrac{\pi}{2}$

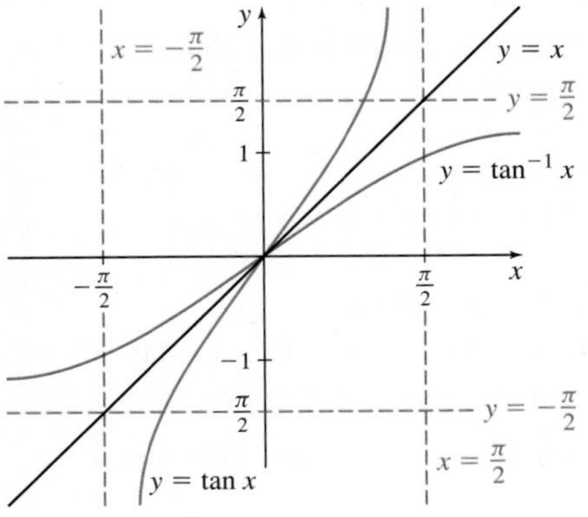

The domain of the function $y = \tan^{-1} x$ is the set of all real numbers and its range is $-\dfrac{\pi}{2} < y < \dfrac{\pi}{2}$. To graph $y = \tan^{-1} x$, reflect the graph of $y = \tan x$, $-\dfrac{\pi}{2} < x < \dfrac{\pi}{2}$, about the line $y = x$, as shown in Figure 94.

$y = \tan^{-1} x, \ -\infty < x < \infty, \ -\dfrac{\pi}{2} < y < \dfrac{\pi}{2}$

Figure 94

NOW WORK   Problem 13.

The Inverse Secant Function

To define the *inverse secant function*, we restrict the domain of the function $y = \sec x$ to the set $\left[0, \dfrac{\pi}{2}\right) \cup \left[\pi, \dfrac{3\pi}{2}\right)$. See Figure 95(a). An equation for the inverse function is obtained by following the steps for finding the inverse of a one-to-one function. By interchanging x and y, we obtain the implicit form of the inverse function: $x = \sec y$, where $0 \le y < \dfrac{\pi}{2}$ or $\pi \le y < \dfrac{3\pi}{2}$. The explicit form is called the *inverse secant* of x and is written $y = \sec^{-1} x$. The graph of $y = \sec^{-1} x$ is given in Figure 95(b).

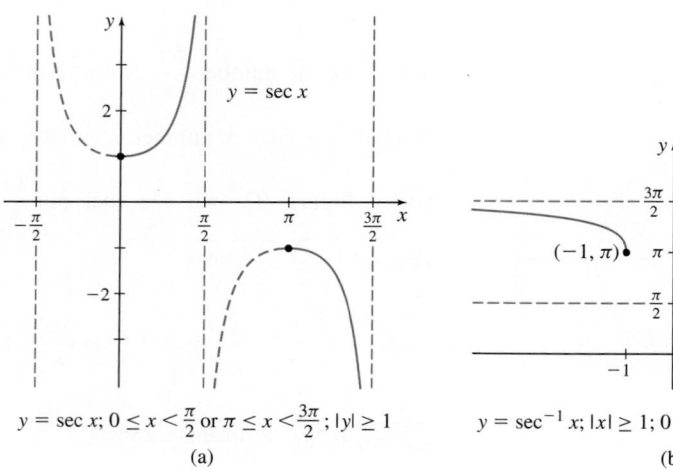

$y = \sec x;\ 0 \le x < \frac{\pi}{2}$ or $\pi \le x < \frac{3\pi}{2}$; $|y| \ge 1$

(a)

$y = \sec^{-1} x;\ |x| \ge 1;\ 0 \le y < \frac{\pi}{2}$ or $\pi \le y < \frac{3\pi}{2}$

(b)

Figure 95

DEFINITION Inverse Secant Function

The **inverse secant function**, symbolized by $y = \sec^{-1} x$, or $y = \operatorname{arcsec} x$, is

$y = \sec^{-1} x$ if and only if $x = \sec y$	
where $|x| \ge 1$ and $0 \le y < \dfrac{\pi}{2}$ or $\pi \le y < \dfrac{3\pi}{2}$	

You may wonder why we chose the restriction we did to define $y = \sec^{-1} x$. The reason is that this restriction not only makes $y = \sec x$ one-to-one, but it also makes $\tan x \ge 0$. With $\tan x \ge 0$, the derivative formula for the inverse secant function is simpler, as we will see in Chapter 3.

The remaining three inverse trigonometric functions are defined as follows.

DEFINITION

- **Inverse Cosine Function** $y = \cos^{-1} x$ if and only if $x = \cos y$,
  where $-1 \le x \le 1$ and $0 \le y \le \pi$

- **Inverse Cotangent Function** $y = \cot^{-1} x$ if and only if $x = \cot y$,
  where $-\infty < x < \infty$ and $0 < y < \pi$

- **Inverse Cosecant Function** $y = \csc^{-1} x$ if and only if $x = \csc y$,
  where $|x| \ge 1$ and $-\pi < y \le -\dfrac{\pi}{2}$ or $0 < y \le \dfrac{\pi}{2}$

The following three identities involve the inverse trigonometric functions. The proofs may be found in books on trigonometry.

THEOREM

• $\cos^{-1} x = \dfrac{\pi}{2} - \sin^{-1} x, \; -1 \le x \le 1$

• $\cot^{-1} x = \dfrac{\pi}{2} - \tan^{-1} x, \; -\infty < x < \infty$

• $\csc^{-1} x = \dfrac{\pi}{2} - \sec^{-1} x, \;\; |x| \ge 1$

These three identities are used whenever the inverse cosine, inverse cotangent, and inverse cosecant functions are needed. The graphs of the inverse cosine, inverse cotangent, and inverse cosecant functions can also be obtained by using these identities to transform the graphs of $y = \sin^{-1} x$, $y = \tan^{-1} x$, and $y = \sec^{-1} x$, respectively.

2 Use the Inverse Trigonometric Functions

CALC CLIP

EXAMPLE 2 **Writing a Trigonometric Expression as an Algebraic Expression**

Write $\sin(\tan^{-1} u)$ as an algebraic expression containing u.

Solution

Let $\theta = \tan^{-1} u$ so that $\tan \theta = u$, where $-\dfrac{\pi}{2} < \theta < \dfrac{\pi}{2}$. We note that in the interval $\left(-\dfrac{\pi}{2}, \dfrac{\pi}{2}\right)$, $\sec \theta > 0$. Then

$$\sin(\tan^{-1} u) = \sin \theta = \sin \theta \cdot \frac{\cos \theta}{\cos \theta} = \tan \theta \cos \theta$$

$$= \frac{\tan \theta}{\sec \theta} \underset{\substack{\uparrow \\ \sec^2 \theta = 1 + \tan^2 \theta; \; \sec \theta > 0}}{=} \frac{\tan \theta}{\sqrt{1 + \tan^2 \theta}} = \frac{u}{\sqrt{1 + u^2}}$$

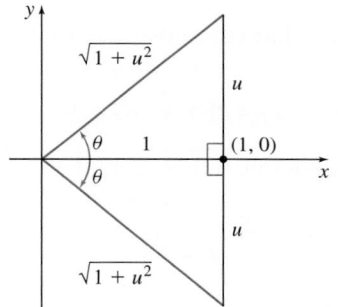

Figure 96 $\tan \theta = u; -\dfrac{\pi}{2} < \theta < \dfrac{\pi}{2}$

An alternate method of obtaining the solution to Example 2 uses right triangles. Let $\theta = \tan^{-1} u$ so that $\tan \theta = u$, $-\dfrac{\pi}{2} < \theta < \dfrac{\pi}{2}$, and label the right triangles drawn in Figure 96. Using the Pythagorean Theorem, the hypotenuse of each triangle is $\sqrt{1 + u^2}$. Then $\sin(\tan^{-1} u) = \sin \theta = \dfrac{u}{\sqrt{1 + u^2}}$. ∎

NOW WORK   Problem **67**.

3 Solve Trigonometric Equations

Inverse trigonometric functions can be used to solve trigonometric equations.

EXAMPLE 3 **Solving Trigonometric Equations**

Solve the equations:

(a) $\sin \theta = \dfrac{1}{2}$ **(b)** $\cos \theta = 0.4$

Give a general formula for all the solutions and list all the solutions in the interval $[-2\pi, 2\pi]$.

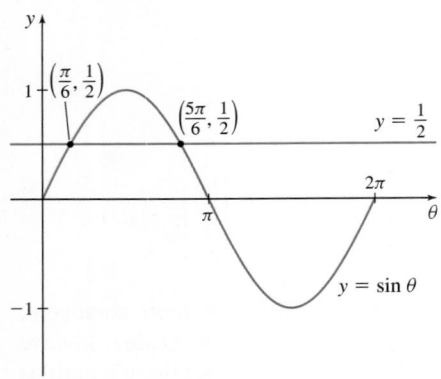

Figure 97

Solution

(a) Use the inverse sine function $y = \sin^{-1} x$, $-\dfrac{\pi}{2} \le y \le \dfrac{\pi}{2}$.

$$\sin \theta = \frac{1}{2}$$

$$\theta = \sin^{-1} \frac{1}{2} \qquad -\frac{\pi}{2} \le \theta \le \frac{\pi}{2}$$

$$\theta = \frac{\pi}{6}$$

Over the interval $[0, 2\pi]$, there are two angles θ for which $\sin \theta = \dfrac{1}{2}$. See Figure 97.

All the solutions of $\sin \theta = \dfrac{1}{2}$ are given by the general formula

$$\theta = \frac{\pi}{6} + 2k\pi \quad \text{or} \quad \theta = \frac{5\pi}{6} + 2k\pi, \qquad \text{where } k \text{ is any integer}$$

The solutions in the interval $[-2\pi, 2\pi]$ are

$$\left\{ -\frac{11\pi}{6}, \ -\frac{7\pi}{6}, \ \frac{\pi}{6}, \ \frac{5\pi}{6} \right\}$$

(b) A calculator must be used to solve $\cos \theta = 0.4$. With your calculator in radian mode,

$$\theta = \cos^{-1}(0.4) \approx 1.159279 \quad 0 \le \theta \le \pi$$

Rounded to three decimal places, $\theta = \cos^{-1} 0.4 = 1.159$ radians. But there is another angle θ in the interval $[0, 2\pi]$ for which $\cos \theta = 0.4$, namely, $\theta \approx 2\pi - 1.159 \approx 5.124$ radians.

Because the cosine function has period 2π, all the solutions of $\cos \theta = 0.4$ are given by the general formulas

$$\theta \approx 1.159 + 2k\pi \quad \text{or} \quad \theta \approx 5.124 + 2k\pi, \qquad \text{where } k \text{ is any integer}$$

The solutions in the interval $[-2\pi, 2\pi]$ are $\{-5.124, -1.159, 1.159, 5.124\}$. ∎

NOW WORK Problem 63.

EXAMPLE 4 Solving a Trigonometric Equation

Solve the equation $\sin(2\theta) = \dfrac{1}{2}$, where $0 \le \theta < 2\pi$.

Solution

In the interval $[0, 2\pi)$, the sine function has the value $\dfrac{1}{2}$ at $\theta = \dfrac{\pi}{6}$ and at $\theta = \dfrac{5\pi}{6}$ as shown in Figure 98. Since the period of the sine function is 2π and the argument in the equation $\sin(2\theta) = \dfrac{1}{2}$ is 2θ, we write the general formula for all the solutions.

$$2\theta = \frac{\pi}{6} + 2k\pi \quad \text{or} \quad 2\theta = \frac{5\pi}{6} + 2k\pi, \qquad \text{where } k \text{ is any integer}$$

$$\theta = \frac{\pi}{12} + k\pi \qquad\qquad \theta = \frac{5\pi}{12} + k\pi$$

The solutions of $\sin(2\theta) = \dfrac{1}{2}$, $0 \le \theta < 2\pi$, are $\left\{ \dfrac{\pi}{12}, \ \dfrac{5\pi}{12}, \ \dfrac{13\pi}{12}, \ \dfrac{17\pi}{12} \right\}$. ∎

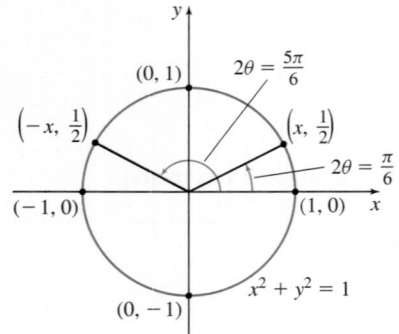

Figure 98

NOW WORK Problem 49.

NEED TO REVIEW? Trigonometric identities are discussed in Appendix Section A.4, pp. A-35 to A-38.

Many trigonometric equations can be solved by applying algebra techniques such as factoring or using the quadratic formula. It is often necessary to begin by using a trigonometric identity to express the given equation in a familiar form.

EXAMPLE 5 **Solving a Trigonometric Equation**

Solve the equation $3 \sin \theta - \cos^2 \theta = 3$, where $0 \le \theta < 2\pi$.

Solution

The equation involves both sine and cosine functions. We use the Pythagorean identity $\sin^2 \theta + \cos^2 \theta = 1$ to rewrite the equation in terms of $\sin \theta$.

$$3 \sin \theta - \cos^2 \theta = 3$$

$$3 \sin \theta - (1 - \sin^2 \theta) = 3 \qquad \cos^2 \theta = 1 - \sin^2 \theta$$

$$\sin^2 \theta + 3 \sin \theta - 4 = 0$$

This is a quadratic equation in $\sin \theta$. Factor the left side and solve for $\sin \theta$.

$$(\sin \theta + 4)(\sin \theta - 1) = 0$$

$$\sin \theta + 4 = 0 \qquad \text{or} \qquad \sin \theta - 1 = 0$$

$$\sin \theta = -4 \qquad \text{or} \qquad \sin \theta = 1$$

The range of the sine function is $-1 \le y \le 1$, so $\sin \theta = -4$ has no solution. Solving $\sin \theta = 1$, we obtain

$$\theta = \sin^{-1} 1 = \frac{\pi}{2}$$

The only solution in the interval $[0, 2\pi)$ is $\dfrac{\pi}{2}$. ∎

NOW WORK Problem 53.

P.7 Assess Your Understanding

Concepts and Vocabulary

1. $y = \sin^{-1} x$ if and only if $x =$ _____,

where $-1 \le x \le 1$ and $-\dfrac{\pi}{2} \le y \le \dfrac{\pi}{2}$.

2. *True or False* $\sin^{-1}(\sin x) = x$, where $-1 \le x \le 1$.

3. *True or False* The domain of $y = \sin^{-1} x$ is $-\dfrac{\pi}{2} \le x \le \dfrac{\pi}{2}$.

4. *True or False* $\sin\left(\sin^{-1} 0\right) = 0$.

5. *True or False* $y = \tan^{-1} x$ means $x = \tan y$,

where $-\infty < x < \infty$ and $-\dfrac{\pi}{2} \le y \le \dfrac{\pi}{2}$.

6. *True or False* The domain of the inverse tangent function is the set of all real numbers.

7. *True or False* $\sec^{-1} 0.5$ is not defined.

8. *True or False* Trigonometric equations can have multiple solutions.

Practice Problems

In Problems 9–20, find the exact value of each expression. Do not use a calculator.

9. $\sin^{-1} 0$ **10.** $\tan^{-1} 1$ **11.** $\sec^{-1}(-1)$

12. $\sec^{-1} \dfrac{2\sqrt{3}}{3}$ **13.** $\tan^{-1}(-1)$ **14.** $\tan^{-1} 0$

15. $\sin^{-1} \dfrac{\sqrt{2}}{2}$ **16.** $\tan^{-1} \dfrac{\sqrt{3}}{3}$ **17.** $\tan^{-1} \sqrt{3}$

18. $\sin^{-1}\left(-\dfrac{\sqrt{3}}{2}\right)$ **19.** $\sec^{-1}(-\sqrt{2})$ **20.** $\sin^{-1}\left(-\dfrac{\sqrt{2}}{2}\right)$

In Problems 21–28, find the exact value of each expression. Do not use a calculator.

21. $\sin^{-1}\left(\sin \dfrac{\pi}{5}\right)$ **22.** $\sin^{-1}\left[\sin\left(-\dfrac{\pi}{10}\right)\right]$

23. $\tan^{-1}\left[\tan\left(-\dfrac{3\pi}{8}\right)\right]$ **24.** $\sin^{-1}\left(\sin \dfrac{3\pi}{7}\right)$

25. $\sin^{-1}\left(\sin \dfrac{9\pi}{8}\right)$ **26.** $\sin^{-1}\left[\sin\left(-\dfrac{5\pi}{3}\right)\right]$

27. $\tan^{-1}\left(\tan \dfrac{4\pi}{5}\right)$ **28.** $\tan^{-1}\left[\tan\left(-\dfrac{2\pi}{3}\right)\right]$

In Problems 29–44, find the exact value, if any, of each composite function. If there is no value, write, "It is not defined." Do not use a calculator.

29. $\sin\left(\sin^{-1}\dfrac{1}{4}\right)$

30. $\sin\left[\sin^{-1}\left(-\dfrac{2}{3}\right)\right]$

31. $\tan\left(\tan^{-1}4\right)$

32. $\tan\left[\tan^{-1}(-2)\right]$

33. $\sin\left(\sin^{-1}1.2\right)$

34. $\sin\left[\sin^{-1}(-2)\right]$

35. $\tan\left(\tan^{-1}\pi\right)$

36. $\sin\left[\sin^{-1}(-1.5)\right]$

37. $\cos\left(\sin^{-1}0.3\right)$

38. $\sin\left(\tan^{-1}2\right)$

39. $\tan\left(\sec^{-1}2\right)$

40. $\cos\left(\tan^{-1}3\right)$

41. $\sin\left(\sin^{-1}0.2+\tan^{-1}2\right)$

42. $\cos\left(\sec^{-1}2+\sin^{-1}0.1\right)$

43. $\tan\left(2\sin^{-1}0.4\right)$

44. $\cos\left(2\tan^{-1}5\right)$

In Problems 45–62, solve each equation on the interval $0 \le \theta < 2\pi$.

45. $\tan\theta = -\dfrac{\sqrt{3}}{3}$

46. $\sec\dfrac{3\theta}{2} = -2$

47. $2\sin\theta + 3 = 2$

48. $1 - \cos\theta = \dfrac{1}{2}$

49. $\sin(3\theta) = -1$ [PAGE 68]

50. $\cos(2\theta) = \dfrac{1}{2}$

51. $4\cos^2\theta = 1$

52. $2\sin^2\theta - 1 = 0$

53. $2\sin^2\theta - 5\sin\theta + 3 = 0$ [PAGE 69]

54. $2\cos^2\theta - 7\cos\theta - 4 = 0$

55. $1 + \sin\theta = 2\cos^2\theta$

56. $\sec^2\theta + \tan\theta = 0$

57. $\sin\theta + \cos\theta = 0$

58. $\tan\theta = \cot\theta$

59. $\cos(2\theta) + 6\sin^2\theta = 4$

60. $\cos(2\theta) = \cos\theta$

61. $\sin(2\theta) + \sin(4\theta) = 0$

62. $\cos(4\theta) - \cos(6\theta) = 0$

In Problems 63–66, use a calculator to solve each equation on the interval $0 \le \theta < 2\pi$. Round answers to three decimal places.

63. $\tan\theta = 5$ [PAGE 68]

64. $\cos\theta = 0.6$

65. $2 + 3\sin\theta = 0$

66. $4 + \sec\theta = 0$

67. Write $\cos(\sin^{-1}u)$ as an algebraic expression containing u, where $|u| \le 1$. [PAGE 67]

68. Write $\tan(\sin^{-1}u)$ as an algebraic expression containing u, where $|u| \le 1$.

69. Show that $y = \sin^{-1}x$ is an odd function. That is, show $\sin^{-1}(-x) = -\sin^{-1}x$.

70. Show that $y = \tan^{-1}x$ is an odd function. That is, show $\tan^{-1}(-x) = -\tan^{-1}x$.

71. (a) On the same set of axes, graph $f(x) = 3\sin(2x) + 2$ and $g(x) = \dfrac{7}{2}$ on the interval $[0, \pi]$.

(b) Solve $f(x) = g(x)$ on the interval $[0, \pi]$, and label the points of intersection on the graph drawn in (a).

(c) Shade the region bounded by $f(x) = 3\sin(2x) + 2$ and $g(x) = \dfrac{7}{2}$ between the points found in (b) on the graph drawn in (a).

(d) Solve $f(x) > g(x)$ on the interval $[0, \pi]$.

72. (a) On the same set of axes, graph $f(x) = -4\cos x$ and $g(x) = 2\cos x + 3$ on the interval $[0, 2\pi]$.

(b) Solve $f(x) = g(x)$ on the interval $[0, 2\pi]$, and label the points of intersection on the graph drawn in (a).

(c) Shade the region bounded by $f(x) = -4\cos x$ and $g(x) = 2\cos x + 3$ between the points found in (b) on the graph drawn in (a).

(d) Solve $f(x) > g(x)$ on the interval $[0, 2\pi]$.

73. Area under a Graph The area A under the graph of $y = \dfrac{1}{1+x^2}$ and above the x-axis between $x = a$ and $x = b$ is given by

$$A = \tan^{-1}b - \tan^{-1}a$$

See the figure.

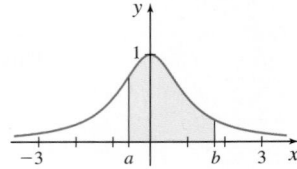

(a) Find the exact area under the graph of $y = \dfrac{1}{1+x^2}$ and above the x-axis between $x = 0$ and $x = \sqrt{3}$.

(b) Find the exact area under the graph of $y = \dfrac{1}{1+x^2}$ and above the x-axis between $x = -\dfrac{\sqrt{3}}{3}$ and $x = 1$.

74. Area under a Graph The area A under the graph of $y = \dfrac{1}{\sqrt{1-x^2}}$ and above the x-axis between $x = a$ and $x = b$ is given by

$$A = \sin^{-1}b - \sin^{-1}a$$

See the figure.

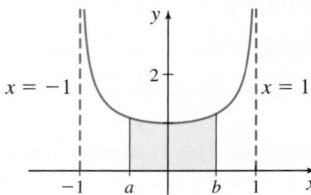

(a) Find the exact area under the graph of $y = \dfrac{1}{\sqrt{1-x^2}}$ and above the x-axis between $x = 0$ and $x = \dfrac{\sqrt{3}}{2}$.

(b) Find the exact area under the graph of $y = \dfrac{1}{\sqrt{1-x^2}}$ and above the x-axis between $x = -\dfrac{1}{2}$ and $x = \dfrac{1}{2}$.

P.8 Sequences; Summation Notation; the Binomial Theorem

OBJECTIVES *When you finish this section you should be able to:*

1 **Write the first several terms of a sequence (p. 71)**
2 **Write the terms of a recursively defined sequence (p. 72)**
3 **Use summation notation (p. 73)**
4 **Find the sum of the first n terms of a sequence (p. 75)**
5 **Use the Binomial Theorem (p. 76)**

When you hear the word *sequence* as used in "a sequence of events," you probably think of something that happens first, then second, and so on. In mathematics, the word *sequence* also refers to outcomes that are first, second, and so on.

> **DEFINITION Sequence**
>
> A **sequence** is a function whose domain is the set of positive integers.

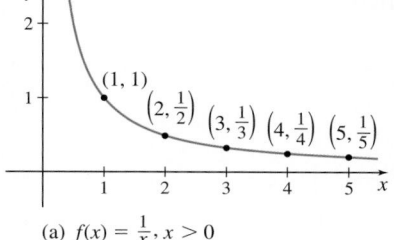

(a) $f(x) = \frac{1}{x}, x > 0$

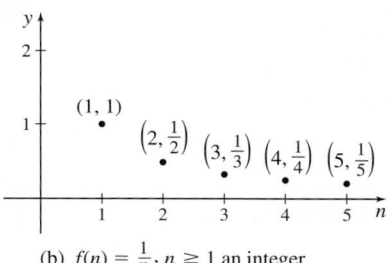

(b) $f(n) = \frac{1}{n}, n \geq 1$ an integer

Figure 99

In a sequence the inputs are $1, 2, 3, \ldots$. Because a sequence is a function, it has a graph. Figure 99(a) shows the graph of the function $f(x) = \frac{1}{x}, x > 0$. If all the points on this graph are removed except those whose x-coordinates are positive integers, that is, if all points are removed except $(1, 1)$, $\left(2, \frac{1}{2}\right)$, $\left(3, \frac{1}{3}\right)$, and so on, the remaining points would be the graph of the sequence $f(n) = \frac{1}{n}$, as shown in Figure 99(b). Notice that we use n to represent the independent variable in a sequence. This is to remind us that n is a positive integer.

1 Write the First Several Terms of a Sequence

A sequence is usually represented by listing its values in order. For example, the sequence whose graph is given in Figure 99(b) might be represented as

$$f(1), f(2), f(3), f(4), \ldots \quad \text{or} \quad 1, \frac{1}{2}, \frac{1}{3}, \frac{1}{4}, \ldots$$

The list never ends, as the dots (**ellipsis**) indicate. The numbers in this ordered list are called the **terms** of the sequence.

In dealing with sequences, we usually use subscripted letters such as a_1 to represent the first term, a_2 for the second term, a_3 for the third term, and so on.

For the sequence $f(n) = \frac{1}{n}$, we write

$$\underbrace{a_1 = f(1) = 1,}_{\text{first term}} \quad \underbrace{a_2 = f(2) = \frac{1}{2},}_{\text{second term}} \quad \underbrace{a_3 = f(3) = \frac{1}{3},}_{\text{third term}} \quad \underbrace{a_4 = f(4) = \frac{1}{4},}_{\text{fourth term}} \cdots \quad \underbrace{a_n = f(n) = \frac{1}{n},}_{n\text{th term}} \cdots$$

In other words, we usually do not use the traditional function notation $f(n)$ for sequences. For this particular sequence, we have a rule for the nth term (sometimes called the **general term**), which is $a_n = \frac{1}{n}$, so it is easy to find any term of the sequence.

When a formula for the nth term of a sequence is known, we usually represent the entire sequence by placing braces around the formula for the nth term. For example, the sequence whose nth term is $b_n = \left(\dfrac{1}{2}\right)^n$ can be represented either as

$$b_1 = \frac{1}{2}, \quad b_2 = \frac{1}{4}, \quad b_3 = \frac{1}{8}, \quad \ldots, \quad b_n = \frac{1}{2^n}, \quad \ldots,$$

or as

$$\{b_n\} = \left\{\left(\frac{1}{2}\right)^n\right\}$$

EXAMPLE 1 Writing the Terms of a Sequence

Write the first six terms of the sequence: $\{S_n\} = \left\{(-1)^{n-1}\dfrac{2}{n}\right\}$

Solution

$$S_1 = (-1)^0\,\frac{2}{1} = 2 \qquad S_2 = (-1)^1\,\frac{2}{2} = -1 \qquad S_3 = (-1)^2\,\frac{2}{3} = \frac{2}{3}$$

$$S_4 = -\frac{1}{2} \qquad\qquad S_5 = \frac{2}{5} \qquad\qquad S_6 = -\frac{1}{3}$$

The first six terms of the sequence are $2,\ -1,\ \dfrac{2}{3},\ -\dfrac{1}{2},\ \dfrac{2}{5},\ -\dfrac{1}{3}.$ ∎

NOW WORK   Problem 17.

EXAMPLE 2 Determining a Sequence from a Pattern

The pattern on the left suggests the nth term of the sequence on the right.

Pattern	nth term	Sequence
$e, \dfrac{e^2}{2}, \dfrac{e^3}{3}, \dfrac{e^4}{4}, \ldots$	$a_n = \dfrac{e^n}{n}$	$\{a_n\} = \left\{\dfrac{e^n}{n}\right\}$
$1, \dfrac{1}{3}, \dfrac{1}{9}, \dfrac{1}{27}, \ldots$	$b_n = \dfrac{1}{3^{n-1}}$	$\{b_n\} = \left\{\dfrac{1}{3^{n-1}}\right\}$
$1,\ 3,\ 5,\ 7, \ldots$	$c_n = 2n - 1$	$\{c_n\} = \{2n - 1\}$
$1,\ 4,\ 9,\ 16,\ 25, \ldots$	$d_n = n^2$	$\{d_n\} = \{n^2\}$
$1, -\dfrac{1}{2}, \dfrac{1}{3}, -\dfrac{1}{4}, \dfrac{1}{5}, \ldots$	$e_n = (-1)^{n+1}\left(\dfrac{1}{n}\right)$	$\{e_n\} = \left\{(-1)^{n+1}\left(\dfrac{1}{n}\right)\right\}$

∎

NOW WORK   Problem 21.

❷ Write the Terms of a Recursively Defined Sequence

A second way of defining a sequence is to assign a value to the first (or the first few) term(s) and specify the nth term by a formula or an equation that involves one or more of the terms preceding it. Such a sequence is defined **recursively**, and the rule or formula is called a **recursive formula**.

EXAMPLE 3   **Writing the Terms of a Recursively Defined Sequence**

Write the first five terms of the recursively defined sequence.

$$s_1 = 1 \qquad s_n = 4s_{n-1} \qquad n \geq 2$$

Solution

The first term is given as $s_1 = 1$. To get the second term, we use $n = 2$ in the formula to obtain $s_2 = 4s_1 = 4 \cdot 1 = 4$. To obtain the third term, we use $n = 3$ in the formula, getting $s_3 = 4s_2 = 4 \cdot 4 = 16$. Each new term requires that we know the value of the preceding term. The first five terms are

$$\begin{aligned} s_1 &= 1 \\ s_2 &= 4s_{2-1} = 4s_1 = 4 \cdot 1 = 4 \\ s_3 &= 4s_{3-1} = 4s_2 = 4 \cdot 4 = 16 \\ s_4 &= 4s_{4-1} = 4s_3 = 4 \cdot 16 = 64 \\ s_5 &= 4s_{5-1} = 4s_4 = 4 \cdot 64 = 256 \end{aligned}$$ ■

NOW WORK   Problem 29.

EXAMPLE 4   **Writing the Terms of a Recursively Defined Sequence**

Write the first five terms of the recursively defined sequence.

$$u_1 = 1 \qquad u_2 = 1 \qquad u_n = u_{n-2} + u_{n-1} \quad n \geq 3$$

Solution

We are given the first two terms. To obtain the third term requires that we know each of the previous two terms.

$$\begin{aligned} u_1 &= 1 \\ u_2 &= 1 \\ u_3 &= u_1 + u_2 = 1 + 1 = 2 \\ u_4 &= u_2 + u_3 = 1 + 2 = 3 \\ u_5 &= u_3 + u_4 = 2 + 3 = 5 \end{aligned}$$ ■

The sequence defined in Example 4 is called a **Fibonacci sequence**; its terms are called **Fibonacci numbers**. These numbers appear in a wide variety of applications.

NOW WORK   Problem 37.

---

❸ Use Summation Notation

It is often important to be able to find the sum of the first n terms of a sequence $\{a_n\}$, namely,

$$a_1 + a_2 + a_3 + \cdots + a_n$$

NOTE Summation notation is called **sigma notation** in some books.

Rather than write down all these terms, we introduce **summation notation**. Using summation notation, the sum is written

$$a_1 + a_2 + a_3 + \cdots + a_n = \sum_{k=1}^{n} a_k$$

The symbol $\sum$ (the uppercase Greek letter sigma, which is an S in our alphabet) is simply an instruction to sum, or add up, the terms. The integer k is called the **index of summation**; it tells us where to start the sum ($k = 1$) and where to end it ($k = n$). The expression $\sum_{k=1}^{n} a_k$ is read, "The sum of the terms a_k (a sub k) from $k = 1$ to $k = n$."

EXAMPLE 5 Expanding Summation Notation

Write out each sum:

(a) $\displaystyle\sum_{k=1}^{n} \frac{1}{k}$ (b) $\displaystyle\sum_{k=1}^{n} k!$

NOTE If $n \geq 0$ is an integer, the **factorial symbol** $n!$ means $0! = 1$, $1! = 1$, and $n! = 1 \cdot 2 \cdot 3 \cdot \cdots \cdot (n-1) \cdot n$, where $n > 1$.

Solution

(a) $\displaystyle\sum_{k=1}^{n} \frac{1}{k} = 1 + \frac{1}{2} + \frac{1}{3} + \cdots + \frac{1}{n}$ (b) $\displaystyle\sum_{k=1}^{n} k! = 1! + 2! + \cdots + n!$ ∎

NOW WORK   Problem 41.

EXAMPLE 6 Writing a Sum in Summation Notation

Express each sum using summation notation:

(a) $1^2 + 2^2 + 3^2 + \cdots + 9^2$ (b) $1 + \dfrac{1}{2} + \dfrac{1}{4} + \dfrac{1}{8} + \cdots + \dfrac{1}{2^{n-1}}$

Solution

(a) The sum $1^2 + 2^2 + 3^2 + \cdots + 9^2$ has 9 terms, each of the form k^2; it starts at $k = 1$ and ends at $k = 9$:

$$1^2 + 2^2 + 3^2 + \cdots + 9^2 = \sum_{k=1}^{9} k^2$$

(b) The sum

$$1 + \frac{1}{2} + \frac{1}{4} + \frac{1}{8} + \cdots + \frac{1}{2^{n-1}}$$

has n terms, each of the form $\dfrac{1}{2^{k-1}}$. It starts at $k = 1$ and ends at $k = n$.

$$1 + \frac{1}{2} + \frac{1}{4} + \frac{1}{8} + \cdots + \frac{1}{2^{n-1}} = \sum_{k=1}^{n} \frac{1}{2^{k-1}}$$ ∎

The index of summation does not need to begin at 1 or end at n. For example, the sum in Example 6(b) could be expressed as

$$\sum_{k=0}^{n-1} \frac{1}{2^k} = 1 + \frac{1}{2} + \frac{1}{4} + \frac{1}{8} + \cdots + \frac{1}{2^{n-1}}$$

Letters other than k can be used as the index. For example,

$$\sum_{j=1}^{n} j! \quad \text{and} \quad \sum_{i=1}^{n} i!$$

each represent the same sum as Example 5(b).

NOW WORK   Problem 51.

④ Find the Sum of the First n Terms of a Sequence

When working with summation notation, the following properties are useful for finding the sum of the first n terms of a sequence.

THEOREM Properties Involving Summation Notation

If $\{a_n\}$ and $\{b_n\}$ are two sequences and c is a real number, then:

$$\sum_{k=1}^{n}(ca_k) = c\sum_{k=1}^{n}a_k \tag{1}$$

$$\sum_{k=1}^{n}(a_k + b_k) = \sum_{k=1}^{n}a_k + \sum_{k=1}^{n}b_k \tag{2}$$

$$\sum_{k=1}^{n}(a_k - b_k) = \sum_{k=1}^{n}a_k - \sum_{k=1}^{n}b_k \tag{3}$$

$$\sum_{k=j+1}^{n}a_k = \sum_{k=1}^{n}a_k - \sum_{k=1}^{j}a_k, \quad \text{where } 0 < j < n \tag{4}$$

The proof of property (1) follows from the distributive property of real numbers. The proofs of properties (2) and (3) are based on the commutative and associative properties of real numbers. Property (4) states that the sum from $j + 1$ to n equals the sum from 1 to n minus the sum from 1 to j. It can be helpful to use this property when the index of summation begins at a number larger than 1.

THEOREM Formulas for Sums of the First n Terms of a Sequence

$$\sum_{k=1}^{n}1 = \underbrace{1 + 1 + 1 + \cdots + 1}_{n \text{ terms}} = n \tag{5}$$

$$\sum_{k=1}^{n}k = 1 + 2 + 3 + \cdots + n = \frac{n(n+1)}{2} \tag{6}$$

$$\sum_{k=1}^{n}k^2 = 1^2 + 2^2 + 3^2 + \cdots + n^2 = \frac{n(n+1)(2n+1)}{6} \tag{7}$$

$$\sum_{k=1}^{n}k^3 = 1^3 + 2^3 + 3^3 + \cdots + n^3 = \left[\frac{n(n+1)}{2}\right]^2 \tag{8}$$

Notice the difference between formulas (5) and (6). In formula (5), the constant 1 is being summed from 1 to n, while in (6) the index of summation k is being summed from 1 to n.

EXAMPLE 7 **Finding Sums**

Find each sum:

(a) $\displaystyle\sum_{k=1}^{100}\frac{1}{2}$ **(b)** $\displaystyle\sum_{k=1}^{5}(3k)$ **(c)** $\displaystyle\sum_{k=1}^{10}(k^3 + 1)$ **(d)** $\displaystyle\sum_{k=6}^{20}(4k^2)$

Solution

(a) $\displaystyle\sum_{k=1}^{100} \frac{1}{2} \underset{\underset{(1)}{\uparrow}}{=} \frac{1}{2} \sum_{k=1}^{100} 1 \underset{\underset{(5)}{\uparrow}}{=} \frac{1}{2} \cdot 100 = 50$

(b) $\displaystyle\sum_{k=1}^{5} (3k) \underset{\underset{(1)}{\uparrow}}{=} 3 \sum_{k=1}^{5} k \underset{\underset{(6)}{\uparrow}}{=} 3 \left(\frac{5(5+1)}{2} \right) = 3 \cdot 15 = 45$

(c) $\displaystyle\sum_{k=1}^{10} (k^3 + 1) \underset{\underset{(2)}{\uparrow}}{=} \sum_{k=1}^{10} k^3 + \sum_{k=1}^{10} 1 \underset{\underset{(8) \text{ and } (5)}{\uparrow}}{=} \left[\frac{10(10+1)}{2} \right]^2 + 10 = 3025 + 10 = 3035$

(d) Notice that the index of summation starts at 6.

$$\sum_{k=6}^{20} (4k^2) \underset{\underset{(1)}{\uparrow}}{=} 4 \sum_{k=6}^{20} k^2 \underset{\underset{(4)}{\uparrow}}{=} 4 \left[\sum_{k=1}^{20} k^2 - \sum_{k=1}^{5} k^2 \right]$$

$$\underset{\underset{(7)}{\uparrow}}{=} 4 \left[\frac{20 \cdot 21 \cdot 41}{6} - \frac{5 \cdot 6 \cdot 11}{6} \right] = 4[2870 - 55] = 11,260 \qquad \blacksquare$$

NOW WORK   Problem 61.

⑤ Use the Binomial Theorem

You already know formulas for expanding $(x + a)^n$ for $n = 2$ and $n = 3$. The *Binomial Theorem* is a formula for the expansion of $(x + a)^n$ for any positive integer n. If $n = 1$, 2 and 3, the expansion of $(x + a)^n$ is straightforward.

- $(x + a)^1 = x + a$ Two terms, beginning with x^1 and ending with a^1

- $(x + a)^2 = x^2 + 2ax + a^2$ Three terms, beginning with x^2 and ending with a^2

- $(x + a)^3 = x^3 + 3ax^2 + 3a^2x + a^3$ Four terms, beginning with x^3 and ending with a^3

Notice that each expansion of $(x + a)^n$ begins with x^n, ends with a^n, and has $n + 1$ terms. As you look at the terms from left to right, the powers of x are decreasing by one, while the powers of a are increasing by one. As a result, we might conjecture that the expansion of $(x + a)^n$ would look like this:

$$(x + a)^n = x^n + \underline{\quad\quad} ax^{n-1} + \underline{\quad\quad} a^2x^{n-2} + \cdots + \underline{\quad\quad} a^{n-1}x + a^n$$

where the blanks are numbers to be found. This, in fact, is the case.

Before filling in the blanks, we introduce the symbol $\dbinom{n}{j}$.

DEFINITION

If j and n are integers with $0 \le j \le n$, the symbol $\dbinom{n}{j}$ is defined as

$$\boxed{\dbinom{n}{j} = \frac{n!}{j!\,(n-j)!}}$$

EXAMPLE 8   Evaluating $\dbinom{n}{j}$

(a) $\dbinom{3}{1} = \dfrac{3!}{1!\,(3-1)!} = \dfrac{3!}{1!\,2!} = \dfrac{3 \cdot 2 \cdot 1}{1(2 \cdot 1)} = \dfrac{6}{2} = 3$

(b) $\dbinom{4}{2} = \dfrac{4!}{2!\,(4-2)!} = \dfrac{4!}{2!\,2!} = \dfrac{4 \cdot 3 \cdot 2 \cdot 1}{(2 \cdot 1)(2 \cdot 1)} = \dfrac{24}{4} = 6$

(c) $\dbinom{8}{7} = \dfrac{8!}{7!\,(8-7)!} = \dfrac{8!}{7!\,1!} = \underset{\substack{\uparrow \\ 8! \,=\, 8 \cdot 7!}}{\dfrac{8 \cdot 7!}{7! \cdot 1!}} = \dfrac{8}{1} = 8$

RECALL $0! = 1$

(d) $\dbinom{n}{0} = \dfrac{n!}{0!\,(n-0)!} = \dfrac{n!}{0!\,n!} = \dfrac{n!}{1 \cdot n!} = 1$

(e) $\dbinom{n}{n} = \dfrac{n!}{n!\,(n-n)!} = \dfrac{n!}{n!\,0!} = \dfrac{n!}{n! \cdot 1} = 1$ ■

NOW WORK   Problem 67.

THEOREM Binomial Theorem

Let x and a be real numbers. For any positive integer n,

$$(x+a)^n = \binom{n}{0}x^n + \binom{n}{1}ax^{n-1} + \binom{n}{2}a^2x^{n-2} + \cdots + \binom{n}{j}a^j x^{n-j}$$

$$+ \cdots + \binom{n}{n}a^n = \sum_{j=0}^{n} \binom{n}{j}a^j x^{n-j}$$

Because of its appearance in the Binomial Theorem, the symbol $\dbinom{n}{j}$ is called a **binomial coefficient**.

EXAMPLE 9   Expanding a Binomial

Use the Binomial Theorem to expand $(x+2)^5$.

Solution

We use the Binomial Theorem with $a = 2$ and $n = 5$. Then

$$(x+2)^5 = \binom{5}{0}x^5 + \binom{5}{1}2x^4 + \binom{5}{2}2^2x^3 + \binom{5}{3}2^3x^2 + \binom{5}{4}2^4x + \binom{5}{5}2^5$$

$$= 1 \cdot x^5 + 5 \cdot 2x^4 + 10 \cdot 4x^3 + 10 \cdot 8x^2 + 5 \cdot 16x + 1 \cdot 32$$

$$= x^5 + 10x^4 + 40x^3 + 80x^2 + 80x + 32$$ ■

NOW WORK   Problem 77.

P.8 Assess Your Understanding

Concepts and Vocabulary

1. A(n) _____ is a function whose domain is the set of positive integers.

2. **True or False** $3! = 6$

3. If $n \geq 0$ is an integer, then $n! = $ _____ when $n \geq 2$.

4. The sequence $a_1 = 5$, $a_n = 3a_{n-1}$ is an example of a(n) _____ sequence.

5. The notation $a_1 + a_2 + a_3 + \cdots + a_n = \sum_{k=1}^{n} a_k$ is an example of _____ notation.

6. **True or False** $\sum_{k=1}^{n} k = 1 + 2 + 3 + \cdots + n = \dfrac{n(n+1)}{2}$.

7. If $s_1 = 1$ and $s_n = 4s_{n-1}$, then $s_3 = $ _____.

8. $\dbinom{n}{0} = $ _____ ; $\dbinom{n}{1} = $ _____

9. **True or False** $\dbinom{n}{j} = \dfrac{j!}{(n-j)!\,n!}$

10. The _____ can be used to expand expressions like $(2x + 3)^6$.

Practice Problems

In Problems 11–18, list the first five terms of each sequence.

11. $\{s_n\} = \{2n\}$

12. $\{s_n\} = \{n^2\}$

13. $\{a_n\} = \left\{\dfrac{n}{n+1}\right\}$

14. $\{a_n\} = \left\{\dfrac{n+1}{2n}\right\}$

15. $\{b_n\} = \{(-1)^n n^2\}$

16. $\{b_n\} = \left\{(-1)^{n-1}\dfrac{n}{n+1}\right\}$

17. $\{s_n\} = \left\{(-1)^{n-1}\dfrac{2^n}{3^n + 1}\right\}$

18. $\{s_n\} = \left\{(-1)^{n-1}\left(\dfrac{4}{3}\right)^n\right\}$

In Problems 19–26, the given pattern continues. Find the nth term of a sequence suggested by the pattern.

19. $2, \dfrac{3}{2}, \dfrac{4}{3}, \dfrac{5}{4}, \ldots$

20. $\dfrac{1}{2 \cdot 3}, \dfrac{1}{3 \cdot 4}, \dfrac{1}{4 \cdot 5}, \dfrac{1}{5 \cdot 6}, \ldots$

21. $1, \dfrac{1}{4}, \dfrac{1}{9}, \dfrac{1}{16}, \dfrac{1}{25}, \ldots$

22. $\dfrac{1}{3}, \dfrac{2}{5}, \dfrac{3}{7}, \dfrac{4}{9}, \dfrac{5}{11}, \ldots$

23. $2, -4, 6, -8, 10, \ldots$

24. $2, -2, 2, -2, 2, \ldots$

25. $2, -3, 4, -5, 6, \ldots$

26. $1, \dfrac{1}{2}, 3, \dfrac{1}{4}, 5, \dfrac{1}{6}, 7, \dfrac{1}{8}, \ldots$

In Problems 27–38, a sequence is defined recursively. Write the first five terms of the sequence.

27. $a_1 = 3$; $a_n = 2 + a_{n-1}$

28. $a_1 = 1$; $a_n = 4 + a_{n-1}$

29. $a_1 = 2$; $a_n = n + a_{n-1}$

30. $a_1 = -1$; $a_n = n + a_{n-1}$

31. $a_1 = 1$; $a_n = 3a_{n-1}$

32. $a_1 = 1$; $a_n = -a_{n-1}$

33. $a_1 = 2$; $a_n = \dfrac{a_{n-1}}{n}$

34. $a_1 = -3$; $a_n = n + a_{n-1}$

35. $a_1 = 1$; $a_2 = -1$; $a_n = a_{n-1} \cdot a_{n-2}$

36. $a_1 = -1$; $a_2 = 2$; $a_n = a_{n-2} + na_{n-1}$

37. $a_1 = A$; $a_n = a_{n-1} + d$

38. $a_1 = A$; $a_n = ra_{n-1}$, $r \neq 0$

In Problems 39–48, write out each sum.

39. $\sum_{k=1}^{n} (k - 2)$

40. $\sum_{k=1}^{n} (2k)$

41. $\sum_{k=1}^{n} \dfrac{k^2}{k+1}$

42. $\sum_{k=1}^{n} \dfrac{(k+1)^2}{k}$

43. $\sum_{k=0}^{n} \dfrac{1}{2^k}$

44. $\sum_{k=0}^{n} \left(\dfrac{2}{3}\right)^k$

45. $\sum_{k=0}^{n} \dfrac{k+1}{2^{k+1}}$

46. $\sum_{k=1}^{n} \dfrac{2k+2}{k}$

47. $\sum_{k=2}^{n} (-1)^k 2^k$

48. $\sum_{k=1}^{n} (-1)^{k+1} \ln k$

In Problems 49–56, express each sum using summation notation.

49. $2 + 4 + 6 + \cdots + 100$

50. $\dfrac{1}{2} + \dfrac{2^3}{2} + \dfrac{3^3}{2} + \cdots + \dfrac{7^3}{2}$

51. $2 + \dfrac{3}{2} + \dfrac{4}{3} + \cdots + \dfrac{35+1}{35}$

52. $\dfrac{3}{2} + \dfrac{5}{2} + \dfrac{7}{2} + \dfrac{9}{2} + \cdots + \dfrac{2(12)+1}{2}$

53. $3 - \dfrac{3}{10} + \dfrac{3}{100} - \dfrac{3}{1000} + \cdots + (-1)^{10}\dfrac{3}{10^{10}}$

54. $-\dfrac{2}{3} + \dfrac{2}{9} - \dfrac{2}{27} + \cdots + (-1)^8 \dfrac{2}{3^8}$

55. $\dfrac{1}{e} - \dfrac{2}{e^2} + \dfrac{3}{e^3} - \cdots + (-1)^{n+1}\dfrac{n}{e^n}$

56. $1 - \dfrac{5}{2} + \dfrac{5^2}{3} - \dfrac{5^3}{4} + \cdots + (-1)^{n-1}\dfrac{5^{n-1}}{n}$

In Problems 57–66, find the sum of each sequence.

57. $\sum_{k=1}^{25} \dfrac{4}{5}$

58. $\sum_{k=1}^{50} 6$

59. $\sum_{k=1}^{30} (3k)$

60. $\sum_{k=1}^{20} \left(-\dfrac{2k}{3}\right)$

61. $\sum_{k=1}^{100} (2k - 1)$

62. $\sum_{k=1}^{18} (4k + 3)$

63. $\sum_{k=1}^{20} (k^2 - 2)$

64. $\sum_{k=0}^{10} (k^2 + 1)$

65. $\sum_{k=6}^{50} \left(\dfrac{k}{2}\right)$

66. $\sum_{k=3}^{20} (3k)$

In Problems 67–72, evaluate each expression.

67. $\dbinom{5}{2}$ **68.** $\dbinom{7}{4}$ **69.** $\dbinom{9}{3}$

70. $\dbinom{8}{4}$ **71.** $\dbinom{50}{1}$ **72.** $\dbinom{100}{2}$

In Problems 73–80, expand each expression using the Binomial Theorem.

73. $(x+1)^4$ **74.** $(x-1)^6$ **75.** $(x-3)^5$ **76.** $(x+2)^4$

77. $(2x+3)^4$ **78.** $(2x-1)^5$ **79.** $(x^2-y^2)^3$ **80.** $(x^2+y^2)^4$

81. Credit Card Debt Jin has a balance of $3000 on his credit card that charges 1% interest per month on any unpaid balance. Jin can afford to pay $100 toward the balance each month. His balance each month after making a $100 payment is given by the recursively defined sequence

$$B_0 = \$3000 \qquad B_n = 1.01B_{n-1} - 100$$

Determine Jin's balance after making the first payment. That is, determine B_1.

82. Approximating $f(x)=e^x$ In calculus, it can be shown that

$$f(x)=e^x = \sum_{k=0}^{\infty} \frac{x^k}{k!}$$

We can approximate the value of $f(x)=e^x$ for any x using the sum

$$f(x)=e^x \approx \sum_{k=0}^{n} \frac{x^k}{k!}$$

for some n.

(a) Approximate $f(1.3)$ with $n=4$.

(b) Approximate $f(1.3)$ with $n=7$.

(c) Use a calculator to approximate $f(1.3)$.

(d) Use trial and error along with a graphing calculator's SEQuence mode to determine the value of n required to approximate $f(1.3)$ correct to eight decimal places.

CHAPTER

1

Limits and Continuity

US Air Force Photo / Alamy Stock Photo

Oil Spills and Dispersant Chemicals

On April 20, 2010, the Deepwater Horizon drilling rig exploded and initiated the worst marine oil spill in recent history. Oil gushed from the well for three months and released millions of gallons of crude oil into the Gulf of Mexico. One technique used to help clean up during and after the spill was the use of the chemical dispersant Corexit. The use of chemical dispersants is debated because some of their ingredients are carcinogens. Further, the use of oil dispersants can increase toxic hydrocarbon levels affecting sea life. Over time, the pollution caused by the oil spill and the dispersants diminish and sea life returns. In the short term, however, pollution raises serious questions about the health of the local sea life and the safety of fish and shellfish for human consumption.

Although the Deepwater Horizon spill remains the worst in U.S. waters, oil and other pollutant spills occur regularly, threatening the environment and the wildlife in the area. In October 2021, a smaller, but still concerning, spill occurred off the coast of Long Beach, California. Just 7 miles from the shore, 26,000 gallons of processed oil poured into the Pacific Ocean, threatening humans and wildlife, as well as beaches, wetlands, and the Santa Ana River. The first response was to limit the size of the spill and keep the spill from spreading. Dispersants were then used to help finish the cleanup.

 Explore a hypothetical lawsuit involving pollution of a lake in the Chapter 1 Project on page 160.

Daphne
Internal Medicine Doctor

AP® Calculus
gave me a leg
up when I
started taking
pre-med classes in college. I was
prepared for the strategic
thinking required in courses like
organic chemistry and had an
understanding of the math
underlying many of the concepts
and equations in biology.

The concept of a limit is central to calculus. To understand calculus, it is essential to know what it means for a function to have a limit, and then how to find a limit of a function. Chapter 1 explains what a limit is, shows how to find a limit of a function, and demonstrates how to prove that limits exist using the definition of limit.

We begin the chapter using numerical and graphical techniques to explore the idea of a limit. Although these methods seem to work well, there are instances in which they fail to identify the correct limit.

In Section 1.2, we use analytic techniques to find limits. Some of the proofs of these techniques are found in Section 1.6, others in Appendix B. A limit found by correctly applying these analytic techniques is precise; there is no doubt that it is correct.

In Sections 1.3–1.5, we continue to study limits and some ways that they are used. For example, we use limits to define *continuity*, an important property of a function.

Section 1.6 provides a precise definition of *limit*, the so-called ε-δ (epsilon-delta) definition, which we use to show when a limit does, and does not, exist. ∎

1.1 Limits of Functions Using Numerical and Graphical Techniques

AP® EXAM TIP

Before starting this chapter, read *What is Calculus*, a short overview of calculus found on pp. iii at the front of this book.

AP® EXAM TIP

BREAK IT DOWN: As you work through the problems in this chapter, remember to follow these steps:

Step 1 Identify the underlying structure and related concepts.

Step 2 Determine the appropriate math rule or procedure.

Step 3 Apply the math rule or procedure.

Step 4 Clearly communicate your answer.

On page 164, see how we've used these steps to solve Section 1.5 AP® Practice Problem 11 on page 149.

OBJECTIVES *When you finish this section, you should be able to:*

1 **Discuss the idea of a limit (p. 83)**
2 **Investigate a limit using a table (p. 84)**
3 **Investigate a limit using a graph (p. 85)**

Calculus can be used to solve certain fundamental questions in geometry. Two of these questions are:

- Given a function f and a point P on its graph, what is the slope of the tangent line to the graph of f at P? See Figure 1.

- Given a nonnegative function f whose domain is the closed interval $[a, b]$, what is the area of the region enclosed by the graph of f, the x-axis, and the vertical lines $x = a$ and $x = b$? See Figure 2.

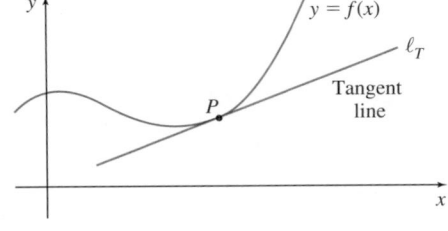

Figure 1   **Figure 2**

These questions, traditionally called the **tangent problem** and the **area problem**, were solved by Gottfried Wilhelm von Leibniz and Sir Isaac Newton during the late seventeenth and early eighteenth centuries. The solutions to the two seemingly different problems are both based on the idea of a limit. Their solutions not only are related to each other, but are also applicable to many other problems in science and geometry. Here, we begin to discuss the tangent problem. The discussion of the area problem begins in Chapter 6.

NEED TO REVIEW? The slope of a line is discussed in Appendix A.3, p. A-20.

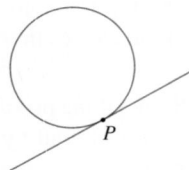

Figure 3 Tangent line to a circle at the point P.

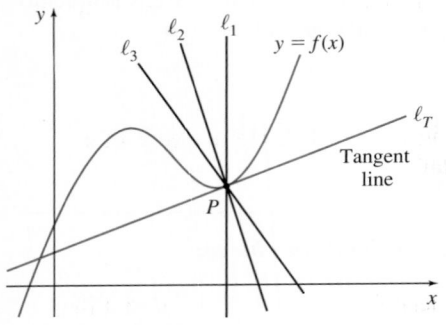

Figure 4

The Slope of the Tangent Line to a Graph

Notice that the line ℓ_T in Figure 1 (page 81) just touches the graph of f at the point P. This unique line is the *tangent line* to the graph of f at P. But how is the tangent line defined?

In plane geometry, a tangent line to a circle is defined as a line having exactly one point in common with the circle, as shown in Figure 3. However, this definition does not work for graphs in general. For example, in Figure 4, the three lines ℓ_1, ℓ_2, and ℓ_3 contain the point P and have exactly one point in common with the graph of f, but they do not meet the requirement of just touching the graph at P. On the other hand, the line ℓ_T just touches the graph of f at P, but it intersects the graph at another point. It is the slope of the tangent line ℓ_T that distinguishes it from all other lines containing P.

So before defining a tangent line, we investigate its slope, which we denote by $m_{\tan}$. We begin with the graph of a function f, a point P on its graph, and the tangent line ℓ_T to f at P, as shown in Figure 5.

The tangent line ℓ_T to the graph of f at P must contain the point P. We denote the coordinates of P by $(c, f(c))$. Since finding a slope requires two points, and we have only one point on the tangent line ℓ_T, we proceed as follows.

Suppose we choose any point $Q = (x, f(x))$, other than P, on the graph of f, as shown in Figure 6. (Q can be to the left or to the right of P; we chose Q to be to the right of P.) The line containing the points $P = (c, f(c))$ and $Q = (x, f(x))$ is called a **secant line** of the graph of f. The slope $m_{\sec}$ of this secant line is

$$m_{\sec} = \frac{f(x) - f(c)}{x - c} \tag{1}$$

Figure 7 shows three different points Q_1, Q_2, and Q_3 on the graph of f that are successively closer to the point P, and three associated secant lines ℓ_1, ℓ_2, and ℓ_3. The closer the point Q is to the point P, the closer the secant line is to the tangent line ℓ_T. The line ℓ_T, the *limiting position* of these secant lines, is the *tangent line to the graph of f at P.*

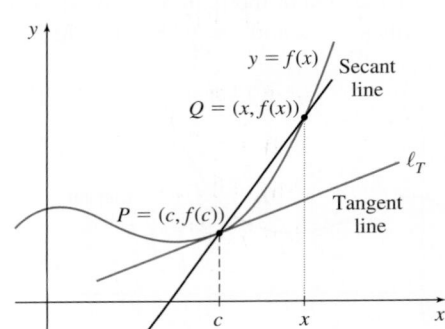

Figure 5 $m_{\tan} = $ slope of the tangent line.

Figure 6 $m_{\sec} = $ slope of a secant line.

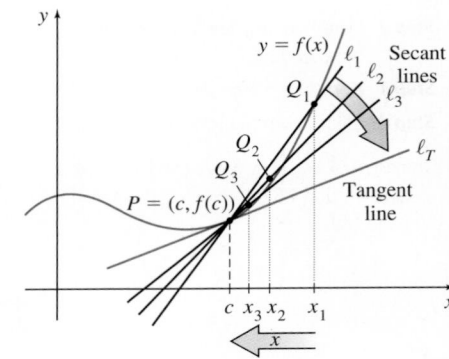

Figure 7

If the limiting position of the secant lines is the tangent line, then the limit of the slopes of the secant lines should equal the slope of the tangent line. Notice in Figure 7 that as the point Q moves closer to the point P, the numbers x get closer to c. So, equation (1) suggests that

$$m_{\tan} = [\text{Slope of the tangent line to } f \text{ at } P]$$

$$= \left[\text{Limit of } \frac{f(x) - f(c)}{x - c} \text{ as } x \text{ gets closer to } c \right]$$

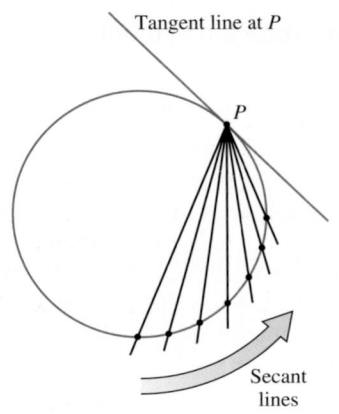

Figure 8

In symbols, we write

$$m_{\tan} = \lim_{x \to c} \frac{f(x) - f(c)}{x - c}$$

The notation $\lim_{x \to c}$ is read, "the limit as x approaches c."

The **tangent line** to the graph of a function f at a point $P = (c, f(c))$ is the line containing the point P whose slope is

$$\boxed{m_{\tan} = \lim_{x \to c} \frac{f(x) - f(c)}{x - c}}$$

provided the limit exists.

As Figure 8 illustrates, this new idea of a tangent line is consistent with the traditional definition of a tangent line to a circle.

NOW WORK Problem 2.

We have begun to answer the tangent problem by introducing the idea of a *limit*. Now we describe the idea of a limit in more detail.

1 Discuss the Idea of a Limit

We begin by asking a question: What does it mean for a function f to have a limit L as x approaches some fixed number c? To answer the question, we need to be more precise about f, L, and c. To have a limit at c, the function f must be defined everywhere in an open interval containing the number c, except possibly at c, and L must be a number. Using these restrictions, we introduce the notation

$$\boxed{\lim_{x \to c} f(x) = L}$$

NEED TO REVIEW? If $a < b$, the open interval (a, b) consists of all real numbers x for which $a < x < b$. Interval notation is discussed in Appendix A.1, p. A-5.

which is read, "the limit as x approaches c of $f(x)$ is equal to the number L."

The notation $\lim_{x \to c} f(x) = L$ can be described as

> The value $f(x)$ can be made as close as we please to L,
> for x sufficiently close to c, but not equal to c.

Figure 9 shows various situations in which $\lim_{x \to c} f(x) = L$.

- In Figure 9(a), $\lim_{x \to c} f(x) = L$ and $f(c) = L$.
- In Figure 9(b), $\lim_{x \to c} f(x) = L$, but $f(c) \neq L$.
- In Figure 9(c) $\lim_{x \to c} f(x) = L$, but f is not defined at c.

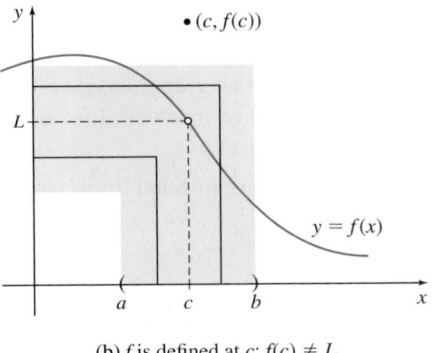

(a) f is defined at c; $f(c) = L$

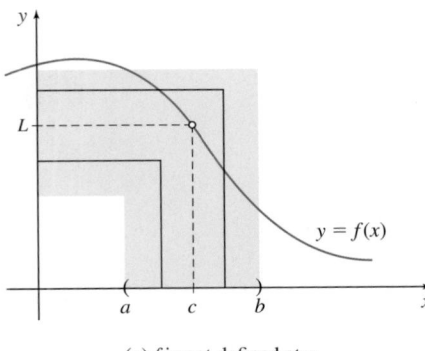

(b) f is defined at c; $f(c) \neq L$

(c) f is not defined at c

Figure 9 $\lim_{x \to c} f(x) = L$

EXAMPLE 1 Reading and Interpreting a Limit Expressed Symbolically

- In words, $\lim\limits_{x \to -1} \dfrac{x^2 - 1}{x + 1} = -2$ is read as, "The limit as x approaches -1 of $\dfrac{x^2 - 1}{x + 1}$ is equal to the number -2."

- The limit may be interpreted as "The value of the function $f(x) = \dfrac{x^2 - 1}{x + 1}$ can be made as close as we please to -2 by choosing x sufficiently close to, but not equal to, -1." ∎

NOW WORK Problem 1 and AP® Practice Problems 3 and 7.

2 Investigate a Limit Using a Table

We can use a table to better understand what it means for a function to have a limit as x approaches a number c.

EXAMPLE 2 Investigating a Limit Using a Table

Investigate $\lim\limits_{x \to 2}(2x + 5)$ using a table.

Solution

We create Table 1 by evaluating $f(x) = 2x + 5$ at values of x near 2, choosing numbers x slightly less than 2 and numbers x slightly greater than 2.

TABLE 1

x	\multicolumn{4}{c}{numbers x slightly less than 2 →}				\multicolumn{4}{c}{← numbers x slightly greater than 2}
x	1.99	1.999	1.9999	1.99999	→	2	←	2.00001	2.0001	2.001	2.01
$f(x) = 2x + 5$	8.98	8.998	8.9998	8.99998		$f(x)$ approaches 9		9.00002	9.0002	9.002	9.02

Table 1 suggests that the value of $f(x) = 2x + 5$ can be made "as close as we please" to 9 by choosing x "sufficiently close" to 2. This suggests that $\lim\limits_{x \to 2}(2x + 5) = 9$. ∎

In creating Table 1, first we used numbers x close to 2 but less than 2, and then we used numbers x close to 2 but greater than 2. When $x < 2$, we say, "x is approaching 2 from the left," and the number 9 is called the **left-hand limit**. When $x > 2$, we say, "x is approaching 2 from the right," and the number 9 is called the **right-hand limit**. Together, these are called the **one-sided limits** of f as x approaches 2.

One-sided limits are symbolized as follows. The left-hand limit, written

$$\lim_{x \to c^-} f(x) = L_{\text{left}}$$

is read, "The limit as x approaches c from the left of $f(x)$ equals L_{left}." It means that the value of f can be made as close as we please to the number L_{left} by choosing $x < c$ and sufficiently close to c.

Similarly, the right-hand limit, written

$$\lim_{x \to c^+} f(x) = L_{\text{right}}$$

is read, "The limit as x approaches c from the right of $f(x)$ equals L_{right}." It means that the value of f can be made as close as we please to the number L_{right} by choosing $x > c$ and sufficiently close to c.

NOW WORK Problem 9.

EXAMPLE 3 **Investigating a Limit Using a Table**

Investigate $\lim_{x\to 0} \dfrac{e^x - 1}{x}$ using a table.

Solution

The domain of $f(x) = \dfrac{e^x - 1}{x}$ is $\{x \mid x \neq 0\}$. So, f is defined everywhere in an open interval containing the number 0, except for 0.

We create Table 2, investigating the left-hand limit $\lim_{x\to 0^-} \dfrac{e^x - 1}{x}$ and the right-hand limit $\lim_{x\to 0^+} \dfrac{e^x - 1}{x}$. First, we evaluate f at numbers less than 0, but close to zero, and then at numbers greater than 0, but close to zero.

TABLE 2

x	-0.01	-0.001	-0.0001	-0.00001	$\to$	0	$\leftarrow$	0.00001	0.0001	0.001	0.01
		x approaches 0 from the left							x approaches 0 from the right		
$f(x) = \dfrac{e^x - 1}{x}$	0.995	0.9995	0.99995	0.999995	$f(x)$ approaches 1			1.000005	1.00005	1.0005	1.005

Table 2 suggests that $\lim_{x\to 0^-} \dfrac{e^x - 1}{x} = 1$ and $\lim_{x\to 0^+} \dfrac{e^x - 1}{x} = 1$.

This suggests $\lim_{x\to 0} \dfrac{e^x - 1}{x} = 1$. ∎

NOW WORK Problem 13 and AP® Practice Problems 2 and 6.

EXAMPLE 4 **Investigating a Limit Using a Table and Technology**

Investigate $\lim_{x\to 0} \dfrac{\sin x}{x}$ using a table.

Solution

The domain of the function $f(x) = \dfrac{\sin x}{x}$ is $\{x \mid x \neq 0\}$. So, f is defined everywhere in an open interval containing 0, except for 0.

We investigate the one-sided limits of $\dfrac{\sin x}{x}$ as x approaches 0 by using a graphing calculator to set up a table. Be sure the mode is set to radians. When making the table, we choose numbers x (in radians) slightly less than 0 and numbers slightly greater than 0. See Figure 10.

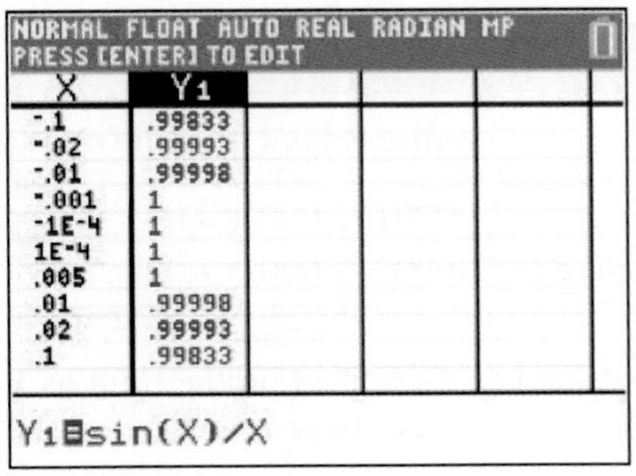

The table in Figure 10 suggests that $\lim_{x\to 0^-} f(x) = 1$ and $\lim_{x\to 0^+} f(x) = 1$. This suggests that $\lim_{x\to 0} \dfrac{\sin x}{x} = 1$. ∎

Figure 10

| NOTE $f(x) = \dfrac{\sin x}{x}$ is an even function, so the Y_1 values in Figure 10 are symmetric about $x = 0$.

NOW WORK Problem 15 and AP® Practice Problem 8.

3 Investigate a Limit Using a Graph

The graph of a function can also help us investigate limits. Figure 11 on page 86 shows the graphs of three different functions f, g, and h. Observe that in each function, *as x gets closer to c, whether from the left or from the right, the value of the function gets closer to the number L.* This is the key idea of a limit.

Notice in Figure 11(b) that the value of g at c does not affect the limit. Notice in Figure 11(c) that h is not defined at c, but the value of h gets closer to the number L for x sufficiently close to c. This suggests that the limit of a function as x approaches c does not depend on the value, if it exists, of the function at c.

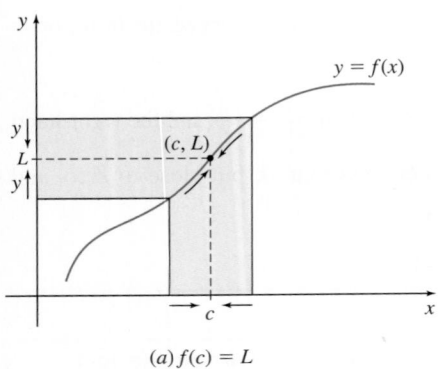

(a) $f(c) = L$

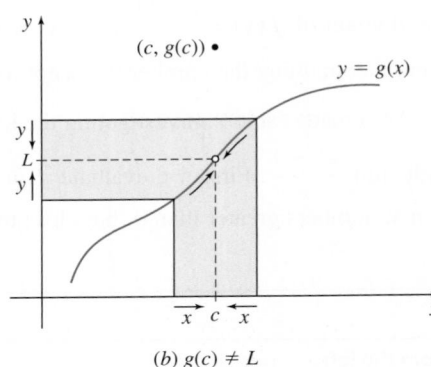

(b) $g(c) \neq L$

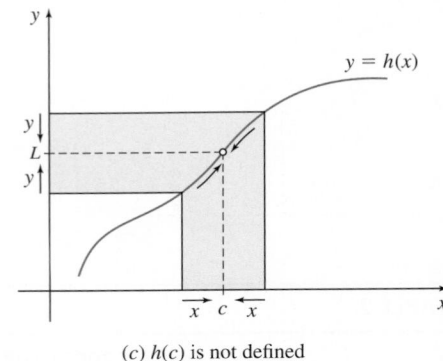

(c) $h(c)$ is not defined

Figure 11

EXAMPLE 5   Investigating a Limit Using a Graph

Use a graph to investigate $\lim\limits_{x \to 2} f(x)$ if $f(x) = \begin{cases} 3x + 1 & \text{if } x \neq 2 \\ 10 & \text{if } x = 2 \end{cases}$.

Solution

The function f is a piecewise-defined function. Its graph is shown in Figure 12. Observe that as x approaches 2 from the left, the value of f is close to 7, and as x approaches 2 from the right, the value of f is close to 7. In fact, we can make the value of f as close as we please to 7 by choosing x sufficiently close to 2 but not equal to 2. This suggests $\lim\limits_{x \to 2} f(x) = 7$. ∎

If we use a table to investigate $\lim\limits_{x \to 2} f(x)$, the result is the same. See Table 3.

NEED TO REVIEW? Piecewise-defined functions are discussed in Section P.1, p. 9.

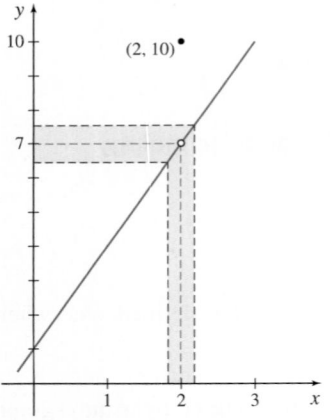

Figure 12 $f(x) = \begin{cases} 3x + 1 & \text{if } x \neq 2 \\ 10 & \text{if } x = 2 \end{cases}$

TABLE 3

	x approaches 2 from the left		x approaches 2 from the right
x	1.99 1.999 1.9999 1.99999	$\to 2 \leftarrow$	2.00001 2.0001 2.001 2.01
$f(x)$	6.97 6.997 6.9997 6.99997	$f(x)$ approaches 7	7.00003 7.0003 7.003 7.03

Figure 12 shows that $f(2) = 10$ but that this value has no impact on the limit as x approaches 2. In fact, $f(2)$ can equal any number, and it would have no effect on the limit as x approaches 2.

We make the following observations:

- The limit L of a function $y = f(x)$ as x approaches a number c does not depend on the value of f at c.
- The limit L of a function $y = f(x)$ as x approaches a number c is unique; that is, a function cannot have more than one limit. (A proof of this property is given in Appendix B.)
- If there is *no single number* that the value of f approaches as x gets close to c, we say that *f* **has no limit as *x* approaches** c, or more simply, that the **limit of *f* does not exist at c.**

NOW WORK   AP® Practice Problem 4.

Examples 6 and 7 illustrate situations in which a limit does not exist.

CALC CLIP

EXAMPLE 6 **Investigating a Limit Using a Graph**

Use a graph to investigate $\lim\limits_{x \to 0} f(x)$ if $f(x) = \begin{cases} x & \text{if} \quad x < 0 \\ 1 & \text{if} \quad x > 0 \end{cases}$.

Solution

Figure 13 shows the graph of f. We first investigate the one-sided limits. The graph suggests that, as x approaches 0 from the left,

$$\lim_{x \to 0^-} f(x) = 0$$

and as x approaches 0 from the right,

$$\lim_{x \to 0^+} f(x) = 1$$

Since there is no single number that the values of f approach when x is close to 0, we conclude that $\lim\limits_{x \to 0} f(x)$ does not exist. ∎

Table 4 uses a numerical technique to support the conclusion that $\lim\limits_{x \to 0} f(x)$ does not exist.

Figure 13 $f(x) = \begin{cases} x & \text{if } x < 0 \\ 1 & \text{if } x > 0 \end{cases}$

TABLE 4

	x approaches 0 from the left						x approaches 0 from the right		
x	−0.01	−0.001	−0.0001	→	0	←	0.0001	0.001	0.01
$f(x)$	−0.01	−0.001	−0.0001	no single number			1	1	1

Examples 5 and 6 lead to the following result.

THEOREM

The limit L of a function $y = f(x)$ as x approaches a number c exists if and only if both one-sided limits exist at c and both one-sided limits are equal. That is,

$$\lim_{x \to c} f(x) = L \quad \text{if and only if} \quad \lim_{x \to c^-} f(x) = \lim_{x \to c^+} f(x) = L$$

NOW WORK Problems 25, 31, and AP® Practice Problem 5.

A one-sided limit is used to describe the behavior of functions such as $f(x) = \sqrt{x-1}$ near $x = 1$. Since the domain of f is $\{x \mid x \geq 1\}$, the left-hand limit, $\lim\limits_{x \to 1^-} \sqrt{x-1}$ makes no sense. But $\lim\limits_{x \to 1^+} \sqrt{x-1} = 0$ suggests how f behaves near and to the right of 1. See Figure 14 and Table 5. They suggest $\lim\limits_{x \to 1^+} f(x) = 0$.

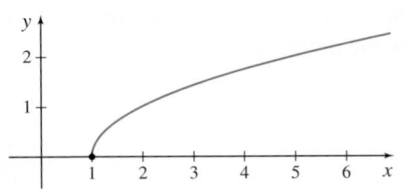

Figure 14 $f(x) = \sqrt{x-1}$

TABLE 5

		x approaches 1 from the right				
x	1 ←	1.0000009	1.000009	1.00009	1.0009	1.009
$f(x) = \sqrt{x-1}$	$f(x)$ approaches 0	0.000949	0.003	0.00949	0.03	0.0949

NOW WORK AP® Practice Problem 1.

Using numeric tables and/or graphs gives us an idea of what a limit might be. That is, these methods suggest a limit, but there are dangers in using these methods, as the following example illustrates.

EXAMPLE 7   Investigating a Limit

Investigate $\lim\limits_{x \to 0} \sin \dfrac{\pi}{x^2}$.

Solution

The domain of the function $f(x) = \sin \dfrac{\pi}{x^2}$ is $\{x \,|\, x \neq 0\}$.

Suppose we let x approach zero in the following way:

TABLE 6

x	$-\dfrac{1}{10}$	$-\dfrac{1}{100}$	$-\dfrac{1}{1000}$	$-\dfrac{1}{10,000}$	$\to \quad 0 \quad \leftarrow$	$\dfrac{1}{10,000}$	$\dfrac{1}{1000}$	$\dfrac{1}{100}$	$\dfrac{1}{10}$
	x approaches 0 from the left →					← *x approaches 0 from the right*			
$f(x) = \sin \dfrac{\pi}{x^2}$	0	0	0	0	$f(x)$ approaches 0	0	0	0	0

Table 6 suggests that $\lim\limits_{x \to 0} \sin \dfrac{\pi}{x^2} = 0$.

Alternatively, suppose we let x approach zero as follows:

TABLE 7

x	$-\dfrac{2}{3}$	$-\dfrac{2}{5}$	$-\dfrac{2}{7}$	$-\dfrac{2}{9}$	$-\dfrac{2}{11}$	$\to \quad 0 \quad \leftarrow$	$\dfrac{2}{11}$	$\dfrac{2}{9}$	$\dfrac{2}{7}$	$\dfrac{2}{5}$	$\dfrac{2}{3}$
	x approaches 0 from the left →						← *x approaches 0 from the right*				
$f(x) = \sin \dfrac{\pi}{x^2}$	0.707	0.707	0.707	0.707	0.707	$f(x)$ approaches 0.707	0.707	0.707	0.707	0.707	0.707

Table 7 suggests that $\lim\limits_{x \to 0} \sin \dfrac{\pi}{x^2} = \dfrac{\sqrt{2}}{2} \approx 0.707$.

In fact, by carefully selecting x, we can make f appear to approach any number in the interval $[-1, 1]$.

Now look at Figure 15, which illustrates that the graph of $f(x) = \sin \dfrac{\pi}{x^2}$ oscillates rapidly as x approaches 0. This suggests that the value of f does not approach a single number and that $\lim\limits_{x \to 0} \sin \dfrac{\pi}{x^2}$ does not exist. ∎

NOW WORK Problem 55.

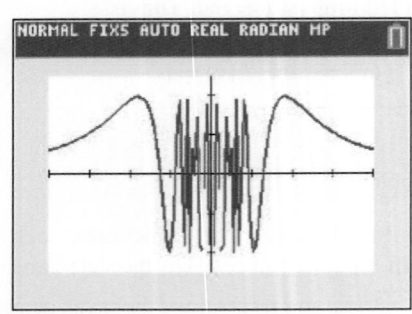

Figure 15 $f(x) = \sin \dfrac{\pi}{x^2}, \; x \neq 0$

So, how do we find a limit with certainty? The answer lies in giving a very precise definition of limit. The next example helps explain the definition.

EXAMPLE 8   Analyzing a Limit

In Example 2, we claimed that $\lim\limits_{x \to 2} (2x + 5) = 9$.

(a) How close must x be to 2, so that $f(x) = 2x + 5$ is within 0.1 of 9?

(b) How close must x be to 2, so that $f(x) = 2x + 5$ is within 0.05 of 9?

Solution

RECALL On the number line, the distance between two points with coordinates a and b is $|a - b|$.

(a) The function $f(x) = 2x + 5$ is within 0.1 of 9, if the distance between $f(x)$ and 9 is less than 0.1 unit. That is, if $|f(x) - 9| \leq 0.1$.

$$|(2x + 5) - 9| \leq 0.1$$
$$|2x - 4| \leq 0.1$$
$$|2(x - 2)| \leq 0.1$$
$$|x - 2| \leq \frac{0.1}{2} = 0.05$$
$$-0.05 \leq x - 2 \leq 0.05$$
$$1.95 \leq x \leq 2.05$$

NEED TO REVIEW? Inequalities involving absolute values are discussed in Appendix A.1, pp. A-7 to A-8.

So, if $1.95 \leq x \leq 2.05$, then $f(x)$ will be within 0.1 of 9.

(b) The function $f(x) = 2x + 5$ is within 0.05 of 9 if $|f(x) - 9| \leq 0.05$. That is,

$$|(2x + 5) - 9| \leq 0.05$$
$$|2x - 4| \leq 0.05$$
$$|x - 2| \leq \frac{0.05}{2} = 0.025$$

So, if $1.975 \leq x \leq 2.025$, then $f(x)$ will be within 0.05 of 9. ∎

Notice that the closer we require f to be to the limit 9, the narrower the interval for x becomes. See Figure 16.

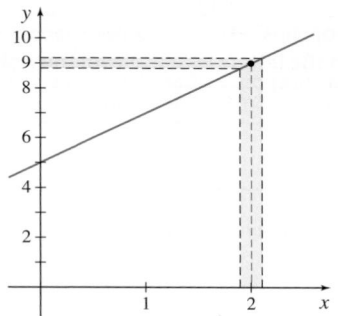

Figure 16 $f(x) = 2x + 5$

NOW WORK Problem 57.

The discussion in Example 8 forms the basis of the definition of a limit. We state the definition here but postpone the details until Section 1.6. It is customary to use the Greek letters ε (epsilon) and δ (delta) in the definition, so we call it the ε-δ *definition of a limit*.

DEFINITION ε-δ **Definition of a Limit**

Suppose f is a function defined everywhere in an open interval containing c, except possibly at c. Then the **limit of the function** f **as** x **approaches** c is the number L, written

$$\boxed{\lim_{x \to c} f(x) = L}$$

if, given any number $\varepsilon > 0$, there is a number $\delta > 0$ so that

$$\boxed{\text{whenever } 0 < |x - c| < \delta \quad \text{then } |f(x) - L| < \varepsilon}$$

Notice in the definition that f is defined everywhere in an open interval containing c except possibly at c. If f is defined at c and there is an open interval containing c that contains no other numbers in the domain of f, then $\lim_{x \to c} f(x)$ does not exist. This conclusion makes sense, since if there are no numbers in the domain of f near c, then it is impossible to find a number that the values of f get closer to.

1.1 Assess Your Understanding

Concepts and Vocabulary

PAGE 84 **1.** *Multiple Choice*   The limit as x approaches c of a function f is written symbolically as
[**(a)** $\lim\limits_{c \to x} f(x)$ **(b)** $\lim f(x)$ **(c)** $\lim\limits_{x \to c} f(x)$]

PAGE 83 **2.** *True or False*   The tangent line to the graph of f at a point $P = (c, f(c))$ is the limiting position of the secant lines passing through P and a point $(x, f(x))$, $x \neq c$, as x moves closer to c.

3. *True or False*   If f is not defined at $x = c$, then $\lim\limits_{x \to c} f(x)$ does not exist.

4. *True or False*   The limit L of a function $y = f(x)$ as x approaches the number c depends on the value of f at c.

5. *True or False*   If $f(c)$ is defined, this suggests that $\lim\limits_{x \to c} f(x)$ exists.

6. *True or False*   The limit of a function $y = f(x)$ as x approaches a number c equals L if at least one of the one-sided limits as x approaches c equals L.

Skill Building

In Problems 7–12, complete each table and investigate the limit.

7. $\lim\limits_{x \to 1} 2x$

	x approaches 1 from the left				x approaches 1 from the right		
x	0.9 0.99 0.999	$\to$	1	$\leftarrow$	1.001 1.01 1.1		
$f(x) = 2x$							

8. $\lim\limits_{x \to 2} (x + 3)$

	x approaches 2 from the left				x approaches 2 from the right		
x	1.9 1.99 1.999	$\to$	2	$\leftarrow$	2.001 2.01 2.1		
$f(x) = x + 3$							

PAGE 84 **9.** $\lim\limits_{x \to 0} (x^2 + 2)$

	x approaches 0 from the left				x approaches 0 from the right		
x	-0.1 -0.01 -0.001	$\to$	0	$\leftarrow$	0.001 0.01 0.1		
$f(x) = x^2 + 2$							

10. $\lim\limits_{x \to -1} (x^2 - 2)$

	x approaches -1 from the left				x approaches -1 from the right		
x	-1.1 -1.01 -1.001	$\to$	-1	$\leftarrow$	-0.999 -0.99 -0.9		
$f(x) = x^2 - 2$							

11. $\lim\limits_{x \to -3} \dfrac{x^2 + 4x + 3}{x + 3}$

	x approaches -3 from the left				x approaches -3 from the right		
x	-3.5 -3.1 -3.01	$\to$	-3	$\leftarrow$	-2.99 -2.9 -2.5		
$f(x) = \dfrac{x^2 + 4x + 3}{x + 3}$							

12. $\lim\limits_{x \to -1} \dfrac{2x^2 + 3x + 1}{x + 1}$

	x approaches -1 from the left				x approaches -1 from the right		
x	-1.1 -1.01 -1.001	$\to$	-1	$\leftarrow$	-0.999 -0.99 -0.9		
$f(x) = \dfrac{2x^2 + 3x + 1}{x + 1}$							

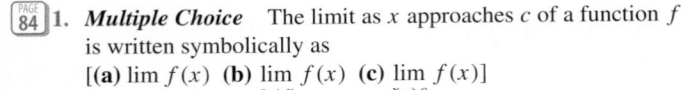 *In Problems 13–16, use technology to complete the table and investigate the limit.*

PAGE 85 **13.** $\lim\limits_{x \to 0} \dfrac{2 - 2e^x}{x}$

	x approaches 0 from the left				x approaches 0 from the right		
x	-0.2 -0.1 -0.01	$\to$	0	$\leftarrow$	0.01 0.1 0.2		
$f(x) = \dfrac{2 - 2e^x}{x}$							

14. $\lim\limits_{x \to 1} \dfrac{\ln x}{x - 1}$

	x approaches 1 from the left				x approaches 1 from the right		
x	0.9 0.99 0.999	$\to$	1	$\leftarrow$	1.001 1.01 1.1		
$f(x) = \dfrac{\ln x}{x - 1}$							

PAGE 85 **15.** $\lim\limits_{x \to 0} \dfrac{1 - \cos x}{x}$, where x is measured in radians

	x approaches 0 from the left				x approaches 0 from the right		
x (in radians)	-0.2 -0.1 -0.01	$\to$	0	$\leftarrow$	0.01 0.1 0.2		
$f(x) = \dfrac{1 - \cos x}{x}$							

16. $\lim\limits_{x \to 0} \dfrac{\sin x}{1 + \tan x}$, where x is measured in radians

	x approaches 0 from the left				x approaches 0 from the right		
x (in radians)	-0.2 -0.1 -0.01	$\to$	0	$\leftarrow$	0.01 0.1 0.2		
$f(x) = \dfrac{\sin x}{1 + \tan x}$							

In Problems 17–20, use the graph to investigate
(a) $\lim\limits_{x \to 2^-} f(x)$, *(b)* $\lim\limits_{x \to 2^+} f(x)$, *(c)* $\lim\limits_{x \to 2} f(x)$.

17.

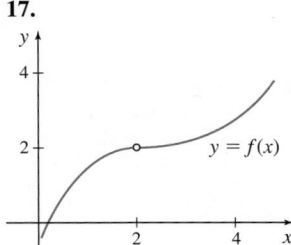

18.

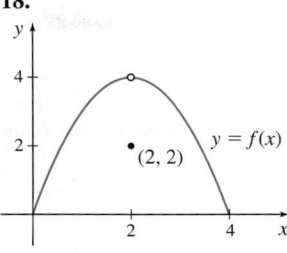

19.

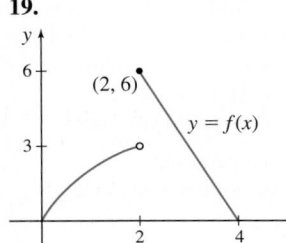

20.

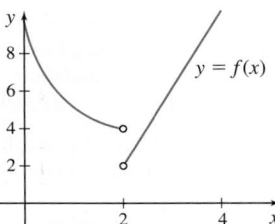

In Problems 21–28, use the graph to investigate $\lim\limits_{x \to c} f(x)$. *If the limit does not exist, explain why.*

21.

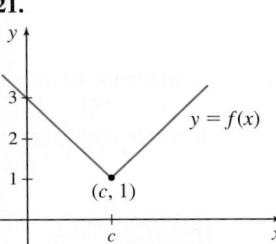

22.

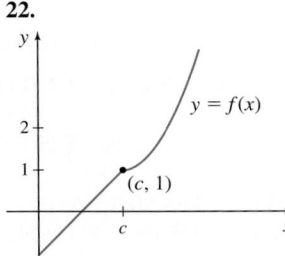

23.

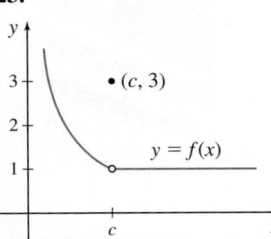

24.

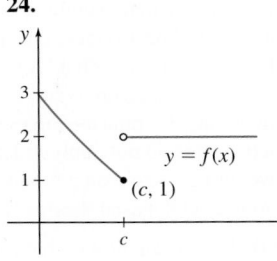

25.

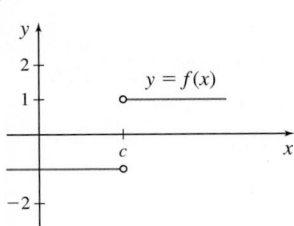

26.

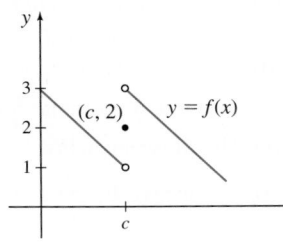

27.

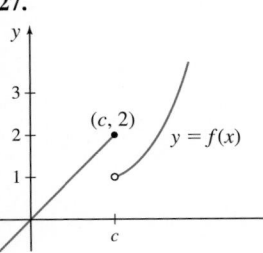

28.

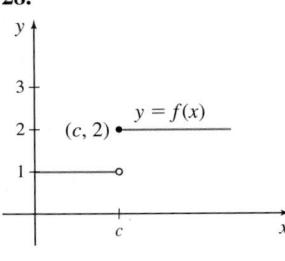

In Problems 29–36, use a graph to investigate $\lim\limits_{x \to c} f(x)$ *at the number c.*

29. $f(x) = \begin{cases} 2x+5 & \text{if } x \le 2 \\ 4x+1 & \text{if } x > 2 \end{cases}$ at $c = 2$

30. $f(x) = \begin{cases} 2x+1 & \text{if } x \le 0 \\ 2x & \text{if } x > 0 \end{cases}$ at $c = 0$

31. $f(x) = \begin{cases} 3x-1 & \text{if } x < 1 \\ 4 & \text{if } x = 1 \\ 4x & \text{if } x > 1 \end{cases}$ at $c = 1$

32. $f(x) = \begin{cases} x+2 & \text{if } x < 2 \\ 4 & \text{if } x = 2 \\ x^2 & \text{if } x > 2 \end{cases}$ at $c = 2$

33. $f(x) = \begin{cases} 2x^2 & \text{if } x < 1 \\ 3x^2-1 & \text{if } x > 1 \end{cases}$ at $c = 1$

34. $f(x) = \begin{cases} x^3 & \text{if } x < -1 \\ x^2-1 & \text{if } x > -1 \end{cases}$ at $c = -1$

35. $f(x) = \begin{cases} x^2 & \text{if } x \le 0 \\ 2x+1 & \text{if } x > 0 \end{cases}$ at $c = 0$

36. $f(x) = \begin{cases} x^2 & \text{if } x < 1 \\ 2 & \text{if } x = 1 \\ -3x+2 & \text{if } x > 1 \end{cases}$ at $c = 1$

Applications and Extensions

In Problems 37–40, sketch a graph of a function with the given properties. Answers will vary.

37. $\lim\limits_{x \to 2} f(x) = 3$; $\lim\limits_{x \to 3^-} f(x) = 3$; $\lim\limits_{x \to 3^+} f(x) = 1$;
  $f(2) = 3$; $f(3) = 1$

38. $\lim\limits_{x \to -1} f(x) = 0$; $\lim\limits_{x \to 2^-} f(x) = -2$; $\lim\limits_{x \to 2^+} f(x) = -2$;
  $f(-1)$ is not defined; $f(2) = -2$

39. $\lim\limits_{x \to 1} f(x) = 4$; $\lim\limits_{x \to 0^-} f(x) = -1$; $\lim\limits_{x \to 0^+} f(x) = 0$;
  $f(0) = -1$; $f(1) = 2$

40. $\lim\limits_{x \to 2} f(x) = 2$; $\lim\limits_{x \to -1} f(x) = 0$; $\lim\limits_{x \to 1} f(x) = 1$;
  $f(-1) = 1$; $f(2) = 3$

In Problems 41–50, use either a graph or a table to investigate each limit.

41. $\lim\limits_{x \to 5^+} \dfrac{|x-5|}{x-5}$

42. $\lim\limits_{x \to 5^-} \dfrac{|x-5|}{x-5}$

43. $\lim\limits_{x \to \left(\frac{1}{2}\right)^-} \lfloor 2x \rfloor$

44. $\lim\limits_{x \to \left(\frac{1}{2}\right)^+} \lfloor 2x \rfloor$

45. $\lim\limits_{x \to \left(\frac{2}{3}\right)^-} \lfloor 2x \rfloor$

46. $\lim\limits_{x \to \left(\frac{2}{3}\right)^+} \lfloor 2x \rfloor$

47. $\lim\limits_{x \to 2^+} \sqrt{|x|-x}$

48. $\lim\limits_{x \to 2^-} \sqrt{|x|-x}$

49. $\lim\limits_{x \to 2^+} \sqrt[3]{\lfloor x \rfloor - x}$

50. $\lim\limits_{x \to 2^-} \sqrt[3]{\lfloor x \rfloor - x}$

51. Slope of a Tangent Line For $f(x) = 3x^2$:

(a) Find the slope of the secant line containing the points $(2, 12)$ and $(3, 27)$.

(b) Find the slope of the secant line containing the points $(2, 12)$ and $(x, f(x))$, $x \ne 2$.

(c) Create a table to investigate the slope of the tangent line to the graph of f at 2 using the result from (b).

(d) On the same set of axes, graph f, the tangent line to the graph of f at the point $(2, 12)$, and the secant line from (a).

52. Slope of a Tangent Line For $f(x) = x^3$:

(a) Find the slope of the secant line containing the points $(2, 8)$ and $(3, 27)$.

(b) Find the slope of the secant line containing the points $(2, 8)$ and $(x, f(x))$, $x \neq 2$.

(c) Create a table to investigate the slope of the tangent line to the graph of f at 2 using the result from (b).

(d) On the same set of axes, graph f, the tangent line to the graph of f at the point $(2, 8)$, and the secant line from (a).

53. Slope of a Tangent Line For $f(x) = \frac{1}{2}x^2 - 1$:

(a) Find the slope m_{sec} of the secant line containing the points $P = (2, f(2))$ and $Q = (2 + h, f(2 + h))$.

(b) Use the result from (a) to complete the following table:

h	-0.5	-0.1	-0.001	0.001	0.1	0.5
m_{sec}						

(c) Investigate the limit of the slope of the secant line found in (a) as $h \to 0$.

(d) What is the slope of the tangent line to the graph of f at the point $P = (2, f(2))$?

(e) On the same set of axes, graph f and the tangent line to the graph of f at $P = (2, f(2))$.

54. Slope of a Tangent Line For $f(x) = x^2 - 1$:

(a) Find the slope m_{sec} of the secant line containing the points $P = (-1, f(-1))$ and $Q = (-1 + h, f(-1 + h))$.

(b) Use the result from (a) to complete the following table:

h	-0.1	-0.01	-0.001	-0.0001	0.0001	0.001	0.01	0.1
m_{sec}								

(c) Investigate the limit of the slope of the secant line found in (a) as $h \to 0$.

(d) What is the slope of the tangent line to the graph of f at the point $P = (-1, f(-1))$?

(e) On the same set of axes, graph f and the tangent line to the graph of f at $P = (-1, f(-1))$.

PAGE 88 **55.** (a) Investigate $\lim\limits_{x \to 0} \cos \dfrac{\pi}{x}$ by using a table and evaluating the

function $f(x) = \cos \dfrac{\pi}{x}$ at

$$x = -\frac{1}{2}, -\frac{1}{4}, -\frac{1}{8}, -\frac{1}{10}, -\frac{1}{12}, \ldots, \frac{1}{12}, \frac{1}{10}, \frac{1}{8}, \frac{1}{4}, \frac{1}{2}.$$

(b) Investigate $\lim\limits_{x \to 0} \cos \dfrac{\pi}{x}$ by using a table and evaluating the

function $f(x) = \cos \dfrac{\pi}{x}$ at

$$x = -1, -\frac{1}{3}, -\frac{1}{5}, -\frac{1}{7}, -\frac{1}{9}, \ldots, \frac{1}{9}, \frac{1}{7}, \frac{1}{5}, \frac{1}{3}, 1.$$

(c) Compare the results from (a) and (b). What do you conclude about the limit? Why do you think this happens? What is your view about using a table to draw a conclusion about limits?

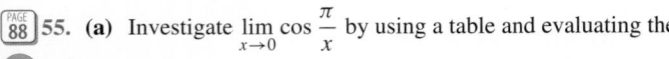 (d) Use technology to graph f. Begin with the x-window $[-2\pi, 2\pi]$ and the y-window $[-1, 1]$. If you were finding $\lim\limits_{x \to 0} f(x)$ using a graph, what would you conclude? Zoom in on the graph. Describe what you see. *Hint*: Be sure your calculator is set to the radian mode.

56. (a) Investigate $\lim\limits_{x \to 0} \cos \dfrac{\pi}{x^2}$ by using a table and evaluating the

function $f(x) = \cos \dfrac{\pi}{x^2}$ at $x = -0.1, -0.01, -0.001,$

$-0.0001, 0.0001, 0.001, 0.01, 0.1$.

(b) Investigate $\lim\limits_{x \to 0} \cos \dfrac{\pi}{x^2}$ by using a table and evaluating the

function $f(x) = \cos \dfrac{\pi}{x^2}$ at

$$x = -\frac{2}{3}, -\frac{2}{5}, -\frac{2}{7}, -\frac{2}{9}, \ldots, \frac{2}{9}, \frac{2}{7}, \frac{2}{5}, \frac{2}{3}.$$

(c) Compare the results from (a) and (b). What do you conclude about the limit? Why do you think this happens? What is your view about using a table to draw a conclusion about limits?

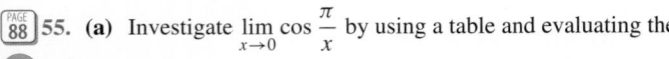 (d) Use technology to graph f. Begin with the x-window $[-2\pi, 2\pi]$ and the y-window $[-1, 1]$. If you were finding $\lim\limits_{x \to 0} f(x)$ using a graph, what would you conclude? Zoom in on the graph. Describe what you see. *Hint*: Be sure your calculator is set to the radian mode.

PAGE 89 **57.** (a) Use a table to investigate $\lim\limits_{x \to 2} \dfrac{x - 8}{2}$.

(b) How close must x be to 2, so that $f(x)$ is within 0.1 of the limit?

(c) How close must x be to 2, so that $f(x)$ is within 0.01 of the limit?

58. (a) Use a table to investigate $\lim\limits_{x \to 2} (5 - 2x)$.

(b) How close must x be to 2, so that $f(x)$ is within 0.1 of the limit?

(c) How close must x be to 2, so that $f(x)$ is within 0.01 of the limit?

59. First-Class Mail As of April 2023, the U.S. Postal Service charged $0.63 postage for first-class letters weighing up to and including 1 ounce, plus a flat fee of $0.24 for each additional or partial ounce up to and including 3.5 ounces. First-class letter rates do not apply to letters weighing more than 3.5 ounces. *Source:* U.S. Postal Service Notice 123.

(a) Find a function C that models the first-class postage charged, in dollars, for a letter weighing w ounces. Assume $w > 0$.

(b) What is the domain of C?

(c) Graph the function C.

(d) Use the graph to investigate $\lim\limits_{w \to 2^-} C(w)$ and $\lim\limits_{w \to 2^+} C(w)$. Do these suggest that $\lim\limits_{w \to 2} C(w)$ exists?

(e) Use the graph to investigate $\lim\limits_{w \to 0^+} C(w)$.

(f) Use the graph to investigate $\lim\limits_{w \to 3.5^-} C(w)$.

60. First-Class Mail   As of April 2023, the U.S. Postal Service charged $1.26 postage for first-class large envelopes weighing up to and including 1 ounce, plus a flat fee of $0.24 for each additional or partial ounce up to and including 13 ounces. First-class rates do not apply to large envelopes weighing more than 13 ounces.

Source: U.S. Postal Service Notice 123.

(a) Find a function C that models the first-class postage charged, in dollars, for a large envelope weighing w ounces. Assume $w > 0$.

(b) What is the domain of C?

(c) Graph the function C.

(d) Use the graph to investigate $\lim\limits_{w \to 1^-} C(w)$ and $\lim\limits_{w \to 1^+} C(w)$. Do these suggest that $\lim\limits_{w \to 1} C(w)$ exists?

(e) Use the graph to investigate $\lim\limits_{w \to 12^-} C(w)$ and $\lim\limits_{w \to 12^+} C(w)$. Do these suggest that $\lim\limits_{w \to 12} C(w)$ exists?

(f) Use the graph to investigate $\lim\limits_{w \to 0^+} C(w)$.

(g) Use the graph to investigate $\lim\limits_{w \to 13^-} C(w)$.

61. Correlating Student Success to Study Time   Professor Smith claims that a student's final exam score is a function of the time t (in hours) that the student studies. He claims that the closer to seven hours a student studies, the closer to 100% the student scores on the final. He claims that studying significantly less than seven hours may cause the student to be underprepared for the test, while studying significantly more than seven hours may cause "burnout."

(a) Write Professor Smith's claim symbolically as a limit.

(b) Write Professor Smith's claim using the ε-δ definition of limit.

Source: Submitted by the students of Millikin University.

62. The definition of the slope of the tangent line to the graph of $y = f(x)$ at the point $(c, f(c))$ is $m_{\tan} = \lim\limits_{x \to c} \dfrac{f(x) - f(c)}{x - c}$.

Another way to express this slope is to define a new variable $h = x - c$. Rewrite the slope of the tangent line $m_{\tan}$ using h and c.

63. If $f(2) = 6$, can you conclude anything about $\lim\limits_{x \to 2} f(x)$? Explain your reasoning.

64. If $\lim\limits_{x \to 2} f(x) = 6$, can you conclude anything about $f(2)$? Explain your reasoning.

65. The graph of $f(x) = \dfrac{x - 3}{3 - x}$ is a straight line with a point punched out.

(a) What straight line and what point?

(b) Use the graph of f to investigate the one-sided limits of f as x approaches 3.

(c) Does the graph suggest that $\lim\limits_{x \to 3} f(x)$ exists? If so, what is it?

66. (a) Use a table to investigate $\lim\limits_{x \to 0} (1 + x)^{1/x}$.

(b) Use graphing technology to graph $g(x) = (1 + x)^{1/x}$.

(c) What do (a) and (b) suggest about $\lim\limits_{x \to 0} (1 + x)^{1/x}$?

[CAS] (d) Find $\lim\limits_{x \to 0} (1 + x)^{1/x}$.

Challenge Problems

For Problems 67–70, investigate each of the following limits.

$$f(x) = \begin{cases} 1 & \text{if } x \text{ is an integer} \\ 0 & \text{if } x \text{ is not an integer} \end{cases}$$

67. $\lim\limits_{x \to 2} f(x)$

68. $\lim\limits_{x \to 1/2} f(x)$

69. $\lim\limits_{x \to 3} f(x)$

70. $\lim\limits_{x \to 0} f(x)$

Preparing for the **AP® Exam**

AP® Practice Problems

Multiple-Choice Questions

[PAGE 87] **1.** To investigate $\lim\limits_{x \to 1^-} \sqrt{1 - x}$ using a table, evaluate $f(x) = \sqrt{1 - x}$, by choosing

(A) numbers close to 0, some slightly less than 0 and some slightly greater than 0.

(B) only numbers slightly less than 1.

(C) only numbers slightly greater than 1.

(D) numbers close to 1, some slightly less than 1 and some slightly greater than 1.

[PAGE 85] **2.** $\lim\limits_{x \to -2^+} x^3 = -8$ is called a

(A) lower limit   (B) negative limit

(C) positive limit   (D) right-hand limit

[PAGE 84] **3.** "The limit as x approaches 0 of the function $f(x) = \cos x$ is equal to the number 1," is written symbolically as

(A) $\lim\limits_{\cos x \to 0} \cos x = 1$ (B) $\lim\limits_{x \to \cos x} \cos x = 1$

(C) $\lim\limits_{x \to 0} \cos x = 1$ (D) $\lim\limits_{x \to 1} \cos x = 0$

[PAGE 86] **4.** The graph of a function f is given below.

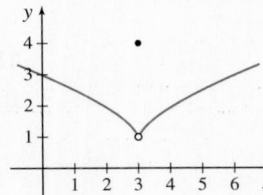

Which statement best describes $\lim\limits_{x \to 3} f(x)$?

(A) $\lim\limits_{x \to 3} f(x) = 1$ (B) $\lim\limits_{x \to 3} f(x) = 3$

(C) $\lim\limits_{x \to 3} f(x) = 4$ (D) $\lim\limits_{x \to 3} f(x)$ doesn't exist.

(continued on the next page)

5. The graph of a piecewise function f is shown.

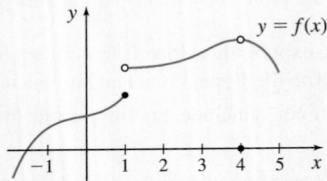

Use the graph to determine which of the following statements is true.

(A) $\displaystyle\lim_{x\to-1^+} f(x)$ does not exist. (B) $\displaystyle\lim_{x\to1^-} f(x) = f(1)$.

(C) $\displaystyle\lim_{x\to1} f(x)$ exists. (D) $\displaystyle\lim_{x\to4} f(x) = f(4)$.

6. The table below gives values of three functions f, g, and h for select numbers x.

x	0.7	0.8	0.9	0.95	1	1.05	1.1	1.2	1.3
$f(x)$	0	0	0	0	0	0.9	0.9	0.9	0.9
$g(x)$	−0.9	−0.95	−0.095	−0.009	undefined	0.009	0.095	0.95	0.995
$h(x)$	1.0	0.08	0.008	0.0008	1	−0.005	−0.025	−0.05	−0.25

For which of these functions does the table suggest that the limit as x approaches 1 is 0?

(A) f only (B) h only (C) f and g only (D) g and h only

Free-Response Questions

7. Interpret $\displaystyle\lim_{x\to2}(x^3 + 3x - 4) = 10$.

8. Use a calculator to create a table to investigate $\displaystyle\lim_{x\to0} \frac{e^x - 1}{x}$.

1.2 Analytic Techniques for Finding Limits of Functions

OBJECTIVES *When you finish this section, you should be able to:*

1 **Find the limit of a sum, a difference, and a product (p. 95)**

2 **Find the limit of a power and the limit of a root (p. 98)**

3 **Find the limit of a polynomial (p. 100)**

4 **Find the limit of a quotient (p. 101)**

5 **Find the limit of an average rate of change (p. 103)**

6 **Find the limit of a difference quotient (p. 104)**

In Section 1.1, we used numerical techniques (tables) and graphical techniques to investigate limits. We saw that these techniques are not always reliable. The only way to be sure a limit is correct is to use the ε-δ definition of a limit. In this section, we state, without proof, analytic techniques based on the ε-δ definition. Some of the results are proved in Section 1.6 and others in Appendix B. As we will see, these analytic techniques involve using algebraic properties of limits.

We begin with two basic limits.

THEOREM The Limit of a Constant

If $f(x) = A$, where A is a constant, then for any real number c,

$$\boxed{\lim_{x\to c} f(x) = \lim_{x\to c} A = A} \qquad (1)$$

The theorem is proved in Section 1.6. See Figure 17 and Table 8 for graphical and numerical support of $\displaystyle\lim_{x\to c} A = A$.

TABLE 8

	x approaches c from the left				x approaches c from the right		
x	$c - 0.01$	$c - 0.001$	$c - 0.0001$	$\to c \leftarrow$	$c + 0.0001$	$c + 0.001$	$c + 0.01$
$f(x) = A$	A	A	A	$f(x)$ remains at A	A	A	A

Figure 17 For x close to c, the value of f remains at A; $\displaystyle\lim_{x\to c} A = A$.

For example,

$$\lim_{x\to5} 2 = 2 \qquad \lim_{x\to\sqrt{2}} \frac{1}{3} = \frac{1}{3} \qquad \lim_{x\to5}(-\pi) = -\pi$$

THEOREM The Limit of the Identity Function

If $f(x) = x$, then for any real number c,

$$\lim_{x \to c} f(x) = \lim_{x \to c} x = c \tag{2}$$

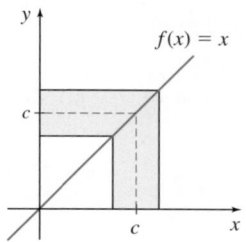

Figure 18 For x close to c, the value of f is just as close to c; $\lim_{x \to c} x = c$.

This theorem is proved in Section 1.6. See Figure 18 and Table 9 for graphical and numerical support of $\lim_{x \to c} x = c$.

TABLE 9

	x approaches c from the left	$\longrightarrow$		x approaches c from the right $\longleftarrow$		
x	$c - 0.01$ $c - 0.001$ $c - 0.0001$	$\to c \leftarrow$		$c + 0.0001$ $c + 0.001$ $c + 0.01$		
$f(x) = x$	$c - 0.01$ $c - 0.001$ $c - 0.0001$	$f(x)$ approaches c	$c + 0.0001$ $c + 0.001$ $c + 0.01$			

For example,

$$\lim_{x \to -5} x = -5 \qquad \lim_{x \to \sqrt{3}} x = \sqrt{3} \qquad \lim_{x \to 0} x = 0$$

① Find the Limit of a Sum, a Difference, and a Product

Many functions are combinations of sums, differences, and products of a constant function and the identity function. The following algebraic properties can be used to find the limit of such functions.

THEOREM Limit of a Sum

If f and g are functions for which $\lim_{x \to c} f(x)$ and $\lim_{x \to c} g(x)$ both exist, then $\lim_{x \to c}[f(x) + g(x)]$ exists and

$$\lim_{x \to c}[f(x) + g(x)] = \lim_{x \to c} f(x) + \lim_{x \to c} g(x)$$

IN WORDS The limit of the sum of two functions equals the sum of their limits.

A proof of this theorem is given in Appendix B.

EXAMPLE 1   Finding the Limit of a Sum

Find $\lim_{x \to -3}(x + 4)$.

Solution

$F(x) = x + 4$ is the sum of two functions $f(x) = x$ and $g(x) = 4$.

From the basic limits given in (1) and (2), we have

$$\lim_{x \to -3} f(x) = \lim_{x \to -3} x = -3 \qquad \text{and} \qquad \lim_{x \to -3} g(x) = \lim_{x \to -3} 4 = 4$$

Then, using the Limit of a Sum, we have

$$\lim_{x \to -3}(x + 4) = \lim_{x \to -3} x + \lim_{x \to -3} 4 = -3 + 4 = 1$$

∎

THEOREM Limit of a Difference

If f and g are functions for which $\lim_{x \to c} f(x)$ and $\lim_{x \to c} g(x)$ both exist, then $\lim_{x \to c}[f(x) - g(x)]$ exists and

$$\lim_{x \to c}[f(x) - g(x)] = \lim_{x \to c} f(x) - \lim_{x \to c} g(x)$$

IN WORDS The limit of the difference of two functions equals the difference of their limits.

EXAMPLE 2   Finding the Limit of a Difference

Find $\lim\limits_{x \to 4}(6 - x)$.

Solution

$F(x) = 6 - x$ is the difference of two functions $f(x) = 6$ and $g(x) = x$.

$$\lim_{x \to 4} f(x) = \lim_{x \to 4} 6 = 6 \qquad \text{and} \qquad \lim_{x \to 4} g(x) = \lim_{x \to 4} x = 4$$

Then, using the Limit of a Difference, we have

$$\lim_{x \to 4}(6 - x) = \lim_{x \to 4} 6 - \lim_{x \to 4} x = 6 - 4 = 2$$

■

THEOREM Limit of a Product

If f and g are functions for which $\lim\limits_{x \to c} f(x)$ and $\lim\limits_{x \to c} g(x)$ both exist, then $\lim\limits_{x \to c}[f(x) \cdot g(x)]$ exists and

$$\boxed{\lim_{x \to c}[f(x) \cdot g(x)] = \lim_{x \to c} f(x) \cdot \lim_{x \to c} g(x)}$$

IN WORDS The limit of the product of two functions equals the product of their limits.

A proof of this theorem is given in Appendix B.

EXAMPLE 3   Finding the Limit of a Product

Find:

(a) $\lim\limits_{x \to 3} x^2$ **(b)** $\lim\limits_{x \to -5}(-4x)$

Solution

(a) $F(x) = x^2$ is the product of two functions, $f(x) = x$ and $g(x) = x$. Then, using the Limit of a Product, we have

$$\lim_{x \to 3} x^2 = \lim_{x \to 3} x \cdot \lim_{x \to 3} x = 3 \cdot 3 = 9$$

(b) $F(x) = -4x$ is the product of two functions, $f(x) = -4$ and $g(x) = x$. Then, using the Limit of a Product, we have

$$\lim_{x \to -5}(-4x) = \lim_{x \to -5}(-4) \cdot \lim_{x \to -5} x = (-4)(-5) = 20$$

■

A *corollary** of the Limit of a Product Theorem is the special case when $f(x) = k$ is a constant function.

COROLLARY Limit of a Constant Times a Function

If g is a function for which $\lim\limits_{x \to c} g(x)$ exists and if k is any real number, then $\lim\limits_{x \to c}[kg(x)]$ exists and

$$\boxed{\lim_{x \to c}[kg(x)] = k \lim_{x \to c} g(x)}$$

IN WORDS The limit of a constant times a function equals the constant times the limit of the function.

You are asked to prove this corollary in Problem 103.

---

*A **corollary** is a theorem that follows directly from a previously proved theorem.

Limit properties often are used in combination.

EXAMPLE 4 Finding a Limit

Find:

(a) $\lim\limits_{x \to 1}[2x(x+4)]$ **(b)** $\lim\limits_{x \to 2^+}[4x(2-x)]$

Solution

(a) $\lim\limits_{x \to 1}[(2x)(x+4)] = \left[\lim\limits_{x \to 1}(2x)\right]\left[\lim\limits_{x \to 1}(x+4)\right]$ **Limit of a Product**

$\qquad\qquad\qquad\qquad = \left[2 \cdot \lim\limits_{x \to 1} x\right] \cdot \left[\lim\limits_{x \to 1} x + \lim\limits_{x \to 1} 4\right]$ **Limit of a Constant Times a Function, Limit of a Sum**

$\qquad\qquad\qquad\qquad = (2 \cdot 1) \cdot (1+4) = 10$ Use (2) and (1), and simplify.

> **NOTE** The limit properties are also true for one-sided limits.

(b) We use properties of limits to find the one-sided limit.

$$\lim\limits_{x \to 2^+}[4x(2-x)] = 4\lim\limits_{x \to 2^+}[x(2-x)] = 4\left[\lim\limits_{x \to 2^+} x\right]\left[\lim\limits_{x \to 2^+}(2-x)\right]$$

$$= 4 \cdot 2\left[\lim\limits_{x \to 2^+} 2 - \lim\limits_{x \to 2^+} x\right] = 4 \cdot 2 \cdot (2-2) = 0 \qquad \blacksquare$$

> **NOW WORK** Problem 13.

We use one-sided limits to find the limit of piecewise-defined functions at numbers where the defining equation changes.

EXAMPLE 5 Finding a Limit for a Piecewise-Defined Function

> **RECALL** The limit *L* of a function $y = f(x)$ as x approaches a number c exists if and only if $\lim\limits_{x \to c^-} f(x) = \lim\limits_{x \to c^+} f(x) = L$.

Find $\lim\limits_{x \to 2} f(x)$, if it exists.

$$f(x) = \begin{cases} 3x+1 & \text{if} \quad x < 2 \\ 2x(x-1) & \text{if} \quad x \geq 2 \end{cases}$$

Solution

Since the rule for f changes at 2, we need to find the one-sided limits of f as x approaches 2.

- For $x < 2$, we use the left-hand limit. Also, because $x < 2$, $f(x) = 3x + 1$.

$$\lim\limits_{x \to 2^-} f(x) = \lim\limits_{x \to 2^-}(3x+1) = \lim\limits_{x \to 2^-}(3x) + \lim\limits_{x \to 2^-} 1 = 3\lim\limits_{x \to 2^-} x + 1 = 3 \cdot 2 + 1 = 7$$

- For $x \geq 2$, we use the right-hand limit. Also, because $x \geq 2$, $f(x) = 2x(x-1)$.

$$\lim\limits_{x \to 2^+} f(x) = \lim\limits_{x \to 2^+}[2x(x-1)] = \lim\limits_{x \to 2^+}(2x) \cdot \lim\limits_{x \to 2^+}(x-1)$$

$$= 2\lim\limits_{x \to 2^+} x \cdot \left[\lim\limits_{x \to 2^+} x - \lim\limits_{x \to 2^+} 1\right] = 2 \cdot 2[2-1] = 4$$

Since $\lim\limits_{x \to 2^-} f(x) = 7 \neq \lim\limits_{x \to 2^+} f(x) = 4$, $\lim\limits_{x \to 2} f(x)$ does not exist. $\blacksquare$

See Figure 19.

Figure 19 $f(x) = \begin{cases} 3x+1 & \text{if} \quad x < 2 \\ 2x(x-1) & \text{if} \quad x \geq 2 \end{cases}$

> **NOW WORK** Problem 73 and AP® Practice Problems 1 and 5.

ORIGINS **Oliver Heaviside** (1850–1925) was a self-taught mathematician and electrical engineer. He developed a branch of mathematics called **operational calculus** in which differential equations are solved by converting them to algebraic equations. Heaviside applied vector calculus to electrical engineering and, perhaps most significantly, he simplified *Maxwell's equations* to the form used by electrical engineers to this day. In 1902 Heaviside claimed there is a layer surrounding Earth from which radio signals bounce, allowing the signals to travel around the Earth. Heaviside's claim was proved true in 1923. The layer, contained in the ionosphere, is named the **Heaviside layer**. The function we discuss here is one of his contributions to mathematics and electrical engineering.

The **Heaviside function**, $u_c(t) = \begin{cases} 0 & \text{if } t < c \\ 1 & \text{if } t \geq c \end{cases}$, is a step function that is used in electrical engineering to model a switch. The switch is off if $t < c$, and it is on if $t \geq c$.

EXAMPLE 6   Finding a Limit of the Heaviside Function

Find $\lim_{t \to 0} u_0(t)$, where $u_0(t) = \begin{cases} 0 & \text{if } t < 0 \\ 1 & \text{if } t \geq 0 \end{cases}$

Solution

Since the Heaviside function changes rules at $t = 0$, we find the one-sided limits as t approaches 0.

For $t < 0$, $\lim_{t \to 0^-} u_0(t) = \lim_{t \to 0^-} 0 = 0$ and   for $t \geq 0$, $\lim_{t \to 0^+} u_0(t) = \lim_{t \to 0^+} 1 = 1$

Since the one-sided limits as t approaches 0 are not equal, $\lim_{t \to 0} u_0(t)$ does not exist. ∎

NOW WORK   Problem 81.

---

2   Find the Limit of a Power and the Limit of a Root

Objective 2 presents three results, Limit of a Power, Limit of a Root, and Limit of a Fractional Power, that are used to find limits of certain types of composite functions. Using the Limit of a Product, if $\lim_{x \to c} f(x)$ exists, then

$$\lim_{x \to c} [f(x)]^2 = \lim_{x \to c} [f(x) \cdot f(x)] = \lim_{x \to c} f(x) \cdot \lim_{x \to c} f(x) = \left[\lim_{x \to c} f(x) \right]^2$$

Repeated use of this property produces the next corollary.

COROLLARY Limit of a Power

If $\lim_{x \to c} f(x)$ exists and if n is a positive integer, then

$$\boxed{\lim_{x \to c} [f(x)]^n = \left[\lim_{x \to c} f(x) \right]^n}$$

EXAMPLE 7   Finding the Limit of a Power

Find: **(a)** $\lim_{x \to 2} x^5$ **(b)** $\lim_{x \to 1} (2x - 3)^3$ **(c)** $\lim_{x \to c} x^n$ n a positive integer

Solution

(a) $\lim_{x \to 2} x^5 = \left(\lim_{x \to 2} x \right)^5 = 2^5 = 32$

(b) $\lim_{x \to 1} (2x - 3)^3 = \left[\lim_{x \to 1} (2x - 3) \right]^3 = \left[\lim_{x \to 1} (2x) - \lim_{x \to 1} 3 \right]^3 = (2 - 3)^3 = -1$

(c) $\lim_{x \to c} x^n = \left[\lim_{x \to c} x \right]^n = c^n$ ∎

The result from Example 7(c) is worth remembering since it is used frequently:

$$\boxed{\lim_{x \to c} x^n = c^n}$$

where c is a number and n is a positive integer.

NOW WORK   Problem 15.

THEOREM Limit of a Root

If $\lim\limits_{x\to c} f(x)$ exists and if $n \geq 2$ is an integer, then

$$\lim_{x\to c} \sqrt[n]{f(x)} = \sqrt[n]{\lim_{x\to c} f(x)}$$

provided $f(x) > 0$ if n is even.

EXAMPLE 8 Finding the Limit of a Root

Find $\lim\limits_{x\to 4} \sqrt[3]{x^2 + 11}$.

Solution

$$\lim_{x\to 4} \sqrt[3]{x^2 + 11} \underset{\substack{\uparrow \\ \text{Limit of a Root}}}{=} \sqrt[3]{\lim_{x\to 4}(x^2 + 11)} \underset{\substack{\uparrow \\ \text{Limit of a Sum}}}{=} \sqrt[3]{\lim_{x\to 4} x^2 + \lim_{x\to 4} 11}$$

$$\underset{\substack{\uparrow \\ \lim\limits_{x\to c} x^2 = c^2}}{=} \sqrt[3]{4^2 + 11} = \sqrt[3]{27} = 3$$

∎

NOW WORK Problem **19** and AP® Practice Problems **6** and **7**.

The Limit of a Power and the Limit of a Root are used together to find the limit of a function with a rational exponent.

THEOREM Limit of a Fractional Power $[f(x)]^{m/n}$

If f is a function for which $\lim\limits_{x\to c} f(x)$ exists and if $[f(x)]^{m/n}$ is defined for positive integers m and n, then

$$\lim_{x\to c} [f(x)]^{m/n} = \left[\lim_{x\to c} f(x)\right]^{m/n}$$

EXAMPLE 9 Finding the Limit of a Fractional Power $[f(x)]^{m/n}$

Find $\lim\limits_{x\to 8}(x+1)^{3/2}$.

Solution

Let $f(x) = x + 1$. Near 8, $x + 1 > 0$, so $(x+1)^{3/2}$ is defined. Then

$$\lim_{x\to 8}[f(x)]^{3/2} = \lim_{x\to 8}(x+1)^{3/2} \underset{\substack{\uparrow \\ \lim\limits_{x\to c}[f(x)]^{m/n} = \left[\lim\limits_{x\to c} f(x)\right]^{m/n}}}{=} \left[\lim_{x\to 8}(x+1)\right]^{3/2} = (8+1)^{3/2} = 9^{3/2} = 27$$

See Figure 20.

∎

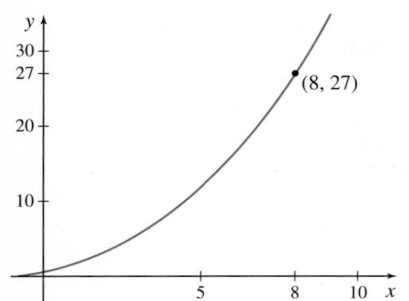

Figure 20 $f(x) = (x+1)^{3/2}$

NOW WORK Problem **23** and AP® Practice Problem **8**.

3 Find the Limit of a Polynomial

Sometimes $\lim_{x \to c} f(x)$ can be found by substituting c for x in $f(x)$. For example,

$$\lim_{x \to 2} (5x^2) = 5 \lim_{x \to 2} x^2 = 5 \cdot 2^2 = 20$$

Since $\lim_{x \to c} x^n = c^n$ if n is a positive integer, we can use the Limit of a Constant Times a Function to obtain a formula for the limit of a monomial $f(x) = ax^n$.

$$\boxed{\lim_{x \to c} (ax^n) = ac^n}$$

where a is any number.

Since a polynomial is the sum of monomials and the limit of a sum is the sum of the limits, we have the following result.

THEOREM Limit of a Polynomial Function

If P is a polynomial function, then

IN WORDS To find the limit of a polynomial as x approaches c, evaluate the polynomial at c.

$$\boxed{\lim_{x \to c} P(x) = P(c)}$$

for any number c.

Proof If P is a polynomial function, that is, if

$$P(x) = a_n x^n + a_{n-1} x^{n-1} + \cdots + a_1 x + a_0$$

where n is a nonnegative integer, then

$$\lim_{x \to c} P(x) = \lim_{x \to c} \left(a_n x^n + a_{n-1} x^{n-1} + \cdots + a_1 x + a_0 \right)$$

$$= \lim_{x \to c} \left(a_n x^n \right) + \lim_{x \to c} \left(a_{n-1} x^{n-1} \right) + \cdots + \lim_{x \to c} (a_1 x) + \lim_{x \to c} a_0$$

$$= a_n c^n + a_{n-1} c^{n-1} + \cdots + a_1 c + a_0 \qquad \text{Limit of a Monomial}$$

$$= P(c) \qquad \blacksquare$$

EXAMPLE 10 Finding the Limit of a Polynomial

Find the limit of each polynomial:

(a) $\lim_{x \to 3} (4x^2 - x + 2) = 4 \cdot 3^2 - 3 + 2 = 35$

(b) $\lim_{x \to -1} (7x^5 + 4x^3 - 2x^2) = 7(-1)^5 + 4(-1)^3 - 2(-1)^2 = -13$

(c) $\lim_{x \to 0} (10x^6 - 4x^5 - 8x + 5) = 10 \cdot 0^6 - 4 \cdot 0^5 - 8 \cdot 0 + 5 = 5$

$\blacksquare$

NOW WORK   Problem 29.

4 Find the Limit of a Quotient

To find the limit of a rational function, which is the quotient of two polynomials, we use the following result.

THEOREM Limit of a Quotient

If f and g are functions for which $\lim\limits_{x \to c} f(x)$ and $\lim\limits_{x \to c} g(x)$ both exist, then $\lim\limits_{x \to c} \left[\dfrac{f(x)}{g(x)} \right]$ exists and

$$\lim_{x \to c} \left[\frac{f(x)}{g(x)} \right] = \frac{\lim\limits_{x \to c} f(x)}{\lim\limits_{x \to c} g(x)}$$

provided $\lim\limits_{x \to c} g(x) \neq 0$.

IN WORDS The limit of the quotient of two functions equals the quotient of their limits, provided that the limit of the denominator is not zero.

NEED TO REVIEW? Rational functions are discussed in Section P.2, p. 23.

COROLLARY Limit of a Rational Function

If the number c is in the domain of a rational function $R(x) = \dfrac{p(x)}{q(x)}$, then

$$\lim_{x \to c} R(x) = R(c) \tag{3}$$

You are asked to prove this corollary in Problem 104.

EXAMPLE 11 Finding the Limit of a Rational Function

Find:

(a) $\lim\limits_{x \to 1} \dfrac{3x^3 - 2x + 1}{4x^2 + 5}$

(b) $\lim\limits_{x \to -2} \dfrac{2x + 4}{3x^2 - 1}$

Solution

(a) Since 1 is in the domain of the rational function $R(x) = \dfrac{3x^3 - 2x + 1}{4x^2 + 5}$,

$$\lim_{x \to 1} R(x) = \underset{\substack{\uparrow \\ \text{Use (3)}}}{R(1)} = \frac{3 - 2 + 1}{4 + 5} = \frac{2}{9}$$

(b) Since -2 is in the domain of the rational function $H(x) = \dfrac{2x + 4}{3x^2 - 1}$,

$$\lim_{x \to -2} H(x) = \underset{\substack{\uparrow \\ \text{Use (3)}}}{H(-2)} = \frac{-4 + 4}{12 - 1} = \frac{0}{11} = 0$$ ∎

NOW WORK Problem 33.

EXAMPLE 12   Finding the Limit of a Quotient

Find $\lim\limits_{x \to 4} \dfrac{\sqrt{3x^2 + 1}}{x - 1}$.

Solution

We seek the limit of the quotient of two functions. Since the limit of the denominator is not zero, $\lim\limits_{x \to 4}(x - 1) \neq 0$, we use the Limit of a Quotient.

$$\lim_{x \to 4} \frac{\sqrt{3x^2 + 1}}{x - 1} = \underset{\underset{\text{Limit of a Quotient}}{\uparrow}}{\frac{\lim\limits_{x \to 4} \sqrt{3x^2 + 1}}{\lim\limits_{x \to 4}(x - 1)}} = \underset{\underset{\text{Limit of a Root}}{\uparrow}}{\frac{\sqrt{\lim\limits_{x \to 4}(3x^2 + 1)}}{\lim\limits_{x \to 4}(x - 1)}} = \frac{\sqrt{3 \cdot 4^2 + 1}}{4 - 1} = \frac{\sqrt{49}}{3} = \frac{7}{3}$$

■

NOW WORK   Problem 31.

Based on these examples, you might be tempted to conclude that finding a limit as x approaches c is simply a matter of substituting the number c into the function. Unfortunately, substitution cannot always be used. For example, consider the rational function $R(x) = \dfrac{p(x)}{q(x)}$. Any number c for which $q(c) = 0$ is not in the domain of R, and $R(c)$ does not exist. So, if c is not in the domain of R, then substitution cannot be used to find $\lim\limits_{x \to c} R(x)$.

The next few examples show that substitution cannot always be used and provide strategies for such cases.

EXAMPLE 13   Finding the Limit of a Rational Function

Find $\lim\limits_{x \to -2} \dfrac{x^2 + 5x + 6}{x^2 - 4}$.

Solution

Since -2 is not in the domain of the rational function, substitution cannot be used. But this does not mean that the limit does not exist! Factoring the numerator and the denominator, we find

$$\frac{x^2 + 5x + 6}{x^2 - 4} = \frac{(x + 2)(x + 3)}{(x + 2)(x - 2)}$$

Since $x \neq -2$, and we are interested in the limit as x approaches -2, the factor $x + 2$ can be divided out. Then

$$\lim_{x \to -2} \frac{x^2 + 5x + 6}{x^2 - 4} = \underset{\underset{\text{Factor}}{\uparrow}}{\lim_{x \to -2} \frac{(x + 2)(x + 3)}{(x + 2)(x - 2)}} = \underset{\underset{\substack{x \neq -2 \\ \text{Divide out } (x + 2)}}{\uparrow}}{\lim_{x \to -2} \frac{x + 3}{x - 2}} = \underset{\underset{\substack{\text{Use the Limit of a} \\ \text{Rational Function}}}{\uparrow}}{\frac{-2 + 3}{-2 - 2}} = -\frac{1}{4}$$

■

NOW WORK   Problem 35 and AP® Practice Problem 2.

The Limit of a Quotient property can only be used when the limit of the denominator of the function is not zero. The next example illustrates a strategy to try if radicals are present.

EXAMPLE 14 Finding the Limit of a Quotient

Find $\displaystyle\lim_{x\to5}\frac{\sqrt{x}-\sqrt{5}}{x-5}$.

Solution

The domain of $h(x)=\dfrac{\sqrt{x}-\sqrt{5}}{x-5}$ is $\{x\,|\,x\geq0,\ x\neq5\}$. Since the limit of the denominator is

$$\lim_{x\to5}(x-5)=0$$

we cannot use the Limit of a Quotient property. A different strategy is necessary. We rationalize the numerator of the quotient.

$$\frac{\sqrt{x}-\sqrt{5}}{x-5}=\frac{(\sqrt{x}-\sqrt{5})}{(x-5)}\cdot\frac{(\sqrt{x}+\sqrt{5})}{(\sqrt{x}+\sqrt{5})}=\frac{x-5}{(x-5)(\sqrt{x}+\sqrt{5})}\underset{\substack{\uparrow\\ x\neq5}}{=}\frac{1}{\sqrt{x}+\sqrt{5}}$$

Do you see why rationalizing the numerator works? It causes the term $x-5$ to appear in the numerator, and since $x\neq5$, the factors $x-5$ can be divided out. Then

$$\lim_{x\to5}\frac{\sqrt{x}-\sqrt{5}}{x-5}=\lim_{x\to5}\frac{1}{\sqrt{x}+\sqrt{5}}=\frac{\displaystyle\lim_{x\to5}1}{\displaystyle\lim_{x\to5}(\sqrt{x}+\sqrt{5})}=\frac{1}{\sqrt{5}+\sqrt{5}}=\frac{1}{2\sqrt{5}}=\frac{\sqrt{5}}{10}$$

Use the Limit of a Quotient

CAUTION When finding a limit, remember to include "$\lim\limits_{x\to c}$" at each step until you let $x\to c$.

NOW WORK Problem 41 and AP® Practice Problem 4.

5 Find the Limit of an Average Rate of Change

The next two examples illustrate limits that we encounter in Chapter 2.

Suppose f is a function whose domain is an interval I. If a and b, where $a\neq b$, are in I, the average rate of change of f from a to b is

NEED TO REVIEW? Average rate of change is discussed in Section P.1, pp. 13–15.

$$\boxed{\frac{\Delta y}{\Delta x}=\frac{f(b)-f(a)}{b-a}\qquad a\neq b}$$

EXAMPLE 15 Finding the Limit of an Average Rate of Change

(a) Find the average rate of change of $f(x)=x^2+3x$ from 2 to x, $x\neq2$.
(b) Find the limit as x approaches 2 of the average rate of change of $f(x)=x^2+3x$ from 2 to x.

Solution

(a) The average rate of change of f from 2 to x is

$$\frac{\Delta y}{\Delta x}=\frac{f(x)-f(2)}{x-2}=\frac{(x^2+3x)-[2^2+3(2)]}{x-2}=\frac{x^2+3x-10}{x-2}=\frac{(x+5)(x-2)}{x-2}$$

(b) The limit as x approaches 2 of the average rate of change is

$$\lim_{x\to2}\frac{f(x)-f(2)}{x-2}=\lim_{x\to2}\frac{(x+5)(x-2)}{x-2}=\lim_{x\to2}(x+5)=7$$

NOW WORK Problem 63 and AP® Practice Problem 3.

6 Find the Limit of a Difference Quotient

In Section P.1, (page 5), we defined the difference quotient of a function f at x as

$$\frac{f(x+h)-f(x)}{h} \qquad h \neq 0$$

EXAMPLE 16 Finding the Limit of a Difference Quotient

(a) For $f(x) = 2x^2 - 3x + 1$, find the difference quotient $\dfrac{f(x+h)-f(x)}{h}$, $h \neq 0$.

(b) Find the limit as h approaches 0 of the difference quotient of $f(x) = 2x^2 - 3x + 1$.

Solution

(a) To find the difference quotient of f, we begin with $f(x+h)$.

$$f(x+h) = 2(x+h)^2 - 3(x+h) + 1 = 2(x^2 + 2xh + h^2) - 3x - 3h + 1$$
$$= 2x^2 + 4xh + 2h^2 - 3x - 3h + 1$$

Now

$$f(x+h) - f(x) = (2x^2 + 4xh + 2h^2 - 3x - 3h + 1) - (2x^2 - 3x + 1) = 4xh + 2h^2 - 3h$$

Then, the difference quotient is

$$\frac{f(x+h)-f(x)}{h} = \frac{4xh + 2h^2 - 3h}{h} = \frac{h(4x + 2h - 3)}{h} = 4x + 2h - 3, \quad h \neq 0$$

(b) $\displaystyle\lim_{h\to 0} \frac{f(x+h)-f(x)}{h} = \lim_{h\to 0}(4x + 2h - 3) = 4x + 0 - 3 = 4x - 3$ ∎

NOW WORK Problem 69 and AP® Practice Problem 9.

Summary

Two Basic Limits

- $\lim_{x\to c} A = A$, where A is a constant

- $\lim_{x\to c} x = c$, c a real number

Properties of Limits

If f and g are functions for which $\lim_{x\to c} f(x)$ and $\lim_{x\to c} g(x)$ both exist, and k is a constant, then

- **Limit of a Sum or a Difference:**
$\lim_{x\to c}[f(x) \pm g(x)] = \lim_{x\to c} f(x) \pm \lim_{x\to c} g(x)$

- **Limit of a Product:** $\lim_{x\to c}[f(x) \cdot g(x)] = \lim_{x\to c} f(x) \cdot \lim_{x\to c} g(x)$

- **Limit of a Constant Times a Function:** $\lim_{x\to c}[kg(x)] = k \lim_{x\to c} g(x)$

- **Limit of a Power:** $\lim_{x\to c}[f(x)]^n = \left[\lim_{x\to c} f(x)\right]^n$
where $n \geq 2$ is an integer

- **Limit of a Root:** $\lim_{x\to c} \sqrt[n]{f(x)} = \sqrt[n]{\lim_{x\to c} f(x)}$
provided $f(x) > 0$ if $n \geq 2$ is even

- **Limit of $[f(x)]^{m/n}$:** $\lim_{x\to c}[f(x)]^{m/n} = \left[\lim_{x\to c} f(x)\right]^{m/n}$
provided $[f(x)]^{m/n}$ is defined for positive integers m and n

- **Limit of a Quotient:** $\lim_{x\to c}\left[\dfrac{f(x)}{g(x)}\right] = \dfrac{\lim_{x\to c} f(x)}{\lim_{x\to c} g(x)}$
provided $\lim_{x\to c} g(x) \neq 0$

- **Limit of a Polynomial Function:** $\lim_{x\to c} P(x) = P(c)$

- **Limit of a Rational Function:** $\lim_{x\to c} R(x) = R(c)$
if c is in the domain of R

1.2 Assess Your Understanding

Concepts and Vocabulary

1. (a) $\lim\limits_{x\to 4} (-3) =$ _____ ; **(b)** $\lim\limits_{x\to 0} \pi =$ _____

2. If $\lim\limits_{x\to c} f(x) = 3$, then $\lim\limits_{x\to c}[f(x)]^5 =$ _____ .

3. If $\lim\limits_{x\to c} f(x) = 64$, then $\lim\limits_{x\to c} \sqrt[3]{f(x)} =$ _____ .

4. (a) $\lim\limits_{x\to -1} x =$ _____ ; **(b)** $\lim\limits_{x\to e} x =$ _____

5. (a) $\lim\limits_{x\to 0} (x-2) =$ _____ ; **(b)** $\lim\limits_{x\to 1/2} (3+x) =$ _____

6. (a) $\lim\limits_{x\to 2} (-3x) =$ _____ ; **(b)** $\lim\limits_{x\to 0} (3x) =$ _____

7. *True or False* If p is a polynomial function, then $\lim\limits_{x\to 5} p(x) = p(5)$.

8. If the domain of a rational function R is $\{x \mid x \neq 0\}$, then $\lim\limits_{x\to 2} R(x) = R(\underline{\quad})$.

9. *True or False* Properties of limits cannot be used for one-sided limits.

10. *True or False* If $f(x) = \dfrac{(x+1)(x+2)}{x+1}$ and $g(x) = x+2$, then $\lim\limits_{x\to -1} f(x) = \lim\limits_{x\to -1} g(x)$.

Skill Building

In Problems 11–44, find each limit using algebraic properties of limits.

11. $\lim\limits_{x\to 3} [2(x+4)]$

12. $\lim\limits_{x\to -2} [3(x+1)]$

[PAGE 97] 13. $\lim\limits_{x\to -2} [x(3x-1)(x+2)]$

14. $\lim\limits_{x\to -1} [x(x-1)(x+10)]$

[PAGE 98] 15. $\lim\limits_{t\to 1} (3t-2)^3$

16. $\lim\limits_{x\to 0} (-3x+1)^2$

17. $\lim\limits_{x\to 4} (3\sqrt{x})$

18. $\lim\limits_{x\to 8} \left(\dfrac{1}{4}\sqrt[3]{x}\right)$

[PAGE 99] 19. $\lim\limits_{x\to 3} \sqrt{5x-4}$

20. $\lim\limits_{t\to 2} \sqrt{3t+4}$

21. $\lim\limits_{t\to 2} [t\sqrt{(5t+3)(t+4)}]$

22. $\lim\limits_{t\to -1} [t\sqrt[3]{(t+1)(2t-1)}]$

[PAGE 99] 23. $\lim\limits_{x\to 3} (\sqrt{x}+x+4)^{1/2}$

24. $\lim\limits_{t\to 2} (t\sqrt{2t}+4)^{1/3}$

25. $\lim\limits_{t\to -1} [4t(t+1)]^{2/3}$

26. $\lim\limits_{x\to 0} (x^2-2x)^{3/5}$

27. $\lim\limits_{t\to 1} (3t^2-2t+4)$

28. $\lim\limits_{x\to 0} (-3x^4+2x+1)$

[PAGE 100] 29. $\lim\limits_{x\to \frac{1}{2}} (2x^4-8x^3+4x-5)$

30. $\lim\limits_{x\to -\frac{1}{3}} (27x^3+9x+1)$

[PAGE 102] 31. $\lim\limits_{x\to 4} \dfrac{x^2+4}{\sqrt{x}}$

32. $\lim\limits_{x\to 3} \dfrac{x^2+5}{\sqrt{3x}}$

[PAGE 101] 33. $\lim\limits_{x\to -2} \dfrac{2x^3+5x}{3x-2}$

34. $\lim\limits_{x\to 1} \dfrac{2x^4-1}{3x^3+2}$

[PAGE 102] 35. $\lim\limits_{x\to 3} \dfrac{x^2-9}{x-3}$

36. $\lim\limits_{x\to -2} \dfrac{x^2-4}{x+2}$

37. $\lim\limits_{x\to -1} \dfrac{x^3-x}{x+1}$

38. $\lim\limits_{x\to -1} \dfrac{x^3+x^2}{x^2-1}$

39. $\lim\limits_{x\to -8} \left(\dfrac{2x}{x+8}+\dfrac{16}{x+8}\right)$

40. $\lim\limits_{x\to 2} \left(\dfrac{3x}{x-2}-\dfrac{6}{x-2}\right)$

[PAGE 103] 41. $\lim\limits_{x\to 2} \dfrac{\sqrt{x}-\sqrt{2}}{x-2}$

42. $\lim\limits_{x\to 3} \dfrac{\sqrt{x}-\sqrt{3}}{x-3}$

43. $\lim\limits_{x\to 4} \dfrac{\sqrt{x+5}-3}{(x-4)(x+1)}$

44. $\lim\limits_{x\to 3} \dfrac{\sqrt{x+1}-2}{x(x-3)}$

In Problems 45–50, find each one-sided limit using properties of limits.

45. $\lim\limits_{x\to 3^-} (x^2-4)$

46. $\lim\limits_{x\to 2^+} (3x^2+x)$

47. $\lim\limits_{x\to 3^-} \dfrac{x^2-9}{x-3}$

48. $\lim\limits_{x\to 3^+} \dfrac{x^2-9}{x-3}$

49. $\lim\limits_{x\to 3^-} (\sqrt{9-x^2}+x)^2$

50. $\lim\limits_{x\to 2^+} (2\sqrt{x^2-4}+3x)$

In Problems 51–58, use the information below to find each limit.

$$\lim_{x\to c} f(x) = 5 \qquad \lim_{x\to c} g(x) = 6 \qquad \lim_{x\to c} h(x) = 0$$

51. $\lim\limits_{x\to c} [f(x)-3g(x)]$

52. $\lim\limits_{x\to c} [5f(x)]$

53. $\lim\limits_{x\to c} [g(x)]^3$

54. $\lim\limits_{x\to c} \dfrac{f(x)}{g(x)-h(x)}$

55. $\lim\limits_{x\to c} \dfrac{h(x)}{g(x)}$

56. $\lim\limits_{x\to c} [4f(x)\cdot g(x)]$

57. $\lim\limits_{x\to c} \left[\dfrac{1}{g(x)}\right]^2$

58. $\lim\limits_{x\to c} \sqrt[3]{5g(x)-3}$

In Problems 59 and 60, use the graphs of the functions and properties of limits to find each limit, if it exists. If the limit does not exist, write, "the limit does not exist," and explain why.

59. (a) $\lim\limits_{x\to 4} [f(x)+g(x)]$

(b) $\lim\limits_{x\to 4} \{f(x)[g(x)-h(x)]\}$

(c) $\lim\limits_{x\to 4} [f(x)\cdot g(x)]$

(d) $\lim\limits_{x\to 4} [2h(x)]$

(e) $\lim\limits_{x\to 4} \dfrac{g(x)}{f(x)}$

(f) $\lim\limits_{x\to 4} \dfrac{h(x)}{f(x)}$

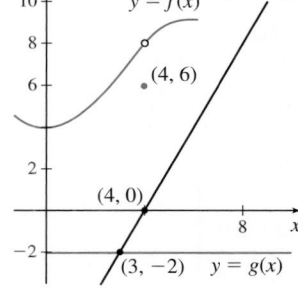

60. (a) $\lim\limits_{x\to 3} \{2[f(x)+h(x)]\}$

(b) $\lim\limits_{x\to 3^-} [g(x)+h(x)]$

(c) $\lim\limits_{x\to 3} \sqrt[3]{h(x)}$

(d) $\lim\limits_{x\to 3} \dfrac{f(x)}{h(x)}$

(e) $\lim\limits_{x\to 3} [h(x)]^3$

(f) $\lim\limits_{x\to 3} [f(x)-2h(x)]^{3/2}$

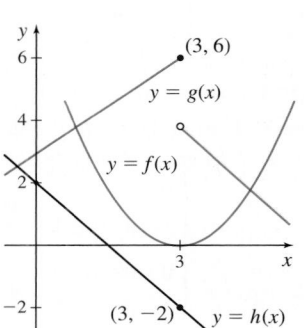

In Problems 61–66, for each function f, find the limit as x approaches c of the average rate of change of f from c to x. That is, find

$$\lim_{x \to c} \frac{f(x) - f(c)}{x - c}$$

61. $f(x) = 3x^2$, $c = 1$

62. $f(x) = 8x^3$, $c = 2$

63. $f(x) = -2x^2 + 4$, $c = 1$

64. $f(x) = 20 - 0.8x^2$, $c = 3$

65. $f(x) = \sqrt{x}$, $c = 1$

66. $f(x) = \sqrt{2x}$, $c = 5$

In Problems 67–72, find the limit of the difference quotient for each function f. That is, find $\lim_{h \to 0} \dfrac{f(x+h) - f(x)}{h}$.

67. $f(x) = 4x - 3$

68. $f(x) = 3x + 5$

69. $f(x) = 3x^2 + 4x + 1$

70. $f(x) = 2x^2 + x$

71. $f(x) = \dfrac{2}{x}$

72. $f(x) = \dfrac{3}{x^2}$

In Problems 73–80, find $\lim\limits_{x \to c^-} f(x)$ and $\lim\limits_{x \to c^+} f(x)$ for the given number c. Based on the results, determine whether $\lim\limits_{x \to c} f(x)$ exists.

73. $f(x) = \begin{cases} 2x - 3 & \text{if } x \le 1 \\ 3 - x & \text{if } x > 1 \end{cases}$ at $c = 1$

74. $f(x) = \begin{cases} 5x + 2 & \text{if } x < 2 \\ 1 + 3x & \text{if } x \ge 2 \end{cases}$ at $c = 2$

75. $f(x) = \begin{cases} 3x - 1 & \text{if } x < 1 \\ 4 & \text{if } x = 1 \\ 2x & \text{if } x > 1 \end{cases}$ at $c = 1$

76. $f(x) = \begin{cases} 3x - 1 & \text{if } x < 1 \\ 2 & \text{if } x = 1 \\ 2x & \text{if } x > 1 \end{cases}$ at $c = 1$

77. $f(x) = \begin{cases} x - 1 & \text{if } x < 1 \\ \sqrt{x - 1} & \text{if } x > 1 \end{cases}$ at $c = 1$

78. $f(x) = \begin{cases} \sqrt{9 - x^2} & \text{if } 0 < x < 3 \\ \sqrt{x^2 - 9} & \text{if } x > 3 \end{cases}$ at $c = 3$

79. $f(x) = \begin{cases} \dfrac{x^2 - 9}{x - 3} & \text{if } x \ne 3 \\ 6 & \text{if } x = 3 \end{cases}$ at $c = 3$

80. $f(x) = \begin{cases} \dfrac{x - 2}{x^2 - 4} & \text{if } x \ne 2 \\ 1 & \text{if } x = 2 \end{cases}$ at $c = 2$

Applications and Extensions

Heaviside Functions *In Problems 81 and 82, find the limit, if it exists, of the given Heaviside function at c.*

81. $u_1(t) = \begin{cases} 0 & \text{if } t < 1 \\ 1 & \text{if } t \ge 1 \end{cases}$ at $c = 1$

82. $u_3(t) = \begin{cases} 0 & \text{if } t < 3 \\ 1 & \text{if } t \ge 3 \end{cases}$ at $c = 3$

In Problems 83–92, find each limit.

83. $\lim\limits_{h \to 0} \dfrac{(x + h)^2 - x^2}{h}$

84. $\lim\limits_{h \to 0} \dfrac{\sqrt{x + h} - \sqrt{x}}{h}$

85. $\lim\limits_{h \to 0} \dfrac{\dfrac{1}{x + h} - \dfrac{1}{x}}{h}$

86. $\lim\limits_{h \to 0} \dfrac{\dfrac{1}{(x + h)^3} - \dfrac{1}{x^3}}{h}$

87. $\lim\limits_{x \to 0} \left[\dfrac{1}{x} \left(\dfrac{1}{4 + x} - \dfrac{1}{4} \right) \right]$

88. $\lim\limits_{x \to -1} \left[\dfrac{2}{x + 1} \left(\dfrac{1}{3} - \dfrac{1}{x + 4} \right) \right]$

89. $\lim\limits_{x \to 7} \dfrac{x - 7}{\sqrt{x + 2} - 3}$

90. $\lim\limits_{x \to 2} \dfrac{x - 2}{\sqrt{x + 2} - 2}$

91. $\lim\limits_{x \to 1} \dfrac{x^3 - 3x^2 + 3x - 1}{x^2 - 2x + 1}$

92. $\lim\limits_{x \to -3} \dfrac{x^3 + 7x^2 + 15x + 9}{x^2 + 6x + 9}$

93. Cost of Water   The Jericho Water District determines quarterly water costs, in dollars, using the following rate schedule:

Water used (in thousands of gallons)	Cost
$0 \le x \le 10$	$12.00
$10 < x \le 30$	$12.00 + 1.26$ for each thousand gallons in excess of 10,000 gallons
$30 < x \le 100$	$37.20 + 2.40$ for each thousand gallons in excess of 30,000 gallons
$x > 100$	$205.20 + 3.18$ for each thousand gallons in excess of 100,000 gallons

Source: Jericho Water District, Syosset, NY.

(a) Find a function C that models the quarterly cost, in dollars, of using x thousand gallons of water.

(b) What is the domain of the function C?

(c) Find each of the following limits. If the limit does not exist, explain why.

$\quad \lim\limits_{x \to 5} C(x) \quad \lim\limits_{x \to 10} C(x) \quad \lim\limits_{x \to 30} C(x) \quad \lim\limits_{x \to 100} C(x)$

(d) What is $\lim\limits_{x \to 0^+} C(x)$?

(e) Graph the function C.

94. Cost of Electricity In January 2023, Florida Power and Light charged customers living in single-family residences for their electric usage according to the following table.

Monthly customer charge for electricity	
$8.99	per household, plus
$0.10858	for each kWH used less than or equal to 1000 kWH, plus
$0.12858	for each kWH used in excess of 1000 kWH

Source: Florida Power and Light, Miami, FL.

(a) Find a function C that models the monthly cost, in dollars, of using x kWH of electricity.

(b) What is the domain of the function C?

(c) Find $\lim\limits_{x \to 1000} C(x)$, if it exists. If the limit does not exist, explain why.

(d) What is $\lim\limits_{x \to 0^+} C(x)$?

(e) Graph the function C.

95. Low-Temperature Physics In thermodynamics, the average molecular kinetic energy (energy of motion) of a gas having molecules of mass m is directly proportional to its temperature T on the absolute (or Kelvin) scale. This can be expressed as $\frac{1}{2}mv^2 = \frac{3}{2}kT$, where $v = v(T)$ is the speed of a typical molecule at time t, and k is a constant, known as the **Boltzmann constant**.

(a) What limit does the molecular speed v approach as the gas temperature T approaches absolute zero (0 K or $-273\,^\circ$C or $-469\,^\circ$F)?

(b) What does this limit suggest about the behavior of a gas as its temperature approaches absolute zero?

96. For the function $f(x) = \begin{cases} 3x+5 & \text{if } x \le 2 \\ 13-x & \text{if } x > 2 \end{cases}$, find

(a) $\lim\limits_{h\to 0^-} \dfrac{f(2+h)-f(2)}{h}$ (b) $\lim\limits_{h\to 0^+} \dfrac{f(2+h)-f(2)}{h}$

(c) Does $\lim\limits_{h\to 0} \dfrac{f(2+h)-f(2)}{h}$ exist?

97. Use the fact that $|x| = \begin{cases} x & \text{if } x \ge 0 \\ -x & \text{if } x < 0 \end{cases}$ to show that $\lim\limits_{x\to 0} |x| = 0$.

98. Use the fact that $|x| = \sqrt{x^2}$ to show that $\lim\limits_{x\to 0} |x| = 0$.

99. Find functions f and g for which $\lim\limits_{x\to c} [f(x)+g(x)]$ may exist even though $\lim\limits_{x\to c} f(x)$ and $\lim\limits_{x\to c} g(x)$ do not exist.

100. Find functions f and g for which $\lim\limits_{x\to c} [f(x)g(x)]$ may exist even though $\lim\limits_{x\to c} f(x)$ and $\lim\limits_{x\to c} g(x)$ do not exist.

101. Find functions f and g for which $\lim\limits_{x\to c}\left[\dfrac{f(x)}{g(x)}\right]$ may exist even though $\lim\limits_{x\to c} f(x)$ and $\lim\limits_{x\to c} g(x)$ do not exist.

102. Find a function f for which $\lim\limits_{x\to c} |f(x)|$ may exist even though $\lim\limits_{x\to c} f(x)$ does not exist.

103. Prove that if g is a function for which $\lim\limits_{x\to c} g(x)$ exists and if k is any real number, then $\lim\limits_{x\to c} [kg(x)]$ exists and $\lim\limits_{x\to c} [kg(x)] = k \lim\limits_{x\to c} g(x)$.

104. Prove that if the number c is in the domain of a rational function $R(x) = \dfrac{p(x)}{q(x)}$, then $\lim\limits_{x\to c} R(x) = R(c)$.

Challenge Problems

105. Find $\lim\limits_{x\to a} \dfrac{x^n - a^n}{x - a}$, n a positive integer.

106. Find $\lim\limits_{x\to -a} \dfrac{x^n + a^n}{x + a}$, n a positive integer.

107. Find $\lim\limits_{x\to 1} \dfrac{x^m - 1}{x^n - 1}$, m, n positive integers.

108. Find $\lim\limits_{x\to 0} \dfrac{\sqrt[3]{1+x} - 1}{x}$.

109. Find $\lim\limits_{x\to 0} \dfrac{\sqrt{(1+ax)(1+bx)} - 1}{x}$.

110. Find $\lim\limits_{x\to 0} \dfrac{\sqrt{(1+a_1x)(1+a_2x)\cdots(1+a_nx)} - 1}{x}$.

111. Find $\lim\limits_{h\to 0} \dfrac{f(h)-f(0)}{h}$ if $f(x) = x|x|$.

Preparing for the AP® Exam

AP® Practice Problems

Multiple-Choice Questions

[PAGE 97] **1.** Consider the piecewise-defined function f given by

$$f(x) = \begin{cases} -x-2 & \text{if } x < -1 \\ x^2 & \text{if } -1 \le x < 2 \\ -4x+12 & \text{if } x \ge 2 \end{cases}$$

Investigate the limits below and decide which limit does NOT exist.

(A) $\lim\limits_{x\to -1^+} f(x)$ (B) $\lim\limits_{x\to 2^-} f(x)$

(C) $\lim\limits_{x\to 2} f(x)$ (D) $\lim\limits_{x\to -1} f(x)$

[PAGE 102] **2.** $\lim\limits_{t\to 5} \dfrac{(5-t)^2}{t-5} =$

(A) -5 (B) 0 (C) 1 (D) 5

[PAGE 103] **3.** Find $\lim\limits_{x\to 2} \dfrac{f(x)-f(2)}{x-2}$ for the function $f(x) = 3x^3 - 4$.

(A) 0 (B) 12 (C) 24 (D) 36

[PAGE 103] **4.** $\lim\limits_{x\to s} \dfrac{x-s}{\sqrt{x} - \sqrt{s}} =$

(A) $2s$ (B) $2\sqrt{s}$ (C) $\sqrt{2s}$ (D) s

[PAGE 97] **5.** For $g(x) = \begin{cases} ax^2 - 5 & \text{if } x < 2 \\ ax+b & \text{if } x > 2 \end{cases}$

find values for a and b so that $\lim\limits_{x\to 2} g(x) = 7$.

(A) $a=1, b=5$ (B) $a=2, b=3$
(C) $a=3, b=1$ (D) $a=6, b=-5$

[PAGE 99] **6.** $\lim\limits_{x\to 4^+} (5\sqrt{x^2 - 16} + 3x) =$

(A) -12 (B) 0 (C) 12 (D) The limit does not exist.

[PAGE 99] **7.** If $\lim\limits_{x\to 2} \sqrt{\dfrac{[f(x)]^2 - 8x + 3}{x+1}} = 9$ and $f(x) \ge 0$ for all x, find $\lim\limits_{x\to 2} f(x)$.

(A) $\sqrt{22}$ (B) $2\sqrt{10}$ (C) 16 (D) 256

[PAGE 99] **8.** $\lim\limits_{x\to 3} [x^{-1/2}(5x-7)^{1/3}] =$

(A) $3^{-1/2}$ (B) $\dfrac{2}{3^{1/2}}$ (C) $\dfrac{8}{3^{1/2}}$ (D) $6^{-1/2}$

[PAGE 104] **9.** For the function

$f(x) = x^2 - 3x + 4$, $\lim\limits_{h\to 0} \dfrac{f(x+h)-f(x)}{h} =$

(A) The limit does not exist. (B) 0 (C) $x-3$ (D) $2x-3$

1.3 Continuity

OBJECTIVES *When you finish this section, you should be able to:*

1 **Determine whether a function is continuous at a number (p. 108)**
2 **Determine intervals on which a function is continuous (p. 111)**
3 **Use properties of continuity (p. 113)**
4 **Use the Intermediate Value Theorem (p. 115)**

Sometimes $\lim\limits_{x \to c} f(x)$ equals $f(c)$ and sometimes it does not. In fact, $f(c)$ may not even be defined and yet $\lim\limits_{x \to c} f(x)$ may exist. In this section, we investigate the relationship between $\lim\limits_{x \to c} f(x)$ and $f(c)$. Figure 21 shows some possibilities.

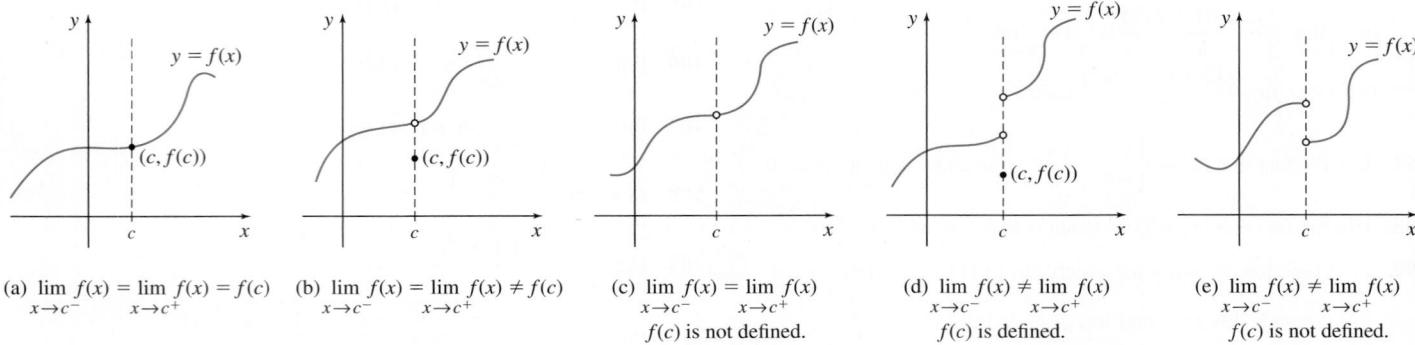

(a) $\lim\limits_{x \to c^-} f(x) = \lim\limits_{x \to c^+} f(x) = f(c)$

(b) $\lim\limits_{x \to c^-} f(x) = \lim\limits_{x \to c^+} f(x) \neq f(c)$

(c) $\lim\limits_{x \to c^-} f(x) = \lim\limits_{x \to c^+} f(x)$
$f(c)$ is not defined.

(d) $\lim\limits_{x \to c^-} f(x) \neq \lim\limits_{x \to c^+} f(x)$
$f(c)$ is defined.

(e) $\lim\limits_{x \to c^-} f(x) \neq \lim\limits_{x \to c^+} f(x)$
$f(c)$ is not defined.

Figure 21

Of these five graphs, the "nicest" one is Figure 21(a). There, $\lim\limits_{x \to c} f(x)$ exists and is equal to $f(c)$. Functions that have this property are said to be *continuous at the number c*. This agrees with the intuitive notion that a function is continuous if its graph can be drawn without lifting the pencil. The functions in Figures 21(b)–(e) are not continuous at c, since each has a break in the graph at c. This leads to the definition of *continuity at a number*.

DEFINITION Continuity at a Number

A function f is **continuous at a number** c if the following three conditions are met:

• $f(c)$ is defined (that is, c is in the domain of f)
• $\lim\limits_{x \to c} f(x)$ exists
• $\lim\limits_{x \to c} f(x) = f(c)$

If a function f is defined on an open interval containing c, except possibly at c, then the function is **discontinuous at** c, whenever any one of the above three conditions is not satisfied.

NOW WORK   AP® Practice Problems 1 and 2.

1 Determine Whether a Function Is Continuous at a Number

EXAMPLE 1   Determining Whether a Function Is Continuous at a Number

(a) Determine whether $f(x) = 3x^2 - 5x + 4$ is continuous at 1.

(b) Determine whether $g(x) = \dfrac{x^2 + 9}{x^2 - 4}$ is continuous at 2.

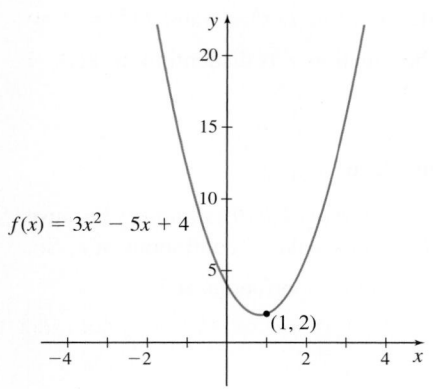

$f(x) = 3x^2 - 5x + 4$

(1, 2)

(a) f is continuous at 1.

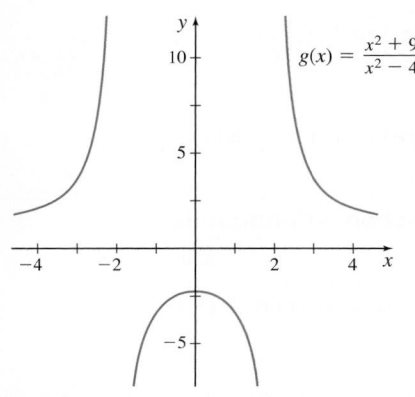

$g(x) = \dfrac{x^2 + 9}{x^2 - 4}$

(b) g is discontinuous at 2.

Figure 22

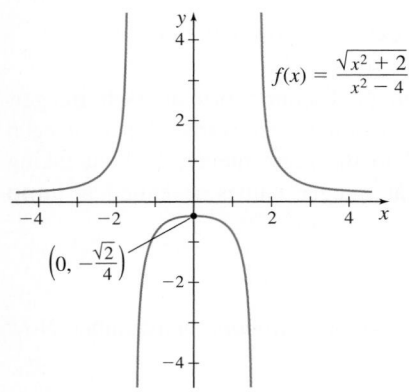

$f(x) = \dfrac{\sqrt{x^2 + 2}}{x^2 - 4}$

$\left(0, -\dfrac{\sqrt{2}}{4}\right)$

Figure 23 f is continuous at 0; f is discontinuous at -2 and 2.

Solution

(a) We begin by checking the conditions for continuity. First, 1 is in the domain of f and $f(1) = 2$. Second, $\displaystyle\lim_{x \to 1} f(x) = \lim_{x \to 1}(3x^2 - 5x + 4) = 2$, so $\displaystyle\lim_{x \to 1} f(x)$ exists. Third, $\displaystyle\lim_{x \to 1} f(x) = f(1)$. Since the three conditions are met, f is continuous at 1.

(b) Since 2 is not in the domain of g, the function g is discontinuous at 2. ∎

Figure 22 shows the graphs of f and g from Example 1. Notice that f is continuous at 1, and its graph is drawn without lifting the pencil. But the function g is discontinuous at 2, and to draw its graph, you must lift your pencil at $x = 2$.

NOW WORK Problem **19** and AP® Practice Problem **5**.

EXAMPLE 2 Determining Whether a Function Is Continuous at a Number

Determine whether $f(x) = \dfrac{\sqrt{x^2 + 2}}{x^2 - 4}$ is continuous at the numbers -2, 0, and 2.

Solution

The domain of f is $\{x \mid x \neq -2,\ x \neq 2\}$. Since f is not defined at -2 and 2, the function f is not continuous at -2 and at 2. The number 0 is in the domain of f. That is, f is defined at 0, and $f(0) = -\dfrac{\sqrt{2}}{4}$. Also,

$$\lim_{x \to 0} f(x) = \lim_{x \to 0} \frac{\sqrt{x^2 + 2}}{x^2 - 4} = \frac{\displaystyle\lim_{x \to 0}\sqrt{x^2 + 2}}{\displaystyle\lim_{x \to 0}(x^2 - 4)} = \frac{\sqrt{\displaystyle\lim_{x \to 0}(x^2 + 2)}}{\displaystyle\lim_{x \to 0} x^2 - \lim_{x \to 0} 4}$$

$$= \frac{\sqrt{0 + 2}}{0 - 4} = -\frac{\sqrt{2}}{4} = f(0)$$

The three conditions of continuity at a number are met. So, the function f is continuous at 0. ∎

Figure 23 shows the graph of f.

NOW WORK Problem **21**.

EXAMPLE 3 Determining Whether a Piecewise-Defined Function Is Continuous

Determine whether the function

$$f(x) = \begin{cases} \dfrac{x^2 - 9}{x - 3} & \text{if } x < 3 \\ 9 & \text{if } x = 3 \\ x^2 - 3 & \text{if } x > 3 \end{cases}$$

is continuous at 3.

Solution

Since $f(3) = 9$, the function f is defined at 3. To check the second condition, we investigate the one-sided limits.

$$\lim_{x \to 3^-} f(x) = \lim_{x \to 3^-} \frac{x^2 - 9}{x - 3} = \lim_{x \to 3^-} \frac{(x - 3)(x + 3)}{x - 3} = \lim_{x \to 3^-}(x + 3) = 6$$

$$\underset{\text{Divide out } x - 3}{\uparrow}$$

$$\lim_{x \to 3^+} f(x) = \lim_{x \to 3^+}(x^2 - 3) = 9 - 3 = 6$$

(continued on the next page)

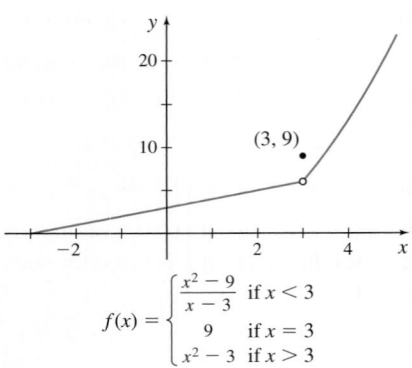

$$f(x) = \begin{cases} \dfrac{x^2 - 9}{x - 3} & \text{if } x < 3 \\ 9 & \text{if } x = 3 \\ x^2 - 3 & \text{if } x > 3 \end{cases}$$

Figure 24 f is discontinuous at 3.

IN WORDS Visually, the graph of a function f with a removable discontinuity has a hole at $(c, \lim\limits_{x \to c} f(x))$. The discontinuity is removable because we can fill in the hole by appropriately defining f. The function will then be continuous at $x = c$.

Since $\lim\limits_{x \to 3^-} f(x) = \lim\limits_{x \to 3^+} f(x)$, then $\lim\limits_{x \to 3} f(x)$ exists. But, $\lim\limits_{x \to 3} f(x) = 6$ and $f(3) = 9$, so the third condition of continuity is not satisfied. The function f is discontinuous at 3. ■

Figure 24 shows the graph of f.

<u>NOW WORK</u>   Problem 25 and AP® Practice Problems 8 and 13.

The discontinuity at $c = 3$ in Example 3 is called a *removable discontinuity* because we can redefine f at the number c to equal $\lim\limits_{x \to c} f(x)$ and make f continuous at c. So, in Example 3, if $f(3)$ is redefined to be 6, then f would be continuous at 3.

DEFINITION Removable Discontinuity

Suppose f is a function that is defined everywhere in an open interval containing c, except possibly at c. The number c is called a **removable discontinuity** of f if the function is discontinuous at c but $\lim\limits_{x \to c} f(x)$ exists. The discontinuity is removed by defining (or redefining) the value of f at c to be $\lim\limits_{x \to c} f(x)$.

<u>NOW WORK</u>   Problems 13 and 35 and AP® Practice Problems 3, 4, and 11.

NEED TO REVIEW? The floor function is discussed in Section P.2, pp. 20–21.

EXAMPLE 4   Determining Whether a Function Is Continuous at a Number

Determine whether the floor function $f(x) = \lfloor x \rfloor$ is continuous at 1.

Solution

The floor function $f(x) = \lfloor x \rfloor =$ the greatest integer $\leq x$. The floor function f is defined at 1 and $f(1) = 1$. But

$$\lim_{x \to 1^-} f(x) = \lim_{x \to 1^-} \lfloor x \rfloor = 0 \qquad \text{and} \qquad \lim_{x \to 1^+} f(x) = \lim_{x \to 1^+} \lfloor x \rfloor = 1$$

So, $\lim\limits_{x \to 1} \lfloor x \rfloor$ does not exist. Since $\lim\limits_{x \to 1} \lfloor x \rfloor$ does not exist, f is discontinuous at 1. ■

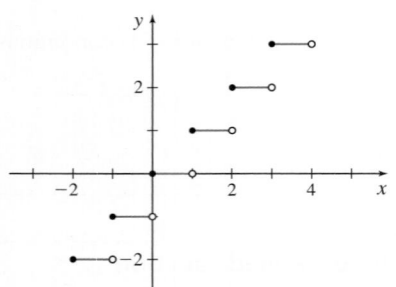

Figure 25 $f(x) = \lfloor x \rfloor$

Figure 25 illustrates that the floor function is discontinuous at each integer. Also, none of the discontinuities of the floor function is removable. Since at each integer the value of the floor function "jumps" to the next integer, without taking on any intermediate values, the discontinuity at integer values is called a **jump discontinuity**.

<u>NOW WORK</u>   Problem 53.

We have defined what it means for a function f to be *continuous* at a number. Now we define *one-sided continuity* at a number.

DEFINITION One-Sided Continuity at a Number

Suppose f is a function defined on the interval $(a, c]$. Then f is **continuous from the left at the number** c if

$$\boxed{\lim_{x \to c^-} f(x) = f(c)}$$

Suppose f is a function defined on the interval $[c, b)$. Then f is **continuous from the right at the number** c if

$$\boxed{\lim_{x \to c^+} f(x) = f(c)}$$

In Example 4, we showed that the floor function $f(x) = \lfloor x \rfloor$ is discontinuous at $x = 1$. But since

$$f(1) = \lfloor 1 \rfloor = 1 \qquad \text{and} \qquad \lim_{x \to 1^+} f(x) = \lfloor x \rfloor = 1$$

the floor function is continuous from the right at 1. In fact, the floor function is discontinuous at each integer n, but it is continuous from the right at every integer n. (Do you see why?)

2 Determine Intervals on Which a Function Is Continuous

So far, we have considered only continuity at a number c. Now, we use one-sided continuity to define continuity on an interval.

DEFINITION Continuity on an Interval

- A function f is **continuous on an open interval** (a, b) if f is continuous at every number in (a, b).
- A function f is **continuous on an interval** $[a, b)$ if f is continuous on the open interval (a, b) and continuous from the right at the number a.
- A function f is **continuous on an interval** $(a, b]$ if f is continuous on the open interval (a, b) and continuous from the left at the number b.
- A function f is **continuous on a closed interval** $[a, b]$ if f is continuous on the open interval (a, b), continuous from the right at a, and continuous from the left at b.

Figure 26 gives examples of graphs over different types of intervals.

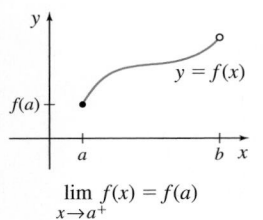
$$\lim_{x \to a^+} f(x) = f(a)$$
(a) f is continuous on $[a, b)$.

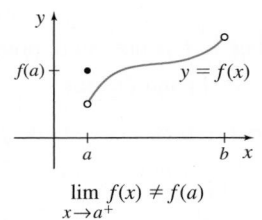
$$\lim_{x \to a^+} f(x) \neq f(a)$$
(b) f is continuous on (a, b).

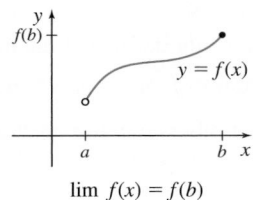
$$\lim_{x \to b^-} f(x) = f(b)$$
(c) f is continuous on $(a, b]$.

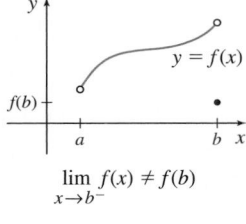
$$\lim_{x \to b^-} f(x) \neq f(b)$$
(d) f is continuous on (a, b).

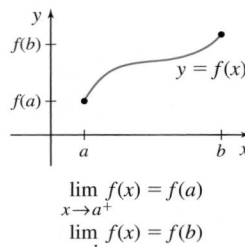
$$\lim_{x \to a^+} f(x) = f(a)$$
$$\lim_{x \to b^-} f(x) = f(b)$$
(e) f is continuous on $[a, b]$.

Figure 26

For example, the graph of the floor function $f(x) = \lfloor x \rfloor$ in Figure 25 illustrates that f is continuous on every interval $[n, n+1)$, n an integer. In each interval, f is continuous from the right at the left endpoint n and is continuous at every number in the open interval $(n, n+1)$.

EXAMPLE 5 Determining Whether a Function Is Continuous on a Closed Interval

Is the function $f(x) = \sqrt{4 - x^2}$ continuous on the closed interval $[-2, 2]$?

Solution

The domain of f is $\{x \mid -2 \leq x \leq 2\}$. So, f is defined for every number in the closed interval $[-2, 2]$.

For any number c in the open interval $(-2, 2)$,

$$\lim_{x \to c} f(x) = \lim_{x \to c} \sqrt{4 - x^2} = \sqrt{\lim_{x \to c} (4 - x^2)} = \sqrt{4 - c^2} = f(c)$$

So, f is continuous on the open interval $(-2, 2)$.

(continued on the next page)

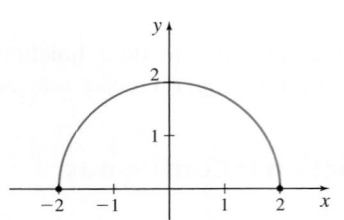

Figure 27 $f(x) = \sqrt{4 - x^2}, -2 \le x \le 2$

To determine whether f is continuous on $[-2, 2]$, we investigate the limit from the right at -2 and the limit from the left at 2. Then,

$$\lim_{x \to -2^+} f(x) = \lim_{x \to -2^+} \sqrt{4 - x^2} = 0 = f(-2)$$

So, f is continuous from the right at -2. Similarly,

$$\lim_{x \to 2^-} f(x) = \lim_{x \to 2^-} \sqrt{4 - x^2} = 0 = f(2)$$

So, f is continuous from the left at 2. We conclude that f is continuous on the closed interval $[-2, 2]$. ∎

Figure 27 shows the graph of f.

NOW WORK Problems **37** and **77**.

DEFINITION Continuity on a Domain

A function f is **continuous on its domain** if it is continuous at every number c in its domain.

EXAMPLE 6 Determining Whether $f(x) = \dfrac{x^2}{x - 1}$ Is Continuous on Its Domain

Determine whether the function $f(x) = \dfrac{x^2}{x - 1}$ is continuous on its domain.

Solution

The domain of $f(x) = \dfrac{x^2}{x - 1}$ is the set $\{x \mid x \ne 1\}$, which is the set of real numbers $(-\infty, 1) \cup (1, \infty)$.

Since f is not defined at the number 1, f is not continuous at 1.

We examine f on the intervals $(-\infty, 1)$ and $(1, \infty)$.

- For all numbers c in the open interval $(-\infty, 1)$, we have $f(c) = \dfrac{c^2}{c - 1}$ and

$$\lim_{x \to c} \frac{x^2}{x - 1} = \frac{\lim_{x \to c} x^2}{\lim_{x \to c}(x - 1)} = \frac{c^2}{c - 1} = f(c)$$

So, f is continuous on the open interval $(-\infty, 1)$.

- For all numbers c in the open interval $(1, \infty)$ we have $f(c) = \dfrac{c^2}{c - 1}$, and

$$\lim_{x \to c} \frac{x^2}{x - 1} = \frac{\lim_{x \to c} x^2}{\lim_{x \to c}(x - 1)} = \frac{c^2}{c - 1} = f(c)$$

So, f is continuous on the open interval $(1, \infty)$.

The function $f(x) = \dfrac{x^2}{x - 1}$ is continuous on its domain. ∎

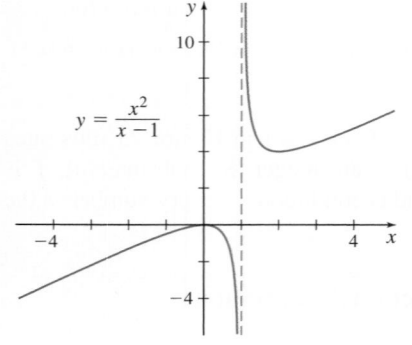

Figure 28 f is continuous on $(-\infty, 1) \cup (1, \infty)$.

Figure 28 shows the graph of f.

When listing the properties of a function in Chapter P, we included the function's domain, its symmetry, and its zeros. Now we add continuity to the list by asking, "Where is the function continuous?" We answer this question here for two important classes of functions: polynomial functions and rational functions.

THEOREM

- A polynomial function is continuous on its domain, all real numbers.
- A rational function is continuous on its domain.

Proof If P is a polynomial function, its domain is the set of real numbers. For a polynomial function,

$$\lim_{x \to c} P(x) = P(c)$$

for any number c. That is, a polynomial function is continuous at every real number.

If $R(x) = \dfrac{p(x)}{q(x)}$ is a rational function, then $p(x)$ and $q(x)$ are polynomials and the domain of R is $\{x \mid q(x) \neq 0\}$. The Limit of a Rational Function (p. 101) states that for all c in the domain of a rational function,

$$\lim_{x \to c} R(x) = R(c)$$

So a rational function is continuous at every number in its domain. ∎

To summarize:

- If a function is continuous *on an interval*, its graph has no holes or gaps on that interval.
- If a function is continuous *on its domain*, it will be continuous at every number in its domain; its graph may have holes or gaps at numbers that are not in the domain.

For example, the function $R(x) = \dfrac{x^2 - 2x + 1}{x - 1}$ is continuous on its domain $\{x \mid x \neq 1\}$ even though the graph has a hole at $(1, 0)$, as shown in Figure 29. The function $f(x) = \dfrac{1}{x}$ is continuous on its domain $\{x \mid x \neq 0\}$, as shown in Figure 30. Notice the behavior of the graph as x goes from negative numbers to positive numbers.

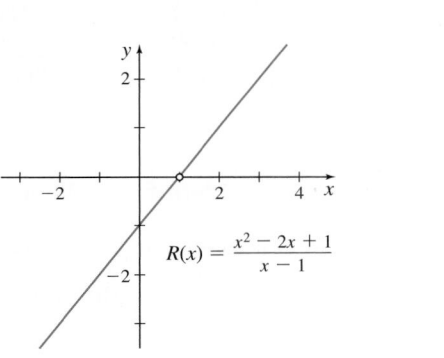

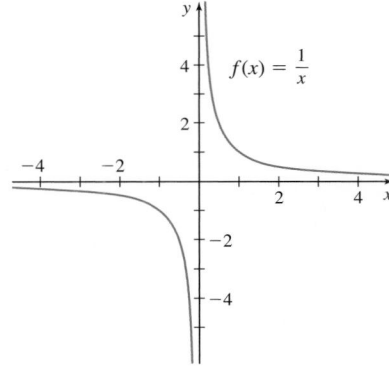

Figure 29 The graph of R has a hole at $(1, 0)$.   **Figure 30** f is not defined at 0.

Notice that the discontinuity at 1 for R is removable, while the discontinuity at 0 for f is not removable.

❸ Use Properties of Continuity

So far we have shown that polynomial and rational functions are continuous on their domains. From these functions, we can build other continuous functions.

THEOREM Continuity of a Sum, Difference, Product, and Quotient

If the functions f and g are continuous at a number c, and if k is a real number, then the functions $f + g$, $f - g$, $f \cdot g$, and kf are also continuous at c. If $g(c) \neq 0$, the function $\dfrac{f}{g}$ is continuous at c.

The proofs of these properties are based on properties of limits explored in Section 1.2. For example, the proof of the continuity of $f + g$ is based on the Limit of a Sum property. That is, if $\lim_{x \to c} f(x)$ and $\lim_{x \to c} g(x)$ exist, then $\lim_{x \to c} [f(x) + g(x)] = \lim_{x \to c} f(x) + \lim_{x \to c} g(x)$.

EXAMPLE 7   Identifying Where Functions Are Continuous

Determine where each function is continuous:

(a) $F(x) = x^2 + 5 - \dfrac{x}{x^2 + 4}$

(b) $G(x) = x^3 + 2x + \dfrac{x^2}{x^2 - 1}$

Solution

First we determine the domain of the function.

(a) F is the difference of the two functions $f(x) = x^2 + 5$ and $g(x) = \dfrac{x}{x^2 + 4}$, each of whose domain is the set of all real numbers. So, the domain of F is the set of all real numbers. Since f and g are continuous on their domains, the difference function F is continuous on its domain.

(b) G is the sum of the two functions $f(x) = x^3 + 2x$, whose domain is the set of all real numbers, and $g(x) = \dfrac{x^2}{x^2 - 1}$, whose domain is $\{x \mid x \neq -1, \ x \neq 1\}$. Since f and g are continuous on their domains, G is continuous on its domain, $\{x \mid x \neq -1, \ x \neq 1\}$. ∎

NOW WORK Problem 45.

NEED TO REVIEW? Composite functions are discussed in Section P.3, pp. 29–31.

The continuity of a composite function depends on the continuity of its components.

THEOREM   Continuity of a Composite Function

If a function g is continuous at c and a function f is continuous at $g(c)$, then the composite function $(f \circ g)(x) = f(g(x))$ is continuous at c. That is,

$$\lim_{x \to c} (f \circ g)(x) = \lim_{x \to c} f(g(x)) = f[\lim_{x \to c} g(x)] = f(g(c))$$

EXAMPLE 8   Identifying Where Functions Are Continuous

Determine where each function is continuous:

(a) $F(x) = \sqrt{x^2 + 4}$

(b) $G(x) = \sqrt{x^2 - 1}$

(c) $H(x) = \dfrac{x^2 - 1}{x^2 - 4} + \sqrt{x - 1}$

Solution

(a) $F = f \circ g$ is the composite of $f(x) = \sqrt{x}$ and $g(x) = x^2 + 4$. f is continuous for $x \geq 0$ and g is continuous for all real numbers. The domain of F is all real numbers and $F = (f \circ g)(x) = \sqrt{x^2 + 4}$ is continuous for all real numbers. That is, F is continuous on its domain.

(b) G is the composite of $f(x) = \sqrt{x}$ and $g(x) = x^2 - 1$. f is continuous for $x \geq 0$ and g is continuous for all real numbers. The domain of G is $\{x \mid x \leq -1\} \cup \{x \mid x \geq 1\}$ and $G = (f \circ g)(x) = \sqrt{x^2 - 1}$ is continuous on its domain.

(c) H is the sum of $f(x) = \dfrac{x^2 - 1}{x^2 - 4}$ and the function $g(x) = \sqrt{x - 1}$. The domain of f is $\{x \mid x \neq -2, x \neq 2\}$; f is continuous on its domain. The domain of g is $x \geq 1$. The domain of H is $\{x \mid 1 \leq x < 2\} \cup \{x \mid x > 2\}$; H is continuous on its domain. ∎

NOW WORK Problem 47.

NEED TO REVIEW? Inverse functions are discussed in Section P.4, pp. 37–41.

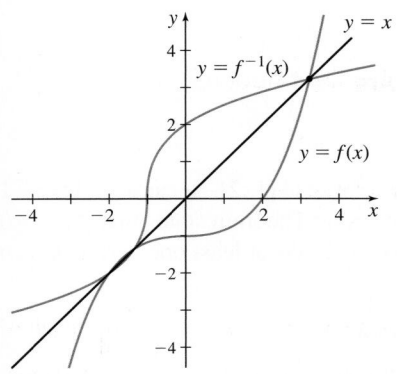

Figure 31

Recall that for any function f that is one-to-one over its domain, its inverse f^{-1} is also a function, and the graphs of f and f^{-1} are symmetric with respect to the line $y = x$. It is intuitive that if f is continuous, then so is f^{-1}. See Figure 31. The following theorem, whose proof is given in Appendix B, confirms this.

> **THEOREM Continuity of an Inverse Function**
>
> If f is a one-to-one function that is continuous on its domain, then its inverse function f^{-1} is also continuous on its domain.

4 Use the Intermediate Value Theorem

Functions that are continuous on a closed interval have many important properties. One of them is stated in the *Intermediate Value Theorem*. The proof of the Intermediate Value Theorem may be found in most books on advanced calculus.

> **THEOREM The Intermediate Value Theorem (IVT)**
>
> Suppose f is a function that is continuous on a closed interval $[a, b]$ and $f(a) \neq f(b)$. If N is any number between $f(a)$ and $f(b)$, then there is at least one number c in the open interval (a, b) for which $f(c) = N$.

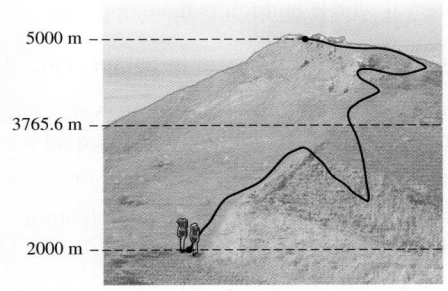

To get a better idea of this result, suppose you climb a mountain, starting at an elevation of 2000 meters and ending at an elevation of 5000 meters. No matter how many ups and downs you take as you climb, at some time your altitude must be 3765.6 meters, or any other number between 2000 and 5000.

In other words, a function f that is continuous on a closed interval $[a, b]$ must take on all values between $f(a)$ and $f(b)$. Figure 32 illustrates this. Figure 33 shows why the continuity of the function is crucial. Notice in Figure 33 that there is a hole in the graph of f at the point (c, N). Because of the discontinuity at c, there is no number c in the open interval (a, b) for which $f(c) = N$.

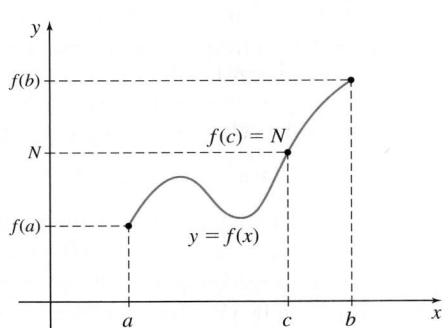

Figure 32 f takes on every value between $f(a)$ and $f(b)$.

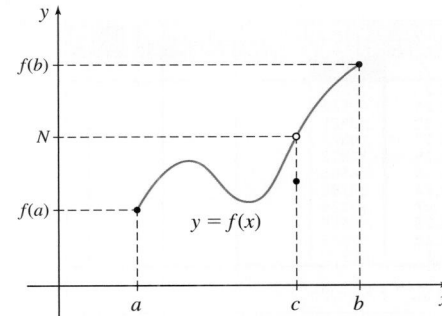

Figure 33 A discontinuity at c results in no number c in (a, b) for which $f(c) = N$.

The Intermediate Value Theorem is an existence theorem. It states that there is at least one number c for which $f(c) = N$, but it does not tell us how to find c. However, we can use the Intermediate Value Theorem to locate an open interval (a, b) that contains c.

An immediate application of the Intermediate Value Theorem involves locating the zeros of a function. Suppose a function f is continuous on the closed interval $[a, b]$ and $f(a)$ and $f(b)$ have opposite signs. Then by the Intermediate Value Theorem, there is at least one number c between a and b for which $f(c) = 0$. That is, f has at least one zero between a and b.

EXAMPLE 9 Using the Intermediate Value Theorem

Use the Intermediate Value Theorem to show that

$$f(x) = x^3 + x^2 - x - 2$$

has a zero between 1 and 2.

Solution

Since f is a polynomial, it is continuous on the closed interval $[1, 2]$. Because $f(1) = -1$ and $f(2) = 8$ have opposite signs, the Intermediate Value Theorem states that $f(c) = 0$ for at least one number c in the interval $(1, 2)$. That is, f has at least one zero between 1 and 2. ■

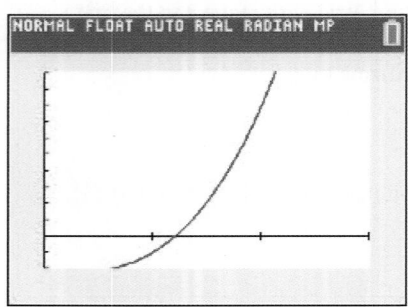

$[0, 3] \times [-2, 10]$

Figure 34 $f(x) = x^3 + x^2 - x - 2$

Figure 34 shows the graph of f on a graphing utility.

NOW WORK Problem 59 and AP® Practice Problems 6, 7, 9, 10, and 12.

The Intermediate Value Theorem can be used to approximate a zero in the interval (a, b) by dividing the interval $[a, b]$ into smaller subintervals. There are two popular methods of subdividing the interval $[a, b]$.

- The **bisection method** bisects $[a, b]$, that is, divides $[a, b]$ into two equal subintervals and compares the sign of $f\left(\frac{a+b}{2}\right)$ to the signs of the previously computed values $f(a)$ and $f(b)$. The subinterval whose endpoints have opposite signs is then bisected, and the process is repeated.
- The second method divides $[a, b]$ into 10 subintervals of equal length and compares the signs of f evaluated at each of the 11 endpoints. The subinterval whose endpoints have opposite signs is then divided into 10 subintervals of equal length and the process is repeated.

We choose to use the second method because it lends itself well to the table feature of a graphing utility. You are asked to use the bisection method in Problems 107–114.

NORMAL FLOAT AUTO REAL RADIAN MP
PRESS [ENTER] TO EDIT

X	Y1
1	-1
1.1	-.559
1.2	-.032
1.3	.587
1.4	1.304
1.5	2.125
1.6	3.056
1.7	4.103
1.8	5.272
1.9	6.569
2	8

Y1=X³+X²−X−2

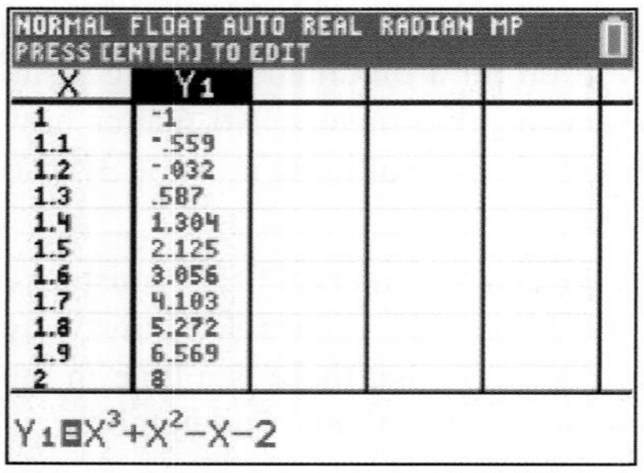

Figure 35

EXAMPLE 10 Using the Intermediate Value Theorem to Approximate a Real Zero of a Function

The function $f(x) = x^3 + x^2 - x - 2$ has a zero in the interval $(1, 2)$. Use the Intermediate Value Theorem to approximate the zero correct to three decimal places.

Solution

Using the TABLE feature on a graphing utility, we subdivide the interval $[1, 2]$ into 10 subintervals, each of length 0.1. Then we find the subinterval whose endpoints have opposite signs, or the endpoint whose value equals 0 (in which case, the exact zero is found). From Figure 35, since $f(1.2) = -0.032$ and $f(1.3) = 0.587$, by the Intermediate Value Theorem, a zero lies in the interval $(1.2, 1.3)$. Correct to one decimal place, the zero is 1.2.

NORMAL FLOAT AUTO REAL RADIAN MP
PRESS [ENTER] TO EDIT

X	Y1
1.2	-.032
1.21	.02566
1.22	.08425
1.23	.14377
1.24	.20422
1.25	.26563
1.26	.32798
1.27	.39128
1.28	.45555
1.29	.52079
1.3	.587

Y1=X³+X²−X−2

Figure 36

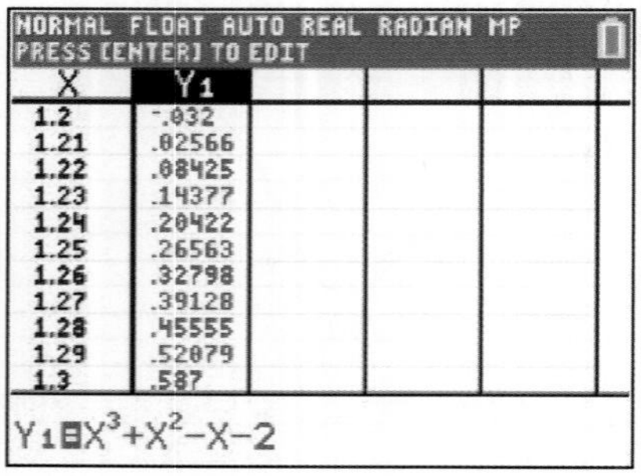

Repeat the process by subdividing the interval $[1.2, 1.3]$ into 10 subintervals, each of length 0.01. See Figure 36. We conclude that the zero is in the interval $(1.20, 1.21)$; so correct to two decimal places, the zero is 1.20.

Now subdivide the interval $[1.20, 1.21]$ into 10 subintervals, each of length 0.001. See Figure 37.

NORMAL FLOAT AUTO REAL RADIAN MP
PRESS [ENTER] TO EDIT

X	Y1
1.2	-.032
1.201	-.0263
1.202	-.0205
1.203	-.0148
1.204	-.009
1.205	-.0033
1.206	.00249
1.207	.00827
1.208	.01405
1.209	.01985
1.21	.02566

Y1=X³+X²−X−2

Figure 37

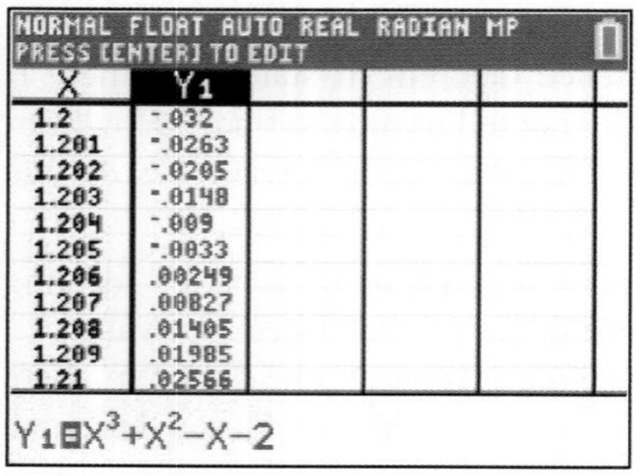

We conclude that the zero of the function f is 1.205, correct to three decimal places. ■

Notice that a benefit of the method used in Example 10 is that each additional iteration results in one additional decimal place of accuracy for the approximation.

NOW WORK Problem 65.

1.3 Assess Your Understanding

Concepts and Vocabulary

1. **True or False** A polynomial function is continuous at every real number.

2. **True or False** Piecewise-defined functions are never continuous at numbers where the function changes equations.

3. The three conditions necessary for a function f to be continuous at a number c are _____, _____, and _____.

4. **True or False** If f is continuous at 0, then $g(x) = \dfrac{1}{4} f(x)$ is continuous at 0.

5. **True or False** If f is a function defined everywhere in an open interval containing c, except possibly at c, then the number c is called a removable discontinuity of f if the function f is not continuous at c.

6. **True or False** If a function f is discontinuous at a number c, then $\lim\limits_{x \to c} f(x)$ does not exist.

7. **True or False** If a function f is continuous on an open interval (a, b), then it is continuous on the closed interval $[a, b]$.

8. **True or False** If a function f is continuous on the closed interval $[a, b]$, then f is continuous on the open interval (a, b).

In Problems 9 and 10, explain whether each function is continuous or discontinuous on its domain.

9. The velocity of a ball thrown up into the air as a function of time, if the ball lands 5 seconds after it is thrown and stops.

10. The temperature of an oven used to bake a potato as a function of time.

11. **True or False** If a function f is continuous on a closed interval $[a, b]$, then the Intermediate Value Theorem guarantees that the function takes on every value between $f(a)$ and $f(b)$.

12. **True or False** If a function f is continuous on a closed interval $[a, b]$ and $f(a) \neq f(b)$, but both $f(a) > 0$ and $f(b) > 0$, then according to the Intermediate Value Theorem, f does not have a zero on the open interval (a, b).

Skill Building

In Problems 13–18, use the graph of $y = f(x)$ (below).

(a) Determine if f is continuous at c.

(b) If f is discontinuous at c, state which condition(s) of the definition of continuity is (are) not satisfied.

(c) If f is discontinuous at c, determine if the discontinuity is removable.

(d) If the discontinuity is removable, define (or redefine) f at c to make f continuous at c.

 13. $c = -3$ 14. $c = 0$

15. $c = 2$ 16. $c = 3$

17. $c = 4$ 18. $c = 5$

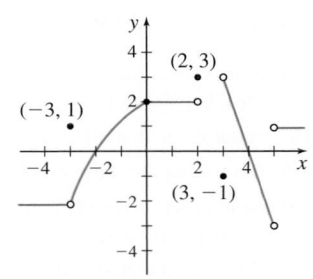

In Problems 19–32, determine whether the function f is continuous at c.

19. $f(x) = x^2 + 3$ at $c = -3$

20. $f(x) = x^3 - 5$ at $c = 4$

21. $f(x) = \dfrac{x}{x^2 + 4}$ at $c = -2$

22. $f(x) = \dfrac{x}{x - 2}$ at $c = 2$

23. $f(x) = \begin{cases} 2x + 5 & \text{if } x \leq 2 \\ 4x + 1 & \text{if } x > 2 \end{cases}$ at $c = 2$

24. $f(x) = \begin{cases} 2x + 1 & \text{if } x \leq 0 \\ 2x & \text{if } x > 0 \end{cases}$ at $c = 0$

25. $f(x) = \begin{cases} 3x - 1 & \text{if } x < 1 \\ 4 & \text{if } x = 1 \\ 2x & \text{if } x > 1 \end{cases}$ at $c = 1$

26. $f(x) = \begin{cases} 3x - 1 & \text{if } x < 1 \\ 2 & \text{if } x = 1 \\ 2x & \text{if } x > 1 \end{cases}$ at $c = 1$

27. $f(x) = \begin{cases} 3x - 1 & \text{if } x < 1 \\ 2x & \text{if } x > 1 \end{cases}$ at $c = 1$

28. $f(x) = \begin{cases} 3x - 1 & \text{if } x < 1 \\ 2 & \text{if } x = 1 \\ 3x & \text{if } x > 1 \end{cases}$ at $c = 1$

29. $f(x) = \begin{cases} x^2 & \text{if } x \leq 0 \\ 2x & \text{if } x > 0 \end{cases}$ at $c = 0$

30. $f(x) = \begin{cases} x^2 & \text{if } x < -1 \\ 2 & \text{if } x = -1 \\ -3x + 2 & \text{if } x > -1 \end{cases}$ at $c = -1$

31. $f(x) = \begin{cases} 4 - 3x^2 & \text{if } x < 0 \\ 4 & \text{if } x = 0 \\ \sqrt{\dfrac{16 - x^2}{4 - x}} & \text{if } 0 < x < 4 \end{cases}$ at $c = 0$

32. $f(x) = \begin{cases} \sqrt{4 + x} & \text{if } -4 \leq x \leq 4 \\ \sqrt{\dfrac{x^2 - 3x - 4}{x - 4}} & \text{if } x > 4 \end{cases}$ at $c = 4$

In Problems 33–36, each function f has a removable discontinuity at c. Define $f(c)$ so that f is continuous at c.

33. $f(x) = \dfrac{x^2 - 4}{x - 2}$, $c = 2$

34. $f(x) = \dfrac{x^2 + x - 12}{x - 3}$, $c = 3$

35. $f(x) = \begin{cases} 1 + x & \text{if } x < 1 \\ 4 & \text{if } x = 1 \\ 2x & \text{if } x > 1 \end{cases}$ $c = 1$

36. $f(x) = \begin{cases} x^2 + 5x & \text{if } x < -1 \\ 0 & \text{if } x = -1 \\ x - 3 & \text{if } x > -1 \end{cases}$ $c = -1$

In Problems 37–40, determine whether each function f is continuous on the given interval. If the answer is no, state the interval(s), if any, on which f is continuous.

37. $f(x) = \dfrac{x^2 - 4}{x - 2}$ on the interval $[-2, 2)$

38. $f(x) = 1 + \dfrac{1}{x}$ on the interval $[-1, 0)$

39. $f(x) = \dfrac{1}{\sqrt{x^2 - 9}}$ on the interval $[-3, 3]$

40. $f(x) = \sqrt{9 - x^2}$ on the interval $[-3, 3]$

In Problems 41–50, determine where each function f is continuous. First determine the domain of the function. Then support your decision using properties of continuity.

41. $f(x) = 2x^2 + 5x - \dfrac{1}{x}$

42. $f(x) = x + 1 + \dfrac{2x}{x^2 + 5}$

43. $f(x) = (x - 1)(x^2 + x + 1)$

44. $f(x) = \sqrt{x}(x^3 - 5)$

45. $f(x) = \dfrac{x - 9}{\sqrt{x} - 3}$

46. $f(x) = \dfrac{x - 4}{\sqrt{x} - 2}$

47. $f(x) = \sqrt{\dfrac{x^2 + 1}{2 - x}}$

48. $f(x) = \sqrt{\dfrac{4}{x^2 - 1}}$

49. $f(x) = (2x^2 + 5x - 3)^{2/3}$

50. $f(x) = (x + 2)^{1/2}$

In Problems 51–56, use the function

$$f(x) = \begin{cases} \sqrt{15 - 3x} & \text{if } x < 2 \\ \sqrt{5} & \text{if } x = 2 \\ 9 - x^2 & \text{if } 2 < x < 3 \\ \lfloor x - 2 \rfloor & \text{if } 3 \le x \end{cases}$$

51. Is f continuous at 0? Why or why not?

52. Is f continuous at 4? Why or why not?

53. Is f continuous at 3? Why or why not?

54. Is f continuous at 2? Why or why not?

55. Is f continuous at 1? Why or why not?

56. Is f continuous at 2.5? Why or why not?

In Problems 57 and 58:

(a) *Use technology to graph f using a suitable scale on each axis.*

(b) *Based on the graph from (a), determine where f is continuous.*

(c) *Use the definition of continuity to determine where f is continuous.*

(d) *What advice would you give a fellow student about using technology to determine where a function is continuous?*

57. $f(x) = \dfrac{x^3 - 8}{x - 2}$ **58.** $f(x) = \dfrac{x^2 - 3x + 2}{3x - 6}$

In Problems 59–64, use the Intermediate Value Theorem to determine which of the functions must have zeros in the given interval. Indicate those for which the theorem gives no information. Do not attempt to locate the zeros.

59. $f(x) = x^3 - 3x$ on $[-2, 2]$

60. $f(x) = x^4 - 1$ on $[-2, 2]$

61. $f(x) = \dfrac{x}{(x + 1)^2} - 1$ on $[10, 20]$

62. $f(x) = x^3 - 2x^2 - x + 2$ on $[3, 4]$

63. $f(x) = \dfrac{x^3 - 8}{x - 2}$ on $[0, 4]$

64. $f(x) = \dfrac{x^2 + 3x + 2}{x^2 - 1}$ on $[-3, 0]$

In Problems 65–72, verify that each function has a zero in the indicated interval. Then use the Intermediate Value Theorem to approximate the zero correct to three decimal places by repeatedly subdividing the interval containing the zero into 10 subintervals.

65. $f(x) = x^3 + 3x - 5$; interval: $[1, 2]$

66. $f(x) = x^3 - 4x + 2$; interval: $[1, 2]$

67. $f(x) = 2x^3 + 3x^2 + 4x - 1$; interval: $[0, 1]$

68. $f(x) = x^3 - x^2 - 2x + 1$; interval: $[0, 1]$

69. $f(x) = x^3 - 6x - 12$; interval: $[3, 4]$

70. $f(x) = 3x^3 + 5x - 40$; interval: $[2, 3]$

71. $f(x) = x^4 - 2x^3 + 21x - 23$; interval: $[1, 2]$

72. $f(x) = x^4 - x^3 + x - 2$; interval: $[1, 2]$

In Problems 73 and 74,

(a) *Use the Intermediate Value Theorem to show that f has a zero in the given interval.*

(b) *Use technology to find the zero rounded to three decimal places.*

73. $f(x) = \sqrt{x^2 + 4x} - 2$ in $[0, 1]$

74. $f(x) = x^3 - x + 2$ in $[-2, 0]$

Applications and Extensions

In Problems 75–78, determine whether each function is (a) continuous from the left (b) continuous from the right at the number c.

75. $f(x) = \begin{cases} x^2 & \text{if } -1 < x < 1 \\ x - 1 & \text{if } |x| \ge 1 \end{cases}$ at $c = -1$

76. $f(x) = \begin{cases} x^2 - 1 & \text{if } -1 < x < 1 \\ |x - 1| & \text{if } |x| \ge 1 \end{cases}$ at $c = -1$

77. $f(x) = \sqrt{(x + 1)(x - 5)}$ at $c = -1$

78. $f(x) = \sqrt{(x - 1)(x - 2)}$ at $c = 1$

79. First-Class Mail As of April 2023, the U.S. Postal Service charged $0.63 postage for first-class letters weighing up to and including 1 ounce, plus a flat fee of $0.24 for each additional or partial ounce up to 3.5 ounces. First-class letter rates do not apply to letters weighing more than 3.5 ounces.

Source: U.S. Postal Service Notice 123.

(a) Find a function C that models the first-class postage charged for a letter weighing w ounces. Assume $w > 0$.

(b) What is the domain of C?

(c) Determine the intervals on which C is continuous.

(d) At numbers where C is not continuous (if any), what type of discontinuity does C have?

(e) What are the practical implications of the answer to (d)?

80. First-Class Mail As of April 2023, the U.S. Postal Service charged $1.26 postage for first-class large envelopes weighing up to and including 1 ounce, plus a flat fee of $0.24 for each additional or partial ounce up to 13 ounces. First-class rates do not apply to large envelopes weighing more than 13 ounces.

Source: U.S. Postal Service Notice 123.

(a) Find a function C that models the first-class postage charged for a large envelope weighing w ounces. Assume $w > 0$.

(b) What is the domain of C?

(c) Determine the intervals on which C is continuous.

(d) At numbers where C is not continuous (if any), what type of discontinuity does C have?

(e) What are the practical implications of the answer to (d)?

81. Cost of Electricity In January 2023, Florida Power and Light charged customers living in single-family residences for their electric usage according to the following table.

Monthly customer charge for electricity:	
$8.99	per household, plus
$0.10858	for each kWH used less than or equal to 1000 kWH, plus
$0.12858	for each kWH used in excess of 1000 kWH

Source: Florida Power and Light, Miami, FL.

(a) Find a function C that models the monthly cost of using x kWH of electricity.

(b) What is the domain of C?

(c) Determine the intervals on which C is continuous.

(d) At numbers where C is not continuous (if any), what type of discontinuity does C have?

(e) What are the practical implications of the answer to (d)?

82. Cost of Water The Jericho Water District determines quarterly water costs, in dollars, using the following rate schedule:

Water used (in thousands of gallons)	Cost
$0 \le x \le 10$	$12.00
$10 < x \le 30$	$12.00 + $1.26 for each thousand gallons in excess of 10,000 gallons
$30 < x \le 100$	$37.20 + $2.40 for each thousand gallons in excess of 30,000 gallons
$x > 100$	$205.20 + $3.18 for each thousand gallons in excess of 100,000 gallons

Source: Jericho Water District, Syosset, NY.

(a) Find a function C that models the quarterly cost of using x thousand gallons of water.

(b) What is the domain of C?

(c) Determine the intervals on which C is continuous.

(d) At numbers where C is not continuous (if any), what type of discontinuity does C have?

(e) What are the practical implications of the answer to (d)?

83. Gravity on Europa

Europa, one of the larger moons of Jupiter, has an icy surface and appears to have oceans beneath the ice. This makes it a candidate for possible extraterrestrial life. Because Europa is much smaller than most planets, its gravity is weaker. If we think of Europa as a sphere with uniform

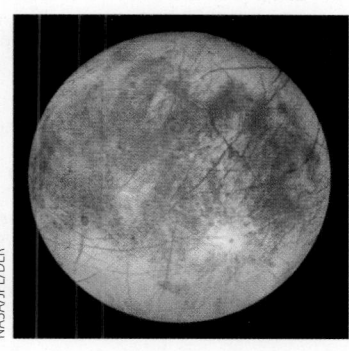

NASA/JPL/DLR

internal density, then inside the sphere, the gravitational field g is given by $g(r) = \dfrac{Gm}{R^3} r$, $0 \le r < R$, where R is the radius of the sphere, r is the distance from the center of the sphere, and G is the universal gravitation constant. Outside a uniform sphere of mass m, the gravitational field g is given by

$$g(r) = \frac{Gm}{r^2}, \quad R < r$$

(a) For the gravitational field of Europa to be continuous at its surface, what must $g(r)$ equal?
Hint: Investigate $\lim\limits_{r \to R} g(r)$.

(b) Determine the gravitational field at Europa's surface. This will indicate the type of gravity environment organisms will experience. Use the following measured values: Europa's mass is 4.8×10^{22} kilograms, its radius is 1.569×10^6 meters, and $G = 6.67 \times 10^{-11}$.

(c) Compare the result found in (b) to the gravitational field on Earth's surface, which is 9.8 meter/second². Is the gravity on Europa less than or greater than that on Earth?

84. Find constants A and B so that the function below is continuous for all x. Graph the resulting function.

$$f(x) = \begin{cases} (x-1)^2 & \text{if } -\infty < x < 0 \\ (A-x)^2 & \text{if } 0 \le x < 1 \\ x+B & \text{if } 1 \le x < \infty \end{cases}$$

85. Find constants A and B so that the function below is continuous for all x. Graph the resulting function.

$$f(x) = \begin{cases} x+A & \text{if } -\infty < x < 4 \\ (x-1)^2 & \text{if } 4 \le x \le 9 \\ Bx+1 & \text{if } 9 < x < \infty \end{cases}$$

86. For the function f below, find k so that f is continuous at 2.

$$f(x) = \begin{cases} \dfrac{\sqrt{2x+5} - \sqrt{x+7}}{x-2} & \text{if } x \ge -\dfrac{5}{2}, x \ne 2 \\ k & \text{if } x = 2 \end{cases}$$

87. Suppose $f(x) = \dfrac{x^2 - 6x - 16}{(x^2 - 7x - 8)\sqrt{x^2 - 4}}$.

(a) For what numbers x is f defined?

(b) For what numbers x is f discontinuous?

(c) Which discontinuities, if any, found in (b) are removable?

88. Intermediate Value Theorem

    (a) Use the Intermediate Value Theorem to show that the function $f(x) = \sin x + x - 3$ has a zero in the interval $[0, \pi]$.

    (b) Approximate the zero rounded to three decimal places.

89. Intermediate Value Theorem

    (a) Use the Intermediate Value Theorem to show that the function $f(x) = e^x + x - 2$ has a zero in the interval $[0, 2]$.

    (b) Approximate the zero rounded to three decimal places.

In Problems 90–93, verify that each function intersects the given line in the indicated interval. Then use the Intermediate Value Theorem to approximate the point of intersection correct to three decimal places by repeatedly subdividing the interval into 10 subintervals.

90. $f(x) = x^3 - 2x^2 - 1$; line: $y = -1$; interval: $(1, 4)$

91. $g(x) = -x^4 + 3x^2 + 3$; line: $y = 3$; interval: $(1, 2)$

92. $h(x) = \dfrac{x^3 - 5}{x^2 + 1}$; line: $y = 1$; interval: $(1, 3)$

93. $r(x) = \dfrac{x - 6}{x^2 + 2}$; line: $y = -1$; interval: $(0, 3)$

94. Graph a function that is continuous on the closed interval $[5, 12]$, is negative at both endpoints, and has exactly three distinct zeros in this interval. Does this contradict the Intermediate Value Theorem? Explain.

95. Graph a function that is continuous on the closed interval $[-1, 2]$, is positive at both endpoints, and has exactly two zeros in this interval. Does this contradict the Intermediate Value Theorem? Explain.

96. Graph a function that is continuous on the closed interval $[-2, 3]$, is positive at -2 and negative at 3, and has exactly two zeros in this interval. Is this possible? Does this contradict the Intermediate Value Theorem? Explain.

97. Graph a function that is continuous on the closed interval $[-5, 0]$, is negative at -5 and positive at 0, and has exactly three zeros in the interval. Is this possible? Does this contradict the Intermediate Value Theorem? Explain.

98. (a) Explain why the Intermediate Value Theorem gives no information about the zeros of the function $f(x) = x^4 - 1$ on the interval $[-2, 2]$.

    (b) Use technology to determine whether or not f has a zero on the interval $[-2, 2]$.

99. (a) Explain why the Intermediate Value Theorem gives no information about the zeros of the function $f(x) = \ln(x^2 + 2)$ on the interval $[-2, 2]$.

    (b) Use technology to determine whether or not f has a zero on the interval $[-2, 2]$.

100. Intermediate Value Theorem

    (a) Use the Intermediate Value Theorem to show that the functions $y = x^3$ and $y = 1 - x^2$ intersect somewhere between $x = 0$ and $x = 1$.

    (b) Use technology to find the coordinates of the point of intersection rounded to three decimal places.

    (c) Use technology to graph both functions on the same set of axes. Be sure the graph shows the point of intersection.

101. Intermediate Value Theorem An airplane is traveling at a speed of 620 miles per hour and then encounters a slight headwind that slows it to 608 miles per hour. After a few minutes, the headwind eases and the plane's speed increases to 614 miles per hour. Explain why the plane's speed is 610 miles per hour on at least two different occasions during the flight.

Source: Submitted by the students of Millikin University.

102. Suppose a function f is defined and continuous on the closed interval $[a, b]$. Is the function $h(x) = \dfrac{1}{f(x)}$ also continuous on the closed interval $[a, b]$? Discuss the continuity of h on $[a, b]$.

103. Given the two functions f and h:

$$f(x) = x^3 - 3x^2 - 4x + 12 \qquad h(x) = \begin{cases} \dfrac{f(x)}{x - 3} & \text{if } x \neq 3 \\ p & \text{if } x = 3 \end{cases}$$

    (a) Find all the zeros of the function f.

    (b) Find the number p so that the function h is continuous at $x = 3$. Justify your answer.

    (c) Determine whether h, with the number found in (b), is even, odd, or neither. Justify your answer.

104. The function $f(x) = \dfrac{|x|}{x}$ is not defined at 0. Explain why it is impossible to define $f(0)$ so that f is continuous at 0.

105. Find two functions f and g that are each continuous at c, yet $\dfrac{f}{g}$ is not continuous at c.

106. Discuss the difference between a discontinuity that is removable and one that is nonremovable. Give an example of each.

Bisection Method for Approximating Zeros of a Function *Suppose the Intermediate Value Theorem indicates that a function f has a zero in the interval (a, b). The bisection method approximates the zero by evaluating f at the midpoint m_1 of the interval (a, b). If $f(m_1) = 0$, then m_1 is the zero we seek and the process ends. If $f(m_1) \neq 0$, then the sign of $f(m_1)$ is opposite that of either $f(a)$ or $f(b)$ (but not both), and the zero lies in that subinterval. Evaluate f at the midpoint m_2 of this subinterval. Continue bisecting the subinterval containing the zero until the desired degree of accuracy is obtained.*

In Problems 107–114, use the bisection method three times to approximate the zero of each function in the given interval.

107. $f(x) = x^3 + 3x - 5$; interval: $[1, 2]$

108. $f(x) = x^3 - 4x + 2$; interval: $[1, 2]$

109. $f(x) = 2x^3 + 3x^2 + 4x - 1$; interval: $[0, 1]$

110. $f(x) = x^3 - x^2 - 2x + 1$; interval: $[0, 1]$

111. $f(x) = x^3 - 6x - 12$; interval: $[3, 4]$

112. $f(x) = 3x^3 + 5x - 40$; interval: $[2, 3]$

113. $f(x) = x^4 - 2x^3 + 21x - 23$; interval: $[1, 2]$

114. $f(x) = x^4 - x^3 + x - 2$; interval: $[1, 2]$

115. Intermediate Value Theorem Use the Intermediate Value Theorem to show that the function $f(x) = \sqrt{x^2 + 4x} - 2$ has a zero in the interval $[0, 1]$. Then approximate the zero correct to one decimal place.

116. Intermediate Value Theorem Use the Intermediate Value Theorem to show that the function $f(x) = x^3 - x + 2$ has a zero in the interval $[-2, 0]$. Then approximate the zero correct to two decimal places.

117. Continuity of a Sum   If f and g are each continuous at c, prove that $f + g$ is continuous at c.
Hint: Use the Limit of a Sum Property.

118. Intermediate Value Theorem   Suppose that the functions f and g are continuous on the interval $[a, b]$. If $f(a) < g(a)$ and $f(b) > g(b)$, prove that the graphs of $y = f(x)$ and $y = g(x)$ intersect somewhere between $x = a$ and $x = b$.
Hint: Define $h(x) = f(x) - g(x)$ and show $h(x) = 0$ for some x between a and b.

Challenge Problems

119. Intermediate Value Theorem   Let $f(x) = \dfrac{1}{x-1} + \dfrac{1}{x-2}$.

Use the Intermediate Value Theorem to prove that there is a real number c between 1 and 2 for which $f(c) = 0$.

120. Intermediate Value Theorem   Prove that there is a real number c between 2.64 and 2.65 for which $c^2 = 7$.

121. Show that the existence of $\lim\limits_{h \to 0} \dfrac{f(a+h) - f(a)}{h}$ implies f is continuous at $x = a$.

122. Find constants A, B, C, and D so that the function below is continuous for all x. Sketch the graph of the resulting function.

$$f(x) = \begin{cases} \dfrac{x^2 + x - 2}{x - 1} & \text{if } -\infty < x < 1 \\ A & \text{if } x = 1 \\ B(x - C)^2 & \text{if } 1 < x < 4 \\ D & \text{if } x = 4 \\ 2x - 8 & \text{if } 4 < x < \infty \end{cases}$$

123. Let f be a function for which $0 \le f(x) \le 1$ for all x in $[0, 1]$. If f is continuous on $[0, 1]$, show that there exists at least one number c in $[0, 1]$ such that $f(c) = c$.
Hint: Let $g(x) = x - f(x)$.

Preparing for the **AP® Exam**

AP® Practice Problems

Multiple-Choice Questions

1. (PAGE 108) The graph of a function f is shown below. Where on the open interval $(-3, 5)$ is f discontinuous?

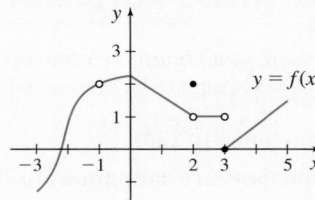

(A) 3 only   (B) −1 and 3 only

(C) 1 only   (D) −1, 2, and 3

2. (PAGE 108) The graph of a function f is shown below.

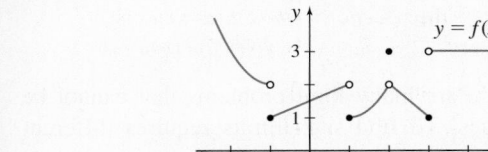

If $\lim\limits_{x \to c} f(x)$ exists and f is not continuous at c, then c equals

(A) −1   (B) 1   (C) 2   (D) 3

3. (PAGE 110) How should the function $f(x) = \dfrac{x^2 - 25}{x + 5}$ be defined at -5 to make it continuous at -5?

(A) −10   (B) −5   (C) 0   (D) 10

4. (PAGE 110) If $f(x) = \begin{cases} \dfrac{\sqrt{2x+5} - \sqrt{x+15}}{x - 10} & \text{if } x \ne 10 \\ k & \text{if } x = 10 \end{cases}$

and if f is continuous at $x = 10$, then $k =$

(A) 0   (B) $\dfrac{1}{10}$ (C) 1   (D) 10

5. (PAGE 109) If $\lim\limits_{x \to c} f(x) = L$, where L is a real number, which of the following must be true?

(A) f is defined at $x = c$.   (B) f is continuous at $x = c$.

(C) $f(c) = L$.   (D) None of the above.

6. (PAGE 116) If $f(x) = x^3 - 2x + 5$ and if $f(c) = 0$ for only one real number c, then c is between

(A) −4 and −2   (B) −2 and −1

(C) −1 and 1   (D) 1 and 3

7. (PAGE 116) The function f is continuous at all real numbers, and $f(-8) = 3$ and $f(-1) = -4$. If f has only one real zero (root), then which number x could satisfy $f(x) = 0$?

(A) −10   (B) −5   (C) 0   (D) 2

8. (PAGE 110) Let f be the function defined by

$$f(x) = \begin{cases} x^2 & \text{if } x < 0 \\ \sqrt{x} & \text{if } 0 \le x < 1 \\ 2 - x & \text{if } 1 \le x < 2 \\ x - 3 & \text{if } x \ge 2 \end{cases}$$

For what numbers x is f NOT continuous?

(A) 1 only   (B) 2 only

(C) 0 and 2 only   (D) 1 and 2 only

9. (PAGE 116) The function f is continuous on the closed interval $[-2, 6]$. If $f(-2) = 7$ and $f(6) = -1$, then the Intermediate Value Theorem guarantees that

(A) $f(0) = 0$.

(B) $f(c) = 2$ for at least one number c between −2 and 6.

(C) $f(c) = 0$ for at least one number c between −1 and 7.

(D) $-1 \le f(x) \le 7$ for all numbers in the closed interval $[-2, 6]$.

(continued on the next page)

10. The function f is continuous on the closed interval $[-2, 2]$. Several values of the function f are given in the table below.

x	-2	0	2
$f(x)$	3	c	2

The equation $f(x) = 1$ must have at least two solutions in the interval $[-2, 2]$ if $c =$

(A) $\dfrac{1}{2}$ (B) 1   (C) 3   (D) 4

11. $f(x) = \dfrac{3x}{x-4}$ if $x \neq 0$, $x \neq 4$, and $f(0) = 5$ and $f(4) = 1$, which of the following must be true?

I. The domain of f is all real numbers.
II. f is continuous for $-2 < x < 2$.
III. f has a removable discontinuity at 0.

(A) II only   (B) III only
(C) I and II only   (D) I and III only

Free-Response Questions

12. (a) Use the Intermediate Value Theorem to show that the graphs of the functions $f(x) = x^3$ and $g(x) = 1 - x^2$ intersect at some point in the interval $0 \leq x \leq 1$.
    (b) Find the point of intersection of the functions f and g.

13. The function f is defined by $f(x) = \begin{cases} x^2 - 2x + 3 & \text{if } x \leq 1 \\ -2x + 5 & \text{if } x > 1 \end{cases}$

(a) Is f continuous at $x = 1$?
(b) Use the definition of continuity to explain your answer.

1.4 Limits and Continuity of Trigonometric, Exponential, and Logarithmic Functions

OBJECTIVES *When you finish this section, you should be able to:*

1 **Use the Squeeze Theorem to find a limit (p. 122)**
2 **Find limits involving trigonometric functions (p. 124)**
3 **Determine where the trigonometric functions are continuous (p. 127)**
4 **Determine where an exponential or a logarithmic function is continuous (p. 129)**

Until now we have found limits using the basic limits

$$\lim_{x \to c} A = A \qquad \lim_{x \to c} x = c$$

and algebraic properties of limits. But there are many limit problems that cannot be found by directly applying these techniques. To find such limits requires different results, such as the *Squeeze Theorem*,* or basic limits involving trigonometric and exponential functions.

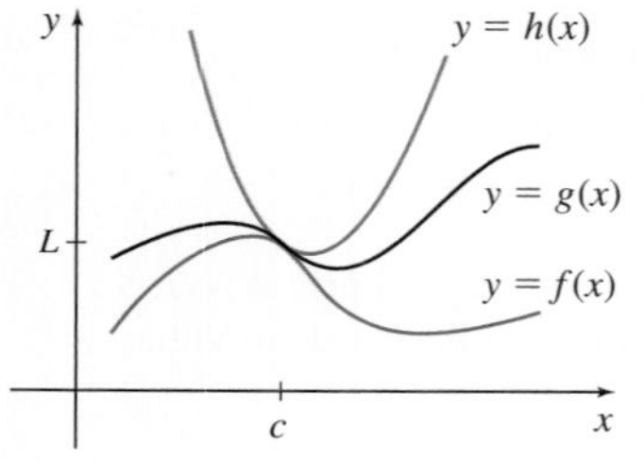

1 Use the Squeeze Theorem to Find a Limit

To use the Squeeze Theorem to find $\lim_{x \to c} g(x)$, we need to know, or be able to find, two functions f and h that "sandwich" the function g between them for all x close to c. That is, in some interval containing c, the functions f, g, and h satisfy the inequality

$$f(x) \leq g(x) \leq h(x)$$

Then if f and h have the same limit L as x approaches c, the function g is "squeezed" to the same limit L as x approaches c. See Figure 38.

$\lim_{x \to c} f(x) = L$, $\lim_{x \to c} h(x) = L$, $\lim_{x \to c} g(x) = L$

Figure 38

---

*The Squeeze Theorem is also known as the Sandwich Theorem and the Pinching Theorem.

We state the Squeeze Theorem here. The proof is given in Appendix B pages B-3 to B-4.

THEOREM   Squeeze Theorem

Suppose the functions f, g, and h have the property that for all x in an open interval containing c, except possibly at c,

$$\boxed{f(x) \leq g(x) \leq h(x)}$$

If

$$\boxed{\lim_{x \to c} f(x) = \lim_{x \to c} h(x) = L}$$

then

$$\boxed{\lim_{x \to c} g(x) = L}$$

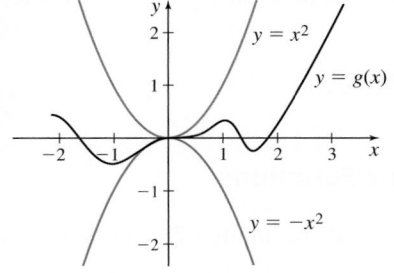

Figure 39

For example, suppose we wish to find $\lim\limits_{x \to 0} g(x)$, and we know that $-x^2 \leq g(x) \leq x^2$ for all $x \neq 0$. Since $\lim\limits_{x \to 0}(-x^2) = 0$ and $\lim\limits_{x \to 0} x^2 = 0$, by the Squeeze Theorem it follows that $\lim\limits_{x \to 0} g(x) = 0$. Figure 39 shows how g is "squeezed" between $y = x^2$ and $y = -x^2$ near 0.

EXAMPLE 1   **Using the Squeeze Theorem to Find a Limit**

Use the Squeeze Theorem to find $\lim\limits_{x \to 0}\left(x \sin \dfrac{1}{x} \right)$.

Solution

If $x \neq 0$, then $g(x) = x \sin \dfrac{1}{x}$ is defined. We seek two functions that "squeeze" $g(x) = x \sin \dfrac{1}{x}$ near 0. Since $-1 \leq \sin x \leq 1$ for all x, we begin with the inequality

$$\left| \sin \frac{1}{x} \right| \leq 1 \qquad x \neq 0$$

Since $x \neq 0$ and we seek to squeeze $g(x) = x \sin \dfrac{1}{x}$, we multiply both sides of the inequality by $|x|$, $x \neq 0$. Since $|x| > 0$, the direction of the inequality is preserved. [Note that if we multiply $\left| \sin \dfrac{1}{x} \right| \leq 1$ by x, we would not know whether the inequality symbol would remain the same or be reversed since we do not know whether $x > 0$ or $x < 0$.]

$$|x| \left| \sin \frac{1}{x} \right| \leq |x| \qquad \text{Multiply both sides by } |x| > 0.$$

$$\left| x \sin \frac{1}{x} \right| \leq |x| \qquad |a| \cdot |b| = |a\,b|.$$

$$-|x| \leq x \sin \frac{1}{x} \leq |x| \qquad |a| \leq b \text{ is equivalent to } -b \leq a \leq b.$$

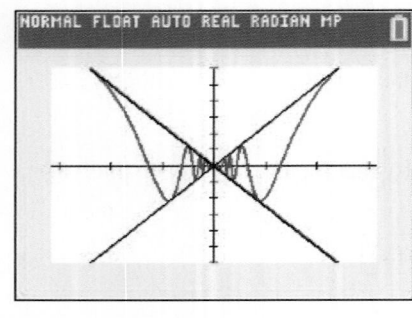

$$\left[-\frac{\pi}{4}, \frac{\pi}{4}\right] \times [-0.6, 0.6]$$

Figure 40 $g(x) = x \sin \dfrac{1}{x}$ is squeezed between $f(x) = -|x|$ and $h(x) = |x|$.

Now use the Squeeze Theorem with $f(x) = -|x|$, $g(x) = x \sin \dfrac{1}{x}$, and $h(x) = |x|$. Since $f(x) \le g(x) \le h(x)$ and

$$\lim_{x \to 0} f(x) = \lim_{x \to 0} (-|x|) = 0 \qquad \text{and} \qquad \lim_{x \to 0} h(x) = \lim_{x \to 0} |x| = 0$$

it follows that

$$\lim_{x \to 0} g(x) = \lim_{x \to 0} \left(x \sin \frac{1}{x} \right) = 0 \qquad\blacksquare$$

Figure 40 illustrates how $g(x) = x \sin \dfrac{1}{x}$ is squeezed between $y = -|x|$ and $y = |x|$.

NOW WORK Problems **5, 47,** and AP® Practice Problem **11.**

2 Find Limits Involving Trigonometric Functions

Knowing $\lim\limits_{x \to c} A = A$ and $\lim\limits_{x \to c} x = c$ helped us to find the limits of many algebraic functions. Knowing several basic trigonometric limits can help to find many limits involving trigonometric functions.

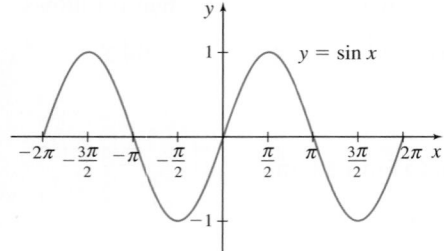

THEOREM Two Basic Trigonometric Limits

$$\lim_{x \to 0} \sin x = 0 \qquad \lim_{x \to 0} \cos x = 1$$

The graphs of $y = \sin x$ and $y = \cos x$ in Figure 41 suggest that $\lim\limits_{x \to 0} \sin x = 0$ and $\lim\limits_{x \to 0} \cos x = 1$. The proofs of these limits use the Squeeze Theorem. Problem 63 provides an outline of the proof that $\lim\limits_{x \to 0} \sin x = 0$.

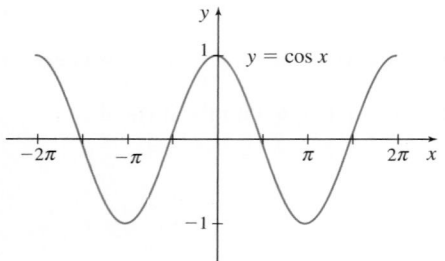

Figure 41

NOW WORK AP® Practice Problem **8.**

A third basic trigonometric limit, $\lim\limits_{\theta \to 0} \dfrac{\sin \theta}{\theta} = 1$, is important in calculus. In Section 1.1, a table suggested that $\lim\limits_{\theta \to 0} \dfrac{\sin \theta}{\theta} = 1$. The function $f(\theta) = \dfrac{\sin \theta}{\theta}$, whose graph is shown in Figure 42, is defined for all real numbers $\theta \ne 0$. The graph suggests $\lim\limits_{\theta \to 0} \dfrac{\sin \theta}{\theta} = 1$.

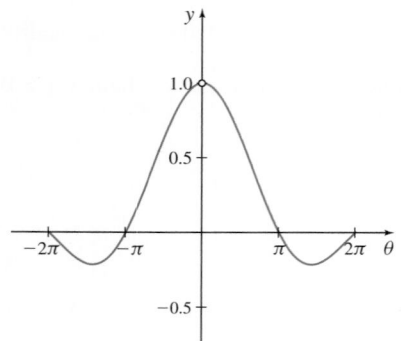

Figure 42 $f(\theta) = \dfrac{\sin \theta}{\theta}$

THEOREM

If θ is measured in radians, then

$$\lim_{\theta \to 0} \frac{\sin \theta}{\theta} = 1$$

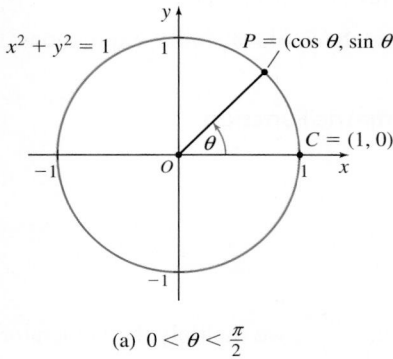

(a) $0 < \theta < \dfrac{\pi}{2}$

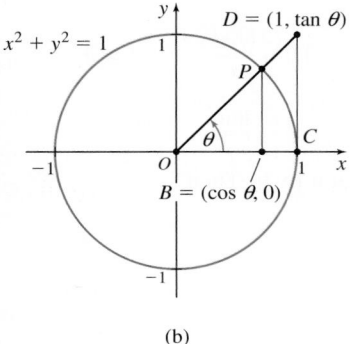

(b)

Figure 43

Proof Although $\dfrac{\sin \theta}{\theta}$ is a quotient, there is no way to divide out θ. To find $\lim\limits_{\theta \to 0^+} \dfrac{\sin \theta}{\theta}$, we let θ be a positive acute central angle of a unit circle, as shown in Figure 43(a). Notice that COP is a sector of the circle.

See Figure 43(b). We add the point $B = (\cos \theta, 0)$ to the graph and form triangle BOP. Next we extend the terminal side of angle θ until it intersects the line $x = 1$ at the point D, forming a second triangle COD. The x-coordinate of D is 1. Since the length of the line segment $\overline{OC}$ is $OC = 1$, then the y-coordinate of D is $CD = \dfrac{CD}{OC} = \tan \theta$. So, $D = (1, \tan \theta)$.

We see from Figure 43(b) that

$$\text{area of triangle } BOP \leq \text{area of sector } COP \leq \text{area of triangle } COD \qquad (1)$$

Each of these areas can be expressed in terms of θ as

- area of triangle $BOP = \dfrac{1}{2} \cos \theta \cdot \sin \theta$ $\qquad$ area of a triangle $= \dfrac{1}{2}$ base $\times$ height

- area of sector $COP = \dfrac{\theta}{2}$ $\qquad$ area of a sector $= \dfrac{1}{2} r^2 \theta; r = 1$

- area of triangle $COD = \dfrac{1}{2} \cdot 1 \cdot \tan \theta = \dfrac{\tan \theta}{2}$ $\qquad$ area of a triangle $= \dfrac{1}{2}$ base $\times$ height

So, for $0 < \theta < \dfrac{\pi}{2}$,

$$\dfrac{1}{2} \cos \theta \cdot \sin \theta \leq \dfrac{\theta}{2} \leq \dfrac{\tan \theta}{2} \qquad \text{From (1)}$$

$$\cos \theta \cdot \sin \theta \leq \theta \leq \dfrac{\sin \theta}{\cos \theta} \qquad \text{Multiply all parts by 2; } \tan \theta = \dfrac{\sin \theta}{\cos \theta}.$$

$$\cos \theta \leq \dfrac{\theta}{\sin \theta} \leq \dfrac{1}{\cos \theta} \qquad \text{Divide by } \sin \theta; \text{ since } 0 < \theta < \dfrac{\pi}{2}, \sin \theta > 0.$$

$$\dfrac{1}{\cos \theta} \geq \dfrac{\sin \theta}{\theta} \geq \cos \theta \qquad \begin{array}{l}\text{Invert each term in the inequality;}\\ \text{reverse the inequality signs.}\end{array}$$

$$\cos \theta \leq \dfrac{\sin \theta}{\theta} \leq \dfrac{1}{\cos \theta} \qquad \text{Rewrite using } \leq.$$

Now use the Squeeze Theorem with $f(\theta) = \cos \theta$, $g(\theta) = \dfrac{\sin \theta}{\theta}$, and $h(\theta) = \dfrac{1}{\cos \theta}$. Since $f(\theta) \leq g(\theta) \leq h(\theta)$, and since

$$\lim_{\theta \to 0^+} f(\theta) = \lim_{\theta \to 0^+} \cos \theta = 1 \quad \text{and} \quad \lim_{\theta \to 0^+} h(\theta) = \lim_{\theta \to 0^+} \dfrac{1}{\cos \theta} = \dfrac{\lim\limits_{\theta \to 0^+} 1}{\lim\limits_{\theta \to 0^+} \cos \theta} = \dfrac{1}{1} = 1$$

then by the Squeeze Theorem, $\lim\limits_{\theta \to 0^+} g(\theta) = \lim\limits_{\theta \to 0^+} \dfrac{\sin \theta}{\theta} = 1$.

Now use the fact that $g(\theta) = \dfrac{\sin \theta}{\theta}$ is an even function, that is,

$$g(-\theta) = \dfrac{\sin(-\theta)}{-\theta} = \dfrac{-\sin \theta}{-\theta} = \dfrac{\sin \theta}{\theta} = g(\theta)$$

So,

$$\lim_{\theta \to 0^-} \dfrac{\sin \theta}{\theta} = \lim_{\theta \to 0^+} \dfrac{\sin(-\theta)}{-\theta} = \lim_{\theta \to 0^+} \dfrac{\sin \theta}{\theta} = 1$$

It follows that $\lim\limits_{\theta \to 0} \dfrac{\sin \theta}{\theta} = 1$. ∎

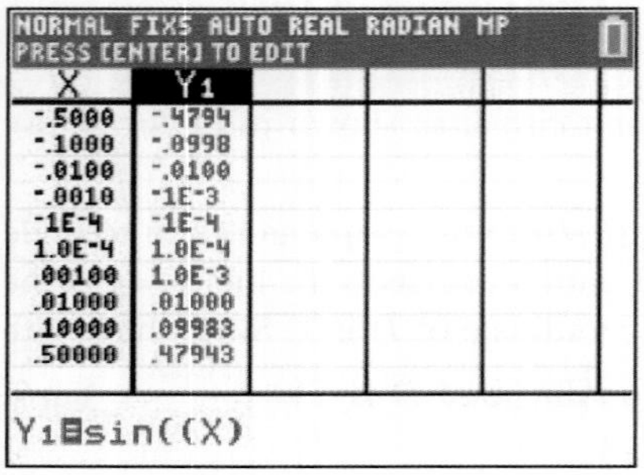

```
NORMAL FIX5 AUTO REAL RADIAN MP        0
PRESS [ENTER] TO EDIT
   X       Y1
 -.5000   -.4794
 -.1000   -.0998
 -.0100   -.0100
 -.0010   -1E-3
 -1E-4    -1E-4
 1.0E-4   1.0E-4
 .00100   1.0E-3
 .01000   .01000
 .10000   .09983
 .50000   .47943

Y1=sin((X)
```

Figure 44

Since $\lim\limits_{\theta \to 0} \dfrac{\sin \theta}{\theta} = 1$, the ratio $\dfrac{\sin \theta}{\theta}$ is close to 1 for values of θ close to 0. That is, $\sin \theta \approx \theta$ for values of θ close to 0. Figure 44 illustrates this property.

The basic limit $\lim\limits_{\theta \to 0} \dfrac{\sin \theta}{\theta} = 1$ can be used to find the limits of similar expressions.

EXAMPLE 2   Finding the Limit of a Trigonometric Function

Find:

(a) $\lim\limits_{\theta \to 0} \dfrac{\sin(3\theta)}{\theta}$ (b) $\lim\limits_{\theta \to 0} \dfrac{\sin(5\theta)}{\sin(2\theta)}$

Solution

(a) Since $\lim\limits_{\theta \to 0} \dfrac{\sin(3\theta)}{\theta}$ is not in the same form as $\lim\limits_{\theta \to 0} \dfrac{\sin \theta}{\theta}$, we multiply the numerator and the denominator by 3, and make the substitution $t = 3\theta$.

$$\lim\limits_{\theta \to 0} \frac{\sin(3\theta)}{\theta} = \lim\limits_{\theta \to 0} \frac{3\sin(3\theta)}{3\theta} \underset{\substack{\uparrow \\ t = 3\theta \\ t \to 0 \text{ as } \theta \to 0}}{=} \lim\limits_{t \to 0} \left(3 \frac{\sin t}{t}\right) = 3 \lim\limits_{t \to 0} \frac{\sin t}{t} \underset{\substack{\uparrow \\ \lim\limits_{t \to 0} \frac{\sin t}{t} = 1}}{=} 3 \cdot 1 = 3$$

(b) Begin by dividing the numerator and the denominator by θ. Then

$$\frac{\sin(5\theta)}{\sin(2\theta)} = \frac{\dfrac{\sin(5\theta)}{\theta}}{\dfrac{\sin(2\theta)}{\theta}}$$

Now follow the approach in (a) on the numerator and on the denominator.

$$\lim\limits_{\theta \to 0} \frac{\sin(5\theta)}{\theta} = \lim\limits_{\theta \to 0} \frac{5\sin(5\theta)}{5\theta} \underset{\substack{\uparrow \\ t = 5\theta \\ t \to 0 \text{ as } \theta \to 0}}{=} \lim\limits_{t \to 0} \frac{5\sin t}{t} = 5 \lim\limits_{t \to 0} \left(\frac{\sin t}{t}\right) = 5$$

$$\lim\limits_{\theta \to 0} \frac{\sin(2\theta)}{\theta} = \lim\limits_{\theta \to 0} \frac{2\sin(2\theta)}{2\theta} \underset{\substack{\uparrow \\ t = 2\theta \\ t \to 0 \text{ as } \theta \to 0}}{=} \lim\limits_{t \to 0} \left(\frac{2\sin t}{t}\right) = 2 \lim\limits_{t \to 0} \frac{\sin t}{t} = 2$$

$$\lim\limits_{\theta \to 0} \frac{\sin(5\theta)}{\sin(2\theta)} = \frac{\lim\limits_{\theta \to 0} \dfrac{\sin(5\theta)}{\theta}}{\lim\limits_{\theta \to 0} \dfrac{\sin(2\theta)}{\theta}} = \frac{5}{2}$$

∎

NOW WORK   Problems 23 and 25 and AP® Practice Problems 1, 2, 4, 6, 7, and 10.

Example 3 establishes an important limit used in Chapter 2.

EXAMPLE 3   Finding a Basic Trigonometric Limit

Establish the limit

$$\boxed{\lim\limits_{\theta \to 0} \frac{\cos \theta - 1}{\theta} = 0}$$

where θ is measured in radians.

Solution

First we rewrite the expression $\dfrac{\cos\theta - 1}{\theta}$ as the product of two terms whose limits are known. For $\theta \neq 0$,

$$\frac{\cos\theta - 1}{\theta} = \frac{\cos\theta - 1}{\theta} \cdot \frac{\cos\theta + 1}{\cos\theta + 1}$$

$$= \underbrace{\frac{\cos^2\theta - 1}{\theta(\cos\theta + 1)}}_{\uparrow} = \frac{-\sin^2\theta}{\theta(\cos\theta + 1)} = \frac{\sin\theta}{\theta} \cdot \frac{(-\sin\theta)}{\cos\theta + 1}$$

$$\sin^2\theta + \cos^2\theta = 1$$

Now we find the limit.

$$\lim_{\theta\to 0}\frac{\cos\theta - 1}{\theta} = \lim_{\theta\to 0}\left[\frac{\sin\theta}{\theta} \cdot \frac{(-\sin\theta)}{\cos\theta + 1}\right] = \left[\lim_{\theta\to 0}\frac{\sin\theta}{\theta}\right] \cdot \left[\lim_{\theta\to 0}\frac{-\sin\theta}{\cos\theta + 1}\right]$$

$$= 1 \cdot \frac{\lim\limits_{\theta\to 0}(-\sin\theta)}{\lim\limits_{\theta\to 0}(\cos\theta + 1)} = \frac{0}{2} = 0 \qquad \blacksquare$$

NOW WORK Problem 31.

---

③ Determine Where the Trigonometric Functions Are Continuous

The graphs of $f(x) = \sin x$ and $g(x) = \cos x$ in Figure 45 suggest that f and g are continuous on their domains, the set of all real numbers.

EXAMPLE 4 **Showing $f(x) = \sin x$ Is Continuous at 0**

- $f(0) = \sin 0 = 0$, so f is defined at 0.
- $\lim\limits_{x\to 0} f(x) = \lim\limits_{x\to 0}\sin x = 0$, so the limit at 0 exists.
- $\lim\limits_{x\to 0}\sin x = \sin 0 = f(0)$.

Since all three conditions of continuity are satisfied, $f(x) = \sin x$ is continuous at 0. ■

In a similar way, we can show that $g(x) = \cos x$ is continuous at 0. That is, $\lim\limits_{x\to 0}\cos x = \cos 0 = 1$.

You are asked to prove the following theorem in Problem 67.

THEOREM

- The sine function $y = \sin x$ is continuous on its domain, all real numbers.
- The cosine function $y = \cos x$ is continuous on its domain, all real numbers.

Based on this theorem the following two limits can be added to the list of basic limits.

$$\lim_{x\to c}\sin x = \sin c \qquad \text{for all real numbers } c$$

$$\lim_{x\to c}\cos x = \cos c \qquad \text{for all real numbers } c$$

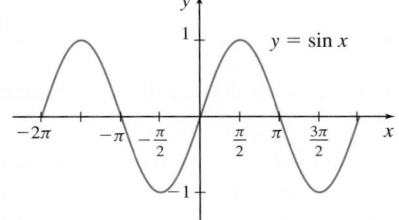

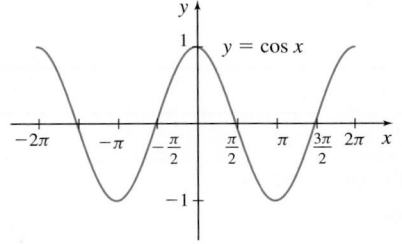

Figure 45

IN WORDS The limit of a sine or cosine function can be found using substitution.

Using the facts that the sine and cosine functions are continuous for all real numbers, we can use basic trigonometric identities to determine where the remaining four trigonometric functions are continuous:

- $y = \tan x$: Since $\tan x = \dfrac{\sin x}{\cos x}$, from the Continuity of a Quotient property, $y = \tan x$ is continuous at all real numbers except those for which $\cos x = 0$. That is, $y = \tan x$ is continuous on its domain, all real numbers except odd multiples of $\dfrac{\pi}{2}$.

- $y = \sec x$: Since $\sec x = \dfrac{1}{\cos x}$, from the Continuity of a Quotient property, $y = \sec x$ is continuous at all real numbers except those for which $\cos x = 0$. That is, $y = \sec x$ is continuous on its domain, all real numbers except odd multiples of $\dfrac{\pi}{2}$.

- $y = \cot x$: Since $\cot x = \dfrac{\cos x}{\sin x}$, from the Continuity of a Quotient property, $y = \cot x$ is continuous at all real numbers except those for which $\sin x = 0$. That is, $y = \cot x$ is continuous on its domain, all real numbers except integer multiples of π.

- $y = \csc x$: Since $\csc x = \dfrac{1}{\sin x}$, from the Continuity of a Quotient property, $y = \csc x$ is continuous at all real numbers except those for which $\sin x = 0$. That is, $y = \csc x$ is continuous on its domain, all real numbers except integer multiples of π.

Recall that a one-to-one function that is continuous on its domain has an inverse function that is continuous on its domain.

NEED TO REVIEW? Inverse trigonometric functions are discussed in Section P.7, pp. 63–69.

Since each of the six trigonometric functions is continuous on its domain, then each is continuous on the restricted domain used to define its inverse trigonometric function. This means the inverse trigonometric functions are continuous on their domains. These results are summarized in Table 10.

TABLE 10 Continuity of the Trigonometric Functions and Their Inverses

Name	Function	Domain	Properties
Sine	$f(x) = \sin x$	all real numbers	continuous on the interval $(-\infty, \infty)$
Cosine	$f(x) = \cos x$	all real numbers	continuous on the interval $(-\infty, \infty)$
Tangent	$f(x) = \tan x$	$\left\{x \mid x \neq \text{odd multiples of } \dfrac{\pi}{2}\right\}$	continuous at all real numbers except odd multiples of $\dfrac{\pi}{2}$
Cosecant	$f(x) = \csc x$	$\{x \mid x \neq \text{multiples of } \pi\}$	continuous at all real numbers except multiples of π
Secant	$f(x) = \sec x$	$\left\{x \mid x \neq \text{odd multiples of } \dfrac{\pi}{2}\right\}$	continuous at all real numbers except odd multiples of $\dfrac{\pi}{2}$
Cotangent	$f(x) = \cot x$	$\{x \mid x \neq \text{multiples of } \pi\}$	continuous at all real numbers except multiples of π
Inverse sine	$f(x) = \sin^{-1} x$	$-1 \leq x \leq 1$	continuous on the closed interval $[-1, 1]$
Inverse cosine	$f(x) = \cos^{-1} x$	$-1 \leq x \leq 1$	continuous on the closed interval $[-1, 1]$
Inverse tangent	$f(x) = \tan^{-1} x$	all real numbers	continuous on the interval $(-\infty, \infty)$
Inverse cosecant	$f(x) = \csc^{-1} x$	$|x| \geq 1$	continuous on the set $(-\infty, -1] \cup [1, \infty)$
Inverse secant	$f(x) = \sec^{-1} x$	$|x| \geq 1$	continuous on the set $(-\infty, -1] \cup [1, \infty)$
Inverse cotangent	$f(x) = \cot^{-1} x$	all real numbers	continuous on the interval $(-\infty, \infty)$

④ Determine Where an Exponential or a Logarithmic Function Is Continuous

The graphs of an exponential function $y = a^x$ and its inverse function $y = \log_a x$ are shown in Figure 46. The graphs suggest that an exponential function and a logarithmic function are continuous on their domains. We state the following theorem without proof.

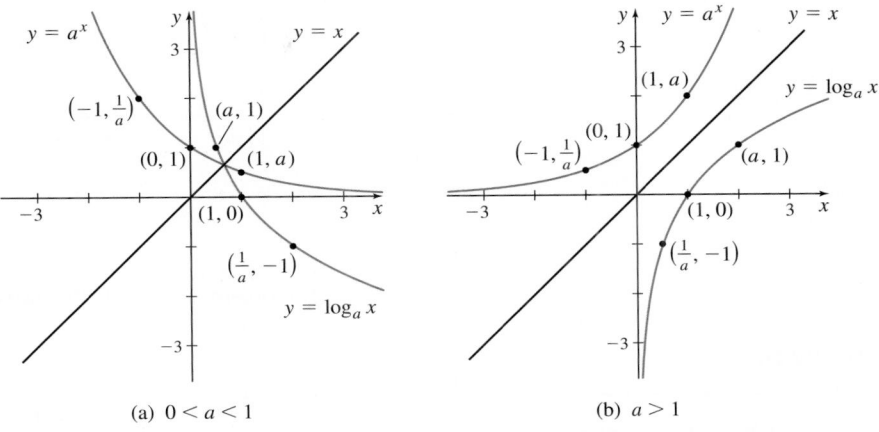

(a) $0 < a < 1$ (b) $a > 1$

Figure 46

THEOREM Continuity of Exponential and Logarithmic Functions

- An exponential function is continuous on its domain, all real numbers.
- A logarithmic function is continuous on its domain, all positive real numbers.

Based on this theorem, the following two limits can be added to the list of basic limits.

IN WORDS The limit of an exponential function or a logarithmic function can be found using substitution.

$$\lim_{x \to c} a^x = a^c \qquad \text{for all real numbers } c, \ a > 0, \ a \neq 1$$

and

$$\lim_{x \to c} \log_a x = \log_a c \qquad \text{for any real number } c > 0, \ a > 0, \ a \neq 1$$

▶ CALC CLIP

EXAMPLE 5 Showing a Composite Function Is Continuous

Show that:

(a) $f(x) = e^{2x}$ is continuous for all real numbers.

(b) $F(x) = \sqrt[3]{\ln x}$ is continuous for $x > 0$.

Solution

(a) The domain of the exponential function is the set of all real numbers, so f is defined for any number c. That is, $f(c) = e^{2c}$. Also for any number c,

$$\lim_{x \to c} f(x) = \lim_{x \to c} e^{2x} = \lim_{x \to c} (e^x)^2 = \left[\lim_{x \to c} e^x\right]^2 = (e^c)^2 = e^{2c} = f(c)$$

Since $\lim_{x \to c} f(x) = f(c)$ for any number c, then f is continuous at all numbers c.

(b) The logarithmic function $f(x) = \ln x$ is continuous on its domain, the set of all positive real numbers. The function $g(x) = \sqrt[3]{x}$ is continuous on its domain, the set of all real numbers. Then for any real number $c > 0$, the composite function $F(x) = (g \circ f)(x) = \sqrt[3]{\ln x}$ is continuous at c. That is, F is continuous at all real numbers $x > 0$. ∎

Figures 47(a) and 47(b) illustrate the graphs of f and F from Example 5.

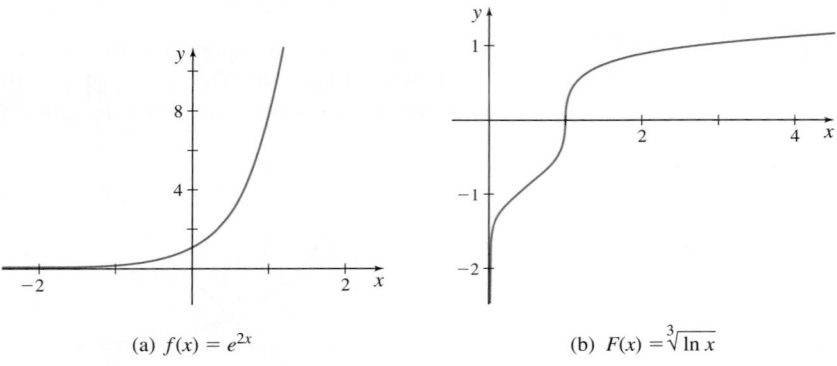

(a) $f(x) = e^{2x}$ (b) $F(x) = \sqrt[3]{\ln x}$

Figure 47

NOW WORK   Problem **45** and AP® Practice Problems **3, 5, 9,** and **12.**

Summary Basic Limits

- $\lim\limits_{x \to c} A = A$, where A is a constant
- $\lim\limits_{x \to c} x = c$
- $\lim\limits_{x \to 0} \sin x = 0$
- $\lim\limits_{x \to 0} \cos x = 1$

- $\lim\limits_{x \to c} \sin x = \sin c$
- $\lim\limits_{x \to c} \cos x = \cos c$
- $\lim\limits_{\theta \to 0} \dfrac{\sin \theta}{\theta} = 1$
- $\lim\limits_{\theta \to 0} \dfrac{\cos \theta - 1}{\theta} = 0$

- $\lim\limits_{x \to c} a^x = a^c, a > 0, a \neq 1$
- $\lim\limits_{x \to c} \log_a x = \log_a c, \ c > 0, a > 0, a \neq 1$

1.4 Assess Your Understanding

Concepts and Vocabulary

1. $\lim\limits_{x \to 0} \sin x = $ _____

2. **True or False** $\lim\limits_{x \to 0} \dfrac{\cos x - 1}{x} = 1$

3. The Squeeze Theorem states that if the functions f, g, and h have the property $f(x) \leq g(x) \leq h(x)$ for all x in an open interval containing c, except possibly at c, and if $\lim\limits_{x \to c} f(x) = \lim\limits_{x \to c} h(x) = L$, then $\lim\limits_{x \to c} g(x) = $ _____.

4. **True or False** $f(x) = \csc x$ is continuous for all real numbers except $x = 0$.

Skill Building

In Problems 5–8, use the Squeeze Theorem to find each limit.

 5. Suppose $-x^2 + 1 \leq g(x) \leq x^2 + 1$ for all x in an open interval containing 0. Find $\lim\limits_{x \to 0} g(x)$.

6. Suppose $-(x - 2)^2 - 3 \leq g(x) \leq (x - 2)^2 - 3$ for all x in an open interval containing 2. Find $\lim\limits_{x \to 2} g(x)$.

7. Suppose $\cos x \leq g(x) \leq 1$ for all x in an open interval containing 0. Find $\lim\limits_{x \to 0} g(x)$.

8. Suppose $-x^2 + 1 \leq g(x) \leq \sec x$ for all x in an open interval containing 0. Find $\lim\limits_{x \to 0} g(x)$.

In Problems 9–22, find each limit.

9. $\lim\limits_{x \to 0} (x^3 + \sin x)$

10. $\lim\limits_{x \to 0} (x^2 - \cos x)$

11. $\lim\limits_{x \to \pi/3} (\cos x + \sin x)$

12. $\lim\limits_{x \to \pi/3} (\sin x - \cos x)$

13. $\lim\limits_{x \to 0} \dfrac{\cos x}{1 + \sin x}$

14. $\lim\limits_{x \to 0} \dfrac{\sin x}{1 + \cos x}$

15. $\lim\limits_{x \to 0} \dfrac{6}{2 + e^x}$

16. $\lim\limits_{x \to 0} \dfrac{e^x - 1}{e^x + 4}$

17. $\lim\limits_{x \to 0} (e^x \sin x)$

18. $\lim\limits_{x \to 0} (e^{-x} \tan x)$

19. $\lim\limits_{x \to 1} \ln \left(\dfrac{e^x}{x} \right)$

20. $\lim\limits_{x \to 1} \ln \left(\dfrac{x}{e^x} \right)$

21. $\lim\limits_{x \to 0} \dfrac{e^{2x}}{1 + e^x}$

22. $\lim\limits_{x \to 0} \dfrac{1 - e^x}{1 - e^{2x}}$

In Problems 23–34, find each limit.

23. $\lim\limits_{x \to 0} \dfrac{\sin(7x)}{x}$

24. $\lim\limits_{x \to 0} \dfrac{\sin \dfrac{x}{3}}{x}$

25. $\lim\limits_{\theta \to 0} \dfrac{\theta + 3 \sin \theta}{2\theta}$

26. $\lim\limits_{x \to 0} \dfrac{2x - 5 \sin(3x)}{x}$

27. $\lim\limits_{\theta \to 0} \dfrac{\sin \theta}{\theta + \tan \theta}$

28. $\lim\limits_{\theta \to 0} \dfrac{\tan \theta}{\theta}$

29. $\lim\limits_{\theta \to 0} \dfrac{5}{\theta \cdot \csc \theta}$

30. $\lim\limits_{\theta \to 0} \dfrac{\sin(3\theta)}{\sin(2\theta)}$

31. $\lim\limits_{\theta \to 0} \dfrac{1 - \cos^2 \theta}{\theta}$

32. $\lim\limits_{\theta \to 0} \dfrac{\cos(4\theta) - 1}{2\theta}$

33. $\lim\limits_{\theta \to 0} (\theta \cdot \cot \theta)$

34. $\lim\limits_{\theta \to 0} \left[\sin \theta \cdot \dfrac{\cot \theta - \csc \theta}{\theta} \right]$

In Problems 35–38, determine whether f is continuous at the number c.

35. $f(x) = \begin{cases} 3\cos x & \text{if } x < 0 \\ 3 & \text{if } x = 0 \\ x+3 & \text{if } x > 0 \end{cases}$ at $c = 0$

36. $f(x) = \begin{cases} \cos x & \text{if } x < 0 \\ 0 & \text{if } x = 0 \\ e^x & \text{if } x > 0 \end{cases}$ at $c = 0$

37. $f(\theta) = \begin{cases} \sin\theta & \text{if } \theta \le \dfrac{\pi}{4} \\ \cos\theta & \text{if } \theta > \dfrac{\pi}{4} \end{cases}$ at $c = \dfrac{\pi}{4}$

38. $f(x) = \begin{cases} \tan^{-1} x & \text{if } x < 1 \\ \ln x & \text{if } x \ge 1 \end{cases}$ at $c = 1$

In Problems 39–46, determine where f is continuous.

39. $f(x) = \cos\left(\dfrac{x^2 - 4x}{x - 4}\right)$ **40.** $f(x) = \sin\left(\dfrac{x^2 - 5x + 1}{2x}\right)$

41. $f(\theta) = \dfrac{1}{1 + \sin\theta}$ **42.** $f(\theta) = \dfrac{1}{1 + \cos^2\theta}$

43. $f(x) = \dfrac{\ln x}{x - 3}$ **44.** $f(x) = \ln(x^2 + 1)$

45. $f(x) = e^x \cos x$ **46.** $f(x) = \dfrac{1 + \cos^2 x}{e^x}$

Applications and Extensions

In Problems 47–50, use the Squeeze Theorem to find each limit.

47. $\lim\limits_{x \to 0}\left(x^2 \sin\dfrac{1}{x}\right)$ **48.** $\lim\limits_{x \to 0}\left[x\left(1 - \cos\dfrac{1}{x}\right)\right]$

49. $\lim\limits_{x \to 0}\left[x^2\left(1 - \cos\dfrac{1}{x}\right)\right]$ **50.** $\lim\limits_{x \to 0}\left[\sqrt{x^3 + 3x^2}\,\sin\dfrac{1}{x}\right]$

In Problems 51–54, show that each statement is true.

51. $\lim\limits_{x \to 0}\dfrac{\sin(ax)}{\sin(bx)} = \dfrac{a}{b};\quad b \ne 0$ **52.** $\lim\limits_{x \to 0}\dfrac{\cos(ax)}{\cos(bx)} = 1$

53. $\lim\limits_{x \to 0}\dfrac{\sin(ax)}{bx} = \dfrac{a}{b};\quad b \ne 0$

54. $\lim\limits_{x \to 0}\dfrac{1 - \cos(ax)}{bx} = 0;\quad a \ne 0, b \ne 0$

55. Show that $\lim\limits_{x \to 0}\dfrac{1 - \cos x}{x^2} = \dfrac{1}{2}$.

56. Squeeze Theorem If $0 \le f(x) \le 1$ for every number x, show that $\lim\limits_{x \to 0}[x^2 f(x)] = 0$.

57. Squeeze Theorem If $0 \le f(x) \le M$ for every x, show that $\lim\limits_{x \to 0}[x^2 f(x)] = 0$.

58. The function $f(x) = \dfrac{\sin(\pi x)}{x}$ is not defined at 0. Decide how to define $f(0)$ so that f is continuous at 0.

59. Define $f(0)$ and $f(1)$ so that the function $f(x) = \dfrac{\sin(\pi x)}{x(1 - x)}$ is continuous on the interval $[0, 1]$.

60. Is $f(x) = \begin{cases} \dfrac{\sin x}{x} & \text{if } x \ne 0 \\ 1 & \text{if } x = 0 \end{cases}$ continuous at 0?

61. Is $f(x) = \begin{cases} \dfrac{1 - \cos x}{x} & \text{if } x \ne 0 \\ 0 & \text{if } x = 0 \end{cases}$ continuous at 0?

62. Squeeze Theorem Show that $\lim\limits_{x \to 0}\left(x^n \sin\dfrac{1}{x}\right) = 0$, where n is a positive integer. *Hint:* Look first at Problem 56.

63. Prove $\lim\limits_{\theta \to 0} \sin\theta = 0$. *Hint:* Use a unit circle as shown in the figure, first assuming $0 < \theta < \dfrac{\pi}{2}$. Then use the fact that $\sin\theta$ is less than the length of the arc AP, and the Squeeze Theorem, to show that $\lim\limits_{\theta \to 0^+} \sin\theta = 0$. Then use a similar argument with $-\dfrac{\pi}{2} < \theta < 0$ to show $\lim\limits_{\theta \to 0^-} \sin\theta = 0$.

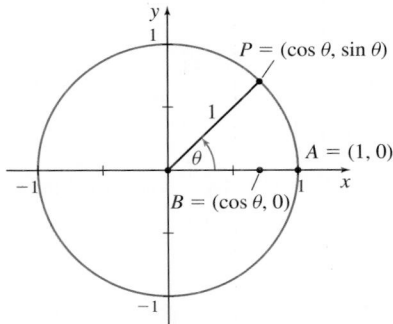

64. Prove $\lim\limits_{\theta \to 0} \cos\theta = 1$. Use either the proof outlined in Problem 63 or a proof using the result $\lim\limits_{\theta \to 0} \sin\theta = 0$ and a Pythagorean identity.

65. Without using limits, explain how you can decide whether $f(x) = \cos(5x^3 + 2x^2 - 8x + 1)$ is continuous.

66. Explain the Squeeze Theorem. Draw a graph to illustrate your explanation.

Challenge Problems

67. Use the Sum Formulas $\sin(a + b) = \sin a \cos b + \cos a \sin b$ and $\cos(a + b) = \cos a \cos b - \sin a \sin b$ to show that the sine function and cosine function are continuous on their domains.

68. Find $\lim\limits_{x \to 0}\dfrac{\sin x^2}{x}$.

69. Squeeze Theorem If $f(x) = \begin{cases} 1 & \text{if } x \text{ is rational} \\ 0 & \text{if } x \text{ is irrational} \end{cases}$ show that $\lim\limits_{x \to 0}[x f(x)] = 0$.

70. Suppose points A and B with coordinates $(0, 0)$ and $(1, 0)$, respectively, are given. Let n be a number greater than 0, and let θ be an angle with the property $0 < \theta < \dfrac{\pi}{1 + n}$. Construct a triangle ABC where $\overline{AC}$ and $\overline{AB}$ form the angle θ, and $\overline{BC}$ and $\overline{BA}$ form the angle $n\theta$ (see the figure below). Let D be the point of intersection of $\overline{AB}$ with the perpendicular from C to $\overline{AB}$. What is the limiting position of D as θ approaches 0?

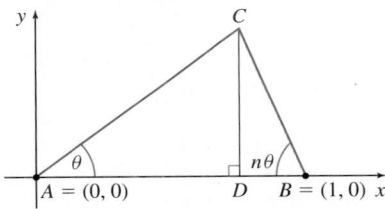

Preparing for the AP® Exam

AP® Practice Problems

Multiple-Choice Questions

PAGE 126 **1.** The function $g(x) = \begin{cases} \dfrac{\sin(2x)}{2x} & \text{if } x \neq 0 \\ k & \text{if } x = 0 \end{cases}$

is continuous at $x = 0$. What is the value of k?

(A) 0   (B) $\dfrac{1}{2}$ (C) 1   (D) 2

PAGE 126 **2.** $\displaystyle\lim_{x \to 0} \frac{\sin(4x)}{2x} =$

(A) 0   (B) $\dfrac{1}{2}$ (C) 1   (D) 2

PAGE 130 **3.** The function $f(x) = \begin{cases} x^3 + 2x^2 & \text{if } x \leq -2 \\ e^{2x+4} & \text{if } x > -2 \end{cases}$.

Find $\displaystyle\lim_{x \to -2} f(x)$ if it exists.

(A) 0   (B) 1   (C) 16   (D) The limit does not exist.

PAGE 126 **4.** $\displaystyle\lim_{x \to 0} \frac{1 - \cos^2(3x)}{x^2} =$

(A) 0   (B) 1   (C) 3   (D) 9

PAGE 130 **5.** Which of the following functions are continuous for all real numbers x?

  I. $f(x) = x^{1/3}$
  II. $g(x) = \sec x$
  III. $h(x) = e^{-x}$

(A) I only
(B) I and II only
(C) I and III only
(D) I, II, and III

PAGE 126 **6.** Find $\displaystyle\lim_{x \to 0} \frac{1}{x \csc x}$ if it exists.

(A) -1 (B) 0   (C) 1   (D) The limit does not exist.

PAGE 126 **7.** $\displaystyle\lim_{x \to \pi/3} \frac{\sin\left(x - \dfrac{\pi}{3}\right)}{x - \dfrac{\pi}{3}} =$

(A) $-\dfrac{\pi}{3}$ (B) 0   (C) 1   (D) $\dfrac{\pi}{3}$

PAGE 124 **8.** $\displaystyle\lim_{x \to 0} \frac{1 - \cos x}{3 \sin^2 x} =$

(A) $\dfrac{1}{6}$ (B) $\dfrac{1}{3}$ (C) $\dfrac{1}{2}$ (D) 1

PAGE 130 **9.** If $f(x) = \begin{cases} \ln x & \text{if } 0 < x < 3 \\ (2x - 3)\ln 3 & \text{if } x \geq 3 \end{cases}$,

then $\displaystyle\lim_{x \to 3} f(x) =$

(A) $\ln 3$ (B) 3   (C) $\ln 9$ (D) The limit does not exist.

PAGE 126 **10.** $\displaystyle\lim_{x \to 0} \frac{\tan(2x)}{3x} =$

(A) $\dfrac{1}{3}$ (B) $\dfrac{1}{2}$ (C) $\dfrac{2}{3}$ (D) 2

PAGE 124 **11.** $\displaystyle\lim_{x \to 0} \left(\sqrt{2} x^2 \sin \frac{1}{x^2} \right) =$

(A) $-\sqrt{2}$ (B) 0   (C) $\sqrt{2}$ (D) The limit does not exist.

PAGE 130 **12.** $\displaystyle\lim_{x \to 0} \frac{\sin(3x)e^{x+1}\ln(x + 2)}{\tan(3x)} =$

(A) 0   (B) $e \ln 2$ (C) $3e \ln 2$ (D) The limit does not exist.

1.5 Infinite Limits; Limits at Infinity; Asymptotes

OBJECTIVES *When you finish this section, you should be able to:*

1 **Investigate infinite limits (p. 133)**
2 **Find the vertical asymptotes of a graph (p. 136)**
3 **Investigate limits at infinity (p. 136)**
4 **Find the horizontal asymptotes of a graph (p. 143)**
5 **Find the asymptotes of the graph of a rational function (p. 144)**

We have described $\displaystyle\lim_{x \to c} f(x) = L$ by saying if a function f is defined everywhere in an open interval containing c, except possibly at c, then the value $f(x)$ can be made as close as we please to L by choosing numbers x sufficiently close to c. Here c and L are real numbers. In this section, we extend the language of limits to allow c to be ∞ or $-\infty$ (*limits at infinity*) and to allow L to be ∞ or $-\infty$ (*infinite limits*). These limits are useful for locating *asymptotes* that aid in graphing some functions.

We begin with infinite limits.

RECALL The symbols ∞ (infinity) and $-\infty$ (negative infinity) are not numbers. The symbol ∞ expresses unboundedness in the positive direction and $-\infty$ expresses unboundedness in the negative direction.

① Investigate Infinite Limits

Consider the function $f(x) = \dfrac{1}{x^2}$. Table 11 lists values of f for selected numbers x that are close to 0 and Figure 48 shows the graph of f.

TABLE 11

x	-0.01	-0.001	-0.0001	$\to 0 \leftarrow$	0.0001	0.001	0.01
	x approaches 0 from the left $\longrightarrow$				$\longleftarrow$ x approaches 0 from the right		
$f(x) = \dfrac{1}{x^2}$	$10{,}000$	10^6	10^8	$f(x)$ becomes unbounded	10^8	10^6	$10{,}000$

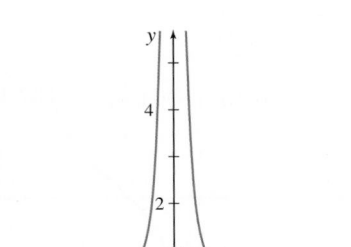

Figure 48 $f(x) = \dfrac{1}{x^2}$

As x approaches 0, the value of $\dfrac{1}{x^2}$ increases without bound. Since the value of $\dfrac{1}{x^2}$ is not approaching a single real number, the *limit* of $f(x)$ as x approaches 0 *does not exist*. However, since the numbers are becoming unbounded, we describe the behavior of the function near zero by writing

$$\lim_{x \to 0} \frac{1}{x^2} = \infty$$

and say that $f(x) = \dfrac{1}{x^2}$ has an **infinite limit** as x approaches 0.

In other words, a function f has an infinite limit at c if f is defined everywhere in an open interval containing c, except possibly at c, and $f(x)$ becomes unbounded when x is sufficiently close to c. (A precise definition of an infinite limit is given in Section 1.6.)

EXAMPLE 1 Investigating $\displaystyle\lim_{x \to 0^-} \frac{1}{x}$ and $\displaystyle\lim_{x \to 0^+} \frac{1}{x}$

Consider the function $f(x) = \dfrac{1}{x}$. Table 12 shows values of f for selected numbers x that are close to 0 and Figure 49 shows the graph of f.

TABLE 12

x	-0.01	-0.001	-0.0001	$\to 0 \leftarrow$	0.0001	0.001	0.01	0.1
	x approaches 0 from the left $\longrightarrow$				$\longleftarrow$ x approaches 0 from the right			
$f(x) = \dfrac{1}{x}$	-100	-1000	$-10{,}000$	$f(x)$ becomes unbounded	$10{,}000$	1000	100	10

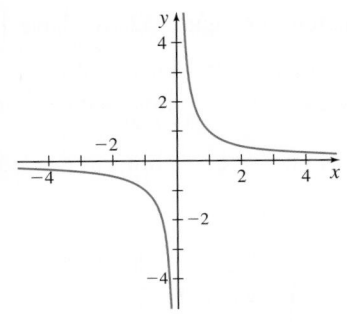

Figure 49 $f(x) = \dfrac{1}{x}$

Here as x gets closer to 0 from the right, the value of $f(x) = \dfrac{1}{x}$ can be made as large as we please. That is, $\dfrac{1}{x}$ becomes unbounded in the positive direction. So,

$$\lim_{x \to 0^+} \frac{1}{x} = \infty$$

Similarly, the notation

$$\lim_{x \to 0^-} \frac{1}{x} = -\infty$$

is used to indicate that $\dfrac{1}{x}$ becomes unbounded in the negative direction as x approaches 0 from the left. ∎

So, there are four possible one-sided infinite limits of a function f at c:

$$\lim_{x \to c^-} f(x) = \infty \qquad \lim_{x \to c^-} f(x) = -\infty \qquad \lim_{x \to c^+} f(x) = \infty \qquad \lim_{x \to c^+} f(x) = -\infty$$

See Figure 50 for illustrations of these possibilities.

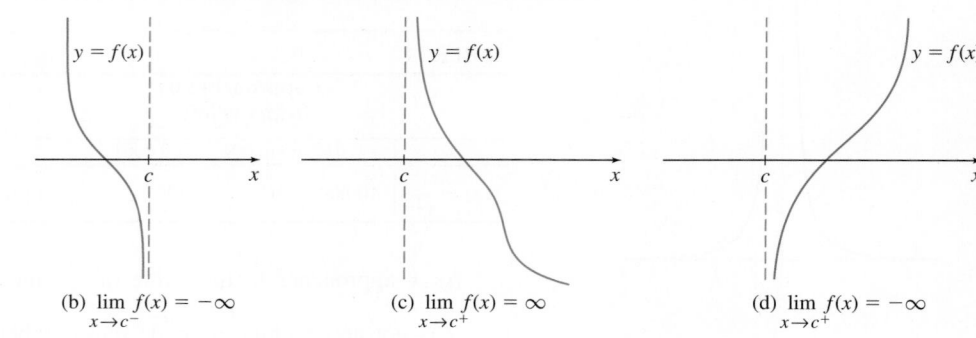

(a) $\lim_{x \to c^-} f(x) = \infty$ (b) $\lim_{x \to c^-} f(x) = -\infty$ (c) $\lim_{x \to c^+} f(x) = \infty$ (d) $\lim_{x \to c^+} f(x) = -\infty$

Figure 50

When we know the graph of a function, we can use it to investigate infinite limits.

EXAMPLE 2   Investigating an Infinite Limit

Investigate $\lim\limits_{x \to 0^+} \ln x$.

Solution

The domain of $f(x) = \ln x$ is $\{x \mid x > 0\}$. See Figure 51. Notice that the graph of $f(x) = \ln x$ becomes unbounded in the negative direction as x approaches 0 from the right. The graph suggests that

$$\boxed{\lim_{x \to 0^+} \ln x = -\infty}$$

∎

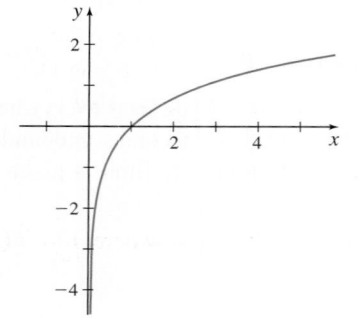

Figure 51 $f(x) = \ln x$

NOW WORK   Problem 11.

Based on the graphs of the trigonometric functions in Figure 52, we have the following infinite limits:

$$\lim_{x \to \pi/2^-} \tan x = \infty \qquad \lim_{x \to \pi/2^+} \tan x = -\infty \qquad \lim_{x \to \pi/2^-} \sec x = \infty \qquad \lim_{x \to \pi/2^+} \sec x = -\infty$$

$$\lim_{x \to 0^-} \csc x = -\infty \qquad \lim_{x \to 0^+} \csc x = \infty \qquad \lim_{x \to 0^-} \cot x = -\infty \qquad \lim_{x \to 0^+} \cot x = \infty$$

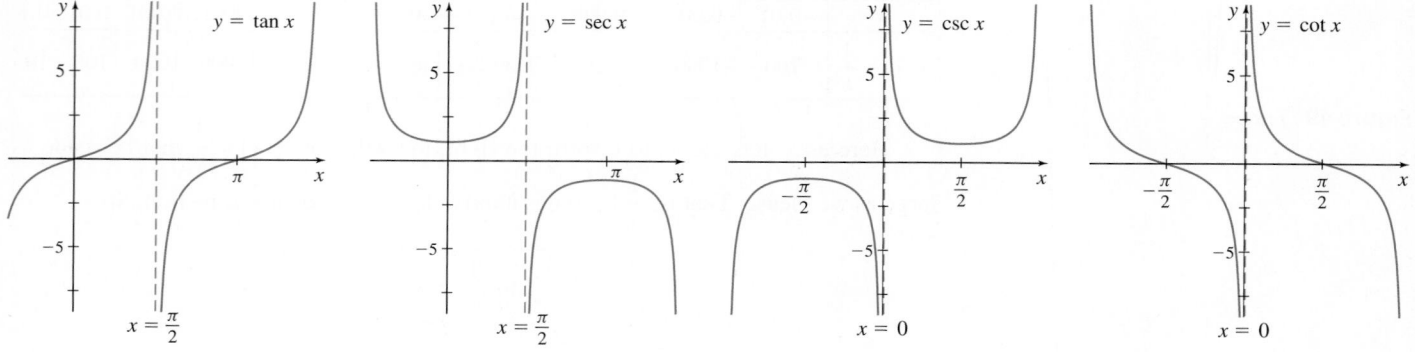

Figure 52

Limits of quotients, in which the limit of the numerator is a nonzero number and the limit of the denominator is 0, often result in infinite limits. (If the limits of both the numerator and the denominator are 0, the quotient is called an *indeterminate form*. We have already discussed some limits of quotients that lead to an indeterminate form. See Examples 13 and 14 on pages 102–103. Limits of indeterminate forms are discussed further in Section 4.4.)

EXAMPLE 3   Investigating an Infinite Limit

Investigate: **(a)** $\lim\limits_{x \to 3^+} \dfrac{2x+1}{x-3}$ **(b)** $\lim\limits_{x \to 1^-} \dfrac{x^2}{\ln x}$

Solution

(a) $\lim\limits_{x \to 3^+} \dfrac{2x+1}{x-3}$ is the right-hand limit of the quotient of two functions, $f(x) = 2x+1$ and $g(x) = x - 3$. Since $\lim\limits_{x \to 3^+} (x-3) = 0$, we cannot use the Limit of a Quotient. So we need a different strategy. As x approaches 3 from the right, we have

$$\lim\limits_{x \to 3^+} (2x+1) = 7 \quad \text{and} \quad \lim\limits_{x \to 3^+} (x-3) = 0$$

Here, the limit of the numerator is 7. The denominator is positive and approaching 0. So the quotient is positive and becoming unbounded in the positive direction. We conclude that

$$\lim\limits_{x \to 3^+} \dfrac{2x+1}{x-3} = \infty$$

(b) As in (a), we cannot use the Limit of a Quotient because $\lim\limits_{x \to 1^-} \ln x = 0$. Here we have

$$\lim\limits_{x \to 1^-} x^2 = 1 \qquad \lim\limits_{x \to 1^-} \ln x = 0$$

We are seeking a left-hand limit, so for numbers close to 1 but less than 1, the numerator, x^2, is positive and the denominator, $\ln x$, is negative. So the quotient is negative and becoming unbounded in the negative direction. We conclude that

$$\lim\limits_{x \to 1^-} \dfrac{x^2}{\ln x} = -\infty$$

■

The discussion below may prove useful when finding $\lim\limits_{x \to c^-} \dfrac{f(x)}{g(x)}$, where $\lim\limits_{x \to c^-} f(x)$ is a nonzero number and $\lim\limits_{x \to c^-} g(x) = 0$.

Investigating $\lim\limits_{x \to c^-} \dfrac{f(x)}{g(x)}$ **when** $\lim\limits_{x \to c^-} f(x) = L$, $L \neq 0$, **and** $\lim\limits_{x \to c^-} g(x) = 0$.

- If $L > 0$ and $g(x) < 0$ for numbers x close to c, but less than c, then

$$\boxed{\lim\limits_{x \to c^-} \dfrac{f(x)}{g(x)} = -\infty}$$

- If $L > 0$ and $g(x) > 0$ for numbers x close to c, but less than c, then

$$\boxed{\lim\limits_{x \to c^-} \dfrac{f(x)}{g(x)} = \infty}$$

- If $L < 0$ and $g(x) < 0$ for numbers x close to c, but less than c, then

$$\boxed{\lim\limits_{x \to c^-} \dfrac{f(x)}{g(x)} = \infty}$$

- If $L < 0$ and $g(x) > 0$ for numbers x close to c, but less than c, then

$$\boxed{\lim\limits_{x \to c^-} \dfrac{f(x)}{g(x)} = -\infty}$$

Similar results can be stated for right-hand limits.

NOW WORK   Problem **27** and AP® Practice Problems **9** and **11**.

2 Find the Vertical Asymptotes of a Graph

Figure 53 illustrates some of the possibilities that can occur when a function has an infinite limit at c. In each case, notice that the graph of f has a vertical asymptote $x = c$.

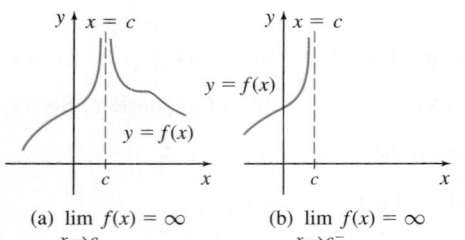
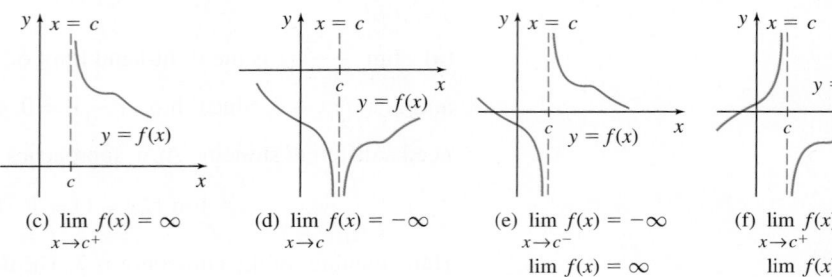

(a) $\lim\limits_{x \to c} f(x) = \infty$

(b) $\lim\limits_{x \to c^-} f(x) = \infty$

(c) $\lim\limits_{x \to c^+} f(x) = \infty$

(d) $\lim\limits_{x \to c} f(x) = -\infty$

(e) $\lim\limits_{x \to c^-} f(x) = -\infty$
$\lim\limits_{x \to c^+} f(x) = \infty$

(f) $\lim\limits_{x \to c^+} f(x) = -\infty$
$\lim\limits_{x \to c^-} f(x) = \infty$

Figure 53

DEFINITION Vertical Asymptote

The line $x = c$ is a **vertical asymptote** of the graph of the function f if any of the following is true:

$$\lim_{x \to c^-} f(x) = \infty \qquad \lim_{x \to c^+} f(x) = \infty \qquad \lim_{x \to c^-} f(x) = -\infty \qquad \lim_{x \to c^+} f(x) = -\infty$$

For rational functions, a vertical asymptote may occur where the denominator equals 0.

EXAMPLE 4 Finding a Vertical Asymptote

Find any vertical asymptote(s) of the graph of $f(x) = \dfrac{x}{(x-3)^2}$.

Solution

The domain of f is $\{x \mid x \neq 3\}$. Since 3 is the only number for which the denominator of f equals 0, we construct Table 13 and investigate the one-sided limits of f as x approaches 3. Table 13 suggests that

$$\lim_{x \to 3} \frac{x}{(x-3)^2} = \infty$$

So, $x = 3$ is a vertical asymptote of the graph of f.

TABLE 13

x	2.9	2.99	2.999	$\to 3 \leftarrow$	3.001	3.01	3.1
	x approaches 3 from the left				*x* approaches 3 from the right		
$f(x) = \dfrac{x}{(x-3)^2}$	290	29,900	2,999,000	$f(x)$ approaches ∞	3,001,000	30,100	310

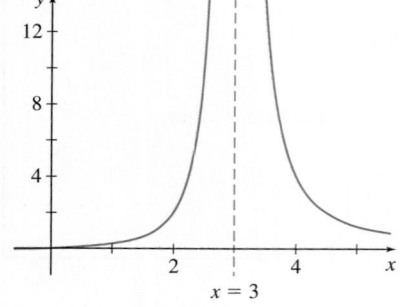

Figure 54 $f(x) = \dfrac{x}{(x-3)^2}$

Figure 54 shows the graph of $f(x) = \dfrac{x}{(x-3)^2}$ and its vertical asymptote. ■

NOW WORK Problems **15** and **63** (find any vertical asymptotes) and AP® Practice Problem **5**.

3 Investigate Limits at Infinity

Now we investigate what happens as x becomes unbounded in either the positive direction or the negative direction. Suppose as x becomes unbounded, the value of a function f approaches some real number L. Then the number L is called the *limit of f at infinity*.

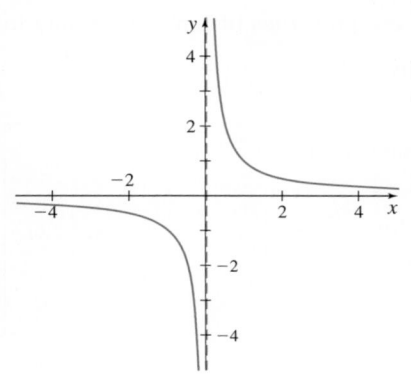

Figure 55 $f(x) = \dfrac{1}{x}$

For example, the graph of $f(x) = \dfrac{1}{x}$ in Figure 55 suggests that as x becomes unbounded in either the positive direction or the negative direction, the values $f(x)$ get closer to 0. Table 14 illustrates this for a few numbers x.

TABLE 14

x	± 100	± 1000	$\pm 10,000$	$\pm 100,000$	x becomes unbounded
$f(x) = \dfrac{1}{x}$	± 0.01	± 0.001	± 0.0001	± 0.00001	$f(x)$ approaches 0

Since f can be made as close as we please to 0 by choosing x sufficiently large, we write

$$\lim_{x \to \infty} \frac{1}{x} = 0$$

and say that the limit as x approaches infinity of f is equal to 0. Similarly, as x becomes unbounded in the negative direction, the function $f(x) = \dfrac{1}{x}$ can be made as close as we please to 0, and we write

$$\lim_{x \to -\infty} \frac{1}{x} = 0$$

Although we do not prove $\lim\limits_{x \to -\infty} \dfrac{1}{x} = 0$ or $\lim\limits_{x \to \infty} \dfrac{1}{x} = 0$ here, they can be proved using the ε-δ Definition of a Limit at Infinity given in Section 1.6.

The limit properties discussed in Section 1.2 hold for infinite limits. Although these properties are stated below for limits as $x \to \infty$, they are also valid for limits as $x \to -\infty$.

THEOREM Properties of Limits at Infinity

If k is a real number, $n \geq 2$ is an integer, and the functions f and g approach real numbers as $x \to \infty$, then the following properties are true:

- $\lim\limits_{x \to \infty} A = A$, where A is a constant

- $\lim\limits_{x \to \infty} [kf(x)] = k \lim\limits_{x \to \infty} f(x)$

- $\lim\limits_{x \to \infty} [f(x) \pm g(x)] = \lim\limits_{x \to \infty} f(x) \pm \lim\limits_{x \to \infty} g(x)$

- $\lim\limits_{x \to \infty} [f(x)g(x)] = \left[\lim\limits_{x \to \infty} f(x) \right] \left[\lim\limits_{x \to \infty} g(x) \right]$

- $\lim\limits_{x \to \infty} \dfrac{f(x)}{g(x)} = \dfrac{\lim\limits_{x \to \infty} f(x)}{\lim\limits_{x \to \infty} g(x)}$, provided $\lim\limits_{x \to \infty} g(x) \neq 0$

- $\lim\limits_{x \to \infty} [f(x)]^n = \left[\lim\limits_{x \to \infty} f(x) \right]^n$

- $\lim\limits_{x \to \infty} \sqrt[n]{f(x)} = \sqrt[n]{\lim\limits_{x \to \infty} f(x)}$, where $f(x) > 0$ if n is even

Since $\lim\limits_{x\to\infty} \dfrac{1}{x}=0$ and $\lim\limits_{x\to-\infty} \dfrac{1}{x}=0$, we can use properties of limits at infinity to obtain two additional properties of limits at infinity.

If $p>0$ is a rational number and k is any real number, then

$$\lim_{x\to\infty} \frac{k}{x^p}=0 \quad\text{and}\quad \lim_{x\to-\infty} \frac{k}{x^p}=0 \tag{1}$$

provided x^p is defined when $x<0$.

You are asked to prove (1) in Problem 92.

EXAMPLE 5 **Finding Limits at Infinity Using the Properties in (1)**

(a) $\lim\limits_{x\to\infty} \dfrac{5}{x^3}=0$ (b) $\lim\limits_{x\to\infty} \dfrac{1}{\sqrt{x}}=0$

(c) $\lim\limits_{x\to-\infty} \dfrac{-6}{x^{1/3}}=0$ (d) $\lim\limits_{x\to-\infty} \dfrac{4}{5x^{8/3}}=0$ ■

EXAMPLE 6 **Finding Limits at Infinity**

Find each limit.

(a) $\lim\limits_{x\to\infty} \left(2+\dfrac{3}{x}\right)$ (b) $\lim\limits_{x\to\infty} \left[\left(2+\dfrac{1}{x}\right)\left(\dfrac{1}{3}-\dfrac{2}{x^3}\right)\right]$

(c) $\lim\limits_{x\to\infty} \left[\dfrac{1}{4}\left(\dfrac{3x^2+5x-1}{x^2}\right)\right]$

Solution

We use properties of limits at infinity.

(a) $\lim\limits_{x\to\infty} \left(2+\dfrac{3}{x}\right) = \lim\limits_{x\to\infty} 2 + \lim\limits_{x\to\infty} \dfrac{3}{x} = 2+0=2$

(b)
$$\lim_{x\to\infty} \left[\left(2+\frac{1}{x}\right)\left(\frac{1}{3}-\frac{2}{x^3}\right)\right] = \lim_{x\to\infty} \left(2+\frac{1}{x}\right)\cdot \lim_{x\to\infty}\left(\frac{1}{3}-\frac{2}{x^3}\right)$$

$$= \left(\lim_{x\to\infty} 2 + \lim_{x\to\infty}\frac{1}{x}\right)\cdot\left(\lim_{x\to\infty}\frac{1}{3} - \lim_{x\to\infty}\frac{2}{x^3}\right)$$

$$= (2+0)\left(\frac{1}{3}-0\right)=\frac{2}{3}$$

(c)
$$\lim_{x\to\infty} \left[\frac{1}{4}\left(\frac{3x^2+5x-1}{x^2}\right)\right] = \frac{1}{4}\lim_{x\to\infty} \frac{3x^2+5x-1}{x^2} = \frac{1}{4}\lim_{x\to\infty}\left(3+\frac{5}{x}-\frac{1}{x^2}\right)$$

$$= \frac{1}{4}\left(\lim_{x\to\infty} 3 + \lim_{x\to\infty}\frac{5}{x} - \lim_{x\to\infty}\frac{1}{x^2}\right) = \frac{1}{4}\cdot 3 = \frac{3}{4}$$ ■

NOW WORK Problem 45 and AP® Practice Problem 3.

For some algebraic functions involving a quotient, we find limits at infinity by dividing the numerator and the denominator by the highest power of x that appears in the denominator.

EXAMPLE 7 **Finding a Limit at Infinity**

CALC CLIP

Find each limit.

(a) $\displaystyle\lim_{x\to\infty}\frac{3x^{3/2}+5x^{1/2}}{x^3-x^{2/3}}$ (b) $\displaystyle\lim_{x\to\infty}\frac{\sqrt{4x^2+10}}{x-5}$

Solution

(a) Divide the numerator and the denominator by x^3, the highest power of x that appears in the denominator.

$$\lim_{x\to\infty}\frac{3x^{3/2}+5x^{1/2}}{x^3-x^{2/3}}=\lim_{x\to\infty}\frac{\dfrac{3x^{3/2}+5x^{1/2}}{x^3}}{\dfrac{x^3-x^{2/3}}{x^3}}=\lim_{x\to\infty}\frac{\dfrac{3x^{3/2}}{x^3}+\dfrac{5x^{1/2}}{x^3}}{1-\dfrac{x^{2/3}}{x^3}}$$

Divide the numerator and
the denominator by x^3.

$$=\lim_{x\to\infty}\frac{\dfrac{3}{x^{3/2}}+\dfrac{5}{x^{5/2}}}{1-\dfrac{1}{x^{7/3}}}=\frac{\displaystyle\lim_{x\to\infty}\left(\dfrac{3}{x^{3/2}}+\dfrac{5}{x^{5/2}}\right)}{\displaystyle\lim_{x\to\infty}\left(1-\dfrac{1}{x^{7/3}}\right)}=\frac{0+0}{1-0}=0$$

Use (1), p. 138

(b) Divide the numerator and the denominator by x, the highest power of x that appears in the denominator. But remember, since $\sqrt{x^2}=|x|=x$ when $x\ge 0$, in the numerator the divisor in the square root will be x^2.

$$\lim_{x\to\infty}\frac{\sqrt{4x^2+10}}{x-5}=\lim_{x\to\infty}\frac{\sqrt{\dfrac{4x^2+10}{x^2}}}{\dfrac{x-5}{x}}=\lim_{x\to\infty}\frac{\sqrt{4+\dfrac{10}{x^2}}}{1-\dfrac{5}{x}}$$

Divide the numerator and the denominator by x.
Remember to divide the numerator by $\sqrt{x^2}=x$.

$$=\frac{\displaystyle\lim_{x\to\infty}\sqrt{4+\dfrac{10}{x^2}}}{\displaystyle\lim_{x\to\infty}\left(1-\dfrac{5}{x}\right)}=\frac{\sqrt{\displaystyle\lim_{x\to\infty}\left(4+\dfrac{10}{x^2}\right)}}{1}=\sqrt{4}=2$$

Use properties of
limits at infinity.

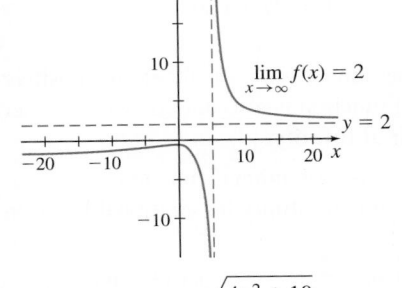

Figure 56 $f(x)=\dfrac{\sqrt{4x^2+10}}{x-5}$

In the graph: $\displaystyle\lim_{x\to\infty}f(x)=2$, $y=2$

The graph of $f(x)=\dfrac{\sqrt{4x^2+10}}{x-5}$ is shown in Figure 56.   ∎

NOW WORK   Problems 47 and 57, and AP® Practice Problem 6.

For polynomial and rational functions, there is an easier approach to finding limits at infinity.

Limits at Infinity of Polynomial and Rational Functions

We begin by investigating limits at infinity of power functions. For any power function $p(x)=x^n$, if n is a positive integer, then p increases without bound as $x\to\infty$. That is,

$$\lim_{x\to\infty}x^n=\infty\qquad n\text{ a positive integer}\qquad(2)$$

To investigate $\displaystyle\lim_{x\to-\infty}x^n$, n a positive integer, we need to consider whether n is even or odd. Remember, as $x\to-\infty$, x is unbounded in the negative direction, so $x<0$.

- If n is positive and even, then $x^n>0$ when $x<0$.
- If n is positive and odd, then $x^n<0$ when $x<0$.

Since in both cases x^n becomes unbounded as $x \to -\infty$, we have

$$\lim_{x \to -\infty} x^n = \infty \text{ if } n \text{ is a positive even integer}$$

$$\lim_{x \to -\infty} x^n = -\infty \text{ if } n \text{ is a positive odd integer}$$

(3)

Now consider a polynomial function, $p(x) = a_n x^n + a_{n-1} x^{n-1} + \cdots + a_1 x + a_0$, where $a_n \neq 0$. To find $\lim_{x \to \pm\infty} p(x)$, factor out x^n. Then

$$\lim_{x \to \pm\infty} p(x) = \lim_{x \to \pm\infty} (a_n x^n + a_{n-1} x^{n-1} + \cdots + a_1 x + a_0)$$

$$= \lim_{x \to \pm\infty} \left[x^n \left(a_n + \frac{a_{n-1}}{x} + \cdots + \frac{a_1}{x^{n-1}} + \frac{a_0}{x^n} \right) \right]$$

Since $\lim_{x \to \pm\infty} \left(a_n + \dfrac{a_{n-1}}{x} + \cdots + \dfrac{a_1}{x^{n-1}} + \dfrac{a_0}{x^n} \right) = a_n$, we have

$$\lim_{x \to \pm\infty} p(x) = a_n \lim_{x \to \pm\infty} x^n$$

(4)

Using (2), (3), and (4), we can find the limit at infinity of any polynomial function.

EXAMPLE 8   **Finding Limits at Infinity of a Polynomial Function**

(a) $\lim_{x \to \infty} (3x^3 - x^2 + 5x + 1) = 3 \lim_{x \to \infty} x^3 = \infty$ Use (4) and (2).

(b) $\lim_{x \to \infty} (x^5 + 2x) = \lim_{x \to \infty} x^5 = \infty$ Use (4) and (2).

(c) $\lim_{x \to -\infty} (2x^4 - x^2 + 2x + 8) = 2 \lim_{x \to -\infty} x^4 = \infty$ Use (4) and (3).

(d) $\lim_{x \to -\infty} (x^3 + 10x + 7) = \lim_{x \to -\infty} x^3 = -\infty$ Use (4) and (3).   ∎

So, for a polynomial function, as x becomes unbounded in either the positive direction or the negative direction, the polynomial function is also unbounded. In other words, polynomial functions have an **infinite limit at infinity**.

Since a rational function R is the quotient of two polynomial functions p and q, where q is not the zero polynomial, we can find a limit at infinity for a rational function using only the leading terms of p and q.

Consider the rational function $R(x) = \dfrac{a_n x^n + a_{n-1} x^{n-1} + \cdots + a_1 x + a_0}{b_m x^m + b_{m-1} x^{m-1} + \cdots + b_1 x + b_0}$, where $a_n \neq 0$, $b_m \neq 0$.

Then we simplify R as follows:

$$R(x) = \frac{x^n \left(a_n + \dfrac{a_{n-1}}{x} + \cdots + \dfrac{a_1}{x^{n-1}} + \dfrac{a_0}{x^n} \right)}{x^m \left(b_m + \dfrac{b_{m-1}}{x} + \cdots + \dfrac{b_1}{x^{m-1}} + \dfrac{b_0}{x^m} \right)}$$

Since $\lim_{x \to \infty} \left(a_n + \dfrac{a_{n-1}}{x} + \cdots + \dfrac{a_1}{x^{n-1}} + \dfrac{a_0}{x^n} \right) = a_n$

and $\lim_{x \to \infty} \left(b_m + \dfrac{b_{m-1}}{x} + \cdots + \dfrac{b_1}{x^{m-1}} + \dfrac{b_0}{x^m} \right) = b_m$, we have

$$\lim_{x \to \infty} R(x) = \lim_{x \to \infty} \frac{x^n}{x^m} \cdot \frac{\lim\limits_{x \to \infty} \left(a_n + \dfrac{a_{n-1}}{x} + \cdots + \dfrac{a_1}{x^{n-1}} + \dfrac{a_0}{x^n} \right)}{\lim\limits_{x \to \infty} \left(b_m + \dfrac{b_{m-1}}{x} + \cdots + \dfrac{b_1}{x^{m-1}} + \dfrac{b_0}{x^m} \right)} = \frac{a_n}{b_m} \lim_{x \to \infty} \frac{x^n}{x^m}$$

Or more generally,

$$\lim_{x \to \pm\infty} R(x) = \lim_{x \to \pm\infty} \frac{a_n x^n + a_{n-1}x^{n-1} + \cdots + a_1 x + a_0}{b_m x^m + b_{m-1}x^{m-1} + \cdots + b_1 x + b_0} = \frac{a_n}{b_m} \lim_{x \to \pm\infty} \frac{x^n}{x^m} \qquad (5)$$

Example 9 below shows the three possibilities that can occur.

EXAMPLE 9   **Finding Limits at Infinity of a Rational Function**

Find each limit.

(a) $\displaystyle\lim_{x \to \infty} \frac{3x^2 - 2x + 8}{x^2 + 1}$ **(b)** $\displaystyle\lim_{x \to \infty} \frac{4x^2 - 5x}{x^3 + 1}$ **(c)** $\displaystyle\lim_{x \to \infty} \frac{5x^4 - 3x^2}{2x^2 + 1}$

Solution

Each of these limits at infinity involve a rational function, so we use (5).

(a) $\displaystyle\lim_{x \to \infty} \frac{3x^2 - 2x + 8}{x^2 + 1} = 3 \lim_{x \to \infty} \frac{x^2}{x^2} = 3$ Here, $n = m$ in (5).

(b) $\displaystyle\lim_{x \to \infty} \frac{4x^2 - 5x}{x^3 + 1} = 4 \lim_{x \to \infty} \frac{x^2}{x^3} = 4 \lim_{x \to \infty} \frac{1}{x} = 0$ Here, $n < m$ in (5).

(c) $\displaystyle\lim_{x \to \infty} \frac{5x^4 - 3x^2}{2x^2 + 1} = \frac{5}{2} \lim_{x \to \infty} \frac{x^4}{x^2} = \frac{5}{2} \lim_{x \to \infty} x^2 = \infty$ Here, $n > m$ in (5). ∎

The graphs of the functions $F(x) = \dfrac{3x^2 - 2x + 8}{x^2 + 1}$ and $G(x) = \dfrac{4x^2 - 5x}{x^3 + 1}$ are shown in Figure 57(a) and (b), respectively. Notice how each graph behaves as x becomes unbounded in the positive direction. The graph of the function $H(x) = \dfrac{5x^4 - 3x^2}{2x^2 + 1}$, which has an infinite limit at infinity, is shown in Figure 57(c).

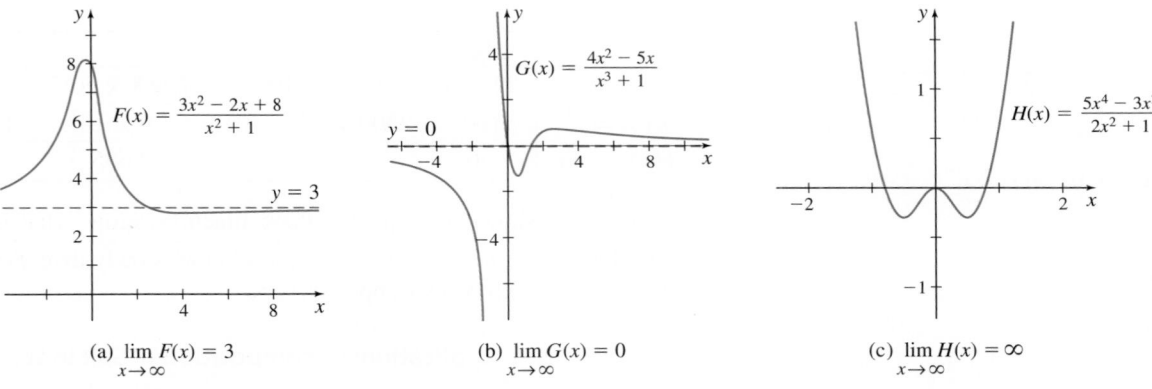

(a) $\displaystyle\lim_{x \to \infty} F(x) = 3$ (b) $\displaystyle\lim_{x \to \infty} G(x) = 0$ (c) $\displaystyle\lim_{x \to \infty} H(x) = \infty$

Figure 57

NOW WORK   Problems **49** and **59**, and AP® Practice Problem **2**.

Other Infinite Limits at Infinity

We have seen that all polynomial functions have infinite limits at infinity, and some rational functions have infinite limits at infinity. Other functions also have an infinite limit at infinity.

For example, consider the function $f(x) = e^x$.

The graph of the exponential function, shown in Figure 58, suggests that

$$\lim_{x \to -\infty} e^x = 0 \qquad \lim_{x \to \infty} e^x = \infty$$

Figure 58 $f(x) = e^x$

These limits are supported by the information in Table 15.

TABLE 15

x	-1	-5	-10	-20	x approaches $-\infty$
$f(x) = e^x$	0.36788	0.00674	0.00005	2×10^{-9}	$f(x)$ approaches 0
x	1	5	10	20	x approaches ∞
$f(x) = e^x$	$e \approx 2.71828$	148.41	22,026	4.85×10^8	$f(x)$ approaches ∞

EXAMPLE 10 **Finding the Limit at Infinity of $g(x) = \ln x$**

Find $\lim_{x \to \infty} \ln x$.

Solution

Table 16 and the graph of $g(x) = \ln x$ in Figure 59 suggest that $g(x) = \ln x$ has an infinite limit at infinity. That is,

$$\lim_{x \to \infty} \ln x = \infty$$

TABLE 16

x	e^{10}	e^{100}	e^{1000}	$e^{10,000}$	$e^{100,000}$	x approaches ∞
$g(x) = \ln x$	10	100	1000	10,000	100,000	$g(x)$ approaches ∞

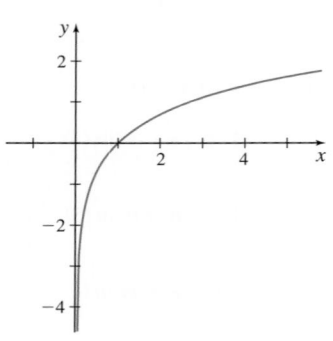

Figure 59 $g(x) = \ln x$

Now let's compare the graph of $f(x) = e^x$ in Figure 58, the graph of $g(x) = \ln x$ in Figure 59, and the graph of $h(x) = x^2$ in Figure 60. As x approaches ∞, all three of the functions approach ∞. But in Table 17, $f(x) = e^x$ approaches infinity more rapidly than $h(x) = x^2$, which approaches infinity more rapidly than $g(x) = \ln x$.

TABLE 17

x	10	50	100	1000	10,000
$f(x) = e^x$	22,026	5.185×10^{21}	2.688×10^{43}	e^{1000}	$e^{10,000}$
$h(x) = x^2$	100	$2500 = 2.5 \times 10^3$	1.0×10^4	1.0×10^6	1.0×10^8
$g(x) = \ln x$	2.303	3.912	4.605	6.908	9.210

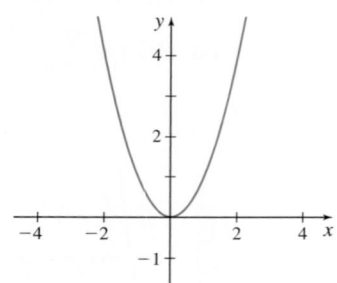

Figure 60 $h(x) = x^2$

Table 17 shows how quickly these functions grow. That is, $g(x) = \ln x$ grows more slowly than $h(x) = x^2$, which grows more slowly than $f(x) = e^x$. Yet, all three functions approach ∞ as x approaches ∞.

EXAMPLE 11 **Application: Decomposition of Salt in Water**

Salt (NaCl) dissolves in water into sodium (Na^+) ions and chloride (Cl^-) ions according to the law of uninhibited decay

$$A(t) = A_0 e^{kt}$$

where $A = A(t)$ is the amount (in kilograms) of undissolved salt present at time t (in hours), A_0 is the original amount of undissolved salt, and k is a negative number that represents the rate of dissolution.

(a) If initially there are 25 kilograms (kg) of undissolved salt and after 10 hours (h) there are 15 kg of undissolved salt remaining, how much undissolved salt is left after one day?

(b) How long will it take until $\dfrac{1}{2}$ kg of undissolved salt remains?

(c) Find $\lim_{t \to \infty} A(t)$.

(d) Interpret the answer found in (c).

Solution

NEED TO REVIEW? Solving exponential equations is discussed in Section P.5, pp. 50–53.

(a) Initially, there are 25 kg of undissolved salt, so $A(0) = A_0 = 25$. To find the number k in $A(t) = A_0 e^{kt}$, we use the fact that at $t = 10$, then $A(10) = 15$. That is,

$$A(10) = 15 = 25e^{10k} \qquad A(t) = A_0 e^{kt},\ A_0 = 25;\ A(10) = 15$$

$$e^{10k} = \frac{3}{5}$$

$$10k = \ln \frac{3}{5}$$

$$k = \frac{1}{10} \ln 0.6$$

So, $A(t) = 25e^{(\frac{1}{10} \ln 0.6)t}$. The amount of undissolved salt that remains after one day (24 h) is

$$A(24) = 25e^{(\frac{1}{10} \ln 0.6)24} \approx 7.337 \text{ kg}$$

(b) We want to find t so that $A(t) = 25e^{(\frac{1}{10} \ln 0.6)t} = \frac{1}{2}$ kg. Then

$$\frac{1}{2} = 25e^{(\frac{1}{10} \ln 0.6)t}$$

$$e^{(\frac{1}{10} \ln 0.6)t} = \frac{1}{50}$$

$$\left(\frac{1}{10} \ln 0.6\right) t = \ln \frac{1}{50}$$

$$t \approx 76.582$$

After approximately 76.6 h, $\frac{1}{2}$ kg of undissolved salt will remain.

(c) Since $\frac{1}{10} \ln 0.6 \approx -0.051$, we have $\lim\limits_{t \to \infty} A(t) = \lim\limits_{t \to \infty} (25e^{-0.051t}) = \lim\limits_{t \to \infty} \frac{25}{e^{0.051t}} = 0$

(d) As t approaches ∞, the amount of undissolved salt in the water approaches 0 kg. Eventually, there will be no undissolved salt. ∎

NOW WORK Problem 79.

---

④ Find the Horizontal Asymptotes of a Graph

Limits at infinity have an important geometric interpretation. When $\lim\limits_{x \to \infty} f(x) = M$, it means that as x approaches ∞, the value of f can be made as close as we please to a number M. That is, the graph of $y = f(x)$ is as close as we please to the horizontal line $y = M$ by choosing x sufficiently large. Similarly, $\lim\limits_{x \to -\infty} f(x) = L$ means that as x approaches $-\infty$, the graph of $y = f(x)$ is as close as we please to the horizontal line $y = L$. These lines are *horizontal asymptotes* of the graph of f.

DEFINITION Horizontal Asymptote

- The line $y = L$ is a **horizontal asymptote** of the graph of a function f if as x approaches $-\infty$, $\lim\limits_{x \to -\infty} f(x) = L$.

- The line $y = M$ is a **horizontal asymptote** of the graph of a function f if as x approaches ∞, $\lim\limits_{x \to \infty} f(x) = M$.

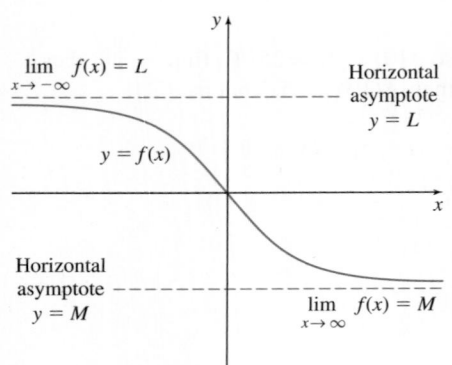

Figure 61

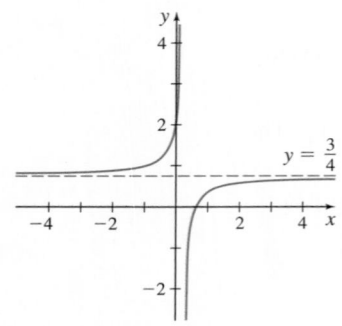

Figure 62 $f(x) = \dfrac{3x-2}{4x-1}$

In Figure 61, $y = L$ is a horizontal asymptote as $x \to -\infty$ because $\lim\limits_{x \to -\infty} f(x) = L$. The line $y = M$ is a horizontal asymptote as $x \to \infty$ because $\lim\limits_{x \to \infty} f(x) = M$. To identify horizontal asymptotes, we find the limits of f at infinity.

EXAMPLE 12 **Finding the Horizontal Asymptotes of a Graph**

Find the horizontal asymptotes, if any, of the graph of $f(x) = \dfrac{3x-2}{4x-1}$.

Solution

We examine the two limits at infinity: $\lim\limits_{x \to -\infty} \dfrac{3x-2}{4x-1}$ and $\lim\limits_{x \to \infty} \dfrac{3x-2}{4x-1}$.

Since $\lim\limits_{x \to -\infty} \dfrac{3x-2}{4x-1} = \dfrac{3}{4}$, the line $y = \dfrac{3}{4}$ is a horizontal asymptote of the graph of f as x approaches $-\infty$.

Since $\lim\limits_{x \to \infty} \dfrac{3x-2}{4x-1} = \dfrac{3}{4}$, the line $y = \dfrac{3}{4}$ is a horizontal asymptote of the graph of f as x approaches ∞. ∎

Figure 62 shows the graph of $f(x) = \dfrac{3x-2}{4x-1}$ and the horizontal asymptote $y = \dfrac{3}{4}$.

NOW WORK Problem 63 (find any horizontal asymptotes) and AP® Practice Problems 1, 4, and 7.

---

5 Find the Asymptotes of the Graph of a Rational Function

In the next example we find the horizontal asymptotes and vertical asymptotes, if any, of the graph of a rational function.

EXAMPLE 13 **Finding the Asymptotes of the Graph of a Rational Function**

Find any asymptotes of the graph of the rational function $R(x) = \dfrac{3x^2 - 12}{2x^2 - 9x + 10}$.

Solution

We begin by factoring R.

$$R(x) = \frac{3x^2 - 12}{2x^2 - 9x + 10} = \frac{3(x-2)(x+2)}{(2x-5)(x-2)}$$

The domain of R is $\left\{ x \mid x \neq \dfrac{5}{2} \text{ and } x \neq 2 \right\}$. Since R is a rational function, it is continuous on its domain, that is, all real numbers except $x = \dfrac{5}{2}$ and $x = 2$.

To check for vertical asymptotes, we find the limits as x approaches $\dfrac{5}{2}$ and 2.

• First we consider $\lim\limits_{x \to 5/2^-} R(x)$.

$$\lim_{x \to 5/2^-} R(x) = \lim_{x \to 5/2^-} \left[\frac{3(x-2)(x+2)}{(2x-5)(x-2)} \right] = \lim_{x \to 5/2^-} \left[\frac{3(x+2)}{(2x-5)} \right] = 3 \lim_{x \to 5/2^-} \frac{x+2}{2x-5} = -\infty$$

That is, as x approaches $\dfrac{5}{2}$ from the left, R approaches $-\infty$. The graph of R has a vertical asymptote on the left at $x = \dfrac{5}{2}$.

To determine the behavior to the right of $x = \dfrac{5}{2}$, we find the right-hand limit.

$$\lim_{x \to 5/2^+} R(x) = \lim_{x \to 5/2^+} \left[\frac{3(x+2)}{(2x-5)} \right] = 3 \lim_{x \to 5/2^+} \frac{x+2}{2x-5} = \infty$$

As x approaches $\dfrac{5}{2}$ from the right, R approaches ∞. The graph of R has a vertical asymptote on the right at $x = \dfrac{5}{2}$.

- Next we consider $\lim\limits_{x \to 2} R(x)$.

$$\lim_{x \to 2} R(x) = \lim_{x \to 2} \frac{3(x-2)(x+2)}{(2x-5)(x-2)} = \lim_{x \to 2} \frac{3(x+2)}{2x-5} = \frac{3(2+2)}{2 \cdot 2 - 5} = \frac{12}{-1} = -12$$

Since the limit is not infinite, the function R does not have a vertical asymptote at 2.

Since 2 is not in the domain of R, the graph of R has a **hole** at the point $(2, -12)$.

To check for horizontal asymptotes, we find the limits at infinity.

$$\lim_{x \to \infty} R(x) = \lim_{x \to \infty} \frac{3x^2 - 12}{2x^2 - 9x + 10} \underset{(5)}{\uparrow} = \lim_{x \to \infty} \frac{3x^2}{2x^2} = \lim_{x \to \infty} \frac{3}{2} = \frac{3}{2}$$

$$\lim_{x \to -\infty} R(x) = \lim_{x \to -\infty} \frac{3x^2 - 12}{2x^2 - 9x + 10} \underset{(5)}{\uparrow} = \lim_{x \to -\infty} \frac{3x^2}{2x^2} = \lim_{x \to -\infty} \frac{3}{2} = \frac{3}{2}$$

The line $y = \dfrac{3}{2}$ is a horizontal asymptote of the graph of R as x approaches $-\infty$ and as x approaches ∞. ∎

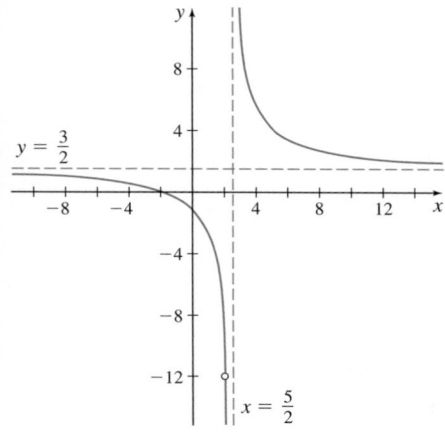

$y = \dfrac{3}{2}$

$x = \dfrac{5}{2}$

Figure 63 $R(x) = \dfrac{3x^2 - 12}{2x^2 - 9x + 10}$

The graph of R and its asymptotes are shown in Figure 63. Notice the hole in the graph at the point $(2, -12)$.

NOW WORK   Problem 69 and AP® Practice Problems 8, 10, and 12.

1.5 Assess Your Understanding

Concepts and Vocabulary

1. **True or False** ∞ is a number.

2. (a) $\lim\limits_{x \to 0^-} \dfrac{1}{x} = $ _____ ; (b) $\lim\limits_{x \to 0^+} \dfrac{1}{x} = $ _____ ;

   (c) $\lim\limits_{x \to 0^+} \ln x = $ _____

3. **True or False**   The graph of a rational function has a vertical asymptote at every number x at which the function is not defined.

4. If $\lim\limits_{x \to 4} f(x) = \infty$, then the line $x = 4$ is a(n) _____ asymptote of the graph of f.

5. (a) $\lim\limits_{x \to \infty} \dfrac{1}{x} = $ _____ ; (b) $\lim\limits_{x \to \infty} \dfrac{1}{x^2} = $ _____ ; (c) $\lim\limits_{x \to \infty} \ln x = $ _____

6. **True or False** $\lim\limits_{x \to -\infty} 5 = 0$.

7. (a) $\lim\limits_{x \to -\infty} e^x = $ _____ ; (b) $\lim\limits_{x \to \infty} e^x = $ _____ ; (c) $\lim\limits_{x \to \infty} e^{-x} = $ _____

8. **True or False**   The graph of a function can have at most two horizontal asymptotes.

Skill Building

In Problems 9–16, use the accompanying graph of $y = f(x)$.

9. Find $\lim\limits_{x \to \infty} f(x)$.

10. Find $\lim\limits_{x \to -\infty} f(x)$.

PAGE 134 11. Find $\lim\limits_{x \to -1^-} f(x)$.

12. Find $\lim\limits_{x \to -1^+} f(x)$.

13. Find $\lim\limits_{x \to 3^-} f(x)$.

14. Find $\lim\limits_{x \to 3^+} f(x)$.

PAGE 136 15. Identify all vertical asymptotes.

16. Identify all horizontal asymptotes.

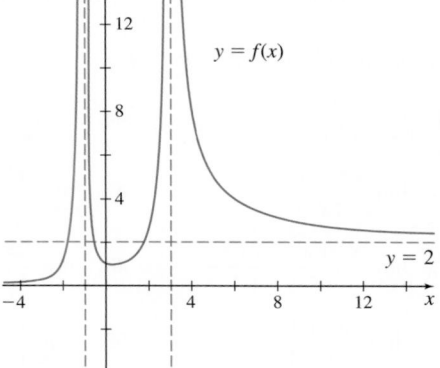

$y = f(x)$

$y = 2$

$x = -1$ $x = 3$

In Problems 17–26, use the graph below of $y = f(x)$.

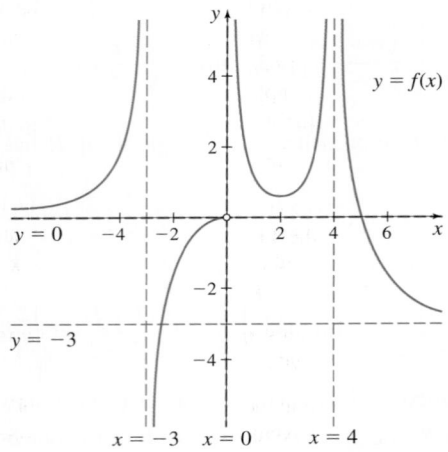

$y = f(x)$

$y = 0$

$y = -3$

$x = -3$ $x = 0$ $x = 4$

17. Find $\lim\limits_{x \to \infty} f(x)$.

18. Find $\lim\limits_{x \to -\infty} f(x)$.

19. Find $\lim\limits_{x \to -3^-} f(x)$.

20. Find $\lim\limits_{x \to -3^+} f(x)$.

21. Find $\lim\limits_{x \to 0^-} f(x)$.

22. Find $\lim\limits_{x \to 0^+} f(x)$.

23. Find $\lim\limits_{x \to 4^-} f(x)$.

24. Find $\lim\limits_{x \to 4^+} f(x)$.

25. Identify all vertical asymptotes.

26. Identify all horizontal asymptotes.

In Problems 27–42, find each limit.

27. $\lim\limits_{x \to 2^-} \dfrac{3x}{x-2}$

28. $\lim\limits_{x \to -4^+} \dfrac{2x+1}{x+4}$

29. $\lim\limits_{x \to 3^+} \dfrac{5}{x^2-9}$

30. $\lim\limits_{x \to 1^-} \dfrac{2x}{x^2-1}$

31. $\lim\limits_{x \to -1^+} \dfrac{5x+3}{x(x+1)}$

32. $\lim\limits_{x \to 0^-} \dfrac{5x+3}{5x(x-1)}$

33. $\lim\limits_{x \to -3^-} \dfrac{1}{x^2-9}$

34. $\lim\limits_{x \to 2^+} \dfrac{x}{x^2-4}$

35. $\lim\limits_{x \to 3} \dfrac{1-x}{(3-x)^2}$

36. $\lim\limits_{x \to -1} \dfrac{x+2}{(x+1)^2}$

37. $\lim\limits_{x \to \pi^-} \cot x$

38. $\lim\limits_{x \to -\pi/2^-} \tan x$

39. $\lim\limits_{x \to \pi/2^+} \csc(2x)$

40. $\lim\limits_{x \to -\pi/2^-} \sec x$

41. $\lim\limits_{x \to -2^+} \ln(x+2)$

42. $\lim\limits_{x \to 4^+} \ln(x-4)$

In Problems 43–60, find each limit.

43. $\lim\limits_{x \to \infty} \dfrac{6}{x^2+9}$

44. $\lim\limits_{x \to -\infty} \dfrac{5}{x^2-4}$

45. $\lim\limits_{x \to \infty} \dfrac{2x+4}{5x}$

46. $\lim\limits_{x \to \infty} \dfrac{x+1}{x}$

47. $\lim\limits_{x \to \infty} \dfrac{x^3+x^2+2x-1}{x^3+x+1}$

48. $\lim\limits_{x \to \infty} \dfrac{2x^2-5x+2}{5x^2+7x-1}$

49. $\lim\limits_{x \to -\infty} \dfrac{x^2+1}{x^3-1}$

50. $\lim\limits_{x \to \infty} \dfrac{x^2-2x+1}{x^3+5x+4}$

51. $\lim\limits_{x \to \infty} \left(\dfrac{3x}{2x+5} - \dfrac{x^2+1}{4x^2+8} \right)$

52. $\lim\limits_{x \to \infty} \left(\dfrac{1}{x^2+x+4} - \dfrac{x+1}{3x-1} \right)$

53. $\lim\limits_{x \to -\infty} \left(2e^x \cdot \dfrac{5x+1}{3x} \right)$

54. $\lim\limits_{x \to -\infty} \left(e^x \cdot \dfrac{x^2+x-3}{2x^3-x^2} \right)$

55. $\lim\limits_{x \to \infty} \dfrac{\sqrt{x}+2}{3x-4}$

56. $\lim\limits_{x \to \infty} \dfrac{\sqrt{3x^3}+2}{x^2+6}$

57. $\lim\limits_{x \to \infty} \sqrt{\dfrac{3x^2-1}{x^2+4}}$

58. $\lim\limits_{x \to \infty} \left(\dfrac{16x^3+2x+1}{2x^3+3x} \right)^{2/3}$

59. $\lim\limits_{x \to -\infty} \dfrac{5x^3}{x^2+1}$

60. $\lim\limits_{x \to -\infty} \dfrac{x^4}{x-2}$

In Problems 61–66, find any horizontal or vertical asymptotes of the graph of f.

61. $f(x) = 3 + \dfrac{1}{x}$

62. $f(x) = 2 - \dfrac{1}{x^2}$

63. $f(x) = \dfrac{x^2}{x^2-1}$

64. $f(x) = \dfrac{2x^2-1}{x^2-1}$

65. $f(x) = \dfrac{\sqrt{2x^2-x+10}}{2x-3}$

66. $f(x) = \dfrac{\sqrt[3]{x^2+5x}}{x-6}$

In Problems 67–72, for each rational function R:

(a) Find the domain of R.
(b) Find any horizontal asymptotes of R.
(c) Find any vertical asymptotes of R.
(d) Discuss the behavior of the graph at numbers where R is not defined.

67. $R(x) = \dfrac{-2x^2+1}{2x^3+4x^2}$

68. $R(x) = \dfrac{x^3}{x^4-1}$

69. $R(x) = \dfrac{x^2+3x-10}{2x^2-7x+6}$

70. $R(x) = \dfrac{x(x-1)^2}{(x+3)^3}$

71. $R(x) = \dfrac{x^3-1}{x-x^2}$

72. $R(x) = \dfrac{4x^5}{x^3-1}$

Applications and Extensions

In Problems 73 and 74:
(a) Sketch a graph of a function f that has the given properties.
(b) Define a function that describes the graph.

73. $f(3) = 0$, $\lim\limits_{x \to \infty} f(x) = 1$, $\lim\limits_{x \to -\infty} f(x) = 1$, $\lim\limits_{x \to 1^-} f(x) = \infty$, $\lim\limits_{x \to 1^+} f(x) = -\infty$

74. $f(2) = 0$, $\lim\limits_{x \to \infty} f(x) = 0$, $\lim\limits_{x \to -\infty} f(x) = 0$, $\lim\limits_{x \to 0} f(x) = \infty$, $\lim\limits_{x \to 5^-} f(x) = -\infty$, $\lim\limits_{x \to 5^+} f(x) = \infty$

75. Newton's Law of Cooling Suppose an object is heated to a temperature u_0. Then at time $t = 0$, the object is put into a medium with a constant lower temperature T causing the object to cool. **Newton's Law of Cooling** states that the temperature u of the object at time t is given by $u = u(t) = (u_0 - T)e^{kt} + T$, where $k < 0$ is a constant.

(a) Find $\lim\limits_{t \to \infty} u(t)$. Is this the value you expected? Explain why or why not.

(b) Find $\lim\limits_{t \to 0^+} u(t)$. Is this the value you expected? Explain why or why not.

Source: Submitted by the students of Millikin University.

76. Environment   A utility company burns coal to generate electricity. The cost C, in dollars, of removing $p\%$ of the pollutants emitted into the air is

$$C = \frac{70{,}000p}{100 - p}, \qquad 0 \le p < 100$$

Find the cost of removing:

(a) 45% of the pollutants.

(b) 90% of the pollutants.

(c) Find $\lim\limits_{p \to 100^-} C$.

(d) Interpret the answer found in (c).

77. Pollution Control   The cost C, in thousands of dollars, to remove a pollutant from a lake is

$$C(x) = \frac{5x}{100 - x}, \qquad 0 \le x < 100$$

where x is the percent of pollutant removed. Find $\lim\limits_{x \to 100^-} C(x)$. Interpret your answer.

78. Population Model   A rare species of insect was discovered in the Amazon Rain Forest. To protect the species, entomologists declared the insect endangered and transferred 25 insects to a protected area. The population P of the new colony t days after the transfer is

$$P(t) = \frac{50(1 + 0.5t)}{2 + 0.01t}$$

(a) What is the projected size of the colony after 1 year (365 days)?

(b) What is the largest population that the protected area can sustain? That is, find $\lim\limits_{t \to \infty} P(t)$.

(c) Graph the population P as a function of time t.

(d) Use the graph from (c) to describe the regeneration of the insect population. Does the graph support the answer to (b)?

79. Population of an Endangered Species   Often environmentalists capture several members of an endangered species and transport them to a controlled environment where they can produce offspring and regenerate their population. Suppose six American bald eagles are captured, tagged, transported to Montana, and set free. Based on past experience, the environmentalists expect the population to grow according to the model

$$P(t) = \frac{500}{1 + 82.3e^{-0.162t}}$$

where t is measured in years.

(a) If the model is correct, how many bald eagles can the environment sustain? That is, find $\lim\limits_{t \to \infty} P(t)$.

(b) Graph the population P as a function of time t.

(c) Use the graph from (b) to describe the growth of the bald eagle population. Does the graph support the answer to (a)?

80. Hailstones   Hailstones typically originate at an altitude of about 3000 meters (m). If a hailstone falls from 3000 m with no air resistance, its speed when it hits the ground would be about 240 meters/second (m/s), which is 540 miles/hour (mi/h)! That would be deadly! But air resistance slows the hailstone considerably. Using a simple model of air resistance, the speed $v = v(t)$ of a hailstone of mass m as a function of time t is given by $v(t) = \dfrac{mg}{k}(1 - e^{-kt/m})$ m/s, where $g = 9.8$ m/s^2 and k is a constant that depends on the size of the hailstone, its mass, and the conditions of the air. For a hailstone with a diameter $d = 1$ centimeter (cm) and mass m $= 4.8 \times 10^{-4}$ kg, k has been measured to be 3.4×10^{-4} kg/s.

(a) Determine the limiting speed of the hailstone by finding $\lim\limits_{t \to \infty} v(t)$. Express your answer in meters per second and miles per hour, using the fact that 1 mi/h ≈ 0.447 m/s. This speed is called the **terminal speed** of the hailstone.

(b) Graph $v = v(t)$. Does the graph support the answer to (a)?

81. Damped Harmonic Motion   The motion of a spring is given by the function

$$x(t) = 1.2e^{-t/2}\cos t + 2.4e^{-t/2}\sin t$$

where x is the distance in meters from the equilibrium position and t is the time in seconds.

(a) Graph $y = x(t)$. What is $\lim\limits_{t \to \infty} x(t)$, as suggested by the graph?

(b) Find $\lim\limits_{t \to \infty} x(t)$.

(c) Compare the results of (a) and (b). Is the answer to (b) supported by the graph in (a)?

82. Decomposition of Chlorine in a Pool   Under certain water conditions, the free chlorine (hypochlorous acid, HOCl) in a swimming pool decomposes according to the law of uninhibited decay, $C = C(t) = C(0)e^{kt}$, where $C = C(t)$ is the amount (in parts per million, ppm) of free chlorine present at time t (in hours) and k is a negative number that represents the rate of decomposition. After shocking his pool, Ben immediately tested the water and found the concentration of free chlorine to be $C_0 = C(0) = 2.5$ ppm. Twenty-four hours later, Ben tested the water again and found the amount of free chlorine to be 2.2 ppm.

(a) What amount of free chlorine will be left after 72 hours?

(b) When the free chlorine reaches 1.0 ppm, the pool should be shocked again. How long can Ben go before he must shock the pool again?

(c) Find $\lim\limits_{t \to \infty} C(t)$.

(d) Interpret the answer found in (c).

83. Decomposition of Sucrose   Reacting with water in an acidic solution at 35 °C, the amount A of sucrose ($C_{12}H_{22}O_{11}$) decomposes into glucose ($C_6H_{12}O_6$) and fructose ($C_6H_{12}O_6$) according to the law of uninhibited decay $A = A(t) = A(0)e^{kt}$, where $A = A(t)$ is the amount (in moles) of sucrose present at time t (in minutes) and k is a negative number that represents the rate of decomposition. An initial amount $A_0 = A(0) = 0.40$ mole of sucrose decomposes to 0.36 mole in 30 minutes.

(a) How much sucrose will remain after 2 hours?

(b) How long will it take until 0.10 mole of sucrose remains?

(c) Find $\lim\limits_{t \to \infty} A(t)$.

(d) Interpret the answer found in (c).

84. Macrophotography The focal length of a camera lens can be approximated by a thin lens. A thin lens of focal length f obeys the thin-lens equation $\dfrac{1}{f} = \dfrac{1}{p} + \dfrac{1}{q}$, where $p > f$ is the distance from the lens to the object being photographed and q is the distance from the lens to the image formed by the lens. See the figure below. To photograph an object, the object's image must be formed on the photo sensors of the camera, which can only occur if q is positive.

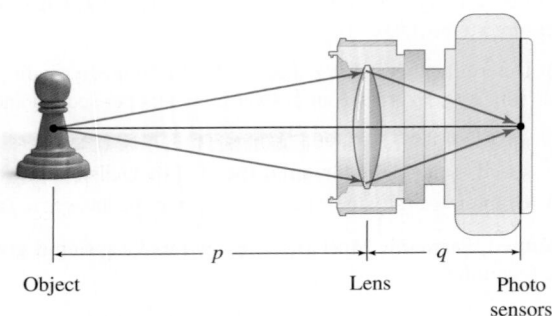

Object   Lens   Photo
                                          sensors

(a) Is the distance q of the image from the lens continuous as the distance of the object being photographed approaches the focal length f of the lens? *Hint*: First solve the thin-lens equation for q and then find $\lim\limits_{p \to f^+} q$.

(b) Use the result from (a) to explain why a camera (or any lens) cannot focus on an object placed close to its focal length.

In Problems 85 and 86, find conditions on a, b, c, and d so that the graph of f has no horizontal or vertical asymptotes.

85. $f(x) = \dfrac{ax^3 + b}{cx^4 + d}$

86. $f(x) = \dfrac{ax + b}{cx + d}$

87. Explain why the following properties are true. Give an example of each.

(a) If n is an even positive integer, then
$$\lim_{x \to c} \frac{1}{(x - c)^n} = \infty.$$

(b) If n is an odd positive integer, then
$$\lim_{x \to c^-} \frac{1}{(x - c)^n} = -\infty.$$

(c) If n is an odd positive integer, then
$$\lim_{x \to c^+} \frac{1}{(x - c)^n} = \infty.$$

88. Explain why a rational function, whose numerator and denominator have no common zeros, will have vertical asymptotes at each point of discontinuity.

89. Explain why a polynomial function of degree 1 or higher cannot have any asymptotes.

90. If P and Q are polynomials of degree m and n, respectively, discuss $\lim\limits_{x \to \infty} \dfrac{P(x)}{Q(x)}$ when:

(a) $m > n$ (b) $m = n$ (c) $m < n$

91. (a) Use a table to investigate $\lim\limits_{x \to \infty} \left(1 + \dfrac{1}{x}\right)^x$.

(CAS) (b) Find $\lim\limits_{x \to \infty} \left(1 + \dfrac{1}{x}\right)^x$.

(c) Compare the results from (a) and (b). Explain the possible causes of any discrepancy.

92. Prove that $\lim\limits_{x \to \pm\infty} \dfrac{k}{x^p} = 0$, for any real number k, and $p > 0$, provided x^p is defined if $x < 0$.

Challenge Problems

93. Kinetic Energy At low speeds the kinetic energy K, that is, the energy due to the motion of an object of mass m and speed v, is given by the formula $K = K(v) = \dfrac{1}{2}mv^2$. But this formula is only an approximation to the general formula and works only for speeds much less than the speed of light, c. The general formula, which holds for all speeds, is

$$K_{\text{gen}}(v) = mc^2 \left[\frac{1}{\sqrt{1 - \dfrac{v^2}{c^2}}} - 1 \right]$$

(a) As an object is accelerated closer and closer to the speed of light, what does its kinetic energy K_{gen} approach?

(b) What does the result suggest about the possibility of reaching the speed of light?

94. $\lim\limits_{x \to \infty} \left(1 + \dfrac{1}{x}\right) = 1$, but $\lim\limits_{x \to \infty} \left(1 + \dfrac{1}{x}\right)^x > 1$. Discuss why the property $\lim\limits_{x \to \infty} [f(x)]^n = \left[\lim\limits_{x \to \infty} f(x)\right]^n$ cannot be used to find the second limit.

AP® Practice Problems

Multiple-Choice Questions

[PAGE 144] **1.** For $x > 0$, the line $y = 1$ is an asymptote of the graph of a function f. Which of the following statements must be true?

(A) $f(x) \neq 1$ for $x > 0$ (B) $\lim\limits_{x \to 1} f(x) = \infty$

(C) $\lim\limits_{x \to \infty} f(x) = 1$ (D) $\lim\limits_{x \to -\infty} f(x) = 1$

[PAGE 141] **2.** $\lim\limits_{x \to \infty} \dfrac{3x^3 + 4x^2 - x + 10}{2x^4 - x^3 + 2x^2 - 2} =$

(A) -5 (B) 0 (C) $\dfrac{3}{2}$ (D) ∞

[PAGE 138] **3.** $\lim\limits_{x \to \infty} \dfrac{5x^3 - x}{8 - x^3} =$

(A) -5 (B) $\dfrac{5}{8}$ (C) 5 (D) ∞

[PAGE 144] **4.** Find all the horizontal asymptotes of the graph of $y = \dfrac{2 + 3^x}{4 - 3^x}$.

(A) $y = -1$ only (B) $y = \dfrac{1}{2}$ only

(C) $y = -1$ and $y = 0$ (D) $y = -1$ and $y = \dfrac{1}{2}$

[PAGE 136] **5.** Find all the vertical asymptotes of the graph of

$$r(x) = \frac{x^2 + 5x + 6}{x^3 - 4x}$$

(A) $x = 0$ and $x = -2$ (B) $x = 0$ and $x = 2$

(C) $x = -2$ and $x = 2$ (D) $x = 0$, $x = -2$ and $x = 2$

[PAGE 139] **6.** $\lim\limits_{x \to -\infty} \dfrac{\sqrt{8x^2 - 4x}}{x + 2} =$

(A) $-\infty$ (B) $-2\sqrt{2}$ (C) 4 (D) $2\sqrt{2}$

[PAGE 144] **7.** The graph of which of the following functions has an asymptote of $y = 1$?

(A) $y = \cos x$ (B) $y = \dfrac{x - 1}{x}$

(C) $y = e^{-x}$ (D) $y = \ln x$

[PAGE 145] **8.** If the graph of $f(x) = \dfrac{ax - b}{x + c}$ has a vertical asymptote $x = -5$ and horizontal asymptote $y = -3$, then $a + c =$

(A) -8 (B) -2 (C) $\dfrac{3}{5}$ (D) 2

[PAGE 135] **9.** $\lim\limits_{x \to 1^-} \dfrac{x}{\ln x} =$

(A) $-\infty$ (B) -1 (C) 1 (D) ∞

[PAGE 145] **10.** The function $f(x) = \dfrac{2x}{|x| - 1}$ has

(A) no vertical asymptote and one horizontal asymptote.

(B) one vertical asymptote and one horizontal asymptote.

(C) two vertical asymptotes and one horizontal asymptote.

(D) two vertical asymptotes and two horizontal asymptotes.

[PAGE 135] **11.** $\lim\limits_{x \to 2^-} \dfrac{5x + 1}{2x - 4} =$

(A) $-\infty$ (B) $-\dfrac{5}{2}$ (C) $\dfrac{5}{2}$ (D) ∞

See the **BREAK IT DOWN** *on page 164 for a stepped out solution to* *AP® Practice Problem 11.*

Free-Response Question

[PAGE 145] **12.** Suppose $R(x) = \dfrac{e^x(x^2 - 5x + 4)}{x^2 - 16}$

(a) For what numbers, if any, is R discontinuous? If possible, define R so R is continuous at these numbers.

(b) Find any horizontal asymptotes of the graph of R.

(c) Find any vertical asymptotes of the graph of R.

1.6 The ε-δ Definition of a Limit

OBJECTIVES *When you finish this section, you should be able to:*

1 Use the ε-δ definition of a limit (p. 152)

Throughout the chapter, we stated that we could be sure a limit was correct only if it was based on the ε-δ definition of a limit. In this section, we examine this definition and how to use it to prove a limit exists, to verify the value of a limit, and to show that a limit does not exist.

Consider the function f defined by

$$f(x) = \begin{cases} 3x + 1 & \text{if } x \neq 2 \\ 10 & \text{if } x = 2 \end{cases}$$

whose graph is given in Figure 64.

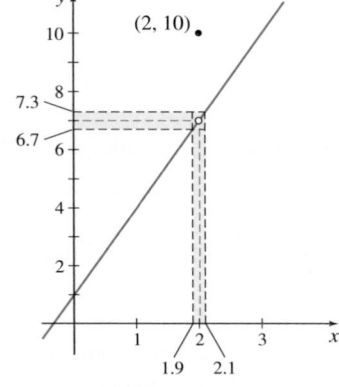

Figure 64 $f(x) = \begin{cases} 3x + 1 & \text{if } x \neq 2 \\ 10 & \text{if } x = 2 \end{cases}$

As x gets closer to 2, the value $f(x)$ gets closer to 7. If in fact, by taking x close enough to 2, we can make $f(x)$ as close to 7 as we please, then $\lim\limits_{x \to 2} f(x) = 7$.

Suppose we want $f(x)$ to differ from 7 by less than 0.3; that is,

$$-0.3 < f(x) - 7 < 0.3$$
$$6.7 < f(x) < 7.3$$

How close must x be to 2? First, we must require $x \neq 2$ because when $x = 2$, then $f(x) = f(2) = 10$, and we obtain $6.7 < 10 < 7.3$, which is impossible.

Then, when $x \neq 2$,

$$-0.3 < f(x) - 7 < 0.3$$
$$-0.3 < (3x + 1) - 7 < 0.3$$
$$-0.3 < 3x - 6 < 0.3$$
$$-0.3 < 3(x - 2) < 0.3$$
$$\frac{-0.3}{3} < x - 2 < \frac{0.3}{3}$$
$$-0.1 < x - 2 < 0.1$$
$$|x - 2| < 0.1$$

That is, whenever $x \neq 2$ and x differs from 2 by less than 0.1, then $f(x)$ differs from 7 by less than 0.3.

Now, generalizing the question, we ask, for $x \neq 2$, how close must x be to 2 to guarantee that $f(x)$ differs from 7 by less than any given positive number ε? (ε might be extremely small.) The statement "$f(x)$ differs from 7 by less than ε" means

$$-\varepsilon < f(x) - 7 < \varepsilon$$
$$7 - \varepsilon < f(x) < 7 + \varepsilon \qquad \text{Add 7 to each expression.}$$

When $x \neq 2$, then $f(x) = 3x + 1$, so

$$7 - \varepsilon < 3x + 1 < 7 + \varepsilon$$

Now we want the middle term to be $x - 2$. This is done as follows:

$$7 - \varepsilon < 3x + 1 < 7 + \varepsilon$$
$$6 - \varepsilon < 3x < 6 + \varepsilon \qquad \text{Subtract 1 from each expression.}$$
$$2 - \frac{\varepsilon}{3} < x < 2 + \frac{\varepsilon}{3} \qquad \text{Divide each expression by 3.}$$
$$-\frac{\varepsilon}{3} < x - 2 < \frac{\varepsilon}{3} \qquad \text{Subtract 2 from each expression.}$$
$$|x - 2| < \frac{\varepsilon}{3}$$

The answer to our question is

"x must be within $\dfrac{\varepsilon}{3}$ of 2, but not equal to 2, to guarantee that $f(x)$ is within ε of 7."

That is, whenever $x \neq 2$ and x differs from 2 by less than $\dfrac{\varepsilon}{3}$, then $f(x)$ differs from 7 by less than ε, which can be restated as

$$\text{whenever} \quad \begin{bmatrix} x \neq 2 \text{ and } x \text{ differs from 2} \\ \text{by less than } \delta = \dfrac{\varepsilon}{3} \end{bmatrix} \quad \text{then} \quad \begin{bmatrix} f(x) \text{ differs from 7} \\ \text{by less than } \varepsilon \end{bmatrix}$$

Notation is used to shorten the statement. We shorten the phrase "ε is any given positive number" by writing $\varepsilon > 0$. Then the statement $f(x)$ differs from 7 by less than ε" is written

$$|f(x) - 7| < \varepsilon$$

Similarly, the statement "x differs from 2 by less than δ" is written $|x - 2| < \delta$. The statement "$x \neq 2$" is handled by writing

$$0 < |x - 2| < \delta$$

So for our example, we write:

given any $\varepsilon > 0$, then there is a number $\delta > 0$ so that

whenever $0 < |x - 2| < \delta$, then $|f(x) - 7| < \varepsilon$

Since the number $\delta = \dfrac{\varepsilon}{3}$ satisfies the inequalities for any number $\varepsilon > 0$, we conclude that $\lim\limits_{x \to 2} f(x) = 7$.

NOW WORK | Problem 41.

This discussion above explains the *definition of the limit of a function.*

DEFINITION Limit of a Function

Suppose f is a function defined everywhere in an open interval containing c, except possibly at c. Then the **limit as x approaches c of $f(x)$ is L**, written

$$\boxed{\lim_{x \to c} f(x) = L}$$

if, given any number $\varepsilon > 0$, there is a number $\delta > 0$ so that

$$\boxed{\text{whenever } \;\; 0 < |x - c| < \delta \qquad \text{then} \qquad |f(x) - L| < \varepsilon}$$

This definition is commonly called the **ε-δ definition of a limit** of a function.

Figure 65 illustrates the definition for three choices of ε. Compare Figures 65(a) and 65(b). Notice that in Figure 65(b), the smaller ε requires a smaller δ. Figure 65(c) illustrates what happens if δ is too large for the choice of ε; here there are values of f, for example, at x_1 and x_2, for which $|f(x) - L| \not< \varepsilon$.

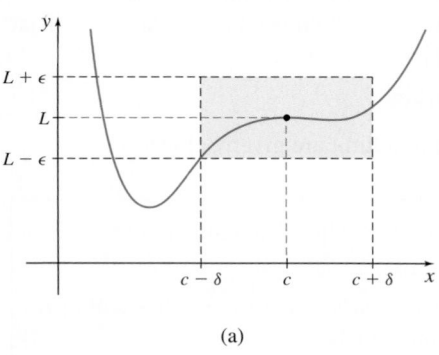

(a)

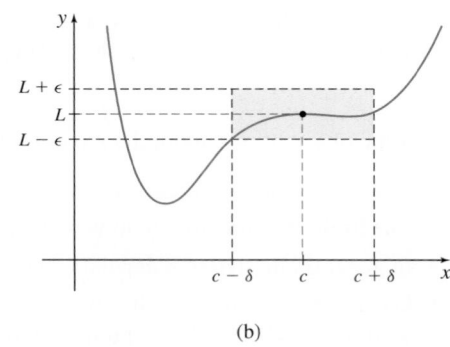

(b)

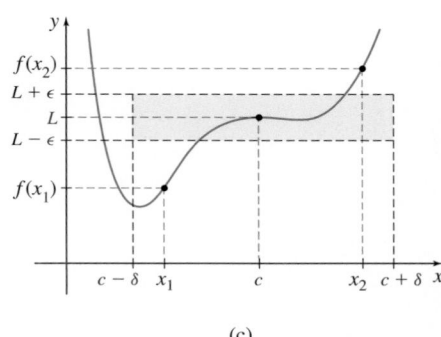

(c)

Figure 65

1 Use the ε-δ Definition of a Limit

EXAMPLE 1 Using the ε-δ Definition of a Limit

Use the ε-δ definition of a limit to prove $\lim\limits_{x \to -1} (1 - 2x) = 3$.

Solution

Given any $\varepsilon > 0$, we must show there is a number $\delta > 0$ so that

$$\text{whenever} \quad 0 < |x - (-1)| = |x + 1| < \delta \quad \text{then} \quad |(1 - 2x) - 3| < \varepsilon$$

The idea is to find a connection between $|x - (-1)| = |x + 1|$ and $|(1 - 2x) - 3|$. Since

$$|(1 - 2x) - 3| = |-2x - 2| = |-2(x + 1)| = |-2| \cdot |x + 1| = 2|x + 1|$$

we see that for any $\varepsilon > 0$,

$$\text{whenever} \quad |x - (-1)| = |x + 1| < \delta = \frac{\varepsilon}{2} \quad \text{then} \quad |(1 - 2x) - 3| = 2|x + 1| < 2\delta = \varepsilon$$

That is, given any $\varepsilon > 0$ there is a δ, $\delta = \dfrac{\varepsilon}{2}$, so that whenever $0 < |x - (-1)| < \delta$, we have $|(1 - 2x) - 3| < \varepsilon$. This proves that $\lim\limits_{x \to -1} (1 - 2x) = 3$. ∎

NOW WORK Problem 21.

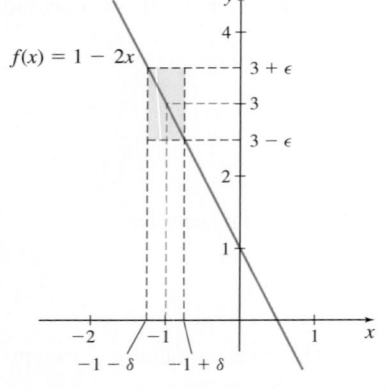

Figure 66 $\lim\limits_{x \to -1} (1 - 2x) = 3$

A geometric interpretation of the ε-δ definition is shown in Figure 66. We see that whenever x on the horizontal axis is between $-1 - \delta$ and $-1 + \delta$, but not equal to -1, then $f(x)$ on the vertical axis is between the horizontal lines $y = 3 + \varepsilon$ and $y = 3 - \varepsilon$. So, $\lim\limits_{x \to -1} f(x) = 3$ describes the behavior of f near -1.

EXAMPLE 2 Using the ε-δ Definition of a Limit

Use the ε-δ definition of a limit to prove that:

(a) $\lim\limits_{x \to c} A = A$, where A and c are real numbers.

(b) $\lim\limits_{x \to c} x = c$, where c is a real number.

Solution

(a) $f(x) = A$ is the constant function whose graph is a horizontal line. Given any $\varepsilon > 0$, we must find $\delta > 0$ so that whenever $0 < |x - c| < \delta$, then $|f(x) - A| < \varepsilon$.

Since $|A - A| = 0$, then $|f(x) - A| < \varepsilon$ no matter what positive number δ is used. That is, any choice of δ guarantees that whenever $0 < |x - c| < \delta$, then $|f(x) - A| < \varepsilon$.

(b) $f(x) = x$ is the identity function. Given any $\varepsilon > 0$, we must find δ so that whenever $0 < |x - c| < \delta$, then $|f(x) - c| = |x - c| < \varepsilon$. The easiest choice is to make $\delta = \varepsilon$. That is, whenever $0 < |x - c| < \delta = \varepsilon$, then $\underset{\underset{\displaystyle f(x) = x}{\uparrow}}{|f(x) - c|} = |x - c| < \varepsilon$. ∎

Some observations about the ε-δ definition of a limit are given below:

- The limit of a function in no way depends on the value of the function at c.
- In general, the size of δ depends on the size of ε.
- For any ε, if a suitable δ has been found, any *smaller positive number* will also work for δ. That is, δ is not uniquely determined when ε is given.

EXAMPLE 3   **Using the ε-δ Definition of a Limit**

Prove: $\lim\limits_{x \to 2} x^2 = 4$

Solution

Given any $\varepsilon > 0$, we must show there is a number $\delta > 0$ so that

whenever $0 < |x - 2| < \delta$ then $|x^2 - 4| < \varepsilon$

To establish a connection between $|x^2 - 4|$ and $|x - 2|$, we write $\left| x^2 - 4 \right|$ as

$$\left| x^2 - 4 \right| = |(x + 2)(x - 2)| = |x + 2| \cdot |x - 2|$$

Now, if we can find a number K for which $|x + 2| < K$, then we can choose $\delta = \dfrac{\varepsilon}{K}$. To find K, we restrict x to some interval centered at 2. For example, suppose the distance between x and 2 is less than 1. Then

$$|x - 2| < 1$$
$$-1 < x - 2 < 1$$
$$1 < x < 3 \qquad \text{Simplify.}$$
$$1 + 2 < x + 2 < 3 + 2 \qquad \text{Add 2 to each part.}$$
$$3 < x + 2 < 5$$

In particular, we have $|x + 2| < 5$. It follows that whenever $|x - 2| < 1$,

$$|x^2 - 4| = |x + 2| \cdot |x - 2| < 5\,|x - 2|$$

If $|x - 2| < \delta = \dfrac{\varepsilon}{5}$, then $\left| x^2 - 4 \right| < 5\,|x - 2| < 5 \cdot \dfrac{\varepsilon}{5} = \varepsilon$, as desired.

But before choosing $\delta = \dfrac{\varepsilon}{5}$, we must remember that there are two constraints on $|x - 2|$. Namely,

$$|x - 2| < 1 \qquad \text{and} \qquad |x - 2| < \frac{\varepsilon}{5}$$

To ensure that both inequalities are satisfied, we select δ to be the smaller of the numbers 1 and $\dfrac{\varepsilon}{5}$, abbreviated as $\delta = \min\left\{ 1, \dfrac{\varepsilon}{5} \right\}$. Now,

whenever $|x - 2| < \delta = \min\left\{ 1, \dfrac{\varepsilon}{5} \right\}$ then $|x^2 - 4| < \varepsilon$

proving $\lim\limits_{x \to 2} x^2 = 4$. ■

NOW WORK   Problem 23.

In Example 3, the decision to restrict x so that $|x - 2| < 1$ was completely arbitrary. However, since we are looking for x close to 2, the interval chosen should be small. In Problem 40, you are asked to verify that if we had restricted x so that $|x - 2| < \dfrac{1}{3}$, then the choice for δ would be less than or equal to the smaller of $\dfrac{1}{3}$ and $\dfrac{3\varepsilon}{13}$; that is, $\delta \le \min\left\{ \dfrac{1}{3}, \dfrac{3\varepsilon}{13} \right\}$.

EXAMPLE 4 Using the ε-δ Definition of a Limit

Prove $\lim\limits_{x\to c}\dfrac{1}{x}=\dfrac{1}{c}$, where $c>0$.

Solution

The domain of $f(x)=\dfrac{1}{x}$ is $\{x\,|\,x\neq 0\}$.

For any $\varepsilon>0$, we need to find a positive number δ so that whenever $0<|x-c|<\delta$, then $\left|\dfrac{1}{x}-\dfrac{1}{c}\right|<\varepsilon$. For $x\neq 0$, and $c>0$, we have

$$\left|\frac{1}{x}-\frac{1}{c}\right|=\left|\frac{c-x}{xc}\right|=\frac{|c-x|}{|x|\cdot|c|}=\frac{|x-c|}{c|x|}$$

The idea is to find a connection between

$$|x-c|\qquad\text{and}\qquad\frac{|x-c|}{c|x|}$$

We proceed as in Example 3. Since we are interested in x near c, we restrict x to a small interval around c, say, $|x-c|<\dfrac{c}{2}$. Then,

$$-\frac{c}{2}<x-c<\frac{c}{2}$$

$$\frac{c}{2}<x<\frac{3c}{2}\qquad\text{Add }c\text{ to each expression.}$$

Since $c>0$, then $x>\dfrac{c}{2}>0$, and $\dfrac{1}{x}<\dfrac{2}{c}$. Now

$$\left|\frac{1}{x}-\frac{1}{c}\right|=\frac{|x-c|}{c|x|}<\frac{2}{c^2}\cdot|x-c|\qquad\text{Substitute }\frac{1}{|x|}<\frac{2}{c}.$$

We can make $\left|\dfrac{1}{x}-\dfrac{1}{c}\right|<\varepsilon$ by choosing $\delta=\dfrac{c^2}{2}\varepsilon$. Then

whenever $|x-c|<\delta=\dfrac{c^2}{2}\varepsilon$, we have $\left|\dfrac{1}{x}-\dfrac{1}{c}\right|<\dfrac{2}{c^2}\cdot|x-c|<\dfrac{2}{c^2}\cdot\left(\dfrac{c^2}{2}\cdot\varepsilon\right)=\varepsilon$

But remember, there are two restrictions on $|x-c|$.

$$|x-c|<\frac{c}{2}\qquad\text{and}\qquad|x-c|<\frac{c^2}{2}\cdot\varepsilon$$

So, given any $\varepsilon>0$, we choose $\delta=\min\left(\dfrac{c}{2},\dfrac{c^2}{2}\cdot\varepsilon\right)$. Then whenever $0<|x-c|<\delta$, we have $\left|\dfrac{1}{x}-\dfrac{1}{c}\right|<\varepsilon$. This proves $\lim\limits_{x\to c}\dfrac{1}{x}=\dfrac{1}{c}$, $c>0$. ∎

NOW WORK Problem 25.

The ε-δ definition of a limit can be used to show that a limit does not exist or that a limit is not equal to a specific number. Example 5 illustrates how the ε-δ definition of a limit is used to show that a limit is not equal to a specific number.

EXAMPLE 5   **Showing a Limit Is Not Equal to a Specific Number**

Use the ε-δ definition of a limit to prove the statement $\lim\limits_{x\to 3}(4x-5) \neq 10$.

Solution

> **NOTE** In a proof by contradiction, we assume that the conclusion is not true and then show this leads to a contradiction.

We use a proof by contradiction. Assume $\lim\limits_{x\to 3}(4x-5)=10$ and choose $\varepsilon = 1$. (Any smaller positive number ε will also work.) Then there is a number $\delta > 0$, so that

$$\text{whenever} \quad 0 < |x-3| < \delta \quad \text{then} \quad |(4x-5)-10| < 1$$

We simplify the right inequality.

$$|(4x-5)-10| = |4x-15| < 1$$
$$-1 < 4x-15 < 1$$
$$14 < 4x < 16$$
$$3.5 < x < 4$$

> **NOTE** For example, if $\delta = \dfrac{1}{4}$, then
> $$3 - \frac{1}{4} < x < 3 + \frac{1}{4}$$
> $$2.75 < x < 3.25$$
> contradicting $3.5 < x < 4$.

According to our assumption, whenever $0 < |x-3| < \delta$, then $3.5 < x < 4$. Regardless of the value of δ, the inequality $0 < |x-3| < \delta$ is satisfied by a number x that is less than 3. This contradicts the fact that $3.5 < x < 4$. The contradiction means that $\lim\limits_{x\to 3}(4x-5) \neq 10$. ∎

NOW WORK   Problem 33.

EXAMPLE 6   **Showing a Limit Does Not Exist**

The **Dirichlet function** is defined by

$$f(x) = \begin{cases} 1 & \text{if } x \text{ is rational} \\ 0 & \text{if } x \text{ is irrational} \end{cases}$$

Prove $\lim\limits_{x\to c} f(x)$ does not exist for any c.

Solution

We use a proof by contradiction. That is, we assume that $\lim\limits_{x\to c} f(x)$ exists and show that this leads to a contradiction.

Assume $\lim\limits_{x\to c} f(x) = L$ for some number c. Now if we are given $\varepsilon = \dfrac{1}{2}$ (or any smaller positive number), then there is a positive number δ, so that

$$\text{whenever} \quad 0 < |x-c| < \delta \quad \text{then} \quad |f(x)-L| < \frac{1}{2}$$

Suppose x_1 is a rational number satisfying $0 < |x_1 - c| < \delta$, and x_2 is an irrational number satisfying $0 < |x_2 - c| < \delta$. Then from the definition of the function f,

$$f(x_1) = 1 \quad \text{and} \quad f(x_2) = 0$$

Using these values in the inequality $|f(x) - L| < \varepsilon$, we get

$$|f(x_1) - L| = |1 - L| < \frac{1}{2} \qquad \text{and} \qquad |f(x_2) - L| = |0 - L| < \frac{1}{2}$$

$$-\frac{1}{2} < 1 - L < \frac{1}{2} \qquad\qquad -\frac{1}{2} < -L < \frac{1}{2}$$

$$-\frac{3}{2} < -L < -\frac{1}{2} \qquad\qquad \frac{1}{2} > L > -\frac{1}{2}$$

$$\frac{1}{2} < L < \frac{3}{2} \qquad\qquad -\frac{1}{2} < L < \frac{1}{2}$$

From the left inequality, we have $L > \frac{1}{2}$, and from the right inequality, we have $L < \frac{1}{2}$. Since it is impossible for both inequalities to be satisfied, we conclude that $\lim\limits_{x \to c} f(x)$ does not exist. ■

EXAMPLE 7   Using the ε-δ Definition of a Limit

Prove that if $\lim\limits_{x \to c} f(x) > 0$, then there is an open interval around c, for which $f(x) > 0$ everywhere in the interval except possibly at c.

Solution

Suppose $\lim\limits_{x \to c} f(x) = L > 0$. Then given any $\varepsilon > 0$, there is a $\delta > 0$ so that

$$\text{whenever} \quad 0 < |x - c| < \delta \quad \text{then} \quad |f(x) - L| < \varepsilon$$

If $\varepsilon = \dfrac{L}{2}$, then from the definition of limit, there is a $\delta > 0$ so that

$$\text{whenever } 0 < |x - c| < \delta \quad \text{then } |f(x) - L| < \frac{L}{2} \quad \text{or equivalently,} \quad \frac{L}{2} < f(x) < \frac{3L}{2}$$

Since $\dfrac{L}{2} > 0$, the last statement proves our assertion that $f(x) > 0$ for all x in the interval $(c - \delta, c + \delta)$ except possibly at c. ■

In Problem 43, you are asked to prove the theorem stating that if $\lim\limits_{x \to c} f(x) < 0$, then there is an open interval around c, for which $f(x) < 0$ everywhere in the interval, except possibly at c.

We close this section with the ε-δ definitions of limits at infinity and infinite limits.

DEFINITION Limit at Infinity

Let f be a function defined on an open interval (b, ∞). Then f has a **limit at infinity**

$$\lim_{x \to \infty} f(x) = L$$

where L is a real number, if given any number $\varepsilon > 0$, there is a positive number M so that whenever $x > M$, then $|f(x) - L| < \varepsilon$.

If f is a function defined on an open interval $(-\infty, a)$, then

$$\lim_{x \to -\infty} f(x) = L$$

if given any number $\varepsilon > 0$, there is a negative number N so that whenever $x < N$, then $|f(x) - L| < \varepsilon$.

Figures 67 and 68 illustrate limits at infinity.

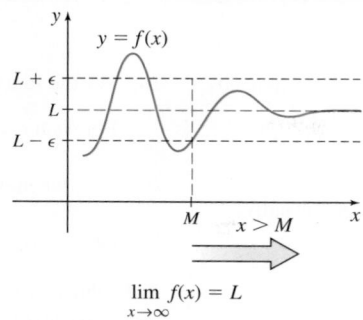

$$\lim_{x\to\infty} f(x) = L$$

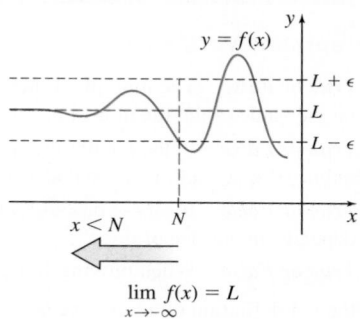

$$\lim_{x\to-\infty} f(x) = L$$

Figure 67 For any $\varepsilon > 0$, there is a positive number M so that whenever $x > M$, then $|f(x) - L| < \varepsilon$.

Figure 68 For any $\varepsilon > 0$, there is a negative number N so that whenever $x < N$, then $|f(x) - L| < \varepsilon$.

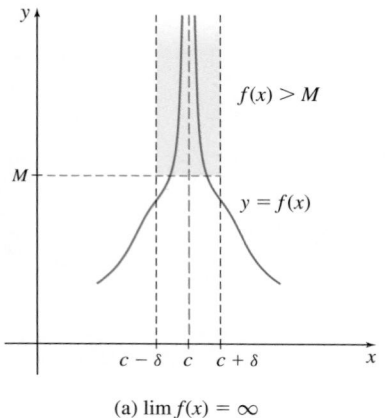

(a) $\lim_{x\to c} f(x) = \infty$

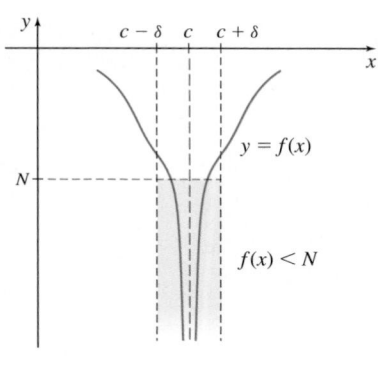

(b) $\lim_{x\to c} f(x) = -\infty$

Figure 69

DEFINITION Infinite Limit

Let f be a function defined everywhere on an open interval containing c, except possibly at c. Then $f(x)$ becomes unbounded in the positive direction (has an infinite limit) as x approaches c, written

$$\lim_{x\to c} f(x) = \infty$$

if, for every positive number M, a positive number δ exists so that

$$\text{whenever } 0 < |x - c| < \delta \text{ then } f(x) > M.$$

Similarly, $f(x)$ becomes unbounded in the negative direction (has an infinite limit) as x approaches c, written

$$\lim_{x\to c} f(x) = -\infty$$

if, for every negative number N, a positive number δ exists so that

$$\text{whenever } 0 < |x - c| < \delta \text{ then } f(x) < N.$$

Figure 69 illustrates infinite limits.

DEFINITION Infinite Limit at Infinity

Let f be a function defined on an open interval (b, ∞). Then f has an **infinite limit at infinity**

$$\lim_{x\to\infty} f(x) = \infty$$

if for any positive number M, there is a corresponding positive number N so that whenever $x > N$, then $f(x) > M$.

A similar definition applies for infinite limits at negative infinity.

1.6 Assess Your Understanding

Concepts and Vocabulary

1. **True or False** The limit of a function as x approaches c depends on the value of the function at c.

2. **True or False** In the ε-δ definition of a limit, we require $0 < |x - c|$ to ensure that $x \neq c$.

3. **True or False** In an ε-δ proof of a limit, the size of δ usually depends on the size of ε.

4. **True or False** When proving $\lim_{x \to c} f(x) = L$ using the ε-δ definition of a limit, we try to find a connection between $|f(x) - c|$ and $|x - c|$.

5. **True or False** Given any $\varepsilon > 0$, suppose there is a $\delta > 0$ so that whenever $0 < |x - c| < \delta$, then $|f(x) - L| < \varepsilon$. Then $\lim_{x \to c} f(x) = L$.

6. **True or False** A function f has a limit L at infinity, if for any given $\varepsilon > 0$, there is a positive number M so that whenever $x > M$, then $|f(x) - L| > \varepsilon$.

Skill Building

In Problems 7–12, for each limit, find the largest δ that "works" for the given ε.

7. $\lim_{x \to 1} (2x) = 2$, $\varepsilon = 0.01$

8. $\lim_{x \to 2} (-3x) = -6$, $\varepsilon = 0.01$

9. $\lim_{x \to 2} (6x - 1) = 11$, $\varepsilon = \dfrac{1}{2}$

10. $\lim_{x \to -3} (2 - 3x) = 11$, $\varepsilon = \dfrac{1}{3}$

11. $\lim_{x \to 2} \left(-\dfrac{1}{2}x + 5\right) = 4$, $\varepsilon = 0.01$

12. $\lim_{x \to \frac{5}{6}} \left(3x + \dfrac{1}{2}\right) = 3$, $\varepsilon = 0.3$

13. For the function $f(x) = 4x - 1$, we have $\lim_{x \to 3} f(x) = 11$.

   For each $\varepsilon > 0$, find a $\delta > 0$ so that

   whenever $0 < |x - 3| < \delta$ then $|(4x - 1) - 11| < \varepsilon$

   (a) $\varepsilon = 0.1$ (b) $\varepsilon = 0.01$

   (c) $\varepsilon = 0.001$ (d) $\varepsilon > 0$ is arbitrary

14. For the function $f(x) = 2 - 5x$, we have $\lim_{x \to -2} f(x) = 12$.

   For each $\varepsilon > 0$, find a $\delta > 0$ so that

   whenever $0 < |x + 2| < \delta$ then $|(2 - 5x) - 12| < \varepsilon$

   (a) $\varepsilon = 0.2$ (b) $\varepsilon = 0.02$

   (c) $\varepsilon = 0.002$ (d) $\varepsilon > 0$ is arbitrary

15. For the function $f(x) = \dfrac{x^2 - 9}{x + 3}$, we have $\lim_{x \to -3} f(x) = -6$.

   For each $\varepsilon > 0$, find a $\delta > 0$ so that

   whenever $0 < |x + 3| < \delta$ then $\left| \dfrac{x^2 - 9}{x + 3} - (-6) \right| < \varepsilon$

   (a) $\varepsilon = 0.1$ (b) $\varepsilon = 0.01$ (c) $\varepsilon > 0$ is arbitrary

16. For the function $f(x) = \dfrac{x^2 - 4}{x - 2}$, we have $\lim_{x \to 2} f(x) = 4$.

   For each $\varepsilon > 0$, find a $\delta > 0$ so that

   whenever $0 < |x - 2| < \delta$ then $\left| \dfrac{x^2 - 4}{x - 2} - 4 \right| < \varepsilon$

   (a) $\varepsilon = 0.1$ (b) $\varepsilon = 0.01$ (c) $\varepsilon > 0$ is arbitrary

In Problems 17–32, write a proof for each limit using the ε-δ definition of a limit.

17. $\lim_{x \to 2} (3x) = 6$

18. $\lim_{x \to 3} (4x) = 12$

19. $\lim_{x \to 0} (2x + 5) = 5$

20. $\lim_{x \to -1} (2 - 3x) = 5$

21. $\lim_{x \to -3} (-5x + 2) = 17$

22. $\lim_{x \to 2} (2x - 3) = 1$

23. $\lim_{x \to 2} (x^2 - 2x) = 0$

24. $\lim_{x \to 0} (x^2 + 3x) = 0$

25. $\lim_{x \to 1} \dfrac{1 + 2x}{3 - x} = \dfrac{3}{2}$

26. $\lim_{x \to 2} \dfrac{2x}{4 + x} = \dfrac{2}{3}$

27. $\lim_{x \to 0} \sqrt[3]{x} = 0$

28. $\lim_{x \to 1} \sqrt{2 - x} = 1$

29. $\lim_{x \to -1} x^2 = 1$

30. $\lim_{x \to 2} x^3 = 8$

31. $\lim_{x \to 3} \dfrac{1}{x} = \dfrac{1}{3}$

32. $\lim_{x \to 2} \dfrac{1}{x^2} = \dfrac{1}{4}$

33. Use the ε-δ definition of a limit to show that the statement $\lim_{x \to 3} (3x - 1) = 12$ is false.

34. Use the ε-δ definition of a limit to show that the statement $\lim_{x \to -2} (4x) = -7$ is false.

Applications and Extensions

35. Show that $\left| \dfrac{1}{x^2 + 9} - \dfrac{1}{18} \right| < \dfrac{7}{234} |x - 3|$ if $2 < x < 4$. Use this to show that $\lim_{x \to 3} \dfrac{1}{x^2 + 9} = \dfrac{1}{18}$.

36. Show that $|(2 + x)^2 - 4| \leq 5 |x|$ if $-1 < x < 1$. Use this to show that $\lim_{x \to 0} (2 + x)^2 = 4$.

37. Show that $\left| \dfrac{1}{x^2+9} - \dfrac{1}{13} \right| \le \dfrac{1}{26} |x-2|$ if $1 < x < 3$. Use this

to show that $\displaystyle\lim_{x \to 2} \dfrac{1}{x^2+9} = \dfrac{1}{13}$.

38. Use the ε-δ definition of a limit to show that $\displaystyle\lim_{x \to 1} x^2 \ne 1.31$.
Hint: Use $\varepsilon = 0.1$.

39. If m and b are any constants, prove that

$$\lim_{x \to c} (mx + b) = mc + b$$

40. Verify that if x is restricted so that $|x - 2| < \dfrac{1}{3}$ in the proof of

$\displaystyle\lim_{x \to 2} x^2 = 4$, then the choice for δ would be less than or equal to

the smaller of $\dfrac{1}{3}$ and $\dfrac{3\varepsilon}{13}$; that is, $\delta \le \min\left\{ \dfrac{1}{3}, \dfrac{3\varepsilon}{13} \right\}$.

41. For $x \ne 3$, how close to 3 must x be to guarantee that $2x - 1$ differs from 5 by less than 0.1?

42. For $x \ne 0$, how close to 0 must x be to guarantee that 3^x differs from 1 by less than 0.1?

43. Prove that if $\displaystyle\lim_{x \to c} f(x) < 0$, then there is an open interval around c for which $f(x) < 0$ everywhere in the interval, except possibly at c.

44. Use the ε-δ definition of a limit at infinity to prove that

$$\lim_{x \to -\infty} \dfrac{1}{x} = 0.$$

45. Use the ε-δ definition of a limit at infinity to prove that

$$\lim_{x \to \infty} \left(-\dfrac{1}{\sqrt{x}} \right) = 0$$

46. For $\displaystyle\lim_{x \to -\infty} \dfrac{1}{x^2} = 0$, find a value of N that satisfies the ε-δ definition of limits at infinity for $\varepsilon = 0.1$.

47. Use the ε-δ definition of limit to prove that no number L exists so

that $\displaystyle\lim_{x \to 0} \dfrac{1}{x} = L$.

48. Explain why in the ε-δ definition of a limit, the inequality $0 < |x - c| < \delta$ has two strict inequality symbols.

49. The ε-δ definition of a limit states, in part, that f is defined everywhere in an open interval containing c, except possibly at c. Discuss the purpose of including the phrase, *except possibly at c*, and why it is necessary.

50. In the ε-δ definition of a limit, what does ε measure? What does δ measure? Give an example to support your explanation.

51. Discuss $\displaystyle\lim_{x \to 0} f(x)$ and $\displaystyle\lim_{x \to 1} f(x)$ if $f(x) = \begin{cases} x^2 & \text{if } x \text{ is rational} \\ 0 & \text{if } x \text{ is irrational} \end{cases}$

52. Discuss $\displaystyle\lim_{x \to 0} f(x)$ if $f(x) = \begin{cases} x^2 & \text{if } x \text{ is rational} \\ \tan x & \text{if } x \text{ is irrational} \end{cases}$

Challenge Problems

53. Use the ε-δ definition of limit to prove that
$$\lim_{x \to 1} (4x^3 + 3x^2 - 24x + 22) = 5.$$

54. If $\displaystyle\lim_{x \to c} f(x) = L$ and $\displaystyle\lim_{x \to c} g(x) = M$, prove that
$$\lim_{x \to c} [f(x) + g(x)] = L + M. \text{ Use the ε-δ definition of limit.}$$

55. For $\displaystyle\lim_{x \to \infty} \dfrac{2 - x}{\sqrt{5 + 4x^2}} = -\dfrac{1}{2}$, find a value of M that satisfies

the ε-δ definition of limits at infinity for $\varepsilon = 0.01$.

56. Use the ε-δ definition of a limit to prove that the linear function $f(x) = ax + b$ is continuous everywhere.

57. Show that the function $f(x) = \begin{cases} 0 & \text{if } x \text{ is rational} \\ x & \text{if } x \text{ is irrational} \end{cases}$

is continuous only at $x = 0$.

58. Suppose that f is defined on an interval (a, b) and there is a number K so that

$$|f(x) - f(c)| \le K|x - c|$$

for all c in (a, b) and x in (a, b). Such a constant K is called a **Lipschitz constant**. Find a Lipschitz constant for $f(x) = x^3$ on $(0, 2)$.

CHAPTER 1 PROJECT

Pollution in Clear Lake

This Project can be done individually or as part of a team.

The Toxic Waste Disposal Company (TWDC) specializes in the disposal of a particularly dangerous pollutant, Agent Yellow (AY). Unfortunately, instead of safely disposing of this pollutant, the company simply dumped AY in (formerly) Clear Lake. Fortunately, they have been caught and are now defending themselves in court.

The facts below are not in dispute. As a result of TWDC's activity, the current concentration of AY in Clear Lake is now 10 ppm (parts per million). Clear Lake is part of a chain of rivers and lakes. Fresh water flows into Clear Lake, and the contaminated water flows downstream from it. The Department of Environmental Protection estimates that the level of contamination in Clear Lake will fall by 20% each year. These facts can be modeled as

$$p(0) = 10 \qquad p(t+1) = 0.80 p(t)$$

where $p = p(t)$, measured in ppm, is the concentration of pollutants in the lake at time t, in years.

1. Explain how the above equations model the facts.
2. Create a table showing the values of t for $t = 0, 1, 2, \ldots, 20$.
3. Show that $p(t) = 10(0.8)^t$.
4. Use technology to graph $p = p(t)$.
5. What is $\lim_{t \to \infty} p(t)$?

Lawyers for TWDC looked at the results in 1–5 above and argued that their client has not done any real damage. They concluded that Clear Lake would eventually return to its former clear and unpolluted state. They even called in a mathematician, who wrote the following on a blackboard:

$$\lim_{t \to \infty} p(t) = 0$$

and explained that this bit of mathematics means, descriptively, that after many years the concentration of AY will, indeed, be close to zero.

Concerned citizens booed the mathematician's testimony. Fortunately, one of them has taken calculus and knows a little bit about limits. She noted that, although "after many years the concentration of AY will approach zero," the townspeople like to swim in Clear Lake and state regulations prohibit swimming unless the concentration of AY is below 2 ppm. She proposed a fine of $100,000 per year for each full year that the lake is unsafe for swimming. She also questioned the mathematician, saying, "Your testimony was correct as far as it went, but I remember from studying calculus that talking about the eventual concentration of AY after many, many years is only a small part of the story. The more precise meaning of your statement $\lim_{t \to \infty} p(t) = 0$ is that given some tolerance T for the concentration of AY, there is some time N (which may be far in the future) so that for all $t > N$, $p(t) < T$."

6. Using a table or a graph for $p = p(t)$, find N so that if $t > N$, then $p(t) < 2$.
7. How much is the fine?

Her words were greeted by applause. The town manager sprang to his feet and noted that although a tolerance of 2 ppm was fine for swimming, the town used Clear Lake for its drinking water and until the concentration of AY dropped below 0.5 ppm, the water would be unsafe for drinking. He proposed a fine of $200,000 per year for each full year the water was unfit for drinking.

8. Using a table or a graph for $p = p(t)$, find N so that if $t > N$, then $p(t) < 0.5$.
9. How much is the fine?
10. How would you find if you were on the jury trying TWDC? If the jury found TWDC guilty, what fine would you recommend? Explain your answers.

Chapter Review

THINGS TO KNOW

1.1 Limits of Functions Using Numerical and Graphical Techniques

- Slope of a secant line: $m_{\sec} = \dfrac{f(x) - f(c)}{x - c}$ (p. 82)

- Slope of a tangent line: $m_{\tan} = \lim_{x \to c} \dfrac{f(x) - f(c)}{x - c}$ (p. 83)

- $\lim_{x \to c} f(x) = L$: read, "The limit as x approaches c of $f(x)$ is equal to the number L." (p. 83)

- $\lim_{x \to c} f(x) = L$: interpreted as, "The value $f(x)$ can be made as close as we please to L, for x sufficiently close to c, but not equal to c." (p. 83)

- One-sided limits (p. 84)

- The limit L of a function $y = f(x)$ as x approaches a number c does not depend on the value of f at c. (p. 86)

- The limit L of a function $y = f(x)$ as x approaches a number c is unique. A function cannot have more than one limit as x approaches c. (p. 86)

- The limit L of a function $y = f(x)$ as x approaches a number c exists if and only if both one-sided limits exist at c and both one-sided limits are equal. That is, $\lim_{x \to c} f(x) = L$ if and only if
$$\lim_{x \to c^-} f(x) = \lim_{x \to c^+} f(x) = L. \quad \text{(p. 87)}$$

1.2 Analytic Techniques for Finding Limits of Functions

Basic Limits

- $\lim_{x \to c} A = A$, A a constant (p. 94)

- $\lim_{x \to c} x = c$, c a real number (p. 95)

US Air Force Photo / Alamy Stock Photo

Properties of Limits   If f and g are functions for which $\lim\limits_{x \to c} f(x)$ and $\lim\limits_{x \to c} g(x)$ both exist and if k is any real number, then:

- $\lim\limits_{x \to c}[f(x) \pm g(x)] = \lim\limits_{x \to c} f(x) \pm \lim\limits_{x \to c} g(x)$ (p. 95)

- $\lim\limits_{x \to c}[f(x) \cdot g(x)] = \lim\limits_{x \to c} f(x) \cdot \lim\limits_{x \to c} g(x)$ (p. 96)

- $\lim\limits_{x \to c}[kg(x)] = k \lim\limits_{x \to c} g(x)$ (p. 96)

- $\lim\limits_{x \to c}[f(x)]^n = \left[\lim\limits_{x \to c} f(x)\right]^n$, $n \geq 2$ is an integer   (p. 98)

- $\lim\limits_{x \to c} \sqrt[n]{f(x)} = \sqrt[n]{\lim\limits_{x \to c} f(x)}$, provided $f(x) > 0$ if n is even (p. 99)

- $\lim\limits_{x \to c}[f(x)]^{m/n} = \left[\lim\limits_{x \to c} f(x)\right]^{m/n}$, provided $[f(x)]^{m/n}$ is defined for positive integers m and n (p. 99)

- $\lim\limits_{x \to c}\left[\dfrac{f(x)}{g(x)}\right] = \dfrac{\lim\limits_{x \to c} f(x)}{\lim\limits_{x \to c} g(x)}$, provided $\lim\limits_{x \to c} g(x) \neq 0$ (p. 101)

- If P is a polynomial function, then $\lim\limits_{x \to c} P(x) = P(c)$. (p. 100)

- If R is a rational function and if c is in the domain of R, then $\lim\limits_{x \to c} R(x) = R(c)$.   (p. 101)

1.3 Continuity

Definitions

- Continuity at a number   (p. 108)
- Removable discontinuity   (p. 110)
- One-sided continuity at a number   (p. 110)
- Continuity on an interval   (p. 111)
- Continuity on a domain   (p. 112)

Properties of Continuity

- A polynomial function is continuous on its domain, all real numbers.   (p. 112)

- A rational function is continuous on its domain.   (p. 112)

- If the functions f and g are continuous at a number c, and if k is a real number, then the functions $f + g$, $f - g$, $f \cdot g$, and kf are also continuous at c. If $g(c) \neq 0$, the function $\dfrac{f}{g}$ is continuous at c.   (p. 113)

- If a function g is continuous at c and a function f is continuous at $g(c)$, then the composite function $(f \circ g)(x) = f(g(x))$ is continuous at c.   (p. 114)

- If f is a one-to-one function that is continuous on its domain, then its inverse function f^{-1} is also continuous on its domain. (p. 115)

The Intermediate Value Theorem   Let f be a function that is continuous on a closed interval $[a, b]$ with $f(a) \neq f(b)$. If N is any number between $f(a)$ and $f(b)$, then there is at least one number c in the open interval (a, b) for which $f(c) = N$.   (p. 115)

1.4 Limits and Continuity of Trigonometric, Exponential, and Logarithmic Functions

Basic Limits

- $\lim\limits_{\theta \to 0} \dfrac{\sin \theta}{\theta} = 1$ (p. 124)

- $\lim\limits_{\theta \to 0} \dfrac{\cos \theta - 1}{\theta} = 0$ (p. 126)

- $\lim\limits_{x \to c} \sin x = \sin c$ (p. 127)

- $\lim\limits_{x \to c} \cos x = \cos c$ (p. 127)

- $\lim\limits_{x \to c} a^x = a^c$; $a > 0$, $a \neq 1$ (p. 129)

- $\lim\limits_{x \to c} \log_a x = \log_a c$; $a > 0$, $a \neq 1$, and $c > 0$ (p. 129)

Squeeze Theorem   If the functions f, g, and h have the property that for all x in an open interval containing c, except possibly at c, $f(x) \leq g(x) \leq h(x)$, and if $\lim\limits_{x \to c} f(x) = \lim\limits_{x \to c} h(x) = L$, then $\lim\limits_{x \to c} g(x) = L$.   (p. 123)

Properties of Continuity

- The six trigonometric functions are continuous on their domains.   (p. 128)

- The six inverse trigonometric functions are continuous on their domains.   (p. 128)

- An exponential function is continuous on its domain, all real numbers.   (p. 129)

- A logarithmic function is continuous on its domain, all positive real numbers.   (p. 129)

1.5 Infinite Limits; Limits at Infinity; Asymptotes

Basic Limits

- $\lim\limits_{x \to 0^-} \dfrac{1}{x} = -\infty$ $\lim\limits_{x \to 0^+} \dfrac{1}{x} = \infty$ (p. 133)

- $\lim\limits_{x \to 0} \dfrac{1}{x^2} = \infty$ (p. 133)

- $\lim\limits_{x \to 0^+} \ln x = -\infty$ (p. 134)

- $\lim\limits_{x \to \infty} \dfrac{1}{x} = 0$ $\lim\limits_{x \to -\infty} \dfrac{1}{x} = 0$ (p. 137)

- $\lim\limits_{x \to \infty} \ln x = \infty$ (p. 142)

- $\lim\limits_{x \to -\infty} e^x = 0$ $\lim\limits_{x \to \infty} e^x = \infty$ (p. 141)

Definitions

- Vertical asymptote   (p. 136)
- Horizontal asymptote   (p. 143)

Properties of Limits at Infinity   If k is a real number, $n \geq 2$ is an integer, and the functions f and g approach real numbers as $x \to \infty$, then:

- $\lim\limits_{x \to \infty} A = A$, where A is a constant

- $\lim\limits_{x \to \infty}[kf(x)] = k \lim\limits_{x \to \infty} f(x)$

- $\lim\limits_{x \to \infty}[f(x) \pm g(x)] = \lim\limits_{x \to \infty} f(x) \pm \lim\limits_{x \to \infty} g(x)$

- $\lim\limits_{x \to \infty}[f(x)g(x)] = \left[\lim\limits_{x \to \infty} f(x)\right]\left[\lim\limits_{x \to \infty} g(x)\right]$

- $\lim\limits_{x \to \infty} \dfrac{f(x)}{g(x)} = \dfrac{\lim\limits_{x \to \infty} f(x)}{\lim\limits_{x \to \infty} g(x)}$ provided $\lim\limits_{x \to \infty} g(x) \neq 0$

- $\lim\limits_{x \to \infty}[f(x)]^n = \left[\lim\limits_{x \to \infty} f(x)\right]^n$

- $\lim\limits_{x \to \infty} \sqrt[n]{f(x)} = \sqrt[n]{\lim\limits_{x \to \infty} f(x)}$, where $f(x) > 0$ if n is even (p. 137)

1.6 The ε-δ Definition of a Limit

Definitions

- Limit of a Function   (p. 151)
- Limit at Infinity   (p. 156)
- Infinite Limit   (p. 157)
- Infinite Limit at Infinity   (p. 157)

Properties of Limits

- If $\lim\limits_{x \to c} f(x) > 0$, then there is an open interval around c, for which $f(x) > 0$ everywhere in the interval, except possibly at c.   (p. 156)

- If $\lim\limits_{x \to c} f(x) < 0$, then there is an open interval around c, for which $f(x) < 0$ everywhere in the interval, except possibly at c.   (p. 156)

OBJECTIVES

Section	You should be able to ...	Example	Review Exercises	AP® Review Problems
1.1	1 Discuss the idea of a limit (p. 83)	1	4	
	2 Investigate a limit using a table (p. 84)	2–4	1	5
	3 Investigate a limit using a graph (p. 85)	5–8	2, 3	
1.2	1 Find the limit of a sum, a difference, and a product (p. 95)	1–6	8, 10, 12, 14, 22, 26, 29, 30, 47, 48	13
	2 Find the limit of a power and the limit of a root (p. 98)	7–9	11, 18, 28, 55	13
	3 Find the limit of a polynomial (p. 100)	10	10, 22	8
	4 Find the limit of a quotient (p. 101)	11–14	13–17, 19–21, 23–25, 27, 56	2, 3, 12, 15
	5 Find the limit of an average rate of change (p. 103)	15	37	
	6 Find the limit of a difference quotient (p. 104)	16	5, 6, 49	
1.3	1 Determine whether a function is continuous at a number (p. 108)	1–4	31–36	11
	2 Determine intervals on which a function is continuous (p. 111)	5, 6	39–42	8
	3 Use properties of continuity (p. 113)	7, 8	39–42	8, 11
	4 Use the Intermediate Value Theorem (p. 115)	9, 10	38, 44–46	6, 14
1.4	1 Use the Squeeze Theorem to find a limit (p. 122)	1	7, 69	15
	2 Find limits involving trigonometric functions (p. 124)	2, 3	9, 51–55	4, 10, 13
	3 Determine where the trigonometric functions are continuous (p. 127)	4	63–65	
	4 Determine where an exponential or a logarithmic function is continuous (p. 129)	5	43	
1.5	1 Investigate infinite limits (p. 133)	1–3	57, 58	
	2 Find the vertical asymptotes of a graph (p. 136)	4	61, 62	1, 7, 9
	3 Investigate limits at infinity (p. 136)	5–10	59, 60	2
	4 Find the horizontal asymptotes of a graph (p. 143)	11	61, 62	1, 7, 9
	5 Find the asymptotes of the graph of a rational function (p. 144)	12	67, 68	7, 9
1.6	1 Use the ε-δ definition of a limit (p. 152)	1–7	50, 66	

REVIEW EXERCISES

1. Use a table of numbers to investigate $\lim\limits_{x \to 0} \dfrac{1 - \cos x}{1 + \cos x}$.

In Problems 2 and 3, use a graph to investigate $\lim\limits_{x \to c} f(x)$.

2. $f(x) = \begin{cases} 2x - 5 & \text{if } x < 1 \\ 6 - 9x & \text{if } x \geq 1 \end{cases}$ at $c = 1$

3. $f(x) = \begin{cases} x^2 + 2 & \text{if } x < 2 \\ 2x + 1 & \text{if } x \geq 2 \end{cases}$ at $c = 2$

4. For $f(x) = x^2 - 3$:

   (a) Find the slope of the secant line joining $(1, -2)$ and $(2, 1)$.

   (b) Find the slope of the tangent line to the graph of f at $(1, -2)$.

In Problems 5 and 6, for each function find the limit of the difference quotient $\lim\limits_{h \to 0} \dfrac{f(x + h) - f(x)}{h}$.

5. $f(x) = \dfrac{3}{x}$

6. $f(x) = 3x^2 + 2x$

7. Find $\lim\limits_{x \to 0} f(x)$ if $1 + \sin x \leq f(x) \leq |x| + 1$.

In Problems 8–22, find each limit.

8. $\lim\limits_{x\to 2}\left(2x-\dfrac{1}{x}\right)$

9. $\lim\limits_{x\to\pi}(x\cos x)$

10. $\lim\limits_{x\to -1}\left(x^3+3x^2-x-1\right)$

11. $\lim\limits_{x\to 0}\sqrt[3]{x(x+2)^3}$

12. $\lim\limits_{x\to 0}[(2x+3)(x^5+5x)]$

13. $\lim\limits_{x\to 3}\dfrac{x^3-27}{x-3}$

14. $\lim\limits_{x\to 3}\left(\dfrac{x^2}{x-3}-\dfrac{3x}{x-3}\right)$

15. $\lim\limits_{x\to 2}\dfrac{x^2-4}{x-2}$

16. $\lim\limits_{x\to -1}\dfrac{x^2+3x+2}{x^2+4x+3}$

17. $\lim\limits_{x\to -2}\dfrac{x^3+5x^2+6x}{x^2+x-2}$

18. $\lim\limits_{x\to 1}\left(x^2-3x+\dfrac{1}{x}\right)^{15}$

19. $\lim\limits_{x\to 2}\dfrac{3-\sqrt{x^2+5}}{x^2-4}$

20. $\lim\limits_{x\to 0}\left\{\dfrac{1}{x}\left[\dfrac{1}{(2+x)^2}-\dfrac{1}{4}\right]\right\}$

21. $\lim\limits_{x\to 0}\dfrac{(x+3)^2-9}{x}$

22. $\lim\limits_{x\to 1}[(x^3-3x^2+3x-1)(x+1)^2]$

In Problems 23–28, find each one-sided limit, if it exists.

23. $\lim\limits_{x\to -2^+}\dfrac{x^2+5x+6}{x+2}$

24. $\lim\limits_{x\to 5^+}\dfrac{|x-5|}{x-5}$

25. $\lim\limits_{x\to 1^-}\dfrac{|x-1|}{x-1}$

26. $\lim\limits_{x\to 3/2^+}\lfloor 2x\rfloor$

27. $\lim\limits_{x\to 4^-}\dfrac{x^2-16}{x-4}$

28. $\lim\limits_{x\to 1^+}\sqrt{x-1}$

In Problems 29 and 30, find $\lim\limits_{x\to c^-}f(x)$ and $\lim\limits_{x\to c^+}f(x)$ for the given c. Determine whether $\lim\limits_{x\to c}f(x)$ exists.

29. $f(x)=\begin{cases}2x+3 & \text{if } x<2\\ 9-x & \text{if } x\geq 2\end{cases}$ at $c=2$

30. $f(x)=\begin{cases}3x+1 & \text{if } x<3\\ 10 & \text{if } x=3 \\ 4x-2 & \text{if } x>3\end{cases}$ at $c=3$

In Problems 31–36, determine whether f is continuous at c.

31. $f(x)=\begin{cases}5x-2 & \text{if } x<1\\ 5 & \text{if } x=1 \\ 2x+1 & \text{if } x>1\end{cases}$ at $c=1$

32. $f(x)=\begin{cases}x^2 & \text{if } x<-1\\ 2 & \text{if } x=-1 \\ -3x-2 & \text{if } x>-1\end{cases}$ at $c=-1$

33. $f(x)=\begin{cases}4-3x^2 & \text{if } x<0\\ 4 & \text{if } x=0 \\ \sqrt{16-x^2} & \text{if } 0<x\leq 4\end{cases}$ at $c=0$

34. $f(x)=\begin{cases}\sqrt{4+x} & \text{if } -4\leq x\leq 4\\ \sqrt{\dfrac{x^2-16}{x-4}} & \text{if } x>4\end{cases}$ at $c=4$

35. $f(x)=\lfloor 2x\rfloor$ at $c=\dfrac{1}{2}$

36. $f(x)=|x-5|$ at $c=5$

37. (a) Find the average rate of change of $f(x)=2x^2-5x$ from 1 to x.

(b) Find the limit as x approaches 1 of the average rate of change found in (a).

38. A function f is defined on the interval $[-1, 1]$ with the following properties: f is continuous on $[-1, 1]$ except at 0, negative at -1, positive at 1, but with no zeros. Does this contradict the Intermediate Value Theorem?

In Problems 39–43, find all numbers x for which f is continuous.

39. $f(x)=\dfrac{x}{x^3-27}$

40. $f(x)=\dfrac{x^2-3}{x^2+5x+6}$

41. $f(x)=\dfrac{2x+1}{x^3+4x^2+4x}$

42. $f(x)=\sqrt{x-1}$

43. $f(x)=2^{-x}$

44. Use the Intermediate Value Theorem to determine whether $2x^3+3x^2-23x-42=0$ has a zero in the interval $[3, 4]$.

In Problems 45 and 46, use the Intermediate Value Theorem to approximate the zero correct to three decimal places.

45. $f(x)=8x^4-2x^2+5x-1$ on the interval $[0, 1]$.

46. $f(x)=3x^3-10x+9$; zero between -3 and -2.

47. Find $\lim\limits_{x\to 0^+}\dfrac{|x|}{x}(1-x)$ and $\lim\limits_{x\to 0^-}\dfrac{|x|}{x}(1-x)$.

What can you say about $\lim\limits_{x\to 0}\dfrac{|x|}{x}(1-x)$?

48. Find $\lim\limits_{x\to 2}\left(\dfrac{x^2}{x-2}-\dfrac{2x}{x-2}\right)$. Then comment on the statement

that this limit is given by $\lim\limits_{x\to 2}\dfrac{x^2}{x-2}-\lim\limits_{x\to 2}\dfrac{2x}{x-2}$.

49. Find $\lim\limits_{h\to 0}\dfrac{f(x+h)-f(x)}{h}$ for $f(x)=\sqrt{x}$.

50. For $\lim\limits_{x\to 3}(2x+1)=7$, find the largest possible δ that "works" for $\varepsilon=0.01$.

In Problems 51–60, find each limit.

51. $\lim\limits_{x\to 0}\cos(\tan x)$

52. $\lim\limits_{x\to 0}\dfrac{\sin\dfrac{x}{4}}{x}$

53. $\lim\limits_{x\to 0}\dfrac{\tan(3x)}{\tan(4x)}$

54. $\lim\limits_{x\to 0}\dfrac{\cos\dfrac{x}{3}-1}{x}$

55. $\lim\limits_{x\to 0}\left(\dfrac{\cos x-1}{x}\right)^{10}$

56. $\lim\limits_{x\to 0}\dfrac{e^{4x}-1}{e^x-1}$

57. $\lim\limits_{x\to \pi/2^+}\tan x$

58. $\lim\limits_{x\to -3}\dfrac{2+x}{(x+3)^2}$

59. $\lim\limits_{x\to \infty}\dfrac{3x^3-2x+1}{x^3-8}$

60. $\lim\limits_{x\to \infty}\dfrac{3x^4+x}{2x^2}$

In Problems 61 and 62, find any vertical and horizontal asymptotes of f.

61. $f(x) = \dfrac{4x-2}{x+3}$

62. $f(x) = \dfrac{2x}{x^2-4}$

63. Let $f(x) = \begin{cases} \dfrac{\tan x}{2x} & \text{if } x \neq 0 \\ \dfrac{1}{2} & \text{if } x = 0 \end{cases}$. Is f continuous at 0?

64. Let $f(x) = \begin{cases} \dfrac{\sin(3x)}{x} & \text{if } x \neq 0 \\ 1 & \text{if } x = 0 \end{cases}$. Is f continuous at 0?

65. The function $f(x) = \dfrac{\cos\left(\pi x + \dfrac{\pi}{2}\right)}{x}$ is not defined at 0. Decide how to define $f(0)$ so that f is continuous at 0.

66. Use the ε-δ definition of a limit to prove $\lim\limits_{x \to -3} (x^2 - 9) \neq 18$.

67. (a) Sketch a graph of a function f that has the following properties:

$$f(-1) = 0 \quad \lim_{x \to \infty} f(x) = 2 \quad \lim_{x \to -\infty} f(x) = 2$$

$$\lim_{x \to 4^-} f(x) = -\infty \quad \lim_{x \to 4^+} f(x) = \infty$$

(b) Define a function that describes your graph.

68. (a) Find the domain and the intercepts (if any) of

$$R(x) = \frac{2x^2 - 5x + 2}{5x^2 - x - 2}.$$

(b) Discuss the behavior of the graph of R at numbers where R is not defined.

(c) Find any vertical or horizontal asymptotes of the function R.

69. If $1 - x^2 \leq f(x) \leq \cos x$ for all x in the interval $-\dfrac{\pi}{2} < x < \dfrac{\pi}{2}$, show that $\lim\limits_{x \to 0} f(x) = 1$.

Break It Down

*Let's take a closer look at **AP® Practice Problem 11** from Section 1.5 on page 149.*

11. $\lim\limits_{x \to 2^-} \dfrac{5x+1}{2x-4} =$

(A) $-\infty$ (B) $-\dfrac{5}{2}$ (C) $\dfrac{5}{2}$ (D) ∞

Step 1	Identify the underlying structure and related concepts.	The problem is to find a **limit**. The limit is a one-sided limit of a quotient.
Step 2	Determine the appropriate math rule or procedure.	The Limit of a Quotient Theorem cannot be used because the limit of the denominator equals 0. That is, $\lim\limits_{x \to 2^-} (2x - 4) = 0$. Since the limit of the numerator is nonzero, $\lim\limits_{x \to 2^-} (5x + 1) = 11 \neq 0$, the limit we seek is infinite.
Step 3	Apply the math rule or procedure.	The limit of the numerator, 11, is positive. As x approaches 2 from the left, x is less than 2 and is close to 2, so the denominator $2x - 4$ is negative and close to 0. So, the quotient is negative and is unbounded. Therefore, $\lim\limits_{x \to 2^-} \dfrac{5x+1}{2x-4} = -\infty$.
Step 4	Clearly communicate your answer.	The answer is (A) $-\infty$.

AP® Review Problems: Chapter 1

Preparing for the AP® Exam

Multiple-Choice Questions

1. Which line is an asymptote to the graph of $f(x) = e^{2x}$?

(A) $x = 0$ (B) $y = 0$ (C) $y = 2$ (D) $y = x$

2. $\lim\limits_{x \to \infty} \dfrac{3x^3 + 4x}{5 - 2x^4} =$

(A) $-\dfrac{3}{2}$ (B) 0 (C) $\dfrac{3}{5}$ (D) $\dfrac{3}{2}$

3. If $f(x) = 5x^3 - 1$, then $\lim\limits_{x \to 0} \dfrac{f(x) - f(0)}{x^3} =$

(A) 0 (B) 1 (C) 5 (D) The limit does not exist.

4. $\lim\limits_{\theta \to 0} \dfrac{\theta^2}{1 - \cos\theta} =$

(A) 0 (B) 1 (C) 2 (D) 4

5. The table gives values of three functions:

x	-0.15	-0.1	-0.05	0	0.05	0.1	0.15
$f(x)$	0.075	0.05	0.025	-4	0.025	0.05	0.075
$g(x)$	-8.3	-8.2	-8.1	undefined	-7.9	-7.8	-7.7
$h(x)$	1.997	1.99	1.9975	1	1.005	1.02	1.045

For which of these functions does the table suggest that the limit as x approaches 0 exists?

(A) f only   (B) h only
(C) f and g only   (D) f and h only

6. If a function f is continuous on the closed interval $[1, 4]$ and if $f(1) = 6$ and $f(4) = -1$, then which of the following must be true?

(A) $f(c) = 0$ for some number c in the open interval $(-1, 6)$.

(B) $f(c) = 1$ for some number c in the open interval $(1, 4)$.

(C) $f(c) = 1$ for some number c in the open interval $(-1, 6)$.

(D) $f(c) \neq -2$ for any number c in the open interval $(1, 4)$.

7. Which are the equations of the asymptotes of the graph of the function $f(x) = \dfrac{x}{x(x^2 - 9)}$?

(A) $x = -3, x = 0, x = 3, y = 0$
(B) $x = -3, x = 0, x = 3, y = 1$
(C) $x = -3, x = 3, y = 0$
(D) $x = -3, x = 3, y = 1$

8. Find the value of k that makes the function

$$f(x) = \begin{cases} x^2 + 2 & \text{if } x \leq -1 \\ kx + 4 & \text{if } x > -1 \end{cases}$$

continuous for all real numbers.

(A) -3 (B) -1 (C) 1 (D) 3

9. An odd function f is continuous for all real numbers. If $\lim\limits_{x \to \infty} f(x) = -2$, then which of the statements must be true?

I. f has no vertical asymptotes.

II. $\lim\limits_{x \to 0} f(x) = 0$

III. The horizontal asymptotes of the graph of f are $y = -2$ and $y = 2$.

(A) I only   (B) III only
(C) I and III only   (D) I, II, and III

10. $\lim\limits_{x \to 0} \dfrac{\sin(2x)}{\tan(3x)} =$

(A) 0 (B) $\dfrac{2}{3}$ (C) 1 (D) $\dfrac{3}{2}$

11. Suppose the function f is continuous for all real numbers. If $f(x) = \dfrac{x^3 + 8}{x + 2}$ when $x \neq -2$, then $f(-2) =$

(A) 0 (B) 4 (C) 8 (D) 12

12. $\lim\limits_{x \to 3^+} \dfrac{(x - 3)^2 - 2x + 6}{x - 3} =$

(A) 0 (B) -2 (C) -3 (D) ∞

13. $\lim\limits_{x \to 0} \sqrt{e^x(4 - \sin x)} \cdot \cos x =$

(A) 0 (B) 1 (C) 2 (D) 4

Free-Response Questions

14. The function $f(x) = x^3 - x^2 + 5$ is defined for $-2 \leq x \leq 1$. Show that there is a number c in the interval $(-2, 1)$ for which $f(c) = 2$.

15. Suppose the functions f, g, and h have the following properties:

(1) $f(x) \leq g(x) \leq h(x)$ except at 2 where $g(2) = 10$.

(2) $\lim\limits_{x \to 2} f(x) = \lim\limits_{x \to 2} h(x) = 5$.

(a) Determine $\lim\limits_{x \to 2} g(x)$. Justify your answer.

(b) Use properties of limits to find

$$\lim\limits_{x \to 2} \dfrac{3f(x) \cdot g(x) + [h(x)]^2}{g(2) \cdot f(x)}.$$

CHAPTER

2

The Derivative and Its Properties

Michael Collins, Apollo 11, NASA

The Apollo Lunar Module
"One Giant Leap for Mankind"

On May 25, 1961, in a special address to Congress, U.S. president John F. Kennedy proposed the goal "before this decade is out, of landing a man on the Moon and returning him safely to the Earth." Roughly eight years later, on July 16, 1969, a Saturn V rocket launched from the Kennedy Space Center in Florida, carrying the *Apollo 11* spacecraft and three astronauts—Neil Armstrong, Buzz Aldrin, and Michael Collins—bound for the Moon.

The *Apollo* spacecraft had three parts: the Command Module with a cabin for the three astronauts; the Service Module that supported the Command Module with propulsion, electrical power, oxygen, and water; and the Lunar Module for landing on the Moon. After its launch, the spacecraft traveled for three days until it entered into lunar orbit. Armstrong and Aldrin then moved into the Lunar Module, which they landed in the flat expanse of the Sea of Tranquility. After more than 21 hours, the first humans to touch the surface of the Moon crawled into the Lunar Module and lifted off to rejoin the Command Module, which Collins had been piloting in lunar orbit. The three astronauts then headed back to Earth, where they splashed down in the Pacific Ocean on July 24.

In 2022, NASA initiated the Artemis I program, designed to pave the way for the first crewed Orion mission and eventually for the return of NASA astronauts to the surface of the Moon and then on to Mars in the 2030s.

 Explore some of the physics at work that allowed engineers and pilots to successfully maneuver the Lunar Module to the Moon's surface in the Chapter 2 Project on page 227.

Ryan Murphy
Athlete

I am a professional swimmer and four-time Olympic gold medalist. The knowledge of velocity and acceleration that I learned in AP® Calculus has helped me better understand and develop the strategy behind my backstroke.

hapter 2 opens by returning to the tangent problem to find an equation of the tangent line to the graph of a function f at a point $P = (c, f(c))$. Remember in Section 1.1 we found that the slope of a tangent line is a limit,

$$m_{\tan} = \lim_{x \to c} \frac{f(x) - f(c)}{x - c}$$

This limit is one of the most significant ideas in calculus, the *derivative*.

In this chapter, we introduce interpretations of the derivative, treat the derivative as a function, and consider some properties of the derivative. By the end of the chapter, you will have a collection of basic derivative formulas and derivative rules that will be used throughout your study of calculus. ∎

2.1 Rates of Change and the Derivative

OBJECTIVES *When you finish this section, you should be able to:*

1 **Find equations for the tangent line and the normal line to the graph of a function (p. 168)**
2 **Find the rate of change of a function (p. 169)**
3 **Find average velocity and instantaneous velocity (p. 170)**
4 **Find the derivative of a function at a number (p. 173)**

AP® EXAM TIP

BREAK IT DOWN: As you work through the problems in this chapter, remember to follow these steps:

Step 1 Identify the underlying structure and related concepts.

Step 2 Determine the appropriate math rule or procedure.

Step 3 Apply the math rule or procedure.

Step 4 Clearly communicate your answer.

On page 231, see how we've used these steps to solve Section 2.2 AP® Problem 12 on page 191.

In Chapter 1, we discussed the tangent problem: *Given a function f and a point P on its graph, what is the slope of the tangent line to the graph of f at P?* See Figure 1, where ℓ_T is the tangent line to the graph of f at the point $P = (c, f(c))$.

The tangent line ℓ_T to the graph of f at P must contain the point P. Since finding the slope requires two points, and we have only one point on the tangent line ℓ_T, we reason as follows.

Suppose we choose any point $Q = (x, f(x))$, other than P, on the graph of f. (Q can be to the left or to the right of P; we chose Q to be to the right of P.) The line containing the points $P = (c, f(c))$ and $Q = (x, f(x))$ is a secant line of the graph of f. The slope $m_{\sec}$ of this secant line is

$$m_{\sec} = \frac{f(x) - f(c)}{x - c} \tag{1}$$

Figure 2 shows three different points Q_1, Q_2, and Q_3 on the graph of f that are successively closer to the point P, and three associated secant lines ℓ_1, ℓ_2, and ℓ_3. The closer the points Q are to the point P, the closer the secant lines are to the tangent line ℓ_T. The line ℓ_T, the *limiting position* of these secant lines, is the *tangent line to the graph* of f at P.

If the limiting position of the secant lines is the tangent line, then the limit of the slopes of the secant lines should equal the slope of the tangent line. Notice in Figure 2 that as the points Q_1, Q_2, and Q_3 move closer to the point P, the numbers x get closer to c. So, equation (1) suggests that

$$m_{\tan} = \text{Slope of the tangent line to } f \text{ at } P$$

$$= \text{Limit of } \frac{f(x) - f(c)}{x - c} \text{ as } x \text{ gets closer to } c$$

$$= \lim_{x \to c} \frac{f(x) - f(c)}{x - c}$$

provided the limit exists.

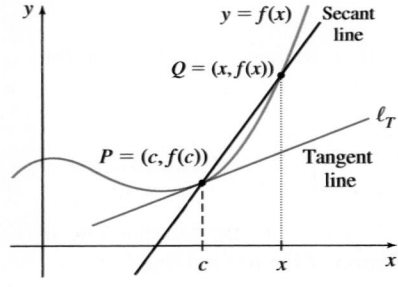

Figure 1 $m_{\sec}$ = slope of the secant line.

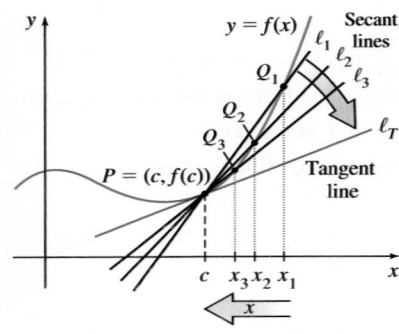

Figure 2 $m_{\tan} = \lim_{x \to c} \dfrac{f(x) - f(c)}{x - c}$

DEFINITION Tangent Line

The **tangent line** to the graph of f at a point P is the line containing the point $P = (c, f(c))$ and having the slope

$$m_{\tan} = \lim_{x \to c} \frac{f(x) - f(c)}{x - c} \qquad (2)$$

provided the limit exists.

NOTE It is possible that the limit in (2) does not exist. The geometric significance of this is discussed in the next section.

The limit in equation (2) that defines the slope of the tangent line occurs so frequently that it is given a special notation $f'(c)$, read, "f prime of c," and called **prime notation**:

$$f'(c) = \lim_{x \to c} \frac{f(x) - f(c)}{x - c} \qquad (3)$$

1 Find Equations for the Tangent Line and the Normal Line to the Graph of a Function

AP® EXAM **TIP**

Problems on the exam often ask about the tangent line.

THEOREM Equation of a Tangent Line

If $m_{\tan} = f'(c)$ exists, then an equation of the tangent line to the graph of a function $y = f(x)$ at the point $P = (c, f(c))$ is

$$y - f(c) = f'(c)(x - c)$$

RECALL Two lines, neither of which is horizontal, with slopes m_1 and m_2, respectively, are perpendicular if and only if

$$m_1 = -\frac{1}{m_2}$$

The line perpendicular to the tangent line at a point P on the graph of a function f is called the **normal line** to the graph of f at P.

THEOREM Equation of a Normal Line

An equation of the normal line to the graph of a function $y = f(x)$ at the point $P = (c, f(c))$ is

$$y - f(c) = -\frac{1}{f'(c)}(x - c)$$

provided $f'(c)$ exists and is not equal to zero. If $f'(c) = 0$, the tangent line is horizontal, the normal line is vertical, and the equation of the normal line is $x = c$.

CALC CLIP **EXAMPLE 1** Finding Equations for the Tangent Line and the Normal Line

(a) Find the slope of the tangent line to the graph of $f(x) = x^2$ at the point $(-2, 4)$.

(b) Use the result from (a) to find an equation of the tangent line at the point $(-2, 4)$.

(c) Find an equation of the normal line to the graph of f at the point $(-2, 4)$.

(d) Graph f, the tangent line to f at $(-2, 4)$, and the normal line to f at $(-2, 4)$ on the same set of axes.

Solution

(a) At the point $(-2, 4)$, the slope of the tangent line is

RECALL One way to find the limit of a quotient when the limit of the denominator is 0 is to factor the numerator and divide out common factors.

$$f'(-2) = \lim_{x \to -2} \frac{f(x) - f(-2)}{x - (-2)} = \lim_{x \to -2} \frac{x^2 - (-2)^2}{x + 2} = \lim_{x \to -2} \frac{x^2 - 4}{x + 2}$$

$$= \lim_{x \to -2} (x - 2) = -4$$

NEED TO REVIEW? The point-slope form of a line is discussed in Appendix A.3, p. A-20.

On the exam, you can leave the equation of a line in point-slope form instead of simplifying to slope-intercept form.

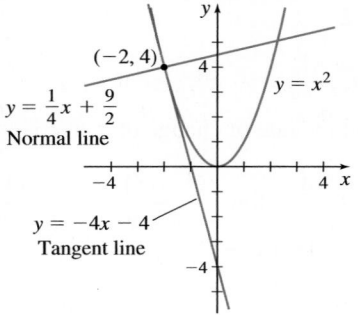

Figure 3 $f(x) = x^2$

(b) We use the result from (a) and the point-slope form of an equation of a line to obtain an equation of the tangent line. An equation of the tangent line containing the point $(-2, 4)$ is

$$y - 4 = f'(-2)[x - (-2)] \qquad \text{Point-slope form of an equation of the tangent line.}$$
$$y - 4 = -4 \cdot (x + 2) \qquad f(-2) = 4; \quad f'(-2) = -4$$
$$y = -4x - 4 \qquad \text{Simplify.}$$

(c) Since the slope of the tangent line to f at $(-2, 4)$ is -4, the slope of the normal line to f at $(-2, 4)$ is $\dfrac{1}{4}$.

Using the point-slope form of an equation of a line, an equation of the normal line is

$$y - 4 = \frac{1}{4}(x + 2)$$
$$y = \frac{1}{4}x + \frac{9}{2}$$

(d) The graphs of f, the tangent line to the graph of f at the point $(-2, 4)$, and the normal line to the graph of f at $(-2, 4)$ are shown in Figure 3. ∎

NOW WORK   Problem 11 and AP® Practice Problems 1 and 5.

2   Find the Rate of Change of a Function

Everything in nature changes. Examples include climate change, change in the phases of the Moon, and change in populations. To describe natural processes mathematically, the ideas of change and rate of change are often used.

Recall that the average rate of change of a function $y = f(x)$ from c to x is given by

$$\boxed{\text{Average rate of change} = \frac{f(x) - f(c)}{x - c} \qquad x \neq c}$$

The average rate of change describes behavior as it changes over an interval. To describe behavior at a specific number, or instant, we use the instantaneous rate of change.

DEFINITION Instantaneous Rate of Change

The **instantaneous rate of change** of f at c is the limit as x approaches c of the average rate of change. Symbolically, the instantaneous rate of change of f at c is

$$\boxed{\lim_{x \to c} \frac{f(x) - f(c)}{x - c}}$$

provided the limit exists.

IN WORDS

- An *average* rate of change describes behavior over an interval.
- An *instantaneous* rate of change describes behavior at a number.

The expression "instantaneous rate of change" is often shortened to *rate of change*. Using prime notation, the **rate of change** of f at c is $f'(c) = \lim\limits_{x \to c} \dfrac{f(x) - f(c)}{x - c}$.

EXAMPLE 2   Finding a Rate of Change

Find the rate of change of the function $f(x) = x^2 - 5x$ at:

(a) $c = 2$

(b) Any real number c

Solution

(a) For $c = 2$,

$$f(x) = x^2 - 5x \qquad \text{and} \qquad f(2) = 2^2 - 5 \cdot 2 = -6$$

The rate of change of f at $c = 2$ is

$$f'(2) = \lim_{x \to 2} \frac{f(x) - f(2)}{x - 2} = \lim_{x \to 2} \frac{(x^2 - 5x) - (-6)}{x - 2} = \lim_{x \to 2} \frac{x^2 - 5x + 6}{x - 2}$$

$$= \lim_{x \to 2} \frac{(x - 2)(x - 3)}{x - 2} = \lim_{x \to 2} (x - 3) = -1$$

(b) If c is any real number, then $f(c) = c^2 - 5c$, and the rate of change of f at c is

$$f'(c) = \lim_{x \to c} \frac{f(x) - f(c)}{x - c} = \lim_{x \to c} \frac{(x^2 - 5x) - (c^2 - 5c)}{x - c} = \lim_{x \to c} \frac{(x^2 - c^2) - 5(x - c)}{x - c}$$

$$= \lim_{x \to c} \frac{(x - c)(x + c) - 5(x - c)}{x - c} = \lim_{x \to c} \frac{(x - c)(x + c - 5)}{x - c}$$

$$= \lim_{x \to c} (x + c - 5) = 2c - 5 \qquad \blacksquare$$

NOW WORK   Problem 17 and AP® Practice Problem 3.

EXAMPLE 3   **Finding the Rate of Change in a Biology Experiment**

In a metabolic experiment, the mass M of glucose decreases according to the function

$$M(t) = 4.5 - 0.03t^2$$

where M is measured in grams (g) and t is the time in hours (h). Find the reaction rate $M'(t)$ at $t = 1$ h.

Solution

The reaction rate at $t = 1$ is $M'(1)$.

$$M'(1) = \lim_{t \to 1} \frac{M(t) - M(1)}{t - 1} = \lim_{t \to 1} \frac{(4.5 - 0.03t^2) - (4.5 - 0.03)}{t - 1}$$

$$= \lim_{t \to 1} \frac{-0.03t^2 + 0.03}{t - 1} = \lim_{t \to 1} \frac{(-0.03)(t^2 - 1)}{t - 1} = \lim_{t \to 1} \frac{(-0.03)(t - 1)(t + 1)}{t - 1}$$

$$= -0.03 \cdot 2 = -0.06$$

The reaction rate at $t = 1$ h is -0.06 g/h. That is, the mass M of glucose at $t = 1$ h is changing at the rate of -0.06 g/h or decreasing at the rate of 0.06 g/h. $\blacksquare$

NOW WORK   Problem 43.

---

❸ Find Average Velocity and Instantaneous Velocity

Average velocity is a physical example of an average rate of change. For example, consider an object moving along a horizontal line with the positive direction to the right, or moving along a vertical line with the positive direction upward. The object's location at time $t = 0$ is called its **initial position**. The initial position is usually marked as the origin O on the line. See Figure 4. We assume the position s at time t of the object from the origin is given by a function $s = f(t)$. Here s is the signed, or directed, distance (using some measure of distance such as centimeters, meters, feet, etc.) of the object from O at time t (in seconds or hours). The function f is usually called the **position function** of the object. Motion along a line is sometimes referred to as **rectilinear motion**.

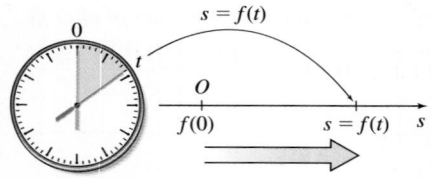

Figure 4 t is the travel time. s is the signed distance of the object from the origin at time t.

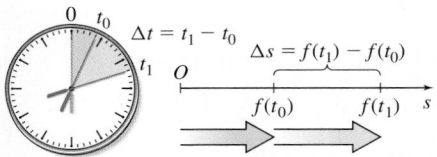

Figure 5 The average velocity is $\dfrac{\Delta s}{\Delta t}$.

DEFINITION Average Velocity

The signed distance s from the origin at time t of an object moving on a line is given by the position function $s = f(t)$. If at time t_0 the object is at $s_0 = f(t_0)$ and at time t_1 the object is at $s_1 = f(t_1)$, then the change in time is $\Delta t = t_1 - t_0$ and the change in position is $\Delta s = s_1 - s_0 = f(t_1) - f(t_0)$. The average rate of change of position with respect to time over the interval $[t_0, t_1]$ is

$$\boxed{\dfrac{\Delta s}{\Delta t} = \dfrac{f(t_1) - f(t_0)}{t_1 - t_0} \qquad t_1 \neq t_0}$$

and is called the **average velocity** of the object over the interval $[t_0, t_1]$. See Figure 5.

EXAMPLE 4   **Finding Average Velocity**

The Mike O'Callaghan–Pat Tillman Memorial Bridge spanning the Colorado River opened on October 16, 2010. Having a length of 1900 ft, it is the longest arch bridge in the Western Hemisphere, and its roadway is 890 ft above the Colorado River.

If a rock falls from the roadway, the function $s = f(t) = 16t^2$ gives the distance s, in feet, that the rock falls after t seconds for $0 \leq t \leq 7.458$. Here 7.458 s is the approximate time it takes the rock to fall 890 ft into the river. The average velocity of the rock during its fall is

$$\dfrac{\Delta s}{\Delta t} = \dfrac{f(7.458) - f(0)}{7.458 - 0} = \dfrac{890 - 0}{7.458} \approx 119.335 \text{ ft/s} \qquad \blacksquare$$

Scott Prokop/Shutterstock

NOW WORK AP® Practice Problem 7.

NOTE Here the motion occurs along a vertical line with the positive direction downward.

The average velocity of the rock in Example 4 approximates the average velocity over the interval $[0, 7.458]$. But the average velocity does not tell us about the velocity at any particular instant of time. That is, it gives no information about the rock's *instantaneous velocity*.

We can investigate the instantaneous velocity of the rock, say, at $t = 3$ s, by computing average velocities for short intervals of time beginning at $t = 3$. First we compute the average velocity for the interval beginning at $t = 3$ and ending at $t = 3.5$. The corresponding distances the rock has fallen are

$$f(3) = 16 \cdot 3^2 = 144 \text{ ft} \qquad \text{and} \qquad f(3.5) = 16 \cdot 3.5^2 = 196 \text{ ft}$$

Then $\Delta t = 3.5 - 3.0 = 0.5$, and during this 0.5-s interval,

$$\text{Average velocity} = \dfrac{\Delta s}{\Delta t} = \dfrac{f(3.5) - f(3)}{3.5 - 3} = \dfrac{196 - 144}{0.5} = 104 \text{ ft/s}$$

Table 1 shows average velocities of the rock for progressively smaller intervals of time.

TABLE 1

Time Interval	Start $t_0 = 3$	End t	Δt	$\dfrac{\Delta s}{\Delta t} = \dfrac{f(t) - f(t_0)}{t - t_0} = \dfrac{16t^2 - 144}{t - 3}$
$[3, 3.1]$	3	3.1	0.1	$\dfrac{\Delta s}{\Delta t} = \dfrac{f(3.1) - f(3)}{3.1 - 3} = \dfrac{16 \cdot 3.1^2 - 144}{0.1} = 97.6$
$[3, 3.01]$	3	3.01	0.01	$\dfrac{\Delta s}{\Delta t} = \dfrac{f(3.01) - f(3)}{3.01 - 3} = \dfrac{16 \cdot 3.01^2 - 144}{0.01} = 96.16$
$[3, 3.0001]$	3	3.0001	0.0001	$\dfrac{\Delta s}{\Delta t} = \dfrac{f(3.0001) - f(3)}{3.0001 - 3} = \dfrac{16 \cdot 3.0001^2 - 144}{0.0001} = 96.0016$

The average velocity of 96.0016 over the time interval [3, 3.0001], ($\Delta t = 0.0001$ s) should be very close to the instantaneous velocity of the rock at $t = 3$ s. As Δt gets closer to 0, the average velocity gets closer to the instantaneous velocity. So, to obtain the instantaneous velocity at $t = 3$ precisely, we use the limit of the average velocity as Δt approaches 0 or, equivalently, as t approaches 3.

$$\lim_{\Delta t \to 0} \frac{\Delta s}{\Delta t} = \lim_{t \to 3} \frac{f(t) - f(3)}{t - 3} = \lim_{t \to 3} \frac{16t^2 - 16 \cdot 3^2}{t - 3} = \lim_{t \to 3} \frac{16(t^2 - 9)}{t - 3}$$

$$= \lim_{t \to 3} \frac{16(t - 3)(t + 3)}{t - 3} = \lim_{t \to 3} [16(t + 3)] = 96$$

The rock's instantaneous velocity at $t = 3$ s is 96 ft/s.

We generalize this result to obtain a definition for instantaneous velocity.

DEFINITION Instantaneous Velocity

If $s = f(t)$ is the position function of an object at time t, the **instantaneous velocity** v of the object at time t_0 is defined as the limit of the average velocity $\dfrac{\Delta s}{\Delta t}$ as Δt approaches 0. That is,

$$v = \lim_{\Delta t \to 0} \frac{\Delta s}{\Delta t} = \lim_{t \to t_0} \frac{f(t) - f(t_0)}{t - t_0}$$

provided the limit exists.

IN WORDS

• Average velocity is measured over an interval of time.
• Instantaneous velocity is measured at a particular instant of time.

We usually shorten "instantaneous velocity" and just use the word "velocity."

NOW WORK Problem 31.

EXAMPLE 5   Finding Velocity

Find the velocity v of the falling rock from Example 4 at:

(a) $t_0 = 1$ s after it begins to fall
(b) $t_0 = 7.4$ s, just before it hits the Colorado River
(c) At any time t_0.

Solution

(a) Use the definition of instantaneous velocity with $f(t) = 16t^2$ and $t_0 = 1$.

$$v = \lim_{\Delta t \to 0} \frac{\Delta s}{\Delta t} = \lim_{t \to 1} \frac{f(t) - f(1)}{t - 1} = \lim_{t \to 1} \frac{16t^2 - 16}{t - 1} = \lim_{t \to 1} \frac{16(t^2 - 1)}{t - 1}$$

$$= \lim_{t \to 1} \frac{16(t - 1)(t + 1)}{t - 1} = \lim_{t \to 1} [16(t + 1)] = 32$$

At 1 s, the velocity of the rock is 32 ft/s.

(b) For $t_0 = 7.4$ s,

$$v = \lim_{\Delta t \to 0} \frac{\Delta s}{\Delta t} = \lim_{t \to 7.4} \frac{f(t) - f(7.4)}{t - 7.4} = \lim_{t \to 7.4} \frac{16t^2 - 16 \cdot (7.4)^2}{t - 7.4}$$

$$= \lim_{t \to 7.4} \frac{16[t^2 - (7.4)^2]}{t - 7.4} = \lim_{t \to 7.4} \frac{16(t - 7.4)(t + 7.4)}{t - 7.4}$$

$$= \lim_{t \to 7.4} [16(t + 7.4)] = 16(14.8) = 236.8$$

NOTE Did you know?
236.8 ft/s is more than 161 mi/h!

At 7.4 s, the velocity of the rock is 236.8 ft/s.

(c)
$$v = \lim_{t \to t_0} \frac{f(t) - f(t_0)}{t - t_0} = \lim_{t \to t_0} \frac{16t^2 - 16t_0^2}{t - t_0} = \lim_{t \to t_0} \frac{16\,(t - t_0)\,(t + t_0)}{t - t_0}$$
$$= 16 \lim_{t \to t_0} (t + t_0) = 32t_0$$

At t_0 seconds, the velocity of the rock is $32t_0$ ft/s. ∎

NOW WORK Problem 33.

4 Find the Derivative of a Function at a Number

Slope of a tangent line, rate of change of a function, and velocity are all found using the same limit,

$$f'(c) = \lim_{x \to c} \frac{f(x) - f(c)}{x - c}$$

The common underlying idea is the mathematical concept of *derivative*.

DEFINITION Derivative of a Function at a Number

If $y = f(x)$ is a function defined on an open interval (a, b), and c is in the interval (a, b), then the **derivative** of f at c, denoted by $f'(c)$, is the number

$$\boxed{f'(c) = \lim_{x \to c} \frac{f(x) - f(c)}{x - c}}$$

provided this limit exists.

EXAMPLE 6 Finding the Derivative of a Function at a Number

Find the derivative of $f(x) = 2x^2 - 3x - 2$ at $x = 2$. That is, find $f'(2)$.

Solution

Using the definition of the derivative, we have

$$f'(2) = \lim_{x \to 2} \frac{f(x) - f(2)}{x - 2} = \lim_{x \to 2} \frac{(2x^2 - 3x - 2) - 0}{x - 2} \quad f(2) = 2 \cdot 4 - 3 \cdot 2 - 2 = 0$$
$$= \lim_{x \to 2} \frac{(x - 2)(2x + 1)}{x - 2}$$
$$= \lim_{x \to 2} (2x + 1) = 5$$

∎

NOW WORK Problem 23 and AP® Practice Problems 2 and 6.

So far we have given three interpretations of the derivative:

- *Geometric interpretation:* If $y = f(x)$, the derivative $f'(c)$ is the slope of the tangent line to the graph of f at the point $(c, f(c))$.
- *Rate of change of a function interpretation:* If $y = f(x)$, the derivative $f'(c)$ is the rate of change of f at c.
- *Physical interpretation:* If the signed distance s from the origin at time t of an object moving along a line is given by the position function $s = f(t)$, the derivative $f'(t_0)$ is the velocity of the object at time t_0.

EXAMPLE 7   **Finding an Equation of a Tangent Line**

(a) Find the derivative of $f(x) = \sqrt{2x}$ at $x = 8$.

(b) Use the derivative $f'(8)$ to find an equation of the tangent line to the graph of f at the point $(8, 4)$.

Solution

(a) The derivative of f at 8 is

$$f'(8) = \lim_{x \to 8} \frac{f(x) - f(8)}{x - 8} = \lim_{\substack{\uparrow x \to 8 \\ f(8) = \sqrt{2 \cdot 8} = 4}} \frac{\sqrt{2x} - 4}{x - 8} = \lim_{\substack{\uparrow x \to 8 \\ \text{Rationalize} \\ \text{the numerator.}}} \frac{(\sqrt{2x} - 4)(\sqrt{2x} + 4)}{(x - 8)(\sqrt{2x} + 4)}$$

$$= \lim_{x \to 8} \frac{2x - 16}{(x - 8)(\sqrt{2x} + 4)} = \lim_{x \to 8} \frac{2(x - 8)}{(x - 8)(\sqrt{2x} + 4)} = \lim_{x \to 8} \frac{2}{\sqrt{2x} + 4} = \frac{1}{4}$$

(b) The slope of the tangent line to the graph of f at the point $(8, 4)$ is $f'(8) = \frac{1}{4}$. Using the point-slope form of a line, we get

$$y - 4 = f'(8)(x - 8) \qquad y - y_1 = m(x - x_1)$$

$$y - 4 = \frac{1}{4}(x - 8) \qquad f'(8) = \frac{1}{4}$$

$$y = \frac{1}{4}x + 2$$

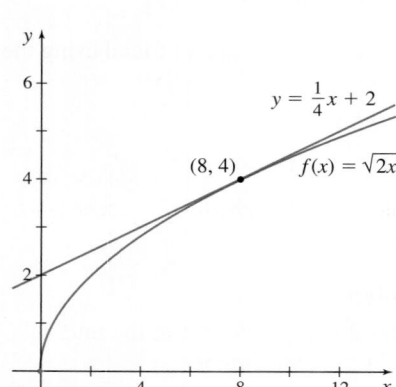

Figure 6

The graphs of f and the tangent line to the graph of f at $(8, 4)$ are shown in Figure 6. ∎

NOW WORK   Problem 15 and AP® Practice Problem 4.

EXAMPLE 8   **Approximating the Derivative of a Function Represented by a Table**

The table below lists several values of a function $y = f(x)$ that is continuous on the interval $[-1, 7]$ and has a derivative at each number in the interval $(-1, 7)$. Approximate the derivative of f at 2.

x	0	1	2	4	6
$f(x)$	0	3	12	32	72

Solution

There are several ways to approximate the derivative of a function defined by a table. Each uses an average rate of change to approximate the rate of change of f at 2, which is the derivative of f at 2.

• Using the average rate of change from 2 to 4, we have

$$\frac{f(4) - f(2)}{4 - 2} = \frac{32 - 12}{2} = 10$$

With this choice, $f'(2)$ is approximately 10.

• Using the average rate of change from 1 to 2, we have

$$\frac{f(2) - f(1)}{2 - 1} = \frac{12 - 3}{1} = 9$$

With this choice, $f'(2)$ is approximately 9.

• A third approximation can be found by averaging the above two approximations.

Then $f'(2)$ is approximately $\dfrac{10 + 9}{2} = \dfrac{19}{2}$.

• **Read the AP® Exam Tip on the left**. Following the tip, we would use the interval of least width containing the number c. So, on the AP® Exam, we would use the interval $[1, 2]$ to approximate the derivative $f'(c)$. Based on the result in the second bullet, we have $f'(2) \approx 9$. ∎

EXAMPLE 9   Approximating the Derivative of a Function Represented by a Table

The table below lists several values of a function $y = f(x)$ that is continuous on the closed interval $[0, 6]$ and has a derivative at each number in the open interval $(0, 6)$. Approximate the derivative of f at 3.4.

x	1	2	3	4	5
$f(x)$	4	6	2	1	6

Solution

Note that 3.4 is not in the table, so we do not know $f(3.4)$. In such cases, approximate the derivative by finding the average rate of change using the interval of least width containing 3.4. In this case, find the average rate of change from 3 to 4, namely,

$$\frac{f(4) - f(3)}{4 - 3} = \frac{1 - 2}{1} = -1$$

Then $f'(3.4)$ is approximately -1. ∎

NOW WORK   Problem 51 and AP® Practice Problems 8 and 9.

2.1 Assess Your Understanding

Concepts and Vocabulary

1. **True or False**   The derivative is used to find instantaneous velocity.

2. **True or False**   The derivative can be used to find the rate of change of a function.

3. The notation $f'(c)$ is read f ____ of c; $f'(c)$ represents the ____ of the tangent line to the graph of f at the point ____ .

4. **True or False**   If it exists, $\lim\limits_{x \to 3} \dfrac{f(x) - f(3)}{x - 3}$ is the derivative of the function f at 3.

5. If $f(x) = 6x - 3$, then $f'(3) =$ ____ .

6. The velocity of an object, the slope of a tangent line, and the rate of change of a function are three different interpretations of the mathematical concept called the ____ .

Skill Building

In Problems 7–16,

   (a) *Find an equation for the tangent line to the graph of each function at the indicated point.*

   (b) *Find an equation of the normal line to each function at the indicated point.*

   (c) *Graph the function, the tangent line, and the normal line at the indicated point on the same set of coordinate axes.*

7. $f(x) = 3x^2$ at $(-2, 12)$

8. $f(x) = x^2 + 2$ at $(-1, 3)$

9. $f(x) = x^3$ at $(-2, -8)$

10. $f(x) = x^3 + 1$ at $(1, 2)$

11. $f(x) = \dfrac{1}{x}$ at $(1, 1)$

12. $f(x) = \sqrt{x}$ at $(4, 2)$

13. $f(x) = \dfrac{1}{x + 5}$ at $\left(1, \dfrac{1}{6}\right)$

14. $f(x) = \dfrac{2}{x + 4}$ at $\left(1, \dfrac{2}{5}\right)$

15. $f(x) = \dfrac{1}{\sqrt{x}}$ at $(1, 1)$

16. $f(x) = \dfrac{1}{x^2}$ at $(1, 1)$

In Problems 17–20, find the rate of change of f at the indicated numbers.

17. $f(x) = 5x - 2$ at **(a)** $c = 0$,   **(b)** $c = 2$

18. $f(x) = x^2 - 1$ at **(a)** $c = -1$,   **(b)** $c = 1$

19. $f(x) = \dfrac{x^2}{x + 3}$ at **(a)** $c = 0$,   **(b)** $c = 1$

20. $f(x) = \dfrac{x}{x^2 - 1}$ at **(a)** $c = 0$,   **(b)** $c = 2$

In Problems 21–30, find the derivative of each function at the given number.

21. $f(x) = 4x + 1$ at 1

22. $f(x) = 5x - 9$ at 2

23. $f(x) = x^2 - 2$ at 0

24. $f(x) = 2x^2 + 4$ at 1

25. $f(x) = 3x^2 + x + 5$ at -1

26. $f(x) = 2x^2 - x - 7$ at -1

27. $f(x) = \sqrt{x}$ at 4

28. $f(x) = \dfrac{1}{x^2}$ at 2

29. $f(x) = \dfrac{2 - 5x}{1 + x}$ at 0

30. $f(x) = \dfrac{2 + 3x}{2 + x}$ at 1

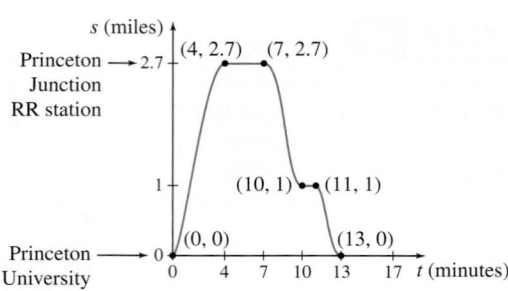

31. **Approximating Velocity**   An object moves on a line according to the position function $s(t) = 10t^2$ (s in centimeters and t in seconds). Approximate the velocity of the object at time $t_0 = 3$ s by letting Δt first equal 0.1 s, then 0.01 s, and finally 0.001 s. What limit does the velocity appear to be approaching? Organize the results in a table.

32. **Approximating Velocity**   An object moves on a line according to the position function $s(t) = 5 - t^2$ (s in centimeters and t in seconds). Approximate the velocity of the object at time $t_0 = 1$ by letting Δt first equal 0.1, then 0.01, and finally 0.001. What limit does the velocity appear to be approaching? Organize the results in a table.

33. **Motion on a Line**   As an object moves on a line, its signed distance s (in meters) from the origin after t seconds is given by the position function $s = f(t) = 3t^2 + 4t$. Find the velocity v at $t_0 = 0$. At $t_0 = 2$. At any time t_0.

34. **Motion on a Line**   As an object moves on a line, its signed distance s (in meters) from the origin after t seconds is given by the position function $s = f(t) = 2t^3 + 4$. Find the velocity v at $t_0 = 0$. At $t_0 = 3$. At any time t_0.

35. **Motion on a Line**   As an object moves on a line, its signed distance s from the origin at time t is given by the position function $s = s(t) = 3t^2 - \dfrac{1}{t}$, where s is in centimeters and t is in seconds. Find the velocity v of the object at $t_0 = 1$ and $t_0 = 4$.

36. **Motion on a Line**   As an object moves on a line, its signed distance s from the origin at time t is given by the position function $s = s(t) = 2\sqrt{t}$, where s is in centimeters and t is in seconds. Find the velocity v of the object at $t_0 = 1$ and $t_0 = 4$.

37. The Princeton Dinky is the shortest rail line in the country. It runs for 2.7 miles, connecting Princeton University to the Princeton Junction railroad station. The Dinky starts from the university and moves north toward Princeton Junction. Its distance from Princeton is shown in the graph (top, right), where the time t is in minutes and the distance s of the Dinky from Princeton University is in miles.

(a) When is the Dinky headed toward Princeton University?

(b) When is it headed toward Princeton Junction?

(c) When is the Dinky stopped?

(d) Find its average velocity on a trip from Princeton to Princeton Junction.

(e) Find its average velocity for the round trip shown in the graph, that is, from $t = 0$ to $t = 13$.

38. Jen walks to the deli, which is six blocks east of her house. After walking two blocks, she realizes she left her phone on her desk, so she runs home. After getting the phone, and closing and locking the door, Jen starts on her way again. At the deli, she waits in line to buy a bottle of **vitaminwater**™, and then she jogs home. The graph below represents Jen's journey. The time t is in minutes, and s is Jen's distance, in blocks, from home.

(a) At what times is she headed toward the deli?

(b) At what times is she headed home?

(c) When is the graph horizontal? What does this indicate?

(d) Find Jen's average velocity from home until she starts back to get her phone.

(e) Find Jen's average velocity from home to the deli after getting her phone.

(f) Find her average velocity from the deli to home.

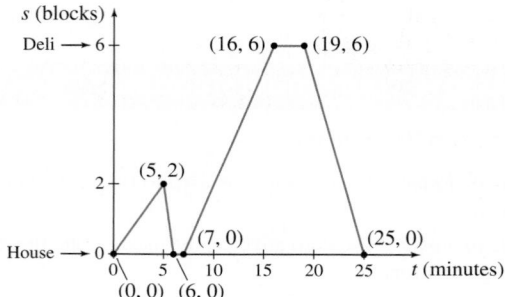

Applications and Extensions

39. **Slope of a Tangent Line**   An equation of the tangent line to the graph of a function f at $(2, 8)$ is $y = -5x + 12$. What is $f'(2)$?

40. **Slope of a Tangent Line**   An equation of the tangent line of a function f at $(3, 4)$ is $y = \dfrac{4}{3}x + 1$. What is $f'(3)$?

41. **Tangent Line**   Does the tangent line to the graph of $y = x^2$ at $(1, 1)$ pass through the point $(2, 5)$?

42. **Tangent Line**   Does the tangent line to the graph of $y = x^3$ at $(1, 1)$ pass through the point $(2, 5)$?

43. **Respiration Rate**   A human being's respiration rate R (in breaths per minute) is given by $R = R(p) = 10.35 + 0.59p$, where p is the partial pressure of carbon dioxide in the lungs. Find the rate of change in respiration when $p = 50$.

44. **Instantaneous Rate of Change**   The volume V of the right circular cylinder of height 5 m and radius r m shown in the figure is $V = V(r) = 5\pi r^2$. Find the instantaneous rate of change of the volume with respect to the radius when $r = 3$ m.

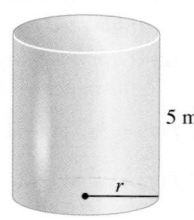

5 m

MancusoMichael/AP Images

45. Market Share During a month-long advertising campaign, the total sales S of a magazine is modeled by the function $S(x) = 5x^2 + 100x + 10,000$, where x, $0 \leq x \leq 30$, represents the number of days since the campaign began.

   (a) What is the average rate of change of sales from $x = 10$ to $x = 20$ days?

   (b) What is the instantaneous rate of change of sales when $x = 10$ days?

46. Demand Equation The demand equation for an item is $p = p(x) = 90 - 0.02x$, where p is the price in dollars and x is the number of units (in thousands) made.

   (a) Assuming all units made can be sold, find the revenue function $R(x) = xp(x)$.

   (b) Marginal Revenue Marginal revenue is defined as the additional revenue earned by selling an additional unit. If we use $R'(x)$ to measure the marginal revenue, find the marginal revenue when 1 million units are sold.

47. Gravity If a ball is dropped from the top of the Empire State Building, 1002 ft above the ground, the distance s (in feet) it falls after t seconds is $s(t) = 16t^2$.

   (a) What is the average velocity of the ball for the first 2 s?

   (b) How long does it take for the ball to hit the ground?

   (c) What is the average velocity of the ball during the time it is falling?

   (d) What is the velocity of the ball when it hits the ground?

48. Velocity A ball is thrown upward. Its height h in feet is given by $h(t) = 100t - 16t^2$, where t is the time elapsed in seconds.

   (a) What is the velocity v of the ball at $t = 0$ s, $t = 1$ s, and $t = 4$ s?

   (b) At what time t does the ball strike the ground?

   (c) At what time t does the ball reach its highest point? *Hint:* At the time the ball reaches its maximum height, it is stationary. So, its velocity $v = 0$.

49. Gravity A rock is dropped from a height of 88.2 m and falls toward Earth in a straight line. In t seconds the rock falls $4.9t^2$ m.

   (a) What is the average velocity of the rock for the first 2 s?

   (b) How long does it take for the rock to hit the ground?

   (c) What is the average velocity of the rock during its fall?

   (d) What is the velocity v of the rock when it hits the ground?

50. Velocity At a certain instant, the speedometer of an automobile reads V mi/h. During the next $\dfrac{1}{4}$ s the automobile travels 20 ft.

Approximate V from this information.

51. The table lists the outside temperature T, in degrees Fahrenheit, in Naples, Florida, on a certain day in January, for selected times x, where x is the number of hours since 12 a.m.

x	5	7	9	12	13	14	16	17
$T(x)$	62	71	74	80	83	84	85	78

   (a) Use the table to approximate $T'(11)$.

   (b) Using appropriate units, interpret $T'(11)$ in the context of the problem.

52. Volume of a Cube A metal cube with each edge of length x centimeters is expanding uniformly as a consequence of being heated.

   (a) Find the average rate of change of the volume of the cube with respect to an edge as x increases from 2.00 to 2.01 cm.

   (b) Find the instantaneous rate of change of the volume of the cube with respect to an edge at the instant when $x = 2$ cm.

53. Rate of Change Show that the rate of change of a linear function $f(x) = mx + b$ is the slope m of the line $y = mx + b$.

54. Rate of Change Show that the rate of change of a quadratic function $f(x) = ax^2 + bx + c$ is a linear function of x.

55. Agriculture The graph represents the diameter d (in centimeters) of a maturing peach as a function of the time t (in days) it is on the tree.

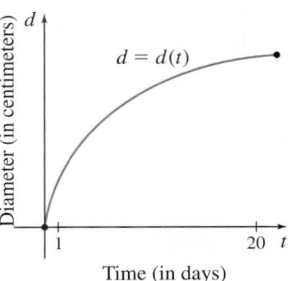

   (a) Interpret the derivative $d'(t)$ as a rate of change.

   (b) Which is larger, $d'(1)$ or $d'(20)$?

   (c) Interpret both $d'(1)$ and $d'(20)$.

56. Business The graph represents the demand d (in gallons) for olive oil as a function of the cost c (in dollars per gallon) of the oil.

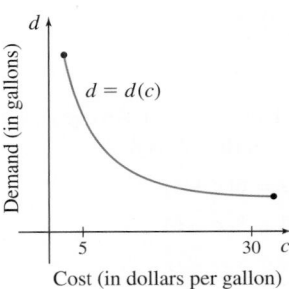

   (a) Interpret the derivative $d'(c)$.

   (b) Which is larger, $d'(5)$ or $d'(30)$? Give an interpretation to $d'(5)$ and $d'(30)$.

AP® Practice Problems

Multiple-Choice Questions

169 **1.** The line $x + y = 5$ is tangent to the graph of $y = f(x)$ at the point where $x = 2$. The values $f(2)$ and $f'(2)$ are:

  (A) $f(2) = 2$; $f'(2) = -1$ (B) $f(2) = 3$; $f'(2) = -1$

  (C) $f(2) = 2$; $f'(2) = 1$ (D) $f(2) = 3$; $f'(2) = 2$

173 **2.** The graph of the function f, given below, consists of three line segments. Find $f'(3)$.

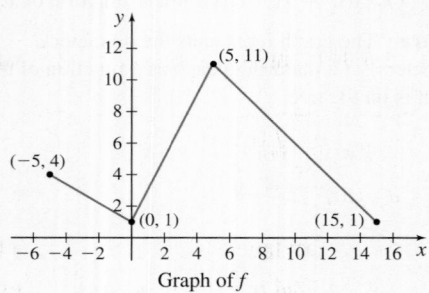

Graph of f

  (A) 1   (B) 2   (C) 3   (D) $f'(3)$ does not exist

170 **3.** What is the instantaneous rate of change of the function $f(x) = 3x^2 + 5$ at $x = 2$?

  (A) 5   (B) 7   (C) 12   (D) 17

174 **4.** The function f is defined on the closed interval $[-2, 16]$. The graph of the derivative of f, $y = f'(x)$, is given below.

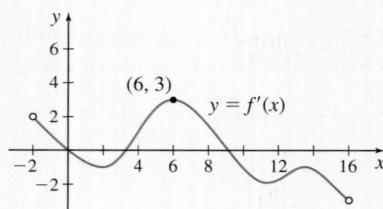

The point $(6, -2)$ is on the graph of $y = f(x)$. An equation of the tangent line to the graph of f at $(6, -2)$ is

  (A) $y = 3$ (B) $y + 2 = 6(x + 3)$

  (C) $y + 2 = 6x$ (D) $y + 2 = 3(x - 6)$

169 **5.** If $x - 3y = -16$ is an equation of the tangent line to the graph of f at the point $(2, 6)$, then $f'(2) =$

  (A) $-\dfrac{1}{3}$ (B) $\dfrac{1}{3}$ (C) -3 (D) $-\dfrac{13}{3}$

173 **6.** If f is a function for which $\lim\limits_{x \to -3} \dfrac{f(x) - f(-3)}{x + 3} = 0$, then which of the following statements must be true?

  (A) $x = -3$ is a vertical asymptote of the graph.

  (B) The derivative of f at $x = -3$ exists.

  (C) The function f is continuous at $x = 3$.

  (D) f is not defined at $x = -3$.

171 **7.** If the position of an object on the x-axis at time t is $4t^2$, then the average velocity of the object over the interval $0 \le t \le 5$ is

  (A) 5   (B) 20   (C) 40   (D) 100

175 **8.** The table below lists several values of a function $y = f(x)$ that is continuous on the closed interval $[-2, 6]$ and has a derivative at each number in the open interval $(-2, 6)$. Approximate the derivative of f at 0.

x	-1	$-1/2$	1	3	5
$f(x)$	-4	-1	2	1	-2

  (A) 0   (B) $\dfrac{1}{3}$ (C) $\dfrac{2}{3}$ (D) 2

Free-Response Question

175 **9.** A tank is filled with 80 liters of water at 7 a.m. ($t = 0$). Over the next 12 hours the water is continuously used and no water is added to replace it. The table below gives the amount of water $A(t)$ (in liters) remaining in the tank at selected times t, where t measures the number of hours after 7 a.m.

t	0	2	5	7	9	12
$A(t)$	80	71	66	60	54	50

Use the table to approximate $A'(6)$. Interpret your answer in the context of the problem.

Retain Your Knowledge

Multiple-Choice Questions

1. Given $f(x) = \dfrac{x^2 + 3x + 2}{x + 1}$, which of the following statements, if any, must be true?

  I. f is continuous at -1.

  II. $\lim\limits_{x \to -1} f(x)$ exists.

  III. The graph of f has a vertical asymptote at $x = -1$.

  (A) II only   (B) I and II only

  (C) II and III only   (D) I, II, and III

2. $\lim\limits_{x \to \frac{\pi}{4}} (4x \cdot \tan x) =$

  (A) 0   (B) $\dfrac{\pi}{4}$ (C) 1   (D) π

3. Suppose $g(x) = x^2 - 3$, $x \ne 4$. Find $\lim\limits_{x \to 4} g(x)$, if it exists.

  (A) 4   (B) 13   (C) 19   (D) The limit does not exist.

Free-Response Question

4. How should the function $r(x) = \dfrac{x^3 - 8}{x - 2}$ be defined at the number 2, so that r is continuous 2?

2.2 The Derivative as a Function; Differentiability

OBJECTIVES *When you finish this section, you should be able to:*

1 **Define the derivative function (p. 179)**
2 **Graph the derivative function (p. 181)**
3 **Identify where a function is not differentiable (p. 182)**
4 **Explain the relationship between differentiability and continuity (p. 184)**

1 Define the Derivative Function

The derivative of a function f at a real number c has been defined as the real number

$$\boxed{\textbf{Form (1)} \quad f'(c) = \lim_{x \to c} \frac{f(x) - f(c)}{x - c}}$$

provided the limit exists. We refer to this representation of the derivative as Form (1). Another way to find the derivative of f at any real number is obtained by rewriting the expression $\dfrac{f(x) - f(c)}{x - c}$ and letting $x = c + h$, $h \neq 0$. Then

$$\frac{f(x) - f(c)}{x - c} = \frac{f(c + h) - f(c)}{(c + h) - c} = \frac{f(c + h) - f(c)}{h}$$

Since $x = c + h$, then as x approaches c, h approaches 0. Form (1) of the derivative with these changes becomes

$$f'(c) = \lim_{x \to c} \frac{f(x) - f(c)}{x - c} = \lim_{h \to 0} \frac{f(c + h) - f(c)}{h}$$

So now, we have an equivalent way to write Form (1) for the derivative of f at a real number c.

$$\boxed{f'(c) = \lim_{h \to 0} \frac{f(c + h) - f(c)}{h}}$$

See Figure 7.

In the expression above for $f'(c)$, c is any real number. That is, the derivative f' is a function, called the *derivative function* of f. Now replace c by x, the independent variable of f.

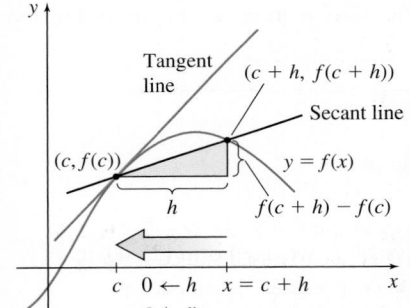

Figure 7 The slope of the tangent line at c is $f'(c) = \lim\limits_{h \to 0} \dfrac{f(c+h) - f(c)}{h}$.

DEFINITION The Derivative Function f'

The **derivative function** f' of a function f is

$$\boxed{\textbf{Form (2)} \quad f'(x) = \lim_{h \to 0} \frac{f(x + h) - f(x)}{h}}$$

provided the limit exists. If f has a derivative, then f is said to be **differentiable**.

IN WORDS In Form (2) the derivative is the limit of a difference quotient.

We refer to this representation of the derivative as Form (2).

EXAMPLE 1 Finding the Derivative Function

Find the derivative of the function $f(x) = x^2 - 5x$ at any real number x using Form (2).

Solution

Using Form (2), we have

$$f'(x) = \lim_{h \to 0} \frac{f(x+h) - f(x)}{h} = \lim_{h \to 0} \frac{[(x+h)^2 - 5(x+h)] - (x^2 - 5x)}{h}$$

$$= \lim_{h \to 0} \frac{[(x^2 + 2xh + h^2) - 5x - 5h] - x^2 + 5x}{h} = \lim_{h \to 0} \frac{2xh + h^2 - 5h}{h}$$

$$= \lim_{h \to 0} \frac{h(2x + h - 5)}{h} = \lim_{h \to 0} (2x + h - 5) = 2x - 5 \qquad \blacksquare$$

NOW WORK AP® Practice Problem 2.

The domain of the function f' is the set of real numbers in the domain of f for which the limit expressed in Form (2) exists. So the domain of f' is a subset of the domain of f.

We can use either Form (1) or Form (2) to find derivatives using the definition. However, if we want the derivative of f at a specified number c, we usually use Form (1) to find $f'(c)$. If we want to find the derivative function of f, we usually use Form (2) to find $f'(x)$. In this section, we use the definitions of the derivative, Forms (1) and (2), to investigate derivatives. In the next section, we begin to develop formulas for finding the derivatives.

EXAMPLE 2 Finding the Derivative Function

NOTE The instruction "differentiate f" means "find the derivative of f."

Differentiate $f(x) = \sqrt{x}$ and determine the domain of f'.

Solution

The domain of f is $\{x \mid x \geq 0\}$. To find the derivative of f, we use Form (2). Then

$$f'(x) = \lim_{h \to 0} \frac{f(x+h) - f(x)}{h} = \lim_{h \to 0} \frac{\sqrt{x+h} - \sqrt{x}}{h}$$

Rationalize the numerator to find the limit.

$$f'(x) = \lim_{h \to 0} \left[\frac{\sqrt{x+h} - \sqrt{x}}{h} \cdot \frac{\sqrt{x+h} + \sqrt{x}}{\sqrt{x+h} + \sqrt{x}} \right] = \lim_{h \to 0} \frac{(x+h) - x}{h(\sqrt{x+h} + \sqrt{x})}$$

$$= \lim_{h \to 0} \frac{h}{h(\sqrt{x+h} + \sqrt{x})} = \lim_{h \to 0} \frac{1}{\sqrt{x+h} + \sqrt{x}} = \frac{1}{2\sqrt{x}}$$

The limit does not exist when $x = 0$. But for all other x in the domain of f, the limit does exist. So, the domain of the derivative function $f'(x) = \dfrac{1}{2\sqrt{x}}$ is $\{x \mid x > 0\}$. $\blacksquare$

In Example 2, notice that the domain of the derivative function f' is a proper subset of the domain of the function f. The graphs of both f and f' are shown in Figure 8.

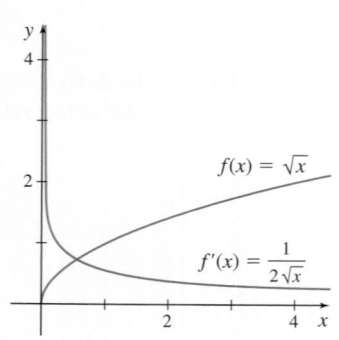

Figure 8

NOW WORK Problem 15 and AP® Practice Problem 8.

EXAMPLE 3   **Interpreting the Derivative as a Rate of Change**

The surface area S (in square meters) of a balloon is expanding as a function of time t (in seconds) according to $S = S(t) = 5t^2$. Find the rate of change of the surface area of the balloon with respect to time. What are the units of $S'(t)$?

Solution

An interpretation of the derivative function $S'(t)$ is the rate of change of $S = S(t)$.

$$S'(t) = \lim_{h \to 0} \frac{S(t+h) - S(t)}{h} = \lim_{h \to 0} \frac{5(t+h)^2 - 5t^2}{h} \qquad \text{Form (2)}$$

$$= \lim_{h \to 0} \frac{5(t^2 + 2th + h^2) - 5t^2}{h}$$

$$= \lim_{h \to 0} \frac{5t^2 + 10th + 5h^2 - 5t^2}{h} = \lim_{h \to 0} \frac{10th + 5h^2}{h}$$

$$= \lim_{h \to 0} \frac{(10t + 5h)h}{h} = \lim_{h \to 0} (10t + 5h) = 10t$$

Since $S'(t)$ is the limit of the quotient of a change in area divided by a change in time, the units of the rate of change are square meters per second (m²/s). The rate of change of the surface area S of the balloon with respect to time is $10t$ m²/s. ∎

NOW WORK   Problem **67** and AP® Practice Problems **11** and **12**.

❷ Graph the Derivative Function

There is a relationship between the graph of a function and the graph of its derivative.

EXAMPLE 4   **Graphing a Function and Its Derivative**

Find f' if $f(x) = x^3 - 1$. Then graph $y = f(x)$ and $y = f'(x)$ on the same set of coordinate axes.

Solution

$f(x) = x^3 - 1$ so

$$f(x + h) = (x + h)^3 - 1 = x^3 + 3hx^2 + 3h^2x + h^3 - 1$$

Using Form (2), we find

$$f'(x) = \lim_{h \to 0} \frac{f(x+h) - f(x)}{h}$$

$$= \lim_{h \to 0} \frac{(x^3 + 3hx^2 + 3h^2x + h^3 - 1) - (x^3 - 1)}{h}$$

$$= \lim_{h \to 0} \frac{3hx^2 + 3h^2x + h^3}{h}$$

$$= \lim_{h \to 0} \frac{h(3x^2 + 3hx + h^2)}{h}$$

$$= \lim_{h \to 0} (3x^2 + 3hx + h^2) = 3x^2$$

The graphs of f and f' are shown in Figure 9. ∎

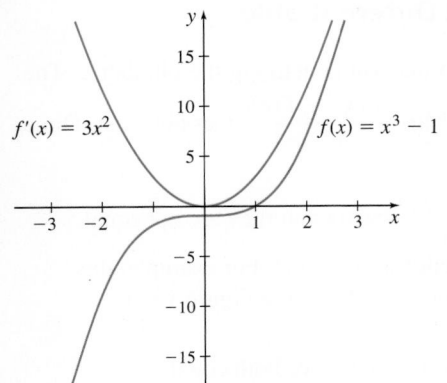

$f'(x) = 3x^2$ $f(x) = x^3 - 1$

Figure 9

NOW WORK   Problem **19**.

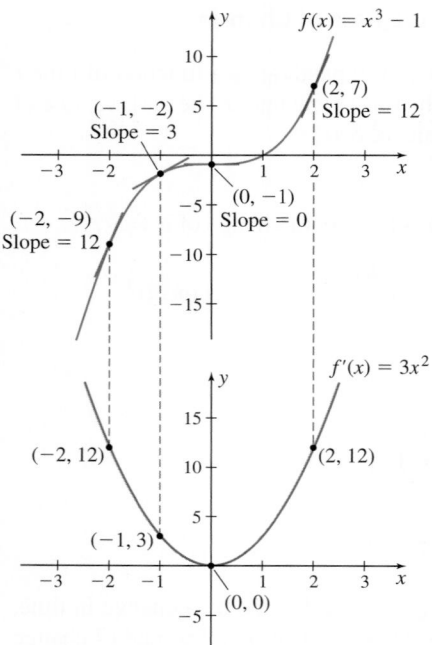

Figure 10

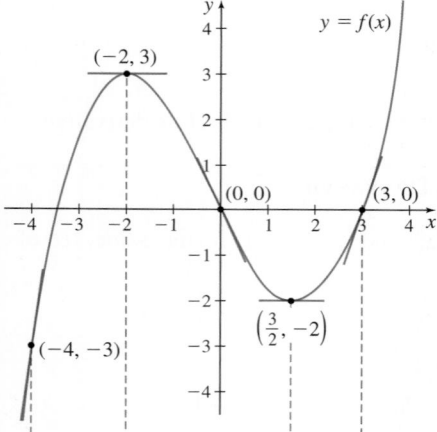

Figure 11

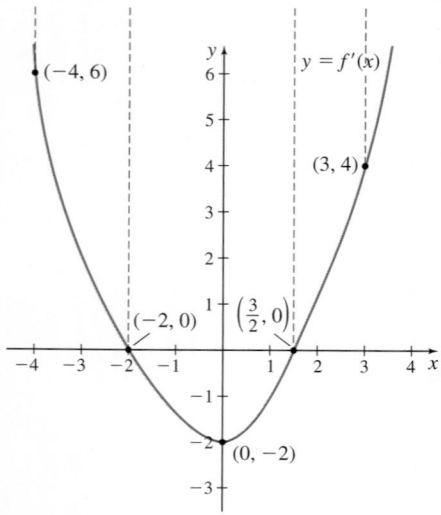

Figure 12

Figure 10 illustrates several tangent lines to the graph of $f(x) = x^3 - 1$. Observe that the tangent line to the graph of f at $(0, -1)$ is horizontal, so its slope is 0. Then $f'(0) = 0$, so the graph of f' contains the point $(0, 0)$. Also notice that every tangent line to the graph of f has a nonnegative slope, so $f'(x) \geq 0$. That is, the range of the function f' is $\{y \mid y \geq 0\}$. Finally, notice that the slope of each tangent line to the graph of f is the y-coordinate of the corresponding point on the graph of the derivative f'.

With these ideas in mind, we can obtain a rough sketch of the derivative function f', even if we know only the graph of the function f.

EXAMPLE 5 Graphing the Derivative Function

Use the graph of the function $y = f(x)$, shown in Figure 11, to sketch the graph of the derivative function $y = f'(x)$.

Solution

We begin by drawing tangent lines to the graph of f at the points shown in Figure 11. At the points $(-2, 3)$ and $\left(\dfrac{3}{2}, -2\right)$ the tangent lines are horizontal, so their slopes are 0.

This means $f'(-2) = 0$ and $f'\left(\dfrac{3}{2}\right) = 0$, so the points $(-2, 0)$ and $\left(\dfrac{3}{2}, 0\right)$ are on the graph of the derivative function. See Figure 12. Now we estimate the slope of the tangent lines to the graph of f at the other selected points. For example, at the point $(-4, -3)$, the slope of the tangent line is positive and the line is rather steep. We estimate the slope to be close to 6, and we plot the point $(-4, 6)$ on the graph in Figure 12. Continue the process and then connect the points with a smooth curve. ■

Notice that at the points where the tangent lines to the graph of f are horizontal (see Figure 11), the graph of the derivative f' intersects the x-axis (see Figure 12). Also notice that wherever the graph of f is increasing, the slopes of the tangent lines are positive, that is, f' is positive, so the graph of f' is above the x-axis. Similarly, wherever the graph of f is decreasing, the slopes of the tangent lines are negative, so the graph of f' is below the x-axis.

NOW WORK   Problem 29.

❸ Identify Where a Function Is Not Differentiable

Suppose a function f is continuous on an open interval containing the number c. The function f is not differentiable at the number c if $\lim\limits_{x \to c} \dfrac{f(x) - f(c)}{x - c}$ does not exist. Three (of several) ways this can happen are:

- $\lim\limits_{x \to c^-} \dfrac{f(x) - f(c)}{x - c}$ exists and $\lim\limits_{x \to c^+} \dfrac{f(x) - f(c)}{x - c}$ exists, but they are not equal. When this happens the graph of f has a **corner** at $(c, f(c))$. For example, the absolute value function $f(x) = |x|$ has a corner at $(0, 0)$. See Figure 13 on page 183.

- The one-sided limits are both infinite and both equal ∞ or both equal $-\infty$. When this happens, the graph of f has a vertical tangent line at $(c, f(c))$. For example, the graph of the cube root function $f(x) = \sqrt[3]{x}$ has a vertical tangent at $(0, 0)$. See Figure 14.

- Both one-sided limits are infinite, but one equals $-\infty$ and the other equals ∞. When this happens, the graph of f has a vertical tangent line at the point $(c, f(c))$. This point is referred to as a **cusp**. For example, the graph of the function $f(x) = x^{2/3}$ has a cusp at $(0, 0)$. See Figure 15.

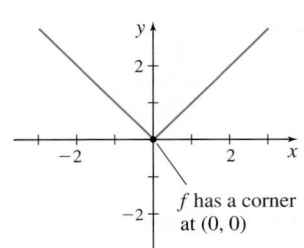

Figure 13 $f(x) = |x|$;
$f'(0)$ does not exist.

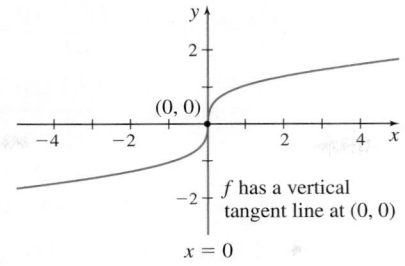

Figure 14 $f(x) = \sqrt[3]{x}$;
$f'(0)$ does not exist.

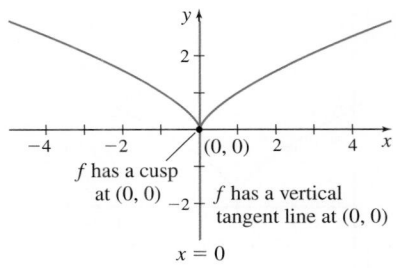

Figure 15 $f(x) = x^{2/3}$;
$f'(0)$ does not exist.

EXAMPLE 6 **Identifying Where a Function Is Not Differentiable**

Given the piecewise defined function $f(x) = \begin{cases} -2x^2 + 4 & \text{if } x < 1 \\ x^2 + 1 & \text{if } x \geq 1 \end{cases}$,
determine whether $f'(1)$ exists.

Solution

Use Form (1) of the definition of a derivative to determine whether $f'(1)$ exists.

$$\lim_{x \to 1} \frac{f(x) - f(1)}{x - 1} = \lim_{x \to 1} \frac{f(x) - 2}{x - 1} \qquad f(1) = 1^2 + 1 = 2$$

If $x < 1$, then $f(x) = -2x^2 + 4$; if $x \geq 1$, then $f(x) = x^2 + 1$. So, it is necessary to find the one-sided limits at 1.

$$\lim_{x \to 1^-} \frac{f(x) - f(1)}{x - 1} = \lim_{x \to 1^-} \frac{(-2x^2 + 4) - 2}{x - 1} = \lim_{x \to 1^-} \frac{-2(x^2 - 1)}{x - 1}$$

$$= -2 \lim_{x \to 1^-} \frac{(x - 1)(x + 1)}{x - 1} = -2 \lim_{x \to 1^-} (x + 1) = -4$$

$$\lim_{x \to 1^+} \frac{f(x) - f(1)}{x - 1} = \lim_{x \to 1^+} \frac{(x^2 + 1) - 2}{x - 1} = \lim_{x \to 1^+} \frac{(x - 1)(x + 1)}{x - 1} = \lim_{x \to 1^+} (x + 1) = 2$$

Since the one-sided limits are not equal, $\lim_{x \to 1} \dfrac{f(x) - f(1)}{x - 1}$ does not exist, and so $f'(1)$ does not exist. ∎

Figure 16 illustrates the graph of the function f from Example 6. At 1, where the derivative does not exist, the graph of f has a corner. We usually say that the graph of f is not *smooth* at a corner.

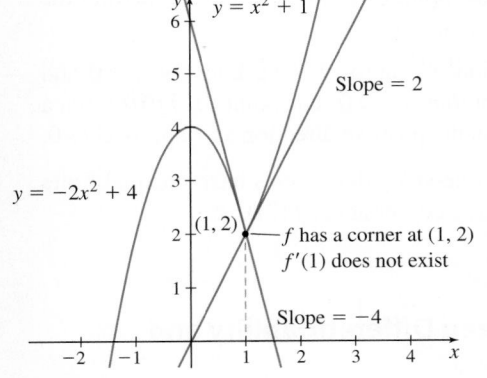

Figure 16 f has a corner at $(1, 2)$.

NOW WORK Problem **39** and AP® Practice Problems **1, 5, 9,** and **10.**

Example 7 illustrates the behavior of the graph of a function f when the derivative at a number c does not exist because $\lim\limits_{x \to c} \dfrac{f(x) - f(c)}{x - c}$ is infinite.

EXAMPLE 7 **Showing That a Function Is Not Differentiable**

CALC CLIP

Show that $f(x) = (x - 2)^{4/5}$ is not differentiable at 2.

Solution

The function f is continuous for all real numbers and $f(2) = (2 - 2)^{4/5} = 0$. Use Form (1) of the definition of the derivative to find the two one-sided limits at 2.

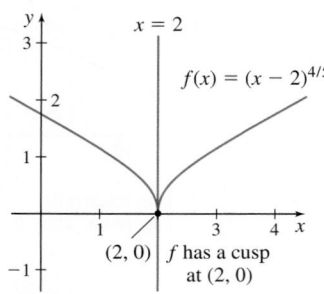

Figure 17 $f'(2)$ does not exist; the point $(2, 0)$ is a cusp of the graph of f.

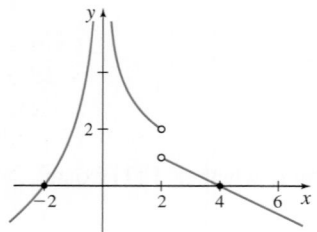

Figure 18 $y = f'(x)$

$$\lim_{x \to 2^-} \frac{f(x) - f(2)}{x - 2} = \lim_{x \to 2^-} \frac{(x-2)^{4/5} - 0}{x - 2} = \lim_{x \to 2^-} \frac{(x-2)^{4/5}}{x - 2} = \lim_{x \to 2^-} \frac{1}{(x-2)^{1/5}} = -\infty$$

$$\lim_{x \to 2^+} \frac{f(x) - f(2)}{x - 2} = \lim_{x \to 2^+} \frac{(x-2)^{4/5} - 0}{x - 2} = \lim_{x \to 2^+} \frac{(x-2)^{4/5}}{x - 2} = \lim_{x \to 2^+} \frac{1}{(x-2)^{1/5}} = \infty$$

Since $\displaystyle\lim_{x \to 2^-} \frac{f(x) - f(2)}{x - 2} = -\infty$ and $\displaystyle\lim_{x \to 2^+} \frac{f(x) - f(2)}{x - 2} = \infty$, we conclude that the function f is not differentiable at 2. The graph of f has a vertical tangent line at the point $(2, 0)$, which is a cusp of the graph. See Figure 17. ∎

NOW WORK   Problem 35.

EXAMPLE 8   **Obtaining Information about $y = f(x)$ from the Graph of Its Derivative Function**

Suppose $y = f(x)$ is continuous for all real numbers. Figure 18 shows the graph of its derivative function f'.

(a)   Does the graph of f have any horizontal tangent lines? If yes, explain why and identify where they occur.

(b)   Does the graph of f have any vertical tangent lines? If yes, explain why, identify where they occur, and determine whether the point is a cusp of f.

(c)   Does the graph of f have any corners? If yes, explain why and identify where they occur.

Solution

(a) Since the derivative f' equals the slope of a tangent line, horizontal tangent lines occur where the derivative equals 0. Since $f'(x) = 0$ for $x = -2$ and $x = 4$, the graph of f has two horizontal tangent lines, one at the point $(-2, f(-2))$ and the other at the point $(4, f(4))$.

(b) As x approaches 0, the derivative function f' approaches ∞ both for $x < 0$ and for $x > 0$. The graph of f has a vertical tangent line at $x = 0$. The point $(0, f(0))$ is not a cusp because both limits become unbounded in the positive direction as x approaches 0.

(c) The derivative is not defined at 2 and the one-sided derivatives have unequal finite limits as x approaches 2. So the graph of f has a corner at $(2, f(2))$. ∎

NOW WORK   Problem 45.

---

4 Explain the Relationship Between Differentiability and Continuity

In Chapter 1, we investigated the continuity of a function. Here we have been investigating the differentiability of a function. An important connection exists between continuity and differentiability.

NEED TO REVIEW? Continuity is discussed in Section 1.3, pp. 108–115.

> **THEOREM**
>
> If a function f is differentiable at a number c, then f is continuous at c.

Proof To show that f is continuous at c, we need to verify that $\displaystyle\lim_{x \to c} f(x) = f(c)$. We begin by observing that if $x \neq c$, then

$$f(x) - f(c) = \left[\frac{f(x) - f(c)}{x - c} \right] (x - c)$$

We take the limit of both sides as $x \to c$, and use the fact that the limit of a product equals the product of the limits (we show later that each limit exists).

$$\lim_{x \to c}[f(x) - f(c)] = \lim_{x \to c} \left\{ \left[\frac{f(x) - f(c)}{x - c} \right] (x - c) \right\} = \left[\lim_{x \to c} \frac{f(x) - f(c)}{x - c} \right] \left[\lim_{x \to c} (x - c) \right]$$

Since f is differentiable at c, we know that

$$\lim_{x \to c} \frac{f(x) - f(c)}{x - c} = f'(c)$$

is a number. Also for any real number c, $\lim_{x \to c}(x - c) = 0$. So

$$\lim_{x \to c}[f(x) - f(c)] = [f'(c)]\left[\lim_{x \to c}(x - c)\right] = f'(c) \cdot 0 = 0$$

That is, $\lim_{x \to c} f(x) = f(c)$, so f is continuous at c. ∎

An equivalent statement of this theorem gives a condition under which a function has no derivative.

COROLLARY

If a function f is discontinuous at a number c, then f is not differentiable at c.

IN WORDS Differentiability implies continuity, but continuity does not imply differentiability.

Let's look at some of the possibilities. In Figure 19(a), the function f is continuous at the number 1 and has a derivative at 1. The function g, graphed in Figure 19(b), is continuous at the number 0, but it has no derivative at 0. So continuity at a number c provides no prediction about differentiability. On the other hand, the function h graphed in Figure 19(c) illustrates the corollary: If h is discontinuous at a number, it is not differentiable at that number.

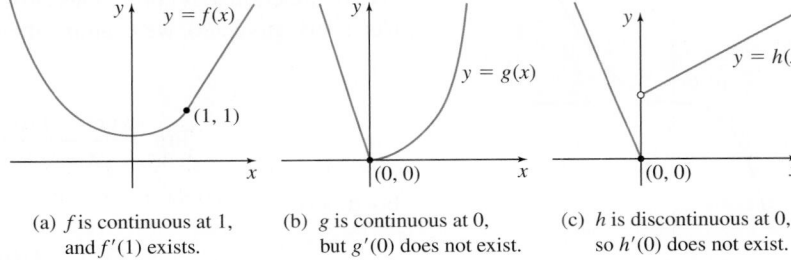

(a) f is continuous at 1, and $f'(1)$ exists.   (b) g is continuous at 0, but $g'(0)$ does not exist.   (c) h is discontinuous at 0, so $h'(0)$ does not exist.

Figure 19

The corollary is useful if we are seeking the derivative of a function f that we suspect is discontinuous at a number c. If we can show that f is discontinuous at c, then the corollary affirms that the function f has no derivative at c. For example, since the floor function $f(x) = \lfloor x \rfloor$ is discontinuous at every integer c, it has no derivative at an integer.

EXAMPLE 9 **Determining Whether a Function Is Differentiable at a Number**

Determine whether the function

$$f(x) = \begin{cases} 2x + 2 & \text{if } x < 3 \\ 5 & \text{if } x = 3 \\ x^2 - 1 & \text{if } x > 3 \end{cases}$$

is differentiable at 3.

Solution

Since f is a piecewise-defined function, it may be discontinuous at 3 and therefore not differentiable at 3. So we begin by determining whether f is continuous at 3.

Since $f(3) = 5$, the function f is defined at 3. Use one-sided limits to check whether $\lim_{x \to 3} f(x)$ exists.

$$\lim_{x \to 3^-} f(x) = \lim_{x \to 3^-}(2x + 2) = 8 \qquad \lim_{x \to 3^+} f(x) = \lim_{x \to 3^+}(x^2 - 1) = 8$$

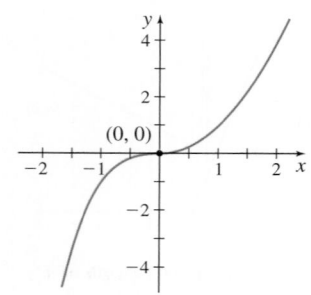

Figure 20 $f(x) = \begin{cases} 2x+2 & \text{if } x < 3 \\ 5 & \text{if } x = 3 \\ x^2 - 1 & \text{if } x > 3 \end{cases}$

Since $\lim\limits_{x\to 3^-} f(x) = \lim\limits_{x\to 3^+} f(x)$, then $\lim\limits_{x\to 3} f(x)$ exists. But $\lim\limits_{x\to 3} f(x) = 8$ and $f(3) = 5$, so f is discontinuous at 3. From the corollary, since f is discontinuous at 3, the function f is not differentiable at 3. ■

Figure 20 shows the graph of f.

NOW WORK   Problem 43.

In Example 9, the function f is discontinuous at 3, so by the corollary, the derivative of f at 3 does not exist. But when a function is continuous at a number c, then sometimes the derivative at c exists and other times the derivative at c does not exist.

EXAMPLE 10   **Determining Whether a Function Is Differentiable at a Number**

Determine whether each piecewise-defined function is differentiable at c. If the function has a derivative at c, find it.

(a) $f(x) = \begin{cases} x^3 & \text{if } x < 0 \\ x^2 & \text{if } x \geq 0 \end{cases}$ $c = 0$ **(b)** $g(x) = \begin{cases} 1 - 2x & \text{if } x \leq 1 \\ x - 2 & \text{if } x > 1 \end{cases}$ $c = 1$

Solution

(a) The function f is continuous at 0, which you should verify. To determine whether f has a derivative at 0, we examine the one-sided limits at 0 using Form (1).

For $x < 0$,

$$\lim_{x\to 0^-} \frac{f(x) - f(0)}{x - 0} = \lim_{x\to 0^-} \frac{x^3 - 0}{x} = \lim_{x\to 0^-} x^2 = 0$$

For $x > 0$,

$$\lim_{x\to 0^+} \frac{f(x) - f(0)}{x - 0} = \lim_{x\to 0^+} \frac{x^2 - 0}{x} = \lim_{x\to 0^+} x = 0$$

Since both one-sided limits are equal, f is differentiable at 0, and $f'(0) = 0$. See Figure 21.

(b) The function g is continuous at 1, which you should verify. To determine whether g is differentiable at 1, examine the one-sided limits at 1 using Form (1).

For $x < 1$,

$$\lim_{x\to 1^-} \frac{g(x) - g(1)}{x - 1} = \lim_{x\to 1^-} \frac{(1 - 2x) - (-1)}{x - 1} = \lim_{x\to 1^-} \frac{2 - 2x}{x - 1}$$
$$= \lim_{x\to 1^-} \frac{-2(x - 1)}{x - 1} = \lim_{x\to 1^-} (-2) = -2$$

For $x > 1$,

$$\lim_{x\to 1^+} \frac{g(x) - g(1)}{x - 1} = \lim_{x\to 1^+} \frac{(x - 2) - (-1)}{x - 1} = \lim_{x\to 1^+} \frac{x - 1}{x - 1} = 1$$

The one-sided limits are not equal, so $\lim\limits_{x\to 1} \dfrac{g(x) - g(1)}{x - 1}$ does not exist. That is, g is not differentiable at 1. See Figure 22. ■

Notice in Figure 21 the tangent lines to the graph of f turn smoothly around the origin. On the other hand, notice in Figure 22 the tangent lines to the graph of g change abruptly at the point $(1, -1)$, where the graph of g has a corner.

Figure 21 $f(x) = \begin{cases} x^3 & \text{if } x < 0 \\ x^2 & \text{if } x \geq 0 \end{cases}$

Figure 22 $g(x) = \begin{cases} 1 - 2x & \text{if } x \leq 1 \\ x - 2 & \text{if } x > 1 \end{cases}$

NOW WORK   Problem 41 and AP® Practice Problems 3, 4, 6, and 7.

2.2 Assess Your Understanding

Concepts and Vocabulary

1. **True or False** The domain of a function f and the domain of its derivative function f' are always equal.

2. **True or False** If a function is continuous at a number c, then it is differentiable at c.

3. **Multiple Choice** If f is continuous at a number c and

   if $\lim\limits_{x \to c} \dfrac{f(x) - f(c)}{x - c}$ is infinite, then the graph of f has

   [(a) a horizontal   (b) a vertical   (c) no]

   tangent line at c.

4. The instruction, "Differentiate f," means to find the _____ of f.

Skill Building

In Problems 5–10, find the derivative of each function f at any real number c. Use Form (1) on page 179.

5. $f(x) = 10$

6. $f(x) = -4$

7. $f(x) = 5x + 8$

8. $f(x) = 6x - 4$

9. $f(x) = 2 - x^2$

10. $f(x) = 2x^2 + 4$

In Problems 11–16, differentiate each function f and determine the domain of f'. Use Form (2) on page 179.

11. $f(x) = 5$

12. $f(x) = -2$

13. $f(x) = 4x^2 + x + 6$

14. $f(x) = 3x^2 - x - 8$

 15. $f(x) = 5\sqrt{x - 1}$

16. $f(x) = 4\sqrt{x + 3}$

In Problems 17–22, differentiate each function f. Graph y = f(x) and y = f'(x) on the same set of coordinate axes.

17. $f(x) = \dfrac{1}{3}x + 1$

18. $f(x) = -4x - 5$

[PAGE 181] 19. $f(x) = 2x^2 - 5x$

20. $f(x) = -3x^2 + 2$

21. $f(x) = x^3 - 8x$

22. $f(x) = -x^3 - 8$

In Problems 23–26, for each figure determine if the graphs represent a function f and its derivative f'. If they do, indicate which is the graph of f and which is the graph of f'.

23.
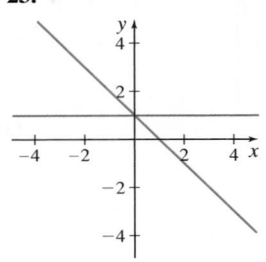

24.
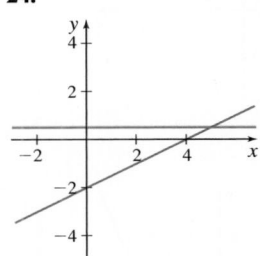

25.
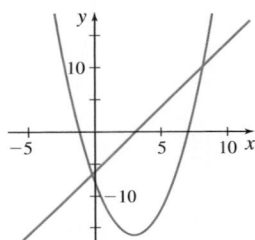

26.
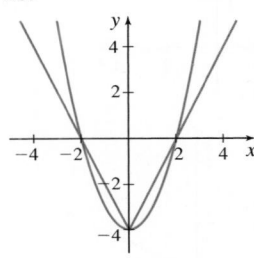

In Problems 27–30, use the graph of f to obtain the graph of f'.

27.
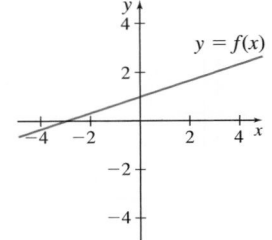

28.
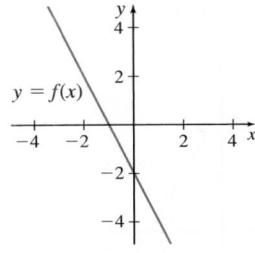

[PAGE 182] 29.
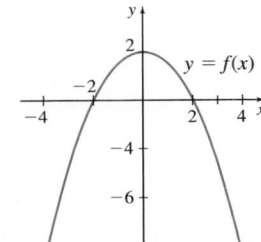

30.
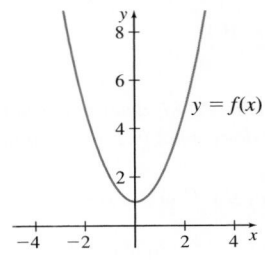

In Problems 31–34, the graph of a function f is given. Match each graph to the graph of its derivative f′ in A–D.

31.

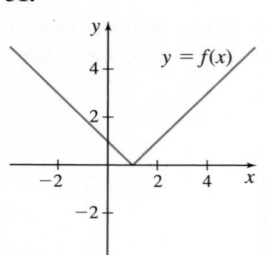

32.

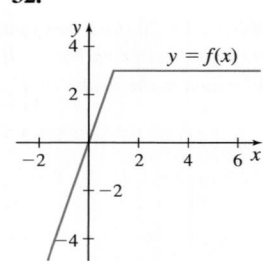

33.

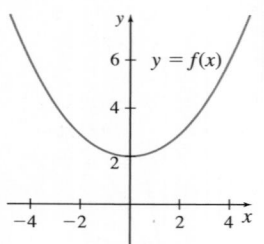

34.

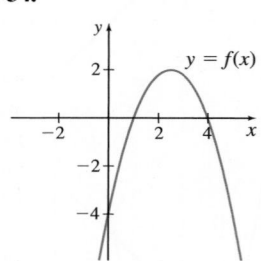

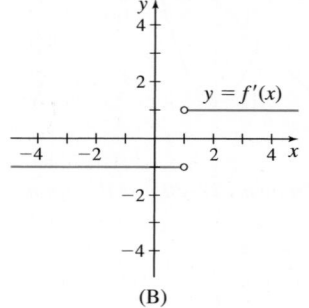

(A)   (B)

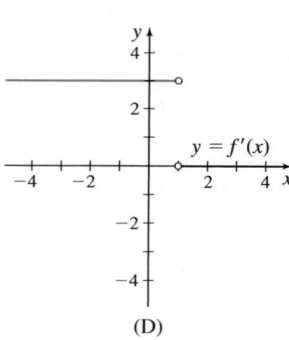

(C)   (D)

In Problems 35–44, determine whether each function f has a derivative at c. If it does, what is f′(c)? If it does not, give the reason why.

[PAGE 184] 35. $f(x) = x^{2/3}$ at $c = -8$

36. $f(x) = 2x^{1/3}$ at $c = 0$

37. $f(x) = |x^2 - 4|$ at $c = 2$

38. $f(x) = |x^2 - 4|$ at $c = -2$

[PAGE 183] 39. $f(x) = \begin{cases} 2x + 3 & \text{if } x < 1 \\ x^2 + 4 & \text{if } x \geq 1 \end{cases}$ at $c = 1$

40. $f(x) = \begin{cases} 3 - 4x & \text{if } x < -1 \\ 2x + 9 & \text{if } x \geq -1 \end{cases}$ at $c = -1$

[PAGE 186] 41. $f(x) = \begin{cases} -4 + 2x & \text{if } x \leq \dfrac{1}{2} \\ 4x^2 - 4 & \text{if } x > \dfrac{1}{2} \end{cases}$ at $c = \dfrac{1}{2}$

42. $f(x) = \begin{cases} 2x^2 + 1 & \text{if } x < -1 \\ -1 - 4x & \text{if } x \geq -1 \end{cases}$ at $c = -1$

[PAGE 186] 43. $f(x) = \begin{cases} 2x^2 + 1 & \text{if } x < -1 \\ 2 + 2x & \text{if } x \geq -1 \end{cases}$ at $c = -1$

44. $f(x) = \begin{cases} 5 - 2x & \text{if } x < 2 \\ x^2 & \text{if } x \geq 2 \end{cases}$ at $c = 2$

In Problems 45–48, each function f is continuous for all real numbers, and the graph of $y = f'(x)$ is given.

(a) Does the graph of f have any horizontal tangent lines? If yes, explain why and identify where they occur.

(b) Does the graph of f have any vertical tangent lines? If yes, explain why, identify where they occur, and determine whether the point is a cusp of f.

(c) Does the graph of f have any corners? If yes, explain why and identify where they occur.

[PAGE 184] 45.

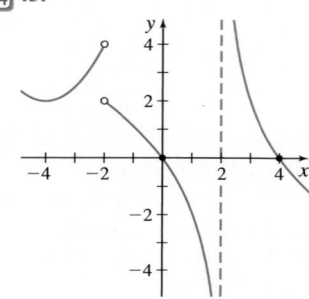

46.

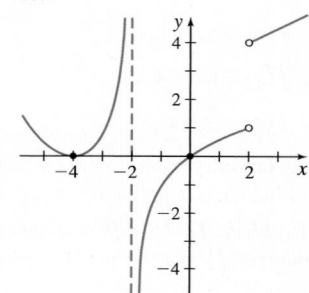

47.

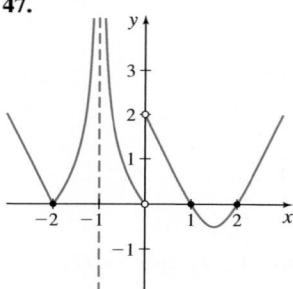

48.

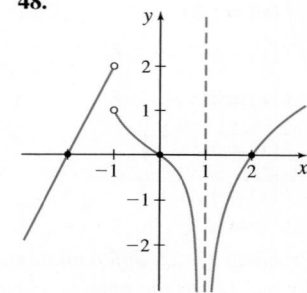

In Problems 49 and 50, use the given points $(c, f(c))$ on the graph of the function f.

(a) For which numbers c does $\lim_{x \to c} f(x)$ exist but f is not continuous at c?

(b) For which numbers c is f continuous at c but not differentiable at c?

49.

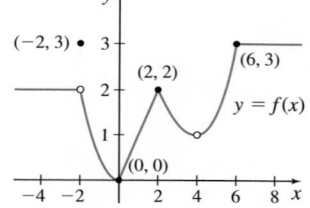

50.

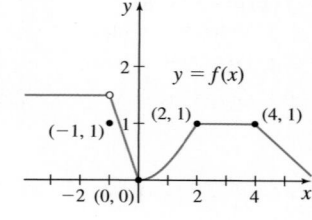

In Problems 51–54, find the derivative of each function.

51. $f(x) = mx + b$

52. $f(x) = ax^2 + bx + c$

53. $f(x) = \dfrac{1}{x^2}$

54. $f(x) = \dfrac{1}{\sqrt{x}}$

Applications and Extensions

In Problems 55–66, each limit represents the derivative of a function f at some number c. Determine f and c in each case.

55. $\lim\limits_{h\to 0} \dfrac{(2+h)^2 - 4}{h}$

56. $\lim\limits_{h\to 0} \dfrac{(2+h)^3 - 8}{h}$

57. $\lim\limits_{x\to 1} \dfrac{x^2 - 1}{x - 1}$

58. $\lim\limits_{x\to 1} \dfrac{x^4 - 1}{x - 1}$

59. $\lim\limits_{h\to 0} \dfrac{\sqrt{9+h} - 3}{h}$

60. $\lim\limits_{h\to 0} \dfrac{(8+h)^{1/3} - 2}{h}$

61. $\lim\limits_{x\to \pi/6} \dfrac{\sin x - \dfrac{1}{2}}{x - \dfrac{\pi}{6}}$

62. $\lim\limits_{x\to \pi/4} \dfrac{\cos x - \dfrac{\sqrt{2}}{2}}{x - \dfrac{\pi}{4}}$

63. $\lim\limits_{x\to 0} \dfrac{2(x+2)^2 - (x+2) - 6}{x}$

64. $\lim\limits_{x\to 0} \dfrac{3x^3 - 2x}{x}$

65. $\lim\limits_{h\to 0} \dfrac{(3+h)^2 + 2(3+h) - 15}{h}$

66. $\lim\limits_{h\to 0} \dfrac{3(h-1)^2 + h - 3}{h}$

67. Units The volume V (in cubic feet) of a balloon is expanding according to $V = V(t) = 4t$, where t is the time (in seconds). Find the rate of change of the volume of the balloon with respect to time. What are the units of $V'(t)$?

68. Units The area A (in square miles) of a circular patch of oil is expanding according to $A = A(t) = 2t$, where t is the time (in hours). At what rate is the area changing with respect to time? What are the units of $A'(t)$?

69. Units A manufacturer of precision digital switches has a daily cost C (in dollars) of $C(x) = 10,000 + 3x$, where x is the number of switches produced daily. What is the rate of change of cost with respect to x? What are the units of $C'(x)$?

70. Units A manufacturer of precision digital switches has daily revenue R (in dollars) of $R(x) = 5x - \dfrac{x^2}{2000}$, where x is the number of switches produced daily. What is the rate of change of revenue with respect to x? What are the units of $R'(x)$?

71. $f(x) = \begin{cases} x^3 & \text{if } x \le 0 \\ x^2 & \text{if } x > 0 \end{cases}$

  (a) Determine whether f is continuous at 0.
  (b) Determine whether $f'(0)$ exists.
  (c) Graph the function f and its derivative f'.

72. For the function $f(x) = \begin{cases} 2x & \text{if } x \le 0 \\ x^2 & \text{if } x > 0 \end{cases}$

  (a) Determine whether f is continuous at 0.
  (b) Determine whether $f'(0)$ exists.
  (c) Graph the function f and its derivative f'.

73. Velocity The distance s (in feet) of an automobile from the origin at time t (in seconds) is given by the position function

$$s = s(t) = \begin{cases} t^3 & \text{if } 0 \le t < 5 \\ 125 & \text{if } t \ge 5 \end{cases}$$

(This could represent a crash test in which a vehicle is accelerated until it hits a brick wall at $t = 5$ s.)

  (a) Find the velocity just before impact (at $t = 4.99$ s) and just after impact (at $t = 5.01$ s).
  (b) Is the velocity function $v = s'(t)$ continuous at $t = 5$?
  (c) How do you interpret the answer to (b)?

74. Population Growth A simple model for population growth states that the rate of change of population size P with respect to time t is proportional to the population size. Express this statement as an equation involving a derivative.

75. Atmospheric Pressure Atmospheric pressure p decreases as the distance x from the surface of Earth increases, and the rate of change of pressure with respect to altitude is proportional to the pressure. Express this law as an equation involving a derivative.

76. Electrical Current Under certain conditions, an electric current I will die out at a rate (with respect to time t) that is proportional to the current remaining. Express this law as an equation involving a derivative.

77. Tangent Line Let $f(x) = x^2 + 2$. Find all points on the graph of f for which the tangent line passes through the origin.

78. Tangent Line Let $f(x) = x^2 - 2x + 1$. Find all points on the graph of f for which the tangent line passes through the point $(1, -1)$.

79. Area and Circumference of a Circle A circle of radius r has area $A = \pi r^2$ and circumference $C = 2\pi r$. If the radius changes from r to $r + \Delta r$, find the:

  (a) Change in area.
  (b) Change in circumference.
  (c) Average rate of change of area with respect to radius.
  (d) Average rate of change of circumference with respect to radius.
  (e) Rate of change of circumference with respect to radius.

80. Volume of a Sphere The volume V of a sphere of radius r is $V = \dfrac{4\pi r^3}{3}$. If the radius changes from r to $r + \Delta r$, find the:

  (a) Change in volume.
  (b) Average rate of change of volume with respect to radius.
  (c) Rate of change of volume with respect to radius.

81. Use the definition of the derivative to show that $f(x) = |x|$ is not differentiable at 0.

82. Use the definition of the derivative to show that $f(x) = \sqrt[3]{x}$ is not differentiable at 0.

83. If f is an even function that is differentiable at c, show that its derivative function is odd. That is, show $f'(-c) = -f'(c)$.

84. If f is an odd function that is differentiable at c, show that its derivative function is even. That is, show $f'(-c) = f'(c)$.

85. Tangent Lines and Derivatives Let f and g be two functions, each with derivatives at c. State the relationship between their tangent lines at c if:

(a) $f'(c) = g'(c)$

(b) $f'(c) = -\dfrac{1}{g'(c)}$ $g'(c) \neq 0$

Challenge Problems

86. Let f be a function defined for all real numbers x. Suppose f has the following properties:

$$f(u+v) = f(u)f(v) \qquad f(0) = 1 \qquad f'(0) \text{ exists}$$

(a) Show that $f'(x)$ exists for all real numbers x.

(b) Show that $f'(x) = f'(0)f(x)$.

87. A function f is defined for all real numbers and has the following three properties:

$$f(1) = 5 \qquad f(3) = 21 \qquad f(a+b) - f(a) = kab + 2b^2$$

for all real numbers a and b where k is a fixed real number independent of a and b.

(a) Use $a = 1$ and $b = 2$ to find k.

(b) Find $f'(3)$.

(c) Find $f'(x)$ for all real x.

88. A function f is **periodic** if there is a positive number p so that $f(x+p) = f(x)$ for all x. Suppose f is differentiable. Show that if f is periodic with period p, then f' is also periodic with period p.

Preparing for the AP® Exam

AP® Practice Problems

Multiple-Choice Questions

PAGE 183

1. The function $f(x) = \begin{cases} x^2 - ax & \text{if } x \leq 1 \\ ax + b & \text{if } x > 1 \end{cases}$, where a and b are constants. If f is differentiable at $x = 1$, then $a + b =$

(A) -3 (B) -2 (C) 0   (D) 2

PAGE 180

2. The graph of the function f, given below, consists of three line segments. Find $\lim\limits_{h \to 0} \dfrac{f(3+h) - f(3)}{h}$.

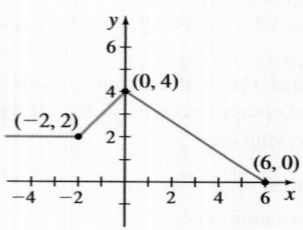

(A) -1 (B) $-\dfrac{2}{3}$ (C) $-\dfrac{3}{2}$ (D) The limit does not exist.

PAGE 186

3. If $f(x) = \begin{cases} \dfrac{x^2 - 25}{x - 5} & \text{if } x \neq 5 \\ 5 & \text{if } x = 5 \end{cases}$

which of the following statements about f are true?

I. $\lim\limits_{x \to 5} f$ exists.

II. f is continuous at $x = 5$.

III. f is differentiable at $x = 5$.

(A) I only   (B) I and II only

(C) I and III only   (D) I, II, and III

PAGE 186

4. Suppose f is a function that is differentiable on the open interval $(-2, 8)$. If $f(0) = 3$, $f(2) = -3$, and $f(7) = 3$, which of the following must be true?

I. f has at least 2 zeros.

II. f is continuous on the closed interval $[-1, 7]$.

III. For some c, $0 < c < 7$, $f(c) = -2$.

(A) I only   (B) I and II only

(C) II and III only   (D) I, II, and III

PAGE 183

5. If $f(x) = |x|$, which of the following statements about f are true?

I. f is continuous at 0.

II. f is differentiable at 0.

III. $f(0) = 0$.

(A) I only   (B) III only

(C) I and III only   (D) I, II, and III

PAGE 186

6. The graph of the function f shown in the figure has horizontal tangent lines at the points $(0, 1)$ and $(2, -1)$ and a vertical tangent line at the point $(1, 0)$. For what numbers x in the open interval $(-2, 3)$ is f not differentiable?

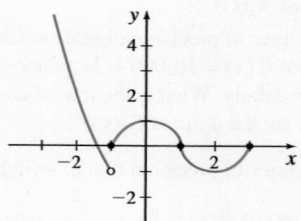

(A) -1 only   (B) -1 and 1 only

(C) $-1, 0$, and 2 only   (D) $-1, 0, 1$, and 2

PAGE 186

7. Let f be a function for which $\lim\limits_{h \to 0} \dfrac{f(1+h) - f(1)}{h} = -3$.

Which of the following must be true?

I. f is continuous at 1.

II. f is differentiable at 1.

III. f' is continuous at 1.

(A) I only   (B) II only

(C) I and II only   (D) I, II, and III

PAGE 180

8. At what point on the graph of $f(x) = x^2 - 4$ is the tangent line parallel to the line $6x - 3y = 2$?

(A) $(1, -3)$ (B) $(1, 2)$ (C) $(2, 0)$ (D) $(2, 4)$

9. At $x = 2$, the function $f(x) = \begin{cases} 4x + 1 & \text{if } x \le 2 \\ 3x^2 - 3 & \text{if } x > 2 \end{cases}$ is

(A) Both continuous and differentiable.
(B) Continuous but not differentiable.
(C) Differentiable but not continuous.
(D) Neither continuous nor differentiable.

 10. The table below lists several values of a function $y = f(x)$.

x	1	2	3	4	5
$f(x)$	4	−6	2	1	6

Suppose f is continuous on the interval $(0, 6)$, except at 3. Suppose f has a derivative at each number in the interval $(0, 6)$ except at 3 and 4. Which of the following statements must be true?

  I. The graph of f has a vertical tangent line at $(4, 1)$.
 II. The graph of f has a corner at the point $(3, 2)$.
III. f has a zero in the interval $(1, 2)$.

(A) I only   (B) II and III only
(C) III only   (D) I, II, and III

Free-Response Questions

 11. A rod of length 12 cm is heated at one end. The table below gives the temperature $T(x)$ in degrees Celsius at selected distances x cm from the heated end.

x	0	2	5	7	9	12
$T(x)$	80	71	66	60	54	50

(a) Use the table to approximate $T'(8)$.
(b) Using appropriate units, interpret $T'(8)$ in the context of the problem.

 12. Oil is leaking from a tank. The amount of oil, in gallons, in the tank is given by $G(t) = 4000 - 3t^2$, where t, $0 \le t \le 24$ is the number of hours past midnight.

(a) Find $G'(5)$ using the definition of the derivative.
(b) Using appropriate units, interpret the meaning of $G'(5)$ in the context of the problem.

See the BREAK IT DOWN on page 231 for a stepped out solution to AP® Practice Problem 12.

Retain Your Knowledge

Multiple-Choice Questions

1. Find $\lim\limits_{x \to 3^+} f(x)$, for the function $f(x) = \begin{cases} 2 & \text{if } x \le 3 \\ x + 1 & \text{if } x > 3 \end{cases}$.

(A) 2   (B) 3   (C) 4   (D) The limit does not exist.

2. $\lim\limits_{x \to 2^-} \dfrac{2x + 6}{2 - x} =$

(A) $-\infty$ (B) -2 (C) -1 (D) ∞

3. Find $\lim\limits_{x \to 0} \dfrac{3x + 2\sin(3x)}{2x}$, if it exists.

(A) $\dfrac{3}{2}$ (B) $\dfrac{5}{2}$ (C) $\dfrac{9}{2}$ (D) The limit does not exist.

Free-Response Question

4. $f(x) = x^3 - 4x^2 + 2x + 1$. Show that there is at least one number c in the interval $[0, 4]$ for which $f(c) = 7$.

2.3 The Derivative of a Polynomial Function; The Derivative of $y = e^x$ and $y = \ln x$

OBJECTIVES *When you finish this section, you should be able to:*

1 **Differentiate a constant function (p. 192)**
2 **Differentiate a power function; the Simple Power Rule (p. 192)**
3 **Differentiate the sum and the difference of two functions (p. 195)**
4 **Differentiate the exponential function $y = e^x$ and the natural logarithm function $y = \ln x$ (p. 197)**

Finding the derivative of a function from the definition can become tedious, especially if the function f is complicated. Just as we did for limits, we derive some basic derivative formulas and some properties of derivatives that make finding a derivative simpler.

Before getting started, we introduce other notations commonly used for the derivative $f'(x)$ of a function $y = f(x)$. The other most commonly used notations are

$$y' \qquad \frac{dy}{dx} \qquad Df(x)$$

Leibniz notation $\dfrac{dy}{dx}$ may be written in several equivalent ways as

$$\frac{dy}{dx} = \frac{d}{dx}y = \frac{d}{dx}f(x)$$

where $\dfrac{d}{dx}$ is an instruction to find the derivative of the function $y = f(x)$ with respect to the independent variable x.

In **operator notation** $Df(x)$, D is said to *operate* on the function, and the result is the derivative of f. To emphasize that the operation is performed with respect to the independent variable x, it is sometimes written $Df(x) = D_x f(x)$.

We use prime notation or Leibniz notation, or sometimes a mixture of the two, depending on which is more convenient. We do not use the notation $Df(x)$ in this book.

❶ Differentiate a Constant Function

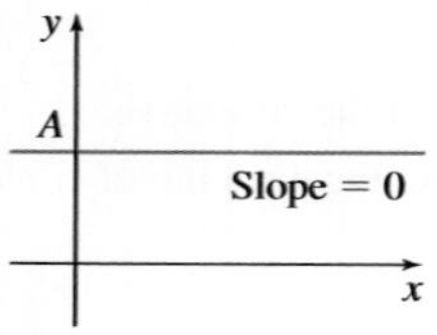

See Figure 23. Since the graph of a constant function $f(x) = A$ is a horizontal line, the tangent line to f at any point is also a horizontal line, whose slope is 0. Since the derivative is the slope of the tangent line, the derivative of f is 0.

> **THEOREM Derivative of a Constant Function**
>
> If f is the constant function $f(x) = A$, then
>
> $$\boxed{f'(x) = 0}$$
>
> That is, if A is a constant, then
>
> $$\boxed{\frac{d}{dx}A = 0}$$

Proof If $f(x) = A$, then its derivative function is given by

$$f'(x) = \lim_{\substack{\uparrow \\ h \to 0}} \frac{f(x+h) - f(x)}{h} = \lim_{\substack{\uparrow \\ h \to 0}} \frac{A - A}{h} = 0$$

The definition of a derivative, Form (2) $f(x) = A$
      $f(x+h) = A$ ∎

EXAMPLE 1 **Differentiating a Constant Function**

(a) If $f(x) = \sqrt{3}$, then $f'(x) = 0$ **(b)** If $f(x) = -\dfrac{1}{2}$, then $f'(x) = 0$

(c) If $f(x) = \pi$, then $\dfrac{d}{dx}\pi = 0$ **(d)** If $f(x) = 0$, then $\dfrac{d}{dx}0 = 0$ ∎

❷ Differentiate a Power Function; the Simple Power Rule

Next we analyze the derivative of a power function $f(x) = x^n$, where $n \geq 1$ is an integer.

When $n = 1$, then $f(x) = x$ is the identity function and its graph is the line $y = x$, as shown in Figure 24.

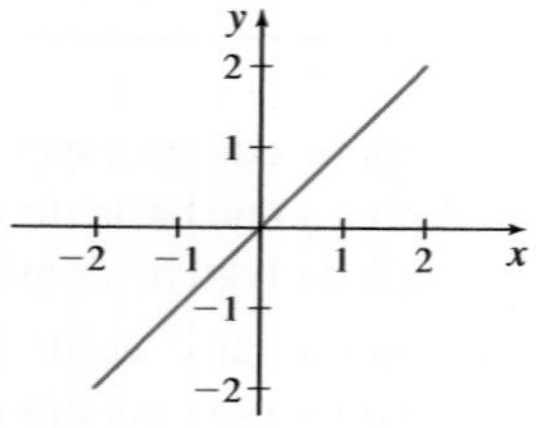

The slope of the line $y = x$ is 1, so we would expect $f'(x) = 1$.

Proof $f'(x) = \dfrac{d}{dx}x = \lim_{h\to 0}\dfrac{f(x+h) - f(x)}{h} = \lim_{\substack{\uparrow \\ h\to 0}}\dfrac{(x+h) - x}{h} = \lim_{h\to 0}\dfrac{h}{h} = 1$

$f(x) = x$, $f(x+h) = x+h$ ∎

(left margin figures)

y axis, A marked, with label **Slope = 0**

Figure 23 $f(x) = A$

Figure 24 $f(x) = x$

THEOREM Derivative of $f(x) = x$

If $f(x) = x$, then

$$f'(x) = \frac{d}{dx}x = 1$$

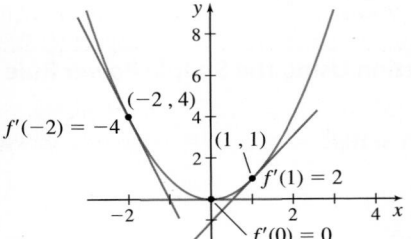

Figure 25 $f(x) = x^2$

When $n = 2$, then $f(x) = x^2$ is the square function. The derivative of f is

$$f'(x) = \frac{d}{dx}x^2 = \lim_{h \to 0} \frac{(x+h)^2 - x^2}{h} = \lim_{h \to 0} \frac{x^2 + 2hx + h^2 - x^2}{h}$$

$$= \lim_{h \to 0} \frac{h(2x + h)}{h} = \lim_{h \to 0}(2x + h) = 2x$$

The slope of the tangent line to the graph of $f(x) = x^2$ is different for every number x. Figure 25 shows the graph of f and several of its tangent lines. Notice that the slope of each tangent line drawn is twice the value of x.

When $n = 3$, then $f(x) = x^3$ is the cube function. The derivative of f is

$$f'(x) = \lim_{h \to 0} \frac{(x+h)^3 - x^3}{h} = \lim_{h \to 0} \frac{x^3 + 3x^2h + 3xh^2 + h^3 - x^3}{h}$$

$$= \lim_{h \to 0} \frac{h(3x^2 + 3xh + h^2)}{h} = \lim_{h \to 0}(3x^2 + 3xh + h^2) = 3x^2$$

Notice that the derivative of each of these power functions is another power function, whose degree is 1 less than the degree of the original function and whose coefficient is the degree of the original function. This rule holds for all power functions as the following theorem, called the *Simple Power Rule*, indicates.

THEOREM Simple Power Rule

The derivative of the power function $y = x^n$, where $n \geq 1$ is an integer, is

IN WORDS The derivative of x raised to an integer power $n \geq 1$ is n times x raised to the power $n - 1$.

$$y' = \frac{d}{dx}x^n = nx^{n-1}$$

NEED TO REVIEW? The Binomial Theorem is discussed in Section P.8, pp. 76–77.

Proof If $f(x) = x^n$ and n is a positive integer, then $f(x + h) = (x + h)^n$. We use the Binomial Theorem to expand $(x + h)^n$. Then

$$f(x+h) = (x+h)^n = x^n + nx^{n-1}h + \frac{n(n-1)}{2}x^{n-2}h^2 + \frac{n(n-1)(n-2)}{6}x^{n-3}h^3 + \cdots + nxh^{n-1} + h^n$$

$$f'(x) = \lim_{h \to 0} \frac{f(x+h) - f(x)}{h}$$

$$= \lim_{h \to 0} \frac{\left[x^n + nx^{n-1}h + \dfrac{n(n-1)}{2}x^{n-2}h^2 + \dfrac{n(n-1)(n-2)}{6}x^{n-3}h^3 + \cdots + nxh^{n-1} + h^n\right] - x^n}{h}$$

$$= \lim_{h \to 0} \frac{nx^{n-1}h + \dfrac{n(n-1)}{2}x^{n-2}h^2 + \dfrac{n(n-1)(n-2)}{6}x^{n-3}h^3 + \cdots + nxh^{n-1} + h^n}{h} \quad \text{Simplify.}$$

(Proof continues on page 194.)

$$= \lim_{h \to 0} \frac{h \left[nx^{n-1} + \frac{n(n-1)}{2} x^{n-2}h + \frac{n(n-1)(n-2)}{6} x^{n-3}h^2 + \cdots + nxh^{n-2} + h^{n-1} \right]}{h}$$

Factor h in the numerator.

$$= \lim_{h \to 0} \left[nx^{n-1} + \frac{n(n-1)}{2} x^{n-2}h + \frac{n(n-1)(n-2)}{6} x^{n-3}h^2 + \cdots + nxh^{n-2} + h^{n-1} \right]$$

Divide out the common h.

$$= nx^{n-1}$$

Take the limit. Only the first term remains.   ■

EXAMPLE 2   Differentiating a Power Function Using the Simple Power Rule

NOTE $\frac{d}{dx} x^n = nx^{n-1}$ is true not only for positive integers n but also for any real number n. But the proof requires future results. As these are developed, we will expand the Simple Power Rule to include an ever-widening set of numbers until we arrive at the fact it is true when n is a real number.

(a) $\frac{d}{dx} x^5 = 5x^4$ **(b)** If $g(x) = x^{10}$, then $g'(x) = 10x^9$.   ■

NOW WORK   Problem 1 and AP® Practice Problem 1.

But what if we want to find the derivative of the function $f(x) = ax^n$ when $a \neq 1$? The next theorem, called the *Constant Multiple Rule,* provides a way.

THEOREM   Constant Multiple Rule

If a function f is differentiable and k is a constant, then $F(x) = kf(x)$ is a function that is differentiable and

$$\boxed{F'(x) = kf'(x)}$$

IN WORDS The derivative of a constant times a differentiable function f equals the constant times the derivative of f.

Proof Use the definition of a derivative, Form (2).

$$F'(x) = \lim_{h \to 0} \frac{F(x+h) - F(x)}{h} = \lim_{h \to 0} \frac{kf(x+h) - kf(x)}{h}$$

$$= \lim_{h \to 0} \frac{k \left[f(x+h) - f(x) \right]}{h} = k \cdot \lim_{h \to 0} \frac{f(x+h) - f(x)}{h} = k \cdot f'(x)$$ ■

Using Leibniz notation, the Constant Multiple Rule takes the form

$$\boxed{\frac{d}{dx} [kf(x)] = k \left[\frac{d}{dx} f(x) \right]}$$

A change in the symbol used for the independent variable does not affect the derivative formula. For example, $\frac{d}{dt} t^2 = 2t$ and $\frac{d}{du} u^5 = 5u^4$.

EXAMPLE 3   Differentiating a Constant Times a Power Function

Find the derivative of each function:

(a) $f(x) = 5x^3$ **(b)** $g(u) = -\frac{1}{2}u^2$ **(c)** $u(x) = \pi^4 x^3$

Solution

Notice that each of these functions involves the product of a constant and a power function. So, we use the Constant Multiple Rule followed by the Simple Power Rule.

(a) $f(x) = 5 \cdot x^3$, so $f'(x) = 5\left[\dfrac{d}{dx} x^3\right] = 5 \cdot 3x^2 = 15x^2$

(b) $g(u) = -\dfrac{1}{2} \cdot u^2$, so $g'(u) = -\dfrac{1}{2} \cdot \dfrac{d}{du} u^2 = -\dfrac{1}{2} \cdot 2u^1 = -u$

(c) $u(x) = \pi^4 x^3$, so $u'(x) = \underset{\underset{\pi \text{ is a constant}}{\uparrow}}{\pi^4} \cdot \dfrac{d}{dx} x^3 = \pi^4 \cdot 3x^2 = 3\pi^4 x^2$ ∎

NOW WORK   Problem 31 and AP® Practice Problem 10.

---

③ Differentiate the Sum and the Difference of Two Functions

We can find the derivative of a function that is the sum of two functions whose derivatives are known by adding the derivatives of each function.

THEOREM Sum Rule

If two functions f and g are differentiable and if $F(x) = f(x) + g(x)$, then F is differentiable and

$$\boxed{F'(x) = f'(x) + g'(x)}$$

IN WORDS The derivative of the sum of two differentiable functions equals the sum of their derivatives. That is, $(f + g)' = f' + g'$.

Proof If $F(x) = f(x) + g(x)$, then

$$F(x + h) - F(x) = [f(x + h) + g(x + h)] - [f(x) + g(x)]$$
$$= [f(x + h) - f(x)] + [g(x + h) - g(x)]$$

So, the derivative of F is

$$F'(x) = \lim_{h \to 0} \frac{[f(x + h) - f(x)] + [g(x + h) - g(x)]}{h}$$

$$= \lim_{h \to 0} \frac{f(x + h) - f(x)}{h} + \lim_{h \to 0} \frac{g(x + h) - g(x)}{h} \qquad \text{The limit of a sum is the sum of the limits.}$$

$$= f'(x) + g'(x) \qquad\qquad\qquad\qquad\qquad\qquad\qquad\qquad\qquad\qquad ∎$$

In Leibniz notation, the Sum Rule takes the form

$$\boxed{\dfrac{d}{dx}[f(x) + g(x)] = \dfrac{d}{dx} f(x) + \dfrac{d}{dx} g(x)}$$

EXAMPLE 4   **Differentiating the Sum of Two Functions**

Find the derivative of $f(x) = 3x^2 + 8$.

Solution

Here f is the sum of $3x^2$ and 8. So, we begin by using the Sum Rule.

$$f'(x) = \frac{d}{dx}(3x^2 + 8) = \underset{\underset{\text{Sum Rule}}{\uparrow}}{\frac{d}{dx}(3x^2)} + \frac{d}{dx}8 = \underset{\underset{\substack{\text{Constant Multiple} \\ \text{Rule}}}{\uparrow}}{3\frac{d}{dx}x^2} + 0 = \underset{\underset{\substack{\text{Simple} \\ \text{Power Rule}}}{\uparrow}}{3 \cdot 2x} = 6x \qquad ∎$$

NOW WORK   Problem 7 and AP® Practice Problem 6.

THEOREM Difference Rule

If the functions f and g are differentiable and if $F(x) = f(x) - g(x)$, then F is differentiable, and $F'(x) = f'(x) - g'(x)$. That is,

$$\frac{d}{dx}[f(x) - g(x)] = \frac{d}{dx}f(x) - \frac{d}{dx}g(x)$$

IN WORDS The derivative of the difference of two differentiable functions is the difference of their derivatives. That is, $(f - g)' = f' - g'$.

The proof of the Difference Rule is left as an exercise. (See Problem 78.)

The Sum and Difference Rules extend to sums (or differences) of more than two functions. That is, if the functions $f_1, f_2, \ldots, f_n$ are all differentiable, and $a_1, a_2, \ldots, a_n$ are constants, then

$$\frac{d}{dx}[a_1 f_1(x) + a_2 f_2(x) + \cdots + a_n f_n(x)] = a_1 \frac{d}{dx}f_1(x) + a_2\frac{d}{dx}f_2(x) + \cdots + a_n\frac{d}{dx}f_n(x)$$

Combining the rules for finding the derivative of a constant, a power function, and a sum or difference allows us to differentiate any polynomial function.

EXAMPLE 5   Differentiating a Polynomial Function

(a) Find the derivative of $f(x) = 2x^4 - 6x^2 + 2x - 3$.

(b) What is $f'(2)$?

(c) Find the slope of the tangent line to the graph of f at the point $(1, -5)$.

(d) Find an equation of the tangent line to the graph of f at the point $(1, -5)$.

(e) Find an equation of the normal line to the graph of f at the point $(1, -5)$.

(f) Use technology to graph f, the tangent line, and the normal line to the graph of f at the point $(1, -5)$ on the same screen.

Solution

(a) $f'(x) = \dfrac{d}{dx}(2x^4 - 6x^2 + 2x - 3) \underset{\underset{\text{Sum/Difference Rules}}{\uparrow}}{=} \dfrac{d}{dx}(2x^4) - \dfrac{d}{dx}(6x^2) + \dfrac{d}{dx}(2x) - \dfrac{d}{dx}3$

$\underset{\underset{\text{Constant Multiple Rule}}{\uparrow}}{=} 2 \cdot \dfrac{d}{dx}x^4 - 6 \cdot \dfrac{d}{dx}x^2 + 2 \cdot \dfrac{d}{dx}x - 0$

$\underset{\underset{\text{Simple Power Rule}}{\uparrow}}{=} 2 \cdot 4x^3 - 6 \cdot 2x + 2 \cdot 1 \underset{\underset{\text{Simplify}}{\uparrow}}{=} 8x^3 - 12x + 2$

(b) $f'(2) = 8 \cdot 2^3 - 12 \cdot 2 + 2 = 64 - 24 + 2 = 42$.

(c) The slope of the tangent line to the graph of f at the point $(1, -5)$ equals $f'(1)$.

$$f'(1) = 8 \cdot 1^3 - 12 \cdot 1 + 2 = 8 - 12 + 2 = -2$$

(d) Use the point-slope form of an equation of a line to find an equation of the tangent line to the graph of f at $(1, -5)$.

$$y - (-5) = -2(x - 1)$$
$$y = -2(x - 1) - 5 = -2x + 2 - 5 = -2x - 3$$

The line $y = -2x - 3$ is tangent to the graph of $f(x) = 2x^4 - 6x^2 + 2x - 3$ at the point $(1, -5)$.

(e) Since the normal line and the tangent line at the point $(1, -5)$ on the graph of f are perpendicular and the slope of the tangent line is -2, the slope of the normal line is $\dfrac{1}{2}$.

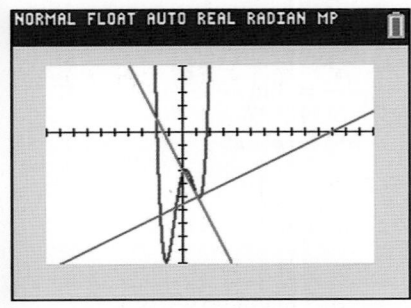

$[-10, 14] \times [-10, 5]$

Figure 26 $f(x)=2x^4-6x^2+2x-3$

Use the point-slope form of an equation of a line to find an equation of the normal line.

$$y-(-5)=\frac{1}{2}(x-1)$$

$$y=\frac{1}{2}(x-1)-5=\frac{1}{2}x-\frac{1}{2}-5=\frac{1}{2}x-\frac{11}{2}$$

The line $y=\frac{1}{2}x-\frac{11}{2}$ is normal to the graph of f at the point $(1, -5)$.

(f) The graphs of f, the tangent line, and the normal line to f at $(1, -5)$ are shown in Figure 26. Because we are graphing the tangent line and the normal line, which are perpendicular to each other, we use a square screen to obtain Figure 26. ∎

NOW WORK   Problem 33 and AP® Practice Problems 2, 5, 8, 11, and 12.

In some applications, we need to solve equations or inequalities involving the derivative of a function.

EXAMPLE 6   **Solving Equations and Inequalities Involving Derivatives**

(a) Find the points on the graph of $f(x)=4x^3-12x^2+2$, where f has a horizontal tangent line.

(b) Where is $f'(x)>0$? Where is $f'(x)<0$?

Solution

(a) The slope of a horizontal tangent line is 0. Since the derivative of f equals the slope of the tangent line, we need to find the numbers x for which $f'(x)=0$.

$$f'(x)=12x^2-24x=12x(x-2)$$

$$12x(x-2)=0 \qquad\qquad\qquad f'(x)=0$$

$$x=0 \text{ or } x=2 \qquad\qquad\qquad \text{Solve.}$$

At the points $(0, f(0))=(0, 2)$ and $(2, f(2))=(2, -14)$, the graph of the function $f(x)=4x^3-12x^2+2$ has horizontal tangent lines.

(b) Since $f'(x)=12x(x-2)$ and we want to solve the inequalities $f'(x)>0$ and $f'(x)<0$, we use the zeros of f', 0 and 2, and form a table using the intervals $(-\infty, 0)$, $(0, 2)$, and $(2, \infty)$. See Table 2.

TABLE 2

Interval	$(-\infty, 0)$	$(0, 2)$	$(2, \infty)$
Sign of $f'(x)=12x(x-2)$	Positive	Negative	Positive

We conclude $f'(x)>0$ on $(-\infty, 0) \cup (2, \infty)$ and $f'(x)<0$ on $(0, 2)$. ∎

Figure 27 shows the graph of f and the two horizontal tangent lines.

NOW WORK   Problem 37 and AP® Practice Problem 3.

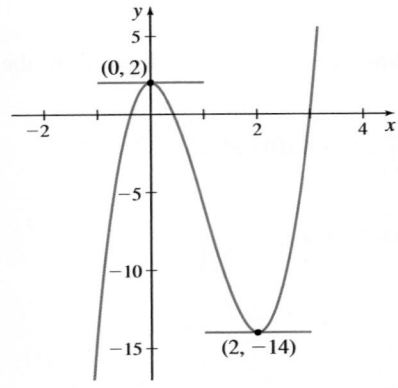

Figure 27 $f(x)=4x^3-12x^2+2$

---

④ Differentiate the Exponential Function $y=e^x$ and the Natural Logarithm Function $y=\ln x$

None of the differentiation rules developed so far allow us to find the derivative of an exponential function. To differentiate $f(x)=a^x$, we need to return to the definition of a derivative.

We begin by making some general observations about the derivative of $f(x)=a^x$, $a>0$ and $a \neq 1$. We then use these observations to find the derivative of the exponential function $y=e^x$.

NEED TO REVIEW? Exponential functions are discussed in Section P.5, pp. 43–46.

Suppose $f(x)=a^x$, where $a > 0$ and $a \neq 1$. The derivative of f is

$$f'(x) = \lim_{h \to 0} \frac{f(x+h) - f(x)}{h} = \lim_{h \to 0} \frac{a^{x+h} - a^x}{h} = \lim_{\substack{\uparrow \\ h \to 0}} \frac{a^x \cdot a^h - a^x}{h}$$

$$a^{x+h} = a^x \cdot a^h$$

$$= \lim_{\substack{\uparrow \\ h \to 0}} \left[a^x \cdot \frac{a^h - 1}{h} \right] = a^x \cdot \lim_{h \to 0} \frac{a^h - 1}{h}$$

Factor out a^x.

provided $\lim_{h \to 0} \dfrac{a^h - 1}{h}$ exists.

Three observations about the derivative of $f(x) = a^x$ are significant:

- $f'(0) = a^0 \lim_{h \to 0} \dfrac{a^h - 1}{h} = \lim_{h \to 0} \dfrac{a^h - 1}{h}$.

- $f'(x)$ is a multiple of a^x. In fact, $\dfrac{d}{dx} a^x = f'(0) \cdot a^x$.

- If $f'(0)$ exists, then $f'(x)$ exists, and the domain of f' is the same as that of $f(x) = a^x$, all real numbers.

NEED TO REVIEW? The number e is discussed in Section P.5, pp. 46–47.

The slope of the tangent line to the graph of $f(x) = a^x$ at the point $(0, 1)$ is $f'(0) = \lim_{h \to 0} \dfrac{a^h - 1}{h}$, and the value of this limit depends on the base a. In Section P.5, the number e was defined as that number for which the slope of the tangent line to the graph of $y = a^x$ at the point $(0, 1)$ equals 1. That is, if $f(x) = e^x$, then $f'(0) = 1$ so that

$$\boxed{\lim_{h \to 0} \frac{e^h - 1}{h} = 1}$$

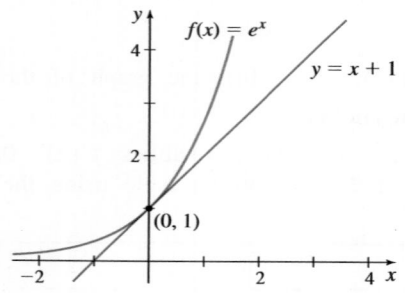

Figure 28

Figure 28 shows $f(x) = e^x$ and the tangent line $y = x + 1$ with slope 1 at the point $(0, 1)$.

Since $\dfrac{d}{dx} a^x = f'(0) \cdot a^x$, if $f(x) = e^x$, then $\dfrac{d}{dx} e^x = f'(0) \cdot e^x = 1 \cdot e^x = e^x$.

THEOREM Derivative of the Exponential Function $y = e^x$

The derivative of the exponential function $y = e^x$ is

$$\boxed{y' = \frac{d}{dx} e^x = e^x} \tag{1}$$

EXAMPLE 7 Differentiating an Expression Involving $y = e^x$

CALC CLIP

Find the derivative of $f(x) = 4e^x + x^3$.

Solution

The function f is the sum of $4e^x$ and x^3. Then

$$f'(x) = \underset{\substack{\uparrow \\ \text{Sum Rule}}}{\frac{d}{dx}(4e^x + x^3)} = \frac{d}{dx}(4e^x) + \frac{d}{dx}x^3 = \underset{\substack{\uparrow \\ \text{Constant Multiple Rule;} \\ \text{Simple Power Rule}}}{4\frac{d}{dx}e^x + 3x^2} = \underset{\substack{\uparrow \\ \text{Use (1).}}}{4e^x + 3x^2}$$

∎

NOW WORK Problem 25 and AP® Practice Problem 4.

NEED TO REVIEW? The natural logarithm function is defined in Section P.5, p. 47.

In Chapter 1, we found that the natural logarithm function $y = \ln x$ is continuous on its domain $\{x \mid x > 0\}$. Below we give the rule for finding the derivative of $y = \ln x$.

THEOREM Derivative of the Natural Logarithm Function $y = \ln x$

The derivative of the natural logarithm function $y = \ln x$, $x > 0$, is

$$\boxed{y' = \frac{d}{dx} \ln x = \frac{1}{x}} \tag{2}$$

We do not have the necessary mathematics to prove (2) now. We will prove the theorem in Chapter 3.

EXAMPLE 8 Differentiating a Function Involving $y = \ln x$

Find the derivative $f(x) = 3 \ln x - 5x^2$.

Solution

The function f is the difference between $3 \ln x$ and $5x^2$. Then using (2), we find that

$$f'(x) = \underset{\underset{\text{Difference Rule}}{\uparrow}}{\frac{d}{dx}(3 \ln x - 5x^2) = \frac{d}{dx}(3 \ln x)} - \underset{\underset{\substack{\text{Constant Multiple Rule;} \\ \text{Simple Power Rule}}}{\uparrow}}{\frac{d}{dx}(5x^2) = 3\frac{d}{dx}(\ln x)} - 5 \cdot 2x = \underset{\underset{\text{Use (2).}}{\uparrow}}{\frac{3}{x}} - 10x \quad\blacksquare$$

NOW WORK   Problem 23 and AP® Practice Problems 7 and 9.

2.3 Assess Your Understanding

Concepts and Vocabulary

  1. $\dfrac{d}{dx} \pi^2 = $ _____ ; $\dfrac{d}{dx} x^3 = $ _____ .

2. When n is a positive integer, the Simple Power Rule states that $\dfrac{d}{dx} x^n = $ _____ .

3. *True or False*   The derivative of a power function of degree greater than 1 is also a power function.

4. If k is a constant and f is a differentiable function, then $\dfrac{d}{dx}[kf(x)] = $ _____ .

5. The derivative of $f(x) = e^x$ is _____ .

6. *True or False*   The derivative of an exponential function $f(x) = a^x$, where $a > 0$ and $a \neq 1$, is always a constant multiple of a^x.

Skill Building

In Problems 7–26, find the derivative of each function using the formulas of this section. (a, b, c, and d, when they appear, are constants.)

  7. $f(x) = 3x + \sqrt{2}$ **8.** $f(x) = 5x - \pi$

9. $f(x) = x^2 + 3x + 4$ **10.** $f(x) = 4x^4 + 2x^2 - 2$

11. $f(u) = 8u^5 - 5u + 1$ **12.** $f(u) = 9u^3 - 2u^2 + 4u + 4$

13. $f(s) = as^3 + \dfrac{3}{2}s^2$ **14.** $f(s) = 4 - \pi s^2$

15. $f(t) = \dfrac{1}{6}(t^6 - 5t)$ **16.** $f(x) = \dfrac{1}{8}(x^8 - 5x^2 + 2)$

17. $f(t) = \dfrac{t^3 + 2}{5}$ **18.** $f(x) = \dfrac{x^7 - 5x}{9}$

19. $f(x) = \dfrac{x^3 + 2x + 1}{7}$ **20.** $f(x) = \dfrac{1}{a}(ax^2 + bx + c), a \neq 0$

21. $f(x) = 4e^x$ **22.** $f(x) = -\dfrac{1}{2}e^x$

23. $f(x) = x - \ln x$ **24.** $f(x) = 5 \ln x + 8$

25. $f(u) = 5 \ln u - 2e^u$ **26.** $f(u) = 3e^u + 10 \ln u$

In Problems 27–32, find each derivative.

27. $\dfrac{d}{dt}\left(\sqrt{3}\, t + \dfrac{1}{2}\right)$ **28.** $\dfrac{d}{dt}\left(\dfrac{2t^4 - 5}{8}\right)$

29. $\dfrac{dA}{dR}$ if $A(R) = \pi R^2$ **30.** $\dfrac{dC}{dR}$ if $C = 2\pi R$

31. $\dfrac{dV}{dr}$ if $V = \dfrac{4}{3}\pi r^3$ **32.** $\dfrac{dP}{dT}$ if $P = 0.2T$

In Problems 33–36:

(a) Find the slope of the tangent line to the graph of each function f at the indicated point.

(b) Find an equation of the tangent line at the point.

(c) Find an equation of the normal line at the point.

(d) Graph f and the tangent line and normal line found in (b) and (c) on the same set of axes.

33. $f(x) = x^3 + 3x - 1$ at $(0, -1)$ **34.** $f(x) = x^4 + 2x - 1$ at $(1, 2)$

35. $f(x) = e^x + 5x$ at $(0, 1)$ **36.** $f(x) = 4 - e^x$ at $(0, 3)$

In Problems 37–42:

(a) *Find the points, if any, at which the graph of each function f has a horizontal tangent line.*

(b) *Find an equation for each horizontal tangent line.*

(c) *Solve the inequality $f'(x) > 0$.*

(d) *Solve the inequality $f'(x) < 0$.*

(e) *Graph f and any horizontal lines found in (b) on the same set of axes.*

(f) *Describe the graph of f for the results obtained in parts (c) and (d).*

37. $f(x) = 3x^2 - 12x + 4$ **38.** $f(x) = x^2 + 4x - 3$

39. $f(x) = x + e^x$ **40.** $f(x) = 2e^x - 1$

41. $f(x) = x^3 - 3x + 2$ **42.** $f(x) = x^4 - 4x^3$

43. Motion on a Line At t seconds, an object moving on a line is s meters from the origin, where $s(t) = t^3 - t + 1$. Find the velocity of the object at $t = 0$ and at $t = 5$.

44. Motion on a Line At t seconds, an object moving on a line is s meters from the origin, where $s(t) = t^4 - t^3 + 1$. Find the velocity of the object at $t = 0$ and at $t = 1$.

Motion on a Line *In Problems 45 and 46, each position function gives the signed distance s from the origin at time t of an object moving on a line:*

(a) *Find the velocity v of the object at any time t.*

(b) *When is the velocity of the object 0?*

45. $s(t) = 2 - 5t + t^2$

46. $s(t) = t^3 - \dfrac{9}{2}t^2 + 6t + 4$

In Problems 47 and 48, use the graphs to find each derivative.

47. Let $u(x) = f(x) + g(x)$ and $v(x) = f(x) - g(x)$.

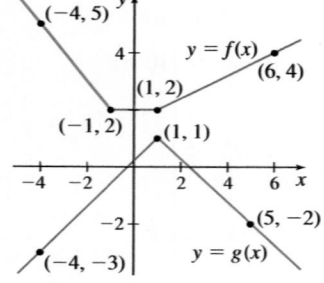

(a) $u'(0)$ (b) $u'(4)$
(c) $v'(-2)$ (d) $v'(6)$
(e) $3u'(5)$ (f) $-2v'(3)$

48. Let $F(t) = f(t) + g(t)$ and $G(t) = g(t) - f(t)$.

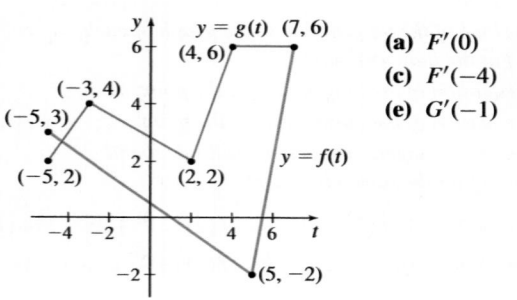

(a) $F'(0)$ (b) $F'(3)$
(c) $F'(-4)$ (d) $G'(-2)$
(e) $G'(-1)$ (f) $G'(6)$

In Problems 49 and 50, for each function f:

(a) *Find $f'(x)$ by expanding $f(x)$ and differentiating the polynomial.*

(CAS) (b) *Find $f'(x)$ using a CAS.*

(c) *Show that the results found in parts (a) and (b) are equivalent.*

49. $f(x) = (2x - 1)^3$ **50.** $f(x) = (x^2 + x)^4$

Applications and Extensions

In Problems 51–56, find each limit.

51. $\displaystyle\lim_{h \to 0} \frac{4\left(\frac{1}{2} + h\right)^8 - 4\left(\frac{1}{2}\right)^8}{h}$

52. $\displaystyle\lim_{h \to 0} \frac{5(2 + h)^5 - 5 \cdot 2^5}{h}$

53. $\displaystyle\lim_{h \to 0} \frac{\sqrt{3}(8 + h)^5 - \sqrt{3} \cdot 8^5}{h}$

54. $\displaystyle\lim_{h \to 0} \frac{\pi(1 + h)^{10} - \pi}{h}$

55. $\displaystyle\lim_{h \to 0} \frac{a(x + h)^3 - ax^3}{h}$

56. $\displaystyle\lim_{h \to 0} \frac{b(x + h)^n - bx^n}{h}$

In Problems 57–62, find an equation of the tangent line(s) to the graph of the function f that is (are) parallel to the line L.

57. $f(x) = 3x^2 - x$; L: $y = 5x$

58. $f(x) = 2x^3 + 1$; L: $y = 6x - 1$

59. $f(x) = e^x$; L: $y - x - 5 = 0$

60. $f(x) = -2e^x$; L: $y + 2x - 8 = 0$

61. $f(x) = 3 \ln x$; L: $y = 3x - 2$

62. $f(x) = \ln x - 2x$; L: $3x - y = 4$

63. Tangent Lines Let $f(x) = 4x^3 - 3x - 1$.

(a) Find an equation of the tangent line to the graph of f at $x = 2$.

(b) Find the coordinates of any points on the graph of f where the tangent line is parallel to $y = x + 12$.

(c) Find an equation of the tangent line to the graph of f at any points found in (b).

(d) Graph f, the tangent line found in (a), the line $y = x + 12$, and any tangent lines found in (c) on the same screen.

64. Tangent Lines Let $f(x) = x^3 + 2x^2 + x - 1$.

(a) Find an equation of the tangent line to the graph of f at $x = 0$.

(b) Find the coordinates of any points on the graph of f where the tangent line is parallel to $y = 3x - 2$.

(c) Find an equation of the tangent line to the graph of f at any points found in (b).

(d) Graph f, the tangent line found in (a), the line $y = 3x - 2$, and any tangent lines found in (c) on the same screen.

65. Tangent Line Show that the line perpendicular to the x-axis and containing the point (x, y) on the graph of $y = e^x$ and the tangent line to the graph of $y = e^x$ at the point (x, y) intersect the x-axis 1 unit apart. See the figure.

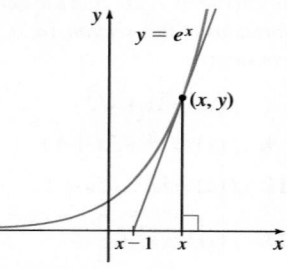

66. Tangent Line Show that the tangent line to the graph of $y = x^n$, $n \geq 2$ an integer, at $(1, 1)$ has y-intercept $1 - n$.

67. Tangent Lines If n is an odd positive integer, show that the tangent lines to the graph of $y = x^n$ at $(1, 1)$ and at $(-1, -1)$ are parallel.

68. Tangent Line If the line $3x - 4y = 0$ is tangent to the graph of $y = x^3 + k$ in the first quadrant, find k.

69. Tangent Line Find the constants a, b, and c so that the graph of $y = ax^2 + bx + c$ contains the point $(-1, 1)$ and is tangent to the line $y = 2x$ at $(0, 0)$.

70. Tangent Line Let T be the tangent line to the graph of $y = x^3$ at the point $\left(\dfrac{1}{2}, \dfrac{1}{8}\right)$. At what other point Q on the graph of $y = x^3$ does the line T intersect the graph? What is the slope of the tangent line at Q?

71. Military Tactics A dive bomber is flying from right to left along the graph of $y = x^2$. When a rocket bomb is released, it follows a path that is approximately along the tangent line. Where should the pilot release the bomb if the target is at $(1, 0)$?

72. Military Tactics Answer the question in Problem 71 if the plane is flying from right to left along the graph of $y = x^3$.

73. Fluid Dynamics The velocity v of a liquid flowing through a cylindrical tube is given by the **Hagen–Poiseuille equation** $v = k(R^2 - r^2)$, where R is the radius of the tube, k is a constant that depends on the length of the tube and the velocity of the liquid at its ends, and r is the variable distance of the liquid from the center of the tube. See the figure below.

(a) Find the rate of change of v with respect to r at the center of the tube.

(b) What is the rate of change halfway from the center to the wall of the tube?

(c) What is the rate of change at the wall of the tube?

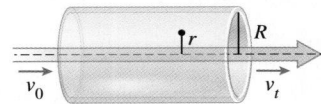

74. Rate of Change Water is leaking out of a swimming pool that measures 20 ft by 40 ft by 6 ft. The amount of water in the pool at a time t is $W(t) = 35{,}000 - 20t^2$ gallons, where t equals the number of hours since the pool was last filled. At what rate is the water leaking when $t = 2$ h?

75. Luminosity of the Sun The luminosity L of a star is the rate at which it radiates energy. This rate depends on the temperature T and surface area A of the star's photosphere (the gaseous surface that emits the light). Luminosity is modeled by the equation $L = \sigma A T^4$, where σ is a constant known as the **Stefan–Boltzmann constant**, and T is expressed in the absolute (Kelvin) scale for which 0 K is absolute zero. As with most stars, the Sun's temperature has gradually increased over the 6 billion years of its existence, causing its luminosity to slowly increase.

(a) Find the rate at which the Sun's luminosity changes with respect to the temperature of its photosphere. Assume that the surface area A remains constant.

(b) Find the rate of change at the present time. The temperature of the photosphere is currently 5800 K (10,000 °F), the radius of the photosphere is $r = 6.96 \times 10^8$ m,

and $\sigma = 5.67 \times 10^{-8}\,\dfrac{\text{W}}{\text{m}^2\,\text{K}^4}$.

(c) Assuming that the rate found in (b) remains constant, how much would the luminosity change if its photosphere temperature increased by 1 K (1 °C or 1.8 °F)? Compare this change to the present luminosity of the Sun.

76. Medicine: Poiseuille's Equation The French physician Poiseuille discovered that the volume V of blood (in cubic centimeters per unit time) flowing through an artery with inner radius R (in centimeters) can be modeled by

$$V(R) = kR^4$$

where $k = \dfrac{\pi}{8\nu l}$ is constant (here ν represents the viscosity of blood and l is the length of the artery).

(a) Find the rate of change of the volume V of blood flowing through the artery with respect to the radius R.

(b) Find the rate of change when $R = 0.03$ and when $R = 0.04$.

(c) If the radius of a partially clogged artery is increased from 0.03 to 0.04 cm, estimate the effect on the rate of change of the volume V with respect to R of the blood flowing through the enlarged artery.

(d) How do you interpret the results found in (b) and (c)?

77. Derivative of an Area Let $f(x) = mx$, $m > 0$. Let $F(x)$, $x > 0$, be defined as the area of the shaded region in the figure. Find $F'(x)$.

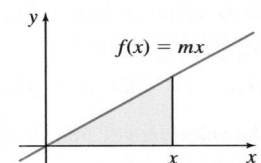

78. The Difference Rule Prove that if f and g are differentiable functions and if $F(x) = f(x) - g(x)$, then

$$F'(x) = f'(x) - g'(x)$$

79. Simple Power Rule Let $f(x) = x^n$, where n is a positive integer. Use a factoring principle to show that

$$f'(c) = \lim_{x \to c} \frac{f(x) - f(c)}{x - c} = nc^{n-1}$$

80. Normal Lines For what nonnegative number b is the line given by $y = -\dfrac{1}{3}x + b$ normal to the graph of $y = x^3$?

81. Normal Lines Let N be the normal line to the graph of $y = x^2$ at the point $(-2, 4)$. At what other point Q does N meet the graph?

Challenge Problems

82. Tangent Line Find a, b, c, d so that the tangent line to the graph of the cubic $y = ax^3 + bx^2 + cx + d$ at the point $(1, 0)$ is $y = 3x - 3$ and at the point $(2, 9)$ is $y = 18x - 27$.

83. Tangent Line Find the fourth degree polynomial that contains the origin and to which the line $x + 2y = 14$ is tangent at both $x = 4$ and $x = -2$.

84. Tangent Lines Find equations for all the lines containing the point $(1, 4)$ that are tangent to the graph of $y = x^3 - 10x^2 + 6x - 2$. At what points do each of the tangent lines touch the graph?

85. The line $x = c$, where $c > 0$, intersects the cubic $y = 2x^3 + 3x^2 - 9$ at the point P and intersects the parabola $y = 4x^2 + 4x + 5$ at the point Q, as shown in the figure below.

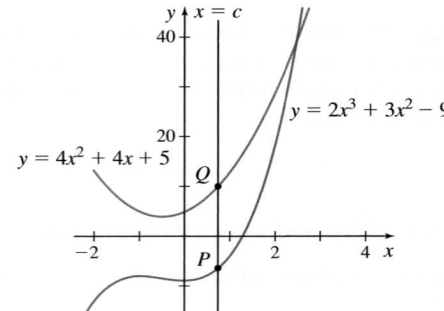

(a) If the tangent line to the cubic at the point P is parallel to the line tangent to the parabola at the point Q, find the number c.

(b) Write an equation for each of the two tangent lines described in (a).

86. $f(x) = Ax^2 + B$, $A > 0$.

(a) Find c, $c > 0$, in terms of A so that the tangent lines to the graph of f at $(c, f(c))$ and $(-c, f(-c))$ are perpendicular.

(b) Find the slopes of the tangent lines in (a).

(c) Find the coordinates, in terms of A and B, of the point of intersection of the tangent lines in (a).

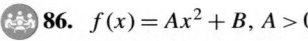

AP® Practice Problems

Multiple-Choice Questions

1. If $g(x) = x$, then $g'(7) =$

(A) 0   (B) 1   (C) 7   (D) $\dfrac{49}{2}$

2. The line $x + y = k$, where k is a constant, is a line tangent to the graph of the function $f(x) = x^2 - 5x + 2$. What is the value of k?

(A) -1 (B) 2   (C) -2 (D) -4

3. An object moves along the x-axis so that its position at time t is $x(t) = 3t^2 - 9t + 7$. For what time t is the velocity of the object zero?

(A) -3 (B) 3   (C) $\dfrac{3}{2}$ (D) 7

4. If $f(x) = e^x$, then $\ln(f'(3)) =$

(A) 3   (B) 0   (C) e^3 (D) $\ln 3$

5. An equation of the line tangent to the graph of $g(x) = x^3 + 2x^2 - 2x + 1$ at the point where $x = -2$ is

(A) $x + 2y = 12$ (B) $y + 2x = 9$
(C) $2x + y = -9$ (D) $y - 2x = 9$

6. The line $9x - 16y = 0$ is tangent to the graph of $f(x) = 3x^3 + k$, where k is a constant, at a point in the first quadrant. Find k.

(A) $\dfrac{3}{32}$ (B) $\dfrac{3}{16}$ (C) $\dfrac{3}{64}$ (D) $\dfrac{9}{64}$

7. If $f(x) = 3x + \ln x$, find $f'(1)$.

(A) 2   (B) 0   (C) 3   (D) 4

8. The cost C (in dollars) of manufacturing x units of a product is $C(x) = 0.3x^2 + 4.02x + 3500$. What is the rate of change of C when $x = 1000$ units?

(A) 307.52   (B) 0.60402   (C) 604.02   (D) 1020

9. Find $f'(1)$ if $f(x) = 3e^x - 5x^3 + 2\ln x - 5$.

(A) -15 (B) $3e - 5$ (C) $3e - 18$ (D) $3e - 13$

10. $\displaystyle\lim_{h \to 0} \dfrac{3(2 + h)^4 - 3 \cdot 16}{h} =$

(A) 0   (B) 32   (C) 48   (D) 96

11. Which is an equation of the line tangent to the graph of $f(x) = x^4 + 3x^2 + 2$ at the point where $f'(x) = 2$?

(A) $y = 2x + 2$ (B) $y = 2x + 2.929$
(C) $y = 2x + 1.678$ (D) $y = 2x - 2.929$

Free-Response Questions

12. For the function $f(x) = x^2 + 4$
(a) Find $f'(1)$.

(b) Find an equation of the line tangent to the graph of f at $x = 1$.

(c) Find $f'(-4)$.

(d) Find an equation of the line tangent to the graph of f at $x = -4$.

(e) Find the point of intersection of the two tangent lines found in (b) and (d).

13. $f(x) = \begin{cases} -x^2 + x + 1 & \text{if } x < 0 \\ 1 & \text{if } x = 0 \\ e^x & \text{if } x > 0 \end{cases}$

(a) Determine whether f is continuous at $x = 0$. Justify your answer.

(b) Find an equation of the line tangent to the graph of f at $x = -1$.

(c) Find an equation of the line tangent to the graph of f at $x = 1$.

Retain Your Knowledge

Multiple-Choice Questions

1. The graph of a piecewise defined function f is shown in the accompanying figure. Use the graph to determine which of the following statements are true.

   I. $\lim\limits_{x \to 2^-} f(x) = 1$

   II. $\lim\limits_{x \to 2^+} f(x)$ does not exist

   III. $\lim\limits_{x \to 4} f(x) = f(4)$

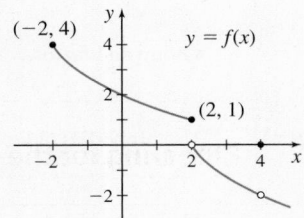

   (A) I only
   (B) I and II only
   (C) I and III only
   (D) I, II, and III

2. $\lim\limits_{x \to 1} \ln \dfrac{x^2}{e^x} =$

   (A) -1 (B) $\dfrac{1}{e}$ (C) 1   (D) 2

3. Given $f(x) = \begin{cases} x^2 & \text{if} & x < 2 \\ 2 & \text{if} & x = 2 \\ x + 1 & \text{if} & x > 2 \end{cases}$,

   find $\lim\limits_{x \to 2} f(x)$, if it exists.

   (A) 2   (B) 3   (C) 4   (D) The limit does not exist.

Free-Response Question

4. Show that $\lim\limits_{x \to 0} x^3 \sin \dfrac{1}{x} = 0$.

   (*Hint*: Use the Squeeze Theorem.)

2.4 Differentiating the Product and the Quotient of Two Functions; Higher-Order Derivatives

OBJECTIVES *When you finish this section, you should be able to:*

1 **Differentiate the product of two functions (p. 203)**
2 **Differentiate the quotient of two functions (p. 206)**
3 **Find higher-order derivatives (p. 208)**
4 **Find the acceleration of an object moving on a line (p. 210)**

In this section, we obtain formulas for differentiating products and quotients of functions. As it turns out, the formulas are not what we might expect. The derivative of the product of two functions is *not* the product of their derivatives, and the derivative of the quotient of two functions is *not* the quotient of their derivatives.

1 Differentiate the Product of Two Functions

Consider the two functions $f(x) = 2x$ and $g(x) = x^3$. Both are differentiable, and their derivatives are $f'(x) = 2$ and $g'(x) = 3x^2$. Form the product

$$F(x) = f(x)g(x) = 2x \cdot x^3 = 2x^4$$

Now find F' using the Constant Multiple Rule and the Simple Power Rule.

$$F'(x) = 2 \cdot 4x^3 = 8x^3$$

Notice that $f'(x)g'(x) = 2 \cdot 3x^2 = 6x^2$ is not equal to $F'(x) = \dfrac{d}{dx}[f(x)g(x)] = 8x^3$. We conclude that the derivative of a product of two functions is *not* the product of their derivatives.

To find the derivative of the product of two differentiable functions f and g, we let $F(x) = f(x)g(x)$ and use the definition of a derivative, namely,

$$F'(x) = \lim_{h \to 0} \frac{[f(x+h)g(x+h)] - [f(x)g(x)]}{h} \qquad \text{Form (2)}$$

We can express F' in an equivalent form that contains the difference quotients for f and g, by subtracting and adding $f(x+h)g(x)$ to the numerator.

$$F'(x) = \lim_{h \to 0} \frac{f(x+h)g(x+h) - f(x+h)g(x) + f(x+h)g(x) - f(x)g(x)}{h}$$

$$= \lim_{h \to 0} \frac{f(x+h)[g(x+h) - g(x)] + [f(x+h) - f(x)]g(x)}{h} \qquad \text{Group and factor.}$$

$$= \left[\lim_{h \to 0} f(x+h) \right] \left[\lim_{h \to 0} \frac{g(x+h) - g(x)}{h} \right] + \left[\lim_{h \to 0} \frac{f(x+h) - f(x)}{h} \right] \left[\lim_{h \to 0} g(x) \right] \qquad \text{Use properties of limits.}$$

$$= \left[\lim_{h \to 0} f(x+h) \right] g'(x) + f'(x) \left[\lim_{h \to 0} g(x) \right] \qquad \text{Definition of a derivative.}$$

$$= f(x)g'(x) + f'(x)g(x) \qquad \begin{array}{l} \lim_{h \to 0} g(x) = g(x) \text{ since } h \text{ is not present.} \\ \text{Since } f \text{ is differentiable, it is} \\ \text{continuous, so } \lim_{h \to 0} f(x+h) = f(x). \end{array}$$

We have proved the following theorem.

> **THEOREM Product Rule**
>
> If f and g are differentiable functions and if $F(x) = f(x)g(x)$, then F is differentiable, and the derivative of the product F is
>
> $$\boxed{F'(x) = [f(x)g(x)]' = f(x)g'(x) + f'(x)g(x)}$$
>
> In Leibniz notation, the Product Rule has the form
>
> $$\boxed{\frac{d}{dx} F(x) = \frac{d}{dx}[f(x)g(x)] = f(x) \left[\frac{d}{dx} g(x) \right] + \left[\frac{d}{dx} f(x) \right] g(x)}$$

IN WORDS The derivative of the product of two differentiable functions equals the first function times the derivative of the second function plus the derivative of the first function times the second function. That is,

$$(fg)' = f \cdot g' + f' \cdot g$$

EXAMPLE 1   **Differentiating the Product of Two Functions**

Find y' if $y = (1 + x^2)e^x$.

Solution

The function y is the product of two functions: a polynomial, $f(x) = 1 + x^2$, and the exponential function, $g(x) = e^x$. By the Product Rule,

$$y' = \frac{d}{dx}[(1+x^2)e^x] = (1+x^2) \left[\frac{d}{dx} e^x \right] + \left[\frac{d}{dx}(1+x^2) \right] e^x = (1+x^2)e^x + 2xe^x$$

↑
Product Rule

At this point, we have found the derivative, but it is customary to simplify the answer. Then

$$y' = \underset{\underset{\text{Factor out } e^x.}{\uparrow}}{(1+x^2+2x)e^x} = \underset{\underset{\text{Factor.}}{\uparrow}}{(x+1)^2 e^x} \qquad \blacksquare$$

NOW WORK Problem **9** and AP® Practice Problem **4**.

Do not use the Product Rule unnecessarily! When one of the factors is a constant, use the Constant Multiple Rule. For example, it is easier to work

$$\frac{d}{dx}[5(x^2+1)] = 5\frac{d}{dx}(x^2+1) = 5 \cdot 2x = 10x$$

than it is to work

$$\frac{d}{dx}[5(x^2+1)] = 5\frac{d}{dx}(x^2+1) + \left[\frac{d}{dx}5\right](x^2+1) = 5 \cdot 2x + 0 \cdot (x^2+1) = 10x$$

Also, it is easier to simplify $f(x) = x^2(4x-3)$ before finding the derivative. That is, it is easier to work

$$\frac{d}{dx}[x^2(4x-3)] = \frac{d}{dx}(4x^3-3x^2) = 12x^2 - 6x$$

than it is to use the Product Rule

$$\frac{d}{dx}[x^2(4x-3)] = x^2\frac{d}{dx}(4x-3) + \left(\frac{d}{dx}x^2\right)(4x-3) = (x^2)(4) + (2x)(4x-3)$$

$$= 4x^2 + 8x^2 - 6x = 12x^2 - 6x$$

EXAMPLE 2 **Differentiating a Product in Two Ways**

CALC CLIP

Find the derivative of $F(v) = (5v^2 - v + 1)(v^3 - 1)$ in two ways:

(a) By using the Product Rule.
(b) By multiplying the factors of the function before finding its derivative.

Solution

(a) F is the product of the two functions $f(v) = 5v^2 - v + 1$ and $g(v) = v^3 - 1$. Using the Product Rule, we get

$$F'(v) = (5v^2 - v + 1)\left[\frac{d}{dv}(v^3 - 1)\right] + \left[\frac{d}{dv}(5v^2 - v + 1)\right](v^3 - 1)$$

$$= (5v^2 - v + 1)(3v^2) + (10v - 1)(v^3 - 1)$$

$$= 15v^4 - 3v^3 + 3v^2 + 10v^4 - 10v - v^3 + 1$$

$$= 25v^4 - 4v^3 + 3v^2 - 10v + 1$$

(b) Here we multiply the factors of F before differentiating.

$$F(v) = (5v^2 - v + 1)(v^3 - 1) = 5v^5 - v^4 + v^3 - 5v^2 + v - 1$$

Then

$$F'(v) = 25v^4 - 4v^3 + 3v^2 - 10v + 1 \qquad \blacksquare$$

Notice that the derivative is the same whether you differentiate and then simplify, or whether you multiply the factors and then differentiate. Use the approach that you find easier.

NOW WORK Problem **13**.

2 Differentiate the Quotient of Two Functions

The derivative of the quotient of two functions is *not* equal to the quotient of their derivatives. Instead, the derivative of the quotient of two functions is found using the *Quotient Rule*.

THEOREM Quotient Rule

If two functions f and g are differentiable and if $F(x) = \dfrac{f(x)}{g(x)}$, $g(x) \neq 0$, then F is differentiable, and the derivative of the quotient F is

$$F'(x) = \left[\frac{f(x)}{g(x)} \right]' = \frac{f'(x)g(x) - f(x)g'(x)}{[g(x)]^2}$$

IN WORDS The derivative of a quotient of two functions is the derivative of the numerator times the denominator, minus the numerator times the derivative of the denominator, all divided by the denominator squared. That is,

$$\left(\frac{f}{g} \right)' = \frac{f'g - fg'}{g^2}$$

In Leibniz notation, the Quotient Rule has the form

$$\frac{d}{dx} F(x) = \frac{d}{dx} \left[\frac{f(x)}{g(x)} \right] = \frac{\left[\dfrac{d}{dx} f(x) \right] g(x) - f(x) \left[\dfrac{d}{dx} g(x) \right]}{[g(x)]^2}$$

Proof We use the definition of a derivative (Form 2) to find $F'(x)$.

$$F'(x) = \lim_{h \to 0} \frac{F(x+h) - F(x)}{h} = \lim_{\substack{\uparrow \\ h \to 0}} \frac{\dfrac{f(x+h)}{g(x+h)} - \dfrac{f(x)}{g(x)}}{h}$$

$$F(x) = \frac{f(x)}{g(x)}$$

$$= \lim_{h \to 0} \frac{f(x+h)g(x) - f(x)g(x+h)}{h[g(x+h)g(x)]}$$

We write F' in an equivalent form that contains the difference quotients for f and g by subtracting and adding $f(x)g(x)$ to the numerator.

$$F'(x) = \lim_{h \to 0} \frac{f(x+h)g(x) - f(x)g(x) + f(x)g(x) - f(x)g(x+h)}{h[g(x+h)g(x)]}$$

Now group and factor the numerator.

$$F'(x) = \lim_{h \to 0} \frac{[f(x+h) - f(x)]g(x) - f(x)[g(x+h) - g(x)]}{h[g(x+h)g(x)]}$$

$$= \lim_{h \to 0} \frac{\left[\dfrac{f(x+h) - f(x)}{h} \right] g(x) - f(x) \left[\dfrac{g(x+h) - g(x)}{h} \right]}{g(x+h)g(x)}$$

$$= \frac{\displaystyle\lim_{h \to 0} \left[\dfrac{f(x+h) - f(x)}{h} \right] \cdot \lim_{h \to 0} g(x) - \lim_{h \to 0} f(x) \cdot \lim_{h \to 0} \left[\dfrac{g(x+h) - g(x)}{h} \right]}{\displaystyle\lim_{h \to 0} g(x+h) \cdot \lim_{h \to 0} g(x)}$$

RECALL Since g is differentiable, it is continuous; so, $\lim\limits_{h \to 0} g(x+h) = g(x)$.

$$= \frac{f'(x)g(x) - f(x)g'(x)}{[g(x)]^2} \qquad \blacksquare$$

EXAMPLE 3   **Differentiating the Quotient of Two Functions**

Find y' if $y = \dfrac{x^2 + 1}{2x - 3}$.

Solution

The function y is the quotient of $f(x) = x^2 + 1$ and $g(x) = 2x - 3$. Using the Quotient Rule, we have

$$y' = \frac{d}{dx}\frac{x^2+1}{2x-3} = \frac{\left[\dfrac{d}{dx}(x^2+1)\right](2x-3) - (x^2+1)\left[\dfrac{d}{dx}(2x-3)\right]}{(2x-3)^2}$$

$$= \frac{(2x)(2x-3) - (x^2+1)(2)}{(2x-3)^2} = \frac{4x^2 - 6x - 2x^2 - 2}{(2x-3)^2} = \frac{2x^2 - 6x - 2}{(2x-3)^2}$$

provided $x \neq \dfrac{3}{2}$. ∎

NOW WORK   Problem 23 and AP® Practice Problems 1, 2, 3, 7, and 8.

COROLLARY Derivative of the Reciprocal of a Function

If a function g is differentiable, then

IN WORDS The derivative of the reciprocal of a function is the negative of the derivative of the denominator divided by the square of the denominator. That is,
$$\left(\frac{1}{g}\right)' = -\frac{g'}{g^2}.$$

$$\boxed{\frac{d}{dx}\frac{1}{g(x)} = -\frac{\dfrac{d}{dx}g(x)}{[g(x)]^2} = -\frac{g'(x)}{[g(x)]^2}}\qquad(1)$$

provided $g(x) \neq 0$.

The proof of the corollary is left as an exercise. (See Problem 98.)

EXAMPLE 4   **Differentiating the Reciprocal of a Function**

(a) $\dfrac{d}{dx}\dfrac{1}{x^2+x} \underset{\substack{\uparrow \\ \text{Use (1).}}}{=} -\dfrac{\dfrac{d}{dx}(x^2+x)}{(x^2+x)^2} = -\dfrac{2x+1}{(x^2+x)^2}$

(b) $\dfrac{d}{dx}e^{-x} = \dfrac{d}{dx}\dfrac{1}{e^x} \underset{\substack{\uparrow \\ \text{Use (1).}}}{=} -\dfrac{\dfrac{d}{dx}e^x}{(e^x)^2} = -\dfrac{e^x}{e^{2x}} = -\dfrac{1}{e^x} = -e^{-x}$ ∎

NOW WORK   Problem 25.

Notice that the derivative of the reciprocal of a function f is *not* the reciprocal of the derivative. That is,

$$\frac{d}{dx}\frac{1}{f(x)} \neq \frac{1}{f'(x)}$$

The rule for the derivative of the reciprocal of a function allows us to extend the Simple Power Rule to all integers. Here is the proof.

Proof Suppose n is a negative integer and $x \neq 0$. Then $m = -n$ is a positive integer, and

$$\frac{d}{dx}x^n = \frac{d}{dx}\frac{1}{x^m} \underset{\substack{\uparrow \\ \text{Use (1).}}}{=} -\frac{\dfrac{d}{dx}x^m}{(x^m)^2} \underset{\substack{\uparrow \\ \text{Simple Power Rule}}}{=} -\frac{mx^{m-1}}{x^{2m}} = -mx^{m-1-2m} = -mx^{-m-1} \underset{\substack{\uparrow \\ \text{Substitute } n = -m.}}{=} nx^{n-1}\quad∎$$

NOTE In Section 2.3, we proved the Simple Power Rule, $\frac{d}{dx}x^n = nx^{n-1}$ where n is a positive integer. Here we have extended the Simple Power Rule from positive integers to all integers. In Chapter 3, we extend the result to include all real numbers.

THEOREM Simple Power Rule

The derivative of $y = x^n$, where n is any integer, is

$$y' = \frac{d}{dx}x^n = nx^{n-1}$$

EXAMPLE 5 Differentiating Using the Simple Power Rule

(a) $\dfrac{d}{dx}x^{-1} = -x^{-2} = -\dfrac{1}{x^2}$

(b) $\dfrac{d}{du}\dfrac{1}{u^2} = \dfrac{d}{du}u^{-2} = -2u^{-3} = -\dfrac{2}{u^3}$

(c) $\dfrac{d}{ds}\dfrac{4}{s^5} = 4\dfrac{d}{ds}s^{-5} = 4 \cdot (-5)\,s^{-6} = -20s^{-6} = -\dfrac{20}{s^6}$ ∎

NOW WORK Problem **31** and AP® Practice Problem **5.**

EXAMPLE 6 Using the Simple Power Rule in Electrical Engineering

Ohm's Law states that the current I running through a wire is inversely proportional to the resistance R in the wire and can be written as $I = \dfrac{V}{R}$, where V is the voltage. Find the rate of change of I with respect to R when $V = 12$ volts.

Solution

The rate of change of I with respect to R is the derivative $\dfrac{dI}{dR}$. We write Ohm's Law with $V = 12$ as $I = \dfrac{V}{R} = 12R^{-1}$ and use the Simple Power Rule.

$$\frac{dI}{dR} = \frac{d}{dR}(12R^{-1}) = 12 \cdot \frac{d}{dR}R^{-1} = 12(-1R^{-2}) = -\frac{12}{R^2}$$

The minus sign in $\dfrac{dI}{dR}$ indicates that the current I decreases as the resistance R in the wire increases. ∎

NOW WORK Problem **91.**

---

③ Find Higher-Order Derivatives

Since the derivative f' is a function, it makes sense to ask about the derivative of f'. The derivative (if there is one) of f' is also a function called the **second derivative** of f and denoted by f'', read "f double prime."

By continuing in this fashion, we can find the **third derivative** of f, the **fourth derivative** of f, and so on, provided that these derivatives exist. Collectively, these are called **higher-order derivatives**.

Leibniz notation also can be used for higher-order derivatives. Table 3 summarizes the notation for higher-order derivatives.

TABLE 3

	Prime Notation		Leibniz Notation	
First Derivative	y'	$f'(x)$	$\dfrac{dy}{dx}$	$\dfrac{d}{dx}f(x)$
Second Derivative	y''	$f''(x)$	$\dfrac{d^2y}{dx^2}$	$\dfrac{d^2}{dx^2}f(x)$
Third Derivative	y'''	$f'''(x)$	$\dfrac{d^3y}{dx^3}$	$\dfrac{d^3}{dx^3}f(x)$
Fourth Derivative	$y^{(4)}$	$f^{(4)}(x)$	$\dfrac{d^4y}{dx^4}$	$\dfrac{d^4}{dx^4}f(x)$
$\vdots$				
nth Derivative	$y^{(n)}$	$f^{(n)}(x)$	$\dfrac{d^ny}{dx^n}$	$\dfrac{d^n}{dx^n}f(x)$

EXAMPLE 7   **Finding Higher-Order Derivatives of a Power Function**

Find the second, third, and fourth derivatives of $y = 2x^3$.

Solution

Use the Simple Power Rule and the Constant Multiple Rule to find each derivative. The first derivative is

$$y' = \frac{d}{dx}(2x^3) = 2 \cdot \frac{d}{dx}x^3 = 2 \cdot 3x^2 = 6x^2$$

The next three derivatives are

$$y'' = \frac{d^2}{dx^2}(2x^3) = \frac{d}{dx}(6x^2) = 6 \cdot \frac{d}{dx}x^2 = 6 \cdot 2x = 12x$$

$$y''' = \frac{d^3}{dx^3}(2x^3) = \frac{d}{dx}(12x) = 12$$

$$y^{(4)} = \frac{d^4}{dx^4}(2x^3) = \frac{d}{dx}12 = 0$$

All derivatives of this function f of order 4 or more equal 0. This result can be generalized.

For a power function f of degree n, where n is a positive integer,

$$\boxed{\begin{aligned}
f(x) &= x^n \\
f'(x) &= nx^{n-1} \\
f''(x) &= n(n-1)x^{n-2} \\
&\;\;\vdots \\
f^{(n)}(x) &= n(n-1)(n-2) \cdot \ldots \cdot 3 \cdot 2 \cdot 1
\end{aligned}}$$

NOTE If $n > 1$ is an integer, the product $n \cdot (n-1) \cdot (n-2) \cdot \ldots \cdot 3 \cdot 2 \cdot 1$ is often written $n!$ and is read, "n factorial." The **factorial symbol** ! means $0! = 1$, $1! = 1$, and $n! = 1 \cdot 2 \cdot 3 \cdot \ldots \cdot (n-1) \cdot n$, where $n > 1$.

The nth-order derivative of $f(x) = x^n$ is a constant, so all derivatives of order greater than n equal 0.

It follows from this discussion that the nth derivative of a polynomial of degree n is a constant and that all derivatives of order $n+1$ and higher equal 0.

NOW WORK   Problem 41.

EXAMPLE 8 Finding Higher-Order Derivatives

Find the second and third derivatives of $y = (1 + x^2)e^x$.

Solution

In Example 1, we found that $y' = (1 + x^2)e^x + 2xe^x = (x^2 + 2x + 1)e^x$. To find y'', use the Product Rule with y'.

$$y'' = \frac{d}{dx}[(x^2 + 2x + 1)e^x] \underset{\substack{\uparrow \\ \text{Product Rule}}}{=} (x^2 + 2x + 1)\left(\frac{d}{dx}e^x\right) + \left[\frac{d}{dx}(x^2 + 2x + 1)\right]e^x$$

$$= (x^2 + 2x + 1)e^x + (2x + 2)e^x = (x^2 + 4x + 3)e^x$$

$$y''' = \frac{d}{dx}[(x^2 + 4x + 3)e^x] \underset{\substack{\uparrow \\ \text{Product Rule}}}{=} (x^2 + 4x + 3)\frac{d}{dx}e^x + \left[\frac{d}{dx}(x^2 + 4x + 3)\right]e^x$$

$$= (x^2 + 4x + 3)e^x + (2x + 4)e^x = (x^2 + 6x + 7)e^x \qquad \blacksquare$$

NOW WORK Problem 45 and AP® Practice Problems 9, 10, and 11.

---

4 Find the Acceleration of an Object Moving on a Line

For an object moving on a line whose signed distance s from the origin at time t is the position function $s = s(t)$, the derivative $s'(t)$ has a physical interpretation as the velocity of the object. The second derivative s'', which is the rate of change of velocity, is called *acceleration*.

DEFINITION Acceleration

For an object moving on a line, its signed distance s from the origin at time t is given by a position function $s = s(t)$. The first derivative $\dfrac{ds}{dt}$ is the velocity $v = v(t)$ of the object at time t.

The **acceleration** $a = a(t)$ of an object at time t is defined as the rate of change of velocity with respect to time. That is,

$$\boxed{a = a(t) = \frac{dv}{dt} = \frac{d}{dt}v = \frac{d}{dt}\left(\frac{ds}{dt}\right) = \frac{d^2s}{dt^2}}$$

IN WORDS Acceleration is the second derivative of a position function with respect to time.

EXAMPLE 9 Analyzing Vertical Motion

A ball is propelled vertically upward from the ground with an initial velocity of 29.4 m/s. The height s (in meters) of the ball above the ground is approximately $s = s(t) = -4.9t^2 + 29.4t$, where t is the number of seconds that elapse from the moment the ball is released.

(a) What is the velocity of the ball at time t? What is its velocity at $t = 1$ s?

(b) When will the ball reach its maximum height?

(c) What is the maximum height the ball reaches?

(d) What is the acceleration of the ball at any time t?

(e) How long is the ball in the air?

(f) What is the velocity of the ball upon impact with the ground? What is its speed?

(g) What is the total distance traveled by the ball?

Solution

(a) Since $s = s(t) = -4.9t^2 + 29.4t$, then

$$v = v(t) = \frac{ds}{dt} = -9.8t + 29.4$$

$$v(1) = -9.8 + 29.4 = 19.6$$

At $t = 1$ s, the velocity of the ball is 19.6 m/s.

(b) As the ball gets higher, its velocity decreases until it reaches its maximum height. Then as the ball drops, its velocity increases. It follows that the ball reaches its maximum height when $v(t) = 0$.

$$v(t) = -9.8t + 29.4 = 0$$

$$9.8t = 29.4$$

$$t = 3$$

The ball reaches its maximum height after 3 s.

(c) The maximum height is

$$s = s(3) = -4.9 \cdot 3^2 + 29.4 \cdot 3 = 44.1$$

The maximum height of the ball is 44.1 m.

(d) The acceleration of the ball at any time t is

$$a = a(t) = \frac{d^2 s}{dt^2} = \frac{dv}{dt} = \frac{d}{dt}(-9.8t + 29.4) = -9.8 \text{ m/s}^2$$

(e) There are two ways to answer the question "How long is the ball in the air?" *First way:* Since it takes 3 s for the ball to reach its maximum height, it follows that it will take another 3 s to reach the ground, for a total time of 6 s in the air. *Second way:* When the ball reaches the ground, $s = s(t) = 0$. Solve for t:

$$s(t) = -4.9t^2 + 29.4t = 0$$

$$t(-4.9t + 29.4) = 0$$

$$t = 0 \qquad \text{or} \qquad t = \frac{29.4}{4.9} = 6$$

The ball is at ground level at $t = 0$ and at $t = 6$, so the ball is in the air for 6 s.

(f) Upon impact with the ground, $t = 6$ s. So the velocity is

$$v(6) = (-9.8)(6) + 29.4 = -29.4$$

Upon impact the direction of the ball is downward, and its speed is 29.4 m/s.

(g) The total distance traveled by the ball is

$$s(3) + s(3) = 2 \, s(3) = 2(44.1) = 88.2 \text{ m}$$

See Figure 29 for an illustration.

NOTE *Speed* and *velocity* are not the same. Speed measures how fast an object is moving and is a nonnegative number. Velocity measures both the speed and the direction of an object and may be a positive number or a negative number or zero.

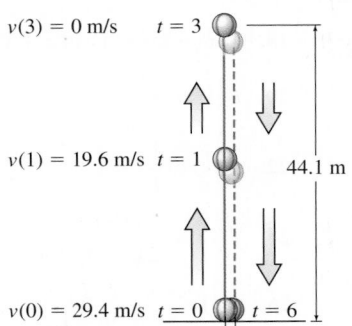

$v(3) = 0$ m/s $t = 3$

$v(1) = 19.6$ m/s $t = 1$ 44.1 m

$v(0) = 29.4$ m/s $t = 0$ $t = 6$

Figure 29

NOWWORK   Problem **83** and AP® Practice Problem **6.**

NOTE The Earth is not perfectly round; it bulges slightly at the equator, and its mass is not distributed uniformly. As a result, the acceleration of a freely falling body varies slightly.

In Example 9, the acceleration of the ball is constant. In fact, acceleration is the same for all falling objects at the same location, provided air resistance is not taken into account. In the sixteenth century, Galileo (1564–1642) discovered this by experimentation.* He also found that all falling bodies obey the law, stating that the distance s they fall when dropped is proportional to the square of the time t it takes to fall that distance, and that the constant of proportionality c is the same for all objects. That is,

$$s = -ct^2$$

The velocity v of the falling object is

$$v = \frac{ds}{dt} = \frac{d}{dt}(-ct^2) = -2ct$$

and its acceleration a is

$$a = \frac{dv}{dt} = \frac{d^2s}{dt^2} = -2c$$

which is a constant. Usually, we denote the constant $2c$ by g so $c = \frac{1}{2}g$. Then

$$\boxed{a = -g \qquad v = -gt \qquad s = -\frac{1}{2}gt^2}$$

The number g is called the **acceleration due to gravity**. For our planet, g is approximately 32 ft/s^2, or 9.8 m/s^2. On the planet Jupiter, $g \approx 26.0$ m/s^2, and on our Moon, $g \approx 1.60$ m/s^2.

2.4 Assess Your Understanding

Concepts and Vocabulary

1. **True or False** The derivative of a product is the product of the derivatives.

2. If $F(x) = f(x)g(x)$, then $F'(x) =$ _____ .

3. **True or False** $\frac{d}{dx}x^n = nx^{n+1}$, for any integer n.

4. If f and $g \neq 0$ are two differentiable functions, then $\frac{d}{dx}\frac{f(x)}{g(x)} =$ _____ .

5. **True or False** $f(x) = \frac{e^x}{x^2}$ can be differentiated using the Quotient Rule or by writing $f(x) = \frac{e^x}{x^2} = x^{-2}e^x$ and using the Product Rule.

6. If $g \neq 0$ is a differentiable function, then $\frac{d}{dx}\frac{1}{g(x)} =$ _____ .

7. If $f(x) = x$, then $f''(x) =$ _____ .

8. When an object moving on a line is modeled by the position function $s = s(t)$, then the acceleration a of the object at time t is given by $a = a(t) =$ _____ .

Skill Building

In Problems 9–40, find the derivative of each function.

[PAGE 205] 9. $f(x) = xe^x$

10. $f(x) = x^2e^x$

11. $f(x) = x^2(x^3 - 1)$

12. $f(x) = x^4(x + 5)$

[PAGE 205] 13. $f(x) = (3x^2 - 5)(3x + 4)$

14. $f(x) = (3x - 2)(5x + 3)$

15. $s(t) = (2t^5 - t)(t^3 - 2t + 1)$

16. $F(u) = (u^4 - 3u^2 + 1)(u^2 - u + 2)$

17. $f(x) = (x^3 + 1)(\ln x + 1)$

18. $f(x) = (x^2 + 1)(\ln x + x)$

19. $g(s) = \frac{2s}{s + 1}$

20. $F(z) = \frac{z + 1}{2z}$

21. $G(u) = \frac{1 - 2u}{1 + 2u}$

22. $f(w) = \frac{1 - w^2}{1 + w^2}$

[PAGE 207] 23. $f(x) = \frac{4x^2 - 2}{3x + 4}$

24. $f(x) = \frac{-3x^3 - 1}{2x^2 + 1}$

[PAGE 207] 25. $f(w) = \frac{1}{w^3 - 1}$

26. $g(v) = \frac{1}{v^2 + 5v - 1}$

*In a famous legend, Galileo dropped a feather and a rock from the top of the Leaning Tower of Pisa, to show that the acceleration due to gravity is constant. He expected them to fall together, but he failed to account for air resistance that slowed the feather. In July 1971, *Apollo 15* astronaut David Scott repeated the experiment on the Moon, where there is no air resistance. He dropped a hammer and a falcon feather from his shoulder height. Both hit the Moon's surface at the same time. A video of this experiment may be found at the NASA website (https://moon.nasa.gov).

27. $s(t) = t^{-3}$

28. $G(u) = u^{-4}$

29. $f(x) = -\dfrac{4}{e^x}$

30. $f(x) = \dfrac{3}{4e^x}$

[PAGE 208] 31. $f(x) = \dfrac{10}{x^4} + \dfrac{3}{x^2}$

32. $f(x) = \dfrac{2}{x^5} - \dfrac{3}{x^3}$

33. $f(x) = 3x^3 - \dfrac{1}{3x^2}$

34. $f(x) = x^5 - \dfrac{5}{x^5}$

35. $s(t) = \dfrac{1}{t} - \dfrac{1}{t^2} + \dfrac{1}{t^3}$

36. $s(t) = \dfrac{1}{t} + \dfrac{1}{t^2} + \dfrac{1}{t^3}$

37. $f(x) = \dfrac{e^x}{x^2}$

38. $f(x) = \dfrac{x^2}{e^x}$

39. $f(x) = \dfrac{x^2 + 1}{xe^x}$

40. $f(x) = \dfrac{xe^x}{x^2 - x}$

In Problems 41–54, find f' and f'' for each function.

[PAGE 209] 41. $f(x) = 5x^2 + x - 8$

42. $f(x) = -4x^2 - 7x$

43. $f(x) = 3 \ln x - 3$

44. $f(x) = x - 4 \ln x$

[PAGE 210] 45. $f(x) = (x + 5)e^x$

46. $f(x) = 3x^4 e^x$

47. $f(x) = (2x + 1)(x^3 + 5)$

48. $f(x) = (3x - 5)(x^2 - 2)$

49. $f(x) = x + \dfrac{1}{x}$

50. $f(x) = x - \dfrac{1}{x}$

51. $f(t) = \dfrac{t^2 - 1}{t}$

52. $f(u) = \dfrac{u + 1}{u}$

53. $f(x) = \dfrac{e^x + x}{x}$

54. $f(x) = \dfrac{e^x}{x}$

55. Find y' and y'' for (a) $y = \dfrac{1}{x}$ and (b) $y = \dfrac{2x - 5}{x}$.

56. Find $\dfrac{dy}{dx}$ and $\dfrac{d^2 y}{dx^2}$ for (a) $y = \dfrac{5}{x^2}$ and (b) $y = \dfrac{2 - 3x}{x}$.

Motion on a Line   *In Problems 57–60, find the velocity $v = v(t)$ and acceleration $a = a(t)$ of an object moving on a line whose signed distance s from the origin at time t is modeled by the position function $s = s(t)$.*

57. $s(t) = 16t^2 + 20t$

58. $s(t) = 16t^2 + 10t + 1$

59. $s(t) = 4.9t^2 + 4t + 4$

60. $s(t) = 4.9t^2 + 5t$

In Problems 61–68, find the indicated derivative.

61. $f^{(4)}(x)$ if $f(x) = x^3 - 6x^2 + 8x - 5$

62. $f^{(5)}(x)$ if $f(x) = 6x^3 + x^2 - 4$

63. $\dfrac{d^8}{dt^8}\left(\dfrac{1}{8}t^8 - \dfrac{1}{7}t^7 + t^5 - t^3\right)$

64. $\dfrac{d^6}{dt^6}(t^6 + 5t^5 - 2t + 4)$

65. $\dfrac{d^7}{du^7}(e^u + u^2)$

66. $\dfrac{d^{10}}{du^{10}}(2e^u)$

67. $\dfrac{d^5}{dx^5}(-e^x)$

68. $\dfrac{d^8}{dx^8}(12x - e^x)$

In Problems 69–72:

(a) Find the slope of the tangent line for each function f at the given point.

(b) Find an equation of the tangent line to the graph of each function f at the given point.

(c) Find the points, if any, where the graph of the function has a horizontal tangent line.

(d) Graph each function, the tangent line found in (b), and any tangent lines found in (c) on the same set of axes.

69. $f(x) = \dfrac{x^2}{x - 1}$ at $\left(-1, -\dfrac{1}{2}\right)$

70. $f(x) = \dfrac{x}{x + 1}$ at $(0, 0)$

71. $f(x) = \dfrac{x^3}{x + 1}$ at $\left(1, \dfrac{1}{2}\right)$

72. $f(x) = \dfrac{x^2 + 1}{x}$ at $\left(2, \dfrac{5}{2}\right)$

In Problems 73–80:

(a) Find the points, if any, at which the graph of each function f has a horizontal tangent line.

(b) Find an equation for each horizontal tangent line.

(c) Solve the inequality $f'(x) > 0$.

(d) Solve the inequality $f'(x) < 0$.

(e) Graph f and any horizontal lines found in (b) on the same set of axes.

(f) Describe the graph of f for the results obtained in (c) and (d).

73. $f(x) = (x + 1)(x^2 - x - 11)$

74. $f(x) = (3x^2 - 2)(2x + 1)$

75. $f(x) = \dfrac{x^2}{x + 1}$

76. $f(x) = \dfrac{x^2 + 1}{x}$

77. $f(x) = xe^x$

78. $f(x) = x^2 e^x$

79. $f(x) = \dfrac{x^2 - 3}{e^x}$

80. $f(x) = \dfrac{e^x}{x^2 + 1}$

In Problems 81 and 82, use the graphs to determine each derivative.

81. Let $u(x) = f(x) \cdot g(x)$ and $v(x) = \dfrac{g(x)}{f(x)}$.

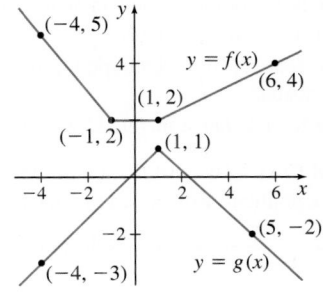

(a) $u'(0)$

(b) $u'(4)$

(c) $v'(-2)$

(d) $v'(6)$

(e) $\dfrac{d}{dx}\dfrac{1}{f(x)}$ at $x = -2$

(f) $\dfrac{d}{dx}\dfrac{1}{g(x)}$ at $x = 4$

82. Let $F(t) = f(t) \cdot g(t)$ and $G(t) = \dfrac{f(t)}{g(t)}$.

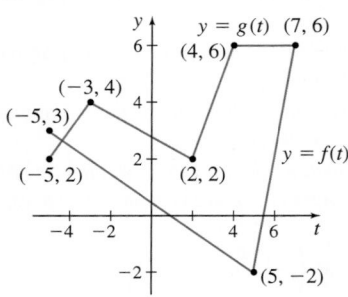

(a) $F'(0)$ (b) $F'(3)$

(c) $F'(-4)$ (d) $G'(-2)$

(e) $G'(-1)$ (f) $\dfrac{d}{dt}\dfrac{1}{f(t)}$ at $t = 3$

Applications and Extensions

 83. Vertical Motion An object is propelled vertically upward from the ground with an initial velocity of 39.2 m/s. The distance s (in meters) of the object from the ground after t seconds is given by the position function $s = s(t) = -4.9t^2 + 39.2t$.

(a) What is the velocity of the object at time t?

(b) When will the object reach its maximum height?

(c) What is the maximum height?

(d) What is the acceleration of the object at any time t?

(e) How long is the object in the air?

(f) What is the velocity of the object upon impact with the ground? What is its speed?

(g) What is the total distance traveled by the object?

84. Vertical Motion A ball is thrown vertically upward from a height of 6 ft with an initial velocity of 80 ft/s. The distance s (in feet) of the ball from the ground after t seconds is given by the position function $s = s(t) = 6 + 80t - 16t^2$.

(a) What is the velocity of the ball after 2 s?

(b) When will the ball reach its maximum height?

(c) What is the maximum height the ball reaches?

(d) What is the acceleration of the ball at any time t?

(e) How long is the ball in the air?

(f) What is the velocity of the ball upon impact with the ground? What is its speed?

(g) What is the total distance traveled by the ball?

85. Environmental Cost The cost C, in thousands of dollars, for the removal of a pollutant from a certain lake is given by the function $C(x) = \dfrac{5x}{110 - x}$, where x is the percent of pollutant removed.

(a) What is the domain of C?

(b) Graph C.

(c) What is the cost to remove 80% of the pollutant?

(d) Find $C'(x)$, the rate of change of the cost C with respect to the amount of pollutant removed.

(e) Find the rate of change of the cost for removing 40%, 60%, 80%, and 90% of the pollutant.

(f) Interpret the answers found in (e).

86. Investing in Fine Art The value V of a painting t years after it is purchased is modeled by the function

$$V(t) = \frac{100t^2 + 50}{t} + 400 \quad 1 \le t \le 5$$

(a) Find the rate of change in the value V with respect to time.

(b) What is the rate of change in value after 2 years?

(c) What is the rate of change in value after 3 years?

(d) Interpret the answers in (b) and (c).

87. Drug Concentration The concentration of a drug in a patient's blood t hours after injection is given by the function $f(t) = \dfrac{0.4t}{2t^2 + 1}$ (in milligrams per liter).

(a) Find the rate of change of the concentration with respect to time.

(b) What is the rate of change of the concentration after 10 min? After 30 min? After 1 hour?

(c) Interpret the answers found in (b).

(d) Graph f for the first 5 hours after administering the drug.

(e) From the graph, approximate the time (in minutes) at which the concentration of the drug is highest. What is the highest concentration of the drug in the patient's blood?

88. Population Growth A population of 1000 bacteria is introduced into a culture and grows in number according to the formula

$$P(t) = 1000\left(1 + \frac{4t}{100 + t^2}\right),$$ where t is measured in hours.

(a) Find the rate of change in population with respect to time.

(b) What is the rate of change in population at $t = 1$, $t = 2$, $t = 3$, and $t = 4$?

(c) Interpret the answers found in (b).

(d) Graph $P = P(t)$, $0 \le t \le 20$.

(e) From the graph, approximate the time (in hours) when the population is the greatest. What is the maximum population of the bacteria in the culture?

89. Economics The price-demand function for a popular e-book is given by $D(p) = \dfrac{100,000}{p^2 + 10p + 50}$, $4 \le p \le 20$, where $D = D(p)$ is the quantity demanded at the price p dollars.

(a) Find $D'(p)$, the rate of change of demand with respect to price.

(b) Find $D'(5)$, $D'(10)$, and $D'(15)$.

(c) Interpret the results found in (b).

90. Intensity of Light The intensity of illumination I on a surface is inversely proportional to the square of the distance r from the surface to the source of light. If the intensity is 1000 units when the distance is 1 m from the light, find the rate of change of the intensity with respect to the distance when the source is 10 meters from the surface.

91. Ideal Gas Law   The Ideal Gas Law, used in chemistry and thermodynamics, relates the pressure p, the volume V, and the absolute temperature T (in Kelvin) of a gas, using the equation $pV = nRT$, where n is the amount of gas (in moles) and $R = 8.31$ is the ideal gas constant. In an experiment, a spherical gas container of radius r meters is placed in a pressure chamber and is slowly compressed while keeping its temperature at 273 K.

(a) Find the rate of change of the pressure p with respect to the radius r of the chamber.

   Hint: The volume V of a sphere is $V = \frac{4}{3}\pi r^3$.

(b) Interpret the sign of the answer found in (a).

(c) If the sphere contains 1.0 mol of gas, find the rate of change of the pressure when $r = \frac{1}{4}$ m.

   Note: The metric unit of pressure is the pascal, Pa.

92. Body Density   The density ρ of an object is its mass m divided by its volume V; that is, $\rho = \frac{m}{V}$. If a person dives below the surface of the ocean, the water pressure on the diver will steadily increase, compressing the diver and therefore increasing body density. Suppose the diver is modeled as a sphere of radius r.

(a) Find the rate of change of the diver's body density with respect to the radius r of the sphere.

   Hint: The volume V of a sphere is $V = \frac{4}{3}\pi r^3$.

(b) Interpret the sign of the answer found in (a).

(c) Find the rate of change of the diver's body density when the radius is 45 cm and the mass is 80,000 g (80 kg).

Jerk and Snap   *Problems 93–96 use the following discussion:* Suppose that an object is moving on a line so that its signed distance s from the origin at time t is given by the position function $s = s(t)$. The velocity $v = v(t)$ of the object at time t is the rate of change of s with respect to time, namely, $v = v(t) = \dfrac{ds}{dt}$. The acceleration $a = a(t)$ of the object at time t is the rate of change of the velocity with respect to time,

$$a = a(t) = \frac{dv}{dt} = \frac{d}{dt}\left(\frac{ds}{dt}\right) = \frac{d^2s}{dt^2}$$

There are also physical interpretations of the third derivative and the fourth derivative of $s = s(t)$. The **jerk** $J = J(t)$ of the object at time t is the rate of change of the acceleration a with respect to time; that is,

$$J = J(t) = \frac{da}{dt} = \frac{d}{dt}\left(\frac{dv}{dt}\right) = \frac{d^2v}{dt^2} = \frac{d^3s}{dt^3}$$

The **snap** $S = S(t)$ of the object at time t is the rate of change of the jerk J with respect to time; that is,

$$S = S(t) = \frac{dJ}{dt} = \frac{d^2a}{dt^2} = \frac{d^3v}{dt^3} = \frac{d^4s}{dt^4}$$

Engineers take jerk into consideration when designing elevators, aircraft, and cars. In these cases, they try to minimize jerk, making for a smooth ride. But when designing thrill rides, such as roller coasters, the jerk is increased, making for an exciting experience.

93. Motion on a Line   As an object moves on a line, its signed distance s from the origin at time t is given by the position function $s = s(t) = t^3 - t + 1$, where s is in meters and t is in seconds.

(a) Find the velocity v, acceleration a, jerk J, and snap S of the object at time t.

(b) When is the velocity of the object 0 m/s?

(c) Find the acceleration of the object at $t = 2$ and at $t = 5$.

(d) Does the jerk of the object ever equal 0 m/s³?

(e) How would you interpret the snap for this object in rectilinear motion?

94. Motion on a Line   As an object moves on a line, its signed distance s from the origin at time t is given by the position function $s = s(t) = \frac{1}{6}t^4 - t^2 + \frac{1}{2}t + 4$, where s is in meters and t is in seconds.

(a) Find the velocity v, acceleration a, jerk J, and snap S of the object at any time t.

(b) Find the velocity of the object at $t = 0$ and at $t = 3$.

(c) Find the acceleration of the object at $t = 0$. Interpret your answer.

(d) Is the jerk of the object constant? In your own words, explain what the jerk says about the acceleration of the object.

(e) How would you interpret the snap for this object in rectilinear motion?

95. Elevator Ride Quality   The ride quality of an elevator depends on several factors, two of which are acceleration and jerk. In a study of 367 persons riding in a 1600-kg elevator that moves at an average speed of 4 m/s, the majority of riders were comfortable in an elevator with vertical motion given by

$$s(t) = 4t + 0.8t^2 + 0.333t^3$$

(a) Find the acceleration that the riders found acceptable.

(b) Find the jerk that the riders found acceptable.

Source: Elevator Ride Quality, January 2007, http://www .lift-report.de/index.php/news/176/368/Elevator-Ride-Quality

96. Elevator Ride Quality   In a hospital, the effects of high acceleration or jerk may be harmful to patients, so the acceleration and jerk need to be lower than in standard elevators. It has been determined that a 1600-kg elevator that is installed in a hospital and that moves at an average speed of 4 m/s should have vertical motion

$$s(t) = 4t + 0.55t^2 + 0.1167t^3$$

(a) Find the acceleration of a hospital elevator.

(b) Find the jerk of a hospital elevator.

Source: Elevator Ride Quality, January 2007, http://www.lift -report.de/index.php/news/176/368/Elevator-Ride-Quality

97. Current Density in a Wire The current density J in a wire is a measure of how much an electrical current is compressed as it flows through a wire and is modeled by the function $J(A) = \dfrac{I}{A}$, where I is the current (in amperes) and A is the cross-sectional area of the wire. In practice, current density, rather than merely current, is often important. For example, superconductors lose their superconductivity if the current density is too high.

(a) As current flows through a wire, it heats the wire, causing it to expand in area A. If a constant current is maintained in a cylindrical wire, find the rate of change of the current density J with respect to the radius r of the wire.

(b) Interpret the sign of the answer found in (a).

(c) Find the rate of change of current density with respect to the radius r when a current of 2.5 amps flows through a wire of radius $r = 0.50$ mm.

98. Derivative of a Reciprocal, Function Prove that if a function g is differentiable, then $\dfrac{d}{dx}\dfrac{1}{g(x)} = -\dfrac{g'(x)}{[g(x)]^2}$, provided $g(x) \neq 0$.

99. Extended Product Rule Show that if f, g, and h are differentiable functions, then

$$\frac{d}{dx}[f(x)g(x)h(x)] = f(x)g(x)h'(x) + f(x)g'(x)h(x)$$
$$+ f'(x)g(x)h(x)$$

From this, deduce that

$$\frac{d}{dx}[f(x)]^3 = 3[f(x)]^2 f'(x)$$

In Problems 100–105, use the Extended Product Rule (Problem 99) to find y'.

100. $y = (x^2 + 1)(x - 1)(x + 5)$

101. $y = (x - 1)(x^2 + 5)(x^3 - 1)$

102. $y = (x^4 + 1)^3$ 　　　　**103.** $y = (x^3 + 1)^3$.

104. $y = (3x + 1)\left(1 + \dfrac{1}{x}\right)(x^{-5} + 1)$

105. $y = \left(1 - \dfrac{1}{x}\right)\left(1 - \dfrac{1}{x^2}\right)\left(1 - \dfrac{1}{x^3}\right)$

106. (Further) Extended Product Rule Write a formula for the derivative of the product of four differentiable functions. That is, find a formula for $\dfrac{d}{dx}[f_1(x)f_2(x)f_3(x)f_4(x)]$. Also find a formula for $\dfrac{d}{dx}[f(x)]^4$.

107. If f and g are differentiable functions, show that if $F(x) = \dfrac{1}{f(x)g(x)}$, then

$$F'(x) = -F(x)\left[\frac{f'(x)}{f(x)} + \frac{g'(x)}{g(x)}\right]$$

provided $f(x) \neq 0$, $g(x) \neq 0$.

108. Higher-Order Derivatives If $f(x) = \dfrac{1}{1-x}$, find a formula for the nth derivative of f. That is, find $f^{(n)}(x)$.

109. Let $f(x) = \dfrac{x^6 - x^4 + x^2}{x^4 + 1}$. Rewrite f in the form $(x^4 + 1)f(x) = x^6 - x^4 + x^2$. Now find $f'(x)$ without using the quotient rule.

110. If f and g are differentiable functions with $f \neq -g$, find the derivative of $\dfrac{fg}{f+g}$.

[CAS] **111.** $f(x) = \dfrac{2x}{x+1}$.

(a) Use technology to find $f'(x)$.

(b) Simplify f' to a single fraction using either algebra or a CAS.

(c) Use technology to find $f^{(5)}(x)$.
Hint: Your CAS may have a method for finding higher-order derivatives without finding other derivatives first.

Challenge Problems

112. Suppose f and g have derivatives up to the fourth order. Find the first four derivatives of the product fg and simplify the answers. In particular, show that the fourth derivative is

$$\frac{d^4}{dx^4}(fg) = f^{(4)}g + 4f^{(3)}g^{(1)} + 6f^{(2)}g^{(2)} + 4f^{(1)}g^{(3)} + fg^{(4)}$$

Identify a pattern for the higher-order derivatives of fg.

113. Suppose $f_1(x), \ldots, f_n(x)$ are differentiable functions.

(a) Find $\dfrac{d}{dx}[f_1(x) \cdot \ldots \cdot f_n(x)]$.

(b) Find $\dfrac{d}{dx}\dfrac{1}{f_1(x) \cdot \ldots \cdot f_n(x)}$.

114. Let a, b, c, and d be real numbers. Define

$$\begin{vmatrix} a & b \\ c & d \end{vmatrix} = ad - bc$$

This is called a 2×2 **determinant** and it arises in the study of linear equations. Let $f_1(x)$, $f_2(x)$, $f_3(x)$, and $f_4(x)$ be differentiable and let

$$D(x) = \begin{vmatrix} f_1(x) & f_2(x) \\ f_3(x) & f_4(x) \end{vmatrix}$$

Show that

$$D'(x) = \begin{vmatrix} f_1'(x) & f_2'(x) \\ f_3(x) & f_4(x) \end{vmatrix} + \begin{vmatrix} f_1(x) & f_2(x) \\ f_3'(x) & f_4'(x) \end{vmatrix}$$

115. Let $f_0(x) = x - 1$

$$f_1(x) = 1 + \frac{1}{x-1}$$

$$f_2(x) = 1 + \cfrac{1}{1 + \cfrac{1}{x-1}}$$

$$f_3(x) = 1 + \cfrac{1}{1 + \cfrac{1}{1 + \cfrac{1}{x-1}}}$$

(a) Write f_1, f_2, f_3, f_4, and f_5 in the form $\dfrac{ax+b}{cx+d}$.

(b) Using the results from (a), write the sequence of numbers representing the coefficients of x in the numerator, beginning with $f_0(x) = x - 1$.

(c) Write the sequence in (b) as a recursive sequence. *Hint:* Look at the sum of consecutive terms.

(d) Find f_0', f_1', f_2', f_3', f_4', and f_5'.

AP® Practice Problems

Multiple-Choice Questions

207 1. What is the instantaneous rate of change at $x = -2$ of the function $f(x) = \dfrac{x-1}{x^2+2}$?

(A) $-\dfrac{1}{6}$ (B) $\dfrac{1}{9}$ (C) $\dfrac{1}{2}$ (D) -1

207 2. An equation of the line tangent to the graph of $f(x) = \dfrac{5x-3}{3x-6}$ at the point $(3, 4)$ is

(A) $7x + 3y = 37$ (B) $7x + 3y = 33$
(C) $7x - 3y = 9$ (D) $13x + 3y = 51$

207 3. If f, g, and h are nonzero differentiable functions of x, then $\dfrac{d}{dx}\left(\dfrac{gh}{f}\right) =$

(A) $\dfrac{fgh' + fg'h - f'gh}{f^2}$ (B) $\dfrac{g'h' - ghf'}{f^2}$

(C) $\dfrac{gh' + g'h}{f'}$ (D) $\dfrac{fgh' + fg'h + f'gh}{f^2}$

205 4. If $y = x^3 e^x + \ln x$, then $\dfrac{dy}{dx} =$

(A) $3x^2 e^x + \dfrac{1}{x}$ (B) $3x^2 + e^x - \dfrac{1}{x}$

(C) $3x^2 e^x (x+1) + x$ (D) $x^2 e^x (x+3) + \dfrac{1}{x}$

208 5. $\dfrac{d}{dt}\left(t^2 - \dfrac{1}{t^2} + \dfrac{1}{t}\right)$ at $t = 2$ is

(A) $\dfrac{7}{2}$ (B) $\dfrac{9}{2}$ (C) $\dfrac{9}{4}$ (D) 4

211 6. The position of an object moving along a line at time t, in seconds, is given by $s(t) = 16t^2 - 5t + 20$ meters. What is the acceleration of the object when $t = 2$?

(A) 32 m/s   (B) 0 m/s²   (C) 32 m/s²   (D) 64 m/s²

207 7. If $y = \dfrac{x-3}{x+3}$, $x \neq -3$, the instantaneous rate of change of y with respect to x at $x = 3$ is

(A) $-\dfrac{1}{6}$ (B) $\dfrac{1}{6}$ (C) $\dfrac{1}{36}$ (D) 1

207 8. Find an equation of the line tangent to the graph of the function $f(x) = \dfrac{x^2}{x+1}$ at $x = 1$.

(A) $8x + 6y = 11$ (B) $-8x + 6y = -5$
(C) $-3x + 4y = -1$ (D) $3x + 4y = 5$

210 9. If $y = xe^x$, then the nth derivative of y is

(A) e^x (B) $(x+n)e^x$ (C) ne^x (D) $x^n e^x$

210 10. $f(x) = \dfrac{x^2 + x + 4}{x}$. Find $f''(c)$ if $f'(c) = 0$ and $c < 0$.

(A) -4 (B) -2 (C) -1 (D) 1

210 11. $f(x) = 3\ln x - 4x + 5e^x$. Find $f''(1)$.

(A) $-4 + 5e$ (B) $-3 + 5e$ (C) $5e - 1$ (D) 4

Retain Your Knowledge

Multiple-Choice Questions

1. $\displaystyle\lim_{x \to 0} \dfrac{\sin(4x)}{\sin x} =$

(A) 0 (B) $\dfrac{1}{4}$ (C) 4 (D) ∞

2. Find $\displaystyle\lim_{t \to 2} \dfrac{(t^2 - 4)^3}{t^2 + t - 6}$ if it exists.

(A) $-\dfrac{8}{5}$ (B) 0 (C) $\dfrac{4}{5}$ (D) The limit does not exist.

3. Find $\displaystyle\lim_{x \to c} \dfrac{f(x) - f(c)}{x - c}$ if $f(x) = \sqrt{3x}$ and $c = 4$.

(A) $\sqrt{3}$ (B) $\dfrac{1}{\sqrt{3}}$ (C) $\dfrac{\sqrt{3}}{4}$ (D) ∞

Free-Response Question

4. Given the function $f(x) = \begin{cases} \sqrt{3x^2 + 4} & \text{if } x < 2 \\ 4 & \text{if } x = 2 \\ 5x - 6 & \text{if } x > 2 \end{cases}$

(a) Determine the domain of f.
(b) Determine whether f is continuous at $x = 2$.
(c) Is f continuous on its domain? Justify your answer.

2.5 The Derivative of the Trigonometric Functions

OBJECTIVE *When you finish this section, you should be able to:*

1 **Differentiate trigonometric functions (p. 218)**

1 Differentiate Trigonometric Functions

To find the derivatives of $y = \sin x$ and $y = \cos x$, we use the limits

$$\lim_{\theta \to 0} \frac{\sin \theta}{\theta} = 1 \quad \text{and} \quad \lim_{\theta \to 0} \frac{\cos \theta - 1}{\theta} = 0$$

that were established in Section 1.4.

> **THEOREM** Derivative of $y = \sin x$
>
> The derivative of $y = \sin x$ is $y' = \cos x$. That is,
>
> $$y' = \frac{d}{dx} \sin x = \cos x$$

Proof

NEED TO REVIEW? The trigonometric functions are discussed in Section P.6, pp. 55–61. Trigonometric identities are discussed in Appendix A.4, pp. A-35 to A-38.

$$y' = \lim_{h \to 0} \frac{\sin(x + h) - \sin x}{h} \qquad \text{The definition of a derivative (Form 2)}$$

$$= \lim_{h \to 0} \frac{\sin x \cos h + \sin h \cos x - \sin x}{h} \qquad \sin(A + B) = \sin A \cos B + \sin B \cos A$$

$$= \lim_{h \to 0} \left[\frac{\sin x \cos h - \sin x}{h} + \frac{\sin h \cos x}{h} \right] \qquad \text{Rearrange terms.}$$

$$= \lim_{h \to 0} \left[\sin x \cdot \frac{\cos h - 1}{h} + \frac{\sin h}{h} \cdot \cos x \right] \qquad \text{Factor.}$$

$$= \left[\lim_{h \to 0} \sin x \right] \left[\lim_{h \to 0} \frac{\cos h - 1}{h} \right] + \left[\lim_{h \to 0} \cos x \right] \left[\lim_{h \to 0} \frac{\sin h}{h} \right] \quad \text{Use properties of limits.}$$

$$= \sin x \cdot 0 + \cos x \cdot 1 = \cos x \qquad \lim_{\theta \to 0} \frac{\cos \theta - 1}{\theta} = 0; \quad \lim_{\theta \to 0} \frac{\sin \theta}{\theta} = 1 \quad \blacksquare$$

The geometry of the derivative $\dfrac{d}{dx} \sin x = \cos x$ is shown in Figure 30. On the graph of $f(x) = \sin x$, horizontal tangent lines are marked as well as the tangent lines that have slopes of 1 and -1. The derivative function is plotted on the second graph, and those points are connected with a smooth curve.

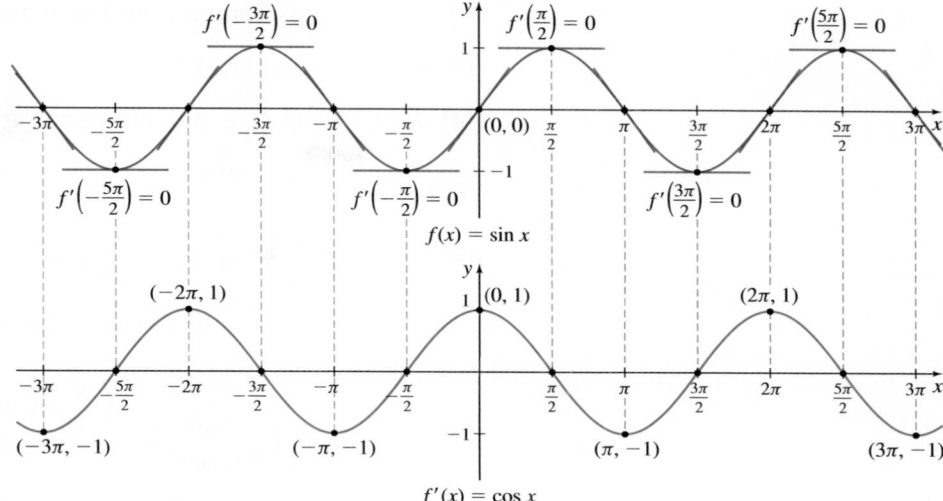

Figure 30

EXAMPLE 1   **Differentiating the Sine Function**

Find y' if:

(a) $y = x + 4\sin x$ **(b)** $y = x^2 \sin x$ **(c)** $y = \dfrac{\sin x}{x}$ **(d)** $y = e^x \sin x$

Solution

(a) Use the Sum Rule and the Constant Multiple Rule.

$$y' = \frac{d}{dx}(x + 4\sin x) = \frac{d}{dx}x + \frac{d}{dx}(4\sin x) = 1 + 4\frac{d}{dx}\sin x = 1 + 4\cos x$$

(b) Use the Product Rule.

$$y' = \frac{d}{dx}(x^2 \sin x) = x^2\left[\frac{d}{dx}\sin x\right] + \left[\frac{d}{dx}x^2\right]\sin x = x^2\cos x + 2x\sin x$$

(c) Use the Quotient Rule.

$$y' = \frac{d}{dx}\left(\frac{\sin x}{x}\right) = \frac{\left[\dfrac{d}{dx}\sin x\right]\cdot x - \sin x \cdot \left[\dfrac{d}{dx}x\right]}{x^2} = \frac{x\cos x - \sin x}{x^2}$$

(d) Use the Product Rule.

$$y' = \frac{d}{dx}(e^x \sin x) = e^x\frac{d}{dx}\sin x + \left(\frac{d}{dx}e^x\right)\sin x$$
$$= e^x \cos x + e^x \sin x = e^x(\cos x + \sin x)$$

NOW WORK   Problems 5, 29, and AP® Practice Problems 1 and 10.

THEOREM Derivative of $y = \cos x$

The derivative of $y = \cos x$ is

$$\boxed{\;y' = \frac{d}{dx}\cos x = -\sin x\;}$$

You are asked to prove this in Problem 75.

CALC CLIP | **EXAMPLE 2**   **Differentiating Trigonometric Functions**

Find the derivative of each function:

(a) $f(x) = x^2 \cos x$ **(b)** $g(\theta) = \dfrac{\cos \theta}{1 - \sin \theta}$ **(c)** $F(t) = \dfrac{e^t}{\cos t}$

Solution

(a) $f'(x) = \dfrac{d}{dx}(x^2 \cos x) = x^2 \dfrac{d}{dx} \cos x + \left(\dfrac{d}{dx} x^2 \right)(\cos x)$

$\qquad\qquad = x^2(-\sin x) + 2x \cos x = 2x \cos x - x^2 \sin x$

(b) $g'(\theta) = \dfrac{d}{d\theta}\left(\dfrac{\cos \theta}{1 - \sin \theta} \right) = \dfrac{\left(\dfrac{d}{d\theta} \cos \theta \right)(1 - \sin \theta) - (\cos \theta)\left[\dfrac{d}{d\theta}(1 - \sin \theta) \right]}{(1 - \sin \theta)^2}$

$\qquad = \dfrac{-\sin \theta\,(1 - \sin \theta) - \cos \theta(-\cos \theta)}{(1 - \sin \theta)^2} = \dfrac{-\sin \theta + \sin^2 \theta + \cos^2 \theta}{(1 - \sin \theta)^2}$

$\qquad = \dfrac{-\sin \theta + 1}{(1 - \sin \theta)^2} = \dfrac{1}{1 - \sin \theta}$

(c) $F'(t) = \dfrac{d}{dt}\left(\dfrac{e^t}{\cos t} \right) = \dfrac{\left(\dfrac{d}{dt} e^t \right)(\cos t) - e^t \left(\dfrac{d}{dt} \cos t \right)}{\cos^2 t} = \dfrac{e^t \cos t - e^t(-\sin t)}{\cos^2 t}$

$\qquad = \dfrac{e^t(\cos t + \sin t)}{\cos^2 t}$ ∎

NOW WORK   Problem 13 and AP Practice® Problems 2, 6, and 8.

EXAMPLE 3   **Identifying Horizontal Tangent Lines**

Find all points on the graph of $f(x) = x + \sin x$ where the tangent line is horizontal.

Solution

Since tangent lines are horizontal at points on the graph of f where $f'(x) = 0$, begin by finding $f'(x) = 1 + \cos x$. Now solve the equation:

$$f'(x) = 1 + \cos x = 0$$
$$\cos x = -1$$
$$x = (2k + 1)\pi$$

where k is an integer.

Since $\sin[(2k + 1)\pi] = 0$, then $f((2k + 1)\pi) = (2k + 1)\pi$. So, at each of the points $((2k + 1)\pi, (2k + 1)\pi)$, the graph of f has a horizontal tangent line. See Figure 31. ∎

Notice in Figure 31 that each of the points with a horizontal tangent line lies on the line $y = x$.

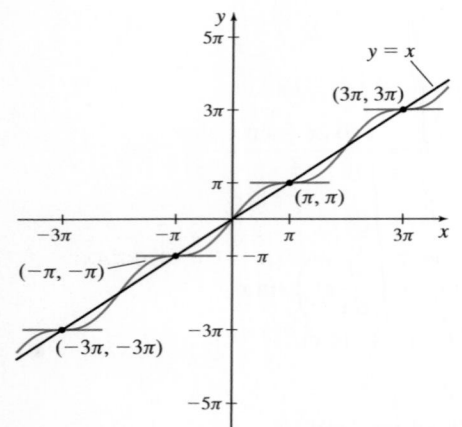

Figure 31 $f(x) = x + \sin x$

NOW WORK   Problem 57 and AP® Practice Problem 9.

The derivatives of the remaining four trigonometric functions are obtained using trigonometric identities and basic derivative rules. We establish the formula for the derivative of $y = \tan x$ in Example 4. You are asked to prove formulas for the derivative of the secant function, the cosecant function, and the cotangent function in the exercises. (See Problems 76–78.)

EXAMPLE 4 **Differentiating $y = \tan x$**

Show that the derivative of $y = \tan x$ is

$$\boxed{y' = \frac{d}{dx}\tan x = \sec^2 x}$$

Solution

$$y' = \frac{d}{dx}\tan x = \frac{d}{dx}\frac{\sin x}{\cos x} = \frac{\left[\dfrac{d}{dx}\sin x\right](\cos x) - (\sin x)\left[\dfrac{d}{dx}\cos x\right]}{\cos^2 x}$$

$$\underset{\text{Identity}}{\uparrow} \quad \underset{\text{Quotient Rule}}{\uparrow}$$

$$= \frac{\cos x \cdot \cos x - \sin x \cdot (-\sin x)}{\cos^2 x} = \frac{\cos^2 x + \sin^2 x}{\cos^2 x} = \frac{1}{\cos^2 x} = \sec^2 x \qquad \blacksquare$$

NOW WORK Problem 15 and AP® Practice Problems 3 and 7.

Table 4 lists the derivatives of the six trigonometric functions along with the domain of each derivative.

TABLE 4

Derivative Function	Domain of the Derivative Function
$\dfrac{d}{dx}\sin x = \cos x$	$(-\infty, \infty)$
$\dfrac{d}{dx}\cos x = -\sin x$	$(-\infty, \infty)$
$\dfrac{d}{dx}\tan x = \sec^2 x$	$\left\{x \,\middle|\, x \neq \dfrac{2k+1}{2}\pi,\, k \text{ an integer}\right\}$
$\dfrac{d}{dx}\cot x = -\csc^2 x$	$\{x \,|\, x \neq k\pi,\, k \text{ an integer}\}$
$\dfrac{d}{dx}\csc x = -\csc x \cot x$	$\{x \,|\, x \neq k\pi,\, k \text{ an integer}\}$
$\dfrac{d}{dx}\sec x = \sec x \tan x$	$\left\{x \,\middle|\, x \neq \dfrac{2k+1}{2}\pi,\, k \text{ an integer}\right\}$

NOTE If the trigonometric function begins with the letter *c*, that is, cosine, cotangent, or cosecant, then its derivative has a minus sign.

NOW WORK Problem 35.

EXAMPLE 5 **Finding the Second Derivative of a Trigonometric Function**

Find $f''\left(\dfrac{\pi}{4}\right)$ if $f(x) = \sec x$.

Solution

If $f(x) = \sec x$, then $f'(x) = \sec x \tan x$ and

$$f''(x) = \frac{d}{dx}(\sec x \tan x) = (\sec x)\left(\frac{d}{dx}\tan x\right) + \left(\frac{d}{dx}\sec x\right)(\tan x)$$

$$\underset{\text{Use the Product Rule.}}{\uparrow}$$

$$= \sec x \cdot \sec^2 x + (\sec x \tan x)\tan x = \sec^3 x + \sec x \tan^2 x$$

$$f''\left(\frac{\pi}{4}\right) = \sec^3\left(\frac{\pi}{4}\right) + \sec\left(\frac{\pi}{4}\right)\tan^2\left(\frac{\pi}{4}\right) = \left(\sqrt{2}\right)^3 + \sqrt{2}\cdot 1^2 = 2\sqrt{2} + \sqrt{2} = 3\sqrt{2}$$

$$\underset{\sec\frac{\pi}{4} = \sqrt{2};\ \tan\frac{\pi}{4} = 1}{\uparrow}$$

$\blacksquare$

NOW WORK Problem 45 and AP® Practice Problems 4, 5, and 11.

Application: Simple Harmonic Motion

Simple harmonic motion is a repetitive motion that can be modeled by a trigonometric function. A swinging pendulum and an oscillating spring are examples of simple harmonic motion.

EXAMPLE 6   Analyzing Simple Harmonic Motion

An object hangs on a spring, making the spring 2 m long in its equilibrium position. See Figure 32. If the object is pulled down 1 m and released, it oscillates up and down. The length l of the spring after t seconds is modeled by the function $l(t) = 2 + \cos t$.

(a) How does the length of the spring vary?

(b) Find the velocity of the object.

(c) At what position is the speed of the object a maximum?

(d) Find the acceleration of the object.

(e) At what position is the acceleration equal to 0?

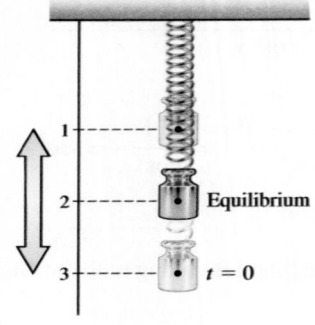

Figure 32

Solution

(a) Since $l(t) = 2 + \cos t$ and $-1 \le \cos t \le 1$, the length of the spring varies between 1 and 3 m.

(b) The velocity v of the object is

$$v = l'(t) = \frac{d}{dt}(2 + \cos t) = -\sin t$$

(c) Speed is the absolute value of velocity. Since $v = -\sin t$, the speed of the object is $|v| = |-\sin t| = |\sin t|$. Since $-1 \le \sin t \le 1$, the object moves the fastest when $|v| = |\sin t| = 1$. This occurs when $\sin t = \pm 1$ or, equivalently, when $\cos t = 0$. So, the speed is a maximum when $l(t) = 2$, that is, when the spring is at the equilibrium position.

(d) The acceleration a of the object is given by

$$a = l''(t) = \frac{d}{dt}l'(t) = \frac{d}{dt}(-\sin t) = -\cos t$$

(e) Since $a = -\cos t$, the acceleration is zero when $\cos t = 0$. So, $a = 0$ when $l(t) = 2$, that is, when the spring is at the equilibrium position. This is the same time at which the speed is maximum. ∎

Figure 33 shows the graphs of the length of the spring $y = l(t)$, the velocity $y = v(t)$, and the acceleration $y = a(t)$.

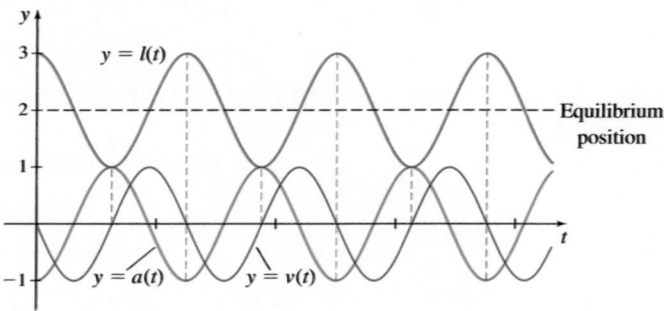

Figure 33 $y = l(t)$(blue), $y = v(t)$(red), $y = a(t)$(green)

NOW WORK   Problem 65.

2.5 Assess Your Understanding

Concepts and Vocabulary

1. **True or False** $\dfrac{d}{dx}\cos x = \sin x$

2. **True or False** $\dfrac{d}{dx}\tan x = \cot x$

3. **True or False** $\dfrac{d^2}{dx^2}\sin x = -\sin x$

4. **True or False** $\dfrac{d}{dx}\sin\dfrac{\pi}{3} = \cos\dfrac{\pi}{3}$

Skill Building

In Problems 5–38, find y'.

5. $y = x - \sin x$

6. $y = \cos x - x^2$

7. $y = \tan x + \sin x$

8. $y = \cos x - \tan x$

9. $y = 3\sin\theta - 2\cos\theta$

10. $y = 4\tan\theta + \sin\theta$

11. $y = \sin x \cos x$

12. $y = \cot x \tan x$

13. $y = t\cos t$

14. $y = t^2\tan t$

15. $y = e^x\tan x$

16. $y = e^x\sec x$

17. $y = \pi\sec u\tan u$

18. $y = \pi u\tan u$

19. $y = \dfrac{\cot x}{x}$

20. $y = \dfrac{\csc x}{x}$

21. $y = x^2\sin x$

22. $y = t^2\tan t$

23. $y = t\tan t - \sqrt{3}\sec t$

24. $y = x\sec x + \sqrt{2}\cot x$

25. $y = \dfrac{\sin\theta}{1-\cos\theta}$

26. $y = \dfrac{x}{\cos x}$

27. $y = \dfrac{\sin t}{1+t}$

28. $y = \dfrac{\tan u}{1+u}$

29. $y = \dfrac{\sin x}{e^x}$

30. $y = \dfrac{\cos x}{e^x}$

31. $y = \dfrac{\sin\theta + \cos\theta}{\sin\theta - \cos\theta}$

32. $y = \dfrac{\sin\theta - \cos\theta}{\sin\theta + \cos\theta}$

33. $y = \dfrac{\sec t}{1+t\sin t}$

34. $y = \dfrac{\csc t}{1+t\cos t}$

35. $y = \csc\theta\cot\theta$

36. $y = \tan\theta\cos\theta$

37. $y = \dfrac{1+\tan x}{1-\tan x}$

38. $y = \dfrac{\csc x - \cot x}{\csc x + \cot x}$

In Problems 39–50, find y''.

39. $y = \sin x$

40. $y = \cos x$

41. $y = \tan\theta$

42. $y = \cot\theta$

43. $y = t\sin t$

44. $y = t\cos t$

45. $y = e^x\sin x$

46. $y = e^x\cos x$

47. $y = 5\sin u - 4\cos u$

48. $y = 6\sin u + 5\cos u$

49. $y = a\sin x + b\cos x$

50. $y = a\sec\theta + b\tan\theta$

In Problems 51–56:
(a) Find an equation of the tangent line to the graph of f at the indicated point.
(b) Graph the function and the tangent line.

51. $f(x) = \sin x$ at $(0,0)$

52. $f(x) = \cos x$ at $\left(\dfrac{\pi}{3}, \dfrac{1}{2}\right)$

53. $f(x) = \tan x$ at $(0,0)$

54. $f(x) = \tan x$ at $\left(\dfrac{\pi}{4}, 1\right)$

55. $f(x) = \sin x + \cos x$ at $\left(\dfrac{\pi}{4}, \sqrt{2}\right)$

56. $f(x) = \sin x - \cos x$ at $\left(\dfrac{\pi}{4}, 0\right)$

In Problems 57–60:
(a) Find all points on the graph of f where the tangent line is horizontal.
(b) Graph the function and the horizontal tangent lines on the interval $[-2\pi, 2\pi]$.

57. $f(x) = 2\sin x + \cos x$

58. $f(x) = \cos x - \sin x$

59. $f(x) = \sec x$

60. $f(x) = \csc x$

Applications and Extensions

In Problems 61 and 62, find the nth derivative of each function.

61. $f(x) = \sin x$

62. $f(\theta) = \cos\theta$

63. What is $\lim\limits_{h\to 0}\dfrac{\cos\left(\dfrac{\pi}{2}+h\right) - \cos\dfrac{\pi}{2}}{h}$?

64. What is $\lim\limits_{h\to 0}\dfrac{\sin(\pi + h) - \sin\pi}{h}$?

65. **Simple Harmonic Motion** The signed distance s (in meters) of an object from the origin at time t (in seconds) is modeled by the position function $s(t) = \dfrac{1}{8}\cos t$.

(a) Find the velocity $v = v(t)$ of the object.
(b) When is the speed of the object a maximum?
(c) Find the acceleration $a = a(t)$ of the object.
(d) When is the acceleration equal to 0?
(e) Graph s, v, and a on the same screen.

66. Simple Harmonic Motion

An object attached to a coiled spring is pulled down a distance $d = 5$ cm from its equilibrium position and then released as shown in the figure. The motion of the object at time t seconds is simple harmonic and is modeled by $d(t) = -5\cos t$.

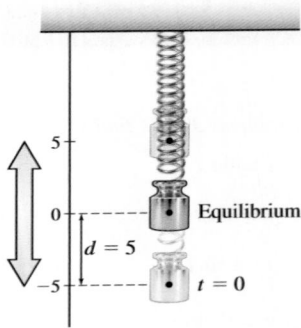

(a) As t varies from 0 to 2π, how does the length of the spring vary?

(b) Find the velocity $v = v(t)$ of the object.

(c) When is the speed of the object a maximum?

(d) Find the acceleration $a = a(t)$ of the object.

(e) When is the acceleration equal to 0?

(f) Graph d, v, and a on the same set of axes.

67. Rate of Change A large, 8-ft-high decorative mirror is placed on a wood floor and leaned against a wall. The weight of the mirror and the slickness of the floor cause the mirror to slip.

(a) If θ is the angle between the top of the mirror and the wall, and y is the distance from the floor to the top of the mirror, what is the rate of change of y with respect to θ?

(b) In feet/radian, how fast is the top of the mirror slipping down the wall when $\theta = \dfrac{\pi}{4}$?

68. Rate of Change The sides of an isosceles triangle are sliding outward. See the figure.

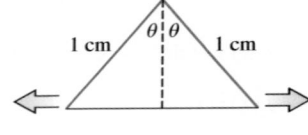

(a) Find the rate of change of the area of the triangle with respect to θ.

(b) How fast is the area changing when $\theta = \dfrac{\pi}{6}$?

69. Sea Waves Waves in deep water tend to have the symmetric form of the function $f(x) = \sin x$. As they approach shore, however, the sea floor creates drag, which changes the shape of the wave. The trough of the wave widens and the height of the wave increases, so the top of the wave is no longer symmetric with the trough. This type of wave can be represented by a function such as

$$w(x) = \frac{4}{2 + \cos x}$$

(a) Graph $w = w(x)$ for $0 \le x \le 4\pi$.

(b) What is the maximum and the minimum value of w?

(c) Find the values of x, $0 < x < 4\pi$, at which $w'(x) = 0$.

(d) Evaluate w' near the peak at π, using $x = \pi - 0.1$, and near the trough at 2π, using $x = 2\pi - 0.1$.

(e) Explain how these values confirm a nonsymmetric wave shape.

70. Swinging Pendulum A simple pendulum is a small-sized ball swinging from a light string. As it swings, the supporting string makes an angle θ with the vertical. See the figure. At an angle θ, the tension in the string is $T = \dfrac{W}{\cos\theta}$, where W is the weight of the swinging ball.

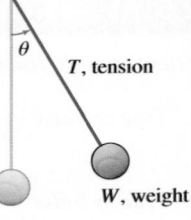

(a) Find the rate of change of the tension T with respect to θ when the pendulum is at its highest point ($\theta = \theta_{\max}$).

(b) Find the rate of change of the tension T with respect to θ when the pendulum is at its lowest point.

(c) What is the tension at the lowest point?

71. Restaurant Sales A restaurant in Naples, Florida, is very busy during the winter months and extremely slow over the summer. But every year the restaurant grows its sales. Suppose over the next two years, the revenue R, in units of \$10,000, is projected to follow the model

$$R = R(t) = \sin t + 0.3t + 1 \qquad 0 \le t \le 12$$

where $t = 0$ corresponds to November 1, 2024; $t = 1$ corresponds to January 1, 2025; $t = 2$ corresponds to March 1, 2025; and so on.

(a) What is the projected revenue for November 1, 2024; March 1, 2025; September 1, 2025; and January 1, 2026?

(b) What is the rate of change of revenue with respect to time?

(c) What is the rate of change of revenue with respect to time for January 1, 2026?

(d) Graph the revenue function and the derivative function $R' = R'(t)$.

(e) Does the graph of R support the facts that every year the restaurant grows its sales and that sales are higher during the winter and lower during the summer? Explain.

72. Polarizing Filters Polarizing filters transmit only light for which the electric field oscillations are in a specific direction. Light is polarized naturally by scattering off the molecules in the atmosphere and by reflecting off many (but not all) types of surfaces. If light of intensity I_0 is already polarized in a certain direction, and the transmission direction of the polarizing filter makes an angle with that direction, then the intensity I of the light after passing through the filter is given by

Malus's Law, $I(\theta) = I_0 \cos^2\theta$.

(a) As you rotate a polarizing filter, θ changes. Find the rate of change of the light intensity I with respect to θ.

(b) Find both the intensity $I(\theta)$ and the rate of change of the intensity with respect to θ, for the angles $\theta = 0°$, $45°$, and $90°$. (Remember to use radians for θ.)

73. If $y = \sin x$ and $y^{(n)}$ is the nth derivative of y with respect to x, find the smallest positive integer n for which $y^{(n)} = y$.

74. Use the identity $\sin A - \sin B = 2 \cos \dfrac{A+B}{2} \sin \dfrac{A-B}{2}$, with $A = x + h$ and $B = x$, to prove that

$$\frac{d}{dx} \sin x = \lim_{h \to 0} \frac{\sin(x+h) - \sin x}{h} = \cos x$$

75. Use the definition of a derivative to prove $\dfrac{d}{dx} \cos x = -\sin x$.

76. Derivative of $y = \sec x$ Use a derivative rule to show that

$$\frac{d}{dx} \sec x = \sec x \tan x$$

77. Derivative of $y = \csc x$ Use a derivative rule to show that

$$\frac{d}{dx} \csc x = -\csc x \cot x$$

78. Derivative of $y = \cot x$ Use a derivative rule to show that

$$\frac{d}{dx} \cot x = -\csc^2 x$$

79. Let $f(x) = \cos x$. Show that finding $f'(0)$ is the same as finding $\displaystyle\lim_{x \to 0} \frac{\cos x - 1}{x}$.

80. Let $f(x) = \sin x$. Show that finding $f'(0)$ is the same as finding $\displaystyle\lim_{x \to 0} \frac{\sin x}{x}$.

81. If $y = A \sin t + B \cos t$, where A and B are constants, show that $y'' + y = 0$.

Challenge Problem

82. For a differentiable function f, let f^* be the function defined by

$$f^*(x) = \lim_{h \to 0} \frac{f(x+h) - f(x-h)}{h}$$

   (a) Find $f^*(x)$ for $f(x) = x^2 + x$.

   (b) Find $f^*(x)$ for $f(x) = \cos x$.

   (c) Write an equation that expresses the relationship between the functions f^* and f', where f' denotes the derivative of f. Justify your answer.

Preparing for the AP® Exam

AP® Practice Problems

Multiple-Choice Questions

PAGE 219 **1.** If $y = x \sin x$, then $\dfrac{dy}{dx} =$

   (A) $x \cos x + \sin x$ (B) $x \cos x - \sin x$

   (C) $\cos x + \sin x$ (D) $(x+1) \cos x$

PAGE 220 **2.** What is $\displaystyle\lim_{h \to 0} \frac{\cos\left(\dfrac{\pi}{3} + h\right) - \cos\dfrac{\pi}{3}}{h}$?

   (A) 0 (B) $\dfrac{1}{2}$ (C) $\dfrac{\sqrt{3}}{2}$ (D) $-\dfrac{\sqrt{3}}{2}$

PAGE 221 **3.** If $f(x) = \tan x$, then $f'\left(\dfrac{\pi}{3}\right)$ equals

   (A) $2\sqrt{3}$ (B) 4 (C) 2 (D) $\dfrac{1}{4}$

PAGE 221 **4.** The position x (in meters) of an object moving along a horizontal line at time t, $0 \le t \le \dfrac{\pi}{2}$, (in seconds) is given by $x(t) = 6 \sin t + \dfrac{3}{2} t^2 + 8$. What is the velocity of the object when its acceleration is zero?

   (A) 6 m/s   (B) $3 + \pi$ m/s

   (C) $\dfrac{6\sqrt{3} + \pi}{2}$ m/s (D) $\left(3\sqrt{3} - \dfrac{\pi}{2}\right)$ m/s

PAGE 221 **5.** If $y = \sin x$, then $\dfrac{d^{50}}{dx^{50}} \sin x$ equals

   (A) $\sin x$ (B) $-\sin x$ (C) $\cos x$ (D) $-\cos x$

PAGE 220 **6.** If $f(x) = \dfrac{x}{\cos x}$, find $f'\left(\dfrac{\pi}{3}\right)$.

   (A) $2 - \dfrac{2\sqrt{3}}{3}\pi$ (B) $1 + \dfrac{\sqrt{3}}{3}\pi$

   (C) $1 - \dfrac{\sqrt{3}}{3}\pi$ (D) $2 + \dfrac{2\sqrt{3}}{3}\pi$

PAGE 221 **7.** If $y = x - \tan x$, then $\dfrac{dy}{dx}$ equals

   (A) $1 - \sec x \tan x$ (B) $-\tan^2 x$

   (C) $\tan^2 x$ (D) $-\sec^2 x$

PAGE 220 **8.** If $g(x) = e^x \cos x + 2\pi$, then $g'(x) =$

   (A) $e^x - \sin x$ (B) $e^x \cos x - e^x \sin x + 3\pi$

   (C) $e^x \cos x - e^x \sin x$ (D) $e^x \cos x + e^x \sin x$

PAGE 220 **9.** At which of the following numbers x, $0 \le x \le 2\pi$, does the graph of $y = x + \cos x$ have a horizontal tangent line?

   (A) 0 only   (B) $\dfrac{\pi}{2}$ only

   (C) $\dfrac{3\pi}{2}$ only (D) 0 and $\dfrac{\pi}{2}$ only

(continued on the next page)

10. An equation of the line tangent to the graph of $f(x) = \sin x$ at $x = \dfrac{2\pi}{3}$ is

(A) $3x + 6y = 4\pi - 3\sqrt{3}$ (B) $3x + 6y = 2\pi + 3\sqrt{3}$

(C) $6y - 3x = 2\pi - 3\sqrt{3}$ (D) $6y - 3x = 4\pi - 3\sqrt{3}$

Free-Response Question

11. If $y = A \sin x + B \cos x$, find $y'' + y$.

Retain Your Knowledge

Multiple-Choice Questions

1. The graph of a function h is shown in the accompanying figure. Use the graph to determine which of the following statements are true.

 I. h is continuous at the number 1.
 II. h is continuous on the interval $[0, 1]$.
 III. h is continuous from the right at the number 1.

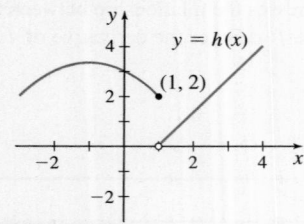

(A) I only   (B) II only

(C) I and II only   (D) II and III only

2. $\lim\limits_{x \to 4^+} \left(\sqrt{x^2 - 16} - x\right)^3 =$

(A) -64 (B) -4 (C) 0 (D) 64

3. $\lim\limits_{x \to 4} \dfrac{x^2 - 5x + 4}{2x - 8} =$

(A) 0 (B) $\dfrac{3}{2}$ (C) 3 (D) ∞

Free-Response Question

4. Find the vertical and horizontal asymptotes, if any, of the graph of the function $f(x) = \dfrac{\cos x}{\ln |x|}$.

CHAPTER 2 PROJECT   **The *Apollo* Lunar Module**

This Project may be done individually or as part of a team.

The Lunar Module (LM) was a small spacecraft that detached from the *Apollo* Command Module and was designed to land on the Moon. Fast and accurate computations were needed to bring the LM from an orbiting speed of about 5500 ft/s to a speed slow enough to land it within a few feet of a designated target on the Moon's surface. The LM carried a 70-lb computer to assist in guiding it successfully to its target. The approach to the target was split into three phases, each of which followed a reference trajectory specified by NASA engineers.* The position and velocity of the LM were monitored by sensors that tracked its deviation from the preassigned path at each moment. Whenever the LM strayed from the reference trajectory, control thrusters were fired to reposition it. In other words, the LM's position and velocity were adjusted by changing its acceleration.

The reference trajectory for each phase was specified by the engineers to have the form

$$r_{ref}(t) = R_T + V_T t + \frac{1}{2}A_T t^2 + \frac{1}{6}J_T t^3 + \frac{1}{24}S_T t^4 \qquad (1)$$

The reference trajectory given in equation (1) is a fourth-degree polynomial, the lowest degree polynomial that has enough free parameters to satisfy all the mission criteria. Now we see that the parameters $R_T = r_{ref}(0)$, $V_T = v_{ref}(0)$, $A_T = a_{ref}(0)$, $J_T = J_{ref}(0)$, and $S_T = S_{ref}(0)$. The five parameters in equation (1) are referred to as the **target parameters**, since they provide the path the LM should follow.

The variable r_{ref} in (1) represents the intended position of the LM at time t before the end of the landing phase. The engineers specified the end of the landing phase to take place at $t = 0$, so that during the phase, t was always negative. Note that the LM was landing in three dimensions, so there were actually three equations like (1). Since each of those equations had this same form, we will work in one dimension, assuming, for example, that r represents the distance of the LM above the surface of the Moon.

1. If the LM follows the reference trajectory, what is the reference velocity $v_{ref}(t)$?

2. What is the reference acceleration $a_{ref}(t)$?
3. The rate of change of acceleration is called **jerk**. Find the reference jerk $J_{ref}(t)$.
4. The rate of change of jerk is called **snap**. Find the reference snap $S_{ref}(t)$.
5. Evaluate $r_{ref}(t)$, $v_{ref}(t)$, $a_{ref}(t)$, $J_{ref}(t)$, and $S_{ref}(t)$ when $t = 0$.

But small variations in propulsion, mass, and countless other variables cause the LM to deviate from the predetermined path. To correct the LM's position and velocity, NASA engineers apply a force to the LM using rocket thrusters. That is, they changed the acceleration. (Remember Newton's second law, $F = ma$.) Engineers modeled the actual trajectory of the LM by

$$r(t) = R_T + V_T t + \frac{1}{2}A_T t^2 + \frac{1}{6}J_A t^3 + \frac{1}{24}S_A t^4 \qquad (2)$$

We know the target parameters for position, velocity, and acceleration. We need to find the actual parameters for jerk and snap to know the proper force (acceleration) to apply.

6. Find the actual velocity $v = v(t)$ of the LM.
7. Find the actual acceleration $a = a(t)$ of the LM.
8. Use equation (2) and the actual velocity found in Problem 6 to express J_A and S_A in terms of R_T, V_T, A_T, $r(t)$, and $v(t)$.
9. Use the results of Problems 7 and 8 to express the actual acceleration $a = a(t)$ in terms of R_T, V_T, A_T, $r(t)$, and $v(t)$.

The result found in Problem 9 provides the acceleration (force) required to keep the LM in its reference trajectory.

10. When riding in an elevator, the sensation one feels just before the elevator stops at a floor is jerk. Would you want jerk to be small or large in an elevator? Explain. Would you want jerk to be small or large on a roller coaster ride? Explain. How would you explain snap?

_____

*A. R. Klumpp, "*Apollo* Lunar-Descent Guidance," MIT Charles Stark Draper Laboratory, R-695, June 1971, http://www.hq.nasa.gov/alsj/ApolloDescentGuidnce.pdf

Chapter Review

THINGS TO KNOW

2.1 Rates of Change and the Derivative

- **Definition** Derivative of a function f at a number c

  Form (1) $f'(c) = \lim\limits_{x \to c} \dfrac{f(x) - f(c)}{x - c}$

  provided the limit exists. (p. 173)

  Three Interpretations of the Derivative

- *Geometric*   If $y = f(x)$, the derivative $f'(c)$ is the slope of the tangent line to the graph of f at the point $(c, f(c))$. (p. 173)

- *Rate of change of a function*   If $y = f(x)$, the derivative $f'(c)$ is the rate of change of f with respect to x at c. (p. 173)
- *Physical*   If the signed distance s from the origin at time t of an object moving on a line is given by the position function $s = f(t)$, the derivative $f'(t_0)$ is the velocity of the object at time t_0. (p. 173)

2.2 The Derivative as a Function
- **Definition of a derivative function**

  Form (2) $f'(x) = \lim\limits_{h \to 0} \dfrac{f(x + h) - f(x)}{h}$

  provided the limit exists. (p. 179)

- **Theorem**   If a function f has a derivative at a number c, then f is continuous at c. (p. 184)
- **Corollary**   If a function f is discontinuous at a number c, then f has no derivative at c. (p. 185)

2.3 The Derivative of a Polynomial Function; The Derivative of $y = e^x$ and $y = \ln x$

- **Leibniz notation** $\dfrac{dy}{dx} = \dfrac{d}{dx} y = \dfrac{d}{dx} f(x)$ (p. 192)

- **Basic derivatives**

$\dfrac{d}{dx} A = 0$ A is a constant (p. 192) $\dfrac{d}{dx} x = 1$ (p. 193)

$\dfrac{d}{dx} e^x = e^x$ (p. 198) $\dfrac{d}{dx} \ln x = \dfrac{1}{x}$ (p. 199)

- *Simple Power Rule* $\dfrac{d}{dx} x^n = nx^{n-1}$, $n \geq 1$, an integer (p. 193)

Properties of Derivatives

- *Sum Rule* (p. 195) $\dfrac{d}{dx}[f + g] = \dfrac{d}{dx} f + \dfrac{d}{dx} g$

$(f + g)' = f' + g'$

- *Difference Rule* (p. 196) $\dfrac{d}{dx}[f - g] = \dfrac{d}{dx} f - \dfrac{d}{dx} g$

$(f - g)' = f' - g'$

- *Constant Multiple Rule* (p. 194)   If k is a constant,

$$\dfrac{d}{dx}[kf] = k \dfrac{d}{dx} f$$
$$(kf)' = k \cdot f'$$

2.4 Differentiating the Product and the Quotient of Two Functions; Higher-Order Derivatives

Properties of Derivatives

- *Product Rule* (p. 204) $\dfrac{d}{dx}(fg) = f\left(\dfrac{d}{dx} g\right) + \left(\dfrac{d}{dx} f\right) g$

$(fg)' = fg' + f'g$

- *Quotient Rule* (p. 206) $\dfrac{d}{dx}\left(\dfrac{f}{g}\right) = \dfrac{\left(\dfrac{d}{dx} f\right) g - f\left(\dfrac{d}{dx} g\right)}{g^2}$

$$\left(\dfrac{f}{g}\right)' = \dfrac{f'g - fg'}{g^2}$$

provided $g(x) \neq 0$

- *Reciprocal Rule* (p. 207) $\dfrac{d}{dx}\left(\dfrac{1}{g}\right) = -\dfrac{\dfrac{d}{dx} g}{g^2}$

$$\left(\dfrac{1}{g}\right)' = -\dfrac{g'}{g^2}$$

provided $g(x) \neq 0$

- *Simple Power Rule* $\dfrac{d}{dx} x^n = nx^{n-1}$, n an integer (p. 208)

- *Higher-order derivatives*   See Table 3 (p. 209)
- *Position Function* $s = s(t)$ (p. 210)
- *Velocity* $v = v(t) = \dfrac{ds}{dt}$ (p. 210)
- *Acceleration* $a = a(t) = \dfrac{dv}{dt} = \dfrac{d^2 s}{dt^2}$ (p. 210)

2.5 The Derivative of the Trigonometric Functions

Basic Derivatives

$\dfrac{d}{dx} \sin x = \cos x$ (p. 218) $\dfrac{d}{dx} \sec x = \sec x \tan x$ (p. 221)

$\dfrac{d}{dx} \cos x = -\sin x$ (p. 219) $\dfrac{d}{dx} \csc x = -\csc x \cot x$ (p. 221)

$\dfrac{d}{dx} \tan x = \sec^2 x$ (p. 221) $\dfrac{d}{dx} \cot x = -\csc^2 x$ (p. 221)

OBJECTIVES

Preparing for the **AP® Exam**

Section	You should be able to ...	Examples	Review Exercises	AP® Review Problems
2.1	1 Find equations for the tangent line and the normal line to the graph of a function (p. 168)	1	67–70	7, 10
	2 Find the rate of change of a function (p. 169)	2, 3	1, 2, 73 (a)	6
	3 Find average velocity and instantaneous velocity (p. 170)	4, 5	71(a), (b); 72(a), (b)	12
	4 Find the derivative of a function at a number (p. 173)	6–9	3–8, 75	5, 11
2.2	1 Define the derivative function (p. 179)	1–3	9–12	2, 13
	2 Graph the derivative function (p. 181)	4, 5	9–12, 15–18	
	3 Identify where a function is not differentiable (p. 182)	6–8	13, 14, 75	4
	4 Explain the relationship between differentiability and continuity (p. 184)	9, 10	13, 14, 75	4
2.3	1 Differentiate a constant function (p. 192)	1		
	2 Differentiate a power function; the simple power rule (p. 192)	2, 3	19–22	
	3 Differentiate the sum and the difference of two functions (p. 195)	4–6	23–26, 33, 34, 40, 51, 52, 67	6, 8, 12
	4 Differentiate the exponential function $y = e^x$ and the natural logarithm function $y = \ln x$ (p. 197)	7, 8	44, 45, 53, 54, 56, 59, 69	6, 7, 9

2.4	1 Differentiate the product of two functions (p. 203)	1, 2	27, 28, 36, 46, 47–50, 53–56, 60	6, 7, 9
	2 Differentiate the quotient of two functions (p. 206)	3–6	29–35, 37–43, 57–59, 68, 73, 74	3, 10
	3 Find higher-order derivatives (p. 208)	7, 8	61–66, 71, 72	8, 12
	4 Find the acceleration of an object moving on a line (p. 210)	9	71, 72	8, 12
2.5	1 Differentiate trigonometric functions (p. 218)	1–6	49–60, 70	1, 9, 11

REVIEW EXERCISES

In Problems 1 and 2, use a definition of the derivative to find the rate of change of f at the indicated numbers.

1. $f(x) = \sqrt{x}$ at **(a)** $c = 1$ **(b)** $c = 4$
    (c) c any positive real number

2. $f(x) = \dfrac{2}{x-1}$ at **(a)** $c = 0$ **(b)** $c = 2$
    (c) c any real number, $c \neq 1$

In Problems 3–8, use a definition of the derivative to find the derivative of each function at the given number.

3. $F(x) = 3x + 6$ at -2 **4.** $f(x) = 8x^2 + 1$ at -1

5. $f(x) = 3x^2 + 5x$ at 0 **6.** $f(x) = \dfrac{3}{x}$ at 1

7. $f(x) = \sqrt{4x+1}$ at 0 **8.** $f(x) = \dfrac{x+1}{2x-3}$ at 1

In Problems 9–12, use a definition of the derivative to find the derivative of each function. Graph f and f' on the same set of axes.

9. $f(x) = x - 6$ **10.** $f(x) = 7 - 3x^2$

11. $f(x) = \dfrac{1}{2x^3}$ **12.** $f(x) = \pi$

In Problems 13 and 14, determine whether the function f has a derivative at c. If it does, find the derivative. If it does not, explain why. Graph each function.

13. $f(x) = |x^3 - 1|$ at $c = 1$

14. $f(x) = \begin{cases} 4 - 3x^2 & \text{if } x \leq -1 \\ -x^3 & \text{if } x > -1 \end{cases}$ at $c = -1$

In Problems 15 and 16, determine whether the graphs represent a function f and its derivative f'. If they do, indicate which is the graph of f and which is the graph of f'.

15.

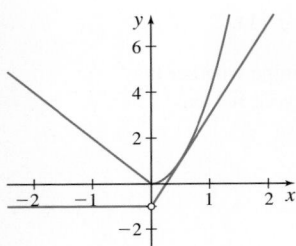

16.

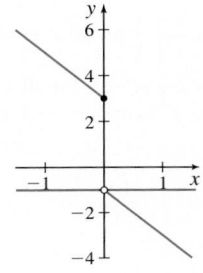

17. Use the information in the graph of $y = f(x)$ to sketch the graph of $y = f'(x)$.

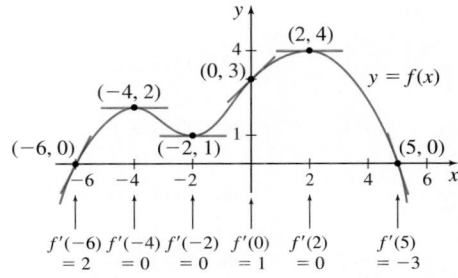

18. Match the graph of $y = f(x)$ with the graph of its derivative.

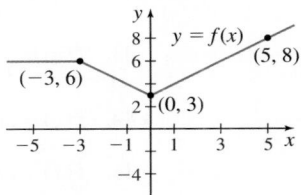

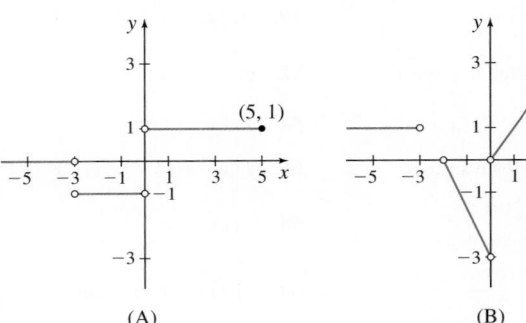

(A)   (B)

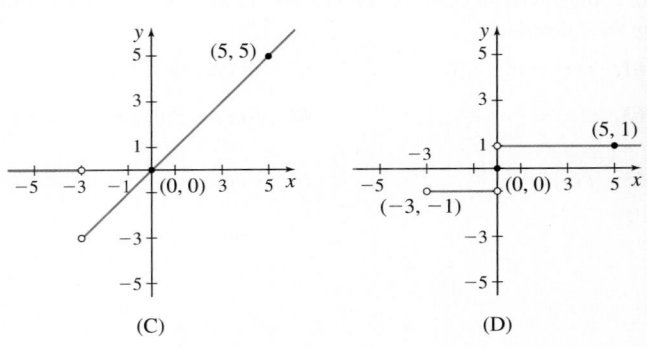

(C)   (D)

In Problems 19–60, find the derivative of each function. Treat a and b, if present, as constants.

19. $f(x) = x^5$

20. $f(x) = ax^3$

21. $f(x) = \dfrac{x^4}{4}$

22. $f(x) = -6x^2$

23. $f(x) = 3x^2 - 4x$

24. $f(x) = 2x^3 + \dfrac{2}{3}x^2 - 6x + 8$

25. $F(x) = 7(x^2 - 4)$

26. $F(x) = \dfrac{5(x+6)}{7}$

27. $f(x) = 5(x^2 - 3x)(x - 6)$

28. $f(x) = (2x^3 + x)(x^2 - 5)$

29. $f(x) = \dfrac{6x^4 - 9x^2}{3x^3}$

30. $f(x) = \dfrac{2x + 2}{5x - 3}$

31. $f(x) = \dfrac{7x}{x - 5}$

32. $f(x) = 2x^{-12}$

33. $f(x) = 2x^2 - 5x^{-2}$

34. $f(x) = 2 + \dfrac{3}{x} + \dfrac{4}{x^2}$

35. $f(x) = \dfrac{a}{x} - \dfrac{b}{x^3}$

36. $f(x) = (x^3 - 1)^2$

37. $f(x) = \dfrac{3}{(x^2 - 3x)^2}$

38. $f(x) = \dfrac{x^2}{x + 1}$

39. $s(t) = \dfrac{t^3}{t - 2}$

40. $f(x) = 3x^{-2} + 2x^{-1} + 1$

41. $F(z) = \dfrac{1}{z^2 + 1}$

42. $f(v) = \dfrac{v - 1}{v^2 + 1}$

43. $g(z) = \dfrac{1}{1 - z + z^2}$

44. $f(x) = 3e^x + x^2$

45. $s(t) = 1 - e^t$

46. $f(x) = ae^x(2x^2 + 7x)$

47. $f(x) = (1 + x)\ln x$

48. $f(x) = 2x \ln x + e^x \tan x$

49. $f(x) = x \sin x$

50. $s(t) = \cos^2 t$

51. $G(u) = \tan u + \sec u$

52. $g(v) = \sin v - \dfrac{1}{3}\cos v$

53. $f(x) = e^x \sin x$

54. $f(x) = e^x \csc x$

55. $f(x) = 2 \sin x \cos x$

56. $f(x) = (e^x + b)\cos x$

57. $f(x) = \dfrac{\sin x}{\csc x}$

58. $f(x) = \dfrac{1 - \cot x}{1 + \cot x}$

59. $f(\theta) = \dfrac{\cos \theta}{2e^\theta}$

60. $f(\theta) = 4\theta \cot \theta \tan \theta$

In Problems 61–66, find the first derivative and the second derivative of each function.

61. $f(x) = (5x + 3)^2$

62. $f(x) = xe^x$

63. $g(u) = \dfrac{u}{2u + 1}$

64. $F(x) = e^x(\sin x + 2 \cos x)$

65. $f(u) = \dfrac{\cos u}{e^u}$

66. $F(x) = \dfrac{\sin x}{x}$

In Problems 67–70, for each function:

(a) Find an equation of the tangent line to the graph of the function at the indicated point.

(b) Find an equation of the normal line to the function at the indicated point.

(c) Graph the function, the tangent line, and the normal line on the same screen.

67. $f(x) = 2x^2 - 3x + 7$ at $(-1, 12)$

68. $y = \dfrac{x^2 + 1}{2x - 1}$ at $\left(2, \dfrac{5}{3}\right)$

69. $f(x) = x^2 - e^x$ at $(0, -1)$

70. $s(t) = 1 + 2 \sin t$ at $(\pi, 1)$

71. Motion on a Line   As an object in moves on a line, its signed distance s (in meters) from the origin at time t (in seconds) is given by the position function

$$s = f(t) = t^2 - 6t$$

(a) Find the average velocity of the object from 0 to 5 s.

(b) Find the velocity at $t = 0$, at $t = 5$, and at any time t.

(c) Find the acceleration at any time t.

72. Motion on a Line   As an object moves on a line, its signed distance s from the origin at time t is given by the position function $s(t) = t - t^2$, where s is in centimeters and t is in seconds.

(a) Find the average velocity of the object from 1 to 3 s.

(b) Find the velocity of the object at $t = 1$ s and $t = 3$ s.

(c) What is its acceleration at $t = 1$ and $t = 3$?

73. Business   The price p in dollars per pound when x pounds of a commodity are demanded is modeled by the function

$$p(x) = \dfrac{10{,}000}{5x + 100} - 5$$

when between 0 and 90 lb are demanded (purchased).

(a) Find the rate of change of price with respect to demand.

(b) What is the revenue function R? (Recall, revenue R equals price times amount purchased.)

(c) What is the marginal revenue R' at $x = 10$ and at $x = 40$ lb?

74. If $f(x) = \dfrac{x - 1}{x + 1}$ for all $x \neq -1$, find $f'(1)$.

75. If $f(x) = 2 + |x - 3|$ for all x, determine whether the derivative f' exists at $x = 3$. Justify your reason.

Break It Down

Let's take a closer look at **AP® Practice Problem 12** *from Section 2.2 on page 191.*

12. Oil is leaking from a tank. The amount of oil, in gallons, in the tank is given by $G(t) = 4000 - 3t^2$, where t, $0 \le t \le 24$ is the number of hours past midnight.

  (a) Find $G'(5)$ using the definition of the derivative.

  (b) Using appropriate units, interpret the meaning of $G'(5)$ in the context of the problem.

Step 1	Identify the underlying structure and the related concepts.	The problem is asking for $G'(5)$, the **derivative** of G at the number 5.
Step 2	Determine the appropriate math rule or procedure.	Because we want the derivative of G at a number, we use Form (1) of the Definition of a Derivative (see p. 179).
Step 3	Apply the math rule or procedure.	Using Form (1), $$G'(5) = \lim_{t \to 5} \frac{G(t) - G(5)}{t - 5} = \lim_{t \to 5} \frac{(4000 - 3t^2) - (4000 - 3 \cdot 5^2)}{t - 5}$$ $$= \lim_{t \to 5} \frac{-3(t^2 - 25)}{t - 5} = \lim_{t \to 5} \frac{-3(t - 5)(t + 5)}{t - 5}$$ $$= \lim_{t \to 5}[-3(t + 5)] = -30$$
Step 4	Clearly communicate your answer.	(a) $G'(5) = -30$ (b) A derivative is a rate of change. In this problem, $G'(t)$ equals the rate of change of G with respect t, that is, the rate of change of the amount of oil, in gallons, with respect to the time in hours. Because $G'(5) = -30$, we say the amount of oil in the tank is decreasing at the rate of 30 gallons per hour when $t = 5$ hours past midnight, or at 5AM.

AP® Review Problems: Chapter 2

Multiple-Choice Questions

▶ **1.** If $f(x) = \sec x$, then $f'\left(\dfrac{\pi}{4}\right) =$

  (A) $\dfrac{\sqrt{2}}{2}$ (B) 2   (C) 1   (D) $\sqrt{2}$

▶ **2.** If a function f is differentiable at c, then $f'(c)$ is given by

  I. $\lim_{x \to c} \dfrac{f(x) - f(c)}{x - c}$

  II. $\lim_{x \to c} \dfrac{f(x + h) - f(x)}{h}$

  III. $\lim_{h \to 0} \dfrac{f(c + h) - f(c)}{h}$

  (A) I only   (B) III only

  (C) I and II only   (D) I and III only

▶ **3.** If $y = \dfrac{3}{x^2 - 5}$, then $\dfrac{dy}{dx} =$

  (A) $\dfrac{6x}{(x^2 - 5)^2}$ (B) $-\dfrac{6x}{(x^2 - 5)^2}$

  (C) $\dfrac{6x}{x^2 - 5}$ (D) $\dfrac{2x}{(x^2 - 5)^2}$

▶ **4.** The graph of the function f is shown below. Which statement about the function is true?

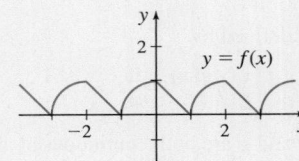

  (A) f is differentiable everywhere.

  (B) $0 \le f'(x) \le 1$, for all real numbers.

  (C) f is continuous everywhere.

  (D) f is an even function.

▶ **5.** The table displays select values of a differentiable function f. What is an approximate value of $f'(2)$?

x	1.996	1.998	2.002	2.004
$f(x)$	3.168	3.181	3.207	3.220

  (A) 6.5   (B) 0.154   (C) 0.013   (D) 1.5

6. If $y = \ln x + xe^x + 6$, what is the instantaneous rate of change of y with respect to x at $x = 5$?

(A) $5 + 6e^5$ (B) $\dfrac{1}{5} + 5e$

(C) $5 + 5e^5$ (D) $6e^5 + \dfrac{1}{5}$

7. An equation of the line tangent to the graph of $f(x) = 3xe^x + 5$ at $x = 0$ is

(A) $y = 3x + 5$ (B) $y = -\dfrac{1}{3}x + 5$

(C) $y = \dfrac{1}{3}x + 5$ (D) $y = -3x + 5$

8. An object moves along a horizontal line so that its position at time t is $x(t) = t^4 - 6t^3 - 2t - 1$. At what time t is the acceleration of the object zero?

(A) at 0 only   (B) at 1 only

(C) at 3 only   (D) at 0 and 3 only

9. If $f(x) = e^x(\sin x + \cos x)$, then $f'(x) =$

(A) $2e^x(\cos x + \sin x)$ (B) $e^x \cos x$

(C) $2e^x \cos x$ (D) $e^x(\cos^2 x - \sin^2 x)$

10. Find an equation of the line tangent to the graph of $f(x) = \dfrac{x+3}{x^2+2}$ at $x = 1$.

(A) $5x + 9y = 17$ (B) $9y - 5x = 7$

(C) $5x + 3y = 9$ (D) $5x + 9y = 7$

11. $\displaystyle\lim_{x \to \frac{\pi}{4}} \dfrac{\tan x - 1}{x - \frac{\pi}{4}} =$

(A) 0   (B) -1

(C) 2   (D) Does not exist.

Free-Response Questions

12. An object moves on a line according to the position function $s = 2t^3 - 15t^2 + 24t + 3$, where t is measured in minutes and s in meters.

(a) When is the velocity of the object 0?

(b) Find the object's acceleration when $t = 3$.

13. Find the value of the limit below and specify the function f for which this is the derivative.

$$\lim_{h \to 0} \frac{[4 - 2(x+h)]^2 - (4-2x)^2}{h}$$

AP® Cumulative Review Problems: Chapters 1–2   Preparing for the AP® Exam

Multiple-Choice Questions

1. $\displaystyle\lim_{x \to 4} \dfrac{x-4}{4-x} =$

(A) -4 (B) -1 (C) 0 (D) The limit does not exist.

2. $\displaystyle\lim_{x \to 0} \dfrac{3x + \sin x}{2x} =$

(A) 0 (B) 1 (C) 2 (D) The limit does not exist.

3. Let h be defined by

$$h(x) = \begin{cases} f(x) \cdot g(x) & \text{if } x \le 1 \\ k + x & \text{if } x > 1 \end{cases}$$

where f and g are both continuous at all real numbers. If $\lim_{x \to 1} f(x) = 2$ and $\lim_{x \to 1} g(x) = -2$, then for what number k is h continuous?

(A) -5 (B) -4 (C) -2 (D) 2

4. Which function has the horizontal asymptotes $y = 1$ and $y = -1$?

(A) $f(x) = \dfrac{2}{\pi} \tan^{-1} x$ (B) $f(x) = e^{-x} + 1$

(C) $f(x) = \dfrac{1 - x^2}{1 + x^2}$ (D) $f(x) = \dfrac{2x^2 - 1}{2x^2 + x}$

5. Suppose the function f is continuous at all real numbers and $f(-2) = 1$ and $f(5) = -3$. Suppose the function g is also continuous at all real numbers and $g(x) = f^{-1}(x)$ for all x. The Intermediate Value Theorem guarantees that

(A) $g(c) = 2$ for at least one c between -3 and 1.

(B) $g(c) = 0$ for at least one c between -2 and 5.

(C) $f(c) = 0$ for at least one c between -3 and 1.

(D) $f(c) = 2$ for at least one c between -2 and 5.

6. The line $x = c$ is a vertical asymptote to the graph of the function f. Which of the following statements cannot be true?

(A) $\displaystyle\lim_{x \to c} f(x) = \infty$ (B) $\displaystyle\lim_{x \to \infty} f(x) = c$

(C) $f(c)$ is not defined.   (D) f is continuous at $x =$

7. The position function of an object moving along a straight line is $s(t) = \dfrac{1}{15}t^3 - \dfrac{1}{2}t^2 + 5t^{-1}$. What is the object's acceleration at $t = 5$?

(A) $-\dfrac{27}{25}$ (B) $-\dfrac{1}{5}$ (C) $\dfrac{1}{5}$ (D) $\dfrac{27}{25}$

▶ 8. If the function

$$f(x) = \begin{cases} 2ax^2 + bx - 1 & \text{if } x \le 3 \\ bx^2 + bx - a & \text{if } x > 3 \end{cases}$$

is continuous for all real numbers x, then

(A) $19a - 15b = 1$ (B) $18a - 9b = 1$

(C) $19a - 9b = 1$ (D) $19a + 15b = 1$

▶ 9. Find the slope of the line tangent to the graph of $f(x) = 2x \ln x$ at the point $(1, 0)$.

(A) 0   (B) 1   (C) 2   (D) 4

▶ 10. An object moving on a line is modeled by the position function

$$s(t) = 3t^4 - 8t^3 - 6t^2 + 24t \qquad t > 0$$

where s is in feet (ft) and t is in seconds (s). Find the acceleration of the object when its velocity is zero.

(A) -24 ft/s^2, 36 ft/s^2, and 72 ft/s^2 only

(B) 36 ft/s^2 only

(C) 36 ft/s^2 and 72 ft/s^2 only

(D) -24 ft/s^2 and 36 ft/s^2 only

Free-Response Question

▶ 11. The velocity of an object moving on a line is modeled by a differentiable function v, where v is measured in meters and $t \ge 0$ is measured in minutes. The velocity of the object is observed at several times t and posted in the table below.

t	0	2	3	6	8
$v(t)$	0	4	13	7	5

(a) Using the data in the table, approximate $v'(5)$.

(b) Interpret the meaning of $v'(5)$. Be sure to include the correct units.

CHAPTER

3

The Derivative of Composite, Implicit, and Inverse Functions

Ayhan Altun/Getty Images

World Population Growth

In the late 1700s, Thomas Malthus predicted that population growth, if left unchecked, would outstrip food resources and lead to mass starvation. His prediction turned out to be incorrect, since in 1800 world population had not yet reached 1 billion and currently world population is in excess of 8 billion. Malthus' most dire predictions did not come true largely because improvements in food production made his models inaccurate. On the other hand, his population growth model is still used today, and population growth remains an important issue in human progress. For example, many people are crowded into growing urban areas, such as Istanbul, Turkey (pictured above), which has a population of over 15.8 million and a density of over 7650 people per square mile.

 In the Chapter 3 Project on page 278, we explore a basic model to predict world population and examine its accuracy.

Luis
Financial Analyst

As a financial
analyst, I use
mathematical
models to
assess investment risk and
return for my firm. The technical
skills I developed from taking
AP® Calculus allow me to
effectively evaluate the financial
health of complex and diverse
companies.

Yolanda Studios

AP® EXAM TIP

BREAK IT DOWN: As you work through the
problems in this chapter, remember to follow
these steps:

Step 1 Identify the underlying structure and
related concepts.

Step 2 Determine the appropriate math rule
or procedure.

Step 3 Apply the math rule or procedure.

Step 4 Clearly communicate your answer.

On page 281, see how we've used these steps
to solve Section 3.4 AP® Problem 13
on page 277.

NEED TO REVIEW? Composite functions
and their properties are discussed in Section
P.3, pp. 29–31.

IN WORDS If you think of the function
$y = f(u)$ as the outside function and the
function $u = g(x)$ as the inside function, then
the derivative of $f \circ g$ is the derivative of the
outside function, evaluated at the inside
function, times the derivative of the inside
function. That is,

$$\frac{d}{dx}(f \circ g)(x) = f'(g(x)) \cdot g'(x)$$

In this chapter, we continue exploring properties of derivatives, beginning with the Chain Rule, which allows us to differentiate composite functions, implicitly defined functions, and inverse functions. The Chain Rule also provides the means to establish derivative formulas for the exponential and logarithmic functions. ∎

3.1 The Chain Rule

OBJECTIVES *When you finish this section, you should be able to:*

1 **Differentiate a composite function (p. 235)**
2 **Differentiate $y = a^x$, $a > 0$, $a \neq 1$ (p. 239)**
3 **Use the Power Rule for Functions to find a derivative (p. 240)**
4 **Use the Chain Rule for multiple composite functions (p. 242)**

Using the differentiation rules developed in Chapter 2, it would be difficult to differentiate the function

$$F(x) = (x^3 - 4x + 1)^{100}$$

We can represent F as the composite function of $f(u) = u^{100}$ and $u = g(x) = x^3 - 4x + 1$; that is, $F(x) = f(g(x)) = (f \circ g)(x) = (x^3 - 4x + 1)^{100}$. In this section, we derive the *Chain Rule*, a result that enables us to find the derivative of a composite function. We use the Chain Rule to find the derivative in applications involving functions such as $A(t) = 102 - 90e^{-0.21t}$ (market penetration, Problem 101) and $v(t) = \frac{mg}{k}(1 - e^{-kt/m})$ (the terminal velocity of a falling object, Problem 103).

① Differentiate a Composite Function

Suppose $y = (f \circ g)(x) = f(g(x))$ is a composite function, where $y = f(u)$ is a differentiable function of u, and $u = g(x)$ is a differentiable function of x. What is the derivative of $(f \circ g)(x)$? It turns out that the derivative of the composite function $f \circ g$ is the product of the derivatives $f'(u) = f'(g(x))$ and $g'(x)$.

THEOREM Chain Rule

If a function g is differentiable at x and a function f is differentiable at $g(x)$, then the composite function $f \circ g$ is differentiable at x and

$$(f \circ g)'(x) = f'(g(x)) \cdot g'(x)$$

For differentiable functions $y = f(u)$ and $u = g(x)$, the Chain Rule, in Leibniz notation, takes the form

$$\frac{dy}{dx} = \frac{dy}{du} \cdot \frac{du}{dx}$$

where in $\frac{dy}{du}$ we substitute $u = g(x)$.

Partial Proof The Chain Rule is proved using the definition of a derivative. First observe that if x changes by a small amount Δx, the corresponding change in $u = g(x)$ is Δu and Δu depends on Δx. Also,

$$g'(x) = \frac{du}{dx} = \lim_{\Delta x \to 0} \frac{\Delta u}{\Delta x}$$

Since $y = f(u)$, the change Δu, which could equal 0, causes a change Δy. Suppose Δu is never 0. Then

$$f'(u) = \frac{dy}{du} = \lim_{\Delta u \to 0} \frac{\Delta y}{\Delta u} \qquad \Delta u \neq 0$$

To find $\dfrac{dy}{dx}$, write

$$\frac{dy}{dx} = \lim_{\Delta x \to 0} \frac{\Delta y}{\Delta x} \underset{\substack{\uparrow \\ \Delta u \neq 0}}{=} \lim_{\Delta x \to 0} \left(\frac{\Delta y}{\Delta x} \cdot \frac{\Delta u}{\Delta u} \right) = \lim_{\Delta x \to 0} \left(\frac{\Delta y}{\Delta u} \cdot \frac{\Delta u}{\Delta x} \right)$$

$$= \left(\lim_{\Delta x \to 0} \frac{\Delta y}{\Delta u} \right) \left(\lim_{\Delta x \to 0} \frac{\Delta u}{\Delta x} \right)$$

Since the differentiable function $u = g(x)$ is continuous, $\Delta u \to 0$ as $\Delta x \to 0$, so in the first factor we can replace $\Delta x \to 0$ by $\Delta u \to 0$. Then

$$\frac{dy}{dx} = \left(\lim_{\Delta u \to 0} \frac{\Delta y}{\Delta u} \right) \left(\lim_{\Delta x \to 0} \frac{\Delta u}{\Delta x} \right) = \frac{dy}{du} \cdot \frac{du}{dx}$$

This proves the Chain Rule if Δu is never 0. To complete the proof, we need to consider the case when Δu may be 0. (This part of the proof is given in Appendix B.) ∎

EXAMPLE 1   **Differentiating a Composite Function**

Find the derivative of:

(a) $y = (x^3 - 4x + 1)^{100}$ **(b)** $y = \cos\left(3x - \dfrac{\pi}{4}\right)$

Solution

(a) In the composite function $y = (x^3 - 4x + 1)^{100}$, let $u = x^3 - 4x + 1$. Then $y = u^{100}$. Now $\dfrac{dy}{du}$ and $\dfrac{du}{dx}$ are

$$\frac{dy}{du} = \frac{d}{du} u^{100} = 100u^{99} = 100(x^3 - 4x + 1)^{99}$$
$$\underset{u = x^3 - 4x + 1}{\uparrow}$$

and

$$\frac{du}{dx} = \frac{d}{dx}(x^3 - 4x + 1) = 3x^2 - 4$$

Now use the Chain Rule to find $\dfrac{dy}{dx}$.

$$\frac{dy}{dx} = \frac{dy}{du} \cdot \frac{du}{dx} = 100(x^3 - 4x + 1)^{99}(3x^2 - 4)$$
$$\underset{\text{Chain Rule}}{\uparrow}$$

(b) In the composite function $y = \cos\left(3x - \dfrac{\pi}{4}\right)$, let $u = 3x - \dfrac{\pi}{4}$. Then $y = \cos u$ and

$$\frac{dy}{du} = \frac{d}{du} \cos u = -\sin u = -\sin\left(3x - \frac{\pi}{4}\right) \quad \text{and} \quad \frac{du}{dx} = \frac{d}{dx}\left(3x - \frac{\pi}{4}\right) = 3$$
$$\underset{u = 3x - \frac{\pi}{4}}{\uparrow}$$

Now use the Chain Rule.

$$\frac{dy}{dx} \underset{\underset{\text{Chain Rule}}{\uparrow}}{=} \frac{dy}{du} \cdot \frac{du}{dx} = -\sin\left(3x - \frac{\pi}{4}\right) \cdot 3 = -3\sin\left(3x - \frac{\pi}{4}\right)$$ ∎

NOW WORK   Problems **9** and **37** and AP® Practice Problem **2.**

EXAMPLE 2   Differentiating a Composite Function

Find y' if:

(a) $y = e^{x^2 - 4}$ **(b)** $y = \sin(4e^x)$

Solution

(a) For $y = e^{x^2 - 4}$, let $u = x^2 - 4$. Then $y = e^u$ and

$$\frac{dy}{du} = \frac{d}{du}e^u = e^u \underset{\underset{u = x^2 - 4}{\uparrow}}{=} e^{x^2 - 4} \quad \text{and} \quad \frac{du}{dx} = \frac{d}{dx}(x^2 - 4) = 2x$$

Using the Chain Rule, we get

$$y' = \frac{dy}{dx} = \frac{dy}{du} \cdot \frac{du}{dx} = e^{x^2 - 4} \cdot 2x = 2xe^{x^2 - 4}$$

(b) For $y = \sin(4e^x)$, let $u = 4e^x$. Then $y = \sin u$ and

$$\frac{dy}{du} = \frac{d}{du}\sin u = \cos u \underset{\underset{u = 4e^x}{\uparrow}}{=} \cos(4e^x) \quad \text{and} \quad \frac{du}{dx} = \frac{d}{dx}(4e^x) = 4e^x$$

Using the Chain Rule, we get

$$y' = \frac{dy}{dx} = \frac{dy}{du} \cdot \frac{du}{dx} = \cos(4e^x) \cdot 4e^x = 4e^x \cos(4e^x)$$ ∎

For composite functions $y = f(u(x))$, where f is an exponential, logarithmic, or trigonometric function, the Chain Rule simplifies finding y'. For example,

- If $y = e^{u(x)}$, where $u = u(x)$ is a differentiable function of x, then by the Chain Rule

$$\frac{dy}{dx} = \frac{dy}{du} \cdot \frac{du}{dx} = e^{u(x)}\frac{du}{dx}$$

That is,

$$\boxed{\frac{d}{dx}e^{u(x)} = e^{u(x)}\frac{du}{dx}}$$

NEED TO REVIEW? The derivatives of the exponential function and the natural logarithm function are discussed on pp. 198 and 199 in Section 2.3.

- If $y = \ln u(x)$, where $u = u(x)$ is a differentiable function of x, then by the Chain Rule,

$$\frac{dy}{dx} = \frac{dy}{du} \cdot \frac{du}{dx} = \frac{1}{u(x)} \cdot \frac{du}{dx}$$

That is,

$$\boxed{\frac{d}{dx}\ln u(x) = \frac{1}{u(x)} \cdot \frac{du}{dx}}$$

• If $u = u(x)$ is a differentiable function,

$$\frac{d}{dx}\sin u(x) = \cos u(x)\frac{du}{dx} \qquad \frac{d}{dx}\sec u(x) = \sec u(x)\tan u(x)\frac{du}{dx}$$

$$\frac{d}{dx}\cos u(x) = -\sin u(x)\frac{du}{dx} \qquad \frac{d}{dx}\csc u(x) = -\csc u(x)\cot u(x)\frac{du}{dx}$$

$$\frac{d}{dx}\tan u(x) = \sec^2 u(x)\frac{du}{dx} \qquad \frac{d}{dx}\cot u(x) = -\csc^2 u(x)\frac{du}{dx}$$

NEED TO REVIEW? The derivatives of the trigonometric functions are discussed on pp. 218 and 219 in Section 2.5.

NOW WORK Problem **41** and AP® Practice Problems **3, 4, 10, 12,** and **15.**

EXAMPLE 3 **Finding an Equation of a Tangent Line**

Find an equation of the tangent line to the graph of $y = 5e^{4x}$ at the point $(0, 5)$.

Solution

The slope of the tangent line to the graph of $y = f(x)$ at the point $(0, 5)$ is $f'(0)$.

$$f'(x) = \frac{d}{dx}(5e^{4x}) = 5\underset{\uparrow}{\frac{d}{dx}}e^{4x} = 5e^{4x}\cdot\underset{\uparrow}{\frac{d}{dx}}(4x) = 5e^{4x}\cdot 4 = 20e^{4x}$$

$$\underset{\substack{\text{Constant}\\\text{Multiple Rule}}}{} \qquad \underset{u=4x;\ \frac{d}{dx}e^u = e^u\frac{du}{dx}}{}$$

$m_{\tan} = f'(0) = 20e^0 = 20$. Using the point-slope form of a line,

$$y - 5 = 20(x - 0) \qquad y - y_0 = m_{\tan}(x - x_0).$$

$$y = 20x + 5$$

The graph of $y = 5e^{4x}$ and the line $y = 20x + 5$ are shown in Figure 1.

[Figure: graph showing $y = 5e^{4x}$ and $y = 20x + 5$, with the point $(0, 5)$ marked; axes labeled from -1 to 1 on x-axis and 10, 20 on y-axis]

Figure 1

NOW WORK Problem **77** and AP® Practice Problems **5, 7,** and **16.**

EXAMPLE 4 **Application: Carbon-14 Dating**

All carbon on Earth contains some carbon-14, which is radioactive and exists in a fixed ratio with some nonradioactive carbon-12. When a living organism dies, the carbon-14 begins to decay at a fixed rate. The formula $P(t) = 100e^{-0.000121t}$ gives the percentage of carbon-14 present at time t years. Notice that when $t = 0$, the percentage of carbon-14 present is 100%. When the preserved bodies of 15-year-old La Doncella and two younger children were found in Argentina in 2005, 93.5% of the carbon-14 remained in their bodies, indicating that the three had died about 550 years earlier.

(a) What is the rate of change of the percentage of carbon-14 present in a 550-year-old fossil?

(b) What is the rate of change of the percentage of carbon-14 present in a 2000-year-old fossil?

Solution

(a) The rate of change of P is given by its derivative

$$P'(t) = \frac{d}{dt}\left(100e^{-0.000121t}\right) = 100\underset{\uparrow}{\left(-0.000121e^{-0.000121t}\right)} = -0.0121e^{-0.000121t}$$

$$\underset{\frac{d}{dx}e^{u(x)} = e^{u(x)}\frac{du}{dx}}{}$$

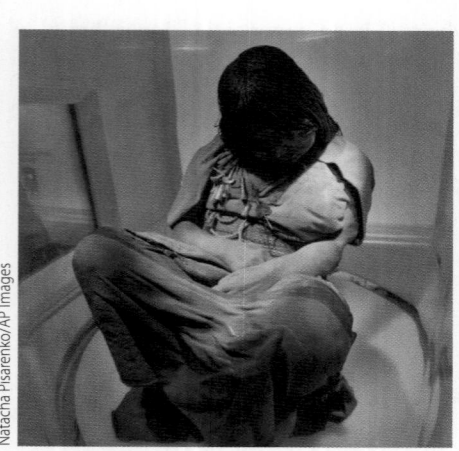

The perfectly preserved mummy of La Doncella, a 15-year-old girl, is displayed in a museum in Salta, Argentina.

At $t = 550$ years,

$$P'(550) = -0.0121e^{-0.000121(550)} \approx -0.0113$$

The percentage of carbon-14 present in a 550-year-old fossil is decreasing at the rate of 0.0113% per year.

(b) When $t = 2000$ years, the rate of change is

$$P'(2000) = -0.0121e^{-0.000121(2000)} \approx -0.0095$$

The percentage of carbon-14 present in a 2000-year-old fossil is decreasing at the rate of 0.0095% per year. ∎

NOW WORK   Problem 99.

When we stated the Chain Rule, we expressed it two ways: using prime notation and using Leibniz notation. In each example so far, we have used Leibniz notation. But when solving numerical problems, using prime notation is often easier. In this form, the Chain Rule states that if a function g is differentiable at x_0 and a function f is differentiable at $g(x_0)$, then

$$\boxed{(f \circ g)'(x_0) = f'(g(x_0))g'(x_0)}$$

EXAMPLE 5   **Differentiating a Composite Function**

Suppose $h = f \circ g$. Use the data in the table to find $h'(1)$.

x	$f(x)$	$f'(x)$	$g(x)$	$g'(x)$
1	2	1	2	-3
2	$\dfrac{1}{2}$	-4	4	5

Solution

Based on the Chain Rule using prime notation, we have

$$h'(x_0) = (f \circ g)'(x_0) = f'(g(x_0))\, g'(x_0)$$

When $x_0 = 1$,

$$h'(1) = f'(g(1)) \cdot g'(1) = f'(2) \cdot (-3) = (-4)(-3) = 12$$
$$\underset{\substack{\uparrow \\ g(1)=2;\ g'(1)=-3}}{} \qquad \underset{\substack{\uparrow \\ f'(2)=-4}}{}$$

∎

NOW WORK   Problem 81 and AP® Practice Problem 13.

② Differentiate $y = a^x$, $a > 0$, $a \neq 1$

NEED TO REVIEW? Properties of logarithms are discussed in Appendix A.1, pp. A-11 to A-12.

The Chain Rule allows us to establish a formula for differentiating an exponential function $y = a^x$ for any base $a > 0$ and $a \neq 1$. We start with the following property of logarithms:

$$a^x = e^{\ln a^x} = e^{x \ln a} \qquad a > 0, a \neq 1$$

Then

$$\frac{d}{dx}a^x = \frac{d}{dx}e^{x \ln a} = e^{x \ln a}\frac{d}{dx}(x \ln a) = e^{x \ln a}\ln a = a^x \ln a$$
$$\underset{\substack{\uparrow \\ \frac{d}{dx}e^{u(x)} = e^{u(x)}\frac{du}{dx}}}{}$$

THEOREM Derivative of $y = a^x$

The derivative of an exponential function $y = a^x$, where $a > 0$ and $a \neq 1$, is

$$\boxed{y' = \frac{d}{dx}a^x = a^x \ln a}$$

If $u = u(x)$ is a differentiable function, then

$$\boxed{\frac{d}{dx}a^{u(x)} = a^{u(x)} \ln a \frac{du}{dx}}$$

IN WORDS The rate of change of any exponential function $y = a^x$ is proportional to a^x, where $\ln a$ is the constant of proportionality.

The second result follows directly from the proof of $\frac{d}{dx}a^x$.

EXAMPLE 6   Differentiating Exponential Functions

Find the derivative of each function:

(a) $f(x) = 2^x$ **(b)** $F(x) = 3^{-x}$ **(c)** $g(x) = \left(\frac{1}{2}\right)^{x^2+1}$

Solution

(a) f is an exponential logarithmic function with base $a = 2$.

$$f'(x) = \frac{d}{dx}2^x = 2^x \ln 2 \qquad \frac{d}{dx}a^x = a^x \ln a$$

(b) $F'(x) = \frac{d}{dx}3^{-x} = 3^{-x} \ln 3 \frac{d}{dx}(-x) \qquad \frac{d}{dx}a^{u(x)} = a^{u(x)} \ln a \frac{du}{dx}$

$$= -3^{-x} \ln 3 \qquad\qquad \frac{d}{dx}(-x) = -1$$

(c) $g'(x) = \frac{d}{dx}\left(\frac{1}{2}\right)^{x^2+1} = \left(\frac{1}{2}\right)^{x^2+1} \ln\frac{1}{2} \cdot \frac{d}{dx}(x^2+1) \qquad \frac{d}{dx}a^{u(x)} = a^{u(x)} \ln a \frac{du}{dx}$

$$= \left(\frac{1}{2}\right)^{x^2+1} \ln\frac{1}{2} \cdot 2x \qquad\qquad \frac{d}{dx}(x^2+1) = 2x$$

$$= \frac{x}{2^{x^2}} \ln\frac{1}{2} \qquad\qquad\qquad \text{Simplify.} \qquad ■$$

·NOW WORK   Problem 47 and AP® Practice Problems 11 and 14.

❸ Use the Power Rule for Functions to Find a Derivative

We use the Chain Rule to establish other derivative formulas, such as a formula for the derivative of a function raised to a power.

THEOREM Power Rule for Functions

If g is a differentiable function and n is an integer, then

$$\boxed{\frac{d}{dx}[g(x)]^n = n[g(x)]^{n-1}g'(x)}$$

Proof If $y=[g(x)]^n$, let $y=u^n$ and $u=g(x)$. Then

$$\frac{dy}{du}=nu^{n-1}=n[g(x)]^{n-1} \quad\text{and}\quad \frac{du}{dx}=g'(x)$$

By the Chain Rule,

$$y'=\frac{d}{dx}[g(x)]^n=\frac{dy}{du}\cdot\frac{du}{dx}=n[g(x)]^{n-1}g'(x) \quad\blacksquare$$

EXAMPLE 7   Using the Power Rule for Functions to Find a Derivative

(a) If $f(x)=(3-x^3)^{-5}$, then

$$f'(x)=\frac{d}{dx}(3-x^3)^{-5}=-5(3-x^3)^{-5-1}\cdot\frac{d}{dx}(3-x^3) \quad \text{Use the Power Rule for Functions.}$$

$$=-5(3-x^3)^{-6}\cdot(-3x^2)=15x^2(3-x^3)^{-6}=\frac{15x^2}{(3-x^3)^6}$$

(b) If $f(\theta)=\cos^3\theta$, then $f(\theta)=(\cos\theta)^3$, and

$$f'(\theta)=\frac{d}{d\theta}(\cos\theta)^3=3(\cos\theta)^{3-1}\cdot\frac{d}{d\theta}\cos\theta \quad\text{Use the Power Rule for Functions.}$$

$$=3\cos^2\theta\cdot(-\sin\theta)=-3\cos^2\theta\sin\theta \quad\blacksquare$$

NOW REWORK Example 7 using the Chain Rule and rework Example 1(a) using the Power Rule for Functions. Decide for yourself which method is easier.

NOW WORK Problem 21.

Often other derivative rules are used along with the Power Rule for Functions.

EXAMPLE 8   Using the Power Rule for Functions with Other Derivative Rules

Find the derivative of:

(a) $f(x)=e^x(x^2+1)^3$ **(b)** $g(x)=\left(\dfrac{3x+2}{4x^2-5}\right)^5$

Solution

(a) The function f is the product of e^x and $(x^2+1)^3$, so we first use the Product Rule.

$$f'(x)=e^x\left[\frac{d}{dx}(x^2+1)^3\right]+\left[\frac{d}{dx}e^x\right](x^2+1)^3 \quad\text{Use the Product Rule.}$$

To complete the solution, we use the Power Rule for Functions to find $\dfrac{d}{dx}(x^2+1)^3$.

$$f'(x)=e^x\left[3(x^2+1)^2\cdot\frac{d}{dx}(x^2+1)\right]+e^x(x^2+1)^3 \quad\text{Use the Power Rule for Functions.}$$

$$=e^x[3(x^2+1)^2\cdot 2x]+e^x(x^2+1)^3=e^x[6x(x^2+1)^2+(x^2+1)^3]$$

$$=e^x(x^2+1)^2[6x+(x^2+1)]=e^x(x^2+1)^2(x^2+6x+1)$$

Factor out $(x^2+1)^2$

(b) g is a function raised to a power, so we begin with the Power Rule for Functions.

$$g'(x) = \frac{d}{dx}\left(\frac{3x+2}{4x^2-5}\right)^5 = 5\left(\frac{3x+2}{4x^2-5}\right)^4\left[\frac{d}{dx}\frac{3x+2}{4x^2-5}\right] \quad \text{Use the Power Rule for Functions.}$$

$$= 5\left(\frac{3x+2}{4x^2-5}\right)^4\left[\frac{(3)(4x^2-5)-(3x+2)(8x)}{(4x^2-5)^2}\right] \quad \text{Use the Quotient Rule.}$$

$$= \frac{5(3x+2)^4[(12x^2-15)-(24x^2+16x)]}{(4x^2-5)^6} \quad \text{Distribute.}$$

$$= \frac{5(3x+2)^4[-12x^2-16x-15]}{(4x^2-5)^6} \quad \text{Simplify.} \quad ■$$

NOW WORK   Problem **29** and AP® Practice Problems **1, 6,** and **9.**

④ Use the Chain Rule for Multiple Composite Functions

The Chain Rule can be extended to multiple composite functions. For example, if the functions

$$y = f(u) \qquad u = g(v) \qquad v = h(x)$$

are each differentiable functions of u, v, and x, respectively, then the composite function $y = (f \circ g \circ h)(x) = f(g(h(x)))$ is a differentiable function of x and

$$\boxed{y' = \frac{dy}{dx} = \frac{dy}{du}\cdot\frac{du}{dv}\cdot\frac{dv}{dx}}$$

where $u = g(v) = g(h(x))$ and $v = h(x)$. This "chain" of factors is the basis for the name Chain Rule.

EXAMPLE 9   Differentiating a Multiple Composite Function

Find y' if:

(a) $y = 5\cos^2(3x+2)$ **(b)** $y = \sin^3\left(\frac{\pi}{2}x\right)$

Solution

(a) For $y = 5\cos^2(3x+2)$, use the Chain Rule with $y = 5u^2$, $u = \cos v$, and $v = 3x+2$. Then $y = 5u^2 = 5\cos^2 v = 5\cos^2(3x+2)$ and

$$\frac{dy}{du} = \frac{d}{du}(5u^2) = 10u = 10\cos(3x+2)$$
$$\uparrow$$
$$u = \cos v$$
$$v = 3x+2$$

$$\frac{du}{dv} = \frac{d}{dv}\cos v = -\sin v = -\sin(3x+2)$$
$$\uparrow$$
$$v = 3x+2$$

$$\frac{dv}{dx} = \frac{d}{dx}(3x+2) = 3$$

Then

$$y' = \frac{dy}{dx} = \frac{dy}{du}\cdot\frac{du}{dv}\cdot\frac{dv}{dx} = [10\cos(3x+2)][-\sin(3x+2)][3]$$
$$\uparrow$$
$$\text{Chain Rule}$$

$$= -30\cos(3x+2)\sin(3x+2)$$

(b) For $y = \sin^3\left(\dfrac{\pi}{2}x\right)$, use the Chain Rule with $y = u^3$, $u = \sin v$, and $v = \dfrac{\pi}{2}x$.

Then $y = u^3 = (\sin v)^3 = \left[\sin\left(\dfrac{\pi}{2}x\right)\right]^3 = \sin^3\left(\dfrac{\pi}{2}x\right)$, and

$$\frac{dy}{du} = \frac{d}{du}u^3 = 3u^2 \underset{\substack{\uparrow \\ u = \sin v \\ v = \frac{\pi}{2}x}}{=} 3\left[\sin\left(\frac{\pi}{2}x\right)\right]^2 = 3\sin^2\left(\frac{\pi}{2}x\right)$$

$$\frac{du}{dv} = \frac{d}{dv}\sin v = \cos v \underset{\substack{\uparrow \\ v = \frac{\pi}{2}x}}{=} \cos\left(\frac{\pi}{2}x\right)$$

$$\frac{dv}{dx} = \frac{d}{dx}\left(\frac{\pi}{2}x\right) = \frac{\pi}{2}$$

Then

$$y' = \frac{dy}{dx} \underset{\substack{\uparrow \\ \text{Chain Rule}}}{=} \frac{dy}{du} \cdot \frac{du}{dv} \cdot \frac{dv}{dx} = 3\sin^2\left(\frac{\pi}{2}x\right) \cdot \cos\left(\frac{\pi}{2}x\right) \cdot \frac{\pi}{2}$$

$$= \frac{3\pi}{2}\sin^2\left(\frac{\pi}{2}x\right)\cos\left(\frac{\pi}{2}x\right) \qquad \blacksquare$$

NOW WORK   Problems **59** and **69** and AP® Practice Problem **8**.

3.1 Assess Your Understanding

Concepts and Vocabulary

1. The derivative of a composite function $(f \circ g)(x)$ can be found using the _____ Rule.

2. **True or False** If $y = f(u)$ and $u = g(x)$ are differentiable functions, then $y = f(g(x))$ is differentiable.

3. **True or False** If $y = f(g(x))$ is a differentiable function, then $y' = f'(g'(x))$.

4. To find the derivative of $y = \tan(1 + \cos x)$, using the Chain Rule, begin with $y = $_____ and $u = $_____.

5. If $y = (x^3 + 4x + 1)^{100}$, then $y' = $_____.

6. If $f(x) = e^{3x^2 + 5}$, then $f'(x) = $_____.

7. **True or False** The Chain Rule can be applied to multiple composite functions.

8. $\dfrac{d}{dx}\sin x^2 = $_____.

Skill Building

In Problems 9–14, write y as a function of x. Find $\dfrac{dy}{dx}$ using the Chain Rule.

PAGE 237 **9.** $y = u^5$, $u = x^3 + 1$

10. $y = u^3$, $u = 2x + 5$

11. $y = \dfrac{u}{u+1}$, $u = x^2 + 1$

12. $y = \dfrac{u-1}{u}$, $u = x^2 - 1$

13. $y = (u+1)^2$, $u = \dfrac{1}{x}$

14. $y = (u^2 - 1)^3$, $u = \dfrac{1}{x+2}$

In Problems 15–32, find the derivative of each function using the Power Rule for Functions.

15. $f(x) = (2x + 7)^2$

16. $f(x) = (3x - 8)^3$

17. $f(x) = (6x - 5)^{-3}$

18. $f(t) = (4t + 1)^{-2}$

19. $g(x) = (x^2 + 5)^4$

20. $F(x) = (x^3 - 2)^5$

PAGE 241 **21.** $f(u) = \left(u - \dfrac{1}{u}\right)^3$

22. $f(x) = \left(x + \dfrac{1}{x}\right)^3$

23. $g(x) = (4x + e^x)^3$

24. $F(x) = (e^x - x^2)^2$

25. $f(x) = \tan^2 x$

26. $f(x) = \sec^3 x$

27. $f(z) = (\tan z + \cos z)^2$

28. $f(z) = (e^z + 2\sin z)^3$

PAGE 242 **29.** $y = (x^2 + 4)^2(2x^3 - 1)^3$

30. $y = (x^2 - 2)^3(3x^4 + 1)^2$

31. $y = \left(\dfrac{\sin x}{x}\right)^2$

32. $y = \left(\dfrac{x + \cos x}{x}\right)^5$

In Problems 33–54, find y'.

33. $y = \cos(3x)$

34. $y = \sin(4x)$

35. $y = 2\sin(x^2 + 2x - 1)$

36. $y = \dfrac{1}{2}\cos(x^3 - 2x + 5)$

PAGE 237 **37.** $y = \cos\dfrac{2}{x}$

38. $y = \sin\dfrac{3}{x}$

39. $y = \ln(x^2 + 4)$

40. $y = \ln\dfrac{1}{x}$

PAGE 238 **41.** $y = e^{1/x}$

42. $y = e^{1/x^2}$

43. $y = \dfrac{1}{(x^4 - 2x + 1)^2}$

44. $y = \dfrac{3}{(x^5 + 2x^2 - 3)^4}$

45. $y = \dfrac{100}{1 + 99e^{-x}}$

46. $y = \dfrac{1}{1 + 2e^{-x}}$

[PAGE 240] **47.** $y = 2^{\sin x}$

48. $y = (\sqrt{3})^{\cos x}$

49. $y = \ln(\sin x)$

50. $y = \sin(\ln x)$

51. $y = 5xe^{3x}$

52. $y = x^3 e^{2x}$

53. $y = x^2 \sin(4x)$

54. $y = x^2 \cos(4x)$

In Problems 55–58, find y'. Treat a and b as constants.

55. $y = e^{-ax} \sin(bx)$

56. $y = e^{ax} \cos(-bx)$

57. $y = \dfrac{e^{ax} - 1}{e^{ax} + 1}$

58. $y = \dfrac{e^{-ax} + 1}{e^{bx} - 1}$

In Problems 59–62, write y as a function of x. Find $\dfrac{dy}{dx}$ using the Chain Rule.

[PAGE 243] **59.** $y = u^3,\ u = 3v^2 + 1,\ v = \dfrac{4}{x^2}$

60. $y = 3u,\ u = 3v^2 - 4,\ v = \dfrac{1}{x}$

61. $y = u^2 + 1,\ u = \dfrac{4}{v},\ v = x^2$

62. $y = u^3 - 1,\ u = -\dfrac{2}{v},\ v = x^3$

In Problems 63–70, find y'.

63. $y = e^{-2x} \cos(3x)$

64. $y = e^{\pi x} \tan(\pi x)$

65. $y = \cos(e^{x^2})$

66. $y = \tan(e^{x^2})$

67. $y = e^{\cos(4x)}$

68. $y = e^{\csc^2 x}$

[PAGE 243] **69.** $y = 4\sin^2(3x)$

70. $y = 2\cos^2(x^2)$

In Problems 71 and 72, find the derivative of each function by:
(a) Using the Chain Rule.
(b) Using the Power Rule for Functions.
(c) Expanding and then differentiating.
(d) Verify the answers from parts (a)–(c) are equal.

71. $y = (x^3 + 1)^2$

72. $y = (x^2 - 2)^3$

In Problems 73–78:
(a) Find an equation of the tangent line to the graph of f at the given point.
(b) Find an equation of the normal line to the graph of f at the given point.
 (c) Use technology to graph f, the tangent line, and the normal line on the same screen.

73. $f(x) = (x^2 - 2x + 1)^5$ at $(1, 0)$

74. $f(x) = (x^3 - x^2 + x - 1)^{10}$ at $(0, 1)$

75. $f(x) = \dfrac{x}{(x^2 - 1)^3}$ at $\left(2, \dfrac{2}{27}\right)$

76. $f(x) = \dfrac{x^2}{(x^2 - 1)^2}$ at $\left(2, \dfrac{4}{9}\right)$

[PAGE 238] **77.** $f(x) = \sin(2x) + \cos\dfrac{x}{2}$ at $(0, 1)$

78. $f(x) = \sin^2 x + \cos^3 x$ at $\left(\dfrac{\pi}{2}, 1\right)$

In Problems 79 and 80, find the indicated derivative.

79. $\dfrac{d^2}{dx^2} \cos(x^5)$

80. $\dfrac{d^3}{dx^3} \sin^3 x$

[PAGE 239] **81.** Suppose $h = f \circ g$. Find $h'(1)$ if $f'(2) = 6$, $f(1) = 4$, $g(1) = 2$, and $g'(1) = -2$.

82. Suppose $h = f \circ g$. Find $h'(1)$ if $f'(3) = 4$, $f(1) = 1$, $g(1) = 3$, and $g'(1) = 3$.

83. Suppose $h = g \circ f$. Find $h'(0)$ if $f(0) = 3$, $f'(0) = -1$, $g(3) = 8$, and $g'(3) = 0$.

84. Suppose $h = g \circ f$. Find $h'(2)$ if $f(1) = 2$, $f'(1) = 4$, $f(2) = -3$, $f'(2) = 4$, $g(-3) = 1$, and $g'(-3) = 3$.

85. If $y = u^5 + u$ and $u = 4x^3 + x - 4$, find $\dfrac{dy}{dx}$ at $x = 1$.

86. If $y = e^u + 3u$ and $u = \cos x$, find $\dfrac{dy}{dx}$ at $x = 0$.

Applications and Extensions

In Problems 87–94, find the indicated derivative.

87. $\dfrac{d}{dx} f(x^2 + 1)$

88. $\dfrac{d}{dx} f(1 - x^2)$

Hint: Let $u = x^2 + 1$.

89. $\dfrac{d}{dx} f\left(\dfrac{x+1}{x-1}\right)$

90. $\dfrac{d}{dx} f\left(\dfrac{1-x}{1+x}\right)$

91. $\dfrac{d}{dx} f(\sin x)$

92. $\dfrac{d}{dx} f(\tan x)$

93. $\dfrac{d^2}{dx^2} f(\cos x)$

94. $\dfrac{d^2}{dx^2} f(e^x)$

95. Motion on a Line   An object moves along a horizontal line so that its position s, in meters, from the origin at time t seconds is given by $s = s(t) = A \cos(\omega t + \phi)$, where A, ω, and ϕ are constants.

(a) Find the velocity v of the object at time t.
(b) When is the velocity of the object 0?
(c) Find the acceleration a of the object at time t.
(d) When is the acceleration of the object 0?

96. Motion on a Line   A bullet is fired horizontally into a bale of paper. The distance s (in meters) the bullet travels into the bale of paper in t seconds is given by

$$s = s(t) = 8 - (2 - t)^3 \quad 0 \le t \le 2$$

(a) Find the velocity v of the bullet at any time t.
(b) Find the velocity of the bullet at $t = 1$.
(c) Find the acceleration a of the bullet at any time t.
(d) Find the acceleration of the bullet at $t = 1$.
(e) How far into the bale of paper did the bullet travel?

97. Motion on a Line   Find the acceleration a of a car if the distance s, in feet, it has traveled along a highway at time $t \ge 0$ seconds is given by

$$s(t) = \dfrac{80}{3}\left[t + \dfrac{3}{\pi} \sin\left(\dfrac{\pi}{6}t\right)\right]$$

98. Motion on a Line An object moves along a line so that at time $t \geq 0$ seconds, its position from the origin is $s(t) = \sin e^t$, in feet.

(a) Find the velocity v and acceleration a of the object at any time t.

(b) At what time does the object first have zero velocity?

(c) What is the acceleration of the object at the time t found in (b)?

 99. Resistance The resistance R (measured in ohms) of an 80-meter-long electric wire of radius x (in centimeters) is given by the formula $R = R(x) = \dfrac{0.0048}{x^2}$. The radius x is given by $x = 0.1991 + 0.000003T$ where T is the temperature in Kelvin. How fast is R changing with respect to T when $T = 320$ K?

100. Pendulum Motion in a Car The motion of a pendulum swinging in the direction of motion of a car moving at a low, constant speed can be modeled by

$$s = s(t) = 0.05 \sin(2t) + 3t \qquad 0 \leq t \leq \pi$$

where s is the distance in meters and t is the time in seconds.

(a) Find the velocity v at $t = \dfrac{\pi}{8}$, $t = \dfrac{\pi}{4}$, and $t = \dfrac{\pi}{2}$.

(b) Find the acceleration a at the times given in (a).

 (c) Graph $s = s(t)$, $v = v(t)$, and $a = a(t)$ on the same screen.

Source: Mathematics students at Trine University, Angola, Indiana.

101. Economics The function $A(t) = 102 - 90\,e^{-0.21t}$ represents the relationship between A, the percentage of the market penetrated by the latest generation smart phones, and t, the time in years, where $t = 0$ corresponds to the year 2025.

(a) Find $\lim\limits_{t \to \infty} A(t)$ and interpret the result.

 (b) Graph the function $A = A(t)$, and explain how the graph supports the answer in (a).

(c) Find the rate of change of A with respect to time.

(d) Evaluate $A'(5)$ and $A'(10)$ and interpret these results.

 (e) Graph the function $A' = A'(t)$, and explain how the graph supports the answers in (d).

102. Meteorology The atmospheric pressure at a height of x meters above sea level is $P(x) = 10^4 e^{-0.00012x}$ kg/m². What is the rate of change of the pressure with respect to the height at $x = 500$ m? At $x = 750$ m?

 103. Hailstones Hailstones originate at an altitude of about 3000 m, although this varies. As they fall, air resistance slows down the hailstones considerably. In one model of air resistance, the speed of a hailstone of mass m as a function of time t is given by $v(t) = \dfrac{mg}{k}(1 - e^{-kt/m})$ m/s,

where $g = 9.8$ m/s² is the acceleration due to gravity and k is a constant that depends on the size of the hailstone and the conditions of the air.

(a) Find the acceleration $a(t)$ of a hailstone as a function of time t.

(b) Find $\lim\limits_{t \to \infty} v(t)$. What does this limit say about the speed of the hailstone?

(c) Find $\lim\limits_{t \to \infty} a(t)$. What does this limit say about the acceleration of the hailstone?

104. Median Earnings The median earnings E, in dollars, of workers 18 years and over are given in the table below:

Year	1980	1985	1990	1995	2000	2005	2010	2015	2020
Median Earnings	12,665	17,181	21,793	26,792	32,604	41,231	49,733	48,000	50,295

Source: U.S. Bureau of the Census, Current Population Survey.

 (a) Find the exponential function of best fit and show that it equals $E = E(t) = 12{,}376\,(1.036)^t$, where t is the number of years since 1979.

(b) Find the rate of change of E with respect to t.

(c) Find the rate of change at $t = 26$ (year 2005).

(d) Find the rate of change at $t = 36$ (year 2015).

(e) Find the rate of change at $t = 41$ (year 2020).

(f) Compare the answers to (c), (d), and (e). Interpret each answer and explain the differences.

105. Motion on a Line An object moves along a line so that at time $t > 0$ its position s from the origin is $s = s(t)$. The velocity v of the object is $v = \dfrac{ds}{dt}$, and its acceleration is $a = \dfrac{dv}{dt} = \dfrac{d^2 s}{dt^2}$. If the velocity $v = v(s)$ is expressed as a function of s, show that the acceleration a can be expressed as $a = v\dfrac{dv}{ds}$.

 106. Student Approval Professor Miller's student approval rating is modeled by the function $Q(t) = 21 + \dfrac{10 \sin \dfrac{2\pi t}{7}}{\sqrt{t} - \sqrt{20}}$,

where $0 \leq t \leq 16$ is the number of weeks since the semester began.

(a) Find $Q'(t)$.

(b) Evaluate $Q'(1)$, $Q'(5)$, and $Q'(10)$.

(c) Interpret the results obtained in (b).

 (d) Use technology to graph $Q(t)$ and $Q'(t)$.

(e) How would you explain the results in (d) to Professor Miller?

Source: Mathematics students at Millikin University, Decatur, Illinois.

107. Angular Velocity If the disk in the figure is rotated about a vertical line through an angle θ, torsion in the wire attempts to turn the disk in the opposite direction. The motion θ at time t (assuming no friction or air resistance) obeys the equation

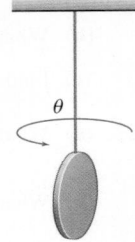

$$\theta(t) = \frac{\pi}{3} \cos \left(\frac{1}{2} \sqrt{\frac{2k}{5}}\,t \right)$$

where k is the coefficient of torsion of the wire.

(a) Find the angular velocity $\omega = \dfrac{d\theta}{dt}$ of the disk at any time t.

(b) What is the angular velocity at $t = 3$?

108. Harmonic Motion   A weight hangs on a spring making it 2 m long when it is stretched out (see the figure). If the weight is pulled down and then released, the weight oscillates up and down, and the length l of the spring after t seconds is given by $l(t) = 2 + \cos(2\pi t)$.

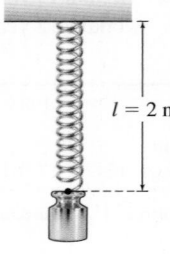

$l = 2$ m

(a) Find the length l of the spring at the times $t = 0, \frac{1}{2}, 1, \frac{3}{2}$, and $\frac{5}{8}$.

(b) Find the velocity v of the weight at time $t = \frac{1}{4}$.

(c) Find the acceleration a of the weight at time $t = \frac{1}{4}$.

109. Find $F'(1)$ if $f(x) = \sin x$ and $F(t) = f(t^2 - 1)$.

110. Normal Line   Find the point on the graph of $y = e^{-x}$ where the normal line to the graph passes through the origin.

111. Use the Chain Rule and the fact that $\cos x = \sin\left(\frac{\pi}{2} - x\right)$ to show that $\dfrac{d}{dx} \cos x = -\sin x$.

112. If $y = e^{2x}$, show that $y'' - 4y = 0$.

113. If $y = e^{-2x}$, show that $y'' - 4y = 0$.

114. If $y = Ae^{2x} + Be^{-2x}$, where A and B are constants, show that $y'' - 4y = 0$.

115. If $y = Ae^{ax} + Be^{-ax}$, where A, B, and a are constants, show that $y'' - a^2 y = 0$.

116. If $y = Ae^{2x} + Be^{3x}$, where A and B are constants, show that $y'' - 5y' + 6y = 0$.

117. If $y = Ae^{-2x} + Be^{-x}$, where A and B are constants, show that $y'' + 3y' + 2y = 0$.

118. If $y = A\sin(\omega t) + B\cos(\omega t)$, where A, B, and ω are constants, show that $y'' + \omega^2 y = 0$.

119. Show that $\dfrac{d}{dx} f(h(x)) = 2xg(x^2)$, if $\dfrac{d}{dx} f(x) = g(x)$ and $h(x) = x^2$.

120. Find the nth derivative of $f(x) = (2x + 3)^n$.

121. Find a general formula for the nth derivative of y.
   (a) $y = e^{ax}$ (b) $y = e^{-ax}$

122. (a) What is $\dfrac{d^{10}}{dx^{10}} \sin(ax)$?

   (b) What is $\dfrac{d^{25}}{dx^{25}} \sin(ax)$?

   (c) Find the nth derivative of $f(x) = \sin(ax)$.

123. (a) What is $\dfrac{d^{11}}{dx^{11}} \cos(ax)$?

   (b) What is $\dfrac{d^{12}}{dx^{12}} \cos(ax)$?

   (c) Find the nth derivative of $f(x) = \cos(ax)$.

124. If $y = e^{-at}[A\sin(\omega t) + B\cos(\omega t)]$, where A, B, a, and ω are constants, find y' and y''.

125. Show that if a function f has the properties:

   a. $f(u + v) = f(u)f(v)$ for all choices of u and v
   b. $f(x) = 1 + xg(x)$, where $\lim\limits_{x \to 0} g(x) = 1$, then $f' = f$.

Challenge Problems

126. Find the nth derivative of $f(x) = \dfrac{1}{3x - 4}$.

127. Let $f_1(x), \ldots, f_n(x)$ be n differentiable functions. Find the derivative of $y = f_1(f_2(f_3(\ldots(f_n(x)\ldots))))$.

128. Let

$$f(x) = \begin{cases} x^2 \sin \dfrac{1}{x} & \text{if } x \neq 0 \\ 0 & \text{if } x = 0 \end{cases}$$

Show that $f'(0)$ exists, but that f' is not continuous at 0.

129. Define f by

$$f(x) = \begin{cases} e^{-1/x^2} & \text{if } x \neq 0 \\ 0 & \text{if } x = 0 \end{cases}$$

Show that f is differentiable on $(-\infty, \infty)$ and find f' for each value of x.
   Hint: To find $f'(0)$, use the definition of the derivative. Then show that $1 < x^2 e^{1/x^2}$ for $x \neq 0$.

130. Suppose $f(x) = x^2$ and $g(x) = |x - 1|$. The functions f and g are continuous on their domains, the set of all real numbers.

   (a) Is f differentiable at all real numbers? If not, where does f' not exist?

   (b) Is g differentiable at all real numbers? If not, where does g' not exist?

   (c) Can the Chain Rule be used to differentiate the composite function $(f \circ g)(x)$ for all x? Explain.

   (d) Is the composite function $(f \circ g)(x)$ differentiable? If so, what is its derivative?

131. Suppose $f(x) = x^4$ and $g(x) = x^{1/3}$. The functions f and g are continuous on their domains, the set of all real numbers.

   (a) Is f differentiable at all real numbers? If not, where does f' not exist?

   (b) Is g differentiable at all real numbers? If not, where does g' not exist?

   (c) Can the Chain Rule be used to differentiate the composite function $(f \circ g)(x)$ for all x? Explain.

   (d) Is the composite function $(f \circ g)(x)$ differentiable? If so, what is its derivative?

132. The function $f(x) = e^x$ has the property $f'(x) = f(x)$. Give an example of another function $g(x)$ such that $g(x)$ is defined for all real x, $g'(x) = g(x)$, and $g(x) \neq f(x)$.

133. Harmonic Motion   The motion of the piston of an automobile engine is approximately simple harmonic. If the stroke of a piston (twice the amplitude) is 10 cm and the angular velocity ω is 60 revolutions per second, then the motion of the piston is given by $s(t) = 5\sin(120\pi t)$ cm.

   (a) Find the acceleration a of the piston at the end of its stroke $\left(t = \dfrac{1}{240} \text{ second}\right)$.

   (b) If the mass of the piston is 1 kg, what resultant force must be exerted on it at this point?
   Hint: Use Newton's Second Law, that is, $F = ma$.

AP® Practice Problems

Multiple-Choice Questions

PAGE 242 **1.** If $f(x) = (x+3)(x^2-2)^4$, then $f'(x) =$

(A) $8x(x^2-2)^3$

(B) $8x(x+3)(x^2-2)^3$

(C) $8x(x^2+x+1)(x^2-2)^3$

(D) $(9x^2+24x-2)(x^2-2)^3$

PAGE 237 **2.** If $f(x) = \sec(4x)$, then $f'\left(\dfrac{\pi}{6}\right) =$

(A) $\dfrac{\sqrt{3}}{2}$ (B) $2\sqrt{3}$ (C) $8\sqrt{3}$ (D) $-8\sqrt{3}$

PAGE 238 **3.** $\dfrac{d}{dx}e^{-3/x} =$

(A) $\dfrac{3e^{-3/x}}{x^2}$ (B) $-3e^{-3/x}$

(C) $-\dfrac{3e^{-3/x}}{x^2}$ (D) $-3x^2e^{-3/x}$

PAGE 238 **4.** $\dfrac{d}{dx}\tan e^{-x} =$

(A) $\sec^2 e^{-x}$ (B) $-x\sec^2 e^{-x}$

(C) $e^{-x}\sec^2 e^{-x}$ (D) $-e^{-x}\sec^2 e^{-x}$

PAGE 238 **5.** An equation of the line tangent to the graph of $f(x) = 2(10-x)^2$ at the point $(9, 2)$ is

(A) $y = -4x + 28$ (B) $y - 2 = -\dfrac{1}{4}(x-9)$

(C) $y - 2 = -4(x-9)$ (D) $y - 2 = 4(x-9)^2$

PAGE 242 **6.** If $y = \left(\dfrac{e^{4x}}{2x}\right)^2$, the instantaneous rate of change of y with respect to x is:

(A) $\dfrac{(4x-1)e^{8x}}{2x^3}$ (B) $\dfrac{(2x-1)e^{8x}}{x^3}$

(C) $\dfrac{(4x-1)e^{8x}}{x^3}$ (D) $\dfrac{(x-1)e^{8x}}{x^3}$

PAGE 238 **7.** An equation of the line tangent to the graph of $g(x) = \sin(2x)$ at $x = \dfrac{\pi}{3}$ is

(A) $y + \dfrac{\sqrt{3}}{2} = -x + \dfrac{\pi}{3}$

(B) $y - \dfrac{\sqrt{3}}{2} = -x + \dfrac{\pi}{3}$

(C) $y - \dfrac{\sqrt{3}}{2} = -1\left(x + \dfrac{\pi}{3}\right)$

(D) $y + \dfrac{\sqrt{3}}{2} = x + \dfrac{\pi}{3}$

PAGE 243 **8.** If $y = \cos^3(x^2)$, then $\dfrac{dy}{dx} =$

(A) $-6x\cos^2(x^2)$ (B) $-6x\cos^2(x^2)\sin x^2$

(C) $6x\cos^2(x^2)\sin x^2$ (D) $-3\cos^2(x^2)\sin x$

PAGE 242 **9.** If $f(x) = e^{\sin^2 x}$, then $f'(x) =$

(A) $2e^{\sin^2 x}\sin x\cos x$ (B) $e^{\sin x(\sin x + 2\cos x)}$

(C) $2e^x\sin x\cos x$ (D) $e^x\sin^2 x$

PAGE 238 **10.** An object is moving along the x-axis. Its position (in kilometers) at time $t \geq 0$ (in hours) is given by $s(t) = \sin(3t) - \cos(4t)$. What is the acceleration of the object at time $t = \dfrac{\pi}{2}$?

(A) 7 km/h^2 (B) 9 km/h^2

(C) -7 km/h^2 (D) 25 km/h^2

PAGE 240 **11.** $\dfrac{d}{dx}5^x =$

(A) $5^x \ln 5$ (B) $(5^{x-1})x$

(C) 5^{x-1} (D) $\dfrac{5^x}{\ln 5}$

PAGE 238 **12.** If $y = e^{kx}$, then $\dfrac{d^k}{dx^k}(e^{kx}) =$

(A) $k^2 e^{kx}$ (B) $k^k e^x$

(C) $k^k e^{kx}$ (D) $k!\, e^{kx}$

PAGE 239 **13.** Suppose $h = g \circ f$. Use the table, to find $h'(2)$

x	f	f'	g	g'
1	2	4	6	-2
2	-3	4	-3	5
-3	1	-2	1	3

(A) 5   (B) 7   (C) 12   (D) 20

PAGE 240 **14.** The velocity v (in meters/second) of an object moving on a line is given by $v(t) = 3 - 1.5^{-t^2}$, $t \geq 0$. What is the acceleration of the object at $t = 4$ seconds?

(A) 0.005 m/s^2 (B) -0.005 m/s^2

(C) 0.012 m/s^2 (D) 2.998 m/s^2

PAGE 238 **15.** If $h(x) = \cos x^3$, then $h''(\sqrt[3]{\pi}) =$

(A) $-9\pi\sqrt[3]{\pi}$ (B) $9\sqrt[3]{\pi}$ (C) $9\pi\sqrt[3]{\pi}$ (D) 0

Free-Response Question

PAGE 238 **16.** $f(x) = \ln(\cos x)$, $-\dfrac{\pi}{2} < x < \dfrac{\pi}{2}$

(a) Find an equation of the line tangent to the graph of f at $x = \dfrac{\pi}{4}$.

(b) Determine where $f'(x) > 0$.

(c) Find $f''\left(\dfrac{\pi}{4}\right)$.

Retain Your Knowledge

Multiple-Choice Questions

1. The function f is continuous on the closed interval $[-5, 5]$. A portion of the graph of its derivative f' is shown below. Based on this graph, which of the following must be true about f?

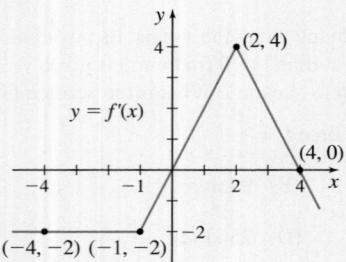

    I. The function f is not differentiable at $x = 2$.
    II. The function f is decreasing for $-4 < x < -1$.
    III. The graph of f has a horizontal tangent line at $(0, f(0))$ and $(4, f(4))$.

(A) I only   (B) I and III only

(C) II and III only   (D) I, II, and III

2. Find the derivative of the function $f(x) = x^2 \sin x$.

(A) $f'(x) = 2x \cos x$

(B) $f'(x) = 2x \sin x - 2x \cos x$

(C) $f'(x) = 2x \sin x - x^2 \cos x$

(D) $f'(x) = x^2 \cos x + 2x \sin x$

3. $\displaystyle \lim_{x \to \infty} \frac{4x^2 - 9}{3x^2 + 2x + 3} =$

(A) 0   (B) 1   (C) $\dfrac{4}{3}$ (D) ∞

Free-Response Question

4. Show that the graphs of the functions $f(x) = x^3$ and $g(x) = 1 - x^2$ intersect for some number x in the open interval $(0, 1)$.

3.2 Implicit Differentiation

OBJECTIVES *When you finish this section, you should be able to:*

1 Find a derivative using implicit differentiation (p. 249)
2 Find higher-order derivatives using implicit differentiation (p. 252)
3 Differentiate functions with rational exponents (p. 253)

NEED TO REVIEW? The implicit form of a function is discussed in Section P.1, p. 5.

So far we have differentiated only functions $y = f(x)$ where the dependent variable y is expressed *explicitly* in terms of the independent variable x. There are functions that are not written in the form $y = f(x)$, but are written in the *implicit* form $F(x, y) = 0$. For example, x and y are related implicitly in the equations

$$xy - 4 = 0 \qquad y^2 + 3x^2 - 1 = 0 \qquad e^{x^2 - y^2} - \cos(xy) = 0$$

The implicit form $xy - 4 = 0$ can easily be written explicitly as the function $y = \dfrac{4}{x}$.

Also, $y^2 + 3x^2 - 1 = 0$ can be written explicitly as the two functions $y_1 = \sqrt{1 - 3x^2}$ and $y_2 = -\sqrt{1 - 3x^2}$. However, for the equation $e^{x^2 - y^2} - \cos(xy) = 0$, it is impossible to express y as an explicit function of x. To find the derivative of a function defined implicitly, we use the technique of *implicit differentiation*.

1 Find a Derivative Using Implicit Differentiation

The method used to differentiate an implicitly defined function is called **implicit differentiation**. It does not require rewriting the function explicitly, but it does require that the dependent variable y is a differentiable function of the independent variable x. So throughout the section, we make the assumption that there is a differentiable function $y = f(x)$ defined by the implicit equation. This assumption made, implicit differentiation consists of differentiating both sides of the implicitly defined function with respect to x and then solving the resulting equation for $\dfrac{dy}{dx}$.

EXAMPLE 1   **Finding a Derivative Using Implicit Differentiation**

Find $\dfrac{dy}{dx}$ if $xy - 4 = 0$.

(a) Use implicit differentiation.

(b) Solve for y and then differentiate.

(c) Verify the results of (a) and (b) are the same.

Solution

(a) To differentiate implicitly, we assume y is a differentiable function of x and differentiate both sides with respect to x.

$$\frac{d}{dx}(xy - 4) = \frac{d}{dx}0 \qquad \text{Differentiate both sides with respect to } x.$$

$$\frac{d}{dx}(xy) - \frac{d}{dx}4 = 0 \qquad \text{Use the Difference Rule.}$$

$$x \cdot \frac{d}{dx}y + \left(\frac{d}{dx}x\right)y - 0 = 0 \qquad \text{Use the Product Rule.}$$

$$x\frac{dy}{dx} + y = 0 \qquad \text{Simplify.}$$

$$\frac{dy}{dx} = -\frac{y}{x} \qquad \text{Solve for } \frac{dy}{dx}. \tag{1}$$

(b) Solve $xy - 4 = 0$ for y, obtaining $y = \dfrac{4}{x} = 4x^{-1}$. Then

$$\frac{dy}{dx} = \frac{d}{dx}(4x^{-1}) = -4x^{-2} = -\frac{4}{x^2} \tag{2}$$

(c) At first glance, the results in (1) and (2) appear to be different. However, since $xy - 4 = 0$, or equivalently, $y = \dfrac{4}{x}$, the result from (1) becomes

$$\frac{dy}{dx} \underset{\underset{(1)}{\uparrow}}{=} -\frac{y}{x} \underset{\underset{y = \frac{4}{x}}{\uparrow}}{=} -\frac{\frac{4}{x}}{x} = -\frac{4}{x^2}$$

which is the same as (2). ∎

In most instances, we will not know the explicit form of the function (as we did in Example 1) and so we will leave the derivative $\dfrac{dy}{dx}$ expressed in terms of x and y [as in (1)].

NOW WORK   Problem 17.

NEED TO REVIEW? The Power Rule for Functions is discussed on pp. 240 and 241.

The Power Rule for Functions is

$$\frac{d}{dx}[f(x)]^n = n[f(x)]^{n-1}f'(x)$$

where n is an integer. If $y = f(x)$, the Power Rule for Functions takes the form

$$\boxed{\frac{d}{dx}y^n = ny^{n-1}\frac{dy}{dx}}$$

This is convenient notation to use with implicit differentiation when y^n appears.

$$\frac{d}{dx}y = 1 \cdot \underset{\underset{n=1}{\uparrow}}{\frac{dy}{dx}} = \frac{dy}{dx} \qquad \frac{d}{dx}y^2 = \underset{\underset{n=2}{\uparrow}}{2y\frac{dy}{dx}} \qquad \frac{d}{dx}y^3 = \underset{\underset{n=3}{\uparrow}}{3y^2\frac{dy}{dx}}$$

To differentiate an implicit function:

- Assume that y is a differentiable function of x.
- Differentiate both sides of the equation with respect to x.
- Solve the resulting equation for $y' = \dfrac{dy}{dx}$.

CALC CLIP

EXAMPLE 2 **Using Implicit Differentiation to Find an Equation of a Tangent Line**

Find an equation of the tangent line to the graph of the ellipse $3x^2 + 4y^2 = 2x$ at the point $\left(\dfrac{1}{2}, -\dfrac{1}{4}\right)$. Graph the equation and the tangent line on the same set of coordinate axes.

Solution

First use implicit differentiation to find the slope of the tangent line.

Assuming y is a differentiable function of x, differentiate both sides with respect to x.

$$\frac{d}{dx}(3x^2+4y^2) = \frac{d}{dx}(2x) \qquad \text{Differentiate both sides with respect to } x.$$

$$\frac{d}{dx}(3x^2) + \frac{d}{dx}(4y^2) = 2 \qquad \text{Use the Sum Rule.}$$

$$3\frac{d}{dx}x^2 + 4\frac{d}{dx}y^2 = 2 \qquad \text{Use the Constant Multiple Rule.}$$

$$6x + 4 \cdot 2y\frac{dy}{dx} = 2 \qquad \frac{d}{dx}y^2 = 2y\frac{dy}{dx}$$

$$6x + 8y\frac{dy}{dx} = 2 \qquad \text{Simplify.}$$

$$\frac{dy}{dx} = \frac{2-6x}{8y} = \frac{1-3x}{4y} \qquad \text{Solve for } \frac{dy}{dx}.$$

provided $y \neq 0$.

Now evaluate $\dfrac{dy}{dx} = \dfrac{1-3x}{4y}$ at $\left(\dfrac{1}{2}, -\dfrac{1}{4}\right)$.

NOTE The notation $\dfrac{dy}{dx}\bigg|_{(x_0,y_0)}$ is used when the derivative $\dfrac{dy}{dx}$ is to be evaluated at a point (x_0, y_0).

$$\frac{dy}{dx}\bigg|_{\left(\frac{1}{2}, -\frac{1}{4}\right)} = \frac{1-3x}{4y}\bigg|_{\left(\frac{1}{2}, -\frac{1}{4}\right)} = \frac{1-3\cdot\frac{1}{2}}{4\cdot\left(-\frac{1}{4}\right)} = \frac{1}{2}$$

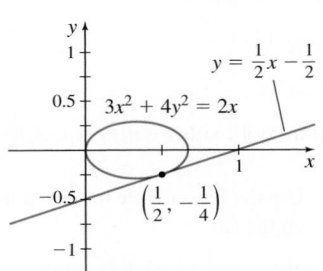

Figure 2

The slope of the tangent line to the graph of $3x^2 + 4y^2 = 2x$ at the point $\left(\dfrac{1}{2}, -\dfrac{1}{4}\right)$ is $\dfrac{1}{2}$. An equation of the tangent line is

$$y + \frac{1}{4} = \frac{1}{2}\left(x - \frac{1}{2}\right)$$

$$y = \frac{1}{2}x - \frac{1}{2}$$

Figure 2 shows the graph of the ellipse $3x^2 + 4y^2 = 2x$ and the graph of the tangent line $y = \dfrac{1}{2}x - \dfrac{1}{2}$ at the point $\left(\dfrac{1}{2}, -\dfrac{1}{4}\right)$. ■

NOW WORK   Problems 15 and 53 and AP® Practice Problems 3, 5, and 12.

EXAMPLE 3 **Using Implicit Differentiation to Find an Equation of a Tangent Line**

Find an equation of the tangent line to the graph of $e^y \cos x = x + 1$ at the point $(0, 0)$.

Solution

Assuming y is a differentiable function of x, use implicit differentiation to find the slope of the tangent line.

$$\frac{d}{dx}(e^y \cos x) = \frac{d}{dx}(x + 1)$$

$$e^y \cdot \frac{d}{dx}(\cos x) + \left(\frac{d}{dx}e^y\right) \cdot \cos x = 1 \qquad \text{Use the Product Rule on the left.}$$

$$e^y(-\sin x) + e^y \frac{dy}{dx} \cdot \cos x = 1 \qquad \frac{d}{dx}e^y = e^y \frac{dy}{dx} \text{ using the Chain Rule.}$$

$$e^y \cos x \frac{dy}{dx} = 1 + e^y \sin x$$

$$y' = \frac{1 + e^y \sin x}{e^y \cos x}$$

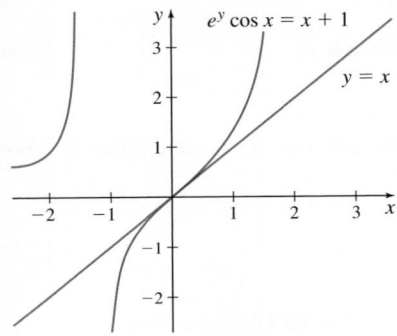

Figure 3

provided $\cos x \neq 0$.

At the point $(0, 0)$, the derivative is

$$\left.\frac{dy}{dx}\right|_{(0,0)} = \frac{1 + e^0 \sin 0}{e^0 \cos 0} = \frac{1 + 1 \cdot 0}{1 \cdot 1} = 1$$

so the slope of the tangent line to the graph at $(0, 0)$ is 1. An equation of the tangent line to the graph at the point $(0, 0)$ is $y = x$. ■

The graph of $e^y \cos x = x + 1$ and the tangent line to the graph at $(0, 0)$ are shown in Figure 3.

NOW WORK   AP® Practice Problem 10.

EXAMPLE 4   Using Implicit Differentiation with the Chain Rule

Find y' if $\tan^2 y = e^x$. Express y' as a function of x alone.

Solution

Assuming y is a differentiable function of x, differentiate both sides with respect to x.

$$\frac{d}{dx}\tan^2 y = \frac{d}{dx}e^x$$
Use the Power Rule for functions on the left.

$$2\tan y \cdot \sec^2 y \frac{dy}{dx} = e^x$$
$\dfrac{d}{dx}\tan y = \sec^2 y \dfrac{dy}{dx}$

$$y' = \frac{dy}{dx} = \frac{e^x}{2\tan y \sec^2 y}$$
Solve for $y' = \dfrac{dy}{dx}$.

$$= \frac{e^x}{2e^{x/2}(1+e^x)} = \frac{e^{x/2}}{2(1+e^x)}$$
$\tan y = e^{x/2}$
$\sec^2 y = 1 + \tan^2 y = 1 + e^x$ ∎

NOW WORK   Problem 25 and AP® Practice Problems 1 and 8.

---

❷ Find Higher-Order Derivatives Using Implicit Differentiation

NEED TO REVIEW? Higher-order derivatives are discussed in Section 2.4, pp. 208–210.

Implicit differentiation can be used to find higher-order derivatives.

EXAMPLE 5   Finding Higher-Order Derivatives

Use implicit differentiation to find y' and y'' if $y^2 - x^2 = 5$. Express y'' in terms of x and y.

Solution

First, we assume there is a twice differentiable function $y = f(x)$ that satisfies $y^2 - x^2 = 5$. Now we find y'.

$$\frac{d}{dx}(y^2 - x^2) = \frac{d}{dx}5$$

$$\frac{d}{dx}y^2 - \frac{d}{dx}x^2 = 0$$

$$2yy' - 2x = 0 \qquad\qquad \frac{d}{dx}y^2 = 2y\frac{dy}{dx} = 2yy' \tag{3}$$

$$y' = \frac{2x}{2y} = \frac{x}{y} \qquad\qquad \text{Solve for } y'. \tag{4}$$

provided $y \neq 0$.

Equations (3) and (4) both involve y'. Either one can be used to find y''. We use (3) because it avoids differentiating a quotient.

$$\frac{d}{dx}(2yy' - 2x) = \frac{d}{dx}0$$

$$\frac{d}{dx}(yy') - \frac{d}{dx}x = 0$$
Use the Difference Rule.

$$y \cdot \frac{d}{dx}y' + \left(\frac{d}{dx}y\right)y' - 1 = 0$$
Use the Product Rule.

$$yy'' + (y')^2 - 1 = 0$$

$$y'' = \frac{1 - (y')^2}{y} \qquad\qquad \text{Solve for } y''. \tag{5}$$

provided $y \neq 0$. To express y'' in terms of x and y, we use (4) and substitute for y' in (5).

$$y'' = \underset{\substack{\uparrow \\ y' = \frac{x}{y}}}{\frac{1 - \left(\frac{x}{y}\right)^2}{y}} = \frac{y^2 - x^2}{y^3} = \underset{\substack{\uparrow \\ y^2 - x^2 = 5}}{\frac{5}{y^3}}$$

∎

NOW WORK Problem **49**.

③ Differentiate Functions with Rational Exponents

In Section 2.4, we showed that the Simple Power Rule

$$\frac{d}{dx} x^n = n x^{n-1}$$

is true if the exponent n is any integer. We use implicit differentiation to extend this result for rational exponents.

> **THEOREM Simple Power Rule for Rational Exponents**
>
> If $y = x^{p/q}$, where $\dfrac{p}{q}$ is a rational number, then
>
> $$\boxed{y' = \frac{d}{dx} x^{p/q} = \frac{p}{q} \cdot x^{(p/q)-1}} \tag{6}$$
>
> provided that $x^{p/q}$ and $x^{(p/q)-1}$ are defined.

Proof We begin with the function $y = x^{p/q}$, where p and $q > 0$ are integers. Now, raise both sides of the equation to the power q to obtain

$$y^q = x^p$$

This is the implicit form of the function $y = x^{p/q}$. Assuming y is differentiable,* we can differentiate implicitly, obtaining

$$\frac{d}{dx} y^q = \frac{d}{dx} x^p$$

$$q y^{q-1} y' = p x^{p-1}$$

Now solve for y'.

$$y' = \underset{\substack{\uparrow \\ y = x^{p/q}}}{\frac{p x^{p-1}}{q y^{q-1}}} = \frac{p}{q} \cdot \frac{x^{p-1}}{(x^{p/q})^{q-1}} = \frac{p}{q} \cdot \frac{x^{p-1}}{x^{p-(p/q)}} = \frac{p}{q} \cdot x^{p-1-[p-(p/q)]} = \frac{p}{q} \cdot x^{(p/q)-1}$$

∎

---

*In Problem 89, you are asked to show that $y = x^{p/q}$ is differentiable.

EXAMPLE 6 **Differentiating Functions with Rational Exponents**

(a) $\dfrac{d}{dx}\sqrt{x} = \dfrac{d}{dx}x^{1/2} = \dfrac{1}{2}x^{(1/2)-1} = \dfrac{1}{2}x^{-1/2} = \dfrac{1}{2x^{1/2}} = \dfrac{1}{2\sqrt{x}} = \dfrac{\sqrt{x}}{2x}$

(b) $\dfrac{d}{du}\sqrt[3]{u} = \dfrac{d}{du}u^{1/3} = \dfrac{1}{3}u^{-2/3} = \dfrac{1}{3\sqrt[3]{u^2}} = \dfrac{\sqrt[3]{u}}{3u}$

(c) $\dfrac{d}{dx}x^{5/2} = \dfrac{5}{2}x^{3/2}$

(d) $\dfrac{d}{ds}s^{-3/2} = -\dfrac{3}{2}s^{-5/2} = -\dfrac{3}{2s^{5/2}}$ ∎

NOW WORK   Problem 31 and AP® Practice Problems 6, 9, and 11.

THEOREM **Power Rule for Functions**

If u is a differentiable function of x and r is a rational number, then

$$\boxed{\dfrac{d}{dx}[u(x)]^r = r[u(x)]^{r-1}u'(x)}$$

provided u^r and u^{r-1} are defined.

EXAMPLE 7 **Differentiating Functions Using the Power Rule for Functions**

(a) $\dfrac{d}{ds}(s^3 - 2s + 1)^{5/3} = \dfrac{5}{3}(s^3 - 2s + 1)^{2/3}\dfrac{d}{ds}(s^3 - 2s + 1) = \dfrac{5}{3}(s^3 - 2s + 1)^{2/3}(3s^2 - 2)$

(b) $\dfrac{d}{dx}\sqrt[3]{x^4 - 3x + 5} = \dfrac{d}{dx}(x^4 - 3x + 5)^{1/3} = \dfrac{1}{3}(x^4 - 3x + 5)^{-2/3}\dfrac{d}{dx}(x^4 - 3x + 5)$

$$= \dfrac{4x^3 - 3}{3(x^4 - 3x + 5)^{2/3}}$$

(c) $\dfrac{d}{d\theta}[\tan(3\theta)]^{-3/4} = -\dfrac{3}{4}[\tan(3\theta)]^{-7/4}\dfrac{d}{d\theta}\tan(3\theta) = -\dfrac{3}{4}[\tan(3\theta)]^{-7/4}\cdot\sec^2(3\theta)\cdot 3$

$$= -\dfrac{9\sec^2(3\theta)}{4[\tan(3\theta)]^{7/4}}$$ ∎

NOW WORK   Problem 39 and AP® Practice Problems 2, 4, and 7.

3.2 Assess Your Understanding

Concepts and Vocabulary

1. True or False Implicit differentiation is a technique for finding the derivative of an implicitly defined function.

2. True or False If $y^q = x^p$ for integers p and q, then $qy^{q-1} = px^{p-1}$.

3. $\dfrac{d}{dx}(3x^{1/3}) = $ _____.

4. If $y = (x^2 + 1)^{3/2}$, then $y' = $ _____.

Skill Building

In Problems 5–30, find $y' = \dfrac{dy}{dx}$ using implicit differentiation.

5. $3x + 2y = 3$

6. $2x - 5y = 7$

7. $x^2 + y^2 = 4x$

8. $y^4 - 4x^2 = 4y$

9. $e^y = \sin x$

10. $e^y = \tan x$

11. $e^{x+y} = y$

12. $e^{x+y} = x^2$

13. $x^2 y = 5$

14. $x^3 y = 8$

251 **15.** $x^2 - y^2 - xy = 2$

16. $x^2 - 4xy + y^2 = y$

249 **17.** $\dfrac{1}{x} + \dfrac{1}{y} = 1$

18. $\dfrac{1}{x} - \dfrac{1}{y} = 4$

19. $x^2 + y^2 = \dfrac{2y}{x}$

20. $x^2 + y^2 = \dfrac{2y^2}{x^2}$

21. $e^x \sin y + e^y \cos x = 4$

22. $e^y \cos x + e^{-x} \sin y = 10$

23. $(x^2 + y)^3 = y$

24. $(x + y^2)^3 = 3x$

252 **25.** $y = \tan(x - y)$

26. $y = \cos(x + y)$

27. $y = x \sin y$

28. $y = x \cos y$

29. $x^2 y = e^{xy}$

30. $ye^x = y - x$

In Problems 31–46, find y'.

254 **31.** $y = x^{2/3} + 4$

32. $y = x^{1/3} - 1$

33. $y = \sqrt[3]{x^2}$

34. $y = \sqrt[4]{x^5}$

35. $y = \sqrt[3]{x} - \dfrac{1}{\sqrt[3]{x}}$

36. $y = \sqrt{x} + \dfrac{1}{\sqrt{x}}$

37. $y = (x^3 - 1)^{3/2}$

38. $y = (x^2 - 1)^{5/3}$

254 **39.** $y = x\sqrt{x^2 - 1}$

40. $y = x\sqrt{x^3 + 1}$

41. $y = e\sqrt{x^2 - 9}$

42. $y = \sqrt{e^x}$

43. $y = (x^2 \cos x)^{3/2}$

44. $y = (x^2 \sin x)^{3/2}$

45. $y = (x^2 - 3)^{3/2}(6x + 1)^{5/3}$

46. $y = \dfrac{(2x^3 - 1)^{4/3}}{(3x + 4)^{5/2}}$

In Problems 47–52, find y' and y''.

47. $x^2 + y^2 = 4$

48. $x^2 - y^2 = 1$

253 **49.** $x^2 - y^2 = 4 + 5x$

50. $4xy = x^2 + y^2$

51. $y = \sqrt{x^2 + 1}$

52. $y = \sqrt{4 - x^2}$

In Problems 53–58, for each implicitly defined equation:

(a) *Find the slope of the tangent line to the graph of the equation at the indicated point.*

(b) *Write an equation for this tangent line.*

(c) *Graph the tangent line on the same axes as the graph of the equation.*

251 **53.** $x^2 + y^2 = 5$ at $(1, 2)$

54. $(x - 3)^2 + (y + 4)^2 = 25$ at $(0, 0)$

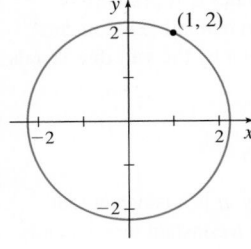

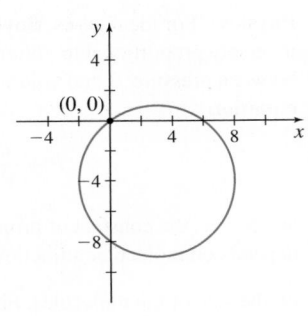

55. $x^2 - y^2 = 8$ at $(3, 1)$

56. $y^2 - 3x^2 = 6$ at $(1, -3)$

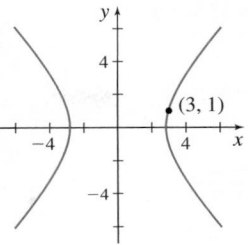

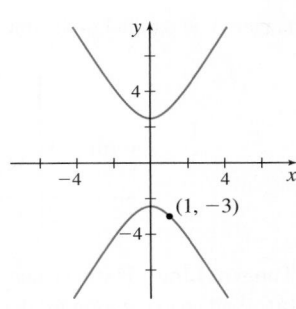

57. $\dfrac{x^2}{4} + \dfrac{y^2}{3} = 1$ at $\left(-1, \dfrac{3}{2}\right)$

58. $x^2 + \dfrac{y^2}{4} = 1$ at $\left(\dfrac{1}{2}, \sqrt{3}\right)$

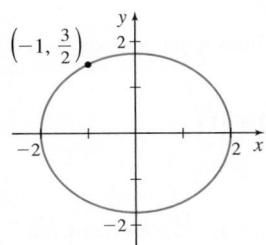

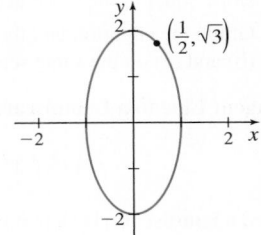

59. Find y' and y'' at the point $(-1, 1)$ on the graph of

$$3x^2 y + 2y^3 = 5x^2$$

60. Find y' and y'' at the point $(0, 0)$ on the graph of

$$4x^3 + 2y^3 = x + y$$

Applications and Extensions

In Problems 61–68, find y'.

Hint: Use the fact that $|x| = \sqrt{x^2}$.

61. $y = |3x|$

62. $y = |x^5|$

63. $y = |2x - 1|$

64. $y = |5 - x^2|$

65. $y = |\cos x|$

66. $y = |\sin x|$

67. $y = \sin |x|$

68. $|x| + |y| = 1$

69. Tangent Line to a Hypocycloid The graph of

$$x^{2/3} + y^{2/3} = 5$$

is called a **hypocycloid**. Part of its graph is shown in the figure. Find an equation of the tangent line to the hypocycloid at the point $(1, 8)$.

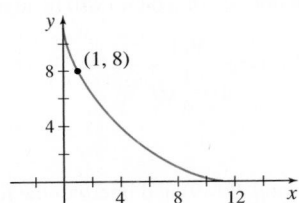

70. Tangent Line   At what point does the graph of $y = \dfrac{1}{\sqrt{x}}$ have a

tangent line parallel to the line $x + 16y = 5$? See the figure.

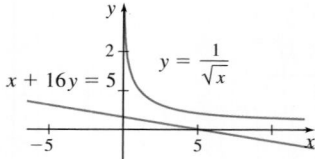

71. Tangent Line   For the equation $x + xy + 2y^2 = 6$:
   (a) Find an expression for the slope of the tangent line at any point (x, y) on the graph.
   (b) Write an equation for the line tangent to the graph at the point $(2, 1)$.
   (c) Find the coordinates of any other point on this graph with slope equal to the slope at $(2, 1)$.
   (CAS) **(d)** Graph the equation and the tangent lines found in parts (b) and (c) on the same screen.

72. Tangent Line to a Lemniscate   The graph of

$$(x^2 + y^2)^2 = x^2 - y^2$$

called a **lemniscate**, is shown in the figure. There are exactly four points at which the tangent line to the lemniscate is horizontal. Find them.

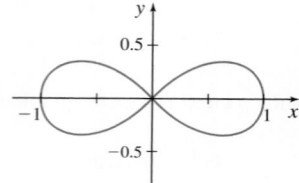

73. Motion Along a Line   An object of mass m moves along a horizontal line so that at time $t > 0$ its position s from the origin and its velocity $v = \dfrac{ds}{dt}$ satisfy the equation

$$m\left(v^2 - v_0^2\right) = k\left(s_0^2 - s^2\right)$$

where k is a positive constant and v_0 and s_0 are the initial velocity and position, respectively, of the object. Show that if $v > 0$, then

$$ma = -ks$$

where $a = \dfrac{d^2 s}{dt^2}$ is the acceleration of the object.

Hint: Differentiate the expression $m\left(v^2 - v_0^2\right) = k\left(s_0^2 - s^2\right)$ with respect to t.

74. Price Function   It is estimated that t months from now the average price (in dollars) of a tablet will be given by

$$P(t) = \frac{300}{1 + \dfrac{1}{6}\sqrt{t}} + 100 \quad 0 \le t \le 60$$

   (a) Find $P'(t)$.
   (b) Find $P'(16)$ and $P'(49)$ and interpret the results.
   (c) Graph $P = P(t)$, and explain how the graph supports the answers in (b).

75. Production Function   The production of commodities sometimes requires several resources such as land, labor, and machinery. If there are two inputs that require amounts x and y, then the output z is given by the function of two variables: $z = f(x, y)$. Here, z is called a **production function**. For example, if x represents land, y represents capital, and z is the amount of a commodity produced, a possible production function is $z = x^{0.5} y^{0.4}$. Set z equal to a fixed amount produced and

show that $\dfrac{dy}{dx} = -\dfrac{5y}{4x}$. This illustrates that the rate of change of capital with respect to land is always negative when the amount produced is fixed.

76. Learning Curve   The psychologist L. L. Thurstone suggested the following function for the time T it takes to memorize a list of n words: $T = f(n) = Cn\sqrt{n - b}$, where C and b are constants depending on the person and the task.

   (a) Find the rate of change of the time T with respect to the number n of words to be memorized.
   (b) Suppose that for a certain person and a certain task, $C = 2$ and $b = 2$. Find $f'(10)$ and $f'(30)$.
   (c) Interpret the results found in (b).

77. The Folium of Descartes   The graph of the equation $x^3 + y^3 = 2xy$ is called the **folium of Descartes**.

   (a) Find y'.
   (b) Find an equation of the tangent line to the folium of Descartes at the point $(1, 1)$.
   (c) Find any points on the graph where the tangent line to the graph is horizontal. Ignore the origin.
   (CAS) **(d)** Graph the equation $x^3 + y^3 = 2xy$. Explain how the graph supports the answers to (b) and (c).

78. If n is an even positive integer, show that the tangent line to the graph of $y = \sqrt[n]{x}$ at $(1, 1)$ is perpendicular to the tangent line to the graph of $y = x^n$ at $(-1, 1)$.

79. At what point(s), if any, is the line $y = x - 1$ parallel to the tangent line to the graph of $y = \sqrt{25 - x^2}$?

80. What is wrong with the following?
If $x + y = e^{x+y}$, then $1 + y' = e^{x+y}(1 + y')$. Since $e^{x+y} > 0$, then $y' = -1$ for all x. Therefore, $x + y = e^{x+y}$ must be a line of slope -1.

81. Show that if a function y is differentiable, and x and y are related by the equation $x^n y^m + x^m y^n = k$, where k is a constant, then

$$\frac{dy}{dx} = -\frac{y(nx^r + my^r)}{x(mx^r + ny^r)} \quad \text{where} \quad r = n - m$$

82. Physics   For ideal gases, **Boyle's law** states that pressure is inversely proportional to volume. A more realistic relationship between pressure P and volume V is given by the **van der Waals equation**

$$P + \frac{a}{V^2} = \frac{C}{V - b}$$

where C is the constant of proportionality, a is a constant that depends on molecular attraction, and b is a constant that depends on the size of the molecules. Find $\dfrac{dV}{dP}$, which measures the compressibility of the gas.

83. Tangent Line Show that an equation for the tangent line at any point (x_0, y_0) on the ellipse $\dfrac{x^2}{a^2} + \dfrac{y^2}{b^2} = 1$ is $\dfrac{xx_0}{a^2} + \dfrac{yy_0}{b^2} = 1$.

84. Tangent Line Show that the slope of the tangent line to a hypocycloid $x^{2/3} + y^{2/3} = a^{2/3}$, $a > 0$, at any point

for which $x \neq 0$, is $-\dfrac{y^{1/3}}{x^{1/3}}$.

85. Tangent Line Use implicit differentiation to show that the tangent line to a circle $x^2 + y^2 = R^2$ at any point P on the circle is perpendicular to OP, where O is the center of the circle.

Challenge Problems

86. Let $A = (2, 1)$ and $B = (5, 2)$ be points on the graph of $f(x) = \sqrt{x - 1}$. A line is moved upward on the graph so that it remains parallel to the secant line AB. Find the coordinates of the last point on the graph of f before the secant line loses contact with the graph.

Orthogonal Graphs *Problems 87 and 88 require the following definition:*

> *The graphs of two functions are said to be **orthogonal** if the tangent lines to the graphs are perpendicular at each point of intersection.*

87. (a) Show that the graphs of $xy = c_1$ and $-x^2 + y^2 = c_2$ are orthogonal, where c_1 and c_2 are positive constants.

[CAS] **(b)** Graph each function on one coordinate system for $c_1 = 1, 2, 3$ and $c_2 = 1, 9, 25$.

88. Find $a > 0$ so that the parabolas $y^2 = 2ax + a^2$ and $y^2 = a^2 - 2ax$ are orthogonal.

89. Show that if p and $q > 0$ are integers, then $y = x^{p/q}$ is a differentiable function of x.

90. We say that y is an **algebraic function** of x if it is a function that satisfies an equation of the form

$$P_0(x)y^n + P_1(x)y^{n-1} + \cdots + P_{n-1}(x)y + P_n(x) = 0$$

where $P_k(x)$, $k = 0, 1, 2, \ldots, n$, are polynomials. For example, $y = \sqrt{x}$ satisfies

$$y^2 - x = 0$$

Use implicit differentiation to obtain a formula for the derivative of an algebraic function.

Preparing for the **AP® Exam**

AP® Practice Problems

Multiple-Choice Questions

[PAGE 252] **1.** If $\sin(x^2 y) = x$, then $\dfrac{dy}{dx}$ equals

(A) $\cos(x^2 y)(2xy + x^2)$ (B) $\dfrac{1}{\cos(x^2 y)(2xy + x^2)}$

(C) $\dfrac{\sec(x^2 y) - 2xy}{x^2}$ (D) $\dfrac{\sec(x^2 y)}{2x}$

[PAGE 254] **2.** If $\dfrac{dy}{dx} = \sqrt{1 - 2y^3}$, then $\dfrac{d^2 y}{dx^2}$ equals

(A) $3y^2$ (B) $-3y^2$

(C) $-\dfrac{3y^2}{\sqrt{x - 2y^3}}$ (D) $-6y^2$

[PAGE 251] **3.** The slope of the line tangent to the graph of $x^2 y - xy^3 = 10$ at the point $(-1, 2)$ is

(A) -12 (B) $-\dfrac{4}{13}$ (C) $-\dfrac{4}{11}$ (D) $\dfrac{12}{13}$

[PAGE 254] **4.** For $f(x) = (x^3 - 4x + 32)^{1/5}$, find $f'(0)$.

(A) $-\dfrac{1}{4}$ (B) $-\dfrac{1}{20}$ (C) $\dfrac{1}{80}$ (D) $\dfrac{1}{20}$

[PAGE 251] **5.** Find y' if $x^4 + 4xy + 6y^2 = 12$.

(A) $-\dfrac{x^3 + y}{x + 3y}$ (B) $-\dfrac{x^3 + 3y}{x + 12y}$

(C) $\dfrac{x^3 + y}{x + 3y}$ (D) $\dfrac{-x^3 + y}{x + 3y}$

[PAGE 254] **6.** Find $\dfrac{dy}{dx}$ at the point $(3, 8)$ on the graph of $5xy^{2/3} - x^2 y = -12$.

(A) 2 (B) 17 (C) -7 (D) $-\dfrac{28}{3}$

[PAGE 254] **7.** What is the domain of the derivative of $f(x) = 3x^{2/3}(x + 5)^{1/3}$?

(A) The set of all real numbers   (B) $\{x \mid x \geq 0\}$

(C) $\{x \mid x \neq 0\}$ (D) $\{x \mid x \neq 0, \ x \neq -5\}$

[PAGE 252] **8.** If $e^y = \tan x$, $0 < x < \dfrac{\pi}{2}$, what is $\dfrac{dy}{dx}$ in terms of x?

(A) $\sec x \csc x$ (B) $\sec^2 x$

(C) $\sec x$ (D) $\sin x \sec x$

[PAGE 254] **9.** $\displaystyle\lim_{h \to 0} \dfrac{\sqrt[3]{x + h} - \sqrt[3]{x}}{h} =$

(A) $-\dfrac{3}{x^{2/3}}, \ x \neq 0$ (B) $\dfrac{1}{3x^{2/3}}, \ x \neq 0$

(C) $\dfrac{1}{3x^{2/3}}, \ x \geq 0$ (D) $\dfrac{x^{2/3}}{3}$

(continued on the next page)

10. The slope of the line tangent to the graph
of $e^{x^2+y^2} = xy + e$ at the point $(0, -1)$ equals

(A) $\dfrac{1}{e}$ (B) $\dfrac{e}{2}$ (C) $\dfrac{1}{2}$ (D) $\dfrac{1}{2e}$

Free-Response Questions

11. (a) Find the slope of the line tangent to the graph
of $x^{2/3} + y^{2/3} = 5$ at the point $(1, 8)$.
(b) Identify any vertical tangent lines of the graph. Just your answer.

12. Consider the equation $x^2 y + 3y^3 = 24$.

(a) Find $\dfrac{dy}{dx}$.

(b) Determine the points, if any, where the line tangent to the graph of the equation is horizontal.

Retain Your Knowledge

Multiple-Choice Questions

1. $\lim\limits_{x\to 0}\left(x^3 \sin\dfrac{1}{x}\right) =$

(A) -1 (B) 0 (C) 1 (D) The limit does not exist.

2. The table below lists the values of a continuous function f for select numbers x. Use the table to approximate $f'(1)$.

x	-1	0	$\dfrac{1}{2}$	2	3	5
$f(x)$	2	1	0	-3	$-\dfrac{3}{2}$	0

(A) -2 (B) -4 (C) $-\dfrac{3}{2}$ (D) $-\dfrac{1}{2}$

3. If $y = x\cos x$, find y''.

(A) $-x\cos x$ (B) $-2\sin x - x\cos x$

(C) $-2\cos x + \sin x$ (D) $-x\cos x$

Free-Response Question

4. Find numbers A and B so that the function is continuous for all numbers x. Justify your answers.

$$f(x) = \begin{cases} 3x - A & \text{if } x \le 2 \\ (2-x)^2 & \text{if } 2 < x < 7 \\ Bx + 4 & \text{if } x \ge 7 \end{cases}$$

3.3 Derivatives of Inverse Functions

OBJECTIVES *When you finish this section, you should be able to:*

1 **Find the derivative of an inverse function (p. 258)**
2 **Find the derivative of the inverse trigonometric functions (p. 260)**

NEED TO REVIEW? Inverse functions are discussed in Section P.4, pp. 38–41.

The Chain Rule is used to find the derivative of an inverse function.

1 Find the Derivative of an Inverse Function

Suppose f is a function and g is its inverse function. Then

$$g(f(x)) = x$$

for all x in the domain of f.

If both f and g are differentiable, we can differentiate both sides with respect to x and use the Chain Rule.

$$\frac{d}{dx}g(f(x)) = \frac{d}{dx}x$$

$$g'(f(x)) \cdot f'(x) = 1 \qquad \text{Use the Chain Rule on the left.}$$

Since the product of the derivatives is never 0, each function has a nonzero derivative on its domain.

Conversely, if a one-to-one function has a nonzero derivative, then its inverse function also has a nonzero derivative as stated in the following theorem.

THEOREM Derivative of an Inverse Function

Suppose $y = f(x)$ and $x = g(y)$ are inverse functions, and suppose f is differentiable on an open interval containing x_0, and $y_0 = f(x_0)$. If $f'(x_0) \neq 0$, then g is differentiable at $y_0 = f(x_0)$ and

$$\left. \frac{d}{dy} g(y) \right|_{y=y_0} = g'(y_0) = \frac{1}{f'(x_0)} \tag{1}$$

where the notation $\left. \dfrac{d}{dy} g(y) \right|_{y=y_0} = g'(y_0)$ means $\dfrac{d}{dy} g(y)$ evaluated at y_0.

We have three comments about this theorem, which is proved in Appendix B:

Suppose $y = f(x)$ and $x = g(y)$ are inverse functions.

- If $y_0 = f(x_0)$, and $f'(x_0) \neq 0$ exists, then $\dfrac{d}{dy} g(y)$ exists at y_0.

- Statement (1) in the Theorem box above can be used to find $\dfrac{d}{dy} g(y)$ at y_0 without knowing a formula for $g = f^{-1}$, provided we can find x_0 and $f'(x_0)$.

- In Leibniz notation, statement (1) is of the form $\dfrac{dy}{dx} = \dfrac{1}{\frac{dx}{dy}}$.

IN WORDS The derivative of a function equals the reciprocal of the derivative of its inverse function.

EXAMPLE 1 Finding the Derivative of an Inverse Function

The function $f(x) = x^5 + x$ has an inverse function g. Find $g'(2)$.

Solution
Using (1) with $y_0 = 2$, we get

$$g'(2) = \frac{1}{f'(x_0)} \qquad \text{where } 2 = f(x_0)$$

To find x_0, we use the equation

$$f(x_0) = x_0^5 + x_0 = 2$$

A solution is $x_0 = 1$. So $g'(2) = \dfrac{1}{f'(1)}$. Since $f'(x) = 5x^4 + 1$, then $f'(1) = 6 \neq 0$ and

$$g'(2) = \frac{1}{f'(1)} = \frac{1}{6}$$

Observe in Example 1 that the derivative of the inverse function g at 2 was obtained without actually knowing a formula for g.

NOW WORK Problem 5 and AP® Practice Problems 3, 5, 8, and 9.

EXAMPLE 2   Finding an Equation of a Tangent Line

The function $f(x) = x^3 + 2x - 4$ is one-to-one and has an inverse function g. Find an equation of the tangent line to the graph of g at the point $(8, 2)$ on g.

Solution

RECALL If the functions f and g are inverses and if $f(a) = b$, then $g(b)$ exists, and $g(b) = a$.

The slope of the tangent line to the graph of g at the point $(8, 2)$ is $g'(8)$. Since f and g are inverse functions, $g(8) = 2$ and $f(2) = 8$. Now use (1) to find $g'(8)$.

$$g'(8) = \frac{1}{f'(2)} \qquad g'(y_0) = \frac{1}{f'(x_0)}; \; x_0 = 2; \; y_0 = 8$$

$$= \frac{1}{14} \qquad f'(x) = 3x^2 + 2; \; f'(2) = 3 \cdot 2^2 + 2 = 14$$

Now use the point-slope form of an equation of a line to find an equation of the tangent line to g at $(8, 2)$.

$$y - 2 = \frac{1}{14}(x - 8)$$
$$14y - 28 = x - 8$$
$$14y - x = 20$$

The line $14y - x = 20$ is tangent to the graph of g at the point $(8, 2)$. ∎

NOW WORK   Problem 51 and AP® Practice Problems 7 and 11.

❷ Find the Derivative of the Inverse Trigonometric Functions

NEED TO REVIEW? Inverse trigonometric functions are discussed in Section P.7, pp. 63–67.

Table 1 lists the inverse trigonometric functions and their domains.

TABLE 1 The Domain of the Inverse Trigonometric Functions

f	Restricted Domain	f^{-1}	Domain
$f(x) = \sin x$	$\left[-\dfrac{\pi}{2}, \dfrac{\pi}{2}\right]$	$f^{-1}(x) = \sin^{-1} x$	$[-1, 1]$
$f(x) = \cos x$	$[0, \pi]$	$f^{-1}(x) = \cos^{-1} x$	$[-1, 1]$
$f(x) = \tan x$	$\left(-\dfrac{\pi}{2}, \dfrac{\pi}{2}\right)$	$f^{-1}(x) = \tan^{-1} x$	$(-\infty, \infty)$
$f(x) = \csc x$	$\left(-\pi, -\dfrac{\pi}{2}\right] \cup \left(0, \dfrac{\pi}{2}\right]$	$f^{-1}(x) = \csc^{-1} x$	$\lvert x \rvert \geq 1$
$f(x) = \sec x$	$\left[0, \dfrac{\pi}{2}\right) \cup \left[\pi, \dfrac{3\pi}{2}\right)$	$f^{-1}(x) = \sec^{-1} x$	$\lvert x \rvert \geq 1$
$f(x) = \cot x$	$(0, \pi)$	$f^{-1}(x) = \cot^{-1} x$	$(-\infty, \infty)$

AP® EXAM TIP

The exam often uses $\arcsin x$ instead of $\sin^{-1} x$, $\arccos x$ instead of $\cos^{-1} x$, and so on. Be familiar with both notations.

CAUTION Keep in mind that $f^{-1}(x)$ represents the inverse function of f and not the reciprocal function. That is, $f^{-1}(x) \neq \dfrac{1}{f(x)}$.

To find the derivative of $y = \sin^{-1} x$, $-1 \leq x \leq 1$, $-\dfrac{\pi}{2} \leq y \leq \dfrac{\pi}{2}$, we write $\sin y = x$ and differentiate implicitly with respect to x.

$$\frac{d}{dx} \sin y = \frac{d}{dx} x$$

NOTE Alternatively, we can use the Derivative of an Inverse Function. If $y = \sin^{-1} x$, then $x = \sin y$, and

$$\frac{dy}{dx} = \frac{1}{\dfrac{dx}{dy}} = \frac{1}{\cos y}$$

provided $\cos y \neq 0$.

$$\cos y \cdot \frac{dy}{dx} = 1 \qquad \frac{d}{dx} \sin y = \cos y \frac{dy}{dx} \text{ using the Chain Rule.}$$

$$\frac{dy}{dx} = \frac{1}{\cos y}$$

provided $\cos y \neq 0$. Since $\cos y = 0$ if $y = -\dfrac{\pi}{2}$ or $y = \dfrac{\pi}{2}$, we exclude these values.

Then $\dfrac{d}{dx}\sin^{-1}x = \dfrac{1}{\cos y}$, $-\dfrac{\pi}{2} < y < \dfrac{\pi}{2}$. Now, if $-\dfrac{\pi}{2} < y < \dfrac{\pi}{2}$, then $\cos y > 0$. Using a Pythagorean identity, we have

$$\cos^2 y = 1 - \sin^2 y$$

$$\cos y = \underset{\substack{\uparrow \\ \cos y > 0}}{\sqrt{1 - \sin^2 y}} = \underset{\substack{\uparrow \\ \sin y = x}}{\sqrt{1 - x^2}}$$

We have proved the following theorem.

$\dfrac{d}{dx}\arcsin x = \dfrac{1}{\sqrt{1-x^2}}$, $-1 < x < 1$

THEOREM Derivative of $y = \sin^{-1}x$

The derivative of the inverse sine function $y = \sin^{-1}x$ is

$$y' = \dfrac{d}{dx}\sin^{-1}x = \dfrac{1}{\sqrt{1-x^2}} \qquad -1 < x < 1$$

EXAMPLE 3 Using the Chain Rule with the Inverse Sine Function

Find y' if:

(a) $y = \sin^{-1}(4x^2)$ **(b)** $y = e^{\sin^{-1}x}$

Solution

(a) If $y = \sin^{-1}u$ and $u = 4x^2$, then $\dfrac{dy}{du} = \dfrac{1}{\sqrt{1-u^2}}$ and $\dfrac{du}{dx} = 8x$. By the Chain Rule,

$$y' = \dfrac{dy}{dx} = \dfrac{dy}{du}\cdot\dfrac{du}{dx} = \dfrac{1}{\sqrt{1-u^2}}\cdot 8x = \dfrac{8x}{\sqrt{1-16x^4}} \qquad u = 4x^2$$

(b) If $y = e^u$ and $u = \sin^{-1}x$, then $\dfrac{dy}{du} = e^u$ and $\dfrac{du}{dx} = \dfrac{1}{\sqrt{1-x^2}}$. By the Chain Rule,

$$y' = \dfrac{dy}{dx} = \dfrac{dy}{du}\cdot\dfrac{du}{dx} = e^u\cdot\underset{\substack{\uparrow \\ u = \sin^{-1}x}}{\dfrac{1}{\sqrt{1-x^2}}} = \dfrac{e^{\sin^{-1}x}}{\sqrt{1-x^2}}$$ ■

NOW WORK   Problem **27** and AP® Practice Problems **2** and **10**.

To derive a formula for the derivative of $y = \tan^{-1}x$, $-\infty < x < \infty$, $-\dfrac{\pi}{2} < y < \dfrac{\pi}{2}$, we write $\tan y = x$ and differentiate implicitly with respect to x. Then

$$\dfrac{d}{dx}\tan y = \dfrac{d}{dx}x$$

$$\sec^2 y\,\dfrac{dy}{dx} = 1 \qquad \text{Use the Chain Rule on the left.}$$

$$\dfrac{dy}{dx} = \dfrac{1}{\sec^2 y}$$

Since $-\dfrac{\pi}{2} < y < \dfrac{\pi}{2}$, then $\sec y \neq 0$.

Now use the Pythagorean identity $\sec^2 y = 1 + \tan^2 y$, and substitute $x = \tan y$. Then

$$y' = \frac{d}{dx} \tan^{-1} x = \frac{1}{1+x^2}$$

AP® EXAM TIP

$\frac{d}{dx} \arctan x = \frac{1}{1+x^2}$

THEOREM Derivative of $y = \tan^{-1} x$

The derivative of the inverse tangent function $y = \tan^{-1} x$ is

$$y' = \frac{d}{dx} \tan^{-1} x = \frac{1}{1+x^2}$$

The domain of y' is all real numbers.

EXAMPLE 4 **Using the Chain Rule with the Inverse Tangent Function**

Find y' if:

(a) $y = \tan^{-1}(4x)$ **(b)** $y = \ln(\tan^{-1} x)$

Solution

(a) Let $y = \tan^{-1} u$ and $u = 4x$. Then $\dfrac{dy}{du} = \dfrac{1}{1+u^2}$ and $\dfrac{du}{dx} = 4$. By the Chain Rule,

$$y' = \frac{dy}{dx} = \frac{dy}{du} \cdot \frac{du}{dx} = \frac{1}{1+u^2} \cdot 4 = \frac{4}{1+16x^2}$$

(b) Let $y = \ln u$ and $u = \tan^{-1} x$. Then $\dfrac{dy}{du} = \dfrac{1}{u}$ and $\dfrac{du}{dx} = \dfrac{1}{1+x^2}$. By the Chain Rule,

$$y' = \frac{dy}{dx} = \frac{dy}{du} \cdot \frac{du}{dx} = \frac{1}{u} \cdot \frac{1}{1+x^2} = \frac{1}{(1+x^2)(\tan^{-1} x)}$$ ∎

NOW WORK Problem 35 and AP® Practice Problems 1, 4, and 6.

Finally, to find the derivative of $y = \sec^{-1} x$, $|x| \geq 1$, $0 \leq y < \dfrac{\pi}{2}$ or $\pi \leq y < \dfrac{3\pi}{2}$, we write $x = \sec y$ and use implicit differentiation.

$$\frac{d}{dx} x = \frac{d}{dx} \sec y$$

$$1 = \sec y \tan y \frac{dy}{dx} \qquad \text{Use the Chain Rule on the right.}$$

We can solve for $\dfrac{dy}{dx}$ provided $\sec y \neq 0$ and $\tan y \neq 0$, or equivalently, $y \neq 0$ and $y \neq \pi$. That is,

$$\frac{dy}{dx} = \frac{1}{\sec y \tan y}, \qquad \text{provided } 0 < y < \frac{\pi}{2} \text{ or } \pi < y < \frac{3\pi}{2}.$$

With these restrictions on y, $\tan y > 0$.* Now we use the Pythagorean identity $1 + \tan^2 y = \sec^2 y$. Then $\tan y = \sqrt{\sec^2 y - 1}$ and

$$\frac{1}{\sec y \tan y} = \frac{1}{\sec y \sqrt{\sec^2 y - 1}} \underset{\substack{\uparrow \\ \sec y = x}}{=} \frac{1}{x\sqrt{x^2 - 1}}$$

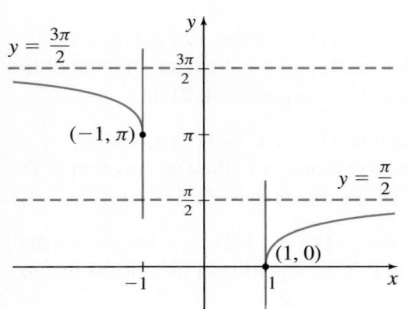

Figure 4 $y = \sec^{-1} x$, $|x| \geq 1$,
$$0 \leq y < \frac{\pi}{2} \text{ or } \pi \leq y < \frac{3\pi}{2}$$

> **THEOREM** Derivative of $y = \sec^{-1} x$
>
> The derivative of the inverse secant function $y = \sec^{-1} x$ is
>
> $$y' = \frac{d}{dx} \sec^{-1} x = \frac{1}{x\sqrt{x^2 - 1}} \qquad |x| > 1$$

Notice that $y = \sec^{-1} x$ is not differentiable when $x = \pm 1$. In fact, as Figure 4 shows, at the points $(-1, \pi)$ and $(1, 0)$, the graph of $y = \sec^{-1} x$ has vertical tangent lines.

NOW WORK Problem **19.**
_____

The formulas for the derivatives of the three remaining inverse trigonometric functions can be obtained using the following identities:

> - Since $\cos^{-1} x = \dfrac{\pi}{2} - \sin^{-1} x$, then $\dfrac{d}{dx} \cos^{-1} x = -\dfrac{1}{\sqrt{1-x^2}}$, where $|x| < 1$.
>
> - Since $\cot^{-1} x = \dfrac{\pi}{2} - \tan^{-1} x$, then $\dfrac{d}{dx} \cot^{-1} x = -\dfrac{1}{1+x^2}$, where $-\infty < x < \infty$.
>
> - Since $\csc^{-1} x = \dfrac{\pi}{2} - \sec^{-1} x$, then $\dfrac{d}{dx} \csc^{-1} x = -\dfrac{1}{x\sqrt{x^2-1}}$, where $|x| > 1$.

_____

*Look back at the definition given for the inverse secant function in Section P.7, p. 66. The restricted domain of $f(\theta) = \sec \theta$ was chosen as $\left\{ \theta \,\middle|\, 0 \leq \theta < \dfrac{\pi}{2} \text{ or } \pi \leq \theta < \dfrac{3\pi}{2} \right\}$ so that $\tan \theta \geq 0$.

3.3 Assess Your Understanding

Concepts and Vocabulary

1. True or False If f is a one-to-one differentiable function and if $f'(x) > 0$, then f has an inverse function whose derivative is positive.

2. True or False $\dfrac{d}{dx} \tan^{-1} x = \dfrac{1}{1+x^2}$, where $-\infty < x < \infty$.

3. True or False If f and g are inverse functions, then

$$g'(y_0) = \frac{1}{f'(y_0)}$$

4. $\dfrac{d}{dx} \sin^{-1} x =$ _____ where $-1 < x < 1$.

Skill Building

In Problems 5–8, the functions f and g are inverse functions.

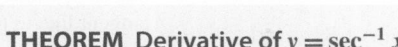

5. If $f(0) = 4$ and $f'(0) = -2$, find $g'(4)$.

6. If $f(1) = -2$ and $f'(1) = 4$, find $g'(-2)$.

7. If $g(4) = -2$ and $g'(4) = \dfrac{1}{2}$, find $f'(-2)$.

8. If $g(-4) = 0$ and $g'(-4) = -\dfrac{1}{3}$, find $f'(0)$.

In Problems 9–16, f and g are inverse functions. For each function f, find $g'(y_0)$.

9. $f(x) = x^5$; $y_0 = 32$

10. $f(x) = x^3$; $y_0 = 27$

11. $f(x) = x^2 + 2$, $x \geq 0$; $y_0 = 6$

12. $f(x) = x^2 - 5$, $x \geq 0$; $y_0 = 4$

13. $f(x) = x^{1/3}$; $y_0 = 2$

14. $f(x) = x^{2/5}$; $y_0 = 4$

15. $f(x) = 3x^{4/3} + 1$; $y_0 = 49$

16. $f(x) = x^{2/3} + 5$; $y_0 = 6$

In Problems 17–42, find the derivative of each function.

17. $f(x) = \sin^{-1}(4x)$

18. $f(x) = \sin^{-1}(3x - 2)$

[263] 19. $g(x) = \sec^{-1}(3x)$

20. $g(x) = \cos^{-1}(2x)$

21. $s(t) = \tan^{-1} \dfrac{t}{2}$

22. $s(t) = \sec^{-1} \dfrac{t}{3}$

23. $f(x) = \tan^{-1}(1 - 2x^2)$

24. $f(x) = \sin^{-1}(1 - x^2)$

25. $f(x) = \sec^{-1}(x^2 + 2)$

26. $f(x) = \cos^{-1} x^2$

[261] 27. $F(x) = \sin^{-1} e^x$

28. $F(x) = \tan^{-1} e^x$

29. $g(x) = \tan^{-1} \dfrac{1}{x}$

30. $g(x) = \sec^{-1} \sqrt{x}$

31. $g(x) = x \sin^{-1} x$

32. $g(x) = x \tan^{-1}(x + 1)$

33. $s(t) = t^2 \sec^{-1} t^3$

34. $s(t) = t^2 \sin^{-1} t^2$

[262] 35. $f(x) = \tan^{-1}(\cos x)$

36. $f(x) = \sin^{-1}(\sin x)$

37. $G(x) = \sin(\tan^{-1} x)$

38. $G(x) = \cos(\tan^{-1} x)$

39. $f(x) = e^{\tan^{-1}(3x)}$

40. $f(x) = e^{\sec^{-1} x^2}$

41. $g(x) = \dfrac{\sin^{-1} x}{\sqrt{1 - x^2}}$

42. $g(x) = \dfrac{\sec^{-1} x}{\sqrt{x^2 - 1}}$

In Problems 43–46, use implicit differentiation to find $\dfrac{dy}{dx}$.

43. $y^2 + \sin^{-1} y = 2x$

44. $xy + \tan^{-1} y = 3$

45. $40 \tan^{-1} y^2 - \pi x^3 y = 2\pi$

46. $\sec^{-1} y - xy = \dfrac{\pi}{3}$

47. The function $f(x) = x^3 + 2x$ has an inverse function g. Find $g'(0)$ and $g'(3)$.

48. The function $f(x) = 2x^3 + x - 3$ has an inverse function g. Find $g'(-3)$ and $g'(0)$.

Applications and Extensions

49. Tangent Line

(a) Find an equation for the tangent line to the graph of $y = \sin^{-1} \dfrac{x}{2}$ at the origin.

(b) Use technology to graph $y = \sin^{-1} \dfrac{x}{2}$ and the tangent line found in (a).

50. Tangent Line

(a) Find an equation for the tangent line to the graph of $y = \tan^{-1} x$ at $x = 1$.

(b) Use technology to graph $y = \tan^{-1} x$ and the tangent line found in (a).

[260] 51. Tangent Line   The function $f(x) = 2x^3 - x$, $x \geq 1$, is one-to-one and has an inverse function g. Find an equation of the tangent line to the graph of g at the point $(14, 2)$ on g.

52. Tangent Line   The function $f(x) = x^5 - 3x$, $x \leq -1$, is one-to-one and has an inverse function g. Find an equation of the tangent line to the graph of g at the point $(-2, 1)$ on g.

53. Normal Line   The function $f(x) = x + 2x^{1/3}$ is one-to-one and has an inverse function g. Find an equation of the normal line to the graph of g at the point $(3, 1)$ on g.

54. Normal Line   The function $f(x) = x^4 + x$, $x > 0$, is one-to-one and has an inverse function g. Find an equation of the normal line to the graph of g at the point $(2, 1)$ on g.

55. Motion on a Line   An object moves along the x-axis so that its position x from the origin (in meters) is given by $x(t) = \sin^{-1} \dfrac{1}{t}$, $t > 0$, where t is the time in seconds.

(a) Find the velocity of the object at $t = 2$ s.

(b) Find the acceleration of the object at $t = 2$ s.

56. If $g(x) = \cos^{-1}(\cos x)$, show that $g'(x) = \dfrac{\sin x}{|\sin x|}$.

57. Show that $\dfrac{d}{dx} \tan^{-1}(\cot x) = -1$.

58. Show that $\dfrac{d}{dx} \cot^{-1} x = \dfrac{d}{dx} \tan^{-1} \dfrac{1}{x}$ for all $x \neq 0$.

59. Show that $\dfrac{d}{dx} \left[\sin^{-1} x - x\sqrt{1 - x^2} \right] = \dfrac{2x^2}{\sqrt{1 - x^2}}$.

Challenge Problem

60. Another way of finding the derivative of $y = \sqrt[n]{x}$ is to use inverse functions. The function $y = f(x) = x^n$, n a positive integer, has the derivative $f'(x) = nx^{n-1}$. So, if $x \neq 0$, then $f'(x) \neq 0$. The inverse function of f, namely, $x = g(y) = \sqrt[n]{y}$, is defined for all y if n is odd and for all $y \geq 0$ if n is even. Since this inverse function is differentiable for all $y \neq 0$, we have

$$g'(y) = \frac{d}{dy} \sqrt[n]{y} = \frac{1}{f'(x)} = \frac{1}{nx^{n-1}}$$

Since $nx^{n-1} = n \left(\sqrt[n]{y} \right)^{n-1} = ny^{(n-1)/n} = ny^{1-(1/n)}$, we have

$$\frac{d}{dy} \sqrt[n]{y} = \frac{d}{dy} y^{1/n} = \frac{1}{ny^{1-(1/n)}} = \frac{1}{n} y^{(1/n)-1}$$

Use the result from above and the Chain Rule to prove the formula

$$\frac{d}{dx} x^{p/q} = \frac{p}{q} x^{(p/q)-1}$$

Preparing for the AP® **Exam**

AP® Practice Problems

Multiple-Choice Questions

PAGE 262 **1.** What is the slope of the line tangent to the graph of $y = \tan^{-1}(2x)$ where $x = -1$?

(A) $\dfrac{2}{5}$ (B) $-\dfrac{2}{5}$ (C) -5 (D) $-\dfrac{5}{2}$

PAGE 261 **2.** $\dfrac{d}{dx}\sin^{-1}(e^{2x}) =$

(A) $-\dfrac{2e^{2x}}{\sqrt{1 - e^{2x}}}$ (B) $\dfrac{2e^{2x}}{\sqrt{1 - e^{4x}}}$

(C) $\dfrac{2e^{2x}}{\sqrt{1 - e^{4x^2}}}$ (D) $\dfrac{2e^{2x}}{\sqrt{1 - e^{2x}}}$

PAGE 259 **3.** The functions f and g are differentiable and g is the inverse of f. If $g(-3) = 2$ and $f'(2) = -\dfrac{1}{3}$, then $g'(-3)$ is

(A) 3 (B) $\dfrac{1}{2}$ (C) -3 (D) $\dfrac{1}{3}$

PAGE 262 **4.** If $y = \tan^{-1}(\cos x)$, then $y' =$

(A) $\dfrac{\sin x}{1 + \cos^2 x}$ (B) $-\dfrac{\cos x}{1 + \sin^2 x}$

(C) $-\dfrac{\sin x}{1 + \cos^2 x}$ (D) $\dfrac{\cos x}{1 + \sin^2 x}$

PAGE 259 **5.** The function g is given by $g(x) = x^3 + x^2 - 3$, $x \geq 0$, so $g(2) = 9$. If the function f is the inverse function of g, find $f'(9)$.

(A) $\dfrac{1}{261}$ (B) -261

(C) 16 (D) $\dfrac{1}{16}$

PAGE 262 **6.** An object is moving along the y-axis. Its position (in centimeters) at time $t > 0$ seconds is given by $y(t) = \tan^{-1} t$. What is the acceleration of the object at $t = 1$ second?

(A) $\dfrac{1}{2}$ cm/s²   (B) $-\dfrac{1}{2}$ cm/s

(C) -2 cm/s²   (D) $-\dfrac{1}{2}$ cm/s²

PAGE 260 **7.** $F(x) = x^3 - x^2 + x - 5$ and g are inverse functions. Find an equation of the line tangent to the graph of g at the point $(-4, 1)$ on g.

(A) $y = \dfrac{1}{2}x - \dfrac{9}{2}$ (B) $y = \dfrac{1}{57}x - \dfrac{53}{57}$

(C) $y = \dfrac{1}{2}x + 3$ (D) $y = 2x + 9$

PAGE 259 **8.** Suppose f and g are inverse functions and g is differentiable. Use the information in the table below to find $f'(4)$.

x	$g(x)$	$g'(x)$
-1	4	5
4	3	-1

(A) -1 (B) $-\dfrac{1}{5}$ (C) $\dfrac{1}{5}$ (D) $\dfrac{1}{4}$

PAGE 259 **9.** The function $f(x) = x^3 + x + 1$ has an inverse function g. Then $g'(3)$ equals

(A) $\dfrac{1}{31}$ (B) $\dfrac{1}{28}$ (C) $\dfrac{1}{4}$ (D) 4

PAGE 261 **10.** $f(x) = \sin^{-1}(4x)$. The equation of the line tangent to the graph of f when $x = \dfrac{1}{8}$ is

(A) $y - \dfrac{\pi}{3} = 8\sqrt{3}\left(x - \dfrac{1}{8}\right)$ (B) $y - \dfrac{\pi}{6} = \dfrac{8\sqrt{3}}{\sqrt{3}}\left(x - \dfrac{1}{8}\right)$

(C) $y - \dfrac{\pi}{6} = \dfrac{8}{3}\left(x - \dfrac{1}{8}\right)$ (D) $y - \dfrac{\pi}{6} = \dfrac{2\sqrt{3}}{\sqrt{3}}\left(x - \dfrac{1}{8}\right)$

Free-Response Question

PAGE 260 **11.** The function $f(x) = x^3 + x^2 - 3$ is one-to-one on the restricted domain $x > 0$, and so has an inverse function g. The graph of f is shown below.

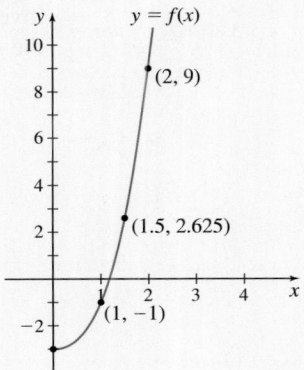

(a) Find $g'(9)$.

(b) Write an equation of the line tangent to the graph of g at the point $(9, 2)$.

Retain Your Knowledge

Multiple-Choice Questions

1. $\lim\limits_{x \to 0} \dfrac{3\cos^2 x}{e^x + x} =$

(A) 0   (B) 1

(C) 3   (D) The limit does not exist.

2. Given $f(x) = \begin{cases} 3x^2 - 10 & \text{if } x < 4 \\ 8x + 6 & \text{if } x \geq 4 \end{cases}$

find $f'(4)$, if it exists.

(A) The derivative does not exist.   (B) $f'(4) = 8$

(C) $f'(4) = 24$ (D) $f'(4) = 38$

3. Find f'' for the function $f(x) = 4x^2 e^x$.

(A) $8xe^x + 4x^2 e^x$ (B) $4e^x(x^2 + 4x + 2)$

(C) $8e^x + 8xe^x + 4x^2 e^x$ (D) $4e^x(x^2 + 4x + 1)$

Free-Response Question

4. The velocity v of an object moving along a line is modeled by

the function $v(t) = \dfrac{3t^2 + t - 7}{t + 2}, 0 \leq t \leq 4$, where t is

measured in hours and distance is measured in km.

(a) Show that there is at least one time t in the interval $[1, 4]$ for which $v(t) = 0$.

(b) What is the acceleration a of the object at time $t = 3$ h?

3.4 Derivatives of Logarithmic Functions

OBJECTIVES *When you finish this section, you should be able to:*

1 **Differentiate logarithmic functions (p. 267)**

2 **Use logarithmic differentiation (p. 270)**

3 **Express *e* as a limit (p. 272)**

NEED TO REVIEW? Logarithmic functions are discussed in Section P.5, pp. 47–49.

Recall that the function $f(x) = a^x$, where $a > 0$ and $a \neq 1$, is one-to-one and is defined for all real numbers. We know from Section 3.1 that $f'(x) = a^x \ln a$ and that the domain of $f'(x)$ is also the set of all real numbers. Moreover, since $f'(x) \neq 0$, the inverse of f, $y = \log_a x$, is differentiable on its domain $(0, \infty)$. See Figure 5 for the graphs of $y = a^x$ and $y = \log_a x$, $a > 0$, $a \neq 1$.

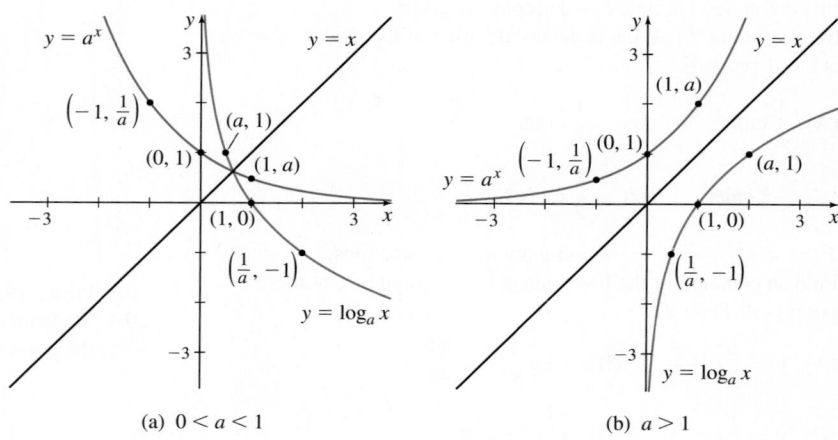

(a) $0 < a < 1$ (b) $a > 1$

Figure 5

❶ Differentiate Logarithmic Functions

To find $y' = \dfrac{d}{dx} \log_a x$, we use the fact that the following statements are equivalent:

$$y = \log_a x \quad \text{if and only if} \quad x = a^y$$

That is, $y = \log_a x$ and $x = a^y$ are inverse functions.

Using the Derivative of an Inverse Function and the fact that $\dfrac{dx}{dy} = \dfrac{d}{dy} a^y = a^y \ln a$, we have

$$\frac{d}{dx} \log_a x = \frac{dy}{dx} = \frac{1}{\dfrac{dx}{dy}} = \frac{1}{a^y \ln a} \underset{\substack{\uparrow \\ x = a^y}}{=} \frac{1}{x \ln a}$$

THEOREM Derivative of a Logarithmic Function

The derivative of the logarithmic function $y = \log_a x$, $x > 0$, $a > 0$, and $a \neq 1$, is

$$\boxed{y' = \frac{d}{dx} \log_a x = \frac{1}{x \ln a} \qquad x > 0} \tag{1}$$

When $a = e$, then $\ln a = \ln e = 1$, and (1) gives the formula from Section 2.3 for the derivative of the natural logarithm function $f(x) = \ln x$.

THEOREM Derivative of the Natural Logarithm Function

The derivative of the natural logarithm function $y = \ln x$, $x > 0$, is

$$\boxed{y' = \frac{d}{dx} \ln x = \frac{1}{x}}$$

EXAMPLE 1 Differentiating Logarithmic Functions

Find y' if:

(a) $y = (\ln x)^2$ **(b)** $y = \dfrac{1}{2} \log x$ **(c)** $\ln x + \ln y = 2x$ **(d)** $y = \log_2 x^3$

Solution

(a) We use the Power Rule for Functions. Then

$$y' = \frac{d}{dx} (\ln x)^2 = 2 \ln x \cdot \underset{\uparrow}{\frac{d}{dx} \ln x} = 2 \ln x \cdot \frac{1}{x} = \frac{2 \ln x}{x}$$

$$\frac{d}{dx}[u(x)]^n = n[u(x)]^{n-1} u'(x)$$

(b) Remember that $\log x = \log_{10} x$. Then $y = \dfrac{1}{2} \log x = \dfrac{1}{2} \log_{10} x$.

$$y' = \frac{1}{2} \left(\frac{d}{dx} \log_{10} x \right) = \frac{1}{2} \cdot \frac{1}{x \ln 10} = \frac{1}{2x \ln 10} \qquad \frac{d}{dx} \log_a x = \frac{1}{x \ln a}$$

(c) We assume y is a differentiable function of x and use implicit differentiation. Then

$$\ln x + \ln y = 2x$$

$$\frac{d}{dx}(\ln x + \ln y) = \frac{d}{dx}(2x)$$

$$\frac{d}{dx}\ln x + \frac{d}{dx}\ln y = 2$$

$$\frac{1}{x} + \frac{1}{y}\frac{dy}{dx} = 2 \qquad \frac{d}{dx}\ln y = \frac{1}{y}\frac{dy}{dx}$$

$$y' = \frac{dy}{dx} = y\left(2 - \frac{1}{x}\right) = \frac{y(2x-1)}{x} \qquad \text{Solve for } y' = \frac{dy}{dx}.$$

(d) We use a property of logarithms and write

$$y = \log_2 x^3 = 3\log_2 x \qquad \log_a u^r = r\log_a u$$

Then,

$$y' = 3\frac{d}{dx}\log_2 x = 3\frac{1}{x\ln 2} = \frac{3}{x\ln 2} \qquad\blacksquare$$

NOW WORK   Problem 9 and AP® Practice Problems 1, 10, and 11.

EXAMPLE 2   Differentiating Logarithmic Functions

Find y' if:

(a) $y = \ln(5x)$ **(b)** $y = x\ln(x^2+1)$ **(c)** $y = \ln|x|$

Solution

(a) $y' = \dfrac{d}{dx}\ln(5x) = \dfrac{\frac{d}{dx}(5x)}{5x} = \dfrac{5}{5x} = \dfrac{1}{x}$ $\dfrac{d}{dx}\ln u(x) = \dfrac{u'(x)}{u(x)}$

(b) $y' = \dfrac{d}{dx}[x\ln(x^2+1)] = x\dfrac{d}{dx}\ln(x^2+1) + \left(\dfrac{d}{dx}x\right)\ln(x^2+1)$ **Product Rule**

$$= x\cdot\underset{\underset{\frac{d}{dx}\ln u(x) = \frac{u'(x)}{u(x)}}{\uparrow}}{\frac{\frac{d}{dx}(x^2+1)}{x^2+1}} + 1\cdot\ln(x^2+1) = \frac{2x^2}{x^2+1} + \ln(x^2+1)$$

(c) Express $y = |x|$ as a piecewise-defined function. Then

$$y = \ln|x| = \begin{cases} \ln x & \text{if } x > 0 \\ \ln(-x) & \text{if } x < 0 \end{cases}$$

$$y' = \frac{d}{dx}\ln|x| = \begin{cases} \dfrac{1}{x} & \text{if } x > 0 \\ \dfrac{1}{-x}(-1) = \dfrac{1}{x} & \text{if } x < 0 \end{cases}$$

That is, $\dfrac{d}{dx}\ln|x| = \dfrac{1}{x}$ for all numbers $x \neq 0$. $\blacksquare$

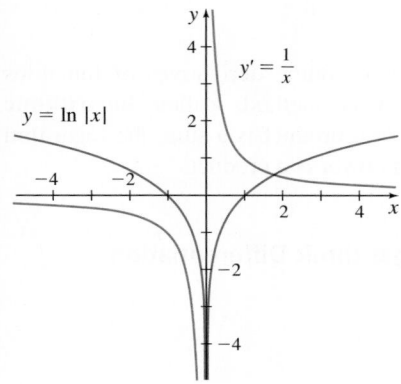

Figure 6

NEED TO REVIEW? Properties of logarithms are discussed in Appendix A.1, pp. A-10 to A-12.

NOTE In the remaining examples, we do not explicitly state the domain of a function containing a logarithm. Instead, we assume that the variable is restricted so all arguments for logarithmic functions are positive.

Another way to find the derivative of $y = \ln |x|$ is to express y as $y = \ln \sqrt{x^2}$ and use the Chain Rule. See Problem 97.

Part (c) proves an important result that is used often:

$$\frac{d}{dx} \ln |x| = \frac{1}{x} \qquad x \neq 0$$

Look at the graph of $y = \ln |x|$ (in blue) in Figure 6. As we move from left to right along the graph, starting where x is unbounded in the negative direction and ending near the y-axis, we see that the tangent lines to $y = \ln |x|$ go from being nearly horizontal with negative slope to being nearly vertical with a very large negative slope. This is reflected in the graph of the derivative $y' = \frac{1}{x}$, where $x < 0$: The graph (in red) starts close to zero and slightly negative and gets more negative as the y-axis is approached.

Similar remarks hold for $x > 0$. The graph of $y' = \frac{1}{x}$ just to the right of the y-axis is unbounded in the positive direction (the tangent lines to $y = \ln |x|$ are nearly vertical with positive slope). As x becomes unbounded in the positive direction, the graph of y' gets closer to zero but remains positive (the tangent lines to $y = \ln |x|$ are nearly horizontal with positive slope).

NOW WORK   Problem 17 and AP® Practice Problems 2, 3, 4, 5, 7, 9, and 13.

Properties of logarithms can sometimes be used to simplify the work needed to find a derivative.

EXAMPLE 3   **Differentiating Logarithmic Functions**

Find y' if $y = \ln \left[\dfrac{(2x-1)^3 \sqrt{2x^4+1}}{x} \right], x > \dfrac{1}{2}$.

Solution

Rather than attempting to use the Chain Rule, Quotient Rule, and Product Rule, we first simplify the right side by using properties of logarithms.

$$y = \ln \left[\frac{(2x-1)^3 \sqrt{2x^4+1}}{x} \right] = \ln(2x-1)^3 + \ln \sqrt{2x^4+1} - \ln x$$

$$= 3\ln(2x-1) + \frac{1}{2}\ln(2x^4+1) - \ln x$$

Now differentiate y.

$$y' = \frac{d}{dx}\left[3\ln(2x-1) + \frac{1}{2}\ln(2x^4+1) - \ln x \right]$$

$$= \frac{d}{dx}[3\ln(2x-1)] + \frac{d}{dx}\left[\frac{1}{2}\ln(2x^4+1) \right] - \frac{d}{dx}\ln x$$

$$= 3 \cdot \frac{2}{2x-1} + \frac{1}{2} \cdot \frac{8x^3}{2x^4+1} - \frac{1}{x} = \frac{6}{2x-1} + \frac{4x^3}{2x^4+1} - \frac{1}{x} \qquad \blacksquare$$

NOW WORK   Problem 27.

2 Use Logarithmic Differentiation

Logarithms and their properties are very useful for finding derivatives of functions that involve products, quotients, or powers. This method, called **logarithmic differentiation**, uses the facts that the logarithm of a product is a sum, the logarithm of a quotient is a difference, and the logarithm of a power is a product.

EXAMPLE 4   **Finding Derivatives Using Logarithmic Differentiation**

Find y' if $y = \dfrac{x^2}{(3x-2)^3}$.

Solution

It is easier to find y' if we take the natural logarithm of each side before differentiating. That is, we write

$$\ln y = \ln\left[\frac{x^2}{(3x-2)^3}\right]$$

and simplify using properties of logarithms.

$$\ln y = \ln x^2 - \ln(3x-2)^3$$
$$= 2\ln x - 3\ln(3x-2)$$

To find y', use implicit differentiation.

$$\frac{d}{dx}\ln y = \frac{d}{dx}[2\ln x - 3\ln(3x-2)]$$

$$\frac{y'}{y} = \frac{d}{dx}(2\ln x) - \frac{d}{dx}[3\ln(3x-2)] \qquad \frac{d}{dx}\ln y = \frac{1}{y}\frac{dy}{dx} = \frac{y'}{y}$$

$$\frac{y'}{y} = \frac{2}{x} - \frac{9}{3x-2}$$

$$y' = y\left[\frac{2}{x} - \frac{9}{3x-2}\right] = \left[\frac{x^2}{(3x-2)^3}\right]\left[\frac{2}{x} - \frac{9}{3x-2}\right]$$ ■

Summarizing these steps, we have the method of Logarithmic Differentiation.

World History Archive/Alamy

ORIGINS Logarithmic differentiation was first used in 1697 by Johann Bernoulli (1667–1748) to find the derivative of $y = x^x$. Johann, a member of a famous family of mathematicians, was the younger brother of Jakob Bernoulli (1654–1705). He was also a contemporary of Newton, Leibniz, the French mathematician Guillaume de l'Hôpital, and the Japanese mathematician Seki Takakazu (1642–1708).

Steps for Using Logarithmic Differentiation

Step 1 If the function $y = f(x)$ consists of products, quotients, and powers, take the natural logarithm of each side. Then simplify using properties of logarithms.

Step 2 Differentiate implicitly, and use the fact that $\dfrac{d}{dx}\ln y = \dfrac{y'}{y}$.

Step 3 Solve for y', and replace y with $f(x)$.

NOW WORK   Problem **51** and AP® Practice Problem **12**.

EXAMPLE 5 **Using Logarithmic Differentiation**

Find y' if $y = x^x$, $x > 0$.

Solution

Notice that x^x is neither x raised to a fixed power a, nor a fixed base a raised to a variable power. We differentiate y using logarithmic differentiation:

Step 1 Take the natural logarithm of each side of $y = x^x$, and simplify using properties of logarithms.

$$\ln y = \ln x^x = x \ln x$$

Step 2 Differentiate implicitly.

$$\frac{d}{dx} \ln y = \frac{d}{dx}(x \ln x)$$

$$\frac{y'}{y} = x \cdot \frac{d}{dx} \ln x + \left(\frac{d}{dx} x\right) \cdot \ln x = x \cdot \frac{1}{x} + 1 \cdot \ln x = 1 + \ln x$$

Step 3 Solve for y': $y' = y(1 + \ln x) = x^x(1 + \ln x)$. ∎

Another way to find the derivative of $y = x^x$ uses the fact that $y = x^x = e^{\ln x^x} = e^{x \ln x}$. See Problem 85.

NOW WORK Problem 57 and AP® Practice Problem 8.

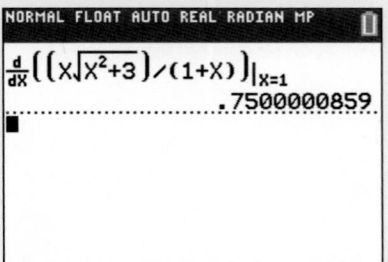
EXAMPLE 6 **Finding an Equation of a Tangent Line**

Find an equation of the tangent line to the graph of $f(x) = \dfrac{x\sqrt{x^2+3}}{1+x}$ at the point $(1, 1)$.

Solution

The slope of the tangent line to the graph of $y = f(x)$ at the point $(1, 1)$ is $f'(1)$. Since the function consists of a product, a quotient, and a power, we use logarithmic differentiation:

Step 1 Take the natural logarithm of each side, and simplify:

$$\ln y = \ln \frac{x\sqrt{x^2+3}}{1+x} = \ln x + \frac{1}{2}\ln(x^2+3) - \ln(1+x)$$

Step 2 Differentiate implicitly.

$$\frac{y'}{y} = \frac{1}{x} + \frac{1}{2} \cdot \frac{2x}{x^2+3} - \frac{1}{1+x} = \frac{1}{x} + \frac{x}{x^2+3} - \frac{1}{1+x}$$

Step 3 Solve for y' and replace y with $f(x)$:

$$y' = f'(x) = y\left(\frac{1}{x} + \frac{x}{x^2+3} - \frac{1}{1+x}\right) = \frac{x\sqrt{x^2+3}}{1+x}\left(\frac{1}{x} + \frac{x}{x^2+3} - \frac{1}{1+x}\right)$$

Now find the slope of the tangent line by evaluating $f'(1)$.

$$f'(1) = \frac{\sqrt{4}}{2}\left(1 + \frac{1}{4} - \frac{1}{2}\right) = \frac{3}{4}$$

(continued on the next page)

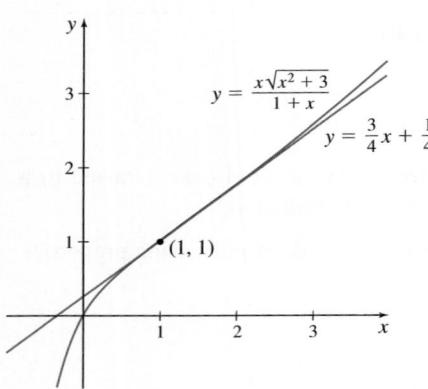

Figure 7

Then an equation of the tangent line to the graph of f at the point $(1, 1)$ is

$$y - 1 = \frac{3}{4}(x - 1)$$

$$y = \frac{3}{4}x + \frac{1}{4}$$

See Figure 7 for the graphs of f and $y = \frac{3}{4}x + \frac{1}{4}$. ∎

NOW WORK   Problem 75 and AP® Practice Problem 6.

We now can prove the Simple Power Rule for finding the derivative of $y = x^a$, where a is any real number.

THEOREM Simple Power Rule

If a is a real number, then

$$\boxed{\frac{d}{dx}x^a = ax^{a-1}}$$

Proof Let $y = x^a$, a real. Then $\ln y = \ln x^a = a \ln x$.

$$\frac{y'}{y} = a\frac{1}{x} = \frac{a}{x}$$

$$y' = y \cdot \frac{a}{x} = \frac{ax^a}{x} = ax^{a-1}$$ ∎

EXAMPLE 7   **Using the Power Rules**

Find the derivative of:

(a) $y = x^{\sqrt{2}}$ **(b)** $y = (x^2 + 1)^\pi$

Solution

(a) We differentiate y using the Simple Power Rule.

$$y' = \frac{d}{dx}x^{\sqrt{2}} = \sqrt{2}x^{\sqrt{2}-1}$$

(b) The Power Rule for Functions also holds when the exponent is any real number. Then

$$y' = \frac{d}{dx}(x^2 + 1)^\pi = \pi(x^2 + 1)^{\pi-1} \cdot 2x = 2\pi x(x^2 + 1)^{\pi-1}$$ ∎

③ Express e as a Limit

In Section P.5, we stated that the number e has this property: $\ln e = 1$. Here we give two equivalent ways of writing e as a limit; both are used in later chapters.

THEOREM The Number e as a Limit

The number e can be expressed by either of the two limits:

$$\boxed{\textbf{(a)} \ \lim_{h \to 0}(1 + h)^{1/h} = e \qquad \textbf{(b)} \ \lim_{n \to \infty}\left(1 + \frac{1}{n}\right)^n = e} \qquad (2)$$

Proof (a) The derivative of $f(x) = \ln x$ is $f'(x) = \dfrac{1}{x}$ so $f'(1) = 1$. Then

$$f'(1) = \lim_{h \to 0} \frac{f(1+h) - f(1)}{h} = \lim_{h \to 0} \frac{\ln(1+h) - \ln 1}{h} = \lim_{h \to 0} \frac{\ln(1+h) - 0}{h}$$

$\uparrow$
Definition of a Derivative

$$= \lim_{h \to 0} \left[\frac{1}{h} \ln(1+h) \right] = \lim_{h \to 0} \ln(1+h)^{1/h}$$

Since $y = \ln x$ is continuous and $f'(1) = 1$,

$$f'(1) = \lim_{h \to 0} \ln(1+h)^{1/h} = \ln \left[\lim_{h \to 0} (1+h)^{1/h} \right] = 1$$

Since $\ln e = 1$,

$$\ln \left[\lim_{h \to 0} (1+h)^{1/h} \right] = \ln e$$

Since $y = \ln x$ is a one-to-one function,

$$\lim_{h \to 0} (1+h)^{1/h} = e$$

TABLE 2

n	$\left(1 + \dfrac{1}{n}\right)^n$
10	2. 593 742 46
1000	2. 716 923 932
10,000	2. 718 145 927
10^6	2. 718 280 469

(b) The limit above is true when $h \to 0^+$. So if we let $n = \dfrac{1}{h}$, then as $h \to 0^+$, we have $n = \dfrac{1}{h} \to \infty$ and

$$e = \lim_{n \to \infty} \left(1 + \frac{1}{n} \right)^n \qquad \blacksquare$$

Table 2 shows the values of $\left(1 + \dfrac{1}{n} \right)^n$ for selected values of n.

EXAMPLE 8 **Expressing a Limit in Terms of e**

Express each limit in terms of the number e:

(a) $\displaystyle\lim_{h \to 0} (1 + 2h)^{1/h}$ **(b)** $\displaystyle\lim_{n \to \infty} \left(1 + \frac{3}{n} \right)^{2n}$

Solution

(a) $\displaystyle\lim_{h \to 0} (1 + 2h)^{1/h}$ resembles $\displaystyle\lim_{h \to 0} (1 + h)^{1/h}$, and with some manipulation, it can be expressed in terms of $\displaystyle\lim_{h \to 0} (1 + h)^{1/h}$.

NOTE Correct to nine decimal places,

$$e = 2.\ 718\ 281\ 828$$

$$(1 + 2h)^{1/h} = [(1 + 2h)^{1/(2h)}]^2$$

Now let $k = 2h$, and note that $h \to 0$ is equivalent to $2h = k \to 0$. So,

$$\lim_{h \to 0} (1 + 2h)^{1/h} = \lim_{k \to 0} \left[(1 + k)^{1/k} \right]^2 = \left[\lim_{k \to 0} (1 + k)^{1/k} \right]^2 = e^2$$

$\uparrow$
(2)

(b) $\lim\limits_{n\to\infty}\left(1+\dfrac{3}{n}\right)^{2n}$ resembles $\lim\limits_{n\to\infty}\left(1+\dfrac{1}{n}\right)^{n}$. We rewrite it as follows:

$$\left(1+\frac{3}{n}\right)^{2n}=\left[\left(1+\frac{3}{n}\right)^{2n}\right]^{3/3}=\left[\left(1+\frac{3}{n}\right)^{n/3}\right]^{6}$$

Let $k=\dfrac{n}{3}$. Since $n\to\infty$ is equivalent to $\dfrac{n}{3}=k\to\infty$, we find that

$$\lim_{n\to\infty}\left(1+\frac{3}{n}\right)^{2n}=\lim_{k\to\infty}\left[\left(1+\frac{1}{k}\right)^{k}\right]^{6}=\left[\lim_{k\to\infty}\left(1+\frac{1}{k}\right)^{k}\right]^{6}=e^{6}\quad\blacksquare$$

NOW WORK Problem 77.

The number e occurs in many applications. For example, in finance, the number e is used to find **continuously compounded interest**. See the discussion preceding Problem 89.

3.4 Assess Your Understanding

Concepts and Vocabulary

1. $\dfrac{d}{dx}\ln x =$ _____.

2. *True or False* $\dfrac{d}{dx}x^{e}=e\,x^{e-1}$.

3. *True or False* $\dfrac{d}{dx}\ln[x\sin^{2}x]=\dfrac{d}{dx}\ln x\cdot\dfrac{d}{dx}\ln\sin^{2}x$.

4. *True or False* $\dfrac{d}{dx}\ln\pi=\dfrac{1}{\pi}$.

5. $\dfrac{d}{dx}\ln|x| =$ ____ for all $x\neq0$.

6. $\lim\limits_{n\to\infty}\left(1+\dfrac{1}{n}\right)^{n} =$ _____.

Skill Building

In Problems 7–44, find the derivative of each function.

7. $f(x)=5\ln x$

8. $f(x)=-3\ln x$

9. $s(t)=\log_{2}t$

10. $s(t)=\log_{3}t$

11. $g(x)=(\cos x)(\ln x)$

12. $g(x)=(\sin x)(\ln x)$

13. $F(x)=\ln(5x)$

14. $F(x)=\ln\dfrac{x}{3}$

15. $s(t)=\ln(e^{t}-e^{-t})$

16. $s(t)=\ln(e^{3t}+e^{-3t})$

17. $f(x)=x\ln(x^{2}+4)$

18. $f(x)=x\ln(x^{2}+5x+1)$

19. $f(x)=x^{2}\log_{5}(3x+5)$

20. $f(x)=x^{3}\log_{4}(2x+3)$

21. $f(x)=\dfrac{1}{2}\ln\dfrac{1+x}{1-x}$

22. $f(x)=\dfrac{1}{2}\ln\dfrac{1+x^{2}}{1-x^{2}}$

23. $f(x)=\ln(\ln x)$

24. $f(x)=\ln\left(\ln\dfrac{1}{x}\right)$

25. $g(x)=\ln\dfrac{x}{\sqrt{x^{2}+1}}$

26. $g(x)=\ln\dfrac{4x^{3}}{\sqrt{x^{2}+4}}$

27. $f(x)=\ln\dfrac{(x^{2}+1)^{2}}{x\sqrt{x^{2}-1}}$

28. $f(x)=\ln\dfrac{x\sqrt{3x-1}}{(x^{2}+1)^{3}}$

29. $F(\theta)=\ln(\sin\theta)$

30. $F(\theta)=\ln(\cos\theta)$

31. $g(x)=\ln\left(x+\sqrt{x^{2}+4}\right)$

32. $g(x)=\ln\left(\sqrt{x+1}+\sqrt{x}\right)$

33. $f(x)=\log_{2}(1+x^{2})$

34. $f(x)=\log_{2}(x^{2}-1)$

35. $f(x)=\tan^{-1}(\ln x)$

36. $f(x)=\sin^{-1}(\ln x)$

37. $s(t)=\ln(\tan^{-1}t)$

38. $s(t)=\ln(\sin^{-1}t)$

39. $G(x)=(\ln x)^{1/2}$

40. $G(x)=(\ln x)^{-1/2}$

41. $f(\theta)=\sin(\ln\theta)$

42. $f(\theta)=\cos(\ln\theta)$

43. $g(x)=(\log_{3}x)^{3/2}$

44. $g(x)=(\log_{2}x)^{4/3}$

In Problems 45–50, use implicit differentiation to find $y'=\dfrac{dy}{dx}$.

45. $x\ln y+y\ln x=2$

46. $\dfrac{\ln y}{x}+\dfrac{\ln x}{y}=2$

47. $\ln(x^{2}+y^{2})=x+y$

48. $\ln(x^{2}-y^{2})=x-y$

49. $\ln\dfrac{y}{x}=y$

50. $\ln\dfrac{y}{x}-\ln\dfrac{x}{y}=1$

In Problems 51–72, use logarithmic differentiation to find y'. Assume that the variable is restricted so that all arguments of logarithmic functions are positive.

[PAGE 270] 51. $y = (x^2 + 1)^2 (2x^3 - 1)^4$

52. $y = (3x^2 + 4)^3 (x^2 + 1)^4$

53. $y = \dfrac{x^2(x^3 + 1)}{\sqrt{x^2 + 1}}$

54. $y = \dfrac{\sqrt{x}(x^3 + 2)^2}{\sqrt[3]{3x + 4}}$

55. $y = \dfrac{x \cos x}{(x^2 + 1)^3 \sin x}$

56. $y = \dfrac{x \sin x}{(1 + e^x)^3 \cos x}$

[PAGE 271] 57. $y = (3x)^x$

58. $y = (x - 1)^x$

59. $y = x^{\ln x}$

60. $y = (2x)^{\ln x}$

61. $y = x^{x^2}$

62. $y = (3x)^{\sqrt{x}}$

63. $y = x^{e^x}$

64. $y = (x^2 + 1)^{e^x}$

65. $y = x^{\sin x}$

66. $y = x^{\cos x}$

67. $y = (\sin x)^x$

68. $y = (\cos x)^x$

69. $y = (\sin x)^{\cos x}$

70. $y = (\sin x)^{\tan x}$

71. $x^y = 7$

72. $y^x = 9$

In Problems 73–76, find an equation of the tangent line to the graph of $y = f(x)$ at the given point.

73. $y = \ln(5x)$ at $\left(\dfrac{1}{5}, 0\right)$

74. $y = x \ln x$ at $(1, 0)$

[PAGE 272] 75. $y = \dfrac{x^2\sqrt{3x - 2}}{(x - 1)^2}$ at $(2, 8)$

76. $y = \dfrac{x(\sqrt[3]{x} + 1)^2}{\sqrt{x + 1}}$ at $(8, 24)$

In Problems 77–80, express each limit in terms of e.

[PAGE 274] 77. $\lim\limits_{n \to \infty} \left(1 + \dfrac{1}{n}\right)^{2n}$

78. $\lim\limits_{n \to \infty} \left(1 + \dfrac{1}{n}\right)^{n/2}$

79. $\lim\limits_{n \to \infty} \left(1 + \dfrac{1}{3n}\right)^n$

80. $\lim\limits_{n \to \infty} \left(1 + \dfrac{4}{n}\right)^n$

Applications and Extensions

81. Find $\dfrac{d^{10}}{dx^{10}}(x^9 \ln x)$.

82. If $f(x) = \ln(x - 1)$, find $f^{(n)}(x)$.

83. If $y = \ln(x^2 + y^2)$, find the value of $\dfrac{dy}{dx}$ at the point $(1, 0)$.

84. If $f(x) = \tan\left(\ln x - \dfrac{1}{\ln x}\right)$, find $f'(e)$.

85. Find y' if $y = x^x$, $x > 0$, by using $y = x^x = e^{\ln x^x}$ and the Chain Rule.

86. If $y = \ln(kx)$, where $x > 0$ and $k > 0$ is a constant, show that $y' = \dfrac{1}{x}$.

In Problems 87 and 88, find y'. Assume that a is a constant.

87. $y = x \tan^{-1} \dfrac{x}{a} - \dfrac{1}{2} a \ln(x^2 + a^2), \quad a \neq 0$

88. $y = x \sin^{-1} \dfrac{x}{a} + a \ln \sqrt{a^2 - x^2}, \quad |a| > |x|, \quad a \neq 0$

Continuously Compounded Interest *In Problems 89 and 90, use the following discussion:*

*Suppose an initial investment, called the **principal** P, earns an annual rate of interest r (expressed as a decimal), which is compounded n times per year. The interest earned on the principal P in the first compounding period is $P\left(\dfrac{r}{n}\right)$, and the resulting amount A of the investment after one compounding period is*

$$A = P + P\left(\dfrac{r}{n}\right) = P\left(1 + \dfrac{r}{n}\right).$$ *After k compounding periods, the amount A of the investment is $A = P\left(1 + \dfrac{r}{n}\right)^k$. Since in t years there are nt compounding periods, the amount A after t years is*

$$A = P\left(1 + \dfrac{r}{n}\right)^{nt}$$

*When interest is compounded so that after t years the accumulated amount is $A = \lim\limits_{n \to \infty} P\left(1 + \dfrac{r}{n}\right)^{nt}$, the interest is said to be **compounded continuously**.*

89. (a) Show that if the annual rate of interest r is compounded continuously, then the amount A after t years is $A = Pe^{rt}$, where P is the initial investment.

(b) If an initial investment of $P = \$5000$ earns 2% interest compounded continuously, how much is the investment worth after 10 years?

(c) How long does it take an investment of $10,000 to double if it is invested at 2.4% compounded continuously?

(d) Show that the rate of change of A with respect to t when the interest rate r is compounded continuously is $\dfrac{dA}{dt} = rA$.

90. A bank offers a certificate of deposit (CD) that matures in 10 years with a rate of interest of 3% compounded continuously. (See Problem 89.) Suppose you purchase such a CD for $2000 in your IRA.

(a) Write an equation that gives the amount A in the CD as a function of time t in years.

(b) How much is the CD worth at maturity?

(c) What is the rate of change of the amount A at $t = 3$? At $t = 5$? At $t = 8$?

(d) Explain the results found in (c).

91. Sound Level of a Leaf Blower   The loudness L, measured in decibels (dB), of a sound of intensity I is defined as $L(x) = 10 \log \dfrac{I(x)}{I_0}$, where x is the distance in meters from the source of the sound and $I_0 = 10^{-12}$ W/m^2 is the least intense sound that a human ear can detect. The intensity I is defined as the power P of the sound wave divided by the area A on which it falls. If the wave spreads out uniformly in all directions, that is, if it is spherical, the surface area is $A(x) = 4\pi x^2$ m^2, and $I(x) = \dfrac{P}{4\pi x^2}$ W/m^2.

(a) If you are 2.0 m from a noisy leaf blower and are walking away from it, at what rate is the loudness L changing with respect to distance x?

(b) Interpret the sign of your answer.

92. Show that $\ln x + \ln y = 2x$ is equivalent to $xy = e^{2x}$. Use $xy = e^{2x}$ to find y'. Compare this result to the solution found in Example 1(c).

93. If $\ln T = kt$, where k is a constant, show that $\dfrac{dT}{dt} = kT$.

94. Graph $y = \left(1 + \dfrac{1}{x}\right)^x$ and $y = e$ on the same screen. Explain how the graph supports the fact that $\lim\limits_{n \to \infty}\left(1 + \dfrac{1}{n}\right)^n = e$.

95. Power Rule for Functions   Show that if u is a function of x that is differentiable and a is a real number, then

$$\frac{d}{dx}[u(x)]^a = a[u(x)]^{a-1}u'(x)$$

provided u^a and u^{a-1} are defined.
Hint: Let $|y| = |[u(x)]^a|$ and use logarithmic differentiation.

96. Show that the tangent lines to the graphs of the family of parabolas $f(x) = -\dfrac{1}{2}x^2 + k$ are always perpendicular to the tangent lines to the graphs of the family of natural logarithms $g(x) = \ln(bx) + c$, where $b > 0$, k, and c are constants.
Source: Mathematics students at Millikin University, Decatur, Illinois.

97. Find the derivative of $y = \ln|x|$ by writing $y = \ln\sqrt{x^2}$ and using the Chain Rule.

Challenge Problem

98. If f and g are differentiable functions, and if $f(x) > 0$, show that

$$\frac{d}{dx}f(x)^{g(x)} = g(x)f(x)^{g(x)-1}f'(x) + f(x)^{g(x)}[\ln f(x)]g'(x)$$

99. Find $h'(x)$ if

$$h(x) = \sqrt[3]{\frac{(4x+3)^2}{(2x^2-1)^4}}$$

Preparing for the AP® Exam

AP® Practice Problems

Multiple-Choice Questions

1. (PAGE 268) If $f(x) = e^x + x^2$, find $\dfrac{d}{dx}[f(\ln x)]$.

(A) $\dfrac{1 + 2\ln x}{x}$ (B) $1 + \dfrac{2\ln x}{x}$

(C) $1 + \dfrac{2}{x^2}$ (D) $x + 2x \ln x$

2. (PAGE 269) $\dfrac{d}{dx}(x^2 e^{\ln x^3}) =$

(A) $2x + 3x^2$ (B) $5x^3$ (C) $5x^4$ (D) $6x^4$

3. (PAGE 269) If $h(x) = \ln(x^2 + 4)$, then $h'(x)$ equals

(A) $\left|\dfrac{2x}{x^2+4}\right|$ (B) $\dfrac{2x}{x^2+4}$ (C) $\dfrac{x^2}{x^2+4}$ (D) $\dfrac{1}{x^2+4}$

4. (PAGE 269) Find the rate of change of y with respect to x when $x = 1$ if $\ln(xy) = x$.

(A) e (B) 0 (C) 1 (D) $e - 1$

5. (PAGE 269) If $f(x) = \ln(e^{x^2 - 3x})$, then $f'(x)$ equals

(A) $\dfrac{1}{e^{x^2-3x}}$ (B) $\dfrac{2x-3}{e^{x^2-3x}}$ (C) $x^2 - 3x$ (D) $2x - 3$

6. (PAGE 272) Find the slope of the line tangent to the graph of $y = \ln(\sec^2 x)$ at $x = \dfrac{\pi}{4}$.

(A) 2 (B) $\dfrac{\sqrt{2}}{2}$ (C) $\sqrt{2}$ (D) $2\sqrt{2}$

7. (PAGE 269) $\dfrac{d}{dx}\ln\left|\sin\dfrac{\pi}{x}\right| =$

(A) $\cot\dfrac{\pi}{x}$ (B) $-\dfrac{\pi}{x^2}\csc\dfrac{\pi}{x}$

(C) $-\dfrac{\pi}{x^2}\cot\dfrac{\pi}{x}$ (D) $\dfrac{\pi}{x}\cot\dfrac{\pi}{x}$

8. (PAGE 271) Find $\dfrac{d}{dx}(x^4 + 2)^x$.

(A) $4x^4(x^4 + 2)^{x-1}$

(B) $\dfrac{4x^4}{x^4 + 2} + \ln(x^4 + 2)$

(C) $(x^4 + 2)^x\left[\dfrac{4x^4}{x^4 + 2} + \ln(x^4 + 2)\right]$

(D) $(x^4 + 2)^x \ln(x^4 + 2)$

9. Find $\dfrac{d^2y}{dx^2}$ for $y = \ln(x\sqrt{x})$.

(A) $-\dfrac{3}{2x^2}$ (B) $\dfrac{3}{2x^2}$ (C) $\dfrac{3}{2x}$ (D) $-\dfrac{3}{x^3}$

10. $\dfrac{d}{dx}\dfrac{\log_2 x}{x} =$

(A) $\dfrac{1 - \log_2 x}{x^2}$ (B) $-\dfrac{\ln 2 + \log_2 x}{x^2}$

(C) $\dfrac{\ln 2 - \log_2 x}{x^2 \ln 2}$ (D) $\dfrac{1 - \ln 2 \log_2 x}{x^2 \ln 2}$

11. If $f(x) = x \ln x$, then $f'(x)$ equals

(A) $x + \ln x$ (B) 1 (C) $1 + \ln x$ (D) $\dfrac{1}{x} + \ln x$

Free-Response Questions

12. Find $f'(0)$, if

$$f(x) = \frac{(x^2 + 2)^4}{(x^2 + 1)^5}$$

13. Suppose $g(x) = \ln(f(x))$, where $f(x) > 0$ for all real numbers and f is differentiable for all real numbers. If $f(4) = 2$ and $f'(4) = -\dfrac{1}{5}$, find $g'(4)$. Show the computations that lead to the answer.

See the **BREAK IT DOWN** *on page 281 for a stepped out solution to AP® Practice Problem 13.*

Retain Your Knowledge

Multiple-Choice Questions

1. Find $\displaystyle\lim_{x \to \pi} \dfrac{\tan x}{\sin x}$.

(A) 0 (B) -1 (C) 1 (D) The limit does not exist.

2. What is the instantaneous rate of change

of $f(x) = \dfrac{\sin x}{1 + \tan x}$ at $x = \dfrac{\pi}{4}$?

(A) -1 (B) 0 (C) $\dfrac{\sqrt{2}}{4}$ (D) 1

3. Find an equation of the line tangent to the graph of $f(x) = 2e^x + 2x - 5$ at $x = 0$.

(A) $y + 6 = 4x$ (B) $y - 3 = 4x - 2$

(C) $y + 3 = 2x$ (D) $y + 3 = 4x$

Free-Response Question

4. Suppose

$$R(x) = \frac{x^2 + 3x + 2}{x^2 - 1}$$

(a) For what numbers, if any, is R discontinuous? If possible, define R so that R is continuous at these numbers.

(b) Find any horizontal asymptotes of the graph of R.

(c) Find any vertical asymptotes of the graph of R.

CHAPTER 3 PROJECT

Ayhan Altun/Getty Images

World Population Growth

This Project can be done individually or as part of a team.

The Law of Uninhibited Growth states that, under certain conditions, the rate of change of a population is proportional to the size of the population at that time. One consequence of this law is that the time it takes for a population to double remains constant.

For example, suppose a culture of bacteria obeys the Law of Uninhibited Growth. Then if it takes five hours for the culture to double from 100 organisms to 200 organisms, in the next five hours it will double again from 200 to 400. We can model this mathematically using the formula

$$P(t) = P_0 \, 2^{t/D}$$

where $P(t)$ is the population at time t, P_0 is the population at time $t = 0$, and D is the doubling time.

If we use this formula to model population growth, a few observations are in order. For example, the model is continuous, but actual population growth is discrete. That is, an actual population would change from 100 to 101 individuals in an instant, as opposed to a model that has a continuous flow from 100 to 101.

The model also produces fractional answers, whereas an actual population is counted in whole numbers. For large populations, however, the growth is continuous enough for the model to match real-world conditions, at least for a short time. In general, as growth continues, there are obstacles to growth at which point the model will fail to be accurate. Situations that follow the model of the Law of Uninhibited Growth vary from the introduction of invasive species into a new environment to the spread of a deadly virus for

which there is no immunization, as was the case of COVID-19 prior to the development of a viable vaccine. Here, we investigate how accurately the model predicts world population.

1. The world population on July 1, 1959, was approximately 2.983435×10^9 persons and had a doubling time of $D = 40$ years. Use these data and the Law of Uninhibited Growth to write a formula for the world population $P = P(t)$. Use this model to solve Problems 2 through 4.

2. Find the rate of change of the world population $P = P(t)$ with respect to time t.

3. Find the rate of change on July 1, 2026 of the world population with respect to time. (Note that $t = 0$ is July 1, 1959.) Round the answer to the nearest whole number.

4. Approximate the world population on July 1, 2026. Round the answer to the nearest person.

5. According to the United Nations, the world population on July 1, 2021 was 7.909295×10^9. Use $t_0 = 2021$, $P_0 = 7.909295 \times 10^9$, and $D = 40$ and find a new formula to model the world population $P = P(t)$.

6. Use the new model from Problem 5 to find the rate of change of the world population on July 1, 2026.

7. Compare the results from Problems 3 and 6. Interpret and explain any discrepancy between the two rates of change.

8. Use the new model to approximate the world population on July 1, 2026. Round the answer to the nearest person.

9. Discuss possible reasons for the discrepancies in the approximations of the 2026 population. Do you think one set of data gives better results than the other?

_____

Source: UN World Population http://population.un.org

Chapter Review

THINGS TO KNOW

3.1 The Chain Rule

Theorems:

- Chain Rule: $(f \circ g)'(x) = f'(g(x)) \cdot g'(x)$
  (p. 235)
  $$\frac{dy}{dx} = \frac{dy}{du} \cdot \frac{du}{dx}$$

- Power Rule for Functions:
  $$\frac{d}{dx}[g(x)]^n = n[g(x)]^{n-1} g'(x), \quad \text{where } n \text{ is an integer}$$
  (p. 240)

Basic Derivative Formulas:

- $\frac{d}{dx} e^{u(x)} = e^{u(x)} \frac{du}{dx}$ (p. 237)

- $\frac{d}{dx} \ln u(x) = \frac{1}{u(x)} \cdot \frac{du}{dx}$

- Derivatives of trigonometric functions (p. 238)

- $\frac{d}{dx} a^x = a^x \ln a \quad a > 0 \text{ and } a \neq 1$ (p. 240)

3.2 Implicit Differentiation

To differentiate an implicit function (p. 249):

- Assume y is a differentiable function of x.
- Differentiate both sides of the equation with respect to x.
- Solve the resulting equation for $y' = \frac{dy}{dx}$.

Theorems:

- Simple Power Rule for rational exponents:
  $$\frac{d}{dx} x^{p/q} = \frac{p}{q} \cdot x^{(p/q)-1},$$
  provided $x^{p/q}$ and $x^{p/q-1}$ are defined. (p. 253)

- Power Rule for Functions: $\frac{d}{dx}[u(x)]^r = r[u(x)]^{r-1} u'(x)$, r a rational number; provided u^r and u^{r-1} are defined. (p. 254)

3.3 Derivatives of Inverse Functions

Theorem: The derivative of an inverse function (p. 259)

Basic Derivative Formulas:

- $\dfrac{d}{dx}\sin^{-1}x = \dfrac{1}{\sqrt{1-x^2}}$ $-1 < x < 1$ (p. 261)

- $\dfrac{d}{dx}\tan^{-1}x = \dfrac{1}{1+x^2}$ (p. 262)

- $\dfrac{d}{dx}\sec^{-1}x = \dfrac{1}{x\sqrt{x^2-1}}$ $|x| > 1$ (p. 263)

3.4 Derivatives of Logarithmic Functions

Basic Derivative Formulas:

- $\dfrac{d}{dx}\log_a x = \dfrac{1}{x\ln a}$, $x > 0, a > 0, a \neq 1$ (p. 267)

- $\dfrac{d}{dx}\ln|x| = \dfrac{1}{x}$, $x \neq 0$ (p. 269)

Steps for Using Logarithmic Differentiation (p. 270):

- **Step 1** If the function $y = f(x)$ consists of products, quotients, and powers, take the natural logarithm of each side. Then simplify using properties of logarithms.
- **Step 2** Differentiate implicitly, and use $\dfrac{d}{dx}\ln y = \dfrac{y'}{y}$.
- **Step 3** Solve for y', and replace y with $f(x)$.

Theorems:

- **Simple Power Rule** If a is a real number, then
$$\frac{d}{dx}x^a = ax^{a-1} \text{ (p. 272)}$$

- The number e can be expressed as
$$\lim_{h \to 0}(1+h)^{1/h} = e \quad\text{or}\quad \lim_{n \to \infty}\left(1+\frac{1}{n}\right)^n = e \text{ (p. 272)}$$

OBJECTIVES

Preparing for the
AP® Exam
AP® Review Problems

Section	You should be able to ...	Examples	Review Exercises	AP® Review Problems
3.1	1 Differentiate a composite function (p. 235)	1–5	1, 13, 24	11, 13
	2 Differentiate $y = a^x$, $a > 0$, $a \neq 1$ (p. 239)	6	19, 22	1, 3
	3 Use the Power Rule for Functions to find a derivative (p. 240)	7, 8	1, 11, 12, 14, 17	10, 13
	4 Use the Chain Rule for multiple composite functions (p. 242)	9	15, 18, 57	2, 4, 13
3.2	1 Find a derivative using implicit differentiation (p. 249)	1–4	39–48, 58, 59	5, 6, 12
	2 Find higher-order derivatives using implicit differentiation (p. 252)	5	45–48	12
	3 Differentiate functions with rational exponents (p. 253)	6, 7	2–8, 15, 16, 53–56	10
3.3	1 Find the derivative of an inverse function (p. 258)	1, 2	49, 50	7
	2 Find the derivative of the inverse trigonometric functions (p. 260)	3, 4	32–38	9
3.4	1 Differentiate logarithmic functions (p. 267)	1–3	20, 21, 23, 25–30, 48	8
	2 Use logarithmic differentiation (p. 270)	4–7	9, 10, 31, 57	
	3 Express e as a limit (p. 272)	8	51, 52	

REVIEW EXERCISES

In Problems 1–38, find the derivative of each function. When a, b, or n appear, they are nonzero constants.

1. $y = (ax + b)^n$

2. $y = \sqrt{2ax}$

3. $y = x\sqrt{1-x}$

4. $y = \dfrac{1}{\sqrt{x^2+1}}$

5. $y = (x^2+4)^{3/2}$

6. $F(x) = \dfrac{x^2}{\sqrt{x^2-1}}$

7. $z = \dfrac{\sqrt{2ax-x^2}}{x}$

8. $y = \sqrt{x} + \sqrt[3]{x}$

9. $y = (e^x - x)^{5x}$

10. $\phi(x) = \dfrac{(x^2-a^2)^{3/2}}{\sqrt{x+a}}$

11. $f(x) = \dfrac{x^2}{(x-1)^2}$

12. $u = (b^{1/2} - x^{1/2})^2$

13. $y = x\sec(2x)$

14. $u = \cos^3 x$

15. $y = \sqrt{a^2 \sin \dfrac{x}{a}}$

16. $\phi(z) = \sqrt{1 + \sin z}$

17. $u = \sin v - \dfrac{1}{3}\sin^3 v$

18. $y = \tan\sqrt{\dfrac{\pi}{x}}$

19. $y = 1.05^x$

20. $v = \ln(y^2 + 1)$

21. $z = \ln(\sqrt{u^2 + 25} - u)$

22. $y = x^2 + 2^x$

23. $y = \ln[\sin(2x)]$

24. $f(x) = e^{-x}\sin(2x + \pi)$

25. $g(x) = \ln(x^2 - 2x)$

26. $y = \ln\dfrac{x^2+1}{x^2-1}$

27. $y = e^{-x}\ln x$

28. $w = \ln\left(\sqrt{x+7} - \sqrt{x}\right)$

29. $y = \dfrac{1}{12}\ln\dfrac{x}{\sqrt{144-x^2}}$

30. $y = \ln(\tan^2 x)$

31. $f(x) = \dfrac{e^x(x^2+4)}{x-2}$

32. $y = \sin^{-1}(x-1) + \sqrt{2x - x^2}$

33. $y = 2\sqrt{x} - 2\tan^{-1}\sqrt{x}$

34. $y = 4\tan^{-1}\dfrac{x}{2} + x$

35. $y = \sin^{-1}(2x - 1)$

36. $y = x^2\tan^{-1}\dfrac{1}{x}$

37. $y = x\tan^{-1}x - \ln\sqrt{1 + x^2}$

38. $y = \sqrt{1 - x^2}(\sin^{-1}x)$

In Problems 39–44, find $y' = \dfrac{dy}{dx}$ *using implicit differentiation.*

39. $x = y^4 + y$

40. $x = \cos^4 y + \cos y$

41. $\ln x + \ln y = x\cos y$

42. $\tan(xy) = x$

43. $y = x + \sin(xy)$

44. $x = \ln(\csc y + \cot y)$

In Problems 45–48, find y' *and* y''.

45. $xy + 3y^2 = 10x$

46. $y^3 + y = x^2$

47. $xe^y = 4x^2$

48. $\ln(x + y) = 8x$

49. The function $f(x) = e^{2x}$ has an inverse function g. Find $g'(1)$.

50. The function $f(x) = \sin x$ defined on the restricted domain $\left[-\dfrac{\pi}{2}, \dfrac{\pi}{2}\right]$ has an inverse function g. Find $g'\left(\dfrac{1}{2}\right)$.

In Problems 51 and 52, express each limit in terms of the number e.

51. $\lim\limits_{n\to\infty}\left(1 + \dfrac{2}{5n}\right)^n$

52. $\lim\limits_{h\to 0}(1 + 3h)^{2/h}$

53. If $f(x) = \sqrt{1 - \sin^2 x}$, find the domain of f'.

54. If $f(x) = x^{1/2}(x - 2)^{3/2}$ for all $x \ge 2$, find the domain of f'.

55. Let f be the function defined by $f(x) = \sqrt{1 + 6x}$.

    (a) What are the domain and the range of f?

    (b) Find the slope of the tangent line to the graph of f at $x = 4$.

    (c) Find the y-intercept of the tangent line to the graph of f at $x = 4$.

    (d) Give the coordinates of the point on the graph of f where the tangent line is parallel to the line $y = x + 12$.

56. **Tangent and Normal Lines** Find equations of the tangent and normal lines to the graph of $y = x\sqrt{x + (x - 1)^2}$ at the point $(2, 2\sqrt{3})$.

57. If $f(x) = (x^2 + 1)^{(2 - 3x)}$, find $f'(1)$.

58. **Tangent Line** Find an equation of the tangent line to the graph of $4xy - y^2 = 3$ at the point $(1, 3)$.

59. Find y' at $x = \dfrac{\pi}{2}$ and $y = \pi$ if $x\sin y + y\cos x = 0$.

Break It Down

Let's take a closer look at AP® Practice Problem 13 from Section 3.4 on page 277.

13. Suppose $g(x) = \ln(f(x))$, where $f(x) > 0$ for all real numbers and f is differentiable for all real numbers. If $f(4) = 2$ and $f'(4) = -\frac{1}{5}$, find $g'(4)$. Show the computations that lead to the answer.

Step 1	Identify the underlying structure and related concepts.	The function $g(x) = \ln(f(x))$ is a composite function, and we are asked to find $g'(4)$, the **derivative** of the function g evaluated at 4.
Step 2	Determine the appropriate math rule or procedure.	To find $g'(x)$, we need to find the derivative of $\ln(f(x))$. First, we notice that $\ln(f(x))$ is a composite function, so we will need to use the chain rule.
Step 3	Apply the math rule or procedure.	$g'(x) = \dfrac{d}{dx}(\ln(f(x))) = \dfrac{1}{f(x)} \cdot f'(x)$ $g'(4) = \dfrac{1}{f(4)} \cdot f'(4)$ $g'(4) = \dfrac{1}{2} \cdot \dfrac{-1}{5} = \dfrac{-1}{10}$
Step 4	Clearly communicate your answer.	$g'(4) = \dfrac{-1}{10}$

AP® Review Problems: Chapter 3

Multiple-Choice Questions

1. Find an equation of the line tangent to the graph of $y = 3^{\sin x} - 4$ when $x = 0$.

(A) $y = (\ln 3)x - 4$ (B) $y = (\ln 3)x - 3$

(C) $y = x - 4$ (D) $y = 3x - 3$

2. If $f(x) = \sin u$, $u = v - \dfrac{1}{v^2}$, and $v = \ln x$, find $f'(e)$.

(A) 0 (B) $\dfrac{3}{e}$ (C) $\dfrac{1}{e} + \dfrac{2}{e \ln 3}$ (D) $\dfrac{2}{e^2}$

3. If $e^{f(x)} = 2 + x^4$, then $f'(x) =$

(A) $\dfrac{4x^3}{e^x}$ (B) $4x^3 e^x$

(C) $\dfrac{4x^3}{2 + x^4}$ (D) $\dfrac{e^x}{2 + x^4}$

4. Find y' if $y = \sin^{4/3}(4x - x^2)$.

(A) $y' = \dfrac{16 - 8x}{3} \sin^{1/3}(4x - x^2)$

(B) $y' = \dfrac{16 - 8x}{3} \sin^{1/3}(4x - x^2) \cdot \cos(4x - x^2)$

(C) $y' = \dfrac{4}{3} \sin^{1/3}(4x - x^2) \cdot \cos(4x - x^2)$

(D) $y' = -\dfrac{8x}{3} \sin^{1/3}(4x - x^2) \cdot \cos(4 - 2x)$

5. The slope of the line tangent to the graph of $x^2 + (xy - 2)^2 = 20$ at the point $(2, -1)$ is

(A) $\dfrac{3}{4}$ (B) -4 (C) $\dfrac{4}{3}$ (D) $-\dfrac{4}{3}$

6. If $y = \sin(3x + 2y)$, find the rate of change of y with respect to x at the origin.

(A) 0 (B) 3 (C) -1 (D) -3

7. The points $(1, 1)$, $(2, 3)$, and $(3, 13)$ are on the graph of the function $f(x) = x^3 - 2x^2 + x + 1$, $x \geq 1$. If the function g is the inverse function of f, then $g'(3) =$

(A) $\dfrac{1}{5}$ (B) $\dfrac{1}{16}$ (C) $\dfrac{1}{3}$ (D) $-\dfrac{1}{5}$

8. Find $\dfrac{d}{dx} \ln(\ln x)$.

(A) 1 (B) $\dfrac{x}{\ln x}$

(C) $\dfrac{1}{x \ln x}$ (D) $\dfrac{1}{(\ln x)^2}$

9. The derivative of $y = \tan^{-1}(xe^x)$ equals

(A) $\dfrac{xe^x + e^x}{\sqrt{1 + x^2 e^{2x}}}$ (B) $\dfrac{e^x + 1}{1 + xe^x}$

(C) $\dfrac{2x^2 e^{2x} + 2xe^x}{1 + x^2 e^{2x}}$ (D) $\dfrac{xe^x + e^x}{1 + x^2 e^{2x}}$

10. The derivative y' of the function $y = \sqrt{x^3 + 2x^2 - 5}$ is given by

(A) $\dfrac{3x^2 + 4x}{2(x^3 + 2x^2 - 5)^{1/2}}$

(B) $\dfrac{3x^2 + 4x}{(x^3 + 2x^2 - 5)^{1/2}}$

(C) $\dfrac{1}{2}(3x^2 + 4x)\sqrt{x^3 + 2x^2 - 5}$

(D) $\dfrac{3x^2 + 4x}{2(x^3 + 2x^2 - 5)}$

11. The functions f and g are differentiable. The function $h(x) = f(g(x)) - [f(x)]^3$. Use the information in the table to find $h'(2)$.

x	$f(x)$	$f'(x)$	$g(x)$	$g'(x)$
2	5	$-\dfrac{1}{3}$	4	-2
4	2	3	-1	$\dfrac{1}{2}$

(A) -78 (B) -28 (C) 17 (D) 19

Free-Response Questions

12. The twice differentiable function $y = f(x)$ is defined implicitly as $xy + y^2 = 4x$.

(a) Find $y' = dy/dx$.

(b) Find y''.

13. An object moves along a horizontal line so that its position x, in meters, at time t, $0 \le t \le 4$, in minutes, i[s] given by

$$x = x(t) = t - \sin^2\left(\frac{\pi}{2}t\right)$$

(a) What is the position of the object at time $t = 2$ minutes?

(b) What is the average rate of change of the position o[f] the object from $t = 0$ to $t = 1$.

(c) What is the velocity of the object at time t?

(d) What is the acceleration of the object at time $t = 1$ minutes.

AP® Cumulative Review Problems: Chapters 1–3   Preparing for the AP® Exam

Multiple-Choice Questions

1. If $f(t) = t^2 \sin t + \ln t$, then $f'(t)$ equals

(A) $2t(\sin t + \ln t) + t^2\left(\cos t + \dfrac{1}{t}\right)$

(B) $2t \sin t - t^2 \cos t + \dfrac{1}{t}$

(C) $2t \sin t + t^2 \cos t + \dfrac{1}{t}\ln t$

(D) $2t \sin t + t^2 \cos t + \dfrac{1}{t}$

2. Find the point(s), if any, where the slope of the line tangent to the graph of $f(x) = \dfrac{x^3}{3} + x^2 - 3x + 1$ is zero.

(A) $(0, 1)$ (B) $(0, -3)$

(C) $(-3, 0)$ and $(1, 0)$ (D) $(-3, 10)$ and $\left(1, -\dfrac{2}{3}\right)$

3. Given $f(x) = e^x(x + \cos x)$, find $f''(0)$.

(A) 0 (B) 2 (C) 3 (D) 5

4. Suppose the function f is continuous on the closed interval $[-2, 6]$, and $f(-2) = 8$ and $f(6) = 1$. By the Intermediate Value Theorem, we know

(A) f has at least one zero on the interval $(-2, 6)$.

(B) $f(x) > 0$ for all x in the interval $[-2, 6]$.

(C) $f(x) = 3$ for at least one number in the interval $(-2, 6)$.

(D) $f'(x)$ is defined for all numbers in the interval $(-2, 6)$.

5. The graph of f is shown in the figure below. Which statement is false?

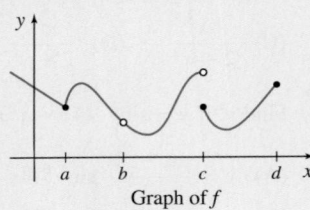

Graph of f

(A) $\lim\limits_{x \to a} f(x)$ exists.   (B) $\lim\limits_{x \to b} f(x)$ exists.

(C) $\lim\limits_{x \to c} f(x)$ exists.   (D) $\lim\limits_{x \to d} f(x)$ does not exist.

6. $\lim\limits_{h\to 0}\dfrac{\ln(e+h)^2-2}{h}=$

(A) $f'(e^2)$ where $f(x)=\ln x$

(B) $f'(e)$ where $f(x)=\ln x^2$

(C) $f'(2)$ where $f(x)=\ln x$

(D) $f'(e)$ where $f(x)=(\ln x)^2$

7. If $f(x)=\dfrac{4x+5}{5x-4}$, then $f'(x)$ equals

(A) $\dfrac{40x+9}{(5x-4)^2}$ (B) $-\dfrac{41}{(5x-4)^2}$

(C) $-\dfrac{9}{(5x-4)^2}$ (D) $\dfrac{9}{(5x-4)^2}$

8. Suppose the function f is defined by

$$f(x)=\begin{cases} 5x+17 & \text{if } x<-3 \\ 2 & \text{if } x=-3 \\ 8-x^2 & \text{if } -3<x<3 \\ x-4 & \text{if } x\geq 3 \end{cases}$$

For what numbers, if any, is f discontinuous?

(A) none   (B) -3 only   (C) 3 only   (D) -3 and 3

9. $\lim\limits_{x\to 4}\dfrac{x^2-16}{\sqrt{x}-2}=$

(A) 2   (B) 4   (C) 32   (D) The limit does not exist.

10. The graph of $y=\dfrac{2x^2+2x+3}{4x^2-4x}$ has

(A) a horizontal asymptote at $y=\dfrac{1}{2}$, but no vertical asymptote.

(B) no horizontal asymptote, but vertical asymptotes at $x=0$ and $x=1$.

(C) a horizontal asymptote at $y=\dfrac{1}{2}$ and vertical asymptotes at $x=0$ and $x=1$.

(D) a horizontal asymptote at $x=2$, but no vertical asymptote.

11. The graph of a differentiable function f is shown below.

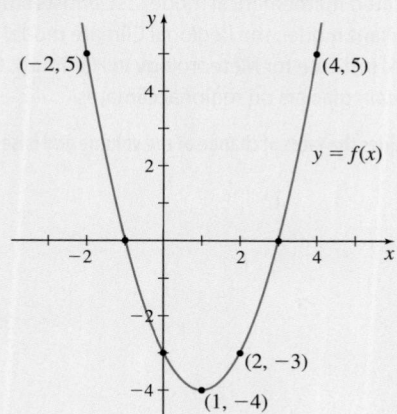

If the function $g(x)=[f(x)]^4$, then which of the choices below best describes $g'(2)$?

(A) $g'(2)=-3$ (B) $g'(2)<0$

(C) $g'(2)=0$ (D) $g'(2)>0$

12. A one-to-one function $y=f(x)$ is continuous on the closed interval $[-3, 5]$ and differentiable on the open interval $(-3, 5)$, except at 0. Also, $f(-3)=-2$ and $f(5)=6$.

Which of the following must be true?

I. f has an inverse function.

II. The graph of f has a vertical tangent line at $x=0$.

III. For at least one number x, $-3<x<5$, $f(x)=-1$.

(A) II only   (B) III only

(C) I and III only   (D) I, II, and III

Free-Response Questions

13. A function f is continuous on the closed interval $-5\leq x\leq 8$. The table below shows x and $f(x)$ for select numbers x in the interval.

x	-5	-2	0	4	7
$f(x)$	4	-1	-2	-3	2

(a) Determine the average rate of change of f from $x=-5$ to $x=7$.

(b) Show that f has at least two zeros for $-5<x<7$.

(c) Determine whether f is decreasing on the open interval $(-2, 4)$? Justify your answer.

14. Suppose $f(x)=\begin{cases} 6x-1 & \text{if} & x<1 \\ x^2+4x & \text{if} & 1\leq x\leq 2 \\ x^3-x^2 & \text{if} & 2<x<10 \end{cases}$

In answering (a)–(d), be sure to justify your conclusions.

(a) Determine whether f is continuous at 1.

(b) Determine whether f is continuous at 2.

(c) Determine whether f is differentiable at 1. If it is, find $f'(1)$.

(d) Determine whether f is differentiable at 2. If it is, find $f'(2)$.

CHAPTER

4

Applications of the Derivative, Part 1

Fotografías Jorge León Cabello/Getty Images

Mountain Glaciers in Retreat

Changes in the size of glaciers significantly impact many aspects of our world, including global climate. Studies have found that melting glaciers affect the temperatures and the amount of precipitation. Using sophisticated mathematical models, scientists study the impact of glaciers on climate. One important model, the Regional Climate model called REMO, created in 1997 at the Max-Planck-Institute for Meteorology in Hamburg, Germany, analyzes the effects of changes in mountain glaciers on regional climates.

 The Chapter 4 Project on page 326 examines the rates of change of the volume and base area of glaciers.

Anna
Engineer

AP® Calc filled a college math credit, allowing me to start taking classes related to my engineering major. It taught me the skills to succeed in later math & science courses and was great practice for the real-life problem solving that I do daily as an engineer.

In Chapter 2, we interpreted the derivative in several ways:

- **Geometric:** If $y = f(x)$, the derivative $f'(c)$ is the slope of the tangent line to the graph of f at the point $(c, f(c))$.
- **Physical:** When $s = f(t)$ represents the position of an object moving on a line, the derivative $f'(t_0)$ is the velocity of the object at time t_0, and $f''(t_0)$ is the acceleration of the object at time t_0.
- **Analytical:** If $y = f(x)$, the derivative $f'(c)$ is the instantaneous rate of change of f at the number c.

In this chapter, we deepen our understanding of these interpretations of the derivative and begin to delve into other applications of the derivative. ∎

4.1 Interpreting a Derivative

OBJECTIVES *When you finish this section, you should be able to:*

1 **Interpret a derivative as the slope of a tangent line to the graph of a function (p. 285)**

2 **Interpret a derivative as an instantaneous rate of change (p. 286)**

3 **Interpret a derivative as velocity or acceleration (p. 287)**

NEED TO REVIEW? The derivative and interpretations of the derivative are discussed in Section 2.1, pp. 168–173.

Suppose f is defined over some open interval containing the number c. Recall that the average rate of change of f from c to any number $x \neq c$ in the interval is given by

$$\frac{\Delta y}{\Delta x} = \frac{f(x) - f(c)}{x - c} \qquad x \neq c$$

where Δy is the change in y and Δx is the change in x.

The derivative of f at c, $f'(c)$, if it exists, equals the limit of the average rate of change of f as x approaches c. That is,

$$f'(c) = \lim_{x \to c} \frac{f(x) - f(c)}{x - c} \qquad x \neq c$$

An average rate of change measures the change in f from c to x; the derivative measures change at c.

1 Interpret a Derivative as the Slope of a Tangent Line to the Graph of a Function

AP® EXAM **TIP**

BREAK IT DOWN: As you work through the problems in this chapter, remember to follow these steps:

Step 1 Identify the underlying structure and related concepts.

Step 2 Determine the appropriate math rule or procedure.

Step 3 Apply the math rule or procedure.

Step 4 Clearly communicate your answer.

On page 329, see how we've used these steps to solve Section 4.2 AP® Practice Problem 9(a) on page 300.

EXAMPLE 1   **Interpreting the Derivative as the Slope of a Tangent Line**

(a) Find the derivative of $f(x) = x \cos x + x - 1$.

(b) Find the slope of the tangent line to the graph of f at the point $(0, f(0))$.

(c) Find an equation of the tangent line to the graph of f at the point $(0, f(0))$.

Solution

(a) Use the Product Rule.

$$f(x) = x \cos x + x - 1$$

$$f'(x) = -x \sin x + \cos x + 1$$

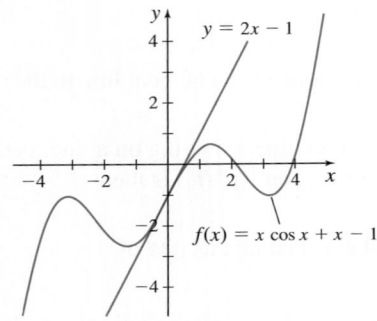

Figure 1

(b) The slope of the tangent line to the graph of f at the point $(0, f(0))$ is the derivative $f'(0)$.

$$f'(0) = -0 \cdot \sin 0 + \cos 0 + 1 = 2$$

(c) Since $f(0) = 0 \cdot \cos 0 + 0 - 1 = -1$, the tangent line contains the point $(0, -1)$. Use the point-slope form of a line with the point $(0, -1)$ and the slope $f'(0) = 2$.

$$y - (-1) = 2(x - 0) \qquad y - y_1 = f'(0)(x - x_1)$$
$$y = 2x - 1$$

The graphs of f and the tangent line $y = 2x - 1$ are shown in Figure 1. ∎

NOW WORK   Problem 7 and AP® Practice Problem 1.

2 Interpret a Derivative as an Instantaneous Rate of Change

EXAMPLE 2   **Interpreting the Derivative as an Instantaneous Rate of Change**

A population of killer bees has migrated to a town in the Midwest. Suppose the population P of killer bees in the town at time t (in weeks) can be modeled by the function

$$P(t) = \frac{1200}{1 + 39e^{-0.24t}}$$

(a) Find the initial number of killer bees that migrated to the town.

(b) How many killer bees are in the town after 1 week ($t = 1$)? Round your answer to the nearest whole number. Find and interpret the rate of change in the killer bee population at $t = 1$.

(c) How many killer bees are in the town after 20 weeks ($t = 20$)? Round your answer to the nearest whole number. Find and interpret the rate of change in the bee population at $t = 20$.

Solution

(a) The initial population is $P(0) = \dfrac{1200}{1 + 39e^{-0.24 \cdot 0}} = \dfrac{1200}{1 + 39 \cdot 1} = 30$

Thirty killer bees migrated to the town.

(b)   After 1 week, the killer bee population is $P(1) = \dfrac{1200}{1 + 39e^{-0.24 \cdot 1}} \approx 37.88$, or approximately 38 bees. The derivative $P'(t)$ equals the rate of change in the killer bee population at time t.

$$P'(t) = \frac{d}{dt}\frac{1200}{1 + 39e^{-0.24t}}$$
$$= -1200 \cdot \frac{-0.24 \cdot 39e^{-0.24t}}{(1 + 39e^{-0.24t})^2} \qquad \frac{d}{dx}\left[\frac{1}{g(x)}\right] = -\frac{g'(x)}{[g(x)]^2}$$
$$= \frac{11{,}232e^{-0.24t}}{(1 + 39e^{-0.24t})^2}$$

After 1 week, the rate of change in the killer bee population is

$$P'(1) = \frac{11{,}232e^{-0.24}}{\left(1 + 39e^{-0.24}\right)^2} \approx 8.804$$

IMPORTANT FACT The units of a derivative always equal the units of the dependent variable divided by the units of the independent variable.

That is, after 1 week, the population is increasing at a rate of approximately 8.8 bees per week.

AP® EXAM TIP

Answers for rate problems must include

- a numerical value,
- the units, and
- the time (or time interval).

(c) After 20 weeks,

$$P(20) = \frac{1200}{1 + 39e^{-0.24 \cdot 20}} \approx 908.43 \quad \text{and} \quad P'(20) = \frac{11{,}232e^{-0.24 \cdot 20}}{(1 + 39e^{-0.24 \cdot 20})^2} \approx 52.974$$

After 20 weeks, there are approximately 908 killer bees in the town, and the killer bee population is increasing at a rate of approximately 53 bees per week. ∎

NOW WORK Problem 13 and AP® Practice Problem 2.

③ Interpret a Derivative as Velocity or Acceleration

NEED TO REVIEW? Acceleration of an object moving on a line is discussed in Section 2.4, Objective 4, pp. 210–212.

Suppose an object moving along a line with position function $s = f(t)$, where f is twice differentiable. The velocity v and the acceleration a of the object are

$$v = v(t) = \frac{ds}{dt} = f'(t) \quad \text{and} \quad a = a(t) = \frac{dv}{dt} = \frac{d^2s}{dt^2} = f''(t)$$

Since acceleration is the rate of change of velocity, we conclude

- The velocity of the object is increasing if $a(t) > 0$.
- The velocity of the object is decreasing if $a(t) < 0$.

EXAMPLE 3 **Interpreting Derivatives as the Velocity and Acceleration of an Object Moving Along a Line**

The position of an object moving on the x-axis is modeled by the function $x(t) = t^3 - 8t^2 + 16t$ kilometers, where t is measured in hours.

(a) Find the average velocity of the object from $t = 2$ to $t = 4$.
(b) Find the velocity of the object at $t = 3$.
(c) Find the average acceleration of the object from $t = 2$ to $t = 4$.
(d) Find the acceleration of the object at $t = 3$.

Solution

(a) The average velocity $\dfrac{\Delta x}{\Delta t}$ of the object from $t = 2$ to $t = 4$ is

$$\frac{\Delta x}{\Delta t} = \frac{x(4) - x(2)}{4 - 2} \qquad \frac{\Delta x}{\Delta t} = \frac{x(t_1) - x(t_0)}{t_1 - t_0} \quad t_1 \neq t_0$$

$$= \frac{(4^3 - 8 \cdot 4^2 + 16 \cdot 4) - (2^3 - 8 \cdot 2^2 + 16 \cdot 2)}{4 - 2}$$

$$= \frac{-8}{2} = -4$$

The average velocity of the object from $t = 2$ to $t = 4$ is -4 km/h.

(b) The velocity $v = v(t)$ of an object whose position function is $x = x(t)$ is

$$v(t) = \frac{dx}{dt} = x'(t) = 3t^2 - 16t + 16$$

At $t = 3$, the velocity is

$$v(3) = 3 \cdot 3^2 - 16 \cdot 3 + 16 = 27 - 48 + 16 = -5 \text{ km/h}$$

The velocity of the object is decreasing at a rate of 5 km/h.

(continued on the next page)

(c) The average acceleration of an object moving on the x-axis is the change in velocity divided by the change in time. From $t = 2$ to $t = 4$, the average acceleration of the object is

$$\frac{\Delta v}{\Delta t} = \frac{v(4) - v(2)}{4 - 2}$$

$$\frac{\Delta v}{\Delta t} = \frac{v(t_1) - v(t_0)}{t_1 - t_0} \quad t_1 \neq t_0$$

$$= \frac{(3 \cdot 4^2 - 16 \cdot 4 + 16) - (3 \cdot 2^2 - 16 \cdot 2 + 16)}{4 - 2} \quad v(t) = 3t^2 - 16t + 16$$

$$= \frac{4}{2} = 2$$

The average acceleration of the object from $t = 2$ to $t = 4$ equals $2 \, \text{km/h}^2$.

(d) The acceleration $a = a(t)$ of an object moving on the x-axis is the rate of change in velocity with respect to time. That is,

$$a = a(t) = v'(t) = \frac{d}{dt}(3t^2 - 16t + 16) = 6t - 16$$

At $t = 3$, $a(3) = 6 \cdot 3 - 16 = 2 \, \text{km/h}^2$. ∎

NOW WORK   Problem 9 and AP® Practice Problems 3 and 4.

4.1 Assess Your Understanding

Concepts and Vocabulary

1. **True or False** The velocity of an object moving on a line equals the change in its position.

2. **Multiple Choice** The acceleration of an object moving on a line equals

   (a) the change in the speed over an interval of time.

   (b) the change in the velocity over an interval of time.

   (c) the rate of change of the position of the object with respect to time.

   (d) the rate of change of the velocity with respect to time.

3. **True or False** Both $f'(x)$ and $f''(x)$ are rates of change.

4. A(n) _____ measures change over an interval; a(n) _____ measures change at an instant.

Skill Building

In Problems 5–8, for each function,

(a) Find the derivative of f.

(b) Find the slope of the tangent line to the graph of f at the point $(c, f(c))$.

(c) Find an equation of the tangent line to the graph of f at the point $(c, f(c))$.

5. $f(x) = x^3 - 4x^2 - 2$ at $c = 3$ 6. $f(x) = 4 - 3x^2$ at $c = -1$

 7. $f(x) = \dfrac{e^x}{x^2 + 2}$ at $c = 0$ 8. $f(x) = e^{2x}(2 - x^2)$ at $c = 0$

In Problems 9–12, the position function of an object moving on a line is given. For each function, find the velocity and acceleration of the object at any time $t \geq 0$.

 9. $s(t) = -16t^2 + 120t + 43$, where s is in feet and t is in seconds

10. $s(t) = 3 - 12t + t^3$, where s is in meters and t is in seconds

11. $s(t) = t \cos^2 t$, where s is in kilometers and t is in hours

12. $s(t) = \dfrac{e^t}{t + 1}$, where s is in inches and t is in minutes

Applications and Extensions

 13. **Population Growth** A colony of fruit flies is growing in a controlled environment. The population is modeled by the function $P(t) = 30e^{0.2t}$, where t is measured in days.

   (a) Find the initial number of fruit flies in the colony.

   (b) How many fruit flies are in the colony after 5 days?

   (c) Find the function that models the rate of change in population of the fruit flies at time t.

   (d) Find and interpret the rate of change in the fruit fly population after 10 days.

 14. **Market Penetration** The percentage A of the market penetrated by the Nintendo Switch gaming system is modeled by the function $A(t) = 100 - 90e^{-0.12t}$, where t expresses the time in years since 2017 when the Nintendo Switch was first introduced.

   (a) Find $A'(t)$ and interpret its meaning.

   (b) Evaluate $A'(5)$ and interpret its meaning,

   (c) Evaluate $A'(15)$ and interpret its meaning.

15. **Units** The surface area S (in cm^2) of a right rectangular cylinder is increasing according to $S = S(t) = 4t - 1$, where t is the time in seconds. At what rate is the surface area changing with respect to time? What are the units of $S'(t)$?

16. **Units** The circumference C (in mi) of an oil spill is expanding according to $C = C(t) = 3t$, where t is measured in days. At what rate is the circumference changing with respect to time? What are the units of $C'(t)$?

17. **Units** The weekly revenue R (in dollars) is given by $R(x) = x^2 + 100x$, where x is the number of items sold. What is the rate of change of R with respect to x? What are the units of R'? Find and interpret $R'(75)$.

18. **Units** The daily cost C (in dollars) of producing x items is modeled by the function $C(x) = 0.1x^2 + 120x + 800$. Find the marginal cost, $C'(x)$, which is the cost of producing one additional item. What are the units of C'? Find and interpret $C'(100)$.

AP® Practice Problems

Multiple-Choice Questions

[PAGE 286] **1.** An equation of the line tangent to the graph of
$f(x) = 2x^3 - 3x + 6$ at the point $(0, f(0))$ is

(A) $y = -3x - 6$ (B) $y = 3x + 6$

(C) $y = -3x - 3$ (D) $y = -3x + 6$

[PAGE 287] **2.** If $y = f(x) = x^2 + \ln x$, what is the rate of change of f at 2?

(A) $\dfrac{9}{2}$ (B) $\dfrac{9}{4}$ (C) $\dfrac{7}{2}$ (D) $\dfrac{7}{4}$

[PAGE 288] **3.** The position function s of an object moving on a line
is $s(t) = \dfrac{t^3}{6} - \dfrac{2t^2}{3} + 4t - 1$. At $t = 3$, the velocity of the object is

(A) increasing.   (B) decreasing.

(C) neither increasing nor decreasing.   (D) stopped.

Free-Response Question

[PAGE 288] **4.** An object is moving along the y-axis, and its velocity at time t
is $v(t) = e^{-t} \sin t^3$ for $0 \le t \le \pi$.

(a) Find the velocity of the object at $t = \dfrac{\pi}{4}$.

(b) Is the velocity increasing or decreasing at $t = \dfrac{\pi}{4}$? Explain your reasoning.

(c) Find the acceleration of the object at $t = \dfrac{\pi}{4}$.

Retain Your Knowledge

Multiple-Choice Questions

1. The table below gives the position s of an object moving on a line at selected times t, $t \ge 0$.

t (seconds)	0	1	3	6	8
s (centimeters)	4	5	12	21	22

Use the data in the table to approximate the velocity of the particle at $t = 4$ s.

(A) -3 cm/s   (B) 3 cm/s   (C) 5 cm/s   (D) $\dfrac{1}{3}$ cm/s

2. Find the horizontal and vertical asymptotes, if any, of the graph of the function

$$f(x) = \frac{x^2 + x - 6}{2x^2 + 9x + 9}$$

(A) vertical asymptotes: $x = -3$, $x = -\dfrac{3}{2}$;

horizontal asymptote: $y = \dfrac{1}{2}$.

(B) no vertical asymptotes; horizontal asymptote: $y = \dfrac{1}{2}$.

(C) vertical asymptote: $x = -\dfrac{3}{2}$; no horizontal asymptote.

(D) vertical asymptote: $x = -\dfrac{3}{2}$; horizontal asymptote: $y = \dfrac{1}{2}$.

3. The function f is differentiable on the open interval $(-6, 6)$. A portion of the graph of f' is shown below.

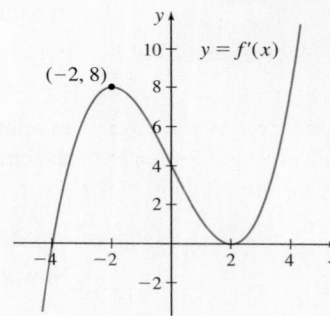

List the numbers x at which f has a horizontal tangent line.

(A) $-4, -2,$ and 2   (B) -4 and 2

(C) 0   (D) -2 and 2

Free-Response Question

4. The function $f(x) = (x + 2)^{1/3}$ is continuous for all real numbers. Find number(s) x, if any, at which the derivative of f does not exist. Explain your reasoning.

4.2 Related Rates

OBJECTIVE *When you finish this section, you should be able to:*

1 Solve related rate problems (p. 290)

In the natural sciences and in many of the social and behavioral sciences, there are quantities that are related to each other but that vary with time. For example, the pressure of an ideal gas of fixed volume is proportional to the temperature, yet each of these variables may change over time. Problems involving the rates of change of related variables are called **related rate problems**. In a related rate problem, we seek the rate at which one of the variables is changing at a certain time when the rates of change of the other variables are known.

1 Solve Related Rate Problems

We approach related rate problems by writing an equation involving time-dependent variables. This equation is often obtained by investigating the geometric and/or physical conditions imposed by the problem. We then differentiate the equation with respect to time and obtain a new equation that involves the variables and their rates of change with respect to time.

For example, suppose an object falling into still water causes a circular ripple, as illustrated in Figure 2. Both the radius and the area of the circle created by the ripple increase with time and their rates of change are related. We use the formula for the area of a circle

$$A = \pi r^2$$

to relate the radius and the area. Both A and r are functions of time t, and so the area of the circle can be expressed as

$$A(t) = \pi [r(t)]^2$$

Now we differentiate both sides with respect to t, obtaining

$$\frac{dA}{dt} = 2\pi r \frac{dr}{dt}$$

The derivatives (rates of change) are related by this equation, so we call them **related rates**. We can solve for one of these rates if the value of the other rate and the values of the variables are known.

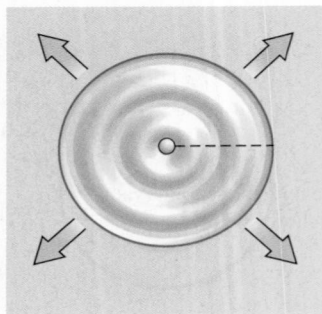

Figure 2

EXAMPLE 1 Solving a Related Rate Problem

A golfer hits a ball into a pond of still water, causing a circular ripple as shown in Figure 3. If the radius of the circle increases at the constant rate of 0.5 m/s, how fast is the area of the circle increasing when the radius of the ripple is 2 m?

Solution

The quantities that are changing, that is, the variables of the problem, are

$$t = \text{the time (in seconds) elapsed from the time when the ball hits the water;}$$

$$r = r(t) = \text{the radius (in meters) of the ripple } t \text{ seconds after the ball hits the water;}$$
and
$$A = A(t) = \text{the area (in square meters) of the circle formed by the ripple } t \text{ seconds after the ball hits the water.}$$

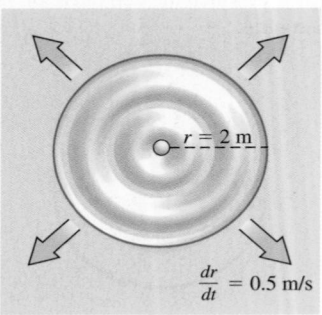

$r = 2$ m

$\frac{dr}{dt} = 0.5$ m/s

Figure 3

The rates of change with respect to time are

$$\frac{dr}{dt} = \text{the rate (in meters per second) at which the radius of the ripple is increasing; and}$$

$$\frac{dA}{dt} = \text{the rate (in meters squared per second) at which the area of the circle is increasing.}$$

It is given that $\dfrac{dr}{dt} = 0.5\,\text{m/s}$. We seek $\dfrac{dA}{dt}$ when $r = 2\,\text{m}$.

The relationship between A and r is given by the formula for the area of a circle:

$$A = \pi r^2$$

Since A and r are functions of t, we differentiate with respect to t to obtain

$$\frac{dA}{dt} = 2\pi r \frac{dr}{dt}$$

Since $\dfrac{dr}{dt} = 0.5\,\text{m/s}$,

$$\frac{dA}{dt} = 2\pi r (0.5) = \pi r$$

When $r = 2\,\text{m}$,

$$\frac{dA}{dt} = 2\pi \approx 6.283$$

When the radius of the circular ripple is $2\,\text{m}$, its area is increasing at a rate of approximately $6.283\,\text{m}^2/\text{s}$. ∎

AP® EXAM TIP

When answering free response questions, be sure to express the answers using precise mathematical language and correct units.

CAUTION Numerical values cannot be substituted (Step 5) until after the equation has been differentiated (Step 4).

Steps for Solving a Related Rate Problem

Step 1 Read the problem carefully, perhaps several times. Pay particular attention to the rate of change you are asked to find. If possible, draw a picture to illustrate the problem.

Step 2 Identify the variables, assign symbols to them, and label the picture. Identify the rates of change as derivatives. Be sure to list the units for each variable and each rate of change. Indicate what is given and what is asked for.

Step 3 Write an equation that relates the variables. It may be necessary to use more than one equation.

Step 4 Differentiate both sides of the equation(s).

Step 5 Substitute numerical values for the variables and the derivatives. Solve for the unknown rate of change.

NOW WORK Problem 7 and AP® Practice Problems 3 and 6.

EXAMPLE 2 Solving a Related Rate Problem

A spherical balloon is inflated at the rate of $10\,\text{m}^3/\text{min}$. Find the rate at which the surface area of the balloon is increasing when the radius of the sphere is $3\,\text{m}$.

Solution

We follow the steps for solving a related rate problem.

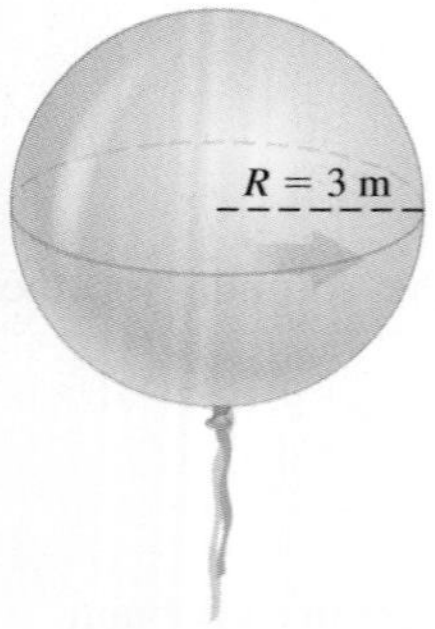

Step 1 Figure 4 shows a sketch of the balloon with its radius labeled.

Step 2 Identify the variables of the problem:

$$t = \text{the time (in minutes) measured from the moment the balloon}$$
$$\text{begins inflating}$$
$$R = R(t) = \text{the radius (in meters) of the balloon at time } t$$
$$V = V(t) = \text{the volume (in meters cubed) of the balloon at time } t$$
$$S = S(t) = \text{the surface area (in meters squared) of the balloon at time } t$$

Identify the rates of change:

$$\frac{dR}{dt} = \text{the rate of change of the radius of the balloon (in meters per minute)}$$

$$\frac{dV}{dt} = \text{the rate of change of the volume of the balloon (in meters cubed per minute)}$$

$$\frac{dS}{dt} = \text{the rate of change of the surface area of the balloon (in meters squared per minute)}$$

We are given $\dfrac{dV}{dt} = 10\,\text{m}^3/\text{min}$, and we seek $\dfrac{dS}{dt}$ when $R = 3\,\text{m}$.

Step 3 Since both the volume V of the balloon (a sphere) and its surface area S can be expressed in terms of the radius R, we use two equations to relate the variables.

$$V = \frac{4}{3}\pi R^3 \quad \text{and} \quad S = 4\pi R^2 \qquad \text{where } V, S, \text{ and } R \text{ are functions of } t$$

Step 4 Use the Chain Rule to differentiate both sides of the equations with respect to time t.

$$\frac{dV}{dt} = 4\pi R^2 \frac{dR}{dt} \qquad \text{and} \qquad \frac{dS}{dt} = 8\pi R \frac{dR}{dt}$$

Combine the equations by solving for $\dfrac{dR}{dt}$ in the equation on the left and substituting the result into the equation for $\dfrac{dS}{dt}$ on the right. Then

$$\frac{dS}{dt} = 8\pi R \left(\frac{\frac{dV}{dt}}{4\pi R^2} \right) = \frac{2}{R}\frac{dV}{dt}$$

Step 5 Substitute $R = 3\,\text{m}$ and $\dfrac{dV}{dt} = 10\,\text{m}^3/\text{min}$.

$$\frac{dS}{dt} = \frac{2}{3} \cdot 10 = \frac{20}{3} \approx 6.667$$

When the radius of the balloon is $3\,\text{m}$, its surface area is increasing at a rate of approximately $6.667\,\text{m}^2/\text{min}$. ∎

Figure 4

NEED TO REVIEW? Geometry formulas are discussed in Appendix A.2, p. A-16.

AP® EXAM TIP

On the exam, approximate answers should be rounded or truncated (your choice) to three decimal places. In this text, we round answers to three decimal places.

NOW WORK Problems 11 and 17 and AP® Practice Problems 1 and 4.

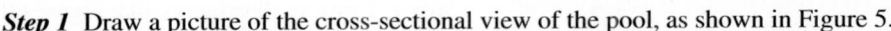

EXAMPLE 3 Solving a Related Rate Problem

A rectangular swimming pool 10 m long and 5 m wide has a depth of 3 m at one end and 1 m at the other end. See Figure 5. If water is pumped into the pool at the rate of 0.3 cubic meters per minute (m³/min), at what rate is the water level rising when it is 1.5 m deep at the deep end?

Solution

Step 1 Draw a picture of the cross-sectional view of the pool, as shown in Figure 5.

Step 2 The width of the pool is 5 m, the water level (measured at the deep end) is h (in meters), the distance from the wall at the deep end to the edge of the water is L (in meters), and the volume of water in the pool is V (in cubic meters, m³). Each of the variables h, L, and V changes with respect to time t.

We are given $\dfrac{dV}{dt} = 300 \text{ l/min}$ and are asked to find $\dfrac{dh}{dt}$ when $h = 1.5$ m.

Step 3 The volume V is related to L and h by the formula

$$V = (\text{Cross-sectional triangular area})(\text{width}) = \frac{1}{2}Lh \cdot 5 = \frac{5}{2}Lh$$

Using similar triangles, L and h are related by the equation

$$\frac{L}{h} = \frac{10}{2} \qquad \text{so} \quad L = 5h$$

Now we can write V as

$$V = \frac{5}{2}Lh = \underset{\underset{L=5h}{\uparrow}}{\frac{5}{2}} \cdot 5h \cdot h = \frac{25}{2}h^2 \tag{1}$$

Both V and h change with time t.

Step 4 Differentiate both sides of equation (1) with respect to t.

$$\frac{dV}{dt} = 25h\frac{dh}{dt}$$

Step 5 Substitute $h = 1.5$ m and $\dfrac{dV}{dt} = 0.3$ m³/min. Then

$$0.3 = 25(1.5)\frac{dh}{dt} \qquad\qquad \frac{dV}{dt} = 25h\frac{dh}{dt}$$

$$\frac{dh}{dt} = \frac{0.3}{25(1.5)} = 0.008$$

When the height of the water at the deep end is 1.5 m, the water level is rising at a rate of 0.008 m/min. ∎

NOW WORK   Problem 23 and AP® Practice Problem 7.

Figure 5

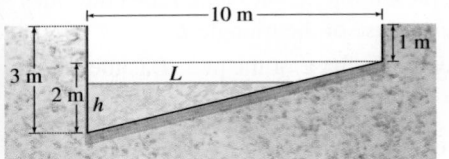

NEED TO REVIEW? Similar triangles are discussed in Appendix A.2, pp. A-14 to A-15.

NOTE 1000 liters = 1 m³.

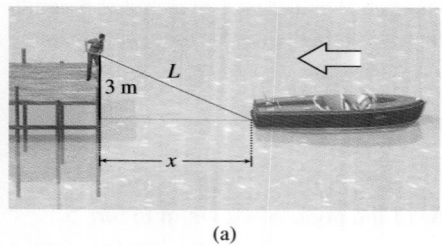

(a)

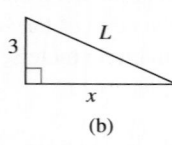

(b)

Figure 6

NEED TO REVIEW? The Pythagorean Theorem is discussed in Appendix A.2, pp. A-12 to A-13.

EXAMPLE 4   Solving a Related Rate Problem

A person standing at the end of a pier is docking a boat by pulling a rope at the rate of 2 m/s. The end of the rope is 3 m above water level. See Figure 6(a). How fast is the boat approaching the base of the pier when 5 m of rope are left to pull in? (Assume the rope never sags and that the rope is attached to the boat at water level.)

Solution

Step 1 Draw an illustration, like Figure 6(b), representing the problem. Label the sides of the triangle 3 and x, and label the hypotenuse of the triangle L.

Step 2 x is the distance (in meters) from the boat to the base of the pier, L is the length of the rope (in meters), and the distance between the water level and the person's hand is 3 m. Both x and L are changing with respect to time, so $\dfrac{dx}{dt}$ is the rate at which the boat approaches the pier, and $\dfrac{dL}{dt}$ is the rate at which the rope is pulled in. Both $\dfrac{dx}{dt}$ and $\dfrac{dL}{dt}$ are measured in meters/second (m/s).

Since the length L of the rope is decreasing, the rate of change of L is $\dfrac{dL}{dt} = -2$ m/s. We seek $\dfrac{dx}{dt}$ when $L = 5$ m.

Step 3 The lengths 3, x, and L are the sides of a right triangle and by the Pythagorean Theorem are related by the equation

$$x^2 + 3^2 = L^2 \qquad (2)$$

Step 4 Differentiate both sides of equation (2) with respect to t.

$$2x \frac{dx}{dt} = 2L \frac{dL}{dt}$$

$$\frac{dx}{dt} = \frac{L}{x} \frac{dL}{dt} \qquad \text{Solve for } \frac{dx}{dt}.$$

Step 5 When $L = 5$, we use equation (2) to find $x = 4$. Since $\dfrac{dL}{dt} = -2$,

$$\frac{dx}{dt} = \frac{5}{4}(-2) \qquad L = 5, x = 4, \frac{dL}{dt} = -2$$

$$\frac{dx}{dt} = -2.5$$

When 5 m of rope are left to be pulled in, the distance of the boat from the pier is decreasing at a rate of − 2.5 m/s. ∎

NOW WORK   Problem 29 and AP® Practice Problems 5 and 8.

EXAMPLE 5   Solving a Related Rate Problem

A revolving light located 5 km from a straight shoreline turns at a constant angular speed of 3 rad/min. With what speed is the spot of light moving along the shore when the beam makes an angle of 60° with the shoreline?

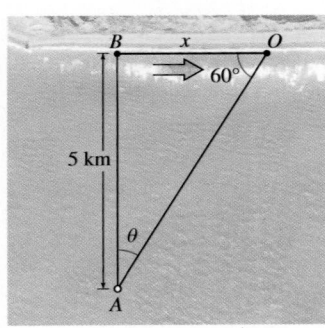

Figure 7

RECALL For motion that is circular, angular speed ω is defined as the rate of change of a central angle θ of the circle with respect to time. That is,

$$\omega = \frac{d\theta}{dt}$$

where θ is measured in radians.

Solution

Figure 7 illustrates the triangle that describes the problem.

$x =$ the distance (in kilometers) of the beam of light from the point B

$\theta =$ the angle (in radians) the beam of light makes with AB

Both variables x and θ change with time t (in minutes). The rates of change are

$$\frac{dx}{dt} = \text{the speed of the spot of light along the shore (in kilometers per minute)}$$

$$\frac{d\theta}{dt} = \text{the angular speed of the beam of light (in radians per minute)}$$

We are given $\dfrac{d\theta}{dt} = 3$ rad/min and we seek $\dfrac{dx}{dt}$ when the angle $AOB = 60°$.

From Figure 7,

$$\tan\theta = \frac{x}{5} \qquad \text{so} \qquad x = 5\tan\theta$$

Then

$$\frac{dx}{dt} = 5\sec^2\theta\,\frac{d\theta}{dt}$$

When $AOB = 60°$, angle $\theta = 30° = \dfrac{\pi}{6}$ radian. Since $\dfrac{d\theta}{dt} = 3$ rad/min,

$$\frac{dx}{dt} = \frac{5}{\cos^2\theta}\frac{d\theta}{dt} = \frac{5}{\left(\cos\dfrac{\pi}{6}\right)^2}\cdot 3 = \frac{15}{\dfrac{3}{4}} = 20$$

When $\theta = 30°$, the light is moving along the shore at a speed of 20 km/min. ∎

NOW WORK Problem 37 and AP® Practice Problems 2 and 9.

4.2 Assess Your Understanding

Concepts and Vocabulary

1. If a spherical balloon of volume V is inflated at a rate of 10 m³/min, where t is the time (in minutes), what is the rate of change of V with respect to t?

2. For the balloon in Problem 1, if the radius r is increasing at the rate of 0.5 m/min, what is the rate of change of r with respect to t?

In Problems 3 and 4, x and y are differentiable functions of t.

Find $\dfrac{dx}{dt}$ when $x = 3$, $y = 4$, and $\dfrac{dy}{dt} = 2$.

3. $x^2 + y^2 = 25$

4. $x^3 y^2 = 432$

Skill Building

5. Suppose h is a differentiable function of t and suppose that

$$\frac{dh}{dt} = \frac{5}{16}\pi \text{ when } h = 8$$

Find $\dfrac{dV}{dt}$ when $h = 8$ if $V = \dfrac{1}{12}\pi h^3$.

6. Suppose x and y are differentiable functions of t and suppose that

when $t = 20$, $\dfrac{dx}{dt} = 5$, $\dfrac{dy}{dt} = 4$, $x = 150$, and $y = 80$. Find $\dfrac{ds}{dt}$

when $t = 20$ if $s^2 = x^2 + y^2$.

PAGE 291 7. Suppose h is a differentiable function of t and suppose that

when $h = 3$, $\dfrac{dh}{dt} = \dfrac{1}{12}$. Find $\dfrac{dV}{dt}$ when $h = 3$ if $V = 80h^2$.

8. Suppose x is a differentiable function of t and suppose

that when $x = 15$, $\dfrac{dx}{dt} = 3$. Find $\dfrac{dy}{dt}$ when $x = 15$

if $y^2 = 625 - x^2$, $y \geq 0$.

9. **Volume of a Cube** If each edge of a cube is increasing at the constant rate of 3 cm/s, how fast is the volume of the cube increasing when the length x of an edge is 10 cm?

10. **Volume of a Sphere** If the radius of a sphere is increasing at 1 cm/s, find the rate of change of its volume when the radius is 6 cm.

PAGE 292 11. **Radius of a Sphere** If the surface area of a sphere is shrinking at the constant rate of 0.1 cm²/h, find the rate of change of its radius when the radius is $\dfrac{20}{\pi}$ cm.

12. **Surface Area of a Sphere** If the radius of a sphere is increasing at the constant rate of 2 cm/min, find the rate of change of its surface area when the radius is 100 cm.

13. Dimensions of a Triangle   Consider a right triangle with hypotenuse of (fixed) length 45 cm and variable legs of lengths x and y, respectively. If the leg of length x increases at the rate of 2 cm/min, at what rate is y changing when $x = 4$ cm?

14. Change in Area   The fixed sides of an isosceles triangle are of length 1 cm. (See the figure.) If the sides slide outward at a speed of 1 cm/min, at what rate is the area enclosed by the triangle changing when $\theta = 30°$?

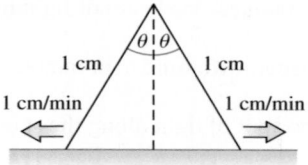

15. Area of a Triangle   An isosceles triangle has equal sides 4 cm long and the included angle is θ. If θ increases at the rate of 2°/min, at what rate is the area of the triangle changing when θ is 30°?

Hint: $\dfrac{d\theta}{dt} = \dfrac{2\pi}{180}$ rad/min.

16. Area of a Rectangle   In a rectangle with a diagonal 15 cm long, one side is increasing at the rate of $2\sqrt{5}$ cm/s. Find the rate of change of the area when that side is 10 cm long.

 17. Change in Surface Area   A spherical balloon filled with gas has a leak that causes the gas to escape at a rate of 1.5 m³/min. At what rate is the surface area of the balloon shrinking when the radius is 4 m?

18. Frozen Snow Ball   Suppose that the volume of a spherical ball of ice decreases (by melting) at a rate proportional to its surface area. Show that its radius decreases at a constant rate.

Applications and Extensions

19. Change in Inclination   A ladder 5 m long is leaning against a wall. If the lower end of the ladder slides away from the wall at the rate of 0.5 m/s, at what rate is the inclination θ of the ladder with respect to the ground changing when the lower end is 4 m from the wall?

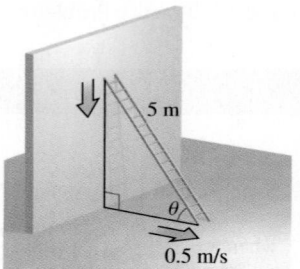

20. Angle of Elevation   A man 2 m tall walks horizontally at a constant rate of 1 m/s toward the base of a tower 25 m tall. When the man is 10 m from the tower, at what rate is the angle of elevation changing if that angle is measured from the horizontal to the line joining the top of the man's head to the top of the tower?

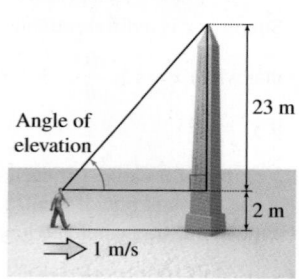

21. Filling a Pool   A public swimming pool is 30 m long and 5 m wide. Its depth is 3 m at the deep end and 1 m at the shallow end. If water is pumped into the pool at the rate of 15 m³/min, how fast is the water level rising when it is 1 m deep at the deep end? Use Figure 5 as a guide.

22. Filling a Tank   Water is flowing into a vertical cylindrical tank of diameter 6 m at the rate of 5 m³/min. Find the rate at which the depth of the water is rising.

23. Fill Rate   A container in the form of a right circular cone (vertex down) has radius 4 m and height 16 m. See the figure. If water is poured into the container at the constant rate of 16 m³/min, how fast is the water level rising when the water is 8 m deep?

Hint: The volume V of a cone of radius r and height h is $V = \dfrac{1}{3}\pi r^2 h$.

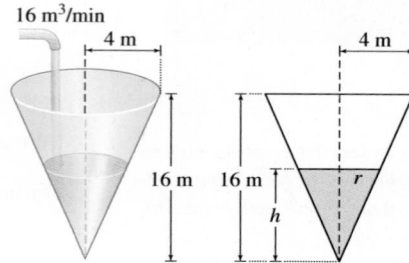

24. Building a Sand Pile   Sand is being poured onto the ground, forming a conical pile whose height equals one-fourth of the diameter of the base. If the sand is falling at the rate of 20 cm³/s, how fast is the height of the sand pile increasing when it is 3 cm high?

25. Is There a Leak?   A cistern in the shape of a cone 4 m deep and 2 m in diameter at the top is being filled with water at a constant rate of 3 m³/min.

(a) At what rate is the water rising when the water in the tank is 3 m deep?

(b) If, when the water is 3 m deep, it is observed that the water rises at a rate of 0.5 m/min, at what rate is water leaking from the tank?

26. Change in Area   The vertices of a rectangle are at $(0, 0)$, $(0, e^x)$, $(x, 0)$, and (x, e^x), $x > 0$. If x increases at the rate of 1 unit per second, at what rate is the area increasing when $x = 10$ units?

27. Distance from the Origin   An object is moving along the parabola $y^2 = 4(3 - x)$. When the object is at the point $(-1, 4)$, its y-coordinate is increasing at the rate of 3 units per second. How fast is the distance from the object to the origin changing at that instant?

28. Funneling Liquid   A conical funnel 15 cm in diameter and 15 cm deep is filled with a liquid that runs out at the rate of 5 cm³/min. At what rate is the depth of liquid changing when its depth is 8 cm?

29. Baseball   A baseball is hit along the third-base line with a speed of 100 ft/s. At what rate is the ball's distance from first base changing when it crosses third base? See the figure.

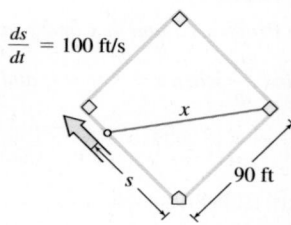

30. Flight of a Falcon   A peregrine falcon flies up from its trainer at an angle of 60° until it has flown 200 ft. It then levels off and continues to fly away. If the constant speed of the bird is 88 ft/s, how fast is the falcon moving away from the falconer after it is 6 s in flight? See the figure.

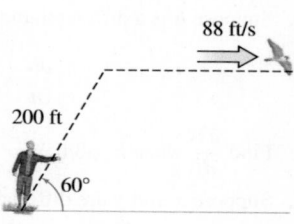

Source: http://www.rspb.org.uk.

31. Boyle's Law A gas is said to be compressed *adiabatically* if there is no gain or loss of heat. When such a gas is diatomic (has two atoms per molecule), it satisfies the equation $PV^{1.4} = k$, where k is a constant, P is the pressure, and V is the volume. At a given instant, the pressure is $20 \, \text{kg/cm}^2$, the volume is $32 \, \text{cm}^3$, and the volume is decreasing at the rate of $2 \, \text{cm}^3/\text{min}$. At what rate is the pressure changing?

32. Heating a Plate When a metal plate is heated, it expands. If the shape of the plate is circular and its radius is increasing at the rate of $0.02 \, \text{cm/s}$, at what rate is the area of the top surface increasing when the radius is $3 \, \text{cm}$?

33. Pollution After a rupture, oil begins to escape from an underwater well. If, as the oil rises, it forms a circular slick whose radius increases at a rate of $0.42 \, \text{ft/min}$, find the rate at which the area of the spill is increasing when the radius is $120 \, \text{ft}$.

34. Flying a Kite A girl flies a kite at a height $30 \, \text{m}$ above her hand. If the kite flies horizontally away from the girl at the rate of $2 \, \text{m/s}$, at what rate is the string being let out when the length of the string released is $70 \, \text{m}$? Assume that the string remains taut.

35. Falling Ladder An 8-m ladder is leaning against a vertical wall. If a person pulls the base of the ladder away from the wall at the rate of $0.5 \, \text{m/s}$, how fast is the top of the ladder moving down the wall when the base of the ladder is

(a) $3 \, \text{m}$ from the wall?

(b) $4 \, \text{m}$ from the wall?

(c) $6 \, \text{m}$ from the wall?

36. Beam from a Lighthouse A light in a lighthouse $2000 \, \text{m}$ from a straight shoreline is rotating at 2 revolutions per minute. How fast is the beam moving along the shore when it passes a point $500 \, \text{m}$ from the point on the shore opposite the lighthouse? *Hint:* One revolution $= 2\pi$ rad.

PAGE 295 **37. Moving Radar Beam** A radar antenna, making one revolution every 5 s, is located on a ship that is $6 \, \text{km}$ from a straight shoreline. How fast is the radar beam moving along the shoreline when the beam makes an angle of $45°$ with the shore?

38. Tracking a Rocket When a rocket is launched, it is tracked by a tracking dish on the ground located a distance D from the point of launch. The dish points toward the rocket and adjusts its angle of elevation θ to the horizontal (ground level) as the rocket rises. Suppose a rocket rises vertically at a constant speed of $2.0 \, \text{m/s}$, with the tracking dish located $150 \, \text{m}$ from the launch point. Find the rate of change of the angle θ of elevation of the tracking dish with respect to time t (tracking rate) for each of the following:

(a) Just after launch.

(b) When the rocket is $100 \, \text{m}$ above the ground.

(c) When the rocket is $1.0 \, \text{km}$ above the ground.

(d) Use the results in (a)–(c) to describe the behavior of the tracking rate as the rocket climbs higher and higher. What limit does the tracking rate approach as the rocket gets extremely high?

39. Lengthening Shadow A child, $1 \, \text{m}$ tall, is walking directly under a street lamp that is $6 \, \text{m}$ above the ground. If the child walks away from the light at the rate of $20 \, \text{m/min}$, how fast is the child's shadow lengthening?

40. Approaching a Pole A boy is walking toward the base of a pole $20 \, \text{m}$ high at the rate of $4 \, \text{km/h}$. At what rate (in meters per second) is the distance from his feet to the top of the pole changing when he is $5 \, \text{m}$ from the pole?

41. Riding a Ferris Wheel A Ferris wheel is $50 \, \text{ft}$ in diameter and its center is located $30 \, \text{ft}$ above the ground. See the image. If the wheel rotates once every 2 min, how fast is a cabin rising when it is $42.5 \, \text{ft}$ above the ground? How fast is it moving horizontally?

Yee Kong Mau/Getty Images

42. Approaching Cars Two cars approach an intersection, one heading east at the rate of $30 \, \text{km/h}$ and the other heading south at the rate of $40 \, \text{km/h}$. At what rate are the two cars approaching each other at the instant when the first car is $100 \, \text{m}$ from the intersection and the second car is $75 \, \text{m}$ from the intersection? Assume the cars maintain their respective speeds.

43. Parting Ways An elevator in a building is located on the fifth floor, which is $25 \, \text{m}$ above the ground. A delivery truck is positioned directly beneath the elevator at street level. If, simultaneously, the elevator goes down at a speed of $5 \, \text{m/s}$ and the truck pulls away at a speed of $8 \, \text{m/s}$, how fast will the elevator and the truck be separating 1 s later? Assume the speeds remain constant at all times.

44. Pulley In order to lift a container $10 \, \text{m}$, a rope is attached to the container and, with the help of a pulley, the container is hoisted. See the figure. If a person holds the end of the rope and walks away from beneath the pulley at the rate of $2 \, \text{m/s}$, how fast is the container rising when the person is $5 \, \text{m}$ away? Assume the end of the rope in the person's hand was originally at the same height as the top of the container.

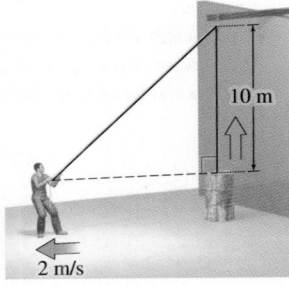

45. Business A manufacturer of precision digital switches has daily cost C and revenue R, in dollars, of $C(x) = 10{,}000 + 3x$ and $R(x) = 5x - \dfrac{x^2}{2000}$, respectively, where x is the daily production of switches. Production is increasing at the rate of 50 switches per day when production is 1000 switches.

(a) Find the rate of increase in cost when production is 1000 switches per day.

(b) Find the rate of increase in revenue when production is 1000 switches per day.

(c) Find the rate of increase in profit when production is 1000 switches per day. *Hint:* Profit = Revenue − Cost.

46. An Enormous Growing Black Hole In 2022, astronomers announced the discovery of the most massive black hole identified to date, which has a mass equal to 100 billion Suns. Huge black holes typically grow by swallowing nearby matter, including whole stars. In this way, the size of the **event horizon** (the distance from the center of the black hole to the position at which no light can escape) increases. From general relativity, the radius R of the event horizon for a black hole of mass m is $R = \dfrac{2Gm}{c^2}$, where G is the gravitational constant and c is the speed of light in a vacuum.

(a) If this huge black hole swallows one Sun-like star per year, at what rate (in kilometers per year) will its event horizon grow? The mass of the Sun is 1.99×10^{30} kg, the speed of light in a vacuum is $c = 3.00 \times 10^8$ m/s, and $G = 6.67 \times 10^{-11}$ m³/(kg · s²).

(b) By what percent does the event horizon change per year?

47. Weight in Space An object that weighs K lb on the surface of Earth weighs approximately

$$W(R) = K \left(\frac{3960}{3960 + R} \right)^2$$

pounds when it is a distance of R mi from Earth's surface. Find the rate at which the weight of an object weighing 1000 lb on Earth's surface is changing when it is 50 mi above Earth's surface and is being lifted at the rate of 10 mi/s.

48. Pistons In a certain piston engine, the distance x in meters between the center of the crank shaft and the head of the piston is given by $x = \cos \theta + \sqrt{16 - \sin^2 \theta}$, where θ is the angle between the crank and the path of the piston head. See the figure below.

(a) If θ increases at the constant rate of 45 rad/s, what is the speed of the piston head when $\theta = \dfrac{\pi}{6}$?

(b) Graph x as a function of θ on the interval $[0, \pi]$. Determine the maximum distance and the minimum distance of the piston head from the center of the crank shaft.

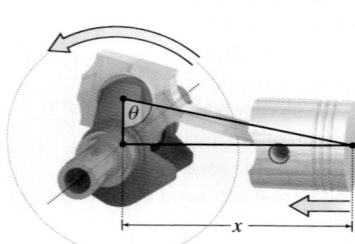

49. Tracking an Airplane A soldier at an anti-aircraft battery observes an airplane flying toward him at an altitude of 4500 ft. See the figure. When the angle of elevation of the battery is 30°, the soldier must increase the angle of elevation by 1°/s to keep the plane in sight. What is the ground speed of the plane?

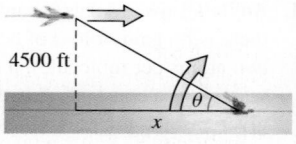

50. Change in the Angle of Elevation A hot air balloon is rising at a speed of 100 m/min. If an observer is standing 200 m from the lift-off point, what is the rate of change of the angle of elevation of the observer's line of sight when the balloon is 600 m high?

51. Rate of Rotation A searchlight is following a plane flying at an altitude of 3000 ft in a straight line over the light; the plane's velocity is 500 mi/h. At what rate is the searchlight turning when the distance between the light and the plane is 5000 ft? See the figure.

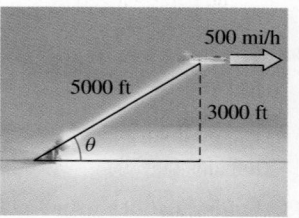

52. Police Chase A police car approaching an intersection at 80 ft/s spots a suspicious vehicle on the cross street. When the police car is 210 ft from the intersection, the policeman turns a spotlight on the vehicle, which is at that time just crossing the intersection at a constant rate of 60 ft/s. See the figure. How fast must the light beam be turning 2 s later for it to follow the suspicious vehicle?

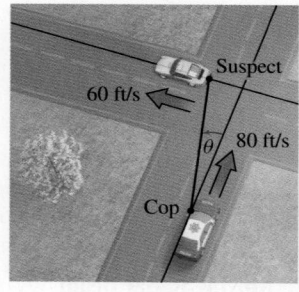

53. Change in Volume The height h and width x of an open box with a square base are related to its volume by the formula $V = hx^2$. Discuss how the volume changes

(a) if h decreases with time, but x remains constant.

(b) if both h and x change with time.

54. Rate of Change Let $y = 2e^{\cos x}$. If both x and y vary with time in such a way that y increases at a steady rate of 5 units per second, at what rate is x changing when $x = \dfrac{\pi}{2}$?

Challenge Problems

 55. Moving Shadows   The dome of an observatory is a hemisphere 60 ft in diameter. A boy is playing near the observatory at sunset. He throws a ball upward so that its shadow climbs to the highest point on the dome. How fast is the shadow moving along the dome $\frac{1}{2}$ s after the ball begins to fall? How did you use the fact that it was sunset in solving the problem? (*Note:* A ball falling from rest covers a distance $s = 16t^2$ ft in t seconds.)

56. Moving Shadows   A railroad train is moving at a speed of 15 mi/h past a station 800 ft long. The track has the shape of the parabola $y^2 = 600x$ as shown in the figure. If the Sun is just rising in the east, find how fast the shadow S of the locomotive L is moving along the wall at the instant it reaches the end of the wall.

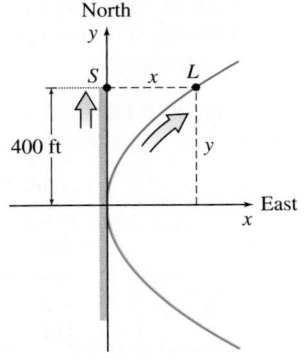

57. Change in Area   The hands of a clock are 2 in. and 3 in. long. See the figure. As the hands move around the clock, they sweep out the triangle *OAB*. At what rate is the area of the triangle changing at 12:10 p.m.?

 58. Distance   A train is traveling northeast at a rate of 25 ft/s along a track that is elevated 20 ft above the ground. The track passes over the street below at an angle of 30°. See the figure. Five seconds after the train passes over the road, a car traveling east passes under the tracks going 40 ft/s. How fast are the train and the car separating after 3 s?

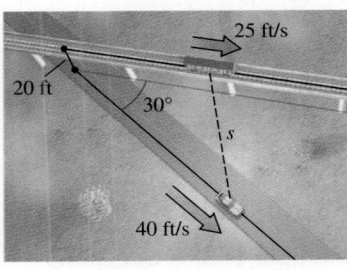

Preparing for the AP® Exam

AP® Practice Problems

Multiple-Choice Questions

`PAGE 292` **1.** A spherical balloon is inflated at the rate of 50 m³/min. Find the rate at which the radius of the balloon is increasing when the diameter is 20 m.

(A) $\dfrac{1}{2\pi}$ m/min   (B) $\dfrac{5}{8\pi}$ m/min

(C) $\dfrac{1}{8\pi}$ m/min   (D) $\dfrac{5}{4\pi}$ m/min

`PAGE 295` **2.** In the right triangle below, θ is changing at the rate of 2 radians per second. At what rate is x changing at the instant when $x = 1$ cm?

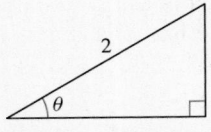

(A) 2 cm/s (B) $2\sqrt{3}$ cm/s (C) $\sqrt{3}$ cm/s (D) $4\sqrt{3}$ cm/s

`PAGE 291` **3.** The radius of a circle is decreasing at a constant rate of 2 in./min. What is the rate of change in the area of the circle when its area is 25π in²?

(A) -20π in²/min   (B) -25π in²/min

(C) 20π in²/min   (D) $20\pi^2$ in²/min

`PAGE 292` **4.** The radius r of a sphere is increasing at a rate of 2 cm/s. At the instant when $r = 12$ cm, what is the rate of change in the surface area S of the sphere? (The surface area S of a sphere with radius r is $S = 4\pi r^2$.)

(A) 96π cm²/s   (B) 1152π cm²/s

(C) 576π cm²/s   (D) 192π cm²/s

`PAGE 294` **5.** The sides of the rectangle shown below are increasing so that the rate of change of y with respect to time t is three times the rate of change of x with respect to t. If $\dfrac{dc}{dt} = 1$, what is the rate of change of x when $x = 6$ and $y = 8$?

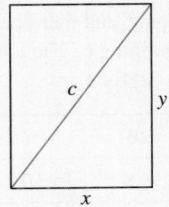

(A) 3   (B) $\dfrac{1}{3}$

(C) 1   (D) $\dfrac{1}{6}$

`PAGE 291` **6.** The area of a circle is increasing at a rate of 48π ft²/h. How fast is the radius of the circle increasing when its area is 36π ft²?

(A) 4 ft/h   (B) 6 ft/h   (C) $4\sqrt{3}$ ft/h   (D) $\dfrac{4}{\pi}$ ft/h

`PAGE 293` **7.** An ice cube in the shape of a sphere is melting. Its volume is decreasing at the constant rate of 8 mm³/min. When the radius of the sphere is 4000 mm, how fast is the surface area decreasing?

(A) 16π mm²/min   (B) 0.002 mm²/min

(C) 0.008 mm²/min   (D) 0.004 mm²/min

(continued on the next page)

8. Two roads cross at right angles. A police officer sits in a car 65 m east of the crossing and observes a car speeding northbound at 84 m/s. At what speed (in meters per second) is the car distancing itself from the police officer 5 seconds after it passes the crossing?

   (A) 166.024 m/s

   (B) 83.012 m/s

   (C) 84 m/s

   (D) 95.859 m/s

Free-Response Question

9. A roofer's 13-meter ladder is placed against the wall of a building with its base on level ground. The top of the ladder slips down the wall as the bottom of the ladder slips away from the building at a constant rate of 5 m/s.

   (a) At what rate is the top of the ladder moving when it is 5 m from the ground?

   (b) At what rate is the area of the triangle formed by the ladder, the wall, and the ground changing when the top of the ladder is 5 m from the ground?

   (c) If θ is the angle formed by the ladder and the ground, what is the rate of change in θ when the top of the ladder is 5 m from the ground?

   *See the **BREAK IT DOWN** on page 329 for a stepped out solution to AP® Practice Problem 9(a).*

Retain Your Knowledge

Multiple-Choice Questions

1. $\dfrac{d}{dx} f\left(\dfrac{x+1}{2x-1}\right) =$

   (A) $\dfrac{-3}{(-2x-1)^2}$

   (B) $f'\left(\dfrac{x+1}{2x-1}\right) \cdot \dfrac{1}{(2x-1)^2}$

   (C) $-f'\left(\dfrac{x+1}{2x-1}\right) \cdot \dfrac{3}{(2x-1)^2}$

   (D) $f'\left(\dfrac{x+1}{2x-1}\right) \cdot \dfrac{3}{(2x-1)^2}$

2. The functions f and g are both differentiable in an open interval containing 6. The table below shows the values of f, g, f', and g' for $x = 6$.

$f(6)$	$f'(6)$	$g(6)$	$g'(6)$
-8	0	3	2

   Use the table to find the derivative of the

   function $h(x) = \dfrac{3\sqrt[3]{f(x)}}{g(x)}$ for $x = 6$.

   (A) $-\dfrac{4}{3}$ (B) $\dfrac{10}{9}$ (C) $\dfrac{4}{3}$ (D) 4

3. Find an equation of the line tangent to the graph of function $f(x) = 3\sin^{-1}(2x)$ at $x = \dfrac{1}{4}$.

   (A) $y - \dfrac{\pi}{2} = \dfrac{6}{\sqrt{3}}\left(x - \dfrac{1}{4}\right)$ (B) $y - \dfrac{\pi}{2} = \dfrac{12}{\sqrt{3}}\left(x - \dfrac{1}{4}\right)$

   (C) $y - \dfrac{\pi}{6} = \dfrac{2}{\sqrt{3}}\left(x - \dfrac{1}{4}\right)$ (D) $y - \dfrac{\pi}{2} = \dfrac{4}{\sqrt{3}}\left(x - \dfrac{1}{4}\right)$

Free-Response Question

4. $f(x) = \begin{cases} 4x + 1 & \text{if} \quad x < 1 \\ 5 & \text{if} \quad x = 1 \\ 2x^2 - 8x + 11 & \text{if} \quad x > 1 \end{cases}$

   (a) Determine whether f is continuous at $x = 1$.

   (b) Find $f'(2)$.

   (c) Determine whether $f'(1)$ exists. Justify the answer.

4.3 Differentials; Linear Approximations; Newton's Method

OBJECTIVES *When you finish this section, you should be able to:*

1 **Find the differential of a function and interpret it geometrically (p. 301)**

2 **Find the linear approximation to a function (p. 303)**

3 **Use differentials in applications (p. 305)**

4 **Use Newton's Method to approximate a real zero of a function (p. 306)**

1 Find the Differential of a Function and Interpret It Geometrically

Recall that for a differentiable function $y = f(x)$, the derivative can be defined as

$$\frac{dy}{dx} = f'(x) = \lim_{\Delta x \to 0} \frac{\Delta y}{\Delta x} = \lim_{\Delta x \to 0} \frac{f(x + \Delta x) - f(x)}{\Delta x}$$

That is, for Δx sufficiently close to 0, we can make $\dfrac{\Delta y}{\Delta x}$ as close as we please to $f'(x)$. This can be expressed as

$$\frac{\Delta y}{\Delta x} \approx f'(x) \qquad \text{when} \qquad \Delta x \approx 0 \quad \Delta x \neq 0$$

Then since $\Delta x \neq 0$,

$$\boxed{\Delta y \approx f'(x)\Delta x \qquad \text{when} \qquad \Delta x \approx 0 \quad \Delta x \neq 0} \tag{1}$$

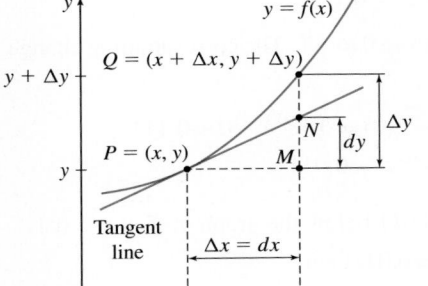

Figure 8

IN WORDS

- Δy equals the change in the height of the graph of f from x to $x + \Delta x$;
- dy equals the change in the height of the tangent line to the graph of f from x to $x + \Delta x$.

DEFINITION Differential

Let $y = f(x)$ be a differentiable function and let Δx denote a change in x.

The **differential of x**, denoted dx, is defined as $dx = \Delta x \neq 0$.

The **differential of y**, denoted dy, is defined as

$$\boxed{dy = f'(x)\,dx} \tag{2}$$

From statement (1), if $\Delta x \approx 0$, then $\Delta y \approx dy = f'(x)dx$. That is, when the change Δx in x is close to 0, then the differential dy is an approximation to the change Δy in y.

To see the geometric relationship between Δy and dy, we use Figure 8. There, $P = (x, y)$ is a point on the graph of $y = f(x)$ and $Q = (x + \Delta x, y + \Delta y)$ is a nearby point also on the graph of f. The change Δy in y

$$\Delta y = f(x + \Delta x) - f(x)$$

is the distance from M to Q. The slope of the tangent line to the graph of f at P is $f'(x) = \dfrac{dy}{dx}$. So numerically, the differential dy measures the distance from M to N. For $\Delta x = dx$ close to 0, $dy = f'(x)dx$ will be close to Δy.

EXAMPLE 1   Finding and Interpreting Differentials Geometrically

For the function $f(x) = xe^x$:

(a) Find an equation of the tangent line to the graph of f at $(0, 0)$.

(b) Find the differential dy.

(c) Compare dy to Δy when $x = 0$ and $\Delta x = 0.5$.

(d) Compare dy to Δy when $x = 0$ and $\Delta x = 0.1$.

(e) Compare dy to Δy when $x = 0$ and $\Delta x = 0.01$.

(f) Interpret the results.

Solution

(a) The slope of the tangent line is

$$f'(x) = xe^x + e^x \qquad \text{Use the Product Rule.}$$

At $(0, 0)$, $f'(0) = 1$, so an equation of the tangent line to the graph of f at $(0, 0)$ is $y = x$.

(b) $dy = f'(x)dx = (xe^x + e^x)dx = (x+1)e^x dx$.

(c) See Figure 9(a). When $x = 0$ and $\Delta x = dx = 0.5$, then

$$dy = (x+1)e^x dx = (0+1)e^0(0.5) = 0.5$$

The tangent line $y = x$ rises by 0.5 as x changes from 0 to 0.5. The corresponding change in the height of the graph f is

$$\Delta y = f(x + \Delta x) - f(x) = f(0.5) - f(0) = 0.5e^{0.5} - 0 \approx 0.824$$
$$|\Delta y - dy| \approx |0.824 - 0.5| = 0.324$$

The graph of the tangent line is approximately 0.324 below the graph of f at $x = 0.5$.

(d) See Figure 9(b). When $x = 0$ and $\Delta x = dx = 0.1$, then

$$dy = (0+1)e^0(0.1) = 0.1$$

The tangent line $y = x$ rises by 0.1 as x changes from 0 to 0.1. The corresponding change in the height of the graph f is

$$\Delta y = f(x + \Delta x) - f(x) = f(0.1) - f(0) = 0.1e^{0.1} - 0 \approx 0.111$$
$$|\Delta y - dy| \approx |0.111 - 0.1| = 0.011$$

The graph of the tangent line is approximately 0.011 below the graph of f at $x = 0.1$.

(e) See Figure 9(c). When $x = 0$ and $\Delta x = dx = 0.01$, then

$$dy = (0+1)e^0(0.01) = 0.01$$

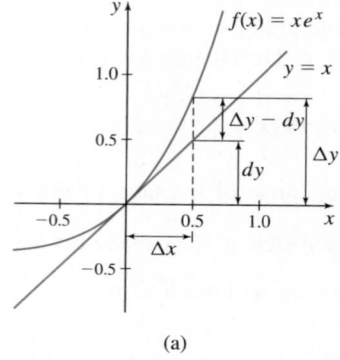

(a)

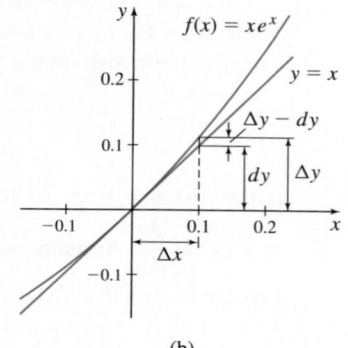

(b)

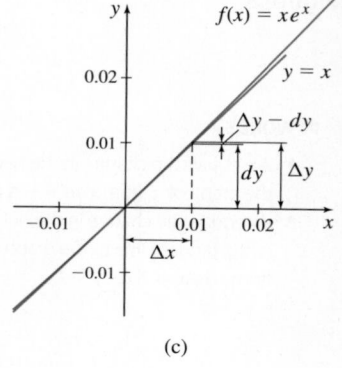

(c)

Figure 9

The tangent line $y = x$ rises by 0.01 as x changes from 0 to 0.01. The corresponding change in the height of the graph of f is

$$\Delta y = f(x + \Delta x) - f(x) = f(0.01) - f(0) = 0.01e^{0.01} - 0 \approx 0.0101$$
$$|\Delta y - dy| \approx |0.0101 - 0.01| = 0.0001$$

The graph of the tangent line is approximately 0.0001 below the graph of f at $x = 0.01$.

(f) The closer Δx is to 0, the closer dy is to Δy. So, we conclude that the closer Δx is to 0, the less the tangent line departs from the graph of the function. That is, we can use the tangent line to f at a point P as a *linear approximation* to f near P. ∎

NOW WORK   Problems 9 and 19.

2 Find the Linear Approximation to a Function

Local linearity is a term used to describe the fact that the tangent line to the graph of a function can be used to approximate the function near the point of tangency.

Suppose $y = f(x)$ is a differentiable function and suppose (x_0, y_0) is a point on the graph of f. Then

$$\Delta y = f(x) - f(x_0) \qquad \text{and} \qquad dy = f'(x_0)dx = f'(x_0)\Delta x = f'(x_0)(x - x_0)$$
$$\underset{dx = \Delta x}{\uparrow}$$

If $dx = \Delta x$ is close to 0, then

$$\Delta y \approx dy$$
$$f(x) - f(x_0) \approx f'(x_0)(x - x_0)$$
$$f(x) \approx f(x_0) + f'(x_0)(x - x_0)$$

See Figure 10. Each figure shows the graph of $y = f(x)$ and the graph of the tangent line at (x_0, y_0). Figures 10(a) and 10(b) show the difference between Δy and dy. Figure 10(c) illustrates that when Δx is close to 0, $\Delta y \approx dy$.

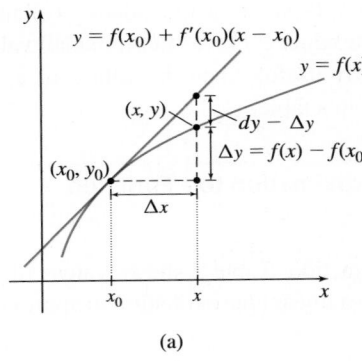

(a)

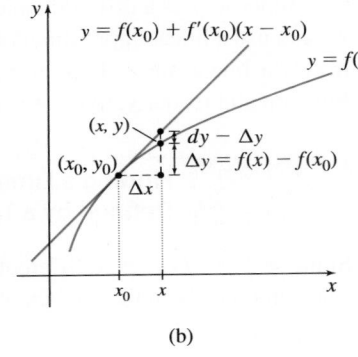

(b)

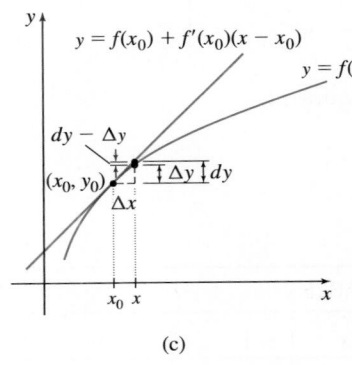

(c)

Figure 10

THEOREM Linear Approximation

The **linear approximation $L(x)$** to a differentiable function f at $x = x_0$ is given by

$$\boxed{L(x) = f(x_0) + f'(x_0)(x - x_0)} \tag{3}$$

The graph of the linear approximation $y = L(x)$ is the tangent line to the graph of f at x_0. The closer x is to x_0, the better the approximation.

- When the tangent line lies above the graph of the function, the linear approximation overestimates the value of the function.
- When the tangent line lies below the graph of the function, the linear approximation underestimates the value of the function.

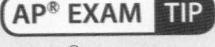

AP® EXAM TIP

The AP® Exam often asks whether a linear approximation underestimates or overestimates the approximation.

EXAMPLE 2 Finding the Linear Approximation to a Function

We call the line L a **local linear approximation of f near x_0**. Although L provides a good approximation of f in an interval centered at x_0, the next example shows that as the interval widens, the accuracy of the approximation of L may decrease.

NOW WORK AP® Practice Problem 7.

EXAMPLE 2 Finding the Linear Approximation to a Function

(a) Find the linear approximation $L(x)$ to $f(x) = \sin x$ near $x = 0$.

(b) Use $L(x)$ to approximate $\sin(-0.3)$, $\sin 0.1$, $\sin 0.4$, $\sin 0.5$, and $\sin \dfrac{\pi}{4}$.

(c) Graph f and L.

Solution

(a) Note that $f(0) = \sin 0 = 0$. Since $f'(x) = \cos x$, then $f'(0) = \cos 0 = 1$. Then, the linear approximation $L(x)$ to f at 0 is

$$L(x) = f(0) + f'(0)(x - 0) = x \qquad \text{Use Equation (3)}.$$

So, for x close to 0, the function $f(x) = \sin x$ can be approximated by the line $L(x) = x$.

(b) Table 1 shows approximate values of $\sin x$ using $L(x) = x$, the true values of $\sin x$, and the absolute error in using the approximation. From Table 1, we see that the farther x is from 0, the worse the line $L(x) = x$ approximates $f(x) = \sin x$.

(c) See Figure 11 for the graphs of $f(x) = \sin x$ and $L(x) = x$. ∎

Observe from Figure 11 that the tangent line $L(x) = x$ lies above the graph of $y = \sin x$ for $x > 0$ and below the graph of $y = \sin x$ for $x < 0$. We conclude that $L(x) = x$ overestimates $f(x) = \sin x$ for $x > 0$ and underestimates f for $x < 0$. Notice in Table 1 that the estimates of 0.1, 0.4, and 0.5 overestimate the true values and the estimate of -0.3 underestimates the true value of the sine function.

NOW WORK Problem 27 and AP® Practice Problems 2–6.

Suppose f is a differentiable function. If we know the values $f(c)$ and $f'(c)$, then we can use a linear approximation to approximate $f(c + \Delta x)$ for small values Δx. This makes a linear approximation particularly useful when the values of a differentiable function and its derivative are expressed in a table.

TABLE 1

$L(x) = x$	$f(x) = \sin x$	Error: $\lvert x - \sin x\rvert$
0.1	0.0998	0.0002
-0.3	-0.2955	0.0045
0.4	0.3894	0.0106
0.5	0.4794	0.0206
$\dfrac{\pi}{4} \approx 0.7854$	0.7071	0.0783

Figure 11

EXAMPLE 3 Finding a Linear Approximation to a Function Defined by a Table

Suppose $y = f(x)$ is a differentiable function. Table 2 shows values of f and f' for select numbers x in the domain of f. Use a linear approximation to approximate $f(2.9)$.

Solution

The linear approximation $L(x)$ to the function f at $x = c$ is

$$L(x) = f(c) + f'(c)(x - c)$$

Since 2.9 is near 3 and 3 is in Table 2, we use $x = 2.9$ and $c = 3$ to find $L(2.9)$.

$$L(2.9) = f(3) + f'(3)(2.9 - 3)$$
$$= -8 - 2(-0.1) = -7.8$$

Using a linear approximation, $f(2.9) \approx -7.8$. ∎

TABLE 2

x	-1	0	1	2	3
$f(x)$	-0	4	0	-6	-8
$f'(x)$	10	-1	-6	-5	-2

NOW WORK Problem 35 and AP® Practice Problems 1 and 8.

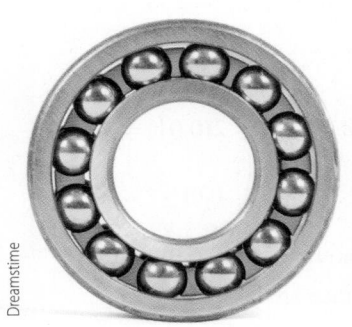

Dreamstime

③ Use Differentials in Applications

The next two examples show applications of differentials. In these examples, we use the differential to approximate the change.

EXAMPLE 4 **Measuring Metal Loss in a Mechanical Process**

A spherical bearing has a radius of 3 cm when it is new. Use differentials to approximate the volume of the metal lost after the bearing wears down to a radius of 2.971 cm. Find the actual volume lost and compare the approximation to the actual volume lost.

Solution

The volume V of a sphere of radius R is $V = \frac{4}{3}\pi R^3$. As a machine operates, friction wears away part of the spherical bearing. The exact volume of metal lost equals the change ΔV in the volume V of the sphere, when the change in the radius of the bearing is $\Delta R = 2.971 - 3 = -0.029$ cm. Since the change ΔR is small, we use the differential dV to approximate the change ΔV. Then

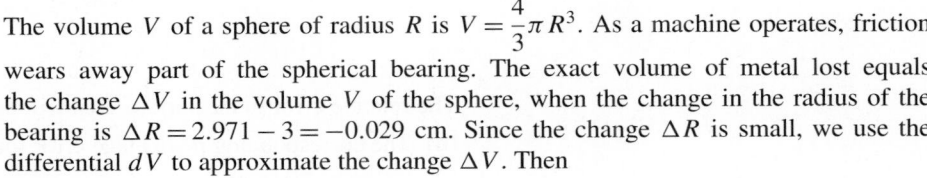

$$\Delta V \approx dV = 4\pi R^2 dR = (4\pi)(3^2)(-0.029) \approx -3.280$$
$$\uparrow \qquad\qquad\qquad \uparrow$$
$$dV = V'(R)\,dR \quad dR = \Delta R = -0.029$$

The approximate loss in volume of the bearing is 3.28 cm³.

The actual loss in volume ΔV is

$$\Delta V = V(R + \Delta R) - V(R) = \frac{4}{3}\pi \cdot 2.971^3 - \frac{4}{3}\pi \cdot 3^3 \approx \frac{4}{3}\pi(-0.7755) \approx -3.248 \text{ cm}^3$$

The approximate change in volume overstates the actual change by 0.032 cm³. ∎

NOW WORK Problem 53.

The use of dy to approximate Δy when $\Delta x = dx$ is small is also helpful in approximating *error*. If Q is the quantity to be measured and if ΔQ is the change in Q, then the

$$\boxed{\begin{array}{l} \textbf{Relative error at } x_0 \textbf{ in } Q = \dfrac{|\Delta Q|}{Q(x_0)} \\[2mm] \textbf{Percentage error at } x_0 \textbf{ in } Q = \dfrac{|\Delta Q|}{Q(x_0)} \cdot 100\% \end{array}}$$

IN WORDS Relative error is the ratio of the change in a quantity to the original quantity.

For example, if $Q = 50$ units and the change ΔQ in Q is measured to be 5 units, then

$$\text{Relative error at 50 in } Q = \frac{5}{50} = 0.10 \qquad \text{Percentage error at 50 in } Q = 10\%$$

When Δx is small, $dQ \approx \Delta Q$. The relative error and percentage error at x_0 in Q can be approximated by $\dfrac{|dQ|}{Q(x_0)}$ and $\dfrac{|dQ|}{Q(x_0)} \cdot 100\%$, respectively.

EXAMPLE 5 **Measuring Error in a Manufacturing Process**

A company manufactures spherical ball bearings of radius 3 cm. The customer will accept a deviation in the radius of at most 1%.

(a) Use differentials to approximate the relative error for the surface area.

(b) Find the corresponding percentage error for the surface area.

(c) Interpret the results in the context of the problem.

Solution

(a) A deviation of at most 1% in the radius R means that the relative error in the radius R must be less than or equal to 0.01. That is, $\dfrac{|\Delta R|}{R} \le 0.01$. The surface area S of a sphere of radius R is given by the formula $S = 4\pi R^2$. We seek the relative error in S, $\dfrac{|\Delta S|}{S}$, which can be approximated by $\dfrac{|dS|}{S}$.

$$\frac{|\Delta S|}{S} \approx \underset{\underset{dS = S'(R)\,dR}{\uparrow}}{\frac{|dS|}{S}} = \frac{8\pi R \cdot |dR|}{4\pi R^2} = \frac{2|dR|}{R} = 2 \cdot \underset{\underset{\frac{|\Delta R|}{R} \le 0.01}{\uparrow}}{\frac{|\Delta R|}{R}} \le 2(0.01) = 0.02$$

The relative error in the surface area will be less than or equal to 0.02.

(b) The corresponding percentage error for the surface area is

$$\frac{|\Delta S|}{S} \cdot 100\% \approx 0.02 \cdot 100\% = 2\%$$

(c) A deviation of at most 1% in the radius results in a possible percentage error of 2% in the surface area. The deviation of 1% in the radius of the ball bearing means the radius of the sphere must be somewhere between $3 - 0.01(3) = 2.97$ cm and $3 + 0.01(3) = 3.03$ cm. The corresponding 2% error in the surface area means the surface area lies within ± 0.02 of $S = 4\pi R^2 = 36\pi$ cm^2. That is, the surface area is between $35.28\pi \approx 110.84$ cm^2 and $36.72\pi \approx 115.36$ cm^2. A rather small error in the radius results in a more significant variation in the surface area! ∎

NOW WORK Problem 65.

4 Use Newton's Method to Approximate a Real Zero of a Function

Newton's Method is used to approximate a real zero of a function. Suppose a function $y = f(x)$ is defined on a closed interval $[a, b]$, and its derivative f' is continuous on the interval (a, b). Also suppose that, from the Intermediate Value Theorem or from a graph or by trial calculations, we know that the function f has a real zero in some open subinterval of (a, b).

Suppose c_1 is some number in the subinterval of (a, b). Draw the tangent line to the graph of f at the point $P_1 = (c_1, f(c_1))$ and label as c_2 the x-intercept of the tangent line. See Figure 12. If c_1 is a *first approximation* to the required zero of f, then c_2 is a *second approximation* to the zero. Repeated use of this procedure often generates increasingly accurate approximations to the actual zero.

Suppose c_1 has been chosen. We seek a formula for finding the approximation c_2. The coordinates of P_1 are $(c_1, f(c_1))$, and the slope of the tangent line to the graph of f at point P_1 is $f'(c_1)$. So, the equation of the tangent line to graph of f at P_1 is

$$y - f(c_1) = f'(c_1)(x - c_1)$$

To find the x-intercept c_2, let $y = 0$. Then c_2 satisfies the equation

$$-f(c_1) = f'(c_1)(c_2 - c_1)$$

$$c_2 = c_1 - \frac{f(c_1)}{f'(c_1)} \qquad \text{if } f'(c_1) \ne 0 \qquad \text{Solve for } c_2.$$

Notice that we are using the linear approximation to f at the point $(c_1, f(c_1))$ to approximate the zero.

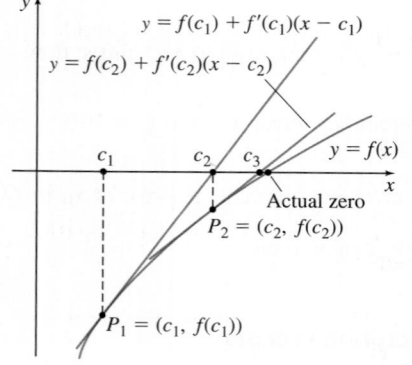

Figure 12

Now repeat the process by drawing the tangent line to the graph of f at the point $P_2 = (c_2, f(c_2))$ and label as c_3 the x-intercept of the tangent line. By continuing this process, we have Newton's Method.

Newton's Method

Suppose a function f is defined on a closed interval $[a, b]$ and its first derivative f' is continuous and not equal to 0 on the open interval (a, b). If $x = c_1$ is a sufficiently close first approximation to a real zero of the function f in (a, b), then the formula

$$c_2 = c_1 - \frac{f(c_1)}{f'(c_1)}$$

gives a second approximation to the zero. The nth approximation to the zero is given by

$$c_n = c_{n-1} - \frac{f(c_{n-1})}{f'(c_{n-1})}$$

ORIGINS When Isaac Newton first introduced his method for finding real zeros in 1669, it was very different from what we did here. Newton's original method (also known as the **Newton–Raphson Method**) applied only to polynomials, and it did not use calculus! In 1690, Joseph Raphson simplified the method and made it recursive, again without calculus. It was not until 1740 that Thomas Simpson used calculus, giving the method the form we use here.

The formula in Newton's Method is a **recursive formula**, because each new approximation is written in terms of the previous one. Since recursive processes are particularly useful in computer programs, variations of Newton's Method are used by many computer algebra systems and graphing utilities to find the zeros of a function.

EXAMPLE 6 **Using Newton's Method to Approximate a Real Zero of a Function**

Use Newton's Method to find a fourth approximation to the positive real zero of the function $f(x) = x^3 + x^2 - x - 2$.

Solution

Since f is continuous for all real numbers and $f(1) = -1$ and $f(2) = 8$, we know from the Intermediate Value Theorem that f has a zero in the interval $(1, 2)$. Also, since f is a polynomial function, both f and f' are differentiable functions. Now

$$f(x) = x^3 + x^2 - x - 2 \qquad \text{and} \qquad f'(x) = 3x^2 + 2x - 1$$

We choose $c_1 = 1.5$ as the first approximation and use Newton's Method. The second approximation to the zero is

$$c_2 = c_1 - \frac{f(c_1)}{f'(c_1)} = 1.5 - \frac{f(1.5)}{f'(1.5)} = 1.5 - \frac{2.125}{8.75} \approx 1.2571429$$

Use Newton's Method again, with $c_2 = 1.2571429$. Then $f(1.2571429) \approx 0.3100644$ and $f'(1.2571429) \approx 6.2555106$. The third approximation to the zero is

$$c_3 = c_2 - \frac{f(c_2)}{f'(c_2)} = 1.2571429 - \frac{0.3100641}{6.2555102} \approx 1.2075763$$

The fourth approximation to the zero is

$$c_4 = c_3 - \frac{f(c_3)}{f'(c_3)} = 1.2075763 - \frac{0.0116012}{5.7898745} \approx 1.2055726$$

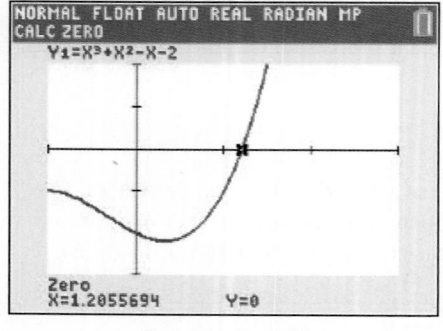

$[-1, 3] \times [-3, 2]$

Figure 13 $f(x) = x^3 + x^2 - x - 2$

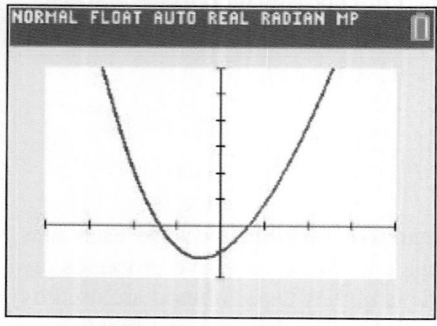

$[-4, 4] \times [-2, 6]$

Figure 14 $f(x) = \sin x + x^2 - 1$

Using graphing technology, the zero of $f(x) = x^3 + x^2 - x - 2$ is given as $x = 1.2055694$. See Figure 13.

NOW WORK Problem 39.

The table feature of a graphing calculator can be used to take advantage of the recursive nature of Newton's Method and speed up the computation.

EXAMPLE 7 **Using Technology with Newton's Method**

The graph of the function $f(x) = \sin x + x^2 - 1$ is shown in Figure 14.

(a) Use the Intermediate Value Theorem to confirm that f has a zero in the interval $(0, 1)$.

(b) Use a graphing calculator with Newton's Method and a first approximation of $c_1 = 0.5$ to find a fourth approximation to the zero.

Solution

(a) The function $f(x) = \sin x + x^2 - 1$ is continuous at all real numbers, so it is continuous on the closed interval $[0, 1]$. Since $f(0) = -1$ and $f(1) = \sin 1 \approx 0.841$ have opposite signs, the Intermediate Value Theorem guarantees that f has a zero in the interval $(0, 1)$.

(b) We begin by finding $f'(x) = \cos x + 2x$. To use Newton's Method with a graphing calculator, enter

$$ x - \frac{\sin x + x^2 - 1}{\cos x + 2x} \qquad x - \frac{f(x)}{f'(x)} $$

into the $Y =$ editor, as Y_1. See Figure 15.

Then create a table (Figure 16) by entering the initial value 0.5 in the X column. The graphing calculator computes $Y_1(0.5)$ and displays 0.64411 in column Y_1 to the right of 0.5. The value Y_1 is the second approximation c_2 that we use in the next iteration. Now enter $Y_1(0.5) = 0.64410789$ in the X column of the next row, and the new entry in column Y_1 is the third approximation c_3. Repeat the process until the desired approximation is obtained. The fourth approximation to the zero of f is 0.63673. ∎

NOW WORK Problem 47.

When Newton's Method Fails

You may wonder, does Newton's Method always work? The answer is no. The list below, while not exhaustive, gives some conditions under which Newton's Method fails.

• Newton's Method fails if the conditions of the theorem are not met

  (a) $f'(c_n) = 0$: Algebraically, division by 0 is not defined. Geometrically, the tangent line is parallel to the x-axis and so has no x-intercept.

  (b) $f'(c)$ is undefined: The process cannot be used.

• Newton's Method fails if the initial estimate c_1 is not "good enough":

  (a) Choosing an initial estimate too far from the required zero could result in approximating a different zero of the function.

  (b) The convergence could approach the zero so slowly that hundreds of iterations are necessary.

• Newton's Method fails if the terms oscillate between two values and so never get closer to the zero.

Problems 80–83 illustrate some of these possibilities.

Figure 15

Figure 16

4.3 Assess Your Understanding

Concepts and Vocabulary

1. **Multiple Choice** If $y = f(x)$ is a differentiable function, the differential $dy =$
   (a) Δy (b) Δx (c) $f(x)dx$ (d) $f'(x)dx$

2. A linear approximation to a differentiable function f near x_0 is given by the function $L(x) =$ _____.

3. **True or False** The difference $|\Delta y - dy|$ measures the departure of the graph of $y = f(x)$ from the graph of the tangent line to f.

4. If Q is a quantity to be measured and ΔQ is the error made in measuring Q, then the relative error in the measurement at x_0 is given by the ratio _____.

5. **True or False** Newton's Method uses tangent lines to the graph of f to approximate the zeros of f.

6. **True or False** Before using Newton's Method, we need a first approximation for the zero.

Skill Building

In Problems 7–18, find the differential dy of each function.

7. $y = x^3 - 2x + 1$

8. $y = e^x + 2x - 1$

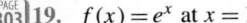 9. $y = 4(x^2 + 1)^{3/2}$

10. $y = \sqrt{x^2 - 1}$

11. $y = 3\sin(2x) + x$

12. $y = \cos^2(3x) - x$

13. $y = e^{-x}$ 14. $y = e^{\sin x}$

15. $y = xe^x$

16. $y = \dfrac{e^{-x}}{x}$ 17. $y = \sin^{-1}(2x)$ 18. $y = \tan^{-1} x^2$

In Problems 19–24:
(a) *Find the differential dy for each function f.*
(b) *Evaluate dy and Δy at the given value of x when*
   (i) $\Delta x = 0.5$, (ii) $\Delta x = 0.1$, and (iii) $\Delta x = 0.01$.
(c) *Find the error $|\Delta y - dy|$ for each choice of $dx = \Delta x$.*

19. $f(x) = e^x$ at $x = 1$

20. $f(x) = e^{-x}$ at $x = 1$

21. $f(x) = x^{2/3}$ at $x = 2$

22. $f(x) = x^{-1/2}$ at $x = 1$

23. $f(x) = \cos x$ at $x = \pi$

24. $f(x) = \tan x$ at $x = 0$

In Problems 25–32:
(a) *Find the linear approximation $L(x)$ to f at x_0.*
(b) *Graph f and L on the same set of axes.*
(c) *Does L overestimate or underestimate the value of $f(x_0)$? Explain.*

25. $f(x) = (x + 1)^5$, $x_0 = 2$

26. $f(x) = x^3 - 1$, $x_0 = 0$

27. $f(x) = \sqrt{x}$, $x_0 = 4$

28. $f(x) = x^{2/3}$, $x_0 = 1$

29. $f(x) = \ln x$, $x_0 = 1$

30. $f(x) = e^x$, $x_0 = 1$

31. $f(x) = \cos x$, $x_0 = \dfrac{\pi}{3}$

32. $f(x) = \sin x$, $x_0 = \dfrac{\pi}{6}$

33. Use differentials to approximate the change in:
   (a) $y = f(x) = x^2$ as x changes from 3 to 3.001.
   (b) $y = f(x) = \dfrac{1}{x + 2}$ as x changes from 2 to 1.98.

34. Use differentials to approximate the change in:
   (a) $y = x^3$ as x changes from 3 to 3.01.
   (b) $y = \dfrac{1}{x - 1}$ as x changes from 2 to 1.98.

In Problems 35–38, use the given information and a linear approximation to approximate the value of the function at c.

35. $f(2) = 8$; $f'(2) = -3$; $c = 2.06$

36. $f(-4) = 3$; $f'(-4) = 2$; $c = -3.6$

37. $f(-1) = 0$; $f'(-1) = \dfrac{3}{2}$; $c = -1.1$

38. $f(5) = \dfrac{1}{2}$; $f'(5) = -3$; $c = 5.2$

In Problems 39–46, for each function:
(a) *Use the Intermediate Value Theorem to confirm that a zero exists in the given interval.*
(b) *Use Newton's Method with the first approximation c_1 to find c_3, the third approximation to the real zero.*

39. $f(x) = x^3 + 3x - 5$, interval: $(1, 2)$. Let $c_1 = 1.5$.

40. $f(x) = x^3 - 4x + 2$, interval: $(1, 2)$. Let $c_1 = 1.5$.

41. $f(x) = 2x^3 + 3x^2 + 4x - 1$, interval: $(0, 1)$. Let $c_1 = 0.5$.

42. $f(x) = x^3 - x^2 - 2x + 1$, interval: $(0, 1)$. Let $c_1 = 0.5$.

43. $f(x) = x^3 - 6x - 12$, interval: $(3, 4)$. Let $c_1 = 3.5$.

44. $f(x) = 3x^3 + 5x - 40$, interval: $(2, 3)$. Let $c_1 = 2.5$.

45. $f(x) = x^4 - 2x^3 + 21x - 23$, interval: $(1, 2)$. Use a first approximation c_1 of your choice.

46. $f(x) = x^4 - x^3 + x - 2$, interval: $(1, 2)$. Use a first approximation c_1 of your choice.

In Problems 47–52, for each function:
(a) *Use the Intermediate Value Theorem to confirm that a zero exists in the given interval.*
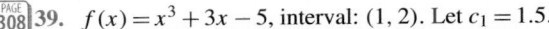 (b) *Use technology with Newton's Method to find c_5, the fifth approximation to the real zero. Use the midpoint of the interval for the first approximation c_1.*

47. $f(x) = x + e^x$, interval: $(-1, 0)$

48. $f(x) = x - e^{-x}$, interval: $(0, 1)$

49. $f(x) = x^3 + \cos^2 x$, interval: $(-1, 0)$

50. $f(x) = x^2 + 2\sin x - 0.5$, interval: $(0, 1)$

51. $f(x) = 5 - \sqrt{x^2 + 2}$, interval: $(4, 5)$

52. $f(x) = 2x^2 + x^{2/3} - 4$, interval: $(1, 2)$

Applications and Extensions

53. **Area of a Disk** A circular plate is heated and expands. If the radius of the plate increases from $R = 10$ cm to $R = 10.1$ cm, use differentials to approximate the increase in the area of the top surface.

54. **Volume of a Cylinder** In a wooden block 3 cm thick, an existing circular hole with a radius of 2 cm is enlarged to a hole with a radius of 2.2 cm. Use differentials to approximate the volume of wood that is removed.

55. Volume of a Balloon   Use differentials to approximate the change in volume of a spherical balloon of radius 3 m as the balloon swells to a radius of 3.1 m.

56. Volume of a Paper Cup   A manufacturer produces paper cups in the shape of a right circular cone with a radius equal to one-fourth the height. Specifications call for the cups to have a top diameter of 4 cm. After production, it is discovered that the diameter measures only 3.8 cm. Use differentials to approximate the loss in capacity of the cup.

57. Volume of a Sphere

(a) Use differentials to approximate the volume of material needed to manufacture a hollow sphere if its inner radius is 2 m and its outer radius is 2.1 m.

(b) Is the approximation overestimating or underestimating the volume of material needed?

(c) Discuss the importance of knowing the answer to (b) if the manufacturer receives an order for 10,000 spheres.

58. Distance Traveled   A bee flies around a circle traced on an equator of a ball with a radius of 7 cm at a constant distance of 2 cm from the ball. An ant travels along the same circle but on the ball.

(a) Use differentials to approximate how many more centimeters the bee travels than the ant in one round-trip journey.

(b) Does the linear approximation overestimate or underestimate the difference in the distances the bugs travel? Explain.

59. Estimating Height   To find the height of a building, the length of the shadow of a 3-m pole placed 9 m from the building is measured. See the figure. This measurement is found to be 1 m, with a percentage error of 1%. The height of the building is estimated to be 30 m. Use differentials to find the percentage error in the estimate.

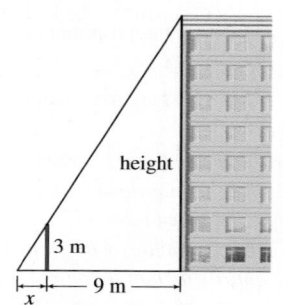

60. Pendulum Length   The period of the pendulum of a grandfather clock is $T = 2\pi\sqrt{\dfrac{l}{g}}$, where l is the length (in meters) of the pendulum, T is the period (in seconds), and g is the acceleration due to gravity (9.8 m/s^2). Suppose an increase in temperature increases the length l of the pendulum, a thin wire, by 1%. What is the corresponding percentage error in the period? How much time will the clock lose (or gain) each day?

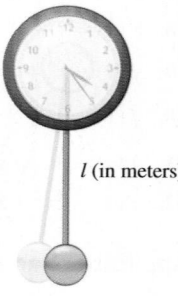

l (in meters)

61. Pendulum Length   Refer to Problem 60. If the pendulum of a grandfather clock is normally 1 m long and the length is increased by 10 cm, use differentials to approximate the number of minutes the clock will lose (or gain) each day.

62. Luminosity of the Sun   The luminosity L of a star is the rate at which it radiates energy. This rate depends on the temperature T (in kelvin where 0 K is absolute zero) and the surface area A of the star's photosphere (the gaseous surface that emits the light). Luminosity at time t is given by the formula $L(t) = \sigma A T^4$, where σ is a constant, known as the **Stefan–Boltzmann constant**.

As with most stars, the Sun's temperature has gradually increased over the 5 billion years of its existence, causing its luminosity to slowly increase. For this problem, we assume that increased luminosity L is due only to an increase in temperature T. That is, we treat A as a constant.

(a) Find the rate of change of the temperature T of the Sun with respect to time t. Write the answer in terms of the rate of change of the Sun's luminosity L with respect to time t.

(b) 4.5 billion years ago, the Sun's luminosity was only 70% of what it is now. If the rate of change of luminosity L with respect to time t is constant, then $\dfrac{\Delta L}{\Delta t} = \dfrac{0.3L_c}{\Delta t} = \dfrac{0.3L_c}{4.5}$, where L_c is the current luminosity. Use differentials to approximate the current rate of change of the temperature T of the Sun in degrees per century.

63. Climbing a Mountain   Weight W is the force on an object due to the pull of gravity. On Earth, this force is given by Newton's Law of Universal Gravitation: $W = \dfrac{GmM}{r^2}$, where m is the mass of the object, $M = 5.974 \times 10^{24}$ kg is the mass of Earth, r is the distance of the object from the center of the Earth, and $G = 6.67 \times 10^{-11}$ m^3/(kg · s^2) is the universal gravitational constant. Suppose a person weighs 70 kg at sea level, that is, when $r = 6370$ km (the radius of Earth). Use differentials to approximate the person's change in weight at the top of Mount Everest, which is 8.8 km above sea level. (In this problem, we treat kilograms as weight.)

64. Body Mass Index   The **body mass index** (**BMI**) is given by the formula

$$\text{BMI} = 703\,\frac{m}{h^2}$$

where m is the person's weight in pounds and h is the person's height in inches. A BMI between 18.5 and 25 indicates that weight is typically healthy for an adult, whereas a BMI greater than 25 indicates that an adult is overweight.

(a) Suppose a man who is 5 ft 6 in. weighs 142 lb in the morning when he first wakes up, but he weighs 148 lb in the afternoon. Calculate his BMI in the morning and use differentials to approximate the change in his BMI in the afternoon. Round both answers to three decimal places.

(b) Did this linear approximation overestimate or underestimate the man's afternoon weight? Explain.

(c) A woman who weighs 165 lb estimates her height at 68 in. with a possible error of ± 1.5 in. Calculate her BMI, assuming a height of 68 in. Then use differentials to approximate the possible error in her calculation of BMI. Round the answers to three decimal places.

(d) In the situations described in (a) and (c), how do you explain the classification of each adult as typically healthy or overweight?

65. Error Estimation The radius of a spherical ball is found by measuring the volume of the sphere (by finding how much water it displaces). It is determined that the volume is 40 cubic centimeters (cm^3), with a tolerance of no more than 1%.

    (a) Use differentials to approximate the relative error in measuring the radius.

    (b) Find the corresponding percentage error in measuring the radius.

    (c) Interpret the results in the context of the problem.

66. Error Estimation The oil pan of a car has the shape of a hemisphere with a radius of 8 cm. The depth h of the oil is measured at 3 cm, with a percentage error of no more than 10%. [*Hint:* The volume V for a spherical segment is

$$V = \frac{1}{3}\pi h^2(3R - h),$$ where R is the radius of the sphere.]

    (a) Use differentials to approximate the relative error in measuring the volume.

    (b) Find the corresponding percentage error in measuring volume.

    (c) Interpret the results in the context of the problem.

67. Error Estimation A closed container is filled with liquid. As the temperature T, in degrees Celsius, increases, the pressure P, in atmospheres, exerted on the container increases according to the ideal gas law, $P = 50T$.

    (a) If the relative error in measuring the temperature is no more than 0.01, what is the relative error in measuring the pressure?

    (b) Find the corresponding percentage error in measuring the pressure.

    (c) Interpret the results in the context of the problem.

68. Error Estimation A container filled with liquid is held at a constant pressure. As the temperature T, in degrees Celsius, decreases, the volume V, in cm^3, of the container decreases according to the ideal gas law, $V = 15T$.

    (a) If the relative error in measuring the temperature is no more than 0.02, what is the relative error in measuring the volume?

    (b) Find the corresponding percentage error in measuring the volume.

    (c) Interpret the results in the context of the problem.

69. Percentage Error If the percentage error in measuring the edge of a cube is 2%, what is the percentage error in computing its volume?

70. Focal Length To photograph an object, a camera's lens forms an image of the object on the camera's photo sensors. A camera lens can be approximated by a thin lens, which obeys the thin-lens equation $\dfrac{1}{f} = \dfrac{1}{p} + \dfrac{1}{q}$, where p is the distance from the lens to the object being photographed, q is the distance from the lens to the image of the object, and f is the focal length of the lens. A camera whose lens has a focal length of 50 mm is being used to photograph a dog. The dog is originally 15 m from the lens but moves 0.33 m (about a foot) closer to the lens. Use differentials to approximate the distance that the image of the dog has moved on the photo sensors.

Using Newton's Method to Solve Equations *In Problems 71–74, use Newton's Method to solve each equation.*

71. $e^{-x} = \ln x$

72. $e^{-x} = x - 4$

73. $e^x = x^2$

74. $e^x = 2\cos x, x > 0$

75. Approximating e Use Newton's Method to approximate the value of e by finding the zero of the equation $\ln x - 1 = 0$. Use $c_1 = 3$ as the first approximation and find the fourth approximation to the zero. Compare the results from this approximation to the value of e obtained with a calculator.

76. Show that the linear approximation of a function $f(x) = (1 + x)^k$, where x is near 0 and k is any number, is given by $y = 1 + kx$.

77. Does it seem reasonable that if a first degree polynomial approximates a differentiable function in an interval near x_0, a higher-degree polynomial should approximate the function over a wider interval? Explain your reasoning.

78. Why does a function need to be differentiable at x_0 for a linear approximation to be used?

79. Newton's Method Suppose you use Newton's Method to solve $f(x) = 0$ for a differentiable function f, and you obtain $x_{n+1} = x_n$. What can you conclude?

80. When Newton's Method Fails Verify that the function $f(x) = -x^3 + 6x^2 - 9x + 6$ has a zero in the interval $(2, 5)$. Show that Newton's Method fails if an initial estimate of $c_1 = 2.9$ is chosen. Repeat Newton's Method with an initial estimate of $c_1 = 3.0$. Explain what occurs for each of these two choices. (The zero is near $x = 4.2$.)

81. When Newton's Method Fails Show that Newton's Method fails for $f(x) = x^3 - 2x + 2$ with an initial estimate of $c_1 = 0$.

82. When Newton's Method Fails Show that Newton's Method fails for $f(x) = x^8 - 1$ if an initial estimate of $c_1 = 0.1$ is chosen. Explain what occurs.

83. When Newton's Method Fails Show that Newton's Method fails for $f(x) = (x - 1)^{1/3}$ with an initial estimate of $c_1 = 2$.

84. Newton's Method

    (a) Use the Intermediate Value Theorem to show that $f(x) = x^4 + 2x^3 - 2x - 2$ has a zero in the interval $(-2, -1)$.

    (b) Use Newton's Method to find c_3, a third approximation to the zero from (a).

    (c) Explain why the initial approximation, $c_1 = -1$, cannot be used in (b).

85. Specific Gravity A solid wooden sphere of diameter d and specific gravity S sinks in water to a depth h, which is determined by the equation $2x^3 - 3x^2 - S = 0$, where $x = \dfrac{h}{d}$. Use Newton's Method to find a third approximation to h for a maple ball of diameter 6 in. for which $S = 0.786$. Use $x = 1.6$ as the first approximation.

Challenge Problem

86. Kepler's Equation The equation $x - p\sin x = M$, called **Kepler's equation**, is used in astronomy to describe the location of an object in elliptical orbit. Use Newton's Method to find a second approximation to x when $p = 0.2$ and $M = 0.85$. Use $c_1 = 1$ as your first approximation.

AP® Practice Problems

Multiple-Choice Questions

 1. Let f be a function for which $f(2) = 6$ and $f'(2) = -3$. If the line tangent to the graph of f at 2 is used to approximate a zero of f, then the approximation is

(A) 0   (B) 4   (C) 6   (D) 12

 2. For small, positive values of h, $\sqrt[3]{8+h}$ is best approximated by

(A) $4 - \dfrac{h}{12}$ (B) $4 + \dfrac{h}{12}$ (C) $2 - \dfrac{h}{12}$ (D) $2 + \dfrac{h}{12}$

 3. A linear approximation to $f(x) = x \sin\left(\dfrac{\pi x}{2}\right) + x^2$ at $x = 3$ is

(A) $y = 5x + 6$ (B) $y = 5x - 9$

(C) $y = 7x - 9$ (D) $y = 7x + 9$

 4. Using the line tangent to the graph of $f(x) = xe^x + 2$ at 0, the approximate value of $f(-0.3)$ is

(A) 2.3   (B) 1.3   (C) 1.7   (D) -2.3

5. If $f'(x) = 2xe^{x^2-1} - 3\pi \sin(\pi x)$ and $f(1) = 4$, approximate $f(1.03)$ using a linear approximation.

(A) 4.06   (B) 5.06   (C) 4   (D) 3.94

 6. The line tangent to the graph of $f(x) = x^3 + 1$ at $x = 1$ is used to approximate $f(x)$ near 1. Which number below is the greatest value of x that results in an error less than or equal to 0.5?

(A) 1.30   (B) 1.35   (C) 1.40   (D) 1.45

 7. A linear approximation L is used to approximate $f(x) = \sqrt{x}$, at $c, c > 0$. The approximation

(A) always underestimates the true value of f at c.

(B) always overestimates the true value of f at c.

(C) sometimes overestimates the true value of f at c.

(D) does not provide enough information to determine whether the true value of f at c is over- or underestimated.

 8. Suppose $y = f(x)$ is a differentiable function. The table below gives values of f and f' for select numbers x in the domain of f. Use a linear approximation to approximate $f(3.1)$.

x	-3	0	1	3	5
$f(x)$	4	4	-1	-2	3
$f'(x)$	1	-1	-2	3	4

(A) 4.1   (B) -2.3 (C) 0.1   (D) -1.7

Retain Your Knowledge

Multiple-Choice Questions

1. The function $f(x) = e^x + x$ has an inverse function g. Then, $g'(1) =$

(A) $\dfrac{1}{2}$ (B) 2   (C) $\dfrac{1}{e+1}$ (D) $\dfrac{1}{e}$

2. Find y' for $e^x \sin y - x^2 \cos y = x$.

(A) $y' = \dfrac{1 + 2x \cos y - e^x \sin y}{x^2 \sin y - e^x \cos y}$

(B) $y' = \dfrac{1 - 2 \cos y - e^x \sin y}{x^2 \sin y + e^x \cos y}$

(C) $y' = 1 + \dfrac{2x \cos y - e^x \sin y}{x^2 \sin y + e^x \cos y}$

(D) $y' = \dfrac{1 + 2x \cos y - e^x \sin y}{x^2 \sin y + e^x \cos y}$

3. If $f(x) = x^4 \ln(2x^2 + 3)$, then $f'(1)$ equals

(A) $\dfrac{4}{\ln 5} + 4 \ln 5$ (B) $4 + 4 \ln 5$

(C) $\dfrac{4}{5} + 4 \ln 5$ (D) $\dfrac{1}{5} + 4 \ln 5$

Free-Response Question

4. The resistance R, in ohms, of a wire of radius x cm is given by the formula $R(x) = \dfrac{0.0048}{x^2}$. The radius x is given by $x = 0.991 + 10^{-5}T$, where T is the temperature in Kelvin. Determine how R is changing with respect to T when $T = 320$ K.

4.4 Indeterminate Forms and L'Hôpital's Rule

OBJECTIVES *When you finish this section, you should be able to:*

1 **Identify indeterminate forms of the type $\dfrac{0}{0}$ and $\dfrac{\infty}{\infty}$ (p. 313)**

2 **Use L'Hôpital's Rule to find a limit (p. 314)**

3 **Find the limit of an indeterminate form of the type $0 \cdot \infty$, $\infty - \infty$, 0^0, 1^∞, or ∞^0 (p. 318)**

NEED TO REVIEW? The limit of a quotient is discussed in Section 1.2, pp. 101–103.

In this section, we reexamine the limit of a quotient and explore what we can do when previously learned strategies cannot be used.

1 Identify Indeterminate Forms of the Type $\dfrac{0}{0}$ and $\dfrac{\infty}{\infty}$

To find $\lim\limits_{x \to c} \dfrac{f(x)}{g(x)}$ we usually first try to use the Limit of a Quotient:

$$\lim_{x \to c} \frac{f(x)}{g(x)} = \frac{\lim\limits_{x \to c} f(x)}{\lim\limits_{x \to c} g(x)} \tag{1}$$

But the quotient property cannot always be used. For example, to find $\lim\limits_{x \to 2} \dfrac{x^2 - 4}{x - 2}$, we cannot use equation (1) since the numerator and the denominator each approach 0 (resulting in the form $\dfrac{0}{0}$). Instead, we use algebra and obtain

$$\lim_{x \to 2} \frac{x^2 - 4}{x - 2} = \lim_{x \to 2} \frac{(x - 2)(x + 2)}{x - 2} = \lim_{x \to 2} (x + 2) = 4$$

NEED TO REVIEW? Limits at infinity and infinite limits are discussed in Section 1.5, pp. 133–143.

To find $\lim\limits_{x \to \infty} \dfrac{3x - 2}{x^2 + 5}$, we cannot use equation (1) since the numerator and the denominator each approach infinity (resulting in the form $\dfrac{\infty}{\infty}$). Instead, we use the end behavior of polynomial functions. Then

$$\lim_{x \to \infty} \frac{3x - 2}{x^2 + 5} = \lim_{x \to \infty} \frac{3}{x} = 0$$

As a third example, to find $\lim\limits_{x \to 0} \dfrac{\sin x}{x}$, we cannot use equation (1) since it leads to the form $\dfrac{0}{0}$. Instead, we use a geometric argument (the Squeeze Theorem) to show that

$$\lim_{x \to 0} \frac{\sin x}{x} = 1$$

When $\lim\limits_{x \to c} \dfrac{f(x)}{g(x)} = \dfrac{\lim\limits_{x \to c} f(x)}{\lim\limits_{x \to c} g(x)}$ leads to the form $\dfrac{0}{0}$ or $\dfrac{\infty}{\infty}$, we say that $\dfrac{f}{g}$ is an

NOTE The word "indeterminate" conveys the idea that the limit cannot be found without additional work.

indeterminate form at c. Then depending on the functions f and g, $\lim\limits_{x \to c} \dfrac{f(x)}{g(x)}$, if it exists, could equal 0, ∞, or some nonzero real number.

DEFINITION Indeterminate Form at c of the Type $\dfrac{0}{0}$ or the Type $\dfrac{\infty}{\infty}$

If the functions f and g are each defined in an open interval containing the number c, except possibly at c, then the quotient $\dfrac{f(x)}{g(x)}$ is called an **indeterminate form at c of the type $\dfrac{0}{0}$** if

$$\lim_{x \to c} f(x) = 0 \qquad \text{and} \qquad \lim_{x \to c} g(x) = 0$$

and an **indeterminate form at c of the type $\dfrac{\infty}{\infty}$** if

$$\lim_{x \to c} f(x) = \pm\infty \qquad \text{and} \qquad \lim_{x \to c} g(x) = \pm\infty$$

NOTE $\dfrac{0}{0}$ and $\dfrac{\infty}{\infty}$ are symbols used to denote an indeterminate form.

These definitions also hold for limits at infinity.

EXAMPLE 1 Identifying an Indeterminate Form of the Type $\dfrac{0}{0}$ or $\dfrac{\infty}{\infty}$

(a) $\dfrac{\cos(3x) - 1}{2x}$ is an indeterminate form at 0 of the type $\dfrac{0}{0}$ since

$$\lim_{x \to 0} [\cos(3x) - 1] = 0 \qquad \text{and} \qquad \lim_{x \to 0} (2x) = 0$$

(b) $\dfrac{x - 1}{x^2 + 2x - 3}$ is an indeterminate form at 1 of the type $\dfrac{0}{0}$ since

$$\lim_{x \to 1} (x - 1) = 0 \qquad \text{and} \qquad \lim_{x \to 1} (x^2 + 2x - 3) = 0$$

(c) $\dfrac{x^2 - 2}{x - 3}$ is not an indeterminate form at 3 of the type $\dfrac{0}{0}$ since $\lim_{x \to 3} (x^2 - 2) \neq 0$.

(d) $\dfrac{x^2}{e^x}$ is an indeterminate form at ∞ of the type $\dfrac{\infty}{\infty}$ since

$$\lim_{x \to \infty} x^2 = \infty \qquad \text{and} \qquad \lim_{x \to \infty} e^x = \infty \qquad \blacksquare$$

NOW WORK Problem 7 and AP® Practice Problem 8.

Classic Image/Alamy

ORIGINS Guillaume François de L'Hôpital (1661–1704) was a French nobleman. When he was 30, he hired Johann Bernoulli to tutor him in calculus. Several years later, he entered into a deal with Bernoulli. L'Hôpital paid Bernoulli an annual sum for Bernoulli to share his mathematical discoveries with him but no one else. In 1696, L'Hôpital published the first textbook on differential calculus. It was immensely popular, the last edition being published in 1781 (seventy-seven years after L'Hôpital's death). The book included the rule we study here. After L'Hôpital's death, Johann Bernoulli made his deal with L'Hôpital public and claimed that L'Hôpital's textbook was his own material. His position was dismissed because he often made such claims. However, in 1921, the manuscripts of Bernoulli's lectures to L'Hôpital were found, showing L'Hôpital's calculus book was indeed largely Johann Bernoulli's work.

2 Use L'Hôpital's Rule to Find a Limit

A theorem, named after the French mathematician Guillaume François de L'Hôpital (pronounced "low-pee-tal"), provides a method for finding the limit of an indeterminate form.

THEOREM L'Hôpital's Rule

Suppose the functions f and g are differentiable on an open interval I containing the number c, except possibly at c, and $g'(x) \neq 0$ for all $x \neq c$ in I. Let L denote either a real number or $\pm\infty$, and suppose $\dfrac{f(x)}{g(x)}$ is an indeterminate form at c of the type $\dfrac{0}{0}$ or $\dfrac{\infty}{\infty}$. If $\lim_{x \to c} \dfrac{f'(x)}{g'(x)} = L$, then $\lim_{x \to c} \dfrac{f(x)}{g(x)} = L$.

L'Hôpital's Rule is also valid for limits at infinity and one-sided limits. A partial proof of L'Hôpital's Rule is given in Appendix B. A limited proof is given here that assumes f' and g' are continuous at c and $g'(c) \neq 0$.

Proof Suppose $\lim_{x \to c} f(x) = 0$ and $\lim_{x \to c} g(x) = 0$. Then $f(c) = 0$ and $g(c) = 0$. Since both f' and g' are continuous at c and $g'(c) \neq 0$, we have

$$\lim_{x \to c} \frac{f'(x)}{g'(x)} = \frac{\lim_{x \to c} f'(x)}{\lim_{x \to c} g'(x)} = \frac{f'(c)}{g'(c)} = \frac{\lim_{x \to c} \dfrac{f(x) - f(c)}{x - c}}{\lim_{x \to c} \dfrac{g(x) - g(c)}{x - c}}$$

↑ Quotient Property   ↑ f', g' continuous

$$= \lim_{x \to c} \frac{\dfrac{f(x) - f(c)}{x - c}}{\dfrac{g(x) - g(c)}{x - c}} = \lim_{x \to c} \frac{f(x) - f(c)}{g(x) - g(c)} = \lim_{x \to c} \frac{f(x)}{g(x)}$$

↑ $f(c) = 0$, $g(c) = 0$

Steps for Finding a Limit Using L'Hôpital's Rule

Step 1 Check that $\dfrac{f}{g}$ is an indeterminate form at c of the type $\dfrac{0}{0}$ or $\dfrac{\infty}{\infty}$.

If it is not, do not use L'Hôpital's Rule.

Step 2 Differentiate f and g separately.

Step 3 Find $\lim_{x \to c} \dfrac{f'(x)}{g'(x)}$. This limit is equal to $\lim_{x \to c} \dfrac{f(x)}{g(x)}$, provided the limit is a number or ∞ or $-\infty$.

Step 4 If $\dfrac{f'}{g'}$ is an indeterminate form at c of the type $\dfrac{0}{0}$ or $\dfrac{\infty}{\infty}$, repeat the process.

IN WORDS If $\lim_{x \to c} \dfrac{f(x)}{g(x)}$ leads to $\dfrac{0}{0}$ or $\dfrac{\infty}{\infty}$, L'Hôpital's Rule states that the limit of the quotient equals the limit of the quotient of their derivatives.

AP® EXAM TIP

On the AP® Exam, it is important that you verify, in writing, that the quotient is an indeterminate form before using L'Hôpital's Rule, as shown in Step 1. Also, do not write, "The limit $= \dfrac{0}{0}$ or $\dfrac{\infty}{\infty}$." Rather write, "The quotient is an indeterminate form of the type $\dfrac{0}{0}$ or $\dfrac{\infty}{\infty}$."

EXAMPLE 2   Using L'Hôpital's Rule to Find a Limit

Find $\lim_{x \to 0} \dfrac{\tan x}{6x}$.

Solution

We follow the steps for finding a limit using L'Hôpital's Rule.

Step 1 Since $\lim_{x \to 0} \tan x = 0$ and $\lim_{x \to 0} (6x) = 0$, the quotient $\dfrac{\tan x}{6x}$ is an indeterminate form at 0 of the type $\dfrac{0}{0}$.

Step 2 $\dfrac{d}{dx} \tan x = \sec^2 x$ and $\dfrac{d}{dx}(6x) = 6$.

Step 3 $\lim_{x \to 0} \dfrac{\frac{d}{dx} \tan x}{\frac{d}{dx}(6x)} = \lim_{x \to 0} \dfrac{\sec^2 x}{6} = \dfrac{1}{6}$.

It follows from L'Hôpital's Rule that $\lim_{x \to 0} \dfrac{\tan x}{6x} = \dfrac{1}{6}$. ∎

CAUTION When using L'Hôpital's Rule, we find the derivative of the numerator and the derivative of the denominator separately. Be sure not to find the derivative of the quotient. Also, using L'Hôpital's Rule to find a limit of an expression that is not an indeterminate form may result in an incorrect answer.

In the solution of Example 2, we were careful to determine that the limit of the ratio of the derivatives, that is, $\lim\limits_{x\to 0}\dfrac{\sec^2 x}{6}$, existed or became infinite before using L'Hôpital's Rule. However, the usual practice is to combine Step 2 and Step 3 as follows:

$$\lim_{x\to 0}\frac{\tan x}{6x}=\lim_{x\to 0}\frac{\dfrac{d}{dx}\tan x}{\dfrac{d}{dx}(6x)}=\lim_{x\to 0}\frac{\sec^2 x}{6}=\frac{1}{6}$$

NOW WORK Problem 29 and AP® Practice Problems 1 and 6.

At times, it is necessary to use L'Hôpital's Rule more than once.

EXAMPLE 3 Using L'Hôpital's Rule to Find a Limit

Find $\lim\limits_{x\to 0}\dfrac{\sin x - x}{x^2}$.

Solution

Use the steps for finding a limit using L'Hôpital's Rule.

Step 1 Since $\lim\limits_{x\to 0}(\sin x - x)=0$ and $\lim\limits_{x\to 0}x^2=0$, the expression $\dfrac{\sin x - x}{x^2}$ is an indeterminate form at 0 of the type $\dfrac{0}{0}$.

Steps 2 and 3 Use L'Hôpital's Rule.

$$\lim_{x\to 0}\frac{\sin x - x}{x^2}\underset{\substack{\uparrow\\ \text{L'Hôpital's Rule}}}{=}\lim_{x\to 0}\frac{\dfrac{d}{dx}(\sin x - x)}{\dfrac{d}{dx}x^2}=\lim_{x\to 0}\frac{\cos x - 1}{2x}=\frac{1}{2}\lim_{x\to 0}\frac{\cos x - 1}{x}$$

Step 4 Since $\lim\limits_{x\to 0}(\cos x - 1)=0$ and $\lim\limits_{x\to 0}x=0$, the expression $\dfrac{\cos x - 1}{x}$ is an indeterminate form at 0 of the type $\dfrac{0}{0}$. So, use L'Hôpital's Rule again.

$$\lim_{x\to 0}\frac{\sin x - x}{x^2}=\frac{1}{2}\lim_{x\to 0}\frac{\cos x - 1}{x}\underset{\substack{\uparrow\\ \text{L'Hôpital's Rule}}}{=}\frac{1}{2}\lim_{x\to 0}\frac{\dfrac{d}{dx}(\cos x - 1)}{\dfrac{d}{dx}x}=\frac{1}{2}\lim_{x\to 0}\frac{-\sin x}{1}=0$$ ∎

NOW WORK Problem 39 and AP® Practice Problems 2 and 4.

EXAMPLE 4 Using L'Hôpital's Rule to Find a Limit at Infinity

Find: **(a)** $\lim\limits_{x\to\infty}\dfrac{\ln x}{x}$ **(b)** $\lim\limits_{x\to\infty}\dfrac{x}{e^x}$ **(c)** $\lim\limits_{x\to\infty}\dfrac{e^x}{x}$

Solution

(a) Since $\lim\limits_{x\to\infty}\ln x=\infty$ and $\lim\limits_{x\to\infty}x=\infty$, $\dfrac{\ln x}{x}$ is an indeterminate form at ∞ of the type $\dfrac{\infty}{\infty}$. Using L'Hôpital's Rule,

$$\lim_{x\to\infty}\frac{\ln x}{x}\underset{\substack{\uparrow\\ \text{L'Hôpital's}\\ \text{Rule}}}{=}\lim_{x\to\infty}\frac{\dfrac{d}{dx}\ln x}{\dfrac{d}{dx}x}=\lim_{x\to\infty}\frac{\dfrac{1}{x}}{1}=\lim_{x\to\infty}\frac{1}{x}=0$$

(b) $\lim\limits_{x\to\infty}x=\infty$ and $\lim\limits_{x\to\infty}e^x=\infty$, so $\dfrac{x}{e^x}$ is an indeterminate form at ∞ of the type $\dfrac{\infty}{\infty}$. Using L'Hôpital's Rule,

$$\lim_{x\to\infty}\frac{x}{e^x}\underset{\substack{\uparrow\\ \text{L'Hôpital's}\\ \text{Rule}}}{=}\lim_{x\to\infty}\frac{\dfrac{d}{dx}x}{\dfrac{d}{dx}e^x}=\lim_{x\to\infty}\frac{1}{e^x}=0$$

(c) From (b), we know that $\dfrac{e^x}{x}$ is an indeterminate form at ∞ of the type $\dfrac{\infty}{\infty}$. Using L'Hôpital's Rule,

$$\lim_{x\to\infty}\frac{e^x}{x}\underset{\substack{\uparrow\\ \text{L'Hôpital's}\\ \text{Rule}}}{=}\lim_{x\to\infty}\frac{\dfrac{d}{dx}e^x}{\dfrac{d}{dx}x}=\lim_{x\to\infty}\frac{e^x}{1}=\infty$$ ∎

NOW WORK Problem 35 and AP® Practice Problems 3 and 5.

The results from Example 4 tell us that $y=x$ grows faster than $y=\ln x$, and that $y=e^x$ grows faster than $y=x$. In fact, the limits found in (a) and (b) are also true if $y=x$ is replaced by $y=x^n$, where n is a positive integer. That is,

$$\lim_{x\to\infty}\frac{\ln x}{x^n}=0 \qquad \lim_{x\to\infty}\frac{x^n}{e^x}=0$$

for $n\ge 1$ an integer. You are asked to verify these results in Problems 93 and 94.

The table in Figure 17 compares the growth rates of $y=\ln x$, $y=x^3$, and $y=e^x$.

Example 5 shows that sometimes simplifying first reduces the effort needed to find the limit.

```
NORMAL FLOAT AUTO REAL RADIAN MP
PRESS + FOR △Tbl
  X       Y1      Y2       Y3
  1       0       1        2.7183
  10      2.3026  1000     22026
  50      3.912   125000   5.2E21
  100     4.6052  1E6      2.7E43
  500     6.2146  1.25E8   ERROR
  1000    6.9078  1E9      ERROR
  5000    8.5172  1.3E11   ERROR
  10000   9.2103  1E12     ERROR
  50000   10.82   1.3E14   ERROR
  100000  11.513  1E15     ERROR
  1E6     13.816  1E18     ERROR

X=1000000
```

Figure 17

EXAMPLE 5 **Using L'Hôpital's Rule to Find a Limit**

Find $\lim\limits_{x\to 0}\dfrac{\tan x-\sin x}{x^2\tan x}$.

Solution

$\dfrac{\tan x-\sin x}{x^2\tan x}$ is an indeterminate form at 0 of the type $\dfrac{0}{0}$. We simplify the expression before using L'Hôpital's Rule. Then it is easier to find the limit.

$$\frac{\tan x-\sin x}{x^2\tan x}=\frac{\dfrac{\sin x}{\cos x}-\sin x}{x^2\cdot\dfrac{\sin x}{\cos x}}=\frac{\dfrac{\sin x-\sin x\cos x}{\cos x}}{\dfrac{x^2\sin x}{\cos x}}=\frac{\sin x(1-\cos x)}{x^2\sin x}=\frac{1-\cos x}{x^2}$$

Since $\dfrac{1-\cos x}{x^2}$ is an indeterminate form at 0 of the type $\dfrac{0}{0}$, we use L'Hôpital's Rule.

$$\lim_{x\to 0}\frac{\tan x-\sin x}{x^2\tan x}=\lim_{x\to 0}\frac{1-\cos x}{x^2}\underset{\text{L'Hôpital's Rule}}{=}\lim_{x\to 0}\frac{\dfrac{d}{dx}(1-\cos x)}{\dfrac{d}{dx}x^2}=\lim_{x\to 0}\frac{\sin x}{2x}$$

$$=\frac{1}{2}\lim_{x\to 0}\frac{\sin x}{x}\underset{\lim\limits_{x\to 0}\frac{\sin x}{x}=1}{=}\frac{1}{2}$$

The next example uses L'Hôpital's Rule to find a one-sided limit.

EXAMPLE 6 Using L'Hôpital's Rule to Find a One-Sided Limit

Find $\displaystyle\lim_{x\to 0^+}\frac{\cot x}{\ln x}$.

Solution

Since $\displaystyle\lim_{x\to 0^+}\cot x=\infty$ and $\displaystyle\lim_{x\to 0^+}\ln x=-\infty$, $\dfrac{\cot x}{\ln x}$ is an indeterminate form at 0^+ of the type $\dfrac{\infty}{\infty}$. Using L'Hôpital's Rule,

$$\lim_{x\to 0^+}\frac{\cot x}{\ln x}\underset{\text{L'Hôpital's Rule}}{=}\lim_{x\to 0^+}\frac{\dfrac{d}{dx}\cot x}{\dfrac{d}{dx}\ln x}=\lim_{x\to 0^+}\frac{-\csc^2 x}{\dfrac{1}{x}}=-\lim_{x\to 0^+}\frac{x}{\sin^2 x}\underset{\text{L'Hôpital's Rule}}{=}-\lim_{x\to 0^+}\frac{\dfrac{d}{dx}x}{\dfrac{d}{dx}\sin^2 x}$$

$$=-\lim_{x\to 0^+}\frac{1}{2\sin x\cos x}=-\frac{1}{2}\cdot\lim_{x\to 0^+}\frac{1}{\sin x}\cdot\lim_{x\to 0^+}\frac{1}{\cos x}$$

$$=-\frac{1}{2}\cdot\lim_{x\to 0^+}\frac{1}{\sin x}\cdot 1=-\infty$$

NOW WORK Problem 61 and AP® Practice Problems 7 and 9.

❸ Find the Limit of an Indeterminate Form of the Type $0\cdot\infty$, $\infty-\infty$, 0^0, 1^∞, or ∞^0

L'Hôpital's Rule can only be used for indeterminate forms of the types $\dfrac{0}{0}$ and $\dfrac{\infty}{\infty}$. We need to rewrite indeterminate forms of the type $0\cdot\infty$, $\infty-\infty$, 0^0, 1^∞, or ∞^0 in the form $\dfrac{0}{0}$ or $\dfrac{\infty}{\infty}$ before using L'Hôpital's Rule to find the limit.

Indeterminate Forms of the Type $0\cdot\infty$

Suppose $\displaystyle\lim_{x\to c}f(x)=0$ and $\displaystyle\lim_{x\to c}g(x)=\infty$. Depending on the functions f and g, $\displaystyle\lim_{x\to c}[f(x)\cdot g(x)]$, if it exists, could equal 0, ∞, or some nonzero number. The product $f\cdot g$ is called an **indeterminate form at c of the type $0\cdot\infty$**.

To find $\lim\limits_{x \to c}(f \cdot g)$, rewrite the product $f \cdot g$ as one of the following quotients:

$$f \cdot g = \frac{f}{\dfrac{1}{g}} \qquad \text{or} \qquad f \cdot g = \frac{g}{\dfrac{1}{f}}$$

The right side of the equation on the left is an indeterminate form at c of the type $\dfrac{0}{0}$; the right side of the equation on the right is of the type $\dfrac{\infty}{\infty}$. Then use L'Hôpital's Rule with one of these quotients, choosing the one for which the derivatives are easier to find. If the first choice does not work, then try the other one.

EXAMPLE 7   **Finding the Limit of an Indeterminate Form of the Type $0 \cdot \infty$**

Find:

(a) $\lim\limits_{x \to 0^+} (x \ln x)$ **(b)** $\lim\limits_{x \to \infty} \left(x \sin \dfrac{1}{x} \right)$

Solution

(a) Since $\lim\limits_{x \to 0^+} x = 0$ and $\lim\limits_{x \to 0^+} \ln x = -\infty$, then $x \ln x$ is an indeterminate form at 0^+ of the type $0 \cdot \infty$. Change $x \ln x$ to an indeterminate form of the type $\dfrac{\infty}{\infty}$ by writing $x \ln x = \dfrac{\ln x}{\dfrac{1}{x}}$. Then use L'Hôpital's Rule.

NOTE We choose to use $\dfrac{\ln x}{\dfrac{1}{x}}$ rather than $\dfrac{x}{\dfrac{1}{\ln x}}$ because it is easier to find the derivatives of $\ln x$ and $\dfrac{1}{x}$ than it is to find the derivatives of x and $\dfrac{1}{\ln x}$.

$$\lim_{x \to 0^+} (x \ln x) = \lim_{x \to 0^+} \frac{\ln x}{\dfrac{1}{x}} = \lim_{x \to 0^+} \frac{\dfrac{d}{dx} \ln x}{\dfrac{d}{dx} \dfrac{1}{x}} = \lim_{x \to 0^+} \frac{\dfrac{1}{x}}{-\dfrac{1}{x^2}} = \lim_{x \to 0^+} (-x) = 0$$

(b) Since $\lim\limits_{x \to \infty} x = \infty$ and $\lim\limits_{x \to \infty} \sin \dfrac{1}{x} = 0$, then $x \sin \dfrac{1}{x}$ is an indeterminate form at ∞ of the type $0 \cdot \infty$. Change $x \sin \dfrac{1}{x}$ to an indeterminate form of the type $\dfrac{0}{0}$ by writing

NOTE Notice in the solution to (b) that we did not need to use L'Hôpital's Rule.

$$\lim_{x \to \infty} x \sin \frac{1}{x} = \lim_{x \to \infty} \frac{\sin \dfrac{1}{x}}{\dfrac{1}{x}} \underset{\substack{\uparrow \\ \text{Let } t = \frac{1}{x}}}{=} \lim_{t \to 0^+} \frac{\sin t}{t} = 1 \qquad \blacksquare$$

NOW WORK   Problem 45.

NOTE The indeterminate form of the type $\infty - \infty$ is a convenient notation for any of the following: $\infty - \infty$, $-\infty - (-\infty)$, $\infty + (-\infty)$. Note that $\infty + \infty = \infty$ and $(-\infty) + (-\infty) = -\infty$ are not indeterminate forms because there is no confusion about these sums.

Indeterminate Forms of the Type $\infty - \infty$

Suppose $\lim\limits_{x \to c} f(x) = \infty$ and $\lim\limits_{x \to c} g(x) = \infty$. Then depending on the functions f and g, $\lim\limits_{x \to c} [f(x) - g(x)]$, if it exists, could equal 0, ∞, or some nonzero number. In such instances the difference $f(x) - g(x)$ is called an **indeterminate form at c of the type $\infty - \infty$.**

If the limit of a function results in the indeterminate form $\infty - \infty$, it is generally possible to rewrite the function as an indeterminate form of the type $\dfrac{0}{0}$ or $\dfrac{\infty}{\infty}$ by using algebra, trigonometry, or properties of logarithms.

EXAMPLE 8 **Finding the Limit of an Indeterminate Form of the Type $\infty - \infty$**

Find $\lim\limits_{x \to 0^+} \left(\dfrac{1}{x} - \dfrac{1}{\sin x} \right)$.

Solution

Since $\lim\limits_{x \to 0^+} \dfrac{1}{x} = \infty$ and $\lim\limits_{x \to 0^+} \dfrac{1}{\sin x} = \infty$, then $\dfrac{1}{x} - \dfrac{1}{\sin x}$ is an indeterminate form at 0^+ of the type $\infty - \infty$. Rewrite the difference as a single fraction.

$$\lim_{x \to 0^+} \left(\frac{1}{x} - \frac{1}{\sin x} \right) = \lim_{x \to 0^+} \frac{\sin x - x}{x \sin x}$$

Then $\dfrac{\sin x - x}{x \sin x}$ is an indeterminate form at 0^+ of the type $\dfrac{0}{0}$. Now use L'Hôpital's Rule.

$$\lim_{x \to 0^+} \left(\frac{1}{x} - \frac{1}{\sin x} \right) = \lim_{x \to 0^+} \frac{\sin x - x}{x \sin x} \underset{\substack{\uparrow \\ \text{L'Hôpital's Rule}}}{=} \lim_{x \to 0^+} \frac{\dfrac{d}{dx}(\sin x - x)}{\dfrac{d}{dx}(x \sin x)} = \lim_{x \to 0^+} \frac{\cos x - 1}{x \cos x + \sin x}$$

$$\underset{\substack{\uparrow \\ \text{Type } \frac{0}{0}; \text{ use L'Hôpital's Rule}}}{=} \lim_{x \to 0^+} \frac{\dfrac{d}{dx}(\cos x - 1)}{\dfrac{d}{dx}(x \cos x + \sin x)} = \lim_{x \to 0^+} \frac{-\sin x}{(-x \sin x + \cos x) + \cos x}$$

$$= \lim_{x \to 0^+} \frac{\sin x}{x \sin x - 2 \cos x} = \frac{0}{-2} = 0 \qquad \blacksquare$$

NOW WORK Problem 47.

Indeterminate Forms of the Type 1^∞, 0^0, or ∞^0

A function of the form $[f(x)]^{g(x)}$ may result in an indeterminate form of the type 1^∞, 0^0, or ∞^0. To find the limit of such a function, we let $y = [f(x)]^{g(x)}$ and take the natural logarithm of each side.

$$\ln y = \ln[f(x)]^{g(x)} = g(x) \ln f(x)$$

The expression on the right will then be an indeterminate form of the type $0 \cdot \infty$, and we find the limit using the method used in Example 7(a).

NEED TO REVIEW? Properties of logarithms are discussed in Appendix A.1, pp. A-11 to A-12.

Steps for Finding $\lim\limits_{x \to c}[f(x)]^{g(x)}$ When $[f(x)]^{g(x)}$ Is an Indeterminate Form at c of the Type 1^∞, 0^0, or ∞^0

Step 1 Let $y = [f(x)]^{g(x)}$ and take the natural logarithm of each side, obtaining $\ln y = g(x) \ln f(x)$.

Step 2 Find $\lim\limits_{x \to c} \ln y$.

Step 3 If $\lim\limits_{x \to c} \ln y = L$, then $\lim\limits_{x \to c} y = e^L$.

These steps also can be used for limits at infinity and for one-sided limits.

EXAMPLE 9   Finding the Limit of an Indeterminate Form of the Type 0^0

Find $\lim\limits_{x \to 0^+} x^x$.

Solution

The expression x^x is an indeterminate form at 0^+ of the type 0^0. Follow the steps for finding $\lim\limits_{x \to c}[f(x)]^{g(x)}$.

Step 1 Let $y = x^x$. Then $\ln y = x \ln x$.

Step 2 $\lim\limits_{x \to 0^+} \ln y = \lim\limits_{x \to 0^+} (x \ln x) = 0$ [from Example 7(a)].

Step 3 Since $\lim\limits_{x \to 0^+} \ln y = 0$, $\lim\limits_{x \to 0^+} y = e^0 = 1$.   ∎

> **CAUTION** Do not stop after finding $\lim\limits_{x \to c} \ln y = L$. Remember, we want to find $\lim\limits_{x \to c} y = e^L$.

NOW WORK   Problem 51.

EXAMPLE 10   Finding the Limit of an Indeterminate Form of the Type 1^∞

Find $\lim\limits_{x \to 0^+} (1 + x)^{1/x}$.

Solution

The expression $(1 + x)^{1/x}$ is an indeterminate form at 0^+ of the type 1^∞.

Step 1 Let $y = (1 + x)^{1/x}$. Then $\ln y = \dfrac{1}{x} \ln(1 + x)$.

Step 2 $\lim\limits_{x \to 0^+} \ln y = \lim\limits_{x \to 0^+} \dfrac{\ln(1 + x)}{x} = \lim\limits_{x \to 0^+} \dfrac{\dfrac{d}{dx} \ln(1 + x)}{\dfrac{d}{dx} x} = \lim\limits_{x \to 0^+} \dfrac{\dfrac{1}{1 + x}}{1} = 1$

Type $\dfrac{0}{0}$; use L'Hôpital's Rule

Step 3 Since $\lim\limits_{x \to 0^+} \ln y = 1$, $\lim\limits_{x \to 0^+} y = e^1 = e$.   ∎

NOW WORK   Problem 85.

4.4 Assess Your Understanding

Concepts and Vocabulary

1. **True or False** $\dfrac{f(x)}{g(x)}$ is an indeterminate form at c of the type $\dfrac{0}{0}$ if $\lim\limits_{x \to c} \dfrac{f(x)}{g(x)}$ does not exist.

2. **True or False** If $\dfrac{f(x)}{g(x)}$ is an indeterminate form at c of the type $\dfrac{0}{0}$, then L'Hôpital's Rule states that $\lim\limits_{x \to c} \dfrac{f(x)}{g(x)} = \lim\limits_{x \to c}\left[\dfrac{d}{dx} \dfrac{f(x)}{g(x)}\right]$.

3. **True or False** $\dfrac{1}{x}$ is an indeterminate form at 0.

4. **True or False** $x \ln x$ is not an indeterminate form at 0^+ because $\lim\limits_{x \to 0^+} x = 0$ and $\lim\limits_{x \to 0^+} \ln x = -\infty$, and $0 \cdot -\infty = 0$.

5. In your own words, explain why $\infty - \infty$ is an indeterminate form, but $\infty + \infty$ is not an indeterminate form.

6. In your own words, explain why $0 \cdot \infty \neq 0$.

Skill Building

In Problems 7–26:

(a) *Determine whether each expression is an indeterminate form at c.*

(b) *If it is, identify the type. If it is not an indeterminate form, state why.*

7. $\dfrac{1 - e^x}{x}$, $c = 0$

8. $\dfrac{1 - e^x}{x - 1}$, $c = 0$

9. $\dfrac{e^x}{x}$, $c = 0$

10. $\dfrac{e^x}{x}$, $c = \infty$

11. $\dfrac{\ln x}{x^2}$, $c = \infty$

12. $\dfrac{\ln(x + 1)}{e^x - 1}$, $c = 0$

13. $\dfrac{\sec x}{x}$, $c = 0$

14. $\dfrac{x}{\sec x - 1}$, $c = 0$

15. $\dfrac{\sin x(1-\cos x)}{x^2}$, $c=0$

16. $\dfrac{\sin x-1}{\cos x}$, $c=\dfrac{\pi}{2}$

17. $\dfrac{\tan x-1}{\sin(4x-\pi)}$, $c=\dfrac{\pi}{4}$

18. $\dfrac{e^x-e^{-x}}{1-\cos x}$, $c=0$

19. $x^2 e^{-x}$, $c=\infty$

20. $x\cot x$, $c=0$

21. $\csc\dfrac{x}{2}-\cot\dfrac{x}{2}$, $c=0$

22. $\dfrac{x}{x-1}+\dfrac{1}{\ln x}$, $c=1$

23. $\left(\dfrac{1}{x^2}\right)^{\sin x}$, $c=0$

24. $(e^x+x)^{1/x}$, $c=0$

25. $(x^2-1)^x$, $c=0$

26. $(\sin x)^x$, $c=0$

In Problems 27–42, identify each quotient as an indeterminate form of the type $\dfrac{0}{0}$ or $\dfrac{\infty}{\infty}$. Then find the limit.

27. $\lim\limits_{x\to 2}\dfrac{x^2+x-6}{x^2-3x+2}$

28. $\lim\limits_{x\to 1}\dfrac{2x^3+5x^2-4x-3}{x^3+x^2-10x+8}$

PAGE 316 29. $\lim\limits_{x\to 1}\dfrac{\ln x}{x^2-1}$

30. $\lim\limits_{x\to 0}\dfrac{\ln(1-x)}{e^x-1}$

31. $\lim\limits_{x\to 0}\dfrac{e^x-e^{-x}}{\sin x}$

32. $\lim\limits_{x\to 0}\dfrac{\tan(2x)}{\ln(1+x)}$

33. $\lim\limits_{x\to 1}\dfrac{\sin(\pi x)}{x-1}$

34. $\lim\limits_{x\to \pi}\dfrac{1+\cos x}{\sin(2x)}$

PAGE 317 35. $\lim\limits_{x\to \infty}\dfrac{x^2}{e^x}$

36. $\lim\limits_{x\to \infty}\dfrac{e^x}{x^4}$

37. $\lim\limits_{x\to \infty}\dfrac{\ln x}{e^x}$

38. $\lim\limits_{x\to \infty}\dfrac{x+\ln x}{x\ln x}$

PAGE 316 39. $\lim\limits_{x\to 0}\dfrac{e^x-1-\sin x}{1-\cos x}$

40. $\lim\limits_{x\to 0}\dfrac{e^x-e^{-x}-2\sin x}{3x^3}$

41. $\lim\limits_{x\to 0}\dfrac{\sin x-x}{x^3}$

42. $\lim\limits_{x\to 0}\dfrac{x^3}{\cos x-1}$

In Problems 43–58, identify each expression as an indeterminate form of the type $0\cdot\infty$, $\infty-\infty$, 0^0, 1^∞, or ∞^0. Then find the limit.

43. $\lim\limits_{x\to 0^+}(x^2\ln x)$

44. $\lim\limits_{x\to \infty}(xe^{-x})$

PAGE 319 45. $\lim\limits_{x\to \infty}[x(e^{1/x}-1)]$

46. $\lim\limits_{x\to \pi/2}[(1-\sin x)\tan x]$

PAGE 320 47. $\lim\limits_{x\to \pi/2}(\sec x-\tan x)$

48. $\lim\limits_{x\to 0}\left(\cot x-\dfrac{1}{x}\right)$

49. $\lim\limits_{x\to 1}\left(\dfrac{1}{\ln x}-\dfrac{x}{\ln x}\right)$

50. $\lim\limits_{x\to 0}\left(\dfrac{1}{x}-\dfrac{1}{e^x-1}\right)$

PAGE 321 51. $\lim\limits_{x\to 0^+}(2x)^{3x}$

52. $\lim\limits_{x\to 0^+}x^{x^2}$

53. $\lim\limits_{x\to \infty}(x+1)^{e^{-x}}$

54. $\lim\limits_{x\to \infty}(1+x^2)^{1/x}$

55. $\lim\limits_{x\to 0^+}(\csc x)^{\sin x}$

56. $\lim\limits_{x\to \infty}x^{1/x}$

57. $\lim\limits_{x\to \pi/2^-}(\sin x)^{\tan x}$

58. $\lim\limits_{x\to 0}(\cos x)^{1/x}$

In Problems 59–88, find each limit.

59. $\lim\limits_{x\to 0^+}\dfrac{\cot x}{\cot(2x)}$

60. $\lim\limits_{x\to \infty}\dfrac{\ln(\ln x)}{\ln x}$

PAGE 318 61. $\lim\limits_{x\to 1/2^-}\dfrac{\ln(1-2x)}{\tan(\pi x)}$

62. $\lim\limits_{x\to 1^-}\dfrac{\ln(1-x)}{\cot(\pi x)}$

63. $\lim\limits_{x\to \infty}\dfrac{x^4+x^3}{e^x+1}$

64. $\lim\limits_{x\to \infty}\dfrac{x^2+x-1}{e^x+e^{-x}}$

65. $\lim\limits_{x\to 0}\dfrac{xe^{4x}-x}{1-\cos(2x)}$

66. $\lim\limits_{x\to 0}\dfrac{x\tan x}{1-\cos x}$

67. $\lim\limits_{x\to 0}\dfrac{\tan^{-1}x}{x}$

68. $\lim\limits_{x\to 0}\dfrac{\tan^{-1}x}{\sin^{-1}x}$

69. $\lim\limits_{x\to 0}\dfrac{\cos x-1}{\cos(2x)-1}$

70. $\lim\limits_{x\to 0}\dfrac{\tan x-\sin x}{x^3}$

71. $\lim\limits_{x\to 0^+}(x^{1/2}\ln x)$

72. $\lim\limits_{x\to \infty}[(x-1)e^{-x^2}]$

73. $\lim\limits_{x\to \pi/2}[\tan x\ln(\sin x)]$

74. $\lim\limits_{x\to 0^+}[\sin x\ln(\sin x)]$

75. $\lim\limits_{x\to 0}[\csc x\ln(x+1)]$

76. $\lim\limits_{x\to \pi/4}[(1-\tan x)\sec(2x)]$

77. $\lim\limits_{x\to a}\left[(a^2-x^2)\tan\left(\dfrac{\pi x}{2a}\right)\right]$

78. $\lim\limits_{x\to 1^+}\left[(1-x)\tan\left(\dfrac{1}{2}\pi x\right)\right]$

79. $\lim\limits_{x\to 1}\left(\dfrac{1}{\ln x}-\dfrac{1}{x-1}\right)$

80. $\lim\limits_{x\to 1}\left(\dfrac{x}{x-1}-\dfrac{1}{\ln x}\right)$

81. $\lim\limits_{x\to \pi/2}\left(x\tan x-\dfrac{\pi}{2}\sec x\right)$

82. $\lim\limits_{x\to \pi}(\cot x-x\csc x)$

83. $\lim\limits_{x\to 1^-}(1-x)^{\tan(\pi x)}$

84. $\lim\limits_{x\to 0^+}x^{\sqrt{x}}$

PAGE 321 85. $\lim\limits_{x\to 0}\left(\dfrac{\sin x}{x}\right)^{1/x}$

86. $\lim\limits_{x\to \infty}\left(1+\dfrac{5}{x}+\dfrac{3}{x^2}\right)^x$

87. $\lim\limits_{x\to (\pi/2)^-}(\tan x)^{\cos x}$

88. $\lim\limits_{x\to 0^+}(x^2+x)^{-\ln x}$

Applications and Extensions

89. **Wolf Population**   In 2024, there were 229 gray wolves in Wyoming. Suppose the population w of wolves in the region at time t follows the logistic growth curve

$$w=w(t)=\dfrac{Ke^{rt}}{\dfrac{K}{40}+e^{rt}-1}$$

where $K=252$, $r=0.283$, and $t=0$ represents the population in the year 2010.

Source: Wyoming Gray Wolf Monitoring and Management: 2021 Annual Report.

(a) Find $\lim\limits_{t\to\infty}w(t)$.

(b) Interpret the answer found in (a) in the context of the problem.

(c) Use technology to graph $w=w(t)$.

90. Skydiving   The downward velocity v of a skydiver with nonlinear air resistance can be modeled by

$$v = v(t) = -A + RA \frac{e^{Bt+C} - 1}{e^{Bt+C} + 1}$$

where t is the time in seconds, and A, B, C, and R are positive constants with $R > 1$.

(a) Find $\lim\limits_{t \to \infty} v(t)$.

(b) Interpret the limit found in (a).

(c) If the velocity v is measured in feet per second, reasonable values of the constants are $A = 108.6$, $B = 0.554$, $C = 0.804$, and $R = 2.62$. Graph the velocity of the skydiver with respect to time.

91. Electricity   The equation governing the amount of current I (in amperes) in a simple RL circuit consisting of a resistance R (in ohms), an inductance L (in henrys), and an electromotive force E (in volts)

is $I = \dfrac{E}{R}(1 - e^{-Rt/L})$.

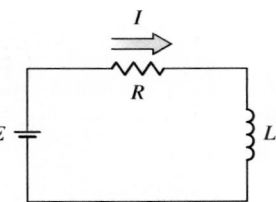

(a) Find $\lim\limits_{t \to \infty} I(t)$ and $\lim\limits_{R \to 0^+} I(t)$.

(b) Interpret these limits.

92. Find $\lim\limits_{x \to 0} \dfrac{a^x - b^x}{x}$, where $a \neq 1$ and $b \neq 1$ are positive real numbers.

93. Show that $\lim\limits_{x \to \infty} \dfrac{\ln x}{x^n} = 0$, for $n \geq 1$ an integer.

94. Show that $\lim\limits_{x \to \infty} \dfrac{x^n}{e^x} = 0$ for $n \geq 1$ an integer.

95. Show that $\lim\limits_{x \to 0^+} (\cos x + 2 \sin x)^{\cot x} = e^2$.

96. Find $\lim\limits_{x \to \infty} \dfrac{P(x)}{e^x}$, where P is a polynomial function.

97. Find $\lim\limits_{x \to \infty} [\ln(x+1) - \ln(x-1)]$.

98. Show that $\lim\limits_{x \to 0^+} \dfrac{e^{-1/x^2}}{x} = 0$.   *Hint:* Write $\dfrac{e^{-1/x^2}}{x} = \dfrac{\frac{1}{x}}{e^{1/x^2}}$.

99. If n is an integer, show that $\lim\limits_{x \to 0^+} \dfrac{e^{-1/x^2}}{x^n} = 0$.

100. Show that $\lim\limits_{x \to \infty} \sqrt[x]{x} = 1$.

101. Show that $\lim\limits_{x \to \infty} \left(1 + \dfrac{a}{x}\right)^x = e^a$, a any real number.

102. Show that $\lim\limits_{x \to \infty} \left(\dfrac{x+a}{x-a}\right)^x = e^{2a}$, $a \neq 0$.

103. (a) Show that the function below has a derivative at 0. What is $f'(0)$?

$$f(x) = \begin{cases} e^{-1/x^2} & \text{if } x \neq 0 \\ 0 & \text{if } x = 0 \end{cases}$$

(b) Graph f.

104. If a, $b \neq 0$ and $c > 0$ are real numbers, show that

$$\lim_{x \to c} \frac{x^a - c^a}{x^b - c^b} = \frac{a}{b} c^{a-b}.$$

105. Prove L'Hôpital's rule when $\dfrac{f(x)}{g(x)}$ is an indeterminate form at $-\infty$ of the type $\dfrac{0}{0}$.

Challenge Problems

106. Explain why L'Hôpital's Rule does not apply to $\lim\limits_{x \to 0} \dfrac{x^2 \sin \frac{1}{x}}{\sin x}$.

107. Find each limit:

(a) $\lim\limits_{x \to \infty} \left(1 + \dfrac{1}{x}\right)^{-x^2}$

(b) $\lim\limits_{x \to \infty} \left(1 + \dfrac{\ln a}{x}\right)^x$, $a > 1$

(c) $\lim\limits_{x \to \infty} \left(1 + \dfrac{1}{x}\right)^{x^2}$

(d) $\lim\limits_{x \to \infty} \left(1 + \dfrac{\sin x}{x}\right)^x$

(e) $\lim\limits_{x \to \infty} (e^x)^{-1/\ln x}$

(f) $\lim\limits_{x \to \infty} \left[\left(\dfrac{1}{a}\right)^x\right]^{-1/x}$, $0 < a < 1$

(g) $\lim\limits_{x \to \infty} x^{1/x}$

(h) $\lim\limits_{x \to \infty} (a^x)^{1/x}$, $a > 1$

(i) $\lim\limits_{x \to \infty} [(2 + \sin x)^x]^{1/x}$

(j) $\lim\limits_{x \to 0^+} x^{-1/\ln x}$

108. Find constants A, B, C, and D so that

$$\lim_{x \to 0} \frac{\sin(Ax) + Bx + Cx^2 + Dx^3}{x^5} = \frac{4}{15}$$

109. A function f has derivatives of all orders.

(a) Find $\lim\limits_{h \to 0} \dfrac{f(x+2h) - 2f(x+h) + f(x)}{h^2}$.

(b) Find $\lim\limits_{h \to 0} \dfrac{f(x+3h) - 3f(x+2h) + 3f(x+h) - f(x)}{h^3}$.

(c) Generalize parts (a) and (b).

110. The formulas in Problem 109 can be used to approximate derivatives. Approximate $f'(2)$, $f''(2)$, and $f'''(2)$ from the table. The data are for $f(x) = \ln x$. Compare the exact values with your approximations.

x	2.0	2.1	2.2	2.3	2.4
$f(x)$	0.6931	0.7419	0.7885	0.8329	0.8755

111. Consider the function $f(t, x) = \dfrac{x^{t+1} - 1}{t + 1}$, where $x > 0$ and $t \neq -1$.

   (a) For x fixed at x_0, show that $\lim\limits_{t \to -1} f(t, x_0) = \ln x_0$.

   (b) For x fixed, define a function $F(t, x)$, where $x > 0$, that is continuous so $F(t, x) = f(t, x)$ for all $t \neq -1$.

   (c) For t fixed, show that $\dfrac{d}{dx} F(t, x) = x^t$, for $x > 0$ and all t.

Source: Michael W. Ecker (2012, September), Unifying Results via L'Hôpital's Rule. *Journal of the American Mathematical Association of Two-Year Colleges,* 4(1) pp. 9–10.

Preparing for the AP® Exam

AP® Practice Problems

Multiple-Choice Questions

[PAGE 316] **1.** $\lim\limits_{x \to 0} \dfrac{e^{4x} - 1}{\sin(2x)} =$

   (A) 0   (B) 2

   (C) 4   (D) does not exist

[PAGE 316] **2.** $\lim\limits_{x \to 0} \dfrac{1 - \cos^2(3x)}{x^2} =$

   (A) 18   (B) 9   (C) 0   (D) 3

[PAGE 317] **3.** Find $\lim\limits_{x \to \infty} \dfrac{x^{-3/2}}{\sin \frac{1}{x}}$.

   (A) $\dfrac{3}{2}$ (B) 1   (C) 0   (D) ∞

[PAGE 316] **4.** $\lim\limits_{x \to 1} \dfrac{\ln x^3}{x^2 - 1} =$

   (A) 0   (B) 1   (C) $\dfrac{3}{2}$ (D) 3

[PAGE 317] **5.** For any positive integer k, $\lim\limits_{x \to \infty} \dfrac{\ln x}{x^k} =$

   (A) 0   (B) 1   (C) $k + 1$ (D) ∞

[PAGE 316] **6.** $\lim\limits_{\theta \to 0} \dfrac{1 - \cos(2\theta)}{3\sin \theta} =$

   (A) -2 (B) $\dfrac{2}{3}$ (C) 0   (D) $-\dfrac{1}{3}$

[PAGE 318] **7.** $\lim\limits_{x \to \frac{\pi}{2}^-} \dfrac{\ln(\cos x)}{\tan x} =$

   (A) $-\infty$ (B) 0   (C) 1   (D) ∞

[PAGE 314] **8.** Which of the following are indeterminate forms at 0?

   I. $\dfrac{x}{\ln(x + 1)}$

   II. $\dfrac{e^x}{x^2 - 2x}$

   III. $\dfrac{x}{1 - \cos(\pi x)}$

   (A) I only   (B) I and III only

   (C) II and III only   (D) I, II, and III

Free-Response Question

[PAGE 318] **9.** For the function $f(x) = \dfrac{e^x + \sin(2x) - 1}{xe^x}$, determine whether there is a value of f at 0 so that f is continuous for all real numbers. Justify your result.

Retain Your Knowledge

Multiple-Choice Questions

1. Given $y = \sin^{-1}(3x)$, find y''.

   (A) $y'' = \dfrac{27x}{(1-9x^2)^{3/2}}$ (B) $y'' = -\dfrac{27x}{(1-9x^2)^{3/2}}$

   (C) $y'' = \dfrac{9x}{2(1-3x^2)^{3/2}}$ (D) $y'' = \dfrac{9x}{(1-9x^2)^{3/2}}$

2. The function $g(x) = -2x^3 + 5x + 3$, $x \geq 1$, is one-to-one; g has an inverse function f. Find an equation of the tangent line to the graph of f at the point $(-3, 2)$ on f.

   (A) $y - 2 = -\dfrac{1}{49}(x+3)$ (B) $y - 2 = \dfrac{1}{49}(x+3)$

   (C) $y - 2 = -\dfrac{1}{19}(x+3)$ (D) $y - 2 = -19(x+3)$

3. $\dfrac{d}{dx}[e^{2x}\tan(3x)] =$

   (A) $3e^{2x}\sec^2(3x)$

   (B) $6e^{2x}\sec^2(3x)$

   (C) $e^{2x}\tan(3x) + 3e^{2x}\sec^2(3x)$

   (D) $2e^{2x}\tan(3x) + 3e^{2x}\sec^2(3x)$

Free-Response Question

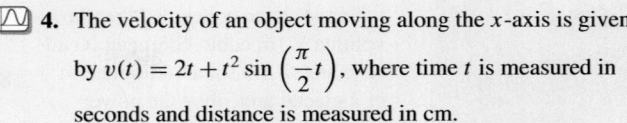 4. The velocity of an object moving along the x-axis is given by $v(t) = 2t + t^2\sin\left(\dfrac{\pi}{2}t\right)$, where time t is measured in seconds and distance is measured in cm.

   (a) Find the velocity of the object at time $t = 3$s.

   (b) Is the velocity increasing or decreasing at $t = 3$s?

   (c) What is the acceleration of the object at $t = 3$s?

   (d) Is the speed of the object increasing or decreasing at $t = 3$s?

CHAPTER 4 PROJECT

Mountain Glaciers in Retreat

This Project can be done individually or as part of a team.

Scientists have concluded that the volume V (in cubic kilometers) and base area A (in square kilometers) of a glacial area obey the power law, $V = 28.5\, A^{1.357}$.

The volume V (in km^3) and base area A (in km^2) of a glacial region each vary with time t (in years).

1. Interpret the derivative of V with respect to t as a rate of change. What are the units of $\dfrac{dV}{dt}$?

2. Interpret the derivative of A with respect to t as a rate of change. What are the units of $\dfrac{dA}{dt}$?

3. Express $\dfrac{dV}{dt}$ in terms of $\dfrac{dA}{dt}$.

Researchers have estimated that the base area of the glaciers in one region of the Swiss Alps in 2008 was approximately 1.836 km^2.

4. What was the volume of these glaciers?

5. Assuming that the base area A decreased at a rate of 0.28 km^2 per year in 2008, what was the rate of change of the volume V of these glaciers in 2008?

6. If these rates of change of volume and base area also occurred in 2009, what was the base area of these Swiss Alps glaciers at the end of 2009? What was the volume of ice?

7. What percentage of volume was lost in 2009?

A study of the glaciers in the Stelvio National Park in the Italian Alps reported the following base areas for the years 1981, 1990, 2003, and 2007.

	1981	1990	2003	2007
Base area (sq km)	42.16	38.23	32.05	29.27

Using the four years in the table, the line of best fit is found to be

$$A = A(t) = -0.505\, t + 42.748 \text{ km}^2$$

where $t = 1$ represents the year 1981.

8. Create a table for the projected base areas in the years 2026, 2030, and 2071.

9. Using the power law, $V = 28.5 A^{1.357}$, add a row to the table that shows the projected volume V of these glaciers for the years 2007, 2026, 2030, and 2071.

10. What is the average rate of change in base area A from 2007 to 2026, from 2007 to 2030, and from 2007 to 2071?

11. What is the average rate of change in volume V of the glaciers from 2007 to 2026, from 2007 to 2030, and from 2007 to 2071?

12. What is the rate of change in the base area A for the years 2026, 2030, and 2071?

13. What is the rate of change in the volume V for the years 2026, 2030, and 2071?

14. Write an essay analyzing the results of this study. What do you conclude about the future size of Earth's glaciers?

References

1. S. Kotlarski, D. Jacob, R. Podzun, and F. Paul, Representing glaciers in a regional climate model, *Climate Dynamics* 34, 27–46 (2010).

2. D.B. Bahr, W.T. Pfeffer, and Georg Kaser A review of volume-area scaling of glaciers, *Reviews of Geophysics* 53, 95–140.

3. D.B. Bahr, M.F. Meier, and S.D. Peckham, The physical basis of glacier volume-area scaling, *Journal of Geophysical Research*, v. 102, n. B9, 355–362 (Sept. 10, 1997).

4. C. D'Agata, D. Bocchiola, D. Maragno, C. Smiraglia, and G.A. Diolaiuti, Glacier shrinkage driven by climate change during half a century (1954–2007) in the Ortles-Cevedale group (Stelvio National Park, Lombardy, Italian Alps), *Theoretical and Applied Climatology* (March 2014).

Chapter Review

THINGS TO KNOW

4.1 Interpreting a Derivative (pp. 285–288)

- The average rate of change of f over the interval $[c, x]$ is given by

$$\frac{\Delta y}{\Delta x} = \frac{f(x) - f(c)}{x - c} \qquad x \neq c$$

where Δy is the change in y and Δx is the change in x.
- The derivative $f'(c)$ equals the limit of the average rate of change of f as x approaches c. That is,

$$f'(c) = \lim_{x \to c} \frac{f(x) - f(c)}{x - c} \qquad x \neq c$$

- The derivative of a function f at a number c is the slope of the tangent line to the graph of f at the point $(c, f(c))$.

- For an object moving along a line,
  - the function $s = s(t)$ is the position of the object at time t.
  - the derivative $v = s'(t)$ is the velocity of the object at time t.
  - the derivative $a = v'(t) = s''(t)$ is the acceleration of the object at time t.

4.2 Related Rates

Steps for solving a related rate problem (p. 291)

4.3 Differentials; Linear Approximations; Newton's Method

- The differential dx of x is defined as $dx = \Delta x \neq 0$, where Δx is the change in x. (p. 301)
- The differential dy of $y = f(x)$ is defined as $dy = f'(x)dx$. (p. 301)

- The **linear approximation** to a differentiable function f at $x = x_0$ is given by $L(x) = f(x_0) + f'(x_0)(x - x_0)$. (p. 303)

- **Relative error** at x_0 in $Q = \dfrac{|\Delta Q|}{Q(x_0)}$ (p. 305)

- **Newton's Method** for finding the zero of a function. (p. 307)

4.4 Indeterminate Forms and L'Hôpital's Rule

Definitions:

- Indeterminate form at c of the type $\dfrac{0}{0}$ or the type $\dfrac{\infty}{\infty}$ (p. 314)

- Indeterminate form at c of the type $0 \cdot \infty$ or $\infty - \infty$ (pp. 318–319)
- Indeterminate form at c of the type 1^∞, 0^0, or ∞^0 (p. 320)

Theorem:

- L'Hôpital's Rule (p. 314)

Procedures:

- Steps for finding a limit using L'Hôpital's Rule (p. 315)
- Steps for finding $\lim_{x \to c}[f(x)]^{g(x)}$, where $[f(x)]^{g(x)}$ is an indeterminate form at c of the type 1^∞, 0^0, or ∞^0 (p. 320)

OBJECTIVES

Section	You should be able to ...	Examples	Review Exercises	AP® Review Problems
4.1	1 Interpret a derivative as the slope of a tangent line to the graph of a function (p. 285)	1	1	
	2 Interpret a derivative as an instantaneous rate of change (p. 286)	2	2	1, 7
	3 Interpret a derivative as velocity or acceleration (p. 287)	3	3	8
4.2	1 Solve related rate problems (p. 290)	1–5	13–15	4, 7
4.3	1 Find the differential of a function and interpret it geometrically (p. 301)	1	4, 5, 9, 10	2
	2 Find the linear approximation to a function (p. 303)	2, 3	6, 7, 8	2, 6
	3 Use differentials in applications (p. 305)	4, 5	11, 12	
	4 Use Newton's Method to approximate a real zero of a function (p. 306)	6, 7	32, 33, 34	
4.4	1 Identify indeterminate forms of the type $\dfrac{0}{0}$ and $\dfrac{\infty}{\infty}$ (p. 313)	1	16–19	5
	2 Use L'Hôpital's Rule to find a limit (p. 314)	2–6	20, 22, 24–27, 30	3, 5
	3 Find the limit of an indeterminate form of the type $0 \cdot \infty$, $\infty - \infty$, 0^0, or ∞^0 (p. 318)	7–10	21, 23, 28, 29, 31	

REVIEW EXERCISES

1. (a) Find the rate of change of $f(x) = 2\sqrt{x}\left[x + (x-1)^2\right]$ at 1.
(b) Find an equation of the tangent line to the graph of f at the point $(1, f(1))$.

2. The cost C of manufacturing x items is modeled by $C(x) = 1500 + 50x + 0.05x^2$ dollars.

(a) Find the cost C of manufacturing 1000 items.
(b) Find $C'(x)$, and interpret it as a rate of change.
(c) Find $C'(1000)$, and interpret it as a rate of change.

3. The position function of an object moving on a line is $s(t) = t^2 - \dfrac{2}{3}t$, where s is in meters and $t \geq 0$ is in seconds.

(a) Find the initial position of the object. Find its position at 3 s.
(b) Find and interpret $s'(3)$ in the context of this problem.
(c) Find and interpret the object's average velocity during the first three seconds.
(d) What is the acceleration of the object after at $t = 3$ s?

4. Find the differential dy if $x^3 + 2y^2 = x^2 y$.

5. Find the differential dy if $y = \ln \dfrac{e^x}{x^2 + 5}$.

6. (a) Find the linear approximation $L(x)$ to $f(x) = x + \ln x$ near $x = 1$.
(b) Use $L(x)$ to approximate $f(1.04)$.

7. (a) Find the linear approximation $L(x)$ to $f(x) = \sin^{-1} x$ near $x = \dfrac{1}{2}$.
(b) Use $L(x)$ to approximate $\sin^{-1} 0.45$.

8. (a) Find the linear approximation $L(x)$ to $f(x) = xe^{2x}$ near $x = 0$.
(b) Use $L(x)$ to approximate $f(-0.2)$.

9. For the function $f(x) = \tan x$:

(a) Find the differential dy and Δy when $x = 0$.
(b) Compare dy to Δy when $x = 0$ and (i) $\Delta x = 0.5$, (ii) $\Delta x = 0.1$, and (iii) $\Delta x = 0.01$.

10. For the function $f(x) = \ln x$:

(a) Find the differential dy and Δy when $x = 1$.
(b) Compare dy to Δy when $x = 1$ and (i) $\Delta x = 0.5$, (ii) $\Delta x = 0.1$, and (iii) $\Delta x = 0.01$.

11. **Measurement Error**   If p is the period of a pendulum of length L, the acceleration due to gravity may be computed by the formula $g = \dfrac{4\pi^2 L}{p^2}$. If L is measured with negligible error but an error of no more than 2% may occur in the measurement of p, what is the approximate percentage error in the computation of g?

12. **Measurement Error**   If the percentage error in measuring the edge of a cube is no more than 5%, what is the percentage error in computing its volume?

13. **Related rates**   A spherical snowball is melting at the rate of $2\,\text{cm}^3/\text{min}$. How fast is the surface area changing when the radius is 5 cm?

14. **Related rates**   A lighthouse is 3 km from a straight shoreline. Its light makes one revolution every 8 s. How fast is the light moving along the shore when it makes an angle of 30° with the shoreline?

15. **Related rates**   Two planes at the same altitude are approaching an airport, one from the north and one from the west. The plane from the north is flying at 250 mi/h and is 30 mi from the airport. The plane from the west is flying at 200 mi/h and is 20 mi from the airport. How fast are the planes approaching each other at that instant?

In Problems 16–19, determine if the expression is an indeterminate form at 0. If it is, identify its type.

16. $\left(\dfrac{1}{x}\right)^{\tan x}$

17. $\dfrac{xe^{3x} - x}{1 - \cos(2x)}$

18. $\dfrac{\tan x - x}{x - \sin x}$

19. $\dfrac{1}{x^2} - \dfrac{1}{x^2 \sec x}$

In Problems 20–31, find each limit.

20. $\displaystyle\lim_{x\to 0}\left[\dfrac{2}{\sin^2 x} - \dfrac{1}{1 - \cos x}\right]$

21. $\displaystyle\lim_{x\to \pi/2} \dfrac{\sec^2 x}{\sec^2(3x)}$

22. $\displaystyle\lim_{x\to 0^-} x\cot(\pi x)$

23. $\displaystyle\lim_{x\to 0} \dfrac{e^x - e^{-x}}{\sin x}$

24. $\displaystyle\lim_{x\to a} \dfrac{ax - x^2}{a^4 - 2a^3 x + 2ax^3 - x^4}$

25. $\displaystyle\lim_{x\to 0} \dfrac{\tan x + \sec x - 1}{\tan x - \sec x + 1}$

26. $\displaystyle\lim_{x\to 0} \dfrac{\tan x - \sin x}{\sin^3 x}$

27. $\displaystyle\lim_{x\to 0} \dfrac{x - \sin x}{x^3}$

28. $\displaystyle\lim_{x\to 1}\left[\dfrac{2}{x^2 - 1} - \dfrac{1}{x - 1}\right]$

29. $\displaystyle\lim_{x\to \infty} (1 + 4x)^{2/x}$

30. $\displaystyle\lim_{x\to 0^+} (\cot x)^x$

31. $\displaystyle\lim_{x\to 4} \dfrac{x^2 - 16}{x^2 + x - 20}$

In Problems 32 and 33, for each function:

(a) *Use the Intermediate Value Theorem to confirm that a zero exists in the given interval.*

(b) *Use Newton's Method to find c_3, the third approximation to the real zero.*

32. $f(x) = 8x^4 - 2x^2 + 5x - 1$, interval: (0, 1). Let $c_1 = 0.5$.

33. $f(x) = 2 - x + \sin x$, interval: $\left(\dfrac{\pi}{2}, \pi\right)$. Let $c_1 = \dfrac{\pi}{2}$.

34. **(a)** Use the Intermediate Value Theorem to confirm that the function $f(x) = 2\cos x - e^x$ has a zero in the interval (0, 1).

    (b) Use technology with Newton's Method to find c_5, the fifth approximation to the real zero. Use the midpoint of the interval for the first approximation c_1.

Break It Down

*Let's take a closer look at **AP® Practice Problem 9 part (a)** from Section 4.2 on page 300.*

9. A roofer's 13-meter ladder is placed against the wall of a building with its base on level ground. The top of the ladder slips down the wall as the bottom of the ladder slips away from the building at a constant rate of 5 m/s.

(a) At what rate is the top of the ladder moving when it is 5 m from the ground?

Step 1	Identify the underlying structure and related concepts.	There are rates changing with respect to time: the rate the top of the ladder slips down the wall, and the rate the bottom of the ladder moves away from the wall. This is a **related rates** question.
Step 2	Determine the appropriate math rule or procedure.	When possible, the first step in solving a related rates problem should be drawing a picture (see the figure). The ladder is 13 m long. Next, identify the variables needed for the given situation. Let h denote the height of the top of the ladder from the ground and let x denote the distance the bottom of the ladder is from the wall. As the ladder slides down the wall, $\dfrac{dh}{dt}$ equals the rate at which the height is changing and $\dfrac{dx}{dt}$ equals the rate at which the distance from the wall is changing. We are told that $\dfrac{dx}{dt} = 5$ m/s and are asked to find the rate of change of the height $\left(\dfrac{dh}{dt}\right)$ when $h = 5$ m.

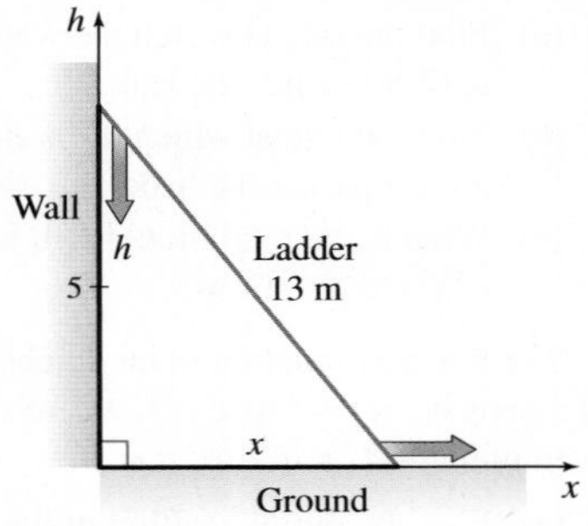

The diagram shows a right triangle with the Wall on the left (vertical side labeled h, with 5 marked), the Ladder (hypotenuse) labeled 13 m, and the Ground (horizontal side labeled x).

Step 3	Apply the math rule or procedure.	Together, the ladder, the height h, and the base x form a right triangle. Using the Pythagorean Theorem, we have $h^2 + x^2 = 13^2$. When $h = 5$ m, we find $x^2 = 169 - 25 = 144$, so $x = 12$ m. The distance x and the height h are each changing over time t, so we will take the derivative of the above expression with respect to t to obtain an equation involving the derivatives of h and x.

$$\frac{d}{dt}(h^2 + x^2) = \frac{d}{dt}(13^2)$$

$$2h\frac{dh}{dt} + 2x\frac{dx}{dt} = 0$$

When $h = 5$ m, $x = 12$ m and $dx/dt = 5$ m/s. After dividing out the 2's, we have

$$5\frac{dh}{dt} + (12)(5) = 0$$

$$\frac{dh}{dt} = -12$$

Step 4	Clearly communicate your answer.	(a) When $x = 5$ m, the height of the ladder is decreasing at a rate of -12 m/s.

AP® Review Problems: Chapter 4

Preparing for the AP® Exam

Multiple-Choice Questions

1. If $y = \tan(3x + 2y)$, find the rate of change of y with respect to x at the origin.

   (A) -3 (B) 1 (C) 3 (D) 5

2. The linear approximation for $f(x) = xe^x$ near $x = 2$ is

   (A) $L(x) = 3e^2(x - 2)$

   (B) $L(x) = xe^x + xe^x(x - 2)$

   (C) $L(x) = 2e^2 + 3e^2(x - 2)$

   (D) $L(x) = 2e^2 + 2e^2(x - 2)$

3. $\displaystyle \lim_{x \to \infty} \frac{x}{\ln x} =$

   (A) 0 (B) 1 (C) e (D) ∞

4. The radius of a circle is increasing at a constant positive rate with respect to time. What is the radius of the circle when the rate of change of the area with respect to time is numerically equal to twice the rate of change of the circumference with respect to time?

   (A) 2 (B) $\dfrac{1}{2}$

   (C) 1 (D) 4

5. $\displaystyle \lim_{x \to 0} \frac{xe^x}{\sin x} =$

   (A) 0 (B) 1 (C) e (D) ∞

6. Use the linear approximation to $f(x) = \tan x$ at $x = 0$ approximate $f(0.2)$.

   (A) -0.2 (B) 0

   (C) 0.2 (D) 0.8

Free-Response Questions

7. Water is being pumped into a cylindrical tank that measures 20 m in height and 4 m in radius at a constant rate of 5 m³/h. Ten meters from the bottom of the tank there is a hole in the tank. Water leaks from that hole at rate of 1 m³/h.

   (a) Find the rate at which the water is rising in the tank until it reaches the leak.

   (b) Find the rate at which the water is rising in the tank after it passes the leak.

   (c) What is the total time it will take for the tank to begin to overflow?

8. The position function of an object moving along the x-axis is $s(t) = t \sin t + 3$, where s in meters and $t \geq 0$ in minutes.

   (a) Find the initial position of the object. Find its position at $t = \dfrac{2\pi}{3}$ min.

   (b) Find and interpret $s'\left(\dfrac{2\pi}{3}\right)$ in the context of this problem.

   (c) Find the average velocity of the object from $t = 0$ to $t = \dfrac{2\pi}{3}$.

AP® Cumulative Review Problems: Chapters 1–4

Preparing for the AP® Exam

Multiple-Choice Questions

1. The graph of f is shown in the figure below. If f is defined at k, but $\lim\limits_{x \to k} f(x)$ does not exist, then k equals

   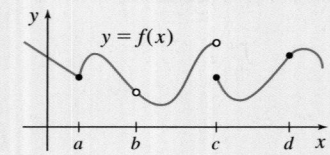

   (A) a (B) b (C) c (D) d

2. The graph of the function $f(x) = \dfrac{x^2 - 9}{2x^2 - 5x - 3}$ has a vertical asymptote at $x = a$ and a horizontal asymptote at $y = b$. What are the values of the numbers a and b?

   (A) $a = -3, b = 2$ (B) $a = -\dfrac{1}{2}, b = \dfrac{1}{2}$

   (C) $a = \dfrac{1}{2}, b = \dfrac{1}{2}$ (D) $a = 3, b = 2$

3. $\dfrac{d}{dx}\tan^{-1}x^2 =$

(A) $\dfrac{2x}{1+x^4}$ (B) $\dfrac{2}{1+x^4}$

(C) $\dfrac{2x}{1+x^2}$ (D) $\dfrac{1}{1+x^4}$

4. Suppose $f(x) = \begin{cases} \dfrac{x^2+x}{x} & \text{if } x \neq 0 \\ 1 & \text{if } x = 0 \end{cases}$.

Which of the following is true?

I $f(0)$ exists.

II $\lim\limits_{x\to 0} f(x)$ exists.

III f is continuous at $x = 0$.

(A) I only   (B) II only

(C) I, II, III   (D) I and II only

5. $\dfrac{d}{dx}\dfrac{\ln x}{x^2} =$

(A) $\dfrac{1-2\ln x}{x}$ (B) $\dfrac{1-2x\ln x}{x^3}$

(C) $\dfrac{1-2\ln x}{x^3}$ (D) $\dfrac{1-2\ln x}{x^4}$

6. If $x^2 - 2xy - y^2 = -1$, then $\dfrac{dy}{dx}$ equals

(A) -1 (B) $\dfrac{-x+y}{x+y}$

(C) $\dfrac{x+y}{-x+y}$ (D) $\dfrac{x-y}{x+y}$

7. If $\lim\limits_{x\to\infty}\dfrac{k}{x^p} = 0$, which combination of p and k is not possible?

(A) $p = -1, k = 1$ (B) $p = 0, k = 0$

(C) $p = 2, k = -1$ (D) $p = 3, k = -2$

8. Suppose the functions f and g are differentiable for all real numbers, $f(0) = \ln 3$, $f'(0) = 2$, $g(0) = -\dfrac{1}{2}$, $g'(0) = \dfrac{1}{2}$. If $h(x) = \dfrac{e^{f(x)}}{[g(x)]^2}$, then $h'(0) =$

(A) -48 (B) 0 (C) 12 (D) 48

9. Suppose the functions f and g are both differentiable, and $y = f(x)$ and $x = g(y)$ are inverse functions. Use the table below, which lists the values of f, f', and g for select numbers x, to find $g'(3)$.

x	$f(x)$	$f'(x)$	$g(x)$
0	3	2	1
2	4	1	3
3	5	3	0

(A) 2   (B) $\dfrac{1}{3}$ (C) $\dfrac{1}{2}$ (D) 1

10. If $\lim\limits_{x\to 1} f(x) = 5$, then $\lim\limits_{x\to 1}\dfrac{(x^3-1)f(x)}{x-1} =$

(A) 5   (B) 15   (C) 0   (D) The limit does not exist.

Free-Response Questions

11. The graph of the derivative of a function f is given below. Also, $f(1) = 3$.

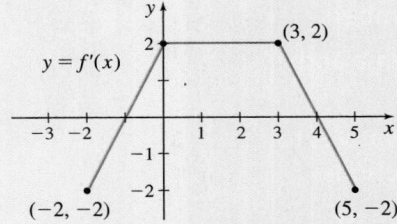

(a) Find an equation of the tangent line to the graph of f at the point $(1, 3)$. Justify your answer.

(b) For what values of x, if any, does the graph of f have a horizontal tangent line? Justify your answer.

(c) Determine whether f is continuous at $x = 0$. Justify your answer.

12. $f(x) = \begin{cases} 2x - 1 & \text{if } -2 \leq x \leq 1 \\ -3x + 4 & \text{if } 1 < x \leq 4 \end{cases}$

(a) Determine whether f is continuous at $x = 1$. Justify your answer.

(b) Determine whether f is differentiable at $x = 1$. Justify your answer.

(c) Find $\lim\limits_{x\to 0}\dfrac{f(x)+\cos x}{e^x + f(x+2)}$. Justify your answer.

CHAPTER

5

Applications of the Derivative, Part 2

Yellow Dog Productions/Getty Images

The U.S. Economy

There are many ways economists collect and report data about the U. S. economy. Two such reports are Unemployment and Gross Domestic Product (GDP). The Unemployment report gives the percentage of people over the age of 16 who are unemployed and are actively looking for a job. The GDP report gives the percent increase (or decrease) in the market value of all goods and services produced over a 3-month period of time, expressed annually. The graphs on the right show Unemployment and GDP in the United States from 2002 to 2021.

 The Chapter 5 Project on page 415 examines the two economic indicators, Unemployment and GDP, and investigates relationships they might have to each other.

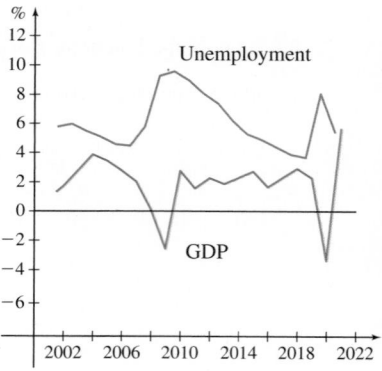

Edwindra
Chief People Officer

AP® Calc taught me how to approach complex problems from many angles and be resilient in finding the answer. I leverage these skills daily as I manage resources and make data-driven decisions about hiring and training.

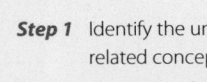

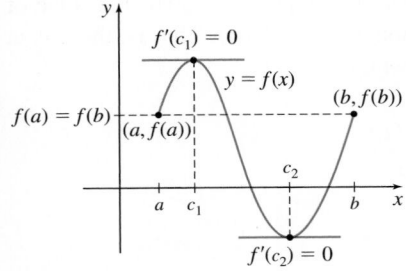

Figure 1

In Chapters 2 and 3, we developed a variety of formulas for finding derivatives. In Chapter 4, we investigated applications of the derivative: giving interpretations to the derivative, finding linear approximations to the graph of a function, analyzing related rate problems, and finding limits using L'Hôpital's Rule.

In this chapter, we continue to explore applications of the derivative. We use the derivative to find optimal (minimum or maximum) values of functions and to investigate properties of the graph of a function. We also introduce the inverse of the derivative operation, the *antiderivative*, and use it to solve problems involving motion on a line.

Before we discuss these applications of the derivative, we need the Mean Value Theorem, the subject of Section 5.1. ∎

5.1 The Mean Value Theorem; The Increasing/Decreasing Function Test

OBJECTIVES *When you finish this section, you should be able to:*

1 Use Rolle's Theorem (p. 333)
2 Use the Mean Value Theorem (p. 334)
3 Determine intervals on which a function is increasing or decreasing (p. 338)

In this section, we prove several theorems. The most significant of these, the *Mean Value Theorem,* is used to develop tests for locating local extreme values. To prove the Mean Value Theorem, we need *Rolle's Theorem.*

1 Use Rolle's Theorem

THEOREM Rolle's Theorem
Suppose f is a function defined on a closed interval $[a, b]$.
If

- f is continuous on $[a, b]$
- f is differentiable on (a, b)
- $f(a) = f(b)$

then there is at least one number c in the open interval (a, b) for which $f'(c) = 0$.

We omit a proof.

Notice that the graph in Figure 1 meets the three conditions of Rolle's Theorem and has at least one number c at which $f'(c) = 0$.

In contrast, the graphs in Figure 2 show that the conclusion of Rolle's Theorem may not hold when one or more of the three conditions are not met.

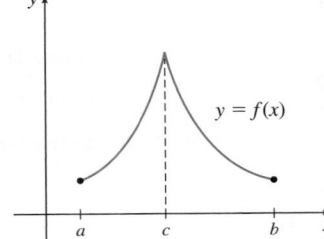

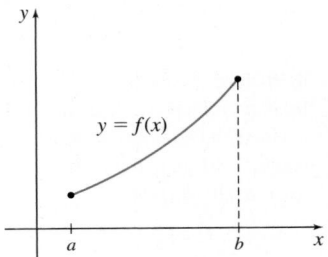

(a) f is defined on $[a, b]$.
f is not continuous at a.
f is differentiable on (a, b).
$f(a) = f(b)$.
No c in (a, b) at which $f'(c) = 0$.

(b) f is defined on $[a, b]$.
f is continuous on $[a, b]$.
f is not differentiable on (a, b), no derivative at c.
$f(a) = f(b)$.
No c in (a, b) at which $f'(c) = 0$.

(c) f is defined on $[a, b]$.
f is continuous on $[a, b]$.
f is differentiable on (a, b).
$f(a) \neq f(b)$.
No c in (a, b) at which $f'(c) = 0$.

Figure 2

ORIGINS Michel Rolle (1652–1719) was a French mathematician whose formal education was limited to elementary school. At age 24, he moved to Paris and married. To support his family, Rolle worked as an accountant, but he also began to study algebra and became interested in the theory of equations. Although Rolle is primarily remembered for the theorem presented here, he was also the first to use the notation $\sqrt[n]{}$ for the nth root.

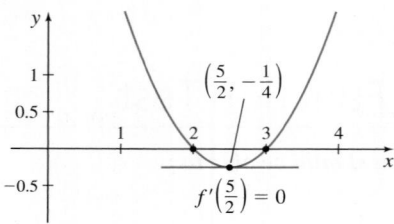

Figure 3 $f(x) = x^2 - 5x + 6$

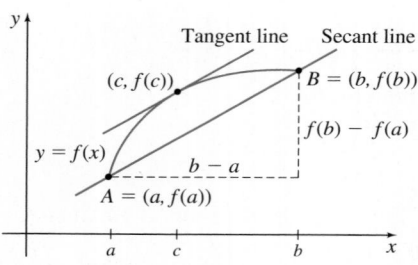

Figure 4 The tangent line at $(c, f(c))$ is parallel to the secant line from A to B.

IN WORDS If a function f is continuous on $[a, b]$ and differentiable on (a, b), then there is a tangent line at c in (a, b) whose slope equals the slope of the secant line from $(a, f(a))$ to $(b, f(b))$.

EXAMPLE 1   Using Rolle's Theorem

Find the x-intercepts of $f(x) = x^2 - 5x + 6$, and show that $f'(c) = 0$ for some number c in the open interval formed by the two x-intercepts. Find c.

Solution

At the x-intercepts, $f(x) = 0$.

$$f(x) = x^2 - 5x + 6 = (x - 2)(x - 3) = 0$$

So, $x = 2$ and $x = 3$ are the x-intercepts of the graph of f, and $f(2) = f(3) = 0$.

Since f is a polynomial, it is continuous on the closed interval $[2, 3]$ formed by the x-intercepts and is differentiable on the open interval $(2, 3)$. The three conditions of Rolle's Theorem are satisfied, guaranteeing that there is a number c in the open interval $(2, 3)$ for which $f'(c) = 0$. Since $f'(x) = 2x - 5$, the number c for which $f'(x) = 0$ is $c = \dfrac{5}{2}$. ∎

See Figure 3 for the graph of f.

NOW WORK Problem 9.

❷ Use the Mean Value Theorem

The importance of Rolle's Theorem is that it can be used to obtain other results, many of which have wide-ranging application. Perhaps the most important of these is the *Mean Value Theorem*, sometimes called the *Theorem of the Mean for Derivatives*.

A geometric example helps explain the Mean Value Theorem. Consider the graph of a function f that is continuous on a closed interval $[a, b]$ and differentiable on the open interval (a, b), as illustrated in Figure 4. Then there is at least one number c between a and b at which the slope $f'(c)$ of the tangent line to the graph of f equals the slope of the secant line joining the points $A = (a, f(a))$ and $B = (b, f(b))$. That is, there is at least one number c in the open interval (a, b) for which

$$f'(c) = \frac{f(b) - f(a)}{b - a}$$

THEOREM Mean Value Theorem

Suppose f is a function defined on a closed interval $[a, b]$.
If

* f is continuous on $[a, b]$
* f is differentiable on (a, b)

then there is at least one number c in the open interval (a, b) for which

$$\boxed{f'(c) = \frac{f(b) - f(a)}{b - a}} \tag{1}$$

Proof The slope $m_{\sec}$ of the secant line containing the points $(a, f(a))$ and $(b, f(b))$ is

$$m_{\sec} = \frac{f(b) - f(a)}{b - a}$$

Then using the point $(a, f(a))$ and the point-slope form of an equation of a line, an equation of this secant line is

$$y - f(a) = \frac{f(b) - f(a)}{b - a}(x - a)$$

$$y = f(a) + \frac{f(b) - f(a)}{b - a}(x - a)$$

NOTE The function g has a geometric significance: Its value equals the vertical distance from the graph of $y = f(x)$ to the secant line joining $(a, f(a))$ to $(b, f(b))$. See Figure 5.

Now construct a function g that satisfies the conditions of Rolle's Theorem, by defining g as

$$g(x) = f(x) - \underbrace{\left[f(a) + \frac{f(b) - f(a)}{b - a}(x - a) \right]}_{\substack{\text{Height of the secant line} \\ \text{containing } (a, f(a)) \text{ and } (b, f(b))}} \quad (2)$$

for all x on $[a, b]$.

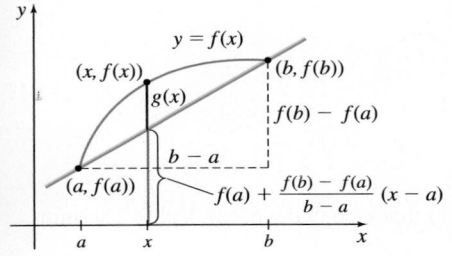

Figure 5

Since f is continuous on $[a, b]$ and differentiable on (a, b), it follows that g is also continuous on $[a, b]$ and differentiable on (a, b). Also,

$$g(a) = f(a) - \left[f(a) + \frac{f(b) - f(a)}{b - a}(a - a) \right] = 0$$

$$g(b) = f(b) - \left[f(a) + \frac{f(b) - f(a)}{b - a}(b - a) \right] = 0$$

Now, since g satisfies the three conditions of Rolle's Theorem, there is a number c in (a,b) at which $g'(c) = 0$. Since $f(a)$ and $\dfrac{f(b) - f(a)}{b - a}$ are constants, the derivative of g, from equation (2) above, is given by

$$g'(x) = f'(x) - \frac{d}{dx}\left[f(a) + \frac{f(b) - f(a)}{b - a}(x - a) \right] = f'(x) - \frac{f(b) - f(a)}{b - a}$$

Now evaluate g' at c and use the fact that $g'(c) = 0$.

$$g'(c) = f'(c) - \left[\frac{f(b) - f(a)}{b - a} \right] = 0$$

$$f'(c) = \frac{f(b) - f(a)}{b - a} \qquad\blacksquare$$

The Mean Value Theorem, like the Intermediate Value Theorem discussed in Section 1.3, p. 115, is an example of an *existence theorem*. It does not tell us how to find the number c; it merely states that at least one number c exists. Often, the number c can be found.

CALC CLIP **EXAMPLE 2** **Verifying the Mean Value Theorem**

(a) Verify that the function $f(x) = x^3 - 3x + 5$, $-1 \leq x \leq 1$ satisfies the conditions of the Mean Value Theorem.

(b) Find the number(s) c guaranteed by the Mean Value Theorem.

(c) Interpret the number(s) geometrically.

Solution

(a) The function f is a polynomial function, so f is continuous on the closed interval $[-1, 1]$ and differentiable on the open interval $(-1, 1)$. The conditions of the Mean Value Theorem are met.

(b) Evaluate f at the endpoints -1 and 1 and then find $f'(x)$.

$$f(-1) = 7 \qquad f(1) = 3 \qquad \text{and} \qquad f'(x) = 3x^2 - 3$$

The number(s) c in the open interval $(-1, 1)$ guaranteed by the Mean Value Theorem satisfy the equation

$$f'(c) = \frac{f(1) - f(-1)}{1 - (-1)}$$

$$3c^2 - 3 = \frac{3 - 7}{1 - (-1)} = \frac{-4}{2} = -2$$

$$3c^2 = 1$$

$$c = \sqrt{\frac{1}{3}} = \frac{\sqrt{3}}{3} \qquad \text{or} \qquad c = -\sqrt{\frac{1}{3}} = -\frac{\sqrt{3}}{3}$$

There are two numbers in the interval $(-1, 1)$ that satisfy the Mean Value Theorem.

(c) At each number, $-\dfrac{\sqrt{3}}{3}$ and $\dfrac{\sqrt{3}}{3}$, the slope of the tangent line to the graph of f equals the slope of the secant line connecting the points $(-1, 7)$ and $(1, 3)$. That is,

$$m_{\tan} = f'\left(-\frac{\sqrt{3}}{3}\right) = f'\left(\frac{\sqrt{3}}{3}\right) = -2 \quad \text{and} \quad m_{\sec} = \frac{f(1) - f(-1)}{1 - (-1)} = \frac{3 - 7}{2} = -2$$

The tangent lines and the secant line are parallel. ∎

See Figure 6.

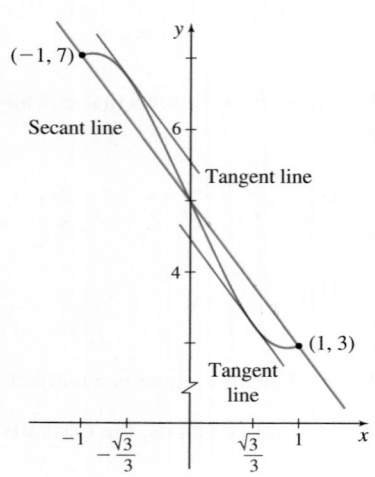

Figure 6 $f(x) = x^3 - 3x + 5$, $-1 \leq x \leq 1$

NOW WORK   Problem 25 and AP® Practice Problems 1 and 3.

The Mean Value Theorem can be applied to motion on a line. Suppose the position function $s = f(t)$ models the signed distance s of an object from the origin at time t. If f is continuous on $[a, b]$ and differentiable on (a, b), the Mean Value Theorem tells us that there is at least one time t at which the velocity of the object equals its average velocity.

Let t_1 and t_2 be two distinct times in $[a, b]$. Then

$$\text{Average velocity over } [t_1, t_2] = \frac{f(t_2) - f(t_1)}{t_2 - t_1}$$

$$\text{Instantaneous velocity } v(t) = \frac{ds}{dt} = f'(t)$$

The Mean Value Theorem states there is a time t_0 in the interval (t_1, t_2) for which

$$f'(t_0) = \frac{f(t_2) - f(t_1)}{t_2 - t_1}$$

That is, the velocity at time t_0 equals the average velocity over the interval $[t_1, t_2]$.

EXAMPLE 3   Applying the Mean Value Theorem to Motion on a Line

Use the Mean Value Theorem to show that a car that travels 110 mi in 2 h must have had a speed of 55 mi/h at least once during the 2 h.

Solution

Let $s = f(t)$ represent the distance s the car has traveled after t hours. Its average velocity during the time period from 0 to 2 h is

$$\text{Average velocity} = \frac{f(2) - f(0)}{2 - 0} = \frac{110 - 0}{2 - 0} = 55 \text{ mi/h}$$

Using the Mean Value Theorem, there is a time t_0, $0 < t_0 < 2$, at which

$$f'(t_0) = \frac{f(2) - f(0)}{2 - 0} = 55$$

That is, the car had a velocity of 55 mi/h at least once during the 2 hour period. ■

NOW WORK   Problem 57.

The following corollaries are consequences of the Mean Value Theorem.

> **COROLLARY**
>
> If a function f is continuous on the closed interval $[a, b]$ and is differentiable on the open interval (a, b), and if $f'(x) = 0$ for all numbers x in (a, b), then f is constant on (a, b).

IN WORDS If $f'(x) = 0$ for all numbers in (a, b), then f is a constant function on the interval (a, b).

Proof To show that f is a constant function, we must show that $f(x_1) = f(x_2)$ for any two numbers x_1 and x_2 in the interval (a, b).

Suppose $x_1 < x_2$. The conditions of the Mean Value Theorem are satisfied on the interval $[x_1, x_2]$, so there is a number c in this interval for which

$$f'(c) = \frac{f(x_2) - f(x_1)}{x_2 - x_1}$$

Since $f'(c) = 0$ for all numbers in (a, b), we have

$$0 = \frac{f(x_2) - f(x_1)}{x_2 - x_1}$$

$$0 = f(x_2) - f(x_1)$$

$$f(x_1) = f(x_2) \qquad ■$$

IN WORDS If an object moving on a line has velocity $v = 0$, then the object is at rest.

Applying the corollary to motion on a line, if the velocity v of an object is zero over an interval (t_1, t_2), then the object does not move during this time interval.

NOW WORK   AP® Practice Problems 7 and 9.

COROLLARY

If the functions f and g are differentiable on an open interval (a, b) and if $f'(x) = g'(x)$ for all numbers x in (a, b), then there is a number C for which $f(x) = g(x) + C$ on (a, b).

IN WORDS If two functions f and g have the same derivative, then the functions differ by a constant, and the graph of f is a vertical shift of the graph of g.

Proof We define the function h so that $h(x) = f(x) - g(x)$. Then h is differentiable since it is the difference of two differentiable functions, and

$$h'(x) = f'(x) - g'(x) = 0$$

Since $h'(x) = 0$ for all numbers x in the interval (a, b), h is a constant function and we can write

$$h(x) = f(x) - g(x) = C \qquad \text{for some number } C$$

That is,

$$f(x) = g(x) + C$$

∎

③ Determine Intervals on Which a Function Is Increasing or Decreasing

NEED TO REVIEW? Increasing and decreasing functions are discussed in Section P.1, pp. 13.

A third corollary that follows from the Mean Value Theorem gives us a way to determine where a function is increasing and where a function is decreasing. Recall that functions are increasing or decreasing on *intervals,* either open or closed or half-open or half-closed.

COROLLARY Increasing/Decreasing Function Test

Suppose f is a function that is differentiable on the open interval (a, b):

- If $f'(x) > 0$ on (a, b), then f is increasing on (a, b).

- If $f'(x) < 0$ on (a, b), then f is decreasing on (a, b).

The proof of the first part of the corollary is given here. The proof of the second part is left as an exercise. See Problem 82.

Proof Since f is differentiable on (a, b), it is continuous on (a, b). To show that f is increasing on (a, b), choose two numbers x_1 and x_2 in (a, b), with $x_1 < x_2$. We need to show that $f(x_1) < f(x_2)$.

Since x_1 and x_2 are in (a, b), f is continuous on the closed interval $[x_1, x_2]$ and differentiable on the open interval (x_1, x_2). By the Mean Value Theorem, there is a number c in the interval (x_1, x_2) for which

$$f'(c) = \frac{f(x_2) - f(x_1)}{x_2 - x_1}$$

Since $f'(x) > 0$ for all x in (a, b), it follows that $f'(c) > 0$. Since $x_1 < x_2$, then $x_2 - x_1 > 0$. As a result,

$$f(x_2) - f(x_1) > 0$$
$$f(x_1) < f(x_2)$$

So, f is increasing on (a, b). ∎

There are a few important observations to make about the Increasing/Decreasing Function Test:

- In using the Increasing/Decreasing Function Test, we determine open intervals on which the derivative is positive or negative.
- Suppose f is continuous on the closed interval $[a, b]$ and differentiable on the open interval (a, b).
  - If $f'(x) > 0$ on (a, b), then f is increasing on the closed interval $[a, b]$.
  - If $f'(x) < 0$ on (a, b), then f is decreasing on the closed interval $[a, b]$.
- The Increasing/Decreasing Function Test is valid if the interval (a, b) is $(-\infty, b)$ or (a, ∞) or $(-\infty, \infty)$.

EXAMPLE 4 Determining Intervals on Which a Function Is Increasing or Decreasing

Determine where the function $f(x) = 2x^3 - 9x^2 + 12x - 5$ is increasing and where it is decreasing.

Solution

The function f is a polynomial, so f is continuous and differentiable at every real number. We find f'.

$$f'(x) = 6x^2 - 18x + 12 = 6(x - 2)(x - 1)$$

The Increasing/Decreasing Function Test states that f is increasing on intervals where $f'(x) > 0$ and that f is decreasing on intervals where $f'(x) < 0$. We solve these inequalities by using the numbers 1 and 2 to form three intervals on the x-axis, as shown in Figure 7. Then we determine the sign of $f'(x)$ on each interval, as shown in Table 1.

NEED TO REVIEW? Solving inequalities is discussed in Appendix A.1, pp. A-6 to A-8.

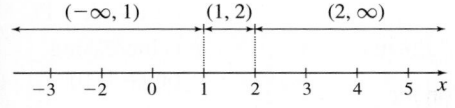

Figure 7

TABLE 1

Interval	Sign of $x - 1$	Sign of $x - 2$	Sign of $f'(x) = 6(x - 2)(x - 1)$	Conclusion
$(-\infty, 1)$	Negative ($-$)	Negative ($-$)	Positive ($+$)	f is increasing
$(1, 2)$	Positive ($+$)	Negative ($-$)	Negative ($-$)	f is decreasing
$(2, \infty)$	Positive ($+$)	Positive ($+$)	Positive ($+$)	f is increasing

We conclude that f is increasing on the intervals $(-\infty, 1)$ and $(2, \infty)$, and f is decreasing on the interval $(1, 2)$. Since f is continuous on its domain, f is increasing on the intervals $(-\infty, 1]$ and $[2, \infty)$ and is decreasing on the interval $[1, 2]$. ∎

Figure 8 shows the graph of f. Notice that the graph of f has horizontal tangent lines at the points $(1, 0)$ and $(2, -1)$.

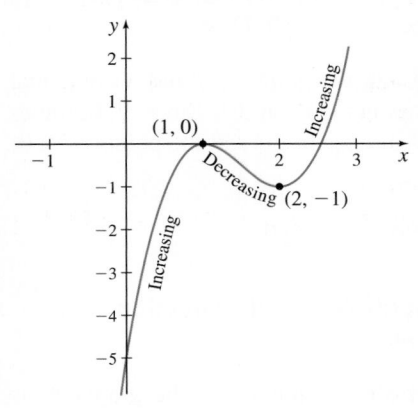

Figure 8 $f(x) = 2x^3 - 9x^2 + 12x - 5$

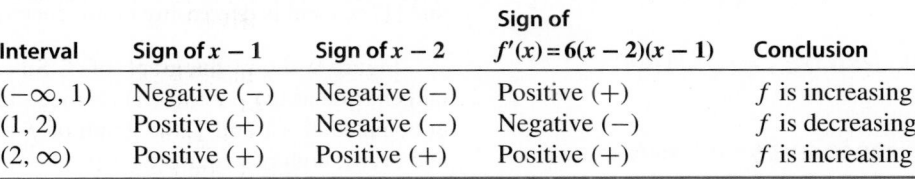

NOW WORK Problem **31** and AP® Practice Problem **5**.

EXAMPLE 5 **Determining Intervals on Which a Function Is Increasing or Decreasing**

Determine where the function $f(x) = (x^2 - 1)^{2/3}$ is increasing and where it is decreasing.

Solution

f is continuous for all real numbers x, and

$$f'(x) = \frac{2}{3}(x^2 - 1)^{-1/3}(2x) = \frac{4x}{3(x^2 - 1)^{1/3}}$$

Note that f' does not exist at -1 and 1, so the domain of f' consists of all real numbers except -1 and 1. Also, $f'(x) = 0$ at $x = 0$.

The Increasing/Decreasing Function Test states that f is increasing on intervals where $f'(x) > 0$ and decreasing on intervals where $f'(x) < 0$. We solve these inequalities by using the numbers at which f' does not exist and at which f' equals 0, namely, -1, 0, and 1, to form four intervals on the x-axis: $(-\infty, -1)$, $(-1, 0)$, $(0, 1)$, and $(1, \infty)$.

Then we determine the sign of f' in each interval. We conclude that f is increasing on the intervals $(-1, 0)$ and $(1, \infty)$ and that f is decreasing on the intervals $(-\infty, -1)$ and $(0, 1)$. See Table 2.

TABLE 2

Interval	Sign of $4x$	Sign of $(x^2 - 1)^{1/3}$	Sign of $f'(x) = \dfrac{4x}{3(x^2 - 1)^{1/3}}$	Conclusion
$(-\infty, -1)$	Negative $(-)$	Positive $(+)$	Negative $(-)$	f is decreasing on $(-\infty, -1)$
$(-1, 0)$	Negative $(-)$	Negative $(-)$	Positive $(+)$	f is increasing on $(-1, 0)$
$(0, 1)$	Positive $(+)$	Negative $(-)$	Negative $(-)$	f is decreasing on $(0, 1)$
$(1, \infty)$	Positive $(+)$	Positive $(+)$	Positive $(+)$	f is increasing on $(1, \infty)$

Since f is continuous on its domain, f is increasing on the intervals $[-1, 0]$ and $[1, \infty)$ and is decreasing on the intervals $(-\infty, -1]$ and $[0, 1]$. ∎

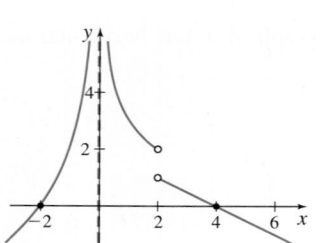

Figure 9 $f(x) = (x^2 - 1)^{2/3}$

NEED TO REVIEW? The term "cusp" is defined in Section 2.2, p. 182.

Figure 9 shows the graph of f. Since $f'(0) = 0$, the graph of f has a horizontal tangent line at the point $(0, 1)$. Notice that f' does not exist at ± 1. Since f' becomes unbounded at -1 and 1, the graph of f has vertical tangent lines at the points $(-1, 0)$ and $(1, 0)$, both of which are cusps.

NOW WORK Problem 37 and AP® Practice Problems 4, 8, 10, and 13.

EXAMPLE 6 **Obtaining Information About the Graph of a Function from the Graph of Its Derivative**

Suppose f is a function that is continuous for all real numbers. The graph of its derivative function f' is shown in Figure 10.

(a) What is the domain of the derivative function f''?

(b) At what numbers x, if any, does the graph of f have a horizontal tangent line?

(c) At what numbers x, if any, does the graph of f have a vertical tangent line?

(d) At what numbers x, if any, does the graph of f have a corner?

(e) On what intervals is the graph of f increasing?

(f) On what intervals is the graph of f decreasing?

Figure 10 The graph of $y = f'(x)$

Solution

(a) The domain of f' is the set of all real numbers except 0 and 2.

(b) The graph of f has a horizontal tangent line when its derivative $f'(x)$ is zero, namely, at $x = -2$ and $x = 4$.

(c) The graph of f has a vertical tangent line where the derivative f' is unbounded. This occurs at $x = 0$.

(d) From the graph of f', we see that the derivative has unequal one-sided limits at $x = 2$. As x approaches 2 from the left, the derivative is 2, and as x approaches 2 from the right, the derivative is 1. This means the graph of f has a corner at $x = 2$.

NEED TO REVIEW? The term "corner" is defined in Section 2.2, p.182.

(e) The graph of f is increasing when $f'(x) > 0$; that is, when the graph of f' is above the x-axis. This occurs on the intervals $(-2, 0)$, $(0, 2)$, and $(2, 4)$. Since f is continuous for all real numbers, f is increasing on $[-2, 4]$.

(f) The graph of f is decreasing when $f'(x) < 0$; that is, when the graph of f' is below the x-axis. This occurs on the intervals $(-\infty, -2)$ and $(4, \infty)$. Since f is continuous for all x, f is decreasing on $(-\infty, -2]$ and $[4, \infty)$. ■

NOW WORK   Problem **47** and AP® Practice Problems **2, 6,** and **14.**

EXAMPLE 7   Investigating Motion on a Line

An object moves along the y-axis with velocity $v = v(t)$, where t is measured in hours and $v(t)$ is measured in meters per hour. The velocity function v is differentiable for $t > 0$, and the object is at the origin when $t = 0$. Table 3 below shows values of $v(t)$ at select times t.

TABLE 3

t	0	0.2	0.8	1.5	2.0	3.8
$v(t)$	0	15	5	-10	5	28

(a) Show that there are at least two times t, $0.2 < t < 3.8$, at which the velocity v of the object is 0.

(b) Show that there is at least one time t, $0.8 < t < 2.0$, at which the acceleration of the object is 0.

Solution

(a) Since $v = v(t)$ is differentiable for $t > 0$, the function v is continuous on the interval $[0.2, 3.8]$. Using the Intermediate Value Theorem, since $v(0.8) > 0$ and $v(1.5) < 0$, there is at least one t in the interval $(0.8, 1.5)$ for which $v(t) = 0$. Similarly, since $v(1.5) < 0$ and $v(2.0) > 0$, there is at least one time t in the interval $(1.5, 2.0)$ for which $v(t) = 0$. So, there are at least two times t, $0.2 < t < 3.8$, at which the velocity of the object is 0.

NOTE Part (b) can also be solved using Rolle's Theorem. Since $v = v(t)$ is continuous on the closed interval $[0.8, 2.0]$ and differentiable on the open interval $(0.8, 2.0)$, and $v(0.8) = v(2.0) = 5$, then $v'(t) = 0$ for some t in $(0.8, 2.0)$.

(b) Since $v = v(t)$ is differentiable for $t > 0$, the function v is continuous on the interval $[0.8, 2.0]$. Since

$$\frac{v(2.0) - v(0.8)}{2.0 - 0.8} = \frac{5 - 5}{1.2} = 0$$

by the Mean Value Theorem, there is at least one time t in the interval $(0.8, 2.0)$ for which the acceleration $a = v'(t) = 0$. ■

NOW WORK   AP® Practice Problems **11** and **12.**

Application: Agricultural Economics

EXAMPLE 8   **Determining Crop Yield**

A variation of the von Liebig* model states that the yield $f(x)$ of a plant, measured in bushels, responds to the amount x of potassium in a fertilizer according to the following square root model:

$$f(x) = -0.057 - 0.417x + 0.852\sqrt{x}$$

For what amounts of potassium will the yield increase? For what amounts of potassium will the yield decrease?

Solution

The yield is increasing when $f'(x) > 0$.

$$f'(x) = -0.417 + \frac{0.426}{\sqrt{x}} = \frac{-0.417\sqrt{x} + 0.426}{\sqrt{x}}$$

Now $f'(x) > 0$ when

$$-0.417\sqrt{x} + 0.426 > 0$$
$$0.417\sqrt{x} < 0.426$$
$$\sqrt{x} < 1.022$$
$$x < 1.044$$

The crop yield is increasing when the amount of potassium in the fertilizer is less than 1.044 and is decreasing when the amount of potassium in the fertilizer is greater than 1.044. ∎

---

*__Source:__ Quirino Paris. (1992), The von Liebig Hypothesis, *American Journal of Agricultural Economics*, 74(4), 1019–1028.

5.1 Assess Your Understanding

Concepts and Vocabulary

1. **True or False**   If a function f is defined and continuous on a closed interval $[a, b]$, differentiable on the open interval (a, b), and if $f(a) = f(b)$, then Rolle's Theorem guarantees that there is at least one number c in the interval (a, b) for which $f(c) = 0$.

2. In your own words, give a geometric interpretation of the Mean Value Theorem.

3. **True or False**   If two functions f and g are differentiable on an open interval (a, b) and if $f'(x) = g'(x)$ for all numbers x in (a, b), then f and g differ by a constant on (a, b).

4. **True or False**   When the derivative f' is positive on an open interval I, then f is positive on I.

Skill Building

In Problems 5–16, verify that each function satisfies the three conditions of Rolle's Theorem on the given interval. Then find all numbers c in (a, b) guaranteed by Rolle's Theorem.

5. $f(x) = x^2 - 3x$ on $[0, 3]$

6. $f(x) = x^2 + 2x$ on $[-2, 0]$

7. $g(x) = x^2 - 2x - 2$ on $[0, 2]$

8. $g(x) = x^2 + 1$ on $[-1, 1]$

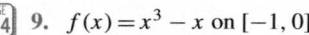 9. $f(x) = x^3 - x$ on $[-1, 0]$

10. $f(x) = x^3 - 4x$ on $[-2, 2]$

11. $f(t) = t^3 - t + 2$ on $[-1, 1]$

12. $f(t) = t^4 - 3$ on $[-2, 2]$

13. $s(t) = t^4 - 2t^2 + 1$ on $[-2, 2]$

14. $s(t) = t^4 + t^2$ on $[-2, 2]$

15. $f(x) = \sin(2x)$ on $[0, \pi]$

16. $f(x) = \sin x + \cos x$ on $[0, 2\pi]$

In Problems 17–20, state why Rolle's Theorem cannot be used with the function f on the given interval.

17. $f(x) = x^2 - 2x + 1$ on $[-2, 1]$

18. $f(x) = x^3 - 3x$ on $[2, 4]$

19. $f(x) = x^{1/3} - x$ on $[-1, 1]$

20. $f(x) = x^{2/5}$ on $[-1, 1]$

In Problems 21–30,

(a) *Verify that each function satisfies the conditions of the Mean Value Theorem on the indicated interval.*

(b) *Find the number(s) c guaranteed by the Mean Value Theorem.*

(c) *Interpret the number(s) c geometrically.*

21. $f(x) = x^2 + 1$ on $[0, 2]$

22. $f(x) = x + 2 + \dfrac{3}{x-1}$ on $[2, 7]$

23. $f(x) = \ln \sqrt{x}$ on $[1, e]$

24. $f(x) = xe^x$ on $[0, 1]$

25. $f(x) = x^3 - 5x^2 + 4x - 2$ on $[1, 3]$

26. $f(x) = x^3 - 7x^2 + 5x$ on $[-2, 2]$

27. $f(x) = \dfrac{x+1}{x}$ on $[1, 3]$

28. $f(x) = \dfrac{x^2}{x+1}$ on $[0, 1]$

29. $f(x) = \sqrt[3]{x^2}$ on $[1, 8]$

30. $f(x) = \sqrt{x-2}$ on $[2, 4]$

In Problems 31–46, determine where each function is increasing and where each is decreasing.

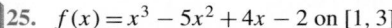

31. $f(x) = x^3 + 6x^2 + 12x + 1$

32. $f(x) = -x^3 + 3x^2 + 4$

33. $f(x) = x^3 - 3x + 1$

34. $f(x) = x^3 - 6x^2 - 3$

35. $f(x) = x^4 - 4x^2 + 1$

36. $f(x) = x^4 + 4x^3 - 2$

37. $f(x) = x^{2/3}(x^2 - 4)$

38. $f(x) = x^{1/3}(x^2 - 7)$

39. $f(x) = |x^3 + 3|$

40. $f(x) = |x^2 - 4|$

41. $f(x) = 3 \sin x$ on $[0, 2\pi]$

42. $f(x) = \cos(2x)$ on $[0, 2\pi]$

43. $f(x) = xe^x$

44. $g(x) = x + e^x$

45. $f(x) = e^x \sin x,\ 0 \le x \le 2\pi$

46. $f(x) = e^x \cos x,\ 0 \le x \le 2\pi$

In Problems 47–50, the graph of the derivative function f' of a function f that is continuous for all real numbers is given.

(a) *What is the domain of the derivative function f'?*

(b) *At what numbers x, if any, does the graph of f have a horizontal tangent line?*

(c) *At what numbers x, if any, does the graph of f have a vertical tangent line?*

(d) *At what numbers x, if any, does the graph of f have a corner?*

(e) *On what intervals is the graph of f increasing?*

(f) *On what intervals is the graph of f decreasing?*

47.

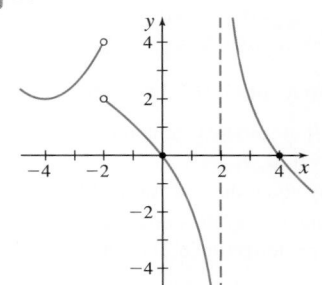

48.

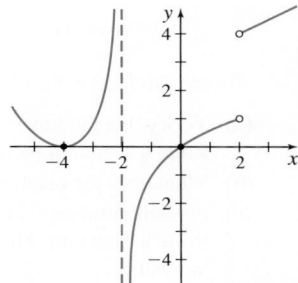

49.

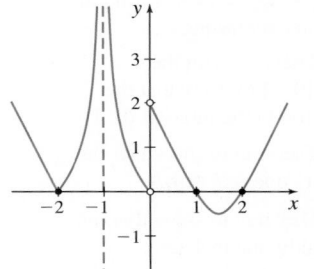

50.

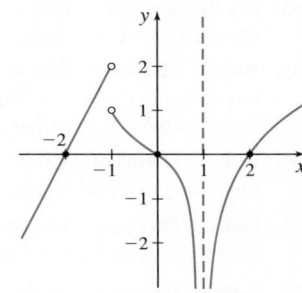

Applications and Extensions

51. Show that the function $f(x) = 2x^3 - 6x^2 + 6x - 5$ is increasing for all x.

52. Show that the function $f(x) = x^3 - 3x^2 + 3x$ is increasing for all x.

53. Show that the function $f(x) = \dfrac{x}{x+1}$ is increasing on any interval not containing $x = -1$.

54. Show that the function $f(x) = \dfrac{x+1}{x}$ is decreasing on any interval not containing $x = 0$.

55. Mean Value Theorem Draw the graph of a function f that is continuous on $[a, b]$ but is not differentiable on (a, b), and for which the conclusion of the Mean Value Theorem does not hold.

56. Mean Value Theorem Draw the graph of a function f that is differentiable on (a, b) but is not continuous on $[a, b]$, and for which the conclusion of the Mean Value Theorem does not hold.

57. Motion on a Line An automobile travels 20 mi down a straight road at an average velocity of 40 mi/h. Show that the automobile must have a velocity of exactly 40 mi/h at some time during the trip. (Assume that the position function is differentiable.)

58. Motion on a Line Suppose a car is traveling on a highway. At 4:00 p.m., the car's speedometer reads 40 mi/h. At 4:12 p.m., it reads 60 mi/h. Show that at some time between 4:00 p.m. and 4:12 p.m., the acceleration was exactly 100 mi/h.

59. Motion on a Line Two stock cars start a race at the same time and finish in a tie. If $f_1(t)$ is the position of one car at time t and $f_2(t)$ is the position of the second car at time t, show that at some time during the race they have the same velocity. *Hint:* Set $f(t) = f_2(t) - f_1(t)$.

60. Motion on a Line Suppose $s = f(t)$ is the position of an object from the origin at time t. If the object is at a specific location at $t = a$, and returns to that location at $t = b$, then $f(a) = f(b)$. Show that there is at least one time $t = c$, $a < c < b$, for which $f'(c) = 0$. That is, show that there is a time c when the velocity of the object is 0.

61. Loaded Beam The vertical deflection d (in feet), of a particular 5-foot-long loaded beam can be approximated by

$$d = d(x) = -\frac{1}{192}x^4 + \frac{25}{384}x^3 - \frac{25}{128}x^2$$

where x (in feet) is the distance from one end of the beam.

(a) Verify that the function $d = d(x)$ satisfies the conditions of Rolle's Theorem on the interval $[0, 5]$.

(b) What does the result in (a) say about the ends of the beam?

(c) Find all numbers c in $(0, 5)$ that satisfy the conclusion of Rolle's Theorem. Then find the deflection d at each number c.

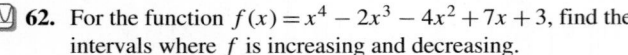 (d) Graph the function d on the interval $[0, 5]$.

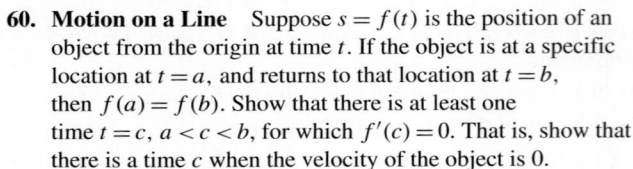 **62.** For the function $f(x) = x^4 - 2x^3 - 4x^2 + 7x + 3$, find the intervals where f is increasing and decreasing.

63. Rolle's Theorem Use Rolle's Theorem with the function $f(x) = (x - 1)\sin x$ on $[0, 1]$ to show that the equation $\tan x + x = 1$ has a solution in the interval $(0, 1)$.

64. Rolle's Theorem Use Rolle's Theorem to show that the function $f(x) = x^3 - 2$ has exactly one real zero.

65. Rolle's Theorem Use Rolle's Theorem to show that the function $f(x) = (x - 8)^3$ has exactly one real zero.

66. Rolle's Theorem Without finding the derivative, show that if $f(x) = (x^2 - 4x + 3)(x^2 + x + 1)$, then $f'(x) = 0$ for at least one number between 1 and 3. Check by finding the derivative and using the Intermediate Value Theorem.

67. Rolle's Theorem Consider $f(x) = |x|$ on the interval $[-1, 1]$. Here $f(1) = f(-1) = 1$ but there is no c in the interval $(-1, 1)$ at which $f'(c) = 0$. Explain why this does not contradict Rolle's Theorem.

68. Mean Value Theorem Consider $f(x) = x^{2/3}$ on the interval $[-1, 1]$. Verify that there is no c in $(-1, 1)$ for which

$$f'(c) = \frac{f(1) - f(-1)}{1 - (-1)}$$

Explain why this does not contradict the Mean Value Theorem.

69. Mean Value Theorem The Mean Value Theorem guarantees that there is a real number N in the interval $(0, 1)$ for which $f'(N) = f(1) - f(0)$ if f is continuous on the interval $[0, 1]$ and differentiable on the interval $(0, 1)$. Find N if $f(x) = \sin^{-1} x$.

70. Mean Value Theorem Show that when the Mean Value Theorem is applied to the function $f(x) = Ax^2 + Bx + C$ in the interval $[a, b]$, the number c referred to in the theorem is the midpoint of the interval.

71. (a) Apply the Increasing/Decreasing Function Test to the function $f(x) = \sqrt{x}$. What do you conclude?

(b) Is f increasing on the interval $[0, \infty)$? Explain.

72. Explain why the function $f(x) = ax^4 + bx^3 + cx^2 + dx + e$ must have a zero between 0 and 1 if

$$\frac{a}{5} + \frac{b}{4} + \frac{c}{3} + \frac{d}{2} + e = 0$$

73. Put It Together If $f'(x)$ and $g'(x)$ exist and $f'(x) > g'(x)$ for all real x, then which of the following statements must be true about the graph of $y = f(x)$ and the graph of $y = g(x)$?

(a) They intersect exactly once.

(b) They intersect no more than once.

(c) They do not intersect.

(d) They could intersect more than once.

(e) They have a common tangent at each point of intersection.

74. Prove that there is no k for which the function

$$f(x) = x^3 - 3x + k$$

has two distinct zeros in the interval $[0, 1]$.

75. Show that $e^x > x^2$ for all $x > 0$. *Hint:* Show that $f(x) = e^x - x^2$ is an increasing function for $x > 0$.

76. Show that $e^x > 1 + x$ for all $x > 0$.

77. Show that $0 < \ln x < x$ for $x > 1$.

78. Show that $\tan \theta \geq \theta$ for all θ in the open interval $\left(0, \frac{\pi}{2}\right)$.

79. Establish the identity $\sin^{-1} x + \cos^{-1} x = \frac{\pi}{2}$ by showing that the derivative of $y = \sin^{-1} x + \cos^{-1} x$ is 0. Then use the fact that when $x = 0$, then $y = \frac{\pi}{2}$.

80. Establish the identity $\tan^{-1} x + \cot^{-1} x = \frac{\pi}{2}$ by showing that the derivative of $y = \tan^{-1} x + \cot^{-1} x$ is 0. Then use the fact that when $x = 1$, then $y = \frac{\pi}{2}$.

81. Let f be a function that is continuous on the closed interval $[a, b]$ and differentiable on the open interval (a, b). If $f(x) = 0$ for three different numbers x in (a, b), show that there must be at least two numbers in (a, b) at which $f'(x) = 0$.

82. Proof of the Increasing/Decreasing Function Test Let f be a function that is differentiable on an open interval (a, b). Show that if $f'(x) < 0$ for all numbers in (a, b), then f is a decreasing function on (a, b). (See the Corollary on p. 338.)

Challenge Problems

83. Find where the general cubic $f(x) = ax^3 + bx^2 + cx + d$ is increasing and where it is decreasing by considering cases depending on the value of $b^2 - 3ac$. *Hint:* $f'(x)$ is a quadratic function; examine its discriminant.

84. Explain why the function $f(x) = x^n + ax + b$, where n is a positive even integer, has at most two distinct real zeros.

85. Explain why the function $f(x) = x^n + ax + b$, where n is a positive odd integer, has at most three distinct real zeros.

86. Explain why the function $f(x) = x^n + ax^2 + b$, where n is a positive odd integer, has at most three distinct real zeros.

87. Explain why the function $f(x) = x^n + ax^2 + b$, where n is a positive even integer, has at most four distinct real zeros.

88. Use Rolle's Theorem to show that between any two real zeros of a polynomial function f, there is a real zero of its derivative function f'.

89. Given $f(x) = \dfrac{ax^n + b}{cx^n + d}$, where $n \geq 2$ is a positive integer and $ad - bc \neq 0$, find the intervals on which f is increasing and decreasing.

90. Mean Value Theorem   Use the Mean Value Theorem to verify that

$$\frac{1}{9} < \sqrt{66} - 8 < \frac{1}{8}$$

Hint: Consider $f(x) = \sqrt{x}$ on the interval $[64,\ 66]$.

91. Show that $a \ln \dfrac{b}{a} \leq b - a < b \ln \dfrac{b}{a}$ if $0 < a < b$.

92. Show that $x \leq \ln(1 + x)^{1 + x}$ for all $x > -1$.

Hint: Consider $f(x) = -x + \ln(1 + x)^{1 + x}$.

93. It can be shown that

$$\frac{d}{dx} \cot^{-1} x = \frac{d}{dx} \tan^{-1} \frac{1}{x}, \quad x \neq 0$$

(See Section 3.3, Assess Your Understanding Problem 58.) You might infer from this that $\cot^{-1} x = \tan^{-1} \dfrac{1}{x} + C$ for all $x \neq 0$, where C is a constant. Show, however, that

$$\cot^{-1} x = \begin{cases} \tan^{-1} \dfrac{1}{x} & \text{if } x > 0 \\ \tan^{-1} \dfrac{1}{x} + \pi & \text{if } x < 0 \end{cases}$$

What is the explanation of the incorrect inference?

94. Show that for any positive integer n, $\dfrac{n}{n+1} < \ln\left(1 + \dfrac{1}{n}\right)^n < 1$.

Preparing for the **AP® Exam**

AP® Practice Problems

Multiple-Choice Questions

1. A function f is continuous on the closed interval $[-2, 5]$, differentiable on the open interval $(-2, 5)$, and $f(-2) = f(5) = 3$. Which of the following statements must be true?
(A) There is a number c in the interval $(-2, 5)$ for which $f'(c) = 0$.
(B) $f'(x) > 0$ for all numbers in the interval $(-2, 5)$.
(C) $f'(x) = 3$ for all numbers in the interval $(-2, 5)$.
(D) None of the above

2. The graph of f' for the interval $[-3, 5]$ is shown below. On what interval(s) is(are) f decreasing?

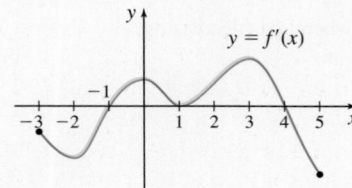

(A) $[-3, -2]$, $[0, 1]$, and $[3, 5]$ (B) $[-3, -1]$ and $[4, 5]$
(C) $[-3, -2]$ and $[3, 5]$ (D) $[-3, -2]$ and $[4, 5]$

3. For the function $f(x) = \sqrt{x}$, find the value of c that satisfies the conclusion of the Mean Value Theorem on the interval $[0, 4]$.

(A) 0   (B) $\dfrac{1}{2}$ (C) 1   (D) 2

4. On what interval(s) is the function $f(x) = e^{3x^3 - x}$ increasing?

(A) $(-\infty, \infty)$ (B) $[0, \infty)$
(C) $\left(-\infty, -\dfrac{1}{3}\right]$ and $\left[\dfrac{1}{3}, \infty\right)$ (D) $\left[-\dfrac{1}{3}, \infty\right)$

5. For which values of x is the function
$f(x) = x^4 - 4x^3 + 4x^2 + 1$ decreasing?

(A) $x < 0$ or $1 < x < 2$ (B) $x < 0$ or $x > 2$
(C) $0 < x < 2$ (D) $x > 2$ only

6. The graph of the derivative of f is shown. Which of the following can be the graph of f?

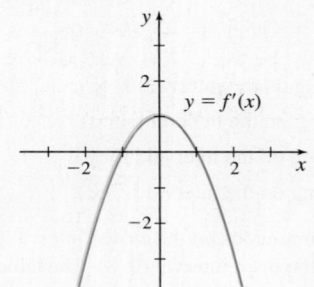

(A)

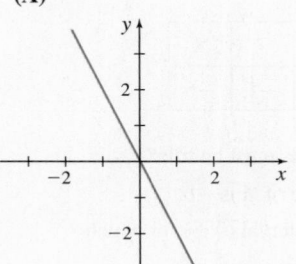

(B)

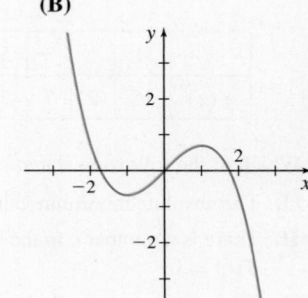

(C)

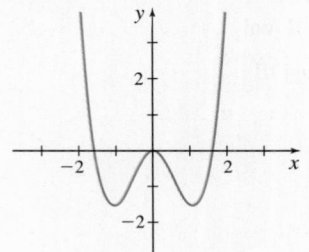

(D)

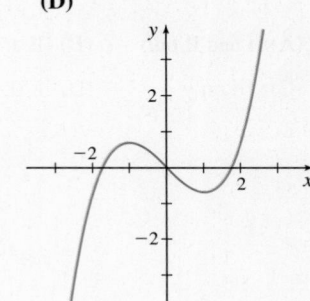

(continued on the next page)

7. Suppose f and g are differentiable functions for which

 • $f(x) > 0$ for all real numbers

 • $g(0) = 4$

If $h(x) = f(x)g(x)$ and $h'(x) = f'(x)g(x)$ for all real numbers, then $g(x)$ equals

(A) $f(x)$ (B) $g'(x)$ (C) 0   (D) 4

8. The domain of the function f is the set of all real numbers.

If $f'(x) = \dfrac{|x^2 - 9|}{x - 3}$, then f is increasing on the interval

(A) $(-\infty, \infty)$.   (B) $[-3, 3]$.

(C) $[3, 9]$.   (D) $[3, \infty)$.

9. An object moves along the x-axis so that its distance from the origin at any time $t \geq 0$ is given

by $x(t) = t^3 + \dfrac{3}{2}t^2 - 18t + 4$. At what times t is the object at rest?

(A) 2 and -3 (B) 2 only

(C) 3 only   (D) 0 and 2

10. Which of the following statements is true for the function

$$f(x) = \frac{\ln x}{x} \quad x > 0$$

(A) f is increasing on the interval $(0, \infty)$.

(B) f is increasing on the interval $[e, \infty)$.

(C) f is decreasing on the interval $[1, e]$.

(D) f is decreasing on the interval $[e, \infty)$.

11. A function f is continuous on the closed interval $[0, 8]$ and differentiable on the open interval $(0, 8)$. The table below shows selected values of $f(x)$.

x	0	1	3	5	7	8
$f(x)$	15	9	7	5	-1	-6

Which of the following statements must be true?

I. The absolute minimum value of f is -6.

II. There is a number c in the interval $(0, 8)$ for which $f(c) = 0$.

III. There is a number c in the open interval $(3, 5)$ for which $f'(c) = -1$.

(A) I and II only   (B) II and III only

(C) III only   (D) I, II, and III

Free-Response Questions

 12. A function $y = f(x)$ is continuous on the closed interval $[0, 9]$ and differentiable on the open interval $(0, 9)$. The derivative function f' is also differentiable on the interval $(0, 9)$. Selected values of f' are given in the table below.

x	1	3	6	7	8
$f'(x)$	2	15	5	0	5

(a) Verify that there is at least one number c in the interval $(1, 8)$ so that $f'(c) = 1$.

(b) Verify that there is at least one number c in the interval $(6, 8)$ so that $f''(c) = 0$.

13. The derivative of f is $f'(x) = e^{x \cos x}$ for $0 < x < \pi$. Determine where f' is increasing.

14. A function f is continuous for $-3 \leq x \leq 4$. The graph of its derivative f' is shown below.

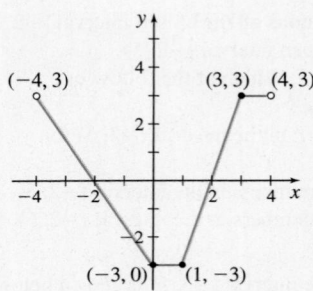

(a) Determine where f is increasing.

(b) Determine where f is decreasing.

Retain Your Knowledge

Multiple-Choice Questions

1. An equation of the line tangent to the graph of the function $f(x) = \dfrac{x}{x^2+4}$ when $x = 3$ is

   (A) $y - \dfrac{3}{13} = \dfrac{7}{13}(x-3)$ (B) $y - \dfrac{3}{13} = -\dfrac{5}{13}(x-3)$

   (C) $y - \dfrac{3}{13} = \dfrac{7}{169}(x-3)$ (D) $y - \dfrac{3}{13} = -\dfrac{5}{169}(x-3)$

2. Suppose $h = g \cdot f$. Use the information in the table below to find $h'(3)$.

x	2	3	4
$f(x)$	7	4	2
$g(x)$	3	2	-4
$f'(x)$	4	-2	3
$g'(x)$	2	6	5

   (A) -10 (B) 7   (C) 1   (D) -2

3. Identify the asymptotes of the function

   $$f(x) = \frac{(2x^2 - x - 15)e^x}{x^2 - 9}.$$

   (A) vertical asymptotes: $x = -3$, $x = 3$; no horizontal asymptote

   (B) vertical asymptotes: $x = -3$, $x = 3$; horizontal asymptote: $y = 0$

   (C) vertical asymptote: $x = -3$; no horizontal asymptote

   (D) vertical asymptote: $x = -3$; horizontal asymptote: $y = 0$

Free-Response Question

4. As an object moves along the x-axis, its signed distance x from the origin at time t is given by the position function

   $$x(t) = \frac{2}{3}t^3 - 3t^2 + 5$$

   where x is in meters and $t \geq 0$ is in minutes.

   (a) What is the position of the function at time 0?

   (b) Find the velocity v of the object as a function of time.

   (c) When is the object's velocity equal to 0? When is the velocity of the object less than 0?

   (d) Find the velocity of the object when the acceleration equals 0. Justify your result.

5.2 Extreme Value Theorem; Global Versus Local Extrema; Critical Numbers

OBJECTIVES *When you finish this section, you should be able to:*

1 **Identify absolute maximum and minimum values and local maximum and minimum values of a function; apply the Extreme Value Theorem (p. 347)**

2 **Find critical numbers (p. 352)**

3 **Find absolute maximum and absolute minimum values; use the Candidates Test (p. 353)**

Often problems in engineering and economics seek to find an optimal, or best, solution. For example, state and local governments try to set tax rates to optimize revenue. If a problem like this can be modeled by a function, then finding the maximum or the minimum values of the function solves the problem.

1 Identify Absolute Maximum and Minimum Values and Local Maximum and Minimum Values of a Function; Apply the Extreme Value Theorem

Figure 11 illustrates the graph of a function f defined on a closed interval $[a, b]$. Pay particular attention to the graph at the numbers x_1, x_2, and x_3. In small open intervals of the x-axis containing x_1 or containing x_3, the value of f is greatest at these numbers. We say that f has *local maxima* at x_1 and x_3, and that $f(x_1)$ and $f(x_3)$ are *local maximum values of* f. Similarly, in small open intervals of the x-axis containing x_2, the value of f is least at x_2. We say f has a *local minimum* at x_2 and $f(x_2)$ is a *local minimum value of* f. ("Maxima" is the plural of "maximum"; "minima" is the plural of "minimum.")

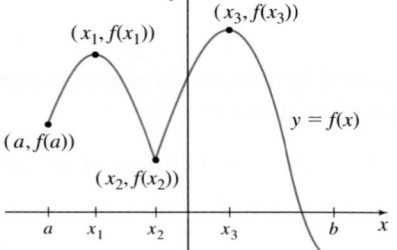

Figure 11

On the closed interval $[a, b]$, the largest value of f is $f(x_3)$, while the smallest value of f is $f(b)$. $f(x_3)$ and $f(b)$ are called, respectively, the *absolute maximum value* and *absolute minimum value* of f on $[a, b]$.

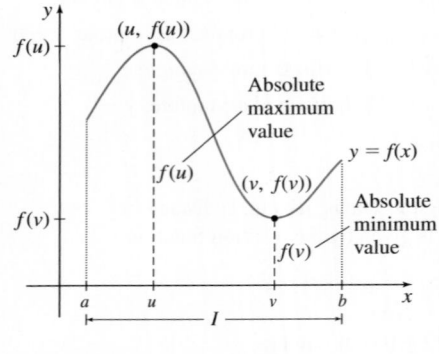

Figure 12 f is defined on an interval I. For all x in I, $f(u) \geq f(x)$ and $f(v) \leq f(x)$. $f(u)$ is the absolute maximum value of f. $f(v)$ is the absolute minimum value of f.

DEFINITION Absolute Maximum; Absolute Minimum

Suppose f is a function defined on an interval I. If there is a number u in the interval for which $f(u) \geq f(x)$ for all x in I, then f has an **absolute maximum** at u and the number $f(u)$ is called the **absolute maximum value** of f on I.

If there is a number v in I for which $f(v) \leq f(x)$ for all x in I, then f has an **absolute minimum** at v and the number $f(v)$ is the **absolute minimum value** of f on I.

The values $f(u)$ and $f(v)$ are sometimes called **global extrema** or the **extreme values** of f on I. ("Extrema" is the plural of the Latin noun *extremum*.)

The *absolute* maximum value and the *absolute* minimum value, if they exist, are the largest and smallest values, respectively, of a function f on the interval I. The interval may be open, closed, or neither. See Figure 12. Contrast this idea with that of a *local* maximum value and a *local* minimum value. These are the largest and smallest values of f in *some open interval* in I. The next definition makes this distinction precise.

DEFINITION Local Maximum; Local Minimum

Suppose f is a function defined on some interval I and u and v are numbers in I. If there is an open interval in I containing u so that $f(u) \geq f(x)$ for all x in this open interval, then f has a **local maximum** (or a **relative maximum**) at u, and the number $f(u)$ is called a **local maximum value**.

Similarly, if there is an open interval in I containing v so that $f(v) \leq f(x)$ for all x in this open interval, then f has a **local minimum** (or a **relative minimum**) at v, and the number $f(v)$ is called a **local minimum value.**

The term **local extreme value** describes either a local maximum value of f or a local minimum value of f.

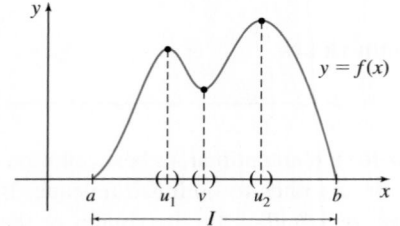

Figure 13 f has a local maximum at u_1 and at u_2. f has a local minimum at v.

Figure 13 illustrates the definition.

Notice in the definition of a local maximum that the interval containing u is required to be *open*. Notice also in the definition of a local maximum that the value $f(u)$ must be greater than or equal to *all* other values of f in this open interval. The word "local" is used to emphasize that this condition holds on *some* open interval containing u. Similar remarks hold for a local minimum. This means that a local extremum cannot occur at the endpoint of an interval.

EXAMPLE 1 Identifying Absolute Maximum and Minimum Values and Local Extreme Values from the Graph of a Function

Figures 14–19 on page 349, show the graphs of six different functions. For each function:

(a) Find the domain.

(b) Determine where the function is continuous.

(c) Identify the absolute maximum value and the absolute minimum value, if they exist.

(d) Identify any local extreme values.

NEED TO REVIEW? Continuity is discussed in Section 1.3, pp. 108–115.

Solution

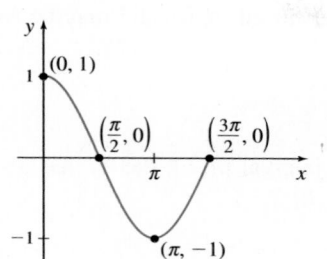

Figure 14

(a) Domain: $\left[0, \dfrac{3\pi}{2}\right]$

(b) Continuous on $\left[0, \dfrac{3\pi}{2}\right]$

(c) Absolute maximum value:
$f(0) = 1$
Absolute minimum value:
$f(\pi) = -1$

(d) No local maximum value;
local minimum value: $f(\pi) = -1$

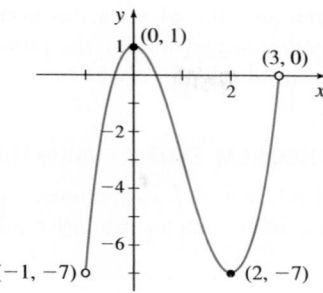

Figure 15

(a) Domain: $(-1, 3)$
(b) Continuous on $(-1, 3)$
(c) Absolute maximum value:
$f(0) = 1$
Absolute minimum value:
$f(2) = -7$
(d) Local maximum value:
$f(0) = 1$
Local minimum value:
$f(2) = -7$

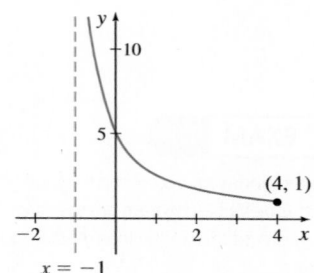

Figure 16

(a) Domain: $(-1, 4]$
(b) Continuous on $(-1, 4]$
(c) No absolute maximum value;
absolute minimum value:
$f(4) = 1$
(d) No local maximum value;
no local minimum value

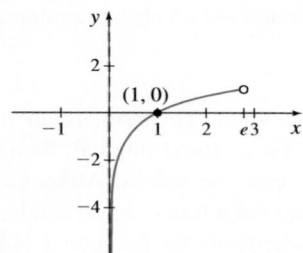

Figure 17

(a) Domain: $(0, e)$
(b) Continuous on $(0, e)$
(c) No absolute maximum value;
no absolute minimum value
(d) No local maximum value;
no local minimum value

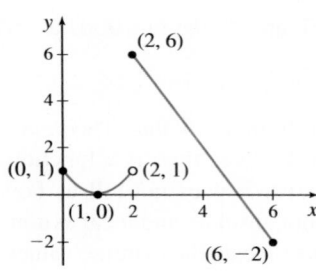

Figure 18

(a) Domain: $[0, 6]$
(b) Continuous on $[0, 6]$ except
at $x = 2$
(c) Absolute maximum value:
$f(2) = 6$
Absolute minimum value:
$f(6) = -2$
(d) Local maximum value: $f(2) = 6$
Local minimum value: $f(1) = 0$

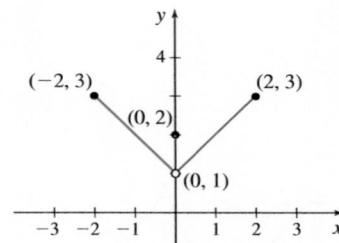

Figure 19

(a) Domain: $[-2, 2]$
(b) Continuous on $[-2, 2]$ except
at $x = 0$
(c) Absolute maximum value:
$f(-2) = 3, \quad f(2) = 3$
No absolute minimum value
(d) Local maximum value: $f(0) = 2$
No local minimum value

NOW WORK   Problem 7.

Example 1 illustrates that a function f can have both an absolute maximum value and an absolute minimum value, can have one but not the other, or can have neither an absolute maximum value nor an absolute minimum value. The next theorem provides conditions for which a function f will always have absolute extrema. Although the theorem seems simple, the proof requires advanced topics and may be found in most advanced calculus books.

THEOREM Extreme Value Theorem

If a function f is continuous on a closed interval $[a, b]$, then f has an absolute maximum and an absolute minimum on $[a, b]$.

Look back at Example 1. Figure 14 illustrates a function that is continuous on a closed interval; it has an absolute maximum and an absolute minimum on the interval. But if a function is not continuous on a closed interval $[a, b]$, then the conclusion of the Extreme Value Theorem may or may not hold.

For example, the functions graphed in Figures 15–17 are all continuous, but not on a closed interval:

- In Figure 15, the function has both an absolute maximum and an absolute minimum.
- In Figure 16, the function has an absolute minimum but no absolute maximum.
- In Figure 17, the function has neither an absolute maximum nor an absolute minimum.

On the other hand, the functions graphed in Figures 18 and 19 are each defined on a closed interval, but neither is continuous on that interval:

- In Figure 18, the function has an absolute maximum and an absolute minimum.
- In Figure 19, the function has an absolute maximum but no absolute minimum.

NOW WORK Problem 9.

The Extreme Value Theorem, like the Mean Value Theorem, is an existence theorem. It states that, if a function is continuous on a closed interval, then extreme values exist. It does not tell us how to find these extreme values. Although we can locate both absolute and local extrema given the graph of a function, we need tools that allow us to locate the extreme values analytically, when only the function f is known.

We begin by examining local extrema more closely. Consider the function $y = f(x)$ whose graph is shown in Figure 20. There is a local maximum at x_1 and a local minimum at x_2. The derivative at these numbers is 0, since the tangent lines to the graph of f at x_1 and x_2 are horizontal. There is also a local maximum at x_3, where the derivative fails to exist. The next theorem provides the details.

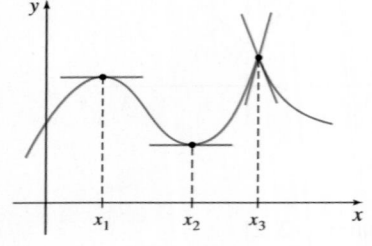

Figure 20

THEOREM Condition for a Local Maximum or a Local Minimum

If a function f has a local maximum or a local minimum at the number c, then either $f'(c) = 0$ or $f'(c)$ does not exist.

Proof Suppose f has a local maximum at c. Then, by definition,

$$f(c) \geq f(x)$$

for all x in some open interval containing c. Equivalently,

$$f(x) - f(c) \leq 0$$

The derivative of f at c is

$$f'(c) = \lim_{x \to c} \frac{f(x) - f(c)}{x - c}$$

provided the limit exists. If this limit does not exist, then $f'(c)$ does not exist and there is nothing further to prove.

If this limit does exist, then

$$\lim_{x \to c^-} \frac{f(x) - f(c)}{x - c} = \lim_{x \to c^+} \frac{f(x) - f(c)}{x - c} \qquad (1)$$

In the limit on the left, $x < c$ and $f(x) - f(c) \le 0$, so $\dfrac{f(x) - f(c)}{x - c} \ge 0$ and

$$\lim_{x \to c^-} \frac{f(x) - f(c)}{x - c} \ge 0 \qquad (2)$$

[Do you see why? If the limit were negative, then there would be an open interval about c, $x < c$, on which $f(x) - f(c) < 0$; refer to Section 1.6, Example 7, p. 156.]

Similarly, in the limit on the right side of (1), $x > c$ and $f(x) - f(c) \le 0$, so $\dfrac{f(x) - f(c)}{x - c} \le 0$ and

$$\lim_{x \to c^+} \frac{f(x) - f(c)}{x - c} \le 0 \qquad (3)$$

Since the limits (2) and (3) must be equal, we have

$$f'(c) = \lim_{x \to c} \frac{f(x) - f(c)}{x - c} = 0$$

The proof when f has a local minimum at c is similar and is left as an exercise (Problem 88). ■

For differentiable functions, the following theorem, often called **Fermat's Theorem**, is simpler.

ORIGINS Pierre de Fermat (1601–1665), a lawyer whose contributions to mathematics were made in his spare time, ranks as one of the great "amateur" mathematicians. Although Fermat is often remembered for his famous "last theorem," he established many fundamental results in number theory and, with Pascal, cofounded the theory of probability. Since his work on calculus was the best done before Newton and Leibniz, he must be considered one of the principal founders of calculus.

THEOREM

If a differentiable function f has a local maximum or a local minimum at c, then $f'(c) = 0$.

In other words, for differentiable functions, local extreme values occur at points where the tangent line to the graph of f is horizontal.

As the theorems show, the numbers at which $f'(x) = 0$ or at which $f'(x)$ does not exist provide a clue for locating where f has local extrema. But unfortunately, knowing that $f'(c) = 0$ or that f' does not exist at c does not guarantee a local extremum occurs at c. For example, in Figure 21, $f'(x_3) = 0$, but f has neither a local maximum nor a local minimum at x_3. Similarly, $f'(x_4)$ does not exist, but f has neither a local maximum nor a local minimum at x_4.

Even though there may be no local extremum found at a number c for which $f'(c) = 0$ or $f'(c)$ does not exist, the collection of all such numbers provides *all* the *possibilities* where f *might* have local extreme values. For this reason, we call these numbers *critical numbers*.

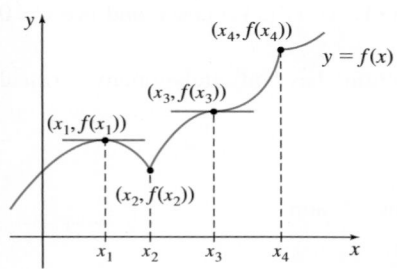

Figure 21

DEFINITION Critical Number

A **critical number** of a function f is a number c in the domain of f for which either $f'(c) = 0$ or $f'(c)$ does not exist.

2 Find Critical Numbers

EXAMPLE 2 Finding Critical Numbers

Find any critical numbers of the following functions:

(a) $f(x) = x^3 - 6x^2 + 9x + 2$ (b) $R(x) = \dfrac{1}{x-2}$

(c) $g(x) = \dfrac{(x-2)^{2/3}}{x}$ (d) $G(x) = \sin x$

Solution

(a) Since f is a polynomial, it is differentiable at every real number. So, the critical numbers occur where $f'(x) = 0$.

$$f'(x) = 3x^2 - 12x + 9 = 3(x-1)(x-3)$$

$f'(x) = 0$ at $x = 1$ and $x = 3$; the numbers 1 and 3 are the critical numbers of f.

(b) The domain of $R(x) = \dfrac{1}{x-2}$ is $\{x \mid x \neq 2\}$, and $R'(x) = -\dfrac{1}{(x-2)^2}$. R' exists for all numbers x in the domain of R (remember, 2 is not in the domain of R). Since R' is never 0, R has no critical numbers.

(c) The domain of $g(x) = \dfrac{(x-2)^{2/3}}{x}$ is $\{x \mid x \neq 0\}$, and the derivative of g is

$$g'(x) = \frac{x \cdot \left[\frac{2}{3}(x-2)^{-1/3}\right] - 1 \cdot (x-2)^{2/3}}{x^2} \underset{\uparrow}{=} \frac{2x - 3(x-2)}{3x^2(x-2)^{1/3}} = \frac{6-x}{3x^2(x-2)^{1/3}}$$

Multiply by $\dfrac{3(x-2)^{1/3}}{3(x-2)^{1/3}}$

Critical numbers occur where $g'(x) = 0$ or where $g'(x)$ does not exist. Since $g'(6) = 0$, $x = 6$ is a critical number. Next, $g'(x)$ does not exist where

$$3x^2(x-2)^{1/3} = 0$$

$$3x^2 = 0 \quad \text{or} \quad (x-2)^{1/3} = 0$$

$$x = 0 \quad \text{or} \quad x = 2$$

We ignore 0 since it is not in the domain of g. The critical numbers of g are 6 and 2.

(d) The domain of $G(x) = \sin x$ is all real numbers, and G is differentiable on its domain, so the critical numbers occur where $G'(x) = 0$. $G'(x) = \cos x$ and $\cos x = 0$ at $x = \pm\dfrac{\pi}{2}, \ \pm\dfrac{3\pi}{2}, \ \pm\dfrac{5\pi}{2}, \ldots$ This function has infinitely many critical numbers. ∎

NOW WORK Problem 25 and AP® Practice Problems 1, 6, and 7.

3 Find Absolute Maximum and Absolute Minimum Values; Use the Candidates Test

The following theorem provides a way to find the absolute extreme values of a function f that is continuous on a closed interval $[a, b]$.

> **THEOREM** Locating Absolute Extreme Values
>
> Suppose f is a function that is continuous on a closed interval $[a, b]$. Then the absolute maximum value and the absolute minimum value of f are the largest and the smallest values, respectively, found among the following:
>
> - The values of f at the critical numbers in the open interval (a, b)
> - $f(a)$ and $f(b)$, the values of f at the endpoints a and b

For any function f satisfying the conditions of the theorem, the Extreme Value Theorem guarantees that extreme values exist, and the theorem tells us how to find them.

> **Steps for Finding the Absolute Extreme Values of a Function f That Is Continuous on a Closed Interval $[a, b]$ (the Candidates Test)**
>
> **Step 1** Locate all critical numbers in the open interval (a, b).
>
> **Step 2** Evaluate f at each critical number and at the endpoints a and b.
>
> **Step 3** The largest value is the absolute maximum value; the smallest value is the absolute minimum value.

EXAMPLE 3 Finding Absolute Maximum and Minimum Values

Find the absolute maximum value and the absolute minimum value of each function:

(a) $f(x) = x^3 - 6x^2 + 9x + 2$ on $[0, 2]$ **(b)** $g(x) = \dfrac{(x - 2)^{2/3}}{x}$ on $[1, 10]$

Solution

(a) The function f, a polynomial function, is continuous on the closed interval $[0, 2]$, so the Extreme Value Theorem guarantees that f has an absolute maximum value and an absolute minimum value on the interval. We use the Candidates Test and follow the steps for finding extreme values.

Step 1 From Example 2(a), the critical numbers of f are 1 and 3. We exclude 3, since it is not in the interval $(0, 2)$.

Step 2 Find the value of f at the critical number 1 and at the endpoints 0 and 2:

$$f(1) = 6 \qquad f(0) = 2 \qquad f(2) = 4$$

Step 3 The largest value 6 is the absolute maximum value of f on the interval $[0, 2]$; the smallest value 2 is the absolute minimum value of f on $[0, 2]$.

(b) The function g is continuous on the closed interval $[1, 10]$, so g has an absolute maximum and an absolute minimum on the interval.

Step 1 From Example 2(c), the critical numbers of g are 2 and 6. Both critical numbers are in the interval $(1, 10)$.

Step 2 Evaluate g at the critical numbers 2 and 6 and at the endpoints 1 and 10:

x	$g(x) = \dfrac{(x-2)^{2/3}}{x}$	
1	$\dfrac{(1-2)^{2/3}}{1} = (-1)^{2/3} = 1$	← absolute maximum value
2	$\dfrac{(2-2)^{2/3}}{2} = 0$	← absolute minimum value
6	$\dfrac{(6-2)^{2/3}}{6} = \dfrac{4^{2/3}}{6} \approx 0.420$	
10	$\dfrac{(10-2)^{2/3}}{10} = \dfrac{8^{2/3}}{10} = 0.4$	

Step 3 The largest value 1 is the absolute maximum value of g on the interval $[1, 10]$; the smallest value 0 is the absolute minimum value of g on $[1, 10]$. ∎

NOW WORK Problem 49 and AP® Practice Problems 2, 3, 8, and 10.

For piecewise-defined functions f, we need to look carefully at the number(s) where the rules for the function change.

▶ CALC CLIP **EXAMPLE 4** **Finding Absolute Maximum and Minimum Values**

Find the absolute maximum value and absolute minimum value of the function

$$f(x) = \begin{cases} 2x - 1 & \text{if } 0 \le x \le 2 \\ x^2 - 5x + 9 & \text{if } 2 < x \le 3 \end{cases}$$

Solution

The function f is continuous on the closed interval $[0, 3]$. (You should verify this.) To find the absolute maximum value and absolute minimum value, we use the Candidates Test.

Step 1 Find the critical numbers in the open interval $(0, 3)$:

- On the open interval $(0, 2)$: $f(x) = 2x - 1$ and $f'(x) = 2$. Since $f'(x) \ne 0$ on the interval $(0, 2)$, there are no critical numbers in $(0, 2)$.

- On the open interval $(2, 3)$: $f(x) = x^2 - 5x + 9$ and $f'(x) = 2x - 5$.

  Solving $f'(x) = 2x - 5 = 0$, we find $x = \dfrac{5}{2}$. Since $\dfrac{5}{2}$ is in the interval $(2, 3)$, $\dfrac{5}{2}$ is a critical number.

- At $x = 2$, the rule for f changes, so we investigate the one-sided limits of $\dfrac{f(x) - f(2)}{x - 2}$ to determine whether $f'(2)$ exists.

$$\lim_{x \to 2^-} \frac{f(x) - f(2)}{x - 2} = \lim_{x \to 2^-} \frac{(2x - 1) - 3}{x - 2} = \lim_{x \to 2^-} \frac{2x - 4}{x - 2}$$

$$= \lim_{x \to 2^-} \frac{2(x - 2)}{x - 2} = 2$$

$$\lim_{x \to 2^+} \frac{f(x) - f(2)}{x - 2} = \lim_{x \to 2^+} \frac{(x^2 - 5x + 9) - 3}{x - 2} = \lim_{x \to 2^+} \frac{x^2 - 5x + 6}{x - 2}$$

$$= \lim_{x \to 2^+} \frac{(x - 2)(x - 3)}{x - 2} = \lim_{x \to 2^+} (x - 3) = -1$$

The one-sided limits are not equal, so the derivative does not exist at 2; 2 is a critical number.

Step 2 Evaluate f at the critical numbers $\dfrac{5}{2}$ and 2 and at the endpoints 0 and 3.

x	$f(x)$	
0	$2 \cdot 0 - 1 = -1$	← absolute minimum value
2	$2 \cdot 2 - 1 = 3$	← absolute maximum value
$\dfrac{5}{2}$	$\left(\dfrac{5}{2}\right)^2 - 5 \cdot \dfrac{5}{2} + 9 = \dfrac{25}{4} - \dfrac{25}{2} + 9 = \dfrac{11}{4}$	
3	$3^2 - 5 \cdot 3 + 9 = 9 - 15 + 9 = 3$	← absolute maximum value

Step 3 The largest value 3 is the absolute maximum value; the smallest value -1 is the absolute minimum value. ∎

The graph of f is shown in Figure 22. Notice that the absolute maximum occurs at 2 and at 3.

NOW WORK   Problem **61** and AP® Practice Problems **4, 5,** and **9.**

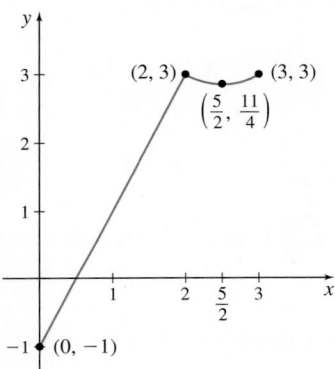

Figure 22

$$f(x) = \begin{cases} 2x - 1 & \text{if } 0 \le x \le 2 \\ x^2 - 5x + 9 & \text{if } 2 < x \le 3 \end{cases}$$

EXAMPLE 5   **Constructing a Rain Gutter**

A rain gutter is to be constructed using a piece of aluminum 12 in. wide. After marking a length of 4 in. from each edge, the piece of aluminum is bent up at an angle θ, as illustrated in Figure 23. The area A of a cross section of the opening, expressed as a function of θ, is

$$A(\theta) = 16 \sin \theta (\cos \theta + 1) \qquad 0 \le \theta \le \frac{\pi}{2}$$

Find the angle θ that maximizes the area A. (This bend will allow the most water to flow through the gutter.)

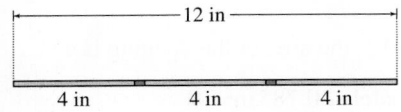

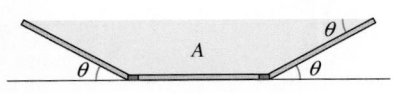

Figure 23

Solution

The function $A = A(\theta)$ is continuous on the closed interval $\left[0, \dfrac{\pi}{2}\right]$. To find the angle θ that maximizes A, follow the three steps of the Candidate's Test.

Step 1 Locate all critical numbers in the open interval $\left(0, \dfrac{\pi}{2}\right)$.

$$
\begin{aligned}
A'(\theta) &= 16 \sin\theta(-\sin\theta) + 16\cos\theta(\cos\theta + 1) \\
&= 16[-\sin^2\theta + \cos^2\theta + \cos\theta] \\
&= 16[(\cos^2\theta - 1) + \cos^2\theta + \cos\theta] \qquad -\sin^2\theta = \cos^2\theta - 1 \\
&= 16[2\cos^2\theta + \cos\theta - 1] = 16(2\cos\theta - 1)(\cos\theta + 1)
\end{aligned}
$$

The critical numbers satisfy the equation $A'(\theta) = 0$, $0 < \theta < \dfrac{\pi}{2}$.

$$16(2\cos\theta - 1)(\cos\theta + 1) = 0$$

$$2\cos\theta - 1 = 0 \qquad \text{or} \qquad \cos\theta + 1 = 0$$

$$\cos\theta = \frac{1}{2} \qquad\qquad\qquad \cos\theta = -1$$

$$\theta = \frac{\pi}{3} \text{ or } \frac{5\pi}{3} \qquad\qquad \theta = \pi$$

Of these solutions, only $\dfrac{\pi}{3}$ is in the interval $\left(0, \dfrac{\pi}{2}\right)$. So, $\dfrac{\pi}{3}$ is the only critical number.

Step 2 Evaluate A at the critical number $\dfrac{\pi}{3}$ and at the endpoints 0 and $\dfrac{\pi}{2}$.

θ	$A(\theta) = 16\sin\theta(\cos\theta + 1)$
0	$16\sin 0(\cos 0 + 1) = 0$
$\dfrac{\pi}{3}$	$16\sin\dfrac{\pi}{3}\left(\cos\dfrac{\pi}{3} + 1\right) = 16\left(\dfrac{\sqrt{3}}{2}\right)\left(\dfrac{1}{2} + 1\right)$
	$\qquad\qquad = 12\sqrt{3} \approx 20.785 \quad \longleftarrow \text{absolute maximum value}$
$\dfrac{\pi}{2}$	$16\sin\dfrac{\pi}{2}\left(\cos\dfrac{\pi}{2} + 1\right) = 16(1)(0 + 1) = 16$

Step 3 If the aluminum is bent at an angle of $\dfrac{\pi}{3}$ (60°), the area of the opening is a maximum. The maximum area is approximately 20.785 in². ∎

NOW WORK   Problem 71.

Application: Physics in Medicine

From physics, the volume V of fluid flowing through a pipe is a function of the radius r of the pipe and the difference in pressure p at each end of the pipe and is given by

$$V = kpr^4$$

where k is a constant.

EXAMPLE 6 **Analyzing a Cough**

Coughing is caused by increased pressure in the lungs and is accompanied by a decrease in the diameter of the windpipe. See Figure 24. The radius r of the windpipe decreases with increased pressure p according to the formula $r_0 - r = cp$, where r_0 is the radius of the windpipe when there is no difference in pressure and c is a positive constant. The volume V of air flowing through the windpipe is

$$V = kpr^4$$

where k is a constant. Find the radius r that allows the most air to flow through the windpipe. Restrict r so that $0 < \dfrac{r_0}{2} \leq r \leq r_0$.

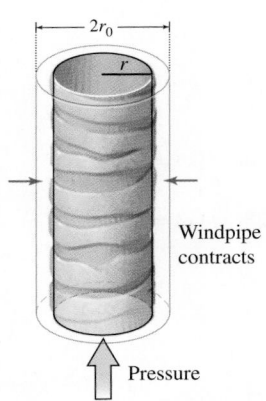

Windpipe contracts

Pressure

Figure 24

Solution

Since $p = \dfrac{r_0 - r}{c}$, we can express V as a function of r:

$$V = V(r) = k\left(\frac{r_0 - r}{c}\right) r^4 = \frac{kr_0}{c}r^4 - \frac{k}{c}r^5 \qquad \frac{r_0}{2} \leq r \leq r_0$$

Now find the absolute maximum of V on the interval $\left[\dfrac{r_0}{2}, r_0\right]$.

$$V'(r) = \frac{4k\,r_0}{c}r^3 - \frac{5k}{c}r^4 = \frac{k}{c}r^3(4r_0 - 5r)$$

The only critical number in the interval $\left(\dfrac{r_0}{2}, r_0\right)$ is $r = \dfrac{4r_0}{5}$.

Evaluate V at the critical number and at the endpoints, $\dfrac{r_0}{2}$ and r_0.

r	$V(r) = k\left(\dfrac{r_0 - r}{c}\right) r^4$
$\dfrac{r_0}{2}$	$k\left(\dfrac{r_0 - \dfrac{r_0}{2}}{c}\right)\left(\dfrac{r_0}{2}\right)^4 = \dfrac{k\,r_0^5}{32c} \approx \dfrac{0.031\,k\,r_0^5}{c}$
$\dfrac{4r_0}{5}$	$k\left(\dfrac{r_0 - \dfrac{4r_0}{5}}{c}\right)\left(\dfrac{4r_0}{5}\right)^4 = \dfrac{k}{c}\cdot\dfrac{4^4\,r_0^5}{5^5} = \dfrac{256\,kr_0^5}{3125c} \approx \dfrac{0.082\,k\,r_0^5}{c}$
r_0	0

The largest of these three values is $\dfrac{256\,kr_0^5}{3125c}$. So, the maximum air flow occurs when the radius of the windpipe is $\dfrac{4r_0}{5}$, that is, when the windpipe contracts by 20%. ∎

5.2 Assess Your Understanding

Concepts and Vocabulary

1. **True or False** Any function f that is defined on a closed interval $[a, b]$ has both an absolute maximum value and an absolute minimum value.

2. **Multiple Choice** A number c in the domain of a function f is called a(n)
   [(**a**) extreme value (**b**) critical number (**c**) local number]
   of f if either $f'(c) = 0$ or $f'(c)$ does not exist.

3. **True or False** At a critical number, there is a local extreme value.

4. **True or False** If a function f is continuous on a closed interval $[a, b]$, then its absolute maximum value is found at a critical number.

5. **True or False** The Extreme Value Theorem tells us where the absolute maximum and absolute minimum can be found.

6. **True or False** If f is differentiable on the interval $(0, 4)$ and $f'(2) = 0$, then f has a local maximum or a local minimum at 2.

Skill Building

In Problems 7 and 8, use the graphs to determine whether the function f has an absolute extremum and/or a local extremum or neither at x_1, x_2, x_3, x_4, x_5, x_6, x_7, and x_8.

 7.
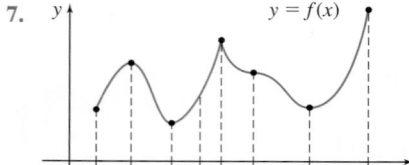

8.
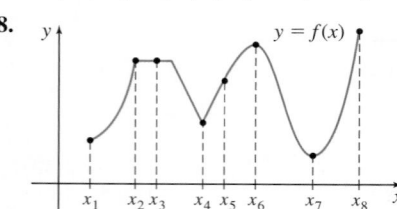

In Problems 9–12, provide a graph of a continuous function f that has the following properties:

 9. domain $[0, 8]$, absolute maximum at 0, absolute minimum at 3, local minimum at 7

10. domain $[-5, 5]$, absolute maximum at 3, absolute minimum at -3

11. domain $[3, 10]$ and has no local extreme points

12. has no absolute extreme values, is differentiable at 4 and has a local minimum at 4, is not differentiable at 0 but has a local maximum at 0

In Problems 13–36, find the critical numbers, if any, of each function.

13. $f(x) = x^2 - 8x$

14. $f(x) = 1 - 6x + x^2$

15. $f(x) = x^3 - 3x^2$

16. $f(x) = x^3 - 6x$

17. $f(x) = x^4 - 2x^2 + 1$

18. $f(x) = 3x^4 - 4x^3$

19. $f(x) = x^{2/3}$

20. $f(x) = x^{1/3}$

21. $f(x) = 2\sqrt{x}$

22. $f(x) = 4 - \sqrt{x}$

23. $f(x) = x + \sin x, \quad 0 \le x \le \pi$

24. $f(x) = x - \cos x, \quad -\dfrac{\pi}{2} \le x \le \dfrac{\pi}{2}$

25. $f(x) = x\sqrt{1 - x^2}$

26. $f(x) = x^2\sqrt{2 - x}$

27. $f(x) = \dfrac{x^2}{x - 1}$

28. $f(x) = \dfrac{x}{x^2 - 1}$

29. $f(x) = (x + 3)^2(x - 1)^{2/3}$

30. $f(x) = (x - 1)^2(x + 1)^{1/3}$

31. $f(x) = \dfrac{(x - 3)^{1/3}}{x - 1}$

32. $f(x) = \dfrac{(x + 3)^{2/3}}{x + 1}$

33. $f(x) = \dfrac{\sqrt[3]{x^2 - 9}}{x}$

34. $f(x) = \dfrac{\sqrt[3]{4 - x^2}}{x}$

35. $f(x) = \begin{cases} 3x & \text{if } 0 \le x < 1 \\ 4 - x & \text{if } 1 \le x \le 2 \end{cases}$

36. $f(x) = \begin{cases} x^2 & \text{if } 0 \le x < 1 \\ 1 - x^2 & \text{if } 1 \le x \le 2 \end{cases}$

In Problems 37–64, find the absolute maximum value and absolute minimum value of each function on the indicated interval. Notice that the functions in Problems 37–58 are the same as those in Problems 13–34 above.

37. $f(x) = x^2 - 8x$ on $[-1, 10]$

38. $f(x) = 1 - 6x + x^2$ on $[0, 4]$

39. $f(x) = x^3 - 3x^2$ on $[1, 4]$

40. $f(x) = x^3 - 6x$ on $[-1, 1]$

41. $f(x) = x^4 - 2x^2 + 1$ on $[0, 2]$

42. $f(x) = 3x^4 - 4x^3$ on $[-2, 0]$

43. $f(x) = x^{2/3}$ on $[-1, 1]$

44. $f(x) = x^{1/3}$ on $[-1, 1]$

45. $f(x) = 2\sqrt{x}$ on $[1, 4]$

46. $f(x) = 4 - \sqrt{x}$ on $[0, 4]$

47. $f(x) = x + \sin x$ on $[0, \pi]$

48. $f(x) = x - \cos x$ on $\left[-\dfrac{\pi}{2}, \dfrac{\pi}{2}\right]$

49. $f(x) = x\sqrt{1 - x^2}$ on $[-1, 1]$

50. $f(x) = x^2\sqrt{2 - x}$ on $[0, 2]$

51. $f(x) = \dfrac{x^2}{x - 1}$ on $\left[-1, \dfrac{1}{2}\right]$

52. $f(x) = \dfrac{x}{x^2 - 1}$ on $\left[-\dfrac{1}{2}, \dfrac{1}{2}\right]$

53. $f(x) = (x+3)^2(x-1)^{2/3}$ on $[-4, 5]$

54. $f(x) = (x-1)^2(x+1)^{1/3}$ on $[-2, 7]$

55. $f(x) = \dfrac{(x-3)^{1/3}}{x-1}$ on $[2, 11]$

56. $f(x) = \dfrac{(x+3)^{2/3}}{x+1}$ on $[-4, -2]$

57. $f(x) = \dfrac{\sqrt[3]{x^2-9}}{x}$ on $[3, 6]$

58. $f(x) = \dfrac{\sqrt[3]{4-x^2}}{x}$ on $[-4, -1]$

59. $f(x) = e^x - 3x$ on $[0, 1]$

60. $f(x) = e^{\cos x}$ on $[-\pi, 2\pi]$

61. $f(x) = \begin{cases} 2x+1 & \text{if } 0 \le x < 1 \\ 3x & \text{if } 1 \le x \le 3 \end{cases}$

62. $f(x) = \begin{cases} x+3 & \text{if } -1 \le x \le 2 \\ 2x+1 & \text{if } 2 < x \le 4 \end{cases}$

63. $f(x) = \begin{cases} x^2 & \text{if } -2 \le x < 1 \\ x^3 & \text{if } 1 \le x \le 2 \end{cases}$

64. $f(x) = \begin{cases} x+2 & \text{if } -1 \le x < 0 \\ 2-x & \text{if } 0 \le x \le 1 \end{cases}$

Applications and Extensions

In Problems 65–68, for each function f :

(a) *Find the derivative f'.*

(b) *Use technology to find the critical numbers of f .*

(c) *Graph f and describe the behavior of f suggested by the graph at each critical number.*

65. $f(x) = 3x^4 - 2x^3 - 21x^2 + 36x$

66. $f(x) = x^2 + 2x - \dfrac{2}{x}$

67. $f(x) = \dfrac{(x^2 - 5x + 2)\sqrt{x+5}}{\sqrt{x^2+2}}$

68. $f(x) = \dfrac{(x^2 - 9x + 16)\sqrt{x+3}}{\sqrt{x^2-4x+6}}$

In Problems 69 and 70, for each function f :

(a) *Find the derivative f'.*

(b) *Use technology to find the absolute maximum value and the absolute minimum value of f on the closed interval [0, 5].*

(c) *Graph f. Are the results from (b) supported by the graph?*

69. $f(x) = x^4 - 12.4x^3 + 49.24x^2 - 68.64x$

70. $f(x) = e^{-x}\sin(2x) + e^{-x/2}\cos(2x)$

71. Cost of Fuel A truck has a top speed of 75 mi/h, and when traveling at the rate of x mi/h, it consumes fuel at the rate

of $\dfrac{1}{200}\left(\dfrac{2500}{x} + x\right)$ gal/mi. If the price of fuel is \$3.60/ gal,

the cost C (in dollars) of driving 200 mi is given by

$$C(x) = 3.60 \cdot \left(\dfrac{2500}{x} + x\right)$$

(a) What is the most economical speed for the truck to travel? Use the interval [10, 75].

(b) Graph the cost function C.

72. Trucking Costs If the driver of the truck in Problem 71 is paid \$28.00 per hour and wages are added to the cost of fuel, what is the most economical speed for the truck to travel?

73. Projectile Motion An object is propelled upward at an angle θ, $45° < \theta < 90°$, to the horizontal with an initial velocity of v_0 ft/s from the base of an inclined plane that makes an angle of $45°$ to the horizontal. See the illustration below. If air resistance is ignored, the distance R, in feet, that the object travels up the inclined plane is given by the function

$$R(\theta) = \dfrac{v_0^2\sqrt{2}}{16}\cos\theta(\sin\theta - \cos\theta)$$

(a) Find the angle θ that maximizes R. What is the maximum value of R?

(b) Graph $R = R(\theta)$, $45° \le \theta \le 90°$, using $v_0 = 32$ ft/s.

(c) Does the graph support the result from (a)?

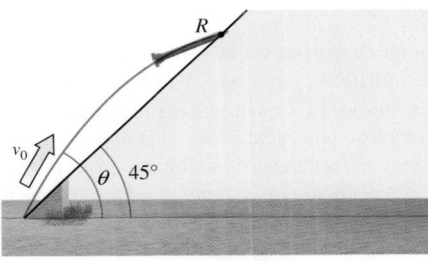

74. Harmonic Motion An object of mass 1 kg moves in simple harmonic motion, with an amplitude $A = 0.24$ m and a period of 4 s. The position s of the object is given by $s(t) = A\cos(\omega t)$, where t is the time in seconds.

(a) Find the position of the object at time t and at time $t = 0.5$ s.

(b) Find the velocity $v = v(t)$ of the object.

(c) Find the velocity of the object when $t = 0.5$ s.

(d) Find the acceleration $a = a(t)$ of the object.

(e) Use Newton's Second Law of Motion, $F = ma$, to find the magnitude and direction of the force acting on the object when $t = 0.5$ s.

(f) Find the minimum time required for the object to move from its initial position to the point where $s = -0.12$ m.

(g) Find the velocity of the object when $s = -0.12$ m.

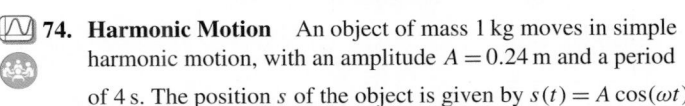

75. Golf Ball Speed   The fastest speed of a golf ball on record is 97.05 m/s (217.1 mi/h). It was hit by Ryan Winther at the Orange County National Driving Range in Orlando, Florida on January 13, 2013. When a ball is hit at an angle θ to the horizontal, $0° \leq \theta \leq 90°$, and lands at the same level from which it was hit, the horizontal range R of the ball is given

by $R = \dfrac{2v_0^2}{g} \sin\theta \cos\theta$, where v_0 is the initial speed of the ball

and $g = 9.8$ m/s^2 is the acceleration due to gravity.

(a) Show that the golf ball achieves its maximum range if the golfer hits it at an angle of $45°$.

(b) What is the maximum range that could be achieved by the record golf ball speed?

Source: http://www.guinnessworldrecords.com

76. Optics   When light goes through a thin slit, it spreads out (diffracts). After passing through the slit, the intensity I

of the light on a distant screen is given by $I = I_0 \left(\dfrac{\sin\alpha}{\alpha} \right)^2$,

where I_0 is the original intensity of the light and α depends on the angle away from the center.

(a) What is the intensity of the light as $\alpha \to 0$?

(b) Show that the bright spots, that is, the places where the intensity has a local maximum, occur when $\tan\alpha = \alpha$.

(c) Is the intensity of the light the same at each bright spot?

Economics   *In Problems 77 and 78, use the following discussion:*

In determining a tax rate on consumer goods, the government is always faced with the question, "What tax rate produces the largest tax revenue?" Imposing a tax may cause the price of the goods to increase and reduce the demand for the product. A very large tax may reduce the demand to zero, with the result that no tax is collected. On the other hand, if no tax is levied, there is no tax revenue at all. (Tax revenue R is the product of the tax rate t times the actual quantity q, in dollars, consumed.)

77. The government has determined that the relationship between the quantity q of a product consumed and the related tax rate t

is $t = \sqrt{27 - 3q^2}$. Find the tax rate that maximizes tax revenue. How much tax is generated by this rate?

78. On a particular product, government economists determine that the relationship between the tax rate t and the quantity q consumed is $t + 3q^2 = 18$. Find the tax rate that maximizes tax revenue and the revenue generated by the tax.

79. (a) Find the critical numbers of

$$f(x) = \frac{x^3}{3} - 0.055x^2 + 0.0028x - 4.$$

(b) Find the absolute extrema of f on the interval $[-1, 1]$.

80. The function $f(x) = Ax^2 + Bx + C$ has a local minimum at 0, and its graph contains the points $(0, 2)$ and $(1, 8)$. Find A, B, and C.

81. (a) Determine the domain of the function

$$f(x) = [(16 - x^2)(x^2 - 9)]^{1/2}.$$

(b) Find the absolute maximum value of f on its domain.

82. Absolute Extreme Values   Without finding them, explain why

the function $f(x) = \sqrt{x(2 - x)}$ must have an absolute maximum value and an absolute minimum value. Then find the absolute extreme values in two ways (one with and one without calculus).

83. Put It Together   If a function f is continuous on the closed interval $[a, b]$, which of the following is necessarily true?

(a) f is differentiable on the open interval (a, b).

(b) If $f(u)$ is an absolute maximum value of f, then $f'(u) = 0$.

(c) $\lim\limits_{x \to c} f(x) = f(\lim\limits_{x \to c} x)$ for $a < c < b$.

(d) $f'(x) = 0$, for some x, $a \leq x \leq b$.

(e) f has an absolute maximum value on $[a, b]$.

84. Write a paragraph that explains the similarities and differences between an absolute extreme value and a local extreme value.

85. Explain in your own words the method for finding the absolute extreme values of a continuous function that is defined on a closed interval.

86. A function f is defined and continuous on the closed interval $[a, b]$. Why can't $f(a)$ be a local extreme value on $[a, b]$?

87. Show that if f has a local minimum at c, then $g(x) = -f(x)$ has a local maximum at c.

88. Show that if f has a local minimum at c, then either $f'(c) = 0$ or $f'(c)$ does not exist.

89. Locating Extreme Values   Find the absolute maximum value

and the absolute minimum value of $f(x) = \sqrt{1 + x^2} + |x - 2|$

on $[0, 3]$, and determine where each occurs.

Challenge Problem ────────────

90. (a) Prove that a rational function of the form $f(x) = \dfrac{ax^{2n} + b}{cx^n + d}$,

$n \geq 1$ an integer, has at most five critical numbers.

(b) Give an example of such a rational function with exactly five critical numbers.

AP® Practice Problems

Multiple-Choice Questions

1. The critical numbers of $g(x) = \sin x + \cos x$, $0 < x < 2\pi$, are

(A) $\dfrac{\pi}{4}$ (B) $\dfrac{3\pi}{4}$ and $\dfrac{7\pi}{4}$

(C) $\dfrac{\pi}{4}$ and $\dfrac{5\pi}{4}$ (D) $\dfrac{\pi}{4}, \dfrac{3\pi}{4}, \dfrac{5\pi}{4}$, and $\dfrac{7\pi}{4}$

2. On the closed interval $[0, 2\pi]$, the absolute minimum of $f(x) = e^{\sin x}$ occurs at

(A) 0 (B) $\dfrac{\pi}{2}$

(C) $\dfrac{3\pi}{2}$ (D) 2π

3. The maximum value of $f(x) = 2x^3 - 15x^2 + 36x$ on the closed interval $[0, 4]$ is

(A) 28 (B) 30 (C) 32 (D) 48

4. For $0 \le x \le 5$, the function $f(x) = 3 - |x - 1|$ has:

(A) both an absolute maximum and an absolute minimum.
(B) an absolute maximum but no absolute minimum.
(C) no absolute maximum but an absolute minimum.
(D) an absolute maximum and two absolute minima.

5. The critical numbers of the function

$$f(x) = \begin{cases} x^2 + 1 & \text{if } -2 \le x \le 1 \\ 3x^2 - 4x + 3 & \text{if } 1 < x \le 3 \end{cases} \quad \text{are}$$

(A) 0 and 1 (B) 0 and $\dfrac{2}{3}$

(C) $0, \dfrac{2}{3}$, and 1 (D) 0

6. $f'(x) = x \sin^2 x - \dfrac{1}{x}$ is the derivative of a function f. How many critical numbers does f have on the open interval $(0, 2\pi)$?

(A) 1 (B) 3 (C) 4 (D) 5

7. A function f is continuous on the closed interval $[0, 8]$ and differentiable on the open interval $(0, 8)$, except at 5. The table below shows selected values of $f(x)$.

x	0	1	3	5	7	8
$f(x)$	15	3	9	-4	-1	4

f has a local maximum at 3 and a local minimum at 5. Which of the following statements must be true?

   I. The absolute minimum value of f is -4.
   II. The function f is decreasing for $0 \le x \le 1$.
   III. $f'(3) = 0$

(A) I and II only   (B) I only

(C) III only   (D) I, II, and III

8. At what value of x does the function $f(x) = xe^{2x}$ for $-1 \le x \le 2$ have an absolute minimum?

(A) $\dfrac{1}{2}$ (B) -1 (C) $-\dfrac{1}{2}$ (D) 2

9. A function f is continuous on the closed interval $[1, 10]$. The table below shows selected values of $f(x)$.

x	1	2	5	7	10
$f(x)$	3	9	-9	-1	4

Which of the following statements must be true?

   I. There is a number c in the interval $[1, 10]$ for which $f(c) = 0$.
   II. There is a number c in the interval $[1, 10]$ for which $f(c) \le f(x)$ for all x in the interval $[1, 10]$.
   III. There is a number c in the interval $(1, 10)$ for which $f'(c) = 0$.

(A) I and II   (B) only I   (C) I and III   (D) I, II, and III

Free-Response Question

10. Find the absolute extreme values of $f(x) = x^2\sqrt{4 - x}$ on the closed interval $[-5, 4]$.

Retain Your Knowledge

Multiple-Choice Questions

1. Given $f(x) = e^{2x}(x^3 + 2x - 7)$, find where $f'(x) = 0$.

(A) 0 (B) 1.087 (C) 1.171 (D) 1.269

2. Find y' at the point $(2, 1)$ on the graph of $5x^2y^2 - 2x^3 = 4y^3$.

(A) -1 (B) $\dfrac{1}{7}$ (C) $\dfrac{1}{2}$ (D) 3

3. Identify the numbers x, if any, where the graph of $f(x) = x^3\sqrt{4 - x}$ has a horizontal tangent line.

(A) 0 and 4 (B) $0, 4$, and $\dfrac{24}{7}$

(C) 0 and $\dfrac{24}{7}$ (D) f has no horizontal tangent lines.

Free-Response Question

4. A right triangle has a hypotenuse of 5 in. The legs, x and y, of the triangle vary with respect to time. If leg x decreases at the rate of 2 in/h, at what rate is leg y changing when $x = 3$ in?

5.3 Local Extrema and Concavity

OBJECTIVES *When you finish this section, you should be able to:*

1 Use the First Derivative Test to find local extrema (p. 362)
2 Use the First Derivative Test with motion on a line (p. 365)
3 Determine the concavity of a function (p. 367)
4 Find inflection points (p. 370)
5 Use the Second Derivative Test to find local extrema (p. 372)

So far we know that if a function f, defined on a closed interval, has a local maximum or a local minimum at a number c in the open interval, then c is a critical number. We are now ready to see how the derivative is used to determine whether a function f has a local maximum, a local minimum, or neither at a critical number.

① Use the First Derivative Test to Find Local Extrema

All local extreme values of a function f occur at critical numbers. While the value of f at each critical number is a candidate for being a local extreme value for f, not every critical number results in a local extreme value.

How do we distinguish critical numbers that result in local extreme values from those that do not? And then how do we determine if a local extreme value is a local maximum value or a local minimum value? Figure 25 provides a clue. If you look from left to right along the graph of f, you see that the graph of f is increasing to the left of x_1, where a local maximum occurs, and is decreasing to its right. The function f is decreasing to the left of x_2, where a local minimum occurs, and is increasing to its right. So, knowing where a function f increases and decreases enables us to find local maximum values and local minimum values.

The next theorem is usually referred to as the *First Derivative Test*, since it relies on information obtained from the first derivative of a function.

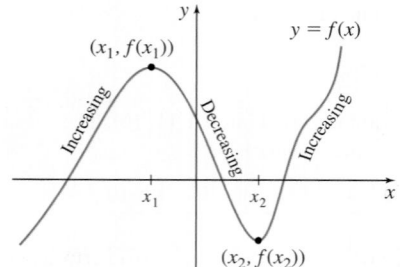

Figure 25

IN WORDS If c is a critical number of f and if f is increasing to the left of c and decreasing to the right of c, then $f(c)$ is a local maximum value. If f is decreasing to the left of c and increasing to the right of c, then $f(c)$ is a local minimum value.

THEOREM First Derivative Test

Suppose f is a function that is continuous on an interval I. Suppose that c is a critical number of f and (a, b) is an open interval in I containing c:

• If $f'(x) > 0$ for $a < x < c$ and $f'(x) < 0$ for $c < x < b$, then $f(c)$ is a local maximum value.
• If $f'(x) < 0$ for $a < x < c$ and $f'(x) > 0$ for $c < x < b$, then $f(c)$ is a local minimum value.
• If $f'(x)$ has the same sign on both sides of c, then $f(c)$ is neither a local maximum value nor a local minimum value.

Partial Proof If $f'(x) > 0$ on the interval (a, c), then f is increasing on (a, c). Also, if $f'(x) < 0$ on the interval (c, b), then f is decreasing on (c, b). So, for all x in (a, b), $f(x) \le f(c)$. That is, $f(c)$ is a local maximum value. ∎

See Figure 26(a).

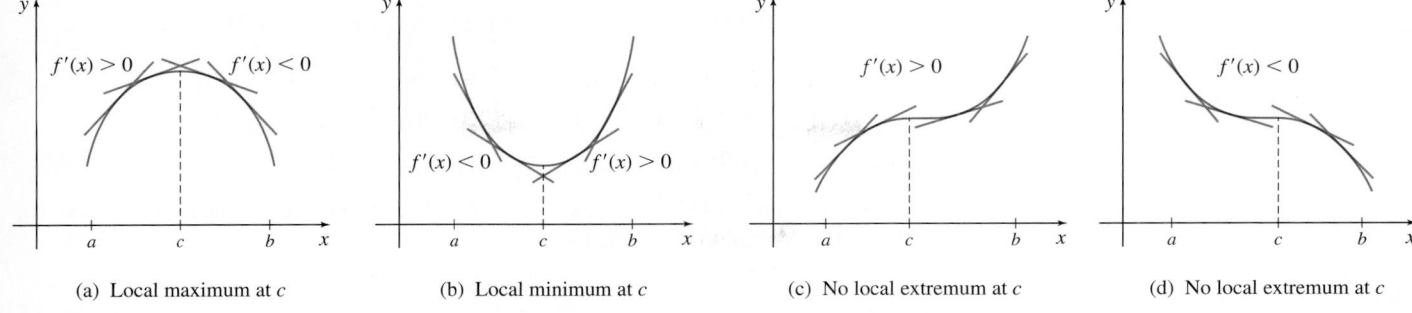

(a) Local maximum at c (b) Local minimum at c (c) No local extremum at c (d) No local extremum at c

Figure 26

Figure 26(b) illustrates the behavior of f' near a local minimum value. In Figures 26(c) and 26(d), f' has the same sign on both sides of c, so f has neither a local maximum nor a local minimum at c.

EXAMPLE 1   Using the First Derivative Test to Find Local Extrema

Find the local extrema of $f(x) = x^4 - 4x^3$.

Solution

Since f is a polynomial function, f is continuous and differentiable at every real number. We begin by finding the critical numbers of f.

$$f'(x) = 4x^3 - 12x^2 = 4x^2(x - 3)$$

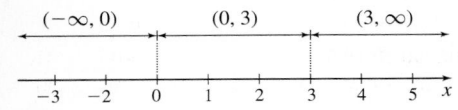

Figure 27

The critical numbers are 0 and 3. We use the critical numbers 0 and 3 to form three intervals on the x-axis, as shown in Figure 27. Then we determine where f is increasing and where it is decreasing by determining the sign of $f'(x)$ in each interval. See Table 4.

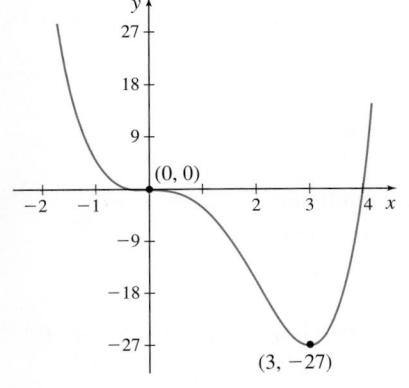

Figure 28 $f(x) = x^4 - 4x^3$

TABLE 4

Interval	Sign of x^2	Sign of $x - 3$	Sign of $f'(x) = 4x^2(x - 3)$	Conclusion
$(-\infty, 0)$	Positive $(+)$	Negative $(-)$	Negative $(-)$	f is decreasing
$(0, 3)$	Positive $(+)$	Negative $(-)$	Negative $(-)$	f is decreasing
$(3, \infty)$	Positive $(+)$	Positive $(+)$	Positive $(+)$	f is increasing

Using the First Derivative Test, f has neither a local maximum nor a local minimum at 0, and f has a local minimum at 3. The local minimum value is $f(3) = -27$. ∎

The graph of f is shown in Figure 28. Notice that the tangent line to the graph of f is horizontal at the points $(0, 0)$ and $(3, -27)$.

NOW WORK   Problem 13 and AP® Practice Problems 3, 10, and 13.

EXAMPLE 2   Using the First Derivative Test to Find Local Extrema

Find the local extrema of $f(x) = x^{2/3}(x - 5)$.

Solution

The domain of f is all real numbers and f is continuous on its domain.

(continued on the next page)

$$f'(x) = \frac{d}{dx}[x^{2/3}(x-5)] = x^{2/3} + \frac{2}{3} \cdot x^{-1/3}(x-5)$$

$$= x^{2/3} + \frac{2(x-5)}{3x^{1/3}} = \frac{3x + 2(x-5)}{3x^{1/3}} = \frac{5(x-2)}{3x^{1/3}}$$

Since $f'(2) = 0$ and $f'(0)$ does not exist, the critical numbers are 0 and 2.

Use the critical numbers 0 and 2 to form three intervals on the x-axis: $(-\infty, 0)$, $(0, 2)$, and $(2, \infty)$ as shown in Table 5.

TABLE 5

Interval	Sign of $x-2$	Sign of $x^{1/3}$	Sign of $f'(x) = \dfrac{5(x-2)}{3x^{1/3}}$	Conclusion
$(-\infty, 0)$	Negative $(-)$	Negative $(-)$	Positive $(+)$	f is increasing
$(0, 2)$	Negative $(-)$	Positive $(+)$	Negative $(-)$	f is decreasing
$(2, \infty)$	Positive $(+)$	Positive $(+)$	Positive $(+)$	f is increasing

The sign of $f'(x)$ on each interval tells us where f is increasing and decreasing. By the First Derivative Test, f has a local maximum at 0 and a local minimum at 2; $f(0) = 0$ is the local maximum value and $f(2) = -3\sqrt[3]{4}$ is the local minimum value. ∎

Since $f'(2) = 0$, the graph of f has a horizontal tangent line at the point $(2, -3\sqrt[3]{4})$. Since $f'(0)$ does not exist and $\lim\limits_{x \to 0^-} f'(x) = \infty$ and $\lim\limits_{x \to 0^+} f'(x) = -\infty$, the graph of f has a vertical tangent line at the point $(0, 0)$, a cusp. The graph of f is shown in Figure 29.

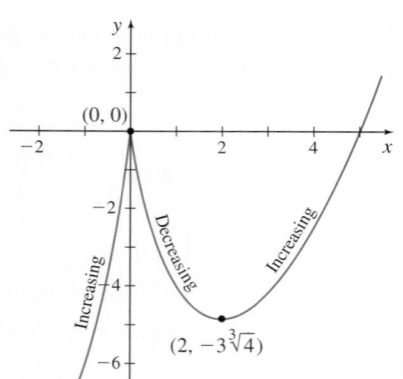

Figure 29 $f(x) = x^{2/3}(x-5)$

NEED TO REVIEW? Corners and cusps are defined in Section 2.2, p.183.

NOW WORK Problems 21 and 51(a) and AP® Practice Problem 12.

EXAMPLE 3 Obtaining Information About the Graph of f from the Graph of Its Derivative

A function f is continuous for all real numbers. Figure 30 shows the graph of its derivative function f'.

(a) Determine the critical numbers of f.
(b) Where is f increasing?
(c) Where is f decreasing?
(d) At what numbers x, if any, does f have a local minimum?
(e) At what numbers x, if any, does f have a local maximum?

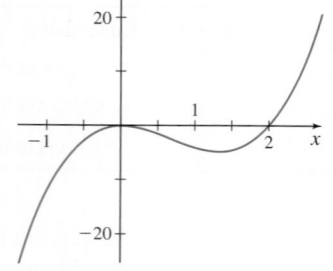

Figure 30 $y = f'(x)$

Solution

(a) The critical numbers of f occur where $f'(x) = 0$ or where $f'(x)$ does not exist. Here f' exists for all real numbers, and the critical numbers of f are 0 and 2.

(b) The function f is increasing where $f'(x) > 0$; that is, where the graph of f' is above the x-axis. This occurs for $x > 2$. So f is increasing on the interval $(2, \infty)$.

(c) The function f is decreasing where $f'(x) < 0$; that is, where the graph of f' is below the x-axis. So f is decreasing on the intervals $(-\infty, 0)$ and $(0, 2)$.

(d) Since $f'(x) < 0$ for $x < 2$ and $f'(x) > 0$ for $x > 2$, by the First Derivative Test, f has a local minimum at $x = 2$.

(e) f has no local maximum. (Do you see why?) There is no number c on the graph of f' where f' changes from positive (increasing) to negative (decreasing). ∎

NOW WORK Problem 35 and AP® Practice Problem 4.

② Use the First Derivative Test with Motion on a Line

Increasing and decreasing functions can be used to investigate the motion of an object as it moves along a horizontal line with positive direction to the right. Suppose the position x of an object from the origin at time t is given by the function $x = f(t)$. The velocity v of the object is $v = \dfrac{dx}{dt}$.

Investigating the Motion of an Object Moving on a Line

- If $v = \dfrac{dx}{dt} > 0$, the position x of the object is increasing with time t, and the object moves to the right.

- If $v = \dfrac{dx}{dt} < 0$, the position x of the object is decreasing with time t, and the object moves to the left.

- If $v = \dfrac{dx}{dt} = 0$, the object is at rest.

This information, along with the First Derivative Test, can be used to find the local extreme values of $x = f(t)$ and to determine at what times t the direction of the motion of the object changes.

Similarly, if the acceleration of the object, $a = \dfrac{dv}{dt} > 0$, then the velocity of the object is increasing, and if $a = \dfrac{dv}{dt} < 0$, then the velocity is decreasing. Again, this information, along with the First Derivative Test, is used to find the local extreme values of the velocity.

Next, recall that the speed of an object moving on a line is the absolute value of the velocity. That is, speed $= |v(t)|$. Then

$$\frac{d}{dt}|v(t)| = \frac{d}{dt}\sqrt{[v(t)]^2} = \frac{v(t)}{(\sqrt{[v(t)]^2})} \cdot v'(t) = \frac{v(t) \cdot a(t)}{|v(t)|}$$

where $a(t) = v'(t)$ is the acceleration of the object. Since $|v(t)|$ is positive, by the Increasing/Decreasing Function Test,

---

- the speed is increasing when $v(t) \cdot a(t) > 0$.
- the speed is decreasing when $v(t) \cdot a(t) < 0$.

In other words,

- the speed is increasing when $v(t)$ and $a(t)$ have the same sign (both positive or both negative).
- the speed is decreasing when $v(t)$ and $a(t)$ are opposite in sign (one positive, one negative).

EXAMPLE 4   **Using the First Derivative Test with Motion Along a Line**

An object moves along a horizontal line so that its position x from the origin at time $t \geq 0$, in seconds, is given by the function

$$x = x(t) = t^3 - 9t^2 + 15t + 3$$

(continued on the next page)

(a) Determine the time intervals when the object is moving to the right.

(b) Determine the time intervals when the object is moving to the left.

(c) When does the object reverse direction?

(d) Draw a figure that illustrates the motion of the object.

(e) When is the velocity of the object increasing and when is it decreasing?

(f) Draw a figure that illustrates the velocity of the object.

(g) When is the acceleration of the object increasing and when is it decreasing?

(h) Draw a figure that illustrates the acceleration of the object.

(i) Determine when the speed of the object is increasing and when the speed is decreasing.

Solution

To investigate the motion, we first find the velocity v.

$$v = \frac{dx}{dt} = \frac{d}{dt}(t^3 - 9t^2 + 15t + 3) = 3t^2 - 18t + 15 = 3(t^2 - 6t + 5) = 3(t-1)(t-5)$$

Solving $\frac{dx}{dt} = 0$, we find the critical numbers are 1 and 5. We use the critical numbers to form three intervals on the nonnegative t-axis: $(0, 1)$, $(1, 5)$, and $(5, \infty)$ as shown in Table 6.

TABLE 6

Time Interval	Sign of $t - 1$	Sign of $t - 5$	Velocity, v	Position x	Motion of the Object
$(0, 1)$	Negative $(-)$	Negative $(-)$	Positive $(+)$	Increasing	To the right
$(1, 5)$	Positive $(+)$	Negative $(-)$	Negative $(-)$	Decreasing	To the left
$(5, \infty)$	Positive $(+)$	Positive $(+)$	Positive $(+)$	Increasing	To the right

We use Table 6 to describe the motion of the object.

(a) The object moves to the right for the first second and again after 5 s.

(b) The object moves to the left from $t = 1$ s to $t = 5$ s.

(c) The object reverses direction at $t = 1$ s and $t = 5$ s.

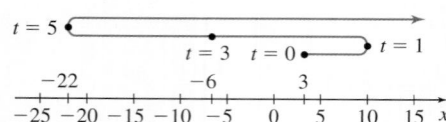

Figure 31 $x(t) = t^3 - 9t^2 + 15t + 3$

(d) Figure 31 illustrates the motion of the object. The motion begins when $t = 0$ at $x = 3$.

(e) To determine when the velocity increases or decreases, we find the acceleration.

$$a = \frac{dv}{dt} = \frac{d}{dt}(3t^2 - 18t + 15) = 6t - 18 = 6(t-3)$$

Since $a < 0$ on the interval $(0, 3)$, the velocity v decreases for the first 3 s. On the interval $(3, \infty)$, $a > 0$, so the velocity v increases from 3 s onward.

Figure 32 $v(t) = 3(t-1)(t-5)$

(f) Figure 32 illustrates the velocity of the object. At $t = 0$ the velocity is 15.

(g) To determine when the acceleration is increasing and decreasing, we need the derivative of a. That is, we need $a'(t)$:

$$a'(t) = \frac{d}{dt}(6t - 18) = 6$$

Since $a'(t) > 0$ for all $t > 0$, the acceleration of the object is increasing for all $t > 0$.

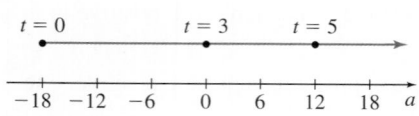

Figure 33 $a(t) = 6t - 8$

(h) Figure 33 illustrates the acceleration of the object. At $t = 0$ the acceleration is -18.

(i) The speed of an object is increasing when $v(t) \cdot a(t) > 0$, and the speed is decreasing when $v(t) \cdot a(t) < 0$.

$$v(t) = 3(t-1)(t-5)$$

Then $v(t) > 0$ for $0 < t < 1$ and $t > 5$, and $v(t) < 0$ for $1 < t < 5$.

$$a(t) = 6(t-3)$$

(continued on the next page)

Then $a(t) > 0$ for $t > 3$, and $a(t) < 0$ for $0 < t < 3$.

The speed of the object is increasing for $1 < t < 3$ seconds and when $t > 5$ seconds. The speed of the object is decreasing for $0 < t < 1$ second and for $3 < t < 5$ seconds. ∎

NOW WORK   Problem **29** and AP® Practice Problems **7** and **15**.

③ Determine the Concavity of a Function

Figure 34 shows the graphs of two familiar functions: $y = x^2$, $x \geq 0$, and $y = \sqrt{x}$. Each graph starts at the origin, passes through the point $(1, 1)$, and is increasing. But there is a noticeable difference in their shapes. The graph of $y = x^2$ bends upward; the function $y = x^2$ is *concave up*. The graph of $y = \sqrt{x}$ bends downward; the function $y = \sqrt{x}$ is *concave down*.

Suppose we draw tangent lines to the graphs of $y = x^2$ and $y = \sqrt{x}$, as shown in Figure 35. Notice that the graph of $y = x^2$ lies above all its tangent lines, and moving from left to right, the slopes of the tangent lines are increasing. That is, the derivative $y' = \dfrac{d}{dx} x^2$ is an increasing function.

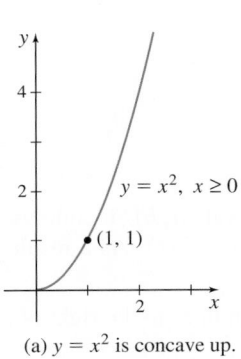

(a) $y = x^2$ is concave up.

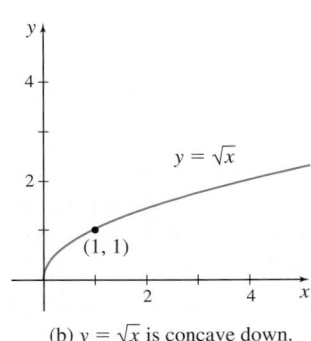

(b) $y = \sqrt{x}$ is concave down.

Figure 34

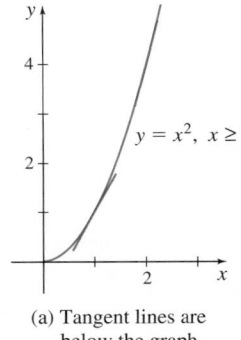

(a) Tangent lines are below the graph.

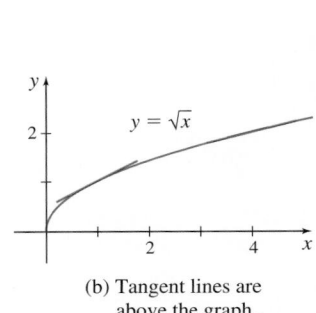

(b) Tangent lines are above the graph.

Figure 35

On the other hand, the graph of $y = \sqrt{x}$ lies below all its tangent lines, and moving from left to right, the slopes of the tangent lines are decreasing. That is, the derivative $y' = \dfrac{d}{dx} \sqrt{x}$ is a decreasing function. This discussion leads to the following definition.

DEFINITION Concave Up; Concave Down

Suppose f is a function that is continuous on a closed interval $[a, b]$ and differentiable on the open interval (a, b).

- f is **concave up** on (a, b) if the graph of f lies above each of its tangent lines throughout (a, b).

- f is **concave down** on (a, b) if the graph of f lies below each of its tangent lines throughout (a, b).

We can formulate a test to determine where a function f is concave up or concave down, provided f'' exists. Since f'' equals the rate of change of f', it follows that if $f''(x) > 0$ on an open interval, then f' is increasing on that interval, and f is concave up on the interval. Similarly, if $f''(x) < 0$ on an open interval, then f' is decreasing on that interval and f is concave down on the interval. Look again at Figures 34 and 35. These observations lead to the *Test for Concavity*.

NOTE When f' and f'' both exist on an open interval (a, b), the function f is called **twice differentiable** on (a, b).

THEOREM Test for Concavity

Suppose f is a function that is continuous on a closed interval $[a, b]$. Suppose f' and f'' exist on the open interval (a, b).

- If $f''(x) > 0$ on the interval (a, b), then f is concave up on (a, b).
- If $f''(x) < 0$ on the interval (a, b), then f is concave down on (a, b).

Proof Suppose $f''(x) > 0$ on the interval (a, b), and c is any fixed number in (a, b). An equation of the tangent line to f at the point $(c, f(c))$ is $y = f(c) + f'(c)(x - c)$

We need to show that the graph of f lies above each of its tangent lines for all x in (a, b). That is, we need to show that

$$f(x) \geq f(c) + f'(c)(x - c) \qquad \text{for all } x \text{ in } (a, b)$$

If $x = c$, then $f(x) = f(c)$ and we are finished.

If $x \neq c$, then by applying the Mean Value Theorem to the function f, there is a number x_1 between c and x, for which $f'(x_1) = \dfrac{f(x) - f(c)}{x - c}$

Now solve for $f(x)$:

$$f(x) = f(c) + f'(x_1)(x - c) \tag{1}$$

There are two possibilities: Either $c < x_1 < x$ or $x < x_1 < c$.

Suppose $c < x_1 < x$. Since $f''(x) > 0$ on the interval (a, b), it follows that f' is increasing on (a, b). For $x_1 > c$, this means that $f'(x_1) > f'(c)$. As a result, from (1), we have $f(x) > f(c) + f'(c)(x - c)$

That is, the graph of f lies above each of its tangent lines to the right of c in (a, b).

Similarly, if $x < x_1 < c$, then $f(x) > f(c) + f'(c)(x - c)$.

In all cases, $f(x) \geq f(c) + f'(c)(x - c)$ so f is concave up on (a, b).

The proof that if $f''(x) < 0$, then f is concave down is left as an exercise. ∎

AP® EXAM TIP

If a function f is twice differentiable, then f and f' are both continuous and the Intermediate Value Theorem and Mean Value Theorem can be applied to both f and f'.

EXAMPLE 5 Determining the Concavity of $f(x) = e^x$

Show that $f(x) = e^x$ is concave up on its domain.

Solution

The domain of $f(x) = e^x$ is all real numbers. The first and second derivatives of f are

$$f'(x) = e^x \qquad f''(x) = e^x$$

Since $f''(x) > 0$ for all real numbers, by the Test for Concavity, f is concave up on its domain. ∎

Figure 36 shows the graph of $f(x) = e^x$ and a selection of tangent lines to the graph. Notice that for any x, the graph of f lies above its tangent lines. So any linear approximation of $f(x) = e^x$ will underestimate the true value of f.

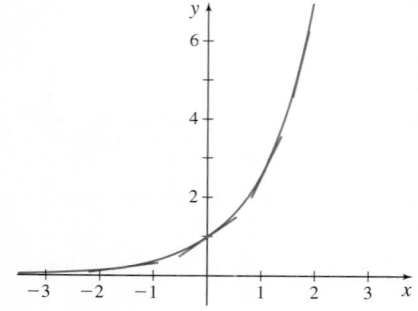

Figure 36 $f(x) = e^x$ is concave up on its domain.

NOW WORK AP® Practice Problem **16.**

▶ **EXAMPLE 6**   **Finding Local Extrema and Determining Concavity**
CALC CLIP

(a) Find any local extrema of the function $f(x) = x^3 - 6x^2 + 9x + 30$.

(b) Determine where $f(x) = x^3 - 6x^2 + 9x + 30$ is concave up and where it is concave down.

Solution

(a) The first derivative of f is $f'(x) = 3x^2 - 12x + 9 = 3(x - 1)(x - 3)$

The function f is a polynomial, so its critical numbers occur when $f'(x) = 0$. So, 1 and 3 are critical numbers of f. Now

- $f'(x) = 3(x - 1)(x - 3) > 0$ if $x < 1$ or $x > 3$.
- $f'(x) < 0$ if $1 < x < 3$.

So f is increasing on $(-\infty, 1)$ and on $(3, \infty)$; f is decreasing on $(1, 3)$.

At 1, f has a local maximum, and at 3, f has a local minimum. The local maximum value is $f(1) = 34$; the local minimum value is $f(3) = 30$.

(b) To determine concavity, we use the second derivative:

$$f''(x) = 6x - 12 = 6(x - 2)$$

Solve the inequalities $f''(x) < 0$ and $f''(x) > 0$ and use the Test for Concavity.

- $f''(x) < 0$ if $x < 2$,   so f is concave down on $(-\infty, 2)$
- $f''(x) > 0$ if $x > 2$,   so f is concave up on $(2, \infty)$ ∎

Figure 37 shows the graph of f. The point $(2, 32)$ where the concavity of f changes is of special importance and is called an *inflection point*.

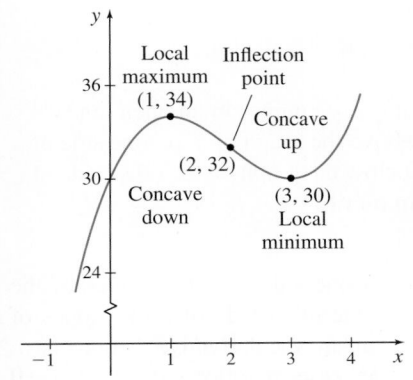

Figure 37 $f(x) = x^3 - 6x^2 + 9x + 30$

NOW WORK   Problem **51(b)** and AP® Practice Problems **9** and **14.**

EXAMPLE 7   **Obtaining Information About an Implicitly-Defined Function**

A twice-differentiable function f is defined implicitly by the equation $y^3 + 3y = x^3 - 4$, and its graph contains the point $(2, 1)$.

(a) Find any points on the graph of f at which f has a horizontal tangent line.

(b) Is the graph of f increasing or decreasing at the point $(2, 1)$?

(c) Is the graph of f concave up or down at the point $(2, 1)$?

Solution

(a) Horizontal tangent lines occur at points where $y' = f'(x) = 0$. We differentiate the equation $y^3 + 3y = x^3 - 4$ implicitly to find y'.

$$3y^2 y' + 3y' = 3x^2$$

$$(y^2 + 1)y' = x^2$$

Now $y' = 0$, if $x = 0$. If $x = 0$, then $y^3 + 3y = -4$, so $y = -1$. The graph of f has a horizontal tangent line at the point $(0, -1)$.

(b) Use the Increasing/Decreasing Function Test. At the point $(2, 1)$, the derivative y' is

$$y' = \frac{x^2}{y^2 + 1} = \frac{4}{2} = 2 \qquad x = 2, \ y = 1$$

Since $y' > 0$ at $(2, 1)$, the graph of f is increasing there.

(continued on the next page)

(c) We need to find y''. Differentiating $3y^2y' + 3y' = 3x^2$ implicitly, we obtain

$$6y(y')^2 + 3y^2y'' + 3y'' = 6x$$

At the point (2, 1), we have $x = 2$, $y = 1$, and $y' = 2$. Substituting, we obtain

$$24 + 6y'' = 12$$
$$y'' = -2$$

Since $y'' < 0$, the graph of f is concave down at the point (2, 1). ∎

NOW WORK AP® Practice Problem **17**.

④ Find Inflection Points

RECALL A function f has a tangent line at a number c when $f'(c)$ exists (a non-vertical tangent line) or when $f'(c)$ is unbounded (a vertical tangent line).

DEFINITION Inflection Point

Suppose f is a function that has a tangent line at every number in an open interval (a, b) containing c. If the tangent lines of f lie above the graph of f on one side of the point $(c, f(c))$ and the tangent lines of f lie below the graph on the other side of the point $(c, f(c))$, then $(c, f(c))$ is an **inflection point** of f.

If $(c, f(c))$ is an inflection point of f, then on one side of c the slopes of the tangent lines are increasing (or decreasing), and on the other side of c the slopes of the tangent lines are decreasing (or increasing). This means the derivative f' must have a local maximum or a local minimum at c. In either case, it follows that $f''(c) = 0$ or $f''(c)$ does not exist.

THEOREM A Condition for an Inflection Point

Suppose f is a function that has tangent lines at every number in an open interval (a, b) containing c. If $(c, f(c))$ is an inflection point of f, then either $f''(c) = 0$ or f'' does not exist at c.

Notice the wording in the theorem. If you *know* that $(c, f(c))$ is an inflection point of f, then the second derivative of f at c is 0 or does not exist. The converse is not necessarily true. In other words, a number at which $f''(x) = 0$ or at which f'' does not exist does not always identify an inflection point.

Steps for Finding the Inflection Points of a Function f

Step 1 Find all numbers in the domain of f at which $f''(x) = 0$ or at which f'' does not exist. Eliminate any number at which f does not have a tangent line.

Step 2 Use the Test for Concavity to determine the concavity of f on both sides of each of the numbers found in Step 1.

Step 3 If the concavity changes, there is an inflection point; otherwise, there is no inflection point.

EXAMPLE 8 **Finding Inflection Points**

Find the inflection points of $f(x) = x^{5/3}$.

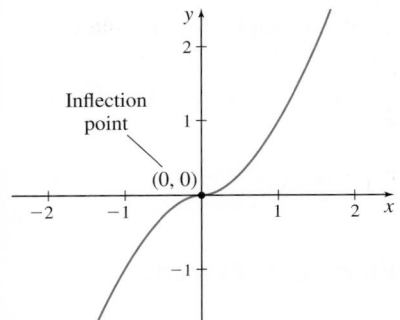

Figure 38 $f(x) = x^{5/3}$

Figure 39 $f(x) = \dfrac{1}{x}$

Solution

We follow the steps for finding an inflection point.

Step 1 The domain of f is all real numbers. The first and second derivatives of f are

$$f'(x) = \frac{5}{3}x^{2/3} \qquad f''(x) = \frac{10}{9}x^{-1/3} = \frac{10}{9x^{1/3}}$$

The second derivative of f does not exist when $x = 0$. Since f' exists for all x, the point $(0, 0)$ is a possible inflection point.

Step 2 Now use the Test for Concavity.

• If $x < 0$ then $f''(x) < 0$ so f is concave down on $(-\infty, 0)$.
• If $x > 0$ then $f''(x) > 0$ so f is concave up on $(0, \infty)$.

Step 3 Since the concavity of f changes at 0, we conclude that $(0, 0)$ is an inflection point of f. ∎

Figure 38 shows the graph of $f(x) = x^{5/3}$ with its inflection point $(0, 0)$ marked.

A change in concavity does not, of itself, guarantee an inflection point. For example, Figure 39 shows the graph of $f(x) = \dfrac{1}{x}$. The first and second derivatives are $f'(x) = -\dfrac{1}{x^2}$ and $f''(x) = \dfrac{2}{x^3}$. Then

$$f''(x) = \frac{2}{x^3} < 0 \quad \text{if } x < 0 \quad \text{and} \quad f''(x) = \frac{2}{x^3} > 0 \quad \text{if } x > 0$$

So, f is concave down on $(-\infty, 0)$ and concave up on $(0, \infty)$, yet f has no inflection point at $x = 0$ because 0 is not in the domain of f.

NOW WORK Problems **49** and **51(c)** and AP® Practice Problems **1, 2, 8,** and **11.**

EXAMPLE 9   **Obtaining Information About the Graph of f from the Graphs of f' and f''**

A function f is continuous for all real numbers. The graphs of f' and f'' are given in Figure 40.

(a) Determine the critical numbers of f.
(b) Where is f increasing? Where is f decreasing?
(c) At what numbers x, if any, does f have a local minimum?
(d) At what numbers x, if any, does f have a local maximum?
(e) Where is f concave up? Where is f concave down?
(f) Find any inflection points of f.

Solution

(a) The critical numbers of f occur where $f'(x) = 0$ or where f' does not exist. See Figure 40(a). Since f' exists for all real numbers and since $f'(0) = 0$ and $f'\left(\dfrac{4}{3}\right) = 0$, the critical numbers are 0 and $\dfrac{4}{3}$.

(b) f is increasing where $f'(x) > 0$; that is, where the graph of f' is above the x-axis. This occurs on the intervals $(-\infty, 0)$ and $\left(\dfrac{4}{3}, \infty\right)$.

f is decreasing where $f'(x) < 0$; that is, where the graph of f' is below the x-axis. This occurs on the interval $\left(0, \dfrac{4}{3}\right)$.

(c) Since f is decreasing for $x < \dfrac{4}{3}$, $(f'(x) < 0)$, and increasing for $x > \dfrac{4}{3}$, $(f'(x) > 0)$, the function f has a local minimum at $x = \dfrac{4}{3}$.

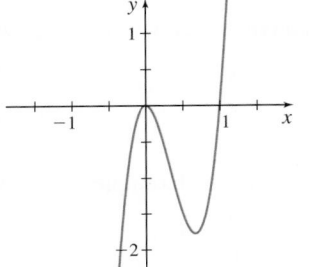

(a) $y = f'(x)$

(b) $y = f''(x)$

Figure 40

(continued on the next page)

(d) Since f is increasing for $x < 0$, $(f'(x) > 0)$, and decreasing for $x > 0$, $(f'(x) < 0)$, the function f has a local maximum at $x = 0$.

(e) f is concave up where $f''(x) > 0$; that is, where the graph of f'' is above the x-axis. See Figure 40(b). So f is concave up on the interval $(1, \infty)$.

  f is concave down where $f''(x) < 0$; that is, where the graph of f'' is below the x-axis. So f is concave down on the intervals $(-\infty, 0)$ and $(0, 1)$.

(f) The only number at which the concavity changes is $x = 1$. Since $f'(1)$ exists, the point $(1, f(1))$ is an inflection point of f. ■

NOW WORK   Problem **63** and AP® Practice Problems **5, 6**, and **19**.

⑤ Use the Second Derivative Test to Find Local Extrema

Suppose c is a critical number of f and $f'(c) = 0$. This means that the graph of f has a horizontal tangent line at the point $(c, f(c))$. If f'' exists on an open interval containing c and $f''(c) > 0$, then the graph of f is concave up on the interval, as shown in Figure 41(a). Intuitively, it would seem that $f(c)$ is a local minimum value of f. If, on the other hand, $f''(c) < 0$, the graph of f is concave down on the interval, and $f(c)$ would appear to be a local maximum value of f, as shown in Figure 41(b). The next theorem, known as the *Second Derivative Test*, confirms our intuition.

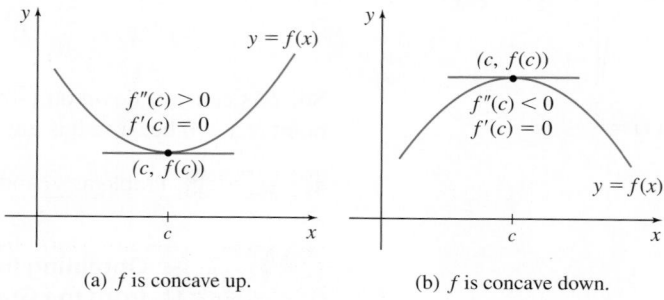

(a) f is concave up.   (b) f is concave down.

Figure 41

THEOREM Second Derivative Test

Suppose f is a function for which f' and f'' exist on an open interval (a, b). Suppose c lies in (a, b) and is a critical number of f.

- If $f''(c) < 0$, then $f(c)$ is a local maximum value.
- If $f''(c) > 0$, then $f(c)$ is a local minimum value.

Proof Suppose $f''(c) < 0$. Then $f''(c) = \lim\limits_{x \to c} \dfrac{f'(x) - f'(c)}{x - c} < 0$

  Now since $\lim\limits_{x \to c} \dfrac{f'(x) - f'(c)}{x - c} < 0$, there is an open interval containing c for which

$$\frac{f'(x) - f'(c)}{x - c} < 0$$

everywhere in the interval, except possibly at c itself (refer to Example 7, p. 156, in Section 1.6). Since c is a critical number, $f'(c) = 0$, so $\dfrac{f'(x)}{x - c} < 0$

  For $x < c$ on this interval, $f'(x) > 0$, and for $x > c$ on this interval, $f'(x) < 0$. By the First Derivative Test, $f(c)$ is a local maximum value. ■

EXAMPLE 10 **Using the Second Derivative Test to Identify Local Extrema**

Use the Second Derivative Test to identify any local extreme values of $f(x) = x - 2\cos x$, $0 \le x \le 2\pi$.

Solution

The first and second derivatives of f are

$$f'(x) = \frac{d}{dx}(x - 2\cos x) = 1 + 2\sin x \qquad \text{and} \qquad f''(x) = \frac{d}{dx}(1 + 2\sin x) = 2\cos x$$

We find the critical numbers by solving the equation $f'(x) = 0$.

$$1 + 2\sin x = 0 \qquad 0 \le x \le 2\pi$$
$$\sin x = -\frac{1}{2}$$
$$x = \frac{7\pi}{6} \qquad \text{or} \qquad x = \frac{11\pi}{6}$$

Using the Second Derivative Test, we get

$$f''\left(\frac{7\pi}{6}\right) = 2\cos\left(\frac{7\pi}{6}\right) = -\sqrt{3} < 0$$

and

$$f''\left(\frac{11\pi}{6}\right) = 2\cos\left(\frac{11\pi}{6}\right) = \sqrt{3} > 0$$

So,

$$f\left(\frac{7\pi}{6}\right) = \frac{7\pi}{6} - 2\cos\frac{7\pi}{6} = \frac{7\pi}{6} + \sqrt{3} \approx 5.397$$

is a local maximum value, and

$$f\left(\frac{11\pi}{6}\right) = \frac{11\pi}{6} - 2\cos\frac{11\pi}{6} = \frac{11\pi}{6} - \sqrt{3} \approx 4.028$$

is a local minimum value. ∎

Figure 42 $y = x - 2\cos x$, $0 \le x \le 2\pi$

See Figure 42 for the graph of f.

Sometimes the Second Derivative Test cannot be used, and sometimes it gives no information.

- If the second derivative of the function does not exist at a critical number, the Second Derivative Test cannot be used.
- If the second derivative exists at a critical number, but equals 0, the Second Derivative Test gives no information.

  In these situations, the First Derivative Test must be used to identify local extreme points.

An example of a function for which the Second Derivative Test cannot be used is $f(x) = x^4$. Both $f'(x) = 4x^3$ and $f''(x) = 12x^2$ exist for all real numbers. Since $f'(0) = 0$, 0 is a critical number of f. But $f''(0) = 0$, so the Second Derivative Test gives no information about the behavior of f at 0.

NOW WORK Problem **59** and AP® Practice Problem **18.**

Application: The Logistic Function in Business

EXAMPLE 11 **Analyzing Monthly Sales**

Unit monthly sales R of a new product over a period of time are expected to follow the logistic function

$$R = R(t) = \frac{20{,}000}{1 + 50e^{-t}} - \frac{20{,}000}{51} \qquad t \geq 0$$

where t is measured in months.

(a) When are the monthly sales increasing? When are they decreasing?

(b) Find the rate of change of sales.

(c) When is the rate of change of sales R' increasing? When is it decreasing?

(d) When is the rate of change of sales a maximum?

(e) Find any inflection points of R.

(f) Interpret the result found in (e) in the context of the problem.

Solution

(a) We find $R'(t)$ and use the Increasing/Decreasing Function Test.

$$R'(t) = \frac{d}{dt}\left(\frac{20{,}000}{1 + 50e^{-t}} - \frac{20{,}000}{51} \right) = 20{,}000 \cdot \frac{50e^{-t}}{(1 + 50e^{-t})^2} = \frac{1{,}000{,}000e^{-t}}{(1 + 50e^{-t})^2}$$

Since $e^{-t} > 0$ for all $t \geq 0$, then $R'(t) > 0$ for $t \geq 0$. The sales function R is an increasing function. So, monthly sales are always increasing.

(b) The rate of change of sales is given by the derivative $R'(t) = \dfrac{1{,}000{,}000e^{-t}}{(1 + 50e^{-t})^2}$, $t \geq 0$.

Note that $R'(t)$ exists for all $t \geq 0$.

(c) Using the Increasing/Decreasing Function Test with R', the rate of change of sales R' is increasing when its derivative $R''(t) > 0$; $R'(t)$ is decreasing when $R''(t) < 0$.

$$R''(t) = \frac{d}{dt} R'(t) = 1{,}000{,}000 \left[\frac{-e^{-t}(1 + 50e^{-t})^2 + 100e^{-2t}(1 + 50e^{-t})}{(1 + 50e^{-t})^4} \right]$$

$$= 1{,}000{,}000e^{-t} \left[\frac{-1 - 50e^{-t} + 100e^{-t}}{(1 + 50e^{-t})^3} \right] = \frac{1{,}000{,}000e^{-t}}{(1 + 50e^{-t})^3}(50e^{-t} - 1)$$

Since $e^{-t} > 0$ for all t, the sign of R'' depends on the sign of $50e^{-t} - 1$.

$$\begin{array}{ll} 50e^{-t} - 1 > 0 & \qquad 50e^{-t} - 1 < 0 \\ 50e^{-t} > 1 & \qquad 50e^{-t} < 1 \\ 50 > e^{t} & \qquad 50 < e^{t} \\ t < \ln 50 & \qquad t > \ln 50 \end{array}$$

Since $R''(t) > 0$ for $t < \ln 50 \approx 3.912$ and $R''(t) < 0$ for $t > \ln 50 \approx 3.912$, the rate of change of sales is increasing for the first 3.9 months and is decreasing from 3.9 months on.

(d) The critical number of R' is $\ln 50 \approx 3.912$. Using the First Derivative Test, the rate of change of sales is a maximum about 3.9 months after the product is introduced.

(e) Since $R''(t) > 0$ for $t < \ln 50$ and $R''(t) < 0$ for $t > \ln 50$, the point $(\ln 50, 9608)$, is the inflection point of R.

(f) The sales function R is an increasing function. Before the inflection point $(\ln 50, 9608)$, the rate of change in sales R' is increasing $(R'' > 0)$, and after the inflection point, R' is decreasing $(R'' < 0)$. ■

See Figure 43 for the graphs of R and R'.

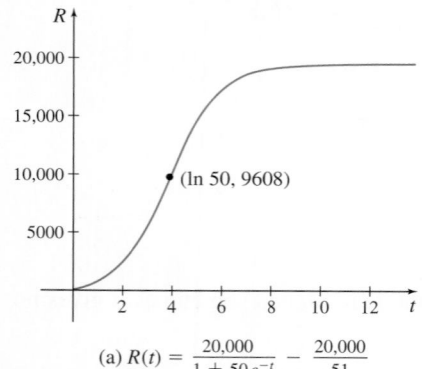

(a) $R(t) = \dfrac{20{,}000}{1 + 50e^{-t}} - \dfrac{20{,}000}{51}$

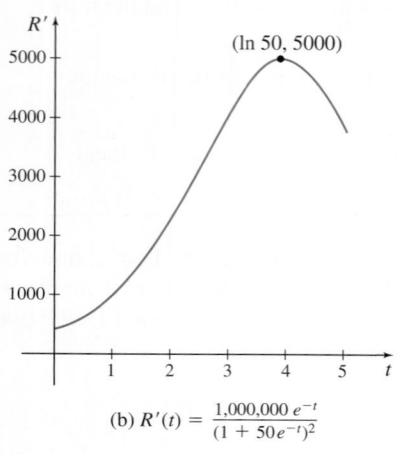

(b) $R'(t) = \dfrac{1{,}000{,}000\, e^{-t}}{(1 + 50e^{-t})^2}$

Figure 43

NOW WORK Problem 93.

5.3 Assess Your Understanding

Concepts and Vocabulary

1. **True or False** If a function f is continuous on the interval $[a, b]$, differentiable on the interval (a, b), and changes from an increasing function to a decreasing function at the point $(c, f(c))$, then $(c, f(c))$ is an inflection point of f.

2. **True or False** Suppose c is a critical number of f and (a, b) is an open interval containing c. If $f'(x)$ is positive on both sides of c, then $f(c)$ is a local maximum value.

3. **Multiple Choice** Suppose a function f is continuous on a closed interval $[a, b]$ and differentiable on the open interval (a, b). If the graph of f lies above each of its tangent lines on the interval (a, b), then on (a, b) f is

   [(**a**) concave up (**b**) concave down (**c**) neither].

4. **Multiple Choice** If the acceleration of an object in moving on a line is negative, then the velocity of the object is

   [(**a**) increasing (**b**) decreasing (**c**) neither].

5. **Multiple Choice** Suppose f is a function that is differentiable on an open interval containing c and the concavity of f changes at the point $(c, f(c))$. Then the point $(c, f(c))$ on the graph of f is a(n)

   [(**a**) inflection point (**b**) critical point (**c**) both (**d**) neither].

6. **Multiple Choice** Suppose a function f is continuous on a closed interval $[a, b]$ and both f' and f'' exist on the open interval (a, b). If $f''(x) > 0$ on the interval (a, b), then on (a, b) f is

   [(**a**) increasing (**b**) decreasing (**c**) concave up (**d**) concave down].

7. **True or False** Suppose f is a function for which f' and f'' exist on an open interval (a, b) and suppose c, $a < c < b$, is a critical number of f. If $f''(c) = 0$, then the Second Derivative Test cannot be used to determine if there is a local extremum at c.

8. **True or False** Suppose a function f is differentiable on the open interval (a, b). If either $f''(c) = 0$ or f'' does not exist at the number c in (a, b), then $(c, f(c))$ is an inflection point of f.

Skill Building

In Problems 9–12, the graph of a function f is given.

(**a**) *Identify the points where each function has a local maximum value, a local minimum value, or an inflection point.*

(**b**) *Identify the intervals on which each function is increasing, decreasing, concave up, or concave down.*

9.

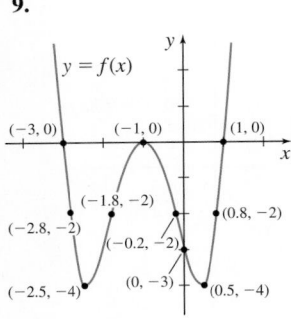

10.

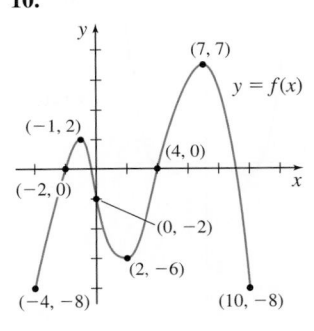

11.

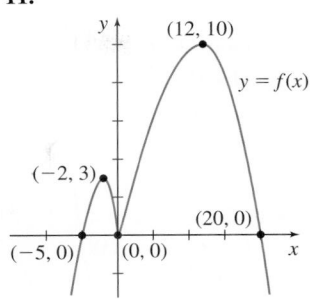

12.

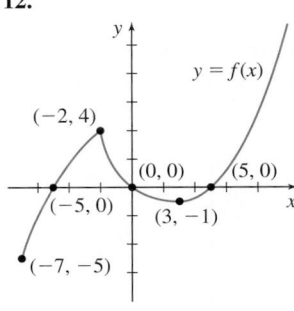

In Problems 13–26, for each function:

(**a**) *Find the critical numbers.*

(**b**) *Use the First Derivative Test to find any local extrema.*

13. $f(x) = x^3 - 6x^2 + 2$

14. $f(x) = x^3 + 6x^2 + 12x + 1$

15. $f(x) = 3x^4 - 4x^3$

16. $h(x) = x^4 + 2x^3 - 3$

17. $f(x) = (5 - 2x)e^x$

18. $f(x) = (x - 8)e^x$

19. $f(x) = x^{2/3} + x^{1/3}$

20. $f(x) = \dfrac{1}{2}x^{2/3} - x^{1/3}$

21. $g(x) = x^{2/3}(x^2 - 4)$

22. $f(x) = x^{1/3}(x^2 - 9)$

23. $f(x) = \dfrac{\ln x}{x^3}$

24. $h(x) = \dfrac{\ln x}{\sqrt{x^3}}$

25. $f(\theta) = \sin \theta - 2\cos \theta$

26. $f(x) = x + 2\sin x$

In Problems 27–34, an object moves along a horizontal line with the positive direction to the right. The position x of the object from the origin at time $t \geq 0$ (in seconds) is given by the function $x = x(t)$.

(**a**) *Determine the intervals during which the object moves to the right and the intervals during which it moves to the left.*

(**b**) *When does the object reverse direction?*

(**c**) *When is the velocity of the object increasing and when is it decreasing?*

(**d**) *Draw a figure to illustrate the motion of the object.*

(**e**) *Draw a figure to illustrate the velocity of the object.*

(**f**) *When is the acceleration of the object increasing, and when is the acceleration decreasing?*

(**g**) *When is the speed of the object increasing, and when is the speed decreasing?*

27. $x = t^2 - 2t + 3$

28. $x = 2t^2 + 8t - 7$

29. $x = 2t^3 + 6t^2 - 18t + 1$

30. $x = 3t^4 - 16t^3 + 24t^2$

31. $x = 2t - \dfrac{6}{t}, \quad t > 0$

32. $x = 3\sqrt{t} - \dfrac{1}{\sqrt{t}}, \quad t > 0$

33. $x = 2\sin(3t), \quad 0 \leq t \leq \dfrac{2\pi}{3}$

34. $x = 3\cos(\pi t), \quad 0 \leq t \leq 2$

In Problems 35–38, the function f is continuous for all real numbers and the graph of its derivative function f′ is given.

(a) *Determine the critical numbers of f.*

(b) *Where is f increasing?*

(c) *Where is f decreasing?*

(d) *At what numbers x, if any, does f have a local minimum?*

(e) *At what numbers x, if any, does f have a local maximum?*

PAGE 364 **35.**

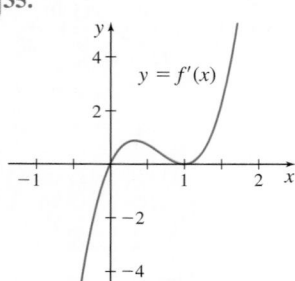

36.

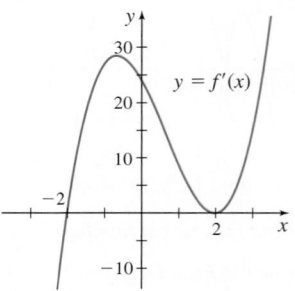

37.

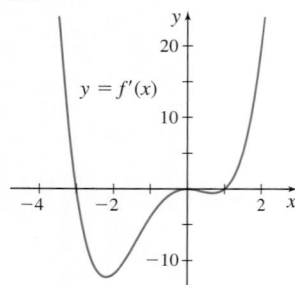

38.

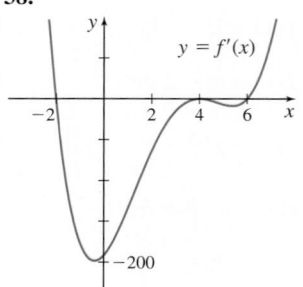

In Problems 39–62:

(a) *Find the local extrema of f.*

(b) *Determine the intervals on which f is concave up and on which it is concave down.*

(c) *Find any points of inflection.*

39. $f(x) = 2x^3 - 6x^2 + 6x - 3$

40. $f(x) = 2x^3 + 9x^2 + 12x - 4$

41. $f(x) = x^4 - 4x$

42. $f(x) = x^4 + 4x$

43. $f(x) = 5x^4 - x^5$

44. $f(x) = 4x^6 + 6x^4$

45. $f(x) = 3x^5 - 20x^3$

46. $f(x) = 3x^5 + 5x^3$

47. $f(x) = x^2 e^x$

48. $f(x) = x^3 e^x$

PAGE 371 **49.** $f(x) = 6x^{4/3} - 3x^{1/3}$

50. $f(x) = x^{2/3} - x^{1/3}$

PAGE 364 **51.** $f(x) = x^{2/3}(x^2 - 8)$

52. $f(x) = x^{1/3}(x^2 - 2)$

53. $f(x) = x^2 - \ln x$

54. $f(x) = \ln x - x$

55. $f(x) = \dfrac{x}{(1 + x^2)^{5/2}}$

56. $f(x) = \dfrac{\sqrt{x}}{1 + x}$

57. $f(x) = x^2 \sqrt{1 - x^2}$

58. $f(x) = x\sqrt{1 - x}$

PAGE 373 **59.** $f(x) = x - 2\sin x, \quad 0 \le x \le 2\pi$

60. $f(x) = x + 2\cos x, \quad 0 \le x \le 2\pi$

61. $f(x) = 2\cos^2 x - \sin^2 x, \quad 0 < x < 2\pi$

62. $f(x) = 2\sin^2 x - \cos^2 x, \quad 0 < x < 2\pi$

In Problems 63–66, the function f is continuous for all real numbers and the graphs of f′ and f″ are given.

(a) *Determine the critical numbers of f.*

(b) *Where is f increasing?*

(c) *Where is f decreasing?*

(d) *At what numbers x, if any, does f have a local minimum?*

(e) *At what numbers x, if any, does f have a local maximum?*

(f) *Where is f concave up?*

(g) *Where is f concave down?*

(h) *List any inflection points.*

PAGE 372 **63.**

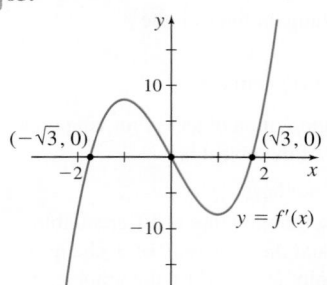 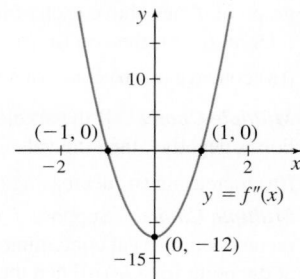

64.

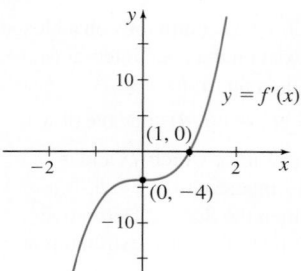 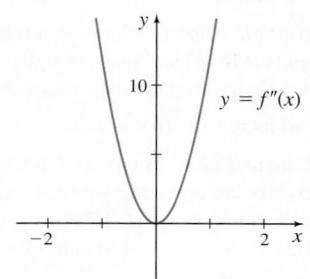

65.

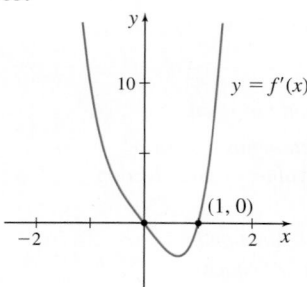 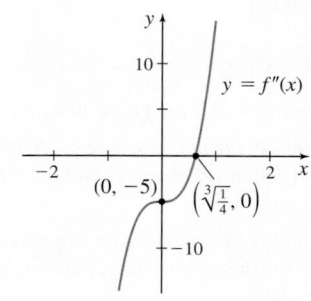

66.

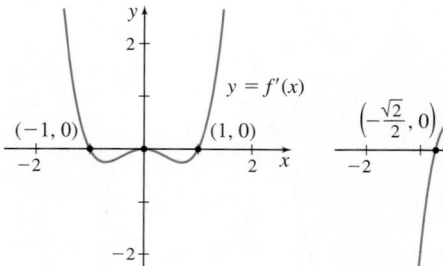 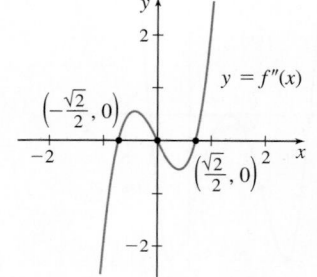

In Problems 67–74, find the local extrema of each function f by:

(a) *Using the First Derivative Test.*

(b) *Using the Second Derivative Test.*

(c) *Discuss which of the two tests you found easier.*

67. $f(x) = -2x^3 + 15x^2 - 36x + 7$

68. $f(x) = x^3 + 10x^2 + 25x - 25$

69. $f(x) = x^4 - 8x^2 - 5$

70. $f(x) = x^4 + 2x^2 + 2$

71. $f(x) = 3x^5 + 5x^4 + 1$

72. $f(x) = 60x^5 + 20x^3$

73. $f(x) = (x - 3)^2 e^x$

74. $f(x) = (x + 1)^2 e^{-x}$

Applications and Extensions

In Problems 75–86, sketch the graph of a continuous function f that has the given properties. Answers will vary.

75. f is concave up on $(-\infty, \infty)$, increasing on $(-\infty, 0)$, decreasing on $(0, \infty)$, and $f(0) = 1$.

76. f is concave up on $(-\infty, 0)$, concave down on $(0, \infty)$, decreasing on $(-\infty, 0)$, increasing on $(0, \infty)$, and $f(0) = 1$.

77. f is concave down on $(-\infty, 1)$, concave up on $(1, \infty)$, decreasing on $(-\infty, 0)$, increasing on $(0, \infty)$, $f(0) = 1$, and $f(1) = 2$.

78. f is concave down on $(-\infty, 0)$, concave up on $(0, \infty)$, increasing on $(-\infty, \infty)$, and $f(0) = 1$ and $f(1) = 2$.

79. $f'(x) > 0$ if $x < 0$; $f'(x) < 0$ if $x > 0$; $f''(x) > 0$ if $x < 0$; $f''(x) > 0$ if $x > 0$ and $f(0) = 1$.

80. $f'(x) > 0$ if $x < 0$; $f'(x) < 0$ if $x > 0$; $f''(x) > 0$ if $x < 0$; $f''(x) < 0$ if $x > 0$ and $f(0) = 1$.

81. $f''(0) = 0$; $f'(0) = 0$; $f''(x) > 0$ if $x < 0$; $f''(x) > 0$ if $x > 0$ and $f(0) = 1$.

82. $f''(0) = 0$; $f'(x) > 0$ if $x \neq 0$; $f''(x) < 0$ if $x < 0$; $f''(x) > 0$ if $x > 0$ and $f(0) = 1$.

83. $f'(0) = 0$; $f'(x) < 0$ if $x \neq 0$; $f''(x) > 0$ if $x < 0$; $f''(x) < 0$ if $x > 0$; $f(0) = 1$.

84. $f''(0) = 0$; $f'(0) = \dfrac{1}{2}$; $f''(x) > 0$ if $x < 0$; $f''(x) < 0$ if $x > 0$ and $f(0) = 1$.

85. $f'(0)$ does not exist; $f''(x) > 0$ if $x < 0$; $f''(x) > 0$ if $x > 0$ and $f(0) = 1$.

86. $f'(0)$ does not exist; $f''(x) < 0$ if $x < 0$; $f''(x) > 0$ if $x > 0$ and $f(0) = 1$.

CAS *In Problems 87–90, for each function:*

(a) *Determine the intervals on which f is concave up and on which it is concave down.*

(b) *Find any points of inflection.*

(c) *Graph f and describe the behavior of f at each inflection point.*

87. $f(x) = e^{-(x-2)^2}$

88. $f(x) = x^2\sqrt{5 - x}$

89. $f(x) = \dfrac{2 - x}{2x^2 - 2x + 1}$

90. $f(x) = \dfrac{3x}{x^2 + 3x + 5}$

91. Inflection Point   For the function $f(x) = ax^3 + bx^2$, find a and b so that the point $(1, 6)$ is an inflection point of f.

92. Inflection Point   For the cubic polynomial function $f(x) = ax^3 + bx^2 + cx + d$, find a, b, c, and d so that 0 is a critical number, $f(0) = 4$, and the point $(1, -2)$ is an inflection point of f.

PAGE 374 **93. Public Health**   In a town of 50,000 people, the number of people at time t who have the flu is $N(t) = \dfrac{10,000}{1 + 9999e^{-t}}$, where t is measured in days. Note that the flu is spread by the one person who has it at $t = 0$.

(a) Find the rate of change of the number of infected people.

(b) When is N' increasing? When is it decreasing?

(c) When is the rate of change of the number of infected people a maximum?

(d) Find any inflection points of N.

(e) Interpret the result found in (d) in the context of the problem.

94. Business   The profit P (in millions of dollars) generated by introducing a new technology is expected to follow the logistic function $P(t) = \dfrac{300}{1 + 50e^{-0.2t}}$, where t is the number of years after its release.

(a) When is annual profit increasing? When is it decreasing?

(b) Find the rate of change in profit.

(c) When is the rate of change in profit increasing? When is it decreasing?

(d) When is the rate of change in profit a maximum?

(e) Find any inflection points of P.

(f) Interpret the result found in (e) in the context of the problem.

95. Population Model   The following data represent the population of the United States:

Year	Population	Year	Population
1900	76,212,168	1970	203,302,031
1910	92,228,486	1980	226,542,203
1920	106,021,537	1990	248,709,873
1930	123,202,624	2000	281,421,906
1940	132,164,569	2010	308,745,538
1950	151,325,798	2020	331,449,281
1960	179,323,175		

Source: U.S. Census Bureau.

An ecologist finds the data fit the logistic function

$$P(t) = \dfrac{656,047,448.3}{1 + 7.47909e^{-0.01708t}}.$$

(a) Draw a scatterplot of the data using the years since 1900 as the independent variable and population as the dependent variable.

(b) Verify that P is the logistic function of best fit.

(c) Find the rate of change in population.

(d) When is P' increasing? When is it decreasing?

(e) When is the rate of change in population a maximum?

(f) Find any inflection points of P.

(g) Interpret the result found in (f) in the context of the problem.

96. Biology The amount of yeast biomass in a culture after t hours is given in the table below.

Time (in hours)	Yeast Biomass	Time (in hours)	Yeast Biomass	Time (in hours)	Yeast Biomass
0	9.6	7	257.3	13	629.4
1	18.3	8	350.7	14	640.8
2	29.0	9	441.0	15	651.1
3	47.2	10	513.3	16	655.9
4	71.1	11	559.7	17	659.6
5	119.1	12	594.8	18	661.8
6	174.6				

Source: Tor Carlson, *Uber Geschwindigkeit und Grosse der Hefevermehrung in Wurze, Biochemische Zeitschrift,* Bd. 57, 1913, pp. 313–334.

The logistic function $y = \dfrac{663.0}{1 + 71.6e^{-0.5470t}}$, where t is time, models the data.

(a) Draw a scatterplot of the data using time t as the independent variable.

(b) Verify that y is the logistic function of best fit.

(c) Find the rate of change in biomass.

(d) When is y' increasing? When is it decreasing?

(e) When is the rate of change in the biomass a maximum?

(f) Find any inflection points of y.

(g) Interpret the result found in (f) in the context of the problem.

97. U.S. Budget The United States budget documents the amount of money (revenue) the federal government takes in (through taxes, for example) and the amount (expenses) it pays out (for social programs, defense, etc.). When revenue exceeds expenses, we say there is a **budget surplus**, and when expenses exceed revenue, there is a **budget deficit**. The function

$$B = B(t) = -4.9t^3 + 21.4t^2 + 204.6t - 1407.2$$

where $0 \le t \le 11$ approximates revenue minus expenses for the years 2010 to 2021, with $t = 0$ representing the year 2010 and B in billions of dollars.

(a) Find all the local extrema of B.

(b) Do the local extreme values represent a budget surplus or a budget deficit?

(c) Find the intervals on which B is concave up or concave down. Identify any inflection points of B.

(d) What does the concavity on either side of the inflection point(s) indicate about the rate of change of the budget? Is it increasing at an increasing rate? Increasing at a decreasing rate?

(e) Graph the function B.

98. If $f(x) = ax^3 + bx^2 + cx + d$, $a \ne 0$, how does the quantity $b^2 - 3ac$ determine the number of potential local extrema?

99. If $f(x) = ax^3 + bx^2 + cx + d$, $a \ne 0$, find a, b, c, and d so that f has a local minimum at 0, a local maximum at 4, and the graph contains the points $(0, 5)$ and $(4, 33)$.

100. Find the local minimum of the function

$$f(x) = \frac{2}{x} + \frac{8}{1-x}, \quad 0 < x < 1.$$

101. Find the local extrema and the inflection points of $y = \sqrt{3} \sin x + \cos x$, $0 \le x \le 2\pi$.

102. If $x > 0$ and $n > 1$, can the expression $x^n - n(x - 1) - 1$ ever be negative?

103. Why must the First Derivative Test be used to find the local extreme values of the function $f(x) = x^{2/3}$?

104. Put It Together Which of the following is true of the function

$$f(x) = x^2 + e^{-2x}$$

(a) f is increasing (b) f is decreasing (c) f is discontinuous at 0

(d) f is concave up (e) f is concave down

105. Put It Together If a function f is continuous for all x and if f has a local maximum at $(-1, 4)$ and a local minimum at $(3, -2)$, which of the following statements must be true?

(a) The graph of f has an inflection point somewhere between $x = -1$ and $x = 3$.

(b) $f'(-1) = 0$.

(c) The graph of f has a horizontal asymptote.

(d) The graph of f has a horizontal tangent line at $x = 3$.

(e) The graph of f intersects both axes.

106. Vertex of a Parabola If $f(x) = ax^2 + bx + c$, $a \ne 0$, prove that f has a local maximum at $-\dfrac{b}{2a}$ if $a < 0$ and has a local minimum at $-\dfrac{b}{2a}$ if $a > 0$.

107. Show that $\sin x \le x$, $0 \le x \le 2\pi$. *Hint:* Let $f(x) = x - \sin x$.

108. Show that $1 - \dfrac{x^2}{2} \le \cos x$, $0 \le x \le 2\pi$. Use the result of Problem 107.

109. Show that $2\sqrt{x} > 3 - \dfrac{1}{x}$, for $x > 1$.

110. Use calculus to show that $x^2 - 8x + 21 > 0$ for all x.

111. Use calculus to show that $3x^4 - 4x^3 - 12x^2 + 40 > 0$ for all x.

112. Show that the function $f(x) = ax^2 + bx + c$, $a \ne 0$, has no inflection points. For what values of a is f concave up? For what values of a is f concave down?

113. Show that every polynomial function of degree 3 $f(x) = ax^3 + bx^2 + cx + d$, $a \ne 0$, has exactly one inflection point.

114. Prove that a polynomial of degree $n \ge 3$ has at most $(n - 1)$ critical numbers and at most $(n - 2)$ inflection points.

115. Show that the function $f(x) = (x - a)^n$, a a constant, has exactly one inflection point if $n \ge 3$ is an odd integer.

116. Show that the function $f(x) = (x - a)^n$, a a constant, has no inflection point if $n \ge 2$ is an even integer.

117. Show that the function $f(x) = \dfrac{ax + b}{ax + d}$ has no critical points and no inflection points.

118. (a) Show that the second derivative of the logistic function

$$f(t) = \frac{c}{1 + ae^{-bt}} \qquad a > 0, c > 0$$

is

$$f''(t) = ab^2 c e^{-bt}(ae^{-bt} - 1)$$

(b) Show that the logistic function f has an inflection point at

$$\left(\frac{\ln a}{b}, \frac{c}{2} \right)$$

Challenge Problems

119. Find the inflection point of $y = (x + 1) \tan^{-1} x$.

120. Bernoulli's Inequality Prove
$(1 + x)^n > 1 + nx$ for $x > -1$, $x \neq 0$, and $n > 1$.
Hint: Let $f(x) = (1 + x)^n - (1 + nx)$.

Preparing for the **AP® Exam**

AP® Practice Problems

Multiple-Choice Questions

1. If $f''(x) = x(x + 1)^2(x - 2)$ is the second derivative of the function f, then list the x-coordinate(s) of the inflection point(s) of f.

(A) $-1, 0$, and 2 (B) 0 only

(C) 2 only   (D) 0 and 2 only

2. If $f(x) = x + \sin x$, find the smallest positive number x at which the function changes concavity.

(A) $\dfrac{\pi}{4}$ (B) $\dfrac{\pi}{2}$ (C) π (D) $\dfrac{3\pi}{2}$

3. The function $g(x) = x^5 + x^3 - 2x - 1$ has a local minimum at

(A) $x = -\dfrac{\sqrt{10}}{5}$ (B) $x = \dfrac{\sqrt{10}}{5}$

(C) $x = \dfrac{2}{5}$ (D) $x = 0$

4. A function f is continuous for all real numbers. The graph of its derivative function f' is shown. The graph has horizontal tangent lines at $(0, 0)$, $(2, 3)$, and $(4.5, -1)$. At which number(s) x does f have a local minimum?

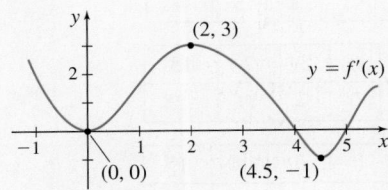

(A) $0, 4$, and 5 (B) 4 only

(C) 5 only   (D) $0, 2$, and 4.5

5. For the function f graphed below, both f' and f'' exist for all numbers x. Which of the following is true?

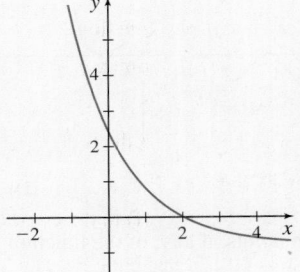

(A) $f(2) < f'(2) < f''(2)$ (B) $f'(2) < f(2) < f''(2)$

(C) $f''(2) < f'(2) < f(2)$ (D) $f''(2) < f(2) < f'(2)$

6. At which point on the graph of f do both f' and f'' have the same sign?

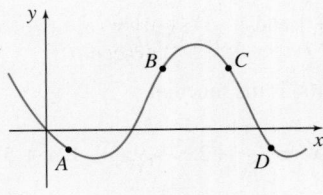

(A) A (B) B (C) C (D) D

(continued on the next page)

367 7. An object moving on a line has position function $s = s(t)$. At time $t = 0$, the object is at the origin. Several values of the velocity v of the object are listed in the table. Which of the following graphs could be the graph of the position of the object for $0 \le t \le 6$?

t	0	1	2	3	4	5	6
$v(t)$	1	−1	2	0	−2	0	1

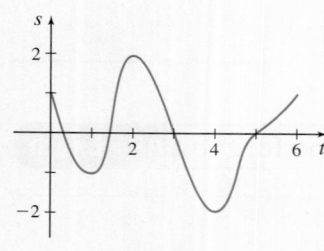

(A)

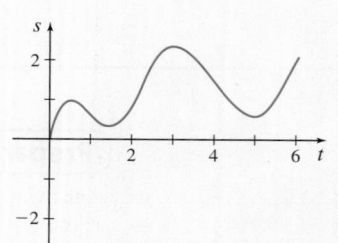

(B)

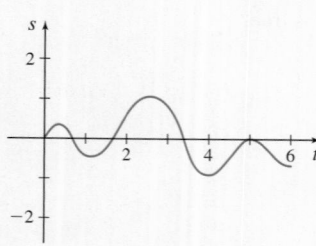

(C)

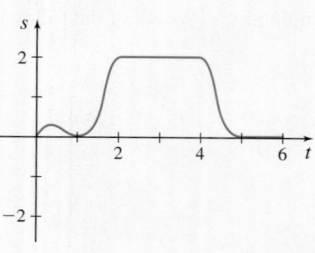

(D)

371 8. Find the inflection points, if any, of the function $f(x) = 3x^4 - 2x^2$.

(A) $(0, 0)$

(B) $\left(-\dfrac{1}{3}, 0\right)$ and $\left(\dfrac{1}{3}, 0\right)$

(C) $\left(-\dfrac{1}{3}, -\dfrac{5}{27}\right)$ and $\left(\dfrac{1}{3}, \dfrac{5}{27}\right)$

(D) $\left(-\dfrac{1}{3}, -\dfrac{5}{27}\right)$ and $\left(\dfrac{1}{3}, -\dfrac{5}{27}\right)$

369 9. For what interval(s) is the function

$$f(x) = \frac{x^4}{2} - 2x^3 - 9x^2 - 12x + 5$$

concave down?

(A) $(-1, 3)$

(B) $(-\infty, -1)$ or $(3, \infty)$

(C) $(-\infty, -3)$ or $(1, \infty)$

(D) Nowhere; the function is always concave up.

363 10. The derivative of a function f is

$$f'(x) = x^2(x - 1)(x + 2)(x + 3)$$

At what number(s) x does f have a relative minimum?

(A) -3 only   (B) 0 and 1

(C) -3 and -2 (D) -3 and 1

371 11. Which statement regarding the function

$$f(x) = (x - 1)^{4/5} - 2$$

is not true?

(A) The point $(1, -2)$ is a local minimum of f.

(B) f is concave down on the intervals $(-\infty, 1)$ and $(1, \infty)$.

(C) The point $(1, -2)$ is an inflection point of f.

(D) f has a vertical tangent line at $(1, -2)$.

364 12. Find the minimum value, if any, of the function $f(x) = x \ln x$.

(A) $-\dfrac{1}{e}$ (B) -1 (C) $\dfrac{1}{e}$ (D) f has no minimum value.

363 13. A polynomial function f has only three critical numbers -5, 0, and 2 and has three local extrema: a local minimum point at $(-5, -3)$, a local maximum point at $(0, -1)$, and a local minimum point at $(2, -6)$. How many zeros does f have?

(A) One   (B) Two   (C) Three   (D) Four

369 14. Determine where the function $f(x) = \dfrac{2x^2}{x + 3}$ is concave up.

(A) $(-3, \infty)$ (B) $(0, \infty)$ (C) $(-2, \infty)$ (D) $(-\infty, -3)$

367 15. An object moves along the x-axis so that its position at time $t \ge 1$ (in seconds) is $x(t) = \dfrac{\ln t}{t}$. At what time t is the object farthest from the origin?

(A) $2e$ seconds   (B) 1 second   (C) 2 seconds   (D) e seconds

368 16. A twice differentiable function $y = f(x)$ has the properties that $f'(x) > 0$ for all x and $f''(x) < 0$ for all x. Which of the following tables could represent the function f?

(A)

x	0	3	6	15
$f(x)$	1	4	7	6

(B)

x	0	3	8	15
$f(x)$	1	2	3	4

(C)

x	0	3	7	15
$f(x)$	0	9	49	125

(D)

x	0	3	10	15
$f(x)$	10	−1	90	115

370 17. A twice differentiable function f is defined implicitly by the equation $x^3 + 3x = y^3 - 4$, and the graph of f contains the point $(1, 2)$. Which of the following must be true about the graph of f at the point $(1, 2)$?

(A) The graph is decreasing and concave down.

(B) The graph is decreasing and concave up.

(C) The graph is increasing and concave down.

(D) The graph is increasing and concave up.

Free-Response Questions

18. A function f is continuous and twice differentiable on the closed interval $[1, 10]$. Also, $f(2) = 5$ and $f(4) = 5$.

    (a) Verify that there is a number c in the interval $(2, 4)$ for which $f'(c) = 0$.

    (b) If $f''(c) = 10$, explain why f has a local minimum at c.

19. A function f is continuous for $-3 \le x \le 4$. The graph of its derivative f' is shown in the accompanying figure.

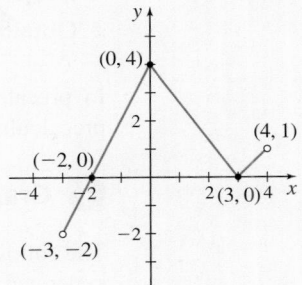

    (a) List the critical numbers of f and indicate whether each locates a local maximum, local minimum, or neither. Justify your answer.

    (b) Where is the function f concave up? Where is the function f concave down? Justify your answers.

    (c) List any numbers at which f has an inflection point.

See the **BREAK IT DOWN** *on page 418 for a stepped out solution to AP® Practice Problem 19(a).*

Retain Your Knowledge

Multiple-Choice Questions

1. Find $\displaystyle\lim_{x \to 0} \frac{\sin(3x^2)}{2x}$

    (A) $\dfrac{1}{2}$ (B) $\dfrac{3}{2}$ (C) 0   (D) The limit does not exist.

2. Identify the graph of $f'(x)$ for the function

$$f(x) = \begin{cases} |x - 1| + 3 & \text{if } x \le 2 \\ x^2 & \text{if } x > 2 \end{cases}$$

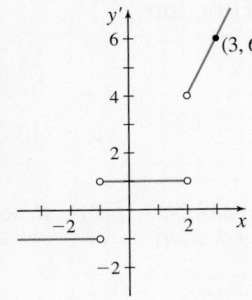

(A)

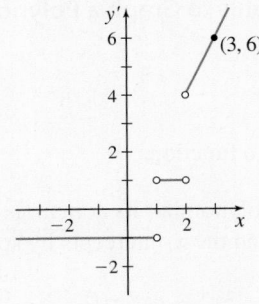

(B)

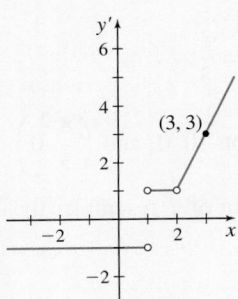

(C)

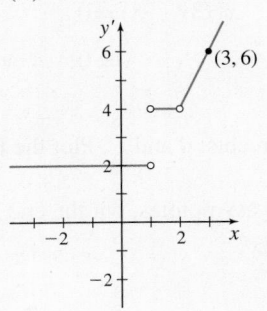

(D)

3. The radius r of a right circular cone is increasing at a rate of 2 cm/h, and the height h of the cone is decreasing at a rate of 3 cm/h. At what rate is the volume V of the cone changing when $r = 6$ cm and $h = 15$ cm? (The volume V of a right circular cone of height h and radius r is $V = \dfrac{1}{3}\pi r^2 h$.)

    (A) -84π cm^3/h   (B) -28π cm^3/h

    (C) 84π cm^3/h   (D) 180π cm^3/h

Free-Response Question

4. If $f(x) = \tan^{-1}(x^3 - 1)$, find $f''(x)$. Justify your answer.

5.4 Using Calculus to Graph Functions

OBJECTIVE *When you finish this section, you should be able to:*

1 Graph a function using calculus (p. 382)
2 Obtain information about a function from its derivatives (p. 390)

In precalculus, algebra was used to graph a function. In this section, we combine precalculus methods with calculus to obtain a more detailed graph.

1 Graph a Function Using Calculus

The following steps are used to graph a function by hand or to analyze a technology-generated graph of a function.

Steps for Graphing a Function $y = f(x)$

Step 1 Find the domain of f and any intercepts.

Step 2 Identify any asymptotes, and examine the end behavior.

Step 3 Find $y = f'(x)$ and use it to find any critical numbers of f. Also find f''.

Step 4 Use the Increasing/Decreasing Function Test to identify intervals on which f is increasing ($f'(x) > 0$) and the intervals on which f is decreasing ($f'(x) < 0$).

Step 5 Use either the First Derivative Test or the Second Derivative Test (if applicable) to identify local maximum values and local minimum values.

Step 6 Use the Test for Concavity to determine intervals where the function is concave up ($f''(x) > 0$) and concave down ($f''(x) < 0$). Identify any inflection points.

Step 7 Graph the function using the information gathered. Plot additional points as needed. It may be helpful to find the slope of the tangent line at such points.

CALC CLIP

EXAMPLE 1   Using Calculus to Graph a Polynomial Function

Graph $f(x) = 3x^4 - 8x^3$.

Solution

Follow the steps for graphing a function.

Step 1 f is a fourth-degree polynomial; its domain is all real numbers. $f(0) = 0$, so the y-intercept is 0. We find the x-intercepts by solving $f(x) = 0$.

$$3x^4 - 8x^3 = 0$$
$$x^3(3x - 8) = 0$$
$$x = 0 \quad \text{or} \quad x = \frac{8}{3}$$

There are two x-intercepts: 0 and $\frac{8}{3}$. Plot the intercepts $(0, 0)$ and $\left(\frac{8}{3}, 0\right)$.

Step 2 Polynomials have no asymptotes, but the end behavior of f resembles the power function $y = 3x^4$.

TIP When graphing a function, piece together a sketch of the graph as you go, to ensure that the information is consistent and makes sense. If it appears to be contradictory, check your work. You may have an error.

Step 3 $f'(x) = 12x^3 - 24x^2 = 12x^2(x-2)$ $f''(x) = 36x^2 - 48x = 12x(3x-4)$

For polynomials, the critical numbers occur when $f'(x) = 0$.

$$12x^2(x-2) = 0$$
$$x = 0 \quad \text{or} \quad x = 2$$

The critical numbers are 0 and 2. At the points $(0, 0)$ and $(2, -16)$, the tangent lines are horizontal. Plot these points.

Step 4 Use the critical numbers 0 and 2 to form three intervals on the x-axis: $(-\infty, 0)$, $(0, 2)$, and $(2, \infty)$. Then determine the sign of $f'(x) = 12x^2(x-2)$ on each interval and whether f is increasing or decreasing on the interval.

Interval	Sign of f'	Conclusion
$(-\infty, 0)$	Negative	f is decreasing on $(-\infty, 0)$
$(0, 2)$	Negative	f is decreasing on $(0, 2)$
$(2, \infty)$	Positive	f is increasing on $(2, \infty)$

Step 5 Use the First Derivative Test. Based on the table above, $f(0) = 0$ is neither a local maximum value nor a local minimum value, and $f(2) = -16$ is a local minimum value.

Step 6 $f''(x) = 36x^2 - 48x = 12x(3x-4)$; the zeros of f'' are $x = 0$ and $x = \dfrac{4}{3}$. Use the numbers 0 and $\dfrac{4}{3}$ to form three intervals on the x-axis: $(-\infty, 0)$, $\left(0, \dfrac{4}{3}\right)$, and $\left(\dfrac{4}{3}, \infty\right)$. Then determine the sign of $f''(x)$ on each interval and whether f is concave up or concave down on the interval.

Interval	Sign of f''	Conclusion
$(-\infty, 0)$	Positive	f is concave up on $(-\infty, 0)$
$\left(0, \dfrac{4}{3}\right)$	Negative	f is concave down on $\left(0, \dfrac{4}{3}\right)$
$\left(\dfrac{4}{3}, \infty\right)$	Positive	f is concave up on $\left(\dfrac{4}{3}, \infty\right)$

The concavity of f changes at 0 and $\dfrac{4}{3}$, so the points $(0, 0)$ and $\left(\dfrac{4}{3}, -\dfrac{256}{27}\right)$ are inflection points. Plot the inflection points.

Step 7 Using the points plotted and the information about the shape of the graph, graph the function. See Figure 44. ∎

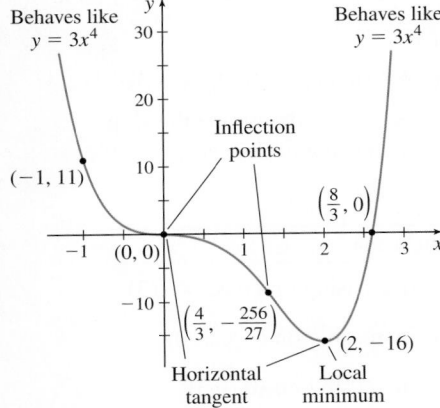

Behaves like $y = 3x^4$ Behaves like $y = 3x^4$
$(-1, 11)$
Inflection points
$\left(\dfrac{8}{3}, 0\right)$
$(0, 0)$
$\left(\dfrac{4}{3}, -\dfrac{256}{27}\right)$
$(2, -16)$
Horizontal tangent   Local minimum

Figure 44 $f(x) = 3x^4 - 8x^3$

NOW WORK Problem 1.

EXAMPLE 2 **Using Calculus to Graph a Rational Function**

Graph $f(x) = \dfrac{6x^2 - 6}{x^3}$.

Solution

Follow the steps for graphing a function.

Step 1 f is a rational function; the domain of f is $\{x \mid x \neq 0\}$. The x-intercepts are -1 and 1. There is no y-intercept. Plot the intercepts $(-1, 0)$ and $(1, 0)$.

(continued on the next page)

Step 2 The degree of the numerator is less than the degree of the denominator, so f has a horizontal asymptote. We find the horizontal asymptotes by finding the limits at infinity.

$$\lim_{x\to\infty}\frac{6x^2-6}{x^3}=\lim_{x\to\infty}\frac{6x^2}{x^3}=\lim_{x\to\infty}\frac{6}{x}=0$$

The line $y=0$ is a horizontal asymptote as $x\to\infty$. It is also a horizontal asymptote as $x\to-\infty$.

We identify vertical asymptotes by checking for infinite limits. [$\lim_{x\to0}f(x)$ is the only one to check—do you see why?]

$$\lim_{x\to0^-}\frac{6x^2-6}{x^3}=\infty \qquad \lim_{x\to0^+}\frac{6x^2-6}{x^3}=-\infty$$

The line $x=0$ is a vertical asymptote. Draw the asymptotes on the graph.

Step 3 $\quad f'(x)=\dfrac{d}{dx}\dfrac{6x^2-6}{x^3}=\dfrac{12x(x^3)-(6x^2-6)(3x^2)}{x^6}$ Use the Quotient Rule.

$$=\frac{-6x^4+18x^2}{x^6}=\frac{6x^2(3-x^2)}{x^6}=\frac{6(3-x^2)}{x^4}$$

$f''(x)=\dfrac{d}{dx}\dfrac{6(3-x^2)}{x^4}=6\cdot\dfrac{(-2x)x^4-(3-x^2)(4x^3)}{x^8}$ Use the Quotient Rule.

$$=6\cdot\frac{2x^5-12x^3}{x^8}=\frac{12(x^2-6)}{x^5}$$

Critical numbers occur where $f'(x)=0$ or where $f'(x)$ does not exist. Since $f'(x)=0$ when $x=\pm\sqrt{3}$, there are two critical numbers, $\sqrt{3}$ and $-\sqrt{3}$. (The derivative does not exist at 0, but, since 0 is not in the domain of f, it is not a critical number.)

Step 4 Use the numbers $-\sqrt{3}$, 0, and $\sqrt{3}$ to form four intervals on the x-axis: $(-\infty,-\sqrt{3})$, $(-\sqrt{3},0)$, $(0,\sqrt{3})$ and $(\sqrt{3},\infty)$. Then determine the sign of $f'(x)$ on each interval and whether f is increasing or decreasing on the interval.

Interval	Sign of f'	Conclusion
$(-\infty,-\sqrt{3})$	Negative	f is decreasing on $(-\infty,-\sqrt{3})$
$(-\sqrt{3},0)$	Positive	f is increasing on $(-\sqrt{3},0)$
$(0,\sqrt{3})$	Positive	f is increasing on $(0,\sqrt{3})$.
$(\sqrt{3},\infty)$	Negative	f is decreasing on $(\sqrt{3},\infty)$

Step 5 Use the First Derivative Test. From the table in Step 4, we conclude

- $f(-\sqrt{3})=-\dfrac{4}{\sqrt{3}}\approx-2.309$ is a local minimum value

- $f(\sqrt{3})=\dfrac{4}{\sqrt{3}}\approx2.309$ is a local maximum value

Plot the local extreme points.

Step 6 Use the Test for Concavity. $f''(x) = \dfrac{12(x^2 - 6)}{x^5} = 0$ when $x = \pm\sqrt{6}$. Now use the numbers $-\sqrt{6}$, 0, and $\sqrt{6}$ to form four intervals on the x-axis: $(-\infty, -\sqrt{6})$, $(-\sqrt{6}, 0)$, $(0, \sqrt{6})$, and $(\sqrt{6}, \infty)$. Then determine the sign of $f''(x)$ on each interval and whether f is concave up or concave down on the interval.

Interval	Sign of f''	Conclusion
$(-\infty, -\sqrt{6})$	Negative	f is concave down on $(-\infty, -\sqrt{6})$
$(-\sqrt{6}, 0)$	Positive	f is concave up on $(-\sqrt{6}, 0)$
$(0, \sqrt{6})$	Negative	f is concave down on $(0, \sqrt{6})$
$(\sqrt{6}, \infty)$	Positive	f is concave up on $(\sqrt{6}, \infty)$

The concavity of the function f changes at $-\sqrt{6}$, 0, and $\sqrt{6}$. So, the points $\left(-\sqrt{6}, -\dfrac{5\sqrt{6}}{6}\right)$ and $\left(\sqrt{6}, \dfrac{5\sqrt{6}}{6}\right)$ are inflection points. Since 0 is not in the domain of f, there is no inflection point at 0. Plot the inflection points.

Step 7 Now use the information about where the function increases/decreases and the concavity to complete the graph. See Figure 45. Notice the apparent symmetry of the graph with respect to the origin. We can verify this symmetry by showing $f(-x) = -f(x)$ and concluding that f is an odd function.

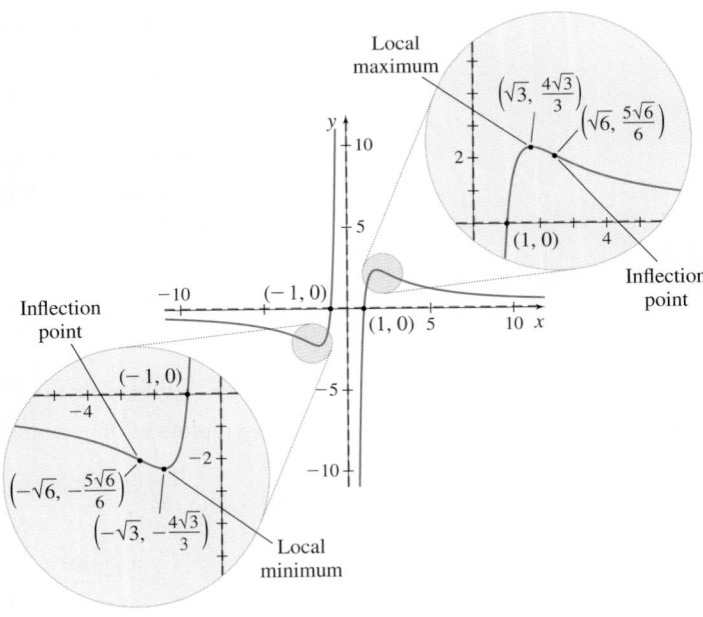

Figure 45 $f(x) = \dfrac{6x^2 - 6}{x^3}$

NOW WORK Problem 11.

Some functions have a graph with an *oblique asymptote*. That is, the graph of the function approaches a line $y = mx + b$, $m \neq 0$, as x becomes unbounded in either the positive or negative direction.

In general, for a function $y = f(x)$, the line $y = mx + b$, $m \neq 0$, is an **oblique asymptote** of the graph of f, if

$$\lim_{x \to -\infty} [f(x) - (mx + b)] = 0 \qquad \text{or} \qquad \lim_{x \to \infty} [f(x) - (mx + b)] = 0$$

In Chapter 1, we saw that rational functions can have vertical or horizontal asymptotes. Rational functions can also have oblique asymptotes, but cannot have both a horizontal asymptote and an oblique asymptote.

- Rational functions for which the degree of the numerator is 1 more than the degree of the denominator have an oblique asymptote.
- Rational functions for which the degree of the numerator is less than or equal to the degree of the denominator have a horizontal asymptote.

EXAMPLE 3 **Using Calculus to Graph a Rational Function**

Graph $f(x) = \dfrac{x^2 - 2x + 6}{x - 3}$.

Solution

Step 1 f is a rational function; the domain of f is $\{x \mid x \neq 3\}$. There are no x-intercepts, since $x^2 - 2x + 6 = 0$ has no real solutions. (Its discriminant is negative.) The y-intercept is $f(0) = -2$. Plot the intercept $(0, -2)$.

Step 2 Identify any vertical asymptotes by checking for infinite limits (3 is the only number to check). Since $\displaystyle\lim_{x \to 3^-} \frac{x^2 - 2x + 6}{x - 3} = -\infty$

and $\displaystyle\lim_{x \to 3^+} \frac{x^2 - 2x + 6}{x - 3} = \infty$, the line $x = 3$ is a vertical asymptote.

The degree of the numerator of f is 1 more than the degree of the denominator, so f will have an oblique asymptote. We divide $x^2 - 2x + 6$ by $x - 3$ to find the line $y = mx + b$.

$$f(x) = \frac{x^2 - 2x + 6}{x - 3} = x + 1 + \frac{9}{x - 3}$$

Since $\displaystyle\lim_{x \to \infty} [f(x) - (x + 1)] = \lim_{x \to \infty} \frac{9}{x - 3} = 0$, then $y = x + 1$ is an oblique asymptote of the graph of f. Draw the asymptotes on the graph.

Step 3
$$f'(x) = \frac{d}{dx} \frac{x^2 - 2x + 6}{x - 3} = \frac{(2x - 2)(x - 3) - (x^2 - 2x + 6)(1)}{(x - 3)^2}$$

$$= \frac{x^2 - 6x}{(x - 3)^2} = \frac{x(x - 6)}{(x - 3)^2}$$

$$f''(x) = \frac{d}{dx} \frac{x^2 - 6x}{(x - 3)^2} = \frac{(2x - 6)(x - 3)^2 - (x^2 - 6x)[2(x - 3)]}{(x - 3)^4}$$

$$= (x - 3) \frac{(2x^2 - 12x + 18) - (2x^2 - 12x)}{(x - 3)^4} = \frac{18}{(x - 3)^3}$$

$f'(x) = 0$ at $x = 0$ and at $x = 6$. So, 0 and 6 are critical numbers. The tangent lines are horizontal at 0 and at 6. Since 3 is not in the domain of f, 3 is not a critical number.

Step 4 Use the numbers 0, 3, and 6 to form four intervals on the x-axis: $(-\infty, 0)$, $(0, 3)$, $(3, 6)$, and $(6, \infty)$. Then determine the sign of $f'(x)$ on each interval and whether f is increasing or decreasing on the interval.

Interval	Sign of $f'(x)$	Conclusion
$(-\infty, 0)$	Positive	f is increasing on $(-\infty, 0)$
$(0, 3)$	Negative	f is decreasing on $(0, 3)$
$(3, 6)$	Negative	f is decreasing on $(3, 6)$
$(6, \infty)$	Positive	f is increasing on $(6, \infty)$

Step 5 Using the First Derivative Test, we find that $f(0) = -2$ is a local maximum value and $f(6) = 10$ is a local minimum value. Plot the points $(0, -2)$ and $(6, 10)$.

Step 6 $f''(x) = \dfrac{18}{(x - 3)^3}$. If $x < 3$, $f''(x) < 0$, and if $x > 3$, $f''(x) > 0$. From the Test for Concavity, we conclude f is concave down on the interval $(-\infty, 3)$ and f is concave up on the interval $(3, \infty)$. The concavity changes at 3, but 3 is not in the domain of f. So, the graph of f has no inflection point.

Step 7 Use the information to complete the graph of the function. See Figure 46. ∎

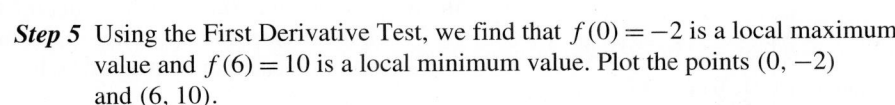

Figure 46 $f(x) = \dfrac{x^2 - 2x + 6}{x - 3}$

(Figure labels: Local minimum; Oblique asymptote $y = x + 1$; $(6, 10)$; $x = 3$ is a vertical asymptote; $(0, -2)$; Local maximum; 12; 8; 4; -4; 4; 8; 12; y; x)

NOW WORK   Problem **13**.

EXAMPLE 4   **Using Calculus to Graph a Function**

Graph $f(x) = 4x^{1/3} - x^{4/3}$.

Solution

Step 1 The domain of f is all real numbers. Since $f(0) = 0$, the y-intercept is 0. Now $f(x) = 0$ when $4x^{1/3} - x^{4/3} = x^{1/3}(4 - x) = 0$ or when $x = 0$ or $x = 4$. So, the x-intercepts are 0 and 4. Plot the intercepts $(0, 0)$ and $(4, 0)$.

Step 2 Since $\lim\limits_{x \to \pm\infty} f(x) = \lim\limits_{x \to \pm\infty} [x^{1/3}(4 - x)] = -\infty$, there is no horizontal asymptote. Since the domain of f is all real numbers, there is no vertical asymptote.

Step 3
$$f'(x) = \frac{d}{dx}(4x^{1/3} - x^{4/3}) = \frac{4}{3}x^{-2/3} - \frac{4}{3}x^{1/3} = \frac{4}{3}\left(\frac{1}{x^{2/3}} - x^{1/3}\right)$$

$$= \frac{4}{3} \cdot \frac{1 - x}{x^{2/3}}$$

$$f''(x) = \frac{d}{dx}\left(\frac{4}{3}x^{-2/3} - \frac{4}{3}x^{1/3}\right) = -\frac{8}{9}x^{-5/3} - \frac{4}{9}x^{-2/3} = -\frac{4}{9}\left(\frac{2}{x^{5/3}} + \frac{1}{x^{2/3}}\right)$$

$$= -\frac{4}{9} \cdot \frac{2 + x}{x^{5/3}}$$

Since $f'(x) = \dfrac{4}{3} \cdot \dfrac{1 - x}{x^{2/3}} = 0$ when $x = 1$ and $f'(x)$ does not exist at $x = 0$, the critical numbers are 0 and 1. At the point $(1, 3)$, the tangent line to the graph is horizontal; at the point $(0, 0)$, the tangent line is vertical. Plot these points.

(continued on the next page)

Step 4 Use the critical numbers 0 and 1 to form three intervals on the x-axis: $(-\infty, 0)$, $(0, 1)$ and $(1, \infty)$. Then determine the sign of $f'(x)$ on each interval and whether f is increasing or decreasing on the interval.

Interval	Sign of f'	Conclusion
$(-\infty, 0)$	Positive	f is increasing on $(-\infty, 0)$
$(0, 1)$	Positive	f is increasing on $(0, 1)$
$(1, \infty)$	Negative	f is decreasing on $(1, \infty)$

Step 5 By the First Derivative Test, $f(1) = 3$ is a local maximum value and $f(0) = 0$ is not a local extreme value.

Step 6 Now test for concavity by using the numbers -2 and 0 to form three intervals on the x-axis: $(-\infty, -2)$, $(-2, 0)$, and $(0, \infty)$. Then determine the sign of $f''(x)$ on each interval and whether f is concave up or concave down on the interval.

Interval	Sign of f''	Conclusion
$(-\infty, -2)$	Negative	f is concave down on the interval $(-\infty, -2)$
$(-2, 0)$	Positive	f is concave up on the interval $(-2, 0)$
$(0, \infty)$	Negative	f is concave down on the interval $(0, \infty)$

The concavity changes at -2 and at 0. Since

$$f(-2) = 4(-2)^{1/3} - (-2)^{4/3} = 4\sqrt[3]{-2} - \sqrt[3]{16} \approx -7.560,$$

the inflection points are $(-2, -7.560)$ and $(0, 0)$. Plot the inflection points.

Step 7 The graph of f is given in Figure 47. ∎

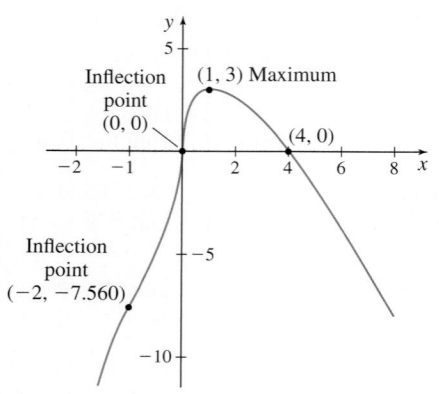

Figure 47 $f(x) = 4x^{1/3} - x^{4/3}$

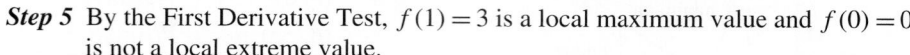
NOW WORK Problem 19.

EXAMPLE 5 **Using Calculus to Graph a Function**

Graph $f(x) = e^x(x^2 - 3)$.

Solution

Step 1 The domain of f is all real numbers. Since $f(0) = -3$, the y-intercept is -3. The x-intercepts occur where

$$f(x) = e^x(x^2 - 3) = 0$$
$$x = \sqrt{3} \quad \text{or} \quad x = -\sqrt{3}$$

Plot the intercepts.

Step 2 Since the domain of f is all real numbers, there are no vertical asymptotes. To determine if there are any horizontal asymptotes, we find the limits at infinity. First we find

$$\lim_{x \to -\infty} f(x) = \lim_{x \to -\infty} [e^x(x^2 - 3)]$$

Since $e^x(x^2 - 3)$ is an indeterminate form at $-\infty$ of the type $0 \cdot \infty$, we write $e^x(x^2 - 3) = \dfrac{x^2 - 3}{e^{-x}}$ and use L'Hôpital's Rule.

$$\lim_{x \to -\infty} f(x) = \lim_{x \to -\infty} [e^x(x^2 - 3)] = \lim_{x \to -\infty} \frac{x^2 - 3}{e^{-x}}$$
$$= \lim_{\substack{\uparrow \\ x \to -\infty}} \frac{2x}{-e^{-x}} = \lim_{\substack{\uparrow \\ x \to -\infty}} \frac{2}{e^{-x}} = 0$$

Type $\dfrac{\infty}{\infty}$; use   Type $\dfrac{\infty}{\infty}$; use

L'Hôpital's Rule   L'Hôpital's Rule

The line $y = 0$ is a horizontal asymptote as $x \to -\infty$.

Since $\lim\limits_{x\to\infty}[e^x(x^2-3)]=\infty$, there is no horizontal asymptote as $x\to\infty$. Draw the asymptote on the graph.

Step 3 $f'(x)=\dfrac{d}{dx}[e^x(x^2-3)]=e^x(2x)+e^x(x^2-3)=e^x(x^2+2x-3)$

$$=e^x(x+3)(x-1)$$

$$f''(x)=\dfrac{d}{dx}[e^x(x^2+2x-3)]=e^x(2x+2)+e^x(x^2+2x-3)$$

$$=e^x(x^2+4x-1)$$

Solving $f'(x)=0$, we find that the critical numbers are -3 and 1.

Step 4   Use the critical numbers -3 and 1 to form three intervals on the x-axis: $(-\infty,-3)$, $(-3,1)$, and $(1,\infty)$. Then determine the sign of $f'(x)$ on each interval and whether f is increasing or decreasing on the interval.

Interval	Sign of f'	Conclusion
$(-\infty,-3)$	Positive	f is increasing on the interval $(-\infty,-3)$
$(-3,1)$	Negative	f is decreasing on the interval $(-3,1)$
$(1,\infty)$	Positive	f is increasing on the interval $(1,\infty)$

Step 5   Use the First Derivative Test to identify the local extrema. From the table in Step 4, there is a local maximum at -3 and a local minimum at 1. Then $f(-3)=6e^{-3}\approx0.30$ is a local maximum value and $f(1)=-2e\approx-5.44$ is a local minimum value. Plot the local extrema.

Step 6   To determine the concavity of f, first we solve $f''(x)=0$. We find $x=-2\pm\sqrt{5}$. Now use the numbers $-2-\sqrt{5}\approx-4.24$ and $-2+\sqrt{5}\approx0.24$ to form three intervals on the x-axis: $(-\infty,-2-\sqrt{5})$, $(-2-\sqrt{5},-2+\sqrt{5})$, and $(-2+\sqrt{5},\infty)$. Then determine the sign of $f''(x)$ on each interval and whether f is concave up or concave down on the interval.

Interval	Sign of f''	Conclusion
$(-\infty,-2-\sqrt{5})$	Positive	f is concave up on the interval $(-\infty,-2-\sqrt{5})$
$(-2-\sqrt{5},-2+\sqrt{5})$	Negative	f is concave down on the interval $(-2-\sqrt{5},-2+\sqrt{5})$
$(-2+\sqrt{5},\infty)$	Positive	f is concave up on the interval $(-2+\sqrt{5},\infty)$

The inflection points are $(-4.24,0.22)$ and $(0.24,-3.73)$. Plot the inflection points.

Step 7   The graph of f is given in Figure 48. ∎

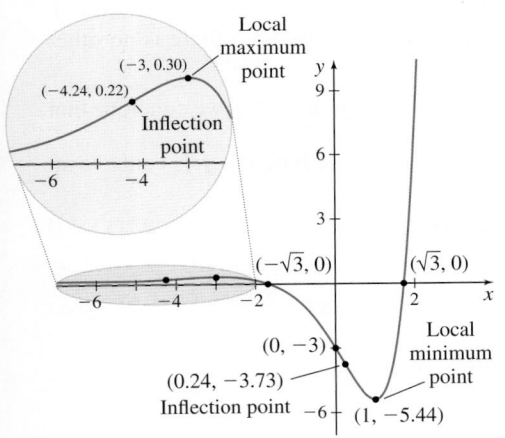

Figure 48 $f(x)=e^x(x^2-3)$

NOW WORK   Problem 39.

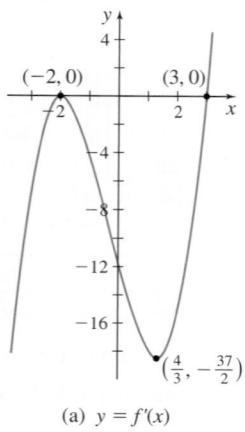

(a) $y = f'(x)$

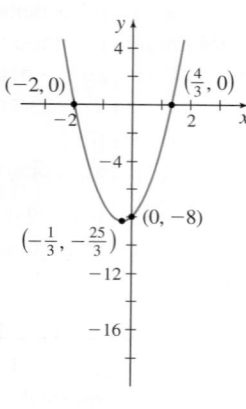

(b) $y = f''(x)$

Figure 49

② Obtain Information About a Function from Its Derivatives

Until now we have been given a function f expressed algebraically and have used calculus to graph the function. We now use the derivatives f' and f'' of a function to obtain information about f and the graph of f.

EXAMPLE 6 **Obtaining Information About a Function f from the Graphs of f' and f''**

A function f is twice differentiable on its domain, the set of real numbers. The graphs of f' and f'' are shown in Figure 49(a) and (b).

(a) Determine the critical numbers of f.

(b) On what intervals is f increasing? On what intervals is f decreasing?

(c) At what numbers x does f have a local minimum? At what numbers x does f have a local maximum?

(d) Where is f concave up? Where is f concave down?

(e) List any inflection points.

Solution

Since f is twice differentiable for all real numbers, both f and f' are continuous on the set of real numbers.

(a) Since f is differentiable at every real number, the critical numbers occur where $f'(x) = 0$. From Figure 49(a), we see that $f'(-2) = 0$ and that $f'(3) = 0$. We conclude that the critical numbers of f are -2 and 3.

(b) Since f is differentiable, the Increasing/Decreasing Function Test can be used. Figure 49(a) shows that $f'(x) > 0$ when $x > 3$. So, f is increasing on the interval $(3, \infty)$. Figure 49(a) also shows that $f'(x) < 0$ on the intervals $(-\infty, -2)$ and $(-2, 3)$, so f is decreasing on $(-\infty, -2)$ and on $(-2, 3)$. Because f is continuous for all real numbers, f is increasing on the interval $[3, \infty)$, and f is decreasing on the interval $(-\infty, 3]$.

(c) $f'(x)$ is negative for $x < 3$ and positive for $x > 3$, so by the First Derivative Test, f has a local minimum at $x = 3$. f has no other local extrema, because there is no other number c where the value of f' changes signs.

(d) Using the Test for Concavity and the graph of f'' (see Figure 49(b)), we conclude that because $f''(x) > 0$ for $-\infty < x < -2$ and for $\dfrac{4}{3} < x < \infty$, the function f is concave up for these values of x. Similarly, f is concave down for $-2 < x < \dfrac{4}{3}$ because $f''(x) < 0$ for these values of x.

(e) The function f has two inflection points $(-2, f(-2))$ and $\left(\dfrac{4}{3}, f\left(\dfrac{4}{3}\right)\right)$, because the concavity of f changes at each of these points. ∎

NOW WORK AP® Practice Problems 1, 2, and 3.

EXAMPLE 7	Obtaining Information About the Graph of a Function f from Numerical Information

A function f is continuous on the closed interval $[-4, 7]$ and is twice differentiable on the open interval $(-4, 7)$ except at $x = 6$. Information about f, f', and f'' is given in Table 7.

TABLE 7

x	-4	$(-4, -2)$	-2	$(-2, 0)$	0	$(0, 4)$	4	$(4, 6)$	6	$(6, 7)$	7
$f(x)$	0	Positive	3	Positive	4	Positive	0	Negative	-5	Negative	-4
$f'(x)$		Positive	1	Positive	0	Negative	-1	Negative	Doesn't exist	Positive	
$f''(x)$		Negative	$-\dfrac{1}{2}$	Negative	$-\dfrac{1}{2}$	Negative	$-\dfrac{1}{2}$	Negative	Doesn't exist	Negative	

(a) Identify the intervals on which the function f is increasing or decreasing.

(b) Identify the critical numbers of f, if any. Determine whether each critical number results in a local maximum, a local minimum, or neither.

(c) Determine the intervals where f is concave up or concave down.

(d) List any inflection points.

(e) Sketch the graph of f.

Solution

(a) The function f is continuous on $[-4, 7]$ and is differentiable on the open intervals $(-4, 6)$ and $(6, 7)$. So, the Increasing/Decreasing Function Test can be used. Since $f'(x)$ is positive on the open intervals $(-4, -2)$, $(-2, 0)$, and $(6, 7)$ and f is continuous, the function f is increasing on the closed intervals $[-4, 0]$ and $[6, 7]$. Since $f'(x)$ is negative on the open intervals $(0, 4)$ and $(4, 6)$, and f is continuous, the function f is decreasing on the closed interval $[0, 6]$.

(b) A number c in the domain of f for which either $f'(c) = 0$ or $f'(c)$ does not exist is a critical number. So, the critical numbers of f are 0 and 6.

- For the critical number 0: Using the Second Derivative Test, $f''(0) = -\dfrac{1}{2}$, so $f(0) = 4$ is a local maximum value.

- For the critical number 6: The Second Derivative Test cannot be used, since the second derivative does not exist at $x = 6$. Using the First Derivative Test, $f'(x) < 0$ for $x < 6$ and $f'(x) > 0$ for $x > 6$. So, $f(6) = -5$ is a local minimum value.

(c) The function f is continuous on $[-4, 7]$ and twice differentiable on the open intervals $(-4, 6)$ and $(6, 7)$. Since $f''(x) < 0$ on the intervals $(-4, 6)$ and $(6, 7)$, then by the Test for Concavity, f is concave down on the intervals $(-4, 6)$ and $(6, 7)$.

(d) Since the concavity of f never changes, f has no inflection point.

(e) Using results from (a)–(d), along with the information about the function f in Table 7, we can draw a sketch of f. See Figure 50. ∎

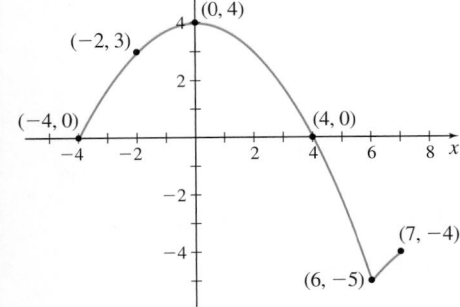

Figure 50 $y = f(x)$

NOW WORK   Problem **53** and AP® Practice Problem **4**.

5.4 Assess Your Understanding

Skill Building

In Problems 1–42, use calculus to graph each function. Follow the steps for graphing a function on page 382.

1. $f(x) = x^4 - 6x^2 + 10$

2. $f(x) = x^4 - 4x$

3. $f(x) = x^5 - 10x^2$

4. $f(x) = x^5 - 3x^3 + 4$

5. $f(x) = 3x^5 + 5x^4$

6. $f(x) = 60x^5 + 20x^3$

7. $f(x) = \dfrac{2}{x^2 - 4}$

8. $f(x) = \dfrac{1}{x^2 - 1}$

9. $f(x) = \dfrac{2x - 1}{x + 1}$

10. $f(x) = \dfrac{x - 2}{x}$

11. $f(x) = \dfrac{x}{x^2 + 1}$

12. $f(x) = \dfrac{2x}{x^2 - 4}$

13. $f(x) = \dfrac{x^2+1}{2x}$ **14.** $f(x) = \dfrac{x^2-1}{2x}$

15. $f(x) = \dfrac{x^2-1}{x^2+2x-3}$ **16.** $f(x) = \dfrac{x^2-x-12}{x^2-9}$

17. $f(x) = \dfrac{x(x^3+1)}{(x^2-4)(x+1)}$ **18.** $f(x) = \dfrac{x(x^3-1)}{(x+1)^2(x-1)}$

19. $f(x) = 1 + \dfrac{1}{x} + \dfrac{1}{x^2}$ **20.** $f(x) = \dfrac{2}{x} + \dfrac{1}{x^2}$

21. $f(x) = \sqrt{3-x}$ **22.** $f(x) = x\sqrt{x+2}$

23. $f(x) = x + \sqrt{x}$ **24.** $f(x) = \sqrt{x} - \sqrt{x+1}$

25. $f(x) = \dfrac{x^2}{\sqrt{x+1}}$ **26.** $f(x) = \dfrac{x}{\sqrt{x^2+2}}$

27. $f(x) = \dfrac{1}{(x+1)(x-2)}$ **28.** $f(x) = \dfrac{1}{(x-1)(x+3)}$

29. $f(x) = x^{2/3} + 3x^{1/3} + 2$ **30.** $f(x) = x^{5/3} - 5x^{2/3}$

31. $f(x) = \sin x - \cos x$ **32.** $f(x) = \sin x + \tan x$

33. $f(x) = \sin^2 x - \cos x$ **34.** $f(x) = \cos^2 x + \sin x$

35. $f(x) = \ln x - x$ **36.** $f(x) = x \ln x$

37. $f(x) = \ln(4-x^2)$ **38.** $f(x) = \ln(x^2+2)$

39. $f(x) = 3e^{3x}(5-x)$ **40.** $f(x) = 3e^{-3x}(x-4)$

41. $f(x) = e^{-x^2}$ **42.** $f(x) = e^{1/x}$

Applications and Extensions

In Problems 43–52, for each function:

(a) Graph the function.

(b) Identify any asymptotes.

(c) Use the graph to identify intervals on which the function increases or decreases and any intervals where the function is concave up or down.

(d) Approximate the local extreme values using the graph.

(e) Compare the approximate local maxima and local minima to the exact local extrema found by using calculus.

(f) Approximate any inflection points using the graph.

43. $f(x) = \dfrac{x^{2/3}}{x-1}$ **44.** $f(x) = \dfrac{5-x}{x^2+3x+4}$

45. $f(x) = x + \sin(2x)$ **46.** $f(x) = x - \cos x$

47. $f(x) = \ln(x\sqrt{x-1})$ **48.** $f(x) = \ln(\tan^2 x)$

49. $f(x) = \sqrt[3]{\sin x}$ **50.** $f(x) = e^{-x} \cos x$

51. $y^2 = x^2(6-x), \ y \geq 0$ **52.** $y^2 = x^2(4-x^2), \ y \geq 0$

In Problems 53–56, graph a function f that is continuous on the interval $[1, 6]$ and satisfies the given conditions.

53. $f'(2)$ does not exist

$f'(3) = -1$

$f''(3) = 0$

$f'(5) = 0$

$f''(x) < 0, \quad 2 < x < 3$

$f''(x) > 0, \quad x > 3$

54. $f'(2) = 0$

$f''(2) = 0$

$f'(3)$ does not exist

$f'(5) = 0$

$f''(x) > 0, \quad 2 < x < 3$

$f''(x) > 0, \quad x > 3$

55. $f'(2) = 0$

$\lim\limits_{x \to 3^-} f'(x) = \infty$

$\lim\limits_{x \to 3^+} f'(x) = \infty$

$f'(5) = 0$

$f''(x) > 0, \quad x < 3$

$f''(x) < 0, \quad x > 3$

56. $f'(2) = 0$

$\lim\limits_{x \to 3^-} f'(x) = -\infty$

$\lim\limits_{x \to 3^+} f'(x) = -\infty$

$f'(5) = 0$

$f''(x) < 0, \quad x < 3$

$f''(x) > 0, \quad x > 3$

57. Graph a function f defined and continuous for $-1 \leq x \leq 2$ that satisfies the following conditions:

$$f(-1) = 1 \quad f(1) = 2 \quad f(2) = 3 \quad f(0) = 0 \quad f\left(\dfrac{1}{2}\right) = 3$$

$$\lim_{x \to -1^+} f'(x) = -\infty \quad \lim_{x \to 1^-} f'(x) = -1 \quad \lim_{x \to 1^+} f'(x) = \infty$$

f has a local minimum at 0. f has a local maximum at $\dfrac{1}{2}$.

58. Graph of a Function Which of the following is true about the graph of $f(x) = \ln|x^2 - 1|$ in the interval $(-1, 1)$?

(a) The graph is increasing.

(b) The graph has a local minimum at $(0, 0)$.

(c) The graph has a range of all real numbers.

(d) The graph is concave down.

(e) The graph has an asymptote $x = 0$.

59. Properties of a Function Suppose $f(x) = \dfrac{1}{x} + \ln x$ is defined only on the closed interval $\dfrac{1}{e} \leq x \leq e$.

(a) Determine the numbers x at which f has its absolute maximum and absolute minimum.

(b) For what numbers x is the graph concave up?

(c) Graph f.

60. Probability The function $f(x) = \dfrac{1}{\sqrt{2\pi}} e^{-x^2/2}$, encountered in probability theory, is called the **standard normal density function**. Determine where this function is increasing and decreasing, find all local maxima and local minima, find all inflection points, and determine the intervals where f is concave up and concave down. Then graph the function.

In Problems 61–64, graph each function. Use L'Hôpital's Rule to find any asymptotes.

61. $f(x) = \dfrac{\sin(3x)}{x\sqrt{4-x^2}}$

62. $f(x) = x^{\sqrt{x}}$

63. $f(x) = x^{1/x}$

64. $f(x) = \dfrac{1}{x} \tan x \quad -\dfrac{\pi}{2} < x < \dfrac{\pi}{2}$

AP® Practice Problems

Multiple-Choice Questions

1. The graphs of a differentiable function f and its first and second derivatives, f' and f'', are shown below. Using the properties of a function and its derivatives, identify each function.

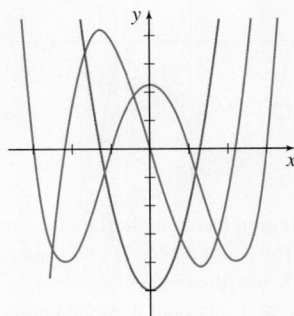

(A) f: blue graph; f': red graph; f'': green graph
(B) f: blue graph; f': green graph; f'': red graph
(C) f: red graph; f': blue graph; f'': green graph
(D) f: green graph; f': red graph; f'': blue graph

2. The function f is twice differentiable on the interval $(-5, 7)$. The graph of f'' is shown below.

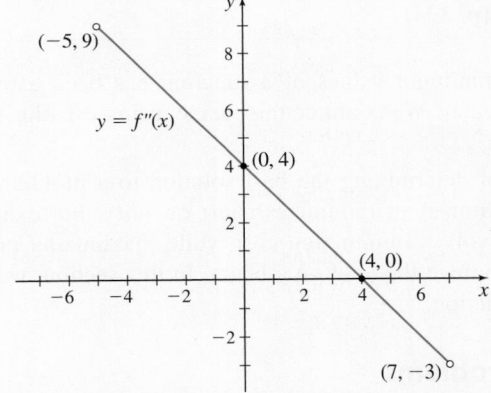

Which of the following statements must be true?
I. f is concave up on the interval $(-5, 4)$.
II. f' is decreasing on the interval $(-5, 4)$.
III. $(4, f(4))$ is an inflection point of f.

(A) II only   (B) III only
(C) I and III only   (D) I, II, and III

Free-Response Questions

3. A function f, defined on $[-0.5, 2.5]$, is twice differentiable. Use the information given in the graphs of f' and f'' below.

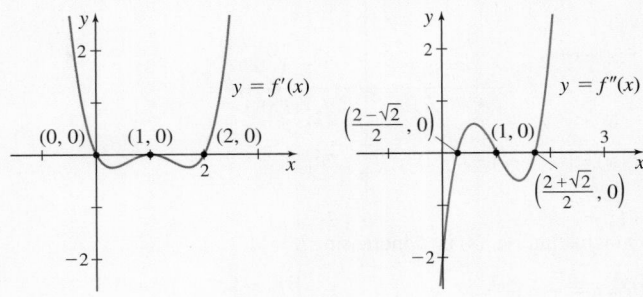

(a) Determine where f is decreasing. Give a reason for your answer.
(b) At what numbers x, if any, does f have a local maximum?
(c) Where is f concave up? Give a reason for your answer.
(d) Identify any numbers x where f has an inflection point.

4. Suppose f is a twice differentiable function that is continuous on the interval $[-10, 10]$. Information about f' and f'' is given in the table below.

x	$x < -\sqrt{3}$	$-\sqrt{3}$	$-\sqrt{3} < x < -1$	-1	$-1 < x < 0$	0
$f'(x)$	Negative	0	Positive	16	Positive	0
$f''(x)$	Positive	48	Positive	0	Negative	-24

x	0	$0 < x < 1$	1	$1 < x < \sqrt{3}$	$\sqrt{3}$	$x > \sqrt{3}$
$f'(x)$	0	Negative	-16	Negative	0	Positive
$f''(x)$	-24	Negative	0	Positive	48	Positive

(a) Identify the interval(s) on which the function f is increasing.
(b) Locate the numbers, if any, for which f has a local maximum. Give a reason for your answer.
(c) Identify any intervals where the graph of f is concave up. Give a reason for your answer.
(d) Identify the numbers x, if any, where f has an inflection point.

Retain Your Knowledge

Multiple-Choice Questions

1. The function g is differentiable. The table below lists values for g for select numbers x in the domain of g.

x	-2	1	3	4	6
$g(x)$	4	6.5	6	3	1

Approximate $g(2.4)$ using a linear approximation.

(A) 6.15   (B) 6   (C) 3.6   (D) -6.15

2. The derivative of $f(x) = \sin^{-1}(3x^2)$ is

(A) $\dfrac{6x}{\sqrt{1 - 9x^2}}$ (B) $\dfrac{1}{\sqrt{1 - 9x^2}}$

(C) $\dfrac{6x}{\sqrt{1 + 3x^4}}$ (D) $\dfrac{6x}{\sqrt{1 - 9x^4}}$

(continued on the next page)

3. A function f is continuous for $-3 \leq x \leq 4$. The graph of its derivative f' is shown below.

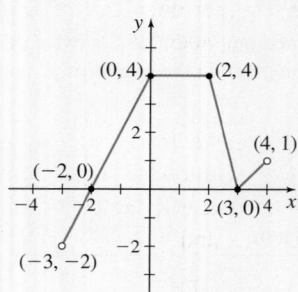

On what interval(s) is f increasing?

(A) $-2 \leq x \leq 4$

(B) $-3 \leq x \leq 0$ and $3 \leq x \leq 4$

(C) $-2 \leq x \leq 0$ and $3 \leq x \leq 4$

(D) $-3 \leq x \leq 2$ and $3 \leq x \leq 4$

Free-Response Question

4. The function f is continuous on $[-4, 4]$. The graph of its derivative function f' is shown below.

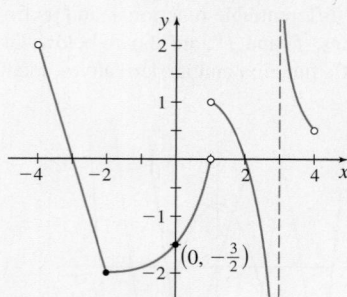

(a) Does the graph of f have any horizontal tangent lines? Give a reason for your answer. If horizontal tangent lines exist, identify where they occur.

(b) Does the graph of f have any vertical tangent lines? Give a reason for your answer.

5.5 Introduction to Optimization

OBJECTIVE *When you finish this section, you should be able to:*

1 Solve optimization problems (p. 394)

Investigating maximum and/or minimum values of a function has been a recurring theme throughout most of this chapter. We continue this theme, using calculus to solve *optimization problems*.

Optimization is a process of determining the best solution to a problem. Often the best solution is one that maximizes or minimizes some quantity. For example, a business owner's goal usually involves minimizing cost while maximizing profit, or an engineer's goal may be to minimize the load on a beam. In this section, we model optimization problems using a function.

1 Solve Optimization Problems

Even though each individual problem has unique features, it is possible to outline a general method for obtaining an optimal solution.

Steps for Solving Optimization Problems

Step 1 Read the problem until you understand it and can identify the quantity for which a maximum or minimum value is to be found. Assign a symbol to represent it.

Step 2 Assign symbols to represent the other variables in the problem. If possible, draw a picture and label it. Determine relationships among the variables.

Step 3 Express the quantity to be maximized or minimized as a function of one of the variables, and determine a meaningful domain for the function.

Step 4 Find the absolute maximum value or absolute minimum value of the function.

Figure 51 Area $A = xy$

EXAMPLE 1 Maximizing an Area

A farmer with 4000 m of fencing wants to enclose a rectangular plot that borders a straight river, as shown in Figure 51. If the farmer does not fence the side along the river, what is the largest rectangular area that can be enclosed?

Solution

Step 1 The quantity to be maximized is the area; we denote it by A.

Step 2 We denote the dimensions of the rectangle by x and y (both in meters), with the length of the side y parallel to the river. The area is $A = xy$. Because there are 4000 m of fence available, the variables x and y are related by the equation

$$x + y + x = 4000$$
$$y = 4000 - 2x$$

Step 3 Now express the area A as a function of x.

$$A = A(x) = x(4000 - 2x) = 4000x - 2x^2 \qquad A = xy, \quad y = 4000 - 2x$$

The domain of A is the closed interval $[0, 2000]$.

Step 4 To find the number x that maximizes $A(x)$, differentiate A with respect to x and find the critical numbers:

$$A'(x) = 4000 - 4x = 0$$

> **NOTE** Since $A = A(x)$ is continuous on the closed interval $[0, 2000]$, the Extreme Value Theorem guarantees A has an absolute maximum.

The only critical number is $x = 1000$. The maximum value of A occurs either at the critical number or at an endpoint of the interval $[0, 2000]$.

$$A(1000) = 2{,}000{,}000 \qquad A(0) = 0 \qquad A(2000) = 0$$

The maximum value is $A(1000) = 2{,}000{,}000 \ \text{m}^2$, which results from fencing a rectangular plot that measures 2000 m along the side parallel to the river and 1000 m along each of the other two sides. ■

We could have reached the same conclusion by noting that $A''(x) = -4 < 0$ for all x, so A is always concave down, and the local maximum at $x = 1000$ must be the absolute maximum of the function.

> **NOW WORK** Problem 3 and AP® Practice Problem 4.

 CALC CLIP

EXAMPLE 2 Maximizing a Volume

From each corner of a square piece of sheet metal 18 cm on a side, remove a small square and turn up the edges to form an open box. What are the dimensions of the box with the largest volume?

Solution

Step 1 The quantity to be maximized is the volume of the box; denote it by V. Denote the length of each side of the small squares by x and the length of each side after the small squares are removed by y, as shown in Figure 52. Both x and y are in centimeters.

Step 2 Then $y = (18 - 2x)$ cm. The height of the box is x cm, and the area of the base of the box is $y^2 \ \text{cm}^2$. So, the volume of the box is $V = xy^2 \ \text{cm}^3$.

Step 3 To express V as a function of one variable, substitute $y = 18 - 2x$ into the formula for V. Then the function to be maximized is

$$V = V(x) = x(18 - 2x)^2$$

Since both $x \geq 0$ and $18 - 2x \geq 0$, we find $x \leq 9$, meaning the domain of V is the closed interval $[0, 9]$. (All other numbers make no physical sense—do you see why?)

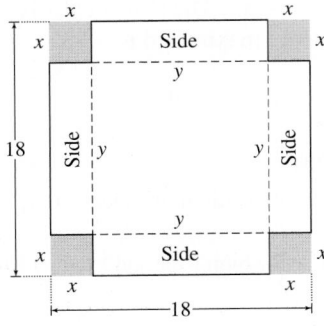

Figure 52 $y = 18 - 2x$

Step 4 To find the value of x that maximizes V, differentiate V and find the critical numbers.

$$V'(x) = \frac{d}{dx}[x(18-2x)^2] = \underset{\underset{\text{Product Rule}}{\uparrow}}{2x(18-2x)(-2) + (18-2x)^2} = \underset{\underset{\text{Simplify and factor.}}{\uparrow}}{(18-2x)(18-6x)}$$

Now solve $V'(x) = 0$ for x. The solutions are

$$x = 9 \qquad \text{or} \qquad x = 3$$

The only critical number in the open interval $(0, 9)$ is 3. Evaluate V at 3 and at the endpoints 0 and 9.

$$V(0) = 0 \qquad V(3) = 3(18-6)^2 = 432 \qquad V(9) = 0$$

The maximum volume is $432\,\text{cm}^3$. The box with the maximum volume has a height of 3 cm. Since $y = 18 - 2 \cdot 3 = 12$ cm, the base of the box is square and measures 12 cm by 12 cm. ∎

NOW WORK Problems 7 and 9.

EXAMPLE 3   **Maximizing an Area**

A manufacturer makes a flexible square play yard, (Figure 53a), with hinges at the four corners. It can be opened at one corner and attached at right angles to a wall or the side of a house, as shown in Figure 53(b). If each side is 3 m in length, the open configuration doubles the available area from $9\,\text{m}^2$ to $18\,\text{m}^2$. Is there a configuration that will more than double the play area?

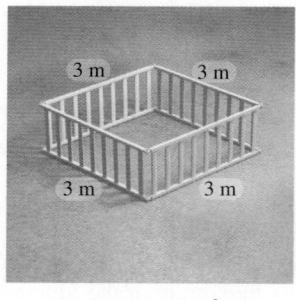
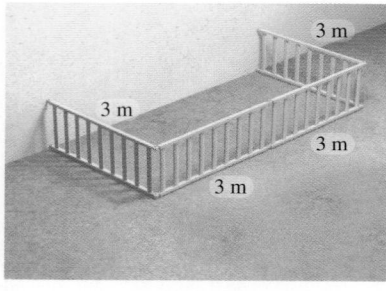

(a) Area $= 9\,\text{m}^2$ (b) Area $= 18\,\text{m}^2$

Figure 53

Solution

Step 1 We want to maximize the play area A.

Step 2 Since the play yard must be attached at right angles to the wall, the possible configurations depend on the amount of wall used as a fifth side, as shown in Figure 54. Use x to represent half the length (in meters) of the wall used as the fifth side.

Step 3 Partition the play area into two sections: a rectangle with area $3 \cdot 2x = 6x$ and a triangle with base $2x$ and height $\sqrt{3^2 - x^2} = \sqrt{9 - x^2}$. The play area A is the sum of the areas of the two sections. The area to be maximized is

$$A = A(x) = 6x + \frac{1}{2} \cdot 2x \sqrt{9 - x^2} = 6x + x\sqrt{9 - x^2}$$

The domain of A is the closed interval $[0, 3]$.

Step 4 To find the maximum area of the play yard, find the critical numbers of A.

$$A'(x) = 6 + x\left[\frac{1}{2}(-2x)(9-x^2)^{-1/2}\right] + \sqrt{9-x^2} \quad \text{Use the Sum Rule and Product Rule.}$$

$$= 6 - \frac{x^2}{\sqrt{9-x^2}} + \sqrt{9-x^2} = \frac{6\sqrt{9-x^2} - 2x^2 + 9}{\sqrt{9-x^2}}$$

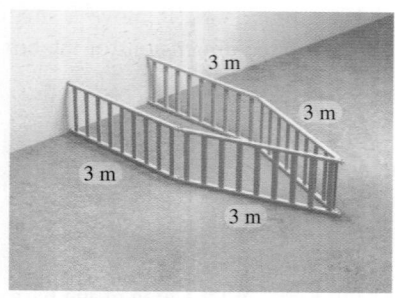
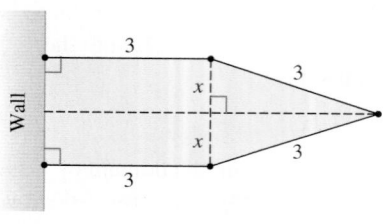

Figure 54

Critical numbers occur when $A'(x) = 0$ or where $A'(x)$ does not exist. A' does not exist at -3 and 3, and $A'(x) = 0$ at

$$6\sqrt{9 - x^2} - 2x^2 + 9 = 0 \qquad\qquad A'(x) = 0$$

$$\sqrt{9 - x^2} = \frac{2x^2 - 9}{6} \qquad\qquad \text{Isolate the radical on the left.}$$

$$9 - x^2 = \left(\frac{2x^2 - 9}{6}\right)^2 \qquad\qquad \text{Square both sides.}$$

$$9 - x^2 = \frac{4x^4 - 36x^2 + 81}{36} \qquad\qquad \text{Solve for } x.$$

$$324 - 36x^2 = 4x^4 - 36x^2 + 81$$

$$324 = 4x^4 + 81$$

$$x^4 = \frac{324 - 81}{4} = \frac{243}{4}$$

$$x = \sqrt[4]{\frac{243}{4}} \approx 2.792$$

The only critical number in the open interval $(0, 3)$ is $\sqrt[4]{\dfrac{243}{4}} \approx 2.792$.

Now evaluate $A(x)$ at the endpoints 0 and 3 and at the critical number $x \approx 2.792$.

$$A(0) = 0 \qquad A(3) = 18 \qquad A(2.792) \approx 19.817$$

Using a wall of length $2x \approx 2(2.792) = 5.584$ m maximizes the area; the configuration shown in Figure 55 increases the play area by approximately 10% (from 18 to 19.817 m²). ∎

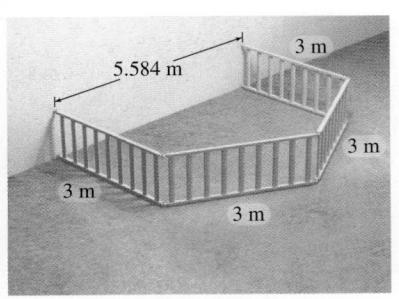

Figure 55 Area ≈ 19.817 m²

EXAMPLE 4   Minimizing Cost

A manufacturer needs to produce a cylindrical container with a capacity of 1000 cm³. The top and bottom of the container are made of material that costs $0.05 per square centimeter, while the side of the container is made of material costing $0.03 per square centimeter. Find the dimensions that will minimize the company's cost of producing the container.

Solution

Figure 56 shows a cylindrical container and the area of its top, bottom, and lateral surface. As shown in the figure, let h denote the height of the container and R denote the radius. The total area of the bottom and top is $2\pi R^2$ cm². The area of the lateral surface of the can is $2\pi Rh$ cm².

The variables h and R are related. Since the capacity (volume) V of the cylinder is 1000 cm³,

$$V = \pi R^2 h = 1000$$

$$h = \frac{1000}{\pi R^2}$$

The cost C, in dollars, of manufacturing the container is

$$C = (0.05)(2\pi R^2) + (0.03)(2\pi Rh) = 0.1\pi R^2 + 0.06\pi Rh$$

↑ The cost of the top and the bottom   ↑ The cost of the side

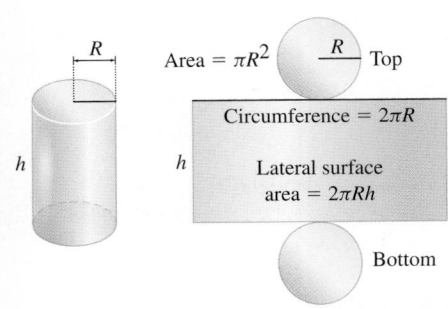

Figure 56

NEED TO REVIEW? Geometry formulas are discussed in Appendix A.2, pp. A-15 to A-16.

By substituting for h, C can be expressed as a function of R.

$$C = C(R) = 0.1\pi R^2 + 0.06\pi R \cdot \frac{1000}{\pi R^2} = 0.1\pi R^2 + \frac{60}{R}$$

This is the function to be minimized. The domain of C is $\{R \mid R > 0\}$.

To find the minimum cost, differentiate C with respect to R.

$$C'(R) = 0.2\pi R - \frac{60}{R^2} = \frac{0.2\pi R^3 - 60}{R^2}$$

Solve $C'(R) = 0$ to find the critical numbers.

$$0.2\pi R^3 - 60 = 0$$

$$R^3 = \frac{300}{\pi}$$

$$R = \sqrt[3]{\frac{300}{\pi}} \approx 4.571 \text{ cm}$$

Now find $C''(R)$ and use the Second Derivative Test.

$$C''(R) = \frac{d}{dR}\left(0.2\pi R - \frac{60}{R^2}\right) = 0.2\pi + \frac{120}{R^3}$$

$$C''\left(\sqrt[3]{\frac{300}{\pi}}\right) = 0.2\pi + \frac{120\pi}{300} = 0.6\pi > 0$$

C has a local minimum at $\sqrt[3]{\dfrac{300}{\pi}}$. Since $C''(R) > 0$ for all R in the domain, the graph of C is always concave up, and the local minimum value is the absolute minimum value. The radius of the container that minimizes the cost is $R \approx 4.571$ cm. The height of the container that minimizes the cost of the material is

> **NOTE** If the costs of the materials for the top, bottom, and lateral surfaces of a cylindrical container are all equal, then the minimum total cost occurs when the surface area is minimum. It can be shown (see Problem 39) that for any fixed volume, the minimum surface area of a cylindrical container is obtained when the height equals twice the radius.

$$h = \frac{1000}{\pi R^2} \approx \frac{1000}{20.892\pi} \approx 15.236 \text{ cm}$$

The minimum cost of the container is

$$C\left(\sqrt[3]{\frac{300}{\pi}}\right) = 0.1\pi \left(\sqrt[3]{\frac{300}{\pi}}\right)^2 + \frac{60}{\sqrt[3]{\dfrac{300}{\pi}}} \approx \$19.69$$

∎

NOW WORK   Problem 11 and AP® Practice Problems 2 and 6.

EXAMPLE 5   Maximizing Area

A rectangle is inscribed in a semicircle of radius 2. Find the dimensions of the rectangle that has the maximum area.

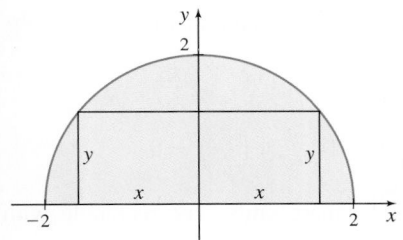

Figure 57

Solution

We present two methods of solution: The first uses analytic geometry, and the second uses trigonometry. To begin, place the semicircle with its diameter along the x-axis and center at the origin. Then inscribe a rectangle in the semicircle, as shown in Figure 57. The length of the inscribed rectangle is $2x$ and its height is y.

Analytic Geometry Method The area A of the inscribed rectangle is $A = 2xy$ and the equation of the semicircle is $x^2 + y^2 = 4$, $y \geq 0$. We solve for y in $x^2 + y^2 = 4$ and obtain $y = \sqrt{4 - x^2}$. Now substitute this expression for y in the area formula for the rectangle to express A as a function of x alone.

$$A = A(x) = \underset{\underset{A = 2xy;\; y = \sqrt{4-x^2}}{\uparrow}}{2x\sqrt{4 - x^2}} \qquad 0 \leq x \leq 2$$

Then using the Product Rule,

$$A'(x) = 2\left[x \cdot \frac{1}{2}(4 - x^2)^{-1/2}(-2x) + \sqrt{4 - x^2}\right] = 2\left[\frac{-x^2}{\sqrt{4 - x^2}} + \sqrt{4 - x^2}\right]$$

$$= 2\left[\frac{-x^2 + 4 - x^2}{\sqrt{4 - x^2}}\right] = 2\left[\frac{-2(x^2 - 2)}{\sqrt{4 - x^2}}\right] = -\frac{4(x^2 - 2)}{\sqrt{4 - x^2}}$$

The only critical number in the open interval $(0, 2)$ is $\sqrt{2}$, where $A'(\sqrt{2}) = 0$. [The numbers $-\sqrt{2}$ and -2 are not in the domain of A, and 2 is not in the open interval $(0, 2)$.] The values of A at the endpoints 0 and 2 and at the critical number $\sqrt{2}$ are

$$A(0) = 0 \qquad A(\sqrt{2}) = 4 \qquad A(2) = 0$$

The maximum area of the inscribed rectangle is 4, and it corresponds to the rectangle whose length is $2x = 2\sqrt{2}$ and whose height is $y = \sqrt{2}$.

Trigonometry Method Using Figure 57, draw the radius $r = 2$ from O to the vertex of the rectangle and place the angle θ in the standard position, as shown in Figure 58. Then $x = 2\cos\theta$ and $y = 2\sin\theta$, $0 \leq \theta \leq \dfrac{\pi}{2}$. The area A of the rectangle is

$$A = 2xy = 2(2\cos\theta)(2\sin\theta) = 8\cos\theta\sin\theta = \underset{\underset{\sin(2\theta) = 2\sin\theta\cos\theta}{\uparrow}}{4\sin(2\theta)}$$

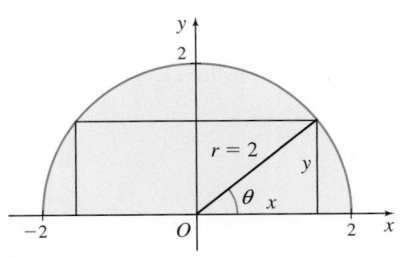

Figure 58

Since the area A is a differentiable function of θ, we obtain the critical numbers by finding $A'(\theta)$ and solving the equation $A'(\theta) = 0$.

$$A'(\theta) = 8\cos(2\theta) = 0 \qquad A(\theta) = 4\sin(2\theta)$$
$$\cos(2\theta) = 0$$
$$2\theta = \cos^{-1}0 = \frac{\pi}{2}$$
$$\theta = \frac{\pi}{4}$$

The only critical number in the open interval $\left(0, \dfrac{\pi}{2}\right)$ is $\dfrac{\pi}{4}$.

Now evaluate $A(\theta) = 4\sin(2\theta)$ at the endpoints 0 and $\dfrac{\pi}{2}$ and the critical number $\dfrac{\pi}{4}$.

$$A(0) = 0 \qquad A\left(\dfrac{\pi}{4}\right) = 4\sin\left(2 \cdot \dfrac{\pi}{4}\right) = 4 \qquad A\left(\dfrac{\pi}{2}\right) = 0$$

The maximum area of the inscribed rectangle is 4 square units. The rectangle with maximum area has

$$\text{length } 2x = 4\cos\dfrac{\pi}{4} = 2\sqrt{2} \text{ and height } y = 2\sin\dfrac{\pi}{4} = \sqrt{2} \qquad \blacksquare$$

NOW WORK Problem **41** and AP® Practice Problems **1, 3,** and **5.**

EXAMPLE 6 Snell's Law of Refraction

Light travels at different speeds in different media (air, water, glass, etc.). Suppose that light travels from a point A in one medium, where its speed is c_1, to a point B in another medium, where its speed is c_2. See Figure 59. We use Fermat's Principle that light always travels along the path that requires the least time to prove Snell's Law of Refraction.

$$\boxed{\dfrac{\sin\theta_1}{c_1} = \dfrac{\sin\theta_2}{c_2}}$$

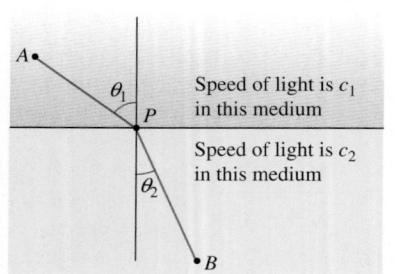

A

θ_1

P

Speed of light is c_1 in this medium

Speed of light is c_2 in this medium

θ_2

B

Figure 59

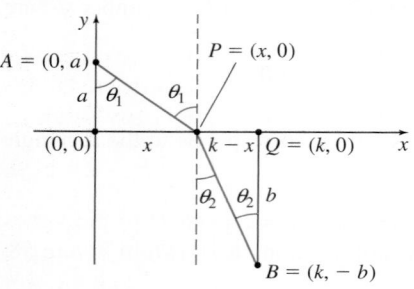

$A = (0, a)$

$P = (x, 0)$

a θ_1 θ_1

$(0, 0)$ x $k - x$ $Q = (k, 0)$ x

θ_2 θ_2 b

$B = (k, -b)$

Figure 60

ORIGINS In 1621 Willebrord Snell (c. 1580–1626) discovered the law of refraction, one of the basic principles of geometric optics. Pierre de Fermat (1601–1665) was able to prove the law mathematically using the principle that light follows the path that takes the least time.

Solution

Position the coordinate system, as illustrated in Figure 60. The light passes from one medium to the other at the point P. Since the shortest distance between two points is a line, the path taken by the light is made up of two line segments—from $A = (0, a)$ to $P = (x, 0)$ and from $P = (x, 0)$ to $B = (k, -b)$, where a, b, and k are positive constants.

Since

$$\text{Time} = \dfrac{\text{Distance}}{\text{Speed}}$$

the travel time t_1 from $A = (0, a)$ to $P = (x, 0)$ is

$$t_1 = \dfrac{\sqrt{x^2 + a^2}}{c_1} \qquad \text{c_1 is the speed of light in medium 1.}$$

and the travel time t_2 from $P = (x, 0)$ to $B = (k, -b)$ is

$$t_2 = \dfrac{\sqrt{(k - x)^2 + b^2}}{c_2} \qquad \text{c_2 is the speed of light in medium 2.}$$

The total time $T = T(x)$ is given by

$$T(x) = t_1 + t_2 = \dfrac{\sqrt{x^2 + a^2}}{c_1} + \dfrac{\sqrt{(k - x)^2 + b^2}}{c_2}$$

To find the least time, find the critical numbers of T.

$$T'(x) = \dfrac{x}{c_1\sqrt{x^2 + a^2}} - \dfrac{k - x}{c_2\sqrt{(k - x)^2 + b^2}} = 0 \qquad (1)$$

From Figure 60,

$$\dfrac{x}{\sqrt{x^2 + a^2}} = \sin\theta_1 \qquad \text{and} \qquad \dfrac{k - x}{\sqrt{(k - x)^2 + b^2}} = \sin\theta_2 \qquad (2)$$

Using the result from (2) in equation (1), we have

$$T'(x) = \frac{\sin\theta_1}{c_1} - \frac{\sin\theta_2}{c_2} = 0$$

$$\frac{\sin\theta_1}{c_1} = \frac{\sin\theta_2}{c_2}$$

Now to be sure that the minimum value of T occurs when $T'(x) = 0$, we show that $T''(x) > 0$. From (1),

$$T''(x) = \frac{d}{dx}\frac{x}{c_1\sqrt{x^2 + a^2}} - \frac{d}{dx}\frac{k - x}{c_2\sqrt{(k - x)^2 + b^2}}$$

$$= \frac{a^2}{c_1(x^2 + a^2)^{3/2}} + \frac{b^2}{c_2[(k - x)^2 + b^2]^{3/2}} > 0$$

Since $T''(x) > 0$ for all x, T is concave up for all x, and the minimum value of T occurs at the critical number. That is, T is a minimum when $\dfrac{\sin\theta_1}{c_1} = \dfrac{\sin\theta_2}{c_2}$. ■

5.5 Assess Your Understanding

Applications and Extensions

1. **Maximizing Area** The owner of a motel has 3000 m of fencing and wants to enclose a rectangular plot of land that borders a straight highway. If she does not fence the side along the highway, what is the largest area that can be enclosed?

2. **Maximizing Area** If the motel owner in Problem 1 decides to also fence the side along the highway, except for 5 m to allow for access, what is the largest area that can be enclosed?

3. **Maximizing Area** Find the dimensions of the rectangle with the largest area that can be enclosed on all sides by L meters of fencing.

4. **Maximizing the Area of a Triangle** An isosceles triangle has a perimeter of fixed length L. What should the dimensions of the triangle be if its area is to be a maximum? See the figure.

5. **Maximizing Area** A gardener with 200 m of available fencing wishes to enclose a rectangular field and then divide it into two plots with a fence parallel to one of the sides, as shown in the figure. What is the largest area that can be enclosed?

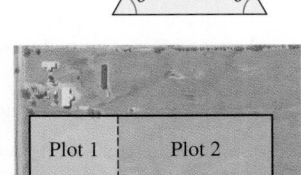

6. **Minimizing Fencing** A realtor wishes to enclose 600 m² of land in a rectangular plot and then divide it into two plots with a fence parallel to one of the sides. What are the dimensions of the rectangular plot that require the least amount of fencing?

7. **Maximizing the Volume of a Box** An open box with a square base is to be made from a square piece of cardboard that measures 12 cm on each side. A square will be cut out from each corner of the cardboard and the sides will be turned up to form the box. Find the dimensions that yield the maximum volume.

8. **Maximizing the Volume of a Box** An open box with a rectangular base is to be made from a piece of tin measuring 30 cm by 45 cm by cutting out a square from each corner and turning up the sides. Find the dimensions that yield the maximum volume.

9. **Minimizing the Surface Area of a Box** An open box with a square base is to have a volume of 2000 cm³. What should be the dimensions of the box if the amount of material used is to be a minimum?

10. **Minimizing the Surface Area of a Box** If the box in Problem 9 is to be closed on top, what should be the dimensions of the box if the amount of material used is to be a minimum?

11. **Minimizing the Cost to Make a Can** A cylindrical container that has a capacity of 10 m³ is to be produced. The top and bottom of the container are to be made of a material that costs $20 per square meter, while the side of the container is to be made of a material costing $15 per square meter. Find the dimensions that minimize the cost of the material.

12. **Minimizing the Cost of Fencing** A builder wishes to fence in 60,000 m² of land in a rectangular shape. For security reasons, the fence along the front part of the land will cost $20 per meter, while the fence for the other three sides will cost $10 per meter. How much of each type of fence should the builder buy to minimize the cost of the fence? What is the minimum cost?

13. **Maximizing Revenue** A car rental agency has 24 cars (each an identical model). The owner of the agency finds that at a price of $18 per day, all the cars can be rented; however, for each $1 increase in rental cost, one of the cars is not rented. What should the agency charge to maximize income?

14. **Maximizing Revenue** A charter flight club charges its members $200 per year. But for each new member in excess of 60, the charge for every member is reduced by $2. What number of members leads to a maximum revenue?

15. **Minimizing Distance** Find the coordinates of the points on the graph of the parabola $y = x^2$ that are closest to the point $\left(2, \dfrac{1}{2}\right)$.

16. **Minimizing Distance** Find the coordinates of the points on the graph of the parabola $y = 2x^2$ that are closest to the point $(1, 4)$.

17. **Minimizing Distance** Find the coordinates of the points on the graph of the parabola $y = 4 - x^2$ that are closest to the point $(6, 2)$.

18. **Minimizing Distance** Find the coordinates of the points on the graph of $y = \sqrt{x}$ that are closest to the point $(4, 0)$.

19. **Minimizing Transportation Cost** A truck has a top speed of 75 mi/h and, when traveling at a speed of x mi/h, consumes gasoline at the rate of $\dfrac{1}{200}\left(\dfrac{1600}{x}+x\right)$ gallons per mile. The truck is to be taken on a 200-mi trip by a driver who is paid at the rate of \$$b$ per hour plus a commission of \$$c$. Since the time required for the trip at x mi/h is $\dfrac{200}{x}$, the cost C of the trip, when gasoline costs \$$a$ per gallon, is

$$C=C(x)=\left(\frac{1600}{x}+x\right)a+\frac{200}{x}b+c$$

Find the speed that minimizes the cost C under each of the following conditions:

(a) $a=\$3.50,\ b=0,\ c=0$

(b) $a=\$3.50,\ b=\$10.00,\ c=\$500$

(c) $a=\$4.00,\ b=\$20.00,\ c=0$

20. **Optimal Placement of a Cable Box** A telephone company is asked to provide cable service to a customer whose house is located 2 km away from the road along which the cable lines run. The nearest cable box is located 5 km down the road. As shown in the figure below, let $5-x$ denote the distance from the box to the connection so that x is the distance from this point to the point on the road closest to the house. If the cost to connect the cable line is \$500 per kilometer along the road and \$600 per kilometer away from the road, where along the road from the box should the company connect the cable line so as to minimize construction cost?

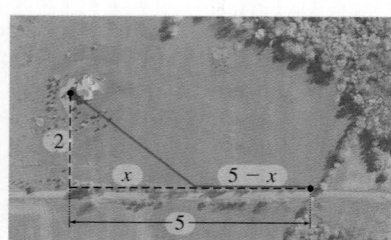

21. **Minimizing a Path** Two houses A and B on the same side of a road are a distance p apart, with distances q and r, respectively, from the center of the road, as shown in the figure on the right. Find the length of the shortest path that goes from A to the road and then on to the other house B.

(a) Use calculus.

(b) Use only elementary geometry.

Hint: Reflect B across the road to a point C that is also a distance r from the center of the road.

22. **Minimizing Travel Time** A small island is 3 km from the nearest point P on the straight shoreline of a large lake. A town is 12 km down the shore from P. See the figure. If a person on the island can row a boat 2.5 km/h and can walk 4 km/h, where should the boat be landed so that the person arrives in town in the shortest time?

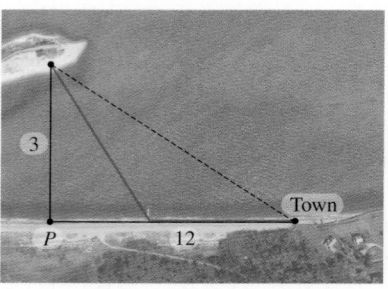

23. **Supporting a Wall** Find the length of the shortest beam that can be used to brace a wall if the beam is to pass over a second wall 2 m high and 5 m from the first wall. See the figure. What is the angle of elevation of the beam?

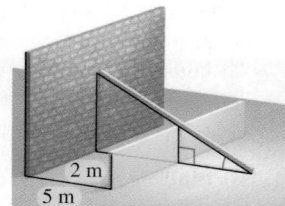

24. **Maximizing the Strength of a Beam** The strength of a rectangular beam is proportional to the product of the width and the cube of its depth. Find the dimensions of the strongest beam that can be cut from a log whose cross section has the form of a circle of fixed radius R. See the figure.

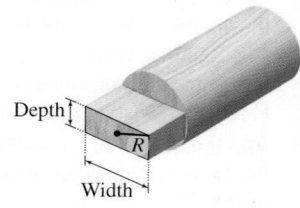

25. **Maximizing the Strength of a Beam** If the strength of a rectangular beam is proportional to the product of its width and the square of its depth, find the dimensions of the strongest beam that can be cut from a log whose cross section has the form of the ellipse $10x^2+9y^2=90$.
Hint: Choose width $=2x$ and depth $=2y$.

26. **Maximizing the Strength of a Beam** The strength of a beam made from a certain wood is proportional to the product of its width and the cube of its depth. Find the dimensions of the rectangular cross section of the beam with maximum strength that can be cut from a log whose original cross section is in the form of the ellipse $b^2x^2+a^2y^2=a^2b^2,\ a\ge b$.

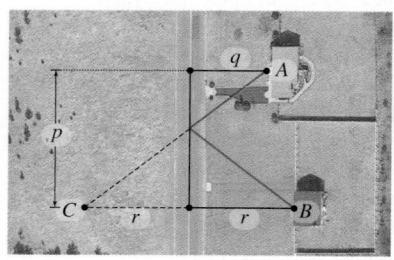

27. Pricing Wine   A winemaker in Walla Walla, Washington, is producing the first vintage for her own label, and she needs to know how much to charge per case of wine. It costs her $132 per case to make the wine. She understands from industry research and an assessment of her marketing list that she can sell $x = 1430 - \dfrac{11}{6}p$ cases of wine, where p is the price of a case in dollars. She can make at most 1100 cases of wine in her production facility. How many cases of wine should she produce, and what price p should she charge per case to maximize her profit P?
Hint: Maximize the profit $P = xp - 132x$, where x equals the number of cases.

28. Optimal Window Dimensions   A Norman window has the shape of a rectangle surmounted by a semicircle of diameter equal to the width of the rectangle, as shown in the figure. If the perimeter of the window is 10 m, what dimensions will admit the most light?

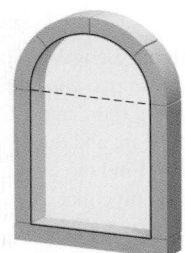

29. Maximizing Volume   The sides of a V-shaped trough are 28 cm wide. Find the angle between the sides of the trough that results in maximum capacity.

30. Maximizing Volume   A metal rain gutter is to have 10-cm sides and a 10-cm horizontal bottom, with the sides making equal angles with the bottom, as shown in the figure. How wide should the opening across the top be for maximum carrying capacity?

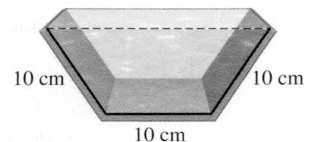

10 cm   10 cm

10 cm

31. Minimizing Construction Cost   A proposed tunnel with a fixed cross-sectional area is to have a horizontal floor, vertical walls of equal height, and a ceiling that is a semicircular cylinder. If the ceiling costs three times as much per square meter to build as the vertical walls and the floor, find the most economical ratio of the diameter of the semicircular cylinder to the height of the vertical walls.

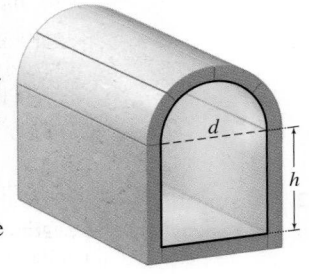

d

h

32. Minimizing Construction Cost   An observatory is to be constructed in the form of a right circular cylinder surmounted by a hemispherical dome. If the hemispherical dome costs three times as much per square meter as the cylindrical wall, what are the most economical dimensions for a given volume? Neglect the floor.

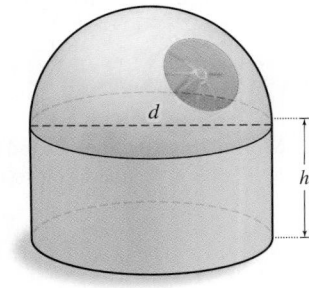

d

h

33. Intensity of Light   The intensity of illumination at a point varies inversely as the square of the distance between the point and the light source. Two lights, one having an intensity eight times that of the other, are 6 m apart. At what point between the two lights is the total illumination least?

34. Drug Concentration   The concentration of a drug in the bloodstream t hours after injection into muscle tissue is given by $C(t) = \dfrac{2t}{16 + t^2}$. When is the concentration greatest?

35. Optimal Wire Length   A wire is to be cut into two pieces. One piece will be bent into a square, and the other piece will be bent into a circle. If the total area enclosed by the two pieces is to be 64 cm^2, what is the minimum length of wire that can be used? What is the maximum length of wire that can be used?

36. Optimal Wire Length   A wire is to be cut into two pieces. One piece will be bent into an equilateral triangle, and the other piece will be bent into a circle. If the total area enclosed by the two pieces is to be 64 cm^2, what is the minimum length of wire that can be used? What is the maximum length of wire that can be used?

37. Optimal Area   A wire 35 cm long is cut into two pieces. One piece is bent into the shape of a square, and the other piece is bent into the shape of a circle.

(a) How should the wire be cut so that the area enclosed is a minimum?

(b) How should the wire be cut so that the area enclosed is a maximum?

(c) Graph the area enclosed as a function of the length of the piece of wire used to make the square. Show that the graph confirms the results of (a) and (b).

38. Optimal Area   A wire 35 cm long is cut into two pieces. One piece is bent into the shape of an equilateral triangle, and the other piece is bent into the shape of a circle.

(a) How should the wire be cut so that the area enclosed is a minimum?

(b) How should the wire be cut so that the area enclosed is a maximum?

(c) Graph the area enclosed as a function of the length of the piece of wire used to make the triangle. Show that the graph confirms the results of (a) and (b).

39. Optimal Dimensions for a Can   Show that a cylindrical container of fixed volume V requires the least material (minimum surface area) when its height is twice its radius.

40. Maximizing Area   Find the triangle of largest area that has two sides along the positive coordinate axes if its hypotenuse is tangent to the graph of $y = 3e^{-x}$.

41. Maximizing Area   Find the largest area of a rectangle with one vertex on the parabola $y = 9 - x^2$, another at the origin, and the remaining two on the positive x-axis and positive y-axis, respectively. See the figure.

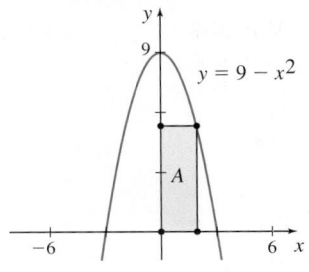

$y = 9 - x^2$

A

42. Maximizing Volume Find the dimensions of the right circular cone of maximum volume having a slant height of 4 ft. See the figure.

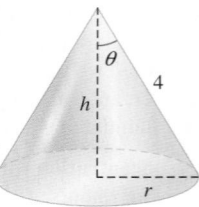

43. Minimizing Distance Let a and b be two positive real numbers. Find the line through the point (a, b) and connecting the points $(0, y_0)$ and $(x_0, 0)$ so that the distance from $(x_0, 0)$ to $(0, y_0)$ is a minimum. (In general, x_0 and y_0 will depend on the line.)

44. Maximizing Velocity An object moves on the x-axis in such a way that its velocity at time t seconds, $t \geq 1$, is given by $v = \dfrac{\ln t}{t}$ cm/s. At what time t does the object attain its maximum velocity?

45. Physics A heavy object of mass m is to be dragged along a horizontal surface by a rope making an angle θ to the horizontal. The force F required to move the object is given by the formula

$$F = \frac{cmg}{c\sin\theta + \cos\theta}$$

where g is the acceleration due to gravity and c is the **coefficient of friction** of the surface. Show that the force is least when $\tan\theta = c$.

46. Chemistry A self-catalytic chemical reaction results in the formation of a product that causes its formation rate to increase. The reaction rate V of many self-catalytic chemicals is given by

$$V = kx(a - x) \qquad 0 \leq x \leq a$$

where k is a positive constant, a is the initial amount of the chemical, and x is the variable amount of the chemical. For what value of x is the reaction rate a maximum?

47. Optimal Viewing Angle A picture 4 m in height is hung on a wall with the lower edge 3 m above the level of an observer's eye. How far from the wall should the observer stand in order to obtain the most favorable view? (That is, the picture should subtend the maximum angle.)

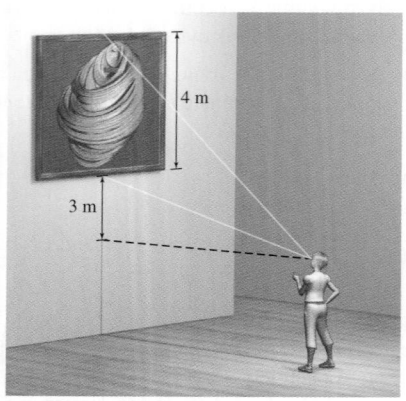

48. Maximizing Transmission Speed Traditional telephone cable is made up of a core of copper wires covered by an insulating material. If x is the ratio of the radius of the core to the thickness of the insulating material, the speed v of signaling is $v = kx^2 \ln \dfrac{1}{x}$, where k is a constant. Determine the ratio x that results in maximum speed.

49. Absolute Minimum If a, b, and c are positive constants, show that the minimum value of $f(x) = ae^{cx} + be^{-cx}$ is $2\sqrt{ab}$.

Challenge Problems

50. Maximizing Length The figure shows two corridors meeting at a right angle. One has width 1 m, and the other, width 8 m. Find the length of the longest pipe that can be carried horizontally from one corridor, around the corner, and into the other corridor.

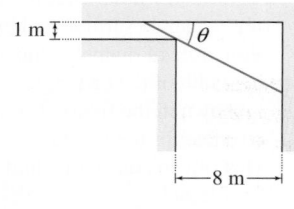

51. Optimal Height of a Lamp In the figure, a circular area of radius 20 ft is surrounded by a walk. A light is placed above the center of the area. What height most strongly illuminates the walk? The intensity of illumination is given by $I = \dfrac{\sin\theta}{s}$, where s is the distance from the source and θ is the angle at which the light strikes the surface.

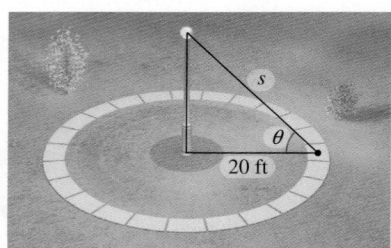

52. Maximizing Area Show that the rectangle of largest area that can be inscribed under the graph of $y = e^{-x^2}$ has two of its vertices at the point of inflection of y. See the figure.

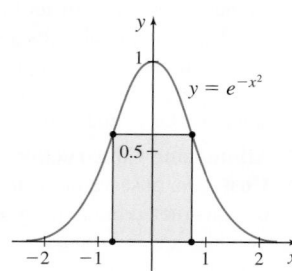

53. Minimizing Distance Find the point (x, y) on the graph of $f(x) = e^{-x/2}$ that is closest to the point $(1, 8)$.

AP® Practice Problems

Multiple-Choice Questions

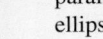 **1.** Find the dimensions of the rectangle with the largest area that can be formed with its base on the x-axis and its upper vertices on the graph of the parabola $y = 9 - x^2$.

    (A) base: $\sqrt{3}$, height: 6   (B) base: 3, height: 6
    (C) base: 5, height: 5   (D) base: $2\sqrt{3}$, height: 6

2. A closed cylindrical aluminum can has a volume of 16π in³. Find the dimensions of the can that minimizes the amount of aluminum used.

    (A) height: 8 in; radius: 2 in   (B) height: 4 in; radius: 2 in
    (C) height: 2 in; radius: 4 in   (D) height: $\dfrac{16}{9}$ in; radius: 3 in

3. What is the area of the largest rectangle with sides parallel to the coordinate axes that can be inscribed in the ellipse $4x^2 + y^2 = 16$?

    (A) 4   (B) $2\sqrt{2}$ (C) $8\sqrt{2}$ (D) 16

 4. If $y = 4x^2 - 3$, what is the local minimum value of the product xy?

    (A) -1 (B) 1   (C) -12 (D) -2

5. What point on the graph of $(x - 1)y = 4$, $x \geq 0$, is closest to the point $(1, 0)$?

    (A) $(0, -4)$ (B) $\left(4, \dfrac{4}{3}\right)$ (C) $(3, 2)$ (D) $(5, 1)$

Free-Response Question

6. A manufacturer needs to produce 10,000,000 cylindrical cans that are to be used for a new product. Each can must hold 8 ft³ of product. For stability, the cost of material for the tops and bottoms of the cans costs \$6.00 per square foot, and the material used for the sides of the cans costs \$4.00 per square foot. Find the dimensions of a can that will minimize production costs.

Retain Your Knowledge

Multiple-Choice Questions

1. Find $\lim\limits_{x \to \infty} \dfrac{x^3 + \ln x}{x \ln x}$

    (A) 0   (B) 9/2   (C) 9   (D) ∞

2. The functions f and g are both differentiable over the interval $0 < x < 10$, and $h = (f + g)^3$. Use the information in the table below to find $h'(4)$.

x	$f(x)$	$g(x)$	$f'(x)$	$g'(x)$
4	2	-3	-1	6

    (A) 3   (B) 15   (C) -15 (D) 5

3. $f(x) = \begin{cases} 2x^3 - 4x + 3 & \text{if } x < 1 \\ \cos(x - 1) & \text{if } x > 1 \end{cases}$. What value of f, if any, at $x = 1$ will make f continuous?

    (A) 1   (B) 0
    (C) -1 (D) No number will make f continuous.

Free-Response Question

4. $\dfrac{x^2}{2} + \dfrac{y^2}{3} = 1$

    (a) Find the slope of the line tangent to the graph of the equation at $\left(-1, \dfrac{\sqrt{6}}{2}\right)$.

    (b) Write an equation for the tangent line containing the point $\left(-1, \dfrac{\sqrt{6}}{2}\right)$.

5.6 Antiderivatives

OBJECTIVES *When you finish this section, you should be able to:*

1 Find antiderivatives (p. 406)

2 Solve motion problems (p. 408)

We have already learned that for each differentiable function f, there is a corresponding derivative function f'. Now consider this question: For a given function f, is there a function F whose derivative is f? That is, is it possible to find a function F so that $F' = \dfrac{dF}{dx} = f$? If such a function F can be found, it is called an *antiderivative of* f.

> **DEFINITION** Antiderivative
>
> A function F is called an **antiderivative** of the function f if $F'(x) = f(x)$ for all x in the domain of f.

① Find Antiderivatives

For example, an antiderivative of the function $f(x) = 2x$ is $F(x) = x^2$, since

$$F'(x) = \frac{d}{dx}x^2 = 2x$$

Another function F whose derivative is $2x$ is $F(x) = x^2 + 3$, since

$$F'(x) = \frac{d}{dx}(x^2 + 3) = 2x$$

This leads us to suspect that the function $f(x) = 2x$ has many antiderivatives. Indeed, any of the functions x^2 or $x^2 + \frac{1}{2}$ or $x^2 + 2$ or $x^2 + \sqrt{5}$ or $x^2 - 1$ has the property that its derivative is $2x$. Any function $F(x) = x^2 + C$, where C is a constant, is an antiderivative of $f(x) = 2x$.

Are there other antiderivatives of $2x$ that are not of the form $x^2 + C$? A corollary of the Mean Value Theorem (p. 338) tells us the answer is no.

> **COROLLARY**
>
> If the functions f and g are differentiable on an open interval (a, b) and if $f'(x) = g'(x)$ for all numbers x in (a, b), then there is a number C for which $f(x) = g(x) + C$ on (a, b).

This result can be stated in the following way.

> **THEOREM The General Form of an Antiderivative**
>
> If a function F is an antiderivative of a function f defined on an interval I, then any other antiderivative of f has the form $F(x) + C$, where C is an (arbitrary) constant.

All the antiderivatives of f can be obtained from the expression $F(x) + C$ by letting C range over all real numbers. For example, all the antiderivatives of $f(x) = x^5$ are of the form $F(x) = \frac{x^6}{6} + C$, where C is a constant. Figure 61 shows the graphs of $F(x) = \frac{x^6}{6} + C$ for some numbers C. The antiderivatives of a function f are a family of functions, each one a vertical translation of the others.

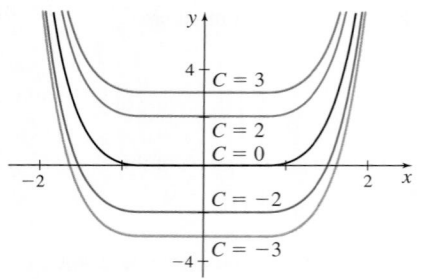

Figure 61 $F(x) = \frac{x^6}{6} + C$

EXAMPLE 1 Finding the Antiderivatives of a Function

Find all the antiderivatives of:

(a) $f(x) = 0$ **(b)** $g(\theta) = -\sin\theta$ **(c)** $h(x) = x^{1/2}$

Solution

(a) Since the derivative of a constant function is 0, all the antiderivatives of f are of the form $F(x) = C$, where C is a constant.

(b) Since $\frac{d}{d\theta}\cos\theta = -\sin\theta$, all the antiderivatives of $g(\theta) = -\sin\theta$ are of the form $G(\theta) = \cos\theta + C$, where C is a constant.

(c) The derivative of $\frac{2}{3}x^{3/2}$ is $\frac{2}{3} \cdot \frac{3}{2}x^{3/2-1} = x^{1/2}$. So, all the antiderivatives of $h(x) = x^{1/2}$ are of the form $H(x) = \frac{2}{3}x^{3/2} + C$, where C is a constant. ■

In Example 1(c), you may wonder how we knew to choose $\frac{2}{3}x^{3/2}$. For any real number a, the Power Rule states $\frac{d}{dx}x^a = ax^{a-1}$. That is, differentiation reduces the exponent by 1. Antidifferentiation is the inverse process, so it suggests we increase the exponent by 1. This is how we obtain the factor $x^{3/2}$ of the antiderivative $\frac{2}{3}x^{3/2}$. The factor $\frac{2}{3}$ is needed so that when we differentiate $\frac{2}{3}x^{3/2}$, the result is $\frac{2}{3}\cdot\frac{3}{2}x^{3/2-1} = x^{1/2}$.

NOW WORK Problems 11 and 17.

Suppose $f(x) = x^a$, where a is a real number and $a \neq -1$.

> The function $F(x)$ defined by
> $$F(x) = \frac{x^{a+1}}{a+1} \quad a \neq -1$$
> is an antiderivative of $f(x) = x^a$. That is, all the antiderivatives of $f(x) = x^a$, $a \neq -1$, are of the form $\frac{x^{a+1}}{a+1} + C$, where C is a constant.

CAUTION Be sure to include the absolute value bars when finding the antiderivative of $\frac{1}{x}$.

Now consider the case when $a = -1$. Then $f(x) = x^{-1} = \frac{1}{x}$. We know that $\frac{d}{dx}\ln|x| = \frac{1}{x} = x^{-1}$, for $x \neq 0$. So, all the antiderivatives of $f(x) = x^{-1}$ are of the form $\ln|x| + C$, provided $x \neq 0$.

Table 8 includes these results along with the antiderivatives of some other common functions. Each listing is based on a derivative formula.

TABLE 8

Function f	Antiderivatives F of f	Function f	Antiderivatives F of f
$f(x) = 0$	$F(x) = C$	$f(x) = \sec^2 x$	$F(x) = \tan x + C$
$f(x) = 1$	$F(x) = x + C$	$f(x) = \sec x \tan x$	$F(x) = \sec x + C$
$f(x) = x^a, \quad a \neq -1$	$F(x) = \frac{x^{a+1}}{a+1} + C$	$f(x) = \csc x \cot x$	$F(x) = -\csc x + C$
		$f(x) = \csc^2 x$	$F(x) = -\cot x + C$
$f(x) = x^{-1} = \frac{1}{x}$	$F(x) = \ln|x| + C$	$f(x) = \frac{1}{\sqrt{1-x^2}}, \quad |x| < 1$	$F(x) = \sin^{-1} x + C$
$f(x) = e^x$	$F(x) = e^x + C$		
$f(x) = a^x$	$F(x) = \frac{a^x}{\ln a} + C, \quad a > 0, a \neq 1$	$f(x) = \frac{1}{1+x^2}$	$F(x) = \tan^{-1} x + C$
$f(x) = \sin x$	$F(x) = -\cos x + C$	$f(x) = \frac{1}{x\sqrt{x^2-1}}, \quad |x| > 1$	$F(x) = \sec^{-1} x + C$
$f(x) = \cos x$	$F(x) = \sin x + C$		

The next two theorems are consequences of properties of derivatives and the relationship between derivatives and antiderivatives.

> **THEOREM Sum Rule and Difference Rule**
>
> If the functions F_1 and F_2 are antiderivatives of the functions f_1 and f_2, respectively, then
>
> - $F_1 + F_2$ is an antiderivative of $f_1 + f_2$, and
> - $F_1 - F_2$ is an antiderivative of $f_1 - f_2$.

The Sum Rule can be extended to any finite sum of functions.

> **THEOREM Constant Multiple Rule**
>
> If k is a real number and if F is an antiderivative of f, then kF is an antiderivative of kf.

IN WORDS
- An antiderivative of the sum of two functions equals the sum of the antiderivatives of the functions.
- An antiderivative of a number times a function equals the number times an antiderivative of the function.

EXAMPLE 2 Finding the Antiderivatives of a Function

Find all the antiderivatives of $f(x) = e^x + \dfrac{6}{x^2} - \sin x$.

Solution

Since f is the sum of three functions, we use the Sum Rule. That is, we find the antiderivatives of each function individually and then add.

- An antiderivative of e^x is e^x.
- An antiderivative of $6x^{-2}$ is $6 \cdot \dfrac{x^{-2+1}}{-2+1} = 6 \cdot \dfrac{x^{-1}}{-1} = -\dfrac{6}{x}$.
- Finally, an antiderivative of $\sin x$ is $-\cos x$.

Then all the antiderivatives of the function f are given by

$$F(x) = e^x - \frac{6}{x} + \cos x + C$$

where C is a constant. ∎

NOTE Since the antiderivatives of e^x are $e^x + C_1$, the antiderivatives of $\dfrac{6}{x^2}$ are $-\dfrac{6}{x} + C_2$, and the antiderivatives of $\sin x$ are $-\cos x + C_3$, in Example 2, the constant $C = C_1 + C_2 + C_3$.

NOW WORK Problem **29** and AP® Practice Problems **2, 3,** and **6.**

❷ Solve Motion Problems

For an object moving along a line, the functions $s = s(t)$, $v = v(t)$, and $a = a(t)$ represent the position s, velocity v, and acceleration a, respectively, of the object at time t. The three quantities s, v, and a are related by the equations

$$\frac{ds}{dt} = v(t) \qquad \frac{dv}{dt} = a(t) \qquad \frac{d^2 s}{dt^2} = a(t)$$

If the acceleration $a = a(t)$ is a known function of the time t, then the velocity can be found by finding an antiderivative of $a = a(t)$. Similarly, if the velocity $v = v(t)$ is a known function of t, then the position s from the origin at time t is an antiderivative of $v = v(t)$.

In physical problems, the values of the velocity v and position s at time $t = 0$ are often given. They are denoted as $v(0) = v_0$ and $s(0) = s_0$ and are referred to as **initial conditions**.

EXAMPLE 3 Solving a Motion Problem

Find the position function $s = s(t)$ of an object at time t if its acceleration a is

$$a(t) = 8t - 3$$

and the initial conditions are $v_0 = v(0) = 4$ and $s_0 = s(0) = 1$.

Solution
The antiderivatives of $a(t) = 8t - 3$ are

$$v(t) = 4t^2 - 3t + C_1$$

Now use the initial condition $v_0 = v(0) = 4$.

$$v_0 = v(0) = 4 \cdot 0^2 - 3 \cdot 0 + C_1 = 4 \quad t=0, \ v_0=4$$
$$C_1 = 4$$

The velocity of the object at time t is $v(t) = 4t^2 - 3t + 4$.

The position function $s = s(t)$ of the object at time t is obtained by finding the antiderivatives of

$$v(t) = 4t^2 - 3t + 4$$

Then

$$s(t) = \frac{4}{3}t^3 - \frac{3}{2}t^2 + 4t + C_2$$

Using the initial condition, $s_0 = s(0) = 1$, we have

$$s_0 = s(0) = 0 - 0 + 0 + C_2 = 1$$
$$C_2 = 1$$

The position function $s = s(t)$ of the object at any time t is

$$s = s(t) = \frac{4}{3}t^3 - \frac{3}{2}t^2 + 4t + 1$$ ∎

NOW WORK Problem 37 and AP® Practice Problems 1, 4, and 5.

EXAMPLE 4 Solving a Motion Problem

When the brakes of a car are applied, the car decelerates at a constant rate of $10\,\text{m/s}^2$. If the car is to stop within $20\,\text{m}$ after the brakes are applied, what is the maximum velocity the car could have been traveling? Express the answer in miles per hour.

Solution

Let the position function $s = s(t)$ represent the distance s in meters the car has traveled t seconds after the brakes are applied. Let v_0 be the initial velocity of the car at the time the brakes are applied ($t = 0$). Since the car decelerates at the rate of $10\,\text{m/s}^2$, its acceleration a, in meters per second squared, is

$$a(t) = \frac{dv}{dt} = -10$$

Then

$$v(t) = -10t + C_1$$

When $t = 0$, $v(0) = v_0$, the initial velocity of the car when the brakes are applied, so $C_1 = v_0$. Then

$$v(t) = \frac{ds}{dt} = -10t + v_0$$

so

$$s(t) = -5t^2 + v_0 t + C_2$$

Since s is the distance the car has traveled after the brakes are applied, the initial condition is $s(0) = 0$. Then $s(0) = -5 \cdot 0 + v_0 \cdot 0 + C_2 = 0$, so $C_2 = 0$. The distance s, in meters, the car travels t seconds after applying the brakes is

$$s(t) = -5t^2 + v_0 t$$

The car stops completely when its velocity equals 0. That is, when

$$v(t) = -10t + v_0 = 0$$

$$t = \frac{v_0}{10}$$

This is the time it takes the car to come to rest. Substituting $\frac{v_0}{10}$ for t in the position function $s = s(t)$, the distance the car has traveled is

$$s\left(\frac{v_0}{10}\right) = -5\left(\frac{v_0}{10}\right)^2 + v_0\left(\frac{v_0}{10}\right) = \frac{v_0^2}{20}$$

If the car is to stop within 20 m, then $s \leq 20$; that is, $\frac{v_0^2}{20} \leq 20$ or equivalently $v_0^2 \leq 400$. The maximum possible velocity v_0 for the car is $v_0 = 20$ m/s.

To express 20 m/s in miles per hour, we proceed as follows:

$$v_0 = 20 \text{ m/s} = \frac{20 \text{ m}}{\text{s}} \cdot \frac{1 \text{ km}}{1000 \text{ m}} \cdot \frac{3600 \text{ s}}{1 \text{ h}}$$

$$= 72 \text{ km/h} \approx \frac{72 \text{ km}}{\text{h}} \cdot \frac{1 \text{ mi}}{1.6 \text{ km}} = 45 \text{ mi/h}$$

The car's maximum possible velocity to stop within 20 m is 45 mi/h. ■

NOW WORK Problem 49.

Freely Falling Objects

An object falling toward Earth is a common example of motion with (nearly) constant acceleration. In the absence of air resistance, all objects, regardless of size, weight, or composition, fall with the same acceleration when released from the same point above Earth's surface, and if the distance fallen is not too great, the acceleration remains constant throughout the fall. This ideal motion, in which air resistance and the small change in acceleration with altitude are neglected, is called **free fall**. The constant acceleration of a freely falling object is called the **acceleration due to gravity** and is denoted by the symbol g. Near Earth's surface, its magnitude is approximately 32 ft/s^2, or 9.8 m/s^2, and its direction is down toward the center of Earth.

EXAMPLE 5 **Solving a Problem Involving Free Fall**

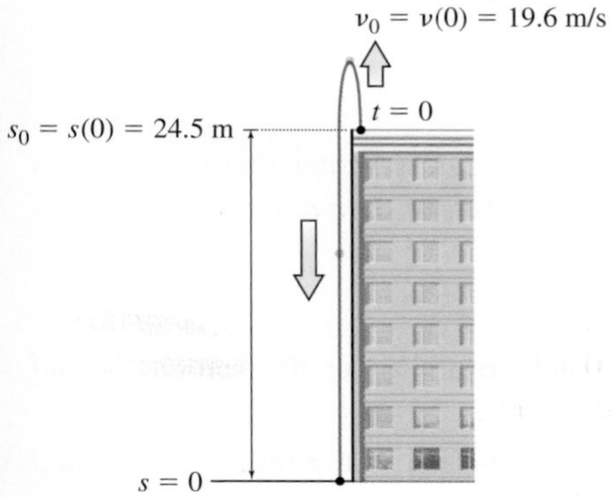

A rock is thrown straight up with an initial velocity of 19.6 m/s from the roof of a building 24.5 m above ground level, as shown in Figure 62.

(a) How long does it take the rock to reach its maximum altitude?

(b) What is the maximum altitude of the rock?

(c) If the rock misses the edge of the building on the way down and eventually strikes the ground, what is the total time the rock is in the air?

$v_0 = v(0) = 19.6$ m/s

$t = 0$

$s_0 = s(0) = 24.5$ m

$s = 0$

Figure 62

Solution

To answer the questions, we need to find the velocity $v = v(t)$ and the position $s = s(t)$ of the rock as functions of time. We begin measuring time when the rock is released. If s is the distance, in meters, of the rock from the ground, then since the rock is released at a height of 24.5 m, $s_0 = s(0) = 24.5$ m.

The initial velocity of the rock is given as $v_0 = v(0) = 19.6$ m/s. If air resistance is ignored, the only force acting on the rock is gravity. Since the acceleration due to gravity is -9.8 m/s^2, the acceleration a of the rock is

$$a = \frac{dv}{dt} = -9.8$$

Then

$$v(t) = -9.8t + v_0 \quad \text{Find the antiderivatives of } a.$$

Using the initial condition, $v_0 = v(0) = 19.6$ m/s, the velocity of the rock at any time t is

$$v(t) = -9.8t + 19.6 \tag{1}$$

Since $\frac{ds}{dt} = v(t) = -9.8t + 19.6$, we find

$$s(t) = -4.9t^2 + 19.6t + s_0$$

Using the initial condition, $s(0) = 24.5$ m, the distance s of the rock from the ground at any time t is

$$s(t) = -4.9t^2 + 19.6t + 24.5 \tag{2}$$

Now we can answer the questions.

(a) The rock reaches its maximum altitude when its velocity is 0.

$$v(t) = -9.8t + 19.6 = 0 \quad \text{From (1)}$$
$$t = 2$$

The rock reaches its maximum altitude at $t = 2$ s.

(b) To find the maximum altitude, evaluate $s(2)$. The maximum altitude of the rock is

$$s(2) = -4.9 \cdot 2^2 + 19.6 \cdot 2 + 24.5 = 44.1 \text{ m} \quad \text{From (2)}$$

(c) We find the total time the rock is in the air by solving $s(t) = 0$.

$$-4.9t^2 + 19.6t + 24.5 = 0$$
$$t^2 - 4t - 5 = 0$$
$$(t - 5)(t + 1) = 0$$
$$t = 5 \quad \text{or} \quad t = -1$$

The only meaningful solution is $t = 5$. The rock is in the air for 5 s. ∎

Now we examine the general problem of freely falling objects.

If F is the weight of an object of mass m, then according to Galileo, assuming air resistance is negligible, a freely falling object obeys the equation

$$F = -mg$$

where g is the acceleration due to gravity. The minus sign indicates that the object is falling. Also, according to **Newton's Second Law of Motion**, $F = ma$, so

$$ma = -mg \qquad \text{or} \qquad a = -g$$

where a is the acceleration of the object. We seek formulas for the velocity v and distance s from Earth of a freely falling object at time t.

Let $t = 0$ be the instant we begin to measure the motion of the falling object, and suppose at this instant the object's vertical distance above Earth is s_0 and its velocity is v_0. Since the acceleration $a = \dfrac{dv}{dt}$ and $a = -g$, we have

$$\boxed{a = \frac{dv}{dt} = -g} \qquad\qquad (3)$$

Using the initial condition $v_0 = v(0)$, we obtain

$$\boxed{v(t) = -gt + v_0}$$

Now since $\dfrac{ds}{dt} = v(t)$, we have

$$\frac{ds}{dt} = -gt + v_0 \qquad\qquad (4)$$

Using the initial condition $s_0 = s(0)$, we obtain

$$\boxed{s(t) = -\frac{1}{2}gt^2 + v_0 t + s_0}$$

NOW WORK   Problem 51.

Newton's First Law of Motion

We close this section with a special case of the *Law of Inertia*, which was originally stated by Galileo.

THEOREM Newton's First Law of Motion

If no force acts on a body, then a body at rest remains at rest and a body in uniform motion (constant velocity) remains in uniform motion.

Proof The force F acting on a body of mass m is given by Newton's Second Law of Motion $F = ma$, where a is the acceleration of the body. If there is no force acting on the body, then $F = 0$. In this case, the acceleration a must be 0. But $a = \dfrac{dv}{dt}$, where v is the velocity of the body. So,

$$\frac{dv}{dt} = 0 \qquad \text{and} \qquad v = C, \text{ a constant}$$

That is, the body is at rest ($v = 0$) or else in a state of uniform motion (v is a nonzero constant). ∎

5.6 Assess Your Understanding

Concepts and Vocabulary

1. A function F is called a(n) _____ of a function f if $F' = f$.
2. *True or False* If F is an antiderivative of f, then $F(x) + C$, where C is a constant, is also an antiderivative of f.
3. All the antiderivatives of $y = x^{-1}$ are _____.
4. *True or False* An antiderivative of $\sin x$ is $-\cos x + \pi$.
5. *True or False* Free fall is an example of motion with constant acceleration.
6. *True or False* If F_1 and F_2 are both antiderivatives of a function f on an interval I, then $F_1 - F_2 = C$, a constant, on I.

Skill Building

In Problems 7–36, find all the antiderivatives of each function.

7. $f(x) = 2$

8. $f(x) = \dfrac{1}{2}$

9. $f(x) = 4x$

10. $f(x) = -6x$

11. $f(x) = 4x^5$

12. $f(x) = x$

13. $f(x) = 5x^{3/2}$

14. $f(x) = x^{5/2} + 2$

15. $f(x) = 2x^{-2}$

16. $f(x) = 3x^{-3}$

17. $f(x) = \sqrt{x}$

18. $f(x) = \dfrac{1}{\sqrt{x}}$

19. $f(x) = 4x^3 - 3x^2 + 1$

20. $f(x) = x^2 - x$

21. $f(x) = (2 - 3x)^2$

22. $f(x) = (3x - 1)^2$

23. $f(x) = \dfrac{3x - 2}{x}$

24. $f(x) = \dfrac{x^2 + 4}{x}$

25. $f(x) = \dfrac{3x^{1/2} - 4}{x}$

26. $f(x) = \dfrac{4x^{3/2} - 1}{x}$

27. $f(x) = 2x - 3\cos x$

28. $f(x) = 2\sin x - \cos x$

29. $f(x) = 4e^x + x$

30. $f(x) = e^{-x} + \sec^2 x$

31. $f(x) = \dfrac{7}{1 + x^2}$

32. $f(x) = x + \dfrac{10}{\sqrt{1 - x^2}}$

33. $f(x) = e^x + \dfrac{1}{x\sqrt{x^2 - 1}}$

34. $f(x) = \sin x + \dfrac{1}{1 + x^2}$

35. $f(x) = \dfrac{1 + xe^x}{x}$

36. $f(x) = \dfrac{2 - xe^x}{x}$

In Problems 37–42, the acceleration of an object moving on a line is given. Find the position function $s = s(t)$ of the object under the given initial conditions.

37. $a(t) = -32\,\text{ft/s}^2$, $s(0) = 0\,\text{ft}$, $v(0) = 128\,\text{ft/s}$

38. $a(t) = -980\,\text{cm/s}^2$, $s(0) = 5\,\text{cm}$, $v(0) = 1980\,\text{cm/s}$

39. $a(t) = 3t\,\text{m/s}^2$, $s(0) = 2\,\text{m}$, $v(0) = 18\,\text{m/s}$

40. $a(t) = 5t - 2\,\text{ft/s}^2$, $s(0) = 0\,\text{ft}$, $v(0) = 8\,\text{ft/s}$

41. $a(t) = \sin t\,\text{ft/s}^2$, $s(0) = 0\,\text{ft}, v(0) = 5\,\text{ft/s}$

42. $a(t) = e^t\,\text{m/s}^2$, $s(0) = 0\,\text{m}, v(0) = 4\,\text{m/s}$

Applications and Extensions

In Problems 43 and 44, find all the antiderivatives of each function.

43. $f(u) = \dfrac{u^2 + 10u + 21}{3u + 9}$

44. $f(t) = \dfrac{t^3 - 5t + 8}{t^5}$

45. Use the fact that

$$\frac{d}{dx}(x\cos x + \sin x) = -x\sin x + 2\cos x$$

to find F if

$$\frac{dF}{dx} = -x\sin x + 2\cos x \qquad \text{and} \qquad F(0) = 1$$

46. Use the fact that

$$\frac{d}{dx}\sin x^2 = 2x\cos x^2$$

to find h if

$$\frac{dh}{dx} = x\cos x^2 \qquad \text{and} \qquad h(0) = 2$$

47. **Motion on a Line** A car decelerates at a constant rate of $10\,\text{m/s}^2$ when its brakes are applied. If the car must stop within 15 m after applying the brakes, what is the maximum allowable velocity for the car? Express the answer in meters per second and in miles per hour.

48. **Motion on a Line** A car can accelerate from 0 to 60 km/h in 10 seconds. If the acceleration is constant, how far does the car travel during this time?

49. **Motion on a Line** A BMW 6 series can accelerate from 0 to 60 mi/h in 5 s. If the acceleration is constant, how far does the car travel during this time?
    Source: BMW USA.

50. **Free Fall** A ball is thrown straight up from ground level, with an initial velocity of 19.6 m/s. How high is the ball thrown? How long will it take the ball to return to ground level?

51. **Free Fall** A child throws a ball straight up. If the ball is to reach a height of 9.8 m, what is the minimum initial velocity that must be imparted to the ball? Assume the initial height of the ball is 1 m.

52. **Free Fall** A ball thrown directly down from a roof 49 m high reaches the ground in 3 s. What is the initial velocity of the ball?

53. **Inertia** A constant force is applied to an object that is initially at rest. If the mass of the object is 4 kg and if its velocity after 6 s is 12 m/s, determine the force applied to it.

54. **Motion on a Line** Starting from rest, with what constant acceleration must a car move to travel 2 km in 2 min? (Give your answer in centimeters per second squared.)

55. **Downhill Speed of a Skier** The down slope acceleration a of a skier is given by $a = a(t) = g\sin\theta$, where t is time, in seconds, $g = 9.8\,\text{m/s}^2$ is the acceleration due to gravity, and θ is the angle of the slope. If the skier starts from rest at the lift, points his skis straight down a $20°$ slope, and does not turn, how fast is he going after 5 s?

56. **Free Fall** A child on top of a building 24 m high drops a rock and then 1 s later throws another rock straight down. What initial velocity must the second rock be given so that the dropped rock and the thrown rock hit the ground at the same time?

57. Free Fall The world's high jump record, set on July 27, 1993, by Cuban jumper Javier Sotomayor, is 2.45 m. If this event were held on the Moon, where the acceleration due to gravity is $1.6 \, \text{m/s}^2$, what height would Sotomayor have attained? Assume that he propels himself with the same force on the Moon as on Earth.

58. Free Fall The 2-m high jump is common today. If this event were held on the Moon, where the acceleration due to gravity is $1.6 \, \text{m/s}^2$, what height would be attained? Assume that an athlete can propel him- or herself with the same force on the Moon as on Earth.

59. Radiation Radiation, such as X-rays or the radiation from radioactivity, is absorbed as it passes through tissue or any other material. The rate of change in the intensity I of the radiation with respect to the depth x of tissue is directly proportional to the intensity I. This proportion can be expressed as an equation by introducing a positive constant of proportionality k, where k depends on the properties of the tissue and the type of radiation.

(a) Show that $\dfrac{dI}{dx} = -kI, \, k > 0.$

(b) Explain why the minus sign is necessary.

(c) Use (a) to find the intensity I as a function of the depth x in the tissue. The intensity of the radiation when it enters the tissue is $I(0) = I_0$.

(d) Find the value of k if the intensity is reduced by 90% of its maximum value at a depth of 2.0 cm.

Challenge Problems

In Problems 60–61, find all the antiderivatives of f.

60. $f(x) = \dfrac{5 + 7\sin x}{2\cos^2 x}$

61. $f(x) = \dfrac{2x^3 + 2x + 3}{1 + x^2}$

62. Moving Shadows A lamp on a post 10 m high stands 25 m from a wall. A boy standing 5 m from the lamp and 20 m from the wall throws a ball straight up with an initial velocity of 19.6 m/s. The acceleration due to gravity is $a = -9.8 \, \text{m/s}^2$. The ball is thrown up from an initial height of 1 m above ground.

(a) How fast is the shadow of the ball moving on the wall 3 s after the ball is released?

(b) Explain if the ball is moving up or down.

(c) How far is the ball above ground at $t = 3$ s?

Preparing for the AP® Exam

AP® Practice Problems

Multiple-Choice Questions

 1. A particle moves along the x-axis with velocity $v(t) = t^2 + t$, $t \geq 0$. If the particle is at $x = -1$, when $t = 0$, then what is the position x of the particle at time $t = 3$?

(A) -1 (B) $\dfrac{29}{2}$ (C) $\dfrac{25}{2}$ (D) 12

 2. If $f(x) = 2e^x - \dfrac{3}{x}$, then the antiderivatives of f are

(A) $2e^x + 3\ln|x| + C$ (B) $e^{2x} - 3\ln|x| + C$
(C) $2e^x - \ln|3x| + C$ (D) $2e^x - 3\ln|x| + C$

 3. Find all the antiderivatives of $f(x) = 2x - \cos x$.

(A) $2x^2 + \sin x + C$ (B) $x^2 + \sin x + C$
(C) $x^2 - \sin x + C$ (D) $x^2 - \cos^2 x \sin x + C$

4. An object is moving along the y-axis so that at any time t its velocity is given by $v(t) = 4t + 1$. If at time $t = 1$, the object is at $y = 3$, then its position y at any time t is

(A) $y(t) = 4t^2 + t - 5$ (B) $y(t) = 2t^2 + t + 3$
(C) $y(t) = 4t - 1$ (D) $y(t) = 2t^2 + t$

5. The acceleration a of an object moving on a line is given by $a(t) = 2 + 12t$, and the velocity v of the object at time $t = 0$ is 5. If $s = s(t)$ is the distance of the object from the origin at time t, find $s(3) - s(1)$.

(A) 86 (B) 70 (C) 30 (D) 52

Free-Response Question

 6. Find the antiderivative F of $f(x) = \dfrac{6x^3 - 3x + 4}{x}$ for which $F(1) = 2$.

Retain Your Knowledge

Multiple-Choice Questions

1. The function $f(x) = x^3 + 5$ has an inverse function g. Find $g'(13)$.

(A) $\dfrac{1}{13^3 + 5}$ (B) $\dfrac{1}{507}$ (C) $\dfrac{1}{13}$ (D) $\dfrac{1}{12}$

2. $\lim\limits_{x \to \frac{\pi}{2}} \dfrac{1 - \sin x}{\cos x} =$

(A) 0 (B) -1 (C) 1 (D) ∞

3. A function $y = f(x)$ is differentiable for $0 < x < 100$. The table below gives values of f for various numbers x in the domain of f. Use a linear approximation to estimate $f(9)$.

x	4	8	11	14
$f(x)$	11	27	9	-9

(A) 11 (B) 21 (C) 15 (D) 18

Free-Response Questions

4. $f(x) = \begin{cases} 3e^x + 1 & \text{if } x \leq 0 \\ (x^2 - 3x + 2)^2 & \text{if } x > 0 \end{cases}$. Determine whether $f'(0)$ exists. Justify your answer.

CHAPTER 5 PROJECT

The U.S. Economy

This Project can be done individually or as part of a team.

Economic data are quite variable and very complicated, making them difficult to model. The models we use here are rough approximations of Unemployment and Gross Domestic Product (GDP).

Suppose Unemployment, defined as the percentage of people over the age of 16 who are actively looking for a job, is modeled by

$$U = U(t) = \frac{5\cos\dfrac{2t}{\pi}}{2 + \sin\dfrac{2t}{\pi}} + 3.3 \qquad 0 \le t \le 20$$

where t is in years and $t = 0$ represents January 1, 2012.

Suppose GDP, defined as the percent increase (or decrease) in the market value of all goods and services produced over a three month period of time, expressed annually, is modeled by

$$G = G(t) = 3.5\cos\left(\frac{2t}{\pi} + 5\right) + 1.2 \qquad 0 \le t \le 20$$

where t is in years and $t = 0$ represents January 1, 2012.

- If GDP > 0, the economy is **expanding**.
- If GDP < 0, the economy is **contracting**.
- An economy is in **recession** if GDP < 0 for two or more consecutive quarters.

1. During what years was the U.S. economy in recession?
2. Find the critical numbers of $U = U(t)$.
3. Determine the intervals on which U is increasing and on which it is decreasing.
4. Find all the local extrema of U.

5. Find the critical numbers of $G = G(t)$.
6. Determine the intervals on which G is increasing and on which it is decreasing.
7. Find all the local extrema of G.
8. (a) Find the inflection points of G.
   (b) Describe in economic terms what the inflection points of G represent.
9. Graph $U = U(t)$ and $G = G(t)$ on the same set of coordinate axes.

Problems 10–12 refer to the graphs on page 332.

Okun's Law states that an increase in unemployment tends to coincide with a decrease in GDP.

10. Using the U.S. data shown in the graphs, during years of recession, what is happening to unemployment? Explain in economic terms why this makes sense.
11. Use your answer to 10 to explain whether or not the graphs generally agree with Okun's Law.
12. One school of economic thought holds the view that when the economy improves (increasing GDP), more jobs are created, resulting in lower unemployment. Others argue that once unemployment improves (decreases), more people are working and this increases GDP. Based on your analysis of the graphs, which position do you support?

_____

U.S. GDP data from:

https://www.macrotrends.net/countries/USA/united-states/gdp-growth-rate

https://fortunly.com/statistics/us-gdp-by-year-guide/

Unemployment data from:

https://www.bls.gov/beta/

To read Arthur Okun's paper, see

http://cowles.econ.yale.edu/P/cp/p01b/p0190.pdf

Chapter Review

THINGS TO KNOW

5.1 The Mean Value Theorem; Increasing/Decreasing Function Test

- **Rolle's Theorem** (p. 333)
- **Mean Value Theorem** Suppose f is a function defined on a closed interval $[a, b]$. If f is continuous on $[a, b]$ and differentiable on (a, b), then there is at least one number c in the open interval (a, b) for which $f'(c) = \dfrac{f(b) - f(a)}{b - a}$
  (p. 334)
- **Corollaries to the Mean Value Theorem**
  - If $f'(x) = 0$ for all numbers x in (a, b), then f is constant on (a, b). (p. 337)
  - If $f'(x) = g'(x)$ for all numbers x in (a, b), then there is a number C for which $f(x) = g(x) + C$ on (a, b). (p. 338)
- **Increasing/Decreasing Function Test** (p. 338)

5.2 Extreme Value Theorem; Global Versus Local Extrema; Critical Numbers

Definitions:

- Absolute maximum; absolute minimum (p. 348)
- Absolute maximum value; absolute minimum value (p. 348)
- Local maximum; local minimum (p. 348)
- Local maximum value; local minimum value (p. 348)
- Critical number (p. 351)

Theorems:

- **Extreme Value Theorem** If a function f is continuous on a closed interval $[a, b]$, then f has an absolute maximum and an absolute minimum on $[a, b]$. (p. 350)
- If a function f has a local maximum or a local minimum at the number c, then either $f'(c) = 0$ or $f'(c)$ does not exist. (p. 350)

Procedure:

- **The Candidates Test**, the steps for finding the absolute extreme values of a function f that is continuous on a closed interval $[a, b]$. (p. 353)

5.3 Local Extrema and Concavity

Definitions:

- Concave up; concave down (p. 367)
- Inflection point (p. 370)

Theorems:

- First Derivative Test (p. 362)
- Test for concavity (p. 368)
- A condition for an inflection point (p. 370)
- Second Derivative Test (p. 372)

Procedure:

- Steps for finding the inflection points of a function (p. 370)

5.4 Using Calculus to Graph Functions

Procedure:

- Steps for graphing a function $y = f(x)$ (p. 382)

5.5 Introduction to Optimization

Procedure:

- Steps for solving optimization problems (p. 394)

5.6 Antiderivatives

Definitions:

- Antiderivative (p. 405)
- Initial condition (p. 408)

Basic Antiderivatives   See Table 8 (p. 407)

Antidifferentiation Properties:

- **Sum Rule and Difference Rule**: If functions F_1 and F_2 are antiderivatives of the functions f_1 and f_2, respectively, then $F_1 + F_2$ is an antiderivative of $f_1 + f_2$ and $F_1 - F_2$ is an antiderivative of $f_1 - f_2$. (p. 408)
- **Constant Multiple Rule**: If k is a real number and if F is an antiderivative of f, then kF is an antiderivative of kf. (p. 408)

Theorems:

- If a function F is an antiderivative of a function f on an interval I, then any other antiderivative of f has the form $F(x) + C$, where C is an (arbitrary) constant. (p. 406)
- Newton's First Law of Motion (p. 412)
- Newton's Second Law of Motion, $F = ma$ (p. 412)

OBJECTIVES

Section	You should be able to …	Examples	Review Exercises	Preparing for the AP® Exam — AP® Review Problems
5.1	1 Use Rolle's Theorem (p. 333)	1	24	
	2 Use the Mean Value Theorem (p. 334)	2, 3	7, 8, 25	7, 14
	3 Determine intervals on which a function is increasing or decreasing (p. 338)	4–7	19(a), 20(a)	1, 2
5.2	1 Identify absolute maximum and minimum values and local maximum and minimum values of a function; apply the Extreme Value Theorem (p. 347)	1	1, 2	
	2 Find critical numbers (p. 352)	2	3, 4(a)	13(a)
	3 Find absolute maximum and absolute minimum values; use the Candidates Test (p. 353)	3–6	5, 6	8, 15(a)
5.3	1 Use the First Derivative Test to find local extrema (p. 362)	1, 2, 3	4(b), 9(a)–11(a)	6, 13(c)
	2 Use the First Derivative Test with motion on a line (p. 365)	4	12	3, 12, 13(b)
	3 Determine the concavity of a function (p. 367)	5, 6, 7	19(b), 20(b)	13(d), 13(f), 15(b)
	4 Find inflection points (p. 370)	8, 9	19(c), 20(c)	4, 5, 13(e)
	5 Use the Second Derivative Test to find local extrema (p. 372)	10	9(b)–11(b)	13(c)
5.4	1 Graph a function using calculus (p. 382)	1–5	13–18	
	2 Obtain information about a function from its derivatives (p. 390)	6, 7	21–23	11, 6
5.5	1 Solve optimization problems (p. 394)	1–6	26, 27, 39–41	9
5.6	1 Find antiderivatives (p. 406)	1, 2	28–35	10
	2 Solve motion problems (p. 408)	3–5	36–38	

REVIEW EXERCISES

In Problems 1 and 2, use the graphs below to determine whether each function has an absolute extremum and/or a local extremum or neither at the indicated points.

1.

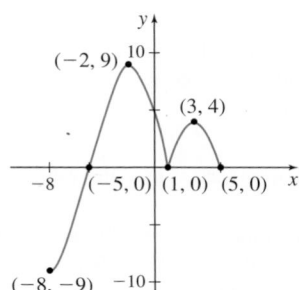

2.

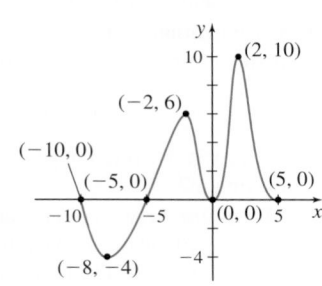

3. Critical Numbers Find all the critical numbers of $f(x) = \cos(2x)$ on the closed interval $[0, \pi]$.

4. Critical Numbers $f(x) = \dfrac{x^2}{2x - 1}$

   (a) Find all the critical numbers of f.

   (b) Find the local extrema of f.

In Problems 5 and 6, find the absolute maximum value and absolute minimum value of each function on the indicated interval.

5. $f(x) = \dfrac{3}{2}x^4 - 2x^3 - 6x^2 + 5$ on $[-2, 3]$

6. $f(x) = x - \sin(2x)$ on $[0, 2\pi]$

7. Mean Value Theorem Verify that the hypotheses for the Mean Value Theorem are satisfied for the function $f(x) = \dfrac{2x - 1}{x}$ on the interval $[1, 4]$. Find a point on the graph of f that has a tangent line whose slope equals the slope of the secant line joining $(1, 1)$ to $\left(4, \dfrac{7}{4}\right)$.

8. Mean Value Theorem Does the Mean Value Theorem apply to the function $f(x) = \sqrt{x}$ on the interval $[0, 9]$? If not, why not? If so, find the number c referred to in the theorem.

In Problems 9–11, find the local extrema of each function:

(a) *Using the First Derivative Test.*

(b) *Using the Second Derivative Test, if possible. If the Second Derivative Test cannot be used, explain why.*

9. $f(x) = x^3 - x^2 - 8x + 1$ **10.** $f(x) = x^2 - 24x^{2/3}$

11. $f(x) = x^4 e^{-2x}$

12. Motion on a Line The position s of an object from the origin at time t is given by $s = s(t) = t^4 + 2t^3 - 36t^2$. Draw figures to illustrate the motion of the object and its velocity.

In Problems 13–18, graph each function. Follow the steps given in Section 5.4.

13. $f(x) = -x^3 - x^2 + 2x$ **14.** $f(x) = x^{1/3}(x^2 - 9)$

15. $f(x) = xe^x$ **16.** $f(x) = \dfrac{x - 3}{x^2 - 4}$

17. $f(x) = x\sqrt{x - 3}$ **18.** $f(x) = x^3 - 3\ln x$

In Problems 19 and 20, for each function:

(a) *Determine the intervals where each function is increasing and decreasing.*

(b) *Determine the intervals on which each function f is concave up and concave down.*

(c) *Identify any inflection points.*

19. $f(x) = x^4 + 12x^2 + 36x - 11$

20. $f(x) = 3x^4 - 2x^3 - 24x^2 - 7x + 2$

21. If y is a function and $y' > 0$ for all x and $y'' < 0$ for all x, which of the following could be part of the graph of $y = f(x)$? See illustrations (A) through (D).

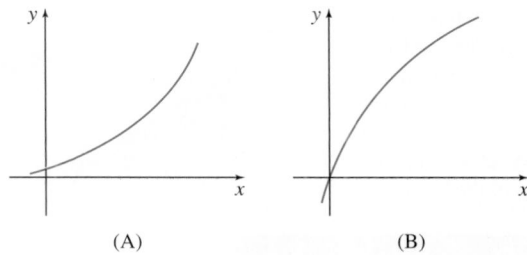

    (A)   (B)

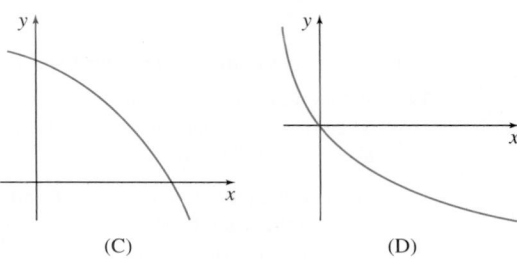

    (C)   (D)

22. Graph a function f that has the following properties:

$$f(-3) = 2; \quad f(-1) = -5; \quad f(2) = -4;$$
$$f(6) = -1; \quad f'(-3) = f'(6) = 0;$$
$$\lim_{x \to 0^-} f(x) = -\infty; \quad \lim_{x \to 0^+} f(x) = \infty;$$
$$f''(x) > 0 \text{ if } x < -3 \text{ or } 0 < x < 4;$$
$$f''(x) < 0 \text{ if } -3 < x < 0 \text{ or } 4 < x$$

23. Graph a function f that has the following properties:

$$f(-2) = 2; \quad f(5) = 1; \quad f(0) = 0;$$
$$f'(x) > 0 \text{ if } x < -2 \text{ or } 5 < x;$$
$$f'(x) < 0 \text{ if } -2 < x < 2 \text{ or } 2 < x < 5;$$
$$f''(x) > 0 \text{ if } x < 0 \text{ or } 2 < x \text{ and } f''(x) < 0 \text{ if } 0 < x < 2;$$
$$\lim_{x \to 2^-} f(x) = -\infty; \quad \lim_{x \to 2^+} f(x) = \infty$$

24. Rolle's Theorem Verify that the hypotheses for Rolle's Theorem are satisfied for the function $f(x) = x^3 - 4x^2 + 4x$ on the interval $[0, 2]$. Find the coordinates of the point(s) at which there is a horizontal tangent line to the graph of f.

25. Mean Value Theorem For the function $f(x) = x\sqrt{x + 1}$, $0 \le x \le b$, the number c satisfying the Mean Value Theorem is $c = 3$. Find b.

26. Minimizing Distance Find the point on the graph of $2y = x^2$ nearest to the point $(4, 1)$.

27. Maximizing Volume An open box is to be made from a piece of cardboard by cutting squares out of each corner and folding up the sides. If the size of the cardboard is 2 ft by 3 ft, what size squares (in inches) should be cut out to maximize the volume of the box?

In Problems 28–35, find all the antiderivatives of each function.

28. $f(x) = x^{1/2}$ **29.** $f(x) = 0$

30. $f(x) = \sec x \tan x$ **31.** $f(x) = \cos x$

32. $f(x) = -2x^{-3}$ **33.** $f(x) = \dfrac{2}{x}$

34. $f(x) = e^x + \dfrac{4}{x}$ **35.** $f(x) = 4x^3 - 9x^2 + 10x - 3$

36. Free Fall Two objects begin a free fall from rest at the same height 1 s apart. How long after the first object begins to fall will the two objects be 10 m apart?

37. Velocity A box moves down an inclined plane with an acceleration $a(t) = t^2(t - 3)\,\text{cm/s}^2$. It covers a distance of 10 cm in 2 s. What was the original velocity of the box?

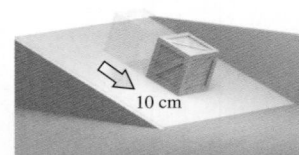

10 cm

38. Motion Problem A motorcycle accelerates at a constant rate from 0 to 72 km/h in 10 s. How far has it traveled in that time?

39. Maximizing Profit A manufacturer has determined that the cost C of producing x items is given by

$$C(x) = 200 + 35x + 0.02x^2 \text{ dollars}$$

Each item produced can be sold for $78. How many items should she produce to maximize profit?

40. Optimization The sales of a new sound system over a period of time are expected to follow the logistic curve

$$f(x) = \frac{5000}{1 + 5e^{-x}} \qquad x \geq 0$$

where x is measured in years. In what year is the sales rate a maximum?

41. Maximum Area Find the area of the rectangle of largest area in the fourth quadrant that has vertices at $(0, 0)$, $(x, 0)$, $x > 0$, and $(0, y)$, $y < 0$. The fourth vertex is on the graph of $y = \ln x$.

Break It Down   Preparing for the AP® Exam

*Let's take a closer look at **AP® Practice Problem 19 part (a)** from Section 5.3 on page 381.*

19. A function f is continuous for $-3 \leq x \leq 4$. The graph of its derivative f' is shown in the accompanying figure.

(a) List the critical numbers of f and indicate whether each locates a local maximum, local minimum, or neither. Justify your answer.

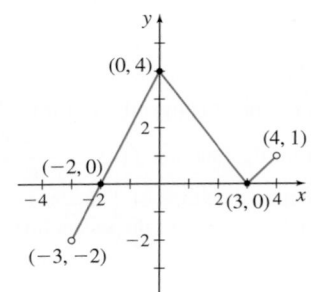

Step 1	Identify the underlying structure and related concepts.	We are asked to find the **critical numbers** of a function, f, and decide whether each one locates a local maximum, a local minimum, or neither given a graph of f' (the derivative of the function f).
Step 2	Determine the appropriate math rule or procedure.	Local extrema occur when a function changes from increasing to decreasing or decreasing to increasing. To find local extrema, we need to identify the critical numbers of f, which occur when $f'(x) = 0$ or where $f'(x)$ does not exist. Then, we can use the First Derivative Test to determine whether the critical numbers correspond to local minima, local maxima, or neither.
Step 3	Apply the math rule or procedure.	From the graph of f' we see that f' exists for all x in the interval $(-3, 4)$. So the critical numbers obey $f'(x) = 0$. Examining the graph of f', we see that $f'(x) = 0$ at $x = -2$ and at $x = 3$. Since $f'(x) < 0$ for $x < -2$, and $f'(x) > 0$ for $x > -2$, by the First Derivative Test, f has a local minimum at $x = -2$. Since $f'(x) > 0$ for $x < 3$ and $f'(x) > 0$ for $x > 3$, $f(3)$ is neither a local maximum value nor a local minimum value.
Step 4	Clearly communicate your answer	The critical numbers of f are $x = -2$ and $x = 3$. Because the function f is decreasing for $x < -2$ and increasing for $x > -2$, the critical number $x = -2$ locates a local minimum. The function f is increasing both for $x < 3$ and $x > 3$, so the critical number 3 does not locate a local extrema.

AP® Review Problems: Chapter 5

Preparing for the AP® Exam

Multiple-Choice Questions

▶ **1.** The graph of a function f is shown on the right. Which of the following could be the graph of the derivative function f'?

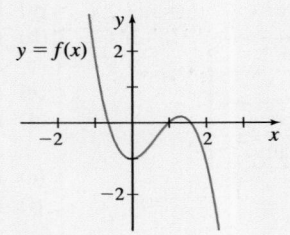

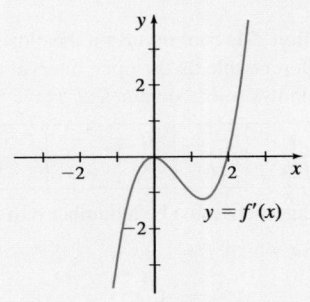

(A)

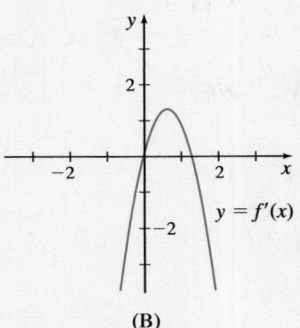

(B)

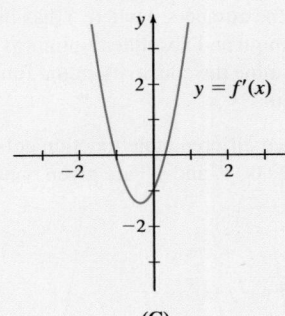

(C)

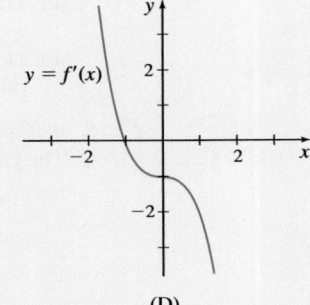

(D)

▶ **2.** On which of the following intervals is the function $f(x) = x^3 - 2x^2 + x$ increasing?

(A) $(-\infty, \infty)$ (B) $(0, \infty)$

(C) $\left(-\infty, \dfrac{1}{3}\right)$ and $(1, \infty)$ (D) $\left(\dfrac{1}{3}, 1\right)$

▶ **3.** An object moves along the y-axis, so that at any time $t \geq 0$, its position is given by $y(t) = te^{-t^2}$. At what time(s) t is the object at rest?

(A) $\dfrac{\sqrt{2}}{2}$ only (B) 0 only

(C) 0 and $\dfrac{\sqrt{2}}{2}$ (D) $\dfrac{\sqrt{2}}{2}$ and $-\dfrac{\sqrt{2}}{2}$

▶ **4.** The polynomial function f has a second derivative function f''. Several values of f'' are given in the table. Which statement must be true?

x	0	1	2	3	4
$f''(x)$	1	−1	2	0	−2

(A) f changes concavity in the interval $(2, 4)$.
(B) f has a point of inflection at $x = 3$.
(C) f has a local maximum at $x = 3$.
(D) f is not increasing on the interval $(0, 2)$.

▶ **5.** The function f is continuous for all real numbers and has a relative maximum at $(-4, 3)$ and a relative minimum at $(1, -8)$. Which of the following statements must be true?

(A) f has a point of inflection between $(-4, 3)$ and $(1, -8)$.
(B) f has a horizontal asymptote at either $(-4, 3)$ or $(1, -8)$.
(C) f has at least one zero.
(D) f is decreasing on the open interval $(-4, 1)$.

▶ **6.** The function $f(x) = 3x^{1/3} - 4x + 1$ has a local minimum at $x =$

(A) 1 (B) $\dfrac{1}{8}$ (C) $-\dfrac{1}{8}$ (D) 0

▶ **7.** Suppose f is a polynomial function of degree greater than 2 and $f(a) = f(b)$ where $a < b$. Then which of the following statements must be true for at least one number c in the interval (a, b)?

  I. $f(c) = 0$

  II. $f'(c) = 0$

  III. $f''(c) = 0$

(A) I only (B) II only

(C) II and III only (D) I, II, and III

▶ **8.** Find the absolute maximum value of the function $f(\theta) = \cos\theta - \cos^2\theta$ on the closed interval $[0, \pi]$.

(A) 1 (B) $\dfrac{1}{2}$

(C) $\dfrac{1}{4}$ (D) 0

▶ **9.** A water storage tank in the shape of a right circular cylinder is constructed so that the sum of the height and the circumference of the tank is 300 m. What are the radius r and the height h of the tank with maximum volume?

(A) radius: 50 m; height 200 m

(B) radius: $\dfrac{100}{\pi}$ m; height 100 m

(C) radius: 100π m; height 100 m

(D) radius: $\dfrac{50}{\pi}$ m; height 200 m

▶ **10.** Find all the antiderivatives of the function $f(x) = 3x^2 + \sec^2 x - 4e^x$.

(A) $x^3 + \tan x - 4e^x + C$

(B) $x^3 + \tan x - 2e^{2x} + C$

(C) $x^3 + \dfrac{\sec^3 x}{3} - 2e^{x^2} + C$

(D) $x^3 + 2\sec^2 x \tan x - 4e^x + C$

11. The table below lists several properties of a continuous function f.

x	$x < 1$	1	$x > 1$
$f(x)$		-2	
$f'(x)$	Positive		Negative
$f''(x)$	Positive		Negative

Which graph can represent f?

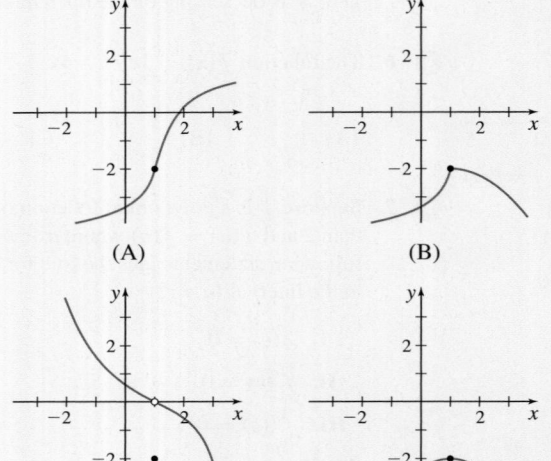

(A)   (B)

(C)   (D)

Free-Response Questions

12. The position $x(t)$ (in feet) of an object moving on a horizontal line at time t, $0 \le t \le 7$ (in seconds), is shown in the graph. The graph has horizontal tangents at time $t = 1$ and $t = 5$ and has an inflection point at $(3, -6)$.

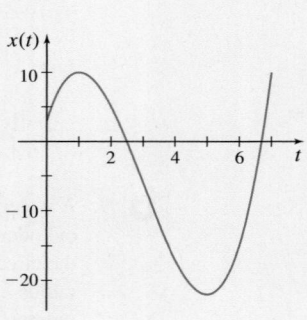

(a) At what time(s) t is the object at rest?
(b) On what interval(s) is the velocity of the object increasing?

13. For the function $f(x) = x^3 + 3x^2 + 2$

(a) Find the critical numbers of f.
(b) Find the intervals where the function is increasing and the intervals where it is decreasing.
(c) Identify the local extreme points.
(d) Find the intervals where the function is concave up and concave down.
(e) Identify any inflection points.
(f) Find an equation of the tangent line to the graph of at its inflection point(s).

14. A function f is continuous on the closed interval $[-2$ and differentiable on the open interval $(-2, 8)$. The ta below shows selected values of $f(x)$.

x	-2	0	3	5	7	8
$f(x)$	6	5	-2	-1	4	6

Show that there must be a number c in the open interv $(3, 5)$ for which $f'(c) = \dfrac{1}{2}$.

15. The function $f(x) = \dfrac{1}{x} + \ln x$ is defined for $\dfrac{1}{e} \le x \le 2$

(a) Find the numbers x where f has its absolute maximum and absolute minimum.
(b) Determine the concavity of the function over its domain.

16. f is a twice differentiable function defined on $[-2, 0.$ The graphs of f' and f'' are given below.

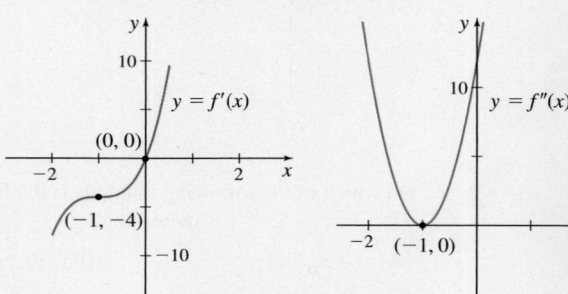

(a) Where is f increasing, and where is f decreasing Give a reason for your answer.
(b) At what numbers x, if any, does f have a local maximum or a local minimum? Give a reason for your answer.
(c) Where is f concave up, and where is f concave down? Give a reason for your answer.

AP® Cumulative Review Problems: Chapters 1–5   Preparing for the AP® Exam

Multiple-Choice Questions

1. $\lim\limits_{h \to 0} \dfrac{\cos\left(\dfrac{\pi}{6} + h\right) - \dfrac{\sqrt{3}}{2}}{h} =$

(A) $\dfrac{1}{2}$ (B) $-\dfrac{1}{2}$ (C) 1   (D) $\dfrac{\sqrt{3}}{2}$

2. The edges of a cube are increasing at a constant rate c 2 cm/min. What is the rate of change in the volume o the cube with respect to time when an edge is 12 cm?

(A) 288 cm^2/min   (B) 432 cm^2/min
(C) 864 cm^3/min   (D) 3456 cm^3/min

3. $\lim\limits_{x\to 0} \dfrac{e^x \sin x \tan x}{x^2} =$

(A) 0   (B) 1   (C) 5   (D) Does not exist

4. If $f(x) = 3x\sqrt{\cos(3x)}$, then $f'(0) =$

(A) 0   (B) 1   (C) 2   (D) 3

5. The critical numbers of $f(x) = 2xe^{-x^2}$ are

(A) 0 only   (B) $\dfrac{\sqrt{2}}{2}$ only

(C) $\dfrac{\sqrt{2}}{2}$ and $-\dfrac{\sqrt{2}}{2}$ only   (D) -1 and 1 only

6. The function $f(x) = x^3 - x^2 - 8x$ is increasing on the interval(s)

(A) $\left(-\infty, -\dfrac{4}{3}\right]$ and $[2, \infty)$

(B) $\left[-\dfrac{4}{3}, 2\right]$

(C) $(-\infty, -2]$ and $\left[\dfrac{4}{3}, \infty\right)$

(D) $\left[\dfrac{1 - \sqrt{33}}{2}, 0\right]$ and $\left[\dfrac{1 + \sqrt{33}}{2}, \infty\right)$

7. Find $f'(x)$ for the function $f(x) = \sin^{-1}(3x)$.

(A) $\dfrac{1}{3\sqrt{1 - 9x^2}}$ (B) $\dfrac{3}{\sqrt{1 - 9x^2}}$

(C) $\dfrac{1}{\sqrt{1 - x^2}}$ (D) $\dfrac{3}{\sqrt{1 - x^2}}$

8. The asymptotes, if any, to the graph of the function
$$f(x) = \dfrac{x^2 + x - 6}{x^4 - 16} \text{ are}$$

(A) Vertical asymptote: $x = -2$,
Horizontal asymptote: $y = -3$

(B) Vertical asymptotes: $x = -2, x = 2$,
Horizontal asymptote: $y = 0$

(C) Vertical asymptotes: $x = -2, x = 2$,
Horizontal asymptotes: None

(D) Vertical asymptote: $x = -2$,
Horizontal asymptote: $y = 0$

9. Find y'' for the twice differentiable function f defined implicitly by the equation
$x^2 - y^2 = 8x + 4y$.

(A) $y'' = \dfrac{(y + 2)^2 - (x - 4)^2}{(y + 2)^3}$

(B) $y'' = \dfrac{(y + 2)^2 - (x - 4)^2}{(y + 2)}$

(C) $y'' = \dfrac{(y + 2) - (x - 4)}{(y + 2)^2}$

(D) $y'' = \dfrac{1 - (x - 4)^2}{(y + 2)^3}$

10. The function f is continuous on the interval $[0, 10]$. The table below gives selected values of f.

x	0	2	4	7	10
$f(x)$	0	5	5	8	18

Which of the following statements must be true about f?

I. There is a number c, $0 < c < 10$, so that $f(c) = 6$.

II. There is a number c, $0 < c < 10$, so that $f'(c) = 0$.

III. There is a number c, $0 \le c \le 10$, so that $f(c) \le f(x)$ for all x in the interval $[0, 10]$.

(A) I and II only   (B) I and III only
(C) II and III only   (D) I, II, III

Free-Response Questions

11. The table below lists several properties of a function f that is continuous and twice differentiable for all real numbers.

x	$x < 6$	6	$6 < x < 12$	12	$12 < x < 18$	18	$x > 18$
$f'(x)$	Positive	0	Negative	8	Negative	0	Positive
$f''(x)$	Negative		Negative		Positive		Positive

(a) At what numbers does f have a local maximum? At what numbers does f have a local minimum?

(b) On what intervals is the graph of f concave up? On what intervals is the graph of f concave down?

(c) List any numbers x at which the graph of f has an inflection point. Justify your answers.

12. The tax on gross income in a small country is computed using the table below.

Gross income, i	Tax, T
$0 up to but not including $30,000	$0
$30,000 up to but not including $50,000	10% of gross income
$50,000 and higher	$5000 + 20% of gross income in excess of $50,000

(a) Use a piecewise-defined function to represent T as a function of i.

(b) Determine the numbers, if any, at which T is continuous.

(c) Determine whether T is differentiable at 30,000. Justify your answer.

(d) Determine whether T is differentiable at 50,000. Justify your answer.

CHAPTER

6

PART 1
The Integral

Brian Handy/500px/Getty Images

Managing the Klamath River

The Klamath River starts in the eastern lava plateaus of Oregon, passes through a farming region of that state, and then crosses Northern California. It is the largest river in the region, it flows through a variety of terrain, and it has different land uses. Historically, the Klamath supported large salmon runs. Now it is used to irrigate agricultural lands and to generate electric power for the region. Downstream, it flows through federal wild lands and is used for fishing, rafting, and kayaking.

If a river is to be well managed for such a variety of uses, its flow must be understood. For that reason, the U.S. Geological Survey (USGS) maintains a number of gauges along the Klamath. These typically measure the depth of the river, which can then be expressed as a rate of flow in cubic feet per second (ft³/s). To understand the river flow fully, we must be able to find the total amount of water that flows down the river over any period of time.

 The Chapter 6, Part 1 Project on page 505 examines ways to obtain the total flow of a river over any period of time from data that provide flow rates.

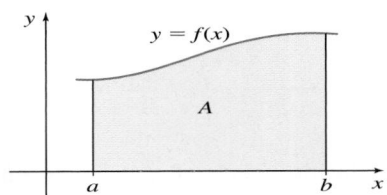

Figure 1 *A* is the area of the region bounded by the graph of f, the x-axis, and the lines $x = a$ and $x = b$.

We began our study of calculus by asking two questions from geometry. The first, the tangent problem —"What is the slope of the tangent line to the graph of a function?"— led to the derivative of a function.

The second question was the area problem, "Given a function f, defined and nonnegative on a closed interval $[a, b]$, what is the area of the region bounded by the graph of f, the x-axis, and the vertical lines $x = a$ and $x = b$?" Figure 1 illustrates this region. Answering the area problem question leads to the *integral* of a function.

Sections 6.1 and 6.2 show how the concept of an integral evolves from the area problem. At first glance, the area problem and the tangent problem look quite dissimilar. However, much of calculus is built on a surprising relationship between the two problems and their associated concepts. This relationship is the basis for the *Fundamental Theorem of Calculus,* discussed in Section 6.3. ∎

6.1 Area

OBJECTIVES *When you finish this section, you should be able to:*

1 **Approximate an area (p. 424)**
2 **Find the area under the graph of a function (p. 428)**

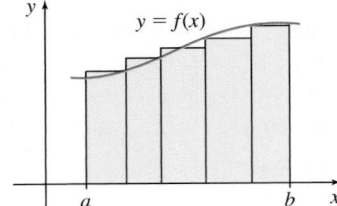

Laura
Screenwriter/playwright

AP® Calculus taught me to take my time coming up with solutions to complex problems in order to solve them thoroughly. As a television writer/producer, I've needed to recognize when more time will help me make important decisions.

The area problem asks, "Given a function f, defined and nonnegative on a closed interval $[a, b]$, what is the area A of the region bounded by the graph of f, the x-axis, and the vertical lines $x = a$ and $x = b$?" To answer the question posed in the area problem, we approximate the area A of the region by adding the areas of a finite number of rectangles, as shown in Figure 2. This approximation of the area A will lead to the *integral*, which gives the *exact* area A of the region. We begin by reviewing the formula for the area of a rectangle.

The area A of a rectangle with height h and width w is given by the geometry formula

$$A = hw$$

See Figure 3. The graph of a constant function $f(x) = h$, for some positive number h, is a horizontal line that lies above the x-axis. The region bounded by this line, the x-axis, and the lines $x = a$ and $x = b$ is the rectangle whose area A is the product of the height h and the width $(b - a)$.

$$A = h(b - a)$$

See Figure 4. If the graph of $y = f(x)$ consists of three horizontal lines, each of positive height, the area A bounded by the graph of f, the x-axis, and the lines $x = a$ and $x = b$ is the sum of the rectangular areas A_1, A_2, and A_3.

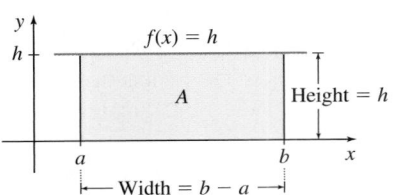

Figure 2 The sum of the areas of the rectangles approximates the area of the region.

Figure 3 $A = h(b - a)$

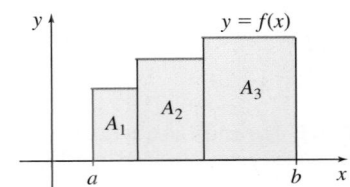

Figure 4 $A = A_1 + A_2 + A_3$

Emily Schwartz

1 Approximate an Area

EXAMPLE 1 Approximating an Area

Approximate the area A of the region bounded by the graph of $f(x) = \frac{1}{2}x + 3$, the x-axis, and the lines $x = 2$ and $x = 4$.

Solution

Figure 5 illustrates the region whose area A is to be approximated.

We begin by drawing a rectangle of height $f(2) = 4$ and width $4 - 2 = 2$. The area of the rectangle, $4 \cdot 2 = 8$, approximates the area A, but it underestimates A, as seen in Figure 6(a).

Alternatively, A can be approximated by a rectangle of height $f(4) = 5$ and width $4 - 2 = 2$. See Figure 6(b). This approximation of the area equals $5 \cdot 2 = 10$, but it overestimates A. We conclude that $8 < A < 10$.

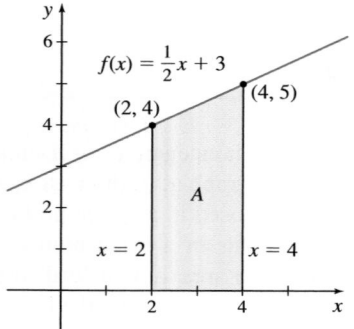

Figure 5

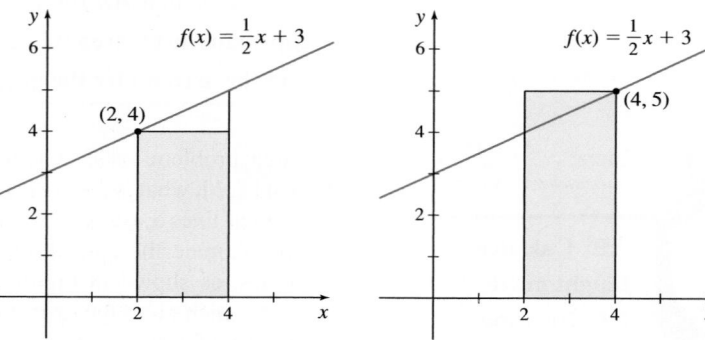

(a) The area A is underestimated. (b) The area A is overestimated.

Figure 6

The approximation of the area A of the region can be improved by dividing the closed interval $[2, 4]$ into two subintervals, $[2, 3]$ and $[3, 4]$. Now draw two rectangles: one rectangle with height $f(2) = \frac{1}{2} \cdot 2 + 3 = 4$ and width $3 - 2 = 1$; the other rectangle with height $f(3) = \frac{1}{2} \cdot 3 + 3 = \frac{9}{2}$ and width $4 - 3 = 1$. As Figure 7(a) on page 425 illustrates, the sum of the areas of the two rectangles

$$4 \cdot 1 + \frac{9}{2} \cdot 1 = \frac{17}{2} = 8.5$$

underestimates the area.

Repeat this process by drawing two rectangles, one of height $f(3) = \frac{9}{2}$ and width 1; the other of height $f(4) = \frac{1}{2} \cdot 4 + 3 = 5$ and width 1. The sum of the areas of these two rectangles,

$$\frac{9}{2} \cdot 1 + 5 \cdot 1 = \frac{19}{2} = 9.5$$

overestimates the area as shown in Figure 7(b). We conclude that

$$8.5 < A < 9.5$$

obtaining a better approximation of the area.

NOTE The actual area in Figure 5 is 9 square units, obtained by using the formula for the area A of a trapezoid with base b and parallel heights h_1 and h_2:

$$A = \frac{1}{2}b(h_1 + h_2) = \frac{1}{2} \cdot 2 \cdot (4 + 5) = 9.$$

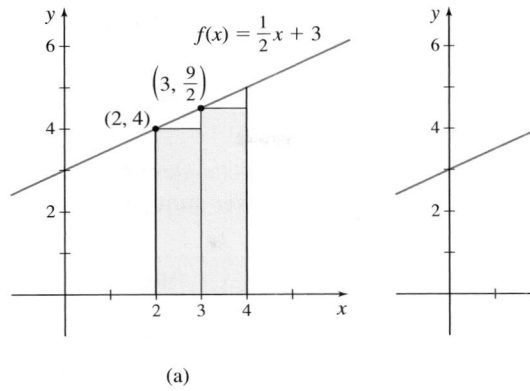

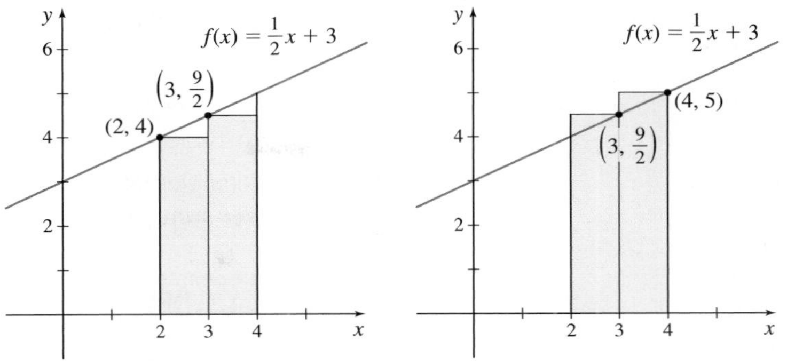

(a)   (b)

Figure 7   ■

Observe that as the number n of subintervals of the interval $[2, 4]$ increases, the approximation of the area A improves. For $n = 1$, the error in approximating A is 1 square unit, but for $n = 2$, the error is only 0.5 square unit.

NOW WORK   Problem 5.

In general, the procedure for approximating the area A is based on the idea of summing the areas of rectangles. We shall refer to the area A of the region bounded by the graph of a function $y = f(x) \geq 0$, the x-axis, and the lines $x = a$ and $x = b$ as the **area under the graph of f from a to b**.

We make two assumptions about the function f:

- f is continuous on the closed interval $[a, b]$.
- f is nonnegative on the closed interval $[a, b]$.

We divide, or **partition**, the interval $[a, b]$ into n nonoverlapping subintervals:

$$[x_0, x_1], [x_1, x_2], \ldots, [x_{i-1}, x_i], \ldots, [x_{n-1}, x_n] \qquad x_0 = a, x_n = b$$

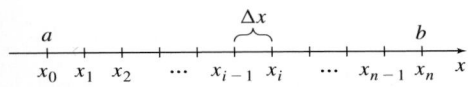

Figure 8

each of the same width Δx. See Figure 8. Since there are n subintervals and the width of the interval $[a, b]$ is $b - a$, the common width Δx of each subinterval is

$$\boxed{\Delta x = \frac{b - a}{n}}$$

NEED TO REVIEW? The Extreme Value Theorem is discussed in Section 5.2, p. 350.

Since f is continuous on the closed interval $[a, b]$, it is continuous on every subinterval $[x_{i-1}, x_i]$ of $[a, b]$. By the Extreme Value Theorem, there is a number in each subinterval where f attains its absolute minimum. Label these numbers $c_1, c_2, c_3, \ldots, c_n$, so that $f(c_i)$ is the absolute minimum value of f in the subinterval $[x_{i-1}, x_i]$. Now construct n rectangles, each having $f(c_i)$ as its height and Δx as its width, as illustrated in Figure 9. This produces n narrow rectangles of heights $f(c_1), f(c_2), \ldots, f(c_n)$ and uniform width $\Delta x = \dfrac{b - a}{n}$, respectively. The areas of the n rectangles are

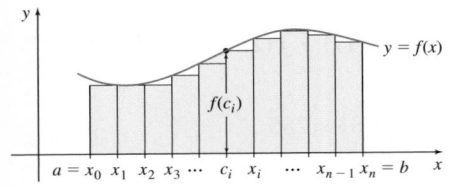

$$\text{Area of the first rectangle} = f(c_1)\Delta x$$
$$\text{Area of the second rectangle} = f(c_2)\Delta x$$
$$\vdots$$
$$\text{Area of the } n\text{th (and last) rectangle} = f(c_n)\Delta x$$

Figure 9 $f(c_i)$ is the absolute minimum value of f on $[x_{i-1}, x_i]$.

NEED TO REVIEW? Summation notation is discussed in Section P.8, pp. 73-74.

The sum s_n of the areas of the n rectangles approximates the area A. That is,

$$A \approx s_n = f(c_1)\Delta x + f(c_2)\Delta x + \cdots + f(c_i)\Delta x + \cdots + f(c_n)\Delta x = \sum_{i=1}^{n} f(c_i)\Delta x$$

Since the rectangles used to approximate the area A lie under the graph of f, the sum s_n, called a **lower sum**, *underestimates* A. That is, $s_n \leq A$.

EXAMPLE 2 Approximating Area Using Lower Sums

Approximate the area A under the graph of $f(x) = x^2$ from 0 to 10 by using lower sums s_n (rectangles that lie under the graph) for:

(a) $n = 2$ subintervals of equal width
(b) $n = 5$ subintervals of equal width
(c) $n = 10$ subintervals of equal width

Solution

(a) For $n = 2$, we partition the closed interval $[0, 10]$ into two subintervals $[0, 5]$ and $[5, 10]$, each of width $\Delta x = \dfrac{10 - 0}{2} = 5$. See Figure 10(a). To compute s_2, we need to know where f attains its minimum value in each subinterval. Since the function f is increasing on $[0, 10]$, the absolute minimum is attained at the left endpoint of each subinterval. So, for $n = 2$, the minimum of f on $[0, 5]$ occurs at 0 and the minimum of f on $[5, 10]$ occurs at 5. The lower sum s_2 is

$$s_2 = \sum_{i=1}^{2} f(c_i)\Delta x = \Delta x \sum_{i=1}^{2} f(c_i) = 5[f(0) + f(5)] = 5(0 + 25) = 125$$

$\Delta x = 5$ $f(0) = 0$
$f(c_1) = f(0); f(c_2) = f(5)$ $f(5) = 25$

(b) For $n = 5$, partition the interval $[0, 10]$ into five subintervals $[0, 2]$, $[2, 4]$, $[4, 6]$, $[6, 8]$, $[8, 10]$, each of width $\Delta x = \dfrac{10 - 0}{5} = 2$. See Figure 10(b). The lower sum s_5 is

$$s_5 = \sum_{i=1}^{5} f(c_i)\Delta x = \Delta x \sum_{i=1}^{5} f(c_i) = 2[f(0) + f(2) + f(4) + f(6) + f(8)]$$

$$= 2(0 + 4 + 16 + 36 + 64) = 240$$

(c) For $n = 10$, partition $[0, 10]$ into 10 subintervals, each of width $\Delta x = \dfrac{10 - 0}{10} = 1$. See Figure 10(c). The lower sum s_{10} is

$$s_{10} = \sum_{i=1}^{10} f(c_i)\Delta x = \Delta x \sum_{i=1}^{10} f(c_i) = 1[f(0) + f(1) + f(2) + \cdots + f(9)]$$

$$= 0 + 1 + 4 + 9 + 16 + 25 + 36 + 49 + 64 + 81 = 285 \qquad \blacksquare$$

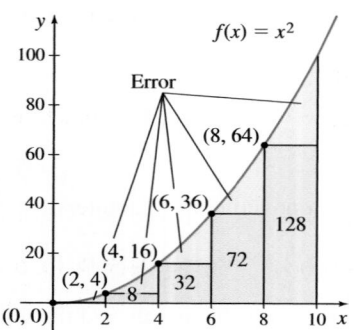

(a) Two subintervals
$s_2 = 125$ sq. units

(b) Five subintervals
$s_5 = 240$ sq. units

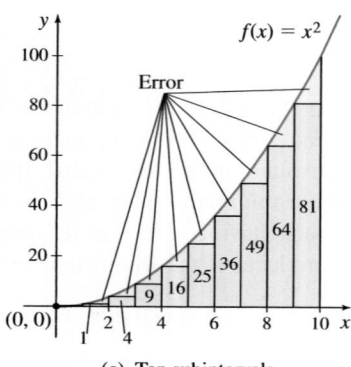

(c) Ten subintervals
$s_{10} = 285$ sq. units

Figure 10

NOW WORK Problem 13(a) and AP® Practice Problem 1.

In general, as Figure 11(a) illustrates, the error due to using lower sums s_n (rectangles that lie below the graph of f) occurs because a portion of the region lies outside the rectangles. To improve the approximation of the area, we increase the number of subintervals. For example, in Figure 11(b), there are four subintervals and the error is reduced; in Figure 11(c), there are eight subintervals and the error is further reduced. By taking a finer and finer partition of the interval $[a, b]$, that is, by increasing the number n of subintervals without bound, we can make the sum of the areas of the rectangles as close as we please to the actual area of the region. (A proof of this statement is usually found in books on advanced calculus.)

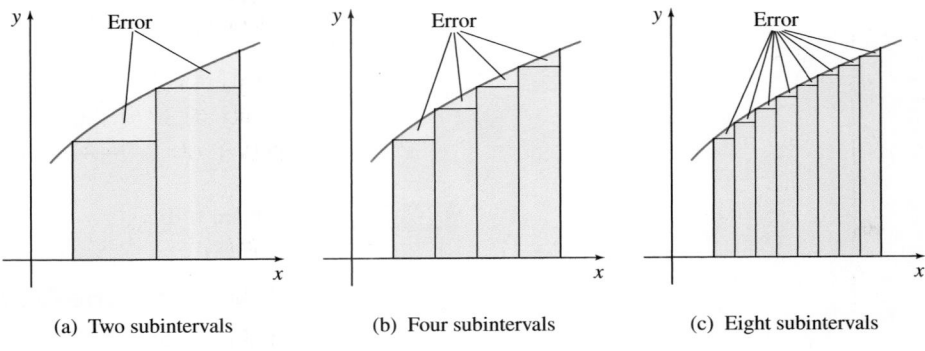

(a) Two subintervals   (b) Four subintervals   (c) Eight subintervals

Figure 11

For functions that are continuous on a closed interval, $\lim\limits_{n \to \infty} s_n$ always exists. With this in mind, we now define the area under the graph of a function f from a to b.

DEFINITION Area A under the Graph of a Function from a to b

Suppose a function f is nonnegative and continuous on a closed interval $[a, b]$. Partition $[a, b]$ into n subintervals $[x_0, x_1], [x_1, x_2], \ldots, [x_{i-1}, x_i], \ldots, [x_{n-1}, x_n]$, each of width

$$\Delta x = \frac{b - a}{n}$$

In each subinterval $[x_{i-1}, x_i]$, let $f(c_i)$ equal the absolute minimum value of f on the subinterval. Form the lower sums

$$s_n = \sum_{i=1}^{n} f(c_i)\Delta x = f(c_1)\Delta x + \cdots + f(c_n)\Delta x$$

The **area A under the graph of f from a to b** is the number

$$A = \lim_{n \to \infty} s_n$$

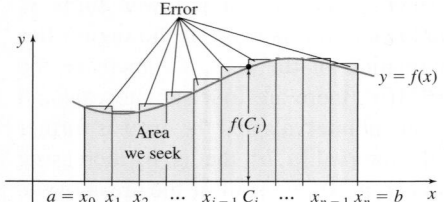

Figure 12 $f(C_i)$ is the absolute maximum value of f on $[x_{i-1}, x_i]$.

The area A is defined using lower sums s_n (rectangles that lie below the graph of f). By a parallel argument, we can choose values $C_1, C_2, \ldots, C_n$ so that the height $f(C_i)$ of the ith rectangle is the absolute maximum value of f on the ith subinterval, as shown in Figure 12. The corresponding **upper sums** S_n are

$$S_n = \sum_{i=1}^{n} f(C_i)\Delta x = f(C_1)\Delta x + \cdots + f(C_n)\Delta x$$

The upper sums S_n (rectangles that extend above the graph of f) *overestimate* the area A, so $S_n \geq A$.

It can be shown that as n increases without bound, the limit of the upper sums S_n equals the limit of the lower sums s_n. That is,

$$\lim_{n \to \infty} s_n = \lim_{n \to \infty} S_n = A$$

NOW WORK   Problem 13(b).

2   Find the Area under the Graph of a Function

In the next example, instead of using a specific number of rectangles to *approximate* area, we partition the interval $[a, b]$ into n subintervals of equal width, obtaining n rectangles. By letting $n \to \infty$, we find the *actual* area under the graph of f from a to b.

EXAMPLE 3   Finding Area Using Upper Sums

Find the area A under the graph of $f(x) = 3x$ from 0 to 10 using upper sums S_n (rectangles that extend above the graph of f). That is, find $A = \lim_{n \to \infty} S_n$.

Solution

Figure 13 illustrates the area A. Partition the closed interval $[0, 10]$ into n subintervals of equal width

$$[x_0, x_1], [x_1, x_2], \ldots, [x_{i-1}, x_i], \ldots, [x_{n-1}, x_n]$$

where $0 = x_0 < x_1 < x_2 < \cdots < x_i < \cdots < x_{n-1} < x_n = 10$ and each subinterval is of width

$$\Delta x = \frac{10 - 0}{n} = \frac{10}{n}$$

As illustrated in Figure 14, coordinates of the endpoints of each subinterval, written in terms of n, are

$$x_0 = 0, \ x_1 = \frac{10}{n}, \ x_2 = 2 \cdot \frac{10}{n}, \ldots, x_{i-1} = (i-1) \cdot \frac{10}{n}, \ x_i = i \cdot \frac{10}{n}, \ldots, x_n = n \cdot \frac{10}{n} = 10$$

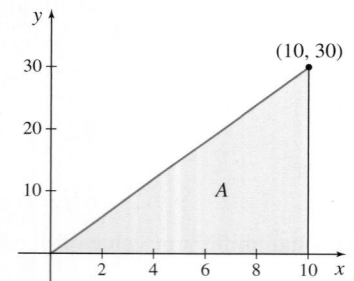

Figure 13 $f(x) = 3x, 0 \leq x \leq 10$

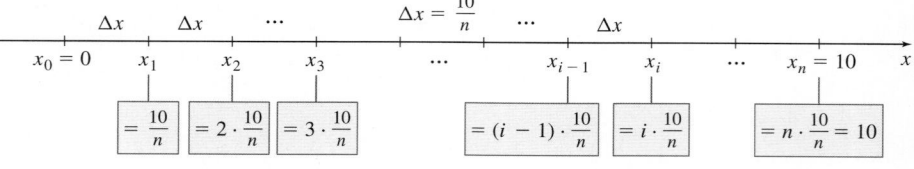

Figure 14

To find A using upper sums S_n (rectangles that extend above the graph of f), we need the absolute maximum value of f in each subinterval. Since the function $f(x) = 3x$ is increasing, the absolute maximum occurs at the right endpoint $x_i = i \cdot \dfrac{10}{n} = \dfrac{10i}{n}$ of each subinterval. So, $C_i = x_i$, $f(C_i) = 3x_i$, and

$$S_n = \sum_{i=1}^{n} f(C_i)\Delta x = \sum_{i=1}^{n} 3x_i \cdot \frac{10}{n} = \sum_{i=1}^{n} \left(3 \cdot \frac{10i}{n} \cdot \frac{10}{n}\right) = \sum_{i=1}^{n} \frac{300}{n^2} i$$

$$\uparrow \qquad\qquad \uparrow$$
$$\Delta x = \frac{10}{n} \qquad x_i = \frac{10i}{n}$$

NEED TO REVIEW? Summation properties are discussed in Section P.8, pp. 75–76.

RECALL $\displaystyle\sum_{i=1}^{n} i = \dfrac{n(n+1)}{2}$ (p. 75)

Using summation properties, we get

$$S_n = \sum_{i=1}^{n} \frac{300}{n^2} i = \frac{300}{n^2} \sum_{i=1}^{n} i = \frac{300}{n^2} \cdot \frac{n(n+1)}{2} = 150 \cdot \frac{n+1}{n} = 150\left(1 + \frac{1}{n}\right)$$

Then

$$A = \lim_{n \to \infty} S_n = \lim_{n \to \infty}\left[150\left(1 + \frac{1}{n}\right)\right] = 150 \lim_{n \to \infty}\left(1 + \frac{1}{n}\right) = 150$$

The area A under the graph of $f(x) = 3x$ from 0 to 10 is 150 square units. ∎

NOTE The region under the graph of $f(x) = 3x$ from 0 to 10 is a triangle. So, we can verify that $A = 150$ by using the geometry formula for the area A of a triangle with base b and height h:

$$A = \frac{1}{2}bh = \frac{1}{2} \cdot 10 \cdot 30 = 150$$

NOW WORK AP® Practice Problem 2.

▶ CALC CLIP **EXAMPLE 4 Finding Area Using Lower Sums**

Find the area A under the graph of $f(x) = 16 - x^2$ from 0 to 4 by using lower sums s_n (rectangles that lie below the graph of f). That is, find $A = \lim_{n \to \infty} s_n$.

Solution

Figure 15 shows the area under the graph of f and a typical rectangle that lies below the graph. Partition the closed interval $[0, 4]$ into n subintervals

$$[x_0, x_1], [x_1, x_2], \ldots, [x_{i-1}, x_i], \ldots, [x_{n-1}, x_n]$$

where $0 = x_0 < x_1 < \cdots < x_i < \cdots < x_{n-1} < x_n = 4$ and each interval is of width

$$\Delta x = \frac{4-0}{n} = \frac{4}{n}$$

As Figure 16 illustrates, the endpoints of each subinterval, written in terms of n, are

$$x_0 = 0, \quad x_1 = 1 \cdot \frac{4}{n}, \quad x_2 = 2 \cdot \frac{4}{n}, \ldots, x_{i-1} = (i-1) \cdot \frac{4}{n}, \quad x_i = i \cdot \frac{4}{n}, \ldots, x_n = n \cdot \frac{4}{n} = 4$$

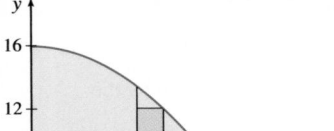

Figure 15 $f(x) = 16 - x^2$, $0 \le x \le 4$

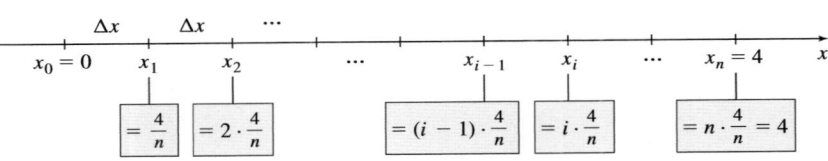

Figure 16

To find A using lower sums s_n (rectangles that lie below the graph of f), we must find the absolute minimum value of f on each subinterval. Since the function f is decreasing, the absolute minimum occurs at the right endpoint of each subinterval.

Since $c_i = i \cdot \dfrac{4}{n} = \dfrac{4i}{n}$ and $\Delta x = \dfrac{4}{n}$,

$$s_n = \sum_{i=1}^{n} f(c_i)\Delta x = \sum_{i=1}^{n}\left[16 - \left(\frac{4i}{n}\right)^2\right]\cdot\frac{4}{n} \qquad c_i = \frac{4i}{n};\ f(c_i) = 16 - c_i^2;\ \Delta x = \frac{4}{n}$$

$$= \sum_{i=1}^{n}\left[\frac{64}{n} - \frac{64i^2}{n^3}\right] = \frac{64}{n}\sum_{i=1}^{n}1 - \frac{64}{n^3}\sum_{i=1}^{n}i^2$$

$$= \frac{64}{n}\cdot n - \frac{64}{n^3}\left[\frac{n(n+1)(2n+1)}{6}\right] = 64 - \frac{32}{3n^2}\left[2n^2 + 3n + 1\right]$$

$$= 64 - \frac{64}{3} - \frac{32}{n} - \frac{32}{3n^2} = \frac{128}{3} - \frac{32}{n} - \frac{32}{3n^2}$$

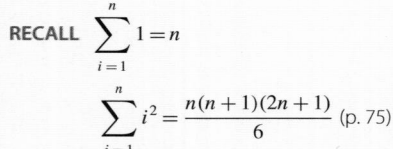

RECALL $\displaystyle\sum_{i=1}^{n}1 = n$

$\displaystyle\sum_{i=1}^{n}i^2 = \frac{n(n+1)(2n+1)}{6}$ (p. 75)

Then

$$A = \lim_{n\to\infty} s_n = \lim_{n\to\infty}\left(\frac{128}{3} - \frac{32}{n} - \frac{32}{3n^2}\right) = \frac{128}{3}$$

The area A under the graph of $f(x) = 16 - x^2$ from 0 to 4 is $\dfrac{128}{3}$ square units. ■

NOW WORK Problem 29.

AP® EXAM TIP

The integral and the Fundamental Theorem of Calculus are important concepts in the AP® Calculus curriculum.

The previous two examples illustrate just how complex it can be to find areas using lower sums and/or upper sums. In the next section, we define a *definite integral* and show how it can be used to find area. Then in Section 6.3, we present the Fundamental Theorem of Calculus, which provides a relatively simple way to find area.

6.1 Assess Your Understanding

Concepts and Vocabulary

1. Explain how rectangles can be used to approximate the area of the region bounded by the graph of a function $y = f(x) \geq 0$, the x-axis, and the lines $x = a$ and $x = b$.

2. **True or False** When a closed interval $[a, b]$ is partitioned into n subintervals each of the same width, the width of each subinterval is $\dfrac{a+b}{n}$.

3. If the closed interval $[-2, 4]$ is partitioned into 12 subintervals, each of the same width, then the width of each subinterval is _____.

4. **True or False** If the area A under the graph of a function f that is continuous and nonnegative on a closed interval $[a, b]$ is approximated using upper sums S_n, then $S_n \geq A$ and $A = \lim\limits_{n\to\infty} S_n$.

Skill Building

5. Approximate the area A of the region bounded by the graph of $f(x) = \dfrac{1}{2}x + 3$, the x-axis, and the lines $x = 2$ and $x = 4$ by dividing the closed interval $[2, 4]$ into four subintervals:

$$\left[2, \frac{5}{2}\right],\ \left[\frac{5}{2}, 3\right],\ \left[3, \frac{7}{2}\right],\ \left[\frac{7}{2}, 4\right]$$

(a) Using the left endpoint of each subinterval, draw four small rectangles that lie below the graph of f and sum the areas of the four rectangles.

(b) Using the right endpoint of each subinterval, draw four small rectangles that extend above the graph of f and sum the areas of the four rectangles.

(c) Compare the answers from parts (a) and (b) to the exact area $A = 9$ and to the estimates obtained in Example 1.

6. Approximate the area A of the region bounded by the graph of $f(x) = 6 - 2x$, the x-axis, and the lines $x = 1$ and $x = 3$ by dividing the closed interval $[1, 3]$ into four subintervals:

$$\left[1, \frac{3}{2}\right], \left[\frac{3}{2}, 2\right], \left[2, \frac{5}{2}\right], \left[\frac{5}{2}, 3\right]$$

(a) Using the right endpoint of each subinterval, draw four small rectangles that lie below the graph of f and sum the areas of the four rectangles.

(b) Using the left endpoint of each subinterval, draw four small rectangles that extend above the graph of f and sum the areas of the four rectangles.

(c) Compare the answers from parts (a) and (b) to the exact area $A = 4$.

In Problems 7 and 8, refer to the graphs below. Approximate the shaded area under the graph of f:

(a) By constructing rectangles using the left endpoint of each subinterval.

(b) By constructing rectangles using the right endpoint of each subinterval.

7.

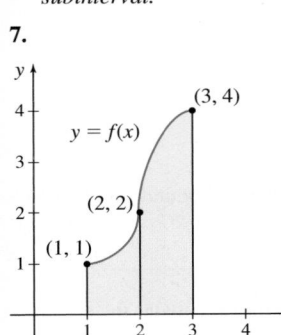

8.

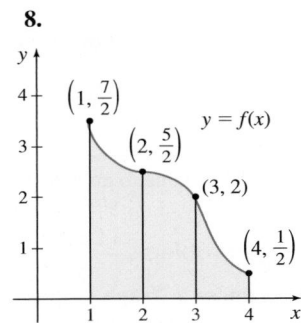

In Problems 9–12, partition each interval into n subintervals each of the same width.

9. $[1, 4]$ with $n = 3$

10. $[0, 9]$ with $n = 9$

11. $[-1, 4]$ with $n = 10$

12. $[-4, 4]$ with $n = 16$

In Problems 13 and 14, refer to the graphs below. Using the indicated subintervals, approximate the shaded area:

(a) By using lower sums s_n (rectangles that lie below the graph of f).

(b) By using upper sums S_n (rectangles that extend above the graph of f).

PAGE 426 **13.**

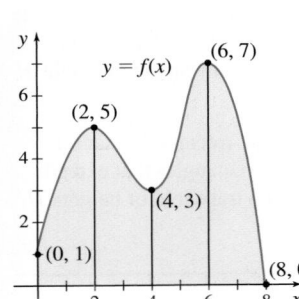

14.

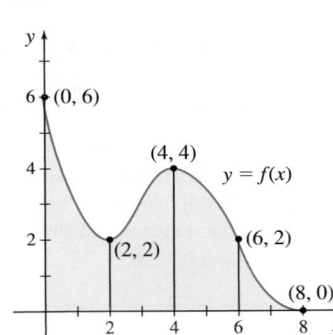

15. Area under a Graph

(a) Graph $y = x$ and indicate the area under the graph from 0 to 3.

(b) Partition the interval $[0, 3]$ into n subintervals each of equal width.

(c) Show that $s_n = \sum_{i=1}^{n} (i - 1) \left(\frac{3}{n}\right)^2$.

(d) Show that $S_n = \sum_{i=1}^{n} i \left(\frac{3}{n}\right)^2$.

(e) Show that $\lim_{n \to \infty} s_n = \lim_{n \to \infty} S_n = \frac{9}{2}$.

16. Area under a Graph

(a) Graph $y = 4x$ and indicate the area under the graph from 0 to 5.

(b) Partition the interval $[0, 5]$ into n subintervals each of equal width.

(c) Show that $s_n = \sum_{i=1}^{n} (i - 1) \frac{100}{n^2}$.

(d) Show that $S_n = \sum_{i=1}^{n} i \frac{100}{n^2}$.

(e) Show that $\lim_{n \to \infty} s_n = \lim_{n \to \infty} S_n = 50$.

In Problems 17–22, approximate the area A under the graph of each function f from a to b for n = 4 and n = 8 subintervals of equal width:

(a) By using lower sums s_n (rectangles that lie below the graph of f).

(b) By using upper sums S_n (rectangles that extend above the graph of f).

17. $f(x) = -x + 10$ on $[0, 8]$

18. $f(x) = 2x + 5$ on $[2, 6]$

19. $f(x) = 16 - x^2$ on $[0, 4]$

20. $f(x) = x^3$ on $[0, 8]$

21. $f(x) = \cos x$ on $\left[-\frac{\pi}{2}, \frac{\pi}{2}\right]$

22. $f(x) = \sin x$ on $[0, \pi]$

23. Rework Example 3 (p. 428) by using lower sums s_n (rectangles that lie below the graph of f).

24. Rework Example 4 (p. 429) by using upper sums S_n (rectangles that extend above the graph of f).

In Problems 25–32, find the area A under the graph of f from a to b:

(a) By using lower sums s_n (rectangles that lie below the graph of f).

(b) By using upper sums S_n (rectangles that extend above the graph of f).

(c) Compare the work required in (a) and (b). Which is easier? Could you have predicted this?

25. $f(x) = 2x + 1$ from $a = 0$ to $b = 4$

26. $f(x) = 3x + 1$ from $a = 0$ to $b = 4$

27. $f(x) = 12 - 3x$ from $a = 0$ to $b = 4$

28. $f(x) = 5 - x$ from $a = 0$ to $b = 4$

[PAGE 430] 29. $f(x) = 4x^2$ from $a = 0$ to $b = 2$

30. $f(x) = \dfrac{1}{2}x^2$ from $a = 0$ to $b = 3$

31. $f(x) = 4 - x^2$ from $a = 0$ to $b = 2$

32. $f(x) = 12 - x^2$ from $a = 0$ to $b = 3$

Applications and Extensions

In Problems 33–38, find the area under the graph of f from a to b. Partition the closed interval $[a, b]$ into n subintervals

$$[x_0, x_1], [x_1, x_2], \ldots, [x_{i-1}, x_i], \ldots, [x_{n-1}, x_n],$$

where $a = x_0 < x_1 < \cdots < x_i < \cdots < x_{n-1} < x_n = b$,

and each subinterval is of width $\Delta x = \dfrac{b-a}{n}$. As the figure below illustrates, the endpoints of each subinterval, written in terms of n, are

$$x_0 = a, \quad x_1 = a + \frac{b-a}{n}, \quad x_2 = a + 2\left(\frac{b-a}{n}\right), \ldots,$$

$$x_{i-1} = a + (i-1)\left(\frac{b-a}{n}\right), \quad x_i = a + i\left(\frac{b-a}{n}\right), \ldots,$$

$$x_n = a + n\left(\frac{b-a}{n}\right) = b$$

$$\Delta x = \frac{b-a}{n}$$

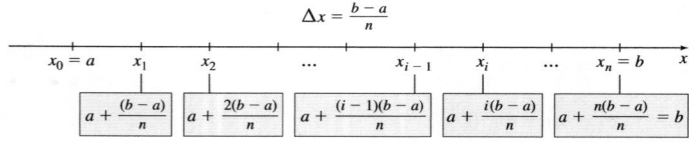

33. $f(x) = x + 3$ from $a = 1$ to $b = 3$

34. $f(x) = 3 - x$ from $a = 1$ to $b = 3$

35. $f(x) = 2x + 5$ from $a = -1$ to $b = 2$

36. $f(x) = 2 - 3x$ from $a = -2$ to $b = 0$

37. $f(x) = 2x^2 + 1$ from $a = 1$ to $b = 3$

38. $f(x) = 4 - x^2$ from $a = 1$ to $b = 2$

In Problems 39–42, approximate the area A under the graph of each function f by partitioning $[a, b]$ into 20 subintervals of equal width and using an upper sum.

39. $f(x) = xe^x$ on $[0, 8]$

40. $f(x) = \ln x$ on $[1, 3]$

41. $f(x) = \dfrac{1}{x}$ on $[1, 5]$

42. $f(x) = \dfrac{1}{x^2}$ on $[2, 6]$

43. (a) Graph $y = \dfrac{4}{x}$ from $x = 1$ to $x = 4$ and shade the area under its graph.

(b) Partition the interval $[1, 4]$ into n subintervals of equal width.

(c) Show that the lower sum s_n is

$$s_n = \sum_{i=1}^{n} \frac{4}{\left(1 + \dfrac{3i}{n}\right)}\left(\frac{3}{n}\right)$$

(d) Show that the upper sum S_n is

$$S_n = \sum_{i=1}^{n} \frac{4}{\left(1 + \dfrac{3(i-1)}{n}\right)}\left(\frac{3}{n}\right)$$

(e) Complete the following table:

n	5	10	50	100
s_n				
S_n				

(f) Use the table to give an upper and lower bound for the area.

Challenge Problems

44. **Area under a Graph** Approximate the area under the graph of $f(x) = x$ from $a \geq 0$ to b by using lower sums s_n and upper sums S_n for a partition of $[a, b]$ into n subintervals, each of width $\dfrac{b-a}{n}$. Show that

$$s_n < \frac{b^2 - a^2}{2} < S_n$$

45. **Area under a Graph** Approximate the area under the graph of $f(x) = x^2$ from $a \geq 0$ to b by using lower sums s_n and upper sums S_n for a partition of $[a, b]$ into n subintervals, each of width $\dfrac{b-a}{n}$. Show that

$$s_n < \frac{b^3 - a^3}{3} < S_n$$

46. **Area of a Right Triangle** Use lower sums s_n (rectangles that lie inside the triangle) and upper sums S_n (rectangles that extend outside the triangle) to find the area of a right triangle of height H and base B.

47. **Area of a Trapezoid** Use lower sums s_n (rectangles that lie inside the trapezoid) and upper sums S_n (rectangles that extend outside the trapezoid) to find the area of a trapezoid of heights H_1 and H_2 and base B.

AP® Practice Problems

Multiple-Choice Question

PAGE 426

1. The approximate area under the graph of $f(x) = x^2 + 1$ from -1 to 3 found by partitioning the interval $[-1, 3]$ into 4 subintervals of equal width and using lower sums s_4 (rectangles that lie under the graph) is

(A) 8   (B) 9   (C) 10   (D) 18

Free-Response Question

PAGE 429

2. The graph of the function $f(x) = 2x$ from 2 to 10 is shown on the right.
   (a) Approximate the area under the graph of f by partitioning the interval $[2, 10]$ into 4 subintervals of equal width and using lower sums s_4 (rectangles that lie under the graph of f).
   (b) Approximate the area under the graph of f by partitioning the interval $[2, 10]$ into 4 subintervals of equal width and using upper sums S_4 (rectangles that extend above the graph of f).
   (c) Find the exact area under the graph using geometry.

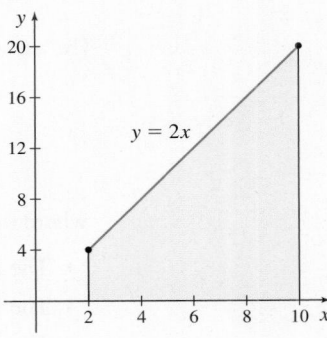

Retain Your Knowledge

Multiple-Choice Questions

1. The graph of a function f is given below. At what number x is f continuous, but not differentiable?

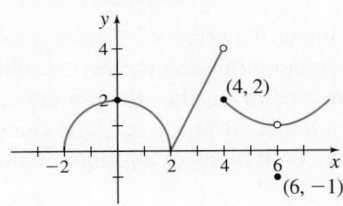

(A) 0   (B) 2   (C) 4   (D) 6

2. If a non-constant function f is continuous on the closed interval $[a, b]$, which of the following statements must be true?

   I. The function has both an absolute maximum and an absolute minimum value in the interval $[a, b]$.
   II. The function has at least one critical number in the interval (a, b).
   III. An absolute extreme value is at one of the endpoints.

(A) I only   (B) I and II only
(C) II and III only (D) I, II, and III

3. If $f(x) = 3^{4x}$, then $f'(x) =$

(A) $4 \cdot 3^{3x}$ (B) $3 \cdot \ln 3^{4x}$

(C) $\dfrac{3^{4x}}{4 \ln 3}$ (D) $3^{4x} \cdot 4 \ln 3$

Free-Response Question

4. An object moves along a horizontal line with velocity v given by $v(t) = \dfrac{t}{4} \sin \dfrac{\pi t}{2}$, in meters/second, for time t, $0 \le t \le 4$.
   (a) Find the acceleration a of the object when $t = 3$.
   (b) What is the speed of the object at time $t = 3$?
   (c) Find all times t in the interval $0 < t < 4$ when the object changes direction. Justify your answer.

6.2 The Definite Integral

OBJECTIVES *When you finish this section, you should be able to:*

1 **Form Riemann sums (p. 435)**

2 **Define a definite integral as the limit of Riemann sums (p. 436)**

3 **Approximate a definite integral using Riemann sums (p. 441)**

4 **Approximate a definite integral using trapezoidal sums (p. 442)**

5 **Know conditions that guarantee a definite integral exists (p. 446)**

The area A under the graph of $y = f(x)$ from a to b was obtained by finding

$$A = \lim_{n \to \infty} s_n = \lim_{n \to \infty} \sum_{i=1}^{n} f(c_i) \Delta x = \lim_{n \to \infty} S_n = \lim_{n \to \infty} \sum_{i=1}^{n} f(C_i) \Delta x$$

where the following assumptions were made:

- The function f is continuous on the closed interval $[a, b]$.
- The function f is nonnegative on the closed interval $[a, b]$.
- The closed interval $[a, b]$ is partitioned into n subintervals, each of width

$$\Delta x = \frac{b - a}{n}$$

- $f(c_i)$ is the absolute minimum value of f on the ith subinterval, $i = 1, 2, \ldots, n$.
- $f(C_i)$ is the absolute maximum value of f on the ith subinterval, $i = 1, 2, \ldots, n$.

In Section 6.1, we found the area A under the graph of f from a to b by choosing either the number c_i, where f has an absolute minimum on the ith subinterval, or the number C_i, where f has an absolute maximum on the ith subinterval. Suppose we arbitrarily choose a number u_i in each subinterval $[x_{i-1}, x_i]$, and draw rectangles of height $f(u_i)$ and width Δx. Then from the definitions of absolute minimum value and absolute maximum value

$$f(c_i) \leq f(u_i) \leq f(C_i)$$

and, since $\Delta x > 0$,

$$f(c_i) \Delta x \leq f(u_i) \Delta x \leq f(C_i) \Delta x$$

Then

$$\sum_{i=1}^{n} f(c_i) \Delta x \leq \sum_{i=1}^{n} f(u_i) \Delta x \leq \sum_{i=1}^{n} f(C_i) \Delta x$$

$$s_n \leq \sum_{i=1}^{n} f(u_i) \Delta x \leq S_n$$

NEED TO REVIEW? The Squeeze Theorem is discussed in Section 1.4, pp. 122–124.

Since $\lim_{n \to \infty} s_n = \lim_{n \to \infty} S_n = A$, by the Squeeze Theorem, we have

$$\lim_{n \to \infty} \sum_{i=1}^{n} f(u_i) \Delta x = A$$

In other words, we can use any number u_i in the ith subinterval to find the area A.

① Form Riemann Sums

We now investigate sums of the form

$$\sum_{i=1}^{n} f(u_i)\Delta x_i$$

using the following more general assumptions:

- The function f is defined on a closed interval $[a, b]$.
- The function f is not necessarily continuous on $[a, b]$.
- The function f is not necessarily nonnegative on $[a, b]$.
- The widths $\Delta x_i = x_i - x_{i-1}$ of the subintervals $[x_{i-1}, x_i]$, $i = 1, 2, \ldots, n$ of $[a, b]$ are not necessarily equal.
- The number u_i may be any number in the subinterval $[x_{i-1}, x_i]$, $i = 1, 2, \ldots, n$.

The sums $\sum_{i=1}^{n} f(u_i)\Delta x_i$, called **Riemann sums** for f on $[a, b]$, are the foundation of integral calculus.

EXAMPLE 1   **Finding Riemann Sums**

For the function $f(x) = x^2 - 3$, $0 \le x \le 6$, partition the interval $[0, 6]$ into four subintervals $[0, 1]$, $[1, 2]$, $[2, 4]$, $[4, 6]$ and find the Riemann sum for which

(a) u_i is the left endpoint of each subinterval.

(b) u_i is the midpoint of each subinterval.

Solution

In forming Riemann sums $\sum_{i=1}^{n} f(u_i)\Delta x_i$, n is the number of subintervals in the partition, $f(u_i)$ is the value of f at the number u_i chosen in the ith subinterval, and Δx_i is the width of the ith subinterval. The four subintervals have width

$$\Delta x_1 = 1 - 0 = 1 \quad \Delta x_2 = 2 - 1 = 1 \quad \Delta x_3 = 4 - 2 = 2 \quad \Delta x_4 = 6 - 4 = 2$$

(a) Figure 17 shows the graph of f, the partition of the interval $[0, 6]$ into the four subintervals, and values of $f(u_i)$ at the left endpoint of each subinterval, namely,

$$f(u_1) = f(0) = -3 \qquad f(u_2) = f(1) = -2$$
$$f(u_3) = f(2) = 1 \qquad f(u_4) = f(4) = 13$$

The Riemann sum is found by adding the products $f(u_i)\Delta x_i$ for $i = 1, 2, 3, 4$.

$$\sum_{i=1}^{4} f(u_i)\Delta x_i = f(u_1)\Delta x_1 + f(u_2)\Delta x_2 + f(u_3)\Delta x_3 + f(u_4)\Delta x_4$$
$$= -3 \cdot 1 + (-2) \cdot 1 + 1 \cdot 2 + 13 \cdot 2 = 23$$

ORIGINS Riemann sums are named after the German mathematician Georg Friedrich Bernhard Riemann (1826–1866). Early in his life, Riemann was home schooled. At age 14, he was sent to a lyceum (high school) and then the University of Göttingen to study theology. Once at Göttingen, he asked for and received permission from his father to study mathematics. He completed his PhD under Karl Friedrich Gauss (1777–1855). In his thesis, Riemann used topology to analyze complex functions. Later he developed a theory of geometry to describe real space. His ideas were far ahead of their time and were not truly appreciated until they provided the mathematical framework for Einstein's Theory of Relativity.

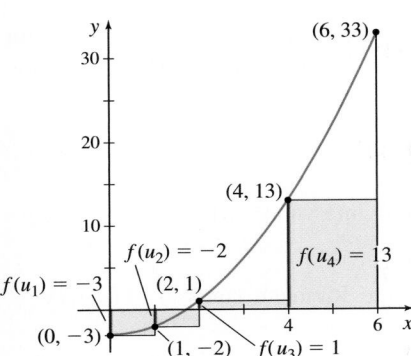

Figure 17 $f(x) = x^2 - 3, 0 \le x \le 6$

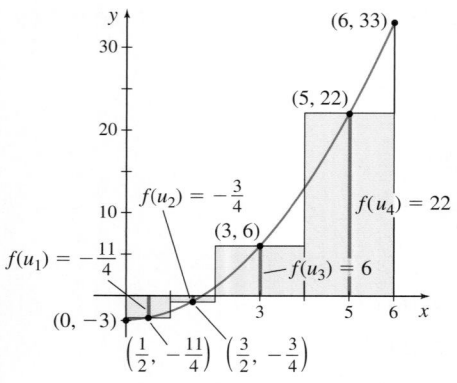

Figure 18 $f(x) = x^2 - 3, 0 \le x \le 6$

(b) See Figure 18. If u_i is chosen as the midpoint of each subinterval, then the value of $f(u_i)$ at the midpoint of each subinterval is

$$f(u_1) = f\left(\frac{1}{2}\right) = -\frac{11}{4} \qquad f(u_2) = f\left(\frac{3}{2}\right) = -\frac{3}{4}$$

$$f(u_3) = f(3) = 6 \qquad f(u_4) = f(5) = 22$$

The Riemann sum is found by adding the products $f(u_i)\Delta x_i$ for $i = 1, 2, 3, 4$.

$$\sum_{i=1}^{4} f(u_i)\Delta x_i = f(u_1)\Delta x_1 + f(u_2)\Delta x_2 + f(u_3)\Delta x_3 + f(u_4)\Delta x_4$$

$$= -\frac{11}{4} \cdot 1 + \left(-\frac{3}{4}\right) \cdot 1 + 6 \cdot 2 + 22 \cdot 2 = \frac{105}{2} = 52.5 \qquad \blacksquare$$

NOW WORK   Problem 9 and AP® Practice Problem 1.

2 Define a Definite Integral as the Limit of Riemann Sums

Suppose a function f is defined on a closed interval $[a, b]$, and we partition the interval $[a, b]$ into n subintervals

$$[x_0, x_1], [x_1, x_2], [x_2, x_3], \ldots, [x_{i-1}, x_i], \ldots, [x_{n-1}, x_n]$$

where

$$a = x_0 < x_1 < x_2 < \cdots < x_{i-1} < x_i < \cdots < x_{n-1} < x_n = b$$

These subintervals are not necessarily of the same width. Denote the width of the first subinterval by $\Delta x_1 = x_1 - x_0$, the width of the second subinterval by $\Delta x_2 = x_2 - x_1$, and so on.

In general, the width of the ith subinterval is

$$\boxed{\Delta x_i = x_i - x_{i-1}}$$

for $i = 1, 2, \ldots, n$. This set of subintervals of the interval $[a, b]$ is called a **partition** of $[a, b]$. The width of the largest subinterval in a partition is called the **norm** of the partition and is denoted by **max Δx_i**.

> **DEFINITION Definite Integral**
>
> Suppose f is a function defined on the closed interval $[a, b]$. Partition $[a, b]$ into n subintervals of width $\Delta x_i = x_i - x_{i-1}$, $i = 1, 2, \ldots, n$. Choose a number u_i in each subinterval, evaluate $f(u_i)$, and form the Riemann sums $\sum_{i=1}^{n} f(u_i)\Delta x_i$.
>
> If $\lim\limits_{\max \Delta x_i \to 0} \sum_{i=1}^{n} f(u_i)\Delta x_i = I$ exists and does not depend on the choice of the partition or on the choice of u_i, then the number I is called the **Riemann integral** or **definite integral** of f from a to b and is denoted by the expression $\int_a^b f(x)\,dx$. That is,
>
> $$\boxed{\int_a^b f(x)\,dx = \lim_{\max \Delta x_i \to 0} \sum_{i=1}^{n} f(u_i)\Delta x_i} \qquad (1)$$

IN WORDS The integral sign $\int$ is an elongated "S" to remind us of a sum. In the boxed equation (1), a and b represent the endpoints of the closed interval $[a, b]$. $\int_a^b$ represents the limit of the sum, and $f(x)\,dx$ represents $f(u_i)$ (the height) times Δx_i (the width).

When the limit exists, then we say that f is **integrable over** $[a, b]$ or that $\int_a^b f(x)\,dx$ **exists**.

In defining the definite integral $\int_a^b f(x)\,dx$, we assumed that $a < b$. To remove this restriction, we give the following definitions.

DEFINITION

- If $f(a)$ is defined, then

$$\int_a^a f(x)\,dx = 0$$

- If $a > b$ and if $\int_b^a f(x)\,dx$ exists, then

$$\int_a^b f(x)\,dx = -\int_b^a f(x)\,dx$$

IN WORDS Interchanging the limits of integration reverses the sign of the definite integral.

As examples,

- $\int_1^1 x^2\,dx = 0$
- $\int_3^2 x^2\,dx = -\int_2^3 x^2\,dx$

NOW WORK Problem 17.

If f is integrable over $[a, b]$, then $\lim\limits_{\max \Delta x_i \to 0} \sum\limits_{i=1}^n f(u_i)\Delta x_i$ exists for any choice of u_i in the ith subinterval, so we are free to choose the u_i any way we please. The choices could be the left endpoint of each subinterval, or the right endpoint, or the midpoint, or any other number in each subinterval. Also, $\lim\limits_{\max \Delta x_i \to 0} \sum\limits_{i=1}^n f(u_i)\Delta x_i$ is independent of the partition of the closed interval $[a, b]$, provided $\max \Delta x_i$ can be made as close as we please to 0. It is this flexibility that makes the definite integral so important in engineering, physics, chemistry, geometry, and economics.

EXAMPLE 2 Expressing the Limit of Riemann Sums as a Definite Integral

The Riemann sums for $f(x) = x^2 - 3$ on the closed interval $[0, 6]$ are

$$\sum_{i=1}^n f(u_i)\Delta x_i = \sum_{i=1}^n \left(u_i^2 - 3\right)\Delta x_i$$

where $[0, 6]$ is partitioned into n subintervals $[x_{i-1}, x_i]$ of width Δx_i and u_i is some number in the subinterval $[x_{i-1}, x_i]$, $i = 1, 2, \ldots, n$. Assuming that the limit of the Riemann sums exists as $\max \Delta x_i \to 0$, express the limit as a definite integral.

Solution

Since $\lim\limits_{\max \Delta x_i \to 0} \sum\limits_{i=1}^n \left(u_i^2 - 3\right)\Delta x_i$ exists, then

$$\lim_{\max \Delta x_i \to 0} \sum_{i=1}^n \left(u_i^2 - 3\right)\Delta x_i = \int_0^6 (x^2 - 3)\,dx$$

NOW WORK Problem 19.

For a definite integral $\int_a^b f(x)\,dx,$

- the number a is called the **lower limit of integration**
- the number b is called the **upper limit of integration**
- the symbol $\int$ is called the **integral sign**
- $f(x)$ is called the **integrand**
- dx is the differential of the independent variable x

The variable used in a definite integral is an *artificial* or a *dummy* variable because it may be replaced by any other symbol. For example,

$$\int_a^b f(x)\,dx \qquad \int_a^b f(t)\,dt \qquad \int_a^b f(s)\,ds \qquad \int_a^b f(\theta)\,d\theta$$

all denote the definite integral of f from a to b, and if any one of them exists, they all exist and equal the same number.

Suppose a function f is integrable over the interval $[a, b]$. Partition $[a, b]$ into n subintervals, each of the same width $\Delta x = \dfrac{b-a}{n}$. Such a partition is called a **regular partition**. For a regular partition, the norm of the partition is

$$\boxed{\max \Delta x_i = \Delta x = \frac{b-a}{n}}$$

So, $\max \Delta x_i \to 0$ implies $\Delta x \to 0$. Furthermore, since $\Delta x = \dfrac{b-a}{n}$, it follows that $\Delta x \to 0$ is equivalent to $n \to \infty$. That is,

$$\boxed{\max \Delta x_i \to 0 \text{ is equivalent to } n \to \infty \text{ and } \Delta x \to 0}$$

As a result, for regular partitions of an interval $[a, b]$, the definite integral can be expressed as

$$\boxed{\int_a^b f(x)\,dx = \lim_{\Delta x \to 0} \sum_{i=1}^n f(u_i)\,\Delta x} \qquad (2a)$$

$$\boxed{\int_a^b f(x)\,dx = \lim_{n \to \infty} \sum_{i=1}^n f(u_i)\frac{b-a}{n}} \qquad (2b)$$

where u_i is a number in the ith subinterval and is expressed in terms of i and Δx (see 2a) or in terms of i and n (see 2b).

EXAMPLE 3 **Expressing a Definite Integral as the Limit of a Riemann Sum**

Express the following definite integrals as the limit of a Riemann sum using a regular partition.

(a) $\int_0^8 \sqrt{x}\,dx$ \qquad (b) $\int_{2\pi}^{8\pi} \sin x\,dx$

Solution

(a) $\Delta x = \dfrac{b-a}{n} = \dfrac{8-0}{n} = \dfrac{8}{n}$. Partition the interval $[0, 8]$ into n subintervals, each of width Δx.

$[0, \Delta x] \; [\Delta x, 2\Delta x] \; [2\Delta x, 3\Delta x] \; \cdots \; [(i-1)\Delta x, i\Delta x] \; \cdots \; [(n-1)\Delta x, n\Delta x]$

(Note that the right endpoint of the last subinterval is $n\Delta x = n \cdot \dfrac{8}{n} = 8$, as it should be.)

Choose $u_i = i\,\Delta x$ (the right endpoint of the ith subinterval). Then, using (2a),

$$\int_0^8 \sqrt{x}\,dx = \lim_{\Delta x \to 0} \sum_{i=1}^n f(u_i)\,\Delta x = \lim_{\Delta x \to 0} \sum_{i=1}^n f(i\Delta x)\,\Delta x = \lim_{\Delta x \to 0} \sum_{i=1}^n \sqrt{i\Delta x}\,\Delta x$$

Using (2b),

$$\int_0^8 \sqrt{x}\,dx = \lim_{n\to\infty} \sum_{i=1}^n f(u_i)\,\frac{b-a}{n} = \lim_{n\to\infty} \sum_{i=1}^n f(i\Delta x)\,\frac{8}{n} = \lim_{n\to\infty} \sum_{i=1}^n \sqrt{i \cdot \frac{8}{n}}\,\frac{8}{n}$$

(b) Here, $\Delta x = \dfrac{b-a}{n} = \dfrac{8\pi - 2\pi}{n} = \dfrac{6\pi}{n}$. Partition the interval $[2\pi, 8\pi]$ into n subintervals, each of width Δx.

$[2\pi, 2\pi + \Delta x] \quad [2\pi + \Delta x, 2\pi + 2\Delta x] \quad [2\pi + 2\Delta x, 2\pi + 3\Delta x] \cdots$
$[2\pi + (i-1)\Delta x, 2\pi + i\Delta x] \quad \cdots \quad [2\pi + (n-1)\Delta x, 2\pi + n\Delta x]$

(Note that the right endpoint of the last subinterval is $2\pi + n\,\Delta x = 2\pi + n \cdot \dfrac{6\pi}{n} = 8\pi$ as it should be.)

Choose $u_i = 2\pi + i\,\Delta x$ (the right endpoint of the ith subinterval). Then, using (2a),

$$\int_{2\pi}^{8\pi} \sin x\,dx = \lim_{\Delta x \to 0} \sum_{i=1}^n f(u_i)\,\Delta x = \lim_{\Delta x \to 0} \sum_{i=1}^n f(2\pi + i\Delta x)\,\Delta x$$

$$= \lim_{\Delta x \to 0} \sum_{i=1}^n \sin(2\pi + i\Delta x)\,\Delta x$$

Using (2b),

$$\int_{2\pi}^{8\pi} \sin x\,dx = \lim_{n\to\infty} \sum_{i=1}^n f(u_i)\,\frac{b-a}{n} = \lim_{n\to\infty} \sum_{i=1}^n f(2\pi + i\Delta x)\,\frac{6\pi}{n}$$

$$= \lim_{n\to\infty} \sum_{i=1}^n \sin\left(2\pi + i \cdot \frac{6\pi}{n}\right)\frac{6\pi}{n} \qquad\blacksquare$$

NOW WORK Problem **33** and AP® Practice Problems **4** and **9**.

EXAMPLE 4 Expressing the Limit of a Riemann Sum as a Definite Integral

Express each limit of a Riemann sum as a definite integral. Assume each limit exists.

(a) $\displaystyle \lim_{n\to\infty} \sum_{i=1}^n \sqrt{\frac{20}{n}i} \cdot \frac{10}{n}$ on the interval $[0, 10]$

(b) $\displaystyle \lim_{n\to\infty} \sum_{i=1}^n \sin\left(\frac{\pi}{4} + \frac{2\pi}{n}i\right) \cdot \frac{2\pi}{n}$ on the interval $\left[\dfrac{\pi}{4}, \dfrac{9\pi}{4}\right]$

(Example continued on the next page)

Solution

(a) For the interval $[0, 10]$, $\Delta x = \dfrac{b-a}{n} = \dfrac{10}{n}$. The partition of the interval $[0, 10]$ begins at 0 and ends at 10.

$$[0, \Delta x] \quad [\Delta x, 2\Delta x] \quad [2\Delta x, 3\Delta x] \quad \cdots \quad [(i-1)\Delta x, i\Delta x] \quad \cdots \quad [(n-1)\Delta x, n\Delta x]$$

(Note that the right endpoint of the last subinterval is $n\Delta x = n\dfrac{10}{n} = 10$, as it should.)

Choose $u_i = i\Delta x$, the right endpoint of the ith subinterval. The expression being summed is

$$\sqrt{\frac{20i}{n}} \cdot \frac{10}{n} = \sqrt{2 \cdot \frac{10}{n} i} \cdot \frac{10}{n} = \sqrt{2 \cdot i\Delta x} \; \Delta x = \sqrt{2u_i} \; \Delta x$$

Then

$$\lim_{n \to \infty} \sum_{i=1}^{n} \sqrt{\frac{20}{n} i} \cdot \frac{10}{n} = \lim_{n \to \infty} \sum_{i=1}^{n} \sqrt{2u_i} \; \Delta x = \int_0^{10} \sqrt{2x} \; dx$$

(b) For the interval $\left[\dfrac{\pi}{4}, \dfrac{9\pi}{4}\right]$, $\Delta x = \dfrac{b-a}{n} = \dfrac{\frac{9\pi}{4} - \frac{\pi}{4}}{n} = \dfrac{2\pi}{n}$. The partition of the interval $\left[\dfrac{\pi}{4}, \dfrac{9\pi}{4}\right]$ begins at $\dfrac{\pi}{4}$ and ends at $\dfrac{9\pi}{4}$.

$$\left[\frac{\pi}{4}, \frac{\pi}{4} + \Delta x\right] \quad \left[\frac{\pi}{4} + \Delta x, \frac{\pi}{4} + 2\Delta x\right] \cdots \left[\frac{\pi}{4} + (i-1)\Delta x, \frac{\pi}{4} + i\Delta x\right]$$

$$\cdots \left[\frac{\pi}{4} + (n-1)\Delta x, \frac{\pi}{4} + n\Delta x\right]$$

(Note that the right endpoint of the last subinterval is

$$\frac{\pi}{4} + n\Delta x = \frac{\pi}{4} + n \cdot \frac{2\pi}{n} = \frac{\pi}{4} + 2\pi = \frac{9\pi}{4}$$

as it should.)

Choose $u_i = \dfrac{\pi}{4} + i\Delta x$, the right endpoint of the ith subinterval. The expression being summed is

$$\sin\left(\frac{\pi}{4} + \frac{2\pi}{n} i\right) \frac{2\pi}{n} = \sin\left(\frac{\pi}{4} + i\Delta x\right) \Delta x = \sin u_i \; \Delta x$$

Then

$$\lim_{n \to \infty} \sum_{i=1}^{n} \sin\left(\frac{\pi}{4} + \frac{2\pi}{n} i\right) \frac{2\pi}{n} = \lim_{n \to \infty} \sum_{i=1}^{n} \sin u_i \; \Delta x = \int_{\pi/4}^{9\pi/4} \sin x \; dx \qquad \blacksquare$$

We can generalize the process used in Example 4 by observing that

$$\int_a^b f(x)\, dx = \lim_{\Delta x \to 0} \sum_{i=1}^{n} f(a + i\Delta x)\, \Delta x \text{ on the interval } [a, b] \text{ and } \Delta x = \frac{b-a}{n}$$

NOW WORK Problem 23 and AP® Practice Problem 6.

AP® EXAM TIP

On the exam,

- Riemann sums formed using the left endpoint of each subinterval are called **Left Riemann sums.**
- Riemann sums formed using the midpoint of each subinterval are called **Midpoint Riemann sums** or **Midpoint sums.**
- Riemann sums formed using the right endpoint of each subinterval are called **Right Riemann sums.**

③ Approximate a Definite Integral Using Riemann Sums

Suppose f is integrable over the closed interval $[a, b]$. Then

$$\int_a^b f(x)\,dx = \lim_{\max \Delta x_i \to 0} \sum_{i=1}^n f(u_i)\Delta x_i$$

To approximate $\int_a^b f(x)\,dx$ using Riemann sums, we usually partition $[a, b]$ into n subintervals, each of the same width $\Delta x = \dfrac{b-a}{n}$. Then

$$\int_a^b f(x)\,dx \approx \sum_{i=1}^n f(u_i)\Delta x = \sum_{i=1}^n f(u_i)\frac{b-a}{n}$$

where u_i is usually chosen as the left endpoint, the right endpoint, or the midpoint of the ith subinterval $[x_{i-1}, x_i]$.

EXAMPLE 5 **Approximating a Definite Integral Using Riemann Sums**

Approximate $\int_1^9 (x-3)^2\,dx$ by partitioning $[1, 9]$ into four subintervals each of width 2

(a) using a Left Riemann sum. (Choose u_i as the left endpoint of the ith subinterval.)

(b) using a Right Riemann sum. (Choose u_i as the right endpoint of the ith subinterval.)

(c) using a Midpoint Riemann sum. (Choose u_i as the midpoint of the ith subinterval.)

Solution

For the given partition $\Delta x = \dfrac{9-1}{4} = 2$ and the four subintervals are

$$[1, 3] \quad [3, 5] \quad [5, 7] \quad [7, 9]$$

AP® EXAM TIP

Be sure to show the approximation as a series of sums of products $f(u_i)\Delta x$ to get full credit on a free-response question.

(a) We choose u_i to be the left endpoint of the ith subinterval.
Then $u_1 = 1, u_2 = 3, u_3 = 5, u_4 = 7$, and

$$\int_1^9 (x-3)^2\,dx \approx \sum_{i=1}^4 f(u_i)\Delta x = f(1)(2) + f(3)(2) + f(5)(2) + f(7)(2)$$
$$= [4 + 0 + 4 + 16] \cdot 2 = 48$$

(b) We choose u_i to be the right endpoint of the ith subinterval.
Then $u_1 = 3, u_2 = 5, u_3 = 7, u_4 = 9$, and

$$\int_1^9 (x-3)^2\,dx \approx \sum_{i=1}^4 f(u_i)\Delta x = f(3)(2) + f(5)(2) + f(7)(2) + f(9)(2)$$
$$= [0 + 4 + 16 + 36] \cdot 2 = 112$$

(c) We choose u_i to be the midpoint of the ith subinterval.
Then $u_1 = 2, u_2 = 4, u_3 = 6, u_4 = 8$, and

$$\int_1^9 (x-3)^2\,dx \approx \sum_{i=1}^4 f(u_i)\Delta x = f(2)(2) + f(4)(2) + f(6)(2) + f(8)(2)$$
$$= [1 + 1 + 9 + 25] \cdot 2 = 72 \qquad ■$$

NOW WORK Problem 39(a) and (b) and AP® Practice Problems 3, 5, 7, and 11.

Depending on the behavior of a function f, it may be possible to determine whether an approximation to the definite integral $\int_a^b f(x)\,dx$ underestimates or overestimates the value of the integral.

EXAMPLE 6 | **Determining Whether a Riemann Sum Underestimates or Overestimates an Integral**

A function f is continuous on a closed interval $[-3, 8]$ and is differentiable on the open interval $(-3, 8)$. Suppose $f'(x) > 0$ for $-3 < x < 5$, $f'(x) = 0$ for $5 \leq x \leq 6$, and $f'(x) < 0$ for $6 < x < 8$.

For each integral, determine whether a Left Riemann sum overestimates the integral, underestimates it, or that there is insufficient information to draw a conclusion.

(a) $\displaystyle\int_{-3}^{5} f(x)dx$ **(b)** $\displaystyle\int_{4}^{6} f(x)dx$ **(c)** $\displaystyle\int_{4}^{8} f(x)dx$

Solution

(a) On the interval $(-3, 5)$, the derivative $f'(x) > 0$, so the function f is increasing on the interval $[-3, 5]$. On every subinterval of $[-3, 5]$, the value of f at the left endpoint is less than that of f at any other number in the subinterval. So a Left Riemann sum underestimates the value of $\int_{-3}^{5} f(x)dx$.

(b) The function f is increasing on the interval $[4, 5]$ and is constant on the interval $[5, 6]$. So the value of f at the left endpoint of any subinterval of $[4, 6]$ is less than or equal to that of f at any other number in the subinterval. So a Left Riemann sum underestimates $\int_{4}^{6} f(x)dx$.

(c) The function f is increasing on the interval $[4, 5]$, is constant on the interval $[5, 6]$, and is decreasing on the interval $[6, 8]$. We cannot draw a conclusion about the relationship between the Left Riemann sum and an arbitrary Riemann sum, so it is impossible to determine whether the Left Riemann sum overestimates or underestimates $\int_{4}^{8} f(x)dx$. ∎

The results found in Example 6 can be generalized.

- Suppose a function f is continuous and increasing on a closed interval $[a, b]$. Then
  - a Left Riemann sum underestimates $\int_{a}^{b} f(x)dx$.
  - a Right Riemann sum overestimates $\int_{a}^{b} f(x)dx$.
- Suppose a function f is continuous and decreasing on a closed interval $[a, b]$. Then
  - a Left Riemann sum overestimates $\int_{a}^{b} f(x)dx$.
  - a Right Riemann sum underestimates $\int_{a}^{b} f(x)dx$.

NOW WORK | AP® Practice Problem 8.

4 Approximate a Definite Integral Using Trapezoidal Sums

Suppose a function f is integrable on the interval $[a, b]$. Partition $[a, b]$ into n subintervals, not necessarily of the same width. Suppose $[x_{i-1}, x_i]$ is the ith subinterval of the partition.

- If $u_i = x_{i-1}$ is the left endpoint of each subinterval, the Left Riemann sums $\sum_{i=1}^{n} f(x_{i-1}) \Delta x_i$ can be used to approximate $\int_{a}^{b} f(x)dx$. That is,

$$\int_{a}^{b} f(x)dx \approx \sum_{i=1}^{n} f(x_{i-1})\Delta x_i \qquad (3)$$

- If $u_i = x_i$ is the right endpoint of each subinterval, the Right Riemann sums $\sum_{i=1}^{n} f(x_i)\Delta x_i$ can be used to approximate $\int_a^b f(x)dx$. That is,

$$\int_a^b f(x)dx \approx \sum_{i=1}^{n} f(x_i)\Delta x_i \qquad (4)$$

We use a Left Riemann sum and a Right Riemann sum to define a *trapezoidal sum*. Begin by adding (3) and (4).

$$2\int_a^b f(x)dx \approx \sum_{i=1}^{n} f(x_{i-1})\Delta x_i + \sum_{i=1}^{n} f(x_i)\Delta x_i$$

or

$$\int_a^b f(x)dx \approx \frac{1}{2}\left[\sum_{i=1}^{n} f(x_{i-1})\Delta x_i + \sum_{i=1}^{n} f(x_i)\Delta x_i\right] \qquad (5)$$

Statement (5) states that if a function f is integrable on a closed interval $[a, b]$, then $\int_a^b f(x)\,dx$ can be approximated by the average of its Left Riemann sums and its Right Riemann sums.

Now look at Figure 19. The shaded region from x_{i-1} to x_i is a trapezoid whose area is

$$\frac{f(x_{i-1}) + f(x_i)}{2} \cdot \Delta x_i$$

and the sum of the trapezoidal areas from $i = 1$ to $i = n$ is

$$\sum_{i=1}^{n} \frac{f(x_{i-1}) + f(x_i)}{2} \cdot \Delta x_i \qquad (6)$$

Statement (6) tells us that $\int_a^b f(x)\,dx$ can be approximated by a sum of trapezoidal areas. Moreover, statements (5) and (6) are equivalent. When (5) or (6) is used to approximate $\int_a^b f(x)\,dx$, it is referred to as a **trapezoidal approximation**.

RECALL The area A of a trapezoid with width Δx and parallel heights l_1 and l_2 is

$$A = \frac{1}{2}(l_1 + l_2)\Delta x.$$

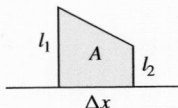

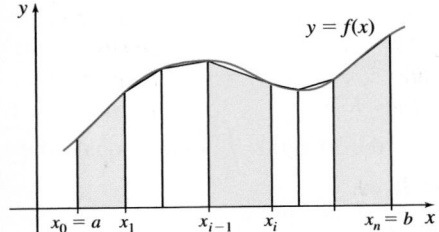

Figure 19 Trapezoidal sums

IN WORDS A trapezoidal sum is the average of a Left Riemann sum and a Right Riemann sum.

THEOREM Trapezoidal Sums

If a function f is integrable on the closed interval $[a, b]$, then for any partition of $[a, b]$ into n subintervals

$$\int_a^b f(x)\,dx \approx \sum_{i=1}^{n} \frac{f(x_{i-1}) + f(x_i)}{2} \cdot \Delta x_i = \frac{1}{2}\left[\sum_{i=1}^{n} f(x_{i-1})\Delta x_i + \sum_{i=1}^{n} f(x_i)\Delta x_i\right]$$

EXAMPLE 7 Approximating $\int_1^5 \dfrac{e^x}{x}\,dx$ Using Trapezoidal Sums

Use trapezoidal sums with four subintervals of equal width to approximate $\int_1^5 \dfrac{e^x}{x}dx$. Express the answer rounded to three decimal places.

Solution

Partition $[1, 5]$ into four subintervals, each of width $\Delta x = \dfrac{5-1}{4} = 1$.

$$[1, 2] \qquad [2, 3] \qquad [3, 4] \qquad [4, 5]$$

The value of $f(x) = \dfrac{e^x}{x}$ corresponding to each endpoint is

$$f(1) = e \qquad f(2) = \frac{e^2}{2} \qquad f(3) = \frac{e^3}{3} \qquad f(4) = \frac{e^4}{4} \qquad f(5) = \frac{e^5}{5}$$

The Left Riemann sum is

$$\sum_{i=2}^{5} f(x_{i-1})\Delta x = f(1) \cdot 1 + f(2) \cdot 1 + f(3) \cdot 1 + f(4) \cdot 1$$

$$= \left[e + \frac{e^2}{2} + \frac{e^3}{3} + \frac{e^4}{4} \right] \cdot 1 = 26.75752$$

The Right Riemann sum is

$$\sum_{i=2}^{5} f(x_{i})\Delta x = f(2) \cdot 1 + f(3) \cdot 1 + f(4) \cdot 1 + f(5) \cdot 1$$

$$= \left[\frac{e^2}{2} + \frac{e^3}{3} + \frac{e^4}{4} + \frac{e^5}{5} \right] \cdot 1 = 53.72187$$

Then the trapezoidal sum that approximates $\int_{1}^{5} \frac{e^x}{x} dx$ is

$$\frac{1}{2} \left[\sum_{i=2}^{5} \frac{e^{i-1}}{x_{i-1}} \cdot 1 + \sum_{i=2}^{5} \frac{e^i}{x_i} \cdot 1 \right] = \frac{1}{2}(26.75752 + 53.72187) \approx 40.240 \qquad \blacksquare$$

NOW WORK   Problem **31** and AP® Practice Problems **2** and **10**.

Since $f(x) = \dfrac{e^x}{x} > 0$ on the interval $[1, 5]$, the integral $\int_{1}^{5} \dfrac{e^x}{x} dx$ equals the area under the graph of f from 1 to 5. Figure 20 shows the graph of $y = \dfrac{e^x}{x}$, the area $A = \int_{1}^{5} \dfrac{e^x}{x} dx$, and the approximation to the area using the trapezoidal sums with four subintervals of equal width.

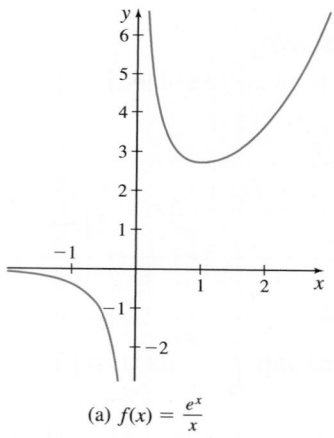

(a) $f(x) = \dfrac{e^x}{x}$

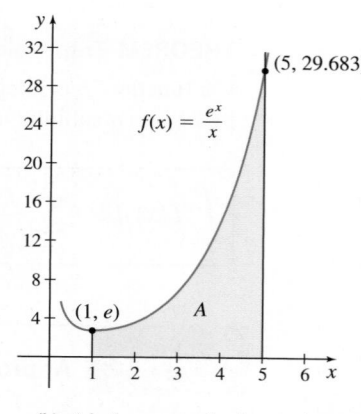

(b) A is the area under the graph of f from 1 to 5.

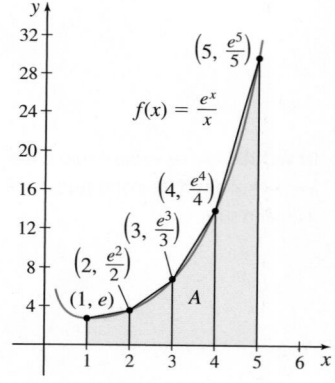

(c) Area approximated with four subintervals of equal width.

Figure 20

Notice in Figure 20(c) that the graph of $y = \dfrac{e^x}{x}$ is concave up on the interval $[1, 5]$. The trapezoids used to approximate $\int_{1}^{5} \dfrac{e^x}{x} dx$ extend above the graph, and so the approximation overestimates the integral. For graphs that are concave down, trapezoidal sums will underestimate the integral.

Suppose f is a function that is continuous on the interval $[a, b]$ and is twice differentiable on the interval (a, b).

- If $f''(x) > 0$ on (a, b), that is, if the graph of f is concave up on (a, b), a trapezoidal sum will overestimate $\int_a^b f(x)\,dx$.
- If $f''(x) < 0$ on (a, b), that is, if the graph of f is concave down on (a, b), a trapezoidal sum will underestimate $\int_a^b f(x)\,dx$.

NOW WORK AP® Practice Problem 12.

EXAMPLE 8 Approximating $\int_1^3 \ln x\,dx$

Approximate $\int_1^3 \ln x\,dx$, rounded to three decimal places, using four subintervals of equal width with

(a) a Left Riemann sum. Does the approximation overestimate or underestimate the integral?

(b) a Right Riemann sum. Does the approximation overestimate or underestimate the integral?

(c) a trapezoidal sum. Does the approximation overestimate or underestimate the integral?

(d) Use a calculator to approximate $\int_1^3 \ln x\,dx$.

Solution

Begin by partitioning $[1, 3]$ into four subintervals, each of width $\Delta x = \dfrac{3-1}{4} = \dfrac{1}{2}$.

$$[1, 1.5] \qquad [1.5, 2] \qquad [2, 2.5] \qquad [2.5, 3]$$

The value of $f(x) = \ln x$ corresponding to each endpoint is

$$f(1) = 0 \qquad f(1.5) = \ln 1.5 \qquad f(2) = \ln 2 \qquad f(2.5) = \ln 2.5 \qquad f(3) = \ln 3$$

See Figure 21.

(a) The Left Riemann sum:

$$\int_1^3 \ln x\,dx \approx \sum_{i=2}^{5} \ln x_{i-1}\Delta x = \ln 1\left(\frac{1}{2}\right) + \ln 1.5\left(\frac{1}{2}\right) + \ln 2\left(\frac{1}{2}\right) + \ln 2.5\left(\frac{1}{2}\right)$$

$$= [0 + \ln 1.5 + \ln 2 + \ln 2.5] \cdot \frac{1}{2} \approx 1.00745 \approx 1.007$$

If $f(x) = \ln x$, then $f'(x) = \dfrac{1}{x} > 0$ on the interval $(1, 3)$, so f is increasing on $[1, 3]$.

The approximation of $\int_1^3 \ln x\,dx$ using a Left Riemann sum underestimates $\int_1^3 \ln x\,dx$ because the function $f(x) = \ln x$ is continuous and increasing on the interval $[1, 3]$.

(b) The Right Riemann sum:

$$\int_1^3 \ln x\,dx \approx \sum_{i=2}^{5} \ln x_i \Delta x = \ln 1.5\left(\frac{1}{2}\right) + \ln 2\left(\frac{1}{2}\right) + \ln 2.5\left(\frac{1}{2}\right) + \ln 3\left(\frac{1}{2}\right)$$

$$= [\ln 1.5 + \ln 2 + \ln 2.5 + \ln 3] \cdot \frac{1}{2} \approx 1.55676 \approx 1.557$$

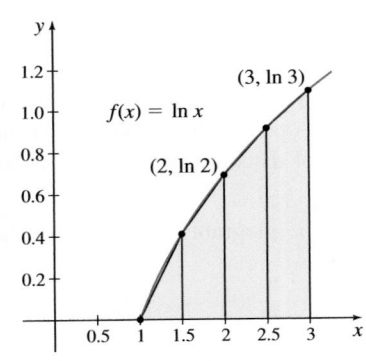

Figure 21 $f(x) = \ln x$, $\Delta x = \dfrac{1}{2}$

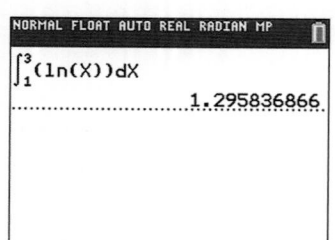

Figure 22

NEED TO REVIEW? Continuity is discussed
in Section 1.3, pp. 108–115.

IN WORDS If a function f is continuous
on $[a, b]$, then it is integrable over $[a, b]$.
But if f is not continuous on $[a, b]$, then f
may or may not be integrable over $[a, b]$.

The approximation of $\int_1^3 \ln x \, dx$ using a Right Riemann sum overestimates $\int_1^3 \ln x \, dx$ because the function $f(x) = \ln x$ is continuous and increasing on the interval $[1, 3]$.

(c) The trapezoidal sum:

$$\int_1^3 \ln x \, dx \approx \frac{1}{2}\left[\sum_{i=2}^5 \ln x_{i-1}\Delta x + \sum_{i=2}^5 \ln x_i \Delta x\right] = \frac{1}{2}(1.00745 + 1.55676) \approx 1.282$$

For $f(x) = \ln x$, $f'(x) = \dfrac{1}{x}$ and $f''(x) = -\dfrac{1}{x^2} < 0$ on the interval $(1, 3)$, so f is concave down on $(1, 3)$. Using a trapezoidal sum, the approximation of $\int_1^3 \ln x \, dx$ underestimates $\int_1^3 \ln x \, dx$ because the function $f(x) = \ln x$ is continuous and concave down on the closed interval $[1, 3]$.

(d) Using a TI-84 Plus CE calculator, $\int_1^3 \ln x \, dx \approx 1.296$. See Figure 22. ∎

NOW WORK Problem 25(a)-(c) and AP® Practice Problem 13.

---

⑤ Know Conditions That Guarantee a Definite Integral Exists

We now state two theorems that give conditions that guarantee a function f is integrable. The proofs of these two results may be found in advanced calculus texts.

> **THEOREM** Existence of the Definite Integral
>
> If a function f is continuous on a closed interval $[a, b]$, then the definite integral $\int_a^b f(x) \, dx$ exists.

The hypothesis of the theorem deserves special attention. First, f is defined on a *closed* interval, and second, f is *continuous* on that interval. There are some functions that are continuous on an open interval (or even a half-open interval) for which the integral does not exist. Also, there are many examples of discontinuous functions for which the integral exists.

The second theorem that guarantees a function f is integrable requires that the function be *bounded*.

> **DEFINITION** Bounded Function
>
> A function f is **bounded on an interval** if there are two numbers m and M for which $m \le f(x) \le M$ for all x in the interval.

For example, the function $f(x) = x^2$ is bounded on the interval $[-1, 3]$ because there are two numbers, $m = 0$ and $M = 9$ for which $0 \le f(x) \le 9$ for all x in the interval.

> **THEOREM** Existence of the Definite Integral
>
> If a function f is bounded on a closed interval $[a, b]$ and has at most a finite number of points of discontinuity on $[a, b]$, then $\int_a^b f(x) \, dx$ exists.

In most of our work with applications of definite integrals, the integrand will be continuous on a closed interval $[a, b]$, guaranteeing the integral exists.

For example, if f is continuous and nonnegative on a closed interval $[a, b]$, the area under the graph of f exists and equals $\int_a^b f(x) \, dx$.

6.2 Assess Your Understanding

Concepts and Vocabulary

1. **True or False** In a Riemann sum $\sum_{i=1}^{n} f(u_i)\Delta x$, u_i is always the left endpoint, the right endpoint, or the midpoint of the ith subinterval, $i = 1, 2, \ldots, n$.

2. **Multiple Choice** In a regular partition of $[0, 40]$ into 20 subintervals, $\Delta x =$ [(**a**) 20 (**b**) 0.5 (**c**) 2 (**d**) 4].

3. **True or False** Trapezoidal sums approximate an integral $\int_a^b f(x)\,dx$ by averaging Left and Right Riemann sums.

4. In the notation for a definite integral $\int_a^b f(x)\,dx$, a is called the _____; b is called the _____; $\int$ is called the _____; and $f(x)$ is called the _____.

5. If $f(a)$ is defined, $\int_a^a f(x)\,dx =$ _____.

6. **True or False** If a function f is integrable over a closed interval $[a, b]$, then $\int_a^b f(x)\,dx = \int_b^a f(x)\,dx$.

7. **True or False** If a function f is continuous on a closed interval $[a, b]$, then the definite integral $\int_a^b f(x)\,dx$ exists.

8. **Multiple Choice** Since $\int_0^2 (3x - 8)\,dx = -10$, then $\int_2^0 (3x - 8)\,dx =$ [(**a**) -10 (**b**) 10 (**c**) 5 (**d**) 0].

Skill Building

In Problems 9 and 10, find the Riemann sum for $f(x) = x$, $0 \le x \le 2$, for the partition and the numbers u_i given below.

 9. Partition: $\left[0, \dfrac{1}{4}\right]$, $\left[\dfrac{1}{4}, \dfrac{1}{2}\right]$, $\left[\dfrac{1}{2}, \dfrac{3}{4}\right]$, $\left[\dfrac{3}{4}, 1\right]$, $[1, 2]$

$$u_1 = \frac{1}{8}, u_2 = \frac{3}{8}, u_3 = \frac{5}{8}, u_4 = \frac{7}{8}, u_5 = \frac{9}{8}$$

10. Partition: $\left[0, \dfrac{1}{2}\right]$, $\left[\dfrac{1}{2}, 1\right]$, $\left[1, \dfrac{3}{2}\right]$, $\left[\dfrac{3}{2}, 2\right]$

$$u_1 = \frac{1}{2}, u_2 = 1, u_3 = \frac{3}{2}, u_4 = 2$$

For Problems 11–14, use the graph below to approximate the area A, where $n = 3$ is the number of subintervals used and each subinterval has the same width.

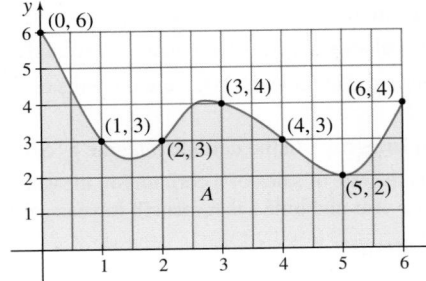

11. Use a Left Riemann sum to approximate A.

12. Use a Right Riemann sum to approximate A.

13. Use a trapezoidal sum to approximate A.

14. Use a Midpoint Riemann sum to approximate A.

15. If $\int_2^5 f(x)\,dx = 10$, find $\int_5^2 f(x)\,dx$.

16. If $\int_{-3}^7 g(x)\,dx = -2$, find $\int_7^{-3} g(x)\,dx$.

In Problems 17–18, find each definite integral.

PAGE 437 17. $\int_4^4 2\theta\,d\theta$ 18. $\int_{-1}^{-1} 8\,dr$

In Problems 19–22, each limit exists. Write the limit of the Riemann sums as a definite integral. Partition each interval $[a, b]$ into n subintervals of equal width $\Delta x = \dfrac{b - a}{n}$.

PAGE 437 19. $\displaystyle\lim_{\Delta x \to 0} \sum_{i=1}^{n} (e^{u_i} + 2)\Delta x$ on $[0, 2]$

20. $\displaystyle\lim_{\Delta x \to 0} \sum_{i=1}^{n} \ln u_i\,\Delta x$ on $[1, 8]$ 21. $\displaystyle\lim_{\Delta x \to 0} \sum_{i=1}^{n} \cos u_i\,\Delta x$ on $[0, 2\pi]$

22. $\displaystyle\lim_{\Delta x \to 0} \sum_{i=1}^{n} (\cos u_i + \sin u_i)\,\Delta x$ on $[0, \pi]$

In Problems 23–26, express each limit as a definite integral.

PAGE 440 23. $\displaystyle\lim_{n \to \infty} \sum_{i=1}^{n} \frac{2}{1 + \frac{3}{n}i} \cdot \frac{3}{n}$ on the interval $[0, 3]$.

24. $\displaystyle\lim_{n \to \infty} \sum_{i=1}^{n} \left(\frac{8}{n}i\right)^{1/3} \cdot \frac{8}{n}$ on the interval $[2, 10]$.

25. $\displaystyle\lim_{n \to \infty} \sum_{i=1}^{n} \left(1 + \frac{e}{n}i\right) \ln\left(1 + \frac{e}{n}i\right) \cdot \frac{e - 1}{n}$ on the interval $[1, e]$.

26. $\displaystyle\lim_{n \to \infty} \sum_{i=1}^{n} \cos\left(\frac{2\pi}{n}i + 3\pi\right) \cdot \frac{2\pi}{n}$ on the interval $[0, 2\pi]$.

In Problems 27–30, the graph of a function is shown. Express the shaded area as a definite integral.

27.
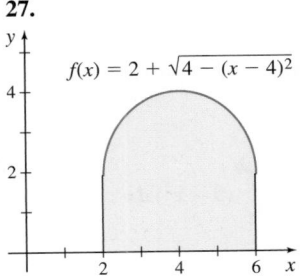
$f(x) = 2 + \sqrt{4 - (x - 4)^2}$

28.
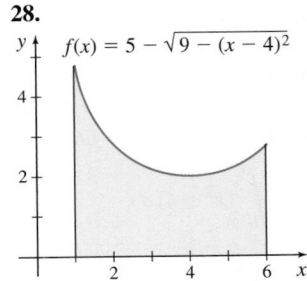
$f(x) = 5 - \sqrt{9 - (x - 4)^2}$

29.
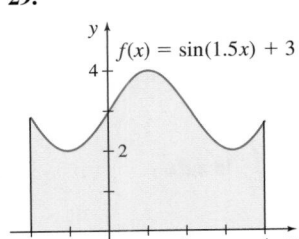
$f(x) = \sin(1.5x) + 3$

30.
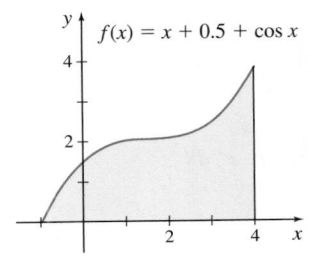
$f(x) = x + 0.5 + \cos x$

In Problems 31 and 32, the graph of a function f defined on an interval [a, b] is given.

(a) *Partition the interval [a, b] into six subintervals (not necessarily of the same width) using the points shown on each graph.*

(b) *Approximate $\int_a^b f(x)\,dx$ using a Left Riemann sum.*

(c) *Approximate $\int_a^b f(x)\,dx$ using a trapezoidal sum.*

[PAGE 443] 31.

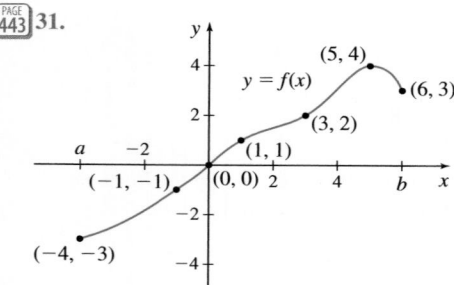

32.

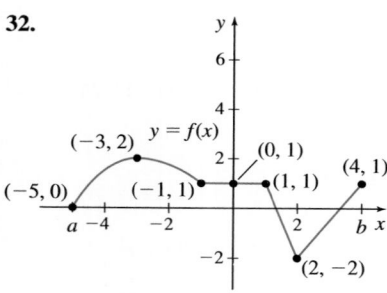

In Problems 33–38, express each definite integral as the limit of a Riemann sum.

[PAGE 439] 33. $\displaystyle\int_0^{\pi} \sin x\,dx$

34. $\displaystyle\int_{-\pi/4}^{\pi/4} \tan x\,dx$

35. $\displaystyle\int_1^4 (x-2)^{1/3}dx$

36. $\displaystyle\int_1^4 (x+2)^{1/3}dx$

37. $\displaystyle\int_1^4 (|x|-2)\,dx$

38. $\displaystyle\int_{-2}^4 |x|\,dx$

In Problems 39–46, approximate each definite integral $\int_a^b f(x)\,dx$ by

(a) *Partitioning [a, b] into two subintervals and choosing u_i as the left endpoint of the ith subinterval.*

(b) *Partitioning [a, b] into four subintervals and choosing u_i as the right endpoint of the ith subinterval.*

(c) *Use technology to find each integral.*

[PAGE 441] 39. $\displaystyle\int_0^8 (x^2-4)\,dx$

40. $\displaystyle\int_1^9 (9-x^2)\,dx$

41. $\displaystyle\int_{-4}^4 (x^2-x)\,dx$

42. $\displaystyle\int_{-2}^6 (x^2-4x)\,dx$

43. $\displaystyle\int_0^{2\pi} \sin^2 x\,dx$

44. $\displaystyle\int_0^{2\pi} \cos^2 x\,dx$

45. $\displaystyle\int_0^8 e^x\,dx$

46. $\displaystyle\int_1^9 \ln x\,dx$

In Problems 47–50,

(a) *Approximate each definite integral by completing the table of Riemann sums using a regular partition of [a, b].*

n	10	50	100
Using a Left Riemann sum			
Using a Right Riemann sum			
Using a trapezoidal sum			

(b) *Use technology to find each definite integral.*

47. $\displaystyle\int_1^5 (2+\sqrt{x})\,dx$

48. $\displaystyle\int_{-1}^3 (e^x + e^{-x})\,dx$

49. $\displaystyle\int_{-1}^1 \frac{3}{1+x^2}\,dx$

50. $\displaystyle\int_0^2 \frac{1}{\sqrt{x^2+4}}\,dx$

Applications and Extensions

In Problems 51–54, for the given function f:

(a) *Graph f.*

(b) *Express the area under the graph of f as an integral.*

(c) *Evaluate the integral using geometry.*

51. $f(x)=\sqrt{9-x^2}, 0\le x\le 3$

52. $f(x)=\sqrt{25-x^2}, -5\le x\le 5$

53. $f(x)=3-\sqrt{6x-x^2}, 0\le x\le 6$

54. $f(x)=\sqrt{4x-x^2}+2, 0\le x\le 4$

55. Units of an Integral In the definite integral $\int_0^5 F(x)\,dx$, F represents a force measured in newtons and x, $0\le x\le 5$, is measured in meters. What are the units of $\int_0^5 F(x)\,dx$?

56. Units of an Integral In the definite integral $\int_0^{50} C(x)\,dx$, C represents the concentration of a drug in grams per liter and x, $0\le x\le 50$, is measured in liters of alcohol. What are the units of $\int_0^{50} C(x)\,dx$?

57. Units of an Integral In the definite integral $\int_a^b v(t)\,dt$, v represents velocity measured in meters per second and time t is measured in seconds. What are the units of $\int_a^b v(t)\,dt$?

58. Units of an Integral In the definite integral $\int_a^b S(t)\,dt$, S represents the rate of sales of a corporation measured in millions of dollars per year and time t is measured in years. What are the units of $\int_a^b S(t)\,dt$?

Challenge Problems

 59. Consider the **Dirichlet function** f, where

$$f(x) = \begin{cases} 1 & \text{if } x \text{ is rational} \\ 0 & \text{if } x \text{ is irrational} \end{cases}$$

Show that $\int_0^1 f(x)\,dx$ does not exist. *Hint*: Evaluate the Riemann sums in two different ways: first by using rational numbers for u_i and then by using irrational numbers for u_i.

60. It can be shown (with a certain amount of work) that if $f(x)$ is integrable on the interval $[a, b]$, then so is $|f(x)|$. Is the converse true?

61. If only regular partitions are allowed, then we could not always partition an interval $[a, b]$ in a way that automatically partitions subintervals $[a, c]$ and $[c, b]$ for $a < c < b$. Why not?

Preparing for the AP® Exam

AP® Practice Problems

Multiple-Choice Questions

 1. For the function $f(x) = 2 - 3x, 0 \le x \le 4$, the interval $[0, 4]$ is partitioned into four subintervals $[0, 1], [1, 2], [2, 3], [3, 4]$. The Riemann sum using the right endpoint of each subinterval equals

(A) -88 (B) -24 (C) -22 (D) -10

 2. A function f is continuous on the closed interval $[0, 30]$ and has the values given in the table.

x	0	10	20	30
$f(x)$	16	12	c	6

Using a trapezoidal sum with three subintervals of equal width, $\int_0^{30} f(x)\,dx \approx 310$. What is the value of c?

(A) 3   (B) 8   (C) 9   (D) 13

 3. Approximate $\int_0^6 (2x - 5)dx$ by partitioning the interval $[0, 6]$ into three subintervals each of width 2 and using a Right Riemann sum.

(A) -10 (B) 9   (C) 18   (D) 22

 4. $\int_0^8 (3 - x^2)dx =$

(A) $\lim\limits_{\Delta x \to 0} \sum\limits_{i=1}^{n} \left(3 - u_i\right)\Delta x$

(B) $\lim\limits_{n \to \infty} \sum\limits_{i=1}^{n} (3 - x^2)\dfrac{8}{n}$

(C) $\lim\limits_{n \to \infty} \sum\limits_{i=1}^{n} \left[3 - \left(\dfrac{8}{n}i\right)^2\right]\dfrac{8}{n}$

(D) $\lim\limits_{n \to \infty} \sum\limits_{i=1}^{n} \left[3 - \left(\dfrac{2}{n}i\right)^2\right]\dfrac{2}{n}$

 5. A function f is continuous on the closed interval $[0, 10]$ and has values

x	0	1	4	8	10
$f(x)$	4	5	10	12	8

Find an approximation to $\int_0^{10} f(x)\,dx$ using a Right Riemann sum with the four subintervals $[0, 1], [1, 4], [4, 8], [8, 10]$.

(A) 70   (B) 99   (C) 83   (D) 62

6. $\lim\limits_{n \to \infty} \dfrac{1}{n}\left[2\left(\dfrac{1}{n}\right)^{2/3} + 2\left(\dfrac{2}{n}\right)^{2/3} + 2\left(\dfrac{3}{n}\right)^{2/3} + \cdots + 2\left(\dfrac{n}{n}\right)^{2/3}\right] =$

(A) $\int_0^1 2x^{2/3}dx$ (B) $\int_0^2 x^{2/3}dx$

(C) $2\int_0^1 \left(\dfrac{1}{x}\right)^{2/3} dx$ (D) $\dfrac{2}{n^{2/3}}\int_0^1 dx$

7. The expression

$$\dfrac{2}{25}\left[\sqrt{\dfrac{2}{25}} + \sqrt{\dfrac{4}{25}} + \sqrt{\dfrac{6}{25}} + \cdots + \sqrt{\dfrac{48}{25}} + \sqrt{\dfrac{50}{25}}\right]$$

is a Riemann sum approximation for

(A) $\dfrac{1}{25}\int_0^2 \sqrt{x}dx$ (B) $\int_0^2 \sqrt{x}dx$

(C) $\int_0^{50} \sqrt{x}dx$ (D) $\int_0^1 \sqrt{2x}dx$

8. The integral $\int_{-2}^4 (x^3 - 4)dx$ is approximated by partitioning the closed interval $[-2, 4]$ into three subintervals of equal width and using a Right Riemann sum. Which of the following is true?

(A) The Right Riemann sum $= -24$; it underestimates $\int_{-2}^4 (x^3 - 4)dx$.

(B) The Right Riemann sum $= 60$; it overestimates $\int_{-2}^4 (x^3 - 4)dx$.

(C) The Right Riemann sum $= 120$; it overestimates $\int_{-2}^4 (x^3 - 4)dx$.

(D) The Right Riemann sum $= 120$; there is not enough information to determine whether it overestimates or underestimates $\int_{-2}^4 (x^3 - 4)dx$.

(continued on the next page)

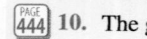

9. Which of the following is equivalent to $\int_4^8 \sqrt[3]{x}\,dx$?

(A) $\displaystyle\lim_{n\to\infty} \sum_{i=1}^{n} \sqrt[3]{4 + \frac{4}{n}i} \cdot \frac{4}{n}$

(B) $\displaystyle\lim_{n\to\infty} \sum_{i=1}^{n} \sqrt[3]{\frac{4}{n}i} \cdot \frac{8}{n}$

(C) $\displaystyle\lim_{n\to\infty} \sum_{i=1}^{n} \sqrt[3]{4 + \frac{4}{n}i} \cdot \frac{8}{n}$

(D) $\displaystyle\lim_{n\to\infty} \sum_{i=1}^{n} \sqrt[3]{\frac{4}{n}i} \cdot \frac{4}{n}$

Free-Response Questions

10. The graph of $f(x) = \sqrt{6x - x^2}$ is shown below.

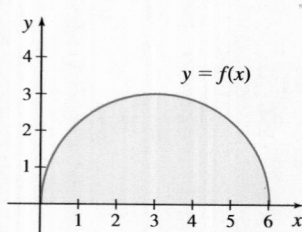

(a) Approximate the area under the graph of $f(x) = \sqrt{6x - x^2}$ using a trapezoidal sum with three subintervals of equal width.

(b) Express the area under the graph of f as a definite integral.

(c) Evaluate the integral using geometry.

11. The graph of $f(x) = (x-2)^3 - 2x + 10$ is shown below.

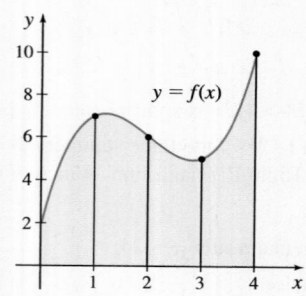

(a) Approximate the area under the graph of f using a Left Riemann sum with $n = 4$ subintervals of width 1.

(b) Express the area under the graph of f as a definite integral.

(c) Use technology to find the area under the graph of f.

12. The table below gives values of a function f that is continuous on the interval $[0, 10]$. The graph of f is increasing and concave up on $(0, 10)$.

x	0	2	5	7	10
$f(x)$	3	4	6	9	15

(a) Use the values in the table to approximate $\int_0^{10} f(x)\,dx$ using a Right Riemann sum with four subintervals.

(b) Does the approximation in (a) overestimate or underestimate $\int_0^{10} f(x)\,dx$? Justify your answer.

(c) Use the values in the table to approximate $\int_0^{10} f(x)\,dx$ using a trapezoidal sum with four subintervals.

(d) Does the approximation in (c) overestimate or underestimate $\int_0^{10} f(x)\,dx$? Justify your answer.

13. Volume In the table below, S is the area in square meters of the cross section of a railroad track cutting through a mountain, and x meters is the corresponding distance along the line. See the picture below. The volume V of earth removed to make the cutting from $x = 0$ to $x = 150$ is given by $V = \int_0^{150} S\,dx$.

x	0	25	50	75	100	125	150
S	105	118	142	120	110	90	78

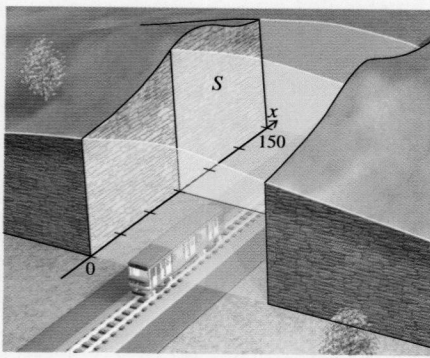

Approximate V using six subintervals of equal width and

(a) a Left Riemann sum.

(b) a Right Riemann sum.

(c) a trapezoidal sum.

Retain Your Knowledge

Multiple-Choice Questions

1. The derivative of $y = x \ln(4x)$ is

(A) $y = \ln(4x)$ (B) $y' = \ln(4x) + \dfrac{1}{4}$

(C) $y' = \dfrac{1}{x}$ (D) $y' = 1 + \ln(4x)$

2. The function f is continuous on the closed interval $[-5, 5]$. The graph of its derivative f' is shown below.

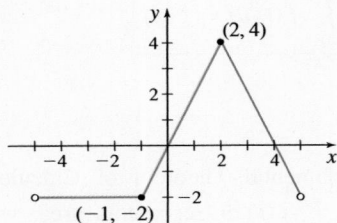

Based on this graph, which of the following must be true about f?

I. The function f is increasing on the interval $-1 \le x \le 2$.

II. The function f has a local minimum at $x = 0$.

III. The graph of f has an inflection point at $x = 2$.

(A) I only   (B) II only

(C) II and III only   (D) I, II, and III

3. An equation of the tangent line to the graph of $f(x) = \cos\left(\dfrac{1}{3}x\right)$ at $x = 2\pi$ is

(A) $y - \dfrac{1}{2} = -\dfrac{\sqrt{3}}{2}(x - 2\pi)$ (B) $y + \dfrac{1}{2} = -\dfrac{\sqrt{3}}{6}(x - 2\pi)$

(C) $y + \dfrac{1}{2} = \dfrac{\sqrt{3}}{6}(x - 2\pi)$ (D) $y + \dfrac{1}{2} = -\dfrac{\sqrt{3}}{2}(x - 2\pi)$

Free-Response Question

4. The graph of f is shown below.

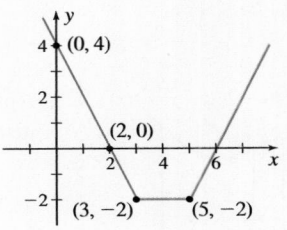

Find $\displaystyle \lim_{x \to 0} \frac{4e^x - f(x)}{\tan x + [f(x+2)]^2}$. Justify your work.

6.3 The Fundamental Theorem of Calculus

OBJECTIVES *When you finish this section, you should be able to:*

1 Use Part 1 of the Fundamental Theorem of Calculus (p. 452)

2 Use Part 2 of the Fundamental Theorem of Calculus (p. 454)

3 Interpret the integral of a rate of change (p. 456)

4 Interpret the integral as an accumulation function (p. 458)

In this section, we discuss the Fundamental Theorem of Calculus, a method for finding integrals more easily, avoiding the need to find the limit of Riemann sums. The Fundamental Theorem is aptly named because it links the two branches of calculus: differential calculus and integral calculus. As it turns out, the Fundamental Theorem of Calculus has two parts, each of which relates an integral to an antiderivative.

NEED TO REVIEW? Antiderivatives are discussed in Section 5.6, pp. 405–408.

We begin with a function f that is continuous on a closed interval $[a, b]$. Then the definite integral $\int_a^b f(x)\, dx$ exists and is equal to a real number. Now if x denotes any number in $[a, b]$, the definite integral $\int_a^x f(t)\, dt$ exists and depends on x. That is, $\int_a^x f(t)\, dt$ is a function of x, which we name I, for "integral."

$$I(x) = \int_a^x f(t)\, dt$$

The domain of I is the closed interval $[a, b]$. The integral that defines I has a *variable upper limit of integration x*. The t that appears in the integrand is a dummy variable. Part 1 of the Fundamental Theorem of Calculus states that if we differentiate I with respect to x, we get back the original function f. That is, $\int_a^x f(t)\, dt$ *is an antiderivative of f.*

THEOREM Fundamental Theorem of Calculus, Part 1

Suppose f is a function that is continuous on a closed interval $[a, b]$. The function I defined by

$$I(x) = \int_a^x f(t)\,dt$$

has the properties that it is continuous on $[a, b]$ and differentiable on (a, b). Moreover,

$$I'(x) = \frac{d}{dx} \int_a^x f(t)\,dt = f(x)$$

for all x in (a, b).

The proof of Part 1 of the Fundamental Theorem of Calculus is given in Appendix B. However, if the integral $\int_a^x f(t)\,dt$ represents area, we can provide justification for the theorem using geometry.

Figure 23 shows the graph of a function f that is nonnegative and continuous on a closed interval $[a, b]$. Then $I(x) = \int_a^x f(t)\,dt$ equals the area under the graph of f from a to x.

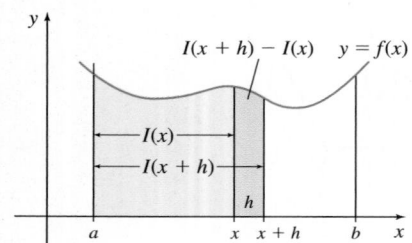

Figure 23

$$I(x) = \int_a^x f(t)\,dt = \text{the area under the graph of } f \text{ from } a \text{ to } x$$

$$I(x+h) = \int_a^{x+h} f(t)\,dt = \text{the area under the graph of } f \text{ from } a \text{ to } x+h$$

$$I(x+h) - I(x) = \text{the area under the graph of } f \text{ from } x \text{ to } x+h$$

$$\frac{I(x+h) - I(x)}{h} = \frac{\text{the area under the graph of } f \text{ from } x \text{ to } x+h}{h} \qquad (1)$$

Based on the definition of a derivative,

$$\lim_{h \to 0} \left[\frac{I(x+h) - I(x)}{h} \right] = I'(x) \qquad (2)$$

Since f is continuous, $\lim_{h \to 0} f(x+h) = f(x)$. As $h \to 0$, the area under the graph of f from x to $x+h$ gets closer to the area of a rectangle with width h and height $f(x)$. That is,

$$\lim_{h \to 0} \frac{\text{the area under the graph of } f \text{ from } x \text{ to } x+h}{h} = \lim_{h \to 0} \frac{h[f(x)]}{h} = f(x) \qquad (3)$$

Combining (1), (2), and (3), it follows that $I'(x) = f(x)$.

1 Use Part 1 of the Fundamental Theorem of Calculus

EXAMPLE 1   Using Part 1 of the Fundamental Theorem of Calculus

(a) $\dfrac{d}{dx} \displaystyle\int_0^x \sqrt{t+1}\,dt = \sqrt{x+1}$ **(b)** $\dfrac{d}{dx} \displaystyle\int_2^x \dfrac{s^3 - 1}{2s^2 + s + 1}\,ds = \dfrac{x^3 - 1}{2x^2 + x + 1}$ ∎

NOW WORK   Problem 5 and AP® Practice Problems 2, 7, 8, and 12.

CALC CLIP **EXAMPLE 2** **Using Part 1 of the Fundamental Theorem of Calculus**

Find $\dfrac{d}{dx} \displaystyle\int_4^{3x^2+1} \sqrt{e^t + t}\, dt$.

NEED TO REVIEW? The Chain Rule is discussed in Section 3.1, pp. 235–243.

Solution

The upper limit of integration is a function of x, so we use the Chain Rule along with Part 1 of the Fundamental Theorem of Calculus.

Let $y = \displaystyle\int_4^{3x^2+1} \sqrt{e^t + t}\, dt$ and $u(x) = 3x^2 + 1$. Then $y = \displaystyle\int_4^{u} \sqrt{e^t + t}\, dt$ and

$$\dfrac{d}{dx} \int_4^{3x^2+1} \sqrt{e^t + t}\, dt = \underbrace{\dfrac{dy}{dx} = \dfrac{dy}{du} \cdot \dfrac{du}{dx}}_{\text{Chain Rule}} = \left[\dfrac{d}{du} \int_4^{u} \sqrt{e^t + t}\, dt\right] \cdot \dfrac{du}{dx}$$

$$= \underbrace{\sqrt{e^u + u} \cdot \dfrac{du}{dx}}_{\substack{\text{Use the Fundamental} \\ \text{Theorem.}}} = \underbrace{\sqrt{e^{(3x^2+1)} + 3x^2 + 1} \cdot 6x}_{u = 3x^2 + 1;\ \frac{du}{dx} = 6x}$$

■

NOW WORK Problem 11 and AP® Practice Problems 6 and 9.

EXAMPLE 3 **Using Part 1 of the Fundamental Theorem of Calculus**

Find $\dfrac{d}{dx} \displaystyle\int_{x^3}^{5} (t^4 + 1)^{1/3}\, dt$.

Solution

To use Part 1 of the Fundamental Theorem of Calculus, the variable must be part of the upper limit of integration. So, we use the fact that $\int_a^b f(x)\, dx = -\int_b^a f(x)\, dx$ to interchange the limits of integration.

$$\dfrac{d}{dx} \int_{x^3}^{5} (t^4 + 1)^{1/3}\, dt = \dfrac{d}{dx}\left[-\int_5^{x^3} (t^4 + 1)^{1/3}\, dt\right] = -\dfrac{d}{dx} \int_5^{x^3} (t^4 + 1)^{1/3}\, dt$$

Now use the Chain Rule. Let $y = \displaystyle\int_5^{x^3} (t^4 + 1)^{1/3} dt$ and $u(x) = x^3$.

$$\dfrac{d}{dx} \int_{x^3}^{5} (t^4 + 1)^{1/3}\, dt = -\dfrac{d}{dx} \int_5^{x^3} (t^4 + 1)^{1/3}\, dt = \underbrace{-\dfrac{dy}{dx} = -\dfrac{dy}{du} \cdot \dfrac{du}{dx}}_{\text{Chain Rule}}$$

$$= -\dfrac{d}{du} \int_5^{u} (t^4 + 1)^{1/3} dt \cdot \dfrac{du}{dx}$$

$$= -(u^4 + 1)^{1/3} \cdot \dfrac{du}{dx} \qquad \text{Use Part 1 of the Fundamental Theorem.}$$

$$= -(x^{12} + 1)^{1/3} \cdot 3x^2 \qquad u = x^3;\ \dfrac{du}{dx} = 3x^2$$

$$= -3x^2(x^{12} + 1)^{1/3}$$

■

NOTE In these examples, the differentiation is with respect to the variable that appears in the upper or lower limit of integration, and the answer is a function of that variable.

NOW WORK Problem 15.

EXAMPLE 4 **Using Part I of the Fundamental Theorem of Calculus**

Functions f and g are defined as

$$f(x) = \int_0^{6x} \sqrt{t^2 + 1}\, dt \qquad g(x) = f(3 \cos x)$$

(a) Find $f'(x)$.

(b) Find $g'\left(\dfrac{\pi}{2}\right)$.

(c) Express the maximum value of f on $0 \le x \le \pi$ as an integral. Do not evaluate.

Solution

(a) $f'(x) = \dfrac{d}{dx} \displaystyle\int_0^{6x} \sqrt{t^2 + 1}\, dt = 6\sqrt{36x^2 + 1}$

(b) $g'(x) = \dfrac{d}{dx} f(3\cos x) = -3\sin x f'(3\cos x)$

$g'\left(\dfrac{\pi}{2}\right) = -3\sin\dfrac{\pi}{2} f'\left(3\cos\dfrac{\pi}{2}\right) = -3 \cdot 1 \cdot f'(0) = -3 \cdot 6 = -18$

(c) From (a), we conclude $f'(x) > 0$ for $0 < x < \pi$. By the Increasing/Decreasing Test, f is increasing for $0 < x < \pi$. So, the maximum value of f occurs at $x = \pi$ and equals $f(\pi) = \displaystyle\int_0^{6\pi} \sqrt{t^2 + 1}\, dt.$ ∎

NOW WORK AP® Practice Problems 14 and 19.

② Use Part 2 of the Fundamental Theorem of Calculus

Part 1 of the Fundamental Theorem of Calculus establishes a relationship between the derivative and the definite integral. Part 2 of the Fundamental Theorem of Calculus provides a method for finding a definite integral without using Riemann sums.

> **THEOREM Fundamental Theorem of Calculus, Part 2**
>
> Suppose f is a function that is continuous on a closed interval $[a, b]$. If F is any antiderivative of f on $[a, b]$, then
>
> $$\boxed{\int_a^b f(x)\, dx = F(b) - F(a)}$$

IN WORDS If $F'(x) = f(x)$, then $\int_a^b f(x)\, dx = F(b) - F(a)$. That is, if you can find an antiderivative of f, you can integrate f without using Riemann sums.

Proof Let $I(x) = \int_a^x f(t)\, dt$. Then from Part 1 of the Fundamental Theorem of Calculus, $I = I(x)$ is continuous for $a \leq x \leq b$ and differentiable for $a < x < b$. So,

$$\frac{d}{dx} \int_a^x f(t)\, dt = f(x) \qquad a < x < b$$

RECALL Any two antiderivatives of a function differ by a constant.

That is, $\int_a^x f(t)\, dt$ is an antiderivative of f. So, if F is any antiderivative of f, then

$$F(x) = \int_a^x f(t)\, dt + C$$

where C is some constant. Since F is continuous on $[a, b]$, we have

$$F(a) = \int_a^a f(t)\, dt + C \qquad F(b) = \int_a^b f(t)\, dt + C$$

NOTE For any constant C, $[F(b) + C] - [F(a) + C] = F(b) - F(a)$. In other words, it does not matter which antiderivative of f is chosen when using Part 2 of the Fundamental Theorem of Calculus, since the same answer is obtained for every antiderivative.

Since, $\int_a^a f(t)\, dt = 0$, subtracting $F(a)$ from $F(b)$ gives

$$F(b) - F(a) = \int_a^b f(t)\, dt$$

Since t is a dummy variable, we can replace t by x and the result follows. ∎

As an aid in computation, we introduce the notation

$$\boxed{\int_a^b f(x)\, dx = \Big[F(x)\Big]_a^b = F(b) - F(a)}$$

The notation $\Big[F(x)\Big]_a^b$ also suggests that to find $\int_a^b f(x)\, dx$, we first find an antiderivative $F(x)$ of $f(x)$. Then we write $\Big[F(x)\Big]_a^b$ to represent $F(b) - F(a)$.

NEED TO REVIEW? Table 8 in Section 5.6, p. 407, provides a list of common antiderivatives.

EXAMPLE 5 Using Part 2 of the Fundamental Theorem of Calculus

Use Part 2 of the Fundamental Theorem of Calculus to find:

(a) $\int_{-2}^{1} x^2\, dx$ **(b)** $\int_{0}^{\pi/6} \cos x\, dx$ **(c)** $\int_{1}^{2} \frac{1}{x}\, dx$ **(d)** $\int_{0}^{\sqrt{3}/2} \frac{1}{\sqrt{1-x^2}}\, dx$

Solution

(a) The function $f(x)=x^2$ is continuous on the closed interval $[-2,1]$. An antiderivative of f is $F(x)=\dfrac{x^3}{3}$. By Part 2 of the Fundamental Theorem of Calculus,

$$\int_{-2}^{1} x^2\, dx = \left[\frac{x^3}{3}\right]_{-2}^{1} = \frac{1^3}{3} - \frac{(-2)^3}{3} = \frac{1}{3} + \frac{8}{3} = \frac{9}{3} = 3$$

(b) The function $f(x)=\cos x$ is continuous on the closed interval $\left[0, \dfrac{\pi}{6}\right]$. An antiderivative of f is $F(x)=\sin x$. By Part 2 of the Fundamental Theorem of Calculus,

$$\int_{0}^{\pi/6} \cos x\, dx = \left[\sin x\right]_{0}^{\pi/6} = \sin \frac{\pi}{6} - \sin 0 = \frac{1}{2}$$

(c) The function $f(x)=\dfrac{1}{x}$ is continuous on the closed interval $[1,2]$. An antiderivative of f is $F(x)=\ln|x|$. By Part 2 of the Fundamental Theorem of Calculus,

$$\int_{1}^{2} \frac{1}{x}\, dx = \left[\ln|x|\right]_{1}^{2} = \ln 2 - \ln 1 = \ln 2 - 0 = \ln 2$$

(d) The function $f(x)=\dfrac{1}{\sqrt{1-x^2}}$ is continuous on the closed interval $\left[0, \dfrac{\sqrt{3}}{2}\right]$. An antiderivative of f is $F(x)=\sin^{-1} x$, provided $|x|<1$. By Part 2 of the Fundamental Theorem of Calculus,

$$\int_{0}^{\sqrt{3}/2} \frac{1}{\sqrt{1-x^2}}\, dx = \left[\sin^{-1} x\right]_{0}^{\sqrt{3}/2} = \sin^{-1}\frac{\sqrt{3}}{2} - \sin^{-1} 0 = \frac{\pi}{3} - 0 = \frac{\pi}{3} \quad\blacksquare$$

NOW WORK Problems 25 and 31 and AP® Practice Problems 1, 3, 4, 5, and 15.

EXAMPLE 6 Finding the Area under a Graph

Find the area under the graph of $f(x)=e^x$ from -1 to 1.

Solution

Figure 24 shows the graph of $f(x)=e^x$ on the closed interval $[-1,1]$.

Since f is continuous on the closed interval $[-1,1]$, the area A under the graph of f from -1 to 1 is given by

$$A = \int_{-1}^{1} e^x\, dx = \left[e^x\right]_{-1}^{1} = e^1 - e^{-1} = e - \frac{1}{e} \quad\blacksquare$$

NOW WORK Problem 53 and AP® Practice Problem 13.

Figure 24 $f(x)=e^x,\ -1\le x\le 1$

3 Interpret the Integral of a Rate of Change

Part 2 of the Fundamental Theorem of Calculus states that if a function f is continuous on a closed interval $[a, b]$, and if F is an antiderivative of f, then

$$\int_a^b f(x)dx = F(b) - F(a)$$

But suppose the integrand is $f'(x)$. Then f is an antiderivative of f', so

$$\int_a^b f'(x)dx = f(b) - f(a)$$

or equivalently,

$$\boxed{f(b) = f(a) + \int_a^b f'(x)dx} \tag{4}$$

This formulation allows us to retrieve all the values of a function when its derivative is known and one value of the function is known.

EXAMPLE 7   **Finding the Value of a Function Given Its Derivative and a Value of the Function**

Suppose f is a differentiable function whose derivative is $f'(x) = 3x^2 + 4$. If $f(1) = 5$, find $f(3)$.

Solution

Since $f(1)$ is known, we use 1 as the lower limit of integration and use $f'(x)$ as the integrand. Since we want $f(3)$, we use 3 as the upper limit of integration. Now use equation (4).

$$f(3) = f(1) + \int_1^3 f'(x)dx = 5 + \int_1^3 (3x^2 + 4)dx$$

$$= 5 + [x^3 + 4x]_1^3$$

$$= 5 + (27 + 12) - (1 + 4)$$

$$= 39 \qquad\blacksquare$$

NOW WORK   AP® Practice Problems **16** and **17**.

EXAMPLE 8   **Using a Definite Integral with Motion on a Line**

The acceleration a of an object moving on a line is given by

$$a = a(t) = 10e^t \, \text{m/s}^2$$

where $t \geq 0$ is the time in seconds.

(a)   If the initial velocity of the object is $v_0 = v(0) = 150 \, \text{m/s}$, what is the velocity of the object at time t?

(b)   If the initial position of the object is $s(0) = 0$, what is the position of the object at $t = 2$?

Solution

(a) If v is the velocity of the object, then

$$v(t) = v(t_0) + \int_{t_0}^{t} v'(u)\, du$$

Since $v'(t) = \dfrac{dv}{dt} = a(t) = 10e^t$, for $t_0 = 0$ we have

$$v(t) = v(0) + \int_0^t 10e^u\, du$$
$$= 150 + 10\,[e^u]_0^t$$
$$= 150 + 10(e^t - 1) = 140 + 10e^t$$

At time t, the velocity of the object is $v(t) = (140 + 10e^t)\,\text{m/s}$.

(b) The position function $s = s(t)$ is given by

$$s(t) = s(t_0) + \int_{t_0}^{t} s'(u)\, du$$

But $s'(t) = \dfrac{ds}{dt} = v(t)$. For $t_0 = 0$, we have

$$s(t) = s(0) + \int_0^t v(u)\, du$$
$$= 0 + \int_0^t (140 + 10e^u)\, du$$
$$= 140\,[u]_0^t + 10\,[e^u]_0^t$$
$$= 140t + 10e^t - 10$$

For $t = 2$,

$$s(2) = 280 + 10e^2 - 10 = 270 + 10e^2$$

After 2 s, the object is $(270 + 10e^2)$ m to the right of 0. ∎

NOW WORK   Problem 65 and AP® Practice Problem 21.

Part 2 of the Fundamental Theorem of Calculus states that if f is a function that is continuous on a closed interval $[a, b]$, and if F is an antiderivative of f, then

$$\int_a^b f(x)\, dx = F(b) - F(a)$$

Since F is an antiderivative of f, then $F' = f$ so

$$\int_a^b F'(x)\, dx = F(b) - F(a)$$

Since the derivative $F'(x)$ is the rate of change of F, we have the following result.

> The integral from a to b of the rate of change of F equals the change in F from a to b. That is,
>
> $$\int_a^b F'(x)\, dx = F(b) - F(a) \qquad (5)$$

EXAMPLE 9 | **Interpreting an Integral Whose Integrand Is a Rate of Change**

The population of the United States is growing at the rate of $P'(t) = 2.867(1.005)^t$ million people per year, where t is the number of years since 2020.

(a) Find $\displaystyle\int_0^5 P'(t)\,dt$.

(b) Interpret $\displaystyle\int_0^5 P'(t)\,dt$ in the context of the problem.

(c) The population of the United States was approximately 336 million people in 2020. What is the projected population in 2025?

Source: U.S. Census Bureau

Solution

(a) We use the Fundamental Theorem of Calculus, Part 2.

$$\int_0^5 2.867(1.005)^t\,dt = 2.867\left[\frac{1.005^t}{\ln 1.005}\right]_0^5 \qquad \text{An antiderivative of } a^t \text{ is } \frac{a^t}{\ln a}.$$

$$= 2.867\left[\frac{1.005^5}{\ln 1.005} - \frac{1.005^0}{\ln 1.005}\right]$$

$$= 2.867\left[\frac{1.005^5 - 1}{\ln 1.005}\right] \approx 14.515$$

> **NOTE** Be aware of the units: $P'(t)$ is measured in population/year and dt is measured in years, so the product $P'(t)\,dt$ is measured in population.

(b) Since $P'(t)$ is the rate of change of the population P (in millions) with respect to time t, in years,

$$\int_0^5 P'(t)\,dt = P(5) - P(0)$$

is the projected change in the population (in millions) of the United States from 2020 ($t=0$) to 2025 ($t=5$). The population of the United States is projected to increase by 14.515 million people from 2020 to 2025.

(c) The projected population of the United States in 2025 is

$$P(5) = P(0) + \int_0^5 P'(t)\,dt = 336 + 14.515 = 350.515 \approx 351 \text{ million people} \quad \blacksquare$$

NOW WORK Problem 63 and AP® Practice Problem 22.

4 Interpret the Integral as an Accumulation Function

If f is a function that is continuous and nonnegative on an interval $[a, b]$, then the definite integral $\int_a^b f(x)\,dx$ exists and equals the area A under the graph of f from a to b. See Figure 25(a).

Suppose we define the function G to be

$$G(x) = \int_a^x f(t)\,dt \quad a \le x \le b$$

Since the independent variable x of G is the upper limit of integration of the definite integral, we can interpret G as the area under the graph of f from a to the number x, as shown in Figure 25(b). Notice that when $x = a$, then $G(a) = \int_a^a f(t)\,dt = 0$ and when $x = b$, then $G(b) = \int_a^b f(t)\,dt = A$. Notice also that as x increases from a to b, the function G also increases, accumulating more area as x goes from a to b. It is for this reason we can describe $G(x) = \int_a^x f(t)\,dt$ as an *accumulation function*.

Now suppose f is not required to be nonnegative on the interval $[a, b]$. Then f may be nonnegative on some subinterval(s) of $[a, b]$ and negative on others. We examine how this affects $G(x) = \int_a^x f(t)\,dt$.

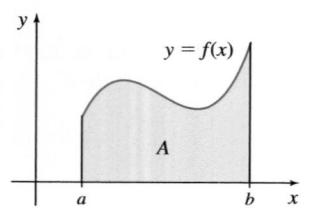

Figure 25 (a) $A = \displaystyle\int_a^b f(x)\,dx$

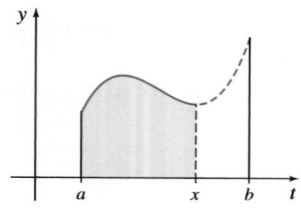

Figure 25 (b) $G(x) = \displaystyle\int_a^x f(t)\,dt,$ $a \le x \le b$

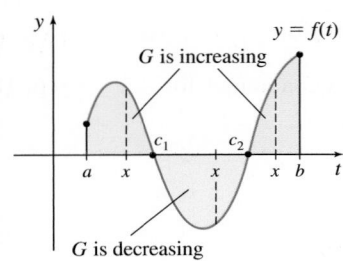

Figure 26 $G(x) = \displaystyle\int_a^x f(t)\,dt$

See Figure 26. On the interval $[a, c_1]$, the function f is nonnegative. As x increases from a to c_1, $G(x) = \int_a^x f(t)\,dt$ accumulates value and is increasing. On the interval $[c_1, c_2]$, $f(t) \leq 0$. As a result, the Riemann sums $\displaystyle\sum_{i=1}^n f(u_i)\Delta t_i$ for f on the interval $[c_1, c_2]$ are negative.

So as x increases from a to c_2, the accumulation function G increases from a to c_1 and then starts to decrease as x goes from c_1 to c_2, possibly becoming negative. Then from c_2 to b, the accumulation function G starts to increase again. The net accumulation from a to b, given by $\int_a^b f(t)\,dt$, can be a positive number, a negative number, or zero.

DEFINITION Accumulation Function

Suppose f is a function that is continuous on the closed interval $[a, b]$. The **accumulation function** G **associated with** f over the interval $[a, b]$ is defined as

$$G(x) = \int_a^x f(t)\,dt \qquad a \leq x \leq b$$

EXAMPLE 10 Interpreting an Integral as an Accumulation Function

The graph of $f(t) = \sin t$, $0 \leq t \leq 2\pi$ is shown in Figure 27.

(a) Find the accumulation function G associated with f over the interval $[0, 2\pi]$.

(b) Find $G\left(\dfrac{\pi}{2}\right)$, $G(\pi)$, $G\left(\dfrac{3\pi}{2}\right)$, and $G(2\pi)$.

(c) Interpret the results found in (b).

Solution

(a) The accumulation function G associated with $f(t) = \sin t$ over the interval $[0, 2\pi]$ is

$$G(x) = \int_0^x \sin t\,dt = \left[-\cos t\right]_0^x = -\cos x - (-\cos 0) = -\cos x + 1$$

(b) Now use $G(x) = -\cos x + 1$ to determine G at the required numbers.

$$G\left(\frac{\pi}{2}\right) = -\cos\frac{\pi}{2} + 1 = 1$$

$$G(\pi) = -\cos\pi + 1 = 1 + 1 = 2$$

$$G\left(\frac{3\pi}{2}\right) = -\cos\frac{3\pi}{2} + 1 = 0 + 1 = 1$$

$$G(2\pi) = -\cos(2\pi) + 1 = -1 + 1 = 0$$

(c) The sine function is positive over the interval $(0, \pi)$ and is negative over the interval $(\pi, 2\pi)$. The accumulation function G of $f(t) = \sin t$ increases from 0 to 1 to 2 as x increases from 0 to π and then decreases from 2 to 1 to 0 as x goes from π to 2π. The net accumulation over the interval $[0, 2\pi]$ equals 0. ∎

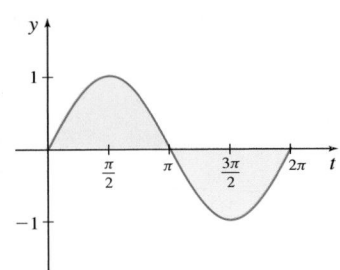

Figure 27 $f(t) = \sin t$, $0 \leq t \leq 2\pi$

NOW WORK   Problem 45 and AP® Practice Problems 11 and 18.

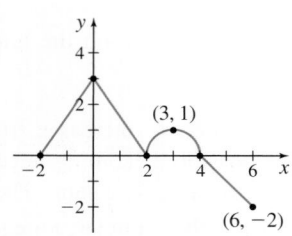

Figure 28 $y = f(x)$

EXAMPLE 11   Analyzing an Accumulation Function

Figure 28 shows the graph of the function f, which is continuous for $-2 \leq x \leq 6$. The function G is defined as $G(x) = \int_{-2}^{x} f(t)\, dt$.

(a) Find $G(0)$, $G(4)$, and $G(6)$.

(b) Determine where G is increasing or decreasing.

(c) Find any local extrema of G.

Solution

(a) $G(0) = \int_{-2}^{0} f(t)\, dt =$ area of the right triangle with base 2 and height $3 = \dfrac{1}{2} \cdot 2 \cdot 3 = 3$.

$G(4) = \int_{-2}^{4} f(t)\, dt =$ area of the triangle with base 4 and height 3 + area of the semicircle with radius 1

$$= \frac{1}{2} \cdot 4 \cdot 3 + \frac{1}{2} \cdot \pi \cdot 1^2 = 6 + \frac{1}{2}\pi$$

$$G(6) = G(4) + \int_{4}^{6} f(t)\, dt$$

The integral from 4 to 6 is represented by a right triangle with base 2 and height 2 that lies below the x-axis. Since the graph is below the x-axis, $\int_{4}^{6} f(t)\,dt$ is negative, so the area of the triangle $\left(\dfrac{1}{2} \cdot 2 \cdot 2 = 2 \right)$ is subtracted from $G(4)$. In other words, $\int_{4}^{6} f(t)\, dt = -\dfrac{1}{2} \cdot 2 \cdot 2 = -2$. Now,

$$G(6) = G(4) + \int_{4}^{6} f(t)\, dt = 6 + \frac{1}{2}\pi - 2 = 4 + \frac{1}{2}\pi$$

(b) We begin by finding $G'(x)$.

$$G'(x) = \frac{d}{dx} \int_{-2}^{x} f(t)\, dt = f(x)$$

Now $G'(x) > 0$ for $-2 < x < 4$ and $G'(x) < 0$ for $4 < x < 6$.

By the Increasing/Decreasing Function Test, G is increasing for $-2 \leq x \leq 4$, and G is decreasing for $4 \leq x \leq 6$.

(c) Using the First Derivative test, G has a local maximum at $x = 4$. From (a), the local maximum value is $6 + \dfrac{1}{2}\pi$. ∎

NOW WORK   AP® Practice Problems 10 and 20.

6.3 Assess Your Understanding

Concepts and Vocabulary

1. According to Part 1 of the Fundamental Theorem of Calculus, if a function f is continuous on a closed interval $[a, b]$, then $\dfrac{d}{dx} \int_{a}^{x} f(t)\, dt = $ _____ for all numbers x in (a, b).

2. *True or False* By Part 2 of the Fundamental Theorem of Calculus, $\int_{a}^{b} x\, dx = b - a$.

3. *True or False* By Part 2 of the Fundamental Theorem of Calculus, $\int_{a}^{b} f(x)\, dx = f(b) - f(a)$.

4. *True or False* $\int_{a}^{b} F'(x)\, dx$ can be interpreted as the rate of change in F from a to b.

Skill Building

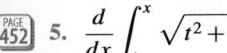

In Problems 5–18, find each derivative using Part 1 of the Fundamental Theorem of Calculus.

5. $\dfrac{d}{dx}\displaystyle\int_1^x \sqrt{t^2+1}\,dt$

6. $\dfrac{d}{dx}\displaystyle\int_3^x \dfrac{t+1}{t}\,dt$

7. $\dfrac{d}{dt}\displaystyle\int_0^t (3+x^2)^{3/2}\,dx$

8. $\dfrac{d}{dx}\displaystyle\int_{-4}^x \left(t^3+8\right)^{1/3}\,dt$

9. $\dfrac{d}{dx}\displaystyle\int_1^x \ln u\,du$

10. $\dfrac{d}{dt}\displaystyle\int_4^t e^x\,dx$

11. $\dfrac{d}{dx}\displaystyle\int_1^{2x^3} \sqrt{t^2+1}\,dt$

12. $\dfrac{d}{dx}\displaystyle\int_1^{\sqrt{x}} \sqrt{t^4+5}\,dt$

13. $\dfrac{d}{dx}\displaystyle\int_2^{x^5} \sec t\,dt$

14. $\dfrac{d}{dx}\displaystyle\int_3^{1/x} \sin^5 t\,dt$

15. $\dfrac{d}{dx}\displaystyle\int_x^5 \sin(t^2)\,dt$

16. $\dfrac{d}{dx}\displaystyle\int_x^3 (t^2-5)^{10}\,dt$

17. $\dfrac{d}{dx}\displaystyle\int_{5x^2}^5 (6t)^{2/3}\,dt$

18. $\dfrac{d}{dx}\displaystyle\int_{x^2}^0 e^{10t}\,dt$

In Problems 19–36, use Part 2 of the Fundamental Theorem of Calculus to find each definite integral.

19. $\displaystyle\int_{-2}^3 dx$

20. $\displaystyle\int_{-2}^3 2\,dx$

21. $\displaystyle\int_{-1}^2 x^3\,dx$

22. $\displaystyle\int_1^3 \dfrac{1}{x^3}\,dx$

23. $\displaystyle\int_0^1 \sqrt{u}\,du$

24. $\displaystyle\int_1^8 \sqrt[3]{y}\,dy$

25. $\displaystyle\int_{\pi/6}^{\pi/2} \csc^2 x\,dx$

26. $\displaystyle\int_0^{\pi/2} \cos x\,dx$

27. $\displaystyle\int_0^{\pi/4} \sec x \tan x\,dx$

28. $\displaystyle\int_{\pi/6}^{\pi/2} \csc x \cot x\,dx$

29. $\displaystyle\int_{-1}^0 e^x\,dx$

30. $\displaystyle\int_{-1}^0 e^{-x}\,dx$

31. $\displaystyle\int_1^e \dfrac{1}{x}\,dx$

32. $\displaystyle\int_e^1 \dfrac{1}{x}\,dx$

33. $\displaystyle\int_0^1 \dfrac{1}{1+x^2}\,dx$

34. $\displaystyle\int_0^{\sqrt{2}/2} \dfrac{1}{\sqrt{1-x^2}}\,dx$

35. $\displaystyle\int_{-1}^8 x^{2/3}\,dx$

36. $\displaystyle\int_0^4 x^{3/2}\,dx$

In Problems 37–42, find $\int_a^b f(x)\,dx$ over the domain of f indicated in the graph.

37.

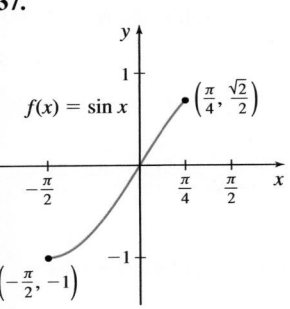

38.

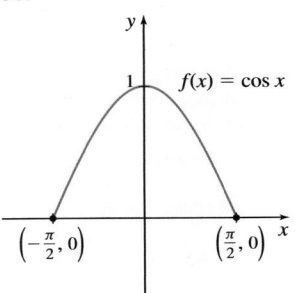

39.

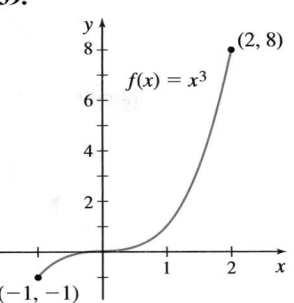

40.

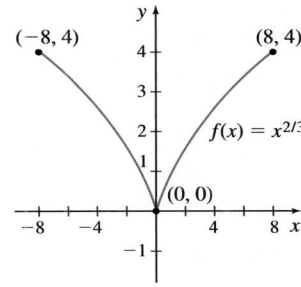

41.

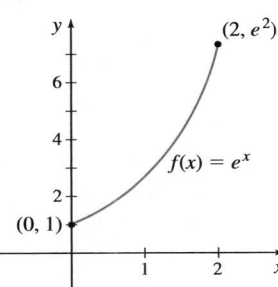

42.

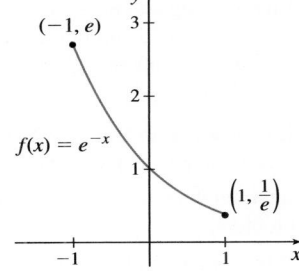

In Problems 43–50,

(a) Write the accumulation function G associated with f over the indicated interval as an integral.

(b) Find G at the given numbers.

(c) Graph G.

43. $f(t)=\cos t,\ 0\le t\le \dfrac{4\pi}{3};\ G\left(\dfrac{\pi}{6}\right),\ G\left(\dfrac{\pi}{2}\right),\ G\left(\dfrac{4\pi}{3}\right)$

44. $f(t)=\sin(2t),\ 0\le t\le \dfrac{3\pi}{2};\ G\left(\dfrac{\pi}{6}\right),\ G\left(\dfrac{\pi}{2}\right),\ G\left(\dfrac{3\pi}{2}\right)$

45. $f(t)=t^3,\ -2\le t\le 3;\ G(-1),\ G(0),\ G(3)$

46. $f(t)=\sqrt[3]{t},\ -8\le t\le 8;\ G(-1),\ G(0),\ G(8)$

47. $f(t)=4-t^2,\ -1\le t\le 3;\ G(0),\ G(2),\ G(3)$

48. $f(t)=8-t^3,\ 0\le t\le 4;\ G(1),\ G(2),\ G(4)$

49. $f(t)=\sqrt{t}-2,\ 0\le t\le 9;\ G(1),\ G(4),\ G(9)$

50. $f(t)=t^2-2t-3,\ 0\le t\le 5;\ G(1),\ G(3),\ G(5)$

Applications and Extensions

51. Given that $f(x)=(2x^3-3)^2$ and $f'(x)=12x^2(2x^3-3)$, find $\int_0^2 [12x^2(2x^3-3)]\,dx$.

52. Given that $f(x)=(x^2+5)^3$ and $f'(x)=6x(x^2+5)^2$, find $\int_{-1}^2 6x(x^2+5)^2\,dx$.

53. Area Find the area under the graph of $f(x)=\dfrac{1}{\sqrt{1-x^2}}$ from 0 to $\dfrac{1}{2}$.

54. Area Find the area under the graph of $f(x)=\sec x \tan x$ from 0 to $\dfrac{\pi}{4}$.

55. Area   Find the area under the graph of $f(x) = \dfrac{1}{x^2+1}$ from 0 to $\sqrt{3}$.

56. Area   Find the area under the graph of $f(x) = \dfrac{1}{1+x^2}$ from 0 to r, where $r > 0$. What happens as $r \to \infty$?

57. Area   Find the area under the graph of $y = \dfrac{1}{\sqrt{x}}$ from $x = 1$ to $x = r$, where $r > 1$. Then examine the behavior of this area as $r \to \infty$.

58. Area   Find the area under the graph of $y = \dfrac{1}{x^2}$ from $x = 1$ to $x = r$, where $r > 1$. Then examine the behavior of this area as $r \to \infty$.

59. Interpreting an Integral   The function $R = R(t)$ models the rate of sales of a corporation measured in millions of dollars per year as a function of the time t in years. Interpret the integral $\int_0^2 R(t)\, dt = 23$ in the context of the problem.

60. Interpreting an Integral   The function $v = v(t)$ models the speed v in meters per second of an object at time t in seconds. Interpret the integral $\int_0^{10} v(t)\, dt = 4.8$ in the context of the problem.

61. Interpreting an Integral   Helium is leaking from a large advertising balloon at a rate of $H(t)$ cubic centimeters per minute, where t is measured in minutes.

 (a) Write an integral that models the change in the amount of helium in the balloon over the interval $a \le t \le b$.
 (b) What are the units of the integral from (a)?
 (c) Interpret $\int_0^{300} H(t)\, dt = -100$ in the context of the problem.

62. Interpreting an Integral   Water is being added to a reservoir at a rate of $w(t)$ kiloliters per hour, where t is measured in hours.

 (a) Write an integral that models the change in the amount of water in the reservoir over the interval $a \le t \le b$.
 (b) What are the units of the integral from (a)?
 (c) Interpret $\int_0^{36} w(t)\, dt = 100$ in the context of the problem.

63. Population Growth   The growth rate of a colony of bacteria is $B'(t) = 3.455(1.259)^t$ grams per hour, where t is the number of hours since time $t = 0$.

 (a) Find $\int_0^6 B'(t)\, dt$.
 (b) Interpret $\int_0^6 B'(t)\, dt$ in the context of the problem.
 (c) If 15 grams of bacteria are present at time $t = 0$, how many grams of bacteria are present after 6 h?

64. Return on Investment   An investment in a hedge fund is growing at a continuous rate of $A'(t) = 1105.17(1.105)^t$ dollars per year.

 (a) Find $\int_0^{10} A'(t)\, dt$.
 (b) Interpret $\int_0^{10} A'(t)\, dt$ in the context of the problem.
 (c) If initially \$1000 is invested, what is the value of the investment after 10 years?

65. Increase in Revenue   The marginal revenue function $R'(x)$ for selling x units of a product is $R'(x) = \sqrt[3]{x}$ (in hundreds of dollars per unit).

 (a) Interpret $\int_a^b R'(x)\, dx$ in the context of the problem.
 (b) How much additional revenue is attained if sales increase from 40 to 50 units?

66. Increase in Cost   The marginal cost function $C'(x)$ of producing x thousand units of a product is $C'(x) = 2x + 6$ thousand dollars.

 (a) Interpret $\int_a^b C'(x)\, dx$ in the context of the problem.
 (b) What does it cost the company to increase production from 10,000 units to 13,000 units?

67. Free Fall   The speed v of an object dropped from rest is given by $v(t) = 9.8t$, where v is in meters per second and time t is in seconds.

 (a) Express the distance traveled in the first 5.2 s as an integral.
 (b) Find the distance traveled in 5.2 s.

68. Area   Find h so that the area under the graph of $y^2 = x^3$, $0 \le x \le 4$, $y \ge 0$, is equal to the area of a rectangle of base 4 and height h.

69. Area   If P is a polynomial that is positive for $x > 0$, and for each $k > 0$, the area under the graph of P from $x = 0$ to $x = k$ is $k^3 + 3k^2 + 6k$, find P.

70. Put It Together   If $f(x) = \displaystyle\int_0^x \dfrac{1}{\sqrt{t^3+2}}\, dt$, which of the following is *false*?

 (a) f is continuous at x for all $x \ge 0$.
 (b) $f(1) > 0$
 (c) $f(0) = \dfrac{1}{\sqrt{2}}$
 (d) $f'(1) = \dfrac{1}{\sqrt{3}}$

In Problems 71–74:
 (a) Use Part 2 of the Fundamental Theorem of Calculus to find each definite integral.
 (b) Determine whether the integrand is an even function, an odd function, or neither.
 (c) Can you make a conjecture about the definite integrals in (a) based on the analysis from (b)?

71. $\displaystyle\int_0^4 x^2\, dx$ and $\displaystyle\int_{-4}^4 x^2\, dx$ **72.** $\displaystyle\int_0^4 x^3\, dx$ and $\displaystyle\int_{-4}^4 x^3\, dx$

73. $\displaystyle\int_0^{\pi/4} \sec^2 x\, dx$ and $\displaystyle\int_{-\pi/4}^{\pi/4} \sec^2 x\, dx$

74. $\displaystyle\int_0^{\pi/4} \sin x\, dx$ and $\displaystyle\int_{-\pi/4}^{\pi/4} \sin x\, dx$

75. Area   Find c, $0 < c < 1$, so that the area under the graph of $y = x^2$ from 0 to c equals the area under the same graph from c to 1.

76. Put It Together   Let A be the area under the graph of $y = \dfrac{1}{x}$ from $x = m$ to $x = 2m$, $m > 0$. Which of the following is true about the area A?

 (a) A is independent of m.
 (b) A increases as m increases.
 (c) A decreases as m increases.
 (d) A decreases as m increases when $m < \dfrac{1}{2}$ and increases as m increases when $m > \dfrac{1}{2}$.
 (e) A increases as m increases when $m < \dfrac{1}{2}$ and decreases as m increases when $m > \dfrac{1}{2}$.

77. If F is a function whose derivative is continuous for all real x, find

$$\lim_{h \to 0} \frac{1}{h} \int_c^{c+h} F'(x)\, dx$$

78. Suppose the closed interval $\left[0, \dfrac{\pi}{2}\right]$ is partitioned into n subintervals, each of length Δx, and u_i is an arbitrary number in the subinterval $[x_{i-1}, x_i]$, $i = 1, 2, \ldots, n$. Explain why

$$\lim_{n \to \infty} \sum_{i=1}^{n} [(\cos u_i)\, \Delta x] = 1$$

79. The interval $[0, 4]$ is partitioned into n subintervals, each of width Δx, and a number u_i is chosen in the subinterval $[x_{i-1}, x_i]$, $i = 1, 2, \ldots, n$. Find

$$\lim_{n \to \infty} \sum_{i=1}^{n} (e^{u_i}\, \Delta x)$$

80. If u and v are differentiable functions and f is a continuous function, find a formula for

$$\frac{d}{dx} \int_{u(x)}^{v(x)} f(t)\, dt$$

81. Suppose that the graph of $y = f(x)$ contains the points $(0, 1)$ and $(2, 5)$. Find $\int_0^2 f'(x)\, dx$. (Assume that f' is continuous.)

82. If f' is continuous on the interval $[a, b]$, show that

$$\int_a^b f(x) f'(x)\, dx = \frac{1}{2}\left\{ [f(b)]^2 - [f(a)]^2 \right\}$$

Hint: Look at the derivative of $F(x) = \dfrac{[f(x)]^2}{2}$.

83. If f'' is continuous on the interval $[a, b]$, show that

$$\int_a^b x f''(x)\, dx = b f'(b) - a f'(a) - f(b) + f(a)$$

Hint: Look at the derivative of $F(x) = x f'(x) - f(x)$.

84. What conditions on f and f' guarantee that

$$f(x) = \int_0^x f'(t)\, dt?$$

85. Suppose that F is an antiderivative of f on the interval $[a, b]$. Partition $[a, b]$ into n subintervals, each of length

$$\Delta x_i = x_i - x_{i-1}, \ i = 1, 2, \ldots, n.$$

  (a) Apply the Mean Value Theorem for derivatives to F in each subinterval $[x_{i-1}, x_i]$ to show that there is a number u_i in the subinterval for which $F(x_i) - F(x_{i-1}) = f(u_i)\Delta x_i$.

  (b) Show that $\displaystyle\sum_{i=1}^{n} [F(x_i) - F(x_{i-1})] = F(b) - F(a)$.

  (c) Use parts (a) and (b) to explain why

$$\int_a^b f(x)\, dx = F(b) - F(a).$$

(In this alternate proof of Part 2 of the Fundamental Theorem of Calculus, the continuity of f is not assumed.)

86. Given $y = \sqrt{x^2 - 1}\,(4 - x)$, $1 \le x \le a$, for what number a will $\int_1^a y\, dx$ have a maximum value?

87. Find $a > 0$, so that the area under the graph of $y = x + \dfrac{1}{x}$ from a to $(a + 1)$ is minimum.

88. If n is a known positive integer, for what number c is

$$\int_1^c x^{n-1}\, dx = \frac{1}{n}$$

89. Let $f(x) = \displaystyle\int_0^x \frac{dt}{\sqrt{1 - t^2}}$, $0 < x < 1$.

  (a) Find $\dfrac{d}{dx} f(\sin x)$.

  (b) Is f one-to-one?

  (c) Does f have an inverse?

AP® Practice Problems

Multiple-Choice Questions

1. $\int_0^{\pi/4} \sec^2 x\, dx =$

(A) $-\dfrac{1}{2}$ (B) $\sqrt{2}$ (C) 1 (D) $\dfrac{1}{2}$

2. If $G(x) = \int_0^x \sin t\, dt$ then $G'\left(\dfrac{\pi}{2}\right)$ is

(A) 0 (B) $\dfrac{1}{2}$ (C) 1 (D) -1

3. $\int_0^x e^t\, dt =$

(A) e^x (B) $e^x - e$ (C) $e^x - 1$ (D) $\dfrac{e^x - 1}{\ln|x|}$

4. $\int_1^e \dfrac{1}{x^3}\, dx =$

(A) $\dfrac{e-1}{2e}$ (B) $\dfrac{e^2-1}{2}$ (C) $\dfrac{1-e^2}{2}$ (D) $\dfrac{e^2-1}{2e^2}$

5. If P is a polynomial function of degree n, what is the degree of the function $Q(x) = \int_0^x P(t)\, dt$?

(A) $n-1$ (B) n (C) $n+1$ (D) $2n$

6. If $F(x) = \int_1^{x^3+1} \sqrt{t^2 - 1}\, dt$, then $F'(x)$ equals

(A) $\sqrt{(x^3+1)^2 - 1}$ (B) $3x^2\sqrt{x^3}$

(C) x^3 (D) $3x^2\sqrt{(x^3+1)^2 - 1}$

7. If $x > 1$, then $\dfrac{d}{dx} \displaystyle\int_e^x \dfrac{1}{t}\, dt =$

(A) $\dfrac{1}{x}$ (B) $\ln x$ (C) $\ln x - 1$ (D) $\dfrac{1}{x} - \dfrac{1}{e}$

8. The graph of a function $y = g(x)$, where $a \le x \le b$, is shown below.

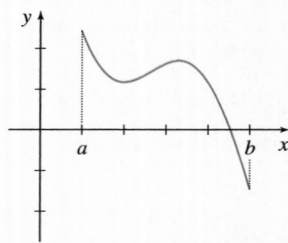

If $g(x) = \displaystyle\int_a^x f(t)\, dt$, then which of the following could be the graph of f on $[a, b]$?

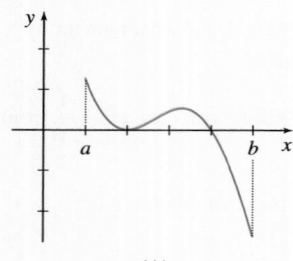

(A)

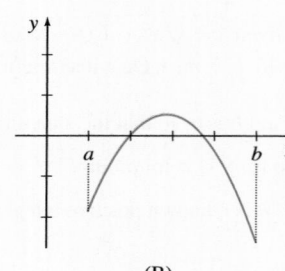

(B)

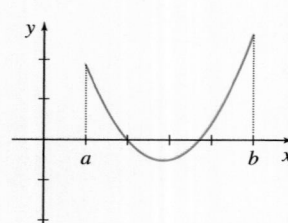

(C)

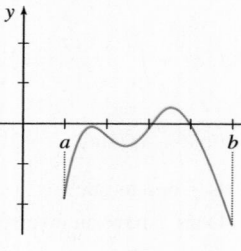

(D)

9. If $f(x) = \displaystyle\int_1^{3x^2} \sqrt{t^2 + 1}\, dt$, then $f'(-2)$ equals

(A) $-12\sqrt{145}$ (B) $-12\sqrt{37}$

(C) $12\sqrt{37}$ (D) $\sqrt{145}$

10. The graph of a function f is shown below. If

$$g(t) = \int_{-3}^t f(x)\, dx$$

for what number t is $g(t)$ the greatest?

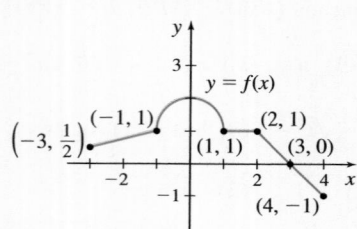

(A) 0   (B) 2   (C) 3   (D) 4

11. Use the function g and the graph shown in Problem 10 to find $g(2)$.

(A) $\dfrac{5 + \pi}{2}$ (B) $\dfrac{9 + \pi}{2}$ (C) $\dfrac{9\pi}{2}$ (D) $\dfrac{9}{2} + \pi$

12. $\dfrac{d}{dx} \displaystyle\int_0^x (\sin t - \cos t)\, dt =$

(A) $\cos x - \sin x$ (B) $-\cos x + \sin x$

(C) $-\cos t - \sin t$ (D) $-\cos x - \sin x$

13. The area under the graph of $f(x) = \sin x$ from 0 to k, $0 \le k \le \dfrac{\pi}{2}$, is 0.2. Then k equals

(A) 1.2   (B) 0.644   (C) 0.356   (D) 0.314

14. Suppose $g(x) = \displaystyle\int_0^x \sin\left(t - \dfrac{\pi}{2}\right) dt$ for $0 \le x \le \dfrac{3\pi}{2}$. On which interval is g increasing?

(A) $\left[0, \dfrac{\pi}{2}\right]$ (B) $[0, \pi]$

(C) $\left[\dfrac{\pi}{2}, \pi\right]$ (D) $\left[\dfrac{\pi}{2}, \dfrac{3\pi}{2}\right]$

15. If $f(x) = \displaystyle\int_0^x \dfrac{t - 5}{t^2 + 1}\, dt$, which of the following statements is false?

(A) f is continuous for all numbers $x \ge 0$.

(B) f is differentiable for all numbers $x > 0$.

(C) $f'(2) = -\dfrac{3}{5}$

(D) $f(0) = -5$

Free-Response Questions

16. The functions f and f' are continuous for all real numbers, and the graph of f contains the points $(0, 8)$ and $(4, 2)$. Find $\int_0^4 f'(x)\, dx$. Justify your answer.

17. f is a differentiable function. The derivative of f is $f'(x) = 5x^2 - 4$. If $f(1) = 5$, find $f(3)$. Justify your result.

18. A chemical compound is filtering through a triple filter and collects in a processing vat at a rate of $f(t)$ cubic centimeters per hour.

(a) Write an integral that models the accumulation of chemical compound in the vat during the time interval $a \le t \le b$.

(b) Interpret $\int_0^{24} f(t)\, dt$ in the context of the problem.

19. Functions f and g are defined as

$$f(x) = \int_0^{4x} \sqrt{t^2 + 9}\, dt \quad \text{and} \quad g(x) = f(3 \sin x)$$

(a) Find $f'(x)$ and $g'(x)$.

(b) Find an equation of the tangent line to the graph of g at $x = \pi$.

(c) Express the maximum value of g for $0 \le x \le \pi$ as an integral. Justify your answer.

20. The graph of a function f is shown below. The function g is defined by $g(x) = \int_{-4}^{x} f(t)\,dt$ for $-4 \le x \le 8$.

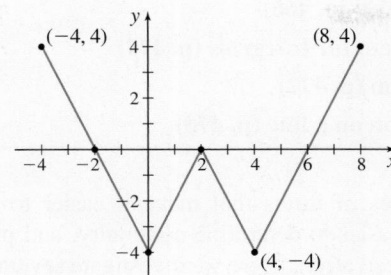

(a) Does g have a relative maximum, a relative minimum, or neither at $x = 2$? Justify your answer.

(b) Does the graph of g have a point of inflection at $x = 4$? Justify your answer.

(c) Find the absolute maximum value and the absolute minimum value of g on the interval $-4 \le x \le 8$. Justify your answers.

21. The acceleration a of an object moving along a line is given by $a = a(t) = 2\sin t + 2$ measured in meters^2 per second, $0 \le t \le 2\pi$.

(a) If the initial velocity v of the object is $v(0) = 20$ m/s, what is the velocity of the object at time t?

(b) If the initial position s of the object is $s(0) = 0$ m, what is the position of the object at time t?

22. A biologist is growing bacteria for a lab experiment. There are 10 mg of bacteria in a controlled environment when she changes the temperature. The amount P of bacteria then grows at a rate of $20(0.95)^t$ per hour, where t is the number of hours since the temperature changed.

(a) Write an integral that models the amount of bacteria at time $t \ge 0$.

(b) Find the amount after 24 hours.

Retain Your Knowledge

Multiple-Choice Questions

1. Suppose $2x^2 + y^2 = 36$, and x and y are differentiable functions of t. Find $\dfrac{dx}{dt}$ when $x = 6$, $y = -3$, and $\dfrac{dy}{dt} = 2$.

(A) $-\dfrac{1}{2}$ (B) $\dfrac{1}{4}$ (C) $\dfrac{1}{2}$ (D) 3

2. $r(x) = \dfrac{(f \circ g)(x)}{h(x)}$. Use the information in the table to determine $r'(3)$.

x	$f(x)$	$f'(x)$	$g(x)$	$g'(x)$	$h(x)$	$h'(x)$
1	5	4	0	-2	2	-3
2	7	$\dfrac{1}{3}$	2	-4	5	-2
3	10	$\dfrac{1}{2}$	1	2	3	0

(A) $\dfrac{20}{9}$ (B) $\dfrac{8}{3}$ (C) $\dfrac{4}{3}$ (D) 8

3. If $y = |5 - x^2|$, then $y' =$

(A) $|-2x|$

(B) $\dfrac{5 - x^2}{|5 - x^2|}$

(C) $\dfrac{-2x(5 - x^2)}{|5 - x^2|}$

(D) $-\dfrac{x}{|5 - x^2|}$

Free-Response Question

4. Suppose the differentiable function $y = f(x)$ is defined implicitly by the equation $x + xy + 2y^2 = 6$.

(a) Find an expression for the slope of the tangent line at any point (x, y) on the graph of f.

(b) Write an equation for the line tangent to the graph of f at the point $(2, 1)$.

(c) Find the coordinates of any other point on the graph of f with a slope equal to the slope at the point $(2, 1)$.

6.4 Applying Properties of the Definite Integral

OBJECTIVES *When you finish this section, you should be able to:*

1 Use properties of the definite integral (p. 466)
2 Work with the Mean Value Theorem for Integrals (p. 471)
3 Find the average value of a function (p. 472)
4 Interpret integrals involving motion on a line (p. 473)

We have seen that there are properties of limits that make it easier to find limits, properties of continuity that make it easier to determine continuity, and properties of derivatives that make it easier to find derivatives. Here we investigate several properties of the definite integral that will make it easier to find integrals.

1 Use Properties of the Definite Integral

Most of the properties in this section require that a function be integrable.

Remember, in Section 6.2 we stated two theorems that guarantee a function f is integrable.

- A function f that is continuous on a closed interval $[a, b]$ is integrable over $[a, b]$.

- A function f that is bounded on a closed interval $[a, b]$ and has at most a finite number of discontinuities on $[a, b]$ is integrable over $[a, b]$.

Proofs of most of the properties in this section require the use of the definition of the definite integral. However, if the condition is added that the integrand has an antiderivative, then the Fundamental Theorem of Calculus can be used to establish the properties. We use this added condition in the proofs included here.

NOTE From now on, we refer to Parts 1 and 2 of the Fundamental Theorem of Calculus simply as the **Fundamental Theorem of Calculus.**

> **THEOREM** The Integral of the Sum of Two Functions
>
> If two functions f and g are integrable on the closed interval $[a, b]$, then
>
> $$\int_a^b [f(x) + g(x)]\, dx = \int_a^b f(x)\, dx + \int_a^b g(x)\, dx \tag{1}$$

IN WORDS The integral of a sum equals the sum of the integrals.

Proof For the proof, we assume that the functions f and g are continuous on $[a, b]$ and that F and G are antiderivatives of f and g, respectively, on (a, b). Then $F' = f$ and $G' = g$ on (a, b).

Also, since $(F + G)' = F' + G' = f + g$, then $F + G$ is an antiderivative of $f + g$ on (a, b). Now

$$\int_a^b [f(x) + g(x)]\, dx = \left[F(x) + G(x) \right]_a^b = [F(b) + G(b)] - [F(a) + G(a)]$$

$$= [F(b) - F(a)] + [G(b) - G(a)]$$

$$= \int_a^b f(x)\, dx + \int_a^b g(x)\, dx \qquad \blacksquare$$

EXAMPLE 1 Using Property (1) of the Definite Integral

$$\int_0^1 (x^2 + e^x)\, dx = \int_0^1 x^2 dx + \int_0^1 e^x\, dx = \left[\frac{x^3}{3}\right]_0^1 + \left[e^x\right]_0^1$$

$$= \left[\frac{1^3}{3} - 0\right] + \left[e^1 - e^0\right] = \frac{1}{3} + e - 1 = e - \frac{2}{3}\qquad\blacksquare$$

NOW WORK   Problem 17 and AP® Practice Problem 6.

THEOREM The Integral of a Constant Times a Function

If a function f is integrable on the closed interval $[a, b]$ and k is a constant, then

$$\int_a^b kf(x)\, dx = k \int_a^b f(x)\, dx \qquad (2)$$

IN WORDS A constant factor can be factored out of an integral.

You are asked to prove this theorem in Problem 111 if f is continuous on the interval $[a, b]$ and f has an antiderivative on the interval (a, b).

EXAMPLE 2 Using Property (2) of the Definite Integral

$$\int_1^e \frac{3}{x}\, dx = 3 \int_1^e \frac{1}{x} dx = 3\left[\ln |x|\right]_1^e = 3(\ln e - \ln 1) = 3(1 - 0) = 3$$
$$\uparrow$$
$$\text{Property (2)}$$
$$\blacksquare$$

NOW WORK   Problem 19 and AP® Practice Problems 3, 10, and 16.

Properties (1) and (2) can be extended as follows:

THEOREM

Suppose each of the functions $f_1, f_2, \ldots, f_n$ is integrable on the closed interval $[a, b]$. If $k_1, k_2, \ldots, k_n$ are constants, then

$$\int_a^b [k_1 f_1(x) + k_2 f_2(x) + \cdots + k_n f_n(x)]\, dx$$

$$= k_1 \int_a^b f_1(x)\, dx + k_2 \int_a^b f_2(x)\, dx + \cdots + k_n \int_a^b f_n(x)\, dx \qquad (3)$$

You are asked to prove this theorem in Problem 104 if each function is continuous on the interval $[a, b]$ and each function has an antiderivative on the interval (a, b).

EXAMPLE 3 Using Division and Property (3) of the Definite Integral

Find $\displaystyle\int_1^2 \frac{3x^3 - 6x^2 - 5x + 4}{2x}\, dx.$

Solution

The function $f(x) = \dfrac{3x^3 - 6x^2 - 5x + 4}{2x}$ is continuous on the closed interval $[1, 2]$ so it is integrable on $[1, 2]$. Divide and use properties of the definite integral.

$$\int_1^2 \frac{3x^3 - 6x^2 - 5x + 4}{2x}dx = \int_1^2 \left[\frac{3}{2}x^2 - 3x - \frac{5}{2} + \frac{2}{x}\right]dx \qquad \text{Divide each term on the numerator by } 2x.$$

$$= \frac{3}{2}\int_1^2 x^2\,dx - 3\int_1^2 x\,dx - \frac{5}{2}\int_1^2 dx + 2\int_1^2 \frac{1}{x}\,dx \qquad \text{Use property (3).}$$

$$= \frac{3}{2}\left[\frac{x^3}{3}\right]_1^2 - 3\left[\frac{x^2}{2}\right]_1^2 - \frac{5}{2}\left[x\right]_1^2 + 2\left[\ln|x|\right]_1^2 \qquad \begin{array}{l}\text{Use the Fundamental}\\\text{Theorem of Calculus.}\end{array}$$

$$= \frac{1}{2}(8 - 1) - \frac{3}{2}(4 - 1) - \frac{5}{2}(2 - 1) + 2(\ln 2 - \ln 1)$$

$$= -\frac{7}{2} + 2\ln 2 \qquad \blacksquare$$

NOW WORK　Problem 29 and AP® Practice Problem 4.

The next property states that a definite integral of a function f from a to b can be evaluated in pieces.

THEOREM

If a function f is integrable on an interval containing the numbers a, b, and c, then

$$\int_a^b f(x)\,dx = \int_a^c f(x)\,dx + \int_c^b f(x)\,dx \qquad (4)$$

Note that there is no restriction on the location of c, except that it is in an interval that contains a and b. In other words, c need not be between a and b.

A proof of this theorem for a function f that is continuous on $[a, b]$ is given in Appendix B. A proof for the more general case can be found in most advanced calculus books.

In particular, if f is continuous and nonnegative on a closed interval $[a, b]$ and if c is a number between a and b, then property (4) has a simple geometric interpretation, as seen in Figure 29.

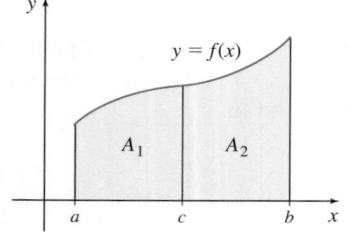

A = Area under the graph of f from a to b
= area A_1 + area A_2

$$\int_a^b f(x)dx = \int_a^c f(x)dx + \int_c^b f(x)dx$$

Figure 29

EXAMPLE 4　**Using Property (4) of the Definite Integral**

(a) If f is integrable on the closed interval $[2, 7]$, then

$$\int_2^7 f(x)\,dx = \int_2^4 f(x)\,dx + \int_4^7 f(x)\,dx$$

(b) If g is integrable on the closed interval $[3, 25]$, then

$$\int_3^{10} g(x)\,dx = \int_3^{25} g(x)\,dx + \int_{25}^{10} g(x)\,dx$$

$\blacksquare$

Example 4(b) illustrates that the number c need not lie between a and b.

NOW WORK　Problem 49 and AP® Practice Problems 1 and 11.

Property (4) is useful when integrating piecewise-defined functions.

EXAMPLE 5

Using Property (4) to Find the Area under the Graph of a Piecewise-defined Function

Find the area A under the graph of

$$f(x) = \begin{cases} x^2 & \text{if} \quad 0 \leq x < 10 \\ 100 & \text{if} \quad 10 \leq x \leq 15 \end{cases}$$

from 0 to 15.

Solution

See Figure 30. Since f is continuous and nonnegative on the closed interval $[0, 15]$, then $\int_0^{15} f(x)\,dx$ exists and equals the area A under the graph of f from 0 to 15. We break the integral at $x = 10$, where the rule for the function changes.

$$\int_0^{15} f(x)\,dx = \int_0^{10} f(x)\,dx + \int_{10}^{15} f(x)\,dx = \int_0^{10} x^2\,dx + \int_{10}^{15} 100\,dx$$

$$= \left[\frac{x^3}{3}\right]_0^{10} + \left[100x\right]_{10}^{15} = \frac{1000}{3} + 500 = \frac{2500}{3}$$

The area under the graph of f is approximately 833.333 square units. ∎

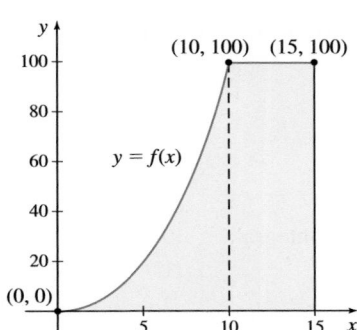

Figure 30 $A = \displaystyle\int_0^{15} f(x)\,dx$

NOW WORK Problem **39** and AP® Practice Problems **7, 8,** and **13.**

Recall that if a function f is bounded and has at most a finite number of points of discontinuity on the closed interval $[a, b]$, then $\int_a^b f(x)\,dx$ exists. We use property (4) to find the integral of a bounded function with discontinuities on $[a, b]$.

EXAMPLE 6

Using Property (4) to Integrate a Bounded Function with Discontinuities

Find $\int_{-1}^{4} f(x)\,dx$ where

$$f(x) = \begin{cases} e^x & \text{if} \quad -1 \leq x \leq 0 \\ x + 3 & \text{if} \quad 0 < x < 2 \\ 4 & \text{if} \quad x = 2 \\ -2x + 9 & \text{if} \quad 2 < x \leq 4 \end{cases}$$

Solution

The graph of f is shown in Figure 31. The function is defined and bounded on the interval $[-1, 4]$ and has two points of discontinuity: a jump discontinuity at $x = 0$ and a removable discontinuity at $x = 2$. So $\int_{-1}^{4} f(x)\,dx$ exists.

(Example continued on the next page)

Figure 31

$$f(x) = \begin{cases} e^x & \text{if} \quad -1 \leq x \leq 0 \\ x + 3 & \text{if} \quad 0 < x < 2 \\ 4 & \text{if} \quad x = 2 \\ -2x + 9 & \text{if} \quad 2 < x \leq 4 \end{cases}$$

Use property (4) to break the integral at the numbers where the function is discontinuous.

$$\int_{-1}^{4} f(x)\,dx = \int_{-1}^{0} f(x)\,dx + \int_{0}^{2} f(x)\,dx + \int_{2}^{4} f(x)\,dx \qquad \text{Use property (4).}$$

$$= \int_{-1}^{0} e^{x}dx + \int_{0}^{2}(x+3)dx + \int_{2}^{4}(-2x+9)dx$$

$$= \left[e^{x}\right]_{-1}^{0} + \left[\frac{x^2}{2} + 3x\right]_{0}^{2} + \left[-x^2 + 9x\right]_{2}^{4}$$

$$= (e^0 - e^{-1}) + [(2+6) - 0] + [(-16 + 36) - (-4 + 18)]$$

$$= 1 - e^{-1} + 8 + 20 - 14 = 15 - \frac{1}{e}$$

∎

NOW WORK Problem **41** and AP® Practice Problem **15**.

The next property establishes bounds on a definite integral.

THEOREM Bounds on an Integral

If a function f is continuous on a closed interval $[a, b]$ and if m and M denote the absolute minimum value and the absolute maximum value, respectively, of f on $[a, b]$, then

$$\boxed{m(b - a) \le \int_{a}^{b} f(x)\,dx \le M\,(b - a)} \qquad (5)$$

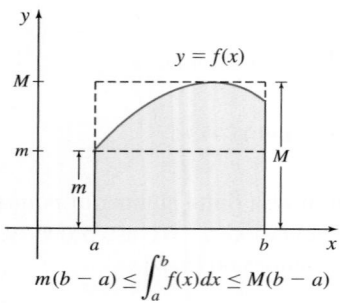

$$m(b - a) \le \int_{a}^{b} f(x)dx \le M(b - a)$$

Figure 32

A proof of this theorem is given in Appendix B.

If f is nonnegative on $[a, b]$, then the inequalities in (5) have a geometric interpretation. In Figure 32, the area of the shaded region is $\int_{a}^{b} f(x)\,dx$. The rectangle of width $b - a$ and height m has area $m(b - a)$. The rectangle of width $b - a$ and height M has area $M(b - a)$. The areas of the three regions are numerically related by the inequalities in the theorem.

EXAMPLE 7 Using Property (5) of Definite Integrals

(a) Find an upper bound and a lower bound for the area A under the graph of $f(x) = \sin x$ from 0 to π.

(b) Find the actual area under the graph.

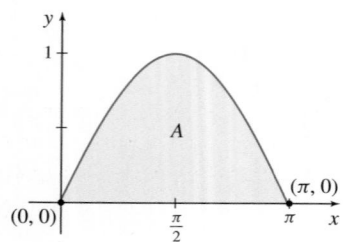

Figure 33 $f(x) = \sin x, 0 \le x \le \pi$

Solution

The graph of f is shown in Figure 33. Since $f(x) \ge 0$ for all x in the closed interval $[0, \pi]$, the area A under its graph is given by the definite integral, $\int_{0}^{\pi} \sin x\,dx$.

NEED TO REVIEW? The Extreme Value Theorem is discussed in Section 5.2, p. 350.

(a) From the Extreme Value Theorem, f has an absolute minimum value and an absolute maximum value on the interval $[0, \pi]$. The absolute maximum of f occurs at $x = \dfrac{\pi}{2}$, and its value is $M = f\left(\dfrac{\pi}{2}\right) = \sin\dfrac{\pi}{2} = 1$. The absolute minimum occurs at $x = 0$ and at $x = \pi$; the absolute minimum value is $m = f(0) = \sin 0 = 0 = f(\pi)$. Using the inequalities in (5), the area under the graph of f from 0 to π is bounded as follows:

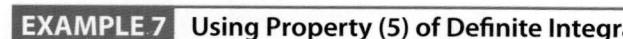

$$0 \le \int_{0}^{\pi} \sin x\,dx \le \pi \qquad b - a = \pi - 0 = \pi;\, m = 0,\, M = 1$$

(b) The actual area under the graph is

$$A = \int_0^\pi \sin x \, dx = \left[-\cos x \right]_0^\pi = -\cos \pi + \cos 0 = 1 + 1 = 2 \text{ square units} \quad \blacksquare$$

NOW WORK   Problem **59** and AP® Practice Problem **14.**

2   Work with the Mean Value Theorem for Integrals

Suppose f is a function that is continuous and nonnegative on a closed interval $[a, b]$. Figure 34 suggests that the area under the graph of f from a to b, $\int_a^b f(x)\, dx$, is equal to the area of some rectangle of width $b - a$ and height $f(u)$ for some choice (or choices) of u in the interval $[a, b]$.

In more general terms, for every function f that is continuous on a closed interval $[a, b]$, there is some number u (not necessarily unique) in the interval $[a, b]$ for which

$$\int_a^b f(x)\, dx = f(u)(b - a)$$

This result is known as the *Mean Value Theorem for Integrals.*

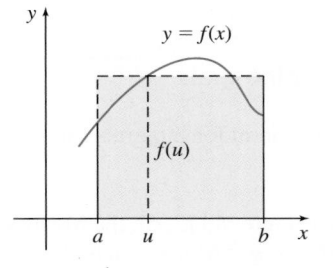

Figure 34 $\displaystyle\int_a^b f(x)\, dx = f(u)(b - a)$

THEOREM   Mean Value Theorem for Integrals

If a function f is continuous on a closed interval $[a, b]$, there is a real number u, $a \le u \le b$, for which

$$\boxed{\int_a^b f(x)\, dx = f(u)(b - a)} \qquad (6)$$

Proof Let f be a function that is continuous on a closed interval $[a, b]$.

- If f is a constant function, say, $f(x) = k$, on $[a, b]$, then

$$\int_a^b f(x)\, dx = \int_a^b k \, dx = k(b - a) = f(u)(b - a)$$

for any choice of u in $[a, b]$.

- If f is not a constant function on $[a, b]$, then by the Extreme Value Theorem, f has an absolute maximum and an absolute minimum on $[a, b]$. Suppose f assumes its absolute minimum at the number c so that $f(c) = m$; and suppose f assumes its absolute maximum at the number C so that $f(C) = M$. Then by the Bounds on an Integral Theorem (5), we have

$$m(b - a) \le \int_a^b f(x)\, dx \le M(b - a) \qquad \text{for all } x \text{ in } [a, b]$$

Divide each part by $(b - a)$ and replace m by $f(c)$ and M by $f(C)$. Then

$$f(c) \le \frac{1}{b - a} \int_a^b f(x)\, dx \le f(C)$$

(Proof continued on the next page)

NEED TO REVIEW? The Intermediate Value Theorem is discussed in Section 1.3, pp. 115–116.

Since $\dfrac{1}{b-a}\displaystyle\int_a^b f(x)\,dx$ is a real number between $f(c)$ and $f(C)$, it follows from the Intermediate Value Theorem that there is a real number u between c and C, for which

$$f(u)=\frac{1}{b-a}\int_a^b f(x)\,dx$$

That is, there is a real number u, $a \le u \le b$, for which

$$\int_a^b f(x)\,dx = f(u)\,(b-a)$$

∎

EXAMPLE 8   Using the Mean Value Theorem for Integrals

Find the number(s) u guaranteed by the Mean Value Theorem for Integrals for $\int_2^6 x^2\,dx$.

Solution

Since the function $f(x)=x^2$ is continuous on the closed interval $[2,6]$, the Mean Value Theorem for Integrals guarantees there is a number u, $2 \le u \le 6$, for which

$$\int_2^6 x^2\,dx = f(u)(6-2)=4u^2 \qquad f(u)=u^2$$

Integrating, we have

$$\int_2^6 x^2\,dx = \left[\frac{x^3}{3}\right]_2^6 = \frac{1}{3}(216-8)=\frac{208}{3}$$

Then

$$4u^2 = \frac{208}{3}$$
$$u^2 = \frac{52}{3} \qquad 2 \le u \le 6$$
$$u = \sqrt{\frac{52}{3}} \approx 4.163 \qquad \text{Disregard the negative solution since } u>0.$$

∎

NOW WORK   Problem 61.

③ Find the Average Value of a Function

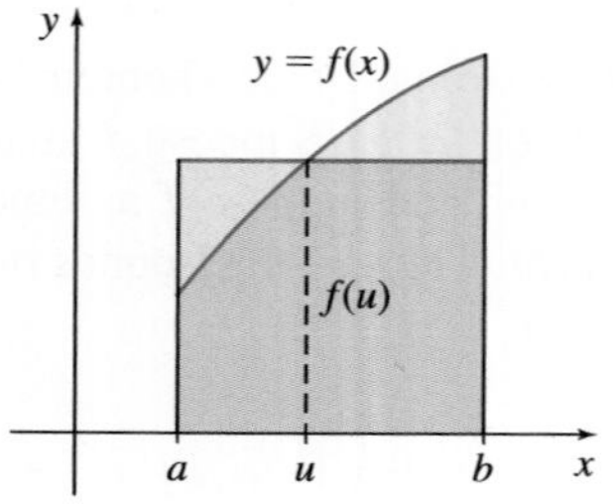

We know from the Mean Value Theorem for Integrals that if a function f is continuous on a closed interval $[a,b]$, there is a real number u, $a \le u \le b$, for which

$$\int_a^b f(x)\,dx = f(u)(b-a)$$

This means that if the function f is also nonnegative on the closed interval $[a,b]$, the area enclosed by a rectangle of height $f(u)$ and width $b-a$ equals the area under the graph of f from a to b. See Figure 35. Consequently, $f(u)$ can be thought of as an *average value*, or *mean value*, of f over $[a,b]$.

This suggests the following definition:

Figure 35 $\displaystyle\int_a^b f(x)\,dx = f(u)(b-a)$

DEFINITION Average Value of a Function over an Interval

Suppose f is a function that is continuous on the closed interval $[a,b]$. The **average value $\bar{y}$ of f over $[a,b]$** is

$$\boxed{\bar{y} = \frac{1}{b-a}\int_a^b f(x)\,dx} \qquad (7)$$

IN WORDS The average value $\bar{y}$ of a function f equals the value $f(u)$ in the Mean Value Theorem for Integrals.

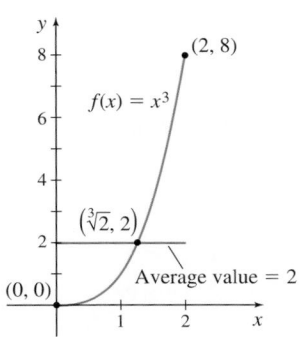

Figure 36 $f(x) = x^3, 0 \le x \le 2$

EXAMPLE 9 **Finding the Average Value of a Function**

Find the average value of $f(x) = x^3$ on the closed interval $[0, 2]$.

Solution

The average value of $f(x) = x^3$ on the closed interval $[0, 2]$ is given by

$$\bar{y} = \frac{1}{b-a} \int_a^b f(x)\, dx = \frac{1}{2-0} \int_0^2 x^3\, dx = \frac{1}{2} \left[\frac{x^4}{4} \right]_0^2 = \frac{1}{2} \cdot \frac{16}{4} = 2$$

The average value of f on $[0, 2]$ is $\bar{y} = 2$. ∎

The function f and its average value are graphed in Figure 36.

NOW WORK   Problem **67** and AP® Practice Problems **2, 5, 9,** and **12.**

④ Interpret Integrals Involving Motion on a Line

For an object moving on a line with velocity $v = v(t)$

- The **net displacement** of an object whose motion begins at $t = a$ and ends at $t = b$ is given by

$$\int_a^b v(t)\, dt$$

- The **total distance** traveled by the object from $t = a$ to $t = b$ is given by

$$\int_a^b |v(t)|\, dt$$

- If $v(t) \ge 0$, for $a \le t \le b$, then the net displacement and the total distance traveled are equal.

Remember speed and velocity are not the same thing. Speed measures how fast an object is moving regardless of direction. Velocity measures both the speed and the direction of an object. Because of this, velocity can be positive, negative, or zero. Speed is the absolute value of velocity. So $\int_a^b v(t)\, dt$ is a sum of signed distances (net displacement) and $\int_a^b |v(t)|\, dt$ is a sum of actual distances traveled (total distance).

EXAMPLE 10 **Interpreting an Integral Involving Motion on a Line**

An object is moving along a horizontal line with velocity $v(t) = 3t^2 - 18t + 15$ (in meters per minute) for time $0 \le t \le 6$ min.

(a) Find the net displacement of the object from $t = 0$ to $t = 6$. Interpret the result in the context of the problem.

(b) Find the total distance traveled by the object from $t = 0$ to $t = 6$ and interpret the result.

Solution

(a) The net displacement of the object from $t = 0$ to $t = 6$ is given by

$$\int_0^6 v(t)\, dt = \int_0^6 (3t^2 - 18t + 15)\, dt = \left[t^3 - 9t^2 + 15t \right]_0^6 = 6^3 - 9 \cdot 6^2 + 15 \cdot 6 - 0 = -18$$

After moving from $t = 0$ to $t = 6$ min, the object is 18 m to the left of where it was at $t = 0$ min.

(b) The total distance traveled by the object from $t=0$ to $t=6$ min is given by $\int_0^6 |v(t)|\,dt$.

Since $v(t) = 3t^2 - 18t + 15 = 3(t-1)(t-5)$

we find that $v(t) \geq 0$ if $0 \leq t \leq 1$ or if $5 \leq t \leq 6$ and $v(t) \leq 0$ if $1 \leq t \leq 5$. Now express $|v(t)|$ as the piecewise-defined function

$$|v(t)| = \begin{cases} v(t) = 3t^2 - 18t + 15 & \text{if } 0 \leq t \leq 1 \\ -v(t) = -(3t^2 - 18t + 15) & \text{if } 1 < t < 5 \\ v(t) = 3t^2 - 18t + 15 & \text{if } 5 \leq t \leq 6 \end{cases}$$

Then

$$\int_0^6 |v(t)|\,dt = \int_0^1 v(t)\,dt + \int_1^5 [-v(t)]\,dt + \int_5^6 v(t)\,dt$$

$$= [t^3 - 9t^2 + 15t]_0^1 - [t^3 - 9t^2 + 15t]_1^5 + [t^3 - 9t^2 + 15t]_5^6$$

$$= (7-0) - (-25-7) + [-18-(-25)] = 7 + 32 + 7 = 46$$

The object traveled a total distance of 46 m from $t=0$ to $t=6$ min. ∎

Figure 37 shows the distinction between the net displacement and the total distance traveled by the object assuming that the object is at the origin when the motion starts.

📖 COMMENT Solving part (b) of Example 10 required evaluating the integral of an absolute value, which requires finding and integrating a piecewise-defined function. Integrals involving absolute value are evaluated more easily using technology. Figure 38 shows the solution to part (b) using a TI-84 Plus CE calculator.

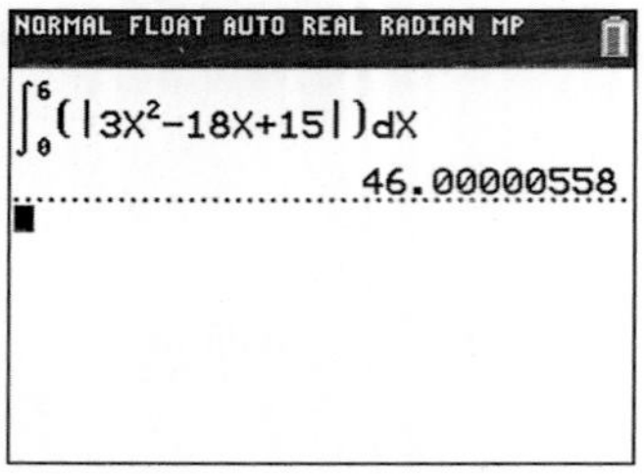

NOW WORK   Problem 81 and AP® Practice Problems 17 and 18.

Total distance: $7 + 32 + 7 = 46$ m

Net displacement: -18 m

Figure 37

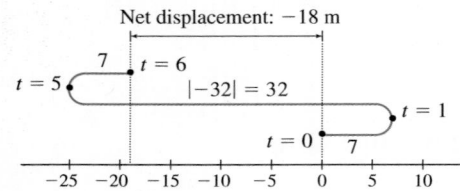

$\int_0^6 (|3X^2 - 18X + 15|)\,dX$

                    46.00000558

Figure 38

6.4 Assess Your Understanding

Concepts and Vocabulary

1. True or False $\int_2^3 (x^2 + x)\,dx = \int_2^3 x^2\,dx + \int_2^3 x\,dx$

2. True or False $\int_0^3 5e^{x^2}\,dx = \int_0^3 5\,dx \cdot \int_0^3 e^{x^2}\,dx$

3. True or False
$$\int_0^5 (x^3 + 1)\,dx = \int_0^{-3} (x^3 + 1)\,dx + \int_{-3}^5 (x^3 + 1)\,dx$$

4. If f is continuous on an interval containing the numbers a, b, and c, and if $\int_a^c f(x)\,dx = 3$ and $\int_c^b f(x)\,dx = -5$, then $\int_a^b f(x)\,dx = $ _____.

5. If a function f is continuous on the closed interval $[a, b]$, then $\bar{y} = \dfrac{1}{b-a} \int_a^b f(x)\,dx$ is the _____ of f over $[a, b]$.

6. True or False If a function f is continuous on a closed interval $[a, b]$ and if m and M denote the absolute minimum value and the absolute maximum value, respectively, of f on $[a, b]$, then

$$m \leq \int_a^b f(x)\,dx \leq M$$

Skill Building

In Problems 7–16, find each definite integral given that
$\int_1^3 f(x)\,dx = 5$, $\int_1^3 g(x)\,dx = -2$, $\int_3^5 f(x)\,dx = 2$, $\int_3^5 g(x)\,dx = 1$.

7. $\int_1^3 [f(x) - g(x)]\,dx$

8. $\int_1^3 [f(x) + g(x)]\,dx$

9. $\int_1^3 [5f(x) - 3g(x)]\,dx$

10. $\int_1^3 [3f(x) + 4g(x)]\,dx$

11. $\int_1^5 [2f(x) - 3g(x)]\,dx$

12. $\int_1^5 [f(x) - g(x)]\,dx$

13. $\int_1^3 [5f(x) - 3g(x) + 7]\,dx$

14. $\int_1^3 [f(x) + 4g(x) + 2x]\,dx$

15. $\int_1^5 [2f(x) - g(x) + 3x^2]\,dx$

16. $\int_1^5 [5f(x) - 3g(x) - 4]\,dx$

In Problems 17–36, find each definite integral using the Fundamental Theorem of Calculus and properties of the definite integral.

17. $\int_0^1 (t^2 - t^{3/2})\, dt$

18. $\int_{-2}^0 (x + x^2)\, dx$

19. $\int_{\pi/2}^{\pi} 4 \sin x\, dx$

20. $\int_0^{\pi/2} 3 \cos x\, dx$

21. $\int_{-\pi/4}^{\pi/4} (1 + 2 \sec x\, \tan x)\, dx$

22. $\int_0^{\pi/4} (1 + \sec^2 x)\, dx$

23. $\int_1^4 (\sqrt{x} - 4x)\, dx$

24. $\int_0^1 (\sqrt[5]{t^2} + 1)\, dt$

25. $\int_{-2}^3 [(x - 1)(x + 3)]\, dx$

26. $\int_0^1 (z^2 + 1)^2\, dz$

27. $\int_1^2 \frac{x^2 - 12}{x^4}\, dx$

28. $\int_1^e \frac{5s^2 + s}{s^2}\, ds$

29. $\int_0^1 \frac{e^{2x} - 1}{e^x}\, dx$

30. $\int_0^1 \frac{e^{2x} + 4}{e^{-x}}\, dx$

31. $\int_{\pi/3}^{\pi/2} \frac{x \sin x + 2}{x}\, dx$

32. $\int_{\pi/6}^{\pi/2} \frac{x \cos x - 4}{x}\, dx$

33. $\int_1^2 \frac{x^2 + 2}{x^2}\, dx$

34. $\int_1^4 \frac{x - \sqrt{x}}{x}\, dx$

35. $\int_0^{1/2} \left(5 + \frac{1}{\sqrt{1 - x^2}}\right) dx$

36. $\int_0^1 \left(1 + \frac{5}{1 + x^2}\right) dx$

In Problems 37–44, use properties of integrals and the Fundamental Theorem of Calculus to find each integral.

37. $\int_{-2}^1 f(x)\, dx$, where $f(x) = \begin{cases} 1 & \text{if } x < 0 \\ x^2 + 1 & \text{if } x \geq 0 \end{cases}$

38. $\int_{-1}^2 f(x)\, dx$, where $f(x) = \begin{cases} x + 1 & \text{if } x < 0 \\ x^2 + 1 & \text{if } x \geq 0 \end{cases}$

39. $\int_{-2}^2 f(x)\, dx$, where $f(x) = \begin{cases} 3x & \text{if } -2 \leq x < 0 \\ 2x^2 & \text{if } 0 \leq x \leq 2 \end{cases}$

40. $\int_0^4 h(x)\, dx$, where $h(x) = \begin{cases} x - 2 & \text{if } 0 \leq x \leq 2 \\ 2 - x & \text{if } 2 < x \leq 4 \end{cases}$

41. $\int_{-4}^2 f(x)\, dx$, where

$f(x) = \begin{cases} 8 - x & \text{if } -4 \leq x \leq -2 \\ \dfrac{x^2 + 7x + 10}{x + 2} & \text{if } -2 < x \leq 2 \end{cases}$

42. $\int_0^5 f(x)\, dx$, where $f(x) = \begin{cases} \dfrac{x^3 - 8}{x - 2} & \text{if } 0 \leq x < 1 \\ 4x - 1 & \text{if } 1 \leq x \leq 5 \end{cases}$

43. $\int_0^{2\pi} f(x)\, dx$, where $f(x) = \begin{cases} \sin x & \text{if } 0 \leq x < \pi \\ \cos x & \text{if } \pi \leq x \leq 2\pi \end{cases}$

44. $\int_{-1}^e f(x)\, dx$, where $f(x) = \begin{cases} x^2 + 4 & \text{if } -1 \leq x < 1 \\ \dfrac{3}{x} & \text{if } 1 \leq x \leq e \end{cases}$

In Problems 45–48, the domain of f is a closed interval [a, b]. Find $\int_a^b f(x)\, dx$.

45. $a = -\dfrac{3}{2}$ and $b = 7$

46. $a = -5$ and $b = 2$

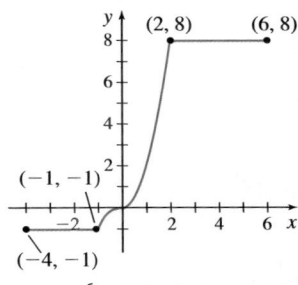

$f(x) = \begin{cases} x^4 & \text{if } -\dfrac{3}{2} \leq x < 1 \\ x & \text{if } 1 \leq x < 4 \\ 4 & \text{if } 4 \leq x \leq 7 \end{cases}$

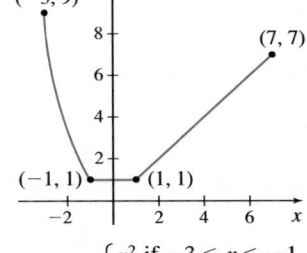

$f(x) = \begin{cases} 4 & \text{if } -5 \leq x < -2 \\ x^2 & \text{if } -2 \leq x \leq 0 \\ -\dfrac{x}{2} & \text{if } 0 < x \leq 2 \end{cases}$

47. $a = -4$ and $b = 6$

48. $a = -3$ and $b = 7$

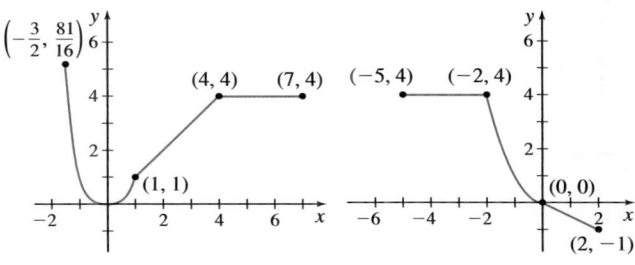

$f(x) = \begin{cases} -1 & \text{if } -4 \leq x \leq -1 \\ x^3 & \text{if } -1 < x < 2 \\ 8 & \text{if } 2 \leq x \leq 6 \end{cases}$

$f(x) = \begin{cases} x^2 & \text{if } -3 \leq x \leq -1 \\ 1 & \text{if } -1 < x \leq 1 \\ x & \text{if } 1 < x \leq 7 \end{cases}$

In Problems 49–52, use properties of definite integrals to verify each statement. Assume that all integrals involved exist.

49. $\int_3^{11} f(x)\, dx - \int_7^{11} f(x)\, dx = \int_3^7 f(x)\, dx$

50. $\int_{-2}^6 f(x)\, dx - \int_3^6 f(x)\, dx = \int_{-2}^3 f(x)\, dx$

51. $\int_0^4 f(x)\, dx - \int_6^4 f(x)\, dx = \int_0^6 f(x)\, dx$

52. $\int_{-1}^3 f(x)\, dx - \int_5^3 f(x)\, dx = \int_{-1}^5 f(x)\, dx$

In Problems 53–60, use the Bounds on an Integral Theorem to obtain a lower estimate and an upper estimate for each integral.

53. $\int_1^3 (5x+1)\,dx$

54. $\int_0^1 (1-x)\,dx$

55. $\int_{\pi/4}^{\pi/2} \sin x\,dx$

56. $\int_{\pi/6}^{\pi/3} \cos x\,dx$

57. $\int_0^1 \sqrt{1+x^2}\,dx$

58. $\int_{-1}^1 \sqrt{1+x^4}\,dx$

[PAGE 471] **59.** $\int_0^1 e^x\,dx$

60. $\int_1^{10} \frac{1}{x}\,dx$

In Problems 61–66, for each integral find the number(s) u guaranteed by the Mean Value Theorem for Integrals.

[PAGE 472] **61.** $\int_0^3 (2x^2+1)\,dx$

62. $\int_0^2 (2-x^3)\,dx$

63. $\int_0^4 x^2\,dx$

64. $\int_0^4 (-x)\,dx$

65. $\int_0^{2\pi} \cos x\,dx$

66. $\int_{-\pi/4}^{\pi/4} \sec x \tan x\,dx$

In Problems 67–76, find the average value of each function f over the given interval.

[PAGE 473] **67.** $f(x)=e^x$ over $[0,1]$

68. $f(x)=\dfrac{1}{x}$ over $[1,e]$

69. $f(x)=x^{2/3}$ over $[-1,1]$

70. $f(x)=\sqrt{x}$ over $[0,4]$

71. $f(x)=\sin x$ over $\left[0,\dfrac{\pi}{2}\right]$

72. $f(x)=\cos x$ over $\left[0,\dfrac{\pi}{2}\right]$

73. $f(x)=1-x^2$ over $[-1,1]$

74. $f(x)=16-x^2$ over $[-4,\ 4]$

75. $f(x)=e^x-\sin x$ over $\left[0,\dfrac{\pi}{2}\right]$

76. $f(x)=x+\cos x$ over $\left[0,\dfrac{\pi}{2}\right]$

In Problems 77–80, find:

(a) *The area under the graph of the function over the indicated interval.*

(b) *The average value of each function over the indicated interval.*

77. $[-1,2]$

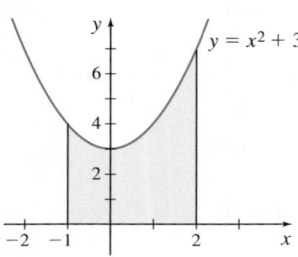

78. $[-2,1]$

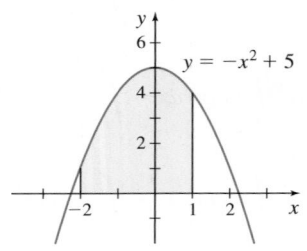

79. $[-1,2]$

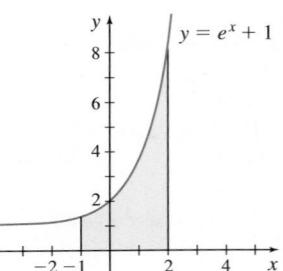

80. $\left[0,\dfrac{3\pi}{4}\right]$

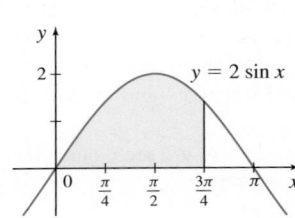

In Problems 81–84, an object is moving along a horizontal line with velocity $v=v(t)$ over the indicated interval.

(a) *Find the net displacement of the object. Interpret the result in the context of the problem.*

(b) *Find the total distance traveled by the object and interpret the result.*

[PAGE 474] **81.** $v(t)=6t^2+12t-18$ m/min; $0\le t\le 3$

82. $v(t)=t^2-2t-8$ m/min; $0\le t\le 6$

83. $v(t)=t^3+2t^2-8t$ m/s; $-2\le t\le 3$

84. $v(t)=t^3+5t^2-6t$ m/s; $-1\le t\le 3$

Applications and Extensions

In Problems 85–88, find each definite integral using the Fundamental Theorem of Calculus and properties of definite integrals.

85. $\int_{-2}^3 (x+|x|)\,dx$

86. $\int_0^3 |x-1|\,dx$

87. $\int_0^2 |3x-1|\,dx$

88. $\int_0^5 |2-x|\,dx$

89. Average Temperature   A rod 3 m long is heated to $25x\ °$C, where x is the distance in meters from one end of the rod. Find the average temperature of the rod.

90. Average Daily Rainfall   The rainfall per day, x days after the beginning of the year, is modeled by the function $r(x)=0.00002(6511+366x-x^2)$, measured in centimeters. Find the average daily rainfall for the first 180 days of the year.

91. Structural Engineering A structural engineer designing a member of a structure must consider the forces that will act on that member. Most often, natural forces like snow, wind, or rain distribute force over the entire member. For practical purposes, however, an engineer determines the distributed force as a single resultant force acting at one point on the member. If the distributed force is given by the function $W = W(x)$, in newtons per meter (N/m), then the magnitude F_R of the resultant force is

$$F_R = \int_a^b W(x)\, dx$$

The position $\bar{x}$ of the resultant force measured in meters from the origin is given by

$$\bar{x} = \frac{\int_a^b x\, W(x)\, dx}{\int_a^b W(x)\, dx}$$

If the distributed force is $W(x) = 0.75x^3$, $0 \le x \le 5$, find:

(a) The magnitude of the resultant force.

(b) The position from the origin of the resultant force.

Source: Problem contributed by the students at Trine University, Avalon, IN.

92. Chemistry: Enthalpy In chemistry, **enthalpy** is a measure of the total energy of a system. For a nonreactive process with no phase change, the change in enthalpy ΔH is given by $\Delta H = \int_{T_1}^{T_2} C_p\, dT$, where C_p is the specific heat of the system in question. The specific heat per mol of the chemical benzene is

$$C_p = 0.126 + (2.34 \times 10^{-6})T,$$

where C_p is in kJ/ (mol °C), and T is in degrees Celsius.

(a) What are the units of the change in enthalpy ΔH?

(b) What is the change in enthalpy ΔH associated with increasing the temperature of 1.0 mol of benzene from 20 °C to 40 °C?

(c) What is the change in enthalpy ΔH associated with increasing the temperature of 1.0 mol of benzene from 20 °C to 60 °C?

(d) Does the enthalpy of benzene increase, decrease, or remain constant as the temperature increases?

Source: Problem contributed by the students at Trine University, Avalon, IN.

93. Average Mass Density The mass density of a metal bar of length 3 m is given by $\rho(x) = 1000 + x - \sqrt{x}$ kilograms per cubic meter, where x is the distance in meters from one end of the bar. What is the average mass density over the length of the entire bar?

94. Average Velocity The velocity at time t, in m/s, of an object moving on a line is given by $v(t) = 4\pi \cos t$. What is the average velocity of the object over the interval $0 \le t \le \pi$?

95. Average Area What is the average area of all circles whose radii are between 1 and 3 m?

96. Area

(a) Use properties of integrals and the Fundamental Theorem of Calculus to find the area under the graph of $y = 3 - |x|$ from -3 to 3. *Hint:* Break the integral at $x = 0$.

(b) Check your answer by using elementary geometry.

97. Area

(a) Use properties of integrals and the Fundamental Theorem of Calculus to find the area under the graph of $y = 1 - \left|\frac{1}{2}x\right|$ from -2 to 2.

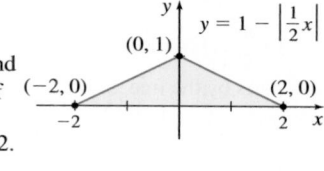

(b) Check your answer by using elementary geometry. See the figure.

98. Area Let A be the area in the first quadrant that is enclosed by the graphs of $y = 3x^2$, $y = \dfrac{3}{x}$, the x-axis, and the line $x = k$, where $k > 1$, as shown in the figure.

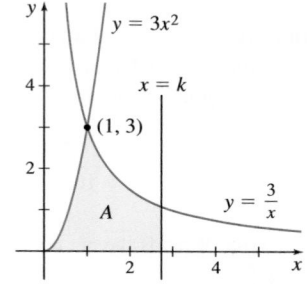

(a) Find the area A as a function of k.

(b) When the area is 7, what is k?

(c) If the area A is increasing at the constant rate of 5 square units per second, at what rate is k increasing when $k = 15$?

99. Motion on a Line A car starting from rest accelerates at the rate of 3 m/s². Find its average speed over the first 8 s.

100. Motion on a Line A car moving at a constant speed of 80 miles per hour begins to decelerate at the rate of 10 mi/h². Find its average speed over the next 10 min.

101. Average Slope

(a) Use the definition of average value of a function to find the *average slope* of the graph of $y = f(x)$, where $a \le x \le b$. (Assume that f' is continuous.)

(b) Give a geometric interpretation.

102. What theorem guarantees that the average slope found in Problem 101 is equal to $f'(u)$ for some u in $[a, b]$? What *different* theorem guarantees the same thing? (Do you see the connection between these theorems?)

103. Prove that if a function f is continuous on a closed interval $[a, b]$ and if k is a constant, then $\int_a^b kf(x)\, dx = k \int_a^b f(x)\, dx$. Assume f has an antiderivative.

104. Prove that if the functions $f_1, f_2, \ldots, f_n$ are continuous on a closed interval $[a, b]$ and if $k_1, k_2, \ldots, k_n$ are constants, then

$$\int_a^b [k_1 f_1(x) + k_2 f_2(x) + \cdots + k_n f_n(x)]\, dx$$

$$= k_1 \int_a^b f_1(x)\, dx + k_2 \int_a^b f_2(x)\, dx + \cdots + k_n \int_a^b f_n(x)\, dx$$

Assume each function f has an antiderivative.

105. Area   The area under the graph of $y = \cos x$ from $-\dfrac{\pi}{2}$ to $\dfrac{\pi}{2}$ is separated into two parts by the line $x = k$, $-\dfrac{\pi}{2} < k < \dfrac{\pi}{2}$, as shown in the figure. If the area under the graph of y from $-\dfrac{\pi}{2}$ to k is three times the area under the graph of y from k to $\dfrac{\pi}{2}$, find k.

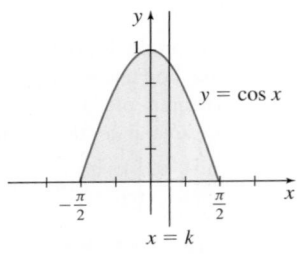

106. Displacement of a Damped Spring   The net displacement x in meters of a damped spring from its equilibrium position at time t seconds is given by

$$x(t) = \frac{\sqrt{15}}{10} e^{-t} \sin\left(\sqrt{15}\,t\right) + \frac{3}{2} e^{-t} \cos\left(\sqrt{15}\,t\right)$$

(a) What is the net displacement of the spring at $t = 0$?

(b) Graph the displacement for the first 2 s of the spring's motion.

(c) Find the average displacement of the spring for the first 2 s of its motion.

107. Area   Let

$$f(x) = |x^4 + 3.44x^3 - 0.5041x^2 - 5.0882x + 1.1523|$$

be defined on the interval $[-3, 1]$. Find the area under the graph of f.

108. If f is continuous on $[a, b]$, show that the functions defined by

$$F(x) = \int_c^x f(t)\,dt \qquad G(x) = \int_d^x f(t)\,dt$$

for any choice of c and d in (a, b) always differ by a constant. Also show that

$$F(x) - G(x) = \int_c^d f(t)\,dt$$

109. Put It Together   Suppose $a < c < b$ and the function f is continuous on $[a, b]$ and differentiable on (a, b). Which of the following is *not* necessarily true?

(I) $\displaystyle\int_a^b f(x)\,dx = \int_a^c f(x)\,dx + \int_c^b f(x)\,dx$

(II) There is a number d in (a, b) for which

$$f'(d) = \frac{f(b) - f(a)}{b - a}$$

(III) $\displaystyle\int_a^b f(x)\,dx \ge 0$

(IV) $\displaystyle\lim_{x \to c} f(x) = f(c)$

(V) If k is a real number, then $\int_a^b kf(x)\,dx = k\int_a^b f(x)\,dx$

110. Minimizing Area   Find $b > 0$ so that the area enclosed in the first quadrant by the graph of $y = 1 + b - bx^2$ and the coordinate axes is a minimum.

111. Area   Find the area enclosed by the graph of $\sqrt{x} + \sqrt{y} = 1$ and the coordinate axes.

Challenge Problems

112. Average Speed   For a freely falling object starting from rest, $v_0 = 0$, find:

(a) The average speed $\bar{v}_t$ with respect to the time t in seconds over the closed interval $[0, 5]$.

(b) The average speed $\bar{v}_s$ with respect to the distance s of the object from its position at $t = 0$ over the closed interval $[0, s_1]$, where s_1 is the distance the object falls from $t = 0$ to $t = 5$ s.

Hint: The derivation of the formulas for freely falling objects is given in Section 5.6, pp. 410–412.

113. Average Speed   If an object falls from rest for 3 s, find:

(a) Its average speed with respect to time.

(b) Its average speed with respect to the distance it travels in 3 s.

Express the answers in terms of g, the acceleration due to gravity.

114. Free Fall   For a freely falling object starting from rest, $v_0 = 0$, find:

(a) The average velocity $\bar{v}_t$ with respect to the time t over the closed interval $[0, t_1]$.

(b) The average velocity $\bar{v}_s$ with respect to the distance s of the object from its position at $t = 0$ over the closed interval $[0, s_1]$, where s_1 is the distance the object falls in time t_1. Assume $s(0) = 0$.

115. Put It Together

(a) What is the domain of $f(x) = 2|x - 1|x^2$?

(b) What is the range of f?

(c) For what values of x is f continuous?

(d) For what values of x is the derivative of f continuous?

(e) Find $\int_0^1 f(x)\,dx$.

116. Probability   A function f that is continuous on the closed interval $[a, b]$, and for which (i) $f(x) \ge 0$ for numbers x in $[a, b]$ and 0 elsewhere and (ii) $\int_a^b f(x)\,dx = 1$, is called a **probability density function**. If $a \le c < d \le b$, the probability of obtaining a value between c and d is defined as $\int_c^d f(x)\,dx$.

(a) Find a constant k so that $f(x) = kx$ is a probability density function on $[0, 2]$.

(b) Find the probability of obtaining a value between 1 and 1.5.

117. Cumulative Probability Distribution   Refer to Problem 116. If f is a probability density function, the **cumulative distribution function** F for f is defined as

$$F(x) = \int_a^x f(t)\,dt \qquad a \le x \le b$$

Find the cumulative distribution function F for the probability density function $f(x) = kx$ of Problem 116(a).

118. For the cumulative distribution function $F(x) = x - 1$, on the interval $[1, 2]$:
  (a) Find the probability density function f corresponding to F.
  (b) Find the probability of obtaining a value between 1.5 and 1.7.

119. Let $f(x) = x^3 - 6x^2 + 11x - 6$. Find $\int_1^3 |f(x)|\,dx$.

120. (a) Prove that if functions f and g are continuous on a closed interval $[a, b]$ and if $f(x) \geq g(x)$ on $[a, b]$, then
$$\int_a^b f(x)\,dx \geq \int_a^b g(x)\,dx$$
  (b) Show that for $x > 1$, $\ln x < 2(\sqrt{x} - 1)$.

121. Prove that the average value of a line segment $y = m(x - x_1) + y_1$ on the interval $[x_1, x_2]$ equals the y-coordinate of the midpoint of the line segment from x_1 to x_2.

122. Prove that if a function f is continuous on a closed interval $[a, b]$ and if $f(x) \geq 0$ on $[a, b]$, then $\int_a^b f(x)\,dx \geq 0$.

123. (a) Prove that if f is continuous on a closed interval $[a, b]$ and $\int_a^b f(x)\,dx = 0$, there is at least one number c in $[a, b]$ for which $f(c) = 0$.
  (b) Give a counterexample to the statement above if f is not required to be continuous.

124. Prove that if f is continuous on $[a, b]$, then
$$\left| \int_a^b f(x)\,dx \right| \leq \int_a^b |f(x)|\,dx$$
Give a geometric interpretation of the inequality.

Preparing for the AP® Exam

AP® Practice Problems

Multiple-Choice Questions

[PAGE 468] 1. If $\int_0^2 f(x)\,dx = -3$ and $\int_0^5 f(x)\,dx = 7$, then $\int_2^5 [4f(x) - 1]\,dx$ equals
(A) 13   (B) 15   (C) 37   (D) 39

[PAGE 473] 2. An object moves along the x-axis with velocity $v = v(t)$ m/s. If $v(t) = 3t^2 + t - 2$, what is the average velocity of the object during the interval $0 \leq t \leq 6$?
(A) 37 m/s   (B) 38 m/s
(C) 41 m/s   (D) 222 m/s

[PAGE 467] 3. If $g(x) = 2f(x) - 4$ on the interval $[-2, 8]$, then $\int_{-2}^8 [f(x) + g(x)]\,dx$ equals
(A) $3\int_{-2}^8 f(x)\,dx - 24$ (B) $3\int_{-2}^8 f(x)\,dx - 40$
(C) $3\int_{-2}^8 f(x)\,dx - 4$ (D) $\int_{-2}^8 f(x)\,dx - 40$

[PAGE 468] 4. $\int_1^e \dfrac{3x^2 + 1}{x}\,dx =$
(A) $\dfrac{3e^2 - 1}{2}$ (B) $\dfrac{3e^2 + 1}{2}$
(C) $e^3 + 1$ (D) $3e^2 - 1$

[PAGE 473] 5. The average value of $f(x) = \sin x$ on the interval $\left[-\dfrac{\pi}{3}, \dfrac{\pi}{2}\right]$ is
(A) $-\dfrac{3}{5\pi}$ (B) $\dfrac{3}{5\pi}$ (C) $\dfrac{1}{2}$ (D) $\dfrac{5\pi}{12}$

[PAGE 467] 6. $\int_1^4 \sqrt{x}\left(x - \dfrac{1}{x}\right)dx =$
(A) $\dfrac{32}{5}$ (B) $\dfrac{44}{5}$ (C) $\dfrac{52}{5}$ (D) $\dfrac{56}{15}$

[PAGE 469] 7. The graph of the piecewise function f is below. What is $\int_{-2}^4 f(x)\,dx$?

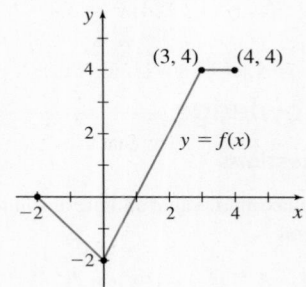

(A) 2   (B) 5   (C) $\dfrac{17}{2}$ (D) 9

[PAGE 469] 8. If $f(x) = \begin{cases} x^3 & \text{if } x \leq 1 \\ \dfrac{1}{x} & \text{if } x > 1 \end{cases}$ then $\int_0^e f(x)\,dx$ equals
(A) 0   (B) $\dfrac{5}{4}$ (C) $\dfrac{1}{4} + e$ (D) $\dfrac{e^4}{4}$

[PAGE 473] 9. What is the average value of $f(x) = \dfrac{1}{x}$ on the closed interval $[1, 4]$?
(A) $\ln \dfrac{4}{3}$ (B) $\dfrac{\ln 3}{3}$ (C) $\dfrac{\ln 4}{3}$ (D) $\ln 4$

[PAGE 467] 10. The area under the graph of $f(x) = x^2(3 - x)$ from 0 to 3 is
(A) 6   (B) $\dfrac{27}{4}$ (C) 7   (D) 7.5

[PAGE 468] 11. If $\int_1^8 f(x)\,dx = 5$ and $\int_8^4 f(x)\,dx = 9$, then $\int_1^4 f(x)\,dx$ equals
(A) -4 (B) 4   (C) 8   (D) 14

[PAGE 473] **12.** What is the average value of the part of the graph of $f(x) = x^3(2 - x)$ that lies in the first quadrant?

(A) $\dfrac{4}{5}$ (B) 1   (C) $\dfrac{8}{5}$ (D) $\dfrac{12}{5}$

[PAGE 469] **13.** $\displaystyle\int_0^6 |x - 4|\, dx =$

(A) 6   (B) 10   (C) 22   (D) 42

[PAGE 471] **14.** The graph of f is shown below.

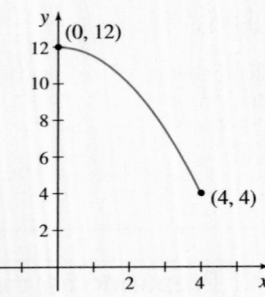

Then $\displaystyle\int_0^4 f(x)\, dx$ must be between

(A) 4 and 12   (B) 20 and 32

(C) 16 and 48   (D) 40 and 60

[PAGE 470] **15.** Find $\displaystyle\int_{-2}^{10} f(x)\, dx$ where $f(x) = \begin{cases} x + 3 & \text{if } -2 \le x < 2 \\ 3x & \text{if } 2 \le x \le 10 \end{cases}$

(A) 144   (B) 150   (C) 156   (D) 306

[PAGE 467] **16.** If $\int_2^6 f(x)\,dx = 10$, then $\int_2^6 3[f(x) + 4]dx =$

(A) 10   (B) 46   (C) 78   (D) 26

Free-Response Questions

[PAGE 474] **17.** An object is moving along a horizontal line with velocity $v(t) = 3t^2 - 6t$, $1 \le t \le 4$ (in meters per second).
  (a) Find the total distance the object moves from $t = 1$ to $t = 4$.
  (b) For what time(s) is the object at rest?
  (c) If at time $t = 1$, the object is 2 m to the right of the origin, what is its position at $t = 4$?
  (d) Find the average velocity of the object from $t = 1$ to $t = 4$.

*See the **BREAK IT DOWN** on page 509 for a stepped out solution to AP® Practice Problem 17(c).*

[PAGE 474] **18.** An object is moving along the y-axis with the position function $y(t) = \sin t^2$, in meters, and time t in minutes, $0 \le t \le \pi$.
  (a) Find the velocity of the object at time t.
  (b) Find the net displacement of the object over the interval $[0, \pi]$.
  (c) Find the total distance traveled by the object over the interval $[0, \pi]$.

Retain Your Knowledge

Multiple-Choice Questions

1. Find the absolute maximum value and the absolute minimum value of the function

$$f(x) = \begin{cases} x^2 - 4 & \text{if } -1 \le x \le 3 \\ -x + 8 & \text{if } 3 < x \le 9 \end{cases}$$

  (A) 9 is the absolute maximum value; -1 is the absolute minimum value
  (B) 5 is the absolute maximum value; -3 is the absolute minimum value
  (C) 5 is the absolute maximum value; -4 is the absolute minimum value
  (D) 3 is the absolute maximum value; -1 is the absolute minimum value

2. A function f is continuous on the closed interval $[-3, 10]$. The table below shows selected values of f.

x	-3	1	3	7	8	10
$f(x)$	0	5	9	0	4	12

Which of the following must be true?
  I. The function has an absolute maximum and an absolute minimum value on $[-3, 10]$.
  II. There is a number c, $-3 < c < 7$, for which $f'(c) = 0$.
  III. There is a number c, $-3 < c < 10$, for which $f'(c) = 2$.

  (A) I only   (B) I and II only
  (C) I and III only   (D) I, II, and III

3. The antiderivatives F of the function $f(x) = x^2 + \sin x$ are

(A) $x^3 + \cos x + C$ (B) $\dfrac{x^3}{3} - \dfrac{1}{2}\sin^2 x \cdot \cos x + C$

(C) $\dfrac{x^3}{3} + \cos x + C$ (D) $\dfrac{x^3}{3} - \cos x + C$

Free-Response Question

4. An object moves along a horizontal line. The velocity of the object in ft/min is given by a differentiable function $v = v(t)$. The velocity is measured at selected times shown in the table.

t	0	1	3	6	8	10
$v(t)$	0	6	8	12	-4	-12

  (a) Approximate the acceleration a of the object at $t = 5$ min.
  (b) Is there a time during the interval $0 < t < 10$ when the velocity v of the object is -6 ft/min? Justify your reasoning.

6.5 The Indefinite Integral; Method of Substitution

OBJECTIVES *When you finish this section, you should be able to:*

1 **Find indefinite integrals (p. 481)**
2 **Use properties of indefinite integrals (p. 483)**
3 **Find an indefinite integral using substitution (p. 484)**
4 **Find a definite integral using substitution (p. 488)**
5 **Integrate even and odd functions (p. 490)**

NEED TO REVIEW? Antiderivatives are discussed in Chapter 5, Section 5.6.

The Fundamental Theorem of Calculus establishes an important relationship between definite integrals and antiderivatives: the definite integral $\int_a^b f(x)\,dx$ can be found easily if an antiderivative of f can be found. Because of this, it is customary to use the integral symbol $\int$ as an instruction to find all the antiderivatives of a function.

1 Find Indefinite Integrals

> **DEFINITION** Indefinite Integral
>
> The expression $\int f(x)\,dx$, called the **indefinite integral of** f, is defined as,
>
> $$\boxed{\int f(x)\,dx = F(x) + C}$$
>
> where F is any function for which $\dfrac{d}{dx}F(x) = f(x)$ and C is a number, called the **constant of integration**.

IN WORDS The indefinite integral $\int f(x)\,dx$ is a symbol for all the antiderivatives of f.

CAUTION In writing an indefinite integral $\int f(x)\,dx$, remember to include the "dx", and in evaluating an indefinite integral, remember to add the constant of integration.

For example,

$$\int (x^2 + 1)\,dx = \frac{x^3}{3} + x + C \qquad \text{Check: } \frac{d}{dx}\left(\frac{x^3}{3} + x + C\right) = \frac{3x^2}{3} + 1 + 0 = x^2 + 1$$

The process of finding either the indefinite integral $\int f(x)\,dx$ or the definite integral $\int_a^b f(x)\,dx$ is called **integration**, and in both cases the function f is called the **integrand**.

It is important to distinguish between the definite integral $\int_a^b f(x)\,dx$ and the indefinite integral $\int f(x)\,dx$. The definite integral is a *number* that depends on the limits of integration a and b. In contrast, the indefinite integral of f is a *family* of functions $F(x) + C$, C a constant, for which $F'(x) = f(x)$. For example,

IN WORDS The definite integral $\int_a^b f(x)\,dx$ is a number; the indefinite integral $\int f(x)\,dx$ is a family of functions.

$$\cdot \quad \int_0^2 x^2\,dx = \left[\frac{x^3}{3}\right]_0^2 = \frac{8}{3} \qquad\qquad \cdot \quad \int x^2\,dx = \frac{x^3}{3} + C$$

We list the indefinite integrals of some important functions in Table 1. Each entry results from a differentiation formula.

TABLE 1 Basic Integral Formulas

- $\displaystyle \int dx = x + C$

- $\displaystyle \int x^a \, dx = \frac{x^{a+1}}{a+1} + C; \quad a \neq -1$

- $\displaystyle \int x^{-1} \, dx = \int \frac{1}{x} \, dx = \ln|x| + C$

- $\displaystyle \int e^x \, dx = e^x + C$

- $\displaystyle \int a^x \, dx = \frac{a^x}{\ln a} + C; \quad a > 0, \, a \neq 1$

- $\displaystyle \int \sin x \, dx = -\cos x + C$

- $\displaystyle \int \cos x \, dx = \sin x + C$

- $\displaystyle \int \sec^2 x \, dx = \tan x + C$

- $\displaystyle \int \sec x \tan x \, dx = \sec x + C$

- $\displaystyle \int \csc x \cot x \, dx = -\csc x + C$

- $\displaystyle \int \csc^2 x \, dx = -\cot x + C$

- $\displaystyle \int \frac{1}{\sqrt{1-x^2}} \, dx = \sin^{-1} x + C, \quad |x| < 1$

- $\displaystyle \int \frac{1}{1+x^2} \, dx = \tan^{-1} x + C$

- $\displaystyle \int \frac{1}{x\sqrt{x^2-1}} \, dx = \sec^{-1} x + C, \quad |x| > 1$

EXAMPLE 1   Finding Indefinite Integrals

Find:

(a) $\displaystyle \int x^4 \, dx$ **(b)** $\displaystyle \int \sqrt{x} \, dx$ **(c)** $\displaystyle \int \frac{\sin x}{\cos^2 x} \, dx$

Solution

(a) The antiderivatives of $f(x) = x^4$ are $F(x) = \dfrac{x^5}{5} + C$, so

$$\int x^4 \, dx = \frac{x^5}{5} + C$$

(b) The antiderivatives of $f(x) = \sqrt{x} = x^{1/2}$ are $F(x) = \dfrac{x^{3/2}}{\frac{3}{2}} + C = \dfrac{2x^{3/2}}{3} + C$, so

$$\int \sqrt{x} \, dx = \frac{2x^{3/2}}{3} + C$$

NEED TO REVIEW? Trigonometric identities are discussed in Appendix A.4, pp. A-35 to A-38.

(c) No integral in Table 1 corresponds to $\displaystyle \int \frac{\sin x}{\cos^2 x} \, dx$, so we begin by using trigonometric identities to rewrite $\dfrac{\sin x}{\cos^2 x}$ in a form whose antiderivative is recognizable.

$$\frac{\sin x}{\cos^2 x} = \frac{\sin x}{\cos x \cdot \cos x} = \frac{1}{\cos x} \cdot \frac{\sin x}{\cos x} = \sec x \tan x$$

Then

$$\int \frac{\sin x}{\cos^2 x} \, dx = \int \sec x \tan x \, dx = \sec x + C$$

The antiderivatives of $\sec x \tan x$ are $\sec x + C$. ∎

NOW WORK   Problems 9 and 11.

2 Use Properties of Indefinite Integrals

Since the definite integral and the indefinite integral are closely related, properties of indefinite integrals are very similar to those of definite integrals:

- **Derivative of an Indefinite Integral**:

$$\boxed{\frac{d}{dx} \int f(x)\,dx = f(x)} \tag{1}$$

Property (1) is a consequence of the definition of $\int f(x)\,dx$. For example,

- $\dfrac{d}{dx} \displaystyle\int \sqrt{x^2 + 1}\,dx = \sqrt{x^2 + 1}$ • $\dfrac{d}{dt} \displaystyle\int e^t \cos t \, dt = e^t \cos t$

NOW WORK AP® Practice Problem 2.

- **Indefinite Integral of the Sum of Two Functions**:

IN WORDS The indefinite integral of a sum of two functions equals the sum of the indefinite integrals.

$$\boxed{\int [f(x) + g(x)]\,dx = \int f(x)\,dx + \int g(x)\,dx} \tag{2}$$

The proof of property (2) follows directly from properties of derivatives and is left as an exercise. See Problem 144.

- **Indefinite Integral of a Constant Times a Function**:
  If k is a constant,

$$\boxed{\int kf(x)\,dx = k \int f(x)\,dx} \tag{3}$$

To prove property (3), differentiate the right side of (3).

IN WORDS To find the indefinite integral of a constant k times a function f, find the indefinite integral of f and then multiply by k.

$$\frac{d}{dx}\left[k \int f(x)\,dx \right] \underset{\substack{\uparrow \\ \text{Constant Multiple Rule}}}{=} k\left[\frac{d}{dx} \int f(x)\,dx \right] \underset{\substack{\uparrow \\ \text{Property (1)}}}{=} kf(x)$$

EXAMPLE 2 **Using Properties of Indefinite Integrals**

(a)
$$\int (2x^{1/3} + 5x^{-1})\,dx = \int 2x^{1/3}\,dx + \int \frac{5}{x}\,dx = 2\int x^{1/3}\,dx + 5\int \frac{1}{x}\,dx$$

$$= 2 \cdot \frac{x^{4/3}}{\frac{4}{3}} + 5\ln|x| + C = \frac{3x^{4/3}}{2} + 5\ln|x| + C$$

(b)
$$\int \left(\frac{12}{x^5} + \frac{1}{\sqrt{x}} \right)\,dx = 12\int \frac{1}{x^5}\,dx + \int \frac{1}{\sqrt{x}}\,dx = 12\int x^{-5}\,dx + \int x^{-1/2}\,dx$$

$$= 12 \cdot \left(\frac{x^{-4}}{-4} \right) + \frac{x^{1/2}}{\frac{1}{2}} + C = -\frac{3}{x^4} + 2\sqrt{x} + C \qquad \blacksquare$$

NOW WORK Problem **17** and AP® Practice Problem **1**.

Sometimes an appropriate algebraic manipulation is required before integrating.

EXAMPLE 3 Using Properties of Indefinite Integrals

$$\int \frac{x^2+6}{x^2+1}\,dx = \int \frac{(x^2+1)+5}{x^2+1}\,dx = \int \left[\frac{x^2+1}{x^2+1}+\frac{5}{x^2+1}\right]dx$$

$$= \int \left[1+\frac{5}{x^2+1}\right]dx = \underset{\underset{\text{Sum Property}}{\uparrow}}{\int dx} + \int \frac{5}{x^2+1}\,dx$$

$$= \int dx + 5\int \frac{1}{x^2+1}\,dx \underset{\uparrow}{=} x+5\tan^{-1}x + C$$

The antiderivatives of $\frac{1}{x^2+1}$ are $\tan^{-1}x+C$.

NOW WORK Problem **13** and AP® Practice Problems **3** and **11**.

③ Find an Indefinite Integral Using Substitution

Indefinite integrals that cannot be found using Table 1 sometimes can be found using the *method of substitution*. In the method of substitution, we change the variable to transform the integrand into a form that is readily integrable.

For example, to find $\int (x^2+5)^3 2x\,dx$, we use the substitution $u=x^2+5$. The differential of $u=x^2+5$ is $du=2x\,dx$. Now we write $(x^2+5)^3 2x\,dx$ in terms of u and du, and integrate the simpler integral.

$$\int \underset{u}{\underbrace{(x^2+5)^3}}\underset{du}{\underbrace{2x\,dx}} = \int u^3\,du = \frac{u^4}{4}+C \underset{\underset{u=x^2+5}{\uparrow}}{=} \frac{(x^2+5)^4}{4}+C$$

We can verify the result by differentiating it using the Power Rule for Functions.

$$\frac{d}{dx}\left[\frac{(x^2+5)^4}{4}+C\right]=\frac{1}{4}\left[4(x^2+5)^3(2x)+0\right]=(x^2+5)^3 2x$$

The method of substitution is based on the Chain Rule, which states that if f and g are differentiable functions, then for the composite function $f\circ g$,

$$\frac{d}{dx}[(f\circ g)(x)]=\frac{d}{dx}f(g(x))=f'(g(x))\,g'(x)$$

The Chain Rule provides a template for finding integrals of the form

$$\int f'(g(x))g'(x)\,dx$$

If we let $u=g(x)$, then the differential $du=g'(x)\,dx$, and we have

$$\int \underset{u}{\underbrace{f'(g(x))}}\underset{du}{\underbrace{g'(x)\,dx}} = \int f'(u)\,du = f(u)+C \underset{\underset{u=g(x)}{\uparrow}}{=} f(g(x))+C$$

Replacing $g(x)$ by u and $g'(x)dx$ by du is called **substitution**. Substitution is a strategy for finding indefinite integrals when the integrand is a composite function.

NOTE Since the integrand is an improper rational expression, polynomial division is an alternative.

NEED TO REVIEW? Differentials are discussed in Section 4.3, pp. 301–303.

NOTE When using the method of substitution, once the substitution u is chosen, write the original integral in terms of u and du.

NEED TO REVIEW? The Chain Rule is discussed in Section 3.1, pp. 235–243.

EXAMPLE 4   **Finding Indefinite Integrals Using Substitution**

Find:

(a) $\displaystyle\int \sin(3x+2)\,dx$ **(b)** $\displaystyle\int x\sqrt{x^2+1}\,dx$ **(c)** $\displaystyle\int \frac{e^{\sqrt{x}}}{\sqrt{x}}\,dx$

Solution

(a) Since $f(x)=\sin(3x+2)$ is a composite function, and we know $\int \sin x\,dx$, we let $u=3x+2$. Then $du=3\,dx$ so $dx=\dfrac{du}{3}$.

$$\int \underbrace{\sin(3x+2)}_{u}\ \underbrace{dx}_{\frac{du}{3}} = \int \sin u\,\frac{du}{3} = \frac{1}{3}\int \sin u\,du$$

$$= \frac{1}{3}(-\cos u)+C \underset{\underset{u=3x+2}{\uparrow}}{=} -\frac{1}{3}\cos(3x+2)+C$$

(b) Let $u=x^2+1$. Then $du=2x\,dx$, so $x\,dx=\dfrac{du}{2}$.

$$\int x\sqrt{x^2+1}\,dx = \int \sqrt{x^2+1}\,x\,dx = \int \sqrt{u}\,\frac{du}{2} = \frac{1}{2}\int u^{1/2}\,du = \frac{1}{2}\cdot\frac{u^{3/2}}{\frac{3}{2}}+C$$

$$= \frac{(x^2+1)^{3/2}}{3}+C$$

(c) Let $u=\sqrt{x}=x^{1/2}$. Then $du=\dfrac{1}{2}x^{-1/2}dx=\dfrac{dx}{2\sqrt{x}}$, so $\dfrac{dx}{\sqrt{x}}=2\,du$.

$$\int \frac{e^{\sqrt{x}}}{\sqrt{x}}\,dx = \int e^{\sqrt{x}}\cdot\frac{dx}{\sqrt{x}} = \int e^u\cdot 2\,du = 2e^u+C = 2e^{\sqrt{x}}+C \qquad\blacksquare$$

NOW.WORK   Problems **21** and **27** and AP® Practice Problem 10.

Substitution is often a good strategy to use when an integrand equals the product of an expression involving a function and its derivative (or a multiple of its derivative). In Example 4(c), $\displaystyle\int \frac{e^{\sqrt{x}}}{\sqrt{x}}dx$, we used the substitution $u=\sqrt{x}$, because

$$\frac{du}{dx}=\frac{d}{dx}\sqrt{x}=\frac{1}{2\sqrt{x}}\ \text{is a multiple of}\ \frac{1}{\sqrt{x}}$$

Similarly, in Example 4(b), $\int x\sqrt{x^2+1}\,dx$, the factor x in the integrand makes the substitution $u=x^2+1$ work. On the other hand, if we try to use this same substitution to integrate $\int \sqrt{x^2+1}\,dx$, then

$$\int \sqrt{x^2+1}\,dx = \int \sqrt{u}\,\frac{du}{2x}\ \underset{\underset{x=\sqrt{u-1}}{\uparrow}}{=}\ \int \frac{\sqrt{u}}{2\sqrt{u-1}}\,du$$

and the resulting integral is *more* complicated than the original integral.

The idea behind substitution is to obtain an integral $\int h(u)\,du$ that is simpler than the original integral $\int f(x)\,dx$. When a substitution does not simplify the integral, try a different substitution. If no substitution works, other integration methods should be tried. Some of these methods are explored in Chapter 6, Part 2.

EXAMPLE 5   Finding Indefinite Integrals Using Substitution

Find:

(a) $\displaystyle\int \frac{e^x}{e^x+4}\,dx$ **(b)** $\displaystyle\int \frac{5x^2\,dx}{4x^3-1}$

Solution

(a) Here the numerator equals the derivative of the denominator. So, we use the substitution $u = e^x + 4$. Then $du = e^x dx$.

$$\int \frac{e^x}{e^x+4}\,dx = \int \frac{1}{e^x+4}\cdot e^x dx = \int \frac{1}{u}\,du = \ln|u| + C = \ln(e^x+4) + C$$
$$\underset{u=e^x+4>0}{\uparrow}$$

(b) Notice that the numerator equals the derivative of the denominator, except for a constant factor. So, we try the substitution $u = 4x^3 - 1$. Then $du = 12x^2 dx$, so $5x^2 dx = \dfrac{5}{12}\,du$.

$$\int \frac{5x^2\,dx}{4x^3-1} = \int \frac{\frac{5}{12}\,du}{u} = \frac{5}{12}\int \frac{du}{u} = \frac{5}{12}\ln|u| + C = \frac{5}{12}\ln\left|4x^3-1\right| + C \quad \blacksquare$$

In Example 5(a) and 5(b), the integrands have a numerator that is the derivative of the denominator. In general, we have the following formula:

$$\boxed{\int \frac{g'(x)}{g(x)}\,dx = \ln|g(x)| + C} \tag{4}$$

IN WORDS If the numerator of the integrand equals the derivative of the denominator, then the integral equals a logarithmic function.

Notice that in formula (4) the integral equals the natural logarithm of the *absolute value of the function g*. The absolute value is necessary since the domain of the logarithm function is the set of positive real numbers. When g is known to be positive, as in Example 5(a), the absolute value is not required.

NOW WORK   Problem **33** and AP® Practice Problems **6** and **7**.

EXAMPLE 6   Using Substitution to Establish an Integration Formula

Show that:

(a)
$$\boxed{\int \tan x\,dx = -\ln|\cos x| + C = \ln|\sec x| + C}$$

(b)
$$\boxed{\int \sec x\,dx = \ln|\sec x + \tan x| + C}$$

Solution

(a) Since $\tan x = \dfrac{\sin x}{\cos x}$, let $u = \cos x$. Then $du = -\sin x\, dx$ and

$$\int \tan x\, dx = \int \frac{\sin x}{\cos x} dx = \int -\frac{du}{u} = -\ln|u| + C = -\ln|\cos x| + C$$

$$= \underset{\underset{r\,\ln x\,=\,\ln x^r}{\uparrow}}{\ln|\cos x|^{-1}} + C = \ln\left|\frac{1}{\cos x}\right| + C = \ln|\sec x| + C$$

(b) To find $\int \sec x\, dx$, multiply the integrand by $\dfrac{\sec x + \tan x}{\sec x + \tan x}$.

$$\int \sec x\, dx = \int \sec x \cdot \frac{\sec x + \tan x}{\sec x + \tan x} dx = \int \frac{\sec^2 x + \sec x \tan x}{\sec x + \tan x} dx$$

Now the numerator equals the derivative of the denominator. So by (4)

$$\int \sec x\, dx = \ln|\sec x + \tan x| + C \qquad \blacksquare$$

NOW WORK   Problem 43.

As we saw in Example 6(b), sometimes we need to manipulate an integral so that a basic integration formula can be used. Unlike differentiation, integration has no prescribed method; some ingenuity and a lot of practice are required. To illustrate, two different substitutions are used to solve Example 7.

EXAMPLE 7   **Finding an Indefinite Integral Using Substitution**

Find $\displaystyle\int x\sqrt{4+x}\, dx$.

Solution

Substitution I   Let $u = 4 + x$. Then $du = dx$. The x term remains. Since $u = 4 + x$, then $x = u - 4$. Substituting gives

$$\int x\sqrt{4+x}\, dx = \int \underset{\underset{x}{\uparrow}}{(u-4)} \underset{\underset{4+x}{\uparrow}}{\sqrt{u}}\ \underset{\underset{dx}{\uparrow}}{du} = \int (u^{3/2} - 4u^{1/2})\, du$$

$$= \frac{u^{5/2}}{\frac{5}{2}} - 4 \cdot \frac{u^{3/2}}{\frac{3}{2}} + C$$

$$= \frac{2(4+x)^{5/2}}{5} - \frac{8(4+x)^{3/2}}{3} + C$$

Substitution II   Let $u = \sqrt{4+x}$, so $u^2 = 4+x$ and $x = u^2 - 4$. Then $dx = 2u\, du$ and

$$\int x\sqrt{4+x}\, dx = \int \underset{\underset{x}{\uparrow}}{(u^2-4)}(u)\underset{\underset{dx}{\uparrow}}{(2u\, du)} = 2\int (u^4 - 4u^2)\, du = 2\left[\frac{u^5}{5} - \frac{4u^3}{3}\right] + C$$

$$= \frac{2}{5}\left(\sqrt{4+x}\right)^5 - \frac{8}{3}\left(\sqrt{4+x}\right)^3 + C = \frac{2(4+x)^{5/2}}{5} - \frac{8(4+x)^{3/2}}{3} + C$$

$\blacksquare$

NOW WORK   Problem 51.

EXAMPLE 8   **Finding Indefinite Integrals Using Substitution**

Find: (a) $\displaystyle\int \frac{dx}{\sqrt{4-x^2}}$ (b) $\displaystyle\int \frac{dx}{9+4x^2}$

Solution

(a) $\displaystyle\int \frac{dx}{\sqrt{4-x^2}}$ resembles $\displaystyle\int \frac{1}{\sqrt{1-x^2}}\, dx = \sin^{-1} x + C$. We begin by rewriting the

integrand as

$$\frac{1}{\sqrt{4-x^2}} = \frac{1}{\sqrt{4\left(1-\dfrac{x^2}{4}\right)}} = \frac{1}{2\sqrt{1-\left(\dfrac{x}{2}\right)^2}}$$

Now let $u = \dfrac{x}{2}$. Then $du = \dfrac{dx}{2}$

$$\int \frac{dx}{\sqrt{4-x^2}} = \int \frac{dx}{2\sqrt{1-\left(\dfrac{x}{2}\right)^2}} \underset{\substack{\uparrow \\ u = \frac{x}{2} \\ du = \frac{dx}{2}}}{=} \int \frac{du}{\sqrt{1-u^2}} = \sin^{-1} u + C$$

$$= \sin^{-1}\left(\frac{x}{2}\right) + C$$

(b) $\displaystyle\int \frac{dx}{9+4x^2}$ resembles $\displaystyle\int \frac{1}{1+x^2}\, dx = \tan^{-1} x + C$. Rewrite the integrand as

$$\frac{1}{9+4x^2} = \frac{1}{9\left(1+\dfrac{4x^2}{9}\right)} = \frac{1}{9\left[1+\left(\dfrac{2x}{3}\right)^2\right]}$$

Now let $u = \dfrac{2x}{3}$. Then $du = \dfrac{2}{3}\, dx$, so $dx = \dfrac{3}{2}\, du$.

$$\int \frac{dx}{9+4x^2} = \int \frac{dx}{9\left[1+\left(\dfrac{2x}{3}\right)^2\right]} = \int \frac{\dfrac{3}{2}du}{9(1+u^2)} = \frac{1}{6}\int \frac{du}{1+u^2}$$

$$= \frac{1}{6}\tan^{-1} u + C = \frac{1}{6}\tan^{-1}\frac{2x}{3} + C$$ ∎

NOW WORK   Problem 55.

---

4 Find a Definite Integral Using Substitution

Two approaches can be used to find a definite integral using substitution:

AP® EXAM TIP

On the exam, either method can be used. Method 2 is often tested in Multiple-Choice Questions. See AP® Practice Problem 4 on p. 496.

- Method 1: Find the related indefinite integral using substitution, and then use the Fundamental Theorem of Calculus.
- Method 2: Find the definite integral directly by making a substitution in the integrand and using the substitution to *change the limits of integration.*

EXAMPLE 9   **Finding a Definite Integral Using Substitution**

Find $\displaystyle\int_0^2 x\sqrt{4-x^2}\, dx$.

Solution

Method 1: Use the related indefinite integral and then use the Fundamental Theorem of Calculus. The related indefinite integral $\int x\sqrt{4-x^2}\,dx$ can be found using the substitution $u = 4 - x^2$. Then $du = -2x\,dx$, so $x\,dx = -\dfrac{du}{2}$.

$$\int x\sqrt{4-x^2}\,dx = \int \sqrt{u}\left(-\frac{du}{2}\right) = -\frac{1}{2}\int u^{1/2}\,du = -\frac{1}{2}\cdot\frac{u^{3/2}}{\frac{3}{2}} + C$$

$$= -\frac{1}{3}(4-x^2)^{3/2} + C$$

Then by the Fundamental Theorem of Calculus,

$$\int_0^2 x\sqrt{4-x^2}\,dx = -\frac{1}{3}\left[(4-x^2)^{3/2}\right]_0^2 = -\frac{1}{3}\left[0 - 4^{3/2}\right] = \frac{8}{3}$$

Method 2: Find the definite integral directly by making a substitution in the integrand and changing the limits of integration. Let $u = 4 - x^2$; then $du = -2x\,dx$. Now use the function $u = 4 - x^2$ to change the limits of integration.

- The lower limit of integration is $x = 0$ so, in terms of u, it changes to $u = 4 - 0^2 = 4$.

- The upper limit of integration is $x = 2$ so the upper limit changes to $u = 4 - 2^2 = 0$.

    Then

$$\int_0^2 x\sqrt{4-x^2}\,dx = \underset{\substack{\uparrow \\ u=4-x^2 \\ x\,dx=-\frac{1}{2}du}}{\int_4^0} \sqrt{u}\left(-\frac{du}{2}\right) = -\frac{1}{2}\int_4^0 \sqrt{u}\,du = -\frac{1}{2}\cdot\left[\frac{u^{3/2}}{\frac{3}{2}}\right]_4^0$$

CAUTION When using substitution to find a definite integral directly, remember to change the limits of integration.

$$= -\frac{1}{3}(0-8) = \frac{8}{3} \qquad \blacksquare$$

NOW WORK   Problem **61** and AP® Practice Problems **4, 5, 8, 9, 12,** and **13.**

EXAMPLE 10   **Finding a Definite Integral Using Substitution**

Find $\displaystyle\int_0^{\pi/2} \frac{1-\cos(2\theta)}{2}\,d\theta$.

Solution

Use properties of integrals to simplify before integrating.

$$\int_0^{\pi/2} \frac{1-\cos(2\theta)}{2}\,d\theta = \frac{1}{2}\int_0^{\pi/2}[1-\cos(2\theta)]\,d\theta$$

$$= \frac{1}{2}\left[\int_0^{\pi/2} d\theta - \int_0^{\pi/2}\cos(2\theta)\,d\theta\right]$$

$$= \frac{1}{2}\int_0^{\pi/2} d\theta - \frac{1}{2}\int_0^{\pi/2}\cos(2\theta)\,d\theta$$

$$= \frac{1}{2}\Big[\theta\Big]_0^{\pi/2} - \frac{1}{2}\int_0^{\pi/2}\cos(2\theta)\,d\theta$$

$$= \frac{\pi}{4} - \frac{1}{2}\int_0^{\pi/2}\cos(2\theta)\,d\theta$$

(Example continued on the next page.)

In the integral on the right, use the substitution $u = 2\theta$. Then $du = 2\,d\theta$ so $d\theta = \dfrac{du}{2}$.
Now change the limits of integration:

- when $\theta = 0$ then $u = 2 \cdot 0 = 0$
- when $\theta = \dfrac{\pi}{2}$ then $u = 2 \cdot \dfrac{\pi}{2} = \pi$

Now

$$\int_0^{\pi/2} \cos(2\theta)\,d\theta = \int_0^{\pi} \cos u\, \frac{du}{2} = \frac{1}{2}\Big[\sin u\Big]_0^{\pi} = \frac{1}{2}\left(\sin \pi - \sin 0\right) = 0$$

Then,

$$\int_0^{\pi/2} \frac{1 - \cos(2\theta)}{2}\,d\theta = \frac{\pi}{4} - \frac{1}{2}\int_0^{\pi/2} \cos(2\theta)\,d\theta = \frac{\pi}{4}$$ ∎

NOW WORK Problem 69.

5 Integrate Even and Odd Functions

RECALL A function f is even if $f(-x) = f(x)$ and is odd if $f(-x) = -f(x)$ for all x in the domain of f.

Integrals of even and odd functions sometimes can be simplified using symmetry. Figure 39 illustrates the conclusions of the theorem that follows.

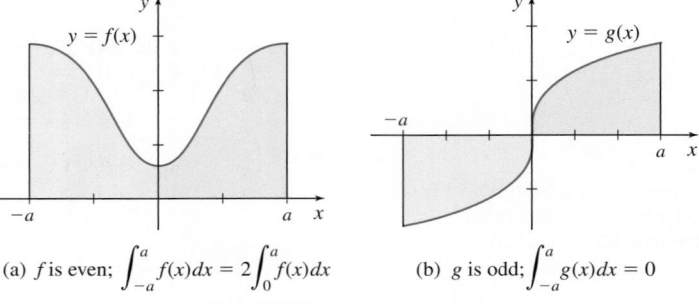

(a) f is even; $\displaystyle\int_{-a}^{a} f(x)dx = 2\int_0^{a} f(x)dx$ (b) g is odd; $\displaystyle\int_{-a}^{a} g(x)dx = 0$

Figure 39

THEOREM The Integrals of Even and Odd Functions

Suppose a function f is continuous on a closed interval $[-a, a]$, $a > 0$.

- If f is an even function, then

$$\boxed{\int_{-a}^{a} f(x)\,dx = 2\int_0^{a} f(x)\,dx}$$

- If f is an odd function, then

$$\boxed{\int_{-a}^{a} f(x)\,dx = 0}$$

The property for even functions is proved here; the proof for odd functions is left as an exercise. See Problem 135.

Proof f is an even function: Since f is continuous on the closed interval $[-a, a]$, $a > 0$, and 0 is in the interval $[-a, a]$, we have

$$\int_{-a}^{a} f(x)\, dx = \int_{-a}^{0} f(x)\, dx + \int_{0}^{a} f(x)\, dx = -\int_{0}^{-a} f(x)\, dx + \int_{0}^{a} f(x)\, dx \quad (5)$$

In $-\int_{0}^{-a} f(x)\, dx$, we use the substitution $u = -x$. Then $du = -dx$. Also, if $x = 0$, then $u = 0$, and if $x = -a$, then $u = a$. Therefore,

$$-\int_{0}^{-a} f(x)\, dx = \int_{0}^{a} f(-u)\, du = \int_{0}^{a} f(u)\, du = \int_{0}^{a} f(x)\, dx \quad (6)$$

$$\uparrow$$
$$f \text{ is even.}$$
$$f(-u) = f(u)$$

Combining (5) and (6), we obtain

$$\int_{-a}^{a} f(x)\, dx = \int_{0}^{a} f(x)\, dx + \int_{0}^{a} f(x)\, dx = 2\int_{0}^{a} f(x)\, dx \quad \blacksquare$$

To use the theorem involving even or odd functions, three conditions must be met:

- The function f must be even or odd.
- The function f must be continuous on the closed interval $[-a, a]$, $a > 0$.
- The limits of integration must be $-a$ and a, $a > 0$.

EXAMPLE 11 **Integrating an Even or Odd Function**

Find:

(a) $\displaystyle\int_{-3}^{3} (x^7 - 4x^3 + x)\, dx$ **(b)** $\displaystyle\int_{-2}^{2} (x^4 - x^2 + 3)\, dx$

Solution

(a) If $f(x) = x^7 - 4x^3 + x$, then

$$f(-x) = (-x)^7 - 4(-x)^3 + (-x) = -(x^7 - 4x^3 + x) = -f(x)$$

Since f is an odd function,

$$\int_{-3}^{3} (x^7 - 4x^3 + x)\, dx = 0$$

(b) If $g(x) = x^4 - x^2 + 3$, then $g(-x) = (-x)^4 - (-x)^2 + 3 = x^4 - x^2 + 3 = g(x)$. Since g is an even function,

$$\int_{-2}^{2} (x^4 - x^2 + 3)\, dx = 2\int_{0}^{2} (x^4 - x^2 + 3)\, dx = 2\left[\frac{x^5}{5} - \frac{x^3}{3} + 3x\right]_{0}^{2}$$

$$= 2\left[\frac{32}{5} - \frac{8}{3} + 6\right] = \frac{292}{15} \quad \blacksquare$$

NOW WORK   Problems **79** and **81.**

EXAMPLE 12   Using Properties of Integrals

If f is an even function and $\int_0^2 f(x)\,dx = -6$ and $\int_{-5}^0 f(x)\,dx = 8$, find $\int_2^5 f(x)\,dx$.

Solution

$\int_2^5 f(x)\,dx = \int_2^0 f(x)\,dx + \int_0^5 f(x)\,dx$

Now $\int_2^0 f(x)\,dx = -\int_0^2 f(x)\,dx = 6$

Since f is even, $\int_0^5 f(x)\,dx = \int_{-5}^0 f(x)\,dx = 8$. Then

$$\int_2^5 f(x)\,dx = \int_2^0 f(x)\,dx + \int_0^5 f(x)\,dx = 6+8 = 14$$

NOW WORK   Problem 87.

6.5 Assess Your Understanding

Concepts and Vocabulary

1. $\dfrac{d}{dx}\displaystyle\int f(x)\,dx = $ _____

2. **True or False** If k is a constant, then $\int kf(x)\,dx = \int k\,dx \cdot \int f(x)\,dx$

3. If a is a real number, $a \neq -1$, then $\int x^a\,dx = $ _____.

4. **True or False** When finding the indefinite integral of a function f, a constant of integration C is added to the result because $\int f(x)\,dx$ denotes all the antiderivatives of f.

5. If the substitution $u = 2x+3$ is used with $\int \sin(2x+3)\,dx$, the result is $\int$ _____ du.

6. **True or False** If the substitution $u = x^2+3$ is used with $\int_0^1 x(x^2+3)^3\,dx$, then $\int_0^1 x(x^2+3)^3\,dx = \dfrac{1}{2}\int_0^1 u^3\,du$.

7. **Multiple Choice** $\int_{-4}^4 x^3\,dx = $ [(a) 128 (b) 4 (c) 0 (d) 64]

8. **True or False** $\int_0^5 x^2\,dx = \dfrac{1}{2}\int_{-5}^5 x^2\,dx$

Skill Building

In Problems 9–20, find each indefinite integral.

9. $\int x^{2/3}\,dx$

10. $\int t^{-4}\,dt$

11. $\int \dfrac{1}{\sqrt{1-x^2}}\,dx$

12. $\int \dfrac{1}{1+x^2}\,dx$

13. $\int \dfrac{5x^2+2xe^x-1}{x}\,dx$

14. $\int \dfrac{xe^x+1}{x}\,dx$

15. $\int \dfrac{\tan x}{\cos x}\,dx$

16. $\int \dfrac{1}{\sin^2 x}\,dx$

17. $\int \dfrac{2}{5\sqrt{1-x^2}}\,dx$

18. $\int -\dfrac{4}{x\sqrt{x^2-1}}\,dx$

19. $\int \dfrac{e^t+e^{-t}}{2}\,dt$

20. $\int \dfrac{e^{2t}-e^{-2t}}{2}\,dt$

In Problems 21–26, find each indefinite integral using the given substitution.

21. $\int e^{3x+1}\,dx$; let $u = 3x+1$

22. $\int \dfrac{dx}{x\ln x}$; let $u = \ln x$

23. $\int (1-t^2)^6 t\,dt$; let $u = (1-t^2)$

24. $\int \sin^5 x \cos x\,dx$; let $u = \sin x$

25. $\int \dfrac{x^2\,dx}{\sqrt{1-x^6}}$; let $u = x^3$

26. $\int \dfrac{e^{-x}}{6+e^{-x}}\,dx$; let $u = 6+e^{-x}$

In Problems 27–60, find each indefinite integral.

27. $\int \sin(3x)\,dx$

28. $\int x\sin x^2\,dx$

29. $\int \sin x \cos^2 x\,dx$

30. $\int \tan^2 x \sec^2 x\,dx$

31. $\int \dfrac{e^{1/x}}{x^2}\,dx$

32. $\int \dfrac{e^{\sqrt[3]{x}}}{\sqrt[3]{x^2}}\,dx$

33. $\int \dfrac{x\,dx}{x^2-1}$

34. $\int \dfrac{5x\,dx}{1-x^2}$

35. $\int \dfrac{e^x}{\sqrt{1+e^x}}\,dx$

36. $\int \dfrac{dx}{x(\ln x)^7}$

37. $\int \dfrac{1}{\sqrt{x}(1+\sqrt{x})^4}\,dx$

38. $\int \dfrac{dx}{\sqrt{x}(1+\sqrt{x})}$

39. $\int \dfrac{3e^x}{\sqrt[4]{e^x-1}}\,dx$

40. $\int \dfrac{[\ln(5x)]^3}{x}\,dx$

41. $\int \dfrac{\cos x\,dx}{2\sin x-1}$

42. $\int \dfrac{\cos(2x)\,dx}{\sin(2x)}$

43. $\int \sec(5x)\,dx$

44. $\int \tan(2x)\,dx$

45. $\displaystyle\int \sqrt{\tan x}\, \sec^2 x\, dx$

46. $\displaystyle\int (2 + 3\cot x)^{3/2}\, \csc^2 x\, dx$

47. $\displaystyle\int \frac{\sin x}{\cos^2 x}\, dx$

48. $\displaystyle\int \frac{\cos x}{\sin^2 x}\, dx$

49. $\displaystyle\int \sin x \cdot e^{\cos x}\, dx$

50. $\displaystyle\int \sec^2 x \cdot e^{\tan x}\, dx$

51. $\displaystyle\int x\sqrt{x+3}\, dx$ [PAGE 487]

52. $\displaystyle\int x\sqrt{4-x}\, dx$

53. $\displaystyle\int [\sin x + \cos(3x)]\, dx$

54. $\displaystyle\int [x^2 + \sqrt{3x+2}]\, dx$

55. $\displaystyle\int \frac{dx}{x^2 + 25}$ [PAGE 488]

56. $\displaystyle\int \frac{\cos x}{1 + \sin^2 x}\, dx$

57. $\displaystyle\int \frac{dx}{\sqrt{9 - x^2}}$

58. $\displaystyle\int \frac{dx}{\sqrt{16 - 9x^2}}$

59. $\displaystyle\int \sin x \cos x\, dx$

60. $\displaystyle\int \sec^2 x \tan x\, dx$

In Problems 61–68, find each definite integral two ways:

(a) By finding the related indefinite integral and then using the Fundamental Theorem of Calculus.

(b) By making a substitution in the integrand and using the substitution to change the limits of integration.

(c) Which method did you prefer? Why?

61. $\displaystyle\int_{-2}^{0} \frac{x}{(x^2 + 3)^2}\, dx$ [PAGE 489]

62. $\displaystyle\int_{-1}^{1} (x^2 - 1)^5 x\, dx$

63. $\displaystyle\int_{0}^{1} x^2 e^{x^3 + 1}\, dx$

64. $\displaystyle\int_{0}^{1} x e^{x^2 - 2}\, dx$

65. $\displaystyle\int_{1}^{6} x\sqrt{x+3}\, dx$

66. $\displaystyle\int_{2}^{6} x^2\sqrt{x-2}\, dx$

67. $\displaystyle\int_{0}^{2} x \cdot 3^{2x^2}\, dx$

68. $\displaystyle\int_{0}^{1} x \cdot 10^{-x^2}\, dx$

In Problems 69–78, find each definite integral.

69. $\displaystyle\int_{1}^{3} \frac{1}{x^2}\sqrt{1 - \frac{1}{x}}\, dx$ [PAGE 490]

70. $\displaystyle\int_{0}^{\pi/4} \frac{\sin(2x)}{\sqrt{5 - 2\cos(2x)}}\, dx$

71. $\displaystyle\int_{0}^{2} \frac{e^{2x}}{e^{2x} + 1}\, dx$

72. $\displaystyle\int_{1}^{3} \frac{e^{3x}}{e^{3x} - 1}\, dx$

73. $\displaystyle\int_{2}^{3} \frac{dx}{x \ln x}$

74. $\displaystyle\int_{2}^{3} \frac{dx}{x(\ln x)^2}$

75. $\displaystyle\int_{0}^{\pi} e^x \cos(e^x)\, dx$

76. $\displaystyle\int_{0}^{\pi} e^{-x} \cos(e^{-x})\, dx$

77. $\displaystyle\int_{0}^{1} \frac{x\, dx}{1 + x^4}$

78. $\displaystyle\int_{0}^{1} \frac{e^x}{1 + e^{2x}}\, dx$

In Problems 79–86, use properties of even and odd functions to find each integral.

79. $\displaystyle\int_{-2}^{2} (x^2 - 4)\, dx$ [PAGE 491]

80. $\displaystyle\int_{-1}^{1} (x^3 - 2x)\, dx$

81. $\displaystyle\int_{-\pi/2}^{\pi/2} \frac{1}{3}\sin\theta\, d\theta$ [PAGE 491]

82. $\displaystyle\int_{-\pi/4}^{\pi/4} \sec^2 x\, dx$

83. $\displaystyle\int_{-1}^{1} \frac{3}{1 + x^2}\, dx$

84. $\displaystyle\int_{-5}^{5} (x^{1/3} + x)\, dx$

85. $\displaystyle\int_{-5}^{5} |2x|\, dx$

86. $\displaystyle\int_{-1}^{1} [|x| - 3]\, dx$

87. If f is an odd function and $\int_0^3 f(x)\, dx = 6$ and $\int_3^5 f(x)\, dx = 2$, find $\int_{-3}^5 f(x)\, dx$. [PAGE 492]

88. If f is an odd function and $\int_{-5}^{10} f(x)\, dx = 8$, find $\int_5^{10} f(x)\, dx$.

89. If f is an even function and $\int_0^4 f(x)\, dx = -2$ and $\int_0^6 f(x)\, dx = 6$, find $\int_{-4}^6 f(x)\, dx$.

90. If f is an even function and $\int_{-3}^0 f(x)\, dx = 4$ and $\int_{-5}^0 f(x)\, dx = 1$, find $\int_3^5 f(x)\, dx$.

Applications and Extensions

In Problems 91 and 92, find each indefinite integral by

(a) Using substitution.

(b) Expanding the integrand.

91. $\displaystyle\int (x + 1)^2\, dx$

92. $\displaystyle\int (x^2 + 1)^2 x\, dx$

In Problems 93 and 94, find each integral three ways:

(a) By using substitution.

(b) By using even-odd properties of the definite integral.

(c) By using trigonometry to simplify the integrand before integrating.

(d) Verify the results are equivalent.

93. $\displaystyle\int_{-\pi/2}^{\pi/2} \cos(2x + \pi)\, dx$

94. $\displaystyle\int_{-\pi/4}^{\pi/4} \sin(7\theta - \pi)\, d\theta$

95. Area Find the area under the graph of $f(x) = \dfrac{x^2}{\sqrt{2x+1}}$ from 0 to 4.

96. Area Find the area under the graph of $f(x) = \dfrac{x}{(x^2 + 1)^2}$ from 0 to 2.

97. Area Find the area under the graph of $y = \dfrac{1}{3x^2 + 1}$ from $x = 0$ to $x = 1$.

98. Area Find the area under the graph of $y = \dfrac{1}{x\sqrt{x^2 - 4}}$ from $x = 3$ to $x = 4$.

99. Area Find the area under the graph of
$$y = a\cos\frac{\pi x}{a}$$
from $x = -\dfrac{a}{2}$ to $x = \dfrac{a}{2}$.

100. Area   Find b so that the area under the graph of

$$y = (x+1)\sqrt{x^2 + 2x + 4}$$

is $\dfrac{56}{3}$ for $0 \le x \le b$.

101. Average Value   Find the average value of $y = \tan x$ on the interval $\left[0, \dfrac{\pi}{4}\right]$.

102. Average Value   Find the average value of $y = \sec x$ on the interval $\left[0, \dfrac{\pi}{4}\right]$.

103. If $\displaystyle\int_0^2 f(x-3)\,dx = 8$, find $\displaystyle\int_{-3}^{-1} f(x)\,dx$.

104. If $\displaystyle\int_{-2}^1 f(x+1)\,dx = \dfrac{5}{2}$, find $\displaystyle\int_{-1}^2 f(x)\,dx$.

105. If $\displaystyle\int_0^4 g\left(\dfrac{x}{2}\right)dx = 8$, find $\displaystyle\int_0^2 g(x)\,dx$.

106. If $\displaystyle\int_0^1 g(3x)\,dx = 6$, find $\displaystyle\int_0^3 g(x)\,dx$.

In Problems 107–112, find each integral.
Hint: Begin by using a Change of Base formula.

107. $\displaystyle\int \dfrac{dx}{x \log_{10} x}$ **108.** $\displaystyle\int \dfrac{dx}{x \log_3 \sqrt[5]{x}}$ **109.** $\displaystyle\int_{10}^{100} \dfrac{dx}{x \log x}$

110. $\displaystyle\int_8^{32} \dfrac{dx}{x \log_2 x}$ **111.** $\displaystyle\int_3^9 \dfrac{dx}{x \log_3 x}$ **112.** $\displaystyle\int_{10}^{100} \dfrac{dx}{x \log_5 x}$

113. If $\displaystyle\int_1^b t^2 (5t^3 - 1)^{1/2}\,dt = \dfrac{38}{45}$, find b.

114. If $\displaystyle\int_a^3 t\sqrt{9 - t^2}\,dt = 6$, find a.

In Problems 115–128, find each integral.

115. $\displaystyle\int \dfrac{x+1}{x^2+1}\,dx$ **116.** $\displaystyle\int \dfrac{2x-3}{1+x^2}\,dx$

117. $\displaystyle\int \left(2\sqrt{x^2+3} - \dfrac{4}{x} + 9\right)^6 \left(\dfrac{x}{\sqrt{x^2+3}} + \dfrac{2}{x^2}\right)dx$

118. $\displaystyle\int \left[\sqrt{(z^2+1)^4 - 3}\right]\left[z(z^2+1)^3\right]dz$

119. $\displaystyle\int \dfrac{x + 4x^3}{\sqrt{x}}\,dx$ **120.** $\displaystyle\int \dfrac{z\,dz}{z + \sqrt{z^2 + 4}}$

121. $\displaystyle\int \sqrt{t}\sqrt{4 + t\sqrt{t}}\,dt$ **122.** $\displaystyle\int_0^1 \dfrac{x+1}{x^2+3}\,dx$

123. $\displaystyle\int 3^{2x+1}\,dx$ **124.** $\displaystyle\int 2^{3x+5}\,dx$

125. $\displaystyle\int \dfrac{\sin x}{\sqrt{4 - \cos^2 x}}\,dx$ **126.** $\displaystyle\int \dfrac{\sec^2 x\,dx}{\sqrt{1 - \tan^2 x}}$

127. $\displaystyle\int_0^1 \dfrac{(z^2+5)(z^3 + 15z - 3)}{196 - (z^3 + 15z - 3)^2}\,dz$

128. $\displaystyle\int_2^{17} \dfrac{dx}{\sqrt{\sqrt{x-1} + (x-1)^{5/4}}}$

129. Use an appropriate substitution to show that

$$\int_0^1 x^m (1-x)^n\,dx = \int_0^1 x^n (1-x)^m\,dx$$

where m, n are positive integers.

130. Find $\displaystyle\int_{-1}^1 f(x)\,dx$, for the function given below

$$f(x) = \begin{cases} x+1 & \text{if } x < 0 \\ \cos(\pi x) & \text{if } x \ge 0 \end{cases}$$

131. If f is continuous on $[a, b]$, show that

$$\int_a^b f(x)\,dx = \int_a^b f(a + b - x)\,dx$$

132. If $\displaystyle\int_0^1 f(x)\,dx = 2$, find

(a) $\displaystyle\int_0^{0.5} f(2x)\,dx$ (b) $\displaystyle\int_0^3 f\left(\dfrac{1}{3}x\right)dx$

(c) $\displaystyle\int_0^{1/5} f(5x)\,dx$

(d) Find the upper and lower limits of integration so that

$$\int_a^b f\left(\dfrac{x}{4}\right)dx = 8$$

(e) Generalize (d) so that $\displaystyle\int_a^b f(kx)\,dx = \dfrac{1}{k}\cdot 2$ for $k > 0$.

133. If $\displaystyle\int_0^2 f(s)\,ds = 5$, find

(a) $\displaystyle\int_{-1}^1 f(s+1)\,ds$ (b) $\displaystyle\int_{-3}^{-1} f(s+3)\,ds$

(c) $\displaystyle\int_4^6 f(s-4)\,ds$

(d) Find upper and lower limits of integration so that

$$\int_a^b f(s-2)\,ds = 5$$

(e) Generalize (d) so that $\displaystyle\int_a^b f(s-k)\,ds = 5$ for $k > 0$.

134. Find $\displaystyle\int_0^b |2x|\,dx$ for any real number b.

135. If f is an odd function, show that $\displaystyle\int_{-a}^a f(x)\,dx = 0$.

136. Find the constant k, where $0 \le k \le 3$, for which

$$\int_0^3 \dfrac{x}{\sqrt{x^2 + 16}}\,dx = \dfrac{3k}{\sqrt{k^2 + 16}}$$

137. If n is a positive integer, for what number $c > 0$ is

$$\int_0^c x^{n-1}\,dx = \dfrac{1}{n}?$$

138. If f is a continuous function defined on the interval $[0, 1]$, show that

$$\int_0^\pi x\,f(\sin x)\,dx = \dfrac{\pi}{2}\int_0^\pi f(\sin x)\,dx$$

139. Prove that $\displaystyle\int \csc x\,dx = \ln|\csc x - \cot x| + C$.
   Hint: Multiply and divide by $(\csc x - \cot x)$.

140. Describe a method for finding $\int_a^b |f(x)|\,dx$ in terms of $F(x) = \int f(x)\,dx$ when f has finitely many zeros.

141. Find $\int \sqrt[n]{a+bx}\,dx$ where a and b are real numbers, $b \neq 0$, and $n \geq 2$ is an integer.

142. If f is continuous for all x, which of the following integrals have the same value?

(a) $\displaystyle\int_a^b f(x)\,dx$ **(b)** $\displaystyle\int_0^{b-a} f(x+a)\,dx$

(c) $\displaystyle\int_{a+c}^{b+c} f(x+c)\,dx$

143. Area Let $f(x) = k\sin(kx)$, where k is a positive constant.

(a) Find the area of the region under one arch of the graph of f.

(b) Find the area of the triangle formed by the x-axis and the tangent lines to one arch of f at the points where the graph of f crosses the x-axis.

144. Prove that $\int [f(x) + g(x)]\,dx = \int f(x)\,dx + \int g(x)\,dx$.

145. Derive the integration formula $\displaystyle\int a^x\,dx = \frac{a^x}{\ln a} + C, a > 0$, $a \neq 1$. *Hint:* Begin with the derivative of $y = a^x$.

146. Use the formula from Problem 145 to find

(a) $\displaystyle\int 2^x\,dx$ **(b)** $\displaystyle\int 3^x\,dx$

147. Electric Potential The electric field strength a distance z from the axis of a ring of radius R carrying a charge Q is given by the formula

$$E(z) = \frac{Qz}{(R^2 + z^2)^{3/2}}$$

If the electric potential V is related to E by $E = -\dfrac{dV}{dz}$, what is $V(z)$?

148. Impulse During a Rocket Launch
The impulse J due to a force F is the product of the force times the amount of time t for which the force acts. When the force varies over time,

$$J = \int_{t_1}^{t_2} F(t)\,dt.$$

We can model the force acting on a rocket during launch by an exponential function $F(t) = Ae^{bt}$, where A and b are constants that depend on the characteristics of the engine. At the instant lift-off occurs ($t = 0$), the force must equal the weight of the rocket.

NASA/Rick Wetherington and Tony Gray

(a) Suppose the rocket weighs 25,000 N (a mass of about 2500 kg or a weight of 5500 lb), and 30 s after lift-off the force acting on the rocket equals twice the weight of the rocket. Find A and b.

(b) Find the impulse delivered to the rocket during the first 30 seconds after the launch.

Challenge Problems

149. Air Resistance on a Falling Object If an object of mass m is dropped, the air resistance on it when it has speed v can be modeled as $F_{\text{air}} = -kv$, where the constant k depends on the shape of the object and the condition of the air. The minus sign is necessary because the direction of the force is opposite to the velocity. Using Newton's Second Law of Motion, this force leads to a downward acceleration $a(t) = ge^{-kt/m}$. (You are asked to prove this in Problem 150.) Using the equation for $a(t)$, find:

(a) $v = v(t)$, if the object starts from rest $v_0 = v(0) = 0$, with the positive direction downward.

(b) $s = s(t)$, if the object starts from the position $s_0 = s(0) = 0$, with the positive direction downward.

(c) What limits do $a(t)$, $v(t)$, and $s(t)$ approach if the object falls for a very long time ($t \to \infty$)? Interpret each result and explain if it is physically reasonable.

(d) Graph $a = a(t)$, $v = v(t)$, $s = s(t)$. Do the graphs support the conclusions obtained in part (c)? Use $g = 9.8$ m/s^2, $k = 5$, and $m = 10$ kg.

150. Air Resistance on a Falling Object (Refer to Problem 149.) If an object of mass m is dropped, the air resistance on it when it has speed v can be modeled as

$$F_{\text{air}} = -kv$$

where the constant k depends on the shape of the object and the condition of the air. The minus sign is necessary because the direction of the force is opposite to the velocity. Using Newton's Second Law of Motion, show that the downward acceleration of the object is

$$a(t) = ge^{-kt/m}$$

where g is the acceleration due to gravity.
Hint: The velocity of the object obeys the equation

$$m\frac{dv}{dt} = mg - kv$$

Solve for v and use the fact that $ma = mg - kv$.

151. Find $\displaystyle\int \frac{x^6 + 3x^4 + 3x^2 + x + 1}{(x^2 + 1)^2}\,dx$.

152. Find $\displaystyle\int \frac{\sqrt[4]{x}}{\sqrt{x} + \sqrt[3]{x}}\,dx$.

153. Find $\displaystyle\int \frac{3x + 2}{x\sqrt{x + 1}}\,dx$.

154. Find $\displaystyle\int \frac{dx}{(x\ln x)[\ln(\ln x)]}$.

155. (a) Find y' if $y = \ln\left[\tan\left(\dfrac{x}{2} + \dfrac{\pi}{4}\right)\right]$.

(b) Use the result to show that

$$\int \sec x\,dx = \ln\left[\tan\left(\frac{x}{2} + \frac{\pi}{4}\right)\right] + C$$

(c) Show that $\ln\left[\tan\left(\dfrac{x}{2} + \dfrac{\pi}{4}\right)\right] = \ln|\sec x + \tan x|$.

156. (a) Find y' if $y = x\sin^{-1}x + \sqrt{1-x^2}$.

(b) Use the result to show that

$$\int \sin^{-1}x\,dx = x\sin^{-1}x + \sqrt{1-x^2} + C$$

157. (a) Find y' if $y = \frac{1}{2}x\sqrt{a^2-x^2} + \frac{1}{2}a^2\sin^{-1}\left(\frac{x}{a}\right), a > 0$.

(b) Use the result to show that

$$\int \sqrt{a^2-x^2}\,dx = \frac{1}{2}x\sqrt{a^2-x^2} + \frac{1}{2}a^2\sin^{-1}\left(\frac{x}{a}\right) + C$$

158. (a) Find y' if $y = \ln|\csc x - \cot x|$.

(b) Use the result to show that

$$\int \csc x\,dx = \ln|\csc x - \cot x| + C$$

159. The formula $\frac{d}{dx}\int f(x)\,dx = f(x)$ states that if a function is integrated and the result is differentiated, the original function is returned. What about the other way around? Is the formula $\int f'(x)\,dx = f(x)$ correct? Be sure to justify your answer.

Preparing for the AP® Exam

AP® Practice Problems

Multiple-Choice Questions

1. $\displaystyle\int \frac{1}{3}\left(e^x - \frac{2}{x}\right)dx =$

(A) $\frac{1}{3}e^x - 2\ln|x| + C$ (B) $\frac{1}{3}\left(e^x - \frac{2}{x^2}\right) + C$

(C) $\frac{1}{3}e^x - \frac{2}{3}\ln|x| + C$ (D) $\frac{1}{3}(e^x - \ln|2x|) + C$

2. The velocity $v = v(t)$ of an object moving on a line is given by $v(t) = \displaystyle\int t\sin\frac{t}{2}\,dt$. What function gives the acceleration of the object?

(A) $a(t) = \frac{1}{2}t\cos\frac{t}{2} + \sin\frac{t}{2}$

(B) $a(t) = \frac{1}{2}t\cos\frac{t}{2}$

(C) $a(t) = t^2\cos\frac{t}{2}$

(D) $a(t) = t\sin\frac{t}{2}$

3. $\displaystyle\int \frac{1+\sin x}{\cos^2 x}\,dx =$

(A) $\tan x + \sec x + C$ (B) $\csc x + \sec^2 x + C$
(C) $x + \sec x + C$ (D) $\sec x + C$

4. Using the substitution $u = \sin x$, $\displaystyle\int_{\pi/6}^{\pi/2}\sin^4 x\cos x\,dx$ becomes

(A) $-\displaystyle\int_{1/2}^1 u^4\,du$ (B) $\displaystyle\int_{\pi/6}^{\pi/2}u^4\,du$

(C) $\displaystyle\int_{\sqrt{3}/2}^0 u^4\,du$ (D) $\displaystyle\int_{1/2}^1 u^4\,du$

5. If the function f is continuous for all real numbers, and if $F'(x) = f(x)$, then $\displaystyle\int_1^4 f(3x)\,dx$ equals

(A) $F(12) - F(3)$ (B) $\frac{1}{3}[F(4) - F(1)]$

(C) $\frac{1}{3}[F(12) - F(3)]$ (D) $3[F(4) - F(1)]$

6. $\displaystyle\int \frac{e^{-3x}}{2 + e^{-3x}}\,dx =$

(A) $-3\ln(2 + e^{-3x}) + C$ (B) $-\frac{1}{3}\ln(2 + e^{-3x}) + C$

(C) $2x - \frac{e^{-3x}}{3} + C$ (D) $\frac{1}{3}\ln|-3e^{-3x}| + C$

7. $\displaystyle\int \frac{x}{1+x^2}\,dx =$

(A) $\arctan x + C$ (B) $\frac{1}{2}\ln|1+x^2| + C$

(C) $\ln|2(1+x^2)| + C$ (D) $2\ln|1+x^2| + C$

8. $\displaystyle\int_2^6 \frac{dx}{\sqrt{2x-3}} =$

(A) 1 (B) 2 (C) 4 (D) 78

9. What is the average value of $f(x) = \sin x\cos x$ on the closed interval $\left[0, \frac{\pi}{2}\right]$?

(A) $\frac{1}{\pi}$ (B) $\frac{2}{\pi}$ (C) $\frac{1}{2}$ (D) $\frac{\pi}{4}$

10. $\int x\sqrt{x^2+3}\,dx =$

(A) $(x^2+3)^{3/2}+C$

(B) $\dfrac{1}{2(x^2+3)^{1/2}}+C$

(C) $\dfrac{1}{3}(x^2+3)^{3/2}+C$

(D) $\dfrac{x^2+3}{2}+C$

11. An object moving along the x-axis has acceleration $a(t)=4t-5$ at time $t\geq 0$. If the object's initial velocity is 3, at what time t does the object first change direction?

(A) 0   (B) 1   (C) $\dfrac{3}{2}$ (D) 3

12. $\displaystyle\int_{e}^{e^4}\dfrac{1}{x\sqrt{\ln x}}\,dx =$

(A) $\dfrac{1}{2}$ (B) 1   (C) 2   (D) $2(e^2-\sqrt{e})$

13. An object moving along the y-axis has velocity $v(t)=2\cos\left(2t+\dfrac{\pi}{2}\right)$, $t\geq 0$. If the object is at $y=5$ when $t=\pi$, find its position when $t=\dfrac{\pi}{2}$.

(A) 1   (B) 3   (C) 4   (D) 5

Retain Your Knowledge

Multiple-Choice Questions

1. A function h is continuous on the closed interval [3, 15]. The table shows selected values of h.

x	3	7	8	10	15
$h(x)$	8	4	3	-2	-4

Which of the following statements must be true?

I. There is a number c in the interval [3, 15] for which $h(c)=7$.

II. h is a decreasing function.

III. There is a number c in the interval (3, 15) for which $h'(c)=0$.

(A) I only   (B) II only

(C) I and II only   (D) I, II, and III

2. As sand falls from above, it forms a right circular cone. The radius r of the cone is increasing at a rate of 4 cm/min, and the height h of the cone is increasing at a rate of 3 cm/min. Which expression gives the rate of change of the volume of the cone with respect to time t? (The volume of a right circular cone is given by $V=\dfrac{1}{3}\pi r^2 h$.)

(A) $\dfrac{1}{3}\pi(2rh+r^2)$ cm³/min   (B) $\dfrac{1}{3}\pi(6r)$ cm³/min

(C) $\dfrac{1}{3}\pi(2rh+3r^2)$ cm³/min   (D) $\dfrac{1}{3}\pi(8rh+3r^2)$ cm³/min

3. Find all the antiderivatives of $f(x)=\cos x+\dfrac{10}{\sqrt{1-x^2}}$.

(A) $F(x)=-\sin x+10\sin^{-1}x+C$

(B) $F(x)=\sin x-20x\sin^{-1}x+C$

(C) $F(x)=\sin x+10\sin^{-1}x+C$

(D) $F(x)=-\sin x+\dfrac{40}{3}x\sin^{-1}x+C$

Free-Response Question

4. The velocity of an object moving on a line is modeled by a differentiable function v, where v is measured in meters/minute and $t\geq 0$ is measured in minutes. The velocity of the object is observed at several times t and posted in the table below.

t	0	2	3	6	8
$v(t)$	0	4	13	7	5

(a) Use a linear approximation to find $v(5)$.

(b) Approximate the acceleration at $t=5$.

6.6 Integrands Containing $ax^2 + bx + c, a \neq 0$; Integration Using Long Division

OBJECTIVE *When you finish this section, you should be able to:*

1 Integrate a function that contains a quadratic expression (p. 498)
2 Integrate an improper rational function using long division (p. 500)

Sometimes an integral that contains a quadratic expression can be found using substitution. For example,

$$\int (x^2 + 5)^{1/2} x \, dx \qquad\qquad \text{Let } u = x^2 + 5. \text{ Then } du = 2x \, dx.$$

$$\int \frac{x - 5}{x^2 - 10x} dx \qquad\qquad \text{Let } u = x^2 - 10x. \text{ Then } du = (2x - 10)dx = 2(x - 5)dx$$

For integrals containing a quadratic expression, always check first to see if substitution can be used. When substitution does not work, often completing the square of the quadratic expression or division can be used.

1 Integrate a Function That Contains a Quadratic Expression

NEED TO REVIEW? Completing the square is discussed in Appendix A.1, pp. A-2 to A-3.

Integrals containing $ax^2 + bx + c$, where $a \neq 0$, b and c are real numbers, can often be integrated by completing the square. We complete the square of $ax^2 + bx + c$ as follows:

$$ax^2 + bx + c = a\left(x^2 + \frac{b}{a}x\right) + c \qquad \text{Factor } a \text{ from the first two terms.}$$

$$= a\left(x^2 + \frac{b}{a}x + \frac{b^2}{4a^2}\right) + c - \frac{b^2}{4a} \qquad \text{Add and subtract } a\left(\frac{1}{2} \cdot \frac{b}{a}\right)^2 = \frac{b^2}{4a}.$$

$$= a\left(x + \frac{b}{2a}\right)^2 + \left(c - \frac{b^2}{4a}\right) \qquad \text{Factor } x^2 + \frac{b}{a}x + \frac{b^2}{4a^2} = \left(x + \frac{b}{2a}\right)^2.$$

After completing the square, use the substitution

$$u = x + \frac{b}{2a}$$

to express the original quadratic expression $ax^2 + bx + c$ in the simpler form $au^2 + r$, where $r = c - \dfrac{b^2}{4a}$.

EXAMPLE 1 Integrating a Function That Contains $x^2 + 6x + 10$

Find $\displaystyle\int \frac{dx}{x^2 + 6x + 10}$.

Solution

The integrand contains the quadratic expression $x^2 + 6x + 10$. So, complete the square.

$$x^2 + 6x + 10 = (x^2 + 6x + 9) + 1 = (x + 3)^2 + 1$$

Now write the integral as

$$\int \frac{dx}{x^2 + 6x + 10} = \int \frac{dx}{(x + 3)^2 + 1}$$

and use the substitution $u = x + 3$. Then $du = dx$, and

$$\int \frac{dx}{x^2 + 6x + 10} = \int \frac{dx}{(x + 3)^2 + 1} = \int \frac{du}{u^2 + 1} = \tan^{-1} u + C = \tan^{-1}(x + 3) + C \quad \blacksquare$$

NOW WORK   Problem 3 and AP® Practice Problem 1.

EXAMPLE 2   **Integrating a Function That Contains $x^2 + x + 1$**

Find **(a)** $\displaystyle\int \frac{dx}{x^2 + x + 1}$ **(b)** $\displaystyle\int \frac{x\,dx}{x^2 + x + 1}$

Solution

(a) The integrand contains the quadratic expression $x^2 + x + 1$. So, complete the square in the denominator.

$$\int \frac{dx}{x^2 + x + 1} \underset{\substack{\uparrow \\ \text{Complete the square.}}}{=} \int \frac{dx}{\left(x^2 + x + \frac{1}{4}\right) + \left(1 - \frac{1}{4}\right)} = \int \frac{dx}{\left(x + \frac{1}{2}\right)^2 + \frac{3}{4}}$$

Now use the substitution $u = x + \dfrac{1}{2}$. Then $du = dx$.

$$\int \frac{dx}{x^2 + x + 1} = \int \frac{dx}{\left(x + \frac{1}{2}\right)^2 + \frac{3}{4}} = \int \frac{du}{u^2 + \frac{3}{4}}$$

$$= \frac{1}{\frac{\sqrt{3}}{2}} \tan^{-1} \frac{u}{\frac{\sqrt{3}}{2}} + C = \frac{2}{\sqrt{3}} \tan^{-1} \frac{2u}{\sqrt{3}} + C$$

$$= \frac{2\sqrt{3}}{3} \tan^{-1} \frac{2x + 1}{\sqrt{3}} + C \qquad\qquad u = x + \frac{1}{2}$$

(b) Here we use a manipulation that causes the derivative of the denominator to appear in the numerator. This helps simplify the integration.

NOTE We could also complete the square of $x^2 + x + 1$ and let $u = x + \dfrac{1}{2}$.

$$\int \frac{x\,dx}{x^2 + x + 1} = \frac{1}{2} \int \frac{2x\,dx}{x^2 + x + 1}$$

$$= \frac{1}{2} \int \frac{[(2x + 1) - 1]\,dx}{x^2 + x + 1} \qquad \text{Add and subtract 1 in the numerator to get } 2x + 1.$$

$$= \frac{1}{2} \int \frac{(2x + 1)\,dx}{x^2 + x + 1} - \frac{1}{2} \int \frac{dx}{x^2 + x + 1} \qquad \text{Write the integral as the sum of two integrals.}$$

(Example continued on the next page.)

RECALL $\int \dfrac{g'(x)}{g(x)} dx = \ln |g(x)| + C$

We found the integral on the right in (a). In the integral on the left, the numerator equals the derivative of the denominator.

$$\frac{1}{2} \int \frac{(2x+1)\,dx}{x^2+x+1} = \frac{1}{2} \ln \left| x^2+x+1 \right|$$

So,

$$\int \frac{x\,dx}{x^2+x+1} = \frac{1}{2} \int \frac{(2x+1)\,dx}{x^2+x+1} - \frac{1}{2} \int \frac{dx}{x^2+x+1}$$

$$= \frac{1}{2} \ln \left| x^2+x+1 \right| - \frac{1}{2} \left[\frac{2\sqrt{3}}{3} \tan^{-1} \frac{2x+1}{\sqrt{3}} \right] + C$$

$$= \frac{1}{2} \ln \left| x^2+x+1 \right| - \frac{\sqrt{3}}{3} \tan^{-1} \frac{2x+1}{\sqrt{3}} + C \qquad ■$$

NOW WORK   Problem 9 and AP® Practice Problem 2.

EXAMPLE 3   **Integrating a Function That Contains $2x - x^2$**

CALC CLIP

Find $\displaystyle\int \frac{dx}{\sqrt{2x - x^2}}$.

Solution

The integrand contains the quadratic expression $2x - x^2$, so complete the square.

$$2x - x^2 = -x^2 + 2x = -(x^2 - 2x) = -(x^2 - 2x + 1) + 1$$
$$= -(x-1)^2 + 1 = 1 - (x-1)^2$$

Then using substitution,

RECALL $\int \dfrac{dx}{\sqrt{1-x^2}} = \sin^{-1} x + C$

$$\int \frac{dx}{\sqrt{2x - x^2}} = \int \frac{dx}{\sqrt{1-(x-1)^2}} \underset{\substack{\uparrow \\ u = x-1 \\ du = dx}}{=} \int \frac{du}{\sqrt{1-u^2}} = \sin^{-1} u + C = \sin^{-1}(x-1) + C \qquad ■$$

NOW WORK   Problem 25.

② Integrate an Improper Rational Function Using Long Division

NEED TO REVIEW? Rational functions are discussed in Section P.2, p. 23.

A rational function R in lowest terms is the ratio of two polynomial functions p and $q \neq 0$, where p and q have no common factors. The domain of R is the set of all real numbers for which $q \neq 0$. The rational function R is called **proper** when the degree of the polynomial p is less than the degree of the polynomial q; otherwise, R is an **improper** rational function.

Using long division, every improper rational function R can be written as the sum of a polynomial function and a proper rational function. Since we know how to integrate polynomial functions, the integral of an improper rational function can be found provided we can integrate the proper rational function that is the remainder after long division.

EXAMPLE 4   **Integrating an Improper Rational Function**

Find $\displaystyle\int \frac{x^2}{x-1}\,dx$.

Solution

The integrand is an improper rational function. Using long division, the integrand can be expressed as the sum of a polynomial and a proper rational function.

$$
\begin{array}{r}
x + 1 + \dfrac{1}{x-1} \\
x-1 \overline{\smash{\big)}\, x^2 } \\
\underline{x^2 - x } \\
x \\
\underline{x - 1} \\
1
\end{array}
$$

The remainder is $\dfrac{1}{x-1}$, a function we know how to integrate. So,

$$
\int \frac{x^2}{x-1}\, dx = \int \left(x + 1 + \frac{1}{x-1} \right) dx = \int x\, dx + \int dx + \int \frac{1}{x-1}\, dx
$$

$$
= \frac{x^2}{2} + x + \ln |x - 1| + C
$$

∎

NOW WORK   Problems **35** and **47** and AP® Practice Problems **3** and **6**.

EXAMPLE 5   **Using Division to Integrate an Improper Rational Function**

Find $\displaystyle\int \frac{x^3 + 5x + 10}{x^2 + 1}\, dx.$

Solution

The integrand $f(x) = \dfrac{x^3 + 5x + 10}{x^2 + 1}$ is an improper rational function. Use long division

to express f as $f(x) = x + \dfrac{4x + 10}{x^2 + 1}$. Then

$$
\int \frac{x^3 + 5x + 10}{x^2 + 1}\, dx = \int \left(x + \frac{4x + 10}{x^2 + 1} \right) dx = \int \left(x + \frac{4x}{x^2 + 1} + \frac{10}{x^2 + 1} \right) dx
$$

$$
= \int x\, dx + \int \frac{4x}{x^2 + 1}\, dx + \int \frac{10}{x^2 + 1}\, dx
$$

Now

• $\displaystyle\int x\, dx = \frac{x^2}{2} + C_1$

• $\displaystyle\int \frac{4x}{x^2 + 1}\, dx = \int \frac{2}{u}\, du = 2 \ln |u| + C_2 = 2 \ln(x^2 + 1) + C_2$
  $\uparrow$
  $u = x^2 + 1,$
  $du = 2x\, dx$

• $\displaystyle\int \frac{10}{x^2 + 1}\, dx = 10 \int \frac{1}{x^2 + 1}\, dx = 10 \tan^{-1} x + C_3$

So, $\displaystyle\int \frac{x^3 + 5x + 10}{x^2 + 1}\, dx = \frac{x^2}{2} + 2 \ln(x^2 + 1) + 10 \tan^{-1} x + C$, where $C = C_1 + C_2 + C_3$.

∎

NOW WORK   Problem **51** and AP® Practice Problems **4** and **5**.

EXAMPLE 6 **Using Division to Integrate an Improper Rational Function**

Find $\displaystyle\int \frac{x^3 - 4x^2 - 2x + 25}{x^2 - 6x + 10} dx$.

Solution

The integrand $f(x) = \dfrac{x^3 - 4x^2 - 2x + 25}{x^2 - 6x + 10}$ is an improper rational function. Use long division. Then,

$$f(x) = \frac{x^3 - 4x^2 - 2x + 25}{x^2 - 6x + 10} = x + 2 + \frac{5}{x^2 - 6x + 10}$$

Now

$$\int \frac{x^3 - 4x^2 - 2x + 25}{x^2 - 6x + 10} dx = \int (x+2)dx + \int \frac{5}{x^2 - 6x + 10} dx$$

- $\displaystyle\int (x+2)dx = \frac{x^2}{2} + 2x + C_1$

- $\displaystyle\int \frac{5}{x^2 - 6x + 10} dx = \underset{\substack{\uparrow \\ \text{Complete the square.}}}{\int \frac{5}{(x-3)^2 + 1} dx} = \underset{\substack{\uparrow \\ u = x - 3, \\ du = dx}}{\int \frac{5}{u^2 + 1} du} = 5\tan^{-1} u + C_2 = 5\tan^{-1}(x-3) + C_2$

Then

$$\int \frac{x^3 - 4x^2 - 2x + 25}{x^2 - 6x + 10} dx = \frac{x^2}{2} + 2x + 5\tan^{-1}(x-3) + C$$

where $C = C_1 + C_2$. ∎

NOW WORK Problem 55 and AP® Practice Problem 7.

6.6 Assess Your Understanding

Concepts and Vocabulary

1. **Multiple Choice** A rational function $R(x) = \dfrac{p(x)}{q(x)}$ is proper when the degree of p is

   (a) less than   (b) equal to   (c) greater than

   the degree of q.

2. **True or False** Every improper rational function can be written as the sum of a polynomial and a proper rational function.

Skill Building

In Problems 3–34, find each integral.

PAGE 499 3. $\displaystyle\int \frac{dx}{x^2 + 4x + 5}$

4. $\displaystyle\int \frac{dx}{x^2 + 2x + 5}$

5. $\displaystyle\int \frac{dx}{x^2 + 4x + 8}$

6. $\displaystyle\int \frac{dx}{x^2 - 6x + 10}$

7. $\displaystyle\int \frac{2\,dx}{3 + 2x + 2x^2}$

8. $\displaystyle\int \frac{3\,dx}{x^2 + 6x + 10}$

PAGE 500 9. $\displaystyle\int \frac{x\,dx}{2x^2 + 2x + 3}$

10. $\displaystyle\int \frac{3x\,dx}{x^2 + 6x + 10}$

11. $\displaystyle\int \frac{dx}{\sqrt{8 + 2x - x^2}}$

12. $\displaystyle\int \frac{dx}{\sqrt{5 - 4x - 2x^2}}$

13. $\displaystyle\int \frac{dx}{\sqrt{4x - x^2}}$

14. $\displaystyle\int \frac{dx}{\sqrt{x^2 - 6x - 10}}$

15. $\displaystyle\int \frac{dx}{(x+1)\sqrt{x^2 + 2x + 2}}$

16. $\displaystyle\int \frac{dx}{(x-4)x^2 - 8x + 17}$

17. $\displaystyle\int \frac{dx}{\sqrt{24 - 2x - x^2}}$

18. $\displaystyle\int \frac{dx}{\sqrt{9x^2 + 6x + 10}}$

19. $\displaystyle\int \frac{x - 5}{\sqrt{x^2 - 2x + 5}} dx$

20. $\displaystyle\int \frac{x + 1}{x^2 - 4x + 3} dx$

21. $\displaystyle\int_1^3 \frac{dx}{\sqrt{x^2 - 2x + 5}}$

22. $\displaystyle\int_{1/2}^1 \frac{x^2\,dx}{\sqrt{2x - x^2}}$

23. $\displaystyle\int \frac{e^x\,dx}{\sqrt{e^{2x} + e^x + 1}}$

24. $\displaystyle\int \frac{\cos x\,dx}{\sqrt{\sin^2 x + 4\sin x + 3}}$

PAGE 500 25. $\displaystyle\int \frac{2x - 3}{\sqrt{4x - x^2 - 3}} dx$

26. $\displaystyle\int \frac{x + 3}{\sqrt{x^2 + 2x + 2}} dx$

27. $\displaystyle\int \frac{dx}{(x^2 - 2x + 10)^{3/2}}$

28. $\displaystyle\int \frac{dx}{\sqrt{x^2 - 2x + 10}}$

29. $\displaystyle\int \frac{dx}{\sqrt{x^2 + 2x - 3}}$

30. $\displaystyle\int x\sqrt{x^2 - 4x - 1}\,dx$

31. $\displaystyle \int \frac{\sqrt{5 + 4x - x^2}}{x - 2}\,dx$ **32.** $\displaystyle \int \sqrt{5 + 4x - x^2}\,dx$

33. $\displaystyle \int \frac{x\,dx}{\sqrt{x^2 + 2x - 3}}$ **34.** $\displaystyle \int \frac{x\,dx}{\sqrt{x^2 - 4x + 3}}$

In Problems 35–44, write each improper rational function as the sum of a polynomial function and a proper rational function.

PAGE 501 **35.** $R(x) = \dfrac{x^2 + 1}{x + 1}$ **36.** $R(x) = \dfrac{x^2 + 4}{x - 2}$

37. $R(x) = \dfrac{x^3 + 3x - 4}{x - 2}$ **38.** $R(x) = \dfrac{x^3 - 3x^2 + 4}{x + 3}$

39. $R(x) = \dfrac{2x^3 + 3x^2 - 17x - 27}{x^2 - 9}$

40. $R(x) = \dfrac{3x^3 - 2x^2 - 3x + 2}{x^2 - 1}$

41. $R(x) = \dfrac{x^4 - 1}{x(x + 4)}$ **42.** $R(x) = \dfrac{x^4 + x^2 + 1}{x(x - 2)}$

43. $R(x) = \dfrac{2x^4 + x^2 - 2}{x^2 + 4}$ **44.** $R(x) = \dfrac{3x^4 + x^2}{x^2 + 9}$

In Problems 45–56, find each integral by first writing the integrand as the sum of a polynomial and a proper rational function.

45. $\displaystyle \int \frac{x^2 + 1}{x + 1}\,dx$ **46.** $\displaystyle \int \frac{x^2 + 4}{x - 2}\,dx$

PAGE 501 **47.** $\displaystyle \int \frac{x^3 + 3x - 4}{x - 2}\,dx$ **48.** $\displaystyle \int \frac{x^3 - 3x^2 + 4}{x + 3}\,dx$

49. $\displaystyle \int \frac{2x^3 + 3x^2 - 17x - 27}{x^2 - 9}\,dx$ **50.** $\displaystyle \int \frac{3x^3 - 2x^2 - 5x + 2}{x^2 - 1}\,dx$

PAGE 501 **51.** $\displaystyle \int \frac{2x^3 - 3x^2 + 8x - 10}{x^2 + 4}\,dx$ **52.** $\displaystyle \int \frac{5x^3 + 2x^2 + 19x + 8}{x^2 + 4}\,dx$

53. $\displaystyle \int \frac{2x^4 + 23x^2 + 3x + 30}{2x^2 + 3}\,dx$

54. $\displaystyle \int \frac{3x^6 + x^4 + 3x^2 - 2x + 3}{3x^2 + 1}\,dx$

PAGE 502 **55.** $\displaystyle \int \frac{2x^3 + 5x^2 - 2x - 16}{x^2 + 4x + 5}\,dx$

56. $\displaystyle \int \frac{3x^4 - 18x^3 + 32x^2 - 12x + 19}{x^2 - 6x + 10}\,dx$

Applications and Extensions

57. Show that if $k > 0$, then

$$\int \frac{dx}{\sqrt{(x + h)^2 + k}} = \ln\left[\sqrt{(x + h)^2 + k} + x + h\right] + C$$

58. Show that if $a > 0$ and $b^2 - 4ac > 0$, then

$$\int \frac{dx}{\sqrt{ax^2 + bx + c}} = \frac{1}{\sqrt{a}} \ln\left|\sqrt{ax^2 + bx + c} + \sqrt{a}\,x + \frac{b}{2\sqrt{a}}\right| + C$$

Challenge Problem

59. Find $\displaystyle \int \sqrt{\frac{a + x}{a - x}}\,dx$, where $a > 0$.

Preparing for the **AP®** Exam

AP® Practice Problems

Multiple-Choice Questions

PAGE 499 **1.** $\displaystyle \int \frac{1}{x^2 + 6x + 13}\,dx =$

(A) $\dfrac{1}{2} \tan^{-1}(x + 3) + C$

(B) $\dfrac{1}{2} \tan^{-1} \dfrac{x + 3}{2} + C$

(C) $\dfrac{1}{4} \tan^{-1} \dfrac{x + 3}{4} + C$

(D) $2 \tan^{-1}(x + 3) + C$

PAGE 500 **2.** $\displaystyle \int_{-1}^{2} \frac{x + 1}{x^2 + 2x + 10}\,dx =$

(A) $\ln 18 - \ln 9$ (B) $\dfrac{1}{2}(\ln 18 - \ln 9)$

(C) $\dfrac{1}{2}\ln 10 - \dfrac{1}{2}\ln 9$ (D) $\ln 9$

PAGE 501 **3.** Find $\displaystyle \int \frac{x^2 + 6}{x + 1}\,dx$.

(A) $\dfrac{x^2}{2} - x + 5\ln|x + 1| + C$

(B) $\dfrac{x^2}{2} + x + 5\ln|x + 1| + C$

(C) $\dfrac{x^2}{2} - x + 7\ln|x + 1| + C$

(D) $6\ln|x + 1| + C$

PAGE 501 **4.** $\displaystyle \int_{0}^{1} \frac{2x^2}{1 + x^2}\,dx =$

(A) $2 + \dfrac{\pi}{4}$ (B) $2 - \dfrac{\pi}{2}$

(C) $2 + \dfrac{\pi}{2}$ (D) $2\ln 2$

501 **5.** $\int \dfrac{x^4 + 3x^2 - 2}{x^2 + 1}\, dx =$

(A) $\dfrac{1}{3}x^3 + 2x + C$

(B) $\dfrac{1}{3}x^3 + 2x - 4\ln(x^2 + 1) + C$

(C) $\dfrac{1}{3}x^3 + 2x + 4\tan^{-1}x + C$

(D) $\dfrac{1}{3}x^3 + 2x - 4\tan^{-1}x + C$

501 **6.** What is the average value of $f(x) = \dfrac{x+2}{x+1}$ over the closed interval $[0, 2]$?

(A) $\dfrac{1}{2}\ln 3$ (B) 1

(C) $1 + \dfrac{1}{2}\ln 3$ (D) $2 + \ln 3$

502 **7.** $\int \dfrac{2x^4 - 8x^3 + 11x^2 - 4x + 6}{x^2 - 4x + 5}\, dx =$

(A) $\dfrac{2}{3}x^3 + x + \ln|x + 1| + C$

(B) $\dfrac{2}{3}x^3 + x + \ln|(x - 2)^2 + 1| + C$

(C) $\dfrac{2}{3}x^3 + x + \tan^{-1}x + C$

(D) $\dfrac{2}{3}x^3 + x + \tan^{-1}(x - 2) + C$

Retain Your Knowledge

Multiple-Choice Questions

1. The function f is continuous on the closed interval $[-5, 5]$. A portion of the graph of its derivative f' is shown below.

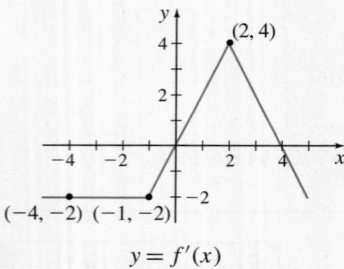

$$y = f'(x)$$

Based on this graph, which of the following must be true?

I. $\int_{-4}^{2} f'(x)\,dx = -3$

II. The function f is decreasing for $2 < x < 4$.

III. The graph of f is concave up on the interval $(-1, 2)$.

(A) III only   (B) I and III only

(C) II and III only   (D) I, II, and III

2. The function f is continuous on the closed interval $[-3, 6]$. The graph of its derivative f' is shown below. At what number c does f have relative maximum?

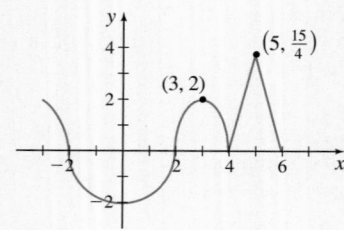

(A) -2 (B) 2 (C) 4 (D) 5

3. $\dfrac{d}{dx}\sin^3(4x) =$

(A) $12\sin^2(4x)$ (B) $12\sin^2(4x)\cos x$

(C) $12\sin^2(4x)\cos(4x)$ (D) $-12\sin^2(4x)\cos(4x)$

Free-Response Question

4. A function $y = f(x)$ is continuous on the closed interval $[-2, 7]$ and differentiable on the open interval $(-2, 7)$. The derivative function f' is also differentiable on the interval $(-2, 7)$. Selected values of f' are given in the table below.

x	-1	0	2	5	6
$f'(x)$	$-\dfrac{3}{2}$	1	3	$-\dfrac{3}{2}$	-5

(a) Show that there is at least one number c in the interval $(-1, 5)$ for which $f''(c) = 0$.

(b) Show that there is at least one number c in the interval $(-1, 6)$ for which $f'(c) = 2$.

CHAPTER 6 PROJECT

Brian Handy/500px/Getty Images

Managing the Klamath River

This Project can be done individually or as part of a team.

There is a gauge on the Klamath River, just downstream from the dam at Keno, Oregon. The U.S. Geological Survey has posted flow rates for this gauge every month since 1930. The averages of these monthly measurements since 1930 are given in Table 2. Notice that the data in Table 2 measure the rate of change of the volume V in cubic feet of water each second over one year; that is, the table gives $\dfrac{dV}{dt} = V'(t)$ in cubic feet per second, where t is in months.

TABLE 2

Month (t)	Flow Rate (ft^3/s)
January (1)	1911.79
February (2)	2045.40
March (3)	2431.73
April (4)	2154.14
May (5)	1592.73
June (6)	945.17
July (7)	669.46
August (8)	851.97
September (9)	1107.30
October (10)	1325.12
November (11)	1551.70
December (12)	1766.33

Source: USGS Surface-Water Monthly Statistics, available at http://waterdata.usgs.gov/or/nwis/monthly

1. Find the factor that will convert the data in Table 2 from ft^3/s to ft^3/day. *Hint:* 1 day = 1 d = 24 hours = 24 h
   1 h = 60 min; 1 min = 60 s

2. If we assume February has 28.25 days, to account for a leap year, then 1 year = 365.25 days. If $V'(t)$ is the rate of flow of water, in cubic feet per day, the total flow of water over 1 year is given by

$$V = V(t) = \int_0^{365.25} V'(t)dt$$

   Approximate the total annual flow using a Riemann sum. *Hint:* Use $\Delta t_1 = 31$, $\Delta t_2 = 28.25$, etc.

3. The solution to Problem 2 finds the sum of 12 rectangles whose widths are Δt_i, $1 \leq i \leq 12$, and whose heights are the flow rate for the ith month. Using the horizontal axis for time and the vertical axis for flow rate in ft^3/day, plot the points of Table 2 as follows: (January 1, flow rate for January), (February 1, flow rate for February), ..., (December 1, flow rate for December) and add the point (December 31, flow rate for January). Beginning with the point at January 1, connect each consecutive pair of points with a line segment, creating 12 trapezoids whose bases are Δt_i, $1 \leq i \leq 12$. Approximate the total annual flow $V = V(t) = \int_0^{365.25} V'(t)\, dt$ by summing the areas of these trapezoids.

4. Using the horizontal axis for time and the vertical axis for flow rate in ft^3/day, plot the points of Table 2 as follows: (January 31, flow rate for January), (February 28, flow rate for February), ..., (December 31, flow rate for December). Then add the point (January 1, flow rate for December) to the left of (January 31, flow rate for January). Connect consecutive points with a line segment, creating 12 trapezoids whose bases are Δt_i, $1 \leq i \leq 12$. Approximate the total annual flow $V = V(t) = \int_0^{365.25} V'(t)\, dt$ by summing the areas of these trapezoids.

5. Why did we add the extra point in Problems 3 and 4? How do you justify the choice?

6. Compare the three approximations. Discuss which might be the most accurate.

7. Another way to approximate $V = V(t) = \int_0^{365.25} V'(t)dt$ is to fit a polynomial function to the data. We could find a polynomial of degree 11 that passes through every point of the data, but a polynomial of degree 6 is sufficient to capture the essence of the behavior. The polynomial function f of degree 6 is

$$f(t) = 2.2434817 \times 10^{-10}t^6 - 2.5288956 \times 10^{-7}t^5$$
$$+ 0.00010598313t^4 - 0.019872628t^3 + 1.557403t^2$$
$$- 39.387734t + 2216.2455$$

   Find the total annual flow using $f(t) = V'(t)$.

8. Use technology to graph the polynomial function f over the closed interval $[0, 12]$. How well does the graph fit the data?

9. A manager could approximate the rate of flow of the river for every minute of every day using the function

$$g(t) = 1529.403 + 510.330 \sin \frac{2\pi t}{365.25} + 489.377 \cos \frac{2\pi t}{365.25}$$
$$- 47.049 \sin \frac{4\pi t}{365.25} - 249.059 \cos \frac{4\pi t}{365.25}$$

   where t represents the day of the year in the interval $[0, 365.25]$. The function g represents the best fit to the data that has the form of a sum of trigonometric functions with the period 1 year. We could fit the data perfectly using more terms, but the improvement in results would not be worth the extra work in handling that approximation. Use g to approximate the total annual flow.

10. Use technology to graph the function g over the closed interval $[0, 12]$. How well does the graph fit the data?

11. Compare the five approximations to the annual flow of the river. Discuss the advantages and disadvantages of using one over another. What method would you recommend to measure the annual flow of the Klamath River?

Chapter Review

THINGS TO KNOW

6.1 Area

Definitions:

- Partition of an interval $[a, b]$ (p. 425)
- Lower sum (p. 426)
- Upper sum (p. 428)
- Area A under the graph of a function f from a to b (p. 427)

6.2 The Definite Integral

Definitions:

- Riemann sums (p. 435)
- Definite Integral (p. 436)
- Regular partition (p. 437)
- $\int_a^a f(x)\, dx = 0$ (p. 437)
- $\int_a^b f(x)\, dx = -\int_b^a f(x)\, dx$ (p. 437)
- Left Riemann sum (p. 437)
- Right Riemann sum (p. 437)
- Midpoint Riemann sum (p. 437)
- Trapezoidal sums (p. 442)
- Bounded Function (p. 445)

Theorems:

- If a function f is continuous on a closed interval $[a, b]$, then the definite integral $\int_a^b f(x)\, dx$ exists. (p. 445)
- If a function f is bounded on a closed interval $[a, b]$ and has at most a finite number of points of discontinuity on $[a, b]$, then $\int_a^b f(x)\, dx$ exists. (p. 445)

6.3 The Fundamental Theorem of Calculus

Fundamental Theorem of Calculus: Suppose f is a function that is continuous on a closed interval $[a, b]$.

- Part 1: The function I defined by $I(x) = \int_a^x f(t)\, dt$ has the properties that it is continuous on $[a, b]$ and differentiable on (a, b). Moreover, $I'(x) = \dfrac{d}{dx} \int_a^x f(t)\, dt = f(x)$, for all x in (a, b). (p. 452)

- Part 2: If F is any antiderivative of f on $[a, b]$, then

$$\int_a^b f(x)\, dx = F(b) - F(a) \quad \text{(p. 454)}$$

- If f is an antiderivative of f' and $f(a)$ is known, then

$$f(b) = f(a) + \int_a^b f'(x)\, dx \quad \text{(p. 456)}$$

- The integral from a to b of the rate of change of F equals the change in F from a to b. That is,

$$\int_a^b F'(x)\, dx = F(b) - F(a) \quad \text{(p. 457)}$$

- The accumulation function G associated with a function f that is continuous on a closed interval $[a, b]$ is

$$G(x) = \int_a^x f(t)\, dt \quad a \le x \le b \quad \text{(p. 459)}$$

6.4 Applying Properties of the Definite Integral

Properties of definite integrals:

If the functions f and g are integrable on the closed interval $[a, b]$ and k is a constant, then

- Integral of a sum:

$$\int_a^b [f(x) + g(x)]\, dx = \int_a^b f(x)\, dx + \int_a^b g(x)\, dx \quad \text{(p. 466)}$$

- Integral of a constant times a function:

$$\int_a^b k f(x)\, dx = k \int_a^b f(x)\, dx \quad \text{(p. 467)}$$

- $\displaystyle\int_a^b [k_1 f_1(x) + k_2 f_2(x) + \cdots + k_n f_n(x)]\, dx$

$$= k_1 \int_a^b f_1(x)\, dx + k_2 \int_a^b f_2(x)\, dx$$

$$+ \cdots + k_n \int_a^b f_n(x)\, dx \quad \text{(p. 467)}$$

- If f is integrable on an interval containing the numbers a, b, and c, then

$$\int_a^b f(x)\, dx = \int_a^c f(x)\, dx + \int_c^b f(x)\, dx \quad \text{(p. 468)}$$

- *Bounds on an Integral:* If a function f is continuous on a closed interval $[a, b]$ and if m and M denote the absolute minimum and absolute maximum values, respectively, of f on $[a, b]$, then

$$m(b - a) \le \int_a^b f(x)\, dx \le M(b - a) \quad \text{(p. 470)}$$

- *Mean Value Theorem for Integrals:* If a function f is continuous on a closed interval $[a, b]$, then there is a real number u, $a \le u \le b$, for which

$$\int_a^b f(x)\, dx = f(u)(b - a) \quad \text{(p. 471)}$$

Definitions:

- Suppose f is a function that is continuous on a closed interval $[a, b]$. The average value $\bar{y}$ of f over $[a, b]$ is

$$\bar{y} = \frac{1}{b - a} \int_a^b f(x)\, dx \quad \text{(p. 472)}$$

- For an object moving on a line with velocity $v = v(t)$
  - net displacement from $t = a$ to $t = b$ is $\int_a^b v(t)\, dt$
  - total distance traveled from $t = a$ to $t = b$ is $\int_a^b |v(t)|\, dt$.

  (p. 473)

6.5 The Indefinite Integral; Method of Substitution

The indefinite integral of f: $\displaystyle\int f(x)\,dx = F(x) + C$ where F is any function for which $\dfrac{d}{dx}F(x) = f(x)$, and C is the constant of integration. (p. 481)

Basic integration formulas: See Table 1.   (p. 482)

Properties of indefinite integrals:

- Derivative of an integral:

$$\frac{d}{dx}\int f(x)\,dx = f(x) \quad \text{(p. 483)}$$

- Integral of a sum:

$$\int [f(x) + g(x)]dx = \int f(x)\,dx + \int g(x)\,dx \quad \text{(p. 483)}$$

- Integral of a constant k times a function:

$$\int kf(x)\,dx = k\int f(x)\,dx \quad \text{(p. 483)}$$

Integration formulas:

- $\displaystyle\int \frac{g'(x)}{g(x)}dx = \ln|g(x)| + C \quad \text{(p. 486)}$

- $\displaystyle\int \tan x\,dx = \ln|\sec x| + C \quad \text{(p. 486)}$

- $\displaystyle\int \sec x\,dx = \ln|\sec x + \tan x| + C \quad \text{(p. 486)}$

Method of substitution (definite integrals):

- **Method 1:** Find the related indefinite integral using substitution. Then use the Fundamental Theorem of Calculus.
- **Method 2:** Find the definite integral directly by making a substitution in the integrand and using the substitution to change the limits of integration. (p. 488)

Theorems:

- If f is an even function, then

$$\int_{-a}^{a} f(x)\,dx = 2\int_{0}^{a} f(x)\,dx \quad \text{(p. 490)}$$

- If f is an odd function, then $\displaystyle\int_{-a}^{a} f(x)\,dx = 0. \quad \text{(p. 490)}$

6.6 Integrands Containing $ax^2 + bx + c$, $a \neq 0$; Integration Using Long Division

Proper and improper rational function   (p. 500)

OBJECTIVES

Preparing for the
AP® Exam
AP® Review Problems

Section	You should be able to …	Examples	Review Exercises	AP® Review Problems
6.1	1 Approximate an area (p. 424)	1, 2	1, 2	
	2 Find the area under the graph of a function (p. 428)	3, 4	3, 4	3, 19, 21
6.2	1 Form Riemann sums (p. 435)	1	5(a), 55	12
	2 Define a definite integral as the limit of Riemann sums (p. 436)	2–4	5(b), 60, 63	17
	3 Approximate a definite integral using Riemann sums (p. 441)	5, 6	58, 59	12
	4 Approximate a definite integral using trapezoidal sums (p. 442)	7, 8	64	14, 16
	5 Know conditions that guarantee a definite integral exists (p. 446)		61	
6.3	1 Use Part 1 of the Fundamental Theorem of Calculus (p. 452)	1–4	7-10, 52, 53	7
	2 Use Part 2 of the Fundamental Theorem of Calculus (p. 454)	5, 6	5(c), 11–13, 17–18, 54	1, 2, 3, 4, 5, 19, 21
	3 Interpret the integral of a rate of change (p. 456)	7–9	6, 21, 22, 57	6, 10, 12
	4 Interpret the integral as an accumulation function (p. 458)	10, 11	62	12, 18, 20
6.4	1 Use properties of the definite integral (p. 466)	1–7	15, 16, 23, 27, 28, 45	2, 3, 4, 5, 11, 19, 20
	2 Work with the Mean Value Theorem for Integrals (p. 471)	8	29, 30	
	3 Find the average value of a function (p. 472)	9	31–34	21
	4 Interpret integrals involving motion on a line (p. 473)	10	56	6
6.5	1 Find indefinite integrals (p. 481)	1	14	11, 13, 15
	2 Use properties of indefinite integrals (p. 483)	2, 3	19, 20, 35, 36	11
	3 Find an indefinite integral using substitution (p. 484)	4–8	37–39, 42, 43, 46, 49	9
	4 Find a definite integral using substitution (p. 488)	9, 10	40, 41, 44, 45, 47, 48 52, 53	2, 8
	5 Integrate even and odd functions (p. 490)	11, 12	25	
6.6	1 Integrate a function that contains a quadratic expression (p. 498)	1–3	26	13, 15
	2 Integrate an improper rational function using long division (p. 500)	4–6	24	13

REVIEW EXERCISES

1. **Area** Approximate the area under the graph of $f(x) = 2x + 1$ from 0 to 4 by finding lower sums s_n and upper sums S_n for $n = 4$ and $n = 8$ subintervals.

2. **Area** Approximate the area under the graph of $f(x) = x^2$ from 0 to 8 by finding lower sums s_n and upper sums S_n for $n = 4$ and $n = 8$ subintervals.

3. **Area** Find the area A under the graph of $y = f(x) = 9 - x^2$ from 0 to 3 by using lower sums s_n (rectangles that lie below the graph of f).

4. **Area** Find the area A under the graph of $y = f(x) = 8 - 2x$ from 0 to 4 using upper sums S_n (rectangles that extend above the graph of f).

5. **Riemann Sums**

   (a) Find the Riemann sum of $f(x) = x^2 - 3x + 3$ on the closed interval $[-1, 3]$ using a regular partition with four subintervals and the numbers $u_1 = -1$, $u_2 = 0$, $u_3 = 2$, and $u_4 = 3$.

   (b) Find the Riemann sums of f by partitioning $[-1, 3]$ into n subintervals of equal width and choosing u_i as the right endpoint of the ith subinterval $[x_{i-1}, x_i]$. Write the limit of the Riemann sums as a definite integral.

   (c) Find the definite integral from (b) using the Fundamental Theorem of Calculus.

6. **Units of an Integral** In the definite integral $\int_a^b a(t)\,dt$, where a represents acceleration measured in meters per second squared and t is measured in seconds, what are the units of $\int_a^b a(t)\,dt$?

In Problems 7–10, find each derivative using the Fundamental Theorem of Calculus.

7. $\dfrac{d}{dx} \int_0^x t^{2/3} \sin t\,dt$

8. $\dfrac{d}{dx} \int_e^x \ln t\,dt$

9. $\dfrac{d}{dx} \int_{x^2}^1 \tan t\,dt$

10. $\dfrac{d}{dx} \int_0^{2\sqrt{x}} \dfrac{t}{t^2+1}\,dt$

In Problems 11–20, find each integral.

11. $\int_1^{\sqrt{2}} x^{-2}\,dx$

12. $\int_1^{e^2} \dfrac{1}{x}\,dx$

13. $\int_0^1 \dfrac{1}{1+x^2}\,dx$

14. $\int \dfrac{1}{x\sqrt{x^2-1}}\,dx$

15. $\int_0^{\ln 2} 4e^x\,dx$

16. $\int_0^2 (x^2 - 3x + 2)\,dx$

17. $\int_1^4 2^x\,dx$

18. $\int_0^{\pi/4} \sec x \tan x\,dx$

19. $\int \dfrac{1 + 2xe^x}{x}\,dx$

20. $\int \dfrac{1}{2}\sin x\,dx$

21. **Interpreting an Integral** The function $v = v(t)$ is the speed v, in kilometers per hour, of a train at a time t, in hours. Interpret the integral $\int_0^{16} v(t)\,dt = 460$.

22. **Interpreting an Integral** The function f is the rate of change of the volume V of oil, in liters per hour, draining from a storage tank at time t (in hours). Interpret the integral $\int_0^2 f(t)\,dt = 100$.

In Problems 23–26, find each integral.

23. $\int_1^2 f(x)\,dx$, where $f(x) = \begin{cases} 3x+2 & \text{if } -2 \le x < 0 \\ 2x^2+2 & \text{if } 0 \le x \le 2 \end{cases}$

24. $\int_1^2 \dfrac{x^3 + 2x - 3}{x}\,dx$

25. $\int_{-\pi/2}^{\pi/2} \sin x\,dx$

26. $\int \dfrac{1}{x^2+9}\,dx$

Bounds on an Integral *In Problems 27 and 28, find lower and upper bounds for each integral.*

27. $\int_0^2 e^{x^2}\,dx$

28. $\int_0^1 \dfrac{1}{1+x^2}\,dx$

In Problems 29 and 30, for each integral find the number(s) u guaranteed by the Mean Value Theorem for Integrals.

29. $\int_0^\pi \sin x\,dx$

30. $\int_{-3}^3 (x^3 + 2x)\,dx$

In Problems 31–34, find the average value of each function over the given interval.

31. $f(x) = \sin x$ over $\left[-\dfrac{\pi}{2}, \dfrac{\pi}{2}\right]$

32. $f(x) = x^3$ over $[1, 4]$

33. $f(x) = e^x$ over $[-1, 1]$

34. $f(x) = 6x^{2/3}$ over $[0, 8]$

35. Find $\dfrac{d}{dx} \int \sqrt{\dfrac{1}{1+4x^2}}\,dx$

36. Find $\dfrac{d}{dx} \int \ln x\,dx$.

In Problems 37–49, find each integral.

37. $\int \dfrac{y\,dy}{(y-2)^3}$

38. $\int \dfrac{x}{(2-3x)^3}\,dx$

39. $\int \sqrt{\dfrac{1+x}{x^5}}\,dx,\ x > 0$

40. $\int_{\pi^2/4}^{4\pi^2} \dfrac{1}{\sqrt{x}} \sin \sqrt{x}\,dx$

41. $\int_1^2 \dfrac{1}{t^4}\left(1 - \dfrac{1}{t^3}\right)^3 dt$

42. $\int \dfrac{e^x+1}{e^x-1}\,dx$

43. $\int \dfrac{dx}{\sqrt{x}\,(1-2\sqrt{x})}$

44. $\int_{1/5}^3 \dfrac{\ln(5x)}{x}\,dx$

45. $\int_{-1}^1 \dfrac{5^{-x}}{2^x}\,dx$

46. $\int e^{x+e^x}\,dx$

47. $\int_0^1 \dfrac{x\,dx}{\sqrt{2-x^4}}$

48. $\int_4^5 \dfrac{dx}{x\sqrt{x^2-9}}$

49. $\int \sqrt[3]{x^3 + 3\cos x}\,(x^2 - \sin x)\,dx$

50. Find $f''(x)$ if $f(x) = \int_0^x \sqrt{1-t^2}\,dt$.

51. Suppose that $F(x)=\int_0^x \sqrt{t}\,dt$ and $G(x)=\int_1^x \sqrt{t}\,dt$. Explain why $F(x)-G(x)$ is constant. Find the constant.

52. If $\int_0^2 f(x+2)\,dx=3$, find $\int_2^4 f(x)\,dx$.

53. If $\int_1^2 f(x-c)\,dx=5$, where c is a constant, find $\int_{1-c}^{2-c} f(x)\,dx$.

54. Area Find the area under the graph of $y=\dfrac{e^x+e^{-x}}{2}$ from $x=0$ to $x=2$.

55. The table below shows the rate of change of area A with respect to time t (in square meters per hour) of an oil spill as a function of time t, in hours.

t	1	2.5	3	4	5
$\dfrac{dA}{dt}$	10	21	30	36	40

(a) Find a Riemann sum for $\dfrac{dA}{dt}$ on the interval $[1,5]$.

(b) What are the units of the Riemann sum?

(c) Interpret the Riemann sum in the context of the problem.

56. An object is moving along a horizontal line with velocity $v(t)=6t-t^2$ m/s over the interval $[0,9]$.

(a) Find the net displacement of the object over the interval $[0,9]$. Interpret the result in the context of the problem.

(b) Find the total distance traveled by the object over the interval $[0,9]$ and interpret the result.

57. Water Supply A sluice gate of a dam is opened and water is released from the reservoir at a rate of $r(t)=100+\sqrt{t}$ gallons per minute, where t measures time in minutes since the gate has been opened. If the gate is opened at 7 a.m. and is left open until 9:24 a.m., how much water is released?

58. Approximate $\int_{-2}^6 \dfrac{2x}{x+3}\,dx$ with a Midpoint Riemann sum formed by partitioning $[-2,6]$ into four subintervals each of width 2.

59. Approximate $\int_0^4 (x^2+1)\,dx$ with a Left Riemann sum formed by partitioning $[0,4]$ into four subintervals each of width 1.

60. Write the definite integral of $f(x)=2x-1$ from 0 to 2 as the limit of a Riemann sum.

61. State a condition that guarantees that a function f is integrable over an interval $[a,b]$.

62. (a) Write the accumulation function F associated with $f(x)=x^3-2x^2-3x$ over the interval $[-2,4]$ as an integral.

(b) Find $F(-1)$, $F(0)$, $F(2)$, $F(3)$, and $F(4)$.

(c) Interpret the results found in (b).

63. Write the limit of the Riemann sums

$$\lim_{n\to\infty}\frac{3}{n}\left[\left(\frac{3}{n}\right)^2+\left(\frac{6}{n}\right)^2+\cdots+\left(\frac{3n}{n}\right)^2\right]$$ as an integral.

The partition is into n subintervals each of the same width.

64. Distance The velocity v (in meters per second) of an object moving in a straight line at time t is given in the table. Use trapezoidal sums with six subintervals of equal width to approximate the distance traveled from $t=1$ to $t=4$.

t (s)	1	1.5	2	2.5	3	3.5	4
v (m/s)	3	4.3	4.6	5.1	5.8	6.2	6.6

Break It Down

Let's take a closer look at AP® Practice Problem 17 part (c) from Section 6.4 on page 480.

17. An object is moving along a horizontal line with velocity $v(t)=3t^2-6t$, $1\le t\le 4$ (in meters per second).

(c) If at time $t=1$, the object is 2 m to the right of the origin, what is its position at $t=4$?

Step 1	Identify the underlying structure and related concepts.	We are asked to find the position of an object moving along a horizontal line from time $t=1$ to $t=4$ given its velocity function $v(t)=3t^2-6t$. We also know that the object is at a position 2 m to the right of the origin at time $t=1$ s.
Step 2	Determine the appropriate math rule or procedure.	We are given $s(1)=2$ and $s'(t)=v(t)=3t^2-6t$, $1\le t\le 4$. So, we can find any value of $s(t)$, $1\le t\le 4$, using $s(b)=s(a)+\int_a^b v(t)\,dt$ where $a=1, b=4$, and $v(t)=3t^2-6t$.
Step 3	Apply the math rule or procedure.	The position of the object at $t=4$ seconds is given by $s(4)=s(1)+\int_1^4 (3t^2-6t)\,dt$. So $s(4)=2+\left[t^3-3t^2\right]_1^4=2+(64-48)-(1-3)=20$.
Step 4	Clearly communicate your answer.	The position of the particle at time $t=4$ is 20 meters to the right of the origin.

AP® Review Problems: Chapter 6 Part 1

Preparing for the AP® Exam

Multiple-Choice Questions

1. The closed interval $[a, b]$ is partitioned into n subintervals each of width $\Delta x = \dfrac{b - a}{n}$. If u_i is any number in the ith subinterval, what is $\displaystyle\lim_{n \to \infty} \sum_{i=1}^{n} \sqrt[3]{u_i}\, \Delta x$?

(A) $\dfrac{3}{2}[b^{2/3} - a^{2/3}]$ (B) $\dfrac{3}{4}[b^{4/3} - a^{4/3}]$

(C) $\dfrac{4}{3}[b^{4/3} - a^{4/3}]$ (D) $\dfrac{1}{3}\left[\dfrac{1}{b^{2/3}} - \dfrac{1}{a^{2/3}}\right]$

2. If $\int_1^8 f(x)\, dx = 3$, find $\int_1^8 f(9 - x)\, dx$.

(A) -3 (B) 6 (C) 3 (D) 9

3. The area under the graph of the function $f(x) = \dfrac{2}{x}$, from $x = k$ to $x = 4k$, $k > 0$ is

(A) $\ln 8$ (B) $2\dfrac{\ln(4k)}{\ln k}$

(C) $2 \ln 4$ (D) $2 \ln(4k)$

4. If $\int_{-3}^{4} f(x)\, dx = 8$ and $\int_{6}^{4} f(x)\, dx = -6$, what is $\int_{-3}^{6} f(x)\, dx$?

(A) -14 (B) 2 (C) 14 (D) 16

5. Find $\int_{-2}^{6} f(x)\, dx$ when

$$f(x) = \begin{cases} -x & \text{if } -2 \le x < 2 \\ x + 3 & \text{if } 2 \le x \le 6 \end{cases}.$$

(A) 24 (B) 28 (C) 32 (D) 44

6. An object is moving along the x-axis. If its velocity v at time t (in minutes) is $v(t) = t^2 - 2t$ (in feet per minute), what is the total distance the object travels between $t = 0$ and $t = 4$ minutes?

(A) 0 ft   (B) $\dfrac{16}{3}$ ft   (C) 8 ft   (D) $\dfrac{40}{3}$ ft

7. The graph of a function f that is differentiable is shown below. If $F(x) = \int_0^x f(t)\, dt$, which of the following is true?

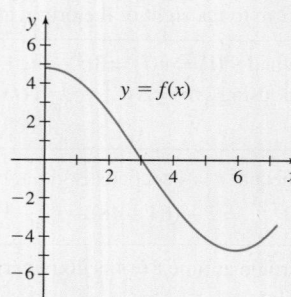

$y = f(x)$

(A) $F(3) < F'(3) < F''(3)$ (B) $F'(3) < F''(3) < F(3)$

(C) $F''(3) < F(3) < F'(3)$ (D) $F''(3) < F'(3) < F(3)$

8. If $\int_0^6 f(3x + 1)\, dx = 9$, then

(A) $\int_0^6 f(u)\, du = 3$ (B) $\int_1^{19} f(u)\, du = 9$

(C) $\int_0^{18} f(u)\, du = 27$ (D) $\int_1^{19} f(u)\, du = 27$

9. $\int x^3 e^{-x^4}\, dx =$

(A) $-\dfrac{1}{4} e^{-x^4} + C$ (B) $-\dfrac{1}{4} e^x + C$

(C) $\dfrac{x^4}{4} e^{-4x^3} + C$ (D) $\dfrac{x^4}{4} e^{-x^4} + C$

10. If at every point (x, y) on the graph of a function f, the slope of the tangent line is given by $3 - 4x$. If the point $(2, 3)$ is on the graph of f, then

(A) $f(x) = -5x + 13$ (B) $f(x) = -2x^2 + 3x - 1$

(C) $f(x) = -2x^2 + 11$ (D) $f(x) = -2x^2 + 3x + 5$

11. $\displaystyle\int \dfrac{x^2 - 3x + 2\sqrt{x} - 1}{x}\, dx =$

(A) $\dfrac{1}{2}x^2 - 3x + 4x^{1/2} - \ln|x| + C$

(B) $\dfrac{1}{3}x^3 - \dfrac{3}{2}x^2 + \dfrac{4}{3}x^{3/2} - \ln|x| + C$

(C) $\dfrac{1}{2}x^2 - 3x + \dfrac{6}{5}x^{5/2} - \ln|x| + C$

(D) $\dfrac{1}{2}x^2 - 3x + x^{1/2} - \ln|x| + C$

12. On a typical day, a dam releases water at a rate of $\dfrac{dA}{dt}$ (hundred thousand gallons per hour) as shown in the graph. Use a Left Riemann sum with four equal subintervals to approximate the total amount A (in hundreds of thousands of gallons) of water released in a day.

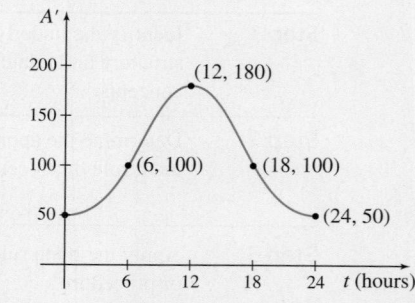

(A) 1720 (B) 2580 (C) 2760 (D) 3660

▶ 13. $\int \dfrac{x^3 + 2x^2 - 3x + 5}{x + 4}\, dx =$

(A) $\dfrac{1}{3}x^3 - x^2 + 5x - 15\ln|x + 4| + C$

(B) $\dfrac{1}{3}x^3 + x^2 + 5x - 15\ln|x + 4| + C$

(C) $\dfrac{1}{3}x^3 - x^2 + 5x - 15\tan^{-1}(x + 4) + C$

(D) $\dfrac{1}{3}x^3 - x^2 + 5x - 15\tan^{-1}(x^2 + 4) + C$

▶ 14. A function f is continuous on its domain, all real numbers. Select values of the function are shown in the table below. Use a trapezoidal sum with three subintervals of equal width to approximate $\int_0^{2.4} f(x)\, dx$.

x	0	0.4	0.8	1.2	1.6	2	2.4
$f(x)$	3	4	4	6	8	10	13

(A) $\dfrac{41}{5}$ (B) 14   (C) 16   (D) 18

▶ 15. $\int \dfrac{dx}{x^2 - 2x + 2} =$

(A) $-\dfrac{1}{x - 1} + C$ (B) $\tan^{-1} x + C$

(C) $\tan^{-1}(x - 1) + C$ (D) $\ln|x^2 - 2x + 2| + C$

▶ 16. A function f is continuous on the closed interval $[0, 10]$ and has values

x	0	1	4	8	10
$f(x)$	4	5	10	12	8

Using trapezoidal sums, $\int_0^{10} f(x)\, dx$ is approximately equal to

(A) 39   (B) 66   (C) 82.5   (D) 91

▶ 17. Which of the following is equivalent to the integral $\int_1^5 (x^3 + x)\, dx$?

(A) $\displaystyle\lim_{n\to\infty} \sum_{i=1}^{n} \left[\left(1 + \dfrac{4}{n}i\right)^3 + \left(1 + \dfrac{4}{n}i\right)\right]\dfrac{4}{n}$

(B) $\displaystyle\lim_{n\to\infty} \sum_{i=1}^{n} \left[\left(2 + \dfrac{4}{n}i\right)\right]\dfrac{4}{n}$

(C) $\displaystyle\lim_{n\to\infty} \sum_{i=1}^{n} \left[\left(1 + \dfrac{4}{n}i\right)^3 + \left(1 + \dfrac{4}{n}i\right)\right]$

(D) $\displaystyle\lim_{n\to\infty} \sum_{i=1}^{n} \left(x^3 + x + \dfrac{4}{n}i\right)\dfrac{4}{n}$

Free-Response Questions

▶ 18. A function is defined and continuous for $-4 \le x \le 4$. The graph of its derivative f' is given below.

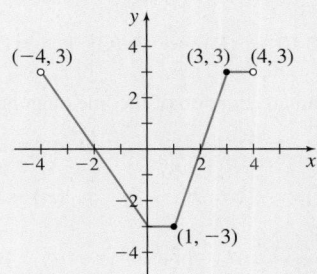

If $G(x) = \int_{-4}^{x} f'(t)\, dt$,

(a) Find $G(0)$.

(b) Find $G(2)$.

(c) Find $G(4)$.

(d) Determine where G is increasing or decreasing.

(e) Find any local maxima of G.

(f) Find any local minima of G.

▶ 19. $f(x) = \begin{cases} x\sin x & \text{if } 0 \le x \le \pi \\ e^x|\sin x| & \text{if } \pi < x \le 2\pi \end{cases}$.

Find the area under the graph of f from 0 to 2π.

▶ 20. Use the graph of the function f shown below to answer the questions.

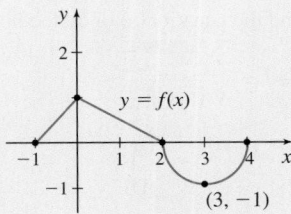

(a) Write $\int_{-1}^{4} f(x)\, dx$ as a sum of three integrals using properties of definite integrals.

(b) Find $\int_{-1}^{4} f(x)\, dx$ using geometry.

(c) Explain why $\int_{-1}^{2} f(x)\, dx > \int_{-1}^{4} f(x)\, dx$.

▶ 21. For the function $f(x) = \dfrac{\sin x}{x^2 + 1}$

(a) Find the derivative of f when $x = 1$.

(b) Find the area under the graph of f from 0 to π.

(c) What is the average value of f over the closed interval $[0, \pi]$?

Multiple-Choice Questions

1. $\lim\limits_{x\to 0^+}\dfrac{(x+4)^2-16}{x}=$

(A) 0   (B) ∞ (C) 8   (D) 1

2. Find an equation of the line tangent to

$f(x)=3x^3-4x^2+1$ at $x=2$.

(A) $y-2=9(x-9)$ (B) $y-9=\dfrac{1}{2}(x-2)$

(C) $y-9=20(x-2)$ (D) $y-20=\dfrac{1}{9}(x-2)$

3. A car's velocity $v=v(t)$ (in ft/s) is measured each second t for $t=0$ to $t=8$ and posted in the table. Use a Right Riemann sum with four subintervals of equal length to approximate the car's average velocity over the interval from 0 to 8 seconds.

t	0	1	2	3	4	5	6	7	8
$v(t)$	0	2	4	6	7	7	8	6	2

(A) 42 ft/s   (B) 7 ft/s   (C) 5 ft/s   (D) 5.25 ft/s

4. Determine the interval(s) on which $f(x)=2xe^{-x}$ is increasing.

(A) $(-\infty,1)$ (B) $\left(-\infty,\dfrac{1}{e}\right)$

(C) $\left(\dfrac{1}{e},1\right)$ (D) $(1,\infty)$

5. Find the equation(s) of all the horizonal asymptotes of

$f(x)=\dfrac{3x}{\sqrt{x^2+4}}$

(A) $y=3$ (B) $y=\dfrac{3}{4}$

(C) $y=\pm 3$ (D) $y=-3$

6. The velocity, in km/h, of an object moving along the x-axis is given by $v(t)=t+\sin t$, for $0\le t\le 2\pi$. Find the average velocity of the object for $\dfrac{\pi}{2}\le t\le\pi$.

(A) $\dfrac{\pi^2}{4}-1$ (B) $\dfrac{3\pi^2}{8}+1$

(C) $\dfrac{2}{\pi}\left(\dfrac{\pi^2}{4}-1\right)$ (D) $\dfrac{2}{\pi}\left(\dfrac{3\pi^2}{8}+1\right)$

7. $\dfrac{d}{dx}\left[\displaystyle\int_4^{\sqrt{x}}\sqrt{t^3+5}\,dt\right]=$

(A) $x^{\frac{1}{2}}\sqrt{x^{\frac{3}{2}}+5}$ (B) $\dfrac{\sqrt{x^{\frac{3}{2}}+5}}{2x^{\frac{1}{2}}}$

(C) $\sqrt{x}\sqrt{x\sqrt{x}+5}-13$ (D) $\sqrt{x^{\frac{3}{2}}+5}$

8. $\displaystyle\int_{\frac{\pi}{6}}^{\frac{2\pi}{3}}\dfrac{x\sin x-6}{x}dx=$

(A) $\dfrac{-1-\sqrt{3}}{2}+\dfrac{189}{2\pi}$

(B) $\dfrac{1+\sqrt{3}}{2}-12\ln 2$

(C) $-1-\dfrac{189}{2\pi^2}$

(D) $\dfrac{1}{2}+\dfrac{\sqrt{3}}{2}+\ln\dfrac{2\pi}{3}-\ln\dfrac{\pi}{6}$

9. Sand is being poured onto the ground forming a conical pile whose height equals $\dfrac{1}{3}$ of the diameter of the base. If the sand is being poured at the rate of $30\,\text{cm}^3/\text{s}$, how fast is the height of the sand pile increasing when the sand pile is 4 cm high?

(A) $\dfrac{5}{6\pi}$ cm/s   (B) $\dfrac{45}{16\pi}$ cm/s

(C) $1,080$ cm/s   (D) $\dfrac{45}{4\pi}$ cm/s

10. During a fire drill at the local high school, students and teachers exit the school building at a rate of $R(t)$ students and teachers per minute, where t is the time in minutes since the fire drill began. Four values for $R(t)$ at the given value of t are provided in the table below. Using a Right Riemann sum with three subintervals and the information from the table, approximate the number of students and teachers who exit the building during the first 9 minutes of the fire drill.

t minutes	0	3	6	9
$R(t)$ teachers and students per minute	90	100	85	50

(A) 605   (B) 705   (C) 975   (D) 825

Free-Response Questions

11. Functions f and g are defined as

$$f(x) = \int_0^{2x} \sqrt{t^2 + 4}\, dt \qquad g(x) = f(5\cos x)$$

(a) Find $f'(x)$.

(b) Find $g'\left(\dfrac{\pi}{2}\right)$.

(c) Express the maximum value of f on $0 \le x \le \pi$ as an integral. Do not evaluate.

12. A function is defined and continuous for $-4 \le x \le 4$. The graph of its derivative, f' is given below.

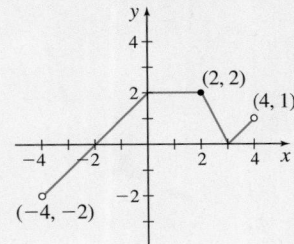

(a) Where is f increasing?

(b) Where is f decreasing?

(c) Where is f concave up?

(d) If $G(x) = \int_{-4}^{x} f'(t)\,dt$, find $G(2)$.

13. $f(x) = |x^2 - x - 6|$

(a) Determine where f is continuous.

(b) Determine where f is differentiable.

(c) Determine the interval(s) on which f is increasing.

(d) Find $\int_{-5}^{6} f(x)\,dx$

14. An object is moving along a horizontal line with velocity $v = v(t) = t \sin t$ (in miles/hour) for $0 \le t \le 2\pi$.

(a) If the initial position of the object is 0, what is its position when $t = \pi$? Justify your answer.

(b) Find the total distance traveled from time $t = 0$ to $t = 2\pi$.

(c) Find the average speed of the object for $0 \le t \le 2\pi$. Justify your answer.

15. Guests arrive at an amusement park from 8 a.m. until 9 p.m. at a rate modeled by the function $P(t) = 200 + 150 \sin\left(\dfrac{\pi}{8}t\right)$. No guests are admitted after 9 p.m. At noon guests begin to depart at a uniform rate of $D(t) = 40$ and continue to depart at this rate until the park closes at midnight, when all remaining guests are ushered out of the park. Both P and D are measured in guests per hour, and t is measured in hours since 8 a.m. ($t = 0$).

(a) How many guests enter the park between 8 a.m. ($t = 0$) and noon ($t = 4$)? Round the answer to the nearest whole number.

(b) What is the average number of guests entering the park per hour between noon ($t = 4$) and 8 p.m. ($t = 12$)?

(c) At what time t, $4 \le t \le 13$, are the greatest number of guests in the park? Justify your answer.

(d) Is the rate of change of the number of guests in the park increasing or decreasing at 1 p.m. ($t = 5$)? Justify your answer.

CHAPTER

6

PART 2
Techniques of Integration

tilialucida/Shutterstock

The Birds of Rügen Island

Where can a variety of shore and grassland birds, such as Redshanks, Lapwings, Pied Avocet, and several species of seagull (pictured above), find an undisturbed habitat? Try Rügen Island, a UNESCO World Heritage Site since June 2011. A department of the Western Pomeranian National Park administers the Western Pomerania Lagoon Area and Jasmund National Park, which is on Rügen Island. This area on the Baltic Sea off the coast of Germany includes some of the most important breeding grounds for water and mud flat birds of the Baltic Sea. How might a scientist model the population of birds that live in a protected area like Rügen Island?

 The Chapter 6, Part 2 Project on page 572 for a discussion of how we can use calculus to model bird populations.

We have developed a sizable collection of basic integration formulas that are listed below. Some of the integrals are the result of simple antidifferentiation. Others, such as $\int \tan x \, dx$, were found using substitution, a *technique of integration*. It is important that you are familiar with each of these integrals.

Integration, unlike differentiation, has no hard and fast rules. It is often difficult, sometimes even impossible, to integrate a function that appears to be simple.

As you study the techniques in this chapter, it is important to recognize the form of the integrand for each technique. Then, with practice, integration becomes easier.

$$\int x^a \, dx = \frac{x^{a+1}}{a+1} + C, a \neq -1$$

$$\int \frac{1}{x} \, dx = \ln |x| + C$$

$$\int e^x \, dx = e^x + C$$

$$\int a^x \, dx = \frac{a^x}{\ln a} + C, a > 0, a \neq 1$$

$$\int \sin x \, dx = -\cos x + C$$

$$\int \cos x \, dx = \sin x + C$$

$$\int \sec^2 x \, dx = \tan x + C$$

$$\int \sec x \tan x \, dx = \sec x + C$$

$$\int \csc^2 x \, dx = -\cot x + C$$

$$\int \csc x \cot x \, dx = -\csc x + C$$

$$\int \tan x \, dx = \ln |\sec x| + C$$

$$\int \cot x \, dx = \ln |\sin x| + C$$

$$\int \sec x \, dx = \ln |\sec x + \tan x| + C$$

$$\int \csc x \, dx = \ln |\csc x - \cot x| + C$$

$$\int \frac{dx}{\sqrt{a^2 - x^2}} = \sin^{-1} \frac{x}{a} + C, \ a > 0$$

$$\int \frac{dx}{x\sqrt{x^2 - a^2}} = \frac{1}{a} \sec^{-1} \frac{x}{a} + C, a > 0$$

$$\int \frac{dx}{a^2 + x^2} = \frac{1}{a} \tan^{-1} \frac{x}{a} + C, a > 0$$

∎

6.7 Integration by Parts

OBJECTIVES *When you finish this section, you should be able to:*

1 **Integrate by parts (p. 516)**
2 **Find a definite integral using integration by parts (p. 520)**
3 **Derive a general formula using integration by parts (p. 522)**

Osy
Software Developer

As a software developer, I am constantly creating graphs and visuals along with coding applications that deal with design and analysis algorithms. AP® Calculus gave me the skills to efficiently utilize datasets to best meet our customers' unique needs.

Integration by parts is a technique of integration based on the Product Rule for derivatives: If $u = f(x)$ and $v = g(x)$ are functions that are differentiable on an open interval (a, b), then

$$\frac{d}{dx}[f(x) \cdot g(x)] = f(x) \, g'(x) + f'(x) \, g(x)$$

Integrating both sides gives

$$\int \frac{d}{dx}[f(x) \cdot g(x)] \, dx = \int [f(x) \, g'(x) + f'(x) \, g(x)] \, dx$$

$$f(x) \cdot g(x) = \int f(x) \, g'(x) \, dx + \int f'(x) g(x) \, dx$$

Solving the above equation for $\int f(x)g'(x) \, dx$ yields

$$\boxed{\int f(x) \, g'(x) \, dx = f(x) \cdot g(x) - \int f'(x) \, g(x) \, dx}$$

which is known as the **integration by parts formula**.

Let $u = f(x)$, $v = g(x)$. Then we can use their differentials, $du = f'(x) \, dx$ and $dv = g'(x) \, dx$, to obtain the integration by parts formula in the form we usually use:

IN WORDS The integral of $u \, dv$ equals uv minus the integral of $v \, du$.

$$\boxed{\int u \, dv = uv - \int v \, du} \tag{1}$$

1 Integrate by Parts

In using the integration by parts formula (1), the goal is to choose u and dv so that $\int v\,du$ is easier to integrate than $\int u\,dv$.

EXAMPLE 1   Using the Integration by Parts Formula $\int u\,dv = u\,v - \int v\,du$

Find $\int x\,e^x\,dx$.

Solution

Choose u and dv so that

$$\int u\,dv = \int x\,e^x\,dx$$

Suppose we choose

$$u = x \quad\text{and}\quad dv = e^x\,dx$$

Then

$$du = dx \quad\text{and}\quad v = \int dv = \int e^x\,dx = e^x$$

Notice that we did not add a constant. Only a particular antiderivative of dv is required at this stage; we add the constant of integration at the end. Using the integration by parts formula (1), we have

$$\int \underbrace{x}_{u}\underbrace{e^x\,dx}_{dv} = \underbrace{x\,e^x}_{uv} - \int \underbrace{e^x}_{v}\underbrace{dx}_{du} = x\,e^x - e^x + C \qquad\blacksquare$$

We intentionally chose $u = x$ and $dv = e^x dx$ so that $\int v\,du$ in the formula is easy to integrate. Suppose, instead, we chose

$$u = e^x \quad\text{and}\quad dv = x\,dx$$

Then

$$du = e^x\,dx \quad\text{and}\quad v = \int x\,dx = \frac{x^2}{2}$$

and the integration by parts formula yields

$$\int x\,e^x\,dx = \int \underbrace{e^x}_{u}\underbrace{x\,dx}_{dv} = \underbrace{e^x\,\frac{x^2}{2}}_{uv} - \int \underbrace{\frac{x^2}{2}}_{v}e^x\,dx\underbrace{}_{du}$$

For this choice of u and v, the integral on the right is more complicated than the original integral, indicating an unwise choice of u and dv.

NOW WORK   Problem 3.

EXAMPLE 2   **Using the Integration by Parts Formula (1)**

Find $\int x \sin x \, dx$.

Solution

Use the integration by parts formula $\int u \, dv = \int x \sin x \, dx$.

Choose

$$u = x \qquad \text{and} \qquad dv = \sin x \, dx$$

Then

$$du = dx \qquad \text{and} \qquad v = \int \sin x \, dx = -\cos x$$

Now

$$\int x \sin x \, dx = -x \cos x + \int \cos x \, dx = -x \cos x + \sin x + C$$
$$\uparrow$$
$$\int u \, dv = uv - \int v \, du$$

∎

NOW WORK   Problem 5 and AP® Practice Problem 6.

EXAMPLE 3   **Using the Integration by Parts Formula (1) to Find $\int \ln x \, dx$**

Derive the formula

$$\boxed{\int \ln x \, dx = x \ln x - x + C}$$

Solution

Use the integration by parts formula (1) with

$$u = \ln x \qquad \text{and} \qquad dv = dx$$

Then

$$du = \frac{1}{x} \, dx \qquad \text{and} \qquad v = \int dx = x$$

Now

> **NOTE** The integral $\int \ln x \, dx$ can be found in the list of integrals at the back of the book and is considered a basic integral.

$$\int \ln x \, dx = x \cdot \ln x - \int x \cdot \frac{1}{x} \, dx = x \ln x - \int dx = x \ln x - x + C$$
$$\uparrow$$
$$\int u \, dv = uv - \int v \, du$$

∎

NOW WORK   Problem 17 and AP® Practice Problem 2.

Unfortunately, there are no exact rules for choosing u and dv. But the following guidelines are helpful:

Integration by Parts: General Guidelines for Choosing u and dv

- dx is always part of dv.
- dv should be easy to integrate.
- u and dv are chosen so that $\int v \, du$ is no more difficult to integrate than the original integral $\int u \, dv$.
- If the new integral is more complicated, try different choices for u and dv.

Table 1 provides additional guidelines to help choose u and dv for several types of integrals that are found using integration by parts. In the table, n is a positive integer.

TABLE 1 Guidelines for Choosing u and dv

Integral; n is a positive integer	u	dv
$\int x^n e^{ax}\,dx$ $\int x^n \sin(ax)\,dx$ $\int x^n \cos(ax)\,dx$ $\int x^n \sec^2(ax)\,dx$ $\int x^n \sec(ax)\tan(ax)\,dx$	$u = x^n$	$dv =$ what remains
$\int x^n \sin^{-1} x\,dx$	$u = \sin^{-1} x$	$dv = x^n\,dx$
$\int x^n \cos^{-1} x\,dx$	$u = \cos^{-1} x$	$dv = x^n\,dx$
$\int x^n \tan^{-1} x\,dx$	$u = \tan^{-1} x$	$dv = x^n\,dx$
$\int x^m (\ln x)^n\,dx;\ \ m$ is a real number, $m \neq -1$	$u = (\ln x)^n$	$dv = x^m\,dx$

EXAMPLE 4   Integrating by Parts to Find $\int \tan^{-1} x\,dx$

Derive the formula

$$\int \tan^{-1} x\,dx = x\,\tan^{-1} x - \frac{1}{2}\ln(1+x^2) + C$$

Solution

Look at Table 1. The entry $\int x^n \tan^{-1} x\,dx$ with $n = 0$ is the integral we seek. So we use the integration by parts formula (1) with

$$u = \tan^{-1} x \qquad \text{and} \qquad dv = dx$$

Then

$$du = \frac{1}{1+x^2}\,dx \qquad \text{and} \qquad v = \int dx = x$$

Now using integration by parts,

$$\int \tan^{-1} x\,dx = \underset{\underset{\int u\,dv = uv - \int v\,du}{\uparrow}}{x \cdot \tan^{-1} x} - \int \frac{x}{1+x^2}\,dx$$

NEED TO REVIEW? The method of substitution is discussed in Section 6.5, pp. 484–490.

To find the integral $\int \dfrac{x}{1+x^2}\,dx$, use the substitution $t = 1 + x^2$. Then $dt = 2x\,dx$, or equivalently, $x\,dx = \dfrac{dt}{2}$.

$$\int \frac{x}{1+x^2}\,dx = \frac{1}{2}\int \frac{dt}{t} = \frac{1}{2}\ln|t| = \frac{1}{2}\ln(1+x^2)$$

NOTE Since $1 + x^2 > 0$, we can drop the absolute value bars and write $\ln(1 + x^2)$.

As a result, $\displaystyle\int \tan^{-1} x\,dx = x\,\tan^{-1} x - \frac{1}{2}\ln(1+x^2) + C.$ ∎

NOW WORK   Problem **9** and AP® Practice Problem **3**.

The next two examples show that sometimes it is necessary to integrate by parts more than once.

EXAMPLE 5   **Integrating by Parts**

Find $\int x^2 e^x\, dx$.

Solution

Use the integration by parts formula with

$$u = x^2 \qquad \text{and} \qquad dv = e^x\, dx$$

Then

$$du = 2x\, dx \qquad \text{and} \qquad v = \int e^x\, dx = e^x$$

Now

$$\int x^2 e^x\, dx = \underset{\underset{\displaystyle \int u\, dv = uv - \int v\, du}{\uparrow}}{x^2 e^x} - 2\int x e^x\, dx$$

The integral on the right is simpler than the original integral. To find it, use integration by parts a second time. (Use the result from Example 1.)

$$\int x e^x\, dx = x e^x - e^x$$

Then

$$\int x^2 e^x\, dx = x^2 e^x - 2(x e^x - e^x) + C = e^x(x^2 - 2x + 2) + C \qquad \blacksquare$$

NOW WORK   Problem **13**.

EXAMPLE 6   **Integrating by Parts**

Find $\int e^{2x} \cos(3x)\, dx$.

Solution

The choice for u and dv is not immediately evident. Both e^{2x} and $\cos(3x)$ are easy to integrate and neither the derivative of e^{2x} nor the derivative of $\cos(3x)$ is simpler to integrate. We try the following and see what happens.

$$u = e^{2x} \qquad \text{and} \qquad dv = \cos(3x)\, dx$$

Then

$$du = 2e^{2x} dx \qquad \text{and} \qquad v = \int \cos(3x)\, dx = \frac{1}{3}\sin(3x)$$

Now

$$\int e^{2x} \cos(3x)\, dx = e^{2x}\cdot\frac{1}{3}\sin(3x) - \int \frac{1}{3}\sin(3x)\cdot 2e^{2x} dx \qquad \scriptstyle \int u\, dv = uv - \int v\, du$$

$$\int e^{2x} \cos(3x)\, dx = \frac{1}{3}e^{2x}\sin(3x) - \frac{2}{3}\int e^{2x}\sin(3x)\, dx \qquad (2)$$

(continued on the next page)

The integral on the right in equation (2) is not simpler than the original integral, but it is no more difficult than the original. To find it, integrate by parts a second time using

$$u = e^{2x} \qquad \text{and} \qquad dv = \sin(3x)\,dx$$

Then

$$du = 2e^{2x}\,dx \quad \text{and} \quad v = \int \sin(3x)\,dx = -\frac{1}{3}\cos(3x)$$

Now

$$\int e^{2x} \sin(3x)\,dx = e^{2x}\left[-\frac{1}{3}\cos(3x)\right] - \int -\frac{1}{3}\cos(3x)\cdot(2e^{2x})\,dx$$

$$= -\frac{1}{3}e^{2x}\cos(3x) + \frac{2}{3}\int e^{2x}\cos(3x)\,dx$$

Then from (2)

$$\int e^{2x} \cos(3x)\,dx = \frac{1}{3}e^{2x}\sin(3x) - \frac{2}{3}\left[-\frac{1}{3}e^{2x}\cos(3x) + \frac{2}{3}\int e^{2x}\cos(3x)\,dx\right]$$

$$\int e^{2x} \cos(3x)\,dx = \frac{1}{3}e^{2x}\sin(3x) + \frac{2}{9}e^{2x}\cos(3x) - \frac{4}{9}\int e^{2x}\cos(3x)\,dx \qquad (3)$$

The integral we are trying to find now appears on both sides of equation (3). By adding $\dfrac{4}{9}\displaystyle\int e^{2x}\cos(3x)\,dx$ to both sides of (3) and simplifying, we obtain

$$\int e^{2x} \cos(3x)\,dx + \frac{4}{9}\int e^{2x}\cos(3x)\,dx = \frac{1}{3}e^{2x}\sin(3x) + \frac{2}{9}e^{2x}\cos(3x)$$

$$\frac{13}{9}\int e^{2x}\cos(3x)\,dx = \frac{1}{3}e^{2x}\sin(3x) + \frac{2}{9}e^{2x}\cos(3x)$$

$$\int e^{2x}\cos(3x)\,dx = \frac{9}{13}\left[\frac{1}{3}e^{2x}\sin(3x) + \frac{2}{9}e^{2x}\cos(3x)\right]$$

Adding the constant of integration,

$$\int e^{2x}\cos(3x)\,dx = \frac{3}{13}e^{2x}\sin(3x) + \frac{2}{13}e^{2x}\cos(3x) + C \qquad \blacksquare$$

We started the solution by choosing $u = e^{2x}$ and $dv = \cos(3x)\,dx$, but we could have started with $u = \cos(3x)$ and $dv = e^{2x}\,dx$. The result would be the same. You should try this approach on your own.

NOW WORK   Problem 31.

❷ Find a Definite Integral Using Integration by Parts

NEED TO REVIEW? Finding a definite integral using substitution is discussed in Section 6.5, pp. 488–490.

There are two approaches for finding a definite integral using integration by parts. They are similar to the approaches used in Section 6.5 for finding a definite integral using substitution.

- Method 1: Find the related indefinite integral using integration by parts, and then use the Fundamental Theorem of Calculus.
- Method 2: Find the definite integral directly by using the limits of integration at each step of the process. That is,

$$\int_a^b u\,dv = [uv]_a^b - \int_a^b v\,du$$

EXAMPLE 7   **Finding a Definite Integral Using Integration by Parts**

Find $\displaystyle\int_0^{\pi/4} x \sec^2 x \, dx$

Solution

We use Method 2 to find the definite integral. Using the guidelines in Table 1, choose

$$u = x \qquad \text{and} \qquad dv = \sec^2 x \, dx$$

Then

$$du = dx \qquad \text{and} \qquad v = \int \sec^2 x \, dx = \tan x$$

Now

$$\int_0^{\pi/4} x \sec^2 x \, dx = [x \tan x]_0^{\pi/4} - \int_0^{\pi/4} \tan x \, dx \qquad\qquad \int_a^b u \, dv = [uv]_a^b - \int_a^b v \, du$$

$$= [x \tan x]_0^{\pi/4} - [\ln |\sec x|]_0^{\pi/4}$$

$$= \left(\frac{\pi}{4} - 0\right) - (\ln\sqrt{2} - 0) = \frac{\pi}{4} - \ln\sqrt{2} \qquad \blacksquare$$

NOW WORK   Problem **43** and AP® Practice Problem **4**.

EXAMPLE 8   **Finding the Area under the Graph of $f(x) = x \ln x$**

Find the area under the graph of $f(x) = x \ln x$ from 1 to 2.

Solution

See Figure 1 for the graph of $f(x) = x \ln x$. The area A under the graph of f from 1 to 2 is $A = \int_1^2 x \ln x \, dx$. We use Method 1 and the integration by parts formula to find $\int x \ln x \, dx$. Using the guidelines in Table 1, choose

$$u = \ln x \qquad \text{and} \qquad dv = x \, dx$$

Then

$$du = \frac{1}{x} \, dx \qquad \text{and} \qquad v = \int x \, dx = \frac{x^2}{2}$$

$$\int x \ln x \, dx = (\ln x) \cdot \frac{x^2}{2} - \int \frac{x^2}{2} \cdot \frac{1}{x} dx = \frac{x^2}{2} \ln x - \frac{1}{2} \int x \, dx = \frac{x^2}{2} \ln x - \frac{x^2}{4}$$

$$\underset{\displaystyle \int u \, dv = uv - \int v \, du}{\uparrow}$$

Now use the Fundamental Theorem of Calculus.

$$A = \int_1^2 x \ln x \, dx = \left[\frac{x^2}{2} \ln x - \frac{x^2}{4}\right]_1^2 = (2\ln 2 - 1) - \left(0 - \frac{1}{4}\right) = 2\ln 2 - \frac{3}{4} \qquad \blacksquare$$

Figure 1 $f(x) = x \ln x$

NOW WORK   Problem **45** and AP® Practice Problems **1** and **5**.

③ Derive a General Formula Using Integration by Parts

Integration by parts is also used to derive general formulas involving integrals.

EXAMPLE 9   Deriving a General Formula

(a) Derive the formula

$$\int e^{ax}\cos(bx)\,dx = \frac{e^{ax}[b\sin(bx)+a\cos(bx)]}{a^2+b^2}+C \qquad b\neq 0 \tag{4}$$

(b) Use formula (4) to find $\int e^{4x}\cos(5x)\,dx$.

NOTE If $b=0$, formula (4) still works provided $a\neq 0$.

$$\int e^{ax}\,dx = \frac{1}{a}e^{ax}+C$$

Solution

(a) Use the integration by parts formula with

$$u=e^{ax} \qquad \text{and} \qquad dv=\cos(bx)\,dx$$

Then

$$du=ae^{ax}\,dx \qquad \text{and} \qquad v=\int \cos(bx)\,dx = \frac{1}{b}\sin(bx)$$

Now

$$\int e^{ax}\cos(bx)\,dx = e^{ax}\frac{\sin(bx)}{b}-\frac{a}{b}\int e^{ax}\sin(bx)\,dx \tag{5}$$

The new integral on the right, $\int e^{ax}\sin(bx)\,dx$, is different from the original integral, but it is essentially of the same form. Use integration by parts again with this integral by choosing

$$u=e^{ax} \qquad \text{and} \qquad dv=\sin(bx)\,dx$$

Then

$$du=ae^{ax}\,dx \qquad \text{and} \qquad v=\int \sin(bx)\,dx = -\frac{1}{b}\cos(bx)$$

Now

$$\int e^{ax}\sin(bx)\,dx = -\frac{1}{b}e^{ax}\cos(bx)+\frac{a}{b}\int e^{ax}\cos(bx)\,dx \tag{6}$$

Substituting the result from (6) into (5), we obtain

$$\int e^{ax}\cos(bx)\,dx = e^{ax}\frac{\sin(bx)}{b}-\frac{a}{b}\left[-\frac{1}{b}e^{ax}\cos(bx)+\frac{a}{b}\int e^{ax}\cos(bx)\,dx\right]$$

$$\int e^{ax}\cos(bx)\,dx = \frac{1}{b}e^{ax}\sin(bx)+\frac{a}{b^2}e^{ax}\cos(bx)-\frac{a^2}{b^2}\int e^{ax}\cos(bx)\,dx$$

Now solve for $\int e^{ax} \cos(bx)\, dx$ and simplify.

$$\int e^{ax} \cos(bx)\, dx + \frac{a^2}{b^2} \int e^{ax} \cos(bx)\, dx = \frac{1}{b} e^{ax} \sin(bx) + \frac{a}{b^2} e^{ax} \cos(bx)$$

$$\left(1 + \frac{a^2}{b^2}\right) \int e^{ax} \cos(bx)\, dx = \frac{1}{b^2} e^{ax}[b \sin(bx) + a \cos(bx)]$$

$$\int e^{ax} \cos(bx)\, dx = \frac{e^{ax}[b \sin(bx) + a \cos(bx)]}{a^2 + b^2} + C$$

(b) To find $\int e^{4x} \cos(5x)\, dx$, use (4) with $a = 4$ and $b = 5$.

$$\int e^{4x} \cos(5x)\, dx = \frac{e^{4x}[5 \sin (5x) + 4 \cos (5x)]}{41} + C \qquad \blacksquare$$

NOW WORK   Problem 77.

EXAMPLE 10   Deriving a General Formula

Derive the formula

$$\boxed{\int \sec^n x\, dx = \frac{\sec^{n-2} x \tan x}{n-1} + \frac{n-2}{n-1} \int \sec^{n-2} x\, dx \qquad n \geq 3} \qquad (7)$$

Solution
We begin by writing $\sec^n x = \sec^{n-2} x \sec^2 x$, and choosing

$$u = \sec^{n-2} x \qquad \text{and} \qquad dv = \sec^2 x\, dx$$

This choice makes $\int dv$ easy to integrate. Then

$$du = [(n-2) \sec^{n-3} x \cdot \sec x \tan x]\, dx = [(n-2) \sec^{n-2} x \tan x]\, dx$$
$$v = \int \sec^2 x\, dx = \tan x$$

Using integration by parts, we get

$$\int \sec^n x\, dx = \sec^{n-2} x \tan x - (n-2) \int \sec^{n-2} x \tan^2 x\, dx$$

To express the integrand on the right in terms of $\sec x$, use the trigonometric identity, $\tan^2 x + 1 = \sec^2 x$, and replace $\tan^2 x$ by $\sec^2 x - 1$, obtaining

$$\int \sec^n x\, dx = \sec^{n-2} x \tan x - (n-2) \int \sec^{n-2} x (\sec^2 x - 1)\, dx$$

$$\int \sec^n x\, dx = \sec^{n-2} x \tan x - (n-2) \int \sec^n x\, dx + (n-2) \int \sec^{n-2} x\, dx$$

Adding $(n-2) \int \sec^n x\, dx$ to each side and simplifying, we obtain

$$(n-1) \int \sec^n x\, dx = \sec^{n-2} x \tan x + (n-2) \int \sec^{n-2} x\, dx$$

Finally, divide both sides by $n - 1$:

$$\int \sec^n x\, dx = \frac{\sec^{n-2} x \tan x}{n-1} + \frac{n-2}{n-1} \int \sec^{n-2} x\, dx \qquad \blacksquare$$

Formula (7) is called a **reduction formula** because repeated applications of the formula eventually lead to an elementary integral. For this reduction formula:

- When n is even, repeated applications lead eventually to

$$\int \sec^2 x \, dx = \tan x + C$$

- When n is odd, repeated applications lead eventually to

$$\int \sec x \, dx = \ln |\sec x + \tan x| + C$$

For example, if $n = 3$,

$$\int \sec^3 x \, dx = \frac{\sec x \, \tan x}{2} + \frac{1}{2} \int \sec x \, dx = \frac{\sec x \, \tan x}{2} + \frac{1}{2} \ln |\sec x + \tan x| + C$$

NOW WORK   Problem 67.

6.7 Assess Your Understanding

Concepts and Vocabulary

1. **True or False** Integration by parts is based on the Product Rule for derivatives.

2. The integration by parts formula states that $\int u \, dv = $ _____.

Skill Building

In Problems 3–34, use integration by parts to find each integral.

3. $\int x e^{2x} \, dx$

4. $\int x e^{-3x} \, dx$

5. $\int x \cos x \, dx$

6. $\int x \sin(3x) \, dx$

7. $\int \sqrt{x} \ln x \, dx$

8. $\int x^{-2} \ln x \, dx$

9. $\int \cot^{-1} x \, dx$

10. $\int \sin^{-1} x \, dx$

11. $\int (\ln x)^2 \, dx$

12. $\int x (\ln x)^2 \, dx$

13. $\int x^2 \sin x \, dx$

14. $\int x^2 \cos x \, dx$

15. $\int x \cos^2 x \, dx$

16. $\int x \sin^2 x \, dx$

17. $\int x^2 \ln x \, dx$

18. $\int \frac{(\ln x)^2}{x^2} \, dx$

19. $\int \frac{x e^x}{(x+1)^2} \, dx$

20. $\int \frac{x e^{3x}}{(3x+1)^2} \, dx$

21. $\int \sin(\ln x) \, dx$

22. $\int \cos(\ln x) \, dx$

23. $\int (\ln x)^3 \, dx$

24. $\int (\ln x)^4 \, dx$

25. $\int x^2 (\ln x)^2 \, dx$

26. $\int x^3 (\ln x)^2 \, dx$

27. $\int x^2 \tan^{-1} x \, dx$

28. $\int x \tan^{-1} x \, dx$

29. $\int 7^x x \, dx$

30. $\int 2^{-x} x \, dx$

31. $\int e^{-x} \cos(2x) \, dx$

32. $\int e^{-2x} \sin(3x) \, dx$

33. $\int e^{2x} \sin x \, dx$

34. $\int e^{3x} \cos(5x) \, dx$

In Problems 35–44, use integration by parts to find each definite integral.

35. $\int_0^\pi e^x \cos x \, dx$

36. $\int_0^{\pi/2} e^{-x} \sin x \, dx$

37. $\int_0^2 x^2 e^{-3x} \, dx$

38. $\int_0^1 x^2 e^{-x} \, dx$

39. $\int_0^{\pi/4} x \sec x \tan x \, dx$

40. $\int_0^{\pi/4} x \tan^2 x \, dx$

41. $\int_1^9 \ln \sqrt{x} \, dx$

42. $\int_{\pi/4}^{3\pi/4} x \csc^2 x \, dx$

43. $\int_1^e (\ln x)^2 \, dx$

44. $\int_0^{\pi/4} x \sec^2 x \, dx$

Applications and Extensions

45. Area under a Graph Find the area under the graph of $y = e^x \sin x$ from 0 to π.

46. Area under a Graph Find the area under the graph of $y = x \cos x$ from $x = 0$ to $x = \dfrac{\pi}{2}$.

47. Area under a Graph Find the area under the graph of $y = xe^{-x}$ from $x = 0$ to $x = 1$.

48. Area under a Graph Find the area under the graph of $y = xe^{3x}$ from $x = 0$ to $x = 2$.

49. Motion on a Line The acceleration $a = a(t)$ of an object moving on a line is given by

$$a(t) = e^{-2t} \sin t \text{ m/s}^2$$

(a) Find the velocity $v = v(t)$ of the object if the initial velocity is $v(0) = 8$ m/s.

(b) Find the position $s = s(t)$ of the object if the initial position is $s(0) = 0$ m.

50. Motion on a Line The acceleration a of an object moving on a line is given by

$$a(t) = t^2 e^{-t} \text{ ft/s}^2$$

(a) Find the velocity $v = v(t)$ of the object if the initial velocity is $v(0) = 5$ ft/s.

(b) Find the position $s = s(t)$ of the object if the initial position is $s(0) = 0$ ft.

51. Volume of a Solid of Revolution The volume V of the solid of revolution generated by revolving the region bounded by the graph of $y = \sin x$ and the x-axis from $x = 0$ to $x = \dfrac{\pi}{2}$ about the y-axis is given by $V = 2\pi \int_0^{\pi/2} x \sin x \, dx$. Find the volume V. See the figure below.

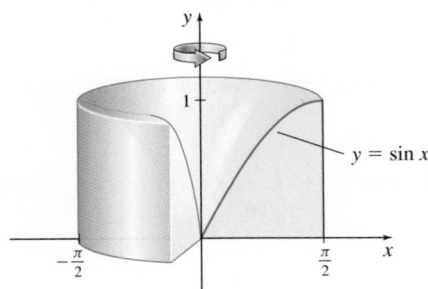

52. Volume of a Solid of Revolution The volume V of the solid of revolution generated by revolving the region bounded by the graph of $y = \cos x$ and the x-axis from $x = 0$ to $x = \dfrac{\pi}{2}$ about the y-axis is given by $V = 2\pi \int_0^{\pi/2} x \cos x \, dx$. Find the volume V. See the figure below.

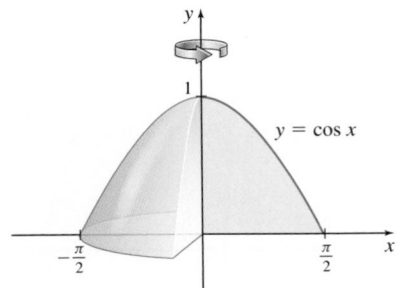

53. A function $y = f(x)$ is continuous and differentiable on the interval $(0, 7)$. If $\int_1^5 f(x)\,dx = 10$, $f(1) = 2$, and $f(5) = -5$, find $\int_1^5 xf'(x)\,dx$.

54. A function $y = f(x)$ is continuous and differentiable on the interval $(2, 6)$. If $\int_3^5 f(x)\,dx = 18$ and $f(3) = 8$ and $f(5) = 11$, then find $\int_3^5 xf'(x)\,dx$.

55. Find the area under the graph of $f(x) = \ln x$ from $x = 1$ to $x = e$.

56. Use integration by parts to show that if f' is continuous on a closed interval $[a, b]$ then

$$\int_a^b f(x)f'(x)\,dx = \frac{1}{2}\left\{[f(b)]^2 - [f(a)]^2\right\}$$

In Problems 57–62, find each integral by first making a substitution and then integrating by parts.

57. $\displaystyle\int \sin\sqrt{x}\,dx$ **58.** $\displaystyle\int e^{\sqrt{x}}\,dx$

59. $\displaystyle\int \cos x \ln(\sin x)\,dx$ **60.** $\displaystyle\int e^x \ln(2 + e^x)\,dx$

61. $\displaystyle\int e^{4x} \cos e^{2x}\,dx$ **62.** $\displaystyle\int \cos x \tan^{-1}(\sin x)\,dx$

63. Find $\displaystyle\int x^3 e^{x^2}\,dx$. *Hint: Let $u = x^2$, $dv = xe^{x^2}\,dx$.*

64. Find $\displaystyle\int x^n \ln x\,dx$; $n \neq -1$, n real.

65. Find $\displaystyle\int xe^x \cos x\,dx$. **66.** Find $\displaystyle\int xe^x \sin x\,dx$.

In Problems 67–70, derive each reduction formula where $n > 1$ is an integer.

67. $\displaystyle\int x^n \sin^{-1} x\,dx = \frac{x^{n+1}}{n+1}\sin^{-1} x - \frac{1}{n+1}\int \frac{x^{n+1}}{\sqrt{1-x^2}}\,dx$

68. $\displaystyle\int \frac{dx}{(x^2+1)^{n+1}} = \left(1 - \frac{1}{2n}\right)\int \frac{dx}{(x^2+1)^n} + \frac{x}{2n(x^2+1)^n}$

69. $\displaystyle\int \sin^n x\,dx = -\frac{\sin^{n-1} x \cos x}{n} + \frac{n-1}{n}\int \sin^{n-2} x\,dx$

70. $\displaystyle\int \sin^n x \cos^m x\,dx = -\frac{\sin^{n-1} x \cos^{m+1} x}{n+m}$
$$+ \frac{n-1}{n+m}\int \sin^{n-2} x \cos^m x\,dx$$

where $m \neq -n$, $m \neq -1$

71. (a) Find $\int x^2 e^{5x}\,dx$.

(b) Using integration by parts, derive a reduction formula for $\int x^n e^{kx}\,dx$, where $k \neq 0$ and $n \geq 2$ is an integer, in which the resulting integrand involves x^{n-1}.

72. (a) Assuming there is a function p for which
$\int x^3 e^x \, dx = p(x)e^x$, show that $p(x) + p'(x) = x^3$.

(b) Use integration by parts to find a polynomial p of degree 3 for which $\int x^3 e^x \, dx = p(x)e^x + C$.

73. (a) Use integration by parts with $u = \sin x$ and $dv = \cos x \, dx$ to find a function f for which $\int \sin x \cos x \, dx = f(x) + C_1$.

(b) Use integration by parts with $u = \cos x$ and $dv = \sin x \, dx$ to find a function g for which $\int \sin x \cos x \, dx = g(x) + C_2$.

(c) Use the trigonometric identity $\sin(2x) = 2 \sin x \cos x$ and substitution to find a function h for which
$$\int \sin x \cos x \, dx = h(x) + C_3$$

(d) Compare the functions f and g. Find a relationship between C_1 and C_2.

(e) Compare the functions f and h. Find a relationship between C_1 and C_3.

74. (a) Graph the functions $f(x) = x^3 e^{-3x}$ and $g(x) = x^2 e^{-3x}$ on the same set of coordinate axes.

(b) Find the area enclosed by the graphs of f and g.

75. Damped Spring The displacement x of a damped spring at time t, $0 \le t \le 5$, is given by
$$x = x(t) = 3e^{-t}\cos(2t) + 2e^{-t}\sin(2t)$$

(a) Graph $x = x(t)$.

(b) Find the least positive number t that satisfies $x(t) = 0$.

(c) Find the area under the graph of $x = x(t)$ from $t = 0$ to the value of t found in (b).

76. Derive the formula
$$\int \ln\left(x + \sqrt{x^2 + a^2}\right) dx$$
$$= x \ln\left(x + \sqrt{x^2 + a^2}\right) - \sqrt{x^2 + a^2} + C$$

77. Derive the formula
$$\int e^{ax} \sin(bx) \, dx = \frac{e^{ax}[a\sin(bx) - b\cos(bx)]}{a^2 + b^2} + C, a > 0, b > 0$$

78. Suppose $F(x) = \int_0^x t\, g'(t) \, dt$ for all $x \ge 0$.
Show that $F(x) = xg(x) - \int_0^x g(t) \, dt$.

79. Use **Wallis' formulas**, given below, to find each definite integral.

$$\int_0^{\pi/2} \sin^n x \, dx = \int_0^{\pi/2} \cos^n x \, dx \quad n > 1 \text{ an integer}$$

$$= \begin{cases} \dfrac{(n-1)(n-3)\cdots(4)(2)}{n(n-2)\cdots(5)(3)(1)} & \text{if } n > 1 \text{ is odd} \\[2ex] \dfrac{(n-1)(n-3)\cdots(5)(3)(1)}{n(n-2)\cdots(4)(2)}\left(\dfrac{\pi}{2}\right) & \text{if } n > 1 \text{ is even} \end{cases}$$

(a) $\displaystyle\int_0^{\pi/2} \sin^6 x \, dx$ **(b)** $\displaystyle\int_0^{\pi/2} \sin^5 x \, dx$

(c) $\displaystyle\int_0^{\pi/2} \cos^8 x \, dx$ **(d)** $\displaystyle\int_0^{\pi/2} \cos^6 x \, dx$

Challenge Problems

80. Derive Wallis' formulas given in Problem 79.
Hint: Use the result of Problem 69.

81. (a) If n is a positive integer, use integration by parts to show that there is a polynomial p of degree n for which
$$\int x^n e^x \, dx = p(x)e^x + C$$

(b) Show that $p(x) + p'(x) = x^n$.

(c) Show that p can be written in the form
$$p(x) = \sum_{k=0}^{n} (-1)^k \frac{n!}{(n-k)!} x^{n-k}$$

82. Show that for any positive integer n, $\int_0^1 e^{x^2} \, dx$ equals
$$e \cdot \left[1 - \frac{2}{3} + \frac{4}{15} - \frac{8}{105} + \cdots + \frac{(-1)^n 2^n}{(2n+1)(2n-1)\cdots 3 \cdot 1}\right]$$
$$+ (-1)^{n+1} \cdot \frac{2^{n+1}}{(2n+1)(2n-1)\cdots 3 \cdot 1} \int_0^1 x^{2n+2} e^{x^2} \, dx$$

83. Use integration by parts to show that if f is a polynomial of degree $n \ge 1$, then $\int f(x)e^x \, dx = g(x)e^x + C$ for some polynomial $g(x)$ of degree n.

84. Start with the identity $f(b) - f(a) = \int_a^b f'(t) \, dt$ and derive the following generalizations of the Mean Value Theorem for Integrals:

(a) $f(b) - f(a) = f'(a)(b-a) - \displaystyle\int_a^b f''(t)(t-b) \, dt$

(b) $f(b) - f(a) = f'(a)(b-a) + \dfrac{f''(a)}{2}(b-a)^2$
$$+ \int_a^b \frac{f'''(t)}{2}(t-b)^2 \, dt$$

85. If $y = f(x)$ has the inverse function given by $x = f^{-1}(y)$, show that
$$\int_a^b f(x) \, dx + \int_{f(a)}^{f(b)} f^{-1}(y) \, dy = bf(b) - af(a)$$

AP® Practice Problems

NOTE: *These problems are BC only.*

Multiple-Choice Questions

1. An object is moving along the x-axis. Its acceleration at any time $t > 0$ is given by $a(t) = \ln t^2$. If the velocity $v = v(t)$ of the object at time $t = 1$ is $v(1) = 2$, then what is its velocity v at time $t = 3$?

(A) $6\ln 3 + 4$ (B) $6\ln 3 - 2$ (C) $6\ln 3$ (D) $6\ln 3 + 6$

2. If $\int xe^{4x}\, dx = f(x)e^{4x} + C$, then $f(x)$ equals

(A) $-\dfrac{1}{4}x - \dfrac{1}{16}$ (B) $\dfrac{1}{4}x - \dfrac{1}{16}$

(C) $\dfrac{1}{4}x - \dfrac{1}{4}$ (D) $\dfrac{1}{4}x - 1$

3. $\int \cos^{-1} x\, dx =$

(A) $\ln|\tan x + \sec x| + C$

(B) $x\cos^{-1} x + \sqrt{1 - x^2} + C$

(C) $x\cos^{-1} x - \sqrt{1 - x^2} + C$

(D) $x\cos^{-1} x - 2\sqrt{1 - x^2} + C$

4. $\int_0^2 xe^{-x}\, dx =$

(A) $1 - 3e^{-2}$ (B) $-1 - e$ (C) $1 + e^{-2}$ (D) $3e^{-2}$

5. A function f is continuous and differentiable on the interval $(-2, 8)$.

If $f(0) = 4$, $f(6) = 10$, and $\int_0^6 f(x)\, dx = 12$,

then $\int_0^6 xf'(x)\, dx =$

(A) 44 (B) 72 (C) 48 (D) -6

Free-Response Question

6. An object is moving along a line with velocity $v(t) = t\,e^t$, in meters/second, for $0 \le t \le 10$. At $t = 0$, the object is at a position 2 meters to the right of 0. What is the position of the object at $t = 4$?

Retain Your Knowledge

Multiple-Choice Questions

1. The graph of a piecewise-defined function f is shown below. What is $\int_{-2}^5 f(x)\, dx$?

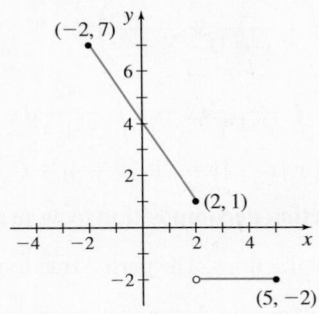

(A) 6 (B) 10 (C) 20 (D) 22

2. The position of an object moving along a straight line is given by $s(t) = 8t - 2t^2$. At what time t is the object at rest?

(A) $t = 0$ (B) $t = 2$

(C) $t = 4$ (D) $t = 0$ and $t = 4$

3. A function f is continuous on $[0, 12]$ and twice differentiable on the open intervals $(0, 6)$ and $(6, 12)$.

x	$0 < x < 4$	4	$4 < x < 6$	6	$6 < x < 9$	9	$9 < x < 12$
$f'(x)$	Negative	0	Negative	Undefined	Positive	0	Negative
$f''(x)$	Positive	0	Negative	Undefined	Negative	0	Negative

From the information provided, f has

(A) a local maximum at 9, no local minimum, and no inflection points.

(B) a local maximum at 9; a local minimum at 6; and inflection points at $(4, f(4))$, $(6, f(6))$, and $(9, f(9))$.

(C) a local maximum at 9, a local minimum at 6, and no inflection points.

(D) a local maximum at 9, a local minimum at 6, and an inflection point at $(4, f(4))$.

Free-Response Question

4. The domain of $\dfrac{\ln x}{x^2 - 1}$ is $x > 0$, $x \ne 1$. If the function $f(x) = \dfrac{\ln x}{x^2 - 1}$ is continuous for all $x > 0$, what is the value of $f(1)$? Justify your answer.

6.8 Integration of Rational Functions Using Partial Fractions

OBJECTIVES *When you finish this section, you should be able to:*

1 **Integrate a proper rational function whose denominator contains only distinct linear factors (p. 528)**

2 **Integrate a proper rational function whose denominator contains a repeated linear factor (p. 531)**

3 **Integrate a proper rational function whose denominator contains a distinct irreducible quadratic factor (p. 532)**

4 **Integrate a proper rational function whose denominator contains a repeated irreducible quadratic factor (p. 533)**

Every improper rational function can be expressed as the sum of a polynomial function and a proper rational function. Since we know how to integrate polynomial functions, once we know how to integrate proper rational functions, we will be able to integrate any rational function.

For this reason, the discussion that follows deals only with the integration of proper rational functions in lowest terms. Such rational functions can be written as the sum of simpler functions, called **partial fractions.**

For example, since $\dfrac{2}{x-1} + \dfrac{3}{x+4} = \dfrac{5x+5}{(x-1)(x+4)}$, the rational function

$$R(x) = \frac{5x+5}{(x-1)(x+4)}$$

can be expressed as the sum $R(x) = \dfrac{2}{x-1} + \dfrac{3}{x+4}$. Then

$$\int \frac{5x+5}{(x-1)(x+4)}\,dx = \int \left(\frac{2}{x-1} + \frac{3}{x+4} \right) dx$$

$$= 2\int \frac{1}{x-1}\,dx + 3\int \frac{1}{x+4}\,dx$$

$$= 2\ln|x-1| + 3\ln|x+4| + C$$

We use a technique called **partial fraction decomposition** to write a proper rational function $R(x) = \dfrac{p(x)}{q(x)}$ as the sum of partial fractions. The partial fractions depend on the nature of the factors of the denominator q. It can be shown that any polynomial q whose coefficients are real numbers can be factored (over the real numbers) into products of linear and/or irreducible quadratic factors. As we shall see, this means *the integral of every rational function can be expressed in terms of algebraic, logarithmic, and/or inverse trigonometric functions.*

① **Integrate a Proper Rational Function Whose Denominator Contains Only Distinct Linear Factors**

Case 1: If the denominator q contains only distinct linear factors, say, $x-a_1$, $x-a_2, \ldots, x-a_n$, then $\dfrac{p}{q}$ can be written as

$$\frac{p(x)}{q(x)} = \frac{A_1}{x-a_1} + \frac{A_2}{x-a_2} + \cdots + \frac{A_n}{x-a_n} \tag{1}$$

where $A_1, A_2, \ldots, A_n$ are real numbers.

To find $\int \dfrac{p(x)}{q(x)} dx$, integrate both sides of (1). Then

$$\int \frac{p(x)}{q(x)} dx = \int \frac{A_1}{x - a_1} dx + \int \frac{A_2}{x - a_2} dx + \cdots + \int \frac{A_n}{x - a_n} dx$$

$$= A_1 \ln |x - a_1| + A_2 \ln |x - a_2| + \cdots + A_n \ln |x - a_n| + C$$

All that remains is to find the numbers $A_1, \ldots, A_n$.

A procedure for finding the numbers $A_1, \ldots, A_n$ is illustrated in Example 1.

NOTE Case 1 type integrands lead to sums of natural logarithms.

EXAMPLE 1 | **Integrating a Proper Rational Function Whose Denominator Contains Only Distinct Linear Factors**

Find $\int \dfrac{x \, dx}{x^2 - 5x + 6}$.

Solution

The integrand is a proper rational function in lowest terms. We begin by factoring the denominator: $x^2 - 5x + 6 = (x - 2)(x - 3)$. Since the factors are linear and distinct, this is a Case 1 type integrand and can be written using the terms $\dfrac{A}{x - 2}$ and $\dfrac{B}{x - 3}$. From (1),

$$\frac{x}{x^2 - 5x + 6} = \frac{x}{(x - 2)(x - 3)} = \frac{A}{x - 2} + \frac{B}{x - 3}$$

Now clear fractions by multiplying both sides of the equation by $(x - 2)(x - 3)$.

$$x = A(x - 3) + B(x - 2)$$
$$x = (A + B)x - (3A + 2B) \qquad \text{Group like terms.}$$

This is an identity in x, so the coefficients of like powers of x must be equal.

$$1 = A + B \qquad \text{The coefficient of } x \text{ equals 1.}$$
$$0 = -3A - 2B \qquad \text{The constant term on the left is 0.}$$

This is a system of two equations containing two variables. Solving the second equation for B, we get $B = -\dfrac{3}{2}A$. Substituting for B in the first equation produces the solution $A = -2$ from which $B = 3$. So,

$$\frac{x}{(x - 2)(x - 3)} \underset{\substack{\uparrow \\ A = -2, \, B = 3}}{=} \frac{-2}{x - 2} + \frac{3}{x - 3}$$

Then

$$\int \frac{x}{(x - 2)(x - 3)} dx = \int \frac{-2}{x - 2} dx + \int \frac{3}{x - 3} dx$$

$$= -2 \ln |x - 2| + 3 \ln |x - 3| + C = \ln \left| \frac{(x - 3)^3}{(x - 2)^2} \right| + C \qquad \blacksquare$$

NOW WORK Problem 5 and AP® Practice Problems 1, 2, 3, and 5.

Alternatively, we can find the unknown numbers in the decomposition of $\dfrac{p}{q}$ by substituting convenient values of x into the identity obtained after clearing fractions.* In Example 1, after clearing fractions, the identity is

$$x = A(x - 3) + B(x - 2)$$

- When $x = 3$, the term involving A drops out, leaving $3 = B \cdot 1$, so that $B = 3$.
- When $x = 2$, the term involving B drops out, leaving $2 = A \cdot (-1)$, so that $A = -2$.

EXAMPLE 2   Using Partial Fractions with a Definite Integral

Find $\displaystyle\int_2^5 \frac{1}{x^2 - 1}\,dx$.

Solution

The factored denominator $x^2 - 1 = (x - 1)(x + 1)$ contains only distinct linear factors. So, $\dfrac{1}{x^2 - 1}$ can be decomposed into partial fractions of the form

$$\frac{1}{x^2 - 1} = \frac{1}{(x - 1)(x + 1)} = \frac{A}{x - 1} + \frac{B}{x + 1}$$

$$1 = A(x + 1) + B(x - 1) \qquad \text{Multiply both sides by } (x - 1)(x + 1).$$

This is an identity in x.

- When $x = 1$, the term involving B drops out. Then $1 = A(2)$, so $A = \dfrac{1}{2}$.
- When $x = -1$, the term involving A drops out. Then $1 = B(-2)$, so $B = -\dfrac{1}{2}$.

Then

$$\frac{1}{x^2 - 1} = \frac{A}{x - 1} + \frac{B}{x + 1} = \frac{1}{2(x - 1)} - \frac{1}{2(x + 1)}$$

$$\int_2^5 \frac{dx}{x^2 - 1} = \frac{1}{2}\int_2^5 \frac{dx}{x - 1} - \frac{1}{2}\int_2^5 \frac{dx}{x + 1} = \frac{1}{2}\left(\int_2^5 \frac{dx}{x - 1} - \int_2^5 \frac{dx}{x + 1}\right)$$

$$= \frac{1}{2}\left\{\left[\ln(x - 1)\right]_2^5 - \left[\ln(x + 1)\right]_2^5\right\}$$

$$= \frac{1}{2}\left[(\ln 4 - \ln 1) - (\ln 6 - \ln 3)\right]$$

$$= \frac{1}{2}(\ln 4 - \ln 6 + \ln 3) \qquad\blacksquare$$

NOW WORK   Problems 31 and 49 and AP® Practice Problems 4 and 6.

---

*This method is discussed in detail in H. J. Straight & R. Dowds (1984, June–July), *American Mathematical Monthly*, *91*(6), 365.

❷ Integrate a Proper Rational Function Whose Denominator Contains a Repeated Linear Factor

> **Case 2**: If the denominator q has a repeated linear factor $(x-a)^n$, $n \geq 2$ an integer, then the decomposition of $\dfrac{p}{q}$ includes the terms
>
> $$\frac{A_1}{x-a}, \frac{A_2}{(x-a)^2}, \ldots, \frac{A_n}{(x-a)^n}$$
>
> where $A_1, A_2, \ldots, A_n$ are real numbers.

| **EXAMPLE 3** | **Integrating a Proper Rational Function Whose Denominator Contains a Repeated Linear Factor** |

Find $\displaystyle\int \frac{dx}{x(x-1)^2}$.

Solution

Since x is a distinct linear factor of the denominator q, and $(x-1)^2$ is a repeated linear factor of the denominator, the decomposition of $\dfrac{1}{x(x-1)^2}$ into partial fractions has the three terms $\dfrac{A}{x}$, $\dfrac{B}{x-1}$, and $\dfrac{C}{(x-1)^2}$.

$$\frac{1}{x(x-1)^2} = \frac{A}{x} + \frac{B}{x-1} + \frac{C}{(x-1)^2} \qquad \text{Write the identity.}$$

$$1 = A(x-1)^2 + B \cdot x(x-1) + C \cdot x \qquad \text{Multiply both sides by } x(x-1)^2.$$

We find A, B, and C by choosing values of x that cause one or more terms to drop out. When $x = 1$, we have $1 = C \cdot 1$, so $C = 1$. When $x = 0$, we have $1 = A(0-1)^2$, so $A = 1$. Now using $A = 1$ and $C = 1$, we have

$$1 = (x-1)^2 + B \cdot x(x-1) + 1 \cdot x$$

Let $x = 2$. (Any choice other than 0 and 1 will also work.) Then

$$1 = 1 + 2B + 2$$
$$B = -1$$

Then

$$\frac{1}{x(x-1)^2} = \frac{1}{x} + \frac{-1}{(x-1)} + \frac{1}{(x-1)^2} \qquad A=1 \quad B=-1 \quad C=1$$

NOTE Case 2 type integrands lead to sums of natural logarithms and rational functions.

So,

$$\int \frac{dx}{x(x-1)^2} = \int \frac{dx}{x} - \int \frac{dx}{x-1} + \int \frac{dx}{(x-1)^2}$$

NOTE To avoid confusion, we use K for the constant of integration whenever C appears in the partial fraction decomposition.

$$= \ln|x| - \ln|x-1| - \frac{1}{x-1} + K \qquad \blacksquare$$

NOW WORK Problem 9.

③ Integrate a Proper Rational Function Whose Denominator Contains a Distinct Irreducible Quadratic Factor

A quadratic polynomial $ax^2 + bx + c$ is called **irreducible** if it cannot be factored into real linear factors. This happens if the discriminant $b^2 - 4ac < 0$. For example, $x^2 + x + 1$ and $x^2 + 4$ are irreducible.

NEED TO REVIEW? The discriminant of a quadratic equation is discussed in Appendix A.1, p. A-3.

Case 3: If the denominator q contains a distinct irreducible quadratic factor $ax^2 + bx + c$, then the decomposition of $\dfrac{p}{q}$ includes the term

$$\boxed{\dfrac{Ax + B}{ax^2 + bx + c}}$$

where A and B are real numbers.

► CALC CLIP

EXAMPLE 4 **Integrating a Proper Rational Function Whose Denominator Contains a Distinct Irreducible Quadratic Factor**

Find $\displaystyle\int \dfrac{3x}{x^3 - 1}\,dx$.

Solution

The denominator, $x^3 - 1 = (x - 1)(x^2 + x + 1)$, contains a nonrepeated linear factor, $x - 1$. Then by Case 1 the decomposition of $\dfrac{3x}{x^3 - 1} = \dfrac{3x}{(x - 1)(x^2 + x + 1)}$ has the term $\dfrac{A}{x - 1}$.

The discriminant of the quadratic expression $x^2 + x + 1$ is negative, so $x^2 + x + 1$ is an irreducible quadratic factor of the denominator $x^3 - 1$, and the decomposition of $\dfrac{3x}{x^3 - 1}$ also contains the term $\dfrac{Bx + C}{x^2 + x + 1}$. Then

$$\dfrac{3x}{x^3 - 1} = \dfrac{A}{x - 1} + \dfrac{Bx + C}{x^2 + x + 1}$$

Clearing the denominators, we have

$$3x = A(x^2 + x + 1) + (Bx + C)(x - 1)$$

This is an identity in x. When $x = 1$, we have $3 = 3A$, so $A = 1$. With $A = 1$, the identity becomes

$$3x = (x^2 + x + 1) + (Bx + C)(x - 1)$$
$$-x^2 + 2x - 1 = (Bx + C)(x - 1)$$
$$-(x - 1)^2 = (Bx + C)(x - 1)$$
$$-(x - 1) = Bx + C$$
$$B = -1 \quad C = 1$$

So,

$$\frac{3x}{x^3-1}=\frac{1}{x-1}+\frac{-x+1}{x^2+x+1}$$

$$\int \frac{3x}{x^3-1}\,dx=\int\left(\frac{1}{x-1}+\frac{-x+1}{x^2+x+1}\right)dx=\int\frac{1}{x-1}\,dx+\int\frac{-x+1}{x^2+x+1}\,dx$$

$$=\ln|x-1|-\int\frac{x-1}{x^2+x+1}\,dx \tag{2}$$

To find the integral on the right, complete the square in the denominator and use substitution.

$$\int\frac{x-1}{x^2+x+1}dx=\int\frac{x-1}{\left(x+\frac{1}{2}\right)^2+\frac{3}{4}}dx \underset{\substack{\uparrow \\ u=x+\frac{1}{2}}}{=}\int\frac{u-\frac{3}{2}}{u^2+\frac{3}{4}}du=\int\frac{u}{u^2+\frac{3}{4}}du-\frac{3}{2}\int\frac{du}{u^2+\frac{3}{4}}$$

$$=\frac{1}{2}\ln\left(u^2+\frac{3}{4}\right)-\frac{3}{2}\left[\frac{2}{\sqrt{3}}\tan^{-1}\left(\frac{2}{\sqrt{3}}u\right)\right]\qquad \int\frac{du}{u^2+a^2}=\frac{1}{a}\tan^{-1}\frac{u}{a}$$

$$=\frac{1}{2}\ln\left(u^2+\frac{3}{4}\right)-\sqrt{3}\tan^{-1}\left(\frac{2}{\sqrt{3}}u\right)$$

$$=\frac{1}{2}\ln(x^2+x+1)-\sqrt{3}\tan^{-1}\frac{2x+1}{\sqrt{3}}\qquad 2u=2x+1$$

NOTE Case 3 type integrands lead to sums of natural logarithms and/or inverse tangent functions.

Then from (2),

$$\int\frac{3x}{x^3-1}dx=\ln|x-1|-\frac{1}{2}\ln(x^2+x+1)+\sqrt{3}\tan^{-1}\frac{2x+1}{\sqrt{3}}+K \qquad ■$$

NOW WORK   Problem 15.

---

④ Integrate a Proper Rational Function Whose Denominator Contains a Repeated Irreducible Quadratic Factor

Case 4: If the denominator q contains a repeated irreducible quadratic polynomial $(x^2+bx+c)^n$, $n\ge 2$ an integer, then the decomposition of $\dfrac{p}{q}$ includes the terms

$$\frac{A_1x+B_1}{x^2+bx+c},\ \frac{A_2x+B_2}{(x^2+bx+c)^2},\ \ldots,\ \frac{A_nx+B_n}{(x^2+bx+c)^n}$$

where $A_1,\ B_1,\ A_2,\ B_2,\ \ldots,\ A_n,\ B_n$ are real numbers.

EXAMPLE 5   **Integrating a Proper Rational Function Whose Denominator Contains a Repeated Irreducible Quadratic Factor**

Find $\displaystyle\int \frac{x^3+1}{(x^2+4)^2}\,dx$.

Solution

The denominator is a repeated, irreducible quadratic, so the decomposition of $\dfrac{x^3+1}{(x^2+4)^2}$ is

$$\frac{x^3+1}{(x^2+4)^2} = \frac{Ax+B}{x^2+4} + \frac{Cx+D}{(x^2+4)^2}$$

Clearing fractions and combining terms give

$$x^3+1 = (Ax+B)(x^2+4) + Cx + D$$
$$x^3+1 = Ax^3 + Bx^2 + (4A+C)x + 4B + D$$

Equating coefficients,

$$A=1 \qquad B=0 \qquad 4A+C=0 \qquad 4B+D=1$$
$$C=-4 \qquad\qquad D=1$$

Then

$$\frac{x^3+1}{(x^2+4)^2} = \frac{x}{x^2+4} + \frac{-4x+1}{(x^2+4)^2}$$

and

$$\int \frac{x^3+1}{(x^2+4)^2}\,dx = \int \frac{x}{x^2+4}\,dx + \int \frac{-4x+1}{(x^2+4)^2}\,dx$$

$$= \int \frac{x}{x^2+4}\,dx - 4\int \frac{x}{(x^2+4)^2}\,dx + \int \frac{dx}{(x^2+4)^2} \qquad (3)$$

In the first two integrals on the right in (3), use the substitution $u=x^2+4$, $x\geq 0$. Then $du=2x\,dx$, and

$$\bullet \quad \int \frac{x}{x^2+4}\,dx = \frac{1}{2}\int \frac{du}{u} = \frac{1}{2}\ln|u| = \frac{1}{2}\ln(x^2+4) \qquad (4)$$

$$\bullet \quad -4\int \frac{x}{(x^2+4)^2}\,dx = -2\int \frac{du}{u^2} = \frac{2}{u} = \frac{2}{x^2+4} \qquad (5)$$

In the third integral on the right in (3), use the trigonometric substitution $x=2\tan\theta$, $-\dfrac{\pi}{2}<\theta<\dfrac{\pi}{2}$. Then $dx=2\sec^2\theta\,d\theta$, and

$$\int \frac{dx}{(x^2+4)^2} = \int \frac{2\sec^2\theta\,d\theta}{\left(4\tan^2\theta+4\right)^2} = \frac{2}{16}\int \frac{\sec^2\theta\,d\theta}{(\sec^2\theta)^2}$$

$$= \frac{1}{8}\int \cos^2\theta\,d\theta = \frac{1}{8}\int \frac{1+\cos(2\theta)}{2}\,d\theta$$

$$= \frac{1}{16}\left[\theta + \frac{1}{2}\sin(2\theta)\right] = \frac{1}{16}(\theta + \sin\theta\cos\theta)$$

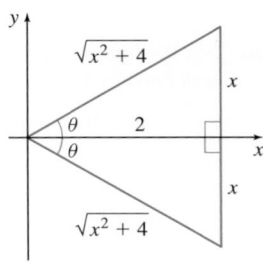

Figure 2 $\tan \theta = \dfrac{x}{2}, -\dfrac{\pi}{2} < \theta < \dfrac{\pi}{2}$

To express the solution in terms of x, either use the triangles in Figure 2 or use trigonometric identities as follows:

$$\sin \theta \cos \theta = \frac{\sin \theta}{\cos \theta} \cdot \cos^2 \theta = \frac{\tan \theta}{\sec^2 \theta} = \frac{\tan \theta}{\tan^2 \theta + 1} = \frac{\dfrac{x}{2}}{\dfrac{x^2}{4} + 1} = \frac{2x}{x^2 + 4}$$

Then

$$\int \frac{dx}{(x^2 + 4)^2} = \frac{1}{16}(\theta + \sin \theta \cos \theta) = \frac{1}{16}\left(\tan^{-1}\frac{x}{2} + \frac{2x}{x^2 + 4}\right) \qquad (6)$$

$$\underset{\uparrow}{} \quad x = 2\tan \theta; \ \theta = \tan^{-1}\frac{x}{2}$$

NOTE Case 4 type integrands lead to sums of natural logarithms, inverse tangents, and/or rational functions.

Now combine the results of (3), (4), (5), and (6):

$$\int \frac{x^3 + 1}{(x^2 + 4)^2}\,dx = \frac{1}{2}\ln(x^2 + 4) + \frac{2}{x^2 + 4} + \frac{1}{16}\tan^{-1}\frac{x}{2} + \frac{x}{8(x^2 + 4)} + K \qquad \blacksquare$$

NOW WORK Problem **17.**

6.8 Assess Your Understanding

Concepts and Vocabulary

1. True or False The integration of a proper rational function always leads to a logarithm.

2. True or False The decomposition of $\dfrac{7x + 1}{(x + 1)^4}$ into partial fractions has three terms: $\dfrac{A}{x + 1} + \dfrac{B}{(x + 1)^2} + \dfrac{C}{(x + 1)^3}$, where A, B, and C are real numbers.

Skill Building

In Problems 3–8, find each integral.
Hint: Each of the denominators contains only distinct linear factors.

3. $\displaystyle\int \frac{dx}{(x - 2)(x + 1)}$

4. $\displaystyle\int \frac{dx}{(x + 4)(x - 1)}$

[PAGE 529] 5. $\displaystyle\int \frac{x\,dx}{(x - 1)(x - 2)}$

6. $\displaystyle\int \frac{3x\,dx}{(x + 2)(x - 4)}$

7. $\displaystyle\int \frac{x\,dx}{(3x - 2)(2x + 1)}$

8. $\displaystyle\int \frac{dx}{(2x + 3)(4x - 1)}$

In Problems 9–12, find each integral.
Hint: Each of the denominators contains a repeated linear factor.

[PAGE 531] 9. $\displaystyle\int \frac{x - 3}{(x + 2)(x + 1)^2}\,dx$

10. $\displaystyle\int \frac{x + 1}{x^2(x - 2)}\,dx$

11. $\displaystyle\int \frac{x^2\,dx}{(x - 1)^2(x + 1)}$

12. $\displaystyle\int \frac{x^2 + x}{(x + 2)(x - 1)^2}\,dx$

In Problems 13–16, find each integral.
Hint: Each of the denominators contains an irreducible quadratic factor.

13. $\displaystyle\int \frac{dx}{x(x^2 + 1)}$

14. $\displaystyle\int \frac{dx}{(x + 1)(x^2 + 4)}$

[PAGE 533] 15. $\displaystyle\int \frac{x^2 + 2x + 3}{(x + 1)(x^2 + 2x + 4)}\,dx$

16. $\displaystyle\int \frac{x^2 - 11x - 18}{x(x^2 + 3x + 3)}\,dx$

In Problems 17–20, find each integral.
Hint: Each of the denominators contains a repeated irreducible quadratic factor.

[PAGE 535] 17. $\displaystyle\int \frac{2x + 1}{(x^2 + 16)^2}\,dx$

18. $\displaystyle\int \frac{x^2 + 2x + 3}{(x^2 + 4)^2}\,dx$

19. $\displaystyle\int \frac{x^3\,dx}{(x^2 + 16)^3}$

20. $\displaystyle\int \frac{x^2\,dx}{(x^2 + 4)^3}$

In Problems 21–30, find each integral.

21. $\displaystyle\int \frac{x\,dx}{x^2 + 2x - 3}$

22. $\displaystyle\int \frac{x^2 - x - 8}{(x + 1)(x^2 + 5x + 6)}\,dx$

23. $\displaystyle\int \frac{10x^2 + 2x}{(x - 1)^2(x^2 + 2)}\,dx$

24. $\displaystyle\int \frac{x + 4}{x^2(x^2 + 4)}\,dx$

25. $\displaystyle\int \frac{7x + 3}{x^3 - 2x^2 - 3x}\,dx$

26. $\displaystyle\int \frac{x^5 + 1}{x^6 - x^4}\,dx$

27. $\displaystyle\int \frac{x^2}{(x - 2)(x - 1)^2}\,dx$

28. $\displaystyle\int \frac{x^2 + 1}{(x + 3)(x - 1)^2}\,dx$

29. $\displaystyle\int \frac{2x + 1}{x^3 - 1}\,dx$

30. $\displaystyle\int \frac{dx}{x^3 - 8}$

In Problems 31–34, find each definite integral.

[PAGE 530] 31. $\displaystyle\int_0^1 \frac{dx}{x^2 - 9}$

32. $\displaystyle\int_2^4 \frac{dx}{x^2 - 25}$

33. $\displaystyle\int_{-2}^3 \frac{dx}{16 - x^2}$

34. $\displaystyle\int_1^2 \frac{dx}{9 - x^2}$

Applications and Extensions

In Problems 35–48, find each integral.
Hint: Make a substitution before using partial fraction decomposition.

35. $\displaystyle\int \frac{\cos\theta}{\sin^2\theta + \sin\theta - 6}\,d\theta$

36. $\displaystyle\int \frac{\sin x}{\cos^2 x - 2\cos x - 8}\,dx$

37. $\displaystyle\int \frac{\sin\theta}{\cos^3\theta + \cos\theta}\,d\theta$

38. $\displaystyle\int \frac{4\cos\theta}{\sin^3\theta + 2\sin\theta}\,d\theta$

39. $\displaystyle\int \frac{e^t}{e^{2t} + e^t - 2}\,dt$

40. $\displaystyle\int \frac{e^x}{e^{2x} + e^x - 6}\,dx$

41. $\displaystyle\int \frac{e^x}{e^{2x} - 1}\,dx$

42. $\displaystyle\int \frac{dx}{e^x - e^{-x}}$

43. $\displaystyle\int \frac{dt}{e^{2t} + 1}$

44. $\displaystyle\int \frac{dt}{e^{3t} + e^t}$

45. $\displaystyle\int \frac{\sin x \cos x}{(\sin x - 1)^2}\,dx$

46. $\displaystyle\int \frac{\cos x \sin x}{(\cos x - 2)^2}\,dx$

47. $\displaystyle\int \frac{\cos x}{(\sin^2 x + 9)^2}\,dx$

48. $\displaystyle\int \frac{\sin x}{(\cos^2 x + 4)^2}\,dx$

49. Area Find the area under the graph of $y = \dfrac{4}{x^2 - 4}$ from $x = 3$ to $x = 5$, as shown in the figure below.

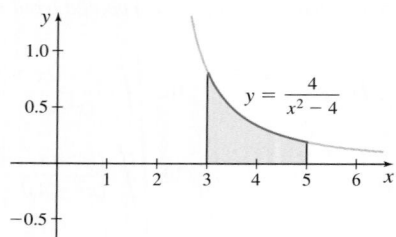

50. Area Find the area under the graph of $y = \dfrac{x - 4}{(x + 3)^2}$ from $x = 4$ to $x = 6$, as shown in the figure below.

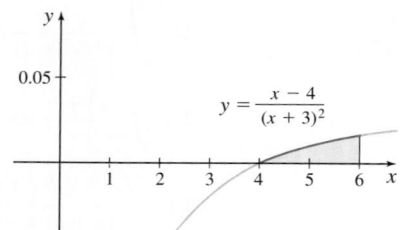

51. Area Find the area under the graph of $y = \dfrac{8}{x^3 + 1}$ from $x = 0$ to $x = 2$, as shown in the figure below.

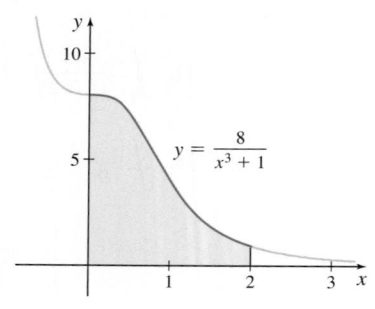

52. Volume of a Solid of Revolution See the figure below. The volume V of the solid of revolution generated by revolving the region bounded by the graph of $y = \dfrac{x}{x^2 - 4}$ and the x-axis from $x = 3$ to $x = 5$ about the x-axis is given by

$$V = \pi \int_3^5 \left(\frac{x}{x^2 - 4}\right)^2 dx$$

Find V.

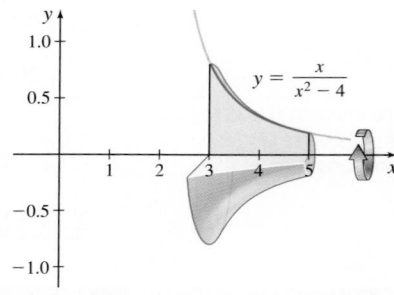

53. (a) Find the zeros of $q(x) = x^3 + 3x^2 - 10x - 24$.
(b) Factor q.
(c) Find the integral $\displaystyle\int \frac{3x - 7}{x^3 + 3x^2 - 10x - 24}\,dx$.

Challenge Problems

In Problems 54–65, simplify each integrand as follows: If the integrand involves fractional powers such as $x^{p/q}$ and $x^{r/s}$, make the substitution $x = u^n$, where n is the least common denominator of $\dfrac{p}{q}$ and $\dfrac{r}{s}$. Then find each integral.

54. $\displaystyle\int \frac{x\,dx}{3 + \sqrt{x}}$

55. $\displaystyle\int \frac{dx}{\sqrt{x} + 2}$

56. $\displaystyle\int \frac{dx}{x - \sqrt[3]{x}}$

57. $\displaystyle\int \frac{x\,dx}{\sqrt[3]{x} - 1}$

58. $\displaystyle\int \frac{dx}{\sqrt{x} + \sqrt[3]{x}}$

59. $\displaystyle\int \frac{dx}{3\sqrt{x} - \sqrt[3]{x}}$

60. $\displaystyle\int \frac{dx}{\sqrt[3]{2} + 3x}$

61. $\displaystyle\int \frac{dx}{\sqrt[4]{1 + 2x}}$

62. $\displaystyle\int \frac{x\,dx}{(1 + x)^{3/4}}$

63. $\displaystyle\int \frac{dx}{(1 + x)^{2/3}}$

64. $\displaystyle\int \frac{\sqrt[3]{x} + 1}{\sqrt[3]{x} - 1}\,dx$

65. $\displaystyle\int \frac{dx}{\sqrt{x}(1 + \sqrt[3]{x})^2}$

Weierstrass Substitution   *In Problems 66–81, use the following substitution, called a **Weierstrass substitution**. If an integrand is a rational expression of* $\sin x$ *or* $\cos x$ *or both, the substitution*

$$z = \tan \frac{x}{2} \qquad -\frac{\pi}{2} < \frac{x}{2} < \frac{\pi}{2}$$

or equivalently,

$$\cos x = \frac{1 - z^2}{1 + z^2} \qquad dx = \frac{2\,dz}{1 + z^2}$$

will transform the integrand into a rational function of z.

66. $\displaystyle\int \frac{dx}{1 - \sin x}$

67. $\displaystyle\int \frac{dx}{1 + \sin x}$

68. $\displaystyle\int \frac{dx}{1 - \cos x}$

69. $\displaystyle\int \frac{dx}{3 + 2\cos x}$

70. $\displaystyle\int \frac{2\,dx}{\sin x + \cos x}$

71. $\displaystyle\int \frac{dx}{1 - \sin x + \cos x}$

72. $\displaystyle\int \frac{\sin x}{3 + \cos x}\,dx$

73. $\displaystyle\int \frac{dx}{\tan x - 1}$

74. $\displaystyle\int \frac{dx}{\tan x - \sin x}$

75. $\displaystyle\int \frac{\sec x}{\tan x - 2}\,dx$

76. $\displaystyle\int \frac{\cot x}{1 + \sin x}\,dx$

77. $\displaystyle\int \frac{\sec x}{1 + \sin x}\,dx$

78. $\displaystyle\int_0^{\pi/2} \frac{dx}{\sin x + 1}$

79. $\displaystyle\int_{\pi/4}^{\pi/3} \frac{\csc x}{3 + 4\tan x}\,dx$

80. $\displaystyle\int_0^{\pi/2} \frac{\cos x}{2 - \cos x}\,dx$

81. $\displaystyle\int_0^{\pi/4} \frac{4\,dx}{\tan x + 1}$

82. Use a Weierstrass substitution to derive the formula

$$\int \csc x\,dx = \ln \sqrt{\frac{1 - \cos x}{1 + \cos x}} + C$$

83. Show that the result obtained in Problem 82 is equivalent to

$$\int \csc x\,dx = \ln |\csc x - \cot x| + C$$

84. Use the methods of this section to find $\displaystyle\int \frac{dx}{1 + x^4}$.

Hint: Factor $1 + x^4$ into irreducible quadratics.

85. Show that the two formulas below are equivalent.

$$\int \sec x\,dx = \ln |\sec x + \tan x| + C$$

$$\int \sec x\,dx = \ln \left| \frac{1 + \tan \dfrac{x}{2}}{1 - \tan \dfrac{x}{2}} \right| + C$$

Hint: $\tan \dfrac{x}{2} = \dfrac{\sin \dfrac{x}{2}}{\cos \dfrac{x}{2}} = \dfrac{\sin^2 \left(\dfrac{x}{2} \right)}{\sin \dfrac{x}{2} \cos \dfrac{x}{2}} = \dfrac{1 - \cos x}{\sin x}$.

Preparing for the AP® Exam

AP® Practice Problems

NOTE: *These problems are BC only.*

Multiple-Choice Questions

PAGE 529 **1.** $\displaystyle\int \frac{12}{x^2 - 9}\,dx =$

(A) $2\ln\left| \dfrac{x - 3}{x + 3} \right| + C$

(B) $2\ln\left| \dfrac{x + 3}{x - 3} \right| + C$

(C) $\dfrac{4}{3}\tan^{-1}\dfrac{x}{3} + C$

(D) $\ln\left| \dfrac{x - 3}{x + 3} \right| + C$

PAGE 529 **2.** $\displaystyle\int \frac{3x}{(x - 2)(x + 1)}\,dx =$

(A) $\dfrac{x^2}{2}[2\ln|x - 2| + \ln|x + 1|] + C$

(B) $\ln|x + 1| - 2\ln|x - 2| + C$

(C) $2\ln|x + 1| + \ln|x - 2| + C$

(D) $2\ln|x - 2| + \ln|x + 1| + C$

PAGE 529 **3.** $\displaystyle\int \frac{x + 6}{x(x + 2)}\,dx =$

(A) $3x - 12\ln|x + 2| + C$

(B) $3x - 2\ln|x + 2| + C$

(C) $3\ln|x| - 2\ln|x + 2| + C$

(D) $3\ln|x| + 2\ln|x + 2| + C$

PAGE 530 **4.** $\displaystyle\int_0^1 \frac{2x - 1}{x^2 + 3x + 2}\,dx =$

(A) $2\ln 2 + 3\ln 3$

(B) $2\ln 2 + 5\ln 3$

(C) $-4\ln 2 + 3\ln 3$

(D) $-8\ln 2 + 5\ln 3$

PAGE 529 **5.** Find a so that

$$\int \frac{4}{(x - a)(x - 2)}\,dx = -4\ln|x - a| + 4\ln|x - 2| + C$$

(A) 0   (B) 2   (C) −1   (D) 1

Free-Response Question

PAGE 530 **6.** The graph of the function

$$f(x) = \frac{8x}{x^2 - 3x + 2}$$

is shown below.

(a) Express the function f as the sum of partial fractions.

(b) Find the area under the graph of f from $x = 3$ to $x = 6$.

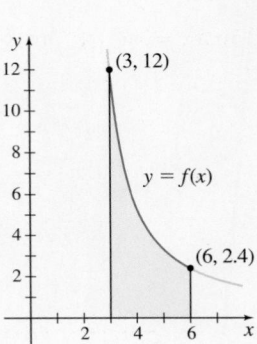

Retain Your Knowledge

Multiple-Choice Questions

1. If $\tan^{-1} x = \ln y$, then $y' =$

(A) $\dfrac{y}{1 + x^2}$ (B) $\dfrac{y}{1 - x^2}$

(C) $\dfrac{2xy}{(1 + x^2)^2}$ (D) $\dfrac{y}{\sqrt{1 - x^2}}$

2. The function f is twice differentiable for all real numbers. A portion of the graph of f is shown below.

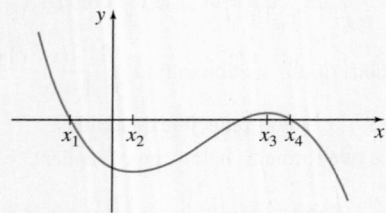

At which number x_i is $f(x_i) > f'(x_i) > f''(x_i)$ always true?

(A) x_1 (B) x_2 (C) x_3 (D) x_4

3. $\int x^3 \cos x^4 \, dx =$

(A) $-\dfrac{1}{4} \sin x^4 + C$ (B) $\dfrac{1}{4} \sin^4 x + C$

(C) $\dfrac{1}{4} \sin x^4 + C$ (D) $\dfrac{x^5}{5} \sin x^4 + C$

Free-Response Question

4. The function g is continuous and concave down for $2 \leq x \leq 12$. The values of g for select numbers x in the domain of g are given in the table below.

x	2	4	9	12
$g(x)$	6	14	17	15

(a) Use a trapezoidal sum with three intervals to approximate $\int_2^{12} g(x) \, dx$.

(b) Does the trapezoidal sum approximation underestimate or overestimate $\int_2^{12} g(x) \, dx$? Justify your answer.

6.9 Improper Integrals

OBJECTIVES *When you finish this section, you should be able to:*

1 Find integrals with an infinite limit of integration (p. 539)
2 Interpret an improper integral geometrically (p. 541)
3 Integrate functions over $[a,\ b]$ that are not defined at an endpoint (p. 542)
4 Use the Comparison Test for improper integrals (p. 545)

In Section 6.2, the definition of the definite integral $\int_a^b f(x)\,dx$ required that both a and b be real numbers. We also required that the function f be defined on the closed interval $[a, b]$. Here we take up instances for which:

• Integrals have infinite limits of integration:

$$\int_1^{\infty} \frac{1}{\sqrt{x}}\,dx \qquad \int_{-\infty}^0 \frac{x-3}{x^3-8}\,dx \qquad \int_{-\infty}^{\infty} \frac{x}{(x^2+1)^2}\,dx$$

• The integrand is not defined at a number in the interval of integration:

$$\int_0^1 \frac{1}{\sqrt{x}}\,dx \qquad \int_0^{\pi/2} \tan x\,dx \qquad \int_{-1}^1 \frac{1}{x^3}\,dx$$

not defined at 0   not defined at $\dfrac{\pi}{2}$ not defined at 0

Integrals like these are called *improper integrals*.

DEFINITION Improper Integral

If a function f is continuous on the interval $[a, \infty)$, then $\int_a^\infty f(x)\,dx$, called an **improper integral**, is defined as

$$\int_a^\infty f(x)\,dx = \lim_{b \to \infty} \int_a^b f(x)\,dx$$

provided the limit exists and is a real number. If $\lim\limits_{b \to \infty} \int_a^b f(x)\,dx$ exists and is a real number, the improper integral $\int_a^\infty f(x)\,dx$ is said to **converge**. If the limit does not exist or if the limit is infinite, the improper integral $\int_a^\infty f(x)\,dx$ is said to **diverge**.

If a function f is continuous on the interval $(-\infty, b]$, then $\int_{-\infty}^b f(x)\,dx$, called an **improper integral**, is defined as

$$\int_{-\infty}^b f(x)\,dx = \lim_{a \to -\infty} \int_a^b f(x)\,dx$$

provided the limit exists and is a real number. If $\lim\limits_{a \to -\infty} \int_a^b f(x)\,dx$ exists and is a real number, the improper integral $\int_{-\infty}^b f(x)\,dx$ **converges**. If the limit does not exist or if the limit is infinite, the improper integral $\int_{-\infty}^b f(x)\,dx$ **diverges**.

1 Find Integrals with an Infinite Limit of Integration

CALC CLIP

EXAMPLE 1 Integrating Functions over Infinite Intervals

Determine whether the following improper integrals converge or diverge:

(a) $\displaystyle\int_1^\infty \frac{1}{x}\,dx$ **(b)** $\displaystyle\int_{-\infty}^0 e^x\,dx$ **(c)** $\displaystyle\int_{\pi/2}^\infty \sin x\,dx$

NEED TO REVIEW? Limits at infinity and infinite limits are discussed in Section 1.5, pp. 136–143.

Solution

(a) To determine whether $\displaystyle\int_1^\infty \frac{1}{x}\,dx$ converges or diverges, we examine the limit:

$$\lim_{b \to \infty} \int_1^b \frac{1}{x}\,dx = \lim_{b \to \infty}\left[\ln|x|\right]_1^b = \lim_{b \to \infty}[\ln b - \ln 1] = \infty$$

CAUTION In (a), it is incorrect to write $\displaystyle\int_1^\infty \frac{1}{x}\,dx = \infty$. Definite integrals, when they exist, are numbers, and ∞ is not a number.

The limit is infinite, so $\displaystyle\int_1^\infty \frac{1}{x}\,dx$ diverges.

(b) To determine whether $\displaystyle\int_{-\infty}^0 e^x\,dx$ converges or diverges, we examine the limit:

$$\lim_{a \to -\infty} \int_a^0 e^x\,dx = \lim_{a \to -\infty}\left[e^x\right]_a^0 = \lim_{a \to -\infty}(1 - e^a) = 1 - 0 = 1$$

Since the limit exists, $\displaystyle\int_{-\infty}^0 e^x\,dx$ converges and equals 1.

(continued on the next page)

(c) To determine whether $\int_{\pi/2}^{\infty} \sin x \, dx$ converges or diverges, we examine the limit:

$$\lim_{b \to \infty} \int_{\pi/2}^{b} \sin x \, dx = \lim_{b \to \infty} \left[-\cos x \right]_{\pi/2}^{b} = \lim_{b \to \infty} [-\cos b + 0] = -\lim_{b \to \infty} \cos b$$

This limit does not exist, since as $b \to \infty$, the value of $\cos b$ oscillates between -1 and 1. So, $\int_{\pi/2}^{\infty} \sin x \, dx$ diverges. ∎

NOW WORK   Problem 17 and AP® Practice Problems 2, 3, 4, and 5.

If both limits of integration are infinite, then the following definition is used.

DEFINITION

If a function f is continuous for all x and if, for any number c, *both* improper integrals $\int_{-\infty}^{c} f(x)\,dx$ and $\int_{c}^{\infty} f(x)\,dx$ converge, then the improper integral $\int_{-\infty}^{\infty} f(x)\,dx$ **converges,** and

$$\int_{-\infty}^{\infty} f(x)\,dx = \int_{-\infty}^{c} f(x)\,dx + \int_{c}^{\infty} f(x)\,dx$$

If *either* or *both* of the integrals on the right diverge, then the improper integral $\int_{-\infty}^{\infty} f(x)\,dx$ **diverges**.

EXAMPLE 2   Integrating Functions over Infinite Intervals

Determine whether $\int_{-\infty}^{\infty} 4x^3 \, dx$ converges or diverges.

Solution

We begin by writing $\int_{-\infty}^{\infty} 4x^3\,dx = \int_{-\infty}^{0} 4x^3\,dx + \int_{0}^{\infty} 4x^3\,dx$ and evaluate each improper integral on the right.

For $\int_{-\infty}^{0} 4x^3\,dx$, we examine the limit

$$\lim_{a \to -\infty} \int_{a}^{0} 4x^3 \, dx = \lim_{a \to -\infty} \left[x^4 \right]_{a}^{0} = \lim_{a \to -\infty} (0 - a^4) = -\infty$$

There is no need to continue. $\int_{-\infty}^{\infty} 4x^3\,dx$ diverges. ∎

NOTE The decision to use 0 to break up the integral is arbitrary. Usually, we choose a number that simplifies finding the integral.

CAUTION   The definition requires that two improper integrals converge in order for $\int_{-\infty}^{\infty} f(x)\,dx$ to converge. Do not set $\int_{-\infty}^{\infty} f(x)\,dx = \lim_{a \to \infty} \int_{-a}^{a} f(x)\,dx$. If this were done in Example 2, the result would have been $\int_{-a}^{a} 4x^3\,dx = \left[x^4 \right]_{-a}^{a} = a^4 - a^4 = 0$, and we would have incorrectly concluded that $\int_{-\infty}^{\infty} f(x)\,dx$ converges and equals 0.

NOW WORK   Problem 23.

② Interpret an Improper Integral Geometrically

If a function f is continuous and nonnegative on the interval $[a, \infty)$, then $\int_a^\infty f(x)\,dx$ can be interpreted geometrically. For each number $b > a$, the definite integral $\int_a^b f(x)\,dx$ represents the area under the graph of $y = f(x)$ from a to b, as shown in Figure 3(a). As $b \to \infty$, this area approaches the area under the graph of $y = f(x)$ over the interval $[a, \infty)$, as shown in Figure 3(b). The area under the graph of $y = f(x)$ to the right of a is defined by $\int_a^\infty f(x)\,dx$, provided the improper integral converges. If the improper integral diverges, there is no area defined.

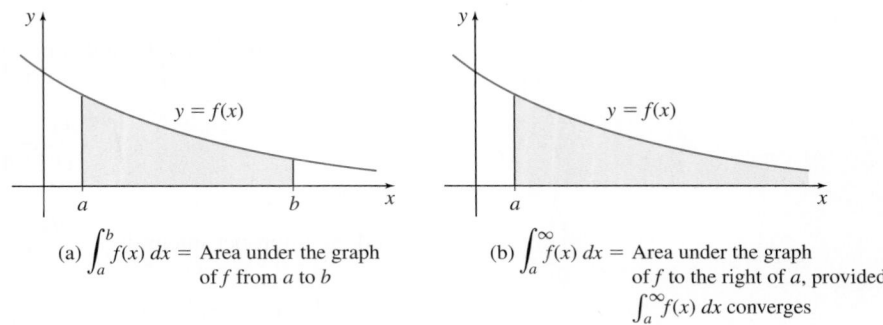

(a) $\displaystyle\int_a^b f(x)\,dx =$ Area under the graph of f from a to b

(b) $\displaystyle\int_a^\infty f(x)\,dx =$ Area under the graph of f to the right of a, provided $\int_a^\infty f(x)\,dx$ converges

Figure 3

EXAMPLE 3 **Determining Whether the Area under a Graph Is Defined**

Determine whether the area under the graph of $y = \dfrac{1}{x^2}$ to the right of $x = 1$ is defined.

Solution

The area, if defined, is given by $\displaystyle\int_1^\infty \frac{1}{x^2}\,dx$.

See Figure 4. To determine whether the area is defined, we examine the limit

$$\lim_{b \to \infty} \int_1^b \frac{1}{x^2}\,dx = \lim_{b \to \infty}\left[-\frac{1}{x}\right]_1^b = \lim_{b \to \infty}\left(-\frac{1}{b} + 1\right) = 1$$

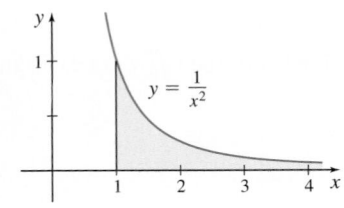

Figure 4

The area under the graph $f(x) = \dfrac{1}{x^2}$ to the right of 1 is defined and equals 1 sq. unit. ■

NOW WORK Problem 73.

The theorem below generalizes the result given in Example 3.

THEOREM

$\displaystyle\int_1^\infty \frac{dx}{x^p}$ converges if $p > 1$ and diverges if $p \le 1$.

Proof We consider two cases: $p = 1$ and $p \neq 1$.

$$p = 1: \quad \int_1^\infty \frac{dx}{x^p} = \int_1^\infty \frac{dx}{x}: \quad \lim_{b \to \infty} \int_1^b \frac{dx}{x} = \lim_{b \to \infty} \ln b = \infty \quad \text{From Example 1(a)}$$

So, for $p = 1$, the integral diverges.

$$p \neq 1: \quad \int_1^\infty \frac{dx}{x^p}: \quad \lim_{b \to \infty} \int_1^b \frac{dx}{x^p} = \lim_{b \to \infty} \left[\frac{x^{-p+1}}{-p+1} \right]_1^b = \lim_{b \to \infty} \left[\frac{1}{1-p} (b^{-p+1} - 1) \right]$$

$$= \frac{1}{1-p} \lim_{b \to \infty} \left(\frac{1}{b^{p-1}} - 1 \right)$$

- If $p > 1$, then $p - 1 > 0$, and $\lim_{b \to \infty} \dfrac{1}{b^{p-1}} = 0$. So,

$$\int_1^\infty \frac{dx}{x^p} = \frac{1}{1-p} (0 - 1) = \frac{1}{p-1}$$

and the improper integral $\displaystyle\int_1^\infty \frac{dx}{x^p}$, $p > 1$, converges.

- If $p < 1$, then $p - 1 < 0$, and $\lim_{b \to \infty} \dfrac{1}{b^{p-1}} = \lim_{b \to \infty} b^{1-p} = \infty$.

So, the improper integral $\displaystyle\int_1^\infty \frac{dx}{x^p}$, $p < 1$, diverges. ∎

NOW WORK AP® Practice Problem 7.

---

❸ Integrate Functions over $[a, b]$ That Are Not Defined at an Endpoint

IN WORDS These improper integrals, sometimes called **improper integrals of the second kind**, have finite limits of integration, but the integrand is not defined at one (but not both) of the limits of integration.

CAUTION Improper integrals of the second kind, as defined in the above IN WORDS, are often not identified as such.

For $\displaystyle\int_a^b f(x)\,dx$, always check the domain of f to see if f is not defined at $x = a$ or at $x = b$.

DEFINITION Improper Integral

If a function f is continuous on $(a, b]$, but is not defined at a, then $\int_a^b f(x)\,dx$ is an **improper integral**, and

$$\boxed{\int_a^b f(x)\,dx = \lim_{t \to a^+} \int_t^b f(x)\,dx}$$

provided the limit exists and is a real number.

If a function f is continuous on $[a, b)$, but is not defined at b, then $\int_a^b f(x)\,dx$ is an **improper integral**, and

$$\boxed{\int_a^b f(x)\,dx = \lim_{t \to b^-} \int_a^t f(x)\,dx}$$

provided the limit exists and is a real number.

When the limit exists and is a real number, the improper integral is said to **converge**; otherwise, it is said to **diverge**.

Be sure to use the correct one-sided limit. If f is not defined at the left endpoint a, then for t to be in the interval of integration, $t > a$ and t must approach a from the right. Similarly, if f is not defined at the right endpoint b, then for t to be in the interval of integration, $t < b$ and t must approach b from the left. Figure 5 may help you remember the correct one-sided limit to use.

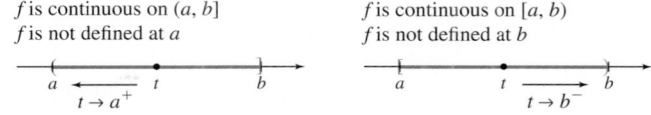

f is continuous on $(a, b]$
f is not defined at a

f is continuous on $[a, b)$
f is not defined at b

Figure 5

If a function f is continuous and nonnegative on the interval $(a, b]$, then the improper integral $\int_a^b f(x)\,dx$ may be interpreted geometrically. For each number t, $a < t \le b$, the integral $\int_t^b f(x)\,dx$ represents the area under the graph of $y = f(x)$ from t to b, as shown in Figure 6(a). As $t \to a^+$, this area approaches the area under the graph of $y = f(x)$ from a to b, as shown in Figure 6(b). That is, if $\int_a^b f(x)\,dx$ converges, its value is defined to be the area A under the graph of f from a to b.

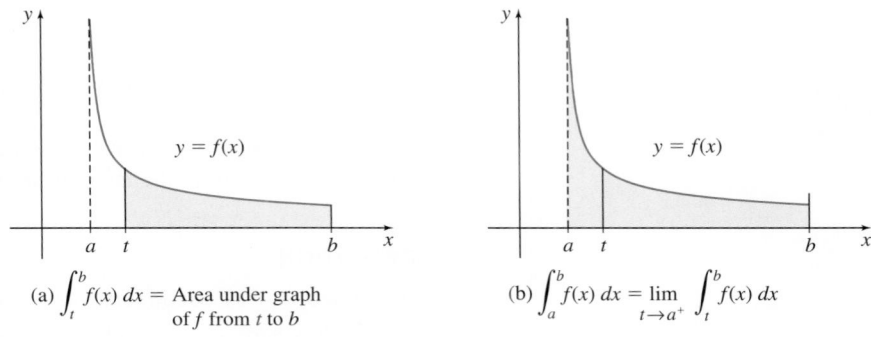

(a) $\int_t^b f(x)\,dx =$ Area under graph of f from t to b

(b) $\int_a^b f(x)\,dx = \lim\limits_{t \to a^+} \int_t^b f(x)\,dx$

Figure 6

EXAMPLE 4 **Determining Whether the Area under a Graph Is Defined**

Determine whether the area under the graph of $y = \dfrac{1}{\sqrt{x}}$ from 0 to 4 is defined.

Solution

Figure 7 shows the graph of $y = \dfrac{1}{\sqrt{x}}$. Since $y = \dfrac{1}{\sqrt{x}}$ is continuous on $(0, 4]$ but is not defined at 0, $\int_0^4 \dfrac{1}{\sqrt{x}}\,dx$ is an improper integral. We determine whether the integral converges or diverges, by examining the limit

$$\lim_{t \to 0^+} \int_t^4 \frac{1}{\sqrt{x}}\,dx = \lim_{t \to 0^+} \int_t^4 x^{-1/2}\,dx = \lim_{t \to 0^+} \left[\frac{x^{1/2}}{\frac{1}{2}} \right]_t^4 = \lim_{t \to 0^+} \left[2\,x^{1/2} \right]_t^4$$

$$= \lim_{t \to 0^+} (2 \cdot 2 - 2\sqrt{t}) = 4 - 2 \lim_{t \to 0^+} \sqrt{t} = 4$$

Since $\int_0^4 \dfrac{1}{\sqrt{x}}\,dx$ converges, the area under the graph of $y = \dfrac{1}{\sqrt{x}}$ from 0 to 4 is defined and equals 4. ∎

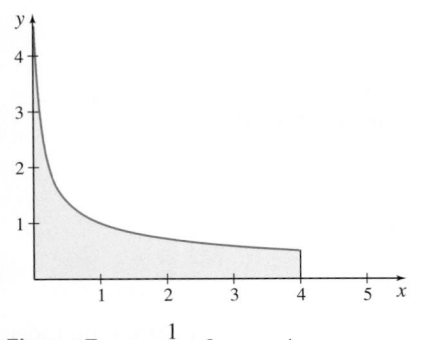

Figure 7 $y = \dfrac{1}{\sqrt{x}}, 0 < x \le 4$

NOW WORK Problem 27.

EXAMPLE 5 | Determining Whether an Improper Integral Converges or Diverges

Determine whether $\displaystyle\int_0^{\pi/2} \tan x \, dx$ converges or diverges.

Solution

The function $f(x) = \tan x$ is continuous on $\left[0, \dfrac{\pi}{2}\right)$ but is not defined at $\dfrac{\pi}{2}$, so $\int_0^{\pi/2} \tan x \, dx$ is an improper integral. To determine whether the integral converges, examine the limit

$$\lim_{t \to (\pi/2)^-} \int_0^t \tan x \, dx = \lim_{t \to (\pi/2)^-} \Big[\ln|\sec x|\Big]_0^t$$

$$= \lim_{t \to (\pi/2)^-} [\ln|\sec t| - \ln|\sec 0|] \qquad \ln|\sec 0| = \ln 1 = 0$$

$$= \lim_{t \to (\pi/2)^-} \ln(\sec t) = \infty$$

So, $\displaystyle\int_0^{\pi/2} \tan x \, dx$ diverges. ∎

NOW WORK | Problem **43** and AP® Practice Problems **1** and **6**.

Another type of improper integral occurs when the integrand is not defined at some number c, $a < c < b$, in the interval $[a, b]$.

DEFINITION

If a function f is continuous on a closed interval $[a, b]$, except at the number c, $a < c < b$, where f is not defined, the integral $\int_a^b f(x)\,dx$ is an **improper integral** and

$$\int_a^b f(x)\,dx = \int_a^c f(x)\,dx + \int_c^b f(x)\,dx$$

provided that both improper integrals on the right converge.

If either improper integral $\int_a^c f(x)\,dx$ or $\int_c^b f(x)\,dx$ diverges, then the improper integral $\int_a^b f(x)\,dx$ also **diverges**.

CAUTION It is a common mistake to look at an improper integral like $\displaystyle\int_0^2 \dfrac{dx}{(x-1)^2}$ and not notice that the integrand is undefined at 1. If that happens, and you use the Fundamental Theorem of Calculus, you obtain

$$\int_0^2 \frac{dx}{(x-1)^2} = \left[\frac{-1}{x-1}\right]_0^2 = -1 - 1 = -2$$

which is an incorrect answer. *Always check the domain of the integrand before attempting to integrate.*

EXAMPLE 6 | Determining Whether an Improper Integral Converges or Diverges

Determine whether $\displaystyle\int_0^2 \frac{1}{(x-1)^2}\,dx$ converges or diverges.

Solution

Since $f(x) = \dfrac{1}{(x-1)^2}$ is not defined at 1, the integral $\displaystyle\int_0^2 \frac{1}{(x-1)^2}\,dx$ is an improper integral on the interval $[0, 2]$. We write the integral as follows:

$$\int_0^2 \frac{1}{(x-1)^2}\,dx = \int_0^1 \frac{1}{(x-1)^2}\,dx + \int_1^2 \frac{1}{(x-1)^2}\,dx$$

and investigate each of the two improper integrals on the right.

$$\lim_{t \to 1^-} \int_0^t \frac{1}{(x-1)^2}\,dx = \lim_{t \to 1^-} \left[\frac{-1}{x-1}\right]_0^t = \lim_{t \to 1^-}\left(\frac{-1}{t-1}-1\right) = \infty$$

$\int_0^1 \frac{1}{(x-1)^2}\,dx$ diverges, so there is no need to investigate the second integral.

The improper integral $\int_0^2 \frac{1}{(x-1)^2}\,dx$ diverges. ∎

NOW WORK   Problem 31.

---

④ Use the Comparison Test for Improper Integrals

So far we have been able to determine whether an improper integral converges or diverges by finding an antiderivative of the integrand. If we cannot find an antiderivative, we may still be able to determine whether the improper integral converges or diverges. Then, if it converges, we can use numerical techniques such as trapezoidal sums, to approximate the integral. One way to determine whether an improper integral converges or diverges is by using the *Comparison Test for Improper Integrals.*

> **THEOREM Comparison Test for Improper Integrals**
>
> Suppose f and g are two functions that are nonnegative and continuous on the interval $[a, \infty)$, and suppose
>
> $$\boxed{f(x) \geq g(x)}$$
>
> for all numbers $x > c$, where $c \geq a$.
>
> • If $\int_a^\infty f(x)\,dx$ converges, then $\int_a^\infty g(x)\,dx$ also converges.
> • If $\int_a^\infty g(x)\,dx$ diverges, then $\int_a^\infty f(x)\,dx$ also diverges.

You are asked to prove the theorem in Problem 96. The proof follows from a property of definite integrals: If the functions f and g are continuous on a closed interval $[a, b]$ and if $f(x) \geq g(x)$ on $[a, b]$, then $\int_a^b f(x)\,dx \geq \int_a^b g(x)\,dx$. (See Section 6.4, Problem 120, p. 479.)

EXAMPLE 7   **Using the Comparison Test for Improper Integrals**

Determine whether $\int_1^\infty e^{-x^2}\,dx$ converges or diverges.

Solution

By definition, $\int_1^\infty e^{-x^2}\,dx = \lim_{b \to \infty} \int_1^b e^{-x^2}\,dx$ converges if the limit exists and equals a real number. Since e^{-x^2} has no antiderivative, we use the Comparison Test for Improper Integrals. We proceed as follows: For $x \geq 1$,

$$x^2 \geq x$$
$$-x^2 \leq -x$$
$$0 < e^{-x^2} \leq e^{-x} \qquad \text{Since } e > 1, \text{ if } a \leq b, \text{ then } e^a \leq e^b.$$

Figure 8 illustrates this.

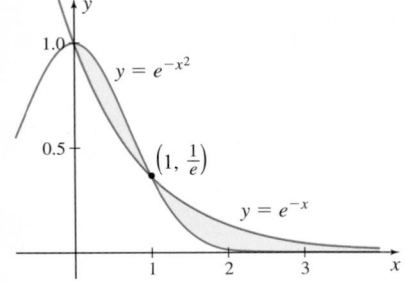

Figure 8

(continued on the next page)

Based on the Comparison Test for Improper Integrals, if $\int_1^\infty e^{-x}dx$ converges, so does $\int_1^\infty e^{-x^2}dx$. We investigate $\int_1^\infty e^{-x}dx$ by examining the limit

$$\lim_{b\to\infty}\int_1^b e^{-x}dx = \lim_{b\to\infty}\left[-e^{-x}\right]_1^b = \lim_{b\to\infty}[-e^{-b}+e^{-1}] = \lim_{b\to\infty}\left[\frac{1}{e}-\frac{1}{e^b}\right]$$

$$= \lim_{b\to\infty}\frac{1}{e} - \lim_{b\to\infty}\frac{1}{e^b} = \frac{1}{e}$$

Since $\int_1^\infty e^{-x}dx$ converges, we conclude that $\int_1^\infty e^{-x^2}dx$ converges. ∎

Notice that the Comparison Test for Improper Integrals does not give the value of $\int_1^\infty e^{-x^2}dx$. To find $\int_1^\infty e^{-x^2}dx$ requires numerical techniques. The Comparison Test does, however, tell us that $0 \le \int_1^\infty e^{-x^2}dx \le \frac{1}{e}$.

NOW WORK   Problem 67.

6.9 Assess Your Understanding

Concepts and Vocabulary

1. **Multiple Choice** If a function f is continuous on the interval $[a, \infty)$, then $\int_a^\infty f(x)\,dx$ is called

   (a) a definite **(b)** an infinite **(c)** an improper **(d)** a proper

   integral.

2. **Multiple Choice** If the $\lim_{b\to\infty}\int_a^b f(x)\,dx$ does not exist, the improper integral $\int_a^\infty f(x)\,dx$

   (a) converges **(b)** diverges **(c)** equals ab.

3. **True or False** If a function f is continuous for all x, then the improper integral $\int_{-\infty}^\infty f(x)\,dx$ always converges.

4. **True or False** If a function f is continuous and nonnegative on the interval $[a, \infty)$, and $\int_a^\infty f(x)\,dx$ converges, then $\int_a^\infty f(x)\,dx$ equals the area under the graph of $y = f(x)$ for $x \ge a$.

5. **True or False** If a function f is continuous for all x, then the improper integral $\int_{-\infty}^\infty f(x)\,dx = \lim_{a\to\infty}\int_{-a}^a f(x)\,dx$.

6. To determine whether the improper integral $\int_a^b f(x)\,dx$ converges or diverges, where f is continuous on $[a, b)$, but is not defined at b, requires finding what limit?

Skill Building

In Problems 7–14, determine whether each integral is improper. For those that are improper, state the reason.

7. $\int_0^\infty x^2\,dx$

8. $\int_0^5 x^3\,dx$

9. $\int_2^3 \dfrac{dx}{x-1}$

10. $\int_1^2 \dfrac{dx}{x-1}$

11. $\int_0^1 \dfrac{1}{x}\,dx$

12. $\int_{-1}^1 \dfrac{x\,dx}{x^2+1}$

13. $\int_0^1 \dfrac{x}{x^2-1}\,dx$

14. $\int_0^\infty e^{-2x}\,dx$

In Problems 15–24, determine whether each improper integral converges or diverges. If it converges, find its value.

15. $\int_1^\infty \dfrac{dx}{x^3}$

16. $\int_{-\infty}^{-10} \dfrac{dx}{x^2}$

PAGE 540 17. $\int_0^\infty e^{2x}\,dx$

18. $\int_0^\infty e^{-x}\,dx$

19. $\int_{-\infty}^{-1} \dfrac{4}{x}\,dx$

20. $\int_1^\infty \dfrac{4}{x}\,dx$

21. $\int_3^\infty \dfrac{dx}{(x-1)^4}$

22. $\int_{-\infty}^0 \dfrac{dx}{(x-1)^4}$

PAGE 540 23. $\int_{-\infty}^\infty \dfrac{dx}{x^2+4}$

24. $\int_{-\infty}^\infty \dfrac{dx}{x^2+1}$

In Problems 25–32, determine whether each improper integral converges or diverges. If it converges, find its value.

25. $\displaystyle\int_0^1 \frac{dx}{x^2}$

26. $\displaystyle\int_0^1 \frac{dx}{x^3}$

27. $\displaystyle\int_0^1 \frac{dx}{x}$

28. $\displaystyle\int_4^6 \frac{dx}{x-4}$

29. $\displaystyle\int_0^4 \frac{dx}{\sqrt{4-x}}$

30. $\displaystyle\int_1^5 \frac{x\,dx}{\sqrt{5-x}}$

31. $\displaystyle\int_{-1}^1 \frac{dx}{\sqrt[3]{x}}$

32. $\displaystyle\int_0^3 \frac{dx}{(x-2)^2}$

In Problems 33–62, determine whether each improper integral converges or diverges. If it converges, find its value.

33. $\displaystyle\int_0^\infty \cos x\,dx$

34. $\displaystyle\int_0^\infty \sin(\pi x)\,dx$

35. $\displaystyle\int_{-\infty}^0 e^x\,dx$

36. $\displaystyle\int_{-\infty}^0 e^{-x}\,dx$

37. $\displaystyle\int_0^{\pi/2} \frac{x\,dx}{\sin x^2}$

38. $\displaystyle\int_0^1 \frac{\ln x\,dx}{x}$

39. $\displaystyle\int_0^1 \frac{dx}{1-x^2}$

40. $\displaystyle\int_1^2 \frac{dx}{\sqrt{x^2-1}}$

41. $\displaystyle\int_0^1 \frac{x}{\sqrt{1-x^2}}\,dx$

42. $\displaystyle\int_0^4 \frac{dx}{\sqrt{4-x}}$

43. $\displaystyle\int_0^{\pi/4} \tan(2x)\,dx$

44. $\displaystyle\int_0^{\pi/2} \csc x\,dx$

45. $\displaystyle\int_0^\infty \frac{x\,dx}{(x+1)^{5/2}}$

46. $\displaystyle\int_2^\infty \frac{dx}{x\sqrt{x^2-1}}$

47. $\displaystyle\int_{-\infty}^\infty \frac{dx}{x^2+4x+5}$

48. $\displaystyle\int_{-\infty}^\infty \frac{dx}{e^x+e^{-x}}$

49. $\displaystyle\int_{-\infty}^2 \frac{dx}{\sqrt{4-x}}$

50. $\displaystyle\int_{-\infty}^1 \frac{x\,dx}{\sqrt{2-x}}$

51. $\displaystyle\int_2^4 \frac{2x\,dx}{\sqrt[3]{x^2-4}}$

52. $\displaystyle\int_0^\pi \frac{1}{1-\cos x}\,dx$

53. $\displaystyle\int_{-1}^1 \frac{1}{x^3}\,dx$

54. $\displaystyle\int_0^2 \frac{dx}{x-1}$

55. $\displaystyle\int_0^2 \frac{dx}{(x-1)^{1/3}}$

56. $\displaystyle\int_{-1}^1 \frac{dx}{x^{5/3}}$

57. $\displaystyle\int_1^2 \frac{dx}{(2-x)^{3/4}}$

58. $\displaystyle\int_0^4 \frac{dx}{\sqrt{8x-x^2}}$

59. $\displaystyle\int_1^3 \frac{2x\,dx}{(x^2-1)^{3/2}}$

60. $\displaystyle\int_0^3 \frac{x\,dx}{(9-x^2)^{3/2}}$

61. $\displaystyle\int_0^\infty xe^{-x^2}\,dx$

62. $\displaystyle\int_0^\infty e^{-x}\sin x\,dx$

In Problems 63–70:

(a) *Use the Comparison Test for Improper Integrals to determine whether each improper integral converges or diverges. Hint: Use the fact that* $\displaystyle\int_1^\infty \frac{dx}{x^p}$ *converges if* $p>1$ *and diverges if* $p\le 1$.

(CAS) **(b)** *If the integral converges, use a CAS to find its value.*

63. $\displaystyle\int_1^\infty \frac{1}{\sqrt{x^2-1}}\,dx$

64. $\displaystyle\int_2^\infty \frac{2}{\sqrt{x^2-4}}\,dx$

65. $\displaystyle\int_1^\infty \frac{1+e^{-x}}{x}\,dx$

66. $\displaystyle\int_1^\infty \frac{3e^{-x}}{x}\,dx$

67. $\displaystyle\int_1^\infty \frac{\sin^2 x}{x^2}\,dx$

68. $\displaystyle\int_1^\infty \frac{\cos^2 x}{x^2}\,dx$

69. $\displaystyle\int_1^\infty \frac{dx}{(x+1)\sqrt{x}}$

70. $\displaystyle\int_1^\infty \frac{dx}{x\sqrt{1+x^2}}$

Applications and Extensions

71. Volume of a Solid of Revolution The volume V, if it is defined, of the solid of revolution generated by revolving the region bounded by the graph of $y=e^{-x}$ and the x-axis to the right of $x=0$ about the x-axis is given by the improper integral $V = \pi \int_0^\infty (e^{-x})^2\,dx$. Find the volume V of the solid, if it exists.

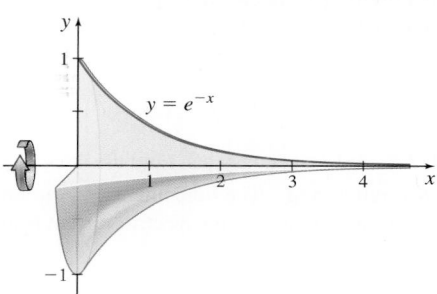

72. Area under a Graph Find the area, if it is defined, under the graph of $y=\dfrac{1}{1+x^2}$ to the right of $x=0$. See the figure below.

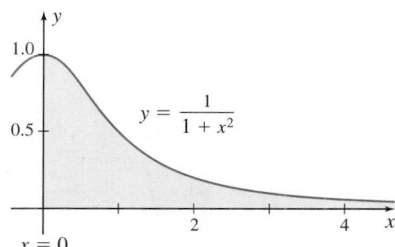

73. Area under a Graph Find the area, if it is defined, bounded by the graph of $y=\dfrac{8a^3}{x^2+4a^2}$, $a>0$, $x\ge 0$ and the x-axis.

74. Drug Reaction The rate of reaction r to a given dose of a drug at time t hours after administration is given by $r(t) = t e^{-t^2}$ (measured in appropriate units).

    (a) Why is it reasonable to define the total *reaction* as the area under the graph of $y = r(t)$ on $[0, \infty)$?

    (b) Find the total reaction to the given dose of the drug.

75. Present Value of Money The present value PV of a capital asset that provides a perpetual stream of revenue that flows continuously at a rate of $R(t)$ dollars per year is given by

$$PV = \int_0^\infty R(t) e^{-rt}\, dt$$

where r, expressed as a decimal, is the annual rate of interest compounded continuously.

    (a) Find the present value of an asset if it provides a constant return of \$100 per year and $r = 8\%$.

    (b) Find the present value of an asset if it provides a return of $R(t) = 1000 + 80t$ dollars per year and $r = 7\%$.

76. Electrical Engineering In a problem in electrical theory, the integral $\int_0^\infty Ri^2\, dt$ occurs, where the current $i = Ie^{-Rt/L}$, t is time, and R, I, and L are positive constants. Find the integral.

77. Magnetic Potential The magnetic potential u at a point on the axis of a circular coil is given by

$$u = \frac{2\pi N I r}{10} \int_x^\infty \frac{dy}{(r^2 + y^2)^{3/2}}$$

where N, I, r, and x are constants. Find the integral.

78. Electrical Engineering The field intensity F around a long ("infinite") straight wire carrying electric current is given by the integral

$$F = \frac{rIm}{10} \int_{-\infty}^\infty \frac{dy}{(r^2 + y^2)^{3/2}}$$

where r, I, and m are constants. Find the integral.

In Problems 79–86, find each improper integral.

79. $\displaystyle\int_0^\infty x e^{-x}\, dx$

80. $\displaystyle\int_0^1 x \ln x\, dx$

81. $\displaystyle\int_0^\infty e^{-x} \cos x\, dx$

82. $\displaystyle\int_0^\infty \tan^{-1} x\, dx$

83. $\displaystyle\int_0^\infty \frac{dx}{(x^2 + 4)^{3/2}}$

84. $\displaystyle\int_4^\infty \frac{dx}{(x^2 - 9)^{3/2}}$

85. $\displaystyle\int_1^\infty \frac{dx}{(x+1)\sqrt{2x + x^2}}$

86. $\displaystyle\int_2^\infty \frac{dx}{(x+2)^2 \sqrt{4x + x^2}}$

87. Show that $\int_0^\infty \sin x\, dx$ and $\int_{-\infty}^0 \sin x\, dx$ each diverge, yet $\lim\limits_{t \to \infty} \int_{-t}^t \sin x\, dx = 0$.

88. Find a function f for which $\int_0^\infty f(x)\, dx$ and $\int_{-\infty}^0 f(x)\, dx$ each diverge, yet $\lim\limits_{t \to \infty} \int_{-t}^t f(x)\, dx = 1$.

89. Use the Comparison Test for Improper Integrals to show that $\displaystyle\int_0^\infty \frac{1}{\sqrt{2 + \sin x}}\, dx$ diverges.

90. Use the Comparison Test for Improper Integrals to show that $\displaystyle\int_2^\infty \frac{\ln x}{\sqrt{x^2 - 1}}\, dx$ diverges.

91. If n is a positive integer, show that:

    (a) $\displaystyle\int_0^\infty x^n e^{-x}\, dx = n \int_0^\infty x^{n-1} e^{-x}\, dx$

    (b) $\displaystyle\int_0^\infty x^n e^{-x}\, dx = n!$

92. Show that $\displaystyle\int_e^\infty \frac{dx}{x(\ln x)^p}$ converges if $p > 1$ and diverges if $p \le 1$.

93. Show that $\displaystyle\int_a^b \frac{dx}{(x - a)^p}$ converges if $0 < p < 1$ and diverges if $p \ge 1$.

94. Show that $\displaystyle\int_a^b \frac{dx}{(b - x)^p}$ converges if $0 < p < 1$ and diverges if $p \ge 1$.

95. Refer to Problems 93 and 94. Discuss the convergence or divergence of the integrals if $p \le 0$. Support your explanation with an example.

96. Comparison Test for Improper Integrals Show that if two functions f and g are nonnegative and continuous on the interval $[a, \infty)$, and if $f(x) \ge g(x)$ for all numbers $x > c$, where $c \ge a$, then

  • If $\int_a^\infty f(x)\, dx$ converges, then $\int_a^\infty g(x)\, dx$ also converges.

  • If $\int_a^\infty g(x)\, dx$ diverges, then $\int_a^\infty f(x)\, dx$ also diverges.

*Laplace transforms are useful in solving a special class of differential equations. The **Laplace transform $L[f(x)]$ of a function f** is defined as*

$$L[f(x)] = \int_0^\infty e^{-sx} f(x)\, dx, \quad x \ge 0, \; s \text{ a complex number}$$

In Problems 97–102, find the Laplace transform of each function.

97. $f(x) = x$

98. $f(x) = \cos x$

99. $f(x) = \sin x$

100. $f(x) = e^x$

101. $f(x) = e^{ax}$

102. $f(x) = 1$

Challenge Problems

103. Find $\displaystyle\int_{-\infty}^\infty e^{(x - e^x)}\, dx$.

104. Find $\displaystyle\int_{-\infty}^a e^{(x - e^x)}\, dx$.

*In Problems 105 and 106, use the following definition: A **probability density function** is a function f, whose domain is the set of all real numbers, with the following properties:*

- $f(x) \geq 0$ for all x

- $\int_{-\infty}^{\infty} f(x)\, dx = 1$

105. Uniform Density Function   Show that the function f below is a probability density function.

$$f(x) = \begin{cases} 0 & \text{if} \quad x < a \\ \dfrac{1}{b-a} & \text{if} \quad a \leq x \leq b, a < b \\ 0 & \text{if} \quad x > b \end{cases}$$

106. Exponential Density Function   Show that the function f is a probability density function for $a > 0$.

$$f(x) = \begin{cases} \dfrac{1}{a} e^{-x/a} & \text{if} \quad x \geq 0 \\ 0 & \text{if} \quad x < 0 \end{cases}$$

*In Problems 107 and 108, use the following definition: The **expected value** or **mean** μ associated with a probability density function f is defined by $\mu = \int_{-\infty}^{\infty} x f(x)\, dx$. The expected value can be thought of as a weighted average of its various probabilities.*

107. Find the expected value μ of the uniform density function f given in Problem 105.

108. Find the expected value μ of the exponential probability density function f defined in Problem 106.

*In Problems 109 and 110, use the following definitions: The **variance** σ^2 of a probability density function f is defined as $\sigma^2 = \int_{-\infty}^{\infty} (x - \mu)^2 f(x)\, dx$. The variance is the average of the squared deviation from the mean. The **standard deviation** σ of a probability density function f is the square root of its variance σ^2.*

109. Find the variance σ^2 and standard deviation σ of the uniform density function defined in Problem 105.

110. Find the variance σ^2 and standard deviation σ of the exponential density function defined in Problem 106.

111. For what numbers a does $\int_0^1 x^a\, dx$ converge?

Preparing for the AP® Exam

AP® Practice Problems

NOTE: *These problems are BC only.*

Multiple-Choice Questions

PAGE 544

1. $\displaystyle\int_0^2 \dfrac{x+2}{x^2 + 4x - 12}\, dx =$

(A) $-\dfrac{\ln 12}{2}$ (B) $\dfrac{1 - \ln 12}{2}$

(C) $\dfrac{\ln 12 - \ln 2}{2}$ (D) diverges

PAGE 540

2. Determine whether $\displaystyle\int_1^{\infty} \dfrac{2}{x^3}\, dx$ converges or diverges. If it converges, find its value.

(A) $\dfrac{1}{2}$ (B) 1   (C) 2   (D) diverges

PAGE 540

3. Determine whether $\displaystyle\int_3^{\infty} \dfrac{8x}{\sqrt[3]{8 - x^2}}\, dx$ converges or diverges. If it converges, find its value.

(A) 0   (B) 18   (C) $6(9^{2/3})$ (D) diverges

PAGE 540

4. Determine whether $\int_{-\infty}^0 x e^{x^2}\, dx$ converges or diverges. If it converges, find its value.

(A) $\dfrac{1}{2}$ (B) 1   (C) $-e$ (D) diverges

PAGE 540

5. Determine whether $\displaystyle\int_2^{\infty} \dfrac{1}{x(\ln x)^2}\, dx$ converges or diverges. If it converges, find its value.

(A) $\dfrac{1}{\ln 2}$ (B) $\dfrac{1}{2}$ (C) $-\dfrac{1}{\ln 2}$ (D) diverges

PAGE 544

6. Determine whether $\displaystyle\int_0^1 \dfrac{1}{\sqrt[4]{x}}\, dx$ converges or diverges. If it converges, find its value.

(A) $\dfrac{3}{4}$ (B) 4   (C) $\dfrac{4}{3}$ (D) diverges

Free-Response Question

PAGE 542

7. When the region bounded by the graph of $y = \dfrac{1}{x^2}$ and the x-axis to the right of the line $x = 1$ is revolved about the x-axis, the volume V, if it is defined, of the solid of revolution that is generated is given by the improper integral

$$\int_1^{\infty} \pi \left(\dfrac{1}{x^2}\right)^2 dx.$$

Determine whether the improper integral converges or diverges. If it converges, find the volume of the solid of revolution.

*See the **BREAK IT DOWN** on page 575 for a stepped out solution to AP® Practice Problem 7.*

Retain Your Knowledge

Multiple-Choice Questions

1. Find all the antiderivatives of $\dfrac{dy}{dx} = e^{3x} + \dfrac{1}{2}\sec x \tan x$

(A) $y = \dfrac{e^{3x}}{3} + \dfrac{1}{2}\tan x + C$ (B) $y = \dfrac{e^{3x}}{3} + \dfrac{1}{2}\sec x + C$

(C) $y = e^{3x} - \dfrac{1}{2}\sec x + C$ (D) $y = 3e^{3x} - \dfrac{1}{2}\tan x + C$

2. The derivative of a function f is $f'(x) = 2x^3 + 4$. If $f(1) = 5$, find $f(2)$.

(A) 24.5   (B) 21   (C) 16.5   (D) 11.5

3. Suppose f is a differentiable, one-to-one function. Several values of f and its derivative f' are given in the table below.

x	$f(x)$	$f'(x)$
4	-2	1
7	4	-3

If g is the inverse function f, then $g'(4) =$

(A) $\dfrac{1}{4}$ (B) $\dfrac{1}{7}$ (C) 1   (D) $-\dfrac{1}{3}$

Free-Response Question

4. A function f is twice differentiable on the interval $-6 < x < 5$. The graph of its derivative $y = f'(x)$ is given below. From the graph, we can read that f' has x-intercepts at -4, 0, and 4, and that the tangent lines to the graph of f' are horizontal at -4, -2, and 2.

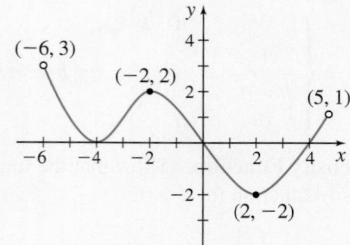

(a) Find the critical numbers of f and determine if each locates a local maximum, local minimum, or neither. Justify your answers.

(b) Find any intervals where the graph of f is both decreasing and concave down. Justify your answer.

(c) At what numbers x does the graph of f have an inflection point? Justify your answers.

6.10 Integrals Containing Trigonometric Functions

OBJECTIVES *When you finish this section, you should be able to:*

1 Find integrals of the form $\int \sin^n x\, dx$ or $\int \cos^n x\, dx$, $n \geq 2$ an integer (p. 550)

2 Find integrals of the form $\int \sin^m x \, \cos^n x\, dx$ (p. 553)

3 Find integrals of the form $\int \tan^m x \, \sec^n x\, dx$ or $\int \cot^m x \, \csc^n x\, dx$ (p. 554)

4 Find integrals of the form $\int \sin(ax) \sin(bx)\, dx$, $\int \sin(ax) \cos(bx)\, dx$, or $\int \cos(ax) \cos(bx)\, dx$ (p. 557)

In this section, we develop techniques to find certain trigonometric integrals. When studying these techniques, concentrate on the strategies used in the examples rather than trying to memorize the results.

① Find Integrals of the Form $\int \sin^n x\, dx$ or $\int \cos^n x\, dx$, $n \geq 2$ an Integer

Although we could use integration by parts to obtain reduction formulas for integrals of the form $\int \sin^n x\, dx$ or $\int \cos^n x\, dx$, $n \geq 2$ an integer, these integrals can be found more easily using trigonometric identities. We consider two cases:

- $n \geq 3$ an odd integer
- $n \geq 2$ an even integer.

$n \geq 3$ **an odd integer** Suppose we want to find $\int \sin^n x\, dx$ or $\int \cos^n x\, dx$ when $n \geq 3$ is an odd integer. We begin by factoring and writing each integral in the form

$$\bullet \quad \int \sin^n x\, dx = \int \sin^{n-1} x \sin x\, dx \qquad \bullet \quad \int \cos^n x\, dx = \int \cos^{n-1} x \cos x\, dx$$

Since n is odd, $(n-1)$ is even. Then use the Pythagorean identity $\sin^2 x + \cos^2 x = 1$.
- For $\int \sin^n x\, dx$, n odd, the substitution $u = \cos x$, $du = -\sin x\, dx$, leads to an integral involving integer powers of u.
- For $\int \cos^n x\, dx$, n odd, the substitution $u = \sin x$, $du = \cos x\, dx$ also leads to an integral involving integer powers of u.

CALC CLIP

| **EXAMPLE 1** | Finding the Integral $\displaystyle\int \sin^5 x\, dx$ |

Find $\displaystyle\int \sin^5 x\, dx$.

Solution

Since the exponent 5 is odd, write $\int \sin^5 x\, dx = \int \sin^4 x \sin x\, dx$, and use the identity $\sin^2 x = 1 - \cos^2 x$.

$$\int \sin^5 x\, dx = \int \sin^4 x \sin x\, dx = \int (\sin^2 x)^2 \sin x\, dx = \int (1 - \cos^2 x)^2 \sin x\, dx$$

$$= \int (1 - 2\cos^2 x + \cos^4 x) \sin x\, dx$$

Now use the substitution $u = \cos x$. Then $du = -\sin x\, dx$, and

$$\int \sin^5 x\, dx = -\int (1 - 2u^2 + u^4)\, du = -u + \frac{2}{3}u^3 - \frac{1}{5}u^5 + C$$

$$= -\cos x + \frac{2}{3}\cos^3 x - \frac{1}{5}\cos^5 x + C \qquad \blacksquare$$

> **NEED TO REVIEW?** The method of substitution is discussed in Section 6.5, pp. 484–490.

| **EXAMPLE 2** | Finding the Integral $\displaystyle\int \cos^3 x\, dx$ |

Find $\displaystyle\int \cos^3 x\, dx$.

Solution

Since the exponent of $\cos x$ is an odd integer, factor and use the Pythagorean identity $\cos^2 x = 1 - \sin^2 x$.

$$\int \cos^3 x\, dx = \int \cos^2 x \cos x\, dx = \int (1 - \sin^2 x) \cos x\, dx$$

$$= \int (1 - u^2)\, du = u - \frac{u^3}{3} + C = \sin x - \frac{\sin^3 x}{3} + C$$

$$\underset{\substack{\uparrow \\ u = \sin x \\ du = \cos x\, dx}}{}$$

$\blacksquare$

NOW WORK Problem 3.

$n \geq 2$ **an even integer** To find $\int \sin^n x\, dx$ or $\int \cos^n x\, dx$ when $n \geq 2$ is an even integer, the preceding strategy does not work. (Try it for yourself.) Instead, we use one of the identities below:

> **NEED TO REVIEW?** Trigonometric identities are discussed in Appendix A.4, pp. A-35 to A-38.

$$\bullet \ \sin^2 x = \frac{1 - \cos(2x)}{2} \qquad \bullet \ \cos^2 x = \frac{1 + \cos(2x)}{2}$$

to obtain a simpler integrand.

| **EXAMPLE 3** | Finding the Integral $\displaystyle\int \sin^2 x\, dx$ |

Find $\displaystyle\int \sin^2 x\, dx$.

Solution

Since the exponent of $\sin x$ is an even integer, we use the identity

$$\sin^2 x = \frac{1 - \cos(2x)}{2}$$

Then

$$\int \sin^2 x \, dx = \frac{1}{2} \int [1 - \cos(2x)] \, dx = \frac{1}{2} \int dx - \frac{1}{2} \int \cos(2x) \, dx$$

$$= \frac{1}{2}x + C_1 - \frac{1}{2} \int \cos u \frac{du}{2} \qquad u = 2x, \, du = 2\,dx$$

$$= \frac{1}{2}x + C_1 - \frac{1}{4} \sin(2x) + C_2$$

Since C_1 and C_2 are constants, we write the solution as

$$\int \sin^2 x \, dx = \frac{1}{2}x - \frac{1}{4} \sin(2x) + C$$

where $C = C_1 + C_2$. ∎

NOTE Usually we will add the constant of integration at the end to avoid letting $C = C_1 + C_2$.

NOW WORK Problem 5.

EXAMPLE 4 Finding the Average Value of a Function

Find the average value $\bar{y}$ of the function $f(x) = \cos^4 x$ over the closed interval $[0, \pi]$.

Solution

NEED TO REVIEW? The average value of a function is discussed in Section 6.4, pp. 472–473.

The average value $\bar{y}$ of a function f over $[a, b]$ is $\bar{y} = \frac{1}{b-a} \int_a^b f(x)\,dx$.

For $f(x) = \cos^4 x$ on $[0, \pi]$, we have

$$\bar{y} = \frac{1}{\pi - 0} \int_0^\pi \cos^4 x \, dx = \frac{1}{\pi} \int_0^\pi (\cos^2 x)^2 \, dx = \frac{1}{4\pi} \int_0^\pi [1 + \cos(2x)]^2 \, dx$$

$$\cos^4 x = (\cos^2 x)^2 \qquad \cos^2 x = \frac{1 + \cos(2x)}{2}$$

$$= \frac{1}{4\pi} \int_0^\pi \left[1 + 2\cos(2x) + \cos^2(2x)\right] dx$$

$$= \frac{1}{4\pi} \left[\int_0^\pi dx + 2 \int_0^\pi \cos(2x)\,dx + \int_0^\pi \cos^2(2x)\,dx \right] \qquad (1)$$

Now

$$\int_0^\pi dx = \pi - 0 = \pi \quad \text{and} \quad \int_0^\pi \cos(2x)\,dx = \int_0^{2\pi} \cos u \frac{du}{2} = \left[\frac{1}{2}\sin u\right]_0^{2\pi} = 0$$

$$u = 2x \qquad du = 2\,dx$$

To find $\int_0^\pi \cos^2(2x)\,dx$, we use the identity $\cos^2\theta = \frac{1 + \cos(2\theta)}{2}$ again to write $\cos^2(2x) = \frac{1 + \cos(4x)}{2}$. Then

$$\int_0^\pi \cos^2(2x)\,dx = \int_0^\pi \frac{1 + \cos(4x)}{2}\,dx = \frac{1}{2}\left[\int_0^\pi dx + \int_0^\pi \cos(4x)\,dx\right]$$

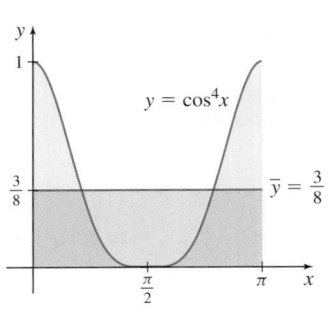

Figure 9 $\bar{y} \cdot \pi = \displaystyle\int_0^\pi \cos^4 x \, dx$

$$= \frac{1}{2}\left[\pi + \int_0^{4\pi} \cos u \, \frac{du}{4}\right] \qquad u = 4x; \, du = 4\,dx$$

$$= \frac{\pi}{2} + \frac{1}{8}[\sin u]_0^{4\pi} = \frac{\pi}{2}$$

So, from (1),

$$\bar{y} = \frac{1}{4\pi}\left[\pi + 0 + \frac{\pi}{2}\right] = \frac{3}{8} \qquad\blacksquare$$

If f is nonnegative on an interval $[a, b]$, the average value of f over the interval $[a, b]$ represents the height of a rectangle with width $b - a$ whose area equals the area under the graph of f from a to b. Figure 9 shows the graph of f from Example 4 and the rectangle of height $\bar{y} = \dfrac{3}{8}$ and width π whose area is equal to the area under the graph of f.

NOW WORK   Problem 59.

Summary   Integrals of the Form $\int \sin^n x \, dx$ or $\int \cos^n x \, dx$, $n \geq 2$ an Integer

	Integral	Use the Identity	u-Substitution
n odd	$\int \sin^n x \, dx = \int \sin^{n-1} x \sin x \, dx$	$\sin^2 x = 1 - \cos^2 x$	$u = \cos x; \quad du = -\sin x \, dx$
	$\int \cos^n x \, dx = \int \cos^{n-1} x \cos x \, dx$	$\cos^2 x = 1 - \sin^2 x$	$u = \sin x; \quad du = \cos x \, dx$
n even	$\int \sin^n x \, dx = \int \left[\dfrac{1 - \cos(2x)}{2}\right]^{n/2} dx$	$\sin^2 x = \dfrac{1 - \cos(2x)}{2}$	$u = 2x; \quad du = 2\,dx$
	$\int \cos^n x \, dx = \int \left[\dfrac{1 + \cos(2x)}{2}\right]^{n/2} dx$	$\cos^2 x = \dfrac{1 + \cos(2x)}{2}$	$u = 2x; \quad du = 2\,dx$

---

2 **Find Integrals of the Form** $\displaystyle\int \sin^m x \cos^n x \, dx$

Integrals of the form $\int \sin^m x \cos^n x \, dx$ are found using variations of previous techniques. We discuss two cases:

- At least one of the exponents m or n is a positive odd integer.
- Both exponents are positive even integers.

If at least one of the exponents is odd, the goal is to write the integral with only one sine factor and the other factors involving the cosine function, or with only one cosine factor and the other factors involving the sine function.

EXAMPLE 5   **Finding** $\displaystyle\int \sin^m x \cos^n x \, dx$, **One of the Exponents Is Odd**

Find $\displaystyle\int \sin^5 x \sqrt{\cos x} \, dx = \int \sin^5 x \, \cos^{1/2} x \, dx$.

Solution
The exponent of $\sin x$ is 5, a positive, odd integer. We factor $\sin x$ from $\sin^5 x$ and write the rest of the integrand in terms of cosines.

$$\int \sin^5 x \cos^{1/2} x \, dx = \int \sin^4 x \, \cos^{1/2} x \sin x \, dx = \int (\sin^2 x)^2 \cos^{1/2} x \sin x \, dx$$

$$= \int (1 - \cos^2 x)^2 \cos^{1/2} x \sin x \, dx$$

(continued on the next page)

Now use the substitution $u = \cos x$.

$$\int \sin^5 x \, \cos^{1/2} x \, dx = \int (1 - \cos^2 x)^2 \, \cos^{1/2} x \, \sin x \, dx = \int (1 - u^2)^2 u^{1/2} (-du)$$

$$\uparrow$$
$$u = \cos x$$
$$du = -\sin x \, dx$$

$$= -\int (1 - 2u^2 + u^4) u^{1/2} \, du$$

$$= -\int (u^{1/2} - 2u^{5/2} + u^{9/2}) \, du = -\frac{2}{3} u^{3/2} + \frac{4}{7} u^{7/2} - \frac{2}{11} u^{11/2} + C$$

$$= u^{3/2} \left[-\frac{2}{3} + \frac{4}{7} u^2 - \frac{2}{11} u^4 \right] + C$$

$$= (\cos x)^{3/2} \left[-\frac{2}{3} + \frac{4}{7} \cos^2 x - \frac{2}{11} \cos^4 x \right] + C \qquad \blacksquare$$

$$\uparrow$$
$$u = \cos x$$

NOW WORK   Problem 11.

If both exponents m and n are positive even integers, then to find $\int \sin^m x \cos^n x \, dx$, use the Pythagorean identity $\sin^2 x + \cos^2 x = 1$ to obtain a sum of integrals, each integral involving only even powers of either $\sin x$ or $\cos x$. For example,

$$\int \sin^2 x \cos^4 x \, dx = \int (1 - \cos^2 x) \cos^4 x \, dx = \int \cos^4 x \, dx - \int \cos^6 x \, dx$$

The two integrals on the right are now of the form $\int \cos^n x \, dx$, n a positive even integer, and we can integrate them using the techniques discussed in Examples 3 and 4.

Summary   Integrals of the Form $\int \sin^m x \cos^n \, dx$

	Integral	Use the Identity	u-Substitution
m odd	$\int \sin^m x \cos^n x \, dx = \int \sin^{m-1} x \cos^n x \sin x \, dx$	$\sin^2 x = 1 - \cos^2 x$	$u = \cos x$ $du = -\sin x \, dx$
n odd	$\int \sin^m x \cos^n x \, dx = \int \sin^m x \cos^{n-1} x \cos x \, dx$	$\cos^2 x = 1 - \sin^2 x$	$u = \sin x$ $du = \cos x \, dx$
m and n both even	$\int \sin^m x \cos^n x \, dx = \int (1 - \cos^2 x)^{m/2} \cos^n x \, dx$ or $\int \sin^m x \cos^n x \, dx = \int \sin^m x (1 - \sin^2 x)^{n/2} \, dx$	$\cos^2 x = \dfrac{1 + \cos(2x)}{2}$ $\sin^2 x = \dfrac{1 - \cos(2x)}{2}$	$u = 2x; \quad du = 2 \, dx$ $u = 2x; \quad du = 2 \, dx$

---

❸ Find Integrals of the Form $\int \tan^m x \, \sec^n x \, dx$ or $\int \cot^m x \, \csc^n x \, dx$

We consider three cases involving integrals of the form $\int \tan^m x \, \sec^n x \, dx$:

- The tangent function is raised to a positive odd integer.
- The secant function is raised to a positive even integer.
- The tangent function is raised to a positive even integer m and the secant function is raised to a positive odd integer n.

The idea is to express the integrand so that we can use either the substitution $u = \tan x$ and $du = \sec^2 x \, dx$ or the substitution $u = \sec x$ and $du = \sec x \tan x \, dx$, while leaving an even power of one of the functions. Then we use a Pythagorean identity to express the integrand in terms of only one trigonometric function.

EXAMPLE 6 Finding the Integral $\int \tan^3 x \sec^4 x\, dx$

Find $\int \tan^3 x \sec^4 x\, dx$.

Solution

Here $\tan x$ is raised to the odd power 3. We factor $\tan x$ from $\tan^3 x$ and use the identity $\tan^2 x = \sec^2 x - 1$.

$$\int \tan^3 x \sec^4 x\, dx = \int \tan^2 x \, \tan x \, \sec^4 x\, dx \qquad \text{Factor } \tan x \text{ from } \tan^3 x.$$

$$= \int (\sec^2 x - 1) \tan x \sec^4 x\, dx \qquad \tan^2 x = \sec^2 x - 1$$

$$= \int (\sec^2 x - 1) \sec^3 x \, \sec x \tan x\, dx \qquad \text{Factor } \sec x \text{ from } \sec^4 x.$$

$$= \int (u^2 - 1) u^3 du \qquad \begin{array}{l}\text{Substitute } u = \sec x;\\ du = \sec x \tan x\, dx.\end{array}$$

$$= \int (u^5 - u^3)\, du = \frac{u^6}{6} - \frac{u^4}{4} + C$$

$$= \frac{\sec^6 x}{6} - \frac{\sec^4 x}{4} + C \qquad u = \sec x \qquad \blacksquare$$

NOW WORK Problem 19.

EXAMPLE 7 Finding the Integral $\int \tan^2 x \sec^6 x\, dx$

Find $\int \tan^2 x \sec^6 x\, dx$.

Solution

Here $\sec x$ is raised to a positive even power. We factor $\sec^2 x$ from $\sec^6 x$ and use the identity $\sec^2 x = 1 + \tan^2 x$. Then

$$\int \tan^2 x \sec^6 x\, dx = \int \tan^2 x \sec^4 x \cdot \sec^2 x\, dx \qquad \text{Factor } \sec^2 x \text{ from } \sec^6 x.$$

$$= \int \tan^2 x (1 + \tan^2 x)^2 \sec^2 x\, dx \qquad \sec^2 x = 1 + \tan^2 x$$

$$= \int u^2 (1 + u^2)^2\, du \qquad \begin{array}{l}\text{Substitute } u = \tan x;\\ du = \sec^2 x\, dx.\end{array}$$

$$= \int (u^2 + 2u^4 + u^6)\, du = \frac{u^3}{3} + \frac{2u^5}{5} + \frac{u^7}{7} + C$$

$$= \frac{\tan^3 x}{3} + 2\frac{\tan^5 x}{5} + \frac{\tan^7 x}{7} + C \qquad u = \tan x \qquad \blacksquare$$

NOW WORK Problem 21.

When the tangent function is raised to a positive even integer m and the secant function is raised to a positive odd integer n, the approach is slightly different. Rather than factoring, we begin by using the identity $\tan^2 x = \sec^2 x - 1$.

EXAMPLE 8   Finding $\int \tan^m x \sec^n x\, dx$ Where m Is Even and n Is Odd

Find $\int \tan^2 x \sec x\, dx$.

Solution

Here $\tan x$ is raised to an even power and $\sec x$ to an odd power. We use the identity $\tan^2 x = \sec^2 x - 1$ to write

$$\int \tan^2 x \sec x\, dx = \int (\sec^2 x - 1)\sec x\, dx = \int (\sec^3 x - \sec x)dx$$

$$= \int \sec^3 x\, dx - \int \sec x\, dx \qquad (2)$$

Next we integrate $\int \sec^3 x\, dx$ by parts. Choose

$$u = \sec x \qquad\qquad \text{and} \qquad dv = \sec^2 x\, dx$$
$$du = \sec x \tan x\, dx \qquad\qquad v = \int \sec^2 x\, dx = \tan x$$

Then

$$\int \sec^3 x\, dx = \sec x\, \tan x - \int \tan^2 x\, \sec x\, dx \qquad\qquad \int u\, dv = uv - \int v\, du$$

$$= \sec x\, \tan x - \int (\sec^2 x - 1)\sec x\, dx \qquad\qquad \tan^2 x = \sec^2 x - 1$$

$$= \sec x\, \tan x - \int \sec^3 x\, dx + \int \sec x\, dx \qquad\text{Write the integral as the sum of two integrals.}$$

$$2\int \sec^3 x\, dx = \sec x\, \tan x + \int \sec x\, dx \qquad\qquad \text{Add } \int \sec^3 x\, dx \text{ to both sides.}$$

$$\int \sec^3 x\, dx = \frac{1}{2}[\sec x\, \tan x + \ln|\sec x + \tan x|] \qquad \text{Solve for } \int \sec^3 x\, dx;\ \int \sec x\, dx = \ln|\sec x + \tan x|.$$

NOTE Substituting $n = 3$ into the reduction formula for $\int \sec^n x\, dx$ (derived in Example 10 of Section 6.7) could also have been used to find $\int \sec^3 x\, dx$.

Now substitute this result in (2) and integrate $\int \sec x\, dx$.

$$\int \tan^2 x \sec x\, dx = \int \sec^3 x\, dx - \int \sec x\, dx$$

$$= \frac{1}{2}[\sec x\, \tan x + \ln|\sec x + \tan x|] - \ln|\sec x + \tan x| + C$$

$$= \frac{1}{2}[\sec x \tan x - \ln|\sec x + \tan x|] + C \qquad\blacksquare$$

NOW WORK   Problem 51.

To find integrals of the form $\int \cot^m x \csc^n x \, dx$, use the same strategies, but with the identity $\csc^2 x = 1 + \cot^2 x$.

Summary   Integrals of the Form $\int \tan^m x \sec^n x \, dx$

	The Integral	Use the Identity	Method
m odd	$\int \tan^m x \sec^n x \, dx = \int \tan^{m-1} x \sec^{n-1} x \sec x \tan x \, dx$	$\tan^2 x = \sec^2 x - 1$	$u = \sec x$ $du = \sec x \tan x \, dx$
n even	$\int \tan^m x \sec^n x \, dx = \int \tan^m x \sec^{n-2} x \sec^2 x \, dx$	$\sec^2 x = 1 + \tan^2 x$	$u = \tan x$ $du = \sec^2 x \, dx$
m even; n odd	$\int \tan^m x \sec^n x \, dx = \int (\sec^2 x - 1)^{m/2} \sec^n x \, dx$	$\tan^2 x = \sec^2 x - 1$	Use integration by parts

4 Find Integrals of the Form $\displaystyle\int \sin(ax)\sin(bx)\,dx,$

$\displaystyle\int \sin(ax)\cos(bx)\,dx,$ **or** $\displaystyle\int \cos(ax)\cos(bx)\,dx$

Trigonometric integrals of the form

$$\int \sin(ax)\sin(bx)\,dx \qquad \int \sin(ax)\cos(bx)\,dx \qquad \int \cos(ax)\cos(bx)\,dx$$

are integrated using the product-to-sum identities:

- $2 \sin A \sin B = \cos(A - B) - \cos(A + B)$
- $2 \sin A \cos B = \sin(A + B) + \sin(A - B)$
- $2 \cos A \cos B = \cos(A - B) + \cos(A + B)$

These identities transform the integrand into a sum of sines and/or cosines.

EXAMPLE 9   **Finding the Integral** $\displaystyle\int \sin(3x)\sin(2x)\,dx$

Find $\displaystyle\int \sin(3x)\sin(2x)\,dx$.

Solution

Use the product-to-sum identity $2 \sin A \sin B = \cos(A - B) - \cos(A + B)$. Then

$$2 \sin(3x)\sin(2x) = \cos(3x - 2x) - \cos(3x + 2x) \qquad A = 3x,\ B = 2x$$

$$\sin(3x)\sin(2x) = \frac{1}{2}[\cos x - \cos(5x)]$$

Then

$$\int \sin(3x)\sin(2x)\,dx = \frac{1}{2}\int [\cos x - \cos(5x)]\,dx = \frac{1}{2}\int \cos x \, dx - \frac{1}{2}\int \cos(5x)\,dx$$

$$\underset{\substack{\uparrow \\ u = 5x \\ du = 5\,dx}}{=}\ \frac{1}{2}\sin x - \frac{1}{2}\int \cos u \, \frac{du}{5} = \frac{1}{2}\sin x - \frac{1}{10}\sin(5x) + C$$

∎

NOW WORK   Problem 27.

6.10 Assess Your Understanding

Concepts and Vocabulary

1. **True or False** To find $\int \cos^5 x\, dx$, factor out $\cos x$ and use the identity $\cos^2 x = 1 - \sin^2 x$.

2. **True or False** To find $\int \sin(2x)\cos(3x)\, dx$, use a product-to-sum identity.

Skill Building

In Problems 3–10, find each integral.

[PAGE 551] 3. $\int \cos^5 x\, dx$

4. $\int \sin^3 x\, dx$

[PAGE 552] 5. $\int \sin^6 x\, dx$

6. $\int \cos^4 x\, dx$

7. $\int \sin^2(\pi x)\, dx$

8. $\int \cos^4(2x)\, dx$

9. $\int_0^\pi \cos^5 x\, dx$

10. $\int_{-\pi/3}^{\pi/3} \sin^3 x\, dx$

In Problems 11–18, find each integral.

[PAGE 554] 11. $\int \sin^3 x \cos^2 x\, dx$

12. $\int \sin^4 x \cos^3 x\, dx$

13. $\int_0^{\pi/2} \sin^3 x(\cos x)^{3/2}\, dx$

14. $\int_0^{\pi/2} \cos^3 x \sqrt{\sin x}\, dx$

15. $\int \sin^3 x \cos^{1/3} x\, dx$

16. $\int \cos^3 x \sin^{1/2} x\, dx$

17. $\int \sin^2\left(\frac{x}{2}\right) \cos^3\left(\frac{x}{2}\right) dx$

18. $\int \sin^3(4x) \cos^3(4x)\, dx$

In Problems 19–26, find each integral.

[PAGE 555] 19. $\int \tan^3 x \sec^3 x\, dx$

20. $\int \tan x \sec^5 x\, dx$

[PAGE 555] 21. $\int \tan^{3/2} x \sec^4 x\, dx$

22. $\int \tan^5 x \sec x\, dx$

23. $\int \tan^3 x(\sec x)^{3/2}\, dx$

24. $\int \sec^4 x \sqrt{\tan x}\, dx$

25. $\int \cot^3 x \csc x\, dx$

26. $\int \cot^3 x \csc^2 x\, dx$

In Problems 27–34, find each integral.

[PAGE 557] 27. $\int \sin(3x) \cos x\, dx$

28. $\int \sin x \cos(3x)\, dx$

29. $\int \cos x \cos(3x)\, dx$

30. $\int \cos(2x) \cos x\, dx$

31. $\int \sin(2x) \sin(4x)\, dx$

32. $\int \sin(3x) \sin x\, dx$

33. $\int_0^{\pi/2} \sin(2x) \sin x\, dx$

34. $\int_0^\pi \cos x \cos(4x)\, dx$

In Problems 35–56, find each integral.

35. $\int_0^{\pi/2} \sin^2 x \cos^5 x\, dx$

36. $\int_0^{\pi/2} \sin^3 x \cos^{1/2} x\, dx$

37. $\int \dfrac{\sin^3 x\, dx}{\cos^2 x}$

38. $\int \dfrac{\cos x\, dx}{\sin^4 x}$

39. $\int_0^{\pi/3} \cos^3(3x)\, dx$

40. $\int_0^{\pi/3} \sin^5(3x)\, dx$

41. $\int_0^\pi \sin^3 x \cos^5 x\, dx$

42. $\int_0^{\pi/2} \sin^3 x \cos^3 x\, dx$

43. $\int \tan^3 x\, dx$

44. $\int \cot^5 x\, dx$

45. $\int \dfrac{\sec^6 x}{\tan^3 x}\, dx$

46. $\int \tan^{1/2} x \sec^2 x\, dx$

47. $\int \csc^4 x \cot^3 x\, dx$

48. $\int \cot^3 x \csc^2 x\, dx$

49. $\int \cot(2x) \csc^4(2x)\, dx$

50. $\int \cot^2(2x) \csc^3(2x)\, dx$

[PAGE 556] 51. $\int_0^{\pi/4} \tan^4 x \sec^3 x\, dx$

52. $\int_0^{\pi/4} \tan^2 x \sec x\, dx$

53. $\int_0^{\pi/2} \sin\dfrac{x}{2} \cos\dfrac{3x}{2}\, dx$

54. $\int_0^{\pi/4} \cos(-x) \sin(4x)\, dx$

55. $\int \sin\dfrac{x}{2} \sin\dfrac{3x}{2}\, dx$

56. $\int \cos(\pi x) \cos(3\pi x)\, dx$

Applications and Extensions

57. **Volume of a Solid of Revolution**   The volume V of the solid of revolution generated by revolving the region bounded by the graph of $y = \sin x$ and the x-axis from $x = 0$ to $x = \pi$ about the x-axis is given by $V = \pi \int_0^\pi \sin^2 x\, dx$. Find the volume of the solid.

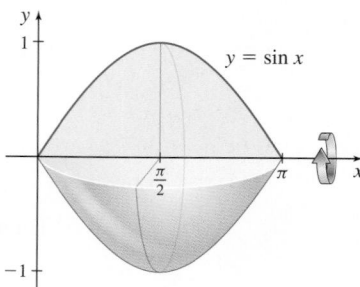

58. **Volume of a Solid of Revolution**
The volume V of the solid of revolution generated by revolving the region bounded by the graphs of $y = \cos x$, $y = \sin x$, and $x = 0$ from $x = 0$ to $x = \dfrac{\pi}{4}$ about the x-axis is given by

$$V = \pi \int_0^{\pi/4} [\cos^2 x - \sin^2 x]\, dx.$$

Find the volume of the solid.

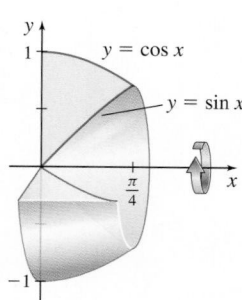

59. Average Value

(a) Find the average value of $f(x) = \sin^3 x$ over the interval $\left[0, \dfrac{\pi}{2}\right]$.

(b) Give a geometric interpretation to the average value.

(c) Use technology to graph f and the average value on the same screen.

60. Average Value

(a) Find the average value of $f(x) = \sin(4x)\cos(2x)$ over the interval $\left[0, \dfrac{\pi}{2}\right]$.

(b) Give a geometric interpretation to the average value.

(c) Use technology to graph f and the average value on the same screen.

61. Motion on a Line   An object is moving along a horizontal line with velocity $v(t) = \sin^2 t \cos^2 t$ (in meters per second) for time t, $0 \le t \le 2\pi$ seconds. Find the net displacement of the object from the origin from $t = 0$ to $t = 2\pi$.

62. Motion on a Line   The acceleration $a = a(t)$ of an object at time t is given by $a(t) = \cos^3 t \sin^2 t$ m/s². At $t = 0$, its velocity is 5 m/s. Find its velocity at any time t.

63. Find $\int \sin^4 x \, dx$.

(a) Using the methods of this section.

(b) Using the reduction formula given in Problem 69 in Section 6.7.

(c) Verify that both results are equivalent.

(d) Use a CAS to find $\int \sin^4 x \, dx$.

64. (a) Use technology to graph the function $f(x) = \sin^n x$, $0 \le x \le \pi$, for $n = 5$, $n = 10$, $n = 20$, and $n = 50$.

(b) Find $\int_0^\pi \sin^n x \, dx$ correct to three decimal places for $n = 5$, $n = 10$, $n = 20$, and $n = 50$.

(c) What does (a) suggest about the shape of the graph of $f(x) = \sin^n x$, $0 \le x \le \pi$, as $n \to \infty$?

(d) Find $\displaystyle\lim_{n\to\infty} \int_0^\pi \sin^n x \, dx$.

65. Derive a formula for $\int \sin(mx)\sin(nx)\,dx$, $m \ne n$.

66. Derive a formula for $\int \sin(mx)\cos(nx)\,dx$, $m \ne n$.

67. Derive a formula for $\int \cos(mx)\cos(nx)\,dx$, $m \ne n$.

Challenge Problems —————————————

68. Use an appropriate substitution to show that

$$\int_0^{\pi/2} \sin^n \theta \, d\theta = \int_0^{\pi/2} \cos^n \theta \, d\theta$$

69. Use the substitution $\sqrt{x} = \sin y$ to find $\displaystyle\int_0^{1/2} \dfrac{\sqrt{x}}{\sqrt{1-x}} \, dx$.

70. (a) What is wrong with the following?

$$\int_0^\pi \cos^4 x \, dx = \int_0^\pi (\cos x)^3 \cos x \, dx = \int_0^\pi (\cos^2 x)^{3/2} \cos x \, dx$$

$$= \int_0^\pi (1 - \sin^2 x)^{3/2} \cos x \, dx = \int_0^0 (1 - u^2)^{3/2} du = 0$$

(b) Find $\int_0^\pi \cos^4 x \, dx$.

6.11 Integration Using Trigonometric Substitution: Integrands Containing $\sqrt{a^2 - x^2}$, $\sqrt{x^2 + a^2}$, or $\sqrt{x^2 - a^2}$, $a > 0$

OBJECTIVES *When you finish this section, you should be able to:*

1 Integrate a function containing $\sqrt{a^2 - x^2}$ (p. 560)

2 Integrate a function containing $\sqrt{x^2 + a^2}$ (p. 561)

3 Integrate a function containing $\sqrt{x^2 - a^2}$ (p. 563)

4 Use trigonometric substitution to find definite integrals (p. 564)

When an integrand contains a square root of the form $\sqrt{a^2 - x^2}$, $\sqrt{x^2 + a^2}$, or $\sqrt{x^2 - a^2}$, $a > 0$, an appropriate trigonometric substitution eliminates the square root and sometimes transforms the integral into a trigonometric integral like those studied earlier.

The substitutions for each of the three types of square roots are given in Table 2.

TABLE 2 Trigonometric Substitution

Integrand Contains	Substitution	Based on the Identity
$\sqrt{a^2 - x^2}$	$x = a \sin\theta,\ -\dfrac{\pi}{2} \le \theta \le \dfrac{\pi}{2}$	$1 - \sin^2\theta = \cos^2\theta$
$\sqrt{x^2 + a^2}$	$x = a \tan\theta,\ -\dfrac{\pi}{2} < \theta < \dfrac{\pi}{2}$	$\tan^2\theta + 1 = \sec^2\theta$
$\sqrt{x^2 - a^2}$	$x = a \sec\theta,\ 0 \le \theta < \dfrac{\pi}{2},\ \pi \le \theta < \dfrac{3\pi}{2}$	$\sec^2\theta - 1 = \tan^2\theta$

NEED TO REVIEW? Right triangle trigonometry is discussed in Appendix A.4, pp. A-30 to A-31.

Although the substitutions can be memorized, it is easier to draw a right triangle and derive them. Each substitution is based on the Pythagorean Theorem. By placing the sides a and x on a right triangle appropriately, the third side of the triangle will represent one of the three types of square roots. See Figure 10.

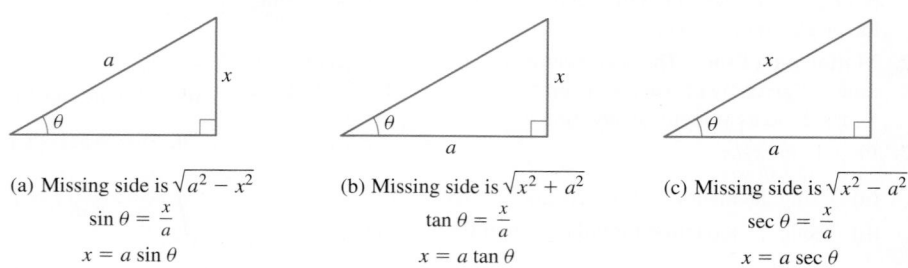

(a) Missing side is $\sqrt{a^2 - x^2}$
$\sin\theta = \dfrac{x}{a}$
$x = a \sin\theta$

(b) Missing side is $\sqrt{x^2 + a^2}$
$\tan\theta = \dfrac{x}{a}$
$x = a \tan\theta$

(c) Missing side is $\sqrt{x^2 - a^2}$
$\sec\theta = \dfrac{x}{a}$
$x = a \sec\theta$

Figure 10

① Integrate a Function Containing $\sqrt{a^2 - x^2}$

When an integrand contains a square root of the form $\sqrt{a^2 - x^2}$, $a > 0$, we use the substitution $x = a \sin\theta$, $-\dfrac{\pi}{2} \le \theta \le \dfrac{\pi}{2}$. This substitution eliminates the square root as follows:

$$\sqrt{a^2 - x^2} = \sqrt{a^2 - a^2\sin^2\theta} \quad \text{Let } x = a\sin\theta;\ -\dfrac{\pi}{2} \le \theta \le \dfrac{\pi}{2}.$$
$$= a\sqrt{1 - \sin^2\theta} \quad \text{Factor out } a^2;\ a > 0.$$
$$= a\sqrt{\cos^2\theta} \quad 1 - \sin^2\theta = \cos^2\theta$$
$$= a\cos\theta \quad \cos\theta \ge 0 \text{ since } -\dfrac{\pi}{2} \le \theta \le \dfrac{\pi}{2}$$

EXAMPLE 1 Integrating a Function That Contains $\sqrt{4 - x^2}$

Find $\displaystyle\int \frac{dx}{x^2\sqrt{4 - x^2}}$.

Solution

The integrand contains the square root $\sqrt{4 - x^2}$ that is of the form $\sqrt{a^2 - x^2}$, where

$a = 2$. Use the substitution $x = 2\sin\theta$, $-\dfrac{\pi}{2} < \theta < \dfrac{\pi}{2}$. Then $dx = 2\cos\theta\, d\theta$. Since

$$\sqrt{4 - x^2} = \underset{\substack{\uparrow \\ x = 2\sin\theta}}{\sqrt{4 - 4\sin^2\theta}} = 2\sqrt{1 - \sin^2\theta} = \underset{\substack{\uparrow \\ \cos\theta > 0 \text{ since } -\frac{\pi}{2} < \theta < \frac{\pi}{2}}}{2\sqrt{\cos^2\theta}} = 2\cos\theta$$

we have

$$\int \frac{dx}{x^2\sqrt{4 - x^2}} = \int \frac{2\cos\theta\, d\theta}{(4\sin^2\theta)(2\cos\theta)} = \int \frac{d\theta}{4\sin^2\theta} = \frac{1}{4}\int \csc^2\theta\, d\theta = -\frac{1}{4}\cot\theta + C$$

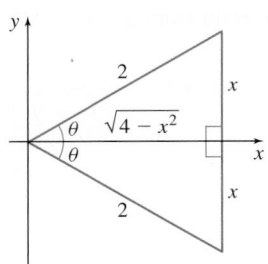

Figure 11 $\sin\theta = \dfrac{x}{2}, -\dfrac{\pi}{2} < \theta < \dfrac{\pi}{2}$

The original integral is a function of x, but the solution we found is a function of θ. To express $\cot\theta$ in terms of x, refer to the right triangles drawn in Figure 11.

Using the Pythagorean Theorem, the third side of each triangle is $\sqrt{4-x^2}$. So,

$$\cot\theta = \frac{\sqrt{4-x^2}}{x} \qquad -\frac{\pi}{2} < \theta < \frac{\pi}{2}$$

Then

$$\int \frac{dx}{x^2\sqrt{4-x^2}} = -\frac{1}{4}\cot\theta + C = -\frac{1}{4}\frac{\sqrt{4-x^2}}{x} + C = -\frac{\sqrt{4-x^2}}{4x} + C \qquad \blacksquare$$

Alternatively, trigonometric identities can be used to express $\cot\theta$ in terms of x. Using identities,

$$\cot\theta = \underset{\substack{\uparrow \\ \cos^2\theta = 1-\sin^2\theta \\ \cos\theta > 0}}{\frac{\cos\theta}{\sin\theta}} = \underset{\substack{\uparrow \\ x = 2\sin\theta \\ \sin\theta = \frac{x}{2}}}{\frac{\sqrt{1-\sin^2\theta}}{\sin\theta}} = \frac{\sqrt{1-\left(\frac{x}{2}\right)^2}}{\frac{x}{2}} = \frac{2\sqrt{1-\frac{x^2}{4}}}{x} = \frac{\sqrt{4-x^2}}{x}$$

NOW WORK   Problem 7.

2 Integrate a Function Containing $\sqrt{x^2 + a^2}$

When an integrand contains a square root of the form $\sqrt{x^2+a^2}$, $a > 0$, use the substitution $x = a\tan\theta$, $-\dfrac{\pi}{2} < \theta < \dfrac{\pi}{2}$. Substituting $x = a\tan\theta$, $-\dfrac{\pi}{2} < \theta < \dfrac{\pi}{2}$, in the expression $\sqrt{x^2+a^2}$ eliminates the square root as follows:

$$\sqrt{x^2+a^2} = \underset{\substack{\uparrow \\ x = a\tan\theta}}{\sqrt{a^2\tan^2\theta + a^2}} = \underset{\substack{\uparrow \\ \text{Factor out } a, \\ a > 0}}{a\sqrt{\tan^2\theta + 1}} = a\sqrt{\sec^2\theta} = \underset{\substack{\uparrow \\ \sec\theta > 0 \\ \text{since } -\frac{\pi}{2} < \theta < \frac{\pi}{2}}}{a\sec\theta}$$

EXAMPLE 2   Integrating a Function That Contains $\sqrt{x^2+9}$

Find $\displaystyle\int \frac{dx}{(x^2+9)^{3/2}}$.

Solution

The integral contains a square root $(x^2+9)^{3/2} = \left(\sqrt{x^2+9}\right)^3$ that is of the form $\sqrt{x^2+a^2}$, where $a = 3$. Use the substitution $x = 3\tan\theta$, $-\dfrac{\pi}{2} < \theta < \dfrac{\pi}{2}$. Then $dx = 3\sec^2\theta\, d\theta$. Since

$$(x^2+9)^{3/2} = \underset{\substack{\uparrow \\ x = 3\tan\theta}}{(9\tan^2\theta + 9)^{3/2}} = 9^{3/2}(\tan^2\theta + 1)^{3/2} = \underset{\substack{\uparrow \\ \tan^2\theta + 1 = \sec^2\theta}}{27(\sec^2\theta)^{3/2}} = \underset{\substack{\uparrow \\ \sec\theta > 0}}{27\sec^3\theta}$$

we have

$$\int \frac{dx}{(x^2+9)^{3/2}} = \int \frac{3\sec^2\theta\, d\theta}{27\sec^3\theta} = \frac{1}{9}\int \frac{d\theta}{\sec\theta} = \frac{1}{9}\int \cos\theta\, d\theta = \frac{1}{9}\sin\theta + C$$

(continued on the next page)

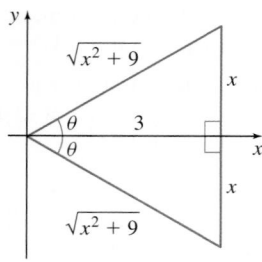

Figure 12 $\tan\theta = \dfrac{x}{3}, \ -\dfrac{\pi}{2} < \theta < \dfrac{\pi}{2}$

To express the solution in terms of x, use either the right triangles in Figure 12 or identities.

From the right triangles, the hypotenuse is $\sqrt{x^2+9}$. So, $\sin\theta = \dfrac{x}{\sqrt{x^2+9}}$. Then

$$\int \frac{dx}{(x^2+9)^{3/2}} = \frac{1}{9}\sin\theta + C = \frac{x}{9\sqrt{x^2+9}} + C \qquad \blacksquare$$

NOW WORK Problem 15.

EXAMPLE 3 **Finding** $\displaystyle\int \frac{dx}{(4x^2+9)^{1/2}}$

Find $\displaystyle\int \frac{dx}{(4x^2+9)^{1/2}}$.

Solution

The integrand contains the expression $4x^2+9$, which equals $(2x)^2 + 3^2$. Then

$$\int \frac{dx}{(4x^2+9)^{1/2}} = \int \frac{dx}{\sqrt{(2x)^2+3^2}}$$

Use the substitution $2x = 3\tan\theta$, $-\dfrac{\pi}{2} < \theta < \dfrac{\pi}{2}$. Then $dx = \dfrac{3}{2}\sec^2\theta\, d\theta$ and

$$\begin{aligned}
\int \frac{dx}{(4x^2+9)^{1/2}} &= \frac{3}{2}\int \frac{1}{\sqrt{9\tan^2\theta + 9}}\cdot\sec^2\theta\, d\theta \\
&= \frac{3}{2}\cdot\frac{1}{3}\int \frac{1}{\sqrt{\tan^2\theta + 1}}\cdot\sec^2\theta\, d\theta \\
&= \frac{1}{2}\int \frac{1}{\sec\theta}\cdot\sec^2\theta\, d\theta = \frac{1}{2}\int \sec\theta\, d\theta \\
&= \frac{1}{2}\ln|\sec\theta + \tan\theta| + C
\end{aligned}$$

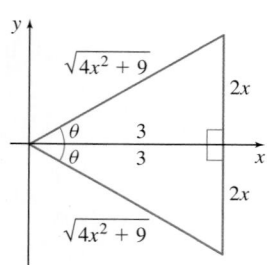

Figure 13 $\tan\theta = \dfrac{2x}{3}, \ -\dfrac{\pi}{2} < \theta < \dfrac{\pi}{2}$

To express the solution in terms of x, refer to the right triangles drawn in Figure 13.

Using the Pythagorean Theorem, the hypotenuse of each triangle is $\sqrt{4x^2+9}$. So,

$$\sec\theta = \frac{\sqrt{4x^2+9}}{3} \quad \text{and} \quad \tan\theta = \frac{2x}{3} \qquad -\frac{\pi}{2} < \theta < \frac{\pi}{2}$$

Then

$$\int \frac{dx}{(4x^2+9)^{1/2}} = \frac{1}{2}\ln\left|\frac{\sqrt{4x^2+9}}{3} + \frac{2x}{3}\right| + C \qquad \blacksquare$$

In general, if the integral contains $\sqrt{b^2x^2+a^2}$, use the substitution $bx = a\tan\theta$ $\left(x = \dfrac{a}{b}\tan\theta\right)$, $-\dfrac{\pi}{2} < \theta < \dfrac{\pi}{2}$.

NOW WORK Problem 45.

❸ Integrate a Function Containing $\sqrt{x^2 - a^2}$

When an integrand contains $\sqrt{x^2 - a^2}$, $a > 0$, use the substitution $x = a \sec \theta$, $0 \leq \theta < \dfrac{\pi}{2}$, $\pi \leq \theta < \dfrac{3\pi}{2}$. Then

$$\underset{\underset{x = a \sec \theta}{\uparrow}}{\sqrt{x^2 - a^2}} = \underset{\underset{a > 0}{\uparrow}}{\sqrt{a^2 \sec^2 \theta - a^2}} = a\sqrt{\sec^2 \theta - 1} = a\sqrt{\tan^2 \theta} = \underset{\underset{\tan \theta \geq 0,\ \text{since } 0 \leq \theta < \frac{\pi}{2},\ \pi \leq \theta < \frac{3\pi}{2}}{\uparrow}}{a \tan \theta}$$

EXAMPLE 4 **Integrating a Function That Contains $\sqrt{x^2 - 4}$**

Find $\displaystyle\int \frac{\sqrt{x^2 - 4}}{x}\, dx$.

Solution

The integrand contains the square root $\sqrt{x^2 - 4}$ which is of the form $\sqrt{x^2 - a^2}$, where $a = 2$. Use the substitution $x = 2 \sec \theta$, $0 \leq \theta < \dfrac{\pi}{2}$, $\pi \leq \theta < \dfrac{3\pi}{2}$.

Then $dx = 2 \sec \theta \tan \theta\, d\theta$, and

$$\underset{\underset{x = 2 \sec \theta}{\uparrow}}{\sqrt{x^2 - 4}} = \sqrt{4 \sec^2 \theta - 4} = 2\sqrt{\sec^2 \theta - 1} = 2\sqrt{\tan^2 \theta} = \underset{\underset{\substack{\tan \theta \geq 0 \\ \text{since } 0 \leq \theta < \frac{\pi}{2},\ \pi \leq \theta < \frac{3\pi}{2}}}{\uparrow}}{2 \tan \theta}$$

Now

$$\int \frac{\sqrt{x^2 - 4}}{x}\, dx = \int \frac{(2 \tan \theta)(2 \sec \theta \tan \theta\, d\theta)}{2 \sec \theta} \qquad \begin{array}{l} x = 2 \sec \theta \\ dx = 2 \sec \theta \tan \theta\, d\theta \end{array}$$

$$= 2 \int \tan^2 \theta\, d\theta = 2 \underset{\underset{\tan^2 \theta = \sec^2 \theta - 1}{\uparrow}}{\int (\sec^2 \theta - 1)\, d\theta}$$

$$= 2 \tan \theta - 2\theta + C$$

To express the solution in terms of x, use either the right triangles in Figure 14 or trigonometric identities.

Using identities, we find

$$\tan \theta = \underset{\underset{\substack{\tan^2 \theta = \sec^2 \theta - 1 \\ \tan \theta \geq 0}}{\uparrow}}{\sqrt{\sec^2 \theta - 1}} = \underset{\underset{\sec \theta = \frac{x}{2}}{\uparrow}}{\sqrt{\frac{x^2}{4} - 1}} = \frac{1}{2}\sqrt{x^2 - 4}$$

Also since $\sec \theta = \dfrac{x}{2}$, $0 \leq \theta < \dfrac{\pi}{2}$, $\pi \leq \theta < \dfrac{3\pi}{2}$, the inverse function $\theta = \sec^{-1}\dfrac{x}{2}$ is defined.

Then

$$\int \frac{\sqrt{x^2 - 4}}{x}\, dx = 2 \tan \theta - 2\theta + C = \sqrt{x^2 - 4} - 2 \sec^{-1}\frac{x}{2} + C \qquad ■$$

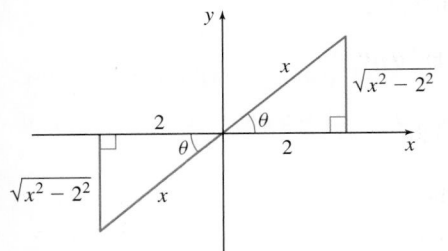

Figure 14 $\sec \theta = \dfrac{x}{2}$,

$0 < \theta < \dfrac{\pi}{2}$, $\pi < \theta < \dfrac{3\pi}{2}$

NOW WORK Problem 29.

4 Use Trigonometric Substitution to Find Definite Integrals

Trigonometric substitution also can be used to find certain types of definite integrals.

EXAMPLE 5   Finding the Area Enclosed by an Ellipse

Find the area A enclosed by the ellipse $\dfrac{x^2}{4} + \dfrac{y^2}{9} = 1$.

Solution

Figure 15 shows the ellipse. Since the ellipse is symmetric with respect to both the x-axis and the y-axis, the total area A of the ellipse is four times the shaded area in the first quadrant, where $0 \le x \le 2$ and $0 \le y \le 3$.

We begin by expressing y as a function of x.

$$\frac{x^2}{4} + \frac{y^2}{9} = 1$$

$$\frac{y^2}{9} = 1 - \frac{x^2}{4} = \frac{4 - x^2}{4}$$

$$y^2 = \frac{9}{4}(4 - x^2)$$

$$y = \frac{3}{2}\sqrt{4 - x^2} \qquad y \ge 0$$

So, the area A of the ellipse is four times the area under the graph of $y = \dfrac{3}{2}\sqrt{4 - x^2}$, $0 \le x \le 2$. That is,

$$A = 4 \int_0^2 \frac{3}{2}\sqrt{4 - x^2}\, dx = 6 \int_0^2 \sqrt{4 - x^2}\, dx$$

Since the integrand contains a square root of the form $\sqrt{a^2 - x^2}$ with $a = 2$, we use the substitution $x = 2\sin\theta$, $-\dfrac{\pi}{2} \le \theta \le \dfrac{\pi}{2}$. Then $dx = 2\cos\theta\, d\theta$. The new limits of integration are:

- When $x = 0$, $2\sin\theta = 0$, so $\theta = 0$.
- When $x = 2$, $2\sin\theta = 2$, so $\sin\theta = 1$ and $\theta = \dfrac{\pi}{2}$.

Then

$$A = 6 \int_0^2 \sqrt{4 - x^2}\, dx = 6 \int_0^{\pi/2} \sqrt{4 - 4\sin^2\theta} \cdot 2\cos\theta\, d\theta$$

$$= 6 \int_0^{\pi/2} 2\sqrt{1 - \sin^2\theta} \cdot 2\cos\theta\, d\theta = 24 \int_0^{\pi/2} \sqrt{\cos^2\theta} \cdot \cos\theta\, d\theta$$

$$\underset{\substack{\uparrow \\ \cos\theta \ge 0}}{=} 24 \int_0^{\pi/2} \cos^2\theta\, d\theta \underset{\substack{\uparrow \\ \cos^2\theta = \frac{1+\cos(2\theta)}{2}}}{=} \frac{24}{2} \int_0^{\pi/2} [1 + \cos(2\theta)]\, d\theta$$

$$= 12\left[\theta + \frac{1}{2}\sin(2\theta)\right]_0^{\pi/2} = 12\left(\frac{\pi}{2} + 0\right) = 6\pi$$

The area of the ellipse is 6π square units. ∎

NOW WORK   Problem 63.

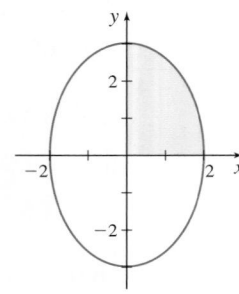

Figure 15 $\dfrac{x^2}{4} + \dfrac{y^2}{9} = 1$

NEED TO REVIEW? The two approaches to finding a definite integral using substitution are discussed in Section 6.5, pp. 488–490.

In Example 5, we changed the limits of integration to find the definite integral. In the next example, we show a second approach to finding a definite integral using substitution.

Using Trigonometric Substitution to Find a Definite Integral

Find the area A under the graph of $y = \dfrac{x^3}{\sqrt{4 - x^2}}$ from 0 to 1. See Figure 16.

Solution

The area $A = \displaystyle\int_0^1 \dfrac{x^3}{\sqrt{4 - x^2}}\, dx$. The integral contains a square root of the form $\sqrt{a^2 - x^2}$, where $a = 2$, so we use the trigonometric substitution $x = 2\sin\theta$, $-\dfrac{\pi}{2} < \theta < \dfrac{\pi}{2}$. Then $dx = 2\cos\theta\, d\theta$.

We find the indefinite integral and then use the Fundamental Theorem of Calculus.

$$\int \frac{x^3}{\sqrt{4 - x^2}}\, dx = \int \frac{8\sin^3\theta}{\sqrt{4 - 4\sin^2\theta}} \cdot 2\cos\theta\, d\theta = \frac{16}{2} \int \frac{\sin^3\theta \cdot \cos\theta}{\sqrt{1 - \sin^2\theta}}\, d\theta$$

$$= 8 \int \frac{\sin^3\theta \cdot \cos\theta}{\cos\theta}\, d\theta = 8 \int \sin^3\theta\, d\theta$$

Since $\sin\theta$ is raised to an odd integer, we write

$$\int \frac{x^3}{\sqrt{4 - x^2}}\, dx = 8 \int \sin^3\theta\, d\theta = 8 \int \sin^2\theta \sin\theta\, d\theta = 8 \int (1 - \cos^2\theta) \sin\theta\, d\theta$$

and use the new substitution $u = \cos\theta$. Then $du = -\sin\theta\, d\theta$.

$$\int \frac{x^3}{\sqrt{4 - x^2}}\, dx = 8 \int (1 - \cos^2\theta) \sin\theta\, d\theta = -8 \int (1 - u^2)\, du$$

$$= -8 \left(u - \frac{u^3}{3} \right)$$

$$= -8 \left(\cos\theta - \frac{\cos^3\theta}{3} \right) = \frac{8\cos^3\theta}{3} - 8\cos\theta$$

To write the integral in terms of x, we use $\cos\theta = \dfrac{\sqrt{4 - x^2}}{2}$.

$$\int \frac{x^3}{\sqrt{4 - x^2}}\, dx = \frac{8}{3} \cdot \left(\frac{\sqrt{4 - x^2}}{2} \right)^3 - 8 \cdot \frac{\sqrt{4 - x^2}}{2}$$

$$= \frac{\left(\sqrt{4 - x^2} \right)^3}{3} - 4\sqrt{4 - x^2}$$

Now using the Fundamental Theorem of Calculus, we find

$$A = \int_0^1 \frac{x^3}{\sqrt{4 - x^2}}\, dx = \left[\frac{\left(\sqrt{4 - x^2} \right)^3}{3} - 4\sqrt{4 - x^2} \right]_0^1$$

$$= \left[\frac{\left(\sqrt{3} \right)^3}{3} - 4\sqrt{3} \right] - \left[\frac{2^3}{3} - 4 \cdot 2 \right]$$

$$= \sqrt{3} - 4\sqrt{3} - \frac{8}{3} + 8 = \frac{16}{3} - 3\sqrt{3} \approx 0.137$$

NOW WORK Problems 53 and 65.

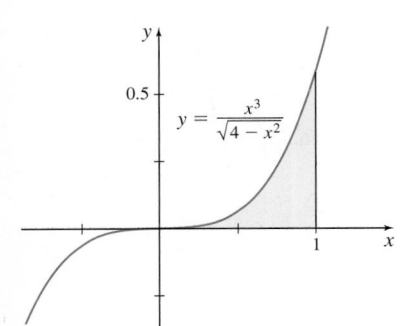

Figure 16 $y = \dfrac{x^3}{\sqrt{4 - x^2}},\ 0 \le x \le 1$

6.11 Assess Your Understanding

Concepts and Vocabulary

1. **True or False**   To find $\int \sqrt{a^2 - x^2}\, dx$, $a > 0$, the substitution $x = a \sin\theta$, $-\dfrac{\pi}{2} \leq \theta \leq \dfrac{\pi}{2}$, can be used.

2. **Multiple Choice**   To find $\int \sqrt{x^2 + 16}\, dx$, use the substitution $x = $ [(**a**) $4 \sin\theta$ (**b**) $\tan\theta$ (**c**) $4 \sec\theta$ (**d**) $4 \tan\theta$].

3. **Multiple Choice**   To find $\int \sqrt{x^2 - 9}\, dx$, use the substitution $x = $ [(**a**) $\sec\theta$ (**b**) $3 \sin\theta$ (**c**) $3 \sec\theta$ (**d**) $3 \tan\theta$].

4. **Multiple Choice**   To find $\int \sqrt{25 - 4x^2}\, dx$, use the substitution $x = \left[(\textbf{a})\ \dfrac{5}{2} \tan\theta \quad (\textbf{b})\ \dfrac{5}{2} \sin\theta \quad (\textbf{c})\ \dfrac{2}{5} \sin\theta \quad (\textbf{d})\ \dfrac{2}{5} \sec\theta\right]$.

Skill Building

In Problems 5–14, find each integral. Each integral contains a term of the form $\sqrt{a^2 - x^2}$.

5. $\displaystyle\int \sqrt{4 - x^2}\, dx$

6. $\displaystyle\int \sqrt{16 - x^2}\, dx$

7. $\displaystyle\int \frac{x^2}{\sqrt{16 - x^2}}\, dx$

8. $\displaystyle\int \frac{x^2}{\sqrt{36 - x^2}}\, dx$

9. $\displaystyle\int \frac{\sqrt{4 - x^2}}{x^2}\, dx$

10. $\displaystyle\int \frac{\sqrt{9 - x^2}}{x^2}\, dx$

11. $\displaystyle\int x^2 \sqrt{4 - x^2}\, dx$

12. $\displaystyle\int x^2 \sqrt{1 - 16x^2}\, dx$

13. $\displaystyle\int \frac{dx}{(4 - x^2)^{3/2}}$

14. $\displaystyle\int \frac{dx}{(1 - x^2)^{3/2}}$

In Problems 15–26, find each integral. Each integral contains a term of the form $\sqrt{x^2 + a^2}$.

15. $\displaystyle\int \sqrt{4 + x^2}\, dx$

16. $\displaystyle\int \sqrt{1 + x^2}\, dx$

17. $\displaystyle\int \frac{dx}{\sqrt{x^2 + 16}}$

18. $\displaystyle\int \frac{dx}{\sqrt{x^2 + 25}}$

19. $\displaystyle\int \sqrt{1 + 9x^2}\, dx$

20. $\displaystyle\int \sqrt{9 + 4x^2}\, dx$

21. $\displaystyle\int \frac{x^2}{\sqrt{4 + 9x^2}}\, dx$

22. $\displaystyle\int \frac{x^2}{\sqrt{x^2 + 16}}\, dx$

23. $\displaystyle\int \frac{dx}{x^2 \sqrt{x^2 + 4}}$

24. $\displaystyle\int \frac{dx}{x^2 \sqrt{4x^2 + 1}}$

25. $\displaystyle\int \frac{dx}{(x^2 + 4)^{3/2}}$

26. $\displaystyle\int \frac{dx}{(x^2 + 1)^{3/2}}$

In Problems 27–36, find each integral. Each integral contains a term of the form $\sqrt{x^2 - a^2}$.

27. $\displaystyle\int \frac{x^2}{\sqrt{x^2 - 25}}\, dx$

28. $\displaystyle\int \frac{x^2}{\sqrt{x^2 - 16}}\, dx$

29. $\displaystyle\int \frac{\sqrt{x^2 - 1}}{x}\, dx$

30. $\displaystyle\int \frac{\sqrt{x^2 - 1}}{x^2}\, dx$

31. $\displaystyle\int \frac{dx}{x^2 \sqrt{x^2 - 36}}$

32. $\displaystyle\int \frac{dx}{x^2 \sqrt{x^2 - 9}}$

33. $\displaystyle\int \frac{dx}{\sqrt{4x^2 - 9}}$

34. $\displaystyle\int \frac{dx}{\sqrt{9x^2 - 4}}$

35. $\displaystyle\int \frac{dx}{(x^2 - 9)^{3/2}}$

36. $\displaystyle\int \frac{dx}{(25x^2 - 1)^{3/2}}$

In Problems 37–48, find each integral.

37. $\displaystyle\int \frac{x^2\, dx}{(x^2 - 9)^{3/2}}$

38. $\displaystyle\int \frac{x^2\, dx}{(x^2 - 4)^{3/2}}$

39. $\displaystyle\int \frac{x^2\, dx}{16 + x^2}$

40. $\displaystyle\int \frac{x^2\, dx}{1 + 16x^2}$

41. $\displaystyle\int \sqrt{4 - 25x^2}\, dx$

42. $\displaystyle\int \sqrt{9 - 16x^2}\, dx$

43. $\displaystyle\int \frac{dx}{(4 - 25x^2)^{3/2}}$

44. $\displaystyle\int \frac{dx}{(1 - 9x^2)^{3/2}}$

45. $\displaystyle\int \sqrt{4 + 25x^2}\, dx$

46. $\displaystyle\int \sqrt{9 + 16x^2}\, dx$

47. $\displaystyle\int \frac{dx}{x^3 \sqrt{x^2 - 16}}$

48. $\displaystyle\int \frac{dx}{x^3 \sqrt{x^2 - 1}}$

In Problems 49–58, find each definite integral.

49. $\displaystyle\int_0^1 \sqrt{1 - x^2}\, dx$

50. $\displaystyle\int_0^{1/2} \sqrt{1 - 4x^2}\, dx$

51. $\displaystyle\int_0^1 \sqrt{1 + x^2}\, dx$

52. $\displaystyle\int_0^2 \frac{x^2}{\sqrt{9 + x^2}}\, dx$

53. $\displaystyle\int_4^5 \frac{x^2}{\sqrt{x^2 - 9}}\, dx$

54. $\displaystyle\int_1^2 \frac{x^2}{\sqrt{4x^2 - 1}}\, dx$

55. $\displaystyle\int_0^2 \frac{x^2\, dx}{(16 - x^2)^{3/2}}$

56. $\displaystyle\int_0^1 \frac{x^2\, dx}{(25 - x^2)^{3/2}}$

57. $\displaystyle\int_0^3 \frac{x^2\, dx}{9 + x^2}$

58. $\displaystyle\int_0^1 \frac{x^2}{25 + x^2}\, dx$

Applications and Extensions

59. **Volume of a Solid of Revolution**
The volume V of the solid of revolution generated by revolving the region bounded by the graph of $y = \dfrac{1}{x^2 + 4}$ and the x-axis from $x = 0$ to $x = 1$ about the x-axis is given by

$$V = \pi \int_0^1 \left(\frac{1}{x^2 + 4}\right)^2 dx.$$

Find the volume V of the solid of revolution.

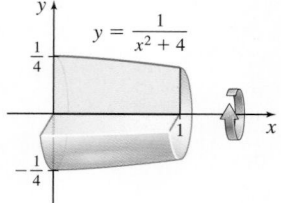

60. Volume of a Solid of Revolution   The volume V of the solid of revolution generated by revolving the region bounded by the graphs of $y = \dfrac{1}{\sqrt{9-x^2}}$, $y=0$, $x=0$, and $x=2$ about the x-axis is given by $V = \pi \displaystyle\int_0^2 \left(\dfrac{1}{\sqrt{9-x^2}} \right)^2 dx$. Find the volume of the solid of revolution.

61. Average Value   Find the average value of the function
$$f(x) = \frac{1}{\sqrt{9-4x^2}} \text{ over the interval } \left[0, \frac{1}{2}\right].$$

62. Average Value   Find the average value of the function $f(x) = \sqrt{x^2 - 4}$ over the interval $[2, 7]$.

63. Area under a Graph   Find the area under the graph of $y = \dfrac{1}{\sqrt{9-x^2}}$ from $x=0$ to $x=2$.

64. Area under a Graph   Find the area under the graph of $y = x^2\sqrt{16-x^2}$, $x \geq 0$.

65. Area under a Graph   Find the area under the graph of $y = \dfrac{x^2}{\sqrt{x^2-1}}$ from $x=3$ to $x=5$.

66. Area under a Graph   Find the area under the graph of $y = \sqrt{16x^2 + 9}$ from $x=0$ to $x=1$.

In Problems 67–72, find each integral.
Hint: Begin with a substitution.

67. $\displaystyle\int \frac{dx}{\sqrt{1-(x-2)^2}}$

68. $\displaystyle\int \sqrt{4-(x+2)^2}\, dx$

69. $\displaystyle\int \frac{dx}{\sqrt{(2x-1)^2 - 4}}$

70. $\displaystyle\int \frac{dx}{(3x-2)\sqrt{(3x-2)^2 + 9}}$

71. $\displaystyle\int e^x \sqrt{25 - e^{2x}}\, dx$

72. $\displaystyle\int e^x \sqrt{4 + e^{2x}}\, dx$

In Problems 73 and 74, use integration by parts and then the methods of this section to find each integral.

73. $\displaystyle\int x \sin^{-1} x\, dx$

74. $\displaystyle\int x \cos^{-1} x\, dx$

In Problems 75–79, use a trigonometric substitution to derive each formula. Assume $a > 0$.

75. $\displaystyle\int \frac{dx}{\sqrt{a^2 - x^2}} = \sin^{-1}\frac{x}{a} + C$

76. $\displaystyle\int \frac{dx}{a^2 + x^2} = \frac{1}{a}\tan^{-1}\frac{x}{a} + C$

77. $\displaystyle\int \frac{dx}{x\sqrt{x^2 - a^2}} = \frac{1}{a}\sec^{-1}\frac{x}{a} + C$

78. $\displaystyle\int \frac{dx}{\sqrt{x^2 - a^2}} = \ln\left| \frac{x + \sqrt{x^2 - a^2}}{a} \right| + C$

79. $\displaystyle\int \frac{dx}{\sqrt{x^2 + a^2}} = \ln\left| x + \sqrt{x^2 + a^2} \right| + C$

80. Area of an Ellipse   Find the area of the region enclosed by the ellipse $\dfrac{x^2}{a^2} + \dfrac{y^2}{b^2} = 1$, where $a > b > 0$.

Challenge Problems

81. Derive the formula
$$\int \sqrt{x^2 - a^2}\, dx = \frac{1}{2}x\sqrt{x^2 - a^2} - \frac{1}{2}a^2 \ln\left| x + \sqrt{x^2 - a^2} \right| + C,$$
$a > 0$.

82. Find $\displaystyle\int \frac{dx}{\sqrt{3x - x^2}}$.

83. Find $\displaystyle\int \frac{\sec^2 x}{\sqrt{\tan^2 x - 6\tan x + 8}}\, dx$.

6.12 Integration Using Tables and Computer Algebra Systems

OBJECTIVES *When you finish this section, you should be able to:*

1 Use a Table of Integrals (p. 567)
2 Use a computer algebra system (p. 569)

While it is important to be able to use techniques of integration, to save time or to check one's work, a Table of Integrals or a computer algebra system (CAS) is often useful.

1 Use a Table of Integrals

The inserts in the back of the book contain a list of integration formulas called a **Table of Integrals**. Many of the integration formulas in the list were derived in this chapter. Although the table seems long, it is far from complete. A more comprehensive table of integrals may be found in Daniel Zwillinger (Ed.), *Standard Mathematical Tables and Formulas*, 33rd ed., Boca Raton, FL: CRC Press, 2018.

EXAMPLE 1   Using a Table of Integrals

Use a Table of Integrals to find $\displaystyle\int \frac{dx}{\sqrt{\left(4x-x^2\right)^3}}$.

Solution

Look through the headings in the Table of Integrals until you locate *Integrals Containing* $\sqrt{2ax-x^2}$. Continue in the subsection until you find a form that closely resembles the integrand given. You should find Integral 74:

$$\int \frac{dx}{\left(2ax-x^2\right)^{3/2}} = \frac{x-a}{a^2\sqrt{2ax-x^2}} + C$$

This is the integral we seek with $a = 2$. So,

$$\int \frac{dx}{\sqrt{\left(4x-x^2\right)^3}} = \frac{x-2}{4\sqrt{4x-x^2}} + C \qquad\blacksquare$$

NOW WORK   Problem 3.

Some integrals in the table are reduction formulas.

EXAMPLE 2   Using a Table of Integrals

Use a Table of Integrals to find $\displaystyle\int x^2 \tan^{-1} x\, dx$.

Solution

Find the subsection of the table titled *Integrals Containing Inverse Trigonometric Functions*. Then look for an integral whose form closely resembles the problem. You should find Integral 106:

$$\int x^n \tan^{-1} x\, dx = \frac{1}{n+1}\left(x^{n+1} \tan^{-1} x - \int \frac{x^{n+1}\, dx}{1+x^2} \right) \qquad n \neq -1$$

This is the integral we seek with $n = 2$.

$$\int x^2 \tan^{-1} x\, dx = \frac{1}{3}\left(x^3 \tan^{-1} x - \int \frac{x^3\, dx}{1+x^2} \right)$$

We find the integral on the right by using the substitution $u = 1+x^2$. Then $du = 2x\, dx$ and

$$\int \frac{x^3}{1+x^2}dx = \int \frac{x^2 x\, dx}{1+x^2} = \int \frac{u-1}{u}\frac{du}{2} = \frac{1}{2}\int\left(1-\frac{1}{u}\right)du = \frac{1}{2}u - \frac{1}{2}\ln|u|$$

$$= \frac{1+x^2}{2} - \frac{\ln(1+x^2)}{2}$$

So,

$$\int x^2 \tan^{-1} x\, dx = \frac{1}{3}\left(x^3 \tan^{-1} x - \int \frac{x^3}{1+x^2}\, dx \right)$$

$$= \frac{1}{3}\left[x^3 \tan^{-1} x - \frac{1+x^2}{2} + \frac{1}{2}\ln(1+x^2) \right] + C \qquad\blacksquare$$

NOW WORK   Problem 11.

Sometimes the given integral is found in the tables after a substitution is made.

CALC CLIP

EXAMPLE 3   Using a Table of Integrals

Use a Table of Integrals to find $\displaystyle\int \frac{3x+5}{\sqrt{3x+6}}\, dx$.

Solution

Find the subsection of the table titled *Integrals Containing* $\sqrt{a+bx}$ (the square root of a linear expression). Then look for an integral whose form closely resembles the problem. The closest one is an integral with x in the numerator and $\sqrt{a+bx}$ in the denominator,

Integral 34: $\qquad \displaystyle\int \frac{x\,dx}{\sqrt{a+bx}} = \frac{2}{3b^2}(bx-2a)\sqrt{a+bx} + C \qquad (1)$

To express the given integral as one with a single variable in the numerator, use the substitution $u = 3x + 5$. Then $du = 3\,dx$. Since

$$\sqrt{3x+6} = \sqrt{(3x+5)+1} = \sqrt{u+1}$$

we find

$$\int \frac{3x+5}{\sqrt{3x+6}}\,dx \underset{\substack{\uparrow \\ u=3x+5,\ \frac{1}{3}du=dx}}{=} \frac{1}{3}\int \frac{u\,du}{\sqrt{u+1}}$$

which is in the form of (1) with $a = 1$ and $b = 1$ So,

$$\int \frac{3x+5}{\sqrt{3x+6}}\,dx = \frac{1}{3}\int \frac{u\,du}{\sqrt{u+1}} \underset{\substack{\uparrow \\ (1)}}{=} \frac{1}{3}\cdot\frac{2}{3}(u-2)\sqrt{1+u} + C$$

$$\underset{\substack{\uparrow \\ u=3x+5}}{=} \frac{2}{9}[(3x+5)-2]\sqrt{1+(3x+5)} + C = \frac{2}{3}(x+1)\sqrt{3x+6} + C \qquad ∎$$

NOW WORK Problem 29.

---

❷ Use a Computer Algebra System

A computer algebra system (or CAS) is computer software that can perform symbolic manipulation of mathematical expressions. Some graphing calculators such as the TI Nspire also have a CAS capability. CAS software packages such as Mathematica, Maple, and Matlab offer more extensive symbolic manipulation capabilities, as well as tools for visualization and numerical approximation. These packages offer intuitive interfaces so that the user can obtain results without needing to write a program. An online CAS, based on Mathematica, can be found at WolframAlpha.com.

A CAS is often used instead of a Table of Integrals. When using a CAS to find an integral, keep in mind:

- A CAS can find indefinite integrals or definite integrals.
- For indefinite integrals, the constant of integration is often omitted.
- For definite integrals, there is often an option to specify whether the solution should be expressed in exact form or as an approximate solution, and to what precision.
- Absolute value symbols are often omitted from logarithmic answers.
- The symbol "log" is often used to mean ln.
- An indefinite integral is sometimes expressed in a different, but equivalent, algebraic form from that found in a table or by hand. With simplification, integrals found by hand, with a Table of Integrals, or with a CAS will be the same.
- When the CAS fails to find a result, either because the integral is infeasible or beyond the capability of the CAS, most systems show this by repeating the integral.
- Many CAS products have the capability to handle improper integrals.

EXAMPLE 4 **Finding an Integral Using a CAS**

Find $\int x^2(2x^3-4)^5 dx$.

Solution

Using WolframAlpha to find the integral, first click on MATH INPUT. Then use the math icons in the purple bar to input the integral. If a symbol is needed that is not on the screen, try clicking on the three dots, followed by "All Math Inputs." Then enter the integral and hit the = icon.

The output is

$$\int x^2(2x^3-4)^5 dx = 32\left(\frac{x^{18}}{18} - \frac{2x^{15}}{3} + \frac{10x^{12}}{3} - \frac{80x^9}{9} + \frac{40x^6}{3} - \frac{32x^3}{3}\right) + C$$

Using Mathematica returns the same result without the constant C. ∎

We can find $\int x^2\left(2x^3-4\right)^5 dx$ using the substitution $u = 2x^3 - 4$.
Then $du = 6x^2\,dx$, and

$$\int x^2(2x^3-4)^5 dx = \frac{1}{6}\int u^5\,du = \frac{1}{6}\cdot\frac{u^6}{6} = \frac{(2x^3-4)^6}{36} + C$$

If this solution is expanded using the Binomial Theorem, it will differ from the CAS answer by a constant.

NOW WORK Problems **35, 43,** and **61** using a CAS and compare your answers to Problems **3, 11,** and **29.**

6.12 Assess Your Understanding

Skill Building

In Problems 1–32, find each integral using the Table of Integrals found at the back of the book.

1. $\int e^{2x}\cos x\,dx$

2. $\int e^{5x}\sin(2x+3)\,dx$

3. $\int x(2+3x)^4\,dx$

4. $\int \frac{x}{(5+2x)^2}\,dx$

5. $\int \frac{x^2\,dx}{\sqrt{6+3x}}$

6. $\int \frac{\sqrt{1+x}}{x}\,dx$

7. $\int \frac{\sqrt{x^2+4}}{x}\,dx$

8. $\int \frac{x^2}{\sqrt{x^2-9}}\,dx$

9. $\int \frac{dx}{(x^2-4)^{3/2}}$

10. $\int (x^2+9)^{3/2}dx$

11. $\int x^3(\ln x)^2 dx$

12. $\int \frac{dx}{x\ln x}$

13. $\int \sqrt{x^2-16}\,dx$

14. $\int \frac{\sqrt{x^2+3}}{x}\,dx$

15. $\int (6-x^2)^{3/2}\,dx$

16. $\int \frac{dx}{x^2\sqrt{10-x^2}}$

17. $\int \sqrt{10x-x^2}\,dx$

18. $\int \frac{dx}{x\sqrt{6x-x^2}}$

19. $\int \cos(3x)\cos(8x)\,dx$

20. $\int \sin(2x)\sin(5x)\,dx$

21. $\int x\tan^{-1}x\,dx$

22. $\int x\sin^{-1}x\,dx$

23. $\int x^4\ln x\,dx$

24. $\int (\ln x)^3 dx$

25. $\int \frac{dx}{x^2\sqrt{8x-x^2}}$

26. $\int \frac{\sqrt{12x-x^2}}{x^2}\,dx$

27. $\int \frac{dx}{(4+x^2)^2}$

28. $\int \frac{dx}{(x^2-25)^3}$

29. $\int (x+1)\sqrt{4x+5}\,dx$

30. $\int \frac{dx}{[(2x+3)^2-1]^{3/2}}$

31. $\int_1^2 \frac{x^3}{\sqrt{3x-x^2}}\,dx$

32. $\int_1^e \frac{1}{x^2\sqrt{x^2+2}}\,dx$

(CAS) In Problems 33–64, redo Problems 1–32 using a CAS. (Answers may vary.)

(CAS) In Problems 65–70, use a CAS to find each integral. (Answers may vary.)

65. $\int \sqrt{1+x^3}\,dx$

66. $\int \sqrt{1+\sin x}\,dx$

67. $\int e^{-x^2}dx$

68. $\int \frac{\cos x}{x}\,dx$

69. $\int x\tan x\,dx$

70. $\int \sqrt{1+e^x}\,dx$

6.13 Mixed Practice

OBJECTIVE *When you finish this section, you should be able to:*

1 **Recognize the form of an integrand and find its integral (p. 571)**

1 Recognize the Form of an Integrand and Find Its Integral

Throughout this chapter, we have discussed many integration techniques. As you now realize, integration, unlike differentiation, is an art that needs practice and often some ingenuity. The techniques studied in each section were based on the form of the integrand. In the *Skill Building* problems at the end of each section, all the integrands followed the forms discussed in the section. Here we have a short section with exercises that "mix up" the integrals.

Always begin by looking at the form of the integrand. Then choose an integration technique that fits that form. Only then attempt to integrate. If the chosen technique does not work, reconsider the form and try again.

6.13 Assess Your Understanding

Skill Building

Find each integral. If the integral is improper, determine whether it converges or diverges. If it converges, find its value.

1. $\displaystyle\int x^2 \sin^2(x^3)\,dx$

2. $\displaystyle\int \cot^3 x \csc x\,dx$

3. $\displaystyle\int (5x-1)\cos x\,dx$

4. $\displaystyle\int \frac{x^2}{(4+x^2)^{3/2}}\,dx$

5. $\displaystyle\int \frac{dx}{\sqrt{4x^2+25}}$

6. $\displaystyle\int_0^1 \frac{dx}{(x+1)(x^2+9)}$

7. $\displaystyle\int_0^1 \frac{4x\,dx}{x^2+5x+6}$

8. $\displaystyle\int x^3 \ln x\,dx$

9. $\displaystyle\int \tan x \sec^3 x\,dx$

10. $\displaystyle\int \frac{dx}{x^2+2x+10}$

11. $\displaystyle\int \tan^2 x \sec^4 x\,dx$

12. $\displaystyle\int_1^\infty \frac{\ln x}{x}\,dx$

13. $\displaystyle\int \frac{dx}{x\sqrt{x^2-16}}$

14. $\displaystyle\int \frac{x+3}{x(x^2+1)^2}\,dx$

15. $\displaystyle\int \tan^{1/3} x \sec^4 x\,dx$

16. $\displaystyle\int \sin(5x)\cos(2x)\,dx$

17. $\displaystyle\int (2x^3+1)\sin(x^4+2x)\,dx$

18. $\displaystyle\int_1^2 \frac{1}{(x-1)^2}\,dx$

19. $\displaystyle\int x^2 \sin^{-1} x\,dx$

20. $\displaystyle\int \sqrt{16x^2-1}\,dx$

21. $\displaystyle\int_0^{\pi/2} \sin(2x)\sin\left(\frac{1}{2}x\right)\,dx$

22. $\displaystyle\int \sin^5(\pi x)\cos^3(\pi x)\,dx$

23. $\displaystyle\int \sqrt{36-x^2}\,dx$

24. $\displaystyle\int \cos^5(2x)\,dx$

25. $\displaystyle\int \frac{x}{x^2+2x+5}\,dx$

26. $\displaystyle\int \frac{x^2-4}{x(x-1)^2}\,dx$

27. $\displaystyle\int_0^2 \frac{dx}{\sqrt{4-x^2}}$

28. $\displaystyle\int x(x+4)^5\,dx$

29. $\displaystyle\int \frac{2x\,dx}{(x-1)(x^2+1)}$

30. $\displaystyle\int_0^1 xe^{-3x}\,dx$

31. $\displaystyle\int \sin^3 x \sqrt[3]{\cos x}\,dx$

32. $\displaystyle\int \frac{\sqrt{4-x^2}}{x}\,dx$

CHAPTER 6 PROJECT   The Birds of Rügen Island

This Project can be done individually or as part of a team.

Let $P = P(t)$ denote the population of a species of rare birds on Rügen Island, where t is the time, in years. Suppose M equals the maximum sustainable number of birds and m equals the minimum population, below which the species becomes extinct. The population P can be modeled by the equation

$$\frac{dP}{dt} = k(M - P)(P - m)$$

where k is a positive constant.

1. Suppose the maximum population M is 1200 birds and the minimum population m is 100 birds. If $k = 0.001$, write the equation that models the population $P = P(t)$.

2. Solve the equation from Problem 1.

3. If the population at time $t = 0$ is 300 birds, how many birds will exist in 5 years?

4. Using technology, graph the solution found in Problem 2.

5. The graph from Problem 4 seems to have an inflection point. Verify this and find it.

6. What conclusions can you draw about the rate of change of population based on your answer to Problem 5?

7. Write an essay about using the given equation to model the bird population. What assumptions are being made? What situations are being ignored?

Chapter Review

THINGS TO KNOW

6.7 Integration by Parts

- Integration by parts formula $\int u\, dv = uv - \int v\, du$ (p. 516)
- Guidelines for choosing u and dv (Table 1, pp. 517–518)

Basic Integral (p. 517):
$$\int \ln x\, dx = x \ln x - x + C$$

6.8 Integration of Rational Functions Using Partial Fractions

Definitions:

- Partial fractions (p. 528)
- Irreducible quadratic factor (p. 532)

Partial Fraction Decomposition $R(x) = \dfrac{p(x)}{q(x)}, q(x) \neq 0$

- Case 1: a proper rational function whose denominator contains only distinct linear factors (p. 528)
- Case 2: a proper rational function whose denominator contains a repeated linear factor $(x - a)^n$, $n \geq 2$ (p. 531)

- Case 3: a proper rational function whose denominator contains a distinct irreducible quadratic factor $x^2 + bx + c$ (p. 532)
- Case 4: a proper rational function whose denominator contains a repeated irreducible quadratic factor $(x^2 + bx + c)^n$, $n \geq 2$ an integer (p. 533)

6.9 Improper Integrals

- Improper integrals of the form $\int_a^\infty f(x)\, dx$; $\int_{-\infty}^b f(x)\, dx$; and $\int_{-\infty}^\infty f(x)\, dx$ (pp. 539, 540)
- Improper integrals $\int_a^b f(x)\, dx$, where $f(a)$, $f(b)$, or $f(c)$, $a < c < b$, is not defined. (pp. 542, 544)

Theorems

- $\displaystyle\int_1^\infty \frac{dx}{x^p}$ converges if $p > 1$ and diverges if $p \leq 1$. (p. 541)
- Comparison Test for Improper Integrals (p. 545)

6.10 Integrals Containing Trigonometric Functions
Integrals of the Form $\int \sin^n x\, dx$ or $\int \cos^n x\, dx$, $n \geq 2$ an Integer (p. 553)

	Integral	Use the Identity	u-substitution
n odd	$\int \sin^n x\, dx = \int \sin^{n-1} x \sin x\, dx$	$\sin^2 x = 1 - \cos^2 x$	$u = \cos x;\quad du = -\sin x\, dx$
	$\int \cos^n x\, dx = \int \cos^{n-1} x \cos x\, dx$	$\cos^2 x = 1 - \sin^2 x$	$u = \sin x;\quad du = \cos x\, dx$
n even	$\int \sin^n x\, dx = \int \left[\dfrac{1 - \cos(2x)}{2}\right]^{n/2} dx$	$\sin^2 x = \dfrac{1 - \cos(2x)}{2}$	$u = 2x;\quad du = 2\, dx$
	$\int \cos^n x\, dx = \int \left[\dfrac{1 + \cos(2x)}{2}\right]^{n/2} dx$	$\cos^2 x = \dfrac{1 + \cos(2x)}{2}$	$u = 2x;\quad du = 2\, dx$

Integrals of the Form $\int \sin^m x \cos^n x\, dx$ (p. 554)

	Integral	Use the Identity	u-substitution
m odd	$\int \sin^m x \cos^n x\, dx = \int \sin^{m-1} x \cos^n x \sin x\, dx$	$\sin^2 x = 1 - \cos^2 x$	$u = \cos x$ $du = -\sin x\, dx$
n odd	$\int \sin^m x \cos^n x\, dx = \int \sin^m x \cos^{n-1} x \cos x\, dx$	$\cos^2 x = 1 - \sin^2 x$	$u = \sin x$ $du = \cos x\, dx$
m and n both even	$\int \sin^m x \cos^n x\, dx = \int (1 - \cos^2 x)^{m/2} \cos^n x\, dx$ or $\int \sin^m x \cos^n x\, dx = \int \sin^m x (1 - \sin^2 x)^{n/2}\, dx$	$\cos^2 x = \dfrac{1 + \cos(2x)}{2}$ $\sin^2 x = \dfrac{1 - \cos(2x)}{2}$	$u = 2x;\quad du = 2\,dx$ $u = 2x;\quad du = 2\,dx$

Integrals of the Form $\int \tan^m x \sec^n x\, dx$ (p. 557)

	Integral	Use the Identity	Method
m odd	$\int \tan^m x \sec^n x\, dx = \int \tan^{m-1} x \sec^{n-1} x \sec x \tan x\, dx$	$\tan^2 x = \sec^2 x - 1$	$u = \sec x$ $du = \sec x \tan x\, dx$
n even	$\int \tan^m x \sec^n x\, dx = \int \tan^m x \sec^{n-2} x \sec^2 x\, dx$	$\sec^2 x = 1 + \tan^2 x$	$u = \tan x$ $du = \sec^2 x\, dx$
m even; n odd	$\int \tan^m x \sec^n x\, dx = \int (\sec^2 x - 1)^{m/2} \sec^n x\, dx$	$\tan^2 x = \sec^2 x - 1$	Use integration by parts

Other Integrals Containing Trigonometric Functions (p. 557)

- $\displaystyle \int \sin(ax) \sin(bx)\, dx = \frac{1}{2} \int [\cos(ax - bx) - \cos(ax + bx)]\, dx$

- $\displaystyle \int \cos(ax) \cos(bx)\, dx = \frac{1}{2} \int [\cos(ax - bx) + \cos(ax + bx)]\, dx$

- $\displaystyle \int \sin(ax) \cos(bx)\, dx = \frac{1}{2} \int [\sin(ax + bx) + \sin(ax - bx)]\, dx$

6.11 Integration Using Trigonometric Substitution: Integrands Containing $\sqrt{a^2 - x^2}$, $\sqrt{x^2 + a^2}$, or $\sqrt{x^2 - a^2}$, $a > 0$

See Table 2 (p. 560):

- Integrands containing $\sqrt{a^2 - x^2}$: Use the substitution $x = a \sin \theta$, $-\dfrac{\pi}{2} \leq \theta \leq \dfrac{\pi}{2}$. (p. 560)

- Integrands containing $\sqrt{a^2 + x^2}$: Use the substitution $x = a \tan \theta$, $-\dfrac{\pi}{2} < \theta < \dfrac{\pi}{2}$. (p. 561)

- Integrands containing $\sqrt{x^2 - a^2}$: Use the substitution $x = a \sec \theta$, $0 \leq \theta < \dfrac{\pi}{2}$ or $\pi \leq \theta < \dfrac{3\pi}{2}$. (p. 563)

6.12 Integration Using Tables and Computer Algebra Systems (p. 567)

6.13 Mixed Practice (p. 571)

OBJECTIVES

Section	You should be able to ...	Examples	Review Exercises	AP® Review Problems
6.7	**1** Integrate by parts (p. 516)	1–6	7, 12, 15, 19	2, 3
	2 Find a definite integral using integration by parts (p. 520)	7, 8	6	1, 7
	3 Derive a general formula using integration by parts (p. 522)	9, 10	29, 30	
6.8	**1** Integrate a proper rational function whose denominator contains only distinct linear factors (p. 528)	1, 2	10, 21, 23	6
	2 Integrate a proper rational function whose denominator contains a repeated linear factor (p. 531)	3	13, 17	
	3 Integrate a proper rational function whose denominator contains a distinct irreducible quadratic factor (p. 532)	4	11	
	4 Integrate a proper rational function whose denominator contains a repeated irreducible quadratic factor (p. 533)	5	16	
6.9	**1** Find integrals with an infinite limit of integration (p. 539)	1, 2	31, 34, 36	4
	2 Interpret an improper integral geometrically (p. 541)	3	39	
	3 Integrate functions over $[a, b]$ that are not defined at an endpoint (p. 542)	4–6	32, 33, 35	5
	4 Use the Comparison Test for improper integrals (p. 545)	7	37, 38	
6.10	**1** Find integrals of the form $\int \sin^n x\, dx$ or $\int \cos^n x\, dx$, $n \geq 2$ an integer (p. 550)	1–4	3, 25	
	2 Find integrals of the form $\int \sin^m x \cos^n x\, dx$ (p. 553)	5	8, 22	
	3 Find integrals of the form $\int \tan^m x \sec^n x\, dx$ or $\int \cot^m x \csc^n x\, dx$ (p. 554)	6–8	1, 2	
	4 Find integrals of the form $\int \sin(ax) \sin(bx)\, dx$, $\int \sin(ax) \cos(bx)\, dx$, or $\int \cos(ax) \cos(bx)\, dx$ (p. 557)	9	26	
6.11	**1** Integrate a function containing $\sqrt{a^2 - x^2}$ (p. 560)	1	4, 9	
	2 Integrate a function containing $\sqrt{x^2 + a^2}$ (p. 561)	2, 3	14, 18, 20	
	3 Integrate a function containing $\sqrt{x^2 - a^2}$ (p. 563)	4	5, 24	
	4 Use trigonometric substitution to find definite integrals (p. 564)	5, 6	27, 28	
6.12	**1** Use a Table of Integrals (p. 567)	1–3	40(a)	
	2 Use a computer algebra system (p. 569)	4	40(b)	

REVIEW EXERCISES

In Problems 1–28, find each integral.

1. $\displaystyle\int \sec^3 \phi \tan \phi \, d\phi$

2. $\displaystyle\int \cot^2 \theta \, \csc \theta \, d\theta$

3. $\displaystyle\int \sin^3 \phi \, d\phi$

4. $\displaystyle\int \frac{x^2}{\sqrt{4 - x^2}} \, dx$

5. $\displaystyle\int \frac{dx}{\sqrt{(x + 2)^2 - 1}}$

6. $\displaystyle\int_0^{\pi/4} x \sin(2x) \, dx$

7. $\displaystyle\int v \csc^2 v \, dv$

8. $\displaystyle\int \sin^2 x \cos^3 x \, dx$

9. $\displaystyle\int (4 - x^2)^{3/2} \, dx$

10. $\displaystyle\int \frac{3x^2 + 1}{x^3 + 2x^2 - 3x} \, dx$

11. $\displaystyle\int \frac{x \, dx}{x^4 - 16}$

12. $\displaystyle\int x^3 e^{x^2} \, dx$

13. $\displaystyle\int \frac{y^2 \, dy}{(y + 1)^3}$

14. $\displaystyle\int \frac{dx}{x^2 \sqrt{x^2 + 25}}$

15. $\displaystyle\int \ln(1 - y) \, dy$

16. $\displaystyle\int \frac{x^3 - 2x - 1}{(x^2 + 1)^2} \, dx$

17. $\displaystyle\int \frac{3x^2 + 2}{x^3 - x^2} \, dx$

18. $\displaystyle\int \frac{dy}{\sqrt{2 + 3y^2}}$

19. $\displaystyle\int x^2 \sin^{-1} x \, dx$

20. $\displaystyle\int \sqrt{16 + 9x^2} \, dx$

21. $\displaystyle\int \frac{dx}{x^2 + 2x}$

22. $\displaystyle\int \sin^4 y \cos^4 y \, dy$

23. $\displaystyle\int \frac{w - 2}{1 - w^2} \, dw$

24. $\displaystyle\int \frac{x}{\sqrt{x^2 - 4}} \, dx$

25. $\displaystyle\int \frac{1}{\sqrt{x}} \cos^2 \sqrt{x} \, dx$

26. $\displaystyle\int \sin\left(\frac{\pi}{2} x\right) \sin(\pi x) \, dx$

27. $\displaystyle\int_0^{\sqrt{3}} \frac{x \, dx}{\sqrt{1 + x^2}}$

28. $\displaystyle\int_0^1 \frac{x^2}{\sqrt{4 - x^2}} \, dx$

In Problems 29 and 30, derive each formula where n > 1 is an integer.

29. $\int x^n \tan^{-1} x\, dx = \dfrac{x^{n+1}}{n+1}\tan^{-1} x - \dfrac{1}{n+1}\int \dfrac{x^{n+1}}{1+x^2}\, dx$

30. $\int x^n(ax+b)^{1/2}dx = \dfrac{2x^n(ax+b)^{3/2}}{(2n+3)a}$
$\qquad -\dfrac{2bn}{(2n+3)a}\int x^{n-1}(ax+b)^{1/2}\, dx$

In Problems 31–34, determine whether each improper integral converges or diverges. If it converges, find its value.

31. $\int_1^\infty \dfrac{e^{-\sqrt{x}}}{\sqrt{x}}\, dx$

32. $\int_0^1 \dfrac{\sin\sqrt{x}}{\sqrt{x}}\, dx$

33. $\int_0^1 \dfrac{x\, dx}{\sqrt{1-x^2}}$

34. $\int_{-\infty}^0 xe^x\, dx$

35. Show that $\int_0^{\pi/2} \dfrac{\sin x}{\cos x}\, dx$ diverges.

36. Show that $\int_1^\infty \dfrac{\sqrt{1+x^{1/8}}}{x^{3/4}}\, dx$ diverges.

In Problems 37 and 38, use the Comparison Test for Improper Integrals to determine whether each improper integral converges or diverges.

37. $\int_1^\infty \dfrac{1+e^{-x}}{x}\, dx$

38. $\int_0^\infty \dfrac{x}{(1+x)^3}\, dx$

39. Area Find the area, if it exists, of the region bounded by the graphs of $y = x^{-2/3}$, $y = 0$, $x = 0$, and $x = 1$.

40. (a) Find $\int \dfrac{\cos^2(2x)dx}{\sin^3(2x)}$ using a Table of Integrals.

$\boxed{\text{CAS}}$ **(b)** Find $\int \dfrac{\cos^2(2x)dx}{\sin^3(2x)}$ using a computer algebra system (CAS).

(c) Verify the results from (a) and (b) are equivalent.

Break It Down

*Let's take a closer look at AP® **Practice Problem 7** from Section 6.9 on page 549.*

7. When the region bounded by the graph of $y = \dfrac{1}{x^2}$ and the x-axis to the right of the line $x = 1$ is revolved about the x-axis, the volume V, if it is defined, of the solid of revolution that is generated is given by the improper integral

$$\int_1^\infty \pi\left(\dfrac{1}{x^2}\right)^2 dx.$$

Determine whether the improper integral converges or diverges. If it converges, find the volume of the solid of revolution.

Step 1	Identify the underlying structure and related concepts.	We are asked to determine whether an improper integral, $\int_1^\infty \pi\left(\dfrac{1}{x^2}\right)^2 dx$, converges or diverges, and to find the value of the integral if it converges.
Step 2	Determine the appropriate math rule or procedure.	Since $\int_1^\infty \pi\left(\dfrac{1}{x^2}\right)^2 dx = \pi\int_1^\infty \dfrac{1}{x^4}dx$ is an improper integral, examine the limit $$\pi\cdot\lim_{b\to\infty}\int_1^b \dfrac{1}{x^4}dx$$
Step 3	Apply the math rule or procedure.	$\pi\cdot\lim_{b\to\infty}\int_1^b \dfrac{1}{x^4}dx = \pi\cdot\lim_{b\to\infty}\left[-\dfrac{1}{3x^3}\right]_1^b = \dfrac{\pi}{3}\cdot\lim_{b\to\infty}\left(-\dfrac{1}{b^3}+1\right) = \dfrac{\pi}{3}$ Since the limit exists, the improper integral converges and has the value $\dfrac{\pi}{3}$.
Step 4	Clearly communicate your answer.	The volume of the solid of revolution is $\dfrac{\pi}{3}$ cubic units.

AP® Review Problems: Chapter 6, Part 2
Preparing for the **AP® Exam**

Multiple-Choice Questions

▶ **1.** A function f is continuous and differentiable on the interval $(-1, 6)$. If $f(1) = -3$, $f(5) = 8$, and $\int_1^5 f(x)dx = 10$, then

$$\int_1^5 xf'(x)dx =$$

(A) 1   (B) 33   (C) 11   (D) −6

▶ **2.** $\int x \cos(\pi x)\, dx =$

(A) $\dfrac{x}{\pi} \sin(\pi x) - \dfrac{1}{\pi^2} \cos(\pi x) + C$

(B) $\dfrac{x}{\pi} \sin(\pi x) + \dfrac{1}{\pi^2} \cos(\pi x) + C$

(C) $\dfrac{x}{\pi} \sin(\pi x) + \pi^2 \cos(\pi x) + C$

(D) $\dfrac{x}{\pi} \sin(\pi x) + \dfrac{1}{\pi^2} \sin^2(\pi x) + C$

▶ **3.** $\int x \csc^2 x\, dx =$

(A) $-x \cot x - \csc^2 x + C$

(B) $-x \cot x - \ln|\sin x| + C$

(C) $-x \cot x - \ln|\csc x| + C$

(D) $-x \cot x + \ln|\sin x| + C$

▶ **4.** Determine whether $\displaystyle\int_1^\infty \dfrac{x}{(x^2 + 1)^3}\, dx$ converges or diverges. If it converges, find its value.

(A) $-\dfrac{1}{16}$ (B) $\dfrac{1}{16}$ (C) $\dfrac{1}{4}$ (D) diverges

▶ **5.** Determine whether $\displaystyle\int_0^{10} \dfrac{\ln x}{x}\, dx$ converges or diverges. If it converges, find its value.

(A) 1   (B) $\dfrac{1}{2} \ln 10$

(C) $\ln 10$ (D) diverges

▶ **6.** $\displaystyle\int \dfrac{2x + 4}{(x - 1)(x - 3)}\, dx =$

(A) $3 \ln|x - 3| - 5 \ln|x - 1| + C$

(B) $5 \ln|x - 3| - 3 \ln|x - 1| + C$

(C) $\ln|x - 3| + \ln|x - 1| + C$

(D) $\ln|x - 3| - \ln|x - 1| + C$

Free-Response Question

▶ **7.** An object is moving along the x-axis with velocity $v(t) = t \sin t$, in meters/second, for $0 \le t \le \pi$. At $t = 0$, the object is at the position $x(0) = -2$.

(a) What is the position of the object at $t = \dfrac{\pi}{2}$?

(b) What is the speed of the object at $t = \dfrac{\pi}{2}$?

(c) When is the object moving to the right?

AP® Cumulative Review Problems: Chapters 1–6, Part 2
Preparing for the **AP® Exam**

Multiple-Choice Questions

▶ **1.** $\displaystyle\lim_{x \to 0^+} \dfrac{\sin(2x)}{x} =$

(A) 2   (B) 1

(C) 0   (D) ∞

▶ **2.** $\dfrac{d}{dx} e^{\sin x} =$

(A) $e^{\cos x}$ (B) $\cos x \cdot e^{\sin x}$

(C) $e^{\sin x} \cdot \ln|\sin x|$ (D) $\dfrac{e^{\sin x}}{\cos x}$

▶ **3.** Apple cider is being poured into a large cylindrical tank at Hassan's Apple Orchard. The amount of apple cider in the tank at time t, $0 \le t \le 6$, is provided by a differentiable function, $A(t)$, where t is measured in minutes. Selected values of $A(t)$, measured in gallons, are provided in the table below:

t minutes	0	1	2	3	4	5	6
$A(t)$ gallons	0	19	21	33	54	75	87

Use trapezoidal sums with three subintervals of equal width to determine the average amount of apple cider in the tank over the time interval $0 \le t \le 6$.

(A) 43.5   (B) 67.5   (C) 79   (D) 39.5

4. $\lim\limits_{x \to 0^+} \dfrac{\ln x}{\cot x} =$

(A) 0   (B) ∞

(C) 1   (D) π

5. A marble rolls along a horizontal line. The velocity of the marble at time t is given by $v(t) = 3t^2 - 2$. If the position of the marble is 3 when $t = 1$, what is the position of the marble when $t = 5$?

(A) 27   (B) 75

(C) 119   (D) 130

6. Let f be a twice-differentiable function of x. If $f(0) = 160$ and $f(15) = 10$, which of the following must be true on the interval, $0 \le x \le 15$?

(A) $f'(x) = 0$ for some x in the given interval
(B) $f'(x) = -10$ for some x in the given interval
(C) $f''(x) = 0$ for some x in the given interval
(D) $f''(x) < 0$ for some x in the given interval

7. $f(x) = \sqrt[4]{x}$. Approximate $f(20)$ by using the tangent line to the graph of $f(x)$ at $(16, 2)$.

(A) $\dfrac{9}{4}$ (B) $\dfrac{7}{3}$

(C) $\dfrac{17}{8}$ (D) $\dfrac{25}{12}$

8. $\displaystyle\int_2^\infty \dfrac{3}{x^3}\,dx =$

(A) diverges   (B) $\dfrac{3}{8}$

(C) $\dfrac{3}{4}$ (D) $\dfrac{3}{16}$

9. Suppose f is a differentiable function for all x. Selected values of f and f' are shown in the table below.

x	2	4	7
$f(x)$	16	10	4
$f'(x)$	-8	-6	-2

If $g(x) = f^{-1}(x)$, find $g'(4)$.

(A) $-\dfrac{1}{6}$ (B) $-\dfrac{1}{2}$

(C) $\dfrac{1}{4}$ (D) $\dfrac{1}{16}$

Free-Response Question

10. A ladder 15 feet long is leaning against the wall of a building 30 feet tall. The base of the ladder is being pulled away from the wall of the building at a rate of 2 feet per second. Assume the wall is vertical to the ground.

(a) How fast is the top of the ladder moving down the wall of the building when the bottom of the ladder is 9 feet from the building?

(b) How fast is the top of the ladder moving down the wall of the building when the bottom of the ladder makes an angle of 45° with the ground?

(c) A triangle is formed by the wall of the building, the ladder, and the ground. Find the rate at which the area of the triangle is changing when the bottom of the ladder is 9 feet from the building.

CHAPTER

7

Differential Equations

Tony Skerl/Shutterstock

The Melting Arctic Ice Cap

The Arctic ice cap is melting. "The average thickness of the Arctic ice cover is declining because it is rapidly losing its thick component, the *multi-year ice*. At the same time, the surface temperature of the Arctic is going up, which results in a shorter ice-forming season," said senior research scientist Josefino Comiso.

Each year during the Arctic winter, new sea ice, called **seasonal ice**, forms. Most of the seasonal ice melts during the summer melt season, but some of it survives. Scientists use this pattern to differentiate among three types of Arctic ice: seasonal ice forms and melts each year, **perennial ice** is defined as ice that has remained through at least one summer melt season, and **multi-year ice** is perennial ice that has survived at least three summer melt seasons. In a recent study, NASA scientists have discovered that the thick, old multi-year ice of the Arctic Ocean is melting at a faster rate than that of the thin seasonal ice and the one- and two-year perennial ice.

 The Chapter 7 Project on page 614 investigates the growth of sea ice.

Sakinah
Geologist

AP® Calculus gave me the tools to think critically, be thorough and precise, and build upon my understanding of advanced math. It greatly prepared me for college as I was able to apply those learned concepts to my science courses and along my career path.

In this chapter, we define a differential equation and investigate methods for finding solutions to differential equations. Included are some important applications such as population growth, radioactive decay, Newton's Law of Cooling, and logistic curves. The analyses and solutions of differential equations are a very useful branch of mathematics, but differential equations are often difficult or impossible to solve algebraically. In practical applications, differential equations are often solved using numerical approximations. ∎

7.1 Classification of Ordinary Differential Equations

OBJECTIVES *When you finish this section, you should be able to:*

1 Classify ordinary differential equations (p. 579)
2 Verify the solution of an ordinary differential equation (p. 580)
3 Find the general solution and a particular solution of a differential equation (p. 581)

1 Classify Ordinary Differential Equations

An **ordinary differential equation** is an equation involving an independent variable x, a dependent variable y, and derivatives of y with respect to x. Examples of ordinary differential equations are:

(a) $y' + y + 5 = 0$ **(b)** $\dfrac{x - y}{x + y} + \dfrac{y + x}{x - y}\dfrac{dy}{dx} = 0$ **(c)** $7y'' - 5y' + 4x^3 y = 0$

(d) $y' + y''' = 1$ **(e)** $\left(\dfrac{dy}{dx}\right)^3 = 8\dfrac{d^2 y}{dx^2}$ **(f)** $5x^2 = 6y^3\left(\dfrac{dy}{dx}\right)^2 + 18$

Notice in the above list that some of the differential equations contain derivatives of different orders. For example, (c) has a first-order derivative y' and a second-order derivative y''. The highest-order derivative that appears in a differential equation is called the **order** of the differential equation. For the differential equations listed above, (a), (b), and (f) are first-order, (c) and (e) are second-order, and (d) is a third-order differential equation.

The exponent of the highest-order derivative in a differential equation is called the **degree** of the equation. Equations (a)–(e) above are of degree 1, and equation (f) is of degree 2.

For example,

$$x^2\frac{d^2 y}{dx^2} + 8\frac{dy}{dx} - 5x^3 y = x + 10$$

is a second-order differential equation of degree 1.

NOTE Other variables besides x and y are often used. If $s = f(t)$, then $\dfrac{d^2 s}{dt^2} + t = 0$ is a second-order differential equation.

EXAMPLE 1 **Classifying a Differential Equation**

What are the order and degree of the differential equation $xe^x\dfrac{d^2 y}{dx^2} - 5e^x\dfrac{dy}{dx} + 4x^3 y = 0$?

Solution

This is a second-order differential equation of degree 1. ∎

NOW WORK Problem 9.

2 Verify the Solution of an Ordinary Differential Equation

A function $y = f(x)$ is called a **solution** to a differential equation if, when y and its derivatives are substituted into the equation, the equation is satisfied for all x for which f and its derivatives are defined.

The **general solution** of a differential equation represents all the solutions to the differential equation and is in the form of an equation with arbitrary constants. The number of arbitrary constants in the general solution agrees with the order of the differential equation.

As an example, consider the second-order differential equation

$$\frac{d^2 y}{dx^2} + y = 0$$

It can be shown that the general solution has the form $y = C_1 \sin x + C_2 \cos x$, where C_1 and C_2 are constants.

Any solution of a differential equation obtained from the general solution by assigning values to the arbitrary constants is called a **particular solution**. So, for this example, choosing $C_1 = 6$ and $C_2 = 9$ yields the particular solution $y = 6 \sin x + 9 \cos x$.

⊳ CALC CLIP

EXAMPLE 2   Verifying Solutions to a Differential Equation

(a) Verify that $y = f(x) = x^4 - 5x + 1$ is a solution of the differential equation $y'' - 12x^2 = 0$.

(b) Verify that $y = g(x) = x^4 + C_1 x + C_2$, where C_1 and C_2 are constants, is a solution of the differential equation $y'' - 12x^2 = 0$.

Solution

(a) For $y = f(x) = x^4 - 5x + 1$, we have

$$y' = 4x^3 - 5 \qquad y'' = 12x^2$$

The function $y = f(x)$ satisfies the differential equation $y'' - 12x^2 = 0$, so $y = f(x)$ is a solution.

(b) For $y = g(x) = x^4 + C_1 x + C_2$, we have

$$y' = 4x^3 + C_1 \qquad y'' = 12x^2$$

The function $y = g(x)$ satisfies the differential equation $y'' - 12x^2 = 0$, so $y = g(x)$ is a solution. ■

Notice in Example 2 that if $C_1 = -5$ and $C_2 = 1$, the solution to (b) results in the solution to (a). It can be shown that $y = x^4 + C_1 x + C_2$ is the general solution of the differential equation $y'' - 12x^2 = 0$. So, $y = x^4 - 5x + 1$ is a particular solution.

NOW WORK   Problem 21 and AP® Practice Problems 1 and 2.

In this chapter, we concentrate on first-order differential equations of the form $\dfrac{dy}{dx} = f(x, y)$.

EXAMPLE 3 **Verifying a Solution to a Differential Equation**

Verify that $x^2 - x^3y + 3y^4 = C$, where C is a constant, is a solution to the differential equation $\dfrac{dy}{dx} = \dfrac{3x^2y - 2x}{12y^3 - x^3}$.

Solution

NEED TO REVIEW? Implicit differentiation is discussed in Section 3.2, pp. 248–254.

Differentiate $x^2 - x^3y + 3y^4 = C$ implicitly with respect to x to find $\dfrac{dy}{dx}$.

$$2x - x^3\frac{dy}{dx} - 3x^2y + 12y^3\frac{dy}{dx} = 0$$

$$(12y^3 - x^3)\frac{dy}{dx} = 3x^2y - 2x$$

$$\frac{dy}{dx} = \frac{3x^2y - 2x}{12y^3 - x^3}$$

The function $y = f(x)$ defined by the equation $x^2 - x^3y + 3y^4 = C$ satisfies the first-order differential equation and so is a solution. ∎

NOW WORK Problem **23** and AP® Practice Problem **6**.

③ Find the General Solution and a Particular Solution of a Differential Equation

AP® EXAM TIP

On the AP® exams, the term *initial condition* is typically used instead of *boundary condition*.

Suppose the general solution of a differential equation has been found. A **boundary condition** or an **initial condition** is a restriction on the variables of the general solution that leads to a particular solution.

EXAMPLE 4 **Finding the General Solution and a Particular Solution of a Differential Equation**

(a) Find the general solution of the differential equation $\dfrac{dy}{dx} = 6x^2 + 5$.

(b) Find the particular solution with the initial condition, if $x = 1$, then $y = 10$.

Solution

NEED TO REVIEW? Antiderivatives are discussed in Section 5.6, pp. 405–408.

(a) The general solution of $\dfrac{dy}{dx} = 6x^2 + 5$ consists of all antiderivatives of $6x^2 + 5$, namely,

$$y = 2x^3 + 5x + C$$

where C is a constant.

(b) The initial condition, if $x = 1$, then $y = 10$, is used to find the constant C.

$$\begin{aligned} y &= 2x^3 + 5x + C && \text{General solution} \\ 10 &= 2 \cdot 1^3 + 5 \cdot 1 + C && \text{Initial condition: } x = 1, \, y = 10 \\ C &= 3 \end{aligned}$$

The particular solution of the differential equation $\dfrac{dy}{dx} = 6x^2 + 5$ satisfying the initial condition, if $x = 1$, then $y = 10$, is

$$y = 2x^3 + 5x + 3$$

∎

NOW WORK Problem **31** and AP® Practice Problem **3**.

EXAMPLE 5 Finding the General Solution and a Particular Solution of a Differential Equation

(a) Find the general solution of the differential equation $\dfrac{dy}{dx} = e^x + 4\sin(2x)$.

(b) Find the particular solution with the initial condition, if $x = 0$, then $y = 4$.

Solution

(a) The general solution of $\dfrac{dy}{dx} = e^x + 4\sin(2x)$ consists of all antiderivatives of $e^x + 4\sin(2x)$, namely,

$$y = e^x - 2\cos(2x) + C$$

where C is a constant.

(b) The initial condition, if $x = 0$, then $y = 4$, is used to find the constant C.

$$\begin{aligned} y &= e^x - 2\cos(2x) + C & \text{General solution}\\ 4 &= e^0 - 2\cos 0 + C & \text{Initial condition: } x = 0,\ y = 4\\ C &= 5 \end{aligned}$$

The particular solution of the differential equation $\dfrac{dy}{dx} = e^x + 4\sin(2x)$ satisfying the initial condition, if $x = 0$, then $y = 4$, is

$$y = e^x - 2\cos(2x) + 5$$

∎

NOW WORK Problem 43 and AP® Practice Problems 4 and 5.

7.1 Assess Your Understanding

Concepts and Vocabulary

1. The differential equation $\dfrac{d^2 y}{dx^2} + 5\left(\dfrac{dy}{dx}\right)^3 - y = 0$ is of order _____ and degree _____.

2. **True or False** An initial condition is a restriction on the variables of the general solution of a differential equation that leads to a particular solution.

3. **True or False** $y = -4x^2 + 100$ is a solution of the differential equation $\dfrac{dy}{dx} = -8x$.

4. **True or False** $y = 8e^{5x}$ is a solution of the differential equation $\dfrac{dy}{dx} + 3 = 8e^{5x}$.

Skill Building

In Problems 5–14, state the order and degree of each equation.

5. $\dfrac{dy}{dx} + x^2 y = xe^x$

6. $\dfrac{d^3 y}{dx^3} + 4\dfrac{d^2 y}{dx^2} - 5\dfrac{dy}{dx} + 3y = \sin x$

7. $\dfrac{d^4 y}{dx^4} + 3\dfrac{d^2 y}{dx^2} + 5y = 0$

8. $\dfrac{d^2 y}{dx^2} + y\sin x = 0$

9. $\dfrac{d^2 y}{dx^2} + x\sin y = 0$

10. $\dfrac{d^6 x}{dt^6} + \dfrac{d^4 x}{dt^4} + \dfrac{d^3 x}{dt^3} + x = t$

11. $\left(\dfrac{dr}{ds}\right)^2 = \left(\dfrac{d^2 r}{ds^2}\right)^3 + 1$

12. $x(y'')^3 + (y')^4 - y = 0$

13. $\dfrac{d^2 y}{ds^2} + 3s\dfrac{dy}{ds} = y$

14. $\dfrac{dy}{dx} = 1 - xy + y^2$

In Problems 15–22, verify that the function y is a solution of the differential equation.

15. $y = e^x + 3e^{-x}$, $\dfrac{d^2 y}{dx^2} - y = 0$

16. $y = 5\sin x + 2\cos x$, $\dfrac{d^2 y}{dx^2} + y = 0$

17. $y = \sin(2x)$, $\dfrac{d^2 y}{dx^2} + 4y = 0$

18. $y = \cos(2x)$, $\dfrac{d^2 y}{dx^2} + 4y = 0$

19. $y = \dfrac{3}{1 - x^3}$, $\dfrac{dy}{dx} = x^2 y^2$

20. $y = 1 + e^{-x^2/2}$, $\dfrac{dy}{dx} + xy = x$

21. $y = e^{-2x}$, $y'' + 3y' + 2y = 0$

22. $y = e^x$, $2y'' + y' - 3y = 0$

In Problems 23–30, use implicit differentiation to verify that the given equation is a solution of the differential equation. C is a constant.

23. $y^2 + 2xy - x^2 = C$, $\dfrac{dy}{dx} = \dfrac{x - y}{x + y}$

24. $x^2 - y^2 = Cx^2 y^2$, $\dfrac{dy}{dx} = \dfrac{y^3}{x^3}$ *Hint:* Solve for C first.

25. $x^2 - y^2 = Cx$, $\dfrac{dy}{dx} = \dfrac{x^2 + y^2}{2xy}$ *Hint:* Solve for C first.

7.2 Separable First-Order Differential Equations; Uninhibited and Inhibited Growth and Decay Models

OBJECTIVES *When you finish this section, you should be able to:*

1 Solve a separable first-order differential equation (p. 585)
2 Solve differential equations involving uninhibited growth and decay (p. 587)
3 Solve differential equations involving inhibited growth and decay (p. 590)
   Application: Newton's Law of Cooling (p. 590)

A **first-order differential equation** is of the form

$$\frac{dy}{dx} = f(x, y) \tag{1}$$

where f is continuous on its domain. Recall, first-order differential equations only contain the first derivative of y with respect to x. In this section we discuss how to solve a certain class of first-order differential equations, called *separable first-order differential equations*, and two applications involving uninhibited growth and decay and inhibited growth and decay.

1 Solve a Separable First-Order Differential Equation

In equation (1), suppose $f(x,y)$ can be written in the form

$$f(x, y) = \frac{M(x)}{N(y)}$$

where M is a function of x alone, N is a function of y alone, and both M and N are continuous on their domains. Then the first-order differential equation (1) is called a separable first-order differential equation.

> **DEFINITION Separable First-Order Differential Equation**
>
> A first-order differential equation is said to be **separable** if it can be written in the form
>
> $$\frac{dy}{dx} = \frac{M(x)}{N(y)} \tag{2}$$
>
> where M is a function of x alone, N is a function of y alone, and both M and N are continuous on their domains. In terms of differentials, equation (2) takes the form
>
> $$N(y)\,dy = M(x)\,dx \tag{3}$$

IN WORDS A differential equation is separable if it can be transformed (using algebra) into an equation with {only y's} dy = {only x's} dx.

For example, the differential equation

$$\frac{dy}{dx} = x \sec y$$

is separable because if we treat dy and dx as differentials, $\frac{dy}{dx} = x \sec y$ can be written as

$$\frac{dy}{\sec y} = x\,dx \qquad \text{Multiply both sides by } dx; \\ \text{divide both sides by } \sec y.$$

The differential equation

$$\frac{dy}{dx} = e^x \sin(xy)$$

is not separable because it cannot be written in the form $N(y)dy = M(x)dx$.

To solve the differential equation $\dfrac{dy}{dx} = \dfrac{M(x)}{N(y)}$, we "*separate*" the variables and express $\dfrac{dy}{dx}$ in terms of differentials as

$$N(y)\,dy = M(x)\,dx$$

Now integrate both sides of the separated equation

$$\int N(y)\,dy = \int M(x)\,dx$$

to obtain the general solution.

CALC CLIP **EXAMPLE 1** **Solving a Separable First-Order Differential Equation**

(a) Find the general solution of the separable first-order differential equation

$$\frac{dy}{dx} = 3xy^2 \qquad y \neq 0$$

(b) Find the particular solution that $y = 2$ if $x = 1$.

Solution

(a) Begin with $\dfrac{dy}{dx} = 3xy^2$ and separate the variables to express the equation using differentials.

$$\frac{dy}{y^2} = 3x\,dx$$

$$\int \frac{dy}{y^2} = \int 3x\,dx \qquad \text{Integrate both sides.}$$

$$-\frac{1}{y} = \frac{3x^2}{2} + C$$

This is the general solution to the differential equation, although it can be written in many equivalent forms. The domain of the general solution is restricted to points (x, y), where $y \neq 0$.

(b) We use the general solution with $x = 1$ and $y = 2$ to find the constant C.

$$-\frac{1}{y} = \frac{3x^2}{2} + C$$

$$-\frac{1}{2} = \frac{3 \cdot 1^2}{2} + C \qquad x = 1, \, y = 2$$

$$-\frac{1}{2} = \frac{3}{2} + C$$

$$C = -2$$

Then the particular solution that $y = 2$ if $x = 1$ is

$$-\frac{1}{y} = \frac{3x^2}{2} - 2 \qquad y \neq 0$$

$\blacksquare$

NOW WORK Problems 3, 15, and AP® Practice Problems 1, 2, 3, 4, 5, 6, and 8.

2 Solve Differential Equations Involving Uninhibited Growth and Decay

There are situations in science, nature, and economics such as radioactive decay, population growth, and interest paid on an investment, in which a quantity A varies with time t in such a way that the rate of change of A with respect to t, $\dfrac{dA}{dt}$, is proportional to A itself. These situations can be modeled by the differential equation

$$\boxed{\dfrac{dA}{dt} = kA} \tag{4}$$

where $k \neq 0$ is a real number.

- If $k > 0$ and $\dfrac{dA}{dt} = kA$, the rate of change of A with respect to t is positive, and the amount A is increasing. (Then $\dfrac{dA}{dt}$ is a **growth model**.)

- If $k < 0$ and $\dfrac{dA}{dt} = kA$, the rate of change of A with respect to t is negative, and the amount A is decreasing. (Then $\dfrac{dA}{dt}$ is a **decay model**.)

Suppose the rate of change of a quantity A is proportional to the amount A. If the initial amount A_0 of the quantity is known, then we have the boundary condition, or initial condition, $A = A(0) = A_0$ when $t = 0$.

We solve the differential equation $\dfrac{dA}{dt} = kA$ by separating the variables.

$$\dfrac{dA}{A} = k\,dt \qquad\qquad \text{Separate the variables.}$$

$$\int \dfrac{1}{A}\,dA = \int k\,dt \qquad\qquad \text{Integrate both sides.}$$

$$\ln|A| = kt + C$$

$$\ln A = kt + C \qquad\qquad A > 0$$

The initial condition requires that $A = A_0$ when $t = 0$. Then $\ln A_0 = C$, so

$$\ln A = kt + \ln A_0$$

$$\ln A - \ln A_0 = kt$$

$$\ln \dfrac{A}{A_0} = kt$$

$$\dfrac{A}{A_0} = e^{kt}$$

$$A = A_0 e^{kt}$$

AP® EXAM TIP

On the exam, you might see this written as "The solution of the differential equation $\dfrac{dy}{dt} = ky$ is $y = y_0\,e^{kt}$, where $y_0 = y(0)$ is the initial condition."

The solution to the differential equation $\dfrac{dA}{dt} = kA$ is

$$\boxed{A = A_0 e^{kt}} \tag{5}$$

where A_0 is the initial amount.

Functions $A = A(t)$, whose rates of change are $\dfrac{dA}{dt} = kA$, are said to follow the **exponential law,** or the **law of uninhibited growth or decay**—or in a business context, **the law of continuously compounded interest**. Figure 1 shows the graphs of the function $A(t) = A_0 e^{kt}$ for both $k > 0$ and $k < 0$.

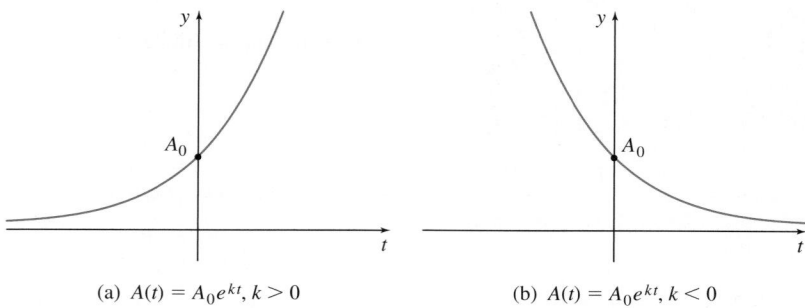

(a) $A(t) = A_0 e^{kt}, k > 0$ (b) $A(t) = A_0 e^{kt}, k < 0$

Figure 1

EXAMPLE 2   Solving a Differential Equation: Uninhibited Growth

NOTE Example 2 is a model of uninhibited growth; it accurately reflects growth in early stages. After a time, bacterial growth no longer continues at a rate proportional to the number present. Factors, such as disease, lack of space, and dwindling food supply, begin to affect the rate of growth.

Assume that a colony of bacteria grows at a rate proportional to the number of bacteria present. If the population of the colony doubles in 5 hours (h), how long does it take for the population of the colony to triple?

Solution

Let $N(t)$ be the number of bacteria present at time t. Then the assumption that this colony of bacteria grows at a rate proportional to the number present can be modeled by

$$\frac{dN}{dt} = kN$$

where k is a positive constant of proportionality. The solution is of the form

$$N(t) = N_0\, e^{kt} \qquad \text{From (5) on page 587}$$

where N_0 is the initial population of the colony. To find k, we use the fact that the initial population N_0 doubles to $2N_0$ in 5 h. That is, $N(5) = 2N_0$.

$$N(5) = N_0\, e^{5k} = 2N_0$$
$$e^{5k} = 2$$
$$5k = \ln 2$$
$$k = \frac{1}{5}\ln 2$$

The time t required for this population to triple satisfies the equation

$$N(t) = 3N_0$$
$$N_0 e^{kt} = 3N_0$$
$$e^{kt} = 3$$
$$kt = \ln 3$$
$$t = \frac{1}{k}\ln 3 = 5\frac{\ln 3}{\ln 2} \approx 8$$
$$\underset{\underset{k = \frac{1}{5}\ln 2}{\uparrow}}{}$$

The population will triple in about 8 h. ∎

NOW WORK Problem 25 and AP® Practice Problems 7 and 9.

Bettmann/Getty Images

ORIGINS Willard F. Libby (1908–1980) grew up in California and went to college and graduate school at UC Berkeley. Libby was a physical chemist who taught at the University of Chicago and later at UCLA. While at Chicago, he developed the methods for using natural carbon-14 to date archaeological artifacts. Libby won the Nobel Prize in Chemistry in 1960 for this work.

bobainsworth/Thinkstock/Getty Images

For a radioactive substance, the **rate of decay** is proportional to the amount of substance present at a given time t. That is, if $A = A(t)$ represents the amount of a radioactive substance at time t, then

$$\frac{dA}{dt} = kA$$

where $k < 0$ and depends on the radioactive substance. The **half-life** of a radioactive substance is the time required for half of the substance to decay.

Carbon dating, a method for determining the age of an artifact, uses the fact that all living organisms contain two kinds of carbon: carbon-12 (a stable carbon) and a small proportion of carbon-14 (a radioactive isotope). When an organism dies, the amount of carbon-12 present remains unchanged, while the amount of carbon-14 begins to decrease. This change in the amount of carbon-14 present relative to the amount of carbon-12 present makes it possible to calculate how long ago the organism died.

EXAMPLE 3 Solving a Differential Equation: Uninhibited Decay

The skull of an animal found in an archaeological dig contains 20% of the original amount of carbon-14. If the half-life of carbon-14 is 5730 years, how long ago did the animal die?

Solution

Let $A = A(t)$ be the amount of carbon-14 present in the skull at time t. Then A satisfies the differential equation $\dfrac{dA}{dt} = kA$, whose solution is

$$A = A_0 e^{kt}$$

where A_0 is the amount of carbon-14 present at time $t = 0$. To determine the constant k, use the fact that when $t = 5730$, half of the original amount A_0 remains.

$$\frac{1}{2} A_0 = A_0 e^{5730k}$$

$$\frac{1}{2} = e^{5730k}$$

$$5730k = \ln \frac{1}{2} = -\ln 2$$

$$k = -\frac{\ln 2}{5730}$$

The relationship between the amount A of carbon-14 present and the time t is

$$A(t) = A_0 e^{(-\ln 2/5730)t}$$

In this skull, 20% of the original amount of carbon-14 remains, so $A(t) = 0.20A_0$.

$$0.20A_0 = A_0 e^{(-\ln 2/5730)t}$$

$$0.20 = e^{(-\ln 2/5730)t}$$

Now take the natural logarithm of both sides.

$$\ln 0.20 = -\frac{\ln 2}{5730} \cdot t$$

$$t = -5730 \cdot \frac{\ln 0.20}{\ln 2} \approx 13{,}300$$

The animal died approximately 13,300 years ago. ∎

NOW WORK   Problem 33.

3 Solve Differential Equations Involving Inhibited Growth and Decay

In uninhibited growth or decay, the amount y either grows without bound or decays to zero. In many situations there is an upper value M that the amount y cannot exceed (inhibited growth), or there is a lower value M that y cannot go below (inhibited decay). For inhibited growth or decay models, the rate of change of y with respect to time t satisfies a different differential equation.

- Inhibited Growth Model:

$$\boxed{\frac{dy}{dt} = k(M - y) \qquad y < M}$$

where $k > 0$ and $M > 0$ are constants. Since $k > 0$ and $y < M$, the derivative $\dfrac{dy}{dt} > 0$ so $y = y(t)$ is increasing. Also, $y < M$, guarantees that the amount y will never be greater than M. Since $y''(t) = -k\dfrac{dy}{dt} < 0$, the graph of $y = y(t)$ is concave down.

- Inhibited Decay Model:

$$\boxed{\frac{dy}{dt} = k(y - M) \qquad y > M}$$

where $k < 0$ and $M > 0$ are constants. Since $k < 0$ and $y > M$, the derivative $\dfrac{dy}{dt} < 0$ so $y = y(t)$ is decreasing. Also, $y > M$ guarantees that the amount y will never be less than M. Since $y''(t) = k\dfrac{dy}{dt} > 0$, the graph of $y = y(t)$ is concave up.

Figures 2 (a) and 2(b) show the graphs of the solutions of the differential equations for inhibited growth and decay. In each graph, the line $y = M$ is a horizontal asymptote.

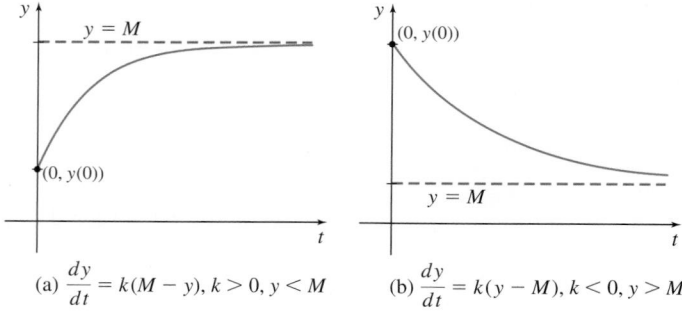

(a) $\dfrac{dy}{dt} = k(M - y), k > 0, y < M$ (b) $\dfrac{dy}{dt} = k(y - M), k < 0, y > M$

Figure 2

Newton's Law of Cooling is an example of an inhibited decay model.

Application: Newton's Law of Cooling

Suppose an object is heated to a temperature u_0. Then at time $t = 0$, the object is put into a medium with a constant lower temperature causing the object to cool. Newton's Law of Cooling states that the rate of change of the temperature of the object with respect to time is continuous and proportional to the difference between the temperature of the object and the constant temperature of the surrounding medium. That is, if $u = u(t)$ is the temperature of the object at time t and if T is the constant temperature, then Newton's Law of Cooling is modeled by the differential equation

$$\boxed{\frac{du}{dt} = k[u(t) - T]} \tag{6}$$

where k is a constant that depends on the object. Since the constant temperature T is lower than $u(0) = u_0$, the object cools and its temperature decreases so that $\dfrac{du}{dt} < 0$. Then, since $T < u(t)$, k is a negative constant.

We find u as a function of t by separating the variables and solving the differential equation $\dfrac{du}{dt} = k\,(u - T)$.

$$\frac{du}{u - T} = k\,dt \qquad \text{Separate the variables.}$$

$$\int \frac{du}{u - T} = \int k\,dt \qquad \text{Integrate both sides.}$$

$$\ln|u - T| = k\,t + C$$

To find C, use the initial condition that at time $t = 0$, the temperature of the object is $u(0) = u_0$. Then

$$\ln|u_0 - T| = k \cdot 0 + C$$

$$C = \ln|u_0 - T|$$

Using this expression for C, we obtain

$$\ln|u - T| = k\,t + \ln|u_0 - T|$$

$$\ln|u - T| - \ln|u_0 - T| = k\,t$$

$$\ln\left|\frac{u - T}{u_0 - T}\right| = k\,t$$

$$\frac{u - T}{u_0 - T} = e^{kt} \qquad T < u_0;\ T < u$$

$$u - T = (u_0 - T)e^{kt}$$

$$\boxed{u(t) = (u_0 - T)\,e^{kt} + T} \qquad (7)$$

EXAMPLE 4   **Using Newton's Law of Cooling**

An object is heated to $90\,°C$ and allowed to cool in a room with a constant temperature of $20\,°C$. If after 10 min the temperature of the object is $60\,°C$, what will its temperature be after 20 min?

Solution

When $t = 0$, $u(0) = 90\,°C$, and when $t = 10$ min, $u(10) = 60\,°C$. Given that the constant temperature $T = 20\,°C$, the differential equation is

$$\frac{du}{dt} = k[u(t) - 20] \qquad \text{Use (6) with } T = 20.$$

$$\frac{du}{u(t) - 20} = k\,dt \qquad \text{Separate the variables.}$$

$$\int \frac{du}{u(t) - 20} = \int k\,dt \qquad \text{Integrate both sides.}$$

$$\boxed{\ln|u(t) - 20| = kt + C} \qquad \text{Evaluate the integrals.} \qquad (8)$$

$$\ln|90 - 20| = k \cdot 0 + C \qquad \text{Substitute } t = 0 \text{ and } u(0) = 90.$$

$$\ln 70 = C \qquad \text{Solve for } C.$$

(continued on the next page)

Then,

$$\ln|u(t) - 20| = kt + \ln 70 \qquad \text{Substitute } C = \ln 70 \text{ in (8).}$$

$$\ln|u(t) - 20| - \ln 70 = kt$$

$$\ln\left|\frac{u(t) - 20}{70}\right| = kt$$

$$\frac{u(t) - 20}{70} = e^{kt} \qquad u(t) > 20$$

$$\boxed{u(t) = 70\, e^{kt} + 20} \qquad \text{Solve for } u(t). \qquad (9)$$

Now use the fact that $u(10) = 60\,^{\circ}\text{C}$ in (9).

$$60 = 70\, e^{10k} + 20$$

$$\frac{40}{70} = e^{10k}$$

$$\frac{1}{10}\ln\frac{4}{7} = k$$

Using $k = 0.1\ln\dfrac{4}{7}$ in (9), the temperature at $t = 20$ min is

$$u(20) = 70\, e^{0.1\ln(4/7)\cdot 20} + 20 \approx 42.857\,^{\circ}\text{C}$$ ∎

See Figure 3 for the graph of $u = u(t)$.

Figure 3 $u(t) = 70\, e^{[0.1\ln(4/7)]t} + 20$, $0 \le t \le 20$

NOW WORK Problem **39** and AP® Practice Problem **10**.

7.2 Assess Your Understanding

Concepts and Vocabulary

1. **True or False** The differential equation $\dfrac{dy}{dx} = y$ can be solved by separating the variables.

2. **True or False** Every first-order differential equation can be solved by separating the variables.

Skill Building

In Problems 3–14, find the general solution of each differential equation.

 3. $\dfrac{dy}{dx} = xy$

4. $\dfrac{dy}{dx} = 5x^2 y^3$

5. $\dfrac{dy}{dx} = \sqrt{xy^2}$

6. $\dfrac{dy}{dx} = \sqrt{xy}$

7. $\dfrac{dy}{dx} = xe^y$

8. $\dfrac{dy}{dx} = 3ye^x$

9. $\dfrac{dy}{dx} = e^x \sin(2y)$

10. $\dfrac{dy}{dx} = e^{3y}\cos x$

11. $\dfrac{dy}{dx} = 5e^{2x - y}$

12. $\dfrac{dy}{dx} = -2e^{5x + y}$

13. $\dfrac{dy}{dx} = x^2(1 + y^2)$

14. $\dfrac{dy}{dx} = x^3\sqrt{1 - y^2}$

In Problems 15–24, find the particular solution of each differential equation using the given initial condition.

586 15. $\dfrac{dy}{dx} = xy^{1/2}$, $y = 1$ if $x = 2$

16. $\dfrac{dy}{dx} = x^{1/2}y$, $y = 1$ if $x = 0$

17. $\dfrac{dy}{dx} = \dfrac{y - 1}{x - 1}$, $y = 2$ if $x = 2$

18. $\dfrac{dy}{dx} = \dfrac{y}{x}$, $y = 2$ if $x = 1$

19. $\dfrac{dy}{dx} = \dfrac{x}{\cos y}$, $y = \pi$ if $x = 2$

20. $\dfrac{dy}{dx} = y\sin x$, $y = e$ if $x = 0$

21. $\dfrac{dy}{dx} = \dfrac{4e^x}{y}$, $y = 2$ if $x = 0$

22. $\dfrac{dy}{dx} = 5ye^x$, $y = 1$ if $x = 0$

23. $\cos x\dfrac{dy}{dx} = y\sin x$, $y = 2$ if $x = \dfrac{\pi}{3}$

24. $(\cos x - 1)\dfrac{dy}{dx} = y\sin x$, $y = 1$ if $x = \dfrac{\pi}{4}$

Applications and Extensions

25. Uninhibited Growth The population of a colony of mosquitoes obeys the uninhibited growth equation $\dfrac{dN}{dt} = kN$. If there are 1500 mosquitoes initially, and there are 2500 mosquitoes after 24 h, what is the mosquito population after 3 days?

26. Radioactive Decay A radioactive substance follows the decay equation $\dfrac{dA}{dt} = kA$. If 25% of the substance disappears in 10 years, what is its half-life?

27. Population Growth The population of a suburb grows at a rate proportional to the population. Suppose the population doubles in size from 4000 to 8000 in an 18-month period and continues at this rate of growth.

(a) Write a differential equation that models the population P at time t in months.

(b) Find the general solution to the differential equation.

(c) Find the particular solution to the differential equation with the initial condition $P(0) = 4000$.

(d) What will the population be in 4 years [$t = 48$]?

28. Uninhibited Growth At any time t in hours, the rate of increase in the area, in millimeters squared (mm^2), of a culture of bacteria is twice the current area A of the culture.

(a) Write a differential equation that models the area of the culture at time t.

(b) Find the general solution to the differential equation.

(c) Find the particular solution to the differential equation if $A = 10 \text{ mm}^2$, when $t = 0$.

29. Radioactive Decay The amount A of the radioactive element radium in a sample decays at a rate proportional to the amount of radium present. The half-life of radium is 1690 years.

(a) Write a differential equation that models the amount A of radium present at time t.

(b) Find the general solution to the differential equation.

(c) Find the particular solution to the differential equation with the initial condition $A(0) = 8$ g.

(d) How much radium will be present in the sample in 100 years?

30. Radioactive Decay Carbon-14 is a radioactive element present in living organisms. After an organism dies, the amount A of carbon-14 present begins to decline at a rate proportional to the amount present at the time of death. The half-life of carbon-14 is 5730 years.

(a) Write a differential equation that models the rate of decay of carbon-14.

(b) Find the general solution to the differential equation.

(c) A piece of fossilized charcoal is found that contains 30% of the carbon-14 that was present when the tree it came from died. How long ago did the tree die?

31. World Population Growth Barring disasters, such as the COVID-19 pandemic, the population P of humans grows at a rate proportional to its current size. According to the U.N. World Population studies, from 2020 to 2025 the population of the more industrialized regions of the world (Europe, North America, Australia, New Zealand, and Japan) is projected to grow at an annual rate of 0.13% per year.

(a) Write a differential equation that models the growth rate of the population.

(b) Find the general solution to the differential equation.

(c) Find the particular solution to the differential equation if in 2020 ($t = 0$), the population of the more industrialized regions of the world was 1.273×10^9.

(d) If the projected rate of growth follows this model, what is the projected population of the more industrialized regions in 2050?

Source: U.N. World Population Prospects: The 2022 Revision.

32. Population Growth in Ecuador Barring disasters, such as the COVID-19 pandemic, the population P of humans grows at a rate proportional to its current size. According to the U.N. World Population studies, from 2021 to 2022 the population of Ecuador grew at an annual rate of 1.257%. Assuming this growth rate continues:

(a) Write a differential equation that models the growth rate of the population.

(b) Find the general solution to the differential equation.

(c) Find the particular solution to the differential equation if in 2021 ($t = 0$), the population of Ecuador was 1.789×10^7.

(d) If the rate of growth continues to follow this model, when will the population of Ecuador reach 20 million persons?

Source: U.N. World Population Prospects: The 2022 Revision.

33. Ötzi the Iceman was found in 1991 by a German couple who were hiking in the Alps near the border of Austria and Italy. Carbon-14 testing determined that Ötzi died 5300 years ago. Assuming the half-life of carbon-14 is 5730 years, what percent of carbon-14 was left in his body? (An interesting note: In September 2010 the complete genome mapping of Ötzi was completed.)

34. Uninhibited Decay Radioactive beryllium is sometimes used to date fossils found in deep-sea sediments. (Carbon-14 dating cannot be used for fossils that lived underwater.) The decay of beryllium satisfies the equation $\dfrac{dA}{dt} = -\alpha A$, where $\alpha = 1.5 \times 10^{-7}$ and t is measured in years. What is the half-life of beryllium?

35. Decomposition of Sucrose Reacting with water in an acidic solution at 35 °C, sucrose ($C_{12}H_{22}O_{11}$) decomposes into glucose ($C_6H_{12}O_6$) and fructose ($C_6H_{12}O_6$) according to the law of uninhibited decay. An initial amount of 0.4 mol of sucrose decomposes to 0.36 mol in 30 min. How much sucrose will remain after 2 h? How long will it take until 0.10 mol of sucrose remains?

36. Chemical Dissociation Salt (NaCl) dissociates in water into sodium (Na^+) and chloride (Cl^-) ions at a rate proportional to its mass. The initial amount of salt is 25 kg, and after 10 h, 15 kg are left.

(a) How much salt will be left after 1 day?

(b) After how many hours will there be less than $\dfrac{1}{2}$ kg of salt left?

37. Voltage Drop The voltage of a certain condenser decreases at a rate proportional to the voltage. If the initial voltage is 20, and 2 s later it is 10, what is the voltage at time t? When will the voltage be 5?

38. Uninhibited Growth The rate of change in the number of bacteria in a culture is proportional to the number present. In a certain laboratory experiment, a culture has 10,000 bacteria initially, 20,000 bacteria at time t_1 minutes, and 100,000 bacteria at $(t_1 + 10)$ minutes.

  (a) In terms of t only, find the number of bacteria in the culture at any time t minutes $(t \geq 0)$.

  (b) How many bacteria are there after 20 min?

  (c) At what time are 20,000 bacteria observed? That is, find the value of t_1.

39. Newton's Law of Heating Newton's Law of Heating states that the rate of change of temperature with respect to time is proportional to the difference between the temperature of the object and the ambient temperature. A thermometer that reads $4\,°C$ is brought into a room that is $30\,°C$.

  (a) Write the differential equation that models the temperature $u = u(t)$ of the thermometer at time t in minutes (min).

  (b) Find the general solution of the differential equation.

  (c) If the thermometer reads $10\,°C$ after 2 min, determine the temperature reading 5 min after the thermometer is first brought into the room.

40. Newton's Law of Cooling A thermometer reading $70\,°F$ is taken outside where the constant temperature is $22\,°F$. Four minutes later, the reading is $32\,°F$.

  (a) Write the differential equation that models the temperature $u = u(t)$ of the thermometer at time t.

  (b) Find the general solution of the differential equation.

  (c) Find the particular solution to the differential equation, using the initial condition that when $t = 0$ min, then $u = 70\,°F$.

  (d) Find the thermometer reading 7 min after the thermometer was brought outside.

  (e) Find the time it takes for the reading to change from $70\,°F$ to within $\dfrac{1}{2}\,°F$ of the air temperature.

41. Forensic Science At 4 p.m., a body was found floating in $12\,°C$ water. When the woman was alive, her body temperature was $37\,°C$ and now it is $20\,°C$. Suppose the rate of change of the temperature $u = u(t)$ of the body with respect to the time t in hours (h) is proportional to $u(t) - T$, where T is the water temperature and the constant of proportionality is -0.159.

  (a) Write a differential equation that models the temperature $u = u(t)$ of the body at time t.

  (b) Find the general solution of the differential equation.

  (c) Find the particular solution of the differential equation, using the initial condition that at the time of death, when $t = 0$ h, her body temperature was $u = 37\,°C$.

  (d) At what time did the woman drown?

  (e) How long does it take for the woman's body to cool to $15\,°C$?

42. Newton's Law of Cooling A pie is removed from a $350\,°F$ oven to cool in a room whose temperature is $72\,°F$.

  (a) Write the differential equation that models the temperature $u = u(t)$ of the pie at time t.

  (b) Find the general solution of the differential equation.

  (c) Find the particular solution to the differential equation, using the initial condition that when $t = 0$ min, then $u = 350\,°F$.

  (d) If $u(5) = 200\,°F$, what is the temperature of the pie after 15 min?

  (e) How long will it take for the pie to be $100\,°F$ and ready to eat?

Preparing for the AP® Exam

AP® Practice Problems

Multiple-Choice Questions

1. Find the solution of the differential equation $\dfrac{dy}{dx} = \dfrac{\cos x}{3y^2}$, with the initial condition $y\left(\dfrac{\pi}{6}\right) = 1$.

  (A) $y^3 = \sin x - \dfrac{1}{2}$ (B) $y = \sin x + \dfrac{1}{2}$

  (C) $y^3 = \sin x + \dfrac{1}{2}$ (D) $y^3 = \sin x + \dfrac{\sqrt{3}}{2}$

2. Which of the following is the solution to the differential equation $\dfrac{dy}{dx} = \dfrac{x}{y}$, with the initial condition $y(0) = 1$?

  (A) $y = \sqrt{x^2 + 1}$ (B) $y = x^2 + 1$

  (C) $y = x + 1$ (D) $y = -\sqrt{x^2 + 1}$

3. Suppose $\dfrac{dy}{dx} = e^y \cos x$, and $y = 0$ when $x = \pi$. Then evaluate y when $x = \dfrac{\pi}{6}$.

  (A) $\ln \dfrac{1}{2}$ (B) $\ln 2$ (C) $\ln \left(1 - \dfrac{\sqrt{3}}{2}\right)^{-1}$ (D) $\dfrac{1}{2}$

4. Solve $\dfrac{dy}{dx} = x^3 y$. Then y equals

  (A) $\dfrac{4}{Cx^4}$ (B) $\dfrac{x^4}{4} + C$ (C) Ce^{3x^2} (D) $Ce^{x^4/4}$

5. If $\dfrac{dy}{dx} = 5y^2$ and $y = 1$ when $x = 3$, then find y when $x = 0$.

(A) $-\dfrac{1}{6}$ (B) $-\dfrac{1}{16}$ (C) $\dfrac{1}{16}$ (D) $\dfrac{1}{6}$

6. If $\dfrac{dy}{dx} = \dfrac{y}{1 + x^2}$ and $y = 1$ if $x = -1$, then y equals

(A) $\dfrac{\pi}{4} e^{\tan^{-1} x}$ (B) $e^{\tan^{-1} x} + \dfrac{\pi}{4}$

(C) $e^{\tan^{-1} x} + e^{\pi/4}$ (D) $e^{\tan^{-1} x + \pi/4}$

7. A population of insects increases according to the uninhibited growth equation $\dfrac{dP}{dt} = kP$, where k is a constant and t is time in days. If the population doubles every 12 days, then k equals

(A) $\dfrac{\ln 2}{12}$ (B) $\dfrac{(\ln 2)^2}{\ln 12}$

(C) $(\ln 2) \ln 12$ (D) $\log_2 12$

8. Find the particular solution of the differential equation $\dfrac{dy}{dx} = e^{x-y}$ if $y = 0$ when $x = 0$.

(A) $y = e^x - 1$ (B) $x + y = 0$
(C) $x = e^y - 1$ (D) $y = x$

9. A colony of bacteria is growing at a rate $\dfrac{dB}{dt} = 6e^{3t/4}$ grams per hour. If initially there are 8 grams of bacteria in the colony, how many grams will be present in 12 hours?

(A) 12 g   (B) 72 g   (C) $6e^9$g   (D) $8e^9$g

Free-Response Question

10. An apple pie is baked to a temperature of 400 °F then placed on a rack to cool in a room with a constant temperature of 70 °F. After 20 min the temperature of the pie is 300 °F. The temperature $u = u(t)$, in °F, of the pie at time t, in minutes, obeys the differential equation

$$\dfrac{du}{dt} = -0.018(u - 70)$$

To the nearest degree, what is the temperature of the pie after 60 min?

Retain Your Knowledge

Multiple-Choice Questions

1. The graph of f is shown below.

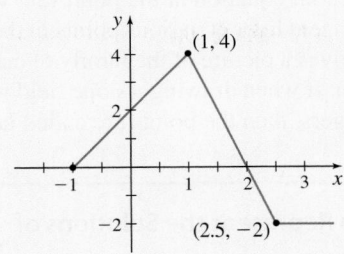

If $g(x) = (4x^2 + 6) \cdot f(x)$, what is $g'(2)$?

(A) 0   (B) 88   (C) −88   (D) −72

2. $\displaystyle\lim_{x \to 0} \dfrac{e^x - \cos x - 3x}{x^2 + 3x} =$

(A) −1   (B) $-\dfrac{2}{3}$ (C) 0   (D) $\dfrac{4}{3}$

3. The limit of the Riemann sum on the interval $\left[\dfrac{\pi}{2}, \dfrac{7\pi}{2}\right]$

$$\lim_{n \to \infty} \sum_{i=1}^{n} \cos\left(\dfrac{\pi}{2} + \dfrac{3\pi}{n} i\right) \dfrac{3\pi}{n} =$$

(A) $\displaystyle\int_{\pi/2}^{7\pi/2} \cos x \, dx$ (B) $\displaystyle\int_{0}^{3\pi} \cos\left(\dfrac{\pi}{2} + x\right) dx$

(C) $\displaystyle\int_{0}^{3\pi} \cos x \, dx$ (D) $\displaystyle\int_{\pi/2}^{7\pi/2} \cos\left(\dfrac{\pi}{2} + x\right) dx$

Free-Response Question

4. The velocity of an object moving on a line is modeled by a differentiable function v, where v is measured in meters and $t \geq 0$ is measured in minutes. The velocity of the object is observed at several times t and posted in the table below.

t	0	3	5	7	8
$v(t)$	0	12	10	0	−8

(a) Using the data in the table, approximate $v'(4)$. Interpret the meaning of $v'(4)$. Be sure to include the correct units.

(b) Show that there is at least one time $0 < t < 8$ at which the acceleration of the object is 0.

7.3 Slope Fields

OBJECTIVE *When you finish this section, you should be able to:*

1 **Use a slope field to represent the solution of a first-order differential equation (p. 596)**

In Sections 7.1 and 7.2 we found the general solution of two kinds of first-order differential equations:

- $\dfrac{dy}{dx} = f(x)$, so $y = \displaystyle\int f(x)\,dx$, where the antiderivative of f can be found.

- $\dfrac{dy}{dx} = \dfrac{M(x)}{N(y)}$, a separable differential equation.

In most instances, finding the general solution of a first-order differential equation $\dfrac{dy}{dx} = f(x, y)$ is not possible. Instead, we use other methods to obtain information about the solution. One method uses *slope fields*.

1 Use a Slope Field to Represent the Solution of a First-Order Differential Equation

In a first-order differential equation $\dfrac{dy}{dx} = f(x, y)$, the function f is continuous on its domain, a set of points in the xy-plane. Although we might not know, nor be able to find, the solution to the differential equation, we can evaluate $f(x_0, y_0)$ at each point (x_0, y_0) in the domain of f. Then $\dfrac{dy}{dx} = f(x_0, y_0)$ equals the slope of the line tangent to the graph of the solution to the differential equation at the point (x_0, y_0). If we draw small line segments representing these tangent lines at various points in the domain of f, the resulting graph, called a **slope field**, gives a picture of the family of curves that make up the solution to the differential equation. If when drawing a slope field we choose only points (x, y) where both x and y are integers, then the points are called **lattice points**.

EXAMPLE 1 Using a Slope Field to Represent the Solutions of $\dfrac{dy}{dx} = y$

(a) Draw a slope field that represents the family of curves that are the solutions of the differential equation $\dfrac{dy}{dx} = y$.

(b) Use the slope field to draw the graphs from the members of the family of solutions of the differential equation that satisfy the initial conditions

    (i) $y = 2$ when $x = 0$,

    (ii) $y = 1$ when $x = 0$,

    (iii) $y = -1$ when $x = 0$, and

    (iv) $y = -2$ when $x = 0$.

Solution

(a) Since $\dfrac{dy}{dx} = y$ does not depend on x, the derivatives $\dfrac{dy}{dx}$ at all points of the form (x, y_0) are equal. For example, at all points $(x, 2)$, the slopes of the lines tangent to the graphs of the solution to the differential equation equal 2. Since the points $(x, 2)$ all lie on the horizontal line $y = 2$, the slope field along $y = 2$ consists of parallel line segments with slope 2. See Figure 4(a). Similarly, along the x-axis, $y = 0$, the slope field consists of parallel line segments with slope 0 (horizontal lines); along $y = -1$, the slope field consists of parallel line segments with slope -1. Figure 4(b) illustrates a slope field of the differential equation $\dfrac{dy}{dx} = y$.

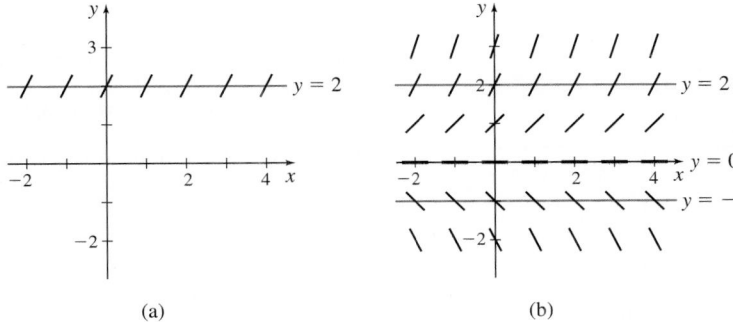

(a)   (b)

Figure 4

(b) (i) Mark the point $(0, 2)$ on the slope field. Then, moving first left and then right from this point, follow the tangent lines and insert a smooth curve. See the blue curve in Figure 4(c).

(ii-iv) Repeat the process used in (i) three more times, using the points $(0, 1)$, $(0, -1)$, and $(0, -2)$. See Figure 4(d), which shows the four graphs of the family of solutions.

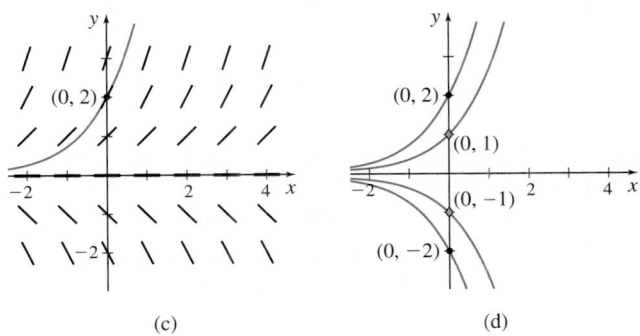

(c)   (d)

Figure 4

Note that the differential equation in Example 1 is separable. From $\dfrac{dy}{y} = dx$, we get $\ln |y| = x + C$ or $y = \pm e^{x+C}$, where C is a constant. This confirms that the graphs in Figure 4(d) are reflections about the x-axis of vertical shifts of the exponential function $y = e^x$.

NOW WORK   Problem 3 and AP® Practice Problem 4.

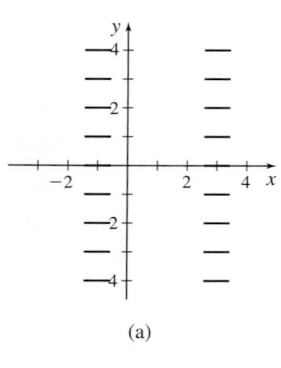

(a)

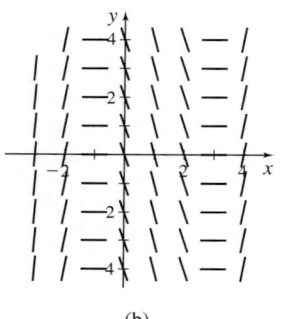

(b)

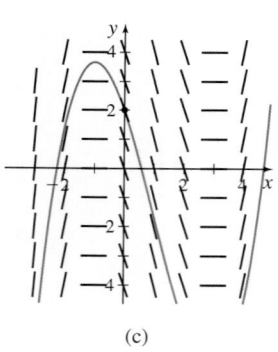

(c)

Figure 5

EXAMPLE 2 **Using a Slope Field to Represent the Solutions**

of $\dfrac{dy}{dx} = x^2 - 2x - 3$

(a) Draw a slope field that represents the solutions of

$$\frac{dy}{dx} = x^2 - 2x - 3$$

(b) Use the slope field to draw the graph of the particular solution of the differential equation with the initial condition, if $x = 0$, then $y = 2$.

Solution

(a) Since $\dfrac{dy}{dx} = x^2 - 2x - 3$ does not depend on y, the derivatives $\dfrac{dy}{dx}$ at all points of the form (x_0, y) are equal. So, for any choice of x_0, the slope field along the vertical line $x = x_0$ consists of short parallel lines. Begin by solving $\dfrac{dy}{dx} = 0$. At these numbers, the tangent lines to the solution of the differential equation are horizontal.

$$\frac{dy}{dx} = x^2 - 2x - 3 = (x - 3)(x + 1)$$

$$(x - 3)(x + 1) = 0$$

$$x = 3 \quad \text{or} \quad x = -1$$

On the vertical lines $x = 3$ and $x = -1$, draw horizontal line segments as illustrated in Figure 5(a). Be sure to extend the lines on both sides of $x = -1$ and $x = 3$.

Next, investigate the behavior of the tangent lines on each side of $x = -1$ and $x = 3$.

• On the line $x = -2$, $\dfrac{dy}{dx} = 5$ (tangent lines have slope 5).

• On the line $x = 0$, $\dfrac{dy}{dx} = -3$ (tangent lines have slope -3).

• On the line $x = 1$, $\dfrac{dy}{dx} = -4$ (tangent lines have slope -4).

• On the line $x = 2$, $\dfrac{dy}{dx} = -3$ (tangent lines have slope -3).

• On the line $x = 4$, $\dfrac{dy}{dx} = 5$ (tangent lines have slope 5).

See Figure 5(b).

(b) The initial condition, if $x = 0$, then $y = 2$, tells us that the particular solution to the differential equation contains the point $(0, 2)$. Plot $(0, 2)$ on the slope field. Then move left and then move right from $(0, 2)$ following the tangent lines to form a smooth curve. See Figure 5(c). ∎

NOW WORK   Problem 13.

EXAMPLE 3 Using a Slope Field to Represent the Solutions of $\dfrac{dy}{dx} = y - x$

Draw a slope field that represents the solutions of the differential equation $\dfrac{dy}{dx} = y - x$.

Solution

The differential equation $\dfrac{dy}{dx} = y - x$ depends on both x and y. It is easiest to investigate the slope field by setting $y - x = k$, or equivalently $y = x + k$, where k is a constant.

- If $k = 0$, then $y = x$, and $\dfrac{dy}{dx} = 0$. Plot horizontal slope lines along the line $y = x$, as shown in Figure 6(a).

- If $k = 1$, then $y = x + 1$, and $\dfrac{dy}{dx} = 1$. Add short lines with slope 1 to the slope field along the line $y = x + 1$, as shown in Figure 6(b).

- If $k = 2$, then $y = x + 2$, and $\dfrac{dy}{dx} = 2$. Add short lines with slope 2 to the slope field along the line $y = x + 2$, as shown in Figure 6(c).

Continue building the slope field by choosing $k = -1$, $k = -2$, and so on, and adding short lines with slopes $\dfrac{dy}{dx} = -1$, $\dfrac{dy}{dx} = -2$ to the graph.

The slope field is shown in Figure 6(d).

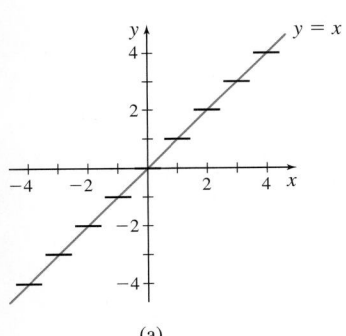
(a)

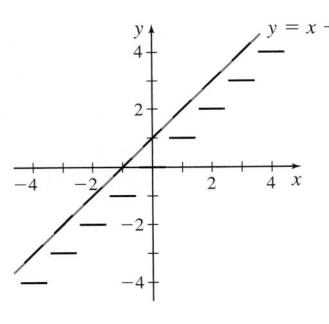
(b)

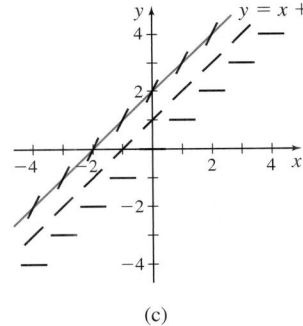
(c)

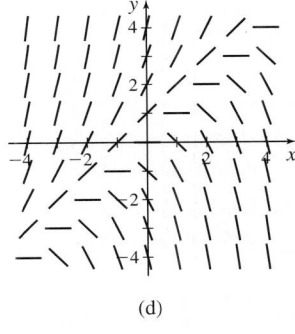
(d)

Figure 6

NOW WORK Problem 11 and AP® Practice Problem 5.

EXAMPLE 4 Interpreting a Slope Field

Which of the following differential equations could have the slope field shown in Figure 7?

(a) $\dfrac{dy}{dx} = -x$ **(b)** $\dfrac{dy}{dx} = x + y$ **(c)** $\dfrac{dy}{dx} = x$ **(d)** $\dfrac{dy}{dx} = x - y$

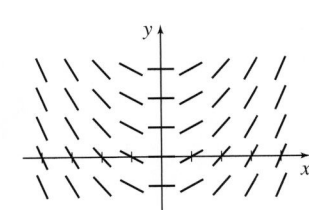

Figure 7

Solution

The tangent lines that lie on a vertical line all have the same slope. That is, for a fixed number x, the slopes $\dfrac{dy}{dx}$ of the tangent lines to the graphs of the solutions of the differential equation are the same. This means $\dfrac{dy}{dx}$ does not depend on y, so the differential equation does not contain y and choices **(b)** and **(d)** can be eliminated. Points in quadrants II and III have negative slopes and points in quadrants I and IV have positive slopes. The differential equation in **(c)** is the only choice that satisfies these conditions. ■

NOW WORK Problem 17 and AP® Practice Problems 1, 2, and 3.

7.3 Assess Your Understanding

Concepts and Vocabulary

1. A(n) _____ gives a picture of the family of curves that make up the solutions to a first-order differential equation.

2. **True or False** A solution to a first-order differential equation equals the slope of the tangent line at each point (x, y) of the solution.

Skill Building

In Problems 3–10,

(a) *Draw a slope field that represents the solutions of each differential equation.*

(b) *Use the slope field to draw some graphs of the family of solutions of the differential equation.*

 3. $\dfrac{dy}{dx} = -2y$ 4. $\dfrac{dy}{dx} = 2x$

5. $\dfrac{dy}{dx} = x^2$ 6. $\dfrac{dy}{dx} = y^2$

7. $\dfrac{dy}{dx} = y^3 - 1$ 8. $\dfrac{dy}{dx} = x^2 - 3x + 2$

9. $\dfrac{dy}{dx} = x + y$ 10. $\dfrac{dy}{dx} = x - y$

In Problems 11–16,

(a) *Draw a slope field that represents the solutions of each differential equation.*

(b) *Use the slope field to draw the particular solution of the differential equation with the given boundary condition.*

 11. $\dfrac{dy}{dx} = \dfrac{1}{2}x + y$; if $x = 0$, then $y = 2$

12. $\dfrac{dy}{dx} = y - \dfrac{1}{2}x$; if $x = 2$, then $y = 0$

 13. $\dfrac{dy}{dx} = x^2 - 1$; if $x = -2$, then $y = 2$

14. $\dfrac{dy}{dx} = y^2 + 1$; if $x = 0$, then $y = -1$

15. $\dfrac{dy}{dx} = e^x$; if $x = 0$, then $y = 0$

16. $\dfrac{dy}{dx} = \dfrac{1}{x}$; if $x = 1$, then $y = 1$

Applications and Extensions

 17. Which of the following differential equations could have the slope field shown below?

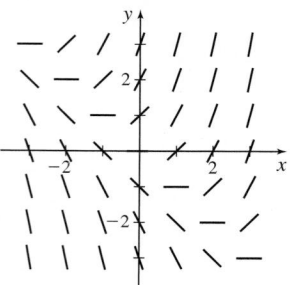

(a) $\dfrac{dy}{dx} = -x$ **(b)** $\dfrac{dy}{dx} = x + y$

(c) $\dfrac{dy}{dx} = x$ **(d)** $\dfrac{dy}{dx} = x - y$

18. Which of the following differential equations could have the slope field shown below?

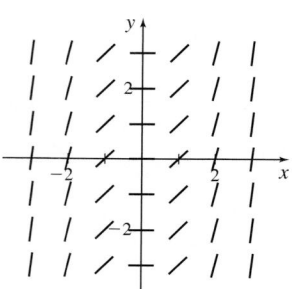

(a) $\dfrac{dy}{dx} = -x$ **(b)** $\dfrac{dy}{dx} = x^2$

(c) $\dfrac{dy}{dx} = 2x + 1$ **(d)** $\dfrac{dy}{dx} = -x^2$

AP® Practice Problems

Multiple-Choice Questions

1. The slope field shown in the figure represents the solution to which differential equation?

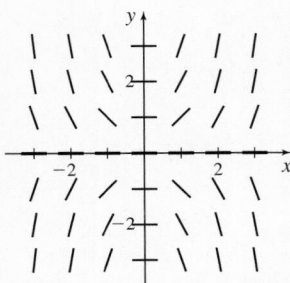

(A) $\dfrac{dy}{dx} = x + y$ (B) $\dfrac{dy}{dx} = xy$

(C) $\dfrac{dy}{dx} = x - y$ (D) $\dfrac{dy}{dx} = \dfrac{x}{y}$

2. Which of the following can be a slope field of $\dfrac{dy}{dx} = \cos x$?

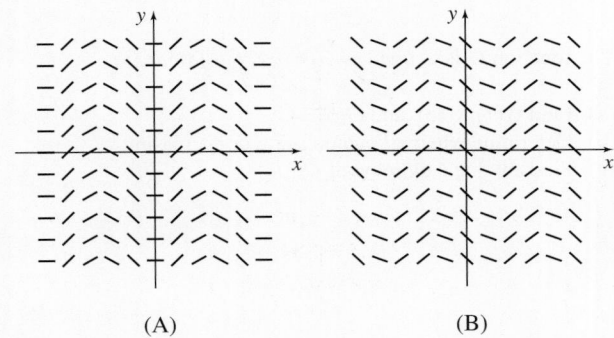

(A)   (B)

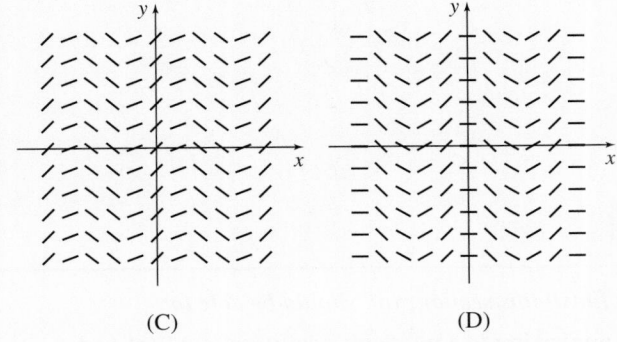

(C)   (D)

3. Which of the following represents the slope field

of $\dfrac{dy}{dx} = x^2 y - 4y$?

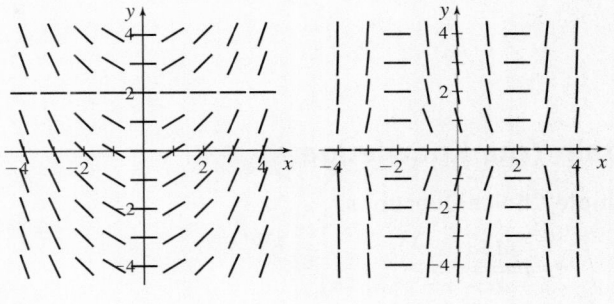

(A)   (B)

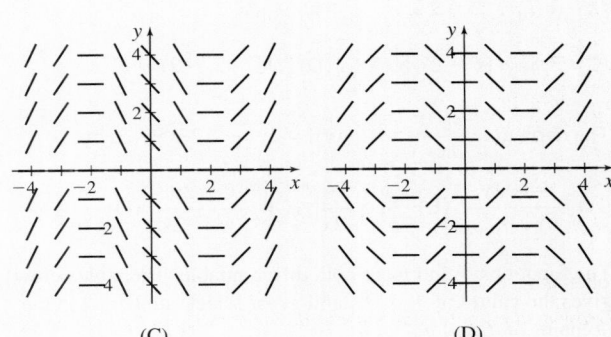

(C)   (D)

Free-Response Questions

4. A slope field for the differential equation $\dfrac{dy}{dx} = y + 2$ is shown below.

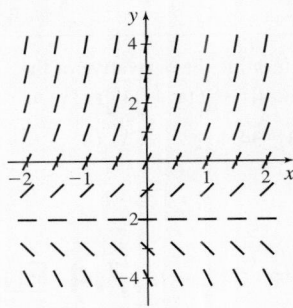

(a) Draw the solution to the differential equation that satisfies the initial condition that $y = -1$ when $x = 0$.

(b) Use separation of variables to find the particular solution

to $\dfrac{dy}{dx} = y + 2$ with the initial condition $y(0) = -1$.

(continued on the next page)

599 **5.** (a) Draw a slope field for the differential equation

$$\frac{dy}{dx} = 2x - y,$$ using the accompanying grid.

(b) Use the slope field in (a) to draw the solution of the differential equation that satisfies the initial condition $y = 0$ when $x = 0$.

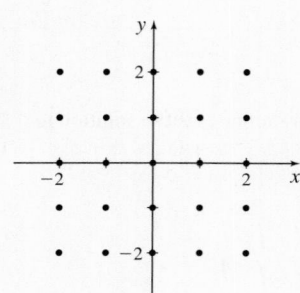

Retain Your Knowledge

Multiple-Choice Questions

1. $\displaystyle\int \frac{3dx}{\sqrt{5 + 4x - x^2}} =$

(A) $3\sin^{-1}\dfrac{x-2}{3} + C$ (B) $\sin^{-1}\dfrac{x-2}{3} + C$

(C) $3\sin^{-1}(x-2) + C$ (D) $\sin^{-1}(x-2) + C$

2. The derivative $\dfrac{dy}{dx}$ of $x^2 + 4xy - 2y^3 = 12$ is given by

(A) $\dfrac{x+2y}{6y^2 - 2x}$ (B) $\dfrac{x+2y}{3y-2x}$ (C) $\dfrac{x+2y}{3y^2 - 2x}$ (D) $\dfrac{2x}{6y^2 - 4x}$

3. The functions f and g are both differentiable. The table below gives the values of f, g, f', and g' for select numbers x in the domains of f and g.

x	$f(x)$	$f'(x)$	$g(x)$	$g'(x)$
2	1	0	-3	2
3	2	4	2	3
6	0	$\dfrac{1}{2}$	8	-1

Use the table to determine the derivative of the function $h(x) = [g(x)]^2 + f(x+1) \cdot g(3x)$ at $x = 2$.

(A) 42   (B) 34   (C) 18   (D) 14

Free-Response Question

4. The graph of a function f, which is continuous for $-3 \le x \le 5$, is shown below.

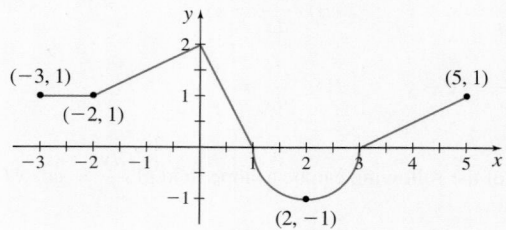

The function G is defined as $G(x) = \displaystyle\int_{-3}^{x} f(t)dt$.

(a) Find $G(1)$, $G(3)$, and $G(5)$.
(b) Determine where G is increasing or decreasing.
(c) Find any local extrema of G.

7.4 Euler's Method

OBJECTIVE *When you finish this section, you should be able to:*

1 Use Euler's method to approximate a particular solution of a first-order differential equation (p. 603)

Euler's method is a numerical algorithm for approximating a particular solution $y = y(x)$ of a first-order differential equation $\dfrac{dy}{dx} = f(x, y)$ with the initial condition if $x = x_0$, then $y = y_0$. It begins with the initial condition (x_0, y_0) and uses a small arbitrary increment h in x often called the **step size**. The first approximation to $y = y(x)$ is given by

$$y_1 = y_0 + hf(x_0, y_0)$$

The second approximation uses y_1 from the first approximation and $x_1 = x_0 + h$. Then

$$y_2 = y_1 + hf(x_1, y_1) \qquad x_1 = x_0 + h$$

Each successive approximation follows the same pattern. The third approximation uses y_2 and $x_2 = x_0 + 2h$, and so on.

$$y_3 = y_2 + hf(x_2, y_2) \qquad x_2 = x_1 + h = x_0 + 2h$$
$$y_4 = y_3 + hf(x_3, y_3) \qquad x_3 = x_2 + h = x_0 + 3h$$
$$\vdots$$
$$y_n = y_{n-1} + hf(x_{n-1}, y_{n-1}) \qquad x_{n-1} = x_0 + (n-1)h$$

Each approximation follows the line tangent to the graph of the solution to the differential equation. That is, it follows the slope field. See Figure 8.

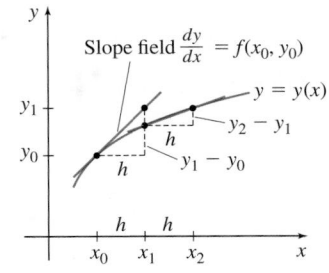

Figure 8 $y_1 = y_0 + hf(x_0, y_0)$

1 Use Euler's Method to Approximate a Particular Solution of a First-Order Differential Equation

EXAMPLE 1 Using Euler's Method

Suppose $y = y(x)$ is the solution to the differential equation

$$\frac{dy}{dx} = xy + x^2$$

with the initial condition if $x = 1$, then $y = 2$. Use Euler's method to approximate $y(1.75)$ using $h = 0.25$ as the step size.

Solution

Begin the first approximation using the initial condition $x_0 = 1$ and $y_0 = 2$, with $h = 0.25$. Then $x_1 = x_0 + h = 1.25$ and

$$\begin{aligned} y(x_1) = y(1.25) \approx y_1 &= y_0 + hf(x_0, y_0) \\ &= 2 + 0.25(1 \cdot 2 + 1^2) \qquad f(x, y) = xy + x^2 \\ &= 2.75 \end{aligned}$$

For the second approximation, we use $x_1 = 1.25$ and $y_1 = 2.75$ with $h = 0.25$. Then $x_2 = x_1 + h = 1.25 + 0.25 = 1.5$, and

$$\begin{aligned} y(x_2) = y(1.5) \approx y_2 &= y_1 + hf(x_1, y_1) \\ &= 2.75 + 0.25[(1.25)(2.75) + 1.25^2] \\ &= 4 \end{aligned}$$

For the third approximation y_3, we use $x_2 = 1.5$ and $y_2 = 4$ with $h = 0.25$. Then $x_3 = x_2 + h = 1.5 + 0.25 = 1.75$, and

$$\begin{aligned} y(x_3) = y(1.75) \approx y_3 &= y_2 + hf(x_2, y_2) \\ &= 4 + 0.25[(1.5)(4) + 1.5^2] \\ &= 6.0625 \end{aligned}$$ ∎

NOW WORK Problem 3 and AP® Practice Problems 1 and 2.

EXAMPLE 2 Using Euler's Method

If $\dfrac{dy}{dx} = yx^2 - y$ and $y(2) = 3$, use Euler's method and two steps of equal size to approximate $y(1.8)$.

Solution

Begin by defining $f(x, y) = yx^2 - y$ and the initial condition $x_0 = 2$ and $y_0 = 3$. Then

$$f(x_0, y_0) = f(2, 3) = 3 \cdot 2^2 - 3 = 9$$

Next, determine the step size h. To get from $x_0 = 2$ to $x = 1.8$ using two steps of equal size requires $h = -0.1$.

Now, use $x_0 = 2$, $y_0 = 3$, and $h = -0.1$. Then $x_1 = x_0 + h = 2 - 0.1 = 1.9$, and the first approximation to $y = y(x)$ is

$$y(x_1) = y(1.9) \approx y_1 = y_0 + hf(x_0, y_0)$$
$$= 3 - 0.1 \cdot 9$$
$$= 2.1$$

The second approximation uses $x_1 = 1.9$, $y_1 = 2.1$, and $h = -0.1$. Then $x_2 = x_1 + h = 1.9 + (-0.1) = 1.8$, and

$$y(x_2) = y(1.8) \approx y_2 = y_1 + hf(x_1, y_1)$$
$$= 2.1 - 0.1(2.1 \cdot 1.9^2 - 2.1)$$
$$= 1.552$$

Using Euler's method with two steps of size -0.1 and the initial condition $y(2) = 3$, an approximation of the particular solution to the differential equation is $y(1.8) \approx 1.552$. ∎

NOW WORK AP® Practice Problems 3 and 4.

7.4 Assess Your Understanding

Concepts and Vocabulary

1. *True or False* Euler's method is used to approximate a particular solution of a first-order differential equation.

2. *True or False* When using Euler's method, the increment h used remains constant for each approximation.

Skill Building

In Problems 3–8, use Euler's method to approximate the particular solution to the differential equation with the given initial condition. Assume $y = y(x)$ is a solution to the differential equation.

3. Approximate $y(0.4)$ if $\dfrac{dy}{dx} = x^2 - y$ and $y = 1$ when $x = 0$. Use $h = 0.2$ as the increment.

4. Approximate $y(0.2)$ if $\dfrac{dy}{dx} = y^2 - x$ and $y = 1$ when $x = 0$. Use $h = 0.1$ as the increment.

5. Approximate $y(1.4)$ if $\dfrac{dy}{dx} = xy^2 - x$ and $y = 2$ when $x = 1$. Use $h = 0.2$ as the step size.

6. Approximate $y(1.2)$ if $\dfrac{dy}{dx} = xy^2 - x$ and $y = 2$ when $x = 1$. Use $h = 0.1$ as the step size.

7. Approximate $y(0.6)$ if $\dfrac{dy}{dx} = 1 - x - 2y$ and $y = 3$ when $x = 1$. Use $h = -0.2$ as the step size.

8. Approximate $y(0.7)$ if $\dfrac{dy}{dx} = 1 - x - 2y$ and $y = 3$ when $x = 1$. Use $h = -0.1$ as the step size.

Applications and Extensions

9. Suppose $\dfrac{dy}{dx} = \cos(\pi x)$ with the initial condition, if $x = 0$, then $y = 1$.

(a) Use Euler's method with step size 0.2 to approximate $y(1)$.

(b) Find the particular solution to the differential equation.

(c) Use the solution from (b) to evaluate $y(1)$.

AP® Practice Problems

Multiple-Choice Question

PAGE 503 1. Suppose $y = f(x)$ is the solution of the differential
   equation $\dfrac{dy}{dx} = x + 2y$ with the initial condition $y(0) = 1$.
   Approximate $f(0.3)$ using Euler's method with
   $(x_0, y_0) = (0, 1)$ and using $h = 0.1$ as the increment.

   (A) 1.20   (B) 1.76   (C) 2.03   (D) 2.78

Free-Response Questions

PAGE 503 2. Suppose $y = f(x)$ is the solution to the differential
   equation $\dfrac{dy}{dx} = 3x - 2y$ with the initial condition $f(1) = 5$.
   What is the approximation for $f(1.2)$ using Euler's method
   starting at $x = 1$ and using two steps of equal size?

PAGE 604 3. Suppose $y = y(t)$ is the particular solution to the differential
   equation

   $$\frac{dy}{dt} = 4y^2 - ty$$

   with the initial condition $y(0) = 1$. Approximate $y(0.4)$ using
   Euler's method and two steps of equal size. Show how you
   reached your conclusion.

PAGE 604 4. A tank contains 81 gal of brine in which 20 lb of salt are
   dissolved. Brine containing 3 lb of dissolved salt per gallon is
   added to the tank at the rate of 5 gal/min. The mixture, kept
   uniform by stirring, drains from the tank at the rate of 2 gal/min.
   The amount s of salt in the tank at time t in minutes is modeled
   by the differential equation

   $$\frac{ds}{dt} = 15 - \frac{2s}{81 + 3t}$$

   Use Euler's method with two steps of equal size to approximate
   the amount of salt in the tank at the end of 4 min, given the
   initial condition $s(0) = 20$.

Retain Your Knowledge

Multiple-Choice Questions

1. $\displaystyle\int_0^\pi x \sin(4x)\, dx =$

   (A) 0   (B) $-\dfrac{1}{4}\pi$

   (C) $\dfrac{1}{4}\pi$ (D) $-\dfrac{1}{4}\pi + \dfrac{1}{16}$

2. Determine whether $\displaystyle\int_{-\infty}^{5} e^{0.001x}\, dx$ converges or diverges.
   If it converges, find its value.

   (A) $1000\, e^{0.005}$ (B) $0.001\, e^{0.001}$

   (C) $1000\,(e^{0.005} - 1)$ (D) diverges

3. If $9xy = x^2 + 2y^2$, then $y'' =$

   (A) $\dfrac{4 - 18y' + 2}{9x - 4y}$ (B) $\dfrac{4(y')^2 + 2}{9x - 4y}$

   (C) $\dfrac{4(y')^2 - 18y' + 2}{9x - 4y}$ (D) $\dfrac{4yy' - 18y' + 2}{9x - 4y}$

Free-Response Question

4. The functions f and g are defined as

   $$f(x) = \int_0^{2x} \sqrt{t^2 + 4}\, dt \qquad g(x) = f(2\sin x)$$

   (a) Find $f'(x)$ and $g'(x)$.
   (b) Find an equation of the tangent line to the graph of g
       at $x = \pi$.

7.5 The Logistic Model

OBJECTIVE *When you finish this section, you should be able to:*

1 Interpret logistic differential equations and logistic functions (p. 606)

The uninhibited growth model $A(t) = A_0 e^{kt}$, $k > 0$, assumes that a population always
grows at a rate proportional to its size. The assumption is valid for the early stages
of growth. But as a population grows and matures, limited resources and other natural
phenomena cause the growth rate to slow and taper off, making the uninhibited growth
model inadequate.

In 1845, Pierre Verhulst proposed a differential equation known as the *logistic differential equation* to model world population. Verhulst's model assumes that a population initially exhibits uninhibited growth. This rate of uninhibited growth is called the **maximum population growth rate** because in Verhulst's model, the population continues to increase over time, but the rate of increase decreases. The model also assumes that the size of the population is bounded from above by some number, called the **carrying capacity** (or *limiting value*).

DEFINITION

The **logistic differential equation** for a population P that follows a logistic model over time t is given by

$$\frac{dP}{dt} = kP\left(1 - \frac{P}{M}\right) \tag{1}$$

where k is the maximum population growth rate and M is the carrying capacity.

1 Interpret Logistic Differential Equations and Logistic Functions

Notice in (1) that when P is small relative to M, then $\frac{P}{M}$ is close to 0 and $\frac{dP}{dt} \approx kP$ so the logistic model approximates the uninhibited growth model. On the other hand, when P is close to the carrying capacity M, then $\frac{P}{M}$ is close to 1 and $\frac{dP}{dt}$ approaches 0, indicating a growth rate close to 0.

Rearrange equation (1) as follows:

$$\frac{dP}{dt} = kP\left(1 - \frac{P}{M}\right) = kP\left(\frac{M-P}{M}\right) = \frac{k}{M}P(M-P)$$

AP® EXAM TIP

The alternate representation of the logistic differential equation

$$\frac{dP}{dt} = \frac{k}{M}P(M-P)$$

is sometimes used on the AP® exam.

DEFINITION (alternate)

An alternate form of the logistic differential equation for a population P that follows a logistic model over time t is

$$\frac{dP}{dt} = \frac{k}{M}P(M-P) \tag{2}$$

In a population in which some members have a disease and some do not, form (2) of the logistic differential equation states that the rate of change of the portion of the population with the disease is proportional to the product of the portion with the disease and the portion of the population without the disease.

EXAMPLE 1 Interpreting a Logistic Differential Equation

Interpret each logistic differential equation:

(a) $\dfrac{dP}{dt} = 0.2P\left(1 - \dfrac{P}{360}\right)$

(b) $\dfrac{dP}{dt} = \dfrac{1}{5000}P(500 - P)$

Solution

(a) To interpret $\dfrac{dP}{dt} = 0.2P\left(1 - \dfrac{P}{360}\right)$, compare it to the logistic differential equation

written in the form $\dfrac{dP}{dt} = kP\left(1 - \dfrac{P}{M}\right)$.

- $M = 360$: The carrying capacity is 360 individuals.
- $k = 0.2$: The maximum growth rate of P is 0.2 or 20%.

(b) To interpret $\dfrac{dP}{dt} = \dfrac{1}{5000}P(500 - P)$, compare it to the logistic differential equation

written in the form $\dfrac{dP}{dt} = \dfrac{k}{M}P(M - P)$.

- $M = 500$: The carrying capacity is 500 individuals.
- $\dfrac{k}{M} = \dfrac{1}{5000}$ so $k = 500 \cdot \dfrac{1}{5000} = \dfrac{1}{10} = 0.10$.

The maximum growth rate of P is 0.10 or 10%. ∎

NOW WORK Problems 11(a), (b), 17(a), (b), and AP® Practice Problems 1, 5, and 6.

The solution of a logistic differential equation is called a **logistic function** and its graph is called a **logistic curve**.

AP® EXAM **TIP**

On the exam, you are expected to be able to identify the carrying capacity M and the maximum growth rate k from either (3) or (4).

THEOREM Logistic Function

The solution to the logistic differential equation

$$\dfrac{dP}{dt} = kP\left(1 - \dfrac{P}{M}\right) \quad \text{or} \quad \dfrac{dP}{dt} = \dfrac{k}{M}P(M - P) \tag{3}$$

is

$$P(t) = \dfrac{M}{1 + ae^{-kt}} \tag{4}$$

where $P(t)$ is the population at time t, M is the carrying capacity, k is the maximum population growth rate, $P_0 = P(0)$, and $a = \dfrac{M - P_0}{P_0}$.

A proof of the theorem is found at the end of the section.

Figure 9 shows the graphs of $P = P(t)$ for a growth model ($k > 0$) and a decay model ($k < 0$).

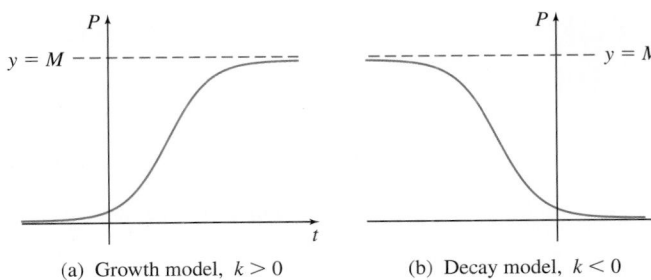

(a) Growth model, $k > 0$ (b) Decay model, $k < 0$

Figure 9 Logistic curves $P(t) = \dfrac{M}{1 + ae^{-kt}}$.

An analysis of the logistic function $P(t) = \dfrac{M}{1 + ae^{-kt}}$ reveals several of its properties.

Properties of the Logistic Function $P(t) = \dfrac{M}{1 + ae^{-kt}}$

- The domain of the logistic function P is the set of all real numbers.
- Since $\dfrac{dP}{dt} = kP\left(1 - \dfrac{P}{M}\right) > 0$ if $k > 0$ and $\dfrac{dP}{dt} < 0$ if $k < 0$, the logistic function $P = P(t)$ increases if $k > 0$ (growth model) and decreases if $k < 0$ (decay model).
- The second derivative of $P = P(t)$ is

$$\frac{d^2 P}{dt^2} = k\left(1 - \frac{2P}{M}\right)\frac{dP}{dt}$$

- $P = P(t)$ has an inflection point when the population equals $\dfrac{M}{2}$.
- If $k > 0$, at the inflection point the rate of change of P is a maximum and P is growing most rapidly.
- If $k < 0$, at the inflection point the rate of change of P is a minimum and P is decreasing most rapidly.
- The function $P = P(t)$ has two horizontal asymptotes $P = 0$ and $P = M$.
- From the information above, we also conclude that $0 < P(t) < M$ for all t.

AP® EXAM TIP

Two important facts:

- The population P is changing most rapidly at the inflection point, that is, at $P = \dfrac{1}{2}M$, where M is the carrying capacity.
- At the inflection point of a growth model, the graph of $P = P(t)$ changes from increasing at an increasing rate to increasing at a decreasing rate.

EXAMPLE 2 **Interpreting a Logistic Function**

For the logistic function $P(t) = \dfrac{650}{1 + 24e^{-0.36t}}$, identify:

(a) the carrying capacity M.

(b) the maximum population growth rate k.

(c) the initial population $P_0 = P(0)$.

(d) the population P when it is growing most rapidly.

Solution

To interpret the logistic function P, compare it to $P(t) = \dfrac{M}{1 + ae^{-kt}}$.

(a) $M = 650$ is the carrying capacity.

(b) $k = 0.36$ is the maximum growth rate of the population.

(c) The initial population $P_0 = P(0)$ is found by evaluating the logistic function at $t = 0$.

$$P(0) = \frac{650}{1 + 24e^{-0.36 \cdot 0}} = \frac{650}{1 + 24 \cdot 1} = \frac{650}{25} = 26$$

Initially there were 26 individuals in the population.

(d) The population is growing most rapidly at the inflection point. This occurs when the population equals $\dfrac{M}{2} = \dfrac{650}{2} = 325$. ∎

NOW WORK Problems **5, 11(c)**, and **17(c)** and AP® Practice Problems **2, 3, 4, 7**, and **8**.

EXAMPLE 3 **Interpreting a Logistic Model: Spread of Disease**

In an experiment involving the spread of disease, one infected mouse is placed in a group of 99 healthy mice. The rate of change of the number of infected mice with respect to time t (in days) follows a logistic model.

(a) If P is the number of infected mice and the maximum daily growth rate of the disease is 10%, write a differential equation that models the experiment.

(b) Write the logistic function that satisfies the differential equation.

(c) Find the time it takes for $\dfrac{1}{2}$ of the population to become infected.

(d) How long does it take for 80% of the population to become infected?

Solution

(a) The total population of mice is 100, the carrying capacity. When $t = 0$, one mouse is infected. The maximum daily growth rate is $k = 0.10$. The differential equation for the experiment is

$$\frac{dP}{dt} = 0.10P \left(1 - \frac{P}{100} \right) \quad P(0) = 1$$

(b) From (a), we know that the carrying capacity M equals 100 mice and the maximum daily growth rate is $k = 0.10$. Since $P(0) = 1$, then

$$a = \frac{M - P_0}{P_0} = \frac{100 - 1}{1} = 99$$

So, the logistic function that satisfies the differential equation is

$$P(t) = \frac{M}{1 + ae^{-kt}} = \frac{100}{1 + 99e^{-0.1t}}$$

(c) Half the population means $P = 50$. Then

$$50 = \frac{100}{1 + 99e^{-0.1t}}$$

$$1 + 99e^{-0.1t} = 2$$

$$-0.1t = \ln \frac{1}{99}$$

$$t = -10(-\ln 99) \approx 45.951$$

Half the population is infected on the 46th day.

(d) 80% of the population is $P = 80$ mice.

$$80 = \frac{100}{1 + 99e^{-0.1t}}$$

$$1 + 99e^{-0.1t} = \frac{5}{4}$$

$$99e^{-0.1t} = \frac{1}{4}$$

$$e^{-0.1t} = \frac{1}{396}$$

$$t = -10 \left(\ln \frac{1}{396} \right) \approx 59.814$$

By 60 days, 80% of the mouse population is infected. ∎

Figure 10 shows the graph of $P = P(t)$

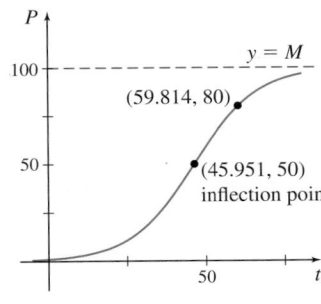

Figure 10 $P(t) = \dfrac{100}{1 + 99e^{-0.1t}}$

NOW WORK Problem 33 and AP® Practice Problem 9.

Proof Logistic Function The logistic function $P = P(t)$ can be found by separating the variables in the logistic differential equation (1).

$$\frac{dP}{dt} = kP\left(1 - \frac{P}{M}\right) \qquad \text{Equation (1)}$$

$$\frac{dP}{P\left(1 - \dfrac{P}{M}\right)} = kdt \qquad \text{Separate the variables.}$$

$$\int \frac{dP}{P\left(1 - \dfrac{P}{M}\right)} = \int kdt \qquad \text{Integrate both sides.} \qquad (5)$$

NEED TO REVIEW? Integrating rational functions using partial fractions is discussed in Section 6.8, pp. 528–535.

The integrand on the left is a rational function with distinct linear factors (Case 1). Express the integrand as

$$\frac{1}{P\left(1 - \dfrac{P}{M}\right)} = \frac{M}{P(M - P)} \qquad \text{Multiply by } \frac{M}{M}.$$

$$\frac{M}{P(M - P)} = \frac{A}{P} + \frac{B}{M - P} \qquad \text{Use Case 1.}$$

$$M = A(M - P) + BP$$

When $P = 0$, the term involving B drops out, leaving $A = 1$. When $P = M$, the term involving A drops out, leaving $B = 1$. Then

$$\frac{M}{P(M - P)} = \frac{1}{P} + \frac{1}{M - P}$$

$$\int \frac{dP}{P\left(1 - \dfrac{P}{M}\right)} = \int \frac{M}{P(M - P)}dP = \int \left(\frac{1}{P} + \frac{1}{M - P}\right)dP$$

$$\int \frac{dP}{P\left(1 - \dfrac{P}{M}\right)} = \ln|P| - \ln|M - P| = \ln\left|\frac{P}{M - P}\right| \qquad (6)$$

Returning to (5), we have

$$\int \frac{dP}{P\left(1 - \dfrac{P}{M}\right)} = \int kdt$$

$$\ln\left|\frac{P}{M - P}\right| = kt + C \qquad \text{From (6)}$$

$$\ln\left|\frac{M - P}{P}\right| = -kt - C$$

$$\frac{M - P}{P} = e^{-kt} \cdot e^{-C}$$

$$\frac{M - P}{P} = ae^{-kt} \qquad (7)$$

where $a = e^{-C}$. Now solve for P.

$$M - P = P(ae^{-kt})$$

$$M = P + P(ae^{-kt})$$

$$P = P(t) = \frac{M}{1 + ae^{-kt}}$$

The value of a is found using the initial condition $P(0) = P_0$ in equation (7). If $t = 0$, then $e^{-kt} = 1$ and

$$a = \frac{M - P_0}{P_0} \qquad \blacksquare$$

7.5 Assess Your Understanding

Concepts and Vocabulary

1. **True or False**   An uninhibited growth model can be used to approximate a logistic growth model when the population P is close to the carrying capacity M.

2. **Multiple Choice**   In a logistic differential equation, $\dfrac{dy}{dx} = \dfrac{k}{M}P(M - P)$, k represents the
   (a) carrying capacity   (b) average growth rate
   (c) maximum growth rate   (d) location of the inflection point

3. **True or False**   All logistic curves have two horizontal asymptotes.

4. **Multiple Choice**   In a logistic model, the population P is growing most rapidly when
   (a) $t = 0$ (b) P is close to 0
   (c) P is close to the carrying capacity M (d) $P = \dfrac{M}{2}$

Skill Building

In Problems 5–10, for each logistic function P, identify

(a) the carrying capacity M.
(b) the maximum population growth rate k.
(c) the initial population $P_0 = P(0)$.
(d) the population P when P is growing most rapidly.

 5. $P(t) = \dfrac{5500}{1 + 99e^{-0.02t}}$

6. $P(t) = \dfrac{120{,}000}{1 + 4999e^{-0.3t}}$

7. $P(t) = \dfrac{660}{1 + 32e^{-0.05t}}$

8. $P(t) = \dfrac{6800}{1 + 16e^{-0.2t}}$

9. $P(t) = \dfrac{150}{1 + 2e^{-0.2t}}$

10. $P(t) = \dfrac{8000}{1 + 63e^{-0.04t}}$

In Problems 11–20, the given logistic differential equation models the rate of change of a population P with respect to time t. For each differential equation, find

(a) the carrying capacity M.
(b) the maximum population growth rate k.
(c) the population when P is growing most rapidly.

 11. $\dfrac{dP}{dt} = 0.12P\left(1 - \dfrac{P}{1000}\right)$ **12.** $\dfrac{dP}{dt} = 0.03P\left(1 - \dfrac{P}{570}\right)$

13. $\dfrac{dP}{dt} = 0.25P\left(1 - \dfrac{P}{4000}\right)$ **14.** $\dfrac{dP}{dt} = 0.08P\left(1 - \dfrac{P}{1600}\right)$

15. $\dfrac{dP}{dt} = 0.002P\left(1 - \dfrac{P}{2800}\right)$ **16.** $\dfrac{dP}{dt} = 0.005P\left(1 - \dfrac{P}{15{,}000}\right)$

 17. $\dfrac{dP}{dt} = 0.00003P(1000 - P)$ **18.** $\dfrac{dP}{dt} = 0.0005P(800 - P)$

19. $\dfrac{dP}{dt} = 0.0002P(1200 - P)$ **20.** $\dfrac{dP}{dt} = 0.00005P(4000 - P)$

In Problems 21 and 22, write the logistic differential equation $\dfrac{dP}{dt}$ whose solution is P.

21. $P(t) = \dfrac{2100}{1 + 29e^{-0.15t}}$ **22.** $P(t) = \dfrac{6000}{1 + 499e^{-0.8t}}$

In Problems 23–28, solve each logistic differential equation for the initial condition $P(0)$.

23. $\dfrac{dP}{dt} = 0.2P\left(1 - \dfrac{P}{800}\right)$; $P(0) = 40$

24. $\dfrac{dP}{dt} = 0.03P\left(1 - \dfrac{P}{1200}\right)$; $P(0) = 25$

25. $\dfrac{dP}{dt} = 0.4P\left(1 - \dfrac{P}{680}\right)$; $P(0) = 20$

26. $\dfrac{dP}{dt} = 0.02P\left(1 - \dfrac{P}{500}\right)$; $P(0) = 8$

27. $\dfrac{dP}{dt} = 0.00075P(400 - P)$; $P(0) = 5$

28. $\dfrac{dP}{dt} = 0.00025P(1000 - P)$; $P(0) = 200$

Applications and Extensions

In Problems 29 and 30, for each logistic function find the population when it is growing most rapidly.

29. $P(t) = \dfrac{100}{1 + 2e^{-0.2t}}$ **30.** $P(t) = \dfrac{60}{1 + 4e^{-0.05t}}$

31. (a) Verify that $P(t) = \dfrac{500}{1 + 4e^{-0.2t}}$ is a solution to the logistic differential equation $\dfrac{dP}{dt} = 0.2P\left(1 - \dfrac{P}{500}\right)$.

   (b) Find the initial condition $P_0 = P(0)$ that yields the solution to (a).

32. Farmers   The number W of farm workers in the United States t years after 1910 follows the logistic decay model

$$\dfrac{dW}{dt} = -0.057W\left(1 - \dfrac{W}{14{,}656{,}248}\right)$$

where $W = 13{,}839{,}705$ when $t = 0$.

   (a) What is the carrying capacity?
   (b) What is the maximum yearly decay rate?
   (c) What is the number of farm workers in the United States at the inflection point?

   Source: U.S. Department of Agriculture

33. Spread of Flu   Suppose one person with the flu is placed in a group of 49 people without the flu. The rate of change of those with the flu with respect to time t (in days) is proportional to the product of the number of those with the flu and the number without flu.

   (a) If P is the number of people with the flu and the maximum daily growth rate is 15%, write a differential equation that models the experiment.
   (b) Write the logistic function that satisfies the differential equation.
   (c) Find the time it takes for $\dfrac{1}{2}$ the population to become infected.
   (d) How long does it take for 80% of the population to become infected?

34. Fruit Fly Population   Four fruit flies are placed in a half-pint bottle with a banana (for food) and yeast (for food and to provide a stimulus to lay eggs). The carrying capacity of the bottle is 230 fruit flies. The rate of change of the fruit fly population P with respect to time t (in days) is proportional to the product of P and $1 - \dfrac{P}{230}$.

  (a) If the maximum daily growth rate is 0.37, write a differential equation that models the experiment.

  (b) Write the logistic function that satisfies the differential equation.

  (c) Find the time it takes for the population to grow to 115 fruit flies.

  (d) How long does it take for the population to grow to 180 fruit flies?

35. In a certain swamp in Florida, 100 insects were released into the swamp. The daily maximum population growth rate is 20%. Entomologists have determined that the swamp can support a colony of 600,000 insects and that the population will follow a logistic growth model.

  (a) Write a logistic differential equation that models the insect population.

  (b) Solve the differential equation.

  (c) Find the time it takes for the population to exceed 100,000 insects.

36. A population P grows according to the logistic differential equation

$$\frac{dP}{dt} = 0.0005\,P(800 - P)$$

  (a) What is the carrying capacity of the population?

  (b) What is the maximum population growth rate of the population?

  (c) How large is the population at the inflection point?

AP® Practice Problems

Multiple-Choice Questions

607 **1.** A population grows according to the logistic differential equation $\dfrac{dP}{dt} = \dfrac{0.01P}{2000}(6000 - P)$. According to the model, the carrying capacity is

  (A) 2000   (B) 4000   (C) 6000   (D) 10,000

608 **2.** A population P grows according to the logistic differential equation $\dfrac{dP}{dt} = 0.08P\left(1 - \dfrac{P}{2000}\right)$, where t is measured in years. What is the population when it is increasing most rapidly?

  (A) 40   (B) 80   (C) 1000   (D) 2000

608 **3.** A population P grows according to the logistic function $P(t) = \dfrac{1200}{1 + 64e^{-0.4t}}$. The maximum growth rate of the population is

  (A) −0.4   (B) 0.4   (C) 0.533   (D) 0.64

608 **4.** Suppose $P(t)$ is the particular solution of the differential equation $\dfrac{dP}{dt} = 0.3P(50 - P)$ with the initial condition if $t = 0$, then $P = 5$. Then $\lim\limits_{t \to \infty} P(t)$ equals

  (A) ∞   (B) 0.3   (C) 15   (D) 50

607 **5.** A virus spreads among a population of N people at a rate proportional to the product of the number of people infected with the virus and the number of people not infected with the virus. Which differential equation can be used to model the situation with respect to time t? Assume k is positive.

  (A) $\dfrac{dP}{dt} = kP$ (B) $\dfrac{dP}{dt} = kN(N - P)$

  (C) $\dfrac{dP}{dt} = kP(P - N)$ (D) $\dfrac{dP}{dt} = \dfrac{kP}{N}(N - P)$

607 **6.** An invasive breed of insect was introduced into a swamp in Florida. Initially 100 insects were present, and their daily maximum population growth rate is 20%. Entomologists have determined that the swamp can support a colony of 600,000 insects and that the population P follows a logistic growth model. Which logistic differential equation models the insect population?

  (A) $\dfrac{dP}{dt} = 20\left(1 - \dfrac{P}{600,000}\right)$

  (B) $\dfrac{dP}{dt} = 100\left(1 - \dfrac{P}{600,000}\right)$

  (C) $\dfrac{dP}{dt} = 0.20P(600,000 - P)$

  (D) $\dfrac{dP}{dt} = 0.20P\left(1 - \dfrac{P}{600,000}\right)$

Free-Response Questions

7. In an attempt to regenerate the American bald eagle population, eagles were captured and released in a federal park in Montana. From previous studies, environmentalists know that the population can be modeled by the logistic function

$$P(t) = \frac{540}{1 + 89e^{-0.162t}}, \text{ where } t \text{ is measured in years.}$$

(a) What is the maximum population growth rate of the American bald eagle?

(b) What is the carrying capacity of the park?

(c) What is the American bald eagle population when it is growing most rapidly?

See the **BREAK IT DOWN** *on page 617 for a stepped out solution to* $AP^{®}$ *Practice Problem 7.*

8. A population P grows according to the logistic differential equation

$$\frac{dP}{dt} = 0.0005P(800 - P)$$

(a) What is the carrying capacity of the population?

(b) What is the maximum population growth rate of the population?

(c) How large is the population when it is increasing most rapidly?

9. The population of hyenas in a game preserve increases according to the logistic function $P(t) = \dfrac{150}{1 + 14e^{-0.125t}}$, where t is measured in years.

(a) How many hyenas were originally in the preserve?

(b) How many hyenas can the preserve maintain?

(c) What is the maximum population growth rate?

(d) How many hyenas are there when the population is growing most rapidly?

Retain Your Knowledge

Multiple-Choice Questions

1. Suppose g is a differentiable function, and

$$\int [g(x) \cdot \cos(2x)]\, dx = \frac{1}{2} g(x) \cdot \sin(2x) - \int x \sin(2x)\, dx$$

Then $g(x)$ could be equal to

(A) x (B) $x^2 + 5$ (C) $\sin(2x)$ (D) $-2x^{1/2}$

2. Find $\displaystyle\int \frac{x+4}{x^2 - x - 2}\, dx$

(A) $2\ln|x - 2| - 3\ln|x + 1| + C$

(B) $2\ln|x - 2| - \ln|x + 1| + C$

(C) $x\ln|x - 2| + 4\ln|x + 1| + C$

(D) $2\ln|x - 2| + \ln|x + 1| + C$

3. Determine whether $\displaystyle\int_0^\infty x^2 e^{-x^3}\, dx$ converges or diverges. If it converges, find its value.

(A) $\dfrac{1}{3}$ (B) $-\dfrac{1}{3}$ (C) -1 (D) diverges

Free-Response Question

4. The function f is differentiable on the interval $(-1, 8)$. A portion of the graph of f is shown below.

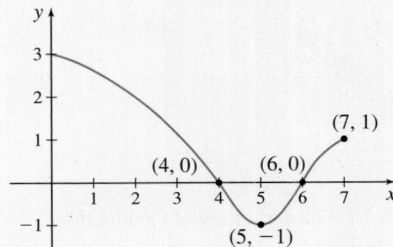

The function $h(x) = \int_0^x f(t)\, dt$, for $0 \le x \le 7$.

(a) On what interval(s) is h increasing? On what interval(s) is h decreasing?

(b) On what interval(s) $0 \le x \le 7$ is h concave up? On what interval(s) is h concave down?

(c) Identify any numbers where h has an inflection point.

(d) Locate any numbers for which h has a relative extreme point. Identify these as relative maximum or relative minimum values.

For (a)–(d), justify your answers.

CHAPTER 7 PROJECT   The Melting Arctic Ice Cap

This Project can be done individually or as part of a team.

In this project, we study the changes in the thickness of sea ice that forms in the Arctic Ocean. There are many factors that affect the behavior of sea ice: abiotic factors such as temperature, both that of the underlying water and air; wind, sea currents, and the motion of the Earth, and biotic factors such as plant and animal life. We examine only one factor: temperature.

We begin with a principle of thermodynamics: The direction of heat transfer is from a warmer to a cooler medium. **Fourier's Law of Conduction** states that the rate of heat flow $\dfrac{dQ}{dt}$ through a homogeneous solid is directly proportional to the area A of the solid perpendicular to the direction of the heat flow, and to the change ΔT in the temperature T, and is inversely proportional to the thickness x of the solid. See Figure 11. The constant of proportionality k depends on the homogeneous solid and is called the **thermal conductivity** of the solid.

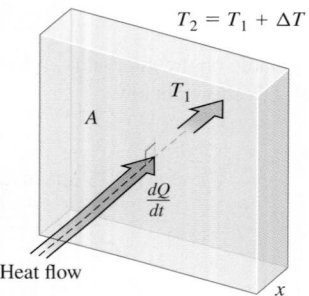

Figure 11 Fourier's Law of Conduction

A simple model depicting sea ice consists of a layer of air, a layer of sea ice, and a layer of seawater. A closer look at this model actually shows another layer directly below the ice. See Figure 12. It is this thin layer of super-cold, almost freezing water that is of interest in this project.

In this model heat flows from the warmer water beneath the ice, through the ice, to the colder air above the ice.

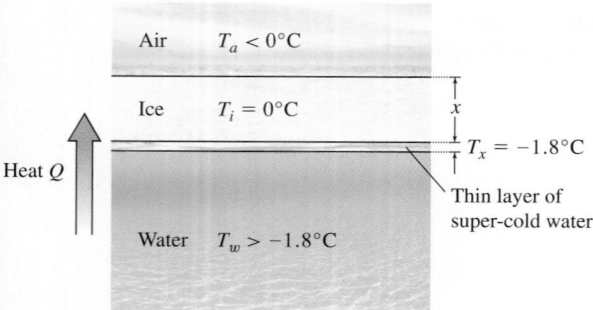

Figure 12 Model of sea ice

1. Write Fourier's Law of Conduction as a differential equation.

2. Using $k = 2.1\ \text{W/m K}$ (the thermal conductivity for sea ice), $A = 1\ \text{m}^2$ as the surface area of the ice, x m as the thickness of the ice, $T_w = -1.8\,°\text{C}$ as the near freezing temperature of the layer of seawater directly beneath the ice, and $T_a = -20\,°\text{C}$ as the (average winter) temperature of Arctic air, write Fourier's Law of Conduction.

Then as the thin layer of seawater directly below the ice loses heat Q, it begins to freeze. The amount of water that freezes satisfies the equation:

$$Q = 334m$$

where 334 J/g is the heat needed to freeze 1 g of water, and m is the mass of the thin layer of water that freezes.

3. Find an expression for the mass m of the layer of water directly below the ice that freezes. (The mass density ρ of freezing seawater is 1.025 g/ml.) Recall that the surface area of the ice is $A = 1\ \text{m}^2$.

4. Find the rate of change in heat Q with respect to the thickness x of the ice.

5. Combine the result from Problem 4 with Fourier's Law of Conductivity (Problem 2) to find the rate of change in thickness of the ice with respect to time.

6. Solve the differential equation found in Problem 5. Then use the initial condition that at time $t = 0$, the thickness of the ice is $x = x_0$.

7. Interpret the solution of the differential equation in Problem 6. Is the rate of change of the thickness of the ice increasing or decreasing over time?

8. Discuss whether this simple model offers a possible explanation for NASA's finding that the multi-year ice is melting faster than seasonal and other perennial ice.

9. Investigate some of the other factors that affect the behavior of sea ice and write a position paper on the impact of at least one these other factors on the depth of ice in the Arctic.

To read more and to see how the Arctic see ice has changed since 1980, visit https://climate.nasa.gov/vital-signs/arctic-sea-ice/

Source: http://www.nasa.gov/topics/earth/features/arctic-antarctic-ice.html

http://www.windows2universe.org/earth/Water/density.html

http://plus.maths.org/content/maths-and-climate-change-melting-arctic

Chapter Review

THINGS TO KNOW

7.1 Classification of Ordinary Differential Equations
- Ordinary differential equation (p. 579)
- Order of a differential equation (p. 579)
- Degree of a differential equation (p. 579)
- General solution; particular solution (p. 580)

7.2 Separable First-Order Differential Equations; Uninhibited and Inhibited Growth and Decay Models

Definitions:
- First-order differential equation $\dfrac{dy}{dx} = f(x, y)$ (p. 585)
- Separable first-order differentiable equation. (p. 585)
- Uninhibited growth model: $\dfrac{dA}{dt} = kA,\ k > 0$ (p. 587)
- Uninhibited decay model: $\dfrac{dA}{dt} = kA,\ k < 0$ (p. 587)
- Half-life of a radioactive substance (p. 589)

- Inhibited growth model:
$$\frac{dy}{dt} = k(M - y),\ y < M, k > 0, M > 0 \quad \text{(p. 590)}$$
- Inhibited decay model:
$$\frac{dy}{dt} = k(y - M),\ y > M, k < 0, M > 0 \quad \text{(p. 590)}$$
- Newton's Law of Cooling (p. 590)

7.3 Slope Fields
A slope field represents the solutions of a first order differential equation. (p. 596)

7.4 Euler's Method
Euler's method is an algorithm used to approximate a particular solution of a first-order differential equation. (p. 603)

7.5 The Logistic Model
- Logistic Differential Equations and the Logistic Function (p. 607)
- Properties of the Logistic Function (p. 608)

OBJECTIVES

Section	You should be able to ...	Examples	Review Exercises	AP® Review Problems
7.1	1 Classify ordinary differential equations (p. 579)	1	1–4	
	2 Verify the solution of an ordinary differential equation (p. 580)	2, 3	5–8	
	3 Find the general solution and a particular solution of a differential equation (p. 581)	4, 5	9, 10, 12, 15	2
7.2	1 Solve a separable first-order differential equation (p. 585)	1	11, 13, 14, 16, 24, 25	1
	2 Solve differential equations involving uninhibited growth and decay (p. 587)	2, 3	22, 23	3
	3 Solve differential equations involving inhibited growth and decay (p. 590)	4		
7.3	1 Use a slope field to represent the solution of a first-order differential equation (p. 596)	1–4	17	5, 7
7.4	1 Use Euler's method to approximate a particular solution of a first-order differential equation (p. 603)	1, 2	21	4
7.5	1 Interpret logistic differential equations and logistic functions (p. 606)	1–3	18, 19, 20	6

REVIEW EXERCISES

In Problems 1–4, state the order and degree of each differential equation.

1. $\dfrac{dy}{dx} + x^2 y = \sin x$

2. $\dfrac{d^2 y}{dx^2} - 5\dfrac{dy}{dx} + 3y = xe^x$

3. $\dfrac{d^3 y}{dx^3} + 4x\dfrac{d^2 y}{dx^2} + \sin^2 x \dfrac{dy}{dx} + 3y = \sin x$

4. $\left(\dfrac{dr}{ds}\right)^3 - r = 1$

In Problems 5–8, verify that the given function is a solution of the differential equation.

5. $y = e^x$, $\quad \dfrac{d^2 y}{dx^2} - y = 0$

6. $y = \dfrac{\ln x}{x^2}$, $\quad \dfrac{d^2 y}{dx^2} + \dfrac{5}{x}\dfrac{dy}{dx} + \dfrac{4}{x^2} y = 0$

7. $y = e^{-3x}$, $\quad \dfrac{d^2 y}{dx^2} + 2\dfrac{dy}{dx} - 3y = 0$

8. $y = x^2 + 3x$, $\quad x\dfrac{dy}{dx} - y = x^2$

In Problems 9–16, (a) find the general solution of each differential equation; (b) find the particular solution for the given initial condition.

9. $\dfrac{dy}{dx} = 3x(4 + e^{x^2})$; if $x = 0$, then $y = 6$

10. $\dfrac{dy}{dx} = \dfrac{3x^3 - 4}{x}$; if $x = e$, then $y = 2$

11. $\dfrac{dy}{dx} = \dfrac{y - 1}{x + 3}$; if $x = -1$, then $y = 0$

12. $\dfrac{dy}{dx} = \cos x - 2\cos x \sin x$; if $x = \dfrac{\pi}{6}$, then $y = -3$

13. $(1 - y)\dfrac{dy}{dx} = y^2 \sin x$; if $x = \pi$, then $y = 1$

14. $\dfrac{dy}{dx} = (3x^2 + 1)\cos^2 y$; if $x = -2$, then $y = \dfrac{\pi}{4}$

15. $\dfrac{dy}{dx} = \dfrac{2x + 1}{3x^2 + 3x + 1}$; if $x = -1$, then $y = 2$

16. $(x^2 - 3)\dfrac{dy}{dx} = -\dfrac{x}{y} \cdot (y^2 + 1)$; if $x = 2$, then $y = -3$

17. (a) Draw a slope field that represents the solutions of the differential equation $\dfrac{dy}{dx} = x^2 + y^2$.

(b) Use the slope field to draw the graph of the particular solution with the initial condition that $y = 1$ if $x = 0$.

18. For the logistic function $P(t) = \dfrac{100}{1 + 24e^{-0.1t}}$, find

(a) the carrying capacity M.

(b) the maximum population growth rate k.

(c) the initial population $P_0 = P(0)$.

(d) the population P when it is increasing most rapidly.

19. The logistic differential equation $\dfrac{dP}{dt} = 0.00003 P(2000 - P)$ models the rate of change of a population P with respect to time t. Find

(a) the carrying capacity M.

(b) the maximum population growth rate k.

(c) the population P when it is increasing most rapidly.

20. Spread of Rumors After a large chip manufacturer posts poor quarterly earnings, the rumor of widespread layoffs spreads among the employees. The rumor spreads throughout the 100,000 employees at a rate that is proportional to the product of the number of employees who have heard the rumor and those who have not. If 10 employees started the rumor, and it has spread to 100 employees after 1 h, how long will it take for 25% of the employees to have heard the rumor?

21. Use Euler's method with increment $h = 0.1$ to approximate the solution $y = y(1.3)$ to the differential equation $\dfrac{dy}{dx} = y + xy^2$ with the boundary condition that $y = 2$ if $x = 1$.

22. Population Growth in China Barring disasters, the population P of humans grows at a rate proportional to its current size. According to the U.N. World Population studies, in 2022 the population of China grew at an annual rate of 0.3% per year.

(a) Write a differential equation that models the growth rate of the population.

(b) Find the general solution of the differential equation.

(c) Find the particular solution of the differential equation if in 2022 ($t = 0$), the population of China was 1.449×10^9.

(d) If the rate of growth continues to follow this model, when will the projected population of China reach 2 billion persons?

Sources: *U.N. World Population Prospects*: The 2022 Revision.

23. Radioactive Decay The amount A of the radioactive element radium in a sample decays at a rate proportional to the amount of radium present. Given the half-life of radium is 1690 years:

(a) Write a differential equation that models the amount A of radium present at time t.

(b) Find the general solution of the differential equation.

(c) Find the particular solution of the differential equation with the initial condition $A(0) = 10$ g.

(d) How much radium will be present in the sample at $t = 300$ years?

In Problems 24 and 25, solve each differential equation using the given initial condition.

24. $\cos y \dfrac{dy}{dx} = \dfrac{\sin y}{x}$; $\quad y = \dfrac{\pi}{3}$ if $x = -1$

25. $\dfrac{dy}{dx} = 3xy$; $\quad y = 4$ if $x = 0$

Break It Down

*Let's take a closer look at parts (a–c) of **AP® Practice Problem 7** from Section 7.5, p. 613.*

7. In an attempt to regenerate the American bald eagle population, eagles were captured and released in a federal park in Montana. From previous studies, environmentalists know that the population can be modeled by the logistic function $P(t) = \dfrac{540}{1 + 89e^{-0.162t}}$, where t is measured in years.

   (a) What is the maximum population growth rate of the American bald eagle?
   (b) What is the carrying capacity of the park?
   (c) What is the American bald eagle population when it is growing most rapidly?

Step 1	Identify the underlying structure and related concepts.	We are given a **logistic function** that models the population P of bald eagles in a park in Montana and are asked to interpret the model.
Step 2	Determine the appropriate math rule or procedure.	The given logistic function is of the form $$P(t) = \dfrac{M}{1 + ae^{-kt}}$$ We know that for a logistic function, M is the carrying capacity, k is the maximum growth rate, and P has an inflection point when $P = \dfrac{M}{2}$.
Step 3	Apply the math rule or procedure.	(a) The maximum population growth rate is $k = 0.162$. (b) The carrying capacity $M = 540$. (c) The population P is growing most rapidly at the inflection point. That is, when the population $= \dfrac{M}{2} = 270$.
Step 4	Clearly communicate your answer.	(a) The maximum population growth rate is 0.162. (b) The carrying capacity is 540 American bald eagles. (c) The population is growing most rapidly when there are 270 bald eagles in the park.

AP® Review Problems: Chapter 7

Multiple-Choice Questions

1. If $\dfrac{dy}{dx} = 2(1 + y^2)x$, then

   (A) $y = x^2 + C$ (B) $\tan^{-1} y = x^2 + C$

   (C) $y = Ce^{x^2}$ (D) $y = \sqrt{e^{x^2}} + C$

2. If the slope of the line tangent to the graph of a function at the point $\left(0, \dfrac{\pi}{6}\right)$ is given by $\dfrac{dy}{dx} = \dfrac{2x + 4}{\cos y}$, then the function is

   (A) $\sin y = x^2 + 4x - \dfrac{1}{2}$

   (B) $\sin y = x^2 + 4x + \dfrac{1}{2}$

   (C) $\cos y = 2x + 4$

   (D) $\sin y = 2x^2 + 4x$

3. A tumor is growing with respect to time t at a rate proportional to its mass m. A differential equation that describes the tumor's growth is

   (A) $\dfrac{dm}{dt} = km, \ k < 0$ (B) $\dfrac{dm}{dt} = km, \ k > 0$

   (C) $\dfrac{dm}{dt} = kt, \ k > 0$ (D) $\dfrac{dm}{dt} = kmt, \ k > 0$

4. Suppose $y = y(x)$ is the solution to the differential equation $\dfrac{dy}{dx} = 2x - y$ with the initial condition $y(0) = 0$. Approximate $y(0.6)$ using Euler's method with $h = 0.2$ as the step size.

   (A) 0.008   (B) 0.224   (C) 0.419   (D) 0.800

▶ **5.** The slope field shown in the figure represents the solutions to which differential equation?

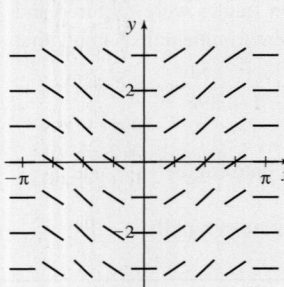

(A) $\dfrac{dy}{dx} = -x^4$ (B) $\dfrac{dy}{dx} = \cos x$

(C) $\dfrac{dy}{dx} = \sin x$ (D) $\dfrac{dy}{dx} = x^3$

Free-Response Questions

▶ **6.** The population P of a community is modeled by the logistic differential equation

$$\frac{dP}{dt} = 0.000025\,P(4000 - P)$$

(a) What is the carrying capacity of the community?

(b) What is the maximum population growth rate?

(c) How large is the population when it is growing most rapidly?

▶ **7.** (a) Using the accompanying grid, draw a slope field for the differential equation

$$\frac{dy}{dx} = \frac{1}{y - 1}$$

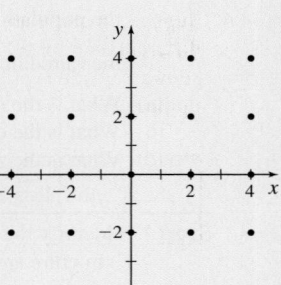

(b) Use the slope field in (a) to draw the solution to the differential equation that satisfies the initial condition $x = 2$ when $y = -2$.

(c) Find the particular solution to $\dfrac{dy}{dx} = \dfrac{1}{y - 1}$ that satisfies the initial condition $x = 2$ when $y = -2$.

AP® Cumulative Review Problems: Chapters 1–7

Preparing for the AP® Exam

Multiple-Choice Questions

▶ **1.** $\displaystyle\lim_{x \to 0} \left(\frac{2e^x - x - 2}{x} \right) =$

(A) 1   (B) $e - 2$ (C) 0   (D) ∞

▶ **2.** If $y^2 - 3x^2 y - 6 = 0$, then $\dfrac{dy}{dx} =$

(A) $\dfrac{6}{y - 3x}$ (B) $\dfrac{6xy}{2y - 3x^2}$

(C) $\dfrac{6 + 3xy}{2y - 3x^2}$ (D) $\dfrac{3xy}{2y + 3x^2}$

▶ **3.** $f'(x) = \dfrac{-3x + 1}{1 + x^2}$, on what interval(s) is f increasing?

(A) $\left(-1, \dfrac{1}{3} \right)$ (B) $\left(\dfrac{1}{3}, 1 \right)$

(C) $\left(-\infty, \dfrac{1}{3} \right)$ (D) $(-\infty, 0)$ and $\left(\dfrac{1}{3}, +\infty \right)$

▶ **4.** $\dfrac{d}{dx} \displaystyle\int_e^{x^2} \ln\left(t^3 + 1\right) dt =$

(A) $\dfrac{1}{x^6 + 1}$ (B) $\ln\left(x^6 + 1\right) - \ln\left(e^3 + 1\right)$

(C) $2x \ln\left(x^6 + 1\right)$ (D) $\ln\left(x^6 + 1\right)$

▶ **5.** $\lim\limits_{h\to 0} \dfrac{(x+h)^2 + 2x + 2h - (x^2 + 2x)}{h} =$

    (A) 0   (B) $4x$ (C) $\dfrac{x^2 + 2x + 1}{2}$ (D) $2x + 2$

▶ **6.** Suppose f and g are functions that are differentiable for all real numbers. The table below shows the values of f, g, f', and g' for the number 4.

x	$f(x)$	$f'(x)$	$g(x)$	$g'(x)$
4	-2	-5	3	7

If $h(x) = \dfrac{f(x)}{g(x)}$, find $h'(4)$.

    (A) $\dfrac{5}{7}$ (B) 7   (C) $-\dfrac{1}{9}$ (D) $-\dfrac{3}{2}$

▶ **7.** A ball is dropped from a height of 1000 ft above the ground. The distance s (in feet) the ball falls after t seconds is $s(t) = 16t^2$. Let $A =$ the average velocity of the ball for the first 3 seconds that it is falling and $B =$ the instantaneous velocity of the ball at 3 seconds. Find $|A - B|$.

    (A) 48   (B) 0   (C) 96   (D) 24

▶ **8.** $\lim\limits_{h\to 0} \dfrac{\cos\left(\dfrac{\pi}{3} + h\right) - \cos\dfrac{\pi}{3}}{h}$

    (A) $\dfrac{1}{2}$ (B) $\dfrac{\sqrt{3}}{2}$ (C) $-\dfrac{\sqrt{3}}{2}$ (D) ∞

▶ **9.** $\lim\limits_{x\to\infty} \dfrac{(\ln x)^2}{\sqrt{x}} =$

    (A) ∞ (B) 0   (C) 2   (D) 4

▶ **10.** $\displaystyle\int_1^2 \dfrac{x^2}{\sqrt{x^3 + 8}}\, dx =$

    (A) $\dfrac{1}{6}$ (B) $\dfrac{2}{3}$ (C) $\dfrac{2}{3}(\sqrt{2} - 1)$ (D) $\dfrac{74}{3}$

Free-Response Questions

▶ **11.** An object moving along a line has velocity $v = v(t) = 3t^2 - 9t + 6$ ft/s, where $t \geq 0$ is the time. Suppose the object begins its motion at the origin.

    (a) Find the acceleration a of the object.

    (b) Find the acceleration when the object is at rest.

    (c) When is the velocity of the object increasing?

▶ **12.** Consider $\displaystyle\int_{\pi/4}^{3\pi/2} \dfrac{\cos x}{x}\, dx$.

    (a) Approximate $\displaystyle\int_{\pi/4}^{3\pi/2} \dfrac{\cos x}{x}\, dx$ using a trapezoidal sum with five subintervals of equal width.

    (b) Approximate $\displaystyle\int_{\pi/4}^{3\pi/2} \dfrac{\cos x}{x}\, dx$ using a graphing calculator.

    (c) Does $\displaystyle\int_{\pi/4}^{3\pi/2} \dfrac{\cos x}{x}\, dx$ represent area? Justify your answer.

▶ **13.** At time $t = 0$, a tank contains 50 gallons of fuel oil. For $0 \leq t \leq 5$, fuel oil is pumped into the tank at a rate of $F(t) = 16t \sin\left(\dfrac{\pi t}{5}\right)$ gallons per minute. Fuel leaks out of the tank through an undetected crack at a rate of $C(t) = \dfrac{5t^2}{t + 3}$ gallons per minute. How many gallons of fuel oil are in the tank at time, $t = 5$? Justify your answer

CHAPTER

8

Applications of the Integral

David R. Frazier Photolibrary, Inc./Alamy Stock Photo

The Cooling Tower at Callaway Nuclear Generating Station

The Callaway Nuclear Generating Station is located in Callaway County, Missouri. Power stations often use cooling towers like the one seen in the photograph above to remove waste and to heat and cool the water that cycles through the system. The Callaway station's energy-producing turbines are powered by steam. Once that steam travels through the turbine system, it passes over a series of tubes carrying cooled water from the tower. As heat from the steam is absorbed by the cooling water, the steam condenses back into water, which is then pumped back into steam generators, and the cycle repeats itself.

The cooling tower at Callaway is 553 ft tall and 430 ft wide at the base and has the capacity to cool 585,000 gallons of water per minute. Water enters the tower with a temperature of about 125 °F and the tower cools it down to 95 °F. Cooling towers like this one are examples of hyperboloid structures.

 The Chapter 8 Project on page 694 investigates the design of cooling towers and determines the quantity of concrete required and the cost of building such a structure.

Charu
Infectious Disease Doctor

I design and oversee clinical trials for vaccines. The principles that I learned in AP® Calculus are instrumental in vaccine development by ensuring that a clinical study appropriately assesses immune response, safety, and effectiveness.

The applications of the definite integral in this chapter rely on two facts:

- If a function f is continuous on a closed interval $[a, b]$, then f is integrable over $[a, b]$. That is,

$$\int_a^b f(x)\,dx = \lim_{n \to \infty} \sum_{i=1}^n f(u_i)\Delta x = \text{a number}, \quad \Delta x = \frac{b-a}{n}$$

- If F is any antiderivative of f on (a, b), then $\int_a^b f(x)\,dx = F(b) - F(a)$.

We begin by using a definite integral to solve problems involving motion on a line. We then use a definite integral to solve geometry problems, finding:

- the area of a region bounded by the graphs of two or more functions;
- the volume of a solid;
- the length of the graph of a function;
- the surface area of a solid.

Finally, we use a definite integral to find:

- the work done by a variable force. ∎

8.1 Applications Involving Motion Along a Line

OBJECTIVES *When you finish this section, you should be able to:*

1 Use an integral to find the total distance traveled and the net displacement of an object moving along a line (p. 621)

2 Determine the average velocity or average acceleration of an object moving along a line (p. 623)

3 Connect position, velocity, and acceleration of an object moving along a line (p. 624)

AP® EXAM TIP

BREAK IT DOWN: As you work through the problems in this chapter, remember to follow these steps:

Step 1 Identify the underlying structure and related concepts.

Step 2 Determine the appropriate math rule or procedure.

Step 3 Apply the math rule or procedure.

Step 4 Clearly communicate your answer.

On page 697, see how we've used these steps to solve Section 8.3 AP® Problem 13(b) on page 653.

1 Use an Integral to Find the Total Distance Traveled and the Net Displacement of an Object Moving Along a Line

How far an object travels and where an object ends in relation to its starting point are two common questions that involve motion on a line. Integration is used to answer both questions. *Total distance traveled* describes how far the object went, and *net displacement* answers how far the object's endpoint is from its starting point. The definitions follow.

DEFINITION Net Displacement; Total Distance Traveled

Suppose an object is moving along a line with velocity $v = v(t)$, where v is continuous for $a \leq t \leq b$.

- The **net displacement** of an object whose motion begins at $t = a$ and ends at $t = b$ is given by

$$\boxed{\int_a^b v(t)dt}$$

- The **total distance** traveled by the object from $t = a$ to $t = b$ is given by

$$\boxed{\int_a^b |v(t)|dt}$$

- If $v(t) \geq 0$, for $a \leq t \leq b$, then the net displacement and the total distance traveled are equal.

Remember speed and velocity are not the same thing. Speed measures how fast an object is moving regardless of its direction. Velocity measures both the speed and the direction of an object. Because of this, velocity can be positive, negative, or zero. So $\int_a^b v(t)\,dt$ is a sum of signed distances (net displacement). Speed is the absolute value of velocity, so $\int_a^b |v(t)|\,dt$ is a sum of actual distances traveled (total distance).

EXAMPLE 1   **Interpreting an Integral Involving Motion Along a Line**

An object is moving along a horizontal line with velocity $v(t) = 3t^2 - 18t + 15$ (in meters per minute) for time t, $0 \le t \le 6$.

(a) Find the net displacement of the object from $t = 0$ to $t = 6$. Interpret the result in the context of the problem.

(b) Find the total distance traveled by the object from $t = 0$ to $t = 6$ and interpret the result.

Solution

(a) The net displacement of the object from $t = 0$ to $t = 6$ is given by

$$\int_0^6 v(t)\,dt = \int_0^6 (3t^2 - 18t + 15)\,dt = \left[t^3 - 9t^2 + 15t\right]_0^6 = 6^3 - 9\cdot 6^2 + 15\cdot 6 - 0 = -18$$

After moving from $t = 0$ to $t = 6$ min, the object is 18 m to the left of where it was at $t = 0$ min.

(b) The total distance traveled by the object from $t = 0$ to $t = 6$ min is given by $\int_0^6 |v(t)|\,dt$. Since $v(t) = 3t^2 - 18t + 15 = 3(t - 1)(t - 5)$, we find that $v(t) \ge 0$ if $0 \le t \le 1$ or if $5 \le t \le 6$, and $v(t) \le 0$ if $1 \le t \le 5$. Now express $|v(t)|$ as the piecewise-defined function:

$$|v(t)| = \begin{cases} v(t) = 3t^2 - 18t + 15 & \text{if } 0 \le t \le 1 \\ -v(t) = -(3t^2 - 18t + 15) & \text{if } 1 < t < 5 \\ v(t) = 3t^2 - 18t + 15 & \text{if } 5 \le t \le 6 \end{cases}$$

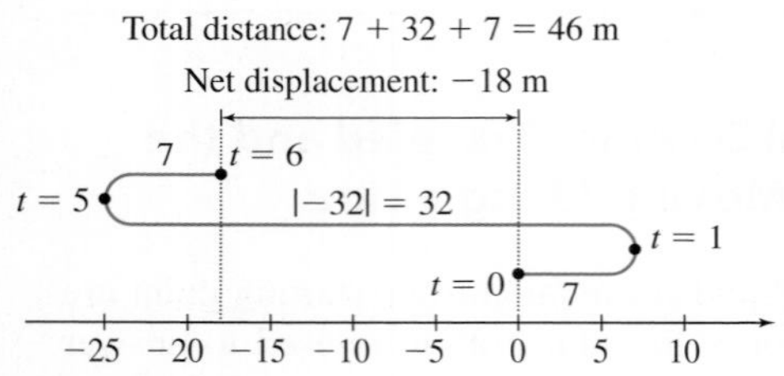

Then

$$\int_0^6 |v(t)|\,dt = \int_0^1 v(t)\,dt + \int_1^5 [-v(t)]\,dt + \int_5^6 v(t)\,dt$$

$$= \left[t^3 - 9t^2 + 15t\right]_0^1 - \left[t^3 - 9t^2 + 15t\right]_1^5 + \left[t^3 - 9t^2 + 15t\right]_5^6$$

$$= (7 - 0) - (-25 - 7) + [-18 - (-25)] = 7 + 32 + 7 = 46$$

The object traveled a total distance of 46 m from $t = 0$ to $t = 6$ min.   ∎

Figure 1 shows the distinction between the net displacement and the total distance traveled by the object, assuming that the object is at the origin when the motion starts.

📐 Solving part (b) of Example 1 required evaluating the integral of an absolute value, which required finding a piecewise-defined function. Integrals involving an absolute value are evaluated more easily using technology. Figure 2 shows the solution to part (b) of Example 1 using a TI-84 Plus CE.

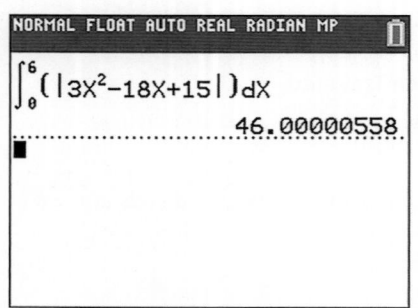

Total distance: 7 + 32 + 7 = 46 m

Net displacement: −18 m

Figure 1

NORMAL FLOAT AUTO REAL RADIAN MP

$$\int_0^6 (|3X^2 - 18X + 15|)\,dX$$

46.00000558

Figure 2

NOW WORK   Problem 5 and AP® Practice Problems 3 and 4.

EXAMPLE 2 **Using Integration with Motion on a Line**

The velocity of an object moving on a line is $v(t) = t^2 - 8t + 5$, for $0 \leq t \leq 6$.

(a) If the position of the object at $t = 0$ is $s(0) = 3$, find the position of the object at $t = 6$.

(b) Find the total distance traveled by the object over the interval $[0, 6]$.

(c) What is the net displacement of the object over the interval $[0, 6]$?

Solution

(a) The position function is $s = s(t) = s(t_0) + \int_{t_0}^{t} v(u)\, du$. When $t_0 = 0$, then

$$s(6) = s(0) + \int_0^6 (u^2 - 8u + 5)\, du$$

$$= 3 + \left[\frac{u^3}{3} - 4u^2 + 5u \right]_0^6 = -39$$

At time $t = 6$, the object is at -39.

(b) The total distance traveled by the object over the interval $[0, 6]$ is given by $\int_0^6 |v(t)|\, dt$. Using a graphing calculator,

$$\int_0^6 |v(t)|\, dt = \int_0^6 |t^2 - 8t + 5|\, dt = 45.310$$

See Figure 3. The total distance traveled by the object over the interval $0 \leq t \leq 6$ is 45.310 units.

(c) Net displacement is given by $\int_0^6 v(t)\, dt$.

$$\int_0^6 v(t)\, dt = \int_0^6 (t^2 - 8t + 5)\, dt = -42$$

The net displacement of the object over the interval $[0, 6]$ is -42 units. ∎

NOW WORK AP® Practice Problem 7.

AP® EXAM TIP

On the AP® Exam, not all parts of a Calculator-Allowed problem need a calculator. When it is possible, it is sometimes faster to do the work by hand.

NORMAL FLOAT AUTO REAL RADIAN MP

$\int_0^6 (|x^2 - 8x + 5|)\, dx$

45.31049852

Figure 3 $v(t) = t^2 - 8t + 5$

2 Determine the Average Velocity or Average Acceleration of an Object Moving Along a Line

NEED TO REVIEW? The average value of a function is discussed in Chapter 6, Part 1 Section 6.4, pp. 472–473.

DEFINITION Average Velocity and Average Acceleration of an Object Moving Along a Line

• Suppose the velocity v of an object moving along a line is given by a function $v = v(t)$ that is continuous on a closed interval $[t_1, t_2]$. The **average velocity** of the object from t_1 to t_2 is given by

$$\frac{1}{t_2 - t_1} \int_{t_1}^{t_2} v(t)\, dt$$

• Suppose the acceleration a of an object moving along a line is given by a function $a = a(t)$ that is continuous on a closed interval $[t_1, t_2]$. The **average acceleration** of the object from t_1 to t_2 is given by

$$\frac{1}{t_2 - t_1} \int_{t_1}^{t_2} a(t)\, dt$$

EXAMPLE 3   **Finding the Average Velocity and Average Acceleration of an Object Moving Along a Line**

Suppose the velocity v, in cm/s, of an object moving along a line is defined by $v = v(t) = t + \sin t$, for $0 \leq t \leq 10$.

(a) What is the average velocity of the object from $t = 0$ to $t = 10$?

(b) What is the average acceleration of the object from $t = 0$ to $t = 10$?

Solution

(a) The velocity v is continuous for $0 \leq t \leq 10$. Using the definition, the average velocity from $t = 0$ to $t = 10$ is

$$\frac{1}{10 - 0} \int_0^{10} v(t)\, dt = \frac{1}{10} \int_0^{10} (t + \sin t)\, dt \approx 5.184 \text{ cm/s}$$

(b) To find the average acceleration, we need to first find the acceleration function $a = a(t)$, which is the derivative of the velocity function.

$$a(t) = \frac{dv}{dt} = 1 + \cos t$$

The function a is continuous for $0 \leq t \leq 10$. Using the definition, the average acceleration from $t = 0$ to $t = 10$ is

$$\frac{1}{10 - 0} \int_0^{10} a(t)\, dt = \frac{1}{10} \int_0^{10} (1 + \cos t)\, dt \approx 0.946 \text{ cm/s}^2$$

∎

NOW WORK   Problem 11 and AP® Practice Problems 2, 5, and 6.

❸ Connect Position, Velocity, and Acceleration of an Object Moving Along a Line

NEED TO REVIEW? See Section 5.3, Objective 2, Use the First Derivative Test with Motion on a Line (p. 365).

Recall that in Chapter 5 we used derivatives to investigate connections among the position, velocity, and acceleration functions of an object moving on a line. The First Derivative Test was often used to investigate properties of the motion.

Suppose an object is moving along a horizontal line and its position x at time t is given by the function $x = x(t)$, for $a \leq t \leq b$. Assuming $x(t)$ is continuous on $[a, b]$ and twice differentiable on (a, b), we have the following results:

- The derivative $\dfrac{dx}{dt}$ of the position function x is the velocity function $v = v(t) = \dfrac{dx}{dt}$ of the object.

  ○ If $v = \dfrac{dx}{dt} > 0$, the object is moving to the right (x is increasing).

  ○ If $v = \dfrac{dx}{dt} < 0$, the object is moving to the left (x is decreasing).

- The derivative of the velocity function v is the acceleration

  function $a = a(t) = \dfrac{dv}{dt}$.

IN WORDS If the velocity and the acceleration are both positive or both negative, the speed of the object is increasing. If the velocity and the acceleration have opposite signs, the speed of the object is decreasing.

- The speed of the object is $|v(t)|$.

  ○ If $v(t) \cdot a(t) > 0$, the speed of the object is increasing.

  ○ If $v(t) \cdot a(t) < 0$, the speed of the object is decreasing.

| EXAMPLE 4 | Connecting the Position, Velocity, and Acceleration of an Object Moving Along a Line |

An object moves along a horizontal line. Its position x, in feet, from the origin at time t, in seconds, is modeled by the function $x = x(t) = t^3 - 9t^2 + 24t - 5$, for $0 \le t \le 6$.

(a) Find the time intervals when the object moves to the right and the intervals when the object moves to the left.

(b) When is the velocity of the object increasing? When is it decreasing?

(c) When is the speed of the object increasing?

(d) What is the average velocity of the object from $t = 0$ to $t = 2$?

Solution

(a) To investigate the motion of the object, it is necessary to find the velocity, v.

$$v = \frac{dx}{dt} = \frac{d}{dt}(t^3 - 9t^2 + 24t - 5) = 3t^2 - 18t + 24$$

Now find the critical numbers of v.

$$3t^2 - 18t + 24 = 3(t^2 - 6t + 8) = 3(t - 2)(t - 4) = 0$$

The critical numbers of v are 2 and 4. Using the critical numbers, partition the interval $[0, 6]$ into three open intervals and determine the times when the velocity is positive (the object is moving to the right) and times when the velocity is negative (the object is moving to the left).

Time interval	Sign of $(t - 2)$	Sign of $(t - 4)$	Sign of $v(t) = 3(t - 2)(t - 4)$	The object moves
(0, 2)	Negative	Negative	Positive	to the right
(2, 4)	Positive	Negative	Negative	to the left
(4, 6)	Positive	Positive	Positive	to the right

The object moves to the right from 0 to 2 seconds and from 4 to 6 seconds. The object moves to the left from 2 to 4 seconds.

(b) To investigate whether the velocity of the object is increasing or decreasing, it is necessary to find $\frac{dv}{dt}$, the acceleration a of the object.

$$a = a(t) = \frac{dv}{dt} = \frac{d}{dt}(3t^2 - 18t + 24) = 6t - 18$$

Because $a < 0$ for $0 < t < 3$, the velocity v decreases on the time interval $(0, 3)$. For $3 < t < 6$, $a > 0$, so the velocity v increases on the time interval $(3, 6)$.

(c) The speed of the object is increasing when $v(t) \cdot a(t) > 0$, and the speed is decreasing when $v(t) \cdot a(t) < 0$.

- $v(t) = 3(t - 2)(t - 4)$
  - $v(t) > 0$ when $0 < t < 2$ or $4 < t < 6$
  - $v(t) < 0$ when $2 < t < 4$.
- $a(t) = 6(t - 3)$
  - $a(t) > 0$ for $3 < t < 6$
  - $a(t) < 0$ for $0 < t < 3$.

The speed of the object is increasing on the time intervals $(2, 3)$ and $(4, 6)$, and the speed of the object is decreasing on the time intervals $(0, 2)$ and $(3, 4)$.

(continued on the next page)

(d) The average velocity of the object from $t = 0$ to $t = 2$ is

$$\frac{1}{2-0}\int_0^2 v(t)dt = \frac{1}{2}\int_0^2 (3t^2 - 18t + 24)\, dt = \frac{1}{2}[t^3 - 9t^2 + 24t]_0^2 = \frac{1}{2}\cdot 20 = 10 \text{ ft/s} \quad\blacksquare$$

NOW WORK AP® Practice Problems **1** and **8**.

📐 **EXAMPLE 5** **Connecting the Position, Velocity, and Acceleration of an Object Moving Along a Line**

The acceleration a of an object moving along a horizontal line is given by $a = a(t) = t^2 e^{-t}$ m/s², for $0 \le t \le 10$. At $t = 0$ s, the velocity of the object is $v(0) = 2$ m/s.

(a) Express $v(t)$ as an integral.

(b) What is the velocity of the object at time $t = 10$?

(c) What is the average velocity of the object over the first 10 seconds, that is, from $t = 0$ to $t = 10$ s?

(d) At what time t is the acceleration a of the object maximum?

Solution

(a) Since the acceleration $a(t) = t^2 e^{-t}$ is a continuous function, it is integrable. Then an expression for the velocity is

$$v(t) = v(0) + \int_0^t u^2 e^{-u}\, du = 2 + \int_0^t u^2 e^{-u}\, du$$

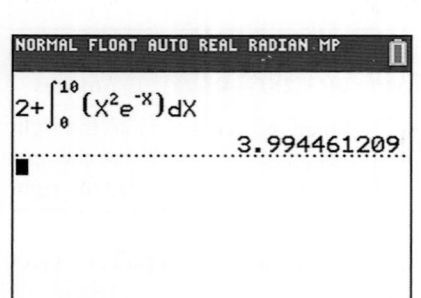

(b) See Figure 4. Using a graphing calculator, the velocity at time $t = 10$ is

$$v(10) = 2 + \int_0^{10} t^2 e^{-t}\, dt \approx 3.994$$

At $t = 10$ s, the velocity of the object is approximately 3.994 m/s.

(c) The average velocity from 0 to 10 seconds is

$$\frac{1}{10-0}\int_0^{10} v(t)dt = \frac{1}{10}\int_0^{10} t^2 e^{-t}\, dt \approx 0.199 \text{ m/s}$$

(d) Since $a = a(t)$ is continuous on the closed interval [0, 10], the Extreme Value Theorem guarantees a has an absolute maximum. To determine when the acceleration a is a maximum, use the Candidates Test. Begin by finding the critical numbers of a in the interval (0, 10).

The derivative of $a(t) = t^2 e^{-t}$ is $\dfrac{da}{dt} = 2te^{-t} - t^2 e^{-t}$. The critical numbers of a satisfy the equation

$$2te^{-t} - t^2 e^{-t} = 0$$
$$(2t - t^2)\, e^{-t} = 0$$
$$t = 0 \text{ or } t = 2 \qquad e^{-t} > 0$$

The only critical number in the interval (0, 10) is $t = 2$. List the values of $a(t)$ at the endpoints $t = 0$ and $t = 10$ and at the critical number $t = 2$.

$$a(0) = 0 \qquad a(2) = 4\, e^{-2} \approx 0.541 \qquad a(10) \approx 0.005$$

The maximum acceleration is ≈ 0.541 m/s², when $t = 2$ s. $\quad\blacksquare$

NOW WORK AP® Practice Problem **9**.

Figure 4 The velocity at $t = 10$ s.

NEED TO REVIEW? The Candidates Test is given in Section 5.2, p. 353.

8.1 Assess Your Understanding

Concepts and Vocabulary

1. **True or False** Net displacement and total distance traveled are equivalent concepts.

2. For an object moving on a line with velocity v, $\int_a^b v(t)\,dt$ measures the _____ of the object from time $t = a$ to time $t = b$.

3. **True or False** An object traveling along a vertical line whose velocity is positive is moving higher.

4. $\dfrac{1}{t_2 - t_1}\displaystyle\int_{t_1}^{t_2} v(t)dt$ defines the _____ from time t_1 to t_2 of an object moving on a line with velocity $v = v(t)$.

Skill Building

In Problems 5–10, an object is moving along a horizontal line with velocity $v = v(t)$ over the indicated interval.

(a) *Find the net displacement of the object. Interpret the result in the context of the problem.*

(b) *Find the total distance traveled by the object and interpret the result.*

 5. $v(t) = 6t^2 + 12t - 18$ m/min; $0 \le t \le 3$

6. $v(t) = t^2 - 2t - 8$ m/min; $0 \le t \le 6$

 7. $v(t) = t^3 + 2t^2 - 8t$ m/s; $-2 \le t \le 3$

8. $v(t) = t^3 + 5t^2 - 6t$ m/s; $-1 \le t \le 3$

9. $v(t) = \dfrac{t}{4}\cos(2t)$ ft/h; $0 \le t \le 5$

10. $v(t) = \dfrac{t}{3}\sin(2t)$ ft/h; $0 \le t \le 4$

Applications and Extensions

11. **Average Velocity** The velocity at time t, in m/s, of an object moving along a line is given by $v(t) = 4\pi \cos t$. What is the average velocity of the object over the interval $0 \le t \le \dfrac{\pi}{2}$?

12. **Average Acceleration** The acceleration at time t in ft/min^2 of an object moving along a line is given by $a(t) = 0.12t^2 - 2$. What is the average acceleration of the object over the interval $0 \le t \le 10$ min?

13. **Average Speed** A car starting from rest accelerates at the rate of 3 m/s^2. Find its average speed over the first 8 s.

14. **Average Speed** A car moving at a constant speed of 80 mi/h begins to decelerate at the rate of 10 mi/h^2. Find its average speed over the next 10 min.

Preparing for the AP® Exam

AP® Practice Problems

Multiple-Choice Questions

 1. Suppose an object is moving along a line and its position at time $t \ge 0$ is given by a twice-differentiable function x. Then which of the following statements must be true?

   I. When the velocity v of the object is less than zero and the acceleration a of the object is less than zero, the speed of the object is decreasing.

   II. The total distance traveled by the object is always greater than or equal to the net displacement of the object over an interval $[t_1,\ t_2]$.

   III. $\int_{t_1}^{t_2} v(t)\,dt$ equals the change in position of the object from time t_1 to time t_2.

   (A) I only   (B) I and II only

   (C) II and III only   (D) I, II, and III

2. The velocity v of an object moving along a line is given by $v(t) = 3\sqrt{t} - \dfrac{1}{\sqrt{t}}$ cm/s. What is the average velocity of the object over the interval $[1, 4]$?

   (A) 12   (B) 4   (C) 6   (D) 3

3. The velocity v of an object moving along a line is given by $v(t) = 0.5t^2 - 2t$, $0 \le t \le 8$, where distance is in meters and time is in minutes. The total distance traveled from $1 \le t \le 6$ is

   (A) 0.83 m   (B) 9.833 m   (C) 21.333 m   (D) 32 m

4. The velocity of an object moving along a horizontal line is given by $v(t) = \dfrac{1}{4}(t^3 - 7t^2 + 10t)$ ft/min. The net displacement of the object from $1 \le t \le 5$ is

   (A) -13.333 ft   (B) -3.333 ft

   (C) -2.604 ft   (D) 4.541 ft

Free-Response Questions

 5. An object moves along a horizontal line and its velocity v, in centimeters per second, is given by $v(t) = 4t^3 - 6t^2 + 12t$ cm/s for $0 \le t \le 10$ s.

   (a) If the position of the object at time $t = 0$ s is -3, what is the position of the object at $t = 2$ s?

   (b) Determine the average velocity of the object from $t = 0$ to $t = 2$ s.

(continued on the next page)

6. The acceleration a of an object moving on a line is given

by $a = a(t) = \dfrac{t}{2} \sin(2t)$ measured in meters per

second^2, $0 \le t \le 2\pi$.

   (a) If the initial velocity v of the object is $v(0) = 20\,\text{m/s}$, what is the velocity of the object at time $t = 4$?

   (b) What is the average acceleration of the object over the interval $[0, 2\pi]$?

7. An object is moving on a line with position

   function $x(t) = e^{t/2} \sin t$, in meters, and time t, in minutes, $0 \le t \le \pi$.

   (a) Find the velocity of the object at time t.

   (b) Find the net displacement of the object over the interval $[0, \pi]$.

   (c) Find the total distance traveled by the object over the interval $[0, \pi]$.

8. An object is moving along a horizontal line. Its position at time t, $0 \le t \le 12$, is given by a differentiable

   function $x = x(t)$. The graph of the derivative $\dfrac{dx}{dt}$ is shown below.

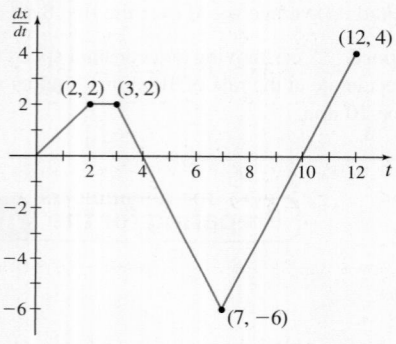

Justify each of your answers.

   (a) For what times t is the object moving to the left?

   (b) What is the acceleration of the object at $t = 11$ s?

   (c) Find the total distance traveled by the object from $t = 0$ to $t = 12$.

   (d) Find the net displacement of the object from time $t = 0$ to $t = 12$.

9. An object is moving along a horizontal line with

   velocity $v(t) = 3t^2 - 6t$, $1 \le t \le 4$ (in meters per second).

   (a) Find the total distance the object moves from $t = 1$ to $t = 4$.

   (b) For what times t is the object at rest?

   (c) If at time $t = 1$ the object is 2 m to the right of the origin, what is its position at $t = 4$?

   (d) Find the average velocity of the object from $t = 1$ to $t = 4$

Retain Your Knowledge

Multiple-Choice Questions

1. The function f is differentiable. The table below lists values of f and f' for select numbers x in the domain of f.

x	-2	0	1	4
$f(x)$	-6	4	6	0
$f'(x)$	7	3	1	-3

Approximate $f(0.3)$ using a linear approximation.

   (A) 3.1   (B) 4.9   (C) 5.7   (D) 6.7

2. $\dfrac{d}{dx} \displaystyle\int_3^{e^{x^2}} \sqrt{\ln t + 2t}\, dt =$

   (A) $\sqrt{\ln e^{x^2} + 2e^{x^2}}$ (B) $e^{x^2}\sqrt{\ln e^{x^2} + 2e^{x^2}}$

   (C) $e^{x^2}\sqrt{x^2 + 2e^{x^2}}$ (D) $2xe^{x^2}\sqrt{x^2 + 2e^{x^2}}$

3. If it is known that the rate of change of f is given by

   $f'(x) = e^x \sin(3x)$ and $f(1) = 4$, then $f(3) =$

   (A) -13.451 (B) 5.472 (C) 9.472 (D) 10.618

Free Response Question

4. The graph of the function $y = f(x) = \begin{cases} \sqrt{x} & \text{if } 0 \le x \le 4 \\ -\dfrac{1}{2}x + 4 & \text{if } 4 < x \le 8 \end{cases}$

is shown below.

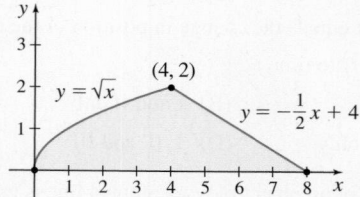

   (a) Approximate the area under the graph of f using a trapezoidal sum with four subintervals of equal width.

   (b) Find the area under the graph of f.

   (c) Find the average value of f.

8.2 Area Between Graphs

OBJECTIVES *When you finish this section, you should be able to:*

1 **Find the area between the graphs of two functions by partitioning the *x*-axis (p. 629)**

2 **Find the area between the graphs of two functions by partitioning the *y*-axis (p. 632)**

When a function f is continuous and nonnegative on a closed interval $[a, b]$, then the definite integral

$$\int_a^b f(x)\, dx$$

equals the area under the graph of $y = f(x)$ from a to b. In this section, we relax the restriction that $f(x)$ is nonnegative and extend the interpretation of the definite integral to find the area of a region bounded by the graphs of two functions.

① Find the Area Between the Graphs of Two Functions by Partitioning the *x*-Axis

Suppose we want to find the area A of the region bounded by the graphs of $y = f(x)$ and $y = g(x)$ and the lines $x = a$ and $x = b$. Occasionally, the area can be found using geometry formulas, and calculus is not needed. But more often the region is irregular, and its area is found using definite integrals.

Assume that the functions f and g are continuous, but not necessarily nonnegative, on the closed interval $[a, b]$ and that $f(x) \geq g(x)$ for all numbers x in $[a, b]$, as illustrated in Figure 5.

To find the area A of the region bounded by the two graphs, partition the interval $[a, b]$ on the *x*-axis into n subintervals:

$$[x_0, x_1], [x_1, x_2], \dots, [x_{i-1}, x_i], \dots, [x_{n-1}, x_n] \qquad x_0 = a \quad x_n = b$$

each of width $\Delta x = \dfrac{b-a}{n}$.

See Figure 6. For each i, $i = 1, 2, \dots, n$, select a number u_i in the subinterval $[x_{i-1}, x_i]$. Then construct n rectangles, each of width Δx and height $f(u_i) - g(u_i)$. The area of the ith rectangle is $[f(u_i) - g(u_i)]\Delta x$.

The sum of the areas of the n rectangles, $\sum\limits_{i=1}^{n} [f(u_i) - g(u_i)]\,\Delta x$, approximates the area A of the region. As the number n of subintervals increases, these sums become better approximations to the area A, and

$$A = \lim_{n \to \infty} \sum_{i=1}^{n} [f(u_i) - g(u_i)]\,\Delta x$$

Since the approximating sums are Riemann sums, and f and g are continuous on the interval $[a, b]$, the limit is a definite integral and

$$A = \int_a^b [f(x) - g(x)]\, dx$$

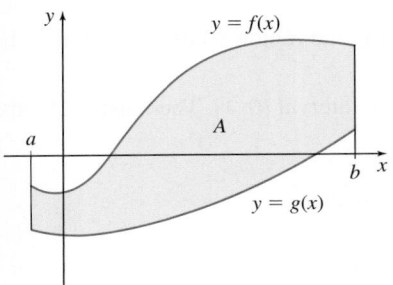

Figure 5 A is the area of the region bounded by the graphs of f and g, $f(x) \geq g(x)$, and the lines $x = a$ and $x = b$.

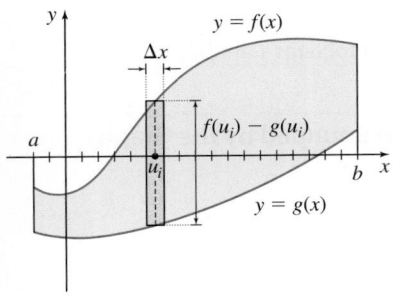

Figure 6 The area of the ith rectangle is $[f(u_i) - g(u_i)]\Delta x$

AREA

The area A of the region bounded by the graphs of $y = f(x)$ and $y = g(x)$, and the lines $x = a$ and $x = b$, where f and g are continuous on the interval $[a, b]$ and $f(x) \geq g(x)$ for all numbers x in $[a, b]$, is

$$A = \int_a^b [f(x) - g(x)]\, dx \qquad (1)$$

Formula (1) for area holds whether the graphs of f and g lie above the x-axis, below the x-axis, or partially above and partially below the x-axis as long as $f(x) \geq g(x)$ on $[a, b]$. It is critical to graph f and g on $[a, b]$ to determine the relationship between the graphs of f and g before setting up the integral. The key is to always subtract the smaller value from the larger value. This ensures that the height of each rectangle is positive.

EXAMPLE 1 | Finding the Area of the Region Bounded by the Graphs of Two Functions

Find the area A of the region bounded by the graphs of $f(x) = e^x$ and $g(x) = \sqrt{x}$ and the lines $x = 0$ and $x = 1$.

Solution

Begin by graphing the two functions and identifying the region whose area A is to be found. See Figure 7.

From the graph, we see that $f(x) > g(x)$ on the interval $[0, 1]$. Then, using (1), the area A is

$$A = \int_a^b [f(x) - g(x)]\, dx = \int_0^1 (e^x - \sqrt{x})\, dx$$

$$= \int_0^1 e^x\, dx - \int_0^1 x^{1/2}\, dx = \left[e^x\right]_0^1 - \left[\frac{x^{3/2}}{\frac{3}{2}}\right]_0^1$$

$$= (e^1 - e^0) - \frac{2}{3}(1 - 0) = e - 1 - \frac{2}{3} = e - \frac{5}{3} \text{ square units} \quad \blacksquare$$

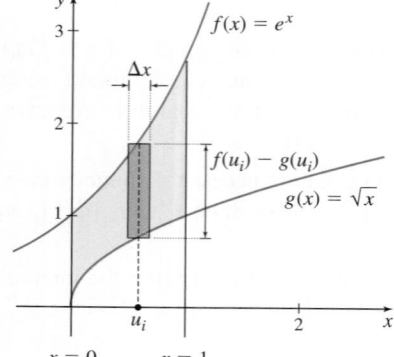

Figure 7 The region bounded by the graphs of $f(x) = e^x$, $g(x) = \sqrt{x}$, and the lines $x = 0$ and $x = 1$.

NOW WORK Problems **3, 7,** and **11** and AP® Practice Problems **1** and **2**.

CALC CLIP

EXAMPLE 2 | Finding the Area of the Region Bounded by the Graphs of Two Functions

Find the area A of the region bounded by the graphs of

$$f(x) = 10x - x^2 \quad \text{and} \quad g(x) = 3x - 8$$

Solution

First graph the two functions. See Figure 8.

The region lies between the graphs of f and g, and the x-coordinates of the points of intersection identify the limits of integration. Solve the equation $f(x) = g(x)$ to find these numbers.

$$10x - x^2 = 3x - 8 \qquad f(x) = g(x)$$
$$x^2 - 7x - 8 = 0$$
$$(x + 1)(x - 8) = 0$$
$$x = -1 \quad \text{or} \quad x = 8$$

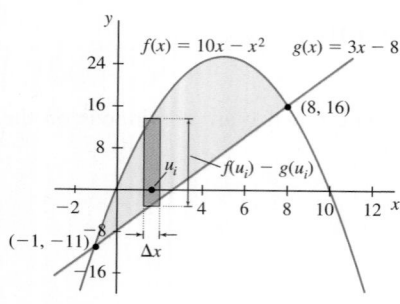

Figure 8 The graphs intersect at $(-1, -11)$ and $(8, 16)$. So, the limits of integration are -1 and 8.

The limits of integration are $a = -1$ and $b = 8$. Since $f(x) \geq g(x)$ on $[-1, 8]$, the area A is given by

$$A = \int_a^b [f(x) - g(x)]\, dx = \int_{-1}^8 [(10x - x^2) - (3x - 8)]\, dx$$

$$= \int_{-1}^8 (-x^2 + 7x + 8)\, dx = \left[\frac{-x^3}{3} + \frac{7x^2}{2} + 8x \right]_{-1}^8$$

$$= \left(-\frac{512}{3} + 224 + 64 \right) - \left(\frac{1}{3} + \frac{7}{2} - 8 \right) = \frac{243}{2} = 121.5 \text{ square units} \quad \blacksquare$$

NOW WORK Problem **21** and AP® Practice Problems **5** and **6**.

EXAMPLE 3 **Finding the Area of the Region Bounded by the Graphs of Two Functions**

Find the area A of the region bounded by the graphs of $y = \sin x$ and $y = \cos x$, $0 \leq x \leq \pi$.

Solution

First we graph the two functions. See Figure 9. The graphs intersect when

$$\sin x = \cos x \qquad 0 \leq x \leq \pi$$

$$\tan x = 1$$

$$x = \frac{\pi}{4}$$

The graphs intersect at the point $\left(\frac{\pi}{4}, \frac{\sqrt{2}}{2} \right)$.

The area A is the sum of area A_1 and area A_2. When $0 \leq x \leq \dfrac{\pi}{4}$, $\cos x \geq \sin x$, so

$$A_1 = \int_0^{\pi/4} (\cos x - \sin x)\, dx = \left[\sin x + \cos x \right]_0^{\pi/4} = 2 \cdot \frac{\sqrt{2}}{2} - 1 = \sqrt{2} - 1$$

When $\dfrac{\pi}{4} \leq x \leq \pi$, $\sin x \geq \cos x$, so

$$A_2 = \int_{\pi/4}^{\pi} (\sin x - \cos x)\, dx = \left[-\cos x - \sin x \right]_{\pi/4}^{\pi} = 1 - 2 \left(-\frac{\sqrt{2}}{2} \right) = 1 + \sqrt{2}$$

The area $A = A_1 + A_2 = \left(\sqrt{2} - 1 \right) + \left(1 + \sqrt{2} \right) = 2\sqrt{2}$. $\blacksquare$

NOW WORK AP® Practice Problems **3** and **10**.

Now suppose $y = g(x) \leq 0$ for all numbers x in the interval $[a, b]$, as illustrated in Figure 10(a). Then by (1), the area A of the region bounded by the graph of $f(x) = 0$ (the x-axis), the graph of g, and the lines $x = a$ and $x = b$, is given by

$$A = \int_a^b [f(x) - g(x)]\, dx = \int_a^b [0 - g(x)]\, dx = -\int_a^b g(x)\, dx$$

By symmetry, this area A is equal to the area under the graph of $y = -g(x) \geq 0$ from a to b. See Figure 10(b).

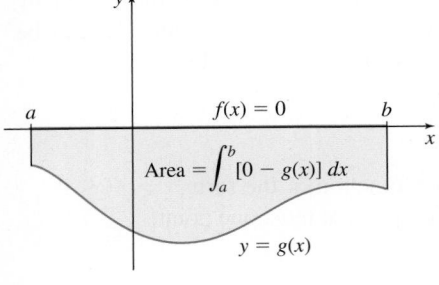

Figure 9

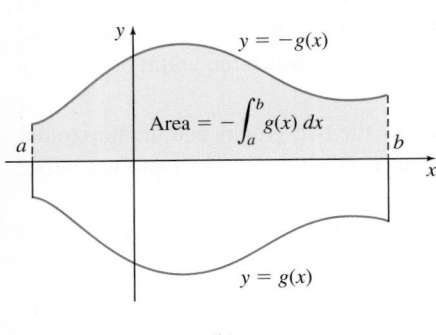

(a)

(b)

Figure 10

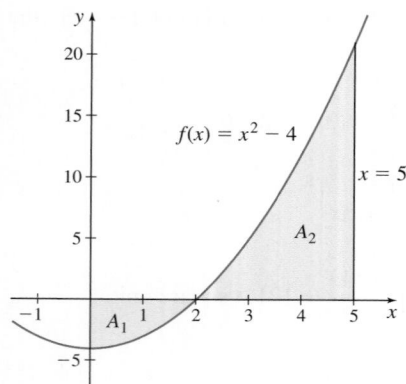

Figure 11 The area A of the region bounded by the graph of f, the x-axis, and the lines $x = 0$ and $x = 5$ is the sum of the areas A_1 and A_2.

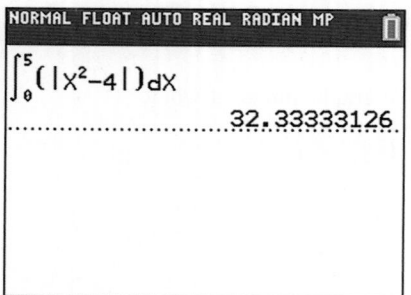

Figure 12

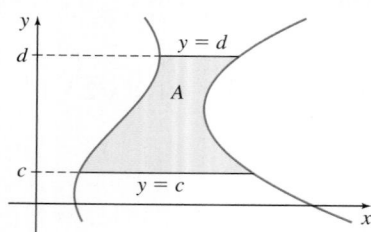

Figure 13

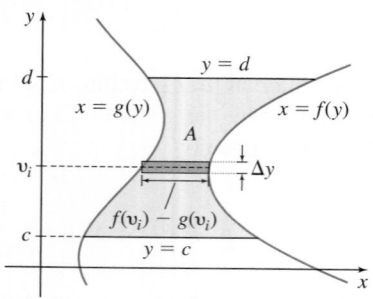

Figure 14

EXAMPLE 4 **Finding the Area Between a Graph and the x-Axis**

Find the area A of the region bounded by the graph of $f(x) = x^2 - 4$, the x-axis, and the lines $x = 0$ and $x = 5$.

Solution

First graph the function f. On the interval $[0, 5]$, the graph of f intersects the x-axis at $x = 2$. As shown in Figure 11, $f(x) \leq 0$ on the interval $[0, 2]$, and $f(x) \geq 0$ on the interval $[2, 5]$. So, the area A of the region is the sum of the areas A_1 and A_2, where

$$A_1 = \int_0^2 [0 - f(x)]\, dx = \int_0^2 -(x^2 - 4)\, dx = \left[-\frac{x^3}{3} + 4x\right]_0^2 = -\frac{8}{3} + 8 = \frac{16}{3}$$

$$A_2 = \int_2^5 f(x)\, dx = \int_2^5 (x^2 - 4)\, dx = \left[\frac{x^3}{3} - 4x\right]_2^5 = \left(\frac{125}{3} - 20\right) - \left(\frac{8}{3} - 8\right)$$

$$= \frac{81}{3} = 27$$

The area is $A = A_1 + A_2 = \dfrac{16}{3} + \dfrac{81}{3} = \dfrac{97}{3} \approx 32.333$ square units. ∎

📓 **COMMENT** Example 4 can also be solved by finding $\int_0^5 |x^2 - 4|\, dx$. See Figure 12.

NOW WORK AP® Practice Problems **7** and **8**.

---

2 **Find the Area Between the Graphs of Two Functions by Partitioning the y-Axis**

In the previous examples, we found the area by partitioning the x-axis. Sometimes it is necessary to partition the y-axis. Look at Figure 13. The region is bounded by the graphs and the horizontal lines $y = c$ and $y = d$.

The graphs that form the left and right borders of the region are not the graphs of functions (they fail the Vertical-line Test). Finding the area A of the region by partitioning the x-axis is not practical. However, both graphs in Figure 13 can be represented as functions of y since each satisfies the Horizontal-line Test.

THEOREM Horizontal-line Test

A set of points in the xy-plane is the graph of a function of the form $x = f(y)$ if and only if every horizontal line intersects the graph in at most one point.

The graphs in Figure 13 satisfy the Horizontal-line Test, so each is of the form x equals a function of y. In Figure 14, the graphs are labeled $x = g(y)$ and $x = f(y)$, where $x = g(y)$ is the horizontal distance from the y-axis to $g(y)$, and $x = f(y)$ is the horizontal distance from the y-axis to $f(y)$. Since the graph of f lies to the right of the graph of g, we know $f(y) \geq g(y)$.

Now to find the area A of the region bounded by the two graphs and the horizontal lines $y = c$ and $y = d$, we partition the interval $[c, d]$ on the y-axis into n subintervals:

$$[y_0, y_1], [y_1, y_2], \ldots, [y_{i-1}, y_i], \ldots, [y_{n-1}, y_n] \qquad y_0 = c \quad y_n = d$$

each of width $\Delta y = \dfrac{d - c}{n}$. For each i, $i = 1, 2, \ldots, n$, we select a number v_i in the subinterval $[y_{i-1}, y_i]$. Then we construct n rectangles, each of width Δy and length $f(v_i) - g(v_i)$. The area of the ith rectangle is $[f(v_i) - g(v_i)]\Delta y$.

The sum of the areas of the n rectangles, $\sum_{i=1}^{n}[f(v_i)-g(v_i)]\Delta y$, approximates the area A. As the number of subintervals increases, the approximation to the area improves, and

$$A = \lim_{n\to\infty}\sum_{i=1}^{n}[f(v_i)-g(v_i)]\,\Delta y$$

The approximating sums are Riemann sums, so if f and g are continuous on the closed interval $[c, d]$, then the limit is a definite integral and

$$A = \int_{c}^{d}[f(y)-g(y)]\,dy$$

AREA (Partitioning the y-Axis)

The area A of the region bounded by the graphs of $x = f(y)$ and $x = g(y)$, and the horizontal lines $y = c$ and $y = d$, where f and g are continuous on the interval $[c, d]$ and $f(y) \geq g(y)$ for all numbers y in $[c, d]$, is

$$A = \int_{c}^{d}[f(y)-g(y)]\,dy \qquad (2)$$

EXAMPLE 5   Finding Area by Partitioning the y-Axis

Find the area A of the region bounded by the graphs of

$$x = f(y) = y+2 \quad \text{and} \quad x = g(y) = y^2$$

Solution

First graph the two functions and identify the region bounded by the two graphs. See Figure 15.

The graphs intersect when $f(y) = g(y)$, that is, when $y+2 = y^2$. Then $y^2 - y - 2 = (y-2)(y+1) = 0$. So, $y = 2$ or $y = -1$. When $y = 2$, $x = 4$; when $y = -1$, $x = 1$. The graphs intersect at the points $(4, 2)$ and $(1, -1)$.

Notice in Figure 15 that the graph of f is to the right of the graph of g; that is, $f(y) \geq g(y)$ for $-1 \leq y \leq 2$. This indicates that we should partition the y-axis and form rectangles from the left graph $(x = y^2)$ to the right graph $(x = y+2)$ as y varies from -1 to 2.

Partitioning the y-axis, the area A of the region is

$$A = \int_{-1}^{2}[f(y)-g(y)]\,dy = \int_{-1}^{2}[(y+2)-y^2]\,dy = \left[\frac{y^2}{2}+2y-\frac{y^3}{3}\right]_{-1}^{2}$$

$$= \left(2+4-\frac{8}{3}\right)-\left(\frac{1}{2}-2+\frac{1}{3}\right) = 4.5 \text{ square units} \qquad \blacksquare$$

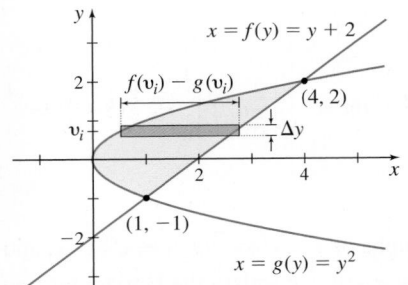

Figure 15

NOW WORK   Problem 15 and AP® Practice Problem 9.

There are times when either the x-axis or the y-axis can be partitioned.

EXAMPLE 6 **Finding the Area Between the Graphs of Two Functions**

Find the area A of the region bounded by the graphs of $y = \sqrt{4 - 4x}$, $y = \sqrt{4 - x}$, and the x-axis:

(a) by partitioning the x-axis.

(b) by partitioning the y-axis.

Solution

(a) Begin by graphing the two equations and identifying the region bounded by the graphs. See Figure 16.

If we partition the x-axis, the area A must be expressed as the sum of the two areas A_1 and A_2 marked in the figure. [Do you see why? The bottom graph changes at $x = 1$ from $y = \sqrt{4 - 4x}$ to $y = 0$ (the x-axis).]

Area A_1 is bounded by the graphs of $y = \sqrt{4 - x}$, $y = \sqrt{4 - 4x}$, and the line $x = 1$. Area A_2 is bounded by the graph $y = \sqrt{4 - x}$, the x-axis, and the line $x = 1$. Then

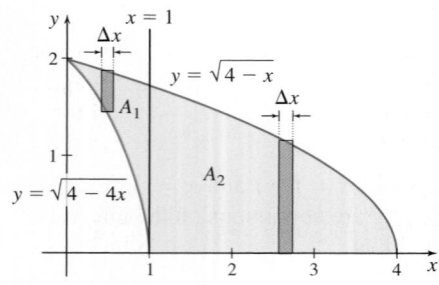

Figure 16 The area A is the sum of the areas A_1 and A_2.

$$A = A_1 + A_2 = \int_0^1 (\sqrt{4 - x} - \sqrt{4 - 4x})\, dx + \int_1^4 \sqrt{4 - x}\, dx$$

$$= \int_0^1 \sqrt{4 - x}\, dx - \int_0^1 \sqrt{4 - 4x}\, dx + \int_1^4 \sqrt{4 - x}\, dx$$

$$= \int_0^4 \sqrt{4 - x}\, dx - \int_0^1 \sqrt{4 - 4x}\, dx \qquad \int_0^1 \sqrt{4 - x}\, dx + \int_1^4 \sqrt{4 - x}\, dx = \int_0^4 \sqrt{4 - x}\, dx$$

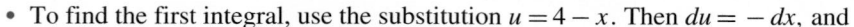

• To find the first integral, use the substitution $u = 4 - x$. Then $du = -\,dx$, and

$$\int_0^4 \sqrt{4 - x}\, dx = -\int_4^0 u^{1/2}\, du = \int_0^4 u^{1/2}\, du = \left[\frac{2}{3}u^{3/2}\right]_0^4 = \frac{2}{3}(8 - 0) = \frac{16}{3}$$

• For the second integral, use the substitution $u = 4 - 4x$. Then $du = -4\, dx$, or equivalently, $dx = -\dfrac{du}{4}$, and

$$\int_0^1 \sqrt{4 - 4x}\, dx = -\frac{1}{4}\int_4^0 u^{1/2}\, du = \frac{1}{4}\int_0^4 u^{1/2}\, du = \frac{1}{4}\left[\frac{2}{3}u^{3/2}\right]_0^4 = \frac{1}{6}(8 - 0) = \frac{4}{3}$$

The area $A = \dfrac{16}{3} - \dfrac{4}{3} = 4$ square units.

(b) Figure 17 shows the region bounded by the graphs of $y = \sqrt{4 - 4x}$, $y = \sqrt{4 - x}$, and the x-axis. Since the graphs of $y = \sqrt{4 - 4x}$ and $y = \sqrt{4 - x}$ satisfy the Horizontal-line Test for $0 \le y \le 2$, we can express $y = \sqrt{4 - 4x}$ as a function $x = f(y)$ and $y = \sqrt{4 - x}$ as a function $x = g(y)$. To find $x = f(y)$, solve $y = \sqrt{4 - 4x}$ for x:

$$y = \sqrt{4 - 4x}$$
$$y^2 = 4 - 4x$$
$$x = \frac{4 - y^2}{4}$$
$$x = f(y) = 1 - \frac{y^2}{4}$$

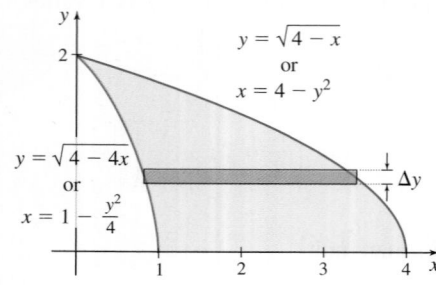

Figure 17

To express $y = \sqrt{4 - x}$ as a function $x = g(y)$, solve for x:

$$y = \sqrt{4 - x}$$
$$y^2 = 4 - x$$
$$x = g(y) = 4 - y^2$$

The graph of $x = g(y) = 4 - y^2$ is to the right of the graph of $x = f(y) = 1 - \dfrac{y^2}{4}$, $0 \le y \le 2$. So, $g(y) \ge f(y)$. Partitioning the y-axis, we have

$$A = \int_0^2 [g(y) - f(y)] \, dy = \int_0^2 \left[(4 - y^2) - \left(1 - \frac{y^2}{4} \right) \right] dy = \int_0^2 \left(3 - \frac{3y^2}{4} \right) dy$$

$$= \left[3y - \frac{y^3}{4} \right]_0^2 = 6 - \frac{8}{4} = 4 \text{ square units} \qquad \blacksquare$$

NOW WORK   Problem **31** and AP® Practice Problem **4**.

When the graphs of functions of x form the top and bottom borders of the region, partitioning the x-axis is usually easier, provided it is easy to evaluate the integrals. If the graphs of functions of y form the left and right borders of the region, partitioning the y-axis is usually easier, provided the integrals with respect to y are easily evaluated.

8.2 Assess Your Understanding

Concepts and Vocabulary

1. Express the area of the region bounded by the graphs of $y = x^2$ and $y = \sqrt{x}$ as an integral using a partition of the x-axis. Do not evaluate the integral.

2. Express the area of the region bounded by the graph of $x = y^2$ and the line $x = 1$ as an integral using a partition of the y-axis. Do not evaluate the integral.

Skill Building

In Problems 3–12, find the area of the region bounded by the graphs of the given equations by partitioning the x-axis.

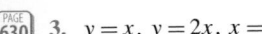 **3.** $y = x$, $y = 2x$, $x = 1$

4. $y = x$, $y = 3x$, $x = 3$

5. $y = x^2$, $y = x$

6. $y = x^2$, $y = 4x$

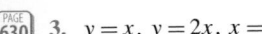 **7.** $y = e^x$, $y = e^{-x}$, $x = \ln 2$

8. $y = e^x$, $y = -x + 1$, $x = 2$

9. $y = x^2$, $y = x^4$

10. $y = x$, $y = x^3$

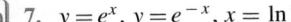 **11.** $y = \cos x$, $y = \dfrac{1}{2}$, $0 \le x \le \dfrac{\pi}{3}$

12. $y = \sin x$, $y = \dfrac{1}{2}$, $\dfrac{\pi}{6} \le x \le \dfrac{5\pi}{6}$

In Problems 13–20, find the area of the region bounded by the graphs of the given equations by partitioning the y-axis.

13. $x = y^2$, $x = 2 - y$

14. $x = y^2$, $x = y + 2$

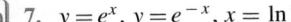 **15.** $x = 9 - y^2$, $x = 5$

16. $x = 16 - y^2$, $x = 7$

17. $x = y^2 + 4$, $y = x - 6$

18. $x = y^2 + 6$, $y = 8 - x$

19. $y = \ln x$, $x = 1$, $y = 2$

20. $y = \ln x$, $x = e$, $y = 0$

In Problems 21–24, find the area of the shaded region in the graph.

 21.

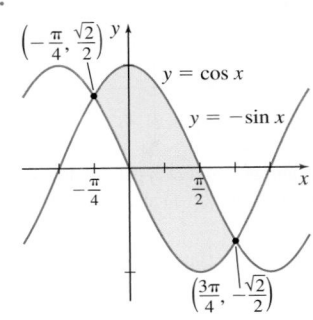

22.

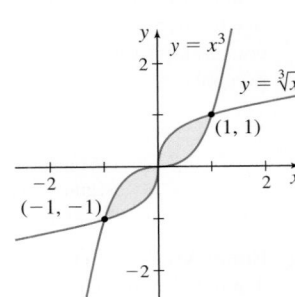

23.

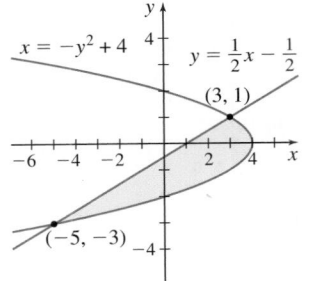

24.

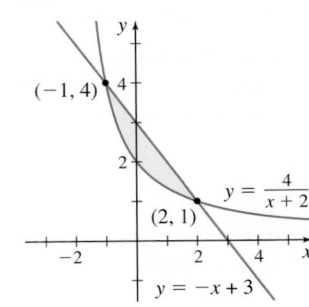

In Problems 25–32, find the area A of the region bounded by the graphs of the given equations:

(a) *by partitioning the x-axis.*
(b) *by partitioning the y-axis.*

25. $y = \sqrt{x}$, $y = x^3$

26. $y = \sqrt{x}$, $y = x^2$

27. $y = x^2 + 1$, $y = x + 1$

28. $y = x^2 + 1$, $y = 4x + 1$

29. $y = \sqrt{9 - x}$, $y = \sqrt{9 - 3x}$, $y = 0$

30. $y = \sqrt{16 - 2x}$, $y = \sqrt{16 - 4x}$, $y = 0$

31. $y = \sqrt{2x - 6}$, $y = \sqrt{x - 2}$, $y = 0$

32. $y = \sqrt{2x - 5}$, $y = \sqrt{4x - 17}$, $y = 0$

In Problems 33–46, find the area of the region bounded by the graphs of the given equations.

33. $y = 4 - x^2$, $y = x^2$

34. $y = 9 - x^2$, $y = x^2$

35. $x = y^2 - 4$, $x = 4 - y^2$

36. $x = y^2$, $x = 16 - y^2$

37. $y = \ln x^2$, $y = 0$, and the line $x = e$

38. $y = \ln x$, $y = 1 - x$, and the line $y = 1$

39. $y = \cos x$, $y = 1 - \dfrac{3}{\pi} x$, $x = \dfrac{\pi}{3}$

40. $y = \sin x$, $y = 1$, $0 \le x \le \dfrac{\pi}{2}$

41. $y = e^{2x}$ and the lines $x = 1$ and $y = 1$

42. $y = e^x$, $y = e^{3x}$, $x = 2$

43. $y^2 = 4x$, $4x - 3y - 4 = 0$

44. $y^2 = 4x + 1$, $x = y + 1$

45. $y = \sin x$, $y = \dfrac{2x}{\pi}$, $x \ge 0$

46. $y = \cos x$, $x \ge 0$, $y = \dfrac{3x}{\pi}$

Applications and Extensions

47. An Archimedean Result Show that the area of the shaded region in the figure is two-thirds of the area of the parallelogram *ABCD*. (This illustrates a result due to Archimedes concerning sectors of parabolas.)

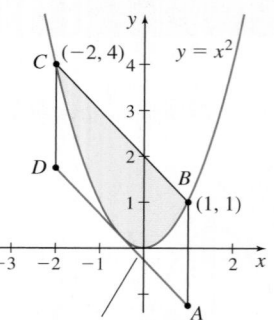

The line segment joining *A* and *D* is parallel to the line segment joining $(-2, 4)$ and $(1, 1)$ and is tangent to the graph of $y = x^2$.

48. Equal Areas Find *h* so that the area of the region bounded by the graphs of $y = x$, $y = 8x$, and $y = \dfrac{1}{x^2}$ is equal to that of an isosceles triangle of base 1 and height *h*. See the figure.

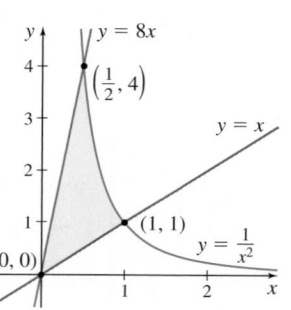

49. The region *R* bounded by the graphs of $y = \cos^2 x$ and $y = \sin^2 x$ is shown below.

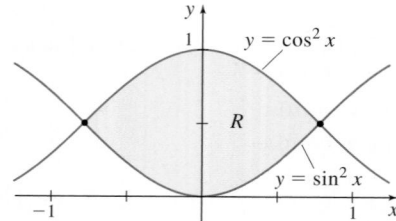

(a) Find the points of intersection of the two graphs.

(b) Find the area of the region in the first quadrant bounded by the graphs of $y = \cos^2 x$ and $y = \sin^2 x$ and the *y*-axis.

50. Area Find the area of the region in the first quadrant bounded by the graphs of $y = \sin(2x)$ and $y = \cos(2x)$, $0 \le x \le \dfrac{\pi}{8}$.

In Problems 51 and 52, find the area of the region bounded by the graphs of f and g.

51. $f(x) = 3 \ln x$ and $g(x) = x \ln x$, $x \ge 1$

52. $f(x) = 4x \ln x$ and $g(x) = x^2 \ln x$, $x \ge 1$

53. Area Find the area of the region bounded by the hyperbola $\dfrac{y^2}{9} - x^2 = 1$ and

(a) the line $y = 4$.   **(b)** the line $y = -7$.

54. Area Find the area of the region bounded by the hyperbola $\dfrac{x^2}{9} - \dfrac{y^2}{16} = 1$ and

(a) the line $x = 6$.   **(b)** the line $x = -8$.

55. Area of a Lune A **lune** is a crescent-shaped region formed when two circles intersect.

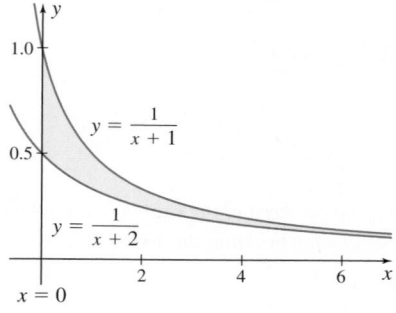

(a) Find the area of the smaller lune formed by the intersection of the two circles $x^2 + y^2 = 4$ and $x^2 + (y - 2)^2 = 1$ (shaded in the figure).

(b) What is the area of the larger lune?

56. Area Between Graphs Find the area, if it is defined, of the region bounded by the graphs of $y = \dfrac{1}{x + 1}$ and $y = \dfrac{1}{x + 2}$ on the interval $[0, \infty)$. See the figure.

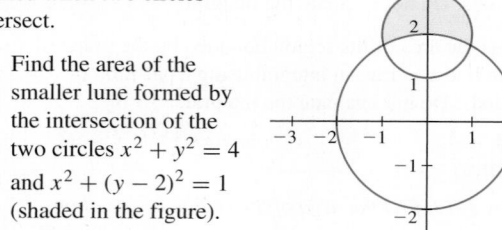

57. Area Find the area of the region bounded by the graphs of

$$y = \sin^{-1} x \quad y = x \quad 0 \le x \le \dfrac{1}{2}$$

Which axis did you choose to partition? Explain your choice.

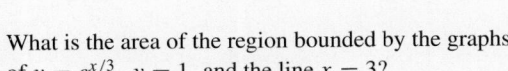

58. Area

(a) Graph $y = x^2$ and $y = \sin x$.

(b) Find the points of intersection of the graphs.

(c) Find the area of the region bounded by the graphs of $y = x^2$ and $y = \sin x$ using a partition of the x-axis.

(d) Find the area of the region bounded by the graphs of $y = x^2$ and $y = \sin x$ using a partition of the y-axis.

59. (a) Graph $y = \sin^{-1} x$, $x + y = 1$, and $y = 0$.

(b) Find the points of intersection of the graphs.

(c) Find the area of the region bounded by the graphs.

60. (a) Graph $y = \cos^{-1} x$, $y = x^3 + 1$, and $y = 0$.

(b) Find the points of intersection of the graphs.

(c) Find the area of the region bounded by the graphs.

61. (a) Graph $2x + y = 3$, $y = \dfrac{1}{1 + x^2}$, and $y = 1$.

(b) Find the points of intersection of the graphs.

(c) Find the area of the region bounded by the graphs.

62. (a) Graph $y = \dfrac{5}{1 + x^2}$, $x = \sqrt{y + 2}$, and $x = 0$.

(b) Find the points of intersection of the graphs.

(c) Find the area of the region bounded by the graphs.

63. (a) Graph $y = \sin^{-1} x$, $y = 1 - x^2$, and $y = 0$.

(b) Find the points of intersection of the graphs.

(c) Find the area of the region bounded by the graphs.

64. Cost of Health Care The cost of health care varies from one country to another. Between 2000 and 2010, the average cost of health insurance for a family of four in the United States was modeled by

$$A(x) = 8020.6596(1.0855^x) \qquad 0 \le x \le 10$$

where $x = 0$ corresponds to the year 2000 and $A(x)$ is measured in U.S. dollars. During the same years, the average cost of health care in Canada was given by

$$C(x) = 4944.6424(1.0711^x) \qquad 0 \le x \le 10$$

where $x = 0$ corresponds to the year 2000 and $C(x)$ is measured in U.S. dollars.

(a) Find the area of the region bounded by the graphs of A and C from $x = 0$ to $x = 10$. Round the answer to the nearest dollar.

(b) Interpret the answer to (a).

Challenge Problems

65. (a) Express the area A of the region in the first quadrant bounded by the y-axis and the graphs of $y = \tan x$ and $y = k$ for $k > 0$ as a function of k.

(b) What is the value of A when $k = 1$?

(c) If the line $y = k$ is moving upward at the rate of $\dfrac{1}{10}$ unit per second, at what rate is A changing when $k = 1$?

66. Find the area bounded by the graph of $y^2 = x^2 - x^4$.

Preparing for the AP® Exam

AP® Practice Problems

Multiple-Choice Questions

1. The graphs of the functions $f(x) = x^2$ and $g(x) = 3x$ are shown below.

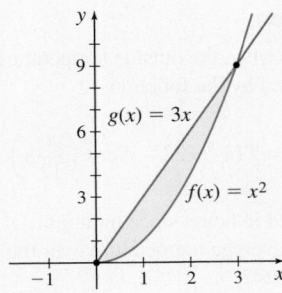

Find the area of the region bounded by the two graphs.

(A) $\dfrac{1}{3}$ (B) $\dfrac{7}{6}$ (C) $\dfrac{9}{2}$ (D) $\dfrac{27}{2}$

2. What is the area of the region bounded by the graphs of $y = e^{x/3}$, $y = 1$, and the line $x = 3$?

(A) $\dfrac{1}{3}(e - 10)$ (B) $e - 4$

(C) $3e - 3$ (D) $3e - 6$

3. Find the area of the region bounded by the graphs of $y = \cos x$ and $y = 1 - \cos x$ over the interval $\left[0, \dfrac{\pi}{2}\right]$.

(A) -2.523 (B) 0.941 (C) 3.988 (D) 1.988

4. The graphs of the functions $f(x) = \sqrt{x - 4}$, $g(x) = -x + 2$, and $y = 2$ are shown in the figure.

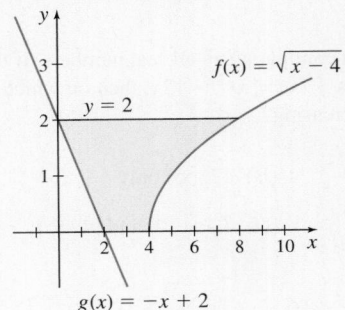

Find the area of the region bounded by the graphs and the x-axis.

(A) $\dfrac{4\sqrt{2} - 14}{3}$ (B) $\dfrac{26}{3}$ (C) $\dfrac{86}{3}$ (D) $\dfrac{112}{3}$

(continued on the next page)

5. The graphs of the functions $y = f(x)$ and $y = g(x)$ intersect at the point (b, c) as shown in the figure below. The area A of the region bounded by the graphs of f, g, and the line $x = 1$ is obtained by finding the integral

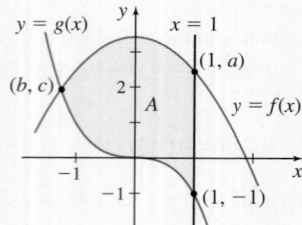

(A) $\int_1^b [f(x) - g(x)]dx$

(B) $\int_b^1 [f(x) - g(x)]dx$

(C) $\int_{-1}^c [f(x) - g(x)]dx$

(D) $\int_b^0 [f(x) - g(x)]dx + \int_0^1 [f(x) - |g(x)|]dx$

6. What is the area of the region bounded by the graphs of $y = x^2 - 4x + 2$ and $y = 7$?

(A) -36 (B) 36 (C) $\dfrac{64}{3}$ (D) 60

7. Find the area of the region bounded by the graphs of the functions $f(x) = e^x + 2$, $g(x) = -2x + 3$, and the line $x = 4$.

(A) $e^4 - 1$ (B) $e^4 + 11$ (C) $e^4 + 12$ (D) $2e^4 + 12$

8. The area of the region bounded by the graphs of $y = \sin x$, $y = \dfrac{1}{\pi}x + 1$, and the y-axis is

(A) $\dfrac{\pi}{2}$ (B) $1 + \dfrac{3\pi}{4}$ (C) $2 + \dfrac{\pi}{2}$ (D) $\dfrac{3\pi}{2} - 1$

9. Find the area of the region bounded by the graph of $x = y^2 + 2y - 3$ and the y-axis.

(A) $-\dfrac{32}{3}$ (B) $-\dfrac{20}{3}$ (C) $\dfrac{20}{3}$ (D) $\dfrac{32}{3}$

Free-Response Question

10. Consider the functions $f(x) = x^3 - 8x^2 + 15x + 2$ and $g(x) = 3x + 2$.

(a) Find the points (x, y) where the graphs of f and g intersect.

(b) Write the integral(s) that represent the area of the region(s) bounded by the graphs of f and g.

(c) Find the area of the region bounded by the graphs.

Retain Your Knowledge

Multiple-Choice Questions

1. Suppose the function f is continuous on the closed interval $[0, 8]$. Then which of the following must be true?

I. f is differentiable at $x = 6$.

II. $\lim\limits_{x \to 6^-} f(x) = \lim\limits_{x \to 6^+} f(x)$

III. There is a number c, $0 \le c \le 8$, so that $f(c) \le f(x)$ for all x in the interval $[0, 8]$.

(A) II only   (B) I and II only

(C) II and III only   (D) I, II, and III

2. The function f is continuous on all real numbers. If the derivative of f is $f'(x) = 3x^2 - 12x$, then on which of the intervals is f increasing?

(A) $[0, \infty)$ only   (B) $[2, \infty)$ only

(C) $[0, 4]$ only   (D) $(-\infty, 0]$ and $[4, \infty)$

3. The graph of the function $f(x) = e^x \cos x$ intersects the x-axis in the interval $[0, 2]$. Find the slope of the tangent line to the graph of f at the point of intersection.

(A) -9.621 (B) -4.810

(C) -1.571 (D) 1.571

Free-Response Question

4. On a certain day in May, the outside temperature T in degrees Celsius was modeled by the function

$$T = T(h) = 22 - 4\cos\left(\frac{\pi}{12}h\right)$$

where h is measured in hours since midnight.

(a) What was the average temperature over the 24-hour period $0 \le h \le 24$?

(b) On that day in May, the sun rose at 6:30 a.m. and set at 8 p.m. What was the average temperature during the daylight hours, that is, over the interval $6.5 \le h \le 20$?

8.3 Volume of a Solid of Revolution: Disks and Washers

OBJECTIVES *When you finish this section, you should be able to:*

1 Use the disk method to find the volume of a solid formed by revolving a region about the x-axis (p. 640)

2 Use the disk method to find the volume of a solid formed by revolving a region about the y-axis (p. 642)

3 Use the washer method to find the volume of a solid formed by revolving a region about the x-axis (p. 644)

4 Use the washer method to find the volume of a solid formed by revolving a region about the y-axis (p. 646)

5 Find the volume of a solid formed by revolving a region about a line parallel to a coordinate axis (p. 647)

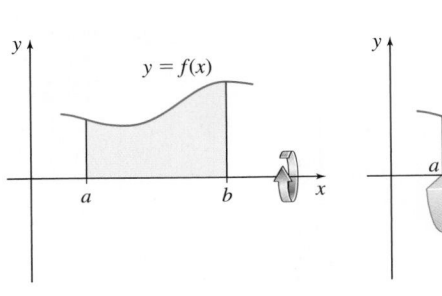

Figure 18 A right circular cylinder

Figure 19 A right circular cone

An example of a solid of revolution is a *right circular cylinder*, which is generated by revolving the region bounded by a horizontal line $y = A$, $A > 0$, the x-axis, and the lines $x = a$ and $x = b$ about the x-axis. See Figure 18. Another familiar example of a solid of revolution, pictured in Figure 19, is a *right circular cone* that is generated by revolving the region in the first quadrant bounded by a line $y = mx$, $m > 0$, the x-axis, and the line $x = h$ about the x-axis.

The volume of these simple solids of revolution can be found using geometry formulas, but finding the volume of most solids of revolution requires calculus.

Suppose a region is bounded by the graph of a function $y = f(x)$ that is continuous and nonnegative on the interval $[a, b]$, the x-axis, and the lines $x = a$ and $x = b$. See Figure 20(a). Now suppose this region is revolved about the x-axis. The result is a **solid of revolution**. See Figure 20(b).

To find the volume of a solid of revolution, we begin by partitioning the interval $[a, b]$ into n subintervals:

$$[a, x_1], [x_1, x_2], \ldots, [x_{i-1}, x_i], \ldots, [x_{n-1}, b]$$

each of width $\Delta x = \dfrac{b - a}{n}$ and selecting a number u_i in each subinterval. Refer to Figure 20(c). The circle obtained by slicing the solid at u_i has radius $r_i = f(u_i)$ and area $A_i = \pi r_i^2 = \pi [f(u_i)]^2$. The volume V_i of the disk obtained from the ith subinterval is $V_i = \pi [f(u_i)]^2 \Delta x$, and an approximation to the volume V of the solid of revolution is

$$V \approx \pi \sum_{i=1}^{n} [f(u_i)]^2 \, \Delta x$$

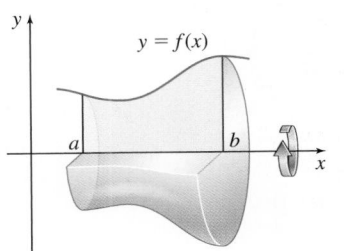

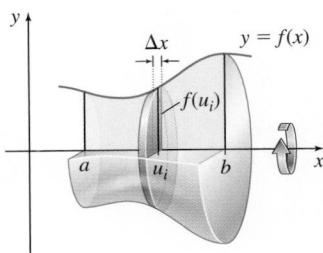

(a) Region to be revolved about the x-axis

(b) Solid of revolution

(c) Radius $= f(u_i)$
Area of circle $= \pi [f(u_i)]^2$
Thickness of disk $= \Delta x$
Volume of disk $= \pi [f(u_i)]^2 \Delta x$

Figure 20

IN WORDS The volume V of a disk is

V = (Area of the circular cross section)
× (Thickness)
= π(Radius)^2(Thickness)

NEED TO REVIEW? Riemann sums and the definite integral are discussed in Section 6.2, pp. 435–446.

As the number n of subintervals increases, the Riemann sums $\pi \sum_{i=1}^{n} [f(u_i)]^2 \Delta x$ become better approximations to the volume V of the solid, and

$$V = \pi \lim_{n \to \infty} \sum_{i=1}^{n} [f(u_i)]^2 \, \Delta x$$

Since the sums are Riemann sums, if f is continuous on the interval $[a, b]$, then the limit is a definite integral and

$$V = \pi \int_a^b [f(x)]^2 \, dx$$

This approach to finding the volume of a solid of revolution is called the **disk method**.

> **Volume of a Solid of Revolution Using the Disk Method**
>
> If a function f is continuous on a closed interval $[a, b]$, then the volume V of the solid of revolution obtained by revolving the region bounded by the graph of f, the x-axis, and the lines $x = a$ and $x = b$ about the x-axis is
>
> $$V = \pi \int_a^b [f(x)]^2 \, dx$$

1 Use the Disk Method to Find the Volume of a Solid Formed by Revolving a Region About the x-Axis

EXAMPLE 1 Using the Disk Method: Revolving About the x-Axis

Find the volume of the solid of revolution generated by revolving the region bounded by the graph of $y = \sqrt{x}$, the x-axis, and the line $x = 5$ about the x-axis.

Solution

We begin by graphing the region to be revolved. See Figure 21(a). Figure 21(b) shows a typical disk, and Figure 21(c) shows the solid of revolution. Using the disk method, the volume V of the solid of revolution is

$$V = \pi \int_0^5 (\sqrt{x})^2 \, dx = \pi \int_0^5 x \, dx = \pi \left[\frac{x^2}{2} \right]_0^5 = \frac{25}{2} \pi \text{ cubic units}$$

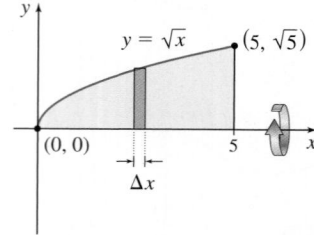

(a) The region to be revolved about the x-axis

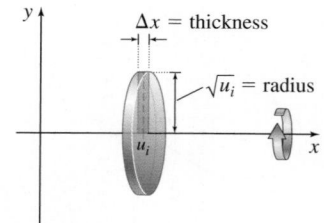

(b) Radius = $\sqrt{u_i}$
Area of circle = $\pi(\sqrt{u_i})^2$
Thickness of disk = Δx
Volume of disk = $\pi(\sqrt{u_i})^2 \Delta x$

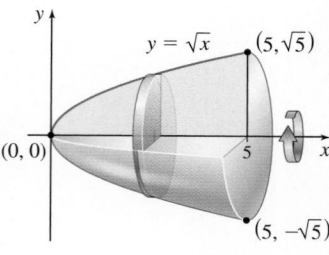

(c) $V = \pi \int_0^5 (\sqrt{x})^2 \, dx$

Figure 21

NOW WORK Problem 5 and AP® Practice Problems 1, 2, and 11.

Notice, the disk method does not require the function f to be nonnegative. Suppose f is a function continuous on the interval $[a, b]$. As Figure 22 illustrates, the volume V of the solid of revolution obtained by revolving the region bounded by the graph of f, the x-axis, and the lines $x = a$ and $x = b$ about the x-axis equals the volume of the solid of revolution obtained by revolving the region bounded by the graph of $y = |f(x)|$, the x-axis, and the lines $x = a$ and $x = b$ about the x-axis. Since $|f(x)|^2 = [f(x)]^2$, the formula for V does not change.

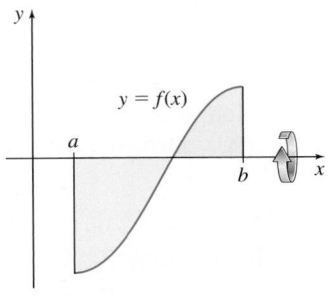

(a) The region to be revolved about the x-axis

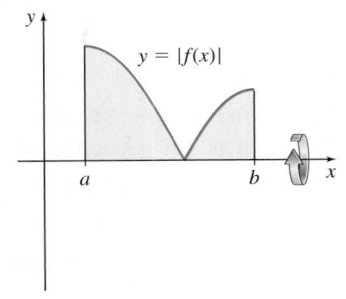

(b) The absolute value of the region to be revolved about the x-axis

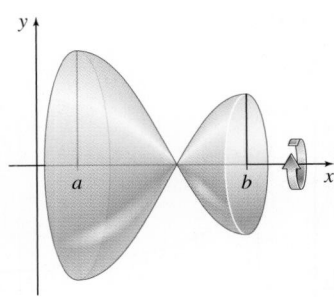

(c) The volume V is the same in both cases:

$$V = \pi \int_a^b [f(x)]^2 \, dx$$

Figure 22

EXAMPLE 2 **Using the Disk Method: Revolving About the x-Axis**

Find the volume of the solid of revolution generated by revolving the region bounded by the graph of $y = x^3$, the x-axis, and the lines $x = -1$ and $x = 2$ about the x-axis.

Solution

Figure 23(a) shows the graph of the region to be revolved about the x-axis. Figure 23(b) illustrates a typical disk, and Figure 23(c) shows the solid of revolution.

Using the disk method, the volume V of the solid of revolution is

$$V = \pi \int_{-1}^{2} x^6 \, dx = \pi \left[\frac{x^7}{7} \right]_{-1}^{2} = \frac{\pi}{7}(128 + 1) = \frac{129}{7}\pi \text{ cubic units}$$

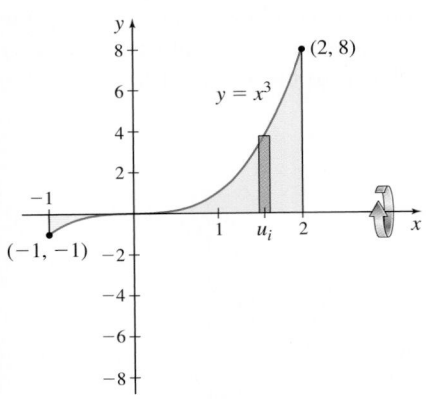

(a) The region to be revolved about the x-axis

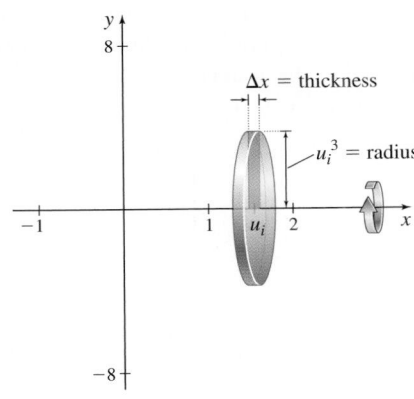

(b) Radius $= u_i^3$
  Area of circle $= \pi(u_i^3)^2 = \pi u_i^6$
  Thickness of disk $= \Delta x$
  Volume of disk $= \pi u_i^6 \, \Delta x$

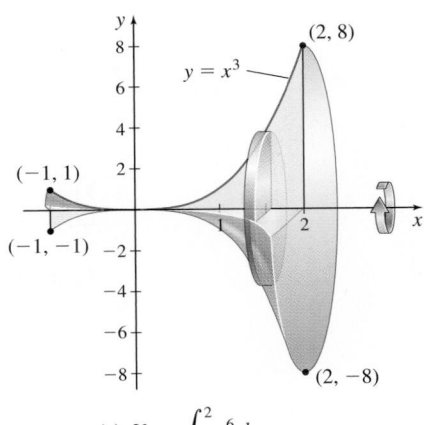

(c) $V = \pi \int_{-1}^{2} x^6 \, dx$

Figure 23

NOW WORK Problem 27 and AP® Practice Problem 12.

EXAMPLE 3 Finding the Volume of Gabriel's Horn

(a) Find the volume of the solid of revolution, called **Gabriel's Horn**, that is generated by revolving the region bounded by the graph of $y = \dfrac{1}{x}$ and the x-axis to the right of 1 about the x-axis.

NEED TO REVIEW? Improper integrals are discussed in Chapter 6, Part 2, Section 6.9, pp. 545–546.

(b) Show that although the volume of Gabriel's Horn is finite, the area A of the region bounded by the graph of $y = \dfrac{1}{x}$ and the x-axis to the right of 1 is not defined.

Solution

(a) Figure 24 illustrates the region being revolved and the solid of revolution that it generates. Using the disk method, the volume V is the improper integral $\pi \displaystyle\int_{1}^{\infty} \left(\dfrac{1}{x}\right)^{2} dx$.

We need to investigate the following limit:

$$\pi \lim_{b \to \infty} \int_{1}^{b} \frac{1}{x^2}\, dx = \pi \lim_{b \to \infty} \left[-\frac{1}{x}\right]_{1}^{b} = \pi \lim_{b \to \infty} \left(-\frac{1}{b} + 1\right) = \pi$$

The volume of Gabriel's Horn is π cubic units.

(b) The area A, if it exists, is defined by the improper integral $\displaystyle\int_{1}^{\infty} \frac{1}{x}\, dx$, so we investigate the limit:

$$\lim_{b \to \infty} \int_{1}^{b} \frac{1}{x}\, dx = \lim_{b \to \infty} \left[\ln x\right]_{1}^{b} = \lim_{b \to \infty} (\ln b - \ln 1) = \lim_{b \to \infty} \ln b = \infty$$

The limit is infinite, so the integral diverges and the area A bounded by the graph of $y = \dfrac{1}{x}$ and the x-axis to the right of 1 is not defined. ∎

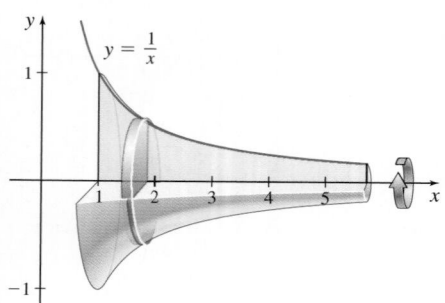

Figure 24 Gabriel's Horn

2 Use the Disk Method to Find the Volume of a Solid Formed by Revolving a Region About the y-Axis

A solid of revolution can be generated by revolving a region about any line. In particular, the volume V of the solid of revolution generated by revolving the region bounded by the graph of $x = g(y)$, where g is continuous on the closed interval $[c, d]$, the y-axis, and the horizontal lines $y = c$ and $y = d$ about the y-axis can be obtained by partitioning the y-axis and taking slices perpendicular to the y-axis of thickness Δy. See Figure 25(a).

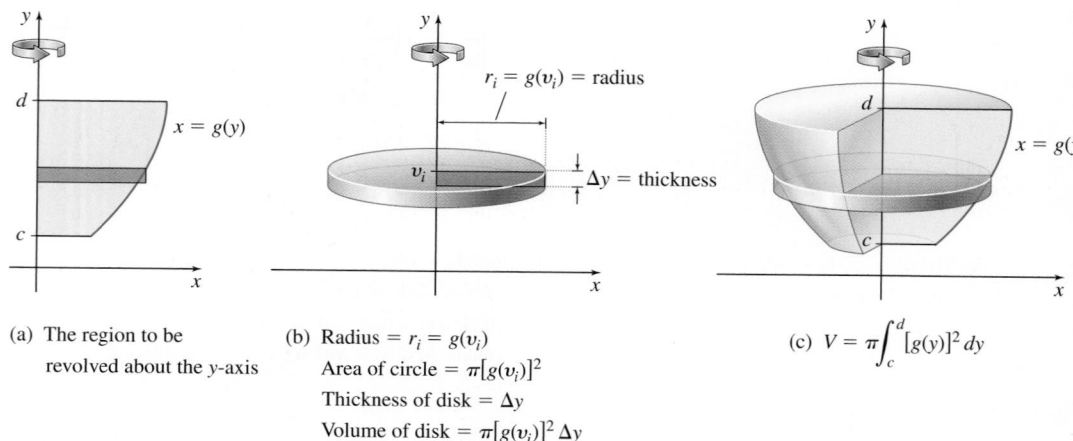

(a) The region to be revolved about the y-axis

(b) Radius $= r_i = g(v_i)$
Area of circle $= \pi[g(v_i)]^2$
Thickness of disk $= \Delta y$
Volume of disk $= \pi[g(v_i)]^2\,\Delta y$

(c) $V = \pi \displaystyle\int_{c}^{d} [g(y)]^2\, dy$

Figure 25

Again, the cross sections are circles, as shown in Figure 25(b). The radius of a typical cross section is $r_i = g(v_i)$, and its area is $A_i = \pi r_i^2 = \pi [g(v_i)]^2$. The volume of the disk of radius r_i and thickness $\Delta y = \dfrac{d-c}{n}$ is $V_i = \pi r_i^2 \Delta y = \pi [g(v_i)]^2 \Delta y$. By summing the volumes of all the disks and taking the limit, the volume V of the solid of revolution is given by

$$V = \pi \int_c^d [g(y)]^2 \, dy$$

See Figure 25(c).

EXAMPLE 4   Using the Disk Method: Revolving About the y-Axis

Use the disk method to find the volume of the solid of revolution generated by revolving the region bounded by the graph of $y = x^3$, the y-axis, and the lines $y = 1$ and $y = 8$ about the y-axis.

Solution

Figure 26(a) shows the region to be revolved. Since the solid is formed by revolving the region about the y-axis, we write $y = x^3$ as $x = \sqrt[3]{y} = y^{1/3}$.

Figure 26(b) illustrates a typical disk, and Figure 26(c) shows the solid of revolution. Using the disk method, the volume V of the solid of revolution is

$$V = \pi \int_1^8 [y^{1/3}]^2 \, dy = \pi \int_1^8 y^{2/3} \, dy = \pi \left[\frac{y^{5/3}}{\dfrac{5}{3}} \right]_1^8$$

$$= \frac{3\pi}{5}(32 - 1) = \frac{93}{5}\pi \text{ cubic units}$$

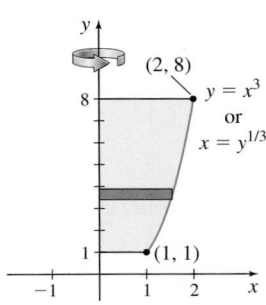

(a) The region to be
   revolved about the y-axis

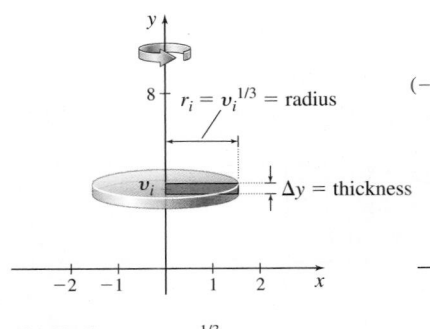

(b) Radius = $r_i = v_i^{1/3}$
   Area of circle = $\pi (v_i^{1/3})^2$
   Thickness of disk = Δy
   Volume of disk = $\pi (v_i^{1/3})^2 \Delta y$

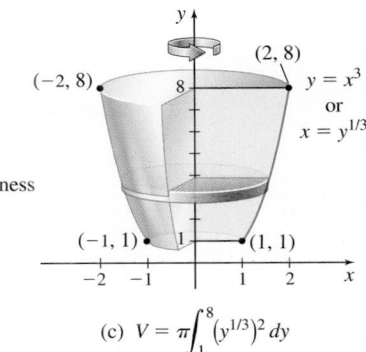

(c) $V = \pi \displaystyle\int_1^8 (y^{1/3})^2 \, dy$

Figure 26

NOW WORK   Problem 15 and AP® Practice Problem 5.

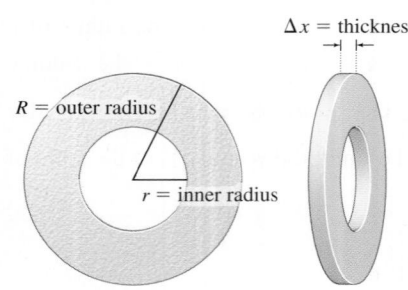

Figure 27 The volume V of the washer is $V = [\pi R^2 - \pi r^2]\Delta x = \pi$ $[(\text{Outer radius})^2 - (\text{Inner radius})^2]$ thickness.

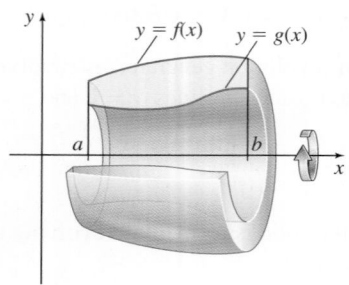

Figure 28 The solid formed by revolving the region bounded by the graphs of f and g and the lines $x = a$ and $x = b$ about the x-axis. Notice that the interior of the solid is hollow.

IN WORDS The volume V of a washer is
$V = \pi[(\text{Outer radius})^2 - (\text{Inner radius})^2]$
$\times (\text{Thickness})$

3 Use the Washer Method to Find the Volume of a Solid Formed by Revolving a Region About the x-Axis

A **washer** is a thin, flat ring with a hole in the middle. It can be represented by two concentric circles, an outer circle of radius R and an inner circle of radius r, as shown in Figure 27. The volume of a washer is found by finding the area of the outer circle, πR^2, subtracting the area of the inner circle, πr^2, then multiplying the result by the thickness of the washer.

Suppose we have two functions $y = f(x)$ and $y = g(x)$, $f(x) \geq g(x) \geq 0$, that are continuous on the closed interval $[a, b]$. If the region bounded by the graphs of f and g and the lines $x = a$ and $x = b$ is revolved about the x-axis, a solid of revolution with a hollow interior is generated, as illustrated in Figure 28. We seek a formula for finding its volume V.

Figure 29(a) shows the region to be revolved about the x-axis. To find the volume V of the resulting solid of revolution using the washer method, we begin by partitioning the interval $[a, b]$ into n subintervals:

$$[a, x_1], [x_1, x_2], \ldots, [x_{i-1}, x_i], \ldots, [x_{n-1}, b]$$

each of width $\Delta x = \dfrac{b - a}{n}$. In each subinterval $[x_{i-1}, x_i]$, $i = 1, 2, \ldots, n$, we select a number u_i. Then form a washer that has outer radius $R_i = f(u_i)$, inner radius $r_i = g(u_i)$, and thickness Δx. A typical washer is shown in Figure 29(b). The area A_i between the concentric circles that form the washer is the difference between the area of the outer circle and the area of the inner circle.

$$A_i = \pi[\text{Outer Radius}]^2 - \pi[\text{Inner radius}]^2 = \pi\{[f(u_i)]^2 - [g(u_i)]^2\}$$

The volume V_i of the washer is

$$V_i = A_i \Delta x = \pi\{[f(u_i)]^2 - [g(u_i)]^2\}\Delta x$$

See Figure 29(c).

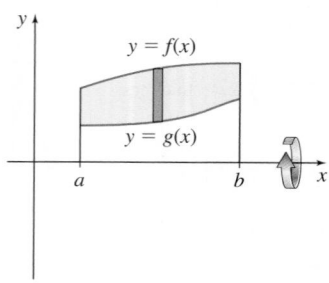

(a) The region to be revolved about the x-axis

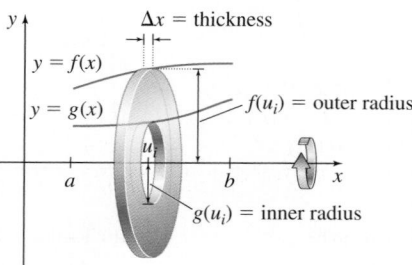

(b) Outer radius $= f(u_i)$
Inner radius $= g(u_i)$
Thickness of washer $= \Delta x$
Volume $= \pi\{[f(u_i)]^2 - [g(u_i)]^2\}\,\Delta x$

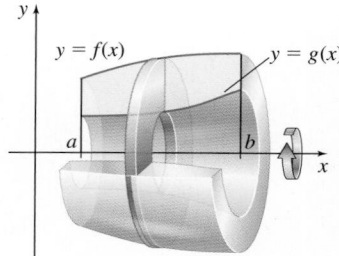

(c) $V = \pi \displaystyle\int_a^b \{[f(x)]^2 - [g(x)]^2\}\,dx$

Figure 29

Volume of a Solid of Revolution Using the Washer Method

Suppose the functions $y = f(x)$ and $y = g(x)$ are continuous on the closed interval $[a, b]$. If $f(x) \geq g(x) \geq 0$ on $[a, b]$, then the volume V of the solid of revolution obtained by revolving the region bounded by the graphs of f and g and the lines $x = a$ and $x = b$ about the x-axis is

$$V = \pi \int_a^b \{[f(x)]^2 - [g(x)]^2\} \, dx \qquad (1)$$

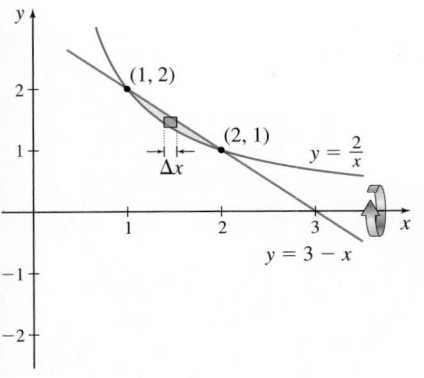

AP® EXAM TIP

Notice that the integrand in (1) is

$$\{[f(x)]^2 - [g(x)]^2\}$$
$$= \{(\text{Outer radius})^2 - (\text{Inner radius})^2\}$$

Be careful not to accidentally write,

$$\{[\text{Outer radius - Inner radius}]^2\}$$

CALC CLIP

▶ **EXAMPLE 5** **Using the Washer Method: Revolving About the x-Axis**

Find the volume V of the solid of revolution generated by revolving the region bounded by the graphs of $y = \dfrac{2}{x}$ and $y = 3 - x$ about the x-axis.

Solution

Begin by graphing the two functions. See Figure 30(a). The x-coordinates of the points of intersection of the graphs satisfy the equation

$$\frac{2}{x} = 3 - x$$

$$x^2 - 3x + 2 = 0$$

$$(x - 1)(x - 2) = 0$$

$$x = 1 \quad \text{or} \quad x = 2$$

So, the region to be revolved lies between the lines $x = 1$ and $x = 2$. Notice that the graph of $y = 3 - x$ lies above the graph of $y = \dfrac{2}{x}$ on the interval $[1, 2]$ so the outer radius is $3 - x$ and the inner radius is $\dfrac{2}{x}$.

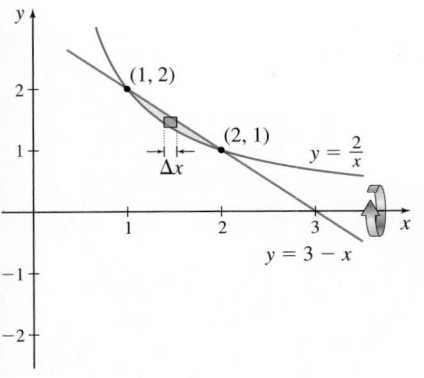

(a) The region to be revolved about the x-axis

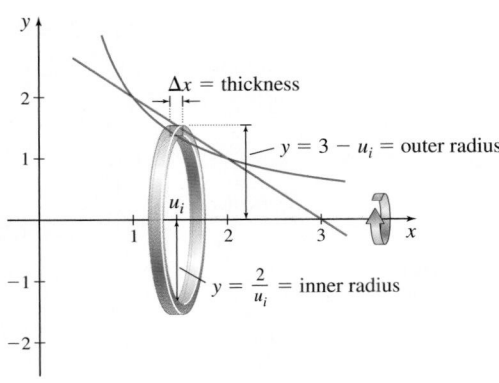

(b) Outer radius $= 3 - u_i$

Inner radius $= \dfrac{2}{u_i}$

Thickness of washer $= \Delta x$

Volume $= \pi \left[\left(3 - u_i\right)^2 - \left(\dfrac{2}{u_i}\right)^2 \right] \Delta x$

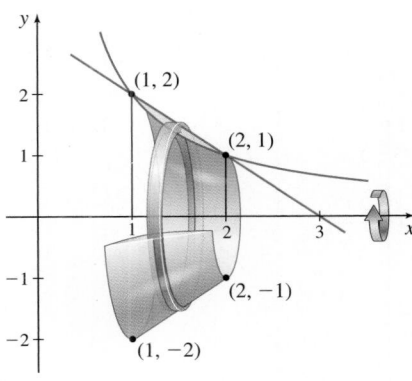

(c) $V = \pi \displaystyle\int_1^2 \left[\left(3 - x\right)^2 - \left(\dfrac{2}{x}\right)^2 \right] dx$

Figure 30

(continued on the next page)

As illustrated in Figure 30(b), if we partition the x-axis, the volume V_i of a typical washer is

$$V_i = \pi \left[(\text{Outer radius})^2 - (\text{Inner radius})^2 \right] \Delta x = \pi \left[(3 - u_i)^2 - \left(\frac{2}{u_i} \right)^2 \right] \Delta x$$

The volume V of the solid of revolution is

$$V = \pi \int_1^2 \left[(3 - x)^2 - \left(\frac{2}{x} \right)^2 \right] dx = \pi \int_1^2 \left(9 - 6x + x^2 - \frac{4}{x^2} \right) dx$$

$$= \pi \left[9x - 3x^2 + \frac{x^3}{3} + \frac{4}{x} \right]_1^2 = \frac{\pi}{3} \text{ cubic units}$$

See Figure 30(c). ■

NOW WORK   Problem **33** and AP® Practice Problem **6**.

④ Use the Washer Method to Find the Volume of a Solid Formed by Revolving a Region About the y-Axis

Suppose the region bounded by the graphs of two functions is revolved about the y-axis. To use the washer method, we partition the y-axis. Then the thickness of a typical washer will be Δy.

EXAMPLE 6   Using the Washer Method: Revolving About the y-Axis

Find the volume V of the solid of revolution generated by revolving the region bounded by the graphs of $y = 2x$ and $y = x^2$ about the y-axis.

Solution

Begin by graphing the two functions. See Figure 31(a). The x-coordinates of the points of intersection of the graphs satisfy the equation

$$2x = x^2$$
$$x^2 - 2x = 0$$
$$x(x - 2) = 0$$
$$x = 0 \quad \text{or} \quad x = 2$$

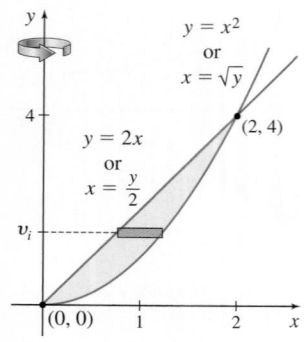

(a) The region to be revolved about the y-axis

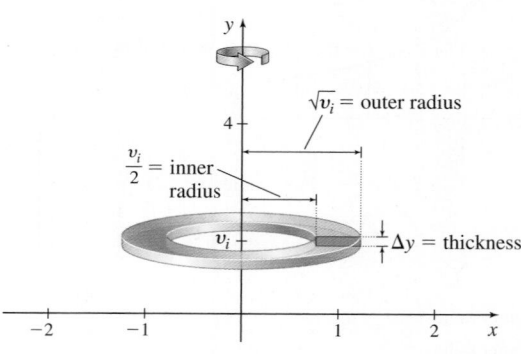
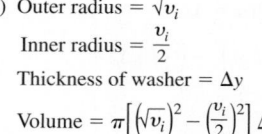

(b) Outer radius = $\sqrt{v_i}$

Inner radius = $\dfrac{v_i}{2}$

Thickness of washer = Δy

Volume = $\pi \left[\left(\sqrt{v_i} \right)^2 - \left(\dfrac{v_i}{2} \right)^2 \right] \Delta y$

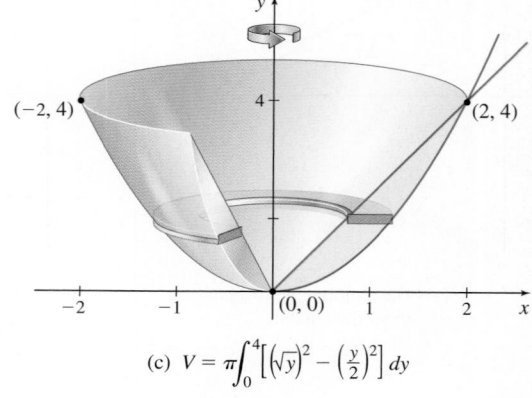

(c) $V = \pi \displaystyle\int_0^4 \left[\left(\sqrt{y} \right)^2 - \left(\dfrac{y}{2} \right)^2 \right] dy$

Figure 31

The points of intersection are $(0, 0)$ and $(2, 4)$. The limits of integration are from $y = 0$ to $y = 4$.

Since the solid is formed by revolving the region about the y-axis from $y = 0$ to $y = 4$, we write $y = 2x$ as $x = \dfrac{y}{2}$ and $y = x^2$ as $x = \sqrt{y}$. The outer radius is $\sqrt{y}$ and the inner radius is $\dfrac{y}{2}$.

As Figure 31(b) illustrates, if we partition the y-axis, the volume V_i of a typical washer is

$$V_i = \pi \left[(\text{Outer radius})^2 - (\text{Inner radius})^2\right] \Delta y = \pi \left[(\sqrt{v_i})^2 - \left(\frac{v_i}{2}\right)^2\right] \Delta y$$

The volume V of the solid of revolution shown in Figure 31(c) is

$$V = \pi \int_0^4 \left[(\sqrt{y})^2 - \left(\frac{y}{2}\right)^2\right] dy = \pi \int_0^4 \left(y - \frac{y^2}{4}\right) dy$$

$$= \pi \left[\frac{y^2}{2} - \frac{y^3}{12}\right]_0^4 = \frac{8}{3}\pi \text{ cubic units} \qquad \blacksquare$$

NOW WORK   Problem **21** and AP® Practice Problems **3** and **4.**

❺ Find the Volume of a Solid Formed by Revolving a Region About a Line Parallel to a Coordinate Axis

Earlier we stated that a solid of revolution can be generated by revolving a region about any line. Now we investigate how to find the volume of a solid formed by revolving a region about a line parallel to either the x-axis or the y-axis.

EXAMPLE 7   **Using the Disk Method: Revolving About a Line Parallel to a Coordinate Axis**

Find the volume V of the solid of revolution generated by revolving the region formed by the function $y = \sqrt{x}$, the x-axis, and the line $x = 4$, shown in Figure 32, about the line $x = 4$.

Solution

Here the solid of revolution is formed by rotating the graph of $y = \sqrt{x}$ about a line parallel to the y-axis. Since the axis of rotation is parallel to the y-axis, we will be partitioning the y-axis and will need to express $y = \sqrt{x}$ as a function of y by solving for x. Then $x = y^2$, $0 \le y \le 2$. In Figure 32, notice that the horizontal distance from the graph of $x = y^2$ to the line $x = 4$ is $4 - x = 4 - y^2$. So, the radius of the solid of revolution is $4 - y^2$.

Using the disk method, the volume V of the solid of revolution is

$$V = \pi \int_0^2 (4 - y^2)^2 dy = \pi \int_0^2 (16 - 8y^2 + y^4) dy = \frac{256}{15}\pi \approx 17.067\pi \qquad \blacksquare$$

NOW WORK   Problem **41** and AP® Practice Problem **10.**

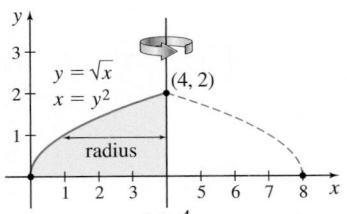

Figure 32 The region to be revolved about the line $x = 4$

EXAMPLE 8 **Using the Washer Method: Revolving About a Horizontal Line**

Find the volume V of the solid of revolution generated by revolving the region bounded by the graphs of $y = 2x$ and $y = x^2$ about the line $y = -5$.

Solution

The region to be revolved is the same as in Example 6, but it is now being revolved about the line $y = -5$, instead of the x-axis. Figure 33 illustrates the region, a typical washer, and the solid of revolution.

In Figure 33(a), notice that the line $y = 2x$ lies $2x + 5$ units above the line $y = -5$ and the parabola $y = x^2$ lies $x^2 + 5$ units above the line $y = -5$. So the outer radius of the washer is $2x + 5$ and the inner radius is $x^2 + 5$. See Figure 33(b). Partitioning the x-axis, the volume V_i of a typical washer is

$$V_i = \pi \left[(2u_i + 5)^2 - (u_i^2 + 5)^2 \right] \Delta x$$

The volume V of the solid of revolution as shown in Figure 33(c) is

$$V = \pi \int_0^2 \left[(2x + 5)^2 - (x^2 + 5)^2 \right] dx = \pi \int_0^2 (-x^4 - 6x^2 + 20x)\, dx$$

$$= \pi \left[-\frac{x^5}{5} - 2x^3 + 10x^2 \right]_0^2 = \frac{88}{5}\pi \text{ cubic units} \quad \blacksquare$$

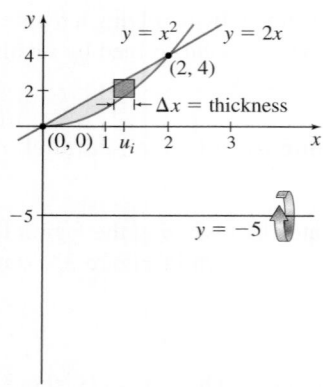

(a) The region to be revolved about the line $y = -5$

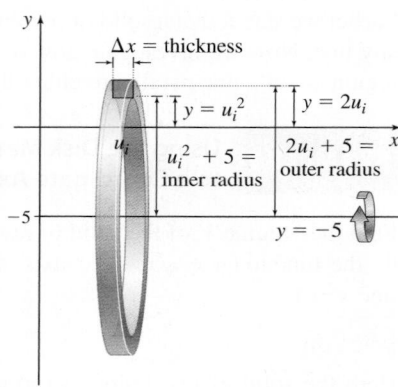

(b) Outer radius $= 2u_i + 5$
Inner radius $= u_i^2 + 5$
Thickness of washer $= \Delta x$
Volume $= \pi[(2u_i + 5)^2 - (u_i^2 + 5)^2]\, \Delta x$

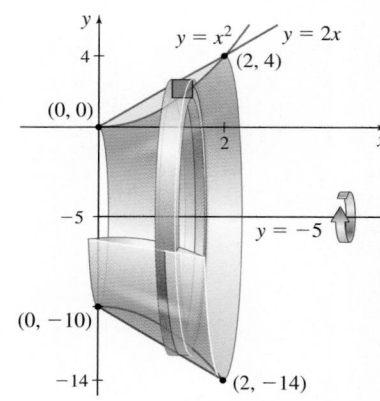

(c) $V = \pi \int_0^2 [(2x + 5)^2 - (x^2 + 5)^2]\, dx$

Figure 33

NOW WORK Problem 39 and AP® Practice Problems 8, 9, 13, and 14.

EXAMPLE 9 **Using the Washer Method: Revolving About a Vertical Line**

Find the volume of the solid of revolution generated by revolving the region bounded by the graphs of $y = 2x$ and $y = x^2$ about the line $x = 2$.

Solution

This example is similar to Example 6 except that the region is revolved about the line $x = 2$, instead of the y-axis. Figure 34 shows the graph of the region, a typical washer, and the solid of revolution.

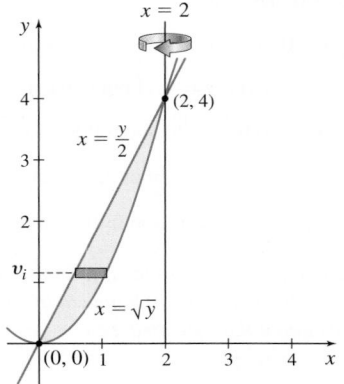

(a) The region to be revolved about the line $x = 2$

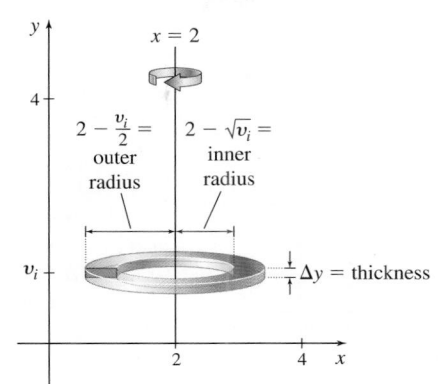

(b) Outer radius $= 2 - \dfrac{v_i}{2}$

Inner radius $= 2 - \sqrt{v_i}$

Thickness of washer $= \Delta y$

Volume $= \pi\left[\left(2 - \dfrac{v_i}{2}\right)^2 - \left(2 - \sqrt{v_i}\right)^2\right]\Delta y$

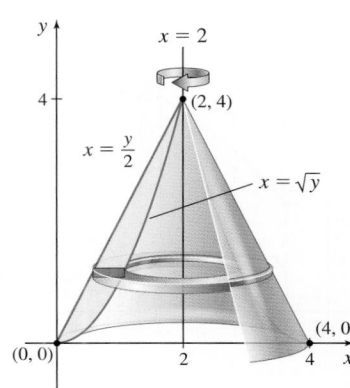

(c) $V = \pi\displaystyle\int_0^4 \left[\left(2 - \dfrac{y}{2}\right)^2 - \left(2 - \sqrt{y}\right)^2\right]dy$

Figure 34

Since the region is revolved about the vertical line $x = 2$, we express $y = 2x$ as $x = \dfrac{y}{2}$ and $y = x^2$ as $x = \sqrt{y}$. In Figure 34(a), notice that the line $x = \dfrac{y}{2}$ lies $2 - \dfrac{y}{2}$ units to the left of the line $x = 2$ and the graph of $x = \sqrt{y}$ lies $2 - \sqrt{y}$ units to the left of the line $x = 2$. So the outer radius is $2 - \dfrac{y}{2}$ and the inner radius is $2 - \sqrt{y}$. The volume V of the solid of revolution is

$$V = \pi \int_0^4 \left[\left(2 - \frac{y}{2}\right)^2 - (2 - \sqrt{y})^2\right]dy$$

$$= \pi \int_0^4 \left[\left(4 - 2y + \frac{y^2}{4}\right) - (4 - 4\sqrt{y} + y)\right]dy$$

$$= \pi \int_0^4 \left(\frac{y^2}{4} - 3y + 4\sqrt{y}\right)dy = \pi\left[\frac{y^3}{12} - \frac{3y^2}{2} + \frac{8y^{3/2}}{3}\right]_0^4 = \frac{8}{3}\pi \text{ cubic units} \quad \blacksquare$$

NOW WORK Problem **43** and AP® Practice Problem **7.**

8.3 Assess Your Understanding

Concepts and Vocabulary

1. If a function f is continuous on a closed interval $[a, b]$, then the volume V of the solid of revolution obtained by revolving the region bounded by the graph of f, the x-axis, and the lines $x = a$ and $x = b$ about the x-axis is found using the formula $V = $ _____.

2. *True or False* When the region bounded by the graphs of the functions f and g, $f(x) \ge g(x) \ge 0$, and the lines $x = a$ and $x = b$ is revolved about the x-axis, the cross section exposed by making a slice at u_i perpendicular to the x-axis is two concentric circles, and the area A_i between the circles is

$$A_i = \pi[f(u_i) - g(u_i)]^2$$

3. *True or False* If the functions f and g are continuous on the closed interval $[a, b]$ and if $f(x) \ge g(x) \ge 0$ on the interval, then the volume V of the solid of revolution obtained by revolving the region bounded by the graphs of f and g and the lines $x = a$ and $x = b$ about the x-axis is

$$V = \pi \int_a^b [f(x) - g(x)]^2\, dx$$

4. *True or False* If the region bounded by the graphs of $y = x^2$ and $y = 2x$ is revolved about the line $y = 6$, the volume V of the solid of revolution generated is found by finding the integral

$$V = \pi \int_6^7 (4x^2 - x^4)\, dx$$

Skill Building

In Problems 5–10, find the volume of the solid of revolution generated by revolving the shaded region about the indicated axis.

5. $y = 2\sqrt{x}$ about the x-axis

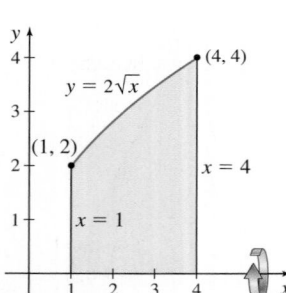

6. $y = x^4$ about the y-axis

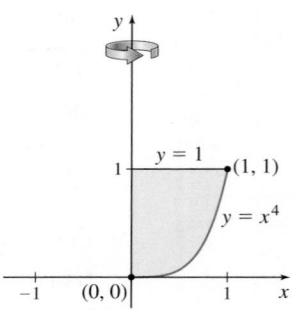

7. $y = \dfrac{1}{x}$ about the y-axis

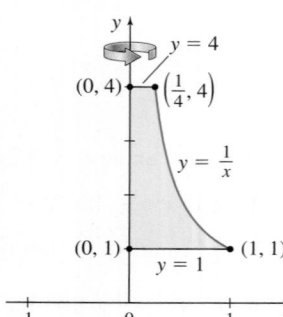

8. $y = x^{2/3}$ about the y-axis

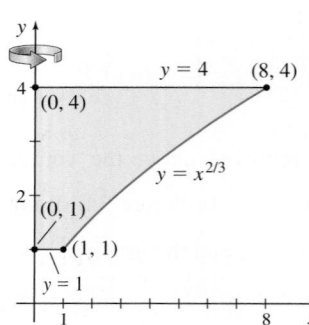

9. $y = \sec x$ about the x-axis

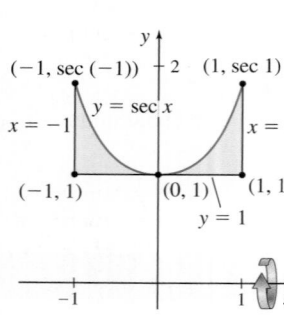

10. $y = x^2$ about the y-axis

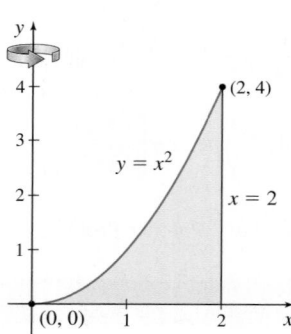

In Problems 11–16, use the disk method to find the volume of the solid of revolution generated by revolving the region bounded by the graphs of the given equations about the indicated axis.

11. $y = 2x^2$, the x-axis, $x = 1$; about the x-axis

12. $y = \sqrt{x}$, the x-axis, $x = 4$, $x = 9$; about the x-axis

13. $y = e^{-x}$, the x-axis, $x = 0$, $x = 2$; about the x-axis

14. $y = e^x$, the x-axis, $x = -1$, $x = 1$; about the x-axis

15. $y = x^2$, $x \ge 0$, $y = 1$, $y = 4$; about the y-axis

16. $y = 2\sqrt{x}$, the y-axis, $y = 4$; about the y-axis

In Problems 17–22, use the washer method to find the volume of the solid of revolution generated by revolving the region bounded by the graphs of the given equations about the indicated axis.

17. $y = x^2$, $x \ge 0$, the y-axis, $y = 4$; about the x-axis

18. $y = 2x^2$, $x \ge 0$, the y-axis, $y = 2$; about the x-axis

19. $y = 2\sqrt{x}$, the y-axis, $y = 4$; about the x-axis

20. $y = x^{2/3}$, the x-axis, $x = 8$; about the y-axis

21. $y = x^3$, the x-axis, $x = 1$; about the y-axis

22. $y = 2x^4$, the x-axis, $x = 1$; about the y-axis

In Problems 23–38, find the volume of the solid of revolution generated by revolving the region bounded by the graphs of the given equations about the indicated axis.

23. $y = \dfrac{1}{x}$, the x-axis, $x = 1$, $x = 2$; about the x-axis

24. $y = \dfrac{1}{x}$, the x-axis, $x = 1$, $x = 2$; about the y-axis

25. $y = \sqrt{x}$, the y-axis, $y = 9$; about the y-axis

26. $y = \sqrt{x}$, the y-axis, $y = 9$; about the x-axis

27. $y = (x - 2)^3$, the x-axis, $x = 0$, $x = 3$; about the x-axis

28. $y = (x - 2)^3$, the x-axis, $x = 0$, $x = 3$; about the y-axis

29. $y = (x + 1)^2$, $x \ge 0$, $y = 16$; about the y-axis

30. $y = (x + 1)^2$, $x \le 0$, $y = 16$; about the x-axis

31. $x = y^4 - 1$, the y-axis; about the y-axis

32. $y = x^4 - 1$, the x-axis; about the x-axis

33. $y = 4x$, $y = x^3$, $x \ge 0$; about the x-axis

34. $y = 2x + 1$, $y = x$, $x = 0$, $x = 3$; about the x-axis

35. $y = 1 - x$, $y = e^x$, $x = 1$; about the x-axis

36. $y = \cos x$, $y = \sin x$, $x = 0$, $x = \dfrac{\pi}{4}$; about the x-axis

37. $y = \csc x$, $y = 0$, $x = \dfrac{\pi}{2}$, $x = \dfrac{3\pi}{4}$; about the x-axis

38. $y = \sec x$, $y = 0$, $x = 0$, $x = \dfrac{\pi}{3}$; about the x-axis

In Problems 39–46, find the volume of the solid of revolution generated by revolving the region bounded by the graphs of the given equations about the indicated line.

39. $y = e^x$, $y = 0$, $x = 0$, $x = 2$; about $y = -1$

40. $y = \dfrac{1}{x}$, $y = 0$, $x = 1$, $x = 4$; about $y = 4$

41. $y = x^2$, the x-axis, $x = 1$; about $x = 1$

42. $y = x^3$, $x = 0$, $y = 1$; about $x = -1$

43. $y = \sqrt{x}$, the x-axis, $x = 4$; about $x = -4$

44. $y = \dfrac{1}{\sqrt{x}}$, the x-axis, $x = 1$, $x = 4$; about $x = 4$

45. $y = \dfrac{1}{x^2}$, $y = 0$, $x = 1$, $x = 4$; about $y = 4$

46. $y = \sqrt{x}$, $y = 0$, $0 \le x \le 4$; about $y = -4$

In Problems 47–50, find the volume of the solid of revolution generated by revolving the indicated region about each line.

47. The region bounded by $y = x^2$, the x-axis, and $x = 3$

   (a) About the x-axis

   (b) About the line $y = -1$

   (c) About the line $y = 10$

   (d) About the line $y = a, \ a \geq 9$

48. The region bounded by $y = x^2$, the x-axis, and $x = 3$

   (a) About the y-axis

   (b) About the line $x = -5$

   (c) About the line $x = 5$

   (d) About the line $x = b, b \geq 3$

49. The region bounded by $y = x^2$, the y-axis, and $y = 4$

   (a) About the y-axis

   (b) About the line $x = -5$

   (c) About the line $x = 5$

   (d) About the line $x = b, b \geq 2$

50. The region bounded by $y = x^2$, the y-axis, and $y = 4$

   (a) About the x-axis

   (b) About the line $y = -1$

   (c) About the line $y = 4$

   (d) About the line $y = a, \ a \geq 4$

Applications and Extensions

51. Volume of a Sphere

   (a) Graph $y = \sqrt{a^2 - x^2}$.

   (b) Revolve the region bounded by $y = \sqrt{a^2 - x^2}$ and the x-axis about the x-axis to generate a sphere of radius a. Use the Disk Method to show that the volume V of the sphere is $\frac{4}{3}\pi a^3$.

52. Volume of a Cone

   (a) Find an equation of the line segment from the point $(0, h)$, $h > 0$, to the point $(a, 0)$, $a > 0$.

   (b) Graph the line segment from (a).

   (c) Revolve the region in the first quadrant bounded by the line segment from (a), the x-axis, and the y-axis, about the y-axis to generate a cone of height h and radius a. Use the Disk Method to show that the volume V of the cone is $\frac{1}{3}\pi a^2 h$.

53. Volume of a Solid of Revolution

The figure shows the solid of revolution generated by revolving the region bounded by the graph of $y = \dfrac{1}{x^2 + 4}$ and the x-axis from $x = 0$ to $x = 1$ about the x-axis.

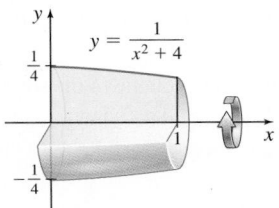

   (a) Express the volume of the solid of revolution as an integral.

   (b) Use technology to find the volume.

54. Volume of a Solid of Revolution A solid of revolution is generated by revolving the region bounded by the graph of $y = \ln x$, the line $x = e$, and the x-axis about the x-axis.

   (a) Express the volume of the solid of revolution as an integral.

   (b) Use technology to find the volume.

55. Volume of an Ellipsoid

   (a) Graph $y = \sqrt{9 - 4x^2}$, the upper half of an ellipse.

   (b) Find the volume of the solid of revolution generated by revolving the region bounded by $y = \sqrt{9 - 4x^2}$ and the x-axis about the x-axis.

56. Volume of a Solid of Revolution

   (a) Graph $y = \sqrt{4x^2 + 1}$, the upper portion of a hyperbola.

   (b) Find the volume of the solid of revolution generated by revolving the region bounded by $y = \sqrt{4x^2 + 1}$, the lines $x = -1$ and $x = 1$, and the x-axis about the x-axis.

57. Mixed Practice A region in the first quadrant is bounded by the x-axis and the graph of $y = kx - x^2$, where $k > 0$.

   (a) In terms of k, find the volume generated when the region is revolved around the x-axis.

   (b) In terms of k, find the area of the region.

58. Mixed Practice

   (a) Find all numbers b for which the graphs of $y = 2x + b$ and $y^2 = 4x$ intersect in two distinct points.

   (b) If $b = -4$, find the area bounded by the graphs of $y = 2x - 4$ and $y^2 = 4x$.

   (c) If $b = 0$, find the volume of the solid generated by revolving about the x-axis the region bounded by the graphs of $y = 2x$ and $y^2 = 4x$.

59. Volume of a Solid of Revolution Find the volume of the solid of revolution generated by revolving the region bounded by the graphs of $y = \cos x$ and $x = 0$ from $x = 0$ to $x = \dfrac{\pi}{2}$ about the line $y = 1$. *Hint:* $\cos^2 x = \dfrac{1 + \cos(2x)}{2}$.

60. Volume of a Solid of Revolution Find the volume of the solid of revolution generated by revolving the region bounded by the graphs of $y = \cos x$ and $x = 0$ from $x = 0$ to $x = \dfrac{\pi}{2}$ about the line $y = -1$. (See the hint in Problem 59.)

61. The Volume of a Solid of Revolution Find the volume of the solid of revolution generated by revolving the region bounded by the graph of $y = \ln x$ and the x-axis from $x = 1$ to $x = e$ about the x-axis.

62. The Volume of a Solid of Revolution Find the volume of the solid of revolution generated by revolving the region bounded by the graph of $y = x\sqrt{\sin x}$ and the x-axis from $x = 0$ to $x = \dfrac{\pi}{2}$ about the x-axis. See the figure below.

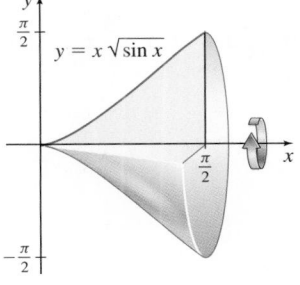

63. The Volume of a Solid of Revolution   Find the volume of the solid of revolution generated by revolving the region bounded by the graph of $y = x\sqrt{\ln x}$, the x-axis, and the lines $x = 1$ and $x = e^2$ about the x-axis.

64. The Volume of a Solid of Revolution   Find the volume, if it is defined, of the solid of revolution generated by revolving the region bounded by the graph of $y = \dfrac{1}{\sqrt{x}}$ and the x-axis to the right of $x = 1$ about the x-axis. See the figure below.

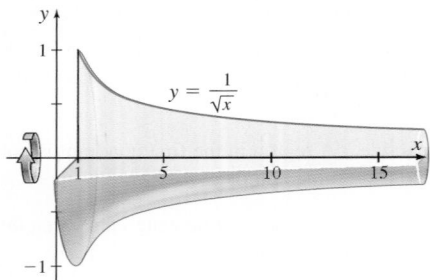

65. Mixed Practice   Let A be the area of the region in the first quadrant bounded by the graph of $y = \tan x$ and the lines $x = 0$ and $y = 1$.

(a) Find A.

(b) Find the volume of the solid of revolution generated by revolving the region about the x-axis.

66. Mixed Practice   Let A be the area of the region in the first quadrant bounded by the graph of $y = \cot x$ and the lines $x = \dfrac{\pi}{4}$ and $x = \dfrac{\pi}{2}$.

(a) Find A.

(b) Find the volume of the solid of revolution generated by revolving the region about the x-axis.

Challenge Problem

 67. Volume of a Solid of Revolution   The graph of the function $P(x) = kx^2$ is symmetric with respect to the y-axis and contains the points $(0, 0)$ and (b, e^{-b^2}), where $b > 0$.

(a) Find k and write an equation for $P = P(x)$.

(b) The region bounded by the graph of P, the y-axis, and the line $y = e^{-b^2}$ is revolved about the y-axis to form a solid. Find its volume.

(c) For what number b is the volume of the solid in (b) a maximum? Justify your answer.

Preparing for the AP® Exam

AP® Practice Problems

Multiple-Choice Questions

PAGE 640 **1.** What is the volume of the solid of revolution generated when the region in the first quadrant bounded by the graph of $y = e^x$, the x-axis, and the line $x = 3$ is revolved about the x-axis?

(A) $2\pi e^6$ (B) $\dfrac{\pi}{2}(e^6 - 1)$

(C) $\pi(e^6 - 1)$ (D) $\dfrac{\pi}{2}(e^3 - 1)$

PAGE 640 **2.** Find the volume of the solid of revolution generated when the region bounded by the graph of $y = \csc x$, the x-axis, and the lines $x = \dfrac{\pi}{4}$ and $x = \dfrac{3\pi}{4}$ is revolved about the x-axis.

(A) 2   (B) π (C) $\sqrt{2}\pi$ (D) 2π

PAGE 647 **3.** The region in the first quadrant bounded by the graph of $y = x^3$, the line $x = 2$, and the x-axis is revolved about the y-axis. The volume of the solid of revolution is

(A) 4π (B) $\dfrac{64}{5}\pi$ (C) $\dfrac{96}{5}\pi$ (D) $\dfrac{128}{7}\pi$

PAGE 647 **4.** The volume of the solid of revolution generated by revolving the region under the graph of $y = \sin x$ from $x = 0$ to $x = \dfrac{\pi}{6}$ about the y-axis is given by

(A) $\pi \displaystyle\int_0^{\pi/6} \left[\left(\dfrac{\pi}{6}\right)^2 - \sin^2 x \right] dx$

(B) $\pi \displaystyle\int_0^{1/2} \left[\left(\dfrac{\pi}{6}\right)^2 - (\arcsin y)^2 \right] dy$

(C) $\pi \displaystyle\int_0^{1/2} (\arcsin y)^2 \, dy$

(D) $\pi \displaystyle\int_0^{1/2} \left[\left(\dfrac{\pi}{6}\right)^2 - \sin^2 y \right] dy$

PAGE 643 **5.** Find the volume of the solid of revolution obtained by revolving the region bounded by the graph of $x = \sqrt{4 - 4y^2}$ and the y-axis about the y-axis.

(A) $\dfrac{4}{3}\pi$ (B) $\dfrac{8}{3}\pi$ (C) $\dfrac{11}{6}\pi$ (D) $\dfrac{16}{3}\pi$

6. The region bounded by the graphs of $y = 1$, $y = e^x$, and the line $x = 2$ is revolved about the x-axis. The volume of the resulting solid of revolution is

(A) $\frac{\pi}{2}(e^4 - 5)$ (B) $2\pi(e^4 - 2)$

(C) $\frac{\pi}{2}(e^4 - 2)$ (D) $\frac{\pi}{2}(e^4 - 4)$

7. The volume of the solid of revolution generated by revolving the region bounded by the graphs of $y = \sqrt{x}$, $y = 2$, and the y-axis about the line $x = 4$ is

(A) $\frac{18}{5}\pi$ (B) $\frac{224}{15}\pi$ (C) $\frac{256}{15}\pi$ (D) $\frac{328}{15}\pi$

8. The volume of the solid of revolution generated by revolving the region bounded by the graphs of $y = x^2$, $x \geq 0$, the y-axis, and $y = 4$ about the line $y = 5$ is given by

(A) $\pi \int_0^2 (5^2 - x^2)\, dx$ (B) $\pi \int_0^5 [25 - (x^2 - 5)]\, dx$

(C) $\pi \int_0^2 [(5 - x^2)^2 - 1]\, dx$ (D) $\pi \int_0^2 [1 - (5 - x^2)]\, dx$

9. The region in the first quadrant bounded by the graphs of $y = x + 2$, $y = 2x$, and the y-axis is revolved about the line $y = -1$. Find the volume of the solid generated.

(A) $\frac{52}{3}\pi$ (B) 12π (C) 28π (D) 32π

10. The integral that represents the volume of the solid of revolution generated by revolving the region in the first quadrant bounded by the graph of $y = x^3$ and the line $y = 8$, about the line $y = 8$, is

(A) $\pi \int_0^2 (8^2 - x^6)\, dx$ (B) $\pi \int_0^2 (8 - x^3)^2\, dx$

(C) $\pi \int_0^8 y^{2/3}\, dy$ (D) $\pi \int_0^8 x^6\, dx$

11. The volume of the solid of revolution generated by revolving the region bounded by the graph of $y = \sec x$, the x-axis, and the lines $x = 0$ and $x = \frac{\pi}{3}$, about the x-axis, equals

(A) $\sqrt{3}\pi$ (B) $\frac{\sqrt{3}}{2}$ (C) $\frac{\sqrt{3}\,\pi}{2}$ (D) $\sqrt{3}$

12. Find the volume of the solid of revolution generated when the region bounded by the graph of $y = \sin x$, the x-axis, and the lines $x = 0$ and $x = 2\pi$ is revolved about the x-axis.

(A) 0   (B) 3.142   (C) 6.283   (D) 9.870

Free-Response Questions

13. The graphs of the functions $f(x) = -x^2 + 4x$ and $g(x) = 3x^2 - 12x + 12$ are shown in the figure below.

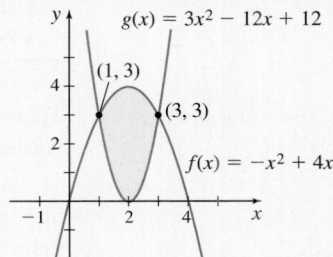

(a) Write, but do not evaluate, an integral that gives the volume of the solid of revolution generated when the region bounded by the graphs of f and g is revolved about the x-axis.

(b) Write, but do not evaluate, an integral that gives the volume of the solid of revolution generated when the regions bounded by the graphs of f and g is revolved about the line $y = -2$.

See the **BREAK IT DOWN** *on page 697 for a stepped out solution to AP® Practice Problem 13(b).*

14. The region R is bounded by the function $y = e^x$, the x-axis, and the lines $x = 0$ and $x = 2$. Find the volume of the solid of revolution generated by revolving R about

(a) the x-axis.

(b) the y-axis.

(c) the line $y = -1$.

In each case, write the integral that represents the volume of the solid.

Retain Your Knowledge

Multiple-Choice Questions

1. The area of a circular disk is increasing at a rate of 4π cm²/s. What is the rate of change of the radius of the disc when its area is $A = 100\pi$ cm²?

(A) $\frac{1}{5}$ cm/s   (B) $\frac{1}{5}\pi$ cm/s   (C) 5 cm/s   (D) 4π cm/s

2. The general solution to the differential equation $\frac{dy}{dx} = \frac{xy}{\sqrt{1 - x^2}}$ is

(A) $y = \sqrt{1 - x^2} + C$ (B) $y = -\sqrt{1 - x^2} + C$

(C) $y = e^{\sqrt{1 - x^2} + C}$ (D) $y = e^{-\sqrt{1 - x^2} + C}$

3. What is the average value of the function $y = \sqrt{\cos x}$ over the interval $\left[-\frac{\pi}{4}, \frac{\pi}{2}\right]$?

(A) 0.300   (B) 4.577   (C) 0.725   (D) 0.824

Free Response Question

4. A differential equation is defined by $\frac{dy}{dx} = 2 - x$.

(a) Use the grid below to draw a slope field for the given differential equation at the points marked on the graph.

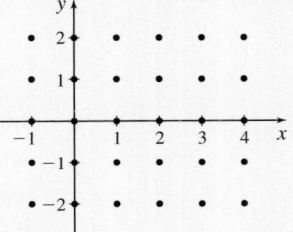

(b) Use the slope field to draw a graph of the particular solution of the differential equation with the initial condition if $x = 2$, then $y = 1$.

(c) Find the general solution to the differential equation.

(d) Find the particular solution to the differential equation with the initial condition if $x = 2$, then $y = 1$.

8.4 Volume of a Solid of Revolution: Cylindrical Shells

OBJECTIVES *When you finish this section, you should be able to:*

1 Use the shell method to find the volume of a solid formed by revolving a region about the *y*-axis (p. 654)

2 Use the shell method to find the volume of a solid formed by revolving a region about the *x*-axis (p. 659)

3 Use the shell method to find the volume of a solid formed by revolving a region about a line parallel to a coordinate axis (p. 661)

NOTE Sewer lines and the water pipes in a house are cylindrical shells.

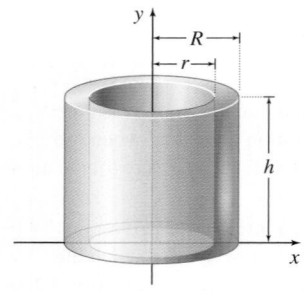

Figure 35 $V = \pi R^2 h - \pi r^2 h$

NEED TO REVIEW? The volume V of a right circular cylinder of height h and radius r is $V = \pi r^2 h$

There are solids of revolution for which the volume is difficult to find using the disk or washer method. In these situations, the volume can often be found using *cylindrical shells*.

A **cylindrical shell** is the solid between two concentric cylinders, as shown in Figure 35. If the inner radius of the cylinder is r and the outer radius is R, the volume V of a cylindrical shell of height h is

$$V = \pi R^2 h - \pi r^2 h$$

That is, the volume of a cylindrical shell equals the volume of the larger cylinder, which has radius R, minus the volume of the smaller cylinder, which has radius r. It is convenient to write this formula as

$$V = \pi(R^2 - r^2)h = \pi(R + r)(R - r)h = 2\pi\left(\frac{R + r}{2}\right)h(R - r)$$

$$V = 2\pi\left(\frac{R + r}{2}\right)h(R - r)$$

$$V = 2\pi \ (\text{Average radius}) \ (\text{Height}) \ (\text{Thickness})$$

1 Use the Shell Method to Find the Volume of a Solid Formed by Revolving a Region About the *y*-Axis

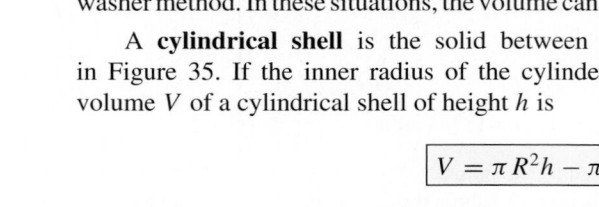

Figure 36

Suppose a function $y = f(x)$ is nonnegative and continuous on the closed interval $[a, b]$, where $a \geq 0$. We seek the volume V of the solid generated by revolving the region bounded by the graph of f, the x-axis, and the lines $x = a$ and $x = b$ about the y-axis. See Figure 36.

We begin by partitioning the interval $[a, b]$ into n subintervals:

$$[a, x_1], [x_1, x_2], \ldots, [x_{i-1}, x_i], \ldots, [x_{n-1}, b]$$

each of width $\Delta x = \dfrac{b - a}{n}$. We concentrate on the rectangle whose base is the subinterval $[x_{i-1}, x_i]$ and whose height is $f(u_i)$, where $u_i = \dfrac{x_{i-1} + x_i}{2}$ is the midpoint of the subinterval. See Figure 37(a) on page 655. When this rectangle is revolved about the y-axis, it generates a cylindrical shell of average radius u_i, height $f(u_i)$, and thickness Δx, as shown in Figure 37(b) on page 655. The volume V_i of this cylindrical shell is

$$V_i = 2\pi(\text{Average radius})(\text{Height})(\text{Thickness}) = 2\pi u_i f(u_i)\Delta x$$

The sum of the volumes of the n cylindrical shells approximates the volume V of the solid generated by revolving the region bounded by the graph of $y = f(x)$, the x-axis, and the lines $x = a$ and $x = b$ about the y-axis. That is,

$$V \approx \sum_{i=1}^{n} [2\pi u_i f(u_i) \Delta x] = 2\pi \sum_{i=1}^{n} [u_i f(u_i) \Delta x]$$

As the number n of subintervals increases, the sums $2\pi \sum_{i=1}^{n} [u_i f(u_i) \Delta x]$ become better approximations to the volume V of the solid. These sums are Riemann sums, and since f is continuous on $[a, b]$, the limit is a definite integral. See Figure 37(c). When cylindrical shells are used to find the volume of a solid of revolution, we refer to the process as the **shell method**.

IN WORDS The volume V of a cylindrical shell is

$V = $ (circumference)(height)(thickness)

$\quad = (2\pi$ radius)(height)(thickness)

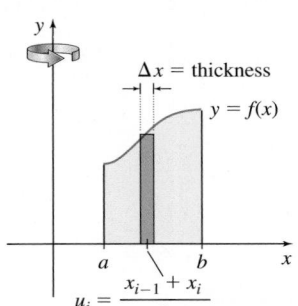

(a) The region to be revolved about the y-axis

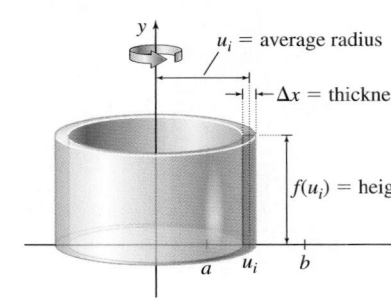

(b) Average radius $= u_i$
Height $= f(u_i)$
Thickness $= \Delta x$
Volume $= 2\pi u_i f(u_i) \Delta x$

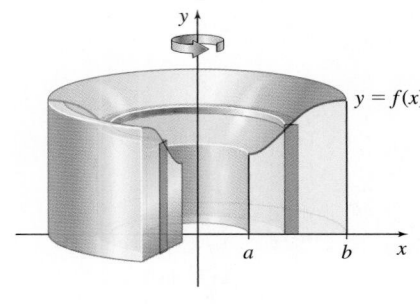

(c) $V = 2\pi \int_a^b x f(x)\, dx$

Figure 37 The shell method

Volume of a Solid of Revolution About the y-Axis: the Shell Method

If $y = f(x)$ is a function that is continuous and nonnegative on the closed interval $[a, b]$, where $a \geq 0$, then the volume V of the solid generated by revolving the region bounded by the graph of f, the x-axis, and the lines $x = a$ and $x = b$ about the y-axis is

$$V = 2\pi \int_a^b x f(x)\, dx$$

It can be shown that the shell method and the washer method of Section 8.3 are equivalent; that is, they both give the same answer.* The advantage of having two equivalent, yet different, formulas is flexibility. There are times when one of the two methods is easier to use, as Example 1 illustrates.

---

*This topic is discussed in detail in an article by Charles A. Cable (February 1984), "The Disk and Shell Method," *American Mathematical Monthly, 91*(2), 139.

EXAMPLE 1 Finding the Volume of a Solid: Revolving About the y-Axis

Find the volume V of the solid generated by revolving the region bounded by the graphs of $f(x) = x^2 + 2x$, the x-axis, and the line $x = 1$ about the y-axis.

Solution

Using the shell method: In the shell method, when a region is revolved about the y-axis, we partition the x-axis and use vertical shells.

Figure 38(a) illustrates the region to be revolved and a typical rectangle of height $f(u_i) = u_i^2 + 2u_i$ and thickness Δx. See Figure 38(b). When the rectangle is revolved about the y-axis, it generates a shell with average radius u_i, whose volume is $V_i = 2\pi$(average radius)(height)(thickness) $= 2\pi u_i f(u_i)\Delta x$. Figure 38(c) illustrates the solid of revolution. The volume V of the solid of revolution is

$$V = 2\pi \int_0^1 xf(x)\, dx = 2\pi \int_0^1 [x(x^2 + 2x)]\, dx = 2\pi \int_0^1 (x^3 + 2x^2)\, dx$$

$$= 2\pi \left[\frac{x^4}{4} + \frac{2x^3}{3} \right]_0^1 = \frac{11}{6}\pi \text{ cubic units}$$

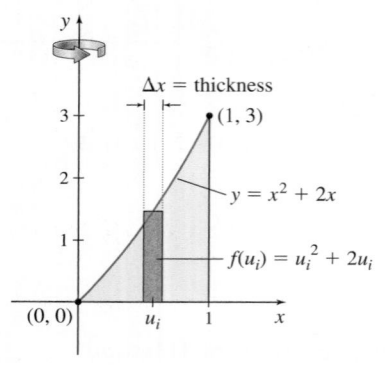

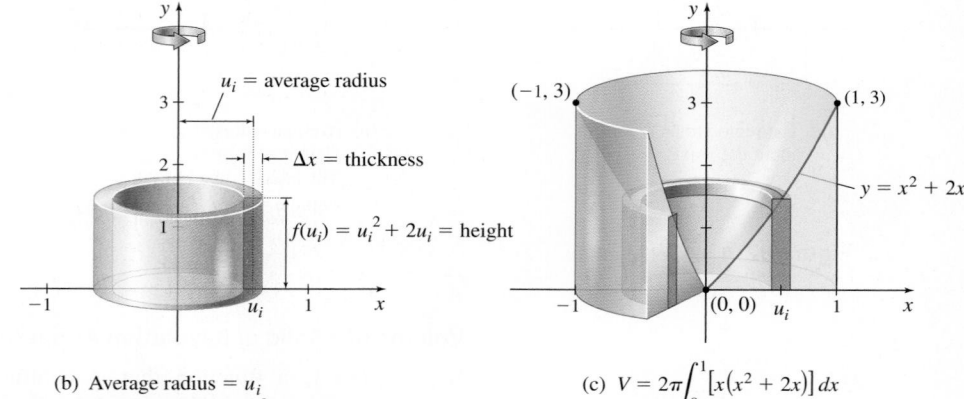

(a) The region to be revolved about the y-axis

(b) Average radius $= u_i$
Height $= f(u_i) = u_i^2 + 2u_i$
Thickness $= \Delta x$
Volume $= 2\pi u_i(u_i^2 + 2u_i)\Delta x$

(c) $V = 2\pi \int_0^1 [x(x^2 + 2x)]\, dx$

Figure 38 The shell method

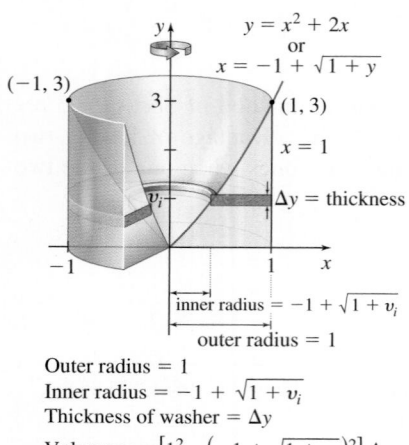

Outer radius $= 1$
Inner radius $= -1 + \sqrt{1 + v_i}$
Thickness of washer $= \Delta y$
Volume $= \pi[1^2 - (-1 + \sqrt{1 + v_i})^2]\Delta y$

Figure 39 The washer method

Using the washer method: Using the washer method, a revolution about the y-axis requires integration with respect to y. This means we need to find the inverse function of $y = f(x)$. We treat $x^2 + 2x - y = 0$ as a quadratic equation in the variable x and use the quadratic formula with $a = 1$, $b = 2$, and $c = -y$ to obtain $x = g(y) = -1 \pm \sqrt{1 + y}$. Since $x \geq 0$, we use the $+$ sign.

See Figure 39. The volume of a typical washer is

$$V_i = \pi[(\text{Outer radius})^2 - (\text{Inner radius})^2] \times (\text{Thickness}).$$

The volume V of the solid of revolution is

$$V = \pi \int_0^3 \left[1^2 - \left(-1 + \sqrt{1 + y} \right)^2 \right] dy = \pi \int_0^3 [1 - (1 - 2\sqrt{1 + y} + 1 + y)]\, dy$$

$$= \pi \int_0^3 [2\sqrt{1 + y} - 1 - y]\, dy = \pi \int_0^3 (2\sqrt{1 + y})\, dy - \pi \int_0^3 (1 + y)\, dy$$

The two integrals are found as follows:

- $\pi \displaystyle\int_0^3 (2\sqrt{1+y})\, dy = 2\pi \int_1^4 u^{1/2}\, du = 2\pi \left[\dfrac{u^{3/2}}{\dfrac{3}{2}}\right]_1^4 = \dfrac{28}{3}\pi$

  Let $u = 1 + y$;
  then $du = dy$

- $\pi \displaystyle\int_0^3 (1+y)\, dy = \pi \left[y + \dfrac{y^2}{2}\right]_0^3 = \dfrac{15}{2}\pi$

The volume V is

$$V = \frac{28}{3}\pi - \frac{15}{2}\pi = \frac{11}{6}\pi \text{ cubic units}$$ ∎

Example 1 gives a clue to when the shell method is preferable to the washer method. When it is difficult, or impossible, to solve $y = f(x)$ for x, we use the shell method. For example, if the function in Example 1 had been $y = f(x) = x^5 + x^2 + 1$, we would not have been able to solve for x, so the practical choice is the shell method.

NOW WORK Problem 5.

The next example illustrates the importance of sketching a graph before using a formula. Notice the limits of integration and how we determined them when we use the washer method.

  EXAMPLE 2 **Finding the Volume of a Solid: Revolving About the *y*-Axis**

CALC CLIP

Find the volume V of the solid generated by revolving the region bounded by the graphs of $f(x) = x^2$ and $g(x) = 12 - x$ to the right of $x = 1$ about the y-axis.

Solution

Using the shell method: Figure 40(a) shows the graph of the region to be revolved and a typical rectangle.

See Figure 40(b). With the shell method, we partition the x-axis and use vertical shells. A typical shell has

- average radius $= u_i$
- height $= g(u_i) - f(u_i) = (12 - u_i) - u_i^2 = 12 - u_i - u_i^2$
- volume $V_i = 2\pi u_i (12 - u_i - u_i^2)\Delta x$

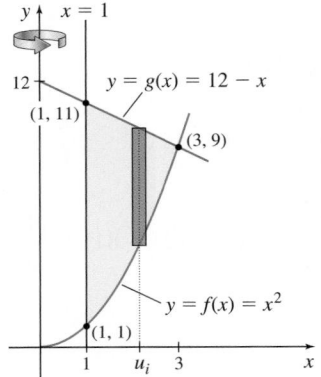

(a) Region to be revolved
about the y-axis

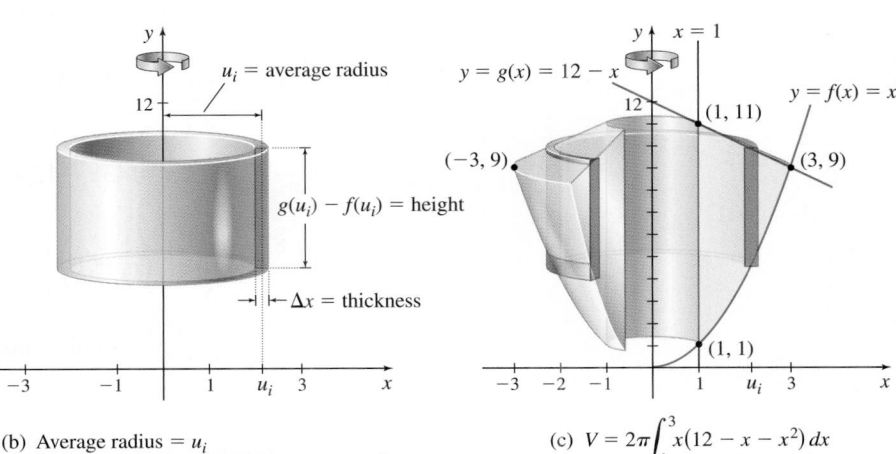

(b) Average radius $= u_i$
Height $= g(u_i) - f(u_i) = (12 - u_i) - u_i^2$
Thickness of shell $= \Delta x$
Volume $= 2\pi u_i (12 - u_i - u_i^2)\, \Delta x$

(c) $V = 2\pi \displaystyle\int_1^3 x(12 - x - x^2)\, dx$

Figure 40 The shell method

(continued on the next page)

Figure 40(c) shows the solid of revolution. Notice that the limits of integration are $x = 1$ and $x = 3$. The volume V of the solid of revolution is

$$V = 2\pi \int_1^3 x(12 - x - x^2)\, dx = 2\pi \int_1^3 (12x - x^2 - x^3)\, dx = 2\pi \left[6x^2 - \frac{x^3}{3} - \frac{x^4}{4} \right]_1^3$$

$$= 2\pi \left[\left(54 - 9 - \frac{81}{4} \right) - \left(6 - \frac{1}{3} - \frac{1}{4} \right) \right] = \frac{116}{3}\, \pi \text{ cubic units}$$

Using the washer method: Figure 41 shows the solid of revolution and typical washers.

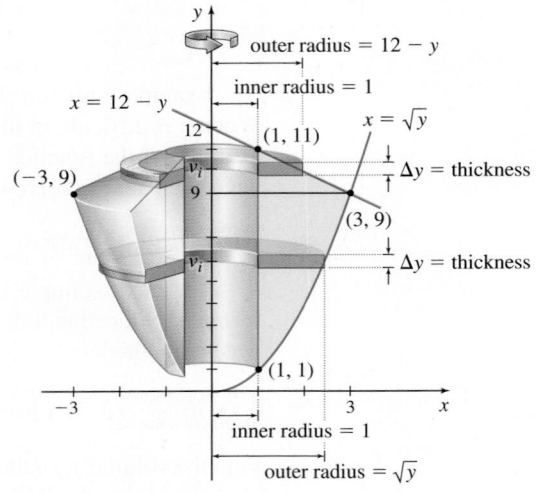

Figure 41   The washer method: $V_i = \pi[(\text{Outer radius})^2 - (\text{Inner radius})^2](\text{Thickness})$

With the washer method, we partition the interval $[1, 11]$ on the y-axis and use horizontal washers. Notice that at $y = 9$, the function on the right changes.

The volume of a typical washer in the interval $[1, 9]$ is

$$V_i = \pi\left[\left(\sqrt{v_i}\right)^2 - 1^2 \right]\Delta y = \pi(v_i - 1)\Delta y$$

The volume of a typical washer in the interval $[9, 11]$ is

$$V_i = \pi\left[(12 - v_i)^2 - 1^2 \right]\Delta y = \pi\left(143 - 24v_i + v_i^2 \right)\Delta y$$

The volume V of the solid of revolution is

$$V = \pi \int_1^9 (y - 1)\, dy + \pi \int_9^{11} (143 - 24y + y^2)\, dy$$

$$= \pi \left[\frac{y^2}{2} - y \right]_1^9 + \pi \left[143y - 12y^2 + \frac{y^3}{3} \right]_9^{11}$$

$$= \pi \left[\left(\frac{81}{2} - 9 \right) - \left(\frac{1}{2} - 1 \right) \right] + \pi \left(143\,(2) - (12)(121 - 81) + \frac{11^3}{3} - \frac{9^3}{3} \right)$$

$$= 32\pi + \frac{20}{3}\pi = \frac{116}{3}\, \pi \text{ cubic units}$$ ∎

NOW WORK   **Problem 15.**

2 Use the Shell Method to Find the Volume of a Solid Formed by Revolving a Region About the x-Axis

Figure 42(a) shows a region that is to be revolved about the x-axis, Figure 42(b) shows a typical shell, and Figure 42(c) shows the solid of revolution.

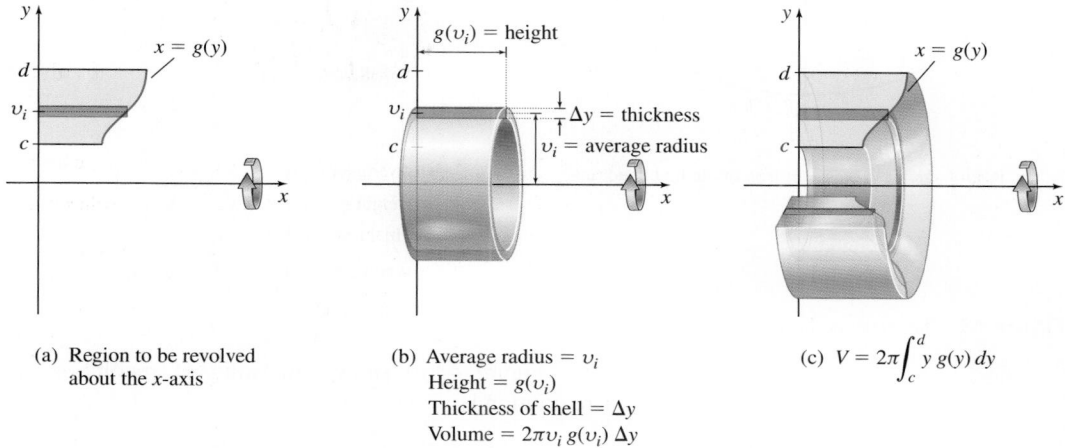

(a) Region to be revolved about the x-axis

(b) Average radius $= v_i$
Height $= g(v_i)$
Thickness of shell $= \Delta y$
Volume $= 2\pi v_i\, g(v_i)\, \Delta y$

(c) $V = 2\pi \displaystyle\int_c^d y\, g(y)\, dy$

Figure 42 The shell method

Volume of a Solid of Revolution About the x-Axis: the Shell Method

If $x = g(y)$ is a function that is continuous and nonnegative on the closed interval $[c, d]$, $c \ge 0$, the volume V of the solid generated by revolving the region bounded by the graphs of $x = g(y)$, the y-axis, and the lines $y = c$ and $y = d$ about the x-axis is

$$V = 2\pi \int_c^d y\, g(y)\, dy$$

EXAMPLE 3 **Using the Shell Method: Revolving About the x-Axis**

Find the volume V of the solid generated by revolving the region bounded by the graph of $\dfrac{x^2}{a^2} + \dfrac{y^2}{b^2} = 1$, $a > 0$, $b > 0$, in the first quadrant, about the x-axis.

Solution

NEED TO REVIEW? The equation of an ellipse is discussed in Appendix A.3, p. A-25.

The equation $\dfrac{x^2}{a^2} + \dfrac{y^2}{b^2} = 1$ defines an ellipse. See Figure 43(a) on page 660. The intercepts of its graph are $(a, 0)$, $(0, b)$, $(-a, 0)$, and $(0, -b)$. The region to be revolved is the shaded region in the first quadrant.

In the shell method, when the region is revolved about the x-axis, partition the y-axis and use horizontal shells. See Figure 43(b). A revolution about the x-axis requires integration with respect to y, so we express the equation of the ellipse as

$$x = g(y) = \frac{a}{b}\sqrt{b^2 - y^2}$$

The volume of a typical shell is

$$V_i = 2\pi \ (\text{Average radius})(\text{Height})(\text{Thickness}) = 2\pi\, v_i\, g(v_i)\Delta y$$

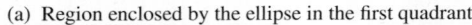
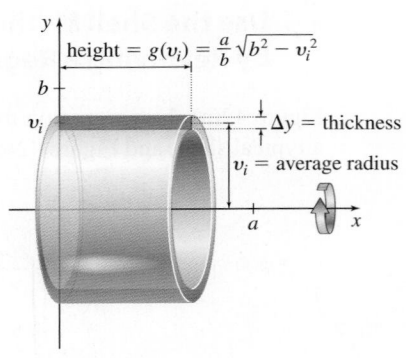
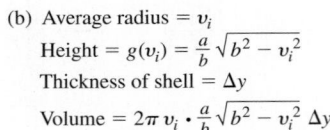
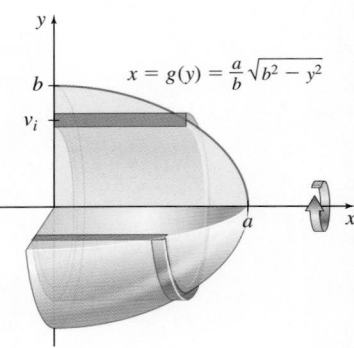

(a) Region enclosed by the ellipse in the first quadrant

(b) Average radius $= v_i$
Height $= g(v_i) = \dfrac{a}{b}\sqrt{b^2 - v_i^2}$
Thickness of shell $= \Delta y$
Volume $= 2\pi\, v_i \cdot \dfrac{a}{b}\sqrt{b^2 - v_i^2}\, \Delta y$

(c) $V = 2\pi \displaystyle\int_0^b y\left(\dfrac{a}{b}\sqrt{b^2 - y^2}\right) dy$

Figure 43 The shell method

Figure 43(c) shows the solid of revolution. The volume V of the solid of revolution is

$$V = 2\pi \int_0^b y\, g(y)\, dy = 2\pi \int_0^b y\left(\frac{a}{b}\sqrt{b^2 - y^2}\right) dy = 2\pi \frac{a}{b}\left(-\frac{1}{2}\right)\int_{b^2}^0 \sqrt{u}\, du$$

$$\underset{\substack{\uparrow \\ \text{Let } u = b^2 - y^2; \\ \text{then } du = -2y\, dy}}{}$$

$$= \frac{\pi a}{b}\int_0^{b^2} u^{1/2}\, du = \frac{\pi a}{b}\left[\frac{u^{3/2}}{\dfrac{3}{2}}\right]_0^{b^2} = \frac{2\pi a}{3b}(b^3) = \frac{2\pi a b^2}{3} \text{ cubic units} \qquad \blacksquare$$

NOW WORK Problem 11.

Using symmetry, the volume V of the **ellipsoid** generated by revolving the region bounded by the graph of $y = \dfrac{b}{a}\sqrt{a^2 - x^2}$, $-a \le x \le a$, about the x-axis is twice the volume found in Example 3. That is,

$$\boxed{V = \frac{4}{3}\pi a b^2}$$

If in Example 3, $a = b = R$, then the solid generated is a **hemisphere** of radius R whose volume V is

$$\boxed{V = \frac{2}{3}\pi R^3}$$

By symmetry, the volume V of a **sphere** of radius R, $(a = b = R)$, is twice the volume of a hemisphere. That is,

$$\boxed{V = \frac{4}{3}\pi R^3}$$

3 Use the Shell Method to Find the Volume of a Solid Formed by Revolving a Region About a Line Parallel to a Coordinate Axis

---

EXAMPLE 4 Using the Shell Method: Revolving About the Line $x = 2$

Find the volume V of the solid of revolution generated by revolving the region bounded by the graph of $y = 2x - 2x^2$ and the x-axis about the line $x = 2$.

Solution

The region bounded by the graph of $y = 2x - 2x^2$ and the x-axis is illustrated in Figure 44(a). A typical shell formed by revolving the region about the line $x = 2$, as shown in Figure 44(b), has an average radius of $2 - u_i$, height $f(u_i) = 2u_i - 2u_i^2$, and thickness Δx. The solid of revolution is depicted in Figure 44(c).

The volume V of the solid is

$$V = 2\pi \int_0^1 (2-x)(2x-2x^2)\,dx = 4\pi \int_0^1 (x^3 - 3x^2 + 2x)\,dx$$

$$= 4\pi \left[\frac{x^4}{4} - x^3 + x^2\right]_0^1 = \pi \text{ cubic units}$$

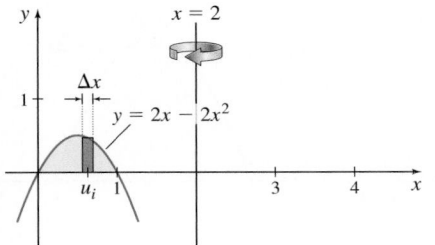
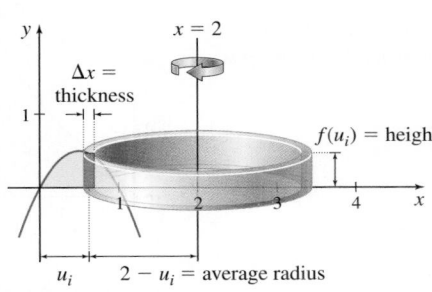
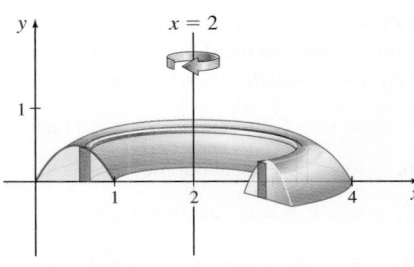

(a) Region to be revolved about the line $x = 2$

(b) Average radius $= 2 - u_i$
Height $= f(u_i) = 2u_i - 2u_i^2$
Thickness of shell $= \Delta x$
Volume $= 2\pi(2 - u_i) f(u_i) \Delta x$

(c) $V = 2\pi \int_0^1 (2-x)(2x-2x^2)\,dx$

Figure 44 The shell method

NOW WORK Problem 17.

Summary

Volume V of the solid obtained by revolving the region R

about the x-axis	about the y-axis
Disk Method	**Shell Method**
$V = \pi \int_a^b [f(x)]^2\,dx$	$V = 2\pi \int_a^b x f(x)\,dx$

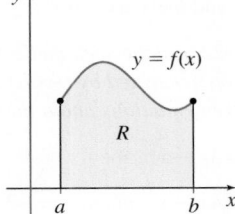

Volume V of the solid obtained by revolving the region R

about the y-axis	about the x-axis
Disk Method	**Shell Method**
$V = \pi \int_c^d [g(y)]^2\,dy$	$V = 2\pi \int_c^d y g(y)\,dy$

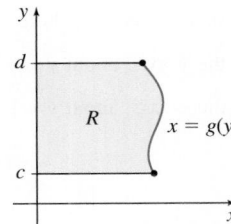

8.4 Assess Your Understanding

Concepts and Vocabulary

1. **True or False**   In using the shell method to find the volume of a solid revolved about the x-axis, the integration is with respect to x.

2. **True or False**   The volume of a cylindrical shell of outer radius R, inner radius r, and height h is given by $V = \pi r^2 h$.

3. **True or False**   The volume of a solid of revolution can be found using either washers or cylindrical shells only if the region is revolved about the y-axis.

4. **True or False**   If $y = f(x)$ is a function that is continuous and nonnegative on the closed interval $[a, b]$, $a \geq 0$, then the volume V of the solid generated by revolving the region bounded by the graph of f and the x-axis from $x = a$ to $x = b$ about the y-axis is

$$V = 2\pi \int_a^b x f(x)\, dx$$

Skill Building

In Problems 5–16, use the shell method to find the volume of the solid of revolution generated by revolving the region bounded by the graphs of the given equations about the indicated axis.

 5. $y = x^2 + 1$, the x-axis, $0 \leq x \leq 1$; about the y-axis

6. $y = x^3$, $y = x^2$; about the y-axis

7. $y = \sqrt{x}$, $y = x^2$; about the y-axis

8. $y = \dfrac{1}{x}$, the x-axis, $x = 1$, $x = 4$; about the y-axis

9. $y = x^3$, the y-axis, $y = 8$; about the x-axis

10. $y = \sqrt{x}$, the y-axis, $y = 2$; about the x-axis

11. $x = \sqrt{y}$, the y-axis, $y = 1$; about the x-axis

12. $x = 4\sqrt{y}$, the y-axis, $y = 4$; about the x-axis

13. $y = x$, $y = x^2$; about the x-axis

14. $y = x$, $y = x^3$ in the first quadrant; about the x-axis

15. $y = e^{-x^2}$ and the x-axis, from $x = 0$ to $x = 2$; about the y-axis

16. $y = x^3 + x$ and the x-axis, $0 \leq x \leq 1$; about the y-axis

In Problems 17–22, use the shell method to find the volume of the solid of revolution generated by revolving the region bounded by the graphs of the given equations about the indicated line.

 17. $y = x^2$, $y = 4x - x^2$; about $x = 4$

18. $y = x^2$, $y = 4x - x^2$; about $x = -3$

19. $y = x^2$, $y = 0$, $x = 1$, $x = 2$; about $x = 1$

20. $y = x^2$, $y = 0$, $x = 1$, $x = 2$; about $x = -2$

21. $x = y - y^2$, the y-axis; about $y = -1$

22. $x = y - y^2$, the y-axis; about $y = 1$

In Problems 23–26, use either the shell method or the disk/washer method to find the volume of the solid of revolution generated by revolving the shaded region in each graph:

(a) *about the x-axis.*
(b) *about the y-axis.*
(c) *Explain why you chose the method you used.*

23.

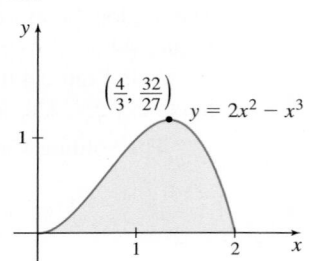

24.

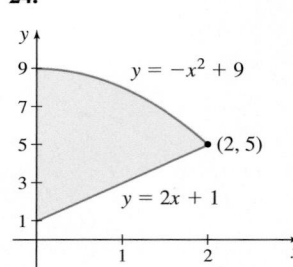

25.

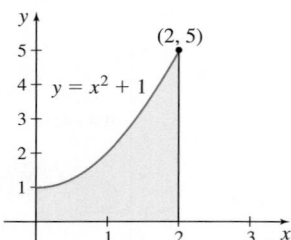

26.

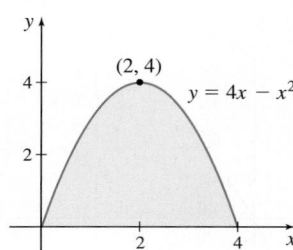

In Problems 27–38, use either the shell method or the disk/washer method to find the volume of the solid of revolution generated by revolving the region bounded by the graphs of the given equations about the indicated axis.

27. $y = \sqrt[3]{x}$, the y-axis, $y = 2$; about the x-axis

28. $y = \sqrt{x} + x$, the x-axis, $x = 1$, $x = 4$; about the y-axis

29. $y = x^3$ and $y = x$ to the right of $x = 0$; about the y-axis

30. $y = x^3$, $y = x^2$; about the x-axis

31. $y = 3x^2$ and $y = 30 - x$ to the right of $x = 1$; about the y-axis

32. $y = 3x^2$ and $y = 30 - x$ to the right of $x = 1$; about the x-axis

33. $y = x^2$ and $y = 8 - x^2$ to the right of $x = 1$; about the x-axis

34. $y = x^2$ and $y = 8 - x^2$ to the right of $x = 1$; about the y-axis

35. $y = x^{2/3}$, $y = x$; about the y-axis

36. $y = x^{2/3} + x^{1/3}$, the x-axis, $x = 1$, $x = 8$; about the y-axis

37. $y = \sqrt{x}$ and $y = 18 - x^2$ to the right of $x = 1$; about the y-axis

38. $y = \sqrt{x}$ and $y = 18 - x^2$ to the right of $x = 1$; about the x-axis

Applications and Extensions

39. Volume of a Solid of Revolution

Find the volume of the solid of revolution generated by revolving the region bounded by the graph of $y = \dfrac{1}{(x^2 + 1)^2}$ and the x-axis from $x = 0$ to $x = 1$ about the y-axis as shown in the figure.

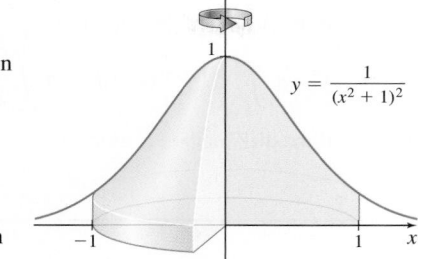

40. Volume of a Solid of Revolution   Find the volume of the solid of revolution generated by revolving the region bounded by the graph of $y = \sqrt{x+1} + x$ and the x-axis from $x = 0$ to $x = 3$ about the y-axis.

41. Volume of a Solid of Revolution   Find the volume of the solid of revolution generated by revolving the region in the first quadrant bounded by the x-axis and the graph of $y = 2x - x^2$
(a) about the x-axis.
(b) about the y-axis.
(c) about the line $x = 3$.
(d) about the line $y = 1$.

42. Volume of a Solid of Revolution   Find the volume of the solid of revolution generated by revolving the region in the first quadrant bounded by the x-axis and the graph of $y = x\sqrt{9 - x}$

(a) about the x-axis.
(b) about the y-axis.
(c) about the line $y = -3$.
(d) about the line $x = -2$.

43. Volume of a Solid of Revolution   Suppose $f(x) \geq 0$ for $x \geq 0$, and the region bounded by the graph of f and the x-axis from $x = 0$ to $x = k$, $k > 0$, is revolved about the x-axis. If the volume of the resulting solid is $\dfrac{1}{5}k^5 + k^4 + \dfrac{4}{3}k^3$, find f.

44. Volume of a Solid of Revolution   Find the volume of the solid generated by revolving about the y-axis the region bounded by the graph of $y = e^{-x^2}$ and the y-axis from $x = 0$ to the positive x-coordinate of the point of inflection of $y = e^{-x^2}$.

45. Volume of a Solid of Revolution   Show that the volume V of the solid generated by revolving the region bounded by the graph of $y = \sin^3 x$ and the x-axis from $x = 0$ to $x = \pi$ about the line $x = -\pi$ is given by $V = 4\pi^2$.

46. Volume of a Cone

(a) Find an equation of the line containing the points $(0, h)$, $h > 0$, and $(a, 0)$, $a > 0$.
(b) Graph the line from (a).
(c) Revolve the region in the first quadrant bounded by the line from (a), and the x- and y-axes about the y-axis to generate a cone of height h and radius a.
(d) Use the Shell Method to show that the volume V of the cone is $\dfrac{1}{3}\pi a^2 h$.

47. Volume of a Solid of Revolution   The figure below shows the solid of revolution generated by revolving the region bounded by the graph of $y = \cos x$ and the x-axis from $x = 0$ to $x = \dfrac{\pi}{2}$ about the y-axis.

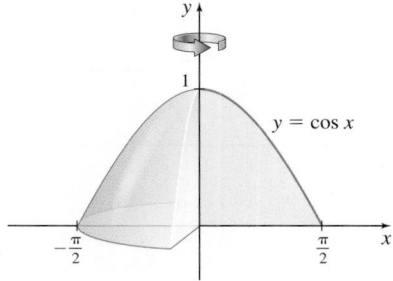

(a) Express the volume of the solid of revolution as an integral.
(b) Use technology to find the volume.

48. Volume of a Solid of Revolution   The figure below shows the solid of revolution generated by revolving the region bounded by the graph of $y = \sin x$ and the x-axis from $x = 0$ to $x = \dfrac{\pi}{2}$ about the y-axis.

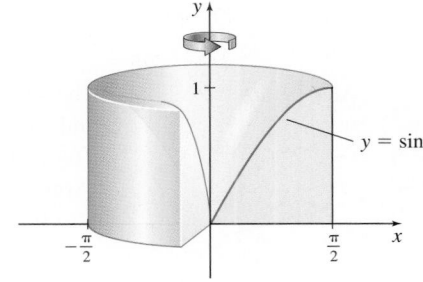

(a) Express the volume of the solid of revolution as an integral.
(b) Use technology to find the volume.

Challenge Problems

49. Suppose a plane region of area A that lies to the right of the y-axis is revolved about the y-axis, generating a solid of volume V. If this same region is revolved about the line $x = -k$, $k > 0$, show that the solid generated has volume $V + 2\pi k A$.

50. Find the volume generated by revolving the region bounded by the parabola $y^2 = 8x$ and its latus rectum about the latus rectum:

(a) by using the disk method.
(b) by using the shell method.

(The **latus rectum** is the chord through the focus perpendicular to the axis of the parabola.) See the figure.

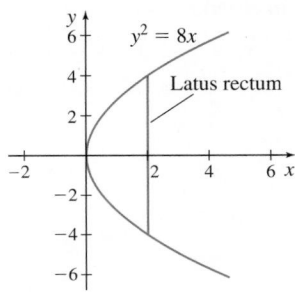

51. Comparing Volume Formulas
In the figure on the right, the region bounded by the graphs of $y = g(x)$, $y = f(x)$, and the y-axis is to be revolved about the y-axis. Show that the resulting volume V is given by:

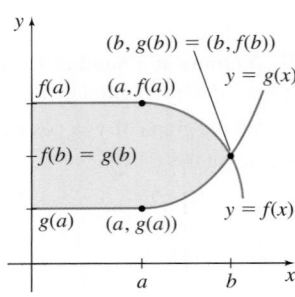

(a) $V = \pi a^2 [f(a) - g(a)]$

$$+ 2\pi \int_a^b x[f(x) - g(x)]\, dx$$

if the shell method is used.

(b) $V = \pi \int_{g(a)}^{g(b)} [g^{-1}(y)]^2 dy + \pi \int_{g(b)}^{f(a)} [f^{-1}(y)]^2 dy$

if the disk method is used.

8.5 Volume of a Solid with a Known Cross Section

OBJECTIVE *When you finish this section, you should be able to:*

1 Find the volume of a solid with a known cross section (p. 664)

In Sections 8.3 and 8.4, we found the volumes of solids of revolution using disks, washers, and cylindrical shells. We now investigate a method to find the volume of a solid that is not necessarily a solid of revolution. This method is referred to as finding the volume of a solid with a known cross section.

1 Find the Volume of a Solid with a Known Cross Section

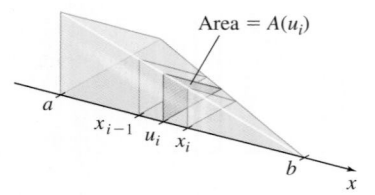

Figure 45 A solid with a triangular cross section

The idea here is to cut a solid into thin slices using planes perpendicular to an axis and then to add the volumes of all the slices to obtain the total volume of the solid. The method relies on the fact that the area of each cross section is known and can be expressed as a function of the position of the slice. See Figure 45.

We begin by partitioning the interval $[a, b]$ on the x-axis into n subintervals:

$$[a, x_1],\ [x_1, x_2],\ \ldots,\ [x_{i-1}, x_i],\ \ldots,\ [x_{n-1}, b]$$

each of width $\Delta x = \dfrac{b - a}{n}$. In the ith subinterval, we select a number u_i and slice through the solid at $x = u_i$ using a plane that is perpendicular to the x-axis. Suppose the area of the cross section that results is known to be $A(u_i)$.

The volume V_i of the thin slice from x_{i-1} to x_i is approximately $V_i = A(u_i)\Delta x$, and the sum of the volumes from all the subintervals is

$$\sum_{i=1}^n V_i = \sum_{i=1}^n [A(u_i)\Delta x]$$

IN WORDS The volume V_i of a slice is
$$V_i = \text{(Area of cross section) (Thickness of the slice)} = A(u_i)\Delta x$$

These sums approximate the volume V of the solid from $x = a$ to $x = b$. As the number n of subintervals increases, the sums $\sum_{i=1}^n V_i = \sum_{i=1}^n [A(u_i)\Delta x]$ become better approximations to the volume V of the solid. These sums are Riemann sums. So, if the area $A(u_i)$ of the cross sections varies continuously with x, then the limit is a definite integral.

Volume of a Solid with a Known Cross Section

If for each x in $[a, b]$, the area $A(x)$ of the cross section of a solid is known and is continuous on $[a, b]$, then the volume V of the solid is

$$\boxed{V = \int_a^b A(x)\, dx} \tag{1}$$

This method for finding the volume of a solid requires that the area of the cross section of the solid at the number x can be expressed as a function of x. For example, if all parallel cross sections of the solid have the same geometric shape (all are semicircles, or triangles, or squares, etc.), then the area of each cross section can be expressed as a function $A(x)$, and the volume of the solid can be found using (1).

NEED TO REVIEW? Geometry formulas are discussed in Appendix A.2, p. A-16.

EXAMPLE 1   **Finding the Volume of a Solid When the Area of a Cross Section Is Known**

The base of a solid is the region R bounded by the x-axis, the function $f(x) = \sqrt{x}$, and the line $x = 4$. Find the volume of the solid if the cross sections perpendicular to the x-axis are rectangles with height x and width $\sqrt{x}$.

Solution

See Figure 46. Each cross section is a rectangle. The area of each cross section is $A = A(x) = (\text{height})(\text{width}) = x\sqrt{x} = x^{3/2}$, where x varies from 0 to 4. The volume V of the solid is

$$V = \int_0^4 A(x)\, dx = \int_0^4 x^{3/2}\, dx = \left[\frac{2}{5}\, x^{5/2} \right]_0^4 = \frac{2}{5}\,(32 - 0) = \frac{64}{5} \text{ cubic units}$$

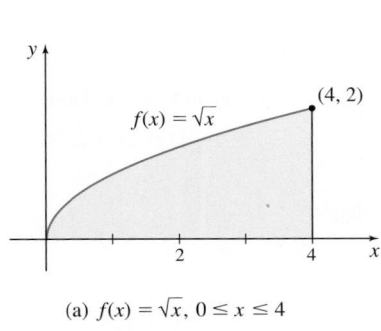

(a) $f(x) = \sqrt{x}$, $0 \le x \le 4$

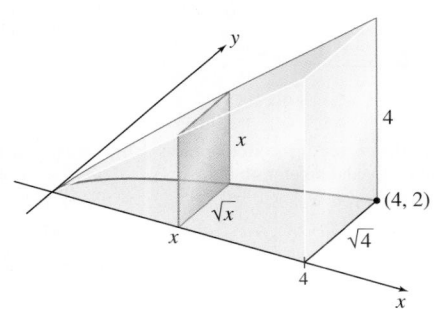

(b) A solid with rectangular cross sections

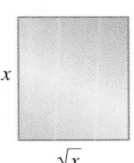

Cross-section
Area $= x\sqrt{x}$

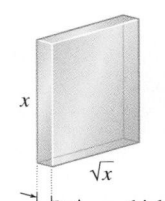

$V = \text{Area} \times \text{Thickness}$
$= x\sqrt{x} \times \Delta x = x^{3/2}\Delta x$

(c) A representative rectangle and the corresponding slice

Figure 46

NOW WORK   Problem 3 and AP® Practice Problems 3 and 4.

The way in which a solid is positioned relative to the x-axis is important if the area A of the cross section is to be easily found, expressed as a function of x, and integrated.

EXAMPLE 2   Finding the Volume of a Solid with a Known Cross Section

A solid has a circular base of radius 3 units. Find the volume V of the solid if every cross section that is perpendicular to a fixed diameter is an equilateral triangle.

Solution

Position the circular base so that its center is at the origin, and the fixed diameter is along the x-axis. See Figure 47(a). Then the equation of the circular base is $x^2 + y^2 = 9$. Each cross section of the solid is an equilateral triangle with sides $= 2y$, height h, and area $A = \sqrt{3}y^2$. See Figure 47(b). Since $y^2 = 9 - x^2$, the volume V_i of a typical slice is

$$V_i = (\text{Area of the cross section})(\text{Thickness}) = A(x_i)\, \Delta x = \sqrt{3}\left(9 - x_i^2\right) \Delta x$$

as shown in Figure 47(c).

The volume V of the solid is

$$V = \int_a^b A(x)\, dx = \int_{-3}^{3} \sqrt{3}(9 - x^2)\, dx = \sqrt{3} \int_{-3}^{3} (9 - x^2)\, dx = \sqrt{3} \left[9x - \frac{x^3}{3} \right]_{-3}^{3}$$

$$= \sqrt{3}\left[(27 - 9) - (-27 + 9)\right]$$

$$= 36\sqrt{3} \text{ cubic units}$$

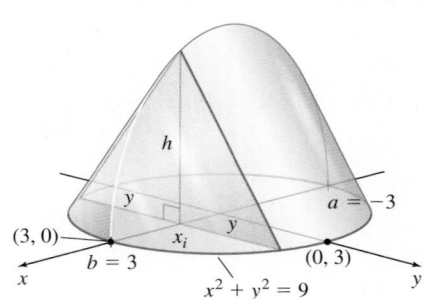
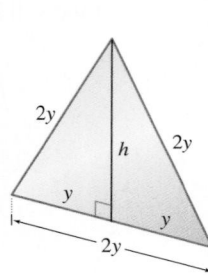
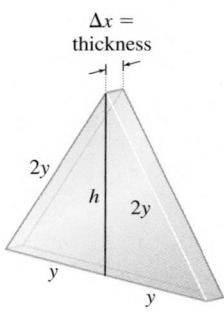

(a) Solid placed with the center of its circular base at the origin and its fixed diameter on the x-axis

(b) Cross section:
$h^2 + y^2 = 4y^2$
$h^2 = 3y^2$
$h = \sqrt{3}y$
Area $= \frac{1}{2}$(Base)(Height)
$= \frac{1}{2} \cdot 2y \cdot \sqrt{3}y = \sqrt{3}y^2$

(c) $V = (\text{Area})(\text{Thickness})$
$= \sqrt{3}y^2\, \Delta x = \sqrt{3}\left(9 - x^2\right) \Delta x$

Figure 47

NOW WORK   Problem 7 and AP® Practice Problem 5.

EXAMPLE 3   Finding the Volume of a Solid with Known Cross Sections

A region R in the xy-plane is bounded by the graphs of $y = \sqrt{x}$ and $y = \frac{1}{8}x^2$.

(a) Find the volume V of a solid whose base is R, if slices perpendicular to the x-axis have cross sections that are squares.

(b) Find the volume V of a solid whose base is R, if slices perpendicular to the y-axis have cross sections that are squares.

Solution

Begin by graphing the region R bounded by the graphs of $y = \sqrt{x}$ and $y = \dfrac{1}{8}x^2$ as shown in Figure 48. The points of intersection of the two graphs satisfy the equation

$$\sqrt{x} = \frac{1}{8}x^2$$

$$x = \frac{1}{64}x^4$$

$$x^4 - 64x = 0$$

$$x(x^3 - 64) = 0$$

$$x = 0 \text{ or } x = 4$$

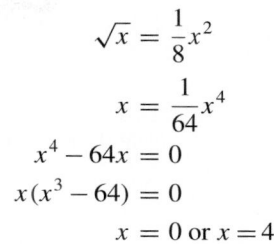

Figure 48

The two graphs intersect at the points (0, 0) and (4, 2).

(a) The solid with slices perpendicular to the x-axis that are squares is shown in Figure 49(a). A slice perpendicular to the x-axis at x_i is a square with side $s_i = \sqrt{x_i} - \dfrac{1}{8}x_i^2$. See Figure 49(b). The area A_i of the cross section at x_i is

$$A_i = s_i^2 = \left(\sqrt{x_i} - \frac{1}{8}x_i^2\right)^2$$

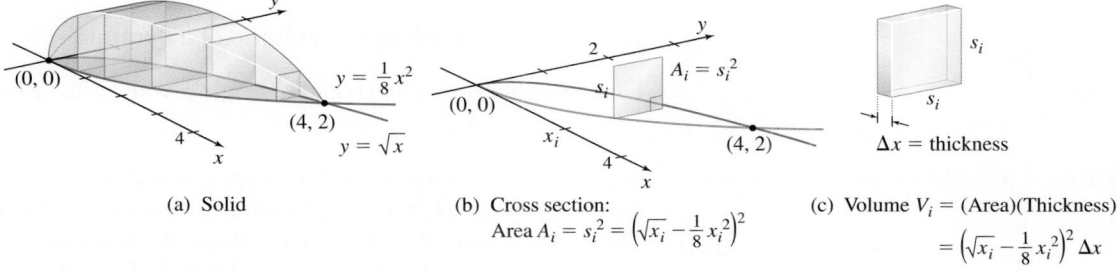

(a) Solid

(b) Cross section:
Area $A_i = s_i^2 = \left(\sqrt{x_i} - \frac{1}{8}x_i^2\right)^2$

(c) Volume V_i = (Area)(Thickness)
$= \left(\sqrt{x_i} - \frac{1}{8}x_i^2\right)^2 \Delta x$

Figure 49

See Figure 49(c). The volume V_i of a typical slice is

$$V_i = (\text{Area of the cross section at } x_i)\,(\text{Thickness of the slice})$$

$$= A(x_i)\Delta x = \left(\sqrt{x_i} - \frac{1}{8}x_i^2\right)^2 \Delta x$$

The volume V of the solid is

$$V = \int_0^4 A(x)\,dx = \int_0^4 \left(\sqrt{x} - \frac{1}{8}x^2\right)^2 dx = \int_0^4 \left(x - \frac{1}{4}x^{5/2} + \frac{1}{64}x^4\right) dx$$

$$= \left[\frac{x^2}{2} - \frac{1}{14}x^{7/2} + \frac{1}{320}x^5\right]_0^4 = 8 - \frac{128}{14} + \frac{1024}{320} = \frac{72}{35} \text{ cubic units}$$

(b) To find the volume of a solid when slices are made perpendicular to the y-axis, follow the same reasoning as in (a), but use a partition of the y-axis. For each y in the closed interval $[c, d]$, the area $A(y)$ of the cross section of the solid is known and is continuous on $[c, d]$. Then the volume V of the solid is

$$\boxed{V = \int_c^d A(y)\,dy} \tag{2}$$

(continued on the next page)

To use (2), we must express $y = \sqrt{x}$ and $y = \dfrac{1}{8}x^2$ as functions of y. That is, we write $y = \sqrt{x}$ as $x = y^2$ and $y = \dfrac{1}{8}x^2$ as $x = \sqrt{8y}$. Then a cross section perpendicular to the y-axis at y_i is a square with side $s_i = \sqrt{8y_i} - y_i^2$. The area A_i of the cross section at y_i is

$$A_i = A(y_i) = s_i^2 = \left(\sqrt{8y_i} - y_i^2\right)^2$$

and the volume V_i of a typical slice is

$$V_i = (\text{Area of the cross section at } y_i)\,(\text{Thickness of the slice})$$

$$= A(y_i)\Delta y = \left(\sqrt{8y_i} - y_i^2\right)^2 \Delta y$$

The volume V of the solid is

$$V = \int_0^2 A(y)\,dy = \int_0^2 \left(\sqrt{8y} - y^2\right)^2 dy = \int_0^2 \left(8y - 4\sqrt{2}\,y^{5/2} + y^4\right)\,dy$$

$$= \left[4y^2 - \frac{8\sqrt{2}}{7}y^{7/2} + \frac{y^5}{5}\right]_0^2 = 16 - \frac{128}{7} + \frac{32}{5} = \frac{144}{35} \text{ cubic units} \quad \blacksquare$$

NOW WORK   Problem 5 and AP® Practice Problems 1, 2, 6, 7, and 8.

EXAMPLE 4   **Finding the Volume of a Pyramid**

Find the volume V of a pyramid of height h with a square base, each side of length b.

Solution

We position the pyramid with its vertex at the origin and its axis along the positive x-axis. Then a typical cross section at x is a square. Let s denote the length of the side of the square at x. See Figure 50(a). We form two triangles: one with height x and side s, the other with height h and side b. These triangles are similar, as shown in Figure 50(b). Then we have

NOTE A **pyramid** is a solid whose base is a polygon and whose sides are triangles with a common vertex.

NEED TO REVIEW? Similar triangles are discussed in Appendix A.2, pp. A–14 to A–15.

$$\frac{x}{s} = \frac{h}{b} \qquad \text{or} \qquad s = \frac{b}{h}x$$

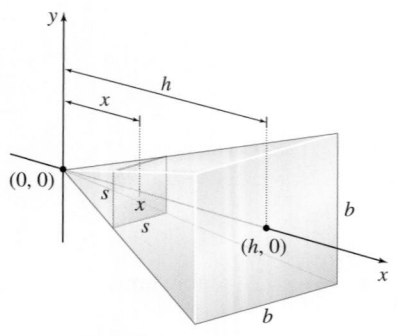

(a) Pyramid placed symmetric to the x-axis and with its vertex at the origin

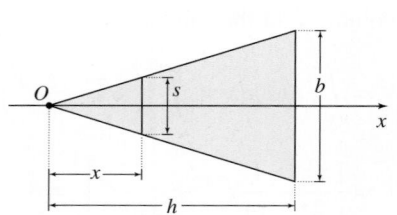

(b) Similar triangles

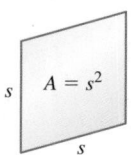

(c) Cross section at x: Area $= A = s^2 = \dfrac{b^2}{h^2}x^2$

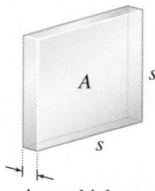

(d) Volume $= V = A\,\Delta x = \dfrac{b^2}{h^2}x^2\Delta x$

Figure 50

So, s is a function of x, and the area A of a typical cross section is $A = s^2 = \dfrac{b^2}{h^2}x^2$. See Figure 50(c). The volume V of a typical cross section is $V = \dfrac{b^2}{h^2}x^2 \Delta x$. See Figure 50(d). Since A is a continuous function of x, where the slice occurred, we have

$$V = \int_0^h A(x)\,dx = \int_0^h \frac{b^2}{h^2}x^2\,dx = \frac{b^2}{h^2}\int_0^h x^2\,dx = \frac{b^2}{h^2}\left[\frac{x^3}{3}\right]_0^h = \frac{1}{3}b^2 h \text{ cubic units} \quad \blacksquare$$

NOW WORK   Problem 11.

EXAMPLE 5   **Finding the Volume of a Cone**

Use circular cross sections to verify that the volume of a right circular cone having radius R and height h is $V = \dfrac{1}{3}\pi R^2 h$.

Solution

The mathematics is simplified if we position the cone with its vertex at the origin and its axis on the x-axis, as shown in Figure 51.

The cone extends from $x = 0$ to $x = h$. The cross section of the cone at any number x is a circle. To obtain its area A, we need the radius r of the circle. Embedded in Figure 51 are two similar right triangles, one with sides x and r; the other with sides h and R, as shown in Figure 52. Because these triangles are similar (equal angles), corresponding sides are in proportion. That is,

$$\frac{r}{x} = \frac{R}{h}$$

$$r = \frac{R}{h}x$$

So, r is a function of x and the area A of the circular cross section is

$$A = A(x) = \pi[r(x)]^2 = \pi\left(\frac{R}{h}x\right)^2 = \frac{\pi R^2}{h^2}x^2$$

Since A is a continuous function of x (where the slice was made), the volume V of the right circular cone is

$$V = \int_0^h \frac{\pi R^2}{h^2}x^2\,dx = \frac{\pi R^2}{h^2}\int_0^h x^2\,dx \qquad V = \int_a^b A(x)\,dx$$

$$= \frac{\pi R^2}{h^2}\left[\frac{x^3}{3}\right]_0^h = \frac{\pi R^2 h}{3} \text{ cubic units} \quad \blacksquare$$

NOTE A right circular cone is a solid of revolution, so the formula for its volume also can be verified using the disk method or the shell method.

NOW WORK   Problem 13.

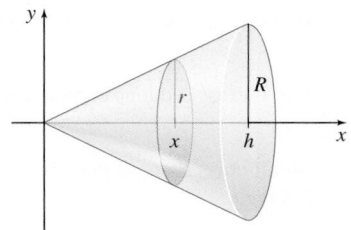

Figure 51

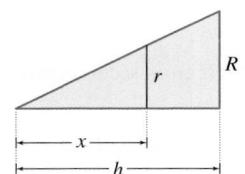

Figure 52

8.5 Assess Your Understanding

Concepts and Vocabulary

1. In your own words, explain how to find the volume of a solid with a known cross section.

2. *True or False*   Finding the volume of a solid using known cross sections works only with solids of revolution.

Skill Building

In Problems 3–10, find the volume of each solid.

 3. Find the volume of the solid whose base is a circle of radius of 2, if cross sections perpendicular to the base are squares. See the figure.

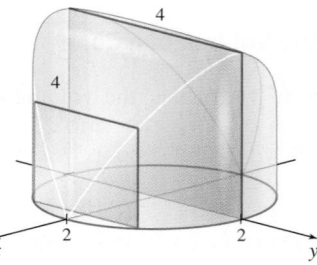

4. Find the volume of the solid whose base is a circle of radius 2, if cross sections perpendicular to the base are isosceles right triangles with one leg on the base.

 5. The base of a solid is the region bounded by the graphs of $y = \sqrt{x}$ and $y = \dfrac{1}{8}x^2$.

   (a) Find the volume V of the solid if the cross sections perpendicular to the x-axis are semicircles. See the figure.

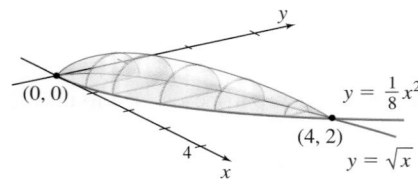

   (b) Find the volume V of the solid if the cross sections perpendicular to the x-axis are equilateral triangles.

6. The base of a solid is the region bounded by the graphs of $y = \sqrt{x}$ and $y = \dfrac{1}{8}x^2$.

   (a) Find the volume V of the solid if the cross sections perpendicular to the y-axis are semicircles.
   (b) Find the volume V of the solid if the cross sections perpendicular to the y-axis are equilateral triangles.

 7. The base of a solid is the region bounded by the graphs of $y = x^2$, $x = 2$, and $y = 0$.

   (a) Find the volume V of the solid if cross sections perpendicular to the x-axis are squares.
   (b) Find the volume V of the solid if cross sections perpendicular to the x-axis are semicircles.
   (c) Find the volume V of the solid if cross sections perpendicular to the x-axis are equilateral triangles.

8. The base of a solid is the region bounded by the graphs of $y = x^2$, $x = 2$, and $y = 0$.

   (a) Find the volume V of the solid if cross sections perpendicular to the y-axis are squares.
   (b) Find the volume V of the solid if cross sections perpendicular to the y-axis are semicircles.
   (c) Find the volume V of the solid if cross sections perpendicular to the y-axis are equilateral triangles.

9. The base of a solid is the region bounded by the graphs of $y = 3\sqrt{3x}$ and $y = x^2$.

   (a) Find the volume V of the solid if cross sections perpendicular to the y-axis are squares.
   (b) Find the volume V of the solid if cross sections perpendicular to the y-axis are semicircles.
   (c) Find the volume V of the solid if cross sections perpendicular to the y-axis are equilateral triangles.

10. The base of a solid is the region bounded by the graphs of $y = 3\sqrt{3x}$ and $y = x^2$.

    (a) Find the volume V of the solid if cross sections perpendicular to the x-axis are squares.
    (b) Find the volume V of the solid if cross sections perpendicular to the x-axis are semicircles.
    (c) Find the volume V of the solid if cross sections perpendicular to the x-axis are equilateral triangles.

Applications and Extensions

In Problems 11 and 12, find the volume of each solid.

 11. The solid is a pyramid 40 m high whose horizontal cross section h meters from the top is a square with sides of length $2h$ meters.

12. The solid is a pyramid 20 m high whose horizontal cross section h meters from the top is a rectangle with sides of length $2h$ and h meters.

13. **Verifying a Geometry Formula**   Use known cross sections to verify that the volume V of a sphere of radius R is $V = \dfrac{4}{3}\pi R^3$.

14. The base of a solid is the region enclosed by the graphs of $y = \sqrt{x}$ and $y = \dfrac{1}{8}x^2$.

    (a) Find the volume V of the solid if cross sections perpendicular to the x-axis are triangles whose base is the distance between the graphs and whose height is three times the base.
    (b) Find the volume V of the solid if cross sections perpendicular to the y-axis are triangles whose base is the distance between the graphs and whose height is three times the base.

In Problems 15 and 16, find the volume of each solid.

15. The solid is horn-shaped; cross sections perpendicular to the x-axis are circles whose diameters extend from the graph of $y = x^{1/2}$ to the graph of $y = \frac{4}{3}x^{1/3}$, $0 \le x \le 1$.

16. The solid is horn-shaped; cross sections perpendicular to the x-axis are circles whose diameters extend from the graph of $y = x^{1/3}$ to $y = \frac{3}{2}x^{1/3}$, $0 \le x \le 1$.

17. **Volume of a Solid**   Find the volume of the cylindrical solid with a bulge in the middle, if cross sections perpendicular to the x-axis are circles whose diameters extend from the graph of $y = e^{-x^2}$ to $y = -e^{-x^2}$, $-1 \le x \le 1$.

Challenge Problems

18. **Volume of Water**   A hemispherical bowl of radius R contains water to the depth h. Find the volume of the water in the bowl.

19. **Volume of Water Left in a Glass**   Suppose a cylindrical glass full of water is tipped until the water level bisects the base and touches the rim. What is the volume of the water remaining?

20. **Volume of a Solid**   Find the volume of a parallelepiped with edge lengths a, b, and c, where the edges having lengths a and b make an acute angle θ with each other, and the edge of length c makes an acute angle of ϕ with the diagonal of the parallelogram formed by a and b.

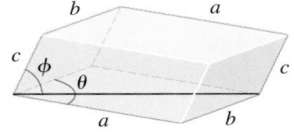

21. **Volume of a Bore**   A hole of radius 2 cm is bored completely through a solid metal sphere of radius 5 cm. If the axis of the hole passes through the center of the sphere, find the volume of the metal removed by the drilling. See the figure.

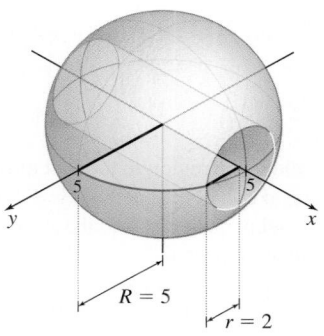

22. **Volume**   The axes of two pipes of equal radii r intersect at right angles. Find their common volume.

23. **Volume of a Cone**   Find the volume of a cone with height h and an elliptical base whose major axis has length $2a$ and minor axis has length $2b$.
Hint: The area of this ellipse is πab.

24. **Volume of a Removed Sector**   Suppose a wedge is cut from a solid right circular cylinder of diameter 10 m (like a wedge cut in a tree by an axe), where one side of the wedge is horizontal and the other is inclined at 30°. See the figure. If the horizontal part of the wedge penetrates 5 m into the cylinder and the two cuts meet along a vertical line through the center of the cylinder, find the volume of the wedge removed.
Hint: Vertical cross sections of the wedge are right triangles.

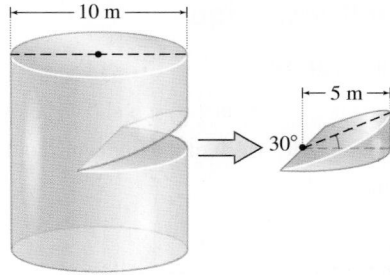

Preparing for the (**AP® Exam**)

AP® Practice Problems

Multiple-Choice Questions

PAGE 668 1. A region R is bounded by the graphs of the functions $f(x) = 3x$ and $g(x) = x^2$, $x \ge 0$. If R forms the base of a solid where cross sections perpendicular to the x-axis are squares, then the volume of the solid is

(A) $\dfrac{9}{2}$ (B) $\dfrac{81}{10}$

(C) $\dfrac{45}{2}$ (D) $\dfrac{162}{5}$

PAGE 668 2. A region R in the first quadrant is bounded by the graphs of $y = e^x$, $y = e^2$, and the y-axis. If R forms the base of a solid where cross sections perpendicular to the x-axis are semicircles, then the volume of the solid is given by

(A) $\dfrac{1}{2}\displaystyle\int_0^e \pi \left(\dfrac{e^2 - e^x}{2}\right)^2 dx$ (B) $\displaystyle\int_0^{e^2} \pi \left(\dfrac{e^2 - e^x}{2}\right)^2 dx$

(C) $\dfrac{1}{2}\displaystyle\int_0^2 \pi \left(\dfrac{e^2 - e^x}{2}\right)^2 dx$ (D) $\dfrac{1}{2}\displaystyle\int_0^2 \pi (e^2 - e^x)^2 dx$

(continued on the next page)

3. The base of a solid is formed by the region bounded by the graphs of $y = e^{-x} + 1$, the x-axis, the y-axis, and the line $x = 4$. If cross sections perpendicular to the x-axis are squares, what is the volume of the solid?

(A) $\frac{1}{2}(5 - e^{-8} - 4e^{-4})$ (B) $8 - e^{-8} - 2e^{-4}$

(C) $\frac{1}{2}(9 - e^{-8})$ (D) $\frac{1}{2}(13 - e^{-8} - 4e^{-4})$

4. The base of a solid is the region in the first quadrant bounded by the coordinate axes and the parabola $y = -x^2 + 9$. Cross sections of the solid perpendicular to the x-axis are squares. What is the volume of the solid?

(A) 18   (B) $\frac{81}{4}$ (C) $\frac{648}{5}$ (D) $\frac{1458}{5}$

5. The base of a solid is the region in the first quadrant bounded by lines $y = 2x$, $x = 3$, and the x-axis. Every cross section of the solid perpendicular to the x-axis is an equilateral triangle. What is the volume of the solid?

(A) 9   (B) $\frac{9\sqrt{3}}{2}$ (C) $9\sqrt{3}$ (D) 27

6. Find the volume of the solid whose base is the region in the first quadrant that is bounded by the graphs of $y = x^2$, $y = 4$, and the y-axis if cross sections perpendicular to the y-axis are squares.

(A) $\frac{32}{5}$ (B) 8   (C) $\frac{256}{15}$ (D) $\frac{1024}{5}$

 7. The base of a solid is the region bounded by the graph of $y = \ln x$, the line $x = e$, and the x-axis. If cross sections perpendicular to the y-axis are semicircles, then the volume of the solid is

(A) $\frac{\pi}{16}\left[7e^2 - 4e - 1\right]$ (B) $\frac{\pi}{16}\left[-e^2 + 4e - 1\right]$

(C) $\frac{\pi}{8}\left[7e^2 - 4e - 1\right]$ (D) $\frac{\pi}{8}\left[-e^2 + 4e - 1\right]$

Free-Response Question

8. The region R bounded by the graphs of $y = (x - 2)^2$, $y = 2\sqrt{x}$, and the line $x = 1$ is shown in the figure below.

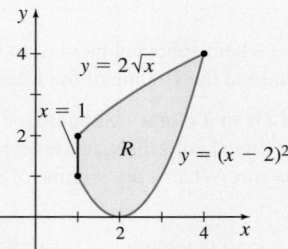

(a) The region forms the base of a solid. If cross sections of the solid perpendicular to the x-axis are squares write an expression that gives the volume of the solid.

(b) Find the volume of the solid.

Retain Your Knowledge

Multiple-Choice Questions

1. The graph of a piecewise function g is shown below. If $f(x) = \int_{-2}^{x} g(t)\, dt$, which of the following could be the greatest value of f?

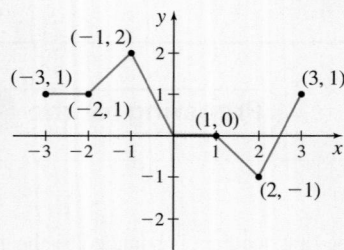

(A) $f(-2)$ (B) $f(-1)$ (C) $f(1)$ (D) $f(3)$

2. The functions f and f' are continuous. The points $(-2, 6)$ and $(5, 0)$ lie on the graph of f. Then $\int_{5}^{-2} f'(x)dx =$

(A) -6 (B) 3   (C) 6   (D) 7

3. Find $\lim\limits_{x \to 0} \dfrac{\cos x - e^x}{e^x \sin x}$ if it exists.

(A) -1 (B) 0   (C) 1   (D) The limit does not exist.

Free-Response Question

4. The graph of the implicit function $x^3 + y^3 = 2xy$ (called a Folium of Descartes) is shown below.

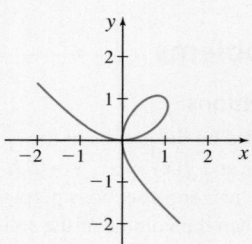

(a) Find the derivative $\dfrac{dy}{dx}$.

(b) Write an equation of the tangent line to the Folium of Descartes at the point $(1, 1)$.

8.6 Arc Length; Surface Area of a Solid of Revolution

OBJECTIVES *When you finish this section, you should be able to:*

1 **Find the arc length of the graph of a function $y = f(x)$ (p. 674)**

2 **Find the arc length of the graph of a function using a partition of the y-axis (p. 677)**

3 **Find the surface area of a solid of revolution (p. 679)**

We have already seen that a definite integral can be used to find the area of a plane region and the volume of certain solids. In this section, we will see that a definite integral can be used to find the length of a graph, and the surface area of a solid of revolution.

The idea of finding the length of a graph had its beginning with Archimedes. Ancient people knew that the ratio of the circumference C of any circle to its diameter d equaled the constant that we call π. That is, $\dfrac{C}{d} = \pi$. But Archimedes is credited as being the first person to analytically investigate the numerical value of π.*

Archimedes drew a circle of diameter 1 unit, then inscribed (and circumscribed) the circle with regular polygons, and computed the perimeters of the polygons. He began with two hexagons (6 sides) and then two dodecagons (12 sides). See Figure 53. He continued until he had two polygons, each with 96 sides. He knew that the circumference C of the circle was larger than the perimeter of the inscribed polygon and smaller than the perimeter of the circumscribed polygon. In this way, he proved that $3\dfrac{10}{71} < \pi < 3\dfrac{1}{7}$. In essence, Archimedes approximated the length of the circle by representing it by smaller and smaller line segments and summing the lengths of the line segments.

ORIGINS Archimedes (287-212 BCE) was a Greek astronomer, physicist, engineer, and mathematician. His contributions include the Archimedes screw, which is still used to lift water or grain, and the Archimedes principle, which states that the weight of an object can be determined by measuring the volume of water it displaces. Archimedes used the principle to measure the amount of gold in the king's crown; modern-day cooks can use it to measure the amount of butter to add to a recipe. Archimedes is also credited with using the Method of Exhaustion to find the area under a parabola and to approximate the value of π.

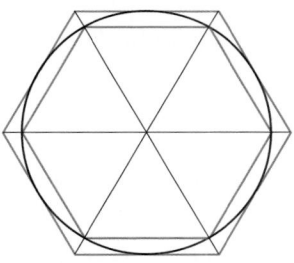
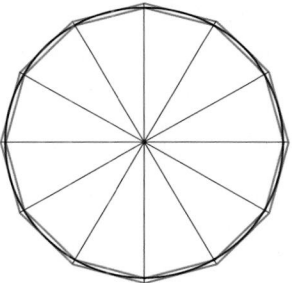

(a) Inscribed and circumscribed hexagons

(b) Inscribed and circumscribed dodecagons

Figure 53

Our approach to finding the length of the graph of a function is similar to that used by Archimedes and follows the ideas we used to find area and volume.

*Petr Beckmann, (1971) *The History of π* (p. 62), New York: St. Martin's Press.

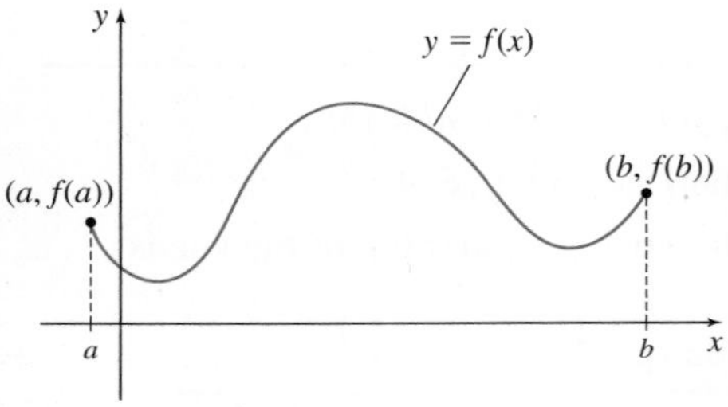

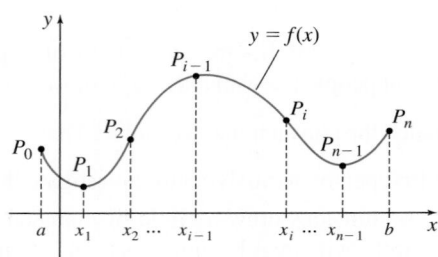

Figure 54 f is continuous on $[a, b]$; f' is continuous on an interval containing a and b.

① Find the Arc Length of the Graph of a Function $y = f(x)$

Suppose we want to find the length of the graph of a function $y = f(x)$ from $x = a$ to $x = b$. We assume f is continuous on $[a, b]$ and the derivative f' of f is continuous on some interval containing a and b. See Figure 54.

We begin by partitioning the closed interval $[a, b]$ into n subintervals:

$$[a, x_1], [x_1, x_2], \ldots, [x_{i-1}, x_i], \ldots, [x_{n-1}, b] \qquad x_0 = a \quad x_n = b$$

each of width $\Delta x = \dfrac{b-a}{n}$. Corresponding to each number $a, x_1, x_2, \ldots, x_{i-1}, x_i, \ldots, x_{n-1}, b$ in the partition, there is a point $P_0, P_1, P_2, \ldots, P_{i-1}, P_i, \ldots, P_{n-1}, P_n$ on the graph of f, as illustrated in Figure 55.

Figure 55 Points $P_0, P_1, P_2, \ldots, P_{n-1}, P_n$ correspond to the numbers $a, x_1, x_2, \ldots, x_{n-1}, b$, respectively.

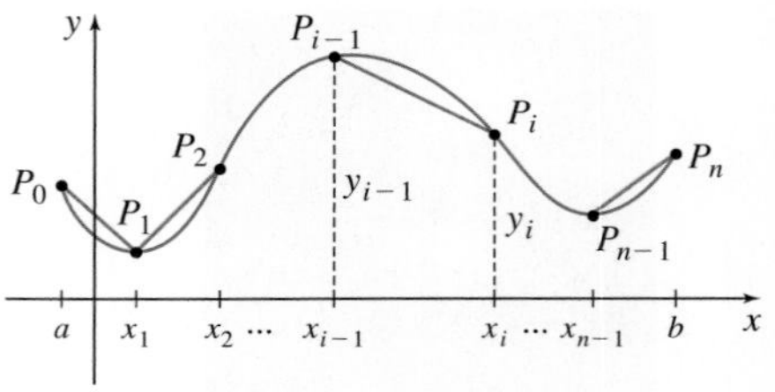

Figure 56 After the interval $[a, b]$ is partitioned, connect consecutive points with line segments.

See Figure 56. When each point $P_{i-1} = (x_{i-1}, y_{i-1})$ is connected to the next point $P_i = (x_i, y_i)$ by a line segment, the sum L of the lengths of the line segments approximates the length of the graph of $y = f(x)$ from $x = a$ to $x = b$. The sum L is given by

$$L = d(P_0, P_1) + d(P_1, P_2) + \cdots + d(P_{n-1}, P_n) = \sum_{i=1}^{n} d(P_{i-1}, P_i)$$

where $d(P_{i-1}, P_i)$ denotes the length of the line segment joining $P_{i-1} = (x_{i-1}, y_{i-1})$ to $P_i = (x_i, y_i)$.

Using the distance formula, the length of the ith line segment is

$$d(P_{i-1}, P_i) = \sqrt{(x_i - x_{i-1})^2 + (y_i - y_{i-1})^2} = \sqrt{(\Delta x)^2 + (\Delta y_i)^2}$$

$$= \sqrt{\frac{(\Delta x)^2 + (\Delta y_i)^2}{(\Delta x)^2}} \cdot \underset{\underset{\Delta x > 0}{\uparrow}}{\sqrt{(\Delta x)^2}} = \sqrt{1 + \left(\frac{\Delta y_i}{\Delta x}\right)^2} \, \Delta x$$

NEED TO REVIEW? The distance formula is discussed in Appendix A.3, p. A-17.

where $\Delta x = \dfrac{b-a}{n}$ and $\Delta y_i = y_i - y_{i-1} = f(x_i) - f(x_{i-1})$. Then the sum L of the lengths of the line segments is

$$\boxed{L = \sum_{i=1}^{n} \sqrt{1 + \left(\frac{\Delta y_i}{\Delta x}\right)^2} \, \Delta x}$$

This sum is not a Riemann sum. To put this sum in the form of a Riemann sum, we need to express $\sqrt{1 + \left(\dfrac{\Delta y_i}{\Delta x}\right)^2}$ in terms of u_i, $x_{i-1} \le u_i \le x_i$. To do this, we first observe that the function f has a derivative f' that is continuous on an interval containing a and b. It follows that f has a derivative on each subinterval (x_{i-1}, x_i) of $[a, b]$. Now use the Mean Value Theorem. Then in each open subinterval (x_{i-1}, x_i), there is a number u_i for which

NEED TO REVIEW? The Mean Value Theorem is discussed in Section 5.1, pp. 334–335.

$$f(x_i) - f(x_{i-1}) = f'(u_i)(x_i - x_{i-1}) \qquad \text{Use the Mean Value Theorem.}$$

$$\Delta y_i = f'(u_i)\, \Delta x \qquad\qquad \Delta y_i = f(x_i) - f(x_{i-1}); \quad \Delta x = x_i - x_{i-1}$$

$$\frac{\Delta y_i}{\Delta x} = f'(u_i) \qquad\qquad \text{Divide both sides by } \Delta x.$$

Now the sum L of the lengths of the line segments can be written as

$$\boxed{L = \sum_{i=1}^{n} \sqrt{1 + [f'(u_i)]^2}\, \Delta x}$$

where u_i is some number in the open subinterval (x_{i-1}, x_i).

This sum L is a Riemann sum. As the number of subintervals increases, that is, as $n \to \infty$, the sum L becomes a better approximation to the arc length of the graph of $y = f(x)$ from $x = a$ to $x = b$. Since f' is continuous on $[a, b]$, the limit of the sum exists and is a definite integral.

Arc Length Formula

Suppose a function $y = f(x)$ is continuous on $[a, b]$ and has a derivative that is continuous on an interval containing a and b. The **arc length** s of the graph of f from $x = a$ to $x = b$ is given by

$$\boxed{s = \int_a^b \sqrt{1 + [f'(x)]^2}\, dx} \tag{1}$$

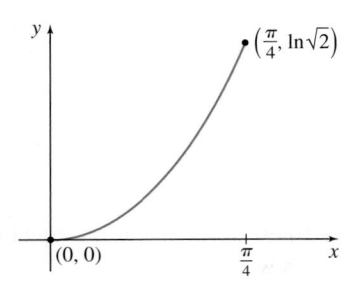

CALC CLIP

EXAMPLE 1 **Finding the Arc Length of a Graph**

Find the arc length of the graph of the function $f(x) = \ln \sec x$ from $x = 0$ to $x = \dfrac{\pi}{4}$.

Solution

The graph of $f(x) = \ln \sec x$ is shown in Figure 57.

The derivative of f is $f'(x) = \dfrac{\sec x \tan x}{\sec x} = \tan x$. Since $f'(x) = \tan x$ is continuous on the open interval $\left(-\dfrac{\pi}{2}, \dfrac{\pi}{2}\right)$, which contains 0 and $\dfrac{\pi}{4}$, we can use the arc length formula (1) to find the arc length s from 0 to $\dfrac{\pi}{4}$.

$$s = \int_0^{\pi/4} \sqrt{1 + [f'(x)]^2}\, dx = \int_0^{\pi/4} \sqrt{1 + \tan^2 x}\, dx = \int_0^{\pi/4} \sqrt{\sec^2 x}\, dx$$

$$= \int_0^{\pi/4} \sec x\, dx = \Big[\ln |\sec x + \tan x|\Big]_0^{\pi/4}$$

$$= \ln \left|\sec \frac{\pi}{4} + \tan \frac{\pi}{4}\right| - \ln |\sec 0 + \tan 0| = \ln |\sqrt{2} + 1| \qquad \blacksquare$$

Figure 57 $f(x) = \ln \sec x,\, 0 \le x \le \dfrac{\pi}{4}$

$\left(\dfrac{\pi}{4}, \ln\sqrt{2}\right)$

$(0, 0)$

$\dfrac{\pi}{4}$

NOW WORK Problem 9 and AP® Practice Problems 1 and 2.

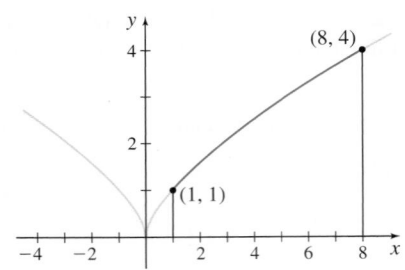

Figure 58 $f(x) = x^{2/3}, 1 \leq x \leq 8$

EXAMPLE 2 Finding the Arc Length of a Graph

Find the arc length of the graph of the function $f(x) = x^{2/3}$ from $x = 1$ to $x = 8$.

Solution

We begin by graphing $f(x) = x^{2/3}$. See Figure 58.

The derivative of $f(x) = x^{2/3}$ is $f'(x) = \dfrac{2}{3}x^{-1/3} = \dfrac{2}{3x^{1/3}}$. Notice that f' is not continuous at 0. However, since f' is continuous on an interval containing 1 and 8 (use $\left[\dfrac{1}{2}, 9\right]$, for example, which avoids 0), we can use the arc length formula (1) to find the arc length s from $x = 1$ to $x = 8$.

$$s = \int_1^8 \sqrt{1 + [f'(x)]^2}\, dx = \int_1^8 \sqrt{1 + \left(\frac{2}{3x^{1/3}}\right)^2}\, dx = \int_1^8 \sqrt{1 + \frac{4}{9x^{2/3}}}\, dx$$

$$= \int_1^8 \sqrt{\frac{9x^{2/3} + 4}{9x^{2/3}}}\, dx = \underset{\substack{\uparrow \\ x > 0 \text{ on } [1, 8], \\ \text{so } \sqrt{x^{2/3}} = x^{1/3}}}{\frac{1}{3} \int_1^8 \left(x^{-1/3}\sqrt{9x^{2/3} + 4}\right) dx}$$

Use the substitution $u = 9x^{2/3} + 4$. Then $du = 6x^{-1/3}dx$ and $x^{-1/3}dx = \dfrac{du}{6}$. The limits of integration change to $u = 9 \cdot 1^{2/3} + 4 = 13$ when $x = 1$, and to $u = 9 \cdot 8^{2/3} + 4 = 40$ when $x = 8$. Then

$$s = \frac{1}{3}\int_1^8 \left[x^{-1/3}\sqrt{9x^{2/3} + 4}\right] dx = \frac{1}{3}\int_{13}^{40} \sqrt{u}\, \frac{du}{6} = \frac{1}{18}\left[\frac{u^{3/2}}{\frac{3}{2}}\right]_{13}^{40}$$

$$= \frac{1}{27}\left(80\sqrt{10} - 13\sqrt{13}\right) \qquad \blacksquare$$

NOW WORK   Problem 17 and AP® Practice Problems 3, 5, and 6.

EXAMPLE 3 Finding the Arc Length of an Ellipse

For the ellipse $x^2 + 4y^2 = 4$,

(a) Set up, but do not evaluate, the integral for the arc length s of the ellipse in the first quadrant from the point $(0,1)$ to the point of intersection of the ellipse and the line $y = x$.

 **(b)** Use technology to evaluate the integral from (a).

Solution

(a) We begin by graphing the ellipse and the line $y = x$. See Figure 59.

The intersection of the ellipse $x^2 + 4y^2 = 4$ and the line $y = x$ satisfies the equation

$$x^2 + 4x^2 = 4 \qquad y = x$$
$$5x^2 = 4$$
$$x = \sqrt{\frac{4}{5}} \qquad x > 0 \text{ in the first quadrant.}$$
$$x = \frac{2\sqrt{5}}{5}$$

In the first quadrant, the ellipse and the line $y = x$ intersect at the point $\left(\dfrac{2\sqrt{5}}{5}, \dfrac{2\sqrt{5}}{5}\right)$.

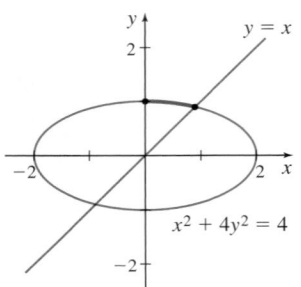

Figure 59

To find the arc length s, we need to express the equation of the ellipse in the explicit form $y = f(x)$ and find the derivative $f'(x)$.

In the first quadrant both $x > 0$ and $y > 0$, so the ellipse $x^2 + 4y^2 = 4$ can be expressed explicitly as

$$4y^2 = 4 - x^2 \qquad\qquad x^2 + 4y^2 = 4$$

$$y^2 = \frac{4 - x^2}{4}$$

$$y = \sqrt{\frac{4 - x^2}{4}} \qquad\qquad y > 0 \text{ in the first quadrant.}$$

$$y = \frac{1}{2}\sqrt{4 - x^2}$$

Then,

$$y' = \frac{d}{dx}\left[\frac{1}{2}\sqrt{4 - x^2}\right] = \frac{d}{dx}\left[\frac{1}{2}(4 - x^2)^{1/2}\right]$$

$$= \frac{1}{2}\left[\frac{1}{2}(4 - x^2)^{-1/2}(-2x)\right] \qquad\qquad \text{Use the Power Rule for Functions.}$$

$$= -\frac{x}{2(4 - x^2)^{1/2}} \qquad\qquad \text{Simplify.}$$

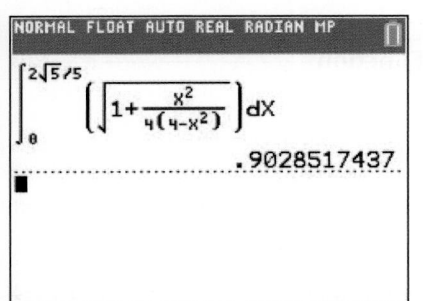

Figure 60

The arc length s of the ellipse from the point $(0, 1)$ to the point $\left(\dfrac{2\sqrt{5}}{5}, \dfrac{2\sqrt{5}}{5}\right)$ is given by

$$s = \int_0^{2\sqrt{5}/5} \sqrt{1 + (y')^2}\,dx = \int_0^{2\sqrt{5}/5} \sqrt{1 + \frac{x^2}{4(4 - x^2)}}\,dx$$

(b) As Figure 60 shows, $s \approx 0.903$. ∎

NOW WORK Problem **29** and AP® Practice Problem **7**.

An observation about Example 3 is noteworthy. Notice that $y' = f'(x)$ is not defined at $x = 2$ so f' is not continuous at 2. The arc length formula (1) cannot be used to compute any arc length that includes $x = 2$ (or $x = -2$). But we can find the arc length of the ellipse from the point $\left(\dfrac{2\sqrt{5}}{5}, \dfrac{2\sqrt{5}}{5}\right)$ to the point $(2, 0)$ by partitioning the y-axis.

② Find the Arc Length of the Graph of a Function Using a Partition of the y-Axis

For a function defined by $x = g(y)$, where g is continuous on $[c, d]$ and g' is continuous on an open interval containing the numbers c and d, we can find the arc length of the graph of g from $y = c$ to $y = d$ by using a partition of the y-axis.

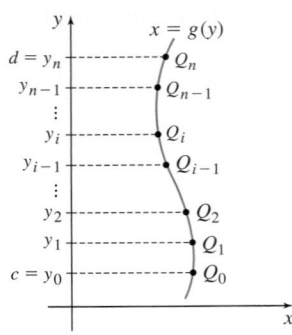

Figure 61 $[c, d]$ is partitioned into

n subintervals of width $\dfrac{d-c}{n}$.

We begin by partitioning the interval $[c, d]$ into n subintervals:

$$[c, y_1], [y_1, y_2], \ldots, [y_{i-1}, y_i], \ldots, [y_{n-1}, d] \qquad y_0 = c \quad y_n = d$$

each of width $\Delta y = \dfrac{d-c}{n}$. Corresponding to each number $c, y_1, y_2, \ldots, y_{i-1}, y_i, \ldots,$
y_{n-1}, d, there are points $Q_0, Q_1, Q_2, \ldots, Q_{i-1}, Q_i, \ldots, Q_{n-1}, Q_n$ on the graph of g,
as shown in Figure 61. By forming line segments connecting each point Q_{i-1} to Q_i,
for $i = 1, 2, 3, \ldots, n$, and following the same process as described for a partition
of the x-axis, we obtain the following result:

Arc Length Formula (Partitioning the y-Axis)

For a function defined by $x = g(y)$, where g is continuous on $[c, d]$ and g' is
continuous on an interval containing c and d, the arc length s of the graph of g
from $y = c$ to $y = d$ is given by

$$s = \int_c^d \sqrt{1 + [g'(y)]^2}\, dy \tag{2}$$

The result in (2) can sometimes be used to find the length of the graph of a function
$y = f(x)$ from a to b when its derivative f' is not continuous at some number in $[a, b]$.

EXAMPLE 4   Finding the Arc Length of a Function

Find the arc length of $f(x) = x^{2/3}$ from $x = -1$ to $x = 8$.

Solution

Since the derivative $f'(x) = \dfrac{2}{3x^{1/3}}$ does not exist at $x = 0$, we cannot use the arc length
formula (1) to find the length of the graph from $x = -1$ to $x = 8$.

From Figure 62, we observe that the length of the graph of f from $x = -1$ to $x = 8$
is the same as the length of the graph of $x = g_1(y) = -y^{3/2}$ from $y = 0$ to $y = 1$ plus the
length of the graph of $x = g_2(y) = y^{3/2}$ from $y = 0$ to $y = 4$.

We investigate $x = g_1(y) = -y^{3/2}$ first. Since $g_1'(y) = -\dfrac{3}{2}y^{1/2}$ is continuous for
all $y \geq 0$, we can use arc length formula (2) to find the arc length s_1 of g_1 from $y = 0$
to $y = 1$.

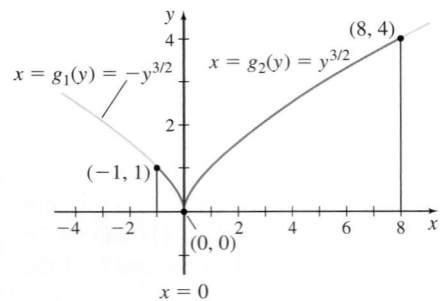

Figure 62 $f(x) = x^{2/3}$, $-1 \leq x \leq 8$

$$s_1 = \int_0^1 \sqrt{1 + [g_1'(y)]^2}\, dy = \int_0^1 \sqrt{1 + \left(-\frac{3}{2}y^{1/2}\right)^2}\, dy$$

$$\underset{g_1'(y) = -\frac{3}{2}y^{1/2}}{\uparrow}$$

$$= \int_0^1 \sqrt{1 + \frac{9}{4}y}\, dy = \frac{1}{2}\int_0^1 \sqrt{4 + 9y}\, dy$$

Use the substitution $u = 4 + 9y$. Then $du = 9\, dy$ or, equivalently, $dy = \dfrac{du}{9}$. The limits of
integration are $u = 4$ when $y = 0$, and $u = 13$ when $y = 1$.

$$s_1 = \frac{1}{2}\int_4^{13} \sqrt{u}\, \frac{du}{9} = \frac{1}{18}\left[\frac{u^{3/2}}{\frac{3}{2}}\right]_4^{13} = \frac{1}{27}\left(13\sqrt{13} - 8\right)$$

Now we investigate $x = g_2(y) = y^{3/2}$. Since $g_2'(y) = \dfrac{3}{2}y^{1/2}$ is continuous for all $y \geq 0$, we can use arc length formula (2) to find the arc length s_2 of g_2 from $y = 0$ to $y = 4$.

$$
s_2 = \int_0^4 \sqrt{1 + [g_2'(y)]^2}\, dy = \int_0^4 \sqrt{1 + \left(\frac{3}{2}y^{1/2}\right)^2}\, dy
$$

$$
= \int_0^4 \sqrt{1 + \frac{9}{4}y}\, dy = \frac{1}{2}\int_0^4 \sqrt{4 + 9y}\, dy \qquad \begin{array}{l}\text{Let } u = 4 + 9y.\\ du = 9\,dy\end{array}
$$

$$
= \frac{1}{2}\int_4^{40} \sqrt{u}\,\frac{du}{9} = \frac{1}{18}\left[\frac{u^{3/2}}{\frac{3}{2}}\right]_4^{40} = \frac{1}{27}\left(80\sqrt{10} - 8\right)
$$

The arc length s of $y = f(x) = x^{2/3}$ from $x = -1$ to $x = 8$ is the sum

$$
s = s_1 + s_2 = \frac{1}{27}\left(13\sqrt{13} - 8\right) + \frac{1}{27}\left(80\sqrt{10} - 8\right)
$$

$$
= \frac{1}{27}\left(80\sqrt{10} + 13\sqrt{13} - 16\right)
$$

∎

NOW WORK Problem **23** and AP® Practice Problem **4**.

3 Find the Surface Area of a Solid of Revolution

The **surface area** of a solid of revolution measures the outer surface of the solid. The formula for the surface area of a solid of revolution depends on the formula for the surface area of a *frustum* of a right circular cone.

Consider a line segment of length L that lies above the x-axis. See Figure 63(a). If the region bounded by the line segment, the x-axis, and the lines $x = a$ and $x = b$, is revolved about an axis, the resulting solid of revolution is a **frustum**. A frustum of a right circular cone has slant height L and base radii r_1 and r_2. See Figure 63(b).

> **NOTE** A frustum is a portion of a solid of revolution that lies between two parallel planes.

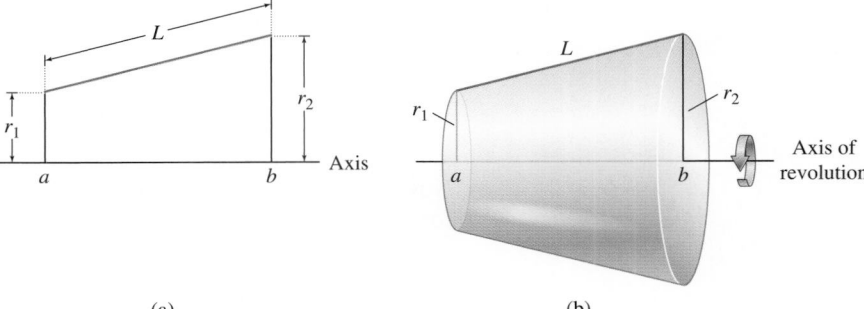

(a)   (b)

Figure 63

The **surface area** S of a frustum of a right circular cone equals the product 2π times the average radius of the frustum times the slant height of the frustum. That is,

$$
\boxed{\, S = 2\pi\left(\frac{r_1 + r_2}{2}\right)L \,}
$$

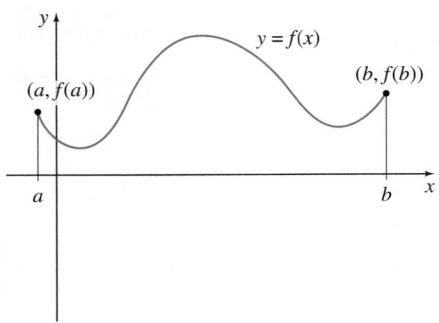

Figure 64

Suppose a solid of revolution is formed by revolving the region shown in Figure 64 bounded by the graph of a function $y = f(x)$, where $f(x) \geq 0$, the x-axis, and the lines $x = a$ and $x = b$, about the x-axis. We assume the derivative f' of f is continuous on some interval containing a and b. To find the surface area of this solid of revolution, we partition the closed interval $[a, b]$ into n subintervals

$$[a, x_1], [x_1, x_2], \ldots, [x_{i-1}, x_i], \ldots, [x_{n-1}, b]$$

each of width $\Delta x = \dfrac{b - a}{n}$. Corresponding to each number $a, x_1, x_2, \ldots, x_{i-1}, x_i, \ldots,$ x_{n-1}, b, there is a point $P_0, P_1, \ldots, P_{i-1}, P_i, \ldots, P_{n-1}, P_n$ on the graph of f. We join each point P_{i-1} to the next point P_i with a line segment and focus on the line segment joining the points P_{i-1} and P_i. See Figure 65(a). When this line segment of length $d(P_{i-1}, P_i)$ is revolved about the x-axis, it generates a frustum of a right circular cone whose surface area S_i is

$$S_i = 2\pi \left[\frac{y_{i-1} + y_i}{2} \right] [d(P_{i-1}, P_i)] \tag{3}$$

See Figure 65(b).

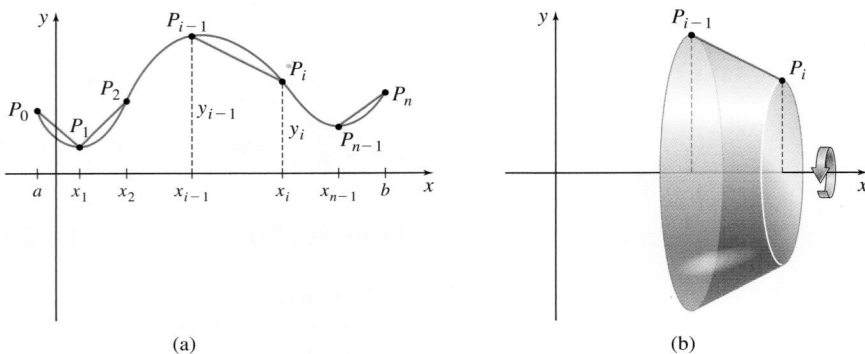

(a)   (b)

Figure 65

We now follow the same reasoning used for finding the arc length of the graph of a function.

Using the distance formula, the length of the ith line segment is

$$d(P_{i-1}, P_i) = \sqrt{(x_i - x_{i-1})^2 + (y_i - y_{i-1})^2}$$

$$= \sqrt{(\Delta x)^2 + (\Delta y_i)^2} = \sqrt{1 + \left(\frac{\Delta y_i}{\Delta x} \right)^2} \, \Delta x$$

Now use the Mean Value Theorem. There is a number u_i in the open interval (x_{i-1}, x_i) for which

$$f(x_i) - f(x_{i-1}) = f'(u_i)\Delta x$$

$$\frac{\Delta y_i}{\Delta x} = f'(u_i)$$

So,

$$d(P_{i-1}, P_i) = \sqrt{1 + [f'(u_i)]^2}\,\Delta x$$

where u_i is a number in the ith subinterval.

Now replace $d(P_{i-1}, P_i)$ in equation (3) with $\sqrt{1 + [f'(u_i)]^2}\,\Delta x$. Then the surface area generated by the sum of the line segments is

$$\sum_{i=1}^{n} S_i = \sum_{i=1}^{n} 2\pi \frac{y_{i-1} + y_i}{2} \sqrt{1 + [f'(u_i)]^2}\,\Delta x$$

These sums approximate the surface area generated by revolving the graph of $y = f(x)$ about the x-axis and lead to the following result.

THEOREM Surface Area of a Solid of Revolution

Suppose a function $y = f(x)$ is nonnegative on the closed interval $[a, b]$ and has a derivative f' that is continuous on an interval containing a and b. The surface area S of the solid of revolution obtained by revolving the region bounded by the graph of $y = f(x)$, the x-axis, and the lines $x = a$ and $x = b$ about the x-axis is given by

$$S = 2\pi \int_a^b f(x)\sqrt{1 + [f'(x)]^2}\, dx \qquad (4)$$

EXAMPLE 5   Finding the Surface Area of a Solid of Revolution

Find the surface area S of the solid generated by revolving the region bounded by the graph of $y = \sqrt{x}$, the x-axis, and the lines $x = 1$ and $x = 4$, about the x-axis.

Solution

We begin with the graph of $y = \sqrt{x}$, $1 \le x \le 4$, and revolve it about the x-axis, as shown in Figure 66.

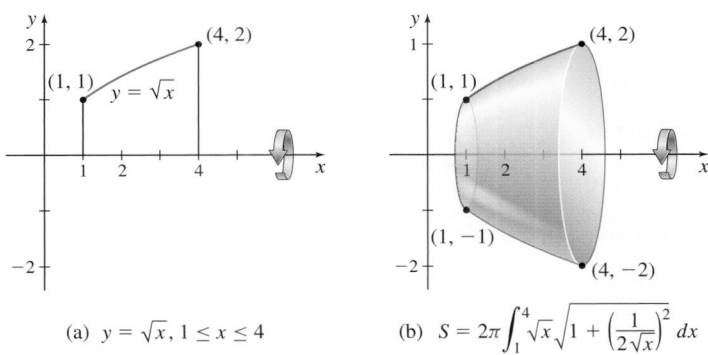

(a) $y = \sqrt{x}$, $1 \le x \le 4$

(b) $S = 2\pi \int_1^4 \sqrt{x}\sqrt{1 + \left(\frac{1}{2\sqrt{x}}\right)^2}\, dx$

Figure 66

Use formula (4) with $f(x) = \sqrt{x}$ and $f'(x) = \dfrac{1}{2\sqrt{x}}$. The surface area S is

$$S = 2\pi \int_a^b f(x)\sqrt{1 + [f'(x)]^2}\, dx = 2\pi \int_1^4 \sqrt{x}\sqrt{1 + \left(\frac{1}{2\sqrt{x}}\right)^2}\, dx$$

$$= 2\pi \int_1^4 \sqrt{x\left(1 + \frac{1}{4x}\right)}\, dx = 2\pi \int_1^4 \frac{1}{2}\sqrt{4x + 1}\, dx$$

$$\underset{\underset{\substack{u = 4x+1;\ \frac{du}{4} = dx \\ \text{when } x = 1,\, u = 5;\ \text{when } x = 4,\, u = 17}}{\uparrow}}{= \pi \int_5^{17} \sqrt{u}\left(\frac{du}{4}\right)} = \frac{\pi}{4}\left[\frac{u^{3/2}}{\frac{3}{2}}\right]_5^{17} = \frac{\pi}{6}(17^{3/2} - 5^{3/2})$$

The surface area of the solid of revolution is $\dfrac{\pi}{6}(17^{3/2} - 5^{3/2}) \approx 30.846$ square units. ∎

NOW WORK   Problem 35.

8.6 Assess Your Understanding

Concepts and Vocabulary

1. **True or False**   If a function f has a derivative that is continuous on an interval containing a and b, the arc length s of the graph of $y = f(x)$ from $x = a$ to $x = b$ is given by the formula $s = \int_a^b \sqrt{1 + [f(x)]^2}\, dx$.

2. **True or False**   If the derivative of a function $y = f(x)$ is not continuous at some number in the interval $[a, b]$, its arc length from $x = a$ to $x = b$ can sometimes be found by partitioning the y-axis.

Skill Building

In Problems 3–6, use the arc length formula to find the length of each line between the points indicated. Verify your answer by using the distance formula.

3. $y = 3x - 1$, from $(1, 2)$ to $(3, 8)$

4. $y = -4x + 1$, from $(-1, 5)$ to $(1, -3)$

5. $2x - 3y + 4 = 0$, from $(1, 2)$ to $(4, 4)$

6. $3x + 4y - 12 = 0$, from $(0, 3)$ to $(4, 0)$

In Problems 7–22, find the arc length of each graph by partitioning the x-axis.

7. $y = x^{2/3} + 1$, from $x = 1$ to $x = 8$

8. $y = x^{2/3} + 6$, from $x = 1$ to $x = 8$

9. $y = x^{3/2}$, from $x = 0$ to $x = 4$

10. $y = x^{3/2} + 4$, from $x = 1$ to $x = 4$

11. $9y^2 = 4x^3$, from $x = 0$ to $x = 1$; $y \geq 0$

12. $y = \dfrac{x^3}{6} + \dfrac{1}{2x}$, from $x = 1$ to $x = 3$

13. $y = \dfrac{2}{3}(x^2 + 1)^{3/2}$, from $x = 1$ to $x = 4$

14. $y = \dfrac{1}{3}(x^2 + 2)^{3/2}$, from $x = 2$ to $x = 4$

15. $y = \dfrac{2}{9}\sqrt{3}(3x^2 + 1)^{3/2}$, from $x = -1$ to $x = 2$

16. $y = (1 - x^{2/3})^{3/2}$, from $x = \dfrac{1}{8}$ to $x = 1$

17. $8y = x^4 + \dfrac{2}{x^2}$, from $x = 1$ to $x = 2$

18. $9y^2 = 4(1 + x^2)^3$, $y \geq 0$, from $x = 0$ to $x = 2\sqrt{2}$

19. $y = \ln(\sin x)$, from $x = \dfrac{\pi}{6}$ to $x = \dfrac{\pi}{3}$

20. $y = \ln(\cos x)$, from $x = \dfrac{\pi}{6}$ to $x = \dfrac{\pi}{3}$

21. $(x + 1)^3 = 4y^2$, $y \geq 0$, from $x = -1$ to $x = 16$

22. $y = x^{3/2} + 8$, from $x = 0$ to $x = 4$

In Problems 23–26, find the arc length of each graph by partitioning the y-axis.

23. $y = x^{2/3}$, from $x = 0$ to $x = 1$

24. $y = x^{2/3}$, from $x = -1$ to $x = 0$

25. $(x + 1)^2 = 4y^3$, $x \geq -1$, from $y = 0$ to $y = 1$

26. $x = \dfrac{2}{3}(y - 5)^{3/2}$, from $y = 5$ to $y = 6$

In Problems 27–32, (a) use the arc length formula (1) to set up the integral for arc length.

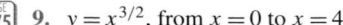 *(b) Use technology to find the arc length.*

27. $y = x^2$, from $x = 0$ to $x = 2$

28. $x = y^2$, from $y = 1$ to $y = 3$

29. $y = \sqrt{25 - x^2}$, from $x = 0$ to $x = 4$

30. $x = \sqrt{4 - y^2}$, from $y = 0$ to $y = 1$

31. $y = \sin x$, from $x = 0$ to $x = \dfrac{\pi}{2}$

32. $x = y + \ln y$, from $y = 1$ to $y = 4$

In Problems 33–42, find the surface area of the solid formed by revolving the region bounded by the graph of $y = f(x)$, the x-axis, and the given lines about the x-axis.

33. $f(x) = 3x + 5$; lines: $x = 0$ and $x = 2$

34. $f(x) = -x + 5$; lines: $x = 1$ and $x = 5$

35. $f(x) = \sqrt{3x}$; lines: $x = 1$ and $x = 2$

36. $f(x) = \sqrt{2x}$; lines: $x = 2$ and $x = 8$

37. $f(x) = 2\sqrt{x}$; lines: $x = 1$ and $x = 4$

38. $f(x) = 4\sqrt{x}$; lines: $x = 4$ and $x = 9$

39. $f(x) = x^3$; lines: $x = 0$ and $x = 2$

40. $f(x) = \dfrac{1}{3}x^3$; lines: $x = 1$ and $x = 3$

41. $f(x) = \sqrt{4 - x^2}$; lines: $x = 0$ and $x = 1$

42. $f(x) = \sqrt{9 - 4x^2}$; lines: $x = 0$ and $x = 1$

In Problems 43–50, (a) set up the integral that represents the surface area of the solid formed by revolving the region bounded by the graph of $y = f(x)$, the x-axis, and the given lines about the x-axis.

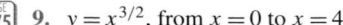 *(b) Use technology to evaluate the integral found in (a).*

43. $f(x) = x^2$; lines: $x = 1$ and $x = 3$

44. $f(x) = 4 - x^2$; lines: $x = 0$ and $x = 2$

45. $f(x) = x^{2/3}$; lines: $x = 1$ and $x = 8$

46. $f(x) = x^{3/2}$; lines: $x = 0$ and $x = 4$

47. $f(x) = e^x$; lines: $x = 0$ and $x = 3$

48. $f(x) = \ln x$; lines: $x = 1$ and $x = e$

49. $f(x) = \sin x$; lines: $x = 0$ and $x = \dfrac{\pi}{2}$

50. $f(x) = \tan x$; lines: $x = 0$ and $x = \dfrac{\pi}{4}$

Applications and Extensions

51. Find the arc length of the graph of $F(x) = \int_0^x \sqrt{16t^2 - 1}\, dt$ from $x = 0$ to $x = 2$.

52. Find the arc length of the graph of $F(x) = \int_0^x \sqrt{4t - 1}\, dt$ from $x = 0$ to $x = 2$.

53. Find the arc length of the graph of $F(x) = \int_0^{3x} \sqrt{e^t - \frac{1}{9}}\, dt$ from $x = 0$ to $x = 4$.

54. Find the arc length of the graph of $F(x) = \int_0^{4x} \sqrt{t^4 - \frac{1}{4}}\, dt$ from $x = 1$ to $x = 2$.

55. Length of a Hypocycloid Find the total length of the **hypocycloid** $x^{2/3} + y^{2/3} = a^{2/3}$, $a > 0$.

56. Distance Along a Curved Path Find the distance between $(1, 1)$ and $(3, 3\sqrt{3})$ along the graph of $y^2 = x^3$.

57. Perimeter Find the perimeter of the region bounded by the graphs of $y^3 = x^2$ and $y = x$.

58. Perimeter Find the perimeter of the region bounded by the graphs of $y = 3(x - 1)^{3/2}$ and $y = 3(x - 1)$.

59. Length of a Graph Find the arc length of the graph of $6xy = y^4 + 3$ from $y = 1$ to $y = 2$.

60. Length of a Graph Find the arc length of the graph of the function $y = \dfrac{e^x + e^{-x}}{2}$ from $x = 0$ to $x = 2$.

61. Length of an Graph Find the arc length of $y = \ln(\csc x)$ from $x = \dfrac{\pi}{4}$ to $x = \dfrac{\pi}{2}$.

62. Length of an Elliptical Arc Set up, but do not attempt to evaluate, the integral for the arc length of the ellipse $\dfrac{x^2}{a^2} + \dfrac{y^2}{b^2} = 1$ from $x = 0$ to $x = \dfrac{a}{2}$ in quadrant I. This integral, which is approximated by numerical techniques (see Chapter 6, Part 1), is called an **elliptical integral of the second kind**.

63. (a) Set up, but do not attempt to evaluate, the integral for the arc length of the ellipse $x^2 + 4y^2 = 4$ in the first quadrant from the point of intersection of the ellipse and the line $y = x$ to the point $(2, 0)$.

   (b) Use technology to evaluate the integral found in (a).

   (c) Use the result obtained in Example 3, p. 676, to find the perimeter of the ellipse $x^2 + 4y^2 = 4$.

64. Surface Area of a Cone

   (a) Find an equation of the line from the origin $(0, 0)$ to the point (h, a), $a > 0$, $h > 0$.

   (b) Graph the line found in (a).

   (c) Revolve the region bounded by the line from (a), the x-axis, and the line $x = h$ about the x-axis to generate a cone of height h and radius a.

   (d) Show that the surface area S of the cone is $S = \pi a\sqrt{a^2 + h^2}$.

65. Search Light The reflector of a search light is formed by revolving an arc of a parabola about its axis. Find the surface area of the reflector, if it measures 1 meter across its widest point and is $\dfrac{1}{4}$ meter deep.

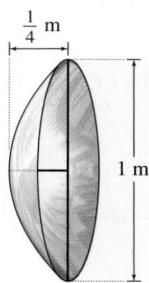

66. Modeling a Ski Slope A ski slope is built on a mountainside and curves upward from ground level to a height h. The shape of the ski slope is modeled by the equation $y = Ax^{3/2}$, where x is the horizontal distance from the bottom of the ski slope measured along the base of the mountain and y is the vertical height of the ski slope at the distance x. See the figure below.

   (a) Find an expression, in terms of A and h, for the length of the ski slope.

   (b) Find A if the ski slope is 150 m high and has a horizontal distance of 250 m along the base.

   (c) If a skier skis directly downhill from the top of the ski slope to the bottom, how far does she travel?

   (d) Describe a simple way to check if the distance obtained in part (c) is reasonable.

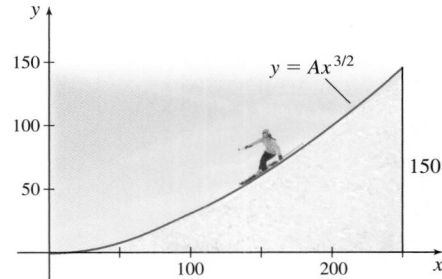

Challenge Problem

67. Inscribe a regular polygon of 2^n sides in a circle of radius 1. The figure illustrates the situation for $n = 3$.

   (a) Show that the indicated angle

   $$\alpha = \frac{45°}{2^{n-2}}.$$

   (b) Show that the formula

   $$s = 2^{n+1}\sin\frac{45°}{2^{n-2}}$$

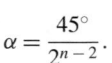
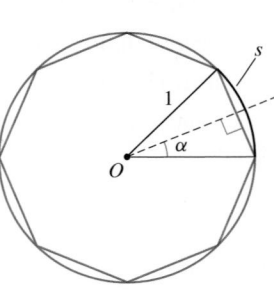

   gives a good approximation to the arc length s (the circumference) of the circle.

   (c) Evaluate this approximation for several values of n, and compare the results with the actual circumference of the circle.

AP® Practice Problems

NOTE: *These problems are BC only.*

Multiple-Choice Questions

1. The arc length of the graph of $f(x) = -x^4 + 2$ from $x = 0$ to $x = 10$ is given by

(A) $\displaystyle\int_0^{10} \sqrt{1 + 16x^3}\, dx$ (B) $\displaystyle\int_0^{10} \sqrt{1 + 4x^3}\, dx$

(C) $\displaystyle\int_0^{10} \sqrt{1 + (-x^4 + 2)^2}\, dx$ (D) $\displaystyle\int_0^{10} \sqrt{1 + 16x^6}\, dx$

2. The arc length of the graph of $y = \tan x$ from $x = a$ to $x = b$, where $-\dfrac{\pi}{2} < a < b < \dfrac{\pi}{2}$ is given by

(A) $\displaystyle\int_{-\pi/2}^{\pi/2} \sqrt{1 + \sec^2 x}\, dx$ (B) $\displaystyle\int_a^b \sqrt{1 + \sec^4 x}\, dx$

(C) $\displaystyle\int_a^b \sqrt{1 + \sec^2 x}\, dx$ (D) $\displaystyle\int_a^b \sqrt{1 + \sec^2 x \tan^2 x}\, dx$

3. The arc length of the graph of $f(x) = \ln \cos x$ from $x = 0$ to $x = \dfrac{\pi}{3}$ is given by

(A) $\displaystyle\int_0^{\pi/3} \sqrt{1 + \sec^2 x}\, dx$

(B) $\displaystyle\int_0^{\pi/3} \sqrt{1 + \tan^2 x}\, dx$

(C) $\displaystyle\int_0^{\pi/3} \sqrt{1 + \tan x}\, dx$

(D) $\displaystyle\int_0^{\pi/3} \sqrt{1 + [\ln |\cos x|]^2}\, dx$

4. The arc length of $y = (x - 8)^{2/3}$ from $x = 0$ to $x = 16$ is given by

(A) $\displaystyle\int_0^{16} \sqrt{1 + \dfrac{4}{9(x - 8)^{2/3}}}\, dx$

(B) $\displaystyle\int_0^4 \sqrt{1 + \dfrac{9}{4}y}\, dy$

(C) $\displaystyle 2\int_0^4 \sqrt{1 + \dfrac{9}{4}y}\, dy$

(D) $\displaystyle\int_0^{16} \sqrt{1 + \dfrac{2}{3(y - 8)^{1/3}}}\, dy$

5. Find the arc length of the graph of $f(x) = x + \ln x$ from $x = 1$ to $x = 10$.

(A) 11.303   (B) 13.507   (C) 14.481   (D) 64.392

6. Find the arc length of the graph of $y = x^2$ from $x = 0$ to $x = 3$.

(A) 4.667   (B) 9.747   (C) 39   (D) 51.600

Free-Response Question

7. (a) The ellipse $4x^2 + y^2 = 4$ and the line $y = 2x$ intersect in the first quadrant. Set up, but do not attempt to evaluate, the integral for the arc length of the ellipse from the point $(1, 0)$ to the point of intersection of the two graphs.
(b) Use technology to evaluate the integral in (a).

Retain Your Knowledge

Multiple-Choice Questions

1. A function $y = f(x)$ is continuous and differentiable on the interval $(0, 10)$. Suppose $\int_2^7 f(x)\, dx = 15$. Values of f and its derivative f' for two numbers x are given in the table below. Find $\int_2^7 x\, f'(x)\, dx$.

x	2	7
$f(x)$	2	6
$f'(x)$	$-\dfrac{3}{2}$	1

(A) -10 (B) 23   (C) 24   (D) 46

2. Which function below can be used to show that a function that is continuous may not be differentiable?

(A) $f(x) = e^{-x}$ (B) $f(x) = x^{4/3}$

(C) $f(x) = x^{2/5}$ (D) $f(x) = \left(\dfrac{1}{e}\right)^{x/2}$

3. The logistic differential equation
$$\dfrac{dP}{dt} = 0.025P\left(1 - \dfrac{P}{20,000}\right)$$
models the rate of change in the population of a newly discovered species of insect. The population is growing most rapidly when the population is

(A) 500 insects   (B) 1000 insects

(C) 10,000 insects   (D) 20,000 insects

Free-Response Question

4. (a) Find the general solution of the differential equation
$$\dfrac{3}{x}\dfrac{dy}{dx} = \dfrac{\cos x}{y^2}.$$

(b) Find the particular solution that satisfies the initial conditions that $y = 4$ if $x = 2\pi$.

8.7 Work

OBJECTIVES *When you finish this section, you should be able to:*

1 **Find the work done by a variable force (p. 685)**
2 **Find the work done by a spring force (p. 687)**
3 **Find the work done to pump a liquid (p. 689)**
   Application: Gravitational Force (p. 691)

In physics, **work** is defined as the energy transferred to or from an object by a force acting on the object. The work W done by a *constant* force F in moving an object a distance x along a straight line in the direction of F is defined to be

$$W = Fx$$

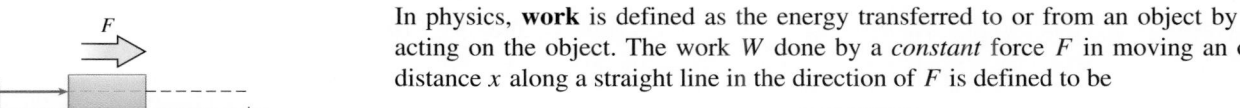

Figure 67 Constant force $W = F(b - a)$

See Figure 67.

One unit of work is the work done by a unit force in moving an object a unit distance in the direction of the force. In the International System of Units (abbreviated SI, for Système International d'Unités), the unit of work is a newton-meter, which is called a **joule** (J). In terms of customary U.S. units, the unit of work is the foot-pound. These terms are summarized in Table 1.

TABLE 1

	Work Units	Force Units	Distance Units
SI	joule (J)	newton (N)	meter (m)
U.S.	foot-pound (ft lb)	pound (lb)	foot (ft)

NOTE 1 N is the force required to accelerate an object of mass 1 kg at 1 m/s². Also, 1 J ≈ 0.7376 ft lb and 1 ft lb ≈ 1.356 J.

For example, the work W required to lift an object weighing 80 lb a distance of 5 ft would be $W = 80 \cdot 5 = 400$ ft lb.

When a force F acts in the same direction as the motion, the work done is positive; if a force F acts in a direction opposite to the motion, the work done is negative.

In some cases, a force F acts along the line of motion of an object, but the magnitude of the force *varies* depending on the position of the object. For example, the force required to raise a cable depends on the length of the cable, and the force required to stretch a spring depends on how far the spring has already been stretched from its normal length. These are examples of **variable forces**.

ORIGINS Joule, the unit of energy, is named for James Prescott Joule, a British physicist and brewer who lived in the nineteenth century.

1 Find the Work Done by a Variable Force

Suppose a variable force $F = F(x)$ acts on an object, where x is the distance of the object from the origin and F is a function that is continuous on a closed interval $[a, b]$. We seek a formula for finding the work done by the force F in moving the object from $x = a$ to $x = b$. We begin by partitioning the interval $[a, b]$ into n subintervals:

$$[a, x_1], [x_1, x_2], \ldots, [x_{i-1}, x_i], \ldots, [x_{n-1}, b] \qquad a = x_0, b = x_n$$

each of width $\Delta x = \dfrac{b - a}{n}$.

Now consider the ith subinterval $[x_{i-1}, x_i]$, and choose a number u_i in $[x_{i-1}, x_i]$. If the width of the subinterval is small, the force $F = F(x)$ acting on the object will not change much over the interval; that is, F can be treated as a constant force. Then the work W done by F to move the object from x_{i-1} to x_i, a distance $\Delta x = x_i - x_{i-1}$, can be approximated by

$$W_i = F(u_i)\,\Delta x$$
$$\uparrow$$
$$W = Fx$$

Summing the work done over all the subintervals approximates the total work W done by the force F in moving the object from a to b. That is, the total work W is approximated by the Riemann sums

$$W \approx F(u_1)\Delta x + F(u_2)\Delta x + \cdots + F(u_n)\Delta x = \sum_{i=1}^{n} F(u_i)\Delta x$$

As the number of subintervals increases, that is, as $n \to \infty$, the approximation improves, and, if $F = F(x)$ is continuous on $[a, b]$, then

$$W = \lim_{n \to \infty} \sum_{i=1}^{n} F(u_i)\Delta x = \int_{a}^{b} F(x)\, dx$$

DEFINITION Work

The **work** W done by a continuously varying force F acting on an object, which moves the object along a straight line in the direction of F from $x = a$ to $x = b$, is

$$\boxed{W = \int_{a}^{b} F(x)\, dx}$$

Consider a freely hanging cable or chain that is being lengthened or shortened. The weight of the cable is a function of its length. As the cable is let out, its weight increases proportionally to its length, and when it is pulled in, it decreases proportionally to its length.

EXAMPLE 1   **Finding the Work Done in Pulling in a Rope**

A 60-ft rope weighing 8 lb per linear foot is used for mooring a cruise ship. As the ship prepares to leave port, the rope is released, and it hangs freely over the side of the ship. How much work is done by the deckhand who winds in the rope?

Solution

We position an x-axis parallel to the side of the ship with the origin O of the axis even with the bottom of the rope and $x = 60$ even with the ship's deck. See Figure 68. The work done by the deckhand depends on the weight of the rope and the length of rope hanging over the edge.

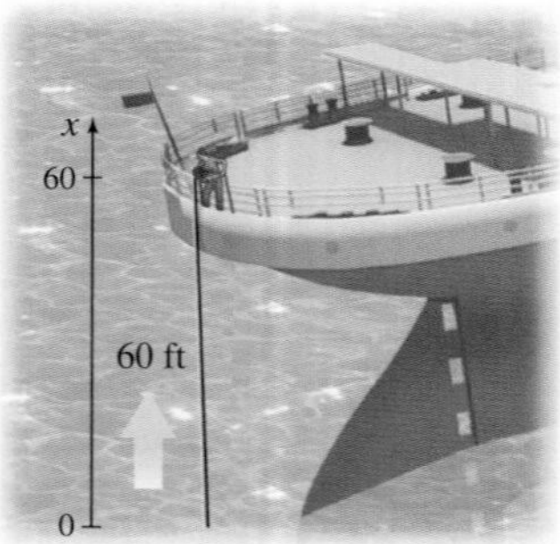

Partition the interval $[0, 60]$ into n subintervals, each of length $\Delta x = \dfrac{60}{n}$, and choose a number u_i in each subinterval. Now think of the rope as n short segments, each of length Δx. Then

$$\text{Weight of the } i\text{th segment} = 8\Delta x \text{ lb} \qquad F$$
$$\text{Distance the } i\text{th segment is lifted} = (60 - u_i) \text{ ft} \qquad x$$
$$\text{Work done in lifting the } i\text{th segment} = 8(60 - u_i)\Delta x \text{ ft lb} \qquad W = Fx$$

The work W required to lift the 60 ft of rope is

$$W = \int_{0}^{60} 8(60 - x)\, dx = \left[480x - 4x^2\right]_{0}^{60} = 14{,}400 \text{ ft lb} \qquad \blacksquare$$

Figure 68

NOW WORK   Problem 11.

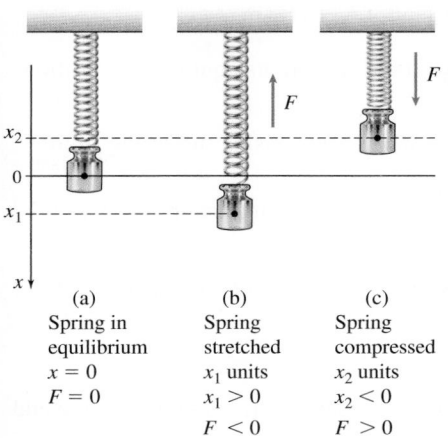

Figure 69

2 Find the Work Done by a Spring Force

A common example of the work done by a variable force is found in stretching or compressing a spring that is fixed at one end. A spring is said to be in **equilibrium** when it is neither extended nor compressed. The **spring force** F needed to extend or compress a spring depends on the *stiffness* of the spring, which is measured by its **spring constant** k, a positive real number. Since a spring always attempts to return to equilibrium, a spring force F is often called a **restoring force**. A spring force F varies with the distance x that the free end of the spring is moved from its equilibrium length and obeys **Hooke's Law**:

$$\boxed{F(x) = -kx}$$

where k is the spring constant measured in newtons per meter in SI units or pounds per foot in U.S. units. The minus sign in Hooke's Law indicates that the direction of a spring force is opposite from the direction of the displacement.

As Figure 69 illustrates, the distance x in Hooke's Law is measured from the equilibrium, or unstretched, position of the spring. This distance x is positive if the spring is stretched from its equilibrium position and is negative if the spring is compressed from its equilibrium position. As a result, a spring force F is negative if the spring is stretched, and F is positive if the spring is compressed.

In applied problems involving springs, the value of k, the spring constant, is often unknown. When we know information about how the spring behaves, the value of k can be found. The next example illustrates this.

EXAMPLE 2   Analyzing a Spring Force

Suppose a spring in equilibrium is 0.8 m long and a spring force of -2 N stretches the spring to a length of 1.2 m.

(a) Find the spring constant k and the spring force F.

(b) What spring force is required to stretch the spring to a length of 3 m?

(c) How much work is done by the spring force in stretching it from equilibrium to 3 m?

Solution

We position an axis parallel to the spring and place the origin at the free end of the spring in equilibrium, as in Figure 70.

(a) When the spring is stretched to a length of 1.2 m due to a spring force of -2 N, then $x = 0.4$ and $F = -2$. Using Hooke's Law, we get

$$-2 = -k(0.4) \qquad \text{Hooke's law: } F(x) = -kx; \ F = -2; \ x = 0.4$$

$$k = \frac{2}{0.4} = 5 \, \text{N/m}$$

The spring constant is $k = 5$. The spring force F is $F = -kx = -5x$.

(b) The spring force F required to stretch the spring to a length of 3 m, that is, a distance $x = 3 - 0.8 = 2.2$ m from equilibrium, is

$$F = -5x = (-5)(2.2) = -11 \, \text{N}$$

(c) The work W done by the spring force F when stretching the spring from equilibrium ($x = 0$) to 3 m ($x = 2.2$) is

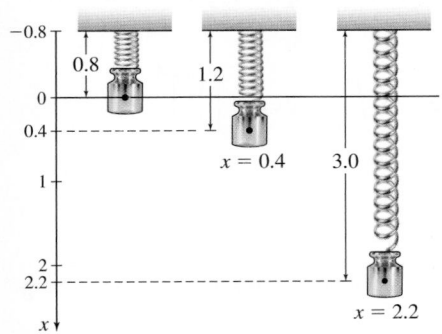

Figure 70

$$W = \int_0^{2.2} F(x) \, dx = -5 \int_0^{2.2} x \, dx = -5 \left[\frac{x^2}{2} \right]_0^{2.2} = -\frac{5}{2}(4.84) = -12.1 \, \text{J}$$
$$\underset{F(x) = -5x}{\uparrow}$$

EXAMPLE 3   **Finding the Work Done by a Spring Force**

Suppose an 0.8 m-long spring has a spring constant of $k = 5$ N/m.

(a) What spring force is required to compress the spring from its equilibrium position to a length of 0.5 m?

(b) How much work is done by the spring force when compressing the spring from equilibrium to a length of 0.5 m?

(c) How much work is done by the spring force when compressing the spring from 1.2 to 0.5 m?

(d) How much work is done by the spring force when compressing the spring from 1 to 0.6 m?

Solution

Begin by positioning an axis parallel to the spring and placing the origin at the free end of the spring in equilibrium. See Figure 71.

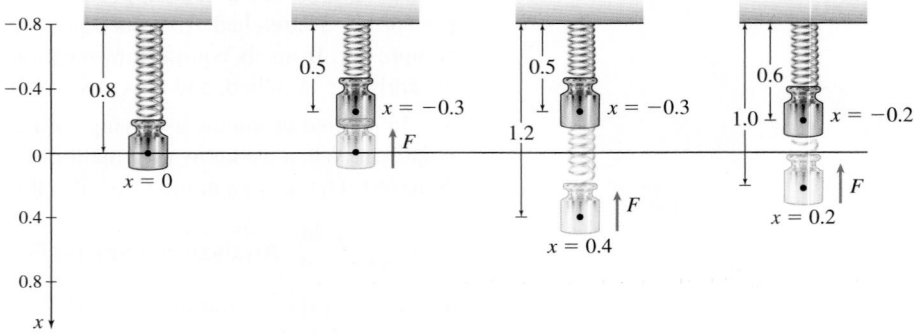

Figure 71

(a) By Hooke's Law, the spring force is $F = -5x$. When the spring is compressed to a length of 0.5 m, then $x = -0.3$. The spring force F required to compress the spring to 0.5 m is

$$F = -kx = -5(-0.3) = 1.5 \text{ N}$$

(b) The work W done by the spring force F when compressing the spring from equilibrium $(x = 0)$ to 0.5 m $(x = -0.3)$ is

$$W = \int_0^{-0.3} F(x)\, dx = \underset{\underset{F(x)\,=\,-5x}{\uparrow}}{\int_0^{-0.3} (-5x)\, dx} = 5 \int_{-0.3}^0 x\, dx = 5\left[\frac{x^2}{2}\right]_{-0.3}^0$$

$$= 0 - \frac{5\,(-0.3)^2}{2} = -0.225 \text{ J}$$

(c) The work W done by the spring force F when compressing the spring from 1.2 m $(x = 0.4)$ to 0.5 m $(x = -0.3)$ is

$$W = \int_{0.4}^{-0.3} (-5x)\, dx = 5 \int_{-0.3}^{0.4} x\, dx = 5\left[\frac{x^2}{2}\right]_{-0.3}^{0.4} = \frac{5}{2}[0.4^2 - (-0.3)^2] = 0.175 \text{ J}$$

(d) The work W done when compressing the spring from 1 m $(x = 0.2)$ to 0.6 m $(x = -0.2)$ is

$$W = \int_{0.2}^{-0.2} (-5x)\, dx = 5 \int_{-0.2}^{0.2} x\, dx = 5\left[\frac{x^2}{2}\right]_{-0.2}^{0.2} = \frac{5}{2}[0.2^2 - (-0.2)^2] = 0 \text{ J}$$ ∎

Observe that the work W done by a spring force can be positive, negative, or zero. If the force brings the spring closer to its equilibrium position $(x = 0)$, then $W > 0$; if the force brings the spring away from its equilibrium position, then $W < 0$. If the spring ends up the same distance from the equilibrium, then no work is done; that is, $W = 0$.

NOW WORK Problem 15.

③ Find the Work Done to Pump a Liquid

Another example of work done by a variable force is found in the work needed to pump a liquid out of a tank. The idea used is that the work needed to lift an object a given distance is the product of the weight (force) of the object and the distance it is lifted, that is,

$$\text{Work} = (\text{Weight of object})(\text{Distance lifted})$$

EXAMPLE 4 Finding the Work Required to Pump Oil Out of a Tank

An oil tank in the shape of a right circular cylinder, with height 30 m and radius 5 m, is two-thirds full of oil. How much work is required to pump all the oil over the top of the tank?

Solution

We position an x-axis parallel to the side of the cylinder with the origin of the axis even with the bottom of the tank and $x = 30$ at the top of the tank, as illustrated in Figure 72.

The work required to pump the oil over the top is the product of the weight of the oil and its distance from the top of the tank. The weight of the oil equals $\rho g V$ newtons, where $\rho \approx 820$ kg/m³ is the mass density of petroleum (mass per unit volume, a constant that depends on the type of liquid involved), $g \approx 9.8$ m/s² (the acceleration due to gravity), and V is the volume of the liquid to be moved. The weight of the oil is

$$\text{Weight} = \rho g V \approx (820)(9.8)V = 8036V \text{ N}$$

The oil fills the tank from $x = 0$ m to $x = 20$ m. We partition the interval $[0, 20]$ into n subintervals, each of width $\Delta x = \dfrac{20}{n}$. Consider the oil in the ith subinterval as a thin layer of thickness Δx. Now choose a number u_i in the ith subinterval. Then

Volume V_i of ith layer = (Area of layer)(Thickness) = $\pi r^2 \Delta x = 25\pi \Delta x$

Weight of ith layer = $\rho g V_i \approx (8036)(25\pi \Delta x) = 200{,}900\pi \Delta x$

Distance ith layer is lifted = $30 - u_i$

Work W_i done in lifting ith layer $\approx (200{,}900\pi \Delta x)(30 - u_i)$

The layers of oil are from 0 to 20 m. So, the work W required to pump all the oil over the top is

$$W = \int_0^{20} 200{,}900\pi (30 - x)\, dx = 200{,}900\pi \int_0^{20} (30 - x)\, dx$$

$$= (200{,}900\,\pi)\left[30x - \frac{x^2}{2}\right]_0^{20}$$

$$= (200{,}900\,\pi)(600 - 200) = 80{,}360{,}000\pi \approx 2.525 \times 10^8 \text{ J} \qquad \blacksquare$$

The choice of $x = 0$ for the position of the bottom of the tank is one of convenience. The work will be the same for other choices of x for the bottom of the tank.

NOW WORK Problem 17(a).

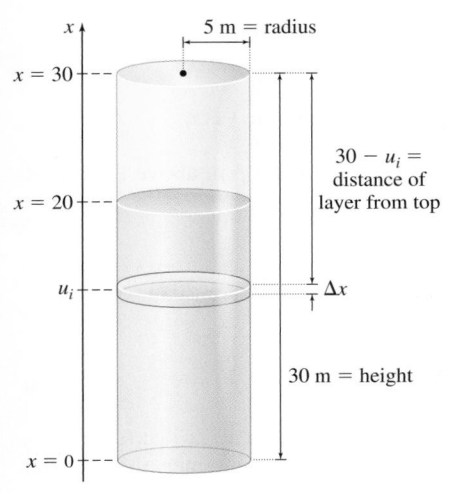

Figure 72

(figure labels: x; 5 m = radius; $x = 30$; $x = 20$; u_i; $x = 0$; $30 - u_i =$ distance of layer from top; Δx; 30 m = height)

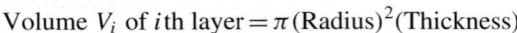

EXAMPLE 5   Finding the Work Required to Pump Water Out of a Tank

A tank in the shape of a hemisphere of radius 2 m is full of water. How much work is required to pump all the water to a level 3 m above the tank? (Density $\rho = 1000 \text{ kg/m}^3$)

Solution

We position an x-axis so the bottom of the tank is at $x = 0$ and the top of the tank is at $x = 2$, as illustrated in Figure 73.

The work required to pump the water to a level 3 m above the top of the tank depends on the weight of the water and its distance from a level 3 m above the tank. The water fills the container from $x = 0$ to $x = 2$.

Partition the interval $[0, 2]$ into n subintervals, each of width $\Delta x = \dfrac{2}{n}$, and choose a number u_i in each subinterval. Now think of the water in the tank as n circular layers, each of thickness Δx. As Figure 74 illustrates, the radius of the circular layer u_i meters from the bottom of the tank is $\sqrt{4u_i - u_i^2}$. Then

$$\text{Volume } V_i \text{ of } i\text{th layer} = \pi (\text{Radius})^2 (\text{Thickness})$$

$$= \pi \left(\sqrt{4u_i - u_i^2} \right)^2 \Delta x = \pi \left(4u_i - u_i^2 \right) \Delta x$$

The density of water is $\rho = 1000 \text{ kg/m}^3$, so

$$\text{Weight of } i\text{th layer} = \rho g V_i = (1000)(9.8)\pi \left(4u_i - u_i^2 \right) \Delta x$$

$$\text{Distance } i\text{th layer is lifted} = 5 - u_i$$

$$\text{Work done in lifting } i\text{th layer} = 9800\,\pi \left(4u_i - u_i^2 \right)(5 - u_i)\, \Delta x$$

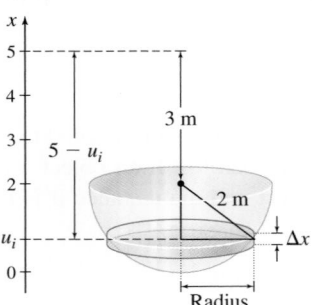

Figure 73

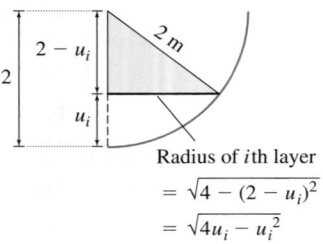

Figure 74

The work W required to lift all the water from the tank to a level 3 m above the top of the tank is given by

$$W = \int_0^2 9800\,\pi(4x - x^2)(5 - x)\,dx = 9800\pi \int_0^2 (x^3 - 9x^2 + 20x)\,dx$$

$$= 9800\,\pi \left[\frac{x^4}{4} - 3x^3 + 10x^2 \right]_0^2 = 196{,}000\pi \approx 615{,}752 \text{ J} \qquad\blacksquare$$

In general, to find the work required to pump liquid from a container, think of the liquid as thin layers of thickness Δx and area $A(x)$. If the liquid is to be lifted a height h above the bottom of the tank, the work required is

$$\boxed{W = \int_a^b \rho g A(x)(h - x)\,dx}$$

where ρ is the mass density of the liquid, g is the acceleration due to gravity, and the liquid to be lifted fills the container from $x = a$ to $x = b$. See Figure 75.

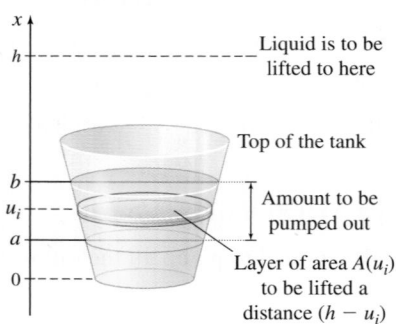

Figure 75

NOW WORK   Problems 17(b) and 31.

Application: Gravitational Force

Denis Kozlenko/iStockphoto

In the mid-1600s, Isaac Newton proposed his theory of gravity. **Newton's Law of Universal Gravitation** states that every body in the universe attracts every other body; he called this force **gravitation**. Newton theorized that the gravitational force F attracting two bodies is proportional to the masses of both bodies and inversely proportional to the square of the distance x between them. That is,

$$F = F(x) = G\frac{m_1 m_2}{x^2}$$

ORIGINS Henry Cavendish (1731-1810) was an English chemist and physicist. Known for his precision and accuracy, Cavendish calculated the density of Earth by measuring the force of the attraction between pairs of lead balls. An immediate result of his experiments was the first calculation of the gravitational constant G.

In a paper appearing in *Science* (2007, January 5), *315*(5808), 74-77, the measurement was reevaluated using atom interferometry to be

$$G = 6.693 \times 10^{-11}\,\frac{N\,m^2}{kg^2}$$

where G is the **gravitational constant**. The widely accepted value of G,

$$G = 6.67 \times 10^{-11}\,\frac{Nm^2}{kg^2}$$

is a result of the work done by Henry Cavendish in 1798.

One of the conclusions of Newton's Law of Universal Gravitation is that the force required to move an object, say, a rocket of mass m kilograms that is at a point x meters above the center of Earth, is $F(x) = G\,\dfrac{Mm}{x^2}$, where M kilograms is the mass of Earth. Then the work required to move an object of mass m from the surface of Earth to a distance r meters from the center of Earth (where R meters is the radius of Earth) is

$$W = \int_R^r \left[G\,\frac{Mm}{x^2} \right] dx = -G\left[\frac{Mm}{x}\right]_R^r = G\,\frac{Mm}{R} - G\,\frac{Mm}{r}$$

$$= GMm\left(\frac{1}{R} - \frac{1}{r}\right) \text{ J}$$

Physicists know that $GM = gR^2$, where $g \approx 9.8\,\text{m/s}^2$ is the acceleration due to gravity of Earth and $R \approx 6.37 \times 10^6$ m. If d is the distance the object is to be moved above the surface of Earth, then $r = R + d$. So, the work required to move an object of mass m to a distance d meters above the surface of Earth is

$$W = gRm\left(1 - \frac{R}{R+d}\right) \tag{1}$$

Observe that although the distance d may be extremely large, the work W required to move an object d meters will never be greater than $gRm \approx (6.24 \times 10^7)m$ J.

8.7 Assess Your Understanding

Concepts and Vocabulary

1. *True or False* Work is the energy transferred to or from an object by a force acting on the object.

2. The work W done by a constant force F in moving an object a distance x along a straight line in the direction of F is _____.

3. A unit of work is called a _____ in SI units and a _____ in the customary U.S. system of units.

4. The work W done by a continuously varying force $F = F(x)$ acting on an object, which moves the object along a straight line in the direction of F from $x = a$ to $x = b$, is given by the definite integral _____.

5. A spring is said to be in _____ when it is neither extended nor compressed.

6. *True or False* The force F required to extend or compress a spring x units is $F = -kx$, where k is the spring constant.

7. *True or False* The mass density ρ of a fluid is defined as mass per unit volume (kg/m^3) and is a constant that depends on the type of fluid.

8. *True or False* Newton's Law of Universal Gravitation affirms that everybody in the universe attracts every other body, and that the force F attracting two bodies is proportional to the product of the masses of both bodies and inversely proportional to the square of the distance between them.

Skill Building

9. How much work is done by a variable force $F(x) = (40 - x)$ N that moves an object along a straight line in the direction of F from $x = 5$ m to $x = 20$ m?

10. How much work is done by a variable force $F(x) = \dfrac{1}{x}$ N that moves an object along a straight line in the direction of F from $x = 1$ m to $x = 2$ m?

11. A 40-m chain weighing 3 kg/m hangs over the side of a bridge. How much work is done by a winch that winds the entire chain in?

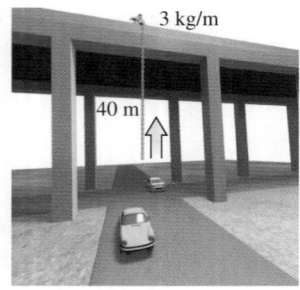

12. A 120-ft chain weighing 240 lb is dangling from the roof of an apartment building. How much work is done in pulling the entire chain up to the roof?

13. A force of 3 N is required to keep a spring extended $\dfrac{1}{4}$ m beyond its equilibrium length. What is its spring constant?

14. A force of 6 lb is required to keep a spring compressed to $\dfrac{1}{2}$ ft shorter than its equilibrium length. What is its spring constant?

15. A spring with spring constant $k = 5$ N/m has an equilibrium length of 0.8 m. How much work is required to stretch the spring to 1.4 m?

16. A spring with spring constant $k = 0.3$ N/m has an equilibrium length of 1.2 m. How much work is required to compress the spring to 1 m?

17. Pumping Water Out of a Pool   A swimming pool in the shape of a right circular cylinder, with height 4 ft and radius 12 ft, is full of water. See the figure.

 (a) How much work is required to pump all the water over the top of the pool?

 (b) How much work is required to pump all the water to a level 5 ft above the pool?

Note: The weight of water is 62.42 lb/ft^3.

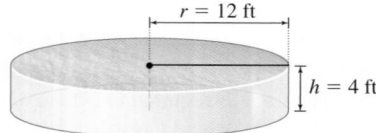

18. Pumping Gasoline Out of a Tank   A storage tank in the shape of a right circular cylinder, with height 10 m and radius 8 m, is full of gasoline.

 (a) How much work is required to pump all the gasoline over the top of the tank?

 (b) How much is required to pump all the gasoline to a level 5 m above the tank?

Note: The density of gasoline is $\rho = 720$ kg/m^3.

19. Pumping Corn Slurry   A container in the shape of an inverted pyramid with a square base of 2 m by 2 m and height of 5 m is filled to a depth of 4 m with corn slurry.

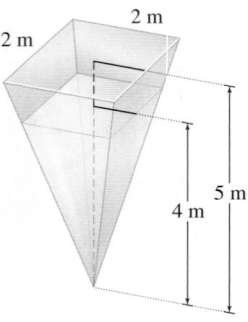

 (a) How much work is required to pump all the slurry over the top of the container?

 (b) How much work is required to pump all the slurry to a level 3 m above the container?

Note: The density of corn slurry is $\rho = 17.9$ kg/m^3.

20. Pumping Olive Oil from a Vat   A vat in the shape of an inverted pyramid with a rectangular base that measures 2 m by 0.5 m and height that measures 4 m is filled to a depth of 2 m with olive oil.

 (a) How much work is required to pump all the olive oil over the top of the vat?

 (b) How much is required to pump all the olive oil to a level 2 m above the vat?

Note: The density of olive oil is $\rho = 0.9$ g/cm^3.

Applications and Extensions

21. Work to Lift an Elevator   How much work is required if six cables, each weighing 0.36 lb/in., lift a 10,000-lb elevator 400 ft? Assume the cables work together and equally share the weight of the elevator.

22. Work by Gravity   A cable with a uniform linear mass density of 9 kg/m is being unwound from a cylindrical drum. If 15 m are already unwound, what is the work done by gravity in unwinding another 60 m of the cable?

23. Work to Lift a Bucket and Chain   A uniform chain 10 m long and with mass 20 kg is hanging from the top of a building 10 m high. If a bucket filled with cement of mass 75 kg is attached to the end of the chain, how much work is required to pull the bucket and chain to the top of the building?

24. Work to Lift a Bucket and Chain   In Problem 23, if a uniform chain 10 m long and with mass 15 kg is used instead, how much work is required to pull the bucket and chain to the top of the building?

25. Work of a Spring   A spring, whose equilibrium length is 1 m, extends to a length of 3 m when a force of 3 N is applied. Find the work needed to extend the spring to a length of 2 m from its equilibrium length.

26. Work of a Spring   A spring, whose equilibrium length is 2 m, is compressed to a length of $\dfrac{1}{2}$ m when a force of 10 N is applied. Find the work required to compress the spring to a length of 1 m.

27. Work of a Spring   A spring, whose equilibrium length is 4 ft, extends to a length of 8 ft when a force of 2 lb is applied. If 9 ft lb of work is required to extend this spring from its equilibrium position, what is its total length?

28. Work of a Spring   If 8 ft lb of work is used on the spring in Problem 27, how far is it extended?

29. Work to Pump Water   A full water tank in the shape of an inverted right circular cone is 8 m across the top and 4 m high. How much work is required to pump all the water over the top of the tank? (The density of water is 1000 kg/m³.)

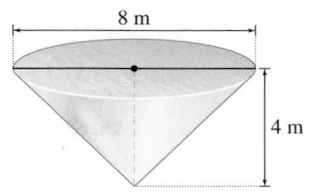

8 m

4 m

30. Work to Pump Water   If the surface of the water in the tank of Problem 29 is 2 m below the top of the tank, how much work is required to pump all the water over the top of the tank?

31. Work to Pump Water   A tank in the shape of a hemispherical bowl of radius 4 m is filled with water to a depth of 2 m. How much work is required to pump all the water over the top of the tank?

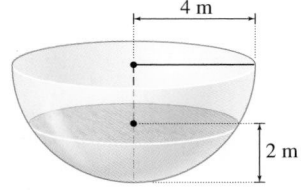

4 m

2 m

32. Work to Pump Water   If the tank in Problem 31 is completely filled with water, how much work is required to pump all the water to a height 2 m above the tank?

33. Work to Pump Water   A cylindrical tank, 4 m in diameter and 6 m high, is full of water. (The density of water is 1000 kg/m³.)

(a) How much work is required to pump half the water over the top of the tank?

(b) How much work is required to pump half the water to a level 3 m above the tank?

34. Work to Pump Water   A swimming pool in the shape of a rectangular parallelepiped 6 ft deep, 30 ft long, and 20 ft wide is filled with water to a depth of 5 ft. How much work is required to pump all the water over the top? (The weight of water is $\rho g = 62.42$ lb/ft³.)

35. Work to Pump Water   A 1 hp motor can do 550 ft lb of work per second. The motor is used to pump the water out of a swimming pool in the shape of a rectangular parallelepiped 5 ft deep, 25 ft long, and 15 ft wide. How long does it take for the pump to empty the pool if the pool is filled to a depth of 4 ft? (The weight of water is 62.42 lb/ft³.)

Newton's Law of Universal Gravitation   *In Problems 36 and 37, use formula (1) on page 691.*

36. The minimum energy required to move an object of mass 30 kg a distance 30 km above the surface of Earth is equal to the work required to accomplish this. Find the work required. (Earth's radius R is approximately 6370 km.)

37. The minimum energy required to move a rocket of mass 1000 kg a distance of 800 km above Earth's surface is equal to the work required to do this. Find the work required.

38. Coulomb's Law   By **Coulomb's Law**, a positive charge m of electricity repels a unit of positive charge at a distance x with the force $\dfrac{m}{x^2}$. What is the work done when the unit charge is carried from $x = 2a$ to $x = a$, $a > 0$?

39. Work to Lift a Leaky Load   In raising a leaky bucket from the bottom of a well 25 ft deep, one-fourth of the water is lost. If the bucket weighs 1.5 lb, the water in the bucket at the start weighs 20 lb, and the amount that has leaked out is assumed to be proportional to the distance the bucket is lifted, find the work done in raising the bucket. Ignore the weight of the rope used to lift the bucket.

40. Work of a Spring   The spring constant on a bumping post in a freight yard is 300,000 N/m. Find the work done in compressing the spring 0.1 m.

41. Work to Pump Water   A container is formed by revolving the region bounded by the graph of $y = x^2$, and the x-axis, $0 \leq x \leq 2$, about the y-axis. How much work is required to fill the container with water from a source 2 units below the x-axis by pumping through a hole in the bottom of the container?

Challenge Problems

42. Work to Move a Piston   The force exerted by a gas in a cylinder on a piston whose area is A, is given by $F = pA$, where p is the force per unit area, or **pressure**. The work W in displacing the piston from x_1 to x_2 is

$$W = \int_{x_1}^{x_2} F\, dx = \int_{x_1}^{x_2} pA\, dx = \int_{V_1}^{V_2} p\, dV$$

where dV is the accompanying infinitesimal change of volume of the gas.

(a) During expansion of a gas at constant temperature (isothermal), the pressure p depends on the volume V according to the relation

$$p = \frac{nRT}{V}$$

where n and R are constants and T is the constant temperature. Calculate the work in expanding the gas isothermally from volume V_1 to volume V_2.

(b) During expansion of a gas at constant entropy (adiabatic), the pressure depends on the volume according to the relation

$$p = \frac{K}{V^\gamma}$$

where K and $\gamma \neq 1$ are constants. Calculate the work in expanding the gas adiabatically from V_1 to V_2.

Expanding Gases   *Problems 43 and 44 use the following discussion. The pressure p (in pounds/square inch, lb/in²) and the volume V (in cubic inches) of an adiabatic expansion of a gas are related by $pV^k = c$, where k and c are constants that depend on the gas. If the gas expands from $V = a$ to $V = b$, the work done (in inch-pounds) is*

$$W = \int_a^b p\, dV$$

43. Work of Expanding Gas   The pressure p of 1 lb of a gas is 100 lb/in.² and the volume V is 2 ft³. Find the work done by the gas in expanding to double its volume according to the law $pV^{1.4} = c$.

44. Work of Expanding Gas   The pressure p and volume V of a certain gas obey the law $pV^{1.2} = 120$ (in inch-pounds). Find the work done when the gas expands from $V = 2.4$ to $V = 4.6$ in.³.

CHAPTER 8 PROJECT

Determining the Amount of Concrete Needed for a Cooling Tower

This Project can be done individually or as part of a team.

A common design for cooling towers is modeled by a branch of a hyperbola rotated about an axis. The design is used because of its strength and efficiency. Not only is the shape stronger than either a cone or cylinder, it takes less material to build. This shape also maximizes the natural upward draft of hot air without the need for fans.*

How much concrete is needed to build a cooling tower that has a base 460 ft wide that is 442 ft below the vertex if the top of the tower is 310 ft wide and 123 ft above the vertex? Assume the walls are a constant 5 in. $= 0.42$ ft thick. See Figure 76.

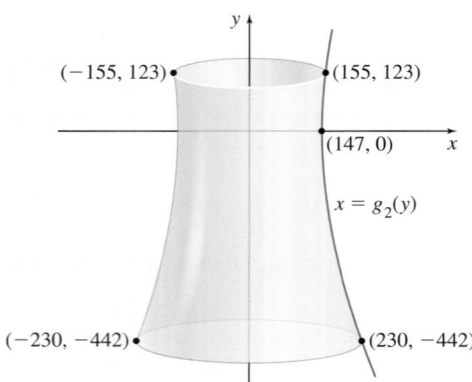

Figure 76

1. Show that the equation of the right branch of the hyperbola in Figure 76 is given by

$$x = g_2(y) = \sqrt{147^2 + 0.16y^2}$$

Verify that the point $(230, -442)$ satisfies $x = g_2(y)$.
Hint: Begin with the equation of a hyperbola

$$\frac{x^2}{a^2} - \frac{y^2}{b^2} = 1$$

Use the points $(155, 123)$ and $(147, 0)$ to find a and b.

---

*The design of an unsupported, reinforced concrete hyperbolic cooling tower was patented in 1918 (UK patent 198,863) by Frederic von Herson and Gerard Kupeers of the Netherlands.

2. Find the volume of the solid of revolution obtained by revolving the area enclosed by $x = g_2(y)$ from $y = -442$ to $y = 123$ about the y-axis.

3. Using the fact that the walls are 0.42 ft thick, show that the equation of the interior hyperbola is given by

$$x = g_1(y) = \sqrt{146.58^2 + 0.16y^2}$$

Verify that the point $(220.58, -442)$ satisfies $g_1(y)$. See Figure 77.

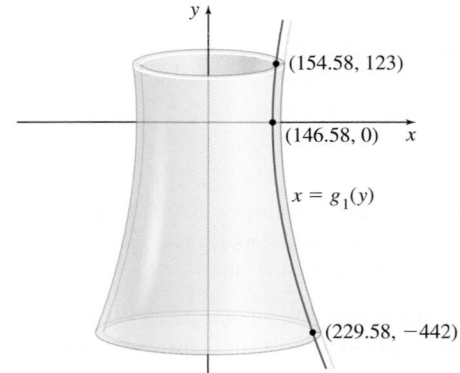

Figure 77

4. Find the volume of the solid of revolution obtained by revolving the area enclosed by $x = g_1(y)$ from $y = -442$ to $y = 123$ about the y-axis.

5. Now determine the volume (in cubic feet) of concrete required for the cooling tower.

6. Do an Internet search to find the cost of concrete. How much will the concrete used in the cooling tower cost?

7. Write a report for an audience that is not familiar with cooling towers, summarizing the findings in Problems 1–6.

Chapter Review

THINGS TO KNOW

8.1 Applications Involving Motion Along a Line

Net displacement: $\int_a^b v(t)\,dt$ (p. 621)

Total distance traveled: $\int_a^b |v(t)|\,dt$ (p. 621)

Average velocity: $\dfrac{1}{b-a}\int_a^b v(t)\,dt$ (p. 623)

Average acceleration: $\dfrac{1}{t_2 - t_1}\int_{t_1}^{t_2} a(t)\,dt$ (p. 623)

8.2 Area Between Graphs

*Area A **between two graphs:***

- (Partitioning the x-axis) $A = \int_a^b [f(x) - g(x)]\,dx$, where $f(x) \ge g(x)$ for all x in the interval $[a, b]$ (p. 630)
- (Partitioning the y-axis) $A = \int_c^d [f(y) - g(y)]\,dy$, where $f(y) \ge g(y)$ for all y in the interval $[c, d]$ (p. 633)

8.3 Volume of a Solid of Revolution: Disks and Washers

Volume V of a solid of revolution: The disk method:

- $V = \pi \int_a^b [f(x)]^2\, dx$, where the region is revolved about the x-axis (p. 640)
- $V = \pi \int_c^d [g(y)]^2\, dy$, where the region is revolved about the y-axis (p. 643)

Volume V of a solid of revolution: The washer method:

- $V = \pi \int_a^b \{[f(x)]^2 - [g(x)]^2\}\, dx$, where the region is revolved about the x-axis (p. 645)
- $V = \pi \int_c^d \{[f(y)]^2 - [g(y)]^2\}\, dy$, where the region is revolved about the y-axis (Example 6, p. 646)

8.4 Volume of a Solid of Revolution: Cylindrical Shells

- Cylindrical shell: The solid region between two concentric cylinders (p. 654)

Volume V of a solid of revolution: The shell method:

- $V = 2\pi \int_a^b x f(x)\, dx$, where the region is revolved about the y-axis (p. 655)
- $V = 2\pi \int_c^d y g(y)\, dy$, where the region is revolved about the x-axis (p. 659)

8.5 Volume of a Solid with a Known Cross Section

Volume V of a solid:

- $V = \int_a^b A(x)\, dx$, where the area $A = A(x)$ of the cross section of a solid is known and is continuous on $[a, b]$ (p. 665)

8.6 Arc Length; Surface Area of a Solid of Revolution

Arc length formulas:

- $s = \int_a^b \sqrt{1 + [f'(x)]^2}\, dx$ (p. 675)
- $s = \int_c^d \sqrt{1 + [g'(y)]^2}\, dy$ (p. 678)

Surface area formula:

- $S = 2\pi \int_a^b f(x)\sqrt{1 + [f'(x)]^2}\,dx$ (p. 681)

8.7 Work

- **Work** is the energy transferred to or from an object by a force acting on the object. (p. 685)

Work formulas:

- F is a continuously varying force that moves an object from a to b: $W = \int_a^b F(x)\, dx$ (p. 686)
- F is a spring force: $F(x) = -kx$ (Hooke's Law) (p. 687)
- W is the work required to pump a liquid out of a container: $W = \int_a^b \rho g A(x)(h-x)dx$ (p. 690)
- F is the attraction between two bodies:

$$F(x) = G\frac{m_1 m_2}{x^2}\quad \text{(p. 691)}$$

OBJECTIVES

Section	You should be able to …	Examples	Review Exercises	AP® Review Problems
8.1	1 Use an integral to find the total distance traveled and the net displacement of an object moving along a line (p. 621)	1, 2	33	9, 15, 16
	2 Determine the average velocity or average acceleration of an object moving along a line (p. 623)	3	35	15, 16
	3 Connect position, velocity, and acceleration of an object moving along a line (p. 624)	4, 5	32	15
8.2	1 Find the area between the graphs of two functions by partitioning the x-axis (p. 629)	1–4	1–5, 19	2, 10, 12, 14
	2 Find the area between the graphs of two functions by partitioning the y-axis (p. 632)	5–6	3, 4, 19	10
8.3	1 Use the disk method to find the volume of a solid formed by revolving a region about the x-axis (p. 640)	1, 2, 3	9, 11	11,14
	2 Use the disk method to find the volume of a solid formed by revolving a region about the y-axis (p. 642)	4	8	8
	3 Use the washer method to find the volume of a solid formed by revolving a region about the x-axis (p. 644)	5	6, 14, 30(a)	5, 7, 10, 13
	4 Use the washer method to find the volume of a solid formed by revolving a region about the y-axis (p. 646)	6	13, 31	6, 10
	5 Find the volume of a solid formed by revolving a region about a line parallel to a coordinate axis (p. 647)	7, 8	10, 12, 15, 30(c), (d)	4, 11
8.4	1 Use the shell method to find the volume of a solid formed by revolving a region about the y-axis (p. 654)	1, 2	7, 8, 13, 31, 30(b)	
	2 Use the shell method to find the volume of a solid formed by revolving a region about the x-axis (p. 659)	3	9, 21	
	3 Use the shell method to find the volume of a solid formed by revolving a region about a line parallel to a coordinate axis (p. 661)	4	12, 15, 30(e), (f)	

REVIEW EXERCISES

Area   *In Problems 1–5, find the area A of the region bounded by the graphs of the given equations.*

1. $y = e^x$, $x = 0$, $y = 4$ **2.** $y = x^2$, $y = 18 - x^2$

3. $x = 2y^2$, $x = 2$ **4.** $y = \dfrac{1}{x}$, $x + y = 4$

5. $y = 4 - x^2$, $y = 3x$

Volume of a Solid of Revolution   *In Problems 6–15, find the volume of the solid of revolution generated by revolving the region bounded by the graphs of the given equations about the given line.*

6. $y = x^2$, $y = 4x - x^2$; about the x-axis

7. $y = x^2 - 5x + 6$, $y = 0$; about the y-axis

8. $x = y^2 - 4$, $x = 0$; about the y-axis

9. $xy = 1$, $x = 1$, $x = 2$, $y = 0$; about the x-axis

10. $y = x^2 - 4$, $y = 0$; about the line $y = -4$

11. $y = 4x - x^2$ and the x-axis; about the x-axis

12. $y^2 = 8x$, $y \geq 0$, and $x = 2$; about the line $x = 2$

13. $y = \dfrac{x^3}{2}$, $y = 0$, $x = 2$; about the y-axis

14. $y = e^x$, $y = 1$, $x = 1$; about the x-axis

15. $y^2 = x^3$, $y = 8$, $x = 0$; about the line $x = 4$

Arc Length   *In Problems 16–18, find the arc length of each graph.*

16. $y = x^{3/2} + 4$ from $x = 2$ to $x = 5$

17. $y = \dfrac{x^3}{6} + \dfrac{1}{2x}$ from $x = 2$ to $x = 6$

18. $2y^3 = x^2$ from $y = 0$ to $y = 2$

19. Area of a Triangle   Use integration to find the area of the triangle formed by the lines $x - y + 1 = 0$, $3x + y - 13 = 0$, and $x + 3y - 7 = 0$.

20. Volume   A solid has a circular base of radius 4 units. Cross sections perpendicular to a fixed diameter are equilateral triangles. Find its volume.

21. Volume   Find the volume of the solid generated by revolving the region bounded by the graph of $4x^2 + 9y^2 = 36$ in the first quadrant about the x-axis.

22. Volume of a Cone   Find the volume of an elliptical cone with base $\dfrac{x^2}{4} + y^2 = 1$ and height 5.

Hint: The area A of an ellipse with semi-axes a and b is $A = \pi ab$.

23. Volume   The base of a solid is bounded by the graphs of $4x + 5y = 20$, $x = 0$, $y = 0$. Every cross section perpendicular to the x-axis is a semicircle. Find the volume of the solid.

24. Arc Length   Find the point P on the graph of $y = \dfrac{2}{3}x^{3/2}$ to the right of the y-axis so that the length of the graph from $(0, 0)$ to P is $\dfrac{52}{3}$.

25. Work   Find the work done in raising an 800-lb anchor 150 ft with a chain weighing 20 lb/ft.

26. Work   Find the work done in raising a container of 1000 kg of silver ore from a mine 1200 m deep with a cable weighing 3 kg/m.

27. Work Pumping Water   A hemispherical water tank has a diameter of 12 m. It is filled to a depth of 4 m. How much work is done in pumping all the water over the edge? (Use $\rho = 1000\,\text{kg/m}^3$.)

28. Work of a Spring   A spring with an unstretched length of 0.6 m requires a force of 4 N to stretch it to 0.8 m. How much work is done in stretching it to 1.4 m?

29. Work of a Spring   Find the unstretched length of a spring if the work required to stretch the spring from 1.0 to 1.4 m is half the work required to stretch it from 1.2 to 1.8 m.

30. Volume of a Solid of Revolution   Find the volume generated when the region bounded by the graphs of $y = 3\sqrt{x}$ and $y = -x^2 + 6x - 2$ is revolved about each of the following lines:

   (a) $y = 0$ **(b)** $x = 0$ **(c)** $y = -2$

   (d) $y = 8$ **(e)** $x = -3$ **(f)** $x = 5$

31. Volume of a Solid of Revolution   Find the volume generated when the triangular region bounded by the lines $x = 3$, $y = 0$, and $2x + y - 12 = 0$ is revolved about the y-axis.

32. Motion Along a Line An object moves along a line so that its position s, in meters, from the origin at time t, $0 \leq t \leq 4$ seconds, is given by the function $s = s(t) = t^4 - 2t^2$.

(a) Find the velocity v of the object.

(b) When is the velocity increasing and when is the velocity decreasing?

(c) Find the acceleration of the object.

(d) Determine when the speed of the object is increasing and when the speed is decreasing.

33. Motion Along a Line An object moves along a line with velocity $v = v(t) = e^{t/2} - 2t$ m/s.

(a) Find the total distance travel by the object from $t = 0$ to $t = 6$ s.

(b) Find the net displacement of the object from $t = 0$ to $t = 6$ s.

34. Surface Area of a Solid of Revolution Find the surface area of the solid formed by revolving the region bounded by the graph of $f(x) = 4\sqrt{2x}$ and the lines $x = 1$ and $x = 4$ about the x-axis.

35. Motion Along a Line An object moves along a line with velocity $v = v(t) = t^2 - 6t + 5$ ft/min.

(a) If at $t = 0$ min the object is 2 ft to the right of the origin, what is the object's position at $t = 3$ min?

(b) What is the average velocity of the object from $0 \leq t \leq 6$ m?

36. Surface Area of a Solid of Revolution Find the surface area of the solid formed by revolving the region bounded by the graph of $f(x) = \dfrac{5}{3}x^3$; $x = 0$; and the line $x = 4$ about the x-axis.

Break It Down

Preparing for the AP® Exam

Let's take a closer look at part (b) of **AP® Practice Problem 13** *from Section 8.3 on page 653.*

13. The graphs of the functions $f(x) = -x^2 + 4x$ and $g(x) = 3x^2 - 12x + 12$ are shown in the figure.

(b) Write, but do not evaluate, an integral that gives the volume of the solid of revolution generated when the regions bounded by the graphs of f and g is revolved about the line $y = -2$.

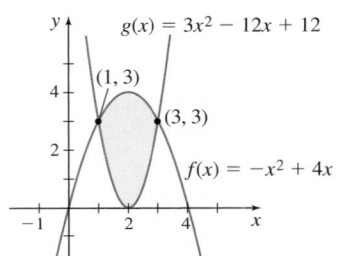

Step 1	Identify the underlying structure and related concepts.	We are asked to find an integral that gives the volume of the solid of revolution generated when the region bounded by two graphs is revolved about a line parallel to a coordinate axis.
Step 2	Determine the appropriate math rule or procedure.	Since the region consists of two graphs, we use the Washer Method. Since the axis of revolution is parallel to the x-axis, the partition will be along the x-axis. The general form of a solid of revolution using washers is $$V = \pi \int_a^b \left[(\text{outside radius})^2 - (\text{inside radius})^2 \right] dx.$$ It is helpful to draw a picture showing the outer radius, the inner radius, and the axis of revolution $y = -2$.
Step 3	Apply the math rule or procedure.	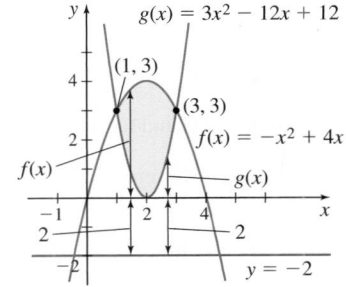 From the drawing, we see the outer radius is $f(x) + 2$ and the inner radius is $g(x) + 2$. The graphs of f and g intersect when $x = 1$ and $x = 3$, so $a = 1$ and $b = 3$.
Step 4	Clearly communicate your answer.	The integral that gives the volume of the solid of revolution generated by revolving the region bounded by the graphs of f and g about the line $y = -2$ is $$\pi \int_1^3 \left[(f(x) + 2)^2 - (g(x) + 2)^2 \right] dx$$ **TIP** When the AP® exam names functions, you can use these function names in your answers. You do not need to write $$\pi \int_1^3 \left[(x^2 + 4x + 2)^2 - (3x^2 - 12x + 12 + 2)^2 \right] dx$$

AP® Review Problems: Chapter 8

Preparing for the AP® **Exam**

Multiple-Choice Questions

1. The base of a solid is the region in the first quadrant bounded by the lines $y = 2x$, $x = 1$, and $x = 4$. Every cross section of the solid perpendicular to the x-axis is a semicircle. What is the volume of the solid?

(A) $\dfrac{21}{2}\pi$ (B) $\dfrac{32}{3}\pi$ (C) 21π (D) 42π

2. The area of the region bounded by the graphs of $y = e^x$ and $y = 2x + 4$, $0 \le x \le 1$, is

(A) $4 - e$ (B) $5 - e$ (C) $6 - e$ (D) $e - 4$

3. The arc length of the graph of $y = \sqrt{x}$ from $x = 2$ to $x = 5$ is given by

(A) $\displaystyle\int_2^5 \sqrt{1 + \dfrac{1}{4x}}\, dx$ (B) $\displaystyle\int_2^5 \sqrt{1 + x}\, dx$

(C) $\displaystyle\int_2^5 \sqrt{1 - \dfrac{1}{4x}}\, dx$ (D) $\displaystyle\int_2^5 \sqrt{1 + \dfrac{x}{4}}\, dx$

4. The region in the first quadrant bounded by the x-axis, the graph of $y = x^2$, and the line $x = 2$ is revolved about the line $y = -1$. The volume of the resulting solid of revolution is

(A) $\dfrac{22}{5}\pi$ (B) $\dfrac{52}{5}\pi$ (C) $\dfrac{176}{15}\pi$ (D) $\dfrac{232}{5}\pi$

5. A solid of revolution is formed by revolving the region bounded by the graphs of $y = \sin x$, $y = \cos x$, $0 \le x \le \dfrac{\pi}{4}$, and the line $x = 0$ about the x-axis. The volume of the solid is

(A) $\dfrac{1}{4}\pi^2$ (B) $\dfrac{1}{2}\pi$ (C) $\dfrac{\sqrt{2}}{4}\pi$ (D) π

6. The region in the first quadrant bounded by the graph of $y = x^3$, the line $x = 2$, and the x-axis, is revolved about the y-axis. The volume of the solid of revolution is

(A) $\dfrac{32\pi}{5}$ (B) $\dfrac{64\pi}{5}$ (C) $\dfrac{256}{7}\pi$ (D) $\dfrac{65,536}{5}\pi$

7. The region in the first quadrant bounded by the graphs of $y = 4 - \dfrac{1}{2}x$, $y = \sqrt{x}$, and the x-axis is revolved about the x-axis. What is the volume of the solid of revolution that is generated?

(A) 26.180   (B) 41.888   (C) 58.643   (D) 100.531

8. Find the volume of the solid of revolution generated when the region in the first quadrant bounded by the graphs of $y = x^2 + 1$, $y = 5$, and the y-axis is revolved about the y-axis.

(A) 8π (B) $\dfrac{15\pi}{2}$ (C) 20π (D) $\dfrac{136\pi}{3}$

9. Find the area of the region bounded by the graphs of $y = \sin(3x)$ and $y = \cos x$, $0 \le t \le \dfrac{\pi}{2}$.

(A) 0.501   (B) 0.667   (C) 0.744   (D) 2.541

Free-Response Questions

10. The graphs of the functions

$$f(x) = \sqrt{x} \text{ and } g(x) = \frac{1}{2}x$$

are shown in the figure.

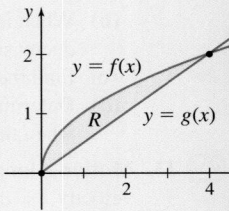

(a) Find the area of the region R bounded by the graph of f and g.

(b) Find the volume of the solid of revolution generated by revolving the region R about the x-axis.

(c) Find the volume of the solid of revolution generated by revolving the region R about the y-axis.

(d) The region R is the base of a solid. Find the volume of the solid if cross sections perpendicular to the base along the x-axis are squares.

11. The region R bounded by the graph of $f(x) = -2x^3 + 4x^2$ and the x-axis is shown.

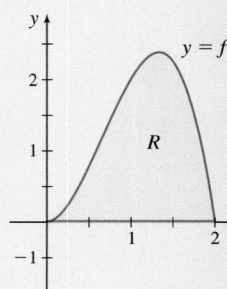

(a) Write, but do not evaluate, an integral to find the volume of the solid of revolution generated by revolving the region R about the x-axis.

(b) Write, but do not evaluate, an integral to find the volume of the solid of revolution generated by revolving the region R about the line $y = -1$.

12. The region R is bounded by the function f, the function $g(x) = x$, and the y-axis and is shown in the graph. Both f and g are twice differentiable for all real numbers.

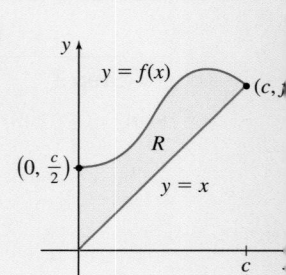

We are told the average value of f on the closed interval $[0, c]$ equals c.

(a) Find the area of the region R in terms of c.

(b) Find the volume V of the solid whose base is R in terms of c, if cross sections perpendicular to the x-axis are rectangles of height $f'(x)$.

(c) Set up, but do not evaluate, the expression (involving one or more integrals) that equals the perimeter of

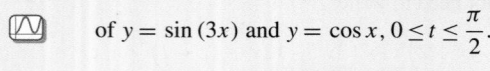

▶ **13.** A region is bounded by $y^2 = 9x$ and a line $y = 3x + b$.

    (a) If $b = -6$, find the area bounded by the graphs of $y = 3x - 6$ and $y^2 = 9x$.

    (b) If $b = 0$, find the volume V of the solid generated when the region R bounded by the graphs of $y = 3x$ and $y^2 = 9x$ is revolved about the x-axis.

    (c) Find the volume of the solid whose base is the region R from part (b) and has known cross sections that are squares.

▶ **14.** The region R bounded by the graph of $f(x) = -x^3 + 4x^2$ and x-axis is shown in the figure below.

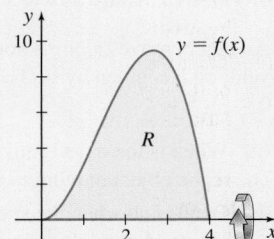

    (a) Find the area under the graph of f from 0 to 4.

    (b) Write, but do not evaluate, an integral for the arc length of the graph of f from $x = 0$ to $x = 4$

    (c) Write, but do not evaluate, an integral to find the volume of the solid of revolution generated by revolving the region R about the x-axis

▶ **15.** The position of an object moving along a horizontal line at time $t \geq 0$ is given by a differentiable function $x = x(t)$. The graph of the derivative $\dfrac{dx}{dt}$ is shown below.

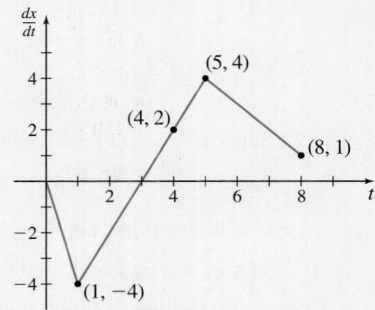

    (a) Find the net displacement of the object over the interval $0 \leq t \leq 8$.

    (b) What is the acceleration of the object at time $t = 4$?

    (c) If the position of the object at time $t = 0$ is -2, what is the position of the object at time $t = 4$?

    (d) What is the average velocity of the object from time $t = 0$ to $t = 8$?

Justify each of your answers.

▶ **16.** An object moves along a horizontal line and its velocity v, in centimeters per second, at time t, $0 \leq t \leq 2$, in seconds, is given by $v(t) = t^3 - 6t^2 + 12t$ cm/s.

    (a) Determine the total distance traveled by the object from 0 to 2 s.

    (b) Determine the average velocity of the object from $t = 0$ to $t = 2$ s.

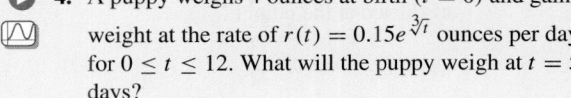

AP® Cumulative Review Problems: Chapters 1–8

Preparing for the AP® Exam

Multiple-Choice Questions

▶ **1.** $\displaystyle\lim_{x \to \infty} \frac{5x - 2}{\sqrt{3x^2 + 2}} =$

    (A) ∞ (B) 0   (C) $\dfrac{5\sqrt{3}}{3}$ (D) $\dfrac{-2}{\sqrt{3}}$

▶ **2.** Find $f'\left(\sqrt[3]{e}\right)$ if $f(x) = \left(\ln x^2\right)^3$.

    (A) $6\sqrt[3]{e}\ln\sqrt[3]{e}$ (B) $\dfrac{8}{3e^{\frac{1}{3}}}$ (C) $6\ln e^{\frac{2}{3}}$ (D) 4

▶ **3.** Suppose f is a twice-differentiable function and $f''(x) = (x - 1)^2 (x + 3)(x - 6)$. What are the x-coordinates of the points of inflection of the graph of f?

    (A) -3 and 1   (B) 1 and 6

    (C) -3 and 6   (D) -3, 1, and 6

▶ **4.** A puppy weighs 4 ounces at birth ($t = 0$) and gains weight at the rate of $r(t) = 0.15e^{\sqrt[3]{t}}$ ounces per day for $0 \leq t \leq 12$. What will the puppy weigh at $t = 5$ days?

    (A) 6.842 oz   (B) 5.846 oz   (C) 8 oz   (D) 7.456 oz

▶ **5.** $f(x) = \begin{cases} \dfrac{x^2 - 9}{x - 3} & \text{if } x \neq 3 \\ 5 & \text{if } x = 3 \end{cases}$

    Which statement about f is true?

    I. $\displaystyle\lim_{x \to 3} f(x)$ exists.

    II. f is continuous at $x = 3$.

    III. f is differentiable at $x = 3$.

    (A) I only   (B) II only   (C) III only   (D) I and II only

▶ **6.** Consider the differential equation $\dfrac{dy}{dx} = \dfrac{3x}{4y}$ where $y \neq 0$. Find an equation of the line tangent to the graph of the solution to the differential equation at the point $(2, 3)$.

(A) $y - 3 = \dfrac{3}{4}(x - 2)$ (B) $y - 2 = \dfrac{1}{2}(x - 3)$

(C) $y - 3 = \dfrac{x}{2} - 1$ (D) $y = \dfrac{-4x}{3} + \dfrac{17}{3}$

▶ **7.** Suppose $y = g(x)$ is a solution of the differential equation $\dfrac{dy}{dx} = 2x - 4y - 2$ with the initial condition $g(1) = k$, k is a constant. Euler's method, starting at $x = 1$ with an increment of $\dfrac{1}{2}$, gives the approximation $g(2) = 0$. Find the value of k.

(A) 2   (B) $\dfrac{1}{2}$ (C) $-\dfrac{1}{2}$ (D) $\dfrac{3}{4}$

▶ **8.** $\displaystyle \int \left(4e^{3x} + \dfrac{2}{x^2} \right) dx =$

(A) $\dfrac{4e^{3x}}{3} + \ln x^2 + C$ (B) $\dfrac{4e^{3x}}{3} - \dfrac{2}{x} + C$

(C) $12e^{3x} - \dfrac{4}{x^3} + C$ (D) $4e^{3x} - \dfrac{1}{x} + C$

▶ **9.** $f(x) = \begin{cases} x^3 \cos(\pi x) & \text{if } x < 3 \\ x^3 - kx - 6 & \text{if } x \geq 3 \end{cases}$ where k is a constant. For what value of k, if any, is f continuous at $x = 3$?

(A) 3   (B) 8

(C) 16   (D) No value of k will work.

▶ **10.** $\displaystyle \lim_{h \to 0} \dfrac{\ln(8 + h) - \ln 8}{h} =$

(A) 0   (B) $\ln 8$

(C) $\dfrac{1}{8}$ (D) The limit does not exist.

▶ **11.** Find the derivative y' of $x + 5xy - y^2 = 11$ evaluated at the point $(2, 1)$.

(A) -2 (B) $-\dfrac{3}{4}$ (C) $\dfrac{1}{2}$ (D) 8

Free-Response Questions

▶ **12.** An object is moving along the x-axis, and its position x is modeled by a twice-differentiable function $x = x(t)$, where t, $0 \leq t \leq 10$, is the time in seconds. The object's position at select times t is given in the table below.

t	0	2	5	7	10
x(meters)	-2	1	4	6	4

(a) Use the data from the table to approximate the velocity of the object when $t = 4$.

(b) Show that there is at least one time when the object is stopped. Explain your reasoning.

(c) Show that there is at least one time when the object is at the origin. Explain your reasoning.

▶ **13.** (a) Approximate the arc length of the graph of $y = \sin x$ from $x = 0$ to $x = \pi$ using trapezoidal sums with four subintervals of equal width.

(b) Does the approximation in (a) overestimate or underestimate the true arc length of the graph? Support your answer.

(c) Find the exact arc length of the graph of $y = \sin x$ from $x = 0$ to $x = \pi$.

▶ **14.** An object moves along a horizontal line and its velocity v is given by $v(t) = t^2 - 4t + 5$, for $0 \leq t \leq 5$ s.

(a) When is the object moving to the left and when is the object moving to the right?

(b) Determine when the speed of the object is increasing and decreasing.

▶ **15.** An object is moving along a horizontal line, and its position at time t, $0 \leq t \leq 12$, is given by a differentiable function $x = x(t)$. The graph of the derivative $\dfrac{dx}{dt}$ is shown below.

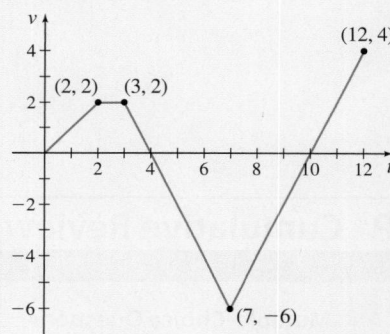

(a) Determine the intervals when the object moves to the right.

(b) When does the object change directions?

(c) What is the acceleration of the object at $t = 10$?

(d) Is the speed increasing or decreasing at $t = 1$?

Justify each of your answers.

▶ **16.** (a) Find all numbers b for which the graphs of $y = 2x + b$ and $y^2 = 4x$ intersect in two distinct points.

(b) If $b = -4$, find the area bounded by the graphs of $y = 2x - 4$ and $y^2 = 4x$.

(c) If $b = 0$, find the volume of the solid generated by revolving about the x-axis the region bounded by the graphs of $y = 2x$ and $y^2 = 4x$.

AP® Practice Exam: Calculus AB

Preparing for the AP® Exam

Section 1: Multiple Choice, Part A

A calculator may not be used for Part A.

1. If $f(x) = e^{4x} + \sin(2x)$, then $f'(0) =$

(A) 1   (B) 2   (C) 4   (D) 6

2. If $\lim\limits_{x \to 1} f(x) = 3$, then $\lim\limits_{x \to 1} \dfrac{(x^2 - 1) f(x)}{x - 1} =$

(A) 0   (B) 4   (C) 6   (D) Does not exist.

3. The graph of the function f is shown below. Which of the following statements is false?

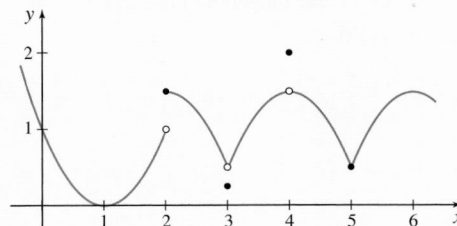

(A) $\lim\limits_{x \to 2^-} f(x) = f(2)$

(B) The function f is discontinuous at $x = 3$.

(C) $\lim\limits_{x \to 4} f(x)$ exists.

(D) The function f is continuous at $x = 5$.

4. Let f be the function $f(x) = x^3 - 15x^2 - 1800x + 2000$. On which interval is the function f both decreasing and concave down?

(A) $x < -20$ (B) $-20 < x < 5$

(C) $5 < x < 30$ (D) $x > 30$

5. If $y = e^{2x} \cos(3x)$, then $\dfrac{dy}{dx} =$

(A) $-6e^{2x} \sin(3x)$

(B) $e^{2x}(2\cos(3x) - 3\sin(3x))$

(C) $e^{2x}(2\cos(3x) + 3\sin(3x))$

(D) $e^{2x}(\cos(3x) - \sin(3x))$

6. Let f be the function defined below. For what value of k is f continuous at $x = 0$?

$$f(x) = \begin{cases} \dfrac{\sin(7x)}{2x} & \text{for } x < 0 \\ k + 2\ln(x + e^{x+1}) & \text{for } x \geq 0 \end{cases}$$

(A) $-\dfrac{3}{2}$ (B) -1 (C) $\dfrac{3}{2}$ (D) $\dfrac{7}{2}$

7. Let f be the function defined by $f(x) = \sqrt[5]{(x-3)^4}$ for all x. Which of the following statements is true?

(A) f is continuous and differentiable at $x = 3$.
(B) f is continuous but not differentiable at $x = 3$.
(C) f is differentiable but not continuous at $x = 3$.
(D) f is not continuous and not differentiable at $x = 3$.

8. The function f is differentiable and its derivative is continuous on the interval $(-4, 5)$. The table below lists several values of f and f' in the interval.

x	-3	-1	0	3
$f(x)$	-16	$\dfrac{2}{3}$	2	2
$f'(x)$	3	-1	0	15

Then $\int_{-1}^{3} f'(x)\, dx =$

(A) -16 (B) $\dfrac{4}{3}$ (C) 12 (D) 16

9. If $y = \sqrt{4x + 6e^{\tan x}}$, then y' equals

(A) $\sqrt{4 + 6e^{\tan x} \sec^2 x}$

(B) $\dfrac{1}{2\sqrt{4x + 6e^{\tan x}}}$

(C) $\dfrac{2 + 3e^{\tan x}}{\sqrt{4x + 6e^{\tan x}}}$

(D) $\dfrac{2 + 3e^{\tan x} \sec^2 x}{\sqrt{4x + 6e^{\tan x}}}$

10. The function f is continuous on $1 \leq x \leq 5$ and differentiable on $1 < x < 5$. If $f(1) = 10$ and $f(5) = 50$, then the Mean Value Theorem guarantees that

(A) f is linear on the interval $1 \leq x \leq 5$.
(B) $f'(c) = 10$ for at least one c between 1 and 5.
(C) $f'(c) = 0$ for at least one c between 1 and 5.
(D) $f(c) = 30$ for at least one c between 1 and 5.

11. The Riemann sum $\dfrac{1}{20}\left(e^{1/20} + e^{2/20} + e^{3/20} + \cdots + e^2\right)$ is an approximation for which integral?

(A) $\int_0^2 e^{x/20}\, dx$ (B) $\int_0^2 e^x\, dx$

(C) $\dfrac{1}{20}\int_0^2 e^x\, dx$ (D) $\dfrac{1}{20}\int_0^2 e^{x/20}\, dx$

12. What is the area of the region in the first quadrant bounded by the graph of $y = 1 + e^{-2x}$ and the line $x = 3$?

(A) $2 - 2e^{-6}$ (B) $3 - \dfrac{1}{2}e^{-6}$

(C) $\dfrac{7}{2} - \dfrac{1}{2}e^{-6}$ (D) $5 - 2e^{-6}$

13. Using the substitution $u = \sqrt{x}$, $\int_1^4 \dfrac{1+x}{1+\sqrt{x}}\, dx$ is equal to which of the following?

(A) $\int_1^4 \dfrac{1+u^2}{1+u}\, du$ (B) $\int_1^2 \dfrac{1+u^2}{1+u}\, du$

(C) $2\int_1^4 \dfrac{u+u^3}{1+u}\, du$ (D) $2\int_1^2 \dfrac{u+u^3}{1+u}\, du$

14. If $f(x) = \sqrt{25 - x^2}$, then the derivative of $f(f(x))$ at $x = -3$ is

(A) -3 (B) -1 (C) 1 (D) 3

15. Let f be the function defined below. What is the value of $\int_{-1}^{1} f(x)\,dx$?

$$f(x) = \begin{cases} e^{2x} & \text{for } x < 0 \\ e^{-4x} & \text{for } x \geq 0 \end{cases}$$

(A) $e^2 - e^{-4}$

(B) $\dfrac{1}{4} - \dfrac{1}{2}e^2 - \dfrac{1}{4}e^{-4}$

(C) $\dfrac{1}{4} - \dfrac{1}{2}e^2 + \dfrac{1}{4}e^{-4}$

(D) $\dfrac{3}{4} - \dfrac{1}{2}e^{-2} - \dfrac{1}{4}e^{-4}$

16. If $f(x) = \sin(2x)$, which of the following is equal to $f'\left(\dfrac{\pi}{4}\right)$?

(A) $\lim\limits_{h \to 0} \dfrac{\sin\left(\dfrac{\pi}{4} + 2h\right) - \sin\left(\dfrac{\pi}{4}\right)}{h}$

(B) $\lim\limits_{h \to 0} \dfrac{\sin\left(\dfrac{\pi}{2} + 2h\right) - \sin\left(\dfrac{\pi}{2}\right)}{h}$

(C) $\lim\limits_{h \to 0} \dfrac{2\sin\left(\dfrac{\pi}{4} + h\right) - 2\sin\left(\dfrac{\pi}{4}\right)}{h}$

(D) $\lim\limits_{h \to 0} \dfrac{\sin\left(\dfrac{\pi}{2} + 2h\right) - 2\sin\left(\dfrac{\pi}{2}\right)}{h}$

17. The graph of a differentiable function f is shown below. Which of the following is true?

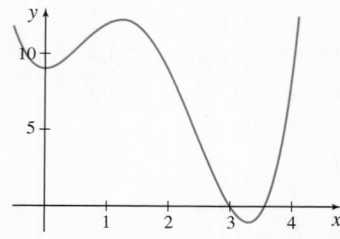

(A) $f(3) < f'(3) < f''(3)$ (B) $f(3) < f''(3) < f'(3)$

(C) $f'(3) < f(3) < f''(3)$ (D) $f''(3) < f(3) < f'(3)$

18. $\lim\limits_{x \to \infty} \dfrac{4x^3 + 3x^2 + 5}{5x^3 + x^2 - x} =$

(A) 0 (B) $\dfrac{4}{5}$ (C) $\dfrac{8}{5}$ (D) ∞

19. A particle moves along the x-axis so that at any time $t > 0$, its acceleration is given by $a(t) = 3$. If the velocity of the particle at time $t = 3$ is -3 and the position at time $t = 2$ is 5, what is the position of the particle when the particle is at rest?

(A) -49 (B) -47 (C) -1 (D) 3.5

20. Let f be the function defined by $f(x) = e^{-0.2x}$. See the figure. A rectangle with two sides along the x-axis and y-axis is inscribed using the graph of f. Find x so that the area of the rectangle is a maximum.

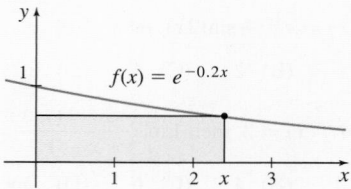

(A) 1.0 (B) 1.2 (C) 2.5 (D) 5.0

21. An equation of the line tangent to the graph of $y = \dfrac{x + 2}{2x - 6}$ at the point $(4, 3)$ is

(A) $y - 3 = -\dfrac{5}{2}(x + 4)$ (B) $y - 3 = -\dfrac{5}{2}(x - 4)$

(C) $y - 4 = -\dfrac{5}{2}(x - 3)$ (D) $y - 3 = \dfrac{5}{2}(x - 4)$

22. Let f be the function $f(x) = e^{kx^2}$, where k is a constant. For what value of k does f have a point of inflection at $x = -2$ and $x = 2$?

(A) $-\dfrac{1}{4}$ (B) $-\dfrac{1}{8}$ (C) $\dfrac{1}{8}$ (D) $\dfrac{1}{4}$

23. $\dfrac{d}{dt}\left(t^2 \displaystyle\int_{2}^{t} \ln(u + 3)\,du\right) =$

(A) $2t \ln(t + 3)$

(B) $2t(\ln(t + 3) - \ln(5))$

(C) $t^2\left(\dfrac{1}{t + 3}\right) + 2t \displaystyle\int_{2}^{t} \ln(u + 3)\,du$

(D) $t^2 \ln(t + 3) + 2t \displaystyle\int_{2}^{t} \ln(u + 3)\,du$

24. If $\dfrac{dy}{dx} = y\cos(2x)$ and $y = 3$ when $x = 0$, then $y =$

(A) $3e^{-0.5\sin(2x)}$ (B) $3e^{0.5\sin(2x)}$

(C) $e^{0.5\sin(2x)} + 3$ (D) $e^{-\sin(2x)} + 3$

25. Let f and g be continuous functions with values given in the table below.

x	$f(x)$	$g(x)$
1	2	1
3	4	5

$\displaystyle\int_{1}^{3} \left(3f'(x) + g(x)g'(x) - 6\right)dx =$

(A) 6 (B) 10 (C) 14 (D) 18

26. At any time $t > 0$, the rate at which a person can memorize a list of M words is proportional to the product of the number of words memorized and the number of words that have not been memorized. If x denotes the number of words memorized at time t, which differential equation models this situation? Assume k is a positive constant.

(A) $\dfrac{dx}{dt} = kx$ (B) $\dfrac{dx}{dt} = kx(x - M)$

(C) $\dfrac{dx}{dt} = kx(M - x)$ (D) $\dfrac{dx}{dt} = kt(M - t)$

27. For what value of x does the function $f(x) = (x - 3)^2(x - 5)$ have a relative minimum?

(A) 3   (B) $\dfrac{7}{3}$ (C) 5   (D) $\dfrac{13}{3}$

28. If $x^2 + 3y^2 = 12$, what is $\dfrac{d^2y}{dx^2}$ at the point $(3, 1)$?

(A) $-\dfrac{3}{2}$ (B) $-\dfrac{4}{3}$ (C) $-\dfrac{1}{2}$ (D) $-\dfrac{1}{3}$

29. $\displaystyle\lim_{x \to 0} \dfrac{3x - \sin(3x)}{1 - \cos(2x)} =$

(A) $-\dfrac{3}{2}$ (B) 0   (C) $\dfrac{9}{4}$ (D) Does not exist.

30. Which of the following pairs of graphs could represent the graph of a function and the graph of its second derivative?

I.

II.

III.

(A) I and II   (B) I and III   (C) II and III   (D) I, II, and III

Section 1: Multiple Choice, Part B

A graphing calculator is required for some of the questions in Part B.

31. If $f(x) = \displaystyle\int_0^x (t - 2) \sin\left(\dfrac{\pi t}{2}\right) dt$, which of the following must be true?

     I. $f(x) < 0$ on the interval $0 < x < 2$.

     II. $f'(x) < 0$ on the interval $0 < x < 2$.

     III. $f''(x) < 0$ on the interval $0 < x < 2$.

(A) I only   (B) I and II   (C) II and III   (D) I and III

32. Let $f(x) = 2x^2 + \sqrt[3]{x^4 - 8}$ be defined for $x \geq 0$ and let g be the inverse of f. What is the value of $g'(10)$?

(A) -0.094 (B) -0.023 (C) 0.023 (D) 0.094

33. What is the slope of the line tangent to the graph of

$$f(x) = e^x - x^e - e$$

at the point where the graph crosses the x-axis?

(A) 0.110   (B) 3.471   (C) 9.106   (D) 55.238

34. Let f be the function given by $f(x) = \ln(x + 1)$ and let g be the function given by $g(x) = \dfrac{1}{\sqrt{x}}$. At what value of x do the graphs of f and g have perpendicular tangent lines?

(A) 0.484   (B) 1.000   (C) 1.358   (D) 2.065

35. A table of values for a continuous function f is shown below. If five subintervals of $[0, 2]$ of equal width are used, what is the approximation of $\int_0^2 f(x)dx$ using trapezoidal sums?

x	0	0.4	0.8	1.2	1.6	2.0
$f(x)$	10	12	13	16	19	20

(A) 18   (B) 30   (C) 36   (D) 60

36. At time $t \geq 0$, the acceleration of a particle moving on the x-axis is $a(t) = 3 - \sqrt{t}$. If the velocity of the particle at time $t = 0$ is 2, what is the velocity at time $t = 4$?

(A) 2   (B) $\dfrac{20}{3}$ (C) $\dfrac{26}{3}$ (D) $\dfrac{58}{3}$

37. What is the area of the region in the first quadrant enclosed by the graphs of $y = \tan x$, $y = 3 - x$, and the y-axis?

(A) 1.905   (B) 2.595   (C) 3.442   (D) 4.132

38. At time $t = 0$, a small animal weighs 3 pounds and grows at the rate of $r(t) = 0.16e^{\sqrt{t}}$ pounds per day, where t is measured in days. How many pounds did the animal gain from $t = 0$ days to $t = 5$ days?

(A) 1.5   (B) 4.0   (C) 4.5   (D) 7.0

39. Let g be the function $g(x) = \dfrac{x \ln x - x}{x + 1}$. Which of the following is a critical number of g?

(A) -1 (B) 0.368 (C) 0.567 (D) 2.718

40. The graph of the function f is shown below.

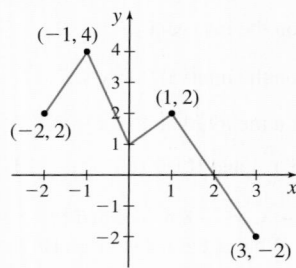

Suppose $g(x) = \int_{-2}^{x} f(t)\, dt$. For which number x is $g(x)$ the greatest?

(A) -1 (B) 1 (C) 2 (D) 3

41. The area of a regular hexagon is

$A = \dfrac{3\sqrt{3}}{2}s^2$ where s is the length of each side. The perimeter of the hexagon is increasing at a constant rate of 30 cm per minute. What is the rate of increase in the area of the hexagon at the instant when the perimeter of the hexagon is 120 cm?

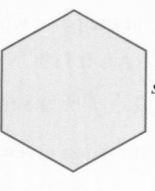

(A) 259.808 (B) 519.615 (C) 1558.846 (D) 3117.691

42. For a continuous function f, let g be the function defined by $g(x) = \int_0^x \left(3f(t) + \sqrt{t^3 + 1}\right) dt$.
If $f(2) = 4$ and $\int_0^2 f(x)dx = 10$, what is $g(2) + g'(2)$?

(A) 30.000 (B) 45.000 (C) 48.241 (D) 78.241

43. The density of a metal rod is given by the function

$f(x) = \sqrt{10 - x^3} \ \dfrac{\text{kg}}{\text{m}}$ for $0 \le x \le 2$ meters.
What is the average density across the length of the rod?

(A) 0.874 (B) 2.288 (C) 2.792 (D) 5.584

44. A particle travels on the x-axis so its velocity at time t is given by $v(t) = e^{0.2t} - 3$ for $0 \le t \le 10$. What is the total distance traveled by the particle from $t = 0$ to $t = 10$?

(A) 1.945 (B) 2.389 (C) 11.945 (D) 14.904

45. Let R be the region bounded by $y = \sin x$ and $y = 1 - \sin x$ as shown below. R is the base of a solid whose cross sections perpendicular to the x-axis are triangles. The height of each triangle is one-half the length of the base. What is the volume of the solid?

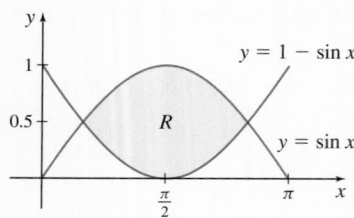

(A) 0.264 (B) 0.272 (C) 0.528 (D) 0.544

Section 2: Free Response, Part A

A graphing calculator is required for Part A.

1. During the workday, which lasts from 7 a.m. to 3 p.m. ($0 \le t \le 8$ hours), raw material is piped into a processing plant at a rate R, where $R = R(t) = 15 \sin \dfrac{t^2}{22}$ kiloliters per hour. The raw material is processed at a rate P, where $P = P(t) = -0.02t^3 + 0.2t^2 + t$ kiloliters per hour. At time $t = 0$, there are 20 kiloliters of raw material already in storage and ready for use.

(a) How many kiloliters of raw material are piped during the workday?

(b) Determine the amount of raw material processed in the first hour of operation.

(c) At what time t, $0 \le t \le 8$, is the raw material available for processing the least? At what time is the amount raw material available the greatest? Justify your answers.

2. Let R be the region in the first quadrant bounded above by the graph of $y = \dfrac{4x}{x^2 + 7}$ and below by the horizontal line $y = \dfrac{1}{2}$.

(a) Find the area of R.

(b) Find the volume of the solid generated when R is revolved about the horizontal line $y = -2$.

(c) The region R is the base of a solid. For this solid, each cross section perpendicular to the x-axis is a rectangle with a height equal to five times the length of the base. Find the volume of the solid.

Section 2: Free Response, Part B

A calculator may not be used for Part B.

3. An object moves along the x-axis with initial position $x(0) = 4$. The velocity of the object at time $t \ge 0$ is given by

$$v(t) = 10e^{-0.1t} \cos\left(\dfrac{\pi t}{4}\right)$$

The graph of $v(t)$ is shown below.

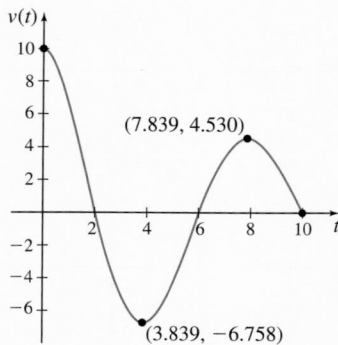

(a) What is the acceleration of the object at time $t = 3.839$?

(b) Write, but do not evaluate, an integral that gives the position of the object at time $t = 5$.

(c) On the interval $0 < t < 6$, when is the object moving to the left? Give a reason for your answer.

(d) Write, but do not evaluate, an integral that gives the total distance traveled by the object over the interval $0 \le t \le 6$.

4. The graph of the function f below consists of polynomials on the intervals $-2 < x < 0$ and $2 < x < 4$ and line segments on the intervals $0 < x < 2$ and $4 < x < 6$.

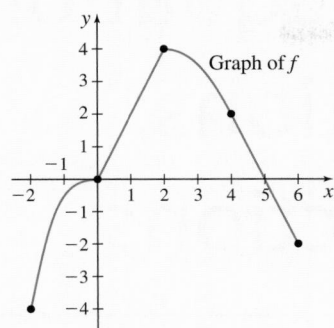

Graph of f

(a) Let g be the function given by $g(x) = \int_0^x f(t)\, dt$.
Find $g(2)$, $g'(2)$, and $g''(2)$, or state that it does not exist.

(b) For the function g defined in part (a), determine where g is increasing.

(c) Let h be the function given by $h(x) = \int_{-2}^x f(t)\, dt$. Find the value of x for which $h(x)$ has a relative maximum value. Justify your answer.

(d) For the function h defined in part (c), if $h(0) = -2$ and $h(6) = \dfrac{26}{3}$, find the value of $\int_2^4 f(t)\, dt$.

5. The elevation of a 6-mile hiking trail, in thousands of feet, is modeled by a twice-differentiable function E of hiking distance x, where x is measured in miles. For $0 < x < 6$, $E'(x) > 0$ and the graph of E is concave up. The table below gives selected values of the elevation, $E(x)$, of the trail over the hiking distance $0 \le x \le 6$.

x (hiking distance)	0	1	3	5	6
E (elevation)	1.0	1.3	2.1	4.1	5.6

(a) Use a trapezoidal sum with the four subintervals given in the table to approximate the value of
$$\frac{1}{6}\int_0^6 E(x)\, dx$$

(b) Using correct units, explain the meaning of
$$\frac{1}{6}\int_0^6 E(x)\, dx$$

(c) For $3 < x < 5$, must there be a distance x for which the elevation increases at a rate of 1000 feet per mile? Justify your answer.

(d) Does a Left Riemann sum under approximate or over approximate $\dfrac{1}{6}\int_0^6 E(x)\, dx$? Give a reason for your answer.

6. Consider the differential equation $\dfrac{dy}{dx} = \dfrac{2(y-1)^2}{\sqrt{x}}$.

(a) Sketch a slope field for the differential equation at the six points indicated in the graph below.

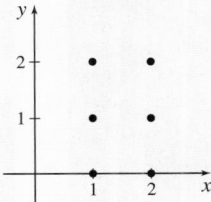

(b) Let $y = f(x)$ be the particular solution of the differential equation with the initial condition $f(1) = 2$.
Find an equation for the tangent line to the graph of f at the point $(1, 2)$.

(c) Find the particular solution $y = f(x)$ of the differential equation with the initial condition $f(1) = 2$.

CHAPTER

9

Parametric Equations; Polar Equations; Vector Functions

paffy/Shutterstock

Polar Graphs and Microphones

If you have ever visited a recording studio or been early to a concert or performance and saw the staff setting up the sound system, you understand a bit about the work that goes into obtaining optimal sound quality. Different microphones are designed to pick up sound from different directions, and sound technicians know that the selection and placement of microphones and monitors (speakers that are aimed back at the musicians so that they can hear the other instruments) vary depending on the specific purpose and venue. The same choices factor into the construction of cell phones and Bluetooth devices, which need to amplify the sound coming out of your mouth and minimize the barking dogs, loud trucks, and random conversations that may be happening around you.

The choice of the right microphone and speaker and their placement to achieve a specific goal requires an understanding of the direction of the sound and the area where it is being picked up.

 The Chapter Project on page 770 examines the design of several different microphones and the directions from where the sounds they amplify originate.

Andy
Director of Instrumental Music

AP® Calculus helped me see the beauty in numbers and to recognize patterns in nature and the world, which I apply when analyzing music as a band director.

Until now, we have worked primarily with explicitly defined functions $y = f(x)$, where x and y represent rectangular coordinates. Because the graphs of such functions satisfy the Vertical-line Test, many graphs, such as circles, cannot be represented using a single, explicitly defined function. To address this issue, we introduce *parametric equations*, a way to represent graphs that are not necessarily those of functions. Additionally, parametric equations are particularly useful because they allow us to model motion along a curve and to determine not only the location of an object (a point) but also the time it is there.

The chapter continues with a discussion of the graph of a polar equation. We use parametric equations to find tangent lines to and the arc length of the graph.

Then we discuss functions whose domain is a subset of real numbers, but whose range consists of vectors. Such functions are called *vector-valued functions* of a real variable, or, more simply, *vector functions*.

Lastly, we work with position, velocity, speed, and acceleration of an object moving along a plane curve. ∎

9.1 Parametric Equations

OBJECTIVES *When you finish this section, you should be able to:*

1 **Graph parametric equations (p. 708)**
2 **Find a rectangular equation for a curve represented parametrically (p. 709)**
3 **Use time as the parameter in parametric equations (p. 712)**
4 **Convert a rectangular equation to parametric equations (p. 712)**

AP® EXAM TIP

BREAK IT DOWN: As you work through the problems in this chapter, remember to follow these steps:

Step 1 Identify the underlying structure and related concepts.

Step 2 Determine the appropriate math rule or procedure.

Step 3 Apply the math rule or procedure.

Step 4 Clearly communicate your answer.

On page 774, see how we've used these steps to solve Section 9.2 AP® Practice Problem 9(b) on page 729.

The graph of an equation of the form $y = f(x)$, where f is a function, is intersected no more than once by any vertical line. There are many graphs, however, such as circles and some parabolas, that do not pass the Vertical-line Test. To study these graphs requires a different model. One model uses a pair of equations and a third variable, called a *parameter*.

DEFINITION

Suppose $x = x(t)$ and $y = y(t)$ are two functions of a third variable t, called the **parameter**, that are defined on the same interval I. Then the equations

$$x = x(t) \qquad y = y(t)$$

where t is in I, are called **parametric equations**, and the graph of the points defined by

$$(x, y) = (x(t), y(t))$$

is called a **plane curve**.

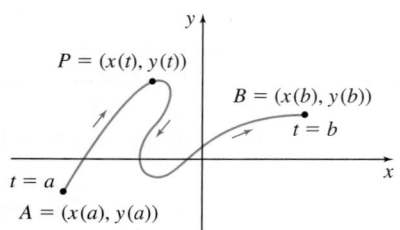

Figure 1 $x = x(t)$, $y = y(t)$, $a \leq t \leq b$

1 Graph Parametric Equations

Parametric equations are particularly useful for describing motion along a curve. Suppose an object moves along a curve represented by the parametric equations

$$x = x(t) \qquad y = y(t)$$

where each function is defined over the interval $a \leq t \leq b$. For a given value of t, the values of $x = x(t)$ and $y = y(t)$ determine a point (x, y) on the curve. In fact, as t varies over the interval from $t = a$ to $t = b$, successive values of t determine the direction of the motion of the object moving along the curve. See Figure 1. The arrows show the direction, or the **orientation**, of the object as it moves from A to B.

EXAMPLE 1 **Graphing a Plane Curve**

Graph the plane curve represented by the parametric equations

$$x(t) = 3t^2 \qquad y(t) = 2t \qquad -2 \leq t \leq 2$$

Indicate the orientation of the curve.

Solution

Corresponding to each number t, $-2 \leq t \leq 2$, there is a number x and a number y that are the coordinates of a point (x, y) on the curve. We form Table 1 by listing various choices of the parameter t and the corresponding values for x and y.

The motion begins when $t = -2$ at the point $(12, -4)$ and ends when $t = 2$ at the point $(12, 4)$. Figure 2 illustrates the plane curve whose parametric equations are $x(t) = 3t^2$ and $y(t) = 2t$. The arrows indicate the orientation of the plane curve for increasing values of the parameter t. ■

TABLE 1

t	$x(t) = 3t^2$	$y(t) = 2t$	(x, y)
-2	12	-4	$(12, -4)$
-1	3	-2	$(3, -2)$
0	0	0	$(0, 0)$
1	3	2	$(3, 2)$
2	12	4	$(12, 4)$

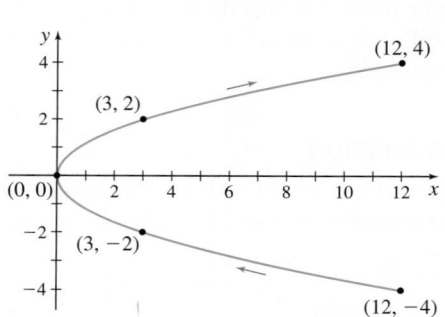

Figure 2 $x(t) = 3t^2$, $y(t) = 2t$, $-2 \leq t \leq 2$

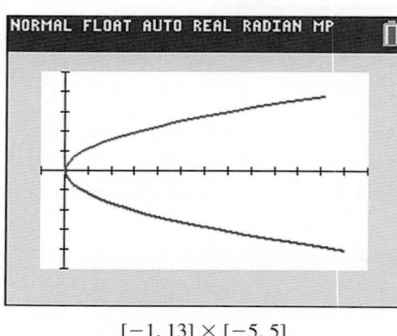

$[-1, 13] \times [-5, 5]$

Figure 3 $x(t) = 3t^2$, $y(t) = 2t$, $-2 \leq t \leq 2$

📝 **COMMENT** Figure 3 shows the graph of the parametric equations in Example 1 on a TI 84 Plus CE. In using a calculator to graph parametric equations with a restricted domain t, be sure to limit t in the graphing window. In Figure 3, t is restricted to Tmin $= -2$; Tmax $= 2$, Tstep $= 0.1$. It is important to choose the step size of T to be small.

② Find a Rectangular Equation for a Curve Represented Parametrically

The plane curve in Figures 2 and 3 should look familiar. To identify it, find the corresponding rectangular equation by eliminating the parameter t from the parametric equations

$$x(t) = 3t^2 \qquad y(t) = 2t \qquad -2 \leq t \leq 2$$

We begin by solving for t in $y = 2t$, obtaining $t = \dfrac{y}{2}$. Then we substitute $t = \dfrac{y}{2}$ in the equation, $x = 3t^2$.

$$x = 3t^2 = 3\left(\frac{y}{2}\right)^2 = \frac{3y^2}{4}$$

The equation $x = \dfrac{3y^2}{4}$ is a parabola with its vertex at the origin and its axis of symmetry along the x-axis. We refer to this equation as the **rectangular equation** of the curve to distinguish it from the parametric equations.

Notice that the plane curve represented by the parametric equations $x(t) = 3t^2$, $y(t) = 2t$, $-2 \leq t \leq 2$, is only part of the parabola $x = \dfrac{3y^2}{4}$. In general, the graph of the rectangular equation obtained by eliminating the parameter will contain more points than the plane curve defined using parametric equations. Therefore, when graphing a plane curve represented by a rectangular equation, we often may have to restrict the graph to match the parametric equations.

EXAMPLE 2	**Finding a Rectangular Equation for a Plane Curve Represented Parametrically**

Find a rectangular equation of the curve whose parametric equations are

$$x(t) = R \cos t \qquad y(t) = R \sin t$$

where $R > 0$ is a constant. Graph the plane curve and indicate its orientation.

Solution

The presence of the sine and cosine functions in the parametric equations suggests using the Pythagorean identity $\cos^2 t + \sin^2 t = 1$. Then

$$\left(\frac{x}{R}\right)^2 + \left(\frac{y}{R}\right)^2 = 1 \qquad \cos t = \frac{x}{R}, \quad \sin t = \frac{y}{R}$$
$$x^2 + y^2 = R^2$$

The graph of the rectangular equation is a circle with center at the origin and radius R. In the parametric equations, as the parameter t increases, the points (x, y) on the circle are traced out in the counterclockwise direction, as shown in Figure 4. ■

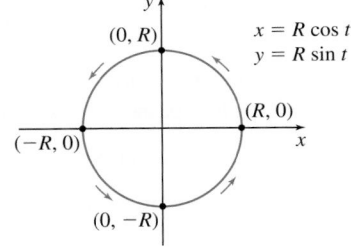

Figure 4 The orientation is counterclockwise.

NOW WORK   Problems **9** and **15**.

When analyzing the parametric equations in Example 2, notice there are no restrictions on t, so t varies from $-\infty$ to ∞. As a result, the circle is repeated each time t increases by 2π.

If we want to describe a curve consisting of exactly one revolution in the counterclockwise direction, the domain must be restricted to an interval of length 2π. For example, we could use

$$x(t) = R \cos t \qquad y(t) = R \sin t \qquad 0 \leq t \leq 2\pi$$

Now the plane curve starts when $t = 0$ at $(R, 0)$, passes through $(0, R)$, $(-R, 0)$, and $(0, -R)$, and ends when $t = 2\pi$ at $(R, 0)$.

If we wanted the curve to consist of exactly three revolutions in the counterclockwise direction, the domain must be restricted to an interval of length 6π. For example, we could use $x(t) = R \cos t$, $y(t) = R \sin t$, with $-2\pi \le t \le 4\pi$, or $0 \le t \le 6\pi$, or $2\pi \le t \le 8\pi$, or any arbitrary interval of length 6π.

EXAMPLE 3 **Finding a Rectangular Equation for a Plane Curve Represented Parametrically**

CALC CLIP

Find rectangular equations for the plane curves represented by each of the parametric equations. Graph each curve and indicate its orientation.

(a) $x(t) = R \cos t$ $y(t) = R \sin t$ $0 \le t \le \pi$ and $R > 0$

(b) $x(t) = R \sin t$ $y(t) = R \cos t$ $0 \le t \le \pi$ and $R > 0$

Solution

(a) We eliminate the parameter t using a Pythagorean identity.

$$\cos^2 t + \sin^2 t = 1$$
$$\left(\frac{x}{R}\right)^2 + \left(\frac{y}{R}\right)^2 = 1 \qquad \cos t = \frac{x}{R}, \quad \sin t = \frac{y}{R}$$
$$x^2 + y^2 = R^2 \qquad\qquad\qquad (1)$$

The rectangular equation represents a circle with radius R and center at the origin. In the parametric equations, $0 \le t \le \pi$, so the curve begins when $t = 0$ at the point $(R, 0)$, passes through the point $(0, R)$ when $t = \dfrac{\pi}{2}$, and ends when $t = \pi$ at the point $(-R, 0)$. The curve is an upper semicircle of radius R with counterclockwise orientation, as shown in Figure 5. If we solve equation (1) for y, we obtain the rectangular equation of the semicircle

$$y = \sqrt{R^2 - x^2} \qquad \text{where } -R \le x \le R$$

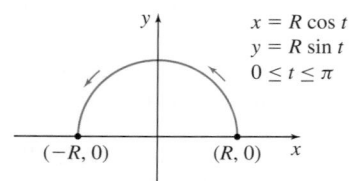

$x = R \cos t$
$y = R \sin t$
$0 \le t \le \pi$

$(-R, 0)$ $(R, 0)$ x

Figure 5 $y = \sqrt{R^2 - x^2}$, $-R \le x \le R$

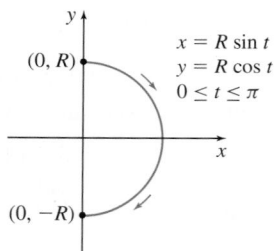

$x = R \sin t$
$y = R \cos t$
$0 \le t \le \pi$

$(0, R)$

$(0, -R)$

Figure 6 $x = \sqrt{R^2 - y^2}$, $-R \le y \le R$

(b) Eliminate the parameter t as in (a), to obtain

$$x^2 + y^2 = R^2 \qquad\qquad (2)$$

The rectangular equation represents a circle with radius R and center at $(0, 0)$. Since $0 \le t \le \pi$, the curve begins when $t = 0$ at the point $(0, R)$, passes through the point $(R, 0)$ when $t = \dfrac{\pi}{2}$, and ends at the point $(0, -R)$ when $t = \pi$. The curve is a right semicircle of radius R with a clockwise orientation, as shown in Figure 6. If we solve equation (2) for x, we obtain the rectangular equation of the semicircle

$$x = \sqrt{R^2 - y^2} \qquad \text{where } -R \le y \le R$$

■

NOTE In Example 3(a) we expressed the rectangular equation as a function $y = f(x)$, and in (b) we expressed the rectangular equation as a function $x = g(y)$.

Parametric equations are not unique; that is, different parametric equations can represent the same graph. Two examples of other parametric equations that represent the graph in Figure 6 are

- $x(t) = R \sin t$ $y(t) = -R \cos(\pi - t)$ $0 \le t \le \pi$
- $x(t) = R \sin(3t)$ $y(t) = R \cos(3t)$ $0 \le t \le \dfrac{\pi}{3}$

There are other examples.

NOW WORK Problem 17.

EXAMPLE 4	**Finding a Rectangular Equation for a Plane Curve Represented Parametrically**

(a) Find a rectangular equation of the plane curve whose parametric equations are

$$x(t) = \cos(2t) \qquad y(t) = \sin t \qquad -\frac{\pi}{2} \le t \le \frac{\pi}{2}$$

(b) Graph the rectangular equation.

(c) Determine the restrictions on x and y so the graph corresponding to the rectangular equation is identical to the plane curve described by

$$x(t) = \cos(2t) \qquad y(t) = \sin t \qquad -\frac{\pi}{2} \le t \le \frac{\pi}{2}$$

(d) Graph the plane curve whose parametric equations are

$$x(t) = \cos(2t) \qquad y(t) = \sin t \qquad -\frac{\pi}{2} \le t \le \frac{\pi}{2}$$

Solution

(a) Here we cannot use a Pythagorean identity to eliminate the parameter t because $\cos(2t)$ and $\sin t$ have different arguments. Instead use a trigonometric identity that involves $\cos(2t)$ and $\sin t$, namely, $\sin^2 t = \dfrac{1 - \cos(2t)}{2}$. Then

$$y^2 = \underset{\underset{y(t)=\sin t}{\uparrow}}{\sin^2 t} = \frac{1 - \cos(2t)}{2} = \underset{\underset{x(t)=\cos(2t)}{\uparrow}}{\frac{1-x}{2}}$$

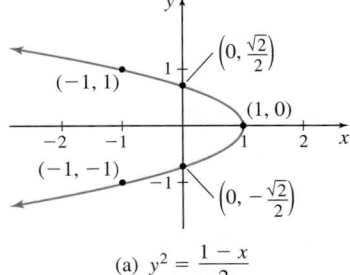

(a) $y^2 = \dfrac{1-x}{2}$

(b) The curve represented by the rectangular equation $y^2 = \dfrac{1-x}{2}$ is the parabola shown in Figure 7(a).

(c) The plane curve represented by the parametric equations does not include all the points on the parabola. Since $x(t) = \cos(2t)$ and $-1 \le \cos(2t) \le 1$, then $-1 \le x \le 1$. Also, since $y(t) = \sin t$, then $-1 \le y \le 1$. Finally, the curve is traced out exactly once in the counterclockwise direction from the point $(-1, -1)$ (when $t = -\dfrac{\pi}{2}$) to the point $(-1, 1)$ (when $t = \dfrac{\pi}{2}$).

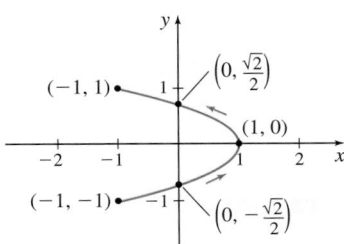

(b) $x(t) = \cos(2t),\ y(t) = \sin t,$
$-\dfrac{\pi}{2} \le t \le \dfrac{\pi}{2}$

Figure 7

(d) The plane curve represented by the given parametric equations is the part of the parabola shown in Figure 7(b). ∎

🖩 **COMMENT** Figures 8(a) and 8(b) show the graphs in Figures 7(a) and 7(b) using a TI 84 Plus CE. Note the MODE chosen and the domain restrictions for each of these screen shots.

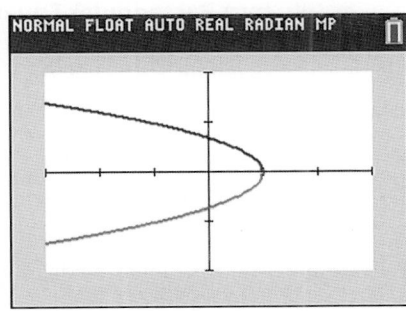

$[-3, 3] \times [-2, 2]$

Figure 8 (a) $y = \begin{cases} -\sqrt{\dfrac{1-x}{2}} \\ \sqrt{\dfrac{1-x}{2}} \end{cases}$

Mode: Function

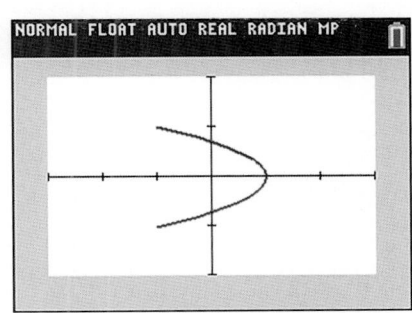

$[-3, 3] \times [-2, 2]$

(b) $x(t) = \cos(2t),\ y(t) = \sin t,\ -\dfrac{\pi}{2} \le t \le \dfrac{\pi}{2}$
Mode: Parametric
Window: Tmin $= -\pi/2$; Tmax $= \pi/2$;
Tstep $= 0.1$

NOW WORK   Problem 21.

❸ Use Time as the Parameter in Parametric Equations

If the parameter t represents time, then the parametric equations $x = x(t)$ and $y = y(t)$ specify how the x- and y-coordinates of an object moving in a plane vary with time. This motion is sometimes referred to as **curvilinear motion**. Describing motion along a plane curve using parametric equations specifies not only *where* the object is, that is, its location (x, y), but also *when* it is there, that is, the time t. A rectangular equation provides only the location of the object.

EXAMPLE 5   **Using Time as the Parameter in Parametric Equations**

Describe the motion of an object that moves along a plane curve so that at time t it has coordinates

$$x(t) = 3 \cos t \qquad y(t) = 4 \sin t \qquad 0 \le t \le 2\pi$$

Solution

Eliminate the parameter t using the Pythagorean identity $\cos^2 t + \sin^2 t = 1$.

$$\frac{x^2}{9} + \frac{y^2}{16} = 1 \qquad \cos t = \frac{x}{3}, \quad \sin t = \frac{y}{4}$$

The plane curve is the ellipse shown in Figure 9. When $t = 0$, the object is at the point $(3, 0)$. As t increases, the object moves around the ellipse in a counterclockwise direction, reaching the point $(0, 4)$ when $t = \dfrac{\pi}{2}$, the point $(-3, 0)$ when $t = \pi$, the point $(0, -4)$ when $t = \dfrac{3\pi}{2}$, and returning to its starting point $(3, 0)$ when $t = 2\pi$. ∎

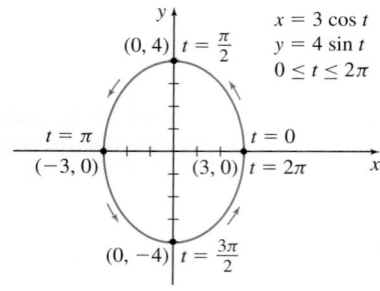

Figure 9 $\dfrac{x^2}{9} + \dfrac{y^2}{16} = 1$

| NOW WORK | Problem 41.

---

❹ Convert a Rectangular Equation to Parametric Equations

Suppose the rectangular equation corresponding to a curve is known. Then the curve can be represented by a variety of parametric equations.

For example, if a curve is defined by the function $y = f(x)$, one way of obtaining parametric equations is to let $x = t$. Then $y = f(t)$ and

$$x(t) = t \qquad y(t) = f(t)$$

where t is in the domain of f, are parametric equations of the curve.

EXAMPLE 6   **Finding Parametric Equations for a Curve Represented by a Rectangular Equation**

Find parametric equations corresponding to the equation $y = x^2 - 4$.

Solution

Let $x = t$. Then parametric equations are

$$x(t) = t \qquad y(t) = t^2 - 4 \qquad -\infty < t < \infty$$ ∎

We can also find parametric equations for $y = x^2 - 4$ by letting $x = t^3$. Then the parametric equations are

$$x(t) = t^3 \qquad y(t) = t^6 - 4 \qquad -\infty < t < \infty$$

Although the choice of x may appear arbitrary, we need to be careful when choosing the substitution for x. The substitution must be a function that allows x to take on all the values in the domain of f. For example, if we let $x(t) = t^2$, then $y(t) = t^4 - 4$. But $x(t) = t^2$, $y(t) = t^4 - 4$ are not parametric equations for all of $y = x^2 - 4$, since only points for which $x \ge 0$ are obtained.

| NOW WORK | Problem 47.

EXAMPLE 7 **Finding Parametric Equations for an Object in Motion**

Find parametric equations that trace out the ellipse

$$\frac{x^2}{4} + y^2 = 1$$

where the parameter t is time (in seconds) and:

(a) The motion around the ellipse is counterclockwise, begins at the point $(2, 0)$, and requires 1 s for a complete revolution.

(b) The motion around the ellipse is clockwise. It begins at the point $(0, 1)$, and requires 2 s for a complete revolution.

Solution

(a) Figure 10 shows the graph of the ellipse $\frac{x^2}{4} + y^2 = 1$.

Since the motion begins at the point $(2, 0)$, we want $x = 2$ and $y = 0$ when $t = 0$. We let

$$x(t) = 2\cos(\omega t) \qquad \text{and} \qquad y(t) = \sin(\omega t)$$

for some constant ω. This choice for $x = x(t)$ and $y = y(t)$ satisfies the equation $\frac{x^2}{4} + y^2 = 1$ and also satisfies the requirement that when $t = 0$, then $x = 2$ and $y = 0$.

For the motion to be counterclockwise, as t increases from 0, the value of x must decrease and the value of y must increase. This requires $\omega > 0$. [Do you see why? If $\omega > 0$, then $x = 2\cos(\omega t)$ is decreasing when $t > 0$ is near zero, and $y = \sin(\omega t)$ is increasing when $t > 0$ is near zero.]

Finally, since 1 revolution takes 1 s, the period is $\dfrac{2\pi}{\omega} = 1$, so $\omega = 2\pi$. Parametric equations that satisfy the conditions given in (a) are

$$x(t) = 2\cos(2\pi t) \qquad y(t) = \sin(2\pi t) \qquad 0 \le t \le 1$$

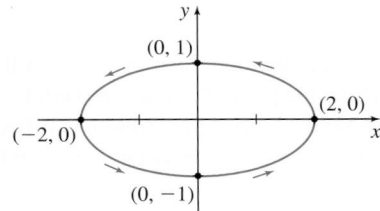

Figure 10 $\dfrac{x^2}{4} + y^2 = 1$

NEED TO REVIEW? The period of a sinusoidal graph is discussed in Section P. 6, p. 58.

(b) Figure 11 shows the graph of the ellipse $\frac{x^2}{4} + y^2 = 1$.

Since the motion begins at the point $(0, 1)$, we want $x = 0$ and $y = 1$ when $t = 0$. We let

$$x(t) = 2\sin(\omega t) \qquad \text{and} \qquad y(t) = \cos(\omega t)$$

for some constant ω. This choice for $x = x(t)$ and $y = y(t)$ satisfies the equation $\frac{x^2}{4} + y^2 = 1$ and also satisfies the requirement that when $t = 0$, then $x = 0$ and $y = 1$.

For the motion to be clockwise, as t increases from 0, the value of x must increase and the value of y must decrease. This requires that $\omega > 0$.

Finally, since 1 revolution takes 2 s, the period is $\dfrac{2\pi}{\omega} = 2$, or $\omega = \pi$. Parametric equations that satisfy the conditions given in (b) are

$$x(t) = 2\sin(\pi t) \qquad y(t) = \cos(\pi t) \qquad 0 \le t \le 2 \qquad ▪$$

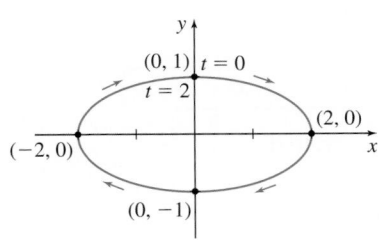

Figure 11 $\dfrac{x^2}{4} + y^2 = 1$

NOW WORK Problem 55.

The Cycloid

Suppose a circle rolls along a horizontal line without slipping. As the circle rolls along the line, a point P on the circle traces out a curve called a **cycloid**, as shown in Figure 12. Deriving the equation of a cycloid in rectangular coordinates is complicated, but the derivation in terms of parametric equations is relatively easy.

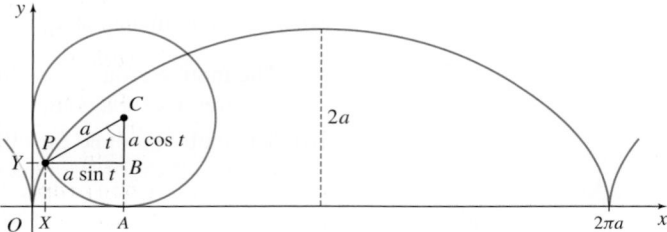

Figure 12 Cycloid

We begin with a circle of radius a. Suppose the fixed line on which the circle rolls is the x-axis. Let the origin be one of the points at which the point P comes into contact with the x-axis. Figure 12 shows the position of this point P after the circle has rolled a bit. The angle t (in radians) measures the angle through which the circle has rolled. Since the circle does not slip, it follows that

$$\text{Arc } AP = d(O, A)$$

RECALL For a circle of radius r, a central angle of θ radians subtends an arc whose length s is $s = r\,\theta$.

The length of the arc AP is

$$at = d(O, A) \qquad s = r\theta, \quad r = a, \quad \theta = t$$

The x-coordinate of the point $P = (x, y)$ is

$$x = d(O, X) = d(O, A) - d(X, A) = at - a\sin t = a(t - \sin t)$$

The y-coordinate of the point P is

$$y = d(O, Y) = d(A, C) - d(B, C) = a - a\cos t = a(1 - \cos t)$$

THEOREM Parametric Equations of a Cycloid

Parametric equations of a cycloid are

$$\boxed{x(t) = a(t - \sin t) \qquad y(t) = a(1 - \cos t) \qquad -\infty < t < \infty}$$

where a is the radius of the circle that traces out the cycloid.

Applications to Mechanics

If $a < 0$ in the parametric equations of the cycloid, the result is an inverted cycloid, as shown in Figure 13(a). The inverted cycloid arises as a result of some remarkable applications in the field of mechanics. We discuss two of them: the *brachistochrone* and *tautochrone*.

NOTE In Greek, *brachistochrone* means "shortest time," and *tautochrone* "same time."

Inverted cycloid as skateboard ramp. Note its resemblance to Figure 13(a).

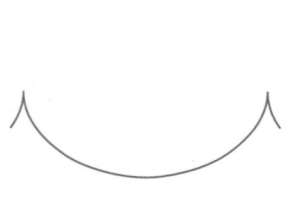

(a) Inverted cycloid

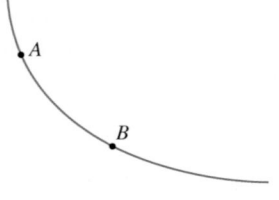

(b) Curve of quickest descent

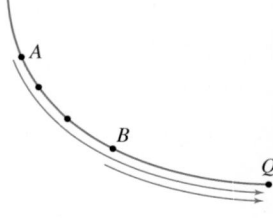

(c) All particles reach Q at the same time.

Figure 13

The **brachistochrone** is the curve of quickest descent. If an object is constrained to follow some path from a point A to a lower point B (not on the same vertical line) and is acted on only by gravity, the time needed to make the descent is minimized if the path is an inverted cycloid. See Figure 13(b). For example, in sliding packages from a loading dock onto a ship, a ramp in the shape of an inverted cycloid might be used so the packages get to the ship in the least amount of time. This discovery, which is attributed to many famous mathematicians (including Johann Bernoulli and Blaise Pascal), was a significant step in creating the branch of mathematics known as the *calculus of variations.*

Suppose Q is the lowest point on an inverted cycloid. If several objects placed at various positions on an inverted cycloid simultaneously begin to slide down the cycloid, they will reach the point Q at the same time, as indicated in Figure 13(c). This is referred to as the **tautochrone property** of the cycloid. It was used by the Dutch mathematician, physicist, and astronomer Christiaan Huygens (1629–1695) to construct a pendulum clock with a bob that swings along an inverted cycloid, as shown in Figure 14. In Huygens's clock, the bob was made to swing along an inverted cycloid by suspending the bob on a thin wire constrained by two plates shaped like cycloids. In a clock of this design, the period of the pendulum is independent of its amplitude.

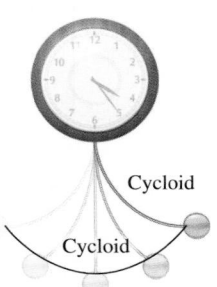

Figure 14 Huygens's pendulum clock.

9.1 Assess Your Understanding

Concepts and Vocabulary

1. Let $x = x(t)$ and $y = y(t)$ be two functions whose common domain is the interval I. The graph of the points defined by $(x, y) = (x(t), y(t))$ is called a(n) _____. The variable t is called a(n) _____.

2. *Multiple Choice* The parametric equations $x(t) = 2 \sin t$, $y(t) = 3 \cos t$ define a(n)
   (a) line **(b)** hyperbola **(c)** ellipse **(d)** parabola.

3. *Multiple Choice* The parametric equations $x(t) = a \sin t$, $y(t) = a \cos t$, $a > 0$, define a
   (a) line **(b)** hyperbola **(c)** parabola **(d)** circle.

4. If a circle rolls along a horizontal line without slipping, a point P on the circle traces out a curve called a(n) _____.

5. *True or False* The parametric equations defining a curve are unique.

6. *True or False* Plane curves represented using parametric equations have an orientation.

Skill Building

In Problems 7–20:

(a) *Find the rectangular equation of each plane curve with the given parametric equations.*

(b) *Graph the plane curve represented by the parametric equations and indicate its orientation.*

7. $x(t) = 2t + 1$, $y(t) = t + 2$; $-\infty < t < \infty$

8. $x(t) = t - 2$, $y(t) = 3t + 1$; $-\infty < t < \infty$

9. $x(t) = 2t + 1$, $y(t) = t + 2$; $0 \le t \le 2$

10. $x(t) = t - 2$, $y(t) = 3t + 1$; $0 \le t \le 2$

11. $x(t) = e^t$, $y(t) = t$; $-\infty < t < \infty$

12. $x(t) = t$, $y(t) = \dfrac{1}{t}$; $-\infty < t < \infty, t \neq 0$

13. $x(t) = \sin t$, $y(t) = \cos t$; $0 \le t \le 2\pi$

14. $x(t) = \cos t$, $y(t) = \sin t$; $0 \le t \le \pi$

15. $x(t) = 2 \sin t$, $y(t) = 3 \cos t$; $0 \le t \le 2\pi$

16. $x(t) = 4 \cos t$, $y(t) = 3 \sin t$; $0 \le t \le 2\pi$

17. $x(t) = 2 \sin t - 3$, $y(t) = 2 \cos t + 1$; $0 \le t \le \pi$

18. $x(t) = 4 \cos t + 1$, $y(t) = 4 \sin t - 3$; $0 \le t \le \pi$

19. $x(t) = t^2$, $y(t) = 2t$; $-\infty < t < \infty$

20. $x(t) = 4t + 1$, $y(t) = 2t$; $-\infty < t < \infty$

In Problems 21–36:

(a) *Find a rectangular equation of each plane curve with the given parametric equations.*

(b) *Graph the rectangular equation.*

(c) *Determine the restrictions on x and y so that the graph corresponding to the rectangular equation is identical to the plane curve.*

(d) *Graph the plane curve represented by the parametric equations.*

21. $x(t) = 2t$, $y(t) = t^2 + 4$; $t > 0$

22. $x(t) = t + 3$, $y(t) = t^3$; $-4 \le t \le 4$

23. $x(t) = t + 5$, $y(t) = \sqrt{t}$; $t \ge 0$

24. $x(t) = 2t^2$, $y(t) = 2t^3$; $0 \le t \le 3$

25. $x(t) = t^{1/2} + 1$, $y(t) = t^{3/2}$; $t \ge 1$

26. $x(t) = 2e^t$, $y(t) = 1 - e^t$; $t \ge 0$

27. $x(t) = \sec t$, $y(t) = \tan t$; $-\dfrac{\pi}{2} < t < \dfrac{\pi}{2}$

28. $x(t) = \cot t, \; y(t) = \csc t; \quad 0 < t < \pi$

29. $x(t) = t^4, \quad y(t) = t^2$

30. $x(t) = t^2, \quad y(t) = t^4$

31. $x(t) = t^2, \quad y(t) = 2t - 1$

32. $x(t) = e^{2t}, \quad y(t) = 2e^t - 1$

33. $x(t) = \left(\dfrac{1}{t}\right)^2, \; y(t) = \dfrac{2}{t} - 1$

34. $x(t) = t^4, \; y(t) = 2t^2 - 1$

35. $x(t) = 3\sin^2 t - 2, \; y(t) = 2\cos t; \quad 0 \le t \le \pi$

36. $x(t) = 1 + 2\sin^2 t, \; y(t) = 2 - \cos t; \quad 0 \le t \le 2\pi$

Using Time as the Parameter *In Problems 37–42, describe the motion of an object that moves along a plane curve so that at time t it has coordinates $(x(t), y(t))$.*

37. $x(t) = \dfrac{1}{t^2}, \quad y(t) = \dfrac{2}{t^2 + 1}; \quad t > 0$

38. $x(t) = \dfrac{3t}{\sqrt{t^2 + 1}}, \quad y(t) = \dfrac{3}{\sqrt{t^2 + 1}}; \quad -\infty < t < \infty$

39. $x(t) = \dfrac{4}{\sqrt{4 - t^2}}, \quad y(t) = \dfrac{4t}{\sqrt{4 - t^2}}; \quad 0 \le t < 2$

40. $x(t) = \sqrt{t - 3}, \quad y(t) = \sqrt{t + 1}; \quad t \ge 3$

41. $x(t) = \sin t - 2, \quad y(t) = 4 - 2\cos t; \quad 0 \le t \le 2\pi$

42. $x(t) = 2 + \tan t, \quad y(t) = 3 - 2\sec t; \quad -\dfrac{\pi}{2} < t < \dfrac{\pi}{2}$

In Problems 43–50, find two different pairs of parametric equations corresponding to each rectangular equation.

43. $y = 4x - 2$

44. $y = -8x + 3$

45. $y = -2x^2 + 1$

46. $y = x^2 + 1$

47. $y = 4x^3$

48. $y = 2x^2$

49. $x = \dfrac{1}{3}\sqrt{y} - 3$

50. $x = y^{3/2}$

In Problems 51–54, find parametric equations that represent the plane curve shown.

51.

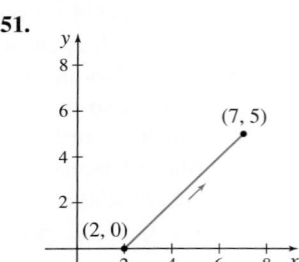

52.

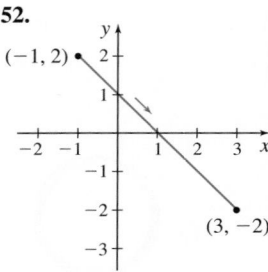

53.

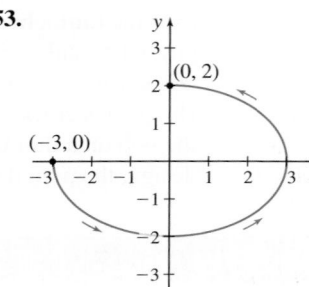

54.

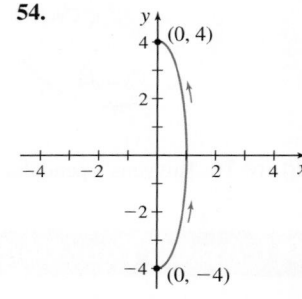

Motion of an Object *In Problems 55–58, find parametric equations for an object that moves along the ellipse $\dfrac{x^2}{9} + \dfrac{y^2}{4} = 1$, where the parameter is time (in seconds) if:*

55. the motion begins at $(3, 0)$, is counterclockwise, and requires 3 s for 1 revolution.

56. the motion begins at $(3, 0)$, is clockwise, and requires 3 s for 1 revolution.

57. the motion begins at $(0, 2)$, is clockwise, and requires 2 s for 1 revolution.

58. the motion begins at $(0, 2)$, is counterclockwise, and requires 1 s for 1 revolution.

In Problems 59 and 60, the parametric equations of four plane curves are given. Graph each curve, indicate its orientation, and compare the graphs.

59. (a) $x(t) = t, \; y(t) = t^2; \quad -4 \le t \le 4$

    (b) $x(t) = \sqrt{t}, \; y(t) = t; \quad 0 \le t \le 16$

    (c) $x(t) = e^t, \; y(t) = e^{2t}; \quad 0 \le t \le \ln 4$

    (d) $x(t) = \cos t, \; y(t) = 1 - \sin^2 t; \quad 0 \le t \le \pi$

60. (a) $x(t) = t, \; y(t) = \sqrt{1 - t^2}; \quad -1 \le t \le 1$

    (b) $x(t) = \sin t, \; y(t) = \cos t; \quad 0 \le t \le 2\pi$

    (c) $x(t) = \cos t, \; y(t) = \sin t; \quad 0 \le t \le 2\pi$

    (d) $x(t) = \sqrt{1 - t^2}, \; y(t) = t; \quad -1 \le t \le 1$

Applications and Extensions

In Problems 61–64, the parametric equations of a plane curve and its graph are given. Match each graph in I–IV to the restricted domain given in (a)–(d). Also indicate the orientation of each graph.

61.

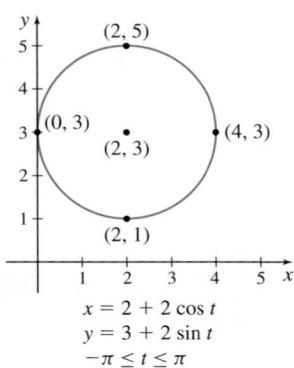

$$x = 2 + 2\cos t$$
$$y = 3 + 2\sin t$$
$$-\pi \le t \le \pi$$

(a) $\left[-\dfrac{\pi}{2}, \dfrac{\pi}{2}\right]$ (b) $\left[-\pi, \dfrac{\pi}{2}\right]$ (c) $[-\pi, 0]$ (d) $[0, \pi]$

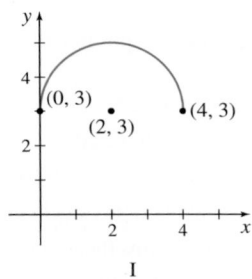

I

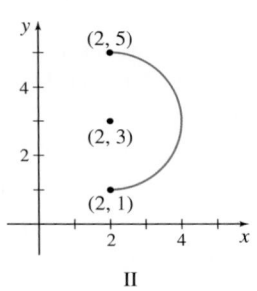

II

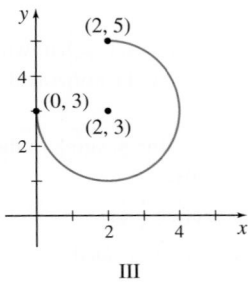

III

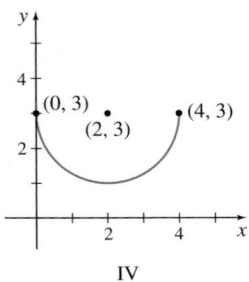

IV

62.

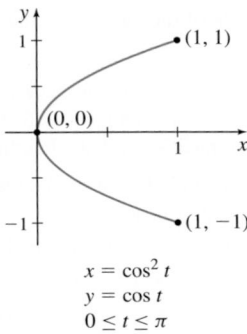

$$x = \cos^2 t$$
$$y = \cos t$$
$$0 \le t \le \pi$$

(a) $\left[\dfrac{\pi}{2}, \pi\right]$ (b) $\left[0, \dfrac{\pi}{3}\right]$ (c) $\left[0, \dfrac{\pi}{2}\right]$ (d) $\left[\dfrac{\pi}{3}, \pi\right]$

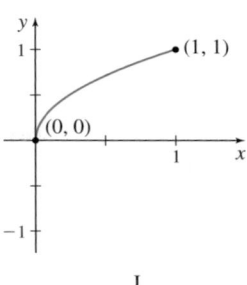

I

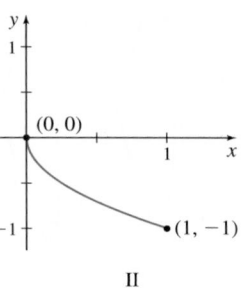

II

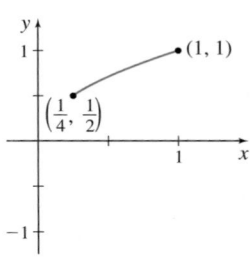

III

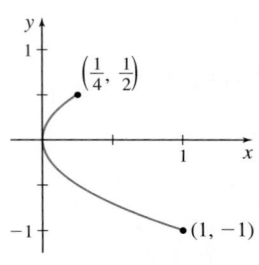

IV

63.

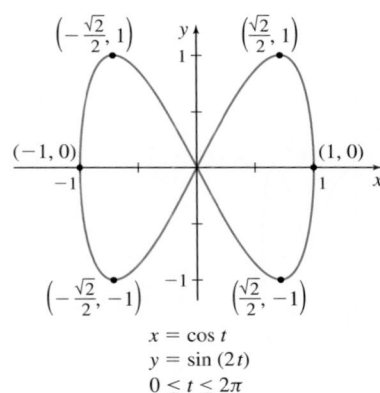

$$x = \cos t$$
$$y = \sin(2t)$$
$$0 \le t \le 2\pi$$

(a) $\left[\dfrac{\pi}{2}, \dfrac{3\pi}{2}\right]$ (b) $[\pi, 2\pi]$ (c) $[0, \pi]$ (d) $\left[\dfrac{5\pi}{4}, 2\pi\right]$

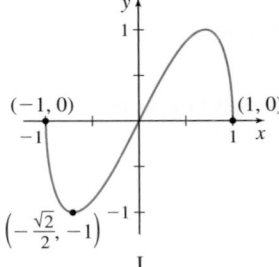

I

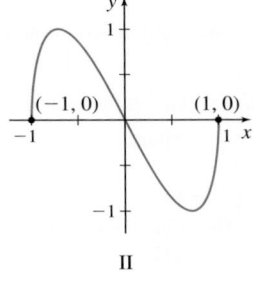

II

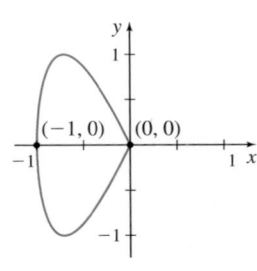

III

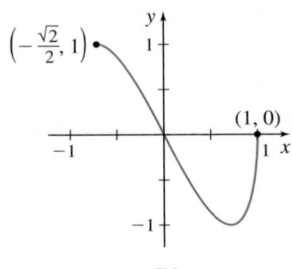

IV

64.

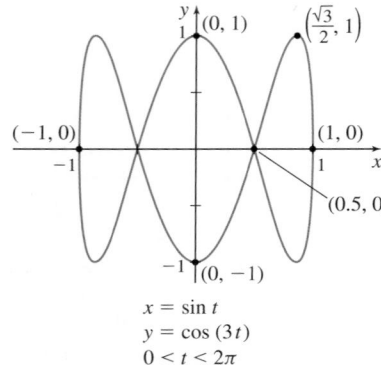

$x = \sin t$
$y = \cos(3t)$
$0 \le t \le 2\pi$

(a) $\left[0, \dfrac{5\pi}{6}\right]$ **(b)** $[\pi, 2\pi]$ **(c)** $\left[0, \dfrac{\pi}{2}\right]$ **(d)** $\left[\dfrac{\pi}{6}, \dfrac{5\pi}{6}\right]$

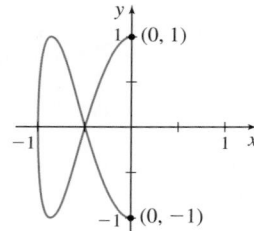

I

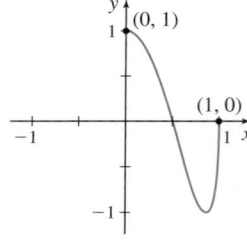

II

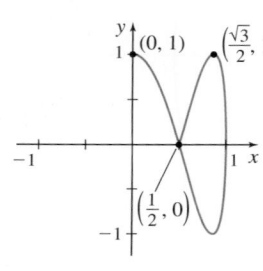

III

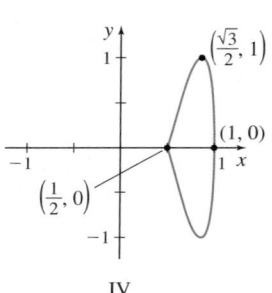

IV

65. (a) Graph the plane curve represented by the parametric equations

$$x(t) = 7(t - \sin t), \; y(t) = 7(1 - \cos t), \; 0 \le t \le 2\pi$$

(b) Find the coordinates of the point (x, y) on the curve when $t = 2.1$.

66. (a) Graph the plane curve represented by the parametric equations

$$x(t) = t - e^t, \; y(t) = 2e^{t/2}, \; -8 \le t \le 2$$

(b) Find the coordinates of the point (x, y) on the curve when $t = 1.5$.

67. Find parametric equations for the ellipse

$$\frac{x^2}{a^2} + \frac{y^2}{b^2} = 1.$$

68. Find parametric equations for the hyperbola

$$\frac{x^2}{a^2} - \frac{y^2}{b^2} = 1.$$

*Problems 69–71 involve projectile motion. When an object is propelled upward at an inclination θ to the horizontal with initial speed v_0, the resulting motion is called **projectile motion**. See the figure. Parametric equations that model the path of the projectile, ignoring air resistance, are given by*

$$x(t) = (v_0 \cos \theta)t \qquad y(t) = -\frac{1}{2}gt^2 + (v_0 \sin \theta)t + h$$

where t is time, g is the constant acceleration due to gravity (approximately 32 ft/s^2 or 9.8 m/s^2), and h is the height from which the projectile is released.

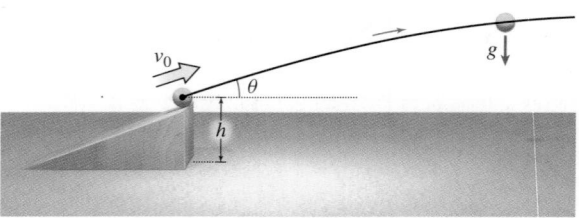

69. Trajectory of a Baseball A baseball is hit with an initial speed of 125 ft/s at an angle of 40° to the horizontal. The ball is hit at a height of 3 ft above the ground.

(a) Find parametric equations that model the position of the ball as a function of the time t in seconds.

(b) What is the height of the baseball after 2 s?

(c) What horizontal distance x has the ball traveled after 2 s?

(d) How long does it take the baseball to travel $x = 300$ ft?

(e) What is the height of the baseball at the time found in (d)?

(f) How long is the ball in the air before it hits the ground?

(g) How far has the baseball traveled horizontally when it hits the ground?

70. Trajectory of a Baseball A pitcher throws a baseball with an initial speed of 145 ft/s at an angle of 20° to the horizontal. The ball leaves his hand at a height of 5 ft.

(a) Find parametric equations that model the position of the ball as a function of the time t in seconds.

(b) What is the height of the baseball after $\dfrac{1}{2}$ s?

(c) What horizontal distance x has the ball traveled after $\dfrac{1}{2}$ s?

(d) How long does it take the baseball to travel $x = 60$ ft?

(e) What is the height of the baseball at the time found in (d)?

71. Trajectory of a Football A quarterback throws a football with an initial speed of 80 ft/s at an angle of 35° to the horizontal. The ball leaves the quarterback's hand at a height of 6 ft.

(a) Find parametric equations that model the position of the ball as a function of time t in seconds.

(b) What is the height of the football after 1 s?

(c) What horizontal distance x has the ball traveled after 1 s?

(d) How long does it take the football to travel $x = 120$ ft?

(e) What is the height of the football at the time found in (d)?

72. The plane curve represented by the parametric equations $x(t) = t$, $y(t) = t^2$, and the plane curve represented by the parametric equations $x(t) = t^2$, $y(t) = t^4$ (where, in each case, time t is the parameter) appear to be identical, but they differ in an important aspect. Identify the difference, and explain its meaning.

73. The circle $x^2 + y^2 = 4$ can be represented by the parametric equations $x(\theta) = 2\cos\theta$, $y(\theta) = 2\sin\theta$, $0 \le \theta \le 2\pi$, or by the parametric equations $x(\theta) = 2\sin\theta$, $y(\theta) = 2\cos\theta$, $0 \le \theta \le 2\pi$. But the plane curves represented by each pair of parametric equations are different. Identify and explain the difference.

74. Uniform Motion   A train leaves a station at 7:15 a.m. and accelerates at the rate of 3 mi/h². Mary, who can run 6 mi/h, arrives at the station 2 s after the train has left. Find parametric equations that model the motion of the train and of Mary as a function of time.
Hint: The position s at time t of an object having acceleration a is $s = \dfrac{1}{2}at^2$.

Challenge Problems

75. Find parametric equations for the circle $x^2 + y^2 = R^2$, using as the parameter the slope m of the line through the point $(-R, 0)$ and a general point $P = (x, y)$ on the circle.

76. Find parametric equations for the parabola $y = x^2$, using as the parameter the slope m of the line joining the point $(1, 1)$ to a general point $P = (x, y)$ on the parabola.

77. Hypocycloid   Let a circle of radius b roll, without slipping, inside a fixed circle with radius a, where $a > b$. A fixed point P on the circle of radius b traces out a curve, called a **hypocycloid**, as shown in the figure. If $A = (a, 0)$ is the initial position of the point P and if t denotes the angle from the positive x-axis to the line segment from the origin to the center of the circle, show that the parametric equations of the hypocycloid are

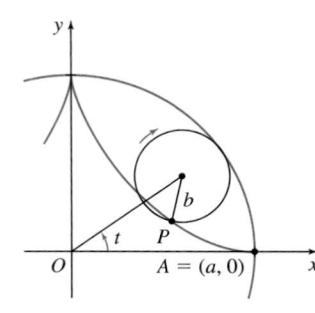

$$x(t) = (a - b)\cos t + b\cos\left(\frac{a-b}{b}t\right)$$

$$y(t) = (a - b)\sin t - b\sin\left(\frac{a-b}{b}t\right) \quad 0 \le t \le 2\pi$$

78. Hypocycloid   Show that the rectangular equation of a hypocycloid with $a = 4b$ is $x^{2/3} + y^{2/3} = a^{2/3}$.

79. Epicycloid   Suppose a circle of radius b rolls on the outside of a second circle, as shown in the figure. Find the parametric equations of the curve, called an **epicycloid**, traced out by the point P.

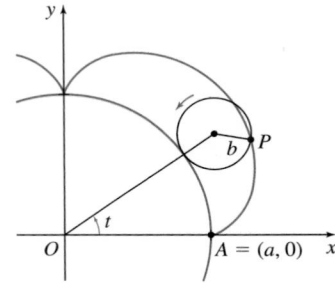

9.2 Tangent Lines; 2nd-Order Derivatives; Arc Length

OBJECTIVES   *When you finish this section, you should be able to:*

1. **Find an equation of the tangent line at a point on a plane curve (p. 721)**
2. **Find 2nd-order derivatives of parametric equations (p. 723)**
3. **Find the arc length of a plane curve (p. 724)**
4. **Investigate the motion of an object moving along a plane curve (p. 725)**

Recall that if a function f is differentiable, the derivative $f'(x) = \dfrac{dy}{dx}$ is the slope of the tangent line to the graph of f at a point (x, y). To obtain a formula for the slope of the tangent line to a curve when it is represented by parametric equations, $x = x(t)$, $y = y(t)$, $a \le t \le b$, we require the curve to be *smooth*.

DEFINITION Smooth Curve

Let C denote a plane curve represented by the parametric equations

$$x = x(t) \qquad y = y(t) \qquad a \le t \le b$$

Suppose each function $x(t)$ and $y(t)$ is continuous on the closed interval $[a, b]$ and differentiable on the open interval (a, b). If both $\dfrac{dx}{dt}$ and $\dfrac{dy}{dt}$ are continuous and are never simultaneously 0 on (a, b), then C is called a **smooth curve** on $[a, b]$.

A smooth curve $x = x(t)$, $y = y(t)$, for which $\dfrac{dx}{dt}$ is never 0, can be represented by the rectangular equation $y = F(x)$, where F is differentiable. (You are asked to prove this in Problem 57.) Then

$$y = F(x)$$

$$y(t) = F(x(t))$$

$$\frac{dy}{dt} = F'(x(t))\frac{dx}{dt} \qquad \text{Use the Chain Rule.}$$

$$\frac{dy}{dt} = \frac{dy}{dx} \cdot \frac{dx}{dt} \qquad\qquad \frac{dy}{dx} = F'(x)$$

$$\frac{dy}{dx} = \frac{\dfrac{dy}{dt}}{\dfrac{dx}{dt}} \qquad\qquad \text{Divide by } \frac{dx}{dt} \neq 0.$$

Since $\dfrac{dy}{dx}$ is the slope of the tangent line to the graph of $y = F(x)$, we have proved the following theorem.

THEOREM Slope of the Tangent Line to a Smooth Curve

For a smooth curve C represented by the parametric equations $x = x(t)$, $y = y(t)$, $a \leq t \leq b$, the slope of the tangent line to C at the point (x, y) is given by

$$\boxed{\frac{dy}{dx} = \frac{\dfrac{dy}{dt}}{\dfrac{dx}{dt}} \qquad \frac{dx}{dt} \neq 0} \tag{1}$$

For a smooth curve, C, $\dfrac{dx}{dt}$ and $\dfrac{dy}{dt}$ are never simultaneously zero. There are three possibilities:

- At a number t where $\dfrac{dx}{dt} = 0$, then $\dfrac{dy}{dt} \neq 0$ and the smooth curve C has a **vertical tangent line**.

- At a number t where $\dfrac{dy}{dt} = 0$, then $\dfrac{dx}{dt} \neq 0$ and the smooth curve C has a **horizontal tangent line**.

- At a number t where neither $\dfrac{dy}{dt}$ nor $\dfrac{dx}{dt}$ equal 0, then the smooth curve C has a tangent line that is neither horizontal nor vertical.

1 Find an Equation of the Tangent Line at a Point on a Plane Curve

NEED TO REVIEW? Equations of tangent lines are discussed in Section 2.1, pp. 167–169.

Now that we have a way to find the slope of the tangent line to a smooth curve C at a point, we can use the point-slope form of a line to obtain an equation of the tangent line.

EXAMPLE 1 Finding an Equation of the Tangent Line to a Smooth Curve

(a) Find an equation of the tangent line to the plane curve with parametric equations $x(t) = 3t^2$, $y(t) = 2t$, when $t = 1$.

(b) Find all the points on the plane curve at which the tangent line is vertical.

Solution

(a) The curve is smooth $\left(\dfrac{dy}{dt} = 2 \text{ is never zero} \right)$, so $\dfrac{dy}{dt}$ and $\dfrac{dx}{dt}$ are never simultaneously zero. Since $\dfrac{dx}{dt} = 6t$ is not 0 at $t = 1$, the slope of the tangent line to the curve is

$$\frac{dy}{dx} = \frac{\dfrac{dy}{dt}}{\dfrac{dx}{dt}} = \frac{\dfrac{d}{dt}(2t)}{\dfrac{d}{dt}(3t^2)} = \frac{2}{6t} = \frac{1}{3t}$$

When $t = 1$, the slope of the tangent line is $\dfrac{1}{3}$. Since $x(1) = 3$ and $y(1) = 2$, an equation of the tangent line is

$$y - 2 = \frac{1}{3}(x - 3) \qquad\quad y - y_1 = m(x - x_1); \quad x_1 = 3,\ y_1 = 2,\ m = \frac{1}{3}$$

$$y = \frac{1}{3}x + 1$$

(b) A vertical tangent line occurs when $\dfrac{dx}{dt} = 0$ and $\dfrac{dy}{dt} \neq 0$. Since $\dfrac{dx}{dt} = 6t = 0$ and $\dfrac{dy}{dt} = 2$ when $t = 0$, there is a vertical tangent line (the y-axis) to the curve at the point $(0, 0)$. ∎

Figure 15 illustrates the results of Example 1.

Figure 15 $x(t) = 3t^2$, $y(t) = 2t$

NOW WORK Problem 17 and AP® Practice Problems 1 and 2.

EXAMPLE 2 Finding the Slope of the Tangent Line to a Cycloid

Consider the cycloid

$$x(t) = a(t - \sin t) \qquad y(t) = a(1 - \cos t) \qquad 0 < t < 2\pi \qquad a > 0$$

(a) Show that the slope of any tangent line to the cycloid is given by $\dfrac{\sin t}{1 - \cos t}$.

(b) Find any points where the tangent line to the cycloid is horizontal.

(c) Find an equation of the tangent line to the cycloid at the point(s) found in (b).

Solution

(a)

$$x(t) = at - a\sin t \qquad y(t) = a - a\cos t$$

$$\frac{dx}{dt} = a - a\cos t \qquad \frac{dy}{dt} = a\sin t$$

For $0 < t < 2\pi$, $\dfrac{dx}{dt} = a(1 - \cos t) \neq 0$, so the cycloid is a smooth curve. Then the slope of any tangent line is

$$\frac{dy}{dx} = \frac{\dfrac{dy}{dt}}{\dfrac{dx}{dt}} = \frac{a\sin t}{a(1 - \cos t)} = \frac{\sin t}{1 - \cos t}$$

(b) The cycloid has a horizontal tangent line when $\dfrac{dy}{dt} = a\sin t = 0$, but $\dfrac{dx}{dt} \neq 0$.

For $0 < t < 2\pi$, $a\sin t = 0$ when $t = \pi$. Since $\dfrac{dx}{dt} \neq 0$ for $0 < t < 2\pi$, the cycloid has a horizontal tangent line when $t = \pi$, at the point $(\pi a, 2a)$.

(c) An equation of the horizontal tangent line when $t = \pi$ is $y = 2a$. ∎

See Figure 16.

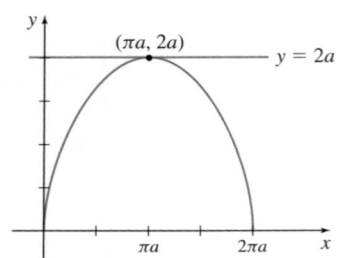

Figure 16 $x(t) = a(t - \sin t)$,
$y(t) = a(1 - \cos t), 0 < t < 2\pi$

NOW WORK Problem 27 and AP® Practice Problem 3.

EXAMPLE 3 Finding an Equation of a Tangent Line to a Smooth Curve

The plane curve represented by the parametric equations $x(t) = t^3 - 4t$, $y(t) = t^2$ crosses itself at the point $(0, 4)$, and so has two tangent lines there, as shown in Figure 17. Find equations for these tangent lines.

Solution

Since $\dfrac{dx}{dt} = 3t^2 - 4 = 0$ only for $t = \dfrac{2\sqrt{3}}{3}$ and for $t = -\dfrac{2\sqrt{3}}{3}$, and $\dfrac{dy}{dt} = 2t = 0$ only for $t = 0$, there is no t for which $\dfrac{dx}{dt}$ and $\dfrac{dy}{dt}$ are simultaneously 0. So the plane curve is a smooth curve.

Next we find the numbers t that correspond to the point $(0, 4)$.

$$
\begin{array}{ll}
x = 0 & y = 4 \\
t^3 - 4t = 0 & t^2 = 4 \\
t(t^2 - 4) = 0 & t = -2 \text{ or } 2 \\
t = 0, -2, \text{ or } 2 &
\end{array}
$$

We exclude $t = 0$, since if $t = 0$, then $y \neq 4$. So we investigate only $t = -2$ and $t = 2$. Since $\dfrac{dx}{dt} = 3t^2 - 4 \neq 0$ for $t = -2$ and $t = 2$, the slopes of the tangent lines are given by

$$\frac{dy}{dx} = \frac{\dfrac{dy}{dt}}{\dfrac{dx}{dt}} = \frac{\dfrac{d}{dt}(t^2)}{\dfrac{d}{dt}(t^3 - 4t)} = \frac{2t}{3t^2 - 4}$$

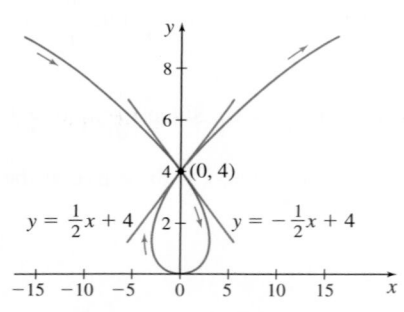

Figure 17 $x(t) = t^3 - 4t, y(t) = t^2$

When $t = -2$, the slope of the tangent line is

$$\frac{dy}{dx} = \frac{2(-2)}{3(-2)^2 - 4} = \frac{-4}{8} = -\frac{1}{2}$$

and an equation of the tangent line at the point $(0, 4)$ is

$$y - 4 = -\frac{1}{2}(x - 0)$$

$$y = -\frac{1}{2}x + 4$$

When $t = 2$, the slope of the tangent line at the point $(0, 4)$ is

$$\frac{dy}{dx} = \frac{2 \cdot 2}{3 \cdot 2^2 - 4} = \frac{4}{8} = \frac{1}{2}$$

and an equation of the tangent line at $(0, 4)$ is

$$y - 4 = \frac{1}{2}(x - 0)$$

$$y = \frac{1}{2}x + 4$$

∎

NOW WORK   Problem 47 and AP® Practice Problem 8.

❷ Find 2nd-Order Derivatives of Parametric Equations

Suppose the parametric equations of a smooth curve are given by $x = x(t)$, $y = y(t)$, for $a \leq t \leq b$. Then $\dfrac{dx}{dt}$ and $\dfrac{dy}{dt}$ are differentiable functions of t, and $\dfrac{dx}{dt}$ and $\dfrac{dy}{dt}$ are never simultaneously zero on (a, b).

If $\dfrac{dx}{dt}$ is never 0 on (a, b), then the curve can be represented by a differentiable function $y = F(x)$, whose derivative is

$$y' = \frac{dy}{dx} = \frac{\dfrac{dy}{dt}}{\dfrac{dx}{dt}} \qquad \frac{dx}{dt} \neq 0$$

Suppose $y = F(x)$ is twice differentiable. Then

$$y'' = \frac{d}{dx}\left(\frac{dy}{dx}\right) = \frac{1}{\dfrac{dx}{dt}} \frac{d}{dt}\left(\frac{\dfrac{dy}{dt}}{\dfrac{dx}{dt}}\right) \qquad \frac{d}{dx} = \frac{1}{\dfrac{dx}{dt}} \frac{d}{dt} \tag{2}$$

EXAMPLE 4   **Finding 2nd-Order Derivatives of Parametric Equations**

(a) Find $y'' = \dfrac{d^2 y}{dx^2}$ of the function $y = F(x)$ defined by the parametric equations

$$x = x(t) = t^3 + 4t + 12 \qquad y = y(t) = t^2$$

for any real number, t.

(b) Evaluate y'' for $t = 1$.

```
NORMAL FLOAT AUTO REAL RADIAN MP

1/ d/dT (T³+4T+12)|_T=1
                       .1428571224
Ans* d/dT (2T/(3T²+4))|_T=1
                       .0058309197
```

Figure 18 $y''(1)$

Solution

(a) Begin by finding $\dfrac{dx}{dt}$ and $\dfrac{dy}{dt}$.

$$\frac{dx}{dt} = 3t^2 + 4 \qquad \frac{dy}{dt} = 2t$$

Then

$$y' = \frac{dy}{dx} = \frac{2t}{3t^2 + 4}$$

$$y'' = \frac{d^2y}{dx^2} = \frac{d}{dx}\left(\frac{dy}{dx}\right) = \frac{1}{\dfrac{dx}{dt}}\frac{d}{dt}\left(\frac{\dfrac{dy}{dt}}{\dfrac{dx}{dt}}\right)$$

$$= \frac{1}{3t^2+4}\frac{d}{dt}\left(\frac{2t}{3t^2+4}\right) = \frac{1}{3t^2+4}\cdot\left[\frac{2(3t^2+4)-2t\cdot 6t}{(3t^2+4)^2}\right] = \frac{8-6t^2}{(3t^2+4)^3}$$

(b) When $t = 1$,

$$y''(1) = \frac{2}{7^3} \approx 0.006$$

∎

NOW WORK Problem 31 and AP® Practice Problem 6.

3 Find the Arc Length of a Plane Curve

For a function $y = f(x)$ that has a derivative that is continuous on some interval containing a and b, the arc length s of the graph of f from a to b is given by $s = \int_a^b \sqrt{1+[f'(x)]^2}\,dx$. For a smooth curve represented by the parametric equations $x = x(t)$, $y = y(t)$, we have the following formula for arc length.

NEED TO REVIEW? Arc length is discussed in Section 8.6, pp. 674–679.

> **THEOREM** Arc Length Formula for Parametric Equations
>
> For a smooth curve C represented by the parametric equations
>
> $$x = x(t) \qquad y = y(t) \qquad a \le t \le b$$
>
> the arc length s of C from $t = a$ to $t = b$ is given by the formula
>
> $$s = \int_a^b \sqrt{\left(\frac{dx}{dt}\right)^2 + \left(\frac{dy}{dt}\right)^2}\,dt$$

CALC CLIP **EXAMPLE 5** Finding Arc Length for Parametric Equations

Find the length s of the smooth curve represented by the parametric equations

$$x(t) = t^3 + 2 \qquad y(t) = 2t^{9/2}$$

from the point where $t = 1$ to the point where $t = 3$. Figure 19 shows the graph of the curve.

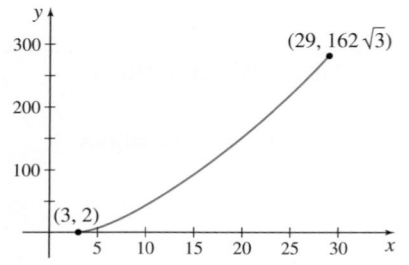

Figure 19 $x(t) = t^3 + 2$, $y(t) = 2t^{9/2}$, $1 \le t \le 3$

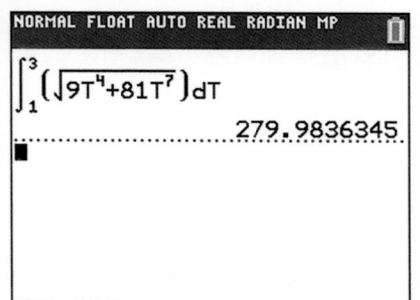

Figure 20 Arc length of the smooth curve traced out by $x(t) = t^3 + 2$, $y(t) = 2t^{9/2}$, $1 \le t \le 3$.

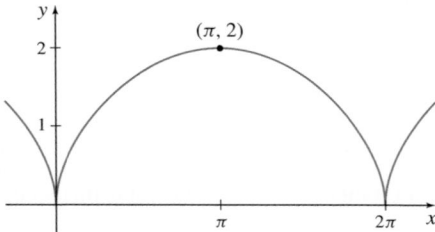

Figure 21 $x(t) = t - \sin t$
$y(t) = 1 - \cos t$

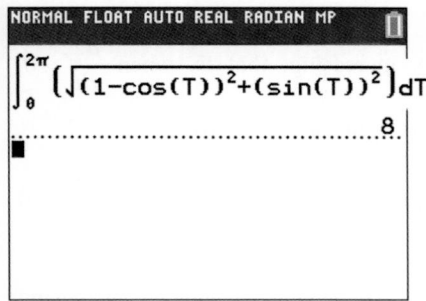

Figure 22 The arc length of one arch of a cycloid.

Solution

We begin by finding the derivatives $\dfrac{dx}{dt}$ and $\dfrac{dy}{dt}$.

$$\frac{dx}{dt} = 3t^2 \quad \text{and} \quad \frac{dy}{dt} = 9t^{7/2}$$

The curve is smooth for $1 \le t \le 3$. Using the arc length formula for parametric equations, we have

$$s = \int_a^b \sqrt{\left(\frac{dx}{dt}\right)^2 + \left(\frac{dy}{dt}\right)^2}\, dt = \int_1^3 \sqrt{(3t^2)^2 + (9t^{7/2})^2}\, dt$$

$$= \int_1^3 \sqrt{9t^4 + 81t^7}\, dt \approx 279.984$$

NOW WORK Problem 35 and AP® Practice Problem 4.

EXAMPLE 6 Finding the Arc Length of a Cycloid

Find the length s of one arch of the cycloid:

$$x(t) = t - \sin t \qquad y(t) = 1 - \cos t$$

Figure 21 shows the graph of the cycloid.

Solution

One arch of the cycloid is obtained when t varies from 0 to 2π. Since

$$\frac{dx}{dt} = 1 - \cos t \quad \text{and} \quad \frac{dy}{dt} = \sin t$$

are both continuous and are never simultaneously 0 on $(0, 2\pi)$, the cycloid is a smooth curve on $[0, 2\pi]$. The arc length s is

$$s = \int_a^b \sqrt{\left(\frac{dx}{dt}\right)^2 + \left(\frac{dy}{dt}\right)^2}\, dt = \int_0^{2\pi} \sqrt{(1 - \cos t)^2 + \sin^2 t} = 8$$

See Figure 22. ■

NOW WORK Problem 39 and AP® Practice Problem 5.

4 Investigate the Motion of an Object Moving Along a Plane Curve

In Section 9.1, we stated that parametric equations are particularly useful for describing motion along a curve. In particular, if t represents time, then the parametric equations

$$x = x(t) \qquad y = y(t) \qquad a \le t \le b$$

model the position of an object moving along a curve, as well as the time t when the object is at that position. In the parametric equations, $x = x(t)$ controls the object's motion right (east) and left (west), and $y = y(t)$ controls the object's motion up (north) and down (south). So, we say that the object is **moving to the right** (x is increasing) when $\dfrac{dx}{dt} > 0$ and **moving to the left** (x is decreasing) when $\dfrac{dx}{dt} < 0$. Similarly, when $\dfrac{dy}{dt} > 0$, y is increasing, so the object is **moving up**, and when $\dfrac{dy}{dt} < 0$, y is decreasing, so the object is **moving down**.

If an object is moving along a smooth curve C, then the arc length of the path traced out by the object as it moves from a to b measures the *total distance traveled* by the object.

DEFINITION Total Distance Traveled

If a smooth curve C is defined by the parametric equations $x = x(t)$, $y = y(t)$ for $a \leq t \leq b$, $a < b$, then the **total distance traveled by the object from $t = a$ to $t = b$** is

$$\int_a^b \sqrt{\left(\frac{dx}{dt}\right)^2 + \left(\frac{dy}{dt}\right)^2}\, dt$$

EXAMPLE 7 **Investigating the Motion of an Object Along a Smooth Curve**

As an object moves along a smooth curve C, its position at time t is $(x(t), y(t))$ for $0 \leq t \leq 10$. The object moves in such a way that

$$\frac{dx}{dt} = t^3 - 8 \qquad \frac{dy}{dt} = t^2 - 9$$

(a) What is the total distance traveled by the object from $t = 0$ to $t = 10$?

(b) For what times t is the object moving to the right?

(c) For what times t is the object moving down?

(d) At what time(s) is the object lowest?

Solution

(a) C is smooth since $\dfrac{dx}{dt}$ and $\dfrac{dy}{dt}$ are never simultaneously zero. The total distance traveled by the object from $t = 0$ to $t = 10$ is

$$\int_0^{10} \sqrt{\left(t^3 - 8\right)^2 + \left(t^2 - 9\right)^2}\, dt = 2467.052 \qquad \text{Use a calculator.}$$

(b) The object moves to the right when x is increasing, that is, when $\dfrac{dx}{dt} > 0$. Since $\dfrac{dx}{dt} > 0$ for $t > 2$, the object is moving to the right for $2 \leq t \leq 10$.

(c) The object moves down when y is decreasing, that is, when $\dfrac{dy}{dt} < 0$. Since $\dfrac{dy}{dt} < 0$ for $t < 3$, the object is moving down for $0 \leq t \leq 3$.

(d) The height of the object is decreasing for $0 \leq t \leq 3$ and is increasing for $3 \leq t \leq 10$ $\left(\dfrac{dy}{dt} > 0\right)$.

NEED TO REVIEW? The First Derivative Test is discussed in Section 5.3, p. 362.

So, by the First Derivative Test, at $t = 3$, the height is a minimum. ∎

NOW WORK Problem 51 and AP® Practice Problems 7, 9, and 10.

9.2 Assess Your Understanding

Concepts and Vocabulary

1. **Multiple Choice** Let C denote a plane curve represented by the parametric equations $x = x(t)$, $y = y(t)$, $a \leq t \leq b$, where each function $x(t)$ and $y(t)$ is continuous on the closed interval $[a, b]$ and differentiable on the open interval (a, b). If both $\dfrac{dx}{dt}$ and $\dfrac{dy}{dt}$ are continuous and never simultaneously 0 on (a, b), then C is called a _____ curve.

  (a) smooth   (b) differentiable   (c) parametric

2. **True or False** If C is a smooth curve, represented by the parametric equations $x = x(t)$ and $y = y(t)$, $a \leq t \leq b$, then the slope of the tangent line to C at the point (x, y) is given by the formula $\dfrac{dy}{dx} = \dfrac{\dfrac{dy}{dt}}{\dfrac{dx}{dt}}$, provided $\dfrac{dx}{dt} \neq 0$.

3. If in the formula for the slope of a tangent line, $\dfrac{dy}{dt} = 0$, but $\dfrac{dx}{dt} \neq 0$, then the curve has a(n) _____ tangent line at the point $(x(t), y(t))$. If $\dfrac{dx}{dt} = 0$, but $\dfrac{dy}{dt} \neq 0$, then the curve has a(n) _____ tangent line at the point $(x(t), y(t))$.

4. **True or False** A smooth curve $x = x(t)$, $y = y(t)$, $a \leq t \leq b$, can be represented by a rectangular equation $y = F(x)$ if $\dfrac{dy}{dt}$ is never 0 on (a, b).

5. **True or False** For a smooth curve C represented by the parametric equations $x = x(t)$, $y = y(t)$, $a \leq t \leq b$, the arc length s of C from $t = a$ to $t = b$ is given by the formula

$$s = \int_a^b \sqrt{\dfrac{d^2x}{dt^2} + \dfrac{d^2y}{dt^2}}\, dt$$

6. **True or False** If a smooth curve is represented by the parametric equations $x = x(t)$, $y = y(t)$, $a \leq t \leq b$, and if $\dfrac{dx}{dt} = 0$ for any number t, $a \leq t \leq b$, then the arc length formula cannot be used.

Skill Building

In Problems 7–14, find $\dfrac{dy}{dx}$. Assume $\dfrac{dx}{dt} \neq 0$.

7. $x(t) = e^t \cos t$, $y(t) = e^t \sin t$

8. $x(t) = 1 + e^{-t}$, $y(t) = e^{3t}$

9. $x(t) = t + \dfrac{1}{t}$, $y(t) = 4 + t$

10. $x(t) = t + \dfrac{1}{t}$, $y(t) = t - \dfrac{1}{t}$

11. $x(t) = \cos t + t \sin t$, $y(t) = \sin t - t \cos t$

12. $x(t) = \cos^3 t$, $y(t) = \sin^3 t$

13. $x(t) = \cot^2 t$, $y(t) = \cot t$ 14. $x(t) = \sin t$, $y(t) = \sec^2 t$

In Problems 15–28, for each pair of parametric equations:

  (a) Find an equation of the tangent line to the curve at the given number.

  (b) Graph the curve and the tangent line.

15. $x(t) = 2t^2$, $y(t) = t$ at $t = 2$

16. $x(t) = t$, $y(t) = 3t^2$ at $t = -2$

17. $x(t) = 3t$, $y(t) = 2t^2 - 1$ at $t = 1$

18. $x(t) = 2t$, $y(t) = t^2 - 2$ at $t = 2$

19. $x(t) = \sqrt{t}$, $y(t) = \dfrac{1}{t}$ at $t = 4$

20. $x(t) = \dfrac{2}{t^2}$, $y(t) = \dfrac{1}{t}$ at $t = 1$

21. $x(t) = \dfrac{t}{t+2}$, $y(t) = \dfrac{4}{t+2}$ at $t = 0$

22. $x(t) = \dfrac{t^2}{1+t}$, $y(t) = \dfrac{1}{1+t}$ at $t = 0$

23. $x(t) = e^t$, $y(t) = e^{-t}$ at $t = 0$

24. $x(t) = e^{2t}$, $y(t) = e^t$ at $t = 0$

25. $x(t) = \sin t$, $y(t) = \cos t$ at $t = \dfrac{\pi}{4}$

26. $x(t) = \sin^2 t$, $y(t) = \cos t$ at $t = \dfrac{\pi}{4}$

In Problems 27–34, an object is moving along the given smooth curve.

  (a) Find any points at which the tangent line is horizontal.

  (b) Find any points at which the tangent line is vertical.

  (c) Find any intervals on which the object is moving up.

  (d) Find any intervals on which the object is moving down.

  (e) Find any intervals on which the object is moving left.

  (f) Find any intervals on which the object is moving right.

27. $x(t) = t^2 - 4t$, $y(t) = t^3 - 3t$, $0 \leq t \leq 4$

28. $x(t) = t^3 - 12t$, $y(t) = 2t^2 - 4t$, $0 \leq t \leq 4$

29. $x(t) = 1 - \cos t$, $y(t) = 1 - \sin t$, $0 \leq t \leq \dfrac{\pi}{2}$

30. $x(t) = 1 + \sin t$, $y(t) = 1 - \cos t$, $0 \leq t \leq 2\pi$

31. $x(t) = -3 \sin t$, $y(t) = 4 \cos t$, $0 \leq t \leq 2\pi$

32. $x(t) = -3 \cos t$, $y(t) = 4 \sin t$, $0 \leq t \leq 2\pi$

33. $x(t) = e^t$, $y(t) = e^{-t}$, $-1 \leq t \leq 4$

34. $x(t) = e^{2t}$, $y(t) = e^t$, $-1 \leq t \leq 4$

In Problems 35–42, find the arc length of each curve on the given interval.

[PAGE 725] 35. $x(t) = t^3$, $y(t) = t^2$; $0 \le t \le 2$

36. $x(t) = 3t^2 + 1$, $y(t) = t^3 - 1$; $0 \le t \le 2$

37. $x(t) = t - 1$, $y(t) = \dfrac{1}{2}t^2$; $0 \le t \le 2$

38. $x(t) = t^2$, $y(t) = 2t$; $1 \le t \le 3$

[PAGE 725] 39. $x(t) = 4 \sin t$, $y(t) = 4 \cos t$; $-\dfrac{\pi}{2} \le t \le \dfrac{\pi}{2}$

40. $x(t) = 6 \sin t$, $y(t) = 6 \cos t$; $-\dfrac{\pi}{2} \le t \le \dfrac{\pi}{2}$

41. $x(t) = 2 \sin t - 1$, $y(t) = 2 \cos t + 1$; $0 \le t \le 2\pi$

42. $x(t) = e^t \sin t$, $y(t) = e^t \cos t$; $0 \le t \le \pi$

In Problems 43–46,

(a) *Use the arc length formula for parametric equations to set up the integral for finding the length of each curve on the given interval.*

(b) *Find the length s of each curve on the given interval.*

(c) *Graph each curve over the given interval.*

43. $x(t) = 2 \cos(2t)$, $y(t) = t^2$, $0 \le t \le 2\pi$

44. $x(t) = t^2$, $y(t) = \sin t$, $0 \le t \le 2\pi$

45. $x(t) = t^2$, $y(t) = \sqrt{t + 2}$, $-2 \le t \le 2$

46. $x(t) = t - \cos t$, $y(t) = 1 - \sin t$, $0 \le t \le \pi$

Applications and Extensions

[PAGE 723] 47. Tangent Lines

(a) Find all the points on the plane curve C represented by

$$x(t) = t^2 + 2 \quad y(t) = t^3 - 4t$$

where the tangent line is horizontal or where it is vertical.

(b) Show that C has two tangent lines at the point $(6, 0)$.

(c) Find equations of these tangent lines.

(d) Graph C.

48. Tangent Lines

(a) Find all the points on the plane curve C represented by

$$x(t) = 2t^2 \quad y(t) = 8t - t^3$$

where the tangent line is horizontal or where it is vertical.

(b) Show that C has two tangent lines at the point $(16, 0)$.

(c) Find equations of these tangent lines.

(d) Graph C.

49. Arc Length

(a) Find the arc length of one arch of the four-cusped hypocycloid

$$x(t) = b \sin^3 t \quad y(t) = b \cos^3 t \quad 0 \le t \le \dfrac{\pi}{2}$$

(b) Graph the portion of the curve for $0 \le t \le \dfrac{\pi}{2}$.

50. Arc Length

(a) Find the arc length of the spiral

$$x(t) = t \cos t \quad y(t) = t \sin t \quad 0 \le t \le \pi$$

(b) Graph the portion of the curve for $0 \le t \le \pi$.

Distance Traveled *In Problems 51–56, find the distance a particle travels along the given path over the indicated time interval.*

[PAGE 726] 51. $x(t) = 3t$, $y(t) = t^2 - 3$; $0 \le t \le 2$

52. $x(t) = t^2$, $y(t) = 3t$; $0 \le t \le 2$

53. $x(t) = \dfrac{t^2}{2} + 1$, $y(t) = \dfrac{1}{3}(2t + 3)^{3/2}$; $0 \le t \le 2$

54. $x(t) = a \cos t$, $y(t) = a \sin t$, $a > 0$; $0 \le t \le \pi$

55. $x(t) = \cos(2t)$, $y(t) = \sin^2 t$; $0 \le t \le \dfrac{\pi}{2}$

56. $x(t) = \dfrac{1}{t}$, $y(t) = \ln t$; $1 \le t \le 2$

Challenge Problems

57. Show that a smooth curve $x = f(t)$, $y = g(t)$, for which $\dfrac{dx}{dt}$ is never 0, can be represented by a rectangular equation $y = F(x)$, where F is differentiable.

Hint: Use the fact that $t = f^{-1}(x)$ exists and is differentiable.

58. Consider the cycloid

$$x(t) = a(t - \sin t) \quad y(t) = a(1 - \cos t) \quad a > 0$$

Discuss the behavior of the tangent line to the cycloid at $t = 0$.

59. Find $\dfrac{d^2 y}{dx^2}$ if $x(\theta) = a \cos^3 \theta$, $y(\theta) = a \sin^3 \theta$.

AP® Practice Problems

Multiple-Choice Questions

[PAGE 721] **1.** The smooth curve C is represented by the parametric equations $x(t) = \tan t$, $y(t) = t^2 - 3t + 8$, $-\dfrac{\pi}{4} \le t < \dfrac{\pi}{2}$.

The slope of the line tangent to C at the point $(0, 8)$ is

(A) -3 (B) $-\dfrac{3}{2}$ (C) 0 (D) $-\dfrac{1}{3}$

[PAGE 721] **2.** Which of the following is an equation of the tangent line to the plane curve represented by the parametric equations $x(t) = 2t - 5$, $y(t) = t^3$ when $t = 2$?

(A) $y = -6x + 14$ (B) $y = 6x$

(C) $y = 6x + 2$ (D) $y = 6x + 14$

[PAGE 722] **3.** For the smooth curve, $x(t) = t^3 - 12t$, $y(t) = 4t^2 + t$, find all the numbers t for which the tangent line is either horizontal or vertical.

(A) Horizontal: $t = \dfrac{1}{8}$; Vertical: $t = -2, 2$

(B) Horizontal: $t = -2, 2$; Vertical: none

(C) Horizontal: $t = -\dfrac{1}{8}$; Vertical: $t = -2, 2$

(D) Horizontal: none, Vertical $t = -\dfrac{1}{8}$

[PAGE 725] **4.** The length of the curve represented by the parametric equations $x(t) = 2 + 4t^3$, $y(t) = 1 + 6t^2$, from $t = 0$ to $t = 1$ is

(A) $8\sqrt{2} - 4$ (B) $2^{7/2} - 1$ (C) 4 (D) $16\sqrt{2} - 8$

[PAGE 725] **5.** The smooth curve C represented by the parametric equations

$$x(t) = 2\sin t, \; y(t) = 1 + e^t, \; -\pi \le t \le 2$$

is shown below.

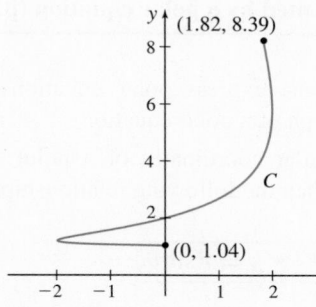

The arc length of C is given by the integral

(A) $\displaystyle\int_{-2}^{1.8} \sqrt{4\cos^2 t + e^{2t}}\, dt$

(B) $\displaystyle\int_{-\pi}^{2} \sqrt{4\sin^2 t + (1 + e^t)^2}\, dt$

(C) $\displaystyle\int_{-\pi}^{2} \sqrt{4\cos^2 t + e^{2t}}\, dt$

(D) $\displaystyle\int_{0}^{1.8} \sqrt{4\cos^2 t + e^{2t}}\, dt$

[PAGE 724] **6.** The function $y = f(x)$ is defined by the parametric equations

$$x = x(t) = t^2 \qquad y = y(t) = 2t^3 + 4t - 3$$

for any real number t. Then $y''(3)$ equals

(A) $\dfrac{25}{54}$ (B) $-\dfrac{25}{54}$ (C) $\dfrac{25}{27}$ (D) $\dfrac{50}{27}$

[PAGE 726] **7.** An object moves along the plane curve C represented by the parametric equations $x(t) = \cos(2t)$, $y(t) = \cos^2 t$. Find the total distance the object travels from $t = \dfrac{\pi}{4}$ to $t = \dfrac{\pi}{3}$.

(A) 0.250 (B) 0.599 (C) 1.118 (D) 1.250

Free-Response Questions

[PAGE 723] **8.** The plane curve represented by the parametric equations

$$x(t) = 3\sin t, \; y(t) = t^2, \; -\dfrac{3}{2}\pi \le t \le \dfrac{3}{2}\pi$$

is shown below. The graph intersects itself at the point $(0, \pi^2)$.

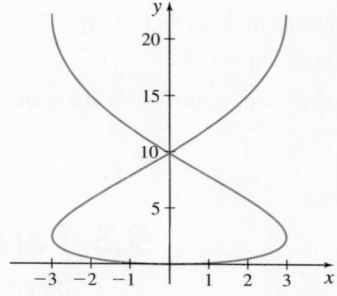

(a) Find the numbers t that correspond to the point $(0, \pi^2)$.

(b) Find $\dfrac{dx}{dt}$ and $\dfrac{dy}{dt}$ and confirm that $\dfrac{dx}{dt} \ne 0$ at the point of intersection.

(c) Find the slope of the tangent lines at the point of intersection.

(d) Find an equation of each tangent line at the point of intersection.

[PAGE 726] **9.** An object moves along a smooth curve C, and its position at time t is $(x(t), y(t))$ for $0 \le t \le 5$. The object moves in such a way that

$$\dfrac{dx}{dt} = 3t^2 - 4 \qquad \dfrac{dy}{dt} = t + 2$$

(a) Express the total distance traveled by the object from $t = 0$ to $t = 4$ as an integral.

(b) For what times t is the object moving to the left?

(c) Does the object ever move down? Justify your answer.

(d) At what time(s) is the object lowest?

See the **BREAK IT DOWN** *on page 774 for a stepped out solution to* AP® *Practice Problem 9(b).*

[PAGE 726] **10.** An object moves along the curve traced out by the parametric equations $x(t) = 3\cos t + \cos(3t)$, $y(t) = 3\sin t - \sin(3t)$, $0 \le t \le 2\pi$. Find the total distance travelled by the object from $t = 0$ to $t = 2\pi$.

Retain Your Knowledge

Multiple-Choice Questions

1. Evaluate $\displaystyle\int_0^3 \frac{x\,dx}{x^2-1}$

   (A) $\dfrac{1}{2}\ln 3$ (B) $\dfrac{1}{2}\ln\dfrac{8}{3}$

   (C) $\ln 8 - \ln 3$ (D) The integral diverges.

2. The graph of a function f, whose domain is $[-2, 8]$, is shown below. Which of the statements below is not true?

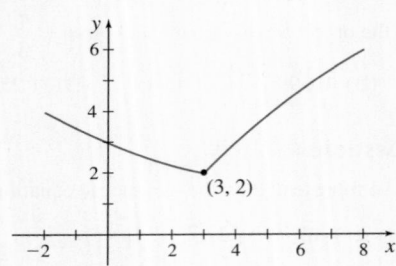

   (A) f is differentiable at all numbers in the interval $(-2, 8)$.

   (B) f is integrable on the interval $[-2, 8]$.

   (C) A critical number of f is 3.

   (D) The graph of f is not continuous at the point $(3, 2)$.

3. The function g is differentiable, and the line tangent to the graph of g at the point $(2, -5)$ contains the point $(6, 0)$. Then $g'(2)$ equals

   (A) 2   (B) $\dfrac{4}{5}$ (C) $-\dfrac{4}{5}$ (D) $\dfrac{5}{4}$

Free-Response Question

4. Functions f and g are defined as

$$f(x) = \int_0^{3x^2} \sqrt{t^3 + 4}\,dt \qquad g(x) = f(\cos x)$$

   (a) Find $f'(x)$.

   (b) Find $g'\left(\dfrac{\pi}{2}\right)$.

   (c) Express the maximum value of f on $0 \le x \le \pi$ as an integral. Do not evaluate.

9.3 Parametric Equations, Tangent Lines, and Arc Length of Polar Equations

OBJECTIVES *When you finish this section, you should be able to:*

1 **Graph a polar equation; find parametric equations and tangent lines (p. 731)**

2 **Investigate the derivative $\dfrac{dr}{d\theta}$ of a polar equation $r = f(\theta)$ (p. 735)**

3 **Find the arc length of a curve represented by a polar equation (p. 736)**

NEED TO REVIEW? Polar coordinates and polar equations are discussed in Appendix A.5, pp. A-41 to A-48.

In this section, we graph polar equations, express polar equations as parametric equations, and find tangent lines to the graph of a polar equation.

To review, if (x, y) are the rectangular coordinates of a point P in the plane and (r, θ) are the polar coordinates of P, then the following relationships hold:

$$\boxed{\bullet\ x = r\cos\theta \quad \bullet\ y = r\sin\theta} \tag{1}$$

The graph of a polar equation $r = f(\theta)$ consists of all points (r, θ) that satisfy the equation. For example, the polar equation $r\sin\theta = 2$ is equivalent to the rectangular equation $y = 2$, so the graph of $r\sin\theta = 2$ is a horizontal line 2 units above the x-axis.

① Graph a Polar Equation; Find Parametric Equations and Tangent Lines

To find parametric equations for a polar equation $r = f(\theta)$, we use θ as the parameter and the relationships given in (1):

$$\bullet \; x = x(\theta) = r\cos\theta = f(\theta)\cos\theta \qquad \bullet \; y = y(\theta) = r\sin\theta = f(\theta)\sin\theta \qquad (2)$$

To find tangent lines to the graph of a polar equation $r = f(\theta)$, where f is a differentiable function of θ, we use the parametric equations (2).

$$\frac{dx}{d\theta} = \frac{d}{d\theta}[f(\theta)\cos\theta] = f'(\theta)\cos\theta - f(\theta)\sin\theta$$

$$\frac{dy}{d\theta} = \frac{d}{d\theta}[f(\theta)\sin\theta] = f'(\theta)\sin\theta + f(\theta)\cos\theta$$

The slope of the tangent line to the graph of $r = f(\theta)$ at a point (r, θ) is given by

$$\frac{dy}{dx} = \frac{\dfrac{dy}{d\theta}}{\dfrac{dx}{d\theta}} = \frac{f'(\theta)\sin\theta + f(\theta)\cos\theta}{f'(\theta)\cos\theta - f(\theta)\sin\theta}$$

TABLE 2

θ	$r = 1 - \sin\theta$	(r, θ)
0	$1 - 0 = 1$	$(1, 0)$
$\dfrac{\pi}{6}$	$1 - \dfrac{1}{2} = \dfrac{1}{2}$	$\left(\dfrac{1}{2}, \dfrac{\pi}{6}\right)$
$\dfrac{\pi}{2}$	$1 - 1 = 0$	$\left(0, \dfrac{\pi}{2}\right)$
$\dfrac{5\pi}{6}$	$1 - \dfrac{1}{2} = \dfrac{1}{2}$	$\left(\dfrac{1}{2}, \dfrac{5\pi}{6}\right)$
π	$1 - 0 = 1$	$(1, \pi)$
$\dfrac{7\pi}{6}$	$1 - \left(-\dfrac{1}{2}\right) = \dfrac{3}{2}$	$\left(\dfrac{3}{2}, \dfrac{7\pi}{6}\right)$
$\dfrac{3\pi}{2}$	$1 - (-1) = 2$	$\left(2, \dfrac{3\pi}{2}\right)$
$\dfrac{11\pi}{6}$	$1 - \left(-\dfrac{1}{2}\right) = \dfrac{3}{2}$	$\left(\dfrac{3}{2}, \dfrac{11\pi}{6}\right)$
2π	$1 - 0 = 1$	$(1, 2\pi)$

NOTE Graphs of polar equations of the form

$r = a(1 + \cos\theta)$	$r = a(1 + \sin\theta)$
$r = a(1 - \cos\theta)$	$r = a(1 - \sin\theta)$

where $a > 0$, are called **cardioids**. A cardioid contains the pole and is heart-shape (giving the curve its name).

EXAMPLE 1   Graphing a Polar Equation (Cardioid); Finding Parametric Equations and Tangent Lines

(a) Graph the polar equation $r = 1 - \sin\theta$, $0 \le \theta \le 2\pi$.

(b) Find parametric equations for $r = 1 - \sin\theta$.

(c) Find the slope of the line tangent to the graph of $r = 1 - \sin\theta$ at the point $(1, \pi)$.

Solution

(a) The polar equation $r = 1 - \sin\theta$ contains $\sin\theta$, which has period 2π. We construct Table 2 using common values of θ that range from 0 to 2π, plot the points (r, θ), and trace out the graph, beginning at the point $(1, 0)$ and ending at the point $(1, 2\pi)$, as shown in Figure 23(a). Figure 23(b) shows the graph using technology.

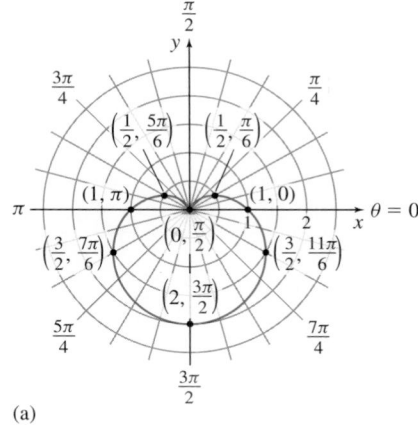

(a)

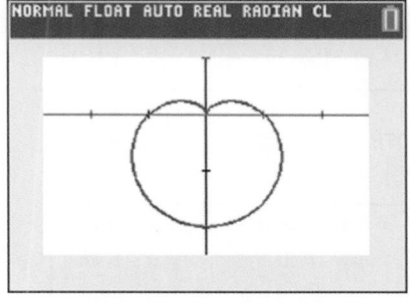

(b) $[-2.5, 2.5] \times [-2.5, 1]$

Figure 23 The cardioid $r = 1 - \sin\theta$.

(b) We obtain parametric equations for $r = 1 - \sin\theta$ by using the conversion formulas $x = r\cos\theta$ and $y = r\sin\theta$:

$$x = r\cos\theta = (1 - \sin\theta)\cos\theta \qquad y = r\sin\theta = (1 - \sin\theta)\sin\theta$$

The parameter is θ and if $0 \le \theta \le 2\pi$, then the graph starts at $(r, \theta) = (1, 0)$ and is traced out exactly once in the counterclockwise direction.

(c) Using the parametric equations found in (b), the slope of the tangent line is

$$\frac{dy}{dx} = \frac{\dfrac{dy}{d\theta}}{\dfrac{dx}{d\theta}} = \frac{-\cos\theta \cdot \sin\theta \ + (1 - \sin\theta) \cdot \cos\theta}{-\cos\theta \cdot \cos\theta \ - (1 - \sin\theta) \cdot \sin\theta}$$

$$= \frac{\cos\theta - 2\sin\theta\cos\theta}{-\sin\theta \ + \sin^2\theta - \cos^2\theta}$$

At the point $(1, \pi)$, the slope m of the tangent line is $m = \dfrac{-1}{-1} = 1$. ∎

COMMENT To graph polar equations on a calculator, be sure the MODE is set to POLAR. Then in the window setting, set θmin and θmax to match the interval for θ given in the problem. Also, be sure θstep is small—for example, 0.1—so the graph appears without corners.

NOW WORK Problem 5(a) and (b) and AP® Practice Problems 1 and 5.

EXAMPLE 2 | **Graphing a Polar Equation (Limaçon with an Inner Loop); Finding Parametric Equations and Vertical Tangent Lines**

(a) Graph the polar equation $r = 1 + 2\cos\theta$, $0 \le \theta \le 2\pi$.

(b) Find parametric equations for $r = 1 + 2\cos\theta$.

(c) Determine any points on the graph where the tangent line is vertical.

Solution

(a) The polar equation $r = 1 + 2\cos\theta$ contains $\cos\theta$, which has period 2π. Construct Table 3 using common values of θ that range from 0 to 2π, plot the points (r, θ), and trace out the graph, beginning at the point $(3, 0)$ and ending at the point $(3, 2\pi)$, as shown in Figure 24(a). Figure 24(b) shows the graph using technology.

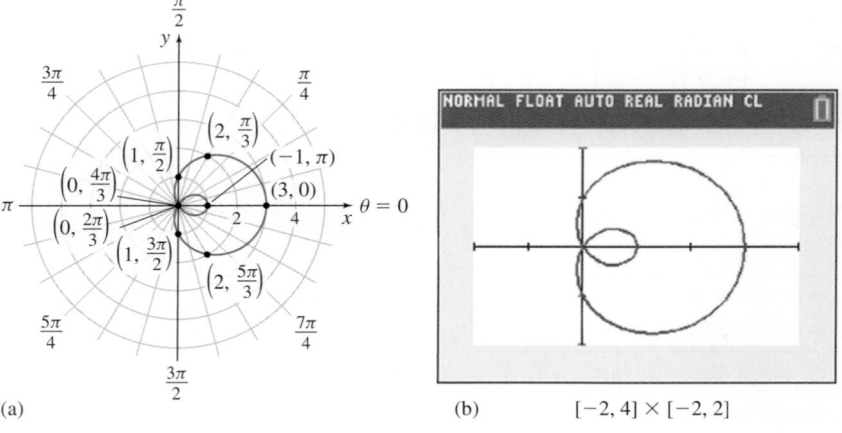

(a)   (b) $[-2, 4] \times [-2, 2]$

Figure 24 The limaçon $r = 1 + 2\cos\theta$.

AP® EXAM TIP

To find the slope of a line tangent to the graph of $r = f(\theta)$, we use

$$\frac{dy}{dx} = \frac{\dfrac{dy}{d\theta}}{\dfrac{dx}{d\theta}}$$

A common mistake is to use $\dfrac{dr}{d\theta}$ to find the slope.

TABLE 3

θ	$r = 1 + 2\cos\theta$	(r, θ)
0	$1 + 2 \cdot 1 = 3$	$(3, 0)$
$\dfrac{\pi}{3}$	$1 + 2 \cdot \dfrac{1}{2} = 2$	$\left(2, \dfrac{\pi}{3}\right)$
$\dfrac{\pi}{2}$	$1 + 2 \cdot 0 = 1$	$\left(1, \dfrac{\pi}{2}\right)$
$\dfrac{2\pi}{3}$	$1 + 2\left(-\dfrac{1}{2}\right) = 0$	$\left(0, \dfrac{2\pi}{3}\right)$
π	$1 + 2(-1) = -1$	$(-1, \pi)$
$\dfrac{4\pi}{3}$	$1 + 2\left(-\dfrac{1}{2}\right) = 0$	$\left(0, \dfrac{4\pi}{3}\right)$
$\dfrac{3\pi}{2}$	$1 + 2 \cdot 0 = 1$	$\left(1, \dfrac{3\pi}{2}\right)$
$\dfrac{5\pi}{3}$	$1 + 2 \cdot \dfrac{1}{2} = 2$	$\left(2, \dfrac{5\pi}{3}\right)$
2π	$1 + 2 \cdot 1 = 3$	$(3, 2\pi)$

NOTE Graphs of polar equations of the form

$r = a + b\cos\theta$	$r = a + b\sin\theta$
$r = a - b\cos\theta$	$r = a - b\sin\theta$
where $0 < a < b$	

are called **limaçons with an inner loop**. A limaçon with an inner loop passes through the pole twice.

(b) We obtain parametric equations for $r = 1 + 2\cos\theta$ by using the conversion formulas $x = r\cos\theta$ and $y = r\sin\theta$:

$$x = r\cos\theta = (1 + 2\cos\theta)\cos\theta \qquad y = r\sin\theta = (1 + 2\cos\theta)\sin\theta$$

Since θ is the parameter and $0 \le \theta \le 2\pi$, the graph starts at $(3, 0)$ and is traced out exactly once in the counterclockwise direction.

(c) Vertical tangent lines occur at points where $\dfrac{dx}{d\theta} = 0$. In (b) we found the parametric equation for $x = x(\theta)$:

$$x = r\cos\theta = (1 + 2\cos\theta)\cos\theta = \cos\theta + 2\cos^2\theta$$

Then

$$\frac{dx}{d\theta} = -\sin\theta - 4\sin\theta\cos\theta = -\sin\theta\,(1 + 4\cos\theta) = 0$$

The solutions to $\dfrac{dx}{d\theta} = 0$ satisfy $\sin\theta = 0$ or $\cos\theta = -\dfrac{1}{4}$. In the interval $[0, 2\pi]$,

$\sin\theta = 0$ when $\theta = 0, \pi$, and 2π; and $\cos\theta = -\dfrac{1}{4}$ when $\theta = \cos^{-1}\left(-\dfrac{1}{4}\right) \approx 1.823$ and ≈ 4.460.

The solution set is $\{0, \pi, 2\pi, 1.823, 4.460\}$. These values of θ correspond to the polar coordinates $(3, 0)$ [or $(3, 2\pi)$], $(-1, \pi)$, $\left(\dfrac{1}{2}, 1.823\right)$, and $\left(\dfrac{1}{2}, 4.460\right)$.

The graph of $r = 1 + 2\cos\theta$ has a vertical tangent line at each of these points. ∎

NOW WORK   Problem 9(a) and (b) and AP® Practice Problem 3.

EXAMPLE 3   **Graphing a Polar Equation (Spiral); Finding Parametric Equations and Tangent Lines**

CALC CLIP

(a) Graph the equation $r = e^{\theta/5}$.

(b) Find parametric equations for $r = e^{\theta/5}$.

(c) Find an equation of the tangent line to the spiral at $\theta = 0$.

Solution

(a) The polar equation $r = e^{\theta/5}$ lacks the symmetry you may have observed in the previous examples. Since there is no number θ for which $r = 0$, the graph does not contain the pole. Also observe that:

- r is positive for all θ.
- r increases as θ increases.
- $r \to 0$ as $\theta \to -\infty$.
- $r \to \infty$ as $\theta \to \infty$.

TABLE 4

θ	$r = e^{\theta/5}$	(r, θ)
$-\dfrac{3\pi}{2}$	0.39	$\left(0.39, -\dfrac{3\pi}{2}\right)$
$-\pi$	0.53	$(0.53, -\pi)$
$-\dfrac{\pi}{2}$	0.73	$\left(0.73, -\dfrac{\pi}{2}\right)$
$-\dfrac{\pi}{4}$	0.85	$\left(0.85, -\dfrac{\pi}{4}\right)$
0	1	$(1, 0)$
$\dfrac{\pi}{4}$	1.17	$\left(1.17, \dfrac{\pi}{4}\right)$
$\dfrac{\pi}{2}$	1.37	$\left(1.37, \dfrac{\pi}{2}\right)$
π	1.87	$(1.87, \pi)$
$\dfrac{3\pi}{2}$	2.57	$\left(2.57, \dfrac{3\pi}{2}\right)$
2π	3.51	$(3.51, 2\pi)$

NOTE Graphs of polar equations of the form $r = e^{\theta/a}$, $a > 0$, are called **logarithmic spirals**, since the equation can be written as $\theta = a \ln r$. A logarithmic spiral spirals infinitely both toward the pole and away from it.

Use a calculator to construct Table 4. Figure 25(a) shows part of the graph $r = e^{\theta/5}$. Figure 25(b) shows the graph for $\theta = -\dfrac{3\pi}{2}$ to $\theta = 2\pi$ using technology.

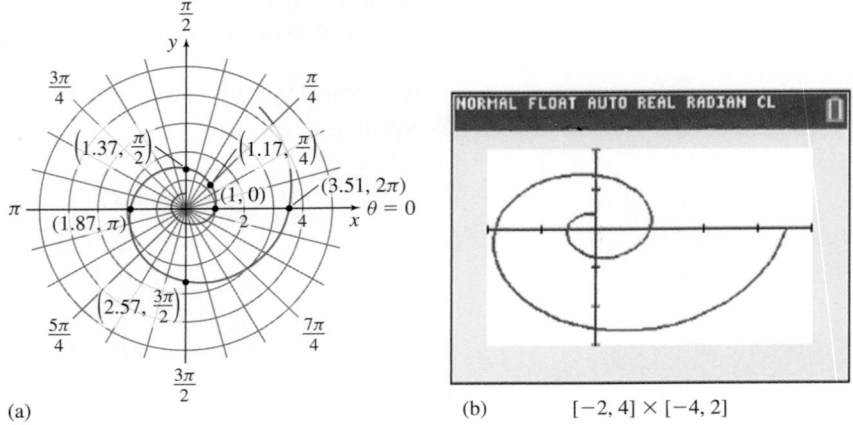

(a)   (b) $[-2, 4] \times [-4, 2]$

Figure 25 The spiral $r = e^{\theta/5}$.

(b) We obtain parametric equations for $r = e^{\theta/5}$ by using the conversion formulas $x = r\cos\theta$ and $y = r\sin\theta$:

$$x = r\cos\theta = e^{\theta/5}\cos\theta \qquad y = r\sin\theta = e^{\theta/5}\sin\theta$$

where θ is the parameter and θ is any real number.

(c) We find the slope of the tangent line using the parametric equations from (b).

$$x = e^{\theta/5}\cos\theta \qquad y = e^{\theta/5}\sin\theta$$

Then

$$\frac{dx}{d\theta} = \frac{d}{d\theta}(e^{\theta/5}\cos\theta) = \frac{1}{5}e^{\theta/5}\cos\theta - e^{\theta/5}\sin\theta$$

$$\frac{dy}{d\theta} = \frac{d}{d\theta}(e^{\theta/5}\sin\theta) = \frac{1}{5}e^{\theta/5}\sin\theta + e^{\theta/5}\cos\theta$$

and

$$\frac{dy}{dx} = \frac{\dfrac{1}{5}e^{\theta/5}\sin\theta + e^{\theta/5}\cos\theta}{\dfrac{1}{5}e^{\theta/5}\cos\theta - e^{\theta/5}\sin\theta} = \frac{\dfrac{1}{5}\sin\theta + \cos\theta}{\dfrac{1}{5}\cos\theta - \sin\theta} = \frac{\sin\theta + 5\cos\theta}{\cos\theta - 5\sin\theta}$$

At $\theta = 0$, the slope of the tangent line is 5. Since $x = 1$ and $y = 0$ at $\theta = 0$, an equation of the tangent line to the spiral is $y = 5(x - 1) = 5x - 5$. ∎

NOW WORK Problem **27** and AP® Practice Problem **4**.

2 Investigate the Derivative $\dfrac{dr}{d\theta}$ of a Polar Equation $r = f(\theta)$

In Objective 1, we discussed the graphs of polar equations and how to find the slope of a line tangent to the graph of $r = f(\theta)$ at a point (r, θ) on the graph. Here we investigate the derivative $\dfrac{dr}{d\theta}$ and its interpretation.

Recall that for a point (r, θ) in the polar coordinate system, $|r|$ is the distance from the pole to the point (r, θ), and θ is an angle whose initial side is the polar axis and whose terminal side is the ray from the pole that contains the point (r, θ). If the function f is differentiable, the derivative $\dfrac{dr}{d\theta}$ of a polar equation $r = f(\theta)$ equals the rate of change of the distance r with respect to θ.

- If $\dfrac{dr}{d\theta} > 0$ for $\alpha \leq \theta \leq \beta, \alpha < \beta$, then r is increasing on the interval $[\alpha, \beta]$. In other words, as θ moves from $\theta = \alpha$ to $\theta = \beta$, the value of r increases so the point (r, θ) is moving farther from the pole.

- If $\dfrac{dr}{d\theta} < 0$ for $\alpha \leq \theta \leq \beta, \alpha < \beta$, then r is decreasing on the interval $[\alpha, \beta]$. In other words, as θ moves from $\theta = \alpha$ to $\theta = \beta$, the value of r decreases so the point (r, θ) is moving closer to the pole.

EXAMPLE 4 Investigating the Derivative $\dfrac{dr}{d\theta}$ of a Polar Equation

Find the derivative $\dfrac{dr}{d\theta}$ and interpret its meaning for the polar equation $r = 4, 0 \leq \theta \leq 2\pi$.

Solution

For $r = 4$, $\dfrac{dr}{d\theta} = \dfrac{d}{d\theta} 4 = 0$. The rate of change of r with respect to θ equals 0, so the distance from the pole to a point on the graph of $r = 4$ is constant for all $\theta, 0 \leq \theta \leq 2\pi$. ∎

That the distance is constant makes sense, since $r = 4$ is a circle of radius 4. See Figure 26.

EXAMPLE 5 Investigating the Derivative $\dfrac{dr}{d\theta}$ of a Polar Equation

Consider the polar equation $r = 1 + \sin\theta, 0 \leq \theta \leq 2\pi$.

(a) Determine where $\dfrac{dr}{d\theta} > 0$ and interpret the result.

(b) Determine where $\dfrac{dr}{d\theta} < 0$ and interpret the result.

(c) List the values of $\theta, 0 < \theta < 2\pi$, if any, at which the graph has a local maximum.

(Example continued on the next page)

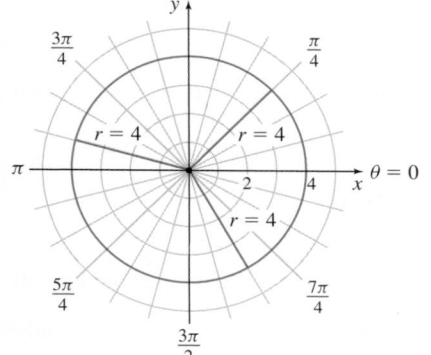

Figure 26 $r = 4$; $\dfrac{dr}{d\theta} = 0, 0 \leq \theta \leq 2\pi$

The distance of r from the pole on the circle is constant.

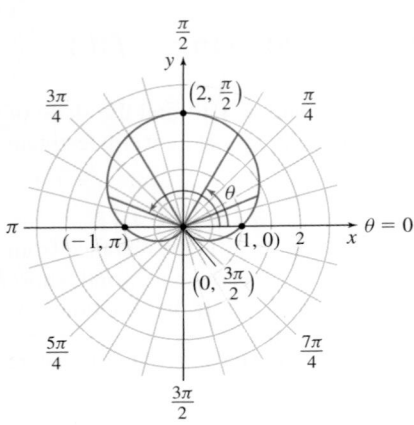

Figure 27 $r = 1 + \sin\theta$, $0 \leq \theta \leq 2\pi$

The distance r from the pole on the curve

increases for $0 \leq \theta < \dfrac{\pi}{2}$ and $\dfrac{3\pi}{2} < \theta \leq 2\pi$

and decreases for $\dfrac{\pi}{2} < \theta < \dfrac{3\pi}{2}$.

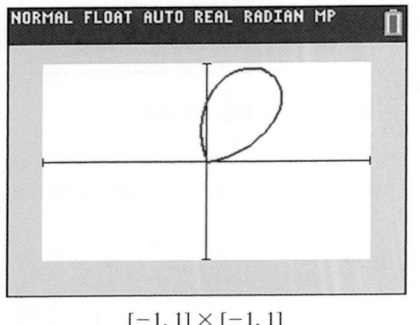

$[-1, 1] \times [-1, 1]$

Figure 28 $r = \sin\theta^2$, $0 \leq \theta \leq \sqrt{\pi}$

Solution

(a) For $r = 1 + \sin\theta$, $\dfrac{dr}{d\theta} = \dfrac{d}{d\theta}(1 + \sin\theta) = \cos\theta$. Then $\dfrac{dr}{d\theta} = \cos\theta > 0$ on the intervals $\left[0, \dfrac{\pi}{2}\right)$ and $\left(3\dfrac{\pi}{2}, 2\pi\right]$. On these intervals, the points (r, θ) are getting farther from the pole.

(b) The derivative $\dfrac{dr}{d\theta} = \cos\theta < 0$ on the interval $\left(\dfrac{\pi}{2}, \dfrac{3\pi}{2}\right)$. On this interval, the points (r, θ) are moving closer to the pole.

(c) Since r is increasing on the interval $\left[0, \dfrac{\pi}{2}\right)$ and decreasing on the interval $\left(\dfrac{\pi}{2}, \dfrac{3\pi}{2}\right)$, by the First Derivative Test, the graph has a local maximum at $\theta = \dfrac{\pi}{2}$. See Figure 27. ∎

NOW WORK AP® Practice Problem 2.

EXAMPLE 6 **Finding the Maximum Distance from $r = f(\theta)$ to the Pole**

Find the maximum distance of $r = \sin\theta^2$, $0 \leq \theta \leq \sqrt{\pi}$, from the pole. See Figure 28.

Solution

Begin by finding the derivative $\dfrac{dr}{d\theta}$ and identifying any critical numbers of r in the open interval $(0, \sqrt{\pi})$.

$$\frac{dr}{d\theta} = \frac{d}{d\theta}\sin\theta^2 = 2\,\theta\cos\theta^2$$

The critical numbers occur where $\dfrac{dr}{d\theta} = 0$. Use a calculator in function mode, and graph the function $\mathbf{Y} = 2x\cos x^2$. Then find any zeros of $\mathbf{Y}$ for $0 < x < \sqrt{\pi}$. There is only one zero, 1.2533141, so $\theta = 1.2533141$ is the only critical number of r in the open interval $(0, \sqrt{\pi})$.

Since the function $r = \sin\theta^2$ is continuous on the closed interval $[0, \sqrt{\pi}]$, we use the Candidates Test and choose the largest value of r at 0, 1.2533141, and $\sqrt{\pi}$.

$$\text{at } \theta = 0, \quad r = \sin 0 = 0$$
$$\text{at } \theta = 1.2533141, \quad r = \sin 1.2533141^2 = 1$$
$$\text{at } \theta = \sqrt{\pi}, \quad r = \sin\left(\sqrt{\pi}^2\right) = 0$$

The maximum distance from the polar curve $r = \sin\theta^2$ to the pole is 1. ∎

NOW WORK AP® Practice Problem 6.

---

❸ Find the Arc Length of a Curve Represented by a Polar Equation

Suppose a curve C is represented by the polar equation $r = f(\theta)$, $\alpha \leq \theta \leq \beta$, where both f and its derivative $f'(\theta) = \dfrac{dr}{d\theta}$ are continuous on an interval containing α and β. Using θ as the parameter, parametric equations for the curve C are

$$x(\theta) = r\cos\theta = f(\theta)\cos\theta \qquad y(\theta) = r\sin\theta = f(\theta)\sin\theta$$

Then

$$\frac{dx}{d\theta} = -f(\theta)\sin\theta + f'(\theta)\cos\theta \qquad \frac{dy}{d\theta} = f(\theta)\cos\theta + f'(\theta)\sin\theta$$

After simplification,

$$\left(\frac{dx}{d\theta}\right)^2 + \left(\frac{dy}{d\theta}\right)^2 = [f(\theta)]^2 + [f'(\theta)]^2 = r^2 + \left(\frac{dr}{d\theta}\right)^2 \qquad r = f(\theta)$$

Since we are using parametric equations, the length s of C from $\theta = \alpha$ to $\theta = \beta$ is

$$s = \int_\alpha^\beta \sqrt{\left(\frac{dx}{d\theta}\right)^2 + \left(\frac{dy}{d\theta}\right)^2}\, d\theta = \int_\alpha^\beta \sqrt{r^2 + \left(\frac{dr}{d\theta}\right)^2}\, d\theta$$

THEOREM Arc Length of the Graph of a Polar Equation

If a curve C is represented by the polar equation $r = f(\theta)$, $\alpha \le \theta \le \beta$, and if $f'(\theta) = \dfrac{dr}{d\theta}$ is continuous on an interval containing α and β, then the arc length s of C from $\theta = \alpha$ to $\theta = \beta$ is

$$s = \int_\alpha^\beta \sqrt{r^2 + \left(\frac{dr}{d\theta}\right)^2}\, d\theta$$

EXAMPLE 7 **Finding the Arc Length of a Logarithmic Spiral**

Find the arc length s of the logarithmic spiral $r = f(\theta) = e^{3\theta}$ from $\theta = 0$ to $\theta = 2$.

Solution

Use the arc length formula $s = \displaystyle\int_\alpha^\beta \sqrt{r^2 + \left(\frac{dr}{d\theta}\right)^2}\, d\theta$ with $r = e^{3\theta}$. Then $\dfrac{dr}{d\theta} = 3e^{3\theta}$ and

$$s = \int_0^2 \sqrt{(e^{3\theta})^2 + (3e^{3\theta})^2}\, d\theta = \int_0^2 \sqrt{10 e^{6\theta}}\, d\theta = \sqrt{10}\int_0^2 e^{3\theta}\, d\theta$$

$$= \sqrt{10}\left[\frac{e^{3\theta}}{3}\right]_0^2 = \frac{\sqrt{10}}{3}(e^6 - 1) \qquad\blacksquare$$

NOW WORK Problem **19** and AP® Practice Problem **7**.

9.3 Assess Your Understanding

Concepts and Vocabulary

1. *True or False* A cardioid passes through the pole.

2. *Multiple Choice* The equations for cardioids and limaçons with an inner loop are very similar. They all have the form

$$r = a \pm b\cos\theta \qquad \text{or} \qquad r = a \pm b\sin\theta, \quad a > 0, b > 0$$

The equations represent a limaçon with an inner loop if

        (a) $a < b$ **(b)** $a > b$ **(c)** $a = b$

a cardioid if

        (a) $a < b$ **(b)** $a > b$ **(c)** $a = b$

3. *True or False* The graph of $r = \sin(4\theta)$ is a rose.

4. The rose $r = \cos(3\theta)$ has _____ petals.

Skill Building

In Problems 5–12, for each polar equation:

(a) Graph the equation.

(b) Find parametric equations that represent the equation.

(c) For $0 \le \theta \le 2\pi$, determine the interval(s) on which r is getting farther from the pole.

PAGE 732   5. $r = 2 + 2\cos\theta$ 6. $r = 3 - 3\sin\theta$

        7. $r = 4 - 2\cos\theta$ 8. $r = 2 + \sin\theta$

PAGE 733   9. $r = 1 + 2\sin\theta$ 10. $r = 2 - 3\cos\theta$

      11. $r = \sin(3\theta)$ 12. $r = 4\cos(4\theta)$

In Problems 13–18, graph each pair of polar equations on the same polar grid. Find polar coordinates of the point(s) of intersection and label the point(s) on the graph.

13. $r = 8\cos\theta$, $r = 2\sec\theta$

14. $r = 8\sin\theta$, $r = 4\csc\theta$

15. $r = \sin\theta$, $r = 1 + \cos\theta$

16. $r = 3$, $r = 2 + 2\cos\theta$

17. $r = 1 + \sin\theta$, $r = 1 + \cos\theta$

18. $r = 1 + \cos\theta$, $r = 3\cos\theta$

In Problems 19–22, find the arc length of each curve.

19. $r = f(\theta) = e^{\theta/2}$ from $\theta = 0$ to $\theta = 2$

20. $r = f(\theta) = e^{2\theta}$ from $\theta = 0$ to $\theta = 2$

21. $r = f(\theta) = \cos^2\dfrac{\theta}{2}$ from $\theta = 0$ to $\theta = \pi$

22. $r = f(\theta) = \sin^2\dfrac{\theta}{2}$ from $\theta = 0$ to $\theta = \pi$

Applications and Extensions

In Problems 23–26, the polar equation for each graph is either $r = a \pm b\cos\theta$ or $r = a \pm b\sin\theta$, $a > 0$, $b > 0$. Select the correct equation and find the values of a and b.

23.
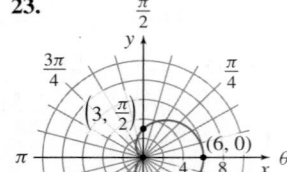

24.
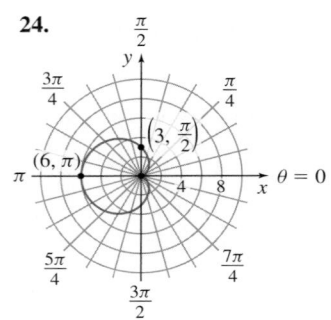

25.
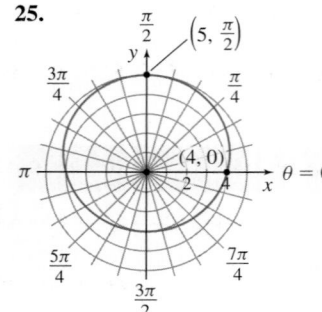

26.
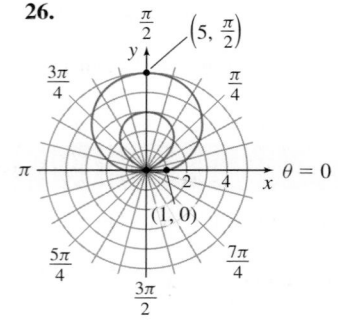

In Problems 27–32, find an equation of the tangent line to each curve at the given value of θ.
Hint: Find parametric equations that represent each polar equation.

27. $r = 2\cos(3\theta)$ at $\theta = \dfrac{\pi}{6}$

28. $r = 3\sin(3\theta)$ at $\theta = \dfrac{\pi}{3}$

29. $r = 2 + \cos\theta$ at $\theta = \dfrac{\pi}{4}$

30. $r = 3 - \sin\theta$ at $\theta = \dfrac{\pi}{6}$

31. $r = 4 + 5\sin\theta$ at $\theta = \dfrac{\pi}{4}$

32. $r = 1 - 2\cos\theta$ at $\theta = \dfrac{\pi}{4}$

Lemniscates Graphs of polar equations of the form $r^2 = a^2\cos(2\theta)$ or $r^2 = a^2\sin(2\theta)$, where $a \neq 0$, are called **lemniscates**. A lemniscate passes through the pole twice and is shaped like the infinity symbol.

In Problems 33–36, for each equation:
(a) Graph the lemniscate.
(b) Find parametric equations that represent the equation.

33. $r^2 = 4\sin(2\theta)$

34. $r^2 = 9\cos(2\theta)$

35. $r^2 = \cos(2\theta)$

36. $r^2 = 16\sin(2\theta)$

In Problems 37–48:
(a) Graph each polar equation.
(b) Find parametric equations that represent each equation.

37. $r = \dfrac{2}{1 - \cos\theta}$ (parabola)

38. $r = \dfrac{1}{1 - \cos\theta}$ (parabola)

39. $r = \dfrac{1}{3 - 2\cos\theta}$ (ellipse)

40. $r = \dfrac{2}{1 - 2\cos\theta}$ (hyperbola)

41. $r = \theta$; $\theta \geq 0$ (spiral of Archimedes)

42. $r = \dfrac{3}{\theta}$; $\theta > 0$ (reciprocal spiral)

43. $r = \csc\theta - 2$; $0 < \theta < \pi$ (conchoid)

44. $r = 3 - \dfrac{1}{2}\csc\theta$ (conchoid)

45. $r = \sin\theta\tan\theta$ (cissoid)

46. $r = \cos\dfrac{\theta}{2}$

47. $r = \tan\theta$ (kappa curve)

48. $r = \cot\theta$ (kappa curve)

49. Show that $r = 4(\cos\theta + 1)$ and $r = 4(\cos\theta - 1)$ have the same graph.

50. Show that $r = 5(\sin\theta + 1)$ and $r = 5(\sin\theta - 1)$ have the same graph.

51. Arc Length Find the arc length of the spiral $r = \theta$ from $\theta = 0$ to $\theta = 2\pi$.

52. Arc Length Find the arc length of the spiral $r = 3\theta$ from $\theta = 0$ to $\theta = 2\pi$.

53. Perimeter Find the perimeter of the cardioid
$$r = f(\theta) = 1 - \cos\theta \quad -\pi \leq \theta \leq \pi$$

54. Exploring Using Technology
(a) Graph $r_1 = 2\cos(4\theta)$. Clear the screen and graph $r_2 = 2\cos(6\theta)$. How many petals does each of the graphs have?
(b) Clear the screen and graph, in order, each on a clear screen, $r_1 = 2\cos(3\theta)$, $r_2 = 2\cos(5\theta)$, and $r_3 = 2\cos(7\theta)$. What do you notice about the number of petals? Do the results support the definition of a rose?

55. Exploring Using Technology Graph $r_1 = 3 - 2\cos\theta$. Clear the screen and graph $r_2 = 3 + 2\cos\theta$. Clear the screen and graph $r_3 = 3 + 2\sin\theta$. Clear the screen and graph $r_4 = 3 - 2\sin\theta$. Describe the pattern.

56. Horizontal and Vertical Tangent Lines Find the horizontal and vertical tangent lines of the cardioid $r = 1 - \sin\theta$ discussed in Example 1.

57. Horizontal and Vertical Tangent Lines Find the horizontal and vertical tangent lines of the cardioid $r = 3 + 3\cos\theta$.

AP® Practice Problems

Multiple-Choice Questions

PAGE 732
1. What is the slope of the tangent line to the cardioid $r = 1 + \cos\theta$ at $\theta = \dfrac{\pi}{6}$?

  (A) -1
  (B) $-\dfrac{\sqrt{3} + 1}{2}$

  (C) $\dfrac{\sqrt{3} + 1}{2}$
  (D) 1

PAGE 736
2. Suppose $r = \theta + 2\sin\theta$, $0 \le \theta \le 2\pi$. Identify the intervals on which r is increasing.

  (A) $\left[0,\ \dfrac{\pi}{2}\right]$ and $\left[\dfrac{3\pi}{2},\ 2\pi\right]$

  (B) $\left[0,\ \dfrac{2\pi}{3}\right]$ and $\left[\dfrac{5\pi}{3},\ 2\pi\right]$

  (C) $[0, 2\pi]$

  (D) $[\pi, 2\pi]$

PAGE 733
3. Parametric equations for the polar equation $r = 5\theta$ are

  (A) $x = 5\theta\cos\theta$; $y = 5\theta\sin\theta$
  (B) $x = 5\cos\theta$; $y = 5\sin\theta$
  (C) $x = 5r\cos\theta$; $y = 5r\sin\theta$
  (D) $x = \cos(5\theta)$; $y = \sin(5\theta)$

PAGE 734
4. Find the line tangent to the rose $r = \sin(3\theta)$ at $\theta = \dfrac{\pi}{4}$.

  (A) $y - \dfrac{1}{2} = \dfrac{1}{2}\left(x - \dfrac{1}{2}\right)$

  (B) $y - \dfrac{1}{2} = -\dfrac{1}{2}\left(x - \dfrac{1}{2}\right)$

  (C) $y - \dfrac{1}{2} = -\dfrac{1}{2}\left(x + \dfrac{1}{2}\right)$

  (D) $y + \dfrac{1}{2} = \dfrac{1}{2}\left(x + \dfrac{1}{2}\right)$

PAGE 732
5. Find the slope of the tangent line to the graph of $r = 4 - \sin\theta$ at $\theta = \pi$.

  (A) -1 (B) 1 (C) 4 (D) -4

Free-Response Questions

PAGE 736
6. The graph of the polar equation $r = \theta e^{-\theta}$, $0 \le \theta \le \dfrac{3\pi}{2}$, is shown below.

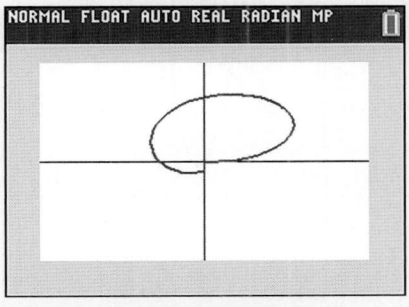

$[-0.5, 0.5] \times [-0.5, 0.5]$

  (a) Determine the maximum distance of the curve from the pole on the interval $\left[0,\ \dfrac{3\pi}{2}\right]$.

  (b) Find the point in the first quadrant at which the tangent line is vertical.

  (c) Find the slope of the line tangent to the graph at $\theta = \dfrac{\pi}{4}$.

PAGE 737
7. A bug travels along the edge of the graph of the polar equation $r = 4 + 2\cos\theta$.

  (a) Write parametric equations for r.

  (b) Find the point (r, θ), $0 < \theta < \dfrac{\pi}{2}$, at which the tangent line is horizontal.

  (c) The bug started its journey when $\theta = 0$ and stopped to rest when $\theta = \dfrac{3\pi}{4}$. What distance did the bug travel?

  (d) After resting for a while, the bug continued its journey, but left the curve and began walking on the line containing the pole. How far did the bug walk on the line to the pole?

Retain Your Knowledge

Multiple-Choice Questions

1. The graphs of $y = f(x) = x^3 - 8x^2 + 16x + 1$ and $y = g(x) = x + 1$ are shown below. Write, but do not evaluate, the integral or integrals that give the area of the region bounded by the graphs.

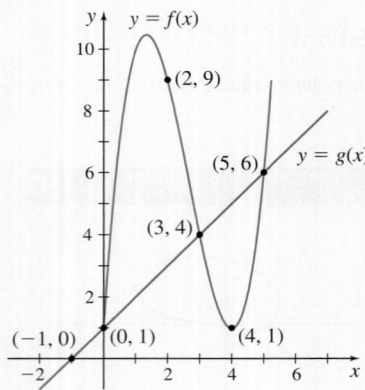

(A) $\displaystyle\int_0^5 [f(x) - g(x)]\,dx$

(B) $\displaystyle\int_1^4 [f(x) - g(x)]\,dx + \int_4^6 [g(x) - f(x)]\,dx$

(C) $\displaystyle\int_0^3 [f(x) - g(x)]\,dx + \int_3^5 [g(x) - f(x)]\,dx$

(D) $\displaystyle\int_0^3 [f(x) - g(x)]\,dx - \int_3^5 [g(x) - f(x)]\,dx$

2. $\displaystyle\int \frac{3x}{(x+4)(x-1)}\,dx =$

(A) $\dfrac{12}{5}\ln|x-1| + \dfrac{3}{5}\ln|x+4| + C$

(B) $12\ln|x+4| + 3\ln|x-1| + C$

(C) $\dfrac{12}{5}\ln|x+4| + \dfrac{3}{5}\ln|x-1| + C$

(D) $\dfrac{12}{5}\ln|x+4| - \dfrac{3}{5}\ln|x-1| + C$

3. Suppose $f(x) = \begin{cases} \dfrac{x^2-9}{x+3} & \text{if } x \neq -3 \\ 3 & \text{if } x = -3 \end{cases}$.

Which of the following statements are true?

I. $\displaystyle\lim_{x \to -3} f(x)$ exists.

II. f is differentiable at -3.

III. f is continuous at -3.

(A) I only   (B) I and III only

(C) II and III only   (D) I, II, and III

Free-Response Question

4. Suppose $x = y^2$, for $-2 \le y \le 2$.

(a) Find the line tangent to the graph of $x = y^2$ at the point $(1, -1)$.

(b) What is the arc length of the curve from $y = -2$ to $y = 2$?

(c) Find the area of the region bounded by the parabola $x = y^2$ and the line $x = 4$.

(d) The region bounded by $x = y^2$ and the line $x = 4$ forms the base of a solid whose cross sections perpendicular to the y-axis are squares. Find the volume of the solid.

9.4 Area in Polar Coordinates

RECALL The area A of the sector of a circle of radius r formed by a central angle of θ radians is $A = \dfrac{1}{2}r^2\theta$. Refer to Appendix A.4, p. A-29.

OBJECTIVES *When you finish this section, you should be able to:*

1 **Find the area of a region bounded by the graph of a polar equation (p. 741)**

2 **Find the area of a region bounded by the graphs of two polar equations (p. 743)**

In this section, we find the area of a region bounded by the graph of a polar equation and two rays that have the pole as a common vertex. The technique used is similar to that used in Chapter 6, Part 1 except, instead of approximating the area using rectangles, we approximate the area using sectors of a circle. Figure 29 illustrates the area of a sector of a circle.

Area $= \dfrac{1}{2}r^2\theta$

Figure 29

1 Find the Area of a Region Bounded by the Graph of a Polar Equation

In Figure 30, $r = f(\theta)$ is a function that is nonnegative and continuous on the interval $\alpha \leq \theta \leq \beta$. Let A denote the area of the region bounded by the graph of $r = f(\theta)$ and the rays $\theta = \alpha$ and $\theta = \beta$, where $0 \leq \alpha < \beta \leq 2\pi$. It is helpful to think of the region as being "swept out" by rays, beginning with the ray $\theta = \alpha$ and continuing to the ray $\theta = \beta$.

Partition the closed interval $[\alpha, \beta]$ into n subintervals:

$$[\alpha, \theta_1], [\theta_1, \theta_2], \ldots, [\theta_{i-1}, \theta_i], \ldots, [\theta_{n-1}, \beta]$$

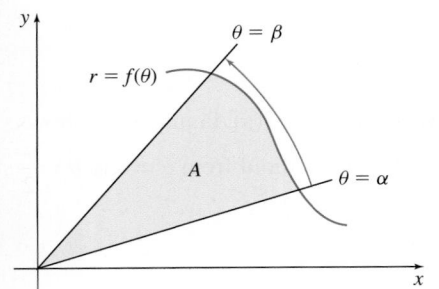

Figure 30

each of width $\Delta\theta = \dfrac{\beta - \alpha}{n}$. As shown in Figure 31, we select an angle $\theta_i{}^*$ in each subinterval $[\theta_{i-1}, \theta_i]$. The quantity $\dfrac{1}{2}[f(\theta_i{}^*)]^2 \Delta\theta$ is the area of the circular sector with radius $r = f(\theta_i{}^*)$ and central angle $\Delta\theta$. The sums of the areas of these sectors $\sum_{i=1}^{n} \dfrac{1}{2}[f(\theta_i{}^*)]^2 \Delta\theta$ are approximations of the area A. As the number n of subintervals increases, the sums $\sum_{i=1}^{n} \dfrac{1}{2} [f(\theta_i{}^*)]^2 \Delta\theta$ become better approximations to A. Since the sums $\sum_{i=1}^{n} \dfrac{1}{2}[f(\theta_i{}^*)]^2 \Delta\theta$ are Riemann sums, and since $r = f(\theta)$ is continuous, the limit of the sums exists and equals a definite integral.

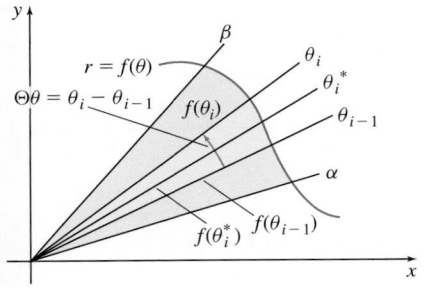

Figure 31

NEED TO REVIEW? The definite integral is discussed in Section 6.2, pp. 436–446.

$$\lim_{n \to \infty} \sum_{i=1}^{n} \frac{1}{2}[f(\theta_i{}^*)]^2 \Delta\theta = \int_{\alpha}^{\beta} \frac{1}{2}[f(\theta)]^2 d\theta = \int_{\alpha}^{\beta} \frac{1}{2} r^2 d\theta$$

THEOREM Area in Polar Coordinates

If $r = f(\theta)$ is nonnegative and continuous on the closed interval $[\alpha, \beta]$, where $\alpha < \beta$ and $\beta - \alpha \leq 2\pi$, then the area A of the region bounded by the graph of $r = f(\theta)$ and the rays $\theta = \alpha$ and $\theta = \beta$ is given by

$$A = \int_{\alpha}^{\beta} \frac{1}{2} r^2 \, d\theta$$

NOTE Be sure to graph the equation $r = f(\theta)$, $\alpha \leq \theta \leq \beta$, before using the area formula. In drawing the graph, include the rays $\theta = \alpha$, indicating the start, and $\theta = \beta$, indicating the end, of the region whose area is to be found. These rays determine the limits of integration in the area formula. Be sure that $r = f(\theta)$ is nonnegative on the interval $[\alpha, \beta]$. Otherwise an incorrect result may be obtained.

EXAMPLE 1 Finding the Area Bounded by Part of a Cardioid

Find the area of the region bounded by the cardioid $r = 1 - \sin\theta$, and the rays $\theta = 0$ and $\theta = \dfrac{\pi}{2}$.

Solution

The graph of the cardioid using technology is shown in Figure 32(a). Figure 32(b) shows the graph drawn by hand with the region bounded by the cardioid from $\theta = 0$ to $\theta = \dfrac{\pi}{2}$ shaded.

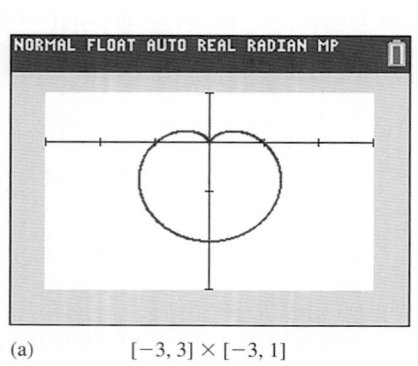

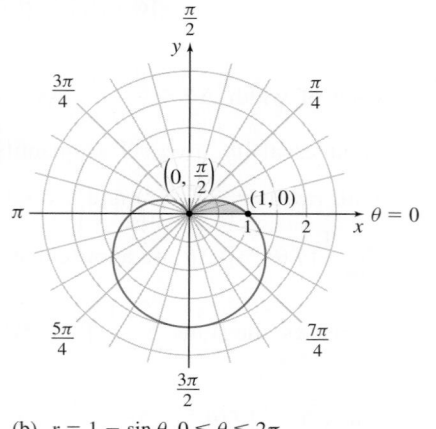

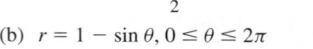

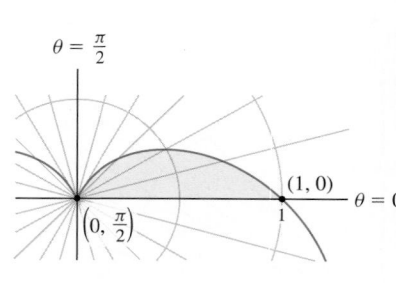

(a) $[-3, 3] \times [-3, 1]$

(b) $r = 1 - \sin\theta,\ 0 \le \theta \le 2\pi$

(c) $r = 1 - \sin\theta,\ 0 \le \theta \le \dfrac{\pi}{2}$

Figure 32

See Figure 32(c). The region is swept out beginning with the ray $\theta = 0$ and ending with the ray $\theta = \dfrac{\pi}{2}$. The limits of integration are 0 and $\dfrac{\pi}{2}$. Since r is nonnegative on the interval $\left[0, \dfrac{\pi}{2}\right]$, the area A is

$$A = \int_{\alpha}^{\beta} \frac{1}{2} r^2 \, d\theta = \int_0^{\pi/2} \frac{1}{2}(1 - \sin\theta)^2 \, d\theta \approx 0.178 \qquad \text{Use a calculator.} \qquad \blacksquare$$

NOW WORK Problem 9 and AP® Practice Problem 1.

EXAMPLE 2 Finding the Area Bounded by a Cardioid

Find the area A bounded by the cardioid $r = 1 - \sin\theta$.

Solution

Look again at the cardioid in Figure 32(b). The region bounded by the cardioid is swept out beginning with the ray $\theta = 0$ and ending with the ray $\theta = 2\pi$. So, the limits of integration are 0 and 2π. Since r is nonnegative on the interval $[0, 2\pi]$, the area A is

$$A = \int_0^{2\pi} \frac{1}{2}(1 - \sin\theta)^2 \, d\theta \approx 4.712 \qquad \text{Use a calculator.} \qquad \blacksquare$$

NOW WORK Problem 13 and AP® Practice Problem 6.

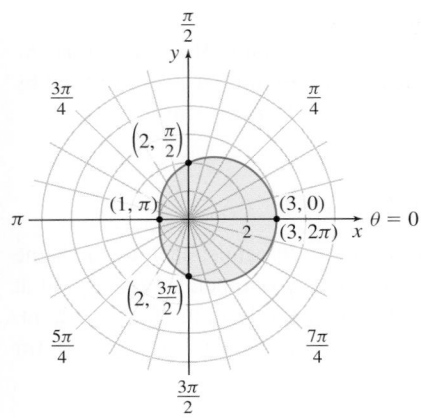

Figure 33 $r = 2 + \cos\theta$

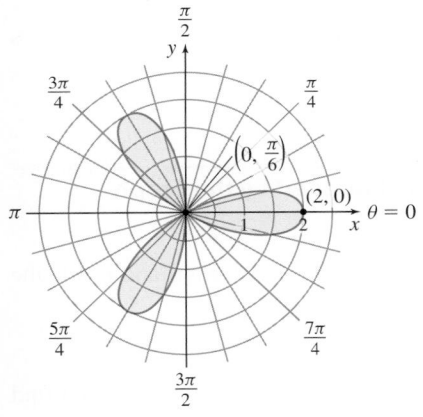

Figure 34 $r = 2\cos(3\theta)$

Figure 35

EXAMPLE 3   Finding the Area Bounded by a Limaçon

Find the area A of the region bounded by the limaçon $r = 2 + \cos\theta$.

Solution

Figure 33 shows the graph of $r = 2 + \cos\theta$, a limaçon without an inner loop. Note that $r = 2 + \cos\theta > 0$ for $0 \le \theta \le 2\pi$.

The area A of the region bounded by the limaçon, the region shaded, is swept out by the rays $\theta = 0$ and $\theta = 2\pi$ [from the point $(3, 0)$ to the point $(3, 2\pi)$].

$$A = \int_0^{2\pi} \frac{1}{2} r^2 d\theta = \int_0^{2\pi} \frac{1}{2}(2 + \cos\theta)^2 \, d\theta \approx 14.137 \qquad \text{Use a calculator.} \qquad \blacksquare$$

NOW WORK   Problem **15** and AP® Practice Problem **4**.

The area formula requires that $r \ge 0$. When there are intervals on which $r < 0$, symmetry can sometimes be used.

EXAMPLE 4   Finding the Area Bounded by a Rose

Find the area A of the region bounded by the rose of $r = 2\cos(3\theta)$, which has three petals.

Solution

Figure 34 shows the rose. Using symmetry, the blue shaded region in quadrant I equals one-sixth of the region bounded by the rose. The blue shaded region is swept out beginning with the ray $\theta = 0$ [the point $(2, 0)$]. It ends at the point $\left(0, \dfrac{\pi}{6}\right)$.

Since $r \ge 0$ on $\left[0, \dfrac{\pi}{6}\right]$, the area of the blue shaded region swept out by the rays $\theta = 0$ and $\theta = \dfrac{\pi}{6}$ is given by $\displaystyle\int_0^{\pi/6} \frac{1}{2} r^2 d\theta$, and the area A of the region bounded by the rose is 6 times the area of the blue shaded region.

$$A = 6\int_0^{\pi/6} \frac{1}{2} r^2 \, d\theta = 3\int_0^{\pi/6} 4\cos^2(3\theta) \, d\theta \approx 3.142 \qquad \text{Use a calculator.} \qquad \blacksquare$$

NOW WORK   Problem **17** and AP® Practice Problem **3**.

❷ Find the Area of a Region Bounded by the Graphs of Two Polar Equations

To find the area A of the region bounded by the graphs of two polar equations, we begin by graphing the equations and finding their points of intersection, if any.

EXAMPLE 5   Finding the Area of the Region Bounded by the Graphs of Two Polar Equations

A sand dollar is a sea urchin with a very distinctive skeleton. Its shape is round with five symmetrically placed slits, called lunules. See Figure 35. The polar equations $r = 4$ (a circle of radius 4 centered at the pole) and $r = 3\sin(5\theta)$ (a rose with 5 petals) roughly approximate the skeleton of a sand dollar with a diameter of 8 cm. Find the area of the skeleton of the sand dollar. That is, find the area A of the region that lies outside the rose $r = 3\sin(5\theta)$ and inside the circle $r = 4$.

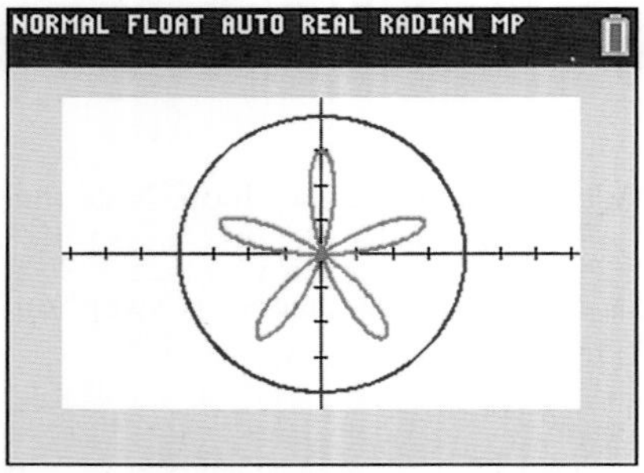

$[-7.2, 7.2] \times [-4.5, 4.5]$

Figure 36

Solution

Begin by graphing both equations on the same screen. See Figure 36. Notice that the graphs do not intersect. So, the area A is the difference between the area A_c enclosed by the circle $r = 4$ cm and the area A_R of the rose with five petals, $r = 3 \sin(5\theta)$.

The area of the circle is

$$A_c = \pi r^2 = 16\pi \approx 50.26548 \text{ cm}^2$$

To find the area A_R of the rose, use symmetry. There are five petals, each of the same area. So, the area of the rose is 5 times the area enclosed by one petal. Look again at Figure 36. We choose to find the area of the petal that lies entirely in the first quadrant. This petal is swept out beginning with the ray $\theta = 0$ and ending when $3 \sin(5\theta) = 0$, for the smallest $\theta > 0$.

$$3 \sin(5\theta) = 0$$
$$5\theta = \sin^{-1} 0 = 0 + k\pi \qquad k = 1 \text{ gives the smallest positive angle } \theta.$$
$$\theta = \frac{\pi}{5}$$

Then

$$A_R = 5 \int_0^{\pi/5} \frac{1}{2} r^2 \, d\theta = \frac{5}{2} \int_0^{\pi/5} 9 \sin^2(5\theta) \, d\theta \approx 7.06853472 \text{ cm}^2$$

The area A of the skeleton of a sand dollar of diameter 8 cm is approximately

$$A = A_c - A_R \approx 43.197 \text{ cm}^2 \qquad \blacksquare$$

NOW WORK Problem 33 and AP® Practice Problem 2.

EXAMPLE 6 **Finding the Area of the Region Bounded by the Graphs of Two Polar Equations**

Find the area A of the region that lies outside the cardioid $r = 1 + \cos\theta$ and inside the circle $r = 3 \cos\theta$.

Solution

Begin by graphing each equation. See Figures 37(a) and 37(b) on page 745. Then find the points of intersection of the two graphs by solving the equation,

$$3 \cos\theta = 1 + \cos\theta$$
$$2 \cos\theta = 1$$
$$\cos\theta = \frac{1}{2}$$
$$\theta = -\frac{\pi}{3} \quad \text{or} \quad \theta = \frac{\pi}{3}$$

The graphs intersect at the points $\left(\frac{3}{2}, -\frac{\pi}{3} \right)$ and $\left(\frac{3}{2}, \frac{\pi}{3} \right)$.

The area A of the region that lies outside the cardioid and inside the circle is shown as the shaded portion in Figure 37(c). Notice that the area A is the difference between the area of the region bounded by the circle $r = 3 \cos\theta$ swept out by the rays $\theta = -\frac{\pi}{3}$ and $\theta = \frac{\pi}{3}$, and the area of the region bounded by the cardioid $r = 1 + \cos\theta$ swept out by the same rays. So,

$$A = \int_{-\pi/3}^{\pi/3} \frac{1}{2} (3\cos\theta)^2 d\theta - \int_{-\pi/3}^{\pi/3} \frac{1}{2} (1 + \cos\theta)^2 d\theta \approx 3.142 \qquad \text{Use a calculator.}$$

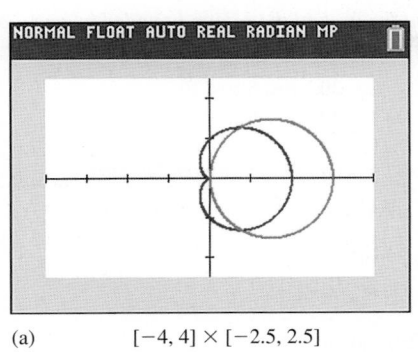

(a) $[-4, 4] \times [-2.5, 2.5]$

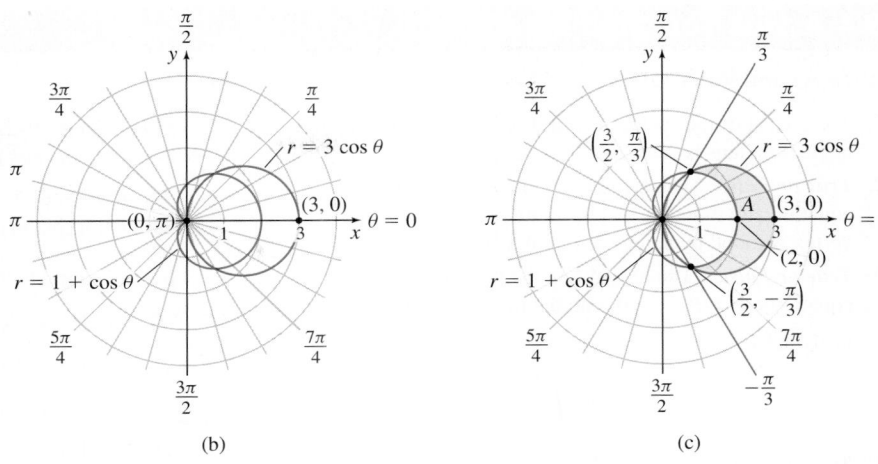

(b)   (c)

Figure 37

CAUTION   The circle and the cardioid shown in Figure 37(b) actually intersect in a third point, the pole. The pole is not identified when we solve the equation to find the points of intersection because the pole has coordinates $(0, \pi)$ on $r = 1 + \cos\theta$, but it has coordinates $\left(0, \dfrac{\pi}{2}\right)$ and $\left(0, \dfrac{3\pi}{2}\right)$ on $r = 3\cos\theta$. This demonstrates the importance of graphing polar equations when looking for their points of intersection. Since the pole presents particular difficulties, let $r = 0$ in each equation to determine whether the graph passes through the pole.

NOW WORK   Problem 23 and AP® Practice Problem 7.

EXAMPLE 7   **Finding the Area of the Region Bounded by the Graphs of Two Polar Equations**

Find the area A of the region common to both the limaçon $r = 3 + 2\cos\theta$ and the circle $r = 3$.

Solution

Begin by graphing the equations. See Figure 38.

The graphs intersect when $\cos\theta = 0$, or when $\theta = \dfrac{\pi}{2}$ and $\theta = \dfrac{3\pi}{2}$, at the points $\left(3, \dfrac{\pi}{2}\right)$ and $\left(3, \dfrac{3\pi}{2}\right)$. Using symmetry, the area A of the region common to the two graphs is 2 times the sum of the area of the quarter circle $r = 3$, $\left[\dfrac{1}{4}\pi r^2\right]$ (shaded in red) and the area of the limaçon, $r = 3 + 2\cos\theta$, $\dfrac{\pi}{2} \le \theta \le \pi$ (shaded in blue). That is,

$$A = 2\left[\frac{1}{4} \cdot 3^2\pi + \int_{\pi/2}^{\pi} \frac{1}{2}(3 + 2\cos\theta)^2\right] d\theta \approx 19.41592654$$

The area of the region common to the graphs of $r = 3 + 2\cos\theta$ and $r = 3$ is approximately 19.416 square units. ■

NOW WORK   Problem 25 and AP® Practice Problem 5.

Figure 38

9.4 Assess Your Understanding

Concepts and Vocabulary

1. The area A of the sector of a circle of radius r and central angle θ radians is $A =$ _____.

2. **True or False** The area bounded by the graph of a polar equation and two rays that have the pole as a common vertex is found by using sectors of a circle.

3. **True or False** The area A bounded by the graph of the equation $r = f(\theta)$, $r \geq 0$, and the rays $\theta = \alpha$ and $\theta = \beta$, is given by

$$A = \int_{\alpha}^{\beta} f(\theta)\, d\theta$$

4. **True or False** When finding the points of intersection of two polar equations $r = f(\theta)$ and $r = g(\theta)$, solving the equation $f(\theta) = g(\theta)$ always identifies every point of intersection.

Skill Building

In Problems 5–8, find the area of the shaded region.

5. $r = \cos(2\theta)$

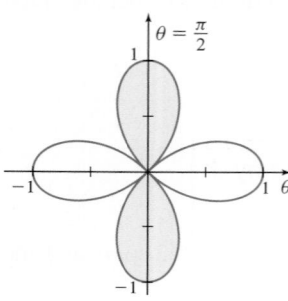

6. $r = 2\sin(3\theta)$

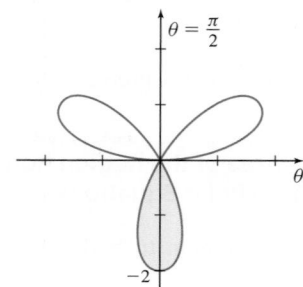

7. $r = 2 + 2\sin\theta$

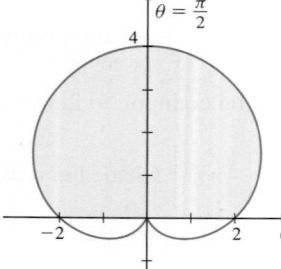

8. $r = 3 - 3\cos\theta$

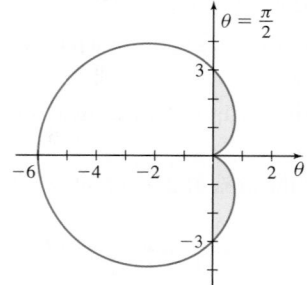

In Problems 9–12, find the area of the region bounded by the graph of each polar equation swept out by the given rays.

9. $r = 3\cos\theta$; $\theta = 0$ to $\theta = \dfrac{\pi}{3}$ [PAGE 742]
10. $r = 3\sin\theta$; $\theta = 0$ to $\theta = \dfrac{\pi}{4}$

11. $r = a\theta$; $\theta = 0$ to $\theta = 2\pi$
12. $r = e^{a\theta}$; $\theta = 0$ to $\theta = \dfrac{\pi}{2}$

In Problems 13–18, find the area of the region bounded by the graph of each polar equation.

13. $r = 1 + \cos\theta$ [PAGE 742]
14. $r = 2 - 2\sin\theta$
15. $r = 3 + \sin\theta$ [PAGE 743]
16. $r = 3(2 - \sin\theta)$
17. $r = 8\sin(3\theta)$ [PAGE 743]
18. $r = \cos(4\theta)$

In Problems 19–22, find the area of the region bounded by one loop of the graph of each polar equation.

19. $r = 4\sin(2\theta)$
20. $r = 5\cos(3\theta)$
21. $r = 4\cos(3\theta)$
22. $r = 6\cos(2\theta)$

In Problems 23–26, find the area of each region described.

23. Inside $r = 2\sin\theta$; outside $r = 1$ [PAGE 744]
24. Inside $r = 4\cos\theta$; outside $r = 2$
25. Inside $r = \sin\theta$; outside $r = 1 - \cos\theta$ [PAGE 745]
26. Inside $r^2 = 4\cos(2\theta)$; outside $r = \sqrt{2}$

Applications and Extensions

In Problems 27–44, find the area of the region:

27. bounded by the small loop of the limaçon $r = 1 + 2\cos\theta$.
28. bounded by the small loop of the limaçon $r = 1 + 2\sin\theta$.
29. bounded by the loop of the graph of $r = 2 - \sec\theta$.
30. bounded by the loop of the graph of $r = 5 + \sec\theta$.
31. bounded by $r = 2\sin^2\dfrac{\theta}{2}$.
32. bounded by $r = 6\cos^2\theta$.
33. inside the circle $r = 8\cos\theta$ and to the right of the line $r = 2\sec\theta$.   [PAGE 744]
34. inside the circle $r = 10\sin\theta$ and above the line $r = 2\csc\theta$.
35. outside the circle $r = 3$ and inside the cardioid $r = 2 + 2\cos\theta$.
36. inside the circle $r = \sin\theta$ and outside the cardioid

$$r = 1 + \cos\theta$$

37. common to the circle $r = \cos\theta$ and the cardioid $r = 1 - \cos\theta$.
38. common to the circles $r = \cos\theta$ and $r = \sin\theta$.
39. common to the inside of the cardioid $r = 1 + \sin\theta$ and the outside of the cardioid $r = 1 + \cos\theta$.
40. common to the inside of the lemniscate $r^2 = 8\cos(2\theta)$ and the outside of the circle $r = 2$.
41. bounded by the rays $\theta = 0$ and $\theta = 1$ and $r = e^{-\theta}$, $0 \leq \theta \leq 1$.
42. bounded by the rays $\theta = 0$ and $\theta = 1$ and $r = e^{\theta}$, $0 \leq \theta \leq 1$.
43. bounded by the rays $\theta = 1$ and $\theta = \pi$ and $r = \dfrac{1}{\theta}$, $1 \leq \theta \leq \pi$.
44. inside the outer loop but outside the inner loop of

$$r = 1 + 2\sin\theta$$

45. **Area** Find the area of the loop of the graph of $r = \sec\theta + 2$.

46. **Area** Find the area bounded by the loop of the **strophoid**

$$r = \sec\theta - 2\cos\theta$$
$$-\dfrac{\pi}{2} < \theta < \dfrac{\pi}{2}$$

as shown in the figure.

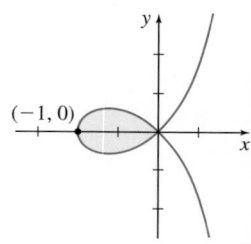

AP® Practice Problems

Multiple-Choice Questions

1. Which integral gives the area in the second quadrant bounded by the graph of the polar equation $r = 2 - \cos\theta$?

(A) $\displaystyle\int_{\pi/2}^{\pi} (2 - \cos\theta)^2 d\theta$

(B) $\displaystyle\int_{\pi/2}^{\pi} (2 - \cos^2\theta) d\theta$

(C) $\displaystyle\frac{1}{2}\int_{\pi/2}^{\pi} (4 - \cos^2\theta) d\theta$

(D) $\displaystyle\frac{1}{2}\int_{\pi/2}^{\pi} (2 - \cos\theta)^2 d\theta$

2. Which integral gives the area of the region common to the graphs of $r = 1 + \sin\theta$ and $r = 3\sin\theta$?

(A) $2\left[\dfrac{1}{2}\displaystyle\int_{0}^{\pi/2} (3\sin\theta)^2 d\theta + \dfrac{1}{2}\displaystyle\int_{0}^{\pi/2} (1 + \sin\theta)^2 d\theta\right]$

(B) $2\left[\dfrac{1}{2}\displaystyle\int_{0}^{\pi/2} (1 + \sin\theta)^2 d\theta - \dfrac{1}{2}\displaystyle\int_{0}^{\pi/2} (3\sin\theta)^2 d\theta\right]$

(C) $2\left[\dfrac{1}{2}\displaystyle\int_{0}^{\pi/2} (1 + \sin\theta)^2 d\theta - \displaystyle\int_{0}^{\pi/6} (3\sin\theta)^2 d\theta\right]$

(D) $2\left[\dfrac{1}{2}\displaystyle\int_{0}^{\pi/6} (3\sin\theta)^2 d\theta + \dfrac{1}{2}\displaystyle\int_{\pi/6}^{\pi/2} (1 + \sin\theta)^2 d\theta\right]$

3. The graph of the polar equation $r = \cos(2\theta)$ is a rose with four petals. Find the area of one petal.

(A) $\dfrac{\pi}{16}$ (B) $\dfrac{\pi}{8}$ (C) $\dfrac{\pi}{4}$ (D) $\dfrac{\pi}{2}$

4. The area of the inner loop of the limaçon $r = 2 - 4\sin\theta$ equals

I. $2 \cdot \dfrac{1}{2}\displaystyle\int_{\pi/6}^{\pi/2} (2 - 4\sin\theta)^2 d\theta$

II. $\dfrac{1}{2}\displaystyle\int_{\pi/6}^{5\pi/6} (2 - 4\sin\theta)^2 d\theta$

III. $2 \cdot \dfrac{1}{2}\displaystyle\int_{\pi/2}^{5\pi/6} (2 - 4\sin\theta)^2 d\theta$

(A) I and II only   (B) I and III only

(C) II and III only   (D) I, II, and III

Free-Response Questions

5. The graphs of the polar equations $r = 2$ and $r = 4\cos\theta$ are shown below.

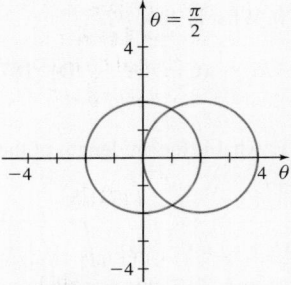

(a) Find the points of intersection of the two graphs.

(b) Find the area of the region that lies outside of the circle $r = 2$ and inside the circle $r = 4\cos\theta$.

(c) Find the area of the region that lies inside the circle $r = 2$ but outside of the circle $r = 4\cos\theta$.

6. Part of the graph of $r = \sqrt{16\sin(2\theta)}$ is shown below.

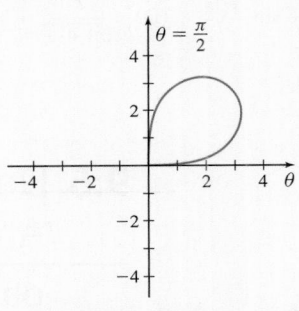

$r = \sqrt{16\sin(2\theta)}, \ 0 \le \theta \le \dfrac{\pi}{2}$

(a) Write an integral to find the area of the region enclosed by the graph.

(b) Find the area of the region.

(c) Find the average distance of the curve from the origin over the interval from $\theta = \dfrac{\pi}{6}$ to $\theta = \dfrac{\pi}{2}$.

7. The graphs of $r_1 = 3 + 2\cos\theta$ and $r_2 = 3$ are shown below.

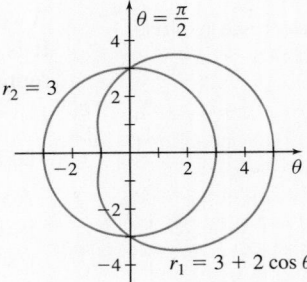

(a) Find the area of the region that lies inside the circle r_2, but outside the graph of r_1.

(b) Set up, but do not evaluate, the integrals to find the area of the region enclosed by the outer perimeter of the combined graphs.

Retain Your Knowledge

Multiple-Choice Questions

1. The rate of change of a population P with respect to time t is modeled by $\dfrac{dP}{dt} = 0.4P\left(1 - \dfrac{P}{750}\right)$, where t is measured in days and $P(0) = 20$. What is $\lim\limits_{t \to \infty} P$?

(A) 300   (B) 375   (C) 750   (D) 1875

2. If $f'(x) = \dfrac{x^3}{3} + 2x$, what is the arc length of the graph of f from $x = 0$ to $x = 4$?

(A) $\displaystyle\int_0^4 \sqrt{1 + \left(x^2 + 2\right)}\, dx$ (B) $\displaystyle\int_0^4 \sqrt{1 + \left(x^2 + 2\right)^2}\, dx$

(C) $\displaystyle\int_0^4 \sqrt{1 - \left(x^2 + 2\right)^2}\, dx$ (D) $\displaystyle\int_0^4 \sqrt{1 + \left(\dfrac{x^3}{3} + 2x\right)^2}\, dx$

3. If $f(x) = x \sin^{-1} x$, then $f'(x)$ equals

(A) $\dfrac{x}{\sqrt{1 - x^2}} + \sin^{-1} x$ (B) $\dfrac{x}{\sqrt{1 + x^2}} - \sin^{-1} x$

(C) $-x \sin^{-2} x + \sin^{-1} x$ (D) $\dfrac{x}{\sqrt{1 + x}} + \sin^{-1} x$

Free-Response Question

4. The graph of the function $f(x) = 12 - x - x^2$ is shown below.

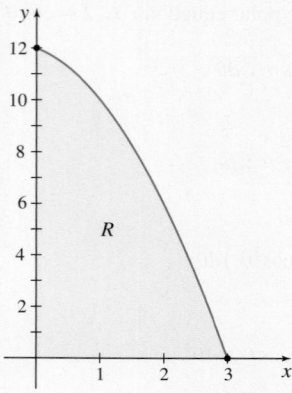

(a) Determine the area of the region R.
(b) Find the volume of the solid generated by revolving the region R about the x-axis.
(c) Find the volume of the solid whose base is R if the cross sections perpendicular to the x-axis are semicircles.
(d) What is the arc length of the graph of f from $x = 0$ to $x = 3$?

9.5 Derivatives and Integrals of Vector Functions; Arc Length

OBJECTIVES *When you finish this section, you should be able to:*

1 **Find the domain of a vector function (p. 749)**
2 **Graph a vector function (p. 749)**
3 **Find the limit and determine the continuity of a vector function (p. 750)**
4 **Find the derivative of a vector function (p. 751)**
5 **Interpret the derivative of a vector function geometrically (p. 752)**
6 **Find the arc length of the curve traced out by a vector function (p. 753)**
7 **Integrate vector functions (p. 755)**

NEED TO REVIEW? Vectors used in calculus are discussed in Appendix A.6, pp. A-51 to A-61.

A *vector function* is denoted by $\mathbf{r} = \mathbf{r}(t)$, where t is a real number and $\mathbf{r}$ is a vector. It is customary to use t as the independent variable, because in applications it often represents time.

DEFINITION Vector Function

A **vector-valued function of a real variable**, or simply, a **vector function**, $\mathbf{r} = \mathbf{r}(t)$, is a function whose domain is a subset of the set of real numbers and whose range is a set of vectors.

NOTE As with vectors and scalars, the symbol for a vector function is written using boldface roman type and the symbol for a real function is written using lightface italic type.

For a vector function $\mathbf{r} = \mathbf{r}(t)$ whose domain is the closed interval $[a, b]$, the components of $\mathbf{r}$ are two real functions $x = x(t)$, $y = y(t)$, each defined for $a \leq t \leq b$. So,

$$\boxed{\mathbf{r} = \mathbf{r}(t) = \langle x(t), y(t) \rangle}$$

If no domain is specified for a vector function, it is assumed that the domain consists of all real numbers for which both of the component functions are defined.

1 Find the Domain of a Vector Function

EXAMPLE 1 Finding the Domain of a Vector Function

If $\mathbf{r}(t) = \langle 2t + 2, \sqrt{t} \rangle$ then the components of $\mathbf{r} = \mathbf{r}(t)$ are

$$x(t) = 2t + 2 \qquad y(t) = \sqrt{t}$$

Since no domain is specified, it is assumed that the domain consists of all real numbers t for which both of the component functions are defined. Since $x = x(t)$ is defined for all real numbers, and $y = y(t)$ is defined for all real numbers $t \geq 0$, the domain of the vector function $\mathbf{r} = \mathbf{r}(t)$ is the intersection of these sets, namely, the set of nonnegative real numbers, $\{t \mid t \geq 0\}$. ∎

NOW WORK   Problem 17.

2 Graph a Vector Function

In general, the graph of a vector function $\mathbf{r} = \mathbf{r}(t)$ with domain $a \leq t \leq b$ is the curve C traced out by the tip of the position vector $\mathbf{r} = \mathbf{r}(t) = \langle x(t), y(t) \rangle$ as t varies over the closed interval $[a, b]$. The components of the vector function are the parametric equations of the curve. The curve C is a **plane curve**, as defined in Section 9.1. So the curve C traced out by a vector function $\mathbf{r} = \mathbf{r}(t)$, $a \leq t \leq b$, has a **direction**, or an **orientation**. We show the orientation of C by placing arrows on its graph in the direction the vector $\mathbf{r} = \mathbf{r}(t)$ moves from $t = a$ to $t = b$. See Figure 39.

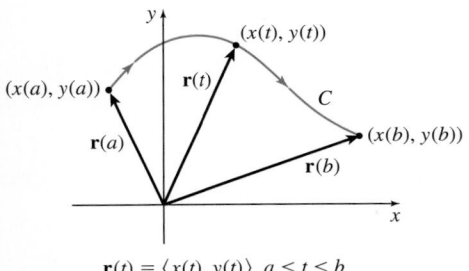

$$\mathbf{r}(t) = \langle x(t), y(t) \rangle, \ a \leq t \leq b$$

Figure 39 A plane curve C.

EXAMPLE 2 Graphing a Vector Function

Discuss the graph of the vector function

$$\mathbf{r}(t) = \langle \cos t, \sin t \rangle \quad 0 \leq t \leq 2\pi$$

Solution

The components of $\mathbf{r} = \mathbf{r}(t)$ are the parametric equations

$$x = x(t) = \cos t \qquad y = y(t) = \sin t \quad 0 \leq t \leq 2\pi$$

Since $x^2 + y^2 = \cos^2 t + \sin^2 t = 1$, the vector function $\mathbf{r} = \mathbf{r}(t)$ traces out a circle with its center at the origin and radius 1. See Figure 40. The curve begins at $\mathbf{r}(0) = \langle \cos 0, \sin 0 \rangle = \langle 1, 0 \rangle$, contains the vectors $\mathbf{r}\left(\dfrac{\pi}{2}\right) = \left\langle \cos \dfrac{\pi}{2}, \sin \dfrac{\pi}{2} \right\rangle = \langle 0, 1 \rangle$ $\mathbf{r}(\pi) = \langle \cos \pi, \sin \pi \rangle = \langle -1, 0 \rangle$, and $\mathbf{r}\left(\dfrac{3\pi}{2}\right) = \langle 0, -1 \rangle$. It ends at $\mathbf{r}(2\pi) = \langle 1, 0 \rangle$, so the circle is traced out in a counterclockwise direction. ∎

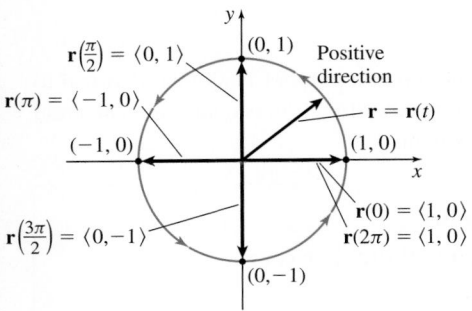

Figure 40 $\mathbf{r}(t) = \langle \cos t, \sin t \rangle \quad 0 \leq t \leq 2\pi$

The domain of the vector function in Example 2 is restricted to $0 \leq t \leq 2\pi$. If t were unrestricted, the domain would be all real numbers. The graph would look the same, as would its orientation. The graph would just keep repeating over intervals of length 2π.

NOW WORK   Problems 11(a) and 29.

EXAMPLE 3 Graphing a Vector Function

Graph the curve C traced out by the vector function

$$\mathbf{r}(t) = \langle 2 + 3t, \, 3 - t \rangle$$

Solution

The components of $\mathbf{r} = \mathbf{r}(t)$ are the parametric equations

$$x(t) = 2 + 3t \qquad y(t) = 3 - t$$

Solve $y = 3 - t$ for t, obtaining $t = 3 - y$, and substitute for t in $x = 2 + 3t$. The result is

$$x = 2 + 3(3 - y)$$
$$x = 2 + 9 - 3y$$
$$x + 3y = 11$$

The curve C traced out by the vector function $\mathbf{r} = \mathbf{r}(t)$ is a line. See Figure 41. ∎

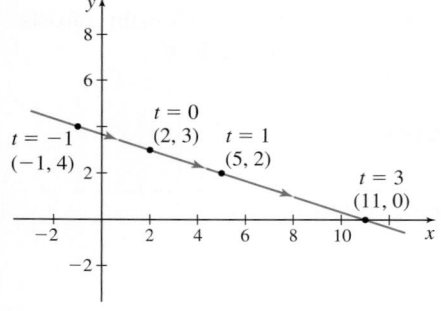

Figure 41 $\mathbf{r}(t) = \langle 2 + 3t, \, 3 - t \rangle$

NOW WORK Problem 34.

3 Find the Limit and Determine the Continuity of a Vector Function

We use the components of a vector function to define limits, continuity, and the derivative of a vector function.

DEFINITION Limit of a Vector Function

For a vector function $\mathbf{r}(t) = \langle x(t), \, y(t) \rangle$

$$\lim_{t \to t_0} \mathbf{r}(t) = \left\langle \lim_{t \to t_0} x(t), \, \lim_{t \to t_0} y(t) \right\rangle$$

provided $\lim_{t \to t_0} x(t)$ and $\lim_{t \to t_0} y(t)$ both exist.

EXAMPLE 4 Finding the Limit of a Vector Function

$$\lim_{t \to \pi/2} \langle 2 \cos t, \, 2 \sin(2t) \rangle = \left\langle \lim_{t \to \pi/2} (2 \cos t), \, \lim_{t \to \pi/2} 2 \sin(2t) \right\rangle = \langle 2 \cdot 0, \, 2 \cdot 0 \rangle = \langle 0, 0 \rangle \quad \blacksquare$$

NOW WORK Problem 11(b).

NEED TO REVIEW? Continuity of a real-valued function is discussed in Section 1.3, pp. 108–116.

The definition of continuity of a vector function at a number is almost identical to the definition for real functions. To determine if a vector function is continuous at a real number, we consider the continuity of its components.

Continuity of a Vector Function at a Number

A vector function $\mathbf{r} = \mathbf{r}(t)$ is **continuous at a real number** t_0 if:

IN WORDS A vector function is continuous at a real number t_0 if, and only if, each of its component functions is continuous at t_0.

- $\mathbf{r}(t_0)$ is defined.
- $\lim_{t \to t_0} \mathbf{r}(t)$ exists.
- $\lim_{t \to t_0} \mathbf{r}(t) = \mathbf{r}(t_0)$.

EXAMPLE 5 Determining Whether a Vector Function Is Continuous

Determine whether the vector function

$$\mathbf{r}(t) = \langle e^t, \sin t \rangle$$

is continuous at 0.

Solution

We check the three conditions for a vector function to be continuous at a number.

- $\mathbf{r}(0) = \langle e^0, \sin 0 \rangle = \langle 1, 0 \rangle$
- $\lim\limits_{t \to 0} \mathbf{r}(t) = \lim\limits_{t \to 0} \langle e^t, \sin t \rangle = \left\langle \lim\limits_{t \to 0} e^t, \lim\limits_{t \to 0} \sin t \right\rangle = \langle 1, 0 \rangle$
- $\mathbf{r}(0) = \lim\limits_{t \to 0} \mathbf{r}(t) = \langle 1, 0 \rangle$

Since all three conditions are met, $\mathbf{r} = \mathbf{r}(t)$ is continuous at 0. ∎

NOW WORK Problem 11(c).

The definition of continuity on an interval for vector functions follows the pattern developed in Chapter 1 for real functions. That is, $\mathbf{r} = \mathbf{r}(t)$ is continuous on an interval I, provided each of its components is continuous on I.

For example, the vector function $\mathbf{r}(t) = \langle e^t, \sin t \rangle$ is continuous for all real numbers since the real functions $x(t) = e^t$ and $y(t) = \sin t$ are continuous for all real numbers.

NOW WORK Problem 37.

4 Find the Derivative of a Vector Function

The definition of a derivative for a vector function is very similar to the definition of a derivative for a real function.

NEED TO REVIEW? The definition of the derivative of a function is discussed in Section 2.2, pp. 179–181.

DEFINITION Derivative of a Vector Function

The **derivative** of the vector function $\mathbf{r} = \mathbf{r}(t)$, denoted by $\mathbf{r}'(t) = \dfrac{d\mathbf{r}}{dt}$, is

$$\boxed{\mathbf{r}'(t) = \frac{d\mathbf{r}}{dt} = \lim_{h \to 0} \frac{\mathbf{r}(t + h) - \mathbf{r}(t)}{h}}$$

---

provided the limit exists. If the limit exists, then we say $\mathbf{r}$ is **differentiable**.

The next theorem provides a simple method for differentiating a vector function.

THEOREM

If $\mathbf{r}(t) = \langle x(t), y(t) \rangle$ and if $x = x(t)$ and $y = y(t)$ are both differentiable real functions, then

$$\boxed{\mathbf{r}'(t) = \frac{d\mathbf{r}}{dt} = \langle x'(t), y'(t) \rangle}$$

IN WORDS To find the derivative of a vector function, differentiate each of its components.

In Leibniz notation, the derivative of a vector function $\mathbf{r}$ has the form

$$\boxed{\frac{d\mathbf{r}}{dt} = \left\langle \frac{dx}{dt}, \frac{dy}{dt} \right\rangle}$$

EXAMPLE 6 **Finding the Derivative of a Vector Function**

Find the derivative of each vector function.

(a) $\mathbf{r}(t) = \langle 2\sin t, 3\cos t \rangle$ (b) $\mathbf{r}(t) = \langle e^t, \ln t \rangle$

Solution

(a) Since $\dfrac{d}{dt}(2\sin t) = 2\cos t$ and $\dfrac{d}{dt}(3\cos t) = -3\sin t$,

$$\mathbf{r}'(t) = \frac{d\mathbf{r}}{dt} = \langle 2\cos t, -3\sin t \rangle$$

(b) $\mathbf{r}'(t) = \dfrac{d\mathbf{r}}{dt} = \left\langle \dfrac{d}{dt}e^t, \dfrac{d}{dt}\ln t \right\rangle = \left\langle e^t, \dfrac{1}{t} \right\rangle$ ∎

NOW WORK   Problem **45** (find $\mathbf{r}'(t)$) and AP® Practice Problem **3**.

Higher-Order Derivatives

Since the derivative $\mathbf{r}'$ is a vector function derived from the vector function $\mathbf{r} = \mathbf{r}(t)$, the derivative of $\mathbf{r}'$ (if it exists) is also a vector function, called the **second derivative** of $\mathbf{r}$, and denoted by $\mathbf{r}''$. If the second derivative exists, then $\mathbf{r}$ is **twice differentiable**. Continuing in this fashion, we can find the **third derivative $\mathbf{r}'''$**, the **fourth derivative** $\mathbf{r}^{(4)}$, and so on, provided these derivatives exist. Collectively, these derivatives are called **higher-order derivatives** of the vector function $\mathbf{r} = \mathbf{r}(t)$.

EXAMPLE 7 **Finding $\mathbf{r}''$ for a Vector Function**

Find $\mathbf{r}''(t)$ for the vector function $\mathbf{r}(t) = \langle e^t, -\cos t \rangle$.

Solution
First find $\mathbf{r}'(t)$

$$\mathbf{r}'(t) = \left\langle \frac{d}{dt}e^t, \frac{d}{dt}(-\cos t) \right\rangle = \langle e^t, \sin t \rangle$$

Then

$$\mathbf{r}''(t) = \left\langle \frac{d}{dt}e^t, \frac{d}{dt}\sin t \right\rangle = \langle e^t, \cos t \rangle$$ ∎

NOW WORK   Problems **45** and **51** and AP® Practice Problem **5**.

⑤ Interpret the Derivative of a Vector Function Geometrically

Suppose a vector function $\mathbf{r} = \mathbf{r}(t)$ is defined on an interval $[a, b]$ and is differentiable on (a, b). Then $\mathbf{r} = \mathbf{r}(t)$ traces out a curve C as t varies over the interval. For a number t_0, $a < t_0 < b$, there is a point P_0 on C given by the vector $\mathbf{r} = \mathbf{r}(t_0)$. If $h \neq 0$, there is a point Q on C different from P_0 given by the vector $\mathbf{r}(t_0 + h)$.

The vector $\mathbf{r}(t_0 + h) - \mathbf{r}(t_0)$ can be thought of as the secant vector from P_0 to Q, as shown in Figure 42. Then the scalar multiple $\dfrac{1}{h}$ of the secant vector from P_0 to Q, $\dfrac{\mathbf{r}(t_0 + h) - \mathbf{r}(t_0)}{h}$, is also a vector in the direction from P_0 to Q.

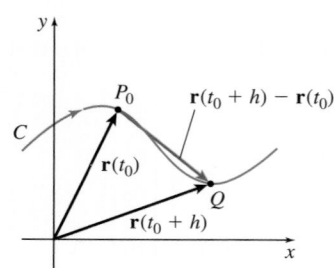

Figure 42 Secant vector.

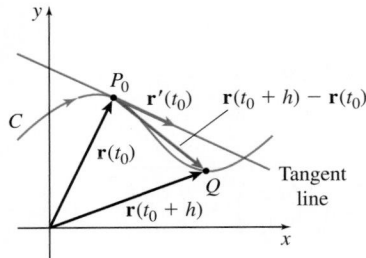

Figure 43 Vector tangent to C at t_0:

$$\mathbf{r}'(t_0) = \lim_{h \to 0} \frac{\mathbf{r}(t_0 + h) - \mathbf{r}(t_0)}{h}$$

NOTE There is no tangent vector defined if $\mathbf{r}'(t_0) = \mathbf{0}$.

As $h \to 0$, the vector $\dfrac{\mathbf{r}(t_0 + h) - \mathbf{r}(t_0)}{h}$ approaches the derivative $\mathbf{r}'(t_0)$. Also, as $h \to 0$, the vectors in the direction from P_0 to Q move along the curve C toward P_0, getting closer and closer to the vector that is tangent to C at P_0. The direction of this vector follows the orientation of C. See Figure 43. This leads to the following definition.

DEFINITION Vector Tangent to a Curve C

Suppose $\mathbf{r} = \mathbf{r}(t)$ is a vector function defined on the interval $[a, b]$ and differentiable on the interval (a, b). Let P_0 be the point on the curve C traced out by $\mathbf{r} = \mathbf{r}(t)$ corresponding to $t = t_0$, $a < t_0 < b$. If $\mathbf{r}'(t_0) \neq \mathbf{0}$, then the vector $\mathbf{r}'(t_0)$ is a **vector tangent** to the curve at t_0. The line containing P_0 in the direction of $\mathbf{r}'(t_0)$ is the **tangent line** to the curve traced out by $\mathbf{r} = \mathbf{r}(t)$ at t_0.

EXAMPLE 8   Finding a Vector Tangent to a Curve

Find the vector tangent to the curve traced out by

$$\mathbf{r}(t) = \langle \cos t, \sin t \rangle \quad 0 \le t \le 2\pi$$

at $t = \dfrac{\pi}{4}$.

Solution

A tangent vector at any point on the curve C traced out by $\mathbf{r} = \mathbf{r}(t)$ is given by

$$\mathbf{r}'(t) = \langle -\sin t, \cos t \rangle$$

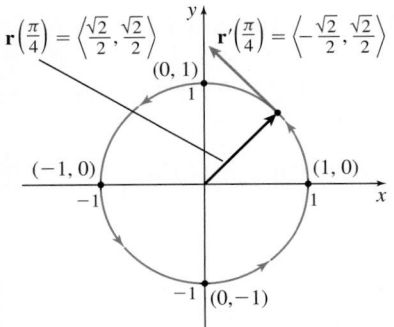

Figure 44 $\mathbf{r}(t) = \langle \cos t, \sin t \rangle$
$0 \le t \le 2\pi$

The vector tangent to C at $t = \dfrac{\pi}{4}$ is $\left\langle -\dfrac{\sqrt{2}}{2}, \dfrac{\sqrt{2}}{2} \right\rangle$. See Figure 44. ∎

NOW WORK   Problems **57** and **63** and AP® Practice Problem **7**.

---

6 Find the Arc Length of the Curve Traced Out by a Vector Function

A formula for the arc length of a smooth plane curve was derived in Section 9.2. The formula for the arc length of a smooth curve traced out by a vector function is given next.

NEED TO REVIEW? The arc length of a smooth curve is discussed in Section 9.2, pp. 724–725.

THEOREM Arc Length of a Vector Function

If a smooth curve C is traced out by the vector function $\mathbf{r} = \mathbf{r}(t)$, $a \le t \le b$, the arc length s along C from $t = a$ to $t = b$ is

$$\boxed{s = \int_a^b \|\mathbf{r}'(t)\|\, dt}$$

NEED TO REVIEW? The magnitude or the length of a vector $\|\mathbf{v}\|$ is discussed in Appendix A.6, p. A-57.

Proof For a smooth plane curve C, traced out by

$$\mathbf{r}(t) = \langle x(t), y(t) \rangle \qquad a \leq t \leq b$$

we have

$$\mathbf{r}'(t) = \left\langle \frac{dx}{dt}, \frac{dy}{dt} \right\rangle \qquad \text{and} \qquad \|\mathbf{r}'(t)\| = \sqrt{\left(\frac{dx}{dt}\right)^2 + \left(\frac{dy}{dt}\right)^2}$$

The arc length of C from a to b is given by

$$s = \int_a^b \sqrt{\left(\frac{dx}{dt}\right)^2 + \left(\frac{dy}{dt}\right)^2}\, dt = \int_a^b \|\mathbf{r}'(t)\|\, dt \qquad \blacksquare$$

EXAMPLE 9 Finding the Total Distance Traveled by an Object

An object travels in the xy-plane along the curve traced out by the vector function

$$\mathbf{r}(t) = \left\langle (1 - t^{2/3})^{3/2}, t \right\rangle \text{ (in meters) for } t \geq 0 \text{ (in seconds)}$$

Find the total distance traveled by the object from $t = 0$ to $t = 1$ second.

Solution

To find the total distance traveled by the object we need to first find $\|\mathbf{r}'(t)\|$.

$$\mathbf{r}'(t) = \left\langle -\frac{(1 - t^{2/3})^{1/2}}{t^{1/3}}, 1 \right\rangle$$

$$\|\mathbf{r}'(t)\| = \sqrt{\frac{1 - t^{2/3}}{t^{2/3}} + 1} = \sqrt{\frac{1}{t^{2/3}}} = t^{-1/3}$$

The total distance traveled from $t = 0$ to $t = 1$ is

$$\int_0^1 \|\mathbf{r}'(t)\|\, dt = \int_0^1 t^{-1/3}\, dt = \left[\frac{3}{2} t^{2/3}\right]_0^1 = \frac{3}{2}$$

The object travels 1.5 meters from $t = 0$ to $t = 1$ second. $\blacksquare$

NOW WORK AP® Practice Problem 8.

For most vector functions, finding arc length requires technology.

EXAMPLE 10 Using Technology to Find the Arc Length of an Ellipse

Use technology to find the arc length s of the ellipse shown in Figure 45 traced out by the vector function $\mathbf{r}(t) = \langle 2\cos t, 3\sin t \rangle$ from $t = 0$ to $t = \dfrac{\pi}{2}$.

Solution

We begin by finding $\mathbf{r}'(t)$ and $\|\mathbf{r}'(t)\|$.

$$\mathbf{r}'(t) = \langle -2\sin t, 3\cos t \rangle \qquad \|\mathbf{r}'(t)\| = \sqrt{4\sin^2 t + 9\cos^2 t}$$

Now use the formula for arc length.

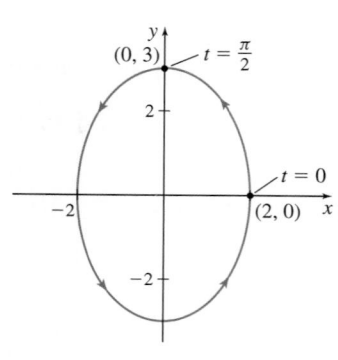

Figure 45 $\mathbf{r}(t) = \langle 2\cos t, 3\sin t \rangle, 0 \leq t \leq \dfrac{\pi}{2}$

$$s = \int_a^b \|\mathbf{r}'(t)\|\, dt = \int_0^{\pi/2} \sqrt{4\sin^2 t + 9\cos^2 t}\, dt$$

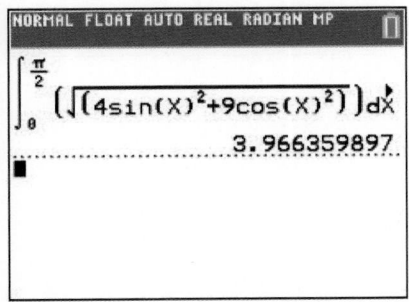

Figure 46

This is an integral that has no antiderivative in terms of elementary functions. To obtain a numerical approximation to the arc length s, we use a graphing calculator. Then

$$s = \int_0^{\pi/2} \sqrt{4 \sin^2 t + 9 \cos^2 t}\, dt \approx 3.966$$

The screen capture is given in Figure 46. ∎

NOW WORK   Problem **73** and AP® Practice Problem **4**.

---

7 Integrate Vector Functions

Let $\mathbf{r} = \mathbf{r}(t)$ be a vector function whose domain is the closed interval $[a, b]$. The **indefinite integral of a vector function** $\mathbf{r} = \mathbf{r}(t)$ is found by integrating each component of the vector function separately. That is, if

$$\mathbf{r}(t) = \langle x(t),\, y(t) \rangle$$

then

$$\int \mathbf{r}(t)\, dt = \left\langle \int x(t)\, dt,\, \int y(t)\, dt \right\rangle$$

When finding the indefinite integral of a vector function, we insert a constant vector of integration after integrating each component.

EXAMPLE 11   **Finding the Indefinite Integral of a Vector Function**

Find $\displaystyle\int \langle e^t,\, \ln t \rangle\, dt$.

Solution

Integrate each component. Remember to insert a constant vector of integration for each component.

$$\int \langle e^t,\, \ln t \rangle\, dt = \left\langle \int e^t dt,\, \int \ln t\, dt \right\rangle$$

$$= \langle e^t + c_1,\, t \ln t - t + c_2 \rangle \quad \text{c_1 and c_2 are constants of integration.}$$

$$= \langle e^t,\, t \ln t - t \rangle + \mathbf{c}$$

where $\mathbf{c} = \langle c_1, c_2 \rangle$ is a constant vector. ∎

NOW WORK   Problem **77** and AP® Practice Problem **2**.

The **definite integral of a vector function** $\mathbf{r} = \mathbf{r}(t)$, $a \le t \le b$, is defined similarly. That is, if

$$\mathbf{r}(t) = \langle x(t),\, y(t) \rangle$$

then

$$\int_a^b \mathbf{r}(t)\, dt = \left\langle \int_a^b x(t)\, dt,\, \int_a^b y(t)\, dt \right\rangle$$

EXAMPLE 12 Integrating a Vector Function

Find each vector integral.

(a) $\displaystyle\int_1^2 \left\langle 6t^2 - t, 4t + \frac{1}{t} \right\rangle dt$ (b) $\displaystyle\int \left\langle e^t, e^{3t} \right\rangle dt$

Solution

(a) Integrate each component.

$$\int_1^2 \left\langle 6t^2 - t, 4t + \frac{1}{t} \right\rangle dt = \left\langle \int_1^2 (6t^2 - t)dt, \int_1^2 \left(4t + \frac{1}{t} \right) dt \right\rangle$$

$$= \left\langle \left[2t^3 - \frac{t^2}{2} \right]_1^2, \; [2t^2 + \ln|t|]_1^2 \right\rangle$$

$$= \left\langle \frac{25}{2}, 6 + \ln 2 \right\rangle$$

(b) Integrate each component, remembering to add a constant vector of integration to each component.

$$\int \left\langle e^t, e^{3t} \right\rangle dt = \left\langle \int e^t dt, \int e^{3t} dt \right\rangle = \left\langle e^t + c_1, \frac{e^{3t}}{3} + c_2 \right\rangle = \left\langle e^t, \frac{e^{3t}}{3} \right\rangle + \mathbf{c}$$

where $\mathbf{c} = \langle c_1, c_2 \rangle$ is a constant vector. ∎

NOW WORK   Problem 89 and AP® Practice Problem 1.

EXAMPLE 13 Finding $\mathbf{r} = \mathbf{r}(t)$ Given $\mathbf{r}'(t)$

Find $\mathbf{r} = \mathbf{r}(t)$ if $\mathbf{r}'(t) = \langle 2t, e^t \rangle$ and $\mathbf{r}(0) = \langle 1, -1 \rangle$.

Solution

$$\mathbf{r}(t) = \int \mathbf{r}'(t)dt = \int \langle 2t, e^t \rangle dt$$

$$= \left\langle \int 2t \, dt, \int e^t \, dt \right\rangle$$

$$= \langle t^2 + c_1, e^t + c_2 \rangle$$

Now use the initial condition $\mathbf{r}(0) = \langle 1, -1 \rangle$.

$$\mathbf{r}(0) = \langle c_1, 1 + c_2 \rangle = \langle 1, -1 \rangle$$

from which we find

$$c_1 = 1 \qquad 1 + c_2 = -1$$
$$c_2 = -2$$

The vector function $\mathbf{r}$ is

$$\mathbf{r}(t) = \langle t^2 + 1, e^t - 2 \rangle$$

∎

NOW WORK   Problem 95 and AP® Practice Problem 6.

We have seen that polar equations can be expressed as parametric equations, and now we see that the components of a vector function are also parametric equations. In the next section, we continue to develop the calculus of vector functions, which leads to further applications involving motion along a curve.

9.5 Assess Your Understanding

Concepts and Vocabulary

1. The domain of $\mathbf{r}(t) = \langle 4t + 1, \sqrt{4-t} \rangle$ is _____.

2. **True or False** A vector function is continuous at a real number t_0 if at least one of its component functions is continuous at t_0.

3. **True or False** To find the derivative of a vector function $\mathbf{r} = \mathbf{r}(t)$, find the derivative of each of its component functions.

4. **True or False** If C is a smooth curve traced out by a vector function $\mathbf{r} = \mathbf{r}(t)$, $a \leq t \leq b$, then $\mathbf{r}'(t)$ is the vector tangent to C at t.

5. **True or False** To integrate a vector function $\mathbf{r} = \mathbf{r}(t) = \langle x(t), y(t) \rangle$, integrate each component individually.

6. **True or False** For a smooth plane curve C traced out by $\mathbf{r} = \mathbf{r}(t)$, $a \leq t \leq b$, the integral $\int_a^b \|\mathbf{r}'(t)\| \, dt$ equals the arc length along C from $t = a$ to $t = b$.

Skill Building

In Problems 7–14,

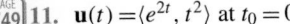

(a) Find the value of each vector function at t_0.

(b) Find the limit of each vector function as $t \to t_0$.

(c) Determine whether each vector function is continuous at t_0.

7. $\mathbf{r}(t) = \langle t^2, -2t \rangle$ at $t_0 = 1$

8. $\mathbf{r}(t) = \langle t^3, 2t \rangle$ at $t_0 = 2$

9. $\mathbf{v}(t) = \langle \sin t, \cos t \rangle$ at $t_0 = \dfrac{\pi}{4}$

10. $\mathbf{v}(t) = \langle \tan t, \cos(2t) \rangle$ at $t_0 = 0$

11. $\mathbf{u}(t) = \langle e^{2t}, t^2 \rangle$ at $t_0 = 0$

12. $\mathbf{u}(t) = \langle \ln t, -t^3 \rangle$ at $t_0 = 1$

13. $\mathbf{g}(t) = \left\langle t, -\cos \dfrac{\pi t}{4} \right\rangle$ at $t_0 = 4$

14. $\mathbf{f}(t) = \left\langle \sin \dfrac{3\pi t}{4}, 3t^2 \right\rangle$ at $t_0 = 1$

In Problems 15–22, find the domain of each vector function.

15. $\mathbf{r}(t) = \langle \cos t, \sin t \rangle$

16. $\mathbf{r}(t) = \langle \cos(2t), \sin t \rangle$

17. $\mathbf{r}(t) = \langle \sqrt{t}, t \rangle$

18. $\mathbf{r}(t) = \left\langle t^2, \dfrac{1}{\sqrt{t}} \right\rangle$

19. $\mathbf{r}(t) = \langle \ln t, t + 1 \rangle$

20. $\mathbf{r}(t) = \langle \sqrt{t}, \ln t \rangle$

21. $\mathbf{r}(t) = \langle \ln(t - 1), 3t \rangle$

22. $\mathbf{r}(t) = \left\langle \sqrt{t^2 - 4}, t \right\rangle$

In Problems 23–32, graph the curve C traced out by each vector function and show its orientation.

23. $\mathbf{r}(t) = \langle 2t, t^2 \rangle \quad t \geq 0$

24. $\mathbf{r}(t) = \langle t^2, -4t \rangle \quad t \leq 0$

25. $\mathbf{r}(t) = \langle t, 0 \rangle \quad -1 \leq t \leq 1$

26. $\mathbf{r}(t) = \langle 0, t \rangle \quad -1 \leq t \leq 1$

27. $\mathbf{r}(t) = \langle 3t, -2t \rangle$

28. $\mathbf{r}(t) = \langle t, t^2 \rangle$

29. $\mathbf{r}(t) = \langle \cos t, -\sin t \rangle \quad 0 \leq t \leq \dfrac{\pi}{2}$

30. $\mathbf{r}(t) = \langle \cos t, \sin t \rangle \quad 0 \leq t \leq \dfrac{\pi}{2}$

31. $\mathbf{r}(t) = \langle \sin^2 t, \cos^2 t \rangle \quad 0 \leq t \leq \dfrac{\pi}{2}$

32. $\mathbf{r}(t) = \langle \sin^2 t, -\cos^2 t \rangle \quad 0 \leq t \leq \dfrac{\pi}{2}$

In Problems 33–36, match each vector function to its plane curve.

33. $\mathbf{r}(t) = \langle t^2, \cos t \rangle \quad 0 \leq t \leq 2\pi$

34. $\mathbf{r}(t) = \langle \cos t, t^2 \rangle \quad 0 \leq t \leq 2\pi$

35. $\mathbf{r}(t) = \langle \sqrt{t}, \cos t \rangle \quad 0 \leq t \leq 2\pi$

36. $\mathbf{r}(t) = \langle \cos t, -t \rangle \quad 0 \leq t \leq 2\pi$

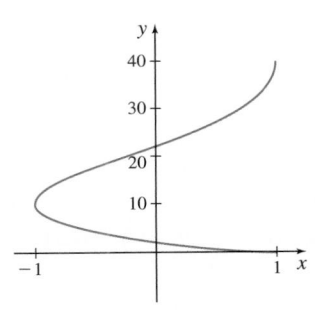

(A)

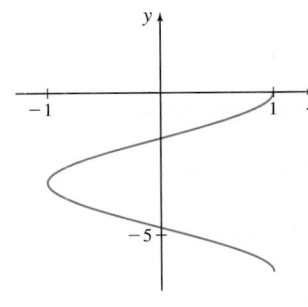

(B)

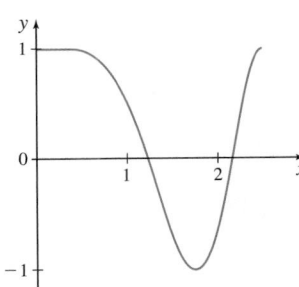

(C)

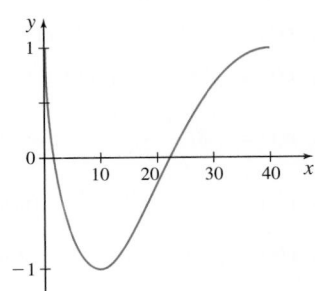

(D)

In Problems 37–42, determine where each vector function is continuous.

37. $\mathbf{r}(t) = \left\langle t^2, \dfrac{2}{1-t^2} \right\rangle$ 　　**38.** $\mathbf{r}(t) = \langle \ln t, e^{-t} \rangle$

39. $\mathbf{r}(t) = \langle \sec t, -\cos t \rangle$ 　　**40.** $\mathbf{r}(t) = \langle \sqrt{t+1}, \tan t \rangle$

41. $\mathbf{r}(t) = \left\langle \dfrac{t}{t+1}, -t \right\rangle$ 　　**42.** $\mathbf{r}(t) = \left\langle \dfrac{t}{t^2+1}, t \right\rangle$

In Problems 43–54, find $\mathbf{r}'(t)$ and $\mathbf{r}''(t)$.

43. $\mathbf{r}(t) = \langle 4t^2, -2t^3 \rangle$ 　　**44.** $\mathbf{r}(t) = \langle 8t, 4t^3 \rangle$

45. $\mathbf{r}(t) = \langle 4\sqrt{t}, 2e^t \rangle$ 　　**46.** $\mathbf{r}(t) = \langle e^{3t}, \sqrt[3]{t} \rangle$

47. $\mathbf{r}(t) = \langle t^2 - t, t^3 \rangle$ 　　**48.** $\mathbf{r}(t) = \langle 1+t, -3t^2 \rangle$

49. $\mathbf{r}(t) = \langle \sin^2 t, -\cos^2 t \rangle$ 　　**50.** $\mathbf{r}(t) = \langle \sin(2t), -\cos(2t) \rangle$

51. $\mathbf{r}(t) = \langle 5e^{t^2+t}, \ln(t^2+t) \rangle$

52. $\mathbf{r}(t) = \langle \sin(3t^3 - t), \cos^2(3t^3 - t) \rangle$

53. $\mathbf{r}(t) = \langle e^t \cos t, e^t \sin t \rangle$

54. $\mathbf{r}(t) = \langle e^{-t} \cos t, e^{-t} \sin t \rangle$

In Problems 55–66,
(a) *Find the vector tangent to each curve at the given t.*
(b) *Graph the curve.*
(c) *Add the graph of the tangent vector to (b).*

55. $\mathbf{r}(t) = \langle t, -t^2 \rangle$　 $t=1$ 　　**56.** $\mathbf{r}(t) = \langle t^2, t+1 \rangle$　 $t=0$

57. $\mathbf{r}(t) = \langle t^2+1, 1-t \rangle$　 $t=1$ 　　**58.** $\mathbf{r}(t) = \langle 2t, t^2-4 \rangle$　 $t=0$

59. $\mathbf{r}(t) = \langle t, -t^3 \rangle$　 $t=-1$ 　　**60.** $\mathbf{r}(t) = \langle t^3, 3t \rangle$　 $t=1$

61. $\mathbf{r}(t) = \langle \cos(2t), \sin(2t) \rangle$　 $t = \dfrac{\pi}{4}$

62. $\mathbf{r}(t) = \langle 2\cos(2t), 2\sin(2t) \rangle$　 $t = \dfrac{\pi}{2}$

63. $\mathbf{r}(t) = \langle 4t, -\sqrt{t} \rangle$　 $t=1$ 　　**64.** $\mathbf{r}(t) = \left\langle \sqrt{t}, \dfrac{1}{2}t \right\rangle$　 $t=4$

65. $\mathbf{r}(t) = \langle e^t, e^{-t} \rangle$　 $t=0$ 　　**66.** $\mathbf{r}(t) = \langle e^{2t}, e^{-t} \rangle$　 $t=0$

In Problems 67–76, find the arc length of each vector function.

67. $\mathbf{r}(t) = \langle t, t^{3/2}+1 \rangle$　 from $t=1$ to $t=8$

68. $\mathbf{r}(t) = \langle t, t^{3/2} \rangle$　 from $t=0$ to $t=4$

69. $\mathbf{r}(t) = \left\langle 8t, \dfrac{t^6+2}{t^2} \right\rangle$　 from $t=1$ to $t=2$

70. $\mathbf{r}(t) = \langle t, (1-t^{3/2})^{3/2} \rangle$　 from $t=\dfrac{1}{8}$ to $t=1$

71. $\mathbf{r}(t) = \langle \sin(2t), \cos(2t) \rangle$　 from $t=0$ to $t=\pi$

72. $\mathbf{r}(t) = \langle \sin t, \cos t \rangle$　 from $t=0$ to $t=\dfrac{\pi}{6}$

73. $\mathbf{r}(t) = \langle t^2, 2(t^2-1) \rangle$　 from $t=0$ to $t=1$

74. $\mathbf{r}(t) = \langle \sin^3 t, \cos^3 t \rangle$　 from $t=0$ to $t=\dfrac{\pi}{2}$

75. $\mathbf{r}(t) = \langle e^t \cos t, e^t \sin t \rangle$　 from $t=0$ to $t=5$

76. $\mathbf{r}(t) = \langle e^{-t} \sin t, e^{-t} \cos t \rangle$　 from $t=0$ to $t=2$

In Problems 77–92, find each integral.

77. $\displaystyle\int \langle \sin t, \cos t \rangle \, dt$ 　　**78.** $\displaystyle\int \langle \cos t, \sin t \rangle \, dt$

79. $\displaystyle\int \langle t^2, -t \rangle \, dt$ 　　**80.** $\displaystyle\int \langle e^t, -\sqrt{t} \rangle \, dt$

81. $\displaystyle\int \langle \ln t, -t \ln t \rangle \, dt$ 　　**82.** $\displaystyle\int \left\langle \ln t, \dfrac{1}{t} \right\rangle \, dt$

83. $\displaystyle\int \langle t-2, -(t-2)^2 \rangle \, dt$ 　　**84.** $\displaystyle\int \langle 3t+1, (3t+1)^2 \rangle \, dt$

85. $\displaystyle\int_3^6 \langle (t-2)^{-3/2}, 2t \rangle dt$ 　　**86.** $\displaystyle\int_1^4 \langle t^{-1/2}, t^{1/2} \rangle dt$

87. $\displaystyle\int_0^{\pi/2} \langle \cos(3t), \sin(2t) \rangle dt$ 　　**88.** $\displaystyle\int_0^{\pi/4} \langle \sin t \cos t, \sec t \tan t \rangle dt$

89. $\displaystyle\int_0^4 \langle 1+3\sqrt{t}, 3t^2+1 \rangle dt$ 　　**90.** $\displaystyle\int_1^8 \langle 2t, \sqrt[3]{t} \rangle dt$

91. $\displaystyle\int_0^1 \langle te^{t^2}, e^t \rangle dt$ 　　**92.** $\displaystyle\int_0^{\pi/4} \langle \sin(2t), \cos(2t) \rangle dt$

In Problems 93–98, find $\mathbf{r}=\mathbf{r}(t)$ given $\mathbf{r}'$ and the initial condition.

93. $\mathbf{r}'(t) = \langle e^t, -\ln t \rangle$　 $\mathbf{r}(1) = \langle 0, 1 \rangle$

94. $\mathbf{r}'(t) = \langle t, e^{-t} \rangle$　 $\mathbf{r}(1) = \langle 1, -1 \rangle$

95. $\mathbf{r}'(t) = \langle 2\sin t, \cos t \rangle$　 $\mathbf{r}(0) = \langle 1, -1 \rangle$

96. $\mathbf{r}'(t) = \langle \cos(2t), \sin(2t) \rangle$　 $\mathbf{r}(0) = \langle 1, 0 \rangle$

97. $\mathbf{r}'(t) = \langle t^{-1}, t \rangle$　 $\mathbf{r}(1) = \langle 1, 1 \rangle$

98. $\mathbf{r}'(t) = \left\langle t^3, \dfrac{1}{t+1} \right\rangle$　 $\mathbf{r}(0) = \langle 1, 1 \rangle$

Applications and Extensions

99. Given the vector function $\mathbf{u}(t) = \langle \cos(\omega t), \sin(\omega t) \rangle$, find $\dfrac{d\mathbf{u}}{dt}$ and $\left\| \dfrac{d\mathbf{u}}{dt} \right\|$.

100. Given the vector function $\mathbf{v}(t) = \langle t^2, t^3 \rangle$, find $\dfrac{d^2\mathbf{v}}{dt^2}$ and $\left\| \dfrac{d^2\mathbf{v}}{dt^2} \right\|$.

101. For $\mathbf{f}(t) = \langle \sin t, \cos t \rangle$, show that $\mathbf{f}(t)$ and $\mathbf{f}''(t)$ are parallel.

102. For $\mathbf{f}(t) = \langle e^{3t}, e^{-3t} \rangle$, show that $\mathbf{f}(t)$ and $\mathbf{f}''(t)$ are parallel.

Preparing for the **AP® Exam**

AP® Practice Problems

Multiple-Choice Questions

[PAGE 756] **1.** Find $\int_0^1 \mathbf{r}(t)\, dt$ where $\mathbf{r}(t) = \left\langle \dfrac{1}{t^2+1}, \dfrac{1}{t+1} \right\rangle$.

(A) $\dfrac{\pi}{4} + \ln 2$ (B) $\left\langle \dfrac{\pi}{4}, \ln 2 \right\rangle$

(C) $\left\langle -\dfrac{1}{2}, \ln 2 \right\rangle$ (D) $\left\langle \tan^{-1} t, \ln(t+1) \right\rangle$

[PAGE 755] **2.** $\int \langle \sec^2 t, \cos t \rangle\, dt =$

(A) $\langle \tan t, \sin t \rangle$ (B) $\langle \tan t, -\sin t \rangle + \mathbf{c}$

(C) $\langle \tan t, \sin t \rangle + \mathbf{c}$ (D) $\left\langle \dfrac{\sec^3 t}{3}, \sin t \right\rangle + \mathbf{c}$

[PAGE 752] **3.** The derivative of the vector function $\mathbf{r}(t) = \langle \sin(3t), -\cos(3t) \rangle$ is

(A) $\mathbf{r}'(t) = \langle \cos(3t), \sin(3t) \rangle$

(B) $\mathbf{r}'(t) = \left\langle \dfrac{1}{3}\cos(3t), \dfrac{1}{3}\sin(3t) \right\rangle$

(C) $\mathbf{r}'(t) = \langle 3\cos(3t), 3\sin(3t) \rangle$

(D) $\mathbf{r}'(t) = \langle 3\cos(3t), -3\sin(3t) \rangle$

[PAGE 755] **4.** The arc length of the curve traced out by $\mathbf{r}(t) = \langle t^3 + 2t, \ln t \rangle$ from $t = 1$ to $t = 5$ is given by

(A) $\displaystyle\int_1^5 \sqrt{(t^3 + 2t)^2 + (\ln t)^2}\, dt$

(B) $\displaystyle\int_1^5 \sqrt{(3t^2 + 2)^2 + \dfrac{1}{t^2}}\, dt$

(C) $\displaystyle\int_1^5 \sqrt{\dfrac{9t^4 + 12t^2 + 5}{t^2}}\, dt$

(D) $\displaystyle\int_1^5 \sqrt{(3t^2 + 2)^2 - \dfrac{1}{t^2}}\, dt$

[PAGE 752] **5.** Which of the following is the second derivative of the vector function $\mathbf{r}(t) = \left\langle 4t, \dfrac{1}{2}t^2 + e^t \right\rangle$?

(A) $\langle 4, t + e^t \rangle$ (B) $\langle 0, 1 + e^t \rangle$

(C) $\langle 1, e^t \rangle$ (D) $\langle 0, e^t \rangle$

[PAGE 756] **6.** What is $\mathbf{r} = \mathbf{r}(t)$ if $\mathbf{r}'(t) = \langle 4e^{4t}, 3t^2 \rangle$ and $\mathbf{r}(0) = \langle 2, -1 \rangle$?

(A) $\langle 1 + e^{4t}, -t^3 \rangle$ (B) $\langle 1 + e^{4t}, t^3 - 1 \rangle$

(C) $\langle 2 + e^{4t}, t^3 - 1 \rangle$ (D) $\langle 3 + e^{4t}, t^3 - 1 \rangle$

[PAGE 753] **7.** The vector tangent to the curve traced out by the vector function $\mathbf{r}(t) = \langle t^2 + 5, 8 - 3t \rangle$ at $t = 2$ is

(A) $\langle 4, -3 \rangle$ (B) $\langle 4, 3 \rangle$

(C) $\langle 4, -6 \rangle$ (D) $\langle 9, 2 \rangle$

[PAGE 754] **8.** If an object travels in the xy-plane along the curve traced out by the vector function $\mathbf{r}(t) = \langle t^{3/2}, -t \rangle$ for $t \geq 0$, then the total distance traveled by the object from $t = 0$ to $t = 4$ is

(A) $\dfrac{16}{3}$ (B) $\dfrac{2}{3}10^{3/2}$

(C) $10^{3/2} - 1$ (D) $\dfrac{8}{27}[10^{3/2} - 1]$

Retain Your Knowledge

Multiple-Choice Questions

1. $\int x^3 \sin x^4\, dx =$

(A) $\dfrac{1}{4}\cos x^4 + C$ (B) $4\cos x^4 + C$

(C) $-\dfrac{1}{4}\cos x^4 + C$ (D) $-\dfrac{1}{4}\cos x^3 + C$

2. Suppose $y = f(x)$ is the solution to the differential equation $\dfrac{dy}{dx} = 2x + y$. Given the initial condition $f(1) = 0$, approximate $f(1.5)$ using Euler's method with step size 0.25.

(A) 1.25   (B) 1.875   (C) 2.312   (D) 3.125

3. An object moving along a horizontal line has velocity $v = v(t)$, where $v = v(t) = \cos\dfrac{t}{2}$ for any time t, $0 \leq t \leq 4\pi$. If the position of the object at time $t = 0$ is $x = 4$, what is the object's position at time $t = \pi$?

(A) 2   (B) 3   (C) 5   (D) 6

Free-Response Question

4. (a) Determine whether $\int_0^\infty e^{-x/2}\, dx$ converges.

(b) If the integral in (a) converges, find the area of the region in the first quadrant that lies between the graph of $y = f(x) = e^{-x/2}$ and the x-axis.

(c) Find the volume of the solid of revolution obtained when the graph of $y = f(x)$ is revolved about the x-axis, $0 \leq x \leq 4$.

9.6 Motion Along a Curve

OBJECTIVES *When you finish this section, you should be able to:*

1 Find the velocity, acceleration, and speed of a particle moving along a smooth curve (p. 760)

2 Find the position, velocity, and speed of a particle with known acceleration (p. 763)

An important physical application of vector functions and their derivatives is the study of moving objects. If we think of the mass of an object as being concentrated at the object's center of gravity, then the object can be thought of as a particle, and it can be represented as a point on a graph. As the particle moves in the plane, its coordinates x and y are each functions of time t, and its position at time t is given by the vector function:

$$\mathbf{r}(t) = \langle x(t), y(t) \rangle$$

1 Find the Velocity, Acceleration, and Speed of a Particle Moving Along a Smooth Curve

If the vector function $\mathbf{r} = \mathbf{r}(t)$ represents the position vector of a particle at time t, we can find the velocity, acceleration, and speed of the particle at any time t.

> **DEFINITIONS** Position; Velocity; Acceleration; Speed
>
> The **position**, **velocity**, **acceleration**, and **speed** of a particle whose motion is along a smooth curve traced out by the twice differentiable vector function
>
> $$\mathbf{r}(t) = \langle x(t), y(t) \rangle \quad a \le t \le b$$
>
> are defined as

Position	$\mathbf{r}(t) = \langle x(t), y(t) \rangle$
Velocity	$\mathbf{v}(t) = \mathbf{r}'(t) = \left\langle \dfrac{dx}{dt}, \dfrac{dy}{dt} \right\rangle$
Acceleration	$\mathbf{a}(t) = \mathbf{r}''(t) = \left\langle \dfrac{d^2x}{dt^2}, \dfrac{d^2y}{dt^2} \right\rangle$
Speed	$v(t) = \|\mathbf{v}(t)\| = \|\mathbf{r}'(t)\| = \sqrt{\left(\dfrac{dx}{dt}\right)^2 + \left(\dfrac{dy}{dt}\right)^2}$

Figure 47 $\mathbf{v} = \mathbf{r}'(t) = \left\langle \dfrac{dx}{dt}, \dfrac{dy}{dt} \right\rangle$

$$v = \|\mathbf{v}\| = \sqrt{\left(\frac{dx}{dt}\right)^2 + \left(\frac{dy}{dt}\right)^2}$$

NOTE Since $\mathbf{v}(t) = \mathbf{r}'(t)$, the velocity vector is directed along the vector tangent to the curve. Also note that speed equals the magnitude of velocity.

See Figure 47.

EXAMPLE 1 Finding Velocity, Acceleration, and Speed

Find the velocity $\mathbf{v}$, acceleration $\mathbf{a}$, and speed v of a particle that is moving along the plane curve $\mathbf{r}(t) = \left\langle \frac{1}{2}t^2 + t, t^3 \right\rangle$ from $t = 0$ to $t = 2$.

Graph the motion of the particle and the vectors $\mathbf{v}(1)$ and $\mathbf{a}(1)$.

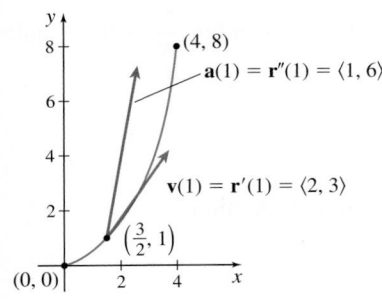

Figure 48 $\mathbf{r}(t) = \left\langle \frac{1}{2}t^2 + t, t^3 \right\rangle$

$0 \le t \le 2$

Solution

The velocity, acceleration, and speed are

$$\mathbf{v}(t) = \mathbf{r}'(t) = \left\langle \frac{d}{dt}\left(\frac{1}{2}t^2 + t \right), \frac{d}{dt}t^3 \right\rangle = \langle t+1, 3t^2 \rangle$$

$$\mathbf{a}(t) = \mathbf{r}''(t) = \frac{d}{dt}\mathbf{r}'(t) = \left\langle \frac{d}{dt}(t+1), \frac{d}{dt}(3t^2) \right\rangle = \langle 1, 6t \rangle$$

$$v(t) = \|\mathbf{v}(t)\| = \sqrt{(t+1)^2 + (3t^2)^2} = \sqrt{9t^4 + t^2 + 2t + 1}$$

At $t = 1$, the velocity is $\mathbf{v}(1) = \langle 2, 3 \rangle$ and the acceleration is $\mathbf{a}(1) = \langle 1, 6 \rangle$. Figure 48 illustrates the graph of $\mathbf{r} = \mathbf{r}(t)$ and the vectors $\mathbf{v}(1)$ and $\mathbf{a}(1)$. ■

NOW WORK   Problems 3 and 5 and AP® Practice Problems 1, 2, 3, 5, and 6.

EXAMPLE 2   Finding the Position, Velocity, and Acceleration of a Particle Moving along a Circle

(a) Find the position $\mathbf{r} = \mathbf{r}(t)$ of a particle that moves counterclockwise along a circle of radius R with a constant speed v_0.

(b) Express the velocity and acceleration of the particle as vector functions of t.

(c) Express the magnitude of the acceleration in terms of v_0 and R.

Solution

For convenience, we place the circle of radius R in the xy-plane, with its center at the origin, and assume that at time $t = 0$ the particle is on the positive x-axis.

(a) The particle is moving counterclockwise along the circle, as shown in Figure 49. Suppose the position vector of the particle is given by $\mathbf{r} = \mathbf{r}(t)$, and $\theta(t)$ is the angle between the positive x-axis and the position vector. Then the vector $\mathbf{r} = \mathbf{r}(t)$ is

$$\mathbf{r}(t) = \langle R\cos[\theta(t)], R\sin[\theta(t)] \rangle \qquad (1)$$

Notice that $\mathbf{r}(0) = \langle R, 0 \rangle$ as required. Also for the motion to be counterclockwise, the function $\theta = \theta(t)$ must be increasing.

(b) The velocity $\mathbf{v}$ of the particle is

$$\mathbf{v}(t) = \frac{d\mathbf{r}}{dt} = \left\langle -R\sin[\theta(t)]\frac{d\theta}{dt}, R\cos[\theta(t)]\frac{d\theta}{dt} \right\rangle \qquad \text{Use the Chain Rule.}$$

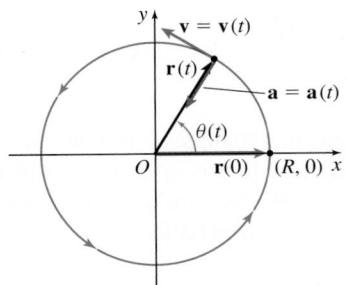

Figure 49

and the speed v of the particle is

$$v(t) = \|\mathbf{v}(t)\| = \sqrt{R^2 \sin^2[\theta(t)]\left(\frac{d\theta}{dt}\right)^2 + R^2 \cos^2[\theta(t)]\left(\frac{d\theta}{dt}\right)^2}$$

$$= \sqrt{R^2\left(\frac{d\theta}{dt}\right)^2} = R\left|\frac{d\theta}{dt}\right|$$

Since we know the speed is constant, $v(t) = v_0$. Also, since the function $\theta = \theta(t)$ is increasing, $\frac{d\theta}{dt} > 0$. As a result,

$$v_0 = R\frac{d\theta}{dt}$$

$$\frac{d\theta}{dt} = \frac{v_0}{R}$$

(Continued on the next page)

Since $\dfrac{d\theta}{dt}$ is the rate at which the angle θ is changing, the quantity $\dfrac{v_0}{R}$ is the angular speed ω of the particle. That is,

$$\frac{d\theta}{dt} = \omega$$

Solve this differential equation, using $\theta(0) = 0$ as the initial condition.

$$d\theta = \omega \, dt$$
$$\theta(t) = \omega t + k$$
$$\theta(0) = k = 0$$
$$\theta(t) = \omega t$$

Substitute $\theta(t) = \omega t$ into the vector function $\mathbf{r} = \mathbf{r}(t)$ [statement (1)] to obtain

$$\mathbf{r}(t) = \langle R\cos(\omega t), \, R\sin(\omega t)\rangle$$

Then the velocity $\mathbf{v}$ and acceleration $\mathbf{a}$ of the particle are

$$\mathbf{v} = \mathbf{v}(t) = \langle -R\omega\sin(\omega t), \, R\omega\cos(\omega t)\rangle$$
$$\mathbf{a} = \mathbf{a}(t) = \langle -R\omega^2\cos(\omega t), \, -R\omega^2\sin(\omega t)\rangle = -\omega^2\langle R\cos(\omega t), \, R\sin(\omega t)\rangle$$
$$= -\omega^2\mathbf{r}(t)$$

(c) Since $\omega = \dfrac{v_0}{R}$, the magnitude of the acceleration is

$$\boxed{\|\mathbf{a}(t)\| = \omega^2\|\mathbf{r}(t)\| = \omega^2 R = \frac{v_0^2}{R}} \tag{2}$$

■

NOTE The force $\mathbf{F}$ acting on the particle in Example 2 is directed toward the center of the circle at any time t. Because of this, it is called a **centripetal** (center-seeking) **force**.

Observe that although the speed of the particle in Example 2 is constant, its velocity $\mathbf{v}$ is not. Furthermore, the direction of the acceleration is opposite that of the vector $\mathbf{r}$ and is directed toward the center of the circle. By Newton's Second Law of Motion, $\mathbf{F} = m\mathbf{a}$, so the force vector $\mathbf{F}$ is an example of a centripetal force.

EXAMPLE 3   Finding the Speed Required for a Near-Earth Circular Orbit

Find the speed required to maintain a satellite in a near-Earth circular orbit. (The gravitational attraction of other bodies is ignored.)

Solution

NOTE Near-Earth orbits are above 100 miles (out of Earth's atmosphere) up to an altitude of approximately 15,000 miles.

Let R be the distance of a satellite from the center of Earth. Then from (2), the magnitude of the acceleration of the satellite whose motion is circular is

$$\|\mathbf{a}(t)\| = \frac{v_0^2}{R}$$

For the satellite to remain in orbit, the magnitude of the acceleration $\|\mathbf{a}(t)\|$ of the satellite at any time t must equal g, the acceleration due to gravity for Earth. As a result,

$$\frac{v_0^2}{R} = g$$
$$v_0 = \sqrt{gR}$$

NOTE The acceleration due to gravity at near-Earth orbits is somewhat less than $g \approx 32.2 \, \text{ft/s}^2 \approx 79{,}036 \, \text{mi/h}^2$, the acceleration due to gravity at Earth's surface.

The speed v_0 required to maintain a near-Earth circular orbit is

$$v_0 = \sqrt{gR}$$

where R is the distance of the satellite from the center of Earth and g is the acceleration due to gravity. ∎

For example, the speed required of a communications satellite whose circular orbit must be 4500 mi from the center of Earth is

$$v_0 = \sqrt{79{,}036 \cdot 4500} \approx 18{,}859 \, \text{mi/h}$$

NOW WORK   Problem 53.

② Find the Position, Velocity, and Speed of a Particle with Known Acceleration

Newton's Second Law of Motion states that $\mathbf{F}(t) = m\mathbf{a}(t)$. Once we know the force $\mathbf{F}(t)$ acting on a particle of mass m, the acceleration vector $\mathbf{a}(t)$ is determined. The velocity vector $\mathbf{v}(t)$ is the integral of $\mathbf{a}(t)$. That is,

$$\mathbf{v}(t) = \int \mathbf{a}(t)\,dt$$

From this we can find the velocity and speed of the particle, provided an initial condition on $\mathbf{v}(t)$ is given. Furthermore, since

$$\mathbf{r}(t) = \int \mathbf{v}(t)\,dt$$

we can find the path of the particle if we know an initial condition on $\mathbf{r}(t)$.

CALC CLIP

EXAMPLE 4   Finding the Velocity and Position of a Particle

Find the velocity vector and position vector of a particle if its acceleration is given by $\mathbf{a}(t) = \langle 0, -9.8 \rangle \, \text{m/s}^2$, $\mathbf{v}(0) = \langle 0, 9.8 \rangle \, \text{m/s}$, and $\mathbf{r}(0) = \langle 0, 0 \rangle \, \text{m}$ for $0 \leq t \leq 2$.

Solution

The velocity vector $\mathbf{v}(t)$ is

$$\mathbf{v}(t) = \int \mathbf{a}(t)\,dt = \left\langle \int 0 \, dt, \int -9.8 \, dt \right\rangle = \langle c_1, -9.8t + c_2 \rangle$$

Since $\mathbf{v}(0) = \langle 0, 9.8 \rangle$, then $c_1 = 0$ and $c_2 = 9.8$, and the velocity vector is

$$\mathbf{v}(t) = \langle 0, -9.8t + 9.8 \rangle$$

The position vector of the particle is given by

$$\mathbf{r}(t) = \int \mathbf{v}(t)\,dt = \left\langle \int 0 \, dt, \int (-9.8t + 9.8) \, dt \right\rangle = \langle d_1, -4.9t^2 + 9.8t + d_2 \rangle$$

Since $\mathbf{r}(0) = \langle 0, 0 \rangle$ then $d_1 = 0$ and $d_2 = 0$, and the position of the particle is given by

$$\mathbf{r}(t) = \langle 0, -4.9t^2 + 9.8t \rangle$$

∎

NOW WORK   Problem 23 and AP® Practice Problems 4 and 7.

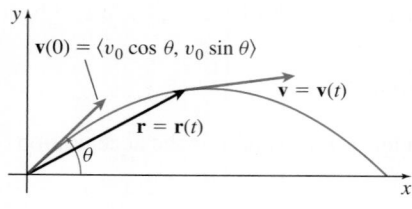

Figure 50

> **NOTE** The assumption $\mathbf{F} = \langle 0, -mg \rangle$ is called the **flat Earth approximation**. It is a good approximation for short-range projectiles. We also ignore air resistance here.

> **RECALL** In SI units, $g \approx 9.8 \, \text{m/s}^2$; in U.S. customary units $g \approx 32 \, \text{ft/s}^2$.

Example 4 is a special case of the projectile problem: Find the path of a projectile fired at an angle θ to the horizontal with initial speed v_0, assuming that the only force acting on the projectile is gravity. We are interested in knowing the maximum height of the projectile, how far it travels, and how long it is in the air. To simplify the analysis, we choose to start the motion at the origin, so $\mathbf{r}(0) = \langle 0, 0 \rangle$. The initial velocity vector $\mathbf{v}(0)$ has magnitude v_0 and is directed at an inclination θ to the horizontal axis, so $\mathbf{v}(0) = v_0 \langle \cos\theta, \sin\theta \rangle = \langle v_0 \cos\theta, v_0 \sin\theta \rangle$.

In Figure 50, $\mathbf{r} = \mathbf{r}(t)$ represents the position of the projectile after time t and $\mathbf{v} = \mathbf{v}(t)$ is the velocity vector. We assume the only force acting on the projectile is $\mathbf{F} = \langle 0, -mg \rangle$ where g is the acceleration due to gravity and m is the mass of the projectile. From Newton's Second Law of Motion, the acceleration $\mathbf{a}(t)$ of the projectile is given by

$$m\mathbf{a}(t) = \langle 0, -mg \rangle \qquad F = ma$$

$$\mathbf{a}(t) = \langle 0, -g \rangle$$

Integrating both sides with respect to t gives

$$\mathbf{v}(t) = \int \mathbf{a}(t)\,dt = \int \langle 0, -g \rangle dt = \left\langle \int 0 \, dt, \int -g\,dt \right\rangle = \langle c_1, -gt + c_2 \rangle = \langle 0, -gt \rangle + \mathbf{c}$$

where $\mathbf{c}$ is a constant vector. When $t = 0$, $\mathbf{v}(0) = \langle v_0 \cos\theta, v_0 \sin\theta \rangle = \mathbf{c}$. So, the velocity vector of the projectile is

$$\boxed{\mathbf{v}(t) = \langle v_0 \cos\theta, \, v_0 \sin\theta - gt \rangle}$$

The position of the projectile is given by

$$\mathbf{r}(t) = \int \mathbf{v}(t)\,dt = \left\langle (v_0 \cos\theta)t, \, (v_0 \sin\theta)t - \frac{1}{2}gt^2 \right\rangle + \mathbf{d}$$

where $\mathbf{d}$ is a constant vector. When $t = 0$, the position of the projectile is at the origin, $\mathbf{r}(0) = \mathbf{0}$, so $\mathbf{d} = \mathbf{0}$. The position of the projectile at time t is

$$\boxed{\mathbf{r}(t) = \left\langle (v_0 \cos\theta)t, \, (v_0 \sin\theta)t - \frac{1}{2}gt^2 \right\rangle}$$

The parametric equations of the motion of the projectile are

$$\boxed{x = (v_0 \cos\theta)t \qquad \text{and} \qquad y = -\frac{1}{2}gt^2 + (v_0 \sin\theta)t}$$

We use $t = \dfrac{x}{v_0 \cos\theta}$ to eliminate t from these equations. Then the rectangular equation of the motion is

$$\boxed{y = -\frac{g}{2v_0^2 \cos^2\theta}x^2 + x \tan\theta}$$

which is the equation of a parabola. See Figure 51.

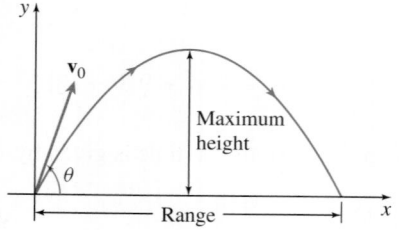

Figure 51 Path of a projectile.

We are interested in knowing how far the projectile travels horizontally, how long it is in the air, and its maximum height.

- As Figure 51 illustrates, the distance between the x-intercepts tells us how far the projectile travels horizontally.
- The x-intercepts of the parabola are found by letting $y = 0$. As a result, the x-intercepts are

$$x = 0 \quad \text{and} \quad x = \frac{2v_0^2 \cos^2 \theta \tan \theta}{g} = \frac{v_0^2 \sin(2\theta)}{g}$$

The number $\dfrac{v_0^2 \sin(2\theta)}{g}$, called the **range** of the projectile, tells us the horizontal distance the projectile travels. In Problem 55, you are asked to show that an inclination of $\theta = \dfrac{\pi}{4}$ gives the maximum range.

- The projectile hits the ground when $y = 0$ and $t > 0$. In the parametric equation of the motion for y, let $y = 0$, and solve for t.

$$y = -\frac{1}{2}gt^2 + (v_0 \sin \theta)t$$

$$0 = -\frac{1}{2}gt^2 + (v_0 \sin \theta)t$$

$$\frac{1}{2}gt = v_0 \sin \theta \qquad t > 0; \ \text{divide out } t$$

$$t = \frac{2v_0 \sin \theta}{g}$$

The other solution to the quadratic equation, $t = 0$, is a result of the initial condition.

The projectile is in the air for $\dfrac{2v_0 \sin \theta}{g}$ seconds.

- The projectile reaches its maximum height when $\dfrac{dy}{dt} = -gt + v_0 \sin \theta = 0$. The maximum height is reached at $t = \dfrac{v_0 \sin \theta}{g}$ seconds. The maximum height is

$$y = -\frac{1}{2}g\left(\frac{v_0^2 \sin^2 \theta}{g^2}\right) + (v_0 \sin \theta)\left(\frac{v_0 \sin \theta}{g}\right) = \frac{v_0^2 \sin^2 \theta}{2g}$$

The formulas for the range of a projectile and its maximum height derived here are valid only for the initial conditions specified by $\mathbf{r}(0) = \mathbf{0}$ and $\mathbf{v}(0) = \langle v_0 \cos \theta, v_0 \sin \theta \rangle$. If different initial conditions are given, the range and maximum height of the projectile will be different. In such instances, merely follow the same pattern of solution, adjusting where needed for different initial conditions.

NOW WORK   Problem 39.

9.6 Assess Your Understanding

Concepts and Vocabulary

1. *True or False*   If a twice differentiable vector function $\mathbf{r}(t) = \langle x(t), y(t) \rangle$ represents the position of a particle moving along a smooth curve, then the velocity vector is $\mathbf{v}(t) = \mathbf{r}'(t)$ and the acceleration vector is $\mathbf{a}(t) = \mathbf{r}''(t)$.

2. *True or False*   If a particle moving along a smooth curve traced out by a twice differentiable vector function $\mathbf{r} = \mathbf{r}(t)$ travels at a constant speed, then the acceleration of the particle is directed along the tangent vector.

Skill Building

In Problems 3–14:
(a) Find the velocity, acceleration, and speed of a particle whose motion is along the plane curve traced out by the vector function $\mathbf{r} = \mathbf{r}(t)$.
(b) For each curve, graph the motion of the particle and the vectors $\mathbf{v}(0)$ and $\mathbf{a}(0)$.

 3. $\mathbf{r}(t) = \langle 2t, t + 1 \rangle$ 4. $\mathbf{r}(t) = \langle 1 - 3t, 2t \rangle$

 5. $\mathbf{r}(t) = \langle t^2 + 1, 1 - t \rangle$ 6. $\mathbf{r}(t) = \langle t, t^2 \rangle$

7. $\mathbf{r}(t) = \langle 4t, -t^3 \rangle$ 8. $\mathbf{r}(t) = \langle t, -t^3 \rangle$

9. $\mathbf{r}(t) = \langle t^3, 3t \rangle$ 10. $\mathbf{r}(t) = \left\langle t^3, \frac{1}{2}t \right\rangle$

11. $\mathbf{r}(t) = \langle 2 \sin t, \cos t \rangle$ 12. $\mathbf{r}(t) = \langle 3 \cos t, 4 \sin t \rangle$

13. $\mathbf{r}(t) = \langle e^t, e^{2t} \rangle$ 14. $\mathbf{r}(t) = \langle e^{-t}, e^{-2t} \rangle$

In Problems 15–22, a particle of mass m is moving along the curve traced out by the vector function $\mathbf{r} = \mathbf{r}(t)$.
(a) Find the velocity of the particle at any time t.
(b) Find the magnitude of the acceleration of the particle.

15. $\mathbf{r}(t) = \langle e^t, e^{-t} \rangle$ 16. $\mathbf{r}(t) = \langle e^{2t}, e^{-t} \rangle$

17. $\mathbf{r}(t) = \langle t, e^t \rangle$ 18. $\mathbf{r}(t) = \langle t, \ln(1 + t) \rangle$

19. $\mathbf{r}(t) = \langle 3 \sin(2t), 3 \cos(2t) \rangle$ 20. $\mathbf{r}(t) = \langle 4 \sin t, -4 \cos t \rangle$

21. $\mathbf{r}(t) = \langle 2 \cos t, -3 \sin t \rangle$ 22. $\mathbf{r}(t) = \langle -\cos(3t), 2 \sin(3t) \rangle$

In Problems 23–26, find the velocity, speed, and position of a particle having the given acceleration, initial velocity, and initial position.

23. $\mathbf{a}(t) = \langle \cos t, \sin t \rangle$ $\mathbf{v}(0) = \langle 1, 0 \rangle$ $\mathbf{r}(0) = \langle 0, 1 \rangle$

24. $\mathbf{a}(t) = \langle \cos t, \sin t \rangle$ $\mathbf{v}(0) = \langle 0, 1 \rangle$ $\mathbf{r}(0) = \langle 1, 0 \rangle$

25. $\mathbf{a}(t) = \langle e^{-t}, 1 \rangle$ $\mathbf{v}(0) = \langle 1, 1 \rangle$ $\mathbf{r}(0) = \langle 1, -1 \rangle$

26. $\mathbf{a}(t) = \langle t^2, -e^{-t} \rangle$ $\mathbf{v}(0) = \langle 1, -1 \rangle$ $\mathbf{r}(0) = \langle 0, 0 \rangle$

Applications and Extensions

27. **Velocity and Acceleration**   Find the velocity and acceleration of a particle moving on the cycloid
$$\mathbf{r}(t) = \langle \pi t - \sin(\pi t), 1 - \cos(\pi t) \rangle$$

28. **Velocity and Acceleration**
   (a) Find the velocity and acceleration of a particle moving on the parabola $\mathbf{r}(t) = \langle t^2 - 2t, t \rangle$.
   (b) At what time t is the speed of the particle 0?

29. **Acceleration** A particle moves along the path $y = 3x^2 - x^3$ with the horizontal component of the velocity equal to $\frac{1}{3}$. Find the acceleration at the points where the velocity $\mathbf{v}$ is horizontal. Graph the motion and indicate $\mathbf{v}$ and $\mathbf{a}$ at these points.

30. Suppose that the vector function $\mathbf{r}(t) = \langle e^t, e^{-t} \rangle$ gives the position of a particle at time t.
   (a) Show that the force on the particle is directed away from the origin.
   (b) What is the minimum speed of the particle and where does it occur?

31. **Centripetal Force**
   (a) If the speed of a motorcycle going around a circular track is increased by 10%, by how much does the frictional force on the tires need to increase to keep the motorcycle from skidding on the track?
   (b) If the radius of the circular track is halved, how much slower must the motorcycle be driven so the frictional force doesn't change?

32. **Frictional Force**   A race car with mass 1000 kg is driven at a constant speed of 200 km/h around a circular track whose radius is 75 m. What frictional force must be exerted by the track on the tires to keep the car from skidding?

33. **Particle Collider**   The Large Hadron Collider (LHC), buried 100 m beneath Earth's surface, is located near Geneva, Switzerland, and spans the border between France and Switzerland. The LHC is 26,659 m in circumference and, at full power, protons travel at a speed of 299,792,455.3 m/s (99.9999991% the speed of light) around the circle. Find the magnitude of the force necessary to keep a proton of mass m moving at this speed.
   Source: http://public.web.cern.ch

34. **Particle Collider**   At the Fermi National Accelerator Laboratory in Batavia, Illinois, protons were accelerated along a circular route with radius 1 km. Find the magnitude of the force necessary to give a proton of mass m a constant speed of 280,000 km/s.

35. **Velocity**   A particle moves on the circle $x^2 + y^2 = 1$ so that at time $t \geq 0$ the position is given by the vector
$$\mathbf{r}(t) = \left\langle \frac{1 - t^2}{1 + t^2}, \frac{2t}{1 + t^2} \right\rangle$$
   (a) Find the velocity vector.
   (b) Is the particle ever at rest? Justify your answer.
   (c) Find the coordinates of the point that the particle approaches as t increases without bound.

36. **Velocity**   The position of a particle at time $t \geq 0$ is given by
$$\mathbf{r}(t) = \langle e^{-t} \cos t, e^{-t} \sin t \rangle$$
   (a) Show that the path of the particle is part of the spiral $r = e^{-\theta}$ (in polar coordinates).
   (b) Graph the path of the particle and indicate its direction of travel.

37. **Magnitude of a Force**   Find the maximum magnitude of the force acting on a particle of mass m whose motion is along the curve $\mathbf{r}(t) = \langle 4 \cos t, -\cos(2t) \rangle$.

38. Oscillating Motion Suppose a particle moves along the curve traced out by $\mathbf{r}(t) = \langle 4\cos t, -2\cos(2t) \rangle$.

 (a) Show that the particle oscillates on an arc of a parabola.

 (b) Graph the path.

 (c) Find the acceleration $\mathbf{a}$ at points of zero velocity.

Applications and Extensions

39. Projectile Motion A projectile is fired at an angle of $30°$ to the horizontal with an initial speed of 520 m/s. What are its range, the time of flight, and the greatest height reached?

40. Projectile Motion A projectile is fired with an initial speed of 200 m/s at an inclination of $60°$ to the horizontal. What are its range, the time of flight, and the greatest height reached?

41. Projectile Motion A projectile is fired with an initial speed of 100 m/s at an inclination of $\tan^{-1}\frac{5}{12}$ to the horizontal.

 (a) Find parametric equations of the path of the projectile.

 (b) Find the range.

 (c) How long is the projectile in the air?

 (d) Graph the trajectory of the projectile.

42. Projectile Motion A projectile is fired with an initial speed of 120 m/s at an inclination of $\tan^{-1}\frac{3}{4}$ to the horizontal.

 (a) Find parametric equations of the path of the projectile.

 (b) Find the range.

 (c) How long is the projectile in the air?

 (d) Graph the trajectory of the projectile.

43. Projectile Motion A projectile is fired up a hill that makes a $30°$ angle to the horizontal. Suppose the projectile is fired at an angle of inclination of $45°$ to the horizontal with an initial speed of 100 ft/s. See the figure.

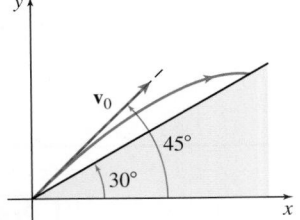

 (a) How far up the hill does the projectile land?

 (b) How long is the projectile in the air?

44. Projectile Motion A projectile is propelled horizontally at a height of 3 m above the ground in order to hit a target 1 m high that is 30 m away. See the figure below. What should the initial velocity $\mathbf{v}_0$ of the projectile be?

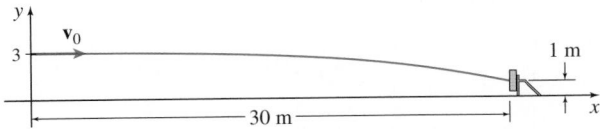

45. A force whose magnitude is 5 N and whose direction is along the positive x-axis is continuously applied to a projectile of mass $m = 1$ kg. At $t = 0$ s, the position of the object is the origin and its velocity is 3 m/s in the direction of the positive y-axis.

 (a) Find the velocity and speed of the projectile after t seconds.

 (b) Find the position after the force has been applied for t seconds.

 (c) How long is the projectile in the air?

 (d) Graph the path of the projectile.

46. An object of mass m is propelled from the point $(1, 2)$ with initial velocity $\mathbf{v}_0 = \langle 3, 4 \rangle$. Thereafter, it is subjected only to the force $\mathbf{F} = \dfrac{m}{\sqrt{2}}\langle -1, -1 \rangle$. Find the vector equation for the position of the object at any time $t > 0$.

47. Projectile Motion: Basketball In a Metro Conference men's basketball game on January 21, 1980, between Florida State University and Virginia Tech, a record was set. Les Henson, who is 6 ft 6 in. tall, made a basket from $89\frac{1}{4}$ ft down court to win the game for Virginia Tech by a score of 79 to 77. Assuming he released the ball at a height of 6 ft 6 in. and threw it at an angle of $45°$ (to maximize distance), what was the initial speed of the ball? See the figure.

48. Projectile Motion: Airplanes A plane is flying at an elevation of 4.0 km with a constant horizontal speed of 400 km/h toward a point directly above its target T. See the figure. At what angle of sight α should a package be released in order to strike the target? *Hint:* $g \approx 127{,}008$ km/h^2.

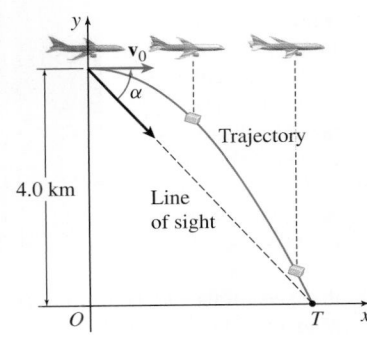

49. Projectile Motion: Baseball A baseball is hit at an angle of $45°$ to the horizontal from an initial height of 3 ft. If the ball just clears the vines in front of the bleachers in Wrigley Field, which are 10 ft high and a distance of 400 ft from home plate, what was the initial speed of the ball? How long did it take the ball to reach the vines?

50. Projectile Motion: Baseball An outfielder throws a baseball at an angle of $45°$ to the horizontal from an initial height of 6 ft. Suppose he can throw the ball with an initial velocity of 100 ft/s.

 (a) What is the farthest the outfielder can be from home plate to ensure that the ball reaches home plate on the fly?

 (b) How long is the ball in flight?

 (c) Use (a) and (b) to determine how fast a player must run to get from third base to home plate on a fly ball and beat the throw to score a run.

Hint: It is 90 ft from third base to home plate, and the runner cannot leave the base until the ball is caught.

51. Projectile Motion: Football   In a field goal attempt on a flat field, a football is kicked at an angle of $30°$ to the horizontal with an initial speed of $65\,\text{ft/s}$.

(a) What horizontal distance does the football travel while it is in the air?

(b) To score a field goal, the ball must clear the cross bar of the goal post, which is 10 ft above the ground. What is the farthest from the goal post the kick can originate and score a field goal?

52. Projectile Motion: Skeet Shooting   A gun, lifted at an angle θ_0 to the horizontal, is aimed at an elevated target T, which is released the moment the gun is fired. See the figure. No matter what the initial speed v_0 of the bullet is, show that it will always hit the falling target.

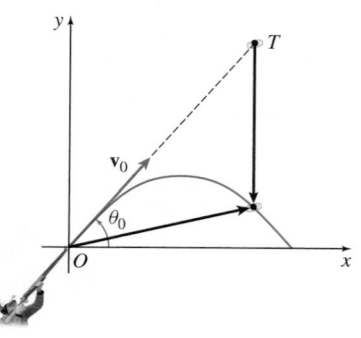

53. Mars Orbit

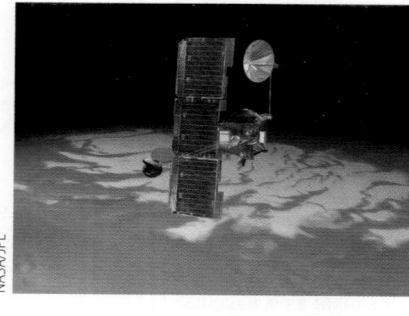

NASA's *Odyssey* orbiter's mission is to relay directly back to Earth the UHF telemetry taken by the rovers on Mars. *Odyssey* is in a "near-Mars" orbit at a height of 400 km above its surface. Find the speed of the *Odyssey* if the acceleration due to gravity on Mars is $3.71\,\text{m/s}^2$ and Mars' radius is 3390 km.

54. Hydrogen Atom   The **Bohr model of an atom** views it as a miniature solar system, with negative electrons in circular orbits around the positive nucleus. A hydrogen atom consists of a single electron in orbit around a single proton. The electric force F_{elec} that the proton of charge $e = 1.60 \times 10^{-19}$ C exerts on the electron, also of charge $e = 1.60 \times 10^{-19}$ C, is $F_{\text{elec}} = \dfrac{ke^2}{r^2}$, where r is the distance in meters between the nucleus and the electron and $k = 9.0 \times 10^9\,\text{N} \cdot \text{m}^3/\text{C}^2$ is a constant. Apply Newton's Second Law of Motion to the electron in a hydrogen atom and show that its speed is $v = \sqrt{\dfrac{ke^2}{mr}}$, where $m = 9.11 \times 10^{-31}$ kg is the mass of the electron.

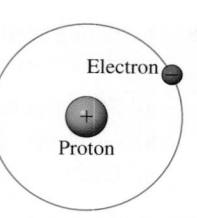

55. (a) Show that the maximum range of the projectile with position vector $\mathbf{r}(t) = \left\langle (v_0 \cos\theta)t, (v_0 \sin\theta)t - \dfrac{1}{2}gt^2 \right\rangle$ occurs when $\theta = \dfrac{\pi}{4}$.

(b) Show that the maximum range is $\dfrac{v_0^2}{g}$.

56. Show that the speed of the projectile whose position vector is given by $\mathbf{r}(t) = \left\langle (v_0 \cos\theta)t, -\dfrac{1}{2}gt^2 + (v_0 \sin\theta)t \right\rangle$ is least when the projectile is at its highest point.

| Preparing for the **AP®** Exam |

AP® Practice Problems

Multiple-Choice Questions

1. If the velocity vector of a particle in motion is

$$\mathbf{v}(t) = \langle t+3, 6t^2 \rangle$$

then the acceleration vector of the particle at $t = 2$ is

(A) $\langle 0, 24 \rangle$ (B) $\langle 1, 24 \rangle$

(C) $\langle 1, 12 \rangle$ (D) $\langle 2, 24 \rangle$

2. A particle moves along the plane curve

$$\mathbf{r}(t) = \langle e^{2t}, t^4 + 2t^2 \rangle$$

What is the acceleration vector of the particle at $t = 0$?

(A) $\mathbf{a}(0) = \langle 1, 4 \rangle$ (B) $\mathbf{a}(0) = \langle 4, 0 \rangle$

(C) $\mathbf{a}(0) = \langle 4, 4 \rangle$ (D) $\mathbf{a}(0) = \langle 4e^2, 4 \rangle$

3. A particle moves in the xy-plane along the curve

$$\mathbf{r}(t) = \langle 2\sqrt{t}, 3t^2 \rangle \quad t > 0$$

What is the velocity of the particle at $t = 9$?

(A) $\left\langle \dfrac{1}{3}, 54 \right\rangle$ (B) $\left\langle \dfrac{2}{3}, 54 \right\rangle$ (C) $\langle 3, 27 \rangle$ (D) $\left\langle \dfrac{1}{3}, 36 \right\rangle$

4. A particle is moving with velocity

$$\mathbf{v}(t) = \langle \pi \cos(\pi t), 3t^2 + 1 \rangle \ \text{m/s}$$

for $0 \le t \le 10$ seconds. Given that the position of the particle at time $t = 2$ s is $\mathbf{r}(2) = \langle 3, -2 \rangle$, the position vector of the particle at t is

(A) $\langle 3, -12 \rangle$ (B) $\langle 3 + \sin(\pi t), t^3 + t + 10 \rangle$

(C) $\langle \sin(\pi t), t^3 + t \rangle$ (D) $\langle 3 + \sin(\pi t), t^3 + t - 12 \rangle$

Free-Response Questions

 5. A particle moves along the plane curve

$$\mathbf{r}(t) = \langle 1 - 2\cos t, 2\sin t \rangle$$

(a) Find the velocity of the particle.
(b) Find the acceleration of the particle.

 6. A particle moves along the plane curve $\mathbf{r}(t) = \langle 2\sin t, 3\cos t \rangle$.

(a) Find the speed of the particle.

(b) What is the speed of the particle at $t = \dfrac{\pi}{2}$?

7. A particle is moving on a smooth curve with acceleration

$$\mathbf{a} = \mathbf{a}(t) = \langle 2t + 1, 3t^2 + 4t \rangle.$$

The particle's velocity at time $t = 0$ is $\mathbf{v}(0) = \langle 1, 2 \rangle$, and its position at time $t = 0$ is $\mathbf{r} = \mathbf{r}(0) = \langle 0, 0 \rangle$.

(a) Find the velocity vector $\mathbf{v}$.

(b) Find the speed of the particle at time $t = 1$.

(c) Find the position vector $\mathbf{r}$.

Retain Your Knowledge

Multiple-Choice Questions

1. $\int x \cos(3x)\, dx =$

(A) $\dfrac{1}{3}x\sin(3x) - \dfrac{1}{9}\cos(3x) + C$

(B) $\dfrac{1}{3}x\sin(3x) + \dfrac{1}{3}\cos(3x) + C$

(C) $x\sin(3x) - \dfrac{1}{3}\cos(3x) + C$

(D) $\dfrac{1}{3}x\sin(3x) + \dfrac{1}{9}\cos(3x) + C$

2. The function f is continuous on the closed interval $[-2, 7]$. The table below contains values of f for select numbers x in the interval.

x	-2	0	1	3	4	7
$f(x)$	-71	1	10	4	1	64

Approximate $\int_{-2}^{7} f(x)\, dx$ using a right Riemann sum with three intervals of width 3.

(A) -180 (B) 75 (C) 207 (D) 225

3. Find $\displaystyle\lim_{x \to 0} \dfrac{2\cos x \sin x}{x}$, if it exists.

(A) 0   (B) 2

(C) -2 (D) The limit does not exist.

Free-Response Question

 4. The velocity v of an object moving along a line is given by $v = v(t) = t^3 - 3t^2 + 2$ m/s, where $0 \le t \le 15$ s.

(a) If at $t = 1$ s, the object is 2 m to the right of the origin, find the position of the object at $t = 5$ s.

(b) At what times t is the object stopped?

(c) Find the acceleration of the object at $t = 2$ s.

(d) When is the speed of the object increasing?

CHAPTER 9 PROJECT

Polar Graphs and Microphones

This Project can be done individually or as part of a team.

Microphones can be configured to have different sensitivities, which are best modeled using various polar patterns. The oldest example is an **omnidirectional microphone**, which records sound equally from all directions. Omnidirectional microphones can be a good option when recording jungle sounds or general ambient noise, or when recording for your cell phone. The recording pattern for this microphone is virtually circular, and a good polar model is

$$r = k$$

where $k > 0$ is the sensitivity parameter. Notice that by adjusting k (say using 1, 4, or 8) we obtain concentric circles. Rotating any one of these circles around a diameter results in a sphere. Notice the microphone receiver is at the center of the sphere and sound from all directions is received equally.

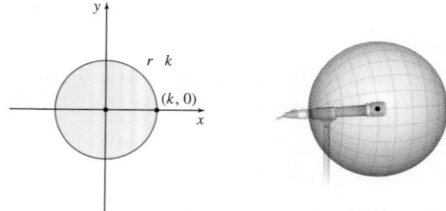

For this model, say when $k = 1$, we can convert the polar equation to parametric equations obtaining

$$x = \cos t \quad y = \sin t \quad 0 \le t \le 2\pi$$

Sound technicians have several things to consider when using microphones, including finding the best location to place the microphone and the best location to place any monitors. Since an omnidirectional microphone collects sound equally from all directions, the microphone should be placed directly in the center of the sounds being recorded with the most important sounds positioned closest to the microphone. Unfortunately, monitors cannot be used with an omnidirectional microphone since there is no place to put them where the microphone does not pick up their sound. To help remedy this and other possible issues, different types of microphone models are used.

A **cardioid microphone** is used when we want to pick up sounds mostly from the front of the microphone. Cardioid microphones are considered "unidirectional." A polar model for a cardioid microphone is

$$r = k(1 + \cos\theta) \quad 0 \le \theta \le 2\pi \quad k > 0$$

Notice that the microphone is placed with the transducer at the pole and pointing toward the polar axis. A cardioid microphone is formed by revolving the cardioid about the polar axis. See the figure.

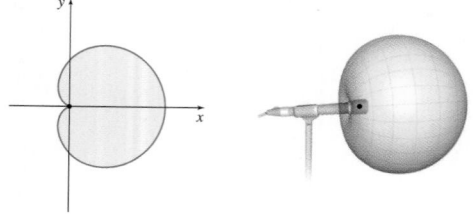

1. Sketch several cardioids using $k = 1, 4, 8$.

2. Explain why any noise created by the person's hand or the microphone stand is unlikely to be picked up by a cardioid microphone.

3. Using $k = 1$, convert the cardioid polar equation to a pair of parametric equations.

4. To determine the percentage of the sound that comes from the "front" of the microphone, we consider the sound coming at the microphone with angles in the polar wedge $-\dfrac{\pi}{4} \le \theta \le \dfrac{\pi}{4}$. With $k = 1$, determine the proportion of the arc length coming in from the front relative to that coming in from all directions.

5. Determine the proportion of the sound that is picked up from behind the microphone in the polar wedge $\dfrac{3\pi}{4} \le \theta \le \dfrac{5\pi}{4}$.

   Since less than 1% of all the sound received comes from the backside of a cardioid microphone, monitors can be placed directly behind the microphone.

   A **hypercardioid microphone** is used to pick up sounds mostly from in front of the microphone, but it also receives some sound from the rear. A polar model for such a hypercardioid microphone is

$$r = \frac{k\,|1 + 2\cos\theta|}{3} \quad 0 \le \theta \le 2\pi \quad k > 0$$

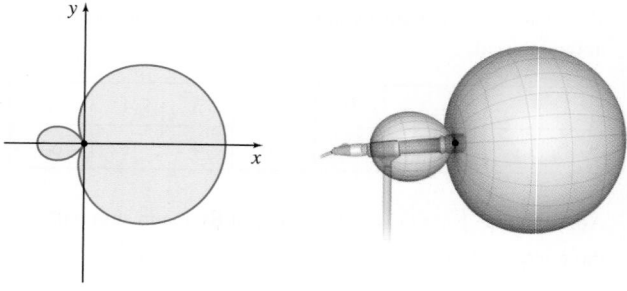

6. Sketch several hypercardioid graphs using $k = 1, 4$, and 8.

7. Express the hypercardioid polar equation with $k = 1$, as a pair of parametric equations.

8. Determine the proportion of the sound that is picked up from the front of a hypercardioid microphone $r = \dfrac{|1 + 2\cos\theta|}{3}$ by finding the percentage of sound coming at the microphone from the polar wedge $-\dfrac{\pi}{4} \le \theta \le \dfrac{\pi}{4}$.

9. Determine the proportion of the sound that is picked up from behind the hypercardioid microphone in the polar wedge $\dfrac{3\pi}{4} \le \theta \le \dfrac{5\pi}{4}$.

10. Compare the results of Problems 4, 5, 8, and 9, and describe how the microphones differ.

Chapter Review

THINGS TO KNOW

9.1 Parametric Equations

- Parametric equations $x = x(t)$, $y = y(t)$, where t is the parameter (p. 707)
- Plane curve: the graph of the points $(x, y) = (x(t), y(t))$ (p. 707)
- Orientation: the direction of the motion along the curve defined for $a \leq t \leq b$ moves as t varies from a to b (p. 708)
- Convert between parametric equations and rectangular equations (p. 712)
- Cycloid: the curve represented by $x(t) = a(t - \sin t)$, $y(t) = a(1 - \cos t)$ (p. 714)

9.2 Tangent Lines; 2nd-Order Derivatives; Arc Length

- Smooth curve (p. 719)
- Slope of a tangent line to a smooth curve

$$\frac{dy}{dx} = \frac{\dfrac{dy}{dt}}{\dfrac{dx}{dt}} \qquad \frac{dx}{dt} \neq 0 \qquad \text{(p. 720)}$$

- Vertical tangent line to a smooth curve

$$\frac{dx}{dt} = 0, \text{ but } \frac{dy}{dt} \neq 0 \qquad \text{(p. 720)}$$

- Horizontal tangent line to a smooth curve

$$\frac{dy}{dt} = 0, \text{ but } \frac{dx}{dt} \neq 0 \qquad \text{(p. 720)}$$

- 2nd-order derivatives of parametric equations (p. 723)
- Arc length formula for parametric equations:

$$s = \int_a^b \sqrt{\left(\frac{dx}{dt}\right)^2 + \left(\frac{dy}{dt}\right)^2} \, dt \qquad \text{(p. 724)}$$

- Total distance traveled (p. 726)

9.3 Parametric Equations, Tangent Lines, and Arc Length of Polar Equations

- Parametric equations for a polar equation $r = f(\theta) : x = f(\theta) \cos \theta$; $y = f(\theta) \sin \theta$ (p. 731)
- Slope of a line tangent to the graph of a polar equation:

$$\frac{dy}{dx} = \frac{\dfrac{dy}{d\theta}}{\dfrac{dx}{d\theta}} \qquad \text{(p. 731)}$$

- $\dfrac{dr}{d\theta}$: The rate of change of the distance r from the pole with respect to θ (p. 735)

- Arc length s of a curve represented by a polar equation from $\theta = \alpha$ to $\theta = \beta$, $\alpha \leq \theta \leq \beta$:

$$s = \int_\alpha^\beta \sqrt{r^2 + \left(\frac{dr}{d\theta}\right)^2} \, d\theta \qquad \text{(p. 737)}$$

9.4 Area in Polar Coordinates

- The area A bounded by the graph of $r = f(\theta)$, $r \geq 0$, and the rays $\theta = \alpha$ and $\theta = \beta$:

$$A = \int_\alpha^\beta \frac{1}{2} r^2 \, d\theta \qquad \text{(p. 741)}$$

9.5 Derivatives and Integrals of Vector Functions; Arc Length

- Vector function $\mathbf{r} = \mathbf{r}(t)$ (p. 748)
- The vector function $\mathbf{r} = \mathbf{r}(t) = \langle x(t), y(t) \rangle$ $a \leq t \leq b$, traces out a **plane curve**. (p. 749)
- Direction of a vector function (p. 749)
- Limit of a vector function (p. 750)
- A vector function $\mathbf{r} = \mathbf{r}(t)$ is **continuous at a real number** t_0 if:
  - $\mathbf{r}(t_0)$ is defined
  - $\lim_{t \to t_0} \mathbf{r}(t)$ exists
  - $\lim_{t \to t_0} \mathbf{r}(t) = \mathbf{r}(t_0)$ (p. 750)
- The **derivative** of the vector function $\mathbf{r} = \mathbf{r}(t)$ is

$$\mathbf{r}'(t) = \frac{d\mathbf{r}}{dt} = \lim_{h \to 0} \frac{\mathbf{r}(t + h) - \mathbf{r}(t)}{h},$$

provided the limit exists. (p. 751)

- Vector tangent to a curve (p. 753)
- Arc length s of a vector function: $s = \int_a^b \|\mathbf{r}'(t)\| dt$ (p. 753)
- If $\mathbf{r}(t) = \langle x(t), y(t) \rangle$, then

$$\int \mathbf{r}(t) dt = \left\langle \int x(t) dt, \int y(t) dt \right\rangle \quad \text{(p. 755)}$$

$$\int_a^b \mathbf{r}(t) dt = \left\langle \int_a^b x(t) dt, \int_a^b y(t) dt \right\rangle \quad \text{(p. 755)}$$

9.6 Motion along a Curve

- Position: $\mathbf{r}(t) = \langle x(t), y(t) \rangle$ (p. 760)
- Velocity: $\mathbf{v}(t) = \mathbf{r}'(t) = \left\langle \dfrac{dx}{dt}, \dfrac{dy}{dt} \right\rangle$ (p. 760)
- Acceleration: $\mathbf{a}(t) = \mathbf{r}''(t) = \left\langle \dfrac{d^2 x}{dt^2}, \dfrac{d^2 y}{dt^2} \right\rangle$ (p. 760)
- Speed: $v(t) = \|\mathbf{v}(t)\| = \|\mathbf{r}'(t)\|$ (p. 760)

OBJECTIVES

Section	You should be able to ...	Examples	Review Exercises	AP® Review Problems
9.1	**1** Graph parametric equations (p. 708)	1	1(b)–6(b)	
	2 Find a rectangular equation for a curve represented parametrically (p. 709)	2–4	1(a)–6(a), 1(c)–6(c)	
	3 Use time as the parameter in parametric equations (p. 712)	5	13, 24, 25	
	4 Convert a rectangular equation to parametric equations (p. 712)	6, 7	11, 12	
9.2	**1** Find an equation of the tangent line at a point on a plane curve (p. 721)	1–3	7–10	4, 5, 23
	2 Find 2nd-order derivatives of parametric equations (p. 723)	4	17–18	1, 8
	3 Find the arc length of a plane curve (p. 724)	5, 6	31–33	23
	4 Investigate the motion of an object moving along a plane curve (p. 725)	7	21–23	2, 6, 7
9.3	**1** Graph a polar equation; find parametric equations and tangent lines (p. 731)	1–3	14–16	3, 9, 26
	2 Investigate the derivative $\dfrac{dr}{d\theta}$ of a polar equation $r = f(\theta)$ (p. 735)	4–6	19, 20	10, 15
	3 Find the arc length of a curve represented by a polar equation (p. 736)	7	34–37	26
9.4	**1** Find the area of a region bounded by the graph of a polar equation (p. 741)	1–4	26, 27	14, 26
	2 Find the area of a region bounded by the graphs of two polar equations (p. 743)	5–7	28–30	11
9.5	**1** Find the domain of a vector function (p. 749)	1	39–42(a)	
	2 Graph a vector function (p. 749)	2, 3	39–42(b)	
	3 Find the limit and determine the continuity of a vector function (p. 750)	4, 5	38, 43, 44	
	4 Find the derivative of a vector function (p. 751)	6, 7	39–42(c), 45, 46	9, 16, 20
	5 Interpret the derivative of a vector function geometrically (p. 752)	8	47–49	13, 24
	6 Find the arc length of the curve traced out by a vector function (p. 753)	9, 10	50–52	19, 25
	7 Integrate vector functions (p. 755)	11–13	59–63	12, 21
9.6	**1** Find the velocity, acceleration, and speed of a particle moving along a smooth curve (p. 760)	1–3	53–58	17, 18, 25
	2 Find the position, velocity, and speed of a particle with known acceleration (p. 763)	4	64	22

REVIEW EXERCISES

In Problems 1–6:

(a) *Find the rectangular equation of each curve.*

(b) *Graph each plane curve whose parametric equations are given and show its orientation.*

(c) *Determine the restrictions on x and y that make the rectangular equation identical to the plane curve.*

1. $x(t) = 4t - 2$, $y(t) = 1 - t$; $-\infty < t < \infty$

2. $x(t) = 2t^2 + 6$, $y(t) = 5 - t$; $-\infty < t < \infty$

3. $x(t) = e^t$, $y(t) = e^{-t}$; $-\infty < t < \infty$

4. $x(t) = \ln t$, $y(t) = t^3$; $t > 0$

5. $x(t) = \sec^2 t$, $y(t) = \tan^2 t$; $0 \le t \le \dfrac{\pi}{4}$

6. $x(t) = t^{3/2}$, $y(t) = 2t + 4$; $t \ge 0$

In Problems 7–10, for the parametric equations below:

(a) *Find an equation of the tangent line to the curve at t.*

(b) *Graph the curve and the tangent line.*

7. $x(t) = t^2 - 4$, $y(t) = t$ at $t = 1$

8. $x(t) = 3\sin t$, $y(t) = 4\cos t + 2$ at $t = \dfrac{\pi}{4}$

9. $x(t) = \dfrac{1}{t^2}$, $y(t) = \sqrt{t^2 + 1}$ at $t = 3$

10. $x(t) = \dfrac{t^2}{1+t}$, $y(t) = \dfrac{t}{1+t}$ at $t = 0$

In Problems 11 and 12, find two different pairs of parametric equations for each rectangular equation.

11. $y = -2x + 4$ **12.** $y = 2x$

13. Describe the motion of an object that moves so that at time t (in seconds) it has coordinates

$$x(t) = 2\cos t \quad y(t) = \sin t \quad 0 \le t \le 2\pi$$

In Problems 14–16, for each polar equation:

(a) *Graph the equation.*
(b) *Find parametric equations that represent the equation.*

14. $r = 1 - \sin\theta$ **15.** $r = 4\cos(2\theta)$

16. $r = e^{0.5\theta}$

17. For the parametric equations $x(t) = t^2$, $y(t) = t^3 - 3t$,

  (a) Find the second derivative $\dfrac{d^2y}{dx^2}$.

  (b) Evaluate the second derivative at time $t = 2$.

18. For the parametric equations $x(t) = 3e^{-t}$, $y(t) = 4 + e^{2t}$,

  (a) Find the second derivative $\dfrac{d^2y}{dx^2}$.

  (b) Evaluate the second derivative at time $t = 0$.

19. Determine the interval(s) on which the polar equation $r = 3\sin\theta + 5$, $0 \le \theta \le 2\pi$, is moving closer to the pole.

20. Determine the interval(s) on which the polar equation $r = 2 - 2\cos\theta$, $0 \le \theta \le 2\pi$, is moving further away from the pole.

📐 *In Problems 21–23,*

  (a) *What is the total distance traveled by the object from $t = 0$ to $t = 10$?*

  (b) *For what times t is the object moving to the left?*

  (c) *For what times t is the object moving up?*

21. $x(t) = t^2$, $y(t) = t^3 - 3t$

22. $x(t) = 3e^{-t}$, $y(t) = 4 + e^{2t}$

23. $x(t) = 2 - t^3$, $y(t) = t^2 - 4t$

In Problems 24 and 25, find parametric equations for an object that moves along the ellipse $\dfrac{x^2}{16} + \dfrac{y^2}{9} = 1$ with the motion described.

24. The motion begins at $(4, 0)$, is counterclockwise, and requires 4 s for a complete revolution.

25. The motion begins at $(0, 3)$, is clockwise, and requires 5 s for a complete revolution.

26. **Area** Find the area of the region bounded by the cardioid $r = 2 + 2\cos\theta$, and the rays $\theta = 0$ and $\theta = \dfrac{\pi}{2}$.

27. **Area** Find the area of the region bounded by the rose $r = 5\cos(2\theta)$.

28. **Area** Find the area of the region inside the circle $r = 4\sin\theta$ and above the line $r = 3\csc\theta$.

29. **Area** Find the area of the region that lies inside the graph of $r^2 = 4\cos(2\theta)$ and outside the circle $r = \sqrt{2}$.

30. **Area** Find the area of the region common to the graphs of $r = \cos\theta$ and $r = 1 - \cos\theta$.

In Problems 31–33, find the arc length of each plane curve.

31. $x(t) = \tan t$, $y(t) = \dfrac{1}{3}(\sec^2 t + 1)$ from $t = 0$ to $t = \dfrac{\pi}{4}$

32. $x(t) = e^t$, $y(t) = \dfrac{1}{2}e^{2t} - \dfrac{1}{4}t$ from $t = 0$ to $t = 2$

33. $x = \dfrac{1}{2}y^2 - \dfrac{1}{4}\ln y$ from $y = 1$ to $y = 2$

In Problems 34–37, find the arc length of each curve represented by a polar equation.

34. $r = 2\sin\theta$ from $\theta = 0$ to $\theta = \pi$

35. $r = e^{-\theta}$ from $\theta = 0$ to $\theta = 2\pi$

36. $r = 3\theta$ from $\theta = 0$ to $\theta = 2\pi$

37. $r = 2\sin^2\dfrac{\theta}{2}$ from $\theta = -\dfrac{\pi}{2}$ to $\theta = \dfrac{\pi}{2}$

38. Find $\lim\limits_{t\to 3}\left\langle (t - 3), \dfrac{t^2 - 3t}{t^2 - 9} \right\rangle$.

In Problems 39–42, for each vector function $\mathbf{r} = \mathbf{r}(t)$:

  (a) *Find the domain.*
  (b) *Graph the curve C traced out by $\mathbf{r} = \mathbf{r}(t)$, indicating its orientation.*
  (c) *Find the value of $\mathbf{r}'(t)$ at the given number t.*

39. $\mathbf{r}(t) = \langle t^2, 3t \rangle$ $t = 2$

40. $\mathbf{r}(t) = \langle 3\cos t, \sin t \rangle$ $0 \le t \le \pi$, $t = \dfrac{\pi}{3}$

41. $\mathbf{r}(t) = \langle t, 2\cos t \rangle$ $t = 0$

42. $\mathbf{r}(t) = \langle t, t \rangle$ $t = 1$

In Problems 43 and 44, determine if the vector function is continuous at each number in the given interval. Identify any numbers where $\mathbf{r} = \mathbf{r}(t)$ is discontinuous.

43. $\mathbf{r}(t) = \left\langle \dfrac{t}{t + 1}, 3t \right\rangle$ interval $[-2, 3]$

44. $\mathbf{r}(t) = \left\langle \dfrac{t}{t - 2}, \sin t \right\rangle$ interval $(0, 2\pi)$

In Problems 45 and 46, find $\mathbf{r}'(t)$ and $\mathbf{r}''(t)$.

45. $\mathbf{r}(t) = \langle 2\cos t, 3\cos t \rangle$ **46.** $\mathbf{r}(t) = \langle 3t, 2e^t \rangle$

In Problems 47–49, for each curve C traced out by $\mathbf{r} = \mathbf{r}(t)$, find a vector tangent to the curve C at $t = 0$.

47. $\mathbf{r}(t) = \langle t, 2t \rangle$ **48.** $\mathbf{r}(t) = \langle e^t, e^{-t} \rangle$

49. $\mathbf{r}(t) = \langle e^t \sin t, e^t \cos t \rangle$

In Problems 50–52, find the arc length of each vector function.

50. $\mathbf{r}(t) = \langle \cos^3 t, \sin^3 t \rangle$ from $t = 0$ to $t = 2\pi$

51. $\mathbf{r}(t) = \langle 3t, (1 + 2t)^{3/2} \rangle$ from $t = 1$ to $t = 7$

52. $\mathbf{r}(t) = \langle \cos t, \sin t \rangle$ from $t = 0$ to $t = \dfrac{2\pi}{3}$

In Problems 53–56, find the velocity **v**, *acceleration* **a**, *and speed v of an object traveling along the curve traced out by the vector function* **r** = **r**(t).

53. $\mathbf{r}(t) = \langle 2\cos t, \sin t \rangle$

54. $\mathbf{r}(t) = \langle e^t \sin t, e^{-t} \rangle$

55. $\mathbf{r}(t) = \langle e^t, e^{-t} \rangle$

56. $\mathbf{r}(t) = \langle e^t \sin t, e^t \cos t \rangle$

57. Find the velocity and acceleration of an object moving on the cycloid $\mathbf{r}(t) = \langle \pi t - \sin(\pi t), 1 - \cos(\pi t) \rangle$.

58. (a) Find the velocity and acceleration of an object moving on the parabola $\mathbf{r}(t) = \langle t^2 - 2t, t \rangle$.

   (b) At what time t is the speed of the object 0?

In Problems 59 and 60, find each integral.

59. $\displaystyle\int \langle t^2 - 2, -(t-2)^2 \rangle\, dt$

60. $\displaystyle\int_0^\pi \left\langle \cos\frac{t}{2}, \sin^2\frac{t}{2} \right\rangle dt$

In Problems 61–63, find **r** = **r**(t) *for the given initial condition.*

61. $\mathbf{r}'(t) = \langle e^{2t}, \ln t \rangle$ $\mathbf{r}(1) = \langle 0, 1 \rangle$

62. $\mathbf{r}'(t) = \langle \sin t \cos t, \tan t \sec t \rangle$ $\mathbf{r}(0) = \langle 1, 0 \rangle$

63. $\dfrac{d\mathbf{r}}{dt} = \langle \cos t, -\sin t \rangle$ $\mathbf{r}(0) = \langle 1, 0 \rangle$

64. Find the speed and the velocity and position vectors at time t for an object whose acceleration at time t is given by

$$\mathbf{a}(t) = \langle \tan t \sec t, \cos t \rangle$$

Assume that the initial velocity and initial position vectors are **0**.

65. Projectile Motion A projectile is fired with a speed of 500 m/s at an inclination of 45° to the horizontal from a point 30 m above level ground. Find the point where the projectile strikes the ground.

Break It Down   Preparing for the AP® Exam

*Let's take a closer look at part (b) of AP® **Practice Problem 9** from Section 9.2.*

9. An object moves along a smooth curve C, and its position at time t is $(x(t), y(t))$ for $0 \le t \le 5$. The object moves in such a way that

$$\frac{dx}{dt} = 3t^2 - 4 \qquad \frac{dy}{dt} = t + 2$$

   (b) For what times t is the object moving to the left?

Step 1	Identify the underlying structure and related concepts.	This problem deals with the motion of an object along a smooth curve defined by parametric equations.
Step 2	Determine the appropriate math rule or procedure.	$x(t)$ and $\dfrac{dx}{dt}$ are the horizontal position and velocity, respectively, of the object. To determine the times for which the object is moving to the left, we need to solve the inequality $\dfrac{dx}{dt} < 0$.
Step 3	Apply the math rule or procedure.	$$\frac{dx}{dt} = 3t^2 - 4 = \left(\sqrt{3}t + 2\right)\left(\sqrt{3}t - 2\right) < 0$$ The factor $\sqrt{3}t + 2 > 0$ for all numbers t, $0 \le t \le 5$. So the derivative $\dfrac{dx}{dt} < 0$ when the factor $\sqrt{3}t - 2 < 0$ or when $t < \dfrac{2}{\sqrt{3}}$. Since $0 \le t \le 5$, $\dfrac{dx}{dt} < 0$ on the interval $\left(0, \dfrac{2}{\sqrt{3}}\right)$.
Step 4	Clearly communicate your answer.	Because $\dfrac{dx}{dt} < 0$ on the interval $\left(0, \dfrac{2}{\sqrt{3}}\right)$, the object is moving to the left for $0 < t < \dfrac{2}{\sqrt{3}}$.

AP® Review Problems: Chapter 9

Multiple-Choice Questions

1. The second derivative $y'' = \dfrac{d^2y}{dx^2}$ of the function defined by the parametric equations

$$x(t) = 2t^3 + t + 4 \qquad y(t) = t^2 - 3$$

equals

(A) $\dfrac{12t^2 + 2}{\left(6t^2 + 1\right)^3}$ (B) $\dfrac{-12t^2 + 2}{\left(6t^2 + 1\right)^3}$

(C) $\dfrac{-12t^2 + 2}{\left(6t^2 + 1\right)^2}$ (D) $\dfrac{-12t^2}{\left(6t^2 + 1\right)^2}$

2. The position of an object moving along a curve C at time t is $(x(t), y(t))$ for $0 \le t \le 10$. If it is known that

$$\frac{dx}{dt} = t^2 - 4 \quad \text{and} \quad \frac{dy}{dt} = t^2 - 4t + 3$$

when is the object moving to the right?

(A) $0 \le t \le 2$ (B) $1 \le t \le 10$

(C) $2 \le t \le 10$ (D) $4 \le t \le 10$

3. What is the slope of the line tangent to the curve of the polar equation $r = 2\cos\theta$ when $\theta = \dfrac{\pi}{3}$?

(A) $-\sqrt{3}$ (B) $\dfrac{\sqrt{3}}{2}$ (C) $\dfrac{\sqrt{3}}{3}$ (D) $-\dfrac{\sqrt{3}}{3}$

4. Find an equation of the line tangent to the plane curve represented by the parametric equations $x(t) = 1 + 2\ln t$, $y(t) = t^3 - 3$, at $t = 1$.

(A) $3x - 2y = 7$ (B) $3x - 2y = 2$

(C) $3x - 2y = -\dfrac{7}{2}$ (D) $2y + 3x = -\dfrac{7}{2}$

5. The plane curve C is represented by the parametric equations $x(t) = 2t^2 + 5$, $y(t) = 3t - t^3$. Find all the points on C where the tangent line is horizontal or vertical.

(A) vertical at $(5, 0)$; horizontal at $(7, 2)$

(B) vertical at $(0, 5)$; horizontal at $(7, 2)$ and $(-7, 2)$

(C) vertical at $(5, 0)$; horizontal at $(7, 2)$ and $(7, -2)$

(D) vertical at $(5, 0)$; horizontal at $(7, 2)$ and $(-7, 2)$

6. Find the distance traveled by an object that moves along the plane curve represented by the parametric equations

$$x(t) = \frac{3}{4}t^2 + 5,\ y(t) = 3 + t^3,\ \text{from } t = 1 \text{ to } t = 4.$$

(A) 65.506   (B) 58.095   (C) 59.492   (D) 64.108

7. An object moves along a curve C so its position at time t is $(x(t), y(t))$ for $1 \le t \le 8$. The object moves in such a way that

$$\frac{dx}{dt} = t^3 - 16t \qquad \frac{dy}{dt} = 3t - t^2$$

At what time is the object at its highest position?

(A) 0   (B) 3   (C) 4   (D) 8

8. The function $y = f(x)$ is defined by the parametric equations

$$x = e^t \qquad y = y(t) = t^2 + 2$$

for any real number t. Then $y''(2)$ equals

(A) -2 (B) -0.271 (C) -0.037 (D) 0.038

9. Parametric equations for $r = 2^{\theta/3}$ are

(A) $x = 2^{\theta/3}\cos\theta$, $y = 2^{\theta/3}\sin\theta$

(B) $x = r\cos 2^{\theta/3}$, $y = r\sin 2^{\theta/3}$

(C) $x = \cos 2^{\theta/3}$, $y = \sin 2^{\theta/3}$

(D) $x = \dfrac{\theta}{3}\cos\theta$, $y = -\dfrac{\theta}{3}\sin\theta$

10. For the polar equation $r = f(\theta)$, where f is differentiable, which of the following statements are true?

  I. The derivative $\dfrac{dr}{d\theta}$ gives the slope of a line tangent to the graph of r.

  II. $|r|$ represents the distance between the polar curve and the pole.

  III. A vertical tangent line occurs when $\dfrac{dx}{d\theta} = 0$

     and $\dfrac{dy}{d\theta} \ne 0$.

(A) I only   (B) I and III only

(C) II and III only   (D) I, II, and III

11. The graphs of the limaçon $r = 2 - \cos\theta$ and the circle $r = \cos\theta$ are shown.

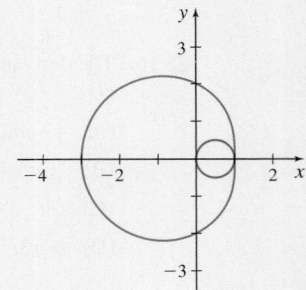

The area inside the limaçon but outside the circle is given by

  I. $\dfrac{1}{2}\displaystyle\int_0^{2\pi} (2 - \cos\theta)^2 d\theta - \int_0^{\pi/2} \cos^2\theta\, d\theta$

  II. $\dfrac{1}{2}\displaystyle\int_0^{2\pi} (2 - \cos\theta)^2 d\theta - \frac{\pi}{4}$

  III. $\dfrac{1}{2}\displaystyle\int_0^{2\pi} (2 - 2\cos\theta)^2 d\theta$

(A) I only   (B) II only

(C) I and II only   (D) I, II, and III

12. Find $\mathbf{r} = \mathbf{r}(t)$ if $\mathbf{r}'(t) = \langle 4t,\ 2t^3 \rangle$ and $\mathbf{r}(0) = \langle 1, 2 \rangle$

(A) $\left\langle 2t^2, t^4 \right\rangle$ (B) $\left\langle 2t^2 - 1,\ \dfrac{1}{2}t^4 - 2 \right\rangle$

(C) $\left\langle 2t^2 + 1,\ \dfrac{1}{2}t^4 + 2 \right\rangle$ (D) $\left\langle 4t + 1,\ 2t^3 + 2 \right\rangle$

13. Which of the following is the vector tangent to the curve traced out by the vector function $\mathbf{r}(t) = \langle e^{2t}, 2\sin(3t)\rangle$ at $t = 0$?

(A) $\left\langle \dfrac{1}{6}, \dfrac{2}{3} \right\rangle$ (B) $\langle 1, 6\rangle$ (C) $\langle 2, 6\rangle$ (D) $\langle 2, -6\rangle$

14. The graph of the polar equation $r = 4\sin(3\theta)$ is shown.

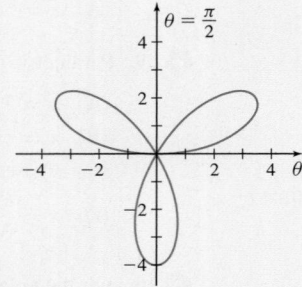

Which integral gives the area enclosed by the loop in the first quadrant?

(A) $\dfrac{1}{2}\int_0^{\pi/2} 4\sin(3\theta)\, d\theta$ (B) $\dfrac{1}{2}\int_0^{2\pi} 16\sin^2(3\theta)\, d\theta$

(C) $\dfrac{1}{2}\int_0^{\pi/2} 16\sin^2(3\theta)\, d\theta$ (D) $\int_0^{\pi/3} 8\sin^2(3\theta)\, d\theta$

15. On what interval(s) is the graph of the polar equation $r = 2 + 2\sin\theta$, $0 \le \theta \le 2\pi$, moving farther from the pole?

(A) $[0,\ \pi]$ (B) $\left[0,\ \dfrac{\pi}{2}\right]$

(C) $\left[\dfrac{\pi}{2},\ \dfrac{3\pi}{2}\right]$ (D) $\left[0,\ \dfrac{\pi}{2}\right]$ and $\left[\dfrac{3\pi}{2},\ 2\pi\right]$

16. The derivative $\mathbf{r}'(t)$ of the vector function $\mathbf{r}(t) = \langle \cos(t^3 + t), \sin(t^3 + t)\rangle$ is

(A) $\langle -\sin(3t^2 + 1), \cos(3t^2 + 1)\rangle$

(B) $\langle (3t^2 + 1)\sin(t^3 + t), (3t^2 + 1)\cos(t^3 + t)\rangle$

(C) $\langle 3t^2 + 1 - \sin(t^3 + t), 3t^2 + 1 + \cos(t^3 + t)\rangle$

(D) $\langle -(3t^2 + 1)\sin(t^3 + t), (3t^2 + 1)\cos(t^3 + t)\rangle$

17. A particle moves along the curve traced out by the vector function $\mathbf{r}(t) = \langle 10\ln t, t^3 + 8\rangle$, $t > 0$. What is the speed of the object at $t = 2$?

(A) $\dfrac{5}{12}$ (B) $\dfrac{12}{5}$ (C) 13 (D) $\langle 5, 12\rangle$

18. A particle moves along the curve traced out by the vector function $\mathbf{r}(t) = \langle e^{3t}, 5t - t^2\rangle$. What is the acceleration of the particle at $t = 0$?

(A) $\langle 9, 0\rangle$ (B) $\langle 6, -2\rangle$ (C) $\langle 9, -2\rangle$ (D) $\langle 9, 2\rangle$

19. The arc length of the smooth curve traced out by $\mathbf{r}(t) = \langle e^t, \sin t\rangle$ from $t = 0$ to $t = \pi$ is given by

(A) $\int_0^\pi \sqrt{(e^t + \sin t)^2}\, dt$

(B) $\int_0^\pi \sqrt{e^{2t} + \cos^2 t}\, dt$

(C) $\int_0^\pi \sqrt{e^{2t} + \sin^2 t}\, dt$

(D) $\int_0^\pi \sqrt{(e^t \cos t)^2 + (e^t \sin t)^2}\, dt$

20. $\dfrac{d}{dt}\langle e^{2t}t^3, e^{2t}(4 - 2t)\rangle =$

(A) $\langle t^2 e^{2t}(2t + 3), 2e^{2t}(3 - 2t)\rangle$

(B) $\langle 6t^2 e^{2t}, -4te^{2t}\rangle$

(C) $\langle 5t^2 e^{2t}, 2e^{2t}\rangle$

(D) $\langle 5t^2 e^{2t}(1 + t), 2e^{2t}(3 - 2t)\rangle$

21. For the vector function $\mathbf{r}(t) = \left\langle \dfrac{3}{t},\ 4t \right\rangle$, find $\int_e^{e^2} \mathbf{r}(t)\, dt$

(A) $\langle 3, 2e^2 - 2e\rangle$ (B) $\langle 3, 2e^4 - e^2\rangle$

(C) $\langle 5, 2e^4 - e^2\rangle$ (D) $\langle 3, 2e^4 - 2e^2\rangle$

22. A particle moves along a plane curve with velocity $\mathbf{v}(t) = \langle 3t + 4, \sqrt{t + 1}\rangle$. If $\mathbf{r}(0) = \langle 3, 1\rangle$, what is the position of the particle at $t = 3$?

(A) $\left\langle \dfrac{57}{2}, \dfrac{17}{3} \right\rangle$ (B) $\left\langle \dfrac{57}{2}, \dfrac{19}{3} \right\rangle$

(C) $\left\langle \dfrac{57}{2}, 6 \right\rangle$ (D) $\langle 171, 34\rangle$

Free-Response Questions

23. The plane curve represented by $x(t) = t - \sin t$, $y(t) = 1 - \cos t$ is a cycloid.

(a) Find the slope of the tangent line to the cycloid for $0 \le t \le 2\pi$.

(b) Find an equation of the tangent line to the cycloid at $t = \dfrac{\pi}{3}$.

(c) Find the length of the cycloid from $t = 0$ to $t = \dfrac{\pi}{2}$.

24. For the vector function $\mathbf{r}(t) = \langle e^t, 2t\rangle$, find the vector tangent to the curve traced out by $\mathbf{r} = \mathbf{r}(t)$ at $t = 0$.

25. The position vector of a particle moving in the xy-plane is

$$\mathbf{r}(t) = \langle 2t^3 + 1, 3\cos(2t - 6)\rangle, t \ge 0$$

(a) Find the velocity vector of the particle at $t = 3$.

(b) Find the acceleration vector of the particle at $t = 3$.

(c) Find the speed of the particle at $t = 3$.

(d) Write an integral that represents the total distance the particle travels from $t = 0$ to $t = 3$.

26. The graph of the polar equation $r = 4 - 3\sin\theta$ is shown.

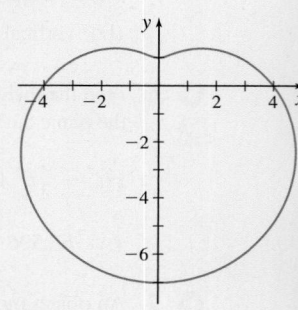

(a) Find the area of the region enclosed by $r = 4 - 3\sin\theta$.

(b) At what points is the line tangent to the graph vertical?

(c) Find the total distance traveled by an object traveling once around the perimeter of the graph.

AP® Cumulative Review Problems: Chapters 1–9

Preparing for the AP® Exam

Multiple-Choice Questions

▶ 1. $\lim\limits_{h \to 0} \dfrac{\cos(\pi + h) + 1}{h} =$

(A) π (B) 0   (C) 1   (D) -1

▶ 2. Find $\dfrac{dy}{dx}$ at $(1, 27)$ for $x^{1/3} + y^{1/3} = 4$

(A) 27   (B) $\dfrac{4}{3}$ (C) -9 (D) -3

▶ 3. Which of the following is the solution to the differential equation $\dfrac{dy}{dx} = \dfrac{x^3}{y^2}$ with the initial condition $y(2) = 3$?

(A) $y = \sqrt[3]{\dfrac{3x^4}{4} + 15}$ (B) $y = \sqrt[3]{\dfrac{3x^4}{4}}$

(C) $y = 3e^{-4 + (x^3/9)}$ (D) $y = x\sqrt{x} + 5$

▶ 4. $\lim\limits_{x \to 0} \dfrac{xe^{3x} + x}{\sin(3x)} =$

(A) 1   (B) 0   (C) e^3 (D) $\dfrac{2}{3}$

▶ 5. What is the slope of the line tangent to the graph of $y = \sin^{-1}(4x)$ at the point where $x = \dfrac{1}{16}$?

(A) $\dfrac{1}{4}$ (B) $\dfrac{16\sqrt{15}}{15}$ (C) $\dfrac{\sqrt{15}}{4}$ (D) 0

▶ 6. If $0 < c < 2$, what is the area of the region bounded by the graphs of $y = x^2 + 1$, $y = 0$, $x = c$, and $x = 2$?

(A) $4 - c^2$ (B) $\dfrac{-c^3}{3} - c + \dfrac{14}{3}$

(C) $\dfrac{c^3}{3} - \dfrac{14}{3}$ (D) $c^2 - 4c + 5$

▶ 7. A particle moves along the x-axis so that at time $t \geq 0$, the acceleration of the particle is $a(t) = 4\sqrt[3]{t}$. The position of the particle is 8 when $t = 0$, and the position is $\dfrac{16}{7}$ when $t = 1$. What is the velocity of the particle at $t = 1$?

(A) $\dfrac{16}{7}$ (B) -4 (C) 8   (D) $-\dfrac{10}{7}$

▶ 8. $\int \cos^3 x \sin x\, dx =$

(A) $\dfrac{-\cos^4 x}{4} + C$ (B) $\dfrac{\sin^2 x}{2} + \dfrac{\sin^4 x}{4} + C$

(C) $-3\cos^3 x \sin x + C$ (D) $\dfrac{\cos^4 x \sin^2 x}{8} + C$

▶ 9. Let f be the function defined by $f(x) = 3\sqrt{x}$. What is the approximation for $f(3)$ found by using the line tangent to the graph of f at the point $(4, 6)$?

(A) $\dfrac{21}{4}$ (B) 6   (C) $\dfrac{27}{4}$ (D) 6.062

▶ 10. A cup of coffee is cooling in a room that has a constant temperature of 72 degrees Fahrenheit (°F). If the initial temperature of the coffee at time $t = 0$ is $190°$F and the temperature of the coffee changes at the rate $R(t) = -5.98e^{-.049t}$ degrees Fahrenheit per minute, what is the temperature, to the nearest degree, of the coffee after 5 minutes?

(A) $120°$F   (B) $163°$F   (C) $73°$F   (D) $45°$F

▶ 11. $\int 4x \ln x\, dx =$

(A) $2x^2 \ln x - x^2 + C$ (B) $4x^2 \ln x - 2\ln x + C$
(C) $4x \ln x - x^2 + C$ (D) $2x^2 \ln x + x^2 + C$

Free-Response Questions

▶ 12. The base of a solid is the region R bounded by the graphs of $y = 8\sqrt{x}$ and $y = x^2$. Express each of the following as integrals.

(a) the area A of the region R.

(b) the volume V of the solid if cross sections perpendicular to the x-axis are squares.

(c) the volume V of the solid if cross sections perpendicular to the x-axis are equilateral triangles.

(d) the volume V of the solid generated by revolving the region R about the x-axis.

▶ 13. The graph of the polar equation $r = 2 + \sin(2\theta)$ is shown.

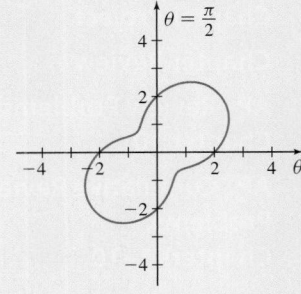

(a) Determine the maximum distance of the curve from the origin on the interval $[0, \pi]$.

(b) Find the area of the region enclosed by the curve.

(c) Find the slope of the line tangent to the curve at $\theta = 0$.

(d) An object is moving along the curve from $\theta = 0$ to $\theta = 2\pi$. Express the velocity of the object at $\theta = \dfrac{\pi}{6}$ as a vector.

UNIT 10   Infinite Sequences and Series

CHAPTER

10

Infinite Series

kelvin38/Shutterstock

How Calculators Calculate

When you want to find the sine of a number, you probably type it into your calculator or computer and use the number that comes out. Have you ever thought about how the calculator/computer obtains the result? Since we are only good at computing integer powers of numbers, we have no direct way to evaluate transcendental functions, such as the exponential function or the sine function. We know that $\sin \pi = 0$, but what is $\sin 3$? Since $\sin 3$ is an irrational number, it is represented by a nonrepeating, nonterminating decimal. So, computers and calculators work with approximations for transcendental functions that involve integer powers.

Most transcendental functions can be expressed as *power series*. As we shall see in Section 10.8, a power series is an infinite sum of monomials with integer exponents. Since we are good at raising numbers to integer powers, we can truncate these series and find good polynomial approximations to transcendental functions on restricted domains. This is how computers evaluate many transcendental functions.

 In the Chapter 10 Project on page 892, we explore how computers obtain lightning-fast approximations.

Miguel
Software engineer

Every time I am confronted with a problem whose solution I need to engineer, I make use of the problem-solving skills I learned in AP® Calculus. In approaching a problem, I know I first have to identify what's being asked of me and then determine which tools I can use to get to the answer.

The idea of adding an infinite collection of numbers has long intrigued mathematicians. In this chapter, we explore the consequences of a definition for an infinite sum. Beginning with sequences, we are led to an *infinite series*, an infinite sum of numbers. We examine various tests to determine whether an infinite series has a sum, *converges*, or whether it does not, *diverges*. Finally, we investigate *power series*, representations of functions by infinite series, and Taylor polynomials, which are used to approximate functions. ■

10.1 Sequences

OBJECTIVES *When you finish this section, you should be able to:*

1 **Write several terms of a sequence (p. 780)**
2 **Find the *n*th term of a sequence (p. 781)**
3 **Use properties of convergent sequences (p. 783)**
4 **Use a related function or the Squeeze Theorem to show a sequence converges (p. 785)**
5 **Determine whether a sequence converges or diverges (p. 787)**

The study of infinite sums of numbers has important applications in physics and engineering since it provides an alternate way of representing functions. In particular, *infinite series* may be used to approximate irrational numbers, such as e, π, and $\ln 2$. The theory of infinite series is developed through the use of a special kind of function called a *sequence*.

NEED TO REVIEW? Sequences are introduced in Section P.8, pp. 71–76.

> **DEFINITION**
>
> A **sequence** is a function whose domain is the set of positive integers and whose range is a subset of the real numbers.

To get an idea of what this means, consider the graph of the function $f(x) = \dfrac{1}{x}$ for $x > 0$, as shown in Figure 1(a). If all the points on the graph are removed *except* the points $(1, 1)$, $\left(2, \dfrac{1}{2}\right)$, $\left(3, \dfrac{1}{3}\right)$, and so on, as shown in Figure 1(b), then these points are the *graph of a sequence*. Notice the graph consists of points, one point for each positive integer.

A sequence is often represented by listing its values in order. For example, the sequence in Figure 1(b) can be written as

$$f(1), f(2), f(3), f(4), f(5), \ldots$$

or as the list

$$1, \frac{1}{2}, \frac{1}{3}, \frac{1}{4}, \frac{1}{5}, \ldots$$

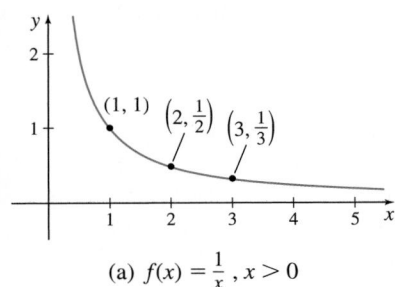

(a) $f(x) = \dfrac{1}{x}$, $x > 0$

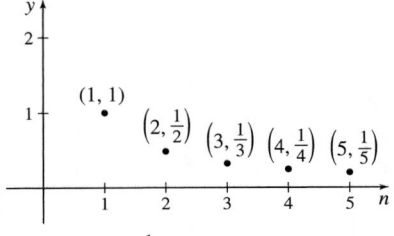

(b) $f(n) = \dfrac{1}{n}$, n is a positive integer

Figure 1

The list never ends, as the three dots at the end (called **ellipsis**) indicate. The numbers in the list are called the **terms** of the sequence. Using subscripted letters to represent the terms of a sequence, this sequence can be written as

$$s_1 = f(1) = 1 \qquad s_2 = f(2) = \frac{1}{2} \qquad s_3 = f(3) = \frac{1}{3} \qquad \cdots \qquad s_n = f(n) = \frac{1}{n} \cdots$$

Chapter 10 • Infinite Series

It is easy to obtain any term of this sequence because $s_n = f(n) = \dfrac{1}{n}$. In general, whenever a rule for the **nth term** of a sequence is known, then any term of the sequence can be found. We also use the nth term to identify the sequence. When the nth term is enclosed in braces, it represents the sequence. The notation $\{s_n\} = \left\{ \dfrac{1}{n} \right\}$

- $\{s_n\}_{n=1}^{\infty} = \left\{ \dfrac{1}{n} \right\}_{n=1}^{\infty}$ and

  $\{s_n\} = \left\{ \dfrac{1}{n} \right\}$ define the same sequence;

- $s_n = \dfrac{1}{n}$ is the nth term of the sequence.

or $\{s_n\}_{n=1}^{\infty} = \left\{ \dfrac{1}{n} \right\}_{n=1}^{\infty}$ both represent the sequence $1, \dfrac{1}{2}, \dfrac{1}{3}, \dfrac{1}{4}, \dfrac{1}{5}, \ldots$.

1 Write Several Terms of a Sequence

BREAK IT DOWN: As you work through the problems in this chapter, remember to follow these steps:

Step 1 Identify the underlying structure and related concepts.

Step 2 Determine the appropriate math rule or procedure.

Step 3 Apply the math rule or procedure.

Step 4 Clearly communicate your answer.

On page 898, see how we've used these steps to solve Section 10.5 AP® Practice Problem 5(c) on page 844.

EXAMPLE 1 Writing Terms of a Sequence

Write the first three terms of each sequence:

(a) $\{b_n\}_{n=1}^{\infty} = \left\{ \dfrac{1}{3n-2} \right\}_{n=1}^{\infty}$ **(b)** $\{c_n\} = \left\{ \dfrac{2n-1}{n^3} \right\}$

Solution

(a) The nth term of this sequence is $b_n = \dfrac{1}{3n-2}$. The first three terms are

$$b_1 = \frac{1}{3 \cdot 1 - 2} = 1 \qquad b_2 = \frac{1}{3 \cdot 2 - 2} = \frac{1}{4} \qquad b_3 = \frac{1}{3 \cdot 3 - 2} = \frac{1}{7}$$

(b) The nth term of this sequence is $c_n = \dfrac{2n-1}{n^3}$. Then

$$c_1 = \frac{2 \cdot 1 - 1}{1^3} = 1 \qquad c_2 = \frac{2 \cdot 2 - 1}{2^3} = \frac{3}{8} \qquad c_3 = \frac{2 \cdot 3 - 1}{3^3} = \frac{5}{27}$$ ■

NOW WORK   Problem 15.

For simplicity, from now on we use the notation $\{s_n\}$, rather than $\{s_n\}_{n=1}^{\infty}$, to represent a sequence.

EXAMPLE 2 Writing Terms of a Sequence

Write the first five terms of the sequence

$$\{a_n\} = \left\{ (-1)^n \left(\frac{1}{2} \right)^n \right\}$$

Solution

The nth term of this sequence is $a_n = (-1)^n \left(\dfrac{1}{2} \right)^n$. So,

$$a_1 = (-1)^1 \left(\frac{1}{2} \right)^1 = -\frac{1}{2} \qquad a_2 = (-1)^2 \left(\frac{1}{2} \right)^2 = \frac{1}{4}$$

$$a_3 = (-1)^3 \left(\frac{1}{2} \right)^3 = -\frac{1}{8} \qquad a_4 = \frac{1}{16} \qquad a_5 = -\frac{1}{32}$$

The first five terms of the sequence $\{a_n\}$ are $-\dfrac{1}{2}, \dfrac{1}{4}, -\dfrac{1}{8}, \dfrac{1}{16}, -\dfrac{1}{32}$. ■

780

Notice the terms of the sequence $\{a_n\}$ given in Example 2 *alternate* between positive and negative due to the factor $(-1)^n$, which equals -1 when n is odd and equals 1 when n is even. Sequences of the form $\{(-1)^n a_n\}$, $\{(-1)^{n-1} a_n\}$, or $\{(-1)^{n+1} a_n\}$, where $a_n > 0$ for all n, are called **alternating sequences**.

NOW WORK   Problem 17.

2 Find the *n*th Term of a Sequence

The rule defining a sequence $\{s_n\}$ is often expressed by an explicit formula for its *n*th term s_n. There are times, however, when a sequence is indicated using the first few terms, suggesting a natural choice for the *n*th term.

EXAMPLE 3   Finding the *n*th Term of a Sequence

Find the *n*th term of each of the following sequences. Assume that the indicated pattern continues.

Sequence	Solution *n*th term
(a) $e, \dfrac{e^2}{2}, \dfrac{e^3}{3}, \dots$	$a_n = \dfrac{e^n}{n}$
(b) $1, \dfrac{1}{3}, \dfrac{1}{9}, \dfrac{1}{27}, \dots$	$b_n = \left(\dfrac{1}{3}\right)^{n-1}$ or $b_n = \dfrac{1}{3^{n-1}}$
(c) $1, 4, 9, 16, 25, \dots$	$c_n = n^2$
(d) $\dfrac{2}{2}, \dfrac{4}{3}, \dfrac{6}{4}, \dfrac{8}{5}, \dots$	$d_n = \dfrac{2n}{n+1}$
(e) $1, -\dfrac{1}{2}, \dfrac{1}{3}, -\dfrac{1}{4}, \dfrac{1}{5}, \dots$	$e_n = \dfrac{(-1)^{n+1}}{n}$
(f) $1, \dfrac{1}{2}, 1, \dfrac{1}{4}, 1, \dfrac{1}{6}, \dots$	$f_n = \begin{cases} 1 & \text{if } n \text{ is odd} \\ \dfrac{1}{n} & \text{if } n \text{ is even} \end{cases}$

The graphs of the sequences (d)–(f) are given in Figure 2.

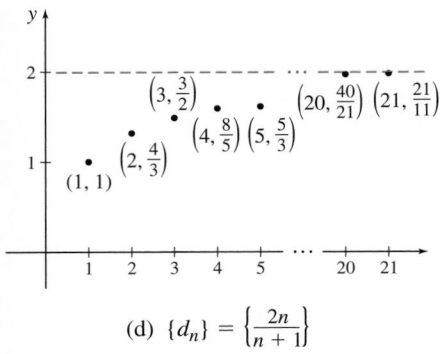

(d) $\{d_n\} = \left\{\dfrac{2n}{n+1}\right\}$

(e) $\{e_n\} = \left\{\dfrac{(-1)^{n+1}}{n}\right\}$

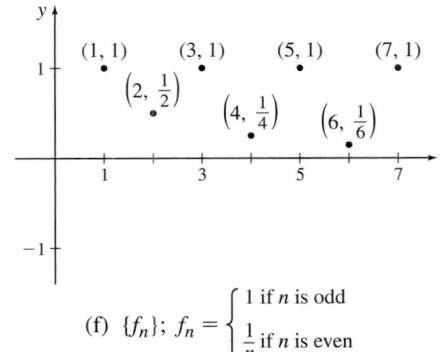

(f) $\{f_n\};\ f_n = \begin{cases} 1 & \text{if } n \text{ is odd} \\ \dfrac{1}{n} & \text{if } n \text{ is even} \end{cases}$

Figure 2

NOW WORK   Problem 25.

Convergent Sequences

As we look from left to right, the points of the graph of the sequence $\{d_n\} = \left\{ \dfrac{2n}{n+1} \right\}$ in Figure 2(d) get closer to the line $y = 2$. If we could look far enough to the right, the points would *appear* to be *on* the line $y = 2$ (although, in reality, the points are always some very small distance below the line). In fact, as Table 1 suggests, $d_n = \dfrac{2n}{n+1}$ can be made as close as we please to 2 by taking n sufficiently large.

TABLE 1

n	1	9	99	999	9999	999,999
$d_n = \dfrac{2n}{n+1}$	1	1.8	1.98	1.998	1.9998	1.999998

We describe this behavior by saying that the sequence $\{d_n\} = \left\{ \dfrac{2n}{n+1} \right\}$ *converges to 2,* and we write $\displaystyle\lim_{n \to \infty} \dfrac{2n}{n+1} = 2$.

DEFINITION Convergent Sequence

Suppose L is a real number and $\{s_n\}$ is a sequence. The sequence $\{s_n\}$ **converges** to the real number L if, for any number $\varepsilon > 0$, there is a positive integer N so that

$$\boxed{|s_n - L| < \varepsilon \qquad \text{for all integers } n > N}$$

If $\{s_n\}$ converges to L, then L is called the **limit of the sequence** and we write $\displaystyle\lim_{n \to \infty} s_n = L$. If a sequence converges, it is said to be **convergent**; otherwise, it is said to be **divergent**.

Figure 3 provides a geometric interpretation of the statement $\displaystyle\lim_{n \to \infty} s_n = L$. The figure also illustrates that the beginning terms of a sequence do not affect the convergence (or divergence) of the sequence. That is, **the beginning terms of a sequence can be ignored when determining convergence or divergence**.

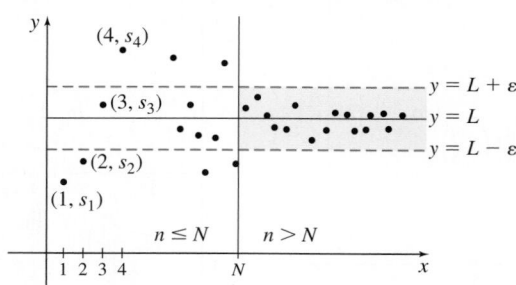

Figure 3

When $n \le N$, the distance between the point (n, s_n) and the line $y = L$ may or may not be less than ε; these terms have no effect on the convergence or divergence of the sequence. But if a sequence $\{s_n\}$ converges, then for any $\varepsilon > 0$, there is an integer N so that for all $n > N$, the distance between the point (n, s_n) and the line $y = L$ remains less than ε, that is, $|s_n - L| < \varepsilon$ for all $n > N$.

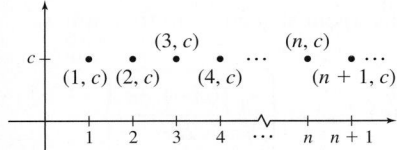

Figure 4 $\{s_n\} = \{c\}$

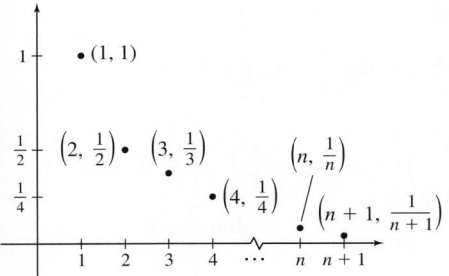

Figure 5 $\{s_n\} = \left\{\dfrac{1}{n}\right\}$

> **NEED TO REVIEW?** Limit properties are listed in Chapter 1, on p. 104.

EXAMPLE 4   Two Limits of Sequences

Use a graph to illustrate the following results.

(a) The sequence $\{s\} = \{c\}$ converges to c. That is,

$$\lim_{n \to \infty} c = c$$

(b) The sequence $\{s\} = \left\{\dfrac{1}{n}\right\}$ converges to 0. That is,

$$\lim_{n \to \infty} \frac{1}{n} = 0$$

Solution

(a) See Figure 4. The graph of the sequence $\{s_n\} = \{c\}$ suggests that $\{s_n\}$ converges to c.

(b) See Figure 5. The graph of the sequence $\{s_n\} = \left\{\dfrac{1}{n}\right\}$ suggests that $\{s_n\}$ converges to 0. ∎

❸ Use Properties of Convergent Sequences

We state, without proof, some properties of convergent sequences. Note the resemblance to the limit properties of functions from Chapter 1.

> **THEOREM**   Properties of Convergent Sequences
>
> If $\{s_n\}$ and $\{t_n\}$ are convergent sequences and if c is a number, then
>
> - **Constant multiple property:**
>
> $$\lim_{n \to \infty} (cs_n) = c \lim_{n \to \infty} s_n \qquad (1)$$
>
> - **Sum and difference properties:**
>
> $$\lim_{n \to \infty} (s_n \pm t_n) = \lim_{n \to \infty} s_n \pm \lim_{n \to \infty} t_n \qquad (2)$$
>
> - **Product property:**
>
> $$\lim_{n \to \infty} (s_n \cdot t_n) = \left(\lim_{n \to \infty} s_n\right)\left(\lim_{n \to \infty} t_n\right) \qquad (3)$$
>
> - **Quotient property:**
>
> $$\lim_{n \to \infty} \frac{s_n}{t_n} = \frac{\lim\limits_{n \to \infty} s_n}{\lim\limits_{n \to \infty} t_n} \quad \text{provided} \quad \lim_{n \to \infty} t_n \neq 0 \qquad (4)$$
>
> - **Power property:**
>
> $$\lim_{n \to \infty} s_n^p = \left[\lim_{n \to \infty} s_n\right]^p \qquad p \geq 2 \text{ is an integer} \qquad (5)$$
>
> - **Root property:**
>
> $$\lim_{n \to \infty} \sqrt[p]{s_n} = \sqrt[p]{\lim_{n \to \infty} s_n} \qquad p \geq 2 \text{ and } s_n > 0 \text{ if } p \text{ is even} \qquad (6)$$

EXAMPLE 5 **Using Properties of Convergent Sequences**

Use the results of Example 4, and properties of convergent sequences to find $\lim_{n\to\infty} s_n$.

(a) $\{s_n\} = \left\{\dfrac{2}{n} + 3\right\}$ **(b)** $\{s_n\} = \left\{\dfrac{4}{n^2}\right\}$ **(c)** $\{s_n\} = \left\{\sqrt[3]{\dfrac{16n^2 + 3n}{2n^2}}\right\}$

Solution

(a) $\lim\limits_{n\to\infty} s_n = \lim\limits_{n\to\infty}\left(\dfrac{2}{n} + 3\right) = \lim\limits_{n\to\infty}\dfrac{2}{n} + \lim\limits_{n\to\infty} 3$

$\qquad = 2\lim\limits_{n\to\infty}\dfrac{1}{n} + \lim\limits_{n\to\infty} 3 = 2\cdot 0 + 3 = 3$ $\lim\limits_{n\to\infty}\dfrac{1}{n} = 0;\ \lim\limits_{n\to\infty} 3 = 3$

(b) $\lim\limits_{n\to\infty} s_n = \lim\limits_{n\to\infty}\dfrac{4}{n^2} = 4\lim\limits_{n\to\infty}\dfrac{1}{n^2} = 4\lim\limits_{n\to\infty}\left(\dfrac{1}{n}\right)^2 \underset{\substack{\uparrow \\ \text{Power} \\ \text{property}}}{=} 4\left(\lim\limits_{n\to\infty}\dfrac{1}{n}\right)^2 \underset{\substack{\uparrow \\ \lim\limits_{n\to\infty}\frac{1}{n} = 0}}{=} 4\cdot 0^2 = 0$

(c) $\lim\limits_{n\to\infty} s_n = \lim\limits_{n\to\infty}\sqrt[3]{\dfrac{16n^2 + 3n}{2n^2}} \underset{\substack{\uparrow \\ \text{Root} \\ \text{property}}}{=} \sqrt[3]{\lim\limits_{n\to\infty}\left(\dfrac{16n^2 + 3n}{2n^2}\right)} = \sqrt[3]{\lim\limits_{n\to\infty}\left(8 + \dfrac{3}{2n}\right)}$

$\qquad = \sqrt[3]{\lim\limits_{n\to\infty} 8 + \lim\limits_{n\to\infty}\dfrac{3}{2n}} \underset{\substack{\uparrow \\ \lim\limits_{n\to\infty} 8 = 8}}{=} \sqrt[3]{8 + \dfrac{3}{2}\lim\limits_{n\to\infty}\dfrac{1}{n}} \underset{\substack{\uparrow \\ \lim\limits_{n\to\infty}\frac{1}{n} = 0}}{=} \sqrt[3]{8 + \dfrac{3}{2}\cdot 0} = \sqrt[3]{8} = 2$ ∎

NOW WORK Problem 37.

The next result is also useful for showing a sequence converges. You are asked to prove it in Problem 142.

THEOREM

Suppose $\{s_n\}$ is a sequence of real numbers. If $\lim\limits_{n\to\infty} s_n = L$ and if f is a function that is continuous at L and is defined for all numbers s_n, then $\lim\limits_{n\to\infty} f(s_n) = f(L)$.

EXAMPLE 6 **Showing a Sequence Converges**

Show $\left\{\ln\left(\dfrac{2}{n} + 3\right)\right\}$ converges and find its limit.

Solution

Since $\lim\limits_{n\to\infty}\left(\dfrac{2}{n} + 3\right) = 3$ [from Example 5(a)], the sequence $\{s_n\} = \left\{\dfrac{2}{n} + 3\right\}$ converges to 3. The function $f(x) = \ln x$ is continuous on its domain, $\{x \mid x > 0\}$, so it is continuous at 3. Then

$$\lim\limits_{n\to\infty} f(s_n) = \lim\limits_{n\to\infty} f\left(\dfrac{2}{n} + 3\right) = \lim\limits_{n\to\infty}\ln\left(\dfrac{2}{n} + 3\right) = \ln\left[\lim\limits_{n\to\infty}\left(\dfrac{2}{n} + 3\right)\right] = \ln 3$$

So, the sequence $\left\{\ln\left(\dfrac{2}{n} + 3\right)\right\}$ converges to $\ln 3$.   ∎

NOW WORK Problem 45.

**4 Use a Related Function or the Squeeze Theorem
to Show a Sequence Converges**

Sometimes a sequence $\{s_n\}$ can be associated with a *related function* f, which can be helpful in determining whether the sequence converges.

DEFINITION Related Function of a Sequence

A **related function** f of the sequence $\{s_n\}$ has the following two properties:

- f is defined on the open interval $(0, \infty)$; that is, the domain of f is the set of positive real numbers.
- $f(n) = s_n$ for all integers $n \geq 1$.

Figure 6 $f(x) = \dfrac{x}{e^x}$; $\{s_n\} = \left\{\dfrac{n}{e^n}\right\}$

EXAMPLE 7 Identifying a Related Function of a Sequence

If $\{s_n\} = \left\{\dfrac{n}{e^n}\right\}$, then a related function is given by $f(x) = \dfrac{x}{e^x}$, where $x > 0$, as shown in Figure 6. ■

There is a connection between the convergence of certain sequences $\{s_n\}$ and the behavior at infinity of a related function f of the sequence $\{s_n\}$. The following result, which we state without proof, explains this connection.

THEOREM

Suppose $\{s_n\}$ is a sequence of real numbers, f is a related function of $\{s_n\}$, and L is a real number.

$$\text{If } \lim_{x \to \infty} f(x) = L \qquad \text{then} \qquad \lim_{n \to \infty} s_n = L \tag{7}$$

NEED TO REVIEW? Limits at infinity are discussed in Section 1.5, pp. 136–143.

EXAMPLE 8 Using a Related Function to Show a Sequence Converges

Show that $\left\{\dfrac{3n^2 + 5n - 2}{6n^2 - 6n + 5}\right\}$ converges and find its limit.

Solution

The function

$$f(x) = \frac{3x^2 + 5x - 2}{6x^2 - 6x + 5} \qquad x > 0$$

is a related function of the sequence $\left\{\dfrac{3n^2 + 5n - 2}{6n^2 - 6n + 5}\right\}$. Since

$$\lim_{x \to \infty} f(x) = \lim_{x \to \infty} \frac{3x^2 + 5x - 2}{6x^2 - 6x + 5} = \lim_{x \to \infty} \frac{3x^2}{6x^2} = \frac{1}{2}$$

the sequence $\left\{\dfrac{3n^2 + 5n - 2}{6n^2 - 6n + 5}\right\}$ converges, and $\displaystyle\lim_{n \to \infty} \frac{3n^2 + 5n - 2}{6n^2 - 6n + 5} = \frac{1}{2}$. ■

Be careful! A related function f can be used to show a sequence $\{s_n\}$ converges only if $\lim\limits_{x \to \infty} f(x) = L$, where L is a *real number*.

Suppose f is the related function of a sequence $\{s_n\}$.

- If $\lim\limits_{x \to \infty} f(x)$ is infinite, then $\{s_n\}$ diverges.
- If $\lim\limits_{x \to \infty} f(x)$ does not exist, then the theorem relating $\lim\limits_{n \to \infty} s_n$ and $\lim\limits_{x \to \infty} f(x)$ provides no information about $\lim\limits_{n \to \infty} s_n$.

NOW WORK Problem 51.

NEED TO REVIEW? L'Hôpital's Rule is discussed in Section 4.4, pp. 314–318.

We can sometimes find the limit of a sequence $\{s_n\}$ by applying L'Hôpital's Rule to its related function f, provided f meets the necessary requirements.

CALC CLIP

EXAMPLE 9   **Using L'Hôpital's Rule to Show a Sequence Converges**

Show that $\left\{ \dfrac{n}{e^n} \right\}$ converges and find its limit.

Solution

We begin with the related function $f(x) = \dfrac{x}{e^x}$, $x > 0$. To find $\lim\limits_{x \to \infty} f(x)$, we note that $\dfrac{x}{e^x}$ is an indeterminate form of the type $\dfrac{\infty}{\infty}$ and use L'Hôpital's Rule.

$$\lim_{x \to \infty} f(x) = \lim_{x \to \infty} \frac{x}{e^x} = \lim_{x \to \infty} \frac{1}{e^x} = 0$$

$$\uparrow$$
$$\text{Use L'Hôpital's Rule}$$

Since $\lim\limits_{x \to \infty} f(x) = 0$, the sequence $\left\{ \dfrac{n}{e^n} \right\}$ converges and $\lim\limits_{n \to \infty} \dfrac{n}{e^n} = 0$. ∎

NOW WORK Problem 57.

NEED TO REVIEW? The Squeeze Theorem for functions is discussed in Section 1.4, pp. 122–124.

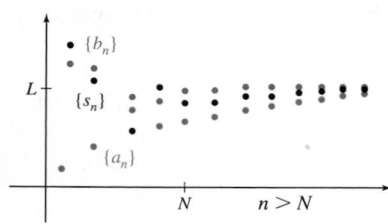

Figure 7 $a_n \le s_n \le b_n$ for $n > N$

THEOREM The Squeeze Theorem for Sequences

Suppose $\{a_n\}$, $\{b_n\}$, and $\{s_n\}$ are sequences and N is a positive integer. If $a_n \le s_n \le b_n$ for every integer $n > N$, and if $\lim\limits_{n \to \infty} a_n = \lim\limits_{n \to \infty} b_n = L$, then $\lim\limits_{n \to \infty} s_n = L$.

Figure 7 illustrates the Squeeze Theorem for sequences.

EXAMPLE 10   **Using the Squeeze Theorem for Sequences**

Show that $\{s_n\} = \left\{ (-1)^n \dfrac{1}{n} \right\}$ converges and find its limit.

Solution

We seek two sequences that "squeeze" $\{s_n\} = \left\{ (-1)^n \dfrac{1}{n} \right\}$ as n becomes large. We begin with $|s_n|$:

$$|s_n| = \left| (-1)^n \frac{1}{n} \right| \le \frac{1}{n}$$

$$-\frac{1}{n} \le (-1)^n \frac{1}{n} \le \frac{1}{n}$$

Notice that s_n is bounded by $\{a_n\} = \left\{ -\dfrac{1}{n} \right\}$ and $\{b_n\} = \left\{ \dfrac{1}{n} \right\}$. Since $a_n \le s_n \le b_n$ for all n and $\displaystyle\lim_{n\to\infty} a_n = \lim_{n\to\infty} \left(-\dfrac{1}{n} \right) = 0$ and $\displaystyle\lim_{n\to\infty} b_n = \lim_{n\to\infty} \dfrac{1}{n} = 0$, then by the Squeeze Theorem for sequences, the sequence $\{s_n\} = \left\{ (-1)^n \dfrac{1}{n} \right\}$ converges and $\displaystyle\lim_{n\to\infty} s_n = 0$. ■

NOW WORK   Problem 59.

⑤ Determine Whether a Sequence Converges or Diverges

A sequence $\{s_n\}$ diverges if $\displaystyle\lim_{n\to\infty} s_n$ does not exist. This can happen if

- there is no single number L that the terms of the sequence approach as $n \to \infty$

or

- $\displaystyle\lim_{n\to\infty} s_n = \infty$.

> **DEFINITION** Divergence of a Sequence to Infinity
>
> The sequence $\{s_n\}$ **diverges to infinity**, that is,
>
> $$\boxed{\lim_{n\to\infty} s_n = \infty}$$
>
> if, given any positive number M, there is a positive integer N so that whenever $n > N$, then $s_n > M$.

EXAMPLE 11   **Showing a Sequence Diverges**

Show that the following sequences diverge:

(a) $\{1 + (-1)^n\}$ **(b)** $\{n\}$

Solution

(a) The terms of the sequence are $0,\ 2,\ 0,\ 2,\ 0,\ 2, \ldots$. See Figure 8. Since the terms alternate between 0 and 2, the terms of the sequence $\{1 + (-1)^n\}$ do not approach a single number L as $n \to \infty$. So, the sequence $\{1 + (-1)^n\}$ is divergent.

(b) The terms of the sequence $\{s_n\} = \{n\}$ are $1,\ 2,\ 3,\ 4, \ldots$. Given any positive number M, we choose a positive integer $N > M$. Then whenever $n > N$, we have $s_n = n > N > M$. That is, the sequence $\{n\}$ diverges to infinity. See Figure 9. ■

NOW WORK   Problem 63.

The next results are useful throughout the chapter.

> **THEOREM** Convergence/Divergence of a Geometric Sequence
>
> The geometric sequence $\{s_n\} = \{r^n\}$, where r is a real number,
>
> - converges to 0, if $-1 < r < 1$.
> - converges to 1, if $r = 1$.
> - diverges for all other numbers.

Problems 138–141 provide an outline of the proof.

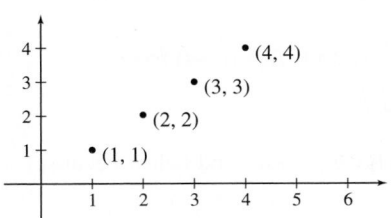

Figure 8 $\{1 + (-1)^n\}$

Figure 9 $\{s_n\} = \{n\}$

EXAMPLE 12 **Determining Whether $\{r^n\}$ Converges or Diverges**

Determine whether the following geometric sequences converge or diverge:

(a) $\{s_n\} = \left\{ \left(\dfrac{3}{4} \right)^n \right\}$ **(b)** $\{t_n\} = \left\{ \left(\dfrac{4}{3} \right)^n \right\}$

Solution

(a) The sequence $\{s_n\} = \left\{ \left(\dfrac{3}{4} \right)^n \right\}$ converges to 0 because $-1 < \dfrac{3}{4} < 1$.

(b) The sequence $\{t_n\} = \left\{ \left(\dfrac{4}{3} \right)^n \right\}$ diverges because $\dfrac{4}{3} > 1$. ∎

NOW WORK Problem 67.

There are other ways to show that a sequence converges or diverges. To explore these, we need to define a *bounded sequence* and a *monotonic sequence*.

Bounded Sequences

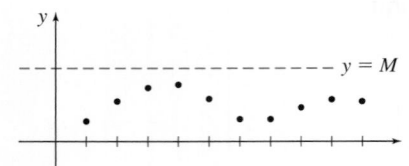

Figure 10 The sequence is bounded from above.

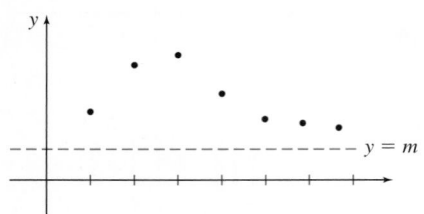

Figure 11 The sequence is bounded from below.

DEFINITION

A sequence $\{s_n\}$ is **bounded from above** if every term of the sequence is less than or equal to some number M. That is,

$$\boxed{s_n \leq M \qquad \text{for all } n}$$

See Figure 10.

Similarly, a sequence $\{s_n\}$ is **bounded from below** if every term of the sequence is greater than or equal to some number m. That is,

$$\boxed{s_n \geq m \qquad \text{for all } n}$$

See Figure 11.

For example, since $s_n = \cos n \leq 1$ for all n, the sequence $\{s_n\} = \{\cos n\}$ is bounded from above by 1, and since $s_n = \cos n \geq -1$ for all n, $\{s_n\}$ is bounded from below by -1.

EXAMPLE 13 **Determining Whether a Sequence Is Bounded from Above or Bounded from Below**

(a) The sequence $\{s_n\} = \left\{ \dfrac{3n}{n+2} \right\}$ is bounded both from above and below because

$$\dfrac{3n}{n+2} = \dfrac{3}{1 + \dfrac{2}{n}} < 3 \qquad \text{and} \qquad \dfrac{3n}{n+2} > 0 \qquad \text{for all } n \geq 1$$

See Figure 12(a).

(b) The sequence $\{a_n\} = \left\{ \dfrac{4n}{3} \right\}$ is bounded from below because $\dfrac{4n}{3} > 1$ for all $n \geq 1$.

It is not bounded from above because $\lim\limits_{n \to \infty} \dfrac{4n}{3} = \dfrac{4}{3} \lim\limits_{n \to \infty} n = \infty$. See Figure 12(b).

(c) The sequence $\{b_n\} = \left\{ (-1)^{n+1} n \right\}$ is neither bounded from above nor bounded from below. If n is odd, $\lim\limits_{n \to \infty} b_n = \lim\limits_{n \to \infty} n = \infty$, and if n is even, $\lim\limits_{n \to \infty} b_n = \lim\limits_{n \to \infty} (-n) = -\infty$. See Figure 12(c).

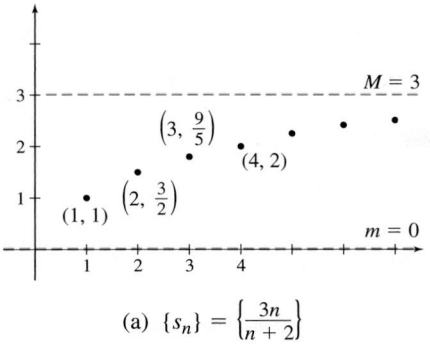

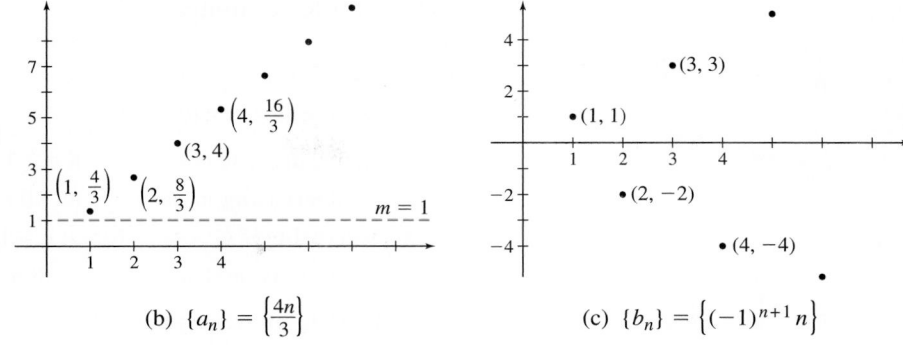

(a) $\{s_n\} = \left\{\dfrac{3n}{n+2}\right\}$ (b) $\{a_n\} = \left\{\dfrac{4n}{3}\right\}$ (c) $\{b_n\} = \left\{(-1)^{n+1} n\right\}$

Figure 12

NOW WORK Problem 73.

DEFINITION

A sequence $\{s_n\}$ is **bounded** if it is bounded both from above and from below. For a bounded sequence $\{s_n\}$, there is a positive number K for which

$$|s_n| \leq K \qquad \text{for all } n \geq 1$$

For example, the sequence $\{s_n\} = \left\{\dfrac{3n}{n+2}\right\}$ [from Example 13(a)] is bounded,

since $\left|\dfrac{3n}{n+2}\right| \leq 3$ for all integers $n \geq 1$.

THEOREM Boundedness Theorem

A convergent sequence is bounded.

CAUTION The converse of the Boundedness Theorem is not true. Bounded sequences may converge, or they may diverge. For example, the sequence $\{1 + (-1)^n\}$ in Example 11(a) is bounded, but it diverges.

NEED TO REVIEW? The Triangle Inequality is discussed in Appendix A.1, p. A-7.

Proof If $\{s_n\}$ is a convergent sequence, there is a number L for which $\lim\limits_{n \to \infty} s_n = L$. We use the definition of the limit of a sequence with $\varepsilon = 1$. Then there is a positive integer N so that

$$|s_n - L| < 1 \qquad \text{for all } n > N$$

Then for all $n > N$,

$$|s_n| = |s_n - L + L| \leq \underset{\underset{\text{Triangle Inequality}}{\uparrow}}{|s_n - L| + |L|} < 1 + |L|$$

If we choose K to be the largest number in the finite collection

$$|s_1|, |s_2|, |s_3|, \ldots, |s_N|, 1 + |L|$$

it follows that $|s_n| \leq K$ for *all* integers $n \geq 1$. That is, the sequence $\{s_n\}$ is bounded. ∎

A restatement of the boundedness theorem provides a test for divergent sequences.

THEOREM Test for Divergence of a Sequence

If a sequence is not bounded from above or if it is not bounded from below, then it diverges.

For example, the sequence $\{-n\}$ is not bounded from below, and the sequences $\left\{\dfrac{4n}{3}\right\}$, $\{2^n\}$, and $\{\ln n\}$ are not bounded from above, so each sequence is divergent.

Monotonic Sequences

DEFINITION

A sequence $\{s_n\}$ is said to be:

- **Increasing** if $s_n < s_{n+1}$ for all $n \geq 1$.
- **Nondecreasing** if $s_n \leq s_{n+1}$ for all $n \geq 1$.
- **Decreasing** if $s_n > s_{n+1}$ for all $n \geq 1$.
- **Nonincreasing** if $s_n \geq s_{n+1}$ for all $n \geq 1$.

If any of the above conditions hold, the sequence is called **monotonic**.

Table 2 lists three ways to show a sequence $\{s_n\}$ is monotonic.

TABLE 2

	To Show $\{s_n\}$ Is Decreasing	To Show $\{s_n\}$ Is Increasing
Algebraic Difference	Show $s_{n+1} - s_n < 0$ for all $n \geq 1$.	Show $s_{n+1} - s_n > 0$ for all $n \geq 1$.
Algebraic Ratio	If $s_n > 0$ for all $n \geq 1$, show $\dfrac{s_{n+1}}{s_n} < 1$ for all $n \geq 1$.	If $s_n > 0$ for all $n \geq 1$, show $\dfrac{s_{n+1}}{s_n} > 1$ for all $n \geq 1$.
Derivative	Show the derivative of a related function f of $\{s_n\}$ is negative for all $x \geq 1$.	Show the derivative of a related function f of $\{s_n\}$ is positive for all $x \geq 1$.

These tests can be extended to show a sequence is nonincreasing or nondecreasing.

EXAMPLE 14 Showing a Sequence Is Monotonic

Show that each of the following sequences is monotonic by determining whether it is increasing, nondecreasing, decreasing, or nonincreasing.

(a) $\{s_n\} = \left\{\dfrac{n}{n+1}\right\}$ **(b)** $\{s_n\} = \left\{\dfrac{e^n}{n!}\right\}$ **(c)** $\{s_n\} = \{\ln n\}$

Solution

(a) We use the algebraic difference test.

$$s_{n+1} - s_n = \frac{n+1}{n+2} - \frac{n}{n+1} = \frac{n^2 + 2n + 1 - n^2 - 2n}{(n+2)(n+1)}$$

$$= \frac{1}{(n+2)(n+1)} > 0 \qquad \text{for all } n \geq 1$$

So, $\{s_n\}$ is an increasing sequence.

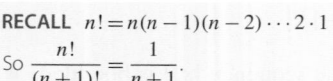

RECALL $n! = n(n-1)(n-2)\cdots 2 \cdot 1$
So $\dfrac{n!}{(n+1)!} = \dfrac{1}{n+1}$.

(b) When the sequence contains a factorial, the algebraic ratio test is usually easiest to use.

$$\frac{s_{n+1}}{s_n} = \frac{\dfrac{e^{n+1}}{(n+1)!}}{\dfrac{e^n}{n!}} = \frac{e^{n+1}}{e^n} \cdot \frac{n!}{(n+1)!} = \frac{e}{n+1} < 1 \qquad \text{for all } n \geq 2$$

After the first term, $\{s_n\} = \left\{\dfrac{e^n}{n!}\right\}$ is a decreasing sequence.

(c) Here we use the derivative of the related function $f(x) = \ln x$ of the sequence $\{s_n\} = \{\ln n\}$. Since $\dfrac{d}{dx}\ln x = \dfrac{1}{x} > 0$ for all $x > 0$, it follows that f is an increasing function and so the sequence $\{\ln n\}$ is an increasing sequence. ∎

NOW WORK Problem 81.

Not every sequence is monotonic. For example, the sequences $\{s_n\} = \left\{ \sin\left(\dfrac{\pi}{2}n\right) \right\}$ and $\{t_n\} = \left\{ 1 + \dfrac{(-1)^n}{n^2} \right\}$, shown in Figures 13 and 14, are not monotonic. Notice that although both $\{s_n\}$ and $\{t_n\}$ are bounded sequences, $\{s_n\}$ diverges and $\{t_n\}$ converges.

The sequence $\{u_n\} = \{n\}$ shown in Figure 15 is monotonic, but it is not bounded from above. So, $\{u_n\}$ is divergent.

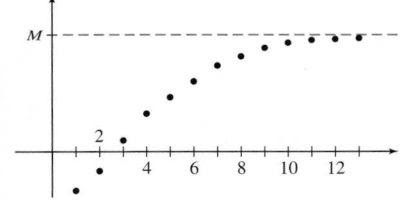

Bounded, not monotonic; diverges

Figure 13 $\{s_n\} = \left\{ \sin\left(\dfrac{\pi}{2}n\right) \right\}$

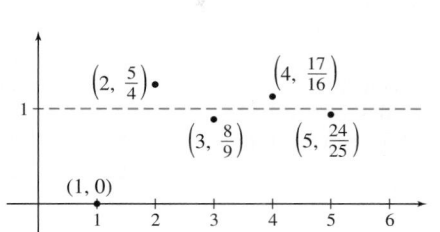

Bounded, not monotonic; converges

Figure 14 $\{t_n\} = \left\{ 1 + \dfrac{(-1)^n}{n^2} \right\}$

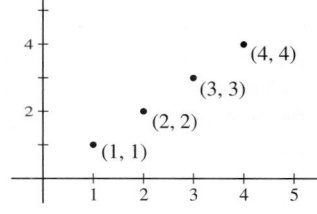

Not bounded, monotonic; diverges

Figure 15 $\{u_n\} = \{n\}$

So, there are examples of monotonic sequences that diverge and examples of bounded sequences that diverge. However, a sequence that is both monotonic and bounded always converges.

> **THEOREM**
>
> An increasing (or nondecreasing) sequence $\{s_n\}$ that is bounded from above converges. A decreasing (or nonincreasing) sequence $\{s_n\}$ that is bounded from below converges.

IN WORDS A bounded, monotonic sequence converges.

Since the convergence of a sequence $\{s_n\}$ is about the behavior of $\{s_n\}$ for large values of n, if a sequence *eventually increases* and is bounded from above, it is convergent. Similar remarks hold for sequences that *eventually decrease* and are bounded from below.

The proof of this theorem can be found in Appendix B. Figure 16 illustrates the theorem for a sequence that is increasing and bounded from above.

Figure 16

EXAMPLE 15   Determining Whether a Sequence Converges or Diverges

Determine whether the sequence $\{s_n\} = \left\{ \dfrac{2^n}{n!} \right\}$ converges or diverges.

Solution

To see if $\left\{ \dfrac{2^n}{n!} \right\}$ is monotonic, find the algebraic ratio $\dfrac{s_{n+1}}{s_n}$:

$$\frac{s_{n+1}}{s_n} = \frac{\dfrac{2^{n+1}}{(n+1)!}}{\dfrac{2^n}{n!}} = \frac{2^{n+1}\, n!}{(n+1)!\, 2^n} = \frac{2}{n+1} \le 1 \qquad \text{for all } n \ge 1$$

Since $s_{n+1} \le s_n$ for $n \ge 1$, the sequence $\{s_n\}$ is nonincreasing.

Next, since each term of the sequence is positive, $s_n > 0$ for $n \ge 1$, the sequence $\{s_n\}$ is bounded from below.

Since $\{s_n\}$ is nonincreasing and bounded from below, it converges. ∎

NOTE Although the sequence $\{s_n\}$ converges, the theorem does not tell us what $\lim\limits_{n \to \infty} s_n$ equals.

NOW WORK   Problem 89.

Summary

To determine whether a sequence converges:

- Look at a few terms of the sequence to see if a trend is developing. For example, the first five terms of the sequence $\left\{1 + \dfrac{(-1)^n}{n^2}\right\}$ are $1-1$, $1+\dfrac{1}{4}$, $1-\dfrac{1}{9}$, $1+\dfrac{1}{16}$, and $1 - \dfrac{1}{25}$. The pattern suggests that the sequence converges to 1.

- Find the limit of the nth term using any available limit technique, including basic limits, limit properties, or a related function (possibly using L'Hôpital's Rule). For example, for the sequence $\left\{\dfrac{\ln n}{n}\right\}$, we examine the limit of the related function $f(x) = \dfrac{\ln x}{x}$.

$$\lim_{x \to \infty} f(x) = \lim_{x \to \infty}\left(\frac{\ln x}{x}\right) = \lim_{x \to \infty} \frac{\frac{1}{x}}{1} = 0$$

and conclude the sequence $\left\{\dfrac{\ln n}{n}\right\}$ converges to 0.

- Show that the sequence is bounded and monotonic.

10.1 Assess Your Understanding

Concepts and Vocabulary

1. True or False A sequence is a function whose domain is the set of positive real numbers.

2. True or False If the sequence $\{s_n\}$ is convergent, then $\lim\limits_{n \to \infty} s_n = 0$.

3. True or False If $f(x)$ is a related function of the sequence $\{s_n\}$ and there is a real number L for which $\lim\limits_{x \to \infty} f(x) = L$, then $\{s_n\}$ converges.

4. Multiple Choice If there is a positive number K for which $|s_n| \le K$ for all integers $n \ge 1$, then $\{s_n\}$ is

 (a) increasing **(b)** bounded **(c)** decreasing **(d)** convergent.

5. True or False A bounded sequence is convergent.

6. True or False An unbounded sequence is divergent.

7. True or False A sequence $\{s_n\}$ is decreasing if and only if $s_n \le s_{n+1}$ for all integers $n \ge 1$.

8. True or False A sequence must be monotonic to be convergent.

9. True or False To use an algebraic ratio to show that the sequence $\{s_n\}$ is increasing, show that $\dfrac{s_{n+1}}{s_n} \ge 0$ for all $n \ge 1$.

10. Multiple Choice If the derivative of a related function f of a sequence $\{s_n\}$ is negative, then the sequence $\{s_n\}$ is

 (a) bounded **(b)** decreasing **(c)** increasing **(d)** convergent.

11. True or False When determining whether a sequence $\{s_n\}$ converges or diverges, the beginning terms of the sequence can be ignored.

12. True or False Sequences that are both bounded and monotonic diverge.

Skill Building

In Problems 13–22, the nth term of a sequence $\{s_n\}$ is given. Write the first four terms of each sequence.

13. $s_n = \dfrac{n+1}{n}$

14. $s_n = \dfrac{2}{n^2}$

15. $s_n = \ln n$

16. $s_n = \dfrac{n}{\ln(n+1)}$

17. $s_n = \dfrac{(-1)^{n+1}}{2n+1}$

18. $s_n = \dfrac{1-(-1)^n}{2}$

19. $s_n = \begin{cases} (-1)^{n+1} & \text{if } n \text{ is even} \\ 1 & \text{if } n \text{ is odd} \end{cases}$

20. $s_n = \begin{cases} n^2 + n & \text{if } n \text{ is even} \\ 4n + 1 & \text{if } n \text{ is odd} \end{cases}$

21. $s_n = \dfrac{n!}{2^n}$

22. $s_n = \dfrac{n!}{n^2}$

In Problems 23–32, the first few terms of a sequence are given. Find an expression for the nth term of each sequence, assuming the indicated pattern continues for all n.

23. $2, 4, 6, 8, 10, \ldots$

24. $1, 3, 5, 7, 9, \ldots$

25. $2, 4, 8, 16, 32, \ldots$

26. $1, 8, 27, 64, 125, \ldots$

27. $\dfrac{1}{2}, -\dfrac{1}{3}, \dfrac{1}{4}, -\dfrac{1}{5}, \dfrac{1}{6}, \ldots$

28. $1, -2, 3, -4, 5, \ldots$

29. $\dfrac{1}{2}, \dfrac{2}{3}, \dfrac{3}{4}, \dfrac{4}{5}, \ldots$

30. $\dfrac{1}{2}, \dfrac{4}{3}, \dfrac{9}{4}, \dfrac{16}{5}, \ldots$

31. $1, 1, 2, 6, 24, 120, 720, \ldots$

32. $1, 1, \dfrac{1}{2}, \dfrac{1}{6}, \dfrac{1}{24}, \dfrac{1}{120}, \ldots$

In Problems 33–44, use properties of convergent sequences to find the limit of each sequence.

33. $\left\{ \dfrac{3}{n} \right\}$ **34.** $\left\{ \dfrac{-2}{n} \right\}$ **35.** $\left\{ 1 - \dfrac{1}{n} \right\}$

36. $\left\{ \dfrac{1}{n} + 4 \right\}$ **37.** $\left\{ \dfrac{4n+2}{n} \right\}$ **38.** $\left\{ \dfrac{2n+1}{n} \right\}$

39. $\left\{ \left(\dfrac{2-n}{n^2} \right)^4 \right\}$ **40.** $\left\{ \left(\dfrac{n^3 - 2n}{n^3} \right)^2 \right\}$

41. $\left\{ \sqrt{\dfrac{n+1}{n^2}} \right\}$ **42.** $\left\{ \sqrt[3]{8 - \dfrac{1}{n}} \right\}$

43. $\left\{ \left(1 - \dfrac{1}{n} \right) \left(1 - \dfrac{1}{n^2} \right) \right\}$

44. $\left\{ \left(1 - \dfrac{1}{n} \right) \left(1 - \dfrac{1}{n^2} \right) \left(1 - \dfrac{1}{n^3} \right) \right\}$

In Problems 45–50, show that each sequence converges. Find its limit.

45. $\left\{ \ln \dfrac{n+1}{3n} \right\}$ **46.** $\left\{ \ln \dfrac{n^2+2}{2n^2+3} \right\}$ **47.** $\left\{ e^{(4/n)-2} \right\}$

48. $\left\{ e^{3+(6/n)} \right\}$ **49.** $\left\{ \sin \dfrac{1}{n} \right\}$ **50.** $\left\{ \cos \dfrac{1}{n} \right\}$

In Problems 51–62, use a related function or the Squeeze Theorem for sequences to show each sequence converges. Find its limit.

51. $\left\{ \dfrac{n^2-4}{n^2+n-2} \right\}$ **52.** $\left\{ \dfrac{n+2}{n^2+6n+8} \right\}$

53. $\left\{ \dfrac{n^2}{2n+1} - \dfrac{n^2}{2n-1} \right\}$ **54.** $\left\{ \dfrac{6n^4-5}{7n^4+3} \right\}$

55. $\left\{ \dfrac{\sqrt{n}+2}{\sqrt{n}+5} \right\}$ **56.** $\left\{ \dfrac{\sqrt{n}}{e^n} \right\}$ **57.** $\left\{ \dfrac{n^2}{3^n} \right\}$

58. $\left\{ \dfrac{(n-1)^2}{e^n} \right\}$ **59.** $\left\{ \dfrac{(-1)^n}{3n^2} \right\}$ **60.** $\left\{ \dfrac{(-1)^n}{\sqrt{n}} \right\}$

61. $\left\{ \dfrac{\sin n}{n} \right\}$ **62.** $\left\{ \dfrac{\cos n}{n} \right\}$

In Problems 63–72, determine whether each sequence converges or diverges.

63. $\{ \cos (\pi n) \}$ **64.** $\left\{ \cos \left(\dfrac{\pi}{2} n \right) \right\}$ **65.** $\{ \sqrt{n} \}$

66. $\{ n^2 \}$ **67.** $\left\{ \left(-\dfrac{1}{3} \right)^n \right\}$ **68.** $\left\{ \left(\dfrac{1}{3} \right)^n \right\}$

69. $\left\{ \left(\dfrac{5}{4} \right)^n \right\}$ **70.** $\left\{ \left(\dfrac{\pi}{2} \right)^n \right\}$ **71.** $\left\{ \dfrac{n+(-1)^n}{n} \right\}$

72. $\left\{ \dfrac{1}{n} + (-1)^n \right\}$

In Problems 73–80, determine whether each sequence is bounded from above, bounded from below, both, or neither.

73. $\left\{ \dfrac{\ln n}{n} \right\}$ **74.** $\left\{ \dfrac{\sin n}{n} \right\}$ **75.** $\left\{ n + \dfrac{1}{n} \right\}$

76. $\left\{ \dfrac{3}{n+1} \right\}$ **77.** $\left\{ \dfrac{n^2}{n+1} \right\}$ **78.** $\left\{ \dfrac{2^n}{n^2} \right\}$

79. $\left\{ \left(-\dfrac{1}{2} \right)^n \right\}$ **80.** $\{ n^{1/2} \}$

In Problems 81–88, determine whether each sequence is monotonic. If the sequence is monotonic, is it increasing, nondecreasing, decreasing, or nonincreasing?

81. $\left\{ \dfrac{3^n}{(n+1)^3} \right\}$ **82.** $\left\{ \dfrac{2n+1}{n} \right\}$ **83.** $\left\{ \dfrac{\ln n}{\sqrt{n}} \right\}$

84. $\left\{ \dfrac{\sqrt{n+1}}{n} \right\}$ **85.** $\left\{ \left(\dfrac{1}{3} \right)^n \right\}$ **86.** $\left\{ \dfrac{n^2}{5^n} \right\}$

87. $\left\{ \dfrac{n!}{3^n} \right\}$ **88.** $\left\{ \dfrac{n!}{n^2} \right\}$

In Problems 89–94, show that each sequence converges by showing it is either increasing (nondecreasing) and bounded from above or decreasing (nonincreasing) and bounded from below.

89. $\{ ne^{-n} \}$ **90.** $\{ \tan^{-1} n \}$ **91.** $\left\{ \dfrac{n}{n+1} \right\}$

92. $\left\{ \dfrac{n}{n^2+1} \right\}$ **93.** $\left\{ 2 - \dfrac{1}{n} \right\}$ **94.** $\left\{ \dfrac{n}{2^n} \right\}$

In Problems 95–114, determine whether each sequence converges or diverges. If it converges, find its limit.

95. $\left\{ \dfrac{3}{n} + 6 \right\}$ **96.** $\left\{ 2 - \dfrac{4}{n} \right\}$

97. $\left\{ \ln \dfrac{n^2+4}{3n^2} \right\}$ **98.** $\left\{ \cos \left(n\pi + \dfrac{\pi}{2} \right) \right\}$

99. $\{ (-1)^n \sqrt{n} \}$ **100.** $\left\{ \dfrac{(-1)^n}{2n} \right\}$

101. $\left\{ \dfrac{3^n+1}{4^n} \right\}$ **102.** $\left\{ n + \sin \dfrac{1}{n} \right\}$

103. $\left\{ \dfrac{\ln(n+1)}{n+1} \right\}$ **104.** $\left\{ \dfrac{\ln(n+1)}{\sqrt{n}} \right\}$

105. $\{ 0.5^n \}$ **106.** $\{ (-2)^n \}$

107. $\left\{ \cos \dfrac{\pi}{n} \right\}$ **108.** $\left\{ \sin \dfrac{\pi}{n} \right\}$

109. $\left\{ \cos \left(\dfrac{n}{e^n} \right) \right\}$ **110.** $\left\{ \sin \left(\dfrac{(n+1)^3}{e^n} \right) \right\}$

111. $\{ e^{1/n} \}$ **112.** $\left\{ \dfrac{1}{ne^{-n}} \right\}$

113. $\left\{ 1 + \left(\dfrac{1}{2} \right)^n \right\}$ **114.** $\left\{ 1 - \left(\dfrac{1}{2} \right)^n \right\}$

In Problems 115 and 116, use a related function to find the limit of each sequence.

115. $\left\{\dfrac{(\ln n)^2}{n}\right\}$

116. $\left\{\sqrt{n}\ln\dfrac{n+1}{n}\right\}$

Applications and Extensions

In Problems 117–126, determine whether each sequence converges or diverges.

117. $\left\{\dfrac{n^2\tan^{-1}n}{n^2+1}\right\}$

118. $\left\{n\sin\dfrac{1}{n}\right\}$

119. $\left\{\dfrac{n+\sin n}{n+\cos(4n)}\right\}$

120. $\left\{\dfrac{n^2}{2n+1}\sin\dfrac{1}{n}\right\}$

121. $\{\ln n-\ln(n+1)\}$

122. $\left\{\ln n^2+\ln\dfrac{1}{n^2+1}\right\}$

123. $\left\{\dfrac{n^2}{\sqrt{n^2+1}}\right\}$

124. $\left\{\dfrac{5^n}{(n+1)^2}\right\}$

125. $\left\{\dfrac{2^n}{(2)(4)(6)\cdots(2n)}\right\}$

126. $\left\{\dfrac{3^{n+1}}{(3)(6)(9)\cdots(3n)}\right\}$

127. The nth term of a sequence is $s_n=\dfrac{1}{n^2+n\cos n+1}$. Does the sequence $\{s_n\}$ converge or diverge? *Hint:* Show that the derivative of $\dfrac{1}{x^2+x\cos x+1}$ is negative for $x>1$.

128. Fibonacci Sequence The famous **Fibonacci sequence** $\{u_n\}$ is defined recursively as

$$u_1=1 \qquad u_2=1 \qquad u_{n+2}=u_n+u_{n+1} \quad n\ge 1$$

(a) Write the first eight terms of the Fibonacci sequence.

(b) Verify that the nth term is given by

$$u_n=\frac{(1+\sqrt{5})^n-(1-\sqrt{5})^n}{2^n\sqrt{5}}$$

Hint: Show that $u_1=1$, $u_2=1$, and $u_{n+2}=u_n+u_{n+1}$.

129. Stocking a Lake Mirror Lake is stocked with rainbow trout. Considering fish reproduction and natural death, along with vigorous efforts by fishermen to decimate the population, managers find that some ratio r, $0<r<1$, of the population persists from one stocking period to the next. If the lake is stocked with h fish each year, the fish population p_n, in year n of the stocking program, is approximately $p_n=rp_{n-1}+h$. If p_0 is 3000, write a general expression for the nth term of the sequence in terms of r and h only. Does this sequence converge?

130. Electronics: A Discharging Capacitor A capacitor is an electronic device that stores an electrical charge. When connected across a resistor, it loses the charge (discharges) in such a way that during a fixed time interval, called the **time constant**, the charge stored in the capacitor is $\dfrac{1}{e}$ of the charge at the beginning of that interval.

(a) Develop a sequence for the charge remaining after n time constants if the initial charge is Q_0.

(b) Does this sequence converge? If yes, to what?

 131. Reflections in a Mirror A highly reflective mirror reflects 95% of the light that falls on it. In a light box having walls made of this mirror, the light will reflect back-and-forth between the mirrors.

(a) If the original intensity of the light is I_0 before it falls on a mirror, develop a sequence to describe the intensity of the light after n reflections.

(b) How many reflections are needed to reduce the light intensity by at least 98%?

132. A Fission Chain Reaction A chain reaction is any sequence of events for which each event causes one or more additional events to occur. For example, in chain-reaction auto accidents, one car rear-ends another car, that car rear-ends another, and so on. In one type of nuclear fission chain reaction, a uranium-235 nucleus is struck by a neutron, causing it to break apart and release several more neutrons. Each of these neutrons strikes another nucleus, causing it to break apart and release additional neutrons, resulting in a chain reaction. In the fission of uranium-235 in nuclear reactors, each fission event releases an average of $2\dfrac{1}{2}$ neutrons, and each of these neutrons causes another fission event. The first fission is triggered by a single free neutron.

(a) Develop a sequence for the average number of neutrons, that is, fission events, that occur at the nth event if we start with one such event.

(b) Does the sequence converge or diverge?

(c) Interpret the answer found in (b).

In Problems 133–136, find the limit of each sequence.

133. $\left\{\left(1+\dfrac{2}{n}\right)^n\right\}$

134. $\left\{\left(1-\dfrac{4}{n}\right)^n\right\}$

135. $\left\{\left(1+\dfrac{1}{n}\right)^{3n}\right\}$

136. $\left\{\left(1+\dfrac{1}{n}\right)^{-2n}\right\}$

137. Use the Squeeze Theorem for sequences to show that the sequence $\left\{(-1)^n\dfrac{1}{n!}\right\}$ converges to 0.

Challenge Problems

138. Show that if $0<r<1$, then $\displaystyle\lim_{n\to\infty}r^n=0$. *Hint:* Let $r=\dfrac{1}{1+p}$, where $p>0$. Then by the Binomial Theorem,

$$r^n=\frac{1}{(1+p)^n}=\frac{1}{1+np+n(n-1)\dfrac{p^2}{2}+\cdots+p^n}<\frac{1}{np}.$$

139. Use the result of Problem 138 to show that if $-1<r<0$, then $\displaystyle\lim_{n\to\infty}r^n=0$.

140. Show that if $r>1$, then $\displaystyle\lim_{n\to\infty}r^n=\infty$. *Hint:* Let $r=1+p$, where $p>0$. Then by the Binomial Theorem,

$$r^n=(1+p)^n=1+np+n(n-1)\frac{p^2}{2}+\cdots+p^n>np.$$

141. Use the result of Problem 140 to show that if $r<-1$, then $\displaystyle\lim_{n\to\infty}r^n$ does not exist.

Hint: r^n oscillates between positive and negative values.

142. Suppose $\{s_n\}$ is a sequence of real numbers. Show that if $\lim_{n\to\infty} s_n = L$ and if f is a function that is continuous at L and is defined for all numbers s_n, then $\lim_{n\to\infty} f(s_n) = f(L)$.

143. Show that if $\lim_{n\to\infty} s_n = L$, then $\lim_{n\to\infty} |s_n|$ exists and $\lim_{n\to\infty} |s_n| = |L|$. Is the converse true?

144. The Limit of a Sequence Is Unique Show that a convergent sequence $\{s_n\}$ cannot have two distinct limits.

145. Review the definition of the limit at infinity of a function from Section 1.6. Write a paragraph that compares and contrasts the limit at infinity of a function f and the limit of a sequence $\{s_n\}$.

146. (a) Show that the sequence $\{\ln n\}$ is increasing.
(b) Show that the sequence $\{\ln n\}$ is unbounded from above.
(c) Conclude $\{\ln n\}$ diverges.
(d) Find the smallest number N so that $\ln N > 20$.
(e) Graph $y = \ln x$ and zoom in for x large.
(f) Does the graph confirm the result in (c)?

147. Let $a_1 > 0$ and $b_1 > 0$ be two real numbers for which $a_1 > b_1$. Define sequences $\{a_n\}$ and $\{b_n\}$ as

$$a_{n+1} = \frac{a_n + b_n}{2} \qquad b_{n+1} = \sqrt{a_n b_n}$$

(a) Show that $b_n < b_{n+1} < a_1$ for all n.
(b) Show that $b_1 < a_{n+1} < a_n$ for all n.
(c) Show that $0 < a_{n+1} - b_{n+1} < \dfrac{a_1 - b_1}{2^n}$.
(d) Show that $\lim_{n\to\infty} a_n$ and $\lim_{n\to\infty} b_n$ each exists and are equal.

In Problems 148–150, determine whether each sequence converges or diverges.

148. $s_n = \dfrac{2^{n-1} \cdot 4^n}{n!}$ **149.** $s_n = \dfrac{n!}{3^n \cdot 4^n}$ **150.** $s_n = \dfrac{n!}{3^n + 8n}$

151. Show that $\{(3^n + 5^n)^{1/n}\}$ converges.

152. Let N be a fixed positive number and define a sequence by $\{a_{n+1}\} = \left\{\dfrac{1}{2}\left[a_n + \dfrac{N}{a_n}\right]\right\}$, where a_1 is a positive number.
(a) Show that the sequence $\{a_n\}$ converges to $\sqrt{N}$.
(b) Use this sequence to approximate $\sqrt{28}$ rounded to three decimal places. How accurate is a_3? a_6?

153. Show that $\{s_n\} = \left\{\dfrac{1\cdot3\cdot5\cdot\,\cdots\,\cdot(2n-1)}{2\cdot4\cdot6\cdot\,\cdots\,\cdot 2n}\right\}$ is bounded and monotonic.

154. Show that $\{s_n\} = \left\{\left(1+\dfrac{1}{n}\right)^n\right\}$ is increasing and bounded from above.
Hint: Use the Binomial Theorem to expand $\left(1+\dfrac{1}{n}\right)^n$.

155. Let $\{s_n\}$ be a convergent sequence, and suppose the nth term of the sequence $\{a_n\}$ is the arithmetic mean (average) of the first n terms of $\{s_n\}$. That is, $a_n = \dfrac{1}{n}(s_1 + s_2 + \cdots + s_n)$. Show that $\{a_n\}$ converges and has the same limit as $\{s_n\}$.

156. Area Let A_n be the area enclosed by a regular n-sided polygon inscribed in a circle of radius R. Show that:
(a) $A_n = \dfrac{n}{2}R^2 \sin\dfrac{2\pi}{n}$.
(b) $\lim_{n\to\infty} A_n = \lim_{n\to\infty}\left(\dfrac{n}{2}R^2 \sin\dfrac{2\pi}{n}\right) = \pi R^2$
(the area of a circle of radius R).

157. Area Let A_n be the area enclosed by a regular n-sided polygon circumscribed around a circle of radius r. Show that:
(a) $A_n = nr^2 \tan\dfrac{\pi}{n}$.
(b) $\lim_{n\to\infty} A_n = \pi r^2$ (the area of a circle of radius r).

Hint: r, called the **apothem** of the polygon, is the perpendicular distance from the center of the polygon to the midpoint of a side.

158. Perimeter Suppose P_n is the perimeter of a regular n-sided polygon inscribed in a circle of radius R. Show that:
(a) $P_n = 2nR \sin\dfrac{\pi}{n}$.
(b) $\lim_{n\to\infty} P_n = 2\pi R$ (the circumference of a circle of radius R).

159. Cauchy Sequence A sequence $\{s_n\}$ is said to be a **Cauchy sequence** if and only if for each $\varepsilon > 0$, there exists a positive integer N for which

$$|s_n - s_m| < \varepsilon \qquad \text{for all } n, m > N$$

Show that every convergent sequence is a Cauchy sequence.

160. (a) Show that the sequence $\{e^{n/(n+2)}\}$ converges.
(b) Find $\lim_{n\to\infty} e^{n/(n+2)}$.
(c) Graph $y = e^{n/(n+2)}$. Does the graph confirm the results of (a) and (b)?

10.2 Infinite Series

1 unit

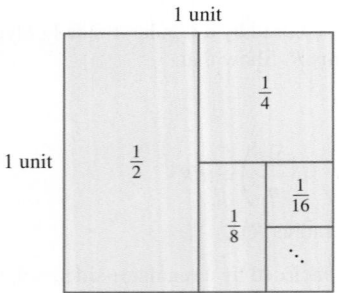

Figure 17

OBJECTIVES *When you finish this section, you should be able to:*

1 **Determine whether a series has a sum (p. 797)**
2 **Determine whether a geometric series converges or diverges (p. 800)**
3 **Show that the harmonic series diverges (p. 804)**
    Application: Using a geometric series in biology (p. 806)

Is it possible for the sum of an infinite collection of nonzero numbers to be finite? Look at Figure 17. The square in the figure has sides of length 1 unit, making its area 1 square unit. If we divide the square into two rectangles of equal area, each rectangle has an area of $\dfrac{1}{2}$ square unit. If one of these rectangles is divided in half, the result is two squares, the area of each equaling $\dfrac{1}{4}$ square unit. If we were to continue the process of dividing one of the smallest regions in half, we would obtain a decomposition of the original area of 1 square unit into regions of area $\dfrac{1}{2}, \dfrac{1}{4}, \dfrac{1}{8}, \dfrac{1}{16}$, and so forth. Therefore,

$$1 = \frac{1}{2} + \frac{1}{4} + \frac{1}{8} + \frac{1}{16} + \cdots \tag{1}$$

Surprised?

Now look at this result from a different point of view by starting with the infinite sum

$$\frac{1}{2} + \frac{1}{4} + \frac{1}{8} + \frac{1}{16} + \cdots$$

One way we might add the fractions is by using *partial sums* to see whether a trend develops. The first five partial sums are

$$S_1 = \frac{1}{2} = 0.5$$

$$S_2 = \frac{1}{2} + \frac{1}{4} = \frac{3}{4} = 0.75$$

$$S_3 = \frac{1}{2} + \frac{1}{4} + \frac{1}{8} = \frac{3}{4} + \frac{1}{8} = \frac{7}{8} = 0.875$$

$$S_4 = \frac{1}{2} + \frac{1}{4} + \frac{1}{8} + \frac{1}{16} = \frac{7}{8} + \frac{1}{16} = \frac{15}{16} = 0.9375$$

$$S_5 = \frac{1}{2} + \frac{1}{4} + \frac{1}{8} + \frac{1}{16} + \frac{1}{32} = \frac{15}{16} + \frac{1}{32} = \frac{31}{32} = 0.96875$$

Each of these sums uses more terms from (1), and each sum seems to be getting closer to 1. The infinite sum in (1) is an example of an *infinite series.*

DEFINITION Infinite Series

If $a_1, a_2, \ldots, a_n, \ldots$ is an infinite collection of numbers, the expression

$$\sum_{k=1}^{\infty} a_k = a_1 + a_2 + \cdots + a_n + \cdots$$

is called an **infinite series** or, simply, a **series**.

NEED TO REVIEW? Sums and summation notation are discussed in Section P.8, pp. 73–76.

The numbers $a_1, a_2, \ldots, a_n, \ldots$ are called the **terms** of the series, and the number a_n is called the **nth term** or **general term** of the series. The symbol $\sum$ stands for summation; k is the **index of summation**. Although the index of summation can begin at any integer, in most of our work with series, it will begin at 1.

① Determine Whether a Series Has a Sum

To define a sum of an infinite series $\sum\limits_{k=1}^{\infty} a_k$, we make use of the sequence $\{S_n\}$ defined by

$$S_1 = a_1$$

$$S_2 = a_1 + a_2 = \sum_{k=1}^{2} a_k$$

$$S_3 = a_1 + a_2 + a_3 = \sum_{k=1}^{3} a_k$$

$$\vdots$$

$$S_n = a_1 + a_2 + \cdots + a_n = \sum_{k=1}^{n} a_k$$

$$\vdots$$

This sequence $\{S_n\}$ is called the **sequence of partial sums** of the series $\sum\limits_{k=1}^{\infty} a_k$.

EXAMPLE 1   **Finding the Sequence of Partial Sums**

Find the sequence of partials sums of the series

$$\sum_{k=1}^{\infty} \frac{1}{2^k} = \frac{1}{2} + \frac{1}{2^2} + \frac{1}{2^3} + \frac{1}{2^4} + \cdots = \frac{1}{2} + \frac{1}{4} + \frac{1}{8} + \frac{1}{16} + \cdots$$

Solution

As it turns out, the partial sums S_n of this series can each be written as 1 minus a power of $\frac{1}{2}$, as follows:

$$S_1 = a_1 = \frac{1}{2} = 1 - \frac{1}{2}$$

$$S_2 = S_1 + a_2 = \left(1 - \frac{1}{2}\right) + \frac{1}{4} = 1 - \frac{1}{4} = 1 - \frac{1}{2^2}$$

$$S_3 = S_2 + a_3 = \left(1 - \frac{1}{4}\right) + \frac{1}{8} = 1 - \frac{1}{8} = 1 - \frac{1}{2^3}$$

$$S_4 = S_3 + a_4 = \left(1 - \frac{1}{8}\right) + \frac{1}{16} = 1 - \frac{1}{16} = 1 - \frac{1}{2^4}$$

$$\vdots$$

$$S_n = 1 - \frac{1}{2^n}$$

$$\vdots$$

∎

The nth partial sum is $S_n = 1 - \dfrac{1}{2^n}$, and as n increases, the sequence $\{S_n\}$ of partial sums approaches a limit. That is,

$$\lim_{n \to \infty} S_n = \lim_{n \to \infty} \left(1 - \frac{1}{2^n} \right) = \lim_{n \to \infty} 1 - \lim_{n \to \infty} \frac{1}{2^n} = 1 - 0 = 1$$

We agree to call this limit the *sum of the series*, and we write

$$\sum_{k=1}^{\infty} \frac{1}{2^k} = \frac{1}{2} + \frac{1}{4} + \frac{1}{8} + \frac{1}{16} + \cdots = 1$$

DEFINITION Convergence/Divergence of an Infinite Series

If the sequence $\{S_n\}$ of partial sums of an infinite series $\displaystyle\sum_{k=1}^{\infty} a_k$ has a limit S, then the series **converges** and is said to have the **sum** S. That is, if $\displaystyle\lim_{n \to \infty} S_n = S$, then

$$\sum_{k=1}^{\infty} a_k = a_1 + a_2 + \cdots + a_n + \cdots = S$$

- If $\displaystyle\lim_{n \to \infty} S_n = \infty$, then the infinite series **diverges**.
- If $\displaystyle\lim_{n \to \infty} S_n$ does not exist, then the infinite series **diverges**.

EXAMPLE 2 Finding the Sum of a Series

Show that

$$\sum_{k=1}^{\infty} \frac{1}{k(k+1)} = \frac{1}{1 \cdot 2} + \frac{1}{2 \cdot 3} + \frac{1}{3 \cdot 4} + \cdots = \frac{1}{2} + \frac{1}{6} + \frac{1}{12} + \cdots = 1$$

Solution

We begin with the sequence $\{S_n\}$ of partial sums,

$$S_1 = \frac{1}{1 \cdot 2}$$

$$S_2 = \frac{1}{1 \cdot 2} + \frac{1}{2 \cdot 3}$$

$$S_3 = \frac{1}{1 \cdot 2} + \frac{1}{2 \cdot 3} + \frac{1}{3 \cdot 4}$$

$$\vdots$$

$$S_n = \frac{1}{1 \cdot 2} + \frac{1}{2 \cdot 3} + \frac{1}{3 \cdot 4} + \cdots + \frac{1}{n(n+1)}$$

$$\vdots$$

We want to express S_n in a simpler form so we can find $\displaystyle\lim_{n \to \infty} S_n$.

NEED TO REVIEW? Partial fractions are discussed in Section 6.8, pp. 528–535.

Using partial fractions, we can write $\dfrac{1}{n(n+1)}$ as

$$\frac{1}{n(n+1)} = \frac{1}{n} - \frac{1}{n+1}$$

Then

$$S_n = \frac{1}{1 \cdot 2} + \frac{1}{2 \cdot 3} + \frac{1}{3 \cdot 4} + \cdots + \frac{1}{n(n+1)}$$

can be written as

$$S_n = \left(\frac{1}{1} - \frac{1}{2}\right) + \left(\frac{1}{2} - \frac{1}{3}\right) + \left(\frac{1}{3} - \frac{1}{4}\right) + \cdots + \left(\frac{1}{n-1} - \frac{1}{n}\right) + \left(\frac{1}{n} - \frac{1}{n+1}\right)$$

$$\frac{1}{1 \cdot 2} = \frac{1}{1} - \frac{1}{2} \qquad \frac{1}{2 \cdot 3} = \frac{1}{2} - \frac{1}{3} \qquad \frac{1}{3 \cdot 4} = \frac{1}{3} - \frac{1}{4}$$

Notice that all the terms except the first and last cancel, so that

$$S_n = 1 - \frac{1}{n+1}$$

Then

$$\lim_{n \to \infty} S_n = \lim_{n \to \infty} \left(1 - \frac{1}{n+1}\right) = 1$$

NOTE Sums for which the middle terms cancel, as in Example 2, are called **telescoping sums**.

The series $\displaystyle\sum_{k=1}^{\infty} \frac{1}{k(k+1)}$ converges, and its sum is 1. ∎

NOW WORK Problem 11.

EXAMPLE 3 **Showing a Series Diverges**

Show that the series $\displaystyle\sum_{k=1}^{\infty} (-1)^k = -1 + 1 - 1 + \cdots$ diverges.

Solution
The sequence $\{S_n\}$ of partial sums is

$$S_1 = -1$$
$$S_2 = -1 + 1 = 0$$
$$S_3 = -1 + 1 - 1 = -1$$
$$S_4 = -1 + 1 - 1 + 1 = 0$$
$$\vdots$$

$$S_n = \begin{cases} -1 & \text{if } n \text{ is odd} \\ 0 & \text{if } n \text{ is even} \end{cases}$$

Since $\displaystyle\lim_{n \to \infty} S_n$ does not exist, the sequence $\{S_n\}$ of partial sums diverges. Therefore, the series diverges. ∎

NOW WORK Problem 49.

EXAMPLE 4 Determining Whether a Series Converges or Diverges

Determine whether the series $\sum\limits_{k=1}^{\infty} k = 1 + 2 + 3 + \cdots$ converges or diverges.

Solution

The sequence $\{S_n\}$ of partial sums is

$$S_1 = 1$$
$$S_2 = 1 + 2$$
$$S_3 = 1 + 2 + 3$$
$$\vdots$$
$$S_n = 1 + 2 + 3 + \cdots + n$$

To express S_n in a way that will make it easy to find $\lim\limits_{n \to \infty} S_n$, we use the formula for the sum of the first n integers:

$$S_n = \sum_{k=1}^{n} k = 1 + 2 + 3 + \cdots + n = \frac{n(n+1)}{2}$$

RECALL

$$\sum_{k=1}^{n} k = 1 + 2 + \cdots + n = \frac{n(n+1)}{2}$$

(Section P.8, p. 75).

Since $\lim\limits_{n \to \infty} S_n = \lim\limits_{n \to \infty} \dfrac{n(n+1)}{2} = \infty$, the sequence $\{S_n\}$ of partial sums diverges.

So, the series $\sum\limits_{k=1}^{\infty} k$ diverges. ∎

NOW WORK Problem 43.

2 Determine Whether a Geometric Series Converges or Diverges

Geometric series occur in a large variety of applications including biology, finance, and probability. They are also useful in analyzing other infinite series.

In a geometric series, the ratio r of any two consecutive terms is a fixed real number.

AP® EXAM TIP

Geometric series are the first, and often the most common, of the three named series encountered on the AP® BC Exam.

DEFINITION Geometric Series

A series of the form

$$\sum_{k=0}^{\infty} ar^k = \sum_{k=1}^{\infty} ar^{k-1} = a + ar + ar^2 + \cdots + ar^{n-1} + \cdots$$

where $a \neq 0$ is called a **geometric series** with ratio r.

To investigate the conditions for convergence of a geometric series, we examine the nth partial sum:

$$S_n = a + ar + ar^2 + \cdots + ar^{n-1} \qquad (2)$$

- If $r = 0$, the nth partial sum is $S_n = a$ and $\lim\limits_{n \to \infty} S_n = a$. The sequence of partial sums converges when $r = 0$.

- If $r = 1$, the series becomes $\sum\limits_{k=1}^{\infty} a = a + a + \cdots + a + \cdots$, and the nth partial sum is

$$S_n = a + a + \cdots + a = na$$

Since $a \neq 0$, $\lim\limits_{n \to \infty} S_n = \infty$ or $-\infty$, so the sequence $\{S_n\}$ of partial sums diverges when $r = 1$.

- If $r = -1$, the series is $\sum\limits_{k=1}^{\infty} a(-1)^{k-1} = a - a + a - a + \cdots$ and the nth partial sum is

$$S_n = \begin{cases} 0 & \text{if } n \text{ is even} \\ a & \text{if } n \text{ is odd} \end{cases}$$

Since $a \neq 0$, $\lim\limits_{n\to\infty} S_n$ does not exist. The sequence $\{S_n\}$ of partial sums diverges when $r = -1$.

- Suppose $r \neq 0$, $r \neq 1$, and $r \neq -1$. Since $r \neq 0$, we multiply both sides of (2) by r to obtain

$$r S_n = ar + ar^2 + \cdots + ar^n$$

Now subtract $r S_n$ from S_n.

$$S_n - r S_n = (a + ar + ar^2 + \cdots + ar^{n-1}) - (ar + ar^2 + \cdots + ar^{n-1} + ar^n)$$

$$= a - ar^n$$

$$S_n(1-r) = a(1-r^n) \qquad \text{Factor.}$$

Since $r \neq 1$, the nth partial sum of the geometric series can be expressed as

$$S_n = \frac{a(1-r^n)}{1-r} = \frac{a - ar^n}{1-r} = \frac{a}{1-r} - \frac{ar^n}{1-r}$$

Now

$$\lim_{n\to\infty} S_n = \lim_{n\to\infty}\left[\frac{a}{1-r} - \frac{ar^n}{1-r}\right] = \lim_{n\to\infty}\frac{a}{1-r} - \lim_{n\to\infty}\frac{ar^n}{1-r} = \frac{a}{1-r} - \frac{a}{1-r}\lim_{n\to\infty} r^n$$

Now use the fact that if $|r| < 1$, then $\lim\limits_{n\to\infty} r^n = 0$ (refer to page 787 in Section 10.1). We conclude that if $|r| < 1$, then $\lim\limits_{n\to\infty} S_n = \frac{a}{1-r}$. So, a geometric series converges to $S = \frac{a}{1-r}$ if $-1 < r < 1$.

If $|r| > 1$, use the fact that $\lim\limits_{n\to\infty} r^n$ does not exist to conclude that a geometric series diverges if $r < -1$ or $r > 1$.

This proves the following theorem:

THEOREM Convergence of a Geometric Series

- If $|r| < 1$, the geometric series $\sum\limits_{k=1}^{\infty} ar^{k-1}$ converges, and its sum is

$$\boxed{\sum_{k=1}^{\infty} ar^{k-1} = \frac{a}{1-r}}$$

- If $|r| \geq 1$, the geometric series $\sum\limits_{k=1}^{\infty} ar^{k-1}$ diverges.

▶ EXAMPLE 5   Determining Whether a Geometric Series Converges
CALC CLIP

Determine whether each geometric series converges or diverges. If it converges, find its sum.

(a) $\displaystyle\sum_{k=1}^{\infty} 8\left(\frac{2}{5}\right)^{k-1}$ (b) $\displaystyle\sum_{k=1}^{\infty}\left(-\frac{5}{9}\right)^{k-1}$ (c) $\displaystyle\sum_{k=1}^{\infty} 3\left(\frac{3}{2}\right)^{k-1}$

(d) $\displaystyle\sum_{k=1}^{\infty}\frac{1}{2^k}$ (e) $\displaystyle\sum_{k=0}^{\infty}\left(\frac{1}{3}\right)^{k-1}$ (f) $\displaystyle\sum_{k=1}^{\infty}\frac{10^{k+1}}{9^{k-2}}$

Solution

Compare each series to $\displaystyle\sum_{k=1}^{\infty} ar^{k-1}$.

(a) $\displaystyle\sum_{k=1}^{\infty} 8\left(\frac{2}{5}\right)^{k-1}$ Here $a=8$ and $r=\frac{2}{5}$. Since $|r|=\frac{2}{5}<1$, the series converges and

$$\sum_{k=1}^{\infty} 8\left(\frac{2}{5}\right)^{k-1} = \frac{8}{1-\dfrac{2}{5}} = 8\cdot\frac{5}{3} = \frac{40}{3}$$

(b) $\displaystyle\sum_{k=1}^{\infty}\left(-\frac{5}{9}\right)^{k-1}$ Here $a=1$ and $r=-\frac{5}{9}$. Since $|r|=\frac{5}{9}<1$, the series converges and

$$\sum_{k=1}^{\infty}\left(-\frac{5}{9}\right)^{k-1} = \frac{1}{1-\left(-\dfrac{5}{9}\right)} = \frac{9}{14}$$

(c) $\displaystyle\sum_{k=1}^{\infty} 3\left(\frac{3}{2}\right)^{k-1}$ Here $a=3$ and $r=\frac{3}{2}$. Since $|r|=\frac{3}{2}>1$, $\displaystyle\sum_{k=1}^{\infty} 3\left(\frac{3}{2}\right)^{k-1}$ diverges.

(d) $\displaystyle\sum_{k=1}^{\infty}\frac{1}{2^k}$ is not in the form $\displaystyle\sum_{k=1}^{\infty} ar^{k-1}$. To place it in this form, we proceed as follows:

$$\sum_{k=1}^{\infty}\frac{1}{2^k} = \sum_{k=1}^{\infty}\left(\frac{1}{2}\right)^k = \sum_{\underset{\uparrow\, k=1}{}}^{\infty}\left[\frac{1}{2}\cdot\left(\frac{1}{2}\right)^{k-1}\right]$$

Write in the form $\displaystyle\sum_{k=1}^{\infty} ar^{k-1}$

So, $\displaystyle\sum_{k=1}^{\infty}\frac{1}{2^k}$ is a geometric series with $a=\frac{1}{2}$ and $r=\frac{1}{2}$. Since $|r|<1$, the series converges, and its sum is

$$\sum_{k=1}^{\infty}\frac{1}{2^k} = \frac{\dfrac{1}{2}}{1-\dfrac{1}{2}} = 1$$

which agrees with the sum we found earlier (p. 798).

(e) $\displaystyle\sum_{k=0}^{\infty}\left(\frac{1}{3}\right)^{k-1}$ starts at 0, so it is not in the form, $\displaystyle\sum_{k=1}^{\infty}ar^{k-1}$. To start the series at 1, write

$$\sum_{k=0}^{\infty}\left(\frac{1}{3}\right)^{k-1}=\left(\frac{1}{3}\right)^{-1}+\sum_{k=1}^{\infty}\left(\frac{1}{3}\right)^{k-1}=3+\sum_{k=1}^{\infty}\left(\frac{1}{3}\right)^{k-1}$$

Now, $\displaystyle\sum_{k=1}^{\infty}\left(\frac{1}{3}\right)^{k-1}$ is a geometric series with $a=1$ and $r=\dfrac{1}{3}$. Since $|r|<1$, the series

converges to $\dfrac{1}{1-\dfrac{1}{3}}=\dfrac{3}{2}$. Then,

$$\sum_{k=0}^{\infty}\left(\frac{1}{3}\right)^{k-1}=3+\frac{3}{2}=\frac{9}{2}$$

(f) We put $\displaystyle\sum_{k=1}^{\infty}\frac{10^{k+1}}{9^{k-2}}$ in the form $\displaystyle\sum_{k=1}^{\infty}ar^{k-1}$ by using the facts that

$$10^{k+1}=10^2\cdot10^{k-1}=100\cdot10^{k-1}\quad\text{and}\quad 9^{k-2}=\frac{1}{9}\cdot9^{k-1}$$

Then

$$\sum_{k=1}^{\infty}\frac{10^{k+1}}{9^{k-2}}=\sum_{k=1}^{\infty}\frac{100\cdot10^{k-1}}{\dfrac{1}{9}\cdot9^{k-1}}=\sum_{k=1}^{\infty}900\cdot\left(\frac{10}{9}\right)^{k-1}$$

Since the ratio $r=\dfrac{10}{9}>1$, the series diverges. ■

NOW WORK   Problems 21, 29, and AP® Practice Problems 1, 2, 3, and 5.

EXAMPLE 6   **Writing a Repeating Decimal as a Fraction**

Express the repeating decimal $0.090909\ldots$ as a quotient of two integers.

Solution
We write the infinite decimal $0.090909\ldots$ as an infinite series:

$$0.090909\ldots=0.09+0.0009+0.000009+0.00000009+\cdots$$
$$=\frac{9}{100}+\frac{9}{10000}+\frac{9}{1000000}+\cdots$$
$$=\frac{9}{100}\left(1+\frac{1}{100}+\frac{1}{10000}+\frac{1}{1000000}+\cdots\right)$$
$$=\sum_{k=1}^{\infty}\frac{9}{100}\left(\frac{1}{100}\right)^{k-1}$$

This is a geometric series with $a=\dfrac{9}{100}$ and $r=\dfrac{1}{100}$. Since $|r|<1$, the series converges, and its sum is

$$\sum_{k=1}^{\infty}\frac{9}{100}\left(\frac{1}{100}\right)^{k-1}=\frac{\dfrac{9}{100}}{1-\dfrac{1}{100}}=\frac{9}{99}=\frac{1}{11}$$

So, $0.090909\ldots=\dfrac{1}{11}$. ■

NOW WORK   Problem 59.

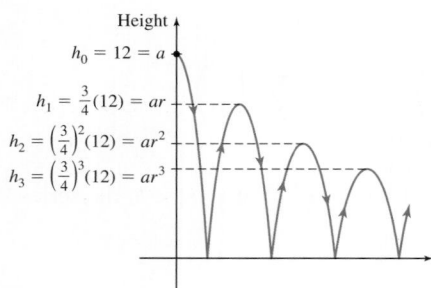

Figure 18

EXAMPLE 7 Using a Geometric Series with a Bouncing Ball

A ball is dropped from a height of 12 m. Each time it strikes the ground, it bounces back to a height three-fourths the distance from which it fell. Find the total distance traveled by the ball. See Figure 18.

Solution

Let h_n denote the height of the ball on the nth bounce. Then

$$h_0 = 12$$
$$h_1 = \frac{3}{4} \cdot 12$$
$$h_2 = \frac{3}{4}\left[\frac{3}{4} \cdot 12\right] = \left(\frac{3}{4}\right)^2 \cdot 12$$
$$\vdots$$
$$h_n = \left(\frac{3}{4}\right)^n \cdot 12$$

After the first bounce, the ball travels up a distance $h_1 = \frac{3}{4} \cdot 12$ and then the same distance back down. Between the first and the second bounce, the total distance traveled is therefore $h_1 + h_1 = 2h_1$. The *total* distance H traveled by the ball is

$$H = h_0 + 2h_1 + 2h_2 + 2h_3 + \cdots = h_0 + \sum_{k=1}^{\infty}(2h_k) = 12 + \sum_{k=1}^{\infty} 2\left[12\left(\frac{3}{4}\right)^k\right]$$
$$= 12 + \sum_{k=1}^{\infty} 24\left[\frac{3}{4}\left(\frac{3}{4}\right)^{k-1}\right]$$
$$= 12 + \sum_{k=1}^{\infty} 18\left(\frac{3}{4}\right)^{k-1}$$

The sum is a geometric series with $a = 18$ and $r = \frac{3}{4}$. The series converges and

$$H = 12 + \sum_{k=1}^{\infty} 18\left(\frac{3}{4}\right)^{k-1} = 12 + \frac{18}{1 - \frac{3}{4}} = 84$$

The ball travels a total distance of 84 m. ∎

NOW WORK Problem 63 and AP® Practice Problem 7.

③ Show that the Harmonic Series Diverges

Another useful series, even though it diverges, is the *harmonic series*.

DEFINITION Harmonic Series

The infinite series

$$\sum_{k=1}^{\infty} \frac{1}{k} = 1 + \frac{1}{2} + \frac{1}{3} + \cdots$$

is called the **harmonic series**.

THEOREM

The harmonic series $\displaystyle\sum_{k=1}^{\infty} \frac{1}{k}$ diverges.

Proof To show that the harmonic series diverges, we look at the partial sums whose indexes of summation are powers of 2. That is, we investigate the sequence S_1, S_2, S_4, S_8, and so on. We can show that this sequence is not bounded as follows:

$$S_1 = 1 > \frac{1}{2} = 1\left(\frac{1}{2}\right)$$

$$S_2 = 1 + \frac{1}{2} > \frac{1}{2} + \frac{1}{2} = 2\left(\frac{1}{2}\right)$$

$$S_4 = 1 + \frac{1}{2} + \left(\frac{1}{3} + \frac{1}{4}\right) > 2\left(\frac{1}{2}\right) + \left(\frac{1}{4} + \frac{1}{4}\right) = 3\left(\frac{1}{2}\right)$$

$$S_8 = 1 + \frac{1}{2} + \left(\frac{1}{3} + \frac{1}{4}\right) + \left(\frac{1}{5} + \frac{1}{6} + \frac{1}{7} + \frac{1}{8}\right)$$

$$> 3\left(\frac{1}{2}\right) + \left(\frac{1}{8} + \frac{1}{8} + \frac{1}{8} + \frac{1}{8}\right) = 4\left(\frac{1}{2}\right)$$

$$\vdots$$

$$S_{2^{n-1}} > n\left(\frac{1}{2}\right)$$

RECALL An unbounded sequence diverges (p. 789).

We conclude that the sequence $\{S_{2^{n-1}}\}$ is not bounded, so the sequence $\{S_n\}$ of partial sums is not bounded. It follows that the sequence $\{S_n\}$ of partial sums diverges. Therefore, the harmonic series $\displaystyle\sum_{k=1}^{\infty} \frac{1}{k}$ diverges. ■

NOW WORK AP® Practice Problems 4 and 6.

Summary

- An infinite series is an expression of the form

$$\sum_{k=1}^{\infty} a_k = a_1 + a_2 + a_3 + \cdots + a_n + \cdots$$

where a_k are numbers. Here a_n is the **nth term** or the **general term** of the series.

- The nth term of the sequence $\{S_n\}$ of partial sums of the infinite series $\displaystyle\sum_{k=1}^{\infty} a_k$ is

$$S_n = \sum_{k=1}^{n} a_k = a_1 + a_2 + a_3 + \cdots + a_n$$

- If $\lim\limits_{n \to \infty} S_n = S$, then the series $\displaystyle\sum_{k=1}^{\infty} a_k$ converges and $\displaystyle\sum_{k=1}^{\infty} a_k = S$.

- If $\lim\limits_{n \to \infty} S_n$ does not exist or if $\lim\limits_{n \to \infty} S_n = \infty$, then the series $\displaystyle\sum_{k=1}^{\infty} a_k$ diverges.

- The geometric series $\displaystyle\sum_{k=1}^{\infty} ar^{k-1} = a + ar + ar^2 + ar^3 + \cdots$

   ○ converges to $\dfrac{a}{1-r}$ if $-1 < r < 1$.

   ○ diverges if $|r| \geq 1$.

- The harmonic series $\displaystyle\sum_{k=1}^{n} \frac{1}{k} = 1 + \frac{1}{2} + \frac{1}{3} + \cdots$ diverges.

Application: Using a Geometric Series in Biology*

This application deals with the rate of occurrence of *retinoblastoma*, a rare type of eye cancer in children. An *allele (allelomorph)* is a gene that gives rise to one of a pair of contrasting characteristics, such as smooth or rough, tall or short. Each person normally has two such genes for each characteristic. An individual may have two "tall" genes, two "short" genes, or one of each. In reproduction, each parent gives one of the two types to the child.

The tendency to develop retinoblastoma apparently depends on the mutation of both copies of a gene, called RB1. The mutation rate from a normal RB1 allele to the mutant RB1 in each generation is approximately $m = 0.00002 = 2 \times 10^{-5}$. In this example, we ignore the very unlikely possibility of mutation from an abnormal RB1 to a normal RB1 gene. At the beginning of the twentieth century, retinoblastoma was nearly always fatal, but by the early 1950s, approximately 70% of children affected with the disease survived, although they usually became blind in one or both eyes. The current (2017) survival rate is about 95%, and the goals of treatment are to prevent the tumor cells from growing and spreading and to preserve vision.

Assume that survivors reproduce at about half the normal rate. (The assumption is based on scientific guesswork.) Then the productive proportion of persons affected with the retinoblastoma in 2017 was $r = (0.5)(0.95) = 0.475$. This rate is remarkable, considering that in 1900, $r \approx 0$, and in 1950 $r \approx 0.35$.

Starting with zero inherited cases in an early generation, for the nth consecutive generation, we obtain a rate of

m	due to mutation in the nth generation
mr	due to mutation in the $(n-1)$st generation
mr^2	due to mutation in the $(n-2)$nd generation
$\vdots$	
mr^n	due to mutation in the zero (original) generation

Then the total rate of occurrence of the disease in the nth generation is

$$p_n = m + mr + \cdots + mr^n = \frac{m(1 - r^{n+1})}{1 - r}$$

from which

$$p = \lim_{n \to \infty} p_n = \frac{m}{1-r} = \frac{2 \times 10^{-5}}{1 - 0.475} = 3.810 \times 10^{-5}$$

indicating that the total rate of persons affected with the disease will be almost twice the mutation rate.

Notice that if $r = 0$, as in 1900, then $p = m = 2 \times 10^{-5}$, and that if $r = 0.35$, as in 1950, then $p = 3.08 \times 10^{-5}$. We see that with better medical care, retinoblastoma has become more frequent. As medical care improves, the rate of occurrence of the disease can be expected to become even greater.

---

*Information from

J. L. Young & M. A. Smith (1999), Retinoblastoma. In L. A. G. Ries, M. A. Smith, J. G. Gurney, M. Linet, T. Tamra, J. L. Young, & G. R. Bunin (Eds.), *Cancer incidence and survival among children and adolescents: United States SEER Program, 1975–1995*, NIH Pub. No. 99-4649, Bethesda, MD: National Cancer Institute.

Children's Hospital of Philadelphia (2017), *Retinoblastoma*, http://www.CHOP.edu.

J. V. Neel & W. J. Schull (1958), *Human heredity*, 3rd ed. (pp. 333–334), Chicago: University of Chicago Press.

10.2 Assess Your Understanding

Concepts and Vocabulary

1. **Multiple Choice** If $a_1, a_2, \ldots, a_n, \ldots$ is an infinite collection of numbers, the expression $\sum\limits_{k=1}^{\infty} a_k = a_1 + a_2 + \cdots + a_n + \cdots$ is called

   (a) an infinite sequence (b) an infinite series (c) a partial sum.

2. **Multiple Choice** If $\sum\limits_{k=1}^{\infty} a_k$ is an infinite series, then the sequence $\{S_n\}$ where $S_n = \sum\limits_{k=1}^{n} a_k$, is called the sequence of

   (a) fractional parts   (b) early terms
   (c) completeness   (d) partial sums

   of the infinite series.

3. **True or False** A series converges if and only if its sequence of partial sums converges.

4. **True or False** A geometric series $\sum\limits_{k=1}^{\infty} ar^{k-1}$, $a \neq 0$, converges if $|r| \leq 1$.

5. The sum of a convergent geometric series $\sum\limits_{k=1}^{\infty} ar^{k-1}$, $a \neq 0$, is $S =$ _____.

6. **True or False** The harmonic series $\sum\limits_{k=1}^{\infty} \dfrac{1}{k}$ converges because $\lim\limits_{n \to \infty} \dfrac{1}{n} = 0$.

Skill Building

In Problems 7–10, find the fourth partial sum of each series.

7. $\sum\limits_{k=1}^{\infty} \left(\dfrac{3}{4}\right)^{k-1}$

8. $\sum\limits_{k=1}^{\infty} \dfrac{(-1)^{k+1}}{3^{k-1}}$

9. $\sum\limits_{k=1}^{\infty} k$

10. $\sum\limits_{k=1}^{\infty} \ln k$

In Problems 11–16, find the sum of each telescoping series.

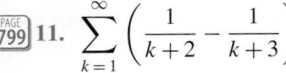 11. $\sum\limits_{k=1}^{\infty} \left(\dfrac{1}{k+2} - \dfrac{1}{k+3}\right)$

12. $\sum\limits_{k=1}^{\infty} \left[\dfrac{1}{k^2} - \dfrac{1}{(k+1)^2}\right]$

13. $\sum\limits_{k=1}^{\infty} \left(\dfrac{1}{3^{k+1}} - \dfrac{1}{3^k}\right)$

14. $\sum\limits_{k=1}^{\infty} \left(\dfrac{1}{4^{k+1}} - \dfrac{1}{4^k}\right)$

15. $\sum\limits_{k=1}^{\infty} \dfrac{1}{4k^2 - 1}$ *Hint:* $\dfrac{1}{4k^2-1} = \dfrac{1}{2}\left(\dfrac{1}{2k-1} - \dfrac{1}{2k+1}\right)$

16. $\sum\limits_{k=2}^{\infty} \dfrac{2}{k^2 - 1}$

In Problems 17–38, determine whether each geometric series converges or diverges. If it converges, find its sum.

17. $\sum\limits_{k=1}^{\infty} (\sqrt{2})^{k-1}$

18. $\sum\limits_{k=1}^{\infty} (0.33)^{k-1}$

19. $\sum\limits_{k=1}^{\infty} 5\left(\dfrac{1}{6}\right)^{k-1}$

20. $\sum\limits_{k=1}^{\infty} 4(1.1)^{k-1}$

21. $\sum\limits_{k=0}^{\infty} 7\left(\dfrac{1}{3}\right)^{k}$

22. $\sum\limits_{k=0}^{\infty} \left(\dfrac{7}{4}\right)^{k}$

23. $\sum\limits_{k=1}^{\infty} (-0.38)^{k-1}$

24. $\sum\limits_{k=1}^{\infty} (-0.38)^{k}$

25. $\sum\limits_{k=0}^{\infty} \dfrac{2^{k+1}}{3^k}$

26. $\sum\limits_{k=0}^{\infty} \dfrac{5^k}{6^{k+1}}$

27. $\sum\limits_{k=0}^{\infty} \dfrac{1}{4^{k+1}}$

28. $\sum\limits_{k=0}^{\infty} \dfrac{4^{k+1}}{3^k}$

29. $\sum\limits_{k=1}^{\infty} \left(\sin\dfrac{\pi}{2}\right)^{k-1}$

30. $\sum\limits_{k=1}^{\infty} \left(\tan\dfrac{\pi}{4}\right)^{k-1}$

31. $\sum\limits_{k=1}^{\infty} \left(-\dfrac{3}{2}\right)^{k-1}$

32. $\sum\limits_{k=1}^{\infty} \left(-\dfrac{2}{3}\right)^{k-1}$

33. $1 + \dfrac{1}{3} + \dfrac{1}{9} + \cdots + \left(\dfrac{1}{3}\right)^{n} + \cdots$

34. $1 + \dfrac{1}{4} + \dfrac{1}{16} + \cdots + \left(\dfrac{1}{4}\right)^{n} + \cdots$

35. $1 + 2 + 4 + \cdots + 2^n + \cdots$

36. $1 - \dfrac{1}{2} + \dfrac{1}{4} - \dfrac{1}{8} + \cdots + \dfrac{(-1)^{n-1}}{2^{n-1}} + \cdots$

37. $\left(\dfrac{1}{7}\right)^{2} + \left(\dfrac{1}{7}\right)^{3} + \cdots + \left(\dfrac{1}{7}\right)^{n} + \cdots$

38. $\left(\dfrac{3}{4}\right)^{5} + \left(\dfrac{3}{4}\right)^{6} + \cdots + \left(\dfrac{3}{4}\right)^{n} + \cdots$

In Problems 39–58, determine whether each series converges or diverges. If it converges, find its sum.

39. $\sum\limits_{k=0}^{\infty} \dfrac{1}{k+1}$

40. $\sum\limits_{k=4}^{\infty} k^{-1}$

41. $\sum\limits_{k=1}^{\infty} \dfrac{1}{100^k}$

42. $\sum\limits_{k=1}^{\infty} e^{-k}$

43. $\sum\limits_{k=1}^{\infty} (-10k)$

44. $\sum\limits_{k=1}^{\infty} \dfrac{3k}{5}$

45. $\sum\limits_{k=1}^{\infty} \left(\cos\dfrac{2\pi}{3}\right)^{k-1}$

46. $\sum\limits_{k=1}^{\infty} \left(\sin\dfrac{\pi}{6}\right)^{k-1}$

47. $\sum\limits_{k=1}^{\infty} \dfrac{\left(\tan\dfrac{\pi}{4}\right)^{k}}{k}$

48. $\sum\limits_{k=1}^{\infty} \dfrac{\left(\sin\dfrac{\pi}{2}\right)^{k}}{k}$

49. $\displaystyle\sum_{k=1}^{\infty} \cos(\pi k)$

50. $\displaystyle\sum_{k=1}^{\infty} \sin\frac{\pi k}{2}$

51. $\displaystyle\sum_{k=1}^{\infty} 2^{-k}3^{k+1}$

52. $\displaystyle\sum_{k=1}^{\infty} 3^{1-k}2^{1+k}$

53. $\displaystyle\sum_{k=1}^{\infty} \left(-\frac{1}{3}\right)^{k}$

54. $\displaystyle\sum_{k=1}^{\infty} \frac{\pi}{3^{k}}$

55. $\displaystyle\sum_{k=1}^{\infty} \ln\frac{k}{k+1}$

56. $\displaystyle\sum_{k=1}^{\infty} \left[e^{2k-1} - e^{2(k+1)^{-1}}\right]$

57. $\displaystyle\sum_{k=1}^{\infty} \left(\sin\frac{1}{k} - \sin\frac{1}{k+1}\right)$

58. $\displaystyle\sum_{k=1}^{\infty} \left(\tan\frac{1}{k} - \tan\frac{1}{k+1}\right)$

In Problems 59–62, express each repeating decimal as a rational number by using a geometric series.

 59. $0.5555\ldots$

60. $0.727272\ldots$

61. $4.28555\ldots$ *Hint:* $4.28555\ldots = 4.28 + 0.00555\ldots$

62. $7.162162\ldots$

Applications and Extensions

63. Distance a Ball Travels A ball is dropped from a height of 18 ft. Each time it strikes the ground, it bounces back to two-thirds of the previous height. Find the total distance traveled by the ball.

64. Diminishing Returns A rich person promises to give you $1000 on January 1, 2025. Each day thereafter you receive $\frac{9}{10}$ of what you got the previous day.

(a) What is the total amount you will receive?

(b) What is the first date on which the amount you receive is less than 1 cent?

65. Stocking a Lake Mirror Lake is stocked periodically with rainbow trout. In year n of the stocking program, the population is given by $p_n = 3000r^n + h\displaystyle\sum_{k=1}^{n} r^{k-1}$, where h is the number of fish added by the program per year and r, $0 < r < 1$, is the percent of fish removed each year.

(a) What does a manager expect the steady rainbow trout population to be as $n \to \infty$?

(b) If $r = 0.5$, how many fish h should be added annually to obtain a steady population of 4000 rainbow trout?

66. Marginal Propensity to Consume Suppose that individuals in the United States spend 90% of every additional dollar that they earn. Then according to economists, an individual's **marginal propensity to consume** is 0.90. For example, if Jane earns an additional dollar, she will spend $0.9(1) = \$0.90$ of it. The individual who earns Jane's $0.90 will spend 90% of it or $0.81. The process of spending continues and results in the series

$$\sum_{k=1}^{\infty} 0.90^{k-1} = 1 + 0.90 + 0.90^2 + 0.90^3 + \cdots$$

The sum of this series is called the **multiplier**. What is the multiplier if the marginal propensity to consume is 90%?

67. Stock Pricing One method of pricing a stock is to discount the stream of future dividends of the stock. Suppose a stock currently pays $\$P$ annually in dividends, and historically, the dividend has increased by $i\%$ annually. If an investor wants an annual rate of return of $r\%$, $r > i$, this method of pricing a stock states that the stock should be priced at the present value of an infinite stream of payments:

$$\text{Price} = P + P\frac{1+i}{1+r} + P\left(\frac{1+i}{1+r}\right)^2 + P\left(\frac{1+i}{1+r}\right)^3 + \cdots$$

(a) Find the price of a stock priced using this method.

(b) Suppose an investor desires a 9% return on a stock that currently pays an annual dividend of $4.00, and, historically, the dividend has been increased by 3% annually. What is the highest price the investor should pay for the stock?

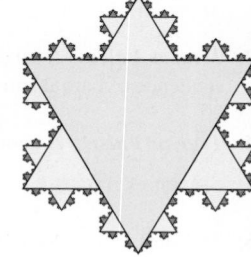 **68. The Koch Snowflake** The area inside the fractal known as the **Koch snowflake** can be described as the sum of the areas of infinitely many equilateral triangles. See the figure. For all but the center (largest) triangle, a triangle in the Koch snowflake is $\frac{1}{9}$ the area of the next largest triangle in the fractal. Suppose the largest (center) triangle has an area of 1 square unit. Then the area of the snowflake is given by the series

$$1 + 3\left(\frac{1}{9}\right) + 12\left(\frac{1}{9}\right)^2 + 48\left(\frac{1}{9}\right)^3 + 192\left(\frac{1}{9}\right)^4 + \cdots$$

Find the area of the Koch snowflake by finding the sum of the series.

69. Zeno's Paradox is about a race between Achilles and a tortoise. The tortoise is allowed a certain lead at the start of the race. Zeno claimed the tortoise must win such a race. He reasoned that for Achilles to overtake the tortoise, at some time he must cover $\frac{1}{2}$ of the distance that originally separated them. Then when he covers another $\frac{1}{4}$ of the original distance separating them, he will still have $\frac{1}{4}$ of that distance remaining, and so on. Therefore by Zeno's reasoning, Achilles never catches the tortoise. Use a series argument to explain this paradox. Assume that the difference in speed between Achilles and the tortoise is a constant v meters per second.

70. Probability A coin-flipping game involves two people who successively flip a coin. The first person to obtain a head is the winner. In probability, it turns out that the person who flips first has the probability of winning given by the series below. Find this probability.

$$\frac{1}{2} + \frac{1}{8} + \frac{1}{32} + \cdots + \frac{1}{2^{2n-1}} + \cdots$$

71. Controlling *Salmonella* *Salmonella* is a common enteric bacterium infecting both humans and farm animals with salmonellosis. Barn surfaces contaminated with *Salmonella* can be the major source of salmonellosis spread in a farm. While cleaning barn surfaces is used as a control measure on pig and cattle farms, the efficiency of cleaning has been a concern. Suppose, on average, there are p kilograms (kg) of feces produced each day and cleaning is performed with the constant efficiency e, $0 < e < 1$. At the end of each day, $(1-e)$ kg of feces from the previous day is added to the amount of the present day. Let $T(n)$ be the total accumulated fecal material on day n. Farmers are concerned when $T(n)$ exceeds the threshold level L.

(a) Express $T(n)$ as a geometric series.

(b) Find $\lim\limits_{n \to \infty} T(n)$.

(c) Determine the minimum cleaning efficiency $e_{\min}$ required to guarantee $T(n) \le L$ for all n.

(d) Suppose that $p = 120$ kg and $L = 180$ kg. Using (b) and (c), find $e_{\min}$ and $T(365)$ for $e = \dfrac{4}{5}$.

Source: R. Gautam, G. Lahodny, M. Bani-Yaghoub, & R. Ivanek. Based on their paper "Understanding the role of cleaning in the control of *Salmonella Typhimurium* in a grower finisher pig herd: a modeling approach."

72. Show that $0.9999\ldots = 1$.

In Problems 73 and 74, use a geometric series to prove the given statement.

73. $\dfrac{x}{x-1} = \sum\limits_{k=1}^{\infty} \dfrac{1}{x^{k-1}}$ for $|x| > 1$

74. $\dfrac{1}{1+x} = \sum\limits_{k=0}^{\infty} (-1)^k x^k$ for $|x| < 1$

75. Find the smallest number n for which $\sum\limits_{k=1}^{n} \dfrac{1}{k} \ge 3$.

76. Find the smallest number n for which $\sum\limits_{k=1}^{n} \dfrac{1}{k} \ge 4$.

77. Show that the series $\sum\limits_{k=1}^{\infty} \dfrac{\sqrt{k+1} - \sqrt{k}}{\sqrt{k(k+1)}}$ converges and has the sum 1.

78. Show that $\sum\limits_{k=1}^{\infty} \dfrac{1}{k(k+2)} = \dfrac{3}{4}$.

79. Show that $\sum\limits_{k=1}^{\infty} \dfrac{1}{k(k+1)(k+2)} = \dfrac{1}{4}$.

80. Show that $\sum\limits_{k=1}^{\infty} \dfrac{1}{k(k+1)(k+2)(k+3)} = \dfrac{1}{18}$.

81. Show that $\sum\limits_{k=1}^{\infty} \dfrac{1}{k(k+1)(k+2)\cdots(k+a)} = \dfrac{1}{a} \cdot \dfrac{1}{a!}$; $a \ge 1$ is an integer.

82. Solve for x: $\dfrac{x}{2+2x} = x + x^2 + x^3 + \cdots$, $|x| < 1$.

83. Show that the sum of any convergent geometric series whose first term and common ratio are rational is rational.

84. The sum S_n of the first n terms of a geometric series is given by the formula

$$S_n = a + ar + ar^2 + \cdots + ar^{n-1} = \frac{a(r^n - 1)}{r - 1} \qquad a > 0, \quad r \ne 1$$

Find $\lim\limits_{r \to 1} \dfrac{a(r^n - 1)}{r - 1}$ and compare the result with a geometric series in which $r = 1$.

Challenge Problems

The following discussion relates to Problems 85 and 86.

An interesting relationship between the nth partial sum of the harmonic series and $\ln n$ was discovered by Euler. In particular, he showed that

$$\gamma = \lim_{n \to \infty} \left(1 + \frac{1}{2} + \frac{1}{3} + \cdots + \frac{1}{n} - \ln n \right)$$

exists and is approximately equal to 0.5772. The **Euler–Mascheroni constant**, as γ is called, appears in many interesting areas of mathematics. For example, it is involved in the evaluation of the exponential integral, $\displaystyle\int_x^{\infty} \dfrac{e^{-t}}{t}\,dt$, which is important in applied mathematics. It is also related to two special functions—the gamma function and Riemann's zeta function (see Challenge Problem 85, Section 10.3). Surprisingly, it is still unknown whether γ is rational or irrational.

 85. The harmonic series diverges quite slowly. For example, the partial sums S_{10}, S_{20}, S_{50}, and S_{100} have approximate values 2.92897, 3.59774, 4.49921, and 5.18738, respectively. In fact, the sum of the first million terms of the harmonic series is about 14.4. With this in mind, what would you conjecture about the rate of convergence of the limit defining γ? Test your conjecture by calculating approximate values for γ by using the partial sums given above.

86. Use $\gamma \approx 0.5772$ to approximate

$$1 + \frac{1}{2} + \frac{1}{3} + \cdots + \frac{1}{1,000,000,000}$$

87. Show that a real number has a repeating decimal if and only if it is rational.

88. (a) Suppose $\sum\limits_{k=1}^{\infty} s_k$ is a series with the property that $s_n \ge 0$ for all integers $n \ge 1$. Show that $\sum\limits_{k=1}^{\infty} s_k$ converges if and only if the sequence $\{S_n\}$ of partial sums is bounded.

(b) Use the result of (a) to show the harmonic series diverges.

AP® Practice Problems

Multiple-Choice Questions

PAGE 803 **1.** For what numbers c does the geometric series

$$\sum_{k=1}^{\infty} \left(\frac{2c + 3}{9} \right)^{k-1}$$

converge?

(A) $-2 < c \le -1$ (B) $c < 3$

(C) $-6 \le c < 3$ (D) $-6 < c < 3$

PAGE 803 **2.** $\displaystyle\sum_{k=1}^{\infty} \frac{7}{3^{k-1}}$

(A) converges and equals $\dfrac{7}{4}$.   (B) converges and equals $\dfrac{14}{3}$.

(C) converges and equals $\dfrac{21}{2}$.   (D) diverges.

PAGE 803 **3.** The sum of the convergent geometric series

$$\sum_{k=1}^{\infty} 4c \left(\frac{c + 3}{9} \right)^{k-1} = 72$$

Determine the number c.

(A) -6 (B) -4 (C) -1 (D) 4

PAGE 805 **4.** Which of the following series diverge?

I. $\displaystyle\sum_{k=1}^{\infty} \left(\sqrt{2} \right)^{k-1}$ **II.** $\displaystyle\sum_{k=1}^{\infty} -\frac{3}{4^k}$ **III.** $\displaystyle\sum_{k=1}^{\infty} \frac{1}{k}$

(A) I and II only   (B) I and III only

(C) II and III only   (D) I, II, and III

PAGE 803 **5.** Determine whether the series $\displaystyle\sum_{k=1}^{\infty} \frac{7^{k-2}}{8^{k+1}}$ converges or diverges. If it converges, find its sum.

(A) converges and equals $\dfrac{1}{64}$ (B) converges and equals $\dfrac{1}{56}$

(C) converges and equals $\dfrac{1}{8}$ (D) The series diverges.

PAGE 805 **6.** Which of the following series diverge?

I. $\displaystyle\sum_{k=1}^{\infty} 3^{-k} \cdot 5^{k+1}$ **II.** $\displaystyle\sum_{k=1}^{\infty} k^{-1}$ **III.** $\displaystyle\sum_{k=1}^{\infty} \pi^{-k}$

(A) I only   (B) III only

(C) I and II only   (D) I, II, and III

Free-Response Question

PAGE 804 **7.** An object hanging on a spring is pulled downward a distance of 100 cm from its equilibrium position (the origin) and released. It recoils upward past the origin to a height 90 cm above the origin. It continues to oscillate up and down about the origin, but with each oscillation the spring travels only $\dfrac{9}{10}$ of the last distance traveled.

(a) Write an infinite series that models the object's movement.
(b) Find the sum of the infinite series.
(c) Interpret the sum of the infinite series in the context of the problem.

Retain Your Knowledge

Multiple-Choice Questions

1. For what numbers p does $\int_1^{\infty} x^{-p/2}\, dx$ converge?

(A) $p > \dfrac{1}{2}$ (B) $p > 1$ (C) $p > 2$ (D) $p > 3$

2. As an object moves along a line, its position at time t is given by $x(t) = \dfrac{5}{2}t^2 - 3t + 7$, for $0 \le t \le 10$. For what time interval is the speed of the object increasing?

(A) $[0, 10]$ (B) $\left[\dfrac{3}{5}, 5 \right]$ (C) $\left[\dfrac{3}{5}, 10 \right]$ (D) $[5, 10]$

3. The graph of the polar equation $r = 3 \sin(2\theta)$ is shown below.

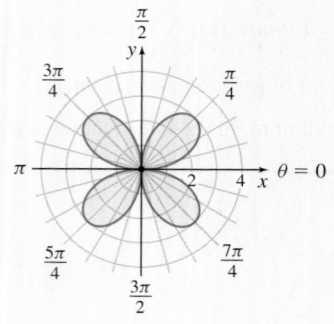

Which integral gives the area enclosed by the polar graph?

(A) $4 \int_0^{\pi/2} 9 \sin^2 (2\theta)\, d\theta$ (B) $2 \int_0^{\pi/4} 9 \sin^2 (2\theta)\, d\theta$

(C) $\dfrac{1}{2} \int_0^{\pi/2} 9 \sin^2 (3\theta)\, d\theta$ (D) $2 \int_0^{\pi/2} 9 \sin^2 (2\theta)\, d\theta$

Free-Response Question

4. The function f is given as $f(x) = \dfrac{(x^2 - 6x + 5)e^x}{x^2 - 25}$.

(a) Determine if there are any numbers where f is discontinuous.
(b) If possible, for each number found in (a), define f so that f is continuous.
(c) Identify any horizontal asymptotes of the graph of f.
(d) Identify any vertical asymptotes of f.

10.3 Properties of Series; Series with Positive Terms; the Integral Test

OBJECTIVES *When you finish this section, you should be able to:*

1. **Use the *n*th Term Test for Divergence (p. 812)**
2. **Work with properties of series (p. 812)**
3. **Use the Integral Test (p. 815)**
4. **Analyze *p*-series (p. 817)**
5. **Approximate the sum of a convergent series (p. 818)**

We have been determining whether a series $\sum_{k=1}^{\infty} a_k$ converges or diverges by finding a single compact expression for the sequence $\{S_n\}$ of partial sums as a function of n, and then examining $\lim_{n \to \infty} S_n$. For most series $\sum_{k=1}^{\infty} a_k$, however, this is not possible. As a result, we develop alternate methods for determining whether $\sum_{k=1}^{\infty} a_k$ is convergent or divergent. Most of the alternate methods only tell us whether a series converges or diverges but provide no information about the sum of a convergent series. Fortunately, in many applications involving series, it is more important to know whether or not a series converges. Knowing the sum of a convergent series, although desirable, is not always necessary.

The next result gives a property of convergent series that is used often.

THEOREM

If the infinite series $\sum_{k=1}^{\infty} a_k$ converges, then $\lim_{n \to \infty} a_n = 0$.

Proof The nth partial sum of $\sum_{k=1}^{\infty} a_k$ is $S_n = \sum_{k=1}^{n} a_k$. Since $S_{n-1} = \sum_{k=1}^{n-1} a_k$, it follows that

$$a_n = S_n - S_{n-1}$$

Since the series $\sum_{k=1}^{\infty} a_k$ converges, the sequence $\{S_n\}$ of partial sums has a limit S. Then $\lim_{n \to \infty} S_n = S$ and $\lim_{n \to \infty} S_{n-1} = S$, so

$$\lim_{n \to \infty} a_n = \lim_{n \to \infty} (S_n - S_{n-1}) = \lim_{n \to \infty} S_n - \lim_{n \to \infty} S_{n-1} = S - S = 0 \qquad \blacksquare$$

IN WORDS

- If $\sum_{k=1}^{\infty} a_k$ converges, then $\lim_{n \to \infty} a_n$ equals 0.
- If $\lim_{n \to \infty} a_n$ equals 0, then the series $\sum_{k=1}^{\infty} a_k$ may converge or diverge.

So if a series $\sum_{k=1}^{\infty} a_k$ converges, then $\lim_{n \to \infty} a_n = 0$. But there are many divergent series $\sum_{k=1}^{\infty} a_k$ for which $\lim_{n \to \infty} a_n = 0$. For example, the limit of the nth term of the harmonic series $\sum_{k=1}^{\infty} \frac{1}{k}$ is $\lim_{n \to \infty} \frac{1}{n} = 0$, and yet it diverges.

By restating the theorem, we obtain a useful test for divergence.

THEOREM nth Term Test for Divergence

If $\lim\limits_{n \to \infty} a_n \neq 0$, the infinite series $\sum\limits_{k=1}^{\infty} a_k$ diverges.

1 Use the nth Term Test for Divergence

EXAMPLE 1   Using the nth Term Test for Divergence

(a) $\sum\limits_{k=1}^{\infty} 87$ diverges, since $\lim\limits_{n \to \infty} 87 = 87 \neq 0$.

(b) $\sum\limits_{k=1}^{\infty} k$ diverges, since $\lim\limits_{n \to \infty} n = \infty \neq 0$.

(c) $\sum\limits_{k=1}^{\infty} (-1)^k$ diverges, since $\lim\limits_{n \to \infty} (-1)^n$ does not exist.

(d) $\sum\limits_{k=1}^{\infty} 2^k$ diverges, since $\lim\limits_{n \to \infty} 2^n = \infty \neq 0$.   ∎

Be careful! In testing a series $\sum\limits_{k=1}^{\infty} a_k$ for convergence/divergence

- if $\lim\limits_{n \to \infty} a_n \neq 0$, the series diverges.
- if $\lim\limits_{n \to \infty} a_n = 0$, no information about the convergence or divergence
  of the series is obtained.

NOW WORK   Problem 17 and AP® Practice Problem 2.

2 Work with Properties of Series

Next we investigate some properties of convergent and divergent series. Knowing these properties can help to determine whether a series converges or diverges.

THEOREM

If two infinite series are identical after a certain term, then either both series converge or both series diverge. If both series converge, they do not necessarily have the same sum.

Proof Consider the two series

$$\sum_{k=1}^{\infty} a_k = a_1 + a_2 + \cdots + a_p + a_{p+1} + \cdots + a_n + \cdots$$

$$\sum_{k=1}^{\infty} b_k = b_1 + b_2 + \cdots + b_p + a_{p+1} + \cdots + a_n + \cdots$$

Notice that after the first p terms, the terms of both series are identical. The nth partial sum S_n of $\sum\limits_{k=1}^{\infty} a_k$ and the nth partial sum T_n of $\sum\limits_{k=1}^{\infty} b_k$ are given by

$$S_n = a_1 + a_2 + \cdots + a_p + a_{p+1} + \cdots + a_n$$

$$T_n = b_1 + b_2 + \cdots + b_p + a_{p+1} + \cdots + a_n$$

After the pth term, the terms are the same.

$$S_n - T_n = (a_1 + \cdots + a_p) - (b_1 + \cdots + b_p)$$

$$S_n = T_n + (a_1 + \cdots + a_p) - (b_1 + \cdots + b_p)$$

For $n > p$

$$\lim_{n \to \infty} S_n = \lim_{n \to \infty} \left[T_n + (a_1 + \cdots + a_p) - (b_1 + \cdots + b_p) \right]$$

$$\lim_{n \to \infty} S_n = \lim_{n \to \infty} T_n + \lim_{n \to \infty} \left[(a_1 + \cdots + a_p) - (b_1 + \cdots + b_p) \right]$$

$$\lim_{n \to \infty} S_n = \lim_{n \to \infty} T_n + k$$

where $k = (a_1 + \cdots + a_p) - (b_1 + \cdots + b_p)$ is some number. Consequently, either both limits exist (both series converge) or neither limit exists (both series diverge). No other possibility can occur. ∎

THEOREM Sum and Difference of Convergent Series

If $\sum\limits_{k=1}^{\infty} a_k = S$ and $\sum\limits_{k=1}^{\infty} b_k = T$ are two convergent series, then the

series $\sum\limits_{k=1}^{\infty} (a_k + b_k)$ and the series $\sum\limits_{k=1}^{\infty} (a_k - b_k)$ also converge. Moreover,

$$\sum_{k=1}^{\infty} (a_k + b_k) = \sum_{k=1}^{\infty} a_k + \sum_{k=1}^{\infty} b_k = S + T$$

$$\sum_{k=1}^{\infty} (a_k - b_k) = \sum_{k=1}^{\infty} a_k - \sum_{k=1}^{\infty} b_k = S - T$$

The proof of the sum part of this theorem is an exercise. See Problem 75.

THEOREM Constant Multiple of a Series

Suppose c is a nonzero real number. If $\sum\limits_{k=1}^{\infty} a_k = S$ is a convergent series, then the

series $\sum\limits_{k=1}^{\infty} (ca_k)$ also converges. Moreover,

$$\sum_{k=1}^{\infty} (ca_k) = c \sum_{k=1}^{\infty} a_k = cS$$

IN WORDS Multiplying each term of a series by the same nonzero constant does not affect the convergence (or divergence) of the series.

If the series $\sum\limits_{k=1}^{\infty} a_k$ diverges, then the series $\sum\limits_{k=1}^{\infty} (ca_k)$ also diverges.

The proof of the convergent part of this theorem is an exercise. See Problem 76.

EXAMPLE 2 **Using Properties of Series**

Determine whether each series converges or diverges. If it converges, find its sum.

(a) $\displaystyle\sum_{k=4}^{\infty} \frac{1}{k}$ **(b)** $\displaystyle\sum_{k=1}^{\infty} \frac{2}{k}$ **(c)** $\displaystyle\sum_{k=1}^{\infty} \left(\frac{1}{2^{k-1}} + \frac{1}{3^{k-1}} \right)$

Solution

(a) Except for the first three terms, the series $\displaystyle\sum_{k=4}^{\infty} \frac{1}{k} = \frac{1}{4} + \frac{1}{5} + \frac{1}{6} + \cdots$ is identical to

the harmonic series, which diverges. So, it follows that $\displaystyle\sum_{k=4}^{\infty} \frac{1}{k}$ also diverges.

(b) Since the harmonic series $\displaystyle\sum_{k=1}^{\infty} \frac{1}{k}$ diverges, the series $\displaystyle\sum_{k=1}^{\infty} \left(2 \cdot \frac{1}{k} \right) = \sum_{k=1}^{\infty} \frac{2}{k}$ diverges.

(c) Since the series $\displaystyle\sum_{k=1}^{\infty} \frac{1}{2^{k-1}}$ and the series $\displaystyle\sum_{k=1}^{\infty} \frac{1}{3^{k-1}}$ are both convergent geometric

series, the series defined by $\displaystyle\sum_{k=1}^{\infty} \left(\frac{1}{2^{k-1}} + \frac{1}{3^{k-1}} \right)$ is also convergent. The sum is

$$\sum_{k=1}^{\infty} \left(\frac{1}{2^{k-1}} + \frac{1}{3^{k-1}} \right) = \sum_{k=1}^{\infty} \frac{1}{2^{k-1}} + \sum_{k=1}^{\infty} \frac{1}{3^{k-1}} = \frac{1}{1 - \dfrac{1}{2}} + \frac{1}{1 - \dfrac{1}{3}} = 2 + \frac{3}{2} = \frac{7}{2} \qquad \blacksquare$$

NOW WORK   Problem **39** and AP® Practice Problem **4.**

Series with Positive Terms

For series that have only positive terms, it is possible to construct tests for convergence that use only the nth term of the series and do not require knowledge of the form of the sequence $\{S_n\}$ of partial sums.

For example, consider an infinite series

$$\sum_{k=1}^{\infty} a_k = a_1 + a_2 + \cdots + a_n + \cdots$$

where each term $a_k > 0$. The general term S_n of the sequence of partial sums $\{S_n\}$ is

$$S_n = (a_1 + a_2 + a_3 + \cdots + a_{n-1}) + a_n = S_{n-1} + a_n > S_{n-1}$$
$$\underset{\substack{\uparrow \\ a_n > 0}}{}$$

That is, the sequence of partial sums is increasing.

There are two possibilities.

- If $\{S_n\}$ is unbounded, then $\{S_n\}$ diverges and the series diverges. This is how we showed that the harmonic series $\displaystyle\sum_{k=1}^{\infty} \frac{1}{k}$ diverges. (p. 805)

RECALL A bounded, monotonic sequence converges.

- If $\{S_n\}$ is bounded, then $\{S_n\}$ converges and the series converges.

We formalize this conclusion as the *General Convergence Test*.

THEOREM General Convergence Test

An infinite series of positive terms converges if and only if its sequence of partial sums is bounded. The sum of such an infinite series will not exceed an upper bound.

We use this theorem to develop other tests for convergence of series with only positive terms; the first one we discuss is the *Integral Test*.

3 Use the Integral Test

THEOREM Integral Test

Suppose f is a function that is continuous, positive, and nonincreasing on the interval $[1, \infty)$. Also suppose $a_k = f(k)$ for all positive integers k. Then

- if the improper integral $\int_1^\infty f(x)\,dx$ converges, the series $\sum\limits_{k=1}^\infty a_k$ converges.

- if the improper integral $\int_1^\infty f(x)\,dx$ diverges, the series $\sum\limits_{k=1}^\infty a_k$ diverges.

NEED TO REVIEW? Improper Integrals are discussed in Section 6.9, pp. 538–546.

A proof of the Integral Test is given at the end of the section.

The Integral Test is used when the nth term of the series is related to a function f that is not only continuous, positive, and nonincreasing on the interval $[1, \infty)$ but also has an antiderivative that can be found readily.

CALC CLIP | **EXAMPLE 3 Using the Integral Test**

Determine whether the series $\sum\limits_{k=1}^\infty a_k = \sum\limits_{k=1}^\infty \dfrac{4}{k^2+1}$ converges or diverges.

Solution

On the interval $[1, \infty)$, the related function $f(x) = \dfrac{4}{x^2+1}$ is continuous, positive, and decreasing. (f is decreasing because $f'(x) = -\dfrac{8x}{(x^2+1)^2} < 0$ for $x \geq 1$.) Also, $a_k = f(k)$ for all positive integers. To use the Integral Test, we need to determine whether the improper integral $\displaystyle\int_1^\infty \dfrac{4}{x^2+1}\,dx$ converges or diverges. We investigate the limit

$$\lim_{b \to \infty} \int_1^b \frac{4}{x^2+1}\,dx = \lim_{b \to \infty}\left[4 \int_1^b \frac{1}{x^2+1}\,dx\right] = 4 \lim_{b \to \infty}\left[\tan^{-1} x\right]_1^b$$

$$= 4 \lim_{b \to \infty}\left[\tan^{-1} b - \frac{\pi}{4}\right] = 4\left[\frac{\pi}{2} - \frac{\pi}{4}\right] = \pi$$

CAUTION In Example 3, the fact that $\displaystyle\int_1^\infty \dfrac{4}{x^2+1}\,dx = \pi$ does not mean that the sum S of the series is π. Do not confuse the value of the improper integral $\int_1^\infty f(x)\,dx$ with the sum S of the series. In general, they are *not* equal.

Since the improper integral converges, the series $\sum\limits_{k=1}^\infty \dfrac{4}{k^2+1}$ converges. ∎

NOW WORK AP® Practice Problem 7.

EXAMPLE 4   Using the Integral Test

Determine whether the series $\displaystyle\sum_{k=1}^{\infty} a_k = \sum_{k=1}^{\infty} \frac{2k}{k^2+1}$ converges or diverges.

Solution

On the interval $[1, \infty)$, the related function $f(x) = \dfrac{2x}{x^2+1}$ is continuous, positive, and

decreasing. (f is decreasing because $f'(x) = \dfrac{2(1-x^2)}{(x^2+1)^2} < 0$ for $x > 1$.) Also, $a_k = f(k)$

for all positive integers k. Now investigate the improper integral $\displaystyle\int_1^{\infty} \frac{2x}{x^2+1}\,dx$ by

examining the limit

$$\lim_{b \to \infty} \int_1^b \frac{2x}{x^2+1}\,dx = \lim_{b \to \infty} \left[\ln(x^2+1)\right]_1^b$$

$$= \lim_{b \to \infty} [\ln(b^2+1) - \ln 2] = \infty$$

Since the improper integral diverges, by the Integral Test the series $\displaystyle\sum_{k=1}^{\infty} \frac{2k}{k^2+1}$ also

diverges. ∎

NOW WORK   Problem 23.

To use the Integral Test, the lower limit of integration does not need to be 1, as we see in the next example.

EXAMPLE 5   Using the Integral Test

Determine whether the series $\displaystyle\sum_{k=2}^{\infty} \frac{1}{k(\ln k)^2}$ converges or diverges.

Solution

On the interval $[2, \infty)$, the related function $f(x) = \dfrac{1}{x(\ln x)^2}$ is continuous, positive,

and decreasing. (Check this for yourself.) To investigate the improper integral

$\displaystyle\int_2^{\infty} \frac{1}{x(\ln x)^2}\,dx$, we first find an antiderivative of $\dfrac{1}{x(\ln x)^2}$. Using the substitution

$u = \ln x$, $du = \dfrac{1}{x}\,dx$, we find

$$\int \frac{dx}{x(\ln x)^2} = \int \frac{du}{u^2} = -\frac{1}{u} + C = -\frac{1}{\ln x} + C$$

Now investigate the limit

$$\lim_{b \to \infty} \left[-\frac{1}{\ln x}\right]_2^b = \lim_{b \to \infty} \left(-\frac{1}{\ln b} + \frac{1}{\ln 2}\right) = \frac{1}{\ln 2}$$

Since the improper integral $\displaystyle\int_2^{\infty} \frac{dx}{x(\ln x)^2}$ converges, by the Integral Test the

series $\displaystyle\sum_{k=2}^{\infty} \frac{1}{k(\ln k)^2}$ converges. ∎

NOW WORK   Problem 27 and AP® Practice Problem 6.

④ Analyze *p*-Series

Another important series of positive terms is a *p-series*.

DEFINITION *p*-Series

A ***p*-series** is an infinite series of the form

$$\sum_{k=1}^{\infty} \frac{1}{k^p} = 1 + \frac{1}{2^p} + \frac{1}{3^p} + \cdots + \frac{1}{n^p} + \cdots$$

where *p* is a positive real number.

A *p*-series is sometimes referred to as a **hyperharmonic series**, since the harmonic series is a special case of a *p*-series when $p = 1$. Some examples of *p*-series are

$$p = 1: \quad \sum_{k=1}^{\infty} \frac{1}{k} = 1 + \frac{1}{2} + \frac{1}{3} + \cdots + \frac{1}{n} + \cdots \qquad \text{The harmonic series}$$

$$p = \frac{1}{2}: \quad \sum_{k=1}^{\infty} \frac{1}{k^{1/2}} = 1 + \frac{1}{2^{1/2}} + \frac{1}{3^{1/2}} + \cdots + \frac{1}{n^{1/2}} + \cdots$$

$$p = 3: \quad \sum_{k=1}^{\infty} \frac{1}{k^3} = 1 + \frac{1}{2^3} + \frac{1}{3^3} + \cdots + \frac{1}{n^3} + \cdots$$

The following theorem establishes the values of *p* for which a *p*-series converges and for which it diverges.

THEOREM Convergence/Divergence of a *p*-Series

The *p*-series

$$\sum_{k=1}^{\infty} \frac{1}{k^p} = 1 + \frac{1}{2^p} + \frac{1}{3^p} + \cdots + \frac{1}{n^p} + \cdots$$

converges if $p > 1$ and diverges if $0 < p \le 1$.

Proof First, notice that if $p = 1$, the series $\sum_{k=1}^{\infty} \frac{1}{k^p} = \sum_{k=1}^{\infty} \frac{1}{k}$ is the harmonic series, which diverges.

Next, on the interval $[1, \infty)$, the function $f(x) = \frac{1}{x^p}$, $p > 0$, is continuous, positive, and decreasing, and $f(k) = \frac{1}{k^p}$ for all positive integers k. By the Integral Test, the series $\sum_{k=1}^{\infty} \frac{1}{k^p}$ converges if and only if the improper integral $\int_1^{\infty} \frac{1}{x^p} dx$ converges. So we investigate $\lim_{b \to \infty} \int_1^b \frac{1}{x^p} dx$, $p > 0$, $p \ne 1$.

NOTE The improper integral $\int_1^{\infty} \frac{1}{x^p} dx$ is discussed in Section 6.9, pp. 541–542.

$$\lim_{b \to \infty} \int_1^b \frac{1}{x^p} dx = \lim_{b \to \infty} \left[\frac{x^{-p+1}}{-p+1} \right]_1^b = \frac{1}{1-p} \lim_{b \to \infty} \left[x^{1-p} \right]_1^b = \frac{1}{1-p} \left[\lim_{b \to \infty} b^{1-p} - 1 \right]$$

- If $0 < p < 1$, then $1 - p > 0$ and $\lim_{b \to \infty} b^{1-p} = \infty$. So $\int_1^{\infty} \frac{1}{x^p} dx$ diverges.

- If $p > 1$, then $1 - p < 0$ and $\lim_{b \to \infty} b^{1-p} = 0$. So $\int_1^{\infty} \frac{1}{x^p} dx$ converges.

So, the *p*-series $\sum_{k=1}^{\infty} \frac{1}{k^p}$ converges if $p > 1$ and diverges if $0 < p \le 1$. ∎

EXAMPLE 6 **Analyzing a p-Series**

(a) The series

$$\sum_{k=1}^{\infty} \frac{1}{k^3} = 1 + \frac{1}{2^3} + \frac{1}{3^3} + \cdots + \frac{1}{n^3} + \cdots$$

converges, since it is a p-series where $p = 3$ ($p > 1$).

(b) The series

$$\sum_{k=1}^{\infty} \frac{1}{\sqrt{k}} = 1 + \frac{1}{\sqrt{2}} + \frac{1}{\sqrt{3}} + \cdots + \frac{1}{\sqrt{n}} + \cdots$$

diverges, since it is a p-series where $p = \frac{1}{2}$ ($0 < p \leq 1$). ∎

NOW WORK Problem 35 and AP® Practice Problems 1, 3, and 5.

5 Approximate the Sum of a Convergent Series

For a convergent p-series, we can find an interval of width 1 that contains the sum of the series.

THEOREM Bounds for the Sum of a Convergent p-Series

If $p > 1$, then

$$\boxed{\frac{1}{p-1} < \sum_{k=1}^{\infty} \frac{1}{k^p} < 1 + \frac{1}{p-1}} \qquad (1)$$

IN WORDS The theorem gives an interval of width 1 that contains the sum of a convergent p-series. But it gives no information about the actual sum.

A proof of this result is given after the solution to Example 8.

EXAMPLE 7 **Finding Bounds for the Sum of a Convergent p-Series**

Find bounds for the sum of each p-series.

(a) $\displaystyle\sum_{k=1}^{\infty} \frac{1}{k^3}$ **(b)** $1 + \dfrac{1}{2^\pi} + \dfrac{1}{3^\pi} + \dfrac{1}{4^\pi} + \cdots$

Solution

(a) $\displaystyle\sum_{k=1}^{\infty} \frac{1}{k^3}$ is a convergent p-series with $p = 3$. Based on (1),

$$\frac{1}{2} < \sum_{k=1}^{\infty} \frac{1}{k^3} < \frac{3}{2} \qquad \frac{1}{p-1} = \frac{1}{3-1} = \frac{1}{2}; \quad 1 + \frac{1}{p-1} = 1 + \frac{1}{2} = \frac{3}{2}$$

The sum lies in the interval $\left(\dfrac{1}{2}, \dfrac{3}{2}\right)$.

(b) $\displaystyle\sum_{k=1}^{\infty} \frac{1}{k^\pi}$, is a convergent p-series with $p = \pi$. Based on (1),

$$\frac{1}{\pi - 1} < \sum_{k=1}^{\infty} \frac{1}{k^\pi} < 1 + \frac{1}{\pi - 1} \quad \text{or} \quad 0.466 < \sum_{k=1}^{\infty} \frac{1}{k^\pi} < 1.467$$

The sum lies in the interval $(0.466, \ 1.467)$. ∎

NOW WORK Problem 55.

HISTORICAL NOTE Since $\displaystyle\sum_{k=1}^{\infty} \frac{1}{k^p}$ converges for $p > 1$, the series has a sum S. However, the exact sum has been found only for positive even integers p. In 1752, Euler was the first to show that $\displaystyle\sum_{k=1}^{\infty} \frac{1}{k^2} = \frac{\pi^2}{6}$. But the sum of other convergent p-series, such as $\displaystyle\sum_{k=1}^{\infty} \frac{1}{k^3}$, is not known.

The Bounds for the Sum of a Convergent p-series Theorem can be extended to any convergent series of positive terms.

THEOREM Bounds for the Sum of a Convergent Series

Consider the series $\sum\limits_{k=1}^{\infty} a_k$, where $a_k > 0$ for all positive integers k. Suppose f is a continuous, positive, decreasing function on the interval $[1, \infty)$ and $a_k = f(k)$ for all positive integers k. If the improper integral $\int_1^{\infty} f(x)\,dx$ converges, then the sum of the series is bounded by

$$\int_1^{\infty} f(x)\,dx < \sum_{k=1}^{\infty} a_k < a_1 + \int_1^{\infty} f(x)\,dx \qquad (2)$$

EXAMPLE 8   Finding Bounds for the Sum of a Convergent Series

Find bounds for the sum of the series $\sum\limits_{k=1}^{\infty} \dfrac{4}{k^2+1}$.

Solution

In Example 3, page 815, we proved that the series $\sum\limits_{k=1}^{\infty} \dfrac{4}{k^2+1}$ converges by using the Integral Test and showing $\int_1^{\infty} \dfrac{4}{x^2+1}\,dx$ converges to π. Based on (2), the sum of the series $\sum\limits_{k=1}^{\infty} \dfrac{4}{k^2+1}$ has the bounds

$$\pi < \sum_{k=1}^{\infty} \frac{4}{k^2+1} < 2 + \pi \qquad a_1 = \frac{4}{2} = 2 \qquad \blacksquare$$

NOW WORK   Problem 63.

Proof of the Bounds for the Sum of a Convergent p-Series Theorem

From Figure 19, we see that the area under the graph of $f(x) = \dfrac{1}{x^p}$ is less than the area of the sums of the rectangles. That is,

$$\left(\text{Area under the graph of } y = \frac{1}{x^p}\right) = \int_1^{\infty} \frac{dx}{x^p} < 1 + \frac{1}{2^p} + \frac{1}{3^p} + \cdots = \sum_{k=1}^{\infty} \frac{1}{k^p} \qquad (3)$$

From Figure 20, we have

$$1 + \left(\text{Area under the graph of } y = \frac{1}{x^p}\right) = 1 + \int_1^{\infty} \frac{dx}{x^p} > 1 + \frac{1}{2^p} + \frac{1}{3^p} + \cdots = \sum_{k=1}^{\infty} \frac{1}{k^p} \qquad (4)$$

Combining (3) and (4), we have

$$\int_1^{\infty} \frac{dx}{x^p} < \sum_{k=1}^{\infty} \frac{1}{k^p} < 1 + \int_1^{\infty} \frac{dx}{x^p} \qquad (5)$$

Since $p > 1$,

$$\int_1^{\infty} \frac{dx}{x^p} = \lim_{b \to \infty} \int_1^{b} \frac{dx}{x^p} = \lim_{b \to \infty} \left[\frac{x^{-p+1}}{-p+1}\right]_1^{b} = \lim_{b \to \infty} \frac{b^{1-p}-1}{1-p} \underset{\underset{p>1}{\uparrow}}{=} \frac{0-1}{1-p} = \frac{1}{p-1}$$

Using this result in (5), we get

$$\frac{1}{p-1} < \sum_{k=1}^{\infty} \frac{1}{k^p} < 1 + \frac{1}{p-1} \qquad \blacksquare$$

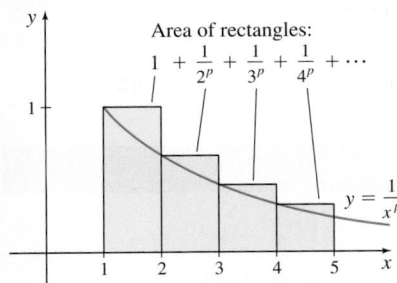

Area of rectangles:
$1 + \dfrac{1}{2^p} + \dfrac{1}{3^p} + \dfrac{1}{4^p} + \cdots$

$y = \dfrac{1}{x^p}$

Figure 19

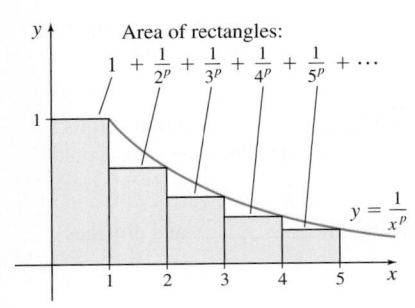

Area of rectangles:
$1 + \dfrac{1}{2^p} + \dfrac{1}{3^p} + \dfrac{1}{4^p} + \dfrac{1}{5^p} + \cdots$

$y = \dfrac{1}{x^p}$

Figure 20

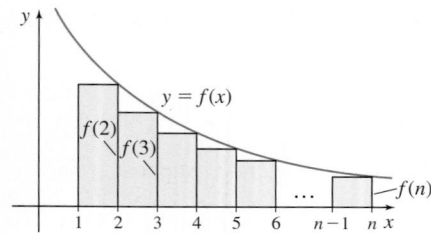

Figure 21

Proof of the Integral Test Suppose $\int_1^\infty f(x)\,dx$ converges. Then $\lim\limits_{n \to \infty} \int_1^n f(x)\,dx$ exists and equals some number L. As Figure 21 shows, $\int_1^n f(x)\,dx$ underestimates the area under the graph of f on $[1, \infty)$. As a result,

$$\int_1^n f(x)\,dx < L \qquad \text{for any number } n$$

But f is decreasing on the interval $[1, n]$. So, the sum of the areas of the $n-1$ rectangles drawn in Figure 21 underestimates the area under the graph of $y = f(x)$ from 1 to n. That is,

$$f(2) + f(3) + \cdots + f(n) < \int_1^n f(x)\,dx < L$$

Now add $f(1) = a_1$ to each part, and use the fact that $a_1 = f(1)$, $a_2 = f(2)$, ..., $a_n = f(n)$. Then

$$a_1 + a_2 + a_3 + \cdots + a_n = f(1) + f(2) + \cdots + f(n) < f(1) + \int_1^n f(x)\,dx < f(1) + L$$

This means that the partial sums of $\sum\limits_{k=1}^\infty a_k$ are bounded from above. Since the partial sums are also increasing, $\sum\limits_{k=1}^\infty a_k$ converges.

Now suppose $\int_1^\infty f(x)\,dx$ diverges. Since f is decreasing on $[1, n]$, the sum of the areas of the $n-1$ rectangles shown in Figure 22 overestimates the area under the graph of $y = f(x)$ from 1 to n. That is,

$$f(1) + f(2) + \cdots + f(n-1) > \int_1^n f(x)\,dx$$

Since $\int_1^n f(x)\,dx \to \infty$ as $n \to \infty$, the nth partial sum S_n must also approach infinity, and it follows that the series $\sum\limits_{k=1}^\infty a_k$ diverges. ∎

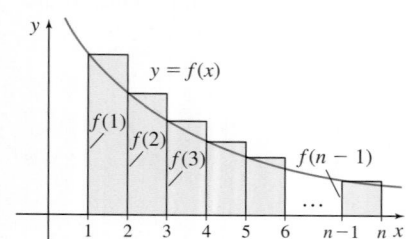

Figure 22

10.3 Assess Your Understanding

Concepts and Vocabulary

1. **Multiple Choice** If the series $\sum\limits_{k=1}^\infty a_k$, $a_k > 0$, converges then $\lim\limits_{n \to \infty} a_n =$
   (a) 0   (b) a_1 (c) a_n (d) ∞.

2. **True or False** If $\lim\limits_{n \to \infty} a_n = 0$, then the series $\sum\limits_{k=1}^\infty a_k$ converges.

3. **True or False** The series $\sum\limits_{k=1}^\infty k^3$ diverges.

4. **True or False** If the first 100 terms of two infinite series are different, but from the 101st term on they are identical, then either both series converge or both series diverge.

5. **True or False** If $\sum\limits_{k=1}^\infty (a_k + b_k)$ converges, then $\sum\limits_{k=1}^\infty a_k$ converges and $\sum\limits_{k=1}^\infty b_k$ converges.

6. **True or False** If $\sum\limits_{k=1}^\infty a_k = S$ is a convergent series and c is a nonzero real number, then $\sum\limits_{k=1}^\infty (ca_k) = cS$.

7. **True or False** Suppose f is a function defined on the interval $[1, \infty)$ that is continuous, positive, and decreasing on its domain, and $a_k = f(k)$ for all positive integers k. Then the series $\sum\limits_{k=1}^\infty a_k$ converges if and only if the improper integral $\int_1^\infty f(x)\,dx$ converges.

8. **True or False** For an infinite series of positive terms, if its sequence of partial sums is not bounded, then the series could converge or diverge.

9. The p-series $\sum\limits_{k=1}^\infty \dfrac{1}{k^p}$ converges if _____ and diverges if _____.

10. Does $\displaystyle\sum_{k=1}^{\infty} \frac{1}{k^{3/2}}$ converge or diverge?

11. Does $\displaystyle\sum_{k=1}^{\infty} \frac{1}{k^{1/2}}$ converge or diverge?

12. **True or False**   If $p > 1$, then the convergent p-series $\displaystyle\sum_{k=1}^{\infty} \frac{1}{k^p}$ is bounded by $\dfrac{1}{p-1} < \displaystyle\sum_{k=1}^{\infty} \frac{1}{k^p} < 1$.

Skill Building

In Problems 13–18, use the nth Term Test for Divergence to show each series diverges.

13. $\displaystyle\sum_{k=1}^{\infty} 16$

14. $\displaystyle\sum_{k=1}^{\infty} \frac{k+9}{k}$

15. $\displaystyle\sum_{k=1}^{\infty} \ln k$

16. $\displaystyle\sum_{k=1}^{\infty} e^k$

[PAGE 812] 17. $\displaystyle\sum_{k=1}^{\infty} \frac{k^2}{k^2+4}$

18. $\displaystyle\sum_{k=1}^{\infty} \frac{k^2+3}{\sqrt{k}}$

In Problems 19–28, use the Integral Test to determine whether each series converges or diverges.

19. $\displaystyle\sum_{k=1}^{\infty} \frac{1}{k^{1.01}}$

20. $\displaystyle\sum_{k=1}^{\infty} \frac{1}{k^{0.9}}$

21. $\displaystyle\sum_{k=1}^{\infty} \frac{\sqrt{\ln k}}{k}$

22. $\displaystyle\sum_{k=2}^{\infty} \frac{1}{k\sqrt{\ln k}}$

[PAGE 816] 23. $\displaystyle\sum_{k=1}^{\infty} ke^{-k^2}$

24. $\displaystyle\sum_{k=1}^{\infty} ke^{-k}$

25. $\displaystyle\sum_{k=1}^{\infty} \frac{1}{k^2+1}$

26. $\displaystyle\sum_{k=2}^{\infty} \frac{1}{k\sqrt{k^2-1}}$

[PAGE 816] 27. $\displaystyle\sum_{k=2}^{\infty} \frac{1}{k \ln k}$

28. $\displaystyle\sum_{k=2}^{\infty} \frac{1}{k(\ln k)^3}$

In Problems 29–38, determine whether each p-series converges or diverges.

29. $\displaystyle\sum_{k=1}^{\infty} \frac{1}{k^2}$

30. $\displaystyle\sum_{k=1}^{\infty} \frac{1}{k^4}$

31. $\displaystyle\sum_{k=1}^{\infty} \frac{1}{k^{1/3}}$

32. $\displaystyle\sum_{k=1}^{\infty} \frac{1}{k^{2/3}}$

33. $\displaystyle\sum_{k=1}^{\infty} \frac{1}{k^e}$

34. $\displaystyle\sum_{k=1}^{\infty} \frac{1}{k^{\sqrt{2}}}$

[PAGE 818] 35. $1 + \dfrac{1}{2\sqrt{2}} + \dfrac{1}{3\sqrt{3}} + \dfrac{1}{4\sqrt{4}} + \cdots$

36. $1 + \dfrac{1}{\sqrt[3]{2}} + \dfrac{1}{\sqrt[3]{3}} + \dfrac{1}{\sqrt[3]{4}} + \cdots$

37. $1 + \dfrac{1}{4\sqrt{2}} + \dfrac{1}{9\sqrt{3}} + \dfrac{1}{16\sqrt{4}} + \cdots$

38. $1 + \dfrac{1}{8} + \dfrac{1}{27} + \dfrac{1}{64} + \cdots$

In Problems 39–54, determine whether each series converges or diverges.

[PAGE 814] 39. $\displaystyle\sum_{k=1}^{\infty} \frac{10}{k}$

40. $\displaystyle\sum_{k=1}^{\infty} \frac{2}{1+k}$

41. $\displaystyle\sum_{k=1}^{\infty} \frac{k^2+1}{4k+1}$

42. $\displaystyle\sum_{k=1}^{\infty} \frac{k^3}{k^3+3}$

43. $\displaystyle\sum_{k=1}^{\infty} \left(k + \frac{1}{k}\right)$

44. $\displaystyle\sum_{k=1}^{\infty} \left(k - \frac{10}{k}\right)$

45. $\displaystyle\sum_{k=1}^{\infty} \left(\frac{1}{3k} - \frac{1}{4k}\right)$

46. $\displaystyle\sum_{k=1}^{\infty} \left(\frac{1}{3^k} - \frac{1}{4^k}\right)$

47. $\displaystyle\sum_{k=1}^{\infty} \sin\left(\frac{\pi}{2}k\right)$

48. $\displaystyle\sum_{k=1}^{\infty} \sec(\pi k)$

49. $\displaystyle\sum_{k=3}^{\infty} \frac{k+1}{k-2}$

50. $\displaystyle\sum_{k=5}^{\infty} \frac{2k^5+3}{k^5-4k^4}$

51. $\displaystyle\sum_{k=2}^{\infty} \frac{1}{k(\ln k)^{1/2}}$

52. $\displaystyle\sum_{k=2}^{\infty} \frac{1}{k(\ln k)^2}$

53. $\displaystyle\sum_{k=3}^{\infty} \frac{2k}{k^2-4}$

54. $\displaystyle\sum_{k=1}^{\infty} \frac{1}{(2k-1)(2k)}$

In Problems 55–64, find bounds for the sum of each convergent series.

[PAGE 818] 55. $\displaystyle\sum_{k=1}^{\infty} \frac{1}{k^2}$

56. $\displaystyle\sum_{k=1}^{\infty} \frac{1}{k^4}$

57. $\displaystyle\sum_{k=1}^{\infty} \frac{1}{k^e}$

58. $\displaystyle\sum_{k=1}^{\infty} \frac{1}{k^{\sqrt{2}}}$

59. $1 + \dfrac{1}{2\sqrt{2}} + \dfrac{1}{3\sqrt{3}} + \dfrac{1}{4\sqrt{4}} + \cdots$

60. $1 + \dfrac{1}{4\sqrt{2}} + \dfrac{1}{9\sqrt{3}} + \dfrac{1}{16\sqrt{4}} + \cdots$

61. $\displaystyle\sum_{k=1}^{\infty} ke^{-k^2}$

62. $\displaystyle\sum_{k=1}^{\infty} k^2 e^{-3k^3}$

[PAGE 819] 63. $\displaystyle\sum_{k=1}^{\infty} \frac{1}{k^2+9}$

64. $\displaystyle\sum_{k=1}^{\infty} \frac{1}{4k^2+1}$

Applications and Extensions

65. **Integral Test**   Use the Integral Test to show that the series $\displaystyle\sum_{k=2}^{\infty} \frac{1}{k(\ln k)^p}$ converges if and only if $p > 1$.

66. **Integral Test**   Use the Integral Test to show that the series $\displaystyle\sum_{k=3}^{\infty} \frac{1}{k(\ln k)\,[\ln(\ln k)^p]}$ converges if and only if $p > 1$.

67. **Faulty Logic**   Suppose $S = 1 + 2 + 4 + 8 + \cdots$. Then

$$2S = 2 + 4 + 8 + 16 + \cdots = -1 + (1 + 2 + 4 + \cdots) = -1 + S$$

Therefore,

$$S = 1 + 2 + 4 + 8 + \cdots = -1$$

What went wrong here?

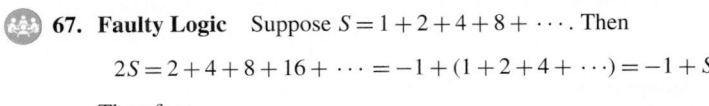

68. Find examples to show that the series $\sum_{k=1}^{\infty}(a_k+b_k)$

and $\sum_{k=1}^{\infty}(a_k-b_k)$ may converge or diverge if $\sum_{k=1}^{\infty}a_k$ and $\sum_{k=1}^{\infty}b_k$ each diverge.

69. Approximating π^2 The p-series $\sum_{k=1}^{\infty}\dfrac{1}{k^2}$ converges to $\dfrac{\pi^2}{6}$.

Use the first hundred terms of the series to approximate π^2.

70. The p-series $\sum_{k=1}^{\infty}\dfrac{1}{k^3}$ converges.

(a) Find an interval of width 1 that contains the sum $\sum_{k=1}^{\infty}\dfrac{1}{k^3}$.

(b) Use the first hundred terms of the series to approximate the sum.

(CAS) *In Problems 71–73, use the Integral Test to determine whether each series converges or diverges.*

71. $\sum_{k=1}^{\infty}\left(k^6 e^{-k}\right)$ **72.** $\sum_{k=1}^{\infty}\dfrac{k+3}{k^2+6k+7}$ **73.** $\sum_{k=2}^{\infty}\dfrac{5k+6}{k^3-1}$

(CAS) **74.** The p-series $\sum_{k=1}^{\infty}\dfrac{1}{k^p}$ converges for $p>1$ and diverges

for $0<p\le 1$. The series $\sum_{k=1}^{\infty}a_k=\sum_{k=1}^{\infty}\dfrac{1}{k^{0.99}}$ diverges and the

series $\sum_{k=1}^{\infty}b_k=\sum_{k=1}^{\infty}\dfrac{1}{k^{1.01}}$ converges.

(a) Find the partial sums S_{10}, S_{1000}, and $S_{100,000}$ for each series.
(b) Explain the results found in (a).

75. Show that the sum of two convergent series is a convergent series.

76. Show that for a nonzero real number c, if $\sum_{k=1}^{\infty}a_k=S$ is a

convergent series, then the series $\sum_{k=1}^{\infty}(ca_k)=cS$.

77. Suppose $\sum_{k=N+1}^{\infty}a_k=S$ and $a_1+a_2+\cdots+a_N=K$.

Prove that $\sum_{k=1}^{\infty}a_k$ converges and its sum is $S+K$.

78. If $\sum_{k=1}^{\infty}a_k$ converges and $\sum_{k=1}^{\infty}b_k$ diverges, then prove

that $\sum_{k=1}^{\infty}(a_k+b_k)$ diverges.

79. Suppose $\sum_{k=1}^{\infty}a_k$ converges, and $a_n>0$ for all n.

Show that $\sum_{k=1}^{\infty}\dfrac{a_k}{1+a_k}$ converges.

Challenge Problems

80. Determine whether the series $\sum_{k=1}^{\infty}\dfrac{1}{k\,\ln\left(1+\dfrac{1}{k}\right)}$ converges.

81. Integral Test Use the Integral Test to show that the

series $\sum_{k=2}^{\infty}\dfrac{1}{(\ln k)^p}$ diverges for all numbers p.

82. For what positive integers p and q does the series $\sum_{k=2}^{\infty}\dfrac{(\ln k)^q}{k^p}$

converge?

83. Find all real numbers x for which $\sum_{k=1}^{\infty}k^x$ converges. Express your answer using interval notation.

84. Consider the finite sum $S_n=\sum_{k=1}^{n}\dfrac{1}{1+k^2}$.

(a) By comparing S_n with an appropriate integral, show that $S_n\le\tan^{-1}n$ for $n\ge 1$.

(b) Use (a) to deduce that $\sum_{k=1}^{\infty}\dfrac{1}{1+k^2}$ converges.

(c) Prove that $\dfrac{\pi}{4}\le\sum_{k=1}^{\infty}\dfrac{1}{1+k^2}\le\dfrac{\pi}{2}$.

85. Riemann's zeta function ζ is defined as

$$\zeta(s)=1+\frac{1}{2^s}+\frac{1}{3^s}+\cdots \qquad \text{for } s>1$$

As mentioned on page 818, Euler showed that $\zeta(2)=\dfrac{\pi^2}{6}$. He also found the value of the zeta function for many other even values of s. As of now, no one knows the value of the zeta function for odd values of s. However, it is not too difficult to approximate these values, as this problem demonstrates.

(a) Find $\sum_{k=1}^{10}\dfrac{1}{k^3}$.

(b) Using integrals in a way analogous to their use in the proof of

the Integral Test, find upper and lower bounds for $\sum_{k=1}^{\infty}\dfrac{1}{k^3}$.

(c) What can you conclude about $\zeta(3)$?

86. (a) By considering graphs like those shown below, show that if f is decreasing, positive, and continuous, then

$$f(n+1)+\cdots+f(m)\le\int_{n}^{m}f(x)\,dx\le f(n)+\cdots+f(m-1)$$

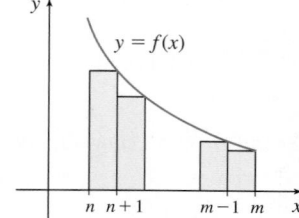
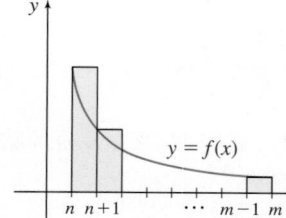

(b) Under the assumption of (a), prove that if $\sum_{k=1}^{\infty}f(k)$

converges, then

$$\sum_{k=n+1}^{\infty}f(k)\le\int_{n}^{\infty}f(x)\,dx\le\sum_{k=n}^{\infty}f(k)$$

(c) Suppose $f(x)=\dfrac{1}{x^2}$. Use the inequality in (b) to determine

exactly how many terms of the series $\sum_{k=1}^{\infty}\dfrac{1}{k^2}$ one must use

to have $|\,\text{Error}\,|<\dfrac{1}{2}\cdot 10^{-2}$. How many terms must one

take to have $|\,\text{Error}\,|<\dfrac{1}{2}\cdot 10^{-10}$?

Source: Based on an article by R. P. Boas, Jr. (April 1977), Partial sums of infinite series, and how they grow, *American Mathematical Monthly,* 84(4), 237–258.

AP® Practice Problems

Multiple-Choice Questions

 1. For what numbers p does the series $\sum\limits_{k=1}^{\infty} \dfrac{1}{k^{p/3}}$ converge?

(A) $p > 1$ (B) $|p| > 1$ (C) $p > 3$ (D) $p > \dfrac{1}{3}$

 2. Which of the following series diverge?

I. $\sum\limits_{k=1}^{\infty} \dfrac{e^{k-1}}{3^{k-1}}$ **II.** $\sum\limits_{k=1}^{\infty} \cos\left(\pi + \dfrac{1}{k}\right)$ **III.** $\sum\limits_{k=1}^{\infty} \dfrac{10}{k}$

(A) I and II only   (B) I and III only
(C) II and III only   (D) I, II, and III

3. For what values of p does the series $\sum\limits_{k=1}^{\infty} k^p$ converge?

(A) $-1 < p < 1$ (B) $p < -1$
(C) $p > 1$ (D) The series diverges.

4. $\sum\limits_{k=1}^{\infty} \dfrac{5^{k-1} - 3^{k-1}}{8^{k-1}} =$

(A) $\dfrac{1}{3}$ (B) $\dfrac{16}{15}$ (C) $\dfrac{49}{24}$ (D) $\dfrac{34}{15}$

5. Which, if any, of the following series converge?

I. $\sum\limits_{k=1}^{\infty} \dfrac{4}{\sqrt[5]{k^2}}$ **II.** $\sum\limits_{k=1}^{\infty} \dfrac{\sqrt{k}+1}{k^3}$ **III.** $\sum\limits_{k=1}^{\infty} \left(\dfrac{8}{k^4} - \dfrac{1}{6^k}\right)$

(A) II only   (B) I and II only
(C) II and III only   (D) I, II, and III

Free-Response Questions

6. (a) Given the infinite series $\sum\limits_{k=3}^{\infty} \dfrac{\ln k}{k}$, find a function f with the property that $f(k) = a_k$ for all positive integers $k \geq 3$.

(b) Show that f is continuous, positive, and decreasing on the interval $[3, \infty)$.

(c) Determine whether the series $\sum\limits_{k=3}^{\infty} \dfrac{\ln k}{k}$ converges or diverges.

7. Show that the infinite series $\sum\limits_{k=1}^{\infty} \dfrac{1}{1 + 9k^2}$ converges.

Retain Your Knowledge

Multiple-Choice Questions

1. $\displaystyle\int \dfrac{2e^x}{1 + 3e^x}\, dx =$

(A) $\dfrac{1}{3} \ln |e^x| + C$ (B) $6 \ln \left|1 + 3e^x\right| + C$

(C) $\dfrac{2}{3} \ln |1 + e^x| + C$ (D) $\dfrac{2}{3} \ln |1 + 3e^x| + C$

2. The general solution to the differential equation $\dfrac{dy}{dx} = \sqrt[3]{x^2 y}$ equals

(A) $y = \dfrac{2x^{5/2}}{5} + C$ (B) $y = \left(\dfrac{3}{5}x^{5/3} + C\right)^{3/2}$

(C) $y = \dfrac{3}{5}x^{5/2} + C$ (D) $y = \left(\dfrac{2x^{5/3}}{5} + C\right)^{3/2}$

3. $\lim\limits_{h \to 0} \dfrac{\tan(x + h) - \tan x}{h} =$

(A) 0 (B) $\sec^2 x$ (C) $\tan^2 x$ (D) $\sec x \cdot \tan x$

Free-Response Question

4. As an object moves along a horizontal line, its velocity is modeled by a differentiable function $v = v(t)$ km/h, $0 \leq t \leq 8$. At time $t = 0$, the object is 1 km to the right of the origin. The table below gives the object's velocity for select times t.

t (hours)	0	2	4	6	8
$v(t)$ (km/h)	20	−10	15	20	35

(a) Show that there is at least one time t, $0 \leq t \leq 8$, when the velocity of the object is 0.

(b) Show that for at least one time t, $4 \leq t \leq 8$, the object's acceleration is 5 km/h².

(c) Approximate $\int_0^8 v(t)\, dt$ using a trapezoidal sum with four subintervals of equal width.

(d) Express the position of the object when $t = 8$ h as an integral.

10.4 Comparison Tests

OBJECTIVES *When you finish this section, you should be able to:*

1 Use Algebraic Comparison Tests for Convergence and Divergence (p. 824)
2 Use the Limit Comparison Test (p. 826)

1 Use Algebraic Comparison Tests for Convergence and Divergence

We can determine whether a series converges or diverges by comparing it to a series whose behavior we already know. Suppose $\sum_{k=1}^{\infty} b_k$ is a series of positive terms that is known to converge to B, and we want to determine whether the series of positive terms $\sum_{k=1}^{\infty} a_k$ converges. If term-by-term $a_k \leq b_k$, that is, if

$$a_1 \leq b_1, \quad a_2 \leq b_2, \ldots, \quad a_n \leq b_n, \ldots$$

then it follows that

$$\left(n\text{th partial sum of } \sum_{k=1}^{\infty} a_k \right) \leq \left(n\text{th partial sum of } \sum_{k=1}^{\infty} b_k \right) < B$$

This shows that the sequence of partial sums $\{S_n\}$ of $\sum_{k=1}^{\infty} a_k$ is bounded, so by the General Convergence Test, $\sum_{k=1}^{\infty} a_k$ must also converge. This proves the *Algebraic Comparison Test for Convergence.*

RECALL The General Convergence Test: An infinite series of positive terms converges if and only if its sequence of partial sums is bounded. (p. 815)

> **THEOREM Algebraic Comparison Test for Convergence**
>
> If $0 < a_k \leq b_k$ for all k and $\sum_{k=1}^{\infty} b_k$ converges, then $\sum_{k=1}^{\infty} a_k$ converges.

It is important to remember that the early terms in a series have no effect on the convergence or divergence of a series. In fact, the Algebraic Comparison Test for Convergence is true if $0 < a_n \leq b_n$ for all $n \geq N$, where N is some suitably selected integer. We use this in the next example by ignoring the first term.

▶ CALC CLIP

EXAMPLE 1 **Using the Algebraic Comparison Test for Convergence**

Show that the series below converges:

$$\sum_{k=1}^{\infty} \frac{1}{k^k} = 1 + \frac{1}{2^2} + \frac{1}{3^3} + \cdots + \frac{1}{n^n} + \cdots$$

Solution

We know that the geometric series $\sum_{k=1}^{\infty} \frac{1}{2^k}$ converges since $|r| = \frac{1}{2} < 1$. Now $\frac{1}{n^n} \leq \frac{1}{2^n}$ for all $n \geq 2$. So, except for the first term, each term of the series $\sum_{k=1}^{\infty} \frac{1}{k^k}$ is less than or equal to the corresponding term of a convergent geometric series. By the Algebraic Comparison Test for Convergence, the series $\sum_{k=1}^{\infty} \frac{1}{k^k}$ converges. ∎

NOTE Do you see why $\frac{1}{n^n} \leq \frac{1}{2^n}$?

$n \geq 2 \Rightarrow n^n \geq 2^n \Rightarrow \frac{1}{n^n} \leq \frac{1}{2^n}$

In Example 1, you may have asked, "How did you know that the given series should be compared to $\sum_{k=1}^{\infty} \frac{1}{2^k}$?" The easy answer is to try a series that you know converges, such as a geometric series with $|r| < 1$ or a p-series with $p > 1$. The honest answer is that only practice and experience will guide you.

There is also an algebraic comparison test for divergence. Suppose $\sum_{k=1}^{\infty} c_k$ is a series of positive terms we know diverges, and $\sum_{k=1}^{\infty} a_k$ is the series of positive terms to be tested. If term-by-term $a_k \geq c_k$, that is, if

$$a_1 \geq c_1, \ a_2 \geq c_2, \ldots, \ a_n \geq c_n, \ldots$$

then it follows that

$$\left(n\text{th partial sum of } \sum_{k=1}^{\infty} a_k \right) \geq \left(n\text{th partial sum of } \sum_{k=1}^{\infty} c_k \right)$$

But we know that the sequence of partial sums of $\sum_{k=1}^{\infty} c_k$ is unbounded. So, the sequence of partial sums of $\sum_{k=1}^{\infty} a_k$ is also unbounded, and $\sum_{k=1}^{\infty} a_k$ diverges.

THEOREM Algebraic Comparison Test for Divergence

If $0 < c_k \leq a_k$ for all k and $\sum_{k=1}^{\infty} c_k$ diverges, then $\sum_{k=1}^{\infty} a_k$ diverges.

EXAMPLE 2   Using the Algebraic Comparison Test for Divergence

Show that the series $\sum_{k=1}^{\infty} \frac{k+3}{k(k+2)}$ diverges.

Solution
Since $\dfrac{n+3}{n+2} > 1$, it follows that

$$\frac{n+3}{n(n+2)} = \frac{n+3}{n+2} \cdot \frac{1}{n} > \frac{1}{n}$$

Term by term, the terms of the series $\sum_{k=1}^{\infty} \frac{k+3}{k(k+2)}$ are greater than the terms of the harmonic series $\sum_{k=1}^{\infty} \frac{1}{k}$, which diverges. It follows from the Algebraic Comparison Test for Divergence that $\sum_{k=1}^{\infty} \frac{k+3}{k(k+2)}$ diverges. ∎

NOW WORK Problems **5, 9,** and AP® Practice Problem **3.**

2 Use the Limit Comparison Test

The Algebraic Comparison Tests for Convergence and Divergence require that certain inequalities hold. Obtaining such inequalities can be challenging. The *Limit Comparison Test* requires certain *conditions* on the limit of a ratio. You may find this comparison test easier to use.

THEOREM Limit Comparison Test

Suppose $\sum_{k=1}^{\infty} a_k$ and $\sum_{k=1}^{\infty} b_k$ are both series of positive terms.

- If $\lim_{n \to \infty} \dfrac{a_n}{b_n} = L, 0 < L < \infty$, then both series converge or both series diverge.

- If $\lim_{n \to \infty} \dfrac{a_n}{b_n} = 0$ and if $\sum_{k=1}^{\infty} b_k$ converges, then $\sum_{k=1}^{\infty} a_k$ converges.

- If $\lim_{n \to \infty} \dfrac{a_n}{b_n} = \infty$ and if $\sum_{k=1}^{\infty} b_k$ diverges, then $\sum_{k=1}^{\infty} a_k$ diverges.

A proof of the Limit Comparison Test is given at the end of the section.

EXAMPLE 3 Using the Limit Comparison Test

Determine whether the series $\sum_{k=1}^{\infty} \dfrac{k+1}{k(k+2)}$ converges or diverges.

Solution

It is tempting to try the same technique that we used in Example 2 and compare $\sum_{k=1}^{\infty} \dfrac{k+1}{k(k+2)}$ to the harmonic series $\sum_{k=1}^{\infty} \dfrac{1}{k}$. Let's see what happens.

$$\text{Since } \frac{k+1}{k+2} < 1, \quad \frac{k+1}{k(k+2)} = \frac{k+1}{k+2} \cdot \frac{1}{k} < \frac{1}{k}$$

The conditions required for the Algebraic Comparison Test for Divergence are not met, so no information about $\sum_{k=1}^{\infty} \dfrac{k+1}{k(k+2)}$ is obtained.

Now use the Limit Comparison Test and compare the series $\sum_{k=1}^{\infty} \dfrac{k+1}{k(k+2)}$ to the harmonic series $\sum_{k=1}^{\infty} \dfrac{1}{k}$. Then with $a_n = \dfrac{n+1}{n(n+2)}$ and $b_n = \dfrac{1}{n}$, we have

$$\lim_{n \to \infty} \frac{a_n}{b_n} = \lim_{n \to \infty} \frac{\dfrac{n+1}{n(n+2)}}{\dfrac{1}{n}} = \lim_{n \to \infty} \frac{n+1}{n+2} = 1$$

Since the limit is a positive number, and $\sum_{k=1}^{\infty} \dfrac{1}{k}$ diverges, by the Limit Comparison Test, $\sum_{k=1}^{\infty} \dfrac{k+1}{k(k+2)}$ also diverges. ∎

NOTE In using the Limit Comparison Test, often the choice of the series to compare can be found by approximating the given series for large k. For example,

$$\frac{k+1}{k(k+2)} \approx \frac{k}{k^2} = \frac{1}{k}$$

for large k.

EXAMPLE 4 **Using the Limit Comparison Test with a p-Series**

Determine whether the series $\displaystyle\sum_{k=1}^{\infty} \frac{1}{2k^{3/2}+5}$ converges or diverges.

Solution

We choose an appropriate p-series to use for comparison by examining the behavior of the series for large values of k:

$$\frac{1}{2k^{3/2}+5} \underset{\underset{\text{for large } k}{\uparrow}}{\approx} \frac{1}{k^{3/2}} \cdot \frac{1}{2}$$

This suggests we choose the p-series $\displaystyle\sum_{k=1}^{\infty} \frac{1}{k^{3/2}}$, which converges, and use the Limit Comparison Test with

$$a_n = \frac{1}{2n^{3/2}+5} \qquad \text{and} \qquad b_n = \frac{1}{n^{3/2}}$$

$$\lim_{n\to\infty} \frac{a_n}{b_n} = \lim_{n\to\infty} \frac{\dfrac{1}{2n^{3/2}+5}}{\dfrac{1}{n^{3/2}}} = \lim_{n\to\infty} \frac{n^{3/2}}{2n^{3/2}+5} = \lim_{n\to\infty} \frac{1}{2+\dfrac{5}{n^{3/2}}} = \frac{1}{2}$$

Since the limit is a positive number and the p-series $\displaystyle\sum_{k=1}^{\infty} \frac{1}{k^{3/2}}$ converges, then by the Limit Comparison Test, $\displaystyle\sum_{k=1}^{\infty} \frac{1}{2k^{3/2}+5}$ also converges. ∎

NOW WORK Problem 15.

EXAMPLE 5 **Using the Limit Comparison Test with a p-Series**

Determine whether the series $\displaystyle\sum_{k=1}^{\infty} \frac{3\sqrt{k}+2}{\sqrt{k^3+3k^2+1}}$ converges or diverges.

Solution

We choose a p-series for comparison by examining how the terms of the series behave for large values of k:

$$\frac{3\sqrt{k}+2}{\sqrt{k^3+3k^2+1}} \approx 3\frac{k^{1/2}}{k^{3/2}} \underset{\underset{\text{for large } k}{\uparrow}}{\approx} \frac{1}{k}$$

So, we compare the series $\displaystyle\sum_{k=1}^{\infty} \frac{3\sqrt{k}+2}{\sqrt{k^3+3k^2+1}}$ to the harmonic series $\displaystyle\sum_{k=1}^{\infty} \frac{1}{k}$, which diverges, and use the Limit Comparison Test with

$$a_n = \frac{3\sqrt{n}+2}{\sqrt{n^3+3n^2+1}} \qquad \text{and} \qquad b_n = \frac{1}{n}$$

$$\lim_{n\to\infty} \frac{a_n}{b_n} = \lim_{n\to\infty} \frac{\dfrac{3\sqrt{n}+2}{\sqrt{n^3+3n^2+1}}}{\dfrac{1}{n}} = \lim_{n\to\infty} \frac{n\left(3\sqrt{n}+2\right)}{\sqrt{n^3+3n^2+1}} = \lim_{n\to\infty} \frac{3n^{3/2}+2n}{\sqrt{n^3+3n^2+1}}$$

$$= \lim_{n\to\infty} \frac{3n^{3/2}}{n^{3/2}} = 3$$

Since the limit is a positive real number and the harmonic series $\displaystyle\sum_{k=1}^{\infty} \frac{1}{k}$ diverges, then by the Limit Comparison Test, $\displaystyle\sum_{k=1}^{\infty} \frac{3\sqrt{k}+2}{\sqrt{k^3+3k^2+1}}$ also diverges. ∎

The Limit Comparison Test is quite versatile for comparing algebraically complex series to a p-series. To find the correct choice of the p-series to use in the comparison, examine the behavior of the terms of the series for large values of k. For example,

- To test $\displaystyle\sum_{k=1}^{\infty} \frac{1}{3k^2 + 5k + 2}$, use the p-series $\displaystyle\sum_{k=1}^{\infty} \frac{1}{k^2}$, because for large k,

$$\frac{1}{3k^2 + 5k + 2} \underset{\substack{\uparrow \\ \text{for large } k}}{\approx} \frac{1}{k^2}$$

- To test $\displaystyle\sum_{k=1}^{\infty} \frac{2k^2 + 5}{3k^3 - 5k^2 + 2}$, use the harmonic series $\displaystyle\sum_{k=1}^{\infty} \frac{1}{k}$, because for large k,

$$\frac{2k^2 + 5}{3k^3 - 5k^2 + 2} \underset{\substack{\uparrow \\ \text{for large } k}}{\approx} \frac{1}{k}$$

- To test $\displaystyle\sum_{k=1}^{\infty} \frac{\sqrt{3k + 1}}{\sqrt{4k^2 - 2k + 1}}$, use the p-series $\displaystyle\sum_{k=1}^{\infty} \frac{1}{k^{1/2}}$, because for large k,

$$\frac{\sqrt{3k + 1}}{\sqrt{4k^2 - 2k + 1}} \underset{\substack{\uparrow \\ \text{for large } k}}{\approx} \frac{1}{\sqrt{k}} = \frac{1}{k^{1/2}}$$

NOW WORK   Problem 17 and AP® Practice Problems 2 and 4.

The geometric series $\displaystyle\sum_{k=1}^{\infty} ar^{k-1}$ is also a valuable series to use when the series to be tested has the index k as an exponent.

EXAMPLE 6   Using the Limit Comparison Test with a Geometric Series

Determine whether the series $\displaystyle\sum_{k=1}^{\infty} \frac{2^k}{3^k + 5}$ converges or diverges.

Solution

The index k of the series $\displaystyle\sum_{k=1}^{\infty} \frac{2^k}{3^k + 5}$ is an exponent. Since for large k,

$$\frac{2^k}{3^k + 5} \approx \frac{2^k}{3^k} = \left(\frac{2}{3}\right)^k$$

compare the series $\displaystyle\sum_{k=1}^{\infty} \frac{2^k}{3^k + 5}$ to the geometric series $\displaystyle\sum_{k=1}^{\infty} \left(\frac{2}{3}\right)^k$, which converges. With

$$a_n = \frac{2^n}{3^n + 5} \quad \text{and} \quad b_n = \left(\frac{2}{3}\right)^n$$

$$\lim_{n \to \infty} \frac{a_n}{b_n} = \lim_{n \to \infty} \frac{\dfrac{2^n}{3^n + 5}}{\left(\dfrac{2}{3}\right)^n} = \lim_{n \to \infty} \left(\frac{2^n}{3^n + 5} \cdot \frac{3^n}{2^n}\right) = \lim_{n \to \infty} \frac{3^n}{3^n + 5} = \lim_{n \to \infty} \frac{1}{1 + \dfrac{5}{3^n}} = 1$$

Since the limit is a positive number and the geometric series $\displaystyle\sum_{k=1}^{\infty} \left(\frac{2}{3}\right)^k$ converges, by the Limit Comparison Test the series $\displaystyle\sum_{k=1}^{\infty} \frac{2^k}{3^k + 5}$ also converges. ∎

NOW WORK   Problem 25 and AP® Practice Problem 1.

Using comparison tests to determine whether a series converges or diverges requires that you know convergent and divergent series to use in the comparison. Table 3 lists some series we have already encountered that are useful for this purpose.

TABLE 3 Summary

Series	Convergent	Divergent
The geometric series $\sum_{k=1}^{\infty} ar^{k-1}$	$\lvert r \rvert < 1$	$\lvert r \rvert \geq 1$
The harmonic series $\sum_{k=1}^{\infty} \frac{1}{k}$		Divergent
The p-series $\sum_{k=1}^{\infty} \frac{1}{k^p}$	$p > 1$	$0 < p \leq 1$
The series $\sum_{k=1}^{\infty} \frac{1}{k^k}$	Convergent	

Proof of the Limit Comparison Test If $\lim_{n \to \infty} \frac{a_n}{b_n} = L > 0$, then for any $\varepsilon > 0$, there is a positive real number N so that $\left\lvert \frac{a_n}{b_n} - L \right\rvert < \varepsilon$, for all $n > N$. If we choose $\varepsilon = \frac{L}{2}$, then

$$\left\lvert \frac{a_n}{b_n} - L \right\rvert < \frac{L}{2} \qquad \text{for all } n > N$$

$$-\frac{L}{2} < \frac{a_n}{b_n} - L < \frac{L}{2} \qquad \text{for all } n > N$$

$$\frac{L}{2} < \frac{a_n}{b_n} < \frac{3L}{2} \qquad \text{for all } n > N$$

Since $b_n > 0$,

$$\frac{L}{2} b_n < a_n < \frac{3L}{2} b_n \qquad \text{for all } n > N$$

RECALL If c is a nonzero real number and if $\sum_{k=1}^{\infty} a_k = S$ is a convergent series, then the series $\sum_{k=1}^{\infty} (ca_k)$ also converges. Furthermore,

$$\sum_{k=1}^{\infty} (ca_k) = c \sum_{k=1}^{\infty} a_k = cS$$

If $\sum_{k=1}^{\infty} a_k$ converges, by the Algebraic Comparison Test for Convergence, the series $\sum_{k=1}^{\infty} \left(\frac{L}{2} b_k \right) = \frac{L}{2} \sum_{k=1}^{\infty} b_k$ also converges. It follows that the series $\sum_{k=1}^{\infty} b_k$ is also convergent.

If $\sum_{k=1}^{\infty} b_k$ converges, then so does $\sum_{k=1}^{\infty} \left(\frac{3L}{2} b_k \right)$. Since $a_n < \frac{3L}{2} b_n$, for all $n > N$, by the Algebraic Comparison Test for Convergence, $\sum_{k=1}^{\infty} a_k$ also converges.

Therefore, $\sum_{k=1}^{\infty} a_k$ and $\sum_{k=1}^{\infty} b_k$ converge together. Since one cannot converge and the other diverge, they must diverge together. ∎

The rest of the proof is left as exercises (Problems 59 and 60).

10.4 Assess Your Understanding

Concepts and Vocabulary

1. Multiple Choice If each term of a series $\sum_{k=1}^{\infty} a_k$ of positive terms is greater than or equal to the corresponding term of a known divergent series $\sum_{k=1}^{\infty} c_k$ of positive terms, then the series $\sum_{k=1}^{\infty} a_k$ is [**(a)** convergent **(b)** divergent].

2. True or False Suppose $\sum_{k=1}^{\infty} a_k$ and $\sum_{k=1}^{\infty} b_k$ are both series of positive terms. The series $\sum_{k=1}^{\infty} a_k$ and the series $\sum_{k=1}^{\infty} b_k$ both converge or both diverge if $\lim_{n \to \infty} \dfrac{a_n}{b_n} = 0$.

3. True or False If $\sum_{k=1}^{\infty} a_k$ and $\sum_{k=1}^{\infty} b_k$ are both series of positive terms and if $\lim_{n \to \infty} \dfrac{a_n}{b_n} = L$, where L is a positive real number, then the series to be tested converges.

4. True or False Since the p-series $\sum_{k=1}^{\infty} \dfrac{1}{k^{3/2}}$ converges and $\lim_{n\to\infty} \dfrac{\dfrac{1}{2n^{3/2}+5}}{\dfrac{1}{n^{3/2}}} = \dfrac{1}{2}$, then by the Limit Comparison Test, the series $\sum_{k=1}^{\infty} \dfrac{1}{2k^{3/2}+5}$ converges to $\dfrac{1}{2}$.

Skill Building

In Problems 5–14, use the Algebraic Comparison Tests for Convergence or Divergence to determine whether each series converges or diverges.

 **5.** $\sum_{k=1}^{\infty} \dfrac{1}{k(k+1)}$: by comparing it with $\sum_{k=1}^{\infty} \dfrac{1}{k^2}$

6. $\sum_{k=1}^{\infty} \dfrac{1}{(k+2)^2}$: by comparing it with $\sum_{k=1}^{\infty} \dfrac{1}{k^2}$

7. $\sum_{k=2}^{\infty} \dfrac{4^k}{7^k+1}$: by comparing it with $\sum_{k=2}^{\infty} \left(\dfrac{4}{7}\right)^k$

8. $\sum_{k=1}^{\infty} \dfrac{1}{(2k-1)(2^k)}$: by comparing it with $\sum_{k=1}^{\infty} \dfrac{1}{2^k}$

 9. $\sum_{k=2}^{\infty} \dfrac{1}{\sqrt{k(k-1)}}$: by comparing it with $\sum_{k=2}^{\infty} \dfrac{1}{k}$

10. $\sum_{k=2}^{\infty} \dfrac{\sqrt{k}}{k-1}$: by comparing it with $\sum_{k=2}^{\infty} \dfrac{1}{\sqrt{k}}$

11. $\sum_{k=1}^{\infty} \dfrac{1}{k(k+1)(k+2)}$

12. $\sum_{k=1}^{\infty} \dfrac{6}{5k-2}$

13. $\sum_{k=1}^{\infty} \dfrac{\sin^2 k}{k^{\pi}}$

14. $\sum_{k=1}^{\infty} \dfrac{\cos^2 k}{k^2+1}$

In Problems 15–36, use the Limit Comparison Test to determine whether each series converges or diverges.

15. $\sum_{k=1}^{\infty} \dfrac{1}{(k+1)(k+2)}$ [PAGE 827]

16. $\sum_{k=1}^{\infty} \dfrac{1}{k^2+1}$

17. $\sum_{k=1}^{\infty} \dfrac{1}{\sqrt{k^2+1}}$ [PAGE 828]

18. $\sum_{k=1}^{\infty} \dfrac{\sqrt{k}}{k+4}$

19. $\sum_{k=1}^{\infty} \dfrac{3\sqrt{k}+2}{2k^2+5}$

20. $\sum_{k=2}^{\infty} \dfrac{3\sqrt{k}+2}{2k-3}$

21. $\sum_{k=2}^{\infty} \dfrac{1}{k\sqrt{k^2-1}}$

22. $\sum_{k=1}^{\infty} \dfrac{k}{(2k-1)^2}$

23. $\sum_{k=1}^{\infty} \dfrac{3^k}{2^k+5}$

24. $\sum_{k=1}^{\infty} \dfrac{8^k}{5+3^k}$

25. $\sum_{k=1}^{\infty} \dfrac{3^k+4}{5^k+3}$ [PAGE 828]

26. $\sum_{k=1}^{\infty} \dfrac{2^k+1}{7^k+4}$

27. $\sum_{k=1}^{\infty} \dfrac{2 \cdot 3^k+5}{4 \cdot 8^k}$

28. $\sum_{k=1}^{\infty} \dfrac{-4 \cdot 5^k}{2 \cdot 6^k}$

29. $\sum_{k=1}^{\infty} \dfrac{8 \cdot 3^k+4 \cdot 5^k}{8^{k-1}}$

30. $\sum_{k=1}^{\infty} \dfrac{2 \cdot 5^k+3^k}{2^{k-1}}$

31. $\sum_{k=1}^{\infty} \dfrac{3k+4}{k2^k}$

32. $\sum_{k=2}^{\infty} \dfrac{k-1}{k2^k}$

33. $\sum_{k=1}^{\infty} \dfrac{1}{2^k+1}$

34. $\sum_{k=1}^{\infty} \dfrac{5}{3^k+2}$

35. $\sum_{k=1}^{\infty} \dfrac{k+5}{k^{k+1}}$

36. $\sum_{k=1}^{\infty} \dfrac{5}{k^k+1}$

In Problems 37–48, use any of the comparison tests to determine whether each series converges or diverges.

37. $\sum_{k=1}^{\infty} \dfrac{6k}{5k^2+2}$

38. $\sum_{k=2}^{\infty} \dfrac{6k+3}{2k^3-2}$

39. $\sum_{k=1}^{\infty} \dfrac{7+k}{(1+k^2)^4}$

40. $\sum_{k=1}^{\infty} \left(\dfrac{7+k}{1+k^2}\right)^4$

41. $\sum_{k=1}^{\infty} \dfrac{e^{1/k}}{k}$

42. $\sum_{k=1}^{\infty} \dfrac{1}{1+e^k}$

43. $\sum_{k=1}^{\infty} \dfrac{\left(1+\dfrac{1}{k}\right)^2}{e^k}$

44. $\sum_{k=1}^{\infty} \dfrac{1}{k \, 2^k}$

45. $\sum_{k=1}^{\infty} \dfrac{1+\sqrt{k}}{k}$

46. $\sum_{k=1}^{\infty} \dfrac{1+3\sqrt{k}}{k^2}$

47. $\sum_{k=1}^{\infty} \left(\dfrac{1}{2}\right)^k \sin^2 k$

48. $\sum_{k=1}^{\infty} \dfrac{\tan^{-1} k}{k^3}$

Applications and Extensions

In Problems 49–56, determine whether each series converges or diverges.

49. $\displaystyle\sum_{k=2}^{\infty} \frac{2}{k^3 \ln k}$

50. $\displaystyle\sum_{k=2}^{\infty} \frac{1}{\sqrt{k}(\ln k)^4}$

51. $\displaystyle\sum_{k=2}^{\infty} \frac{\ln k}{k+3}$

52. $\displaystyle\sum_{k=2}^{\infty} \frac{(\ln k)^2}{k^{5/2}}$

53. $\displaystyle\sum_{k=1}^{\infty} \sin \frac{1}{k}$

54. $\displaystyle\sum_{k=1}^{\infty} \tan \frac{1}{k}$

55. $\displaystyle\sum_{k=1}^{\infty} \frac{1}{k!}$

56. $\displaystyle\sum_{k=1}^{\infty} \frac{k!}{k^k}$

57. It is known that $\displaystyle\sum_{k=1}^{\infty} \frac{1}{k^2}$ is a convergent *p*-series.

   (a) Use the Algebraic Comparison Test for Convergence with $\displaystyle\sum_{k=1}^{\infty} \frac{1}{k^2}$ to show that $\displaystyle\sum_{k=1}^{\infty} \frac{1}{k^2 + 1}$ converges.

   (b) Explain why the Algebraic Comparison Test for Convergence and $\displaystyle\sum_{k=2}^{\infty} \frac{1}{k^2}$ cannot be used to show $\displaystyle\sum_{k=2}^{\infty} \frac{1}{k^2 - 1}$ converges.

   (c) Can the Limit Comparison Test be used to show $\displaystyle\sum_{k=2}^{\infty} \frac{1}{k^2 - 1}$ converges? If it cannot, explain why.

58. Show that any series of the form $\displaystyle\sum_{k=1}^{\infty} \frac{d_k}{10^k}$, where the d_k are digits (0, 1, 2, ..., 9), converges.

59. Suppose the series $\displaystyle\sum_{k=1}^{\infty} a_k$ of positive terms is to be tested for convergence or divergence, and the series $\displaystyle\sum_{k=1}^{\infty} b_k$ of positive terms diverges. Show that if $\displaystyle\lim_{n \to \infty} \frac{a_n}{b_n} = \infty$, then $\displaystyle\sum_{k=1}^{\infty} a_k$ diverges.

60. Suppose the series $\displaystyle\sum_{k=1}^{\infty} a_k$ of positive terms is to be tested for convergence or divergence, and the series $\displaystyle\sum_{k=1}^{\infty} b_k$ of positive terms converges. Show that if $\displaystyle\lim_{n \to \infty} \frac{a_n}{b_n} = 0$, then $\displaystyle\sum_{k=1}^{\infty} a_k$ converges.

In Problems 61–64, use the results of Problems 59 and 60 to determine whether each of the following series converges or diverges.

61. $\displaystyle\sum_{k=2}^{\infty} \frac{1}{\ln k}$

62. $\displaystyle\sum_{k=2}^{\infty} \left(\frac{1}{\ln k} \right)^2$

63. $\displaystyle\sum_{k=2}^{\infty} \frac{\ln k}{k^2}$

64. $\displaystyle\sum_{k=2}^{\infty} \frac{1}{(k \ln k)^2}$

65. (a) Show that $\displaystyle\sum_{k=2}^{\infty} \frac{\ln k}{k^p}$ converges for $p > 1$.

   (b) Show that $\displaystyle\sum_{k=2}^{\infty} \frac{(\ln k)^r}{k^p}$, where r is a positive number, converges for $p > 1$.

66. (a) Show that $\displaystyle\sum_{k=2}^{\infty} \frac{\ln k}{k^p}$ diverges for $0 < p \leq 1$.

   (b) Show that $\displaystyle\sum_{k=2}^{\infty} \frac{(\ln k)^r}{k^p}$, where r is a positive real number and $0 < p \leq 1$, diverges.

In Problems 67–70, use the results of Problems 65 and 66, to determine whether each of the following series converges or diverges.

67. $\displaystyle\sum_{k=2}^{\infty} \frac{\sqrt{\ln k}}{\sqrt{k}}$

68. $\displaystyle\sum_{k=2}^{\infty} \frac{\ln k}{k}$

69. $\displaystyle\sum_{k=2}^{\infty} \frac{(\ln k)^2}{\sqrt{k^3}}$

70. $\displaystyle\sum_{k=2}^{\infty} \frac{\ln k}{k^3}$

71. Use the Algebraic Comparison Tests for Convergence or Divergence to show that the *p*-series $\displaystyle\sum_{k=1}^{\infty} \frac{1}{k^p}$:

   (a) converges if $p > 1$.   **(b)** diverges if $0 < p \leq 1$.

72. If the series $\displaystyle\sum_{k=1}^{\infty} a_k$ of positive terms converges, show that the series $\displaystyle\sum_{k=1}^{\infty} \frac{a_k}{k}$ also converges.

73. Show that $\displaystyle\sum_{k=1}^{\infty} \frac{1}{1 + 2^k}$ converges.

Challenge Problems

In Problems 74 and 75, determine whether each series converges or diverges.

74. $\displaystyle\sum_{k=2}^{\infty} \frac{\ln(2k+1)}{\sqrt{k^2 - 2}\sqrt{k^3 - 2k - 3}}$

75. $\displaystyle\sum_{k=1}^{\infty} \frac{\sqrt{k}}{\sqrt{(k^3 - k + 1)\ln(2k + 1)}}$

76. Show that the series $\displaystyle\sum_{k=1}^{\infty} \frac{1 + \sin k}{4^k}$ converges.

77. Show that the series $\displaystyle\sum_{k=1}^{\infty} \frac{1}{k^{1 + 1/k}}$ diverges.

78. (a) Determine whether the series $\displaystyle\sum_{k=4}^{\infty} \frac{k^2 - 3k - 2}{k^2(k+1)^2}$ converges or diverges.

   (b) If it converges, find its sum.

Preparing for the AP® Exam

AP® Practice Problems

Multiple-Choice Questions

828 1. Which of the following series converge?

I. $\displaystyle\sum_{k=1}^{\infty} \frac{2\pi^k}{3^k \pi}$ II. $\displaystyle\sum_{k=1}^{\infty} \frac{k^2}{2k^3 + 1}$ III. $\displaystyle\sum_{k=1}^{\infty} \frac{k^2 + 3\sqrt[3]{k}}{2k^5}$

(A) I only   (B) III only

(C) I and III only   (D) II and III only

828 2. Which of the following series diverge?

I. $\displaystyle\sum_{k=1}^{\infty} \frac{7k - 5}{k^3}$ II. $\displaystyle\sum_{k=1}^{\infty} \frac{k^2}{2k^3 + 1}$ III. $\displaystyle\sum_{k=1}^{\infty} \frac{k + 3}{(k - 3)^2 + 1}$

(A) II only   (B) III only

(C) I and II only   (D) II and III only

Free-Response Questions

825 3. Determine whether the series $\displaystyle\sum_{k=1}^{\infty} \frac{\sin^2 k}{k^2 + 2k + 1}$ converges or diverges. Justify your answer.

828 4. Determine whether the series $\displaystyle\sum_{k=1}^{\infty} \frac{2k^2 - 1}{k(k^2 + 3)}$ converges or diverges. Justify your answer.

Retain Your Knowledge

Multiple-Choice Questions

1. Find $\displaystyle\lim_{\theta \to 0} \frac{\sin(2\theta)}{\sin(6\theta)}$, if it exists.

(A) $\sin\left(\dfrac{1}{3}\theta\right)$ (B) $\dfrac{1}{3}$

(C) 1   (D) The limit does not exist.

2. At what point on the graph of the function $f(x) = x^3 - 8x$, $x > 0$, is the line tangent to the graph of f parallel to the line $2y - 8x = 1$?

(A) $(2, -8)$ (B) $(2, 4)$

(C) $\left(2, \dfrac{17}{2}\right)$ (D) $(4, 32)$

3. The arc length of the graph of $f(x) = \ln x^2$ from $x = 1$ to $x = 5$ is given by

(A) $\displaystyle\int_1^5 \sqrt{1 + \left(\frac{2}{x}\right)^2}\, dx$ (B) $\displaystyle\int_1^5 \sqrt{1 - \left(\frac{2}{x}\right)^2}\, dx$

(C) $\displaystyle\int_1^5 \sqrt{1 + \frac{2}{x}}\, dx$ (D) $\displaystyle\int_1^5 \sqrt{1 + \left(\frac{1}{x^2}\right)^2}\, dx$

Free-Response Question

4. (a) Sketch a slope field for the differential equation $\dfrac{dy}{dx} = x - y$ on the grid below.

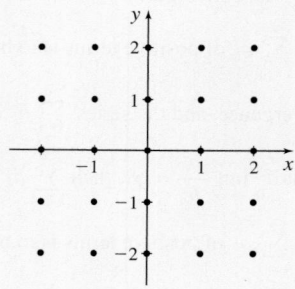

(b) Suppose $y = f(x)$ is the particular solution to the differential equation with the initial condition $f(0) = -2$. Use Euler's method starting at $x = 0$ with two steps of equal size to approximate the solution to $\dfrac{dy}{dx} = x - y$ at $x = 1$.

10.5 Alternating Series; Absolute and Conditional Convergence

OBJECTIVES *When you finish this section, you should be able to:*

1 Determine whether an alternating series converges or diverges (p. 833)
2 Approximate the sum of a convergent alternating series (p. 836)
3 Use the Absolute Convergence Test (p. 838)

Sections 10.3 and 10.4 deal with series whose terms are all positive. We now investigate series that have some positive and some negative terms. The properties of series with positive terms often help to explain the properties of series that include terms with mixed signs.

The most common series with both positive and negative terms are *alternating series*, in which the terms alternate between positive and negative.

DEFINITION Alternating Series

An **alternating series** is a series either of the form

$$\sum_{k=1}^{\infty}(-1)^{k+1}a_k = a_1 - a_2 + a_3 - a_4 + \cdots$$

or of the form

$$\sum_{k=1}^{\infty}(-1)^{k}a_k = -a_1 + a_2 - a_3 + a_4 - \cdots$$

where $a_k > 0$ for all integers $k \geq 1$.

Two examples of alternating series are

$$1 - \frac{1}{2} + \frac{1}{3} - \frac{1}{4} + \cdots = \sum_{k=1}^{\infty}\frac{(-1)^{k+1}}{k} \quad \text{and} \quad -1 + \frac{1}{3!} - \frac{1}{5!} + \frac{1}{7!} - \cdots = \sum_{k=1}^{\infty}\frac{(-1)^{k}}{(2k-1)!}$$

1 Determine Whether an Alternating Series Converges or Diverges

To determine whether an alternating series converges, we use the *Alternating Series Test*.

THEOREM Alternating Series Test

If the numbers a_k, where $a_k > 0$, of an alternating series

$$\sum_{k=1}^{\infty}(-1)^{k+1}a_k \quad \text{or} \quad \sum_{k=1}^{\infty}(-1)^{k}a_k$$

satisfy the two conditions:

- $\lim_{n \to \infty} a_n = 0$
- the a_k are nonincreasing; that is, $a_1 \geq a_2 \geq a_3 \geq \cdots \geq a_n \geq a_{n+1} \geq \cdots$

then the alternating series converges.

NOTE The Alternating Series Test is credited to Leibniz and is often called the **Leibniz Test** or **Leibniz Criterion**.

The Alternating Series Test also works if the a_k are *eventually* nonincreasing.

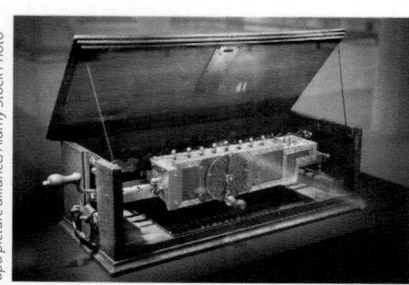

ORIGINS Gottfried Wilhelm von Leibniz (1646–1716) was a German mathematician, philosopher, and inventor. To calculus students, Leibniz is most famous for the derivative notation $\dfrac{dy}{dx}$ and integral notation $\int f(x)\, dx$. He also developed the binary system of mathematics on which modern computers are based, and invented the first mechanical calculator (shown here) that could perform addition, subtraction, multiplication and division quickly and accurately—he called it an arithmetic machine, or a stepped reckoner.

Proof For the alternating series $\sum\limits_{k=1}^{\infty} (-1)^{k+1} a_k$, $a_k > 0$, consider the partial sums with an even number of terms, and group them into pairs:

$$S_{2n} = a_1 - a_2 + a_3 - a_4 + \cdots + a_{2n-1} - a_{2n}$$
$$= (a_1 - a_2) + (a_3 - a_4) + \cdots + (a_{2n-1} - a_{2n})$$

Since $a_1 \geq a_2 \geq a_3 \geq a_4 \geq \cdots \geq a_{2n-1} \geq a_{2n} \geq a_{2n+1} \geq \cdots$, the difference within each pair of parentheses is either 0 or positive. So, the sequence $\{S_{2n}\}$ of partial sums is nondecreasing. That is,

$$S_2 \leq S_4 \leq S_6 \leq \cdots \leq S_{2n} \leq S_{2n+2} \leq \cdots \tag{1}$$

For partial sums with an odd number of terms, group the terms a little differently, and write

$$S_{2n+1} = a_1 - a_2 + a_3 - a_4 + a_5 - \cdots - a_{2n} + a_{2n+1}$$
$$= a_1 - (a_2 - a_3) - (a_4 - a_5) - \cdots - (a_{2n} - a_{2n+1})$$

Again, the difference within each pair of parentheses is either 0 or positive. But this time the difference is subtracted from the previous term, so the sequence $\{S_{2n+1}\}$ of partial sums is nonincreasing. That is,

$$S_1 \geq S_3 \geq S_5 \geq \cdots \geq S_{2n-1} \geq S_{2n+1} \geq \cdots \tag{2}$$

Now since $S_{2n+1} - S_{2n} = a_{2n+1} > 0$, we have

$$S_{2n} < S_{2n+1} \tag{3}$$

Combine (1), (2), and (3) as follows:

$$\underbrace{S_2 \leq S_4 \leq S_6 \leq \cdots \leq S_{2n}}_{(1)} < \underbrace{S_{2n+1}}_{(3)} \leq \underbrace{\cdots \leq S_5 \leq S_3 \leq S_1}_{(2)}$$

Figure 23 shows the sequences $\{S_{2n}\}$ and $\{S_{2n+1}\}$.

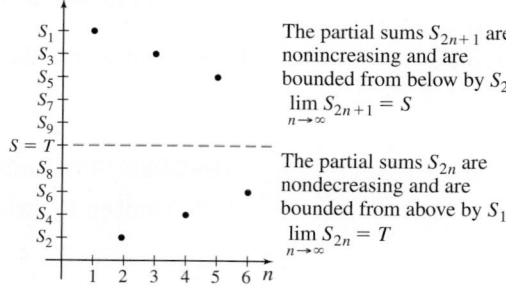

The partial sums S_{2n+1} are nonincreasing and are bounded from below by S_2.
$$\lim_{n \to \infty} S_{2n+1} = S$$

The partial sums S_{2n} are nondecreasing and are bounded from above by S_1.
$$\lim_{n \to \infty} S_{2n} = T$$

Figure 23

RECALL A bounded, monotonic sequence converges.

The sequence S_2, S_4, S_6, ... is nondecreasing and bounded above by S_1. So, this sequence has a limit; call it T. Similarly, the sequence S_1, S_3, S_5, ... is nonincreasing and is bounded from below by S_2, so it, too, has a limit; call it S. Since $S_{2n+1} - S_{2n} = a_{2n+1}$, we have

$$S - T = \lim_{n \to \infty} S_{2n+1} - \lim_{n \to \infty} S_{2n} = \lim_{n \to \infty} (S_{2n+1} - S_{2n}) = \lim_{n \to \infty} a_{2n+1}$$

Since $\lim_{n \to \infty} a_{2n+1} = 0$, then $S - T = 0$, so $S = T$. The sequences $\{S_{2n}\}$ and $\{S_{2n+1}\}$ both converge to S. It follows that

$$\lim_{n \to \infty} S_n = S$$

and so the alternating series converges. ∎

EXAMPLE 1   **Showing the Alternating Harmonic Series Converges**

Show that the **alternating harmonic series** $\sum\limits_{k=1}^{\infty} \dfrac{(-1)^{k+1}}{k} = 1 - \dfrac{1}{2} + \dfrac{1}{3} - \dfrac{1}{4} + \cdots$ converges.

Solution

NOTE In Section 10.8, we show that the sum of the alternating harmonic series is **ln 2**.

We check the two conditions of the Alternating Series Test. Since $\lim\limits_{n \to \infty} a_n = \lim\limits_{n \to \infty} \dfrac{1}{n} = 0$, the first condition is met. Since $a_{n+1} = \dfrac{1}{n+1} < \dfrac{1}{n} = a_n$, the a_n are nonincreasing, so the second condition is met. By the Alternating Series Test, the series converges. ∎

NOW WORK   Problem 7.

When using the Alternating Series Test, check the limit of the nth term first. If $\lim\limits_{n \to \infty} a_n \neq 0$, the nth Term Test for Divergence tells us immediately that the series diverges. For example, in testing the alternating series

$$\sum_{k=2}^{\infty} (-1)^k \frac{k}{k-1} = 2 - \frac{3}{2} + \frac{4}{3} - \frac{5}{4} + \cdots$$

we find $\lim\limits_{n \to \infty} a_n = \lim\limits_{n \to \infty} \dfrac{n}{n-1} = 1$. There is no need to look further. By the nth Term Test for Divergence, the series diverges.

EXAMPLE 2   **Using the Alternating Series Test**

Determine whether the series $\sum\limits_{k=1}^{\infty} (-1)^k \dfrac{2k^2 + 9}{k^2 + 1}$ converges or diverges.

Solution

The series $\sum\limits_{k=1}^{\infty} (-1)^k \dfrac{2k^2 + 9}{k^2 + 1}$ is an alternating series, where

$$a_n = \frac{2n^2 + 9}{n^2 + 1}$$

We check $\lim\limits_{n \to \infty} a_n$.

$$\lim_{n \to \infty} a_n = \lim_{n \to \infty} \frac{2n^2 + 9}{n^2 + 1} = \lim_{n \to \infty} \frac{2 + \dfrac{9}{n^2}}{1 + \dfrac{1}{n^2}} = \frac{2}{1} = 2$$

Since $\lim\limits_{n \to \infty} a_n \neq 0$, by the nth Term Test for Divergence, the series diverges. ∎

NOW WORK   Problem 9.

For the alternating series $\sum_{k=1}^{\infty} (-1)^{k+1} a_k$, $a_k > 0$, suppose $\lim_{n \to \infty} a_n = 0$. To continue with the Alternating Series Test, check whether the terms a_k are nonincreasing.

Three Ways to Verify That the a_k Are Nonincreasing

- Use the Algebraic Difference test and show that $a_{n+1} - a_n \leq 0$ for all $n \geq 1$.

- Use the Algebraic Ratio test and show that $\dfrac{a_{n+1}}{a_n} \leq 1$ for all $n \geq 1$.

- Use the derivative of the related function f, defined for $x > 0$ and for which $a_n = f(n)$ for all n, and show that $f'(x) \leq 0$ for all $x > 0$.

NOTE If $\lim_{n \to \infty} a_n = 0$, for an alternating series and if the terms are increasing, the Alternating Series Test gives no information. In these cases, the alternating series may converge (Problem 61) or diverge (Problem 62).

EXAMPLE 3 Showing an Alternating Series Converges

Show that the alternating series $\sum_{k=0}^{\infty} \dfrac{(-1)^k}{(2k)!} = 1 - \dfrac{1}{2} + \dfrac{1}{24} - \dfrac{1}{720} + \cdots$ converges.

Solution

First, $\lim_{n \to \infty} a_n = \lim_{n \to \infty} \dfrac{1}{(2n)!} = 0$, so the first condition is met. Next, using the Algebraic Ratio test, we verify that the terms $a_k = \dfrac{1}{(2k)!}$ are nonincreasing. Since

$$\frac{a_{n+1}}{a_n} = \frac{\dfrac{1}{[2(n+1)]!}}{\dfrac{1}{(2n)!}} = \frac{(2n)!}{(2n+2)!} = \frac{(2n)!}{(2n+2)(2n+1)(2n)!}$$

$$= \frac{1}{(2n+2)(2n+1)} < 1 \qquad \text{for all } n \geq 1$$

the terms a_k are nonincreasing. By the Alternating Series Test, the series converges. ∎

NOW WORK   Problem 13 and AP® Practice Problems 1, 5(a), and 5(b).

---

❷ Approximate the Sum of a Convergent Alternating Series

An important property of a convergent alternating series that satisfies the two conditions of the Alternating Series Test is that the error made by using a partial sum S_n to approximate the sum S of the series is no larger than the $(n+1)$st term of the series.

THEOREM Error Bound for a Convergent Alternating Series

For a convergent alternating series $\sum_{k=1}^{\infty} (-1)^{k+1} a_k$ that satisfies the two conditions of the Alternating Series Test, the error E_n in using the sum S_n of the first n terms as an approximation to the sum S of the series is numerically less than or equal to the $(n+1)$st term of the series. That is,

$$\boxed{|E_n| \leq a_{n+1}}$$

NOTE The error bound given here is only applicable to alternating series that satisfy the two conditions required in the Alternating Series Test.

Proof We follow the ideas used in the proof of the Alternating Series Test.

If n is even, $S_n < S$, and

$$0 < S - S_n = a_{n+1} - (a_{n+2} - a_{n+3}) - \cdots \leq a_{n+1}$$

If n is odd, $S_n > S$, and

$$0 < S_n - S = a_{n+1} - (a_{n+2} - a_{n+3}) - \cdots \leq a_{n+1}$$

In either case,

$$|E_n| = |S - S_n| \leq a_{n+1} \qquad \blacksquare$$

In Example 1, we found that the alternating harmonic series $\displaystyle\sum_{k=1}^{\infty} \frac{(-1)^{k+1}}{k}$ converges. Table 4 shows how slowly the series converges. For example, using an additional 900 terms from 99 to 999 reduces the estimated error by only $|E_{999} - E_{99}| = 0.009$.

TABLE 4

Series	Number of Terms	Estimate of the Sum	Bound Error
$\displaystyle\sum_{k=1}^{\infty} \frac{(-1)^{k+1}}{k}$	$n = 3$	$S_3 = 1 - \dfrac{1}{2} + \dfrac{1}{3} \approx 0.833$	$\|E_3\| \leq \dfrac{1}{4} = 0.25$
	$n = 9$	$S_9 = \displaystyle\sum_{k=1}^{9} \frac{(-1)^{k+1}}{k} \approx 0.746$	$\|E_9\| \leq \dfrac{1}{10} = 0.1$
	$n = 99$	$S_{99} = \displaystyle\sum_{k=1}^{99} \frac{(-1)^{k+1}}{k} \approx 0.698$	$\|E_{99}\| \leq \dfrac{1}{100} = 0.01$
	$n = 999$	$S_{999} = \displaystyle\sum_{k=1}^{999} \frac{(-1)^{k+1}}{k} \approx 0.694$	$\|E_{999}\| \leq \dfrac{1}{1000} = 0.001$

EXAMPLE 4 **Approximating the Sum of a Convergent Alternating Series**

Approximate the sum S of the alternating series

$$\sum_{k=0}^{\infty} \frac{(-1)^k}{(2k)!} = 1 - \frac{1}{2!} + \frac{1}{4!} - \frac{1}{6!} + \frac{1}{8!} - \cdots$$

so that the error is less than or equal to 0.0001.

Solution

We proved in Example 3 that the series $\displaystyle\sum_{k=0}^{\infty} \frac{(-1)^k}{(2k)!}$ converges by showing that it satisfies the conditions of the Alternating Series Test. The fifth term of the series, $\dfrac{1}{8!} = \dfrac{1}{40{,}320} \approx 0.000025$, is the first term less than or equal to 0.0001. This term represents an upper bound to the error when the sum S of the series is approximated by adding the first four terms. So,

$$S \approx \sum_{k=0}^{3} \frac{(-1)^k}{(2k)!} = 1 - \frac{1}{2!} + \frac{1}{4!} - \frac{1}{6!} = 1 - \frac{1}{2} + \frac{1}{24} - \frac{1}{720} = \frac{389}{720} \approx 0.540$$

approximates the sum with an error less than or equal to 0.0001. $\blacksquare$

NOW WORK Problems 23, 29, and AP® Practice Problems 2, 3, and 5(c).

Absolute and Conditional Convergence

The ideas of *absolute convergence* and *conditional convergence* are used to describe the convergence or divergence of a series $\sum_{k=1}^{\infty} a_k$ in which the terms a_k are sometimes positive and sometimes negative (not necessarily alternating).

> **DEFINITION** Absolute Convergence
>
> A series $\sum_{k=1}^{\infty} a_k$ is **absolutely convergent** if the series
>
> $$\sum_{k=1}^{\infty} |a_k| = |a_1| + |a_2| + \cdots + |a_n| + \cdots$$
>
> is convergent.

③ Use the Absolute Convergence Test

> **THEOREM** Absolute Convergence Test
>
> If a series $\sum_{k=1}^{\infty} a_k$ is absolutely convergent, then it is convergent.

Proof For any n,

$$-|a_n| \le a_n \le |a_n|$$

Adding $|a_n|$ yields

$$0 \le a_n + |a_n| \le 2|a_n|$$

Since $\sum_{k=1}^{\infty} |a_k|$ converges, then $\sum_{k=1}^{\infty} (2|a_k|) = 2\sum_{k=1}^{\infty} |a_k|$ also converges. By the Algebraic Comparison Test for Convergence, $\sum_{k=1}^{\infty} (a_k + |a_k|)$ converges. But $a_n = (a_n + |a_n|) - |a_n|$. Since $\sum_{k=1}^{\infty} a_k$ is the difference of two convergent series, it also converges. ∎

EXAMPLE 5   Using the Absolute Convergence Test

Determine whether the series $1 - \dfrac{1}{2} - \dfrac{1}{4} + \dfrac{1}{8} - \dfrac{1}{16} - \dfrac{1}{32} + \dfrac{1}{64} - \cdots$ converges.

Solution

First note that the given series is not an alternating series, so the Alternating Series Test cannot be used. Next, the series of the absolute values of each term,

$$1 + \frac{1}{2} + \frac{1}{4} + \cdots + \frac{1}{2^{n-1}} + \cdots = \sum_{k=1}^{\infty} \left(\frac{1}{2}\right)^{k-1}$$

is a geometric series with $r = \dfrac{1}{2}$, which converges. So by the Absolute Convergence Test, the series $1 - \dfrac{1}{2} - \dfrac{1}{4} + \dfrac{1}{8} - \dfrac{1}{16} - \dfrac{1}{32} + \dfrac{1}{64} - \cdots$ converges. ∎

NOW WORK   Problem 37.

EXAMPLE 6 **Using the Absolute Convergence Test**

Determine whether the series $\displaystyle\sum_{k=1}^{\infty} \frac{\sin k}{k^2} = \frac{\sin 1}{1^2} + \frac{\sin 2}{2^2} + \frac{\sin 3}{3^2} + \cdots$ converges.

Solution

This series has both positive and negative terms, but it is not an alternating series. We investigate the series of the absolute values of each term $\displaystyle\sum_{k=1}^{\infty} \left| \frac{\sin k}{k^2} \right|$. Since $|\sin n| \le 1$,

$$\left| \frac{\sin n}{n^2} \right| \le \frac{1}{n^2}$$

for all n, and since $\displaystyle\sum_{k=1}^{\infty} \frac{1}{k^2}$ is a convergent p-series, then by the Algebraic Comparison Test for Convergence, the series $\displaystyle\sum_{k=1}^{\infty} \left| \frac{\sin k}{k^2} \right|$ converges. Since $\displaystyle\sum_{k=1}^{\infty} \frac{\sin k}{k^2}$ is absolutely convergent, it follows that $\displaystyle\sum_{k=1}^{\infty} \frac{\sin k}{k^2}$ is convergent. ∎

NOW WORK Problem **39** and AP® Practice Problem **1**.

The converse of the Absolute Convergence Test is not true. That is, if a series is convergent, the series may or may not be absolutely convergent. There are convergent series that are not absolutely convergent. For instance, we have shown that the alternating harmonic series

$$\sum_{k=1}^{\infty} \frac{(-1)^{k+1}}{k} = 1 - \frac{1}{2} + \frac{1}{3} - \frac{1}{4} + \frac{1}{5} - \cdots$$

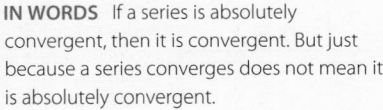

IN WORDS If a series is absolutely convergent, then it is convergent. But just because a series converges does not mean it is absolutely convergent.

converges. The series of absolute values is the harmonic series $\displaystyle\sum_{k=1}^{\infty} \frac{1}{k}$, which is divergent.

DEFINITION Conditional Convergence

A series that is convergent without being absolutely convergent is called **conditionally convergent**.

CALC CLIP

EXAMPLE 7 **Determining Whether a Series Is Absolutely Convergent, Conditionally Convergent, or Divergent**

Determine the numbers p for which the series $\displaystyle\sum_{k=1}^{\infty} \frac{(-1)^{k+1}}{k^p}$ is absolutely convergent, conditionally convergent, or divergent.

Solution

We begin by testing the series for absolute convergence. The series of absolute values is $\displaystyle\sum_{k=1}^{\infty} \left| \frac{(-1)^{k+1}}{k^p} \right| = \sum_{k=1}^{\infty} \frac{1}{k^p}$. This is a p-series, which converges if $p > 1$ and diverges if $p \le 1$. So, $\displaystyle\sum_{k=1}^{\infty} \frac{(-1)^{k+1}}{k^p}$ is absolutely convergent if $p > 1$.

(Example continued on the next page)

It remains to determine what happens when $p \le 1$. We use the Alternating Series Test, and begin by investigating $\lim\limits_{n \to \infty} a_n$:

- $\lim\limits_{n \to \infty} \dfrac{1}{n^p} = 0 \quad 0 < p \le 1$ • $\lim\limits_{n \to \infty} \dfrac{1}{n^p} = 1 \quad p = 0$ • $\lim\limits_{n \to \infty} \dfrac{1}{n^p} = \infty \quad p < 0$

By the nth Term Test for Divergence, $\sum\limits_{k=1}^{\infty} \dfrac{(-1)^{k+1}}{k^p}$ diverges if $p \le 0$.

Continuing, we check the second condition of the Alternating Series Test when $0 < p \le 1$. Using the related function $f(x) = \dfrac{1}{x^p}$, for $x > 0$, we have $f'(x) = -\dfrac{p}{x^{p+1}}$. Since $f'(x) < 0$ for $0 < p \le 1$, the second condition of the Alternating Series Test is satisfied. We conclude that $\sum\limits_{k=1}^{\infty} \dfrac{(-1)^{k+1}}{k^p}$ is conditionally convergent if $0 < p \le 1$. ∎

To summarize, the alternating p-series, $\sum\limits_{k=1}^{\infty} \dfrac{(-1)^{k+1}}{k^p}$ is absolutely convergent if $p > 1$, conditionally convergent if $0 < p \le 1$, and divergent if $p \le 0$.

NOW WORK   Problem **43** and AP® Practice Problems **4** and **6**.

As illustrated in Figure 24, a series $\sum\limits_{k=1}^{\infty} a_k$ is either convergent or divergent. If it is convergent, it is either absolutely convergent or conditionally convergent. The flowchart in Figure 25 can be used to determine whether a series is absolutely convergent, conditionally convergent, or divergent.

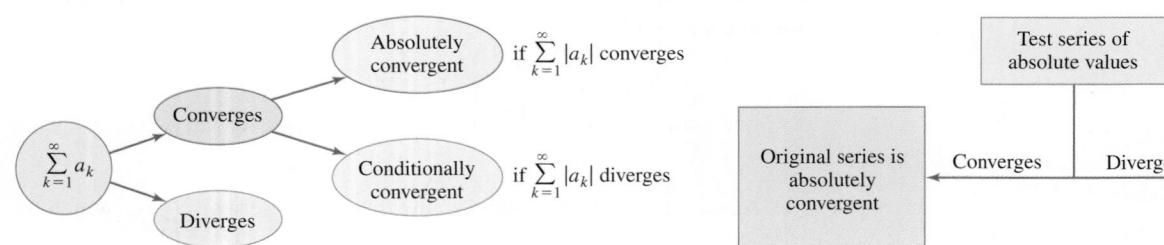

Figure 24   **Figure 25**

Absolute convergence of a series is stronger than conditional convergence, and absolutely convergent series and conditionally convergent series have different properties. We state, but do not prove, some of these properties now.

Properties of Absolutely Convergent and Conditionally Convergent Series

- If a series $\sum\limits_{k=1}^{\infty} a_k$ is absolutely convergent, any rearrangement of its terms results in a series that is also absolutely convergent. Both series will have the same sum. See Problem 78.

- If two series $\sum\limits_{k=1}^{\infty} a_k$ and $\sum\limits_{k=1}^{\infty} b_k$ are absolutely convergent, the series $\sum\limits_{k=1}^{\infty} (a_k + b_k)$ is also absolutely convergent. Moreover,

$$\sum_{k=1}^{\infty} (a_k + b_k) = \sum_{k=1}^{\infty} a_k + \sum_{k=1}^{\infty} b_k$$

See Problem 74.

(Continued on the next page)

- If two series $\sum_{k=1}^{\infty} a_k$ and $\sum_{k=1}^{\infty} b_k$ are absolutely convergent, then the series

$$\sum_{k=1}^{\infty} c_k = \sum_{k=1}^{\infty} (a_1 b_k + a_2 b_{k-1} + \cdots + a_k b_1)$$

is also absolutely convergent. The sum of the series $\sum_{k=1}^{\infty} c_k$ equals the product of the sums of the two original series.

- If a series is absolutely convergent, then the series consisting of just the positive terms converges, as does the series consisting of just the negative terms. See Problem 75.

- If a series is conditionally convergent, then the series consisting of just the positive terms diverges, as does the series consisting of just the negative terms. See Problems 63, 64, and 76.

- If a series $\sum_{k=1}^{\infty} a_k$ is conditionally convergent, its terms may be rearranged to form a new series that converges to *any* sum we like. In fact, they may be rearranged to form a *divergent series*. See Problems 65–67.

10.5 Assess Your Understanding

Concepts and Vocabulary

1. True or False The series $\sum_{k=1}^{\infty} (-1)^k \cos(k\pi)$ is an alternating series.

2. True or False $\sum_{k=1}^{\infty} [1 + (-1)^k]$ is an alternating series.

3. True or False In an alternating series, $\sum_{k=1}^{\infty} (-1)^k a_k$, $a_k > 0$, if $\lim_{n \to \infty} a_n = 0$, then the series is convergent.

4. True or False If the alternating series $\sum_{k=1}^{\infty} (-1)^{k+1} a_k$ satisfies the two conditions of the Alternating Series Test, then the error E_n in using the sum S_n of the first n terms as an approximation of the sum S of the series is $|E_n| \le a_n$.

5. True or False A series that is not absolutely convergent is divergent.

6. True or False If a series is absolutely convergent, then the series converges.

Skill Building

In Problems 7–22, use the Alternating Series Test to determine whether each alternating series converges or diverges.

 7. $\sum_{k=1}^{\infty} (-1)^{k+1} \dfrac{1}{k^2}$

8. $\sum_{k=1}^{\infty} (-1)^{k+1} \dfrac{1}{2\sqrt{k}}$

 9. $\sum_{k=1}^{\infty} (-1)^{k+1} \dfrac{k}{2k+1}$

10. $\sum_{k=1}^{\infty} (-1)^{k+1} \dfrac{k+1}{k}$

11. $\sum_{k=1}^{\infty} (-1)^{k+1} \dfrac{k^2}{5k^2+2}$

12. $\sum_{k=1}^{\infty} (-1)^{k+1} \dfrac{k+1}{k^2}$

 13. $\sum_{k=1}^{\infty} \dfrac{(-1)^{k+1}}{(k+1)2^k}$

14. $\sum_{k=2}^{\infty} (-1)^k \dfrac{1}{k \ln k}$

15. $\sum_{k=2}^{\infty} (-1)^k \dfrac{1}{1+2^{-k}}$

16. $\sum_{k=0}^{\infty} (-1)^k \dfrac{1}{k!}$

17. $\sum_{k=1}^{\infty} (-1)^{k+1} \left(\dfrac{k}{k+1} \right)^k$

18. $\sum_{k=1}^{\infty} (-1)^{k+1} \dfrac{k^2}{(k+1)^3}$

19. $\sum_{k=1}^{\infty} (-1)^k e^{-k}$

20. $\sum_{k=1}^{\infty} (-1)^k k e^{-k}$

21. $\sum_{k=1}^{\infty} (-1)^k \tan^{-1} k$

22. $\sum_{k=1}^{\infty} (-1)^k e^{2/k}$

In Problems 23–28:

(a) *Show that each alternating series satisfies the conditions of the Alternating Series Test.*

(b) *How many terms must be added so the error in using the sum S_n of the first n terms as an approximation to the sum S of the series is less than or equal to 0.001?*

(c) *Approximate the sum S of the series so that the error is less than or equal to 0.001.*

 **23.** $\sum_{k=1}^{\infty} \dfrac{(-1)^{k+1}}{k^5}$

24. $\sum_{k=1}^{\infty} \dfrac{(-1)^{k+1}}{k^4}$

25. $\sum_{k=1}^{\infty} \dfrac{(-1)^{k+1}}{k!}$

26. $\sum_{k=1}^{\infty} \dfrac{(-1)^{k+1}}{k^k}$

27. $\sum_{k=1}^{\infty} \dfrac{(-1)^{k+1}}{2^k}$

28. $\sum_{k=1}^{\infty} \dfrac{(-1)^{k+1}}{3^k}$

In Problems 29–36, approximate the sum of each series using the first three terms and find a bound to the error in using this approximation.

29. $\displaystyle\sum_{k=1}^{\infty}(-1)^{k+1}\frac{1}{k^2}$ $\qquad$ **30.** $\displaystyle\sum_{k=0}^{\infty}(-1)^{k}\frac{1}{k!}$

31. $\displaystyle\sum_{k=1}^{\infty}(-1)^{k+1}\frac{1}{k^4}$ $\qquad$ **32.** $\displaystyle\sum_{k=1}^{\infty}(-1)^{k+1}\left(\frac{1}{\sqrt{k}}\right)^{k}$

33. $\displaystyle\sum_{k=0}^{\infty}(-1)^{k}\frac{1}{k!}\left(\frac{1}{3}\right)^{k}$ $\qquad$ **34.** $\displaystyle\sum_{k=0}^{\infty}(-1)^{k}\frac{1}{k!}\left(\frac{1}{2}\right)^{k}$

35. $\displaystyle\sum_{k=0}^{\infty}(-1)^{k}\frac{1}{2k+1}\left(\frac{1}{3}\right)^{2k+1}$ $\qquad$ **36.** $\displaystyle\sum_{k=1}^{\infty}(-1)^{k+1}\frac{1}{k^k}$

In Problems 37–40, use the Absolute Convergence Test to show that each series converges.

37. $\displaystyle\sum_{k=1}^{\infty}\frac{(-1)^{k+1}}{k^2}$ $\qquad$ **38.** $\displaystyle\sum_{k=1}^{\infty}\frac{(-1)^{k+1}}{k^3}$

39. $\displaystyle\sum_{k=1}^{\infty}(-1)^{k+1}\frac{\sin k}{k^2+1}$ $\qquad$ **40.** $\displaystyle\sum_{k=1}^{\infty}(-1)^{k+1}\frac{\cos k}{k^2}$

In Problems 41–52, determine whether each series is absolutely convergent, conditionally convergent, or divergent.

41. $\displaystyle\sum_{k=1}^{\infty}(-1)^{k+1}\left(\frac{1}{5}\right)^{k}$ $\qquad$ **42.** $\displaystyle\sum_{k=1}^{\infty}(-1)^{k+1}\frac{5^k}{6^{k+1}}$

43. $\displaystyle\sum_{k=1}^{\infty}(-1)^{k+1}\frac{e^k}{k}$ $\qquad$ **44.** $\displaystyle\sum_{k=1}^{\infty}\frac{(-1)^{k+1}2^k}{k^2}$

45. $\displaystyle\sum_{k=1}^{\infty}\frac{(-1)^{k+1}}{k(k+1)}$ $\qquad$ **46.** $\displaystyle\sum_{k=1}^{\infty}\frac{(-1)^{k+1}}{k\sqrt{k+3}}$

47. $\displaystyle\sum_{k=1}^{\infty}\frac{(-1)^{k+1}\sqrt{k}}{k^2+1}$ $\qquad$ **48.** $\displaystyle\sum_{k=1}^{\infty}\frac{(-1)^{k+1}\sqrt{k}}{k+1}$

49. $\displaystyle\sum_{k=2}^{\infty}(-1)^{k+1}\frac{\ln k}{k}$ $\qquad$ **50.** $\displaystyle\sum_{k=2}^{\infty}\frac{(-1)^{k+1}\ln k}{k^3}$

51. $\displaystyle\sum_{k=1}^{\infty}(-1)^{k+1}\frac{1}{k\,e^k}$ $\qquad$ **52.** $\displaystyle\sum_{k=1}^{\infty}\frac{(-1)^{k+1}}{e^k}$

Applications and Extensions

In Problems 53–60, determine whether each series is absolutely convergent, conditionally convergent, or divergent.

53. $\displaystyle\sum_{k=2}^{\infty}\frac{(-1)^{k}}{k\,\ln k}$ $\qquad$ **54.** $\displaystyle\sum_{k=2}^{\infty}\frac{(-1)^{k}}{k\,(\ln k)^2}$

55. $\displaystyle\sum_{k=1}^{\infty}\frac{\cos k}{3^{k-1}}$ $\qquad$ **56.** $\displaystyle\sum_{k=1}^{\infty}\frac{(-1)^{k+1}\tan^{-1}k}{k}$

57. $\displaystyle\sum_{k=1}^{\infty}\frac{(-1)^{k+1}}{k^{1/k}}$ $\qquad$ **58.** $\displaystyle\sum_{k=1}^{\infty}(-1)^{k+1}\left(\frac{k}{k+1}\right)^{k}$

59. $1-\dfrac{1}{2!}+\dfrac{1}{3!}-\dfrac{1}{4!}+\dfrac{1}{5!}-\cdots$ $\qquad$ **60.** $1-\dfrac{1}{3^2}+\dfrac{1}{5^2}-\dfrac{1}{7^2}+\cdots$

61. (a) Show that the Alternating Series Test is not applicable to the series
$$\frac{1}{2}-\frac{1}{3}+\frac{1}{2^2}-\frac{1}{3^2}+\frac{1}{2^3}-\frac{1}{3^3}+\cdots$$
(b) Show the series converges.
(c) Find the sum of the series.

62. (a) Show that the Alternating Series Test is not applicable to the series
$$\sum_{k=1}^{\infty}(-1)^{k}a_k \quad\text{where } a_k=\begin{cases}\dfrac{1}{k} & \text{if } k \text{ is odd}\\[2mm]\dfrac{1}{k^2} & \text{if } k \text{ is even}\end{cases}$$
(b) Show the series diverges.

63. Show that the positive terms of $\displaystyle\sum_{k=1}^{\infty}\frac{(-1)^{k+1}}{k}$ diverge.

64. Show that the negative terms of $\displaystyle\sum_{k=1}^{\infty}\frac{(-1)^{k+1}}{k}$ diverge.

65. Show that the terms of the series $\displaystyle\sum_{k=1}^{\infty}\frac{(-1)^{k+1}}{k}$ can be rearranged so the resulting series converges to 0.

66. Show that the terms of the series $\displaystyle\sum_{k=1}^{\infty}\frac{(-1)^{k+1}}{k}$ can be rearranged so the resulting series converges to 2.

67. Show that the terms of the series $\displaystyle\sum_{k=1}^{\infty}\frac{(-1)^{k+1}}{k}$ can be rearranged so the resulting series diverges.

68. Show that the series
$$e^{-x}\cos x+e^{-2x}\cos(2x)+e^{-3x}\cos(3x)+\cdots$$
is absolutely convergent for all positive values of x.
Hint: Use the fact that $|\cos\theta|\le1$.

69. Determine whether the series below converges (absolutely or conditionally) or diverges.
$$1+r\cos\theta+r^2\cos(2\theta)+r^3\cos(3\theta)+\cdots$$

70. What is wrong with the following argument?
$$A=1-\frac{1}{2}+\frac{1}{3}-\frac{1}{4}+\frac{1}{5}-\frac{1}{6}+\frac{1}{7}-\frac{1}{8}+\cdots$$
$$\frac{1}{2}A=\frac{1}{2}-\frac{1}{4}+\frac{1}{6}-\frac{1}{8}+\cdots$$
So,
$$A+\frac{1}{2}A=1+\frac{1}{3}-\frac{1}{2}+\frac{1}{5}+\frac{1}{7}-\frac{1}{4}+\cdots$$
The series on the right is a rearrangement of the terms of the series A. So its sum is A, meaning
$$A+\frac{1}{2}A=A$$
$$A=0$$
But,
$$A=\left(1-\frac{1}{2}\right)+\left(\frac{1}{3}-\frac{1}{4}\right)+\left(\frac{1}{5}-\frac{1}{6}\right)+\cdots>0$$

Bernoulli's Error *In Problems 71–73, consider an incorrect argument given by Jakob Bernoulli to prove that*

$$\sum_{k=1}^{\infty} \frac{1}{k(k+1)} = \frac{1}{2} + \frac{1}{6} + \frac{1}{12} + \cdots = 1$$

Bernoulli's argument went as follows:

Let $N = 1 + \dfrac{1}{2} + \dfrac{1}{3} + \dfrac{1}{4} + \cdots$. *Then*

$$N - 1 = \frac{1}{2} + \frac{1}{3} + \frac{1}{4} + \frac{1}{5} + \cdots.$$

Now subtract term-by-term to get

$$N - (N-1) = \left(1 - \frac{1}{2}\right) + \left(\frac{1}{2} - \frac{1}{3}\right) + \left(\frac{1}{3} - \frac{1}{4}\right) + \cdots \qquad (4)$$

$$1 = \frac{1}{2} + \frac{1}{6} + \frac{1}{12} + \cdots$$

71. What is wrong with Bernoulli's argument?

72. In general, what can be said about the convergence or divergence of a series formed by taking the term-by-term difference (or sum) of two divergent series? Support your answer with examples.

73. Although the method is wrong. Bernoulli's conclusion is correct; that is, it is true that $\sum\limits_{k=1}^{\infty} \dfrac{1}{k(k+1)} = 1$. Prove it!

Hint: Look at the partial sums using the form of the series in (4).

74. Show that if two series $\sum\limits_{k=1}^{\infty} a_k$ and $\sum\limits_{k=1}^{\infty} b_k$ are absolutely convergent, the series $\sum\limits_{k=1}^{\infty} (a_k + b_k)$ is also absolutely convergent.

Moreover, $\sum\limits_{k=1}^{\infty} (a_k + b_k) = \sum\limits_{k=1}^{\infty} a_k + \sum\limits_{k=1}^{\infty} b_k$.

75. Show that if a series is absolutely convergent, then the series consisting of just the positive terms converges, as does the series consisting of just the negative terms.

76. Show that if a series is conditionally convergent, then the series consisting of just the positive terms diverges, as does the series consisting of just the negative terms.

77. Determine whether the series $\sum\limits_{k=1}^{\infty} c_k$, where

$$c_k = \begin{cases} \dfrac{1}{a^k} & \text{if } k \text{ is even} \\[2mm] -\dfrac{1}{b^k} & \text{if } k \text{ is odd} \end{cases} \qquad a > 1, \ b > 1,$$

converges absolutely, converges conditionally, or diverges.

78. Prove that if a series $\sum\limits_{k=1}^{\infty} a_k$ is absolutely convergent, any rearrangement of its terms results in a series that is also absolutely convergent.

Hint: Use the triangle inequality: $|a + b| \le |a| + |b|$.

Challenge Problems

In Problems 79–84, determine whether each series is absolutely convergent, conditionally convergent, or divergent.

79. $\displaystyle\sum_{k=2}^{\infty} \frac{(-1)^k}{\sqrt[p]{k^3 + 1}}$, $p > 2$

80. $\displaystyle\sum_{k=2}^{\infty} (-1)^k \ln \frac{k+1}{k}$

81. $\displaystyle\sum_{k=2}^{\infty} \frac{(-1)^k}{(\ln k)^{\ln k}}$

82. $\displaystyle\sum_{k=2}^{\infty} (-1)^k k^{(1-k)/k}$

83. $\displaystyle\sum_{k=1}^{\infty} (-1)^{k+1} \frac{k}{(k+1)^2}$

84. $\displaystyle\sum_{k=1}^{\infty} (-1)^{k+1} \frac{\sqrt{k}}{(k+1)}$

85. Let $\{a_n\}$ be a sequence that is decreasing and is bounded from below by 0. Define

$$R_n = \sum_{k=n+1}^{\infty} (-1)^{k+1} a_k \qquad \text{and} \qquad \Delta a_k = a_k - a_{k+1}$$

Suppose that the sequence $\{\Delta a_k\}$ decreases.

(a) Show that the series $\sum\limits_{k=1}^{\infty} (-1)^{k+1} \Delta a_k$ is a convergent alternating series.

(b) Show that

$$|R_n| = \frac{a_n}{2} + \frac{1}{2} \sum_{k=1}^{\infty} (-1)^k \Delta a_{k+n-1}$$

and $\sum\limits_{k=1}^{\infty} (-1)^k \Delta a_{k+n-1} < 0$. Deduce $|R_n| < \dfrac{1}{2} a_n$.

(c) Show that

$$|R_n| = \frac{a_{n+1}}{2} + \frac{1}{2} \sum_{k=1}^{\infty} (-1)^{k+1} \Delta a_{k+n}$$

(d) Conclude that $\dfrac{a_{n+1}}{2} < |R_n|$.

Source: Based on R. Johnsonbaugh. (1979). Summing an alternating series, *American Mathematical Monthly*, Vol. 86, No. 8, pp. 637–648.

AP® Practice Problems

Multiple-Choice Questions

 1. Which of the following series converge?

$$\textbf{I. } \sum_{k=1}^{\infty} (-1)^k \frac{1}{k^2} \qquad \textbf{II. } \sum_{k=1}^{\infty} (-1)^k \left(\frac{5}{3}\right)^k$$

$$\textbf{III. } \sum_{k=1}^{\infty} (-1)^k \frac{1}{\sqrt{k}}$$

(A) I only   (B) I and II only

(C) I and III only   (D) I, II, and III

 2. The alternating series $\sum_{k=1}^{\infty} \frac{(-1)^k k}{10^k}$ converges. What is the maximum error incurred by using the first three nonzero terms to approximate the sum of the series?

(A) −0.083   (B) 0.003   (C) 0.0004   (D) 0.0826

3. What is the fewest number of terms of the series $\sum_{k=1}^{\infty} \frac{(-1)^k}{k^3}$ that must be added to approximate the sum so that the error is less than or equal to 0.001?

(A) 7   (B) 9   (C) 10   (D) 11

 4. Which of the following series converge conditionally, but not absolutely?

$$\textbf{I. } \sum_{k=1}^{\infty} (-1)^{k+1} \frac{3}{k} \qquad \textbf{II. } \sum_{k=1}^{\infty} (-1)^{k+1} \left(\frac{1}{k}\right)^{4/3}$$

$$\textbf{III. } \sum_{k=0}^{\infty} (-1)^k \left(\frac{3}{4}\right)^k$$

(A) I only   (B) I and II only

(C) I and III only   (D) I, II, and III

Free-Response Questions

 5. (a) Write out the first five terms of the series $\sum_{k=0}^{\infty} \frac{(-1)^k}{k!}$.

(b) Show the series $\sum_{k=0}^{\infty} \frac{(-1)^k}{k!}$ converges.

(c) How many terms of the series are necessary to approximate the sum S with an error less than or equal to 0.0001?

See the **BREAK IT DOWN** *on page 898 for a stepped out solution to AP® Practice Problem 5(c).*

 6. Determine whether the series $\sum_{k=1}^{\infty} \frac{\cos(2k)}{4^k}$ converges absolutely, converges conditionally, or diverges. Show your work.

Retain Your Knowledge

Multiple-Choice Questions

1. The functions f and g are both differentiable for $0 < x < 10$, and the function h is defined as $h(x) = (g \cdot f)(x) + 6x \cdot g(x)$. Values for the functions f, g, and their derivatives for select numbers are given in the table below.

x	$f(x)$	$f'(x)$	$g(x)$	$g'(x)$
−2	1	5	0	3
4	−2	$\frac{1}{3}$	8	1

Using the data in the table, $h'(4)$ equals

(A) 24   (B) 76   (C) $76\frac{1}{3}$ (D) 73

2. Find $\lim_{x \to 0} \frac{\ln(x+1)}{\tan(3x)}$, if it exists.

(A) $\frac{1}{3}$ (B) 0   (C) 1   (D) The limit does not exist.

3. Suppose $\int_1^4 f(x)\,dx = 8$. Then $\int_1^4 2[f(x) - 6]\,dx =$

(A) −20   (B) −2   (C) 4   (D) 34

Free-Response Question

4. The graphs of the polar equation $r = 4 + 3\cos\theta$ (in blue) and the circle $r = 4$ (in red) for $0 \le \theta \le 2\pi$ are shown.

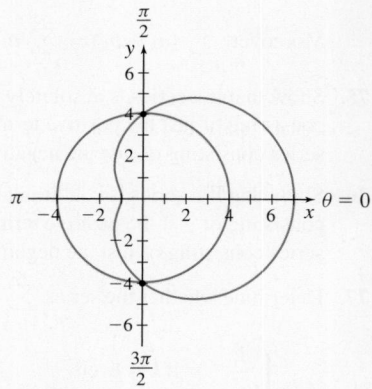

(a) Find an equation of the line tangent to the graph of $r = 4 + 3\cos\theta$ at $\theta = \frac{\pi}{2}$.

(b) Find the area of the region bounded by the graph of $r = 4 + 3\cos\theta$.

(c) Determine the average distance from the pole to the curve $r = 4 + 3\cos\theta$ for $0 \le \theta \le \pi$.

(d) The circle $r = 4$ is graphed on the same polar axis as $r = 4 + 3\cos\theta$. The two graphs intersect at $\theta = \frac{\pi}{2}$ and $\frac{3\pi}{2}$. What is the area of the region that lies outside the circle and inside the graph $r = 4 + 3\cos\theta$?

10.6 Ratio Test; Root Test

OBJECTIVES *When you finish this section, you should be able to:*

1 Use the Ratio Test (p. 845)
2 Use the Root Test (p. 848)

One of the most practical tests for convergence of a series of nonzero terms makes use of the ratio of two consecutive terms.

1 Use the Ratio Test

THEOREM Ratio Test

Suppose $\sum\limits_{k=1}^{\infty} a_k$ is a series of nonzero terms.

1. Suppose $\lim\limits_{n\to\infty} \left| \dfrac{a_{n+1}}{a_n} \right| = L$, a number.

   • If $L < 1$, then the series $\sum\limits_{k=1}^{\infty} a_k$ converges absolutely and

     so $\sum\limits_{k=1}^{\infty} a_k$ is convergent.

   • If $L = 1$, the test provides no information about whether the series converges or diverges.

   • If $L > 1$, then the series $\sum\limits_{k=1}^{\infty} a_k$ diverges.

2. If $\lim\limits_{n\to\infty} \left| \dfrac{a_{n+1}}{a_n} \right| = \infty$, then the series $\sum\limits_{k=1}^{\infty} a_k$ diverges.

A proof of the Ratio Test is given at the end of the section.

> ▶ CALC CLIP
> **EXAMPLE 1** **Using the Ratio Test**

Use the Ratio Test to determine whether each series converges or diverges.

(a) $\sum\limits_{k=1}^{\infty} \dfrac{k}{4^k}$ **(b)** $\sum\limits_{k=1}^{\infty} \dfrac{2^k}{k}$ **(c)** $\sum\limits_{k=1}^{\infty} \dfrac{3k+1}{k^2}$ **(d)** $\sum\limits_{k=1}^{\infty} \dfrac{1}{k!}$

Solution

(a) $\sum\limits_{k=1}^{\infty} \dfrac{k}{4^k}$ is a series of nonzero terms; $a_{n+1} = \dfrac{n+1}{4^{n+1}}$ and $a_n = \dfrac{n}{4^n}$. The absolute value of their ratio is

$$\left| \frac{a_{n+1}}{a_n} \right| = \frac{\dfrac{n+1}{4^{n+1}}}{\dfrac{n}{4^n}} = \frac{n+1}{4^{n+1}} \cdot \frac{4^n}{n} = \frac{n+1}{4n}$$

Then

$$\lim_{n \to \infty} \left| \frac{a_{n+1}}{a_n} \right| = \lim_{n \to \infty} \frac{n+1}{4n} = \frac{1}{4} < 1$$

Since the limit is less than 1, the series converges.

(Example continued on the next page)

(b) $\sum\limits_{k=1}^{\infty} \dfrac{2^k}{k}$ is a series of nonzero terms; $a_{n+1} = \dfrac{2^{n+1}}{n+1}$ and $a_n = \dfrac{2^n}{n}$. The absolute value of their ratio is

$$\left|\frac{a_{n+1}}{a_n}\right| = \frac{2^{n+1}}{n+1} \cdot \frac{n}{2^n} = \frac{2n}{n+1}$$

Then

$$\lim_{n \to \infty} \left|\frac{a_{n+1}}{a_n}\right| = \lim_{n \to \infty} \frac{2n}{n+1} = 2 > 1$$

Since the limit is greater than 1, the series diverges.

(c) $\sum\limits_{k=1}^{\infty} \dfrac{3k+1}{k^2}$ is a series of nonzero terms; $a_{n+1} = \dfrac{3(n+1)+1}{(n+1)^2} = \dfrac{3n+4}{(n+1)^2}$ and $a_n = \dfrac{3n+1}{n^2}$. The absolute value of their ratio is

$$\left|\frac{a_{n+1}}{a_n}\right| = \frac{3n+4}{(n+1)^2} \cdot \frac{n^2}{3n+1} = \frac{3n+4}{n^2+2n+1} \cdot \frac{n^2}{3n+1} = \frac{3n^3+4n^2}{3n^3+7n^2+5n+1}$$

Then

$$\lim_{n \to \infty} \left|\frac{a_{n+1}}{a_n}\right| = \lim_{n \to \infty} \frac{3n^3+4n^2}{3n^3+7n^2+5n+1} = 1$$

The Ratio Test provides no information about this series. Another test must be used. $\left(\text{You can show that the series diverges by comparing it to the harmonic series } \sum\limits_{k=1}^{\infty} \dfrac{1}{k}.\right)$

(d) $\sum\limits_{k=1}^{\infty} \dfrac{1}{k!}$ is a series of nonzero terms; $a_{n+1} = \dfrac{1}{(n+1)!}$ and $a_n = \dfrac{1}{n!}$. The absolute value of their ratio is

$$\left|\frac{a_{n+1}}{a_n}\right| = \frac{n!}{(n+1)!} = \frac{1}{n+1}$$

Then

$$\lim_{n \to \infty} \left|\frac{a_{n+1}}{a_n}\right| = \lim_{n \to \infty} \frac{1}{n+1} = 0$$

NOTE In Section 10.9, we show that $\sum\limits_{k=1}^{\infty} \dfrac{1}{k!} = e$.

Since the limit is less than 1, the series $\sum\limits_{k=1}^{\infty} \dfrac{1}{k!}$ converges. ∎

NOW WORK   Problem 7 and AP® Practice Problem 2.

As Example 1 illustrates, the Ratio Test is useful in determining whether a series containing factorials and/or powers converges or diverges.

EXAMPLE 2   Using the Ratio Test

Use the Ratio Test to determine whether the series $\sum\limits_{k=1}^{\infty} \dfrac{k!}{k^k}$ converges or diverges.

Solution

$\sum\limits_{k=1}^{\infty} \dfrac{k!}{k^k}$ is a series of nonzero terms. Since $a_{n+1} = \dfrac{(n+1)!}{(n+1)^{n+1}}$ and $a_n = \dfrac{n!}{n^n}$, the absolute value of their ratio is

$$\left| \frac{a_{n+1}}{a_n} \right| = \frac{\dfrac{(n+1)!}{(n+1)^{n+1}}}{\dfrac{n!}{n^n}} = \frac{(n+1)!}{(n+1)^{n+1}} \cdot \frac{n^n}{n!} = \frac{(n+1) \cdot n^n}{(n+1)^{(n+1)}} = \frac{n^n}{(n+1)^n}$$

$$= \left(\frac{n}{n+1} \right)^n = \left(\frac{1}{1 + \dfrac{1}{n}} \right)^n = \frac{1}{\left(1 + \dfrac{1}{n} \right)^n}$$

So,

$$\lim_{n \to \infty} \left| \frac{a_{n+1}}{a_n} \right| = \lim_{n \to \infty} \frac{1}{\left(1 + \dfrac{1}{n} \right)^n} = \frac{\lim\limits_{n \to \infty} 1}{\lim\limits_{n \to \infty} \left(1 + \dfrac{1}{n} \right)^n} = \frac{1}{e}$$

Since $\dfrac{1}{e} < 1$, the series converges. ∎

NEED TO REVIEW? The number e expressed as a limit is discussed in Section 3.4, pp. 272–274.

CAUTION In Example 2, the ratio $\left| \dfrac{a_{n+1}}{a_n} \right|$ converges to $\dfrac{1}{e}$. This does not mean that $\sum\limits_{k=1}^{\infty} \dfrac{k!}{k^k}$ converges to $\dfrac{1}{e}$. In fact, the sum of the series $\sum\limits_{k=1}^{\infty} \dfrac{k!}{k^k}$ is *not* known; all that is known is that the series converges.

NOW WORK   Problem 15 and AP® Practice Problem 4.

EXAMPLE 3   Using the Ratio Test

Use the Ratio Test to determine whether the series $\sum\limits_{k=1}^{\infty} (-1)^k \dfrac{k!}{10^k}$ converges or diverges.

Solution

$\sum\limits_{k=1}^{\infty} (-1)^k \dfrac{k!}{10^k}$ is a series of nonzero terms.

$$a_{n+1} = (-1)^{n+1} \frac{(n+1)!}{10^{n+1}} \qquad a_n = (-1)^n \frac{n!}{10^n}$$

$$\lim_{n \to \infty} \left| \frac{a_{n+1}}{a_n} \right| = \lim_{n \to \infty} \frac{\dfrac{(n+1)!}{10^{n+1}}}{\dfrac{n!}{10^n}} = \lim_{n \to \infty} \frac{(n+1)! \, 10^n}{n! \, 10^{n+1}} = \lim_{n \to \infty} \frac{n+1}{10} = \infty$$

By the Ratio Test, the series $\sum\limits_{k=1}^{\infty} (-1)^k \dfrac{k!}{10^k}$ diverges. ∎

NOW WORK   Problem 19 and AP® Practice Problems 1 and 3.

We conclude the discussion of the Ratio Test with these observations:

- To test $\sum\limits_{k=1}^{\infty} a_k$ for convergence, it is important to check whether the *limit of the ratio* $\left| \dfrac{a_{n+1}}{a_n} \right|$, not the ratio itself, is less than 1.

  For example, for the harmonic series $\sum\limits_{k=1}^{\infty} \dfrac{1}{k}$, which diverges,

  the ratio $\left| \dfrac{a_{n+1}}{a_n} \right| = \left| \dfrac{n}{n+1} \right| < 1$, but $\lim\limits_{n \to \infty} \left| \dfrac{a_{n+1}}{a_n} \right| = \lim\limits_{n \to \infty} \dfrac{n}{n+1} = 1$.

- For divergence, it is sufficient to show that the ratio $\left| \dfrac{a_{n+1}}{a_n} \right| > 1$ for all n.

- The Ratio Test provides no information if $\lim\limits_{n \to \infty} \left| \dfrac{a_{n+1}}{a_n} \right| = 1$.

- If the general term a_n of an infinite series involves n, either exponentially or factorially, the Ratio Test often answers the question of convergence or divergence.

② Use the Root Test

The *Root Test* works well for series of nonzero terms whose nth term involves an nth power.

THEOREM Root Test

Suppose $\sum\limits_{k=1}^{\infty} a_k$ is a series of nonzero terms and $\lim\limits_{n \to \infty} \sqrt[n]{|a_n|} = L$, a number.

- If $L < 1$, then $\sum\limits_{k=1}^{\infty} a_k$ is absolutely convergent, so the series $\sum\limits_{k=1}^{\infty} a_k$ converges.

- If $L > 1$, then $\sum\limits_{k=1}^{\infty} a_k$ diverges.

- If $L = 1$, the test provides no information.

The proof of the Root Test is similar to the proof of the Ratio Test. It is left as an exercise (Problem 67).

EXAMPLE 4 Using the Root Test

Use the Root Test to determine whether the series $\sum\limits_{k=1}^{\infty} \dfrac{e^k}{k^k}$ converges or diverges.

Solution

$\sum\limits_{k=1}^{\infty} \dfrac{e^k}{k^k}$ is a series of nonzero terms. The nth term is $a_n = \dfrac{e^n}{n^n} = \left(\dfrac{e}{n} \right)^n$. Since a_n involves an nth power, we use the Root Test.

$$\lim\limits_{n \to \infty} \sqrt[n]{|a_n|} = \lim\limits_{n \to \infty} \sqrt[n]{\left(\dfrac{e}{n} \right)^n} = \lim\limits_{n \to \infty} \dfrac{e}{n} = 0 < 1$$

The series $\sum\limits_{k=1}^{\infty} \dfrac{e^k}{k^k}$ converges. ∎

NOW WORK Problem 33.

EXAMPLE 5   **Using the Root Test**

Use the Root Test to determine whether the series $\displaystyle\sum_{k=1}^{\infty} \left(\frac{8k+3}{5k-2}\right)^{k}$

converges or diverges.

Solution

$\displaystyle\sum_{k=1}^{\infty} \left(\frac{8k+3}{5k-2}\right)^{k}$ is a series of nonzero terms. The nth term is $a_n = \left(\frac{8n+3}{5n-2}\right)^{n}$. Since a_n involves an nth power, we use the Root Test.

$$\lim_{n \to \infty} \sqrt[n]{|a_n|} = \lim_{n \to \infty} \sqrt[n]{\left(\frac{8n+3}{5n-2}\right)^{n}} = \lim_{n \to \infty} \frac{8n+3}{5n-2} = \frac{8}{5} > 1$$

The series diverges. ∎

NOW WORK   Problem 29.

Proof of the Ratio Test   Suppose $\displaystyle\sum_{k=1}^{\infty} a_k$ is a series of nonzero terms.

Case 1: $\displaystyle\lim_{n \to \infty} \left|\frac{a_{n+1}}{a_n}\right| = L$, a number.

- $0 \le L < 1$ Let r be any number for which $L < r < 1$. Since $\displaystyle\lim_{n \to \infty} \left|\frac{a_{n+1}}{a_n}\right| = L$ and $L < r$, then by the definition of the limit of a sequence, we can find a number N so that for any number $n > N$, the ratio $\left|\dfrac{a_{n+1}}{a_n}\right|$ can be made as close as we please to L and be less than r. Then

$$\left|\frac{a_{N+1}}{a_N}\right| < r \qquad \text{or} \qquad |a_{N+1}| < r \cdot |a_N|$$

$$\left|\frac{a_{N+2}}{a_{N+1}}\right| < r \qquad \text{or} \qquad |a_{N+2}| < r \cdot |a_{N+1}| < r^2 \cdot |a_N|$$

$$\left|\frac{a_{N+3}}{a_{N+2}}\right| < r \qquad \text{or} \qquad |a_{N+3}| < r \cdot |a_{N+2}| < r^3 \cdot |a_N|$$

Each term of the series $|a_{N+1}| + |a_{N+2}| + \cdots$ is less than the corresponding term of the geometric series $|a_N| r + |a_N| r^2 + |a_N| r^3 + \cdots$. Since $|r| < 1$, the geometric series converges. By the Algebraic Comparison Test for Convergence, the series $|a_{N+1}| + |a_{N+2}| + \cdots$ also converges. So, the series $\displaystyle\sum_{k=1}^{\infty} |a_k|$ converges. By the Absolute Convergence Test, the series $\displaystyle\sum_{k=1}^{\infty} a_k$ converges.

- $L = 1$ To show that the test provides no information for $L = 1$, we exhibit two series, one that diverges and another that converges, to show that no conclusion can be drawn. Consider $\displaystyle\sum_{k=1}^{\infty} \frac{1}{k}$ and $\displaystyle\sum_{k=1}^{\infty} \frac{1}{k^2}$. The first is the harmonic series, which diverges. The second is a p-series with $p > 1$, which converges. It is left to you to show that $\displaystyle\lim_{n \to \infty} \left|\frac{a_{n+1}}{a_n}\right| = 1$ in each case. (See Problems 57 and 58.)

• $L > 1$ Let r be any number for which $1 < r < L$. Since $\lim\limits_{n \to \infty} \left| \dfrac{a_{n+1}}{a_n} \right| = L$, there

is a number N so that for any number $n > N$, the ratio $\left| \dfrac{a_{n+1}}{a_n} \right|$ can be made as close

as we please to L and will be greater than r. That is, for all numbers $n > N$, the

ratio $\left| \dfrac{a_{n+1}}{a_n} \right| > r > 1$ so that $|a_{n+1}| > |a_n|$. After the Nth term, the absolute value

of the terms are positive and increasing. So, $\lim\limits_{n \to \infty} |a_n| \neq 0$ and, therefore,

$\lim\limits_{n \to \infty} a_n \neq 0$. By the nth Term Test for Divergence, the series diverges.

Case 2: $\lim\limits_{n \to \infty} \left| \dfrac{a_{n+1}}{a_n} \right| = \infty$ The proof that this series diverges is left as an exercise (Problem 66). ∎

10.6 Assess Your Understanding

Concepts and Vocabulary

1. *True or False*   The Ratio Test can be used to show that the series $\sum\limits_{k=1}^{\infty} \cos(k\pi)$ diverges.

2. *True or False*   In using the Ratio Test, if $\lim\limits_{n \to \infty} \left| \dfrac{a_{n+1}}{a_n} \right| = L < 1$, then the sum of the series $\sum\limits_{k=1}^{\infty} a_k$ equals L.

3. *True or False*   In using the Ratio Test, if $\lim\limits_{n \to \infty} \left| \dfrac{a_{n+1}}{a_n} \right| = 1$, then the Ratio Test indicates that the series $\sum\limits_{k=1}^{\infty} a_k$ converges.

4. *True or False*   The Root Test works well if the nth term of a series of nonzero terms involves an nth root.

Skill Building

In Problems 5–28, use the Ratio Test to determine whether each series converges or diverges or state that it provides no information.

5. $\sum\limits_{k=1}^{\infty} \dfrac{4k^2 - 1}{2^k}$

6. $\sum\limits_{k=1}^{\infty} \dfrac{1}{(2k+1)2^k}$

[PAGE 846] 7. $\sum\limits_{k=1}^{\infty} k \left(\dfrac{2}{3} \right)^k$

8. $\sum\limits_{k=1}^{\infty} \dfrac{5^k}{k^2}$

9. $\sum\limits_{k=1}^{\infty} \dfrac{10^k}{(2k)!}$

10. $\sum\limits_{k=1}^{\infty} \dfrac{(2k)!}{5^k 3^{k-1}}$

11. $\sum\limits_{k=1}^{\infty} \dfrac{k}{(2k-2)!}$

12. $\sum\limits_{k=1}^{\infty} \dfrac{(k+1)!}{3^k}$

13. $\sum\limits_{k=1}^{\infty} \dfrac{2^k}{k(k+1)}$

14. $\sum\limits_{k=1}^{\infty} \dfrac{k!}{k^2(k+1)^2}$

[PAGE 847] 15. $\sum\limits_{k=1}^{\infty} \dfrac{k^3}{k!}$

16. $\sum\limits_{k=1}^{\infty} \dfrac{k!}{k^{k+1}}$

17. $\sum\limits_{k=1}^{\infty} (-1)^k \dfrac{k^2}{2^k}$

18. $\sum\limits_{k=1}^{\infty} (-1)^k \dfrac{k}{e^k}$

[PAGE 847] 19. $\sum\limits_{k=2}^{\infty} (-1)^k \dfrac{\ln k}{3^k}$

20. $\sum\limits_{k=2}^{\infty} (-1)^k \dfrac{\ln k}{k!}$

21. $\sum\limits_{k=1}^{\infty} (-1)^k \dfrac{5^k}{3^{k-1}}$

22. $\sum\limits_{k=1}^{\infty} (-1)^k \dfrac{3^k + 4}{2^{k-1}}$

23. $\sum\limits_{k=1}^{\infty} \dfrac{3^{k-1}}{k \cdot 2^k}$

24. $\sum\limits_{k=1}^{\infty} \dfrac{k(k+2)}{3^k}$

25. $\sum\limits_{k=1}^{\infty} \dfrac{k}{e^k}$

26. $\sum\limits_{k=1}^{\infty} \dfrac{e^k}{k^3}$

27. $\sum\limits_{k=1}^{\infty} k \cdot 2^k$

28. $\sum\limits_{k=1}^{\infty} \dfrac{4^k}{k}$

In Problems 29–46, use the Root Test to determine whether each series converges or diverges or state that it provides no information.

[PAGE 849] 29. $\sum\limits_{k=1}^{\infty} \left(\dfrac{2k+1}{5k+1} \right)^k$

30. $\sum\limits_{k=1}^{\infty} \left(\dfrac{3k-1}{2k+1} \right)^k$

31. $\sum\limits_{k=1}^{\infty} \left(\dfrac{k}{5} \right)^k$

32. $\sum\limits_{k=1}^{\infty} \dfrac{\pi^{2k}}{k^k}$

[PAGE 848] 33. $\sum\limits_{k=2}^{\infty} \left(\dfrac{\ln k}{k} \right)^k$

34. $\sum\limits_{k=2}^{\infty} \left(\dfrac{1}{\ln k} \right)^k$

35. $\sum\limits_{k=2}^{\infty} (-1)^k \dfrac{(\ln k)^k}{2^k}$

36. $\sum\limits_{k=1}^{\infty} (-1)^k \dfrac{2^{k+1}}{e^k}$

37. $\sum\limits_{k=1}^{\infty} (-1)^k \left(\dfrac{k+1}{k} \right)^k$

38. $\sum\limits_{k=1}^{\infty} (-1)^k \left(\dfrac{2k+3}{k+1} \right)^k$

39. $\displaystyle\sum_{k=1}^{\infty} (-1)^k \frac{k^k}{e^{2k}}$

40. $\displaystyle\sum_{k=2}^{\infty} (-1)^k \frac{k \ln k}{e^k}$

41. $\displaystyle\sum_{k=1}^{\infty} \left(\frac{\sqrt{k^2+1}}{3k}\right)^k$

42. $\displaystyle\sum_{k=1}^{\infty} \left(\frac{\sqrt{4k^2+1}}{k}\right)^k$

43. $\displaystyle\sum_{k=1}^{\infty} \frac{k^2}{2^k}$

44. $\displaystyle\sum_{k=1}^{\infty} \frac{k^3}{3^k}$

45. $\displaystyle\sum_{k=1}^{\infty} \frac{k^4}{5^k}$

46. $\displaystyle\sum_{k=1}^{\infty} \frac{k}{3^k}$

In Problems 47–56, use the Ratio Test or Root Test to determine whether each series converges or diverges or state that it provides no information.

47. $\displaystyle\sum_{k=1}^{\infty} \frac{10}{(3k+1)^k}$

48. $\displaystyle\sum_{k=1}^{\infty} \left(1+\frac{1}{k}\right)^{k^2}$

49. $\displaystyle\sum_{k=1}^{\infty} \frac{(k+1)(k+2)}{k!}$

50. $\displaystyle\sum_{k=1}^{\infty} \frac{k!}{(3k+1)!}$

51. $\displaystyle\sum_{k=1}^{\infty} \frac{k \ln k}{2^k}$

52. $\displaystyle\sum_{k=1}^{\infty} \left[\ln\left(e^3+\frac{1}{k}\right)\right]^k$

53. $\displaystyle\sum_{k=1}^{\infty} \sin^k\left(\frac{1}{k}\right)$

54. $\displaystyle\sum_{k=1}^{\infty} \frac{k^k}{2^{k^2}}$

55. $\displaystyle\sum_{k=1}^{\infty} \frac{\left(1+\frac{1}{k}\right)^{2k}}{e^k}$

56. $\displaystyle\sum_{k=2}^{\infty} \frac{2^k(k+1)}{k^2(k+2)}$

Applications and Extensions

57. For the divergent series $\displaystyle\sum_{k=1}^{\infty} \frac{1}{k}$, show that $\displaystyle\lim_{n\to\infty}\left|\frac{a_{n+1}}{a_n}\right|=1$.

58. For the convergent series $\displaystyle\sum_{k=1}^{\infty} \frac{1}{k^2}$, show that $\displaystyle\lim_{n\to\infty}\left|\frac{a_{n+1}}{a_n}\right|=1$.

59. Give an example of a convergent series $\displaystyle\sum_{k=1}^{\infty} a_k$ for which $\displaystyle\lim_{n\to\infty}\left|\frac{a_{n+1}}{a_n}\right|\neq\infty$ does not exist.

60. Give an example of a divergent series $\displaystyle\sum_{k=1}^{\infty} a_k$ for which $\displaystyle\lim_{n\to\infty}\left|\frac{a_{n+1}}{a_n}\right|\neq\infty$ does not exist.

61. (a) Show that the series $\displaystyle\sum_{k=1}^{\infty} \frac{(-1)^k 3^k}{k!}$ converges.

[CAS] **(b)** Use technology to find the sum of the series.

62. Determine whether the following series is convergent or divergent:

$$\frac{1}{3} - \frac{2^3}{3^2} + \frac{3^3}{3^3} - \frac{4^3}{3^4} + \cdots + \frac{(-1)^{n-1}n^3}{3^n} + \cdots$$

63. Show that $\displaystyle\lim_{n\to\infty} \frac{n!}{n^n}=0$, where n denotes a positive integer.

64. Show that the Root Test provides no information for $\displaystyle\sum_{k=1}^{\infty} \frac{1}{k}$ and $\displaystyle\sum_{k=1}^{\infty} \frac{1}{k^2}$.

65. Determine whether the series $\displaystyle\sum_{k=1}^{\infty} \frac{e^k}{k^k}$ converges or diverges.

66. Prove that $\displaystyle\sum_{k=1}^{\infty} a_k$ diverges if $\displaystyle\lim_{n\to\infty}\left|\frac{a_{n+1}}{a_n}\right|=\infty$.

67. Prove the Root Test.

68. The terms of the series

$$\frac{1}{4} + \frac{1}{2} + \frac{1}{8} + \frac{1}{4} + \frac{1}{16} + \frac{1}{8} + \frac{1}{32} + \cdots$$

are $a_{2k}=\dfrac{1}{2^k}$ and $a_{2k-1}=\dfrac{1}{2^{k+1}}$.

(a) Show that using the Ratio Test to determine whether the series converges provides no information.

(b) Show that using the Root Test to determine whether the series converges is conclusive.

(c) Does the series converge?

Challenge Problems

69. Show that the following series converges:

$$1+\frac{2}{2^2}+\frac{3}{3^3}+\frac{1}{4^4}+\frac{2}{5^5}+\frac{3}{6^6}+\cdots$$

70. Show that $\displaystyle\sum_{k=1}^{\infty} \frac{(k+1)^2}{(k+2)!}$ converges. *Hint:* Use the Limit Comparison Test and the convergent series $\displaystyle\sum_{k=1}^{\infty} \frac{1}{k!}$.

71. Suppose $0<a<b<1$. Use the Root Test to show that the series

$$a+b+a^2+b^2+a^3+b^3+\cdots$$

converges.

72. Show that if the Ratio Test indicates a series converges, then so will the Root Test. The converse is not true. Refer to Problem 68.

Preparing for the **AP® Exam**

AP® Practice Problems

Multiple-Choice Questions

 1. Which of the following series converge?

   I. $\displaystyle\sum_{k=1}^{\infty} \frac{k^2}{(3k+1)!}$

   II. $\displaystyle\sum_{k=1}^{\infty} \frac{k^k}{k!}$

   III. $\displaystyle\sum_{k=1}^{\infty} k\left(\frac{2}{3}\right)^k$

   (A) II only   (B) III only
   (C) I and III only   (D) II and III only

2. For which of the following series does the Ratio Test provide no information?

   I. $\displaystyle\sum_{k=1}^{\infty} \frac{(-1)^k \sqrt{k}}{k+1}$

   II. $\displaystyle\sum_{k=2}^{\infty} \frac{1}{k \ln k}$

   III. $\displaystyle\sum_{k=1}^{\infty} \frac{k^2 + 3k}{k+1}$

   (A) I only   (B) II only
   (C) I and II only   (D) I, II, and III

3. Which of the following series diverge?

   I. $\displaystyle\sum_{k=1}^{\infty} \frac{k!}{100^k}$

   II. $\displaystyle\sum_{k=1}^{\infty} \frac{20^k}{2^{k^2}}$

   III. $\displaystyle\sum_{k=1}^{\infty} \frac{\sqrt{k}}{k^3 + 1}$

   (A) I only   (B) I and II only
   (C) I and III only (D) I, II, and III

Free-Response Question

4. Find the real numbers x for which $\displaystyle\sum_{k=1}^{\infty} \frac{x^k}{k^2}$ converges or diverges.

Retain Your Knowledge

Multiple-Choice Questions

1. Suppose $y = x^2 e^x - x \ln x$. Then y'' equals

   (A) $(x^2 + 4x + 2)e^x + \dfrac{1}{x}$ (B) $(x^2 + 4x + 2)e^x - \dfrac{1}{x}$

   (C) $\dfrac{\left(x^2 + 4x + 2\right)e^x}{x}$ (D) $6x^2 e^x + 2e^x - \dfrac{1}{x}$

2. $f(x) = x^3 - 6x^2 + 9x$ is defined for $0 \le x \le 2$. Find the absolute maximum value of f.

   (A) 1   (B) 2   (C) 4   (D) 50

3. The population growth of a town follows the logistic function $P(t) = \dfrac{2500}{1 + 49e^{-0.25t}}$. What was the initial population?

   (A) 25   (B) 49   (C) 50   (D) 1250

Free-Response Question

4. The table below gives several values of a differentiable function f for select numbers x.

x	−1	0	1	2	3
$f(x)$	2	−1	−3	−3	6

   (a) Show that f has a zero in the closed interval [0, 3].

   (b) Use a linear approximation to estimate the zero.

   (c) Show for some number c, $0 < c < 1$, that $f'(c) = -2$.

   (d) Approximate $\int_0^3 f(x)\,dx$ using a Right Riemann sum with three subintervals.

10.7 Summary of Tests

OBJECTIVE *When you finish this section, you should be able to:*

1 **Choose an appropriate test to determine whether a series converges (p. 853)**

In the previous sections, we discussed a variety of tests that can be used to determine whether a series converges or diverges. In the exercises following each section, the series under consideration used the tests for convergence/divergence discussed in that section. In practice, such information is not provided. This section summarizes the tests that we have discussed and gives some clues as to what test has the best chance of answering the fundamental question, "Does the series converge or diverge?"

1 Choose an Appropriate Test to Determine Whether a Series Converges

The following outline is a guide to help you choose a test to use when determining the convergence or divergence of a series. Table 5 (on page 854) lists the tests we have discussed. Table 6 (on page 855) describes important series we have analyzed.

Guide to Choosing a Test to Determine Whether a Series Is Convergent

1. Check to see if the series is a geometric series or a *p*-series. If yes, then use the conclusion given for these series in Table 6.

2. Find $\lim_{n \to \infty} a_n$ of the series $\sum_{k=1}^{\infty} a_k$. If $\lim_{n \to \infty} a_n \neq 0$, then by the *n*th Term Test for Divergence, the series diverges.

3. If the series $\sum_{k=1}^{\infty} a_k$ has only positive terms and meets the conditions of the Integral Test, find the related function f. Use the Integral Test if $\int_1^{\infty} f(x)\, dx$ is easy to find.

4. If the series $\sum_{k=1}^{\infty} a_k$ has only positive terms and the *n*th term is a quotient of sums or differences of powers of *n*, the Limit Comparison Test with an appropriate *p*-series will usually work.

5. If the series $\sum_{k=1}^{\infty} a_k$ has only positive terms and the preceding attempts fail, then try the Algebraic Comparison Test for Convergence or the Algebraic Comparison Test for Divergence.

6. *Series with some negative terms.*

   • For an alternating series, use the Alternating Series Test. It is sometimes better to use the Absolute Convergence Test first.

   • For other series containing negative terms, always use the Absolute Convergence Test first.

7. If the series $\sum_{k=1}^{\infty} a_k$ has nonzero terms that involve products, factorials, or powers, the Ratio Test is a good choice.

8. If the series $\sum_{k=1}^{\infty} a_k$ has nonzero terms and the *n*th term involves an *n*th power, try the Root Test.

TABLE 5 Tests for Convergence and Divergence of Series

Test Name	Description	Comment
nth Term Test for Divergence for all series (p. 812)	$\sum_{k=1}^{\infty} a_k$ diverges if $\lim_{n \to \infty} a_n \neq 0$.	No information is obtained about convergence if $\lim_{n \to \infty} a_n = 0$.
Integral Test for series of positive terms (p. 815)	$\sum_{k=1}^{\infty} a_k$ converges (diverges) if $\int_1^{\infty} f(x)\, dx$ converges (diverges), where f is continuous, positive, and nonincreasing for $x \geq 1$; and $f(k) = a_k$ for all k.	Good to use if f is easy to integrate.
Algebraic Comparison Test for Convergence for series of positive terms (p. 824)	$\sum_{k=1}^{\infty} a_k$ converges if $0 < a_k \leq b_k$ and the series $\sum_{k=1}^{\infty} b_k$ converges.	$\sum_{k=1}^{\infty} b_k$ must have positive terms and be convergent.
Algebraic Comparison Test for Divergence for series of positive terms (p. 825)	$\sum_{k=1}^{\infty} a_k$ diverges if $a_k \geq c_k > 0$ and the series $\sum_{k=1}^{\infty} c_k$ diverges.	$\sum_{k=1}^{\infty} c_k$ must have positive terms and be divergent.
Limit Comparison Test for series of positive terms (p. 826)	$\sum_{k=1}^{\infty} a_k$ converges (diverges) if $\sum_{k=1}^{\infty} b_k$ converges (diverges), and $\lim_{n \to \infty} \dfrac{a_n}{b_n} = L$, a positive real number.	$\sum_{k=1}^{\infty} b_k$ must have positive terms, whose convergence (divergence) can be determined.
Alternating Series Test (p. 833)	$\sum_{k=1}^{\infty} (-1)^{k+1} a_k$, $a_k > 0$, converges if • $\lim_{n \to \infty} a_n = 0$ and • the a_k are nonincreasing.	The error made by using the nth partial sum to approximate the sum S of the series is less than or equal to $|a_{n+1}|$.
Absolute Convergence Test (p. 838)	If $\sum_{k=1}^{\infty} |a_k|$ converges, then $\sum_{k=1}^{\infty} a_k$ converges.	The converse is not true. That is, if $\sum_{k=1}^{\infty} |a_k|$ diverges, $\sum_{k=1}^{\infty} a_k$ may or may not converge.
Ratio Test for series with nonzero terms (p. 845)	$\sum_{k=1}^{\infty} a_k$ converges if $\lim_{n \to \infty} \left| \dfrac{a_{n+1}}{a_n} \right| < 1$. $\sum_{k=1}^{\infty} a_k$ diverges if $\lim_{n \to \infty} \left| \dfrac{a_{n+1}}{a_n} \right| > 1$ or if $\lim_{n \to \infty} \left| \dfrac{a_{n+1}}{a_n} \right| = \infty$.	Good to use if a_n includes factorials or powers. The Ratio Test provides no information if $\lim_{n \to \infty} \left| \dfrac{a_{n+1}}{a_n} \right| = 1$.
Root Test for series with nonzero terms (p. 848)	$\sum_{k=1}^{\infty} a_k$ converges if $\lim_{n \to \infty} \sqrt[n]{|a_n|} < 1$. $\sum_{k=1}^{\infty} a_k$ diverges if $\lim_{n \to \infty} \sqrt[n]{|a_n|} > 1$ or if $\lim_{n \to \infty} \sqrt[n]{|a_n|} = \infty$.	Good to use if a_n involves nth powers. The Root Test provides no information if $\lim_{n \to \infty} \sqrt[n]{|a_n|} = 1$.

TABLE 6 Important Series

Series Name	Series Description	Comments
Geometric series (pp. 800–801)	$\sum\limits_{k=1}^{\infty} ar^{k-1} = a + ar + ar^2 + \cdots,\ a \neq 0$	Converges to $\dfrac{a}{1-r}$ if $|r| < 1$; diverges if $|r| \geq 1$.
Harmonic series (p. 804)	$\sum\limits_{k=1}^{\infty} \dfrac{1}{k} = 1 + \dfrac{1}{2} + \dfrac{1}{3} + \cdots$	Diverges.
p-series (p. 817)	$\sum\limits_{k=1}^{\infty} \dfrac{1}{k^p} = 1 + \dfrac{1}{2^p} + \dfrac{1}{3^p} + \cdots$	Converges if $p > 1$; diverges if $0 < p \leq 1$.
k-to-the-k series (p. 824)	$\sum\limits_{k=1}^{\infty} \dfrac{1}{k^k} = 1 + \dfrac{1}{2^2} + \dfrac{1}{3^3} + \dfrac{1}{4^4} + \cdots$	Converges.
Alternating harmonic series (p. 835)	$\sum\limits_{k=1}^{\infty} \dfrac{(-1)^{k+1}}{k} = 1 - \dfrac{1}{2} + \dfrac{1}{3} - \dfrac{1}{4} + \cdots$	Converges.

10.7 Assess Your Understanding

Concepts and Vocabulary

1. *True or False* The series $\sum\limits_{k=1}^{\infty} \dfrac{1}{k^p}$ converges if $p \geq 1$.

2. *True or False* According to the *n*th Term Test for Divergence, an infinite series $\sum\limits_{k=1}^{\infty} a_k$ converges if $\lim\limits_{n \to \infty} a_n = 0$.

3. *True or False* If a series is absolutely convergent, then it is convergent.

4. *True or False* If a series is not absolutely convergent, then it is divergent.

5. *True or False* According to the Ratio Test, a series $\sum\limits_{k=1}^{\infty} a_k$ of nonzero terms converges if $\left| \dfrac{a_{n+1}}{a_n} \right| < 1$.

6. To use the Algebraic Comparison Test for Convergence to show that a series $\sum\limits_{k=1}^{\infty} a_k$ of positive terms converges, find a series $\sum\limits_{k=1}^{\infty} b_k$ that is known to converge and show that $0 < $ _____ $\leq$ _____ for all k.

Skill Building

In Problems 7–39, determine whether each series converges (absolutely or conditionally) or diverges. Use any applicable test.

7. $\sum\limits_{k=1}^{\infty} \dfrac{9k^3 + 5k^2}{k^{5/2} + 4}$

8. $\sum\limits_{k=1}^{\infty} \dfrac{(-1)^{k+1}}{\sqrt{2k+1}}$

9. $6 + 2 + \dfrac{2}{3} + \dfrac{2}{9} + \dfrac{2}{27} + \cdots$

10. $\sum\limits_{k=1}^{\infty} \dfrac{1}{k^2} \sin \dfrac{\pi}{k}$

11. $\sum\limits_{k=1}^{\infty} \dfrac{3k+2}{k^3+1}$

12. $1 + \dfrac{2^2+1}{2^3+1} + \dfrac{3^2+1}{3^3+1} + \dfrac{4^2+1}{4^3+1} + \cdots$

13. $\sum\limits_{k=1}^{\infty} \dfrac{k+4}{k\sqrt{3k-2}}$

14. $\sum\limits_{k=1}^{\infty} \dfrac{\sin k}{k^3}$

15. $\sum\limits_{k=1}^{\infty} \dfrac{3^{2k-1}}{k^2+2k}$

16. $\sum\limits_{k=1}^{\infty} \dfrac{5^k}{k!}$

17. $\sum\limits_{k=1}^{\infty} \left(1 + \dfrac{2}{k}\right)^k$

18. $\sum\limits_{k=1}^{\infty} \dfrac{k^2+4}{e^k}$

19. $\dfrac{2}{3} - \dfrac{3}{4} \cdot \dfrac{1}{2} + \dfrac{4}{5} \cdot \dfrac{1}{3} - \dfrac{5}{6} \cdot \dfrac{1}{4} + \cdots$

20. $2 + \dfrac{3}{2} \cdot \dfrac{1}{4} + \dfrac{4}{3} \cdot \dfrac{1}{4^2} + \dfrac{5}{4} \cdot \dfrac{1}{4^3} + \cdots$

21. $1 + \dfrac{1 \cdot 3}{2!} + \dfrac{1 \cdot 3 \cdot 5}{3!} + \dfrac{1 \cdot 3 \cdot 5 \cdot 7}{4!} + \cdots$

22. $\dfrac{1}{\sqrt{1 \cdot 2 \cdot 3}} + \dfrac{1}{\sqrt{2 \cdot 3 \cdot 4}} + \dfrac{1}{\sqrt{3 \cdot 4 \cdot 5}} + \cdots$

23. $\sum\limits_{k=1}^{\infty} \dfrac{k!}{(2k)!}$

24. $\sum\limits_{k=1}^{\infty} k^3 e^{-k^4}$

25. $\sum\limits_{k=1}^{\infty} \dfrac{1}{\sqrt{k} + 100}$

26. $\sum\limits_{k=1}^{\infty} \dfrac{k^2 + 5k}{3 + 5k^2}$

27. $\sum\limits_{k=1}^{\infty} \dfrac{1}{\sqrt[3]{k^4 + 4}}$

28. $\sum\limits_{k=1}^{\infty} \dfrac{1}{11} \left(\dfrac{-3}{2}\right)^k$

29. $\dfrac{1}{3} - \dfrac{2}{4} + \dfrac{3}{5} - \dfrac{4}{6} + \cdots$

30. $\sum\limits_{k=1}^{\infty} \dfrac{k(-4)^{3k}}{5^k}$

31. $\sum\limits_{k=1}^{\infty} \left(-\dfrac{1}{k}\right)^k$

32. $\sum\limits_{k=1}^{\infty} \dfrac{5}{2^k + 1}$

33. $\sum\limits_{k=1}^{\infty} e^{-k^2}$

34. $\dfrac{\sin \sqrt{1}}{1^{3/2}} + \dfrac{\sin \sqrt{2}}{2^{3/2}} + \dfrac{\sin \sqrt{3}}{3^{3/2}} + \cdots$

35. $\sum\limits_{k=2}^{\infty} \dfrac{(-1)^{k-1}}{k(\ln k)^3}$

36. $\sum\limits_{k=1}^{\infty} \dfrac{1}{(2k)^k}$

37. $\sum\limits_{k=2}^{\infty} \left(\dfrac{\ln k}{1000}\right)^k$

38. $\sum\limits_{k=1}^{\infty} \dfrac{1}{e^k}$

39. $\sum\limits_{k=1}^{\infty} \dfrac{\tan^{-1} k}{k^2}$

In Problems 40–42, determine whether each series converges or diverges. If it converges, find its sum.

40. $\displaystyle\sum_{k=1}^{\infty} \left(\sqrt{k+1} - \sqrt{k} \right)$

41. $\displaystyle\sum_{k=4}^{\infty} \left(\frac{1}{k-3} - \frac{1}{k} \right)$

42. $\displaystyle\sum_{k=2}^{\infty} \ln \frac{k}{k+1}$

43. Determine whether $1 + \dfrac{1 \cdot 2}{1 \cdot 3} + \dfrac{1 \cdot 2 \cdot 3}{1 \cdot 3 \cdot 5} + \dfrac{1 \cdot 2 \cdot 3 \cdot 4}{1 \cdot 3 \cdot 5 \cdot 7} + \cdots$ converges or diverges.

44. (a) Show that the series $\displaystyle\sum_{k=1}^{\infty} \left[\left(\frac{2}{3} \right)^k - \frac{2}{k^2 + 2k} \right]$ converges.

(b) Find the sum of the series.

45. (a) Show that the series $\displaystyle\sum_{k=1}^{\infty} \left[\left(-\frac{1}{4} \right)^k + \frac{3}{k(k+1)} \right]$ converges.

(b) Find the sum of the series.

Challenge Problems

46. (a) Determine whether the series
$$1 - 1 - \frac{1}{2} + \frac{1}{3} + \frac{1}{3} - \frac{1}{9} - \frac{1}{4} + \frac{1}{27} + \frac{1}{5} - \frac{1}{81} - \cdots \text{ converges}$$
or diverges.

(b) Find the sum of the series if it converges.

In Problems 47 and 48, determine whether each series converges or diverges.

47. $\displaystyle\sum_{k=2}^{\infty} \frac{\ln k}{2k^3 - 1}$

48. $\displaystyle\sum_{k=1}^{\infty} \sin^3 \left(\frac{1}{k} \right)$

10.8 Power Series

OBJECTIVES *When you finish this section, you should be able to:*

1 **Determine whether a power series converges (p. 857)**
2 **Find the interval of convergence of a power series (p. 860)**
3 **Define a function using a power series (p. 862)**
4 **Use properties of power series (p. 864)**

In this section, we study series with variable terms, called *power series*. Just as a polynomial is the sum of a *finite* number of monomials, a power series is the sum of an *infinite* number of monomials.

DEFINITION Power Series

If x is a variable, then a series of the form

$$\sum_{k=0}^{\infty} a_k x^k = a_0 + \sum_{k=1}^{\infty} a_k x^k = a_0 + a_1 x + a_2 x^2 + \cdots$$

where the coefficients $a_0,\ a_1,\ a_2, \ldots$ are constants, is called a **power series in x** or a **power series centered at 0**.

A series of the form

$$\sum_{k=0}^{\infty} a_k (x - c)^k = a_0 + \sum_{k=1}^{\infty} a_k (x - c)^k = a_0 + a_1 (x - c) + a_2 (x - c)^2 + \cdots$$

where c is a constant, is called a **power series in $(x - c)$** or a **power series centered at c**.

IN WORDS A power series $\displaystyle\sum_{k=0}^{\infty} a_k x^k$ is a sum of an infinite number of monomials.

If $x = 0$ in a power series $\displaystyle\sum_{k=0}^{\infty} a_k x^k$, or if $x = c$ in a power series $\displaystyle\sum_{k=0}^{\infty} a_k (x - c)^k$, then the power series equals a_0.

① Determine Whether a Power Series Converges

For a particular value of x, a power series in x reduces to a series of real numbers like the series studied so far. For example, $\sum_{k=1}^{\infty} \dfrac{x^k}{k}$ is a power series in x. The series converges (to 0) if $x = 0$. If $x = 1$, it becomes the harmonic series $\sum_{k=1}^{\infty} \dfrac{1}{k}$, which is divergent.

If $x = -1$, it becomes the alternating harmonic series $\sum_{k=1}^{\infty} (-1)^k \dfrac{1}{k}$, which is convergent.

To find all numbers x for which a power series in x is convergent, the Ratio Test (p. 845) or the Root Test (p. 848) is usually used since, in a power series, x is raised to a power.

EXAMPLE 1 **Determining Whether a Power Series Converges**

Find all numbers x for which each power series in x converges.

(a) $\displaystyle\sum_{k=0}^{\infty} \frac{x^k}{k!} = 1 + x + \frac{x^2}{2!} + \frac{x^3}{3!} + \cdots$

(b) $\displaystyle\sum_{k=0}^{\infty} \frac{kx^k}{4^k} = \frac{x}{4} + \frac{2x^2}{4^2} + \frac{3x^3}{4^3} + \cdots$

(c) $\displaystyle\sum_{k=0}^{\infty} k!\, x^k = 1 + x + 2!\, x^2 + 3!\, x^3 + \cdots$

Solution

(a) For the series $\displaystyle\sum_{k=0}^{\infty} \frac{x^k}{k!}$, we use the Ratio Test with

$$a_n = \frac{x^n}{n!} \qquad \text{and} \qquad a_{n+1} = \frac{x^{n+1}}{(n+1)!}$$

Then

$$\lim_{n \to \infty} \left| \frac{a_{n+1}}{a_n} \right| = \lim_{n \to \infty} \left| \frac{\dfrac{x^{n+1}}{(n+1)!}}{\dfrac{x^n}{n!}} \right| = \lim_{n \to \infty} \frac{|x|^{n+1}\, n!}{(n+1)!\, |x|^n} = |x| \lim_{n \to \infty} \frac{1}{n+1} = 0$$

Since the limit is less than 1 for every number x, it follows from the Ratio Test that the power series $\displaystyle\sum_{k=0}^{\infty} \frac{x^k}{k!}$ is absolutely convergent for all real numbers.

NOTE Since $\displaystyle\sum_{k=0}^{\infty} \frac{x^k}{k!}$ converges absolutely for every number x, the limit of the nth term equals 0. That is, $\displaystyle\lim_{n \to \infty} \frac{x^n}{n!} = 0$ for every number x.

(b) For $\displaystyle\sum_{k=0}^{\infty} \frac{kx^k}{4^k}$, we use the Ratio Test with $a_n = \dfrac{nx^n}{4^n}$ and $a_{n+1} = \dfrac{(n+1)x^{n+1}}{4^{n+1}}$. Then

$$\lim_{n \to \infty} \left| \frac{a_{n+1}}{a_n} \right| = \lim_{n \to \infty} \frac{\dfrac{(n+1)\,|x|^{n+1}}{4^{n+1}}}{\dfrac{n\,|x|^n}{4^n}} = \lim_{n \to \infty} \frac{(n+1)\,|x|^{n+1} \cdot 4^n}{4^{n+1} \cdot n\,|x|^n}$$

$$= |x| \lim_{n \to \infty} \frac{n+1}{4n} = \frac{|x|}{4}$$

(Example continued on the next page)

By the Ratio Test, the series converges absolutely if $\dfrac{|x|}{4} < 1$, or equivalently if $|x| < 4$.

It diverges if $\dfrac{|x|}{4} > 1$ or equivalently if $|x| > 4$. The Ratio Test gives no information when $\dfrac{|x|}{4} = 1$, that is, when $x = -4$ or $x = 4$. However, we can check these values directly by replacing x by 4 and -4.

For $x = 4$, the series becomes

$$\sum_{k=0}^{\infty} \frac{k 4^k}{4^k} = \sum_{k=1}^{\infty} k = 1 + 2 + \cdots$$

which diverges.

For $x = -4$, the series becomes

$$\sum_{k=0}^{\infty} \frac{k(-4)^k}{4^k} = \sum_{k=0}^{\infty} \frac{(-1)^k k (4^k)}{4^k} = \sum_{k=0}^{\infty} (-1)^k k = -1 + 2 - 3 + \cdots$$

which also diverges. (Look at the sequence of partial sums.)

The series $\displaystyle\sum_{k=0}^{\infty} \frac{k x^k}{4^k}$ converges absolutely for $-4 < x < 4$ and diverges for $|x| \geq 4$.

(c) For $\displaystyle\sum_{k=0}^{\infty} k! x^k$, we use the Ratio Test with $a_n = n! x^n$ and $a_{n+1} = (n+1)! x^{n+1}$. Then

$$\lim_{n \to \infty} \left| \frac{a_{n+1}}{a_n} \right| = \lim_{n \to \infty} \frac{(n+1)! |x|^{n+1}}{n! \, |x|^n} = |x| \lim_{n \to \infty} (n+1) = \begin{cases} 0 & \text{if } x = 0 \\ \infty & \text{if } x \neq 0 \end{cases}$$

We conclude that the power series $\displaystyle\sum_{k=0}^{\infty} k! \, x^k$ converges only when $x = 0$. For any other number x, the power series diverges. ∎

NOW WORK Problem 15 and AP® Practice Problem 5.

The next theorem gives more information about the numbers for which a power series converges or diverges.

THEOREM Convergence/Divergence of a Power Series

(a) If the power series $\displaystyle\sum_{k=0}^{\infty} a_k x^k$ converges for a number $x_0 \neq 0$, then it converges absolutely for all numbers x for which $|x| < |x_0|$.

(b) If the power series $\displaystyle\sum_{k=0}^{\infty} a_k x^k$ diverges for a number x_1, then it diverges for all numbers x for which $|x| > |x_1|$.

RECALL If a series $\displaystyle\sum_{k=0}^{\infty} a_k$ converges, then $\displaystyle\lim_{n \to \infty} a_n = 0$.

NOTE The definition of the limit of a sequence is found on page 782.

Proof Part (a) Assume that $\displaystyle\sum_{k=0}^{\infty} a_k x_0^k$ converges. Then

$$\lim_{n \to \infty} \left(a_n x_0^n \right) = 0$$

Using the definition of the limit of a sequence and choosing $\varepsilon = 1$, there is a positive integer N for which $|a_n x_0^n| < 1$ for all $n > N$. Now for any number x for which $|x| < |x_0|$, we have

$$|a_n x^n| = \left| \frac{a_n x^n x_0^n}{x_0^n} \right| = |a_n x_0^n| \left| \frac{x}{x_0} \right|^n \underset{\substack{\uparrow \\ |a_n x_0^n| < 1}}{<} \left| \frac{x}{x_0} \right|^n \qquad \text{for } n > N$$

Since $|x| < |x_0|$, then $\left| \dfrac{x}{x_0} \right| < 1$, so the series $\displaystyle\sum_{k=0}^{\infty} \left| \dfrac{x}{x_0} \right|^k$ is a convergent geometric series.

Therefore, by the Algebraic Comparison Test for Convergence, the series $\displaystyle\sum_{k=0}^{\infty} \left| a_k x^k \right|$ converges, and so the power series $\displaystyle\sum_{k=0}^{\infty} a_k x^k$ converges absolutely for all numbers x for which $|x| < |x_0|$.

Part (b) Suppose the series converges for some number x, $|x| > |x_1|$. Then it must converge for x_1 [by Part (a)], which contradicts the hypothesis of the theorem. Therefore, the series diverges for all x for which $|x| > |x_1|$. ∎

EXAMPLE 2 **Using the Convergence/Divergence of a Power Series Theorem**

(a) If the power series $\displaystyle\sum_{k=0}^{\infty} a_k x^k$ converges for $x = 6$, then it converges absolutely for $x = 3$.

(b) If the power series $\displaystyle\sum_{k=0}^{\infty} a_k x^k$ diverges for $x = 4$, then it also diverges for $x = -5$. ∎

NOW WORK Problems 9, 51, and AP® Practice Problems 1, 2 and 7.

The next result is a consequence of the previous theorem. It states that every power series belongs to one of three categories.

THEOREM

For a power series $\displaystyle\sum_{k=0}^{\infty} a_k (x - c)^k$, exactly one of the following is true:

- The series converges only if $x = c$.
- The series converges absolutely for all x.
- There is a positive number R for which the series converges absolutely for all x, $|x - c| < R$, and diverges for all x, $|x - c| > R$. The behavior of the series at $|x - c| = R$ must be determined separately.

In the theorem, the number R is called the **radius of convergence**. If the series converges only for $x = c$, then $R = 0$; if the series converges absolutely for all x, then $R = \infty$. If the series converges absolutely for $|x - c| < R$, $0 < R < \infty$, we call the set of all numbers x for which the power series converges the **interval of convergence** of the power series. Once the radius R of convergence is determined, we test the endpoints $x = c - R$ and $x = c + R$ to find the interval of convergence.

As Example 1 illustrates, the Ratio Test is a useful method for determining the radius of convergence of a power series. However, the test gives no information about convergence or divergence at the endpoints of the interval of convergence. At an endpoint, a power series may be convergent or divergent.

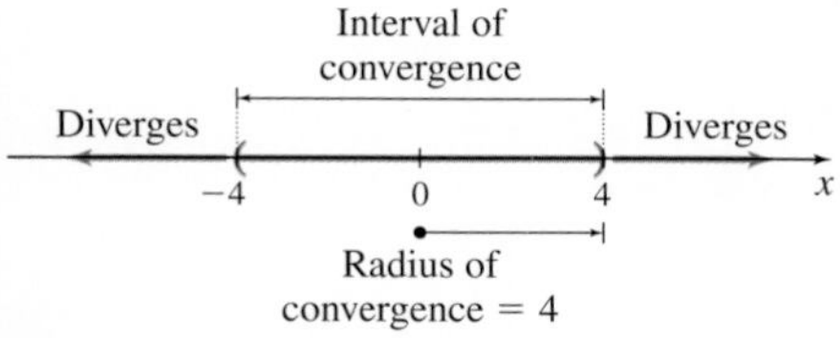

For example, the series $\displaystyle\sum_{k=0}^{\infty} \dfrac{k x^k}{4^k}$ in Example 1(b) converges absolutely for $|x| < 4$ and diverges for $|x| \geq 4$. So, the radius of convergence is $R = 4$, and the interval of convergence is the open interval $(-4, 4)$, as shown in Figure 26.

NOW WORK Problems 4 and 11 and AP® Practice Problem 8.

Figure 26 Convergence/divergence of $\displaystyle\sum_{k=0}^{\infty} \dfrac{k\, x^k}{4^k}$

2 Find the Interval of Convergence of a Power Series

EXAMPLE 3 Finding the Interval of Convergence of a Power Series

Find the radius of convergence and the interval of convergence of the power series

$$\sum_{k=1}^{\infty} \frac{x^{2k}}{k}$$

Solution

We use the Ratio Test with $a_n = \dfrac{x^{2n}}{n}$ and $a_{n+1} = \dfrac{x^{2(n+1)}}{n+1}$. Then

$$\lim_{n \to \infty} \left| \frac{a_{n+1}}{a_n} \right| = \lim_{n \to \infty} \left| \frac{\dfrac{x^{2(n+1)}}{n+1}}{\dfrac{x^{2n}}{n}} \right| = \lim_{n \to \infty} \left| \frac{x^{2n+2}}{n+1} \cdot \frac{n}{x^{2n}} \right| = x^2 \lim_{n \to \infty} \frac{n}{n+1} = x^2$$

The series converges absolutely if $x^2 < 1$, or equivalently, if $-1 < x < 1$.
The radius of convergence is $R = 1$.

To find the interval of convergence, we test the endpoints: $x = -1$ and $x = 1$.
When $x = 1$ or $x = -1$, $\dfrac{x^{2k}}{k} = \dfrac{(x^2)^k}{k} = \dfrac{1^k}{k} = \dfrac{1}{k}$, so the series reduces to the harmonic series $\sum_{k=1}^{\infty} \dfrac{1}{k}$, which diverges. Consequently, the interval of convergence is $(-1, 1)$, as shown in Figure 27. ∎

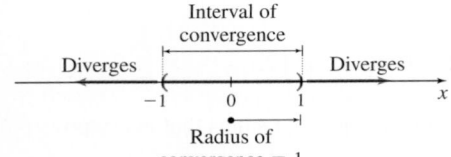

Interval of convergence

Diverges   Diverges

Radius of convergence = 1

Figure 27 Convergence/divergence of $\sum_{k=0}^{\infty} \dfrac{x^{2k}}{k}$

NOW WORK Problem 19(a) and AP® Practice Problem 4.

The power series discussed next converges for all real numbers.

EXAMPLE 4 Finding the Interval of Convergence of a Power Series

Find the radius of convergence R and the interval of convergence of the power series

$$\sum_{k=0}^{\infty} \frac{x^k}{(k+2)^{2k}}$$

Solution

We use the Root Test with $a_n = \dfrac{x^n}{(n+2)^{2n}}$. Then

$$\lim_{n \to \infty} \sqrt[n]{\left| \frac{x^n}{(n+2)^{2n}} \right|} = \lim_{n \to \infty} \frac{|x|}{(n+2)^2} = |x| \lim_{n \to \infty} \frac{1}{(n+2)^2} = 0$$

The series converges absolutely for all x. The radius of convergence is $R = \infty$, and the interval of convergence is $(-\infty, \infty)$. ∎

NOW WORK Problems 19(b) and 21.

The methods used in Examples 3 and 4 can also be used to find the radius of convergence of a power series in $x - c$.

EXAMPLE 5   Finding the Interval of Convergence of a Power Series

Find the radius of convergence R and the interval of convergence of the power series

$$\sum_{k=0}^{\infty}(-1)^k \frac{(x-2)^k}{k+1}$$

Solution

$\sum_{k=0}^{\infty}(-1)^k \dfrac{(x-2)^k}{k+1}$ is a power series centered at 2. Use the Ratio Test

with $a_n = (-1)^n \dfrac{(x-2)^n}{n+1}$ and $a_{n+1} = (-1)^{n+1} \dfrac{(x-2)^{n+1}}{n+2}$. Then

$$\lim_{n \to \infty}\left|\frac{a_{n+1}}{a_n}\right| = \lim_{n \to \infty}\left|\frac{\dfrac{(-1)^{n+1}(x-2)^{n+1}}{n+2}}{\dfrac{(-1)^n(x-2)^n}{n+1}}\right| = \lim_{n \to \infty}\left|\frac{(n+1)(x-2)}{n+2}\right|$$

$$= |x-2| \lim_{n \to \infty}\frac{n+1}{n+2} = |x-2|$$

The series converges absolutely if $|x-2| < 1$, or equivalently if $1 < x < 3$. The radius of convergence is $R = 1$.

To find the interval of convergence, test the endpoints.

- If $x = 1$,

$$\sum_{k=0}^{\infty}(-1)^k \frac{(x-2)^k}{k+1} = \sum_{k=0}^{\infty}(-1)^k \frac{(-1)^k}{k+1} = \sum_{k=0}^{\infty}\frac{(-1)^{2k}}{k+1} = \sum_{k=0}^{\infty}\frac{1}{k+1}$$

$$= 1 + \frac{1}{2} + \frac{1}{3} + \cdots + \frac{1}{n+1} + \cdots$$

which is the divergent harmonic series.

- If $x = 3$,

$$\sum_{k=0}^{\infty}(-1)^k \frac{(x-2)^k}{k+1} = \sum_{k=0}^{\infty}(-1)^k \frac{1}{k+1} = 1 - \frac{1}{2} + \frac{1}{3} - \frac{1}{4} + \cdots + \frac{(-1)^n}{n+1} + \cdots$$

which is the convergent alternating harmonic series.

The interval of convergence of the series $\sum_{k=0}^{\infty}(-1)^k \dfrac{(x-2)^k}{k+1}$ is $(1, 3]$. See Figure 28. ∎

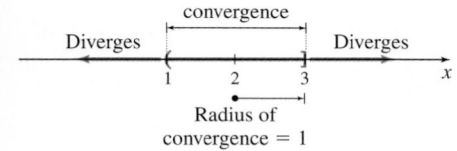

Diverges ← | Interval of convergence | → Diverges

1   2   3 x

Radius of convergence = 1

Figure 28 Convergence/divergence of $\sum_{k=0}^{\infty}(-1)\dfrac{(x-2)^k}{k+1}$

NOW WORK   Problem 29 and AP® Practice Problems 3 and 9.

***STEPS* for finding the radius of convergence R and the interval of convergence of a power series**

STEP 1 Use the Ratio Test (or Root Test) to find the radius R of convergence.

STEP 2 If $R > 0$, the interval of convergence is $(c - R, c + R)$ or $(c - R, c + R]$ or $[c - R, c + R)$ or $[c - R, c + R]$.

STEP 3 Test the endpoints $c - R$ and $c + R$ to determine the correct interval of convergence.

- If the power series is centered at 0, the interval of convergence is also centered at 0.

- If the power series is centered at a number c, $c \neq 0$, the interval of convergence is centered at c.

3 Define a Function Using a Power Series

A power series $\sum\limits_{k=0}^{\infty} a_k x^k$ defines a function

$$f(x) = a_0 + a_1 x + a_2 x^2 + \cdots + a_n x^n + \cdots$$

The domain of f is the interval of convergence of the power series.

If f is defined by the power series $\sum\limits_{k=0}^{\infty} a_k x^k$, whose interval of convergence is I, and if x_0 is a number in I, then f can be evaluated at x_0 by finding the sum of the series

$$f(x_0) = \sum_{k=0}^{\infty} a_k x_0^k = a_0 + a_1 x_0 + a_2 x_0^2 + \cdots + a_n x_0^n + \cdots$$

EXAMPLE 6   Analyzing a Function Defined by a Power Series

A function f is defined by the power series $f(x) = \sum\limits_{k=0}^{\infty} x^k$.

(a) Find the domain of f.

(b) Evaluate $f\left(\dfrac{1}{2}\right)$ and $f\left(-\dfrac{1}{3}\right)$.

(c) Find f by summing the series.

Solution

(a) $\sum\limits_{k=0}^{\infty} x^k$ is a power series centered at 0 with $a_k = 1$. Then

$$f(x) = 1 + x + x^2 + x^3 + x^4 + \cdots$$

The domain of f equals the interval of convergence of the power series.

The series $\sum\limits_{k=0}^{\infty} x^k$ is a geometric series, so it converges for $|x| < 1$. The radius R of convergence is $R = 1$, and the interval of convergence is $(-1, 1)$. The domain of f is the open interval $(-1, 1)$.

(b) The numbers $\dfrac{1}{2}$ and $-\dfrac{1}{3}$ are in the interval $(-1, 1)$, so they are in the domain of f.

Then $f\left(\dfrac{1}{2}\right)$ is a geometric series with $r = \dfrac{1}{2}$, $a = 1$, and

$$f\left(\frac{1}{2}\right) = 1 + \frac{1}{2} + \left(\frac{1}{2}\right)^2 + \left(\frac{1}{2}\right)^3 + \cdots = \frac{a}{1-r} = \frac{1}{1 - \dfrac{1}{2}} = 2$$

Similarly,

$$f\left(-\frac{1}{3}\right) = 1 - \frac{1}{3} + \left(-\frac{1}{3}\right)^2 + \left(-\frac{1}{3}\right)^3 + \cdots = \frac{1}{1 + \dfrac{1}{3}} = \frac{3}{4}$$

(c) Since f is defined by a geometric series, we can find f by summing the series.

$$f(x) = \sum_{k=0}^{\infty} x^k = 1 + x + x^2 + \cdots + x^n + \cdots = \underset{\underset{a=1;\, r=x}{\uparrow}}{\frac{1}{1-x}} \quad -1 < x < 1$$

∎

In Example 6, the function f defined by the power series $\sum_{k=0}^{\infty} x^k$ was found to be

$$f(x) = \frac{1}{1-x} = 1 + x + x^2 + x^3 + \cdots = \sum_{k=0}^{\infty} x^k \quad -1 < x < 1 \tag{1}$$

We say that the function f is **represented** by the power series.

NOW WORK Problem 45.

Many other functions can be represented by power series formed from variations of the function $f(x) = \dfrac{1}{1-x} = \sum_{k=0}^{\infty} x^k, -1 < x < 1.$

EXAMPLE 7 **Representing a Function by a Power Series Centered at 0**

Each of the following functions is a variation of the function $f(x) = \dfrac{1}{1-x}$. Represent each function by a power series centered at 0:

(a) $h(x) = \dfrac{1}{1 - 2x^2}$ **(b)** $g(x) = \dfrac{1}{3+x}$ **(c)** $F(x) = \dfrac{x^2}{1-x}$

Solution

(a) In the power series representation for $f(x) = \dfrac{1}{1-x}$, statement (1), replace x by $2x^2$. The resulting series

$$\sum_{k=0}^{\infty} (2x^2)^k = 1 + 2x^2 + (2x^2)^2 + \cdots + (2x^2)^n + \cdots$$

converges if $|2x^2| < 1$, or equivalently if $-\dfrac{\sqrt{2}}{2} < x < \dfrac{\sqrt{2}}{2}$. Then on the open interval $\left(-\dfrac{\sqrt{2}}{2}, \dfrac{\sqrt{2}}{2}\right)$, the function $h(x) = \dfrac{1}{1-2x^2}$ is represented by the power series

$$h(x) = \frac{1}{1-2x^2} = 1 + (2x^2) + (2x^2)^2 + (2x^2)^3 + \cdots$$

$$= 1 + 2x^2 + 4x^4 + 8x^6 + \cdots + 2^n x^{2n} + \cdots = \sum_{k=0}^{\infty} (2x^2)^k = \sum_{k=0}^{\infty} 2^k x^{2k}$$

(b) We begin by writing

$$g(x) = \frac{1}{3+x} = \frac{1}{3}\left(\frac{1}{1+\dfrac{x}{3}}\right) = \frac{1}{3}\left[\frac{1}{1-\left(-\dfrac{x}{3}\right)}\right]$$

Now in the power series representation for $f(x) = \dfrac{1}{1-x}$, replace x by $-\dfrac{x}{3}$. The resulting series

$$\sum_{k=0}^{\infty}\left(-\frac{x}{3}\right)^k = 1 + \left(-\frac{x}{3}\right) + \left(-\frac{x}{3}\right)^2 + \cdots + \left(-\frac{x}{3}\right)^n + \cdots$$

converges if $\left|-\dfrac{x}{3}\right| < 1$, or equivalently if $-3 < x < 3$. Then in the open interval $(-3, 3)$, $g(x) = \dfrac{1}{3+x}$ is represented by the power series

$$g(x) = \frac{1}{3+x} = \frac{1}{3}\left[1 + \left(-\frac{x}{3}\right) + \left(-\frac{x}{3}\right)^2 + \left(-\frac{x}{3}\right)^3 + \cdots\right] = \frac{1}{3}\sum_{k=0}^{\infty}(-1)^k\left(\frac{x}{3}\right)^k$$

$$= \sum_{k=0}^{\infty}\frac{(-1)^k x^k}{3^{k+1}}$$

(Example continued on the next page)

(c) $F(x) = \dfrac{x^2}{1-x} = x^2 \left(\dfrac{1}{1-x} \right)$. Now for all numbers in the interval $(-1,\ 1)$,

$$\frac{1}{1-x} = 1 + x + x^2 + \cdots + x^n + \cdots$$

So for any number x in the interval of convergence, $-1 < x < 1$, we have

$$F(x) = x^2 \left(1 + x + x^2 + \cdots + x^n + \cdots \right) = x^2 + x^3 + x^4 + \cdots + x^{n+2} + \cdots = \sum_{k=2}^{\infty} x^k$$

> **NOTE** Notice that the power series representation for F begins at $k = 2$.

∎

NOW WORK Problem 55.

---

4 Use Properties of Power Series

The function f represented by a power series has properties similar to those of a polynomial. We state three of these properties without proof.

> **THEOREM** Properties of Power Series
>
> Suppose $\displaystyle\sum_{k=0}^{\infty} a_k x^k$ is a power series in x having a nonzero radius of convergence R.
>
> Define the function f as
>
> $$f(x) = \sum_{k=0}^{\infty} a_k x^k = a_0 + a_1 x + a_2 x^2 + \cdots + a_n x^n + \cdots \qquad -R < x < R$$
>
> • *Continuity property:*
>
> $$\boxed{\lim_{x \to x_0} \left(\sum_{k=0}^{\infty} a_k x^k \right) = \sum_{k=0}^{\infty} \left(\lim_{x \to x_0} a_k x^k \right) = \sum_{k=0}^{\infty} a_k x_0^k \qquad -R < x_0 < R}$$
>
> • *Differentiation property:*
>
> $$\boxed{\frac{d}{dx} \left(\sum_{k=0}^{\infty} a_k x^k \right) = \sum_{k=0}^{\infty} \left(\frac{d}{dx} a_k x^k \right) = \sum_{k=1}^{\infty} k a_k x^{k-1}}$$
>
> • *Integration property:*
>
> $$\boxed{\int_0^x \left(\sum_{k=0}^{\infty} a_k t^k \right) dt = \sum_{k=0}^{\infty} \left(\int_0^x a_k t^k\, dt \right) = \sum_{k=0}^{\infty} \frac{a_k x^{k+1}}{k+1}}$$
>
> The power series obtained by differentiating or integrating $\displaystyle\sum_{k=0}^{\infty} a_k x^k$ is convergent and has the same radius of convergence R as $\displaystyle\sum_{k=0}^{\infty} a_k x^k$.

> **CAUTION** The theorem states that the *radii of convergence* of the three series
>
> $$\sum_{k=0}^{\infty} a_k x^k,\ \sum_{k=1}^{\infty} k a_k x^{k-1},\ \text{and}\ \sum_{k=0}^{\infty} \frac{a_k x^{k+1}}{k+1}$$
>
> are the same. This does not imply that the *intervals of convergence* are the same; the endpoints of the interval must be checked separately for each series. For example, the interval of convergence of $\displaystyle\sum_{k=1}^{\infty} \frac{x^k}{k}$ is $[-1,\ 1)$, but the interval of convergence of its derivative $\displaystyle\sum_{k=2}^{\infty} x^{k-1}$ is $(-1,\ 1)$.

The differentiation and integration properties of power series state that a power series can be differentiated and integrated term-by-term and that the resulting series represent the derivative and integral, respectively, of the function represented by the original power series. Moreover, it can be shown that the power series obtained by differentiating (or integrating) a power series whose radius of convergence is R, converges and has the same radius of convergence R as the original power series. (See Problem 86.)

The differentiation and integration properties can be used to obtain new functions defined by a power series.

EXAMPLE 8   **Using the Differentiation Property of Power Series**

Use the differentiation property on the power series $f(x) = \dfrac{1}{1-x} = \sum\limits_{k=0}^{\infty} x^k$ to find a power series representation for $f'(x) = \dfrac{1}{(1-x)^2}$.

Solution

The function $f(x) = \dfrac{1}{1-x}$, defined on the open interval $(-1, \ 1)$, is represented by the power series

$$f(x) = \frac{1}{1-x} = 1 + x + x^2 + \cdots + x^n + \cdots = \sum_{k=0}^{\infty} x^k$$

Using the differentiation property,

$$f'(x) = \frac{1}{(1-x)^2} = 1 + 2x + 3x^2 + \cdots + nx^{n-1} + \cdots = \sum_{k=1}^{\infty} kx^{k-1}$$

whose radius of convergence is 1. ∎

NOW WORK   Problems 59(a), 65, and AP® Practice Problem 6.

EXAMPLE 9   **Finding the Power Series Representation for $\ln\dfrac{1}{1-x}$**

(a) Find the power series representation for $\ln\dfrac{1}{1-x}$.

(b) Find its interval of convergence.

(c) Find $\ln 2$.

Solution

(a) If $y = \ln\dfrac{1}{1-x}$, then $y' = \dfrac{1}{1-x}$. That is, y' is represented by the geometric series $\sum\limits_{k=0}^{\infty} x^k$, which has a radius of convergence 1. So, if we use the integration property of power series for $y' = \dfrac{1}{1-x}$, we obtain a series for $y = \ln\dfrac{1}{1-x}$, which will also have a radius of convergence of 1.

$$y' = \frac{1}{1-x} = 1 + x + x^2 + \cdots + x^n + \cdots = \sum_{k=0}^{\infty} x^k$$

$$\int_0^x \frac{1}{1-t}\, dt = \int_0^x (1 + t + t^2 + \cdots + t^n + \cdots)\, dt$$

$$\ln\frac{1}{1-x} = x + \frac{x^2}{2} + \frac{x^3}{3} + \cdots + \frac{x^{n+1}}{n+1} + \cdots = \sum_{k=0}^{\infty} \frac{x^{k+1}}{k+1}$$

The radius of convergence of this series is 1.

(Example continued on the next page)

(b) To find the interval of convergence, we investigate the endpoints. When $x = 1$,

$$x + \frac{x^2}{2} + \frac{x^3}{3} + \cdots + \frac{x^{n+1}}{n+1} + \cdots = 1 + \frac{1}{2} + \frac{1}{3} + \cdots$$

the harmonic series, which diverges.

When $x = -1$,

$$x + \frac{x^2}{2} + \frac{x^3}{3} + \cdots + \frac{x^{n+1}}{n+1} + \cdots = -1 + \frac{1}{2} - \frac{1}{3} + \cdots + (-1)^{n+1}\frac{1}{n+1} + \cdots$$

an alternating harmonic series, which converges.

The interval of convergence of the power series $\sum\limits_{k=0}^{\infty} \dfrac{x^{k+1}}{k+1}$ is $[-1, 1)$. So,

$$\boxed{\ln \frac{1}{1-x} = x + \frac{x^2}{2} + \frac{x^3}{3} + \cdots = \sum_{k=0}^{\infty} \frac{x^{k+1}}{k+1} \qquad -1 \le x < 1} \qquad (2)$$

(c) To find $\ln 2$, notice that when $x = -1$, we have $\ln \dfrac{1}{1-x} = \ln \dfrac{1}{2} = -\ln 2$. In the power series (2), let $x = -1$. Then

$$\ln \frac{1}{2} = -\ln 2 = -1 + \frac{1}{2} - \frac{1}{3} + \cdots$$

$$\boxed{\ln 2 = 1 - \frac{1}{2} + \frac{1}{3} \cdots = \sum_{k=0}^{\infty} \frac{(-1)^k}{k+1}}$$

The sum of the alternating harmonic series is $\ln 2$. ∎

NOW WORK   Problem 59(b) and AP® Practice Problems 7 and 10.

EXAMPLE 10   Finding the Power Series Representation for $\tan^{-1} x$

Show that the power series representation for $\tan^{-1} x$ is

$$\boxed{\tan^{-1} x = x - \frac{x^3}{3} + \frac{x^5}{5} - \frac{x^7}{7} + \cdots + (-1)^n \frac{x^{2n+1}}{2n+1} + \cdots = \sum_{k=0}^{\infty} \frac{(-1)^k x^{2k+1}}{2k+1}}$$

Find the radius of convergence and the interval of convergence.

Solution

If $y = \tan^{-1} x$, then $y' = \dfrac{1}{1+x^2}$, which is the sum of the geometric series $\sum\limits_{k=0}^{\infty} (-1)^k x^{2k}$. That is,

$$\frac{1}{1+x^2} = \sum_{k=0}^{\infty} (-1)^k x^{2k} = 1 - x^2 + x^4 - x^6 + \cdots$$

This series converges when $|x^2| < 1$, or equivalently for $-1 < x < 1$. Now use the integration property of power series to integrate $y' = \dfrac{1}{1+x^2}$ and obtain a series for $y = \tan^{-1} x$.

$$\int_0^x \frac{dt}{1+t^2} = \int_0^x (1 - t^2 + t^4 - \cdots)\, dt$$

$$\tan^{-1} x = x - \frac{x^3}{3} + \frac{x^5}{5} - \frac{x^7}{7} + \cdots + (-1)^n \frac{x^{2n+1}}{2n+1} + \cdots = \sum_{k=0}^{\infty} (-1)^k \frac{x^{2k+1}}{2k+1}$$

ORIGINS The power series representation for $\tan^{-1} x$ is called **Gregory's series** (or Gregory–Leibniz series or Madhava–Gregory series). James Gregory (1638–1675) was a Scottish mathematician. His mother, Janet Anderson, was his teacher and taught him geometry. Gregory, like many other mathematicians of his time, was searching for a good way to approximate π, a result of which was Gregory's series. Gregory was also a major contributor to the theory of optics, and he is credited with inventing the reflective telescope.

The radius of convergence is 1. To find the interval of convergence, we check $x = -1$ and $x = 1$. For $x = -1$,

$$-1 + \frac{1}{3} - \frac{1}{5} + \frac{1}{7} - \cdots$$

For $x = 1$, we get

$$1 - \frac{1}{3} + \frac{1}{5} - \frac{1}{7} + \cdots$$

Both of these series satisfy the two conditions of the Alternating Series Test, and so each one converges. The interval of convergence is the closed interval $[-1, \ 1]$. ∎

Since $\tan^{-1} 1 = \dfrac{\pi}{4}$, we can use Gregory's series to approximate π. Then

$$\frac{\pi}{4} = \sum_{k=0}^{\infty} (-1)^k \frac{1^{2k+1}}{2k+1} = \sum_{k=0}^{\infty} \frac{(-1)^k}{2k+1} = 1 - \frac{1}{3} + \frac{1}{5} - \cdots$$

$$\pi = 4\left(1 - \frac{1}{3} + \frac{1}{5} - \cdots\right)$$

While we now have an approximation for π, unfortunately the series converges very slowly, requiring many terms to get close to π. See Problem 82.

COMMENT The approximation for π is an alternating harmonic series, which is conditionally convergent. Any rearrangement of the terms can lead to a different sum.

NOW WORK Problem 69.

10.8 Assess Your Understanding

Concepts and Vocabulary

1. **True or False** Every power series $\sum\limits_{k=0}^{\infty} a_k (x - c)^k$ converges for at least one number.

2. **True or False** Let b_n denote the nth term of the power series $\sum\limits_{k=0}^{\infty} a_k x^k$. If $\lim\limits_{n \to \infty} \left| \dfrac{b_{n+1}}{b_n} \right| < 1$ for every number x, then $\sum\limits_{k=0}^{\infty} a_k x^k$ is absolutely convergent on the interval $(-\infty, \infty)$.

3. **True or False** If the radius of convergence of a power series $\sum\limits_{k=0}^{\infty} a_k x^k$ is 0, then the power series converges only for $x = 0$.

4. **True or False** If a power series converges at one endpoint of its interval of convergence, then it must converge at its other endpoint.

5. **True or False** The power series $\sum\limits_{k=0}^{\infty} a_k x^k$ and $\sum\limits_{k=0}^{\infty} a_k (x - 3)^k$ have the same radius of convergence.

6. **True or False** The power series $\sum\limits_{k=0}^{\infty} a_k x^k$ and $\sum\limits_{k=0}^{\infty} a_k (x - 3)^k$ have the same interval of convergence.

7. **True or False** If the power series $\sum\limits_{k=0}^{\infty} a_k x^k$ converges for $x = 8$, then it converges for $x = -8$.

8. **True or False** If the power series $\sum\limits_{k=0}^{\infty} a_k x^k$ converges for $x = 3$, then it converges for $x = 1$.

9. **True or False** If the power series $\sum\limits_{k=0}^{\infty} a_k x^k$ converges for $x = -4$, then it converges for $x = 3$.

10. **True or False** If the power series $\sum\limits_{k=0}^{\infty} a_k x^k$ converges for $x = 3$, then it diverges for $x = 5$.

11. **True or False** A possible interval of convergence for the power series $\sum\limits_{k=0}^{\infty} a_k x^k$ is $[-2, 4]$.

12. **True or False** If the power series $\sum\limits_{k=0}^{\infty} a_k x^k$ diverges for a number x_1, then it converges for all numbers x for which $|x| < |x_1|$.

Skill Building

In Problems 13–16, find all numbers x for which each power series converges.

13. $\sum\limits_{k=0}^{\infty} k x^k$

14. $\sum\limits_{k=0}^{\infty} \dfrac{k x^k}{3^k}$

15. $\sum\limits_{k=0}^{\infty} \dfrac{(x+1)^k}{3^k}$

16. $\sum\limits_{k=1}^{\infty} \dfrac{(x-2)^k}{k^2}$

In Problems 17–26:

(a) *Use the Ratio Test to find the radius of convergence and the interval of convergence of each power series.*

(b) *Use the Root Test to find the radius of convergence and the interval of convergence of each power series.*

(c) *Which test, the Ratio Test or the Root Test, did you find easier to use? Give the reasons why.*

17. $\displaystyle\sum_{k=0}^{\infty} \frac{x^k}{2^k(k+1)}$

18. $\displaystyle\sum_{k=0}^{\infty} (-1)^k \frac{x^k}{2^k(k+1)}$

[PAGE 860] **19.** $\displaystyle\sum_{k=0}^{\infty} \frac{x^k}{k+5}$

20. $\displaystyle\sum_{k=0}^{\infty} \frac{x^k}{1+k^2}$

[PAGE 860] **21.** $\displaystyle\sum_{k=0}^{\infty} \frac{k^2 x^k}{3^k}$

22. $\displaystyle\sum_{k=0}^{\infty} \frac{2^k x^k}{3^k}$

23. $\displaystyle\sum_{k=0}^{\infty} \frac{kx^k}{2k+1}$

24. $\displaystyle\sum_{k=0}^{\infty} (6x)^k$

25. $\displaystyle\sum_{k=0}^{\infty} (x-3)^k$

26. $\displaystyle\sum_{k=0}^{\infty} \frac{k(2x)^k}{3^k}$

In Problems 27–44, find the radius of convergence and the interval of convergence of each power series.

27. $\displaystyle\sum_{k=1}^{\infty} \frac{x^k}{k^3}$

28. $\displaystyle\sum_{k=2}^{\infty} \frac{x^k}{\ln k}$

[PAGE 861] **29.** $\displaystyle\sum_{k=1}^{\infty} \frac{(x-2)^k}{k^3}$

30. $\displaystyle\sum_{k=0}^{\infty} \frac{k(x-2)^k}{3^k}$

31. $\displaystyle\sum_{k=0}^{\infty} \frac{(-1)^k}{(2k+1)!} x^{2k+1}$

32. $\displaystyle\sum_{k=1}^{\infty} (kx)^k$

33. $\displaystyle\sum_{k=1}^{\infty} \frac{kx^k}{\ln(k+1)}$

34. $\displaystyle\sum_{k=1}^{\infty} \frac{x^k}{\ln(k+1)}$

35. $\displaystyle\sum_{k=0}^{\infty} \frac{k(k+1)x^k}{4^k}$

36. $\displaystyle\sum_{k=1}^{\infty} \frac{(-1)^k(x-5)^k}{k(k+1)}$

37. $\displaystyle\sum_{k=0}^{\infty} (-1)^k \frac{(x-3)^{2k}}{9^k}$

38. $\displaystyle\sum_{k=0}^{\infty} \frac{x^k}{e^k}$

39. $\displaystyle\sum_{k=0}^{\infty} (-1)^k \frac{(2x)^k}{k!}$

40. $\displaystyle\sum_{k=0}^{\infty} \frac{(x+1)^k}{k!}$

41. $\displaystyle\sum_{k=0}^{\infty} (-1)^k \frac{(x-1)^{4k}}{k!}$

42. $\displaystyle\sum_{k=1}^{\infty} \frac{(x+1)^k}{k(k+1)(k+2)}$

43. $\displaystyle\sum_{k=1}^{\infty} \frac{k^k x^k}{k!}$

44. $\displaystyle\sum_{k=0}^{\infty} \frac{3^k(x-2)^k}{k!}$

[PAGE 863] **45.** A function f is defined by the power series $f(x) = \displaystyle\sum_{k=0}^{\infty} \frac{x^k}{3^k}$.

    (a) Find the domain of f.

    (b) Evaluate $f(2)$ and $f(-1)$.

    (c) Find f by summing the series.

46. A function f is defined by the power series

$$f(x) = \sum_{k=0}^{\infty} (-1)^k \left(\frac{x}{2}\right)^k$$

    (a) Find the domain of f.

    (b) Evaluate $f(0)$ and $f(1)$.

    (c) Find f by summing the series.

47. A function f is defined by the power series

$$f(x) = \sum_{k=0}^{\infty} \frac{(x-2)^k}{2^k}$$

    (a) Find the domain of f.

    (b) Evaluate $f(1)$ and $f(2)$.

    (c) Find f by summing the series.

48. A function f is defined by the power series

$$f(x) = \sum_{k=0}^{\infty} (-1)^k(x+3)^k$$

    (a) Find the domain of f.

    (b) Evaluate $f(-3)$ and $f(-2.5)$.

    (c) Find f by summing the series.

49. If $\displaystyle\sum_{k=0}^{\infty} a_k x^k$ converges for $x = 3$, what, if anything, can be said about the convergence at $x = 2$? Can anything be said about the convergence at $x = 5$?

50. If $\displaystyle\sum_{k=0}^{\infty} a_k(x-2)^k$ converges for $x = 6$, at what other numbers x must the series necessarily converge?

[PAGE 859] **51.** If the series $\displaystyle\sum_{k=0}^{\infty} a_k x^k$ converges for $x = 6$ and diverges for $x = -8$, what, if anything, can be said about the truth of the following statements?

    (a) The series converges for $x = 2$.

    (b) The series diverges for $x = 7$.

    (c) The series is absolutely convergent for $x = 6$.

    (d) The series converges for $x = -6$.

    (e) The series diverges for $x = 10$.

    (f) The series is absolutely convergent for $x = 4$.

52. If the radius of convergence of the power series $\displaystyle\sum_{k=0}^{\infty} a_k(x-3)^k$ is $R = 5$, what, if anything, can be said about the truth of the following statements?

    (a) The series converges for $x = 2$.

    (b) The series diverges for $x = 7$.

    (c) The series diverges for $x = 8$.

    (d) The series converges for $x = -6$.

    (e) The series converges for $x = -2$.

In Problems 53–58:

(a) *Use a geometric series to represent each function as a power series centered at 0.*

(b) *Determine the radius of convergence and the interval of convergence of each series.*

53. $f(x) = \dfrac{1}{1+x^3}$

54. $f(x) = \dfrac{1}{1-x^2}$

[PAGE 864] **55.** $f(x) = \dfrac{1}{6-2x}$

56. $f(x) = \dfrac{4}{x+2}$

57. $f(x) = \dfrac{x}{1+x^3}$

58. $f(x) = \dfrac{4x^2}{x+2}$

In Problems 59–62:

(a) Use the differentiation property of power series to find a power series representation for $f'(x)$.

(b) Use the integration property of power series to find a power series representation for the indefinite integral of f.

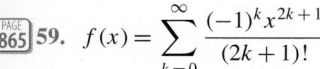 **59.** $f(x) = \sum_{k=0}^{\infty} \frac{(-1)^k x^{2k+1}}{(2k+1)!}$ **60.** $f(x) = \sum_{k=0}^{\infty} \frac{(-1)^k x^{2k}}{(2k)!}$

61. $f(x) = \sum_{k=0}^{\infty} \frac{x^k}{k!}$ **62.** $f(x) = \sum_{k=0}^{\infty} \frac{(-1)^k x^k}{k!}$

In Problems 63–70, find a power series representation of f. Use a geometric series and properties of a power series.

63. $f(x) = \dfrac{1}{(1+x)^2}$ **64.** $f(x) = \dfrac{1}{(1-x)^3}$

65. $f(x) = \dfrac{2}{3(1-x)^2}$ **66.** $f(x) = \dfrac{1}{(1-x)^4}$

67. $f(x) = \ln \dfrac{1}{1+x}$ **68.** $f(x) = \ln(1-2x)$

69. $f(x) = \ln(1-x^2)$ **70.** $f(x) = \ln(1+x^2)$

Applications and Extensions

In Problems 71–78, find all x for which each power series converges.

71. $\sum_{k=1}^{\infty} \dfrac{x^k}{k}$ **72.** $\sum_{k=1}^{\infty} \dfrac{(x-4)^k}{k}$

73. $\sum_{k=1}^{\infty} \dfrac{x^k}{2k+1}$ **74.** $\sum_{k=1}^{\infty} \dfrac{x^k}{k^2}$

75. $\sum_{k=0}^{\infty} x^{k^2}$ **76.** $\sum_{k=1}^{\infty} \dfrac{k^a}{a^k}(x-a)^k, \quad a \neq 0$

77. $\sum_{k=0}^{\infty} \dfrac{(k!)^2}{(2k)!}(x-1)^k$ **78.** $\sum_{k=0}^{\infty} \dfrac{\sqrt{k!}}{(2k)!} x^k$

79. (a) In the geometric series $\dfrac{1}{1-x} = \sum_{k=0}^{\infty} x^k, \ -1 < x < 1$,

replace x by x^2 to obtain the power series representation for $\dfrac{1}{1-x^2}$.

(b) What is its interval of convergence?

80. (a) Integrate the power series found in Problem 79 for $\dfrac{1}{1-x^2}$

to obtain the power series representation for $\dfrac{1}{2} \ln \dfrac{1+x}{1-x}$.

(b) What is its interval of convergence?

81. Use the power series found in Problem 80 to get an approximation for $\ln 2$ correct to three decimal places.

82. Use the first 1000 terms of Gregory's series to approximate $\dfrac{\pi}{4}$.

What is the approximation for π?

83. If $R > 0$ is the radius of convergence of $\sum_{k=1}^{\infty} a_k x^k$, show

that $\lim_{n \to \infty} \left| \dfrac{a_{n+1}}{a_n} \right| = \dfrac{1}{R}$, provided this limit exists.

84. If R is the radius of convergence of $\sum_{k=1}^{\infty} a_k x^k$, show that the

radius of convergence of $\sum_{k=1}^{\infty} a_k x^{2k}$ is $\sqrt{R}$.

85. Prove that if a power series is absolutely convergent at one endpoint of its interval of convergence, then the power series is absolutely convergent at the other endpoint.

86. Suppose $\sum_{k=0}^{\infty} a_k x^k$ converges for $|x| < R$ and that $\lim_{n \to \infty} \left| \dfrac{a_{n+1}}{a_n} \right|$

exists. Show that $\sum_{k=1}^{\infty} k a_k x^{k-1}$ and $\sum_{k=0}^{\infty} \dfrac{a_k}{k+1} x^{k+1}$ also converge

for $|x| < R$.

Challenge Problems

87. Consider the differential equation

$$(1+x^2)\, y'' - 4x y' + 6y = 0$$

Assuming there is a solution $y(x) = \sum_{k=0}^{\infty} a_k x^k$, substitute and

obtain a formula for a_k. Your answer should have the form

$$y(x) = a_0(1 - 3x^2) + a_1 \left(x - \dfrac{1}{3} x^3 \right) \qquad a_0, \ a_1 \text{ real numbers}$$

88. If the series $\sum_{k=0}^{\infty} a_k 3^k$ converges, show that the series $\sum_{k=1}^{\infty} k a_k 2^k$

also converges.

89. Find the interval of convergence of the series $\sum_{k=1}^{\infty} \dfrac{(x-2)^k}{k(3^k)}$.

90. Let a power series $S(x)$ be convergent for $|x| < R$. Assume

that $S(x) = \sum_{k=0}^{\infty} a_k x^k$ with partial sums $S_n(x) = \sum_{k=0}^{n} a_k x^k$.

Suppose for any number $\varepsilon > 0$, there is a number N so that

when $n > N$, $|S(x) - S_n(x)| < \dfrac{\varepsilon}{3}$ for all $|x| < R$.

Show that $S(x)$ is continuous for all $|x| < R$.

91. Find the power series in x, denoted by $f(x)$, for which $f''(x) + f(x) = 0$ and $f(0) = 0$, $f'(0) = 1$. What is the radius of convergence of the series?

92. The **Bessel function of order** m of the first kind, where m is a nonnegative integer, is defined as

$$J_m(x) = \sum_{k=0}^{\infty} (-1)^k \dfrac{1}{(k+m)! \ k!} \left(\dfrac{x}{2} \right)^{2k+m}$$

Show that:

(a) $J_0(x) = x^{-1} \dfrac{d}{dx}(x J_1(x))$

(b) $J_1(x) = x^{-2} \dfrac{d}{dx}(x^2 J_2(x))$

AP® Practice Problems

Multiple-Choice Questions

[PAGE 859] **1.** If $\sum\limits_{k=0}^{\infty} a_k x^k$ converges for $x = -8$, then which of the following must be true?

I. $\sum\limits_{k=0}^{\infty} a_k x^k$ converges for $x = 8$

II. $\sum\limits_{k=0}^{\infty} a_k x^k$ converges for $x = 0$

III. $\sum\limits_{k=0}^{\infty} a_k x^k$ converges for $x = -6$

(A) I only   (B) II only
(C) II and III only   (D) I, II, and III

[PAGE 859] **2.** If the radius of convergence of the series $\sum\limits_{k=0}^{\infty} a_k (x-3)^k$ is 2, then which of the following must be true?

I. The series converges for $x = 1$.
II. The series converges for $x = 2$.
III. The series converges for $x = 5$.

(A) I only   (B) II only
(C) I and III only   (D) I, II, and III

[PAGE 861] **3.** The interval of convergence of the power series $\sum\limits_{k=1}^{\infty} \dfrac{(x+3)^k}{k}$ is

(A) $(-4, -2)$ (B) $[-4, -2)$
(C) $[-4, -2]$ (D) $x = -3$ only

[PAGE 860] **4.** Which series has an interval of convergence $[-1, 1]$?

I. $\sum\limits_{k=1}^{\infty} k x^k$ **II.** $\sum\limits_{k=1}^{\infty} \dfrac{x^k}{k}$ **III.** $\sum\limits_{k=1}^{\infty} \dfrac{x^k}{k^3}$

(A) I only   (B) III only
(C) I and III only   (D) II and III only

[PAGE 858] **5.** Find all numbers x for which the power series $\sum\limits_{k=0}^{\infty} \dfrac{x^k}{3^k}$ converges.

(A) $\left[-\dfrac{1}{3}, \dfrac{1}{3}\right]$ (B) $[-1, 1]$

(C) $(-3, 3)$ (D) $[-3, 3]$

[PAGE 865] **6.** The power series representation of a function f is $\sum\limits_{k=0}^{\infty} \dfrac{x^k}{k!}$. Then the power series for $f'(x)$ is

(A) $\sum\limits_{k=1}^{\infty} \dfrac{x^{k-1}}{(k-1)!}$

(B) $\sum\limits_{k=0}^{\infty} \dfrac{x^k}{(k+1)!}$

(C) $\sum\limits_{k=1}^{\infty} \dfrac{k x^{k-1}}{(k+1)!}$

(D) $\sum\limits_{k=1}^{\infty} \dfrac{k x^{k-1}}{(k-1)!}$

[PAGE 859] **7.** If the series $\sum\limits_{k=1}^{\infty} a_k x^k$ converges for $x = 2$, then which of the following must be true?

I. The radius of convergence R is greater than 2.
II. The radius of convergence R is centered at 0.
III. The series converges for $x = -1$.

(A) I only   (B) II only
(C) II and III only   (D) I, II, and III

[PAGE 859] **8.** If the series $\sum\limits_{k=1}^{\infty} a_k x^k$ converges for $x = 6$, then the interval of convergence could be

(A) $(-6, 6]$ (B) $[-4, 8)$ (C) $[0, 6)$ (D) $(-7, 6)$

Free-Response Questions

[PAGE 861] **9.** (a) Find the radius of convergence for the power series

$$\sum\limits_{k=1}^{\infty} \dfrac{(x-2)^k}{3^k}$$

(b) What is the interval of convergence for the power series?

[PAGE 866] **10.** (a) Use the power series

$$\ln(1+x) = \sum\limits_{k=0}^{\infty} \dfrac{(-1)^k x^{k+1}}{k+1}, \quad -1 < x \le 1,$$

to find the first four terms and the general term for the power series representation for $f(x) = \ln(1+x^2)$.

(b) Use properties of power series to find the derivative of f.

(c) What is the interval of convergence for f'?

Retain Your Knowledge

Multiple-Choice Questions

1. Which differential equation has the slope field shown below?

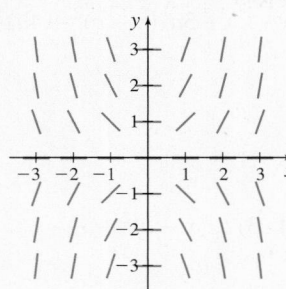

(A) $\dfrac{dy}{dx} = x$ (B) $\dfrac{dy}{dx} = xy$

(C) $\dfrac{dy}{dx} = y$ (D) $\dfrac{dy}{dx} = x - y$

2. The slope of the line tangent to the graph of the polar equation $r = 4\theta$ when $\theta = \pi$ is

(A) $-\pi$ (B) 4   (C) π (D) 4π

3. $\displaystyle\int \dfrac{8}{x^2 - 16}\, dx =$

(A) $2\tan^{-1}\dfrac{x}{4} + C$ (B) $\dfrac{1}{8}\tan^{-1}\dfrac{x}{16} + C$

(C) $\ln\left|\dfrac{x+4}{x-4}\right| + C$ (D) $\ln\left|\dfrac{x-4}{x+4}\right| + C$

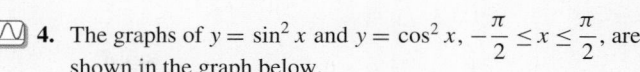 **4.** The graphs of $y = \sin^2 x$ and $y = \cos^2 x$, $-\dfrac{\pi}{2} \le x \le \dfrac{\pi}{2}$, are shown in the graph below.

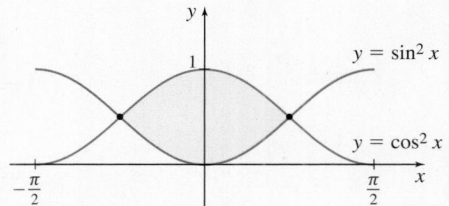

(a) Find the points where the graphs intersect.
(b) Find the area A of the region bounded by the two graphs.
(c) Find the volume V of the solid generated when the region bounded by the graphs is revolved about the x-axis.

10.9 Taylor Series; Maclaurin Series

OBJECTIVES *When you finish this section, you should be able to:*

1 **Express a function as a Taylor series or a Maclaurin series (p. 873)**
2 **Determine the convergence of a Taylor/Maclaurin series (p. 874)**
3 **Find Taylor/Maclaurin expansions (p. 878)**

We saw in Section 10.8 that it is often possible to obtain a power series representation for a function by starting with a power series whose sum is known, and differentiating, integrating, or substituting. But what if the sum is not known? In other words, so far we have seen that functions can be represented by power series, and if we know the sum of the power series, then we know the function. In this section, we investigate what the power series representation of a function must look like *if it has a power series representation.*

Consider the power series in $(x - c)$:

$$\sum_{k=0}^{\infty} a_k(x - c)^k = a_0 + a_1(x - c) + a_2(x - c)^2 + \cdots + a_n(x - c)^n + \cdots$$

and suppose its interval of convergence is the open interval $(c - R,\ c + R)$, $R > 0$. We define the function f as the series

$$f(x) = \sum_{k=0}^{\infty} a_k(x - c)^k = a_0 + a_1(x - c) + a_2(x - c)^2 + \cdots + a_n(x - c)^n + \cdots \qquad (1)$$

The coefficients a_0, a_1, ... can be expressed in terms of f and its derivatives in the following way. Repeatedly differentiate the function using the differentiation property of a power series,

$$f'(x) = \sum_{k=1}^{\infty} k\, a_k (x-c)^{k-1} = a_1 + 2a_2(x-c) + 3a_3(x-c)^2 + 4a_4(x-c)^3 + \cdots$$

$$f''(x) = \sum_{k=2}^{\infty} [k(k-1)]\, a_k (x-c)^{k-2}$$

$$= (2\cdot 1)\, a_2 + (3\cdot 2)\, a_3(x-c) + (4\cdot 3)\, a_4(x-c)^2 + \cdots$$

$$f'''(x) = \sum_{k=3}^{\infty} [k(k-1)(k-2)]a_k(x-c)^{k-3}$$

$$= (3\cdot 2\cdot 1)\, a_3 + (4\cdot 3\cdot 2)\, a_4(x-c) + \cdots$$

and for any positive integer n,

$$f^{(n)}(x) = \sum_{k=n}^{\infty} k(k-1)(k-2)\ldots(k-n+1)a_k(x-c)^{k-n}$$

$$= [n\cdot(n-1)\cdot \ldots \cdot 1]\, a_n + [(n+1)\cdot n\cdot(n-1)\cdot \ldots \cdot 2]\, a_{n+1}(x-c) + \cdots$$

$$= n!\, a_n + (n+1)!\, a_{n+1}(x-c) + \cdots$$

Now let $x = c$ in each derivative and solve for a_k.

$$f(c) = a_0 \qquad\qquad a_0 = f(c)$$

$$f'(c) = a_1 \qquad\qquad a_1 = f'(c)$$

$$f''(c) = 2a_2 \qquad\qquad a_2 = \frac{f''(c)}{2!}$$

$$f'''(c) = 3!\, a_3 \qquad\qquad a_3 = \frac{f'''(c)}{3!}$$

$$\vdots \qquad\qquad\qquad \vdots$$

$$f^{(n)}(c) = n!\, a_n \qquad\qquad a_n = \frac{f^{(n)}(c)}{n!}$$

If we substitute for a_k in (1), we obtain

$$f(x) = f(c) + f'(c)(x-c) + \frac{f''(c)}{2!}(x-c)^2 + \cdots + \frac{f^{(n)}(c)}{n!}(x-c)^n + \cdots$$

and have proved the following result.

THEOREM Taylor Series; Maclaurin Series

Suppose f is a function that has derivatives of all orders on the open interval $(c - R, c + R)$, $R > 0$. If f can be represented by the power series $\sum_{k=0}^{\infty} a_k (x - c)^k$, whose radius of convergence is R, then

$$
f(x) = f(c) + f'(c)(x - c) + \frac{f''(c)}{2!}(x - c)^2 + \cdots + \frac{f^{(n)}(c)}{n!}(x - c)^n + \cdots
$$

$$
= \sum_{k=0}^{\infty} \frac{f^{(k)}(c)}{k!}(x - c)^k \qquad\qquad (2)
$$

for all numbers x in the open interval $(c - R, c + R)$. A power series that has the form of equation (2) is called a **Taylor series** of the function f.

When $c = 0$, the Taylor series

$$
f(x) = f(0) + f'(0)\, x + \frac{f''(0)}{2!}x^2 + \cdots + \frac{f^{(n)}(0)}{n!}x^n + \cdots = \sum_{k=0}^{\infty} \frac{f^{(k)}(0)}{k!}x^k
$$

is called a **Maclaurin series**.

The Taylor series in $(x - c)$ of a function f is referred to as the **Taylor expansion of f about c**; the Maclaurin series of a function f is called the **Maclaurin expansion of f about 0**.

In a Taylor series, all the derivatives are evaluated at c, and the interval of convergence has its center at c. In a Maclaurin series, all the derivatives are evaluated at 0, and the interval of convergence has its center at 0.

❶ Express a Function as a Taylor Series or a Maclaurin Series

The next example shows what a Maclaurin expansion of $f(x) = e^x$ must look like (if there is one).

EXAMPLE 1   **Expressing a Function as a Maclaurin Series**

Assuming that $f(x) = e^x$ can be represented by a power series in x, find its Maclaurin series.

Solution

To express a function f as a Maclaurin series, we begin by evaluating f and its derivatives at 0.

$$
\begin{aligned}
f(x) &= e^x & f(0) &= 1 \\
f'(x) &= e^x & f'(0) &= 1 \\
f''(x) &= e^x & f''(0) &= 1 \\
&\;\;\vdots & &\;\;\vdots
\end{aligned}
$$

Then use the definition of a Maclaurin series.

$$
f(x) = \sum_{k=0}^{\infty} \frac{f^{(k)}(0)}{k!}x^k = 1 + x + \frac{x^2}{2!} + \frac{x^3}{3!} + \cdots + \frac{x^n}{n!} + \cdots = \sum_{k=0}^{\infty} \frac{x^k}{k!} \qquad (3)
$$

$$
\underset{f^{(k)}(0) = 1}{\uparrow}
$$

ORIGINS Colin Maclaurin (1698–1746) was a Scottish mathematician. An orphan, he and his brother were raised by an uncle. Maclaurin was 11 years old when he entered the University of Glasgow, and within a year he had taught himself from Euclid's *Elements* (geometry). Maclaurin was 19 when he was appointed professor of Mathematics at the University of Aberdeen, Scotland. Maclaurin was an avid supporter of Newton's mathematics. In response to a claim that Newton's calculus lacked rigor, Maclaurin wrote the first systematic treatise on Newton's methods. In it he used Taylor series centered about 0, which are now known as Maclaurin series.

NOW WORK   Problem 3 and AP® Practice Problem 5.

But how can we be sure the Maclaurin series (3) converges to e^x? We know [Example 1(a), p. 857] that the power series $\sum\limits_{k=0}^{\infty} \dfrac{x^k}{k!}$ converges absolutely for all numbers x. But does the series $\sum\limits_{k=0}^{\infty} \dfrac{x^k}{k!}$ converge to e^x? To answer these questions, we need to investigate the convergence of a Taylor series.

② Determine the Convergence of a Taylor/Maclaurin Series

Suppose a function f has derivatives of all orders on the open interval $(c - R, c + R)$, $R > 0$. If f can be represented by a power series $\sum\limits_{k=0}^{\infty} a_k(x - c)^k$ whose radius of convergence is R, then

$$f(x) = f(c) + f'(c)(x - c) + \frac{f''(c)}{2!}(x - c)^2 + \cdots + \frac{f^{(n)}(c)}{n!}(x - c)^n + \cdots = \sum_{k=0}^{\infty} \frac{f^{(k)}(c)}{k!}(x - c)^k \quad (4)$$

for all x in the interval $(c - R, c + R)$. We seek conditions on f that guarantee the power series (4) converges to f.

We rewrite (4) as follows

$$f(x) = P_n(x) + R_n \quad (5)$$

> **IN WORDS** Statement (4) is the Taylor Series for f. Statement (5) expresses f as the sum of a polynomial (called a Taylor polynomial) plus a term called the remainder.
>
> Taylor Series = Taylor Polynomial + Remainder

where $P_n(x) = f(c) + f'(c)(x - c) + \dfrac{f''(c)}{2!}(x - c)^2 + \cdots + \dfrac{f^{(n)}(c)}{n!}(x - c)^n$, called the **Taylor polynomial** $P_n(x)$ of f of degree n, and R_n is the **remainder**.

THEOREM Taylor's Formula with the Lagrange Form of the Remainder

Suppose f is a function whose first $n + 1$ derivatives are continuous on an open interval I containing the number c. Then for every x in I, there is a number u between c and x for which

$$\boxed{\begin{aligned} f(x) &= f(c) + f'(c)(x - c) + \frac{f''(c)}{2!}(x - c)^2 + \cdots + \frac{f^{(n)}(c)}{n!}(x - c)^n + R_n(x) \\ &= P_n(x) + R_n(x) \end{aligned}}$$

(6)

where

$$\boxed{R_n(x) = \frac{f^{(n+1)}(u)}{(n+1)!}(x - c)^{n+1}} \quad (7)$$

is the remainder after $n + 1$ terms and $P_n(x)$ is the Taylor polynomial of degree n of f at c.

> **NOTE** The remainder $R_n(x)$ contains the $(n + 1)$st derivative of f evaluated at u, a number in the interval I that is not equal to c.

The proof of this theorem is in Appendix B.

The remainder

$$R_n(x) = \frac{f^{n+1}(u)(x - c)^{n+1}}{(n+1)!}$$

where u is a number between c and x, is the **Lagrange form of the remainder**.

ORIGINS The Taylor polynomial is named after the English mathematician Brook Taylor (1685–1731). Taylor grew up in an affluent, but strict, home. He was home-schooled in the arts and the classics until he attended Cambridge University in 1703, where he studied mathematics. He was an accomplished musician and painter as well as a mathematician.

In Section 10.10, we focus on Equations (6) and (7) as follows:

- In Equation (6), we use the Taylor polynomial $P_n(x)$ and its graph to approximate the function $y = f(x)$ and its graph.
- Then we use the remainder $R_n(x)$, Equation (7), to establish an error bound on the approximation.

For the rest of this section we examine the convergence of the Taylor series

$$\sum_{k=0}^{\infty} \frac{f^k(x)(x-c)^k}{k!}$$

to the function f.

The Taylor series in $x - c$ of a function f is $\sum_{k=0}^{\infty} \frac{f^{(k)}(c)}{k!}(x-c)^k$. Notice that the nth partial sum of the Taylor series in $x - c$ of f is precisely the Taylor polynomial $P_n(x)$ of f at c. If the Taylor series $\sum_{k=0}^{\infty} \frac{f^{(k)}(c)}{k!}(x-c)^k$ converges to $f(x)$, it follows that

$$f(x) = \lim_{n \to \infty} P_n(x)$$

But Taylor's Formula with the Lagrange form of the remainder, Equation (6), states that

$$f(x) = P_n(x) + R_n(x)$$

So, if the Taylor series converges, we must have

$$f(x) = \lim_{n \to \infty} [P_n(x) + R_n(x)] = \lim_{n \to \infty} P_n(x) + \lim_{n \to \infty} R_n(x) = f(x) + \lim_{n \to \infty} R_n(x)$$

That is, $\lim_{n \to \infty} R_n(x) = 0$.

THEOREM Convergence of a Taylor Series

If a function f has derivatives of all orders in an open interval $I = (c - R, c + R)$, $R > 0$, centered at c, and if

$$\lim_{n \to \infty} R_n(x) = \lim_{n \to \infty} \frac{f^{(n+1)}(u)}{(n+1)!}(x-c)^{n+1} = 0$$

for all numbers x in I, where u is a number between c and x, then

$$f(x) = \sum_{k=0}^{\infty} \frac{f^{(k)}(c)}{k!}(x-c)^k = f(c) + f'(c)(x-c) + \frac{f''(c)}{2!}(x-c)^2 + \cdots + \frac{f^{(n)}(c)}{n!}(x-c)^n + \cdots$$

for all numbers x in I.

At first glance, the convergence theorem appears simple, but in practice it is not always easy to show that $\lim_{n \to \infty} R_n(x) = 0$. One reason is that the term $f^{(n+1)}(u)$, which appears in $R_n(x)$, depends on u, making the limit difficult to find.

For many familiar functions we can prove that its Maclaurin and Taylor series converge for all numbers in their domain. We consider $f(x) = e^x$ first.

EXAMPLE 2	Determining the Convergence of the Maclaurin Series for $f(x) = e^x$

Show that $1 + x + \dfrac{x^2}{2!} + \dfrac{x^3}{3!} + \cdots + \dfrac{x^n}{n!} + \cdots$ converges to e^x for every number x. That is, prove that

$$e^x = 1 + \frac{x}{1!} + \frac{x^2}{2!} + \frac{x^3}{3!} + \cdots + \frac{x^n}{n!} + \cdots = \sum_{k=0}^{\infty} \frac{x^k}{k!} \qquad (8)$$

for all real numbers.

Solution

To prove that $1 + \dfrac{x}{1!} + \dfrac{x^2}{2!} + \dfrac{x^3}{3!} + \cdots = e^x$ for every number x, we need to show that $\lim\limits_{n \to \infty} R_n(x) = 0$. Since $f^{(n+1)}(x) = e^x$, we have

$$R_n(x) = \frac{f^{(n+1)}(u)x^{n+1}}{(n+1)!} = \frac{e^u x^{n+1}}{(n+1)!}$$

where u is between 0 and x.

To show that $\lim\limits_{n \to \infty} R_n(x) = 0$, we consider two cases: $x > 0$ and $x < 0$.

Case 1: When $x > 0$, then $0 < u < x$, so that $1 < e^u < e^x$ and, for every positive integer n,

$$0 < R_n(x) = \frac{e^u x^{n+1}}{(n+1)!} < \frac{e^x x^{n+1}}{(n+1)!} \qquad (9)$$

By the Ratio Test, the series $\sum\limits_{k=0}^{\infty} \dfrac{x^{k+1}}{(k+1)!}$ converges for all x. It follows that $\lim\limits_{n \to \infty} \dfrac{x^{n+1}}{(n+1)!} = 0$, and, therefore,

$$\lim_{n \to \infty} \frac{e^x x^{n+1}}{(n+1)!} = e^x \lim_{n \to \infty} \frac{x^{n+1}}{(n+1)!} = 0$$

NEED TO REVIEW? The Squeeze Theorem is discussed in Section 1.4, pp. 122–124.

Using the Squeeze Theorem with (9), $\lim\limits_{n \to \infty} R_n(x) = 0$.

Case 2: When $x < 0$, then $x < u < 0$ and $e^x < e^u < 1$, so that

$$0 \le |R_n(x)| = \frac{e^u \cdot |x|^{n+1}}{(n+1)!} < \frac{|x|^{n+1}}{(n+1)!}$$

Since $\lim\limits_{n \to \infty} \dfrac{|x|^{n+1}}{(n+1)!} = 0$, by the Squeeze Theorem, $\lim\limits_{n \to \infty} R_n(x) = 0$.

So for all x, $\lim\limits_{n \to \infty} R_n(x) = 0$. As a result,

$$e^x = 1 + x + \frac{x^2}{2!} + \frac{x^3}{3!} + \cdots + \frac{x^n}{n!} + \cdots = \sum_{k=0}^{\infty} \frac{x^k}{k!}$$

for all numbers x. ∎

EXAMPLE 3 | **Finding the Maclaurin Expansion That Converges to $f(x) = \sin x$**

(a) Find the Maclaurin expansion for $f(x) = \sin x$.

(b) Show that it converges to $\sin x$ for all numbers x.

Solution

(a) The value of f and its derivatives at 0 are

$$
\begin{aligned}
f(x) &= \sin x & f(0) &= 0 \\
f'(x) &= \cos x & f'(0) &= 1 \\
f''(x) &= -\sin x & f''(0) &= 0 \\
f'''(x) &= -\cos x & f'''(0) &= -1
\end{aligned}
$$

Higher-order derivatives follow this same pattern, so the Maclaurin expansion for $f(x) = \sin x$ is

$$
\sin x = f(0) + \frac{f'(0)}{1!}x + \frac{f''(0)}{2!}x^2 + \frac{f'''(0)}{3!}x^3 + \cdots = x - \frac{x^3}{3!} + \frac{x^5}{5!} - \frac{x^7}{7!} + \cdots
$$

$$
= \sum_{k=0}^{\infty} (-1)^k \frac{x^{2k+1}}{(2k+1)!}
$$

(b) To prove that the series actually converges to $\sin x$ for all x, we need to show that the remainders $R_{2n+1}(x)$ and $R_{2n}(x)$ approach zero. For $R_{2n+1}(x)$ we have

$$
R_{2n+1}(x) = (-1)^{n+1} \frac{f^{(2n+2)}(u) \cdot x^{2n+2}}{(2n+2)!}
$$

where u is between 0 and x. Since $|f^{(2n+2)}(u)| = |\sin u| \leq 1$ for every number u, then

$$
0 \leq |R_{2n+1}(x)| = \frac{|f^{(2n+2)}(u)|}{(2n+2)!}|x|^{2n+2} \leq \frac{|x|^{2n+2}}{(2n+2)!}
$$

By the Ratio Test, the series $\displaystyle\sum_{k=0}^{\infty} \frac{|x|^{2k+2}}{(2k+2)!}$ converges for all x, so

$$
\lim_{n\to\infty} \frac{|x|^{2n+2}}{(2n+2)!} = 0
$$

By the Squeeze Theorem, $\displaystyle\lim_{n\to\infty} |R_{2n+1}(x)| = 0$ and therefore $\displaystyle\lim_{n\to\infty} R_{2n+1}(x) = 0$ for all x.

A similar argument holds for the remainder $|R_{2n}(x)| = \dfrac{|\cos u| \, x^{2n+1}}{(2n+1)!}$.

We conclude that the series $\displaystyle\sum_{k=0}^{\infty} (-1)^k \frac{x^{2k+1}}{(2k+1)!}$ converges to $\sin x$ for all x. That is,

$$
\sin x = x - \frac{x^3}{3!} + \frac{x^5}{5!} - \cdots + (-1)^n \frac{x^{2n+1}}{(2n+1)!} + \cdots = \sum_{k=0}^{\infty} (-1)^k \frac{x^{2k+1}}{(2k+1)!} \qquad (10)
$$

∎

3 Find Taylor/Maclaurin Expansions

The task of finding the Taylor expansion for a function f by taking successive derivatives and then showing that $\lim_{n \to \infty} R_n(x) = 0$ can be challenging. Consequently, it is usually easier to use a known series and properties of power series, such as the Differentiation Property and the Integration Property (see page 864), to find the Taylor expansion of a function f.

For example, we can find the Maclaurin expansion for $f(x) = \cos x$ by differentiating the Maclaurin series for $f(x) = \sin x$.

EXAMPLE 4 **Finding the Maclaurin Expansion for $f(x) = \cos x$**

Find the Maclaurin expansion for $f(x) = \cos x$.

Solution

Use the differentiation property of power series and the Maclaurin expansion for $\sin x$ [equation (10)].

$$\frac{d}{dx} \sin x = \frac{d}{dx}\left(x - \frac{x^3}{3!} + \frac{x^5}{5!} - \cdots + (-1)^n \frac{x^{2n+1}}{(2n+1)!} + \cdots \right)$$

$$= \frac{d}{dx} \sum_{k=0}^{\infty} (-1)^k \frac{x^{2k+1}}{(2k+1)!}$$

Then

$$\boxed{\cos x = 1 - \frac{x^2}{2!} + \frac{x^4}{4!} - \cdots + (-1)^n \frac{x^{2n}}{(2n)!} + \cdots = \sum_{k=0}^{\infty} (-1)^k \frac{x^{2k}}{(2k)!}} \quad (11)$$

for all numbers x. ∎

NOW WORK AP® Practice Problem **6.**

We can use known Maclaurin expansions to obtain the power series representations of other functions. For example, if in the Maclaurin expansion for e^x, the variable x is replaced by $-x$, then

$$f(x) = e^{-x} = 1 - x + \frac{x^2}{2!} - \frac{x^3}{3!} + \cdots + (-1)^n \frac{x^n}{n!} + \cdots$$

for all numbers x.

EXAMPLE 5 **Finding the Maclaurin Expansion for $f(x) = \sin(2x)$**

Find the Maclaurin expansion for $f(x) = \sin(2x)$.

Solution

Use the Maclaurin expansion for $\sin x$, Equation (10), and substitute $2x$ for x.

$$\sin x = x - \frac{x^3}{3!} + \frac{x^5}{5!} - \cdots + (-1)^n \frac{x^{2n+1}}{(2n+1)!} + \cdots$$

$$\sin(2x) = 2x - \frac{(2x)^3}{3!} + \frac{(2x)^5}{5!} - \cdots + (-1)^n \frac{(2x)^{2n+1}}{(2n+1)!} + \cdots$$

$$= \sum_{k=0}^{\infty} (-1)^k \frac{(2x)^{2k+1}}{(2k+1)!}$$

∎

NOW WORK Problems 9, 13, and AP® Practice Problem 2.

NOTE The Maclaurin expansion for $f(x) = \cos x$ could also have been found by integrating the Maclaurin expansion for $f(x) = \sin x$.

EXAMPLE 6 Finding the Maclaurin Expansion for $f(x) = e^x \cos x$

Find the first five terms of the Maclaurin expansion for $f(x) = e^x \cos x$.

Solution

The Maclaurin expansion for $f(x) = e^x \cos x$ is obtained by multiplying the Maclaurin expansion for e^x by the Maclaurin expansion for $\cos x$. That is,

$$e^x \cos x = \left(1 + x + \frac{x^2}{2!} + \frac{x^3}{3!} + \frac{x^4}{4!} + \frac{x^5}{5!} + \cdots\right)\left(1 - \frac{x^2}{2!} + \frac{x^4}{4!} - \cdots\right)$$

Then for all x,

$$e^x \cos x = 1\left(1 - \frac{x^2}{2!} + \frac{x^4}{4!} - \cdots\right) + x\left(1 - \frac{x^2}{2!} + \frac{x^4}{4!} - \cdots\right)$$

$$+ \frac{x^2}{2!}\left(1 - \frac{x^2}{2!} + \frac{x^4}{4!} - \cdots\right) + \frac{x^3}{3!}\left(1 - \frac{x^2}{2!} + \frac{x^4}{4!} - \cdots\right)$$

$$+ \frac{x^4}{4!}\left(1 - \frac{x^2}{2!} + \frac{x^4}{4!} - \cdots\right) + \frac{x^5}{5!}\left(1 - \frac{x^2}{2!} + \frac{x^4}{4!} - \cdots\right) + \cdots$$

$$= \left(1 - \frac{x^2}{2} + \frac{x^4}{24}\right) + \left(x - \frac{x^3}{2} + \frac{x^5}{24}\right) + \left(\frac{x^2}{2} - \frac{x^4}{4}\right) + \left(\frac{x^3}{6} - \frac{x^5}{12}\right)$$

$$+ \frac{x^4}{24} + \frac{x^5}{120} + \cdots$$

$$= 1 + x + \left(-\frac{1}{2} + \frac{1}{2}\right)x^2 + \left(-\frac{1}{2} + \frac{1}{6}\right)x^3 + \left(\frac{1}{24} - \frac{1}{4} + \frac{1}{24}\right)x^4$$

$$+ \left(\frac{1}{24} - \frac{1}{12} + \frac{1}{120}\right)x^5 + \cdots$$

$$= 1 + x - \frac{1}{3}x^3 - \frac{1}{6}x^4 - \frac{1}{30}x^5 + \cdots$$ ∎

NOW WORK Problem **21** and AP® Practice Problems **1** and **3**.

EXAMPLE 7 Finding the Taylor Expansion for $f(x) = \cos x$ about $\dfrac{\pi}{2}$

Find the Taylor expansion for $f(x) = \cos x$ about $\dfrac{\pi}{2}$.

Solution

To express $f(x) = \cos x$ as a Taylor expansion about $\dfrac{\pi}{2}$, we evaluate f and its derivatives at $\dfrac{\pi}{2}$.

$$f(x) = \cos x \qquad f\left(\frac{\pi}{2}\right) = 0$$

$$f'(x) = -\sin x \qquad f'\left(\frac{\pi}{2}\right) = -1$$

$$f''(x) = -\cos x \qquad f''\left(\frac{\pi}{2}\right) = 0$$

$$f'''(x) = \sin x \qquad f'''\left(\frac{\pi}{2}\right) = 1$$

(Example continued on the next page)

For derivatives of odd order, $f^{(2n+1)}\left(\dfrac{\pi}{2}\right) = (-1)^{n+1}$. For derivatives of even order, $f^{(2n)}\left(\dfrac{\pi}{2}\right) = 0$. The Taylor expansion for $f(x) = \cos x$ about $\dfrac{\pi}{2}$ is

$$f(x) = \cos x = f\left(\frac{\pi}{2}\right) + f'\left(\frac{\pi}{2}\right)\left(x - \frac{\pi}{2}\right) + \frac{f''\left(\frac{\pi}{2}\right)}{2!}\left(x - \frac{\pi}{2}\right)^2$$

$$+ \frac{f'''\left(\frac{\pi}{2}\right)}{3!}\left(x - \frac{\pi}{2}\right)^3 + \cdots$$

$$= -\left(x - \frac{\pi}{2}\right) + \frac{1}{3!}\left(x - \frac{\pi}{2}\right)^3 - \frac{1}{5!}\left(x - \frac{\pi}{2}\right)^5 + \cdots$$

$$= \sum_{k=0}^{\infty} \frac{(-1)^{k+1}}{(2k+1)!}\left(x - \frac{\pi}{2}\right)^{2k+1}$$

AP® EXAM TIP

On the exam, you are expected to know the first four series shown in Table 7.

The radius of convergence is ∞; the interval of convergence is $(-\infty, \infty)$. ∎

NOW WORK Problem **29** and AP® Practice Problem **4**.

TABLE 7 Summary

Series	Comment
$e^x = \sum\limits_{k=0}^{\infty} \dfrac{x^k}{k!} = 1 + \dfrac{x}{1!} + \dfrac{x^2}{2!} + \dfrac{x^3}{3!} + \cdots$	Converges for all real numbers x.
$\sin x = \sum\limits_{k=0}^{\infty} \dfrac{(-1)^k x^{2k+1}}{(2k+1)!} = x - \dfrac{x^3}{3!} + \dfrac{x^5}{5!} - \dfrac{x^7}{7!} + \cdots$	Converges for all real numbers x.
$\cos x = \sum\limits_{k=0}^{\infty} \dfrac{(-1)^k x^{2k}}{(2k)!} = 1 - \dfrac{x^2}{2!} + \dfrac{x^4}{4!} - \dfrac{x^6}{6!} + \cdots$	Converges for all real numbers x.
$\dfrac{1}{1-x} = \sum\limits_{k=0}^{\infty} x^k = 1 + x + x^2 + x^3 + \cdots$	Converges for $\lvert x \rvert < 1$.
$\ln \dfrac{1}{1-x} = \sum\limits_{k=1}^{\infty} \dfrac{x^k}{k} = x + \dfrac{x^2}{2} + \dfrac{x^3}{3} + \cdots + \dfrac{x^{n+1}}{n+1} + \cdots$	Converges on the interval $[-1, 1)$.
$\ln(1+x) = \sum\limits_{k=0}^{\infty} \dfrac{(-1)^k x^{k+1}}{k+1}$	Converges on the interval $(-1, 1]$.
$\tan^{-1} x = \sum\limits_{k=0}^{\infty} (-1)^k \dfrac{x^{2k+1}}{2k+1} = x - \dfrac{x^3}{3} + \dfrac{x^5}{5} - \cdots$	Converges on the interval $[-1, 1]$.

10.9 Assess Your Understanding

Concepts and Vocabulary

1. The series representation of a function f given by the power series $f(x) = f(c) + f'(c)(x - c) + \dfrac{f''(c)(x - c)^2}{2!} + \cdots + \dfrac{f^{(n)}(c)(x - c)^n}{n!} + \cdots$ is called a(n) _____ about c.

2. If $c = 0$ in the Taylor expansion of a function f, then the expansion is called a(n) _____ expansion.

3. Write the first four terms of the Maclaurin expansion of f if

$$f(0) = 2,\ f'(0) = -4,\ f''(0) = 3,\ f'''(0) = -2.$$

4. Write the first four terms of the Taylor expansion of f centered at 5, if

$$f(5) = 3,\ f'(5) = 2,\ f''(5) = -1,\ f'''(5) = -4$$

Skill Building

In Problems 5–8, find the Taylor expansion of each function about the given number c. Comment on the result.

5. $f(x) = 3x^3 + 2x^2 + 5x - 6; \quad c = 0$

6. $f(x) = 4x^4 - 2x^3 - x; \quad c = 0$

7. $f(x) = 3x^3 + 2x^2 + 5x - 6; \quad c = 1$

8. $f(x) = 4x^4 - 2x^3 + x; \quad c = 1$

In Problems 9–20, find the first four nonzero terms and the general term of the Maclaurin expansion for each function using series from Table 7 on page 880.

 9. $f(x) = e^{x^2}$

10. $f(x) = e^{-x^2}$

11. $f(x) = xe^x$

12. $f(x) = xe^{-x}$

 13. $f(x) = x^2 \sin x$

14. $f(x) = x^2 \cos x$

15. $f(x) = \dfrac{1}{1 - x^2}$

16. $f(x) = \dfrac{1}{1 + x^2}$

17. $f(x) = \ln \dfrac{1}{1 - 3x}$

18. $f(x) = \ln \dfrac{1}{1 + x}$

19. $f(x) = \ln(1 + x^2)$

20. $f(x) = \ln(1 - 2x)$

In Problems 21–28, find the first five nonzero terms of the Maclaurin expansion.

 21. $f(x) = e^{-x} + \cos x$

22. $f(x) = e^x + \sin x$

23. $f(x) = \dfrac{e^x}{1 - x}$

24. $f(x) = \dfrac{e^x}{1 + x}$

25. $f(x) = \dfrac{\cos x}{1 - x}$

26. $f(x) = \dfrac{\sin x}{1 + x}$

27. $f(x) = e^{-x} \sin x$

28. $f(x) = e^{-x} \cos x$

In Problems 29–38, find the Taylor expansion of each function about the given number c.

29. $f(x) = \sqrt{x}; \quad c = 1$

30. $f(x) = \sqrt[3]{x}; \quad c = 1$

31. $f(x) = \ln x; \quad c = 1$

32. $f(x) = \ln(x - 1); \quad c = 2$

33. $f(x) = \dfrac{1}{x}; \quad c = 1$

34. $f(x) = \dfrac{1}{\sqrt{x}}; \quad c = 4$

35. $f(x) = \sin x; \quad c = \dfrac{\pi}{6}$

36. $f(x) = \cos x; \quad c = -\dfrac{\pi}{2}$

37. $f(x) = e^{2x}; \quad c = 1$

38. $f(x) = e^{-3x}; \quad c = 2$

Applications and Extensions

In Problems 39–42, use a Maclaurin expansion to find each integral.

39. $\displaystyle\int e^{x^2}\, dx$

40. $\displaystyle\int e^{\sqrt{x}}\, dx$

41. $\displaystyle\int \sin x^2\, dx$

42. $\displaystyle\int \cos x^2\, dx$

43. Find the first five nonzero terms of the Maclaurin expansion for $f(x) = \sec x$.

44. Probability The **standard normal distribution**
$p(x) = \dfrac{1}{\sqrt{2\pi}} e^{-x^2/2}$ is important in probability and statistics. If a random variable Z has a standard normal distribution, then the probability that an observation of Z is between $Z = a$ and $Z = b$ is given by

$$P(a \le Z \le b) = \frac{1}{\sqrt{2\pi}} \int_a^b e^{-x^2/2}\, dx$$

(a) Find the Maclaurin expansion for $p(x) = \dfrac{1}{\sqrt{2\pi}} e^{-x^2/2}$.

(b) Use the properties of a power series to find a power series representation for P.

(c) Use the first four terms of the series representation for P to approximate $P(-0.5 \le Z \le 0.3)$.

(d) Use technology to approximate $P(-0.5 \le Z \le 0.3)$.

45. Even Functions Show that if f is an even function, then the Maclaurin expansion for f has only even powers of x.

46. Odd Functions Show that if f is an odd function, then the Maclaurin expansion for f has only odd powers of x.

47. Show that the interval of convergence of the Maclaurin expansion for $f(x) = \sin^{-1} x$ is $[-1, 1]$.

48. Euler's Error Euler believed $\dfrac{1}{2} = 1 - 1 + 1 - 1 + 1 - 1 + \cdots$. He based his argument to support this equation on his belief in the identification of a series and the values of the function from which it was derived.

(a) Write the Maclaurin expansion for $\dfrac{1}{1 + x}$. Do this without calculating any derivatives.

(b) Evaluate both sides of the equation you derived in (a) at $x = 1$ to arrive at the formula above.

(c) Criticize the procedure used in (b).

Challenge Problems

49. Find the exact sum of the infinite series:

$$\frac{x^3}{1 \cdot 3} - \frac{x^5}{3 \cdot 5} + \frac{x^7}{5 \cdot 7} - \frac{x^9}{7 \cdot 9} + \cdots \quad \text{for } x = 1$$

50. Find an elementary expression for $\displaystyle\sum_{k=1}^{\infty} \frac{x^{k+1}}{k(k+1)}$.

Hint: Integrate the series for $\ln \dfrac{1}{1 - x}$.

51. Show that $\displaystyle\sum_{k=1}^{\infty} \frac{k}{(k+1)!} = 1$.

52. Let $s_n = \dfrac{1}{1!} + \dfrac{1}{2!} + \cdots + \dfrac{1}{n!}$, $\quad n = 1, 2, 3, \ldots$.

  (a) Show that $n! \geq 2^{n-1}$.

  (b) Show that $0 < s_n \leq 1 + \dfrac{1}{2} + \left(\dfrac{1}{2}\right)^2 + \cdots + \left(\dfrac{1}{2}\right)^{n-1}$.

  (c) Show that $0 < s_n < s_{n+1} < 2$. Then conclude that $S = \lim\limits_{n \to \infty} s_n$ and $S \leq 2$.

  (d) Let $t_n = \left[1 + \dfrac{1}{n}\right]^n$. Show that

$$t_n = 1 + 1 + \dfrac{1}{2!}\left[1 - \dfrac{1}{n}\right] + \dfrac{1}{3!}\left[1 - \dfrac{1}{n}\right]\left[1 - \dfrac{2}{n}\right] + \cdots$$

$$+ \dfrac{1}{n!}\left[1 - \dfrac{1}{n}\right]\left[1 - \dfrac{2}{n}\right]\cdots\left[1 - \dfrac{n-1}{n}\right] < s_n + 1$$

  (e) Show that $0 < t_n < t_{n+1} < 3$. Then conclude that $e = \lim\limits_{n \to \infty} t_n \leq 3$.

53. Show that $\left[1 + \dfrac{1}{n}\right]^n < e$ for all $n > 0$.

54. From the fact that $\sin t \leq t$ for all $t \geq 0$, use integration repeatedly to prove

$$1 - \dfrac{x^2}{2!} \leq \cos x \leq 1 - \dfrac{x^2}{2!} + \dfrac{x^4}{4!} \qquad \text{for all } x \geq 0$$

55. Find the first four nonzero terms of the Maclaurin expansion for $f(x) = (1 + x)^x$.

56. Show that $f(x) = \begin{cases} e^{-1/x^2} & x \neq 0 \\ 0 & x = 0 \end{cases}$ has a Maclaurin expansion at $x = 0$. Then show that the Maclaurin series does not converge to f.

Preparing for the AP® Exam

AP® Practice Problems

Multiple-Choice Questions

[879] **1.** The coefficient of x^8 in the Maclaurin expansion for $f(x) = (3x)^2 \cos x$ is

  (A) $-\dfrac{1}{240}$ (B) $-\dfrac{1}{80}$ (C) $\dfrac{1}{80}$ (D) $\dfrac{3}{8}$

[878] **2.** The Maclaurin expansion for $f(x) = e^{x/3}$ is

  (A) $\displaystyle\sum_{k=0}^{\infty} \dfrac{x^k}{3^k}$ (B) $\displaystyle\sum_{k=0}^{\infty} \dfrac{x^k}{3k!}$

  (C) $\displaystyle\sum_{k=0}^{\infty} \dfrac{x^k}{3^k k!}$ (D) $\displaystyle\sum_{k=0}^{\infty} \dfrac{x^k}{(3k)!}$

[879] **3.** The Maclaurin expansion of $f(x) = \dfrac{e^{2x} - 1}{x}$ is

  (A) $\displaystyle\sum_{k=0}^{\infty} \dfrac{2^k x^k}{k!}$ (B) $\displaystyle\sum_{k=0}^{\infty} \dfrac{2^{k+1} x^k}{(k+1)!}$

  (C) $\displaystyle\sum_{k=0}^{\infty} \dfrac{(2x)^k}{(k+1)!}$ (D) $\displaystyle\sum_{k=0}^{\infty} \dfrac{2x^k}{(k+1)!}$

[880] **4.** The Taylor expansion for $\int_0^x \cos t\, dt$ about $\dfrac{\pi}{2}$ is

  (A) $\displaystyle\sum_{k=0}^{\infty} (-1)^k \dfrac{\left(x - \dfrac{\pi}{2}\right)^{2k}}{(2k)!}$ (B) $\displaystyle\sum_{k=0}^{\infty} (-1)^k \dfrac{\left(x - \dfrac{\pi}{2}\right)^{2k+1}}{2k + 1}$

  (C) $\displaystyle\sum_{k=0}^{\infty} (-1)^k \dfrac{\left(x - \dfrac{\pi}{2}\right)^{k}}{k!}$ (D) $\displaystyle\sum_{k=0}^{\infty} (-1)^k \dfrac{\left(x - \dfrac{\pi}{2}\right)^{2k+1}}{(2k + 1)!}$

[873] **5.** What is the first nonzero term of the Maclaurin expansion of $f(x) = \ln(2x^3 + 1)$?

  (A) $\ln(3x^2 + 1)$ (B) $6x^2$ (C) $2x^3$ (D) $3x^2$

Free-Response Question

[878] **6.** The Maclaurin expansion of $f(x) = \dfrac{1}{(1-x)^2}$ equals $\displaystyle\sum_{k=0}^{\infty} a_k x^k$.

  (a) Find the coefficients of the first four terms of the Maclaurin expansion.

  (b) Given the Maclaurin expansion of $g(x) = \dfrac{2}{(1-x)^3}$ equals $\displaystyle\sum_{k=0}^{\infty} b_k x^k$, express b_k in terms of a_k.

Retain Your Knowledge

Multiple-Choice Questions

1. An object is moving along the x-axis. Its position (in ft.) at time $t \geq 0$ minutes is given by $x(t) = \sin^2(3t)$. What is the acceleration of the object at time $t = \dfrac{\pi}{6}$ min?

   (A) 0 ft/min²   (B) -18 ft/min²
   (C) 18 ft/min²   (D) 36 ft/min²

2. The function f is differentiable for all x in the open interval $(0, 7)$. Use the information in the table below to choose the correct statement.

x	$0 < x < 2$	2	$2 < x < 5$	5	$5 < x < 7$
$f''(x)$	negative	0	negative	does not exist	positive

   (A) $(2, f(2))$ is an inflection point; f is concave down on the interval $(0, 5)$.
   (B) $(2, f(2))$ is an inflection point because $f''(2) = 0$.
   (C) $(5, f(5))$ is not an inflection point because $f''(5)$ doesn't exist.
   (D) $(5, f(5))$ is an inflection point because the concavity of f changes at $(5, f(5))$.

3. A particle moves along a smooth curve traced out by the parametric equations $x(t) = \dfrac{t^3}{3} - t^2 - 6t + 4$ and $y(t) = 3t^2$, $-2 \leq t \leq 2$. At what point, if any, is the slope of the tangent line equal to 2?

   (A) $(4, 108)$ (B) $(4, 0)$ (C) $\left(\dfrac{26}{3}, 3\right)$ (D) No point

Free-Response Question

4. The graphs of $y_1 = x^3$ and $y_2 = 4x$ are shown below.

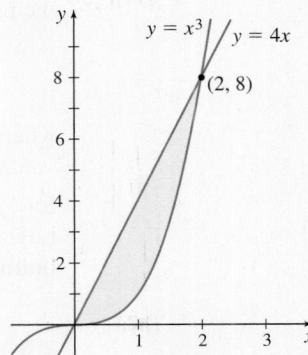

   (a) Find the area of the region bounded by the two graphs.
   (b) Find the length of the perimeter of the region bounded by the two graphs.
   (c) Determine the volume of the solid generated by revolving the region bounded by the two graphs about the y-axis.
   (d) The region forms the base of a solid. If cross sections of the solid perpendicular to the x-axis are rectangles whose heights are twice their widths, express the volume of the solid as an integral.

10.10 Taylor Polynomial Approximations and the Lagrange Error Bound

OBJECTIVES *When you finish this section, you should be able to:*

1 **Approximate a function and its graph using a Taylor polynomial (p. 884)**
2 **Obtain Lagrange error bounds for Taylor polynomial approximations (p. 887)**
3 **Obtain an error bound for a Taylor polynomial approximation with alternating terms (p. 889)**

We begin by restating Taylor's Formula with Remainder. Suppose f is a function whose first $n + 1$ derivatives are continuous on an open interval I containing the number c. Then for every x in I, there is a number u between c and x for which

$$f(x) = f(c) + f'(c)(x - c) + \frac{f''(c)}{2!}(x - c)^2 + \cdots + \frac{f^{(n)}(c)}{n!}(x - c)^n + R_n(x) = P_n(x) + R_n(x) \tag{1}$$

where

$$P_n(x) = f(c) + f'(c)(x - c) + \frac{f''(c)}{2!}(x - c)^2 + \cdots + \frac{f^{(n)}(c)}{n!}(x - c)^n \tag{2}$$

is the Taylor polynomial of degree n at c and

$$R_n(x) = \frac{f^{(n+1)}(u)}{(n+1)!}(x - c)^{n+1} \tag{3}$$

is the Lagrange form of the remainder after $n + 1$ terms.

If f has derivatives of all orders in the open interval I and if $\lim\limits_{n \to \infty} R_n(x) = 0$, then the infinite series $\sum\limits_{k=0}^{\infty} \dfrac{f^{(k)}(c)}{k!}(x-c)^k$ converges to $f(x)$ and we can use $P_n(x)$ to approximate both the function f and its graph.

The accuracy of using $P_n(x)$ as an approximation to f for x close to c depends on the remainder $R_n(x)$. The error involved in the approximation equals

$$|R_n(x)| = \frac{|f^{n+1}(u)||x-c|^{n+1}}{(n+1)!}$$

where x is close to c and u is between c and x. Since we don't know what u equals, the value of $|R_n(x)|$ cannot be determined. However, we can usually find an upper estimate for $|f^{n+1}(u)|$ and obtain an **error bound E** for which $|R_n(x)| \le E$. Since we use the Lagrange form of the remainder, we refer to this error bound as the **Lagrange error bound**.

① Approximate a Function and Its Graph Using a Taylor Polynomial

In Section 4.3, we found that if a function f is differentiable on an interval (a, b), then we can approximate f at a number c in the interval using the linear approximation

$$L(x) = f(c) + f'(c)(x-c)$$

where $y = L(x)$ is the tangent line to the graph of f at the point $(c, f(c))$. This linear approximation for f is good if x is close enough to c, but as the interval widens, the accuracy of the estimate decreases.

Notice that $y = L(x)$ is a first degree polynomial whose terms are identical to the first two terms of the Taylor polynomial evaluated at c. If f is twice differentiable on the interval (a, b), we can improve the approximation by using a second degree polynomial P_2 whose graph has the same tangent line as f at c and whose second derivative $P''(c) = f''(c)$. This polynomial at c is

$$P_2(x) = f(c) + f'(c)(x-c) + \frac{f''(c)}{2}(x-c)^2$$

Continuing in this fashion, if the function f has derivatives of all orders at c, and if the Taylor series converges to f, then the Taylor polynomial

$$P_n(x) = f(c) + f'(c)(x-c) + \frac{f''(c)}{2!}(x-c)^2 + \cdots + \frac{f^{(n)}(c)}{n!}(x-c)^n$$

can be used to approximate f near c.

EXAMPLE 1   Finding a Taylor Polynomial for $f(x) = e^x$

Find the Taylor polynomial $P_n(x)$ for $f(x) = e^x$ at 0.

Solution

The Taylor polynomial $P_n(x)$ at 0 is a polynomial of degree n. Since the Maclaurin expansion of $f(x) = e^x$ is

$$f(x) = \sum_{k=0}^{\infty} \frac{f^{(k)}(0)}{k!}x^k = 1 + x + \frac{x^2}{2!} + \frac{x^3}{3!} + \cdots + \frac{x^n}{n!} + \cdots = \sum_{k=0}^{\infty} \frac{x^k}{k!}$$

the Taylor polynomial $P_n(x)$ at 0 for $f(x) = e^x$ is

$$P_n(x) = 1 + x + \frac{x^2}{2!} + \frac{x^3}{3!} + \cdots + \frac{x^n}{n!}$$

∎

Figure 29 illustrates how the graphs of the Taylor polynomials $P_1(x)$, $P_2(x)$, and $P_3(x)$ for $f(x) = e^x$ approximate the graph of $f(x) = e^x$ near 0.

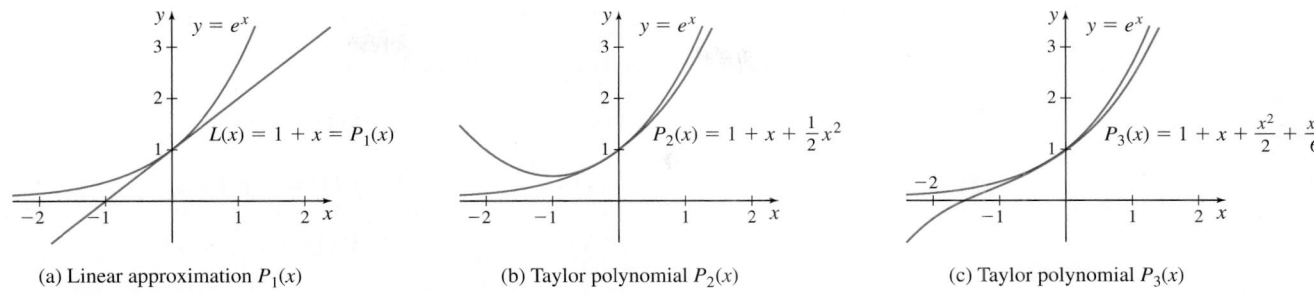

(a) Linear approximation $P_1(x)$ (b) Taylor polynomial $P_2(x)$ (c) Taylor polynomial $P_3(x)$

Figure 29

NOW WORK   AP® Practice Problem 2.

As the next example shows, we can find a Taylor polynomial for a function without using its Taylor expansion.

EXAMPLE 2   Finding a Taylor Polynomial Without Using a Taylor Expansion

Find the Taylor polynomial $P_3(x)$ for $f(x) = \sqrt{x}$ at 1.

Solution
The first three derivatives of $f(x) = \sqrt{x}$ are:

$$f'(x) = \frac{1}{2\sqrt{x}} \qquad f''(x) = \frac{d}{dx} \frac{x^{-1/2}}{2} = -\frac{1}{4x^{3/2}} \qquad f'''(x) = \frac{d}{dx} \left(-\frac{x^{-3/2}}{4} \right) = \frac{3}{8x^{5/2}}$$

Then

$$f(1) = 1 \quad f'(1) = \frac{1}{2} \quad f''(1) = -\frac{1}{4} \quad f'''(1) = \frac{3}{8}$$

The Taylor polynomial $P_3(x)$ for $f(x) = \sqrt{x}$ at 1 is

$$P_3(x) = f(1) + f'(1)(x-1) + \frac{f''(1)}{2!}(x-1)^2 + \frac{f'''(1)}{3!}(x-1)^3$$

$$= 1 + \frac{x-1}{2} - \frac{(x-1)^2}{8} + \frac{(x-1)^3}{16}$$ ∎

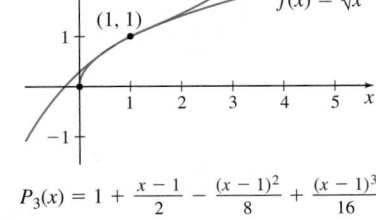

$P_3(x) = 1 + \frac{x-1}{2} - \frac{(x-1)^2}{8} + \frac{(x-1)^3}{16}$

Figure 30

The graphs of $y = \sqrt{x}$ and $y = P_3(x)$ are shown in Figure 30. Notice how the graphs of $y = P_3(x)$ and $y = f(x)$ are virtually identical for values of x near 1.

NOW WORK   Problem 3 and AP® Practice Problem 3.

EXAMPLE 3   Finding a Taylor Polynomial for $f(x) = \ln x$

Find the Taylor polynomial $P_4(x)$ for $f(x) = \ln x$ at 1.

Solution

f and the first four derivatives of $f(x) = \ln x$ at 1 are:

$$
\begin{aligned}
f(x) &= \ln x & f(1) &= \ln 1 = 0 \\[4pt]
f'(x) &= \frac{1}{x} & f'(1) &= \frac{1}{1} = 1 \\[4pt]
f''(x) &= -\frac{1}{x^2} & f''(1) &= -\frac{1}{1^2} = -1 \\[4pt]
f'''(x) &= \frac{2}{x^3} & f'''(1) &= \frac{2}{1^3} = 2 \\[4pt]
f^{(4)}(x) &= -\frac{6}{x^4} & f^{(4)}(1) &= -\frac{6}{1^4} = -6
\end{aligned}
$$

The Taylor polynomial $P_4(x)$ for $f(x) = \ln x$ at 1 is

$$
\begin{aligned}
P_4(x) &= f(1) + f'(1)(x-1) + \frac{f''(1)}{2!}(x-1)^2 + \frac{f'''(1)}{3!}(x-1)^3 + \frac{f^{(4)}(1)}{4!}(x-1)^4 \\[6pt]
&= 0 + 1(x-1) + \frac{(-1)}{2!}(x-1)^2 + \frac{2}{3!}(x-1)^3 + \frac{(-6)}{4!}(x-1)^4 \\[6pt]
&= (x-1) - \frac{(x-1)^2}{2} + \frac{(x-1)^3}{3} - \frac{(x-1)^4}{4}
\end{aligned}
$$

■

The graphs of $y = \ln x$ and $y = P_4(x)$ are shown in Figure 31. Notice how the graphs of $y = P_4(x)$ and $y = \ln x$ are virtually identical for values of x near 1.

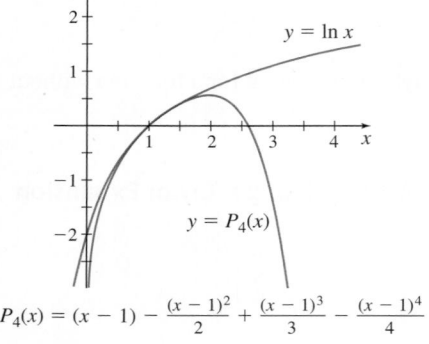

$P_4(x) = (x-1) - \frac{(x-1)^2}{2} + \frac{(x-1)^3}{3} - \frac{(x-1)^4}{4}$

Figure 31

NOW WORK   Problem 5.

EXAMPLE 4   Finding a Taylor Polynomial of Degree Three

Find the Taylor polynomial $P_3(x)$ for $f(x) = xe^x$ at 0.

Solution

We begin with the Maclaurin expansion for e^x.

$$
e^x = \sum_{k=0}^{\infty} \frac{f^{(k)}(0)}{k!} x^k = 1 + x + \frac{x^2}{2!} + \frac{x^3}{3!} + \cdots = \sum_{k=0}^{\infty} \frac{x^k}{k!}
$$

Then the Maclaurin expansion for $f(x) = xe^x$ is

$$
xe^x = \sum_{k=0}^{\infty} \frac{f^{(k)}(0)}{k!} x^{k+1} = x + x^2 + \frac{x^3}{2!} + \frac{x^4}{3!} + \cdots = \sum_{k=0}^{\infty} \frac{x^{k+1}}{k!}
$$

The Taylor polynomial $P_3(x)$ at 0 for $f(x) = xe^x$ is the polynomial of degree three in the Maclaurin expansion for xe^x, namely

$$
P_3(x) = x + x^2 + \frac{x^3}{2!}
$$

■

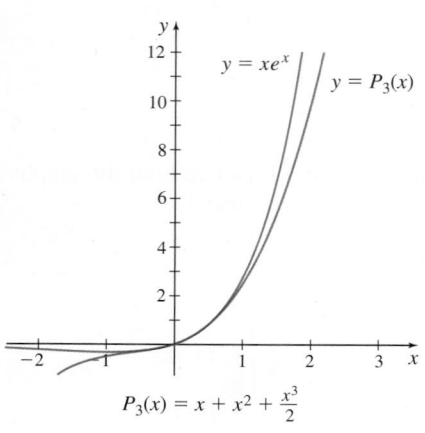

$P_3(x) = x + x^2 + \frac{x^3}{2}$

Figure 32 Taylor polynomial $P_3(x)$

Figure 32 shows how the graph of the Taylor polynomial $P_3(x)$ approximates the graph of $f(x) = xe^x$ near 0.

② Obtain Lagrange Error Bounds for Taylor Polynomial Approximations

EXAMPLE 5   Approximating $y = \sin x$ Using a Taylor Polynomial

(a) Approximate $y = \sin x$ using the Taylor polynomial $P_7(x)$ of $y = \sin x$ at 0.

(b) Graph $y = \sin x$ and the Taylor polynomial $P_7(x)$.

(c) Use (a) to approximate $\sin 0.1$.

(d) What is the Lagrange error bound in using this approximation?

Solution

(a) The Maclaurin expansion for $y = \sin x$ found in Example 3 (p. 877) of Section 10.9 is

$$y = \sin x = x - \frac{x^3}{3!} + \frac{x^5}{5!} - \frac{x^7}{7!} + \cdots = \sum_{k=0}^{\infty} (-1)^k \frac{x^{2k+1}}{(2k+1)!}$$

The Taylor polynomial $P_7(x)$ of degree 7 for $y = \sin x$ at 0 is

$$P_7(x) = x - \frac{x^3}{3!} + \frac{x^5}{5!} - \frac{x^7}{7!} \tag{4}$$

So, $\sin x \approx P_7(x)$ for x close to 0.

(b) The graphs of $y = \sin x$ and $y = P_7(x)$ are given in Figure 33.

(c) Using (4), we can approximate $\sin 0.1$.

$$\sin 0.1 \approx 0.1 - \frac{0.1^3}{3!} + \frac{0.1^5}{5!} - \frac{0.1^7}{7!} \approx 0.0998$$

(d) The Lagrange error bound to the approximation for $\sin 0.1$ is given by

$$|R_n(x)| = \frac{|f^{n+1}(u)||x - c|^{n+1}}{(n+1)!}$$

where u is between $c = 0$ and $x = 0.1$ and $n = 7$. Since $|f^{(8)}(u)| = |\sin u| \leq 1$, we have

$$|R_7(x)| = \frac{|f^{(8)}(u)||x - c|^8}{8!} = \frac{|\sin u|(0.1)^8}{8!} \underset{\underset{|\sin u| \leq 1}{\uparrow}}{\leq} \frac{10^{-8}}{8!} \approx 2.48 \cdot 10^{-13} \qquad ■$$

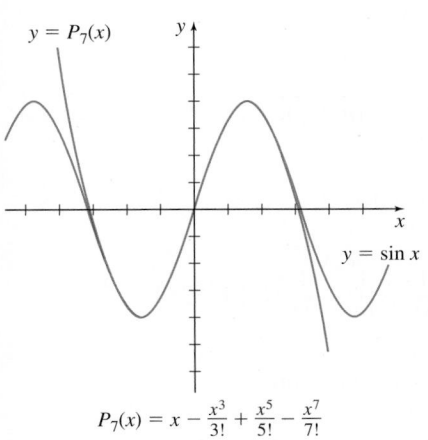

$y = P_7(x)$

$y = \sin x$

$P_7(x) = x - \frac{x^3}{3!} + \frac{x^5}{5!} - \frac{x^7}{7!}$

Figure 33

NOW WORK   Problem **13** and AP® Practice Problems **1** and **5**.

EXAMPLE 6   Approximating e

Use a Taylor polynomial to approximate the number e so that the Lagrange error bound is less than 0.0001.

Solution

Express e^x as a Maclaurin expansion with the Lagrange form of the remainder. Then

$$e^x = P_n(x) + R_n(x)$$

where

$$P_n(x) = 1 + \frac{x}{1!} + \frac{x^2}{2!} + \cdots + \frac{x^n}{n!}$$

is the Taylor polynomial of degree n at 0 of $f(x) = e^x$, and the Lagrange remainder is

$$R_n(x) = \frac{e^u}{(n+1)!} x^{n+1} \qquad \frac{d^{n+1}}{dx^{n+1}} e^x = e^x$$

where u is a number between 0 and x.

(Example continued on the next page)

If $x = 1$,

$$e^1 = P_n(1) + R_n(1)$$

$$e = 1 + \frac{1}{1!} + \frac{1}{2!} + \cdots + \frac{1}{n!} + \frac{e^u}{(n+1)!} \qquad 0 < u < 1$$

We need to find an integer n so that the remainder $\dfrac{e^u}{(n+1)!}$ is less than 0.0001. Our problem in finding this integer n is that we do not know the exact value of e^u. However, since $x = 1$, we do know that $0 < u < 1$. Also, since $f(x) = e^x$ is an increasing function, $e^u < e^1 = e$. Finally, $e < 3$, so

$$0 < \frac{e^u}{(n+1)!} < \frac{e}{(n+1)!} < \frac{3}{(n+1)!}$$

The smallest n for which $\dfrac{3}{(n+1)!} < 0.0001$ is $n = 7$.

$$P_7(1) = 1 + 1 + \frac{1}{2} + \frac{1}{3!} + \frac{1}{4!} + \frac{1}{5!} + \frac{1}{6!} + \frac{1}{7!} \approx 2.718254$$

and the Lagrange error bound is less than 0.0001. In other words,

$$|e - 2.718254| < 0.0001 \text{ or } 2.718154 < e < 2.718354 \qquad \blacksquare$$

NOW WORK Problem 23 and AP® Practice Problem 6.

EXAMPLE 7 Approximating a Logarithm Using a Taylor Polynomial

(a) Use a Taylor polynomial of degree 4 to approximate $\ln 1.1$.

(b) Find the Lagrange error bound for this approximation.

Solution

Because 1.1 is closer to 1 than to 0, the Maclaurin expansion for $f(x) = \ln(x + 1)$ is a better choice than using a Maclaurin expansion for $y = \ln x$.

$$f(x) = \ln(1 + x) \qquad\qquad f(0) = \ln(1 + 0) = 0$$

$$f'(x) = (1 + x)^{-1} \qquad\qquad f'(0) = (1 + 0)^{-1} = 1$$

$$f''(x) = -(1 + x)^{-2} \qquad\qquad f''(0) = -(1 + 0)^{-2} = -1$$

$$f'''(x) = 2(1 + x)^{-3} \qquad\qquad f'''(0) = 2(1 + 0)^{-3} = 2$$

$$f^{(4)}(x) = -6(1 + x)^{-4} \qquad\qquad f^{(4)}(0) = -6(1 + 0)^{-4} = -6$$

The Taylor polynomial for $f(x) = \ln(1 + x)$ of degree 4 is

$$P_4(x) = f(0) + f'(0)x + \frac{f''(0)}{2!}x^2 + \frac{f'''(0)}{3!}x^3 + \frac{f^{(4)}(0)}{4!}x^4$$

$$= 0 + 1x - \frac{1}{2!}x^2 + \frac{2}{3!}x^3 - \frac{6}{4!}x^4 = x - \frac{1}{2}x^2 + \frac{1}{3}x^3 - \frac{1}{4}x^4$$

If $x = 0.1$, then $f(0.1) = \ln(1.1) \approx P_4(0.1) = 0.1 - \dfrac{0.1^2}{2} + \dfrac{0.1^3}{3} - \dfrac{0.1^4}{4} \approx 0.095308$.

(b) The Lagrange error bound for this approximation is $|R_4(x)| = \left| \dfrac{f^{(5)}(u)}{5!} x^5 \right|$,

where $0 < u < 0.1$. Since $f^{(5)}(x) = 24(1+x)^{-5} = \dfrac{24}{(1+x)^5}$, then $\left| f^{(5)}(u) \right| < 24$. So,

$$|R_4(x)| < \left| \dfrac{24}{5!} \cdot 0.1^5 \right| = 2 \times 10^{-6}$$

The Lagrange error bound for approximating ln 1.1 using a Taylor polynomial of degree 4 is less than 0.000002. ∎

NOW WORK AP® Practice Problems **4** and **7**.

③ Obtain an Error Bound for a Taylor Polynomial Approximation with Alternating Terms

Suppose a Taylor expansion of a function has alternating terms. Then it is often easier to calculate the error bound of a Taylor polynomial approximation of f using the error bound for a convergent alternating series rather than using a Lagrange error bound.

Recall an alternating series $\sum\limits_{k=1}^{\infty} (-1)^{k+1} a_k$ for which the $a_k > 0$ are nonincreasing and for which $\lim\limits_{n \to \infty} a_n = 0$ is convergent. Then the error E_n in using the sum S_n of the first n terms as an approximation to the sum of the series is numerically less than or equal to the $(n+1)$st term of the series. That is, $|E_n| \le a_{n+1}$.

EXAMPLE 8 **Approximating $y = \dfrac{1}{\sqrt{e}}$ Using a Taylor Polynomial**

(a) Find the Taylor polynomial $P_5(x)$ for $f(x) = e^{-x}$ at 0.

(b) Use $P_5(x)$ to approximate $f\left(\dfrac{1}{2}\right) = \dfrac{1}{\sqrt{e}} = e^{-1/2}$.

(c) Use the alternating series error bound to approximate error in using the Taylor polynomial $P_5(x)$ to approximate $\dfrac{1}{\sqrt{e}}$.

Solution

(a) The derivatives of $f(x) = e^{-x}$ are:

$$f'(x) = -e^{-x} \quad f''(x) = e^{-x} \quad f'''(x) = -e^{-x} \quad f^{(4)}(x) = e^{-x} \quad f^{(5)}(x) = -e^{-x}$$

So,

$$f'(0) = -1 \quad f''(0) = 1 \quad f'''(0) = -1 \quad f^{(4)}(0) = 1 \quad f^{(5)}(0) = -1$$

The Taylor polynomial $P_5(x)$ of $f(x) = e^{-x}$ at 0 is

$$P_5(x) = 1 - x + \dfrac{x^2}{2!} - \dfrac{x^3}{3!} + \dfrac{x^4}{4!} - \dfrac{x^5}{5!}$$

(b)

$$e^{-1/2} \approx P_5\left(\dfrac{1}{2}\right) = 1 - \dfrac{1}{2} + \dfrac{\tfrac{1}{4}}{2!} - \dfrac{\tfrac{1}{8}}{3!} + \dfrac{\tfrac{1}{16}}{4!} - \dfrac{\tfrac{1}{32}}{5!} \approx 0.607$$

(Example continued on the next page)

(c) The Maclaurin expansion for $f(x) = e^{-x}$ is

$$e^{-x} = 1 - x + \frac{x^2}{2!} - \frac{x^3}{3!} + \cdots + (-1)^n \frac{x^n}{n!} + \cdots$$

Since the Maclaurin expansion satisfies the conditions of a convergent alternating series, the error bound for $P_5 \left(\dfrac{1}{2} \right)$ is numerically less than or equal to

$$\frac{\frac{1}{2^6}}{6!} \approx 0.0000217$$

The error in using the Taylor polynomial of degree 5 to approximate $e^{-1/2}$ is at most 2.17×10^{-5}. ∎

NOW WORK   AP® Practice Problem **8.**

10.10 Assess Your Understanding

Skill Building

In Problems 1–12, for each function f find the Taylor Polynomial $P_5(x)$ for f at c.

1. $f(x) = \ln x^2$ at $c = 1$

2. $f(x) = \ln(1 + x)$ at $c = 0$

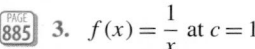 **3.** $f(x) = \dfrac{1}{x}$ at $c = 1$

4. $f(x) = \dfrac{1}{x^2}$ at $c = 1$

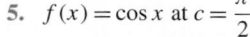 **5.** $f(x) = \cos x$ at $c = \dfrac{\pi}{2}$

6. $f(x) = \sin x$ at $c = \dfrac{\pi}{4}$

7. $f(x) = e^{2x}$ at $c = 0$

8. $f(x) = e^{-x}$ at $c = 0$

9. $f(x) = \dfrac{1}{1 - 2x}$ at $c = 0$

10. $f(x) = \dfrac{1}{1 + x}$ at $c = 0$

11. $f(x) = x \ln x$ at $c = 1$

12. $f(x) = xe^x$ at $c = 1$

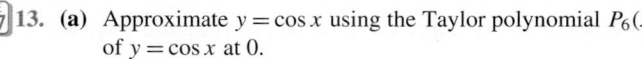 **13. (a)** Approximate $y = \cos x$ using the Taylor polynomial $P_6(x)$ of $y = \cos x$ at 0.

(b) Graph $y = \cos x$ and $y = P_6(x)$.

(c) Use $P_6(x)$ to approximate $\cos \dfrac{\pi}{90}$, $(2°)$.

(d) What is the Lagrange error bound in using this approximation?

(e) If $P_6(x)$ is used to approximate $y = \cos \dfrac{\pi}{90}$, for what values of x is the error less than or equal to 0.0001?

14. (a) Approximate $y = e^x$ using the Taylor polynomial $P_3(x)$ of $y = e^x$ at 0.

(b) Graph $y = e^x$ and $y = P_3(x)$.

(c) Use $P_3(x)$ to approximate $e^{1/2}$.

(d) What is the Lagrange error bound in using this approximation?

(e) If $P_3(x)$ is used to approximate $y = e^{1/2}$, for what values of x is the error less than or equal to 0.0001?

15. (a) Approximate $y = \sqrt[3]{1 + x}$ using the Taylor polynomial $P_4(x)$ of $y = \sqrt[3]{1 + x}$ at 0.

(b) Graph $y = \sqrt[3]{1 + x}$ and $y = P_4(x)$.

(c) Comment on the graphs in (b) and the result of the approximation.

(d) Use $P_4(x)$ to approximate $\sqrt[3]{0.9}$. What is the error in using this approximation?

16. (a) Approximate $y = \dfrac{1}{\sqrt{4 + x}}$ using the Taylor polynomial $P_4(x)$ of $y = \dfrac{1}{\sqrt{4 + x}}$ at 0.

(b) Graph $y = \dfrac{1}{\sqrt{4 + x}}$ and $y = P_4(x)$.

(c) Comment on the graphs in (b) and the result of the approximation.

(d) Use $P_4(x)$ to approximate $\dfrac{1}{\sqrt{4.2}}$. What is the error in using this approximation?

17. (a) Approximate $y = \tan^{-1} x$ using the Taylor polynomial $P_9(x)$ of $y = \tan^{-1} x$ at 0.

(b) Graph $y = \tan^{-1} x$ and $y = P_9(x)$.

(c) Comment on the graphs in (b) and the result of the approximation.

18. (a) Approximate $y = \dfrac{1}{1 - x}$ using the Taylor polynomial $P_3(x)$ of $y = \dfrac{1}{1 - x}$ at 0.

(b) Graph $y = \dfrac{1}{1 - x}$ and $y = P_3(x)$.

(c) Comment on the graphs in (b) and the result of the approximation.

In Problems 19–22, write each polynomial function f as a polynomial function in $(x - c)$.

19. $f(x) = 3x^3 + 2x^2 - 6x + 5$ $c = 1$

20. $f(x) = 4x^3 - 2x^2 - 4$ $c = 1$

21. $f(x) = 2x^4 - 6x^3 + x$ $c = -1$

22. $f(x) = -3x^4 + 2x^2 - 5$ $c = -1$

In Problems 23–24, use a Taylor polynomial to approximate each number so that the Lagrange error bound is less than the number shown. What is the degree of the Taylor polynomial?

23. $\sqrt{e}$; Error < 0.001

24. $\ln 1.03$; Error < 0.00001

Applications and Extensions

In Problems 25–28, approximate each integral using the first four terms of a Maclaurin series.

25. $\displaystyle\int_0^1 \sin x^2\, dx$

26. $\displaystyle\int_0^1 \cos x^2\, dx$

27. $\displaystyle\int_0^{0.1} \frac{e^x}{1-x}\, dx$

28. $\displaystyle\int_0^{0.1} \frac{\cos x}{1-x}\, dx$

29. The graphs of $y = \sin x$ and $y = \lambda x$ intersect near $x = \pi$ if λ is small. Let $f(x) = \sin x - \lambda x$. Find the Taylor polynomial $P_2(x)$ for f at π, and use it to show that an approximate solution of the equation $\sin x = \lambda x$ is $x = \dfrac{\pi}{1+\lambda}$.

30. The graphs of $y = \cot x$ and $y = \lambda x$ intersect near $x = \dfrac{\pi}{2}$ if λ is small. Let $f(x) = \cot x - \lambda x$. Find the Taylor polynomial $P_2(x)$ for f at $\dfrac{\pi}{2}$, and use it to find an approximate solution of the equation $\cot x = \lambda x$.

Preparing for the **AP® Exam**

AP® Practice Problems

Multiple-Choice Questions

 1. Use the Taylor polynomial $P_4(x)$ of $y = \cos x$ at 0 to approximate $\cos 0.2$.

    (A) 1   (B) 0.980   (C) 0.803   (D) 0.801

 2. If a function f can be represented by the Taylor polynomial

$$P_3(x) = 4 + 2(x-2) + 3(x-2)^2 + \frac{1}{2}(x-2)^3$$

then $f''(2) =$

    (A) 1   (B) 2   (C) 3   (D) 6

 3. If $f(-1) = 4$, $f'(-1) = -3$, $f''(-1) = 3$, and $f'''(-1) = 2$ then the Taylor polynomial $P_3(x)$ of degree 3 of f at -1 is

    (A) $P_3(x) = 4 - 3(x+1) + \dfrac{3}{2}(x+1)^2 + \dfrac{2}{3}(x+1)^3$

    (B) $P_3(x) = 4 - 3(x+1) + \dfrac{3}{2}(x+1)^2 - \dfrac{1}{3}(x+1)^3$

    (C) $P_3(x) = 4 - 3(x-1) + \dfrac{3}{2}(x-1)^2 + \dfrac{1}{3}(x-1)^3$

    (D) $P_3(x) = 4 - 3(x+1) + \dfrac{3}{2}(x+1)^2 + \dfrac{1}{3}(x+1)^3$

 4. The Taylor polynomial $P_3(x)$ for the function $g(x) = \ln x$ centered at 1 is given by

    (A) $1 + (x-1) - \dfrac{1}{2}(x-1)^2 + \dfrac{1}{3}(x-1)^3$

    (B) $(x-1) - \dfrac{1}{2}(x-1)^2 + \dfrac{1}{3}(x-1)^3$

    (C) $(x-1) - \dfrac{1}{2!}(x-1)^2 + \dfrac{1}{3!}(x-1)^3$

    (D) $(x-1) + \dfrac{1}{2}(x-1)^2 + \dfrac{1}{3}(x-1)^3$

 5. The Lagrange error bound in using a Taylor polynomial of degree 5 for $f(x) = \sin x$ at $\dfrac{2\pi}{3}$ to approximate $\sin 2$ is no more than

    (A) $\dfrac{\left|\frac{2\pi}{3} - 2\right|^5}{5!}$

    (B) $\dfrac{\left|\frac{2\pi}{3} - 2\right|^6}{6!}$

    (C) $\dfrac{\left|\cos \frac{2\pi}{3}\right|\left|\frac{2\pi}{3} - 2\right|^6}{6!}$

    (D) $\dfrac{\left|\sin \frac{2\pi}{3}\right|\left|\frac{2\pi}{3} - 2\right|^6}{6!}$

Free-Response Questions

 6. (a) Write the Maclaurin series for $f(x) = e^{-x}$.

    (b) Approximate $e^{-0.1}$ with the Taylor polynomial $P_2(x)$ of $f(x) = e^{-x}$.

    (c) How many terms are necessary to approximate $e^{-0.1}$ with an error less than or equal to 0.00001?

7. The function $f(x) = \dfrac{1}{\sqrt{2\pi}} e^{-x^2/2}$ is defined for all real numbers.

    (a) Write the first four nonzero terms of the Maclaurin expansion for f.

    (b) Use the Maclaurin expansion from (a) and properties of a power series to approximate $\displaystyle\int_0^1 \frac{1}{\sqrt{2\pi}} e^{-x^2/2}\, dx$.

    (c) Find the maximum error in using the first four terms of the Maclaurin expansion to approximate $\displaystyle\int_0^1 \frac{1}{\sqrt{2\pi}} e^{-x^2/2}\, dx$.

    (d) Justify the method used to find the error in (c).

 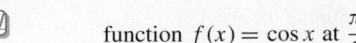 **8.** (a) Find the Taylor polynomial $P_7(x)$ for the function $f(x) = \cos x$ at $\dfrac{\pi}{2}$.

    (b) Use $P_7(x)$ to approximate $f(1.5) = \cos 1.5$.

    (c) Find the maximum error in using $P_7(x)$ to approximate $\cos 1.5$.

Retain Your Knowledge

Multiple-Choice Questions

1. Oil is pumped into a cylindrical tank at a rate of 40 m³/min. If the diameter of the tank is 20 m, find the rate the depth of the oil in the tank is rising.

   (A) $\dfrac{2}{5}$ m/min   (B) $\dfrac{2}{5\pi}$ m/min

   (C) $\dfrac{2\pi}{5}$ m/min   (D) 40 m/min

2. A function $y = f(x)$ is continuous and differentiable for all real numbers. Suppose $\int_2^{10} f(x)\, dx = 16$. The table below gives values of f and f' for two numbers.

x	2	10
$f(x)$	-3	5
$f'(x)$	1	$\dfrac{1}{2}$

   Evaluate $\displaystyle\int_2^{10} x f'(x)\, dx$.

   (A) 20   (B) 34   (C) 40   (D) 72

3. Determine whether $\displaystyle\int_0^1 \dfrac{\ln x}{x}\, dx$ converges or diverges. If it converges, find its value.

   (A) $-\dfrac{1}{2}$ (B) $\dfrac{1}{2}$

   (C) 1   (D) The integral diverges.

Free-Response Question

4. An object moves along a smooth plane curve so that at time t, $\dfrac{\pi}{6} \le t \le 2\pi$, its position is given by the parametric equations $x = x(t)$, $y = y(t)$. We are told that $\dfrac{dx}{dt} = 12t$ and $\dfrac{dy}{dt} = -6\cos(2t)$.

   (a) Find any times t, $\dfrac{\pi}{6} \le t \le 2\pi$, at which the line tangent to the plane curve is horizontal.

   (b) Find the second derivative $\dfrac{d^2 y}{dx^2}$ of the function $y = f(x)$ represented by the parametric equations.

   (c) Find the total distance traveled by the object from $t = \dfrac{\pi}{6}$ to $t = 2\pi$.

   (d) Find the speed of the object at time $t = \pi$.

CHAPTER 10 PROJECT   ## How Calculators Calculate

This Project can be done individually or as part of a team.

The sine function is used in many scientific applications, so a calculator/computer must be able to evaluate it with lightning-fast speed.

While we know how to find the exact value of the sine function for many numbers, such as 0, $\dfrac{\pi}{6}$, $\dfrac{\pi}{2}$, and so on, we have no method for finding the exact value of $\sin 3$ (which should be close to $\sin \pi$) or $\sin 1.5$ (which should be close to $\sin \dfrac{\pi}{2}$). Since the sine function can be evaluated at any real number, we first use some of its properties to restrict its domain to something more manageable.

1. Explain why we can evaluate $\sin x$ for any x using only the interval $\left[-\dfrac{\pi}{2}, \dfrac{\pi}{2} \right]$. (We could restrict that domain further, but this will work for now. See Problem 6 below.)

2. Use the Maclaurin expansion for $\sin x$ to find an approximation for $\sin \dfrac{1}{2}$ correct to within 10^{-5}. Compare your approximation to the one your calculator/computer provides. How many terms of the series do you need to obtain this accuracy?

3. Use the Maclaurin expansion for $\sin x$ to find an approximation for $\sin \dfrac{3}{2}$ correct to within 10^{-5}. How many terms of the series do you need to obtain this accuracy?

4. Explain why the approximation in Problem 3 requires more terms than that of Problem 2.

5. Represent $\sin x$ as a Taylor expansion about $\dfrac{\pi}{4}$.

6. Explain why we can evaluate $\sin x$ for any x using only the interval $\left[0, \dfrac{\pi}{2} \right]$.

7. Use the result of Problem 5 to find an approximation for $\sin \dfrac{1}{2}$ correct to within 10^{-5}. Compare the result with the values your calculator/computer supplies for $\sin \dfrac{1}{2}$, as well as with the result from the Maclaurin approximation obtained in Problem 2. How many terms of the series do you need for the approximation?

8. Use the Taylor expansion for $\sin x$ at $\dfrac{\pi}{2}$ to find an approximation for $\sin \dfrac{3}{2}$ correct to within 10^{-5}. Compare the result with the value your calculator/computer supplies for $\sin \dfrac{3}{2}$, as well as with the result from the Maclaurin approximation obtained in Problem 3. How many terms of the series do you need for the approximation?

   The answers to Problems 3 and 8 reveal why Maclaurin series or Taylor series are not used to approximate the value of most functions. But often the methods used are similar. For example, a Chebyshev polynomial approximation to the sine function on the interval $\left[-\dfrac{\pi}{2}, \dfrac{\pi}{2}\right]$ still has the form of a Maclaurin series, but it was designed to converge more uniformly than the Maclaurin series, so that it can be expected to give answers near $\dfrac{\pi}{2}$ that are roughly as accurate as those near zero.

Chebyshev polynomials are commonly found in mathematical libraries for calculators/computers. For example, the widely used GNU Compiler Collection* uses Chebyshev polynomials to evaluate trigonometric functions. The Chebyshev polynomial approximation of degree 7 for the sine function is

$$S_7(x) = 0.9999966013x - 0.1666482357x^3$$
$$+ 0.008306286146x^5 - 0.1836274858 \times 10^{-3}x^7 \quad (1)$$

The Chebyshev polynomials are designed to remain close to a function across an entire closed interval. They seek to keep the approximation within a specified distance of the function being approximated at every point of that interval. If S_n is a Chebyshev approximation of degree n to the sine function on $\left[-\dfrac{\pi}{2}, \dfrac{\pi}{2}\right]$, then the error estimate in using $S_n(x)$ is given by

$$\max_{-\pi/2 \le x \le \pi/2} |\sin x - S_n(x)| \le \frac{\left(\dfrac{\pi}{2}\right)^{n+1}}{2^n(n+1)!} \quad (2)$$

Like most error estimates of this type, it gives an upper bound to the error.

8. Use the Chebyshev polynomial approximation in (1) for $x = \dfrac{1}{2}$ and $x = \dfrac{3}{2}$. Compare the results with the values your calculator/computer supplies for $\sin \dfrac{1}{2}$ and $\sin \dfrac{3}{2}$, as well as with the results from the Maclaurin approximations and the Taylor series approximations obtained in Problems 2, 3, 7, and 8.

10. Define $E_7(x) = |\sin x - S_7(x)|$. Use technology to graph E_7 on $\left[-\dfrac{\pi}{2}, \dfrac{\pi}{2}\right]$.

11. Find the local maximum and local minimum values of E_7 on $\left[-\dfrac{\pi}{2}, \dfrac{\pi}{2}\right]$. Compare these numbers with the error estimate in equation (2), and discuss the characteristics of the error.

12. Which approximation for the sine function would be preferable: the Maclaurin approximation, the Taylor approximation, or the Chebyshev approximation? Why?

---

*For more information on the GNU Compiler Collection (GCC), go to https://www.gcc.gnu.org.

Chapter Review

THINGS TO KNOW

10.1 Sequences
Definitions:

- A sequence is a function whose domain is the set of positive integers and whose range is a subset of real numbers. (p. 779)
- nth term of a sequence (p. 780)
- Alternating sequence (p. 781)
- Limit of a sequence (p. 782)
- Convergence; divergence of a sequence (p. 782)
- Related function of a sequence (p. 785)
- Divergence of a sequence to infinity (p. 787)
- Bounded sequence (p. 789)
- Monotonic sequence (p. 790)

Properties of a Convergent Sequence: (p. 783)

If $\{s_n\}$ and $\{t_n\}$ are convergent sequences and if c is a number, then

- Constant multiple property: $\lim\limits_{n \to \infty} (cs_n) = c \lim\limits_{n \to \infty} s_n$
- Sum and difference properties:

$$\lim_{n \to \infty} (s_n \pm t_n) = \lim_{n \to \infty} s_n \pm \lim_{n \to \infty} t_n$$

- Product property: $\lim\limits_{n \to \infty} (s_n \cdot t_n) = \left(\lim\limits_{n \to \infty} s_n\right)\left(\lim\limits_{n \to \infty} t_n\right)$

- Quotient property: $\lim\limits_{n \to \infty} \dfrac{s_n}{t_n} = \dfrac{\lim\limits_{n \to \infty} s_n}{\lim\limits_{n \to \infty} t_n}$

provided $\lim\limits_{n \to \infty} t_n \ne 0$

- Power property: $\lim\limits_{n \to \infty} s_n^p = \left[\lim\limits_{n \to \infty} s_n\right]^p$
  where $p \ge 2$ is an integer
- Root property: $\lim\limits_{n \to \infty} \sqrt[p]{s_n} = \sqrt[p]{\lim\limits_{n \to \infty} s_n}$
  where $p \ge 2$ and $s_n > 0$ if p is even

Theorems:

- Suppose $\{s_n\}$ is a sequence of real numbers. If $\lim\limits_{n \to \infty} s_n = L$ and if f is a function that is continuous at L and is defined for all numbers s_n, then $\lim\limits_{n \to \infty} f(s_n) = f(L)$. (p. 784)
- Suppose $\{s_n\}$ is a sequence and f is a related function of $\{s_n\}$. Suppose L is a real number. If $\lim\limits_{x \to \infty} f(x) = L$, then $\lim\limits_{n \to \infty} s_n = L$. (p. 786)
- The Squeeze Theorem for sequences (p. 786)
- The geometric sequence $\{r^n\}$, where r is a real number,
  - converges to 0, for $-1 < r < 1$.
  - converges to 1, for $r = 1$.
  - diverges for all other numbers. (p. 787)
- A convergent sequence is bounded (p. 789)
- If a sequence is not bounded from above or if it is not bounded from below, then it diverges. (p. 789)
- An increasing (or nondecreasing) sequence $\{s_n\}$ that is bounded from above converges. (p. 791)
- A decreasing (or nonincreasing) sequence $\{s_n\}$ that is bounded from below converges. (p. 791)

Procedure: Ways to show a sequence is monotonic: Table 2 (p. 790)

Summary: How to determine whether a sequence converges (p. 792)

10.2 Infinite Series

- If $a_1, a_2, \ldots, a_n, \ldots$ is an infinite collection of numbers, the expression $\sum_{k=1}^{\infty} a_k = a_1 + a_2 + \cdots + a_n + \cdots$ is called an infinite series or, simply, a series. (p. 796)
- nth term or general term of a series (p. 797)
- Partial sum $S_n = \sum_{k=1}^{n} a_k$, where S_n is the sum of the first n terms of the series $\sum_{k=1}^{\infty} a_k$ (p. 797)
- Convergence, divergence of a series (p. 798)
- Geometric series $\sum_{k=1}^{\infty} ar^{k-1} = a + ar + ar^2 + \cdots, a \neq 0$ (p. 800)
  - $\sum_{k=1}^{\infty} ar^{k-1}$ converges if $|r| < 1$, and its sum is $\dfrac{a}{1-r}$
  - $\sum_{k=1}^{\infty} ar^{k-1}$ diverges if $|r| \geq 1$. (p. 801)
- Harmonic series $\sum_{k=1}^{\infty} \dfrac{1}{k} = 1 + \dfrac{1}{2} + \dfrac{1}{3} + \cdots$ (p. 804)
  The harmonic series diverges. (p. 804)

Summary: Series and convergence of series (p. 805)

10.3 Properties of Series; Series with Positive Terms; the Integral Test

- If the series $\sum_{k=1}^{\infty} a_k$ converges, then $\lim_{n \to \infty} a_n = 0$. (p. 811)
- The nth Term Test for Divergence:

  If $\lim_{n \to \infty} a_n \neq 0$, the infinite series $\sum_{k=1}^{\infty} a_k$ diverges. (p. 812)
- If two infinite series are identical after a certain term, then either both series converge or both series diverge. If both series converge, they do not necessarily have the same sum. (p. 812)
- The General Convergence Test (p. 815)
- The Integral Test (p. 815)
  Suppose f is a function that is continuous, positive, and nonincreasing on the interval $[1, \infty)$ and $a_k = f(k)$ for all positive integers k.
  - If the improper integral $\int_1^{\infty} f(x)\,dx$ converges, the series $\sum_{k=1}^{\infty} a_k$ converges.
  - If the improper integral $\int_1^{\infty} f(x)\,dx$ diverges, the series $\sum_{k=1}^{\infty} a_k$ diverges.
- A p-series $\sum_{k=1}^{\infty} \dfrac{1}{k^p} = 1 + \dfrac{1}{2^p} + \dfrac{1}{3^p} + \cdots + \dfrac{1}{n^p} + \cdots$, where p is a positive real number, converges if $p > 1$ and diverges if $0 < p \leq 1$. (p. 817)
- Bounds for the sum of a convergent p-series: (p. 818)

  If $p > 1$, then $\dfrac{1}{p-1} < \sum_{k=1}^{\infty} \dfrac{1}{k^p} < 1 + \dfrac{1}{p-1}$.
- Bounds for the sum of a convergent series: (p. 819)

  For the series $\sum_{k=1}^{\infty} a_k, a_k > 0$, suppose f is a function that is continuous, positive, and decreasing on $[1, \infty)$, and $a_k = f(k)$ for all positive integers k. If the improper integral $\int_1^{\infty} f(x)\,dx$ converges, the sum of the series $\sum_{k=1}^{\infty} a_k$ is bounded by

  $$\int_1^{\infty} f(x)\,dx < \sum_{k=1}^{\infty} a_k < a_1 + \int_1^{\infty} f(x)\,dx$$

Properties of Convergent Series: If $\sum_{k=1}^{\infty} a_k$ and $\sum_{k=1}^{\infty} b_k$ are two convergent series and if $c \neq 0$ is a number, then

- Sum and difference properties:
  $$\sum_{k=1}^{\infty} (a_k \pm b_k) = \sum_{k=1}^{\infty} a_k \pm \sum_{k=1}^{\infty} b_k \quad \text{(p. 813)}$$
- Constant multiple property:
  $$\sum_{k=1}^{\infty} (ca_k) = c \sum_{k=1}^{\infty} a_k \quad c \neq 0 \quad \text{(p. 813)}$$
- If $\sum_{k=1}^{\infty} a_k$ diverges, then $\sum_{k=1}^{\infty} (ca_k)$ also diverges. (p. 813)

10.4 Comparison Tests
Theorems:

- Algebraic Comparison Test for Convergence: If $0 < a_k \leq b_k$ for all k and $\sum_{k=1}^{\infty} b_k$ converges, then $\sum_{k=1}^{\infty} a_k$ converges. (p. 824)
- Algebraic Comparison Test for Divergence: If $0 < c_k \leq a_k$ for all k and $\sum_{k=1}^{\infty} c_k$ diverges, then $\sum_{k=1}^{\infty} a_k$ diverges. (p. 825)
- Limit Comparison Test: (p. 826)

  Suppose $\sum_{k=1}^{\infty} a_k$ and $\sum_{k=1}^{\infty} b_k$ are both series of positive terms.
  - If $\lim_{n \to \infty} \dfrac{a_n}{b_n} = L, 0 < L < \infty$, then both series converge or both diverge.
  - If $\lim_{n \to \infty} \dfrac{a_n}{b_n} = 0$ and if $\sum_{k=1}^{\infty} b_k$ converges, then $\sum_{k=1}^{\infty} a_k$ converges.
  - If $\lim_{n \to \infty} \dfrac{a_n}{b_n} = \infty$ and if $\sum_{k=1}^{\infty} b_k$ diverges, then $\sum_{k=1}^{\infty} a_k$ diverges.

Summary: Table 3: Series often used for comparisons (p. 829)

10.5 Alternating Series; Absolute and Conditional Convergence
Definitions:

- Alternating series (p. 833)
- A series $\sum_{k=1}^{\infty} a_k$ is absolutely convergent if the series $\sum_{k=1}^{\infty} |a_k|$ is convergent. (p. 838)
- A series that is convergent without being absolutely convergent is conditionally convergent. (p. 839)

Theorems:

- Alternating Series Test: (p. 833)
- The alternating harmonic series $\sum_{k=1}^{\infty} \dfrac{(-1)^{k+1}}{k}$ converges. (p. 835)
- Error bound for a convergent alternating series (p. 836)
- Absolute Convergence Test: If a series $\sum_{k=1}^{\infty} a_k$ is absolutely convergent, then it is convergent. (p. 838)

Properties of Absolutely Convergent and Conditionally Convergent Series: (pp. 840–841)

10.6 Ratio Test; Root Test
- Ratio Test (p. 845)
- Root Test (p. 848)

10.7 Summary of Tests
- Guide to choosing a test (p. 853)
- Tests for convergence and divergence (Table 5; p. 854)
- Important Series (Table 6; p. 855)

10.8 Power Series
Definitions:

- Power series: $\displaystyle\sum_{k=0}^{\infty} a_k x^k$ or $\displaystyle\sum_{k=0}^{\infty} a_k (x-c)^k$, where c is a constant. (p. 856)
- Radius of convergence (p. 859)
- Interval of convergence (p. 859)

Theorems:

- If a power series centered at 0 converges for a number $x_0 \neq 0$, then it converges absolutely for all numbers x for which $|x| < |x_0|$. (p. 858)
- If a power series centered at 0 diverges for a number x_1, then it diverges for all numbers x for which $|x| > |x_1|$. (p. 858)
- For a power series centered at c, exactly one of the following is true (p. 859):
  - The series converges for only $x = c$.
  - The series converges absolutely for all x.
  - There is a positive number R for which the series converges absolutely for all x, $|x - c| < R$, and diverges for all x, $|x - c| > R$.

Properties of Power Series: (p. 864)

Let $f(x) = \displaystyle\sum_{k=0}^{\infty} a_k x^k$ be a power series in x having a nonzero radius of convergence R.

- *Continuity property*: If $-R < x_0 < R$,
$$\lim_{x \to x_0} \left(\sum_{k=0}^{\infty} a_k x^k \right) = \sum_{k=0}^{\infty} \left(\lim_{x \to x_0} a_k x^k \right) = \sum_{k=0}^{\infty} a_k x_0^k$$

- *Differentiation property*:
$$\frac{d}{dx}\left(\sum_{k=0}^{\infty} a_k x^k \right) = \sum_{k=0}^{\infty} \left(\frac{d}{dx} a_k x^k \right) = \sum_{k=1}^{\infty} k\, a_k\, x^{k-1}$$

- *Integration property*:
$$\int_0^x \left(\sum_{k=0}^{\infty} a_k t^k \right) dt = \sum_{k=0}^{\infty} \left(\int_0^x a_k t^k\, dt \right) = \sum_{k=0}^{\infty} \frac{a_k x^{k+1}}{k+1}$$

The power series obtained by differentiating or integrating a convergent power series have the same radius of convergence R as the original power series.

10.9 Taylor Series; Maclaurin Series
Theorems:

- Taylor series: (p. 873)
$$f(x) = f(c) + f'(c)(x-c) + \frac{f''(c)}{2!}(x-c)^2$$
$$+ \cdots + \frac{f^{(n)}(c)}{n!}(x-c)^n + \cdots = \sum_{k=0}^{\infty} \frac{f^{(k)}(c)}{k!}(x-c)^k$$

- Maclaurin series (p. 873)
$$f(x) = f(0) + f'(0)\,x + \frac{f''(0)\,x^2}{2!} + \cdots + \frac{f^{(n)}(0)\,x^n}{n!} + \cdots$$
$$= \sum_{k=0}^{\infty} \frac{f^{(k)}(0)}{k!} x^k$$

- Taylor's formula with the Lagrange form of remainder (p. 874)
- Convergence of a Taylor series (p. 875)

Table 7:   A list of important Maclaurin expansions (p. 880)

10.10 Taylor Polynomial Approximations and the Lagrange Error Bound

- Taylor polynomial $P_n(x)$ of degree n of f at c. (p. 883)

$$P_n(x) = f(c) + f'(c)(x-c) + \frac{f''(c)}{2!}(x-c)^2 + \cdots + \frac{f^{(n)}(c)}{n!}(x-c)^n$$

- Lagrange Error Bound (p. 883)
$$R_n(x) = \frac{f^{(n+1)}(u)}{(n+1)!}(x-c)^{n+1} \text{ where } u \text{ is between } c \text{ and } x$$

- Under certain conditions, the error bound for a convergent alternating series can be used to approximate the error of a Taylor polynomial approximation. (p. 889)

OBJECTIVES

Section	You should be able to ...	Examples	Review Exercises	AP Review Problems
10.1	1 Write several terms of a sequence (p. 780)	1, 2	1, 2	
	2 Find the nth term of a sequence (p. 781)	3, 4	3	
	3 Use properties of convergent sequences (p. 783)	5, 6	4, 5	
	4 Use a related function or the Squeeze Theorem to show a sequence converges (p. 785)	7–10	6, 7	
	5 Determine whether a sequence converges or diverges (p. 787)	11–15	8–13	
10.2	1 Determine whether a series has a sum (p. 797)	1–4	14, 15	14
	2 Determine whether a geometric series converges or diverges (p. 800)	5–7	17–20	4, 13, 14
	3 Show that the harmonic series diverges (p. 804)		16	
10.3	1 Use the nth Term Test for Divergence (p. 812)	1	21	2
	2 Work with properties of series (p. 812)	2	25–27	9
	3 Use the Integral Test (p. 815)	3–5	22, 23	7, 17
	4 Analyze p-series (p. 817)	6	24	3
	5 Approximate the Sum of a Convergent Series (p. 818)	7, 8	24, 53, 54	

REVIEW EXERCISES

In Problems 1 and 2, the nth term of a sequence $\{s_n\}$ is given. Write the first five terms of each sequence.

1. $s_n = \dfrac{(-1)^{n+1}}{n^4}$ **2.** $s_n = \dfrac{2^n}{3^n}$

3. Find an expression for the *n*th term of the sequence,

$$2, -\frac{3}{2}, \frac{9}{8}, -\frac{27}{32}, \frac{81}{128}, \cdots$$

assuming the indicated pattern continues for all *n*.

In Problems 4 and 5, use properties of convergent sequences to find the limit of each sequence.

4. $\left\{ 1 + \dfrac{n}{n^2 + 1} \right\}$ **5.** $\left\{ \ln \dfrac{n+2}{n} \right\}$

In Problems 6 and 7, use a related function or the Squeeze Theorem for sequences to show each sequence converges. Find its limit.

6. $\left\{ \tan^{-1} n \right\}$ **7.** $\left\{ \dfrac{(-1)^n}{(n+1)^2} \right\}$

8. Determine if the sequence $\left\{ \dfrac{e^n}{(n+2)^2} \right\}$ is monotonic. If it is monotonic, is it increasing, nondecreasing, decreasing, or nonincreasing? Is it bounded from above and/or from below? Does it converge?

In Problems 9–12, determine whether each sequence converges or diverges. If it converges, find its limit.

9. $\{n!\}$ **10.** $\left\{ \left(\dfrac{5}{8} \right)^n \right\}$

11. $\left\{ \left(-\dfrac{1}{2} \right)^n \right\}$ **12.** $\left\{ (-1)^n + e^{-n} \right\}$

13. Show that sequence $\left\{ 1 + \dfrac{2}{n} \right\}$ converges by showing it is either bounded from above and increasing or is bounded from below and decreasing.

14. Find the fifth partial sum of $\displaystyle\sum_{k=1}^{\infty} \dfrac{(-1)^k}{4^{k-1}}$.

15. Find the sum of the telescoping series $\displaystyle\sum_{k=1}^{\infty} \left(\dfrac{4}{k+4} - \dfrac{4}{k+5} \right)$.

In Problems 16–19, determine whether each series converges or diverges. If it converges, find its sum.

16. $\displaystyle\sum_{k=1}^{\infty} \dfrac{\cos^2(k\pi)}{k}$ **17.** $\displaystyle\sum_{k=1}^{\infty} -(\ln 2)^k$

18. $\displaystyle\sum_{k=0}^{\infty} \dfrac{e}{3^k}$ **19.** $\displaystyle\sum_{k=1}^{\infty} (4^{1/3})^k$

20. Express 0.123123123 . . . as a rational number using a geometric series.

21. Show that the series $\sum_{k=1}^{\infty} \dfrac{3k-2}{k}$ diverges.

In Problems 22 and 23, use the Integral Test to determine whether each series converges or diverges.

22. $\sum_{k=1}^{\infty} \dfrac{\ln k}{k^2}$

23. $\sum_{k=1}^{\infty} \dfrac{1}{4k^2+9}$

24. Determine whether the *p*-series $\sum_{k=1}^{\infty} \dfrac{1}{k^{5/2}}$ converges or diverges. If it converges, approximate the sum.

In Problems 25–27, determine whether each series converges or diverges.

25. $\sum_{k=5}^{\infty} \left[\dfrac{1}{k^5} \cdot \dfrac{1}{2^k} \right]$

26. $\sum_{k=1}^{\infty} \left[\dfrac{3}{5^k} - \left(\dfrac{2}{3} \right)^{k-1} \right]$

27. $\sum_{k=1}^{\infty} \dfrac{3}{k^5}$

In Problems 28–30, use a Comparison Test to determine whether each series converges or diverges.

28. $\sum_{k=1}^{\infty} \dfrac{1}{\sqrt{k+1}}$

29. $\sum_{k=1}^{\infty} \dfrac{k+1}{k^{k+1}}$

30. $\sum_{k=1}^{\infty} \dfrac{4}{k\,3^k}$

In Problems 31–33, determine whether each alternating series converges or diverges. If it converges and satisfies the conditions of the Alternating Series Test, approximate the sum of the series so that the error is less than or equal to 0.001.

31. $\sum_{k=1}^{\infty} (-1)^{k+1} \dfrac{k+2}{k(k+1)}$

32. $\sum_{k=1}^{\infty} (-1)^{k+1} \dfrac{k^2}{e^k}$

33. $\sum_{k=1}^{\infty} (-1)^k \dfrac{3}{\sqrt[3]{k}}$

In Problems 34–37, determine whether each series converges (absolutely or conditionally) or diverges.

34. $\sum_{k=1}^{\infty} \sin\left(\dfrac{\pi}{2} k \right)$

35. $\sum_{k=1}^{\infty} \dfrac{(-1)^{k+1}}{\sqrt{k}}$

36. $\sum_{k=1}^{\infty} \dfrac{\cos k}{k^3}$

37. $\dfrac{1}{2} - \dfrac{4}{2^3+1} + \dfrac{9}{3^3+1} - \dfrac{16}{4^3+1} + \cdots$

In Problems 38 and 39, use the Ratio Test to determine whether each series converges or diverges.

38. $\sum_{k=1}^{\infty} \dfrac{2^k}{k!}$

39. $\sum_{k=1}^{\infty} \dfrac{k!}{e^{k^2}}$

In Problems 40 and 41, use the Root Test to determine whether each series converges or diverges.

40. $\sum_{k=1}^{\infty} \dfrac{2^k}{(k+3)^{k+1}}$

41. $\sum_{k=1}^{\infty} \left[\ln\left(e^4 + \dfrac{1}{k^2} \right) \right]^k$

In Problems 42–52, determine whether each series converges or diverges.

42. $\sum_{k=1}^{\infty} (-1)^{k+1} \dfrac{2^{k+1}}{3^k}$

43. $\sum_{k=1}^{\infty} \ln\left(1 + \dfrac{1}{k} \right)$

44. $\sum_{k=5}^{\infty} \dfrac{3}{k\sqrt{k-4}}$

45. $\sum_{k=1}^{\infty} \dfrac{1}{\left(1 + \dfrac{k^2+1}{k^2} \right)^k}$

46. $\sum_{k=1}^{\infty} \dfrac{2 \cdot 4 \cdot 6 \cdots (2k)}{1 \cdot 3 \cdot 5 \cdots (2k-1)}$

47. $\sum_{k=1}^{\infty} \dfrac{k^2}{(1+k^3) \ln \sqrt[3]{1+k^3}}$

48. $\sum_{k=1}^{\infty} \dfrac{k^{10}}{2^k}$

49. $\sum_{k=1}^{\infty} \dfrac{\left(1 + \dfrac{1}{k^2} \right)^{k^2}}{2^k}$

50. $\sum_{k=1}^{\infty} \left(\dfrac{k^2+1}{k} \right)^k$

51. $\sum_{k=1}^{\infty} \dfrac{k!}{3^{k^k}}$

52. $\sum_{k=1}^{\infty} (-1)^{k+1} \dfrac{k+2}{3k-2}$

53. Find bounds for the sum of the convergent series $\sum_{k=2}^{\infty} \dfrac{6}{k(\ln k)^3}$.

54. Find bounds for the sum of the convergent series $\sum_{k=1}^{\infty} \dfrac{2}{k^{4/3}}$.

In Problems 55–60,

(a) Find the radius of convergence of each power series.

(b) Find the interval of convergence of each power series.

55. $\sum_{k=1}^{\infty} \dfrac{(x-3)^{3k-1}}{k^2}$

56. $\sum_{k=1}^{\infty} \dfrac{x^k}{\sqrt[3]{k}}$

57. $\sum_{k=0}^{\infty} (-1)^k \dfrac{1}{k!\,(k+1)} \left(\dfrac{x}{2} \right)^{2k+1}$

58. $\sum_{k=1}^{\infty} \dfrac{k^k}{(k!)^2} x^k$

59. $\sum_{k=1}^{\infty} \dfrac{(x-1)^k}{k}$

60. $\sum_{k=0}^{\infty} \dfrac{3^k x^k}{5^k}$

In Problems 61 and 62, express each function as a power series centered at 0.

61. $f(x) = \dfrac{2}{x+3}$

62. $f(x) = \dfrac{1}{1-3x}$

63. (a) Use properties of a power series to find the power series representation for $\displaystyle\int \dfrac{1}{1-3x^2}\,dx$.

(b) Use (a) to approximate $\displaystyle\int_0^{1/2} \dfrac{1}{1-3x^2}\,dx$ correct to within 0.001.

64. Find the Taylor expansion of $f(x) = \dfrac{1}{1-2x}$ about $c = 1$.

65. Find the Taylor expansion of $f(x) = e^{x/2}$ about $c = 1$.

66. Find the Maclaurin expansion of $f(x) = 2x^3 - 3x^2 + x + 5$. Comment on the result.

67. Find the Taylor expansion of $f(x) = \tan x$ about $c = \dfrac{\pi}{4}$.

68. Find the first five terms of the Maclaurin expansion for

$$f(x) = e^{-x}\sin x$$

69. Use the Maclaurin expansion for $f(x) = e^x$ to approximate $e^{0.3}$ so that the Lagrange error bound is less than 0.0001.

70. (a) Approximate $y = \cos x$ using the Taylor polynomial $P_8(x)$ of $y = \cos x$ at $\dfrac{\pi}{2}$.

(b) Graph $y = \cos x$ and $y = P_8(x)$.

(c) Use $P_8(x)$ to approximate $\cos 88°$.

(d) What is the error bound in using this approximation?

(e) If the approximation in (a) is used, for what values of x is the Lagrange error bound less than or equal to 0.0001?

Break It Down

*Let's take a closer look at part (c) of AP® **Practice Problem 5** from Section 10.5.*

5. (a) Write out the first five terms of the series $\displaystyle\sum_{k=0}^{\infty} \dfrac{(-1)^k}{k!}$.

(b) Show the series $\displaystyle\sum_{k=0}^{\infty} \dfrac{(-1)^k}{k!}$ converges.

(c) How many terms of the series are necessary to approximate the sum S with an error less than or equal to 0.0001?

Step 1	Identify the underlying structure.	Since $\dfrac{1}{k!} > 0$ for all integers $k \geq 0$, the series $\displaystyle\sum_{k=0}^{\infty} \dfrac{(-1)^k}{k!}$ is an alternating series.
Step 2	Identify an appropriate math rule or procedure.	Part (b) established that $$\sum_{k=0}^{\infty} \dfrac{(-1)^k}{k!}$$ converges using the Alternating Series Test. So the error in using the sum of the first n terms of the series is numerically less than or equal to the $(n+1)$st term of the series.
Step 3	Apply the math rule or procedure.	The $(n+1)$st term is $\dfrac{(-1)^n}{n!}$. For the error to be less than 0.0001, we need to solve the inequality $\dfrac{1}{n!} \leq \dfrac{1}{10000}$. Solving the inequality $n! \geq 10{,}000$ results in an equivalent solution. Using trial and error, $6! = 720$, $7! = 7 \cdot 720 = 5040$, $8! = 8 \cdot 5040 = 40{,}320$. Since $(n+1)! = 8! > 10{,}000$ is the smallest factorial to exceed $10{,}000$, $n = 7$. But because the series begins at $k = 0$, $n = 7$ is the 8th term of the series.
Step 4	Clearly communicate your answer.	Eight terms of the series $\displaystyle\sum_{k=0}^{\infty} \dfrac{(-1)^k}{k!}$ are needed to approximate the sum S with an error less than or equal to 0.0001.

AP® Review Problems: Chapter 10

Preparing for the AP® Exam

Multiple-Choice Questions

1. Suppose $f(2) = 3$; $f'(2) = 0$; $f''(2) = 5$; $f'''(2) = -4$; and $f^{(4)}(2) = -2$. Then the Taylor polynomial $P_4(x)$ of degree 4 of f at 2 is

(A) $P_4(x) = 3(x - 2) + \frac{5}{2}(x - 2)^2 - \frac{4}{6}(x - 2)^3 - \frac{2}{24}(x - 2)^4$

(B) $P_4(x) = 3 + \frac{5}{2}(x - 2)^2 - \frac{2}{3}(x - 2)^3 - \frac{1}{12}(x - 2)^4$

(C) $P_4(x) = 3 + (x - 2) + \frac{5}{2}(x - 2)^2$
$\qquad\qquad - \frac{2}{3}(x - 2)^3 + \frac{1}{12}(x - 2)^4$

(D) $P_4(x) = 3 + \frac{5}{2!}(x - 2)^2 - \frac{4}{3!}(x - 2)^3 + \frac{2}{4!}(x - 2)^4$

2. If $0 < a_k \le b_k$ for all k and $\sum_{k=1}^{\infty} a_k$ diverges, then which statement must be false?

(A) $\lim_{n \to \infty} a_n = 0$ (B) $\lim_{n \to \infty} b_n = 0$

(C) $\sum_{k=1}^{\infty} b_k = 1$ (D) $\sum_{k=1}^{\infty} (-1)^k a_k$ diverges

3. Which of the following series converge?

   I. $\sum_{k=1}^{\infty} k^{-3/2}$ **II.** $\sum_{k=1}^{\infty} k^{-1}$ **III.** $\sum_{k=1}^{\infty} 2^{1-k}$

(A) I only   (B) III only

(C) I and III only   (D) I, II, and III

4. Determine whether the infinite series $\sum_{k=1}^{\infty} \left(-\frac{5}{6}\right)^{k-1}$ converges or diverges. If the series converges, find its sum S.

(A) converges; $S = -\frac{5}{11}$ (B) converges; $S = \frac{6}{11}$

(C) converges; $S = 6$ (D) The series diverges.

5. The interval of convergence of $\sum_{k=1}^{\infty} \frac{(x - 2)^k}{k^2}$ is

(A) $-1 < x < 1$ (B) $-1 < x \le 1$

(C) $1 \le x < 3$ (D) $1 \le x \le 3$

6. For which integer n do all three infinite series converge?

$$\sum_{k=1}^{\infty} k^{-n/3} \qquad \sum_{k=1}^{\infty} \frac{(-1)^{nk}}{k} \qquad \sum_{k=1}^{\infty} \left(\frac{n}{6}\right)^k$$

(A) 2   (B) 3   (C) 4   (D) 5

7. If a function f is continuous, positive, and decreasing on the interval $[1, \infty)$ and if $a_k = f(k)$ for all positive integers k, then the infinite series

(A) $\sum_{k=1}^{\infty} a_k$ converges.

(B) $\sum_{k=1}^{\infty} a_k = \int_1^{\infty} f(x)\,dx$ if $\int_1^{\infty} f(x)\,dx$ converges.

(C) $\sum_{k=1}^{\infty} a_k = f(x)$

(D) $\sum_{k=1}^{\infty} a_k$ converges if $\int_1^{\infty} f(x)\,dx$ converges.

8. $\sum_{k=1}^{\infty} \frac{3^{k+1} + 4k!}{3^k \cdot k!} =$

(A) $e + 2$ (B) $3e - 1$

(C) $3e + \frac{8}{3}$ (D) $3e + 2$

9. $\sum_{k=1}^{\infty} \frac{4 - 2^{k-1}}{5^{k-1}} =$

(A) $\frac{10}{3}$ (B) $\frac{40}{7}$

(C) $\frac{20}{3}$ (D) The series diverges.

10. Which of the series diverge?

   I. $2 - 1 + \frac{1}{2} - \frac{1}{4} + \frac{1}{8} - \cdots$

   II. $2 + \frac{4}{2^{3/2}} + \frac{6}{3^{3/2}} + \frac{8}{4^{3/2}} + \cdots$

   III. $6 - 4 - \frac{8}{3} - \frac{16}{9} - \frac{32}{27} - \cdots$

(A) I only   (B) II only

(C) I and II only   (D) I and III only

11. The interval of convergence of the power series $\sum_{k=0}^{\infty} \left(\frac{x - 1}{4}\right)^k$ is

(A) $-1 \le x \le 1$ (B) $-4 < x < 4$

(C) $-4 < x \le 4$ (D) $-3 < x < 5$

▶ **12.** Which series is the Maclaurin expansion of $f(x) = x^2 \sin x$?

(A) $x^2 - \dfrac{x^4}{2!} + \dfrac{x^6}{4!} - \dfrac{x^8}{6!} + \cdots$

(B) $x^2 - \dfrac{x^6}{3!} + \dfrac{x^{10}}{5!} - \dfrac{x^{14}}{7!} + \cdots$

(C) $x^3 - \dfrac{x^5}{3!} + \dfrac{x^7}{5!} - \dfrac{x^9}{7!} + \cdots$

(D) $\dfrac{x^3}{3!} - \dfrac{x^5}{5!} + \dfrac{x^7}{7!} - \dfrac{x^9}{9!} + \cdots$

▶ **13.** Determine whether the series $\displaystyle\sum_{k=1}^{\infty} \dfrac{4^{k+2}}{5^k}$ converges.

If it converges, its sum equals

(A) converges; 5   (B) converges; 64

(C) converges; 80   (D) The series diverges.

▶ **14.** Determine whether the series $\displaystyle\sum_{k=0}^{\infty} \dfrac{12^2}{3^{k-1}}$ converges.

If the series converges, find the sum.

(A) converges; 63   (B) converges; 216

(C) converges; 648   (D) The series diverges.

▶ **15.** A Maclaurin polynomial for $f(x) = e^x$ is used to approximate $e^{1/3}$. What is the degree of the Maclaurin polynomial needed to ensure that the Lagrange error bound is less than 0.00001?

(A) 3   (B) 5   (C) 7   (D) 9

▶ **16.** A fourth degree Taylor polynomial for $f(x) = \sin x$ at $\dfrac{\pi}{6}$ is used to approximate $\sin 35°$. The Lagrange error bound guarantees the error in using the approximation is less than

(A) $\dfrac{1}{4!}\left(\dfrac{\pi}{36}\right)^4$ (B) $\dfrac{1}{5!}\left(\dfrac{\pi}{36}\right)^5$

(C) $\dfrac{1}{5!}\left(\dfrac{\pi}{6}\right)^5$ (D) $\dfrac{1}{7!}\left(\dfrac{\pi}{6}\right)^7$

Free-Response Questions

▶ **17.** Determine whether the infinite series $\displaystyle\sum_{k=1}^{\infty} \dfrac{(\ln k)^2}{k}$ converges or diverges. Be sure to show all your work.

▶ **18.** Determine whether the infinite series $\displaystyle\sum_{k=1}^{\infty} (-1)^{k+1} \dfrac{\sqrt{k}}{k+1}$ is absolutely convergent, conditionally convergent, or divergent. Show your work.

▶ **19.** (a) Write the first five nonzero terms of the Maclaurin expansion of $f(x) = \tan^{-1} x$.

(b) Use properties of power series to obtain the Maclaurin series for $g(x) = \int_0^x \tan^{-1} t \, dt$.

(c) Using the fact that the radius of convergence R of the Maclaurin series representation for $f(x) = \tan^{-1} x$ is 1, find the interval of convergence of the series representation of g.

(d) Use the result from (b) to approximate $\int_0^{1/4} \tan^{-1} x \, dx$ so the error is less than 0.001.

▶ **20.** The function f has derivatives of all orders, and the Maclaurin series for f is given by

$$\sum_{k=1}^{\infty} (-1)^{k+1} \frac{x^{2k}}{2k+1} = \frac{x^2}{3} - \frac{x^4}{5} + \frac{x^6}{7} - \frac{x^8}{9} + \cdots$$

(a) Using the Ratio Test, determine the interval of convergence of the Maclaurin series for f.

(b) Find the first three nonzero terms and the general term of the Maclaurin series for f'.

(c) Write the first four nonzero terms of the Maclaurin series for $\cos x$. Use the Maclaurin series for $\cos x$ write the Taylor polynomial of degree 4 for $h(x) = f(x) \cos x$ about $x = 0$.

AP® Cumulative Review Problems: Chapters 1–10   **Preparing for the** AP® Exam

Multiple-Choice Questions

▶ 1. Find the area of the region bounded by $y = \dfrac{9}{x^2 + 9}$, the x-axis, and the vertical lines, $x = -3$ and $x = 3$.

  (A) $\dfrac{3\pi}{2}$ (B) $\dfrac{\pi}{4}$ (C) $\dfrac{3\pi}{4}$ (D) 1

▶ 2. If the function $f(x) = \dfrac{e^x \tan x}{x + \sin(2x)}$ is continuous for all real numbers, what is the value of $f(0)$?

  (A) -1 (B) 0   (C) $\dfrac{1}{3}$ (D) $\dfrac{1}{2}$

▶ 3. Find the volume of the solid formed by revolving the trapezoid with vertices $(3, 0)$, $(3, 3)$, $(5, 0)$, and $(5, 5)$ about the x-axis.

  (A) $\dfrac{5\pi}{3}$ (B) $\dfrac{96\pi}{5}$ (C) $\dfrac{25\pi}{9}$ (D) $\dfrac{98\pi}{3}$

▶ 4. $\displaystyle\int \dfrac{e^x}{4 + e^{2x}}\, dx =$

  (A) $\ln |e^{2x} + 4| + C$ (B) $\dfrac{e^{3x}}{3} + 4x + C$

  (C) $e^x + \dfrac{1}{(4 + e^x)^2} + C$ (D) $\dfrac{1}{2} \tan^{-1}\left(\dfrac{e^x}{2}\right) + C$

▶ 5. $\displaystyle\lim_{x\to 0} \dfrac{xe^x}{\tan x} =$

  (A) 1   (B) ∞ (C) 0   (D) 2

▶ 6. $\dfrac{d}{dx} \displaystyle\int_3^{2x^2} (2t^2 + 5t)^2\, dt =$

  (A) $16x^5(4x^2 + 5)^2$ (B) $8x^2(4x^2 + 5)^2$

  (C) $(2x^2 + 5x)^2$ (D) $16x^5(4x^2 + 5)$

▶ 7. Find the average value of $f(x) = \sin^2(3x)$ on the interval $\left[\dfrac{\pi}{4}, \dfrac{\pi}{2}\right]$.

  (A) $\dfrac{\pi}{8}$ (B) $\dfrac{3\pi - 2}{6\pi}$

  (C) $\dfrac{6\pi - 1}{12}$ (D) $\dfrac{6\pi - 2}{3\pi}$

▶ 8. If a bowling ball is dropped from a height of 200 feet, its height s at time t is given by the position function $s(t) = -16t^2 + 200$, where s is measured in feet and t is measured in seconds. Find the average velocity of the falling bowling ball over the time interval $[1, 2]$.

  (A) -48 ft/sec   (B) -60 ft/sec

  (C) -32 ft/sec   (D) -16 ft/sec

▶ 9. If $y = x^5 - x^4 + x^3 - x^2 + x + 1$, then $\dfrac{d^3 y}{dx^3}$ evaluated at $x = -1$ is

  (A) 3375   (B) -1 (C) -3 (D) 90

▶ 10. Find the length of the graph of $y = \ln(\sin x)$ from $x = \dfrac{\pi}{4}$ to $x = \dfrac{\pi}{2}$.

  (A) $\dfrac{\pi}{4}$ (B) $-\ln(\sqrt{2} + 1)$

  (C) $\ln |\sqrt{2} - 1|$ (D) $-\ln |\sqrt{2} - 1|$

▶ 11. Find the area of one petal of the rose curve given by $r = 4\cos(3\theta)$.

  (A) $\dfrac{9\pi}{4}$ (B) 4π (C) $\dfrac{4\pi}{3}$ (D) $\dfrac{3\sqrt{3}}{2}$

Free-Response Questions

▶ 12. A particle moving in the xy-plane has position $\langle x(t), y(t)\rangle$ at time $t \geq 0$, where $\dfrac{dx}{dt} = t^2 + 1$ and $\dfrac{dy}{dt} = t^3$. At time $t = 0$ the particle is at the position $(2, 3)$.

  (a) Find the velocity vector at time t.
  (b) Find the position vector of the particle at time t.
  (c) Find the speed of the particle at time $t = 2$.
  (d) Find the acceleration vector of the particle at time t.
  (e) Find the slope of the line tangent to the particle's path at time $t = 2$.

▶ 13. For $t \geq 0$, a toy rocket is launched from the position $(5, 10)$ and moves along a curve so that its position at time t is $(x(t), y(t))$. It is known that $x'(t) = 6t - 3t^2$ and $y'(t) = 6 - 2t$.

  (a) Find the slope of the path of the rocket at $t = 1$. Find the speed of the rocket at $t = 1$. Find the acceleration vector of the rocked at $t = 1$.
  (b) Write but do not evaluate an integral expression to find the distance traveled by the rocket from $t = 1$ to $t = 4$.
  (c) The rocket is programmed to strike a target located at the position $(-45, 15)$. Will the rocket strike the target? If so, when will the rocket strike the target? Justify your answer.

▶ 14. Suppose $f(x) = \dfrac{2x}{x^2 - 3x + 2}$.

  (a) Identify any vertical asymptotes of the function f.
  (b) Find $\displaystyle\int \dfrac{2x}{x^2 - 3x + 2}\, dx$.
  (c) Find the average value of $f(x) = \dfrac{2x}{x^2 - 3x + 2}$ on the interval $[3, 5]$.

AP® Practice Exam: Calculus BC   Preparing for the AP® Exam

Section 1: Multiple Choice, Part A

A calculator may not be used for Part A.

1. $\displaystyle\int \frac{3x^2 - 7x - 4}{x - 3}\, dx =$

(A) $\dfrac{3}{2}x^2 + 2x + 2\ln|x - 3| + C$

(B) $\dfrac{3}{2}x^2 + 2x - 10\ln|x - 3| + C$

(C) $\dfrac{1}{2}(3x + 2)^2 + 2\ln|x - 3| + C$

(D) $3x + 2 + 2\ln|x - 3| + C$

2. $\displaystyle\int_0^{x^2} e^{3t}\, dt =$

(A) $\dfrac{1}{3}\left(e^{3x^2} - 1\right)$ (B) $3\left(e^{3x^2} - 1\right)$

(C) $2xe^{3x^2}$ (D) $\dfrac{1}{3}(2x)\left(e^{3x^2} - 1\right)$

3. What is the slope of the tangent line to the graph of $y = e^{6x} - \sin(4x)$ at $x = 0$?

(A) 0   (B) 1   (C) 2   (D) 10

4. $\displaystyle\lim_{x \to 0} \frac{2e^{5x} - 5e^{2x} + 3}{x^2 - 2\cos(2x) + 2} =$

(A) 0   (B) $\dfrac{3}{4}$ (C) $\dfrac{3}{2}$ (D) 3

5. Which of the following series converges?

I. $\displaystyle\sum_{k=1}^{\infty} \frac{1}{k^{2/3}}$

II. $\displaystyle\sum_{k=1}^{\infty} \frac{k}{k^3 + 5k + 1}$

III. $\displaystyle\sum_{k=1}^{\infty} (-1)^k \left(\frac{1}{k}\right)$

(A) I and II   (B) I and III

(C) II and III   (D) III only

6. $\displaystyle\int \frac{4x - 1}{2x^2 - x - 1}\, dx =$

(A) $\ln|(x - 1)(2x + 1)| + C$

(B) $\ln\left|(x - 1)(2x + 1)^2\right| + C$

(C) $2\ln|(x - 1)(2x + 1)| + C$

(D) $2\ln\left|\dfrac{x - 1}{2x + 1}\right| + C$

7. Find y' at the point $(3, 6)$ on the graph of $y^2 = 2x^2 + xy$.

(A) $\dfrac{1}{6}$ (B) $\dfrac{6}{5}$ (C) $\dfrac{7}{4}$ (D) 2

8. The function f has a second derivative given by $f''(x) = x^2(x - 1)\sqrt{x + 1}$. At what values of x does f have a point of inflection?

(A) 0 only   (B) 1 only

(C) 0 and 1 only   (D) $-1, 0$, and 1

9. If the function f is continuous and if $F'(x) = f(x)$ for all real numbers x, then $\int_2^5 f(3x + 2)\, dx$ equals

(A) $3F(5) - 3F(2)$ (B) $\dfrac{1}{3}F(5) - \dfrac{1}{3}F(2)$

(C) $\dfrac{1}{3}F(17) - \dfrac{1}{3}F(8)$ (D) $3F(17) - 3F(8)$

10. What is the sum of the series $\displaystyle\sum_{k=1}^{\infty} \frac{3^{k+2}}{5^{k+1}}$?

(A) $\dfrac{3}{2}$ (B) 2   (C) $\dfrac{5}{2}$ (D) $\dfrac{27}{10}$

11. The base of a solid S is the region enclosed by the graph of $y = \sqrt{x}$, the line $x = 3$, the line $x = 1$, and the x-axis. If the cross sections of S perpendicular to the x-axis are squares, then the volume of S is

(A) 2   (B) $2\sqrt{3}$ (C) 4   (D) $\dfrac{4}{3}\left(3\sqrt{3} - 1\right)$

12. Let f be the function given by $f(x) = x^x$. If three subintervals of equal length are used, what is the value of the Left Riemann sum approximation for $\int_{1.2}^{1.8} x^x\, dx$?

(A) $5.0(1.2^{1.2} + 1.4^{1.4} + 1.6^{1.6})$

(B) $0.2(1.2^{1.2} + 1.5^{1.5} + 1.8^{1.8})$

(C) $0.2(1.2^{1.2} + 1.4^{1.4} + 1.6^{1.6})$

(D) $0.2(1.4^{1.4} + 1.6^{1.6} + 1.8^{1.8})$

13. The volume of a box with a square top and bottom is to be k cubic inches. If a minimum amount of cardboard is to be used to construct the box, what must be the area, in square inches, of the top of the box?

(A) $\sqrt[3]{k}$ (B) $2\sqrt[3]{k}$ (C) $\sqrt[3]{k^2}$ (D) $4\sqrt[3]{k^2}$

14. Let $P(x) = a + b(x - 1) + c(x - 1)^2 + d(x - 1)^3$ be the third-degree Taylor polynomial for the function f about $x = 1$. A table of values for $f^{(k)}(1)$ is given below. Find $c + d$.

k	0	1	2	3
$f^{(k)}(1)$	4	4	-2	3

(A) $-\dfrac{3}{2}$ (B) $-\dfrac{1}{2}$ (C) 0   (D) 1

15. Suppose $\dfrac{dy}{dx} = \dfrac{x+y}{x-y}$, where y is a twice differentiable function of x and $x = 2$ when $y = 4$.

What is the value of $\dfrac{d^2y}{dx^2}$ when $x = 2$?

(A) -20 (B) -5 (C) -2 (D) $-\dfrac{1}{2}$

16. Which of the following gives the area of the region enclosed by the loop of the graph of the polar curve $r = 2\sin(6\theta)$ shown in the figure below?

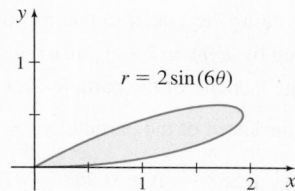

(A) $4\displaystyle\int_0^{\pi/3} \sin^2(6\theta)\,d\theta$ (B) $4\displaystyle\int_0^{\pi/6} \sin^2(6\theta)\,d\theta$

(C) $2\displaystyle\int_0^{\pi/6} \sin^2(6\theta)\,d\theta$ (D) $2\displaystyle\int_0^{\pi/6} \sin(6\theta)\,d\theta$

17. For $p > 0$, $\displaystyle\int_3^\infty \dfrac{1}{(x-2)^p}\,dx$ diverges for

(A) $0 < p \le 1$ (B) $0 < p < 1$

(C) $p \ge 1$ (D) $p \ge 2$

18. Let g be the function given by $g(x) = \displaystyle\int_0^{x^2} e^{-t^2}\,dt$.
The function g has a point of inflection at $x =$

(A) 0 (B) $-\dfrac{\sqrt{2}}{2}, \dfrac{\sqrt{2}}{2}$ (C) $-\sqrt{2}, \sqrt{2}$ (D) $-2, 2$

19. Let g be the function defined by $g(x) = \displaystyle\int_1^x \sqrt{9t^2-1}\,dt$.
Which integral represents the length of the graph of g on the interval $[2, 5]$?

(A) $\displaystyle\int_2^5 \sqrt{9x^2-1}\,dx$ (B) $\displaystyle\int_2^5 \left(1 + \sqrt{9x^2-1}\right)dx$

(C) $\displaystyle\int_2^5 \left(1 + \sqrt{9x^2-1}\right)^2 dx$ (D) $\displaystyle\int_2^5 3x\,dx$

20. Which of the following is the Maclaurin series for $\cos(\sqrt{x})$?

(A) $1 - \dfrac{x}{2} + \dfrac{x^2}{4!} - \dfrac{x^3}{6!} + \cdots$

(B) $1 - \dfrac{x}{2} + \dfrac{x^2}{4} - \dfrac{x^3}{6} + \cdots$

(C) $1 + \dfrac{x}{2} + \dfrac{x^2}{4} + \dfrac{x^3}{6} + \cdots$

(D) $1 - \dfrac{\sqrt{x}}{2} + \dfrac{x}{4!} - \dfrac{x\sqrt{x}}{6!} + \cdots$

21. What is the radius of convergence of the power series

$$\sum_{k=1}^{\infty} \dfrac{(x+2)^k}{5^k k^2}?$$

(A) 3 (B) 4 (C) 5 (D) 7

22. In an experiment involving the spread of disease, one infected mouse is placed in a closed population of healthy mice. The rate of change of the number of infected mice with respect to time t (in days) follows the model $\dfrac{dP}{dt} = 0.05P\left(1 - \dfrac{P}{500}\right)$ where $P(t)$ is the number of infected mice at $t > 0$. Which of the following statements are true?

   I. The carrying capacity is 500.

   II. The population is growing most rapidly when there are 250 mice.

   III. $\displaystyle\lim_{P \to 500} \dfrac{dP}{dt} = 1$

(A) I only   (B) I and II only
(C) II and III only   (D) I, II, and III

23. A function f is continuous on the closed interval $[0, 8]$ and differentiable on the open interval $(0, 8)$. If $f(0) = 5$ and $f(8) = -2$, which of the following MUST be true?

   I. $f(x) = 0$ for some c, $0 < c < 8$.

   II. $f'(x) = 0$ for some c, $0 < c < 8$.

   III. $f'(x) < 0$ for some c, $0 < c < 8$.

(A) I and II only   (B) I and III only
(C) II and III only   (D) I, II, and III

24. $\displaystyle\int_1^2 x^3 \ln x\,dx =$

(A) $12\ln 2 - \dfrac{9}{2}$ (B) $4\ln 2 - \dfrac{17}{16}$

(C) $4\ln 2 - 1$ (D) $4\ln 2 - \dfrac{15}{16}$

25. What is the coefficient of x^3 in the Taylor series for $f(x) = e^x \sin(x)$ about $x = 0$?

(A) $\dfrac{1}{6}$ (B) $\dfrac{1}{3}$ (C) $\dfrac{1}{2}$ (D) $\dfrac{5}{6}$

26. Given $f(x) = \begin{cases} e^{2x} & \text{if } x \le 0 \\ \cos x & \text{if } x > 0 \end{cases}$, then $\displaystyle\int_{-1}^{\pi/2} f(x)\,dx =$

(A) $-\dfrac{1}{2} - \dfrac{1}{2}e^{2x}$ (B) $2 - e^{-2}$

(C) $\dfrac{3}{2} - \dfrac{1}{2}e^{-2}$ (D) $\dfrac{3}{2} - \dfrac{1}{2}e^2$

27. A curve in the plane is defined parametrically by the equations

$$x = t + 4\sqrt{t} \quad \text{and} \quad y = 4t^2 - 2t$$

An equation of the line tangent to the curve at $t = 1$ is

(A) $y = 2x - 8$ (B) $y = 2x$
(C) $y = 6x - 28$ (D) $y = 6x - 4$

28. If $\dfrac{dy}{dt} = ky^2$ and k is a nonzero constant, then y could be

(A) $y = \dfrac{-1}{3 + kt}$ (B) $y = 3 - \dfrac{1}{kt}$

(C) $y = e^{kt} + 3$ (D) $y = 3e^{kt}$

29. The graph of a differentiable function f is shown below. Given $g(x) = f(x) + 3$ and $h(x) = \dfrac{f(x)}{g(x)}$, which of the following statements about $h'(1)$ is true?

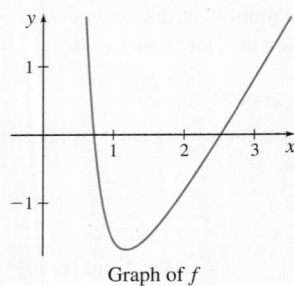

Graph of f

(A) $h'(1) < 0$ (B) $h'(1) = 0$

(C) $h'(1) > 0$ (D) $h'(1)$ is undefined.

30. Refer to the slope field pictured below. This slope field could belong to which of the following differential equations?

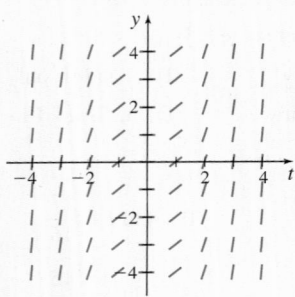

(A) $\dfrac{dy}{dt} = t^2 y^4$ (B) $\dfrac{dy}{dt} = t^2$

(C) $\dfrac{dy}{dt} = y$ (D) $\dfrac{dy}{dt} = t \cdot y$

Section 1: Multiple Choice, Part B

A graphing calculator is required for some of the questions in Part B.

31. Assume that f and f' have the values given in the table below. Use Euler's method with two equal steps to approximate the value of $f(2.5)$.

x	2	2.25	2.50
$f'(x)$	-0.4	-0.2	-0.1
$f(x)$	10		

(A) 9.80   (B) 9.85   (C) 9.90   (D) 9.95

32. Sand is poured onto a pile at a rate of $r(t) = \dfrac{25t}{t + 1}$ cubic feet per minute, where t is the number of minutes since the pouring process began. If the pile contained 100 cubic feet of sand when the process began, how much sand, to the nearest cubic foot, is in the pile after 30 minutes?

(A) 564   (B) 664   (C) 764   (D) 864

33. What is the area bounded by the graphs of $y = e^{-x^2}$ and $8y - x = 4$?

(A) 0.049   (B) 0.473   (C) 0.530   (D) 2.244

34. A particle moves along the x-axis so that its position at time $t \geq 0$ is given by $P(t) = 2 + \int_0^t \ln(u^2 + 1)\, du$. Let a denote the average velocity of the particle over the interval $[1, 5]$ and let b denote the speed of the particle at $t = 3$. Find $\dfrac{a}{b}$.

(A) 0.762   (B) 0.858   (C) 0.953   (D) 1.170

35. If $0 < a_n < \dfrac{2n}{n + 1}$ for $n \geq 1$, which of the following statements must be true?

(A) $\displaystyle\sum_{n=1}^{\infty} \dfrac{2n}{n + 1}$ converges. Therefore, $\displaystyle\sum_{n=1}^{\infty} a_n$ will also converge.

(B) $\displaystyle\sum_{n=1}^{\infty} \dfrac{2n}{n + 1}$ diverges. Therefore, $\displaystyle\sum_{n=1}^{\infty} a_n$ will also diverge.

(C) $\displaystyle\sum_{n=1}^{\infty} \dfrac{2n}{n + 1}$ converges. However, $\displaystyle\sum_{n=1}^{\infty} a_n$ may or may not converge.

(D) $\displaystyle\sum_{n=1}^{\infty} \dfrac{2n}{n + 1}$ diverges. However, $\displaystyle\sum_{n=1}^{\infty} a_n$ may or may not diverge.

36. Let g be a twice differentiable function and $G(x) = \int_0^x g(t)\, dt$. The table below gives selected values of G as well as the first derivative of G. Find the value of $\int_0^3 x g'(x)\, dx$.

x	1	3
$G(x)$	3	4
$G'(x)$	6	2

(A) 3   (B) 4   (C) 2   (D) $\dfrac{22}{3}$

37. The table below gives selected values of a differentiable function f defined on the interval $[1, 5]$. The number $x = 3.2$ satisfies the conclusion of the Mean Value Theorem on the interval $[1, 5]$. What is the slope of the line tangent to f at $x = 3.2$?

x	1	2	3	4	5
$f(x)$	29	27	24	17	7

(A) -7 (B) $-\dfrac{11}{2}$ (C) -5 (D) $-\dfrac{22}{5}$

38. Which of the following differential equations for a population P could model the logistic growth function shown in the figure below?

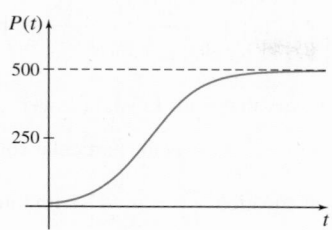

(A) $\dfrac{dP}{dt} = 2P - 0.004P^2$ (B) $\dfrac{dP}{dt} = P - 0.004P^2$

(C) $\dfrac{dP}{dt} = 2P^2 - 0.004P$ (D) $\dfrac{dP}{dt} = P^2 - 0.004P$

39. Find the volume of the solid obtained by revolving the curve
$y = \sqrt{x} + \dfrac{3\cos x}{x}$ from $x = \dfrac{\pi}{2}$ to $x = 3\pi$ about the x-axis.

(A) 41.120   (B) 52.127   (C) 129.183   (D) 142.093

40. A particle moves in the xy-plane so that its position at any time t is given by $x(t) = t^2 + 3t$ and $y(t) = 2t^3 - 21t^2 + 60t$. What is the speed of the particle when $t = 3$?

(A) 9   (B) 12   (C) 15   (D) 21

41. Let g be a differentiable function and $G(x) = \int_0^x g(t)\,dt$. The table below gives a value of G and values of the first two derivatives of G. Use a second-degree Taylor polynomial for the function G about $x = 1$ to approximate $\int_0^{0.8} g(t)\,dt$.

x	1
$G(x)$	3
$G'(x)$	6
$G''(x)$	5

(A) 1.9   (B) 2.0   (C) 4.3   (D) 9.4

42. A particle travels along a straight line with a velocity
of $v(t) = \left(\dfrac{5}{t - 10}\right)\sin(2\pi\sqrt{t})$ feet per second. What is the
total distance, in feet, traveled by the particle during the time interval $0 \leq t \leq 9$ seconds?

(A) 2.945   (B) 5.497   (C) 7.006   (D) 10.994

43. A particle travels along a straight line with velocity $v(t)$. The table below gives the values for $v(t)$ and $v'(t)$ for selected times. At which of the given values of t is the speed of the particle increasing?

t	2	4	6	8
$v(t)$	-5	6	-9	18
$v'(t)$	-3	-8	6	10

(A) 8 only   (B) 4 and 6   (C) 4 and 8   (D) 2 and 8

44. Sand is poured through a chute onto a conical pile at the rate of 25 cubic feet per minute. The bottom radius of the conical pile is always half the height. How fast does the radius of the base change when the pile is 10 feet high? The volume of a cone is

$$V = \frac{1}{3}\pi r^2 h$$

(A) 0.010   (B) 0.040   (C) 0.159   (D) 0.318

45. The function f is continuous on the closed interval $[1, 3]$. If $f(1) = -2$ and $f(3) = 4$, then the Extreme Value Theorem guarantees that

(A) $f(c) = 0$ for at least one c in the interval $[1, 3]$.

(B) $f'(c) = 0$ for at least one c in the interval $[1, 3]$.

(C) $f'(c) = 3$ for at least one c in the interval $[1, 3]$.

(D) f attains an absolute maximum value $f(c) \geq f(x)$ for all x in the interval $[1, 3]$.

Section 2: Free Response, Part A

A graphing calculator is required for Part A.

1. The figure below shows the graphs of
$r_1(\theta) = 2 + 3\sin\theta + 4\cos\theta$ and $r_2(\theta) = 2 + 6\cos\theta$
on the interval $0 \leq \theta \leq \dfrac{\pi}{2}$. Let S be the shaded region
bounded by the two graphs, the x-axis, and the y-axis. The two curves intersect at point P.

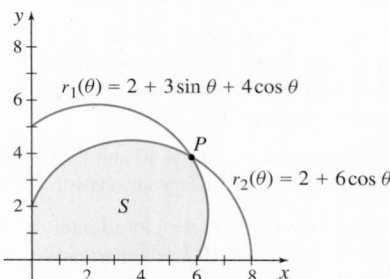

(a) Write, but do not evaluate, an expression involving one or more integrals that gives the area of S.

(b) Find the angle θ in the interval $0 \leq \theta \leq \dfrac{\pi}{2}$ that corresponds to the point on the curve
$$r_1(\theta) = 2 + 3\sin\theta + 4\cos\theta$$
that is furthest from the origin. Justify your answer.

(c) The radial distance between the two curves changes for $0 \leq \theta \leq \dfrac{\pi}{2}$. Suppose θ changes at a constant rate of
$\dfrac{d\theta}{dt} = \dfrac{1}{3}$ for all times $t \geq 0$. Find the rate at which the
distance between the two curves changes for $\theta = \dfrac{\pi}{6}$.

2. At a large distribution center, workers process orders during the five-hour shift from noon to 5:00 p.m. Orders are received at a rate of $r(t) = 400\sqrt{9 + 4t - t^2}$ orders per hour for $0 \leq t \leq 5$ where t is the number of hours since noon. During the same time interval, orders are processed and packaged by the workers at a constant rate of 1200 orders per hour. At the beginning of the shift, there are 2000 orders that had not been processed and packaged.

(a) Using correct units, find and explain the meaning of $\int_3^4 r(t)\,dt$. Express your answer to the nearest integer.

(b) Write an equation involving an integral expression of $r(t)$ that gives $N(t)$, the number of unprocessed orders at time t.

(c) At what time t, for $0 \leq t \leq 5$, is the number of unprocessed orders a maximum?

Section 2: Free Response Part B

No calculator is allowed for Part B.

3. Water is dripping from a faucet into a 2-gallon bucket that has a height h of 24 in. The height of the water in the bucket at time t, $0 \le t \le 60$, can be modeled by a differentiable function h, where h is measured in inches and t in minutes. The table below shows the height of the water at select times t.

t (minutes)	0	10	20	30	40	50	60
h (inches)	0	9.4	15.2	18.6	20.8	22.0	22.8

(a) Using the data above, approximate $h'(25)$. Show the computations and explain the result, including the units.

(b) Approximate $\int_0^{60} h(t)\, dt$ with a trapezoidal sum with three intervals of equal length. Use the result to calculate $\frac{1}{60} \int_0^{60} h(t)\, dt$, and interpret the meaning of the result. Be sure to use correct units.

(c) If the function that models h is $h(t) = 24 - 24e^{-0.05t}$, determine the concavity of h. Does the trapezoidal sum found in (b) overestimate or underestimate the height of the water in the bucket?

(d) Is there a time t between $t = 40$ and $t = 60$ minutes at which $h'(t) = 0.10$? Justify your answer.

4. Let f be a function defined on the closed interval $0 \le x \le 16$ with $f(0) = -3$. The graph of f', the derivative of f, consists of 4 line segments as shown below. Also shown is the graph of $y = \frac{4}{x}$ which intersects the graph of f' at points P and Q.

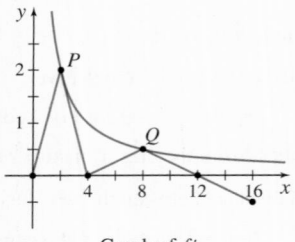

Graph of f'

(a) Determine whether f has a relative minimum, relative maximum, or neither at $x = 12$. Justify your answer.

(b) Let g be a function given by $g(x) = (f(x))^2$. Write an equation of the line tangent to the graph of g at $x = 8$.

(c) Let h be a function given by $h(x) = xe^{-0.25f(x)}$. Find all the critical numbers of h.

(d) Find $\int_4^8 x f''(x)\, dx$.

5. Consider the graphs of the function $y = f(x)$ and the line tangent to the graph of f at $x = 0$ pictured below.

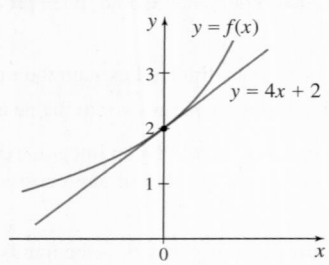

(a) Let g be a function given by $g(x) = \int_0^x f(t)\, dt$. Find

$$\lim_{x \to 0} \left(\frac{\sin(2x) - g(x)}{e^{3x} - \cos(2x) - 3x} \right)$$

Show the work that leads to your answer.

(b) The function $y = f(x)$ is the particular solution to the differential equation $\frac{dy}{dx} = \frac{y^2}{x+1}$ with initial condition $f(0) = 2$. Use Euler's method, starting at $x = 0$ with two steps of equal size, to approximate $f\left(\frac{1}{2}\right)$. Show the computations that lead to your answer.

(c) Use separation of variables to find $y = f(x)$, the particular solution to the differential equation $\frac{dy}{dx} = \frac{y^2}{x+1}$ with initial condition $f(0) = 2$.

6. The function f is defined by $f(x) = \frac{1}{(1+x^2)^2}$. The Maclaurin series for f is given by

$$1 - 2x^2 + 3x^4 - 4x^6 + \cdots + (-1)^n(n+1)x^{2n} + \cdots$$

(a) Use the series to find each of the following:

   i. the value of $f^{(4)}(0)$

   ii. the coefficient of the x^{22} term

(b) Find the radius of convergence for the Maclaurin series for f. Show the work that leads to your answer.

(c) Let $P(x)$ be the fourth-degree Maclaurin polynomial for $f(x)$. Use the graph of $f^{(5)}$ shown below together with the Lagrange error bound to show that

$$|f(0.5) - P(0.5)| < \frac{1}{10}$$

Show the work that leads to your answer.

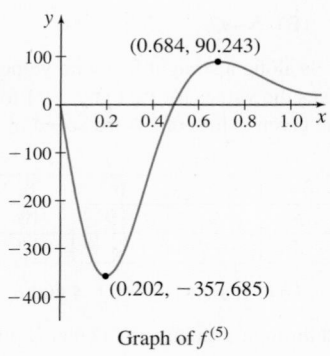

Graph of $f^{(5)}$

(d) Find the first four nonzero terms and the general term for the Maclaurin series representing $\int_0^x \frac{1}{(1+t^2)^2}\, dt$.

Precalculus Used in Calculus

The topics reviewed here are not exhaustive of the precalculus used in calculus. However, they do represent a large body of the material you will see in calculus. If you encounter difficulty with any of this material, consult a textbook in precalculus for more detail and explanation.

A.1 Algebra Used in Calculus

OBJECTIVES *When you finish this section, you should be able to:*

1 Factor and simplify algebraic expressions (p. A-1)
2 Complete the square (p. A-2)
3 Solve equations (p. A-3)
4 Use interval notation; solve inequalities (p. A-5)
5 Work with exponents (p. A-8)
6 Work with logarithms (p. A-10)

1 Factor and Simplify Algebraic Expressions

EXAMPLE 1 Factoring Algebraic Expressions

Factor each expression completely:

(a) $2(x+3)(x-2)^3 + (x+3)^2 \cdot 3 \cdot (x-2)^2$

(b) $\dfrac{4}{3}x^{1/3}(2x+1) + 2x^{4/3}$

Solution

(a) In expression (a), $(x+3)$ and $(x-2)^2$ are **common factors**, factors found in each term. Factor them out.

$$2(x+3)(x-2)^3 + (x+3)^2 \cdot 3 \cdot (x-2)^2$$

$\qquad = (x+3)(x-2)^2[2(x-2) + 3(x+3)]$ Factor out $(x+3)(x-2)^2$.

$\qquad = (x+3)(x-2)^2(5x+5)$ Simplify.

$\qquad = 5(x+3)(x-2)^2(x+1)$ Factor out 5.

Kathleen
Director of Gene Therapy Process Development

I design and run labs that develop manufacturing processes for novel gene therapies. The principles that I learned in AP® Calculus are instrumental in planning the processes that allow us to consistently manufacture the same product at small and large scales.

(b) We begin by writing the term $2x^{4/3}$ as a fraction with a denominator of 3.

$$\frac{4}{3}x^{1/3}(2x+1) + 2x^{4/3} = \frac{4x^{1/3}(2x+1)}{3} + \frac{6x^{4/3}}{3}$$

$$= \frac{4x^{1/3}(2x+1) + 6x^{4/3}}{3} \qquad \text{Add the two fractions.}$$

$$= \frac{2x^{1/3}[2(2x+1) + 3x]}{3} \qquad \begin{array}{l}\text{2 and } x^{1/3} \text{ are common factors.}\\ \text{Factor them out.}\end{array}$$

$$= \frac{2x^{1/3}(7x+2)}{3} \qquad \text{Simplify.} \qquad \blacksquare$$

EXAMPLE 2   Simplifying Algebraic Expressions

(a) Simplify $\dfrac{(x^2+1)\cdot 3 - (3x+4)\cdot 2x}{(x^2+1)^2}$.

(b) Write the expression $(x^2+1)^{1/2} + x \cdot \dfrac{1}{2}(x^2+1)^{-1/2} \cdot 2x$ as a single quotient in which only positive exponents appear.

Solution

(a) $\dfrac{(x^2+1)\cdot 3 - (3x+4)\cdot 2x}{(x^2+1)^2} = \dfrac{3x^2+3-(6x^2+8x)}{(x^2+1)^2} = \dfrac{3x^2+3-6x^2-8x}{(x^2+1)^2}$

$$= \frac{-3x^2-8x+3}{(x^2+1)^2} \underset{\underset{\text{Factor}}{\uparrow}}{=} \frac{-(3x-1)(x+3)}{(x^2+1)^2}$$

(b) $(x^2+1)^{1/2} + x \cdot \dfrac{1}{2}(x^2+1)^{-1/2} \cdot 2x = (x^2+1)^{1/2} + \dfrac{x^2}{(x^2+1)^{1/2}}$

$$= \frac{(x^2+1)^{1/2}(x^2+1)^{1/2}}{(x^2+1)^{1/2}} + \frac{x^2}{(x^2+1)^{1/2}}$$

$$= \frac{(x^2+1)+x^2}{(x^2+1)^{1/2}} = \frac{2x^2+1}{(x^2+1)^{1/2}} \qquad \blacksquare$$

---

② Complete the Square

We complete the square in one variable by modifying an expression of the form $x^2 + bx$ to make it a perfect square. Perfect squares are trinomials of the form

$$x^2 + 2ax + a^2 = (x+a)^2 \quad \text{or} \quad x^2 - 2ax + a^2 = (x-a)^2$$

For example, $x^2 + 6x + 9$ is a perfect square because $x^2 + 6x + 9 = (x+3)^2$. And $p^2 - 12p + 36$ is a perfect square because $p^2 - 12p + 36 = (p-6)^2$.

To make $x^2 + 6x$ a perfect square, we must add 9. The number to be added is chosen by dividing the coefficient of the first-degree term, which is 6, by 2 and squaring the result $\left[\left(\dfrac{6}{2}\right)^2 = 9\right]$.

EXAMPLE 3   **Completing the Square**

Determine the number that must be added to each expression to complete the square. Then factor.

Solution

Expression	Add	Result	Factored Form
$y^2 + 8y$	$\left(\dfrac{1}{2} \cdot 8\right)^2 = 16$	$y^2 + 8y + 16$	$(y + 4)^2$
$a^2 - 20a$	$\left(\dfrac{1}{2} \cdot (-20)\right)^2 = 100$	$a^2 - 20a + 100$	$(a - 10)^2$
$p^2 - 5p$	$\left(\dfrac{1}{2} \cdot (-5)\right)^2 = \dfrac{25}{4}$	$p^2 - 5p + \dfrac{25}{4}$	$\left(p - \dfrac{5}{2}\right)^2$
$2x^2 + 6x = 2(x^2 + 3x)$	$\left(\dfrac{1}{2} \cdot 3\right)^2 = \dfrac{9}{4}$	$2\left(x^2 + 3x + \dfrac{9}{4}\right)$	$2\left(x + \dfrac{3}{2}\right)^2$

∎

CAUTION   The original expression $x^2 + bx$ and the perfect square $x^2 + bx + \left(\dfrac{b}{2}\right)^2$ are not equal. So when completing the square within an equation or an inequality, we must not only add $\left(\dfrac{b}{2}\right)^2$, we must also subtract it. That is,

$$x^2 + bx = x^2 + bx + \underbrace{\left(\frac{b}{2}\right)^2 - \left(\frac{b}{2}\right)^2}_{=\,0} = \left(x + \frac{b}{2}\right)^2 - \left(\frac{b}{2}\right)^2$$

③ Solve Equations

To solve a **quadratic equation** $ax^2 + bx + c = 0$, $a \neq 0$, the *quadratic formula* can be used.

THEOREM Quadratic Formula

Consider the quadratic equation

$$ax^2 + bx + c = 0 \quad a \neq 0$$

- If $b^2 - 4ac < 0$, the equation has no real solution.
- If $b^2 - 4ac \geq 0$, the real solution(s) of the equation is (are) given by the **quadratic formula**:

$$\boxed{x = \frac{-b \pm \sqrt{b^2 - 4ac}}{2a}}$$

The expression $b^2 - 4ac$ is called the **discriminant** of the quadratic equation.

EXAMPLE 4 Solving Quadratic Equations

Solve each equation: **(a)** $3x^2 - 5x + 1 = 0$ **(b)** $x^2 + x = -1$

Solution

(a) The discriminant is $b^2 - 4ac = 25 - 12 = 13$. We use the quadratic formula.

$$x = \frac{-b \pm \sqrt{b^2 - 4ac}}{2a} = \frac{5 \pm \sqrt{13}}{6} \qquad a = 3, b = -5, c = 1$$

The solutions are $x = \dfrac{5 - \sqrt{13}}{6} \approx 0.232$ and $x = \dfrac{5 + \sqrt{13}}{6} \approx 1.434$.

(b) The quadratic equation in standard form is $x^2 + x + 1 = 0$. Since its discriminant is $b^2 - 4ac = 1 - 4 = -3$, this equation has no real solution. ∎

EXAMPLE 5 Solving Equations

Solve each equation:

(a) $\dfrac{3}{x-2} = \dfrac{1}{x-1} + \dfrac{7}{(x-1)(x-2)}$ **(b)** $x^3 - x^2 - 4x + 4 = 0$

(c) $\sqrt{x-1} = x - 7$ **(d)** $|1 - x| = 2$

Solution

NOTE The set of real numbers that a variable can assume is called the **domain of the variable**.

(a) First, notice that the domain of the variable is $\{x \mid x \neq 1,\ x \neq 2\}$. Now clear the equation of rational expressions by multiplying both sides by $(x-1)(x-2)$.

$$\frac{3}{x-2} = \frac{1}{x-1} + \frac{7}{(x-1)(x-2)}$$

$$(x-1)(x-2)\frac{3}{x-2} = (x-1)(x-2)\left[\frac{1}{x-1} + \frac{7}{(x-1)(x-2)}\right] \quad \text{Multiply both sides by } (x-1)(x-2).$$

$$3x - 3 = (x-1)(x-2)\frac{1}{x-1} + (x-1)(x-2)\frac{7}{(x-1)(x-2)} \quad \text{Distribute.}$$

$$3x - 3 = (x-2) + 7 \quad \text{Simplify.}$$

$$3x - 3 = x + 5$$

$$2x = 8$$

$$x = 4$$

Since 4 is in the domain of the variable, the solution is 4.

(b) Group the terms of $x^3 - x^2 - 4x + 4 = 0$, and factor by grouping.

$$x^3 - x^2 - 4x + 4 = 0$$

$$(x^3 - x^2) - (4x - 4) = 0 \quad \text{Group the terms.}$$

$$x^2(x-1) - 4(x-1) = 0 \quad \text{Factor out the common factor from each group.}$$

$$(x^2 - 4)(x-1) = 0 \quad \text{Factor out the common factor } (x-1).$$

$$(x-2)(x+2)(x-1) = 0 \quad x^2 - 4 = (x-2)(x+2)$$

$$x - 2 = 0 \ \text{ or } \ x + 2 = 0 \ \text{ or } \ x - 1 = 0 \quad \text{Set each factor equal to 0.}$$

$$x = 2 \qquad\qquad x = -2 \qquad\qquad x = 1 \quad \text{Solve.}$$

The solutions are -2, 1, and 2.

CAUTION Squaring both sides of an equation may lead to **extraneous solutions** (answers that do not satisfy the equation). Check all apparent solutions.

(c) Square both sides of the equation since the index of a square root is 2.

$$\sqrt{x-1} = x-7$$
$$(\sqrt{x-1})^2 = (x-7)^2 \qquad \text{Square both sides.}$$
$$x - 1 = x^2 - 14x + 49$$
$$x^2 - 15x + 50 = 0 \qquad \text{Put in standard form.}$$
$$(x-10)(x-5) = 0 \qquad \text{Factor.}$$
$$x = 10 \text{ or } x = 5 \qquad \text{Set each factor equal to 0 and solve.}$$

Check: $x = 10$: $\sqrt{x-1} = \sqrt{10-1} = \sqrt{9} = 3$ and $x - 7 = 10 - 7 = 3$

$x = 5$: $\sqrt{x-1} = \sqrt{5-1} = \sqrt{4} = 2$ and $x - 7 = 5 - 7 = -2$

The apparent solution 5 is extraneous; the only solution of the equation is 10.

RECALL $|a| = a$ if $a \geq 0$
$|a| = -a$ if $a < 0$
If $|x| = b, b \geq 0$, then $x = b$ or $x = -b$.

(d) $|1 - x| = 2$

$$1 - x = 2 \quad \text{or} \quad 1 - x = -2 \qquad \text{The expression inside the absolute value bars equals 2 or } -2.$$
$$-x = 1 \qquad\qquad -x = -3 \qquad \text{Simplify.}$$
$$x = -1 \qquad\qquad x = 3 \qquad \text{Simplify.}$$

The solutions are -1 and 3. ∎

④ Use Interval Notation; Solve Inequalities

In expressing the solution to an inequality, *interval notation* is often used.

NOTE Every interval of real numbers contains both rational numbers (numbers that can be expressed as the quotient of two integers), and irrational numbers (numbers that are not rational).

DEFINITION Interval Notation

Suppose a and b represent two real numbers with $a < b$.

A **closed interval**, denoted by $[a, b]$, consists of all real numbers x for which $a \leq x \leq b$.

An **open interval**, denoted by (a, b), consists of all real numbers x for which $a < x < b$.

The **half-open**, or **half-closed**, **intervals** are:

- $(a, b]$, consisting of all real numbers x for which $a < x \leq b$, and
- $[a, b)$, consisting of all real numbers x for which $a \leq x < b$.

In each of these definitions, a is called the **left endpoint** and b the **right endpoint** of the interval.

The symbol ∞ (read "infinity") is *not* a real number, but a notational device used to indicate unboundedness in the positive direction. The symbol $-\infty$ (read "negative infinity") also is not a real number, but a notational device used to indicate unboundedness in the negative direction. Using the symbols ∞ and $-\infty$, we define five other kinds of intervals:

- $[a, \infty)$, consisting of all real numbers x for which $x \geq a$
- (a, ∞), consisting of all real numbers x for which $x > a$
- $(-\infty, a]$, consisting of all real numbers x for which $x \leq a$
- $(-\infty, a)$, consisting of all real numbers x for which $x < a$
- $(-\infty, \infty)$, consisting of all real numbers x

Notice that the endpoints ∞ and $-\infty$ are never included in the interval, since neither is a real number.

Table 1 summarizes interval notation, corresponding inequality notation, and their graphs.

TABLE 1

Interval	Inequality	Graph
The open interval (a, b)	$a < x < b$	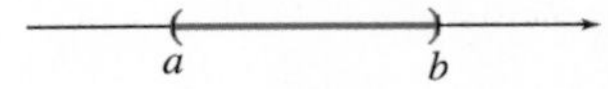
The closed interval $[a, b]$	$a \leq x \leq b$	
The half-open interval $[a, b)$	$a \leq x < b$	
The half-open interval $(a, b]$	$a < x \leq b$	
The interval $[a, \infty)$	$a \leq x < \infty$	
The interval (a, ∞)	$a < x < \infty$	
The interval $(-\infty, a]$	$-\infty < x \leq a$	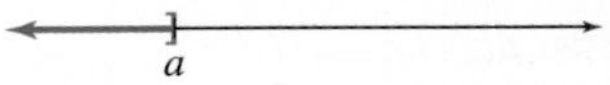
The interval $(-\infty, a)$	$-\infty < x < a$	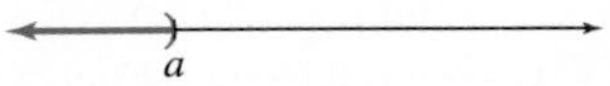

EXAMPLE 6 Solving Inequalities

Solve each inequality and graph the solution.

(a) $4x + 7 \geq 2x - 3$ **(b)** $x^2 - 4x + 3 > 0$ **(c)** $x^2 + x + 1 < 0$ **(d)** $\dfrac{1+x}{1-x} > 0$

Solution

(a) $4x + 7 \geq 2x - 3$

$\qquad 4x \geq 2x - 10$ Subtract 7 from both sides.

$\qquad 2x \geq -10$ Subtract $2x$ from both sides.

$\qquad x \geq -5$ Divide both sides by 2.

$\qquad\qquad\qquad$ (The direction of the inequality symbol is unchanged.)

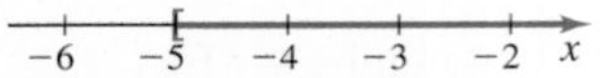

The solution using interval notation is $[-5, \infty)$. See Figure 1 for the graph of the solution.

Figure 1 $x \geq -5$

(b) This is a quadratic inequality. The related quadratic equation

$$x^2 - 4x + 3 = (x - 1)(x - 3) = 0$$

has two solutions, 1 and 3. We use these numbers to partition the number line into three open intervals. Now select a test number in each interval, and determine the value of $x^2 - 4x + 3$ at the test number. See Table 2.

NOTE The test number can be any real number in the interval, but it cannot be an endpoint.

TABLE 2

Interval	Test Number	Value of $x^2 - 4x + 3$	Sign of $x^2 - 4x + 3$
$(-\infty, 1)$	0	3	Positive
$(1, 3)$	2	-1	Negative
$(3, \infty)$	4	3	Positive

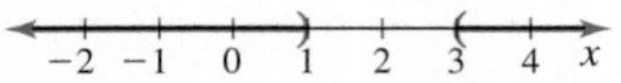

Figure 2 $x < 1$ or $x > 3$

We conclude that $x^2 - 4x + 3 > 0$ for any number in the set $(-\infty, 1) \cup (3, \infty)$. See Figure 2 for the graph of the solution.

(c) The quadratic equation $x^2 + x + 1 = 0$ has no real solution, since its discriminant is negative. [See Example 4(b).] When this happens, the quadratic inequality is either positive for all real numbers or negative for all real numbers. To see which is true, evaluate $x^2 + x + 1$ at some number, say, 0. At 0, $x^2 + x + 1 = 1$, which is positive. So, $x^2 + x + 1 > 0$ for all real numbers x. The inequality $x^2 + x + 1 < 0$ has no solution.

(d) The only solution of the rational equation $\dfrac{1+x}{1-x} = 0$ is $x = -1$; also, the expression $\dfrac{1+x}{1-x}$ is not defined for $x = 1$. We use the solution -1 and the value 1, at which the expression is undefined, to partition the real number line into three open intervals. Now select a test number in each interval, and evaluate the rational expression $\dfrac{1+x}{1-x}$ at each test number. See Table 3.

TABLE 3

Interval	Test Number	Value of $\dfrac{1+x}{1-x}$	Sign of $\dfrac{1+x}{1-x}$
$(-\infty, -1)$	-2	$-\dfrac{1}{3}$	Negative
$(-1, 1)$	0	1	Positive
$(1, \infty)$	2	-3	Negative

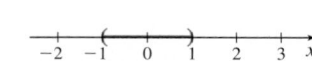

Figure 3 $-1 < x < 1$

We conclude that $\dfrac{1+x}{1-x} > 0$ for any number in the interval $(-1, 1)$. See Figure 3 for the graph of the solution. ∎

The *Triangle Inequality* is an important relationship involving the absolute value of a sum.

THEOREM Triangle Inequality

If x and y are real numbers, then

$$|x + y| \leq |x| + |y|$$

Use the following theorem as a guide to solving inequalities involving absolute values.

THEOREM Inequalities Involving Absolute Value

If a is a positive real number and if u is an algebraic expression, then

• $\|u\| < a$ is equivalent to $-a < u < a$	(1)
• $\|u\| > a$ is equivalent to $u < -a \quad$ or $\quad u > a$	(2)

Similar relationships hold for the nonstrict inequalities $|u| \leq a$ and $|u| \geq a$.

EXAMPLE 7 **Solving Inequalities Involving Absolute Value**

Solve each inequality and graph the solution:

(a) $|3 - 4x| < 11$ **(b)** $|2x + 4| - 1 \leq 9$ **(c)** $\left| \dfrac{4x + 1}{2} - \dfrac{3}{5} \right| > 1$

Solution

(a) The absolute value is less than the number 11, so statement (1) applies.

$$|3 - 4x| < 11$$

$-11 < 3 - 4x < 11$	Use statement (1).
$-14 < -4x < 8$	Subtract 3 from each part.
$\dfrac{-14}{-4} > x > \dfrac{8}{-4}$	Divide each part by -4, which reverses the inequality signs.
$-2 < x < \dfrac{7}{2}$	Simplify and rearrange the ordering.

RECALL Multiplying (or dividing) an inequality by a negative quantity reverses the direction of the inequality sign.

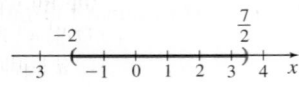

Figure 4 $-2 < x < \dfrac{7}{2}$

The solutions are all the numbers in the open interval $\left(-2, \dfrac{7}{2}\right)$. See Figure 4 for the graph of the solutions.

(b) First put $|2x + 4| - 1 \le 9$ into the form $|u| \le a$.

$|2x + 4| - 1 \le 9$	
$|2x + 4| \le 10$	Add 1 to each side.
$-10 \le 2x + 4 \le 10$	Use statement (1) with $\le$.
$-14 \le 2x \le 6$	
$-7 \le x \le 3$	

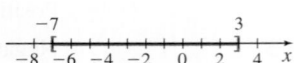

Figure 5 $-7 \le x \le 3$

The solutions are all the numbers in the closed interval $[-7, 3]$. See Figure 5 for the graph of the solutions.

(c) $\left|\dfrac{4x + 1}{2} - \dfrac{3}{5}\right| > 1$ is in the form of statement (2). So, begin by simplifying the expression inside the absolute value.

$$\left|\frac{4x + 1}{2} - \frac{3}{5}\right| = \left|\frac{5(4x + 1)}{10} - \frac{2 \cdot 3}{10}\right| = \left|\frac{20x + 5 - 6}{10}\right| = \left|\frac{20x - 1}{10}\right|$$

The original inequality is equivalent to the inequality

$$\left|\frac{20x - 1}{10}\right| > 1$$

$\dfrac{20x - 1}{10} < -1$	or	$\dfrac{20x - 1}{10} > 1$	Use statement (2).
$20x - 1 < -10$	or	$20x - 1 > 10$	
$20x < -9$	or	$20x > 11$	
$x < -\dfrac{9}{20}$	or	$x > \dfrac{11}{20}$	

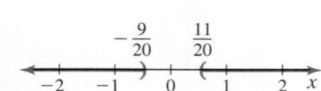

Figure 6 $x < -\dfrac{9}{20}$ or $x > \dfrac{11}{20}$

The solutions are all the numbers in the set $\left(-\infty, -\dfrac{9}{20}\right) \cup \left(\dfrac{11}{20}, \infty\right)$. See Figure 6 for the graph of the solutions. ∎

⑤ Work with Exponents

Integer exponents provide a shorthand notation for repeated multiplication. For example,

$$2^3 = 2 \cdot 2 \cdot 2 = 8 \quad \text{or} \quad \left(\frac{1}{3}\right)^4 = \frac{1}{3} \cdot \frac{1}{3} \cdot \frac{1}{3} \cdot \frac{1}{3} = \frac{1}{81}$$

DEFINITION

If a is a real number and n is a positive integer, then the symbol a^n represents the product of n factors of a. That is,

$$a^n = \underbrace{a \cdot a \cdot \ldots \cdot a}_{n \text{ factors}}$$

Here, it is understood that $a^1 = a$.

Then $a^2 = a \cdot a$ and $a^3 = a \cdot a \cdot a$, and so on. In the expression a^n, the number a is called the **base** and the number n is called the **exponent** or **power**. We read a^n as "a raised to the power n" or as "a to the nth power." We usually read a^2 as "a squared" and a^3 as "a cubed."

DEFINITION

If $a \neq 0$, then

$$a^0 = 1$$

If n is a positive integer, then

$$a^{-n} = \frac{1}{a^n} \qquad a \neq 0$$

The following properties, called the *Laws of Exponents*, can be proved using the preceding definitions.

THEOREM Laws of Exponents

In each of these properties a and b are real numbers, and u and v are integers.

$$
\begin{aligned}
&\bullet\ a^u a^v = a^{u+v} \quad \bullet\ (a^u)^v = a^{uv} \quad \bullet\ (ab)^u = a^u b^u \\[2mm]
&\bullet\ \frac{a^u}{a^v} = a^{u-v} = \frac{1}{a^{v-u}} \quad \text{if } a \neq 0 \\[2mm]
&\bullet\ \left(\frac{a}{b}\right)^u = \frac{a^u}{b^u} \quad \text{if } b \neq 0
\end{aligned}
$$

DEFINITION Principal nth Root

The **principal nth root of a real number** a, where $n \geq 2$ is an integer, symbolized by $\sqrt[n]{a}$, is defined as the solution of the equation $b^n = a$.

$$\sqrt[n]{a} = b \text{ is equivalent to } a = b^n$$

IN WORDS The symbol $\sqrt[n]{a}$ means "the number that, when raised to the nth power, equals a."

If n is even, then both $a \geq 0$ and $b \geq 0$, and if n is odd, then a and b are any real numbers and both have the same sign.

If a is negative and n is even, then $\sqrt[n]{a}$ is not defined. When $\sqrt[n]{a}$ is defined, the principal nth root of a number is unique.

The symbol $\sqrt[n]{a}$ for the principal nth root of a is called a **radical**; the integer n is called the **index**, and a is called the **radicand**. If the index of a radical is 2, we call $\sqrt[2]{a}$ the **square root** of a and omit the index 2 by writing $\sqrt{a}$. If the index is 3, we call $\sqrt[3]{a}$ the **cube root** of a.

Radicals are used to define rational exponents.

DEFINITION $a^{1/n}$

If a is a real number and $n \geq 2$ is an integer, then

$$a^{1/n} = \sqrt[n]{a}$$

provided that $a^{1/n} = \sqrt[n]{a}$ exists.

DEFINITION $a^{m/n}$

If a is a real number and m and n are integers with $n \geq 2$, then

$$a^{m/n} = (a^{1/n})^m$$

provided that $a^{1/n} = \sqrt[n]{a}$ exists.

From the two definitions, we have

$$a^{m/n} = \sqrt[n]{a^m} = (\sqrt[n]{a})^m$$

In simplifying a rational expression $a^{m/n}$, either $\sqrt[n]{a^m}$ or $(\sqrt[n]{a})^m$ can be used. The choice depends on which is easier to simplify. Generally, taking the root first, as in $(\sqrt[n]{a})^m$, is easier.

But does a^x have meaning, where the base a is a positive real number and the exponent x is an irrational number? The answer is yes, and although a rigorous definition requires calculus, the basis for the definition is easy to follow: Select a rational number r that is formed by truncating (removing) all but a finite number of digits from the irrational number x. Then it is reasonable to expect that

$$a^x \approx a^r$$

For example, take the irrational number $\pi = 3.14159 \ldots$. Then an approximation to a^π is

$$a^\pi \approx a^{3.14}$$

where the digits after the hundredths position have been truncated from the value for π. A better approximation would be

$$a^\pi \approx a^{3.14159}$$

where the digits after the hundred-thousandths position have been truncated. Continuing in this way, we can obtain approximations to a^π to any desired degree of accuracy.

It can be shown that the Laws of Exponents hold for real number exponents u and v.

6 Work with Logarithms

The definition of a *logarithm* is based on an exponential relationship.

DEFINITION Logarithm

Suppose $y = a^x$, $a > 0$, $a \neq 1$, and x is a real number. The **logarithm with base a of y**, symbolized by $\log_a y$, is the exponent to which a must be raised to obtain y. That is,

$$\log_a y = x \text{ is equivalent to } y = a^x$$

IN WORDS A logarithm is an exponent. That is, $\log_a y$ equals the exponent x that makes $a^x = y$.

EXAMPLE 8   Working with Logarithms

(a) If $x = \log_3 y$, then $y = 3^x$. For example, $4 = \log_3 81$ is equivalent to $81 = 3^4$.

(b) If $x = \log_5 y$, then $y = 5^x$. For example,

$$-1 = \log_5 \frac{1}{5} \text{ is equivalent to } \frac{1}{5} = 5^{-1}$$ ∎

THEOREM Properties of Logarithms

Suppose u and a are positive real numbers, $a \neq 1$, and r is any real number. The number $\log_a u$ is the exponent to which a must be raised to obtain u. That is,

$$\boxed{a^{\log_a u} = u}$$

The logarithm with base a of a raised to a power r equals that power. That is,

$$\boxed{\log_a a^r = r}$$

EXAMPLE 9   Using Properties of Logarithms

(a) $2^{\log_2 \pi} = \pi$

(b) $\log_{0.2} 0.2^{(-\sqrt{2})} = -\sqrt{2}$

(c) $\log_{1/5} \left(\frac{1}{5}\right)^{kt} = kt$ ∎

THEOREM Properties of Logarithms

In the following properties, u, v, and a are positive real numbers, $a \neq 1$, and r is any real number:

- **The Log of a Product Equals the Sum of the Logs**

$$\boxed{\log_a (uv) = \log_a u + \log_a v} \tag{3}$$

- **The Log of a Quotient Equals the Difference of the Logs**

$$\boxed{\log_a \left(\frac{u}{v}\right) = \log_a u - \log_a v} \tag{4}$$

- **$\log_a u^r$ Equals the Product of r and $\log_a u$**

$$\boxed{\log_a u^r = r \log_a u} \tag{5}$$

CAUTION In using properties (3), (4), and (5), be careful about the values that the variable may assume. For example, the domain of the variable for $\log_a x$ is $x > 0$, and for $\log_a (x - 1)$ it is $x > 1$. So, the equality $\log_a x + \log_a (x - 1) = \log_a [x(x - 1)]$ is true only for $x > 1$.

EXAMPLE 10   Using Properties (3), (4), and (5) of Logarithms

(a) $\log_a \left(x\sqrt{x^2 + 1}\right) = \log_a x + \log_a \sqrt{x^2 + 1}$ $\log_a(uv) = \log_a u + \log_a v$

$$= \log_a x + \log_a (x^2 + 1)^{1/2}$$

$$= \log_a x + \frac{1}{2} \log_a (x^2 + 1) \qquad \log_a u^r = r \log_a u$$

(b) $\log_a \dfrac{x^2}{(x - 1)^3} = \log_a x^2 - \log_a (x - 1)^3 = 2 \log_a x - 3 \log_a (x - 1)$

$\log_a \left(\dfrac{u}{v}\right) = \log_a u - \log_a v$ $\qquad \log_a u^r = r \log_a u$

(c) $\log_a x + \log_a 9 + \log_a(x^2+1) - \log_a 5 = \log_a(9x) + \log_a(x^2+1) - \log_a 5$

$$= \log_a[9x(x^2+1)] - \log_a 5$$

$$= \log_a \frac{9x(x^2+1)}{5} \qquad \blacksquare$$

CAUTION Common errors made by some students include:

• Expressing the sum of logarithms as the logarithm of a sum

$$\log_a u + \log_a v \quad \text{is } not \text{ equal to} \quad \log_a(u+v)$$

Correct statement: $\log_a u + \log_a v = \log_a(uv)$ Property (3)

• Expressing the difference of logarithms as the quotient of logarithms

$$\log_a u - \log_a v \quad \text{is } not \text{ equal to} \quad \frac{\log_a u}{\log_a v}$$

Correct statement: $\log_a u - \log_a v = \log_a\left(\dfrac{u}{v}\right)$ Property (4)

• Expressing a logarithm raised to a power as the product of the power and the logarithm

$$(\log_a u)^r \quad \text{is } not \text{ equal to} \quad r\log_a u$$

Correct statement: $\log_a u^r = r\log_a u$ Property (5)

Since most calculators can calculate only logarithms with base 10, called **common logarithms** (abbreviated log), and logarithms with base $e \approx 2.718$, called **natural logarithms** (abbreviated ln), it is often useful to be able to change the bases of logarithms.

THEOREM Change-of-Base Formula

If $a \neq 1$, $b \neq 1$, and u are positive real numbers, then

$$\boxed{\log_a u = \frac{\log_b u}{\log_b a}}$$

For example, to approximate $\log_2 15$, we use the change-of-base formula to change the logarithm with base 2 to a common logarithm. Then we use a calculator.

$$\log_2 15 = \underset{\substack{\uparrow \\ \text{Change-of-Base Formula}}}{\frac{\log 15}{\log 2}} \approx 3.907$$

A.2 Geometry Used in Calculus

OBJECTIVES *When you finish this section, you should be able to:*

1 **Use properties of triangles and the Pythagorean Theorem (p. A-12)**

2 **Work with congruent triangles and similar triangles (p. A-14)**

3 **Use geometry formulas (p. A-16)**

① Use Properties of Triangles and the Pythagorean Theorem

A **triangle** is a three-sided polygon. The lengths of the sides are labeled with lowercase letters, such as a, b, and c, and the angles are labeled with uppercase letters, such as A, B, and C, with angle A opposite side a, angle B opposite side b, and angle C opposite side c, as shown in Figure 7.

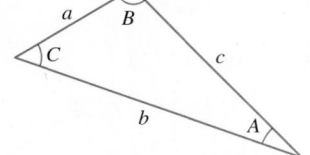

Figure 7

Refer to the triangle in Figure 7. The sum of the lengths of any two sides of a triangle is always greater than the length of the remaining side. The sum of the measures of the three angles of a triangle, when measured in degrees, equals 180°. That is,

$$a + b > c \qquad a + c > b \qquad b + c > a \qquad A + B + C = 180°$$

An **isosceles triangle** is a triangle with two equal sides. In an isosceles triangle, the angles opposite the two equal sides are equal.

An **equilateral triangle** is a triangle with three equal sides. In an equilateral triangle, each angle measures 60°.

The *Pythagorean Theorem* is a statement about *right triangles*. A **right triangle** contains a **right angle**, that is, an angle measuring 90°. The side of the triangle opposite the 90° angle is called the **hypotenuse**; the remaining two sides are called **legs**. In Figure 8, c represents the length of the hypotenuse, and a and b represent the lengths of the legs.

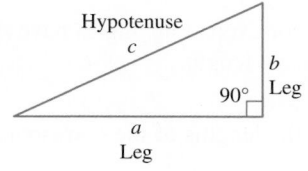

Figure 8 A right triangle

> **THEOREM** Pythagorean Theorem
>
> In a right triangle, the square of the length of the hypotenuse is equal to the sum of the squares of the lengths of the two legs. That is, in the right triangle shown in Figure 8,
>
> $$\boxed{c^2 = a^2 + b^2}$$

EXAMPLE 1   **Finding the Hypotenuse of a Right Triangle**

In a right triangle, one leg has length 4 and the other has length 3. What is the length of the hypotenuse?

Solution

Since the triangle is a right triangle, we use the Pythagorean Theorem with $a = 4$ and $b = 3$ to find the length c of the hypotenuse.

$$c^2 = a^2 + b^2$$
$$c^2 = 4^2 + 3^2 = 16 + 9 = 25$$
$$c = \sqrt{25} = 5$$

The converse of the Pythagorean Theorem is also true.

> **THEOREM** Converse of the Pythagorean Theorem
>
> In a triangle, if the square of the length of one side equals the sum of the squares of the lengths of the other two sides, the triangle is a right triangle. The 90° angle is opposite the longest side.

EXAMPLE 2   **Using the Converse of the Pythagorean Theorem**

Show that a triangle whose sides have lengths 5, 12, and 13 is a right triangle. Identify the hypotenuse.

Solution

We square the lengths of the sides.

$$5^2 = 25 \qquad 12^2 = 144 \qquad 13^2 = 169$$

Notice that the sum of 25 and 144 equals 169. So, the triangle is a right triangle. The longest side, 13, is the hypotenuse. ■

See Figure 9.

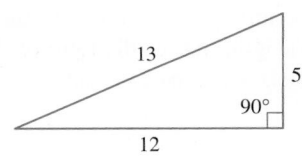

Figure 9

② Work with Congruent Triangles and Similar Triangles

The word "congruent" means "coinciding when superimposed." For example, two angles are congruent if they have the same measure, and two line segments are congruent if they have the same length.

> **IN WORDS** Two triangles are congruent if they are the same size and shape.

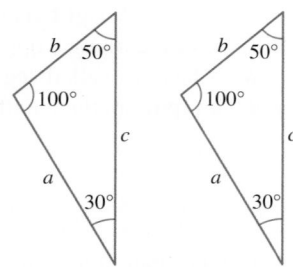

Figure 10 Congruent triangles.

DEFINITION Congruent Triangles

Two triangles are **congruent** if in each triangle, the corresponding angles have the same measure and the corresponding sides have the same length.

In Figure 10, corresponding angles are equal and the lengths of the corresponding sides are equal. So, these triangles are congruent.

It is not necessary to verify that all three angles and all three sides have the same measure to determine whether two triangles are congruent.

Determining Congruent Triangles

- **Angle-Side-Angle Case (ASA)** Two triangles are congruent if the measures of two angles from each triangle are equal and the lengths of the corresponding sides between the two angles are equal.

For example, in Figure 11, the two triangles are congruent because both triangles have an angle measuring 40°, an angle measuring 80°, and the sides between these angles are both 10 units in length.

- **Side-Side-Side Case (SSS)** Two triangles are congruent if the lengths of the sides of one triangle and the lengths of the sides of the other triangle are the same.

For example, in Figure 12, the two triangles are congruent because both triangles have sides with lengths 8, 15, and 20 units.

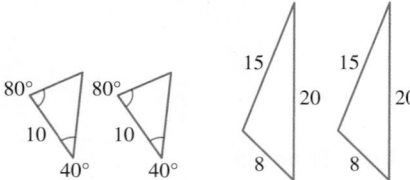

Figure 11 ASA   **Figure 12** SSS

- **Side-Angle-Side Case (SAS)** Two triangles are congruent if the lengths of two corresponding sides of the triangles are the same and the angles between the two sides have the same measure.

For example, in Figure 13, the two triangles are congruent because both triangles have sides of length 7 units and 8 units, and the angle between the two congruent sides in each triangle measures 40°.

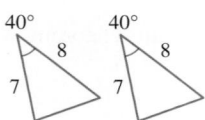

Figure 13 SAS

- **Angle-Angle-Side Case (AAS)** Two triangles are congruent if the measures of two angles and the length of a nonincluded side of one triangle are the same as the corresponding parts of the other triangle.

For example, in Figure 14, the two triangles are congruent because both triangles have angles measuring 35° and 96°, and the nonincluded side adjacent to the 35° angle has length 10 units.

CAUTION Knowing that two triangles have equal angles is not sufficient to conclude that the triangles are congruent. Similarly, knowing that the lengths of two corresponding sides and the measure of a nonincluded angle of the triangles are the same is not sufficient to conclude that the triangles are congruent.

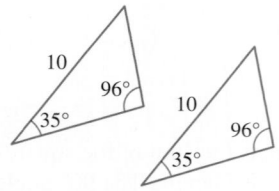

Figure 14 AAS

We contrast congruent triangles with *similar* triangles.

> **IN WORDS** Two triangles are similar if they have the same shape, but (possibly) different sizes.

DEFINITION Similar Triangles

Two triangles are **similar** if in each triangle corresponding angles have the same measure and corresponding sides are proportional in length, that is, the ratio of the lengths of the corresponding sides of each triangle equals the same constant.

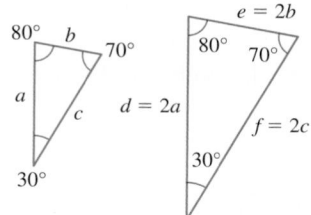

Figure 15

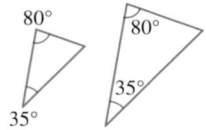

Figure 16 AA

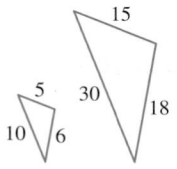

Figure 17 SSS

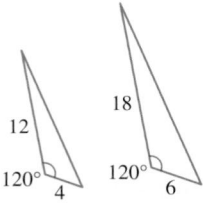

Figure 18 SAS

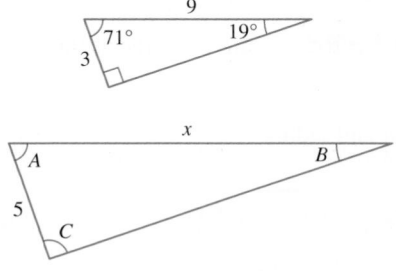

Figure 19

For example, the triangles in Figure 15 are similar because each of the corresponding angles of the two triangles has the same measure. Also, the lengths of the corresponding sides are proportional: each side of the triangle on the right is twice as long as the corresponding side of the triangle on the left. That is, the ratio of the corresponding sides is a constant: $\dfrac{d}{a} = \dfrac{e}{b} = \dfrac{f}{c} = 2$.

It is not necessary to verify that all three angles are equal and all three sides are proportional to determine whether two triangles are similar.

Determining Similar Triangles

- **Angle-Angle Case (AA)** Two triangles are similar if the measures of two angles from each triangle are the same.

  For example, the two triangles in Figure 16 are similar because each triangle has an angle measuring 35° and an angle measuring 80°.

- **Side-Side-Side Case (SSS)** Two triangles are similar if the lengths of the sides of one triangle are proportional to the lengths of the sides of the second triangle.

  For example, Figure 17 shows two triangles, one with sides of lengths 5, 6, 10 and the other with sides of lengths 15, 18, 30. These two triangles are similar because

  $$\frac{5}{15} = \frac{6}{18} = \frac{10}{30} = \frac{1}{3}$$

- **Side-Angle-Side Case (SAS)** Two triangles are similar if the lengths of two sides of one triangle are proportional to the lengths of two sides of the second triangle, and the angles between the corresponding two sides have the same measure.

  See Figure 18. The two triangles are similar because $\dfrac{4}{6} = \dfrac{12}{18} = \dfrac{2}{3}$ and the angle between sides 4 and 12 and the angle between sides 6 and 18 each measures 120°.

EXAMPLE 3   **Using Similar Triangles**

Given that the triangles in Figure 19 are similar, find the missing length x and angles A, B, and C.

Solution

Because the triangles are similar, corresponding angles have the same measure. So, $A = 71°$, $B = 19°$, and $C = 90°$. Also corresponding sides are proportional. That is, $\dfrac{x}{9} = \dfrac{5}{3}$. We solve this equation for x.

$$\frac{x}{9} = \frac{5}{3}$$

$$9 \cdot \frac{x}{9} = 9 \cdot \frac{5}{3} \qquad \text{Multiply both sides by 9.}$$

$$x = 15 \qquad \text{Simplify.}$$

The missing length is 15 units. ∎

❸ Use Geometry Formulas

Certain formulas from geometry are useful in solving calculus problems. Some of these are listed here.

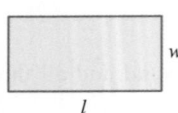

For a rectangle of length l and width w,

$$\boxed{\text{Area} = lw \qquad \text{Perimeter} = 2l + 2w}$$

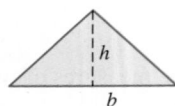

For a triangle with base b and height h,

$$\boxed{\text{Area} = \frac{1}{2}bh}$$

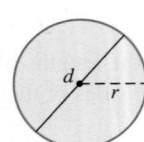

For a circle of radius r (diameter $d = 2r$),

$$\boxed{\text{Area} = \pi r^2 \qquad \text{Circumference} = 2\pi r = \pi d}$$

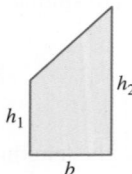

For a trapezoid with base b and parallel heights h_1 and h_2,

$$\boxed{\text{Area} = \frac{1}{2}b(h_1 + h_2)}$$

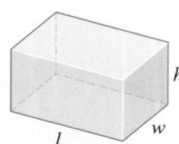

For a closed rectangular box of length l, width w, and height h,

$$\boxed{\text{Volume} = lwh \qquad \text{Surface area} = 2lh + 2wh + 2lw}$$

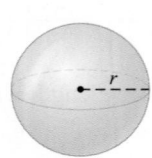

For a sphere of radius r,

$$\boxed{\text{Volume} = \frac{4}{3}\pi r^3 \qquad \text{Surface area} = 4\pi r^2}$$

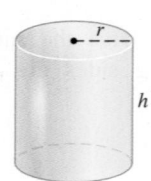

For a closed, right circular cylinder of height h and radius r,

$$\boxed{\text{Volume} = \pi r^2 h \qquad \text{Surface area} = 2\pi r^2 + 2\pi rh}$$

If the cylinder has no top or bottom, the surface area is $2\pi rh$.

For a right circular cone of height h, radius r, and slant height $l = \sqrt{h^2 + r^2}$ that is open on top,

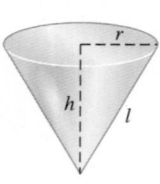

$$\boxed{\text{Volume} = \frac{1}{3}\pi r^2 h \quad \text{Lateral surface area} = \pi rl = \pi r\sqrt{h^2 + r^2}}$$

If the cone is closed on top, the surface area is $\pi r^2 + \pi rl$.

A.3 Analytic Geometry Used in Calculus

OBJECTIVES *When you finish this section, you should be able to:*

1 Use the distance formula (p. A-17)
2 Graph equations, find intercepts, and test for symmetry (p. A-17)
3 Work with equations of a line (p. A-20)
4 Work with the equation of a circle (p. A-22)
5 Graph parabolas, ellipses, and hyperbolas (p. A-24)

1 Use the Distance Formula

NOTE If (x, y) are the coordinates of a point P, then x is called the **x-coordinate** of P and y is called the **y-coordinate** of P.

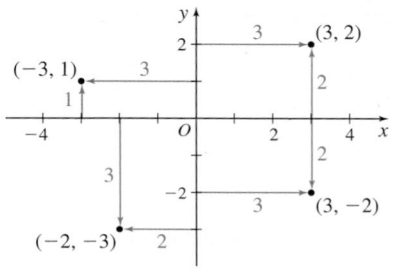

Figure 20

Any point P in the xy-plane can be located using an **ordered pair** (x, y) of real numbers. The ordered pair (x, y), called the **coordinates** of P, gives enough information to locate the point P in the plane. For example, to locate the point with coordinates $(-3, 1)$, move 3 units along the x-axis to the left of O, and then move straight up 1 unit. Then **plot** the point by placing a dot at this location. See Figure 20 in which the points with coordinates $(-3, 1)$, $(-2, -3)$, $(3, -2)$, and $(3, 2)$ are plotted.

> **THEOREM** Distance Formula
>
> The distance between two points $P_1 = (x_1, y_1)$ and $P_2 = (x_2, y_2)$, denoted by $d(P_1, P_1)$, is
>
> $$d(P_1, P_2) = \sqrt{(x_2 - x_1)^2 + (y_2 - y_1)^2}$$

EXAMPLE 1   Using the Distance Formula

Find the distance d between the points $(-3, 5)$ and $(3, 2)$.

Solution

Use the distance formula with $P_1 = (x_1, y_1) = (-3, 5)$ and $P_2 = (x_2, y_2) = (3, 2)$.

$$d = \sqrt{[3 - (-3)]^2 + (2 - 5)^2} = \sqrt{6^2 + (-3)^2} = \sqrt{36 + 9} = \sqrt{45} = 3\sqrt{5} \approx 6.708$$

■

2 Graph Equations, Find Intercepts, and Test for Symmetry

An **equation in two variables**, say, x and y, is a statement in which two expressions involving x and y are equal. The expressions are called the **sides** of the equation. Since an equation is a statement, it may be true or false, depending on the value of the variables. Any values of x and y that result in a true statement are said to **satisfy** the equation.

For example, the following are equations in two variables x and y:

$$x^2 + y^2 = 5 \qquad 2x - y = 6 \qquad y = 2x + 5 \qquad x^2 = y$$

The equation $x^2 + y^2 = 5$ is satisfied for $x = 1$ and $y = 2$, since $1^2 + 2^2 = 1 + 4 = 5$. Other choices of x and y, such as $x = -1$ and $y = -2$, also satisfy this equation. It is not satisfied for $x = 2$ and $y = 3$, since $2^2 + 3^2 = 4 + 9 = 13 \neq 5$.

The **graph of an equation in two variables** x and y consists of the set of points in the xy-plane whose coordinates (x, y) satisfy the equation.

EXAMPLE 2   Graphing an Equation by Plotting Points

Graph the equation $y = x^2$.

Solution

Table 4 lists several points on the graph. In Figure 21(a) the points are plotted and then they are connected with a smooth curve to obtain the graph (a *parabola*). Figure 21(b) shows the graph using graphing technology.

TABLE 4

x	$y = x^2$	(x, y)
-4	16	$(-4, 16)$
-3	9	$(-3, 9)$
-2	4	$(-2, 4)$
-1	1	$(-1, 1)$
0	0	$(0, 0)$
1	1	$(1, 1)$
2	4	$(2, 4)$
3	9	$(3, 9)$
4	16	$(4, 16)$

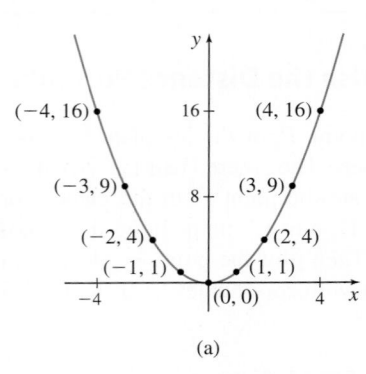

(a)

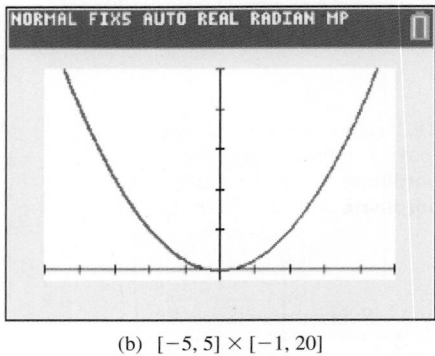

(b) $[-5, 5] \times [-1, 20]$

Figure 21 $y = x^2$

The graphs in Figure 21 do not show all the points whose coordinates (x, y) satisfy the equation $y = x^2$. For example, the point $(6, 36)$ satisfies the equation $y = x^2$, but it is not shown. It is important when graphing to present enough of the graph so that any viewer will "see" the rest of the graph as an obvious continuation of what is actually there.

The points, if any, at which a graph crosses or touches a coordinate axis are called the **intercepts** of the graph. See Figure 22. The x-coordinate of a point where a graph crosses or touches the x-axis is an **x-intercept**. At an x-intercept, the y-coordinate equals 0. The y-coordinate of a point where a graph crosses or touches the y-axis is a **y-intercept**. At a y-intercept, the x-coordinate equals 0. When graphing an equation, all its intercepts should be displayed or be easily inferred from the part of the graph that is displayed.

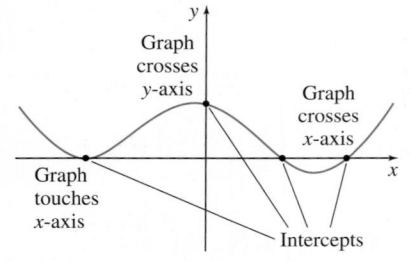

Figure 22

EXAMPLE 3   Finding the Intercepts of a Graph

Find the x-intercept(s) and the y-intercept(s) of the graph of $y = x^2 - 4$.

Solution

To find the x-intercept(s), we let $y = 0$ and solve the equation

$$x^2 - 4 = 0$$

$$(x + 2)(x - 2) = 0 \qquad \text{Factor.}$$

$$x = -2 \quad \text{or} \quad x = 2 \qquad \text{Set each factor equal to 0 and solve.}$$

The equation has two solutions, -2 and 2. The x-intercepts are -2 and 2.

To find the y-intercept(s), we let $x = 0$ in the equation.

$$y = 0^2 - 4 = -4$$

The y-intercept is -4. ∎

DEFINITION Symmetry

- A graph is **symmetric with respect to the x-axis** if, for every point (x, y) on the graph, the point $(x, -y)$ is also on the graph. See Figure 23.
- A graph is **symmetric with respect to the y-axis** if, for every point (x, y) on the graph, the point $(-x, y)$ is also on the graph. See Figure 24.
- A graph is **symmetric with respect to the origin** if, for every point (x, y) on the graph, the point $(-x, -y)$ is also on the graph. See Figure 25.

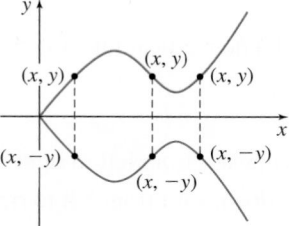

Figure 23 Symmetry with respect to the x-axis.

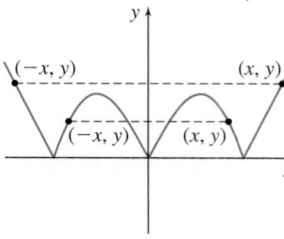

Figure 24 Symmetry with respect to the y-axis.

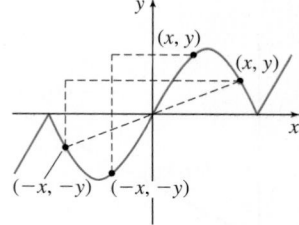

Figure 25 Symmetry with respect to the origin.

EXAMPLE 4 **Testing for Symmetry**

Test the graph of $y = \dfrac{4x^2}{x^2 + 1}$ for symmetry.

Solution

- x-axis: To test for symmetry with respect to the x-axis, we replace y with $-y$. Since $-y = \dfrac{4x^2}{x^2 + 1}$ is not equivalent to $y = \dfrac{4x^2}{x^2 + 1}$, the graph of the equation is not symmetric with respect to the x-axis.
- y-axis: To test for symmetry with respect to the y-axis, we replace x with $-x$. Since $y = \dfrac{4(-x)^2}{(-x)^2 + 1} = \dfrac{4x^2}{x^2 + 1}$ is equivalent to $y = \dfrac{4x^2}{x^2 + 1}$, the graph of the equation is symmetric with respect to the y-axis.
- Origin: To test for symmetry with respect to the origin, we replace x with $-x$ and y with $-y$.

$$-y = \frac{4(-x)^2}{(-x)^2 + 1} \qquad \text{Replace } x \text{ by } -x \text{ and } y \text{ by } -y.$$

$$-y = \frac{4x^2}{x^2 + 1} \qquad \text{Simplify.}$$

$$y = -\frac{4x^2}{x^2 + 1} \qquad \text{Multiply both sides by } -1.$$

Since the result is not equivalent to the original equation, the graph of the equation is not symmetric with respect to the origin. ∎

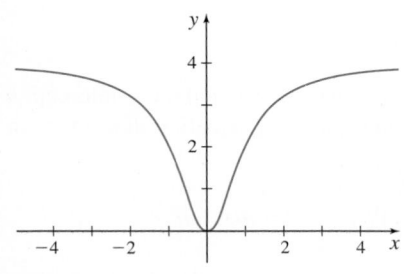

Figure 26 $y = \dfrac{4x^2}{x^2 + 1}$

Figure 26 shows the graph of $y = \dfrac{4x^2}{x^2 + 1}$ and confirms the symmetry with respect to the y-axis.

3 Work with Equations of a Line

DEFINITION Slope

Let $P = (x_1, y_1)$ and $Q = (x_2, y_2)$ be two distinct points. If $x_1 \neq x_2$, the **slope** m of the nonvertical line L containing P and Q is defined by the formula

$$m = \frac{y_2 - y_1}{x_2 - x_1} \qquad x_1 \neq x_2$$

If $x_1 = x_2$, then L is a **vertical line** and the slope m of L is **undefined**.

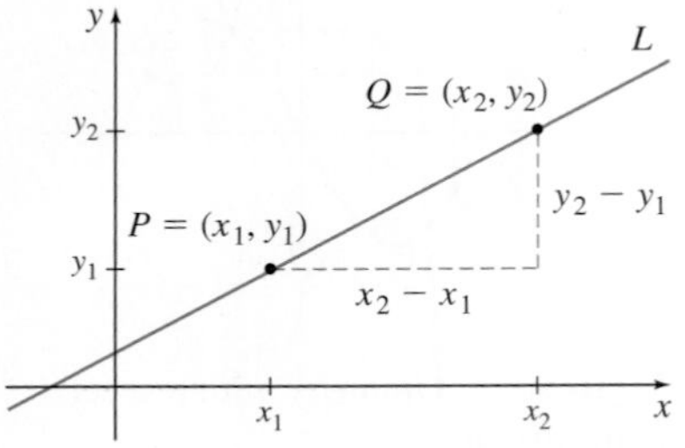

Figure 27(a) provides an illustration of the slope of a nonvertical line; Figure 27(b) illustrates a vertical line.

From the definition of slope, we conclude:

- When the slope of a line is positive, the line slants upward from left to right.
- When the slope of a line is negative, the line slants downward from left to right.
- When the slope of a line is 0, the line is horizontal.
- When the slope of a line is undefined, the line is vertical.

THEOREM Equation of a Vertical Line

A vertical line is given by an equation of the form

$$x = a$$

where a is the x-intercept.

THEOREM Point-Slope Form of an Equation of a Line

An equation of a nonvertical line with slope m that contains the point (x_1, y_1) is

$$y - y_1 = m(x - x_1)$$

(a) Slope of L is $m = \dfrac{y_2 - y_1}{x_2 - x_1}$.

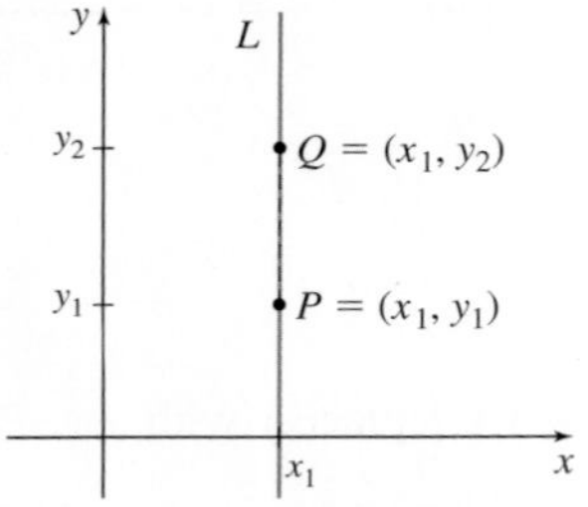

(b) L is vertical; the slope is undefined.

Figure 27

EXAMPLE 5   Using the Point-Slope Form of a Line

An equation of the line with slope 4 and containing the point $(1, 2)$ can be found by using the point-slope form with $m = 4$, $x_1 = 1$, and $y_1 = 2$.

$$y - y_1 = m(x - x_1) \qquad \text{Point-slope form of a line.}$$
$$y - 2 = 4(x - 1) \qquad m = 4, x_1 = 1, y_1 = 2$$

See Figure 28 for the graph. ■

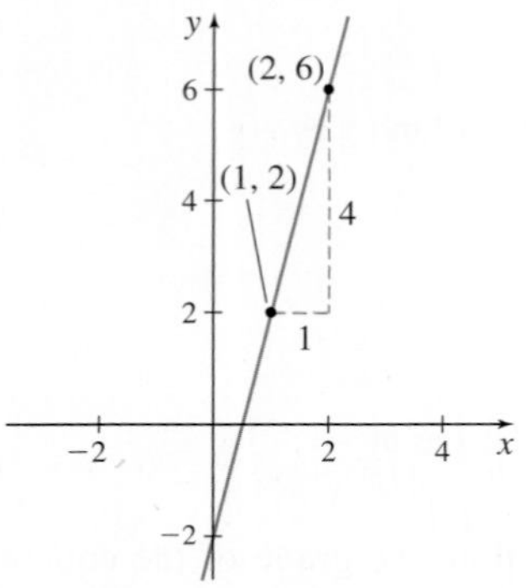

Another useful equation of a line is obtained when the slope m and the y-intercept b are known. The point-slope form is used to obtain the following equation of a line with slope m, containing the point $(0, b)$:

$$y - b = m(x - 0) \qquad \text{or equivalently} \qquad y = mx + b$$

Figure 28 $y - 2 = 4(x - 1)$

For example, if the equation $y - 2 = 4(x - 1)$ from Example 5 is solved for y, we find that $y = 4x - 2$, the slope intercept form of the line.

THEOREM Slope-Intercept Form of an Equation of a Line

An equation of a line with slope m and y-intercept b is

$$\boxed{y = mx + b}$$

For a horizontal line, the slope m is 0.

THEOREM Equation of a Horizontal Line

A horizontal line is given by an equation of the form

$$\boxed{y = b}$$

where b is the y-intercept.

EXAMPLE 6 Finding the Slope and y-Intercept from an Equation of a Line

Find the slope m and the y-intercept of the equation $2x + 4y = 8$. Graph the equation.

Solution

To obtain the slope and y-intercept, we write the equation in slope-intercept form by solving for y.

$$2x + 4y = 8$$
$$4y = -2x + 8$$
$$y = -\frac{1}{2}x + 2 \qquad y = mx + b$$

The coefficient of x is the slope. The slope is $-\dfrac{1}{2}$. The constant 2 is the y-intercept, so the point $(0, 2)$ is on the graph. Now use the slope $-\dfrac{1}{2}$. Starting at the point $(0, 2)$, we move 2 units to the right and then 1 unit down to the point $(2, 1)$. We plot this point and draw a line through the two points. See Figure 29. ■

When two lines in the plane do not intersect, they are **parallel**.

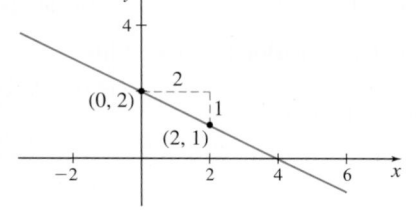

Figure 29 $2x + 4y = 8$

THEOREM Criterion for Parallel Lines

Two nonvertical lines are parallel if and only if their slopes are equal and they have different y-intercepts.

EXAMPLE 7 Finding an Equation of a Line Parallel to a Given Line

Find an equation of the line that contains the point $(1, 2)$ and is parallel to the line $y = 5x$. Write the equation in slope-intercept form.

Solution

Since the two lines are parallel, the slope of the line we seek equals the slope of the line $y = 5x$, which is 5. Since the line we seek also contains the point $(1, 2)$, we use the point-slope form to obtain the equation.

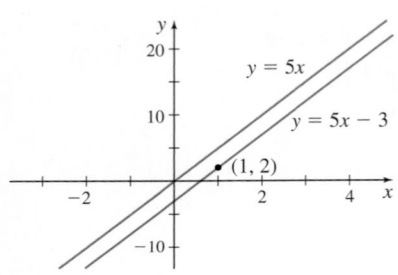

Figure 30 Parallel lines.

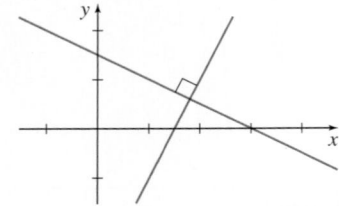

Figure 31 Perpendicular lines.

$$y - y_1 = m(x - x_1) \qquad \text{Point-slope form}$$
$$y - 2 = 5(x - 1)$$
$$y - 2 = 5x - 5$$
$$y = 5x - 3 \qquad \text{Slope-intercept form}$$

The line $y = 5x - 3$ contains the point $(1, 2)$ and is parallel to the line $y = 5x$. ∎

See Figure 30.

When two lines intersect at a right angle (90°), they are **perpendicular**. See Figure 31.

THEOREM Criterion for Perpendicular Lines

Two nonvertical lines are perpendicular if and only if the product of their slopes equals -1.

EXAMPLE 8 **Finding an Equation of a Line Perpendicular to a Given Line**

Find an equation of the line that contains the point $(-1, 3)$ and is perpendicular to the line $4x + y = -1$. Write the equation in slope-intercept form.

Solution

We begin by writing the equation of the given line in slope-intercept form to find its slope.

$$4x + y = -1$$
$$y = -4x - 1$$

This line has a slope of -4. Every line perpendicular to this line has slope $\dfrac{1}{4}$. Because the line we seek contains the point $(-1, 3)$, we use the point-slope form of a line.

$$y - y_1 = m(x - x_1) \qquad \text{Point-slope form}$$
$$y - 3 = \frac{1}{4}[x - (-1)]$$
$$y - 3 = \frac{1}{4}x + \frac{1}{4}$$
$$y = \frac{1}{4}x + \frac{13}{4} \qquad \text{Slope-intercept form}$$

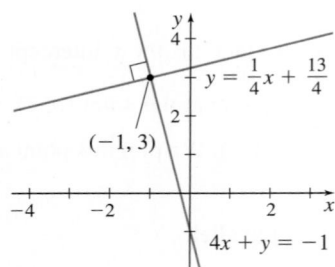

Figure 32

Figure 32 shows the graphs of $4x + y = -1$ and $y = \dfrac{1}{4}x + \dfrac{13}{4}$. ∎

④ Work with the Equation of a Circle

An advantage of a coordinate system is that it enables a geometric statement to be translated into an algebraic statement, and vice versa. Consider, for example, the following geometric statement that defines a circle.

DEFINITION Circle

A **circle** is the set of points in the xy-plane that is a fixed distance r from a fixed point (h, k). The fixed distance r is called the **radius**, and the fixed point (h, k) is called the **center** of the circle.

Figure 33 shows the graph of a circle. To find an equation that describes a circle, let (x, y) represent the coordinates of any point on the circle with radius r and

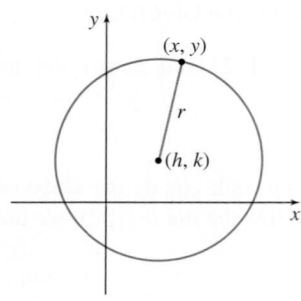

Figure 33 $(x - h)^2 + (y - k)^2 = r^2$

center (h, k). Then the distance between the point (x, y) and (h, k) equals r. That is, by the distance formula,

$$\sqrt{(x - h)^2 + (y - k)^2} = r$$

Now we square both sides to obtain

$$\boxed{(x - h)^2 + (y - k)^2 = r^2}$$

This form of the equation of a circle is called the **standard form**.

THEOREM

The standard form of a circle with radius r and center at the origin $(0, 0)$ is

$$\boxed{x^2 + y^2 = r^2}$$

DEFINITION Unit Circle

The circle with radius $r = 1$ and center at the origin is called the **unit circle** and has the equation

$$\boxed{x^2 + y^2 = 1}$$

See Figure 34.

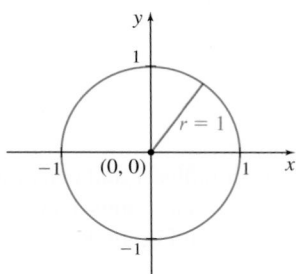

Figure 34 The unit circle: $x^2 + y^2 = 1$.

EXAMPLE 9   Graphing a Circle

Graph the equation $(x + 3)^2 + (y - 2)^2 = 16$.

Solution

This is the standard form of an equation of a circle. To graph the circle, first identify the center and the radius of the circle.

$$(x + 3)^2 + (y - 2)^2 = 16 \qquad (x - h)^2 + (y - k)^2 = r^2$$
$$(x - (-3))^2 + (y - 2)^2 = 4^2$$
$$\begin{array}{ccc} \uparrow & \uparrow & \uparrow \\ h & k & r^2 \end{array}$$

The circle has its center at the point $(-3, 2)$; the radius of the circle is 4 units. To graph the circle, first plot the center $(-3, 2)$. Then locate the four points on the circle that are 4 units to the left, to the right, above, and below the center. These four points are used as guides to obtain the graph. See Figure 35. ∎

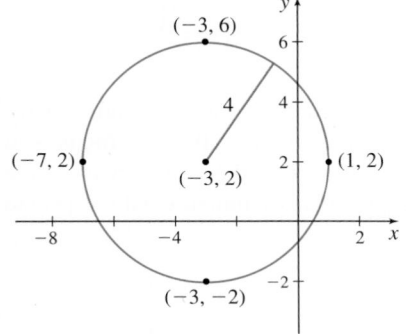

Figure 35 $(x + 3)^2 + (y - 2)^2 = 16$

DEFINITION General Form of the Equation of a Circle

The equation

$$\boxed{x^2 + y^2 + ax + bx + c = 0}$$

is called the **general form of the equation of a circle**.

NEED TO REVIEW? Completing the square is discussed in Section A.1, p. A-2.

If the equation of a circle is given in the general form, use the method of completing the square to put the equation in standard form. Then the center and radius can be easily identified.

EXAMPLE 10 Graphing a Circle Whose Equation Is in General Form

Graph the equation

$$x^2 + y^2 + 4x - 6y + 12 = 0$$

Solution

Complete the square in both x and y to put the equation in standard form. First, group the terms involving x, group the terms involving y, and put the constant on the right side of the equation. The result is

$$(x^2 + 4x) + (y^2 - 6y) = -12$$

Next, complete the square of each expression in parentheses. Remember that any number added to the left side of an equation must be added to the right side.

$$(x^2 + 4x + 4) + (y^2 - 6y + 9) = -12 + 4 + 9$$

$$\text{Add } \left(\frac{4}{2}\right)^2 = 4 \quad \text{Add } \left(\frac{-6}{2}\right)^2 = 9$$

$$(x + 2)^2 + (y - 3)^2 = 1$$

This is the standard form of the equation of a circle with radius 1 and center at the point $(-2, 3)$. To graph the circle, we plot the center $(-2, 3)$ and points 1 unit to the right, to the left, above, and below the point $(-2, 3)$, as shown in Figure 36. ∎

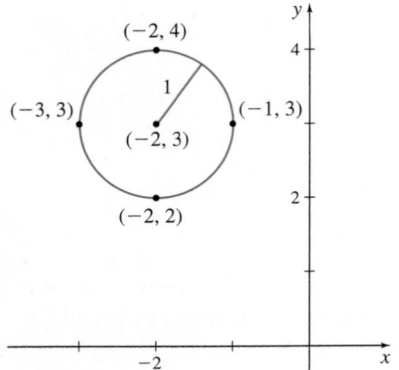

Figure 36 $x^2 + y^2 + 4x - 6y + 12 = 0$

5 Graph Parabolas, Ellipses, and Hyperbolas

The graph of the equation $y = x^2$ is an example of a *parabola*. See Figure 21 on page A-18. In fact, every parabola has a shape like the graph of $y = x^2$.

A **parabola** is the graph of an equation in one of the four forms:

$$\boxed{y^2 = 4ax \qquad y^2 = -4ax \qquad x^2 = 4ay \qquad x^2 = -4ay}$$

where $a > 0$.

In each of the parabolas in Figure 37, the **vertex** V is at the origin $(0, 0)$. When the equation is of the form $y^2 = 4ax$, the point $(a, 0)$ is called the **focus**, and the parabola is symmetric with respect to the x-axis. When the equation is of the form $x^2 = 4ay$, the point $(0, a)$ is the focus, and the parabola is symmetric with respect to the y-axis.

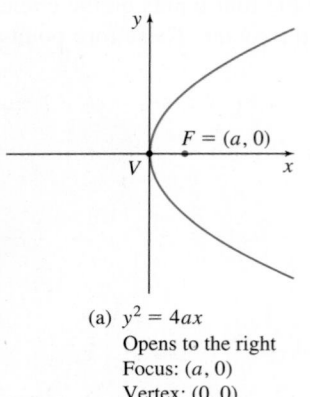
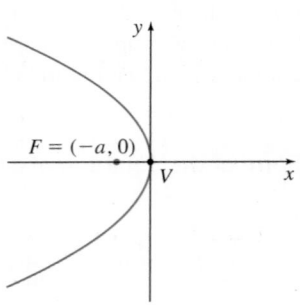
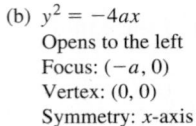
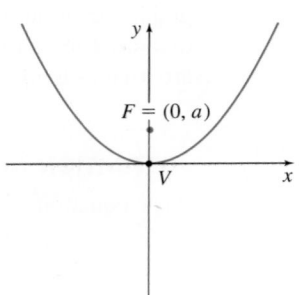
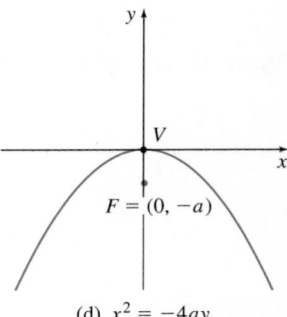

(a) $y^2 = 4ax$
Opens to the right
Focus: $(a, 0)$
Vertex: $(0, 0)$
Symmetry: x-axis

(b) $y^2 = -4ax$
Opens to the left
Focus: $(-a, 0)$
Vertex: $(0, 0)$
Symmetry: x-axis

(c) $x^2 = 4ay$
Opens up
Focus: $(0, a)$
Vertex: $(0, 0)$
Symmetry: y-axis

(d) $x^2 = -4ay$
Opens down
Focus: $(0, -a)$
Vertex: $(0, 0)$
Symmetry: y-axis

Figure 37

EXAMPLE 11 **Graphing a Parabola with Vertex at the Origin**

Graph the equation $x^2 = 12y$.

Solution

The graph of $x^2 = 12y$ is a parabola of the form

$$x^2 = 4ay$$

where $a = 3$. Look at Figure 37(c). The graph will open up, with focus at $F = (0, 3)$. The vertex is at $(0, 0)$. To graph the parabola, we let $y = a$; that is, let $y = 3$. (Any other positive number will also work.)

$$x^2 = 12y = 12 \cdot 3 = 36$$
$$\uparrow$$
$$y = a = 3$$

$$x = -6 \qquad \text{or} \qquad x = 6$$

The points $(-6, 3)$ and $(6, 3)$ are on the parabola and establish its opening. See Figure 38. ∎

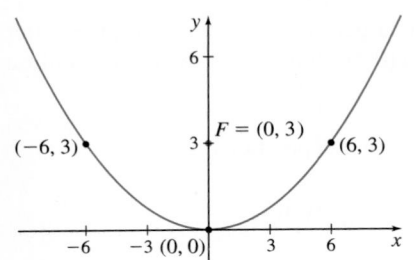

Figure 38 $x^2 = 12y$

An **ellipse** is the graph of an equation in one of the two forms:

$$\frac{x^2}{a^2} + \frac{y^2}{b^2} = 1 \qquad \frac{x^2}{b^2} + \frac{y^2}{a^2} = 1$$

where $a \geq b > 0$.

An ellipse is oval-shaped. The line segment dividing the ellipse in half the long way is the **major axis**; its length is $2a$. The two points of intersection of the ellipse and the major axis are the **vertices** of the ellipse. The midpoint of the vertices is the **center** of the ellipse. Along the major axis c units from the center of the ellipse, where $c^2 = a^2 - b^2$, $c > 0$, are two points, called **foci**, that determine the shape of the ellipse. The line segment through the center of the ellipse and perpendicular to the major axis is the **minor axis**. See Figure 39.

NOTE If $a = b$, then $\dfrac{x^2}{a^2} + \dfrac{y^2}{a^2} = 1$ is the equation of a circle of radius a.

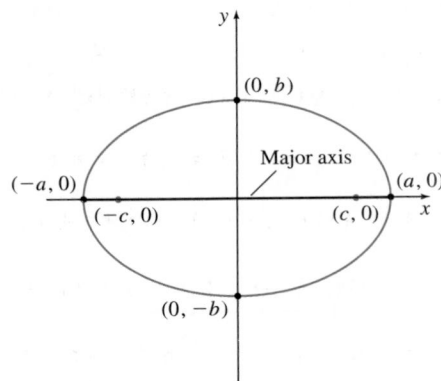

(a) $\dfrac{x^2}{a^2} + \dfrac{y^2}{b^2} = 1, a \geq b > 0$
Major axis: along the x-axis
Center: $(0, 0)$
Vertices: $(-a, 0), (a, 0)$
Foci: $(-c, 0), (c, 0)$, where $c^2 = a^2 - b^2, c > 0$
Symmetry: x-axis, y-axis, origin

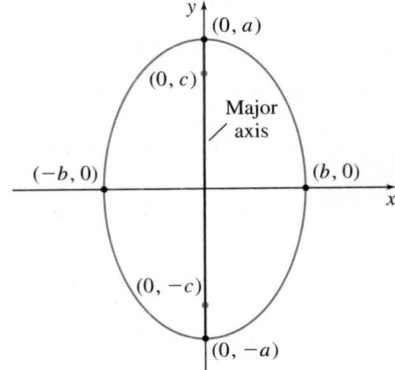

(b) $\dfrac{x^2}{b^2} + \dfrac{y^2}{a^2} = 1, a \geq b > 0$
Major axis: along the y-axis
Center: $(0, 0)$
Vertices: $(0, -a), (0, a)$
Foci: $(0, -c), (0, c)$, where $c^2 = a^2 - b^2, c > 0$
Symmetry: x-axis, y-axis, origin

Figure 39

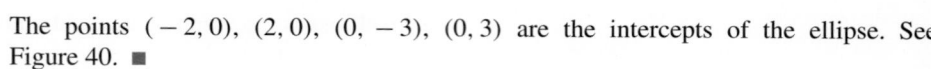

EXAMPLE 12 **Graphing an Ellipse with Center at the Origin**

Graph the equation: $9x^2 + 4y^2 = 36$

Solution

To put the equation in standard form, divide each side by 36.

$$\frac{x^2}{4} + \frac{y^2}{9} = 1$$

The graph of this equation is an ellipse. Since the larger number is under y^2, the major axis is along the y-axis. The center is at the origin $(0, 0)$. It is easiest to graph an ellipse by finding its intercepts:

x - intercepts: Let $y = 0$

$$\frac{x^2}{4} + \frac{0^2}{9} = 1$$

$$x^2 = 4$$

$$x = -2 \text{ and } x = 2$$

y - intercepts: Let $x = 0$

$$0^2 + \frac{y^2}{9} = 1$$

$$y^2 = 9$$

$$y = -3 \text{ and } y = 3$$

The points $(-2, 0)$, $(2, 0)$, $(0, -3)$, $(0, 3)$ are the intercepts of the ellipse. See Figure 40. ∎

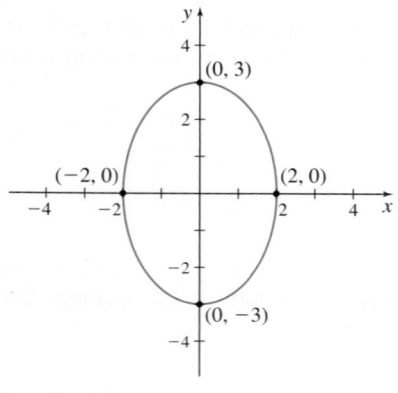

Figure 40 $9x^2 + 4y^2 = 36$

A **hyperbola** is the graph of an equation in one of the two forms:

$$\frac{x^2}{a^2} - \frac{y^2}{b^2} = 1 \qquad \frac{y^2}{a^2} - \frac{x^2}{b^2} = 1$$

where $a > 0$ and $b > 0$.

Two points, called **foci**, determine the shape of the hyperbola. The midpoint of the line segment containing the foci is called the **center** of the hyperbola. The foci are located c units from the center, where $c^2 = a^2 + b^2$, $c > 0$. The line containing the foci is called the **transverse axis**. The **vertices** of a hyperbola are its intercepts.

See Figure 41. The graph of a hyperbola consists of two branches.

- For the hyperbola $\dfrac{x^2}{a^2} - \dfrac{y^2}{b^2} = 1$, the branches of the graph open left and right.

- For the hyperbola $\dfrac{y^2}{a^2} - \dfrac{x^2}{b^2} = 1$, the branches of the graph open up and down.

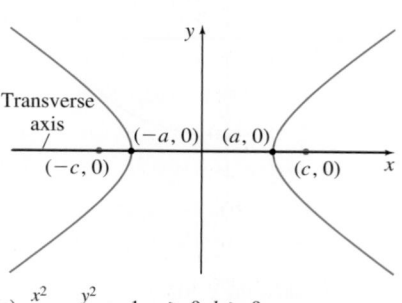

(a) $\frac{x^2}{a^2} - \frac{y^2}{b^2} = 1, a > 0, b > 0$
Branches open to left and right
Center: $(0, 0)$
Vertices: $(-a, 0)$, $(a, 0)$
Foci: $(-c, 0)$, $(c, 0)$, where $c^2 = a^2 + b^2, c > 0$
Symmetry: x-axis, y-axis, origin

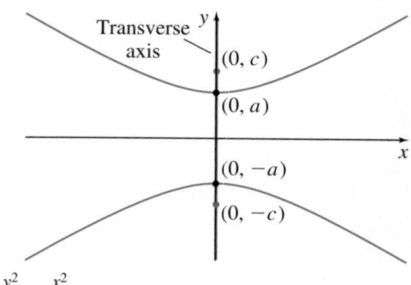

(b) $\frac{y^2}{a^2} - \frac{x^2}{b^2} = 1, a > 0, b > 0$
Branches open up and down
Center: $(0, 0)$
Vertices: $(0, -a)$, $(0, a)$
Foci: $(0, -c)$, $(0, c)$, where $c^2 = a^2 + b^2, c > 0$
Symmetry: x-axis, y-axis, origin

Figure 41

EXAMPLE 13 **Graphing a Hyperbola with Center at the Origin**

Graph the equation $\dfrac{y^2}{4} - \dfrac{x^2}{5} = 1$.

Solution

The graph of $\dfrac{y^2}{4} - \dfrac{x^2}{5} = 1$ is a hyperbola. The hyperbola consists of two branches, one opening up, the other opening down, like the graph in Figure 41(b). The hyperbola has no x-intercepts. To find the y-intercepts, we let $x = 0$ and solve for y.

$$\frac{y^2}{4} = 1$$
$$y^2 = 4$$
$$y = -2 \quad \text{or} \quad y = 2$$

The y-intercepts are -2 and 2, so the vertices are $(0, -2)$ and $(0, 2)$. The transverse axis is the vertical line $x = 0$. To graph the hyperbola, let $y = \pm 3$ (or any two numbers one >2 and the other <-2).

Then

$$\frac{y^2}{4} - \frac{x^2}{5} = 1$$
$$\frac{9}{4} - \frac{x^2}{5} = 1 \qquad y = \pm 3$$
$$\frac{x^2}{5} = \frac{5}{4}$$
$$x^2 = \frac{25}{4}$$
$$x = -\frac{5}{2} \text{ or } x = \frac{5}{2}$$

The points $\left(-\dfrac{5}{2}, 3\right)$, $\left(-\dfrac{5}{2}, -3\right)$, $\left(\dfrac{5}{2}, 3\right)$, and $\left(\dfrac{5}{2}, -3\right)$ are on the hyperbola. See Figure 42 for the graph. ∎

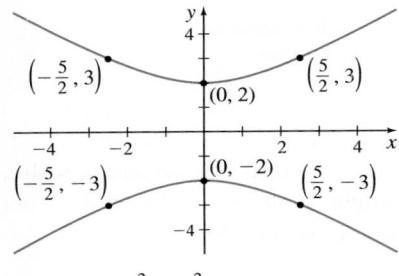

Figure 42 $\dfrac{y^2}{4} - \dfrac{x^2}{5} = 1$

A.4 Trigonometry Used in Calculus

OBJECTIVES *When you finish this section, you should be able to:*

1 **Work with angles, arc length of a circle, and circular motion (p. A-27)**
2 **Define and evaluate trigonometric functions (p. A-30)**
3 **Determine the domain and the range of the trigonometric functions (p. A-34)**
4 **Use basic trigonometry identities (p. A-35)**
5 **Use sum and difference, double-angle and half-angle, and sum-to-product and product-to-sum formulas (p. A-37)**
6 **Solve triangles using the Law of Sines and the Law of Cosines (p. A-38)**

① Work with Angles, Arc Length of a Circle, and Circular Motion

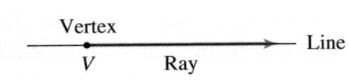

Figure 43

A **ray**, or **half-line**, is the portion of a line that starts at a point V on the line and extends indefinitely in one direction. The point V of a ray is called its **vertex**. See Figure 43.

If two rays are drawn with a common vertex, they form an **angle**. We call one ray of an angle the **initial side** and the other ray the **terminal side**. The angle formed is identified by showing the direction and amount of rotation from the initial side to the terminal side. If the rotation is in the counterclockwise direction, the angle is **positive**; if the rotation is clockwise, the angle is **negative**. See Figure 44.

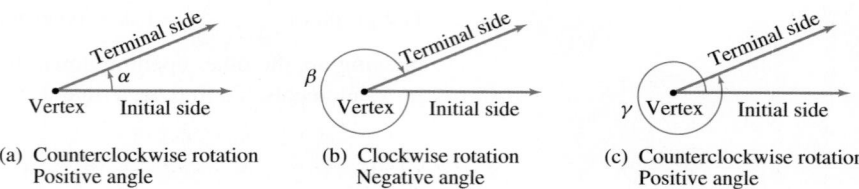

(a) Counterclockwise rotation
    Positive angle

(b) Clockwise rotation
    Negative angle

(c) Counterclockwise rotation
    Positive angle

Figure 44

An angle θ is in **standard position** if its vertex is at the origin of a rectangular coordinate system and its initial side is on the positive x-axis. See Figure 45.

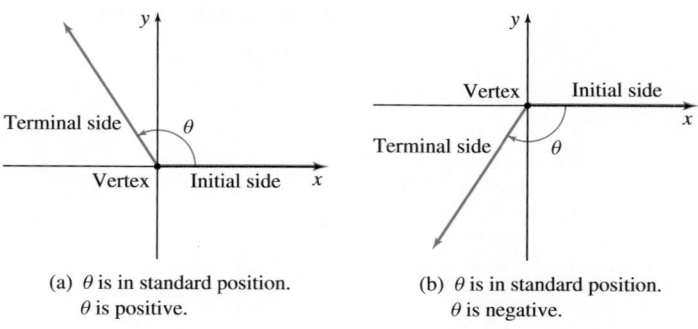

(a) θ is in standard position.
    θ is positive.

(b) θ is in standard position.
    θ is negative.

Figure 45

Suppose an angle θ is in standard position. If its terminal side coincides with a coordinate axis, we say that θ is a **quadrantal angle**. If its terminal side does not coincide with a coordinate axis, we say that θ **lies in a quadrant**. See Figure 46.

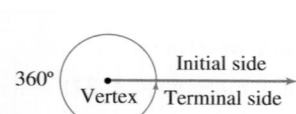

Figure 47 One revolution counterclockwise is 360°.

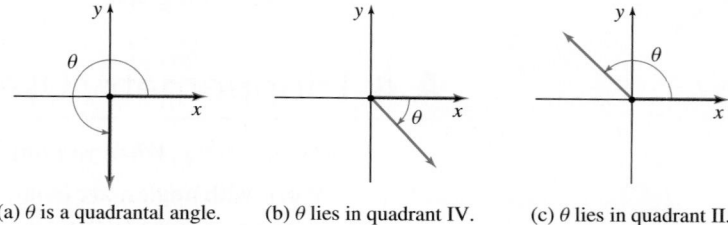

(a) θ is a quadrantal angle.

(b) θ lies in quadrant IV.

(c) θ lies in quadrant II.

Figure 46

Angles are measured by determining the amount of rotation needed for the initial side to coincide with the terminal side. The two commonly used measures for angles are *degrees* and *radians*.

See Figure 47. The angle formed by rotating the initial side exactly once in the counterclockwise direction until it coincides with itself (one revolution) measures 360 degrees, abbreviated 360°.

A **central angle** is a positive angle whose vertex is at the center of a circle. The rays of a central angle subtend (intersect) an arc on the circle. If the radius of the circle is r and the arc subtended by the central angle is also of length r, then the measure of the angle is **1 radian**. See Figure 48.

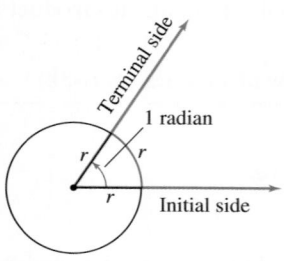

Figure 48 1 radian.

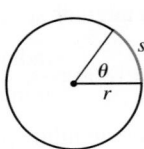

Figure 49 $s = r\theta$

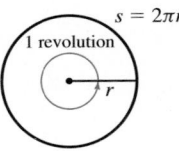

Figure 50 1 revolution $= 2\pi$ radians.

THEOREM Arc Length

For a circle of radius r, a central angle of θ radians subtends an arc whose length s is

$$s = r\theta$$

See Figure 49.

Consider a circle of radius r. A central angle of one revolution will subtend an arc equal to the circumference of the circle, as shown in Figure 50. Because the circumference of a circle equals $2\pi r$, we use $s = 2\pi r$ in the formula for arc length to find the radian measure of an angle of one revolution.

$$s = r\theta$$
$$2\pi r = r\theta \qquad \text{For one revolution: } s = 2\pi r.$$
$$\theta = 2\pi \text{ radians} \qquad \text{Solve for } \theta.$$

Since one revolution is equivalent to $360°$, we have $360° = 2\pi$ radians so that

$$180° = \pi \text{ radians}$$

In calculus, radians are generally used to measure angles, unless degrees are specifically mentioned. Table 5 lists the radian and degree measures for some common angles.

NOTE If the measure of an angle is given as a number, it is understood to mean radians. If the measure of an angle is given in degrees, then it will be marked either with the symbol ° or the word "degrees."

TABLE 5

Radians	0	$\dfrac{\pi}{6}$	$\dfrac{\pi}{4}$	$\dfrac{\pi}{3}$	$\dfrac{\pi}{2}$	$\dfrac{2\pi}{3}$	$\dfrac{3\pi}{4}$	$\dfrac{5\pi}{6}$	π	$\dfrac{3\pi}{2}$	2π
Degrees	0°	30°	45°	60°	90°	120°	135°	150°	180°	270°	360°

THEOREM Area of a Sector

The area A of the sector of a circle of radius r formed by a central angle of θ radians is

$$A = \frac{1}{2}r^2\theta$$

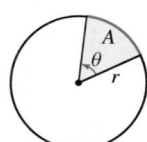

Figure 51 $A = \dfrac{1}{2}r^2\theta$

See Figure 51.

DEFINITION Linear Speed, Angular Speed

Suppose an object moves around a circle of radius r at a constant speed. If s is the distance traveled along the circle in time t, then the **linear speed** v of the object is

$$v = \frac{s}{t}$$

The **angular speed** ω (the Greek letter omega) of an object moving at a constant speed around a circle of radius r is

$$\omega = \frac{\theta}{t}$$

where θ is the angle (measured in radians) swept out in time t.

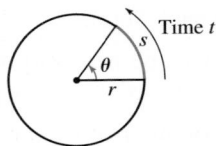

Figure 52 $v = \dfrac{s}{t}$; $\omega = \dfrac{\theta}{t}$

See Figure 52.

Angular speed is used to describe the turning rate of an engine. For example, an engine idling at 900 rpm (revolutions per minute) rotates at an angular speed of

$$900 \; \frac{\text{revolutions}}{\text{minute}} = 900 \; \frac{\text{revolutions}}{\text{minute}} \cdot 2\pi \; \frac{\text{radians}}{\text{revolution}} = 1800\pi \; \frac{\text{radians}}{\text{minute}}$$

> **NOTE** In the formula $s = r\theta$ for the arc length of a circle, the angle θ must be measured in radians.

There is an important relationship between linear speed and angular speed:

$$\text{linear speed} \;=\; v = \frac{s}{t} = \frac{r\,\theta}{t} = r \cdot \frac{\theta}{t} = r\omega$$
$$\uparrow \qquad \uparrow$$
$$s = r\theta \qquad \omega = \frac{\theta}{t}$$

So,

$$\boxed{v = r\omega}$$

where ω is measured in radians per unit time.

When using the equation $v = r\omega$, remember that $v = \frac{s}{t}$ has the dimensions of length per unit of time (such as meters per second or miles per hour), r has the same length dimension as s, and ω has the dimensions of radians per unit of time. If the angular speed is given in terms of *revolutions* per unit of time (as is often the case), be sure to convert it to *radians* per unit of time, using the fact that one revolution $= 2\pi$ radians, before using the formula $v = r\omega$.

② Define and Evaluate Trigonometric Functions

There are two common approaches to trigonometry: One uses right triangles and the other uses the unit circle. We suggest you first review the approach you are most familiar with and then read the other approach. The two approaches are given side-by-side on pages A-30 and A-31.

Right Triangle Approach

Suppose θ is an **acute angle**; that is, $0 < \theta < \frac{\pi}{2}$, as shown in Figure 53(a). Using the angle θ, we form a right triangle, like the one illustrated in Figure 53(b), with hypotenuse of length c and legs of lengths a and b. The three sides of the right triangle can be used to form exactly six ratios:

$$\frac{b}{c} \qquad \frac{a}{c} \qquad \frac{b}{a} \qquad \frac{c}{b} \qquad \frac{c}{a} \qquad \frac{a}{b}$$

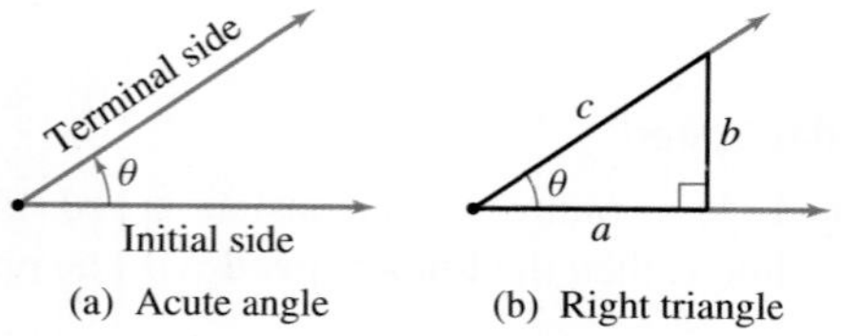

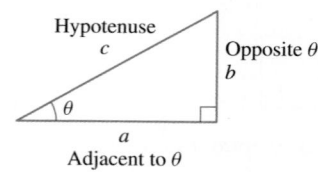

(a) Acute angle   (b) Right triangle

Figure 53

Figure 54 shows the relationship of the sides to the angle θ.

Hypotenuse c
Opposite θ b
θ
a
Adjacent to θ

Figure 54

Unit Circle Approach

Let t be any real number. We position a t-axis perpendicular to the x-axis so that $t = 0$ coincides with the point $(1, 0)$ of the xy-plane, and positive values of t are above the x-axis. See Figure 56(a).

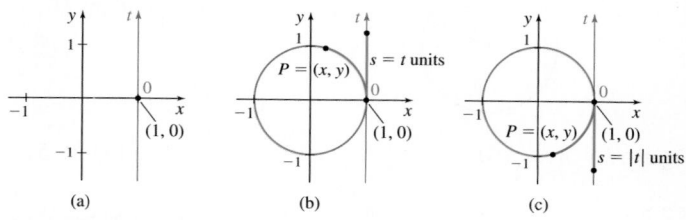

(a)   (b)   (c)

Figure 56

Now look at the unit circle in Figure 56(b). If $t > 0$, we begin at the point $(1, 0)$ on the unit circle and travel $s = t$ units in the counterclockwise direction along the circle to arrive at the point $P = (x, y)$. In this sense, the length $s = t$ units on the t-axis is being **wrapped** around the unit circle.

If $t < 0$, we begin at the point $(1, 0)$ on the unit circle and travel $s = |t|$ units in the clockwise direction along the circle to the point $P = (x, y)$, as shown in Figure 56(c).

If $t > 2\pi$ or if $t < -2\pi$, it will be necessary to travel around the circle more than once before arriving at the point P.

This discussion tells us that, for any real number t, there is a unique point $P = (x, y)$ on the unit circle, called **the point on the unit circle that corresponds to t**. The coordinates of

Because the ratios depend only on the angle θ and not on the triangle itself, each ratio is given a name that involves θ.

DEFINITION   Trigonometric Functions

The six ratios of a right triangle are called the **trigonometric functions of an acute angle** θ and are defined as follows:

Function name	Abbreviation	Value	Relation using words
sine of θ	$\sin \theta$	$\dfrac{b}{c}$	$\dfrac{\text{opposite}}{\text{hypotenuse}}$
cosine of θ	$\cos \theta$	$\dfrac{a}{c}$	$\dfrac{\text{adjacent}}{\text{hypotenuse}}$
tangent of θ	$\tan \theta$	$\dfrac{b}{a}$	$\dfrac{\text{opposite}}{\text{adjacent}}$
cosecant of θ	$\csc \theta$	$\dfrac{c}{b}$	$\dfrac{\text{hypotenuse}}{\text{opposite}}$
secant of θ	$\sec \theta$	$\dfrac{c}{a}$	$\dfrac{\text{hypotenuse}}{\text{adjacent}}$
cotangent of θ	$\cot \theta$	$\dfrac{a}{b}$	$\dfrac{\text{adjacent}}{\text{opposite}}$

To extend the definition of the trigonometric functions to include angles that are not acute, the angle is placed in standard position in a rectangular coordinate system, as shown in Figure 55.

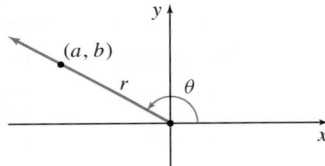

Figure 55

DEFINITION   Trigonometric Functions of θ

Suppose θ is any angle in standard position and $P = (a, b)$ is any point, except the origin, on the terminal side of θ. If $r = \sqrt{a^2 + b^2}$ denotes the distance of the point P from the origin, then the **six trigonometric functions of** θ are defined as the ratios

$$\sin \theta = \frac{b}{r} \qquad \cos \theta = \frac{a}{r} \qquad \tan \theta = \frac{b}{a}$$
$$\csc \theta = \frac{r}{b} \qquad \sec \theta = \frac{r}{a} \qquad \cot \theta = \frac{a}{b}$$

provided no denominator equals 0. If a denominator equals 0, that trigonometric function of the angle θ is not defined.

the point $P = (x, y)$ are used to define the *six trigonometric functions of* t.

DEFINITION   Trigonometric Functions of a Number t

Suppose t is a real number and $P = (x, y)$ is the point on the unit circle that corresponds to t.

- The **sine function** is defined as

$$\sin t = y$$

- The **cosine function** is defined as

$$\cos t = x$$

- If $x \neq 0$, the **tangent function** is defined as

$$\tan t = \frac{y}{x} \qquad x \neq 0$$

- If $y \neq 0$, the **cosecant function** is defined as

$$\csc t = \frac{1}{y} \qquad y \neq 0$$

- If $x \neq 0$, the **secant function** is defined as

$$\sec t = \frac{1}{x} \qquad x \neq 0$$

- If $y \neq 0$, the **cotangent function** is defined as

$$\cot t = \frac{x}{y} \qquad y \neq 0$$

Notice in these definitions that, if $x = 0$, that is, if the point P is on the y-axis, then the tangent function and the secant function are not defined. Also, if $y = 0$, that is, if the point P is on the x-axis, then the cosecant function and the cotangent function are not defined.

Because the unit circle is used in these definitions, the trigonometric functions are sometimes called **circular functions**.

Suppose θ is an angle in standard position, whose terminal side is the ray from the origin through point P. Since the unit circle has radius $r = 1$ unit, then, using the formula for arc length, we find that

$$s = r\theta = \theta$$

So, if $s = |t|$ units, then $\theta = t$ radians and the trigonometric functions of the angle θ are defined as

$$\sin \theta = \sin t \qquad \cos \theta = \cos t \qquad \tan \theta = \tan t$$
$$\csc \theta = \csc t \qquad \sec \theta = \sec t \qquad \cot \theta = \cot t$$

Evaluating Trigonometric Functions

Using properties of right triangles, we can find the values of the six trigonometric functions of $\frac{\pi}{6} = 30°$, $\frac{\pi}{4} = 45°$, and $\frac{\pi}{3} = 60°$ given in Table 6.

TABLE 6

θ (Radians)	θ (Degrees)	$\sin\theta$	$\cos\theta$	$\tan\theta$	$\csc\theta$	$\sec\theta$	$\cot\theta$
$\frac{\pi}{6}$	30°	$\frac{1}{2}$	$\frac{\sqrt{3}}{2}$	$\frac{\sqrt{3}}{3}$	2	$\frac{2\sqrt{3}}{3}$	$\sqrt{3}$
$\frac{\pi}{4}$	45°	$\frac{\sqrt{2}}{2}$	$\frac{\sqrt{2}}{2}$	1	$\sqrt{2}$	$\sqrt{2}$	1
$\frac{\pi}{3}$	60°	$\frac{\sqrt{3}}{2}$	$\frac{1}{2}$	$\sqrt{3}$	$\frac{2\sqrt{3}}{3}$	2	$\frac{\sqrt{3}}{3}$

The values of the trigonometric functions at the quadrantal angles

$$0 = 0° \qquad \frac{\pi}{2} = 90° \qquad \frac{3\pi}{2} = 270° \qquad 2\pi = 360°$$

are given in Table 7.

TABLE 7

Radians	Degrees	$\sin\theta$	$\cos\theta$	$\tan\theta$	$\csc\theta$	$\sec\theta$	$\cot\theta$
0	0°	0	1	0	Not defined	1	Not defined
$\frac{\pi}{2}$	90°	1	0	Not defined	1	Not defined	0
π	180°	0	-1	0	Not defined	-1	Not defined
$\frac{3\pi}{2}$	270°	-1	0	Not defined	-1	Not defined	0
2π	360°	0	1	0	Not defined	1	Not defined

Table 8 lists the signs of the six trigonometric functions for each quadrant.

TABLE 8

Quadrant	$\sin\theta$, $\csc\theta$	$\cos\theta$, $\sec\theta$	$\tan\theta$, $\cot\theta$	Conclusion
I	Positive	Positive	Positive	All trigonometric functions are positive.
II	Positive	Negative	Negative	Only sine and its reciprocal, cosecant, are positive.
III	Negative	Negative	Positive	Only tangent and its reciprocal, cotangent, are positive.
IV	Negative	Positive	Negative	Only cosine and its reciprocal, secant, are positive.

For nonacute angles θ that lie in a quadrant, the acute angle formed by the terminal side of θ and the x-axis, called a **reference angle**, is often used.

Figure 57 illustrates the reference angle for some general angles θ. Note that a reference angle is always an acute angle.

Although there are formulas for calculating reference angles, it is usually easier to find the reference angle for a given angle by making a quick sketch of the angle. The advantage of using reference angles is that, except for the correct sign, the values of the trigonometric functions of a general angle θ equal the values of the trigonometric functions of its reference angle.

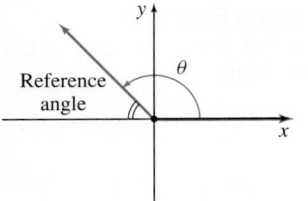
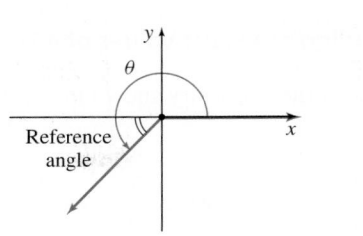
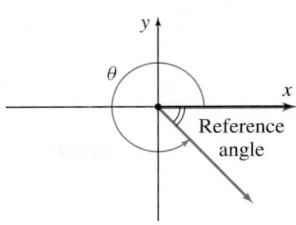
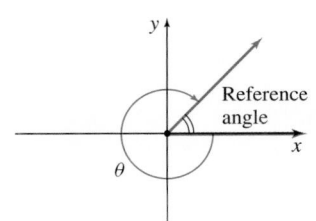

Figure 57

THEOREM Reference Angles

If θ is an angle that lies in a quadrant and if A is its reference angle, then

$$
\begin{array}{ccc}
\sin \theta = \pm \sin A & \cos \theta = \pm \cos A & \tan \theta = \pm \tan A \\
\csc \theta = \pm \csc A & \sec \theta = \pm \sec A & \cot \theta = \pm \cot A
\end{array}
$$

where the $+$ or $-$ sign depends on the quadrant in which θ lies.

EXAMPLE 1   **Using Reference Angles**

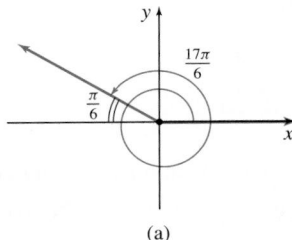

(a)

Find the exact value of: **(a)** $\cos \dfrac{17\pi}{6}$ **(b)** $\tan \left(-\dfrac{\pi}{3}\right)$

Solution

(a) Refer to Figure 58(a). The angle $\dfrac{17\pi}{6}$ is in quadrant II, where the cosine function is negative. The reference angle for $\dfrac{17\pi}{6}$ is $\dfrac{\pi}{6}$. Since $\cos \dfrac{\pi}{6} = \dfrac{\sqrt{3}}{2}$,

$$
\cos \frac{17\pi}{6} = - \cos \frac{\pi}{6} = - \frac{\sqrt{3}}{2}
$$

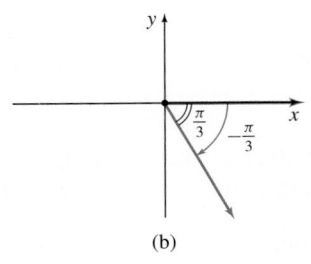

(b)

Figure 58

(b) Refer to Figure 58(b). The angle $-\dfrac{\pi}{3}$ is in quadrant IV, where the tangent function is negative. The reference angle for $-\dfrac{\pi}{3}$ is $\dfrac{\pi}{3}$. Since $\tan \dfrac{\pi}{3} = \sqrt{3}$,

$$
\tan \left(-\frac{\pi}{3}\right) = - \tan \frac{\pi}{3} = - \sqrt{3}
$$ ∎

The use of symmetry provides information about integer multiples of the angles $\dfrac{\pi}{4} = 45°$, $\dfrac{\pi}{6} = 30°$, and $\dfrac{\pi}{3} = 60°$. See the unit circles in Figure 59(a)–(c).

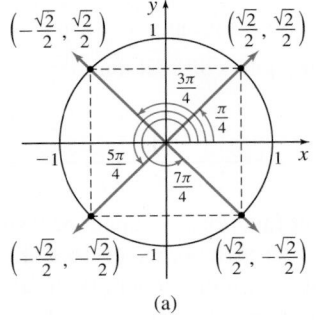

(a)

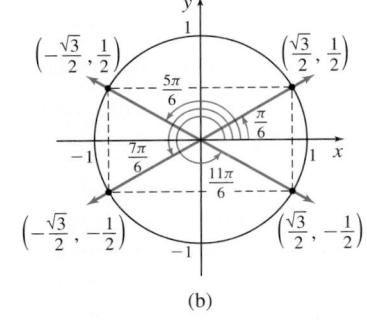

(b)

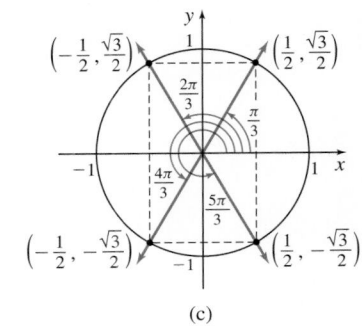

(c)

Figure 59

EXAMPLE 2 **Finding the Exact Values of a Trigonometry Function**

Use reference angles or the symmetry shown in Figure 59 to obtain:

(a) $\sin \dfrac{7\pi}{4} = -\dfrac{\sqrt{2}}{2}$ (b) $\cos \dfrac{7\pi}{6} = -\dfrac{\sqrt{3}}{2}$ (c) $\tan \dfrac{2\pi}{3} = \dfrac{\dfrac{\sqrt{3}}{2}}{-\dfrac{1}{2}} = -\sqrt{3}$

(d) $\cos\left(-\dfrac{3\pi}{4}\right) = -\dfrac{\sqrt{2}}{2}$ (e) $\sin\left(-\dfrac{\pi}{6}\right) = -\dfrac{1}{2}$ (f) $\sin \dfrac{7\pi}{3} = \dfrac{\sqrt{3}}{2}$ ∎

---

③ Determine the Domain and the Range of the Trigonometric Functions

Suppose θ is an angle in standard position and $P = (x, y)$ is the point on the unit circle corresponding to θ. Then the six trigonometric functions are

$$
\begin{array}{lll}
\sin \theta = y & \cos \theta = x & \tan \theta = \dfrac{y}{x} \\[2mm]
\csc \theta = \dfrac{1}{y} & \sec \theta = \dfrac{1}{x} & \cot \theta = \dfrac{x}{y}
\end{array}
$$

For $\sin \theta$ and $\cos \theta$, there is no concern about dividing by 0, so θ can be any angle. It follows that the domain of the sine and cosine functions is all real numbers.

> • The domain of the sine function is all real numbers.
> • The domain of the cosine function is all real numbers.

For the tangent and secant functions, x cannot be 0 since this results in division by 0. On the unit circle, there are two such points, $(0, 1)$ and $(0, -1)$. These two points correspond to $\dfrac{\pi}{2} = 90°$ and $\dfrac{3\pi}{2} = 270°$ or, more generally, to any angle that is an odd integer multiple of $\dfrac{\pi}{2}$, such as $\pm\dfrac{\pi}{2} = \pm90°$, $\pm\dfrac{3\pi}{2} = \pm270°$, $\pm\dfrac{5\pi}{2} = \pm450°$, and so on. These angles must be excluded from the domain of the tangent function and the domain of the secant function.

> • The domain of the tangent function is all real numbers, except odd integer multiples of $\dfrac{\pi}{2} = 90°$.
> • The domain of the secant function is all real numbers, except odd integer multiples of $\dfrac{\pi}{2} = 90°$.

For the cotangent and cosecant functions, y cannot be 0 since this results in division by 0. On the unit circle, there are two such points, $(1, 0)$ and $(-1, 0)$. These two points correspond to $0 = 0°$ and $\pi = 180°$ or, more generally, to any angle that is an integer multiple of π, such as $0 = 0°$, $\pm\pi = \pm180°$, $\pm2\pi = \pm360°$, $\pm3\pi = \pm540°$, and so on. These angles must be excluded from the domain of the cotangent function and the domain of the cosecant function.

- The domain of the cotangent function is all real numbers, except integer multiples of $\pi = 180°$.
- The domain of the cosecant function is all real numbers, except integer multiples of $\pi = 180°$.

Now we investigate the range of the six trigonometric functions. If $P = (x, y)$ is the point on the unit circle corresponding to the angle θ, then it follows that $-1 \leq x \leq 1$ and $-1 \leq y \leq 1$. Since $\sin \theta = y$ and $\cos \theta = x$, we have

$$-1 \leq \sin \theta \leq 1 \qquad \text{and} \qquad -1 \leq \cos \theta \leq 1$$

The range of both the sine function and the cosine function is the closed interval $[-1, 1]$. Using absolute value notation, we have $|\sin \theta| \leq 1$ and $|\cos \theta| \leq 1$.

If θ is not an integer multiple of $\pi = 180°$, then $\csc \theta = \dfrac{1}{y}$. Since $y = \sin \theta$ and $|y| = |\sin \theta| \leq 1$, it follows that $|\csc \theta| = \dfrac{1}{|\sin \theta|} = \dfrac{1}{|y|} \geq 1$. That is, $\dfrac{1}{y} \leq -1$ or $\dfrac{1}{y} \geq 1$. Then the range of the cosecant function consists of all real numbers in the set $(-\infty, -1] \cup [1, \infty)$. That is,

$$\csc \theta \leq -1 \qquad \text{or} \qquad \csc \theta \geq 1$$

If θ is not an odd integer multiple of $\dfrac{\pi}{2} = 90°$, then $\sec \theta = \dfrac{1}{x}$. Since $x = \cos \theta$ and $|x| = |\cos \theta| \leq 1$, it follows that $|\sec \theta| = \dfrac{1}{|\cos \theta|} = \dfrac{1}{|x|} \geq 1$. That is, $\dfrac{1}{x} \leq -1$ or $\dfrac{1}{x} \geq 1$. Then the range of the secant function consists of all real numbers in the set $(-\infty, -1] \cup [1, \infty)$. That is,

$$\sec \theta \leq -1 \qquad \text{or} \qquad \sec \theta \geq 1$$

The range of both the tangent function and the cotangent function is all real numbers:

$$-\infty < \tan \theta < \infty \qquad \text{and} \qquad -\infty < \cot \theta < \infty$$

4 Use Basic Trigonometry Identities

DEFINITION Identity

Two expressions a and b involving a variable x are **identically equal** if

$$a = b$$

for every value of x for which both expressions are defined. Such an equation is referred to as an **identity**. An equation that is not an identity is called a **conditional equation**.

For example, the equation $2x + 3 = x$ is a conditional equation because it is true only for $x = -3$. However, the equation $x^2 + 2x = (x + 1)^2 - 1$ is true for any value of x, so it is an identity.

The basic trigonometry identities listed on page A-36 are consequences of the definitions of the six trigonometric functions.

Basic Trigonometry Identities

- **Quotient Identities**

$$\tan\theta = \frac{\sin\theta}{\cos\theta} \qquad \cot\theta = \frac{\cos\theta}{\sin\theta}$$

- **Reciprocal Identities**

$$\csc\theta = \frac{1}{\sin\theta} \qquad \sec\theta = \frac{1}{\cos\theta} \qquad \cot\theta = \frac{1}{\tan\theta}$$

- **Pythagorean Identities**

$$\sin^2\theta + \cos^2\theta = 1 \qquad \tan^2\theta + 1 = \sec^2\theta \qquad \cot^2\theta + 1 = \csc^2\theta$$

- **Even/Odd Identities**

$$\cos(-\theta) = \cos\theta \qquad \sin(-\theta) = -\sin\theta \qquad \tan(-\theta) = -\tan\theta$$

EXAMPLE 3   Using Basic Trigonometry Identities

(a) $\tan\dfrac{\pi}{9} - \dfrac{\sin\dfrac{\pi}{9}}{\cos\dfrac{\pi}{9}} = \tan\dfrac{\pi}{9} - \tan\dfrac{\pi}{9} = 0$

$$\uparrow$$
$$\frac{\sin\theta}{\cos\theta} = \tan\theta$$

(b) $\sin^2\dfrac{\pi}{12} + \dfrac{1}{\sec^2\dfrac{\pi}{12}} = \sin^2\dfrac{\pi}{12} + \cos^2\dfrac{\pi}{12} = 1$

$$\uparrow \qquad\qquad \uparrow$$
$$\cos\theta = \frac{1}{\sec\theta} \qquad \sin^2\theta + \cos^2\theta = 1$$

■

EXAMPLE 4   Using Basic Trigonometry Identities

Given that $\sin\theta = \dfrac{1}{3}$ and $\cos\theta < 0$, find the exact value of each of the remaining five trigonometric functions.

Solution

Begin by solving the identity $\sin^2\theta + \cos^2\theta = 1$ for $\cos\theta$.

$$\sin^2\theta + \cos^2\theta = 1$$
$$\cos^2\theta = 1 - \sin^2\theta$$
$$\cos\theta = \pm\sqrt{1 - \sin^2\theta}$$

Because $\cos\theta < 0$, we choose the negative value of the radical and use the fact that $\sin\theta = \dfrac{1}{3}$.

$$\cos\theta = -\sqrt{1 - \sin^2\theta} = -\sqrt{1 - \frac{1}{9}} = -\sqrt{\frac{8}{9}} = -\frac{2\sqrt{2}}{3}$$

$$\uparrow$$
$$\sin\theta = \frac{1}{3}$$

The values of $\sin\theta$ and $\cos\theta$ are now known, so we use the quotient and reciprocal identities to obtain

$$\tan\theta = \frac{\sin\theta}{\cos\theta} = \frac{\frac{1}{3}}{-\frac{2\sqrt{2}}{3}} = -\frac{1}{2\sqrt{2}} = -\frac{\sqrt{2}}{4} \qquad \cot\theta = \frac{1}{\tan\theta} = -2\sqrt{2}$$

$$\sec\theta = \frac{1}{\cos\theta} = \frac{1}{-\frac{2\sqrt{2}}{3}} = -\frac{3}{2\sqrt{2}} = -\frac{3\sqrt{2}}{4} \qquad \csc\theta = \frac{1}{\sin\theta} = \frac{1}{\frac{1}{3}} = 3 \qquad \blacksquare$$

5 Use Sum and Difference, Double-Angle and Half-Angle, and Sum-to-Product and Product-to-Sum Formulas

THEOREM Sum and Difference Formulas for Sine, Cosine, and Tangent

- $\sin(A+B) = \sin A\cos B + \cos A\sin B$
- $\sin(A-B) = \sin A\cos B - \cos A\sin B$
- $\cos(A+B) = \cos A\cos B - \sin A\sin B$
- $\cos(A-B) = \cos A\cos B + \sin A\sin B$
- $\tan(A+B) = \dfrac{\tan A + \tan B}{1 - \tan A\tan B}$
- $\tan(A-B) = \dfrac{\tan A - \tan B}{1 + \tan A\tan B}$

EXAMPLE 5 Using Trigonometry Identities

If $\sin A = \dfrac{4}{5}$, $\dfrac{\pi}{2} < A < \pi$, and $\sin B = -\dfrac{2\sqrt{5}}{5}$, $\pi < B < \dfrac{3\pi}{2}$, find the exact value of:

(a) $\cos A$ **(b)** $\cos B$ **(c)** $\cos(A+B)$ **(d)** $\sin(A+B)$

Solution

(a) Use a Pythagorean identity to find $\cos A$.

$$\cos A = -\sqrt{1-\sin^2 A} = -\sqrt{1-\left(\frac{4}{5}\right)^2} = -\sqrt{1-\frac{16}{25}} = -\sqrt{\frac{9}{25}} = -\frac{3}{5}$$

$\uparrow$
A is in quadrant II;
$\cos A < 0$.

(b) We also find $\cos B$ using a Pythagorean identity.

$$\cos B = -\sqrt{1-\sin^2 B} = -\sqrt{1-\frac{4}{5}} = -\sqrt{\frac{1}{5}} = -\frac{\sqrt{5}}{5}$$

(c) Now use the results from (a) and (b) and a sum formula to find $\cos(A+B)$.

$$\cos(A+B) = \cos A\cos B - \sin A\sin B$$

$$= \left(-\frac{3}{5}\right)\left(-\frac{\sqrt{5}}{5}\right) - \left(\frac{4}{5}\right)\left(-\frac{2\sqrt{5}}{5}\right) = \frac{11\sqrt{5}}{25}$$

(d) Again use a sum formula to find $\sin(A + B)$.

$$\sin(A + B) = \sin A \cos B + \cos A \sin B = \left(\frac{4}{5}\right)\left(-\frac{\sqrt{5}}{5}\right) + \left(-\frac{3}{5}\right)\left(-\frac{2\sqrt{5}}{5}\right) = \frac{2\sqrt{5}}{25}$$

∎

THEOREM Double-Angle Formulas

- $\sin(2\theta) = 2 \sin \theta \cos \theta$ • $\cos(2\theta) = \cos^2 \theta - \sin^2 \theta$ • $\tan(2\theta) = \dfrac{2 \tan \theta}{1 - \tan^2 \theta}$
- $\cos(2\theta) = 1 - 2 \sin^2 \theta$ • $\cos(2\theta) = 2 \cos^2 \theta - 1$

The double-angle formulas in the first row are derived directly from the sum formulas by letting $\theta = A = B$. A Pythagorean identity is used to obtain those in the second row.

By rearranging the double-angle formulas, other identities can be obtained that are used in calculus.

- $\sin^2 \theta = \dfrac{1 - \cos(2\theta)}{2}$ • $\cos^2 \theta = \dfrac{1 + \cos(2\theta)}{2}$

If, in the double-angle formulas, we replace θ with $\frac{1}{2}\theta$, we obtain these identities.

- $\sin^2 \dfrac{\theta}{2} = \dfrac{1 - \cos \theta}{2}$ • $\cos^2 \dfrac{\theta}{2} = \dfrac{1 + \cos \theta}{2}$ • $\tan^2 \dfrac{\theta}{2} = \dfrac{1 - \cos \theta}{1 + \cos \theta}$

Taking the square root produces the **half-angle formulas**.

- $\sin \dfrac{\theta}{2} = \pm\sqrt{\dfrac{1 - \cos \theta}{2}}$ • $\cos \dfrac{\theta}{2} = \pm\sqrt{\dfrac{1 + \cos \theta}{2}}$ • $\tan \dfrac{\theta}{2} = \pm\sqrt{\dfrac{1 - \cos \theta}{1 + \cos \theta}}$

where the $+$ or $-$ sign is determined by the quadrant of the angle $\dfrac{\theta}{2}$.

The **product-to-sum** and **sum-to-product identities** are also used in calculus. They are a result of combining the sum and difference formulas.

- $\sin A \, \cos B = \dfrac{1}{2}[\sin(A + B) + \sin(A - B)]$

- $\cos A \, \cos B = \dfrac{1}{2}[\cos(A + B) + \cos(A - B)]$

- $\sin A \, \sin B = \dfrac{1}{2}[\cos(A - B) - \cos(A + B)]$

6 Solve Triangles Using the Law of Sines and the Law of Cosines

If no angle of a triangle is a right angle, the triangle is called **oblique**. An oblique triangle either has three acute angles or two acute angles and one obtuse angle (an angle between $\dfrac{\pi}{2}$ and π). See Figure 60.

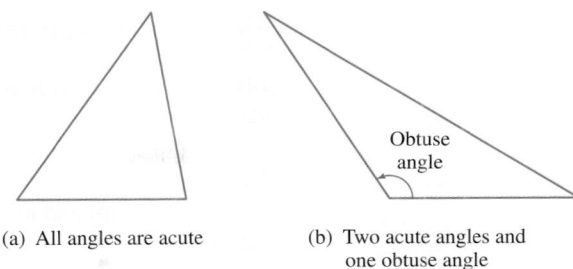

(a) All angles are acute   (b) Two acute angles and
                                      one obtuse angle

Figure 60 Two oblique triangles.

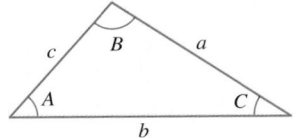

Figure 61

> **NOTE** To solve an oblique triangle, you must know the length of at least one side. Knowing only angles results in a family of similar triangles.

In the discussion that follows, we label an oblique triangle so that side a is opposite angle A, side b is opposite angle B, and side c is opposite angle C, as shown in Figure 61.

To **solve an oblique triangle** means to find the lengths of its sides and the measures of its angles. To do this, we need to know the length of one side along with (i) two angles; (ii) one angle and one other side; or (iii) the other two sides. There are four possibilities to consider:

Case 1: One side and two angles are known (ASA or SAA).

Case 2: Two sides and the angle opposite one of them are known (SSA).

Case 3: Two sides and the included angle are known (SAS).

Case 4: Three sides are known (SSS).

Figure 62 illustrates the four cases.

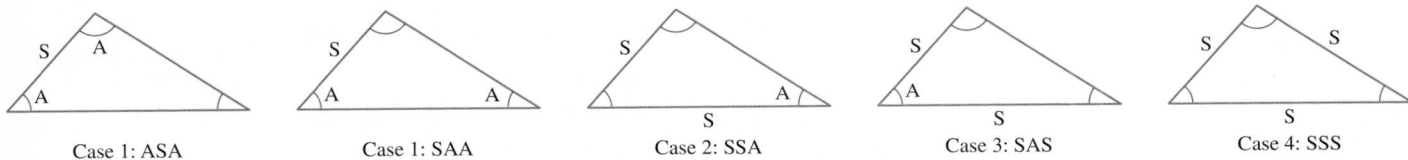

Case 1: ASA   Case 1: SAA   Case 2: SSA   Case 3: SAS   Case 4: SSS

Figure 62

The *Law of Sines* is used to solve triangles for which Case 1 or 2 holds.

> **THEOREM Law of Sines**
>
> For a triangle with sides a, b, c and opposite angles A, B, C, respectively,
>
> $$\frac{\sin A}{a} = \frac{\sin B}{b} = \frac{\sin C}{c} \qquad (1)$$

The Law of Sines actually consists of three equalities:

$$\frac{\sin A}{a} = \frac{\sin B}{b} \qquad \frac{\sin A}{a} = \frac{\sin C}{c} \qquad \frac{\sin B}{b} = \frac{\sin C}{c}$$

Formula (1) is a compact way to write these three equations.

For Case 1 (ASA or SAA), two angles are given. The third angle can be found using the fact that the sum of the angles in a triangle measures $180°$. Then use the Law of Sines (twice) to find the two missing sides.

Case 2 (SSA), which applies to triangles for which two sides and the angle opposite one of them are known, is referred to as the **ambiguous case**, because the information can result in one triangle, two triangles, or no triangle at all.

EXAMPLE 6   **Using the Law of the Sines to Solve a Triangle**

Solve the triangle with side $a = 6$, side $b = 8$, and angle $A = 35°$, which is opposite side a.

Solution

Because two sides and an opposite angle are known (SSA), we use the Law of Sines to find angle B.

$$\frac{\sin A}{a} = \frac{\sin B}{b}$$

Since $a = 6$, $b = 8$, and $A = 35°$, we have

$$\frac{\sin 35°}{6} = \frac{\sin B}{8}$$

$$\sin B = \frac{8 \sin 35°}{6} \approx 0.765$$

$\sin B \approx 0.765$ has two solutions

$$B_1 \approx 49.9° \quad \text{or} \quad B_2 \approx 180° - 49.9° = 130.1°$$

For both choices of B, we have $A + B < 180°$. So, there are two triangles, one containing the angle $B_1 \approx 49.9°$ and the other containing the angle $B_2 \approx 130.1°$. The third angle C is either

$$C_1 = 180° - A - B_1 \approx 95.1° \qquad \text{or} \qquad C_2 = 180° - A - B \approx 14.9°$$
$$\uparrow \qquad\qquad\qquad\qquad\qquad\qquad \uparrow$$
$$A = 35° \qquad\qquad\qquad\qquad\qquad A = 35°$$
$$B_1 = 49.9° \qquad\qquad\qquad\qquad B_2 = 130.1°$$

The third side c satisfies the Law of Sines, so

$$\frac{\sin A}{a} = \frac{\sin C_1}{c_1} \qquad\qquad\qquad \frac{\sin A}{a} = \frac{\sin C_2}{c_2}$$

$$\frac{\sin 35°}{6} = \frac{\sin 95.1°}{c_1} \qquad\qquad\qquad \frac{\sin 35°}{6} = \frac{\sin 14.9°}{c_2}$$

$$c_1 = \frac{6 \sin 95.1°}{\sin 35°} \approx 10.42 \qquad\qquad c_2 = \frac{6 \sin 14.9°}{\sin 35°} \approx 2.69$$

The two solved triangles are illustrated in Figure 63. ∎

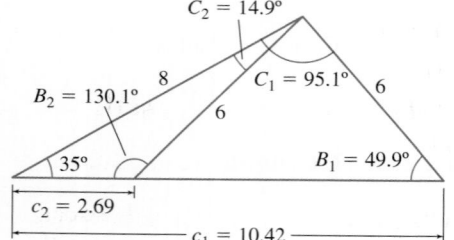

Figure 63

Cases 3 (SAS) and 4 (SSS) are solved using the *Law of Cosines*.

THEOREM   **Law of Cosines**

For a triangle with sides a, b, c and opposite angles A, B, C, respectively,

$$\boxed{\begin{aligned} & \bullet \;\; c^2 = a^2 + b^2 - 2ab \cos C \\ & \bullet \;\; b^2 = a^2 + c^2 - 2ac \cos B \\ & \bullet \;\; a^2 = b^2 + c^2 - 2bc \cos A \end{aligned}}$$

NOTE If the triangle is a right triangle, the Law of Cosines reduces to the Pythagorean Theorem.

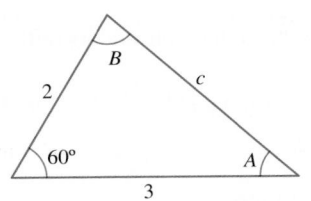

Figure 64

EXAMPLE 7 Using the Law of Cosines to Solve a Triangle

Solve the triangle: $a = 2$, $b = 3$, $C = 60°$ shown in Figure 64.

Solution

Because two sides a and b and the included angle $C = 60°$ (SAS) are known, use the Law of Cosines to find the third side c.

$$c^2 = a^2 + b^2 - 2ab \cos C = 2^2 + 3^2 - 2 \cdot 2 \cdot 3 \cdot \cos 60° = 13 - \left(12 \cdot \frac{1}{2} \right) = 7$$

$$c = \sqrt{7}$$

Side c is of length $\sqrt{7}$. To find either the angle A or B, use the Law of Cosines. For A,

$$a^2 = b^2 + c^2 - 2bc \cos A$$

$$2bc \cos A = b^2 + c^2 - a^2$$

$$\cos A = \frac{b^2 + c^2 - a^2}{2bc} = \frac{9 + 7 - 4}{2 \cdot 3\sqrt{7}} = \frac{12}{6\sqrt{7}} = \frac{2\sqrt{7}}{7}$$

$$A \approx 40.9°$$

Then to find the third angle, use the fact that the sum of the angles of a triangle, using degrees, equals $180°$. That is,

$$40.9° + B + 60° = 180°$$

$$B = 79.1° \quad \blacksquare$$

A.5 Polar Coordinates Used in Calculus: AP® Calculus, BC Only

AP® EXAM TIP

Appendix A.5 contains precalculus mathematics needed for Polar Equations, a BC Calculus only topic, discussed in Chapter 9 (Unit 9).

OBJECTIVES *When you finish this section, you should be able to:*

1 **Plot points using polar coordinates (p. A-41)**
2 **Convert between rectangular coordinates and polar coordinates (p. A-44)**
3 **Identify and graph polar equations (p. A-47)**

In a rectangular coordinate system, there are two perpendicular axes, one horizontal (the x-axis) and one vertical (the y-axis). The point of intersection of the axes is the origin and is labeled O. A point is represented by an ordered pair of numbers (x, y), where x and y equal the signed distance of the point from the y-axis and the x-axis, respectively.

A **polar coordinate system** is constructed by selecting a point O, called the **pole,** and a ray with its vertex at the pole, called the **polar axis**. It is customary to have the pole coincide with the origin and the polar axis coincide with the positive x-axis of the rectangular coordinate system, as shown in Figure 65.

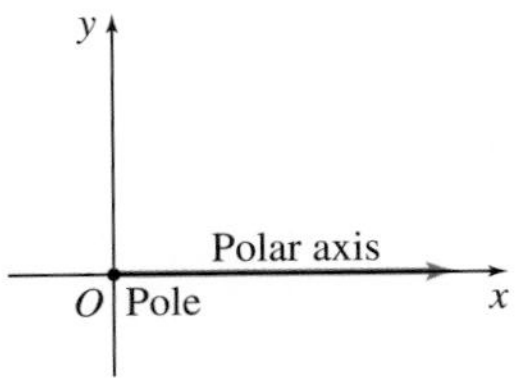

Figure 65

① Plot Points Using Polar Coordinates

A point P in the polar coordinate system is represented by an ordered pair of numbers (r, θ), called the **polar coordinates** of P. If $r > 0$, then r is the distance of the point from the pole (the origin), and θ is an angle (measured in radians or degrees) whose initial side is the polar axis (the positive x-axis) and whose terminal side is a ray from the pole through the point P. See Figure 66.

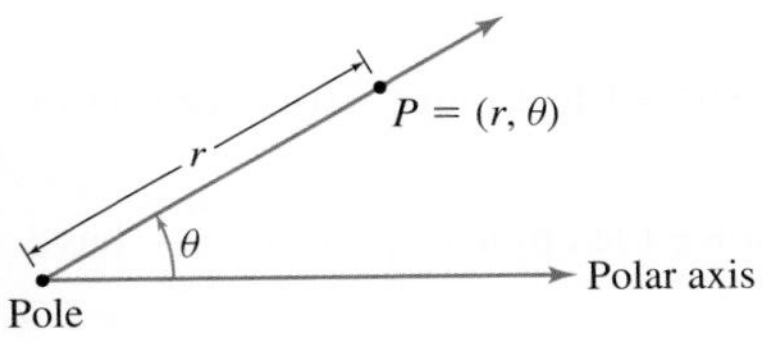

Figure 66 Polar coordinates.

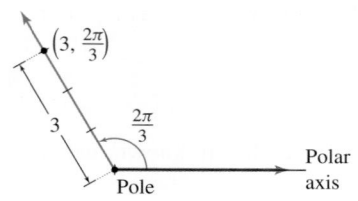

Figure 67

NOTE The points (r, θ) and $(-r, \theta)$ are reflections about the pole.

As an example, suppose the polar coordinates of a point are $\left(3, \dfrac{2\pi}{3}\right)$. We locate the point by drawing an angle of $\dfrac{2\pi}{3}$ radians with its vertex at the pole and its initial side along the polar axis. Then $\left(3, \dfrac{2\pi}{3}\right)$ is the point on the terminal side of the angle that is 3 units from the pole. See Figure 67.

- If $r = 0$, then the point $P = (0, \theta)$ is at the pole for any θ.
- If $r < 0$, the point $P = (r, \theta)$ is *not* on the terminal side of θ. Instead, we extend the terminal side of θ in the opposite direction. The point P is on this extended ray a distance $|r|$ from the pole. See Figure 68.

So, the point $\left(-3, \dfrac{2\pi}{3}\right)$ is located 3 units from the pole on the extension in the opposite direction of the ray that forms the angle $\dfrac{2\pi}{3}$ with the polar axis. See Figure 69.

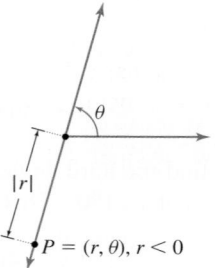

Figure 68

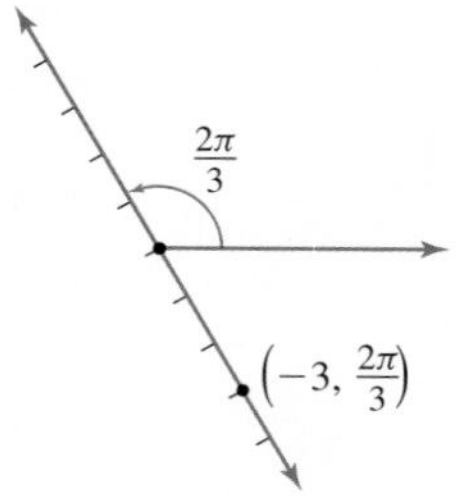

Figure 69

EXAMPLE 1   Plotting Points Using Polar Coordinates

Plot the points with the following polar coordinates:

(a) $\left(3, \dfrac{5\pi}{3}\right)$ (b) $\left(2, -\dfrac{\pi}{4}\right)$ (c) $(3, 0)$ (d) $\left(-2, \dfrac{\pi}{4}\right)$

Solution

Figure 70 shows the points.

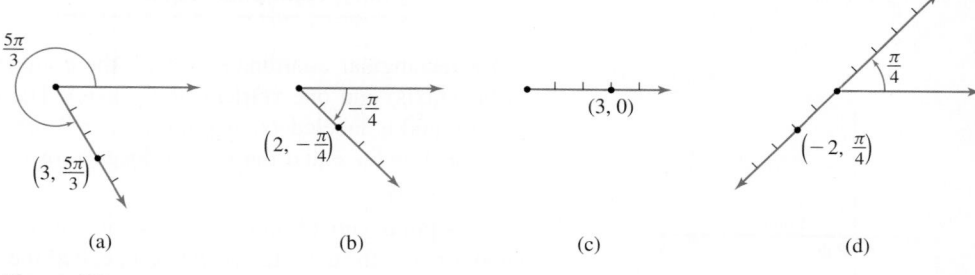

(a)   (b)   (c)   (d)

Figure 70

NOW WORK   Problems 13 and 17.

The polar coordinates $\left(3, \dfrac{2\pi}{3}\right)$ of the point P shown in Figure 71(a) are one of many possible polar coordinates of P. For example, since the angles $\dfrac{8\pi}{3}$ and $-\dfrac{4\pi}{3}$ have the same terminal side as the angle $\dfrac{2\pi}{3}$, the point P can be represented by any of the polar coordinates $\left(3, \dfrac{2\pi}{3}\right)$, $\left(3, \dfrac{8\pi}{3}\right)$, and $\left(3, -\dfrac{4\pi}{3}\right)$, as shown in Figures 71(a)–(c).

The point $\left(3, \dfrac{2\pi}{3}\right)$ can also be represented by the polar coordinates $\left(-3, -\dfrac{\pi}{3}\right)$, as illustrated in Figure 71(d).

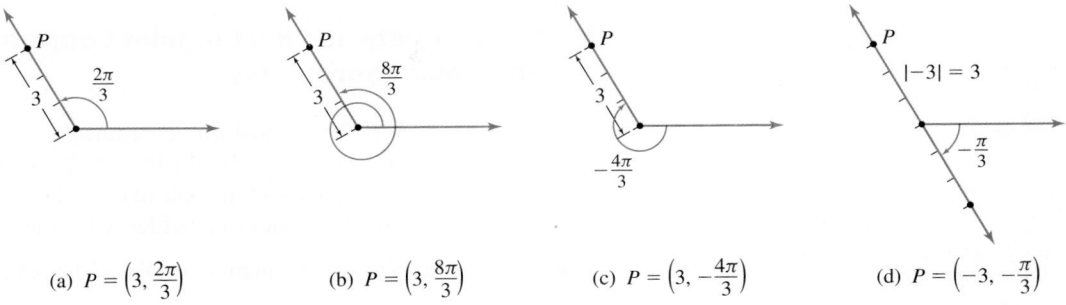

(a) $P = \left(3, \frac{2\pi}{3}\right)$ (b) $P = \left(3, \frac{8\pi}{3}\right)$ (c) $P = \left(3, -\frac{4\pi}{3}\right)$ (d) $P = \left(-3, -\frac{\pi}{3}\right)$

Figure 71

So, there is a major difference between rectangular and polar coordinate systems. In a rectangular system, each point in the plane corresponds to exactly one pair of rectangular coordinates; in a polar coordinate system, every point in the plane can be represented by infinitely many polar coordinates.

EXAMPLE 2   Plotting a Point Using Polar Coordinates

Plot the point P whose polar coordinates are $\left(-2, -\dfrac{3\pi}{4}\right)$. Then find three other polar coordinates of the same point with the properties:

(a) $r > 0, \quad 0 < \theta < 2\pi$ **(b)** $r > 0, \quad -2\pi < \theta < 0$ **(c)** $r < 0, \quad 0 < \theta < 2\pi$

Solution

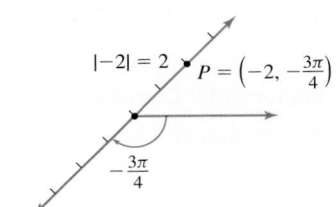

Figure 72

The point $\left(-2, -\dfrac{3\pi}{4}\right)$ is located by first drawing the angle $-\dfrac{3\pi}{4}$. Then P is on the extension of the terminal side of θ in the opposite direction a distance 2 units from the pole, as shown in Figure 72.

(a) The point $P = (r, \theta)$, $r > 0$, $0 < \theta < 2\pi$ is $\left(2, \dfrac{\pi}{4}\right)$, as shown in Figure 73(a).

(b) The point $P = (r, \theta)$, $r > 0$, $-2\pi < \theta < 0$ is $\left(2, -\dfrac{7\pi}{4}\right)$, as shown in Figure 73(b).

(c) The point $P = (r, \theta)$, $r < 0$, $0 < \theta < 2\pi$ is $\left(-2, \dfrac{5\pi}{4}\right)$, as shown in Figure 73(c).

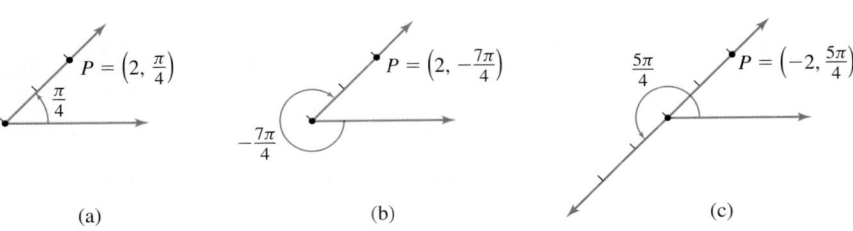

(a)   (b)   (c)

Figure 73   ■

NOW WORK   Problem 19.

Summary

- A point P with polar coordinates (r, θ) can be represented by either

  $(r, \theta + 2n\pi)$ or $(-r, \theta + (2n+1)\pi)$ n an integer

- The polar coordinates of the pole are $(0, \theta)$, where θ is any angle.

2 Convert between Rectangular Coordinates and Polar Coordinates

It is sometimes useful to transform coordinates or equations in rectangular form into polar form, or vice versa. To do this, recall that the origin in the rectangular coordinate system coincides with the pole in the polar coordinate system and the positive x-axis in the rectangular system coincides with the polar axis in the polar system. The positive y-axis in the rectangular system is the ray $\theta = \dfrac{\pi}{2}$ in the polar system.

Suppose a point P has the polar coordinates (r, θ) and the rectangular coordinates (x, y), as shown in Figure 74. If $r > 0$, then P is on the terminal side of θ, and

$$\cos \theta = \frac{x}{r} \qquad \text{and} \qquad \sin \theta = \frac{y}{r}$$

If $r < 0$, then $P = (r, \theta)$ can be represented as $(-r, \pi + \theta)$, where $-r > 0$. Since

$$\cos(\theta + \pi) = -\cos\theta = \frac{x}{-r} \qquad \text{and} \qquad \sin(\theta + \pi) = -\sin\theta = \frac{y}{-r}$$

then whether $r > 0$ or $r < 0$,

$$x = r \cos \theta \qquad \text{and} \qquad y = r \sin \theta$$

If $r = 0$, then P is the pole and the same relationships hold.

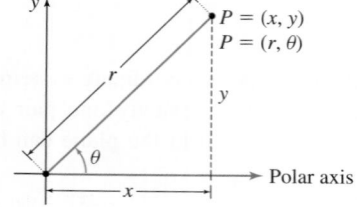

Figure 74

NEED TO REVIEW? The definitions of the trigonometric functions and their properties are discussed in Appendix A.4, pp. A-27–A-41.

> **THEOREM Conversion from Polar Coordinates to Rectangular Coordinates**
>
> If P is a point with polar coordinates (r, θ), then the rectangular coordinates (x, y) of P are given by
>
> $$x = r \cos \theta \qquad \text{and} \qquad y = r \sin \theta$$

EXAMPLE 3 Converting from Polar Coordinates to Rectangular Coordinates

Find the rectangular coordinates of each point whose polar coordinates are:

(a) $\left(4, \dfrac{\pi}{3}\right)$ **(b)** $\left(-2, \dfrac{3\pi}{4}\right)$ **(c)** $\left(-3, -\dfrac{5\pi}{6}\right)$

Solution

(a) Use the equations $x = r \cos \theta$ and $y = r \sin \theta$ with $r = 4$ and $\theta = \dfrac{\pi}{3}$.

$$x = 4 \cos \frac{\pi}{3} = 4 \cdot \frac{1}{2} = 2 \qquad \text{and} \qquad y = 4 \sin \frac{\pi}{3} = 4 \cdot \frac{\sqrt{3}}{2} = 2\sqrt{3}$$

The rectangular coordinates are $(2, 2\sqrt{3})$.

(b) Use the equations $x = r\cos\theta$ and $y = r\sin\theta$ with $r = -2$ and $\theta = \dfrac{3\pi}{4}$.

$$x = -2\cos\frac{3\pi}{4} = -2\left(-\frac{\sqrt{2}}{2}\right) = \sqrt{2} \quad \text{and} \quad y = -2\sin\frac{3\pi}{4} = -2\cdot\frac{\sqrt{2}}{2} = -\sqrt{2}$$

The rectangular coordinates are $(\sqrt{2},\, -\sqrt{2})$.

(c) Use the equations $x = r\cos\theta$ and $y = r\sin\theta$ with $r = -3$ and $\theta = -\dfrac{5\pi}{6}$.

$$x = -3\cos\left(-\frac{5\pi}{6}\right) = -3\left(-\frac{\sqrt{3}}{2}\right) = \frac{3\sqrt{3}}{2}$$

$$y = -3\sin\left(-\frac{5\pi}{6}\right) = -3\left(-\frac{1}{2}\right) = \frac{3}{2}$$

The rectangular coordinates are $\left(\dfrac{3\sqrt{3}}{2},\, \dfrac{3}{2}\right)$. ∎

The points $\left(4,\, \dfrac{\pi}{3}\right)$, $\left(-2,\, \dfrac{3\pi}{4}\right)$, and $\left(-3,\, -\dfrac{5\pi}{6}\right)$ and the points $(2,\, 2\sqrt{3})$, $(\sqrt{2},\, -\sqrt{2})$, and $\left(\dfrac{3\sqrt{3}}{2},\, \dfrac{3}{2}\right)$ are graphed in their respective coordinate systems in Figure 75.

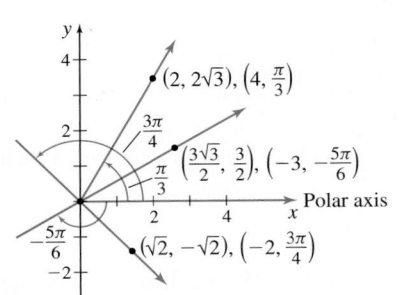

Figure 75

NOW WORK   Problems **25** and **29**.

Now suppose P has the rectangular coordinates (x, y). To represent P in polar coordinates, use the fact that $x = r\cos\theta$, $y = r\sin\theta$. Then

$$x^2 + y^2 = r^2\cos^2\theta + r^2\sin^2\theta = r^2$$

- If $x \neq 0$, then $\dfrac{y}{x} = \dfrac{r\sin\theta}{r\cos\theta} = \tan\theta$.

- If $x = 0$, the point $P = (x, y)$ is on the y-axis. So, $r = y$ and $\theta = \dfrac{\pi}{2}$.

THEOREM Conversion from Rectangular Coordinates to Polar Coordinates

If P is any point in the plane with rectangular coordinates (x, y), the polar coordinates (r, θ) of P are given by

$$\boxed{\; r = \sqrt{x^2 + y^2} \quad \text{and} \quad \tan\theta = \frac{y}{x} \quad \text{if } x \neq 0 \;}$$

$$\boxed{\; r = y \quad \text{and} \quad \theta = \frac{\pi}{2} \quad \text{if } x = 0 \;}$$

NEED TO REVIEW? The inverse tangent function is discussed in Section P.7, p. 64.

Be careful when using this theorem. If $x \neq 0$, $\tan\theta = \dfrac{y}{x}$, so $\theta = \tan^{-1}\dfrac{y}{x}$, $-\dfrac{\pi}{2} < \theta < \dfrac{\pi}{2}$, placing θ in quadrant I or IV. If the point lies in quadrant II or III, we must find $\tan^{-1}\dfrac{y}{x}$ in quadrant I or IV, respectively, and then add π to the result. It is advisable to plot the point (x, y) at the start to identify the quadrant that contains the point.

| EXAMPLE 4 | Converting from Rectangular Coordinates to Polar Coordinates |

Find polar coordinates of each point whose rectangular coordinates are:

(a) $(4, -4)$ (b) $(-1, -\sqrt{3})$ (c) $(4, -1)$

Solution

(a) The point $(4, -4)$, plotted in Figure 76, is in quadrant IV. The distance from the pole to the point $(4, -4)$ is

$$r = \sqrt{x^2 + y^2} = \sqrt{4^2 + (-4)^2} = \sqrt{32} = 4\sqrt{2}$$

Since the point $(4, -4)$ is in quadrant IV, $-\dfrac{\pi}{2} < \theta < 0$. So,

$$\theta = \tan^{-1}\frac{y}{x} = \tan^{-1}\left(\frac{-4}{4}\right) = \tan^{-1}(-1) = -\frac{\pi}{4}$$

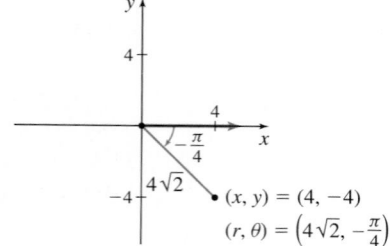

Figure 76

A pair of polar coordinates for this point is $\left(4\sqrt{2}, -\dfrac{\pi}{4}\right)$. Other possible representations include $\left(-4\sqrt{2}, \dfrac{3\pi}{4}\right)$ and $\left(4\sqrt{2}, \dfrac{7\pi}{4}\right)$.

(b) The point $(-1, -\sqrt{3})$, plotted in Figure 77, is in quadrant III. The distance from the pole to the point $(-1, -\sqrt{3})$ is

$$r = \sqrt{(-1)^2 + (-\sqrt{3})^2} = \sqrt{1 + 3} = 2$$

Since the point $(-1, -\sqrt{3})$ lies in quadrant III and the inverse tangent function gives an angle in quadrant I, we add π to $\tan^{-1}\dfrac{y}{x}$ to obtain an angle in quadrant III.

$$\theta = \tan^{-1}\left(\frac{-\sqrt{3}}{-1}\right) + \pi = \tan^{-1}\sqrt{3} + \pi = \frac{\pi}{3} + \pi = \frac{4\pi}{3}$$

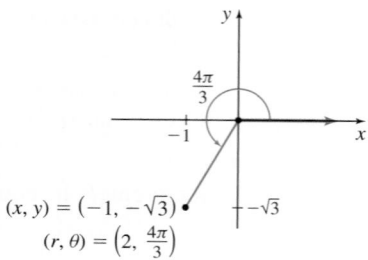

Figure 77

A pair of polar coordinates for the point is $\left(2, \dfrac{4\pi}{3}\right)$. Other possible representations include $\left(-2, \dfrac{\pi}{3}\right)$ and $\left(2, -\dfrac{2\pi}{3}\right)$.

(c) The point $(4, -1)$, plotted in Figure 78, lies in quadrant IV. The distance from the pole to the point $(4, -1)$ is

$$r = \sqrt{x^2 + y^2} = \sqrt{4^2 + (-1)^2} = \sqrt{17}$$

Since the point $(4, -1)$ is in quadrant IV, $-\dfrac{\pi}{2} < \theta < 0$. So,

$$\theta = \tan^{-1}\frac{y}{x} = \tan^{-1}\left(\frac{-1}{4}\right) \approx -0.245 \text{ radians}$$

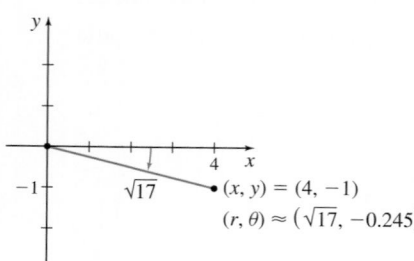

Figure 78

A pair of polar coordinates for this point is $(\sqrt{17}, -0.245)$. Other possible representations for the point include $\left(\sqrt{17}, \tan^{-1}\left(-\dfrac{1}{4}\right) + 2\pi\right) \approx (\sqrt{17}, 6.038)$ and $\left(-\sqrt{17}, \tan^{-1}\left(-\dfrac{1}{4}\right) + \pi\right) \approx (-\sqrt{17}, 2.897)$. ∎

NOW WORK   Problems 31 and 35.

3 Identify and Graph Polar Equations

Just as a rectangular grid is used to plot points given by rectangular coordinates, a *polar grid* is used to plot points given by polar coordinates. A **polar grid** consists of concentric circles centered at the pole and of rays with vertices at the pole, as shown in Figure 79. An equation whose variables are polar coordinates is called a **polar equation**, and the **graph of a polar equation** is the set of all points for which at least one of the polar coordinate representations satisfies the equation.

Polar equations of circles with their center at the pole, lines containing the pole, horizontal and vertical lines, and circles containing the pole have simple polar equations. In Section 9.3, we graph other important polar equations.

There are occasions when geometry is all that is needed to graph a polar equation. But usually other methods are required. One method used to graph polar equations is to convert the equation to rectangular coordinates. In the discussion that follows, (x, y) represents the rectangular coordinates of a point P, and (r, θ) represents polar coordinates of the point P.

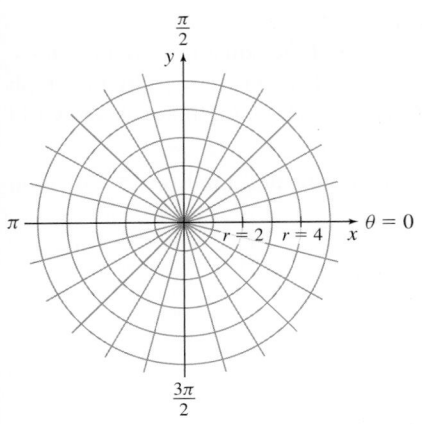

Figure 79 Polar grid.

EXAMPLE 5   Identifying and Graphing a Polar Equation

Identify and graph each equation:

  (a) $r = 3$ **(b)** $\theta = \dfrac{\pi}{4}$

Solution

(a) If r is fixed at 3 and θ is allowed to vary, the graph is a circle with its center at the pole and radius 3, as shown in Figure 80. To confirm this, we convert the polar equation $r = 3$ to a rectangular equation.

$$r = 3$$
$$r^2 = 9 \qquad \text{Square both sides.}$$
$$x^2 + y^2 = 9 \qquad r^2 = x^2 + y^2$$

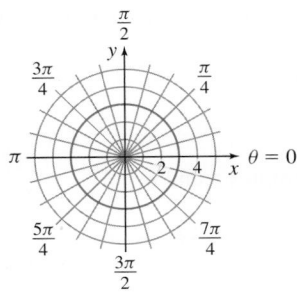

Figure 80 $r = 3$ or $x^2 + y^2 = 9$

(b) If θ is fixed at $\dfrac{\pi}{4}$ and r is allowed to vary, the result is a line containing the pole, making an angle of $\dfrac{\pi}{4}$ with the polar axis. That is, the graph of $\theta = \dfrac{\pi}{4}$ is a line containing the pole with slope $\tan \theta = \tan \dfrac{\pi}{4} = 1$, as shown in Figure 81. To confirm this, convert the polar equation to a rectangular equation.

$$\theta = \frac{\pi}{4}$$
$$\tan \theta = \tan \frac{\pi}{4}$$
$$\frac{y}{x} = 1$$
$$y = x$$

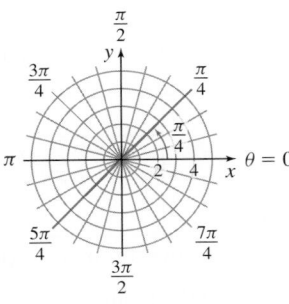

Figure 81 $\theta = \dfrac{\pi}{4}$ or $y = x$

NOW WORK   Problems 39 and 49.

▶ **EXAMPLE 6**   Identifying and Graphing Polar Equations

CALC CLIP

Identify and graph the equations:

  (a) $r \sin \theta = 2$ **(b)** $r = 4 \sin \theta$

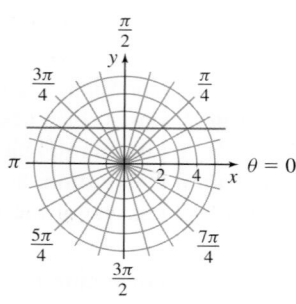

Figure 82 $r \sin \theta = 2$ or $y = 2$

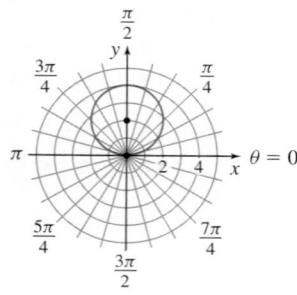

Figure 83 $r = 4 \sin \theta$ or $x^2 + (y - 2)^2 = 4$

Solution

(a) Here both r and θ are allowed to vary, so the graph of the equation is not obvious. If we use the fact that $y = r \sin \theta$, the equation $r \sin \theta = 2$ becomes $y = 2$. So, the graph of $r \sin \theta = 2$ is the horizontal line $y = 2$ that lies 2 units above the pole, as shown in Figure 82.

(b) To convert the equation $r = 4 \sin \theta$ to rectangular coordinates, multiply the equation by r to obtain

$$r^2 = 4r \sin \theta$$

Since $r^2 = x^2 + y^2$ and $y = r \sin \theta$, we have

$$\begin{aligned} r^2 &= 4r \sin \theta \\ x^2 + y^2 &= 4y && r^2 = x^2 + y^2, \quad r \sin \theta = y \\ x^2 + (y^2 - 4y) &= 0 \\ x^2 + \left(y^2 - 4y + 4\right) &= 4 && \text{Complete the square in } y. \\ x^2 + (y - 2)^2 &= 4 && \text{Factor.} \end{aligned}$$

This is the standard form of the equation of a circle with its center at the point (0, 2) and radius 2 in rectangular coordinates. See Figure 83. Notice that the circle passes through the pole. ∎

NOW WORK Problems 53 and 55.

Table 9 summarizes and extends the results of Examples 5 and 6.

TABLE 9

Description	Line passing through the pole making an angle α with the polar axis	Vertical line	Horizontal line
Rectangular equation	$y = (\tan \alpha)x$	$x = a$	$y = b$
Polar equation	$\theta = \alpha$	$r \cos \theta = a$	$r \sin \theta = b$
Typical graph	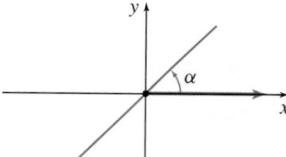	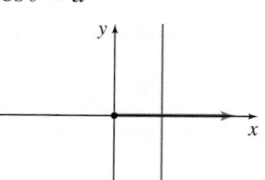	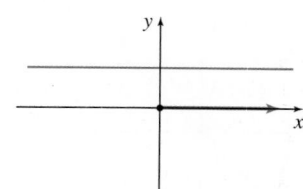

Description	Circle, center at the pole, radius a	Circle, passing through the pole, tangent to the line $\theta = \dfrac{\pi}{2}$, center on the polar axis, radius a	Circle, passing through the pole, tangent to the polar axis, center on the line $\theta = \dfrac{\pi}{2}$, radius a
Rectangular equation	$x^2 + y^2 = a^2, \quad a > 0$	$x^2 + y^2 = \pm 2ax, \quad a > 0$	$x^2 + y^2 = \pm 2ay, \quad a > 0$
Polar equation	$r = a, \quad a > 0$	$r = \pm 2a \cos \theta, \quad a > 0$	$r = \pm 2a \sin \theta, \quad a > 0$
Typical graph	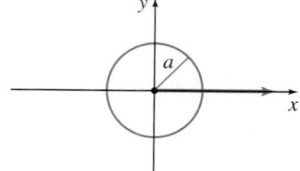	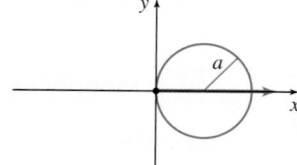	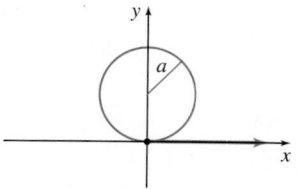

A.5 Assess Your Understanding

Concepts and Vocabulary

1. In a polar coordinate system, the origin is called the _____, and the _____ coincides with the positive x-axis of the rectangular coordinate system.

2. **True or False** Another representation in polar coordinates for the point $\left(2, \frac{\pi}{3}\right)$ is $\left(2, \frac{4\pi}{3}\right)$.

3. **True or False** In a polar coordinate system, each point in the plane has exactly one pair of polar coordinates.

4. **True or False** In polar coordinates (r, θ), the number r can be negative.

5. **True or False** If (r, θ) are the polar coordinates of the point P, then $|r|$ is the distance of the point P from the pole.

6. To convert the point (r, θ) in polar coordinates to a point (x, y) in rectangular coordinates, use the formulas $x = $ _____ and $y = $ _____.

Practice Problems

In Problems 7–12, match each point in polar coordinates with A, B, C, or D on the graph.

7. $\left(2, -\frac{11\pi}{6}\right)$ 8. $\left(-2, -\frac{\pi}{6}\right)$

9. $\left(-2, \frac{\pi}{6}\right)$ 10. $\left(2, \frac{7\pi}{6}\right)$

11. $\left(2, \frac{5\pi}{6}\right)$ 12. $\left(-2, \frac{5\pi}{6}\right)$

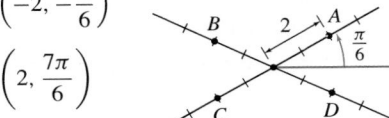

In Problems 13–18, polar coordinates of a point are given. Plot each point in a polar coordinate system.

13. $\left(4, \frac{\pi}{3}\right)$ 14. $\left(-4, \frac{\pi}{3}\right)$ 15. $\left(\sqrt{2}, \frac{\pi}{4}\right)$

16. $\left(7, -\frac{7\pi}{4}\right)$ 17. $\left(-6, \frac{4\pi}{3}\right)$ 18. $\left(5, \frac{\pi}{2}\right)$

In Problems 19–24, polar coordinates of a point are given. Find other polar coordinates (r, θ) of the point for which:

(a) $r > 0, -2\pi \leq \theta < 0$
(b) $r < 0, 0 \leq \theta < 2\pi$
(c) $r > 0, 2\pi \leq \theta < 4\pi$

19. $\left(5, \frac{2\pi}{3}\right)$ 20. $\left(4, \frac{3\pi}{4}\right)$ 21. $\left(1, \frac{\pi}{2}\right)$

22. $(2, \pi)$ 23. $\left(-3, -\frac{\pi}{4}\right)$ 24. $\left(-2, -\frac{2\pi}{3}\right)$

In Problems 25–30, polar coordinates of a point are given. Find the rectangular coordinates of each point.

25. $\left(6, \frac{\pi}{6}\right)$ 26. $\left(-6, \frac{\pi}{6}\right)$ 27. $\left(5, \frac{\pi}{2}\right)$

28. $\left(8, \frac{\pi}{4}\right)$ 29. $\left(2\sqrt{2}, -\frac{\pi}{4}\right)$ 30. $\left(-5, -\frac{\pi}{3}\right)$

In Problems 31–38, rectangular coordinates of a point are given. Plot the point. Find polar coordinates (r, θ) of each point for which $r > 0$ and $0 \leq \theta < 2\pi$.

31. $(5, 0)$ 32. $(2, -2)$ 33. $(-2, 2)$ 34. $(-2, -2\sqrt{3})$

35. $(\sqrt{3}, 1)$ 36. $(0, -3)$ 37. $(3, 2)$ 38. $(-6.5, 1.2)$

In Problems 39–46, match each of the graphs (A) through (H) to one of the following polar equations.

39. $r = 2$ 40. $\theta = \frac{\pi}{4}$ 41. $r = 2\cos\theta$

42. $r\cos\theta = 2$ 43. $r = -2\cos\theta$ 44. $r = 2\sin\theta$

45. $\theta = \frac{3\pi}{4}$ 46. $r\sin\theta = 2$

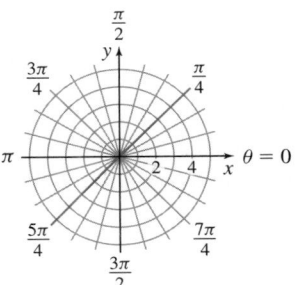

(A)

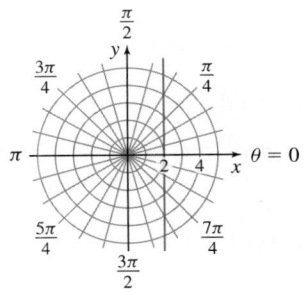

(B)

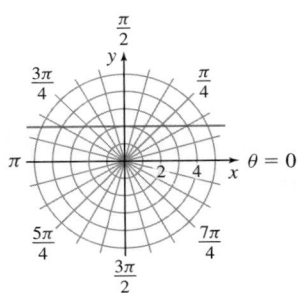

(C)

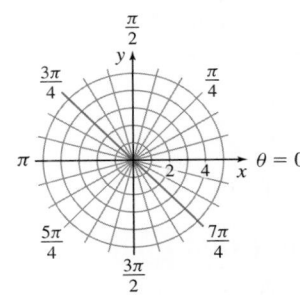

(D)

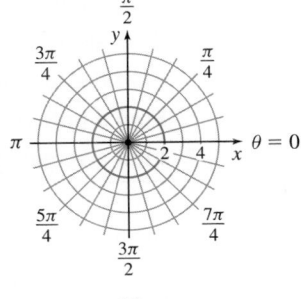

(E)

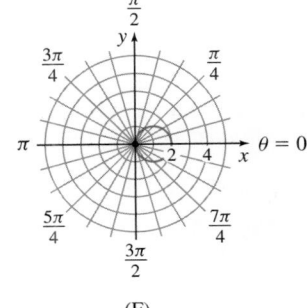

(F)

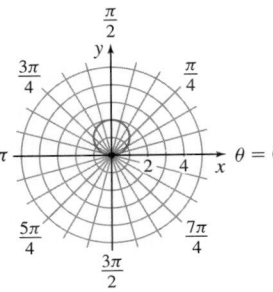

(G)

(H)

In Problems 47–58, identify and graph each polar equation. Convert to a rectangular equation if necessary.

47. $r = 4$ **48.** $r = 2$ PAGE A-47 **49.** $\theta = \dfrac{\pi}{3}$

50. $\theta = -\dfrac{\pi}{4}$ **51.** $r \sin \theta = 4$ **52.** $r \cos \theta = 4$

PAGE A-48 **53.** $r \cos \theta = -2$ **54.** $r \sin \theta = -2$ PAGE A-48 **55.** $r = 2 \cos \theta$

56. $r = 2 \sin \theta$ **57.** $r = -4 \sin \theta$ **58.** $r = -4 \cos \theta$

In Problems 59–68, the letters r and θ represent polar coordinates. Write each equation in rectangular coordinates x and y.

59. $r = \cos \theta$ **60.** $r = 2 + \cos \theta$ **61.** $r^2 = \sin \theta$

62. $r^2 = 1 - \sin \theta$ **63.** $r = \dfrac{4}{1 - \cos \theta}$ **64.** $r = \dfrac{3}{3 - \cos \theta}$

65. $r^2 = \theta$ **66.** $\theta = -\dfrac{\pi}{4}$ **67.** $r = 2$ **68.** $r = -5$

69. Horizontal Line   Show that the graph of the equation $r \sin \theta = a$ is a horizontal line a units above the pole if $a > 0$, and $|a|$ units below the pole if $a < 0$.

70. Vertical Line   Show that the graph of the equation $r \cos \theta = a$ is a vertical line a units to the right of the pole if $a > 0$, and $|a|$ units to the left of the pole if $a < 0$.

71. Circle   Show that the graph of the equation $r = 2a \sin \theta$, $a > 0$, is a circle of radius a with its center at the rectangular coordinates $(0, a)$.

72. Circle   Show that the graph of the equation $r = -2a \sin \theta$, $a > 0$, is a circle of radius a with its center at the rectangular coordinates $(0, -a)$.

73. Circle   Show that the graph of the equation $r = 2a \cos \theta$, $a > 0$, is a circle of radius a with its center at the rectangular coordinates $(a, 0)$.

74. Circle   Show that the graph of the equation $r = -2a \cos \theta$, $a > 0$, is a circle of radius a with its center at the rectangular coordinates $(-a, 0)$.

75. Exploring Using Technology
   (a) Use a square screen to graph $r_1 = \sin \theta$, $r_2 = 2 \sin \theta$, and $r_3 = 3 \sin \theta$.
   (b) Describe how varying the constant a, $a > 0$, alters the graph of $r = a \sin \theta$.
   (c) Graph $r_1 = -\sin \theta$, $r_2 = -2 \sin \theta$, and $r_3 = -3 \sin \theta$.
   (d) Describe how varying the constant a, $a < 0$, alters the graph of $r = a \sin \theta$.

76. Exploring Using Technology
   (a) Use a square screen to graph $r_1 = \cos \theta$, $r_2 = 2 \cos \theta$, and $r_3 = 3 \cos \theta$.
   (b) Describe how varying the constant a, $a > 0$, alters the graph of $r = a \cos \theta$.
   (c) Graph $r_1 = -\cos \theta$, $r_2 = -2 \cos \theta$, and $r_3 = -3 \cos \theta$.
   (d) Describe how varying the constant a, $a < 0$, alters the graph of $r = a \cos \theta$.

A.6 Vectors Used in Calculus: AP® Calculus, BC Only

AP® EXAM TIP

Appendix A.6 contains precalculus mathematics needed for Vector Functions, a BC Calculus only topic discussed in Chapter 9 (Unit 9).

ORIGINS   The word "vector" comes from the Latin word meaning "to carry."

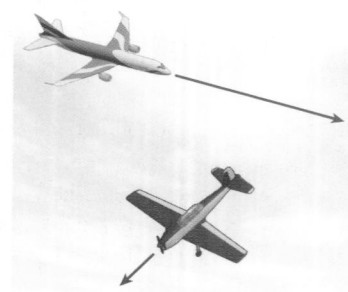

Figure 84

OBJECTIVES *When you finish this section, you should be able to:*

1 **Represent vectors geometrically (p. A-51)**
2 **Use properties of vectors (p. A-53)**
3 **Represent a vector algebraically (p. A-54)**
4 **Add, subtract, and find scalar multiples of vectors (p. A-55)**
5 **Find the magnitude of a vector (p. A-57)**
6 **Find a unit vector (p. A-58)**
7 **Find a vector from its direction and magnitude (p. A-59)**

Geometrically, a **vector** is an entity that has both magnitude and direction, and is often represented by an arrow pointing in the *direction* of the vector. The length of the arrow represents the *magnitude* of the vector.

Velocity is a vector. The velocity of an airplane can be represented by an arrow that points in the direction of its flight; the length of the arrow represents its speed (the magnitude of the velocity). See Figure 84. If the airplane speeds up, the length of the arrow increases.

When working with vectors, we refer to real numbers as *scalars*. **Scalars** are quantities that have magnitude but no direction. Physical examples of scalar quantities are time, mass, temperature, and speed. In print, boldface Roman letters are used to denote vectors, to distinguish them from the lightface, italic letters used for scalars. For handwritten work, an arrow is placed over a letter to denote a vector.

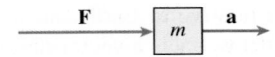

Figure 85 $\mathbf{F} = m\mathbf{a}$

Force and *acceleration* are also vectors. If a force $\mathbf{F}$ is exerted on an object of mass m, the force causes the object to accelerate in the direction of the force. See Figure 85. If $\mathbf{a}$ is the acceleration, then $\mathbf{F}$ and $\mathbf{a}$ are related by the equation $\mathbf{F} = m\mathbf{a}$, Newton's Second Law of Motion.

① Represent Vectors Geometrically

Vectors are closely related to the geometric idea of a *directed line segment*. If P and Q are distinct points, there is exactly one line containing both P and Q. See Figure 86(a). The points P and Q, and those on the line between P and Q, form the **line segment** $\overline{PQ}$. See Figure 86(b). If the points are ordered so they begin with P and end with Q, then we have a **directed line segment from** P **to** Q, denoted by $\overrightarrow{PQ}$. For the directed line segment $\overrightarrow{PQ}$, the point P is the **initial point** and the point Q is the **terminal point**. See Figure 86(c).

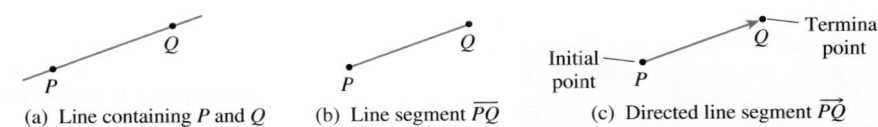

(a) Line containing P and Q (b) Line segment $\overline{PQ}$ (c) Directed line segment $\overrightarrow{PQ}$

Figure 86

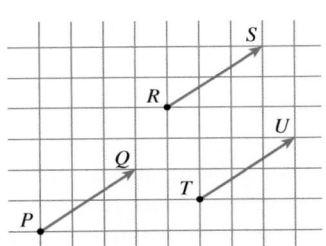

Figure 87 The vectors $\overrightarrow{PQ}$, $\overrightarrow{RS}$, and $\overrightarrow{TU}$ are equal.

The **magnitude** of the directed line segment $\overrightarrow{PQ}$ is the distance from the point P to the point Q; that is, it is the length of the line segment $\overline{PQ}$. The **direction** of $\overrightarrow{PQ}$ is from P to Q. If a vector $\mathbf{v}$ has the same magnitude and the same direction as the directed line segment $\overrightarrow{PQ}$, then

$$\mathbf{v} = \overrightarrow{PQ}$$

The vector whose magnitude is 0 is called the **zero vector 0**. The zero vector is assigned no direction.

Two vectors $\mathbf{v}$ and $\mathbf{w}$ are **equal**, written $\mathbf{v} = \mathbf{w}$, if they have the same magnitude and the same direction. For example, the vectors in Figure 87 are equal even though they have different initial points and terminal points.

As a result, it is useful to think of a vector as an arrow, keeping in mind that two arrows (vectors) are equal if they have the same direction and the same length (magnitude).

Since location does not affect a vector, we **add** two vectors $\mathbf{v}$ and $\mathbf{w}$ by positioning vector $\mathbf{w}$ so that its initial point coincides with the terminal point of $\mathbf{v}$. The **vector sum** $\mathbf{v} + \mathbf{w}$ is the unique vector represented by the arrow from the initial point of $\mathbf{v}$ to the terminal point of $\mathbf{w}$, as shown in Figure 88.

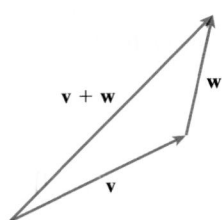

Figure 88 The sum of vectors $\mathbf{v}$ and $\mathbf{w}$ is found by positioning the initial point of $\mathbf{w}$ at the terminal point of $\mathbf{v}$.

We said that forces can be represented by vectors. But how do we know that forces "add" the same way vectors do? Well, physicists say they do, and laboratory experiments bear it out. So if $\mathbf{F}_1$ and $\mathbf{F}_2$ are two forces simultaneously acting on an object, the vector sum $\mathbf{F}_1 + \mathbf{F}_2$ is equal to the force that produces the same effect as that obtained when the forces $\mathbf{F}_1$ and $\mathbf{F}_2$ are applied at the same time to the same object. The force $\mathbf{F}_1 + \mathbf{F}_2$ is called the **resultant force**. See Figure 89.

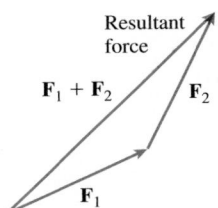

Figure 89 $\mathbf{F}_1 + \mathbf{F}_2$ is the resultant force.

If **w** is a vector describing the velocity of the wind relative to Earth and if **v** is a vector describing the velocity of an airplane in the air, then **w** + **v** is a vector describing the velocity of the airplane relative to Earth. See Figure 90.

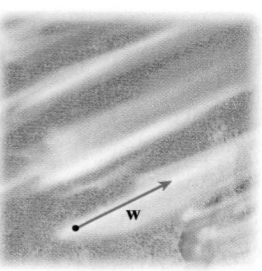

(a) Velocity of the wind relative to Earth

(b) Velocity of an airplane relative to the air

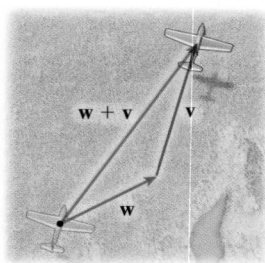

(c) Resultant equals the velocity of an airplane relative to Earth

Figure 90

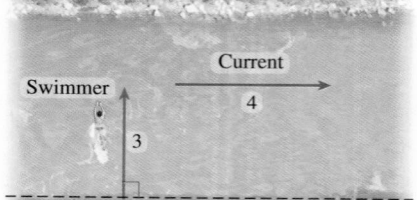

Figure 91

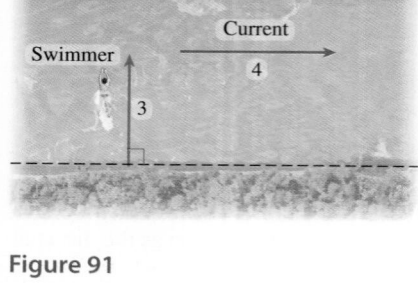

Figure 92

EXAMPLE 1   Graphing a Resultant Vector

A swimmer swims at a constant speed of 3 miles per hour (mi/h) and heads directly across a river whose current is 4 mi/h.

(a) Draw vectors representing the swimmer and the current.

(b) Draw the resultant vector.

(c) Interpret the result.

Solution

(a) See Figure 91. We represent the swimmer by a vector pointing directly across the river and the current by a vector pointing parallel to a straight shore line. Since the swimmer's speed is 3 mi/h, and the current's speed is 4 mi/h, the length of the vector representing the swimmer is 3 units, and the length of the vector representing the current is 4 units.

(b) The resultant vector is the sum of the vectors in (a). See Figure 92.

(c) The resultant vector shows the true speed of the swimmer and true direction of the swim. In fact, the true speed is 5 mi/h, and the true direction can be described in terms of the angle θ as $\theta = \tan^{-1} \dfrac{4}{3} \approx 53.1°$. ■

NOW WORK   Problem 27.

Next, we examine properties of vectors. Many of these properties are similar to properties of real numbers.

Properties of Vector Addition

Let **u**, **v**, and **w** represent any three vectors.

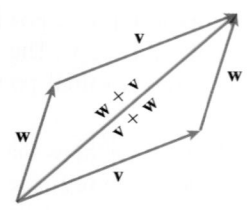

Figure 93 **w** + **v** = **v** + **w**

- **Commutative property of vector addition** (Figure 93):

$$\boxed{\mathbf{w} + \mathbf{v} = \mathbf{v} + \mathbf{w}}$$

- **Associative property of vector addition** (Figure 94):

$$\boxed{\mathbf{u} + (\mathbf{v} + \mathbf{w}) = (\mathbf{u} + \mathbf{v}) + \mathbf{w}}$$

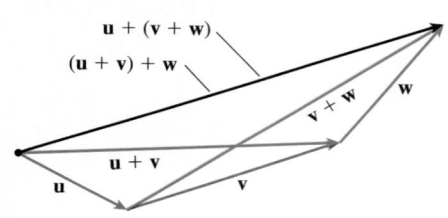

Figure 94 **u** + (**v** + **w**) = (**u** + **v**) + **w**

- **Additive identity property** The zero vector **0** is called the **additive identity**. The sum of any vector **v** and the zero vector **0** equals **v**.

$$\mathbf{v} + \mathbf{0} = \mathbf{0} + \mathbf{v} = \mathbf{v}$$

- **Additive inverse property** For any vector **v**, the vector −**v**, called the **additive inverse** of **v**, has the same magnitude as **v**, but the opposite direction, as shown in Figure 95. The sum of **v** and its additive inverse −**v** is the zero vector **0**.

$$\mathbf{v} + (-\mathbf{v}) = (-\mathbf{v}) + \mathbf{v} = \mathbf{0}$$

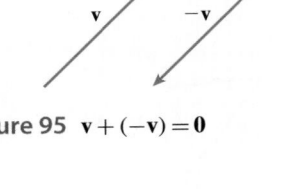

Figure 95 $\mathbf{v} + (-\mathbf{v}) = \mathbf{0}$

DEFINITION Difference of Two Vectors

The **difference** of two vectors **v** − **w** is defined as

$$\mathbf{v} - \mathbf{w} = \mathbf{v} + (-\mathbf{w})$$

The difference **v** − **w** between two vectors can be obtained by positioning the initial point of −**w** with the terminal point of **v**. Then the vector from the initial point of **v** to the terminal point of −**w** equals the vector **v** − **w**, as shown in Figure 96.

The product of a vector **v** and a scalar *a* produces a vector called the *scalar multiple* of **v**.

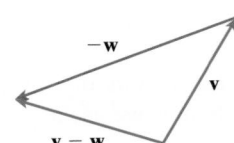

Figure 96 $\mathbf{v} - \mathbf{w} = \mathbf{v} + (-\mathbf{w})$

DEFINITION Scalar Multiple

If *a* is a scalar and **v** is a vector, the product *a***v**, called a **scalar multiple** of **v**, is the vector with magnitude |*a*| times the magnitude of **v**, in the same direction as **v** when *a* > 0 and in the opposite direction from **v** when *a* < 0.

If either *a* = 0 or **v** = **0**, then *a***v** = **0**.

See Figure 97 for illustrations of scalar multiples of **v**. Notice that the additive inverse of **v** can be represented as the scalar multiple (−1)**v**.

We saw earlier that if **a** is the acceleration of an object of mass *m* due to a force **F** being exerted, then by Newton's second law of motion, **F** = *m***a**. Here *m***a** is the product of the scalar *m* and the vector **a**.

Scalar multiples have the following properties:

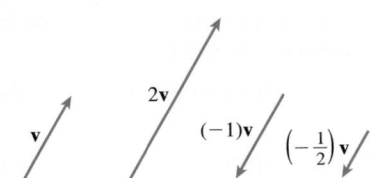

Figure 97 Scalar multiples of **v**.

Properties of Scalar Multiplication

If **v** and **w** are any two vectors and **0** is the zero vector, then:

- $0\mathbf{v} = \mathbf{0}$
- $1\mathbf{v} = \mathbf{v}$
- $(-1)\mathbf{v} = -\mathbf{v}$
- $(a + b)\mathbf{v} = a\mathbf{v} + b\mathbf{v}$
- $a(\mathbf{v} + \mathbf{w}) = a\mathbf{v} + a\mathbf{w}$
- $a(b\mathbf{v}) = (ab)\mathbf{v}$

❷ Use Properties of Vectors

EXAMPLE 2 Using Properties of Vectors

Use the vectors illustrated in Figure 98 on page A-54 to graph each of the following vectors:

(a) $\mathbf{v} - \mathbf{w}$ **(b)** $2\mathbf{v} - \mathbf{w} + \mathbf{u}$ **(c)** $\frac{2}{3}\mathbf{u} + \frac{1}{2}(\mathbf{v} - \mathbf{w})$

Solution

We reposition the vectors in Figure 98 as shown in Figure 99 to graph the desired vectors.

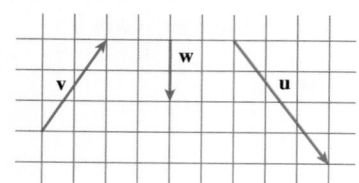

Figure 98

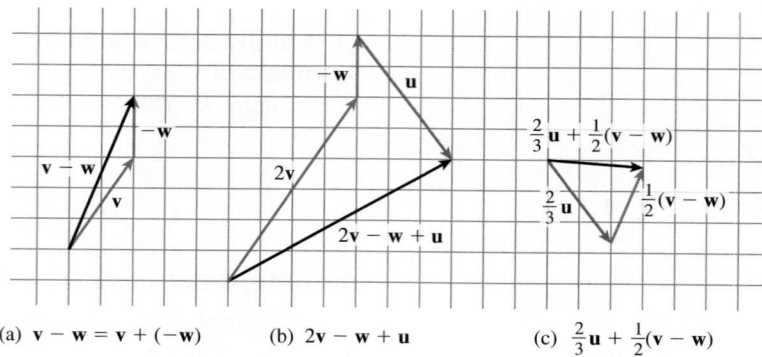

(a) $\mathbf{v} - \mathbf{w} = \mathbf{v} + (-\mathbf{w})$ (b) $2\mathbf{v} - \mathbf{w} + \mathbf{u}$ (c) $\frac{2}{3}\mathbf{u} + \frac{1}{2}(\mathbf{v} - \mathbf{w})$

Figure 99

■

NOW WORK Problems 11 and 15.

We have been treating vectors geometrically, but to use vectors in a meaningful way, we need to look at vectors algebraically as well. To do this, we use a rectangular coordinate system.

3 Represent a Vector Algebraically

If $\mathbf{v}$ is a vector, and if its initial point is at the origin and its terminal point is at the point (v_1, v_2), then we may represent $\mathbf{v}$ by the ordered pair

$$\mathbf{v} = \langle v_1, v_2 \rangle$$

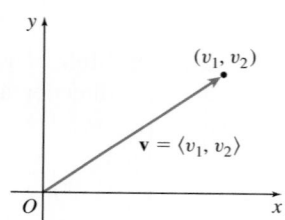

Figure 100

The numbers v_1 is the **x-component** and v_2 is the **y-component** of the vector $\mathbf{v}$. Together, the numbers v_1 and v_2 are called the **components** of $\mathbf{v}$. See Figure 100.

The vectors defined here, as well as any other vector whose initial point is at the origin, are called **position vectors**.

Any vector whose initial point is not at the origin is equal to a unique position vector. If $\mathbf{v}$ is a vector with initial point $P_1 = (x_1, y_1)$ and terminal point $P_2 = (x_2, y_2)$, then $\mathbf{v}$ equals the position vector

IN WORDS Every vector can be represented by a unique position vector.

$$\mathbf{v} = \langle x_2 - x_1, y_2 - y_1 \rangle$$

As Figure 101 illustrates, this feature allows us to move vectors around in applications.

If both the initial point and the terminal point of a vector $\mathbf{v}$ are at the origin, then $\mathbf{v}$ is the **zero vector**. As a result,

$$\mathbf{0} = \langle 0, 0 \rangle$$

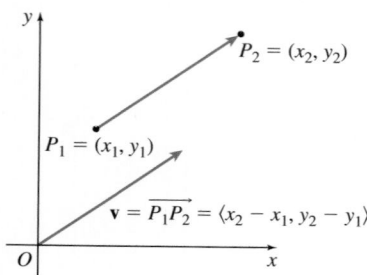

Figure 101

EXAMPLE 3 **Finding a Position Vector**

Find the unique position vector $\mathbf{v}$ of a vector whose initial point is $P_1 = (-1, 2)$ and whose terminal point is $P_2 = (4, 6)$.

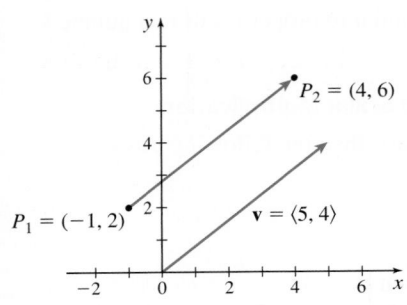

Figure 102

IN WORDS Two position vectors are equal if and only if their x-components are equal and their y-components are equal.

Solution

To find $\mathbf{v}$, subtract corresponding components.

$$\mathbf{v} = \langle 4 - (-1), 6 - 2 \rangle = \langle 5, 4 \rangle$$

See Figure 102. ∎

NOW WORK Problem 29.

Two position vectors $\mathbf{v}$ and $\mathbf{w}$ are equal if and only if the terminal point of $\mathbf{v}$ is the same as the terminal point of $\mathbf{w}$.

> $\mathbf{v} = \langle v_1, v_2 \rangle$ and $\mathbf{w} = \langle w_1, w_2 \rangle$ are equal if and only if $v_1 = w_1$ and $v_2 = w_2$.

④ Add, Subtract, and Find Scalar Multiples of Vectors

We have described the operations of addition, subtraction, and scalar multiplication geometrically. Here we present them algebraically.

Vectors are added or subtracted and scalar multiples are formed using the components of the vectors.

> **DEFINITION Operations with Vectors**
>
> If $\mathbf{v} = \langle v_1, v_2 \rangle$ and $\mathbf{w} = \langle w_1, w_2 \rangle$ are two vectors and if a is a scalar, then
>
> - $\mathbf{v} + \mathbf{w} = \langle v_1 + w_1, v_2 + w_2 \rangle$
> - $\mathbf{v} - \mathbf{w} = \langle v_1 - w_1, v_2 - w_2 \rangle$
> - $a\,\mathbf{v} = \langle av_1, av_2 \rangle$

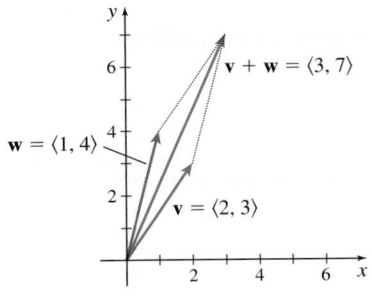

Figure 103

For example, if $\mathbf{v} = \langle 2, 3 \rangle$ and $\mathbf{w} = \langle 1, 4 \rangle$, then $\mathbf{v} + \mathbf{w} = \langle 2 + 1, 3 + 4 \rangle = \langle 3, 7 \rangle$. See Figure 103.

Notice that the algebraic sum $\mathbf{v} + \mathbf{w}$ is the same as the resultant vector obtained geometrically. In other words, the algebraic representation of vectors agrees with the geometric representation discussed earlier.

EXAMPLE 4 Adding and Subtracting Vectors Algebraically

If $\mathbf{v} = \langle 2, 3 \rangle$ and $\mathbf{w} = \langle -1, -2 \rangle$, find:

(a) $\mathbf{v} + \mathbf{w}$ (b) $\mathbf{w} - \mathbf{v}$ (c) $\dfrac{1}{2}\mathbf{w}$ (d) $2\mathbf{v} + 3\mathbf{w}$

Solution

(a) $\mathbf{v} + \mathbf{w} = \langle 2, 3 \rangle + \langle -1, -2 \rangle = \langle 2 + (-1), 3 + (-2) \rangle = \langle 1, 1 \rangle$

(b) $\mathbf{w} - \mathbf{v} = \langle -1, -2 \rangle - \langle 2, 3 \rangle = \langle -1-2, -2 - 3 \rangle = \langle -3, -5 \rangle$

(c) $\dfrac{1}{2}\mathbf{w} = \dfrac{1}{2} \langle -1, -2 \rangle = \left\langle \dfrac{1}{2}(-1), \dfrac{1}{2}(-2) \right\rangle = \left\langle -\dfrac{1}{2}, -1 \right\rangle$

(d) $2\mathbf{v} + 3\mathbf{w} = 2 \langle 2, 3 \rangle + 3 \langle -1, -2 \rangle = \langle 4, 6 \rangle + \langle -3, -6 \rangle = \langle 1, 0 \rangle$ ∎

NOW WORK Problem 35.

The properties listed below for vectors are similar to properties of real numbers.

THEOREM Properties of Vector Addition and Scalar Multiplication

If **u**, **v**, and **w** are vectors and if a and b are scalars, then the following properties hold:

- **Commutative property of vector addition:**

$$\mathbf{u} + \mathbf{v} = \mathbf{v} + \mathbf{u}$$

- **Associative property of vector addition:**

$$\mathbf{u} + (\mathbf{v} + \mathbf{w}) = (\mathbf{u} + \mathbf{v}) + \mathbf{w}$$

- **Additive identity 0:**

$$\mathbf{v} + \mathbf{0} = \mathbf{0} + \mathbf{v} = \mathbf{v}$$

- **Additive inverse property:**

$$\mathbf{v} + (-\mathbf{v}) = \mathbf{0}$$

- **Distributive property of scalar multiplication over vector addition:**

$$a(\mathbf{u} + \mathbf{v}) = a\,\mathbf{u} + a\,\mathbf{v}$$

- **Distributive property of scalar multiplication over scalar addition:**

$$(a + b)\mathbf{v} = a\,\mathbf{v} + b\,\mathbf{v}$$

- **Linearity property of scalar multiplication:**

$$a(b\,\mathbf{v}) = (ab)\mathbf{v}$$

- **Scalar multiplication identity:**

$$1\mathbf{v} = \mathbf{v}$$

We proved some of these properties geometrically. For example, earlier in this section we gave a geometric proof of the commutative property of addition. Here we algebraically prove the commutative property and the scalar multiplication identity property.

Proof Commutative property of vector addition: If $\mathbf{u} = \langle u_1, u_2 \rangle$ and if $\mathbf{v} = \langle v_1, v_2 \rangle$, then

$$\mathbf{u} + \mathbf{v} = \langle u_1, u_2 \rangle + \langle v_1, v_2 \rangle = \langle u_1 + v_1, u_2 + v_2 \rangle = \langle v_1 + u_1, v_2 + u_2 \rangle$$

$$= \langle v_1, v_2 \rangle + \langle u_1, u_2 \rangle = \mathbf{v} + \mathbf{u}$$

Scalar multiplication identity: If $\mathbf{v} = \langle v_1, v_2 \rangle$, then

$$1\mathbf{v} = \langle 1v_1, 1v_2 \rangle = \langle v_1, v_2 \rangle = \mathbf{v} \qquad \blacksquare$$

DEFINITION Parallel Vectors

Let **v** and **w** be two distinct nonzero vectors. If there is a nonzero scalar a for which $\mathbf{v} = a\,\mathbf{w}$, then **v** and **w** are called **parallel vectors**.

EXAMPLE 5   **Showing Two Vectors Are Parallel**

(a) The vectors

$$\mathbf{v} = \langle 2, 3 \rangle \qquad \text{and} \qquad \mathbf{w} = \langle 4, 6 \rangle$$

are parallel, since $\mathbf{v} = \dfrac{1}{2}\mathbf{w}$. In this case, **v** and **w** have the same direction.

(b) The vectors

$$\mathbf{v} = \langle 1, 2 \rangle \qquad \text{and} \qquad \mathbf{w} = \langle -3, -6 \rangle$$

are parallel, since $\mathbf{w} = -3\,\mathbf{v}$. In this case, **v** and **w** have opposite directions.   ∎

NOW WORK   Problem 53.

---

⑤ Find the Magnitude of a Vector

The representation of **v** as the vector $\langle v_1, v_2 \rangle$ gives the direction of **v**. The definition below gives the magnitude of **v**.

DEFINITION

The **magnitude** or **length** of a vector **v**, denoted by the symbol $\|\mathbf{v}\|$, is the distance from the initial point to the terminal point of **v**. $\|\mathbf{v}\|$ is also called the **norm** of the vector **v**.

THEOREM Magnitude of a Vector

If $\mathbf{v} = \langle v_1, v_2 \rangle$ is a vector, the magnitude of **v** is

$$\boxed{\|\mathbf{v}\| = \sqrt{v_1^2 + v_2^2}}$$

The proof follows from the Distance Formula.

EXAMPLE 6   **Finding the Magnitude of a Vector**

Find the magnitude of $\mathbf{v} = \langle 3, -4 \rangle$

Solution

$$\|\mathbf{v}\| = \sqrt{v_1^2 + v_2^2} = \sqrt{3^2 + (-4)^2} = \sqrt{9 + 16} = 5$$ ∎

NOW WORK   Problem 37.

THEOREM Magnitude of the Scalar Multiple of a Vector

If a is scalar and **v** is a vector, then

$$\boxed{\|a\mathbf{v}\| = |a|\,\|\mathbf{v}\|}$$

Proof Suppose $\mathbf{v} = \langle v_1, v_2 \rangle$. Then the scalar multiple of a and $\mathbf{v}$ is $a\mathbf{v} = \langle av_1, av_2 \rangle$ and

$$\|a\mathbf{v}\| = \sqrt{(av_1)^2 + (av_2)^2} = \sqrt{a^2 v_1^2 + a^2 v_2^2} = \sqrt{a^2 (v_1^2 + v_2^2)}$$
$$= \sqrt{a^2}\sqrt{v_1^2 + v_2^2} = |a|\,\|\mathbf{v}\|$$

∎

6 Find a Unit Vector

DEFINITION

A vector with magnitude 1 is called a **unit vector.**

In many applications, it is necessary to find the unit vector with the same direction as a given vector $\mathbf{v}$.

THEOREM Unit Vector in the Direction of v

For any nonzero vector $\mathbf{v}$, the vector

$$\mathbf{u} = \frac{\mathbf{v}}{\|\mathbf{v}\|}$$

is the unit vector that has the same direction as $\mathbf{v}$.

Proof Since $\dfrac{1}{\|\mathbf{v}\|}$ is a positive scalar, the vector $\mathbf{u} = \dfrac{\mathbf{v}}{\|\mathbf{v}\|} = \dfrac{1}{\|\mathbf{v}\|}\mathbf{v}$ has the same direction as $\mathbf{v}$. Next we show that the magnitude of $\mathbf{u}$ is 1.

$$\|\mathbf{u}\| = \left\| \frac{\mathbf{v}}{\|\mathbf{v}\|} \right\| = \left\| \frac{1}{\|\mathbf{v}\|}\mathbf{v} \right\| = \frac{1}{\|\mathbf{v}\|}\|\mathbf{v}\| = 1$$

Since $\|\mathbf{u}\| = 1$, $\mathbf{u}$ is a unit vector in the direction of $\mathbf{v}$. ∎

Multiplying a nonzero vector $\mathbf{v}$ by $\dfrac{1}{\|\mathbf{v}\|}$ to obtain a unit vector that has the same direction as $\mathbf{v}$ is called **normalizing v**.

EXAMPLE 7 Normalizing a Vector

Normalize each vector. That is, find a unit vector $\mathbf{u}$ that has the same direction as:

(a) $\mathbf{v} = \langle 3, -4 \rangle$ (b) $\mathbf{v} = \langle -1, 2 \rangle$

Solution

(a) Since $\mathbf{v} = \langle 3, -4 \rangle$, then $\|\mathbf{v}\| = \sqrt{9 + 16} = 5$. The unit vector $\mathbf{u}$ in the same direction as $\mathbf{v}$ is

$$\mathbf{u} = \frac{\mathbf{v}}{\|\mathbf{v}\|} = \frac{\mathbf{v}}{5} = \left\langle \frac{3}{5}, -\frac{4}{5} \right\rangle$$

(b) Since $\mathbf{v} = \langle -1, 2 \rangle$, then $\|\mathbf{v}\| = \sqrt{1 + 4} = \sqrt{5}$. The unit vector $\mathbf{u}$ in the same direction as $\mathbf{v}$ is

$$\mathbf{u} = \frac{\mathbf{v}}{\|\mathbf{v}\|} = \frac{\mathbf{v}}{\sqrt{5}} = \left\langle -\frac{1}{\sqrt{5}}, \frac{2}{\sqrt{5}} \right\rangle = \left\langle -\frac{\sqrt{5}}{5}, \frac{2\sqrt{5}}{5} \right\rangle$$

∎

NOW WORK   Problem 45.

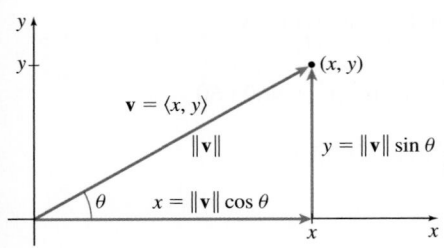

Figure 104 $\mathbf{v} = \langle \|\mathbf{v}\| \cos \theta, \|\mathbf{v}\| \sin \theta \rangle$

7 Find a Vector from Its Direction and Magnitude

Often a vector is described by its magnitude and direction, rather than by its components. For example, a weather report warning of 40 mi/h southwesterly winds (from the southwest) describes a velocity vector, but to work with it requires finding its components.

A nonzero vector $\mathbf{v}$ can be described by specifying the angle θ, $0 \le \theta < 2\pi$, between $\mathbf{v}$ and the positive x-axis, which gives its direction, and by specifying its magnitude $\|\mathbf{v}\|$. See Figure 104. If $\mathbf{v} = \langle x, y \rangle$, then $x = \|\mathbf{v}\| \cos \theta$ and $y = \|\mathbf{v}\| \sin \theta$. So, $\mathbf{v} = \langle \|\mathbf{v}\| \cos \theta, \|\mathbf{v}\| \sin \theta \rangle$.

THEOREM

If $\mathbf{v}$ is a nonzero vector, and if θ is the angle between $\mathbf{v}$ and the positive x-axis, then

$$\mathbf{v} = \langle \|\mathbf{v}\| \cos \theta, \|\mathbf{v}\| \sin \theta \rangle$$

EXAMPLE 8   Using Vectors to Model a Problem

NOTE Wind direction is measured from where it blows.

An airplane has an air speed of 400 km/h and is headed east. There is a northwesterly wind of 80 km/h. (Northwesterly winds blow toward the southeast.)

(a) Find a vector representing the velocity of the airplane in the air.

(b) Find a vector representing the wind velocity.

(c) Find a vector representing the velocity of the airplane relative to the ground.

(d) Find the true speed of the airplane.

Solution

We use a coordinate system with the direction north along the positive y-axis. Then the direction east is along the positive x-axis. See Figure 105.

Using a scale of 1 unit = 1 km/h, define

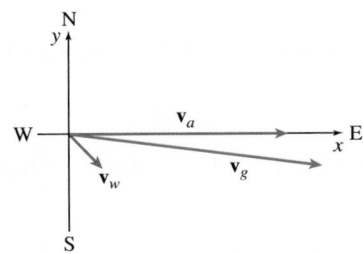

Figure 105

$$\mathbf{v}_a = \text{Velocity of the airplane in the air}$$

$$\mathbf{v}_w = \text{Velocity of the wind}$$

$$\mathbf{v}_g = \text{Velocity of the airplane relative to the ground}$$

Figure 105 shows the vectors $\mathbf{v}_a$, $\mathbf{v}_w$, $\mathbf{v}_g$.

(a) The vector $\mathbf{v}_a$ has magnitude 400 and direction east, so $\theta = 0$ and

$$\mathbf{v}_a = \langle 400 \cos 0, 400 \sin 0 \rangle = \langle 400, 0 \rangle$$

(b) The vector $\mathbf{v}_w$ has magnitude 80 and makes an angle of $\dfrac{7\pi}{4}$ with the positive x-axis, as shown in Figure 106. Since the magnitude of $\mathbf{v}_w$ is 80,

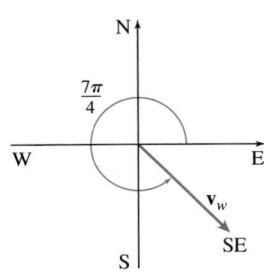

Figure 106

$$\mathbf{v}_w = \left\langle 80 \cos \frac{7\pi}{4}, 80 \sin \frac{7\pi}{4} \right\rangle = \left\langle 80 \cdot \frac{\sqrt{2}}{2}, 80 \left(-\frac{\sqrt{2}}{2} \right) \right\rangle = \langle 40\sqrt{2}, -40\sqrt{2} \rangle$$

(c) The velocity $\mathbf{v}_g$ of the airplane relative to the ground is the resultant of $\mathbf{v}_a$ and $\mathbf{v}_w$.

$$\mathbf{v}_g = \mathbf{v}_a + \mathbf{v}_w = \langle 400, 0 \rangle + \left\langle 40\sqrt{2}, -40\sqrt{2} \right\rangle = \left\langle 400 + 40\sqrt{2}, -40\sqrt{2} \right\rangle$$

(d) The true speed of the airplane is the magnitude of the vector $\mathbf{v}_g$.

$$\|\mathbf{v}_g\| = \sqrt{(400 + 40\sqrt{2})^2 + (40\sqrt{2})^2} \approx 460.060$$

The true speed of the airplane is approximately 460.060 km/h. ∎

NOW WORK Problem 63.

When multiple forces $\mathbf{F}_1, \mathbf{F}_2, \ldots, \mathbf{F}_n$ simultaneously act on an object, the vector sum $\mathbf{F}_1 + \mathbf{F}_2 + \cdots + \mathbf{F}_n$ is the **resultant force**. The resultant force produces the same effect on the object as the effect obtained when all the forces $\mathbf{F}_1, \mathbf{F}_2, \ldots, \mathbf{F}_n$ act on the same object at the same time. An object is in **static equilibrium** if the object is at rest and the resultant force is $\mathbf{0}$.

EXAMPLE 9 Finding the Tension in a Cable

A commercial air conditioner weighing 8600 lb is suspended from two cables, as shown in Figure 107. What is the tension in each cable?

Solution

We represent the cables and the air conditioner with a force diagram, as shown in Figure 108.

The tension in each cable is the magnitude $\|\mathbf{F}_1\|$ and $\|\mathbf{F}_2\|$, respectively. The weight of the air conditioner is $\|\mathbf{F}_3\| = 8600$ lb.

$$\mathbf{F}_1 = \langle \|\mathbf{F}_1\| \cos 45°, \|\mathbf{F}_1\| \sin 45° \rangle = \left\langle \frac{\sqrt{2}}{2}\|\mathbf{F}_1\|, \frac{\sqrt{2}}{2}\|\mathbf{F}_1\| \right\rangle$$

$$\mathbf{F}_2 = \langle \|\mathbf{F}_2\| \cos 140°, \|\mathbf{F}_2\| \sin 140° \rangle = \langle \cos 140° \|\mathbf{F}_2\|, \sin 140° \|\mathbf{F}_2\| \rangle$$

$$\mathbf{F}_3 = \langle 8600 \cos 270°, 8600 \sin 270° \rangle = \langle 0, -8600 \rangle$$

For the air conditioner to be in static equilibrium, the sum of the forces acting on it must equal $\mathbf{0}$. That is,

$$\mathbf{F}_1 + \mathbf{F}_2 + \mathbf{F}_3 = \left\langle \frac{\sqrt{2}}{2}\|\mathbf{F}_1\|, \frac{\sqrt{2}}{2}\|\mathbf{F}_1\| \right\rangle + \langle \cos 140° \|\mathbf{F}_2\|, \sin 140° \|\mathbf{F}_2\| \rangle + \langle 0, -8600 \rangle = \langle 0, 0 \rangle$$

Since the x-component and the y-component equal 0, we have a system of two equations:

$$\begin{cases} \dfrac{\sqrt{2}}{2}\|\mathbf{F}_1\| + \cos 140° \|\mathbf{F}_2\| = 0 & \text{(1)} \\[2mm] \dfrac{\sqrt{2}}{2}\|\mathbf{F}_1\| + \sin 140° \|\mathbf{F}_2\| - 8600 = 0 & \text{(2)} \end{cases}$$

Solve equation (1) for $\|\mathbf{F}_1\|$:

$$\|\mathbf{F}_1\| = -\frac{2 \cos 140° \|\mathbf{F}_2\|}{\sqrt{2}} \qquad\qquad \text{(3)}$$

and substitute the result into equation (2).

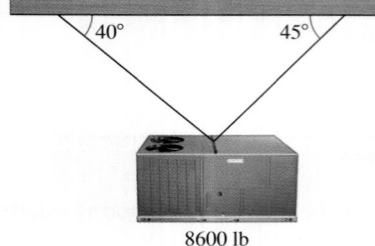

Figure 107

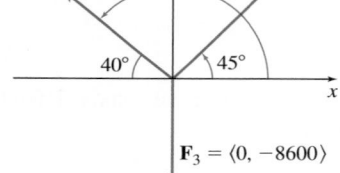

Figure 108

$$\frac{\sqrt{2}}{2}\left(-\frac{2\cos 140°\,\|\mathbf{F}_2\|}{\sqrt{2}}\right)+\sin 140°\,\|\mathbf{F}_2\|-8600=0$$

$$\left[-\cos 140°+\sin 140°\right]\|\mathbf{F}_2\|=8600$$

$$\|\mathbf{F}_2\|=\frac{8600}{\sin 140°-\cos 140°}\approx 6104$$

Then from (3)

$$\|\mathbf{F}_1\|=-\frac{2\cos 140°\,\|\mathbf{F}_2\|}{\sqrt{2}}\approx-\frac{2\cos 140°}{\sqrt{2}}(6104)\approx 6613$$

The left cable has a tension of approximately 6104 lb, and the right cable has a tension of approximately 6613 lb. ∎

NOW WORK   Problem 65.

A.6 Assess Your Understanding

Concepts and Vocabulary

1. True or False   A vector is an entity that has magnitude and direction.

2. Scalars are quantities that have only _____.

3. Multiple Choice   The vectors −**v** and **v** have

    (a) the same   **(b)** different

magnitude and the

    (c) same   **(d)** opposite

direction.

4. A vector whose magnitude is 1 is called a(n) _____ vector.

5. v_1 is called the _____ of $\mathbf{v}=\langle v_1, v_2\rangle$.

6. True or False   If $\mathbf{w}=4\mathbf{v}$, then **w** has the same direction as **v**.

7. True or False   The magnitude of a vector is found by adding its components.

8. True or False $\|-2\,\mathbf{v}\|=-2\|\mathbf{v}\|$

Practice Problems

9. State which of the following are scalars and which are vectors:

    (a) Volume   **(b)** Speed   **(c)** Force   **(d)** Work   **(e)** Mass

    (f) Distance   **(g)** Age   **(h)** Velocity   **(i)** Acceleration

In Problems 10–18, use the vectors in the figure below to graph each of the following vectors:

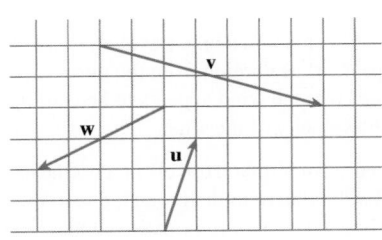

10. $2\mathbf{v}$

11. $-2\mathbf{v}$ [PAGE A-54]

12. $\mathbf{v}+\mathbf{w}$

13. $\mathbf{v}-\mathbf{w}$

14. $\mathbf{w}-\mathbf{v}$

15. $\mathbf{v}-2\mathbf{w}$ [PAGE A-54]

16. $(\mathbf{v}+\mathbf{w})+3\mathbf{u}$

17. $\mathbf{v}+(\mathbf{w}+3\mathbf{u})$

18. $2\mathbf{u}-\dfrac{1}{3}(\mathbf{v}-\mathbf{w})$

In Problems 19–26, use the vectors in the figure below.

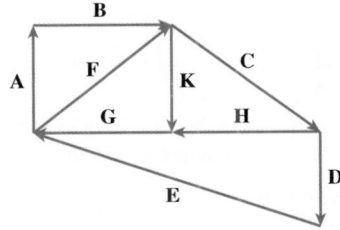

19. Find the vector **x** if $\mathbf{x}+\mathbf{B}=\mathbf{F}$.

20. Find the vector **x** if $\mathbf{x}+\mathbf{K}=\mathbf{C}$.

21. Write **C** in terms of **E**, **D**, and **F**.

22. Write **G** in terms of **C**, **D**, **E**, and **K**.

23. Write **E** in terms of **G**, **H**, and **D**.

24. Write **E** in terms of **A**, **B**, **C**, and **D**.

25. What is $\mathbf{A}+\mathbf{B}+\mathbf{K}+\mathbf{G}$?

26. What is $\mathbf{A}+\mathbf{B}+\mathbf{C}+\mathbf{H}+\mathbf{G}$?

27. Resultant Force   Two forces $\mathbf{F}_1$ and $\mathbf{F}_2$ act on an object as shown in the figure. Graph the vector representing the resultant force; that is, find $\mathbf{F}_1+\mathbf{F}_2$.   [PAGE A-52]

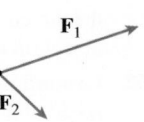

28. Resultant Force   Two forces $\mathbf{F}_1$ and $\mathbf{F}_2$ act on an object as shown in the figure. Graph the vector representing the resultant force; that is, find $\mathbf{F}_1+\mathbf{F}_2$.

*In Problems 29–32, the vector **v** has initial point P_1 and terminal point P_2.*

(a) Write each vector **v** as a position vector.

(b) Graph the vector **v** and the corresponding position vector.

 29. $P_1 = (2, 3)$ and $P_2 = (6, -2)$

30. $P_1 = (4, -1)$ and $P_2 = (-3, 2)$

31. $P_1 = (0, 5)$ and $P_2 = (-1, 6)$

32. $P_1 = (-2, 0)$ and $P_2 = (3, 2)$

In Problems 33–40, use $\mathbf{v} = \langle 3, -1 \rangle$ and $\mathbf{w} = \langle -3, 2 \rangle$ to find each quantity.

33. $2\mathbf{v} - \mathbf{w}$ **34.** $\mathbf{v} + 5\mathbf{w}$ **35.** $\dfrac{1}{3}\mathbf{v} + \dfrac{1}{2}\mathbf{w}$ **36.** $\dfrac{2}{3}\mathbf{v} - \dfrac{1}{2}\mathbf{w}$

**37.** $\|\mathbf{v}\|$ **38.** $\|\mathbf{w}\|$ **39.** $\|2\mathbf{v} - \mathbf{w}\|$ **40.** $\|\mathbf{v} + \mathbf{w}\|$

*In Problems 41–44, find the magnitude of each vector **v**.*

41. $\mathbf{v} = \langle 4, -3 \rangle$ **42.** $\mathbf{v} = \langle 1, -12 \rangle$ **43.** $\mathbf{v} = \langle 1, 1 \rangle$

44. $\mathbf{v} = \langle \cos\theta, \sin\theta \rangle$

*In Problems 45–50, normalize each vector **v**. What is the unit vector in the opposite direction of **v**?*

45. $\mathbf{v} = \langle 5, 12 \rangle$ **46.** $\mathbf{v} = \langle 3, 4 \rangle$ **47.** $\mathbf{v} = \langle 2, 1 \rangle$

48. $\mathbf{v} = \langle 1, -1 \rangle$ **49.** $\mathbf{v} = \left\langle \dfrac{1}{2}, \dfrac{\sqrt{3}}{2} \right\rangle$ **50.** $\mathbf{v} = \left\langle \dfrac{\sqrt{2}}{2}, -\dfrac{\sqrt{2}}{2} \right\rangle$

*In Problems 51–54, determine whether the vectors **v** and **w** are parallel. If the vectors are parallel, are their directions the same or opposite?*

51. $\mathbf{v} = \langle 2, 6 \rangle$, $\mathbf{w} = \langle -1, -3 \rangle$ **52.** $\mathbf{v} = \langle \sqrt{2}, 1 \rangle$, $\mathbf{w} = \langle 2, 1 \rangle$

53. $\mathbf{v} = \langle 12, 3 \rangle$, $\mathbf{w} = \langle 4, -1 \rangle$ **54.** $\mathbf{v} = \langle -6, 8 \rangle$, $\mathbf{w} = \langle -3, 4 \rangle$

55. If $\mathbf{v} = \langle 2, 1 \rangle$ and $\mathbf{w} = \langle x, 3 \rangle$, find all numbers x so that $\|\mathbf{v} + \mathbf{w}\| = 5$.

56. A vector **v** has initial point $P_1 = (-3, 1)$ and terminal point $P_2 = (x, 4)$. Find all numbers x so that **v** has length 5.

In Problems 57–60, find the vector $\mathbf{v} = \langle a, b \rangle$ that has the indicated properties.

57. Magnitude of **v** is 4; **v** makes an angle of $30°$ with the positive x-axis.

58. Magnitude of **v** is 2; **v** makes an angle of $45°$ with the positive x-axis.

59. The y component of **v** is 1; **v** makes an angle of $135°$ with the positive x-axis.

60. The x component of **v** is -3; **v** makes an angle of $210°$ with the positive x-axis.

61. Acceleration A force $\mathbf{F} = \langle 10, -15 \rangle$, measured in newtons (N), is applied to an object of mass 10 kg, causing it to accelerate in the direction of **F**. Find the magnitude of the acceleration of the object (in newtons per kilogram).

62. Ground Speed of an Airplane An airplane has an air speed of 500 km/h in an easterly direction. If there is a southeast wind of 60 km/h, find the speed of the airplane relative to the ground.

63. Position of an Airplane An airplane travels 200 mi due west and then travels 150 mi in a direction $60°$ north of west. Find the vector that gives the position of the airplane.

64. Average Air Speed of an Airplane An airplane, after 1 h of flying, arrives at a point 200 mi due south of the departure point. If, during the flight, there was a steady northwest wind of 20 mi/h, what was the airplane's average air speed?

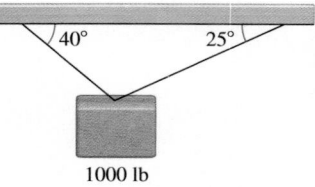
65. Static Equilibrium A weight of 1000 lb is suspended from two cables, as shown in the figure. What is the tension in each cable?

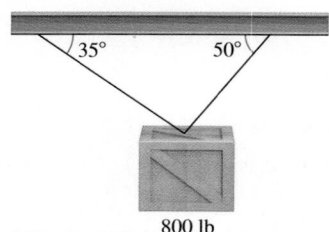
66. Static Equilibrium A weight of 800 lb is suspended from two cables attached to a horizontal steel beam, as shown in the figure. What is the tension in each cable?

67. Tug of War Three pieces of rope are knotted together for use in a three-team tug of war. Each rope is pulled parallel to the ground. As viewed from above, Team A pulls with a force of 1100 lb at an angle of $0°$, Team B pulls with a force of 1050 lb at an angle of $120°$, and Team C pulls with a force of 1000 lb at an angle of $240°$.

(a) Draw a force diagram representing the forces acting on the knot.

(b) Find the resultant force acting on the knot.

68. Pushing a Wheelbarrow A man pushes a wheelbarrow up an incline of $20°$ to the ground with a force of 100 lb. Represent graphically this force.

69. Pulling a Wagon A child pulls a wagon with a force of 40 lb. The handle of the wagon makes an angle of $30°$ with the ground. Represent graphically this force.

70. Orthopedics A patient with a broken tibia has the lower half of his leg suspended horizontally by a cable attached to the center of his foot. See the figure. The tension in the cable needs to exert a force of 25 N horizontally for traction and 40 N vertically upward to support the leg. Find the tension in the cable. *Hint:* Put the center of the patient's foot at the origin of an xy-plane.

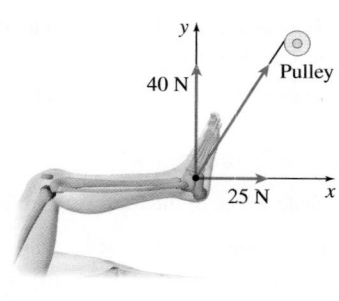

71. Wind Velocity A woman on a bike traveling east at 8 mi/h finds that the wind appears to be coming from the north. Upon doubling her speed, she finds that the wind appears to be coming from the northeast. Find the velocity of the wind.

72. Theoretical Speed An airplane travels in a northeasterly direction at 250 km/h relative to the ground, due to the fact that there is an easterly wind of 50 km/h relative to the ground. How fast would the plane be going if there were no wind?

Theorems and Proofs

B.1 Limit Theorems and Proofs

Uniqueness of a Limit

The limit of a function f, if it exists, is unique; that is, a function can only have one limit.

THEOREM A Limit Is Unique

If a function f is defined on an open interval containing the number c, except possibly at c itself, and if $\lim_{x \to c} f(x) = L_1$ and $\lim_{x \to c} f(x) = L_2$, then $L_1 = L_2$.

NOTE An indirect proof (a proof by contradiction) begins by assuming the conclusion is false. Then we show that this assumption leads to a contradiction.

Proof Assume that $L_1 \neq L_2$. We will show that this assumption leads to a contradiction. By the definition of the limit of a function f, $\lim_{x \to c} f(x) = L_1$ if, for any given number $\varepsilon > 0$, there is a number $\delta_1 > 0$, so that

$$\text{whenever } 0 < |x - c| < \delta_1 \quad \text{then } |f(x) - L_1| < \varepsilon \tag{1}$$

Similarly, $\lim_{x \to c} f(x) = L_2$ if, for any given number $\varepsilon > 0$, there is a number $\delta_2 > 0$, so that

$$\text{whenever } 0 < |x - c| < \delta_2 \quad \text{then } |f(x) - L_2| < \varepsilon \tag{2}$$

Now

$$L_1 - L_2 = L_1 - f(x) + f(x) - L_2$$

so, by applying the Triangle Inequality (see Appendix A.1, p. A-7), we get

$$|L_1 - L_2| = |L_1 - f(x) + f(x) - L_2| \leq |L_1 - f(x)| + |f(x) - L_2| \tag{3}$$

For any given number $\varepsilon > 0$, let δ be the smaller of δ_1 and δ_2. Then from (1) to (3), we can conclude that whenever $0 < |x - c| < \delta \leq \delta_1$ and $0 < |x - c| < \delta \leq \delta_2$, we have

$$|L_1 - L_2| < \varepsilon + \varepsilon = 2\varepsilon \tag{4}$$

In particular, (4) is true for $\varepsilon = \dfrac{1}{2}|L_1 - L_2| > 0$. (Remember $L_1 \neq L_2$.) Then from (4),

$$|L_1 - L_2| < 2\varepsilon = |L_1 - L_2|$$

which is a contradiction. Therefore, $L_1 = L_2$, and the limit, if it exists, is unique. ∎

Algebra of Limits

THEOREM Limit of a Sum

If f and g are functions for which $\lim\limits_{x \to c} f(x)$ and $\lim\limits_{x \to c} g(x)$ both exist, then $\lim\limits_{x \to c} [f(x) + g(x)]$ exists and

$$\lim_{x \to c}[f(x) + g(x)] = \lim_{x \to c} f(x) + \lim_{x \to c} g(x)$$

Proof Suppose $\lim\limits_{x \to c} f(x) = L$ and $\lim\limits_{x \to c} g(x) = M$. We need to show that for any number $\varepsilon > 0$, there is a number $\delta > 0$, so that

$$\text{whenever } 0 < |x - c| < \delta \quad \text{then } |[f(x) + g(x)] - [L + M]| < \varepsilon$$

Since $\lim\limits_{x \to c} f(x) = L$, by the definition of a limit, given the number $\dfrac{\varepsilon}{2} > 0$, there is a number $\delta_1 > 0$, so that

$$\text{whenever } 0 < |x - c| < \delta_1 \quad \text{then } |f(x) - L| < \frac{\varepsilon}{2}$$

Since $\lim\limits_{x \to c} g(x) = M$, for this same number $\dfrac{\varepsilon}{2}$, there is a number $\delta_2 > 0$, so that

$$\text{whenever } 0 < |x - c| < \delta_2 \quad \text{then } |g(x) - M| < \frac{\varepsilon}{2}$$

Let δ be the smaller of δ_1 and δ_2. Then $\delta \le \delta_1$ and $\delta \le \delta_2$. Using this δ,

$$\text{whenever } 0 < |x - c| < \delta \quad \text{then } |f(x) - L| < \frac{\varepsilon}{2}$$

$$\text{whenever } 0 < |x - c| < \delta \quad \text{then } |g(x) - M| < \frac{\varepsilon}{2}$$

That is, whenever $0 < |x - c| < \delta$,

$$|[f(x) + g(x)] - [L + M]| = |[f(x) - L] + [g(x) - M]|$$
$$\le |f(x) - L| + |g(x) - M| \quad \text{Use the Triangle Inequality.}$$
$$< \frac{\varepsilon}{2} + \frac{\varepsilon}{2} = \varepsilon$$

So, $\lim\limits_{x \to c}[f(x) + g(x)] = L + M$. ∎

THEOREM Limit of a Product

If f and g are functions for which $\lim\limits_{x \to c} f(x)$ and $\lim\limits_{x \to c} g(x)$ both exist, then $\lim\limits_{x \to c} [f(x) \cdot g(x)]$ exists and

$$\lim_{x \to c} [f(x) \cdot g(x)] = \lim_{x \to c} f(x) \cdot \lim_{x \to c} g(x)$$

Proof Suppose $\lim\limits_{x \to c} f(x) = L$ and $\lim\limits_{x \to c} g(x) = M$. We need to show that for any number $\varepsilon > 0$, there is a number $\delta > 0$, so that

$$\text{whenever } 0 < |x - c| < \delta \quad \text{then } |f(x) \cdot g(x) - L \cdot M| < \varepsilon$$

Subtracting and adding $f(x) \cdot M$ in the expression $f(x) \cdot g(x) - L \cdot M$ result in terms involving $g(x) - M$ and $f(x) - L$:

$$
\begin{aligned}
|f(x) \cdot g(x) - L \cdot M| &= |f(x) \cdot g(x) - f(x) \cdot M + f(x) \cdot M - L \cdot M| \\
&= |f(x) \cdot [g(x) - M] + [f(x) - L] \cdot M| \\
&\le |f(x)| \cdot |g(x) - M| + |f(x) - L| \cdot |M|
\end{aligned}
\tag{5}
$$
$$\uparrow$$

Use the Triangle Inequality.

Since $\lim\limits_{x \to c} f(x) = L$, there is a number $\delta_1 > 0$, so that whenever $0 < |x - c| < \delta_1$, then

$$
|f(x) - L| < 1, \qquad \text{from which} \qquad |f(x)| < 1 + |L| \tag{6}
$$

Also since $\lim\limits_{x \to c} g(x) = M$, given a number $\varepsilon > 0$, there is a number δ_2, so that whenever $0 < |x - c| < \delta_2$, then

$$
|g(x) - M| < \frac{\varepsilon}{1 + |L| + |M|} \tag{7}
$$

Given a number $\varepsilon > 0$, there is a number δ_3, so that whenever $0 < |x - c| < \delta_3$, then

$$
|f(x) - L| < \frac{\varepsilon}{1 + |L| + |M|} \tag{8}
$$

Choose δ to be the minimum of δ_1, δ_2, and δ_3 and combine (5) to (8). Then for any given $\varepsilon > 0$, there is a $\delta > 0$, so that whenever $0 < |x - c| < \delta$, we have

$$
\begin{aligned}
|f(x) \cdot g(x) - L \cdot M| &< [1 + |L|]\frac{\varepsilon}{1 + |L| + |M|} + |M|\frac{\varepsilon}{1 + |L| + |M|} \\
&< [1 + |L| + |M|]\frac{\varepsilon}{1 + |L| + |M|} = \varepsilon
\end{aligned}
$$

That is, $\lim\limits_{x \to c} [f(x) \cdot g(x)] = \lim\limits_{x \to c} f(x) \cdot \lim\limits_{x \to c} g(x)$. ∎

THEOREM Squeeze Theorem

If the functions f, g, and h have the property that for all x in an open interval containing c, except possibly at c itself,

$$\boxed{f(x) \le g(x) \le h(x)}$$

and if

$$\boxed{\lim\limits_{x \to c} f(x) = \lim\limits_{x \to c} h(x) = L}$$

then

$$\boxed{\lim\limits_{x \to c} g(x) = L}$$

Proof Since $\lim\limits_{x \to c} f(x) = \lim\limits_{x \to c} h(x) = L$, then for any number $\varepsilon > 0$, there are positive numbers δ_1 and δ_2, so that

$$
\text{whenever } 0 < |x - c| < \delta_1 \quad \text{then } |f(x) - L| < \varepsilon
$$
$$
\text{whenever } 0 < |x - c| < \delta_2 \quad \text{then } |h(x) - L| < \varepsilon
$$

Choose δ to be the smaller of the numbers δ_1 and δ_2. Then $0 < |x - c| < \delta$ implies that both $|f(x) - L| < \varepsilon$ and $|h(x) - L| < \varepsilon$. In other words, $0 < |x - c| < \delta$ implies that both

$$L - \varepsilon < f(x) < L + \varepsilon \qquad \text{and} \qquad L - \varepsilon < h(x) < L + \varepsilon$$

Since $f(x) \le g(x) \le h(x)$ for all $x \ne c$ in the open interval, it follows that whenever $0 < |x - c| < \delta$ and x is in the open interval, we have

$$L - \varepsilon < f(x) \le g(x) \le h(x) < L + \varepsilon$$

Then for any given number $\varepsilon > 0$, there is a positive number δ, so that whenever $0 < |x - c| < \delta$, then $L - \varepsilon < g(x) < L + \varepsilon$, or equivalently, $|g(x) - L| < \varepsilon$. That is, $\lim_{x \to c} g(x) = L$. ∎

B.2 Theorems and Proofs Involving Inverse Functions

THEOREM Continuity of the Inverse Function

If f is a one-to-one function that is continuous on its domain, then its inverse function f^{-1} is also continuous on its domain.

Proof Let (a, b) be the largest open interval included in the domain of f. If f is continuous on its domain, then it is continuous on (a, b). Since f is one-to-one, then f is either increasing on (a, b) or decreasing on (a, b).

Suppose f is increasing on (a, b). Then f^{-1} is also increasing. Let $y_0 = f(x_0)$. We need to show that f^{-1} is continuous at y_0, given that f is continuous at x_0. The number $f^{-1}(y_0) = x_0$ is in the open interval (a, b). Choose $\varepsilon > 0$ sufficiently small, so that $f^{-1}(y_0) - \varepsilon$ and $f^{-1}(y_0) + \varepsilon$ are also in (a, b). Then choose δ so that

$$f[f^{-1}(y_0) - \varepsilon] < y_0 - \delta \qquad \text{and} \qquad y_0 + \delta < f[f^{-1}(y_0) + \varepsilon]$$

Then whenever $y_0 - \delta < y < y_0 + \delta$, we have

$$f[f^{-1}(y_0) - \varepsilon] < y < f[f^{-1}(y_0) + \varepsilon]$$

This means whenever $0 < |y - y_0| < \delta$, then

$$f^{-1}(y_0) - \varepsilon < f^{-1}(y) < f^{-1}(y_0) + \varepsilon \qquad \text{or equivalently,} \qquad |f^{-1}(y) - f^{-1}(y_0)| < \varepsilon$$

That is, $\lim_{y \to y_0} f^{-1}(y) = f^{-1}(y_0)$, so f^{-1} is continuous at y_0.

The case where f is decreasing on (a, b) is proved in a similar way. ∎

THEOREM Derivative of the Inverse Function

Let $y = f(x)$ and $x = g(y)$ be inverse functions. If f is differentiable on an open interval containing x_0 and if $f'(x_0) \ne 0$, then g is differentiable at $y_0 = f(x_0)$ and

$$\frac{d}{dy} g(y_0) = \frac{1}{f'(x_0)}$$

where the notation $f'(x_0)$ means the value of $f'(x)$ at x_0 and the notation $\frac{d}{dy} g(y_0)$ means the value of $\frac{d}{dy} g(y)$ at y_0.

Proof Since f and g are inverses of one another, then $f(x) = y$ if and only if $x = g(y)$. So, we have the following identity, where $g(y_0) = x_0$:

$$\frac{g(y) - g(y_0)}{y - y_0} = \frac{x - x_0}{f(x) - f(x_0)} = \frac{1}{\dfrac{f(x) - f(x_0)}{x - x_0}}$$

By the continuity of an inverse function, the continuity of f at x_0 implies the continuity of g at y_0, and $y \to y_0$ as $x \to x_0$.

Now take the limits of both sides of the above identity. Since $f'(x_0) \neq 0$, we have

$$g'(y_0) = \lim_{y \to y_0} \frac{g(y) - g(y_0)}{y - y_0} = \frac{1}{\displaystyle\lim_{x \to x_0} \frac{f(x) - f(x_0)}{x - x_0}} = \frac{1}{f'(x_0)}$$ ∎

B.3 Derivative Theorems and Proofs

A proof of the Chain Rule when Δu is never 0 appears in Chapter 3. Here, we consider the case when Δu may be 0.

THEOREM   Chain Rule

If a function g is differentiable at x_0, and a function f is differentiable at $g(x_0)$, then the composite function $f \circ g$ is differentiable at x_0 and

$$\boxed{(f \circ g)'(x_0) = f'(g(x_0)) \cdot g'(x_0)}$$

Using Leibniz notation, if $y = f(u)$ and $u = g(x)$, then

$$\boxed{\frac{dy}{dx} = \frac{dy}{du} \cdot \frac{du}{dx}}$$

where $\dfrac{dy}{du}$ is evaluated at $u_0 = g(x_0)$ and $\dfrac{du}{dx}$ is evaluated at x_0.

Proof For the fixed number x_0, let $\Delta u = g(x_0 + \Delta x) - g(x_0)$. Since the function $u = g(x)$ is differentiable at x_0, it is also continuous at x_0, and therefore $\Delta u \to 0$ as $\Delta x \to 0$.

For the fixed number $u_0 = g(x_0)$, let $\Delta y = f(u_0 + \Delta u) - f(u_0)$. Since the function $y = f(u)$ is differentiable at the number u_0, we can write $\displaystyle\lim_{\Delta u \to 0} \frac{\Delta y}{\Delta u} = \frac{dy}{du}(u_0)$. This implies that for any $\Delta u \neq 0$:

$$\frac{\Delta y}{\Delta u} = \frac{dy}{du}(u_0) + \alpha \tag{1}$$

where $\alpha = \alpha(\Delta u)$ is a function of Δu for which $\alpha \to 0$ as $\Delta u \to 0$. Now define $\alpha = 0$ when $\Delta u = 0$, so that $\alpha = \alpha(\Delta u)$ is continuous at 0. Multiplying (1) by $\Delta u \neq 0$, we obtain

$$\Delta y = \frac{dy}{du}(u_0) \cdot \Delta u + \alpha(\Delta u) \cdot \Delta u \tag{2}$$

Notice that equation (2) is true for all Δu:

- If Δu is not equal to 0, then (2) is a consequence of (1).
- If Δu equals 0, then the left-hand side of (2) is

$$\Delta y = f(u_0 + \Delta u) - f(u_0) = f(u_0) - f(u_0) = 0$$

and the right-hand side of (2) is also 0.

Now divide (2) by $\Delta x \neq 0$:

$$\frac{\Delta y}{\Delta x} = \frac{dy}{du}(u_0) \cdot \frac{\Delta u}{\Delta x} + \alpha(\Delta u) \cdot \frac{\Delta u}{\Delta x} \qquad (3)$$

Since the function $u = g(x)$ is differentiable at x_0, $\lim\limits_{\Delta x \to 0} \dfrac{\Delta u}{\Delta x} = \dfrac{du}{dx}(x_0)$. Also, since $\Delta u \to 0$ when $\Delta x \to 0$ and $\alpha(\Delta u)$ is continuous at $\Delta u = 0$, we conclude that $\alpha(\Delta u) \to 0$ as $\Delta x \to 0$. So we can take the limit of (3) as $\Delta x \to 0$, which proves that the derivative $\dfrac{dy}{dx}(x_0)$ exists and is equal to

$$\frac{dy}{dx}(x_0) = \lim_{\Delta x \to 0} \frac{\Delta y}{\Delta x} = \lim_{\Delta x \to 0}\left[\frac{dy}{du}(u_0) \cdot \frac{\Delta u}{\Delta x} + \alpha(\Delta u) \cdot \frac{\Delta u}{\Delta x}\right]$$

$$= \frac{dy}{du}(u_0) \cdot \left[\lim_{\Delta x \to 0} \frac{\Delta u}{\Delta x}\right] + \left[\lim_{\Delta x \to 0} \alpha(\Delta u)\right] \cdot \left[\lim_{\Delta x \to 0} \frac{\Delta u}{\Delta x}\right]$$

$$= \frac{dy}{du}(u_0) \cdot \frac{du}{dx}(x_0) + 0 \cdot \frac{du}{dx}(x_0) = \frac{dy}{du}(u_0) \cdot \frac{du}{dx}(x_0) \qquad \blacksquare$$

Partial Proof of L'Hôpital's Rule

To prove L'Hôpital's Rule requires an extension of the Mean Value Theorem, called *Cauchy's Mean Value Theorem.*

THEOREM Cauchy's Mean Value Theorem

If the functions f and g are continuous on the closed interval $[a, b]$ and differentiable on the open interval (a, b), and if $g'(x) \neq 0$ on (a, b), then there is a number c in (a, b) for which

$$\boxed{\frac{f'(c)}{g'(c)} = \frac{f(b) - f(a)}{g(b) - g(a)}}$$

ORIGINS The theorem was named after the French mathematician Augustin Cauchy (1789–1857).

Notice that under the conditions of Cauchy's Mean Value Theorem, $g(b) \neq g(a)$, because otherwise, by Rolle's Theorem, $g'(c) = 0$ for some c in the interval (a, b).

Proof Define the function h as

$$h(x) = [g(b) - g(a)][f(x) - f(a)] - [g(x) - g(a)][f(b) - f(a)] \qquad a \leq x \leq b$$

Then h is continuous on $[a, b]$ and differentiable on (a, b) and $h(a) = h(b) = 0$. So by Rolle's Theorem, there is a number c in the interval (a, b) for which $h'(c) = 0$. That is,

$$h'(c) = [g(b) - g(a)]f'(c) - g'(c)[f(b) - f(a)] = 0$$

$$[g(b) - g(a)]f'(c) = g'(c)[f(b) - f(a)]$$

$$\frac{f'(c)}{g'(c)} = \frac{f(b) - f(a)}{g(b) - g(a)} \qquad \blacksquare$$

NOTE A special case of Cauchy's Mean Value Theorem is the Mean Value Theorem. To get the Mean Value Theorem, let $g(x) = x$. Then $g'(x) = 1$, $g(b) = b$, and $g(a) = a$, giving the Mean Value Theorem.

THEOREM L'Hôpital's Rule

Suppose the functions f and g are differentiable on an open interval I containing the number c, except possibly at c, and $g'(x) \neq 0$ for all $x \neq c$ in I. Let L denote either a real number or $\pm\infty$, and suppose $\dfrac{f(x)}{g(x)}$ is an indeterminate form at c of the type $\dfrac{0}{0}$ or $\dfrac{\infty}{\infty}$. If $\lim\limits_{x \to c} \dfrac{f'(x)}{g'(x)} = L$, then $\lim\limits_{x \to c} \dfrac{f(x)}{g(x)} = L$.

Partial Proof   Suppose $\dfrac{f(x)}{g(x)}$ is an indeterminate form at c of the type $\dfrac{0}{0}$, and

suppose $\lim\limits_{x \to c} \dfrac{f'(x)}{g'(x)} = L$, where L is a real number. We need to prove $\lim\limits_{x \to c} \dfrac{f(x)}{g(x)} = L$.
First define the functions F and G as follows:

$$F(x) = \begin{cases} f(x) & \text{if } x \neq c \\ 0 & \text{if } x = c \end{cases} \qquad G(x) = \begin{cases} g(x) & \text{if } x \neq c \\ 0 & \text{if } x = c \end{cases}$$

Both F and G are continuous at c, since $\lim\limits_{x \to c} F(x) = \lim\limits_{x \to c} f(x) = 0 = F(c)$
and $\lim\limits_{x \to c} G(x) = \lim\limits_{x \to c} g(x) = 0 = G(c)$. Also,

$$F'(x) = f'(x) \qquad \text{and} \qquad G'(x) = g'(x)$$

for all x in the interval I, except possibly at c. Since the conditions for Cauchy's Mean Value Theorem are met by F and G in either $[x, c]$ or $[c, x]$, there is a number u between c and x for which

$$\frac{F(x) - F(c)}{G(x) - G(c)} = \frac{F'(u)}{G'(u)} = \frac{f'(u)}{g'(u)}$$

Since $F(c) = 0$ and $G(c) = 0$, this simplifies to $\dfrac{f(x)}{g(x)} = \dfrac{f'(u)}{g'(u)}$.
Since u is between c and x, it follows that

$$\lim_{x \to c} \frac{f(x)}{g(x)} = \lim_{u \to c} \frac{f'(u)}{g'(u)} = L$$

A similar argument is used if L is infinite. The proof when $\dfrac{f(x)}{g(x)}$ is an indeterminate
form at ∞ of the type $\dfrac{\infty}{\infty}$ is omitted here, but it may be found in books on advanced
calculus. ∎

The use of L'Hôpital's Rule when $c = \infty$ for an indeterminate form of the
type $\dfrac{0}{0}$ is justified by the following argument. In $\lim\limits_{x \to \infty} \dfrac{f(x)}{g(x)}$, let $x = \dfrac{1}{u}$. Then
as $x \to \infty$, $u \to 0^+$, and

$$\lim_{x \to \infty} \frac{f(x)}{g(x)} = \lim_{u \to 0^+} \frac{f\left(\dfrac{1}{u}\right)}{g\left(\dfrac{1}{u}\right)} = \lim_{u \to 0^+} \frac{\dfrac{d}{du} f\left(\dfrac{1}{u}\right)}{\dfrac{d}{du} g\left(\dfrac{1}{u}\right)} = \lim_{u \to 0^+} \frac{-\dfrac{1}{u^2} f'\left(\dfrac{1}{u}\right)}{-\dfrac{1}{u^2} g'\left(\dfrac{1}{u}\right)} = \lim_{x \to \infty} \frac{f'(x)}{g'(x)} = L$$

$$\uparrow \text{ Chain Rule} \qquad\qquad\qquad \uparrow \; x = \frac{1}{u}$$

B.4 Integral Theorems and Proofs

THEOREM

If a function f is continuous on an interval containing the numbers a, b, and c, then

$$\int_a^b f(x)\,dx = \int_a^c f(x)\,dx + \int_c^b f(x)\,dx$$

Proof Since f is continuous on an interval containing a, b, and c, the three integrals above exist.

Part 1 Assume $a < b < c$. Since f is continuous on $[a, b]$ and on $[b, c]$, given any $\varepsilon > 0$, there is a number $\delta_1 > 0$, so that

$$\left| \sum_{i=1}^{k} f(u_i) \Delta x_i - \int_a^b f(x)\, dx \right| < \frac{\varepsilon}{2} \tag{1}$$

for every Riemann sum $\sum_{i=1}^{k} f(u_i) \Delta x_i$ for f on $[a, b]$, where $x_{i-1} \le u_i \le x_i$, $i = 1, 2, \ldots, k$, and whose partition P_1 of $[a, b]$ has norm $\| P_1 \| < \delta_1$. There is also a number $\delta_2 > 0$ for which

$$\left| \sum_{i=k+1}^{n} f(u_i) \Delta x_i - \int_b^c f(x)\, dx \right| < \frac{\varepsilon}{2} \tag{2}$$

for every Riemann sum $\sum_{i=k+1}^{n} f(u_i) \Delta x_i$ for f on $[b, c]$, where $x_{i-1} \le u_i \le x_i$, $i = k+1, k+2, \ldots, n$, and whose partition P_2 of $[b, c]$ has norm $\| P_2 \| < \delta_2$.

NOTE The partition P_1 is $x_0 = a, \ldots, x_k = b$. The partition P_2 is $x_k = b, \ldots, x_n = c$.

Let δ be the smaller of δ_1 and δ_2. Then (1) and (2) hold, with δ replacing δ_1 and δ_2. If (1) and (2) are added and if $\| P_1 \| < \delta$ and $\| P_2 \| < \delta$, then

$$\left| \sum_{i=1}^{k} f(u_i) \Delta x_i - \int_a^b f(x)\, dx \right| + \left| \sum_{i=k+1}^{n} f(u_i) \Delta x_i - \int_b^c f(x)\, dx \right| < \frac{\varepsilon}{2} + \frac{\varepsilon}{2} = \varepsilon$$

Using the Triangle Inequality, this result implies that for $\| P_1 \| < \delta$ and $\| P_2 \| < \delta$,

$$\left| \sum_{i=1}^{k} f(u_i) \Delta x_i - \int_a^b f(x)\, dx + \sum_{i=k+1}^{n} f(u_i) \Delta x_i - \int_b^c f(x)\, dx \right| < \varepsilon \tag{3}$$

Denote $P_1 \cup P_2$ by P^*. Then P^* is a partition of $[a, c]$ having the number $b = x_k$ as an endpoint of the kth subinterval. So,

$$\sum_{i=1}^{k} f(u_i) \Delta x_i + \sum_{i=k+1}^{n} f(u_i) \Delta x_i = \sum_{i=1}^{n} f(u_i) \Delta x_i$$

are Riemann sums for f on P^*. Since $\| P^* \| < \delta$ implies that $\| P_1 \| < \delta$ and $\| P_2 \| < \delta$, it follows from (3) that

$$\left| \sum_{i=1}^{n} f(u_i) \Delta x_i - \left[\int_a^b f(x)\, dx + \int_b^c f(x)\, dx \right] \right| < \varepsilon$$

for every Riemann sum $\sum_{i=1}^{n} f(u_i) \Delta x_i$ for f on $[a, c]$ whose partition P^* of $[a, c]$ has b as an endpoint of a subinterval of the partition and has norm $\| P^* \| < \delta$. Therefore,

$$\int_a^c f(x)\, dx = \int_a^b f(x)\, dx + \int_b^c f(x)\, dx$$

Part 2 There are six possible orderings (permutations) of the numbers a, b, and c:

$$a < b < c \quad a < c < b \quad b < a < c \quad b < c < a \quad c < a < b \quad c < b < a$$

In Part 1, we showed that the theorem is true for the order $a < b < c$. Now consider any other order, say, $b < c < a$. From Part 1,

$$\int_b^c f(x)\,dx + \int_c^a f(x)\,dx = \int_b^a f(x)\,dx \qquad (4)$$

But,

$$\int_c^a f(x)\,dx = -\int_a^c f(x)\,dx \quad \text{and} \quad \int_b^a f(x)\,dx = -\int_a^b f(x)\,dx$$

Now we substitute this into (4).

$$\int_b^c f(x)\,dx - \int_a^c f(x)\,dx = -\int_a^b f(x)\,dx$$

$$\int_a^b f(x)\,dx + \int_b^c f(x)\,dx = \int_a^c f(x)\,dx$$

proving the theorem for $b < c < a$.

The proofs for the remaining four permutations of a, b, and c are similar. ∎

THEOREM Fundamental Theorem of Calculus, Part 1

Let f be a function that is continuous on a closed interval $[a, b]$. The function I defined by

$$\boxed{I(x) = \int_a^x f(t)\,dt}$$

has the property that it is continuous on $[a, b]$ and differentiable on (a, b). Moreover,

$$\boxed{I'(x) = \frac{d}{dx}\left[\int_a^x f(t)\,dt\right] = f(x)}$$

for all x in (a, b).

Proof Let x and $x + h$, $h \neq 0$, be in the interval (a, b). Then

$$I(x) = \int_a^x f(t)\,dt \qquad I(x + h) = \int_a^{x+h} f(t)\,dt$$

and

$$I(x + h) - I(x) = \int_a^{x+h} f(t)\,dt + \int_x^a f(t)\,dt \qquad \int_x^a f(t)\,dt = -\int_a^x f(t)\,dt$$

$$= \int_x^a f(t)\,dt + \int_a^{x+h} f(t)\,dt = \int_x^{x+h} f(t)\,dt$$

Dividing both sides by $h \neq 0$, we get

$$\frac{I(x + h) - I(x)}{h} = \frac{1}{h}\int_x^{x+h} f(t)\,dt \qquad (5)$$

NEED TO REVIEW? The Mean Value Theorem for Integrals is discussed in Section 6.4.

Now we use the Mean Value Theorem for Integrals in the integral on the right. There are two possibilities: either $h > 0$ or $h < 0$.

If $h > 0$, there is a number u, where $x \le u \le x + h$, for which

$$\int_x^{x+h} f(t)\,dt = f(u)h$$

$$\frac{1}{h} \int_x^{x+h} f(t)\,dt = f(u)$$

$$\frac{I(x+h) - I(x)}{h} = f(u) \qquad \text{From (5)}$$

Since $x \le u \le x + h$, as $h \to 0^+$, u approaches x^+, so

$$\lim_{h \to 0^+} \frac{I(x+h) - I(x)}{h} = \lim_{h \to 0^+} f(u) = \lim_{u \to x^+} f(u) = f(x)$$
$$\underset{f \text{ is continuous.}}{\uparrow}$$

Using a similar argument for $h < 0$, we obtain

$$\lim_{h \to 0^-} \frac{I(x+h) - I(x)}{h} = f(x)$$

Since the two one-sided limits are equal,

$$\lim_{h \to 0} \frac{I(x+h) - I(x)}{h} = f(x)$$

The limit is the derivative of the function I, meaning $I'(x) = f(x)$ for all x in (a, b). ∎

THEOREM Bounds on an Integral

If a function f is continuous on a closed interval $[a, b]$ and if m and M denote the absolute minimum and absolute maximum values of f on $[a, b]$, respectively, then

$$\boxed{m(b - a) \le \int_a^b f(x)\,dx \le M(b - a)}$$

The Bounds on an Integral Theorem is proved using an indirect proof. That is, it is proved by contradiction.

Proof Part 1 $m(b - a) \le \displaystyle\int_a^b f(x)\,dx$

Assume

$$m(b - a) > \int_a^b f(x)\,dx \tag{6}$$

Since f is continuous on $[a, b]$,

$$\lim_{\|P\| \to 0} \sum_{i=1}^n f(u_i)\Delta x_i = \int_a^b f(x)\,dx$$

By (6), $m(b - a) - \displaystyle\int_a^b f(x)\,dx > 0$. We choose ε, so that

$$\varepsilon = m(b - a) - \int_a^b f(x)\,dx > 0 \tag{7}$$

Then there is a number $\delta > 0$, so that for all partitions P of $[a, b]$ with norm $\|P\| < \delta$, we have

$$\left| \sum_{i=1}^{n} f(u_i) \Delta x_i - \int_{a}^{b} f(x)\, dx \right| < \varepsilon$$

which is equivalent to

$$\int_{a}^{b} f(x)\, dx - \varepsilon < \sum_{i=1}^{n} f(u_i) \Delta x_i < \int_{a}^{b} f(x)\, dx + \varepsilon$$

By (7), the right inequality can be expressed as

$$\sum_{i=1}^{n} f(u_i) \Delta x_i < \int_{a}^{b} f(x)\, dx + \varepsilon = \int_{a}^{b} f(x)\, dx + \left[m(b-a) - \int_{a}^{b} f(x)\, dx \right]$$
$$= m(b-a)$$

Consequently,

$$\sum_{i=1}^{n} f(u_i) \Delta x_i < m(b-a) = \sum_{i=1}^{n} m\, \Delta x_i$$

implying that for every partition P of $[a, b]$ with norm $\|P\| < \delta$,

$$f(u_i) < m$$

for some u_i in $[a, b]$. But this is impossible because m is the absolute minimum of f on $[a, b]$. Therefore, the assumption $m(b-a) > \int_{a}^{b} f(x)\, dx$ is false. That is,

$$m(b-a) \leq \int_{a}^{b} f(x)\, dx$$

Part 2 To prove $\int_{a}^{b} f(x)\, dx \leq M(b-a)$, use a similar argument. ∎

B.5 A Bounded Monotonic Sequence Converges

THEOREM

An increasing (or nondecreasing) sequence $\{s_n\}$ that is bounded from above converges.

A decreasing (or nonincreasing) sequence $\{s_n\}$ that is bounded from below converges.

To prove this theorem, we need the following property of real numbers. The set of real numbers is defined by a collection of axioms. One of these axioms is the *Completeness Axiom*.

Completeness Axiom of Real Numbers

If S is a nonempty set of real numbers that has an upper bound, then it has a least upper bound. Similarly, if S has a lower bound, then it has a greatest lower bound.

As an example, consider the set S: $\{x \mid x^2 < 2, \ x > 0\}$. The set of upper bounds to S is the set $\{x \mid x^2 \geq 2, \ x > 0\}$.

(a) If our universe is the set of rational numbers, the set of upper bounds has no minimum (since $\sqrt{2}$ is not rational).

(b) If our universe is the set of real numbers, then by the Completeness Axiom, the set of upper bounds has a minimum ($\sqrt{2}$). That is, this axiom completes the set of real numbers by incorporating the set of irrational numbers with the set of rational numbers to form the set of real numbers.

We prove the theorem for a nondecreasing sequence $\{s_n\}$. The proofs for the other three cases are similar.

Proof Suppose $\{s_n\}$ is a nondecreasing sequence that is bounded from above. Since $\{s_n\}$ is bounded from above, there is a positive number K (an upper bound), so that $s_n \leq K$ for every n. From the Completeness Axiom, the set $\{s_n\}$ has a least upper bound L. That is, $s_n \leq L$ for every n.

Then for any $\varepsilon > 0$, $L - \varepsilon$ is not an upper bound of $\{s_n\}$. That is, $L - \varepsilon < s_N$ for some integer N. Since $\{s_n\}$ is nondecreasing, $s_N \leq s_n$ for all $n > N$. Then for all $n > N$,

$$L - \varepsilon < s_n \leq L < L + \varepsilon$$

That is, $|s_n - L| < \varepsilon$ for all $n > N$, so the sequence $\{s_n\}$ converges to L. ■

B.6 Taylor's Formula with the Lagrange Form of the Remainder

THEOREM Taylor's Formula with the Lagrange Form of the Remainder

Let f be a function whose first $n + 1$ derivatives are continuous on an open interval I containing the number c. Then for every x in the interval, there is a number u between x and c for which

$$f(x) = f(c) + f'(c)(x - c) + \frac{f''(c)}{2!}(x - c)^2 + \cdots + \frac{f^{(n)}(c)}{n!}(x - c)^n + R_n(x)$$

where

$$R_n(x) = \frac{f^{(n+1)}(u)}{(n+1)!}(x - c)^{n+1}$$

NOTE If $n \geq 0$ is an integer, the **factorial symbol** $n!$ means $0! = 1$, $1! = 1$, and $n! = 1 \cdot 2 \cdot 3 \cdot \cdots \cdot (n-1) \cdot n$, where $n > 1$.

Proof For a fixed number $x \neq c$ in the open interval I, there is a number L (depending on x) for which

$$f(x) = f(c) + \frac{f'(c)}{1!}(x - c) + \frac{f''(c)}{2!}(x - c)^2 + \cdots + \frac{f^{(n)}(c)}{n!}(x - c)^n + \frac{L}{(n+1)!}(x - c)^{n+1} \quad (1)$$

Define the function F to be

$$F(t) = f(x) - f(t) - \frac{f'(t)}{1!}(x - t) - \frac{f''(t)}{2!}(x - t)^2 - \cdots - \frac{f^{(n)}(t)}{n!}(x - t)^n - \frac{L}{(n+1)!}(x - t)^{n+1} \quad (2)$$

The domain of F is $c \le t \le x$ if $x > c$ and $x \le t \le c$ if $x < c$. Since $f(t)$, $f'(t)$, $f''(t)$, $\ldots$, $f^{(n)}(t)$ are each continuous, then F is continuous on its domain. Furthermore, F is differentiable and

$$\frac{dF}{dt} = F'(t) = -f'(t) + \left[f'(t) - \frac{f''(t)}{1!}(x-t) \right] + \left[\frac{f''(t)}{1!}(x-t) - \frac{f'''(t)}{2!}(x-t)^2 \right]$$

$$+ \cdots + \left[\frac{f^{(n)}(t)}{(n-1)!}(x-t)^{n-1} - \frac{f^{(n+1)}(t)}{n!}(x-t)^n \right] + \frac{L}{n!}(x-t)^n$$

$$= -\frac{f^{(n+1)}(t)}{n!}(x-t)^n + \frac{L}{n!}(x-t)^n$$

for all t between x and c. From (1) and (2), we have

$$F(c) = f(x) - f(c) - \frac{f'(c)}{1!}(x-c) - \frac{f''(c)}{2!}(x-c)^2 - \cdots - \frac{f^{(n)}(c)}{n!}(x-c)^n - \frac{L}{(n+1)!}(x-c)^{n+1} = 0$$

Then

$$F(x) = f(x) - f(x) - \frac{f'(x)}{1!}(x-x) - \cdots - \frac{f^{(n)}(x)}{n!}(x-x)^n - \frac{L}{(n+1)!}(x-x)^{n+1} = 0$$

Now apply Rolle's Theorem to F. Then there is a number u between c and x for which

$$F'(u) = -\frac{f^{(n+1)}(u)}{n!}(x-u)^n + \frac{L}{n!}(x-u)^n = 0$$

Solving for L, we find $L = f^{(n+1)}(u)$. Now let $t = c$ and $L = f^{(n+1)}(u)$ in (2) and solve for $f(x)$. Then

$$f(x) = f(c) + f'(c)(x-c) + \frac{f''(c)}{2!}(x-c)^2 + \cdots + \frac{f^{(n)}(c)}{n!}(x-c)^n + R_n(x)$$

where

$$R_n(x) = \frac{f^{(n+1)}(u)}{(n+1)!}(x-c)^{n+1}$$

∎

Using Calculators on the AP® Calculus Exam

OBJECTIVES *When you finish this section, you should be able to use a graphing calculator to:*

1 **Graph an explicitly defined function (p. C-2)**

2 **Solve an equation (p. C-2)**

3 **Approximate the derivative of a function at a number (p. C-4)**

4 **Approximate a definite integral (p. C-4)**

Most modern applications of calculus involve real-world data collected in a particular context and use calculators and other technology to make calculations that would otherwise be tedious or impossible. Although technology removes many of the burdens of calculations, it does not lessen the importance of knowing and understanding the underlying mathematics. One still must understand the concepts, definitions, theorems, and notation of calculus before using technology to interpret the results.

On the AP® Calculus AB and BC exams, approved graphing calculators can be used on one-third of the exam (the last 15 of 45 multiple-choice questions and the first 2 of 6 free-response questions). While a majority of schools use Texas Instruments (TI) calculators, this guide has been written to assist you in navigating the skills required with any calculator, provided that you are familiar with the specific syntax and keystrokes required of your machine.

In *Calculus for the AP® Course*, 4th edition, you will see screen captures generated by a TI-84 Plus CE graphing calculator and exercises marked with the icon 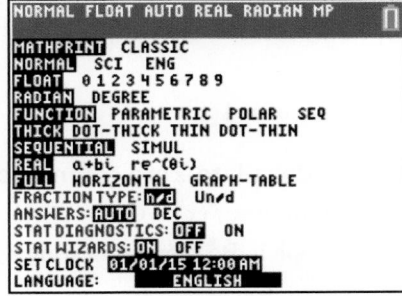, which alert you that a graphing calculator is recommended.

Here are some tips for using a graphing calculator in AP® Calculus.

• Set the calculator to radian mode (not degrees). See Figure 1. Calculus uses radian measure. Assume all angles are measured in radians, unless degrees are specifically stated.

Figure 1

- Become familiar with the power of the **Y= editor**. The Y= editor is used for graphing functions, but it can be used for much more. Once a function has been entered into the Y= editor, it can be recalled later, avoiding the need to reenter the function.
- Use the store command **STO→**, which places numbers in the memory of the calculator. Saving decimals here and rounding or truncating at the end of a solution not only provides better approximations but also provides the approximations expected by the AP® graders. On the AP® Calculus exams, approximations are expressed using three decimal places, either truncated or rounded.

1 Graph an Explicitly Defined Function

Recall that an explicitly defined function is one that is written in the form $y = f(x)$. Any function that is not expressed explicitly must be solved for y before graphing it. Once a function is expressed explicitly, it can be graphed by following the steps below.

- Express the function explicitly, if it is not already in the form $y = f(x)$.
- Open the Y= editor and enter the expression $f(x)$.
- Graph the function.
- Be sure the graph shows properties of the function such as its intercepts, its maximum and minimum points, and its end behavior. If any of these are not visible, adjust the viewing window.

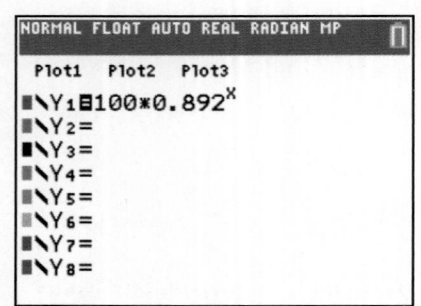

(a)

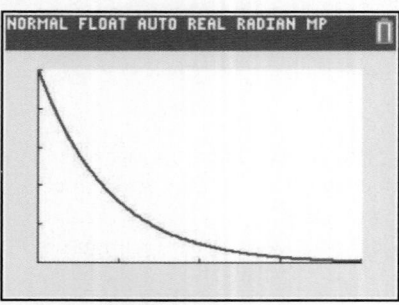

(b) $f(x) = 100 \cdot 0.892^x$
Window: $[0, 40] \times [0, 100]$

Figure 2

EXAMPLE 1   **Graphing a Function**

Graph the function $f(x) = 100 \cdot 0.892^x$, where $x \geq 0$.

Solution

Enter $\backslash Y_1 = 100 \cdot 0.892^x$ in the Y= editor. See Figure 2(a). If we graph f using the standard viewing window: $-10 \leq x \leq 10, -10 \leq y \leq 10$, the screen is blank. To set the viewing window, it is important to analyze the function being graphed.

- Since $x \geq 0$, we do not need negative numbers x in the viewing window.
- Since f is an exponential function, the range of f is $y > 0$.
- Since the base of the exponential function $f = f(x)$ is less than 1, f is decreasing.
- When $x = 0$, $f(0) = 100 \cdot 0.892^0 = 100$. Since f is decreasing, the Ymax of the viewing window is 100.

In Figure 2(b) we use the viewing window to $0 \leq x \leq 40$, Xscl $= 10$; $0 \leq y \leq 100$, Yscl $= 10$. ∎

2 Solve an Equation

The second approved use of a graphing calculator is to solve equations. Here we explore two methods for solving equations:

- Using the *Calculate-Zero* command
- Using the *Calculate-Intersect* command

Solving an equation using the Calculate-Zero command

- First write the equation in the form $y = f(x) = 0$.
- Enter y in the Y= editor, and graph it. Be sure the viewing window shows the x-intercepts of the graph.
- Use the **Calculate-Zero** command to find any x-intercepts of the graph of the function. These are the solutions of the equation.

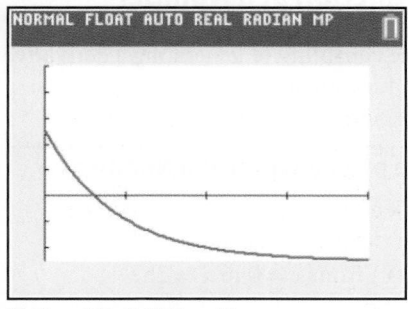

(a) $Y_2 = 100 \cdot 0.892^x - 50$
Window: $[0, 40] \times [-50, 100]$

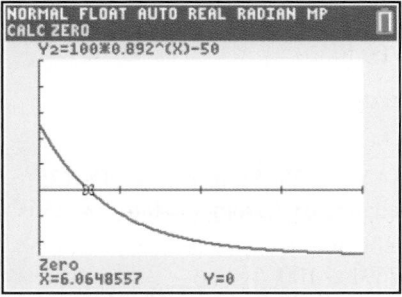

(b) $x \approx 6.065$

Figure 3

EXAMPLE 2 Solving an Equation Using the Calculate-Zero Command

The function f is given by $f(x) = 100 \cdot 0.892^x$. For what number x is $f(x) = 50$?

Solution

Begin by writing $100 \cdot 0.892^x = 50$ in the form $f(x) = 100 \cdot 0.892^x - 50 = 0$. Then enter the function in the Y= editor as \Y$_2$ and graph it. See Figure 3(a). Be sure to use a viewing window that shows the x-intercepts. Use the Calculate-Zero command to locate the x-intercept as shown in Figure 3(b). The solution to the equation is $x \approx 6.065$. ∎

A second method of approximating a solution to an equation using a graphing calculator is with the Calculate-Intersect command. Here we graph the left side of the equation and the right side of the equation separately and approximate the x-value(s) of the intersection(s) of the two graphs.

Solving an equation using the Calculate-Intersect command

- In the Y= editor, enter the left side of the equation in \Y$_1$= and the right side of the equation in \Y$_2$=.
- Graph the two equations in the same screen.
- Use the **Calculate-Intersect** command to find the point(s) of intersection of the two graphs.
- The x-value(s) of each point of intersection is(are) the solution(s) of the equation.

EXAMPLE 3 Solving an Equation Using the Calculate-Intersect Command

Solve Example 2 using the Calculate-Intersect command.

Solution

To solve the equation $100 \cdot 0.892^x = 50$ using a graphing calculator and the Calculate-Intersect command, use \Y$_1$ = $100 \cdot 0.892^x$ and enter \Y$_3$ = 50. See Figure 4(a). Notice that \Y$_2$= has been turned off. Graph \Y$_1$ and \Y$_3$ and use the Calculate-Intersect Command to find the x-value of the point of intersection. See Figures 4(b) and 4(c).

The solution to the equation is $x \approx 6.065$. ∎

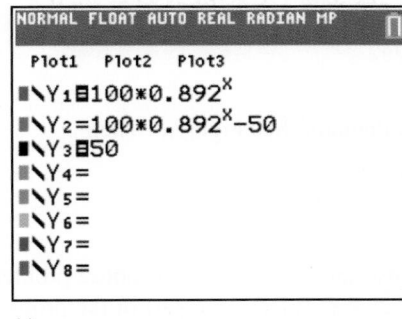

(a)

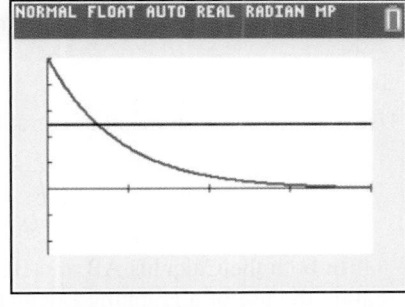

(b) $Y_1 = 100 \cdot 0.892^x$; $Y_3 = 50$
Window: $[0, 40] \times [0, 100]$

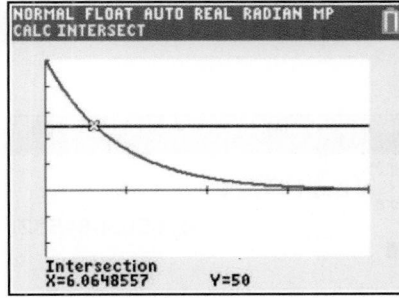

(c) $x \approx 6.065$

Figure 4

③ Approximate the Derivative of a Function at a Number

Although most graphing calculators do not have the capability of generating a derivative function, they can approximate the derivative of a function at a number.

EXAMPLE 4   Approximating the Derivative of a Function at a Number

Given the function $f(x) = 100 \cdot 0.892^x$,

(a)   What is the average rate of change of $y = f(x)$ from $x = 0$ to $x = 25$?

(b)   What is the value of $f'(20)$?

Solution

(a) The average rate of change of $y = f(x)$ from $x = 0$ to $x = 25$ is given by

$$\frac{f(25) - f(0)}{25 - 0}$$

NEED TO REVIEW? The average rate of change of a function is discussed in Section P.1, pp. 13–15.

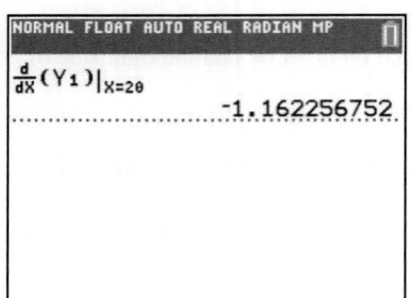

Figure 5

Use the function $\backslash Y_1$ in the Y= editor. Evaluate $\backslash Y_1$ at 25. Store it as C. $[Y_1(25) = 5.7427691; \text{ANS} \rightarrow \text{C.}]$ Then the average rate of change of f from $x = 0$ to $x = 25$ is

$$\frac{f(25) - f(0)}{25 - 0} \underset{\underset{f(0)\,=\,100}{\uparrow}}{=} \frac{C - 100}{25} = \frac{5.7427691 - 100}{25} \approx -3.770$$

(b) Using the Derivative command with the function Y_1, we find that $f'(20) \approx -1.162$ rounded to three decimal places. See Figure 5. ∎

④ Approximate a Definite Integral

The last graphing-calculator tool required for the AP® Calculus exams is to approximate a definite integral with a calculator. Recall that to use the Fundamental Theorem of Calculus to integrate a function f, we must find an antiderivative of f. If this is not possible, or if the antiderivative is difficult to work with, a graphing calculator can provide an approximation to the integral.

EXAMPLE 5   Approximating a Definite Integral with a Graphing Calculator

Use a graphing calculator to approximate $\int_0^{30} 0.215^x \, dx$.

Solution

Use the Integrate command of a graphing calculator. See Figure 6. Then

$$\int_0^{30} 0.215^x \, dx \approx 0.651$$

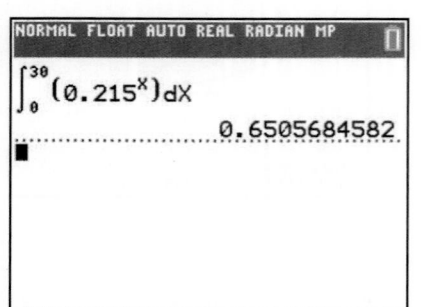

Figure 6

∎

In both the Calculus AB and BC exams there are two free-response problems that require the use of a graphing calculator to answer at least one part of the problem. It is most important that in answering these questions you show all your work. That is, the solution only earns points if the equation, the derivative, or the integral is correctly and completely written in the answer book.

For example, if a calculator enabled question asks for the volume of a solid of revolution, and you decide to use the washer method to find the volume, your solution must include the correct definite integral followed by an approximately equal sign and then the calculator answer.

Answers

Chapter P

Section P.1 (AWV = Answers will vary.)

1. Independent, dependent **2.** True **3.** False **4.** False **5.** False **6.** Vertical **7.** $5, -3$ **8.** -2 **9.** (a) **10.** (a) **11.** False **12.** 8
13. (a) -4 **(b)** $3x^2 - 2x - 4$ **(c)** $-3x^2 - 2x + 4$ **(d)** $3x^2 + 8x + 1$ **(e)** $3x^2 + 6xh + 3h^2 + 2x + 2h - 4$ **15. (a)** 4 **(b)** $|x| + 4$ **(c)** $-|x| - 4$
(d) $|x + 1| + 4$ **(e)** $|x + h| + 4$ **17.** $(-\infty, \infty)$ **19.** $(-\infty, -3] \cup [3, \infty)$ **21.** $\{x \mid x \neq -2, 0, 2\}$ **23.** -3 **25.** $\dfrac{1}{\sqrt{x + h + 7} + \sqrt{x + 7}}$

27. $2x + h + 2$ **29.** It is not a function. **31. (a)** Domain: $[-\pi, \pi]$, range: $[-1, 1]$ **(b)** Intercepts: $\left(-\dfrac{\pi}{2}, 0\right), \left(\dfrac{\pi}{2}, 0\right), (0, 1)$

(c) Symmetric with respect to the y-axis only.

33. (a) $f(-1) = 2,$ **(b)** **(c)** Domain: $[-2, \infty)$, **35. (a)** $f(-1) = 0,$ **(b)** **(c)** Domain:
$f(0) = 3,$ range: $(-\infty, 4) \cup \{5\}$, $f(0) = 0,$ $(-\infty, \infty)$, range:
$f(1) = 5,$ intercepts: $(0, 3)$, $f(1) = 1,$ $(-\infty, \infty)$, intercepts:
$f(8) = -6$ $(2, 0)$ $f(8) = 64$ $(-1, 0), (0, 0)$

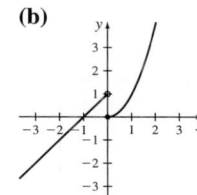

37. $f(0) = 3, f(-6) = -3$ **39.** Negative **41.** $(-3, 6) \cup (10, 11]$ **43.** $[-3, 3]$ **45.** 3 **47.** Once **49.** $-5, 8$ **51.** $[4, 8]$ **53.** $[0, 8]$
55. $\{x \mid x \neq 6\}$ **57.** $-3, (4, -3)$ **59.** -2 **61.** Odd; symmetric with respect to the origin only.
63. Neither; not symmetric to the x-axis, y-axis, or the origin. **65. (a)** -6 **(b)** -8 **(c)** -10 **(d)** $-2(x + 1), x \neq 1$

67. $f(x) = \begin{cases} -x & \text{if } -1 \leq x < 0 \\ \dfrac{1}{2}x & \text{if } 0 \leq x \leq 2 \end{cases}$ or $f(x) = \begin{cases} -x & \text{if } -1 \leq x \leq 0 \\ \dfrac{1}{2}x & \text{if } 0 < x \leq 2 \end{cases}$ domain: $[-1, 2]$, range: $[0, 1]$

69. $f(x) = \begin{cases} -1 & \text{if } x < 1 \\ 0 & \text{if } x = 1 \\ 2 - x & \text{if } 1 < x \leq 2 \end{cases}$ domain: $(-\infty, 2]$, range: $\{-1\} \cup [0, 1)$

71. $f(x) = \begin{cases} -x^3 & \text{if } -2 < x < 1 \\ 0 & \text{if } x = 1 \\ x^2 & \text{if } 1 < x \leq 3 \end{cases}$ domain: $(-2, 3]$, range: $(-1, 9]$ **73. (a)** \$250.60 **(b)** \$2655.40 **(c)** AWV.

75. (a) The independent variable is a; the dependent variable is P. **(b)** $P(20) = 245.331$ million Americans. AWV.
(c) $P(0) = 359.571$ million Americans. AWV.

Section P.2

1. (a) **2.** True **3.** True **4.** False **5.** False **6.** Zero **7.** (b) **8.** True **9.** True **10.** False **11.** D **13.** F **15.** C **17.** G
19. (a) 2 **(b)** 3 **(c)** -4 **21. (a)** 7 with multiplicity 1; -4 with multiplicity 3 **(b)** x-intercepts: $7, -4$, y-intercept: -1344
(c) Crosses at 7 and at -4 **(d)** The graph of f will resemble the graph of $y = 3x^4$. **23.** (c), (e), (f)

25. Domain: $\{x \mid x \neq -3\}$, intercept: $(0, 0)$ **27.** Domain: $(-\infty, \infty)$, intercepts: $(0, 0), \left(\dfrac{1}{3}, 0\right)$ **29. (a)** $A(x) = 16x - x^3$ **(b)** $[0, 4]$

31. (a) The pattern in the scatter plot suggests a quadratic relationship. **(b)** 283.8 feet
(c) 15.63 feet

Section P.3 (AWV = Answers will vary.)

1. $\{x \mid 0 \leq x \leq 5\}$ **2.** False **3.** True **4.** False **5.** True **6.** False **7.** True **8.** Horizontal, right **9.** y **10.** $-5, -2, 2$
11. (a) $(f + g)(x) = 5x + 1$, domain: $(-\infty, \infty)$ **(b)** $(f - g)(x) = x + 7$, domain: $(-\infty, \infty)$ **(c)** $(f \cdot g)(x) = 6x^2 - x - 12$,

domain: $(-\infty, \infty)$ **(d)** $\left(\dfrac{f}{g}\right)(x) = \dfrac{3x + 4}{2x - 3}$, domain: $\left\{x \mid x \neq \dfrac{3}{2}\right\}$ **13. (a)** $(f + g)(x) = \sqrt{x + 1} + \dfrac{2}{x}$, domain: $\{x \mid x \geq -1, x \neq 0\}$

(b) $(f - g)(x) = \sqrt{x + 1} - \dfrac{2}{x}$, domain: $\{x \mid x \geq -1, x \neq 0\}$ **(c)** $(f \cdot g)(x) = \dfrac{2\sqrt{x + 1}}{x}$, domain: $\{x \mid x \geq -1, x \neq 0\}$

(d) $\left(\dfrac{f}{g}\right)(x) = \dfrac{1}{2}x\sqrt{x+1}$, domain: $\{x|x \geq -1, x \neq 0\}$ **15. (a)** $(f \circ g)(4) = 98$ **(b)** $(g \circ f)(2) = 49$ **(c)** $(f \circ f)(1) = 4$

(d) $(g \circ g)(0) = 4$ **17. (a)** $(f \circ g)(1) = -1$ **(b)** $(f \circ g)(-1) = -1$ **(c)** $(g \circ f)(-1) = 8$ **(d)** $(g \circ f)(1) = 8$ **(e)** $(g \circ g)(-2) = 8$

(f) $(f \circ f)(-1) = -7$ **19. (a)** $(g \circ f)(-1) = 4$ **(b)** $(g \circ f)(6) = 2$ **(c)** $(f \circ g)(6) = 1$ **(d)** $(f \circ g)(4) = -2$

21. (a) $(f \circ g)(x) = 24x + 1$, domain: $(-\infty, \infty)$ **(b)** $(g \circ f)(x) = 24x + 8$, domain: $(-\infty, \infty)$ **(c)** $(f \circ f)(x) = 9x + 4$, domain: $(-\infty, \infty)$

(d) $(g \circ g)(x) = 64x$, domain: $(-\infty, \infty)$ **23. (a)** $(f \circ g)(x) = x$, domain: $\{x|x \geq 1\}$ **(b)** $(g \circ f)(x) = |x|$, domain: $(-\infty, \infty)$

(c) $(f \circ f)(x) = x^4 + 2x^2 + 2$, domain: $(-\infty, \infty)$ **(d)** $(g \circ g)(x) = \sqrt{\sqrt{x-1}-1}$, domain: $\{x|x \geq 2\}$ **25. (a)** $(f \circ g)(x) = \dfrac{2}{2-x}$,

domain: $\{x|x \neq 0, x \neq 2\}$ **(b)** $(g \circ f)(x) = \dfrac{2(x-1)}{x}$, domain: $\{x|x \neq 0, x \neq 1\}$ **(c)** $(f \circ f)(x) = x$, domain: $\{x|x \neq 1\}$

(d) $(g \circ g)(x) = x$, domain: $\{x|x \neq 0\}$ **27.** $f(x) = x^4, g(x) = 2x + 3$ **29.** $f(x) = \sqrt{x}, g(x) = x^2 + 1$ **31.** $f(x) = |x|, g(x) = 2x + 1$

33.   **35.**   **37.**   **39.**

41.   **43.**   **45.**

47. (a)   **(b)**   **(c)**   **(d)**

(e)   **(f)**   **(g)**

49. (a)   **(b)**   **(c)** AWV.   **(d)**   **(e)** AWV.

Section P.4   (SSM = See Student Solutions Manual; AWV = Answers will vary.)

1. False   **2.** $[4, \infty)$ **3.** False   **4.** True   **5.** False   **6.** False   **7.** AWV.   **8.** AWV.   **9.** One-to-one

11. Not one-to-one   **13.** One-to-one   **15.** SSM.   **17.** SSM.

19. (a) One-to-one   **(b)** $\{(5, -3), (9, -2), (2, -1), (11, 0), (-5, 1)\}$,

(c) Domain of f: $\{-3, -2, -1, 0, 1\}$, range of f: $\{-5, 2, 5, 9, 11\}$, domain of f^{-1}: $\{-5, 2, 5, 9, 11\}$, range of f^{-1}: $\{-3, -2, -1, 0, 1\}$

21. (a) Not one-to-one   **(b)** Does not apply   **(c)** Domain of f: $\{-10, -3, -2, 1, 2\}$, range of f: $\{0, 1, 2, 9\}$

23. 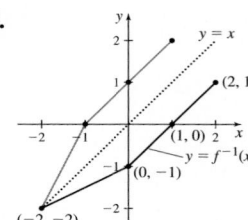   **25.** 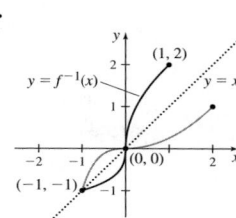   **27.**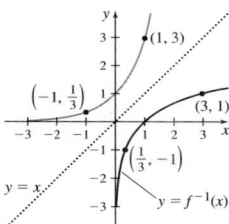

29. (a) $f^{-1}(x) = \dfrac{x-2}{4}$ **(b)** Both the domain and range of f are $(-\infty, \infty)$. Both the domain and range of f^{-1} are $(-\infty, \infty)$.

31. (a) $f^{-1}(x) = x^3 - 10$ **(b)** Both the domain and range of f are $(-\infty, \infty)$. Both the domain and range of f^{-1} are $(-\infty, \infty)$.

33. (a) $f^{-1}(x) = 2 + \dfrac{1}{x}$ **(b)** Domain of f: $\{x \mid x \neq 2\}$, range of f: $\{y \mid y \neq 0\}$, domain of f^{-1}: $\{x \mid x \neq 0\}$, range of f^{-1}: $\{y \mid y \neq 2\}$

35. (a) $f^{-1}(x) = \dfrac{3 - 2x}{x - 2}$ **(b)** Domain of f: $\{x \mid x \neq -2\}$, range of f: $\{y \mid y \neq 2\}$, domain of f^{-1}: $\{x \mid x \neq 2\}$, range of f^{-1}: $\{y \mid y \neq -2\}$

37. (a) $f^{-1}(x) = \sqrt{x - 4}$ **(b)** Domain of f: $\{x \mid x \geq 0\}$, range of f: $\{y \mid y \geq 4\}$, domain of f^{-1}: $\{x \mid x \geq 4\}$, range of f^{-1}: $\{y \mid y \geq 0\}$

Section P.5   (AWV = Answers will vary.)

1. $\left(-1, \dfrac{1}{a}\right), (0, 1), (1, a)$ **2.** False   **3.** 4   **4.** 3   **5.** False   **6.** False   **7.** 1   **8.** $\{x \mid x > 0\}$ **9.** $\left(\dfrac{1}{a}, -1\right), (1, 0), (a, 1)$ **10. (a)**   **11.** False

12. True   **13.** True   **14.** 1   **15.** AWV.   **16.** 0   **17. (a)** $g(-1) = \dfrac{9}{4}, \left(-1, \dfrac{9}{4}\right)$ **(b)** $x = 3, (3, 66)$ **19. (a)**   **21. (c)**   **23. (b)**

25. Domain: $(-\infty, \infty)$, range: $(0, \infty)$ **27.** Domain: $(-\infty, \infty)$, range: $(0, \infty)$ **29.** Domain: $(-\infty, \infty)$, range: $(0, \infty)$

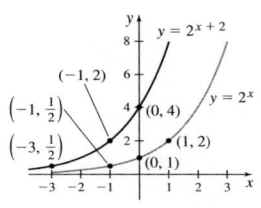   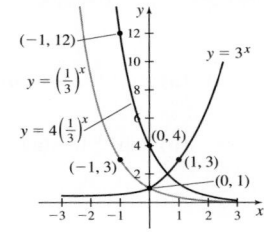   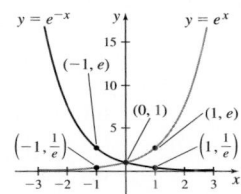

31. $\{x \mid x \neq 0\}$ **33.** $\{x \mid x > 1\}$ **35. (b)**   **37. (f)**   **39. (c)**

41. (a) $\{x \mid x > -4\}$ **(b and f)**   **43. (a)** $(-\infty, \infty)$ **(b and f)**
   (c) $(-\infty, \infty)$ **(c)** $\{y \mid y > 2\}$
   (d) $f^{-1}(x) = e^x - 4$ **(d)** $f^{-1}(x) = \ln\left(\dfrac{x-2}{3}\right)$
   (e) $(-\infty, \infty)$ **(e)** $\{y \mid y > 2\}$

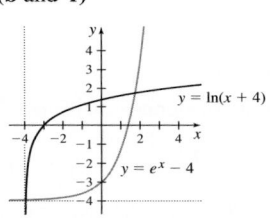   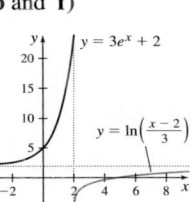

45. It shifts the x-intercept c units to the left.   **47.** $x = 0, x = 2$ **49.** $x = \dfrac{1}{2}$ **51.** $x = \dfrac{1 - \ln 4}{2}$ **53.** $x = \dfrac{\ln\left(\dfrac{9}{5}\right)}{3 \ln 2}$ **55.** $x = \dfrac{\ln 3}{\ln 36}$

57. $x = \dfrac{7}{2}$ **59.** $x = \dfrac{1}{2}$ **61.** $x = 3$ **63.** $x \approx 2.787$ **65.** $x \approx 1.315$

67. (a)   **(b)** $\left(\dfrac{3}{2}, \ln\dfrac{11}{2}\right)$ **(c)** $\left\{x \mid -\dfrac{1}{3} < x < \dfrac{3}{2}\right\}$

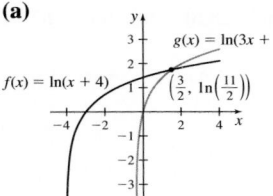

Section P.6   (AWV = Answers will vary.)

1. $2\pi, \pi$ **2.** All real numbers except odd multiples of $\dfrac{\pi}{2}$.   **3.** $\{y \mid -1 \leq y \leq 1\}$ **4.** The period of $\tan x$ is π.   **5.** False   **6.** $3, \dfrac{\pi}{3}$ **7.** True

8. False   **9.** True   **10.** Origin   **11.** y-axis   **12.** AWV.   **13.** -1 **15.** $-\dfrac{2\sqrt{3}}{3}$ **17.** 0   **19.** -1 **21. (f)**   **23. (a)**   **25. (d)**

27.

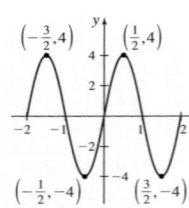

29.

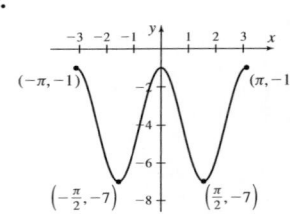

31.

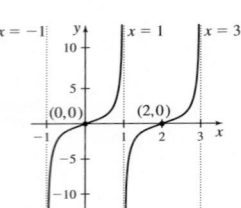

33. Amplitude: $\dfrac{1}{2}$, Period: 2   **35.** Amplitude: 3, Period: 2π **37.** $f(x) = 2\sin(2x)$ **39.** $f(x) = \dfrac{1}{2}\cos(2x)$ **41.** $f(x) = -\sin\left(\dfrac{3}{2}x\right)$

43. $f(x) = 1 - \cos\left(\dfrac{4\pi}{3}x\right)$ **45.** $f(x) = \cot x$ **47.** $f(x) = \tan\left(x - \dfrac{\pi}{2}\right)$

Section P.7   (SSM = See Student Solutions Manual.)

1. $\sin y$ **2.** False   **3.** False   **4.** True   **5.** False   **6.** True   **7.** True   **8.** True   **9.** 0   **11.** π **13.** $-\dfrac{\pi}{4}$ **15.** $\dfrac{\pi}{4}$ **17.** $\dfrac{\pi}{3}$ **19.** $\dfrac{5\pi}{4}$ **21.** $\dfrac{\pi}{5}$

23. $-\dfrac{3\pi}{8}$ **25.** $-\dfrac{\pi}{8}$ **27.** $-\dfrac{\pi}{5}$ **29.** $\dfrac{1}{4}$ **31.** 4   **33.** It is not defined.   **35.** π **37.** $\sqrt{0.91}$ **39.** $\sqrt{3}$ **41.** $\dfrac{1+4\sqrt{6}}{5\sqrt{5}}$ **43.** $\dfrac{4\sqrt{21}}{17}$

45. $\left\{\dfrac{5\pi}{6}, \dfrac{11\pi}{6}\right\}$ **47.** $\left\{\dfrac{7\pi}{6}, \dfrac{11\pi}{6}\right\}$ **49.** $\left\{\dfrac{\pi}{2}, \dfrac{7\pi}{6}, \dfrac{11\pi}{6}\right\}$ **51.** $\left\{\dfrac{\pi}{3}, \dfrac{2\pi}{3}, \dfrac{4\pi}{3}, \dfrac{5\pi}{3}\right\}$ **53.** $\left\{\dfrac{\pi}{2}\right\}$ **55.** $\left\{\dfrac{\pi}{6}, \dfrac{5\pi}{6}, \dfrac{3\pi}{2}\right\}$ **57.** $\left\{\dfrac{3\pi}{4}, \dfrac{7\pi}{4}\right\}$

59. $\left\{\dfrac{\pi}{3}, \dfrac{2\pi}{3}, \dfrac{4\pi}{3}, \dfrac{5\pi}{3}\right\}$ **61.** $\left\{0, \dfrac{\pi}{3}, \dfrac{\pi}{2}, \dfrac{2\pi}{3}, \pi, \dfrac{4\pi}{3}, \dfrac{3\pi}{2}, \dfrac{5\pi}{3}\right\}$ **63.** $\{1.373, 4.515\}$ **65.** $\{3.871, 5.553\}$

67. $\cos(\sin^{-1} u) = \sqrt{1 - u^2}, |u| \le 1$ **69.** SSM.

71. (a and c)

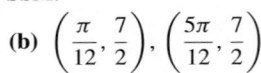

(b) $\left(\dfrac{\pi}{12}, \dfrac{7}{2}\right), \left(\dfrac{5\pi}{12}, \dfrac{7}{2}\right)$

(c) See (a).

(d) $\left\{x \,\middle|\, \dfrac{\pi}{12} < x < \dfrac{5\pi}{12}\right\}$

73. (a) $\tan^{-1}\sqrt{3} - \tan^{-1} 0 = \dfrac{\pi}{3}$

(b) $\tan^{-1} 1 - \tan^{-1}\left(-\dfrac{\sqrt{3}}{3}\right) = \dfrac{5\pi}{12}$

Section P.8

1. sequence   **2.** True   **3.** $1 \cdot 2 \cdot 3 \cdot \dots \cdot (n-1) \cdot n$ **4.** recursively defined   **5.** summation (or sigma)   **6.** True   **7.** 16   **8.** $1; n$ **9.** False

10. Binomial Theorem   **11.** 2, 4, 6, 8, 10   **13.** $\dfrac{1}{2}, \dfrac{2}{3}, \dfrac{3}{4}, \dfrac{4}{5}, \dfrac{5}{6}$ **15.** $-1, 4, -9, 16, -25$ **17.** $\dfrac{1}{2}, -\dfrac{2}{5}, \dfrac{2}{7}, -\dfrac{8}{41}, \dfrac{8}{61}$ **19.** $a_n = \dfrac{n+1}{n}$

21. $a_n = \dfrac{1}{n^2}$ **23.** $a_n = (-1)^{n+1}(2n)$ **25.** $a_n = (-1)^{n+1}(n+1)$ **27.** $3, 5, 7, 9, 11$ **29.** $2, 4, 7, 11, 16$ **31.** $1, 3, 9, 27, 81$

33. $2, 1, \dfrac{1}{3}, \dfrac{1}{12}, \dfrac{1}{60}$ **35.** $1, -1, -1, 1, -1$ **37.** $A, A+d, A+2d, A+3d, A+4d$ **39.** $-1 + 0 + 1 + 2 + \dots + (n-2)$

41. $\dfrac{1}{2} + \dfrac{4}{3} + \dfrac{9}{4} + \dfrac{16}{5} + \dots + \dfrac{n^2}{n+1}$ **43.** $1 + \dfrac{1}{2} + \dfrac{1}{4} + \dfrac{1}{8} + \dots + \dfrac{1}{2^n}$ **45.** $\dfrac{1}{2} + \dfrac{1}{2} + \dfrac{3}{8} + \dfrac{1}{4} + \dots + \dfrac{n+1}{2^{n+1}}$ **47.** $4 - 8 + 16 - 32 + \dots + (-1)^n 2^n$

49. $\displaystyle\sum_{k=1}^{50} 2k$ **51.** $\displaystyle\sum_{k=1}^{35} \dfrac{k+1}{k}$ **53.** $\displaystyle\sum_{k=0}^{10} (-1)^k \left(\dfrac{3}{10^k}\right)$ **55.** $\displaystyle\sum_{k=1}^{n} (-1)^{k-1} \dfrac{k}{e^k}$ **57.** 20   **59.** 1395   **61.** 10,000   **63.** 2830   **65.** 630   **67.** 10

69. 84   **71.** 50   **73.** $x^4 + 4x^3 + 6x^2 + 4x + 1$ **75.** $x^5 - 15x^4 + 90x^3 - 270x^2 + 405x - 243$ **77.** $16x^4 + 96x^3 + 216x^2 + 216x + 81$

79. $x^6 - 3x^4 y^2 + 3x^2 y^4 - y^6$ **81.** After the first payment, Jin's balance is $2930.

Chapter 1

Section 1.1   (SSM = See Student Solutions Manual; AWV = Answers will vary.)

1. (c)   2. True   **3.** False   **4.** False   **5.** False   **6.** False

7. $\lim\limits_{x \to 1} 2x = 2$

	x approaches 1 from the left				x approaches 1 from the right		
x	0.9	0.99	0.999	$\to 1 \leftarrow$	1.001	1.01	1.1
$f(x) = 2x$	1.8	1.98	1.998	$f(x)$ approaches 2	2.002	2.02	2.2

9. $\lim\limits_{x \to 0} (x^2 + 2) = 2$

	x approaches 0 from the left				x approaches 0 from the right		
x	-0.1	-0.01	-0.001	$\to 0 \leftarrow$	0.001	0.01	0.1
$f(x) = x^2 + 2$	2.01	2.0001	2.000001	$f(x)$ approaches 2	2.000001	2.0001	2.01

11. $\lim\limits_{x \to -3} \dfrac{x^2 + 4x + 3}{x + 3} = -2$

	x approaches -3 from the left				x approaches -3 from the right		
x	-3.5	-3.1	-3.01	$\to -3 \leftarrow$	-2.99	-2.9	-2.5
$f(x) = \dfrac{x^2 + 4x + 3}{x + 3}$	-2.5	-2.1	-2.01	$f(x)$ approaches -2	-1.99	-1.90	-1.5

13. $\lim\limits_{x \to 0} \dfrac{2 - 2e^x}{x} = -2$

	x approaches 0 from the left				x approaches 0 from the right		
x	-0.2	-0.1	-0.01	$\to 0 \leftarrow$	0.01	0.1	0.2
$f(x) = \dfrac{2 - 2e^x}{x}$	-1.8127	-1.9033	-1.9900	$f(x)$ approaches -2	-2.0100	-2.1034	-2.2140

15. $\lim\limits_{x \to 0} \dfrac{1 - \cos x}{x} = 0$

	x approaches 0 from the left				x approaches 0 from the right		
x	-0.2	-0.1	-0.01	$\to 0 \leftarrow$	0.01	0.1	0.2
$f(x) = \dfrac{1 - \cos x}{x}$	-0.09967	-0.04996	-0.00500	$f(x)$ approaches 0	0.00500	0.04996	0.09967

17. (a) 2   **(b)** 2   **(c)** 2   **19. (a)** 3   **(b)** 6   **(c)** The limit does not exist.   **21.** 1   **23.** 1   **25.** The limit does not exist because the two one-sided limits are not equal.   **27.** The limit does not exist because the two one-sided limits are not equal. **29.** 9
31. The limit does not exist.   **33.** 2   **35.** The limit does not exist.   **37.** AWV.   **39.** AWV.   **41.** 1   **43.** 0   **45.** 1   **47.** 0   **49.** 0

51.   (a) $m_{\text{sec}} = 15$
(b) $m_{\text{sec}} = 3(x + 2)$
(c) $\lim\limits_{x \to 2} m_{\text{sec}} = 12$
(d)

53. (a) $m_{\text{sec}} = 2 + \dfrac{1}{2}h$ for $h \neq 0$
(b)

h	-0.5	-0.1	-0.001	0.001	0.1	0.5
m_{sec}	1.75	1.95	1.9995	2.0005	2.05	2.25

(c) $\lim\limits_{h \to 0} m_{\text{sec}} = 2$ **(d)** $m_{\text{tan}} = 2$
(e)

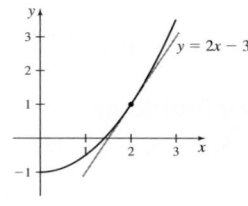

55. (a) These values suggest that $\lim\limits_{x \to 0} \cos \dfrac{\pi}{x}$ may be 1.   **(b)** These values suggest that $\lim\limits_{x \to 0} \cos \dfrac{\pi}{x}$ may be -1.
(c) $\lim\limits_{x \to 0} \cos \dfrac{\pi}{x}$ does not exist. AWV.   **(d)** AWV.

57. (a) $\lim\limits_{x \to 2} \dfrac{x - 8}{2} = -3$ **(b)** $1.8 \le x \le 2.2$ **(c)** $1.98 \le x \le 2.02$

	x approaches 2 from the left				x approaches 2 from the right		
x	1.9	1.99	1.999	$\to 2 \leftarrow$	2.001	2.01	2.1
$f(x) = \dfrac{x - 8}{2}$	-3.05	-3.005	-3.0005	$f(x)$ approaches -3	-2.9995	-2.995	-2.95

59. (a) $C(w) = \begin{cases} 0.63 & \text{if } 0 < w \le 1 \\ 0.87 & \text{if } 1 < w \le 2 \\ 1.11 & \text{if } 2 < w \le 3 \\ 1.35 & \text{if } 3 < w \le 3.5 \end{cases}$ **(b)** $\{w | 0 < w \le 3.5\}$ **(c)** See the graph below.

(d) $\lim\limits_{w \to 2^-} C(w) = 0.87$
$\lim\limits_{w \to 2^+} C(w) = 1.11$
$\lim\limits_{w \to 2} C(w)$ does not exist.
(e) $\lim\limits_{w \to 0^+} C(w) = 0.63$
(f) $\lim\limits_{w \to 3.5^-} C(w) = 1.35$

61. (a) $\lim\limits_{t \to 7} S(t) = 100$ **(b)** Given any $\varepsilon > 0$, there is a number $\delta > 0$ so that whenever $0 < |t - 7| < \delta$ then $|S(t) - 100| < \varepsilon$.
63. No, the value of the function at $x = 2$ has no bearing on $\lim\limits_{x \to 2} f(x)$.   **65. (a)** The graph of f is the horizontal line $y = -1$ excluding
the point $(3, -1)$.   **(b)** $\lim\limits_{x \to 3^-} f(x) = \lim\limits_{x \to 3^+} f(x) = -1$.   **(c)** The graph suggests $\lim\limits_{x \to 3} f(x) = -1$.   **67.** 0   **69.** 0

AP® Practice Problems   (AWV = Answers will vary.)

1. B   **3.** C   **5.** B   **7.** AWV.

Section 1.2   (SSM = See Student Solutions Manual; AWV = Answers will vary.)

1. (a) -3 **(b)** π **2.** 243   **3.** 4   **4. (a)** -1 **(b)** e **5. (a)** -2 **(b)** $\dfrac{7}{2}$ **6. (a)** -6 **(b)** 0   **7.** True   **8.** 2   **9.** False   **10.** True   **11.** 14

13. 0   **15.** 1   **17.** 6   **19.** $\sqrt{11}$ **21.** $2\sqrt{78}$ **23.** $\sqrt{7 + \sqrt{3}}$ **25.** 0   **27.** 5   **29.** $-\dfrac{31}{8}$ **31.** 10   **33.** $\dfrac{13}{4}$ **35.** 6   **37.** 2   **39.** 2   **41.** $\dfrac{\sqrt{2}}{4}$

43. $\dfrac{1}{30}$ **45.** 5   **47.** 6   **49.** 9   **51.** -13 **53.** 216   **55.** 0   **57.** $\dfrac{1}{36}$ **59. (a)** 6   **(b)** -16 **(c)** -16 **(d)** 0   **(e)** $-\dfrac{1}{4}$ **(f)** 0   **61.** 6   **63.** -4

65. $\dfrac{1}{2}$ **67.** 4   **69.** $6x + 4$ **71.** $-\dfrac{2}{x^2}$ **73.** $\lim\limits_{x \to 1^-} f(x) = -1$, $\lim\limits_{x \to 1^+} f(x) = 2$, $\lim\limits_{x \to 1} f(x)$ does not exist   **75.** $\lim\limits_{x \to 1^-} f(x) = 2$, $\lim\limits_{x \to 1^+} f(x) = 2$,
$\lim\limits_{x \to 1} f(x) = 2$ **77.** $\lim\limits_{x \to 1^-} f(x) = 0$, $\lim\limits_{x \to 1^+} f(x) = 0$, $\lim\limits_{x \to 1} f(x) = 0$ **79.** $\lim\limits_{x \to 3^-} f(x) = 6$, $\lim\limits_{x \to 3^+} f(x) = 6$, $\lim\limits_{x \to 3} f(x) = 6$

81. The limit does not exist.   **83.** $2x$ **85.** $-\dfrac{1}{x^2}$ **87.** $-\dfrac{1}{16}$ **89.** 6   **91.** 0

93. (a) $C(x) = \begin{cases} 12.00 & \text{if } 0 \le x \le 10 \\ 12.00 + 1.26(x - 10) & \text{if } 10 < x \le 30 \\ 37.20 + 2.40(x - 30) & \text{if } 30 < x \le 100 \\ 205.20 + 3.18(x - 100) & \text{if } x > 100 \end{cases}$ **(d)** 12.00   **(e)**

(b) $\{x | x \ge 0\}$ **(c)** $\lim\limits_{x \to 5} C(x) = 12.00$, $\lim\limits_{x \to 10} C(x) = 12.00$,
$\lim\limits_{x \to 30} C(x) = 37.20$, $\lim\limits_{x \to 100} C(x) = 205.20$

95. (a) 0   **(b)** As the temperature of a gas approaches zero, the molecules in the gas stop moving.
97. $\lim\limits_{x \to 0} |x| = 0$ since $\lim\limits_{x \to 0^-} |x| = 0$ and $\lim\limits_{x \to 0^+} |x| = 0$ **99.** AWV.   **101.** AWV.   **103.** SSM.
105. na^{n-1} **107.** $\dfrac{m}{n}$ **109.** $\dfrac{a + b}{2}$ **111.** 0

AP® Practice Problems

1. D   **3.** D   **5.** C   **7.** C   **9.** D

Section 1.3   (SSM = See Student Solutions Manual; AWV = Answers will vary.)

1. True   **2.** False   **3.** $f(c)$ is defined, $\lim\limits_{x \to c} f(x)$ exists, $\lim\limits_{x \to c} f(x) = f(c)$ **4.** True   **5.** False   **6.** False   **7.** False   **8.** True
9. The function is discontinuous.   **10.** The function is continuous.   **11.** True   **12.** False   **13. (a)** The function is not continuous at $c = -3$.
    (b) $\lim\limits_{x \to -3} f(x) \ne f(-3)$ **(c)** The discontinuity is removable.   **(d)** $f(-3) = -2$ **15. (a)** The function is not continuous at $c = 2$.
    (b) $\lim\limits_{x \to 2} f(x)$ does not exist.   **(c)** The discontinuity is not removable.   **17. (a)** The function is continuous at $c = 4$.
19. The function is continuous at $c = -3$.   **21.** The function is continuous at $c = -2$.   **23.** The function is continuous at $c = 2$.
25. The function is not continuous at $c = 1$.   **27.** The function is not continuous at $c = 1$.   **29.** The function is continuous at $c = 0$.
31. The function is not continuous at $c = 0$.   **33.** $f(2) = 4$ **35.** $f(1) = 2$ **37.** The function is continuous on the given interval.
39. The function is not continuous on the given interval. The function is continuous on $(-\infty, -3)$ and $(3, \infty)$.
41. The function is continuous on $\{x | x \ne 0\}$.   **43.** The function is continuous on the set of all real numbers.
45. The function is continuous on $\{x | x \ge 0, x \ne 9\}$.   **47.** The function is continuous on the set $\{x | x < 2\}$.
49. The function is continuous on the set of all real numbers.   **51.** f is continuous at 0 because $\lim\limits_{x \to 0} f(x) = f(0)$.

53. f is not continuous at 3 because $\lim\limits_{x\to 3} f(x)$ does not exist. **55.** f is continuous at 1 because $\lim\limits_{x\to 1} f(x) = f(1)$.

57. (a)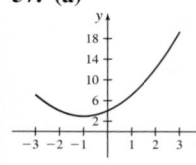
(b) Based on the graph, it appears that the function is continuous for all real numbers.
(c) f is actually continuous at all real numbers except $x = 2$.
(d) AWV.

59. Yes. A zero exists on the given interval. **61.** The IVT gives no information. **63.** The IVT gives no information. **65.** 1.154 **67.** 0.211
69. 3.134 **71.** 1.157 **73. (a)** Since f is continuous on $[0, 1]$, $f(0) < 0$, and $f(1) > 0$, the IVT guarantees that f has a zero on the interval $(0, 1)$. **(b)** 0.828 **75. (a)** The function is continuous from the left at -1. **(b)** The function is not continuous from the right at -1.
77. (a) The function is continuous from the left at -1. **(b)** The function is not continuous from the right at -1.

79. (a) $C(w) = \begin{cases} 0.63 & \text{if } 0 < w \le 1 \\ 0.87 & \text{if } 1 < w \le 2 \\ 1.11 & \text{if } 2 < w \le 3 \\ 1.35 & \text{if } 3 < w \le 3.5 \end{cases}$ **(b)** $\{w \mid 0 < w \le 3.5\}$

(c) The function is continuous on the intervals $(0, 1]$, $(1, 2]$, $(2, 3]$, and $(3, 3.5]$.

(d) At $w = 1$, $w = 2$, and $w = 3$, the function has a jump discontinuity. **(e)** AWV.

81. (a) $C(x) = \begin{cases} 8.99 + 0.10858x & \text{if } 0 \le x \le 1000 \\ 117.57 + 0.12858(x - 1000) & \text{if } x > 1000 \end{cases}$ **(b)** $\{x \mid x \ge 0\}$ **(c)** C is continuous for all $x \ge 0$.

(d) See (c). **(e)** AWV. **83. (a)** $\dfrac{Gm}{R^2}$ **(b)** 1.3 m/s² **(c)** The gravity on Europa is less than the gravity on Earth.
85. $A = 5$, $B = 7$

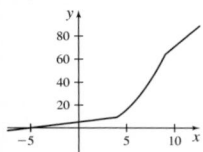

87. (a) $\{x \mid x < 2\} \cup \{x \mid x > 2, x \ne 8\}$ **(b)** f is discontinuous at $x = 8$ and on the interval $[-2, 2]$. **(c)** The discontinuity at $x = 8$ is removable.
89. (a) Since f is continuous on $[0, 2]$, $f(0) < 0$ and $f(2) > 0$, the IVT guarantees that f must have a zero on the interval $(0, 2)$. **(b)** $x \approx 0.443$
91. 1.732 **93.** 1.561 **95.** This does not contradict the IVT. $f(-1)$ and $f(2)$ are both positive, so there is no guarantee that the function has a zero in the interval $(-1, 2)$.
97. This does not contradict the IVT. The IVT guarantees that f must have at least one zero on the interval $(-5, 0)$.
99. (a) Since $f(-2)$ and $f(2)$ are both positive, there is no guarantee that the function has a zero in the interval $(-2, 2)$. **(b)** The graph below indicates that f does not have a zero in the interval $(-2, 2)$.

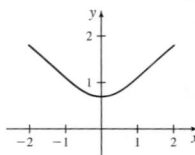

101. SSM. **103. (a)** $x = -2$, $x = 2$, $x = 3$ **(b)** $p = 5$ **(c)** h is an even function. **105.** AWV.
107. $(1.125, 1.25)$ **109.** $(0.125, 0.25)$ **111.** $(3.125, 3.25)$ **113.** $(1.125, 1.25)$ **115.** Since f is continuous on $[0, 1]$, $f(0) < 0$, and $f(1) > 0$, the IVT guarantees that f has a zero on the interval $(0, 1)$. Correct to one decimal place, the zero is $x = 0.8$ **117.** SSM.
119. SSM. **121.** SSM. **123.** SSM.

AP® Practice Problems (SSM = See Student Solutions Manual.)

1. D **3.** A **5.** D **7.** B **9.** B **11.** D **13.** SSM.

Section 1.4 (SSM = See Student Solutions Manual; AWV = Answers will vary.)

1. 0 **2.** False **3.** L **4.** False **5.** 1 **7.** 1 **9.** 0 **11.** $\dfrac{\sqrt{3}}{2} + \dfrac{1}{2}$ **13.** 1 **15.** 2 **17.** 0 **19.** 1 **21.** $\dfrac{1}{2}$ **23.** 7 **25.** 2 **27.** $\dfrac{1}{2}$ **29.** 5
31. 0 **33.** 1 **35.** f is continuous at $c = 0$. **37.** f is continuous at $c = \dfrac{\pi}{4}$. **39.** f is continuous on $\{x \mid x \ne 4\}$.
41. f is continuous on $\left\{ x \mid x \ne \dfrac{3\pi}{2} + 2k\pi \right\}$, k an integer. **43.** f is continuous on $\{x \mid x > 0, x \ne 3\}$.
45. f is continuous on the set of all real numbers. **47.** 0 **49.** 0 **51.** SSM. **53.** SSM. **55.** SSM. **57.** SSM.
59. $f(0) = \pi$, $f(1) = \pi$ **61.** Yes **63.** SSM. **65.** AWV. **67.** SSM. **69.** SSM.

AP® Practice Problems

1. C **3.** D **5.** C **7.** C **9.** D **11.** B

Section 1.5 (SSM = See Student Solutions Manual; AWV = Answers will vary.)

1. False **2. (a)** $-\infty$ **(b)** ∞ **(c)** $-\infty$ **3.** False **4.** Vertical **5. (a)** 0 **(b)** 0 **(c)** ∞ **6.** False **7. (a)** 0 **(b)** ∞ **(c)** 0 **8.** True

9. 2 **11.** ∞ **13.** ∞ **15.** $x = -1, x = 3$ **17.** -3 **19.** ∞ **21.** 0 **23.** ∞ **25.** $x = -3, x = 0, x = 4$ **27.** $-\infty$ **29.** ∞ **31.** ∞

33. ∞ **35.** $-\infty$ **37.** $-\infty$ **39.** $-\infty$ **41.** $-\infty$ **43.** 0 **45.** $\dfrac{2}{5}$ **47.** 1 **49.** 0 **51.** $\dfrac{5}{4}$ **53.** 0 **55.** 0 **57.** $\sqrt{3}$ **59.** $-\infty$

61. $x = 0$ is a vertical asymptote. $y = 3$ is a horizontal asymptote.

63. $x = -1$ and $x = 1$ are vertical asymptotes. $y = 1$ is a horizontal asymptote.

65. $x = \dfrac{3}{2}$ is a vertical asymptote. $y = \dfrac{\sqrt{2}}{2}$ and $y = -\dfrac{\sqrt{2}}{2}$ are horizontal asymptotes.

67. (a) $\{x \mid x \neq -2, x \neq 0\}$ **(b)** $y = 0$ is a horizontal asymptote. **(c)** $x = 0$ and $x = -2$ are vertical asymptotes. **(d)** SSM.

69. (a) $\left\{x \mid x \neq \dfrac{3}{2}, x \neq 2\right\}$ **(b)** $y = \dfrac{1}{2}$ is a horizontal asymptote. **(c)** $x = \dfrac{3}{2}$ is a vertical asymptote. **(d)** SSM.

71. (a) $\{x \mid x \neq 0, x \neq 1\}$ **(b)** There are no horizontal asymptotes. **(c)** $x = 0$ is a vertical asymptote. **(d)** SSM. **73.** AWV.

75. (a) T **(b)** u_0 **77.** ∞

79. (a) 500 **(b)**

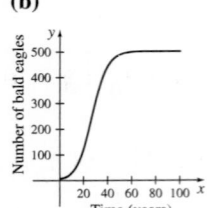

(c) AWV. **81. (a)**

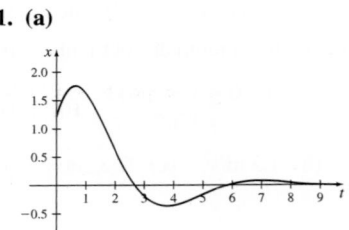

The graph suggests that $\lim\limits_{t \to \infty} x(t) = 0$.

(b) 0 **(c)** AWV.

83. (a) 0.26 moles **(b)** 395 min **(c)** 0 moles **(d)** SSM. **85.** SSM. **87.** SSM. **89.** SSM.

91. (a)

x	100	10,000	1,000,000	1,000,000,000	$\to \infty$
$f(x) = \left(1 + \dfrac{1}{x}\right)^x$	2.70481	2.718146	2.718280	2.718282	2.718282

(b) $e \approx 2.718281828$

(c) SSM.

93. (a) ∞ **(b)** SSM.

AP® Practice Problems

1. C **3.** A **5.** B **7.** B **9.** A **11.** A

Section 1.6 (SSM = See Student Solutions Manual.)

1. False **2.** True **3.** True **4.** False **5.** True **6.** False **7.** $\delta = 0.005$ **9.** $\delta = \dfrac{1}{12}$ **11.** $\delta = 0.02$

13. (a) $\delta \leq 0.025$ **(b)** $\delta \leq 0.0025$ **(c)** $\delta \leq 0.00025$ **(d)** $\delta \leq \dfrac{\varepsilon}{4}$ **15. (a)** $\delta \leq 0.1$ **(b)** $\delta \leq 0.01$ **(c)** $\delta \leq \varepsilon$

17. Given any $\varepsilon > 0$, let $\delta \leq \dfrac{\varepsilon}{3}$. SSM for the complete proof. **19.** Given any $\varepsilon > 0$, let $\delta \leq \dfrac{\varepsilon}{2}$. SSM for the complete proof.

21. Given any $\varepsilon > 0$, let $\delta \leq \dfrac{\varepsilon}{5}$. SSM for the complete proof. **23.** Given any $\varepsilon > 0$, let $\delta \leq \min\left\{1, \dfrac{\varepsilon}{3}\right\}$. SSM for the complete proof.

25. Given any $\varepsilon > 0$, let $\delta \leq \min\left\{1, \dfrac{2\varepsilon}{7}\right\}$. SSM for the complete proof. **27.** Given any $\varepsilon > 0$, let $\delta \leq \varepsilon^3$. SSM for the complete proof.

29. Given any $\varepsilon > 0$, let $\delta \leq \min\left\{1, \dfrac{\varepsilon}{3}\right\}$. SSM for the complete proof. **31.** Given any $\varepsilon > 0$, let $\delta \leq \min\{1, 6\varepsilon\}$. SSM for the complete proof.

33. SSM. **35.** Given any $\varepsilon > 0$, let $\delta \leq \min\left\{1, \dfrac{234}{7}\varepsilon\right\}$. SSM for the complete proof.

37. Given any $\varepsilon > 0$, let $\delta \leq \min\{1, 26\varepsilon\}$. SSM for the complete proof. **39.** Given any $\varepsilon > 0$, let $\delta \leq \dfrac{\varepsilon}{1 + |m|}$. SSM for the complete proof.

41. x must be within 0.05 of 3. **43.** SSM. **45.** SSM. **47.** SSM. **49.** SSM. **51.** SSM.

53. Given any $\varepsilon > 0$, let $\delta \leq \min\left\{1, \dfrac{\varepsilon}{47}\right\}$. SSM for the complete proof. **55.** $M = 101$ **57.** SSM.

Review Exercises   (SSM = See Student Solutions Manual; AWV = Answers will vary.)

1. $\lim\limits_{x \to 0} \dfrac{1 - \cos x}{1 + \cos x} = 0$.

		x approaches 0 from the left				x approaches 0 from the right		
x		-0.1	-0.01	-0.001	$\to 0 \leftarrow$	0.001	0.01	0.1
$f(x) = \dfrac{1 - \cos x}{1 + \cos x}$		0.002504	0.000025	0.00000025	$f(x)$ approaches 0	0.00000025	0.000025	0.002504

3. $\lim\limits_{x \to 2} f(x)$ does not exist.   **5.** $-\dfrac{3}{x^2}$ **7.** 1   **9.** $-\pi$ **11.** 0   **13.** 27   **15.** 4   **17.** $\dfrac{2}{3}$ **19.** $-\dfrac{1}{6}$ **21.** 6   **23.** 1   **25.** -1 **27.** 8

29. $\lim\limits_{x \to 2^-} f(x) = 7$, $\lim\limits_{x \to 2^+} f(x) = 7$, $\lim\limits_{x \to 2} f(x) = 7$ **31.** f is not continuous at $c = 1$.   **33.** f is continuous at $c = 0$.

35. f is not continuous at $c = \dfrac{1}{2}$.   **37. (a)** $2x - 3$ **(b)** -1 **39.** f is continuous on the set $\{x \,|\, x \neq 3\}$.

41. f is continuous on the set $\{x \,|\, x \neq -2, x \neq 0\}$.   **43.** f is continuous on the set of all real numbers.   **45.** 0.215

47. $\lim\limits_{x \to 0^+} \dfrac{|x|}{x}(1 - x) = 1$, $\lim\limits_{x \to 0^-} \dfrac{|x|}{x}(1 - x) = -1$, $\lim\limits_{x \to 0} \dfrac{|x|}{x}(1 - x)$ does not exist.   **49.** $\dfrac{1}{2\sqrt{x}}$ **51.** 1   **53.** $\dfrac{3}{4}$ **55.** 0   **57.** $-\infty$ **59.** 3

61. $x = -3$ is a vertical asymptote. $y = 4$ is a horizontal asymptote.   **63.** Yes   **65.** $f(0) = -\pi$

67. (a) AWV. One possibility is: **(b)** $f(x) = \dfrac{2x + 2}{x - 4}$ **69.** SSM.

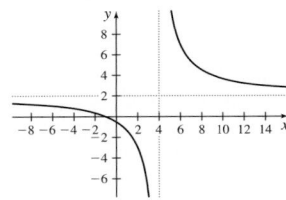

AP® Review Problems: Chapter 1   (SSM = See Student Solutions Manual.)

1. B   **3.** C   **5.** C   **7.** C   **9.** D   **11.** D   **13.** C   **15.** SSM.

Chapter 2

Section 2.1   (SSM = See Student Solutions Manual; AWV = Answers will vary.)

1. True   **2.** True   **3.** Prime, slope, $(c, f(c))$ **4.** True   **5.** 6   **6.** derivative

7. (a) $y = -12x - 12$ **9. (a)** $y = 12x + 16$ **11. (a)** $y = -x + 2$ **13. (a)** $y = -\dfrac{1}{36}x + \dfrac{7}{36}$ **15. (a)** $y = -\dfrac{1}{2}x + \dfrac{3}{2}$

(b) $y = \dfrac{1}{12}x + \dfrac{73}{6}$ **(b)** $y = -\dfrac{1}{12}x - \dfrac{49}{6}$ **(b)** $y = x$ **(b)** $y = 36x - \dfrac{215}{6}$ **(b)** $y = 2x - 1$

(c)   **(c)**   **(c)**   **(c)**   **(c)**

17. (a) 5   **(b)** 5   **19. (a)** 0   **(b)** $\dfrac{7}{16}$ **21.** $f'(1) = 4$ **23.** $f'(0) = 0$ **25.** $f'(-1) = -5$ **27.** $f'(4) = \dfrac{1}{4}$ **29.** $f'(0) = -7$

31.

Time Interval	Δt	$\dfrac{\Delta s}{\Delta t}$
$[3, 3.1]$	0.1	61
$[3, 3.01]$	0.01	60.1
$[3, 3.001]$	0.001	60.01

The velocity appears to approach 60 cm/s.

33. $v = 4$ m/s at $t_0 = 0$, $v = 16$ m/s at $t_0 = 2$, $v = 6t_0 + 4$ m/s at any t_0

35. $v = 7$ cm/s at $t_0 = 1$, $v = \dfrac{385}{16}$ cm/s at $t_0 = 4$

37. (a) $7 \leq t \leq 10$ and $11 \leq t \leq 13$ **(b)** $0 \leq t \leq 4$ **(c)** $4 \leq t \leq 7$ and $10 \leq t \leq 11$ **(d)** Average speed ≈ 0.675 mi/min

(e) 0.415 mi/min   **39.** -5 **41.** No   **43.** $R'(50) = 0.59$ **45. (a)** 250 sales per day   **(b)** 200 sales per day

47. (a) 32 ft/s   **(b)** $t \approx 7.914$ s   **(c)** Average velocity ≈ 126.618 ft/s   **(d)** $v(7.914) \approx 253.235$ ft/s

49. (a) 9.8 m/s   **(b)** $t \approx 4.243$ s   **(c)** Average velocity ≈ 20.789 m/s   **(d)** $v(4.243) \approx 41.578$ m/s   **51. (a)** $T'(11) \approx 2°$F/hr.   **(b)** AWV.

53. SSM. **55. (a)** $d'(t)$ is the rate of change of the diameter (in centimeters) with respect to time (in days).
(b) $d'(1) > d'(20)$ **(c)** $d'(1)$ is the instantaneous rate of change of the peach's diameter on day 1 and $d'(20)$ is the instantaneous rate of change of the peach's diameter on day 20.

AP® Practice Problems (SSM = See Student Solutions Manual.)   Retain Your Knowledge

1. B   **3.** C   **5.** B   **7.** B   **9.** SSM.   **1.** A   **3.** B

Section 2.2   (SSM = See Student Solutions Manual; AWV = Answers will vary.)

1. False   **2.** False   **3. (b)** a vertical   **4.** derivative   **5.** 0   **7.** 5   **9.** $-2c$ **11.** $f'(x) = 0$, all real numbers

13. $f'(x) = 8x + 1$, all real numbers   **15.** $f'(x) = \dfrac{5}{2\sqrt{x-1}}, x > 1$

17. $f'(x) = \dfrac{1}{3}$ **19.** $f'(x) = 4x - 5$ **21.** $f'(x) = 3x^2 - 8$

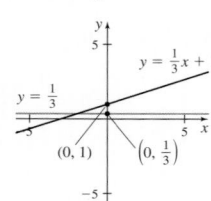   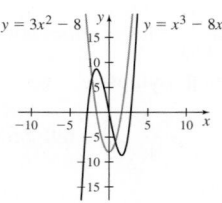

23. Not a graph of f and f'.   **25.** Graph of f and f'. The blue curve is the graph of f; the green curve is the graph of f'.
27.   **29.**

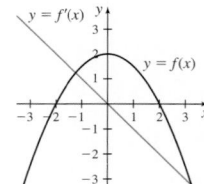

31. (B)   **33.** (A)   **35.** $f'(-8) = -\dfrac{1}{3}$ **37.** $f'(2)$ does not exist.   **39.** $f'(1) = 2$ **41.** $f'\left(\dfrac{1}{2}\right)$ does not exist.

43. $f'(-1)$ does not exist.   **45. (a)** Yes, at $x = 0$ and $x = 4$ **(b)** Yes, a cusp at $x = 2$ **(c)** Yes, at $x = -2$
47. (a) Yes, at $x = -2, x = 1, x = 2$ **(b)** Yes, at $x = -1$, not a cusp   **(c)** Yes, at $x = 0$

49. (a) -2 and 4   **(b)** 0, 2, 6   **51.** $f'(x) = m$ **53.** $f'(x) = -\dfrac{2}{x^3}$ **55.** $f(x) = x^2, c = 2$ **57.** $f(x) = x^2, c = 1$

59. $f(x) = \sqrt{x}, c = 9$ **61.** $f(x) = \sin x, c = \dfrac{\pi}{6}$ **63.** $f(x) = 2(x+2)^2 - (x+2), c = 0$ **65.** $f(x) = x^2 + 2x, c = 3$

67. $V'(t) = 4$ ft³/s   **69.** $C'(x) = 3$; dollars/switch   **71. (a)** Continuous at 0   **(b)** $f'(0) = 0$ **(c)**

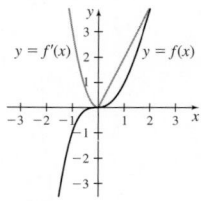

73. (a) $s'(4.99) = 74.7003$ ft/s; $s'(5.01) = 0$ ft/s   **(b)** Not continuous   **(c)** AWV.   **75.** $p'(x) = -kp, k > 0$
77. $(\sqrt{2}, 4)$ and $(-\sqrt{2}, 4)$ **79. (a)** $2\pi r(\Delta r) + \pi(\Delta r)^2$ **(b)** $2\pi\,\Delta r$ **(c)** $2\pi r + \pi\,\Delta r$ **(d)** 2π **(e)** 2π
81. SSM.   **83.** SSM.   **85. (a)** Parallel   **(b)** Perpendicular   **87. (a)** $k = 4$ **(b)** $f'(3) = 12$ **(c)** $f'(x) = 4x$

AP® Practice Problems   (SSM = See Student Solutions Manual.)   Retain Your Knowledge

1. C   **3.** A   **5.** C   **7.** C   **9.** B   **11.** SSM.   **1.** C   **3.** C

Section 2.3　(SSM = See Student Solutions Manual; AWV = Answers will vary.)

1. $0; 3x^2$　**2.** nx^{n-1}　**3.** True　**4.** $k\left[\dfrac{d}{dx}f(x)\right]$　**5.** e^x　**6.** True　**7.** $f'(x)=3$　**9.** $f'(x)=2x+3$　**11.** $f'(u)=40u^4-5$

13. $f'(s)=3as^2+3s$　**15.** $f'(t)=t^5-\dfrac{5}{6}$　**17.** $f'(t)=\dfrac{3}{5}t^2$　**19.** $f'(x)=\dfrac{3x^2+2}{7}$　**21.** $f'(x)=4e^x$　**23.** $f'(x)=1-\dfrac{1}{x}$

25. $f'(u)=\dfrac{5}{u}-2e^u$　**27.** $\sqrt{3}$　**29.** $2\pi R$　**31.** $4\pi r^2$

33. **(a)** 3　**(b)** $y=3x-1$　　**(c)** $y=-\dfrac{1}{3}x-1$　　　　**35.** **(a)** 6　**(b)** $y=6x+1$　**(c)** $y=-\dfrac{1}{6}x+1$

(d)
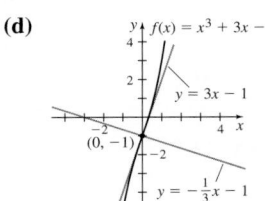

(d)

37. **(a)** $(2,-8)$　**(e)**　　　　**(f)** f is increasing when $x>2$ and decreasing when $x<2$.
(b) $y=-8$
(c) $x>2$
(d) $x<2$

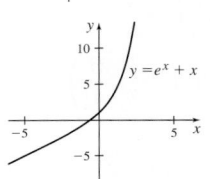

39. **(a)** None　**(e)**　　　　**(f)** f is increasing for all x.
(b) None
(c) All real numbers
(d) None

41. **(a)** $(1,0),(-1,4)$　**(e)**　　　　**(f)** f is increasing when $x<-1$ or $x>1$
(b) $y=0, y=4$　　　　　　　　and decreasing when $-1<x<1$.
(c) $x<-1$ or $x>1$
(d) $-1<x<1$

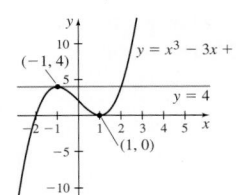

43. $v(0)=-1$ m/s and $v(5)=74$ m/s　**45.** **(a)** $v(t)=2t-5$　**(b)** $t=\dfrac{5}{2}$

47. **(a)** $\dfrac{4}{5}$　**(b)** $-\dfrac{7}{20}$　**(c)** $-\dfrac{9}{5}$　**(d)** $\dfrac{23}{20}$　**(e)** $-\dfrac{21}{20}$　**(f)** $-\dfrac{23}{10}$　**49.** **(a)** $f'(x)=24x^2-24x+6$　**(b)** $f'(x)=6(2x-1)^2$　**(c)** SSM.

51. $\dfrac{1}{4}$　**53.** $20{,}480\sqrt{3}$　**55.** $3ax^2$　**57.** $y=5x-3$　**59.** $y=x+1$　**61.** $y=3(x-1)$

63. **(a)** $y=45x-65$　　　　　　　　　　　　　　　**(d)**

(b) $\left(\dfrac{\sqrt{3}}{3}, -\dfrac{5\sqrt{3}}{9}-1\right), \left(-\dfrac{\sqrt{3}}{3}, \dfrac{5\sqrt{3}}{9}-1\right)$

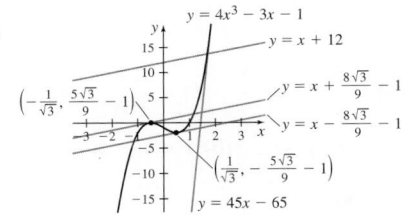

(c) At $x=\dfrac{\sqrt{3}}{3}, y=x-\dfrac{8\sqrt{3}}{9}-1$ and at $x=-\dfrac{\sqrt{3}}{3}, y=x+\dfrac{8\sqrt{3}}{9}-1$

65. SSM.　**67.** SSM.　**69.** $a=3, b=2, c=0$　**71.** $(2,4)$　**73.** **(a)** 0　**(b)** $-kR$　**(c)** $-2kR$

75. **(a)** $\dfrac{dL}{dT}=4\sigma AT^3$　**(b)** $2.694\times10^{23}\ \dfrac{\text{W}}{\text{K}}$　**(c)** 2.694×10^{23} W　**77.** $F'(x)=mx$　**79.** SSM.　**81.** $Q=\left(\dfrac{9}{4}, \dfrac{81}{16}\right)$

83. $y=-\dfrac{7}{64}x^4+\dfrac{7}{16}x^3+\dfrac{21}{16}x^2-4x$　**85.** **(a)** $c=1$　**(b)** $y=12x-16$ and $y=12x+1$

AP® Practice Problems

1. B　**3.** C　**5.** D　**7.** D　**9.** D　**11.** C　**13.** SSM.

Retain Your Knowledge

1. A　**3.** D

Section 2.4 (SSM = See Student Solutions Manual; AWV = Answers will vary.)

1. False **2.** $f(x)g'(x) + f'(x)g(x)$ **3.** False **4.** $\dfrac{\left[\dfrac{d}{dx}f(x)\right]g(x) - f(x)\left[\dfrac{d}{dx}g(x)\right]}{[g(x)]^2}$ **5.** True **6.** $-\dfrac{\dfrac{d}{dx}g(x)}{[g(x)]^2}$ **7.** 0 **8.** $\dfrac{d^2s}{dt^2}$

9. $f'(x) = e^x(x+1)$ **11.** $f'(x) = 5x^4 - 2x$ **13.** $f'(x) = 27x^2 + 24x - 15$ **15.** $s'(t) = 16t^7 - 24t^5 + 10t^4 - 4t^3 + 4t - 1$

17. $f'(x) = 4x^2 + 3x^2\ln x + \dfrac{1}{x}$ **19.** $g'(s) = \dfrac{2}{(s+1)^2}$ **21.** $G'(u) = -\dfrac{4}{(1+2u)^2}$ **23.** $f'(x) = \dfrac{2(6x^2 + 16x + 3)}{(3x+4)^2}$

25. $f'(w) = -\dfrac{3w^2}{(w^3 - 1)^2}$ **27.** $s'(t) = -\dfrac{3}{t^4}$ **29.** $f'(x) = \dfrac{4}{e^x}$ **31.** $f'(x) = -\dfrac{40}{x^5} - \dfrac{6}{x^3}$ **33.** $f'(x) = 9x^2 + \dfrac{2}{3x^3}$

35. $s'(t) = -\dfrac{1}{t^2} + \dfrac{2}{t^3} - \dfrac{3}{t^4}$ **37.** $f'(x) = \dfrac{(x-2)e^x}{x^3}$ **39.** $f'(x) = -\dfrac{(x^3 - x^2 + x + 1)}{x^2 e^x}$ **41.** $f'(x) = 10x + 1$, $f''(x) = 10$

43. $f'(x) = \dfrac{3}{x}$, $f''(x) = -\dfrac{3}{x^2}$ **45.** $f'(x) = e^x(x+6)$, $f''(x) = e^x(x+7)$ **47.** $f'(x) = 8x^3 + 3x^2 + 10$, $f''(x) = 24x^2 + 6x$

49. $f'(x) = 1 - \dfrac{1}{x^2}$, $f''(x) = \dfrac{2}{x^3}$ **51.** $f'(t) = 1 + \dfrac{1}{t^2}$, $f''(t) = -\dfrac{2}{t^3}$ **53.** $f'(x) = e^x\left(\dfrac{1}{x} - \dfrac{1}{x^2}\right)$; $f''(x) = e^x\left(\dfrac{2}{x^3} - \dfrac{2}{x^2} + \dfrac{1}{x}\right)$

55. (a) $y' = -\dfrac{1}{x^2}$, $y'' = \dfrac{2}{x^3}$ (b) $y' = \dfrac{5}{x^2}$, $y'' = -\dfrac{10}{x^3}$ **57.** $v(t) = 32t + 20$, $a(t) = 32$

59. $v(t) = 9.8t + 4$, $a(t) = 9.8$ **61.** $f^{(4)}(x) = 0$ **63.** 5040 **65.** e^u **67.** $-e^x$

69. (a) $\dfrac{3}{4}$ (d)

(b) $y = \dfrac{3}{4}x + \dfrac{1}{4}$

(c) $(0, 0)$, $(2, 4)$

71. (a) $\dfrac{5}{4}$ (d)

(b) $y = \dfrac{5}{4}x - \dfrac{3}{4}$

(c) $(0, 0)$, $\left(-\dfrac{3}{2}, \dfrac{27}{4}\right)$

73. (a) $(-2, 5)$, $(2, -27)$ (e)

(b) $y = 5$, $y = -27$

(c) $x < -2$ or $x > 2$

(d) $-2 < x < 2$

(f) AWV.

75. (a) $(0, 0)$, $(-2, -4)$ (e)

(b) $y = 0$, $y = -4$

(c) $x < -2$ or $x > 0$

(d) $-2 < x < -1$ or $-1 < x < 0$

(f) AWV.

77. (a) $(-1, -e^{-1})$ (e)

(b) $y = -\dfrac{1}{e}$

(c) $x > -1$

(d) $x < -1$

(f) AWV. **79.** (a) $(-1, -2e)$, $\left(3, \dfrac{6}{e^3}\right)$ (e)

(b) $y = -2e$, $y = \dfrac{6}{e^3}$

(c) $-1 < x < 3$

(d) $x < -1$ or $x > 3$

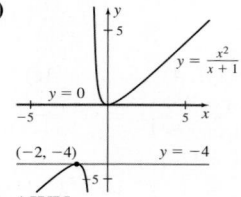

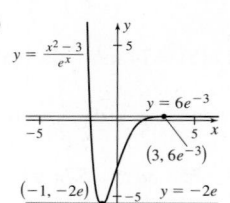

(f) AWV.

81. (a) $\dfrac{8}{5}$ (b) $-\dfrac{29}{10}$ (c) $\dfrac{1}{9}$ (d) $-\dfrac{19}{160}$ (e) $\dfrac{1}{9}$ (f) $\dfrac{12}{25}$

83. (a) $v(t) = -9.8t + 39.2$ (b) $t = 4$ s (c) 78.4 m (d) -9.8 m/s^2 (e) $t = 8$ s (f) $v = -39.2$ m/s (g) 156.8 m

85. (a) $0 \le x \le 100$ (b)

(c) $13,333.33 (d) $C'(x) = \dfrac{550}{(110 - x)^2}$

(e) $C'(40) = 0.112 = \$112/\%$, $C'(60) = 0.220 = \$220/\%$, $C'(80) = 0.611 = \$611/\%$, $C'(90) = 1.375 = \$1,375/\%$

(f) AWV.

87. (a) $f'(t) = \dfrac{0.4 - 0.8t^2}{(2t^2 + 1)^2}$ mg/L/h

(b) $f'\left(\dfrac{1}{6}\right) \approx 0.339$ mg/L/h, $f'\left(\dfrac{1}{2}\right) \approx 0.089$ mg/L/h,

$f'(1) \approx -0.044$ mg/L/h

(c) AWV.

(d)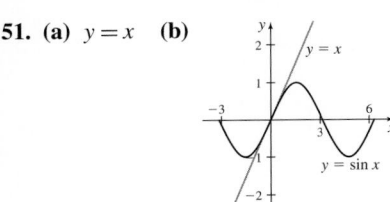

(e) Concentration is highest at approximately 45 min and is about 0.14 mg/L.

89. (a) $-100{,}000 \cdot \dfrac{2p + 10}{(p^2 + 10p + 50)^2}$ books/dollar **(b)** $D'(5) = -128$, $D'(10) = -48$, $D'(15) = -22.145$ **(c)** AWV.

91. (a) $\dfrac{dp}{dr} = -\dfrac{9nRT}{4\pi r^4}$ **(b)** As the radius increases, pressure within the container decreases. **(c)** $p'\left(\dfrac{1}{4}\right) \approx -415{,}945.358$ Pa/m

93. (a) $v(t) = 3t^2 - 1$ m/s, $a(t) = 6t$ m/s^2, $J(t) = 6$ m/s^3, $S(t) = 0$ m/s^4 **(b)** $t = \pm\dfrac{\sqrt{3}}{3}$ s **(c)** $a(2) = 12$ m/s^2, $a(5) = 30$ m/s^2

(d) No **(e)** AWV. **95. (a)** $a(t) = 1.6 + 1.998t$ m/s^2 **(b)** $J(t) = 1.998$ m/s^3

97. (a) $\dfrac{dJ}{dr} = -\dfrac{2I}{\pi r^3}$ **(b)** As radius increases, current density decreases. **(c)** -1.273×10^{10} amps/m^3

99. SSM. **101.** $y' = 6x^5 - 5x^4 + 20x^3 - 18x^2 + 2x - 5$ **103.** $y' = 9x^2(x^3 + 1)^2$ **105.** $y' = \dfrac{1}{x^2} + \dfrac{2}{x^3} - \dfrac{4}{x^5} - \dfrac{5}{x^6} + \dfrac{6}{x^7}$

107. SSM. **109.** $f'(x) = \dfrac{6x^5 - 4x^3 + 2x}{x^4 + 1} - 4x^3 \left[\dfrac{x^6 - x^4 + x^2}{(x^4 + 1)^2}\right]$

111. (a) $f'(x) = \dfrac{2}{x+1} - \dfrac{2x}{(x+1)^2}$ **(b)** $f'(x) = \dfrac{2}{(x+1)^2}$ **(c)** $f^{(5)}(x) = \dfrac{240}{(x+1)^6}$

113. (a) $[f_1'(x) \cdot f_2(x) \cdots f_n(x)] + [f_1(x) \cdot f_2'(x) \cdot f_3(x) \cdots f_n(x)] + \cdots + [f_1(x) \cdots f_{n-1}(x) \cdot f_n'(x)]$

(b) $-\dfrac{1}{f_1(x) \cdots f_n(x)}\left[\dfrac{f_1'(x)}{f_1(x)} + \cdots + \dfrac{f_n'(x)}{f_n(x)}\right]$

115. (a) $f_1(x) = \dfrac{x}{x-1}$, $f_2(x) = \dfrac{2x-1}{x}$, $f_3(x) = \dfrac{3x-1}{2x-1}$, $f_4(x) = \dfrac{5x-2}{3x-1}$, $f_5(x) = \dfrac{8x-3}{5x-2}$

(b) $a_0 = 1$, $a_1 = 1$, $a_2 = 2$, $a_3 = 3$, $a_4 = 5$, $a_5 = 8$, ... or $\{1, 1, 2, 3, 5, 8, \ldots\}$ **(c)** $a_{n+2} = a_{n+1} + a_n$

(d) $f_0'(x) = 1$, $f_1'(x) = -\dfrac{1}{(x-1)^2}$, $f_2'(x) = \dfrac{1}{x^2}$, $f_3'(x) = -\dfrac{1}{(2x-1)^2}$, $f_4'(x) = \dfrac{1}{(3x-1)^2}$, $f_5'(x) = -\dfrac{1}{(5x-2)^2}$

AP® Practice Problems

1. A **3.** A **5.** D **7.** B **9.** B **11.** B

Retain Your Knowledge

1. C **3.** C

Section 2.5 (SSM = See Student Solutions Manual; AWV = Answers will vary.)

1. False **2.** False **3.** True **4.** False **5.** $y' = 1 - \cos x$ **7.** $y' = \sec^2 x + \cos x$ **9.** $y' = 3\cos\theta + 2\sin\theta$

11. $y' = \cos^2 x - \sin^2 x = \cos(2x)$ **13.** $y' = \cos t - t\sin t$ **15.** $y' = e^x(\sec^2 x + \tan x)$ **17.** $y' = \pi\sec u(\sec^2 u + \tan^2 u)$

19. $y' = -\dfrac{x\csc^2 x + \cot x}{x^2}$ **21.** $y' = x(x\cos x + 2\sin x)$ **23.** $y' = t\sec^2 t + \tan t - \sqrt{3}\sec t\tan t$ **25.** $y' = -\dfrac{1}{1 - \cos\theta}$

27. $y' = \dfrac{\cos t + t\cos t - \sin t}{(1+t)^2}$ **29.** $y' = \dfrac{\cos x - \sin x}{e^x}$ **31.** $y' = -\dfrac{2}{(\sin\theta - \cos\theta)^2}$ **33.** $y' = \dfrac{\sec t\tan t + t\tan^2 t - t - \tan t}{(1 + t\sin t)^2}$

35. $y' = -\csc\theta(\csc^2\theta + \cot^2\theta)$ **37.** $y' = \dfrac{2\sec^2 x}{(1 - \tan x)^2}$ **39.** $y'' = -\sin x$ **41.** $y'' = 2\tan\theta\sec^2\theta$ **43.** $y'' = 2\cos t - t\sin t$

45. $y'' = 2e^x \cos x$ **47.** $y'' = -5\sin u + 4\cos u$ **49.** $y'' = -(a\sin x + b\cos x)$

51. (a) $y = x$ **(b)**

53. (a) $y = x$ **(b)**

55. (a) $y = \sqrt{2}$ **(b)**

57. (a) $\left(\tan^{-1} 2 + n\pi, (-1)^n \sqrt{5} \right)$ **(b)**
where n is an integer

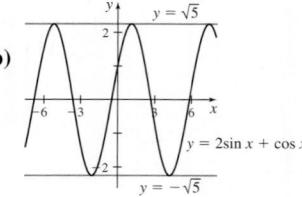

59. (a) $(n\pi, (-1)^n)$ **(b)**
where n is an integer

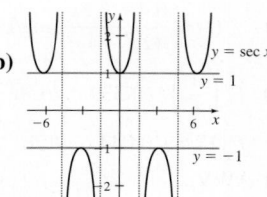

61. $f^{(n)}(x) = \begin{cases} (-1)^{n/2} \sin x & \text{if } n \text{ is even} \\ (-1)^{(n-1)/2} \cos x & \text{if } n \text{ is odd} \end{cases}$

63. -1 **65. (a)** $v(t) = -\dfrac{1}{8} \sin t$ m/s

(b) $\dfrac{\pi}{2} + n\pi$, where n is an integer

(c) $a(t) = -\dfrac{1}{8} \cos t$ m/s^2

(d) $\dfrac{\pi}{2} + n\pi$, where n is an integer

(e)

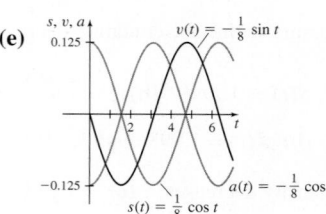

67. (a) $\dfrac{dy}{d\theta} = -8 \sin \theta$

(b) $\dfrac{dy}{d\theta} = -4\sqrt{2}$ ft/radian

69. (a)

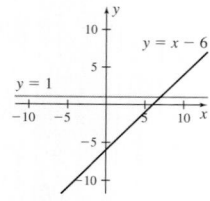

(b) Max $= 4$, Min $= \dfrac{4}{3}$

(c) $x = \pi, 2\pi, 3\pi$

(d) $w'(\pi - 0.1) \approx 0.395$, $w'(2\pi - 0.1) \approx -0.045$

(e) A symmetric wave would give slopes having equal magnitudes but opposite signs.

71. (a) $\$10{,}000$, $\$25{,}093$, $\$15{,}411$, $\$37{,}570$ **(d)**

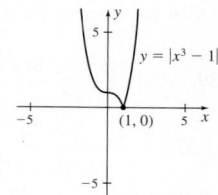

(b) $R'(t) = \cos t + 0.3$

(c) $\$10{,}539$ per month

(e) AWV.

73. $n = 4$ **75.** SSM. **77.** SSM. **79.** SSM. **81.** SSM.

AP® Practice Problems (SSM = See Student Solutions Manual.)

1. A   **3.** B   **5.** B   **7.** B   **9.** B   **11.** SSM.

Retain Your Knowledge

1. B   **3.** B

Review Exercises

1. (a) $\dfrac{1}{2}$ **(b)** $\dfrac{1}{4}$ **(c)** $\dfrac{1}{2\sqrt{c}}$ **3.** 3 **5.** 5 **7.** 2

9. $f'(x) = 1$

11. $f'(x) = -\dfrac{3}{2x^4}$

13. f does not have a derivative at $c = 1$. AWV.

15. Not the graph of a function and its derivative.

17.

19. $f'(x) = 5x^4$ **21.** $f'(x) = x^3$ **23.** $f'(x) = 6x - 4$ **25.** $F'(x) = 14x$ **27.** $f'(x) = 15(x^2 - 6x + 6)$ **29.** $f'(x) = 2 + \dfrac{3}{x^2}$

31. $f'(x) = -\dfrac{35}{(x-5)^2}$ **33.** $f'(x) = 4x + \dfrac{10}{x^3}$ **35.** $f'(x) = -\dfrac{a}{x^2} + \dfrac{3b}{x^4}$ **37.** $f'(x) = -\dfrac{6(2x-3)}{(x^2-3x)^3}$ **39.** $s'(t) = \dfrac{2t^2(t-3)}{(t-2)^2}$

41. $F'(z) = -\dfrac{2z}{(z^2+1)^2}$ **43.** $g'(z) = -\dfrac{2z-1}{(1-z+z^2)^2}$ **45.** $s'(t) = -e^t$ **47.** $f'(x) = \dfrac{1}{x} + 1 + \ln x$ **49.** $f'(x) = x\cos x + \sin x$

51. $G'(u) = \sec u(\sec u + \tan u)$ **53.** $f'(x) = e^x(\cos x + \sin x)$ **55.** $f'(x) = 2(\cos^2 x - \sin^2 x) = 2\cos(2x)$

57. $f'(x) = 2\cos x \sin x = \sin(2x)$ **59.** $f'(\theta) = -\dfrac{\sin\theta + \cos\theta}{2e^\theta}$ **61.** $f'(x) = 50x + 30$, $f''(x) = 50$

63. $g'(u) = \dfrac{1}{(2u+1)^2}$, $g''(u) = -\dfrac{4}{(2u+1)^3}$ **65.** $f'(u) = -\dfrac{\sin u + \cos u}{e^u}$, $f''(u) = \dfrac{2\sin u}{e^u}$

67. (a) $y = -7x + 5$ **(c)**   **69. (a)** $y = -x - 1$ **(c)**
(b) $y = \dfrac{1}{7}x + \dfrac{85}{7}$ **(b)** $y = x - 1$

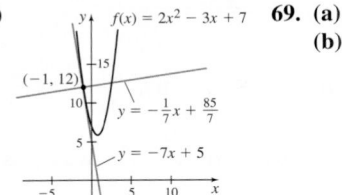

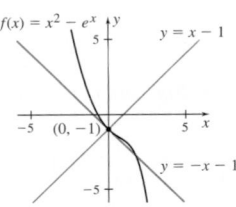

71. (a) -1 m/s **(b)** $v(0) = -6$ m/s, $v(5) = 4$ m/s, $v(t) = 2t - 6$ m/s   **(c)** $a(t) = 2$ m/s^2

73. (a) $\dfrac{dp}{dx} = -\dfrac{50{,}000}{(5x+100)^2}$ **(b)** $R(x) = \dfrac{10{,}000x}{5x+100} - 5x$ **(c)** $R'(10) = \dfrac{355}{9} \approx \$39.44$/lb, $R'(40) = \dfrac{55}{9} \approx \$6.11$/lb

75. $f'(3)$ does not exist.

AP® Review Problems: Chapter 2   (SSM = See Student Solutions Manual.)

1. D   **3.** B   **5.** A   **7.** A   **9.** C   **11.** C   **13.** SSM.

AP® Cumulative Review Problems: Chapters 1–2   (SSM = See Student Solutions Manual.)

1. B   **3.** D   **5.** A   **7.** D   **9.** C   **11.** SSM.

Chapter 3

Section 3.1   (SSM = See Student Solutions Manual.)

1. Chain   **2.** True   **3.** False   **4.** $\tan u$; $1 + \cos x$ **5.** $100(x^3 + 4x + 1)^{99}(3x^2 + 4)$ **6.** $6xe^{3x^2+5}$ **7.** True   **8.** $2x\cos x^2$

9. $y = (x^3 + 1)^5$; $\dfrac{dy}{dx} = 15x^2(x^3+1)^4$ **11.** $y = \dfrac{x^2+1}{x^2+2}$; $\dfrac{dy}{dx} = \dfrac{2x}{(x^2+2)^2}$ **13.** $y = \left(\dfrac{1}{x}+1\right)^2$; $\dfrac{dy}{dx} = -2\left(\dfrac{1}{x}+1\right)\left(\dfrac{1}{x^2}\right)$

15. $f'(x) = 4(2x+7)$ **17.** $f'(x) = -\dfrac{18}{(6x-5)^4}$ **19.** $g'(x) = 8x(x^2+5)^3$ **21.** $f'(u) = 3\left(u - \dfrac{1}{u}\right)^2\left(1 + \dfrac{1}{u^2}\right)$

23. $g'(x) = 3(4x + e^x)^2(4 + e^x)$ **25.** $f'(x) = 2\tan x \sec^2 x$ **27.** $f'(z) = 2(\tan z + \cos z)(\sec^2 z - \sin z)$

29. $y' = 4x(x^2+4)(2x^3-1)^3 + 18x^2(x^2+4)^2(2x^3-1)^2 = 2x(x^2+4)(2x^3-1)^2(13x^3 + 36x - 2)$

31. $y' = \dfrac{2\sin x(x\cos x - \sin x)}{x^3}$ **33.** $y' = -3\sin(3x)$ **35.** $y' = 4(x+1)\cos(x^2+2x-1)$ **37.** $y' = \dfrac{2}{x^2}\sin\dfrac{2}{x}$

39. $y' = \dfrac{2x}{x^2+4}$ **41.** $y' = -\dfrac{1}{x^2}e^{1/x}$ **43.** $y' = -\dfrac{2(4x^3-2)}{(x^4-2x+1)^3}$ **45.** $y' = \dfrac{9900e^{-x}}{(1+99e^{-x})^2}$

47. $y' = 2^{\sin x} \ln 2 \cdot \cos x$ **49.** $y' = \cot x$ **51.** $y' = 5e^{3x} + 15xe^{3x}$

53. $y' = 2x\sin(4x) + 4x^2\cos(4x)$ **55.** $y' = -ae^{-ax}\sin(bx) + be^{-ax}\cos(bx)$ **57.** $y' = \dfrac{2ae^{ax}}{(e^{ax}+1)^2}$

59. $y = \left(\dfrac{48}{x^4}+1\right)^3$; $\dfrac{dy}{dx} = -\dfrac{576}{x^5}\left(\dfrac{48}{x^4}+1\right)^2$ **61.** $y = \dfrac{16}{x^4}+1$; $\dfrac{dy}{dx} = -\dfrac{64}{x^5}$ **63.** $y' = -2e^{-2x}\cos(3x) - 3e^{-2x}\sin(3x)$

65. $y' = -2xe^{x^2}\sin(e^{x^2})$ **67.** $y' = -4\sin(4x)e^{\cos(4x)}$ **69.** $y' = 24\sin(3x)\cos(3x)$

71. (a) $y' = 6x^2(x^3+1)$ **(b)** $y' = 6x^2(x^3+1)$ **(c)** $y' = 6x^5 + 6x^2$ **(d)** The answers are equal.

73. (a) $y = 0$

(b) $x = 1$

(c)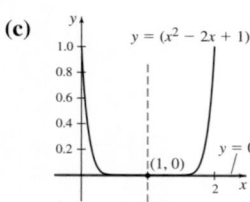

75. (a) $y = -\dfrac{7}{27}x + \dfrac{16}{27}$

(b) $y - \dfrac{2}{27} = \dfrac{27}{7}(x - 2)$

(c)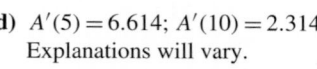

77. (a) $y = 2x + 1$

(b) $y = -\dfrac{1}{2}x + 1$

(c)

79. $\dfrac{d^2 y}{dx^2}\cos x^5 = -25x^8 \cos(x^5) - 20x^3 \sin(x^5)$ **81.** $h'(1) = -12$ **83.** $h'(0) = 0$ **85.** 78 **87.** $\dfrac{df}{dx} = f'(x^2 + 1)\cdot(2x)$

89. $\dfrac{df}{dx} = f'\left(\dfrac{x+1}{x-1}\right)\cdot\dfrac{-2}{(x-1)^2}$ **91.** $\dfrac{df}{dx} = f'(\sin x)\cdot\cos x$ **93.** $\dfrac{d^2 f}{dx^2} = f''(\cos x)\cdot\sin^2 x - f'(\cos x)\cdot\cos x$

95. (a) $v(t) = -A\omega\sin(\omega t + \phi)$ **(b)** $t = \dfrac{k\pi - \phi}{\omega}$, k an integer **(c)** $a(t) = -A\omega^2\cos(\omega t + \phi)$

(d) $t = \dfrac{(2k-1)\dfrac{\pi}{2} - \phi}{\omega}$, k an integer **97.** $a(t) = \dfrac{-20\pi}{9}\sin\dfrac{\pi t}{6}$ **99.** $\dfrac{dR}{dT} = -3.597 \times 10^{-6}$ ohms/K

101. (a) 102 **(c)** $\dfrac{dA}{dt} = 18.9e^{-0.21t}$ **(e)** Explanations will vary.

(b) 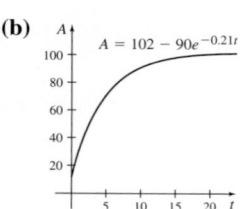 **(d)** $A'(5) = 6.614$; $A'(10) = 2.314$ Explanations will vary.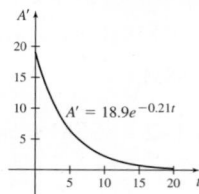

103. (a) $a(t) = ge^{-kt/m}$ m/s^2 **(b)** $\dfrac{mg}{k}$ m/s; explanations will vary. **(c)** 0 m/s^2; explanations will vary.

105. SSM. **107. (a)** $\omega(t) = -\dfrac{\pi}{6}\sqrt{\dfrac{2k}{5}}\sin\left(\dfrac{1}{2}\sqrt{\dfrac{2k}{5}}t\right)$ **(b)** $\omega(3) = -\dfrac{\pi}{6}\sqrt{\dfrac{2k}{5}}\sin\left(\dfrac{3}{2}\sqrt{\dfrac{2k}{5}}\right)$

109. $F'(1) = 2$ **111.** SSM. **113.** SSM. **115.** SSM. **117.** SSM. **119.** SSM.

121. (a) $y^{(n)}(x) = a^n e^{ax}$ **(b)** $y^{(n)}(x) = (-1)^n a^n e^{-ax}$ **123. (a)** $\dfrac{d^{11}}{dx^{11}}\cos(ax) = a^{11}\sin(ax)$ **(b)** $\dfrac{d^{12}}{dx^{12}}\cos(ax) = a^{12}\cos(ax)$

(c) $f^{(n)}(x) = -a^n\sin(ax)$, $n = 1 + 4k$, $k = 0, 1, 2, 3, \ldots$; $f^{(n)}(x) = -a^n\cos(ax)$, $n = 2 + 4k$, $k = 0, 1, 2, 3, \ldots$;
$f^{(n)}(x) = a^n\sin(ax)$, $n = 3 + 4k$, $k = 0, 1, 2, 3, \ldots$; $f^{(n)}(x) = a^n\cos(ax)$, $n = 4 + 4k$, $k = 0, 1, 2, 3, \ldots$

125. SSM. **127.** SSM. **129.** SSM. **131.** SSM. **133. (a)** $-72{,}000\pi^2$cm/s^2 **(b)** $-72{,}000\pi^2$kg $\cdot$ cm/s^2

AP® Practice Problems

1. D **3.** A **5.** C **7.** B **9.** A **11.** A **13.** C **15.** C

Retain Your Knowledge

1. C **3.** C

Section 3.2 (SSM = See Student Solutions Manual; AWV = Answers will vary.)

1. True **2.** False **3.** $\dfrac{1}{x^{2/3}}$ **4.** $3x(x^2+1)^{1/2}$ **5.** $y' = -\dfrac{3}{2}$ **7.** $y' = \dfrac{2-x}{y}$ **9.** $y' = \dfrac{\cos x}{e^y}$ **11.** $y' = -\dfrac{e^{x+y}}{e^{x+y}-1}$ **13.** $y' = -\dfrac{2y}{x}$

15. $y' = \dfrac{2x-y}{2y+x}$ **17.** $y' = -\dfrac{y^2}{x^2}$ **19.** $y' = \dfrac{x^3+y}{x(1-xy)}$ **21.** $y' = \dfrac{e^y\sin x - e^x\sin y}{e^x\cos y + e^y\cos x}$ **23.** $y' = -\dfrac{6x(x^2+y)^2}{3(x^2+y)^2-1}$

25. $y' = \dfrac{\sec^2(x-y)}{1+\sec^2(x-y)}$ **27.** $y' = \dfrac{\sin y}{1-x\cos y}$ **29.** $y' = \dfrac{ye^{xy}-2xy}{x^2-xe^{xy}}$ **31.** $y' = \dfrac{2}{3x^{1/3}}$ **33.** $y' = \dfrac{2}{3x^{1/3}}$ **35.** $y' = \dfrac{1}{3x^{2/3}} + \dfrac{1}{3x^{4/3}}$

37. $y' = \dfrac{9x^2(x^3-1)^{1/2}}{2}$ **39.** $y' = (x^2-1)^{1/2} + \dfrac{x^2}{(x^2-1)^{1/2}}$ **41.** $y' = \dfrac{xe^{(x^2-9)^{1/2}}}{(x^2-9)^{1/2}}$ **43.** $y' = \dfrac{3}{2}(x^2\cos x)^{1/2}(2x\cos x - x^2\sin x)$

45. $y' = 3x(x^2-3)^{1/2}(6x+1)^{5/3} + 10(x^2-3)^{3/2}(6x+1)^{2/3} = (6x+1)^{2/3}(x^2-3)^{1/2}(28x^2+3x-30)$

47. $y' = -\dfrac{x}{y}$, $y'' = -\dfrac{x^2+y^2}{y^3}$ **49.** $y' = \dfrac{2x-5}{2y}$, $y'' = \dfrac{4y^2-(2x-5)^2}{4y^3}$ **51.** $y' = \dfrac{x}{(x^2+1)^{1/2}}$, $y'' = \dfrac{1}{(x^2+1)^{3/2}}$

53. (a) $-1/2$ **(b)** $y = -\dfrac{x}{2} + \dfrac{5}{2}$ **55. (a)** 3 **(b)** $y = 3x - 8$ **57. (a)** $1/2$ **(b)** $y = \dfrac{x}{2} + 2$

(c)   **(c)**   **(c)**

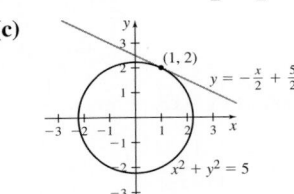 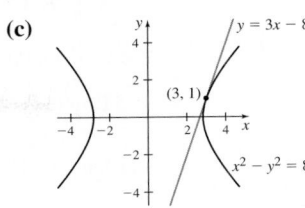 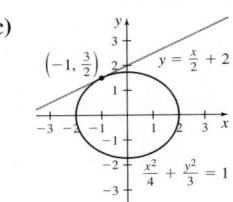

59. $y' = -\dfrac{4}{9}; \; y'' = -\dfrac{100}{243}$ **61.** $y' = \dfrac{3x}{|x|}$ **63.** $y' = \dfrac{2(2x-1)}{|2x-1|}$ **65.** $y' = \dfrac{-\sin x \cos x}{|\cos x|}$ **67.** $y' = \cos|x| \cdot \dfrac{|x|}{x}$ **69.** $y = -2x + 10$

71. (a) $y' = \dfrac{-y-1}{x+4y}$ **(b)** $y = -\dfrac{1}{3}x + \dfrac{5}{3}$ **(c)** $(6, -3), (2, 1)$ **(d)**

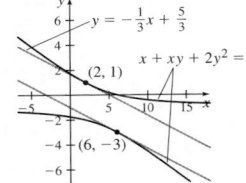

73. SSM. **75.** SSM.

77. (a) $y' = \dfrac{3x^2 - 2y}{2x - 3y^2}$ **(b)** $y = -x + 2$ **(c)** $\left(\dfrac{2^{4/3}}{3}, \dfrac{2^{5/3}}{3}\right)$ **(d)**

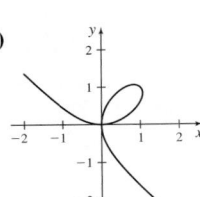

Explanations will vary.

79. $\left(\dfrac{-5}{\sqrt{2}}, \dfrac{5}{\sqrt{2}}\right)$ **81.** SSM. **83.** SSM. **85.** SSM.

87. (a) $\dfrac{dy}{dx} = -\dfrac{y}{x}$ for the first and $\dfrac{dy}{dx} = \dfrac{x}{y}$ for the second. **(b)**

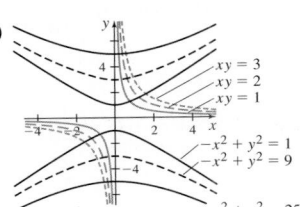

89. SSM.

AP® Practice Problems (SSM = See Student Solutions Manual.)

1. C **3.** D **5.** A **7.** D **9.** B **11.** SSM.

Retain Your Knowledge

1. B **3.** B

Section 3.3 (SSM = See Student Solutions Manual.)

1. True **2.** True **3.** False **4.** $\dfrac{1}{\sqrt{1-x^2}}$ **5.** $g'(4) = -\dfrac{1}{2}$ **7.** $f'(-2) = 2$ **9.** $g'(32) = \dfrac{1}{80}$ **11.** $g'(6) = \dfrac{1}{4}$ **13.** $g'(2) = 12$

15. $g'(49) = \dfrac{1}{8}$ **17.** $f'(x) = \dfrac{4}{\sqrt{1-16x^2}}$ **19.** $g'(x) = \dfrac{1}{x\sqrt{9x^2-1}}$ **21.** $s'(t) = \dfrac{2}{4+t^2}$ **23.** $f'(x) = -\dfrac{2x}{2x^4 - 2x^2 + 1}$

25. $f'(x) = \dfrac{2x}{(x^2+2)\sqrt{(x^2+2)^2 - 1}}$ **27.** $F'(x) = \dfrac{e^x}{\sqrt{1-e^{2x}}}$ **29.** $g'(x) = -\dfrac{1}{x^2+1}$ **31.** $g'(x) = \dfrac{x}{\sqrt{1-x^2}} + \sin^{-1} x$

33. $s'(t) = \dfrac{3t}{\sqrt{t^6 - 1}} + 2t\sec^{-1} t^3$ **35.** $f'(x) = \dfrac{-\sin x}{1+\cos x}$ **37.** $G'(x) = \dfrac{\cos(\tan^{-1} x)}{1+x^2}$ **39.** $f'(x) = \dfrac{3e^{\tan^{-1}(3x)}}{1+9x^2}$

41. $g'(x) = \dfrac{1}{1-x^2} + \dfrac{x\sin^{-1} x}{(1-x^2)^{3/2}}$ **43.** $\dfrac{dy}{dx} = \dfrac{2\sqrt{1-y^2}}{2y\sqrt{1-y^2} + 1}$ **45.** $\dfrac{dy}{dx} = \dfrac{3\pi x^2 y(1+y^4)}{80y - \pi x^3(1+y^4)}$ **47.** $g'(0) = \dfrac{1}{2}, \; g'(3) = \dfrac{1}{5}$

49. (a) $y = \dfrac{1}{2}x$ **51.** $y - 2 = \dfrac{1}{23}(x - 14)$ **53.** $5x + 3y = 18$ **55. (a)** $-\dfrac{\sqrt{3}}{6}$ m/s **(b)** $\dfrac{7\sqrt{3}}{36}$ m/s² **57.** SSM. **59.** SSM.

(b)

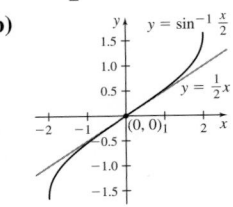

AP® Practice Problems   (SSM = See Student Solutions Manual.)

1. A   **3.** C   **5.** D   **7.** C   **9.** C   **11.** SSM.

Retain Your Knowledge

1. C   **3.** B

Section 3.4   (SSM = See Student Solutions Manual; AWV = Answers will vary.)

1. $\dfrac{1}{x}$ **2.** True   **3.** False   **4.** False   **5.** $\dfrac{1}{x}$ **6.** e **7.** $f'(x) = \dfrac{5}{x}$ **9.** $s'(t) = \dfrac{1}{t \ln 2}$ **11.** $g'(x) = \dfrac{1}{x}\cos x - \sin x \ln x$ **13.** $F'(x) = \dfrac{1}{x}$

15. $s'(t) = \dfrac{e^t + e^{-t}}{e^t - e^{-t}}$ **17.** $f'(x) = \dfrac{2x^2}{x^2 + 4} + \ln(x^2 + 4)$ **19.** $f'(x) = \dfrac{3x^2}{(3x+5)\ln 5} + 2x \log_5(3x+5)$ **21.** $f'(x) = \dfrac{1}{1 - x^2}$

23. $f'(x) = \dfrac{1}{x \ln x}$ **25.** $g'(x) = \dfrac{1}{x(x^2 + 1)}$ **27.** $f'(x) = \dfrac{4x}{x^2 + 1} - \dfrac{1}{x} - \dfrac{x}{x^2 - 1}$ **29.** $F'(\theta) = \cot \theta$ **31.** $g'(x) = \dfrac{1}{\sqrt{x^2 + 4}}$

33. $f'(x) = \dfrac{2x}{(1 + x^2)\ln 2}$ **35.** $f'(x) = \dfrac{1}{x[1 + (\ln x)^2]}$ **37.** $s'(t) = \dfrac{1}{\tan^{-1} t} \cdot \dfrac{1}{1 + t^2}$ **39.** $G'(x) = \dfrac{1}{2x(\ln x)^{1/2}}$

41. $f'(\theta) = \cos(\ln \theta) \cdot \dfrac{1}{\theta}$ **43.** $g'(x) = \dfrac{3(\log_3 x)^{1/2}}{2x \ln 3}$ **45.** $\dfrac{dy}{dx} = -\dfrac{y^2 + xy \ln y}{x^2 + xy \ln x}$ **47.** $\dfrac{dy}{dx} = \dfrac{x^2 + y^2 - 2x}{2y - x^2 - y^2}$ **49.** $\dfrac{dy}{dx} = \dfrac{y}{x(1 - y)}$

51. $y' = (x^2 + 1)^2 (2x^3 - 1)^4 \left(\dfrac{4x}{x^2 + 1} + \dfrac{24x^2}{2x^3 - 1} \right)$ **53.** $y' = \dfrac{x^2(x^3 + 1)}{\sqrt{x^2 + 1}} \left(\dfrac{2}{x} + \dfrac{3x^2}{x^3 + 1} - \dfrac{x}{x^2 + 1} \right)$

55. $y' = \dfrac{x \cos x}{(x^2 + 1)^3 \sin x} \left(\dfrac{1}{x} - \tan x - \dfrac{6x}{x^2 + 1} - \cot x \right)$ **57.** $y' = (3x)^x [\ln(3x) + 1]$ **59.** $y' = 2x^{\ln x} \dfrac{\ln x}{x}$ **61.** $y' = x^{x^2}(2x \ln x + x)$

63. $y' = x e^x \left(e^x \ln x + \dfrac{e^x}{x} \right)$ **65.** $y' = x^{\sin x} \left(\cos x \ln x + \dfrac{\sin x}{x} \right)$ **67.** $y' = (\sin x)^x (\ln(\sin x) + x \cot x)$

69. $y' = (\sin x)^{\cos x} (-\sin x \ln(\sin x) + \cos x \cot x)$ **71.** $y' = -\dfrac{y}{x \ln x}$ **73.** $y = 5x - 1$ **75.** $y = -5x + 18$

77. e^2 **79.** $e^{1/3}$ **81.** $\dfrac{d^{10} y}{dx^{10}} = \dfrac{9!}{x} = \dfrac{362,880}{x}$ **83.** $\dfrac{dy}{dx} = 2$ **85.** $y' = x^x(\ln x + 1)$ **87.** $y' = \tan^{-1}\dfrac{x}{a}$

89. (a) SSM.   **(b)** \$6107.01   **(c)** 28.881 years   **(d)** SSM.   **91. (a)** $-\dfrac{10}{\ln 10}$ dB/m   **(b)** AWV.   **93.** SSM.   **95.** SSM.   **97.** SSM.

99. $-\dfrac{8(6x^2 + 6x + 1)}{3(4x + 3)^{1/3}(2x^2 - 1)^{7/3}}$

AP® Practice Problems   (SSM = See Student Solutions Manual.)

1. B   **3.** B   **5.** D   **7.** C   **9.** A   **11.** C   **13.** SSM.

Retain Your Knowledge

1. B   **3.** D

Review Exercises

1. $an(ax + b)^{n-1}$ **3.** $(1 - x)^{1/2} - \dfrac{x}{2(1 - x)^{1/2}}$ **5.** $3x(x^2 + 4)^{1/2}$ **7.** $-\dfrac{a}{x\sqrt{2ax - x^2}}$ **9.** $(e^x - x)^{5x} \left(5\ln(e^x - x) + \dfrac{5x(e^x - 1)}{e^x - x} \right)$

11. $-\dfrac{2x}{(x - 1)^3}$ **13.** $\sec(2x) + 2x \sec(2x) \tan(2x)$ **15.** $\dfrac{a \cos \dfrac{x}{a}}{2 \left(a^2 \sin \dfrac{x}{a} \right)^{1/2}}$ **17.** $\cos v - \sin^2 v \cos v = \cos^3 v$ **19.** $1.05^x \ln 1.05$

21. $-\dfrac{1}{\sqrt{u^2 + 25}}$ **23.** $2\cot(2x)$ **25.** $\dfrac{2x - 2}{x^2 - 2x}$ **27.** $e^{-x}\left(-\ln x + \dfrac{1}{x} \right)$ **29.** $\dfrac{12}{x(144 - x^2)}$ **31.** $\dfrac{e^x(x^2 + 4)}{(x - 2)} \left[1 + \dfrac{2x}{x^2 + 4} - \dfrac{1}{x - 2} \right]$

33. $\dfrac{\sqrt{x}}{1 + x}$ **35.** $\dfrac{1}{\sqrt{x - x^2}}$ **37.** $\tan^{-1}(x)$ **39.** $\dfrac{1}{4y^3 + 1}$ **41.** $\dfrac{y(x \cos y - 1)}{x(1 + xy \sin y)}$ **43.** $\dfrac{1 + y \cos(xy)}{1 - x \cos(xy)}$

45. $y' = \dfrac{10 - y}{6y + x}$; $y'' = \dfrac{2(y - 10)(3y + x + 30)}{(6y + x)^3}$ **47.** $y' = \dfrac{8x - e^y}{xe^y}$; $y'' = \dfrac{-64x^2 + 8xe^y + e^{2y}}{x^2 e^{2y}}$ **49.** $\dfrac{1}{2}$ **51.** $e^{2/5}$

53. $x \neq (2k - 1)\dfrac{\pi}{2}$, k an integer   **55. (a)** Domain: $\left\{ x \mid x \geq -\dfrac{1}{6} \right\}$; Range: $\{ y \mid y \geq 0 \}$ **(b)** $\dfrac{3}{5}$ **(c)** $\left(0, \dfrac{13}{5} \right)$ **(d)** $\left(\dfrac{4}{3}, 3 \right)$

57. $-\dfrac{1}{2}(1 + \ln 8)$ **59.** -2

AP® Review Problems: Chapter 3   (SSM = See Student Solutions Manual.)

1. B   **3.** C   **5.** A   **7.** A   **9.** D   **11.** D   **13.** SSM.

AP® Cumulative Review: Chapters 1–3   (SSM = See Student Solutions Manual.)

1. D   **3.** B   **5.** C   **7.** B   **9.** C   **11.** B   **13.** SSM.

Chapter 4

Section 4.1 (AWV = Answers will vary.)

1. False **2.** d **3.** True **4.** average rate of change; derivative **5. (a)** $f'(x) = 3x^2 - 8x$ **(b)** $f'(3) = 3$ **(c)** $y = 3x - 20$

7. (a) $f'(x) = \dfrac{e^x(x^2 - 2x + 2)}{(x^2 + 2)^2}$ **(b)** $f'(0) = \dfrac{1}{2}$ **(c)** $y = \dfrac{1}{2}x + \dfrac{1}{2}$ **9.** $v(t) = -32t + 120$ ft/s; $a(t) = -32$ ft/s^2

11. $v(t) = \cos^2 t - 2t\cos t \sin t$ km/h; $a(t) = -4\cos t \sin t - 2t(\cos^2 t - \sin^2 t)$ km/h^2

13. (a) $P(0) = 30$. Originally there were 30 fruit flies in the colony. **(b)** $P(5) = 30e \approx 81.548$. After five days there were approximately 82 fruit flies in the colony. **(c)** $P'(t) = 6e^{0.2t}$ **(d)** $P'(10) = 6e^2 \approx 44.334$. AWV. **15.** $S'(t) = 4$ cm^2/s

17. The rate of change of R with respect to x is $R'(x) = 2x + 100$. The units are dollars per item sold. AWV.

AP® Practice Problems   ### Retain Your Knowledge

1. D **3.** A   **1.** B **3.** B

Section 4.2 (AWV = Answers will vary.)

1. 10 m^3/min **2.** 0.5 m/min **3.** $-\dfrac{8}{3}$ **4.** -1 **5.** $5\pi^2$ **7.** 40 **9.** 900 cm^3/s **11.** -0.000625 cm/h

13. $-\dfrac{8}{\sqrt{2009}} \approx -0.178$ cm/min **15.** $\dfrac{2\sqrt{3}\pi}{45} \approx 0.242$ cm^2/min **17.** shrinking at a rate of 0.75 m^2/min

19. decreasing at a rate of $\dfrac{1}{6}$ rad/s **21.** 0.2 m/min **23.** $\dfrac{4}{\pi} \approx 1.273$ m/min **25. (a)** $\dfrac{16}{3\pi} \approx 1.698$ m/min **(b)** $\dfrac{96 - 9\pi}{32} \approx 2.116$ m^3/min

27. $\dfrac{18}{\sqrt{17}} \approx 4.366$ units/s **29.** $50\sqrt{2}$ ft/s **31.** 1.75 kg/cm^2/min **33.** $100.8\pi \approx 316.673$ ft^2/min

35. (a) $\dfrac{1.5}{\sqrt{55}} \approx 0.202$ m/s **(b)** $\dfrac{0.5}{\sqrt{3}} \approx 0.289$ m/s **(c)** $\dfrac{1.5}{\sqrt{7}} \approx 0.567$ m/s **37.** $\dfrac{24\pi}{5} \approx 15.08$ km/s **39.** 4 m/min

41. The cabin is rising at $\dfrac{25\sqrt{3}\pi}{2} \approx 68.017$ ft/min. It is moving horizontally at $-\dfrac{25\pi}{2} \approx -39.270$ ft/min. **43.** $-\dfrac{9\sqrt{29}}{29} \approx -1.671$ m/s

45. (a) \$150/day **(b)** \$200/day **(c)** \$50/day **47.** ≈ -4.864 lb/s **49.** $100\pi \approx 314$ ft/s **51.** -316.8 rad/h **53. (a)** AWV. **(b)** AWV.

55. ≈ 32.071 ft/s; AWV. **57.** ≈ 0.165 in^2/min

AP® Practice Problems (SSM = See Student Solutions Manual.)   ### Retain Your Knowledge

1. C **3.** A **5.** B **7.** D **9.** SSM.   **1.** C **3.** B

Section 4.3 (SSM = See Student Solutions Manual; AWV = Answers will vary.)

1. (d) **2.** $f(x_0) + f'(x_0)(x - x_0)$ **3.** True **4.** $\dfrac{|\Delta Q|}{Q(x_0)}$ **5.** True **6.** True **7.** $dy = (3x^2 - 2)\,dx$ **9.** $dy = 12x(x^2 + 1)^{1/2}\,dx$

11. $dy = (6\cos(2x) + 1)\,dx$ **13.** $dy = -e^{-x}\,dx$ **15.** $dy = (e^x + xe^x)\,dx$ **17.** $dy = \dfrac{2\,dx}{\sqrt{1 - 4x^2}}$

19. (a) $dy = e^x\,dx$ **(b) (i)** $dy = \dfrac{1}{2}e \approx 1.359$, $\Delta y = e^{1.5} - e \approx 1.763$ **(ii)** $dy = \dfrac{1}{10}e \approx 0.272$, $\Delta y = e^{1.1} - e \approx 0.286$

  (iii) $dy = \dfrac{1}{100}e \approx 0.027$, $\Delta y = e^{1.01} - e \approx 0.027$; **(c) (i)** 0.404 **(ii)** 0.014 **(iii)** 0.000136

21. (a) $dy = \dfrac{2}{3\sqrt[3]{x}}\,dx$ **(b) (i)** $dy \approx 0.265$, $\Delta y \approx 0.255$ **(ii)** $dy \approx 0.0529$, $\Delta y \approx 0.0525$ **(iii)** $dy \approx 0.0053$, $\Delta y \approx 0.0053$;

  (c) (i) 0.00995, **(ii)** 0.000431, **(iii)** 4×10^{-6} **23. (a)** $dy = -\sin x\,dx$ **(b) (i)** $dy = 0$, $\Delta y \approx 0.122$

  (ii) $dy = 0$, $\Delta y \approx 0.005$ **(iii)** $dy = 0$, $\Delta y \approx 0.00005$; **(c) (i)** 0.122, **(ii)** 0.005 **(iii)** 0.00005

25. (a) $L(x) = 405x - 567$ **27. (a)** $L(x) = \dfrac{x}{4} + 1$ **29. (a)** $L(x) = x - 1$ **31. (a)** $L(x) = -\dfrac{\sqrt{3}x}{2} + \dfrac{1}{2} + \dfrac{\sqrt{3}\pi}{6}$

(b)

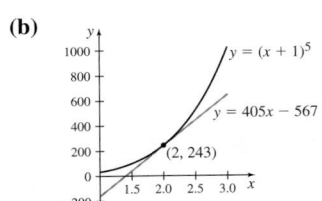

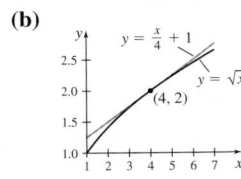

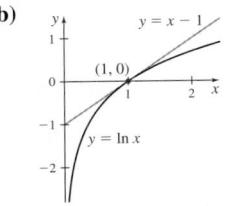

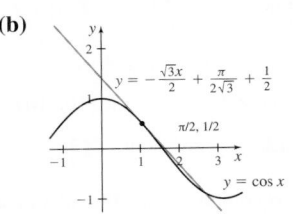

(c) Underestimates, AWV.   **(c)** Overestimates. AWV.   **(c)** Overestimates. AWV.   **(c)** Overestimates. AWV.

33. (a) 0.006 **(b)** 0.00125 **35.** 7.82 **37.** −0.15 **39. (a)** SSM. **(b)** $c_3 = 1.155$ **41. (a)** SSM. **(b)** $c_3 = 0.214$ **43. (a)** SSM. **(b)** $c_3 = 3.1$

45. (a) SSM. **(b)** AWV. **47. (a)** SSM. **(b)** $c_5 = -0.567$ **49. (a)** SSM. **(b)** $c_5 = -0.791$ **51. (a)** SSM. **(b)** $c_5 = 4.796$ **53.** 2π cm^2

55. 3.6π m^3 **57. (a)** 1.6π m^3 **(b)** Underestimate **(c)** AWV. **59.** 0.9% **61.** 72 min **63.** −0.0019 kg **65. (a)** The relative error in the

radius is no more than $\dfrac{1}{300}$. **(b)** $\dfrac{1}{3}$% **(c)** AWV. **67. (a)** The relative error in the pressure is no more than 0.01. **(b)** 1% **(c)** AWV.

69. 6% **71.** 1.310 **73.** −0.703 **75.** 2.718 **77.** AWV. **79.** SSM. **81.** SSM. **83.** SSM. **85.** 9.871 in.

AP® Practice Problems   Retain Your Knowledge

1. B **3.** B **5.** A **7.** B   **1.** A **3.** C

Section 4.4   (SSM = See Student Solutions Manual; AWV = Answers will vary.)

1. False **2.** False **3.** False **4.** False **5.** AWV. **6.** AWV. **7. (a)** Yes **(b)** $\dfrac{0}{0}$ **9. (a)** No **(b)** AWV. **11. (a)** Yes **(b)** $\dfrac{\infty}{\infty}$

13. (a) No **(b)** AWV. **15. (a)** Yes **(b)** $\dfrac{0}{0}$ **17. (a)** Yes **(b)** $\dfrac{0}{0}$ **19. (a)** Yes **(b)** $0 \cdot \infty$ **21. (a)** Yes **(b)** $\infty - \infty$

23. (a) Yes **(b)** ∞^0 **25. (a)** No **(b)** AWV. **27.** $\dfrac{0}{0}, 5$ **29.** $\dfrac{0}{0}, \dfrac{1}{2}$ **31.** $\dfrac{0}{0}, 2$ **33.** $\dfrac{0}{0}, -\pi$ **35.** $\dfrac{\infty}{\infty}, 0$ **37.** $\dfrac{\infty}{\infty}, 0$ **39.** $\dfrac{0}{0}, 1$ **41.** $\dfrac{0}{0}, -\dfrac{1}{6}$

43. $0 \cdot \infty, 0$ **45.** $0 \cdot \infty, 1$ **47.** $\infty - \infty, 0$ **49.** $\infty - \infty, -1$ **51.** $0^0, 1$ **53.** $\infty^0, 1$ **55.** $\infty^0, 1$ **57.** $1^\infty, 1$ **59.** 2 **61.** 0 **63.** 0 **65.** 2

67. 1 **69.** $\dfrac{1}{4}$ **71.** 0 **73.** 0 **75.** 1 **77.** $\dfrac{4a^2}{\pi}$ **79.** $\dfrac{1}{2}$ **81.** −1 **83.** 1 **85.** 1 **87.** 1

89. (a) 252 wolves **(b)** AWV. **(c)**

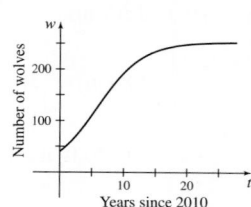

91. (a) $\dfrac{E}{R}, \dfrac{Et}{L}$ **(b)** AWV. **93.** SSM. **95.** SSM. **97.** 0 **99.** SSM.

101. SSM. **103. (a)** SSM. $f'(0) = 0$ **(b)**

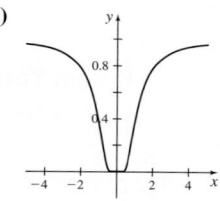

105. SSM.

107. (a) 0 **(b)** a **(c)** ∞ **(d)** Does not exist **(e)** 0 **(f)** a **(g)** 1 **(h)** a **(i)** Does not exist **(j)** e^{-1}

109. (a) $f''(x)$ **(b)** $f'''(x)$ **(c)** $\displaystyle\lim_{h\to 0} \dfrac{\displaystyle\sum_{k=0}^{n} (-1)^k \dfrac{n!}{k!\,(n-k)!} f(x + (n-k)h)}{h^n} = f^{(n)}(x)$ **111.** SSM.

AP® Practice Problems   (SSM = See Student Solutions Manual.)   Retain Your Knowledge

1. B **3.** C **5.** A **7.** B **9.** SSM.   **1.** A **3.** D

Review Exercises

1. (a) $f'(1) = 3$ **(b)** $y = 3x - 1$ **3. (a)** $s(0) = 0$ m; $s(3) = 7$ m **(b)** $s'(3) = \dfrac{16}{3}$; The velocity of the object at time $t = 3$ s is $\dfrac{16}{3}$ m/s.

(c) The average velocity of the object during the first three seconds is $\dfrac{7}{3}$ m/s. **(d)** The acceleration after 3 s is 2 m/s^2.

5. $dy = \left(1 - \dfrac{2x}{x^2 + 5}\right) dx$ **7. (a)** $L(x) = \dfrac{\pi}{6} + \dfrac{2\sqrt{3}}{3}\left(x - \dfrac{1}{2}\right)$ **(b)** $L(0.45) \approx 0.466$

9. (a) $dy = dx$; $\Delta y = \tan(\Delta x)$

(b) (i) $dy = 0.5$, $\Delta y \approx 0.546$; **(ii)** $dy = 0.1$, $\Delta y \approx 0.100$; **(iii)** $dy = 0.01$, $\Delta y \approx 0.010$

11. 4% **13.** $-\dfrac{4}{5}$ cm^2/min **15.** 318.953 mi/h **17.** Yes, $\dfrac{0}{0}$ **19.** Yes, $\infty - \infty$ **21.** 9 **23.** 2 **25.** 1 **27.** $\dfrac{1}{6}$ **29.** 1 **31.** $\dfrac{8}{9}$

33. (a) f is continuous at all real numbers, and $f\left(\dfrac{\pi}{2}\right) = 1.429$ and $f(\pi) = -1.142$ have opposite signs. So f has a zero on the interval $\left(\dfrac{\pi}{2}, \pi\right)$.

(b) 2.568

AP® Review Problems: Chapter 4   (SSM = See Student Solutions Manual.)
1. A   **3.** D   **5.** B   **7.** SSM.

AP® Cumulative Review Problems: Chapters 1–4   (SSM = See Student Solutions Manual.)
1. C   **3.** A   **5.** C   **7.** A   **9.** C   **11.** SSM.

Chapter 5
Section 5.1   (SSM = See Student Solutions Manual; AWV = Answers will vary.)

1. False   **2.** AWV.   **3.** True   **4.** False   **5.** $\dfrac{3}{2}$ **7.** 1   **9.** $-\dfrac{\sqrt{3}}{3}$ **11.** $\pm\dfrac{\sqrt{3}}{3}$ **13.** $\pm1, 0$ **15.** $\dfrac{\pi}{4}, \dfrac{3\pi}{4}$ **17.** $f(-2) \neq f(1)$

19. f is not differentiable at $x = 0$, so f is not differentiable on $(-1, 1)$.   **21. (a)** SSM.   **(b)** 1   **(c)** AWV.   **23. (a)** SSM.   **(b)** $e - 1$

(c) AWV.   **25. (a)** SSM.   **(b)** $\dfrac{7}{3}$ **(c)** AWV.   **27. (a)** SSM.   **(b)** $\sqrt{3}$ **(c)** AWV.   **29. (a)** SSM.   **(b)** $\dfrac{2744}{729}$ **(c)** AWV.

31. f is increasing on $(-\infty, \infty)$. **33.** f is increasing on $(-\infty, -1]$ and $[1, \infty)$ and decreasing on $[-1, 1]$.

35. f is increasing on $[-\sqrt{2}, 0]$ and $[\sqrt{2}, \infty)$ and decreasing on $(-\infty, -\sqrt{2}]$ and $[0, \sqrt{2}]$.

37. f is increasing on $[-1, 0]$ and $[1, \infty)$ and decreasing on $(-\infty, -1]$ and $[0, 1]$.

39. f is increasing on $[-\sqrt[3]{3}, \infty)$ and decreasing on $(-\infty, -\sqrt[3]{3}]$.   **41.** f is increasing on $\left[0, \dfrac{\pi}{2}\right]$ and $\left[\dfrac{3\pi}{2}, 2\pi\right]$ and decreasing on $\left[\dfrac{\pi}{2}, \dfrac{3\pi}{2}\right]$.

43. f is increasing on $[-1, \infty)$ and decreasing on $(-\infty, -1]$.   **45.** f is increasing on $\left[0, \dfrac{3\pi}{4}\right]$ and $\left[\dfrac{7\pi}{4}, 2\pi\right]$ and decreasing on $\left[\dfrac{3\pi}{4}, \dfrac{7\pi}{4}\right]$.

47. (a) $\{x \mid x \neq -2, x \neq 2\}$ **(b)** 0 and 4   **(c)** 2   **(d)** -2 **(e)** $(-\infty, 0], (2, 4]$ **(f)** $[0, 2), [4, \infty)$

49. (a) $\{x \mid x \neq -1, x \neq 0\}$ **(b)** $-2, 1, 2$ **(c)** -1 **(d)** 0   **(e)** $(-\infty, -1], [2, \infty)$ **(f)** $[1, 2]$

51. SSM.   **53.** SSM.   **55.** AWV.   **57.** SSM.   **59.** SSM.

61. (a) SSM.   **63.** SSM.   **65.** SSM.   **67.** AWV.   **69.** $\sqrt{1 - \dfrac{4}{\pi^2}}$

(b) $d(0) = d(5) = 0$

(c) ≈ 2.892 ft; $d(2.892) \approx -0.423$ ft.

(d)

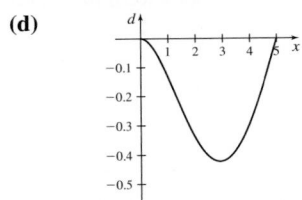

71. (a) The function f is increasing on the interval $(0, \infty)$.   **(b)** Yes.   **73. (b)**   **75.** SSM.   **77.** SSM.   **79.** SSM.   **81.** SSM.

83. For $b^2 - 3ac \leq 0$, f is increasing on $(-\infty, \infty)$ for $a > 0$ and decreasing on $(-\infty, \infty)$ for $a < 0$.

For $b^2 - 3ac > 0$ and $a > 0$, f is increasing on $(-\infty, x_1]$ and $[x_2, \infty)$ and decreasing on $[x_1, x_2]$.

For $b^2 - 3ac > 0$ and $a < 0$, f is increasing on $[x_1, x_2]$ and decreasing on $(-\infty, x_1]$ and $[x_2, \infty)$,

where $x_1 = \min\left\{\dfrac{-b \pm \sqrt{b^2 - 3ac}}{3a}\right\}$ and $x_2 = \max\left\{\dfrac{-b \pm \sqrt{b^2 - 3ac}}{3a}\right\}$.

85. SSM.   **87.** SSM.

89. If n is odd and $ad - bc > 0$, then f is increasing on the domain. If n is odd and $ad - bc < 0$, then f is decreasing on the domain. If n is even and $ad - bc > 0$, then f is increasing on the domain where $x > 0$ and decreasing on the domain where $x < 0$. If n is even and $ad - bc < 0$, then f is increasing on the domain where $x < 0$ and decreasing on the domain where $x > 0$.

91. SSM.   **93.** SSM.

AP® Practice Problems   (SSM = See Student Solutions Manual.)
1. A   **3.** C   **5.** A   **7.** D   **9.** B   **11.** B   **13.** SSM.

Retain Your Knowledge
1. D   **3.** D

Section 5.2 (SSM = See Student Solutions Manual; AWV = Answers will vary.)

1. False   **2.** (b)   **3.** False   **4.** False   **5.** False   **6.** False   **7.** x_1: neither, x_2: local maximum,

x_3: local minimum and absolute minimum, x_4: neither, x_5: local maximum, x_6: neither, x_7: local minimum, x_8: absolute maximum

9. AWV.   **11.** AWV.   **13.** 4   **15.** 0, 2   **17.** $-1, 0, 1$ **19.** 0   **21.** 0   **23.** π **25.** $-\dfrac{\sqrt{2}}{2}, -1, 1, \dfrac{\sqrt{2}}{2}$ **27.** 0, 2

29. $-3, 0, 1$ **31.** 3, 4   **33.** $-3\sqrt{3}, -3, 3, 3\sqrt{3}$ **35.** 1   **37.** Absolute maximum 20 at $x = 10$, absolute minimum -16 at $x = 4$

39. Absolute maximum 16 at $x = 4$, absolute minimum -4 at $x = 2$ **41.** Absolute maximum 9 at $x = 2$, absolute minimum 0 at $x = 1$

43. Absolute maximum 1 at $x = -1, 1$, absolute minimum 0 at $x = 0$ **45.** Absolute maximum 4 at $x = 4$, absolute minimum 2 at $x = 1$

47. Absolute maximum π at $x = \pi$, absolute minimum 0 at $x = 0$ **49.** Absolute maximum $\dfrac{1}{2}$ at $x = \dfrac{\sqrt{2}}{2}$, absolute minimum $-\dfrac{1}{2}$ at $x = -\dfrac{\sqrt{2}}{2}$

51. Absolute maximum 0 at $x = 0$, absolute minimum $-\dfrac{1}{2}$ at $x = -1$ and $x = \dfrac{1}{2}$

53. Absolute maximum $128\sqrt[3]{2}$ at $x = 5$, absolute minimum 0 at $x = -3$ and $x = 1$

55. Absolute maximum $\dfrac{1}{3}$ at $x = 4$, absolute minimum -1 at $x = 2$ **57.** Absolute maximum $\dfrac{\sqrt[3]{18}}{3\sqrt{3}}$ at $x = 3\sqrt{3}$, absolute minimum 0 at $x = 3$

59. Absolute maximum 1 at $x = 0$, absolute minimum $e - 3$ at $x = 1$ **61.** Absolute maximum 9 at $x = 3$, absolute minimum 1 at $x = 0$

63. Absolute maximum 8 at $x = 2$, absolute minimum 0 at $x = 0$

65. (a) $f'(x) = 12x^3 - 6x^2 - 42x + 36$ (b) $-2, 1, \dfrac{3}{2}$ (c)
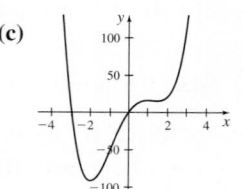
Absolute minimum at $x = -2$,
local maximum at $x = 1$,
local minimum at $x = \dfrac{3}{2}$

67. (a) $f'(x) = \dfrac{3x^4 + 5x^3 + 8x^2 - 10x - 96}{2\sqrt{x+5}(x^2+2)^{3/2}}$ (b) $-5, \approx -2.364, \approx 1.977$ (c)
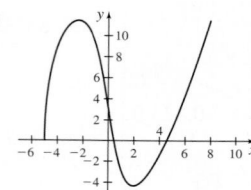
Local maximum at $x \approx -2.364$, absolute minimum at $x \approx 1.977$, neither a local minimum nor a local maximum at $x = -5$

69. (a) $f'(x) = 4x^3 - 37.2x^2 + 98.48x - 68.64$ **71.** (a) 50 mi/h   **73.** (a) $67.5°, 0.0183\, v_0^2$ ft

(b) Absolute maximum 0 at $x = 0$, absolute minimum -37.2 at $x = 5$

(c)
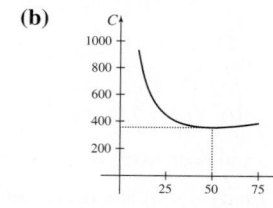

(b)
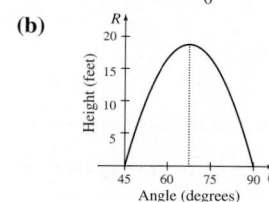

(c) AWV.

75. (a) SSM.   (b) ≈ 961.1 m   **77.** A tax rate of $\dfrac{3\sqrt{6}}{2} \approx 3.674$, maximizes revenue. The maximum revenue generated is $\dfrac{9\sqrt{3}}{2} \approx 7.794$.

79. (a) 0.04, 0.07   (b) Absolute maximum -3.719 at $x = 1$, absolute minimum -4.391 at $x = -1$ **81.** (a) $[-4, -3] \cup [3, 4]$ (b) $\dfrac{7}{2}$

83. (c) (e)   **85.** SSM.   **87.** SSM.   **89.** Absolute maximum $\sqrt{10} + 1$ at $x = 3$, absolute minimum $\sqrt{5}$ at $x = 2$

AP® Practice Problems   ## Retain Your Knowledge

1. C   **3.** C   **5.** D   **7.** C   **9.** A   **1.** C   **3.** C

Section 5.3 (SSM = See Student Solutions Manual; AWV = Answers will vary.)

1. False **2.** False **3.** (a) **4.** (b) **5.** (a) **6.** (c) **7.** True **8.** False

9. (a) Local maximum at $(-1, 0)$, local minima at $(-2.5, -4)$ and $(0.5, -4)$, inflection points $(-1.8, -2)$ and $(-0.2, -2)$
  (b) Increasing on $[-2.5, -1]$ and $[0.5, \infty)$, decreasing on $(-\infty, -2.5]$ and $[-1, 0.5]$, concave up on $(-\infty, -1.8)$ and $(-0.2, \infty)$, concave down on $(-1.8, -0.2)$

11. (a) Local maxima at $(-2, 3)$ and $(12, 10)$, local minimum at $(0, 0)$, no inflection point (b) Increasing on $(-\infty, -2]$ and $[0, 12]$, decreasing on $[-2, 0]$ and $[12, \infty)$, never concave up, concave down on $(-\infty, 0)$ and $(0, \infty)$

13. (a) $0, 4$ (b) Local maximum 2 at 0, local minimum -30 at 4 **15.** (a) $0, 1$ (b) Local minimum -1 at 1

17. (a) $\dfrac{3}{2}$ (b) Local maximum $2e^{3/2}$ at $\dfrac{3}{2}$ **19.** (a) $-\dfrac{1}{8}, 0$ (b) Local minimum $-\dfrac{1}{4}$ at $-\dfrac{1}{8}$

21. (a) $-1, 0, 1$ (b) Local maximum 0 at 0, local minimum -3 at -1 and 1 **23.** (a) $\sqrt[3]{e}$ (b) Local maximum $\dfrac{1}{3e}$ at $\sqrt[3]{e}$

25. (a) $k\pi - \tan^{-1}\dfrac{1}{2}$, k an integer (b) Local maximum $\sqrt{5}$ at $(2k+1)\pi - \tan^{-1}\dfrac{1}{2}$, local minimum $-\sqrt{5}$ at $2k\pi - \tan^{-1}\dfrac{1}{2}$, k an integer

27. (a) The object moves right on $(1, \infty)$, left on $(0, 1)$.
  (b) $t = 1$ (c) The velocity is increasing on $(0, \infty)$.
  (f) The acceleration is neither increasing nor decreasing on $(0, \infty)$.
  (g) The speed is decreasing on $(0, 1)$ and is increasing on $(1, \infty)$.
  (d) (e)

29. (a) The object moves right on $(1, \infty)$, left on $(0, 1)$.
  (b) $t = 1$ (c) The velocity is increasing on $(0, \infty)$.
  (f) The acceleration is increasing on $(0, \infty)$.
  (g) The speed is decreasing on $(0, 1)$ and is increasing on $(1, \infty)$.
  (d) (e)

31. (a) The object moves right on $(0, \infty)$.
  (b) No direction reverse (c) The velocity is decreasing on $(0, \infty)$.
  (f) The acceleration is increasing on $(0, \infty)$.
  (g) The speed is decreasing on $(1, \infty)$.
  (d) (e)

33. (a) The object moves right on $\left(0, \dfrac{\pi}{6}\right)$ and $\left(\dfrac{\pi}{2}, \dfrac{2\pi}{3}\right)$, left on $\left(\dfrac{\pi}{6}, \dfrac{\pi}{2}\right)$. (b) $t = \dfrac{\pi}{6}$ and $t = \dfrac{\pi}{2}$

  (c) The velocity is increasing on $\left(\dfrac{\pi}{3}, \dfrac{2\pi}{3}\right)$, decreasing on $\left(0, \dfrac{\pi}{3}\right)$. (d) (e)

  (f) The acceleration is increasing $\left(\dfrac{\pi}{6}, \dfrac{\pi}{2}\right)$, and the acceleration is decreasing on the intervals $\left(0, \dfrac{\pi}{6}\right)$ and $\left(\dfrac{\pi}{2}, \dfrac{2\pi}{3}\right)$.

  (g) The speed is decreasing on the intervals $\left(0, \dfrac{\pi}{6}\right)$ and $\left(\dfrac{\pi}{3}, \dfrac{\pi}{2}\right)$, and is increasing on the intervals $\left(\dfrac{\pi}{6}, \dfrac{\pi}{3}\right)$ and $\left(\dfrac{\pi}{2}, \dfrac{2\pi}{3}\right)$.

35. (a) $0, 1$ (b) $[0, 1], [1, \infty)$ (c) $(-\infty, 0]$ (d) 0 (e) None

37. (a) $-3, 0, 1$ (b) $(-\infty, -3], [1, \infty)$ (c) $[-3, 0], [0, 1]$ (d) 1 (e) -3

39. (a) No local extrema (b) Concave up on $(1, \infty)$, concave down on $(-\infty, 1)$ (c) $(1, -1)$

41. (a) Local minimum -3 at $x = 1$ (b) Concave up on $(-\infty, 0)$ and $(0, \infty)$ (c) No inflection point

43. (a) Local maximum 256 at $x = 4$, local minimum 0 at $x = 0$
  (b) Concave up on $(-\infty, 0)$ and $(0, 3)$, concave down on $(3, \infty)$ (c) $(3, 162)$

45. (a) Local maximum 64 at $x = -2$, local minimum -64 at $x = 2$
  (b) Concave up on $(-\sqrt{2}, 0)$ and $(\sqrt{2}, \infty)$, concave down on $(-\infty, -\sqrt{2})$ and $(0, \sqrt{2})$ (c) $(0, 0)$, $(\sqrt{2}, -28\sqrt{2})$, $(-\sqrt{2}, 28\sqrt{2})$

47. (a) Local maximum $4e^{-2}$ at $x = -2$, local minimum 0 at $x = 0$ (b) Concave up on $(-\infty, -2 - \sqrt{2})$ and $(-2 + \sqrt{2}, \infty)$, concave down on $(-2 - \sqrt{2}, -2 + \sqrt{2})$ (c) $\left(-2 - \sqrt{2}, \left(6 + 4\sqrt{2}\right)e^{-2 - \sqrt{2}}\right)$, $\left(-2 + \sqrt{2}, \left(6 - 4\sqrt{2}\right)e^{-2 + \sqrt{2}}\right)$

49. (a) Local minimum $-\dfrac{9}{8}$ at $x = \dfrac{1}{8}$ (b) Concave up on $\left(-\infty, -\dfrac{1}{4}\right)$ and $(0, \infty)$, concave down on $\left(-\dfrac{1}{4}, 0\right)$ (c) $\left(-\dfrac{1}{4}, \dfrac{9\sqrt[3]{2}}{4}\right)$, $(0, 0)$

51. (a) Local maximum 0 at $x = 0$, local minimum $-6\sqrt[3]{2}$ at $x = -\sqrt{2}$ and $\sqrt{2}$ (b) Concave up on $(-\infty, 0)$ and $(0, \infty)$ (c) No inflection point

53. (a) Local minimum $\dfrac{1}{2} + \dfrac{1}{2}\ln 2$ at $x = \dfrac{\sqrt{2}}{2}$ (b) Concave up on $(0, \infty)$ (c) No inflection point

55. (a) Local maximum $\dfrac{16\sqrt{5}}{125}$ at $x=\dfrac{1}{2}$, local minimum $-\dfrac{16\sqrt{5}}{125}$ at $x=-\dfrac{1}{2}$ **(b)** Concave up on $\left(-\dfrac{\sqrt{3}}{2},0\right)$ and $\left(\dfrac{\sqrt{3}}{2},\infty\right)$,

concave down on $\left(-\infty,-\dfrac{\sqrt{3}}{2}\right)$ and $\left(0,\dfrac{\sqrt{3}}{2}\right)$ **(c)** $\left(-\dfrac{\sqrt{3}}{2},-\dfrac{16\sqrt{3}}{7^{5/2}}\right),\left(\dfrac{\sqrt{3}}{2},\dfrac{16\sqrt{3}}{7^{5/2}}\right),(0,0)$

57. (a) Local maximum $\dfrac{2\sqrt{3}}{9}$ at $x=\pm\dfrac{\sqrt{6}}{3}$, local minimum 0 at $x=0$ **(b)** Concave up on $\left(-\dfrac{1}{6}\sqrt{27-3\sqrt{33}},\dfrac{1}{6}\sqrt{27-3\sqrt{33}}\right)$,

concave down on $\left(-1,-\dfrac{1}{6}\sqrt{27-3\sqrt{33}}\right)$ and $\left(\dfrac{1}{6}\sqrt{27-3\sqrt{33}},1\right)$ **(c)** $\left(\pm\dfrac{1}{6}\sqrt{27-3\sqrt{33}},\dfrac{9-\sqrt{33}}{12}\sqrt{\dfrac{3+\sqrt{33}}{12}}\right)$

59. (a) Local maximum $\dfrac{5\pi}{3}+\sqrt{3}$ at $x=\dfrac{5\pi}{3}$, local minimum $\dfrac{\pi}{3}-\sqrt{3}$ at $x=\dfrac{\pi}{3}$ **(b)** Concave up on $(0,\pi)$, concave down on $(\pi,2\pi)$
(c) (π,π)

61. (a) Local maximum: 2 at $x=\pi$; local minima: -1 at $x=\dfrac{\pi}{2}$ and -1 at $x=\dfrac{3\pi}{2}$ **(b)** Concave down on $\left(0,\dfrac{\pi}{4}\right),\left(\dfrac{3\pi}{4},\dfrac{5\pi}{4}\right)$,

and $\left(\dfrac{7\pi}{4},2\pi\right)$. Concave up on $\left(\dfrac{\pi}{4},\dfrac{3\pi}{4}\right)$ and $\left(\dfrac{5\pi}{4},\dfrac{7\pi}{4}\right)$. **(c)** Inflection points: $\left(\dfrac{\pi}{4},\dfrac{1}{2}\right),\left(\dfrac{3\pi}{4},\dfrac{1}{2}\right),\left(\dfrac{5\pi}{4},\dfrac{1}{2}\right)$, and $\left(\dfrac{7\pi}{4},\dfrac{1}{2}\right)$

63. (a) $-\sqrt{3},0,\sqrt{3}$ **(b)** $[-\sqrt{3},0],[\sqrt{3},\infty)$ **(c)** $(-\infty,-\sqrt{3}],[0,\sqrt{3}]$ **(d)** $-\sqrt{3},\sqrt{3}$ **(e)** 0
(f) $(-\infty,-1),(1,\infty)$ **(g)** $(-1,1)$ **(h)** $(-1,f(-1)),(1,f(1))$

65. (a) $0,1$ **(b)** $(-\infty,0],[1,\infty)$ **(c)** $[0,1]$ **(d)** 1 **(e)** 0 **(f)** $\left(\sqrt[3]{\dfrac{1}{4}},\infty\right)$ **(g)** $\left(-\infty,\sqrt[3]{\dfrac{1}{4}}\right)$ **(h)** $\left(\sqrt[3]{\dfrac{1}{4}},f\left(\sqrt[3]{\dfrac{1}{4}}\right)\right)$

67. (a) and (b) Local maximum -20 at $x=3$, local minimum -21 at $x=2$ **(c)** AWV.

69. (a) and (b) Local minima -21 at $x=-2$ and $x=2$; local maximum of -5 at $x=0$ **(c)** AWV.

71. (a) Local minimum of 1 at $x=0$; local maximum of $\dfrac{337}{81}$ at $x=-\dfrac{4}{3}$

(b) No information about $x=0$; local maximum of $\dfrac{337}{81}$ at $x=-\dfrac{4}{3}$ **(c)** AWV.

73. (a) and (b) Local maximum $4e$ at $x=1$, local minimum 0 at $x=3$ **(c)** AWV.

75. AWV.
One possible answer below.

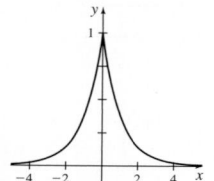

77. AWV.
One possible answer below.

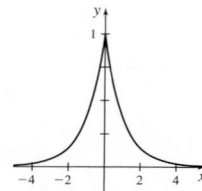

79. AWV.
One possible answer below.

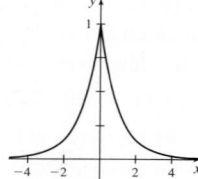

81. AWV.
One possible answer below.

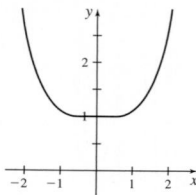

83. AWV.
One possible answer below.

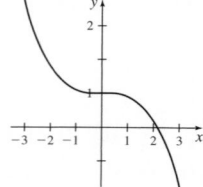

85. AWV.
One possible answer below.

87. (a) Concave up on $\left(-\infty,2-\dfrac{\sqrt{2}}{2}\right)$ and $\left(2+\dfrac{\sqrt{2}}{2},\infty\right)$, concave down on $\left(2-\dfrac{\sqrt{2}}{2},2+\dfrac{\sqrt{2}}{2}\right)$

(b) $\left(2-\dfrac{\sqrt{2}}{2},\dfrac{1}{\sqrt{e}}\right),\left(2+\dfrac{\sqrt{2}}{2},\dfrac{1}{\sqrt{e}}\right)$ **(c)**

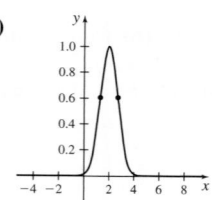

89. (a) Concave up on $(-\infty, 0.135)$ and $(0.721, 5.144)$, concave down on $(0.135, 0.721)$ and $(5.144, \infty)$
(b) $(0.135, 2.433), (0.721, 2.139), (5.144, -0.072)$ **(c)**

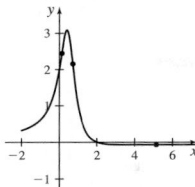

91. $a = -3, b = 9$ **93. (a)** $N'(t) = \dfrac{99,990,000e^{-t}}{(1 + 9999e^{-t})^2}$ **(b)** $N'(t)$ is increasing on the interval $(0, \ln 9999)$ and decreasing on the

interval $(\ln 9999, \infty)$. **(c)** $\ln 9999$ **(d)** $(\ln 9999, 5000)$ **(e)** AWV.

95. (a)

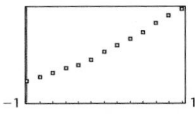

(b) AWV **(c)** $P'(t) = \dfrac{8.3054 \times 10^7 \cdot e^{-0.01708t}}{(1 + 7.47909 \cdot e^{-0.01708t})^2}$

(d) $P'(t)$ is increasing on the interval $[0, 117.804]$ and decreasing on the interval $[117.804, \infty)$.
(e) The rate of change in population is maximum when $t = 117.804$.

(f) $(117.804, 328\,020\,639.6)$ **(g)** AWV.
97. (a) There is no local minimum, the local maximum occurs at approximately $(5.46, -450)$. That is, local maximum occurs in year 2015 and was
approximately -450 billion dollars. **(b)** It represents a budget deficit.
(c) Concave up from $0 < t < 1.456$, concave down on from $1.456 < t < 11$; $(1.456, -1.079)$ is a point of inflection. **(d)** AWV.

(e)

Xscl = 1, Yscl = 1000

99. $a = -\dfrac{7}{8}, b = \dfrac{21}{4}, c = 0,$ and $d = 5$

101. Local maximum 2 at $x = \dfrac{\pi}{3}$, local minimum -2 at $x = \dfrac{4\pi}{3}$. Inflection points $\left(\dfrac{5\pi}{6}, 0\right), \left(\dfrac{11\pi}{6}, 0\right)$.

103. $f''(x)$ does not exist at the critical number $x = 0$. **105.** (e) **107.** SSM. **109.** SSM. **111.** SSM. **113.** SSM.
115. SSM. **117.** SSM. **119.** SSM.

AP® Practice Problems (SSM = See Student Solutions Manual.) **Retain Your Knowledge**
1. D **3.** B **5.** B **7.** B **9.** A **11.** C **13.** B **15.** D **17.** D **19.** SSM. **1.** C **3.** C

Section 5.4 (AWV = Answers will vary.)

1.

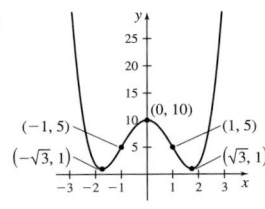

3.

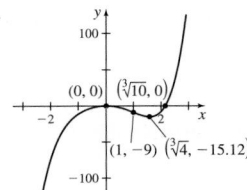

5.

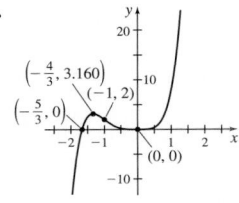

7.

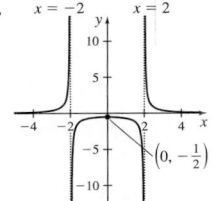

9.

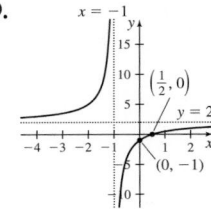

11.

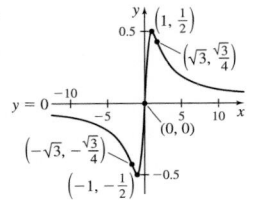

13.

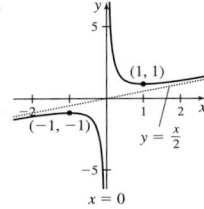

15.

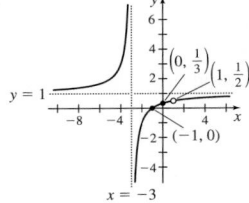

17.

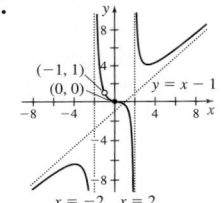

19.

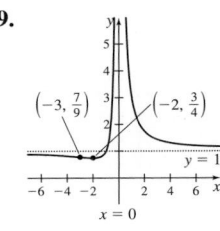

21.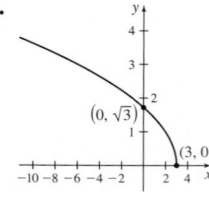

23.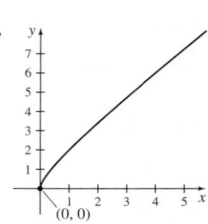

25.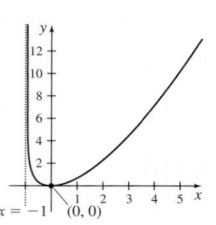

27.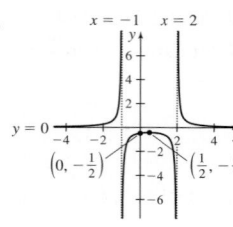

29.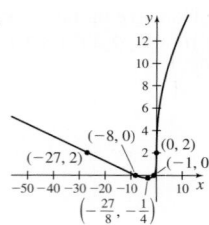

31.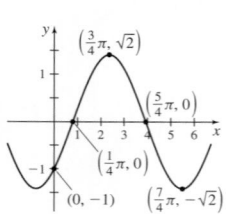

33.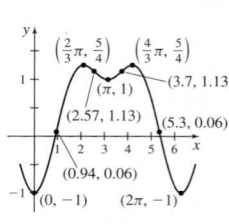

35.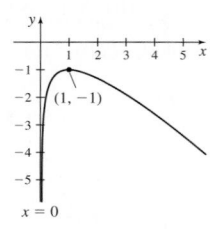

37.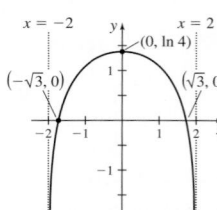

39.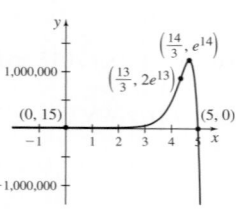

41.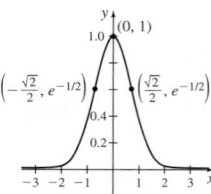

43. (a)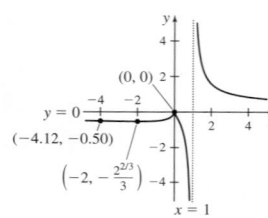

(b) Vertical asymptote: $x = 1$, horizontal asymptote: $y = 0$
(c) Decreasing on approximately $(-\infty, -2)$ and $(0, 1)$ and $(1, \infty)$ and increasing on approximately $(-2, 0)$, concave down on approximately $(-\infty, -4)$ and $(0.1, 1)$ and concave up on approximately $(-4, 0)$ and $(0, 0.1)$ and $(1, \infty)$
(d) Local minimum at approximately -2 and local maximum at approximately 0
(e) They are the same.
(f) Points of inflection are approximately $\left(-4, -\dfrac{2\sqrt[3]{2}}{5}\right)$ and $(0.1, -0.239)$.

45. (a)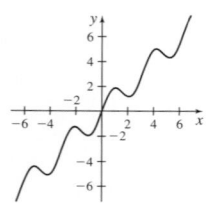

(b) No asymptotes
(c) Increasing on $\left(-\dfrac{\pi}{3} + k\pi, \dfrac{\pi}{3} + k\pi\right)$ and decreasing on $\left(\dfrac{\pi}{3} + k\pi, \dfrac{2\pi}{3} + k\pi\right)$, k an integer, concave up on $\left(-\dfrac{\pi}{2} + k\pi, k\pi\right)$ and concave down on $\left(k\pi, \dfrac{\pi}{2} + k\pi\right)$, k an integer
(d) Local maximum at approximately $\dfrac{\pi}{3} + k\pi$ and local minimum at approximately $\dfrac{2\pi}{3} + k\pi$, k an integer
(e) They are the same.   **(f)** Points of inflection at approximately $\left(\dfrac{k\pi}{2}, \dfrac{k\pi}{2}\right)$, k an integer

47. (a)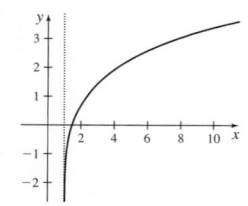

(b) Vertical asymptote: $x = 1$
(c) Increasing and concave down on approximately $(1, \infty)$
(d) No local extrema
(e) They are the same.
(f) No points of inflection

49. (a)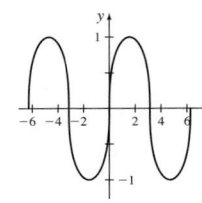

(b) No asymptotes

(c) Increasing approximately on $\left(-\dfrac{\pi}{2}+2k\pi,\ \dfrac{\pi}{2}+2k\pi\right)$ and decreasing on $\left(\dfrac{\pi}{2}+2k\pi,\ \dfrac{3\pi}{2}+2k\pi\right)$, concave up approximately on $(-\pi+2k\pi,\ 2k\pi)$ and concave down approximately on $(2k\pi,\ \pi+2k\pi)$, k an integer

(d) Local maximum at approximately $\dfrac{\pi}{2}+2k\pi$ and local minimum at approximately $\dfrac{3\pi}{2}+2k\pi$, k an integer

(e) They are the same. **(f)** Points of inflection at approximately $(k\pi,\ 0)$, k an integer

51. (a)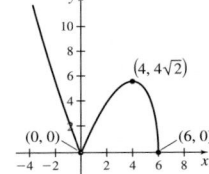

(b) There are no asymptotes.

(c) The function is increasing on $[0, 4]$, it is decreasing on $(-\infty, 0]$ and $[4, 6]$. It is concave up on $(-\infty, 0)$ and is concave down on $(0, 6)$.

(d) There is a local minimum at $(0, 0)$ and a local maximum at $(4, 4\sqrt{2})$.

(e) They are the same.

(f) $(0, 0)$ is an inflection point.

53. AWV. **55.** AWV. **57.** AWV.

59. (a) Absolute maximum at $x=\dfrac{1}{e}$ and absolute minimum at $x=1$

(b) Concave up on $\left(\dfrac{1}{e},\ 2\right)$

(c)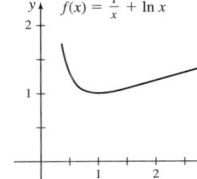

61.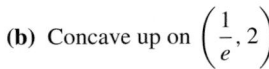

$x=\pm 2$ are two vertical asymptotes.

63.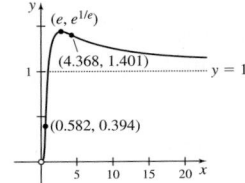

$y=1$ is a horizontal asymptote.

AP® Practice Problems

1. A **3.** SSM.

Retain Your Knowledge

1. A **3.** A

Section 5.5 (SSM = See Student Solutions Manual.)

1. $1{,}125{,}000$ m² **3.** $\dfrac{L}{4}$ m $\times \dfrac{L}{4}$ m **5.** $\dfrac{5000}{3}$ m² **7.** 8 cm × 8 cm × 2 cm **9.** $10\sqrt[3]{4}\times 10\sqrt[3]{4}\times 5\sqrt[3]{4}\approx 15.874$ cm × 15.874 cm × 7.937 cm

11. Radius $\sqrt[3]{\dfrac{15}{4\pi}}\approx 1.061$ m, height $\dfrac{8}{3}\sqrt[3]{\dfrac{15}{4\pi}}\approx 2.829$ m **13.** \$21 per day **15.** $(1, 1)$ **17.** $(1.784, 0.817)$

19. (a) 40 mi/h **(b)** ≈ 46.6 mi/h **(c)** ≈ 51.0 mi/h **21. (a)** and **(b)** $\sqrt{p^2+(q+r)^2}$ **23.** ≈ 9.582 m, $\approx 36.383°$

25. Width $2\sqrt{3}$, depth $\dfrac{4\sqrt{15}}{3}$ **27.** 594 cases, \$456 per case **29.** $\dfrac{\pi}{2}$ **31.** $\dfrac{2}{\pi+1}$ **33.** 2 m from the weak light source

35. Minimum length: 28.359 cm, maximum length: 42.758 cm

37. (a) Cut 19.603 cm for a square and 15.397 cm for a circle.

(b) Cut 0 cm for a square and 35 cm for a circle.

(c)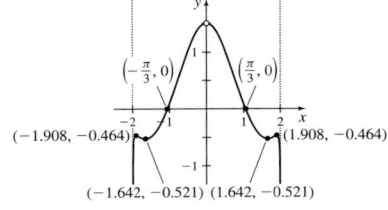

39. SSM.   **41.** $6\sqrt{3}$ **43.** $y = -x\sqrt[3]{\dfrac{b}{a}} + b + a\sqrt[3]{\dfrac{b}{a}}$ **45.** SSM.

47. Observer should stand approximately 4.583 m from the wall.   **49.** SSM.   **51.** 20 ft   **53.** $(-3.758, 6.547)$

AP® Practice Problems   Retain Your Knowledge
1. D   **3.** D   **5.** C   **1.** D   **3.** A

Section 5.6   (SSM = See Student Solutions Manual; AWV = Answers will vary.)

1. Antiderivative   **2.** True   **3.** $\ln|x| + C$ **4.** True   **5.** True   **6.** True   **7.** $2x + C$ **9.** $2x^2 + C$ **11.** $\dfrac{2}{3}x^6 + C$

13. $2x^{5/2} + C$ **15.** $-\dfrac{2}{x} + C$ **17.** $\dfrac{2}{3}x^{3/2} + C$ **19.** $x^4 - x^3 + x + C$ **21.** $3x^3 - 6x^2 + 4x + C$ **23.** $3x - 2\ln|x| + C$

25. $6x^{1/2} - 4\ln|x| + C$ **27.** $x^2 - 3\sin x + C$ **29.** $4e^x + \dfrac{1}{2}x^2 + C$ **31.** $7\tan^{-1}x + C$ **33.** $e^x + \sec^{-1}x + C$ **35.** $\ln|x| + e^x + C$

37. $s(t) = -16t^2 + 128t$ **39.** $s(t) = \dfrac{1}{2}t^3 + 18t + 2$ **41.** $s(t) = 6t - \sin t$ **43.** $\dfrac{1}{6}u^2 + \dfrac{7}{3}u + C$ **45.** $F(x) = x\cos x + \sin x + 1$

47. $v = 10\sqrt{3}$ m/s ≈ 38.74 mi/h   **49.** 220 ft   **51.** 13.133 m/s   **53.** 8 N   **55.** ≈ 16.759 m/s   **57.** 15.00625 m

59. (a) SSM.   **(b)** AWV.   **(c)** $I(x) = I_0 e^{-kx}$ **(d)** $-\dfrac{1}{2}\ln 0.1$ cm^{-1} **61.** $x^2 + 3\tan^{-1}x + C$

AP® Practice Problems   Retain Your Knowledge
1. C   **3.** C   **5.** B   **1.** D   **3.** B

Review Exercises   (AWV = Answers will vary.)

1. $(-8, -9)$ is the absolute minimum, $(-5, 0)$ is neither, $(-2, 9)$ is a local maximum and the absolute maximum, $(1, 0)$ is a local minimum, $(3, 4)$ is a local maximum, $(5, 0)$ is neither   **3.** $0, \dfrac{\pi}{2}, \pi$

5. Absolute maximum value of 21 at $x = -2$ and absolute minimum value of -11 at $x = 2$ **7.** $\left(2, \dfrac{3}{2}\right)$

9. (a) Local maximum $\dfrac{203}{27}$ at $x = -\dfrac{4}{3}$ and local minimum -11 at $x = 2$ **(b)** Same as (a)

11. (a) Local maximum $16e^{-4}$ at $x = 2$ and local minimum 0 at $x = 0$

(b) Local maximum $16e^{-4}$ at $x = 2$; $f''(0) = 0$ so Second Derivative Test gives no information

13. 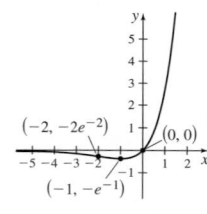   **15.** 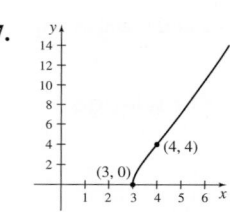   **17.**   **19. (a)** Increasing on $(-1.207, \infty)$ and decreasing on $(-\infty, -1.207)$
(b) Concave up on $(-\infty, \infty)$
(c) No point of inflection

21. (B)   **23.** AWV.   **25.** $\dfrac{105}{16}$ **27.** $10 - 2\sqrt{7}$ in.   **29.** $F(x) = C$ **31.** $F(x) = \sin x + C$ **33.** $F(x) = 2\ln|x| + C$

35. $F(x) = x^4 - 3x^3 + 5x^2 - 3x + C$ **37.** $\dfrac{31}{5}$ cm/s   **39.** 1075 items   **41.** $\dfrac{1}{e}$

AP® Review Problems: Chapter 5   (SSM = See Student Solutions Manual.)
1. B   **3.** A   **5.** C   **7.** B   **9.** B   **11.** B   **13.** SSM.   **15.** SSM.

AP® Cumulative Review Problems: Chapters 1–5   (SSM = See Student Solutions Manual.)
1. B   **3.** B   **5.** C   **7.** B   **9.** A   **11.** SSM.

Chapter 6, Part 1

Section 6.1 (SSM = See Student Solutions Manual; AWV = Answers will vary.)

1. AWV. **2.** False **3.** $\dfrac{1}{2}$ **4.** True **5. (a)** Area is $\dfrac{35}{4}$. **(b)** Area is $\dfrac{37}{4}$. **(c)** $\dfrac{35}{4} < 9 < \dfrac{37}{4}$

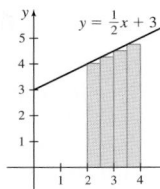 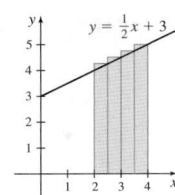

7. (a) 3 **(b)** 6 **9.** [1, 2], [2, 3], [3, 4]

11. $\left[-1, -\dfrac{1}{2}\right], \left[-\dfrac{1}{2}, 0\right], \left[0, \dfrac{1}{2}\right], \left[\dfrac{1}{2}, 1\right], \left[1, \dfrac{3}{2}\right], \left[\dfrac{3}{2}, 2\right], \left[2, \dfrac{5}{2}\right], \left[\dfrac{5}{2}, 3\right], \left[3, \dfrac{7}{2}\right], \left[\dfrac{7}{2}, 4\right]$ **13. (a)** 14 **(b)** 48

15. (a)

(b) $\left[0, \dfrac{3}{n}\right], \left[\dfrac{3}{n}, 2 \cdot \dfrac{3}{n}\right], \ldots, \left[(n-1) \cdot \dfrac{3}{n}, 3\right]$ **(c)** SSM.

(d) SSM. **(e)** SSM.

17. (a) $s_4 = 40, s_8 = 44$ **(b)** $S_4 = 56, S_8 = 52$ **19. (a)** $s_4 = 34, s_8 = \dfrac{77}{2}$ **(b)** $S_4 = 50, S_8 = \dfrac{93}{2}$

21. (a) $s_4 = \dfrac{\sqrt{2}}{4}\pi \approx 1.111, s_8 \approx 1.582$ **(b)** $S_4 = \dfrac{\sqrt{2}+2}{4}\pi \approx 2.682, S_8 \approx 2.367$

23. $s_n = \displaystyle\sum_{i=1}^{n} \left(3(i-1)\dfrac{10}{n}\right)\dfrac{10}{n} = 150 - \dfrac{150}{n}; \lim_{n \to \infty} s_n = 150$

25. (a) $A = \displaystyle\lim_{n \to \infty} s_n = \lim_{n \to \infty}\left(20 - \dfrac{16}{n}\right) = 20$ **(b)** $A = \displaystyle\lim_{n \to \infty} S_n = \lim_{n \to \infty}\left(20 + \dfrac{16}{n}\right) = 20$ **(c)** AWV.

27. (a) $A = \displaystyle\lim_{n \to \infty} s_n = \lim_{n \to \infty}\left(24 - \dfrac{24}{n}\right) = 24$ **(b)** $A = \displaystyle\lim_{n \to \infty} S_n = \lim_{n \to \infty}\left(24 + \dfrac{24}{n}\right) = 24$ **(c)** AWV.

29. (a) $A = \displaystyle\lim_{n \to \infty} s_n = \lim_{n \to \infty}\left(\dfrac{32}{3} - \dfrac{16}{n} + \dfrac{16}{3n^2}\right) = \dfrac{32}{3}$ **(b)** $A = \displaystyle\lim_{n \to \infty} S_n = \lim_{n \to \infty}\left(\dfrac{32}{3} + \dfrac{16}{n} + \dfrac{16}{3n^2}\right) = \dfrac{32}{3}$ **(c)** AWV.

31. (a) $A = \displaystyle\lim_{n \to \infty} s_n = \lim_{n \to \infty}\left(\dfrac{16}{3} - \dfrac{4}{n} - \dfrac{4}{3n^2}\right) = \dfrac{16}{3}$ **(b)** $A = \displaystyle\lim_{n \to \infty} S_n = \lim_{n \to \infty}\left(\dfrac{16}{3} + \dfrac{4}{n} - \dfrac{4}{3n^2}\right) = \dfrac{16}{3}$ **(c)** AWV.

33. 10 **35.** 18 **37.** $\dfrac{58}{3}$ **39.** $A \approx 25{,}994$ **41.** $A \approx 1.693$

43. (a)

(b) $\left[1, 1 + \dfrac{3}{n}\right], \left[1 + \dfrac{3}{n}, 1 + 2 \cdot \dfrac{3}{n}\right], \ldots, \left[1 + (n-1) \cdot \dfrac{3}{n}, 4\right]$.

(c) SSM. **(d)** SSM.

(e)

n	5	10	50	100
s_n	4.754	5.123	5.456	5.500
S_n	6.554	6.023	5.636	5.590

(f) $5.500 \leq A \leq 5.590$

45. SSM. **47.** SSM.

AP® Practice Problems
1. B

Retain Your Knowledge
1. B **3.** D

Section 6.2 (SSM = See Student Solutions Manual.)

1. False   **2.** (c) 2   **3.** True   **4.** Lower limit of integration; upper limit of integration; integral sign; integrand   **5.** 0

6. False   **7.** True   **8.** (b) 10   **9.** $\dfrac{13}{8}$ **11.** 24   **13.** 22   **15.** -10 **17.** 0   **19.** $\int_0^2 (e^x + 2)\, dx$ **21.** $\int_0^{2\pi} \cos x\, dx$

23. $\displaystyle\int_0^3 \dfrac{2}{1+x}\, dx$ **25.** $\int_1^e x \ln x\, dx$ **27.** $\int_2^6 \left(2 + \sqrt{4 - (x-4)^2}\right) dx$ **29.** $\int_{-2}^4 [\sin(1.5x) + 3]\, dx$

31. (a) $[-4, -1], [-1, 0], [0, 1], [1, 3], [3, 5], [5, 6]$ (b) 0   (c) 6.5   **33.** $\displaystyle\lim_{\Delta x \to 0} \sum_{i=1}^n \sin(i\,\Delta x)\, \Delta x$

35. $\displaystyle\lim_{\Delta x \to 0} \sum_{i=1}^n (i\,\Delta x - 1)^{1/3} \cdot \Delta x$ **37.** $\displaystyle\lim_{\Delta x \to 0} \sum_{i=1}^n (|1 + i\,\Delta x| - 2)\, \Delta x$ **39.** (a) 32   (b) 208   (c) ≈ 138.667

41. (a) 80   (b) 40   (c) ≈ 42.667 **43.** (a) 0   (b) π (c) ≈ 3.142 **45.** (a) $4 + 4e^4$ (b) $(e^2 + e^4 + e^6 + e^8) \cdot 2$ (c) ≈ 2979.958

47. (a)

n	10	50	100
Left	14.536	14.737	14.762
Right	15.030	14.836	14.812
Trap. sum	14.783	14.787	14.787

49. (a)

n	10	50	100
Left	4.702	4.712	4.712
Right	4.702	4.712	4.712
Trap. sum	4.702	4.712	4.712

(b) ≈ 14.787

(b) $\dfrac{3\pi}{2} \approx 4.712$

51. (a) 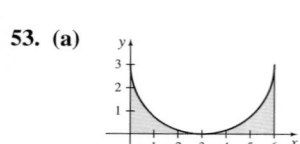   (b) $A = \displaystyle\int_0^3 \sqrt{9 - x^2}\, dx$ (c) $A \approx 7.069$ (d) $A = \dfrac{9\pi}{4} \approx 7.069$

53. (a)   (b) $A = \displaystyle\int_0^6 (3 - \sqrt{6x - x^2})\, dx$ (c) $A \approx 3.863$ (d) $A = 18 - \dfrac{9\pi}{2} \approx 3.863$

55. newton-meters   **57.** meters   **59.** SSM.   **61.** SSM.

AP® Practice Problems   (SSM = See Student Solutions Manual.)   Retain Your Knowledge
1. C   **3.** C   **5.** B   **7.** B   **9.** A   **11.** SSM.   **13.** SSM.   **1.** D   **3.** B

Section 6.3   (SSM = See Student Solutions Manual; AWV = Answers will vary.)

1. $f(x)$ **2.** False   **3.** False   **4.** False   **5.** $\sqrt{x^2 + 1}$ **7.** $(3 + t^2)^{3/2}$ **9.** $\ln x$ **11.** $6x^2 \sqrt{4x^6 + 1}$

13. $5x^4 \sec x^5$ **15.** $-\sin x^2$ **17.** $-10x(30x^2)^{2/3}$ **19.** 5   **21.** $\dfrac{15}{4}$ **23.** $\dfrac{2}{3}$ **25.** $\sqrt{3}$

27. $\sqrt{2} - 1$ **29.** $\dfrac{e-1}{e}$ **31.** 1   **33.** $\dfrac{\pi}{4}$ **35.** $\dfrac{99}{5}$ **37.** $-\dfrac{\sqrt{2}}{2}$ **39.** $\dfrac{15}{4}$ **41.** $e^2 - 1$

43. (a) $G(x) = \displaystyle\int_0^x \cos t\, dt$ (b) $G\left(\dfrac{\pi}{6}\right) = \dfrac{1}{2}; G\left(\dfrac{\pi}{2}\right) = 1; G\left(\dfrac{4\pi}{3}\right) = -\dfrac{\sqrt{3}}{2}$ (c)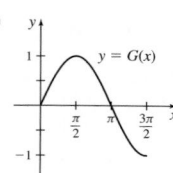

45. (a) $G(x) = \displaystyle\int_{-2}^x t^3\, dt$ (b) $G(-1) = -\dfrac{15}{4}; G(0) = -4; G(3) = \dfrac{65}{4}$ (c)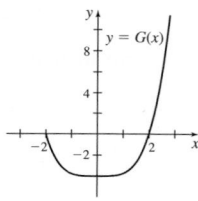

47. (a) $G(x) = \int_{-1}^{x} (4 - t^2)\, dt$ **(b)** $G(0) = \dfrac{11}{3}$; $G(2) = 9$; $G(3) = \dfrac{20}{3}$ **(c)**

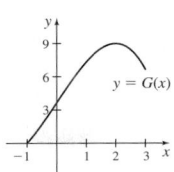

49. (a) $G(x) = \int_{0}^{x} (\sqrt{t} - 2)\, dt$ **(b)** $G(1) = -\dfrac{4}{3}$; $G(4) = -\dfrac{8}{3}$; $G(9) = 0$ **(c)**

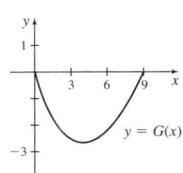

51. 160 **53.** $\dfrac{\pi}{6}$ **55.** $\dfrac{\pi}{3}$ **57.** $2\sqrt{r} - 2$, which goes to infinity as $r \to \infty$

59. The corporation has $23 million in sales over two years.

61. (a) $\int_{a}^{b} H(t)\, dt$ **(b)** cm^3 **(c)** $100\ \text{cm}^3$ of helium leaked out in the first 5 h. **63. (a)** $44.740\ \text{g}$ **(b)** AWV. **(c)** $59.740\ \text{g}$

65. (a) AWV. **(b)** $\dfrac{3}{4}[50^{4/3} - 40^{4/3}] \approx 35.553$ hundreds of dollars in additional revenue

67. (a) $\int_{0}^{5.2} 9.8t\, dt$ **(b)** $132.5\ \text{m}$ **69.** $P(x) = 3x^2 + 6x + 6$ **71. (a)** $\dfrac{64}{3}$ and $\dfrac{128}{3}$ **(b)** Even **(c)** AWV.

73. (a) 1 and 2 **(b)** Even **(c)** AWV. **75.** $\dfrac{1}{\sqrt[3]{2}}$ **77.** $F'(c)$ **79.** $e^4 - 1$ **81.** 4 **83.** SSM.

85. (a) SSM. **(b)** SSM. **(c)** AWV. **87.** $a = \dfrac{\sqrt{5} - 1}{2}$ **89. (a)** $\dfrac{\cos x}{\sqrt{1 - \sin^2 x}} = \dfrac{\cos x}{|\cos x|}$ **(b)** Yes **(c)** Yes

AP® Practice Problems (SSM = See Student Solutions Manual.)

1. C **3.** C **5.** C **7.** A **9.** A **11.** B **13.** B **15.** D **17.** SSM. **19.** SSM. **21.** SSM.

Retain Your Knowledge

1. C **3.** C

Section 6.4 (SSM = See Student Solutions Manual; AWV = Answers will vary.)

1. True **2.** False **3.** True **4.** -2 **5.** Average value **6.** False **7.** 7 **9.** 31 **11.** 17 **13.** 45 **15.** 139 **17.** $-\dfrac{1}{15}$

19. 4 **21.** $\dfrac{\pi}{2}$ **23.** $-\dfrac{76}{3}$ **25.** $\dfrac{5}{3}$ **27.** -3 **29.** $e + \dfrac{1}{e} - 2$ **31.** $\dfrac{1}{2} + 2\ln\dfrac{3}{2}$ **33.** 2 **35.** $\dfrac{\pi + 15}{6}$ **37.** $\dfrac{10}{3}$ **39.** $-\dfrac{2}{3}$

41. 42 **43.** 2 **45.** $\dfrac{679}{32}$ **47.** $\dfrac{131}{4}$ **49.** SSM. **51.** SSM. **53.** 12 to 32 **55.** $\dfrac{\sqrt{2}\pi}{8}$ to $\dfrac{\pi}{4}$ **57.** 1 to $\sqrt{2}$ **59.** 1 to e

61. $u = \sqrt{3}$ **63.** $u = \dfrac{4\sqrt{3}}{3}$ **65.** $u = \dfrac{\pi}{2}, \dfrac{3\pi}{2}$ **67.** $e - 1$ **69.** $\dfrac{3}{5}$ **71.** $\dfrac{2}{\pi}$ **73.** $\dfrac{2}{3}$ **75.** $\dfrac{2}{\pi}\left(e^{\pi/2} - 2\right)$

77. (a) 12 **(b)** 4 **(c)** AWV. **79. (a)** $e^2 + 3 - \dfrac{1}{e}$ **(b)** $\dfrac{1}{3}\left(e^2 + 3 - \dfrac{1}{e}\right)$ **(c)** AWV. **81. (a)** 54 m **(b)** 74 m

83. (a) $\dfrac{235}{12}\,\text{m}$ **(b)** $\dfrac{395}{12}\ \text{m}$ **85.** 9 **87.** $\dfrac{13}{3}$ **89.** $37.5\ °\text{C}$ **91. (a)** $\dfrac{1875}{16} = 117.188\ \text{N}$ **(b)** 4 m

93. $\approx 1000.345\ \text{kg/m}^3$ **95.** $\dfrac{13\pi}{3}\ \text{m}^2$ **97. (a)** 2 **(b)** $\dfrac{1}{2}bh = 2$ **99.** 12 m/s **101. (a)** $\dfrac{f(b) - f(a)}{b - a}$ **(b)** AWV.

103. SSM. **105.** $k = \dfrac{\pi}{6}$ **107.** ≈ 8.296 **109.** Only III need not be true. **111.** $\dfrac{1}{6}$ **113. (a)** $\dfrac{3}{2}g$ **(b)** $2g$

115. (a) $(-\infty, \infty)$ **(b)** $[0, \infty)$ **(c)** $(-\infty, \infty)$ **(d)** All $x \neq 1$ **(e)** $\dfrac{1}{6}$ **117.** $F(x) = \dfrac{x^2}{4}$ **119.** $\dfrac{1}{2}$

121. SSM. **123. (a)** SSM. **(b)** AWV.

AP® Practice Problems (SSM = See Student Solutions Manual.)   ## Retain Your Knowledge

1. C **3.** B **5.** B **7.** B **9.** C **11.** D **13.** B **15.** C **17.** SSM.   **1.** C **3.** D

Section 6.5 (SSM = See Student Solutions Manual; AWV = Answers will vary.)

1. $f(x)$ **2.** False **3.** $\dfrac{x^{a+1}}{a+1}+C$ **4.** True **5.** $\dfrac{1}{2}\sin u$ **6.** False **7.** (c) **8.** True **9.** $\dfrac{3}{5}x^{5/3}+C$ **11.** $\sin^{-1}x+C$

13. $\dfrac{5}{2}x^2+2e^x-\ln|x|+C$ **15.** $\sec x+C$ **17.** $\dfrac{2}{5}\sin^{-1}x+C$ **19.** $\dfrac{e^t-e^{-t}}{2}+C$ **21.** $\dfrac{1}{3}e^{3x+1}+C$ **23.** $-\dfrac{1}{14}(1-t^2)^7+C$

25. $\dfrac{1}{3}\sin^{-1}x^3+C$ **27.** $-\dfrac{1}{3}\cos(3x)+C$ **29.** $-\dfrac{1}{3}\cos^3 x+C$ **31.** $-e^{1/x}+C$ **33.** $\dfrac{1}{2}\ln|x^2-1|+C$ **35.** $2\sqrt{1+e^x}+C$

37. $-\dfrac{2}{3(\sqrt{x}+1)^3}+C$ **39.** $4(e^x-1)^{3/4}+C$ **41.** $\dfrac{1}{2}\ln|2\sin x-1|+C$ **43.** $\dfrac{1}{5}\ln|\sec(5x)+\tan(5x)|+C$ **45.** $\dfrac{2}{3}(\tan x)^{3/2}+C$

47. $\sec x+C$ **49.** $-e^{\cos x}+C$ **51.** $\dfrac{2}{5}(x+3)^{5/2}-2(x+3)^{3/2}+C$ **53.** $\dfrac{1}{3}\sin(3x)-\cos x+C$ **55.** $\dfrac{1}{5}\tan^{-1}\dfrac{x}{5}+C$

57. $\sin^{-1}\dfrac{x}{3}+C$ **59.** $\dfrac{1}{2}\sin^2 x+C$ **61.** (a) (b) $-\dfrac{2}{21}$ (c) AWV. **63.** (a) (b) $\dfrac{1}{3}(e^2-e)$ (c) AWV.

65. (a) (b) $\dfrac{232}{5}$ (c) AWV. **67.** (a) (b) $\dfrac{1640}{\ln 3}$ (c) AWV. **69.** $\left(\dfrac{2}{3}\right)^{5/2}$ **71.** $\dfrac{1}{2}\ln\dfrac{e^4+1}{2}$ **73.** $\ln\left(\dfrac{\ln 3}{\ln 2}\right)$

75. $\sin e^{\pi}-\sin 1$ **77.** $\dfrac{\pi}{8}$ **79.** $-\dfrac{32}{3}$ **81.** 0 **83.** $\dfrac{3\pi}{2}$ **85.** 50 **87.** 2 **89.** 4 **91.** (a) $\dfrac{1}{3}(x+1)^3+C$ (b) $\dfrac{1}{3}x^3+x^2+x+C$

93. (a) 0 (b) 0 (c) 0 (d) SSM. **95.** $\dfrac{124}{15}$ **97.** $\dfrac{\sqrt{3}\pi}{9}$ **99.** $\dfrac{2a^2}{\pi}$ **101.** $\dfrac{2\ln 2}{\pi}$

103. 8 **105.** 4 **107.** $(\ln 10)\ln|\ln x|+C$ **109.** $(\ln 10)\ln 2$ **111.** $(\ln 2)\ln 3$ **113.** $b=\sqrt[3]{2}$ **115.** $\dfrac{1}{2}\ln(x^2+1)+\tan^{-1}x+C$

117. $\dfrac{1}{14}(2\sqrt{x^2+3}-\dfrac{4}{x}+9)^7+C$ **119.** $\dfrac{2}{3}x^{3/2}+\dfrac{8}{7}x^{7/2}+C$ **121.** $\dfrac{4}{9}(4+t^{3/2})^{3/2}+C$ **123.** $\dfrac{3^{2x+1}}{2\ln 3}+C$

125. $-\sin^{-1}\dfrac{\cos x}{2}+C$ **127.** $\dfrac{1}{6}\ln\dfrac{187}{27}$ **129.** SSM. **131.** SSM.

133. (a) 5 (b) 5 (c) 5 (d) $a=2, b=4$ (e) $a=k, b=k+2$ **135.** Hint: Split the integral at 0; SSM.

137. $c=1$ **139.** SSM. **141.** $\dfrac{n}{b(n+1)}(a+bx)^{(n+1)/n}+C$ **143.** (a) 2 (b) $\dfrac{\pi^2}{4}$ **145.** SSM. **147.** $V(z)=\dfrac{Q}{\sqrt{R^2+z^2}}+C$

149. (a) $v(t)=\dfrac{mg}{k}(1-e^{-kt/m})$ (b) $s(t)=\dfrac{mg}{k}\left[\dfrac{m}{k}(e^{-kt/m}-1)+t\right]$ (c) As $t\to\infty, a\to 0, v\to\dfrac{mg}{k}, s\to\infty$

  (d) SSM. **151.** $\dfrac{x^3}{3}+x-\dfrac{1}{2(x^2+1)}+C$ **153.** $6\sqrt{x+1}+2\ln|\sqrt{x+1}-1|-2\ln|\sqrt{x+1}+1|+C$

155. (a) $y'=\sec x$ (b) SSM. (c) SSM. **157.** (a) $y'=\sqrt{a^2-x^2}$ (b) SSM. **159.** No, AWV.

AP® Practice Problems   Retain Your Knowledge

1. C **3.** A **5.** C **7.** B **9.** A **11.** B **13.** B   **1.** A **3.** C

Section 6.6 (SSM = See Student Solutions Manual.)

1. (a) **2.** True **3.** $\tan^{-1}(x+2)+C$ **5.** $\dfrac{1}{2}\tan^{-1}\dfrac{x+2}{2}+C$ **7.** $\dfrac{2\sqrt{5}}{5}\tan^{-1}\dfrac{2x+1}{\sqrt{5}}+C$

9. $\dfrac{1}{4}\ln(2x^2+2x+3)-\dfrac{\sqrt{5}}{10}\tan^{-1}\dfrac{2x+1}{\sqrt{5}}+C$ **11.** $\sin^{-1}\dfrac{x-1}{3}+C$ **13.** $\sin^{-1}\dfrac{x-2}{2}+C$ **15.** $\ln\left|\dfrac{\sqrt{x^2+2x+2}-1}{x+1}\right|+C$

17. $\sin^{-1}\dfrac{x+1}{5}+C$ **19.** $\sqrt{x^2-2x+5}-4\ln\left|\dfrac{x-1+\sqrt{x^2-2x+5}}{2}\right|+C$ **21.** $\ln(\sqrt{2}+1)+C$

23. $\ln\left(e^x+\dfrac{1}{2}+\sqrt{e^{2x}+e^x+1}\right)+C$ **25.** $-2\sqrt{4x-x^2-3}+\sin^{-1}(x-2)+C$ **27.** $\dfrac{x-1}{9\sqrt{x^2-2x+10}}+C$

29. $\ln|x+1+\sqrt{x^2+2x-3}|+C$ **31.** $\sqrt{5+4x-x^2}-3\ln|3+\sqrt{5+4x-x^2}|+3\ln|x-2|+C$

33. $\sqrt{x^2+2x-3}-\ln|x+1+\sqrt{x^2+2x-3}|+C$ **35.** $x-1+\dfrac{2}{x+1}$ **37.** $x^2+2x+7+\dfrac{10}{x-2}$ **39.** $2x+3+\dfrac{x}{x^2-9}$

41. $x^2-4x+16-\dfrac{64x+1}{x^2+4x}$ **43.** $2x^2-7+\dfrac{26}{x^2+4}$ **45.** $\dfrac{x^2}{2}-x+2\ln|x+1|+C$ **47.** $\dfrac{x^3}{3}+x^2+7x+10\ln|x-2|+C$

49. $x^2 + 3x + \dfrac{1}{2}\ln|x^2 - 9| + C$ **51.** $x^2 - 3x + \dfrac{1}{2}\tan^{-1}\dfrac{x}{2} + C$ **53.** $\dfrac{x^3}{3} + 10x + \dfrac{3}{4}\ln|2x^2 + 3| + C$ **55.** $x^2 - 3x - \tan^{-1}(x + 2) + C$

57. SSM. **59.** $a\sin^{-1}\dfrac{x}{a} - \sqrt{a^2 - x^2} + C$

AP® Practice Problems   Retain Your Knowledge

1. B **3.** C **5.** D **7.** D   **1.** B   **3.** C

Review Exercises   (AWV = Answers will vary.)

1. $s_4 = 16$, $S_4 = 24$, $s_8 = 18$, $S_8 = 22$ **3.** $A = \lim\limits_{n \to \infty} s_n = 18$

5. (a) 14 **(b)** $\lim\limits_{n \to \infty} \sum\limits_{i=1}^{n} \dfrac{4}{n}\left[\left(-1 + \dfrac{4i}{n}\right)^2 - 3\left(-1 + \dfrac{4i}{n}\right) + 3\right] = \displaystyle\int_{-1}^{3}(x^2 - 3x + 3)\,dx$ **(c)** $\dfrac{28}{3}$ **7.** $x^{2/3}\sin x$

9. $-2x\tan x^2$ **11.** $1 - \dfrac{\sqrt{2}}{2}$ **13.** $\dfrac{\pi}{4}$ **15.** 4 **17.** $\dfrac{14}{\ln 2}$ **19.** $2e^x + \ln|x| + C$ **21.** The train traveled a distance of 460 km in 16 h.

23. $\dfrac{22}{3}$ **25.** 0 **27.** $2 \le \int_0^2 e^{x^2}\,dx \le 2e^4$ **29.** $\sin^{-1}\dfrac{2}{\pi}$ and $\pi - \sin^{-1}\dfrac{2}{\pi}$ **31.** 0 **33.** $\dfrac{e^2 - 1}{2e}$ **35.** $\sqrt{\dfrac{1}{1 + 4x^2}}$

37. $-\dfrac{1}{y - 2} - \dfrac{1}{(y - 2)^2} + C$ **39.** $-\dfrac{2}{3}\left(\dfrac{1 + x}{x}\right)^{3/2} + C$ **41.** $\dfrac{1}{12}\left(\dfrac{7}{8}\right)^4$ **43.** $-\ln|1 - 2\sqrt{x}| + C$ **45.** $\dfrac{99}{10\ln 10}$ **47.** $\dfrac{\pi}{8}$

49. $\dfrac{1}{4}(x^3 + 3\cos x)^{4/3} + C$ **51.** $\dfrac{2}{3}$ **53.** 5 **55. (a)** Using a Left Riemann sum: 91.5 **(b)** square meters **(c)** AWV.

57. 15,552 gallons **59.** 18 **61.** AWV. **63.** $\displaystyle\int_0^3 x^2\,dx$

AP® Review Problems: Chapter 6, Part 1   (SSM = See Student Solutions Manual.)

1. B **3.** C **5.** B **7.** D **9.** A **11.** A **13.** A **15.** C **17.** A **19.** SSM. **21.** SSM.

AP® Cumulative Review: Chapters 1–6, Part 1   (SSM = See Student Solutions Manual.)

1. C **3.** D **5.** C **7.** B **9.** A **11.** SSM. **13.** SSM. **15.** SSM.

Chapter 6, Part 2

Section 6.7   (SSM = See Student Solutions Manual.)

1. True **2.** $uv - \int v\,du$ **3.** $\dfrac{1}{2}xe^{2x} - \dfrac{1}{4}e^{2x} + C$ **5.** $x\sin x + \cos x + C$ **7.** $\dfrac{2}{3}x^{3/2}\ln x - \dfrac{4}{9}x^{3/2} + C$

9. $x\cot^{-1}x + \dfrac{1}{2}\ln(x^2 + 1) + C$ **11.** $x(\ln x)^2 - 2x\ln x + 2x + C$ **13.** $-x^2\cos x + 2x\sin x + 2\cos x + C$

15. $\dfrac{1}{2}x\cos x\sin x + \dfrac{1}{4}x^2 - \dfrac{1}{4}\sin^2 x + C$ **17.** $\dfrac{x^3\ln x}{3} - \dfrac{x^3}{9} + C$ **19.** $\dfrac{e^x}{x + 1} + C$ **21.** $-\dfrac{\ln x}{3x^3} + \dfrac{1}{9x^3} + C$

23. $x(\ln x)^3 - 3x(\ln x)^2 + 6x\ln x - 6x + C$ **25.** $\dfrac{x^3}{27}[9(\ln x)^2 - 6\ln x + 2] + C$ **27.** $\dfrac{1}{3}x^3\tan^{-1}x - \dfrac{1}{6}x^2 + \dfrac{1}{6}\ln(1 + x^2) + C$

29. $7^x\left[\dfrac{x}{\ln 7} - \dfrac{1}{(\ln 7)^2}\right] + C$ **31.** $\dfrac{e^{-x}}{5}[2\sin(2x) - \cos(2x)] + C$ **33.** $\dfrac{e^{2x}}{5}(2\sin x - \cos x) + C$ **35.** $-\dfrac{e^\pi + 1}{2}$

37. $\dfrac{2}{27}\left(1 - \dfrac{25}{e^6}\right)$ **39.** $\dfrac{\sqrt{2}}{4}\pi - \ln(\sqrt{2} + 1)$ **41.** $9\ln 3 - 4$ **43.** $e - 2$ **45.** $\dfrac{e^\pi + 1}{2}$ **47.** $1 - \dfrac{2}{e}$

49. (a) $v(t) = -\dfrac{1}{5}e^{-2t}(\cos t + 2\sin t) + \dfrac{41}{5}$ **(b)** $s(t) = \dfrac{1}{25}e^{-2t}(3\sin t + 4\cos t) + \dfrac{41}{5}t - \dfrac{4}{25}$ **51.** 2π

53. -37 **55.** 1 **57.** $2\sin\sqrt{x} - 2\sqrt{x}\cos\sqrt{x} + C$ **59.** $(\sin x)\ln(\sin x) - \sin x + C$ **61.** $\dfrac{1}{2}(e^{2x}\sin e^{2x} + \cos e^{2x}) + C$

63. $\dfrac{1}{2}e^{x^2}(x^2 - 1) + C$ **65.** $\dfrac{1}{2}e^x(x\sin x + x\cos x - \sin x) + C$ **67.** SSM. **69.** SSM.

71. (a) $\dfrac{1}{5}e^{5x}\left(x^2 - \dfrac{2}{5}x + \dfrac{2}{25}\right) + C$ **(b)** $\displaystyle\int x^n e^{kx}\,dx = \dfrac{1}{k}\left(x^n e^{kx} - n\int x^{n-1}e^{kx}\,dx\right)$

73. (a) $f(x) = \dfrac{1}{2}\sin^2 x$ **(b)** $g(x) = -\dfrac{1}{2}\cos^2 x$ **(c)** $h(x) = -\dfrac{1}{4}\cos(2x)$ **(d)** $C_2 = \dfrac{1}{2} + C_1$ **(e)** $C_3 = \dfrac{1}{4} + C_1$

75. (a) **(b)** ≈ 1.079 **(c)** ≈ 1.890

77. SSM. **79. (a)** $\dfrac{5\pi}{32}$ **(b)** $\dfrac{8}{15}$ **(c)** $\dfrac{35\pi}{256}$ **(d)** $\dfrac{5\pi}{32}$ **81.** SSM. **83.** SSM. **85.** SSM.

AP® Practice Problems
1. B **3.** C **5.** C

Retain Your Knowledge
1. B **3.** D

Section 6.8 (SSM = See Student Solutions Manual.)

1. False **2.** False **3.** $\dfrac{1}{3}\ln|x-2|-\dfrac{1}{3}\ln|x+1|+C$ **5.** $-\ln|x-1|+2\ln|x-2|+C$

7. $\dfrac{2}{21}\ln|3x-2|+\dfrac{1}{14}\ln|2x+1|+C$ **9.** $-5\ln|x+2|+5\ln|x+1|+\dfrac{4}{x+1}+C$

11. $\dfrac{1}{4}\ln|x+1|+\dfrac{3}{4}\ln|x-1|-\dfrac{1}{2(x-1)}+C$ **13.** $\ln|x|-\dfrac{1}{2}\ln(x^2+1)+C$ **15.** $\dfrac{1}{6}\ln(x^2+2x+4)+\dfrac{2}{3}\ln|x+1|+C$

17. $\dfrac{x-32}{32(x^2+16)}+\dfrac{1}{128}\tan^{-1}\dfrac{x}{4}+C$ **19.** $-\dfrac{1}{2(x^2+16)}+\dfrac{4}{(x^2+16)^2}+C$ **21.** $\dfrac{1}{4}\ln|x-1|+\dfrac{3}{4}\ln|x+3|+C$

23. $\dfrac{14}{3}\ln|x-1|-\dfrac{4}{x-1}-\dfrac{7}{3}\ln(x^2+2)+\dfrac{2\sqrt{2}}{3}\tan^{-1}\dfrac{\sqrt{2}x}{2}+C$ **25.** $-\ln|x|-\ln|x+1|+2\ln|x-3|+C$

27. $4\ln|x-2|-3\ln|x-1|+\dfrac{1}{x-1}+C$ **29.** $\ln|x-1|-\dfrac{1}{2}\ln|x^2+x+1|+\dfrac{\sqrt{3}}{3}\tan^{-1}\dfrac{\sqrt{3}(2x+1)}{3}+C$ **31.** $-\dfrac{1}{6}\ln 2$

33. $\dfrac{1}{8}\ln 21$ **35.** $\dfrac{1}{5}\ln(2-\sin\theta)-\dfrac{1}{5}\ln(\sin\theta+3)+C$ **37.** $-\ln|\cos\theta|+\dfrac{1}{2}\ln(\cos^2\theta+1)+C$

39. $\dfrac{1}{3}\ln|e^t-1|-\dfrac{1}{3}\ln(e^t+2)+C$ **41.** $\dfrac{1}{2}\ln|e^x-1|-\dfrac{1}{2}\ln(e^x+1)+C$ **43.** $t-\dfrac{1}{2}\ln(e^{2t}+1)+C$

45. $\ln|\sin x-1|-\dfrac{1}{\sin x-1}+C$ **47.** $\dfrac{1}{54}\tan^{-1}\dfrac{\sin x}{3}+\dfrac{\sin x}{18(\sin^2 x+9)}+C$ **49.** $\ln\dfrac{15}{7}$ **51.** $\dfrac{4\pi\sqrt{3}}{3}+\dfrac{4}{3}\ln 3$

53. (a) $-4,-2,3$ **(b)** $(x+4)(x+2)(x-3)$ **(c)** $\dfrac{13}{10}\ln|x+2|+\dfrac{2}{35}\ln|x-3|-\dfrac{19}{14}\ln|x+4|+C$ **55.** $2\sqrt{x}-4\ln(\sqrt{x}+2)+C$

57. $\dfrac{3}{5}x^{5/3}+\dfrac{3}{4}x^{4/3}+\dfrac{3}{2}x^{2/3}+x+3x^{1/3}+3\ln|x^{1/3}-1|+C$ **59.** $\dfrac{2}{3}x^{1/2}+\dfrac{1}{3}x^{1/3}+\dfrac{2}{9}x^{1/6}+\dfrac{2}{27}\ln|3x^{1/6}-1|+C$

61. $\dfrac{2}{3}(2x+1)^{3/4}+C$ **63.** $3(1+x)^{1/3}+C$ **65.** $-\dfrac{3x^{1/6}}{x^{1/3}+1}+3\tan^{-1}x^{1/6}+C$

67. $-\dfrac{2}{1+\tan\dfrac{x}{2}}+C$ **69.** $\dfrac{2\sqrt{5}}{5}\tan^{-1}\left(\dfrac{\sqrt{5}}{5}\tan\dfrac{x}{2}\right)+C$ **71.** $-\ln\left|1-\tan\dfrac{x}{2}\right|+C$

73. $\dfrac{1}{2}\ln\left|\tan^2\dfrac{x}{2}+2\tan\dfrac{x}{2}-1\right|-\dfrac{1}{2}\ln\left(\tan^2\dfrac{x}{2}+1\right)-\dfrac{x}{2}+C$ **75.** $\dfrac{\sqrt{5}}{5}\ln\left|\tan\dfrac{x}{2}-\dfrac{\sqrt{5}}{2}+\dfrac{1}{2}\right|-\dfrac{\sqrt{5}}{5}\ln\left|\tan\dfrac{x}{2}+\dfrac{\sqrt{5}}{2}+\dfrac{1}{2}\right|+C$

77. $-\dfrac{1}{2}\ln\left|\tan\dfrac{x}{2}-1\right|+\dfrac{1}{2}\ln\left|\tan\dfrac{x}{2}+1\right|+\dfrac{1}{\tan\dfrac{x}{2}+1}-\dfrac{1}{\left(\tan\dfrac{x}{2}+1\right)^2}+C$

79. $\dfrac{1}{3}\ln\dfrac{\sqrt{3}}{3}-\dfrac{1}{3}\ln\left(\sqrt{2}-1\right)-\dfrac{4}{15}\ln\left(\sqrt{3}+1\right)-\dfrac{4}{15}\ln\left(4-\sqrt{2}\right)+\dfrac{4}{15}\ln(3\sqrt{2}-2)+\dfrac{4}{15}\ln\left(3-\dfrac{\sqrt{3}}{3}\right)$ **81.** $\dfrac{\pi}{2}+\ln 2$

83. SSM. **85.** SSM.

AP® Practice Problems
1. A **3.** C **5.** D

Retain Your Knowledge
1. A **3.** C

Section 6.9 (SSM = See Student Solutions Manual.)

1. (c) **2.** (b) **3.** False **4.** True **5.** False **6.** $\displaystyle\lim_{t\to b^-}\int_a^t f(x)\,dx$ **7.** Yes, ∞ is a limit of integration. **9.** No.

11. Yes, the function is undefined at 0. **13.** Yes, the function is undefined at 1. **15.** Converges to $\dfrac{1}{2}$. **17.** Diverges

19. Diverges **21.** Converges to $\dfrac{1}{24}$ **23.** Converges to $\dfrac{\pi}{2}$ **25.** Diverges **27.** Diverges **29.** Converges to 4 **31.** Converges to 0

33. Diverges **35.** Converges to 1 **37.** Diverges **39.** Diverges **41.** Converges to 1 **43.** Diverges **45.** Converges to $\dfrac{4}{3}$

47. Converges to π **49.** Diverges **51.** Converges to $\sqrt[3]{486}$ **53.** Diverges **55.** Converges to 0 **57.** Converges to 4

59. Diverges **61.** Converges to $\dfrac{1}{2}$ **63.** Diverges **65.** Diverges **67.** (a) Converges (b) ≈ 0.673 **69.** (a) Converges (b) $\dfrac{\pi}{2}$

71. $\dfrac{\pi}{2}$ **73.** $2\pi a^2$ **75.** (a) \$1250 (b) \$30,612 **77.** $\dfrac{\pi N I}{5r}\left(1-\dfrac{x}{\sqrt{r^2+x^2}}\right)$ **79.** 1 **81.** $\dfrac{1}{2}$ **83.** $\dfrac{1}{4}$

85. $\dfrac{\pi}{2}-\sec^{-1}2=\dfrac{\pi}{6}$ **87.** SSM. **89.** SSM. **91.** SSM. **93.** SSM. **95.** SSM. **97.** $\dfrac{1}{s^2}$ **99.** $\dfrac{1}{1+s^2}$

101. $\dfrac{1}{s-a}$ **103.** 1 **105.** SSM. **107.** $\dfrac{a+b}{2}$ **109.** $\sigma^2=\dfrac{(b-a)^2}{12}$, $\sigma=\dfrac{b-a}{2\sqrt{3}}$ **111.** $a>-1$

AP® Practice Problems (SSM = See Student Solutions Manual.)   Retain Your Knowledge

1. D **3.** D **5.** A **7.** SSM.   **1.** B **3.** D

Section 6.10 (SSM = See Student Solutions Manual; AWV = Answers will vary.)

1. True **2.** True **3.** $\dfrac{\sin^5 x}{5}-\dfrac{2\sin^3 x}{3}+\sin x+C$ **5.** $\dfrac{5}{16}x-\dfrac{1}{4}\sin(2x)+\dfrac{3}{64}\sin(4x)+\dfrac{1}{48}\sin^3(2x)+C$

7. $\dfrac{x}{2}-\dfrac{\sin(2\pi x)}{4\pi}+C$ **9.** 0 **11.** $\dfrac{\cos^5 x}{5}-\dfrac{\cos^3 x}{3}+C$ **13.** $\dfrac{8}{45}$ **15.** $\dfrac{3}{10}(\cos x)^{10/3}-\dfrac{3}{4}(\cos x)^{4/3}+C$

17. $\dfrac{2}{3}\sin^3\dfrac{x}{2}-\dfrac{2}{5}\sin^5\dfrac{x}{2}+C$ **19.** $\dfrac{\sec^5 x}{5}-\dfrac{\sec^3 x}{3}+C$ **21.** $\dfrac{2}{5}\tan^{5/2}x+\dfrac{2}{9}\tan^{9/2}x+C$ **23.** $\dfrac{2}{7}(\sec x)^{7/2}-\dfrac{2}{3}(\sec x)^{3/2}+C$

25. $\csc x-\dfrac{\csc^3 x}{3}+C$ **27.** $-\dfrac{1}{8}\cos(4x)-\dfrac{1}{4}\cos(2x)+C$ **29.** $\dfrac{1}{8}\sin(4x)+\dfrac{1}{4}\sin(2x)+C$ **31.** $\dfrac{1}{4}\sin(2x)-\dfrac{1}{12}\sin(6x)+C$

33. $\dfrac{2}{3}$ **35.** $\dfrac{8}{105}$ **37.** $\sec x+\cos x+C$ **39.** 0 **41.** 0 **43.** $\dfrac{1}{2}\tan^2 x-\ln|\sec x|+C$ **45.** $-\dfrac{1}{2}\cot^2 x+2\ln|\tan x|+\dfrac{1}{2}\tan^2 x+C$

47. $\dfrac{\csc^4 x}{4}-\dfrac{\csc^6 x}{6}+C$ or $-\dfrac{\cot^4 x}{4}-\dfrac{\cot^6 x}{6}+C$ **49.** $-\dfrac{1}{8}\csc^4(2x)+C$ **51.** $\dfrac{7\sqrt{2}}{48}+\dfrac{\ln(1+\sqrt{2})}{16}$ **53.** 0

55. $\dfrac{1}{2}\sin x-\dfrac{1}{4}\sin(2x)+C$ **57.** $\dfrac{\pi^2}{2}$ **59.** (a) $\bar{y}=\dfrac{4}{3\pi}$ (b) AWV. (c)

61. $\dfrac{\pi}{4}$

63. (a) $\dfrac{3x}{8}-\dfrac{1}{4}\sin(2x)+\dfrac{1}{32}\sin(4x)+C$ (b) $-\dfrac{1}{4}\sin^3 x\cos x-\dfrac{3}{8}\sin x\cos x+\dfrac{3}{8}x+C$ (c) SSM. (d) Same as (a).

65. $\dfrac{1}{2(m-n)}\sin[(m-n)x]-\dfrac{1}{2(m+n)}\sin[(m+n)x]+C$

67. $\dfrac{1}{2(m-n)}\sin[(m-n)x]+\dfrac{1}{2(m+n)}\sin[(m+n)x]+C$ **69.** $\dfrac{\pi}{4}-\dfrac{1}{2}$

Section 6.11 (SSM = See Student Solutions Manual.)

1. True **2.** (d) **3.** (c) **4.** (b) **5.** $\dfrac{x}{2}\sqrt{4-x^2}+2\sin^{-1}\dfrac{x}{2}+C$ **7.** $-\dfrac{x}{2}\sqrt{16-x^2}+8\sin^{-1}\dfrac{x}{4}+C$

9. $-\dfrac{1}{x}\sqrt{4-x^2}-\sin^{-1}\dfrac{x}{2}+C$ **11.** $\dfrac{x(x^2-2)}{4}\sqrt{4-x^2}+2\sin^{-1}\dfrac{x}{2}+C$ **13.** $\dfrac{1}{4}\dfrac{x}{\sqrt{4-x^2}}+C$

15. $\dfrac{1}{2}x\sqrt{4+x^2}+2\ln\left|\dfrac{\sqrt{4+x^2}+x}{2}\right|+C$ **17.** $\ln\left|\dfrac{\sqrt{x^2+16}+x}{4}\right|+C$ **19.** $\dfrac{x}{2}\sqrt{1+9x^2}+\dfrac{1}{6}\ln|\sqrt{1+9x^2}+3x|+C$

21. $\dfrac{1}{18}x\sqrt{4+9x^2}-\dfrac{2}{27}\ln\left|\dfrac{\sqrt{4+9x^2}+3x}{2}\right|+C$ **23.** $-\dfrac{\sqrt{x^2+4}}{4x}+C$ **25.** $\dfrac{x}{4\sqrt{x^2+4}}+C$

27. $\dfrac{x}{2}\sqrt{x^2-25}+\dfrac{25}{2}\ln\left|\dfrac{x+\sqrt{x^2-25}}{5}\right|+C$ **29.** $\sqrt{x^2-1}-\sec^{-1}x+C$ **31.** $\dfrac{\sqrt{x^2-36}}{36x}+C$

33. $\dfrac{1}{2}\ln\left|\dfrac{2x+\sqrt{4x^2-9}}{3}\right|+C$ **35.** $-\dfrac{x}{9\sqrt{x^2-9}}+C$ **37.** $-\dfrac{x}{\sqrt{x^2-9}}+\ln\left|\dfrac{x+\sqrt{x^2-9}}{3}\right|+C$ **39.** $x-4\tan^{-1}\dfrac{x}{4}+C$

41. $\dfrac{x}{2}\sqrt{4-25x^2}+\dfrac{2}{5}\sin^{-1}\dfrac{5x}{2}+C$ **43.** $\dfrac{x}{4\sqrt{4-25x^2}}+C$ **45.** $\dfrac{1}{2}x\sqrt{4+25x^2}+\dfrac{2}{5}\ln\left|\dfrac{\sqrt{4+25x^2}+5x}{2}\right|+C$

47. $\dfrac{1}{128}\sec^{-1}\dfrac{x}{4}+\dfrac{\sqrt{x^2-16}}{32x^2}+C$ **49.** $\dfrac{\pi}{4}$ **51.** $\dfrac{\sqrt{2}+\ln(1+\sqrt{2})}{2}$ **53.** $10-2\sqrt{7}+9\ln 3-\dfrac{9}{2}\ln(4+\sqrt{7})$ **55.** $\dfrac{\sqrt{3}}{3}-\dfrac{\pi}{6}$ **57.** $3-\dfrac{3\pi}{4}$

59. $\dfrac{\pi}{40}+\dfrac{\pi}{16}\tan^{-1}\dfrac{1}{2}$ **61.** $\bar{y}=\sin^{-1}\dfrac{1}{3}\approx 0.340$ **63.** $\sin^{-1}\dfrac{2}{3}$ **65.** $\dfrac{1}{2}\ln(2\sqrt{6}+5)-\dfrac{1}{2}\ln(2\sqrt{2}+3)-3\sqrt{2}+5\sqrt{6}$ **67.** $\sin^{-1}(x-2)+C$

69. $\dfrac{1}{2}\ln\left|\dfrac{2x-1}{2}+\dfrac{\sqrt{4x^2-4x-3}}{2}\right|+C$ **71.** $\dfrac{e^x}{2}\sqrt{25-e^{2x}}+\dfrac{25}{2}\sin^{-1}\dfrac{e^x}{5}+C$ **73.** $\dfrac{1}{2}x^2\sin^{-1}x-\dfrac{1}{4}\sin^{-1}x+\dfrac{1}{4}x\sqrt{1-x^2}+C$

75. SSM. **77.** SSM. **79.** SSM. **81.** SSM. **83.** $\ln\left|\tan x-3+\sqrt{\tan^2 x-6\tan x+8}\right|+C$

Section 6.12 (AWV = Answers will vary.)

1. $\dfrac{e^{2x}}{5}(\sin x+2\cos x)+C$ **3.** $\dfrac{(2+3x)^5}{9}\left(\dfrac{2+3x}{6}-\dfrac{2}{5}\right)+C$ **5.** $\dfrac{2}{15}(32-8x+3x^2)\sqrt{6+3x}+C$

7. $\sqrt{x^2+4}-2\ln\left|\dfrac{2+\sqrt{x^2+4}}{x}\right|+C$ **9.** $-\dfrac{x}{4\sqrt{x^2-4}}+C$ **11.** $\dfrac{x^4(\ln x)^2}{4}-\dfrac{x^4}{8}\left(\ln x-\dfrac{1}{4}\right)+C$

13. $\dfrac{x}{2}\sqrt{x^2-16}-8\ln\left|x+\sqrt{x^2-16}\right|+C$ **15.** $\dfrac{x}{4}(6-x^2)^{3/2}+\dfrac{9}{4}x\sqrt{6-x^2}+\dfrac{27}{2}\sin^{-1}\dfrac{\sqrt{6}x}{6}+C$

17. $\dfrac{x-5}{2}\sqrt{10x-x^2}+\dfrac{25}{2}\cos^{-1}\dfrac{5-x}{5}+C$ **19.** $\dfrac{\sin(11x)}{22}+\dfrac{\sin(5x)}{10}+C$ **21.** $\dfrac{x^2+1}{2}\tan^{-1}x-\dfrac{x}{2}+C$

23. $\dfrac{x^5}{5}\left(\ln x-\dfrac{1}{5}\right)+C$ **25.** $-\dfrac{\sqrt{8x-x^2}}{12x^2}-\dfrac{\sqrt{8x-x^2}}{48x}+C$ **27.** $\dfrac{1}{8}\left[\dfrac{x}{4+x^2}+\dfrac{1}{2}\tan^{-1}\dfrac{x}{2}\right]+C$

29. $\dfrac{1}{60}(6x+5)(4x+5)^{3/2}+C$ **31.** $\dfrac{135}{8}\sin^{-1}\dfrac{1}{3}-\dfrac{9\sqrt{2}}{4}$ **33.** $\dfrac{e^{2x}}{5}(\sin x+2\cos x)$ **35.** $\dfrac{27}{2}x^6+\dfrac{216}{5}x^5+54x^4+32x^3+8x^2$

37. $\dfrac{2}{45}\sqrt{3}\sqrt{x+2}\,(3x^2-8x+32)$ **39.** $\sqrt{x^2+4}-2\,\text{arctanh}\dfrac{2}{\sqrt{x^2+4}}$ **41.** $-\dfrac{1}{4}\dfrac{x}{\sqrt{x^2-4}}$ **43.** $\dfrac{1}{32}x^4(8\ln^2 x-4\ln x+1)$

45. $\dfrac{1}{2}x\sqrt{x^2-16}-8\ln(x+\sqrt{x^2-16})$ **47.** $\dfrac{9}{4}x\sqrt{6-x^2}-\dfrac{27}{2}i\ln\left(ix+\sqrt{6-x^2}\right)+\dfrac{1}{4}x(6-x^2)^{3/2}$

49. $\dfrac{1}{2}x\sqrt{-x(x-10)}-\dfrac{25}{2}i\ln(ix+\sqrt{-x(x-10)}-5i)-\dfrac{5}{2}\sqrt{-x(x-10)}$ **51.** $\dfrac{1}{10}\sin 5x+\dfrac{1}{22}\sin 11x$

53. $\dfrac{1}{2}\arctan x-\dfrac{1}{2}x-\dfrac{1}{4}\pi+\dfrac{1}{2}x^2\arctan x$ **55.** $\dfrac{1}{5}x^5\ln x-\dfrac{1}{25}x^5$ **57.** $-\dfrac{(x+4)\sqrt{-(x-8)x}}{48x^2}$

59. $\dfrac{1}{32(x^2+4)}\left(4x-4\pi+8\arctan\dfrac{1}{2}x+2x^2\arctan\dfrac{1}{2}x-\pi x^2\right)$ **61.** $\dfrac{1}{60}(4x+5)^{3/2}(6x+5)$

63. $-\dfrac{9\sqrt{2}}{4}+\dfrac{135}{8}\cos^{-1}\dfrac{2\sqrt{2}}{3}$ **65.** AWV. **67.** AWV. **69.** AWV.

Section 6.13

1. $\dfrac{1}{12}[2x^3 - \sin(2x^3)] + C$ **3.** $(5x - 1)\sin x + 5\cos x + C$ **5.** $\dfrac{1}{2}\ln\left|\dfrac{\sqrt{4x^2 + 25}}{5} + \dfrac{2x}{5}\right| + C$ **7.** $4(8\ln 2 - 5\ln 3)$

9. $\dfrac{1}{3}\sec^3 x + C$ **11.** $\dfrac{1}{5}\tan^5 x + \dfrac{1}{3}\tan^3 x + C$ **13.** $\dfrac{1}{4}\sec^{-1}\dfrac{x}{4} + C$ **15.** $\dfrac{3}{10}(\tan x)^{10/3} + \dfrac{3}{4}(\tan x)^{4/3} + C$

17. $-\dfrac{1}{2}\cos(x^4 + 2x) + C$ **19.** $\dfrac{1}{3}x^3\sin^{-1} x - \dfrac{1}{9}(1 - x^2)^{3/2} + \dfrac{1}{3}(1 - x^2)^{1/2} + C$ **21.** $\dfrac{4\sqrt{2}}{15}$

23. $18\sin^{-1}\left(\dfrac{x}{6}\right) + \dfrac{x\sqrt{36 - x^2}}{2} + C$ **25.** $\dfrac{1}{2}\ln(x^2 + 2x + 5) - \dfrac{1}{2}\tan^{-1}\left(\dfrac{x + 1}{2}\right) + C$ **27.** Converges; $\dfrac{\pi}{2}$

29. $\ln|x - 1| + \tan^{-1} x - \dfrac{1}{2}\ln(x^2 + 1) + C$ **31.** $\dfrac{3}{10}(\cos x)^{10/3} - \dfrac{3}{4}(\cos x)^{4/3} + C$

Review Exercises (SSM = See Student Solutions Manual.)

1. $\dfrac{1}{3}\sec^3 \phi + C$ **3.** $\dfrac{1}{3}\cos^3 \phi - \cos \phi + C$ **5.** $\ln\left|x + 2 + \sqrt{(x + 2)^2 - 1}\right| + C$ **7.** $-v\cot v + \ln|\sin v| + C$

9. $6\sin^{-1}\dfrac{x}{2} + 2x\sqrt{4 - x^2} + \dfrac{1}{4}x\sqrt{4 - x^2}\left(2 - x^2\right) + C$ **11.** $\dfrac{1}{16}\ln\left|\dfrac{x^2 - 4}{x^2 + 4}\right| + C$ **13.** $\ln|y + 1| + \dfrac{2}{y + 1} - \dfrac{1}{2(y + 1)^2} + C$

15. $y\ln|1 - y| - y - \ln|1 - y| + C$ **17.** $-2\ln|x| + \dfrac{2}{x} + 5\ln|x - 1| + C$ **19.** $\dfrac{1}{3}x^3\sin^{-1} x + \dfrac{1}{3}\sqrt{1 - x^2} - \dfrac{1}{9}\left(1 - x^2\right)^{3/2} + C$

21. $\dfrac{1}{2}\ln\left|\dfrac{x}{x + 2}\right| + C$ **23.** $\dfrac{1}{2}\ln|1 - w| - \dfrac{3}{2}\ln|w + 1| + C$ **25.** $\sqrt{x} + \sin\sqrt{x}\cos\sqrt{x} + C$ **27.** 1 **29.** SSM. **31.** Converges to $\dfrac{2}{e}$.
33. Converges to 1. **35.** SSM. **37.** Diverges **39.** 3

AP® Review Problems: Chapter 6, Part 2 (SSM = See Student Solutions Manual.)

1. B **3.** D **5.** D **7.** SSM.

AP® Cumulative Review Problems: Chapters 1–6, Part 2

1. A **3.** D **5.** C **7.** C **9.** B

Chapter 7

Section 7.1 (SSM = See Student Solutions Manual.)

1. 2; 1 **2.** True **3.** True **4.** False **5.** This is a first-order differential equation of degree 1.
7. This is a fourth-order differential equation of degree 1. **9.** This is a second-order differential equation of degree 1.
11. This is a second-order differential equation of degree 3. **13.** This is a second-order differential equation of degree 1.
15. SSM. **17.** SSM. **19.** SSM. **21.** SSM. **23.** SSM. **25.** SSM. **27.** SSM. **29.** SSM.
31. (a) $y = x^3 - 3x^2 + x + C$ (b) $y = x^3 - 3x^2 + x + 3$ **33.** (a) $y = 2x^5 + 3x^2 + C$ (b) $y = 2x^5 + 3x^2 - 1$
35. (a) $y = e^x + \cos x + C$ (b) $y = e^x + \cos x + 1$ **37.** (a) $y = 3x^{4/3} + x^2 + C$ (b) $y = 3x^{4/3} + x^2 - 6$
39. (a) $y = \ln|x| - \dfrac{x^3}{3} + C$ (b) $y = \ln|x| - \dfrac{x^3}{3} + \dfrac{7}{3}$ **41.** (a) $y = 2\sin(x - 1) - \dfrac{x^2}{2} + C$ (b) $y = 2\sin(x - 1) - \dfrac{x^2}{2} + \dfrac{9}{2}$
43. (a) $y = \dfrac{1}{4}(3x - 1)^{4/3} + C$ (b) $y = \dfrac{1}{4}(3x - 1)^{4/3} + 1$ **45.** SSM. **47.** SSM. **49.** $n = -3, 2$
51. The Schrödinger equation is a second-order differential equation of degree 1.

AP® Practice Problems   Retain Your Knowledge

1. B **3.** D **5.** B   **1.** D **3.** D

Section 7.2

1. True **2.** False **3.** $\ln|y| = \dfrac{x^2}{2} + C$ **5.** $\ln|y| = \dfrac{2x^{3/2}}{3} + C$ **7.** $e^{-y} = -\dfrac{x^2}{2} + C$ **9.** $\ln|\csc(2y) - \cot(2y)| = 2e^x + C$

11. $e^y = \dfrac{5}{2}e^{2x} + C$ **13.** $\tan^{-1} y = \dfrac{x^3}{3} + C$ **15.** $y = \dfrac{x^4}{16}$ **17.** $|y - 1| = |x - 1|$ **19.** $\sin y = \dfrac{x^2}{2} - 2$ **21.** $y^2 = 8e^x - 4$

23. $y = \sec x$ **25.** ≈ 6944 mosquitoes **27.** (a) $\dfrac{dP}{dt} = kP$ (b) $P = P_0 e^{kt}$ (c) $P = 4000 e^{\frac{\ln 2}{18}t}$ (d) $\approx 25{,}398$ people

29. (a) $\dfrac{dA}{dt} = kA$ **(b)** $A = A_0 e^{kt}$ **(c)** $A = 8e^{-\frac{\ln 2}{1690}t}$ **(d)** ≈ 7.679 g

31. (a) $\dfrac{dP}{dt} = 1.0013P$ **(b)** $P = P(t) = P_0 e^{\ln 1.0013 t}$ **(c)** $P(t) = (1.273 \times 10^9)\, e^{1.0013t}$, where t is the number of years since 2020.

(d) If the population continues to follow this model, the projected population in 2050 will be approximately 1.324×10^9 people.

33. $\approx 52.67\%$ remains **35.** ≈ 0.262 mol and ≈ 394.7 min **37.** $V = 20e^{-\frac{\ln 2}{2}t}$ and $t = 4$ s

39. (a) $\dfrac{du}{dt} = k[u(t) - 30],\, k < 0$ **(b)** $u = -26e^{kt} + 30$ **(c)** $\approx 16.507\,°C$

41. (a) $\dfrac{du}{dt} = -0.159[u(t) - 12]$ **(b)** $u = (u_0 - 12)e^{-0.159t} + 12$ **(c)** $u = 25e^{-0.159t} + 12$ **(d)** $\approx 8{:}50$ a.m.

(e) It takes about 13 h, 20 min to cool to $15\,°C$.

AP® Practice Problems

1. C **3.** B **5.** C **7.** A **9.** D

Retain Your Knowledge

1. C **3.** A

Section 7.3 (SSM = See Student Solutions Manual; AWV = Answers will vary.)

1. slope field **2.** False

3. (a) 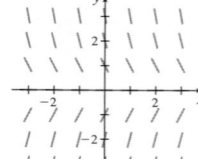 **(b)** 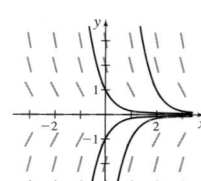 **5. (a)** 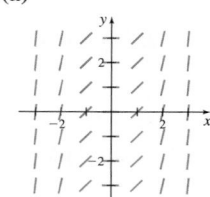 **(b)**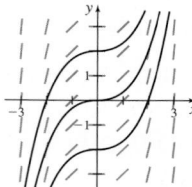

7. (a) 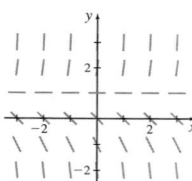 **(b)** 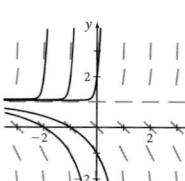 **9. (a)** 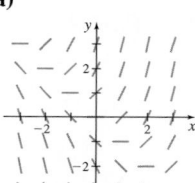 **(b)**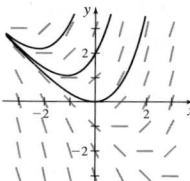

11. 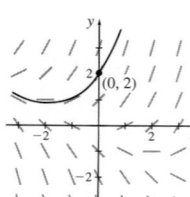 **13.** 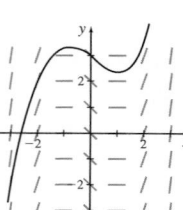 **15.** 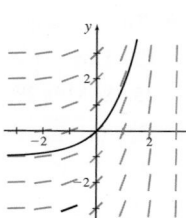 **17. (b)**

AP® Practice Problems (SSM = See Student Solutions Manual.)

1. B **3.** B **5.** SSM.

Retain Your Knowledge

1. A **3.** D

Section 7.4 (SSM = See Student Solutions Manual; AWV = Answers will vary.)

1. True **2.** True **3.** $y(0.4) \approx 0.648$ **5.** $y(1.4) \approx 3.982$ **7.** $y(0.6) \approx 5.84$

9. (a) $y(1) \approx 1.2$ **(b)** $y = \dfrac{1}{\pi}\sin(\pi x) + 1$ **(c)** $y(1) = 1$

AP® Practice Problems (SSM = See Student Solutions Manual.)

1. B **3.** SSM.

Retain Your Knowledge

1. B **3.** C

Section 7.5 (SSM = See Student Solutions Manual; AWV = Answers will vary.)

1. False **2.** (c) **3.** True **4.** (d) **5. (a)** $M = 5500$ **(b)** $k = 0.02$ **(c)** $P(0) = 55$ **(d)** 2750

7. (a) $M = 660$ **(b)** $k = 0.05$ **(c)** $P(0) = 20$ **(d)** 330 **9. (a)** $M = 150$ **(b)** $k = 0.2$ **(c)** $P(0) = 50$ **(d)** 75

11. (a) $M = 1000$ **(b)** $k = 0.12$ **(c)** 500 **13. (a)** $M = 4000$ **(b)** $k = 0.25$ **(c)** 2000

15. (a) $M = 2800$ **(b)** $k = 0.002$ **(c)** 1400 **17. (a)** $M = 1000$ **(b)** $k = 0.03$ **(c)** 500

19. (a) $M = 1200$ **(b)** $k = 0.24$ **(c)** 600 **21.** $\dfrac{dP}{dt} = 0.15P\left(1 - \dfrac{P}{2100}\right)$ **23.** $P(t) = \dfrac{800}{1 + 19e^{-0.2t}}$

25. $P(t) = \dfrac{680}{1 + 33e^{-0.4t}}$ **27.** $P(t) = \dfrac{400}{1 + 79e^{-0.3t}}$ **29.** 50 **31. (a)** SSM. **(b)** $P_0 = 100$

33. (a) $\dfrac{dP}{dt} = 0.15P\left(1 - \dfrac{P}{50}\right)$, $P(0) = 1$ **(b)** $P(t) = \dfrac{50}{1 + 49e^{-0.15t}}$ **(c)** $t = \dfrac{-\ln 49}{-0.15} \approx 25.945$. By the 26th day half the

population is infected. **(d)** $t = \dfrac{-\ln 196}{-0.15} \approx 35.187$. On the 35th day 80% of the population is infected.

35. (a) $\dfrac{dP}{dt} = 0.20P\left(1 - \dfrac{P}{600{,}000}\right)$, $P(0) = 100$ **(b)** $P(t) = \dfrac{600{,}000}{1 + 5999e^{-0.20t}}$

(c) If $P = 100{,}000$, then $t = -5 \cdot \ln \dfrac{5}{5999} \approx 35.450$. The population exceeds 100,000 on the 35th day.

AP® Practice Problems (SSM = See Student Solutions Manual.)

1. C **3.** B **5.** D **7.** SSM. **9.** SSM.

Retain Your Knowledge
1. B **3.** A

Review Exercises (SSM = See Student Solutions Manual; AWV = Answers will vary.)

1. This is a first-order differential equation of degree 1. **3.** This is a third-order differential equation of degree 1. **5.** SSM. **7.** SSM.

9. (a) $y = 6x^2 + \dfrac{3}{2}e^{x^2} + C$ **(b)** $y = 6x^2 + \dfrac{3}{2}e^{x^2} + \dfrac{9}{2}$ **11. (a)** $y = C(x + 3) + 1$ **(b)** $y = -\dfrac{1}{2}x - \dfrac{1}{2}$

13. (a) $-\cos x = -\dfrac{1}{y} - \ln|y| + C$ **(b)** $\cos x = \dfrac{1}{y} + \ln|y| - 2$

15. (a) $y = \dfrac{1}{3}\ln|3x^2 + 3x + 1| + C$ **(b)** $y = \dfrac{1}{3}\ln|3x^2 + 3x + 1| + 2$

17. (a)   **(b)**

19. (a) $M = 2000$ **(b)** $k = 0.06$ **(c)** 1000

21. $y(1.3) \approx 5.522$ **23. (a)** $\dfrac{dA}{dt} = kA$, $k < 0$ **(b)** $A = A_0 e^{kt}$ **(c)** $A = 10e^{-(\ln 2/1690)t}$ **(d)** ≈ 8.842 g **25.** $y = 4e^{3x^2/2}$

AP® Review Problems: Chapter 7 (SSM = See Student Solutions Manual.)

1. B **3.** B **5.** C **7.** SSM.

AP® Cumulative Review Problems: Chapters 1–7 (SSM = See Student Solutions Manual.)

1. A **3.** C **5.** D **7.** A **9.** B **11.** SSM. **13.** SSM.

Chapter 8

Section 8.1 (AWV = Answers will vary.)

1. False **2.** net displacement **3.** True **4.** average velocity **5. (a)** 54 m **(b)** 74 m

7. (a) 2.25 m; AWV. **(b)** 15.583 m; AWV. **9. (a)** −0.445 **(b)** 2.907 **11.** 8 m/s **13.** 12 m/s

AP® Practice Problems (SSM = See Student Solutions Manual.)

1. C **3.** B **5.** SSM. **7.** SSM. **9.** SSM.

Retain Your Knowledge
1. B **3.** C

Section 8.2 (SSM = See Student Solutions Manual.)

1. $\int_0^1 \left(\sqrt{x} - x^2\right) dx$ **2.** $\int_{-1}^1 \left(1 - y^2\right) dy$ **3.** $\dfrac{1}{2}$ **5.** $\dfrac{1}{6}$ **7.** $\dfrac{1}{2}$ **9.** $\dfrac{4}{15}$ **11.** $\dfrac{\sqrt{3}}{2} - \dfrac{\pi}{6}$ **13.** $\dfrac{9}{2}$ **15.** $\dfrac{32}{3}$ **17.** $\dfrac{9}{2}$

19. $e^2 - 3$ **21.** $2\sqrt{2}$ **23.** $\dfrac{32}{3}$ **25.** (a) $\int_0^1 \left(\sqrt{x} - x^3\right) dx = \dfrac{5}{12}$ (b) $\int_0^1 \left(\sqrt[3]{y} - y^2\right) dy = \dfrac{5}{12}$

27. (a) $\int_0^1 \left((x+1) - (x^2+1)\right) dx = \dfrac{1}{6}$ (b) $\int_1^2 \left(\sqrt{y-1} - (y-1)\right) dy = \dfrac{1}{6}$

29. (a) $\int_0^3 \left(\sqrt{9-x} - \sqrt{9-3x}\right) dx + \int_3^9 \sqrt{9-x}\, dx = 12$ (b) $\int_0^3 \left((9 - y^2) - \left(3 - \dfrac{y^2}{3}\right)\right) dy = 12$

31. (a) $\int_2^3 \sqrt{x-2}\, dx + \int_3^4 \left(\sqrt{x-2} - \sqrt{2x-6}\right) dx = \dfrac{2\sqrt{2}}{3}$ (b) $\int_0^{\sqrt{2}} \left(\left(\dfrac{y^2}{2} + 3\right) - (y^2 + 2)\right) dy = \dfrac{2\sqrt{2}}{3}$

33. $\dfrac{16\sqrt{2}}{3}$ **35.** $\dfrac{64}{3}$ **37.** 2   **39.** $\dfrac{\sqrt{3}}{2} - \dfrac{\pi}{6}$ **41.** $\dfrac{e^2 - 3}{2}$ **43.** $\dfrac{125}{24}$ **45.** $1 - \dfrac{\pi}{4}$ **47.** SSM.

49. (a) $\left(-\dfrac{\pi}{4}, \dfrac{1}{2}\right), \left(\dfrac{\pi}{4}, \dfrac{1}{2}\right)$ (b) $\dfrac{1}{2}$ **51.** $\dfrac{9}{2} \ln 3 - 4$ **53.** (a) $\dfrac{4\sqrt{7}}{3} - 3 \ln \dfrac{(4 + \sqrt{7})}{3}$ (b) $\dfrac{14\sqrt{10}}{3} - 3 \ln \dfrac{(7 + 2\sqrt{10})}{3}$

55. (a) $\pi - \left[4 \sin^{-1} \dfrac{\sqrt{15}}{8} + \sin^{-1} \dfrac{\sqrt{15}}{4} - \dfrac{\sqrt{15}}{2}\right]$ (b) $4\pi - \left[4 \sin^{-1} \dfrac{\sqrt{15}}{8} + \sin^{-1} \dfrac{\sqrt{15}}{4} - \dfrac{\sqrt{15}}{2}\right]$ **57.** $\dfrac{\sqrt{3}}{2} + \dfrac{\pi}{12} - \dfrac{9}{8}$

59. (a)
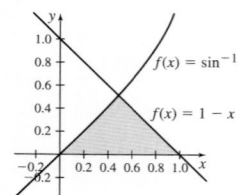
(b) $(0, 0), \approx (0.489, 0.511), (1, 0)$
(c) $A \approx 0.253$

61. (a)
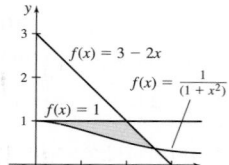
(b) $(0, 1), (1, 1)$ and $\approx (1.317, 0.366)$
(c) $A \approx 0.295$

63. (a)
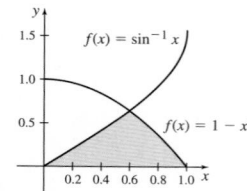
(b) $(0, 0), (1, 0)$, and $\approx (0.599, 0.642)$ (c) $A \approx 0.325$

65. (a) $A(k) = k \tan^{-1} k - \ln \sec(\tan^{-1} k)$ (b) $A(1) = \dfrac{\pi}{4} - \dfrac{\ln 2}{2}$ (c) $\dfrac{\pi}{40}$

AP® Practice Problems   Retain Your Knowledge

1. C   **3.** B   **5.** B   **7.** B   **9.** D   **1.** C   **3.** B

Section 8.3 (SSM = See Student Solutions Manual; AWV = Answers will vary.)

1. $\pi \int_a^b [f(x)]^2\, dx$ **2.** False   **3.** False   **4.** False   **5.** $V = 30\pi$ **7.** $\dfrac{3\pi}{4}$ **9.** $2\pi (\tan 1 - 1)$ **11.** $\dfrac{4\pi}{5}$ **13.** $\dfrac{1}{2}\pi \left(1 - \dfrac{1}{e^4}\right)$

15. $\dfrac{15\pi}{2}$ **17.** $\dfrac{128\pi}{5}$ **19.** 32π **21.** $\dfrac{2\pi}{5}$ **23.** $\dfrac{\pi}{2}$ **25.** $\dfrac{59{,}049\pi}{5}$ **27.** $\dfrac{129\pi}{7}$ **29.** $\dfrac{117\pi}{2}$ **31.** $\dfrac{64\pi}{45}$ **33.** $\dfrac{512\pi}{21}$

35. $\left(\dfrac{e^2}{2} - \dfrac{5}{6}\right)\pi$ **37.** π **39.** $\left(\dfrac{e^4}{2} + 2e^2 - \dfrac{5}{2}\right)\pi$ **41.** $\dfrac{\pi}{6}$ **43.** $\dfrac{1024\pi}{15}$ **45.** $\dfrac{363\pi}{64}$

47. (a) $\dfrac{243}{5}\pi$ (b) $\dfrac{333}{5}\pi$ (c) $\dfrac{657}{5}\pi$ (d) $\dfrac{90a - 243}{5}\pi$ **49.** (a) 8π (b) $\dfrac{184}{3}\pi$ (c) $\dfrac{136}{3}\pi$ (d) $\dfrac{32b - 24}{3}\pi$

51. (a)

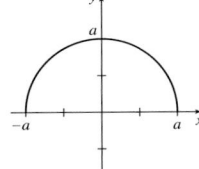

(b) SSM. **53. (a)** $\pi \int_0^1 \dfrac{1}{(x^2+4)^2}dx$ **(b)** ≈ 0.170

55. (a)

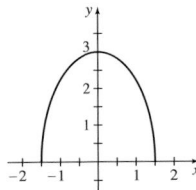

(b) 18π **57. (a)** $\dfrac{k^5\pi}{30}$ **(b)** $\dfrac{k^3}{6}$ **59.** $\left(2-\dfrac{\pi}{4}\right)\pi$ **61.** $\pi e - 2\pi$ **63.** $\dfrac{\pi}{9}(5e^6+1)$

65. (a) $A=\dfrac{\pi}{4}-\ln\sqrt{2}$ **(b)** $V=\dfrac{\pi^2}{2}-\pi$ **67. (a)** $k=\dfrac{e^{-b^2}}{b^2}$, $P(x)=\dfrac{e^{-b^2}}{b^2}x^2$ **(b)** $V=\dfrac{b^2e^{-b^2}\pi}{2}$ **(c)** $b=1$; AWV.

AP® Practice Problems (SSM = See Student Solutions Manual.) **Retain Your Knowledge**

1. B **3.** B **5.** D **7.** B **9.** B **11.** A. **13.** SSM.   **1.** A **3.** D

Section 8.4 (SSM = See Student Solutions Manual; AWV = Answers will vary.)

1. False **2.** False **3.** False **4.** True **5.** $\dfrac{3\pi}{2}$ **7.** $\dfrac{3\pi}{10}$ **9.** $\dfrac{768\pi}{7}$ **11.** $\dfrac{4\pi}{5}$ **13.** $\dfrac{2\pi}{15}$ **15.** $\pi\left(1-\dfrac{1}{e^4}\right)$ **17.** 16π **19.** $\dfrac{17\pi}{6}$

21. $\dfrac{\pi}{2}$ **23. (a)** $\dfrac{128\pi}{105}$ **(b)** $\dfrac{16\pi}{5}$ **(c)** AWV. **25. (a)** $\dfrac{206\pi}{15}$ **(b)** 12π **(c)** AWV. **27.** $\dfrac{64\pi}{5}$ **29.** $\dfrac{4\pi}{15}$ **31.** $\dfrac{308\pi}{3}$

33. $\dfrac{80\pi}{3}$ **35.** $\dfrac{\pi}{12}$ **37.** $\dfrac{1177\pi}{10}$ **39.** $\dfrac{\pi}{2}$ **41. (a)** $\dfrac{16}{15}\pi$ **(b)** $\dfrac{8\pi}{3}$ **(c)** $\dfrac{16\pi}{3}$ **(d)** $\dfrac{8}{5}\pi$ **43.** $f(x)=\dfrac{x^2+2x}{\sqrt{\pi}}$ **45.** SSM.

47. (a) $2\pi\int_0^{\pi/2}x\cos x\,dx$ or $\pi\int_0^1(\cos^{-1}y)^2\,dy$ **(b)** $\pi(\pi-2)\approx 3.586$ **49.** SSM. **51.** SSM.

Section 8.5 (SSM = See Student Solutions Manual; AWV = Answers will vary.)

1. AWV. **2.** False **3.** $\dfrac{128}{3}$ **5. (a)** $\dfrac{9}{35}\pi$ **(b)** $\dfrac{18\sqrt{3}}{35}$ **7. (a)** $\dfrac{32}{5}$ **(b)** $\dfrac{4}{5}\pi$ **(c)** $\dfrac{8\sqrt{3}}{5}$

9. (a) $\dfrac{729}{70}$ **(b)** $\dfrac{729}{560}\pi$ **(c)** $\dfrac{729\sqrt{3}}{280}$ **11.** $\dfrac{256{,}000}{3}$ m³ **13.** SSM. **15.** $\dfrac{37}{1320}\pi$ **17.** ≈ 3.758

19. $\dfrac{2}{3}hr^2$, where h is the height of the glass and r is its radius **21.** $\left(\dfrac{500}{3}-28\sqrt{21}\right)\pi$ **23.** $\dfrac{\pi}{3}bah$

AP® Practice Problems **Retain Your Knowledge**

1. B **3.** D **5.** C **7.** B   **1.** C **3.** A

Section 8.6 (SSM = See Student Solutions Manual; AWV = Answers will vary.)

1. False **2.** True **3.** $2\sqrt{10}$ **5.** $\sqrt{13}$ **7.** $\dfrac{80\sqrt{10}-13\sqrt{13}}{27}$ **9.** $\dfrac{80\sqrt{10}-8}{27}$ **11.** $\dfrac{4\sqrt{2}-2}{3}$ **13.** 45 **15.** 21 **17.** $\dfrac{33}{16}$

19. $\ln\dfrac{\sqrt{3}}{3}-\ln(2-\sqrt{3})$ **21.** $\dfrac{79}{2}$ **23.** $\dfrac{13\sqrt{13}-8}{27}$ **25.** $\dfrac{20\sqrt{10}-2}{27}$ **27. (a)** $\int_0^2\sqrt{1+(2x)^2}\,dx$ **(b)** $s\approx 4.647$

29. (a) $\int_0^4\dfrac{5}{\sqrt{25-x^2}}\,dx$ **(b)** $s\approx 4.636$ **31. (a)** $\int_0^{\pi/2}\sqrt{1+\cos^2 x}\,dx$ **(b)** $s\approx 1.910$

33. $32\sqrt{10}\pi$ **35.** $\dfrac{\sqrt{3}}{6}\pi(11\sqrt{11}-7\sqrt{7})$ **37.** $\dfrac{8}{3}\pi(5\sqrt{5}-2\sqrt{2})$ **39.** $\dfrac{\pi}{27}(145^{3/2}-1)$ **41.** 4π

43. (a) $S=2\pi\int_1^3 x^2\sqrt{1+4x^2}\,dx$ **(b)** $S\approx 257.508$ **45. (a)** $S=2\pi\int_1^8 x^{2/3}\sqrt{1+\dfrac{4}{9x^{2/3}}}\,dx$ **(b)** $S\approx 126.220$

47. (a) $S = 2\pi \int_0^3 e^x \sqrt{1 + e^{2x}}\, dx$ **(b)** $S \approx 1273.371$ **49.** (a) $S = 2\pi \int_0^{\pi/2} \sin x \sqrt{1 + \cos^2 x}\, dx$ **(b)** $S \approx 7.212$ **51.** 8

53. $2(e^6 - 1)$ **55.** $6a$ **57.** $\sqrt{2} + \dfrac{1}{27}\left(13\sqrt{13} - 8\right)$ **59.** $\dfrac{17}{12}$ **61.** $-\ln(\sqrt{2} - 1)$

63. (a) $s = \displaystyle\int_0^{2\sqrt{5}/5} \sqrt{\dfrac{1 + 3y^2}{1 - y^2}}\, dy$ **(b)** ≈ 1.519 **(c)** $P \approx 9.688$ **65.** $\dfrac{\pi}{6}(2\sqrt{2} - 1)\, \mathrm{m}^2$ **67.** (a) SSM. **(b)** SSM. **(c)** AWV.

AP® Practice Problems (SSM = See Student Solutions Manual.)

1. D **3.** B **5.** C **7.** SSM.

Retain Your Knowledge

1. B **3.** C

Section 8.7

1. True **2.** $W = Fx$ **3.** joule; foot-pound **4.** $W = \int_a^b F(x)\, dx$ **5.** Equilibrium **6.** True **7.** True **8.** True

9. $\dfrac{825}{2}$ J **11.** 23,520 J **13.** $k = 12$ N/m **15.** -0.9 J **17.** (a) $(62.42)(1152)\pi \approx 2.259 \times 10^5$ ft lb

(b) $(62.42)(4032)\pi \approx 7.907 \times 10^5$ ft lb **19.** (a) $\dfrac{512}{75}\rho g \approx 1197.534$ J **(b)** $\dfrac{256}{15}\rho g \approx 2993.835$ J **21.** 6.074×10^6 ft lb

23. 8330 J **25.** $-\dfrac{3}{4}$ J **27.** 10 ft **29.** $\dfrac{64}{3}\pi\rho g \approx 6.568 \times 10^5$ J **31.** $36\pi\rho g \approx 1.108 \times 10^6$ J **33.** (a) $18\pi\rho g \approx 5.542 \times 10^5$ J

(b) $54\pi\rho g \approx 1.663 \times 10^6$ J **35.** $t \approx 8.5$ m **37.** 6.965×10^9 J **39.** 475 ft lb **41.** $\dfrac{112}{3}\pi\rho g$ **43.** 2.092×10^5 in. lb

Review Exercises

1. $4\ln 4 - 3$ **3.** $\dfrac{8}{3}$ **5.** $\dfrac{125}{6}$ **7.** $\dfrac{5\pi}{6}$ **9.** $\dfrac{\pi}{2}$ **11.** $\dfrac{512\pi}{15}$ **13.** $\dfrac{32\pi}{5}$ **15.** $\dfrac{3456\pi}{35}$ **17.** $\dfrac{209}{6}$ **19.** 4 **21.** 8π **23.** $\dfrac{10\pi}{3}$

25. 345,000 ft lb **27.** $256\pi\rho g \approx 7.882 \times 10^6$ J **29.** 0.3 m **31.** 72π **33.** (a) 9.508 m **(b)** 2.171 m **35.** (a) -1 ft **(b)** -1 ft/min

AP® Review Problems: Chapter 8 (SSM = See Student Solutions Manual.)

1. A **3.** A **5.** B **7.** B **9.** SSM. **11.** SSM. **13.** SSM. **15.** SSM.

AP® Cumulative Review Problems: Chapters 1–8 (SSM = See Student Solutions Manual.)

1. C **3.** C **5.** A **7.** C **9.** C **11.** B **13.** SSM. **15.** SSM.

AP® Practice Exam: Calculus AB

Section 1: Multiple Choice:

1. D **2.** C **3.** A **4.** B **5.** B **6.** C **7.** B **8.** B **9.** D **10.** B **11.** B **12.** C **13.** D **14.** B **15.** D **16.** B **17.** C **18.** B
19. C **20.** D **21.** B **22.** B **23.** D **24.** B **25.** A **26.** C **27.** D **28.** B **29.** B **30.** D **31.** B **32.** D **33.** C **34.** A
35. B **36.** C **37.** A **38.** B **39.** C **40.** C **41.** B **42.** C **43.** C **44.** D **45.** B

Section 2: Free Response (SSM = See Student Solutions Manual.)

1. SSM. **2.** SSM. **3.** SSM. **4.** SSM. **5.** SSM. **6.** SSM.

Chapter 9

Section 9.1 (SSM = See Student Solutions Manual; AWV = Answers will vary.)

1. Plane curve; parameter **2.** (c) **3.** (d) **4.** Cycloid **5.** False **6.** True

7. (a) $y = \dfrac{1}{2}x + \dfrac{3}{2}$
(b)
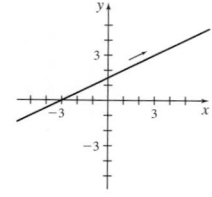

9. (a) $x = 2y - 3,\ 1 \le x \le 5$
(b)
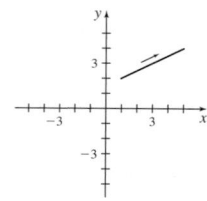

11. (a) $x = e^y$
(b)
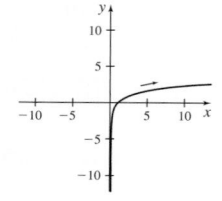

13. (a) $x^2 + y^2 = 1$
(b)
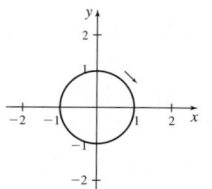

15. (a) $\dfrac{x^2}{4} + \dfrac{y^2}{9} = 1$ **17. (a)** $\dfrac{(x+3)^2}{4} + \dfrac{(y-1)^2}{4} = 1$ **19. (a)** $x = \dfrac{y^2}{4}$ **21. (a)** $y = \dfrac{x^2}{4} + 4$ **(c)** $x > 0, y > 4$

(b) 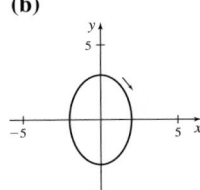   **(b)** 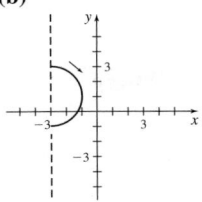   **(b)** 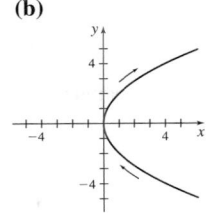   **(b)** 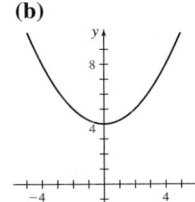   **(d)**

23. (a) $x = y^2 + 5$ **(c)** $x \geq 5, y \geq 0$ **25. (a)** $y = (x-1)^3$ **(c)** $x \geq 2, y \geq 1$

(b) 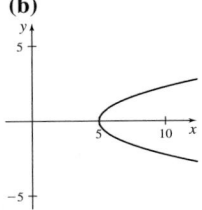   **(d)** 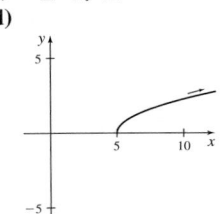   **(b)** 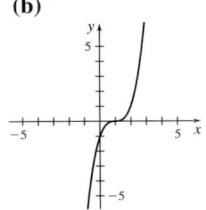   **(d)**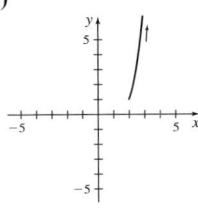

27. (a) $x = \sec \tan^{-1} y$ **(c)** $x \geq 1$ **29. (a)** $x = y^2$ **(c)** $x \geq 0, y \geq 0$

(b) 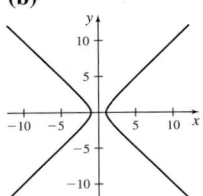   **(d)** 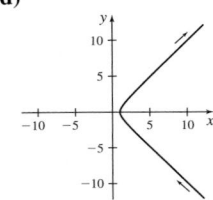   **(b)** 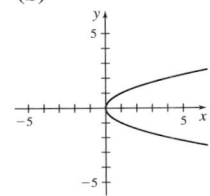   **(d)**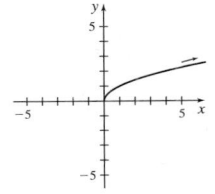

31. (a) $x = \left(\dfrac{y+1}{2}\right)^2$ **(c)** $x \geq 0$ **33. (a)** $x = \left(\dfrac{y+1}{2}\right)^2$ **(c)** $x > 0, y \neq -1$

(b)   **(d)**   **(b)**   **(d)**

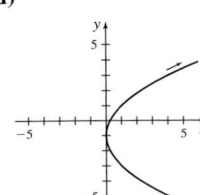   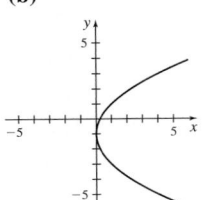

35. (a) $\dfrac{x+2}{3} + \dfrac{y^2}{4} = 1$ **(c)** $-2 \leq x \leq 1, -2 \leq y \leq 2$

(b)   **(d)**

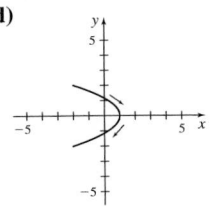

37. $y = \dfrac{2x}{1+x}, x \geq 0, 0 \leq y < 2$; AWV.   **39.** $y = 2\sqrt{x^2 - 4}, x \geq 2, y \geq 0$; AWV.

41. $(x+2)^2 + \left(\dfrac{4-y}{2}\right)^2 = 1, -3 \leq x \leq -1, 2 \leq y \leq 6$; AWV.   **43.** AWV.   **45.** AWV.   **47.** AWV.   **49.** AWV.   **51.** AWV.

53. AWV.   **55.** $x(t) = 3\cos\left(\dfrac{2\pi}{3}t\right), y(t) = 2\sin\left(\dfrac{2\pi}{3}t\right), 0 \leq t \leq 3$ **57.** $x(t) = 3\sin(\pi t), y(t) = 2\cos(\pi t), 0 \leq t \leq 2$

59. (a) 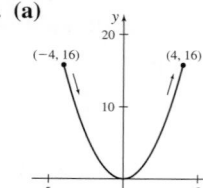 **(b)** 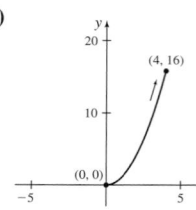 **(c)** 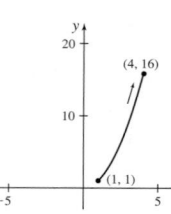 **(d)**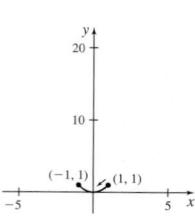

61. $I \rightarrow (d)$ counterclockwise, $II \rightarrow (a)$ counterclockwise, $III \rightarrow (b)$ counterclockwise, $IV \rightarrow (c)$ counterclockwise

63. $I \rightarrow (c)$ from $(1, 0)$ to $(-1, 0)$, $II \rightarrow (b)$ from $(-1, 0)$ to $(1, 0)$, $III \rightarrow (a)$ clockwise, $IV \rightarrow (d)$ from $\left(-\dfrac{\sqrt{2}}{2}, 1\right)$ to $(1, 0)$

65. (a) 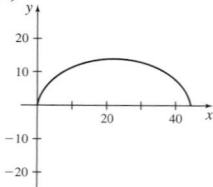 **(b)** $x \approx 8.66$, $y \approx 10.53$ **67.** AWV.

69. (a) $x(t) = (125 \cos 40°)t$, $y(t) = -16t^2 + (125 \sin 40°)t + 3$ **(b)** $y \approx 99.7$ ft
(c) $x \approx 191.5$ ft **(d)** $t \approx 3.13$ s **(e)** $y \approx 97.7$ ft **(f)** $t \approx 5.06$ s **(g)** $x \approx 484.5$ ft

71. (a) $x(t) = (80 \cos 35°)t$, $y(t) = -16t^2 + (80 \sin 35°)t + 6$ **(b)** $y \approx 35.9$ ft **(c)** $x \approx 65.5$ ft **(d)** $t \approx 1.83$ s **(e)** $y \approx 36.4$ ft

73. The first circle is counterclockwise from $(2, 0)$; the second is clockwise from $(0, 2)$. AWV.

75. $x = \dfrac{R(1 - m^2)}{1 + m^2}$, $y = \dfrac{2Rm}{1 + m^2}$ **77.** SSM. **79.** $x(t) = (a + b) \cos t - b \cos\left(\dfrac{a + b}{b}t\right)$, $y(t) = (a + b) \sin t - b \sin\left(\dfrac{a + b}{b}t\right)$

Section 9.2 (SSM = See Student Solutions Manual.)

1. (a) **2.** True **3.** Horizontal; vertical **4.** False **5.** False **6.** False **7.** $\dfrac{dy}{dx} = \dfrac{\sin t + \cos t}{\cos t - \sin t}$ **9.** $\dfrac{dy}{dx} = \dfrac{1}{1 - \dfrac{1}{t^2}}$

11. $\dfrac{dy}{dx} = \tan t$ **13.** $\dfrac{dy}{dx} = \dfrac{1}{2 \cot t}$

15. (a) $y = \dfrac{1}{8}x + 1$ **17. (a)** $y = \dfrac{4}{3}x - 3$ **19. (a)** $y = -\dfrac{1}{4}x + \dfrac{3}{4}$ **21. (a)** $y = -2x + 2$
(b) 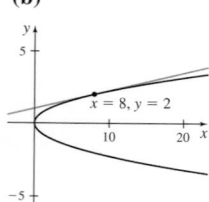 **(b)** 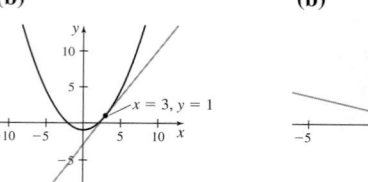 **(b)** **(b)**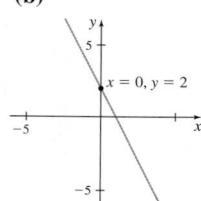

23. (a) $y = -x + 2$ **25. (a)** $y = -x + \sqrt{2}$
(b) 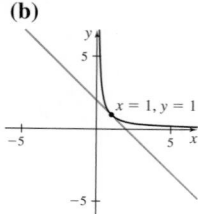 **(b)**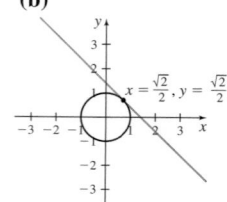

27. (a) $(-3, -2)$ **(b)** $(-4, 2)$ **(c)** $1 \le t \le 4$ **(d)** $0 \le t \le 1$ **(e)** $0 \le t \le 2$ **(f)** $2 \le t \le 4$

29. (a) $(1, 0)$ **(b)** $(0, 1)$ **(c)** The object never moves up. **(d)** $0 \le t \le \dfrac{\pi}{2}$ **(e)** The object never moves left. **(f)** $0 \le t \le \dfrac{\pi}{2}$

31. (a) $(0, 4), (\pi, -4), (2\pi, 4)$ **(b)** $(3, 0), (-3, 0)$ **(c)** $\pi \le t \le 2\pi$ **(d)** $0 \le t \le \pi$ **(e)** $\dfrac{\pi}{2} \le t \le \dfrac{3\pi}{2}$ **(f)** $0 \le t \le \dfrac{\pi}{2}$ and $\dfrac{3\pi}{2} \le t \le 2\pi$

33. (a) There are no horizontal tangent lines. **(b)** There are no vertical tangent lines. **(c)** The object never moves up. **(d)** $-1 \le t \le 4$
(e) The object never moves to the left. **(f)** $-1 \le t \le 4$

35. $\dfrac{8}{27}\left(10\sqrt{10}-1\right)$ **37.** $\sqrt{5}+\dfrac{1}{2}\ln(2+\sqrt{5})$ **39.** 4π **41.** 4π

43. (a) $s=\displaystyle\int_{0}^{2\pi}\sqrt{[-4\sin(2t)]^2+(2t)^2}\,dt$ **45. (a)** $s=\displaystyle\int_{-2}^{2}\sqrt{(2t)^2+\left(\dfrac{1}{2\sqrt{t+2}}\right)^2}\,dt$

(b) $s\approx 44.528$ **(b)** $s\approx 8.429$

(c) 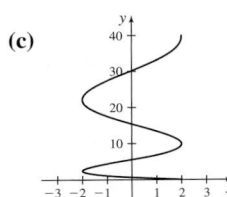 **(c)**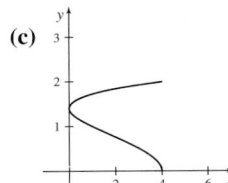

47. (a) Horizontal at $\left(\dfrac{10}{3},-\dfrac{16\sqrt{3}}{9}\right),\left(\dfrac{10}{3},\dfrac{16\sqrt{3}}{9}\right)$; vertical at $(2,0)$ **49. (a)** $s=\dfrac{3b}{2}$

(b) $t=2, t=-2$; SSM. **(b)** graphed using $b=1$

(c) $y=2x-12,\ y=-2x+12$

(d)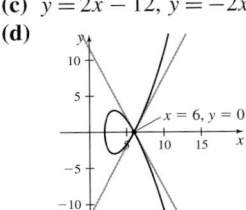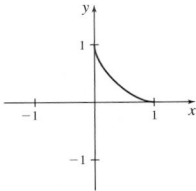

51. $s=5+\dfrac{9}{4}\ln 3$ **53.** $s=\dfrac{3\sqrt{11}-\sqrt{3}}{2}+\ln\left(\dfrac{3+\sqrt{11}}{1+\sqrt{3}}\right)$ **55.** $s=\sqrt{5}$ **57.** SSM. **59.** $\dfrac{1}{3a\cos^4\theta\sin\theta}$

AP® Practice Problems (SSM = See Student Solutions Manual.) **Retain Your Knowledge**

1. A **3.** C **5.** C **7.** B **9.** SSM. **1.** D **3.** D

Section 9.3 (SSM = See Student Solutions Manual.)

1. True **2.** (a), (c) **3.** True **4.** 3

5. (a) 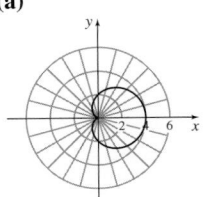 **7. (a)** 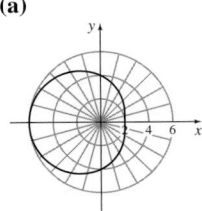 **9. (a)** 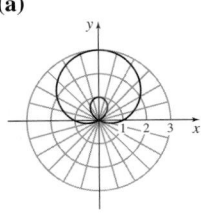 **11. (a)**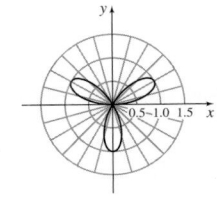

(b) $x=2(1+\cos\theta)\cos\theta,$ **(b)** $x=2(2-\cos\theta)\cos\theta,$ **(b)** $x=(1+2\sin\theta)\cos\theta,$ **(b)** $x=\sin(3\theta)\cos\theta,$
$y=2(1+\cos\theta)\sin\theta$ $y=2(2-\cos\theta)\sin\theta$ $y=(1+2\sin\theta)\sin\theta$ $y=\sin(3\theta)\sin\theta$

(c) $[\pi,2\pi]$ **(c)** $[0,\pi]$ **(c)** $\left[0,\dfrac{\pi}{2}\right],\left[\dfrac{3\pi}{2},2\pi\right]$ **(c)** $\left[0,\dfrac{\pi}{6}\right],\left[\dfrac{\pi}{2},\dfrac{5\pi}{6}\right]$
$\left[\dfrac{7\pi}{6},\dfrac{3\pi}{2}\right],\left[\dfrac{11\pi}{6},2\pi\right]$

13. 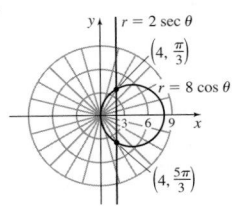 **15.** 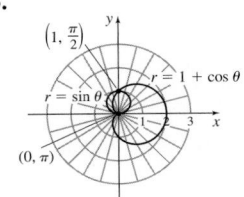 **17.**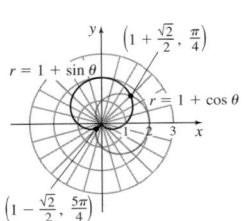

Points of intersection: Points of intersection: Points of intersection:

$\left(4,\dfrac{\pi}{3}\right),\left(4,\dfrac{5\pi}{3}\right)$ $(0,\pi),\left(1,\dfrac{\pi}{2}\right)$ $\left(1+\dfrac{\sqrt{2}}{2},\dfrac{\pi}{4}\right),\left(1-\dfrac{\sqrt{2}}{2},\dfrac{5\pi}{4}\right),(0,\theta)$ (the pole)

19. $\sqrt{5}(e-1)$ **21.** 2 **23.** $r=3+3\cos\theta$ **25.** $r=4+\sin\theta$ **27.** $y=\dfrac{\sqrt{3}}{3}x$

29. $y=\left(\sqrt{2}-2\right)x+\dfrac{5\sqrt{2}}{2}-\dfrac{1}{2}$ **31.** $y=-\left(1+\dfrac{5\sqrt{2}}{4}\right)x+\dfrac{57\sqrt{2}}{8}+10$

33. (a) 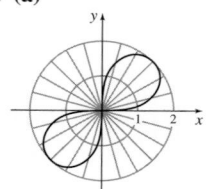 **35. (a)** 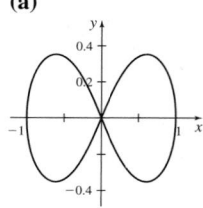 **37. (a)** 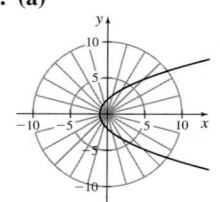 **39. (a)** 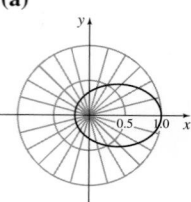 **41. (a)**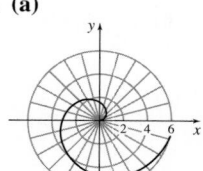

(b) $x=2\sqrt{\sin(2\theta)}\cos\theta$, $y=2\sqrt{\sin(2\theta)}\sin\theta$

(b) $x=\sqrt{\cos(2\theta)}\cos\theta$, $y=\sqrt{\cos(2\theta)}\sin\theta$

(b) $x=\dfrac{2\cos\theta}{1-\cos\theta}$, $y=\dfrac{2\sin\theta}{1-\cos\theta}$

(b) $x=\dfrac{\cos\theta}{3-2\cos\theta}$, $y=\dfrac{\sin\theta}{3-2\cos\theta}$

(b) $x=\theta\cos\theta$, $y=\theta\sin\theta$

43. (a) 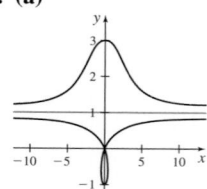 **45. (a)** 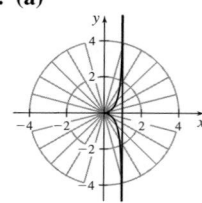 **47. (a)**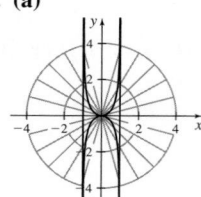

(b) $x=(\csc\theta-2)\cos\theta$, $y=(\csc\theta-2)\sin\theta$

(b) $x=\sin^2\theta$, $y=\sin^2\theta\tan\theta$

(b) $x=\sin\theta$, $y=\sin\theta\tan\theta$

49. SSM. **51.** $\pi\sqrt{1+4\pi^2}+\dfrac{1}{2}\ln\left(2\pi+\sqrt{4\pi^2+1}\right)$ **53.** 8 **55.** AWV.

57. Horizontal: $y=\dfrac{9\sqrt{3}}{4}$, $y=-\dfrac{9\sqrt{3}}{4}$; vertical: $x=-\dfrac{3}{4}$, $x=6$

AP® Practice Problems (SSM = See Student Solutions Manual.) Retain Your Knowledge

1. A **3.** A **5.** C **7.** SSM. **1.** C **3.** A

Section 9.4

1. $\dfrac{1}{2}r^2\theta$ **2.** True **3.** False **4.** True **5.** $\dfrac{\pi}{4}$ **7.** $8+3\pi$ **9.** $\dfrac{3}{16}\left(3\sqrt{3}+4\pi\right)$ **11.** $\dfrac{4\pi^3a^2}{3}$ **13.** $\dfrac{3\pi}{2}$ **15.** $\dfrac{19\pi}{2}$ **17.** 16π

19. 2π **21.** $\dfrac{4}{3}\pi$ **23.** $\dfrac{\sqrt{3}}{2}+\dfrac{\pi}{3}$ **25.** $1-\dfrac{\pi}{4}$ **27.** $\pi-\dfrac{3\sqrt{3}}{2}$ **29.** $\sqrt{3}+\dfrac{4\pi}{3}-4\ln\left(2+\sqrt{3}\right)$ **31.** $\dfrac{3\pi}{2}$ **33.** $4\sqrt{3}+\dfrac{32\pi}{3}$

35. $\dfrac{9\sqrt{3}}{2}-\pi$ **37.** $\dfrac{7\pi}{12}-\sqrt{3}$ **39.** $2\sqrt{2}$ **41.** $\dfrac{1-e^{-2}}{4}$ **43.** $\dfrac{\pi-1}{2\pi}$ **45.** $\sqrt{3}+\dfrac{4\pi}{3}-4\ln\left(2+\sqrt{3}\right)$

AP® Practice Problems (SSM = See Student Solutions Manual.) Retain Your Knowledge

1. D **3.** B **5.** SSM. **7.** SSM. **1.** C **3.** A

Section 9.5

1. $\{t\,|\,t\le 4\}$ **2.** False **3.** True **4.** True **5.** True **6.** True

7. (a) $\langle 1,-2\rangle$ **(b)** $\langle 1,-2\rangle$ **(c)** $\mathbf{r}$ is continuous at $t_0=1$. **9. (a)** $\left\langle\dfrac{\sqrt{2}}{2},\dfrac{\sqrt{2}}{2}\right\rangle$ **(b)** $\left\langle\dfrac{\sqrt{2}}{2},\dfrac{\sqrt{2}}{2}\right\rangle$ **(c)** $\mathbf{v}$ is continuous at $t_0=\dfrac{\pi}{4}$.

11. (a) $\langle 1,0\rangle$ **(b)** $\langle 1,0\rangle$ **(c)** $\mathbf{u}$ is continuous at $t_0=0$. **13. (a)** $\langle 4,1\rangle$ **(b)** $\langle 4,1\rangle$ **(c)** $\mathbf{g}$ is continuous at $t_0=4$.

15. All real numbers **17.** $\{t\,|\,t\ge 0\}$ **19.** $\{t\,|\,t>0\}$ **21.** $\{t\,|\,t>1\}$

23.
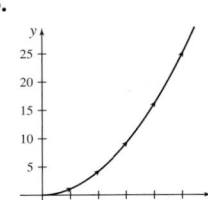

25.
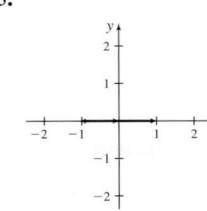

27.
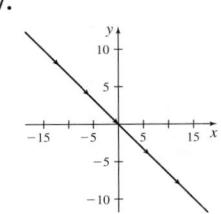

29.
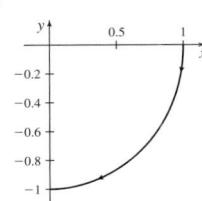

31.
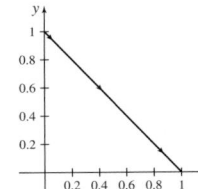

33. D **35.** C **37.** $\{t\,|\,t \neq -1, 1\}$ **39.** $\left\{t\,|\,t \neq (2k+1)\dfrac{\pi}{2}\right\}$, k an integer **41.** $\{t\,|\,t \neq -1\}$

43. $\mathbf{r}'(t) = \langle 8t, -6t^2 \rangle$ and $\mathbf{r}''(t) = \langle 8, -12t \rangle$ **45.** $\mathbf{r}'(t) = \left\langle \dfrac{2\sqrt{t}}{t}, 2e^t \right\rangle$ and $\mathbf{r}''(t) = \left\langle -\dfrac{1}{t^{3/2}}, 2e^t \right\rangle$

47. $\mathbf{r}'(t) = \langle 2t - 1, 3t^2 \rangle$ and $\mathbf{r}''(t)\,\langle 2, 6t \rangle$ **49.** $\mathbf{r}'(t) = \langle \sin(2t), \sin(2t) \rangle$ and $\mathbf{r}''(t) = \langle 2\cos(2t), 2\cos(2t) \rangle$

51. $\mathbf{r}'(t) = \left\langle 5(2t+1)\,e^{t^2+t}, \dfrac{2t+1}{t^2+t} \right\rangle$; $\mathbf{r}''(t) = \left\langle [10 + 5(2t+1)^2]\,e^{t^2+t}, -\dfrac{2t^2+2t+1}{t^2(t+1)^2} \right\rangle$

53. $\mathbf{r}'(t) = \langle e^t \cos t - e^t \sin t, e^t \sin t + e^t \cos t \rangle$ and $\mathbf{r}''(t) = \langle -2e^t \sin t, 2e^t \cos t \rangle$

55. (a) $\mathbf{r}'(1) = \langle 1, -2 \rangle$ **57. (a)** $\mathbf{r}'(1) = \langle 2, -1 \rangle$ **59. (a)** $\mathbf{r}'(-1) = \langle 1, -3 \rangle$ **61. (a)** $\mathbf{r}'\left(\dfrac{\pi}{4}\right) = \langle -2, 0 \rangle$

(b) and **(c)** **(b)** and **(c)** **(b)** and **(c)** **(b)** and **(c)**

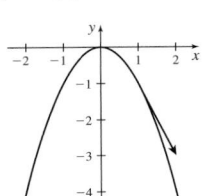
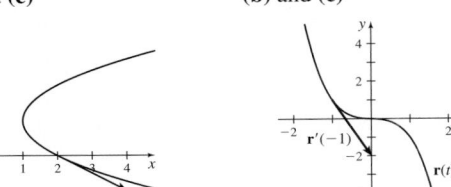
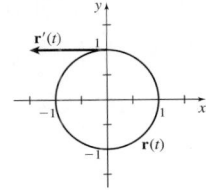

63. (a) $\mathbf{r}'(1) = \left\langle 4, -\dfrac{1}{2} \right\rangle$ **65. (a)** $\mathbf{r}'(0) = \langle 1, -1 \rangle$ **67.** $s = \dfrac{1}{27}\left(76^{3/2} - 13^{3/2}\right)$ **69.** $s = \dfrac{33}{2}$ **71.** $s = 2\pi$

(b) and **(c)** **(b)** and **(c)** **73.** $s = \sqrt{5}$ **75.** $s = \sqrt{2}(e^5 - 1)$

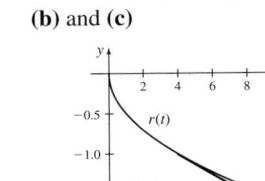
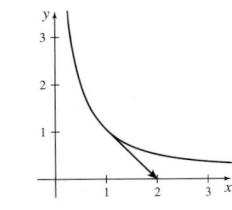

77. $\langle -\cos t, \sin t \rangle + \mathbf{c}$ **79.** $\left\langle \dfrac{1}{3}t^3, -\dfrac{1}{2}t^2 \right\rangle + \mathbf{c}$ **81.** $\left\langle t\ln t - t, \dfrac{t^2}{4} - \dfrac{t^2}{2}\ln t \right\rangle + \mathbf{c}$

83. $\left\langle \dfrac{t^2}{2} - 2t, -\dfrac{1}{3}(t-2)^3 \right\rangle + \mathbf{c}$ **85.** $\langle 1, 27 \rangle$ **87.** $\left\langle -\dfrac{1}{3}, 1 \right\rangle$ **89.** $\langle 20, 68 \rangle$ **91.** $\left\langle \dfrac{1}{2}(e-1), e-1 \right\rangle$

93. $\mathbf{r}(t) = \langle e^t - e, t - t\ln t \rangle$ **95.** $\mathbf{r}(t) = \langle 3 - 2\cos t, \sin t - 1 \rangle$ **97.** $\mathbf{r}(t) = \left\langle \ln t + 1, \dfrac{1}{2}(t^2 + 1) \right\rangle$

99. $\dfrac{d\mathbf{u}}{dt} = \langle -\omega \sin(\omega t), \omega \cos(\omega t) \rangle$ and $\left\|\dfrac{d\mathbf{u}}{dt}\right\| = |\omega|$ **101.** SSM.

AP® Practice Problems **Retain Your Knowledge**

1. B **3.** C **5.** B **7.** A **1.** C **3.** D

Section 9.6 (SSM = See Student Solutions Manual.)

1. True **2.** False

3. (a) $\mathbf{v}(t) = \langle 2, 1 \rangle$, $\mathbf{a}(t) = \langle 0, 0 \rangle$,
$v(t) = \sqrt{5}$

(b)

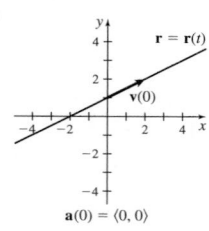

$\mathbf{a}(0) = \langle 0, 0 \rangle$

5. (a) $\mathbf{v}(t) = \langle 2t, -1 \rangle$, $\mathbf{a}(t) = \langle 2, 0 \rangle$, and
$v(t) = \sqrt{4t^2 + 1}$

(b)

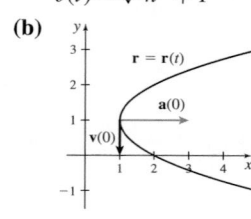

7. (a) $\mathbf{v}(t) = \langle 4, -3t^2 \rangle$, $\mathbf{a}(t) = \langle 0, -6t \rangle$, and
$v(t) = \sqrt{16 + 9t^4}$

(b)

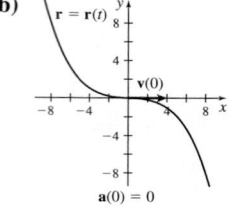

$\mathbf{a}(0) = 0$

9. (a) $\mathbf{v}(t) = \langle 3t^2, 3 \rangle$,
$\mathbf{a}(t) = \langle 6t, 0 \rangle$,
$v(t) = 3\sqrt{t^4 + 1}$

(b)

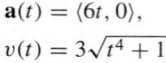

$\mathbf{a}(0) = \langle 0, 0 \rangle$

11. (a) $\mathbf{v}(t) = \langle 2\cos t, -\sin t \rangle$,
$\mathbf{a}(t) = \langle -2\sin t, -\cos t \rangle$,
$v(t) = \sqrt{3\cos^2 t + 1}$

(b)

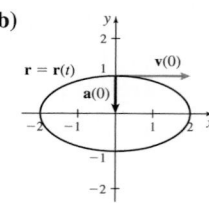

13. (a) $\mathbf{v}(t) = \langle e^t, 2e^{2t} \rangle$,
$\mathbf{a}(t) = \langle e^t, 4e^{2t} \rangle$,
$v(t) = e^t \sqrt{1 + 4e^{2t}}$

(b)

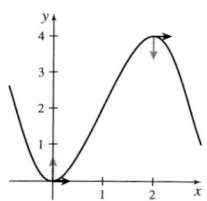

$\mathbf{v}(0) = \langle 1, 2 \rangle$ $\mathbf{a}(0) = \langle 1, 4 \rangle$

15. (a) $\mathbf{v}(t) = \langle e^t, -e^{-t} \rangle$ **(b)** $\|\mathbf{a}(t)\| = \sqrt{e^{2t} + e^{-2t}}$ **17. (a)** $\mathbf{v}(t) = \langle 1, e^t \rangle$ **(b)** $\|\mathbf{a}(t)\| = e^t$

19. (a) $\mathbf{v}(t) = \langle 6\cos(2t), -6\sin(2t) \rangle$ **(b)** $\|\mathbf{a}(t)\| = 12$ **21. (a)** $\mathbf{v}(t) = \langle -2\sin t, -3\cos t \rangle$ **(b)** $\|\mathbf{a}(t)\| = \sqrt{5\sin^2 t + 4}$

23. $\mathbf{v}(t) = \langle \sin t + 1, 1 - \cos t \rangle$, $v(t) = \sqrt{3 + 2\sin t - 2\cos t}$, and $\mathbf{r}(t) = \langle t - \cos t + 1, t - \sin t + 1 \rangle$

25. $\mathbf{v}(t) = \langle 2 - e^{-t}, t + 1 \rangle$, $v(t) = \sqrt{e^{-2t} - 4e^{-t} + t^2 + 2t + 5}$, and $\mathbf{r}(t) = \left\langle 2t + e^{-t}, \frac{1}{2}t^2 + t - 1 \right\rangle$

27. $\mathbf{v}(t) = \langle \pi - \pi\cos(\pi t), \pi\sin(\pi t) \rangle$, $\mathbf{a}(t) = \langle \pi^2 \sin(\pi t), \pi^2 \cos(\pi t) \rangle$ **29.** $\mathbf{a}(0) = \left\langle 0, \frac{2}{3} \right\rangle$ and $\mathbf{a}(6) = \left\langle 0, -\frac{2}{3} \right\rangle$

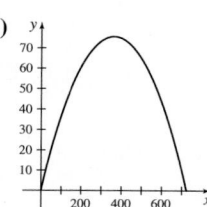

31. (a) 350 N **(b)** 35 km/h slower **33.** 2.118×10^{13} m N **35. (a)** $\mathbf{v}(t) = \left\langle -\dfrac{4t}{(t^2 + 1)^2}, -\dfrac{2(t^2 - 1)}{(t^2 + 1)^2} \right\rangle$ **(b)** No. AWV. **(c)** $(-1, 0)$

37. $4m\sqrt{2}$ **39.** The range $\approx 23{,}895$ m, the time of flight ≈ 53 s, and the greatest height reached ≈ 3449 m.

41. (a) $x = \dfrac{1200t}{13}$ and $y = -4.9t^2 + \dfrac{500t}{13}$ **(b)** Range ≈ 724.6 m. **(c)** Time of flight ≈ 7.8 s. **(d)**

43. (a) The projectile travels approximately 152.5 ft up the hill. **(b)** The projectile is in the air for approximately 1.9 s.

45. (a) $\mathbf{v}(t) = \langle 5t, -9.8t + 3 \rangle$ and $v(t) = \sqrt{121.04t^2 - 58.8t + 9}$ **(b)** $\mathbf{r}(t) = \left\langle \dfrac{5}{2}t^2, -4.9t^2 + 3t \right\rangle$

(c) ≈ 0.612 s **(d)**

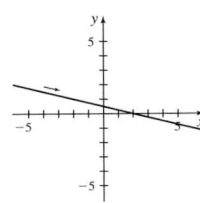

47. $v_0 \approx 54.521$ ft/s **49.** The initial speed of the ball is approximately 114 ft/s. It took the ball approximately 5 s to reach the vines.

51. (a) Range ≈ 114.342 ft **(b)** ≈ 93 ft **53.** $v \approx 3750$ m/s **55.** SSM.

AP® Practice Problems (SSM = See Student Solutions Manual.)

1. B **3.** A **5.** SSM. **7.** SSM.

Retain Your Knowledge

1. D **3.** B

Review Exercises (SSM = See Student Solutions Manual.)

1. (a) $x = -4y + 2$ **3. (a)** $y = \dfrac{1}{x}$ **5. (a)** $y + 1 = x$ **7. (a)** $y = \dfrac{1}{2}x + \dfrac{5}{2}$ **9. (a)** $y = -\dfrac{81}{2\sqrt{10}}x + \dfrac{9}{2\sqrt{10}} + \sqrt{10}$

(b) **(b)** **(b)** **(b)** **(b)**

(c) No restrictions **(c)** $x > 0, y > 0$ **(c)** $1 \le x \le 2,$ $0 \le y \le 1$

11. AWV. **13.** $\dfrac{x^2}{4} + y^2 = 1$ **15. (a)** **(b)** $x = 4\cos(2\theta)\cos\theta,\ y = 4\cos(2\theta)\sin\theta$

17. (a) $\dfrac{d^2y}{dx^2} = \dfrac{3t^2 + 3}{4t^3}$ **(b)** $\dfrac{15}{32}$ **19.** The object moves closer to the pole when $\dfrac{\pi}{2} \le \theta \le \dfrac{3\pi}{2}$.

21. (a) 901.196 units **(b)** The object never moves to the left. **(c)** The object moves up when $1 \le t \le 10$.

23. (a) 1004.574 units **(b)** The object is always moving to the left. **(c)** The object moves up when $2 \le t \le 10$.

25. $x = 4\sin\left(\dfrac{2\pi t}{5}\right),\ y = 3\cos\left(\dfrac{2\pi t}{5}\right),\ 0 \le t \le 5$ **27.** $\dfrac{25\pi}{2}$ **29.** $2\sqrt{3} - \dfrac{2\pi}{3}$

31. $\dfrac{\sqrt{13}}{6} + \dfrac{3}{4}\ln\dfrac{\sqrt{13}+2}{3}$ **33.** $\dfrac{3}{2} + \dfrac{\ln 2}{4}$ **35.** $\sqrt{2}(1 - e^{-2\pi})$ **37.** $2\left(4 - 2\sqrt{2}\right)$

39. (a) All real numbers. **(b)** **41. (a)** All real numbers. **(b)**
(c) $\mathbf{r}'(2) = \langle 4, 3 \rangle$ **(c)** $\mathbf{r}'(0) = \langle 1, 0 \rangle$

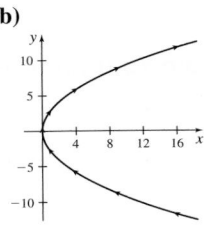

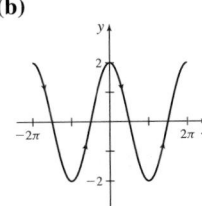

43. Discontinuous at $t = -1$ **45.** $\mathbf{r}'(t) = \langle -2\sin t, -3\sin t \rangle$ and $\mathbf{r}''(t) = \langle -2\cos t, -3\cos t \rangle = -\mathbf{r}(t)$ **47. (a)** $\mathbf{r}'(0) = \langle 1, 2 \rangle$

49. $\langle 1, 1 \rangle$ **51.** 56 **53.** $\mathbf{v}(t) = \langle -2\sin t, \cos t \rangle$; $\mathbf{a}(t) = \langle -2\cos t, -\sin t \rangle$; $v(t) = \sqrt{3\sin^2 t + 1}$

55. $\mathbf{v}(t) = \langle e^t, -e^{-t} \rangle$; $\mathbf{a}(t) = \langle e^t, e^{-t} \rangle$; $v(t) = \sqrt{e^{2t} + e^{-2t}}$ **57.** $\mathbf{v}(t) = \langle \pi - \pi \cos(\pi t), \pi \sin(\pi t) \rangle$; $\mathbf{a}(t) = \langle \pi^2 \sin(\pi t), \pi^2 \cos(\pi t) \rangle$

59. $\left\langle \dfrac{1}{3}t^3 - 2t, -\dfrac{(t-2)^3}{3} \right\rangle + \mathbf{c}$ **61.** $\mathbf{r}(t) = \left\langle \dfrac{1}{2}e^{2t} - \dfrac{1}{2}e^2, t \ln t - t + 2 \right\rangle$ **63.** $\mathbf{r}(t) = \langle \sin t + 1, \cos t - 1 \rangle$ **65.** $\approx 25{,}540$ m away

AP® Review Problems: Chapter 9   (SSM = See Student Solutions Manual.)

1. B   **3.** C   **5.** C   **7.** B   **9.** A   **11.** C   **13.** C   **15.** D   **17.** C   **19.** B   **21.** D   **23.** SSM.   **25.** SSM.

AP® Cumulative Review Problems: Chapters 1–9   (SSM = See Student Solutions Manual.)

1. B   **3.** A   **5.** B   **7.** B   **9.** A   **11.** A   **13.** SSM.

Chapter 10

Section 10.1   (SSM = See Student Solutions Manual; AWV = Answers will vary.)

1. False   **2.** False   **3.** True   **4.** (b)   **5.** False   **6.** True   **7.** False   **8.** False   **9.** False   **10.** (b)   **11.** True   **12.** False

13. $2, \dfrac{3}{2}, \dfrac{4}{3}, \dfrac{5}{4}$ **15.** $0, \ln 2, \ln 3, \ln 4$ **17.** $\dfrac{1}{3}, -\dfrac{1}{5}, \dfrac{1}{7}, -\dfrac{1}{9}$ **19.** $1, -1, 1, -1$ **21.** $\dfrac{1}{2}, \dfrac{1}{2}, \dfrac{3}{4}, \dfrac{3}{2}$ **23.** $a_n = 2n$ **25.** $a_n = 2^n$

27. $a_n = \dfrac{(-1)^{n+1}}{n+1}$ **29.** $a_n = \dfrac{n}{n+1}$ **31.** $a_n = (n-1)!$ **33.** 0   **35.** 1   **37.** 4   **39.** 0   **41.** 0   **43.** 1   **45.** $\ln \dfrac{1}{3}$ **47.** e^{-2}

49. 0   **51.** 1   **53.** $-\dfrac{1}{2}$ **55.** 1   **57.** 0   **59.** 0   **61.** 0   **63.** Diverges   **65.** Diverges   **67.** Converges   **69.** Diverges

71. Converges   **73.** Bounded from above and from below   **75.** Bounded from below   **77.** Bounded from below

79. Bounded from above and from below   **81.** Nonmonotonic   **83.** Nonmonotonic   **85.** Monotonic decreasing

87. Nonmonotonic   **89.** Decreasing, bounded from below   **91.** Increasing, bounded from above

93. Increasing, bounded from above   **95.** Converges; 6   **97.** Converges; $\ln \dfrac{1}{3}$ **99.** Diverges   **101.** Converges; 0

103. Converges; 0   **105.** Converges; 0   **107.** Converges; 1   **109.** Converges; 1   **111.** Converges; 1   **113.** Converges; 1

115. 0   **117.** Converges   **119.** Converges   **121.** Converges   **123.** Diverges   **125.** Converges   **127.** Converges

129. $p_n = 3000r^n + h\dfrac{r^n - 1}{r - 1}$; Converges; $\lim\limits_{n \to \infty} p_n = \dfrac{h}{1-r}$ **131.** (a) $I_n = 0.95^n I_0$ (b) 77   **133.** e^2 **135.** e^3

137. SSM.   **139.** SSM.   **141.** SSM.   **143.** SSM.   **145.** AWV.   **147.** SSM.   **149.** Diverges

151. SSM.   **153.** SSM.   **155.** SSM.   **157.** SSM.   **159.** SSM.

Section 10.2   (SSM = See Student Solutions Manual.)

1. (b)   **2.** (d)   **3.** True   **4.** False   **5.** $\dfrac{a}{1-r}$ **6.** False   **7.** $\dfrac{175}{64}$ **9.** 10   **11.** $\dfrac{1}{3}$ **13.** $-\dfrac{1}{3}$ **15.** $\dfrac{1}{2}$ **17.** Diverges

19. Converges; 6   **21.** Converges; $\dfrac{21}{2}$ **23.** Converges; $\dfrac{50}{69}$ **25.** Converges; 6   **27.** Converges; $\dfrac{1}{3}$ **29.** Diverges

31. Diverges   **33.** Converges; $\dfrac{3}{2}$ **35.** Diverges   **37.** Converges; $\dfrac{1}{42}$ **39.** Diverges   **41.** Converges; $\dfrac{1}{99}$ **43.** Diverges

45. Converges; $\dfrac{2}{3}$ **47.** Diverges   **49.** Diverges   **51.** Diverges   **53.** Converges; $-\dfrac{1}{4}$ **55.** Diverges   **57.** Converges; $\sin 1$

59. $\dfrac{5}{9}$ **61.** $\dfrac{3857}{900}$ **63.** 90 ft   **65.** (a) $\dfrac{h}{1-r}$ (b) $h = 2000$ **67.** (a) $P\dfrac{1+r}{r-i}$ (b) \$6.67   **69.** SSM.

71. (a) $T(n) = p \sum\limits_{k=1}^{n} (1-e)^{k-1}$ (b) $\lim\limits_{n \to \infty} T(n) = \dfrac{p}{e}$ (c) $e_{min} = \dfrac{p}{L}$ (d) $e_{min} = \dfrac{2}{3}$; $T(365) = 150$ kg

73. SSM.   **75.** $n = 11$ **77.** SSM.   **79.** SSM.   **81.** SSM.   **83.** SSM.   **85.** SSM.   **87.** SSM.

AP® Practice Problems   (SSM = See Student Solutions Manual.)   ### Retain Your Knowledge

1. D   **3.** D   **5.** B   **7.** SSM.   **1.** C   **3.** D

Section 10.3   (SSM = See Student Solutions Manual; AWV = Answers will vary.)

1. (a)   **2.** False   **3.** True   **4.** True   **5.** False   **6.** True   **7.** True   **8.** False   **9.** $p > 1; 0 < p \le 1$ **10.** Converges

11. Diverges   **12.** False   **13.** Diverges   **15.** Diverges   **17.** Diverges   **19.** Converges   **21.** Diverges   **23.** Converges

25. Converges   **27.** Diverges   **29.** Converges   **31.** Diverges   **33.** Converges   **35.** Converges   **37.** Converges

39. Diverges   **41.** Diverges   **43.** Diverges   **45.** Diverges   **47.** Diverges   **49.** Diverges   **51.** Diverges   **53.** Diverges

55. $1 < \sum\limits_{k=1}^{\infty} \dfrac{1}{k^2} < 2$ **57.** $\dfrac{1}{e-1} < \sum\limits_{k=1}^{\infty} \dfrac{1}{k^e} < \dfrac{e}{e-1}$ **59.** $2 < 1 + \dfrac{1}{2\sqrt{2}} + \dfrac{1}{3\sqrt{3}} + \dfrac{1}{4\sqrt{4}} + \cdots < 3$ **61.** $\dfrac{1}{2e} < \sum\limits_{k=1}^{\infty} k e^{-k^2} < \dfrac{3}{2e}$

63. $\dfrac{1}{3}\left(\dfrac{\pi}{2} - \tan^{-1}\dfrac{1}{3}\right) < \sum\limits_{k=1}^{\infty} \dfrac{1}{k^2+9} < \dfrac{1}{10} + \dfrac{1}{3}\left(\dfrac{\pi}{2} - \tan^{-1}\dfrac{1}{3}\right)$ **65.** SSM. **67.** SSM. **69.** $\pi^2 \approx 9.8099$ **71.** Converges

73. Converges **75.** SSM. **77.** SSM. **79.** SSM. **81.** SSM. **83.** $(-\infty, -1)$

85. **(a)** 1.1975 **(b)** Upper is 1.5, lower is 0.5. **(c)** AWV.

AP® Practice Problems   (SSM = See Student Solutions Manual.)   Retain Your Knowledge

1. C **3.** B **5.** C **7.** SSM.   **1.** D **3.** B

Section 10.4   (SSM = See Student Solutions Manual.)

1. **(b)** **2.** False **3.** False **4.** False **5.** Converges **7.** Converges **9.** Diverges **11.** Converges **13.** Converges

15. Converges **17.** Diverges **19.** Converges **21.** Converges **23.** Diverges **25.** Converges **27.** Converges **29.** Converges

31. Converges **33.** Converges **35.** Converges **37.** Diverges **39.** Converges **41.** Diverges **43.** Converges **45.** Diverges

47. Converges **49.** Converges **51.** Diverges **53.** Diverges **55.** Converges **57.** SSM. **59.** SSM. **61.** Diverges

63. Converges **65.** SSM. **67.** Diverges **69.** Converges **71.** SSM. **73.** SSM. **75.** Diverges **77.** SSM.

AP® Practice Problems   (SSM = See Student Solutions Manual.)   Retain Your Knowledge

1. B **3.** SSM.   **1.** B **3.** A

Section 10.5   (SSM = See Student Solutions Manual.)

1. False **2.** False **3.** False **4.** False **5.** False **6.** True **7.** Converges **9.** Diverges **11.** Diverges **13.** Converges

15. Diverges **17.** Diverges **19.** Converges **21.** Diverges **23.** **(a)** SSM. **(b)** 3 terms **(c)** $\dfrac{7565}{7776} \approx 0.973$

25. **(a)** SSM. **(b)** 6 terms **(c)** $\dfrac{91}{144} \approx 0.632$ **27.** **(a)** SSM. **(b)** 9 terms **(c)** $\dfrac{171}{512} \approx 0.334$

29. 0.8611; upper estimate to the error is 0.0625 **31.** 0.9498; upper estimate to the error is 0.0039

33. 0.7222; upper estimate to the error is 0.00617 **35.** 0.3218; upper estimate to the error is 6.53×10^{-5}

37. SSM. **39.** SSM. **41.** Absolutely convergent **43.** Diverges **45.** Absolutely convergent

47. Absolutely convergent **49.** Conditionally convergent **51.** Absolutely convergent **53.** Conditionally convergent

55. Absolutely convergent **57.** Diverges **59.** Absolutely convergent **61.** **(a)** SSM. **(b)** SSM. **(c)** $\dfrac{1}{2}$

63. SSM. **65.** SSM. **67.** SSM. **69.** Absolutely convergent for $|r| < 1$; divergent if $|r| \ge 1$ **71.** SSM. **73.** SSM.

75. SSM. **77.** Absolutely convergent **79.** Absolutely convergent if $2 < p < 3$; conditionally convergent if $p \ge 3$

81. Absolutely convergent **83.** Conditionally convergent **85.** SSM.

AP® Practice Problems   (SSM = See Student Solutions Manual.)   Retain Your Knowledge

1. C **3.** B **5.** SSM.   **1.** D **3.** A

Section 10.6   (SSM = See Student Solutions Manual; AWV = Answers will vary.)

1. False **2.** False **3.** False **4.** False **5.** Converges **7.** Converges **9.** Converges **11.** Converges **13.** Diverges

15. Converges **17.** Converges **19.** Converges **21.** Diverges **23.** Diverges **25.** Converges **27.** Diverges

29. Converges **31.** Diverges **33.** Converges **35.** Diverges **37.** The Root Test provides no information.

39. Diverges **41.** Converges **43.** Converges **45.** Converges **47.** Converges **49.** Converges **51.** Converges

53. Converges **55.** Converges **57.** SSM. **59.** AWV. **61.** **(a)** SSM. **(b)** $\dfrac{1-e^3}{e^3}$ **63.** SSM.

65. Converges. **67.** SSM. **69.** SSM. **71.** SSM.

AP® Practice Problems   Retain Your Knowledge

1. C **3.** A.   **1.** B **3.** C

Section 10.7 (SSM = See Student Solutions Manual.)

1. False **2.** False **3.** True **4.** False **5.** False **6.** a_k, b_k **7.** Diverges by the Test for Divergence

9. Absolutely convergent by the Geometric Series Test **11.** Absolutely convergent by the Limit Comparison Test

13. Diverges by the Comparison Test for Divergence **15.** Diverges by the Ratio Test

17. Diverges by the Test for Divergence **19.** Conditionally convergent by the Alternating Series Test

21. Diverges by the Ratio Test **23.** Absolutely convergent by the Ratio Test **25.** Diverges by the Limit Comparison Test

27. Absolutely convergent by the Comparison Test for Convergence **29.** Diverges by the Test for Divergence

31. Absolutely convergent by the Root Test **33.** Absolutely convergent by the Root Test

35. Absolutely convergent by the Integral Test **37.** Diverges by the Root Test

39. Absolutely convergent by the Comparison Test for Convergence **41.** Converges; $\dfrac{11}{6}$ **43.** Converges by the Ratio Test

45. (a) SSM. **(b)** $\dfrac{14}{5}$ **47.** Converges by the Comparison Test

Section 10.8 (SSM = See Student Solutions Manual; AWV = Answers will vary.)

1. True **2.** True **3.** True **4.** False **5.** True **6.** False **7.** False **8.** True **9.** True **10.** False **11.** False **12.** False

13. $-1 < x < 1$ **15.** $-4 < x < 2$ **17. (a, b)** $R = 2; [-2, 2)$ **(c)** AWV. **19. (a, b)** $R = 1; [-1, 1)$ **(c)** AWV.

21. (a, b) $R = 3; (-3, 3)$ **(c)** AWV. **23. (a, b)** $R = 1; (-1, 1)$ **(c)** AWV.

25. (a, b) $R = 1; (2, 4)$ **(c)** AWV. **27.** $R = 1; [-1, 1]$ **29.** $R = 1; [1, 3]$

31. $R = \infty; (-\infty, \infty)$ **33.** $R = 1; (-1, 1)$ **35.** $R = 4; (-4, 4)$ **37.** $R = 3; (0, 6)$ **39.** $R = \infty; (-\infty, \infty)$

41. $R = \infty; (-\infty, \infty)$ **43.** $R = \dfrac{1}{e}; \left[-\dfrac{1}{e}, \dfrac{1}{e} \right)$; series does converge at $x = -\dfrac{1}{e}$

45. (a) $-3 < x < 3$ **(b)** $f(2) = 3, f(-1) = \dfrac{3}{4}$ **(c)** $f(x) = \dfrac{3}{3-x}$ **47. (a)** $0 < x < 4$ **(b)** $f(1) = \dfrac{2}{3}, f(2) = 1$ **(c)** $f(x) = \dfrac{2}{4-x}$

49. Converges at $x = 2$; no information about $x = 5$ **51. (a)** True **(b)** False **(c)** False **(d)** False **(e)** True **(f)** True

53. (a) $f(x) = \displaystyle\sum_{k=0}^{\infty} (-1)^k x^{3k}$ **(b)** $R = 1; (-1, 1)$ **55. (a)** $f(x) = \dfrac{1}{6} \displaystyle\sum_{k=0}^{\infty} \left(\dfrac{x}{3} \right)^k$ **(b)** $R = 3; (-3, 3)$

57. (a) $f(x) = \displaystyle\sum_{k=0}^{\infty} (-1)^k x^{3k+1}$ **(b)** $R = 1; (-1, 1)$ **59. (a)** $f'(x) = \displaystyle\sum_{k=0}^{\infty} \dfrac{(-1)^k x^{2k}}{(2k)!}$ **(b)** $\displaystyle\int_0^x f(t)\,dt = \displaystyle\sum_{k=0}^{\infty} \dfrac{(-1)^k x^{2k+2}}{(2k+2)!}$

61. (a) $f'(x) = \displaystyle\sum_{k=1}^{\infty} \dfrac{x^{k-1}}{(k-1)!}$ **(b)** $\displaystyle\int_0^x f(t)\,dt = \displaystyle\sum_{k=0}^{\infty} \dfrac{x^{k+1}}{(k+1)!}$ **63.** $f(x) = \displaystyle\sum_{k=1}^{\infty} (-1)^{k-1} k x^{k-1}, \ -1 < x < 1$

65. $f(x) = \dfrac{2}{3} \displaystyle\sum_{k=1}^{\infty} k x^{k-1}, \ -1 < x < 1$ **67.** $f(x) = \displaystyle\sum_{k=0}^{\infty} \dfrac{(-1)^{k+1} x^{k+1}}{k+1}, \ -1 < x \le 1$ **69.** $f(x) = -\displaystyle\sum_{k=0}^{\infty} \dfrac{x^{2k+2}}{k+1}, \ -1 < x < 1$

71. $-1 \le x < 1$ **73.** $-1 \le x < 1$ **75.** $-1 < x < 1$ **77.** $-3 < x < 5$ **79. (a)** $\displaystyle\sum_{k=0}^{\infty} x^{2k}$ **(b)** $(-1, 1)$

81. 0.693 **83.** SSM. **85.** SSM. **87.** SSM. **89.** $[-1, 5)$ **91.** $\displaystyle\sum_{k=0}^{\infty} \dfrac{(-1)^k x^{2k+1}}{(2k+1)!}; R = \infty$

AP® Practice Problems (SSM = See Student Solutions Manual.)
1. C **3.** B **5.** C **7.** C **9.** SSM.

Retain Your Knowledge
1. B **3.** D

Section 10.9 (SSM = See Student Solutions Manual.)

1. Taylor expansion **2.** Maclaurin **3.** $2 - 4x + \dfrac{3}{2}x^2 - \dfrac{1}{3}x^3$ **4.** $3 + 2(x-5) - \dfrac{1}{2}(x-5)^2 - \dfrac{2}{3}(x-5)^3$

5. $f(x) = -6 + 5x + 2x^2 + 3x^3$ **7.** $f(x) = 4 + 18(x-1) + 11(x-1)^2 + 3(x-1)^3$

9. $1 + \dfrac{x^2}{1} + \dfrac{x^4}{2} + \dfrac{x^6}{3!}; \displaystyle\sum_{k=0}^{\infty} \dfrac{x^{2k}}{k!}$ **11.** $x + \dfrac{x^2}{1} + \dfrac{x^3}{2} + \dfrac{x^4}{3!}; \displaystyle\sum_{k=0}^{\infty} \dfrac{x^{k+1}}{k!}$ **13.** $x^3 - \dfrac{x^5}{3!} + \dfrac{x^7}{5!} - \dfrac{x^9}{7!}; \displaystyle\sum_{k=0}^{\infty} \dfrac{(-1)^k x^{2k+3}}{(2k+1)!}$

15. $1 + x^2 + x^4 + x^6$; $\displaystyle\sum_{k=0}^{\infty} x^{2k}$ **17.** $3x + \dfrac{9x^2}{2} + \dfrac{27x^3}{3} + \dfrac{81x^4}{4}$; $\displaystyle\sum_{k=1}^{\infty} \dfrac{(3x)^k}{k}$ **19.** $x^2 - \dfrac{x^4}{2} + \dfrac{x^6}{3} - \dfrac{x^8}{4}$; $\displaystyle\sum_{k=0}^{\infty} (-1)^k \dfrac{x^{2k+2}}{k+1}$

21. $2 - x - \dfrac{x^3}{3!} + \dfrac{2x^4}{4!} - \dfrac{x^5}{5!}$ **23.** $1 + 2x + \dfrac{5}{2}x^2 + \dfrac{8}{3}x^3 + \dfrac{65}{24}x^4$ **25.** $1 + x + \dfrac{x^2}{2} + \dfrac{x^3}{2} + \dfrac{13}{24}x^4$

27. $x - x^2 + \dfrac{x^3}{3} - \dfrac{x^5}{30} + \dfrac{x^6}{90}$ **29.** $f(x) = \sqrt{x} = 1 + \dfrac{1}{2}(x-1) - \dfrac{1}{8}(x-1)^2 + \dfrac{1}{16}(x-1)^3 - \dfrac{5}{128}(x-1)^4 \cdots$

31. $f(x) = \ln x = (x-1) - \dfrac{(x-1)^2}{2} + \dfrac{(x-1)^3}{3} - \dfrac{(x-1)^4}{4} + \cdots$ **33.** $f(x) = \displaystyle\sum_{k=0}^{\infty} (-1)^k (x-1)^k$

35. $f(x) = \displaystyle\sum_{k=0}^{\infty} \dfrac{\sin\left(\dfrac{1}{6}(\pi + 3k\pi)\right)\left(x - \dfrac{\pi}{6}\right)^k}{k!}$ **37.** $e^2 \displaystyle\sum_{k=0}^{\infty} \dfrac{2^k(x-1)^k}{k!}$ **39.** $x + \dfrac{x^3}{3} + \dfrac{x^5}{10} + \dfrac{x^7}{42} + \dfrac{x^9}{216} + \cdots$

41. $\dfrac{x^3}{3} - \dfrac{x^7}{42} + \dfrac{x^{11}}{1320} - \dfrac{x^{15}}{15 \cdot 7!} + \cdots$ **43.** $\sec x = 1 + \dfrac{x^2}{2} + \dfrac{5x^4}{24} + \dfrac{61x^6}{720} + \dfrac{277x^8}{8064}$ **45.** SSM.

47. SSM. **49.** $\dfrac{\pi}{4} - \dfrac{1}{2}$ **51.** SSM. **53.** SSM. **55.** $1 + x^2 - \dfrac{x^3}{2} + \dfrac{5x^4}{6} + \cdots$

AP® Practice Problems

1. B **3.** B **5.** C

Retain Your Knowledge

1. B **3.** C

Section 10.10 (SSM = See Student Solutions Manual; AWV = Answers will vary.)

1. $P_5(x) = 2(x-1) - (x-1)^2 + \dfrac{2}{3}(x-1)^3 - \dfrac{1}{2}(x-1)^4 + \dfrac{2}{5}(x-1)^5$

3. $P_5(x) = 1 - (x-1) + (x-1)^2 - (x-1)^3 + (x-1)^4 - (x-1)^5$

5. $P_5(x) = -\left(x - \dfrac{\pi}{2}\right) + \dfrac{\left(x - \dfrac{\pi}{2}\right)^3}{3!} - \dfrac{\left(x - \dfrac{\pi}{2}\right)^5}{5!}$ **7.** $P_5(x) = 1 + 2x + 2x^2 + \dfrac{4}{3}x^3 + \dfrac{2}{3}x^4 + \dfrac{4}{15}x^5$

9. $P_5(x) = 1 + 2x + 4x^2 + 8x^3 + 16x^4 + 32x^5$ **11.** $P_5(x) = (x-1) + \dfrac{1}{2}(x-1)^2 - \dfrac{1}{6}(x-1)^3 + \dfrac{1}{12}(x-1)^4 - \dfrac{1}{20}(x-1)^5$

13. (a) $P_6(x) = 1 - \dfrac{x^2}{2!} + \dfrac{x^4}{4!} - \dfrac{x^6}{6!}$ **15. (a)** $P_4(x) = 1 + \dfrac{x}{3} - \dfrac{x^2}{9} + \dfrac{5x^3}{81} - \dfrac{10x^4}{243}$ **17. (a)** $P_9(x) = x - \dfrac{x^3}{3} + \dfrac{x^5}{5} - \dfrac{x^7}{7} + \dfrac{x^9}{9}$

(b)

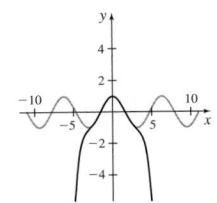

(b)

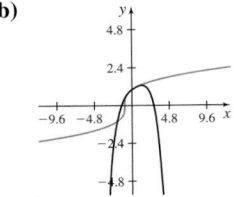

(b)

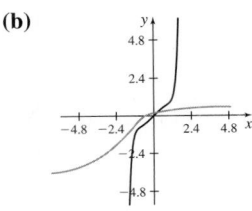

(c) $\cos \dfrac{\pi}{90} \approx 0.999$

(d) $|E| \leq 5.47 \times 10^{-17}$

(e) $|x| \leq 1.190$

(c) AWV.

(d) ≈ 0.965; the error is $\leq 3.018 \times 10^{-7}$

(c) AWV.

19. $f(x) = 4 + 7(x-1) + 11(x-1)^2 + 3(x-1)^3$ **21.** $f(x) = 7 - 25(x+1) + 30(x+1)^2 - 14(x+1)^3 + 2(x+1)^4$

23. degree 4 **25.** 0.310 **27.** 0.111 **29.** $P_2(x) = -(1 + \lambda)x + \pi$; SSM.

AP® Practice Problems (SSM = See Student Solutions Manual.)

1. B **3.** D **5.** B **7.** SSM.

Retain Your Knowledge

1. B **3.** D

Review Exercises (SSM = See Student Solutions Manual.)

1. $1, -\dfrac{1}{16}, \dfrac{1}{81}, -\dfrac{1}{256}, \dfrac{1}{625}$ **3.** $a_n = 2\left(-\dfrac{3}{4}\right)^{n-1}$ **5.** 0 **7.** 0 **9.** Diverges **11.** Converges; 0 **13.** SSM. **15.** $\dfrac{4}{5}$

17. Converges; $\dfrac{\ln 2}{\ln 2 - 1}$ **19.** Diverges **21.** SSM. **23.** Converges **25.** Converges **27.** Converges **29.** Converges

31. Converges; 1.079 **33.** Converges; -1.715 **35.** Conditionally convergent **37.** Conditionally convergent

39. Converges **41.** Diverges **43.** Diverges **45.** Converges **47.** Diverges **49.** Converges **51.** Converges

53. $6.244 < \displaystyle\sum_{k=2}^{\infty} \dfrac{6}{k(\ln k)^3} < 15.252$ **55. (a)** $R = 1$ **(b)** $[2, 4]$ **57. (a)** $R = \infty$ **(b)** $(-\infty, \infty)$

59. (a) $R = 1$ **(b)** $[0, 2)$ **61.** $f(x) = \dfrac{2}{3}\displaystyle\sum_{k=0}^{\infty}(-1)^k\left(\dfrac{x}{3}\right)^k, -3 < x < 3$ **63. (a)** $\displaystyle\sum_{k=0}^{\infty} \dfrac{3^k}{2k+1}x^{2k+1}$ **(b)** 0.758

65. $\displaystyle\sum_{k=0}^{\infty} \dfrac{\sqrt{e}(x-1)^k}{2^k k!}$ **67.** $1 + 2\left(x - \dfrac{\pi}{4}\right) + 2\left(x - \dfrac{\pi}{4}\right)^2 + \dfrac{8}{3}\left(x - \dfrac{\pi}{4}\right)^3 + \dfrac{10}{3}\left(x - \dfrac{\pi}{4}\right)^4 + \dfrac{64}{15}\left(x - \dfrac{\pi}{4}\right)^5 + \cdots$ **69.** 1.348

AP® Review Problems: Chapter 10 (SSM = See Student Solutions Manual.)

1. B **3.** C **5.** D **7.** D **9.** A **11.** D **13.** B **15.** B **17.** SSM. **19.** SSM.

AP® Cumulative Review Problems: Chapters 1–10 (SSM = See Student Solutions Manual.)

1. A **3.** D **5.** A **7.** B **9.** D **11.** C **13.** SSM.

AP® Practice Exam: Calculus BC
Section 1: Multiple Choice

1. A **2.** A **3.** C **4.** D **5.** C **6.** A **7.** D **8.** B **9.** C **10.** D **11.** C **12.** C **13.** C **14.** B **15.** B **16.** C **17.** A **18.** B

19. D **20.** A **21.** C **22.** B **23.** B **24.** D **25.** B **26.** C **27.** A **28.** A **29.** A **30.** B **31.** B **32.** C **33.** C **34.** C

35. D **36.** C **37.** B **38.** A **39.** C **40.** C **41.** A **42.** C **43.** D **44.** C **45.** D

Section 2: Free Response (SSM = See Student Solutions Manual.)

1. SSM. **2.** SSM. **3.** SSM. **4.** SSM. **5.** SSM. **6.** SSM.

Appendix A.5 (SSM = See Student Solutions Manual; AWV = Answers will vary.)

1. Pole, polar axis **2.** False **3.** False **4.** True **5.** True **6.** $x = r\cos\theta, y = r\sin\theta$ **7.** A **9.** C **11.** B

13. **15.** **17.**

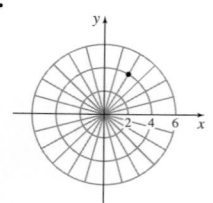 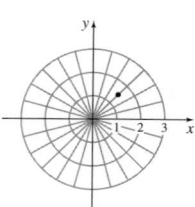 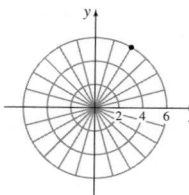

19. (a) $\left(5, -\dfrac{4\pi}{3}\right)$ **(b)** $\left(-5, \dfrac{5\pi}{3}\right)$ **(c)** $\left(5, \dfrac{8\pi}{3}\right)$ **21. (a)** $\left(1, -\dfrac{3\pi}{2}\right)$ **(b)** $\left(-1, \dfrac{3\pi}{2}\right)$ **(c)** $\left(1, \dfrac{5\pi}{2}\right)$

23. (a) $\left(3, -\dfrac{5\pi}{4}\right)$ **(b)** $\left(-3, \dfrac{7\pi}{4}\right)$ **(c)** $\left(3, \dfrac{11\pi}{4}\right)$ **25.** $(3\sqrt{3}, 3)$ **27.** $(0, 5)$ **29.** $(2, -2)$

31.
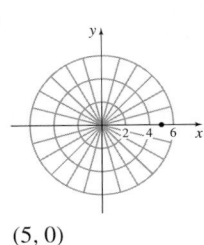
$(5, 0)$

33.
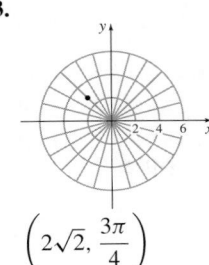
$\left(2\sqrt{2}, \dfrac{3\pi}{4}\right)$

35.
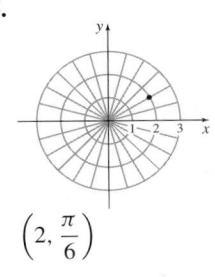
$\left(2, \dfrac{\pi}{6}\right)$

37.
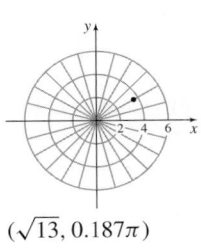
$(\sqrt{13}, 0.187\pi)$

39. E **41.** F **43.** H **45.** D

47. A circle centered at (0, 0) of radius 4.
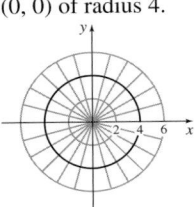

49. A line through the origin.
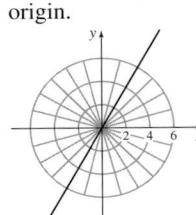

51. The line $y = 4$.
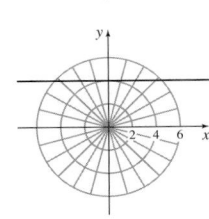

53. The line $x = -2$.
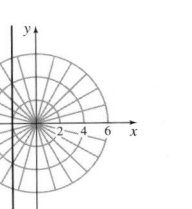

55. The circle centered at (1, 0) of radius 1.
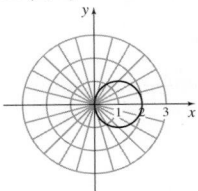

57. The circle centered at (0, −2) of radius 2.
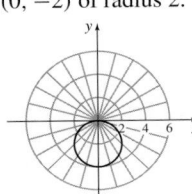

59. $\left(x - \dfrac{1}{2}\right)^2 + y^2 = \dfrac{1}{4}$ **61.** $y = (x^2 + y^2)^{3/2}$ **63.** $\sqrt{x^2 + y^2} - x = 4$ **65.** $y = x\tan(x^2 + y^2)$

67. $y = \sqrt{4 - x^2}$ **69.** SSM. **71.** SSM. **73.** SSM. **75. (a)**
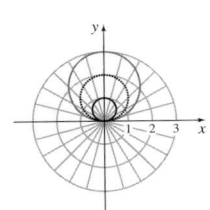
(b) AWV. **(c)**
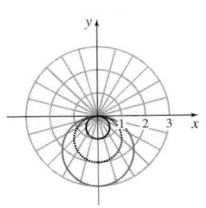
(d) AWV.

Appendix A.6

1. True **2.** Magnitude **3. (a); (d)** **4.** Unit **5.** Components **6.** True **7.** False **8.** False
9. Scalars: **(a), (b), (d), (e), (f), (g)**; Vectors: **(c), (h), (i)**
11. $-2\mathbf{v}$
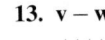

13. $\mathbf{v} - \mathbf{w}$
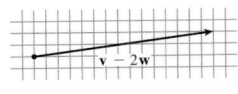

15. $\mathbf{v} - 2\mathbf{w}$
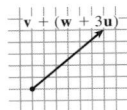

17. $\mathbf{v} + (\mathbf{w} + 3\mathbf{u})$ **19.** $\mathbf{x} = \mathbf{A}$ **21.** $\mathbf{C} = -\mathbf{D} - \mathbf{E} - \mathbf{F}$

23. $\mathbf{E} = \mathbf{G} + \mathbf{H} - \mathbf{D}$ **25.** $\mathbf{0}$ **27.**
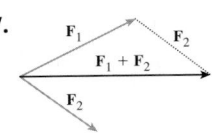

29. (a) $\langle 4, -5 \rangle$
(b)
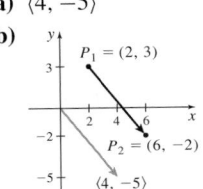

31. (a) $\langle -1, 1 \rangle$
(b)
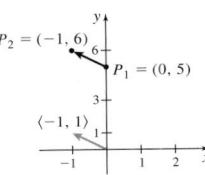

33. $\langle 9, -4 \rangle$ **35.** $\left\langle -\frac{1}{2}, \frac{2}{3} \right\rangle$ **37.** $\sqrt{10}$ **39.** $\sqrt{97}$ **41.** 5 **43.** $\sqrt{2}$ **45.** $\left\langle \frac{5}{13}, \frac{12}{13} \right\rangle, \left\langle -\frac{5}{13}, -\frac{12}{13} \right\rangle$ **47.** $\left\langle \frac{2}{\sqrt{5}}, \frac{1}{\sqrt{5}} \right\rangle; \left\langle -\frac{2}{\sqrt{5}}, -\frac{1}{\sqrt{5}} \right\rangle$

49. $\left\langle \frac{1}{2}, \frac{\sqrt{3}}{2} \right\rangle; \left\langle -\frac{1}{2}, -\frac{\sqrt{3}}{2} \right\rangle$ **51.** Parallel; opposite **53.** Not parallel **55.** $x = -5, 1$ **57.** $\langle 2\sqrt{3}, 2 \rangle$ **59.** $\langle -1, 1 \rangle$ **61.** $\sqrt{3.25}$ N/kg

63. $\langle -275, 75\sqrt{3} \rangle$ **65.** Left: ≈ 1000 lb; right: ≈ 845.237 lb.

67. (a) $\mathbf{A} = \langle 1100, 0 \rangle$, $\mathbf{B} = \langle -525, 525\sqrt{3} \rangle$, $\mathbf{C} = \langle -500, -500\sqrt{3} \rangle$ **(b)** $\langle 75, 25\sqrt{3} \rangle$ **69.** $\mathbf{F} = \langle 20\sqrt{3}, 20 \rangle$

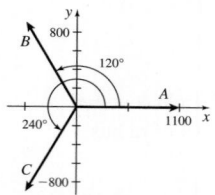

71. $8\sqrt{2}$ mi/h from northwest

Index of Applications

Note: *Italics* indicate examples; **bold** indicates chapter projects

IA-1

Subject Index

Note: **Boldface** indicates a definition; *italics* indicates an exercise; n indicates a footnote; *t* indicates a table.

FORMULAS/EQUATIONS

Distance Formula

If $P_1 = (x_1, y_1)$ and $P_2 = (x_2, y_2)$, the distance from P_1 to P_2 is

$$d(P_1, P_2) = \sqrt{(x_2 - x_1)^2 + (y_2 - y_1)^2}$$

Standard Equation of a Circle

The standard equation of a circle of radius r with center at (h, k) is

$$(x - h)^2 + (y - k)^2 = r^2$$

Slope Formula

The slope m of the line containing the points $P_1 = (x_1, y_1)$ and $P_2 = (x_2, y_2)$ is

$$m = \frac{y_2 - y_1}{x_2 - x_1} \quad \text{if } x_1 \neq x_2$$

$$m \text{ is undefined} \quad \text{if } x_1 = x_2$$

Point-Slope Equation of a Line

The equation of a line with slope m containing the point (x_1, y_1) is

$$y - y_1 = m(x - x_1)$$

Slope-Intercept Equation of a Line

The equation of a line with slope m and y-intercept b is

$$y = mx + b$$

Quadratic Formula

The solutions of the quadratic equation $ax^2 + bx + c = 0$, $a \neq 0$, are

$$x = \frac{-b \pm \sqrt{b^2 - 4ac}}{2a}$$

If the discriminant $b^2 - 4ac > 0$, there are two unequal real solutions.

If the discriminant $b^2 - 4ac = 0$, there is a repeated real solution.

If the discriminant $b^2 - 4ac < 0$, there are two complex solutions that are not real.

GEOMETRY FORMULAS

Circle

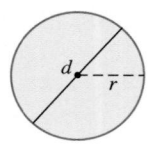

r = Radius, A = Area, C = Circumference

$$A = \pi r^2 \qquad C = 2\pi r$$

Triangle

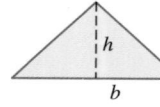

b = Base, h = Altitude (Height), A = Area

$$A = \frac{1}{2}bh$$

Rectangle

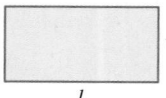

l = Length, w = Width, A = Area, P = Perimeter

$$A = lw \qquad P = 2l + 2w$$

Rectangular Box (closed)

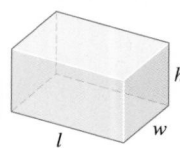

l = Length, w = Width, h = Height, V = Volume, S = Surface area

$$V = lwh \qquad S = 2lw + 2lh + 2wh$$

Sphere

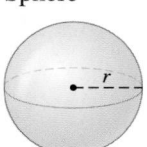

r = Radius, V = Volume, S = Surface area

$$V = \frac{4}{3}\pi r^3 \qquad S = 4\pi r^2$$

Right Circular Cylinder (closed)

r = Radius, h = Height, V = Volume, S = Surface area

$$V = \pi r^2 h \qquad S = 2\pi r^2 + 2\pi rh$$

Right Circular Cone

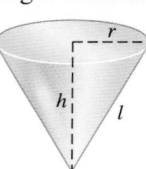

For a right circular cone of height h, radius r, and slant height $l = \sqrt{h^2 + r^2}$ that is open on top,

$$\text{Volume } V = \frac{1}{3}\pi r^2 h$$

$$\text{Lateral surface area} = \pi rl = \pi r\sqrt{h^2 + r^2}$$

If the cone is closed on top,

$$\text{Surface area} = \pi rl + \pi r^2$$

CONICS

Parabola

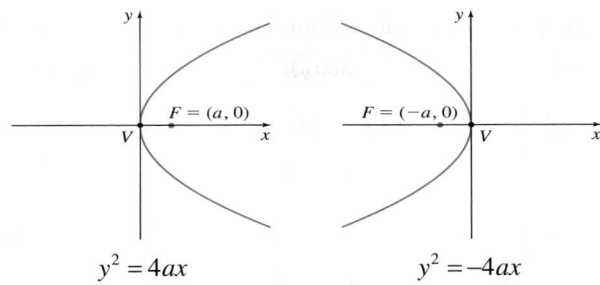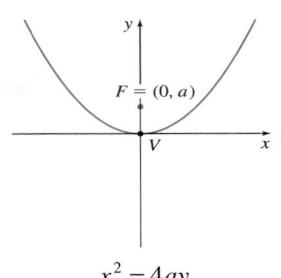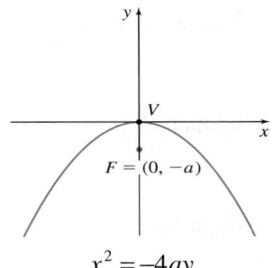

$$y^2 = 4ax \qquad y^2 = -4ax \qquad x^2 = 4ay \qquad x^2 = -4ay$$

Ellipse

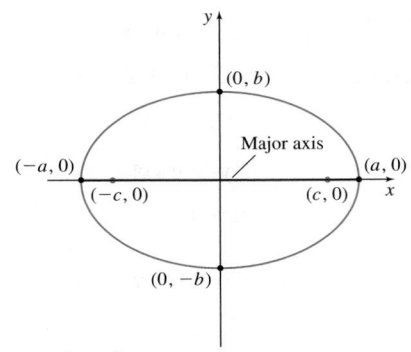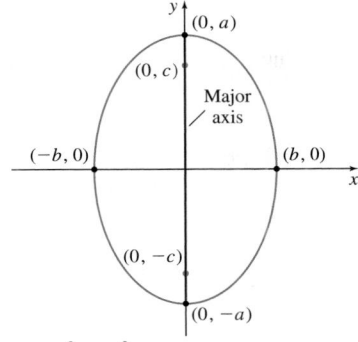

$$\frac{x^2}{a^2} + \frac{y^2}{b^2} = 1, \quad a \geq b > 0, \quad c^2 = a^2 - b^2$$

$$\frac{x^2}{b^2} + \frac{y^2}{a^2} = 1, \quad a \geq b > 0, \quad c^2 = a^2 - b^2$$

Hyperbola

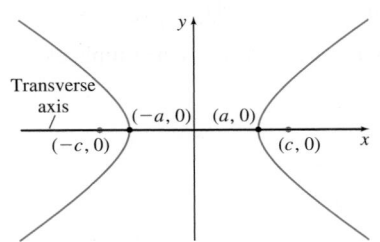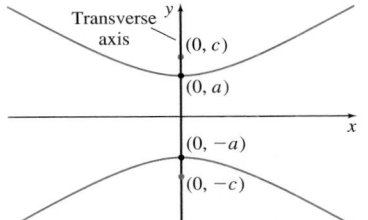

$$\frac{x^2}{a^2} - \frac{y^2}{b^2} = 1, \quad a > 0, b > 0, \quad c^2 = a^2 + b^2$$

Asymptotes: $y = \dfrac{b}{a}x, \quad y = -\dfrac{b}{a}x$

$$\frac{y^2}{a^2} - \frac{x^2}{b^2} = 1, \quad a > 0, b > 0, \quad c^2 = a^2 + b^2$$

Asymptotes: $y = \dfrac{a}{b}x, \quad y = -\dfrac{a}{b}x$

PROPERTIES OF LOGARITHMS

$\log_a(MN) = \log_a M + \log_a N \quad a > 0, a \neq 1$

$\log_a\left(\dfrac{M}{N}\right) = \log_a M - \log_a N \quad a > 0, a \neq 1$

$\log_a M^r = r \log_a M \quad a > 0, a \neq 1$

$\log_a M = \dfrac{\log M}{\log a} = \dfrac{\ln M}{\ln a} \quad a > 0, a \neq 1$

FACTORIAL

$0! = 1 \quad 1! = 1 \qquad n! = n(n-1) \cdot \ldots \cdot 3 \cdot 2 \cdot 1$

BINOMIAL THEOREM

$$(a+b)^n = a^n + \binom{n}{1}ba^{n-1} + \binom{n}{2}b^2 a^{n-2} + \cdots + \binom{n}{n-1}b^{n-1}a + b^n$$

where $\dbinom{n}{r} = \dfrac{n!}{(n-r)!r!}$

LIBRARY OF FUNCTIONS

Identity Function
$$f(x) = x$$

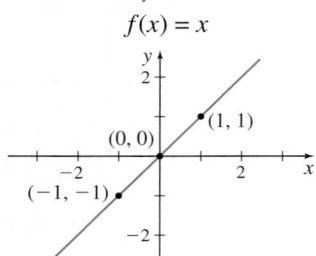

Square Function
$$f(x) = x^2$$

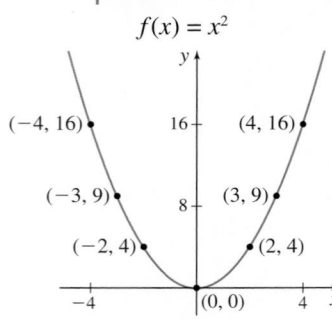

Cube Function
$$f(x) = x^3$$

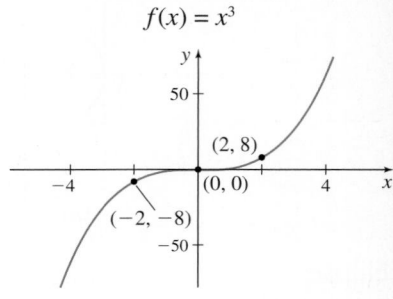

Square Root Function
$$f(x) = \sqrt{x}$$

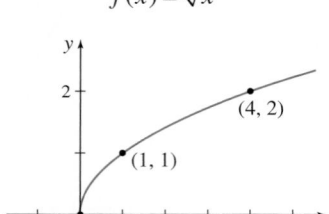

Reciprocal Function
$$f(x) = \frac{1}{x}$$

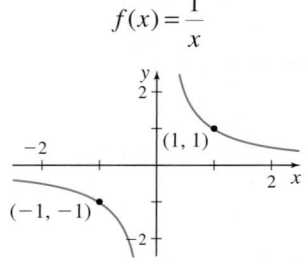

Cube Root Function
$$f(x) = \sqrt[3]{x}$$

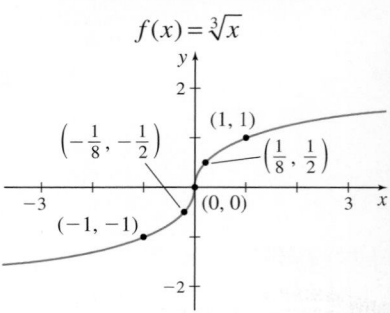

Absolute Value Function
$$f(x) = |x|$$

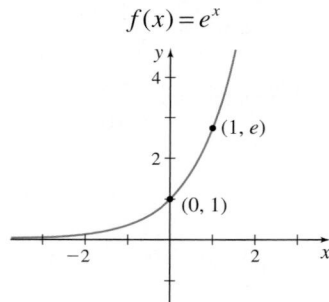

Exponential Function
$$f(x) = e^x$$

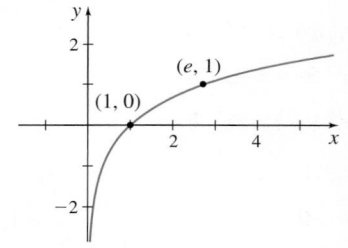

Natural Logarithm Function
$$f(x) = \ln x$$

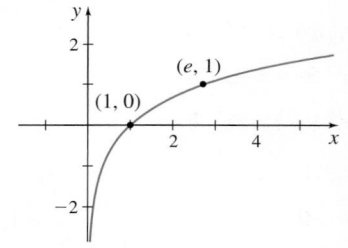

Sine Function
$$f(x) = \sin x$$

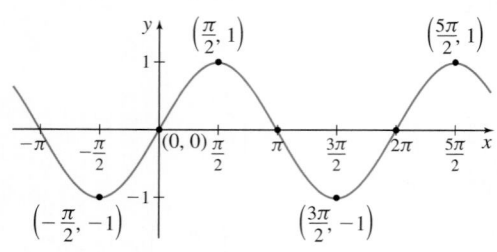

Cosine Function
$$f(x) = \cos x$$

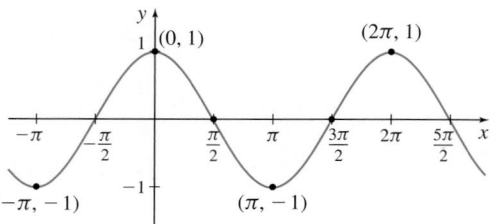

Tangent Function
$$f(x) = \tan x$$

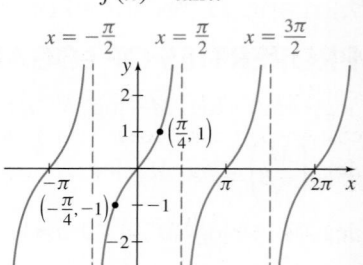

Cosecant Function
$$f(x) = \csc x$$

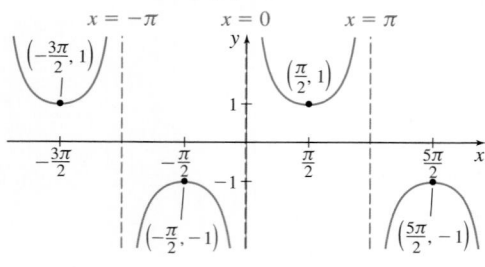

Secant Function
$$f(x) = \sec x$$

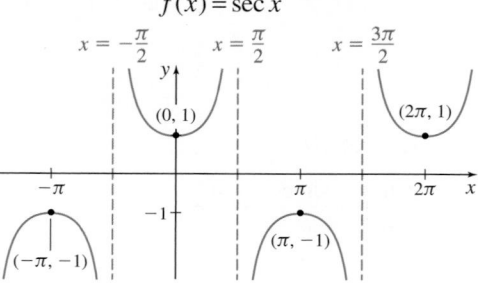

Cotangent Function
$$f(x) = \cot x$$

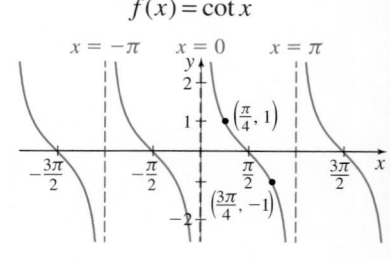

TRIGONOMETRIC FUNCTIONS

Let t be a real number and let $P = (x, y)$ be the point on the unit circle that corresponds to t.

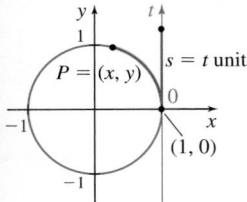 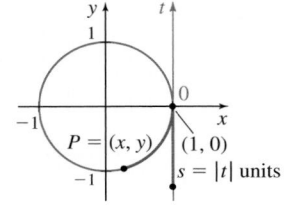

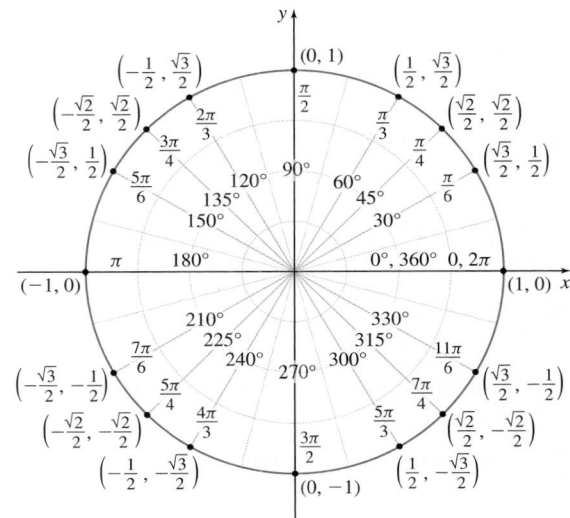

$$\sin t = y \qquad \cos t = x \qquad \tan t = \frac{y}{x}, \ x \neq 0$$

$$\csc t = \frac{1}{y}, \ y \neq 0 \qquad \sec t = \frac{1}{x}, \ x \neq 0 \qquad \cot t = \frac{x}{y}, \ y \neq 0$$

TRIGONOMETRIC IDENTITIES

Fundamental Identities

$$\tan\theta = \frac{\sin\theta}{\cos\theta} \qquad \cot\theta = \frac{\cos\theta}{\sin\theta}$$

$$\csc\theta = \frac{1}{\sin\theta} \qquad \sec\theta = \frac{1}{\cos\theta} \qquad \cot\theta = \frac{1}{\tan\theta}$$

$$\sin^2\theta + \cos^2\theta = 1$$

$$\tan^2\theta + 1 = \sec^2\theta$$

$$\cot^2\theta + 1 = \csc^2\theta$$

Even-Odd Identities

$$\sin(-\theta) = -\sin\theta \qquad \csc(-\theta) = -\csc\theta$$

$$\cos(-\theta) = \cos\theta \qquad \sec(-\theta) = \sec\theta$$

$$\tan(-\theta) = -\tan\theta \qquad \cot(-\theta) = -\cot\theta$$

Sum and Difference Formulas

$$\sin(A+B) = \sin A \cos B + \cos A \sin B$$

$$\sin(A-B) = \sin A \cos B - \cos A \sin B$$

$$\cos(A+B) = \cos A \cos B - \sin A \sin B$$

$$\cos(A-B) = \cos A \cos B + \sin A \sin B$$

$$\tan(A+B) = \frac{\tan A + \tan B}{1 - \tan A \tan B}$$

$$\tan(A-B) = \frac{\tan A - \tan B}{1 + \tan A \tan B}$$

Half-Angle Formulas

$$\sin\frac{\theta}{2} = \pm\sqrt{\frac{1-\cos\theta}{2}}$$

$$\cos\frac{\theta}{2} = \pm\sqrt{\frac{1+\cos\theta}{2}}$$

$$\tan\frac{\theta}{2} = \frac{1-\cos\theta}{\sin\theta}$$

Product-to-Sum Formulas

$$\sin A \sin B = \frac{1}{2}[\cos(A-B) - \cos(A+B)]$$

$$\cos A \cos B = \frac{1}{2}[\cos(A-B) + \cos(A+B)]$$

$$\sin A \cos B = \frac{1}{2}[\sin(A+B) + \sin(A-B)]$$

Sum-to-Product Formulas

$$\sin A + \sin B = 2\sin\frac{A+B}{2}\cos\frac{A-B}{2}$$

$$\sin A - \sin B = 2\sin\frac{A-B}{2}\cos\frac{A+B}{2}$$

$$\cos A + \cos B = 2\cos\frac{A+B}{2}\cos\frac{A-B}{2}$$

$$\cos A - \cos B = -2\sin\frac{A+B}{2}\sin\frac{A-B}{2}$$

Double-Angle Formulas

$$\sin(2\theta) = 2\sin\theta\cos\theta$$

$$\cos(2\theta) = \cos^2\theta - \sin^2\theta$$

$$\cos(2\theta) = 2\cos^2\theta - 1$$

$$\cos(2\theta) = 1 - 2\sin^2\theta$$

$$\tan(2\theta) = \frac{2\tan\theta}{1 - \tan^2\theta}$$

SOLVING TRIANGLES

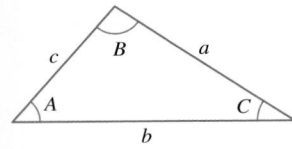

Law of Sines

$$\frac{\sin A}{a} = \frac{\sin B}{b} = \frac{\sin C}{c}$$

Law of Cosines

$$a^2 = b^2 + c^2 - 2bc\cos A$$

$$b^2 = a^2 + c^2 - 2ac\cos B$$

$$c^2 = a^2 + b^2 - 2ab\cos C$$

TABLE OF DERIVATIVES

1. $\dfrac{d}{dx}A = 0,\quad A$ is a constant

2. $\dfrac{d}{dx}[ku(x)] = k\dfrac{d}{dx}u(x),\ k$ a constant

3. $\dfrac{d}{dx}[u(x) + v(x)] = \dfrac{d}{dx}u(x) + \dfrac{d}{dx}v(x)$

4. $\dfrac{d}{dx}[u(x) - v(x)] = \dfrac{d}{dx}u(x) - \dfrac{d}{dx}v(x)$

5. $\dfrac{d}{dx}[u(x)v(x)] = u(x)\left[\dfrac{d}{dx}v(x)\right] + \left[\dfrac{d}{dx}u(x)\right]v(x)$

6. $\dfrac{dy}{dx}\left[\dfrac{u(x)}{v(x)}\right] = \dfrac{\left[\dfrac{d}{dx}u(x)\right]v(x) - u(x)\left[\dfrac{d}{dx}v(x)\right]}{[v(x)]^2}$

7. $\dfrac{dy}{dx} = \dfrac{dy}{du}\cdot\dfrac{du}{dx},\quad y = f(u),\ \ u = g(x)$

8. $\dfrac{d}{dx}[u(x)]^a = a[u(x)]^{a-1}u'(x)$

9. $\dfrac{d}{dx}e^{u(x)} = e^{u(x)}u'(x)$

10. $\dfrac{d}{dx}\ln u(x) = \dfrac{u'(x)}{u(x)}$

11. $\dfrac{d}{dx}a^{u(x)} = a^{u(x)}u'(x)\ln a$

12. $\dfrac{d}{dx}\log_a u(x) = \dfrac{u'(x)}{u(x)\ln a},\ a > 0,\ a \neq 1$

13. $\dfrac{d}{dx}\sin u(x) = \cos u(x)\,u'(x)$

14. $\dfrac{d}{dx}\cos u(x) = -\sin u(x)\,u'(x)$

15. $\dfrac{d}{dx}\tan u(x) = \sec^2 u(x)\,u'(x)$

16. $\dfrac{d}{dx}\cot u(x) = -\csc^2 u(x)\,u'(x)$

17. $\dfrac{d}{dx}\sec u(x) = \sec u(x)\tan u(x)\,u'(x)$

18. $\dfrac{d}{dx}\csc u(x) = -\csc u(x)\cot u(x)\,u'(x)$

19. $\dfrac{d}{dx}\sin^{-1}u(x) = \dfrac{u'(x)}{\sqrt{1-[u(x)]^2}}$

20. $\dfrac{d}{dx}\tan^{-1}u(x) = \dfrac{u'(x)}{1+[u(x)]^2}$

21. $\dfrac{d}{dx}\sec^{-1}u(x) = \dfrac{u'(x)}{u(x)\sqrt{[u(x)]^2-1}}$

22. $\dfrac{d}{dx}\cos^{-1}u(x) = -\dfrac{u'(x)}{\sqrt{1-[u(x)]^2}}$

23. $\dfrac{d}{dx}\cot^{-1}u(x) = -\dfrac{u'(x)}{1+[u(x)]^2}$

24. $\dfrac{d}{dx}\csc^{-1}u(x) = -\dfrac{u'(x)}{u(x)\sqrt{[u(x)]^2-1}}$

TABLE OF INTEGRALS

General Formulas

1. $\displaystyle\int [f(x)+g(x)]\,dx = \int f(x)\,dx + \int g(x)\,dx$

2. $\displaystyle\int [f(x)-g(x)]\,dx = \int f(x)\,dx - \int g(x)\,dx$

3. $\displaystyle\int kf(x)\,dx = k\int f(x)\,dx,\quad k$ a constant

4. Substitution: $\displaystyle\int f'(g(x))\,g'(x)\,dx = \int f'(u)\,du$

5. Integration by Parts Formula: $\displaystyle\int u\,dv = uv - \int v\,du$

Basic Integrals

1. $\displaystyle\int x^a\,dx = \dfrac{x^{a+1}}{a+1} + C,\quad a \neq -1$

2. $\displaystyle\int \dfrac{1}{x}\,dx = \ln|x| + C$

3. $\displaystyle\int e^x\,dx = e^x + C$

4. $\displaystyle\int a^x\,dx = \dfrac{a^x}{\ln a} + C,\ a > 0, a \neq 1$

5. $\displaystyle\int \ln x\,dx = x\ln x - x + C$

6. $\displaystyle\int \sin x\,dx = -\cos x + C$

7. $\displaystyle\int \cos x\,dx = \sin x + C$

8. $\displaystyle\int \sec^2 x\,dx = \tan x + C$

9. $\displaystyle\int \sec x\tan x\,dx = \sec x + C$

10. $\displaystyle\int \csc x\cot x\,dx = -\csc x + C$

11. $\displaystyle\int \csc^2 x\,dx = -\cot x + C$

12. $\displaystyle\int \tan x\,dx = \ln|\sec x| + C$

13. $\displaystyle\int \sec x\,dx = \ln|\sec x + \tan x| + C$

14. $\displaystyle\int \cot x\,dx = \ln|\sin x| + C$

15. $\displaystyle\int \csc x\,dx = \ln|\csc x - \cot x| + C$

16. $\displaystyle\int \dfrac{dx}{\sqrt{a^2-x^2}} = \sin^{-1}\dfrac{x}{a} + C,\ a > 0$

17. $\displaystyle\int \dfrac{dx}{a^2+x^2} = \dfrac{1}{a}\tan^{-1}\dfrac{x}{a} + C,\ a > 0$

18. $\displaystyle\int \dfrac{dx}{x\sqrt{x^2-a^2}} = \dfrac{1}{a}\sec^{-1}\dfrac{x}{a} + C,\ a > 0$

Integrals Involving $a + bx$ $\quad a \neq 0, b \neq 0$

19. $\displaystyle \int \frac{dx}{a+bx} = \frac{1}{b} \ln |a+bx| + C$

20. $\displaystyle \int \frac{x\,dx}{a+bx} = \frac{1}{b^2}(a+bx - a \ln |a+bx|) + C$

21. $\displaystyle \int \frac{x\,dx}{(a+bx)^2} = \frac{a}{b^2(a+bx)} + \frac{1}{b^2} \ln |a+bx| + C$

22. $\displaystyle \int \frac{x^2\,dx}{(a+bx)^2} = \frac{1}{b^3}\left(a+bx - \frac{a^2}{a+bx} - 2a \ln |a+bx| \right) + C$

23. $\displaystyle \int \frac{dx}{x(a+bx)^2} = \frac{1}{a(a+bx)} - \frac{1}{a^2} \ln \left| \frac{a+bx}{x} \right| + C$

24. $\displaystyle \int \frac{dx}{x^2(a+bx)} = -\frac{1}{ax} + \frac{b}{a^2} \ln \left| \frac{a+bx}{x} \right| + C$

25. $\displaystyle \int x(a+bx)^n \, dx = \frac{(a+bx)^{n+1}}{b^2}\left(\frac{a+bx}{n+2} - \frac{a}{n+1} \right) + C, \quad n \neq -1, -2$

26. $\displaystyle \int \frac{x\,dx}{(a+bx)(c+dx)} = \frac{1}{bc-ad}\left(-\frac{a}{b} \ln |a+bx| + \frac{c}{d} \ln |c+dx| \right) + C, \quad bc - ad \neq 0$

27. $\displaystyle \int \frac{x\,dx}{(a+bx)^2(c+dx)} = \frac{1}{bc-ad}\left[\frac{a}{b(a+bx)} + \frac{c}{bc-ad} \ln \left| \frac{a+bx}{c+dx} \right| \right] + C, \quad bc - ad \neq 0$

Integrals Involving $a^2 - x^2$, $a^2 + x^2$, $x^2 - a^2$ $\quad a > 0$

28. $\displaystyle \int \frac{dx}{a^2 - x^2} = \frac{1}{2a} \ln \left| \frac{x+a}{x-a} \right| + C$

29. $\displaystyle \int \frac{dx}{x^2 - a^2} = \frac{1}{2a} \ln \left| \frac{x-a}{x+a} \right| + C$

30. $\displaystyle \int \frac{dx}{(a^2 \pm x^2)^n} = \frac{1}{2(n-1)a^2}\left[\frac{x}{(a^2 \pm x^2)^{n-1}} + (2n-3) \int \frac{dx}{(a^2 \pm x^2)^{n-1}} \right], \quad n \neq 1$

31. $\displaystyle \int \frac{dx}{(x^2 - a^2)^n} = \frac{1}{2(n-1)a^2}\left[-\frac{x}{(x^2 - a^2)^{n-1}} - (2n-3) \int \frac{dx}{(x^2 - a^2)^{n-1}} \right], \quad n \neq 1$

Integrals Containing $\sqrt{a+bx}$ $a \neq 0, b \neq 0$

32. $\displaystyle \int x\sqrt{a+bx}\,dx = \frac{2}{15b^2}(3bx-2a)(a+bx)^{3/2}+C$

33. $\displaystyle \int x^n\sqrt{a+bx}\,dx = \frac{2}{b(2n+3)}\left[x^n(a+bx)^{3/2}-na\int x^{n-1}\sqrt{a+bx}\,dx\right]$

34. $\displaystyle \int \frac{x\,dx}{\sqrt{a+bx}} = \frac{2}{3b^2}(bx-2a)\sqrt{a+bx}+C$

35. $\displaystyle \int \frac{x^2 dx}{\sqrt{a+bx}} = \frac{2}{15b^2}(8a^2-4abx+3b^2x^2)\sqrt{a+bx}+C$

36. $\displaystyle \int \frac{x^n dx}{\sqrt{a+bx}} = \frac{2x^n\sqrt{a+bx}}{b(2n+1)} - \frac{2na}{b(2n+1)}\int \frac{x^{n-1}dx}{\sqrt{a+bx}}$

37. $\displaystyle \int \frac{dx}{x\sqrt{a+bx}} = \begin{cases} \dfrac{1}{\sqrt{a}}\ln\left|\dfrac{\sqrt{a+bx}-\sqrt{a}}{\sqrt{a+bx}+\sqrt{a}}\right|+C, & a>0 \\[3ex] \dfrac{2}{\sqrt{-a}}\tan^{-1}\sqrt{\dfrac{a+bx}{-a}}+C, & a<0 \end{cases}$

38. $\displaystyle \int \frac{dx}{x^n\sqrt{a+bx}} = -\frac{\sqrt{a+bx}}{a(n-1)x^{n-1}} - \frac{b(2n-3)}{2a(n-1)}\int \frac{dx}{x^{n-1}\sqrt{a+bx}}$

39. $\displaystyle \int \frac{\sqrt{a+bx}}{x}\,dx = 2\sqrt{a+bx}+a\int \frac{dx}{x\sqrt{a+bx}}$

40. $\displaystyle \int \frac{\sqrt{a+bx}}{x^2}\,dx = -\frac{\sqrt{a+bx}}{x} + \frac{b}{2}\int \frac{dx}{x\sqrt{a+bx}}$

Integrals Containing $\sqrt{x^2 \pm a^2}$ $a > 0$

41. $\displaystyle \int \sqrt{x^2\pm a^2}\,dx = \frac{x}{2}\sqrt{x^2\pm a^2} \pm \frac{a^2}{2}\ln\left|x+\sqrt{x^2\pm a^2}\right|+C$

42. $\displaystyle \int x\sqrt{x^2\pm a^2}\,dx = \frac{1}{3}(x^2\pm a^2)^{3/2}+C$

43. $\displaystyle \int x^2\sqrt{x^2\pm a^2}\,dx = \frac{x}{8}(2x^2\pm a^2)\sqrt{x^2\pm a^2} - \frac{a^4}{8}\ln\left|x+\sqrt{x^2\pm a^2}\right|+C$

44. $\displaystyle \int \frac{\sqrt{x^2+a^2}}{x}\,dx = \sqrt{x^2+a^2} - a\ln\left|\frac{a+\sqrt{x^2+a^2}}{x}\right|+C$

45. $\displaystyle \int \frac{\sqrt{x^2-a^2}}{x}\,dx = \sqrt{x^2-a^2} - a\sec^{-1}\frac{x}{a}+C$

46. $\displaystyle \int \frac{\sqrt{x^2\pm a^2}}{x^2}\,dx = -\frac{\sqrt{x^2\pm a^2}}{x} + \ln\left|x+\sqrt{x^2\pm a^2}\right|+C$

47. $\displaystyle \int \frac{dx}{\sqrt{x^2\pm a^2}} = \ln\left|x+\sqrt{x^2\pm a^2}\right|+C$

48. $\displaystyle \int \frac{x^2 dx}{\sqrt{x^2\pm a^2}} = \frac{x}{2}\sqrt{x^2\pm a^2} \mp \frac{a^2}{2}\ln\left|x+\sqrt{x^2\pm a^2}\right|+C$

49. $\displaystyle \int \frac{dx}{x\sqrt{x^2+a^2}} = -\frac{1}{a}\ln\left|\frac{a+\sqrt{x^2+a^2}}{x}\right|+C$

50. $\displaystyle \int \frac{dx}{x\sqrt{x^2-a^2}} = \frac{1}{a}\sec^{-1}\frac{x}{a}+C$

51. $\displaystyle \int \frac{dx}{x^2\sqrt{x^2\pm a^2}} = \mp\frac{\sqrt{x^2\pm a^2}}{a^2 x}+C$

52. $\displaystyle \int (x^2\pm a^2)^{3/2}dx = \frac{x}{8}(2x^2\pm 5a^2)\sqrt{x^2\pm a^2} + \frac{3a^4}{8}\ln\left|x+\sqrt{x^2\pm a^2}\right|+C$

53. $\displaystyle \int \frac{dx}{(x^2\pm a^2)^{3/2}} = \pm\frac{x}{a^2\sqrt{x^2\pm a^2}}+C$

Integrals Containing $\sqrt{a^2 - x^2}$ $\quad a > 0$

54. $\displaystyle\int \sqrt{a^2 - x^2}\, dx = \frac{x}{2}\sqrt{a^2 - x^2} + \frac{a^2}{2}\sin^{-1}\frac{x}{a} + C$

55. $\displaystyle\int x^2\sqrt{a^2 - x^2}\, dx = \frac{x}{8}(2x^2 - a^2)\sqrt{a^2 - x^2} + \frac{a^4}{8}\sin^{-1}\frac{x}{a} + C$

56. $\displaystyle\int \frac{\sqrt{a^2 - x^2}}{x}\, dx = \sqrt{a^2 - x^2} - a\ln\left|\frac{a + \sqrt{a^2 - x^2}}{x}\right| + C$

57. $\displaystyle\int \frac{\sqrt{a^2 - x^2}}{x^2}\, dx = -\frac{\sqrt{a^2 - x^2}}{x} - \sin^{-1}\frac{x}{a} + C$

58. $\displaystyle\int \frac{x^2}{\sqrt{a^2 - x^2}}\, dx = -\frac{x}{2}\sqrt{a^2 - x^2} + \frac{a^2}{2}\sin^{-1}\frac{x}{a} + C$

59. $\displaystyle\int \frac{dx}{x\sqrt{a^2 - x^2}} = -\frac{1}{a}\ln\left|\frac{a + \sqrt{a^2 - x^2}}{x}\right| + C$

60. $\displaystyle\int \frac{dx}{x^2\sqrt{a^2 - x^2}} = -\frac{\sqrt{a^2 - x^2}}{a^2 x} + C$

61. $\displaystyle\int (a^2 - x^2)^{3/2}\, dx = \frac{x}{4}(a^2 - x^2)^{3/2} + \frac{3a^2 x}{8}\sqrt{a^2 - x^2} + \frac{3a^4}{8}\sin^{-1}\frac{x}{a} + C$

62. $\displaystyle\int \frac{dx}{(a^2 - x^2)^{3/2}} = \frac{x}{a^2\sqrt{a^2 - x^2}} + C$

Integrals Containing $\sqrt{2ax - x^2}$ $\quad a > 0$

63. $\displaystyle\int \sqrt{2ax - x^2}\, dx = \frac{x - a}{2}\sqrt{2ax - x^2} + \frac{a^2}{2}\cos^{-1}\left(\frac{a - x}{a}\right) + C$

64. $\displaystyle\int x\sqrt{2ax - x^2}\, dx = \frac{2x^2 - ax - 3a^2}{6}\sqrt{2ax - x^2} + \frac{a^3}{2}\cos^{-1}\left(\frac{a - x}{a}\right) + C$

65. $\displaystyle\int \frac{\sqrt{2ax - x^2}}{x}\, dx = \sqrt{2ax - x^2} + a\cos^{-1}\left(\frac{a - x}{a}\right) + C$

66. $\displaystyle\int \frac{\sqrt{2ax - x^2}}{x^2}\, dx = -\frac{2\sqrt{2ax - x^2}}{x} - \cos^{-1}\left(\frac{a - x}{a}\right) + C$

67. $\displaystyle\int \frac{dx}{\sqrt{2ax - x^2}} = \cos^{-1}\left(\frac{a - x}{a}\right) + C$

68. $\displaystyle\int \frac{x\, dx}{\sqrt{2ax - x^2}} = -\sqrt{2ax - x^2} + a\cos^{-1}\left(\frac{a - x}{a}\right) + C$

69. $\displaystyle\int \frac{x^2\, dx}{\sqrt{2ax - x^2}} = -\frac{x + 3a}{2}\sqrt{2ax - x^2} + \frac{3a^2}{2}\cos^{-1}\left(\frac{a - x}{a}\right) + C$

70. $\displaystyle\int \frac{dx}{x\sqrt{2ax - x^2}} = -\frac{\sqrt{2ax - x^2}}{ax} + C$

71. $\displaystyle\int \frac{\sqrt{2ax - x^2}}{x^n}\, dx = \frac{(2ax - x^2)^{3/2}}{(3 - 2n)ax^n} + \frac{n - 3}{(2n - 3)a}\int \frac{\sqrt{2ax - x^2}}{x^{n-1}}\, dx, \quad n \neq \frac{3}{2}$

72. $\displaystyle\int \frac{x^n\, dx}{\sqrt{2ax - x^2}} = -\frac{x^{n-1}\sqrt{2ax - x^2}}{n} + \frac{a(2n - 1)}{n}\int \frac{x^{n-1}}{\sqrt{2ax - x^2}}\, dx$

73. $\displaystyle\int \frac{dx}{x^n\sqrt{2ax - x^2}} = \frac{\sqrt{2ax - x^2}}{a(1 - 2n)x^n} + \frac{n - 1}{(2n - 1)a}\int \frac{dx}{x^{n-1}\sqrt{2ax - x^2}}$

74. $\displaystyle\int \frac{dx}{(2ax - x^2)^{3/2}} = \frac{x - a}{a^2\sqrt{2ax - x^2}} + C$

75. $\displaystyle\int \frac{x\, dx}{(2ax - x^2)^{3/2}} = \frac{x}{a\sqrt{2ax - x^2}} + C$

Integrals Containing Trigonometric Functions

76. $\displaystyle\int \sin^2 x \, dx = \frac{x}{2} - \frac{\sin(2x)}{4} + C$

77. $\displaystyle\int \cos^2 x \, dx = \frac{x}{2} + \frac{\sin(2x)}{4} + C$

78. $\displaystyle\int \tan^2 x \, dx = \tan x - x + C$

79. $\displaystyle\int \cot^2 x \, dx = -\cot x - x + C$

80. $\displaystyle\int \sec^3 x \, dx = \frac{1}{2} \sec x \tan x + \frac{1}{2} \ln|\sec x + \tan x| + C$

81. $\displaystyle\int \csc^3 x \, dx = -\frac{1}{2} \csc x \cot x + \frac{1}{2} \ln|\csc x - \cot x| + C$

82. $\displaystyle\int \sin^n x \, dx = -\frac{1}{n} \sin^{n-1} x \cos x + \frac{n-1}{n} \int \sin^{n-2} x \, dx$

83. $\displaystyle\int \cos^n x \, dx = \frac{1}{n} \cos^{n-1} x \sin x + \frac{n-1}{n} \int \cos^{n-2} x \, dx$

84. $\displaystyle\int \tan^n x \, dx = \frac{1}{n-1} \tan^{n-1} x - \int \tan^{n-2} x \, dx$

85. $\displaystyle\int \cot^n x \, dx = \frac{-1}{n-1} \cot^{n-1} x - \int \cot^{n-2} x \, dx$

86. $\displaystyle\int \sec^n x \, dx = \frac{1}{n-1} \tan x \sec^{n-2} x + \frac{n-2}{n-1} \int \sec^{n-2} x \, dx$

87. $\displaystyle\int \csc^n x \, dx = \frac{-1}{n-1} \cot x \csc^{n-2} x + \frac{n-2}{n-1} \int \csc^{n-2} x \, dx$

88. $\displaystyle\int \sin(mx) \sin(nx) \, dx = -\frac{\sin[(m+n)x]}{2(m+n)} + \frac{\sin[(m-n)x]}{2(m-n)} + C, \ m^2 \neq n^2$

89. $\displaystyle\int \cos(mx) \cos(nx) \, dx = \frac{\sin[(m+n)x]}{2(m+n)} + \frac{\sin[(m-n)x]}{2(m-n)} + C, m^2 \neq n^2$

90. $\displaystyle\int \sin(mx) \cos(nx) \, dx = -\frac{\cos[(m+n)x]}{2(m+n)} - \frac{\cos[(m-n)x]}{2(m-n)} + C, \ m^2 \neq n^2$

91. $\displaystyle\int x \sin x \, dx = \sin x - x \cos x + C$

92. $\displaystyle\int x \cos x \, dx = \cos x + x \sin x + C$

93. $\displaystyle\int x^2 \sin x \, dx = 2x \sin x + (2 - x^2) \cos x + C$

94. $\displaystyle\int x^2 \cos x \, dx = 2x \cos x + (x^2 - 2) \sin x + C$

95. $\displaystyle\int x^n \sin x \, dx = -x^n \cos x + n \int x^{n-1} \cos x \, dx$

96. $\displaystyle\int x^n \cos x \, dx = x^n \sin x - n \int x^{n-1} \sin x \, dx$

97. (a) $\displaystyle\int \sin^m x \cos^n x \, dx = -\frac{\sin^{m-1} x \cos^{n+1} x}{m+n} + \frac{m-1}{m+n} \int \sin^{m-2} x \cos^n x \, dx$

(b) $\displaystyle\int \sin^m x \cos^n x \, dx = \frac{\sin^{m+1} x \cos^{n-1} x}{m+n} + \frac{n-1}{m+n} \int \sin^m x \cos^{n-2} x \, dx$

(if $m = -n$ use formula 84 or 85)

Integrals Containing Inverse Trigonometric Functions

98. $\displaystyle\int \sin^{-1}x\ dx = x\sin^{-1}x + \sqrt{1-x^2} + C$

99. $\displaystyle\int \cos^{-1}x\ dx = x\cos^{-1}x - \sqrt{1-x^2} + C$

100. $\displaystyle\int \tan^{-1}x\ dx = x\tan^{-1}x - \frac{1}{2}\ln(1+x^2) + C$

101. $\displaystyle\int x\sin^{-1}x\ dx = \frac{2x^2-1}{4}\sin^{-1}x + \frac{x\sqrt{1-x^2}}{4} + C$

102. $\displaystyle\int x\cos^{-1}x\ dx = \frac{2x^2-1}{4}\cos^{-1}x - \frac{x\sqrt{1-x^2}}{4} + C$

103. $\displaystyle\int x\tan^{-1}x\ dx = \frac{x^2+1}{2}\tan^{-1}x - \frac{x}{2} + C$

104. $\displaystyle\int x^n \sin^{-1}x\ dx = \frac{1}{n+1}\left(x^{n+1}\sin^{-1}x - \int \frac{x^{n+1}dx}{\sqrt{1-x^2}} \right),\quad n\neq -1$

105. $\displaystyle\int x^n \cos^{-1}x\ dx = \frac{1}{n+1}\left(x^{n+1}\cos^{-1}x + \int \frac{x^{n+1}dx}{\sqrt{1-x^2}} \right),\quad n\neq -1$

106. $\displaystyle\int x^n \tan^{-1}x\ dx = \frac{1}{n+1}\left(x^{n+1}\tan^{-1}x - \int \frac{x^{n+1}dx}{1+x^2} \right),\quad n\neq -1$

Integrals Containing Exponential and Logarithmic Functions

107. $\displaystyle\int xe^{ax}\ dx = \frac{1}{a^2}(ax-1)e^{ax} + C$

108. $\displaystyle\int x^n e^{ax}\ dx = \frac{1}{a}x^n e^{ax} - \frac{n}{a}\int x^{n-1}e^{ax}\ dx$

109. $\displaystyle\int \frac{e^x}{x^n}\ dx = -\frac{e^x}{(n-1)x^{n-1}} + \frac{1}{n-1}\int \frac{e^x}{x^{n-1}}\ dx$

110. $\displaystyle\int (\ln x)^n\ dx = x(\ln x)^n - n\int (\ln x)^{n-1}\ dx$

111. $\displaystyle\int x^n \ln x\ dx = \left(\frac{x^{n+1}}{n+1} \right)\left(\ln x - \frac{1}{n+1} \right) + C$

112. $\displaystyle\int \frac{(\ln x)^n}{x}\ dx = \frac{(\ln x)^{n+1}}{n+1} + C,\quad n\neq -1$

113. $\displaystyle\int \frac{dx}{x\ln x} = \ln|\ln x| + C$

114. $\displaystyle\int x^m (\ln x)^n\ dx = \frac{x^{m+1}(\ln x)^n}{m+1} - \frac{n}{m+1}\int x^m (\ln x)^{n-1}\ dx$

115. $\displaystyle\int \frac{x^m}{(\ln x)^n}\ dx = -\frac{x^{m+1}}{(n-1)(\ln x)^{n-1}} + \frac{m+1}{n-1}\int \frac{x^m}{(\ln x)^{n-1}}\ dx$

116. $\displaystyle\int e^{ax}\sin(bx)\ dx = \frac{e^{ax}}{a^2+b^2}[a\sin(bx) - b\cos(bx)] + C$

117. $\displaystyle\int e^{ax}\cos(bx)\ dx = \frac{e^{ax}}{a^2+b^2}[a\cos(bx) + b\sin(bx)] + C$